Combined Heating, Cooling & Power Handbook: Technologies & Applications

Second Edition

An Integrated Approach to Energy Resource Optimization

By Neil Petchers

Combined Heating, Cooling & Power Handbook: Technologies & Applications

Second Edition

An Integrated Approach to Energy Resource Optimization

By Neil Petchers

LONDON AND NEW YORK

First published by Fairmont Press in 2012.

Published 2020 by River Publishers
River Publishers
Alsbjergvej 10, 9260 Gistrup, Denmark
www.riverpublishers.com

Distributed exclusively by Routledge
2 Park Square, Milton Park, Abingdon, Oxon OX14 4RN
605 Third Avenue, New York, NY 10017

First issued in paperback 2023

Routledge is an imprint of the Taylor & Francis Group, an informa business

Library of Congress Cataloging-in-Publication Data

Petchers, Neil.
Combined heating, cooling & power handbook : technologies & applications an integrated approach to energy resource optimization / by Neil Petchers. -- 2nd ed.
p. cm.
Includes index.
ISBN 0-88173-689-9 (alk. paper) -- ISBN 978-8-7702-2300-3 (electronic) -- ISBN 978-1-4665-5334-7 (Taylor & Francis distribution : alk. paper) 1. Energy conversion. 2. Power plants--Energy conservation. 3. Cogeneration of electric power and heat. I. Title. II. Title: Combined heating, cooling, and power handbook.

TK1041.P48 2012
621.1'99--dc23

2012004109

Combined heating, cooling & power handbook : technologies & applications an integrated approach to energy resource optimization, 2nd edition / Neil Petchers

ISBN 13: 978-0-88173-689-2 (The Fairmont Press, Inc.)
ISBN 13: 978-1-4665-5334-7 (hbk)
ISBN 13: 978-87-7022-913-5 (pbk)
ISBN 13: 978-8-7702-2300-3 (online)
ISBN 13: 978-1-0031-5169-2 (ebk)

Dedication

In loving memory of my parents and greatest supporters, Ben and Esther, and for my children Brian and Adam.

Table of Contents

Foreword

The best unit of fuel is the one that is not used and the best unit of pollution is the one that is not created.

Since the initial publication of this book, we have seen perhaps the most dramatic fluctuations in fuel and electricity prices in history. While prices recently receded from peaks reached a few years ago, overall energy cost inflation over the past 10 years has been 7 percent – more than double that of the consumer price index. Rising energy costs, combined with increased environmental compliance costs, make improvements in energy and environmental efficiency an increasingly valuable investment. Still, we have not seen sufficient public policy support to bolster the value proposition for energy efficiency as a cornerstone of our national energy, security, and environmental protection policy.

This book focuses on how to make sound technology application and energy management decisions based on current and prudently anticipated actual conditions. Numerous commercially available technologies and mechanisms are in place to achieve improved economic performance. This is essentially a technical "how to" book. Despite the numerous complex graphics, tables, and equations, it has been written in a practical and straightforward manner, supported by data to provide value to scientists and engineers and to any individual interested in management of energy and environmental resources. Very few conclusions are offered, since so many factors change over time and each technology application depends on site-specific conditions. While there are myriad examples and calculations provided, their purpose is not to provide answers, but to show how to go about asking the questions and doing the analysis to find the answer. In fact, while some general rules of thumb often hold true, changing just one important variable can produce a different result and, in turn, a different technical and economic decision. Solid information and critical thinking are the keys to success in ever-changing circumstances.

This foreword is a discussion of the bigger picture – the conditions that could and should exist to allow these technology applications to reach their potential. We need a broader vision of societal objectives, the reasoning behind their significance, the options available, and the actions that can be taken to achieve them. The work described in this book takes place in a larger economic, political, and social context than a campus of buildings or the savings on a single balance sheet. The work of the installers, engineers, and project professionals that implement energy efficiency improvements is directly affected – positively or negatively – by public policies. Those policies influence the energy marketplace in countless ways, including energy prices and incentives for installation of more efficient equipment. The economics of the marketplace determine the value proposition for any given efficiency, renewable, or distributed generation project, but public policies both shape the energy market and, in turn, are shaped by it. Current and future professionals in this field should know that each project is part of a much larger economic and political dynamic. That work, our work, could reach farther and accomplish more with the right set of policies. A larger role for energy and environmental efficiency in the energy economy would have far-reaching beneficial effects.

Since publication of the first edition a decade ago, we have heard much talk and political promise of a green revolution, green economy, and aggressively tackling the threat of climate change. Concerns over environmental protection, national energy security, dependence on imported oil, and risks associated with nuclear power have spawned some energy regulations, restarted some utility-sponsored demand-side management programs, and led to other public policy initiatives. The environmental regulatory trend has been toward increasing regulations or incentives to reduce pollution, notably air emissions associated with energy conversion and consumption. Advancements in technology have and will continue to provide solutions. Examples include a four-fold reduction in the cost of photovoltaics, dry low-nitrous oxides (NO_X) combustion, hybrid vehicles, the modern wind turbine, advancements in biofuels, and steady overall efficiency increases in the fleets of almost every energy consuming technology. However, progress has been outpaced by growing challenges. Moreover, the real obstacles are perhaps as much political as technical – the lack of political will of the people and our leaders to drive the changes needed to reach our national energy, security, and environmental protection objectives. One encompassing way to overcome this obstacle is adoption of energy externalities in our public policy.

Externalities

In economics terms, externalities refers to activities of individuals (or firms) that affect other non-involved individuals, either positively or negatively. When a company dumps untreated waste into a river or pumps soot into the air and someone else has to pay to clean it up, the cost of that cleanup is not included in the product price. This is a negative externality. Much of environmental regulation is aimed at reducing negative externality costs to taxpayers and forcing those costs back to their source. This is called internalization. A positive externality is an action that imposes a positive benefit on a third party. An example is when a facility reduces its energy usage or polluting emissions. This saves money for the person, business, or institution, but simultaneously provides value and saves costs to society, for example by reducing health care costs and lost production from pollution related illness.

The study of energy externalities is quite mature. Great efforts have been made to determine the externality costs of various energy sources. This can be integrated into a national accounting system for energy pricing that considers and reallocates the true cost otherwise borne by taxpayers. Resources such as energy, water, and clean air remain undervalued because the full range of external cost factors, including security, public health, and environmental damage mitigation, have not been captured, or monetized, through the combined impact of market price and regulation. Hence, price signals still do not match the true cost benefits of resource conservation, security, and environmental protection. Consequently, much of the potential for increased energy efficiency and conservation of resources and decreased pollution still lies ahead.

Increased proliferation of vehicular efficiency, expansion of renewable resource utilization, fuel switching, cogeneration technology, and a litany of other energy efficiency improvements are well within our grasp, requiring only reasonable additional shifts forward in public policy. Governments usually do not have the political will (or desire) to force the market to internalize these costs. Yet, if energy corporations were forced to pay for rigorous safety measures, more costly insurance policies, stringent penalties for polluting, or taxes from which they were previously exempt, more of the costs of energy resources would be internalized and market behaviors would be adjusted accordingly. If, simultaneously, the implementation of energy efficiency measures or the switching to more societal beneficial energy sources were fully rewarded with public policy support, reflective of the benefits being created to taxpayers, market behaviors would also be adjusted. The combination of requirement, penalty, and reward would then be a powerful force in moving the nation toward key public policy objectives.

The U.S. constitutes about 5 percent of the world's population, but consumes nearly 25 percent of the world's fossil fuel and produces 25 percent of the world's greenhouse gas (GHG) emissions. The two most prominent areas of energy related externality costs are 1) those associated with U.S. dependence on foreign oil and 2) the environmental impact associated with the production and usage of fossil fuels. Elimination of dependence on foreign oil and a reversal in the trend of increased use of all fossil fuels should be among the top priorities in U.S. energy policy, since they affect almost all aspects of our way of life, present and future.

Dependence on foreign oil sends the wealth of the U.S. overseas, increases our national security risks, compromises our foreign policy, leads to instability in the oil producing regions of the world, increases military expenditures, raises our national debt, and lowers domestic employment. The U.S. currently consumes about 20 million barrels of oil per day, which nearly equals the total exports of all Middle East countries. Despite recent downward trends, the U.S. still imports nearly half of the oil it uses, which is about one-third of all of the oil that is exported around the world. Thus, we are the principle contributor to all of the wealth generated by oil producing countries. About 70 percent of U.S. oil consumption is for the transportation sector, with 71 percent of that, or half the nation's oil consumption, going to automotive transport vehicles. At the current market price of $100 per barrel, we spend about $767 billion per year on purchasing crude oil, with half of this sent overseas. This hurts the economy by reducing domestic investment, employment, and tax revenues. It also has a tremendously negatively effect on our balance of trade and growing annual foreign debt, directly or indirectly contributes financially to many nations the U.S. considers enemies or unfriendly, and pays for massive military build-ups, particularly in the Middle East.

The true cost of dependence on foreign oil cannot be understood without examining the associated military cost for mitigating or protecting against military build-ups and creation of nuclear superpowers and associated regional instability, the mitigation of which currently must be paid for by all taxpayers. Each year, we spend roughly $100 billion in military expenditures to maintain stability and protect oil shipping lanes. This does not consider the cost to amortize the most recent two wars in the Middle East (which many believe were motivated, at least partially, by protecting our access to oil exports) or the very real potential for future widescale military engagement for the same reason. Reducing our reliance on oil to a level that is at or below domestic production would result in a corresponding reduction in many of these expenditures.

The military cost of stabilizing oil producing regions is a classic example of an externality; neither the oil companies nor those who use the oil directly pay for this huge military expense. The American people pay for it in their tax bill. Moreover, since it is dissociated from the price paid for oil, there is no reflection of the real price of oil to the consumer. For every barrel of imported oil, the buyer pays about $100, plus the U.S. taxpayer contributes an additional $25 for military support (which does not include consideration of costs for amortizing the past Gulf wars). If we consider the negative impact on our balance of trade, national debt, and jobs lost to other countries, the number would be significantly higher, perhaps equal to the $100 per barrel that is paid directly for oil. This would mean that the real price of imported oil (before consideration of added costs associated with environmental impacts) is about twice the market cost and the U.S. taxpayer pays for half of every gallon.

Environmental externalities represent the most significant costs associated with fossil fuel usage (i.e., oil, natural gas, and coal). There are several types of environmental externalities – those that occur as raw materials are mined and processed and those that occur when fossil fuels are consumed. Environmental costs include traditionally identified criteria pollutants, e.g., NO_X, sulfur oxide (SO_X), carbon monoxide (CO), and particulate matter (PM). The annual cost to U.S. taxpayers for increased health care costs and pollution cleanup is in the tens of billions of dollars. It also includes the more recently identified additional costs associated with addressing climate change. Externality assessment for environmental costs should include the present and future annualized costs of GHG emissions, predominantly carbon dioxide (CO_2), methane, and NO_X. Note that since CO_2 represents the preponderance of GHGs, they are commonly combined and represented in terms of CO_2 equivalent.

Levels of GHGs have increased by about 25 percent since large-scale industrialization began, and over the past few decades, 85 percent of CO_2 emissions have resulted primarily from fossil fuels. This has produced an imbalance between emissions and absorption, resulting in the growth of GHG in the atmosphere. While estimates for an emerging carbon emissions market vary widely, as do the estimated costs for various means of GHG reduction, elimination, or sequestration, $30 per ton of GHG may be considered a reasonable proxy. As GHG emissions continue to increase and accumulate in the atmosphere, demand for more GHG mitigation will be required and the cost per ton of mitigation is expected to increase to perhaps double or triple current levels. This means that the present value of the lifecycle climate change impact of all fossil fuel use in the U.S. is estimated to be in the tens of billions, if not as high as $100 billion, annually and is expected to rise significantly in the future.

Considering all of the costs associated with fuel use, externality costs that are borne by the taxpayer, not internalized by the seller or buyer, results in costs that can approach the total cost of these energy sources themselves. Charging the true externality cost to the buyer or seller of the fossil fuel instead of the taxpayer would result in a tremendous general tax reduction, even if this expense was redistributed and charged to the seller or buyer. This would also send out appropriate market signals based on the full cost of using fossil fuels and place associated accountability for it on the user. Should political will emerge to drive a national commitment to recognize these very real costs, an important question is which type of accounting system makes the most sense to incorporate externalities – taxes, regulations, or incentives.

One of the great ironies is that the strongest free market advocates do not understand or acknowledge that the current model is actually anti free market in that federal and state taxes are assessed to the public at large for problems that can be quantifiably tied to specific sellers or users. The real cost of using natural resources is not apparent to the consumer, since the U.S. taxpayer continues to unknowingly foot a large portion of the bill. As a result, energy pricing does not encourage the consumer to alter behavior. There is general public agreement on goals of national security, reducing dependence on foreign oil, protecting the environment, reducing energy costs, creating jobs, and improving the economy. The path to these goals is in taking the necessary steps to eliminate dependence on foreign oil and dramatically reduce the environmental impact of fossil fuel usage. If this is accomplished, increased national security, reduced energy costs, job creation, and an improved economy will be the long-term results. Wealth will not be exported from this nation, but will instead be invested domestically, jobs will be created, and products will be exported, thereby reducing the national debt and increasing wealth. When we stop pouring wealth into nations in volatile regions that use it for increased militarization, we will increase national security, lower the need for military spending, and support foreign policy that will no longer be compromised by oil dependence. Finally, when we reverse the trends of increased pollution and global warming, hundreds of billions of dollars, both present and future, will be saved by our taxpayers in combating the dangerous impacts on health and safety and untold catastrophes in the event that global warming reaches a critical point.

Solutions to Mitigating Externality Costs

Many solutions have been offered to eliminate U.S. dependence on foreign oil and reduce pollution. One popular suggestion is to increase drilling for domestic oil. However, it will take years for significant yields to come to market and drilling can endanger pristine wilderness and provide various safety risks to local areas. It would also entail expanding into areas with higher cost and risk to drill and produce. This, in turn, may generate additional externalities and require additional subsidies from the American taxpayers, both in tax breaks and cleanup costs when spills and damages occur. The catastrophic 2010 Gulf of Mexico oil spill illustrated the inherent risks associated with deep water drilling, which are exacerbated when safety regulations are softened to make oil production cheaper. Moreover, this solution would offer no progress toward the other great challenge of environmental protection.

Taxes

A simple but far reaching approach to address both major areas of externalities would be a tax on product sales, adding the negative externality cost to each gallon of oil, cubic foot of natural gas, pound of coal (or kWh generated from it), etc. This would go directly to the source of the problem. Those who use more would pay more and those who use less would pay less. As an example, there is precedent for taxes on gasoline in the U.S., but this is generally less than $0.50 per gallon and is used primarily to cover road use and other infrastructure costs. A more aggressive example is found in Europe, where gasoline prices include taxes as high as $3.50 per gallon. On the surface, this might seem like more taxes are being charged, but the reverse is actually true. When companies and individuals pay their fair share for externalities, the high gasoline taxes, in this example, offset the need to increase general taxes while sending strong, accurate market signals. Under this scenario, it costs less to assess the correct externalities and let the market react to these prices than it does to tax people for things they are not responsible for. One measure of the success of this approach is that most European countries use far less transportation energy per person than the U.S. While still plagued by the same challenges as the U.S., namely dependence on oil and the threat of global warming, they are taking far more aggressive steps to remediate both.

When considering the purchase of a new vehicle, for example, a gasoline price of $7 or $8 per gallon would make a more efficient vehicle far more attractive. When considering a hybrid or one of the many new breeds of higher fuel efficiency and alternative energy source powered vehicles, the return on the incremental investment would be substantially improved. In many cases, the annual fuel cost savings would be greater than the annual financing cost payment on the incremental investment, resulting in more cash in the consumer's pocket each year versus in the scenario of a purchase of a conventional vehicle.

Forms of externality taxes have long been applied through U.S. government programs. The Gas Guzzler tax is an example of one intended to lower gasoline usage. Enacted in 1978, it is assessed on new cars that do not meet required fuel economy levels. It is only marginally effective because it has a very low miles per gallon (mpg) threshold, does not apply to trucks, minivans, and sport utility vehicles, and is not tied to actual usage. While a step in the right direction, an externality tax that aligns gasoline prices with true costs would be far more comprehensive and effective.

In contrast to taxing energy unit purchase, taxing carbon emissions directly may be superior in capturing the environmental costs from fossil fuels. This aligns costs with environmental damage done. While this has much merit, it is harder to measure and would require a massive effort to track and enforce. It also does not cover externality costs such as military expenditures required to support our dependence on foreign oil. Hence, even if it is a more accurate measure of environmental impact, it may not be as widely applicable and cost-effective as an externality tax on buying or selling energy.

Incentives/Subsidies

On the other side of the ledger, instead of taxes, positive incentives such as subsidies can reduce negative externality costs. Subsidies take several forms, such as rebates, tax credits, or startup funding. They are intended to support the development and commercialization of promising technologies, or developing infrastructure for new technologies such as electric vehicle charging stations and fueling stations for biofuels or natural gas. They can also be used to impact consumer behavior in a manner similar to externality taxes. For example, the U.S. government initially offered subsidies for hybrid vehicles to help launch the technology. This was met with good success in their commercialization. However, once the subsidies were eliminated, the market stagnated. It can be argued that subsidies used to offset a portion of the incremental cost for hybrids and other vehicles that lower fuel usage should be ongoing because they reduce externality costs that would otherwise be charged to the general taxpayer. The equivalent value of the avoided externality cost can be determined over the lifecycle of the incremental investment, using the government's cost of capital as a discount rate to set a maximum subsidy level. If lower levels are sufficient to move the market, then additional benefits will accrue to taxpayers.

The recent Cash for Clunkers program is an example. About 700,000 cars were purchased under the program. At an average rebate of $4,000 and an average 9.2 mpg increase in fuel efficiency, fuel savings total 277 gallons, or $1,246 at $4.50/gallon. In addition, the program is credited with saving tens of thousands of jobs and increasing safety due to the improved technology of the newer cars. While the program was perhaps more focused on bailing out automotive manufacturers and employment creation, it provided insight into how subsidies can effectively drive consumer behavior toward positive goals and be a cost-effective use of taxpayer money. The benefits may have been even greater had the program been implemented in conjunction with efforts to drive support for higher and timelier vehicular efficiency standards and further advancement in wide-scale deployment of other higher fuel efficiency and alternative energy source powered vehicles.

Regulations

The establishment of an externality tax or increased subsidies would not eliminate the need for regulation, but instead dovetail with it. Aggressive regulations are commonly found in the form of rigorous standards, with penalties for non-compliance. Examples of existing regulations include minimum efficiency standards for a wide range of appliances, home insulation levels, and the air permitting process for stationary emissions sources. For automotive vehicles, Corporate Average Fuel Economy (CAFE) regulations were first enacted by the U.S. Congress in 1975 and intended to improve the average fuel economy of automotive vehicles.

As part of the Energy Independence and Security Act of 2007 (EISA 2007), a CAFE standard was established to be 35 mpg by 2020, a 40 percent increase, which is expected to save the U.S. billions of gallons of fuel each year. One major challenge is closing the various loopholes that provide work-arounds for manufacturers to meet the requirements without accomplishing the goals. For example, certain types and sizes of vehicles have been exempt and a vehicle capable of operating on up to 85 percent ethanol gets counted as only 15 percent of its actual fuel usage on gasoline even if it never runs on ethanol, so car manufacturers can show far higher CAFE standard attainment than will ever be realized on the road. Enforcement of CAFE standards is also a challenge. Because penalties have been relatively modest, some manufacturers have found it more economical to continue to pay the penalties rather than comply. Another problem is that the latest changes call for action to be taken a decade into the future, after which point it will take another decade for the slow integration of new vehicles into the national mix. This is a very long time to wait when the needs for fuel reduction are so pressing in the present.

Since 1972, some of the most rigorous regulations on energy usage and environmental impact have come under the Clean Air Act (CAA). Over the past several years, there has been much political and legal debate over the Environmental Protection Agency's (EPA) position that GHGs endanger human health and welfare, and therefore could be regulated under the CAA. Recently, the EPA proposed the first-ever Carbon Pollution Standard rule for power plants. The rule would establish a new source performance standard (NSPS) for CO_2 emissions from fossil fuel electric generating units at a rate of 1,000 lbm per MWh. About 95 percent of all natural gas combined cycle power plants already meet the standard, but even the most efficient coal power plants exceed this threshold with an average of 1,800 lbm CO_2 per MWh.

If it remains in effect, this rule would result in no new investment in coal electric generation unless it had extremely advanced CO_2 controls, such as Carbon Capture and Storage (CCS). As discussed in Chapter 17, CCS is a process in which CO_2 is separated from emission streams and injected into geologic formations, avoiding its release into the atmosphere. CCS is considered to be in its technological infancy, with significant commercialization more than a decade away. Numerous barriers to the technology remain, such as the cost and feasibility of permanently storing CO_2 underground and the difficulty of constructing sufficient infrastructure to transport CO_2 to injection sites. Such challenges would have to (and are expected to) be overcome to enable widespread deployment. In the interim, this EPA rule would likely lead to the deployment of more natural gas-fired plants, more on-site distributed generation, and more large-scale renewable technology applications. It may also rekindle utility support of customer efficiency and demand-side management programs as a means to reduce load and avoid the need for new capacity. Even if it endures, the rule does not address existing sources and emissions from coal-fired power plants, which are already the largest single source of GHGs in the U.S. They are also a major source of criteria pollutants, notably SOx, NOx, and PM, as well as hazardous air pollutants (HAPs), notably mercury and acid gases, as well as other byproducts.

Another option for addressing the nation's energy challenges is to create incentive-based regulation that has mandatory compliance but uses the economic behavior of businesses and individuals to attain desired goals. The Cap and Trade program for carbon emissions is an example of this approach that has been suggested to Congress but not acted upon. Under such a program, an emissions cap would be established nationally for the pollution that sources can emit. Allowances equal to the cap would be distributed and could be traded freely. Emissions sources would have flexibility to choose how to meet their limits, either by reducing their own emissions or purchasing allowances from other sources. Those that can over-perform in emissions reduction can sell credits and earn profits.

The flexibility of such a program is that everyone can determine the cheapest way to reduce their emissions while government determines the overall emissions cap at a level that guarantees air quality and that environmental goals are met. This was very effective in addressing the acid rain problem and certainly can provide great value when applied effectively to electric generation and other major energy users. The approach, however, seems administratively complex and less viable for the transportation sector and smaller energy users, which combine to represent a high percent of fossil fuel usage and pollution generation.

Automotive Vehicles

Given that automotive transport vehicles consume half of all U.S. oil, the challenges of reducing oil dependence and environmental protection must include a virtual transformation of this sector. While transportation vehicles are not addressed in this book, they are predominantly powered by the same type of reciprocating engines discussed in Chapter 9. Of course, transportation vehicles must carry their engine with them, and the engines are less efficient than stationary engines and have no potential for heat recovery, but the same comparisons apply.

The wide range of alternative vehicles currently on the market include hybrids and those powered by biofuels, natural gas, and electricity. These vehicles can operate exclusively on their intended energy source, or with a flex-fuel design that enables operation with gasoline or diesel fuel. Of the 266 million vehicles in the U.S. today, only 10 million are alternative and flex-fuel vehicles. While the numbers keep growing, it is unlikely that they will achieve the scale necessary to help eliminate national dependence on foreign oil in the near future without support from the use of externalities taxes or subsidies as discussed above.

Alternative Fuels

For the U.S. to reduce to insignificance its dependence on foreign oil while simultaneously addressing its pressing environmental challenges, efficiency improvements must be combined with increased usage of alternative energy sources. A massive transformation in the vehicular sector alone would have a dramatic impact on these objectives. However, a combination of energy sources and technologies is needed to practically and economically achieve these objectives.

The practical major fossil fuel alternatives to fuel oil (and coal) are natural gas, liquid petroleum gas (LGP), and electricity (derived from fossil fuels). The major non-fossil fuel alternatives are biomass and biomass-derived fuels, such as ethanol, biodiesel, and methanol, and electricity produced from renewable sources. These may be used separately or in combination with oil. Numerous other energy sources are available that may have potential in the future, such as hydrogen, but the focus here is on near term, cost-effective options.

Biomass

Biomass is biological material derived from living or recently living organisms. In principle, recycling the carbon associated with biomass is a long-term process under which CO_2 captured by the plants is released back to the atmosphere. Considering the theoretical carbon cycle as a whole, there are no net CO_2 emissions from burning the biomass. In reality, there are many imperfections in how the cycle operates, such as the need for transportation, energy used in processing, and the fact that not all of the biomass harvested will continually be renewed. Still, compared to fossil fuels, biomass will result in far lower GHG emissions.

Biofuels, such as ethanol, methanol, and biodiesel, hold great potential for reducing oil usage and overall GHG emissions and improving the U.S. economy. Their promise is predicated upon appropriate standards and incentives being used to shape the nascent bio-energy industry to provide these benefits in a sound and sustainable manner.

Ethanol is the most widely produced alternative liquid biofuel in the U.S., with 14 billion gallons produced in 2011. It is produced mainly from crops that contain sugar (e.g., sugar cane or sugar beet) by pretreating starch crops (corn or wheat) or cellulose to produce sugars. The Ford Model T, which revolutionized auto manufacturing in 1908, was designed to run on ethanol or gasoline. Current primary sources of ethanol are corn and sugar cane, with corn predominating in the U.S. About 90 percent of all gasoline sold in the U.S. has a 10 percent ethanol blend, called E10, which is used primarily to replace the toxic MBTE as an octane enhancer.

Until 2012, the U.S. government stimulated ethanol usage through the Volumetric Ethanol Excise Tax Credit, which offered refiners a credit of $0.45 per gallon. Its termination will likely place additional limitations on the growth potential of ethanol as a mainstream alternative fuel. Unlike startup type subsidies, in this case an argument can be made for providing continuous subsidies for biofuels such as ethanol, if needed, to keep them market competitive, to the extent that it balances the absence of externality costs imposed upon oil. The debate is complicated further by uncertainties about the demand that ethanol production places on usable farm land and the impact on corn and other agricultural costs. There is near-term potential to increase the yields of corn-based ethanol and significantly lower the negative environmental impact through improved farming and production methods. This would need to be demonstrated to garner broader based support for expanded ethanol production and is an area that could benefit from government research and development support. The benefits of improved agricultural processes are certainly not limited to improved corn production alone. Given the extensive use of fossil fuels for fertilizer, irrigation, drying and transportation, and the various other inefficiencies in farming and land management, production efficiency improvements in the agricultural sector present significant opportunities for the reduction in fossil fuel usage and pollution.

Another potentially massive source is cellulosic ethanol produced from wood, grasses, or non-edible parts of plants such as corn stover, switchgrass, miscanthus, woodchips, and byproducts of lawn and tree maintenance. These have advantages over corn and sugar as a feedstock for ethanol because of their high productivity per acre, low agricultural effort, far greater GHG reduction potential, and less sulfur, CO, and particulate emissions. Methanol, produced from natural gas, coal, crude oil, and biomass, is another widely used alcohol similar to ethanol. When produced from biomass, it is a renewable resource. Methanol can be used to replace gasoline or converted to dimethyl ether to replace diesel fuel. Producing ethanol or methanol from cellulose promises up to a tenfold increase in the volume of commercially available biofuel that can be produced in the U.S. and abroad. This alone would reduce nearly half of the amount of total imported oil while producing corresponding reductions in GHG emissions. Still, the massive conversion to alternative energy sources requires production and vehicular technologies not currently on the market. However, the costs and production time would be relatively modest given genuine market demand stimulated by public policy support.

Biodiesel, a cleaner burning replacement for petroleum-based diesel fuel, is produced from renewable sources such as used and unused vegetable oils and animal fats. It is nontoxic and biodegradable. Just as Henry Ford had focused on ethanol for his Model T, Rudolph Diesel originally used peanut oil for his engines. As the cost of petroleum distillate has continued to rise, there has been renewed interest in biodiesel. The most common sources of oil for biodiesel production in the U.S. are soybean oil and yellow grease (primarily recycled cooking oils). Production costs remain higher than petroleum-based diesel and a subsidy of a $1 per gallon tax credit for refiners for crop-based biodiesel (not waste oil), which had been driving increased production, was recently repealed. Given the limited availability of yellow grease and the high cost of soybean oil, it does not appear that biodiesel can compete with petroleum-based diesel in large quantities unless subsidies are restored.

An effective biofuels policy must recognize that the choice of feedstock is just one of many factors that influence the environmental impact of biofuel production and all forms of biomass. Other key factors to consider are carbon emissions from converting land from other uses to feedstock production, tillage methods, energy use for irrigation, fertilizer application rates, the source of thermal energy and electricity at the refinery, and whether CO_2 produced during fermentation is sequestered or released to the atmosphere. The various carbon cycle imperfections noted above make biomass less than 100 percent renewable, but still superior to fossil fuels.

Biomass has historically been a major fuel source around the world and is now being used increasingly as a direct replacement for fossil fuels in the U.S. Co-firing of biomass with a fossil fuel (commonly coal) in industrial combustion applications is also on the rise and its usage is considered largely renewable. In other cases, such as biomass usage for cooking fires throughout much of the world, notably in Africa, South America, and large parts of Asia, it is not considered fully renewable, as wood and other gathered fuels are not replaced in equal quantity. Billions of people throughout the world still cook on traditional three-stone fires, which are an extremely inefficient use of wood and other biomass fuels. They deplete forests and cause severe indoor air pollution, which is responsible for millions of deaths and untold cases of respiratory illness. Field testing in Maasai villages in Tanzania reveals smoke and CO levels that are 35 times World Health Organization standards.

Inexpensive high-efficiency stoves are available that can reduce wood and other biomass fuels used by 50 to 70 percent, reducing smoke and CO levels by 90 percent and having a significant impact on GHG emissions world-wide. One hundred dollars invested in a new, more efficient stove in an African village, for example, can produce more GHG emission reductions than a typical $1,000 efficiency investment in a European or American factory. Given the world-wide nature of climate change, it may be more cost-effective for nations such as the U.S. to invest in such technology applications abroad to achieve a higher rate of return on the investment (or create more offsets) in eliminating GHG than can be achieved through domestic public policy investment. With deforesting being one of the major sources of GHG emissions, forest protection (in addition to reforesting) is seen as among the most powerful actions in combating climate change.

Natural Gas

Of the relatively clean fossil fuels that can play a large role in reducing U.S. dependence on foreign oil and improving environmental conditions, natural gas is considered the blossoming giant. Whether for vehicles, homes, buildings, industrial processes, or electric generation stations, natural gas should be favored over oil and coal, given consideration of all externalities. After a pronounced spike in natural gas price several years ago, there has been a steady and overall dramatic decrease in price, making it less costly than oil. New exploration techniques have vastly increased available reserves and supply is believed to continue to be abundant for the foreseeable future. It can replace oil in almost every vehicular and stationary application, it can replace coal as a primary source of electric generation in major power plants, and it offers on-site conversion from electric technologies, cogeneration, and other distributed generation capacity.

For vehicular applications, liquefied petroleum gas (LPG) and natural gas are considered among the most promising alternatives. Although not prevalent in the U.S., there are more than 20 million LPG or natural gas vehicles (NGVs) worldwide. Both fuels are available in large quantity throughout the U.S. and have superior environmental characteristics versus oil. They can be deployed economically in large scale and used in dedicated or flex-fuel vehicles. LPG, made primarily of propane with butane, is sometimes referred to internationally as Autogas. NGV applications include liquefied natural gas (LNG) or compressed natural gas (CNG). With only slight modifications required, primarily in the fuel storage tank, engine, and chassis, flex-fuel designs allow drivers to take advantage of widespread availability of gasoline, but use a cleaner, more economical, domestically produced alternative when natural gas is available. Rebates/tax credits have been available to help stimulate the market, though with only a modicum of success to date. Home refueling stations are just now being commercialized. These are the slow-fill variety requiring car owners to refuel overnight. The compression/filling system adds cost, but offers the convenience of filling at home and generally will have a far lower operating cost than a commercial filling station. The largest challenge, however, is to build the infrastructure with enough "quick fill" stations so it can become a regular way of life. There are only about 1,000 CNG filling stations in the U.S. versus about 160,000 gasoline stations. While CNG stations are being added, until a critical mass is reached and home filling station technology becomes commercially and economically available, the advancement of CNG will be delayed.

Natural gas is often described as the cleanest fossil fuel, producing less CO_2 per energy unit delivered than either coal or oil and far fewer criteria pollutants than other hydrocarbon fuels. However, it still contributes substantially to global GHG emissions. Methane is a primary GHG and further exploration, production, and transportation of natural gas and LPG adds methane release from leakage and losses. There are also several potential environmental issues associated with horizontal drilling and hydraulic fracturing, or "fracking," used in the production of shale gas, which is the largest source of new reserves. The drilling and fracturing of wells requires large amounts of water and produces large amounts of wastewater. If mismanaged, hydraulic fracturing fluid can be released by spills, leaks, or other exposure pathways. The use of potentially hazardous chemicals in the fracturing fluid has raised concerns of contamination of drinking water sources and negative impacts on natural habitats. Hydraulic fracturing has come under international scrutiny due to environmental, health, and safety concerns and has been suspended or banned in some countries. Advanced technologies to scrub emissions and purify water produced at the well-head are available and ready to deploy. Many believe that fracking can be cleaner and safer with the right technology and increased monitoring and regulatory oversight.

While it could well hold the key to eliminating U.S. dependence on oil and does offer a degree of improved environmental characteristics, natural gas may not provide significant enough benefit in the battle against global warming to be a comprehensive solution. In fact, since more and more of the U.S. natural gas reserves are harvested from shale gas, environmental concerns and, in turn, externality costs are increasing.

Electricity

The U.S. uses approximately 20 percent of the world's electricity, or about 4 billion GWh per year. Of this, an estimated 70 percent is produced by fossil fuels (45 percent coal, 24 percent natural gas, and 1 percent oil), 19 percent is produced from nuclear energy, 6 percent from hydro, and 4 percent from renewables. While the electricity generation sector has virtually no impact on U.S. dependence on foreign oil, it could play a significant role in reducing oil dependance to the extent electric vehicles become a mainstream technology. Correspondingly, electricity production plays a monumental role in creating environmental externalities, most notably due to coal usage. It is somewhat of a paradox that increased electric usage could help solve one major source of externalities – oil dependency – while exacerbating another – environmental impact.

Among the largest potential options for oil usage reduction are hybrids, electric vehicles (EVs), and plug-in hybrid electric vehicles (PHEVs). Hybrids use a standard engine and an electric motor assist powered by a battery pack that is recharged as braking energy is used to spin an electric generator instead of being dissipated as heat. A PHEV is a hybrid vehicle with larger battery capacity that can operate in all-electric mode until its batteries are exhausted and then engage the engine. EVs and PHEVs can use household current to charge the battery pack. Depending on driving range, PHEVs can use from zero gasoline or diesel fuel to, at worst, that of a traditional hybrid. For most drivers, the all-electric range of a fully charged PHEV will enable them to complete their daily commute with electricity use alone.

Both EVs and PHEVs are just entering the market and can eventually become market leaders. To expand their market penetration, they will require improved infrastructure for wide-scale charging stations. The combination of savings or rebates/tax credits on the incremental cost of an EV or PHEV, however, must be sufficient to overcome the increased first cost. Current tax credits for new EVs and PHEVs are $2,500 to $7,500, depending on battery pack capacity. Additional state incentives are also available, which seem to be adequate to drive the market. This is still less than the incremental cost of these recently introduced vehicles. Research and development continues to be critical to improve battery technology, making it lighter, longer lasting, and less costly. Learning from past experience, subsidies should be ongoing long after commercialization to ensure a sustainably high market share.

Since most vehicles will be charged at night, an increase in electric generation and distribution infrastructure would not be required because nighttime systems loads are so much lower. This is, therefore, very attractive load for electric utilities, many of which offer special off-peak rates, thus reducing the cost of powering an EV or PHEV. EVs and PHEVs also remove vehicle exhaust from congested cities with high pollution levels.

The environmental impacts/benefits of the various alternative fuels and of simple hybrids are fairly predictable. However, the impacts of EVs and PHEVs are complicated because the sources of electricity vary widely. If the electricity is from a renewable source, then the environmental impacts will be very positive. If the source of the electric generation is natural gas, there will be environmental benefit in reducing criteria pollutants and GHG emissions as well. However, if the source is coal, then there is a tradeoff being made, since swapping coal for oil will result in greater negative environmental impact across the board. This is a very important consideration and should be the only real impediment to a massive switch to EVs and PHEVs.

Burning 1 million Btu (MMBtu) of coal produces about 220 lbm of CO_2, compared with 163 lbm and 117 lbm of CO_2 for oil and natural gas, respectively. Typical coal plants produce about 2.1 lbm CO_2 per kWh. Typical natural gas-fired combined cycle power plants produce about 0.9 lbm CO_2 per kWh, or well below half that of a coal plant. At a \$30 per ton proxy rate for GHG credits, this correlates to \$0.032 per coal-generated kWh and \$0.013 per kWh for a natural gas-powered combined cycle plant. This means that the conversion of a coal plant to a combined cycle gas plant would create environmental externality value of about \$0.019/kWh in CO_2 value alone. If we consider all environmental impacts between coal and gas options as externalities, the difference is even greater. If we consider complete elimination of electricity usage from a coal plant through energy efficiency or use of renewable generation, that same \$0.032 of value is created. If we add additional externality costs – pollution from coal from mining to generation – this value could well be doubled. Add further to this the value provided to the utility system in avoided capacity and transmission requirements and the value of efficiency grows still higher. In fact, the combined value of benefits created by the elimination of electric usage may well be double the value the customer currently saves when investing in efficiency, distributed generation, or renewables at their facility. If this were applied as a subsidy incentive, it would then double the return on investment for these technologies, resulting in a far greater number and size of projects being implemented, more jobs created, and far more elimination of fossil fuel usage and its associated externality costs.

Nuclear plant expansion has been suggested as a means of reducing dependence on foreign oil and GHG emissions. Of course, since only 1 percent of all electricity generated in the U.S. is from oil, this offers little potential. As a major energy source used for electric generation, nuclear power is somewhat enigmatic. It scores well in several operating cost and environmental categories. For example, switching all power plants from coal to nuclear would drastically slash GHG emissions. The principle problems with nuclear are its high capital cost, its relatively low job creation rate, the potential for catastrophic accidents, and the unresolved challenge of dealing with nuclear fuel waste, both from a security and environmental safety perspective. The latter two are difficult to quantify, so it is challenging to assign an accurate externality price on nuclear power. Yet, they give us good cause to seek superior options for electric power production.

The heart of this challenge is to strike a balance between oil independence and environmental protection. We must unravel the paradox that increased electricity usage could reduce one source of externalities while increasing another. If electric technology is to serve as a replacement for oil usage or any other fossil fuel usage application, then the continued transformation toward safe, highly efficient, and renewable sources of electric generation, as well as the reduction of electricity requirement through intense user efficiency efforts, must be considered as part of the evaluation of the overall solution from a national perspective.

Efficiency Improvements in Buildings

While significantly reducing oil use in vehicles must be the major focus in eliminating our dependence on foreign oil and improving environmental protection, an equally important component of any serious effort to achieve these goals will be efficiency and energy source switching in buildings. While not a major oil consuming sector, residential and commercial buildings account for 40 percent of energy consumed in the U.S., and fossil fuels constitute 77 percent of that energy to heat, cool, illuminate, and operate those buildings. Hundreds of potential facility energy efficiency opportunities are available – turning lights off during unoccupied periods, daylighting, sophisticated energy management control systems, advanced heating and cooling technologies, complex internal heat pump designs, on-site power generation – many applications of which are presented in this book. Typical energy usage reductions achievable through efficiency improvements in buildings range from 10 to 50 percent. With sufficient investment in special designs or the use of renewables, fossil fuel use can be completely eliminated.

Government (or public) facilities and many institutional facilities have long investment horizons and will often be willing to invest in energy efficiency, distributed generation, and renewables projects that offer a 10-year simple payback or longer. This will typically generate energy usage reductions of 25 to 35 percent. However, with restrictions on capital budgets, these efficiency opportunities are often foregone. As discussed in Chapter 43, a viable option for these facilities is the use of energy savings performance contracting (ESPC). Under such an arrangement, an energy service company (ESCO) will develop, design, implement, and, in some cases, maintain a series of energy (and often water) conservation measures. The ESCO will either finance the project or arrange for financing and provide a guarantee that the savings will be realized. The savings generated from the project are sufficient to pay for the annual finance payments plus maintenance of the installed equipment so that no capital expenditure is required by the customer. At current interest rates, projects with simple paybacks of 15 years or longer can be financed over a 20-year term. In some cases, this increases the energy reduction potential for these facilities to 40 percent or more.

For commercial and industrial facilities, investment horizons are far shorter, typically requiring paybacks of 2 to 4 years. This limits the energy reduction potential to 10 to 20 percent. Unfortunately, programs like ESPC cannot be employed because accounting procedures in these sectors do not recognize this financing as off balance sheet, and hence it is treated as any other capital investment. There is still a wealth of potentially economical technology applications available. However, given the massive size of this energy consuming sector, externality charges to fossil fuel costs (including electricity generated from fossil fuels) or the use of subsidies for implementation of measures that eliminate their use could essentially double the rate of return on investment, making so many more projects economically viable and, in turn, causing a tremendous reduction in fuel usage and pollution.

Summary

It is entirely practical and feasible to dramatically limit the U.S. dependence on foreign oil within this decade and simultaneously reduce the amount of general pollution and GHG emissions produced from today's levels. This can and should be largely driven by an accounting system that blends regulations, externality cost charges, and subsidies in a logical manner. Discouraging oil and coal use, continued legislative increase in CAFE and other equipment and facility standards, expansion of biofuels, including the commercialization of cellulosic biofuels, and switching to LPG vehicles, NGV vehicles, hybrids, EVs, and PHEVs can move the needle close enough. The rest can be accomplished through facility fuel switches away from oil and coal to natural gas and renewable energy sources and intensified efficiency initiatives, many of which are presented in this book. Similarly, with a broader national policy and accounting system that targets all fossil fuel usage, it is possible to arrest the trend of growing GHG emissions, which are affecting climate change.

Through the Cash for Clunkers program, the U.S. was able to reduce oil use at a cost of $15 per gallon per year ($4,000 incentive for 277 gallons per year). With a broader, more well-developed and ongoing program, the needed public policy investment for such accomplishment could be reduced by one-third or one-half. Accordingly, with a subsidy of perhaps $10 per gallon per year to support mass transformation to one of the alternatives presented above, the U.S. can replace all foreign oil usage for an investment of $1.5 trillion. Assuming for simplicity a 10-year vehicle or facility efficiency measure service life, this investment would have to be renewed every 10 years or simply supported by $150 billion per year on an ongoing basis. At $3.00 per gallon, the externality cost avoidance of 10 million barrels per day (153 billion gallons per year) would be $460 billion per year, or $4.6 trillion over 10 years. This means that the return on a public policy investment of $1.5 trillion in subsides would be $4.6 trillion over 10 years. Even at a more conservative externality cost estimate of only $2 per gallon, the return over 10 years would be $3 trillion. This is still a winning investment for the U.S., as it would rid the use of foreign oil, reverse the trend of growing pollution and GHG emissions, create jobs, increase national security, and earn a good return for taxpayers.

There has been a long history of military conflict surrounding oil, particularly in the oil-rich Middle East. If we declared war on usage of foreign oil itself, the goals of this war would be better understood and the casualties would be oil cartel profits, not American lives. Along the way, we would make great progress in our war against climate change as well. Finally, this type of war would provide a healthy return on investment for the U.S. taxpayer through massive reductions in life-cycle externality costs. This could be accomplished by applying an externality tax on fossil fuels or a subsidy program for incentivizing their elimination while at the same time strengthening regulations. This presents the U.S. with a great financial investment opportunity.

Most of the American public is resigned to notions that we cannot eliminate our dependence on foreign oil, as it would destroy our economy and cause too much hardship, or that we have to wait decades for technology advancements for this to be possible. Many people also believe that there is little that can be done to more strongly protect the environment and combat climate change. The realization that the damages might come to be so severe that the time lost now can never be recovered (at least not without extraordinary cost and hardship) is almost too depressing to contemplate. Instead, many people are satisfied with the notion of tax cuts and lowering the price at the pump through any means, no matter how much they are mortgaging our future. What must be understood is that the very opposite approach – one that uses the combination of externalities tax, subsidies, and stronger regulations in pursuit of the goal of 1) eliminating foreign oil dependence and 2) reducing environmental externalities – will ultimately cost taxpayers less while sustainably increasing jobs, improving the economy and strengthening the security of our nation.

When the Soviet Union successfully launched Sputnik in 1957, the U.S. public became convinced that space travel was a reality and was motivated to compete and achieve a similar feat. This raised the bar of possibility, motivation, and commitment. When John F. Kennedy declared, in 1961, that the U.S. would send an American safely to the Moon before the end of the decade, he inspired a nation by giving it not only a goal, but belief that the goal was achievable. Had he suggested we would land on the moon within 30 or 40 years or perhaps get close to the moon, it would have had little impact on the hearts and minds of the people. We have become accustomed to similar deflating proclamations regarding foreign oil dependence and the environment. Yet, with similar belief and motivation as inspired by President Kennedy and achieved ahead of schedule in 1969, this great nation can fully sever its dependence on foreign oil and successfully protect the environment within a similar period. However, with each passing day of reluctance and inaction, precious time is lost that can only be made up with future sacrifice that will make the task at hand today seem easy in retrospect. It is my hope that the technologies and methodologies presented in this book will arm you with many of the tools and weapons needed in this noble war. The rest must come from our political will to accomplish the challenging but readily achievable and wonderfully beneficial objectives.

Preface to Second Edition

This second edition represents a major update to the first edition. Nearly every chapter has been amended to reflect the latest technology advancements and many changes and developments in the areas of utility and environmental regulation. While these will always be subject to change, fuel and utility rates were updated in the various examples to reflect current conditions.

In the technology section, new or modified technologies were included and equipment and system efficiency, performance, and pricing were updated to reflect the latest developments. Advancements and developing cycle modifications were included for reciprocating engines, combustion gas turbines, and combined cycles. Updates on various gasification methods were included in several chapters. The chapter on renewable and alternative power technologies has been updated with focus on advancements and market penetration in wind and solar power technologies.

The utility section was updated to reflect the current state of deregulation and market pricing, and includes a discussion on smart grid and other emerging programs. Updates to the Public Utility Regulatory Policies Act and the evolution of the Energy Policy Act of 2005 were provided along with updates on the independent system operator function and retail wheeling. Discussion on the extreme volatility in oil and gas prices and changes in electric utility rates and rebate programs since the first edition has been provided.

The environmental sections were revised to reflect the current emphasis on climate change and greenhouse gas emissions and carbon management. Carbon dioxide emissions data and climate change considerations were also added to the chapters on heating value and combustion of fuel. Other environmental updates include Clean Air Act amendments, regulations, and timelines and the latest developments in the use and control of refrigerants since the issuance of the Montreal Protocols.

The technologies addressed in the application sections were also updated to reflect latest developments, including improved efficiency and performance and cost data. The mechanical drive chapter discusses improvements and maturity of various technologies, notably electric motor ratings and variable speed controls. Cooling and refrigeration technologies were also updated, including changes in refrigerant types and phase-outs and overall performance.

Numerous examples were re-crafted to reflect updated equipment performance, energy rate pricing, and equipment. A discussion on the volatility of utility rates was added to the technical analysis chapter. Case studies were modified accordingly and a general update of currently available financing vehicles was included.

Acknowledgements to Second Edition

I would not have undertaken the task of producing this second edition without the commitment of my friend and colleague of more than two decades, Scott Silver, for without his talent and support I would not have completed it. Everlasting thanks to my friend and faithful companion through so many of my journeys and endeavors, John "Joker" O'Keefe. I am also very grateful for the assistance provided by Scott Hutchins, John Brown, Sarah Wade, and Steve Allenby. Thank you to my many colleagues who, over the past 30 years, have taught me, inspired me, supported me, and shared the burden of making a meaningful contribution to this field. Finally, thank you to all the readers who took the time to write me notes of support or who found use in the book to put into practice.

I have also included again the listing of acknowledgements to the first edition, as their contributions live on.

Acknowledgements to First Edition

In writing this book over the past decade, I sought to bring the collective knowledge of an industry to bear with solid engineering documentation covering a wide range of topics. Yet, I felt compelled to tell this story of energy, environmental, and cost efficiency with a single voice for consistency of purpose. In pursuing these objectives, I sought out leading experts in the field to review, edit, confirm, or challenge every chapter, formula, and statement. They were also an invaluable source of photographs and graphics and a wealth of performance data and technical descriptions. More than 100 manufacturers, vendors, public agencies, trade associations, utilities, engineering firms, and colleagues contributed to this effort, providing over 1,000 graphics and countless technical descriptions, edits, and revisions so that the story could be told with knowledge and fact beyond the capabilities of any single author. At the end of this long and difficult endeavor, my hope is that the reader will hear a single voice that really is a choir, with thousands of notes contributed by hundreds of individual voices.

I ask the reader to consider the source of each graphic and data table as a footnote and a part of the book's bibliography. I extend my appreciation to each of these contributing companies, agencies, associations, and individuals. In particular, I would like to acknowledge Gary Melickian and the American Gas Association and its member companies for their early support and technical assistance.

I warmly thank many of my colleagues for their editorial and technical review contributions, notably Paul Pimentel, Jerry Reilley, Scott Silver, and Phil Zacuto, who have been beacons of light in support of this effort over many years.

I acknowledge the contributions of several other individuals, including Pentti Aalto, Anthony Bobelis, Anthony Catner, Wasi Choundhury, Tim Costello, James Daley, Robert Dawson, Linda Factor, Manfred Grove, Kevin Harper, Robert Jorgenson, Fred Jones, Naveen Kapur, Jim Moore, Glenn Petty, Joe Singer, Steve Stultz, Sarah Wade, and Neil Zobler.

I thank Anne Turner for designing and executing the entire layout and for her artistic contributions, including numerous illustrations, Teri Sharpe for line editing, and John O'Keefe and Bryan Gianninoto for endless administrative support.

Finally, I thank my parents for their encouragement and my wife Lori and sons Brian and Adam for their support and patience.

This is at best a partial list and I hope that those I have omitted will forgive the oversight and know that all contributions and supporting efforts have been gratefully received.

Despite all care to avoid error, a work of this magnitude cannot be expected to be without flaw. I apologize for any such oversight or omission and will be indebted to those who discover mistakes and take the trouble to let me know so that they will be corrected at the earliest opportunity.

Neil Petchers

INTRODUCTION

While intended as a road map for overall energy resource optimization, the principal theme of this book is to evaluate energy and resource requirements interactively with an integrated approach to energy and cost-efficiency project development, seeking a match between power production and heating/cooling requirements. Therefore, there is a strong interrelationship between most of the chapters. The parts and sections are built upon each other in a sequential flow that weaves building blocks of information along the way into one overall fabric.

The book has been organized into four parts. Part 1 provides the basic building blocks, Part 2 sets the stage, Part 3 provides potential solutions, and Part 4 guides the reader through steps and best practices to choose and implement such solutions. While each chapter is a thread in this fabric, they have also been developed to stand on their own. The author hopes that much of the specific subject content will prove useful independently from the rest of the book. Hence, a balance was sought between the redundancy of fully developed, independent chapters with all necessary background provided in one place versus weaving cross-references throughout chapters in an integrated tapestry.

Part 1, Theory And Technology, provides a theoretical basis for understanding the interrelations of heat and power resources. It provides an introduction to basic heat and power thermodynamics and includes sections on heat and power generation technologies and equipment.

Section I, *Optimizing Heat and Power Resources*, presents thermodynamic theory on heat and power resources. It includes an introduction to various power cycles — simple, cogeneration, and combined cycles — and basic power cycle performance expressions. There is also a comparative discussion on localized and central station power generation and a brief overview on selection of power-generating systems.

Section II, *Thermal Technologies*, presents processes and equipment used to generate useful thermal energy streams, such as steam and hot water. It starts with background theory on fuel characteristics, the combustion process, and the properties and values of steam, and proceeds with chapters on the main thermal technologies. This includes components and systems used in the main boiler technologies — conventional fuel-fired firetube and watertube designs. Details are provided on systems designed for operation on renewable and waste recovery energy sources, and heat recovery heat exchangers used with reciprocating engines and gas turbines to generate hot water and steam.

Section III, *Prime Mover Technologies*, includes a series of chapters on the main classes of prime movers used to generate shaft power — reciprocating engines, combustion gas turbines, and steam turbines — and on combined and steam injection cycles that use recovered heat to augment power generating capacity and performance. The chapters focus on how shaft power is generated and made available as a driver for various applications, rather than on a particular application, such as electric power generation or mechanical drive service. Additional chapters are included on control technologies for prime mover operation and alternative power generation technologies — hydropower, wind power, solar photovoltaic, and hydrogen-powered fuel cell technologies.

Part 2, Operating Environment, describes the infrastructure in which the theories and technologies described in Part 1 must be applied. Having learned of the theory and available technologies, applications cannot be effectively devised, analyzed for cost-effectiveness, and implemented without knowledge of environmental factors and utility rate structures.

Section IV, *Environmental Considerations*, covers the regulatory status of air pollution programs and prescribes ways to permit projects and control emissions, with chapters on the national framework of air pollution and climate change regulatory programs, the permitting process, permitting strategies, and techniques for controlling air emissions. Information is also provided on environmental regulations for refrigerants, with emphasis on chlorofluorocarbon (CFC) and hydrochlorofluorocarbon (HCFC) phase-out and the use of current and evolving, more environmentally benign refrigerants.

Section V, *Utility Industry and Energy Rates*, presents extensive overviews of the gas and electric utility industry. It describes the role of the pipelines, merchants, and utilities in the evolving deregulated environment of the natural gas industry and parallel aspects of the electric utility industry, with treatment of utility integrated resource planning and interaction with non-utility generators. Detail is also included on utility rate structures — reasoning behind their construction and how they work and on utility bill analysis — to determine discrete and weighted average costs for operation on specific load profiles. Since energy rates vary so much over time and from location to location, the focus is on how pricing and utility structures work as opposed to current price levels. These must also be specifically evaluated for each location and application, with sensitivities developed to predict the range of anticipated or potential change.

Part 3, Applications, presents detail on a series of different types of applications and discusses how opportunities can be identified and successfully exploited. It builds on the understanding of the infrastructure and the technologies developed in the first two parts of the book. Whereas in Part 1, the thermal and prime power technologies were described generically, in this part they are combined with secondary technologies such as electric generators and mechanical drive equipment in specific site applications. Additional theories and technologies are introduced as they relate specifically to these applications.

Section VI, *Localized Electric Generation*, focuses on non-utility electric generation applications for commercial, industrial, and institutional facilities, as well as district systems and independent power production. This largely consists of applying traditional prime movers to electric generators to produce electricity in simple cycle, cogeneration cycle, combined cycle, and steam injection cycle applications, but also includes renewable and alternative power production technologies. Chapter topics include a basic introduction to electricity and electric generators and an extensive review of generator driver applications. It covers application and equipment selection processes and provides an example of a detailed electric cogeneration system feasibility study. Also included are chapters on generator switchgear, controls, grid interconnection, and safety considerations.

Section VII, *Mechanical Drive Services*, focuses on applying electric motor and prime mover drivers to mechanical equipment. An overview of mechanical drive applications is provided with detail on electric motors and prime movers as mechanical equipment drivers, with performance information and guidance for application-specific driver selection. Additional chapter topics include air (and gas) compressors, pumps, and fans, providing technology descriptions and detail on performance and application compatibility for different driver and equipment combinations.

Section VIII, *Refrigeration and Air Conditioning*, focuses on various space and process cooling applications. Theory in refrigeration cycles and psychrometrics is provided to support a more complete understanding of application requirements and options. Technical detail is provided on heat extraction — evaporators, chilled water systems, economizers, and thermal storage — and on heat rejection — condensers, cooling towers, heat pumps, and heat recovery. Additional chapter topics include vapor compression cycle, absorption cycle, and desiccant dehumidification system technologies and applications.

Part 4, Analysis and Implementation, includes only one section. Section IX, *Integrated Approach to Energy Resource Optimization Projects*, puts the information presented in the first three parts to practical use. Knowing the application options available and the theory and technology behind them, as well as the energy and environmental infrastructure in which they are to be applied, the next steps involve project development, implementation, and operation. The first chapter provides an overview of the development and implementation of multi-technology application projects using an integrated approach that considers a facility as a dynamic entity with interrelated systems. The next chapter provides detail on technical analysis for identifying project opportunities and analyzing their technical merit, with a step-by-step multi-phased development approach. Ensuing chapters include detail on financial analysis techniques to evaluate project financial performance potential and contract vehicles, and funding sources to secure and support project implementation. The final chapter covers project implementation and operation, with approaches to project design engineering, construction, and long-term operations, maintenance, and repair.

This book should not be considered an engineering reference manual, as much as a preliminary source of information and, hopefully, inspiration to pursue development of integrated energy and environmentally efficient projects that will provide solid financial returns. The author has chosen a presentation style and a "show and tell" format to add real-world perspective. For most topics, the text is heavily supplemented with graphic support from credible sources. The book includes in excess of 1,000 graphics, including photographs, cutaway drawings, layout schematics, performance curves, and data tables. When reading about a technology or application, one can find numerous photographs of equipment from a wide range of manufacturers. To facilitate a more in-depth understanding, components are featured both independently and with labeled cutaway and schematic drawings of entire systems. Building upon this, a wide range of performance information is provided based on manufacturer's data and contributions from various independent engineering sources.

Numerous examples are provided of actual field applications, with supporting documentation of system layouts and performance. Many comparative analyses are also provided showing both simplified and more complex examples of how equipment and systems are selected for various applications. A recurring warning to the reader is not to take the conclusions of these examples as definitive and not to convert them into firmly preconceived notions. Applications must be evaluated against actual site conditions and with consideration of local infrastructure, energy rates, and environmental regulations. The past decade has evidenced great change and volatility of energy rates and regulations, and no less should be expected going forward. Technologies and their performance also continue to change and evolve. Hence, the purpose of these examples is to show different ways in which to analyze opportunities; it is the approach that is paramount, not the conclusions. Equipped with this information, the reader must then evaluate actual situations and prevailing conditions and solicit more detailed design information to make the necessary adjustments for each potential application under consideration.

PART 1
Theory and Technology

Section I
Optimizing Heat and Power Resources

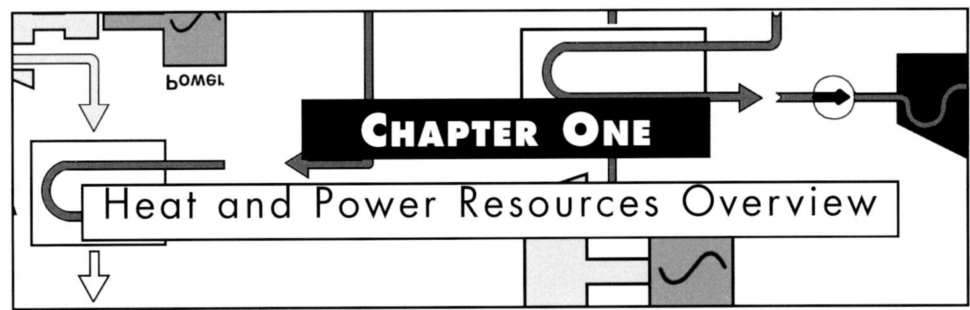

Chapter One

Heat and Power Resources Overview

Virtually every facility requires energy conversion for both power and heat. Power may be purchased from an electric utility or private provider, or it may be produced on site. Power is used as electricity for lights and computers and to drive equipment via electric motors. It is also used as mechanical energy in the form of a rotating shaft that directly drives equipment. Heat – or thermal energy – is usually produced on site from purchased fuel through various types of energy conversion devices. Heat is used to raise steam, hot water, or hot air for space heating or process use, or to produce a cooling effect through certain heat-driven cycles.

Power is generally produced by application of prime movers, either on site or at centralized electric generation plants. Prime movers are devices that convert fuel or heat energy into mechanical energy, which in turn can be used to drive virtually any type of shaft-powered equipment, including electric generators and motor vehicles. Due to the laws of thermodynamics, heat is produced as a necessary by-product of power production.

Much of the technology discussed in this book involves three major types of prime movers: reciprocating engines, combustion gas turbine engines, and steam turbine engines. Most of the applications in this book involve strategic deployment of prime mover and certain heat-cycle technologies in commercial, industrial, and institutional facilities. This chapter introduces a number of terms used to describe and compare the application of these technologies.

Facilities rarely have a consistent requirement for power and heat. Generally, these requirements vary based on the time of use or outside ambient conditions. The portion of a facility's power or heat requirements that is constant is referred to as baseload. The portion that varies is referred to as intermittent load. Maximum intermittent requirements are referred to as the peak load.

If thermal requirements are not considered, baseload power requirements are usually met most economically through purchased power from central utility power plants rather than localized on-site production. Advantages associated with centralized power production include: economy of scale, preferential fuel purchase opportunity, lower staffing levels per unit output, diversity, and reserve capacity. These advantages are usually sufficient to overcome inherent disadvantages of centralized power production, such as system efficiency losses associated with power transmission and distribution, as well as an assortment of regulatory obligations.

Intermittent and peak load requirements, on the other hand, are usually served by centralized utility systems with lower economic efficiency. In some cases, these requirements can be served more economically by strategic application of on-site power production technologies. Examples are on-site peak shaving electric generation, which is the on-site production of electricity during peak usage and/or cost periods, and various types of mechanical drive services.

If thermal energy requirements are taken into consideration, on-site power production has a significant thermodynamic efficiency advantage over centralized power production, because heat energy rejected from the power production process can be used. Centralized power plants usually have no use for this heat energy and must liberate it to the environment at an economic loss. When a facility can recover and use this heat energy, the thermodynamic efficiency advantage translates into an economic advantage that may exceed the economic advantages of centralized power production.

Comparison of life-cycle costs determines the degree to which it is economical to produce shaft power on site, rather than purchase power from an electric utility or private power producer. Such decisions involve analysis of an entire facility's energy usage characteristics, including concurrent requirements for both power and heat, since on-site prime movers can provide both.

The life-cycle cost elements of an on-site prime mover are primarily capital, fuel, and operations and maintenance costs, which are also the primary constituents of electric utility and other centralized power producer costs. Electric utility rates assign different portions of these capital and operational costs to different time periods based on the utility's cost to serve. Rate designs, which often include demand charges and seasonal and time-of-use rates, send price signals that influence consumer behavior. The relationship between these price signals and on-site energy load characteristics will largely determine which portion of electricity requirements can be provided more economically by on-site prime movers than by electricity purchased from a utility or other centralized source.

Additionally, the marginal, or incremental, cost of utility power production will largely determine whether it is economical to produce more power than is required on site and export the excess to other sellers or users.

Investment in an on-site prime mover shifts many additional cost factors onto the individual facility. These costs are capital, fuel procurement, and operation and maintenance, as well as costs associated with reserve capacity, emissions control, space considerations, and insurance. The potential payoff for absorbing these added cost factors is lower operating costs and increased economic performance.

Prime Mover Cycle Terminology

There are numerous terms used to describe application of prime mover cycles. These include topping, bottoming, simple, combined, and cogeneration cycles. Definitions for these terms are flexible. Commonly, topping and bottoming are applied to cycles to indicate the stage at which an energy stream is used to produce power. Alternatively, the term base unit may be used to indicate the primary system to which a topping or bottoming cycle is applied.

The **topping cycle** may be characterized as one that uses a high temperature working fluid to generate power followed by use of recovered heat. In contrast, in a **bottoming cycle,** the working fluid is used as a high-temperature heat source before being used for power generation.

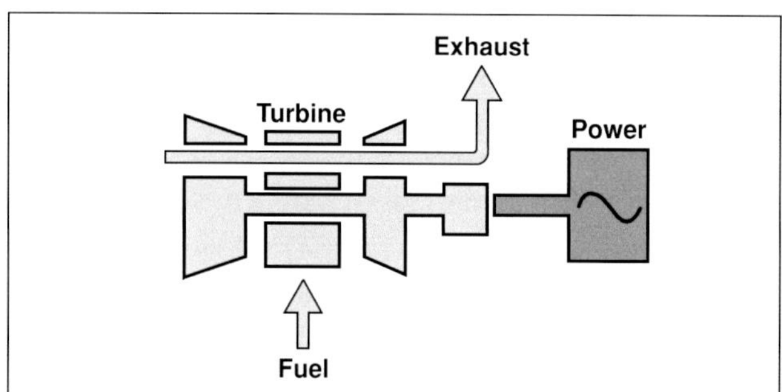

Fig. 1-1 Application of Simple-Cycle Gas Turbine Base Unit.
Source: ABB

Under these definitions, topping cycles have the capacity to independently deliver mechanical or electrical energy from the conversion of fuel or heat energy. Bottoming cycles cannot operate without a preceding energy conversion cycle or process. Bottoming cycles tend to be physically large and relatively expensive due to the low quality of the energy input.

In the context of large power generation plants, the terms topping and bottoming cycle may be applied to the systems that are added to base systems in order to enhance overall thermal efficiency. The base system may be a pre-existing system or a new one, and it normally is the major power producer. For example, if a gas turbine is fitted with a heat recovery boiler that supplies a steam turbine, then the steam turbine plant becomes a bottoming unit on the base gas turbine. Figure 1-1 illustrates a simple-cycle gas turbine base unit. Figure 1-2 illustrates the addition of a heat recovery system to this base unit.

Simple cycle refers to the conventional application of a single prime mover cycle. As shown in Figure 1-3, a **combined cycle**, as the name implies, is the sequential linking of any topping and bottoming cycle, or two simple cycles. The classic combined cycle is a gas turbine in conjunction with a steam turbine. The gas turbine generates shaft power at the upper range of the energy stream. Its exhaust heat is converted to steam in a heat recovery steam generator, and then passed through a steam turbine to generate additional power. The entire plant would be referred to as a combined-cycle plant.

From the power plant perspective, a reciprocating engine applied in mechanical drive or power generation service may be considered as a base unit since it would be the major power producer. A combustion gas turbine might be the base unit or the topping cycle, and the steam turbine might be the base unit or the bottoming cycle. Where the base demand is for process steam, conventional steam cycles with extraction steam turbines might be considered as a topping cycle.

Figure 1-4 shows several variations on application concepts for repowering a power plant with 150 megawatt (MW) gas turbines. In these examples, approximately

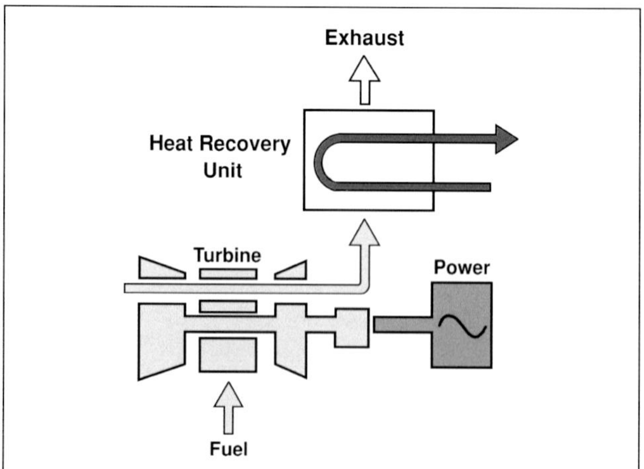

Fig. 1-2 Application of Heat Recovery Unit to Gas Turbine Base Unit.
Source: ABB

75 MW can be recovered from its exhaust energy. The basic combined-cycle unit provides a total output of 225 MW and offers the highest thermal fuel efficiency of the various options. The fully fired boiler concept is sometimes referred to as **hot windbox refiring**. From the power plant perspective, it is considered a topping cycle in that the gas turbine provides only 150 MW of a total plant output of 600 MW. The gas turbine can also be applied in a topping cycle as shown in the feedwater heat exchanger repowering, which can be well utilized in a wide plant capacity range. Note that plant efficiency increases as the repowered plant size becomes smaller relative to the capacity of the gas turbine. Figure 1-5 shows the potential perfor-mance achieved with each of these options as a function of repowered plant output.

Cogeneration is the sequential use of fuel energy to produce more than one finished energy product, such as electric power, steam, refrigeration, thermal drying, air heating, or a host of others. While two finished products can be made by splitting the output of a single-boiler steam supply between a steam turbine power cycle and a heating application, this is not cogeneration. What distinguishes cogeneration and the thermodynamic efficiency benefits it produces is the operative concept of "sequential use" and production of both power and usable thermal energy. Cogeneration may be applied to both simple and combined cycles.

Heat recovery turns a relatively inefficient simple-cycle power generation process into a more efficient cogeneration or combined-cycle process. Heat recovery is the effective capture and use of heat rejected from the power cycles. Rejected heat is the energy associated with streams of air, exhaust gasses, and liquids that exit the system and enter the environment as waste products.

As shown in Figure 1-2, heat recovery is applied to the simple-cycle gas turbine for the purpose of sequentially providing thermal energy to a process, thereby transforming that system into a cogeneration cycle. As shown in Figure 1-3, heat recovery is applied to a simple-cycle gas turbine for the purpose of powering a steam turbine, thereby transforming the system into a combined-cycle system. Figure 1-6 shows the application of heat recovery twice. Heat is first recovered from the simple-cycle gas turbine, transforming the system into a combined cycle. Heat is then recovered again from the steam-turbine cycle and used for a process application, transforming the entire system into a cogeneration combined cycle.

Regulatory Terminology Applied to Prime Mover Cycles

To establish federal regulations and Qualifying Facility (QF) cogeneration system efficiency standards, the Federal Energy Regulatory Commission (FERC) defined cogeneration as "the combined production of electric power and useful thermal energy by sequential use of energy from one source of fuel." As defined by FERC, a topping cycle first uses thermal energy to produce electricity, and then uses the remaining energy for thermal process. In a bottoming cycle, the process is reversed.

Figures 1-7 through 1-11 are diagrammatic examples of cogeneration topping, bottoming, and combined cycles, consistent with commonly used regulatory definitions.

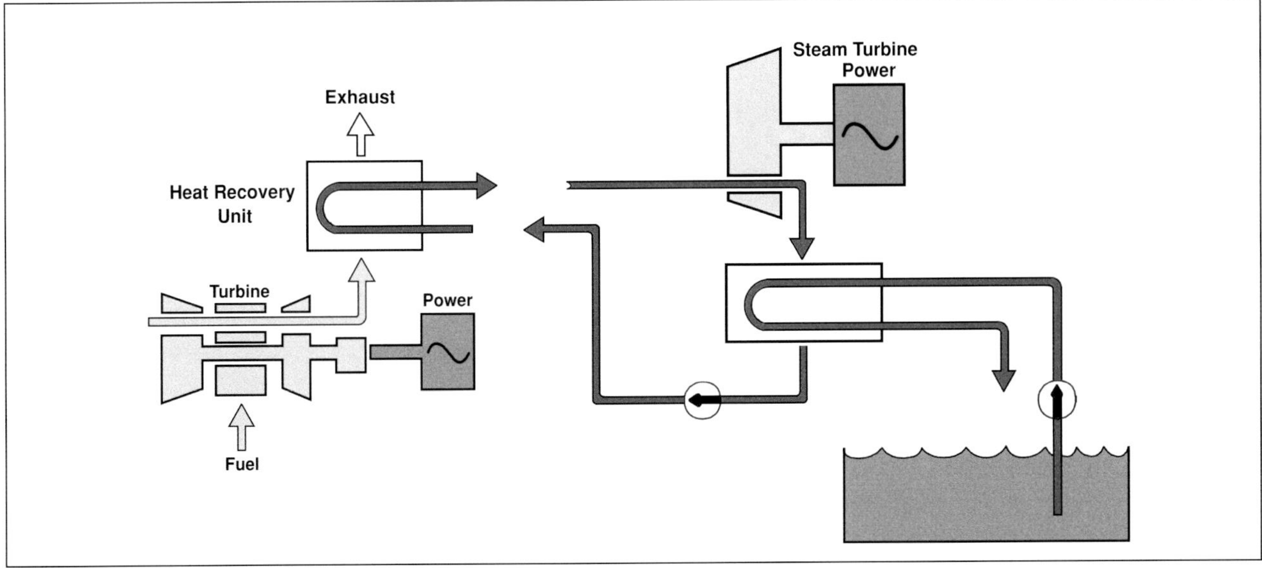

Fig. 1-3 Combined-Cycle System Featuring Gas Turbine, Heat Recovery Unit and Steam Turbine. Source: ABB

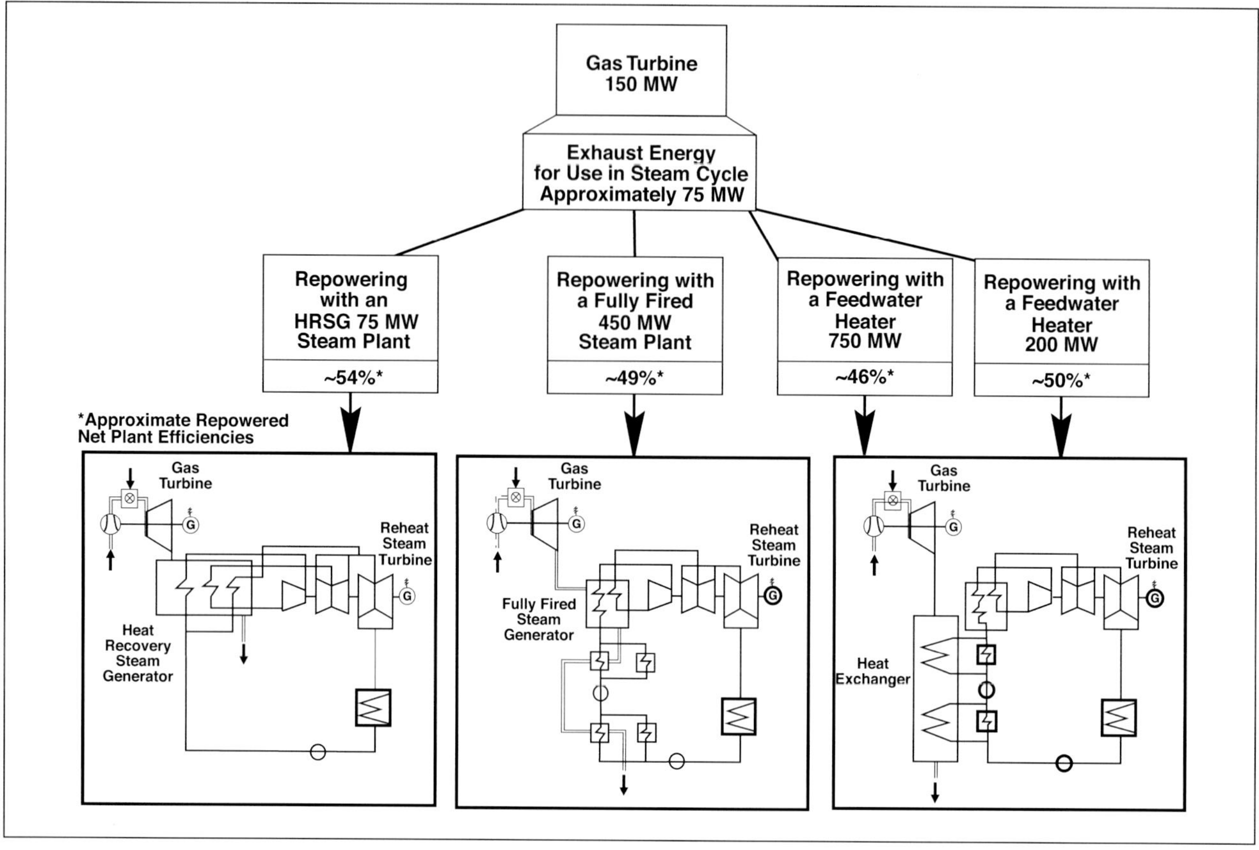

Fig. 1-4 Various Repowering Options Featuring Addition of 150 MW Gas Turbine. Source: Siemens Power Corp.

SUMMARY

Facilities have numerous options for meeting their power and heat resource requirements. Electricity will most commonly be purchased from electric utilities or other centralized power producers, which will employ simple- or combined-cycle systems to generate the power. Facilities may also employ simple or combined power cycles to generate their own power in the form of electricity or direct shaft power output, thereby reducing or eliminating their purchase of electricity. Facilities may also employ cogeneration cycles, which sequentially serve both power and heat requirements.

To determine if it is economical to apply prime mover technology on site, a facility should perform a life-cycle cost-benefit analysis. Terms such as thermal efficiency, fuel rate, fuel credit, net fuel rate, and fuel and cost chargeable-to-power are useful in this pursuit. Still, the repeated statement that mechanical or electrical energy is more valuable than heat energy must be considered within the context of available market alternatives. The relative values of heat and power are not fixed, but are ever changing along with the energy market and the technologies available.

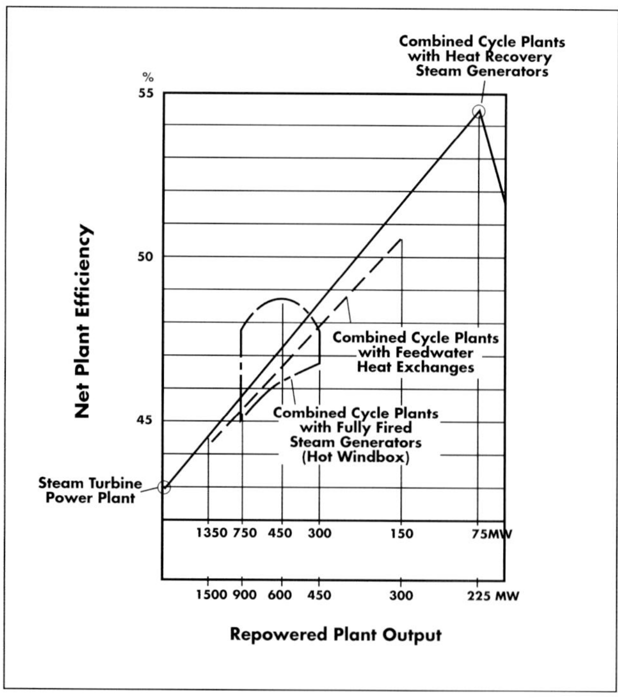

Fig. 1-5 Efficiency Range of Repowered Plant with 150 MW Gas Turbine as a Function of Repowered Plant Output. Source: Siemens Power Corp.

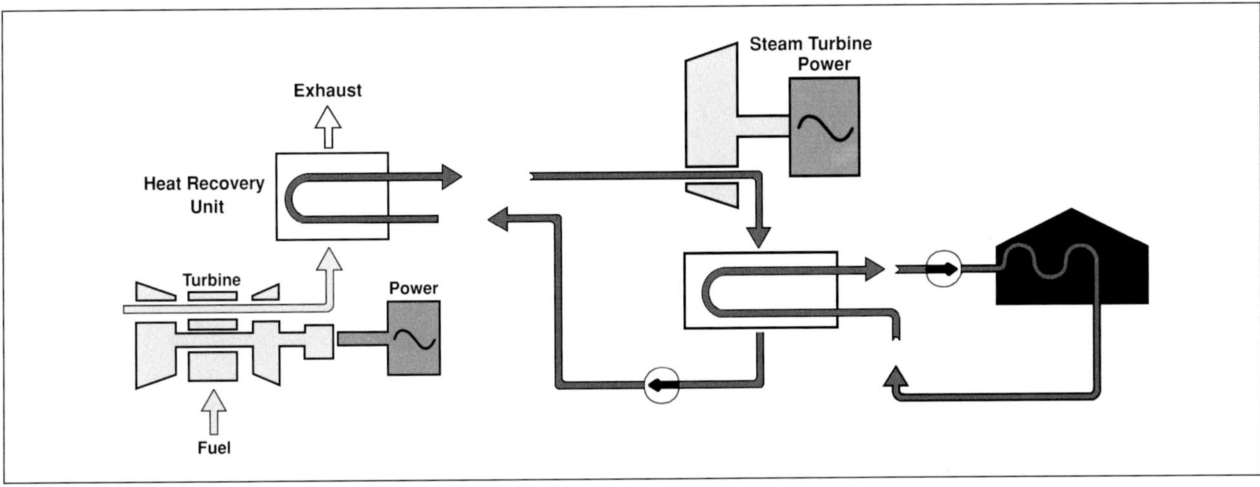

Fig. 1-6 Cogeneration Combined Cycle with Heat Recovered from Gas Turbine and Steam Turbine Cycles. Source: ABB

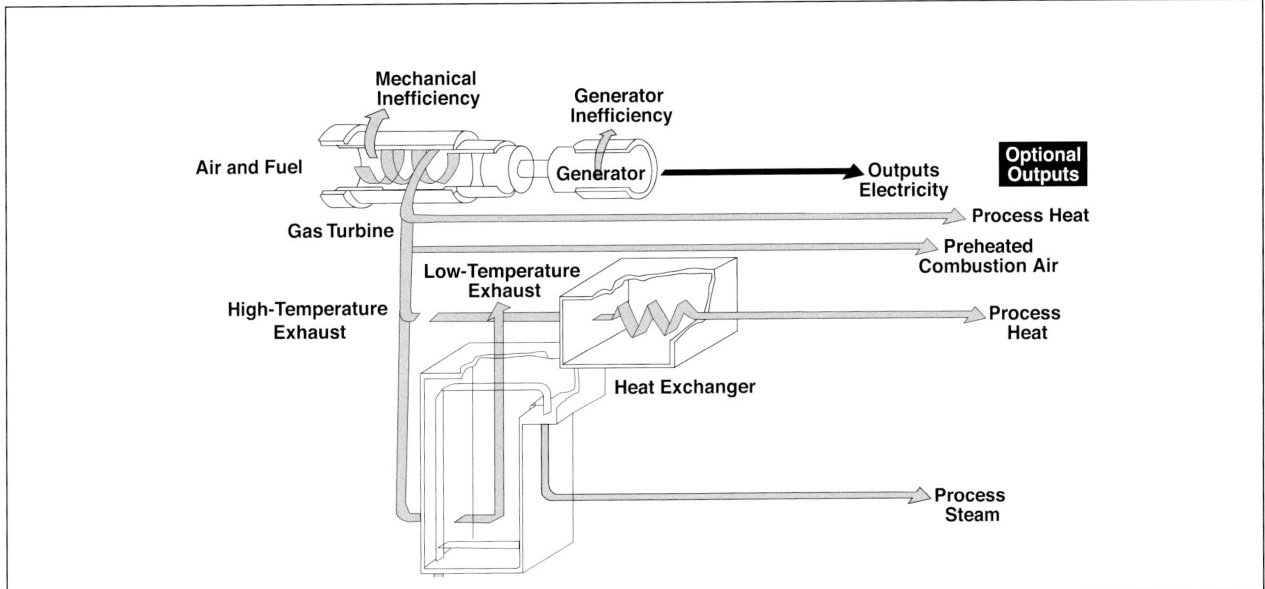

Fig. 1-7 Gas-Turbine Topping Cycle. Source: U.S. DoE

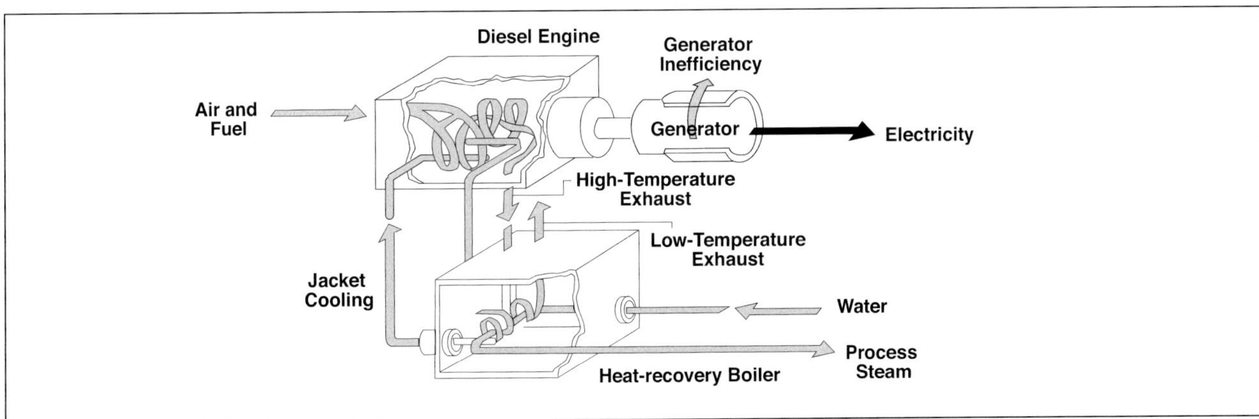

Fig. 1-8 Reciprocating Engine Topping Cycle. Source: U.S. DoE

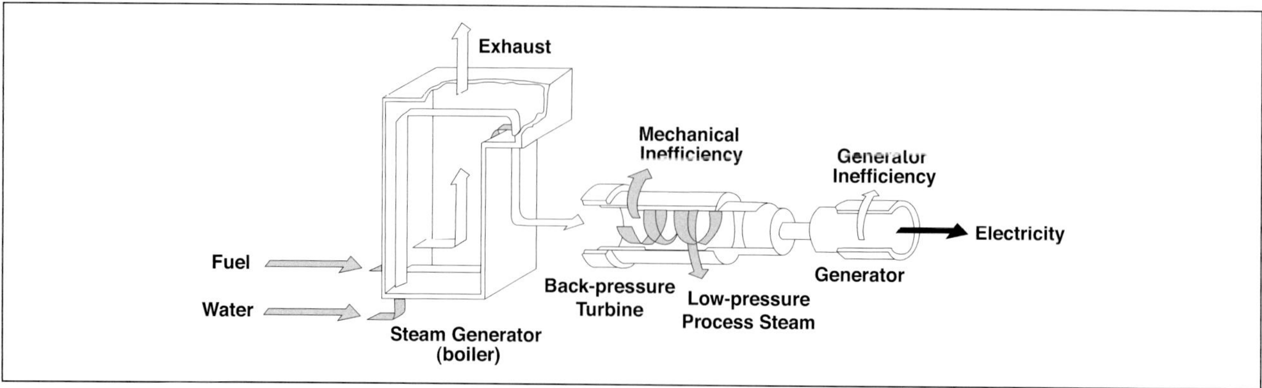

Fig. 1-9 Steam-Turbine Topping Cycle. Source: U.S. DoE

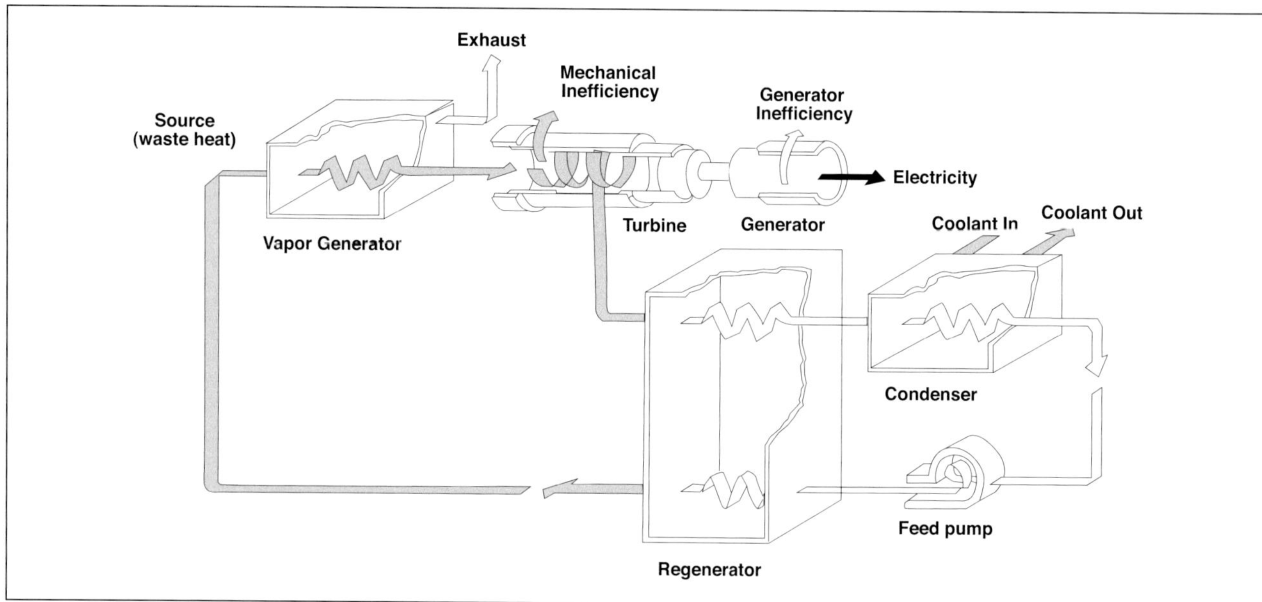

Fig. 1-10 Steam Turbine Rankine Bottoming Cycle. Source: U.S. DoE

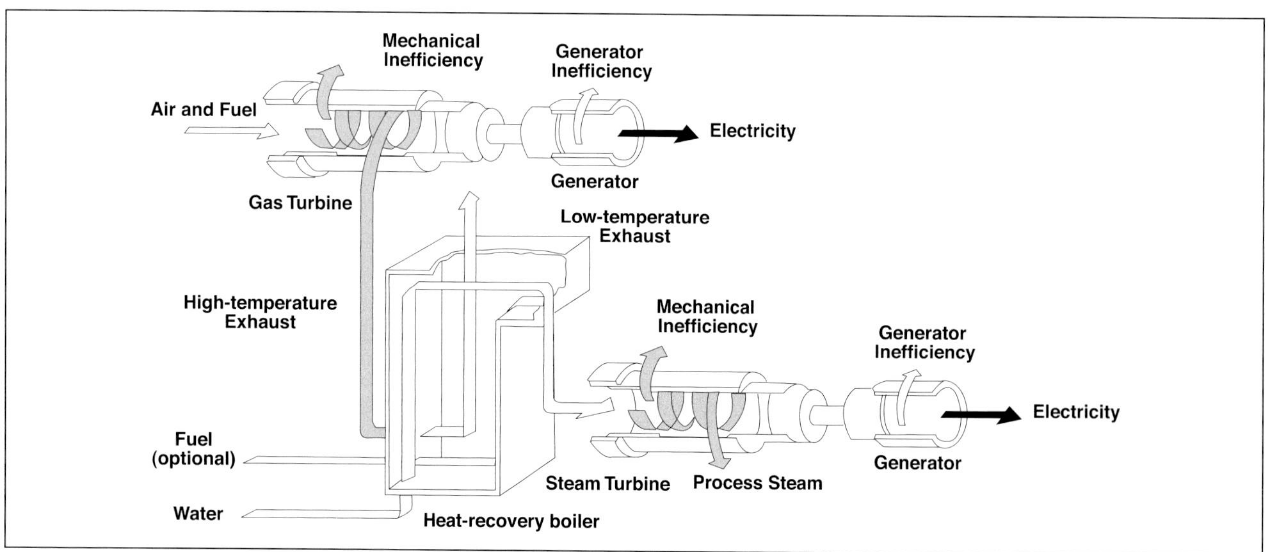

Fig. 1-11 Combined-Cycle. Source: U.S. DoE

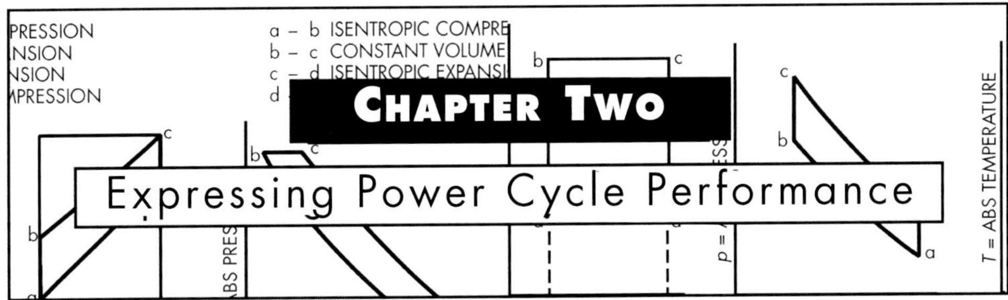

Chapter Two

Expressing Power Cycle Performance

The thermodynamic and economic performance of heat and power systems can be evaluated and expressed in many ways. This chapter presents some of the concepts and equations that may be used to determine the various performance characteristics of systems featuring simple cycles, cogeneration cycles, and combined cycles. General definitions are first provided, followed by equations, presented in generic format. These are followed by examples featuring commonly used power and thermal energy units.

Energy

There are two general types of energy: kinetic and potential. **Kinetic energy (KE)** is the energy of a given mass of material due to its motion relative to another body. **Potential energy (PE)** is the energy of a given mass of material as a result of the material's position in a force field. Generally, potential forms can be easily converted into kinetic forms and vice versa.

Energy can be transformed from one form to another. For example, a rock resting on the top of a hill has the potential to roll down, or fuel has the potential to be combusted and liberate heat energy. The more common classifications of energy include: mechanical, electrical, electromagnetic, chemical, nuclear, gravitational, and thermal.

Thermal energy is associated with atomic and molecular vibration. It is considered a basic energy form because all other energy forms can be completely converted into thermal energy. But the second law of thermodynamics limits conversion of thermal energy into other forms.

Enthalpy (H or h) is the thermodynamic property defined as:

$$H = U + PV \qquad (2\text{-}1)$$

Where:
U = Internal energy of system
P = System pressure
V = System volume

Enthalpy is a general measure of the internally stored energy per unit mass. Commonly used units for expressing enthalpy are Btu/lbm and kJ/kg. **Internal energy (U)** includes all forms of energy of a given system, except its gross kinetic and potential energy. It is associated with the thermodynamic state of the system.

Two key properties that are used to measure energy levels are temperature and pressure:

Temperature is a property that is the measure of the average kinetic energy possessed by the molecules of a substance. The higher the temperature, the greater the kinetic energy or molecular activity of the substance. The common scales of temperature are called the Fahrenheit (F) and Celsius or Centigrade (C) temperatures. These are defined by using the ice point and the boiling point of water at atmospheric pressure at sea level. The ice point is 32°F (0°C) and the boiling point is 212°F (100°C). Fahrenheit temperature can be determined from Centigrade temperature as follows: F = (C x 9/5) + 32. Centigrade temperature can be determined from Fahrenheit temperature as follows: C = (F – 32) x 5/9.

The scales of absolute temperature are called the Rankine (R) and Kelvin (K) temperature scales. These temperature scales are commonly used with steam applications. Both of these temperature scales use absolute zero as a point of origin. Since absolute zero is measured as –459.59°F, Rankine temperature can be determined by adding 459.59 to the Fahrenheit temperature. Similarly, since absolute zero is measured as –273.15°C, Kelvin temperature can be determined by adding 273.15 to the Celsius temperature.

Pressure (p) is the force per unit area exerted on or by a fluid. Pressure, expressed in pounds per square inch (psi), Pascal, or bar, is typically expressed as either absolute pressure, e.g., psi absolute (lbf/in^2 abs, or psia), or gauge pressure, e.g., psi gauge (lbf/in^2 g or psig).

- **Absolute pressure (psia)**, which is generally used in steam tables and most fluid and thermodynamic equations, is the true force per unit of area expressed as pounds per square inch exerted by a fluid on the wall of the vessel containing it. Standard atmospheric pressure is 14.696 lbf/in^2 abs, or 29.92 in. of mercury atmospheric (in. Hg atm) at sea level, or 406.8 inches of water (in. wg). In SI units, standard atmospheric pressure is 1.013 Bar, 101,325 Pascal (P), or 76.0 centimeters of mercury (cm HgA).
- **Gauge pressure (psig)** is the difference between absolute pressure of a fluid and ambient atmospheric pres-

sure. Since atmospheric pressure at sea level is about 14.7 psia (101,325 P), absolute or true pressure is determined, using approximate values, simply by adding 14.7 (101,325) to gauge pressure.

- **Vacuum, or negative gauge pressure,** is pressure below atmospheric pressure.

HEAT AND WORK

In more general terms, all energy in a thermodynamic system can be classified as either heat or work.

Heat is thermal energy that is transferred across the boundary of systems with differing temperatures, always in the direction of the lower temperature. Heat transfer occurs when two adjacent bodies of mass are not in equilibrium due to a difference in temperature. Heat is commonly expressed in calories or British thermal units.

- **British thermal unit (Btu)** is the amount of energy required to raise the temperature of 1 lbm of water by 1.0°F, standardized at 60°F (i.e., from 59.5°F to 60.5°F).
- **Calorie (c)** is the amount of heat necessary to raise the temperature of 1 gram of water from 14.5°C to 15.5°C.

Other commonly used heat units, which are defined below under the definition of work, are the Joule (J) and the Watt-hour (Wh). In SI units, kilo (k) and mega (M) are used to indicate 1,000 and 1,000,000 units, respectively (e.g., kcal, kJ, kWh, MJ, and MWh). Correspondingly, in English units, M and MM are used to indicate 1,000 and 1,000,000 units, respectively (e.g., MBtu and MMBtu).

Work (w), or mechanical energy, is done when a force acts through a distance. Work is the product of force and the displacement along the line of force. Doing work requires the expenditure of energy. Ft-lbf, Joule (J), horsepower-hour (hp-h), and kWh are the common measures of work. A Joule is the amount of energy equal to the work done by a force of 1 Newton (N) when the point at which the force is applied is displaced 1 meter in the direction of the force. This is also known as 1 Newton-meter of energy.

The following terms are useful in more fully understanding the concept of work.

Force (F) is the action that will cause acceleration of a mass. Change in velocity of an object is caused by force. If no force acts on an object, it moves at constant velocity. Newton's first law of motion states that a body at rest will stay at rest and a body in uniform motion will continue its uniform motion unless acted upon by a force. Force, expressed in pound-force (lbf) or in Newton (N), is the product of the **mass** of an object (m) and its **acceleration** (a) caused by that force. Mass, expressed in lbm slugs, or kilograms (kg), is a measure of the quantity of matter of which an object is composed. Acceleration, expressed in feet per second per second (ft/s^2) or in meters per second per second (m/s^2), is a measure of the time rate of change of velocity. Thus:

$$F = ma \tag{2-2}$$

Torque is the force applied at a distance from an axis of rotation and is expressed in lbf-ft or Newton-meter (N-m). It is the product of, for example, the force applied to a lever, or crank arm, and the perpendicular distance from the line of action of the force to the axis of rotation. In rotating machines, the forces required to accelerate (or decelerate) the speed of rotation add up to a torque that is proportional to the angular acceleration times the moment of inertia of the machine.

Power (P) is the rate of doing work and is given by the formula:

$$P = \frac{dE}{dt} \tag{2-3}$$

where power (P) is equal to the rate of energy expenditure (dE) over a given time interval (dt). Thus, power refers to the rate of mechanical energy expenditure over a given time interval or the rate of doing work.

In rotating machines, power is equal to torque times the speed of rotation. Therefore, if speed is held constant, torque and power are proportional. Commonly used units for expressing power are Btu per hour (Btu/h), Joule per second (J/s), ft-lbf per minute (ft-lbf/m), meter-kg per minute (m-kg/m), horsepower (hp), and Watt (W).

- **Horsepower (hp)** is an English system unit based on the power needed to raise a weight of 550 lbm through a height of 1 foot in 1 second (550 ft-lbm of energy per second) and was originally derived from the estimated power of one horse pulling a load. One hp is approximately equal to 2,545 Btu/h and 745.7 Watt. **Brake power** is a measure of the power generated by a prime mover. It can be measured at the crankshaft or the flywheel. The term brake refers to a mechanical arrangement used to measure the output torque, which, when multiplied by rotational speed, yields the power.

- **Watt (W)** is the unit of power equal to 1 J/s and approximately 3.413 Btu/h. 1 kilowatt (kW) is approximately equal to 1.34 hp.

- **Metric horsepower (PS)** is the power that raises a mass of 75 kg through a height of 1 meter in 1 second. One PS is approximately equal to 0.9863 hp and 735.5 Watt.

Given that power is the rate of doing work, applying a unit of time to a unit of power yields a measure of work. Hence, hp-h or Wh are the common work units. Given that power already has a unit of time, work expressions can be reduced to basic energy units such as Btu, calorie, or Joule. In many thermodynamic applications, Btu is considered to be a more convenient English system unit than ft-lbf. In terms of potential work energy, it is defined by the relationship 1 Btu = 778.16 ft-lbf. One hp is approximately equal to 2,545 Btu. The Joule, which may be considered the mechanical equivalent to heat, is equal to 4.1855 calories.

Power Cycles

Energy cycles are a series of thermodynamic processes during which the working fluid undergoes changes involving energy transitions and is then returned to its original state. In this process, the working fluid undergoes changes involving pressure, temperature, and energy levels while producing a usable transfer of energy.

The purpose of any practical thermodynamic cycle is to convert energy from one form to another more useful form. Thus, the practical goals are either to convert heat into work or, in the reverse, to use work to remove heat from a cold to hot region. The classic example used to demonstrate thermodynamic cycles is the **heat engine**. In the heat engine, only heat and work flow across the operating system's boundaries. The engine can deliver work to external devices or receive work from an external device and cause heat to flow from a low temperature level to a high temperature level.

Power cycles are processes in which heat energy is converted to work energy. During these processes, the energy of a fuel is converted into heat energy, which in turn is used to produce electrical or mechanical energy in the form of shaft power. The processes in the cycle are governed by the principle of the first law of thermodynamics, also known as the law of the conservation of energy.

First Law of Thermodynamics

The first law of thermodynamics, or conservation of energy, states that energy can be neither created nor destroyed, but only converted from one form to another. The motion of the molecules causes thermal or internal energy embodied within a system. Transient forms of energy include heat and work.

The first law of thermodynamics for a system undergoing a cycle is:

$$\int \delta Q = \int \delta W \tag{2-4}$$

For the non-cycle process, the law is:

$$\delta Q = dE + \delta W \tag{2-4a}$$

Where:

Q = The heat transferred to the system during the process
W = The work done by the system during the process
E = Total energy of the system (internal, kinetic, and potential) in the given state (this energy may be associated with the motion and position of the molecules, with the structure of the atoms, with chemical reactions, with gravity, or with any of a number of other forms of interaction)

Second Law of Thermodynamics

The second law of thermodynamics states that energy always seeks a lower level, or, in a manner of speaking, only runs downhill. Energy is only useful when it moves through a device from a higher level to a lower level. The level of obtainable benefit is proportional to the decrease in level that is available. The decrease in level is denoted by an increase in molecular disorder, or entropy (s).

Entropy is the thermodynamic property of the system held constant in a reversible adiabatic process. The entropy is a useful parameter in accounting of energy conversions that result in production of work as well as energy unavailable to do work (i.e., heat loss from the system and loss due to internal irreversibilities). It measures the relative molecular disorder of a given system. Entropy can be expressed as:

$$ds = \left(\frac{\delta Q}{T}\right) \tag{2-5}$$

Where:

s = Entropy
Q = Heat
T = Temperature at the boundary

Commonly used units for expressing entropy are Btu/lbm per °F and kJ/kg per °K.

This increase in entropy is unavoidable because energy never develops enough work to restore itself to the original level. The conversion of energy is, therefore, never completely reversible, hence, there are no perpetual motion machines.

The second law of thermodynamics can be stated in a number of equivalent ways. One statement is: no heat engine, either actual or ideal, when operating in a cycle, can convert all the heat energy supplied to it into work. Some of the heat energy must be transferred to a heat sink at a temperature lower than the temperature at which the heat energy is supplied. A consequence of the second law is that for any actual process,

$$ds_{system} \geq 0 \qquad (2\text{-}6)$$

Where:
s = Entropy

The principle of the increase of entropy, based on the second law of thermodynamics, is that the only processes that can occur are those in which the net change in the entropy of system plus the surrounding region increases. Therefore,

$$d_{system} + d_{surrounding} \geq 0 \qquad (2\text{-}7)$$

The important concepts operating in a thermo-dynamic system are then energy and entropy. A unit of energy represents a certain potential to do work. The entropy change of the system is a measure of the irreversibility of that process and the degree to which energy has not been made available to perform work.

SUMMARY OF FIRST AND SECOND LAWS

From the second law, it follows that with each process, interaction, reaction, or exchange of a stream of energy, some of its potential to do work is lost due to process irreversibilities, characterized by an increase in molecular disorder (increased entropy). In practice, in addition to these losses, processes may experience energetic losses, which are essentially heat losses (i.e., radiation, conduction, and convection).

When a fuel is combusted, it automatically loses energy. From the moment heat is collected on a solar plate, or rushing water is used to generate power, the stream of energy continues to decline in its total useful value to do work. The first law shows that on a global level, no energy is lost. It is only lost from within a given system to another. The second law states that even on a global level, availability to do work is lost, since entropy increases throughout the universe with no compensation. Whereas work and all other forms of energy can be wholly converted to heat, the converse is not true.

When these concepts are applied to a heat engine, there is a range of possibilities. At one end of the spectrum, a large portion of the input energy can be converted to prime power (useful shaft work) and most of the rest can be converted to useful thermal energy. At the other end of the spectrum, a small fraction of the input may be converted to prime power and excess energy may be a costly nuisance requiring rejection through the use of additional energy in such forms as pumps, fans, and cooling equipment. From an economic perspective, the same quantity of energy resulting from the same process can have a positive value on a cold day as a heat source and a negative value on a hot day as an excess heat load.

Power producing equipment (i.e., prime movers) operate on several different cycles. Generally, reciprocating engines operate on either the Otto cycle or the Diesel cycle, combustion turbines operate on the Brayton cycle, and steam turbines operate on the Rankine cycle.

In each of these power cycles, the initial and final states of the system are identical. At the end of the cycle, all of the properties have the same value they had at the beginning of the cycle, except that heat has been added, heat has been rejected, and work has been done. An analysis of the power cycle involves an accounting of all of the energy exchanges occurring at each of the processes, so that the sum of all energy inputs equals the sum of all outputs.

POWER CYCLE THERMAL EFFICIENCY

Efficiency is a general concept used to describe the effectiveness of energy conversion from one state or form to another. Thermodynamic efficiency and coefficient of performance (COP) are terms used for expressing efficiency for devices that operate in cycles or for individual system components that operate in processes. The term COP is usually reserved for refrigeration systems. The thermal efficiency of a cycle is the ratio of output energy to input energy. In a heat engine, the desired form of energy is the work; the form of energy that costs money is heat from the high-temperature source (directly or indirectly, the cost of the fuel).

CARNOT CYCLE

The **Carnot cycle** is of particular importance because its thermal efficiency represents the maximum value obtainable for any heat engine. The Carnot cycle represents a theoretical efficiency that serves as a standard for cycle efficiencies. It is based on the concept that the thermal efficiency of a reversible engine is a function solely of the upper and lower temperatures of the cycle and not a function of the working substances.

The Carnot cycle for both vapor and gas cycles typifies ideal performance for power producing cycles. In the Carnot cycle, heat is taken from an infinite reservoir, at temperature T_H, isothermally (without temperature change) and reversibly. The energy received produces work by expanding a working fluid reversibly and adiabatically in an ideal frictionless engine. During expansion, the engine produces a net work output. The working fluid, at temperature T_L, rejects heat at constant temperature reversibly and isothermally to an infinite sink and is then adiabatically and reversibly compressed to its initial state.

Figure 2-1 shows a temperature-entropy (T-s) diagram of the Carnot cycle for a heat engine. The four basic steps or processes are indicated below.

Step A-B	Adiabatic isentropic compression (no change in entropy)
Step B-C	Isothermal addition of heat enters cycle at constant temperature T_H with an increase in entropy
Step C-D	Adiabatic isentropic expansion with no change in entropy (ds = 0); with expansion work and an equivalent decrease in enthalpy
Step D-A	Isothermal rejection of heat; heat is rejected at constant temperature T_L with a reduction of entropy

In Step B-C, the heat added to the cycle is indicated by Q_{in}. In Step D-A, the heat rejected is indicated by

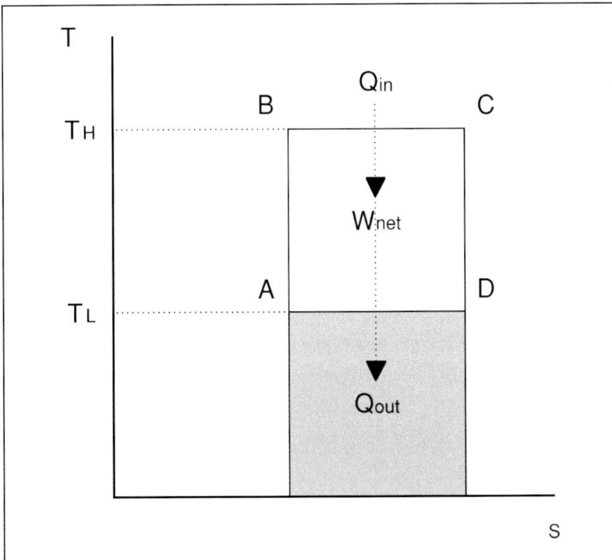

Fig. 2-1 Temperature-Entropy Diagram for Carnot Heat Engine.

Q_{out}. In Steps B-C-D-A, the net work of the cycle is indicated by W_{net}. From these values, cycle efficiency (h_{th}) can be expressed as follows:

$$\eta_{th} = \frac{(T_H - T_L)}{T_H} = \frac{(Q_{in} - Q_{out})}{Q_{in}} \qquad (2\text{-}8)$$

Where:

η_{th}	=	Cycle efficiency
T_H	=	Initial temperature at which heat is added to cycle
T_L	=	Final temperature at which heat is rejected from cycle
Q_{in}	=	Heat energy input
Q_{out}	=	Heat energy rejected

The following example shows ideal cycle efficiency based on Equation 2-8, given an initial temperature, T_H = 540°F, and a final temperature, T_L = 40°F. Temperatures are expressed in degrees R (°F + 460 = °R).

$$\eta_{th} = \frac{(T_H - T_L)}{T_L} x\ 100\% = \frac{1{,}000 - 500}{1{,}000} x\ 100\% = 50\%$$

Figure 2-2 shows the Carnot cycle with steam as the working fluid based on the temperatures shown in the above example. The figure includes both a pressure-volume (PV) diagram and a temperature-entropy (Ts) diagram, indicating the relationship between pressure and temperature as measures of energy. As shown, the four basic steps are similar to that of the Carnot heat engine.

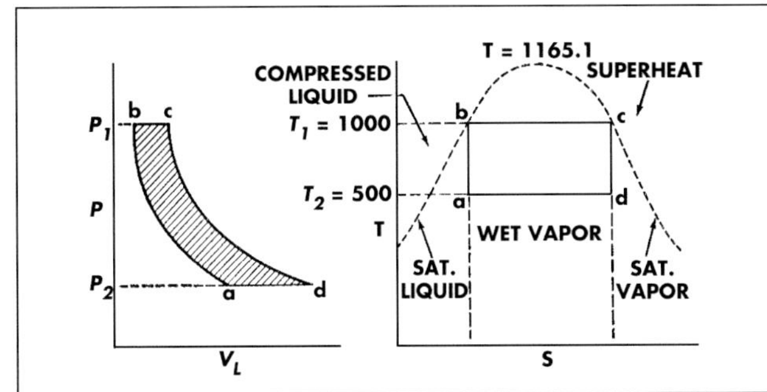

Fig. 2-2 Carnot Cycle with Steam as Working Fluid.
Source: Babcock &Wilcox

Step A-B	Isentropic compression
Step B-C	Constant pressure (hence, constant temperature heat addition)
Step C-D	Isentropic expansion
Step D-A	Constant pressure (hence, constant temperature heat rejection)

Figure 2-3 shows the pressure-volume and temperature-entropy diagrams for Carnot, Otto, Diesel, and Brayton gas

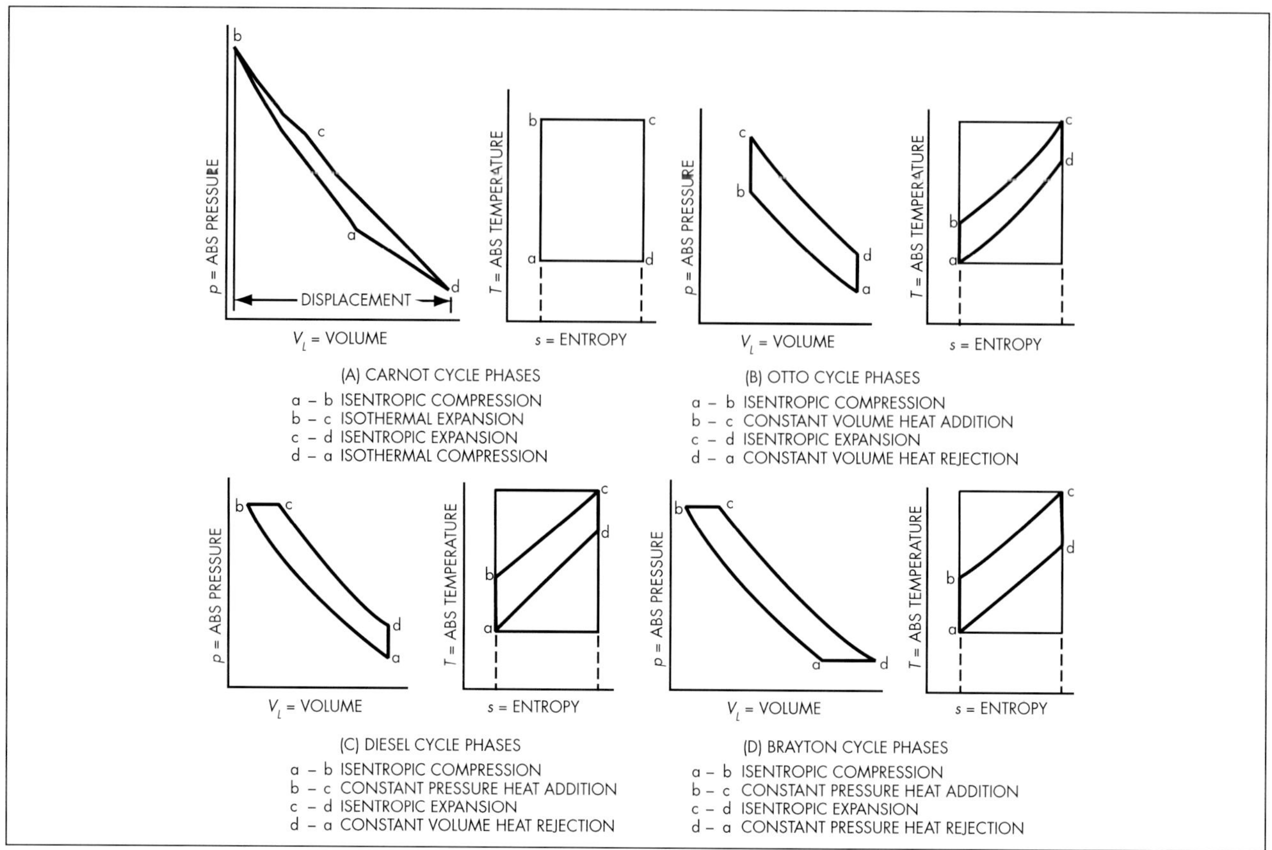

Fig. 2-3 P-V and T-s Diagrams for Carnot, Otto, Diesel, and Brayton Gas Cycles. Source: Babcock & Wilcox

cycles. Work for each cycle is again represented by area A-B-C-D. Note that mean effective pressure (MEP) is the work of the cycle divided by the displacement.

Practice Versus Theory

Given the first law of thermodynamics, all of the energy that goes into an operating system is work or heat. The first law treats heat and work as being interchangeable, though some qualifications must apply. Work and all other forms of energy can be wholly converted to heat, but the converse is not generally true.

The second law of thermodynamics shows that given a source of heat, only a portion of the heat can be converted to work in a heat-work cycle. The rest must be rejected to a heat sink. Thus, it is impossible to have a heat engine that is 100% efficient and even the most perfect cycle must be less than perfectly efficient. Further, the ideal performance of a given power cycle is always less than the Carnot efficiency. During an ideal (reversible) process, it is theoretically possible to reach the maximum potential efficiency for the specific process. Cycle design and operational improvements are initiated in an effort to approach the theoretical limits of the Carnot cycle. There are, however, factors that render the process irreversible.

The practical limit of power cycle efficiencies is set by metallurgical limits or strength of available materials (ability to operate under high temperatures and pressures) and by the ambient temperature of the heat sink. In addition, all practical applications of cycles and any other thermal process will be subject to energetic or heat losses resulting from friction, sustained expansion, convection, and conduction. Thus, reversible thermodynamic processes exist in theory only, defining only the limiting case for heat flow and work processes.

From the comparison of reversible and irreversible processes and cycles, it becomes clear that a critical concern in evaluating various cycles is theoretical and practical cycle efficiency. Theoretical efficiency shows the maximum efficiency that could, in theory, be attained from any given cycle. Practical efficiency shows what can be expected from a particular system operating on a given cycle. In the ensuing chapters covering prime mover technologies, thermal efficiency and other expressions of performance are presented first in theory and then in a practical context that can be applied to actual applications.

The following series of expressions are used throughout the various chapters that present combined heat and power system technologies and applications. They are used to present practical thermodynamic and economic performance measurements, based primarily on fuel and heat input and work and heat output.

Power Cycle Performance Expressions

Thermodynamic (or thermal) efficiency is obtained using the first law of thermodynamics and is calculated as the net work produced divided by the heat energy consumed. It is generically expressed in percent as:

$$\eta_{th} = \frac{Net\ work\ output}{Heat\ energy\ input} \ x\ 100\% \qquad (2\text{-}9)$$

In the English system, work output is commonly expressed as either hp (shaft output) or kW_e (electric generator output). Heat energy input is typically given as Btu/h. Alternative English and SI expressions for thermal efficiency are listed in Table 2-1.

To express efficiency in terms of fuel energy input (e.g., to a steam boiler serving turbine-driven equipment), boiler efficiency must be included in the expression. Thus, the fuel utilization efficiency, η_{TF}, would be expressed as:

$$\eta_{TF} = \eta_{th}\ x\ \eta_{Boiler} \qquad (2\text{-}10)$$

Heat rate is the inverse of thermal efficiency or the amount of energy input required (heat added) to generate shaft work. It is expressed as:

$$Heat\ rate = \frac{Heat\ energy\ input}{Net\ work\ output} \qquad (2\text{-}11)$$

In English units, heat rate is commonly expressed as Btu per horsepower-hour (Btu/hp-h) or Btu per kilowatt-hour (Btu/kWh). Alternative English unit and SI expressions are listed in Table 2-1.

Fuel rate is the heat rate expressed in terms of units of fuel added (or input) per unit of work unit output. Fuel units are commonly expressed on a mass-unit basis as pounds (lbm), tons, or kilograms (kg) or on a volumetric-unit basis as gallons (gal), liters (L), barrels, cubic feet (ft^3 or cf), or cubic meters (m^3 or cm). Fuel rate is expressed as:

$$Fuel\ rate = \frac{Fuel\ input}{Net\ work\ output} \qquad (2\text{-}12)$$

For operation on natural gas, Equation 2-12 becomes:

$$Fuel\ (gas)\ rate = \frac{Btu_{input}/hp\text{-}h}{Btu/cf} = cf/hp\text{-}h \qquad (2\text{-}12a)$$

In SI units, the corresponding expression would be in m^3/kWh_m.

Fuel-Specific Efficiency and Heat Rate

Conversion from heat rate to fuel rate requires a knowledge of the energy density, or heating value, of the particular fuel. Liquid fuels generally are expressed on an energy per unit mass or energy per unit volume basis (Btu/lbm or Btu/gal). For gaseous fuels, values are expressed on a volume basis, and the reference conditions for the fuel volume measurements (pressure, temperature, and degree saturation with water vapor), as well as test conditions, have to be stated explicitly.

Care must be taken when converting heat rates to fuel rates to identify whether the fuel energy density refers to the higher heating value (HHV) or lower heating value (LHV) of fuel. During the combustion of hydrocarbon fuels, some of the oxygen is combined with hydrogen, forming water vapor that may leave the combustion device either in vapor or condensed to liquid state. When the latent heat of vaporization is extracted from the flue products, causing the water to become liquid, the fuel's energy density is identified as HHV. When the equipment used allows the water to remain in the vapor state, the energy density is identified as LHV. Since different fuels have varying amounts of hydrogen, the numerical relationship between LHV and HHV varies.

Prime mover (reciprocating engine and combustion gas turbine) performance is usually based on LHV, while fuel energy is often expressed in HHV. Thus, in order to convert heat rates specified in LHV to purchasable fuel units, one must know the energy density of the fuel and the ratio of LHV to HHV. Refer to chapter 5 for additional details.

To express true fuel-specific thermal efficiency, when heat added is expressed on an LHV basis, Equation 2-9 becomes:

$$\eta_{th}(HHV) = \frac{Net\ work}{Heat\ added\ (LHV)}\ x\ (LHV/HHV)\ x\ 100\% \qquad (2\text{-}13)$$

To express true fuel-specific heat rate, when heat added is expressed on an LHV basis, Equation 2-11 becomes:

$$Heat\ rate\ \ (HHV) = \frac{Heat\ added\ (LHV)}{Work\ units\ of\ output}\ x\ (HHV/LHV) \qquad (2\text{-}14)$$

Consideration of Recovered Heat

Overall thermal efficiency is the ratio of work output plus heat recovered to the heat input. Equation 2-9 thus becomes:

$$\eta_{th} = \left(\frac{Net\ work + Recovered\ heat}{Heat\ added}\right) x\ 100\% \qquad (2\text{-}15)$$

Since mechanical or electrical power is usually more valuable than heat energy, combined product values are not highly informative and can result in misleading conclusions about the relative thermodynamic and economic value of a cogeneration cycle. Though not fully definitive, the terms net thermal efficiency and net heat rate are often more revealing.

Net thermal efficiency reflects incremental energy usage for power generation by subtracting recovered heat from the total energy input, assuming recovered heat replaces other energy usage. Equation 2-9 thus becomes:

$$\eta_{th} = \left(\frac{Net\ work\ output}{Heat\ added - Heat\ recovered}\right) x\ 100\% \qquad (2\text{-}16)$$

Net heat rate is the energy input required to generate shaft work minus the amount of heat recovered:

$$Net\ heat\ rate = \frac{Heat\ added - Heat\ recovered}{Net\ work\ output} \qquad (2\text{-}17)$$

Energy-, Fuel-, and Cost-Chargeable-to-Power

An important distinction needs to be made in comparing prime mover applications, which use rejected heat (or pass on thermal energy to useful processes), with typical simple-cycle processes. Whereas simple cycle thermal efficiencies refer to a single energy product, thermal efficiency measurements of cogeneration-type applications refer to combined production of prime energy and thermal energy.

When a facility is considering an investment in some type of prime mover system, the investment analysis is based on economic performance. The following indices are useful in expressing the performance of cogeneration systems.

Energy Credit

The value of heat recovery can be measured by the cost avoided in using recovered thermal energy (or heat) for a specific purpose, as opposed to using another source of energy. Most commonly, recovered heat replaces thermal energy output from some type of fuel-burning equipment, usually a boiler or furnace. In these cases, the value of recovered thermal energy is equivalent to the cost of fuel energy that would have otherwise been consumed. This displaced energy is commonly referred to as an **energy credit** or **fuel credit**. The amount of energy displaced by recovered heat is a function of the efficiency of the displaced boiler (or other energy conversion equipment).

Displaced energy resulting from heat recovery can be expressed as:

$$Energy\ credit = \frac{Heat\ recovered}{Efficiency\ of\ displaced\ boiler} \qquad (2\text{-}18)$$

Examples

If prime mover heat recovery provides 100,000 Btu of usable thermal energy, reducing boiler operation at 84% conversion efficiency, Equation 2-18 becomes:

$$Energy\ credit = \frac{100{,}000\ Btu}{0.84} = 119{,}000\ Btu \qquad (2\text{-}18a)$$

It is important to note that the displaced energy is not necessarily fuel energy. If, for example, recovered heat were used to displace an electric heat source with an efficiency of 98%, the displaced energy would be expressed as displaced electricity and Equation 2-18 would become:

$$\begin{matrix}Displaced\\ electricity\end{matrix} = \frac{100{,}000\ Btu}{3{,}413\ Btu/kWh\ x\ 0.98} = 29.9\ kWh \qquad (2\text{-}18b)$$

Alternatively, if recovered heat were used to generate chilled water via a double-effect absorption chiller (at a rate of 10,000 Btu/ton-h), the displaced energy could be expressed as a chilled water or cooling credit:

$$\begin{matrix}Displaced\\ cooling\end{matrix} = \frac{100{,}000\ Btu}{10{,}000\ Btu/ton\text{-}h} = 10.0\ ton\text{-}h \qquad (2\text{-}18c)$$

If operation of the absorption chiller resulted in the displacement of an electric chiller, there would be an electricity credit calculated by considering the efficiency of the original chiller (kW/ton) and related auxiliaries.

Energy-Chargeable-to-Power

The difference between the total energy input to the prime mover and the energy credit represents the actual amount of fuel energy consumed for the purpose of generating shaft power. **Energy-chargeable-to-power (ECP)** provides a useful means of expressing the net heat rate of prime mover applications that include heat recovery:

$$ECP = \frac{Total\ heat\ added - Displaced\ fuel\ energy}{Net\ power\ output} \qquad (2\text{-}19)$$

where the displaced fuel energy is computed from Equation 2-18. Calculation of ECP using English and SI units is illustrated in Table 2-1.

For a back-pressure steam turbine with exhaust heat recovery, ECP can be expressed using Equation 2-19 based on boiler input, heat recovery fuel savings, and turbine power output. Alternatively, ECP can be calculated based on steam flow and turbine inlet and exhaust conditions. When steam flow is expressed in lbm/h and steam energy content (enthalpy) in Btu/lbm, Equation 2-19 becomes:

$$ECP = \frac{lbm/h\ (Btu/lbm_{Input} - Btu/lbm_{Discharge})}{hp \times \eta_{Boiler}} = Btu/hp\text{-}h \qquad (2\text{-}19a)$$

Applying ECP to Combined Cycles

In the case of combined-cycle operation, the net power output of the system includes the output of the topping cycle (typically a gas turbine or reciprocating engine) and the bottoming cycle (typically a steam turbine). When recoverable heat is used both to generate additional power and to displace other fuel usage, Equation 2-19 becomes:

$$ECP = \frac{Total\ energy\ input - Displaced\ energy}{TC_{Output} + BC_{Output} - Auxiliary\ power\ input} \qquad (2\text{-}20)$$

Where:
TC =Topping cycle
BC =Bottoming cycle

Fuel-Chargeable-to-Power

Fuel-chargeable-to-power (FCP) provides an alternative means of expressing the net fuel rate of prime mover (cogeneration) applications that effectively use rejected heat. It expresses the total fuel rate of power generation minus a fuel credit (if rejected heat is used), divided by power production minus any auxiliary power requirements. When energy input and displaced fuel energy are replaced with fuel input and fuel displaced (or fuel savings), Equation 2-19 becomes:

$$FCP = \frac{Total\ fuel\ input - Fuel\ displaced}{Net\ power\ input} \qquad (2\text{-}21)$$

When work rate is expressed in English units as hp and fuel rate as cf of natural gas, Equation 2-21 becomes:

$$FCP = \frac{(cf_{Input} - cf_{Displaced})}{hp\text{-}h_{Net\ output}} = cf/hp\text{-}h \qquad (2\text{-}21a)$$

Operating Cost-Chargeable-to-Power

The operating **cost-chargeable-to-power (CCP)** provides a means of expressing economic performance, or the net cost of prime mover (cogeneration) applications that effectively use rejected heat. The CCP expression follows the same format as the ECP and FCP expressions. In addition to energy operating costs, the CCP expression may be extended to include all operating costs associated with the system, such as operations and maintenance (O&M) costs, water costs, and environmental compliance costs. Equation 2-21 becomes:

$$CCP = \frac{Fuel\ cost - Fuel\ credit\ value + Other\ operating\ costs}{Net\ power\ output} \qquad (2\text{-}22)$$

Relationship of ECP, FCP, and CCP

If the type and per-unit cost of fuel is the same for both the prime mover and the fuel use displaced by heat recovery, FCP can be derived from ECP, and CCP can be expressed as a function of ECP or FCP:

$$FCP = \frac{ECP}{Energy\ units\ per\ unit\ of\ fuel} \qquad (2\text{-}23)$$

$$CCP = ECP \times Cost\ per\ energy\ unit \qquad (2\text{-}24)$$

$$CCP = FCP \times Cost\ per\ fuel\ unit \qquad (2\text{-}25)$$

Commonly, however, the fuel used by the prime mover may be different than the fuel used by the heat energy source that is displaced. Alternatively, the same fuel type may be used but at different prices, due to contract specifics, such as the ability to operate on an alternative fuel. In either of these cases, CCP cannot be expressed as a function of ECP or FCP. Instead, a separate fuel credit term must be used. The displaced fuel energy resulting from heat recovery is expressed by Equation 2-18. By assigning a separate financial value or avoided cost to this, Equation 2-18 becomes:

$$Fuel\ credit\ value = Fuel\ credit \times Avoided\ cost\ per\ fuel\ unit \qquad (2\text{-}26)$$

The CCP expression can be manipulated to express performance for a variety of practical applications. The displaced energy use, for example, may in fact be electricity as opposed to fuel. Still, the basic form of Equation

2-22 applies—the value of the energy credit is subtracted from the cost of the total energy input, plus any additional operating costs, and this net cost is divided by the net power output to yield the CCP.

Total and Life-Cycle CCP

The CCP may also be extended to include capital costs as well as operating costs by including an annualized capital cost component based on an assumed time-valued cost of capital. The change (D) in operating costs represents any increases or decreases in related costs, such as O&M, chemical treatment, water use, or sewer use. The total annual CCP can then be expressed as:

$$CCP_{Total} = \frac{\textit{Fuel cost} - \textit{Energy credit} + \Delta\ \textit{Operating costs} + \textit{Annualized capital cost}}{\textit{Net power output}} \quad (2\text{-}27)$$

This expression can be extended further to reflect life-cycle CCP. Based on an assumed financial discount rate, the life cycle CCP would be the sum of the present values of the annualized CCP over the life of the systems. To perform this calculation, a present value factor is assigned to each year, based on the assumed discount rate. The present value factor serves to reduce CCP in each future year to an equivalent cost in the present based on a given discount rate.

Summary Examples

Following are four examples of how the terms and formulae provided above are applied to specific representative prime mover systems. All values are given in English units and the assumed fuel is pipeline grade natural gas with an HHV of 1,020 Btu/cf and LHV of 918 Btu/cf. Fuel costs are given for each example on the basis of 1,000 cf (1 Mcf).

For simplicity, the performance expressions are applied under idealized full load operating conditions. In actual applications, prime mover performance will be directly related to continually varying operating conditions. For example, operation under varying load conditions or at varying inlet air temperatures usually results in a change in heat rate from rated full load operation. Numerous other factors will also influence actual performance.

Fuel input is calculated on an HHV basis, using Equation 2-14 as follows:

$$6{,}758\ Btu/hp \times 1.11 = 7{,}507\ Btu/hp\text{-}h$$

The **thermal efficiency** (HHV basis) of the system for simple-cycle operation, based on Equation 2-9, is:

Term	English Units	SI Units
Thermal efficiency (Eq. 2-9) $\frac{\textit{Net work output}}{\textit{Heat energy input}} \times 100\%$	$\frac{hp \times 2{,}545\ Btu/hp\text{-}h}{Btu/h} \times 100\%$	$\frac{kW_m \times 3{,}600\ kJ/kWh}{kJ/h} \times 100\%$ or $\frac{kW_m}{kWh_h/h} \times 100\%$
Heat rate (Eq. 2-11) $\frac{\textit{Heat energy input}}{\textit{Net work input}}$	$\frac{Btu/h}{hp} = Btu/hp\text{-}h$	$\frac{kJ/h}{kW_m} = kJ/kWh_m$ or $\frac{kWh_h/h}{kW_m} = kWh_h/kWh_m$
Net thermal efficiency (Eq. 2-16) $\left(\frac{\textit{Net work output}}{\textit{Heat added} - \textit{Heat recovered}}\right) \times 100\%$	$\left(\frac{hp \times 2{,}545\ Btu/hp\text{-}h}{Btu_{Heat\ added}/h - Btu_{Heat\ recovered}/h}\right) \times 100\%$	$\left(\frac{kW_m \times 3{,}600\ kJ/kWh_m}{kJ_{Heat\ added}/h - kJ_{Heat\ recovered}/h}\right) \times 100\%$
Net heat rate (Eq. 2-17) $\frac{\textit{Heat added} - \textit{Heat recovered}}{\textit{Net work output}}$	$\frac{Btu_{Energy\ added}/h - Btu_{Heat\ recovered}/h}{hp}$	$\frac{kJ_{Energy\ added}/h - kJ_{Heat\ recovered}/h}{kW_m}$
Energy-chargeable-to-power (Eq. 2-19) $\frac{\textit{Total heat added} - \textit{Displaced fuel energy}}{\textit{Net power output}}$	$\frac{Btu_{Heat\ added} - Btu_{Heat\ recovered} \times \eta_{Boiler}}{hp\text{-}h_{Output} - hp\text{-}h_{Auxilliary\ input}}$	$\frac{kJ_{Heat\ added} - kJ_{Heat\ recovered} \times \eta_{Boiler}}{kWh_{mOutput} - kWh_{m\ Auxiliary\ input}}$

Table 2-1 English and SI Units Commonly Used in Performance Terms.

Example 1
Reciprocating Engine Applied to Mechanical Drive Service
Full-load Performance Values

Total fuel input (LHV) per hp-h output	= 6,758 Btu (LHV)
Ratio of fuel HHV to LHV	= 1.11
Total energy recovered per hp-h output	= 3,003 Btu (40%)
Efficiency of displaced boiler	= 82%
O&M cost for total system per hp-h output	= $0.008/hp-h
Fuel cost (natural gas)	= $6.00/Mcf
Horsepower to Btu conversion factor	= 2,545 Btu/hp-h

$$\frac{2{,}545\ Btu_{Shaft}/hp\text{-}h}{7{,}507\ Btu_{Input}/hp\text{-}h} \times 100\% = 33.9\%$$

The **fuel rate** for simple-cycle operation when operating on natural gas, using Equation 2-12, is:

$$\frac{7{,}507\ Bth/hp\text{-}h}{1{,}020\ Btu/cf} = 7.360\ cf/hp\text{-}h$$

The **overall thermal efficiency** (HHV basis), based on Equation 2-15, is:

$$\left(\frac{2{,}545\ Btu_{Shaft}/hp\text{-}h + 3{,}003\ Btu_{Heat}/hp\text{-}h}{7{,}507\ Btu_{Energy\ input}/hp\text{-}h}\right) \times 100\% = 73.9\%$$

The **ECP,** based on Equation 2-19, is:

$$7{,}507\ Btu/hp\text{-}h\quad \frac{3{,}003\ Btu/hp\text{-}h}{0.082} = 3{,}845\ Btu/hp\text{-}h$$

The **FCP,** based on Equation 2-21, is:

$$\frac{3{,}845\ Btu/hp\text{-}h}{1{,}020\ Btu/hp\text{-}h} = 3.770\ cf/hp\text{-}h$$

Gas rates increased to more typical current values; same firm rate use for all examples:

Firm gas rate	$3.75/Mcf	→	$6.00/Mcf
Interruptible gas rate	$2.70/Mcf	→	$4.00/Mcf

O&M costs inflated 2.5% for 7 years:

O&M cost (engine)	$0.007/hp-h	→	$0.008/hp-h
O&M cost (BP ST)	$0.003/hp-h	→	$0.004/hp-h
O&M cost (GT)	$0.008/hp-h	→	$0.010/hp-h
O&M cost (combined)	$0.006/hp-h	→	$0.007/hp-h

The **CCP,** based on Equation 2-22 is:

$$3.770\ cf/hp\text{-}h \times \$6.00/Mcf \times 1\ Mcf/1{,}000\ cf + \$0.008/hp\text{-}h = \$0.023/hp\text{-}h + \$0.008/hp\text{-}h = \$0.031/hp\text{-}h$$

Example 2
Back-Pressure Steam Turbine Applied to Mechanical Drive Service
Full-load Performance Values

Steam turbine inlet enthalpy (@ 250 psig, saturated)	= 1,201.7 Btu/lbm
Steam turbine exhaust enthalpy	= 1,129.0 Btu/lbm
Condensate return enthalpy	= 170.3 Btu/lbm
Total lbm of steam input per hp-h output	= 35.0 lbm/hp-h
Boiler efficiency	= 83%
Actual enthalpy drop in turbine per lbm of steam	= 72.7 Btu/lbm
Incremental O&M cost for turbine per hp-h output	= $0.004/hp-h
Fuel cost (natural gas)	= $6.00/Mcf

The **thermal efficiency** of the system for simple-cycle operation (full steam use without use of back-pressure steam), based on Equation 2-9, is:

$$\frac{2{,}545\ Btu_{shaft}/hp\text{-}h}{35.0\ lbm/hp\text{-}h \times (1{,}201.7 - 170.3)\ Btu/bm} \times 100\% = 7.05\%$$

The **fuel energy efficiency** for simple-cycle operation, based on Equation 2-10, is:

$$7.05\% \times 0.83 = 5.86\%$$

The **fuel rate** for simple-cycle operation on natural gas, based on Equation 2-12, is:

$$\frac{35.0\ lbm/hp\text{-}h \times (1{,}201.7 - 170.3)\ Btu/lbm}{1{,}020\ Btu/df \times 0.83} = 42.640\ cf/hp\text{-}h$$

The **ECP** (assuming full use of back-pressure steam), based on Equation 2-19, is:

$$35.0\ lbm/hp\text{-}h \times \frac{(1{,}201.7 - 1{,}129.0)\ Btu/lbm}{1{,}020\ Btu/cf \times 0.83} = 3{,}066\ Btu/hp\text{-}h$$

The **FCP**, based on Equation 2-21, is:

$$\frac{3{,}066\ Bth/hp\text{-}h}{1{,}020\ Btu/cf} = 3.006\ cf/hp\text{-}h$$

The **CCP,** based on Equation 2-22, is:

Example 3
Combustion Gas Turbine Applied to Electric Power Cogeneration
Full-load Performance Values

Total fuel input (LHV) per kWh output	= 11,045 Btu (LHV)
Total energy recovered per kWh output	= 6,380 Btu (52%)
Efficiency of displaced boiler	= 82%
O&M cost for total system per kWh output	= $0.010/kWh
Fuel cost (natural gas) for the gas turbine	= $6.00/Mcf
Fuel cost (natural gas) for the displaced boiler	= $4.00/Mcf

Example 4
Combined-Cycle Cogeneration System Applied to Electric Power Generation (Gas Turbine System with Back-Pressure Steam Turbine)
Full-load Performance Values

Total fuel input (LHV) per kWh gas turbine output	= 11,300 Btu (LHV)
Steam turbine kW output per kW gas turbine output	= 0.175 kW (597 Btu/kWh)
Energy recovered per kWh combined cycle output	= 4,540 Btu
Efficiency of displaced boiler	= 83%
O&M cost for total system per kWh output	= $0.007/kWh
Fuel cost (natural gas)	= $6.00/Mcf

3.006 cf/hp-h x $6.00/Mcf x 1 Mcf/1,000 cf + $0.004/hp-h = $0.018/hp-h + $0.004/hp-h = $0.022/hp-h

Fuel input (HHV Basis) = 11,045 Btu/hp-h x 1.11 = 12,279 Btu/hp-h

The **thermal efficiency** (HHV Basis) for simple-cycle operation, based on Equation 2-9, is:

$$\left(\frac{3{,}413 \text{ Btu/kWh}}{12{,}270 \text{ Btu/kWh}}\right) \times 100\% = 27.8\%$$

The **fuel rate**, on an HHV basis, for simple-cycle operation, based on Equation 2-12, is:

$$\frac{12{,}270 \text{ Btu/kWh}}{1{,}020 \text{ Btu/hp-h}} = 12.029 \text{ cf/kWh}$$

The **ECP** (HHV Basis), based on Equation 2-19, is:

$$12{,}270 \text{ Btu/kWh} - \frac{6{,}380 \text{ Btu/kWh}}{0.82} = 4{,}486 \text{ Btu/kWh}$$

In this case, the CCP will not be computed as a function of ECP or FCP because the cost of fuel input to the gas turbine and the cost of fuel that is being displaced by use of recovered heat are different. Therefore, the fuel credit must be calculated separately and then applied to the CCP expression.

The **fuel credit,** based on Equation 2-18, is:

$$12.029 \text{ cf/kWh} \times \frac{0.52}{0.82} = 7.628 \text{ cf/kWh}$$

To apply the CCP expression, the cost of the total fuel input and the value of the fuel credit must be calculated separately as follows:

Total fuel input cost = 12.029 cf/kWh x $6.00/Mcf x 1 Mcf/1,000 cf = $0.072/kWh

Fuel credit value = 7.628 cf/kWh x $4.00/Mcf x 1 Mcf/1,000 cf = $0.031/kWh

The **CCP**, based on Equation 2-22, is:

$fuel cost – $fuel credit + $O&M cost = $Operating cost
$0.072/kWh – $0.031/kWh + $0.010/kWh = $0.051/kWh

Fuel input (HHV basis) = 11,300 Btu/kWh x 1.11 = 12,553 Btu/kWh

The **heat rate** of the combined-cycle system, based on Equation 2-11, is:

$$\frac{12{,}553 \text{ Btu}}{1{,}175 \text{ kWh}} = 10.683 \text{ Btu/kWh (HHV Basis)}$$

The **fuel rate**, on an HHV basis, based on Equation 2-12, is:

$$\frac{10{,}683 \text{ Btu/kWh}}{1{,}020 \text{ Btu/cf}} = 10.474 \text{ cf/kWh}$$

The **energy credit** for recovered heat, on an HHV basis, based on Equation 2-18, is:

$$\frac{4{,}540 \text{ Btu/kWh}_{Total}}{0.83} = 5{,}470 \text{ Btu/kWh}_{Total}$$

and the **fuel credit** is:

$$\frac{5{,}470 \text{ Btu/kWh}}{1{,}020 \text{ Btu/cf}} = 5.363 \text{ cf/kWh}$$

The **ECP**, on an HHV basis, based on Equation 2-20, is:

$$10{,}683\ Btu/kWh - \frac{4{,}540\ Btu/kWh}{0.83} = 5{,}213\ Btu/kWh$$

The **FCP**, based on Equation 2-23, is:

$$\frac{5{,}213\ Btu/kWh}{1{,}020\ Btu/cf} = 5.111\ cf/kWh$$

The **CCP**, based on Equations 2-22 and 2-24, is:

$$5.111\ cf/kWh \times \$6.00/Mcf \times 1\ Mcf/1{,}000\ cf + \$0.007/kWh$$
$$= \$0.031/kWh + \$0.007/kWh = \$0.038/kWh$$

Beyond ECP, FCP, and CCP

Absent from the above discussion of performance indicators are several essential factors that affect actual project economic performance for any prime mover application. When considering an investment in a prime mover system, the analysis is based not on the potential output and efficiency of the system, but the match of that potential output with the facility's internal requirements or ability to sell the outputs. While the CCP of a given system provides useful and compelling information, it provides only a limited perspective from which to make an investment decision.

In most cases, higher cogeneration heat recovery indicates a higher overall heat and fuel rate, even if the ECP or FCP remains the same. This is because as more input energy is converted to thermal energy, less is converted into power, and more fuel must be used per unit of power output. Depending on the available load or rejected energy sink of displaceable thermal energy, it is often better to consider a more thermally efficient power producing system with less heat recovery.

With an unlimited sink of thermal energy requirement, power generating thermal efficiency becomes less of a consideration. A relatively inefficient power system may be considered because decreased capital costs and increased thermal energy generation efficiency compensate for decreased power output. However, if there is a limited amount of thermal energy that can be used, the emphasis shifts to more thermally efficient prime movers that produce a higher percentage of shaft power and a lower percentage of recoverable thermal energy.

Each prime mover system considered will have a different heat rate and different combination of power and thermal outputs. Available temperatures and pressures of thermal outputs will also vary. There will be different emissions rates for different pollutants, different space requirement, and different expected service lives. Moreover, these performance characteristics will vary as operating conditions vary.

Additionally, each of the prime mover systems will require different levels of maintenance and operating attention, and they will have different capital costs. All of these factors will vary, even for the same equipment, depending on site conditions. Translating performance into cost and values requires a detailed analysis of the site conditions, the equipment performance under these conditions, and the interaction of all factors for each alternative.

Chapter Three

Localized vs Central Station Power Generation

For a given application, an on-site prime mover possesses several potential cost advantages over an electrically driven end-use device. These advantages are due largely to a higher utilization of input energy when compared with a central electric generation plant. Potential advantages include recovery of high-grade heat (as opposed to rejection of low-grade heat by central electric generation plants) and reduction of electrical transmission and end-use device inefficiencies.

Economics do not always come out in favor of a more efficient end-use. Large centralized power plants also possess several advantages over the smaller user, including economy of scale, preferential fuel cost, relatively low staffing levels, and reserve capacity.

Central Utility Plants

The electric utility industry, which still produces the majority of the nation's power, has traditionally relied on two basic thermodynamic cycles to generate electric power: the conventional steam turbine **Rankine cycle** for baseload and intermediate load applications, and the gas turbine **Brayton cycle** for peaking applications. While these cycles have proven reliable for the utility industry, the use of natural gas or oil fuels and heat recovery technology opens up a number of other cycles that offer significantly better thermodynamic performance for large central power plants, as well as smaller facilities. These newer gas turbine based combined-cycle central power power plants have thermodynamic efficiencies over 50% as compared to the older traditional Rankine cycle based plants with efficiencies of 30%-to-40%. As is customary in the United States, this chapter presents boiler efficiencies and overall plant efficiencies and heat rates based on Higher Heating Value (HHV), whereas the prime mover heat rates and efficiencies presented in Chapter 9 and 10 are on a Lower Heating Value (LHV). Refer to Chapter 5 for a discussion of the difference in these ratings.

Conventional Utility Rankine Steam Cycle Plants

Most large utility plants rely on the conventional Rankine steam cycle, in which high-pressure steam is expanded through a set of turbine wheels to generate mechanical energy, which then drives a generator to make electric power. The high-pressure steam is produced in a steam generator, or boiler, which may be fueled by coal, oil, natural gas, wood, refuse, or nuclear fission.

Figure 3-1 is a simplified schematic representation of a large utility supercritical steam power plant, featuring a high-pressure and a low-pressure turbine, with regeneration and reheat. The boiler section may be divided into three functional components: an **economizer,** which heats incoming feedwater nearly to its boiling point; an **evaporator,** which adds energy to convert the feedwater into steam; and a **superheater**, which raises the temperature of the steam prior to its use in the steam turbine. Large utility boilers may employ several stages of economizer, evaporator, and superheater, as well as other components, in order to maximize the thermodynamic efficiency of the system. These features

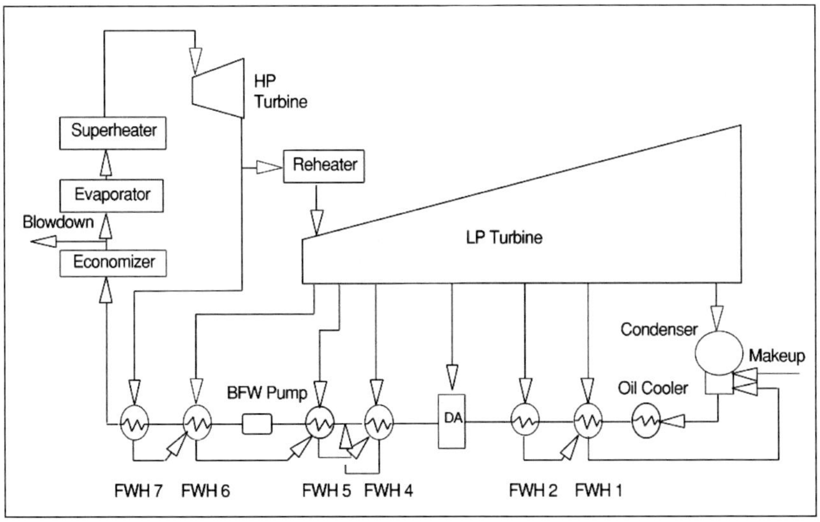

Fig. 3-1 Simplified Diagram of Large Utility Steam Generation Reheat Cycle. Source: Cogen Designs, Inc.

are generally uneconomical in smaller capacity plants.

In the basic cycle, superheated steam exits the boiler and enters the first stage of the steam turbine. Most of the steam passes completely through the steam turbine and is exhausted, drawn to a partial vacuum at a condenser served by external cooling water.

The lower the pressure at the steam exhaust, which is set by the condenser operating temperature, the greater the amount of mechanical energy that can be obtained from the expanding steam, and the higher the power cycle efficiency. Typical condenser operating temperature and pressure are 90°F (32°C) and 1.5 in (3.8 cm) of Hg (about 0.75 psi), respectively. The condenser returns the exhaust steam to liquid state by removing its latent heat, which is rejected to the outside environment in the condensing part of the cycle. Liquid condensate is returned to the boiler through the deaerator to restart the process.

Although boiler feedwater is chemically treated to remove impurities that contribute to scaling and fouling of boiler surfaces, concentrations of dissolved materials, which tend to build up over time, must be eliminated through a blowdown process that removes a small percentage of boiler feedwater from the system. To replace this blowdown loss, fresh make-up water is added before the feedwater enters the deaerator. In the deaerator, extraction steam is used to strip dissolved air from the feedwater.

Larger utility plants may also use a reheat cycle to optimize cycle efficiency while avoiding excessive moisture in the low-pressure stages of the turbine. In the reheat cycle, the steam is withdrawn after partial expansion, re-superheated at constant pressure in the boiler, and introduced into a low-pressure turbine.

A **regenerative cycle** is an additional efficiency enhancement that reduces condenser losses by using a portion of available latent heat for deaeration and feedwater heating. Steam is extracted between high-pressure and low-pressure turbines and at intermediate points in the low-pressure turbine for use by the deaerator and feedwater heating heat exchangers.

Figure 3-2 is a simplified schematic representation of a large utility subcritical steam power plant. It features a high-pressure and a low-pressure turbine, with regeneration and reheat.

Table 3-1 lists historical average heat rates for traditional electric utility steam-electric power plants utilizing the Rankine thermodynamic cycle. Recent fossil-fueled heat rates of 10,300 Btu/kWh (10,860 kJ/kWh) correspond to a thermal efficiency of 33%, with heat rates for nuclear steam-electric plants being slightly higher. Cycle efficiency is proportional to the difference between the temperature at which energy is added (at the boiler) and rejected (at the condenser). However, increases in system pressure and temperature are limited by the tolerance of component materials and decreases in condenser pressure and temperature are limited by ambient environmental conditions. As such, efficiency levels of conventional plants have been relatively constant over time. The only substantial change in this data set has been the addition of combined heating and power plants, with improved efficiencies due to the higher utilization of the fuel source, with heat rates averaging approximately 8,700 Btu/kWh.

Individual steam plant thermal efficiencies can range from under 30% to more than 40%. At the higher end, the more efficient plants use supercritical steam conditions and several stages of reheat. Due to cost and technical considerations, these types of plants have not been widely used in the United States, where, currently, the best steam cycles offer a thermal efficiency of about 34%. Based on these typical thermal efficiencies, a conventional steam cycle releases almost two-thirds of the energy in the fuel into the environment. Figure 3-3 illustrates how this occurs.

The left stacked bar, entitled "Boiler," represents the energy added to a pound of feedwater as it passes into and through the three sections of the steam generator. Since not all of the energy content in the fuel is available to increase the steam energy content, and since the boiler itself has energy losses, in total, 1,458 Btu (1,538 kJ) of fuel energy must be added per pound of steam to increase the energy content of the steam from approximately 200 Btu/lbm (465 kJ/kg) to 1,500 Btu/lbm (3,488 kJ/kg), a 1,300 Btu/lbm (3,023 kJ/kg) increase. The majority of these energy losses, which, in this example, amount to 12% of the fuel energy content, are rejected to the outside environment through the boiler stack.

Expansion of steam through the steam turbine is represented by the diagonal arrow running from the boiler to the stacked bar on the right entitled "condenser". In this example, expansion releases about 452 Btu of useful mechanical work per pound (1,046 kJ per kg) of steam, which is converted by the generator into electric power. Efficiency losses in the generator itself, and other parasitic and auxiliary electric demands of the generating station, slightly reduce the electric energy available for export from the plant.

By far, the largest source of inefficiency in the conventional steam cycle is fundamental to the thermodynamic principles underlying the cycle. The latent heat

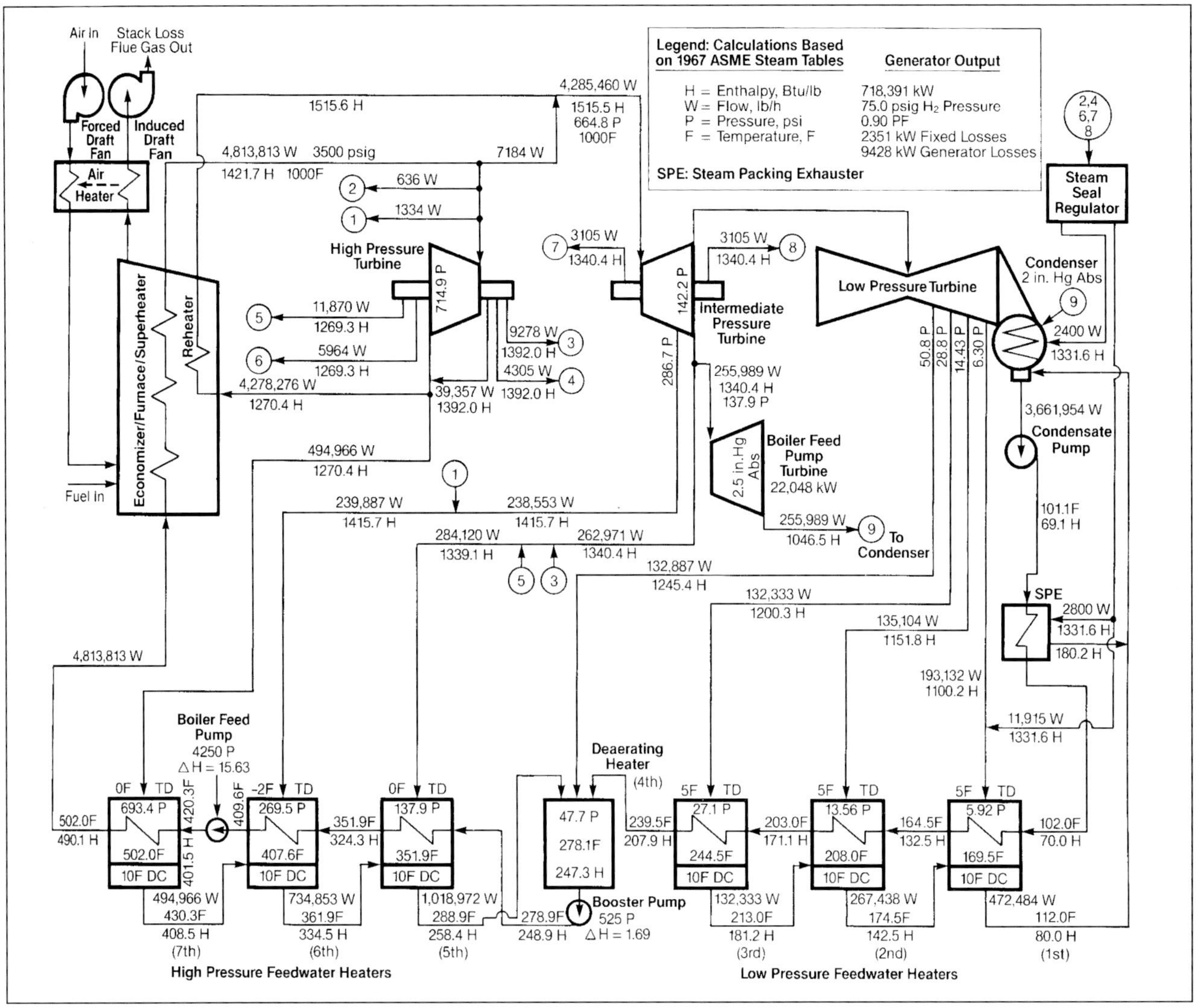

Fig. 3-2 Subcritical Pressure, 2,400 psig (166 bar) Steam Turbine Cycle Heat Balance Diagram. Source: Babcock & Wilcox

Period	Fossil-Fueled (Btu/kWh)	Nuclear (Btu/kWh)
1965-69	10,429	11,463
1970-74	10,436	10,934
1975-79	10,386	10,930
1980-84	10,451	10,952
1985-89	10,391	10,771
1990-93	10,323	10,694

Table 3-1 Average Heat Rates for Utility Steam-Electric Plants, 1965-1993. Source: U.S. DoE/EIA

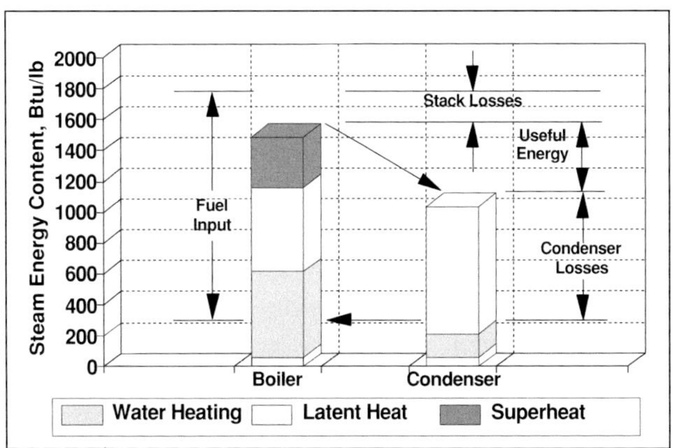

Fig. 3-3 Conventional Power Cycle Energy Utilization. Source: Cogen Designs, Inc.

released through the condenser to the environment from the condensing steam after expansion in the steam turbine serves no useful purpose in the power cycle. In this example, roughly 830 Btu/lbm (1,930 kJ/kg) of steam is lost through the condenser. An additional 150 Btu/lbm (349 kJ/kg) of steam is described as "water heating" on the condenser bar below the condenser losses. This energy represents the steam extracted from the steam turbine and recovered for deaeration and feedwater heating.

In summary, this representative steam cycle example indicates a fuel energy input requirement of 1,458 Btu (1,538 kJ) to produce 452 Btu (479 kJ) of useful work, for an overall power cycle efficiency of 31%. On a heat rate basis, this cycle requires about 11,010 Btu/kWh (11,613 kJ/kWh) of electric energy produced.

Conventional Utility Gas Turbine Cycle Peaking Plants

The other basic thermodynamic cycle used by electric utility plants to generate electric power is the open-cycle gas turbine, which operates on the Brayton cycle. Whereas steam cycle plants are used for base and intermediate load applications, gas turbine cycle plants are used by many utilities to serve peaking power requirements. Whereas the steam cycle operates essentially on a closed loop, the gas turbine uses air in a once-through open cycle. Since air is essentially a non-condensable fluid at normal operating temperatures, the gas turbine cycle does not use a boiler and condenser. Instead, a combustor burns fuels such as natural gas or oil in direct contact with compressed air, directing the mixture into a power turbine, which then expands the gases and exhausts them directly into the atmosphere. While the steam cycle uses boiler feed pumps to elevate feedwater to the required operating pressures, the gas turbine uses a multiple-stage compressor to elevate the pressure of the incoming air stream to operating pressure. The power turbine, through a common shaft, supplies the energy required for this air compression.

There are numerous other differences between the conventional gas turbine cycle and the conventional steam turbine cycle described above. Whereas the steam turbine operates at high pressure of up to 2,000 psig (139 bar) and a relatively moderate temperature of about 1,000°F (538°C), the gas turbine operates at a relatively moderate pressure of up to 460 psi (33 bar) and a high temperature of about 2,000°F (1,093°C). Whereas the steam turbine exhausts to a deep vacuum of about 0.75 psia (5 kPa) at a low temperature, the gas turbine exhausts to atmospheric pressure (about 14.7 psi or 101.4 kPa), but at a temperature of about 1,000°F (538°C). Despite these differences, the thermodynamic efficiency of the two cycles are relatively close, though historically, the basic gas turbine cycle has been slightly less thermally efficient than the steam cycle.

Figure 3-4 is a basic heat balance diagram for a simple-cycle gas turbine system. Currently, some of the newest gas turbine plants offer thermodynamic efficiencies superior to the conventional steam cycle plants with simple-cycle heat rates below 10,000 Btu/kWh (10,548 kJ/kWh) and thermal efficiencies greater than 34%. Still, gas turbine plants often do not match the operating cost-efficiency of conventional steam plants because they operate on higher-cost natural gas and distillate oil fuels, as opposed to lower-cost boiler fuels such as coal. However, due to their relative simplicity, gas turbine-cycle plants have much lower capital costs than steam-cycle plants with full environmental control systems. As a result, the gas turbine plants offer distinct advantages over conventional steam-cycle plants for low load factor peaking service, where the impact of higher operating cost is less significant than capital cost.

Combined-Cycles

The fact that a gas turbine exhausts large quantities of air/gases at temperatures around those of a fairly efficient steam cycle (1,000°F or 538°C) allows for significant enhancements in overall cycle efficiency. As shown in Figure 3-5, if the exhaust from a gas turbine is fed to a **heat recovery steam generator (HRSG)**, the steam that is raised can drive a steam-powered cycle. The steam cycle converts about 15 to 20% of the rejected heat energy into additional electric power, resulting in a combined-cycle thermal efficiency of about 43 to 53%. On a heat rate basis, this plant will require only about 6,400 to 8,000 Btu/kWh (6,750 to 8,440 kJ/kWh), compared with 10,000 to 11,000 Btu/kWh (10,550 to 11,600 kJ/kWh)

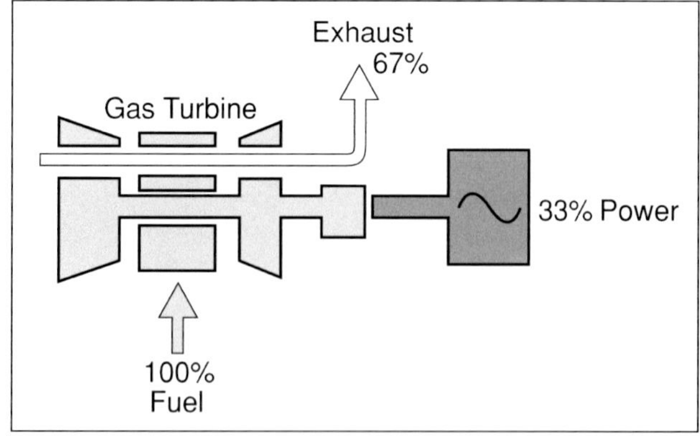

Fig. 3-4 Basic Heat balance Diagram for a Simple-Cycle Gas Turbine System.

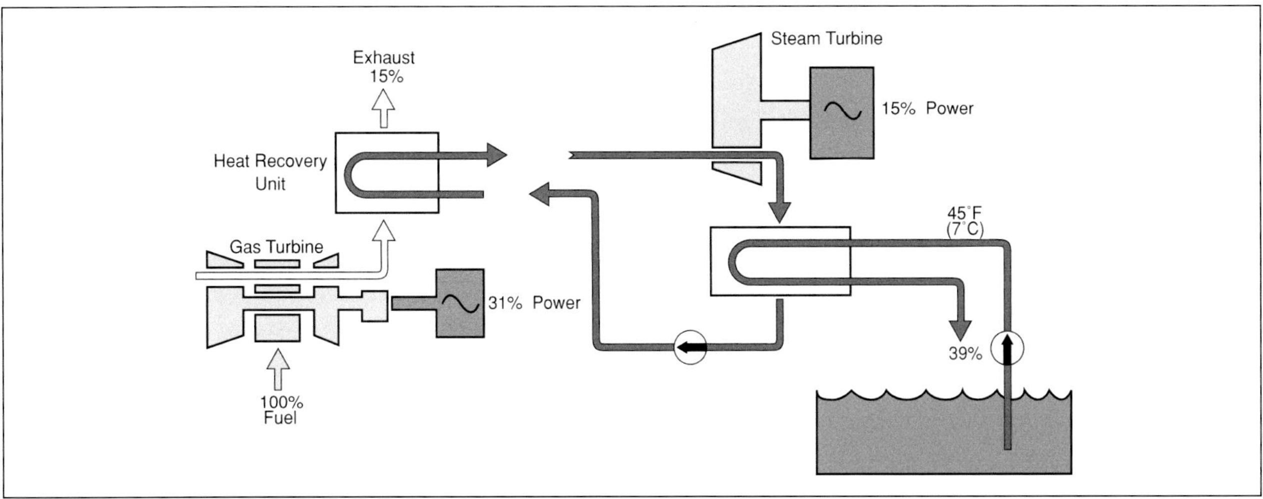

Fig. 3-5 Combined Cycle System Featuring Gas Turbine, Heat Recovery Steam Generator and Condensing Steam Turbine.

for conventional-cycle plants.

The combined-cycle plant capital cost per kW falls between that of a simple gas turbine cycle and a conventional steam cycle. Its exceptional thermal efficiency ratings allow it to compete favorably for intermediate and baseload service.

Table 3-2 shows the Average Operating Heat Rate for Selected Energy Sources from 2001-through-2008. The decreasing heat rates for natural gas fueled plants are a result of more gas turbine based combined-cycle power plants coming on-line.

Other Advanced Cycles

There are a number of other advanced cycles that can be applied to gas turbine or combined cycle systems:

An **integrated gasification combined cycle (IGCC)** is a technology that turns coal (or petrolem based by-products) into synthesis gas (syngas) by heating the fuel in a reduced oxidation state. It then removes impurities from the syngas before it is combusted in a convential gas-turbine generator. This results in lower emissions of sulfur dioxide, particulates and mercury. Excess heat from the primary gas turbine is then passed to a steam cycle, similarly to a combined cycle gas turbine. The technology has been successfully applied to petroleum byproducts in the refinery industry but has not been demonstrated to be commercially viable for large scale central electric plants fueled with coal.

The **steam injection cycle** is similar in many respects to a combined-cycle plant. These plants generate steam

Operating Heat Rate for Selected Energy Sources, 2001 through 2008				
(Btu per kWh)				
Year	**Coal**	**Petroleum**	**Natural Gas**	**Nuclear**
2001	10,378	10,742	10,051	10,443
2002	10,314	10,641	9,533	10,442
2003	10,297	10,610	9,207	10,421
2004	10,331	10,571	8,647	10,427
2005	10,373	10,631	8,551	10,436
2006	10,351	10,809	8,471	10,436
2007.	10,375	10,794	8,403	10,485
2008	10,378	11,015	8,305	10,453

Table 3-2 Average Operating Heat Rate for Selected Energy Sources, 2001 through 2008 (Btu per kWh)

from recovered heat from the basic gas turbine cycle. However, instead of using the steam to drive a second power cycle, steam injection-cycle plants inject the steam directly into the gas turbine. The increased total mass flow and energy input to the gas turbine result in significantly enhanced system capacity and overall cycle efficiency. Since steam is used in an open cycle as opposed to a closed condensing steam turbine cycle, the overall efficiency will be slightly lower than that achieved with a combined-cycle. Water usage is also considerably greater.

A **regenerator cycle** uses a heat exchanger, or **recuperator,** which transfers heat from turbine exhaust to compressor discharge air prior to combustion of fuel. Recovered heat displaces a portion of fuel that would otherwise be required, thereby enhancing overall cycle efficiency.

The **reheat cycle** uses an additional combustor or reheat element in which additional fuel is combusted using the oxygen present in the exhaust gas. The reheat cycle increases the thermal efficiency of the turbine cycle by increasing the average temperature of the gases doing expansion work in the turbine section.

The **intercooling cycle** is used to decrease the work of compression required by the gas turbine cycle by cooling the air in the middle of its compression cycle. For this purpose, two or more compressor sections are used. The intercooler is a heat exchanger through which air exiting the low-pressure compressor passes prior to entering the high-pressure compressor. Intercooling results in lower high-pressure compressor exit temperature, which allows for higher pressure ratios and, therefore, a significant increase in turbine capacity.

The **humid air turbine (HAT)** cycle is currently under development and is expected to be in production within a few years. In the HAT cycle, exhaust heat from the gas turbine is used to heat and humidify the combustion air. The HAT cycle will operate with intercooling and high-pressure ratios and is expected to offer a thermal efficiency of 45% (LHV) or greater. The HAT cycle is also being designed to operate with fuel from a coal gasifier.

There have also been advancements in coal-fired steam cycle plants. Development has been driven by the demand to minimize air emissions and waste production as much as for improved cycle efficiency. **Fluidized-bed combustion (FBC)** is a technology that has widespread appeal because of its low emission characteristics.

In an FBC unit, solid, liquid, or gaseous fuels, together with inert materials such as sand, silica, or alumina, and/or sorbents such as limestone are kept suspended through the action of primary air distributed below the combustor floor. Fluidization promotes turbulence, which makes the mass of solids behave more like a liquid. The results of FBC are lower and more uniform distribution of temperature. Fluidized-bed configurations include bubbling-bed and circulating-fluidized-bed designs, with atmospheric- or elevated-pressure operations.

Research and development for enhancing cycle efficiency of both conventional steam and gas turbine cycle plants is largely focused on improved materials. Given that specific capacity and cycle efficiency are closely tied to increased firing temperatures, materials technology has become a limiting factor. Component development technology has centered on ceramics and advanced alloys for high temperature. Increasing pressure and temperature tolerance due to improved component material strength and turbine blade cooling will allow for improved cycle efficiency.

Power Plant Costs

Table 3-3 shows comparative characteristics of current and developing fossil-fueled generating cycle technologies by one nationally recognized source. Included are comparative heat rates, capital (overnight), and operations and maintenance (O&M) costs in 2008 $/MWh. O&M costs are differentiated between fixed and variable components. Overnight costs do not include financing costs during construction and other peripheral costs, such as site purchase and development. Actual turnkey costs can, therefore, be significantly higher. Additionally, the labor and material costs associated with centralized power plant construction, as well as localized on-site systems, vary from region to region. Regional multipliers for new construction costs that may be applied to the various new fossil-fueled generating technologies can be found in standard construction estimating references.

On-Site Application of Conventional and Advanced Cycle Plants

Use of the conventional and advanced steam turbine Rankine cycle and gas turbine Brayton cycle are not limited to large utility applications. Along with Otto- and Diesel-cycle reciprocating engines, they have been effectively applied on-site at industrial, commercial, and institutional facilities for generating electric power or for mechanical drive service. Figure 3-6 shows representative electric generation efficiencies on a lower heating value (LHV) basis of small capacity simple-cycle gas turbines and reciprocating engines. Reciprocating engine categories included are naturally aspirated and turbocharged spark ignition, Otto-cycle engines

Technology	Overnight Costs 2008 $/kW	Heat Rate Btu/kWh	Fixed O & M 2008 $/kW	Variable O & M 2008 $/MWh
Base and Intermediate Loaded Technologies:				
Pulverized coal (super critical)	2,485	9,118	28.1	4.68
Coal Gasification (IGCC)	3,359	8,528	39.4	2.98
Natural Gas Combined cycle	1,186	6,647	11.9	2.05
Nuclear (generation III/III+)	3,682	10,400	69.2	2.6
Peaking Plant Technologies:				
Combustion turbine	304	11,456	1.9	3.4
Advanced combustion turbine	488	9,149	32.2	2.8

Source: CRS Report to Congress November, 2008 for Base and Intermediate Loaded Technologies; Peaking technologies were extrapolated from 1987 values using the price escalation for Combined Cycles

Table 3-3. Cost and Performance for Typical Electric Power Generation Technologies

and Diesel-cycle engines.

LOCALIZED ON-SITE ELECTRIC POWER GENERATION VS. CENTRALIZED POWER PLANT

In applications of conventional-cycle steam plants and advanced combined-cycle gas turbine plants for intermediate and baseload operation, the centralized plants enjoy several distinct advantages over smaller systems. As noted previously, these include economy of scale, preferential fuel costs, relatively lower staffing levels, potential for higher generation efficiency, and reserve capacity.

However, localized on-site conventional and advanced cycle plants also enjoy certain benefits over centralized electric generating plants. One significant advantage of prime movers applied on site is the lack of the significant transmission and distribution inefficiencies. High voltage transmission systems typically exhibit efficiency losses of 4 to 10%. Under extreme load and temperature conditions, this may increase to more than 15%. Step-down transformers may contribute between 2 and 10% additional efficiency loss to delivered electric power. However, new efficiency standards should limit transformer losses to 1% for new installations.

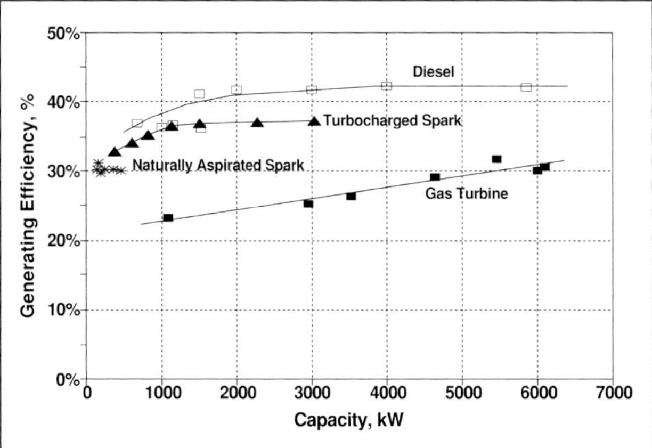

Fig. 3-6 Gas Turbine and Reciprocating Engine-Generator Set Efficiency as a Function of Capacity. Source: Cogen Designs, Inc.

For example, a conventional steam plant featuring a cycle efficiency of 32% with transmission and step-down transformer losses totaling 12% will provide a delivered efficiency to the consumer of only 28%. An advanced combined-cycle system featuring a total cycle efficiency of 43% will provide a delivered efficiency of only 38%.

Despite these losses, the cost-efficiency benefits enjoyed by central electric generation plants usually favor purchased electricity options over on-site electric generation options for providing baseload electric power. Exceptions include circumstances in which centrally generated electric power has been unavailable or too costly to deliver due to the remote location of a facility, or for providing emergency back-up power in the event of a utility outage. Other exceptions include conditions in which a substation or other electric service upgrade necessary to accommodate additional electric power requirements are cost-prohibitive, or when electric distribution system reliability is poor.

For intermittent and peak loads, there are numerous conditions under which on-site electric generation, absent heat recovery, can be economically applied. Strategic peak shaving electric generation can prove cost-effective in areas that face steep demand charges or high on-peak usage components in time-of-use (TOU) differentiated rates. For example, high-speed reciprocating engines are commonly applied to peak shaving generation duty. Required emergency or standby generation capacity can also be used to purchase on-peak power on a less expensive non-firm or interruptible basis.

On-site Fuel Generated Heat vs. On-site Conversion of Electricity for Heat

On-site fuel combustion to generate thermal energy in the form of hot air, hot water, or steam can typically be provided at an efficiency of 75 to 85%. In comparison, electric resistance heating is provided at a delivered efficiency of 28 to 38% from a centralized power plant. While the end-use conversion efficiency of electricity to heat might be as high as 99%, source energy utilization for on-site fuel combustion shows an overwhelming efficiency advantage in most cases.

In a limited range of applications, electric-drive heat pump technologies provide efficient transfer of heat from ambient or low-temperature sources to meet low-grade heating loads. Heat pumps are commonly used, for example, for space heating applications in areas subject to mild winter weather conditions.

On-Site Powered Mechanical Drive Service vs. Centralized Power Plant

Prime movers are sometimes applied in simple-cycle operation for mechanical drive service. In such cases, generator, motor, and electric interconnection costs and associated efficiency losses are avoided, reducing capital and operating costs. These advantages, along with several other benefits, can produce more attractive project economics than on-site electric generation.

An on-site prime mover may compete with an electric motor to drive devices such as compressors, pumps, and fans. In this case, electrically driven end-use equipment includes motor efficiency losses ranging from 5 to 15%. When motor efficiency is considered, a delivered electric service efficiency of 28% provides mechanical service shaft power at an efficiency of less than 26%.

Another advantage of many prime movers applied on-site for mechanical drive services is the ability to track, or follow, load with variable speed operation. **Load following** is matching the amount of work a device is performing with the amount of energy a prime mover is supplying the device. Perfectly efficient load following exactly matches equipment's work load with its energy consumption.

When constant speed drive, such as that produced by a conventional electric motor, is used in a system with variable output, equipment must be throttled to match production output. This yields inefficiency. With variable speed operation (by a prime mover or variable speed electric motor), throttling is not required over a wide range of operating conditions. This allows for operation at, or close to, full-load efficiency under varying load conditions. Figure 3-7 shows a representative comparison of cycle efficiency of variable or optimal (O) speed versus constant (C) speed operation for a reciprocating engine.

Given these advantages, there are numerous conditions under which prime movers without the benefit of heat recovery may be economically applied on-site for mechanical drive service, even though electric motor drives enjoy a capital cost advantage and are less complex than prime mover drives. Circumstances that promote economic performance advantages include seasonal variations in fuel costs and TOU-differentiated electric rates.

When load profiles are highly variable, fixed costs in the form of electric rate demand charges drive up the average cost of purchased electricity. For example, steam turbine and reciprocating engine-driven air-conditioning chillers are fairly common due to contrasting trends of low summer fuel costs and high summer electric rates. Another case where prime movers may be particularly economical for mechanical drive service is when facilities must upgrade electric service capacity to accommodate added loads that include seasonal or daily spikes in electric demand.

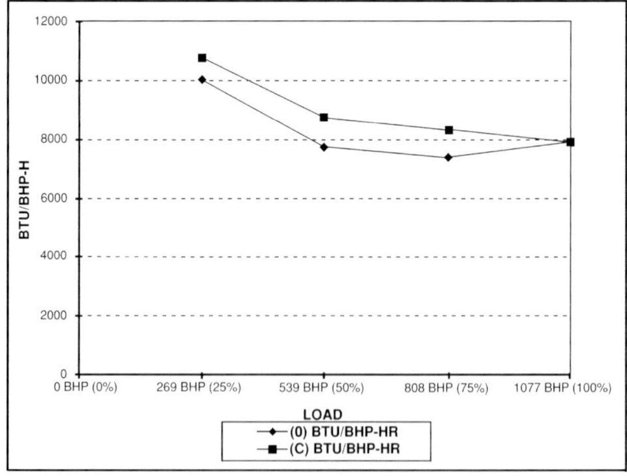

Fig. 3-7 Performance Comparison of Constant and Variable Speed Reciprocating Engine Operation. Source: Caterpillar Engine Division

Applied Cogeneration Cycles

Cogeneration increases power cycle net efficiency by using rejected heat (thermal energy) to produce other useful products. These products range from hot water for heating to process steam to chilled water for cooling. Most importantly, recovering the prime mover's thermal energy can eliminate the energy otherwise consumed to conventionally produce the same useful products. Generally, cogeneration cycles cannot be applied in centralized electric power generation plants because there is no host demand for the available thermal energy — the largest exceptions being large-scale city-wide district heating systems.

Figure 3-8 is a basic heat balance diagram for a cogeneration-cycle system featuring a gas turbine and heat recovery unit. The recovery of the thermal energy to displace these energy sources provides an added financial value to an on-site prime mover application in the form of an energy credit. Compared with a conventional or combined-cycle central electric generation plant, an on-site cogeneration cycle not only achieves higher overall and net energy efficiencies, but also provides other environmental benefits associated with air quality and non-renewable resource conservation.

Cogeneration Cycle Efficiency Example

Figures 3-9 and 3-10 illustrate the potential efficiency improvement provided by cogeneration in meeting local power and thermal loads. In Figure 3-9, independent power and thermal energy conversions are shown. In the steam power cycle on the left side of the figure, roughly 57% of fuel energy input is lost in condenser heat rejection. The right side of the figure illustrates heat generation for a process steam load, such as paper drying, steam distillation, or absorption chilling. In this application, a fired boiler generates steam at a pressure of about 150 psig (11 bar), which is delivered to the process and returned to the boiler as liquid condensate at about 200°F (93°C). Condensate, containing about 167 Btu/lbm (388 kJ/kg), is heated, evaporated, and superheated by the boiler, adding roughly 1,080 Btu of energy to each pound (2,511 kJ to each kg) of condensate. Stack losses result in a fuel energy requirement of about 1,300 Btu/lbm (3,020 kJ/kg) of steam. The process can use all of the latent heat energy in the steam, and the application has a thermal efficiency of about 83%.

Figure 3-10 illustrates energy use by a cogeneration system in which steam is fed sequentially through a steam turbine, followed by the process application. The result is that the heat rejected by the condenser in the conventional cycle is

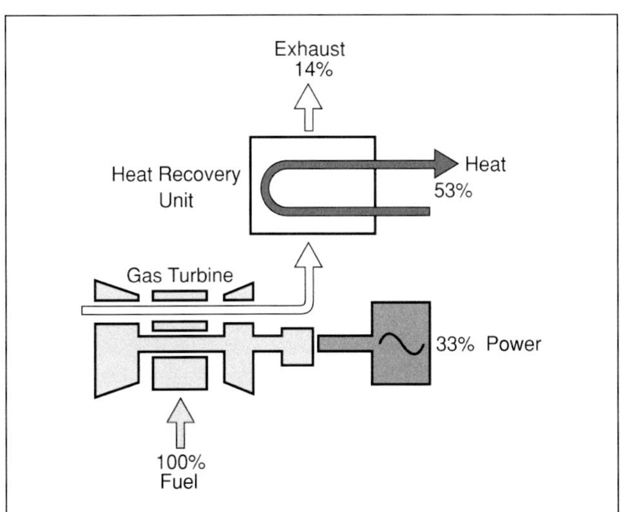

Fig. 3-8 Basic Heat Balance for Cogeneration System.

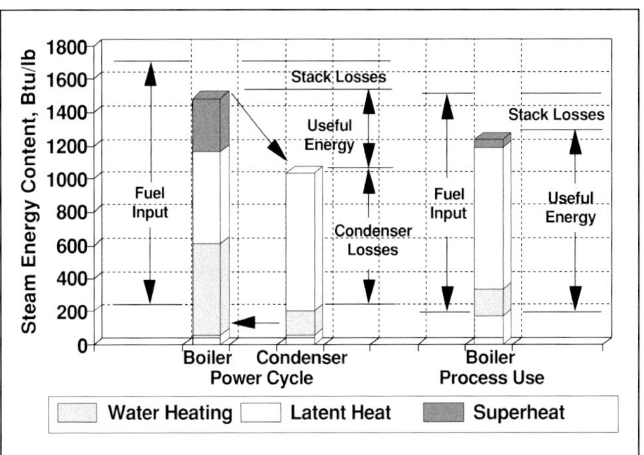

Fig. 3-9 Energy Utilization by Power Cycle and Process Loads. Source: Cogen Designs, Inc.

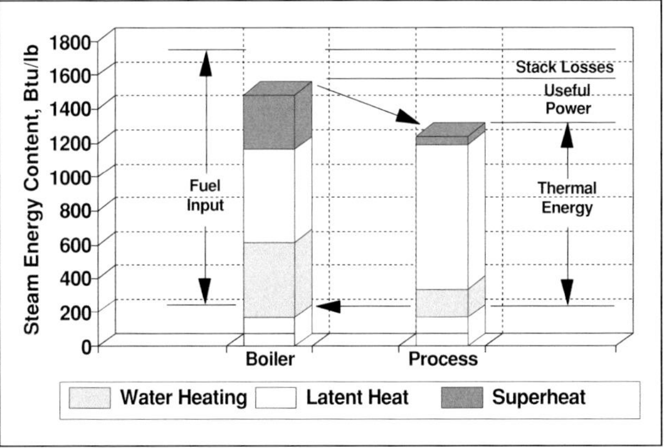

Fig. 3-10 Cogeneration System Energy Utilization. Source: Cogen Designs, Inc.

directed instead to the process heating load. This approach also eliminates the process boiler and its stack loses.

While the power production capabilities of the system are reduced as a result of raising the steam turbine exhaust pressure from vacuum to 150 psi (10 bar), the gains due to latent heat utilization and elimination of one set of stack losses result in an overall thermal efficiency of greater than 85%. In this example, for each pound of process steam delivered, the steam turbine generator would also produce 0.05 kWh of electrical energy.

The separate conventional steam cycle power plants and process boilers shown in Figure 3-9 require about 1,850 Btu (1,950 kJ) of fuel energy, while the cogeneration facility shown in Figure 3-10 requires only 1,420 Btu (1,498 kJ) of fuel energy, giving a 23% energy savings overall. This is the benefit of the "sequential use" concept of cogeneration. Similar benefits also accrue to gas turbine and reciprocating engine cogeneration cycles.

COGENERATION COMBINED-CYCLES

Cogeneration can also enhance the thermal efficiency of advanced power cycles, such as the combined cycle. Figure 3-11 is a simplified heat balance diagram for combined-cycle cogeneration, in which steam turbine exhaust serves a process heating load. The result is a significant upgrade in the net efficiency of the power cycle, from 43 to 50% without cogeneration, to 65% or greater. A comparison of Figure 3-11 with Figure 3-5 shows that only 10% of the fuel energy input is converted to power in the steam turbine in the cogeneration combined cycle as opposed to 15% with the standard combined cycle. However, this shortfall in power generation is more than balanced by delivery of 44% of input energy to a process application, as opposed to losing 39% of input energy to the condenser in the standard combined-cycle system.

The performance of this type of cogeneration cycle can be compared with other power cycles through the concept of a heat rate based on **fuel-chargeable-to-power** (FCP). The combined cycle cogeneration plant would have an FCP heat rate of 6,500 Btu/kWh (6,856 kJ/kWh) or less, compared with 10,000 to 11,000 Btu/kWh (10,550 to 11,600 kJ/kWh) for a conventional steam cycle and 7,000 to 8,000 Btu/kWh (7,380 to 8,440 kJ/kWh) for a combined-cycle power plant.

If the cogeneration combined-cycle is designed from the outset to maximize energy recovery for the power and process steam applications, still further efficiency gains are possible. For example, adding a second evaporator section to the boiler, operating at the process steam application pressure, can reduce stack losses and produce additional process steam without the addition of more fuel. The result is an overall cycle efficiency of nearly 72% and an FCP heat rate of only 6,180 Btu/kWh (6,520 kJ/kWh). This cycle is twice as efficient as the average conventional steam cycle and nearly three times as efficient as several of the conventional steam cycle plants currently being operated across the country.

COGENERATION CYCLE COMPARISON

Figure 3-12 is a comparison of FCP heat rate for three advanced cogeneration cycle alternatives, each utilizing supplemental duct firing to enhance the available energy to the HRSG:

Alternative-1: A simple-cycle-cogeneration system featuring a gas turbine and HRSG with bypass stack and supplementary firing (ref Fig. 3-8).

Alternative-2: A combined-cycle system featuring a gas turbine, HRSG, back-pressure steam turbine cycle with bypass stack, and supplementary firing (ref Fig. 3-11).

Alternative-3: A combined-cycle system featuring a gas turvine, HRSG, condensing-extraction steam turvine cycle with supplementary firing, but no bypass stack (ref Fig. 3-5).

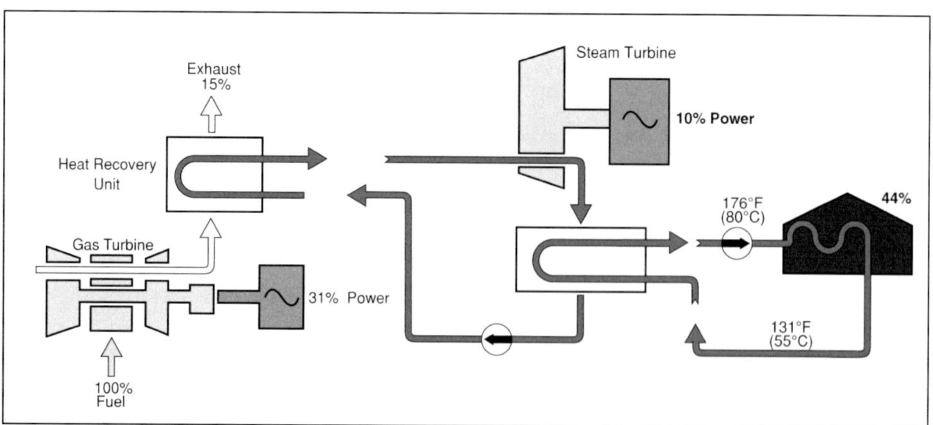

Fig. 3-11 Basic Heat Balance Diagram for Combined-Cycle Cogeneration System Featuring Gas Turbine, Heat Recovery Steam Generator and Non-Condensing Steam Turbine.

The comparisons are based on a basic power cycle featuring steam operating conditions of 900 psig/900°F (63 bar/482°C) and 150 psig (11.4 bar) saturated process steam, 2.5 in. (6.4 cm) Hg condenser pressure, 2% blowdown, and 180°F (82°C) condensate return. The displaced boiler energy conversion efficiency taken into account in the FCP expression is assumed to be 83%.

For each system, the FCP heat rate varies as the amount of steam sent to process (actually steam per MWh of power generated) varies. In the first two alternatives, zero steam sales represents a gas turbine with all exhaust vented to the atmosphere (a peaking turbine), and the indicated 11,000 Btu/kWh (11,600 kJ/kWh) heat rate is similar to those of conventional steam cycle plants.

The simple cycle (Alternative 1) has the highest steam/power ratio break point of 3.5 Mlb/MWh, because this HRSG is producing steam at the process application pressure (in this example 150 psig or 11.4 bar) rather than at the turbine pressure (900 psig or 63 bar). With firing, this cycle continues to improve until it reaches its maximum cycle efficiency at an FCP of just below 6,000 Btu/kWh (6,330 kJ/kWh). The overall cycle heat rate continues to improve as duct firing is increased. The maximum duct firing rate is typically limited by a maximum HRSG design temperature.

The back-pressure cycle (Alternative 2) break point is at a slightly greater steam export than Alternative 3 before firing, since all steam passes completely through the steam turbine. None of the fired steam ever goes to condenser, so this cycle keeps getting more efficient as supplemental firing increases. As with Alternative 1, the overall cycle heat rate continues to improve as duct firing is increased.

Alternative 3 at 8,000 Btu/kWh (8,440 kJ/kWh), is roughly 25% more efficient than the average conventional steam cycle plant, since the listed condensing-extraction steam turbine remains in operation at zero steam sales. The FCP heat rate drops with increased process steam use, until a ratio of about 3 Mlb/MWh (1,360 kg/MWh) is reached. This is the point at which no more steam can be taken from the condensing section of the steam turbine (below the extraction port) without overheating, and additional steam must come from supplemental firing with a duct burner or from a fired heat recovery boiler. Because this "fired" steam is used less efficiently, the heat rate curve starts to climb back slowly with greater firing, but the cycle remains roughly twice as efficient as conventionally produced power purchased from the utility grid.

Emissions Comparison of Cogeneration vs. No Cogeneration

Paramount to the concept of controlling pollution, and air emissions in particular, is the understanding that, in addition to the use of less polluting fuels and more effective pollution control technologies, the best solution is often simply to use less fuel by improving efficiency. The least environmentally harmful fuel unit is the one that is not consumed.

Figure 3-13 compares the NO_X and CO_2 (greenhouse gas) emissions of a representative combined-cycle cogeneration system applied on site against the comparable emissions of a utility-built combined cycle plus an industrial (gas-fired) boiler producing steam for industry. The point of

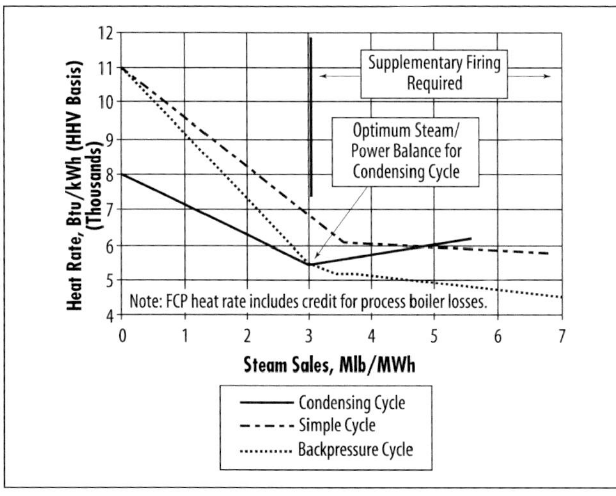

Fig. 3-12 Fuel-Chargeable-to-Power Comparison of Three Cogeneration Cycles. Source: Cogen Designs, Inc.

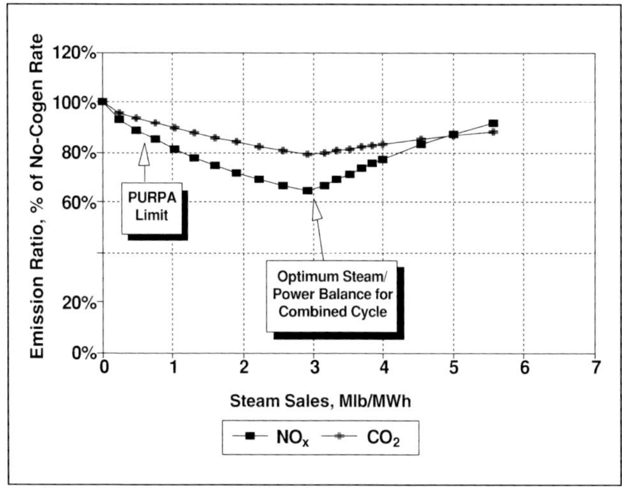

Fig. 3-13 Comparison of Cogeneration Cycle Emissions with No Cogeneration. Source: Cogen Designs, Inc.

comparison in the figure (100% Emissions Ratio) is based on a total emissions by the central utility and the local industrial boiler. Relative emissions at varying steam sales are indicated for a combined-cycle cogeneration plant equivalent to Alternative 3 in the previous FCP comparison.

The assumptions are that the combined cycle is equipped with low-NO_X combustors (25 ppm) and supplemental burners (0.15 lbm NO_X/MMBtu or 0.068 gram/MJ), and that the industrial boiler was also equipped with low-NO_X burners (0.15 lbm NO_X/MMBtu or 0.068 gram/MJ). In this case, the lowest emissions (both NO_X and CO_2) fall at the point where supplemental firing would just start to be needed to produce more steam. Figure 3-14 shows specific CO_2 emissions from different applied power plant technologies as a function of net station efficiency.

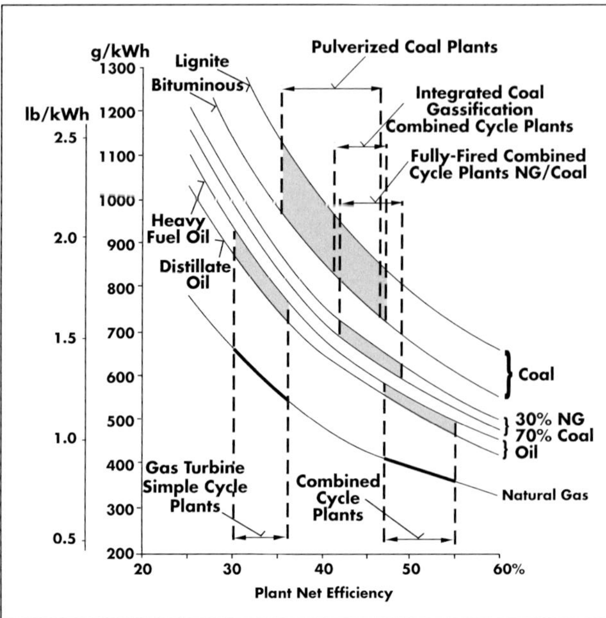

Fig. 3-14 Specific CO_2 emissions from different applied power plant technologies as a function of net station efficiency. Source: Siemens Power Corp.

Application of Cogeneration Cycles On-site vs. Purchased Electricity

The benefits of avoided transmission and distribution losses associated with electricity purchased from a centralized plant and the ability to employ cogeneration cycles often swings life-cycle costs in the favor of on-site prime mover applications. Although centralized systems can employ combined-cycles and other advanced cycles for electric power production, the FCP thermal efficiency comparison always favors cogeneration.

When a facility can use recovered heat for thermal processes, the cogeneration cycle efficiency benefits may shift the life-cycle cost advantage away from the purchased electricity option. When cogeneration cycles can be used for on-site mechanical drive service, the additional benefits of load tracking and avoided electric motor efficiency losses may further improve life-cycle costs relative to purchased electricity options.

Comparison of On-site Electric Generation vs. Mechanical Drive Services

Although mechanical drive applications may enjoy capital cost and efficiency advantages, they are more limited since project size and economics are tied to the location, characteristics, hours of operation, and load factor of only one application. There are thus more opportunities for electric generation and the applications are more flexible. In addition, on-site electric generation systems often can be sized larger than individual mechanical drive systems. One larger system may often be less costly to install and more efficient to use than several smaller ones.

On-site electric generation offers the ability to produce power that can be exported to the grid for sale to the local utility, to another utility via wholesale wheeling, or possibly retail purchases via retail wheeling. On-site generation also allows the facility to choose whether to generate power or purchase power at any given time, depending on the market price. In most cases, mechanical drive end-use outputs, such as chilled water, are less amenable to export for sale outside the facility.

CHAPTER FOUR

Selection of Power Generation Systems

Determining the technical and financial feasibility for on-site power generation requires knowledge of numerous site-specific conditions and the characteristics of alternative prime mover systems. Considerations include facility load profile, main and secondary fuel sources, main and secondary useful products, environmental impacts, physical space requirements, auxiliary components and their parasitic energy consumption, noise levels, and aesthetics. Typically, the most difficult decisions are the ones subject to negotiable values and services, such as prime mover manufacturer, installation contractors and maintenance services.

This chapter discusses prime mover selection considerations, including application and data requirements and evaluation of prime mover performance ratings. This is followed by a comparative discussion of reciprocating engine, combustion gas turbine, and steam turbine systems.

APPLICATION AND SYSTEM DATA REQUIREMENTS

Site-specific prime mover and system data required in virtually all prime mover evaluations are:

- Type of application and current supply system
- Metered or engineering estimate of loads versus time, outdoor temperature, occupancy, or production
- Reliability requirements
- Operating cost of the current supply systems
- Current on-site utility-derived system capacity, service entry characteristics, and reliability
- Site-specific conditions, such as elevation and ambient temperature ranges, that will impact the performance of individual prime movers being considered
- Planned growth of loads
- Types of fuels and fuel storage available to the site
- Field experience under continuous-duty applications of potential power generation systems, including availability, O&M costs, reliability, and compliance with environmental regulations
- Air quality regulations and the facility's existing permit parameters
- Potential siting locations and physical space and zoning limitations
- Federal, state, local, and facility design and installation codes and regulations
- Current energy supplier pricing rates and expected trends

The importance of accurate data collection cannot be overstated. High-quality baseline data is needed to ensure the feasibility of the selected approach and to ensure an accurate and comprehensive analysis of real energy and financial performance.

While field experience is the best indicator of system performance, it is not a guarantee. A broad array of factors, such as site installation design, operation and maintenance procedures, fuel quality, air contamination, and water treatment practices, can influence the success of any installation.

Three of the more important factors that must be considered when evaluating prime movers are performance ratings, environmental regulations, and operations and maintenance costs. These are discussed below.

PERFORMANCE RATINGS

Prime mover performance ratings include data on capacity, thermal efficiency, and other operating characteristics, such as air emissions. [It is a standard across many energy-related industries to interchangeably use the terms energy efficiency and performance.]

Ratings may be expressed as a function of power (e.g., hp) or brake power (e.g., bhp). Brake power is the power that does useful work, measured at the crankshaft or flywheel. The term "brake" is used because typically the force can be measured by a dynamometer based on the braking power necessary to stop the engine. Related ratings such as specific fuel consumption (scf) or mean effective pressure (mep) are expressed as brake specific fuel consumption (bsfc) and brake mean effective pressure (bmep) when related to brake power.

Fuel consumption is sometimes expressed as specific fuel consumption or brake specific fuel consumption, where "specific" refers to the consumption on a per unit of power output basis. For a natural gas-fueled prime mover, fuel consumed may be expressed in terms such as cubic feet (cf) or cubic meter (m^3), or 100 cubic feet (Ccf) of gas consumed. In most cases, cf (m^3) actually refers to standard cubic feet (scf) or standard cubic meter (sm^3), with measurement at standard conditions of atmospheric pressure of about 14.7 psia (101.3 kPa) and an ambient temperature of 60°F (16°C).

The natural evolution of competing products and

industries, as well as customer demands, has created various specialized performance ratings for prime movers and their competing technologies. For instance, in a chilled water application, a natural gas fueled prime mover's performance may be measured by the number of cf consumed in one hour to produce 1 ton (kW_r in SI units with the subscript r indicating refrigeration output) of chilled water, or cf/ton-h (m^3/kWh_r). In an electric generation application, a natural gas-fueled prime mover's performance may be represented as cf/h per kW (m^3/h per kW) or cf/kWh (m^3/kWh).

Manufacturers do not always emphasize parasitic energy usage by sub-systems that offset a portion of the power produced by the prime mover. Depending upon the type of prime mover, sub-systems such as compressors, fans, hydraulic pumps, and controls may account for significant amounts of energy. Only when all of these parasitic uses are identified and normalized to the useful product will an accurate picture of the prime mover's performance appear.

Regardless of the type of prime mover, it is important to understand two important points about performance ratings. First, prime mover capacity and thermal efficiency ratings are based upon various assumed conditions that affect performance. Second, each prime mover manufacturer selects the assumptions used in published performance ratings. Therefore, an accurate comparison of prime movers is not possible without knowing the conditions at which they were rated. It is critical to know if the rating conditions are similar to the actual conditions in which the system would operate and, if not, how to adjust published ratings to fit actual site conditions. The following is a partial list of variables that may affect prime mover performance ratings:

- Methodology used to determine the rating
- Fuel's heat content (or energy density)
- Whether heat content is specified in higher or lower heating value (HHV or LHV)
- Ambient environmental conditions, such as temperature, humidity, atmospheric pressure, etc.
- Power factor
- Integral accessories, such as turbochargers
- Non-integral parasitics (non-integral power losses)
- Location of take-off (i.e., flywheel, generator terminals, etc.)
- Continuous rating, peak rating, or other

Consider, for example, a case in which a particular facility is comparing cogeneration alternatives that include a 1.5 MW reciprocating engine generator set and a similarly sized gas turbine engine generator set. After manufacturers' published data is corrected to the same conditions, each unit generates electricity at 27% efficiency (i.e., at generator terminals). However, the reciprocating engine only requires 35 psig (3.4 bar) natural gas, while the gas turbine engine requires natural gas at 250 psig (18.3 bar). The local natural gas distributor supplies 20 psig (2.4 bar) natural gas at the metering header. Under these conditions, the reciprocating engine needs a 40 kW compressor, thus reducing its system electrical efficiency rating to 26%. Under the same conditions, the gas turbine engine needs a 100 kW compressor, thus reducing its system electrical efficiency rating to 24%. At a displaced electricity price of $0.06/kWh and 7,500 operating hours per year, this 2% efficiency advantage is responsible for an additional $27,000 per year in savings.

In some cases, performance data use standards adopted from various institutions. An example is the International Standards Organization (ISO). ISO standards are used extensively in Europe and in the United States. Numerous other American Standards are also used. Regardless of the standard used, it is critical to identify the measurement tolerances allowed. This will indicate the allowable margin of error between published data and actual measurements after all site-specific adjustments are made.

The Impact of Emissions Control Requirements on Prime Mover Selection

The Clean Air Act (CAA), first enacted in 1970, establishes the nation's air regulatory priorities and a basic regulatory framework. It has been amended several times over the years. The Clean Air Act Amendment of 1990 (CAAA) expanded significantly the scope and detail of the national air regulatory program. Federal and state regulations require emission source owners to minimize or reduce air emissions under a number of regulatory circumstances. The type of regulation influences the preferred control approach, as does the pollutant to be controlled and the stringency of the requirement.

In order to install and operate prime movers or other fuel burning equipment, permits must be acquired to confirm compliance with environmental regulations. There are two general types of permits under state and federal air regulations: permits allowing the construction or modification of air pollution sources, generally referred to as a new source or preconstruction permit, and permits authorizing the operation of air emissions sources. The requirements for obtaining these permits vary dramatically from state to state.

The ability to control regulated air emissions from prime movers and related equipment is an important component in the selection and operation of the prime mover control system.

Natural gas-fired Otto-cycle engines, for example, may be favored over a more thermally efficient or less costly liquid-fueled Diesel engine, largely on the basis of the need to minimize emission of certain pollutants. Steam turbines may be selected over direct fuel-burning reciprocating engines and gas turbines in some cases, because the fuel-burning component (i.e., the boiler) has already been permitted.

Comparison of Prime Mover Characteristics

Once all of the required data for a potential prime mover application is gathered, various prime mover system options can be compared. Once the field of options has been limited by preliminary screening, more detailed analyses are performed on remaining options that are close in economic and environmental performance.

During the preliminary screening phase, certain generalizations may be applied to different categories of prime movers (i.e., reciprocating engines, gas turbines and steam turbines) and to types within these categories, such as condensing or non-condensing steam turbines, single- or multi-shaft gas turbines, and Otto- or Diesel-cycle reciprocating engines. It is important, however, to maintain a broad perspective at the outset so as not to eliminate a prime mover class or type that could, surprisingly, show better economic performance than would be indicated by preconceived generalizations.

Comparison of Combustion Gas Turbines and Reciprocating Engines

Generally, when comparing reciprocating engines and combustion gas turbines, gas turbine economic performance will improve under the following conditions:

- Power production maintained continuously at full load
- High-temperature thermal energy required by on-site processes
- A high-capacity system

Conversely, reciprocating engine economic performance tends to improve relative to gas turbine performance when operation is not continuous or loads vary, low-temperature thermal energy is needed, and as capacity requirements decrease.

In most cases, reciprocating engines offer higher full- and part-load simple-cycle thermal efficiencies than do gas turbines. Reciprocating engines operate on a batch-type combustion cycle, which permits higher peak temperatures and, therefore, promotes high cycle efficiencies. Gas turbines, which are characterized by continuous combustion, have component temperature limitations that reduce cycle efficiency.

Reciprocating engines also tend to be more efficient over a broader range of load and ambient conditions than gas turbines. Due to the relatively high compressor power requirement of a gas turbine, even very small changes away from full-load design conditions produces degraded efficiency in both the compressor and turbine sections. For the same reason, gas turbines are more sensitive to changes in ambient air conditions than reciprocating engines.

In small capacities, both capital cost and thermal efficiency tend to favor reciprocating engines. As equipment capacities increase, these differences tend to decrease. Gas turbines are available in capacities of up to several 100,000 hp (kW), while the largest reciprocating engines are about 75,000 hp (56 MW). Stationary applications featuring reciprocating engines of capacities greater than 30,000 hp (22 MW) are not common, though multiple smaller capacity engines may also be applied in large capacity applications.

Gas turbines offer greater power density than reciprocating engines, since they are physically smaller and lighter per unit of power output. This can result in easier installation. However, when considering total system size, inclusive of heat recovery units and other auxiliary components, systems featuring gas turbines will not necessarily be smaller than systems featuring reciprocating engines.

All recoverable thermal energy from gas turbines is in the form of high-temperature exhaust that can be used to raise high-pressure steam. The high oxygen content of gas turbine exhaust also allows for supplementary duct firing, providing added production of high-pressure steam at very high efficiency. In contrast, only a portion of the recoverable energy from reciprocating engines is in the form of high-temperature exhaust. The remaining portion is low-temperature heat, recoverable from engine coolant systems. In cases where all or most of a facility's thermal energy requirements is in the form of high-pressure steam, with limited application for lower temperature recovered energy, gas turbines have an advantage. Reciprocating engines have an advantage where low-temperature recoverable energy in the form of hot water or low-pressure steam can be effectively used. Also, while most combined-cycle applications include gas turbines with heat recovery steam generation, reciprocating engines may also be effectively used.

Gas turbines generally offer lower maintenance cost requirements and greater reliability than reciprocating engines. In practice, this will depend on the type of application and specific units being compared. Maintenance costs for smaller capacity, high-speed reciprocating engines may be significantly greater than with gas turbines. However, with decreased engine operating speeds and increased capacity, the difference tends to diminish. Very large capacity low-speed reciprocating engines may be equally reliable, or, in some cases, more reliable than comparable capacity gas turbines.

Comparison of Steam Turbines with Gas Turbines and Reciprocating Engines

Relatively low capital and maintenance costs and high net thermal efficiencies characterize steam turbines. However, comparisons with reciprocating engines and gas turbines will vary widely depending on applications and the characteristics of the host facility. Generally, it is not economical to install a complete steam generation and distribution system in a facility solely for the purpose of applying steam turbine technology. However, if a central high-pressure steam generation system is required or is already in place, there are a variety of applications that tend to favor steam turbines over reciprocating engines and gas turbines. A classic example is a facility that requires high-pressure steam to serve a portion of the total load and low-pressure steam to serve the balance of the load. The steam turbine can serve as a pressure-reducing station, producing power as the steam pressure is dropped. If the facility has a low-pressure steam distribution system in place, a back-pressure or extraction turbine can be applied, perhaps far more cost-effectively than any other prime mover option.

Other applications that favor steam turbines involve heat recovery from process applications and, of course, combined-cycle applications. Condensing steam turbine applications that do not utilize recovered thermal energy are usually less efficient than other prime mover alternatives. In addition, these applications usually require more costly turbines than do back-pressure applications and involve additional auxiliary equipment, such as surface condensers and cooling towers. Still, they may prove cost-effective for certain applications. Steam turbine-driven chillers, for example, are designed around excess boiler capacity in non-heating season months. Incremental maintenance costs are low (for an existing steam system) and project capital costs may be lower than with the other prime mover options.

A potential advantage of steam turbines is that the cost per unit of fuel may be lower because boilers have more fuel choice flexibility than do most reciprocating engines and gas turbines. Another potential advantage of steam turbines is that if a steam generation system is already in place, environmental permitting requirements may be avoided. This would depend on the prevailing local environmental regulations and whether or not the facility has taken restrictions on boiler operation.

In summary, the preceding generalizations should be applied with caution. They are by no means hard and fast rules and are subject to debate within the energy industry. While many of these generalizations may often prove true, there are many variables involved in the prime mover selection process. A site-specific investigation of all available options is required to determine the best strategy for meeting facility energy requirements. This involves consideration of prevailing fuel and electricity purchase options, environmental regulations, and the characteristics of alternative systems with respect to power and thermal load profiles.

Comparison of Alternative Prime Mover Options

In addition to the three main types of prime movers discussed, four other power-producing technologies warrant consideration in the selection process under certain circumstances: wind turbines, water turbines, photovoltaic cells, and fuel cells. Wind, water, and photovoltaic systems rely exclusively on renewable resources. While fuel cells currently rely on traditional fossil fuel energy sources, they are being promoted in today's market as "green technologies," along with the other three renewable sources, due to their nearly air emissions-free operation. Public policy support, in the form of financial incentives, the ability to avoid air permitting and emissions control costs, and other strategic advantages have elevated each of these technologies to a market position that merits consideration in the power generation source selection process.

Wind and water (hydro) turbines are traditional prime movers in every sense of the term and, in fact, predate, by centuries, reciprocating engines and gas and steam turbines. Driven by wind or water instead of combustion gases or steam, these prime movers produce rotational power in much the same way as a gas or steam turbine. Their distinguishing characteristic is that they rely on no-cost and emission-free renewable resources as their energy sources. Their ability to effectively operate on an essentially no-cost energy source, with only moderately higher initial capital costs, allows them to be quite market-competitive in many circumstances. However, applications are limited to locations in which this free energy source is available at sufficient levels to cost-effectively produce power.

It may be somewhat misleading to refer to hydropower as an alternative energy source, since it provides nearly 20% of the world's electricity. In some countries, such as Canada or New Zealand, hydropower is the major source of electricity and in the United States it provides more than 10% of the nation's electricity and is the dominant electricity source in the Northwest. However, where once it was a fundamental tenant of industry (initially for mechanical power production and later for electricity production), today hydropower applications are greatly limited for individual commercial, industrial, and institutional (CI&I) facilities. Instead most applications reside in the domain of major utilities, regional power authorities, and independent power producers (IPPs) that locate facilities at remote sites and feed electricity to the grid.

Still, in certain limited cases, facilities located on or near rivers can effectively apply hydropower systems for mechanical drive or electricity production applications. Today, standardized "micro-turbine" systems are available for applications ranging from a few MW down to a few kW. These can be applied in "run-of-the-river" type systems that use the power in river water as it passes through the plant without causing an appreciable change in the river flow. These systems can be built on small dams that impound very little water or, in some cases, do not even require a reservoir or dam. Hence, given unique access to a flowing river, hydropower merits consideration as a prime mover system of choice for a CI&I facility.

While a far smaller contributor to the world energy mix (well below 1%) than hydropower, wind turbines, or wind-energy conversion systems (WECSs) as they are now commonly referred, have had a similar historic evolution. These systems were once more common-place for mechanical service applications for pumping or grain production processes; WECSs are also now most commonly found in utility or IPP remote site applications that feed electricity to the grid. Today, several thousand MW of capacity are being installed each year, largely in Europe and the United States, mostly in large wind-farm installations and located where their tall towers are unobtrusive and accessible to steady wind flow at sufficient speed and frequency to be cost-effective.

While these grid-connected systems require wind speed of about 12 or 13 miles per hour (5.4 to 5.8 m/s) for operation, smaller mechanical drive and non-grid connected electric generator applications can effectively operate at wind speed as low as 7 or 8 miles per hour (3.1 to 3.6 m/s). Given their energy cost- and emission-free characteristics, WECSs can be life cycle cost-competitive in today's market, with the appropriate location for siting and steady minimum wind speed availability. In fact, in areas where it is difficult and costly to permit new combustion sources, such as California, WECSs may demonstrate superior economic performance over traditional fuel/steam-driven prime mover systems. They can be particularly attractive for remote locations for which grid connection is either costly or inaccessible. In these cases, they can be used in mixed (hybrid) system applications with other prime mover systems (e.g., Diesel engines) and/or various forms of storage (e.g., batteries or pumped water storage) to provide a steady, reliable power source.

Photovoltaic and fuel cell systems are uniquely distinct from traditional prime movers. Unlike conventional power generation systems that operate on power cycles, these technologies rely on photoelectric or electrochemical reactions to produce direct current (dc) electricity, which, through an inverter, can be converted to alternating current (ac) electricity. Compared with hydropower and even with wind power, these electric power generation technologies have experienced extremely small market penetration to date and, except in very specialized market niches, have not been considered cost-competitive as power producing alternatives due primarily to very high initial capital costs. These costs can be partially offset by elimination of air-permitting and emissions controls costs, which can be substantial in areas with the strictest air permitting limitations (e.g., Severe and Extreme Non-attainment areas) and, in many cases, government-sponsored financial incentives. Hence, they do merit consideration for strategic application under a limited set of circumstances.

While still extremely capital cost intensive, the production costs of photovoltaics have continued to decrease while their electric generation efficiency has increased. For maximum effectiveness and associated financial return, they must be sited in locations with continuous or near continuous sunlight, such as the Southwestern United States. In areas with limited or high-cost access to the electric grid, photovoltaics have proven moderately cost-effective, though usually as part of a mixed or hybrid system, typically with a Diesel engine backup, to provide for high reliability and full load-tracking capability.

Fuel cell costs have also continued to decrease due to increased production and technology advancements, while increasing in thermal efficiency. Because they are not power-cycle limited, it is believed that over time, thermal efficiency will surpass that of traditional combustion and steam-driven prime movers. Capital and long-term maintenance costs (largely the periodic cost of stack replacement) remain significant obstacles to wide-scale market penetration. They have found somewhat of a niche market in what is referred to as "clean power" applications, where the avoidance of power conditioners, uninterrupted power supply (UPS) systems, and standby generators can offset much of the first-cost premium.

While each of these alternative technologies is limited for wide-scale application in CI&I facilities, each offers advantages, which, under the proper conditions, can make them life cycle cost competitive. Moreover, increasing environmental control costs and energy costs, and continual advances in design and production effectiveness, provide the expectation that these "green" technologies will continue to penetrate the energy market place at an increasing rate.

Section II

Thermal Technologies

Chapter Five

Heating Value and Combustion of Fuel

All fuels contain energy. To harness this energy for productive purposes, however, the fuel must undergo a chemical reaction. The most common reaction is the heat liberating process of combustion. **Combustion** is the rapid combination of oxygen with a fuel in the presence of a source of ignition, resulting in the release of heat. The heating value of the fuel is a measure of the amount of heat liberated during combustion. The heating value of the fuel must be known in order to perform thermodynamic and cost analyses of any energy system. The heating values are calculated based on the chemical constituents of the fuels – usually fossil fuel.

The primary constituents of fossil fuels are carbon, hydrogen, and sometimes sulfur. Combined with oxygen from air, carbon forms carbon dioxide and heat, hydrogen forms water vapor and heat, and sulfur forms sulfur dioxide and heat. The products, or chemical compounds, resulting from the three basic combustion processes are combinations of these elements in specific fixed proportions. Heat resulting from combustion is the result of excess energy that the newly formed molecules are forced to liberate due to their internal composition. This type of chemical reaction, in which there is a net release of energy, is termed exothermic.

Perfect Combustion

Perfect combustion is obtained by mixing and burning the correct proportions of fuel and oxygen so that nothing other than the combustion product is left over. The chemically correct ratio of oxygen (or air) to fuel that produces perfect combustion is referred to as the **stoichiometric ratio,** or stoichiometric air-fuel ratio (AFR).

- If more oxygen than the stoichiometric amount (excess air) is used in the process, the air-fuel mixture is considered **"lean"** and the fire is **"oxidizing."** The flame produced tends to be shorter and clearer.
- If more fuel than the stoichiometric quantity (not enough oxygen) is used in the process, the mixture is considered **"rich"** and the fire is **"reducing."** This results in a flame that tends to be longer and sometimes smoky. It is sometimes referred to as incomplete combustion because not all of the fuel gets enough oxygen to burn completely. One of the results can be the formation of carbon monoxide.

Lambda (λ) is the ratio of actual combustion AFR to the stoichiometric AFR. At the stoichiometric point, lambda is 1.0. At fuel-rich conditions, lambda is less than 1.0, while at fuel-lean conditions, lambda is greater than 1.0. The inverse of lambda is the fuel-air **equivalence ratio** (ϕ). This expresses the ratio of the actual fuel-air ratio to the stoichiometric ratio. An equivalence ratio for fuel-lean combustion is less than 1.0, and an equivalence ratio for fuel-rich combustion is greater than 1.0.

Oxidation

Oxidation is a chemical reaction that increases the oxygen content of a chemical compound. Oxidation of a fossil fuel, in the presence of an ignition source, results in combustion. Oxygen for the combustion process usually comes from air, but because oxygen comprises only about 20% of air, the volume of air required for combustion is much greater than the volume of pure oxygen required. Almost all of the remaining 80% of air is composed of nitrogen, which does not take part in the combustion reaction. Nitrogen does, however, absorb some of the heat, resulting in lower flame temperature. Nitrogen in combustion air may also produce nitrogen oxides (NO_X), which are considered harmful emissions. Compared with the complete combustion with pure oxygen, the presence of nitrogen in combustion results in lower efficiency.

Many of the fossil fuels are mixtures of chemical compounds called **hydrocarbons**, which consist primarily of hydrogen and carbon. During the combustion process, a series of hydrogen and carbon compounds is formed and reformed before ultimately combining to create carbon dioxide and water. These compounds vary depending on the type of hydrocarbon burned, as well as the temperature, pressure, amount of oxygen, and degree of mixing. Optimal combustion results from a proper fuel-air ratio, thorough mixing, initial and sustained ignition of the mixture, and proper flame positioning.

Fuels generally are burned as gases. Liquid fuels must be evaporated to form combustible vapors. Some liquid fuels, such as heavy oil, which are not readily vaporized, are commonly atomized. **Atomizing** speeds up evaporation by producing tiny particles of liquid, which increases the surface area.

When burning solid carbon fuels, such as coal, oxygen contacts the carbon surface and forms carbon monoxide gas, which then must be moved away from the surface to allow more oxygen to react with the fuel. Solid fuel is often pulverized to increase its surface area and speed up mass contact.

Ignition is typically accomplished by increasing the rate of the natural oxidation reaction by adding another heat source. Heat is added until the reaction itself releases heat at a rate faster than heat lost to sustain the reaction. Minimum ignition temperature varies with each fuel and is influenced by several variables, such as pressure, velocity, air-fuel mixture uniformity, and ignition source. Ignition temperature usually decreases with increased pressure of the mixture and increases with increasing air moisture content. Table 5-1 shows approximate values or ranges of minimum ignition temperature of fuels and for combustible constituents of fuels commonly used in air at atmospheric pressure. Note that this tabulation is only a guide, since actual values will vary widely.

Combustible	Formula	Temperature, F
Sulfur	S	470
Charcoal	C	650
Fixed carbon (bituminous coal)	C	765
Fixed carbon (semi-anthracite)	C	870
Fixed carbon (anthracite)	C	840 to 1115
Acetylene	C_2H_2	580 to 825
Ethane	C_2H_6	880 to 1165
Ethylene	C_2H_4	900 to 1020
Hydrogen	H_2	1065 to 1095
Methane	CH_4	1170 to 1380
Carbon monoxide	CO	1130 to 1215
Kerosene	--	490 to 560
Gasoline	--	500 to 800

Table 5-1 Representative Ignition Temperature of Fuels in Air. Source: Babcock & Wilcox

Upon ignition, the incoming fuel-air mixture (including non-combustibles) heats up. Radiation or direct contact with surrounding gases or solids transfers the heat. The flame temperature is highest when the losses to the surroundings are lowest. Addition of excess air or fuel provides more material to absorb the heat of combustion, and therefore lowers the flame temperature.

Efficiency increases with increased flame temperature. Beyond approximately 3,500°F (1,927°C), dissociation, or reverse combustion, occurs and heat that was originally liberated is absorbed by the breakdown of combustion products into combustibles and oxygen, slowing the rise of flame temperature.

Higher and Lower Heating Values of Fuels

The energy density — or the heating value — of any fuel depends on its specific chemical composition. At one time, heating value was empirically determined by reacting a known quantity of fuel in a calorimeter with enough oxygen to ensure its complete combustion and measuring the heat quantity generated during the combustion. The amount of information available on the chemical compounds comprising the various fuels today allows the heating values to be calculated from fuel composition.

The results of such calculations for liquids generally are expressed on energy per unit mass or energy per unit volume basis — Btu/lbm (kJ/kg) or Btu/gallon (kJ/liter). For gaseous fuels, values are expressed on volume basis, and the reference conditions for the fuel volume measurements (pressure, temperature, and degree saturation with water vapor), as well as test conditions, have to be stated explicitly. The alternative "shortcut" to such measurement is to use the criteria established in various national standards as much as possible.

Thus, the ideal energy density for methane is about 1,012 Btu/cf (37,710 kJ/m^3) of dry gas. This is based on the volume determined at 14.73 psia (101.6 kPa) and 60°F (15.55°C) through application of the ideal gas law. The water produced during combustion is condensed to liquid state, and all combustion products are cooled to the initial temperature of 60°F at the start of the test. Real energy density takes into account the large number of methane molecules in the real volume and differs slightly from the ideal density. The real energy density of methane, which compensates for the compressibility factor, is 1,014 Btu per standard cf (37,784 kJ/Nm^3) on dry volume basis.

In addition to the compressibility of gas, there is one more important factor. Most fuels, in addition to carbon, contain hydrogen and are known as hydrocarbons. During the combustion of such fuels, oxygen is combined with carbon, forming carbon dioxide, and with hydrogen, forming water vapor. While the carbon dioxide remains in the gaseous state at normal ambient temperatures, the water vapor in the flue products may leave the combustion device either in vapor or condensed to liquid state. Condensation takes place when the finite quantity of heat, known as the **latent heat of vaporization**, is extracted from the water formed during condensation.

When this additional amount of energy is extracted from the flue products, causing the water to become liquid, the fuel's energy density is identified as **higher heating value (HHV)**. When the equipment used allows the water to remain in the vapor state, the energy density is identified as **lower heating value (LHV)**. Thus,

$$LHV = HHV - h_{fg}\, mH_2 0 \qquad (5\text{-}1)$$

Where:

h_{fg} = Heat of vaporization of water per unit mass

$mH_2 0$ = Mass of water

The relationship between HHV and LHV can be derived from the basic combustion equation. The combustion equation for pure methane (CH_4), which is the primary constituent of natural gas, is:

$$CH_4 + 2O_2 \rightarrow CO_2 + 2H_2 0 \qquad (5\text{-}2)$$

This formula states that one molecule of methane combined with two molecules of oxygen forms one molecule of carbon dioxide and two molecules of water. Since the number of moles of a substance is identical to the number of molecules, the above equation can be expressed in molar mass as well. A mole is defined as the mass equal in numerical amount to the molecular weight of a substance. Thus, one mole of CH_4 equals 16.04 lbm (7.28 kg). [Note that given this one-to-one relationship during combustion, all calculations have to be done on ideal rather than real volume basis — compensation for compressibility would produce misleading results.] Equation 5-2 now has the following form in English system mass units:

$$16.04\ lbm\ (CH_4) + 64.00\ lbm\ (O_2) = 44.01\ lbm\ (CO_2) + 36.03\ lbm\ (H_2 0)$$

Dividing both sides by molar mass of methane (16.04 lbm), the Equation yields all quantities on a mass per unit mass of methane basis:

$$1\ (CH_4) + 3.990\ (O_2) = 2.744\ (CO_2) + 2.246\ (H_2 0)$$

Thus, 2.246 lbm (kg) of water is produced for each lbm (kg) of methane burned.

If, for example, methane is at the standard reference conditions of 14.73 psia and 60°F applicable to natural gas per ANSI Z132.1, it will have an ideal density of 0.04237 lbm/cf. The mass of water per unit volume of methane is 0.09516 lbm/cf.

Steam tables give the latent heat of vaporization, in English system units, as 1059.8 Btu/lbm for water at 60°F and 14.73 psia. Therefore, ideal LHV for methane on standard volume basis is given by Equation 5-1 as:

$$LHV = HHV - h_{fg}\, mH_2 0 = 1012.0 - 1059.8 \times 0.09516 = 911.14\ Btu$$

As noted earlier, natural gas consists primarily of methane, with various but small amounts of ethane, butane, pentane, and other inert gases, such as carbon dioxide and nitrogen. Although the energy density of such mixtures may vary significantly, the LHV/HHV ratio remains reasonably close to 0.901 for most pipeline gases in the United States. For example, the above results for methane show LHV/HHV = 911.14/1012 = 0.9003. In calculations used for most applications, an average factor such as this affects the accuracy of the results very little and, therefore, is acceptable.

In most commercial transactions, the energy density of a fuel is specified using its HHV. LHV is accounted for by the test procedures employed. For example, combustion efficiency of a boiler is generally computed by measuring losses and net output is determined by subtracting losses from the input calculated on HHV.

One notable exception is internal combustion in which the net output is determined by direct measurement of torque at the flywheel, rather than subtraction of losses. For this reason, the performance of engines and other similar equipment is defined on an LHV basis, as this provides more reliable fuel consumption data. While reference temperature for automotive fuel energy density determination is 77°F (25°C), the natural gas industry uses 60°F (15.55°C). The difference in values will be minor on a LHV basis, but can be significant on an HHV basis.

Pipeline Natural Gas

Natural gas is a mixture of several gases, most of which are hydrocarbons, i.e., molecules consisting of carbon and hydrogen atoms. These hydrocarbons are also known as paraffins (or alkenes) belonging to the alkyl group. The chemical formula of paraffins has the format of C_nH_{2n+2}, which includes methane (CH_4), ethane (C_2H_6), propane (C_3H_8), etc. Each succeeding member of the series has one more carbon (C) atom with two more hydrogen (H) atoms. Hydrocarbons deficient in one hydrogen atom take the names methyl, ethyl, etc.

Although the composition of natural gas varies from gas field to gas field, it is always a mixture, predominantly of methane (CH_4), ethane (C_2H_6), propane (C_3H_8), and butane (C_4H_{10}), and usually small amounts of helium (He), carbon dioxide (CO_2), and nitrogen (N_2). In some cases, hydrogen sulfide (H_2S), pentane (C_5H_{12}), or hexane (C_6H_{14}) may also be present.

In its original state, as produced at the wellhead, natural gas may be referred to as field gas or wet gas. The term refers to the presence of liquefied hydrocarbons, such as butane and pentane. Prior to distribution, compounds containing sulfur are removed, as are helium, water, and the bulk of the wet ends (gases). The remaining constituents are sometimes adjusted to produce what is commonly known as "pipeline quality" natural gas. Generally, it contains less than 7 lbm of water vapor per million cf (MMcf) and is lower still in the winter. For most applications, this amount is insignificant and the gas is considered to be dry.

The remaining mixture, which is transported through pipelines, is typically composed of methane, ethane, propane, nitrogen, and carbon dioxide. Whereas field gas may be composed of anywhere from 70 to 99% methane, pipeline gas will generally constitute between 80 and 95% methane.

Natural gas is non-toxic, colorless, and odorless. Because natural gas cannot be smelled, an odorant is added to allow for detection for safety. The odorant most commonly used, Mercaptan, is a hydrocarbon with a sulfur atom added. Roughly, 0.002% of sulfur is added. Although Mercaptan can be corrosive, this is too small an amount to cause corrosion in the pipelines or contribute to air pollution.

Natural gas composition variability in most applications is of minor concern, because such variations are usually insignificant. The **Wobbe Index** is a useful measure of energy delivered to a burner, since the energy varies directly with the heating value of the gas and inversely with the square root of the relative density (also known as specific gravity, or SG) as it flows through an orifice. Thus, two gases with different heating values and different relative densities, but with the same Wobbe Index, will deliver the same heat at the burner and, hence, yield the same performance. Since the Wobbe Index allows comparison of the volumetric energy content of different gas fuels at different temperatures, it is key to defining which fuels can be run in the same fuel system and when multiple manifolds or gas systems are required. It is generally accepted that a difference between two Wobbe Indices of up to 5% will allow full interchangeability of such fuel gases. But, note that the Wobbe Index may be based on higher or lower heating values, depending on application. The internationally used symbols for Wobbe Index, according to International Gas Union, are W_S based on HHV and W_i based on LHV.

The composition of pipeline natural gas is determined by gas chromatography and summarized in a gas quality report. The data generally includes the mole percent of each component present on dry volume basis, real specific gravity and compressibility of the gas at reference pressure and temperature, and adjusted HHV in Btu/scf. The adjusted HHV value in Btu/scf is obtained by dividing the calculated ideal values in Btu/cf by the compressibility factor, which indicates the degree to which the real gas mixture departs from results, calculated using the ideal gas law. There are several computer programs capable of doing this, that are available through the American Gas Association (A.G.A.).

For the composition of the pipeline gas, the computer program based on A.G.A. Transmission Measurement Committee Report No. 8 (A.G.A. Cat. #8806) defines the key gas parameters as shown in Tables 5-2 and 5-3:

Table 5-2 Key Gas Parameters

Component		Mole%		Molar Mass	
Nitrogen		1.80		28.0134 lbm	(12.7060 kg)
Carbon dioxide		0.50		44.0100 lbm	(19.9620 kg)
Methane		92.20		16.0403 lbm	(7.2758 kg)
Ethane		5.00		30.0700 lbm	(13.6390 kg)
Propane		0.30		44.0970 lbm	(20.0020 kg)
N-butane		0.05		58.1230 lbm	(26.3640 kg)
I-butane		0.05		58.1230 lbm	(26.3640 kg)
N-pentane		0.05		72.1500 lbm	(32.7270 kg)
I-pentane		0.05		72.1500 lbm	(32.7270 kg)
Total	=	100.00	M_{NG} =	17.2820 lbm	(7.8390 kg)

Table 5-3 Real Properties at 14.73 psia and 60°F

Density	0.04574	lbm/scf	(0.7327 kg/Nm^3)
Relative density (SG)	0.59772		
Base compressibility	0.99780		
Higher heating value	1038.9	Btu/scf	(38,710.7 kJ/Nm^3)
Lower heating value	937.0	Btu/scf	(34,913.8 kJ/Nm^3)
Wobbe Index$_{HHV}$	1343.8		

It should be noted that the computer program expresses properties based on standard volume basis by dividing the appropriate ideal quantities by the gas compressibility. Such division does not produce real gas heating value, for example, but only allows calculation of custody transfer rates on energy basis using the real gas flow rate rather than the ideal gas flow rate. The HHV and

LHV values for combustion calculations are recovered by multiplying the standard volume basis values by the base compressibility. Thus, for the above composition, the ideal HHV is 1036.6 Btu/cf (38,626.4 kJ/m^3) and the ideal LHV is 934.9 Btu/cf (34,836.8 kJ/m^3), based on a LHV/HHV ratio of 0.9019.

The computer program recalculates the density and certain other parameters to any reference conditions specified. For natural gas vehicle (NGV) applications, one could easily compute the density values at EPA's reference conditions of 20°C and 101.325 kilopascal (kPa), which is approximately 14.6969 psia. By resetting the default values, the relative density, compressibility, HHV, LHV, and Wobbe Index can be recalculated to international standard pressure (101.325 kPa) and temperature (15°C) reference conditions with quantities in customary units referred to A.G.A. standard reference conditions. Density will be calculated at any pressure and temperature specified.

Alternative programming methods may use American Society for Testing and Materials (ASTM) D 3588, *Standard Practice for Calculating Heat Value, Compressibility Factor, and Relative Density (Specific Gravity) of Gaseous Fuels.* While the compressibility factor calculation is not as rigorous as that in the A.G.A. program, the results are quite adequate and computed data are applicable to all common types of utility gaseous fuels on the basis of HHV and LHV. Although the base conditions used are 14.696 psia (101.325 kPa) and 60°F (15.6°C), calculation procedures for other conditions are given as well.

As previously mentioned, the LHV and HHV of pipeline gas, as well as the LHV/HHV ratio, vary depending on the mass percent of hydrogen and the amount of inert gas. As the proportions of propane and other heavier constituents increase, the energy density of the gas increases. Whereas the LHV of methane (CH_4) is 90.03% of its HHV, the LHV of ethane (C_2H_6) is 91.47% and that of propane (C_3H_8) is 92.00% of HHV. Reducing the amount of non-hydrocarbons, of course, will increase the energy density of the gas.

Table 5-4 shows sample analyses of natural gas from several U.S. fields. In these samples, the methane constituent ranges from 83.4 to 93.3% by volume. Notice that the heating value, in Btu/cf, and specific gravity (relative to air) are lowest for Sample 3, which features the highest methane content. This is to be expected due to the absence of heavier, higher energy density constituents such as ethane. A comparison of Sample 5 with Samples 1 and 2 reveals a significant difference in heating value, even though the methane content and specific gravity is almost identical. This is also to be expected since the non-methane constituents of Samples 1 and 2 are primarily ethane (about 15% by volume), while Sample 5 features more than 8% nitrogen, which does not contribute to heating value.

Given the varying energy density levels, natural gas sales today are conducted using therm as the base unit. A therm is defined by the U.S. Secretary of Commerce as "a natural gas energy unit equal to 105,480,400 Joules" (*Federal Register Vol. 33, No. 146, July 27, 1968*). Since natural gas transactions are based on higher heating value, quantities expressed in therms always have a connotation that is not apparent when Btu is used.

Sample No. Source:	1 PA	2 S.C.	3 Ohio	4 LA	5 OK
Analyses:					
Constituents, % by vol.					
H_2, Hydrogen	--	--	1.82	--	--
CH_4, Methane	83.40	84.00	93.33	90.00	84.10
C_2H_4, Ethylene	--	0.25	--	--	--
C_2H_6, Ethane	15.80	14.80	--	5.00	6.70
CO, Carbon monoxide	--	--	0.45	--	--
CO_2, Carbon dioxide	--	0.70	0.22	--	0.80
N_2, Nitrogen	0.80	0.50	3.40	5.00	8.40
O_2, Oxygen	--	--	0.35	--	--
H_2S, Hydrogen sulfide	--	--	0.18	--	--
Ultimate, % by wt.					
S, Sulfur	--	--	0.34	--	--
H_2, Hydrogen	23.53	23.30	23.20	22.68	20.85
C, Carbon	75.25	74.72	69.12	69.26	64.84
N_2, Nitrogen	1.22	0.76	5.76	8.06	12.90
O_2, Oxygen	--	1.22	1.58	--	1.41
Specific gravity (rel. to air)	0.636	0.636	0.567	0.600	0.630
HHV					
Btu/ft^3 at 60°F and 30 in. Hg	1,129	1,116	964	1,022	974
(kJ/m^3 at 16°C and 102 kPa)	(42,065)	(41,581)	(35,918)	(38,079)	(36,290)
Btu/lb(kJ/kg) of fuel	23,170 (53,893)	22,904 (53,275)	22,077 (51,351)	21,824 (50,763)	20,160 (46,892)

Table 5-4 Selected Samples of Natural Gas from U.S. Fields. Source: Babcock & Wilcox

However, when the Btu is defined as the amount of energy required to raise the temperature of one lbm of pure water from 59°F to 60°F, it has a value of 1054.804 Joules (J). This particular Btu value is somewhat smaller than the International Table Btu (1 Btu_{IT} = 1,055.056 J), which is slowly becoming the standard Btu for all applications. It should be noted, however, that only when Btu_{59} is used, the legally defined therm is equivalent to 100,000 Btu. In many less rigorous applications, the therm is taken to represent the energy content of 100 cf (1 Ccf) of natural gas; similarly, the energy content of 1,000 cf (1 Mcf) is equated to 1 million Btu computed on an HHV basis, or 1 decatherm.

The heat content of natural gas production has remained fairly constant over the past few decades. The heat content of production (dry) has ranged from 1,026 to 1,032 Btu/cf and the heat content of production marketed (wet) has ranged from 1,098 to 1,115 Btu/cf (both as reported by the DoE EIA). Note that the heat content of production marketed (wet) is calculated annually by adding the heat content of dry natural gas production and the total heat content of natural gas plant liquids production and dividing this sum by the total quantity of marketed (wet) natural gas production.

Other Fuel Gases

There are several other types of fuel gas that can be used in both boiler and engine prime-mover applications. These include liquefied petroleum gases (LPG or LP), manufactured or mixed gases, digester gas, and sour gas, among others. These fuel gases are usually used when pipeline gas is not available, as a reserve supply for natural gas, or, in some cases, as a primary engine fuel. As reserve fuels, the role of these gases, along with liquefied natural gas (LNG), is important. These gases often allow facilities to withstand interruption of natural gas supplies and permit cost-effective interruptible natural gas contracts. Spark ignition engines are also more compatible with some of these gases as alternative fuels than with oil.

Liquefied Petroleum and Natural Gases

Liquefied petroleum gases are propane with an HHV of 2,522 Btu/cf (93,976 kJ/m^3), butane with an HHV of 3,260 Btu/cf (121,476 kJ/m^3), or a mixture of the two. These fuel gases are obtained from natural gas or produced as a by-product from oil refining. In the refinement of wet, wellhead gas, the methane and ethane (and sometimes a small portion of the propane) are used for dry pipeline gas, while most of the propane and butane are used for LP gases. Commercially distributed LP mixtures can be easily liquefied, reduced in volume by moderate pressure, and transported and stored in tanks.

A propane-air mixture can sometimes be used in the same equipment as natural gas without significant burner adjustments. In internal combustion engines, propane is sometimes used as a back-up fuel to natural gas. Both propane and butane are heavier than air, while natural gas is lighter than air. Therefore, special mechanical ventilation precautions must be taken when switching fuels. In many states, liquid propane is prohibited within the confines of a building.

Manufactured Gases

There are several gases made from either coal or oil that are classified as **manufactured gas**. Generally, these gases are of relatively low heating values and have a high percentage of free hydrogen. These gases are either produced for use as fuel or result as by-products of other processes.

All manufactured gases must be cleaned to reduce dust and solid impurities. Tar and ammonia also must be removed by washing or scrubbing the gases. Sulfur is often present and is sometimes removed by passing the gases through iron oxide beds.

Synthetic natural gas (SNG) is made from coal or petroleum naphtha and has a low energy density. Though generally not commercially competitive at current costs, given the continued improvements in coal gasification technology and the vast amounts of coal reserves in this country, SNG may play an important role in the future. Modern SNG made from coal is often upgraded to high energy density pipeline quality, which has an HHV of 1,000 Btu/cf (37,263 kJ/m^3).

Producer gas is made when coal or coke is burned with a deficiency of air and a controlled amount of steam. The resulting product consists primarily of nitrogen, carbon monoxide, and hydrogen after removal of entrained ash and sulfur compounds. It has a very low energy density of under 150 Btu/cf (5,589 kJ/m^3), which generally restricts its use to locations near its source.

Illuminating gas includes gases made by a number of processes. Passing steam through a hot bed of coke makes blue-water gas. Carbon in the coke combines with the steam in an endothermic reaction to form H_2 and CO. Carbureted-water gas is an enriched water-gas, formed by passing the gas through a checkerwork of hot bricks sprayed with oil. The oil, in turn, is cracked to a gas by the heat. Coal and oil gases are formed by applying heat to coal and oil to drive off the hydrogen, methane, carbon monoxide, and ethylene. Approximate HHVs of these gases are: **blue-water gas**, just under 300 Btu/cf (11,200 kJ/m^3); **carbureted-water gas**, about 500 Btu/cf (18,600 kJ/m^3); **coal gas**, just under 600 Btu/cf (22,400 kJ/m^3); and **oil gas**, just under 800 Btu/cf (29,800 kJ/m^3).

Coke-oven gas is similar to standard coal gas, but is obtained as a by-product of coke production processes. Several valuable products are recovered from the process of converting coal to coke, including ammonium sulfate, oils, and tars. The noncondensible portion is the coke oven gas.

Blast-furnace gas is a by-product of steel mills and is similar to producer gas. Both blast-furnace and coke-oven gases offer a beneficial application of waste products.

Biomass Gases

Biomass gases may be produced from anaerobic digestion of any type of biomass from methane-producing organisms. Common feedstocks for biomass gases are wood waste, landfill, and sewer sludge. Types of biomass gases include sanitary landfill gas and digester gas.

Sanitary landfill gas is produced from digesting landfill. Biological degradation of the organic material produces large quantities of methane, which can be captured under a non-permeable cap over the site, pumped out of the landfill, filtered, and used commercially. This gas has a moderate heating value of about 500 to 600 Btu/cf (18,600 to 22,400 kJ/m^3) due to a high content of CO_2 (about 35 to 45%).

The presence of hydrogen sulfide (H_2S) can damage equipment, even in small quantities. Levels as low as 0.1% of H_2S must be treated to avoid corrosion of certain equipment metals. The term "sweet" refers to sulfur-free gas and "sour" refers to gas containing a large proportion of sulfur compounds.

Digester (sewage or sludge) gas is a mixture of several component gases produced in digester tanks where biodegrading takes place. It can be made from sewage, animal waste, and liquid effluent from vegetable oil mills and alcohol mills. Generally, these gases consist of about two-thirds methane and one-third carbon dioxide, with a few percent nitrogen and minute quantities of other gases such as oxygen, hydrogen, and hydrogen sulfide.

Methane is the primary combustible component that makes this gas compatible with equipment designed for natural gas usage. At 66% methane, the HHV is 668 Btu/cf (24,891 kJ/m^3). Of course, if a significant amount of H_2S (0.01%) is present, the gas must be treated to avoid corrosion of certain equipment metals.

Another technique for producing combustible gas from biomass is similar to the process used for producer gas. In this case, biomass, most commonly wood waste materials, are used instead of coal or coke.

Oil

As with natural gas, there is no specific heating value that can be assigned to oil because each tanker or barrel has a different composition of constituent elements. Determination of HHV and LHV is done in the same way as with gas, although the composition of oil is far more complex and constituents are more difficult to identify and calculate. Empirical analysis yields specific values for different types of oil. The ratio of LHV to HHV of oil is generally greater than that of natural gas because oil has a lower hydrogen-to-carbon content and, thus, produces less water when combusted.

Due to the methods used in their production, fuel oils fall into two broad classifications: distillates and residuals. Both **distillate oils** and **residual oils** are parts of the crude oil. As heat is applied to crude oil, the light, more volatile oil vaporizes and leaves the crude base. The vapors are then captured and condensed. The condensed liquid is the distillate. The distillates, therefore, consist of overhead or distilled fractions; the residuals are the bottoms remaining from the distillation, or blends of these bottoms with distillates. The American Society for Testing and Materials (ASTM) classifies the fuel oils by grade number.

Grade No. 1 is a light distillate intended for use in vaporizing type burners in which the oil is converted to a vapor by contact with a heated surface or by radiation. High volatility is necessary to ensure that evaporation proceeds with a minimum of residue.

Grade No. 2 is a heavier distillate than Grade No. 1 and is intended for use in atomizing burners, i.e., the oil is sprayed into a combustion chamber where the tiny droplets burn while in suspension. This grade of oil is used in most domestic burners and in many medium capacity commercial-industrial burners where its ease of handling and ready availability sometimes justifies its higher cost over the residuals.

Grade No. 4 (Light) is a heavy distillate fuel or distillate-residual fuel blend. It is intended for use both in pressure-atomizing commercial-industrial burners not requiring higher cost distillates and in burners equipped to atomize oils of higher viscosity. Its permissible viscosity range allows it to be pumped and atomized at relatively low storage temperatures.

Grade No. 4 is usually a heavy distillate/residual fuel blend, but can be a heavy distillate fuel. It is intended for use in burners equipped with devices that atomize oils of higher viscosity than domestic burners can handle. Its permissible viscosity range allows it to be pumped and atomized at relatively low storage temperatures. Thus, in all but extremely cold weather, it requires no preheating for handling.

Grade No. 5 (Light) is residual fuel of intermediate viscosity for burners capable of handling fuel more viscous than Grade No. 4 without preheating. Preheating can be necessary in some types of equipment for burning and in colder climates for handling.

Grade No. 5 (Heavy) is a residual oil more viscous than Grade No. 5 (Light) and is intended for use in similar service. Preheating may be necessary in some types of equipment for burning and in colder climates for handling.

Grade No. 6, sometimes referred to as Bunker C, is a high-viscosity oil used mostly in commercial and industrial heating. It requires preheating in the storage tank to permit pumping and additional preheating at the burner to permit atomizing. The extra equipment and maintenance required to handle this fuel usually preclude its use in small installations.

Note that the sulfur content (percent by weight) will vary widely within each grade of oil. Typically, it will be lowest with Grade No. 1 and increase with increasing grade number. Typical sulfur content will range from 0.01 to 0.5% by weight for Grade No. 1 and will range from 0.7 to 3.5% for Grade No. 6. Hydrogen content shows the reverse trend, and is highest for Grade No. 1 and decreases with increasing grade number.

The petroleum industry generally uses the **American Petroleum Institute (API) gravity scale** to determine the relative density of oil. The scale was devised jointly by the API and the National Bureau of Standards. The relationship between degree API gravity and the specific gravity is given by the following formula:

$$\textit{Deg. API gravity} = \frac{141.5}{\textit{Specific gravity at 60/60°F}} - 131.5 \qquad (5\text{-}4)$$

where specific gravity at 60/60°F (16/16°C) means that both oil and water are at 60°F (16°C).

Given this relationship, heavier liquid fuels are denoted by lower API gravity values. The specific gravity is lowest for Grade No. 1 and increases with increasing grade number. Conversely, the Degree API gravity is highest for Grade No. 1 and decreases with increasing grade number. The gross heating value in Btu/lbm (kJ/kg) is generally highest for Grade No. 1 and decreases with increasing grade number. However, since the density, in lbm/gallon (kg/liter), increases significantly with increasing grade number, so does the heat value per gallon (liter).

There are several ASTM standards for determining the heating values of liquid fuels, depending on the precision of results desired. When the heating value is determined by calorimetry, the procedure most frequently used is described in the ASTM D 240, *Standard Test Method for Heat of Combustion of Liquid Hydrocarbon Fuels by Bomb Calorimeter.*

For fuel oils, the gross heat of combustion is reported in preference to the net, which is the quantity required in practical engineering applications. Since high accuracy requirements for energy density determination are rarely a high priority for these fuels, the procedures described in ASTM D 4868, *Standard Test Method for Estimation of Net and Gross Heat of Combustion of Burner and Diesel Fuels,* suffice in most instances. Calculations are based on density, sulfur content, water content, and ash content (the method is not applicable to pure hydrocarbons).

Table 5-5 includes the HHV and LHV, in English system units, of some commonly used liquid fuels. Table 5-6 lists approximate heat content of various petroleum products in million Btu per barrel, as adopted by the U.S. Energy Information Administration (EIA) per the Bureau of Mines. Table 5-7 lists the chemical elements and compounds found in commonly used fuels for combustion.

COMPARISON OF FUEL HEATING VALUES

Table 5-5 HHV and LHV of Liquid Fuels by Weight and Volume

Fuels	HHV		LHV		LHV/HHV
	Btu/lbm	Btu/gal	Btu/lbm	Btu/gal	
No. 2 oil	19,580	142,031	18,421	133,623	0.9408
No. 4 oil	18,890	146,476	17,804	138,055	0.9425
No. 6 oil	18,270	150,808	17,312	142,901	0.9476
Propane (1)	21,653		19,922		0.9201
Butane (1)	21,266		19,623		0.9227
Gasoline (2) (3)	19,657	121,808	18,434	114,235	0.9378
Reformulated gasoline (2) (4)	19,545	120,103	18,304	112,477	0.9365
Methanol (M-85) (2) (5)	11,274	73,882	10,115	66,289	0.8972

(1) At API reference conditions of 14.696 psia and 60°F, propane and butane are gases.Therefore, values are not for Btu/gal.
(2) Calculated values based on raw data from Clean Fuel Test Fleet, Report No. 5.
(3) Based on sample of Indolene No. 3, which is a gasoline test fuel. Note that average unleaded would be only 110,500 Btu/gal.
(4) Reformulated gasoline has 10% rather than 15% MTBE.
(5) M-85 with 15% gasoline.

Table 5-6 Approximate Heat Content of Petroleum Product (MMBtu/Barrel)

Butane	4.326
Distillate fuel oil	5.825
Ethane	3.082
Isobutane	3.974
Jet fuel, kerosene type	5.670
Jet fuel, naphtha type	5.355
Kerosene	5.670
Motor gasoline	5.253
Natural gasoline	4.620
Petroleum coke	6.024
Propane	3.836
Residual fuel oil	6.287
Special naphtha	5.248

AIR AND COMBUSTION

Oxygen reacts with many materials. When such reaction proceeds at a slow rate, it is generally referred to as corrosion or oxidation. When the reaction proceeds at a rapid pace, producing large amounts of heat and usually accompanied by light, it is called combustion. Three ingredients have to be present for combustion to occur: a source of ignition (heat), fuel, and oxygen. Remove any one of these from this triangle and combustion will not take place.

The source of oxygen in most combustion processes is atmospheric air. Air comprises oxygen (20.946 mol%), nitrogen (78.102 mol%), argon (0.916 mol%), carbon dioxide (0.033 mol%), and a variety of other gases in smaller amounts. In rough calculations, the air composition is frequently taken to be 21 vol.% oxygen and 79 vol.% inert gases lumped under the name of atmospheric nitrogen. Based on National Bureau of Standards 1978 data, a more exact approach indicates that each mole of oxygen involved in combustion will carry with it:

$$\frac{(1-0.20946)}{0.20946} = 3.774 \textit{ moles of atmospheric nitrogen}$$

The molar mass of air, in English system units, is 28.9625, i.e., each mole of air has a mass of 28.9625 lbm. However, because atmospheric nitrogen contains

						Heat of Combustion[c]				ft³ per ft³ of Combustible						lb per lb of Combustible						Theor
				Specific		Btu per ft³		Btu per lb		Required for Combustion			Flue Products			Required for Combustion			Flue Gas Products			air lb/
No. Substance	Formula	Molecular Weight[a]	Density,[b] lb per ft³	Volume[b] ft³ per lb	Specific Gravity[b] (air=1)	Gross	Net[d]	Gross	Net[d]	O_2	N_{2a}	Air	CO_2	H_2O	N_{2a}	O_2	N_{2a}	Air	CO_2	H_2O	N_{2a}	10,000 Btu
1 Carbon	C	12.0110	—	—	—	—	—	14,093	14,093	1.0	3.773	4.773	1.0	—	3.773	2.664	8.846	11.510	3.664	—	8.846	8.167
2 Hydrogen	H_2	2.0159	0.0053	187.970	0.0695	325.0	274.6	61,095	51,625	0.5	1.887	2.387	—	1.0	1.887	7.936	26.353	34.290	—	8.937	26.353	5.613
3 Oxygen	O_2	31.9988	0.0846	11.819	1.1053	—	—	—	—	—	—	—	—	—	—	—	—	—	—	—	—	—
4 Nitrogen	N_2	28.0135	0.0744	13.443	0.9717	—	—	—	—	—	—	—	—	—	—	—	—	—	—	—	—	—
4 Nitrogen (atm.)	N_{2a}	28.1610	0.0748	13.372	0.9769	—	—	—	—	—	—	—	—	—	—	—	—	—	—	—	—	—
5 Carbon Monoxide	CO	28.0104	0.0740	13.506	0.9672	321.9	321.9	4,347	4,347	0.5	1.887	2.387	1.0	—	1.887	0.571	1.897	2.468	1.571	—	1.897	5.677
6 Carbon Dioxide	CO_2	44.0098	0.1170	8.547	1.5284	—	—	—	—	—	—	—	—	—	—	—	—	—	—	—	—	—
Parafin series C_nH_{2n+2}																						
7 Methane	CH_4	16.0428	0.0424	23.574	0.5541	1013	912	23,875	21,495	2.0	7.547	9.547	1.0	2.0	7.547	3.989	13.246	17.235	2.743	2.246	13.246	7.219
8 Ethane	C_2H_6	30.0697	0.0803	12.455	1.0488	1792	1639	22,323	20,418	3.5	13.206	16.706	2.0	3.0	13.206	3.724	12.367	16.092	2.927	1.797	12.367	7.209
9 Propane	C_3H_8	44.0966	0.1196[e]	8.361[e]	1.5624	2592	2385	21,669	19,937	5.0	18.866	23.866	3.0	4.0	18.866	3.628	12.047	15.676	2.994	1.634	12.047	7.234
10 n-Butane	C_4H_{10}	58.1235	0.1582[e]	6.321[e]	2.0666	3373	3113	21,321	19,679	6.5	24.526	31.026	4.0	5.0	24.526	3.578	11.882	15.460	3.029	1.550	11.882	7.251
11 Isobutane	C_4H_{10}	58.1235	0.1582[e]	6.321[e]	2.0666	3365	3105	21,271	19,629	6.5	24.526	31.026	4.0	5.0	24.526	3.578	11.882	15.460	3.029	1.550	11.882	7.268
12 n-Pentane	C_5H_{12}	72.1504	0.1904[e]	5.252[e]	2.4872	4017	3714	21,095	19,507	8.0	30.186	38.186	5.0	6.0	30.186	3.548	11.781	15.329	3.050	1.498	11.781	7.267
13 Isopentane	C_5H_{12}	72.1504	0.1904[e]	5.252[e]	2.4872	4007	3705	21,047	19,459	8.0	30.186	38.186	5.0	6.0	30.186	3.548	11.781	15.329	3.050	1.498	11.781	7.283
14 Neopentane	C_5H_{12}	72.1504	0.1904[e]	5.252[e]	2.4872	3994	3692	20,978	19,390	8.0	30.186	38.186	5.0	6.0	30.186	3.548	11.781	15.329	3.050	1.498	11.781	7.307
15 n-Hexane	C_6H_{14}	86.1773	0.2274[e]	4.398[e]	2.9702	4767	4415	20,966	19,415	9.5	35.846	45.346	6.0	7.0	35.846	3.527	11.713	15.240	3.064	1.463	11.713	7.269
Olefin series C_nH_{2n}																						
16 Ethylene	C_2H_4	28.0538	0.0746	13.412	0.9740	1613	1512	21,636	20,275	3.0	11.320	14.320	2.0	2.0	11.320	3.422	11.362	14.784	3.138	1.284	11.362	6.833
17 Propylene	C_3H_6	42.0807	0.1110[e]	9.009	1.4500	2336	2185	21,048	19,687	4.5	16.980	21.480	3.0	3.0	16.980	3.422	11.362	14.784	3.138	1.284	11.362	7.024
18 n-Butene (Butylene)	C_4H_8	56.1076	0.1480[e]	6.757[e]	1.9333[e]	3086	2885	20,854	19,493	6.0	22.640	28.640	4.0	4.0	22.640	3.422	11.362	14.784	3.138	1.284	11.362	7.089
19 Isobutene	C_4H_8	56.1076	0.1480[e]	6.757[e]	1.9333	3069	2868	20,737	19,376	6.0	22.640	28.640	4.0	4.0	22.640	3.422	11.362	14.784	3.138	1.284	11.362	7.129
20 n-Pentene	C_5H_{10}	70.1345	0.1852[e]	5.400[e]	2.4191	3837	3585	20,720	19,359	7.5	28.300	35.800	5.0	5.0	28.300	3.422	11.362	14.784	3.138	1.284	11.362	7.135
Aromatic series C_nH_{2n-6}																						
21 Benzene	C_6H_6	78.1137	0.2060[e]	4.854[e]	2.6912[e]	3746	3595	18,184	17,451	7.5	28.300	35.800	6.0	3.0	28.300	3.072	10.201	13.274	3.380	0.692	10.201	7.300
22 Toluene	C_7H_8	92.1406	0.2431[e]	4.114[e]	3.1753[e]	4497	4296	18,501	17,672	9.0	33.959	42.959	7.0	4.0	33.959	3.125	10.378	13.504	3.343	0.782	10.378	7.299
23 Xlyene	C_8H_{10}	106.1675	0.2803	3.568[e]	3.6612[e]	5222	4970	18,633	17,734	10.5	39.619	50.119	8.0	5.0	39.619	3.164	10.508	13.673	3.316	0.848	10.508	7.338
Miscellaneous																						
24 Acetylene	C_2H_2	26.0379	0.0697	14.345	0.9106	1499	1448	21,502	20,769	2.5	9.433	11.933	2.0	1.0	9.433	3.072	10.201	13.274	3.380	0.692	10.201	6.173
25 Naphthalene	$C_{10}H_8$	128.1736	0.3384[e]	2.955[e]	4.4206[e]	5855	5654	17,303	16,707	12.0	45.279	57.279	10.0	4.0	45.279	2.995	9.947	12.943	3.434	0.562	9.947	7.480
26 Methyl alcohol	CH_3OH	32.0422	0.0846[e]	11.820	1.1052	868	767	10,258	9,066	1.5	5.660	7.160	1.0	2.0	5.660	1.498	4.974	6.472	1.373	1.124	4.974	6.309
27 Ethyl alcohol	C_2H_5OH	46.0691	0.1216[e]	8.224[e]	1.5884[e]	1600	1449	13,161	11,918	3.0	11.320	14.320	2.0	3.0	11.320	2.084	6.919	9.003	1.911	1.173	6.919	6.841
28 Ammonia	NH_3	17.0306	0.0456[e]	21.930[e]	0.5957[e]	441	364	9,667	7,986	0.75	2.830	3.580	—	1.5	3.330	1.409	4.679	6.088	—	1.587	5.502	6.298
29 Sulfur	S	32.0660	—	—	—	—	—	3,980	3,980	1.0	3.773	4.773	SO_2 1.0	—	3.773	1.000	3.320	4.320	SO_2 1.998	—	3.320	10.854
30 Hydrogen sulfide	H_2S	34.0819	0.0911	10.978[e]	1.1899[e]	646	595	7,097	6,537	1.5	5.660	7.160	SO_2 1.0	1.0	5.660	1.410	4.682	6.093	SO_2 1.880	0.529	4.682	8.585
31 Sulfur dioxide	SO_2	64.0648	0.1733	5.770	2.2640	—	—	—	—	—	—	—	—	—	—	—	—	—	—	—	—	—
32 Water vapor	H_2O	18.0153	0.0476	21.017	0.6215	—	—	—	—	—	—	—	—	—	—	—	—	—	—	—	—	—
33 Air	—	28.9660	0.0766	13.063	1.000	—	—	—	—	—	—	—	—	—	—	—	—	—	—	—	—	—

All gas volumes corrected to 60F and 30 in. Hg dry.

[a] 1987 Atomic Weights: C=12.011, H=1.00794, 0=15.9994, N=14.0067, S=32.066.

[b] Densities calculated from values given in grams per liter at 0C and 760 mm in International Critical Tables, allowing for known deviations from gas laws. Where no densities were available, the volume of the mole was taken as 22.415 liters. The ideal values for specific gravity may be found in Reference 2.

[c] For gases saturated with water at 60F, 1.74% of the Btu value must be deducted. Rossini, F.D. and others[3]

[d] Correction from gross to net heating value determined by deducting 1059.7 Btu/lb of water in products of combustion. ASME Steam Tables, 1983[4]

[e] Either the density or the coefficient of expansion has been assumed. Some of the materials cannot exist as gases at 60F and 30 in. Hg, in which case the values are theoretical ones. Under the actual concentrations in which these materials are present, their partial pressure is low enough to keep them as gases.

Table 5-7 Chemical Elements and Compounds Found in Commonly Used Fuels. Source: Babcock & Wilcox

traces of other species, its molar mass is slightly different from the 28.0134 lbm value of pure nitrogen. Its mass is 28.1580 lbm/mol.

Table 5-8 summarizes the molecular and weight relationships between fuel and oxygen for constituents commonly involved in combustion. The heat of combustion for each constituent is also tabulated.

There are many ways to compute the amount of air needed for complete combustion of various fuels. As a general rule, roughly 10 cf of air is required for every 1,000 Btu (or 1 m^3 of air for every 3.77 MJ) of fuel. However, there is usually a need to perform more precise, fuel-specific computations.

Since the highest efficiency is achieved in most applications by minimizing the amount of hot flue products leaving the gas-fired device at temperatures above the ambient, the idea of using the bare minimum of air to achieve complete combustion is attractive. When the flue products contain no oxygen or fuel, with all reactants having been adjusted to chemically correct proportions to produce water and carbon dioxide only, the process is called stoichiometric combustion. Thus, there is a specific air-fuel ratio that results in complete combustion for every combustible hydrocarbon.

One common method computes the amount of oxygen required to achieve stoichiometry by determining the amount of oxygen required for complete oxidation of each combustible in the fuel. It sums these amounts, and then determines the amount of air, in moles, by dividing this sum by 0.20946 (the molar fraction of oxygen in the air). Referring to methane combustion illustrated by Equation 5-2, each carbon atom will combine with two atoms of oxygen to form one atom of carbon dioxide and two atoms of hydrogen, which will combine with one atom of oxygen to form a molecule of water. Thus, the total oxygen requirement is 2 moles for complete combustion under ideal conditions. Since the source of this oxygen is atmospheric air, the burner will have to handle 2/0.20946 = 9.548 moles of air, consisting of two moles of oxygen and 7.548 moles of atmospheric nitrogen.

The stoichiometric combustion equation for methane with air thus becomes:

$$CH_4 + 2\,(O_2 + 3.774\,N_2) \rightarrow CO_2 + 2\,H_2O + 7.548\,N_2$$

Some combustion applications, however, require even a higher precision. This may be done by converting the components (such as those listed on a gas quality report) into elemental — or atomic — equivalents and expressing the chemical make-up using an equivalent formula, i.e., formula in the form $C_aH_bO_cN_d$. The subscript values are determined by multiplying the mole fraction of each component present in the natural gas mixture by the number of atoms of the four primary elements. This can be illustrated using the example of pipeline quality natural gas with the composition given in Table 5-9, which lists all hydrocarbons, as well as inert constituents.

Thus, this particular natural gas composition may be represented by C1.045 H4.034 O0.010 N0.036, with C1.040 H4.034 representing the hydrocarbon portion of the fuel. The mass of hydrocarbons is 16.557 lbm in each mole of fuel having a molar mass of 17.282 lbm (previously computed using A.G.A. Cat. #8806 computer program), easily computed from relative atomic masses components shown in Table 5-10.

Noting that any free oxygen present in the fuel will react, reducing the need for combustion air, the stoichiometric combustion equation of any mixture of sulfur-free hydrocarbons represented by $C_aH_bO_cN_d$, including natural gas, may be written as:

$$C_aH_bO_cN_d + [(4a + b - 2c)/4] \times (O_2 + 3.774\,N_2) \rightarrow CO_2 + (b/2)\,H_2O + 0.9435 \times [(4a + b - 2c)]\,N_2 + (d/2)N_2 \quad (5\text{-}6)$$

Combustible	Reaction	Moles	Mass or weight, lb	Heat of Combustion (High) Btu/lb of Fuel
Carbon (to CO)	$2C + O_2 = 2CO$	2 + 1 = 2	24 + 32 = 56	3,950
Carbon (to CO_2)	$C + O_2 = CO_2$	1 + 1 = 1	12 + 32 = 44	14,093
Carbon monoxide	$2CO + O_2 = 2CO_2$	2 + 1 = 2	56 + 32 = 88	4,347
Hydrogen	$2H_2 + O_2 = 2H_2O$	2 + 1 = 2	4 + 32 = 36	61,095
Sulfur (to SO_2)	$S + O_2 = SO_2$	1 + 1 = 1	32 + 32 = 64	3,980
Methane	$CH_4 + 2O_2 = CO_2 + 2H_2O$	1 + 2 = 1 + 2	16 + 64 = 80	23,875
Acetylene	$2C_2H_2 + 5O_2 = 4CO_2 + 2H_2O$	2 + 5 = 4 + 2	52 + 160 = 212	21,502
Ethylene	$C_2H_4 + 3O_2 = 2CO_2 + 2H_2O$	1 + 3 = 2 + 2	28 + 96 = 124	21,636
Ethane	$2C_2H_6 + 7O_2 = 4CO_2 + 6H_2O$	2 + 7 = 4 + 6	60 + 224 = 284	22,323
Hydrogen sulfide	$2H_2S + 3O_2 = 2SO_2 + 2H_2O$	2 + 3 = 2 + 2	68 + 96 = 164	7,097

Table 5-8 Molecular and Weight Relationships Between Fuel and Oxygen for Common Combustion Constituents. Source: Babcock & Wilcox

Table 5-9 Composition of Sample Pipeline Gas

Pipeline Gas Component	Mole Fraction	Atoms Present C	H	O	N
Methane, CH_4	0.9220	0.92200	3.68800		
Ethane, C_2H_6	0.0500	0.10000	0.30000		
Propane, C_3H_8	0.0030	0.00900	0.02400		
N-butane, C_4H_{10}	0.0005	0.00200	0.00500		
I-butane, C_4H_{10}	0.0005	0.00200	0.00500		
N-pentane, C_5H_{12}	0.0005	0.00250	0.00600		
I-pentane, C_5H_{12}	0.0005	0.00250	0.00600		
HC Total	0.9770	1.04000	4.03400		
Nitrogen, N_2	0.0180				0.03600
Carbon Dioxide, CO_2	0.0050	0.00500	0.01000		
Natural Gas Totals	1.0000	1.04500	4.03400	0.01000	0.03600

Table 5-10 Relative Atomic Masses Components

Carbon, C	1.045 x 12.011	=	12.55	1.040 x 12.011	= 12.491
Hydrogen, H	4.034 x 1.00794	=	4.066		= 4.066
Oxygen, O	0.010 x 15.9994	=	0.160		
Nitrogen, N	0.036 x 14.0067	=	0.504		
Mass per fuel mole	NG	=	17.282 lbm	HC	=16.557 lbm

Based on the given example of pipeline quality gas, the above equation becomes:

$$C_{1.045}H_{4.034}O_{0.010}N_{0.036} + 2.0485\ O_2 + 7.731\ N_2 \rightarrow$$
$$1.045\ CO_2 + 2.017\ H_2O + 7.731\ N_2 + 0.018\ N_2$$

In many applications, combustion air requirements may be estimated with reasonable accuracy. For each volume of the fuel above, for example, roughly 9.78 volumes of air (2.049 volumes of oxygen and 7.731 volumes of nitrogen) have to be added to achieve theoretical complete combustion.

Table 5-11 shows theoretical air requirements for various fuels. In the first column, theoretical air has been tabulated on a mass per mass of fuel basis. The wide variation of values, however, is of little significance when comparing the various fuels. But, when the theoretical air is converted to a mass per unit heat input from fuel basis, as shown in Columns 3 and 4, the theoretical air varies little among fuels. Note that values are listed in lbm per 10,000 Btu and that carbon and hydrogen, the principal combustible fuel elements, are shown for reference. Table 5-12 provides a fuel analysis and shows the theoretical air, fuel, and moisture content for representative samples of heavy oil and natural gas fuels. Note that constituents are shown by weight for the oil sample and by volume for the natural gas sample.

Since perfect mixing of fuel and air is never achieved in practice and a portion of the combustion air volume may be occupied by water vapor, the fuel has to be burned with slight excess air. Humid air will contain varying percentages of moisture, depending on temperature and degree of saturation — typically the proportion will be about 1 vol%, but may rise to about 4% under extreme conditions. Thus, the rule of thumb is that each cf of gaseous fuel (1,000 Btu/cf HHV) will require about 10 cf combustion air. For oil and coal, the theoretical air value is 9.7cf per 1,000 Btu HHV; however, much more excess air is required to avoid smoke and unburned hydrocarbons.

While the general combustion equation expressed in molar units is exact, conversion from moles to volume units is approximate unless compressibilities of the fuel gas and the air are taken into account. They are different for each composition and are not linear with pressure. Thus, in internal combustion work — especially in which the air-fuel ratio has to be controlled precisely to minimize emission of pollutants — the air/fuel ratio is generally based on mass. For any hydrocarbon mixture described by $C_aH_bO_cN_d$, all the terms needed for air-fuel determination are given in the reactants side of the general combustion equation. The stoichiometric air-fuel ratio can be shown to be:

$$(A/F)s = \frac{34.567\ (4a + b - 2c)}{(12.011a + 1.00794b + 15.9994c + 14.0067d)} \qquad (5\text{-}7)$$

Fuel	Theoretical Air, lb/lb Fuel	HHV Btu/lb	Theoretical Air Typical lb/10^4 Btu	Theoretical Air Range lb/10^4 Btu
Bituminous coal (VM* >30%)	9.07	12,000	7.56	7.35 to 7.75
Subbituminous coal (VM* >30%)	6.05	8,000	7.56	7.35 to 7.75
Oil	13.69	18,400	7.46	7.35 to 7.55
Natural gas	15.74	21,800	7.22	7.15 to 7.35
Wood	3.94	5,831	6.75	6.60 to 6.90
MSW*and RDF*	4.13	5,500	7.50	7.20 to 7.80
Carbon	11.50	14,093	8.16	–
Hydrogen	34.28	61,100	5.61	–

*VM = volatile matter, moisture and ash free basis
MSW = municipal waste
RDF = refuse-derived fuel

Table 5-11 Theoretical Air Required for Various Fuels. Source: Babcock & Wilcox

Heavy Fuel Oil, % by wt		Natural Gas, % by vol	
S	1.16	CH_4	85.3
H_2	10.33	C_2H_6	12.6
C	87.87	CO_2	0.1
N_2	0.14	N_2	1.7
O_2	0.50	O_2	0.3
		Sp Gr	0.626
		Btu/ft^3, as-fired	1090
Btu/lb, as-fired	18,400	Btu/lb, as-fired	22,379
Theoretical Air, Fuel and Moisture			
Theoretical air, lb/10,000 Btu	7.437	Theoretical air, lb/10,000 Btu	7.206
Fuel, lb/ 10,000 Btu	0.543	Fuel, lb/ 10,000 Btu	0.440
Moisture, lb/ 10,000 Btu	0.502	Moisture, lb/ 10,000 Btu	0.912

Table 5-12 Fuel Analysis Comparison for Heavy Oil and Natural Gas Fuels. Source: Babcock & Wilcox

For the natural gas described as $C_{1.045}H_{4.034}O_{0.010}N_{0.036}$, the $(A/F)_S$ is 16.39 (283.242/17.282), which can be interpreted as 16.39 lbm dry air requirement for each lbm of gas burned (16.39 lbm-air/lbm-gas, or any other equivalent mass units).

To determine the actual air-fuel ratio for lean-burn conditions, an excess air factor λ is defined as:

$$\lambda = \frac{(A/F)\ actual}{(A/F)s} \tag{5-8}$$

Thus, the actual ratio is obtained by multiplying the stoichiometric ratio by λ.

The quantity of excess O_2 is a nearly exact indication of excess air. Fig. 5-1 is a dry flue gas volumetric combustion chart that is used in field testing. It relates O_2, CO_2, and N_2 (by difference). For complete uniform combustion of a specific fuel, all points should lie along a straight line drawn through the pivot point. This line, which ranges for various fuels, is referred to as the combustion line. Lines indicating constant excess air have been superimposed on the volumetric combustion chart. Note that excess air is essentially constant for a given O_2 level of a wide range of fuels. Since the calculated excess air result is insensitive to variations in moisture for specific types/sources of fuel, O_2 is a constant indicator of excess air when the gas is sampled on a wet or in-situ basis.

Heating Value of Air-Fuel Mixtures

Once the heating value of the specific fuel being burned and the amount of air required for complete combustion of the flammables is determined, the heating value of the air-fuel mixture, or air-fuel charge as it is sometimes called, can be determined as follows:

$$\text{Heating value of air-fuel mixture} = \frac{\text{Heating value of fuel}}{(1\ cf\ fuel + cf\ air\ required)} = Btu/cf \tag{5-9}$$

The hypothetical pipeline gas fuel mixture (neglecting compressibility) yields the following:

$$\text{HHV of air-fuel mixture} = \frac{1{,}036.6}{(1 + 9.78)} = 96.16\ Btu/cf$$

$$\text{LHV of air-fuel mixture} = \frac{934.9}{(1 + 9.78)} = 86.73\ Btu/cf$$

Adjustments of the HHV and LHV of the air-fuel mixture are made by recalculating the values with more or less air. Excess air (lean) will result in decreased Btu/cf (kJ/m^3) values and excess fuel (rich) will result in increased Btu/cf (kJ/m^3) values.

All of these computations are based on complete stoichiometric combustion. Actual combustion may deviate from this for several reasons. In boiler applications, for example, it is common to use a leaner mixture. In the event of imperfect mixing of fuel and air, or other such imperfections, this provides a margin of error on the side of excess air rather than excess fuel. While both result in decreased efficiency, air is less costly than fuel. Another consideration is exhaust emissions. Rich burn may result in increased emission of excess carbon monoxide (CO) as well as the accumulation of soot on heat exchange surfaces, while lean burn may result in higher NO_X emissions. Chapter 17 contains a detailed discussion on methods and technologies to control emissions from combustion of fossil fuels such as NOx and particulate matter that are released as unwanted by-products of combustion.

Although a rich mixture is often used to improve performance in internal combustion engine applications, emissions requirements have led to a movement toward lean-burn rather than rich-burn engines. For both boilers and engines, however, increasingly stringent emissions requirements have produced a carefully planned richer or leaner operation.

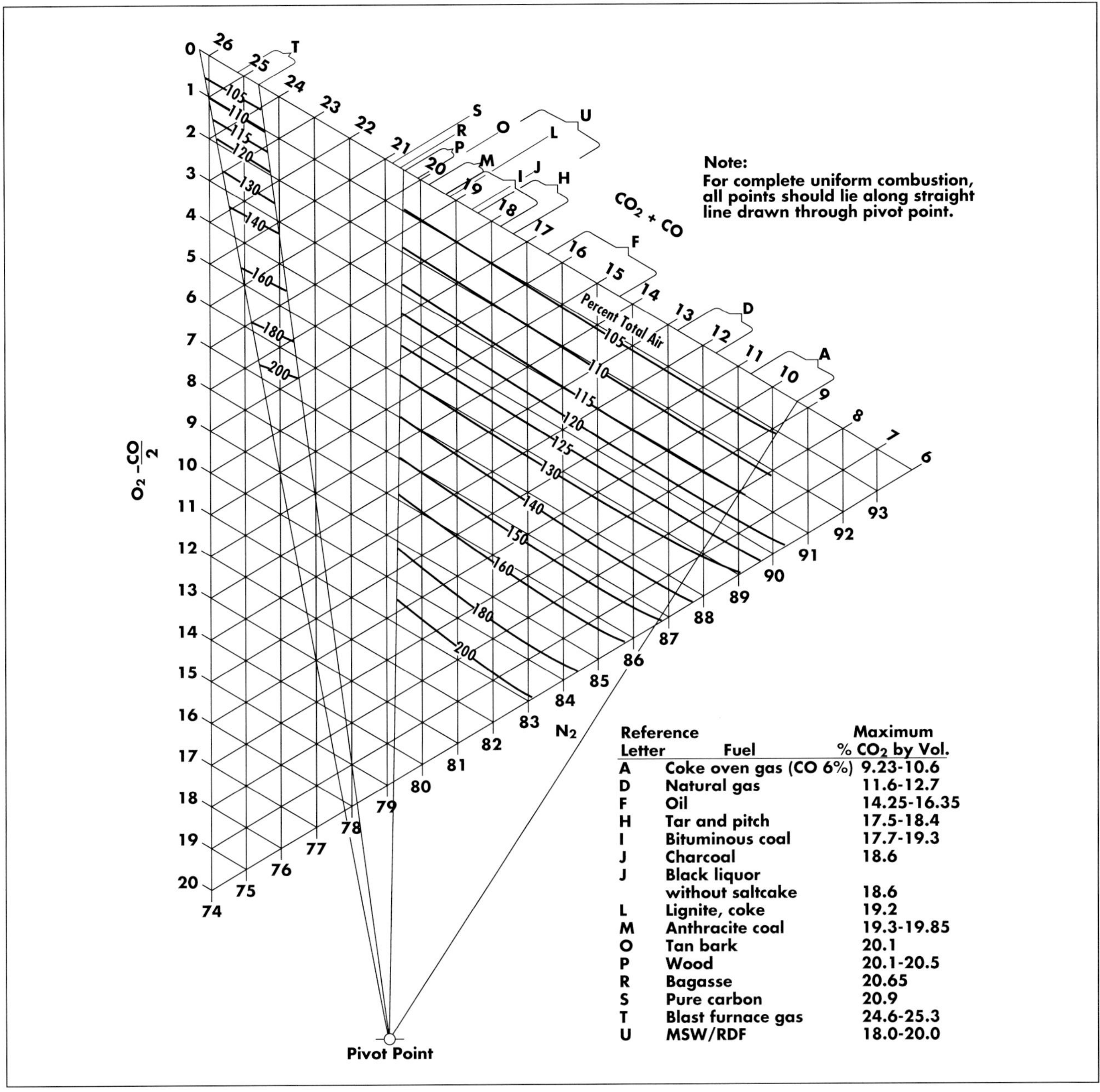

Reference Letter	Fuel	Maximum % CO_2 by Vol.
A	Coke oven gas (CO 6%)	9.23-10.6
D	Natural gas	11.6-12.7
F	Oil	14.25-16.35
H	Tar and pitch	17.5-18.4
I	Bituminous coal	17.7-19.3
J	Charcoal	18.6
J	Black liquor without saltcake	18.6
L	Lignite, coke	19.2
M	Anthracite coal	19.3-19.85
O	Tan bark	20.1
P	Wood	20.1-20.5
R	Bagasse	20.65
S	Pure carbon	20.9
T	Blast furnace gas	24.6-25.3
U	MSW/RDF	18.0-20.0

Fig. 5-1 Dry Fuel Gas Volumetric Combustion Chart. Source: Babcock & Wilcox

Flue Gas and Combustion Analysis

Flue gas analysis is used for any fuel burning equipment, including engines and boilers, to indicate the air-fuel ratio and completeness of combustion and to measure the concentration and amount of all emission elements, including pollutants such as CO, CO_2, NO_X and SO_2.

Carbon dioxide measurement in exhaust gas, for example, can be used to determine combustion efficiency. Optimum, or stoichiometrically correct, combustion results in the formation of the maximum amount of CO_2 (about 12%) and the minimum amount of O_2 and CO (0%). For the given example of pipeline quality gas, the stoichiometric CO_2% volume basis is at maximum: 9.67% on wet flue and 11.88% on dry flue basis.

Several methods are used to measure flue gas composition. These include:

- **O_2 analyzers**, which are commonly designed to measure the difference in O_2 partial pressures on the two sides of a zirconium or palladium wafer or oxide cell. An air-powered aspirator draws gas samples past sensors providing independent and continuous percent-by-volume measurement of O_2 and in some cases CO.
- **CO spectroscopic analyzers**, which sample the entire gas stream by shining a beam of infrared light across the stack and averaging CO concentration across most of the stack and measuring it. A CO-sensitive filter and a non-CO-sensitive filter are set on a rotating wheel and pass alternately across the source to allow alternate measurement and reference infrared impulses.
- **CO_2 analyzers**, which are commonly used for equipment of all sizes. In the smallest applications, a flue gas sample is pumped into a container with a fluid, such as pyrite, that expands in proportion to the amount of CO_2 present. Scaled to the exhaust gas temperature this measurement indicates combustion efficiency. Large boiler systems use systems such as non-dispersive-infrared photometers.

Flue-gas analyzers are either extractive or in-situ. These terms refer to the manner in which the gas sample is delivered to the analyzer. Extractive units draw a sample out the stack, condition, and then transport the sample for remote analysis. In-situ analyzers expose the sampling apparatus directly to the flue gas and measurement combines a light source shining across the stack with a receiver/analyzer. They use absorption spectroscopy and measure ultraviolet, visible, and infrared portions of the optical spectrum. Detection of the absorbed frequencies in the spectrum from a narrow-band source identifies the constituents and their concentrations.

Additional Energy Sources

In addition to the gaseous and liquid fuels discussed in this chapter, there are a host of other energy sources that may be productively used in providing for heating, cooling, process use, and power generation at commercial, industrial, and institutional (CI&I) facilities. These range from fossil fuel sources, such as coal, shale, and tar sands, to renewable sources, such as solar, geothermal, wind, water, and biomass. Other diverse sources range from hydrogen gas to a wide variety of energy sources congealed in common trash. Below is a brief review of several of these alternative sources.

Coal

While mentioned earlier only as a feedstock for various coal-derived fuel gases, coal itself is the world's second leading energy source, providing nearly one-third of the world's energy. It is also still the most commonly used fuel for large-scale utility electric power generation plants, which consume about one third of the energy used in the United States. While far less costly on a heat content basis, relative to gas and oil, its use is still limited at most CI&I facilities in the United States due to increased costs for transportation, handling, combustion equipment, and environmental compliance. Except in very large applications, these facilities typically do not enjoy the economies of scale achieved in central utility applications, where the many peripheral cost obstacles can be overcome. Still, coal is found in use in some industrial facilities and in numerous homes throughout the country. One of the biggest deterrents to increased use of coal as a fuel source is its contribution to climate change, since combustion of coal results in higher rates of release of CO_2 into the atmosphere than with other fossil fuels. Research is underway to address this problem through large-scale carbon capture and storage technology. This technology is discussed in Chapter 17.

Coal is formed from plants by chemical and geological processes that occur over millions of years. Layers of plant debris are deposited in wet or swampy regions under conditions that prevent exposure to air and complete decay as the debris accumulates. Bacterial action, pressure, and temperature act on the organic matter over time to form coal. This geochemical process is called coalification. In the early stages, peat is formed. It is progressively transformed into lignite, which eventually can become anthracite, given the proper progression of geological changes.

The four basic types of coal are as follows:

- **Lignite** is the lowest rank coal. Lignite coals are soft and brownish and have heating values of less than 8,300 Btu/lbm (6,980 kJ/kg), with a moisture content of about 30%. Lignite coals dry out when exposed to air, so spontaneous combustion during storage is a concern. Long distance shipment is often not economical due to their high moisture content and low energy density.
- **Subbituminous** coals have a reasonably high heating value of 8,300 to 11,500 Btu/lbm (19,300 to 26,750 kJ/kg), with moisture content ranging from 15 to 30%. They have become increasing attrac-

tive as fuel sources due to their relatively low ash and sulfur content. With a sulfur content of 1% or less, switching to subbituminous coals has become an attractive option for many power plants to limit SO_2 emissions. Like lignite coals, they are high in volatile matter content and can easily ignite.

- **Bituminous** coals are the most commonly used coals in electric utility boilers. They have heating values of 10,500 to 14,000 Btu/lbm (24,420 to 36,050 kJ/kg) and lower moisture content and higher fixed carbon content (69 to 86%) than subbituminous coals. With a higher energy density, lower moisture content, and lower volatility, these coals are easier to transport and store than lignite and subbituminous coals. They are still fairly volatile, which, combined with their high energy density, enables bituminous coals to burn easily when pulverized to a fine powder. Some types of bituminous coal, when heated in the absence of air, soften and release volatiles to form the porous, hard, black product known as coke. Coke, referred to earlier in the chapter, is used as fuel in blast furnaces to make iron and can also serve as a feedstock for various types of fuel gases.
- **Anthracite** is the highest rank of coal. It is shinny, hard, and brittle and has the highest fixed content of carbon (86 to 98%) and the lowest moisture content (about 3%) of all coals. Heating values are about 15,000 Btu/lbm (34,900 kJ/kg), which is similar or sometimes a bit lower than the best quality bituminous coals. Anthracite is low in sulfur and volatiles and burns with a hot, clean flame, making it a premium coal for domestic heating.

There are a variety of criteria for ranking coal, given its wide characteristic variability. Some of the various ASTM analyses include determination of moisture, volatile matter, fixed carbon, and ash. *ASTM D 3176* analysis includes measurements of carbon, hydrogen, nitrogen, and sulfur content and the calculation of oxygen content. *ASTM D 2015* analysis is used to determine the gross calorific value of coal on various bases (i.e., dry, moisture and ash free, etc.). These constituents vary tremendously amongst different types of coal and even different variations of the same classes of coals. Carbon content may vary anywhere from less than 20% to more than 95% and sulfur content may vary from less than 0.5% to more than 8%. HHV is defined as the heat released from combustion of a unit mass of fuel, with the products in the form of ash, gaseous CO_2, SO_2, nitrogen and liquid water, exclusive of any water added as vapor. LHV is calculated from HHV (*ASTM Standard D 407*) by deducting 1,030 Btu/lbm (2,396 kJ/kg) of water derived from the fuel, including the water originally present as moisture and that formed by combustion.

Coal-Derived and Related Fossil Fuels

As noted earlier, there are a variety of fuels derived from coal. These include coke, char, and liquids and a wide array of gases from coal gasification, including coke oven gas, blast furnace gas, water gas, and producer gas. There are also several new gasification processes as sources of synthetic natural gas that are currently under development.

There is variety of coal liquefaction processes that are under development. The three basic processes used for liquefaction are direct and indirect liquefaction and pyrolysis. In the direct liquefaction process, coal is slurried in a process-derived oil and then reacted with hydrogen at high pressure and moderate temperature to form liquid hydrocarbons. Ash and unreacted coal are removed from the product and the unused coal is gasified to produce the required hydrogen. In the indirect process, coal is first gasified to a synthetic gas and then catalytically converted to methanol or other liquid hydrocarbons. The direct method has received more development focus because it offers higher conversion efficiencies (about 65 to 70%). Pyrolysis requires heat to drive off volatile portions of the coal, producing gases, liquids, and solid products.

The ability to transform coal to useful gases and liquids has a great appeal in that these by-products can be more easily used in various combustion applications and can be used with reduced pollution control expense. While some of the synthetic gases are already commercially used, other products may become increasingly prominent in the future, depending on the status and market price of natural gas reserves.

In addition to the vast global reserve of coal, other fossil fuels in large reserve include shale and tar sands. Shale and tar sands are amenable to producing oil. Based on current technologies and oil prices, no extensive commercial production of shale oil is expected for many years. Tar sands, however, have already been economically produced and remain a potentially valuable energy source, depending on technology advances and world energy prices.

Hydrogen

Hydrogen, the simplest element, is composed of one proton and one electron. It is an abundant resource, making up more than 90% of the composition of the universe and is the third most abundant element in the earth's surface. More than 30% of the mass of the sun is atomic hydrogen and water is mostly composed of hydrogen. Under ordinary (earthly) conditions, hydrogen is a colorless, odorless, tasteless, and nonpoisonous gas composed of diatomic molecules (H_2). It is a clean-burning fuel with an energy density of about 61,000 Btu/lbm (140,000 kJ/kg) on an HHV basis. With a specific volume of 190 ft^3/lbm (12 m^3/kg), this corresponds to a volumetric energy density of 320 Btu/cf (11,920 kJ/m^3) on an HHV basis and 270 Btu/cf (10,060 kJ/m^3) on an LHV basis. Hydrogen produces no smoke or particulate products and, if burned with oxygen, produces only water vapor as its sole by-product. If burned conventionally with ambient combustion air, it will, however, produce some NO_X emissions. While its specific gravity is one-tenth that of natural gas and its energy density nearly three times that of natural gas on a mass basis, hydrogen is a gas at all normal, reasonable temperatures. Hence, its lower energy density on a volumetric basis is a clear deficiency.

Hydrogen is perhaps more appropriately termed an energy carrier, not a fuel source, since it must be manufactured. It is produced through steam reforming or through electrolysis. **Steam reforming** involves the use of high-temperature steam to separate hydrogen from carbon atoms in a feedstock fuel (i.e., natural gas, oil, etc.). Currently, this process is generally only used in the manufacture of fertilizer and chemicals or to upgrade the quality of certain petroleum products. **Electrolysis** involves the splitting of water into its basic elements, hydrogen and oxygen. This is done by passing an electric current through water to separate the atoms ($2H_2O \rightarrow 2H_2 + O_2$). Hydrogen collects at the negatively charged cathode and oxygen collects at the positively charged anode. The energy required to produce hydrogen via electrolysis (assuming a typical 1.23 volt source) is about 51,000 Btu/lbm (33 kWh/kg). Other experimental methods to produce hydrogen include photoelectrolysis, which uses sunlight to split the water molecules and other reforming processes using various types of biomass as feeder fuels.

The high-energy input requirement and other process costs have limited hydrogen's use as a fuel to very specific applications, notably as a rocket fuel and for electricity production in fuel cells. In a fuel cell, the electrolysis process is reversed by combining hydrogen and oxygen, through an electrochemical process, which produces electricity, heat, and water. In a typical fuel cell application, hydrogen-rich gas is produced through the steam reforming process from a fossil fuel source, such as natural gas. It is then combined with oxygen to generate direct current electric power and thermal energy. The hydrogen-rich fuel is oxidized at the anode to form hydrogen ions and release electrons, while air flows over a cathode surface. The hydrogen ions move from the anode through the electrolyte solution to the cathode. The electrons flow through an external electrical circuit, generating direct current output.

Hydrogen is also now being contemplated as a fuel additive for its beneficial emissions characteristics. In the future, it may play a more critical role as an energy resource as the efficiency of the production process is increased and/or cost reduced.

Renewable Energy Resources

There is a wide range of renewable energy sources currently in use and abundantly available for future use as mainstream energy sources. These include wood, wind, tidal power, hydropower, geothermal energy, hydrogen, and, the most abundant of all, solar energy.

Solar

Solar energy is the source of almost all of the earth's energy. About one two-billionth of the sun's radiation impinges on the earth, about half of which is not reflected back into interstellar space by the earth's atmosphere. The total solar-radiation intensity on a surface normal to the sun's rays is about 434.6 Btu/ft^2-h (1,370 W/m^2 or 4,932 kJ/m^2-h). This quantity, called the **solar constant**, undergoes small (1%) variations that primarily affect the short-wave portion of the spectrum. Also, since the earth-sun distance varies as does the angle between the earth's equatorial plane and the sun line, seasonally, there are corresponding variations in terrestrial solar-radiation intensity. Additionally, in passing through the earth's atmosphere, the sun's radiation is partially and selectively absorbed, scattered, and reflected by ozone and water vapor, air molecules, natural dust, clouds, and air-borne pollutants.

The total solar radiation intensity on a horizontal surface at sea level varies from zero at sunrise and sunset to a maximum of about 340 Btu/ft^2-h (1,070 W/m^2 or 3,852 kJ/m^2-h) on perfectly clear summer days at noon.

In addition to the natural chemical reactions that sustain growth of plants and animals, convert carbon dioxide into oxygen through photosynthesis, among other reactions, when applied through various energy conversion devices, the sun's radiation can provide for numerous thermal and electrical reactions, the by-products of which can be used productively.

- Thermal energy is produced when the sun's radiation is absorbed and converted into heat, through various passive and active systems. This conversion can be used for heating air or water. It can also serve to evaporate seawater to produce salt or distill it into potable water. While traditional flat-plate collectors typically heat air, water, or other fluids to temperatures of 100 to 180°F (38 to 82°C), when converted in specially designed solar furnaces, temperatures as high as 6,500°F (3,600°C) can be produced. These furnaces employ precise concentrators that focus the sun's rays on small areas.
- Electrical energy is produced from the sun by photovoltaic (PV) cells. A single PV cell is a thin semiconductor wafer, generally made of highly purified silicon. The wafer is doped on one side with atoms that produce a surplus of electrons and the other side with atoms that produce a deficit of electrons. This establishes a voltage difference between the two sides of the wafer. Metallic contacts are made to both sides of the wafer. When the wafer is bombarded by photons from solar radiation, electrons are knocked off the silicon atoms and are drawn to one side of the wafer by the voltage difference and can flow through an external circuit attached to the metal contacts on each side of the wafer.

Geothermal

Geothermal energy is heat (thermal) derived from the earth (geo). It is the thermal energy contained in the rock and fluid (that fills the fractures and pores within the rock) in the earth's crust. In most areas, this heat reaches the surface in a very diffuse state. However, due to a variety of geological processes, some areas, including substantial portions of the Western United States, are underlain by relatively shallow geothermal resources. Below the crust of the earth, the top layer of the mantle is hot, liquid rock called magma. The crust of the earth floats on this liquid magma mantle. When magma breaks through the surface of the earth in a volcano, it is called lava.

For every 330 ft (100 m) you go below ground, the temperature of the rock increases about 5.5°F (3°C). Deep under the surface, water sometimes makes its way close to the hot rock and turns into hot water or into steam. The hot water can reach temperatures of more than 300°F (148°C). This is higher than the boiling point for water; it is only because the water is contained under pressure that it does not turn to steam. When this hot water comes up through a crack in the earth, it immediately flashes to steam and we call it a geyser or hot spring, the most famous of which is the Geysers in California. Sometimes, people use the hot water in swimming pools or in health spas.

The current production of geothermal energy from all uses places third among renewables, following hydroelectricity and biomass, and ahead of solar and wind. Geothermal resources can be classified as low temperature, less than 194°F (90°C), moderate temperature, 194 to 302°F (90 to 150°C), and high temperature, greater than 302°F (150°C). The application of these resource is influenced by temperature, pressure, location, and accessibility. The highest temperature resources are generally used only for electric power generation. Current U.S. geothermal electric power generation totals more than 2,200 MW. Uses for low and moderate temperature resources can be divided into two categories: direct use and ground-source heat pumps.

- Direct use, as the name implies, involves using the heat in the water directly (without a heat pump or power plant) for such applications as heating of buildings, industrial processes, greenhouses, aquaculture (growing of fish), and resorts. Direct-use projects generally operate at temperatures between 100 to 300°F (38 to 149°C) from hot water found near the earth's surface. Current U.S. installed capacity of direct use systems totals more than 1,600 MMBtu/h (470 MW), or enough to heat 40,000 average-sized houses.
- Using resource temperatures of 40 to 100°F (4 to 38°C), the geothermal heat pump, a device which moves heat from one place to another with the addition of mechanical energy, uses the relatively constant temperature of soil or surface (ground) water as a heat source and sink for the heat pump, which provides heating and/or cooling for buildings.

Historically, the extraction and use of geothermal energy for electric generation systems has focused on dry or vapor-dominated and wet or liquid-dominated (partially flashed steam from hot water at or near saturation temperature) sources. Bottom-hole temperatures as high as 800°F (427°C) have been reported at a depth of 8,100 ft (2,469 m).

Under-developed resource types of geothermal energy include hot dry rock and geopressured zones. The hot dry rock category, which is believed to offer much promise as a useful energy source, is a geological formation that possesses a high heat content, but no meteoric or magmatic waters to provide a heat-transport medium. Hence, the injection of water is required in order to carry heat to the surface. In these areas, the thermal gradient exceeds 45°F per mile (15°C per kilometer), so very deep drilling is not required. While the thermal conductivity of rock may limit energy transfer, there remains vast untapped energy stored in this form throughout many mountain ranges in the country. Geopressed zones, which are found at depths of 5,000 to 20,000 ft (1,500 to 6,100 m), contain water at temperatures ranging from 140 to 375°F (60 to 190°C) at fluid pressures from 3,000 to 14,000 psi (207 to 966 bar). There is a high degree of potential energy value available in this high-pressure fluid that can be applied to electric generation. The potential of this resource is enormous, but so far, no technology exists to extract energy in a commercially useable way.

Note that geothermal energy does pollute the environment with moderate amounts of sulfur and carbon emissions. However, it is believed that since the tapped sources are natural, much of this would occur anyway.

Wind and Water

Two other wide-spread renewable energy sources are wind and water, which both possess energy that can be converted into power and produce useful work through wind and water turbines, respectively. Both have provided useful mechanical power for centuries and have, in recent times, become valuable, pollution-free components of the world's usable energy resource mix.

- A **wind turbine** obtains its power input by converting the force of the wind into torque acting on the rotor blades. The amount of energy that the wind transfers to the rotor depends on the density of the air, the rotor area, the wind speed, and the conversion efficiency of the turbine. About 1 to 2% of the energy coming to the earth from the sun is converted to wind energy. The kinetic energy of any moving body is proportional to its mass. Hence, the kinetic energy in the wind depends on the density of the air; the heavier the air, the more energy potential it possesses. Consistent with the affinity laws discussed in Section VII, while mass flow rate varies proportionally with speed, power varies with the cube of speed. Hence, a two-fold increase in wind speed will produce an eight-fold increase in power production potential. Wind turbines can produce useful power to drive mechanical devices or non-interconnected electric generators at speeds starting at about 7 miles per hour (3 m/s), and for interconnected electric generators at speeds of about 12 miles per hour (5.4 m/s). In accordance with the cubed relationship of speed and power, as available wind speed increases, power production potential increases at a dramatic rate.

  The use of wind power for productive purposes is growing at a rapid rate, with several thousand MW of installed capacity being added each year, notably in the United States and Europe. The driving forces are its wholly renewable character that can provide power with no energy component cost (equivalent to free fuel, though obviously not without capital and maintenance cost requirements), no air emissions, and no depletion of non-renewable energy resources.
- **Hydropower** is currently the world's largest renewable source of power, providing about 6% of worldwide energy supply or nearly 20% of the world's electricity. A water, or hydraulic, turbine obtains its power input as the potential energy of a volume of water at a given height is converted to kinetic energy as it flows from high points to low points (falls through a distance) because of the force of gravity. The power available in the water that can be transferred to the turbine is thus a function of two factors: 1) the head (which is essentially the elevation difference over which the water flows, less friction); and 2) the mass flow, which is the density and velocity rate of the water, both of which are subject to modest deviation depending on gravitational force variation with altitude. The turbine is driven by water under pressure from a penstock or forebay. The energy in the flowing water is converted to mechanical energy by revolving a wheel fitted with blades, buckets, or vanes, and then, in most cases, to electrical energy through an electric generator.

  Hydropower installations do alter their natural surroundings, sometimes with negative and sometimes positive environmental impact. Hence, the effects of such a plant must be carefully studied before a decision is made if a location or project approach is appropriate. However, the positive economic and environmental benefits of this no-cost (for the energy resource component), pollution-free, and fossil fuel depletion-free renewable resource are driving forces in its proliferation as a fundamental component of the world's power production energy resources.

Biomass

Biomass and related waste contribute several percent of the nation's current annual fuel resource mix. While commonly used as feedstock for production of biogasses, as discussed previously, biomass is commonly used as a solid fuel, often in the proximity of its production. Wood waste is the primary component of biomass fuel, though peat and other agriculture by-products are also common components.

- Wood contributes more than three-quarters of the total biomass energy. Where there is ample wood in supply, home heating and some large central steam generation (utility, industrial, and institutional) plants use wood or wood waste as a primary fuel source. Wood waste provides about half of the energy consumed by the forest products industry, as the conversion of logs to lumber results in about 50% waste in the form of bark, shavings, and sawdust. Fresh timber typically contains 30 to 50% moisture or more, mostly in the cell structure of the wood. This may be reduced by half after a year of air drying. When moisture content is high, it may be necessary to mix the wood with low-moisture fuel to ensure that there is enough energy density, on an LHV basis, entering the boiler to support proper combustion. Dry wood has an energy density range of 8,500 to 9,100 Btu/lbm (19,800 to 21,200 kJ/kg). This is reduced to about 4,375 Btu/lbm (10,170 kJ/kg) at 50% moisture and 1,750 Btu/lbm (4,070 kJ/kg) at 80% moisture content. In addition to biogasses, wood can be transformed into several other fuel products. Wood charcoal, for example, is made by heating wood to a high temperature in the absence of air. Weight and volume are reduced by 75 and 50%, respectively, producing a product that has a higher energy density and is easier to transport.
- Peat is another major biomass type fuel source. Discussed above as an early stage in the metamorphosis of vegetable matter into coal, peat is a complex mixture of carbon, hydrogen, and oxygen in a ratio similar to that of cellulose and lignin. It is generally low in sulfur, nitrogen, and ash from the parent vegetation. Dry, ash-free ultimate analysis ranges from 50 to 65% carbon, 5.5 to 7% hydrogen, 30 to 40% oxygen, 1 to 2% nitrogen and less than 1% sulfur. While used largely in the United States for soil improvement, mulch, filler for fertilizers, and litter for domestic animals, it is also a viable fuel source. Its energy density, depending on the source and method of processing, may range from 3,700 to 8,000 Btu/lbm (8,600 to 18,600 kJ/kg), with an average of about 6,000 Btu/lbm (14,950 kJ/kg) with standard air-dried processed material.
- There are a number of other biomass type fuels used for generation of industrial steam and power. Aside from their value as fuel, the burning of wastes minimizes disposal requirements. These benefits, however, are often limited by environmental regulations, which vary by region. Typically, these by-products are cellulosic in character and offer energy densities of 7,000 to 9,500 Btu/lbm (16,300 to 22,100 kJ/kg), depending on the level of resinous material present. These may include bagasse (which is a fibrous by-product of sugarcane processing), corncobs, rice straw or hulls, wheat straw, and pine bark. Cattle manure is another useful by-product type fuel. Other related materials are discussed below under refuse-derived fuels.

Refuse-Derived Fuels

In addition to the various waste-related biomass sources used for methane-based gas production, solid waste is a useful direct fuel source that allows for the production of useful thermal energy to be combined with the elimination of the solid waste. Typically, three-quarters of most solid waste is combustible, composing as much as 90% of the waste volume. Solid waste is a commonly used fuel source in centralized plants for steam production. The steam is used for central distribution as a district thermal energy source, electricity production or both and the ash product of combustion is typically landfilled, but at greatly reduced volumes than with conventional disposal. The basic categories of solid waste include domestic, industrial scrap and plastic, and agricultural refuse. Domestic refuse, typically consisting of garbage, grass, leaves, rags, paper cardboard, wood scraps, etc., offers energy densities ranging from 1,500 to 8,000 Btu/lbm (3,500 to 18,600 kJ/kg). Industrial scrap and plastic refuse, typically consisting of leather, cellophane, rubber, polyethylene, oil waste, etc., offers greater energy densities ranging from 8,000 to 20,000 Btu/lbm (18,600 to 46,500 kJ/kg).

The waste-to-energy industry (WTE) has been a

growing energy source producer (mostly electric generation) for many decades as waste in the United States continues to increase. Waste production has more than doubled over the past four decades. Municipal solid waste (MSW), defined by the EPA to include durable goods, containers and packaging, food wastes, yard wastes, and miscellaneous inorganic waste, has increased from 88 million tons (80 billion kg) in 1960 to current levels of over 200 million tons (181.5 billion kg). Of this total, paper and paperboard account for 39%, yard wastes 31%, plastics 10%, metals 8%, food 7%, and glass 6%.

In 1960, about 27 million tons (24.5 billion kg) or 30% of the MSW generated was incinerated, most without energy recovery or air pollution controls. This declined to half that sum by 1980. Then, following the enactment of the Public Utilities Regulatory Act (PURPA) and related legislation and the subsequent emergence of energy markets, combustion of MSW increased to 32 million tons (29 billion kg) by 1990. Most of these are waste-to-energy plants that produce steam, electricity, or both for sale or campus distribution, with air pollution controls.

The MSW industry currently has four components: recycling, composting, landfilling, and combustion. Refuse can be mass-burned in as-received form or separated, classified, shredded and reclaimed to form a higher energy density, homogeneous product known as refuse-derived fuel (RDF).

While beneficial from the perspective of accomplishing two objectives simultaneously (i.e., the production of energy and the elimination of solid waste), there are several obstacles and limitations on the growth of WTE combustion plants. The movement toward recycling, composting, and landfilling has reduced the growth rate of refuse available for direct combustion. Target rates of 25% reduction in the volume of refuse to be landfilled have been achieved in many municipal recycling programs, which have become commonplace. Moreover, the need for very strict environmental control of air emissions and solid waste greatly impact project economics. Finally, siting is a long-standing political and logistical difficulty. Even so, when combusted properly with full environmental control and recovery of heat, refuse can be a valuable fuel energy resource.

Climate Change And Fuel Combustion

As noted above in the section on flue gas and combustion analysis, burning of fossil fuels creates CO_2 as a by-product. CO_2 is known to contribute to the "greenhouse gas" effect, which causes climate change. It is, therefore, important to consider CO_2 emissions when comparing and selecting fuel types. Table 5-13 provides CO_2 emissions factors for the most common fuel types, expressed as lbm of CO_2 per unit volume of fuel and per million Btu (MMBtu) of fuel consumed. A conversion factor of 2204.6 lbm per metric ton can be used to convert the resulting emissions quantities into metric tons, the most widespread unit of measure for this greenhouse gas.

Electricity

While electricity is not a fuel source in the classic sense, its climate change contribution should also be considered in heat and power resource evaluation. This is because electricity generation typically involves fuel combustion in central utility generating plants, which emit CO_2. Likewise, if local power production systems that burn fossil fuels will offset electricity consumption, then they will affect CO_2 emissions at the host facility. Therefore, the net CO_2 emissions impact of electricity generation or electric savings projects should be included in the evaluation of those projects.

Determination of CO_2 emissions impact from electricity is often not as straightforward as fuel emissions, because the mix of source energy used to generate electricity varies by state. Some states rely heavily on coal-fired power plants, with their correspondingly high CO_2 emissions, while others have a higher reliance on hydropower and nuclear sources, which operate with no CO_2 emissions. Table 5-14 provides values for CO_2 emissions per kWh of electricity by state. For compliance with state regulations or for participation in various state programs, these factors can be used to calculate the impact of electricity that is generated locally, and therefore reduced from grid-purchased power. As with fuel, a conversion factor of 2204.6 lbm per metric ton can be used to convert the resulting emissions quantities into metric tons.

Fuel	Emission Coefficients		
	lbm CO_2 per Unit Volume or Mass		lbm CO_2 per MMBtu
Petroleum Products			
Aviation Gasoline	18	per gallon	153
Distillate Fuel (No. 1, No. 2, No. 4 Fuel Oil and Diesel)	22	per gallon	161
Jet Fuel	21	per gallon	156
Kerosene	22	per gallon	160
Liquified Petroleum Gases (LPG)	13	per gallon	139
Motor Gasoline	20	per gallon	156
Petroleum Coke	32	per gallon	225
Natural Gas and Other Gaseous Fuels			
Methane	116	per 1000 ft3	115
Landfill Gas[1]		per 1000 ft3	115
Flare Gas	134	per 1000 ft3	121
Natural Gas (Pipeline)	121	per 1000 ft3	117
Propane	13	per gallon	139
Coal			
Anthracite	3852	per short ton	227
Bituminous	4931	per short ton	205
Subbituminous	3716	per short ton	213
Lignite	2792	per short ton	215
Renewable Sources			
Biomass	Varies depending on composition of biomass		
Geothermal Energy	0		0
Wind	0		0
Photovoltaic and Solar Thermal	0		0
Hydropower	0		0
Tires/Tire-Derived Fuel	6160	short tons	190
Wood and Wood Waste[2]	3814	per short ton	222
Municipal Solid Waste[2]	1999	per short ton	200

Table 5-13. Fuel and Energy Source Codes and Emission Coefficients Source: U.S. Environmental Protection Agency

Notes:

1. For a landfill gas coefficient per thousand standard cubic foot, multiply the methane factor by the share of the landfill gas that is methane.
2. These biofuels contain "biogenic" carbon. Under international greenhouse gas accounting methods developed by the Intergovernmental Panel on Climate Change, biogenic carbon is part of the natural carbon balance and it will not add to atmospheric concentrations of carbon dioxide. Reporters may wish to use an emission factor of zero for wood, wood waste, and other biomass fuels in which the carbon is entirely biogenic. Municipal solid waste, however, normally contains inorganic materials principally plastics that contain carbon that is not biogenic. The proportion of plastics in municipal solid waste varies considerably depending on climate, season, socio-economic factors, and waste management practices. As a result, EIA does not estimate a non-biogenic carbon dioxide emission factor for municipal solid waste. The U.S. Environmental Protection Agency estimates that, in 1997, municipal solid waste in the United States contained 15.93 percent plastics and the carbon dioxide emission factor for these materials was 5,771 lbs per ton. Using this information, a proxy for a national average non-biogenic emission factor of 919 lbs carbon dioxide per short ton of municipal solid waste can be derived. This represents 91.9 lbs carbon dioxide per million Btu, assuming the average energy content of municipal solid waste is 5,000 Btu/lb.

State Name	lbm CO_2 per kWh	State Name	lbm CO_2 per kWh
Alabama	1.34	Montana	1.59
Alaska	1.09	Nebraska	1.61
Arizona	1.16	Nevada	1.44
Arkansas	1.23	New Hampshire	0.79
California	0.54	New Jersey	0.72
Colorado	1.91	New Mexico	1.94
Connecticut	0.80	New York	0.83
Delaware	2.02	North Carolina	1.22
District of Columbia	2.43	North Dakota	2.33
Florida	1.34	Ohio	1.77
Georgia	1.40	Oklahoma	1.56
Hawaii	1.73	Oregon	0.40
Idaho	0.13	Pennsylvania	1.24
Illinois	1.13	Rhode Island	0.96
Indiana	2.09	South Carolina	0.89
Iowa	1.91	South Dakota	1.18
Kansas	1.89	Tennessee	1.26
Kentucky	2.06	Texas	1.36
Louisiana	1.18	Utah	2.10
Maine	0.74	Vermont	0.00
Maryland (*)	1.35	Virginia	1.20
Massachusetts	1.26	Washington	0.33
Michigan	1.35	West Virginia	1.93
Minnesota	1.59	Wisconsin	1.72
Mississippi	1.23	Wyoming	2.25
Missouri	1.85		

Table 5-14. CO_2 Emissions Factors for Electricity, by State Source: U.S. Environmental Protection Agency

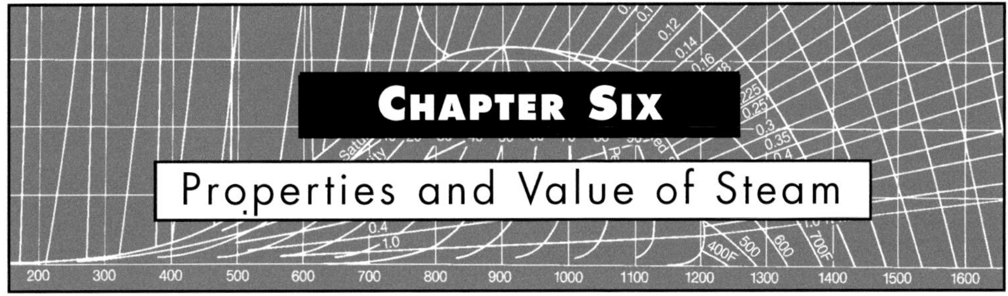

Chapter Six
Properties and Value of Steam

Nearly half of the energy consumed by industry goes into the production of steam, most of which is used for process applications. Steam is one of the least expensive yet most effective heat transfer media. Because its temperature is fixed by pressure — an easily controllable parameter — it allows for excellent control of process temperature.

The conversion of a lbm or kg of water from a liquid to a vapor absorbs large quantities of heat and relatively small quantities of steam can move large amounts of heat inexpensively. High heat transfer rates can be achieved with relatively small pieces of equipment and correspondingly low investments of space and capital.

Steam is water in a gaseous state. Steam generation is the process of converting water from a liquid into a vapor by application of heat. It is classified as either saturated or superheated, dry or wet. The condition of steam, its energy level, its specific volume, and its ability to do work are functions of its pressure, temperature, and quality.

The data frequently used in the United States for identifying the properties of steam and water are the Steam Tables published by the American Society of Mechanical Engineers (ASME). These tables provide data on saturated water, steam, and superheated vapor, and both pressure and temperature organize them. For each specific condition, enthalpy, entropy, specific volume, and internal energy are given. All tabulated properties are specific properties and are tabulated per unit mass. The nomenclature commonly used in steam tables is shown in Figure 6-1.

Fig. 6-1 Commonly Used Nomenclature in Steam Tables

Variable	English Units	Metric Units
t = temperature	°F	°C
p = absolute pressure or bar	psia or in. Hg	megapascals (MPa)
v = specific volume	ft³/lbm	m³/kg
h = enthalpy	Btu/lbm	kJ/kg
s = entropy	Btu/lbm °F	kJ/kg °K
u = internal energy	Btu/lbm	kJ/kg

subscripts:
f refers to a property of the saturated liquid
g refers to a property of the saturated vapor
fg refers to a property change due to evaporation

Selected data from the ASME 1983 Steam Tables in English units are summarized in Tables 1, 2 and 3 at the conclusion of this chapter, with corresponding SI tabulations summarized in Tables 4, 5, 6, and 7. The first two columns of Tables 1, 2, 4, and 5 define the unique relationship between pressure and temperature referred to as saturated conditions, where liquid and vapor phases of water can coexist at thermodynamic equilibrium. For a given pressure, steam heated above the saturation temperature is referred to as superheated steam, while water cooled below the saturation temperature is referred to as subcooled or compressed water. Properties of superheated steam are provided in Tables 3 and 7. Under superheated or subcooled conditions, fluid properties, such as enthalpy, entropy, and specific volume, are unique functions of temperature as well as pressure. At saturated conditions, the quality, or mass fraction, of steam (x) is an additional parameter required for definition.

Definitions

The following definitions are useful for understanding the properties of steam and the use of steam tables in performing various computations that relate to steam applications. These definitions build on some of the basic thermodynamic terms presented in Chapter 2.

The temperature of a system (t) indicates the index of the kinetic energy possessed by the molecules of the substance(s) in the system. At standard atmospheric pressure at sea level, the boiling point of water is 212° in the F scale and 100° in the C scale. The boiling point of water will be greater than 212°F at pressures greater than atmospheric and lower than 212°F at pressures less than atmospheric. The absolute temperature scales — Rankine (R) and Kelvin (K) — are often used with steam applications. Since absolute zero is measured as -459.59°F (-273°C), the boiling point of water at atmospheric pressure at sea level is 671.59°R (373°K).

Steam pressure (p) is the force exerted per unit area by the steam against the containing vessel. Steam pressure is typically measured in terms of either absolute pressure or gauge pressure. **Absolute pressure**, which is generally used in steam tables, is the actual internal force per unit of area expressed as pounds per square inch exerted by the steam on the wall of a vessel in which it was contained.

Gauge pressure is the difference between absolute pressure and ambient (typically taken at atmospheric) pressure. The letters a or g are often affixed to the pressure term to indicate atmospheric and gauge, respectively (e.g., psia and psig).

Because steam is usually measured with sensing devices that are exposed to the atmosphere, these devices register the difference between the actual steam pressure and the ambient pressure exerted by the atmosphere. This difference is referred to as gauge pressure. Since atmospheric pressure at sea level, in English system units, is about 14.7 psia, absolute pressure is determined simply by adding 14.7 to gauge pressure. Tables of steam properties more commonly list absolute pressures.

Vacuum, or negative gauge pressure, is pressure below atmospheric pressure. This should not be confused with absolute pressure, which indicates pressure above absolute zero pressure. For example, 28 in. Hg absolute is roughly equal to 2 in. Hg vacuum.

Specific volume of steam (v) is the amount of space occupied per unit mass of steam and is expressed as ft^3/lbm, or m^3/kg. It is the reciprocal of density, which is expressed as lbm/ft^3, or kg/m^3. At 250 psig (1.825 MPa), for example, the specific volume is 1.74 ft^3/lbm (0.11 m^3/kg). At 0 psig (0.10 MPa), the specific volume of steam is approximately 15 times greater, or 26.8 ft^3/lbm (1.7 m^3/kg).

Internal energy (u), a thermodynamic property, includes all forms of energy in a system, excluding the kinetic and potential energy of the system as a whole, such as center of mass, velocity or elevation in a gravitational field. It is associated with the atomic and molecular sources of energy of the system.

Enthalpy of steam (h or H), is a thermodynamic property of the steam at a given condition, equal to u + pv, and expressed as Btu/lbm or kJ/kg. The change in enthalpy includes heat absorbed or evolved during a change of state (latent heat) and heat absorbed or evolved during a change of temperature without a change of state (sensible heat). The enthalpy is a useful parameter in accounting of energy changes in a system as it undergoes thermal, mechanical, and other energy related changes.

Entropy (s), a thermodynamic property, is the property of steam held constant in a reversible adiabatic process. The entropy is a useful parameter in the accounting of energy conversions that result in the production of work as well as energy unavailable to do work, such as heat loss from the system and loss due to internal irreversibilities.

Steam Properties

When heating water and steam at constant pressure, as in a boiler, one of five conditions exist:

1) Water only
2) Water and steam
3) Saturated steam only
4) Superheated steam
5) Supercritical steam

When heat is added to water in a boiler at constant pressure, the temperature of the water increases. The continued addition of heat will cause the temperature to increase until the pressure of the vapor generated is equal to the boiler pressure (the pressure boost of the boiler feed pumps). At this point, the liquid and vapor are said to be saturated.

Boiling point is the temperature at which water begins to boil. It is a function of pressure: as pressure increases the boiling point increases. At atmospheric pressure at sea level, for example, the boiling point is 212°F (100°C). As more heat is added at constant pressure after the boiling point is reached, the saturated liquid starts to vaporize. This vaporization of liquid is known as a phase change from liquid to vapor. The boiling liquid is called a saturated liquid and the vapor that is generated is called a saturated steam. The enthalpy of saturated liquid refers to the sensible heat required to raise the temperature of 1 lbm (or kg) of water from 32°F (0°C) to the saturation temperature, in this case, the boiling temperature. The heat required to change 1 lbm (or kg) of water from liquid to vapor is called the latent heat or heat of vaporization.

Saturation temperature is the temperature at which both the water and steam are in equilibrium for any given pressure. As long as liquid is present, vaporization of a two-phase mixture of liquid and vapor will continue at constant saturation temperature as heat is added. Whereas at atmospheric pressure, the saturation temperature is 212°F (100°C), at 300 psia (20.7 bar), for example, the saturation temperature is about 417°F (214°C).

A mixture of vapor and liquid is at a state that is somewhere between the saturated liquid and the saturated vapor states. This is referred to as the wet region. As long as liquid is present, the mixture is said to be wet. After all of the liquid is vaporized and only steam is present, it is said to be dry saturated steam. The enthalpy of the saturated steam is the enthalpy of the saturated liquid, plus the heat of vaporization. This is the total amount of heat that must be added to convert water initially at 32°F (0°C) to 100% steam. At 250 psia (17.2 bar), for example, the sensible and latent heat content of dry saturated steam are about 376 and 826 Btu/

lbm (875 and 1,921 kJ/kg), respectively. The total enthalpy is the sum of these two, which is about 1,202 Btu/lbm (2,796 kJ/kg).

The thermodynamic state of steam properties in this saturated region requires specification of two independent variables, and pressure and temperature are not independent in this region. Typically, one of these (temperature or pressure) along with vapor (steam) quality is used to specify conditions. **Steam quality** is a measure of the amount of vapor in the two-phase liquid-vapor mixture. It is defined as the ratio of the mass of vapor present to the total mass of the mixture and may be represented as a percent ranging from 0 to 100%:

$$Quality = x = \frac{mass_{vapor}}{mass_{liquid} + mass_{vapor}} \qquad (6\text{-}1)$$

The further addition of heat to saturated dry steam at constant pressure causes the temperature of the vapor to increase. This state is called the superheat state and the vapor is said to be superheated. In the superheat region, steam quality remains at 100%.

Superheated steam is thus steam heated to a state where it has a higher enthalpy than is associated with its saturation temperature. Unlike saturated steam, which has only a single associated temperature for any given pressure, superheated steam may exist at any temperature above the saturation temperature. Since pressure and temperature are independent in this superheat region, their specification uniquely defines the thermodynamic state of the vapor. Superheated steam properties must, therefore, be tabulated as a function of both temperature and pressure.

The **critical point** is the point at which water turns to steam without boiling and their states are indistinguishable. This occurs at a pressure of 3208.2 psia (22.09 MPa or 221.25 bar) and a temperature of 705.5°F (374.2°C or 647.3°K). As the pressure and temperature of water approach the critical point, the value of the heat of vaporization decreases and becomes zero at the critical point.

Figure 6-2 is a temperature-enthalpy diagram for steam at pressures ranging from 0 to 5,500 psia (0 to 382.8 bar) and temperatures ranging from 300 to 1,200°F (422 to 922°K). Curves are provided showing steam by weight (SBW), or quality, in percent. Figures 6-3 (English units) and 6-4 (SI units)

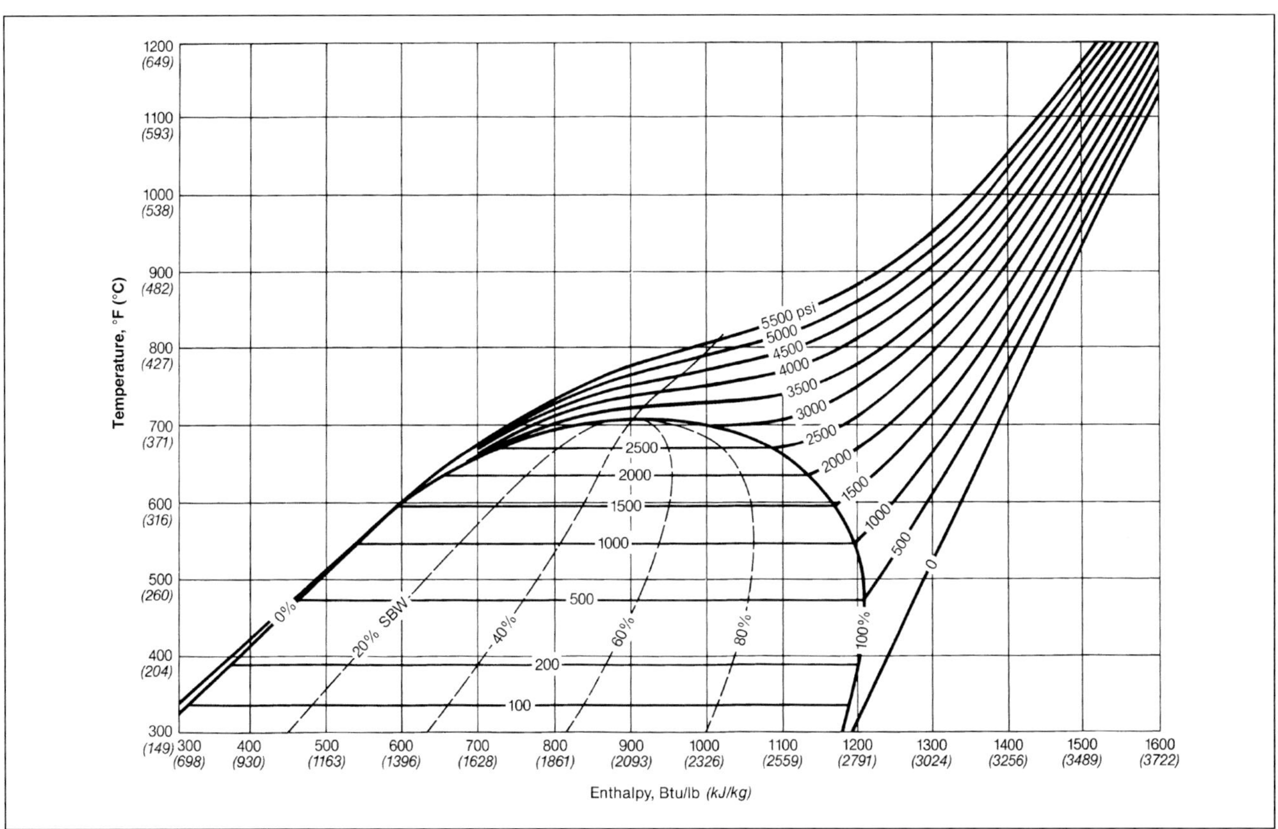

Fig. 6-2 Temperature-Enthalpy Diagram. Source: Babcock & Wilcox

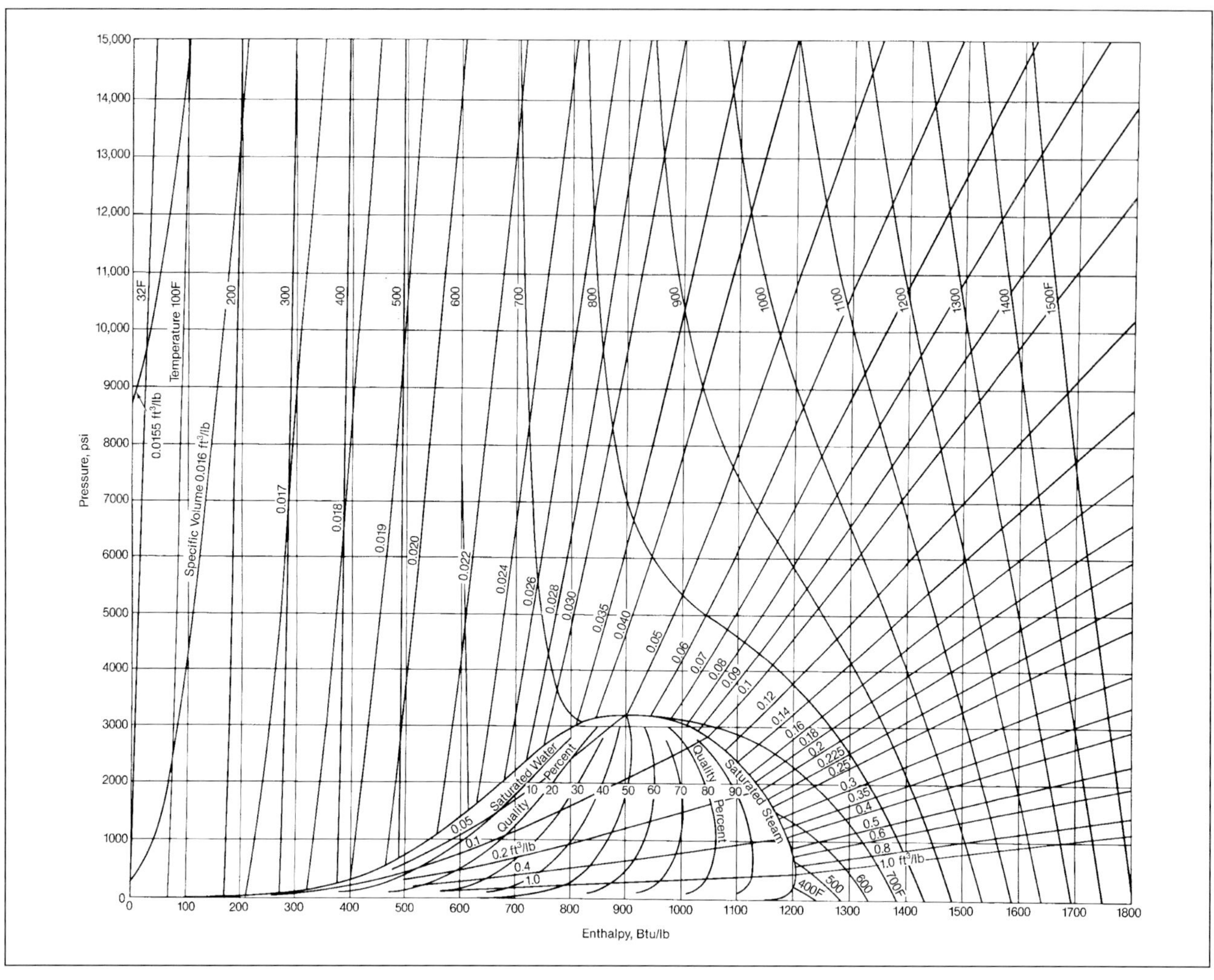

Fig. 6-3 Pressure-Enthalpy Chart for Steam, English Units. Source: Babcock & Wilcox

show the values of enthalpy and specific volume for steam and water over a wide range of pressure and temperatures.

Using Steam Tables

Using steam table nomenclature, the properties of a unit mass of a mixture of liquid and vapor (saturation state properties) of a quality, x, are given by the following expressions:

$$v_x = v_f + x v_{fg} \tag{6-2}$$

$$h_x = h_f + x h_{fg} \tag{6-3}$$

$$u_x = u_f + x u_{fg} \tag{6-4}$$

$$s_x = s_f + s_{fg} \tag{6-5}$$

Where:

f = Liquid

g = Vapor

fg property = g value minus the f value

Note that the change in property going from saturated liquid to saturated vapor (subscript fg) is not given in the steam tables for specific volume. That can be determined by subtracting v_f from v_g. The energy equation applied to the vaporization process may be expressed as:

$$h_{fg} = u_{fg} + p v_{fg} \tag{6-6}$$

Where:

h_{fg} = Heat of vaporization, or the heat required to vaporize a unit mass of liquid at constant pressure and temperature

u_{fg} = Internal energy of the mixture

pv_{fg} = The work performed during vaporization

The following sets of example problems are easily solved with values found in the English units steam tables. Similarly,

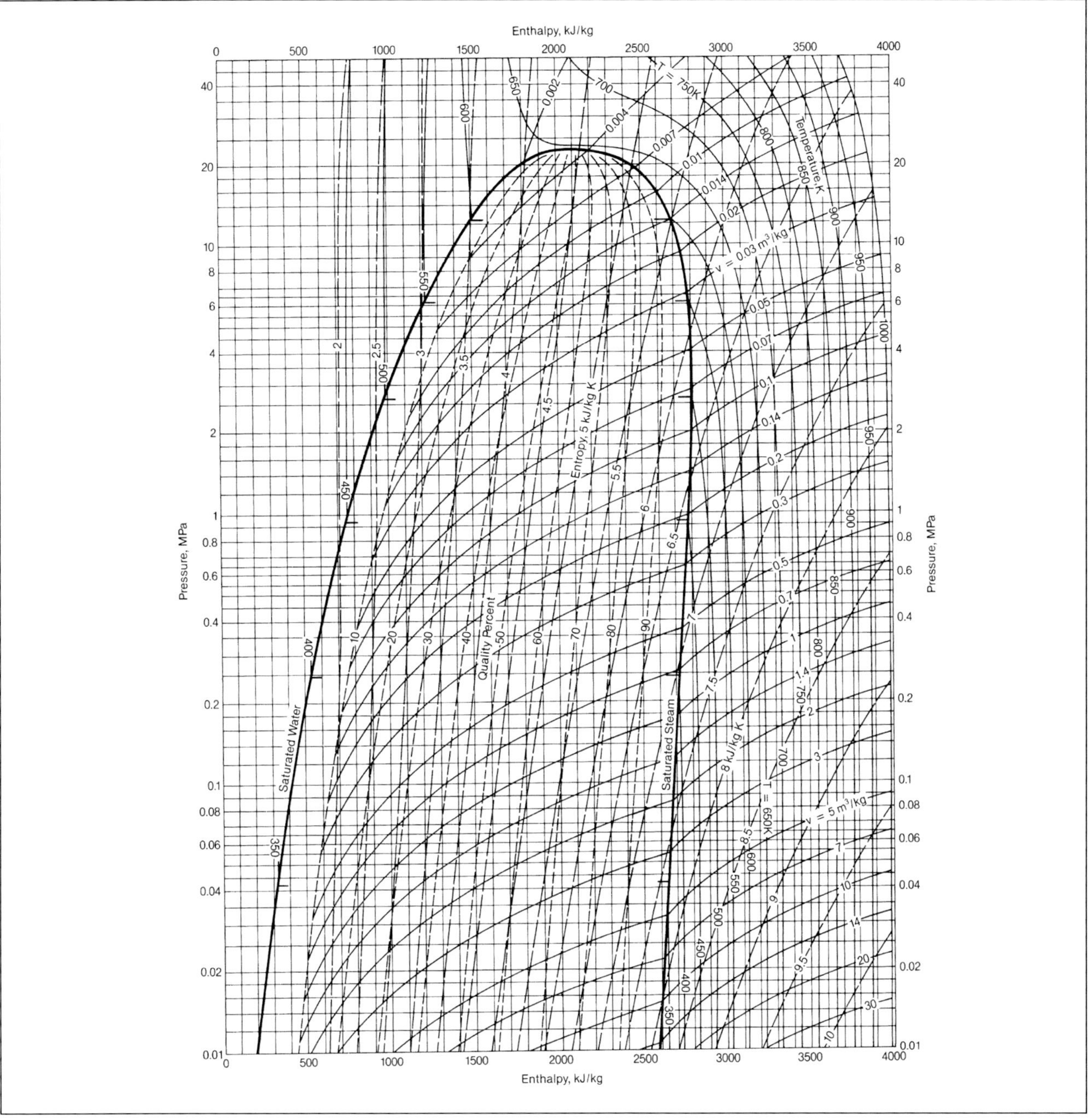

Fig. 6-4 Pressure-Enthalpy Chart for Steam, SI Units. Source: Babcock & Wilcox

the SI units steam tables will provide the values necessary to solve these same problems when expressed in SI units.

(1) To determine how much heat is required to raise the temperature of 3,000 lbm of saturated water from 70°F to 180°F, first determine the enthalpy difference at each condition:

147.9 Btu/lbm – 38.0 Btu/lbm = 109.9 Btu/lbm

Then multiply the difference in enthalpy per lbm by the total lbm of water being heated to determine the total heat required:

109.9 Btu/lbm x 3,000 lbm = 329,700 Btu

To determine the fuel required to heat the water, divide the total heat required by the heating unit's efficiency and by the Btu content per unit of fuel. If the heating unit had an efficiency of 80%, and the fuel used was natural gas, with a Btu content of 1,000 Btu/cf, the fuel requirement would be:

$$\frac{329{,}700\ Btu}{(0.80)\ (1{,}000\ Btu/cf)} = 412.13\ cf$$

(2) To determine the specific volume, enthalpy, internal energy and entropy, of a wet steam mixture at 120 psia with a steam quality of 80%, first identify the values from the steam tables for the saturation pressure of 120 psia. Then insert these values along with the steam quality into Equations 6-2 through 6-5:

$$v_x = v_f + xv_{fg} = 0.017886 + 0.8(3.730 - 0.017886) = 2.988\ cf/lbm$$

$$h_x = h_f + xh_{fg} = 312.67 + 0.8(878.5) = 1015.47\ Btu/lbm$$

$$u_x = u_f + xu_{fg} = 312.27 + 0.8(796.0) = 949.07\ Btu/lbm$$

$$s_x = s_f + s_{fg} = 0.49201 + 0.8(1.0966) = 1.3693\ Btu/lbm \cdot {}^\circ F$$

(3) To determine how much heat is required to raise (or convert) 3,000 lbm of feedwater at 200°F to dry saturated steam at 135 psig (approximately 150 psia), first determine the enthalpy difference per lbm:

1194.1 Btu/lbm – 168.0 Btu/lbm = 1026.1 Btu/lbm

Then multiply the difference in enthalpy per lbm by the total lbm of water being heated to determine the total heat required:

1,026.1 Btu/lbm x 3,000 lbm = 3,078,300 Btu

(4) To determine how much heat would be required if the steam were 95% dry (or 5% wet), using Problem (3), first determine the enthalpy per lbm of wet steam at 135 psig:

1194.1 Btu/lbm – (0.05 x 863.6 Btu/lbm) =
1194.1 Btu/lbm – 43.18 Btu/lbm = 1150.92 Btu/lbm

Then determine the difference in enthalpy per lbm at each condition:

1150.92 Btu/lbm – 168.0 Btu/lbm = 982.92 Btu/lbm

Then multiply the difference in enthalpy per lbm by the total pounds of water being heated to determine the total heat required:

982.92 Btu/lbm x 3,000 lbm = 2,948,760 Btu

(5) To determine how much heat is required to raise (or convert) 10,000 lbm of dry saturated steam at 200 psia to superheated steam at 500°F, first identify the enthalpy of 200 psia dry saturated steam. This may be found under the saturation pressure or under the properties of superheated steam for 200 psia. This is 1,199.3 Btu/lbm at its saturation temperature of 381.86°F. Then locate the enthalpy under properties of superheated steam for 200 psia at 500°F (this is 1,268.8 Btu/lbm). Next, determine the enthalpy difference:

1268.8 Btu/lbm – 1199.3 Btu/lbm = 69.5 Btu/lbm

Then multiply the difference in enthalpy per lbm by the total lbm of steam being superheated (by about 118°F) to determine the total heat required.

69.5 Btu/lbm x 10,000 lbm = 695,000 Btu

MOLLIER DIAGRAM

A convenient alternative to using the steam tables is to use a **Mollier diagram.** A Mollier diagram is a graphic representation of the thermodynamic properties of steam. The ordinate "h" is enthalpy and the abscissa "s" is entropy. Using the graphical representation of the relationships of various steam conditions, computations made for analysis of various cycles and processes can be performed quickly with a fairly high degree of accuracy. Figures 6-5 (English units) and 6-6 (SI units) are Mollier diagrams for steam.

The following are some example problems, in English units, using a Mollier Diagram:

(6) From Example Problem (2), to determine the enthalpy of a wet steam mixture at 120 psia with a steam quality of 80%, first read the lines of constant moisture in the wet region that corresponds to (1-x). Thus, at 20% moisture (80% steam quality) and 120 psia, the enthalpy is shown to be 1,015 Btu/lbm, which agrees with the values taken from the steam tables.

(7) To solve Example Problem (5) using the Mollier Diagram, first locate 200 psia saturated steam. This is found by moving down the 200 psia line of constant pressure until reaching the saturation line. This shows about 1,200 Btu/lbm (1,199.3). Then move up the 200 psia constant pressure line until reaching the constant temperature line for 500°F. This shows about 1,269 Btu/lbm (1,268.8).

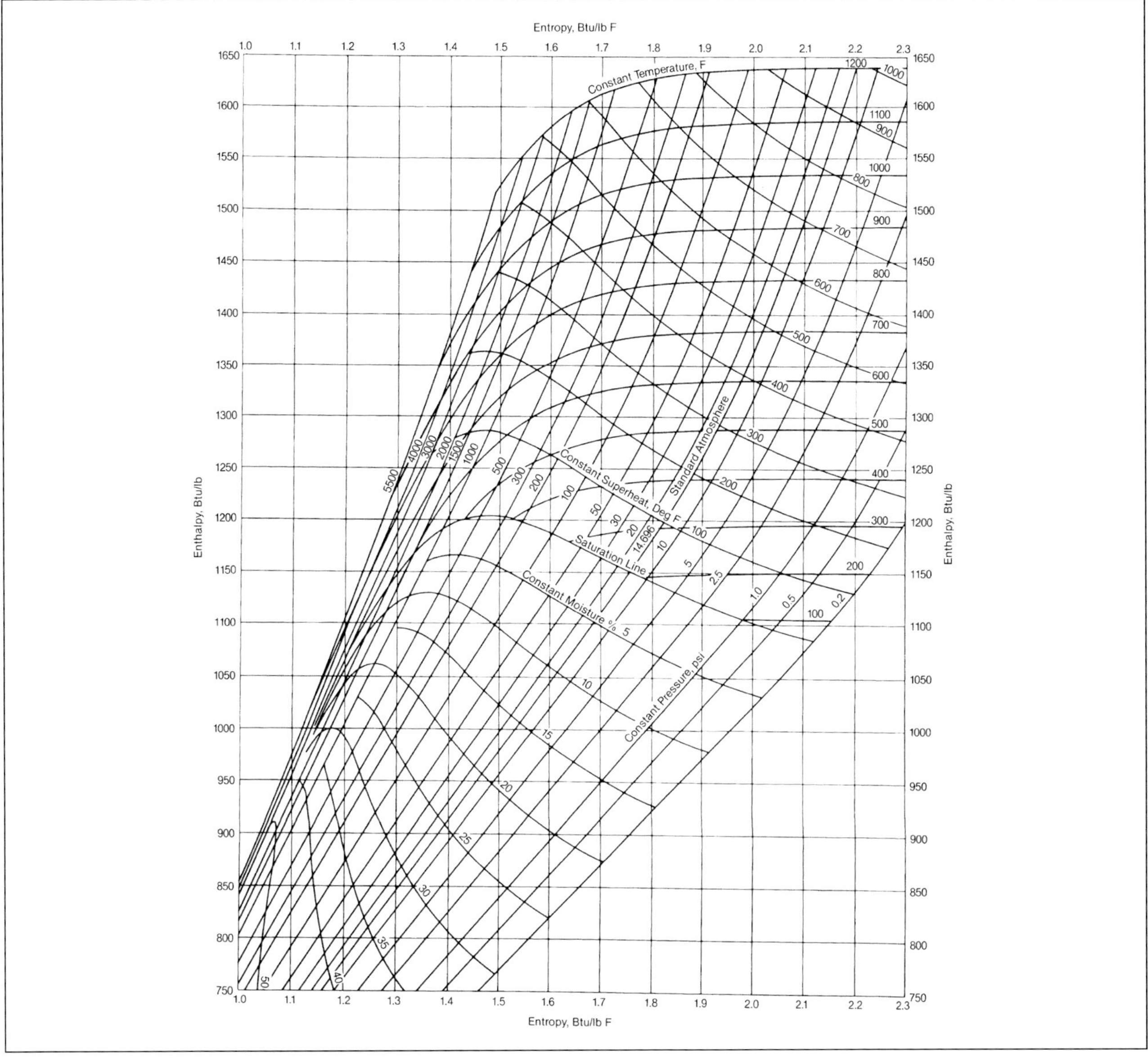

Fig. 6-5 Mollier Diagram for Steam (English Units). Source: Babcock & Wilcox

First determine the enthalpy difference per lbm at each condition:

1269 Btu/lbm – 1200 Btu/lbm = 69 Btu/lbm

Then multiply the difference in enthalpy per lbm by the total lbm of steam being superheated (by about 118°F) to determine the total heat required.

69 Btu/lbm x 10,000 lbm = 690,000 Btu

Notice that this result is off by 5,000 Btu, or about 0.7%, versus the computation based on the steam tables.

THE ECONOMIC VALUE OF STEAM

The positive value of steam is derived from two basic features:

- It can act as a heat transfer medium, conveying thermal energy from the heat released in a boiler from combustion of fuel to various processes and services requiring heat; and
- It can act as a working fluid to produce useful shaft work by expanding from higher to lower pressure through a turbine or engine.

Steam has negative value when it cannot be used

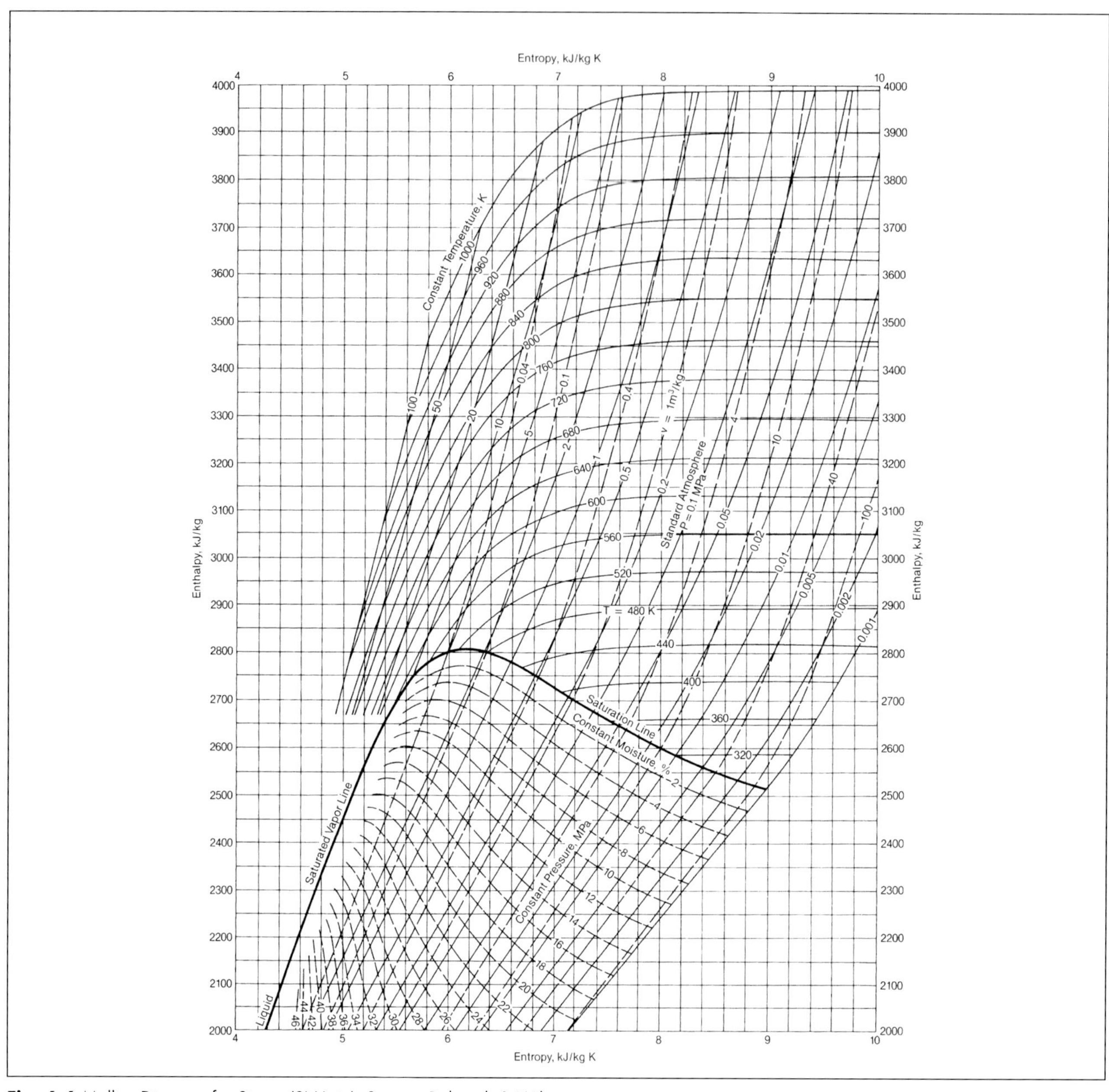

Fig. 6-6 Mollier Diagram for Steam (SI Units). Source: Babcock & Wilcox

productively at a facility and must be discarded. It may also provide unwanted heat in a facility, which then requires work either to exhaust or to cool air or equipment.

The variable cost of steam is primarily a function of fuel cost, boiler efficiency and the difference in enthalpy between incoming feedwater temperatures and exiting steam conditions. There are also a series of ancillary costs for steam, such as makeup water, water treatment, maintenance, blowdown losses, feedwater and condensate pumping, and combustion air supply.

Value of Steam as a Heat Transfer Medium

Steam usage is generally expressed, in English units, in lbm or thousand lbm (Mlbm) and, in SI units, in kg. Rudimentary steam balances are quantified in these terms and steam is typically sold in these quantities. However, as can be seen by steam tables and thermodynamic analysis, steam only has value based on what can be done with it. A lbm (or kg) of steam at 0 gauge pressure may have a negative value. Even though it still has heat value, it often cannot be used productively and, therefore, must be condensed and eliminated.

A more useful way to express the value of steam, however, is based on its Btu (kJ) content. Energy accounting systems often refer to energy usage in terms of dollars per million Btu (or kJ) of steam or million Btu (or kJ) of steam per unit of production. From the steam tables, it can be seen that a lbm (kg) of saturated steam at 250 psig (18.3 bar), which has a saturation temperature of 406°F (208°C) and an enthalpy of 1,202 Btu/lbm (2,795 kJ/kg), has more value than a lbm (kg) of saturated steam at 25 psig (2.7 bar), which has a saturation temperature of 267°F (131°C) and an enthalpy of 1,170 Btu/lbm (2,722 kJ/kg). As a heat transfer media, the difference in value is more than the proportionality of the total enthalpy, which is only a few percent — 32 Btu/lbm (74 kJ/kg). This is because the saturation temperature of the higher pressure steam is about one and one half times greater — 139°F (59°C) — than the lower pressure steam.

When heating a fluid entering at 150°F (66°C), for example, the difference in saturation temperatures versus the temperature of the fluid to be heated is more than double — 256°F vs. 127°F (124°C vs. 53°C). The lower pressure steam can provide the same amount of heat to the fluid, but a much larger heat exchanger will be required. The maximum fluid temperature attainable is limited to some temperature lower than the saturation temperature of the steam.

Value of Steam for Producing Useful Shaft Work

A comprehensive accounting system for the value of steam energy must evaluate the opportunities for generating prime power as well as for providing for heat transfer. A look at a simple pressure-reducing valve (PRV) demonstrates the inadequacy of any type of simple cost per Btu ratio.

The expansion of steam by throttling it through a valve is referred to as adiabatic expansion. Since no work is performed and there is no change in kinetic or potential energy, the entering enthalpy equals the leaving enthalpy. If, for example, 1,000 lbm (454 kg) of steam at 475 psig and 700°F (33.8 bar and 371°C), which has an enthalpy of 1,357 Btu/lbm (3,156 kJ/kg), is expanded to atmospheric pressure through a PRV, the resulting steam at 0 psig and 646°F (101 kPa and 341°C) would also have an enthalpy of 1,357 Btu/lbm (3,156 kJ/kg). Using enthalpy as the sole measurement of value, this steam, after undergoing a 100% efficient expansion, would have the same economic value as in its prior state.

However, there is still an important loss. The theoretical maximum amount of work that could be extracted from the steam in its original state is approximately 204 hp (152 kW). After adiabatic expansion through the PRV, this is reduced to about 126 hp (94 kW). This results in a 38% loss of ability to do theoretical work.

Both levels of work production are, however, unrealistic. In fact, it is practically impossible for steam to do work starting at 0 psig (101 kPa). From a theoretical maximum shaft work-producing level of availability, the adiabatic expansion is only 62% efficient. Actually, it would likely be 0% efficient, as no shaft work would be done after the expansion. In many respects, the PRV, efficient as it seems, is a major source of loss in the economic value of steam.

In addition to the loss of ability to do work resulting from the combustion of fuel in a boiler, the approach temperature between the flame and the steam temperature is also a loss of work potential. Ironically, viewed from the standpoint of enthalpy, most of the potential work loss from steam turbine power generation is in the condenser because most of the heat is lost at that stage. However, at that point, the majority of the work-producing potential of the steam has already been used.

This points out the critical difference between heat loss and loss of work potential. The steam turbine allows for the production of prime power with a relatively minor reduction in enthalpy. To determine the economic value of steam, its full potential for producing useful shaft work and providing thermal energy transfer must be identified with respect to all aspects of the "task at hand" and compared with all of the alternatives for providing the same values.

The process of determining value includes the efficiency of alternative equipment and the cost of energy to power that alternative equipment during any given hour. Given the tremendous real-world differences in energy costs, the relative value of the shaft work of steam may be as much as 10 or even 50 times greater during one hour than another.

Sometimes it will be most economical to produce every last unit of power possible from high-pressure steam in a steam turbine. Sometimes it will be most economical to use the same pressure steam in a topping cycle. And, sometimes, when compared with off-peak electric rates, steam will be most economically used solely for the purpose of thermal energy transfer.

Properties of Saturated Steam and Saturated Water (Temperature)

Temp F	Press. psia	Volume, ft³/lb Water v_f	Volume Evap v_{fg}	Volume Steam v_g	Enthalpy,[2] Btu/lb Water H_f	Enthalpy Evap H_{fg}	Enthalpy Steam H_g	Entropy, Btu/lb F Water s_f	Entropy Evap s_{fg}	Entropy Steam s_g	Temp F
32	0.08859	0.01602	3305	3305	−0.02	1075.5	1075.5	0.0000	2.1873	2.1873	32
35	0.09991	0.01602	2948	2948	3.00	1073.8	1076.8	0.0061	2.1706	2.1767	35
40	0.12163	0.01602	2446	2446	8.03	1071.0	1079.0	0.0162	2.1432	2.1594	40
45	0.14744	0.01602	2037.7	2037.8	13.04	1068.1	1081.2	0.0262	2.1164	2.1426	45
50	0.17796	0.01602	1704.8	1704.8	18.05	1065.3	1083.4	0.0361	2.0901	2.1262	50
60	0.2561	0.01603	1207.6	1207.6	28.06	1059.7	1087.7	0.0555	2.0391	2.0946	60
70	0.3629	0.01605	868.3	868.4	38.05	1054.0	1092.1	0.0745	1.9900	2.0645	70
80	0.5068	0.01607	633.3	633.3	48.04	1048.4	1096.4	0.0932	1.9426	2.0359	80
90	0.6981	0.01610	468.1	468.1	58.02	1042.7	1100.8	0.1115	1.8970	2.0086	90
100	0.9492	0.01613	350.4	350.4	68.00	1037.1	1105.1	0.1295	1.8530	1.9825	100
110	1.2750	0.01617	265.4	265.4	77.98	1031.4	1109.3	0.1472	1.8105	1.9577	110
120	1.6927	0.01620	203.25	203.26	87.97	1025.6	1113.6	0.1646	1.7693	1.9339	120
130	2.2230	0.01625	157.32	157.33	97.96	1019.8	1117.8	0.1817	1.7295	1.9112	130
140	2.8892	0.01629	122.98	123.00	107.95	1014.0	1122.0	0.1985	1.6910	1.8895	140
150	3.718	0.01634	97.05	97.07	117.95	1008.2	1126.1	0.2150	1.6536	1.8686	150
160	4.741	0.01640	77.27	77.29	127.96	1002.2	1130.2	0.2313	1.6174	1.8487	160
170	5.993	0.01645	62.04	62.06	137.97	996.2	1134.2	0.2473	1.5822	1.8295	170
180	7.511	0.01651	50.21	50.22	148.00	990.2	1138.2	0.2631	1.5480	1.8111	180
190	9.340	0.01657	40.94	40.96	158.04	984.1	1142.1	0.2787	1.5148	1.7934	190
200	11.526	0.01664	33.62	33.64	168.09	977.9	1146.0	0.2940	1.4824	1.7764	200
210	14.123	0.01671	27.80	27.82	178.15	971.6	1149.7	0.3091	1.4509	1.7600	210
212	14.696	0.01672	26.78	26.80	180.17	970.3	1150.5	0.3121	1.4447	1.7568	212
220	17.186	0.01678	23.13	23.15	188.23	965.2	1153.4	0.3241	1.4201	1.7442	220
230	20.779	0.01685	19.364	19.381	198.33	958.7	1157.1	0.3388	1.3902	1.7290	230
240	24.968	0.01693	16.304	16.321	208.45	952.1	1160.6	0.3533	1.3609	1.7142	240
250	29.825	0.01701	13.802	13.819	218.59	945.4	1164.0	0.3677	1.3323	1.7000	250
260	35.427	0.01709	11.745	11.762	228.76	938.6	1167.4	0.3819	1.3043	1.6862	260
270	41.856	0.01718	10.042	10.060	238.95	931.7	1170.6	0.3960	1.2769	1.6729	270
280	49.200	0.01726	8.627	8.644	249.17	924.6	1173.8	0.4098	1.2501	1.6599	280
290	57.550	0.01736	7.443	7.460	259.4	917.4	1176.8	0.4236	1.2238	1.6473	290
300	67.005	0.01745	6.448	6.466	269.7	910.0	1179.7	0.4372	1.1979	1.6351	300
310	77.67	0.01755	5.609	5.626	280.0	902.5	1182.5	0.4506	1.1726	1.6232	310
320	89.64	0.01766	4.896	4.914	290.4	894.8	1185.2	0.4640	1.1477	1.6116	320
340	117.99	0.01787	3.770	3.788	311.3	878.8	1190.1	0.4902	1.0990	1.5892	340
360	153.01	0.01811	2.939	2.957	332.3	862.1	1194.4	0.5161	1.0517	1.5678	360
380	195.73	0.01836	2.317	2.335	353.6	844.5	1198.0	0.5416	1.0057	1.5473	380
400	247.26	0.01864	1.8444	1.8630	375.1	825.9	1201.0	0.5667	0.9607	1.5274	400
420	308.78	0.01894	1.4808	1.4997	396.9	806.2	1203.1	0.5915	0.9165	1.5080	420
440	381.54	0.01926	1.1976	1.2169	419.0	785.4	1204.4	0.6161	0.8729	1.4890	440
460	466.9	0.0196	0.9746	0.9942	441.5	763.2	1204.8	0.6405	0.8299	1.4704	460
480	566.2	0.0200	0.7972	0.8172	464.5	739.6	1204.1	0.6648	0.7871	1.4518	480
500	680.9	0.0204	0.6545	0.6749	487.9	714.3	1202.2	0.6890	0.7443	1.4333	500
520	812.5	0.0209	0.5386	0.5596	512.0	687.0	1199.0	0.7133	0.7013	1.4146	520
540	962.8	0.0215	0.4437	0.4651	536.8	657.5	1194.3	0.7378	0.6577	1.3954	540
560	1133.4	0.0221	0.3651	0.3871	562.4	625.3	1187.7	0.7625	0.6132	1.3757	560
580	1326.2	0.0228	0.2994	0.3222	589.1	589.9	1179.0	0.7876	0.5673	1.3550	580
600	1543.2	0.0236	0.2438	0.2675	617.1	550.6	1167.7	0.8134	0.5196	1.3330	600
620	1786.9	0.0247	0.1962	0.2208	646.9	506.3	1153.2	0.8403	0.4689	1.3092	620
640	2059.9	0.0260	0.1543	0.1802	679.1	454.6	1133.7	0.8686	0.4134	1.2821	640
660	2365.7	0.0277	0.1166	0.1443	714.9	392.1	1107.0	0.8995	0.3502	1.2498	660
680	2708.6	0.0304	0.0808	0.1112	758.5	310.1	1068.5	0.9365	0.2720	1.2086	680
700	3094.3	0.0366	0.0386	0.0752	822.4	172.7	995.2	0.9901	0.1490	1.1390	700
705.5	3208.2	0.0508	0	0.0508	906.0	0	906.0	1.0612	0	1.0612	705.5

1. In the balance of *Steam*, enthalpy is denoted by H in place of h to avoid confusion with heat transfer coefficient.

Courtesy of Department of Mechanical Engineering, Stanford University

Table 6-1 Properties of Saturated Steam and Saturated Water (Temperature), English Units. Source: Babcock & Wilcox

Properties of Saturated Steam and Saturated Water (Pressure)

Press. psia	Temp F	Volume, ft^3/lb			Enthalpy,[2] Btu/lb			Entropy, Btu/lb F			Energy, Btu/lb		Press. psia
		Water v_f	Evap v_{fg}	Steam v_g	Water H_f	Evap H_{fg}	Steam H_g	Water s_f	Evap s_{fg}	Steam s_g	Water u_f	Steam u_g	
0.0886	32.018	0.01602	3302.4	3302.4	0.00	1075.5	1075.5	0	2.1872	2.1872	0	1021.3	**0.0886**
0.10	35.023	0.01602	2945.5	2945.5	3.03	1073.8	1076.8	0.0061	2.1705	2.1766	3.03	1022.3	**0.10**
0.15	45.453	0.01602	2004.7	2004.7	13.50	1067.9	1081.4	0.0271	2.1140	2.1411	13.50	1025.7	**0.15**
0.20	53.160	0.01603	1526.3	1526.3	21.22	1063.5	1084.7	0.0422	2.0738	2.1160	21.22	1028.3	**0.20**
0.30	64.484	0.01604	1039.7	1039.7	32.54	1057.1	1089.7	0.0641	2.0168	2.0809	32.54	1032.0	**0.30**
0.40	72.869	0.01606	792.0	792.1	40.92	1052.4	1093.3	0.0799	1.9762	2.0562	40.92	1034.7	**0.40**
0.5	79.586	0.01607	641.5	641.5	47.62	1048.6	1096.3	0.0925	1.9446	2.0370	47.62	1036.9	**0.5**
0.6	85.218	0.01609	540.0	540.1	53.25	1045.5	1098.7	0.1028	1.9186	2.0215	53.24	1038.7	**0.6**
0.7	90.09	0.01610	466.93	466.94	58.10	1042.7	1100.8	0.3	1.8966	2.0083	58.10	1040.3	**0.7**
0.8	94.38	0.01611	411.67	411.69	62.39	1040.3	1102.6	0.1117	1.8775	1.9970	62.39	1041.7	**0.8**
0.9	98.24	0.01612	368.41	368.43	66.24	1038.1	1104.3	0.1264	1.8606	1.9870	66.24	1042.9	**0.9**
1.0	101.74	0.01614	333.59	333.60	69.73	1036.1	1105.8	0.1326	1.8455	1.9781	69.73	1044.1	**1.0**
2.0	126.07	0.01623	173.74	173.76	94.03	1022.1	1116.2	0.1750	1.7450	1.9200	94.03	1051.8	**2.0**
3.0	141.47	0.01630	118.71	118.73	109.42	1013.2	1122.6	0.2009	1.6854	1.8864	109.41	1056.7	**3.0**
4.0	152.96	0.01636	90.63	90.64	120.92	1006.4	1127.3	0.2199	1.6428	1.8626	120.90	1060.2	**4.0**
5.0	162.24	0.01641	73.515	73.53	130.20	1000.9	1131.1	0.2349	1.6094	1.8443	130.18	1063.1	**5.0**
6.0	170.05	0.01645	61.967	61.98	138.03	996.2	1134.2	0.2474	1.5820	1.8294	138.01	1065.4	**6.0**
7.0	176.84	0.01649	53.634	53.65	144.83	992.1	1136.9	0.2581	1.5587	1.8168	144.81	1067.4	**7.0**
8.0	182.86	0.01653	47.328	47.35	150.87	988.5	1139.3	0.2676	1.5384	1.8060	150.84	1069.2	**8.0**
9.0	188.27	0.01656	42.385	42.40	156.30	985.1	1141.4	0.2760	1.5204	1.7964	156.28	1070.8	**9.0**
10	193.21	0.01659	38.404	38.42	161.26	982.1	1143.3	0.2836	1.5043	1.7879	161.23	1072.3	**10**
14.696	212.00	0.01672	26.782	26.80	180.17	970.3	1150.5	0.3121	1.4447	1.7568	180.12	1077.6	**14.696**
15	213.03	0.01673	26.274	26.29	181.21	969.7	1150.9	0.3137	1.4415	1.7552	181.16	1077.9	**15**
20	227.96	0.01683	20.070	20.087	196.27	960.1	1156.3	0.3358	1.3962	1.7320	196.21	1082.0	**20**
30	250.34	0.01701	13.7266	13.744	218.9	945.2	1164.1	0.3682	1.3313	1.6995	218.8	1087.9	**30**
40	267.25	0.01715	10.4794	10.497	236.1	933.6	1169.8	0.3921	1.2844	1.6765	236.0	1092.1	**40**
50	281.02	0.01727	8.4967	8.514	250.2	923.9	1174.1	0.4112	1.2474	1.6586	250.1	1095.3	**50**
60	292.71	0.01738	7.1562	7.174	262.2	915.4	1177.6	0.4273	1.2167	1.6440	262.0	1098.0	**60**
70	302.93	0.01748	6.1875	6.205	272.7	907.8	1180.6	0.4411	1.1905	1.6316	272.5	1100.2	**70**
80	312.04	0.01757	5.4536	5.471	282.1	900.9	1183.1	0.4534	1.1675	1.6208	281.9	1102.1	**80**
90	320.28	0.01766	4.8777	4.895	290.7	894.6	1185.3	0.4643	1.1470	1.6113	290.4	1103.7	**90**
100	327.82	0.01774	4.4133	4.431	298.5	888.6	1187.2	0.4743	1.1284	1.6027	298.2	1105.2	**100**
120	341.27	0.01789	3.7097	3.728	312.6	877.8	1190.4	0.4919	1.0960	1.5879	312.2	1107.6	**120**
140	353.04	0.01803	3.2010	3.219	325.0	868.0	1193.0	0.5071	1.0681	1.5752	324.5	1109.6	**140**
160	363.55	0.01815	2.8155	2.834	336.1	859.0	1195.1	0.5206	1.0435	1.5641	335.5	1111.2	**160**
180	373.08	0.01827	2.5129	2.531	346.2	850.7	1196.9	0.5328	1.0215	1.5543	345.6	1112.5	**180**
200	381.80	0.01839	2.2689	2.287	355.5	842.8	1198.3	0.5438	1.0016	1.5454	354.8	1113.7	**200**
250	400.97	0.01865	1.8245	1.8432	376.1	825.0	1201.1	0.5679	0.9585	1.5264	375.3	1115.8	**250**
300	417.35	0.01889	1.5238	1.5427	394.0	808.9	1202.9	0.5882	0.9223	1.5105	392.9	1117.2	**300**
350	431.73	0.01913	1.3064	1.3255	409.8	794.2	1204.0	0.6059	0.8909	1.4968	408.6	1118.1	**350**
400	444.60	0.0193	1.14162	1.1610	424.2	780.4	1204.6	0.6217	0.8630	1.4847	422.7	1118.7	**400**
450	456.28	0.0195	1.01224	1.0318	437.3	767.5	1204.8	0.6360	0.8378	1.4738	435.7	1118.9	**450**
500	467.01	0.0198	0.90787	0.9276	449.5	755.1	1204.7	0.6490	0.8148	1.4639	447.7	1118.8	**500**
550	476.94	0.0199	0.82183	0.8418	460.9	743.3	1204.3	0.6611	0.7936	1.4547	458.9	1118.6	**550**
600	486.20	0.0201	0.74962	0.7698	471.7	732.0	1203.7	0.6723	0.7738	1.4461	469.5	1118.2	**600**
700	503.08	0.0205	0.63505	0.6556	491.6	710.2	1201.8	0.6928	0.7377	1.4304	488.9	1116.9	**700**
800	518.21	0.0209	0.54809	0.5690	509.8	689.6	1199.4	0.7111	0.7051	1.4163	506.7	1115.2	**800**
900	531.95	0.0212	0.47968	0.5009	526.7	669.7	1196.4	0.7279	0.6753	1.4032	523.2	1113.0	**900**
1000	544.58	0.0216	0.42436	0.4460	542.6	650.4	1192.9	0.7434	0.6476	1.3910	538.6	1110.4	**1000**
1100	556.28	0.0220	0.37863	0.4006	557.5	631.5	1189.1	0.7578	0.6216	1.3794	553.1	1107.5	**1100**
1200	567.19	0.0223	0.34013	0.3625	571.9	613.0	1184.8	0.7714	0.5969	1.3683	566.9	1104.3	**1200**
1300	577.42	0.0227	0.30722	0.3299	585.6	594.6	1180.2	0.7843	0.5733	1.3577	580.1	1100.9	**1300**
1400	587.07	0.0231	0.27871	0.3018	598.8	576.5	1175.3	0.7966	0.5507	1.3474	592.9	1097.1	**1400**
1500	596.20	0.0235	0.25372	0.2772	611.7	558.4	1170.1	0.8085	0.5288	1.3373	605.2	1093.1	**1500**
2000	635.80	0.0257	0.16266	0.1883	672.1	466.2	1138.3	0.8625	0.4256	1.2881	662.6	1068.6	**2000**
2500	668.11	0.0286	0.10209	0.1307	731.7	361.6	1093.3	0.9139	0.3206	1.2345	718.5	1032.9	**2500**
3000	695.33	0.0343	0.05073	0.0850	801.8	218.4	1020.3	0.9728	0.1891	1.1619	782.8	973.1	**3000**
3208.2	705.47	0.0508	0	0.0508	906.0	0	906.0	1.0612	0	1.0612	875.9	875.9	**3208.2**

Courtesy of Department of Mechanical Engineering, Stanford University

Table 6-2 Properties of Saturated Steam and Saturated Water (Pressure), English Units. Source: Babcock & Wilcox

Properties of Superheated Steam and Compressed Water (Temperature and Pressure)

Press.,psia (sat. temp)		Temperature, F 100	200	300	400	500	600	700	800	900	1000	1100	1200	1300	1400	1500
1	v	0.0161	392.5	452.3	511.9	571.5	631.1	690.7								
	H	68.00	1150.2	1195.7	1241.8	1288.6	1336.1	1384.5								
(101.74)	s	0.1295	2.0509	2.1152	2.1722	2.2237	2.2708	2.3144								
5	v	0.0161	78.14	90.24	102.24	114.21	126.15	138.08	150.01	161.94	173.86	185.78	197.70	209.62	221.53	233.45
	H	68.01	1148.6	1194.8	1241.3	1288.2	1335.9	1384.3	1433.6	1483.7	1534.7	1586.7	1639.6	1693.3	1748.0	1803.5
(162.24)	s	0.1295	1.8716	1.9369	1.9943	2.0460	2.0932	2.1369	2.1776	2.2159	2.2521	2.2866	2.3194	2.3509	2.3811	2.4101
10	v	0.0161	38.84	44.98	51.03	57.04	63.03	69.00	74.98	80.94	86.91	92.87	98.84	104.80	110.76	116.72
	H	68.02	1146.6	1193.7	1240.6	1287.8	1335.5	1384.0	1433.4	1483.5	1534.6	1586.6	1639.5	1693.3	1747.9	1803.4
(193.21)	s	0.1295	1.7928	1.8593	1.9173	1.9692	2.0166	2.0603	2.1011	2.1394	2.1757	2.2101	2.2430	2.2744	2.3046	2.3337
15	v	0.0161	0.0166	29.899	33.963	37.985	41.986	45.978	49.964	53.946	57.926	61.905	65.882	69.858	73.833	77.807
	H	68.04	168.09	1192.5	1239.9	1287.3	1335.2	1383.8	1433.2	1483.4	1534.5	1586.5	1639.4	1693.2	1747.8	1803.4
(213.03)	s	0.1295	0.2940	1.8134	1.8720	1.9242	1.9717	2.0155	2.0563	2.0946	2.1309	2.1653	2.1982	2.2297	2.2599	2.2890
20	v	0.0161	0.0166	22.356	25.428	28.457	31.466	34.465	37.458	40.447	43.435	46.420	49.405	52.388	55.370	58.352
	H	68.05	168.11	1191.4	1239.2	1286.9	1334.9	1383.5	1432.9	1483.2	1534.3	1586.3	1639.3	1693.1	1747.8	1803.3
(227.96)	s	0.1295	0.2940	1.7805	1.8397	1.8921	1.9397	1.9836	2.0244	2.0628	2.0991	2.1336	2.1665	2.1979	2.2282	2.2572
40	v	0.0161	0.0166	11.036	12.624	14.165	15.685	17.195	18.699	20.199	21.697	23.194	24.689	26.183	27.676	29.168
	H	68.10	168.15	1186.6	1236.4	1285.0	1333.6	1382.5	1432.1	1482.5	1533.7	1585.8	1638.8	1992.7	1747.5	1803.0
(267.25)	s	0.1295	0.2940	1.6992	1.7608	1.8143	1.8624	1.9065	1.9476	1.9860	2.0224	2.0569	2.0899	2.1224	2.1516	2.1807
60	v	0.0161	0.0166	7.257	8.354	9.400	10.425	11.438	12.446	13.450	14.452	15.452	16.450	17.448	18.445	19.441
	H	68.15	168.20	1181.6	1233.5	1283.2	1332.3	1381.5	1431.3	1481.8	1533.2	1585.3	1638.4	1692.4	1747.1	1802.8
(292.71)	s	0.1295	0.2939	1.6492	1.7134	1.7681	1.8168	1.8612	1.9024	1.9410	1.9774	2.0120	2.0450	2.0765	2.1068	2.1359
80	v	0.0161	0.0166	0.0175	6.218	7.018	7.794	8.560	9.319	10.075	10.829	11.581	12.331	13.081	13.829	14.577
	H	68.21	168.24	269.74	1230.5	1281.3	1330.9	1380.5	1430.5	1481.1	1532.6	1584.9	1638.0	1692.0	1746.8	1802.5
(312.04)	s	0.1295	0.2939	0.4371	1.6790	1.7349	1.7842	1.8289	1.8702	1.9089	1.9454	1.9800	2.0131	2.0446	2.0750	2.1041
100	v	0.0161	0.0166	0.0175	4.935	5.588	6.216	6.833	7.443	8.050	8.655	9.258	9.860	10.460	11.060	11.659
	H	68.26	168.29	269.77	1227.4	1279.3	1329.6	1379.5	1429.7	1480.4	1532.0	1584.4	1637.6	1691.6	1746.5	1802.2
(327.82)	s	0.1295	0.2939	0.4371	1.6516	1.7088	1.7586	1.8036	1.8451	1.8839	1.9205	1.9552	1.9883	2.0199	2.0502	2.0794
120	v	0.0161	0.0166	0.0175	4.0786	4.6341	5.1637	5.6831	6.1928	6.7006	7.2060	7.7096	8.2119	8.7130	9.2134	9.7130
	H	68.31	168.33	269.81	1224.1	1277.4	1328.1	1378.4	1428.8	1479.8	1531.4	1583.9	1637.1	1691.3	1746.2	1802.0
(341.27)	s	0.1295	0.2939	0.4371	1.6286	1.6872	1.7376	1.7829	1.8246	1.8635	1.9001	1.9349	1.9680	1.9996	2.0300	2.0592
140	v	0.0161	0.0166	0.0175	3.4661	3.9526	4.4119	4.8585	5.2995	5.7364	6.1709	6.6036	7.0349	7.4652	7.8946	8.3233
	H	68.37	168.38	269.85	1220.8	1275.3	1326.8	1377.4	1428.0	1479.1	1530.8	1583.4	1636.7	1690.9	1745.9	1801.7
(353.04)	s	0.1295	0.2939	0.4370	1.6085	1.6686	1.7196	1.7652	1.8071	1.8461	1.8828	1.9176	1.9508	1.9825	2.0129	2.0421
160	v	0.0161	0.0166	0.0175	3.0060	3.4413	3.8480	4.2420	4.6295	5.0132	5.3945	5.7741	6.1522	6.5293	6.9055	7.2811
	H	68.42	168.42	269.89	1217.4	1273.3	1325.4	1376.4	1427.2	1478.4	1530.3	1582.9	1636.3	1690.5	1745.6	1801.4
(363.55)	s	0.1294	0.2938	0.4370	1.5906	1.6522	1.7039	1.7499	1.7919	1.8310	1.8678	1.9027	1.9359	1.9676	1.9980	2.0273
180	v	0.0161	0.0166	0.0174	2.6474	3.0433	3.4093	3.7621	4.1084	4.4505	4.7907	5.1289	5.4657	5.8014	6.1363	6.4704
	H	68.47	168.47	269.92	1213.8	1271.2	1324.0	1375.3	1426.3	1477.7	1529.7	1582.4	1635.9	1690.2	1745.3	1801.2
(373.08)	s	0.1294	0.2938	0.4370	1.5743	1.6376	1.6900	1.7362	1.7784	1.8176	1.8545	1.8894	1.9227	1.9545	1.9849	2.0142
200	v	0.0161	0.0166	0.0174	2.3598	2.7247	3.0583	3.3783	3.6915	4.0008	4.3077	4.6128	4.9165	5.2191	5.5209	5.8219
	H	68.52	168.51	269.96	1210.1	1269.0	1322.6	1374.3	1425.5	1477.0	1529.1	1581.9	1635.4	1689.8	1745.0	1800.9
(381.80)	s	0.1294	0.2938	0.4369	1.5593	1.6242	1.6776	1.7239	1.7663	1.8057	1.8426	1.8776	1.9109	1.9427	1.9732	2.0025
250	v	0.0161	0.0166	0.0174	0.0186	2.1504	2.4662	2.6872	2.9410	3.1909	3.4382	3.6837	3.9278	4.1709	4.4131	4.6546
	H	68.66	168.63	270.05	375.10	1263.5	1319.0	1371.6	1423.4	1475.3	1527.6	1580.6	1634.4	1688.9	1744.2	1800.2
(400.97)	s	0.1294	0.2937	0.4368	0.5667	1.5951	1.6502	1.6976	1.7405	1.7801	1.8173	1.8524	1.8858	1.9177	1.9482	1.9776
300	v	0.0161	0.0166	0.0174	0.0186	1.7665	2.0044	2.2263	2.4407	2.6509	2.8585	3.0643	3.2688	3.4721	3.6746	3.8764
	H	68.79	168.74	270.14	375.15	1257.7	1315.2	1368.9	1421.3	1473.6	1526.2	1579.4	1633.3	1688.0	1743.4	1799.6
(417.35)	s	0.1294	0.2937	0.4307	0.5665	1.5703	1.6274	1.6758	1.7192	1.7591	1.7964	1.8317	1.8652	1.8972	1.9278	1.9572
350	v	0.0161	0.0166	0.0174	0.0186	1.4913	1.7028	1.8970	2.0832	2.2652	2.4445	2.6219	2.7980	2.9730	3.1471	3.3205
	H	68.92	168.85	270.24	375.21	1251.5	1311.4	1366.2	1419.2	1471.8	1524.7	1578.2	1632.3	1687.1	1742.6	1798.9
(431.73)	s	0.1293	0.2936	0.4367	0.5664	1.5483	1.6077	1.6571	1.7009	1.7411	1.7787	1.8141	1.8477	1.8798	1.9105	1.9400
400	v	0.0161	0.0166	0.0174	0.0162	1.2841	1.4763	1.6499	1.8151	1.9759	2.1339	2.2901	2.4450	2.5987	2.7515	2.9037
	H	69.05	168.97	270.33	375.27	1245.1	1307.4	1363.4	1417.0	1470.1	1523.3	1576.9	1631.2	1686.2	1741.9	1798.2
(444.60)	s	0.1293	0.2935	0.4366	0.5663	1.5282	1.5901	1.6406	1.6850	1.7255	1.7632	1.7988	1.8325	1.8647	1.8955	1.9250
500	v	0.0161	0.0166	0.0174	0.0186	0.9919	1.1584	1.3037	1.4397	1.5708	1.6992	1.8256	1.9507	2.0746	2.1977	2.3200
	H	69.32	169.19	270.51	375.38	1231.2	1299.1	1357.7	1412.7	1466.6	1520.3	1574.4	1629.1	1684.4	1740.3	1796.9
(467.01)	s	0.1292	0.2934	0.4364	0.5660	1.4921	1.5595	1.6123	1.6578	1.6990	1.7371	1.7730	1.8069	1.8393	1.8702	1.8998

1. See Note 1 Table 1.

Courtesy of Department of Mechanical Engineering, Stanford University

Table 6-3a Properties of Superheated Steam and Compressed Water (Temperature and Pressure), English Units. Source: Babcock & Wilcox

Properties of Superheated Steam and Compressed Water (Temperature and Pressure)

Press., psia (sat. temp)		100	200	300	400	500	600	700	800	900	1000	1100	1200	1300	1400	1500
		Temperature, F														
600	v	0.0161	0.0166	0.0174	0.0186	0.7944	0.9456	1.0726	1.1892	1.3008	1.4093	1.5160	1.6211	1.7252	1.8284	1.9309
	H	69.58	169.42	270.70	375.49	1215.9	1290.3	1351.8	1408.3	1463.0	1517.4	1571.9	1627.0	1682.6	1738.8	1795.6
(486.20)	s	0.1292	0.2933	0.4362	0.5657	1.4590	1.5329	1.5844	1.6351	1.6769	1.7155	1.7517	1.7859	1.8184	1.8494	1.8792
700	v	0.0161	0.0166	0.0174	0.0186	0.0204	0.7928	0.9072	1.0102	1.1078	1.2023	1.2948	1.3858	1.4757	1.5647	1.6530
	H	69.84	169.65	270.89	375.61	487.93	1281.0	1345.6	1403.7	1459.4	1514.4	1569.4	1624.8	1680.7	1737.2	1794.3
(503.08)	s	0.1291	0.2932	0.4360	0.5655	0.6889	1.5090	1.5673	1.6154	1.6580	1.6970	1.7335	1.7679	1.8006	1.8318	1.8617
800	v	0.0161	0.0166	0.0174	0.0186	0.0204	0.6774	0.7828	0.8759	0.9631	1.0470	1.1289	1.2093	1.2885	1.3669	1.4446
	H	70.11	169.88	271.07	375.73	487.88	1271.1	1339.2	1399.1	1455.8	1511.4	1566.9	1622.7	1678.9	1735.0	1792.9
(518.21)	s	0.1290	0.2930	0.4358	0.5652	0.6885	1.4869	1.5484	1.5980	1.6413	1.6807	1.7175	1.7522	1.7851	1.8164	1.8464
900	v	0.0161	0.0166	0.0174	0.0186	0.0204	0.5869	0.6858	0.7713	0.8504	0.9262	0.9998	1.0720	1.1430	1.2131	1.2825
	H	70.37	170.10	271.26	375.84	487.83	1260.6	1332.7	1394.4	1452.2	1508.5	1564.4	1620.6	1677.1	1734.1	1791.6
(531.95)	s	0.1290	0.2929	0.4357	0.5649	0.6881	1.4659	1.5311	1.5822	1.6263	1.6662	1.7033	1.7382	1.7713	1.8028	1.8329
1000	v	0.0161	0.0166	0.0174	0.0186	0.0204	0.5137	0.6080	0.6875	0.7603	0.8295	0.8966	0.9622	1.0266	1.0901	1.1529
	H	70.63	170.33	271.44	375.96	487.79	1249.3	1325.9	1389.6	1448.5	1504.4	1561.9	1618.4	1675.3	1732.5	1790.3
(544.58)	s	0.1289	0.2928	0.4355	0.5647	0.6876	1.4457	1.5149	1.5677	1.6126	1.6530	1.6905	1.7256	1.7589	1.7905	1.8207
1100	v	0.0161	0.0166	0.0174	0.0185	0.0203	0.4531	0.5440	0.6188	0.6865	0.7505	0.8121	0.8723	0.9313	0.9894	1.0468
	H	70.90	170.56	271.63	376.08	487.75	1237.3	1318.8	1384.7	1444.7	1502.4	1559.4	1616.3	1673.5	1731.0	1789.0
(556.28)	s	0.1289	0.2927	0.4353	0.5644	0.6872	1.4259	1.4996	1.5542	1.6000	1.6410	1.6787	1.7141	1.7475	1.7793	1.8097
1200	v	0.0161	0.0166	0.0174	0.0185	0.0203	0.4016	0.4905	0.5615	0.6250	0.6845	0.7418	0.7974	0.8519	0.9055	0.9584
	H	71.16	170.78	271.82	376.20	487.72	1224.2	1311.5	1379.7	1440.9	1499.4	1556.9	1614.2	1671.6	1729.4	1787.6
(567.19)	s	0.1288	0.2926	0.4351	0.5642	0.6868	1.4061	1.4851	1.5415	1.5883	1.6298	1.6679	1.7035	1.7371	1.7691	1.7996
1400	v	0.0161	0.0166	0.0174	0.0185	0.0203	0.3176	0.4059	0.4712	0.5282	0.5809	0.6311	0.6798	0.7272	0.7737	0.8195
	H	71.68	171.24	272.19	376.44	487.65	1194.1	1296.1	1369.3	1433.2	1493.2	1551.8	1609.9	1668.0	1726.3	1785.0
(587.07)	s	0.1287	0.2923	0.4348	0.5636	0.6859	1.3652	1.4575	1.5182	1.5670	1.6096	1.6484	1.6845	1.7185	1.7508	1.7815
1600	v	0.0161	0.0166	0.0173	0.0185	0.0202	0.0236	0.3415	0.4032	0.4555	0.5031	0.5482	0.5915	0.6336	0.6748	0.7153
	H	72.21	171.69	272.57	376.69	487.60	616.77	1279.4	1358.5	1425.2	1486.9	1546.6	1605.6	1664.3	1723.2	1782.3
(604.87)	s	0.1286	0.2921	0.4344	0.5631	0.6851	0.8129	1.4312	1.4968	1.5478	1.5916	1.6312	1.6678	1.7022	1.7344	1.7657
1800	v	0.0160	0.0165	0.0173	0.0185	0.0202	0.0235	0.2906	0.3500	0.3988	0.4426	0.4836	0.5229	0.5609	0.5980	0.6343
	H	72.73	172.15	272.95	376.93	487.56	615.58	1261.1	1347.2	1417.1	1480.6	1541.1	1601.2	1660.7	1720.1	1779.7
(621.02)	s	0.1284	0.2918	0.4341	0.5626	0.6843	0.8109	1.4054	1.4768	1.5302	1.5753	1.6156	1.6528	1.6876	1.7204	1.7516
2000	v	0.0160	0.0165	0.0173	0.0184	0.0201	0.0233	0.2488	0.3072	0.3534	0.3942	0.4320	0.4680	0.5027	0.5365	0.5695
	H	73.26	172.60	273.32	377.19	487.53	614.48	1240.9	1353.4	1408.7	1474.1	1536.2	1596.9	1657.0	1717.0	1777.1
(635.80)	s	0.1283	0.2916	0.4337	0.5621	0.6834	0.8091	1.3794	1.4578	1.5138	1.5603	1.6014	1.6391	1.6743	1.7075	1.7389
2500	v	0.0160	0.0165	0.0173	0.0184	0.0200	0.0230	0.1681	0.2293	0.2712	0.3068	0.3390	0.3692	0.3980	0.4259	0.4529
	H	74.57	173.74	274.27	377.82	487.50	612.08	1176.7	1303.4	1386.7	1457.5	1522.9	1585.9	1647.8	1709.2	1770.4
(668.11)	s	0.1280	0.2910	0.4329	0.5609	0.6815	0.8048	1.3076	1.4129	1.4766	1.5269	1.5703	1.6094	1.6456	1.6796	1.7116
3000	v	0.0160	0.0165	0.0172	0.0183	0.0200	0.0228	0.0982	0.1759	0.2161	0.2484	0.2770	0.3033	0.3282	0.3522	0.3753
	H	75.88	174.88	275.22	378.47	487.52	610.08	1060.5	1267.0	1363.2	1440.2	1509.4	1574.8	1638.5	1701.4	1761.8
(695.33)	s	0.1277	0.2904	0.4320	0.5597	0.6796	0.8009	1.1966	1.3692	1.4429	1.4976	1.5434	1.5841	1.6214	1.6561	1.6888
3200	v	0.0160	0.0165	0.0172	0.0183	0.0199	0.0227	0.0335	0.1588	0.1987	0.2301	0.2576	0.2327	0.3065	0.3291	0.3510
	H	76.4	175.3	275.6	378.7	487.5	609.4	800.8	1250.9	1353.4	1433.1	1503.8	1570.3	1634.8	1698.3	1761.2
(705.08)	s	0.1276	0.2902	0.4317	0.5592	0.6788	0.7994	0.9708	1.3515	1.4300	1.4866	1.5335	1.5749	1.6126	1.6477	1.6806
3500	v	0.0160	0.0164	0.0172	0.0183	0.0199	0.0225	0.0307	0.1364	0.1764	0.2066	0.2326	0.2563	0.2784	0.2995	0.3198
	H	77.2	176.0	276.2	379.1	487.6	608.4	779.4	1224.6	1338.2	1422.2	1495.5	1563.3	1629.2	1693.6	1757.2
	s	0.1274	0.2899	0.4312	0.5585	0.6777	0.7973	0.9508	1.3242	1.4112	1.4709	1.5194	1.5618	1.6002	1.6358	1.6691
4000	v	0.0159	0.0164	0.0172	0.0182	0.0198	0.0223	0.0287	0.1052	0.1463	0.1752	0.1994	0.2210	0.2411	0.2601	0.2783
	H	78.5	177.2	277.1	379.8	487.7	606.9	763.0	1174.3	1311.6	1403.6	1481.3	1552.2	1619.8	1685.7	1750.6
	s	0.1271	0.2893	0.4304	0.5573	0.6760	0.7940	0.9343	1.2754	1.3807	1.4461	1.4976	1.5417	1.5812	1.6177	1.6516
5000	v	0.0159	0.0164	0.0171	0.0181	0.0196	0.0219	0.0268	0.0591	0.1038	0.1312	0.1529	0.1718	0.1890	0.2050	0.2203
	H	81.1	179.5	279.1	381.2	488.1	604.6	746.0	1042.9	1252.9	1364.6	1452.1	1529.1	1600.9	1670.0	1737.4
	s	0.1265	0.2881	0.4287	0.5550	0.6726	0.7880	0.9153	1.1593	1.3207	1.4001	1.4582	1.5061	1.5481	1.5863	1.6216
6000	v	0.0159	0.0163	0.0170	0.0180	0.0195	0.0216	0.0256	0.0397	0.0757	0.1020	0.1221	0.1391	0.1544	0.1684	0.1817
	H	83.7	181.7	281.0	382.7	488.6	602.9	736.1	945.1	1188.8	1323.6	1422.3	1505.9	1582.0	1654.2	1724.2
	s	0.1258	0.2870	0.4271	0.5528	0.6693	0.7826	0.9026	1.0176	1.2615	1.3574	1.4229	1.4748	1.5194	1.5593	1.5962
7000	v	0.0158	0.0163	0.0170	0.0180	0.0193	0.0213	0.0248	0.0334	0.0573	0.0816	0.1004	0.1160	0.1298	0.1424	0.1542
	H	86.2	184.4	283.0	384.2	489.3	601.7	729.3	901.8	1124.9	1281.7	1392.2	1482.6	1563.1	1638.6	1711.1
	s	0.1252	0.2859	0.4256	0.5507	0.6663	0.7777	0.8926	1.0350	1.2055	1.3171	1.3904	1.4466	1.4938	1.5355	1.5735

Courtesy of Department of Mechanical Engineering, Stanford University

Table 6-3b Properties of Superheated Steam and Compressed Water (Temperature and Pressure), English Units. Source: Babcock & Wilcox

SI Properties of Saturated Steam and Saturated Water (Temperature)

Temp K	Press. MPa	Volume, m^3/kg Water v_f	Steam v_g	Enthalpy, kJ/kg Water H_f	Evap H_{fg}	Steam H_g	Entropy, kJ/(kg K) Water s_f	Evap s_{fg}	Steam s_g
273.16	0.0006113	0.001000	206.1	0.0	2500.9	2500.9	0.0	9.1555	9.1555
275	0.0006980	0.001000	181.7	7.5	2496.8	2504.3	0.0274	9.0792	9.1066
280	0.0009912	0.001000	130.3	28.1	2485.4	2513.5	0.1015	8.8765	8.9780
285	0.001388	0.001001	94.67	48.8	2473.9	2522.7	0.1749	8.6803	8.8552
290	0.001919	0.001001	69.67	69.7	2462.2	2531.9	0.2475	8.4903	8.7378
295	0.002620	0.001002	51.90	90.7	2450.3	2541.0	0.3193	8.3061	8.6254
300	0.003536	0.001004	39.10	111.7	2438.4	2550.1	0.3900	8.1279	8.5179
305	0.004718	0.001005	29.78	132.8	2426.3	2559.1	0.4598	7.9551	8.4149
310	0.006230	0.001007	22.91	153.9	2414.3	2568.2	0.5285	7.7878	8.3163
315	0.008143	0.001009	17.80	175.1	2402.0	2577.1	0.5961	7.6255	8.2216
320	0.01054	0.001011	13.96	196.2	2389.8	2586.0	0.6626	7.4682	8.1308
325	0.01353	0.001013	11.04	217.3	2377.6	2594.9	0.7280	7.3156	8.0436
330	0.01721	0.001015	8.809	238.4	2365.3	2603.7	0.7924	7.1675	7.9599
335	0.02171	0.001018	7.083	259.4	2353.0	2612.4	0.8557	7.0236	7.8793
340	0.02718	0.001021	5.737	280.5	2340.5	2621.0	0.9180	6.8838	7.8018
345	0.03377	0.001024	4.680	301.5	2328.0	2629.5	0.9793	6.7479	7.7272
350	0.04166	0.001027	3.844	322.5	2315.4	2637.9	1.0397	6.6156	7.6553
355	0.05105	0.001030	3.178	343.4	2302.9	2646.3	1.0991	6.4869	7.5860
360	0.06215	0.001034	2.643	364.4	2290.1	2654.5	1.1577	6.3615	7.5192
365	0.07521	0.001037	2.211	385.3	2277.3	2662.6	1.2155	6.2391	7.4546
370	0.09047	0.001041	1.860	406.3	2264.3	2670.6	1.2725	6.1198	7.3923
375	0.1082	0.001045	1.573	427.3	2251.2	2678.5	1.3288	6.0032	7.3320
380	0.1288	0.001049	1.337	448.3	2237.9	2686.2	1.3843	5.8894	7.2737
385	0.1524	0.001053	1.142	469.3	2224.5	2693.8	1.4393	5.7779	7.2172
390	0.1795	0.001058	0.9800	490.4	2210.9	2701.3	1.4936	5.6688	7.1624
395	0.2104	0.001062	0.8445	511.5	2197.0	2708.5	1.5473	5.5621	7.1094
400	0.2456	0.001067	0.7308	532.7	2182.9	2715.6	1.6005	5.4573	7.0578
405	0.2854	0.001072	0.6349	554.0	2168.6	2722.6	1.6532	5.3546	7.0078
410	0.3302	0.001077	0.5537	575.3	2154.0	2729.3	1.7054	5.2537	6.9591
415	0.3806	0.001082	0.4846	596.7	2139.1	2735.8	1.7572	5.1545	6.9117
420	0.4370	0.001087	0.4256	618.2	2123.9	2742.1	1.8085	5.0570	6.8655
425	0.4999	0.001093	0.3750	639.8	2108.4	2748.2	1.8594	4.9611	6.8205
430	0.5699	0.001099	0.3314	661.4	2092.7	2754.1	1.9099	4.8667	6.7766
435	0.6474	0.001104	0.2938	683.1	2076.6	2759.7	1.9599	4.7737	6.7336
440	0.7332	0.001110	0.2612	705.0	2060.0	2765.0	2.0096	4.6820	6.6916
445	0.8277	0.001117	0.2328	726.9	2043.2	2770.1	2.0590	4.5914	6.6504
450	0.9315	0.001123	0.2080	749.0	2025.9	2774.9	2.1080	4.5020	6.6100
455	1.045	0.001130	0.1864	771.1	2008.2	2779.3	2.1567	4.4136	6.5703
460	1.170	0.001137	0.1673	793.4	1990.1	2783.5	2.2050	4.3263	6.5313
465	1.306	0.001144	0.1506	815.7	1971.6	2787.3	2.2530	4.2399	6.4929
470	1.454	0.001152	0.1358	838.2	1952.6	2790.8	2.3007	4.1544	6.4551
475	1.615	0.001159	0.1227	860.8	1933.0	2793.8	2.3482	4.0695	6.4177
480	1.789	0.001167	0.1111	883.5	1913.0	2796.5	2.3953	3.9855	6.3808
490	2.181	0.001184	0.09150	929.3	1871.4	2800.7	2.4887	3.8193	6.3080
500	2.637	0.001202	0.07585	975.6	1827.5	2803.1	2.5813	3.6550	6.2363
510	3.163	0.001222	0.06323	1022.6	1781.0	2803.6	2.6731	3.4921	6.1652
520	3.766	0.001244	0.05296	1070.4	1731.7	2802.1	2.7644	3.3301	6.0945
530	4.453	0.001267	0.04454	1119.1	1679.1	2798.2	2.8555	3.1681	6.0236
540	5.233	0.001293	0.03758	1168.9	1622.9	2791.8	2.9466	3.0055	5.9521
550	6.112	0.001322	0.03179	1219.9	1562.7	2782.6	3.0382	2.8413	5.8795
560	7.100	0.001355	0.02694	1272.5	1497.8	2770.3	3.1306	2.6746	5.8052
570	8.206	0.001391	0.02284	1326.9	1427.5	2754.4	3.2241	2.5044	5.7285
580	9.439	0.001433	0.01934	1383.3	1350.9	2734.2	3.3193	2.3291	5.6484
590	10.81	0.001482	0.01635	1442.3	1266.6	2708.9	3.4167	2.1468	5.5635
600	12.33	0.001540	0.01375	1504.6	1172.5	2677.1	3.5174	1.9543	5.4717
610	14.02	0.001611	0.01146	1571.1	1065.6	2636.7	3.6231	1.7468	5.3699
620	15.88	0.001704	0.009422	1644.3	939.6	2583.9	3.7370	1.5154	5.2524
630	17.95	0.001837	0.007532	1729.3	781.4	2510.7	3.8671	1.2404	5.1075
640	20.25	0.002076	0.005626	1842.9	550.5	2393.4	4.0389	0.8602	4.8991
647.29	22.089	0.003155	0.003155	2098.8	0.0	2098.8	4.4289	0.0	4.4289

Courtesy of Department of Mechanical Engineering, Stanford University

Table 6-4 Properties of Saturated Steam and Saturated Water (Temperature), SI Units. Source: Babcock & Wilcox

SI Properties of Saturated Steam and Saturated Water (Pressure)

Press. MPa	Temp K	Volume, m³/kg Water v_f	Steam v_g	Enthalpy, kJ/kg Water H_f	Evap H_{fg}	Steam H_g	Entropy, kJ/(kg K) Water s_f	Evap s_{fg}	Steam s_g
0.00080	276.92	0.001000	159.7	15.4	2492.4	2507.8	0.0559	9.0007	9.0566
0.0010	280.13	0.001000	129.2	28.6	2485.1	2513.7	0.1034	8.8714	8.9748
0.0012	282.81	0.001000	108.7	39.7	2479.0	2518.7	0.1429	8.7654	8.9083
0.0014	285.13	0.001001	93.92	49.3	2473.6	2522.9	0.1768	8.6754	8.8522
0.0016	287.17	0.001001	82.76	57.8	2468.9	2526.7	0.2065	8.5972	8.8037
0.0018	288.99	0.001001	74.03	65.5	2464.5	2530.0	0.2330	8.5280	8.7610
0.0020	290.65	0.001002	67.00	72.4	2460.6	2533.0	0.2569	8.4659	8.7228
0.0025	294.23	0.001002	54.25	87.5	2452.1	2539.6	0.3083	8.3340	8.6423
0.0030	297.23	0.001003	45.67	100.1	2445.0	2545.1	0.3510	8.2258	8.5768
0.0040	302.12	0.001004	34.80	120.7	2433.2	2553.9	0.4197	8.0541	8.4738
0.0050	306.03	0.001005	28.19	137.2	2423.8	2561.0	0.4740	7.9203	8.3943
0.0060	309.31	0.001007	23.74	151.0	2415.9	2566.9	0.5191	7.8105	8.3296
0.0080	314.66	0.001009	18.10	173.7	2402.8	2576.5	0.5915	7.6364	8.2279
0.010	318.96	0.001010	14.67	191.8	2392.4	2584.2	0.6488	7.5006	8.1494
0.012	322.57	0.001012	12.36	207.1	2383.5	2590.6	0.6964	7.3891	8.0855
0.014	325.70	0.001013	10.69	220.3	2375.8	2596.1	0.7371	7.2946	8.0317
0.016	328.47	0.001015	9.433	231.9	2369.1	2601.0	0.7728	7.2124	7.9852
0.018	330.96	0.001016	8.445	242.4	2362.9	2605.3	0.8045	7.1397	7.9442
0.020	333.22	0.001017	7.649	251.9	2357.4	2609.3	0.8332	7.0745	7.9077
0.025	338.12	0.001020	6.204	272.6	2345.1	2617.7	0.8947	6.9359	7.8306
0.030	342.26	0.001022	5.229	289.9	2334.9	2624.8	0.9458	6.8220	7.7678
0.040	349.02	0.001026	3.993	318.3	2318.0	2636.3	1.0279	6.6413	7.6692
0.050	354.48	0.001030	3.240	341.3	2304.1	2645.4	1.0930	6.5001	7.5931
0.060	359.09	0.001033	2.732	360.6	2292.4	2653.0	1.1471	6.3841	7.5312
0.080	366.65	0.001038	2.087	392.3	2273.0	2665.3	1.2344	6.1994	7.4338
0.10	372.78	0.001043	1.694	418.0	2257.0	2675.0	1.3038	6.0548	7.3586
0.101325	373.14	0.001043	1.673	419.5	2256.1	2675.6	1.3079	6.0462	7.3541
0.12	377.96	0.001047	1.428	439.7	2243.4	2683.1	1.3617	5.9356	7.2973
0.14	382.46	0.001051	1.237	458.6	2231.4	2690.0	1.4115	5.8341	7.2456
0.16	386.47	0.001054	1.091	475.5	2220.5	2696.0	1.4553	5.7456	7.2009
0.18	390.09	0.001058	0.9775	490.8	2210.6	2701.4	1.4945	5.6670	7.1615
0.20	393.38	0.001061	0.8857	504.7	2201.5	2706.2	1.5300	5.5963	7.1263
0.25	400.59	0.001067	0.7187	535.2	2181.3	2716.5	1.6068	5.4451	7.0519
0.30	406.70	0.001073	0.6058	561.2	2163.7	2724.9	1.6710	5.3201	6.9911
0.40	416.78	0.001084	0.4625	604.3	2133.8	2738.1	1.7755	5.1196	6.8951
0.50	425.01	0.001093	0.3749	639.8	2108.4	2748.2	1.8594	4.9611	6.8205
0.60	432.00	0.001101	0.3157	670.1	2086.3	2756.4	1.9299	4.8293	6.7592
0.80	443.59	0.001115	0.2404	720.7	2048.0	2768.7	2.0451	4.6169	6.6620
1.0	453.06	0.001127	0.1944	762.5	2015.1	2777.6	2.1378	4.4479	6.5857
1.2	461.14	0.001139	0.1633	798.5	1985.9	2784.4	2.2160	4.3065	6.5225
1.4	468.22	0.001149	0.1408	830.2	1959.4	2789.6	2.2838	4.1847	6.4685
1.6	474.56	0.001159	0.1238	858.8	1934.8	2793.6	2.3440	4.0770	6.4210
1.8	480.30	0.001168	0.1104	884.9	1911.8	2796.7	2.3981	3.9805	6.3786
2.0	485.57	0.001176	0.09963	908.9	1890.2	2799.1	2.4474	3.8927	6.3401
2.5	497.15	0.001197	0.07998	962.4	1840.2	2802.6	2.5549	3.7018	6.2567
3.0	507.05	0.001216	0.06668	1008.7	1795.0	2803.7	2.6461	3.5400	6.1861
4.0	523.55	0.001252	0.04978	1087.6	1713.4	2801.0	2.7968	3.2725	6.0693
5.0	537.14	0.001286	0.03944	1154.5	1639.4	2793.9	2.9206	3.0520	5.9726
6.0	548.79	0.001319	0.03244	1213.7	1570.2	2783.9	3.0271	2.8613	5.8884
7.0	559.03	0.001352	0.02737	1267.4	1504.3	2771.7	3.1216	2.6909	5.8125
8.0	568.22	0.001385	0.02352	1317.0	1440.5	2757.5	3.2073	2.5351	5.7424
10.	584.22	0.001453	0.01803	1407.9	1316.4	2724.3	3.3600	2.2533	5.6133
11.	591.30	0.001489	0.01599	1450.2	1255.0	2705.2	3.4296	2.1224	5.5520
12.	597.90	0.001527	0.01426	1491.2	1193.2	2684.4	3.4960	1.9956	5.4916
13.	604.09	0.001567	0.01278	1531.1	1130.7	2661.8	3.5599	1.8717	5.4316
14.	609.90	0.001610	0.01149	1570.4	1066.8	2637.2	3.6220	1.7490	5.3710
16.	620.59	0.001710	0.009307	1648.9	931.3	2580.2	3.7441	1.5007	5.2448
18.	630.22	0.001840	0.007492	1731.4	777.4	2508.8	3.8703	1.2336	5.1039
20.	638.96	0.002041	0.005836	1828.5	581.0	2409.5	4.0172	0.9093	4.9265
22.089	647.29	0.003155	0.003155	2098.8	0.0	2098.8	4.4289	0.0	4.4289

Courtesy of Department of Mechanical Engineering, Stanford University

Table 6-5 Properties of Saturated Steam and Saturated Water (Pressure), SI Units. Source: Babcock & Wilcox

SI Properties of Compressed Water (Temperature and Pressure)										
		Temperature, K (Sat. Pressure, MPa)								
Press. MPa		400 (0.2456)	425 (0.4999)	450 (0.9315)	475 (1.615)	500 (2.637)	525 (4.098)	550 (6.112)	575 (8.806)	600 (12.33)
sat	ρ,kg/m³	937.35	915.08	890.25	862.64	831.71	796.64	756.18	708.38	649.40
	H,kJ/kg	532.69	639.71	748.98	860.80	975.65	1094.63	1219.93	1354.82	1504.56
	s,kJ/(kg K)	1.60049	1.85933	2.10801	2.34815	2.58128	2.80995	3.03821	3.27144	3.51742
0.50	ρ,kg/m³	937.51	915.08							
	H,kJ/kg	532.82	639.71							
	s,kJ/(kg K)	1.60020	1.85933							
0.70	ρ,kg/m³	937.62	915.22							
	H,kJ/kg	532.94	639.84							
	s,kJ/(kg K)	1.59999	1.85914							
1.00	ρ,kg/m³	937.79	915.41	890.30						
	H,kJ/kg	533.12	640.02	749.01						
	s,kJ/(kg K)	1.59968	1.85884	2.10793						
1.40	ρ,kg/m³	938.01	915.66	890.58						
	H,kJ/kg	533.37	640.27	749.20						
	s,kJ/(kg K)	1.59928	1.85843	2.10740						
2.00	ρ,kg/m³	938.33	916.03	890.99	862.93					
	H,kJ/kg	533.76	640.64	749.50	860.92					
	s,kJ/(kg K)	1.59868	1.85779	2.10661	2.34748					
3.00	ρ,kg/m³	938.86	916.63	891.66	863.70	832.04				
	H,kJ/kg	534.42	641.26	750.01	861.24	975.68				
	s,kJ/(kg K)	1.59771	1.85672	2.10529	2.34578	2.58049				
5.00	ρ,kg/m³	939.90	917.80	892.99	865.23	833.88	797.71			
	H,kJ/kg	535.77	642.49	751.04	861.95	975.97	1094.50			
	s,kJ/(kg K)	1.59579	1.85454	2.10267	2.34248	2.57633	2.80757			
7.00	ρ,kg/m³	940.93	918.96	894.30	866.74	835.70	800.06	757.63		
	H,kJ/kg	537.13	643.73	752.08	862.71	976.34	1094.30	1219.41		
	s,kJ/(kg K)	1.59390	1.85237	2.10006	2.33927	2.57233	2.80247	3.03517		
10.00	ρ,kg/m³	942.45	920.66	896.23	868.97	838.39	803.48	762.36	711.24	
	H,kJ/kg	539.18	645.59	753.67	863.91	976.99	1094.16	1217.91	1353.16	
	s,kJ/(kg K)	1.59110	1.84912	2.09618	2.33455	2.56651	2.79513	3.02530	3.26566	
14.00	ρ,kg/m³	944.45	922.89	898.75	871.89	841.87	807.86	768.28	720.13	656.02
	H,kJ/kg	541.93	648.10	755.82	865.57	977.99	1094.20	1216.32	1348.37	1499.43
	s,kJ/(kg K)	1.58742	1.84484	2.09110	2.32843	2.55903	2.78578	3.01295	3.24764	3.50461
20.00	ρ,kg/m³	947.40	926.15	902.43	876.12	846.90	814.10	776.50	731.86	675.64
	H,kJ/kg	546.09	651.88	759.11	868.19	979.70	1094.61	1214.62	1342.80	1485.18
	s,kJ/(kg K)	1.58199	1.83852	2.08365	2.31954	2.54829	2.77251	2.99579	3.22363	3.46589
30.00	ρ,kg/m³	952.15	931.39	908.31	882.86	854.83	823.75	788.76	748.41	700.23
	H,kJ/kg	553.10	658.28	764.75	872.87	983.09	1096.17	1213.36	1336.77	1469.95
	s,kJ/(kg K)	1.57321	1.82824	2.07165	2.30546	2.53159	2.75225	2.97029	3.18966	3.41631
50.00	ρ,kg/m³	961.23	941.29	919.30	895.30	869.22	840.82	809.63	774.92	735.63
	H,kJ/kg	567.28	671.32	776.46	882.95	991.08	1101.29	1214.33	1331.34	1454.04
	s,kJ/(kg K)	1.55640	1.80868	2.04904	2.27933	2.50117	2.71624	2.92655	3.13458	3.34342
100.00	ρ,kg/m³	982.01	963.53	943.46	921.97	899.14	875.01	849.47	822.32	793.27
	H,kJ/kg	603.30	705.01	807.50	910.95	1015.43	1121.02	1227.92	1336.47	1447.18
	s,kJ/(kg K)	1.51781	1.76443	1.99876	2.22248	2.43683	2.64289	2.84180	3.03479	3.22326

Courtesy of Department of Mechanical Engineering, Stanford University

Table 6-6 Properties of Compressed Water (Temperature and Pressure), SI Units. Source: Babcock & Wilcox

SI Properties of Superheated Steam (Temperature and Pressure)
(Pressure from 0.0010 to 0.0400 MPa)

Press. MPa (Sat Temp, K)		Temperature, K								
		Sat.	350	400	450	500	550	600	650	700
0.0010	v,m³/kg	129.2	161.5	184.6	207.7	230.7	253.8	276.9	300.0	323.1
(280.1)	H,kJ/kg	2513.7	2644.4	2739.0	2834.7	2931.8	3030.2	3130.1	3231.7	3334.8
	s,kJ/(kg K)	8.9748	9.3913	9.6439	9.8693	10.0737	10.2614	10.4353	10.5978	10.7506
0.0020	v,m³/kg	67.00	80.73	92.28	103.8	115.4	126.9	138.4	150.0	161.5
(290.7)	H,kJ/kg	2533.0	2644.3	2738.9	2834.7	2931.7	3030.2	3130.1	3231.6	3334.8
	s,kJ/(kg K)	8.7228	9.0710	9.3238	9.5493	9.7538	9.9414	10.1153	10.2779	10.4307
0.0040	v,m³/kg	34.80	40.35	46.13	51.91	57.68	63.45	69.22	74.99	80.76
(302.1)	H,kJ/kg	2553.9	2644.0	2738.8	2834.6	2931.7	3030.1	3130.1	3231.6	3334.8
	s,kJ/(kg K)	8.4738	8.7504	9.0035	9.2292	9.4338	9.6215	9.7954	9.9579	10.1108
0.0070	v,m³/kg	20.53	23.04	26.35	29.66	32.96	36.25	39.55	42.85	46.15
(312.2)	H,kJ/kg	2572.0	2643.5	2738.5	2834.4	2931.5	3030.0	3130.0	3231.6	3334.7
	s,kJ/(kg K)	8.2750	8.4911	8.7447	8.9707	9.1753	9.3631	9.5371	9.6996	9.8525
0.0100	v,m³/kg	14.67	16.12	18.44	20.75	23.07	25.38	27.69	29.99	32.30
(319.0)	H,kJ/kg	2584.2	2643.0	2738.2	2834.2	2931.4	3030.0	3130.0	3231.5	3334.7
	s,kJ/(kg K)	8.1494	8.3254	8.5796	8.8058	9.0106	9.1984	9.3724	9.5349	9.6878
0.0200	v,m³/kg	7.649	8.044	9.210	10.37	11.53	12.68	13.84	14.99	16.15
(333.2)	H,kJ/kg	2609.3	2641.4	2737.3	2833.7	2931.0	3029.7	3129.7	3231.3	3334.5
	s,kJ/(kg K)	7.9077	8.0019	8.2579	8.4849	8.6901	8.8781	9.0522	9.2148	9.3678
0.0400	v,m³/kg	3.993	4.005	4.595	5.179	5.759	6.339	6.917	7.495	8.073
(349.0)	H,kJ/kg	2636.3	2638.2	2735.5	2832.5	2930.3	3029.1	3129.3	3231.0	3334.3
	s,kJ/(kg K)	7.6692	7.6747	7.9344	8.1631	8.3690	8.5574	8.7318	8.8945	9.0476

Press. MPa (Sat Temp, K)		Temperature, K								
		750	800	850	900	950	1000	1050	1100	1150
0.0010	v,m³/kg	346.1	369.2	392.3	415.4	438.4	461.5	484.6	507.7	530.7
(280.1)	H,kJ/kg	3439.6	3546.1	3654.3	3764.3	3876.0	3989.5	4104.7	4221.7	4340.5
	s,kJ/(kg K)	10.8952	11.0327	11.1639	11.2896	11.4104	11.5268	11.6392	11.7481	11.8536
0.0020	v,m³/kg	173.1	184.6	196.1	207.7	219.2	230.8	242.3	253.8	265.4
(290.7)	H,kJ/kg	3439.6	3546.1	3654.3	3764.3	3876.0	3989.5	4104.7	4221.7	4340.5
	s,kJ/(kg K)	10.5753	10.7128	10.8440	10.9697	11.0905	11.2069	11.3193	11.4282	11.5337
0.0040	v,m³/kg	86.53	92.30	98.07	103.8	109.6	115.4	121.1	126.9	132.7
(302.1)	H,kJ/kg	3439.6	3546.1	3654.3	3764.3	3876.0	3989.5	4104.7	4221.7	4340.5
	s,kJ/(kg K)	10.2554	10.3929	10.5241	10.6498	10.7706	10.8870	10.9994	11.1083	11.2138
0.0070	v,m³/kg	49.45	52.74	56.04	59.34	62.63	65.93	69.23	72.52	75.82
(312.2)	H,kJ/kg	3439.5	3546.0	3654.3	3764.3	3876.0	3989.5	4104.7	4221.7	4340.5
	s,kJ/(kg K)	9.9971	10.1346	10.2658	10.3915	10.5123	10.6287	10.7412	10.8500	10.9556
0.0100	v,m³/kg	34.61	36.92	39.23	41.54	43.84	46.15	48.46	50.77	53.07
(319.0)	H,kJ/kg	3439.5	3546.0	3654.3	3764.2	3876.0	3989.5	4104.7	4221.7	4340.4
	s,kJ/(kg K)	9.8324	9.9699	10.1011	10.2269	10.3477	10.4641	10.5765	10.6854	10.7910
0.0200	v,m³/kg	17.30	18.46	19.61	20.77	21.92	23.07	24.23	25.38	26.54
(333.2)	H,kJ/kg	3439.4	3545.9	3654.2	3764.2	3875.9	3989.4	4104.7	4221.7	4340.4
	s,kJ/(kg K)	9.5124	9.6499	9.7812	9.9069	10.0277	10.1442	10.2566	10.3655	10.4710
0.0400	v,m³/kg	8.650	9.228	9.805	10.38	10.96	11.54	12.11	12.69	13.27
(349.0)	H,kJ/kg	3439.2	3545.7	3654.0	3764.0	3875.8	3989.3	4104.6	4221.6	4340.3
	s,kJ/(kg K)	9.1923	9.3299	9.4611	9.5869	9.7077	9.8242	9.9366	10.0455	10.1511

Courtesy of Department of Mechanical Engineering, Stanford University

Table 6-7a Properties of Superheated Steam and Compressed Water (Temperature and Pressure), SI Units. Source: Babcock & Wilcox

SI Properties of Superheated Steam (Temperature and Pressure)
(Pressure from 0.070 to 2.000 MPa)

Press. MPa (Sat. Temp, K)		Temperature, K								
		Sat.	400	450	500	550	600	650	700	750
0.070	$v, m^3/kg$	2.365	2.617	2.953	3.287	3.619	3.950	4.281	4.612	4.942
(363.1)	H, kJ/kg	2659.6	2732.7	2830.8	2929.1	3028.3	3128.7	3230.5	3333.8	3438.8
	s, kJ/(kg K)	7.4789	7.6707	7.9019	8.1090	8.2980	8.4727	8.6357	8.7889	8.9337
0.101325	$v, m^3/kg$	1.673	1.802	2.036	2.268	2.498	2.727	2.956	3.185	3.413
(373.1)	H, kJ/kg	2675.6	2729.7	2829.0	2927.9	3027.4	3128.0	3229.9	3333.4	3438.4
	s, kJ/(kg K)	7.3541	7.4942	7.7281	7.9365	8.1261	8.3012	8.4644	8.6177	8.7627
0.200	$v, m^3/kg$	0.8857	0.9024	1.025	1.144	1.262	1.379	1.495	1.612	1.728
(393.4)	H, kJ/kg	2706.2	2720.2	2823.2	2924.0	3024.6	3125.8	3228.2	3332.0	3437.3
	s, kJ/(kg K)	7.1263	7.1616	7.4044	7.6168	7.8085	7.9847	8.1486	8.3024	8.4477
0.400	$v, m^3/kg$	0.4625		0.5053	0.5671	0.6273	0.6866	0.7455	0.8040	0.8624
(416.8)	H, kJ/kg	2738.1		2811.0	2916.0	3018.8	3121.5	3224.8	3329.2	3435.0
	s, kJ/(kg K)	6.8951		7.0634	7.2848	7.4808	7.6594	7.8248	7.9795	8.1255
0.700	$v, m^3/kg$	0.2729		0.2822	0.3197	0.3552	0.3899	0.4240	0.4579	0.4915
(438.1)	H, kJ/kg	2763.1		2791.3	2903.4	3010.0	3114.8	3219.5	3325.0	3431.5
	s, kJ/(kg K)	6.7072		6.7709	7.0073	7.2104	7.3928	7.5605	7.7168	7.8637
1.000	$v, m^3/kg$	0.1944			0.2206	0.2464	0.2712	0.2955	0.3194	0.3432
(453.1)	H, kJ/kg	2777.6			2890.2	3000.9	3108.0	3214.2	3320.7	3428.0
	s, kJ/(kg K)	6.5857			6.8223	7.0333	7.2198	7.3898	7.5476	7.6956
2.000	$v, m^3/kg$	0.09963			0.1044	0.1191	0.1326	0.1454	0.1578	0.1701
(485.6)	H, kJ/kg	2799.1			2840.5	2968.4	3084.5	3196.1	3306.2	3416.1
	s, kJ/(kg K)	6.3401			6.4242	6.6683	6.8704	7.0491	7.2122	7.3639

Press. MPa (Sat. Temp, K)		Temperature, K								
		800	850	900	950	1000	1050	1100	1150	1200
0.070	$v, m^3/kg$	5.272	5.602	5.932	6.262	6.592	6.922	7.251	7.581	7.911
(363.1)	H, kJ/kg	3545.4	3653.8	3763.8	3875.6	3989.1	4104.4	4221.5	4340.2	4460.7
	s, kJ/(kg K)	9.0713	9.2027	9.3284	9.4493	9.5658	9.6783	9.7872	9.8927	9.9953
0.101325	$v, m^3/kg$	3.641	3.870	4.098	4.326	4.554	4.781	5.009	5.237	5.465
(373.1)	H, kJ/kg	3545.1	3653.5	3763.6	3875.4	3989.0	4104.3	4221.3	4340.1	4460.6
	s, kJ/(kg K)	8.9004	9.0317	9.1576	9.2785	9.3950	9.5075	9.6164	9.7220	9.8245
0.200	$v, m^3/kg$	1.844	1.959	2.075	2.191	2.306	2.422	2.537	2.653	2.768
(393.4)	H, kJ/kg	3544.2	3652.7	3762.9	3874.8	3988.5	4103.8	4220.9	4339.7	4460.3
	s, kJ/(kg K)	8.5856	8.7172	8.8432	8.9642	9.0808	9.1933	9.3023	9.4079	9.5105
0.400	$v, m^3/kg$	0.9206	0.9787	1.037	1.095	1.153	1.210	1.268	1.326	1.384
(416.8)	H, kJ/kg	3542.2	3651.1	3761.5	3873.6	3987.4	4102.9	4220.1	4339.0	4459.6
	s, kJ/(kg K)	8.2639	8.3959	8.5221	8.6433	8.7601	8.8728	8.9818	9.0875	9.1901
0.700	$v, m^3/kg$	0.5250	0.5584	0.5917	0.6249	0.6581	0.6912	0.7244	0.7575	0.7905
(438.1)	H, kJ/kg	3539.3	3648.6	3759.4	3871.8	3985.8	4101.5	4218.9	4337.9	4458.6
	s, kJ/(kg K)	8.0029	8.1354	8.2621	8.3836	8.5006	8.6135	8.7226	8.8284	8.9312
1.000	$v, m^3/kg$	0.3668	0.3902	0.4137	0.4370	0.4603	0.4836	0.5068	0.5300	0.5532
(453.1)	H, kJ/kg	3536.4	3646.1	3757.3	3870.0	3984.3	4100.1	4217.6	4336.8	4457.6
	s, kJ/(kg K)	7.8356	7.9686	8.0957	8.2175	8.3348	8.4478	8.5571	8.6631	8.7659
2.000	$v, m^3/kg$	0.1821	0.1941	0.2060	0.2178	0.2295	0.2413	0.2530	0.2646	0.2763
(485.6)	H, kJ/kg	3526.5	3637.8	3750.2	3863.9	3979.0	4095.5	4213.5	4333.1	4454.2
	s, kJ/(kg K)	7.5064	7.6413	7.7698	7.8928	8.0108	8.1245	8.2343	8.3406	8.4437

Courtesy of Department of Mechanical Engineering, Stanford University

Table 6-7b Properties of Superheated Steam and Compressed Water (Temperature and Pressure), SI Units. Source: Babcock & Wilcox

SI Properties of Superheated Steam (Temperature and Pressure)
(Pressure from 3.0 to 100 MPa)

Press. MPa (Sat. Temp, K)		Temperature, K Sat.	600	650	700	750	800	900	1000	1100
3.0	v,m³/kg	0.06668	0.08628	0.09532	0.1040	0.1124	0.1206	0.1367	0.1526	0.1684
(507.1)	H,kJ/kg	2803.7	3059.6	3177.3	3291.3	3404.0	3516.5	3743.1	3973.7	4209.4
	s,kJ/(kg K)	6.1861	6.6516	6.8402	7.0091	7.1646	7.3098	7.5767	7.8196	8.0442
4.0	v,m³/kg	0.04978	0.06304	0.07024	0.07699	0.08849	0.08981	0.1021	0.1142	0.1261
(523.6)	H,kJ/kg	2801.0	3033.0	3157.8	3276.1	3391.6	3506.3	3735.8	3968.3	4205.3
	s,kJ/(kg K)	6.0693	6.4847	6.6846	6.8600	7.0194	7.1674	7.4378	7.6827	7.9085
6.0	v,m³/kg	0.03244	0.03958	0.04507	0.04998	0.05459	0.05901	0.06750	0.07571	0.08375
(548.8)	H,kJ/kg	2783.9	2973.8	3116.3	3244.4	3366.3	3485.5	3721.2	3957.5	4197.0
	s,kJ/(kg K)	5.8884	6.2201	6.4486	6.6385	6.8067	6.9605	7.2382	7.4872	7.7154
8.0	v,m³/kg	0.02352	0.02759	0.03239	0.03643	0.04011	0.04359	0.05018	0.05648	0.06260
(568.2)	H,kJ/kg	2757.5	2904.1	3071.1	3210.9	3340.0	3464.0	3706.2	3946.6	4188.7
	s,kJ/(kg K)	5.7424	5.9938	6.2617	6.4691	6.6472	6.8073	7.0927	7.3459	7.5766
10.0	v,m³/kg	0.01803	0.02008	0.02468	0.02825	0.03141	0.03434	0.03979	0.04494	0.04992
(584.2)	H,kJ/kg	2724.3	2818.3	3021.4	3175.6	3312.6	3442.0	3691.0	3935.6	4180.3
	s,kJ/(kg K)	5.6133	5.7722	6.0983	6.3270	6.5162	6.6833	6.9767	7.2344	7.4676
15.0	v,m³/kg	0.01034		0.01404	0.01723	0.01975	0.02196	0.02593	0.02956	0.03301
(615.4)	H,kJ/kg	2610.1		2868.5	3077.4	3239.6	3384.3	3652.0	3907.7	4159.4
	s,kJ/(kg K)	5.3090		5.7192	6.0295	6.2536	6.4404	6.7559	7.0254	7.2653
20.0	v,m³/kg	0.00584		0.00790	0.01156	0.01386	0.01575	0.01900	0.02188	0.02456
(639.0)	H,kJ/kg	2409.5		2625.1	2961.0	3159.4	3323.0	3611.7	3879.3	4138.5
	s,kJ/(kg K)	4.9265		5.2616	5.7622	6.0363	6.2477	6.5881	6.8702	7.1172
30.0	v,m³/kg	0.00543	0.00786	0.00951	0.01087	0.01208	0.01318	0.01421	0.01518	0.01612
	H,kJ/kg	2633.3	2973.5	3189.4	3367.6	3528.0	3678.1	3821.6	3960.6	4096.5
	s,kJ/(kg K)	5.1776	5.6490	5.9280	6.1442	6.3276	6.4900	6.6373	6.7729	6.8993
40.0	v,m³/kg	0.00260	0.00483	0.00640	0.00760	0.00863	0.00954	0.01039	0.01117	0.01192
	H,kJ/kg	2221.6	2752.8	3043.7	3258.4	3441.6	3607.8	3763.3	3911.6	4054.6
	s,kJ/(kg K)	4.5363	5.2720	5.6483	5.9089	6.1185	6.2982	6.4578	6.6025	6.7357
50.0	v,m³/kg	0.00203	0.00323	0.00459	0.00567	0.00658	0.00738	0.00811	0.00878	0.00941
	H,kJ/kg	2074.8	2535.1	2893.2	3147.2	3354.5	3537.3	3705.1	3862.8	4013.2
	s,kJ/(kg K)	4.2943	4.9293	5.3926	5.7009	5.9380	6.1358	6.3080	6.4619	6.6019
60.0	v,m³/kg	0.00183	0.00251	0.00350	0.00444	0.00526	0.00597	0.00661	0.00720	0.00775
	H,kJ/kg	2013.6	2387.9	2754.9	3039.5	3269.1	3468.0	3647.8	3814.7	3972.3
	s,kJ/(kg K)	4.1795	4.6954	5.1697	5.5152	5.7779	5.9930	6.1776	6.3405	6.4872
70.0	v,m³/kg	0.00172	0.00217	0.00287	0.00364	0.00435	0.00498	0.00556	0.00609	0.00658
	H,kJ/kg	1977.8	2301.9	2644.3	2941.8	3188.5	3401.4	3592.2	3767.9	3932.5
	s,kJ/(kg K)	4.1031	4.5498	4.9920	5.3530	5.6353	5.8656	6.0615	6.2329	6.3861
80.0	v,m³/kg	0.00164	0.00197	0.00248	0.00310	0.00371	0.00427	0.00479	0.00527	0.00571
	H,kJ/kg	1953.8	2248.4	2563.0	2858.7	3115.3	3339.0	3539.4	3722.9	3893.9
	s,kJ/(kg K)	4.0448	4.4510	4.8571	5.2159	5.5094	5.7514	5.9571	6.1362	6.2953
100.0	v,m³/kg	0.00153	0.00176	0.00207	0.00247	0.00291	0.00335	0.00377	0.00416	0.00453
	H,kJ/kg	1923.7	2186.1	2461.2	2736.7	2995.5	3230.7	3444.2	3640.0	3821.8
	s,kJ/(kg K)	3.9567	4.3186	4.6735	5.0077	5.3037	5.5581	5.7772	5.9684	6.1375

Courtesy of Department of Mechanical Engineering, Stanford University

Table 6-7c Properties of Superheated Steam and Compressed Water (Temperature and Pressure), SI Units. Source: Babcock & Wilcox

Chapter Seven

Boilers

A **boiler** is a water-filled pressure vessel that, with the addition of heat, produces hot water or steam. Commonly, the term boiler refers to a system that transfers water into the pressure vessel, provides a source of controlled heat, and transfers resulting steam to a piping system for delivery to points of use. Boiler is sometimes used interchangeably with **steam generator**, although the latter is usually reserved for large industrial-type boilers. The term boiler may also refer specifically to the section of a steam generator in which water is evaporated. Functionally speaking, a boiler is a type of **heat exchanger**, a device used to transfer heat from one fluid to another. **Water heaters** or **hot water boilers** are systems that provide a temperature rise without evaporation.

Boiler Operation

Water is the working fluid in the boiler, as well as the product, and provides the heat sink for cooling boiler components. The overall flow-path of water in the boiler is an open system. A regulating valve is used to maintain the required water level in the boiler and control entering feedwater. When heat is added, the water temperature rises and its volume increases slightly. At the boiling temperature, the water begins to boil isothermally (at constant temperature) while the volume increases greatly — this is **evaporation.** Finally, after all of the water has evaporated to steam, the volume continues to increase and the temperature rises — this is **superheating**.

The heat transfer sections of a conventional steam generator may be divided into three functional components: an economizer, which uses recovered energy to heat the feedwater nearly to its boiling point, an evaporator, and a superheater. Very large boilers may use several stages of superheaters, evaporators, and economizers to maximize the thermal efficiency of boiler and power generating systems.

In the basic large boiler system, saturated or superheated steam at a set pressure and temperature exits the boiler into a central steam header and then passes to various branches in the distribution system. Steam pressure may be reduced at local **pressure reducing stations** to meet the requirements of various loads. At the point of use, the steam condenses to liquid water with a transfer of heat to the designated load. Liquid **condensate** is returned to the boiler through the deaerator to start the process over again.

Boiler feedwater may consist of return condensate, fresh makeup water or a mixture of both. Where nearly all condensate is returned, only a small percentage of makeup water is required. But in systems that include large steam and condensate losses, makeup may comprise a high proportion of boiler feedwater.

Impurities that contribute to scaling and fouling of boiler surfaces are removed through treatment of boiler feedwater and through a blowdown process. Impurities are broadly classified as suspended or dissolved organic and inorganic matter and dissolved gases. The concentration of impurities is typically expressed in terms of the parts by weight of the constituent per million parts of water (ppm).

Water treatment may include filtering to remove suspended solids and chemical treatment for removal of hardness, or dissolved minerals. Calcium, magnesium, and silica are the principal scale-forming impurities that must be chemically removed. In the **blowdown** process, a continuous bleed removes a small percentage of boiler feedwater from the system (usually just before the boiler evaporator section) to prevent impurities contained in makeup water from concentrating in the system to damaging levels.

Treated makeup water is added before the feedwater enters the **deaerator,** where steam is used to strip dissolved oxygen from the feedwater by direct contact, typically at 224°F (107°C) or above, the point of minimum solubility of oxygen in water. At the deaerator discharge, boiler feed pumps pressurize the feedwater to boiler operating pressure. To avoid cavitation (the flashing of water to steam in the impeller of the feedwater pumps), the deaerator must be located at an elevation greater than the feedwater pumps. Recirculating lines are used to protect the pumps from overheating by allowing a minimum flow to be maintained at all times.

Figures 7-1 through 7-3 illustrate typical water conditioning, feedwater pumping, and deaeration systems.

Fig. 7-1 Boiler Water Conditioner Module on Skid.
Source: Cleaver Brooks

Fig. 7-2 Packaged Boiler Feedwater System.
Source: Cleaver Brooks

Consider the operation of an industrial gas-fired boiler designed to provide superheated steam at 600 psia (41.4 bar) and 850°F (454°C). Figure 7-4 provides a cutaway view and Figure 7-5 a sectional view of such a boiler. Combustion of natural gas yields a combustion gas stream at a temperature of about 3,600°F (1,982°C) and the minimum gas outlet temperature is set at about 310°F (154°C).

Figure 7-6 graphs steam-water temperature versus enthalpy (total heat) for the example boiler. As shown, the feedwater is supplied at a temperature of 280°F (138°C). About 20% of the total heat absorbed in the process is required to raise the water from its inlet temperature to its saturation temperature of 490°F (254°C). Another 60% is used to evaporate the water to produce saturated steam. The remaining 20% of absorbed heat is used to raise the steam to the superheated temperature of 850°F (454°C).

In general, boiler input refers to the total amount of fuel energy input to the burners, typically expressed in

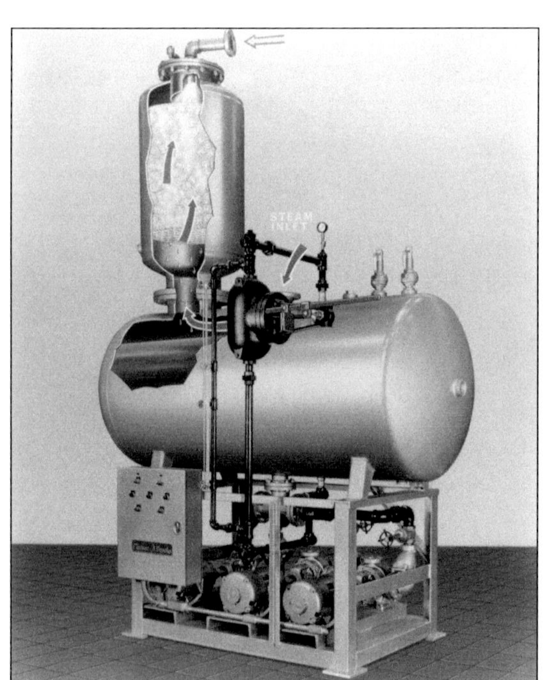

Fig. 7-3 Cutaway View of Packaged Deaerator.
Source: Cleaver Brooks

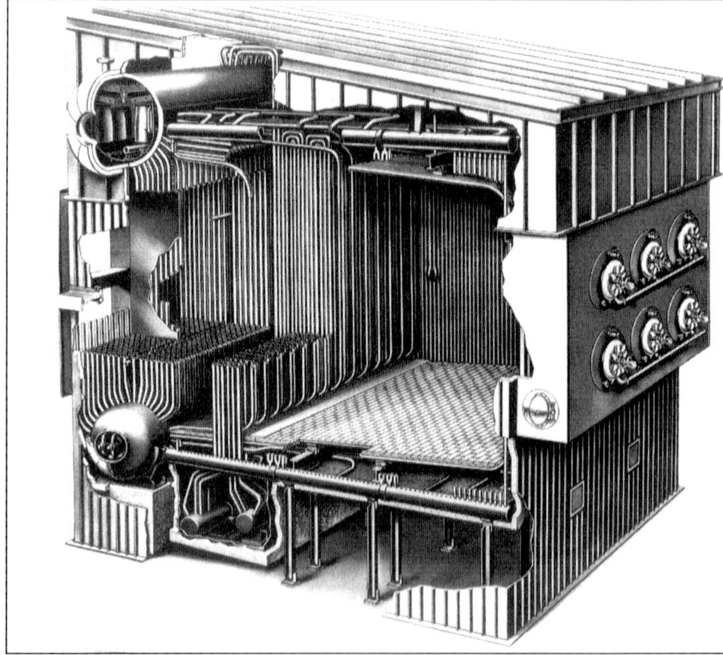

Fig. 7-4 Cutaway View of Gas-Fired Industrial Boiler.
Source: Babcock & Wilcox

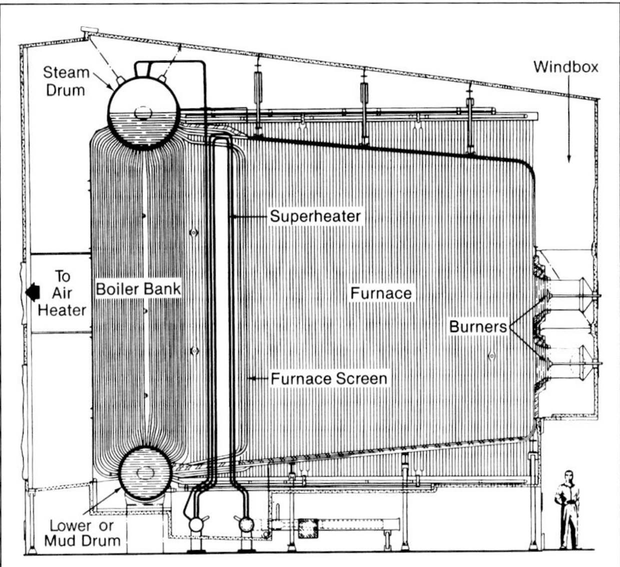

Fig. 7-5 Section View of Gas-Fired Industrial Boiler.
Source: Babcock & Wilcox

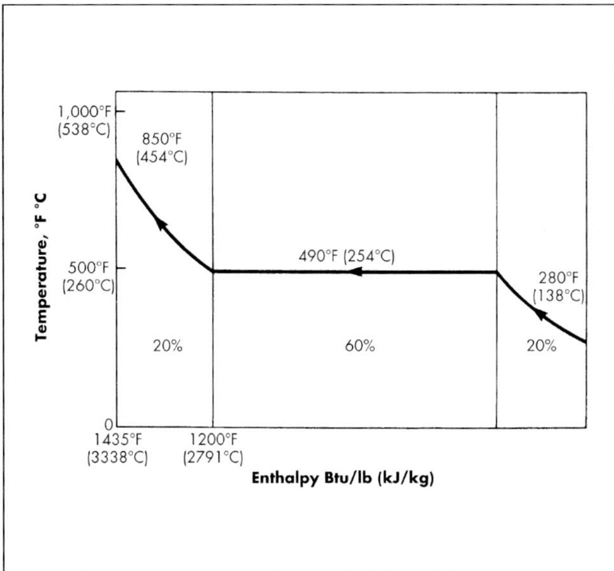

Fig. 7-6 Steam-Water Temperature Profile for Gas-Fired Industrial Boiler.
Source: Babcock & Wilcox

Btu/h or kJ/h (HHV basis). Boiler capacity or output may be specified as heat (Btu or kJ per hour), as steam flow (lbm or kg of steam per hour) or as boiler horsepower (BHP). Boiler horsepower is defined differently from mechanical horsepower; the following relationships are commonly assumed (at 212°F or 100°C):

1 BHP	=	33,472 Btu/h
1 BHP	=	34.5 lbm/h
1 BHP	=	9.8 kW/h
1 BHP	=	35,291 kJ/h

HEAT TRANSFER

Boilers are devices used to transfer heat. As defined in Chapter 2, **heat** is thermal energy that is transferred across the boundary of systems with differing temperatures, always in the direction of the lower temperature. Heat transfer occurs when two adjacent bodies of mass are not in equilibrium due to a difference in temperature.

There are three general modes of heat transfer: conduction, convection, and radiation. One or more of these accounts for heat transfer in all applications. All three methods are important in the process of steam generation.

Conduction is the transfer of heat energy through a material due to a temperature difference across it. This is consistent with the second law of thermodynamics statement that heat flows in the direction of decreasing temperature. For fluids, conduction takes place by molecular impact; for solid nonconductors, by molecular vibration; and for metals, primarily by electronic movement. The rate of heat transfer through materials, due to a temperature difference for one-dimensional conduction, is given by Fourier's law, and is expressed as:

$$Q_k = -kA\frac{dT}{dx} \quad (7\text{-}1)$$

Where:

Q_k = Rate of heat transfer by conduction
A = Area normal to the direction of heat flow
T = Temperature
x = Distance along the direction of heat flow
k = A property of the material called thermal conductivity
dT/dx = Temperature gradient

Consistent with the second law, the flow of heat is positive when the temperature gradient, $\Delta T/L$, is negative. The minus sign indicates that heat flow is in the direction of decreasing temperature.

The discrete form of Fourier's law derived by integrating the differential form with a constant temperature gradient is shown as Figure 7-7 and is expressed as:

$$Q_k = \frac{kA}{L}(T_1 - T_2) \quad (7\text{-}2)$$

Where:

L = Length in the direction of heat flow
T_1 & T_2 = Temperatures of the two surfaces

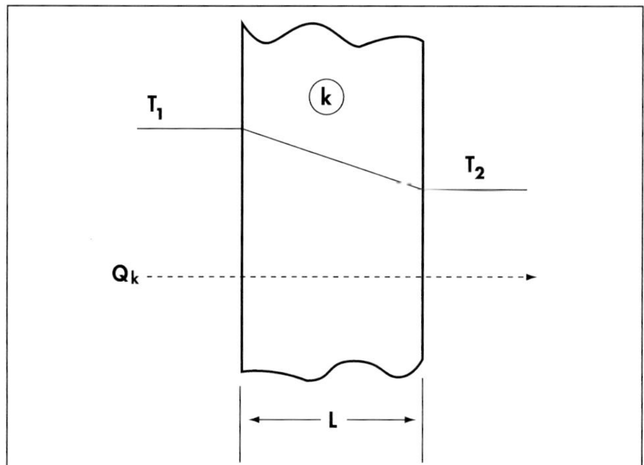

Fig. 7-7 Conduction Heat Transfer.

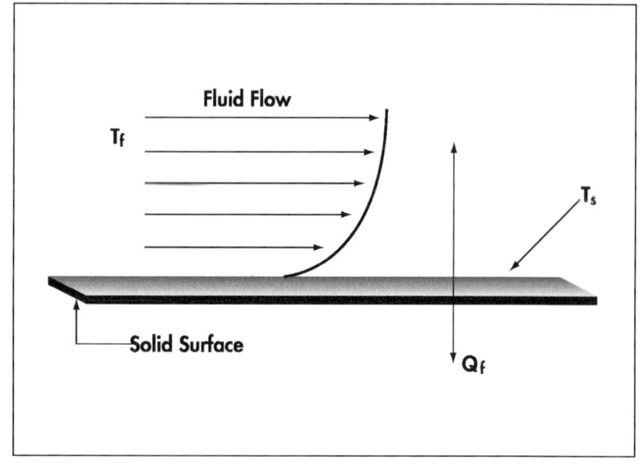

Fig. 7-8 Convection Heat Transfer.

Thermal conductivity (k), a property of the material, quantifies its ability to conduct heat. It is expressed as Btu/h • ft • °F (W/m • °K). Boilers are designed to optimize conduction of heat from combustion gases to liquids or vapors via a metal medium. Generally, thermal conductivities are highest for solids, lower for liquid, and lower still for gases. For metals, k is extremely high: k for pure metals ranges from 30 to 240 (52 to 415) and for alloys from 8 to 70 (14 to 121). Copper, for example, has a k value of 223 (386), while iron and carbon steel are 42 and 25 (73 and 43), respectively. For gases, k is extremely small. Air, for example, has a k value of 0.014 (0.024). For nonmetallic liquids, k ranges from 0.05 to 0.40 (0.087 to 0.69). Water, for example, has a k value of 0.32 (0.55).

Convection is the transfer of heat within a fluid (gas or liquid) due to a combination of molecular conduction and macroscopic fluid motion. Convection occurs adjacent to heated surfaces as a result of fluid motion past the surfaces. As a fluid is heated, its density typically decreases. If part of a fluid mass is heated, the cooler, denser portion acts to displace the heated portion. A convection current is the continuous flow of cooler fluid to, and heated fluid from, the heated area. Natural convection occurs when the fluid motion is due to these local density differences alone. Forced convection occurs when mechanical forces from devices such as fans give motion to the fluids. The rate of heat transfer by convection can be shown as Figure 7-8 and expressed as:

$$Q_f = hA\,(T_s - T_f) \qquad (7\text{-}3)$$

Where:

Q_f = Rate of heat transfer by convection

h = Local heat transfer coefficient

A = Surface area

T_s = Surface temperature

T_f = Fluid temperature

Like thermal conductivity, the heat transfer coefficient (h) ranges widely between substances and may depend on the temperature of the fluid and surface. It is expressed as Btu per hour per square foot per °F (Btu/h • ft^2 • °F) or Watt per square meter per °K (W/m^2 • °K). For air (free convection), h ranges from 1 to 5 (6 to 28) and for forced convection, h typically ranges from 5 to 50 (28 to 280). For steam (forced convection), h ranges from 300 to 800 (1,700 to 4,500). For water (forced convection), h ranges from 50 to 2,000 (280 to 11,000) and for water (boiling), h ranges from 500 to 20,000 (2,800 to 114,000).

Radiation is the transfer of energy between bodies by electromagnetic waves. Solely the temperature of a body produces electromagnetic radiation and, unlike conduction and convection, radiative transfer requires no intervening medium. All surfaces whose temperatures are above absolute zero continuously emit thermal radiation. Electromagnetic waves that result from conversion of energy at the body's surface emanate from the surface and strike another body. Some of the thermal radiation is absorbed by the receiving body and converted into internal energy. The remaining portion is reflected from or transmitted through the body. For a perfect radiator, radiant energy transfer is expressed, based on the Stefan-Boltzmann law, as:

$$Q_R = A\,\sigma\,T_s^4 \qquad (7\text{-}4)$$

Where:

Q_R = Rate of heat transfer by radiation

A = Surface area

σ = Stefan-Boltzmann constant (in English engineering units σ = 0.1713 x 10^{-8} Btu/h • ft² • °R⁴ and in SI units σ = 5.67 x 10^{-8} W/m² • °K⁴)

T_S^4 = Absolute surface temperature

The rate of radiant energy transfer from a body is a function of the temperature as given above and the nature of the surface. A perfect radiator, or **blackbody**, absorbs all of the radiant energy reaching its surface and emits radiant energy at the maximum theoretical rate. The net radiation heat transfer between two blackbody surfaces in a vacuum or non-interactive gas is expressed as:

$$Q_R = AF\sigma(T_1^4 - T_2^4) \tag{7-5}$$

Where F is the configuration factor and represents the fraction of radiant energy leaving surface 1 that directly strikes surface 2, and T_1 and T_2 are the absolute surface temperatures. Radiation heat transfer is shown diagrammatically in Figure 7-9.

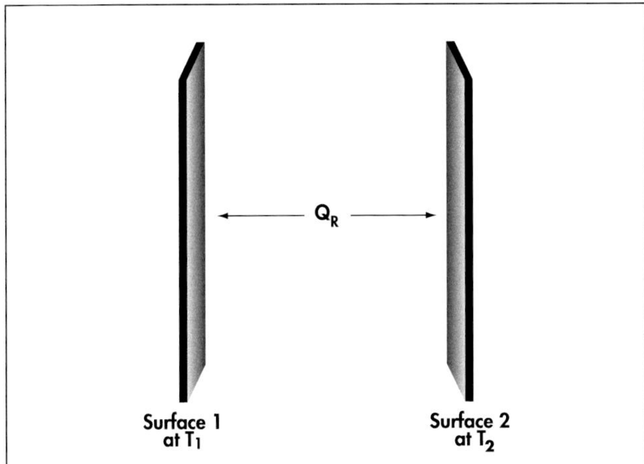

Fig. 7-9 Radiation Heat Transfer.

A real radiator, or graybody, absorbs less than 100% of the energy incident on it. The heat transfer by radiation of a graybody is expressed as:

$$Q_R = \varepsilon A \sigma T_S^4 \tag{7-6}$$

where ε is emissivity, or the emittance of the surface, and can be expressed as:

$$\varepsilon = \frac{\textit{Energy actually radiated by the system}}{\textit{Energy radiated if the system were a blackbody}} \tag{7-7}$$

Like k and h, ε values also vary widely among materials. For polished metals ε ranges from 0.01 to 0.08. For oxidized metals, ε ranges from 0.25 to 0.7. For special paints, ε can be 0.98 or even higher, though it is always less than 1.

Heat Exchangers

As noted above, a boiler is a type of heat exchanger, a device in which energy exchange occurs between two different fluids. There are three basic heat transfer arrangements: parallel flow, counterflow, and crossflow. In parallel flow, both fluids enter at the same relative location and flow in parallel paths over the heating surface. In counterflow, the two fluids enter at opposite ends and flow in opposite directions over the surface. In crossflow, the paths of the two fluids are generally perpendicular to one another. Counterflow is the most efficient heat exchanger arrangement, producing the greatest temperature difference and requiring the least heating surface.

The heat transfer rate, Q, can be related to the characteristics of the heat exchanger by the equation

$$Q = UAF\Delta T_{LMTD} \tag{7-8}$$

Where:

U = Overall heat transfer coefficient

A = Surface area

F = Arrangement factor

ΔT_{LMTD} = Mean temperature difference

The overall heat transfer coefficient, U, expressed in Btu/h • ft² • °F (W/m² • °K), is the coefficient used to express the aggregate effect of all the conductances. The U value of a pane of glass, for example, is about 1. The U value for a water-to-water heat exchanger typically ranges from 150 to 300 (850 to 1,700). The U value for steam-to-gas ranges from 5 to 50 (28 to 280) and for water-to-gas from 10 to 20 (57 to 114).

The log mean temperature difference is defined as:

$$\Delta T_{LMTD} = \frac{\Delta T_1 - \Delta T_2}{ln\,(\Delta T_1 / \Delta T_2)} \tag{7-9}$$

Where:

ΔT_1 = Initial temperature difference between the hot and cold fluids

ΔT_2 = Final temperature difference between these media

ln = Natural logarithm

The log mean temperature difference depends on the relative directions of flow of the fluids as they pass through or over the surface; that is, fluids may flow counter to, parallel to, or across one another. Figure 7-10 shows the three flow arrangements and presents Equation 7-9 written specifically for each case.

Fouling factors must be considered in the deter-

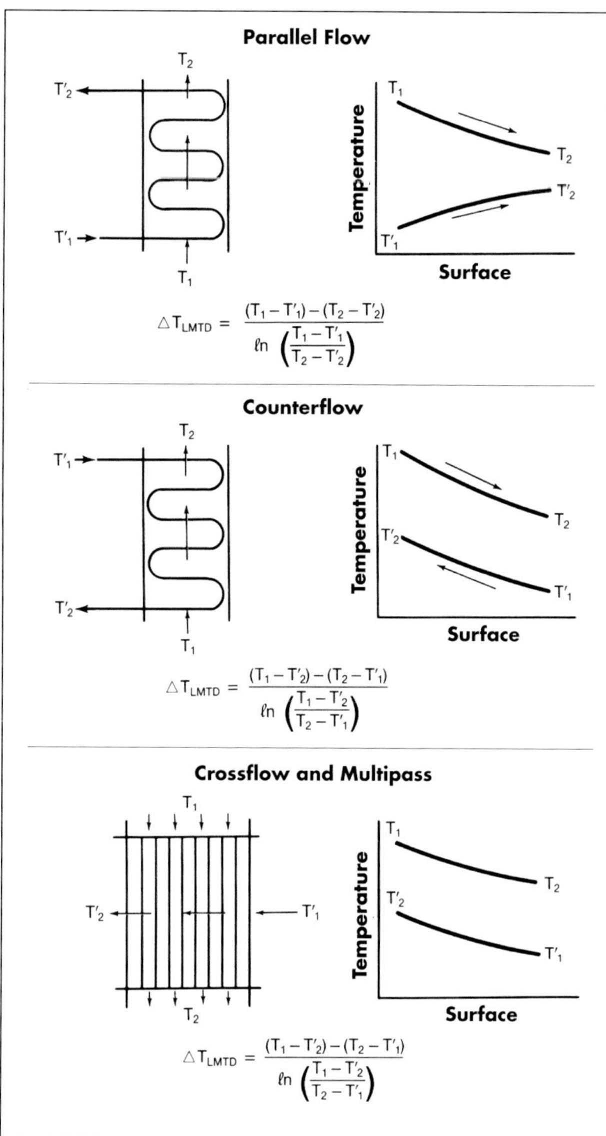

Fig. 7-10 Log Mean Temperature Difference for Basic Heat Exchanger Arrangements. Source: Babcock & Wilcox

mination of the size of heat transfer (exchange) materials. Fouling is a site- and material-specific heat exchange parameter related to the condition of the surface of the heat exchange material. Fouling can result from external tube deposits, such as oil films or solids from a liquid or gaseous stream, internal tube-surface scaling due to precipitation of solid compounds from solution, or corrosion of surfaces, external or internal. Fouling reduces the effectiveness of heat transfer, because it reduces the value of the overall heat transfer coefficient as shown by the equation:

$$\frac{1}{U_f} = R_f + \frac{1}{U} \qquad (7\text{-}10)$$

Where:

U_f = Overall heat transfer coefficient after fouling has occurred

U = Overall heat transfer coefficient of the clean exchanger

R_f = Fouling factor

The fouling factor is a measure of the decrease in heat transfer versus a clean surface resulting from fouling. It is a design parameter — with typical values ranging from 0.001 to 0.01 hr • ft^2 • °F/Btu (0.0002 to 0.002 hr • m^2 • °K/Wh) — included in the specification of all heat exchangers. Important considerations are the chemical composition of the exhaust stream and the exit temperature.

Boiler Design

Boiler selection and design are affected by many factors. Capacity, operating temperatures and pressures, and fuel type are among the most significant. Factors such as duty cycle and anticipated range of operating loads, as well as water quality and emissions control requirements, must also be carefully considered.

A number of codes, standards, laws, and regulations cover boilers and related equipment. The boiler industry is largely regulated by the American Society of Mechanical Engineers (ASME) codes. These codes govern boiler design, inspection, and quality assurance. Numerous other codes and requirements may be mandated by individual insurance companies and by federal and state regulations. These include environmental standards, operating standards, and sewer discharge temperature regulations, to name a few. The selection process must therefore include a review of all requirements to ensure safe, proper, and legal operation.

The following are descriptions of boiler types and design features. Both firetube and watertube boilers are discussed, with a more detailed focus on watertube designs. While the descriptions relate to fuel-fired boilers, many of the same design features apply to non-fired and partially fired heat recovery steam generators. Heat recovery is discussed in more detail in the following chapter.

Firetube Boilers

A firetube boiler directs hot combustion gases through tubes surrounded by water within a boiler shell. The firetube design provides abundant heat transfer surface and helps distribute steam formation uniformly throughout the mass of water. During operation, firetube boilers respond fairly quickly to load swings because they contain a large reservoir of water at the saturation point.

Most firetube boilers are designed to burn liquid and gaseous fuels, but some burn solid fuels.

Firetube boilers can be designed with one or more passes of combustion gas through the length of the boiler. Increasing the number of passes increases overall thermodynamic efficiency. However, as with any heat exchanger design, a point of diminishing returns is reached when the cost of added surface area outweighs the efficiency benefits. Four passes are usually the economical limit. Figure 7-11 shows a cutaway view of a typical firetube package boiler, featuring three passes.

Fig. 7-11 Cutaway View of Three-Pass Packaged Firetube Boiler. Source: Donlee Technologies

Due to size, cost, and other design limitations, firetube boilers usually do not exceed capacities of 50,000 lbm (23,000 kg) of steam per hour and saturated pressures are generally limited to below 250 psig (18.3 bar). They are commonly used for hot water, as well as low-pressure and high-pressure steam applications. Disadvantages of firetube boilers include the inability to generate superheated steam, prolonged startup and shutdown time, and possible inefficiency due to the inherent thermal inertia of the design.

Water Flow Path in Firetube Boilers

The natural circulation flow path of water in the firetube boiler is developed by the positioning of internal baffles, which direct the movement of the water around the tubes, and the location of the feedwater supply. Internal flow is largely the result of the water temperature rise. The water changes density when it is heated from relatively cold feedwater to the high-temperature required to change state and become steam. The warm water begins to rise across the tubes, coming in contact with the second pass firetubes and then the first pass tube.

The energy required to complete the phase change of the water to steam is absorbed between the second and first passes. The boiling action becomes increasingly violent as the water/steam mixture rises to the top of the boiler. As the mixture passes around the third pass firetubes, steam bubbles break free from the water into the boiler steam space. Steam rising through the hot water is scrubbed, reducing the moisture content to less than 1%. The steam then moves through the baffled dry pan, eliminating most of the entrained liquid water.

Fuel Flow Path in Firetube Boilers

Since the fuel supplied to the fired boiler contains the chemical energy required for heat generation, the regulation of fuel flow dictates the output of the boiler. The burner control system is designed to regulate firing rate to ensure that the amount of energy input is equal to the amount of energy removed from the boiler in the steam.

When operating on natural gas, the gas is supplied to the gas pressure regulator, which reduces the gas pressure to the level that is required by the burner. In a typical system, the reduced-pressure fuel gas passes through dual isolation valves designed to automatically interrupt gas flow under unsafe conditions. The gas then passes through a butterfly valve, which is controlled to automatically modulate fuel flow to the burner. Upon entering the burner, the gas is distributed into the gas ring, where it is injected into the combustion air.

When operating on fuel oil, the oil is supplied under pressure to the burner by an oil pump. The oil is sprayed into the furnace through a nozzle that atomizes the oil into very small droplets. Oil atomization may be accomplished in several ways, such as forcing the oil through a nozzle at high pressure, use of a spinning nozzle, or through injection into a high-speed stream of air or steam.

modulates to maintain required pressurization to ensure proper combustion gas flow out of the boiler. Figure 7-12 is a diagram showing the path followed in a three-pass firetube boiler.

Figure 7-13 shows a cutaway view of a four-pass firetube boiler. Figure 7-14 shows a firetube boiler, featuring a cyclonic burner, designed to enhance heat transfer due to the intimate burner flame contact on the firetube wall.

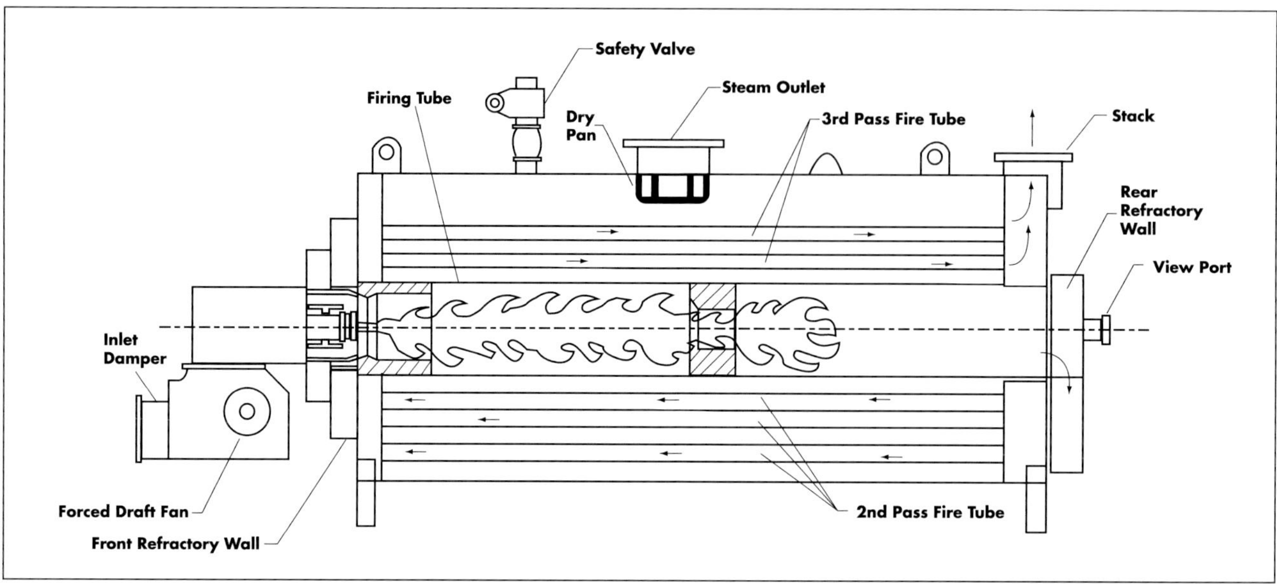

Fig. 7-12 Labeled Diagram of Packaged Firetube Boiler. Source: Donlee Technologies

Combustion Air in Firetube Boilers

The combustion air system supplies oxygen to the burner for burning the fuel. The air in a typical system is controlled by a series of dampers and fans. An inlet damper usually controls the volume of air supplied to a forced draft fan, which provides the driving force for the air injection into the boiler furnace. The air is compressed and forced into the burner tube assembly and whirl assembly, which forms a vortex pattern to promote proper mixing. The resulting air-fuel mixture combusts in the furnace and the hot gases and combustion byproducts are forced through the firetubes.

In a typical three-pass firetube boiler, the first pass is the furnace tube through the center of the boiler to the rear refractory wall, which then redirects the hot gas flow toward the second-pass tubes on the bottom of the boiler shell. The second-pass tubes carry the hot gases forward to the turning box area, which directs them into the third-pass firetubes and to the exhaust breeching and stack. The breeching contains a draft control unit, composed of a pressure sensing device and a damper. The draft damper

Watertube Boilers

Watertube boilers are generally larger than firetube units and have a completely different construction. They generally contain about 75% less water than comparable firetube boilers, can handle higher pressures (well in excess of 1,000 psi or 69 bar), and generally range in capacity from 10,000 lbm (4,500 kg) to several million lbm (or kg) of steam per hour.

Watertube boilers use individual water and steam drums with a multitude of connecting tubes through which a steam and water mixture circulates. Basic sections of the watertube boiler include the following:

Insulating walls house the boiler components. They contain the hot flue gases and are insulated to minimize heat losses from the boiler.

In a fired boiler, the **burner** and **furnace area** receive fuel supplied by the burner in a condition to burn and begin the

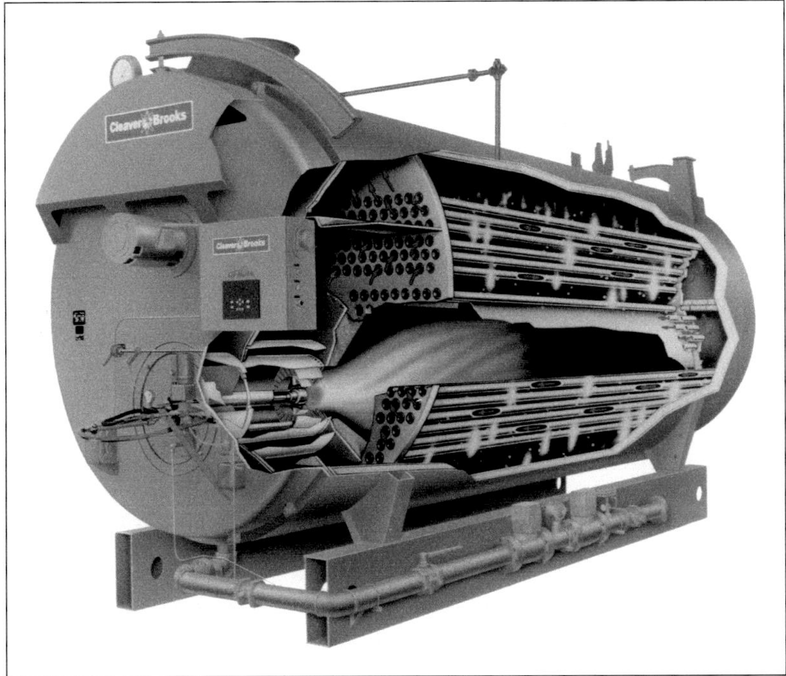

Fig. 7-13 Cutaway View of Four-Pass Firetube Boiler. Source: Cleaver Brooks

steam generation process. The furnace is generally located near the bottom of the boiler to allow the heat to rise, giving maximum exposure to the watertubes. Furnaces are designed to promote complete burning of fuel and must be capable of withstanding high temperatures and pressures.

The **steam drum** is the vessel in which the saturated steam is separated from the steam-water mixture and into which feedwater is introduced. The water separated from the mixture is recirculated with feedwater back to the heat-absorbing surfaces.

Waterwalls are located immediately inside the insulating walls. They are composed of the watertubes directing the feedwater in the boiler. In a typical large industrial boiler, the furnace area is surrounded by as much heat-absorbing tubing as possible. The waterwall surface constitutes only a small portion of the total heat transfer surface, but accounts for a large portion (as much as half) of the total heat absorption. Exposure to radiant heat in the region of highest temperature results in the highest rates of heat transfer of all of the boiler sections.

The waterwalls are composed of two types of watertubes: risers (heated legs) and downcomers (unheated legs). As steam forms in the risers, the density of the steam-water mix decreases, allowing it to be displaced by the heavier water in the downcomers. This results in a continuous circulation flow of water from the steam drum, through the downcomer, and up the riser back to the drum.

The **superheater** section is where saturated steam from the upper steam drum is further heated. The superheater is usually located in the upper portion of the furnace space, where temperatures are still quite high and where there is still some effective radiation. A superheater is a single-phase heat exchanger with steam flowing inside the tubes and combustion gases passing outside. The superheater may have nearly the same amount of heat transfer surface as the waterwalls, but accounts for a much smaller portion of the total heat absorbed.

Reheaters are used in very large boilers designed for power generation plants to reheat steam that has already been partially expanded in a steam turbine.

The **convective section (boiler)**, which is the main convective steaming surface (boiler section), is generally located after the superheater section, where it is exposed to lower temperatures. This section normally accounts for a high proportion of the total heat transfer surface in the boiler.

Fig. 7-14 Firetube Boiler Featuring Cyclonic Burner. Source: Donlee Technologies

Economizer sections are used for transfer of lower-level heat to feedwater using combustion gas near the boiler outlet. This section comprises a small percentage of the total heat transfer surface (maybe 10 %) and accounts for an even lower percentage of total heat absorption. Still, it is critical for maximizing efficiency. Economizers may have bare or finned tubes to facilitate convective heat transfer, with bare tubes used for dirtier fuels. Tubing is usually low-carbon steel. Some smaller, low-pressure boilers use cast-iron economizers.

Air heaters may be used downstream of the economizer to further extract energy from the combustion gases before they pass through the stack. The air heater requires a very large heat transfer area because of the lower temperature gas stream and the poor gas-to-gas heat transfer rate. Generally, air heaters are used with large solid-fuel boilers because they provide the added benefit of using hot air to evaporate moisture from the solid fuel, thereby allowing for more rapid and efficient combustion. Air heaters do not necessarily have to be pressure vessel type heat exchangers with large surface area requirements. One type of air heater is a heat wheel in which rotating metal surfaces contact the exhaust gas stream, where they absorb heat, and the incoming fresh air stream, where they discharge heat.

An additional component that may be added to a steam generation system is the **desuperheater**. A desuperheater may be used if the superheater, which is designed for full-load conditions, produces excess temperature during operation at partial loads. Desuperheaters cool superheated steam by spraying water into the piping either ahead of or behind a superheater (or reheater) section. If located downstream of the superheater, the desuperheater will condition the steam to remove entrained moisture before it is passed along to process or to the steam turbine.

Figure 7-15 is a temperature-enthalpy diagram for a very large subcritical boiler used to supply steam for power generation. This example illustrates the relative heat absorption for each boiler section. Heat absorption for water preheating, evaporation, and superheating are 30%, 32%, and 38%, respectively.

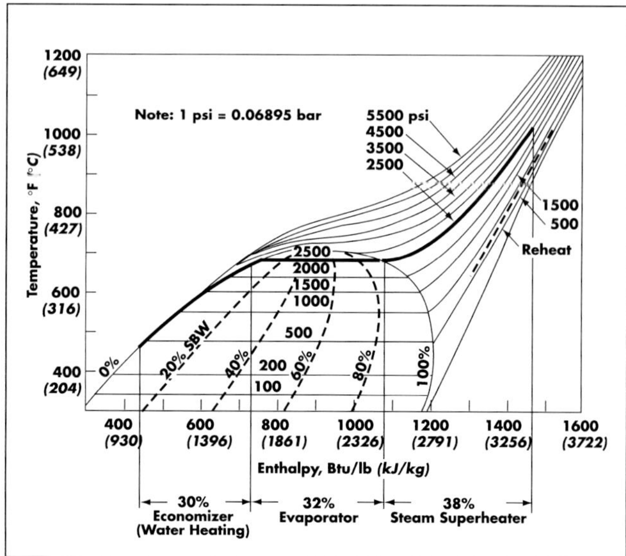

Fig. 7-15 Temperature-Enthalpy Diagram for Large Boiler Used for Power Generation Plant. Source: Babcock & Wilcox

Watertube Boiler Circulation

The purpose of steam-water flow circuitry is to provide the required steam output at the specified temperature and pressure. The circuitry flow also provides cooling of the tube walls under expected operating conditions. Four of the most common systems are illustrated in Figure 7-16. These systems are typically classified as either recirculating or once-through. In recirculating systems, water is only partially evaporated into steam in the boiler tubes. The residual water, plus the makeup water supply, are then recirculated to the boiler tube inlet for further heating and steam generation. Once-through systems provide for continuous evaporation of slightly subcooled water to 100% steam without steam-water separation.

Natural-circulation boilers depend on the difference in fluid density to drive circulation. Cold water is forced to the bottom of the boiler and hotter

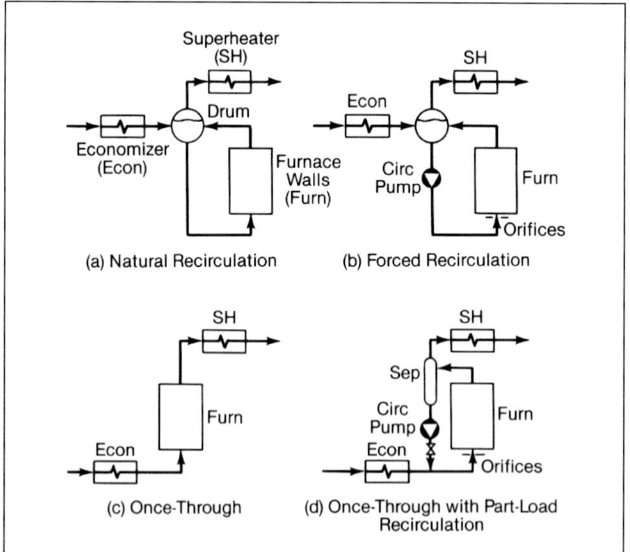

Fig. 7-16 Common Watertube Boiler Circulation Systems. Source: Babcock & Wilcox

water or steam-water mixtures to the top and into the steam separation drum. Forced circulation boilers use pumps to force water to flow through the tubes. These are commonly used when space limitations require a more compact design, or when tubes must be placed horizontally rather than vertically. Because forced circulation units can include subdivisions of pressure parts, they can be used to handle large capacities and high pressures. Figure 7-17 provides elevation and plan view illustrations of convection tubes in an industrial watertube boiler.

The total circulation rate potential depends on several factors: operating pressure, boiler height, heat input rate, and cross-sectional (free flow) area. The force that produces circulation is equal to the difference in weight between the equal columns of water and steam-water mixture. The greater the pressure, the greater the density, and the force-producing circulation is diminished. As pressure rises, natural circulation becomes increasingly more difficult and forced circulation is more likely to be required. Greater boiler height generally produces a greater total flow rate, because it produces a greater total pressure difference between the risers and downcomers.

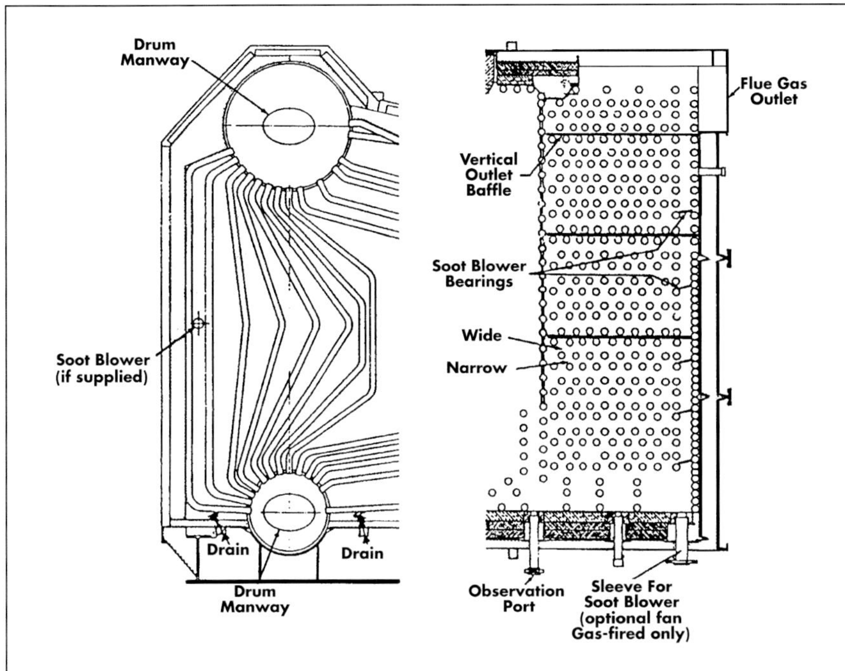

Fig. 7-17 Elevation and Plan View of Convection Tubes in Industrial Watertube Boiler. Source: Cleaver Brooks

Figure 7-18 shows simple natural and forced circulation loops for watertube boilers. In the natural circulation diagram, no steam is present in the unheated tube segment A-B. Heat addition generates a steam-water mixture in segment B-C. Gravity causes the more dense water to flow downward in segment A-B, which causes the steam-water mixture in segment B-C to move upward into the steam drum. In the forced circulation diagram, a mechanical pump is added to the simple flow loop and the pressure difference created by the pump controls the water flow rate.

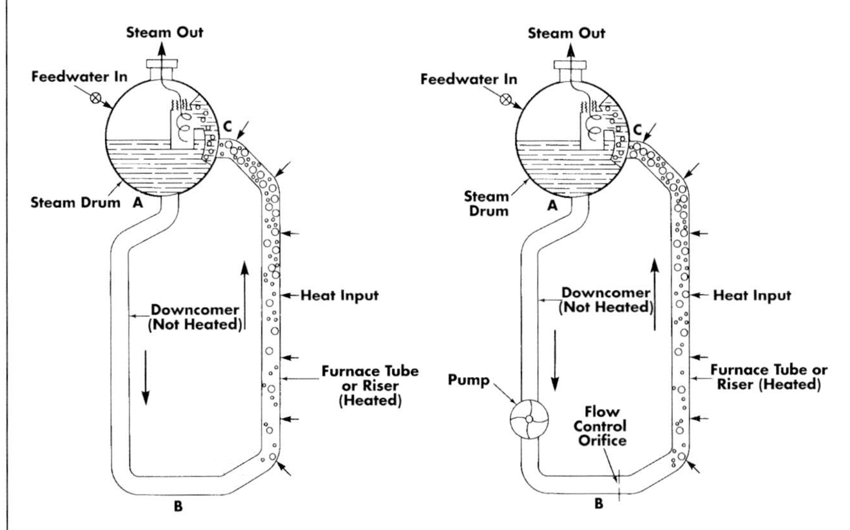

Fig. 7-18 Simple Natural and Forced Circulation Loops for Watertube Boiler. Source: Babcock & Wilcox

Figure 7-19 shows a packaged industrial watertube boiler featuring a convective steam drum and a cyclonic burner. This design is suitable for applications in the range of 25,000 to 100,000 lbm (11,000 to 45,000 kg) of steam per hour at pressures of up to 850 psi (59 bar). A boiler efficiency in excess of 84% is maintained over the operating range from full-rated capacity down to

Fig. 7-19 Packaged Watertube Boiler Featuring Cyclonic Burner and Convective Drum. Source: Donlee Technologies

10% of rated capacity. Figure 7-20 shows a cutaway diagram of this steam generator with labeled components.

Figure 7-21 shows four 500 Bhp forced circulation steam generators with a common exhaust. These are somewhat uncommon boilers that feature a coil-type design. Figure 7-22 provides a labeled illustration of the design of these units.

Once-Through Boilers

In natural- and forced-circulation boiler designs, much more water is circulated than steam is generated, and the drum serves to collect and release steam. The once-through boiler design consists of a single or several parallel serpentine tubes through which feedwater enters and saturated or superheated steam exits. Once-through boilers often have a separator to deliver saturated steam to the superheater and to return entrained liquid water to the feed pump section.

Once-through boilers are especially attractive for applications requiring pressures above the critical point at which water converts to steam without boiling. In once-through boilers, water flow is roughly proportional to the firing rate and is, therefore, inadequate to provide waterwall-tube protection at start-up and while operating at low loads. Bypass systems are used to allow for sufficient flow, or pumps are used to increase circulation.

Packaged Versus Field-Erected Boilers

Shop-assembled packaged boilers are available in capacities as large as several hundred thousand lbm (or kg) of steam per hour. Sizes and mechanical features are

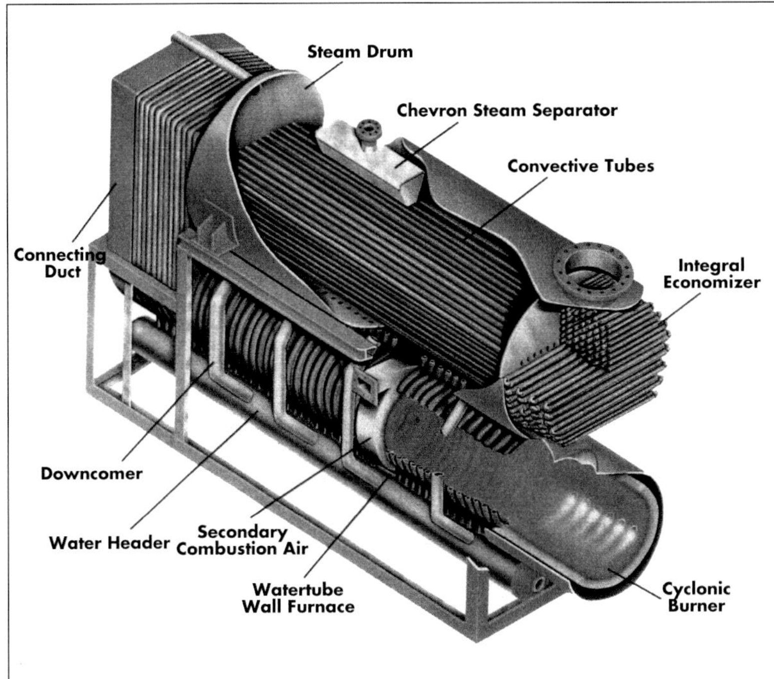

Fig. 7-20 Labeled Cutaway View of Packaged Watertube Boiler Shown in Fig. 7-19. Source: Donlee Technologies

often limited by transportation considerations. Generally, they are designed to burn liquid or gaseous fuels and are less expensive than field-erected units.

Packaged boiler designs usually seek to maximize the use of vertical or near-vertical waterwall tubes in the radiant section surrounding the furnace, as well as in the convective section. Economizers are commonly used and an emphasis is placed on maximizing the use of natural circulation. Most packaged units are designed for pressures ranging from 125 to 1,000 psig (9.6 to 70.0 bar), with temperatures of up to 950°F (510°C) in the larger units. As size, pressure, and temperature requirements increase, field-erected boilers are more practical than shop assembled units. Field-erected units are common for solid fuels in capacities ranging down to about 50,000 lbm (23,000 kg) of steam per hour. Figure 7-23 shows a basic commercial/industrial

Figure 7-21 Multiple 500 Bhp Coil-Type Forced Circulation Steam Generators. Source: Clayton Industries

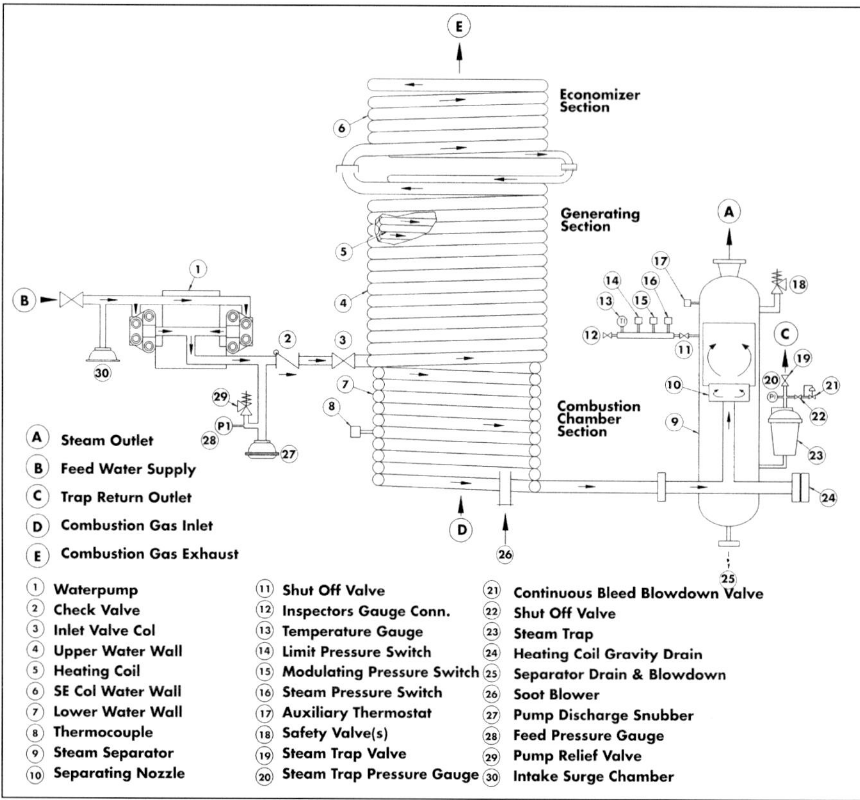

Fig. 7-22 Cutaway Diagram of 500 Bhp Coil-Type Forced Circulation Steam Generators. Source: Clayton Industries

Fig. 7-23 Basic Commercial/Industrial Packaged Watertube Boiler. Source: Cleaver Brooks

Fig. 7-24 Cutaway View of Integral Furnace Boiler–Membrane Wall Construction. Source: Babcock & Wilcox

packaged watertube boiler.

Figure 7-24 illustrates a shop-assembled two-drum bottom-supported boiler design. This is a forced draft unit capable of pressures of up to 1,800 psi (124 bar) and temperatures of 1,000°F (538°C). The design features the furnace on one side and boiler bank on the other, separated by a baffle wall. Burners are directed toward the rear wall where the gas turns 180 degrees and flows frontward to the gas outlet. The unit can be configured with an economizer or air heater to enhance thermal efficiency. Shop-assembled packaged watertube boilers, such as the one shown in Figure 7-25, are available for low-pressure steam and hot water applications in low capacities.

Fluidized-Bed Combustion

Fluidized-bed combustion (FBC) is a technology that has widespread appeal because of its low emission characteristics, particularly for systems that use sulfur-bearing fuels, such as coal, and its potential for improved heat transfer. Fluidized-bed configurations include bubbling-bed and circulating fluidized bed designs, with atmospheric- or elevated-pressure operation.

In an FBC unit, solid, liquid, or gaseous fuels, together with inert materials such as sand, silica, or alumina and/or sorbents, such as limestone, are kept suspended through the action of primary air distributed below the combustor floor. Fluidization promotes turbulence, which makes the mass of solids behave more like a liquid. Lower, more uniform distribution of heat results. Depending on the size of the combustion chamber,

FBC can release the same amount of heat as a conventional boiler, but at a lower temperature. The lower peak temperatures result in lower rates of formation of NOX emissions. The turbulence in the combustion zone allows for good combustion of solid fuels that are difficult to burn, including high ash waste coals, biomass and lignite.

The FBC boiler includes a specialized firing system that is surrounded by conventional waterwalls. In the bubbling-bed design, shown in Figure 7-26, steam-generating tubes are often set in the bed to achieve the desired heat balance and bed operating temperature. Steaming is controlled by adjusting the firing rate and height, temperature, and other primary bed parameters.

Circulating fluidized-bed designs do not have in-bed tubes, but may use an external heat exchanger and some type of hot-solids (cyclone) separator. The heat exchanger is a refractory-lined box with an immersed tube bundle designed to cool the solids returning from the separator. This compensates for variations in heat-absorption resulting from varying load conditions and fuel properties.

While the largest current and future use of FBC is for solid fuel-burning utility boilers, FBC is increasingly being used for industrial boilers. Growth in the industrial sector is due in part to the need to comply with Clean Air Act Amendment (CAAA) regulations.

Fig. 7-25 Low-Capacity Packaged Watertube Boiler. Source: Cleaver Brooks

Boiler Controls and Combustion Systems

Instrumentation and controls are essential to all systems to promote safe, economic, and reliable operation. Boiler control may range from simple manual devices to sophisticated automatic control systems that operate the boiler and all associated equipment. When steam turbines are applied, the overall control system may be extended to include the turbine controls as well.

Small-capacity boilers may include relatively simple packaged control systems that purge the furnace, start and stop the burner, maintain the required steam pressure and water level, and provide required safeties. More sophisticated distributed control systems serving large-capacity boilers can include the following:

- Boiler instrumentation
- Combustion (fuel and air) control
- Drum level and feedwater flow control
- Fuel-firing system and burner sequence control
- Start-up and bypass control
- Steam flow and pressure control
- Steam temperature control for superheater and reheater outlet
- Data processing, recording, and display
- Performance calculation and analysis
- Emissions measurement, calculation, and analysis

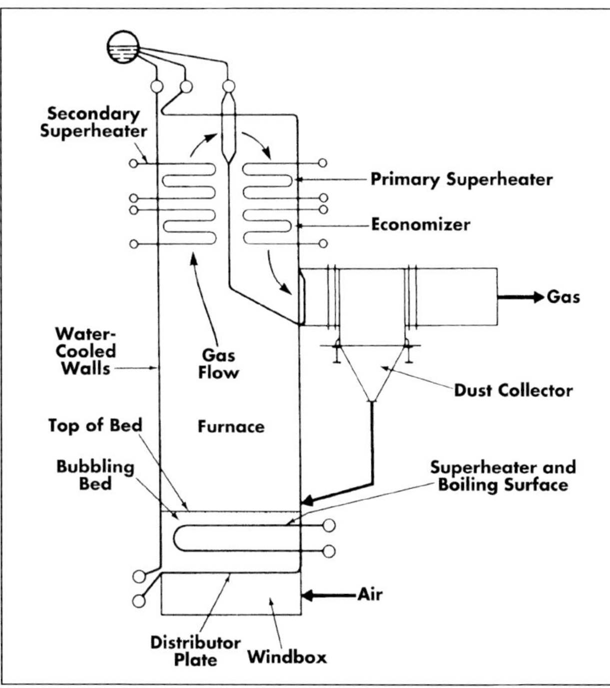

Fig. 7-26 Main Features of a Bubbling Fluidized-Bed Boiler. Source: Babcock & Wilcox

- Unit trip and alarm annunciation system
- Steam turbine and electric generator controls (where applicable)

Combustion and Fuel-Firing Systems

The combustion control system regulates the fuel and air input, or firing rate, to the furnace in response to load demand. Fuel-firing systems provide safe, controlled conversion of the chemical energy into heat energy that is transferred to the heat-absorbing surfaces of the boiler. To accomplish this, the fuel-burning system introduces the fuel and air for combustion, mixes these reactants, ignites the combustible mixture, and distributes the flame envelope and the products of combustion. The overall combustion system consists of the means of delivering, measuring, and regulating fuel and air to the furnace, as well as the burners, ignitors, and flame safety equipment.

Variations in boiler outlet pressure are often used as an index of balance between fuel-energy input and energy output demand. As the system uses more steam energy than the boiler is producing, the pressure in the boiler will start to drop. The burner control will sense this pressure drop and interpret it as a need for higher input. The controller will increase the amount of air and fuel being supplied to the burner to increase the firing rate. This increases the amount of energy added to the boiler, causing the boiler to produce steam at a faster rate. This increase in steam production matches the additional steam load, holding the firing rate at the new higher value. If the system has a reduction in steam demand, the pressure will increase and the controller will reduce the firing rate.

Boilers may be equipped with single or multiple burners. Burner design determines the mixing characteristics of fuel and air, fuel particle size and distribution, and the size and shape of the flame envelope. Figure 7-27 shows a burner assembly for a packaged watertube boiler. Figure 7-28 shows a gas and oil combination burner for a very large capacity boiler, with separate combustion air inlets to the inner and outer firing zones.

Figure 7-29 illustrates a control system for burning gas and oil, either separately or together in a large boiler system. The fuel and air flows are controlled from steam pressure through the boiler master. For single-burner systems, the oil or gas header pressure can be used as an index of air flow. For multiple-burner systems, a flow measuring instrument, such as a pitot tube, is used to provide more accurate air flow measurement and control. In all cases, the fuel delivery system must be able to regulate fuel pressure and flow to the burners and must be safeguarded in accordance with applicable codes.

Combustion Air Systems

The combustion air system supplies the oxygen to the burner for burning the fuel and removes the spent gases from the boiler. An adequate flow of air and combustion gases is required for the complete and effective combustion of fuel. Either the stack alone, or a combination of the stack and fans, produces the pressure differential necessary to maintain the required flow.

Draft is the difference between ambient atmospheric pressure and the static pressure of the combustion gases in the furnace, gas passage, flue, or stack. Four types of draft control are possible:

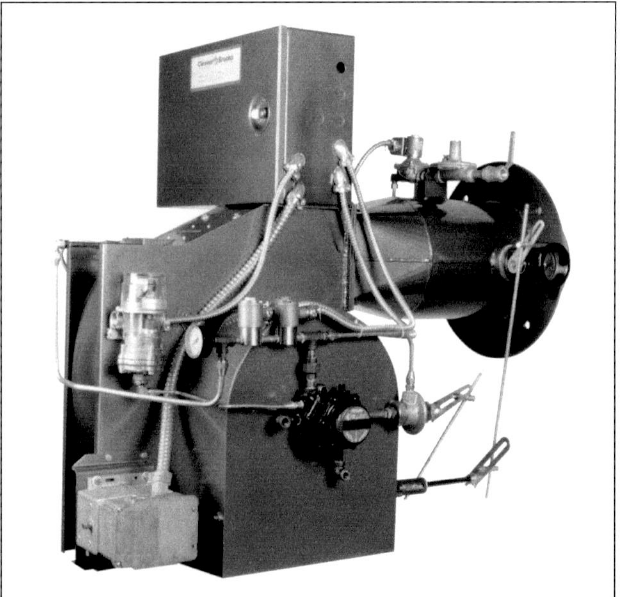

Fig. 7-27 Burner Assembly for Packaged Watertube Boiler. Source: Cleaver Brooks

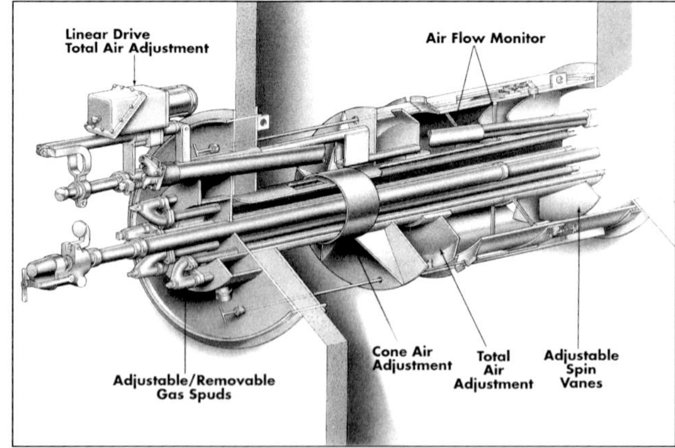

Fig. 7-28 Gas and Oil Combination Burner for Very Large Capacity Boiler. Source: Babcock & Wilcox

- Forced draft boilers operate with the air and combustion products maintained above atmospheric pressure. A fan or fans at the inlet to the boiler system provide the necessary pressure to force the air and flue gas through the system.
- Induced draft boilers maintain the air and combustion products below atmospheric pressure due to the suction imposed by an induced draft fan at the boiler exit.
- Natural draft boilers achieve the required flow through the boiler by the stack alone when the boiler pressure drop is small or the stack is tall. However, most solid fuel boilers and other large modern boilers require a fan at the boiler outlet to draw the flow through the boiler. A modulating damper at the boiler outlet typically operates to maintain proper boiler draft conditions.
- Balanced draft boilers have a forced draft fan at the system inlet and an induced draft fan near the system outlet. The static pressure is above atmospheric at the forced draft fan outlet and decreases below atmospheric pressure at the inlet to the induced draft fan.

To meet the output requirements of the system, a means of varying fan output is required, such as variable speed control, inlet vanes, and/or damper control. Dampers are commonly used to control the flow and temperature of air and flue gas. They can also be used to isolate equipment in the air or flue gas stream when such equipment is out of service or requires maintenance.

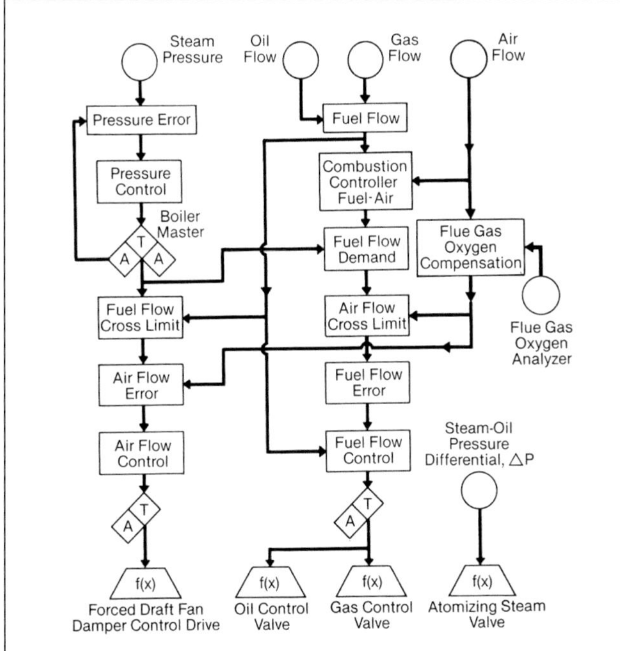

Fig. 7-29 Diagram of Combustion Control for a Gas- and Oil-Fired Boiler. Source: Babcock & Wilcox

Feedwater Control Systems

Feedwater control regulates the flow of water to the boiler to maintain the level in the drum within the desired limits. The control system will vary with the type and capacity of the boiler and the load characteristics. While small boilers may utilize a constant-speed pump with on/off control, a pneumatic or electrically operated feedwater control system is used by most larger units. Control sequences are classified as one-, two-, or three-element feedwater systems.

In simple level, or single-element control, the drum level is the only indicator used to control feedwater flow. Drum level is sensed by a transducer and fed back to the control system, which modulates a feedwater control valve. If the drum level rises above its normal point, the modulating valve is closed. If drum level decreases below the normal point, the valve is opened to increased feedwater flow.

Typically, the steam drum is maintained at half water and half steam. Under steady-state conditions, steam flow will be relatively constant, indicating a constant requirement for feedwater and a constant firing rate. Under this condition single-element control may be effective. However, if steam demand varies, a single-element control can provide inadequate response.

A rapid change in pressure changes the size of all steam bubbles, which causes an increase or decrease in total water volume. **Swell** refers to an increase in water volume in the steam drum, corresponding to a increase in firing rate. **Shrink** is the opposite event, referring to a decrease in steam drum volume, corresponding to a decrease in firing rate. A single-element control responds to swell by decreasing feedwater flow rate. However, since swell is actually an indication that more feedwater is required, the level control will then have to catch up to establish proper feedwater flow. In large systems, this delay results in unstable feedwater control.

Two-element control allows for response to sudden load changes. A steam flow transducer senses changes in load and compensates for the wrong-way reaction of the drum-level controller. With a sudden increase in demand, feedwater flow will remain initially unchanged. After swell occurs and the drum level begins to drop back to the normal point, feedwater flow is gradually increased to match the additional mass flow requirements.

In large, complex systems, a third element, a feedwater flow transducer, may be used to maintain water flow equal to actual steam demand. The three-element control loop considers drum level and the difference between steam and

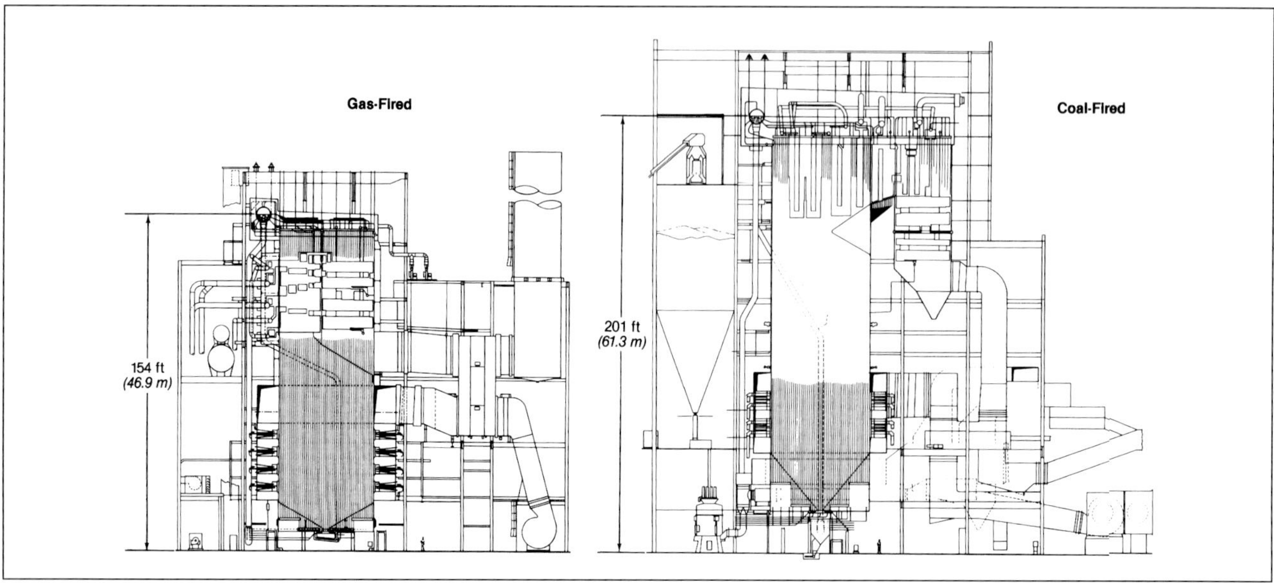

Fig. 7-30 Comparison of Gas and Coal-Fired Steam Generator. Source: Babcock & Wilcox

feedwater flow in determining the correct control response. During low-load operation (0 to 20%), it is difficult to measure steam and water flows accurately; therefore, the sequence may convert to single-element control using drum level as the only control variable. Three-level control allows for optimization of boiler system efficiency, but it is typically only economical in very large systems.

Boiler Safety

The safe operation of any steam generation system is dependent on effective safety control equipment and on the operator's ability to control boiler parameters. Operators must understand the operation of the plant, how to avoid emergency situations, and how to respond in any potential emergency. Since boilers are a source of numerous potential hazards, the importance of boiler safety cannot be overemphasized.

The **safety valve(s)** is the most critical valve on the boiler system. Its purpose is to limit the internal boiler pressure to a point below its safe operating level. For safe operation, one or more safety valves must be installed in an approved manner on the boiler pressure parts so that they cannot be isolated from the steam space. The valve must be set to activate at an approved set-point pressure and must be able to carry all of the steam that the boiler can generate without exceeding the specified pressure rise.

Various elements of the fuel firing, combustion air, and feedwater systems are critical for safe operation. Perhaps the most important burner control function is to prevent furnace explosions that could threaten the safety of operating personnel and damage the boiler and auxiliary equipment. Draft control systems must maintain furnace pressure at a desired set point and prevent furnace overpressurization or implosion. Drum water level measurement is another critical safety feature. An insufficient head of water may cause a reduction in watertube circulation, causing tubes to overheat and fail.

Identification of and compliance with all safety issues and prevailing operating codes, along with rigorous safety training for all operators, is paramount and must override all other issues. Safety instructions should also be posted in environmentally protected casings so that they are clearly visible to all operators and personnel in the boiler operating area.

Impact of Fuel Type and Environmental Regulations

Fuel type has a significant impact on steam generator system configuration and design. Depending on the fuel type, widely differing provisions must be made for fuel handling and preparation, fuel combustion, fouling, corrosion, heat recovery, and emissions control. Figure 7-30 shows a comparison of very large gas- and coal-fired steam generators of the same capacity. Notice the gas system is far more compact due to minimal needs for fuel storage and handling. A much smaller furnace is needed for combustion; due to the absence of ash deposits, lower excess air requirements and closely spaced heat transfer surfaces may be used. The coal system is much more complex and requires more extensive air emissions controls and equipment to collect and remove solid wastes.

Fuel type is perhaps the most significant factor with

respect to environmental regulations. Coal systems present the most serious challenges and require the highest costs for compliance. Generally, natural gas presents the fewest difficulties, though with some technologies, uncontrolled NO_X emissions may be close to that of distillate oil-fired systems.

There are numerous environmental considerations associated with boiler operation. Environmental protection and the control of solid, liquid, and gaseous effluents or emissions are key elements in the design and operation of all boilers. Combustion system emissions are strictly regulated by federal, state, and local governments, with requirements typically more stringent for larger capacity systems. Currently, the most significant emission control requirements relating to boiler combustion systems are for sulfur dioxide (SO_2), oxides of nitrogen (NO_X), volatile organic compounds (VOC), and airborne particulates. With coal-firing systems, numerous additional environmental regulations apply to fugative dust from coal piles and related fuel-handling equipment and for solid wastes arising from the collection of coal ash and from byproducts of flue gas desulfurization (FGD).

Additional environmental controls are required for aqueous discharges, which arise from a number of sources. These sources include boiler blowdown, boiler chemical cleaning solutions, cooling tower blowdown (which may be applicable to plants featuring condensing steam turbines), oil leakage, and a variety of other low volume wastes. Refer to section IV for additional detail on environmental regulations and compliance.

Recently the Environmental Protection Agency (EPA) has proposed limits on Hazardous Air Pollutants (HAPs) for solid fuel and fuel oil fired units. The predominant HAP of concern is mercury emissions from coal fired units.

Design and Provisions for Refuse, Wood, Biomass and By-Product Fuel Boilers

As discussed in Chapter 5, a wide range of gaseous, liquid, and solid by-product, refuse, and biomass fuel sources can be productively used. These fuels may be burned in conjunction with conventional fossil fuels or independently to produce hot water or steam. Many gaseous and liquid recovery type fuels can be burned in standard or modified versions of the same conventional boilers (i.e., front-fired and tangential-fired) used for natural gas and fuel oil combustion. Solid refuse and biomass boilers are generally similar to those used for coal burning. These may be spreader/stoker-fired, suspension-fired, or a combination of both. Other combustion system designs that can be used for certain applications include Dutch ovens, fluidized beds, and gasification systems.

For boilers designed for burning refuse, wood, and biomass, special provisions must be made for delivery, storage, handling, combustion, emissions control, and disposal. In some cases, special boiler designs are required. In other cases, modifications can be made to conventional boilers, notably the addition of fuel feeding mechanisms.

Gaseous and Liquid Fuels

Relatively clean biomass-derived and other by-product gaseous fuels with high energy density of 500 Btu/cf (18.6 MJ/m^3) or more may require only minor changes in firing system design from conventional natural gas burning systems. With lower energy density gases, firing systems with high volumetric capacity, such as tangential fired units, are commonly used. Difficulties may arise when multiple types of gases are used and/or when there is frequent variation in fuel energy density. The furnace and heat exchange surfaces must be specified with sufficient size to generate necessary load output with the lowest energy density fuel expected and must also have the cleaning and pollution control devices necessary for the full range of gaseous fuel types contemplated.

Care must be taken to prevent damage from highly corrosive elements present in many of these gases. For example, if hydrogen sulfide (H_2S) is present in even very small quantities, it can cause serious equipment problems because it is readily converted to forms of sulfuric acid. Gas treatment prior to combustion may be needed using lime baths or other neutralizing agents to avoid corrosion of certain boiler metals. In many cases, special design features, including use of corrosion resistant materials, must be incorporated for protection in the furnace walls and tubes. Other gaseous fuels, such as blast-furnace gas, which carries with it iron oxide dust, can foul gas mains, firing systems, boiler furnaces, and convection heating surfaces. Multiple stages of cleaning (i.e., mechanical or electrostatic separation and extensive washing) may be required. In some cases, boiler design must include as many soot-cleaning devices as needed in a pulverized coal-fired unit.

Many liquid recovery-type fuels can be burned with conventional atomization in standard boilers designed for fuel-oil combustion. Many refuse-type liquids can be burned in suspension, in combination with coal or other solid fuels, when introduced through fuel nozzles in the furnace walls above the solid fuel on the stoker grate.

Some by-product gases and liquids are burned in specially designed recovery-type boilers. A wide variety of liquors (e.g., sodium, calcium, magnesium, ammonium, etc.) are burned in recovery furnaces. Recovery boilers are designed to allow for the recovery of valuable chemicals that can be recycled and the generation of usable thermal output, while operating in an environmentally responsible manner. Unlike the firing of conventional fossil fuels, recovery boiler combustion goes through several distinct stages.

Black liquor (which is a mixture of inorganic and organic solid by-products from paper and pulp industry processes that is partially dissolved in an aqueous solution) is a common by-product burned in recovery boilers. Black liquor firing is composed of drying, volatile burning, char burning, and smelt coalescence. Operating problems resulting from plugging and aggressive corrosion can be particularly prevalent. Plugging results from condensation of the fume or normal gases given off by black liquor combustion, as well as from the carryover of smelt, or unburned black liquor, into convection heat transfer sections of the boiler. With proper design and operation, condensation is limited to the furnace area where deposits can be controlled by soot blowers. Corrosion is mitigated through special furnace floor and lower wall designs featuring studded carbon steel tubes covered with refractory or bimetallic composite tubing. Additionally, special provisions must be made for emergency shut down if water is suspected to have entered the furnace of an operating boiler. This is critical to ensure avoidance of a potentially explosive smelt-water reaction.

Design and Provisions for Solid Refuse, Wood and Biomass Fuel Boilers

For use of solid refuse, wood, and biomass fuels, boiler component design varies considerably from traditional fossil fuel units, but is generally more similar to systems designed for coal than for gas or oil. Similar to coal units, they are larger, more costly, and more maintenance intensive than standard gas or oil burning units. Provisions for fuel handling, combustion, emissions control, and disposal are key design considerations with these fuels. The wide variation in energy density and moisture content requires careful design consideration and there is generally a need to modify heat transfer surfaces to suit the intended fuel composition. Selection of alternative grate technologies can also be critical in designing a proper furnace configuration for these fuels.

Refuse Fuel Boilers

Two main techniques used for combustion of refuse fuel are **mass burning** or prepared **refuse-derived fuel (RDF) burning**. Mass burning uses the refuse in its as-received, unprepared state. Only large or non-combustible items are discarded. In a typical mass burning **municipal solid waste (MSW)** steam generator operation, refuse collection vehicles dump the refuse directly into storage pits. Overhead cranes with grapples first mix the fuel to create consistency and move the refuse from the pit to a stoker-charging hopper. Hydraulic rams move the refuse onto the stoker grates. The combustible portion of the refuse is burned off and the noncombustible portion passes through and drops into an ash pit for reclamation or disposal. To effectively burn this heterogeneous MSW fuel, special designs are required that allow for long furnace residence time prior to flue-gas contact with heat-transfer surfaces in order to completely oxidize the fuel and reduce the potential for corrosive attack of the heating surfaces. Lower heating surface metal temperatures are also generally maintained to limit corrosion.

With RDF combustion, the refuse is first separated, classified, and reclaimed to yield recyclable products. The balance is then moved to the boiler through multiple feeders onto a traveling grate stoker. In addition to shredding to reduce the size and create a more homogeneous fuel, an assortment of separation processes are used to eliminate materials such as stones, grit, and dirt and recover materials such as ferrous metals and aluminum cans. Overhead magnets are used to recover ferrous metals and can yield a recovery rate of up to 90%. An eddy current separator can be used for the removal of aluminum cans. Other devices such as a rotating trommel screen (a perforated drum) and air density separators are used to further sort and separate the various materials. In addition to the recycling benefits, the resulting homogeneous RDF has a higher energy density and burns more efficiently than mass-burn fuel, produces less than half the ash, and can be more effectively metered to match heat input demand. These benefits provide an incremental return on the additional capital and operating cost requirements for RDF systems. Figure 7-31 shows the layout of a typical RDF refuse-to-energy system. Notice the extensive relative space requirement for fuel storage, processing, and feeding.

The RDF is generally burned partially in suspension and partially on a stoker. In large boilers typically used for power generation plants, more finely shredded prepared refuse (largely consisting of light plastics and paper) can also be fired in suspension to supplement conventional fuels. In dedicated RDF-fired boilers, RDF may be used exclusively or a wide range of alternative fuels (e.g., natural gas, oil, coal, wood, and biomass) may be used as supplemental or back-up fuels. In addition to this flexibility, RDF provides for more thermally efficient combustion because of lower excess air requirements than mass burn units and higher steam generation rates per lbm (kg) of fuel consumed as compared to mass-burn operation.

Generally, production of a given quantity of steam requires that nearly 3 times as much MSW be burned as coal (by weight). Therefore, a major consideration in boiler design is the size needed to handle the physical amount of refuse that is delivered to the plant (regardless of the heating value). The maximum expected heat input must also be considered. Different refuse sources yield different heating values, so some analysis of the range of energy density of the anticipated refuse mix should be performed. If the energy density is greater than anticipated, the boiler capacity must be reduced to avoid overheating and the system will not be able to process all the available refuse delivered on a daily basis.

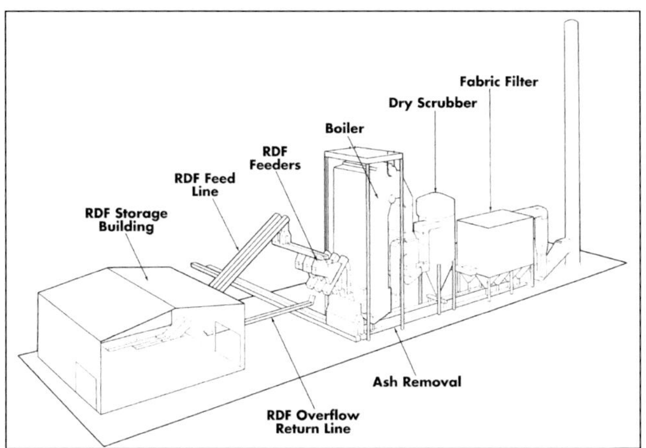

Fig. 7-31 Typical Layout of RDF Refuse-to-Energy System. Source: Babcock and Wilcox

The combustion of MSW requires a rugged, reliable stoker to successfully convey and burn the various types of refuse. Most stokers use some variation of a reciprocating grate action, with either forward moving or reverse acting grate movement. Some combination of stationary and moving grates is used to move the refuse through the furnace area, while allowing time for complete combustion.

Combustion air requirements include a primary source, or under-grate air, and a secondary source, or over-fire air. Under-grate air is typically fed to individual air plenums beneath each grate. Grate surface and damper action are designed to meter the primary air to the burning refuse uniformly. Because refuse contains a high percentage of volatiles, over-fire air should compose a large portion (25 to 50%) of the overall combustion air. Over-fire air enters through ports in the front and rear furnace walls. It serves to provide the quantity of air and turbulence needed to mix the furnace gases with combustion air and provide the oxygen necessary for complete combustion of volatiles in the lower furnace. Excess air levels of 80 to 100% are typically maintained to ensure that the heterogeneous MSW has sufficient air to efficiently oxidize the available carbon and hydrogen. To aid with wet fuels, the air system may include steam coil air heaters to dry the fuel and maintain proper furnace temperature. Figure 7-32 shows the fuel feeding and combustion air system for a refuse boiler.

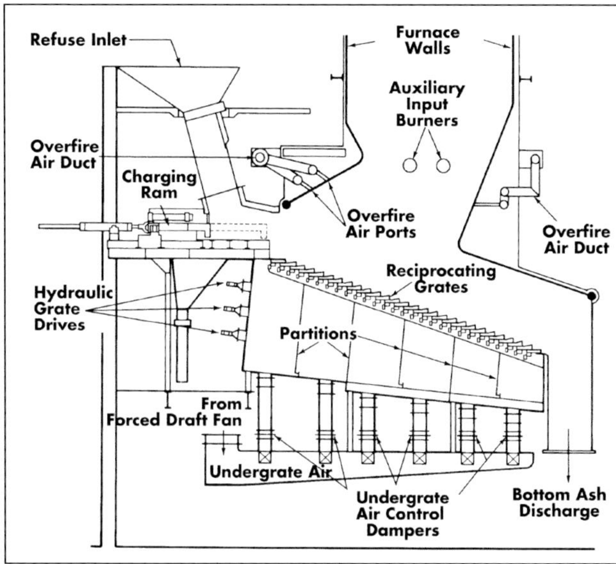

Fig. 7-32 Fuel Feeding and Combustion Air System for Refuse Boiler. Source: Babcock and Wilcox

Because refuse is a high fouling fuel, more rigorous and costly routine maintenance procedures are necessary versus conventional fossil fuel units. Designs should include good access to convection sections for frequent inspection and cleaning. To prevent plugging of gas passages, it is necessary to remove ash and slag deposits from external tube surfaces. Steam or air soot blowers are most commonly used. Corrosion prevention is an important consideration with refuse-fuel combustion. In addition to the corrosive materials present in conventional fossil fuels (such as sodium and sulfur chlorides, among other additional chemical elements), are other persistent corrosives that deposit on the various steam generator sections. Special corrosion resistant materials are used for protection in the furnace walls and tubes. Strict water treatment programs are also particularly important to minimize corrosion.

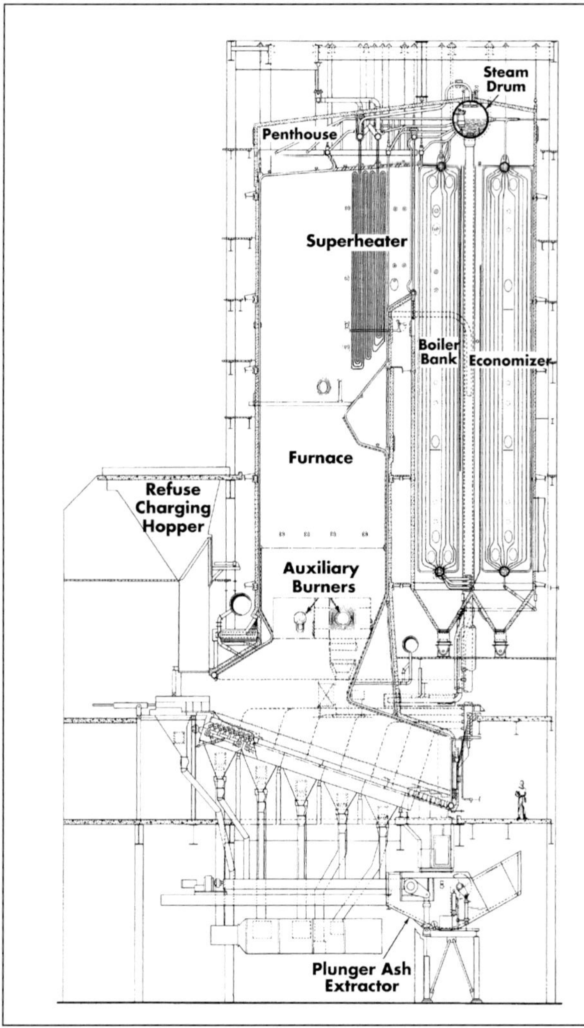

Fig. 7-33 Large-Capacity Mass Burning Unit. Source: Babcock and Wilcox

Ash products consist of light (fly) ash and coarse (or stoker) ash. Fly ash is entrained in the gas stream until it is removed in the particulate collection devise or falls out into the boiler, economizer, or air heater hoppers. Coarse ash, which comes from the fuel and slag deposits on the grates, walls, and tubes, is discharged through the stoker discharge chute and from the stoker siftings hoppers. The ash can be quenched through water spray nozzles or discharged into a water bath and then dewatered and squeezed for extraction and delivery to landfill sites.

Due to their cool-burning temperature characteristics and the low level of fuel bound nitrogen, refuse boilers tend to produce relatively low levels of NO_X emissions. Still, in larger capacity systems, some type of catalytic reduction system may be required. Refuse fuel is also fairly low in sulfur content, resulting in relatively low SO_2 emissions. However, they do produce a host of other pollutants in larger quantities than conventional fossil fuels. Dry scrubbers, used in coal plants for SO_2 emissions control, can provide that same function and are also effective in controlling HCL, dioxin, furan, and heavy metal emissions. Electrostatic precipitator or baghouse are used to control particulate emissions. Finally, tight combustion control of both under-grate and over-fire air is required to limit CO emissions.

Figure 7-33 shows a typical large-capacity mass burning unit. Figure 7-34 is an RDF-fired power plant located in South Dakota. It can extract up to 700 kWh of electricity per ton of processed solid waste.

Wood and Biomass Boilers

As with refuse burning, wood and biomass boiler component design varies considerably from traditional units, but is more similar to systems design for coal than for gas or oil. Provisions for fuel handling, combustion, emissions control, and waste disposal are also similar to those with refuse boilers. Moisture content and energy density variation are also key design and operation challenges. There are several types of combustion systems used for wood and biomass fuels, including Dutch ovens, traveling or varying grates, fluidized beds, and fuel gasification systems.

A traditional **Dutch oven** is a refractory-walled cell, connected to a conventional boiler setting, that is commonly used for wood waste combustion. Wood waste is introduced through an opening in the roof of the Dutch oven and burns in a pile on its floor. Over-fire air is introduced around the periphery through rows of holes or nozzles in the refractory walls. Because of its high amount of refractory surface, only a small portion of the energy

Fig. 7-34 RDF-Fired Power Plant. Source: Philip Shepherd, DoE/NREL.

released in combustion is absorbed. This allows the Dutch oven to burn fuels with moisture content of up to 60%. A disadvantage of this traditional design is that it does not respond quickly to varying load demand and changes in fuel consistency. The refractory is also subject to damage from spalling and erosion caused by rocks or tramp metal introduced with the fuel and from rapid cooling or overheating resulting from rapid changes in fuel moisture content. It also requires regular shutdown or low-load operation (with multiple oven units) for manual removal of accumulated ash. Figure 7-35 shows a large-capacity traditional Dutch oven furnace featuring two ovens, each with their own hogged fuel feeder.

Traveling grate and **vibrating grate** are variant designs that allow for automatic ash discharge. The traveling grate design, which was adopted from a spreader-stoker coal-firing system, features cast iron grate bars attached to chains that are driven by a slow moving sprocket drive system. The grate bars have holes in them to admit under-grate air, which also serves to cool the grate bar castings. This design requires a high degree of under-grate air

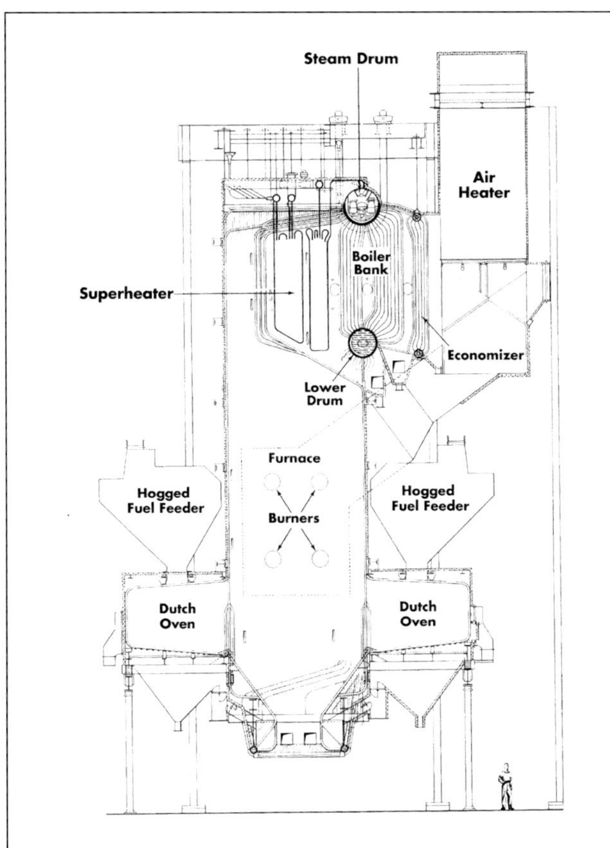

Fig. 7-35 Large-Capacity Dutch Oven Furnace. Source: Babcock and Wilcox

(60 to 85%) for sufficient cooling. The vibrating grate is a modern modification of the traveling grate design that reduces the number of moving parts and associated maintenance costs. Its iron grate bars are attached to a frame that vibrates on an intermittent basis, controlled by an adjustable timer. These can be air- or water-cooled. Because a large proportion of the combustible content of wood and biomass fuels is volatile, a relatively large portion of combustion air requirement above the fuel, in the form of over-fire air, is required. Water-cooled designs are beneficial because they can tolerate higher temperatures under the grate, which allows for a higher percentage of over-fire air and a relatively thin fuel bed. Figure 7-36 shows a large capacity modern water-cooled vibrating grate biomass unit.

Fuel distribution systems for wood and biomass fuels are designed to spread the fuels as evenly as possible over the grate surface. The two most common devices used for introducing wood-product and biomass fuels into the furnace for semi-suspension firing are mechanical distributors and wind-swept spouts. Mechanical distributors use a rotating paddle wheel to distribute the fuel. These can be designed for variable speed operation to enhance even fuel distribution. Wind-swept spouts use high-pressure air, which is continuously varied by a rotary damper. The ramp at the bottom of the spout and the air pressure can be varied to optimize fuel distribution.

Stable combustion can be maintained in most water-cooled furnaces at fuel moisture contents as high as 65% by weight. The use of preheated combustion air reduces the time required for fuel drying prior to ignition. With high-moisture content biomass fuel, it may be economical to dry the fuel with boiler flue gas before firing it in the boiler furnace, rather than pressing or air-drying the fuel for a long time to remove moisture.

Because biomass-fired boilers are susceptible to ash and carbon carryover, the convective heat transfer surfaces must be designed to accommodate soot blowers. Like coal and refuse boilers, ash handling is an important design consideration in biomass-fired boilers. In addition to the fly ash, bottom ash, which consists mainly of sand and stones, is the ash that is raked or conveyed off the grate, plus the ash that falls through the grate bar holes into the undergrate hopper (also called siftings). This ash can be collected using a conveyer with a dewatering incline at the discharge end. Since the vast majority of the ash residue produced by wood and biomass-fired boilers is in the form of gas born particulate, particulate emissions control is a primary environmental consideration.

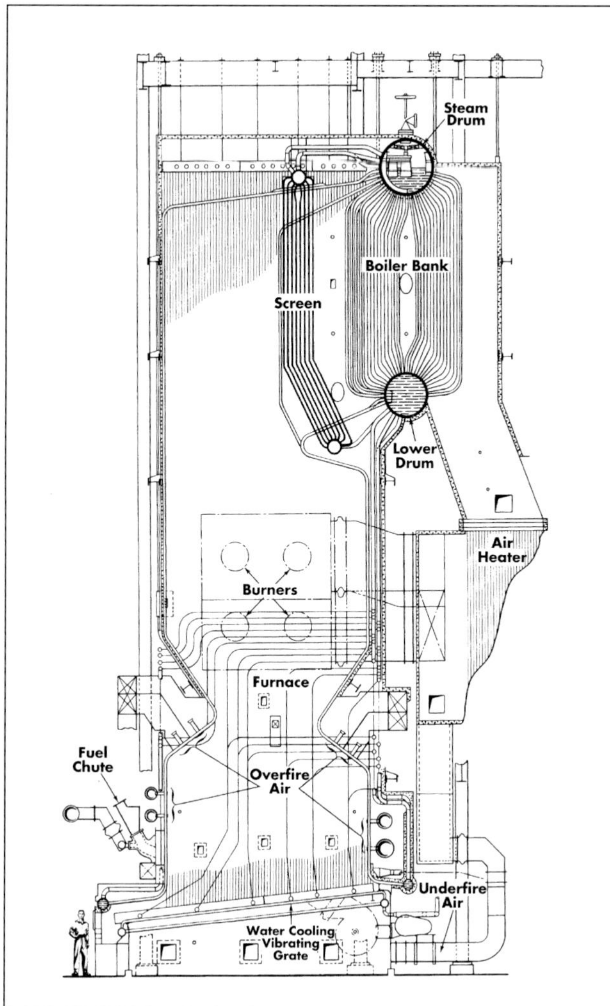

Fig. 7-36 Large-Capacity Water-Cooled, Vibrating Grade Biomass Unit. Source: Babcock and Wilcox

Mechanical dust collectors can be used after the last heat trap on the boiler to collect the larger size fly ash particles. To meet air emissions standards, electrostatic precipitators can be used after the mechanical collector to reduce the particulate concentration in the flue gas. As with refuse burning, the low combustion temperatures produce little thermal NO_X emissions. Hence, NO_X emissions will mostly be a function of the fuel nitrogen content, which can vary widely. SO_2 emissions are generally quite low with wood and biomass firing. CO and VOC emissions will vary widely based on excess air levels and consistency of both fuel heating value and fuel distribution, and on fuel moisture content.

Fluidized bed combustion has been successfully applied to a wide range of wood and biomass fuels. The large percentage of inert material (sand) has a positive dampening effect on brief fluctuations in biomass and wood fuel heating value on steam generation. The lower operating temperatures of 1,400 to 1,600°F (760 to 870°C) versus 2,200°F (1,200°C) for conventional spreader stoker units results in lower NO_X emissions production, which is a particularly important benefit for high nitrogen content wood and biomass fuels. Bubbling bed technology is typically selected for fuels of lower energy density (lower calorific value), while circulating fluid bed design is more suitable for high energy density wood and biomass fuels.

Figure 7-37 is a large biomass-fired power plant located in California. It operates on residues produced by nearby companies in the forest products industry. The system currently burns wood chips prepared from wood waste directly and produces 50 MW of electricity.

Gasification

Gasification is a process that converts fuels such as coal, coke, waste oils, refuse and biomass to gaseous fuel through partial oxidation. Its is applicable to a very wide range of energy sources and is also considered a viable alternative to recovery boilers operation such as that described above for black liquor. In the process, undesirable substances, such as sulfur and ash, can be removed, producing some thermal energy output along with a clean, transportable gaseous energy source. In many cases, usable chemical components can be recovered.

In contrast to combustion where chemical reactions take place in an oxygen-rich, excess air environment, with gasification, chemical reactions take place in an oxygen-lean reducing atmosphere. This results in less heat being released and new gaseous by-products, such as carbon monoxide, hydrogen, carbon dioxide, and methane, being produced. These gaseous by-products contain sufficient potential chemical energy to be used as a fuel source. In some cases, however, the by-products are intended to be used for chemical synthesis rather than combustion.

There are many types of gasification processes, including moving- or fixed-bed, fluidized-bed, and entrained-flow systems. Figure 7-38 shows these three generic processes applied to coal gasification, with corresponding temperature profiles. Process and system selection will depend largely on the type of solid fuel available and the intended gaseous by-products. In general, the feedstock fuel is prepared and fed to the gasifier in either dry or slurried form. The feedstock reacts in the gasifier with steam and oxygen at high temperature and pressure in a reducing (oxygen starved) atmosphere. This produces the by-product gas (also called synthetic or producer gas). In the process, a portion of the fuel undergoes partial oxidation by precisely controlling the amount of oxygen fed to the gasifier. The heat released in this first exothermic reaction provides the necessary energy for the primary endothermic gasification reaction to proceed rapidly.

The high temperature in the gasifier converts the inorganic materials in the feedstock (such as ash and metals) to a vitrified material resembling coarse sand. With some feedstocks, valuable metals are concentrated and recovered for reuse. The by-product gas must then generally be treated to remove various unwanted and/or polluting elements. Trace elements or other impurities are removed from the gas and are either recirculated to the gasifier or recovered.

Figure 7-39 shows five stages associated with the biomass gasification process, with the last stage indicating end-use options for the synthetic or producer gas. Figure 7-40 is a biomass gasifier plant located in Hawaii. It is fueled by bagasse, a fibrous residue by-product of sugar cane, that can be seen piled in the foreground.

Gasification offers two very important advantages over conventional excess air combustion. First, the syn-gas that is produced can be utilized to fuel combustion turbines and reciprocating engines, for an overall improved efficiency in converting fuel to electricity. Secondly, the air pollutants emitted from combusting syn-gas are less than from traditional combustion of solid fuels.

Solar Thermal Energy Systems

Thermal energy generation from solar energy can be accomplished with a wide range of conversion devices, from passive and active space and water heating to high-pressure steam generation. Hot water is commonly generated using traditional solar plates, while high-pressure steam is raised in specially designed solar furnaces. Both systems operate by concentrating heat and transferring it to a fluid. A solar collector differs in several respects from more conventional heat exchangers. The latter usually accomplish a fluid-to-fluid exchange with high heat

Fig. 7-37 Large Capacity Biomass-Fired Power Plant. Source: Warren Gretz, DoE/NREL

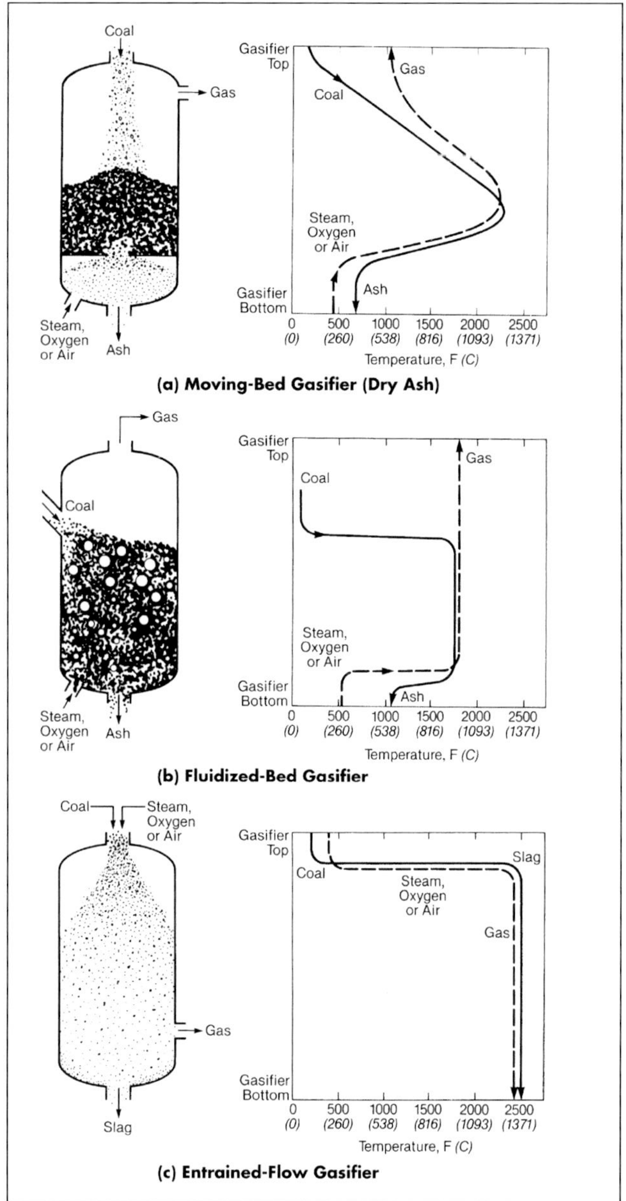

Fig. 7-38 Three Generic Coal Gasification Reactors and their Temperature Profiles. Source: Babcock and Wilcox (Courtesy of the Electric Power Research Institute)

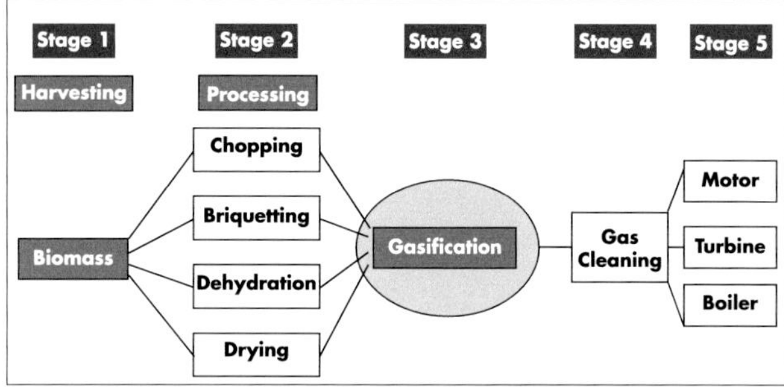

Fig. 7-39 Five Stages of Biomass Gasification Process.

Fig. 7-40 Biomass (Bagasse) Fuel Gasifier and Fuel Supply. Source: Warren Gretz, DoE/NREL

transfer rates and with radiation as an unimportant factor. In the solar collector, radiation is the primary mode of heat transfer. This radiant energy is then transferred to a fluid such as water or a water-glycol mixture.

Solar thermal collectors can be divided into three categories:

- **Low-temperature collectors** provide low-grade heat, less than 110°F (43°C), through either metallic or non-metallic absorbers. These are used for applications such as swimming pool heating and low-grade water and space heating.
- **Medium-temperature collectors** provide heat at temperatures ranging from 110 to 180°F (43 to 82°C), either through glazed flat-plate collectors using air or liquid as the heat transfer medium, or through concentrator collectors that concentrate the heat to elevated levels. These typically feature evacuated tube collectors and are most commonly used for domestic water heating or a variety of low-to-medium temperature process, space, and domestic water heating applications.
- **High-temperature systems** feature heliostat, parabolic dish, or trough type collectors that intensely concentrate solar energy, allowing for generation of extremely high fluid temperatures sufficient to produce high-pressure steam.

Low- and medium-temperature passive and active systems are the predominate solar thermal energy technology applied throughout the world for residential and CI&I facilities. Other than in passive construction design, the major applications are in solar water heating and building space heating. The basic components of a typical flat-plate solar collector are: the solar energy-absorbing surface, with means for transferring the absorbed energy to a fluid; and the envelopes, or enclosures that are transparent to solar

Fig. 7-41 Typical Solar Flat Plate Collector.
Source: Roch A. Ducey, DoE/NREL

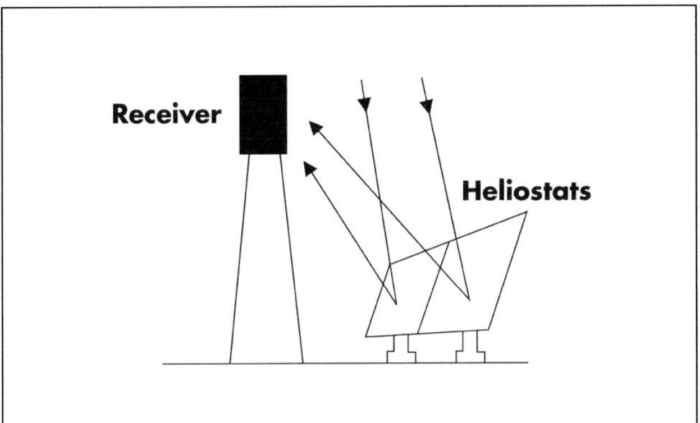

Fig. 7-42 Illustration of Power Tower System.

radiation, but reduce convection and radiation losses to the surroundings. As shown in Figure 7-41, flat-plate collectors are usually mounted in a stationary position with an orientation (tilt and azimuth angle) optimized for the particular location and time of year in which the device is intended to operate. A complete system would also include a mechanism for circulating the heat transfer fluid, either by mechanical pumping or natural convection, a storage tank, and controls.

High-temperature systems are most typically applied in large-scale independent power production (IPP) projects that produce steam to drive a conventional steam-turbine generator and provide electricity to the grid. They may also be used for more localized district steam systems or other specialized heat process applications. Concen-trating solar thermal systems can absorb and concentrate 1,000 times or more normal intensity of the sun.

Currently these systems use three types of concentrators:

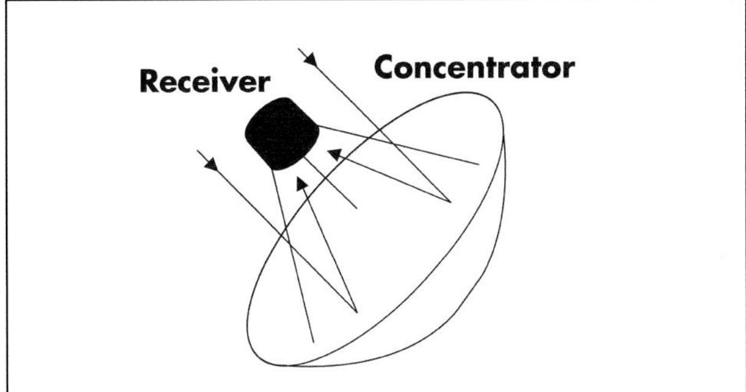

Fig. 7-43 Illustration of Parabolic Dish System.

- **Central receiver** (or **power tower) systems** (Figure 7-42) that use a circular field array of heliostats (individually tracking, highly reflective mirrors) that track the sun and focus it on a central receiver.
- **Parabolic dish systems** (Figure 7-43) that use an array of parabolic dish-shaped reflectors (mirrors) to focus/concentrate solar energy onto a receiver located at the focal point of the dish. Fluid in the receiver is heated to a typical temperature of about 1,380°F (750°C). This can be used to produce superheated steam, or, in some systems currently under prototype development, operate an integral, small-capacity engine.
- **Parabolic trough systems** (Figure 7-44) that use parabolic trough-shaped reflectors (mirrors) to focus/concentrate sunlight on thermally efficient receiver tubes, running the length of the trough, that contain a heat transfer fluid. The fluid is heated to a typical temperature of about 735°F (390°C) and pumped through a series of heat exchangers to produce superheated steam. The troughs are situated in parallel rows, aligned on a north-south axis that enables the single-axis troughs to track the sun continuously from east to west.

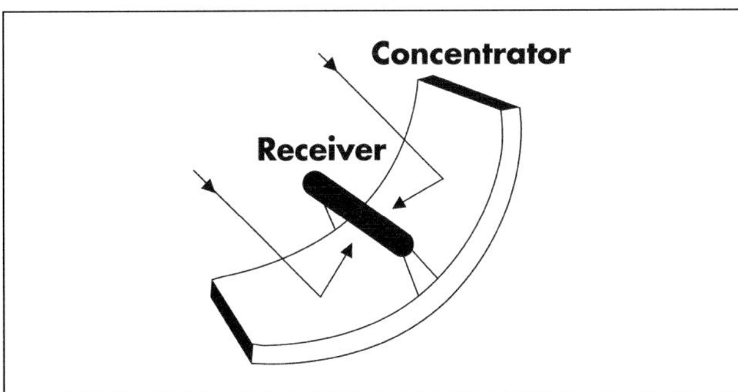

Fig. 7-44 Illustration of Parabolic Trough System.

When applied to electric power generation, the solar trough and power tower rely on conventional steam turbine generators, while the parabolic dish systems use integral Stirling or Brayton-cycle type engines. Currently, the parabolic trough systems are the most commercially advanced high-temperature systems and can be practical in a wide range of applications and capacities. Figure 7-45 shows a parabolic trough-based solar hot water system applied at a correctional facility in Colorado. This system provides about 20,000 gallons (75,700 liters) of hot water per day for laundry and other domestic uses. The power tower has been proven effective in demonstration projects for grid-feeding electric power generation, while the parabolic dish systems are still in prototype development with models in operation that can generate up to 25 kW of electricity.

Figure 7-46 shows a solar thermal collector that utilizes a Fresnel reflector optic to obtain 25 times sun concentration. The optic comprises of lightweight highly reflective aluminum mirrors, which pivot in unison to follow the sun. Solar energy is collected from the mirrors in a selectively coated stainless steel receiver tube capable of operating at up to 428°F (220°C) and 580 psia (40 bar) pressure. A wide variety of fluids can be heated, including thermal oil, water, saturated steam, brines/salines, and slurries. The entire optic is enclosed in a sealed glazed canopy to protect the mirrors from wind, rain and dirt. The collector is mounted with racking systems similar to conventional solar thermal collectors. Mounting orientations can be east/west or north/south, and mounting inclination can be optimized to favor either winter or summer thermal output.

Figure 7-47 shows a parabolic dish concentrator system featuring stretched membrane heliostats. With this self-contained system design, the receiver absorbs energy reflected by the concentrator and transfers it to the engine's working fluid. In addition to the dish module itself, continued research and development is focused on the engine technology and associated interface issues.

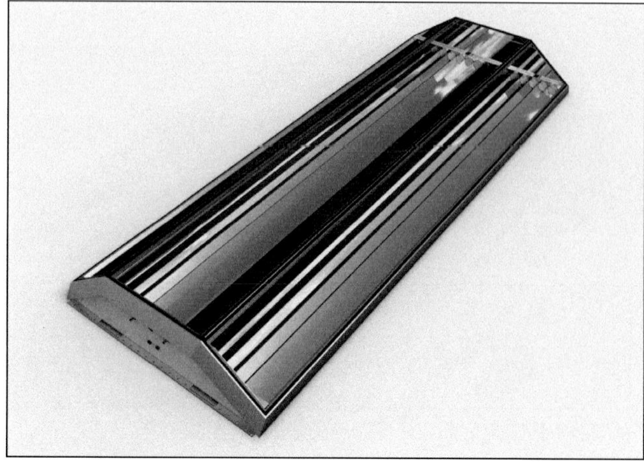

Figure 7-46 Solar Thermal Collector with Fresnel Reflector Optic. Source: Chromosun

Figure 7-47 Parabolic Dish Engine System Prototype. Source: Warren Gretz, DoE/NREL

Figure 7-45 Parabolic Trough-Based Solar Hot Water System Source: Warren Gretz, DoE/NREL

These solar thermal generation systems are limited to about 25% availability, even when located in the most ideal sites. Availability and reliability can be enhanced through a variety of means, including the addition of thermal storage and fossil fuel boilers for steam generation or battery storage and back-up combustion engines for electricity production. The addition of thermal storage can greatly increase output availability. One currently used thermal storage system features two storage tanks. Liquid salt is pumped from a “cold” storage tank through the receiver, where it is heated and then pumped to a "hot" tank for storage. When steam is needed, hot salt is pumped to a steam generation system. From the steam generator, the salt is returned to the cold tank, where it is stored and eventually reheated in the receiver. These tanks contain molten nitrate salt, which is a clear liquid with properties similar to water at temperatures above its 464°F (240°C) melting phase point. The salt is pumped from the cold storage tank to the receiver, where it is heated in the tubes to about 1,050°F (565°C). The salt is then pumped to the hot storage tank, where is remains available for dispatch to the steam generator. After producing steam it is then pumped back to the cold storage tank at a temperature of about 545°F (285°C). While such a high-temperature storage system allows availability to be increased dramatically, it remains very costly.

In addition to steam generation, high-powered solar furnaces can be used directly for process applications, replacing conventional furnaces or very costly laser furnaces. Current prototype applications range from high-temperature coatings on metals and ceramics (where it is advantageous to heat only the surface of the material without affecting the base material) to providing effective decontaminating of hazardous wastes. The solar furnace’s ability to quickly generate temperatures in excess of 5,500°F (3,000°C), focus it with great precision, and selectively heat the surface of a sample have led to studies of phase transformation hardening, thin-film deposition, rapid thermal tempering, and a wide variety of treatments for metal, ceramic, and composite materials to obtain higher-value materials with desired properties, such as superconductivity or greater resistance to corrosion, friction, and oxidation. Such solar furnaces are also well suited to the destruction of hazardous wastes. Focusing a beam of concentrated light onto hazardous wastes breaks down numerous toxic chemicals, including dioxin and polychlorobiphenyls (PCBs). The ultraviolet portion of the solar radiation breaks the bonds holding together the hazardous components.

All power cycles reject heat to ambient air, exhaust gasses, and/or liquid coolants. **Heat recovery** is the capture of a portion of the rejected thermal energy and its use in an economic manner. Recovered heat that is harnessed for effective use replaces purchased energy, reducing fuel consumption and cost, as well as emissions of harmful air or thermal pollutants. When heat recovery serves as a reliable source of supply, it can sometimes offset capital costs by reducing investment in conventional thermal energy-generating equipment.

Heat recovery is a fundamental element in many prime mover applications due to the low efficiency (20 to 40%) of power generation processes. For example, closed-cycle condensing steam turbines give up most of their exhaust energy to the environment through the condensing process; open-cycle internal combustion engines pass their exhaust energy directly to the environment. A large portion of this remaining energy may, in some cases, be captured and used in a secondary power cycle or to meet a variety of thermal loads.

PRIME MOVER HEAT BALANCE

Heat recovery from each of the main types of prime movers (i.e., reciprocating engines, gas turbines, and steam turbines) presents a different set of conditions. A **heat balance** is an accounting of the energy flows involved in the process, including input fuel energy, power output, and rejected heat. The amount of rejected heat that can be used will depend on the form and condition in which it is rejected and the form and condition in which it can be recovered and applied for a given application.

For example, reciprocating engine-generated heat can be recovered in substantial quantity from both the coolant system and the exhaust gas. Most of the heat of friction can be recovered in reciprocating engine operations. But some of the heat is lost to the ambient surroundings as radiation or convection. Some energy is also lost due to inefficiency in the heat exchange process. Exhaust heat recovery is limited by the allowable final exhaust temperature and the required temperature of the available heat sink. Coolant temperatures are often too low to be useful for heat recovery.

The heat balance for a reciprocating engine can be summarized as:

Total heat input = work output + exhaust heat + radiation + jacket water + oil cooler + inter/after cooler (8-1)

Gas turbines reject a high proportion of input energy to exhaust gases, most of which can be recovered. Exhaust gas heat recovery is limited mainly by process temperature requirement and the practical limitations of reducing final exhaust gas exit temperatures below acid dewpoint levels.

Steam turbine heat recovery requires power output to be purposefully reduced in order to improve overall system efficiency. Steam is extracted or discharged from the turbine at a higher than optimal pressure for secondary uses.

In a large condensing turbine, approximately one third of input energy can be used for generating shaft power; the rest is dissipated at relatively low temperature to the environment. By reducing power output to one quarter of input energy, output steam can be provided at a high enough temperature to be effectively used for process heating.

Natural gas offers significant advantages as an engine fuel for exhaust gas heat recovery applications. In particular, gas use allows increased effectiveness due to reduced fouling of heat exchange surfaces. Heat recovery capacity is also increased because the products of gas combustion can be cooled to lower temperatures than those of sulfur-bearing fuels.

HEAT RECOVERY APPLICATION OVERVIEW

Recovered heat can be used for many purposes. Applications include air or water heating, steam production for thermal loads or for additional power production, absorption chilling, and regeneration of desiccant dehumidification systems. The most common heat recovery applications are described below:

- Exhaust gas can be used directly for process air heating, serving industrial ovens and dryers, or for the preheating of combustion air.
- Fluid heating can be accomplished with several types of **heat exchangers** that use exhaust gas from a gas turbine or reciprocating engine, or circulating water from engine coolant systems.
- Steam production at a wide range of pressures can be accomplished by passing exhaust gas through a **heat**

recovery steam generator (HRSG). Output and temperature can be increased through supplementary firing.

- **Absorption refrigeration cycles** can provide chilled water using exhaust gas, steam, or hot water. Engine coolant system hot water or low-pressure steam can only be used for single-stage absorption, while exhaust gas or high-pressure steam can be used for high-efficiency two-stage absorption chilling.

A schematic representation of heat recovery application options for a gas turbine cogeneration system is shown in Figure 8-1.

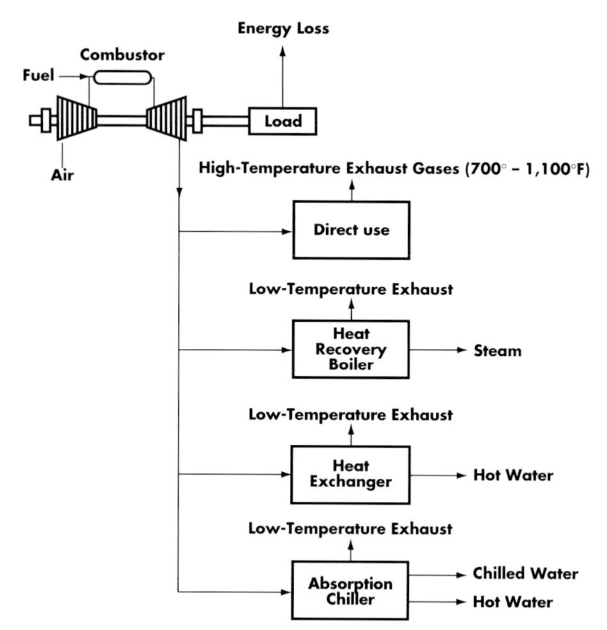

Fig. 8-1 Representative Heat Recovery Application Options for Gas Turbine.

QUANTIFYING HEAT RECOVERY

The following elements are critical to the cost-effectiveness of heat recovery applications:

- The temperature, or **quality**, of the available rejected heat
- The match between available heat rejection and facility thermal loads
- The capital and operating cost of heat recovery equipment
- The cost of supply alternatives, i.e., the cost of fuel

Quality of Rejected Heat

Combustion engines produce a combination of high- and low-temperature heat outputs. Temperature has a strong relationship to the usefulness of recoverable heat. The term "quality" is sometimes used to define usefulness along these lines.

High-temperature exhaust gases, typically at 600 to 1,200°F (300 to 650°C), can be used for generating high-pressure steam. Low-temperature heat is available from reciprocating engine water jacket, lubricating oil, valve cage, and charge air cooling systems at temperatures in the range of 100 to 260°F (38 to 127°C). Heat recovery from these systems can be used to generate hot water or low-pressure steam only.

Low-pressure steam from **ebullient** systems or high-temperature forced circulation cooling systems is limited to 15 psig/250°F (2 bar/121°C). Typical temperatures for hot water raised by engine coolant systems are 160 to 220°F (71 to 104°C), though certain engines are designed to operate at temperatures as high as 260°F (127°C).

Recovered heat from back-pressure or extraction steam turbines can be of varying temperatures and is a function of the specified exhaust pressure. Common outlet pressures are 15 to 50 psig (2 to 4.4 bar), although other pressures between 5 psig (1.4 bar) and 250 psig (18.3 bar) or even higher may be used.

Quantifying Available Rejected Heat

The quantity of available rejected heat can be expressed in terms of the total enthalpy of the rejected heat stream:

$$Q = \dot{m}\,(h_2 - h_1) \tag{8-2}$$

Where:

Q = Total available heat of the rejected heat stream, Btu/h (kJ/h)
$\dot{m}$ = Mass flow rate of the rejected stream, lbm/h (kg/h)
h_1 = Inlet enthalpy of the rejected stream, Btu/lbm (kJ/kg)
h_2 = Outlet enthalpy of the rejected stream, Btu/lbm (kJ/kg)

The equation used by manufacturers to calculate the amount of thermal energy that can be recovered from the combustion engine exhaust gas is expressed as:

$$Q = \dot{m}c_p\,(T_1 - T_2) \tag{8-3}$$

Where:

Q = Total heat available, Btu/h (kJ/h)
$\dot{m}$ = Exhaust gas mass flow rate through system, lbm/h (kg/h)

c_p = Specific heat at constant pressure of the exhaust gas, Btu/lbm per °F (kJ/kg per °C); N_2 constitutes the major part — around 75% of the total exhaust gases volume and the specific heat for N_2 ranges from 0.2495 to 0.2942 Btu/lbm per °F (1.045 to 1.232 kJ/kg per °C) for the correspond-ing temperature range 200 to 2,000°F (93 to 1,093°C)
T_1 = Exhaust gas inlet temperature, in °F (°C)
T_2 = Exhaust gas exit temperature, in °F (°C)

Rejected Heat — Storage or Rejection

The demands for prime power and thermal energy are not always perfectly matched. When configuring a system, it is usually most economical to satisfy non-varying **baseloads.** In theory, this means that available power and recovered thermal energy can be used all of the time. This perfect relationship, however, is not always attainable.

If available heat cannot be used when it is recovered, it must either be stored for later use or rejected to the environment. The most common types of **storage** associated with recovered heat from prime movers are hot water or chilled water generated by a heat recovery absorption chiller. A need for storage may indicate an imbalance in supply and demand that results in increased costs. Sometimes, however, storage is an economic enhancement technique that balances a system and improves efficiency, as when systems are purposely sized to use stored hot or chilled water during an electric utility's on-peak billing period.

Figure 8-2 is a schematic layout of a reciprocating engine-generator application featuring domestic hot water heating with series recovery of exhaust and engine coolant heat. In this application, two 10,000 gal (37,850 l) hot water storage tanks are used to balance the thermal load. The total heat recovered from the 300 kW system can produce about 1,500 gal (5,678 l) per hour of 180°F (82°C) water, charging storage tanks in about 13 hours of continuous operation. The system is appropriate for a facility with a continuous electric load, but only an intermittent (e.g., daytime) hot water load. For example, during an 11 hour period of hot water demand (at up to 3,300 gph or 12,491 lph), the system could provide 1,500 gph (5,678 lph) heat recovery output, plus 1,800 gph (6,813 lph) from storage.

Quantifying Storage

The quantity of thermal energy that can be stored is dependent on the temperature difference that can be achieved and on the thermal qualities of the storage materials. The quantity of heat stored is expressed as:

$$q = V\rho c(T_2 - T_1) \tag{8-4}$$

Where:

q = Quantity of heat stored in Btu (kJ) (note that lower case q is used for quantity of heat, as opposed to upper case Q, which is used for rate of heat flow)
V = Volume of the storage material in ft^3 (m^3)
ρ = Density of the storage material in lbm/ft^3 (kg/m^3)
c = Specific heat of the storage material in Btu/lbm per °F (kJ/kg per °C)
T_2 = Final temperature in °F (°C)
T_1 = Initial temperature in °F (°C)

For a simple closed loop water-to-water heat exchange application, a simplified expression for the Btu/lbm • °F value of thermal energy storage potential is:

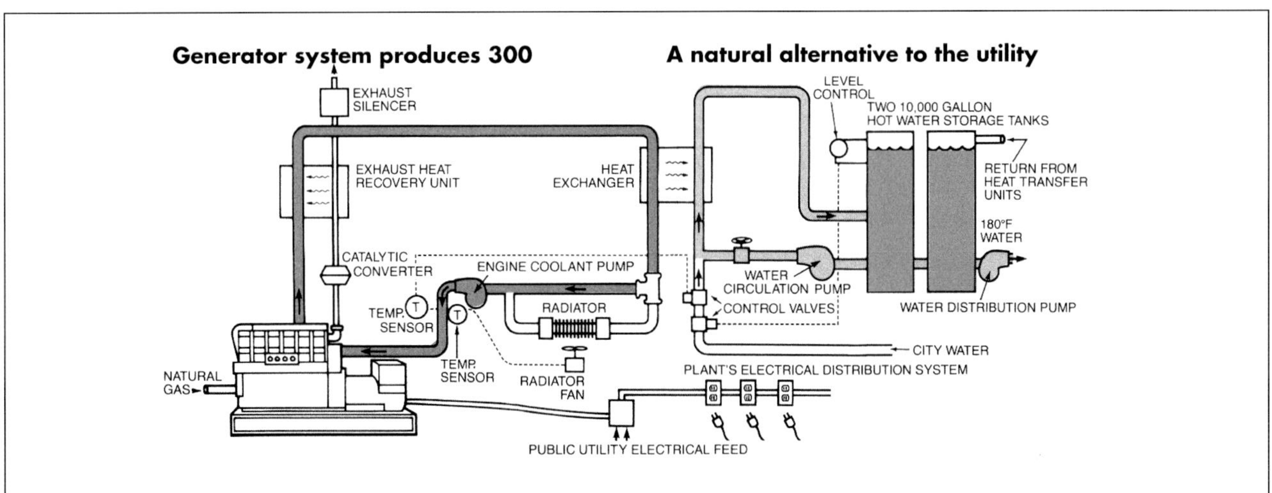

Fig. 8-2 300 kW Engine-Generator Set with Hot Water Heat Recovery System and 20,000 Gallon Storage Capacity. Source: Waukesha Engine Division

$$Q = V_f(gpm)\ (60\ min/h)\ (8.34\ lbm/gal)\ (1.0\ Btu/lbm\ °F) \qquad (8\text{-}5)$$

or

$$Q = 500.4V_f(T_2 - T_1) \qquad (8\text{-}6)$$

Where:

Q	=	Heat available in Btu/h
V_f	=	Volumetric flow rate in gpm
T_2	=	Final temperature in °F
T_1	=	Initial temperature in °F
1.0 Btu/lbm • °F	=	Specific heat of water
8.34 lbm/gal	=	Density of water

Dumping Rejected Heat

When heat recovery is used, there are two factors that are critical to the integrity of the heat recovery system:

- The ability to provide for critical thermal energy requirements in the event of an interruption of the supply of recovered heat.
- The ability to effectively dump the rejected heat in the event of an interruption of the demand for the thermal energy.

A means of dumping rejected heat on an emergency basis may be critical to ensuring required equipment cooling. In addition, it may be economical to operate prime movers on occasion without full utilization of available rejected heat.

A reciprocating engine requires an alternate means of heat rejection for both engine exhaust and coolant systems. When heat recovery is not required, exhaust must either be bypassed around the heat recovery system and discharged directly to the atmosphere, or the recovered heat must be subsequently dumped. Engine cooling systems may utilize an independent fan-powered dump radiator or an interconnection with the facility's cooling system, typically a cooling tower.

An integral component of a gas turbine/HRSG system is a **diverter**. The diverter is installed upstream of the HRSG and vents to atmosphere the turbine excess exhaust gas stream before it enters the HRSG. Diverter valves also protect the heat recovery boiler in the event of a malfunction of the feedwater system and permit start-up of the gas turbine before the boiler.

Unless specified otherwise, large diverter valves for gas turbine applications have a leakage rate of 3% to 5% when the diverting exhaust gasses around the exhaust heat exchangers. For this reason, most HRSG manufactures require that the HRSG be equipped with high water level and low water level alarms to protect it when the gas turbine is in operation.

Heat Recovery Heat Exchangers

There are many types of heat exchangers, although their common function is the transfer of heat from one medium to another. This usually requires both fluids to flow past a separating membrane that provides conductive transfer from the high-temperature fluid to the low-temperature fluid. The rate of heat transfer depends on the logarithmic mean temperature difference, ΔT_{LMTD}, between the fluids.

Flow arrangements may be characterized as parallel flow, counterflow, crossflow, and mixed flow. Counterflow is the most efficient heat exchanger arrangement, producing the greatest temperature difference and requiring the least heating surface. Heat exchangers can also be classified as being of either direct-contact or indirect-contact design.

Direct-contact applications involve some type of mixing of fluid streams: a deaerator might be considered a type of heat exchanger in that steam and condensate are mixed to produce a heated feedwater stream. Another example is the heat wheel, in which a rotating medium is heated by direct contact with one gas or air stream and then moved to contact and heat a second air stream.

Indirect-contact heat exchangers include separate circuits for each fluid stream, allowing fluids to be at different pressures and preventing mixing or contamination. Important types of indirect-contact heat exchangers include the following:

- **Shell-and-tube heat exchangers** include a cylinder, or shell, containing a tube bundle. Generally, the higher-pressure fluid flows through the tubes with the lower-pressure fluid in the shell. When the heating fluid is steam, flow will usually be through the shell. There are a wide variety of shell-and-tube configurations, sizes, and materials of varying cost and effectiveness.
- **Concentric tube (double pipe** or **multi-tube) heat exchangers** can be used for handling high-pressure fluids. Designs include a single tube or bundles of tubes of a single length, a spiral coil, or a tube bundle with hairpin bends. The flow arrangement can be either parallel or counterflow. Concentric tube heat exchangers typically offer the benefit of easy disassembly.
- **Plate-and-frame liquid/liquid heat exchangers** include contoured plates with gasketed edges that are clamped tightly together in a frame. Hot and cold fluids are directed through the spaces between the plates, allowing heat transfer at high effectiveness.
- **Finned-tube HRSGs** are commonly used in large pow-

er generation systems. The design, which resembles a shell-and-tube heat exchanger, is used to generate steam from high-temperature exhaust gas.

Heat Recovery Systems

Small heat recovery systems, including required heat exchangers, controls, piping and ductwork, pumps, silencers, diverters, expansion tanks, safety devices, etc., can be supplied as a single package. In some cases, heat recovery is provided as an integral part of a packaged prime mover system. This is particularly common in small reciprocating engine systems. In other cases, the heat recovery system is a separately packaged unit that is interconnected with the prime mover-driven system. Larger systems are entirely field assembled.

Heat recovery systems used in cogeneration applications include the following:

1. Air heating systems. These include systems in which exhaust gas is used directly for process heating or combustion air preheating, as well as systems in which a heat exchanger is used to separate the energy streams. More elaborate heat recovery systems may use jacket water separately, or in combination with exhaust gas, for ventilation air heating or other purposes.

2. Fluid heating systems. In many cases, particularly in smaller packaged reciprocating engine-driven systems, both exhaust gas and engine jacket heat are used to produce hot water. In a typical configuration, water is circulated through engine jacket and exhaust gas heat exchangers in series. Pressurized systems are required for high-temperature applications to prevent boiling. Alternatively, single-phase fluid heating is done by pumping water or some type of organic liquid through a closed-loop heat recovery unit where it is heated by the hot exhaust gas to the required temperature and distributed for process use.

Many modern high efficiency reciprocating engines limit the jacket water temperature to approximately 203°F (95°C). This limit can significantly reduce the opportunity to recover useful heat from the jacket water system because the incoming cooling water that is being returned from the process heating system where it has been used must be as slow as 166°F (74°C) with the engine manufacturer's standard heat exchanger design. In addition to jacket water, heat recovery from a reciprocating engine can be enhanced by also recovering heat from the lube oil cooling system and the first stage of the intercooler. With the addition of these systems, the cooling water returning from the process heating system must be as low as 159.5°F (71°C) to fully recover the thermal energy from the lube oil system and 140 °F (60°C) to recover all of the available thermal energy from the First Stage of the Intercooler. Figure 8-3 shows a typical heat balance for a 4,366-kW GE/Jenbacher recip-

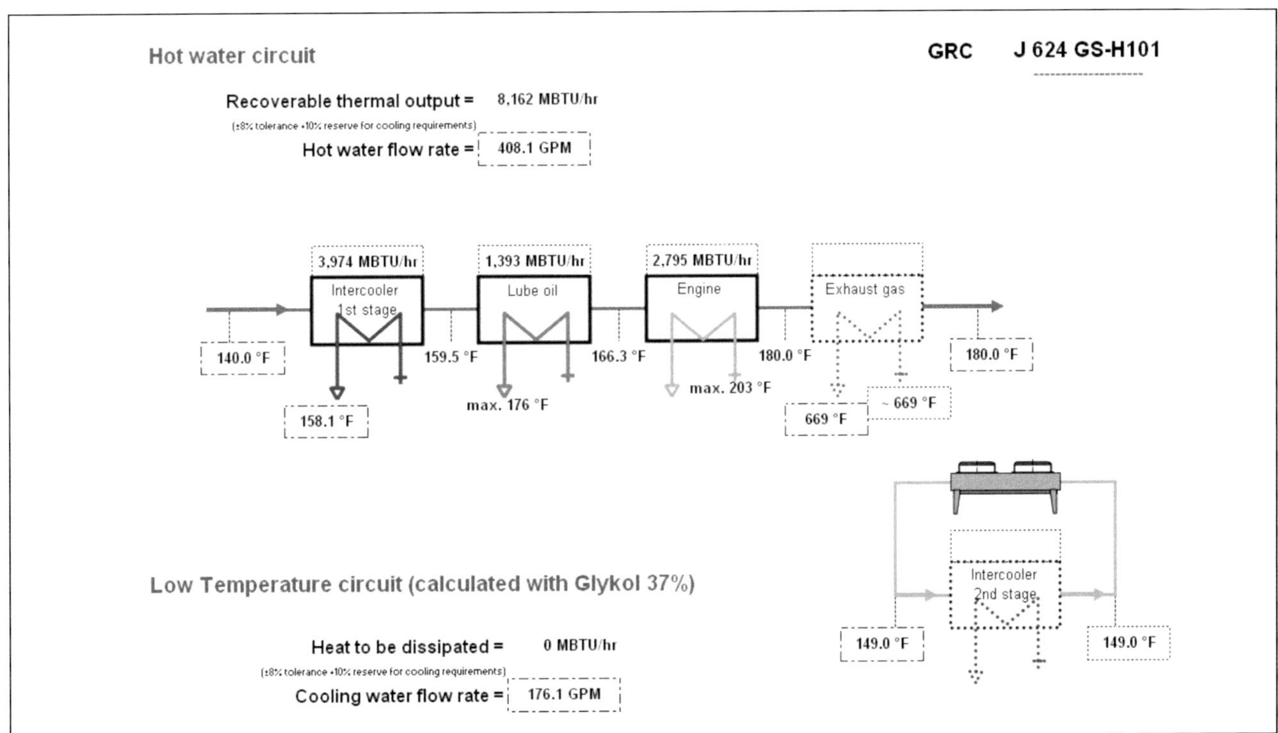

Fig. 8-3 Representative Heat Recovery Application for a Reciprocating Engine.

rocating engine-generator. In this example, the maximum exit temperature of the hot water thermal recovery system is 180°F (82°C) because the thermal energy from the exhaust flue gas was used to produce high pressure steam rather than to boost the final hot water temperature.

3. Low-pressure steam-generating systems using exhaust gas and/or engine jacket water heat. In mid- to large-capacity engines, low-pressure steam at pressures of up to 15 psig (2 bar) can be produced from engine coolant system rejected heat. This can be accomplished using high-temperature forced circulation systems or ebulliently cooled systems.

In an ebullient system (Figures 8-8 and 8-9), the heat of vaporization removes rejected heat from the engine. As hot water passes through the engine cooling passages, small steam bubbles are formed as the jacket water absorbs heat from the engine. The mixture rises to a steam separator above the engine where the steam is discharged and the water is recirculated back to the engine. No jacket water pump is required, since flow is assured by change in coolant density as it gains heat. The temperature differential available for engine cooling is usually quite low (2 to 3°F or 1.1 to 1.7°C). Two or more engines are often connected to one boiler, usually with separate gas passages to prevent exhaust gas from condensing on a shut-down engine. The boiler must be above the engine to provide sufficient head for recirculation.

Because of aggressive cooling required by major engine components, many modern gas engine designs tend to be incompatible with ebullient cooling. High coolant velocities and the associated small flow areas do not allow for adequate gravity flow. However, some engines are capable of operating with coolant temperatures as high as 260°F (127°C), allowing jacket water cooling systems to be used to generate 15 psig (2 bar) steam. As jacket water outlet temperatures are increased with forced high-temperature systems, the quantity of heat rejection to the jacket water will decrease and the heat rejection to lube oil, ambient air, and exhaust gas will increase.

Whether steam is raised by ebullient cooling or forced circulation cooling, steam pressure available from a reciprocating engine jacket water heat recovery system is limited to about 15 psig (2 bar). Where thermal loads require high-pressure steam, it is sometimes cost-effective to employ a steam compressor. In essence, this approach reduces net power generation by the prime mover plant to maximize the use of available rejected heat.

An alternative to providing high-pressure steam (or high-temperature hot water) to an existing process or load could be to redesign the process to operate on lower quality heat. This might involve, for example, changing distribution piping or heat exchange coils.

4. High-pressure steam generation systems using combustion engine exhaust heat. Heat recovery steam generators (HRSGs) are commonly used to recover heat from gas turbine and reciprocating engine exhaust. Elaborate systems use multiple-pressure boilers and can include feedwater heating economizers and superheaters. Steam output can be controlled either by a pressure regulator, which bypasses excess steam to a condenser, or by diverting the exhaust gas around the heat recovery unit. Sometimes a small condenser is included with a bypass valve setup to ensure that exhaust gas leakage around the diverter (which generates small quantities of unwanted steam) does not result in damage from condensation and corrosion.

Reciprocating Engine Coolant System Heat Recovery

Figures 8-4 through 8-8 present alternative heat recovery configurations that may be applied to reciprocating engine cogeneration systems. Note that each of these configurations features a load balancing heat exchanger (or condenser) that must be included in the engine loop, not the load loop. Figure 8-4 illustrates a standard temperature water system with series exhaust heat recovery. A muffler is included in series with the engine system.

Figures 8-5 and 8-6 show standard and high-temperature water system designs. The high-temperature water system employs elevated jacket water temperatures of 210 to 260°F (99 to 127°C). The standard thermostat and bypass are removed and replaced by an external control. A static head must be provided in the engine coolant circuit to assure a pressure of 4 to 5 psig (129 to 136 kPa) above the pressure at which steam forms. The source of this pressure may be a static head imposed by an elevated expansion tank or controlled air pressure in the expansion tank.

Figure 8-7 illustrates a high-temperature water steam system. A circulation pump forces water through the cylinder block to the steam separator. In the steam separator, some of the water flashes to steam at up to 14 psig (2 bar) and the water returns to the engine. Figures 8-8 and 8-9 illustrate ebullient systems. Notice the location of the pressure control valve, excess steam valve, and load balancing condenser in the ebullient systems.

Effective water treatment is essential in all of the above configurations to minimize deposits and avoid corrosion. In many applications, process water (e.g., from space heating systems) is of inferior quality and should not be circulated direct-

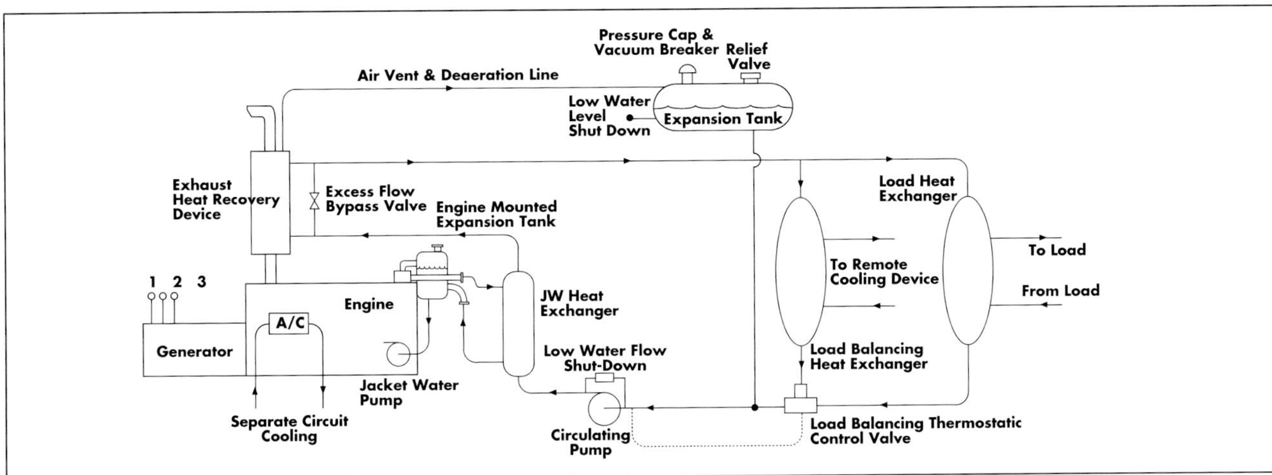

Fig. 8-4 Schematic of Reciprocating Engine Standard Temperature Water System with Series Exhaust Heat Recovery. Source: Caterpillar Engine Division

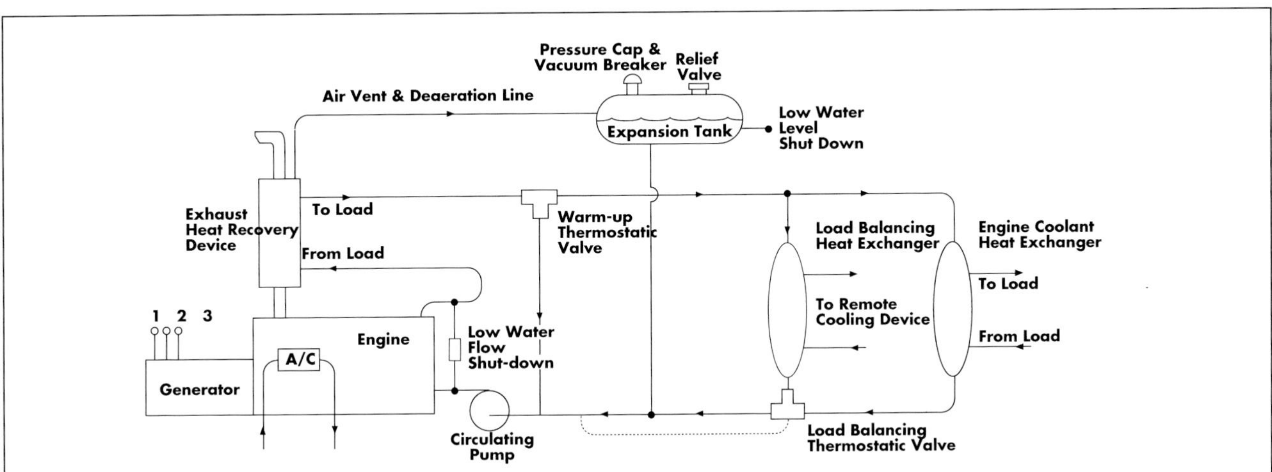

Fig. 8-5 Schematic of Reciprocating Engine Standard Temperature Water Heat Recovery System. Source: Caterpillar Engine Division

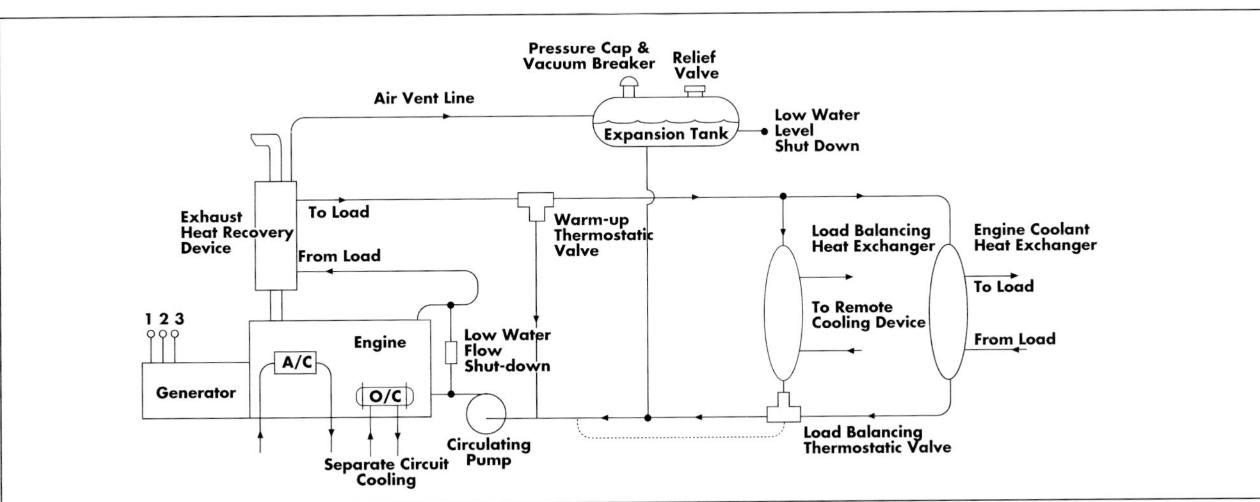

Fig. 8-6 Schematic of Reciprocating Engine High-Temperature Heat Recovery System. Source: Caterpillar Engine Division

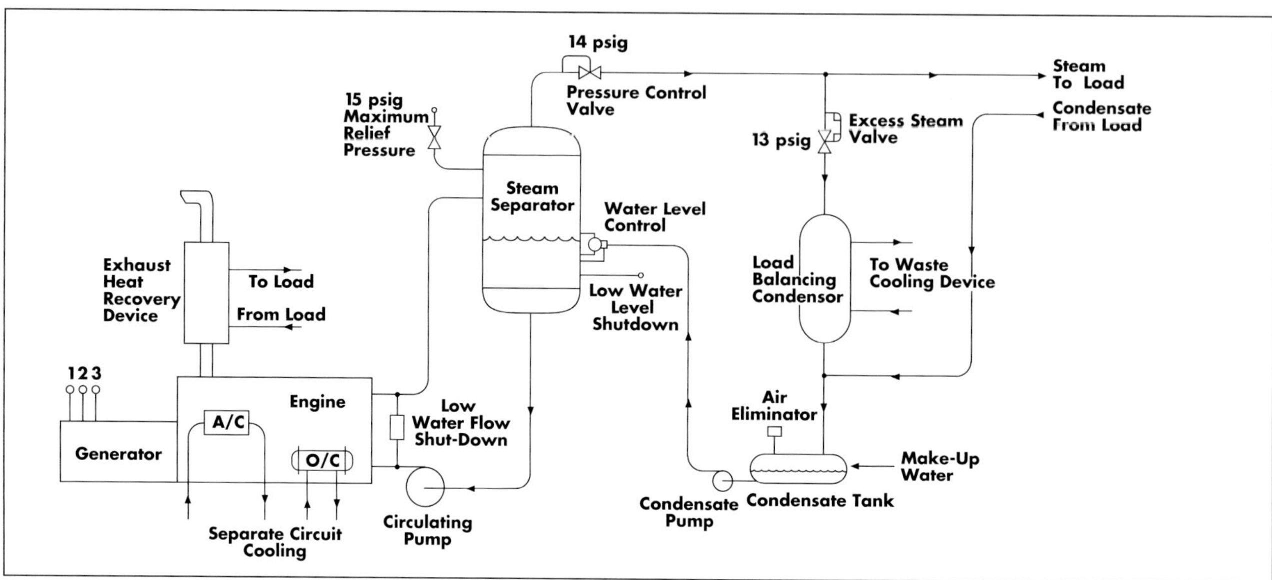

Fig. 8-7 Schematic of Reciprocating Engine High-Temperature Water-Steam Heat Recovery System. Source: Caterpillar Engine Division

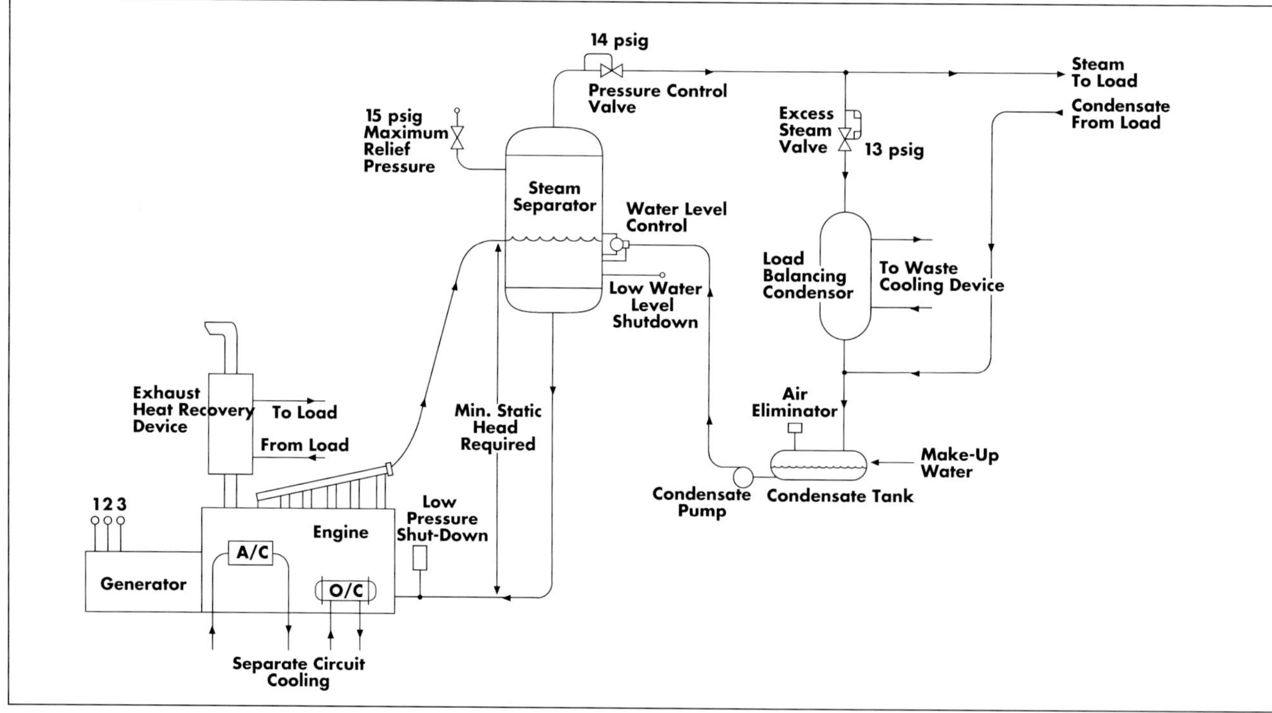

Fig. 8-8 Schematic of Reciprocating Engine Ebullient Cooling Heat Recovery System Producing 14 psig Steam. Source: Caterpillar Engine Division

ly to engines. It is important to carefully follow manufacturers' specifications for water quality, as well as for safety.

Figure 8-10 shows a reciprocating engine heat recovery system featuring intercooler, lube oil, jacket water and exhaust heat exchangers piped in parallel. At full load the engine-generator system produces 1,203 kW and the heat recovery system delivers 5.95 million Btu (6,271 MJ) per hour of recovered heat in the form of hot water. In one circuit, 119 gpm (450 lpm) of incoming water at 85°F (29°C) is heated to 132.8°F (56°C) as it passes through the coolant system heat exchangers, absorbing 2.88 million Btu (3,036 MJ) per hour. The flow is then com-

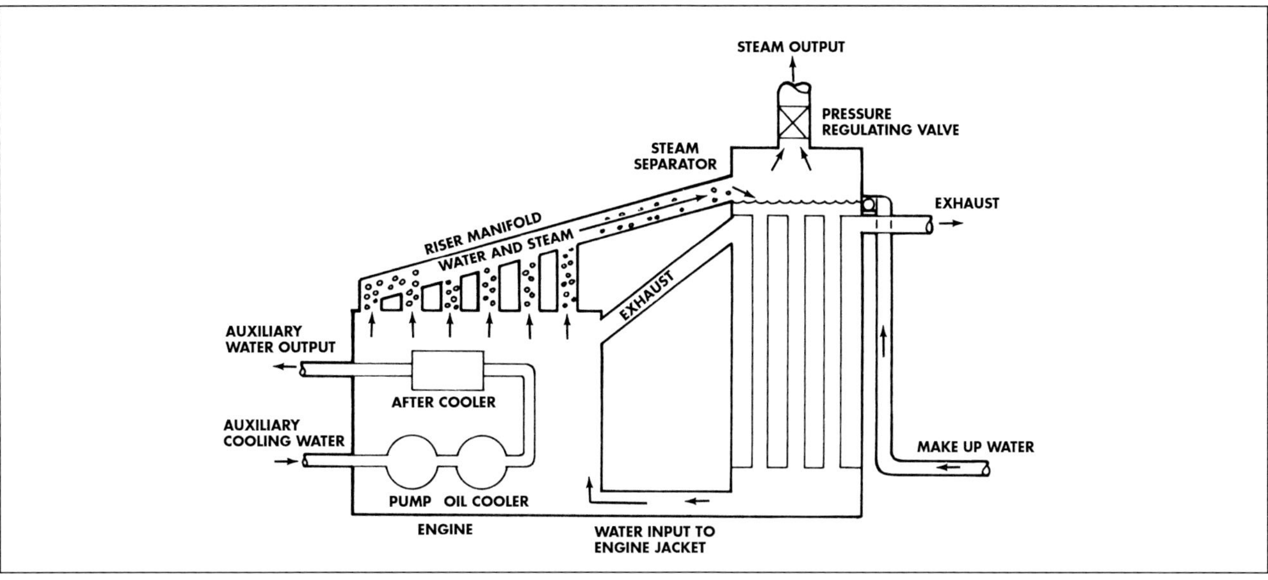

Fig. 8-9 Diagram of Ebullient Cooling Heat Recovery System. Source: Caterpillar Engine Division

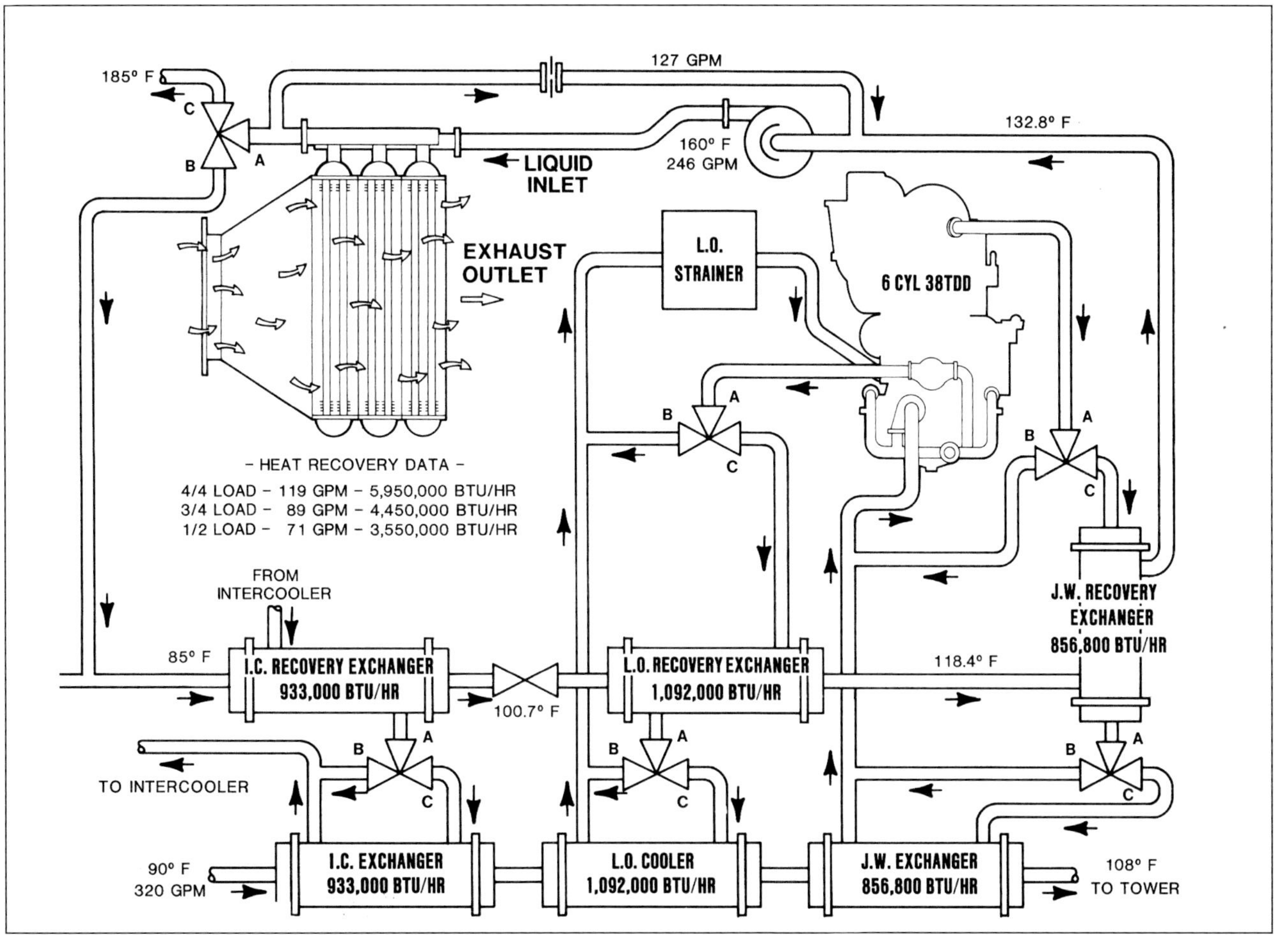

127 GPM
185° F
C
B
A
160° F
246 GPM
132.8° F
LIQUID INLET
EXHAUST OUTLET
L.O. STRAINER
6 CYL 38TDD

- HEAT RECOVERY DATA -
4/4 LOAD - 119 GPM - 5,950,000 BTU/HR
3/4 LOAD - 89 GPM - 4,450,000 BTU/HR
1/2 LOAD - 71 GPM - 3,550,000 BTU/HR

FROM INTERCOOLER
85° F
I.C. RECOVERY EXCHANGER 933,000 BTU/HR
100.7° F
L.O. RECOVERY EXCHANGER 1,092,000 BTU/HR
118.4° F
J.W. RECOVERY EXCHANGER 856,800 BTU/HR
TO INTERCOOLER
90° F
320 GPM
I.C. EXCHANGER 933,000 BTU/HR
L.O. COOLER 1,092,000 BTU/HR
J.W. EXCHANGER 856,800 BTU/HR
108° F
TO TOWER

Fig. 8-10 Schematic Diagram of Reciprocating Engine Hot Water Heat Recovery System. Source: Fairbanks Morse Engine Division

bined with the exhaust heat recovery circuit, elevating the temperature from 160 to 185°F (71 to 85°C). Of the combined flow, 127 gpm (481 lpm) is recirculated and 119 gpm (450 lpm) is passed on to process at 185°F (85°C). In the absence of a thermal load, the 2.88 million Btu (3,036 MJ) per hour coolant system load would be rejected to a cooling tower loop.

Heat Recovery Boilers

Heat recovery boilers are used to produce hot water or steam from the heat of passing exhaust gases. The **HRSG** is essentially a counterflow heat exchanger composed of a series of superheater, boiler, and economizer sections positioned from gas inlet to gas outlet to maximize heat recovery and supply the rated steam flow at the proper temperature and pressure. In some cases, exhaust gas heat recovery may also provide for deaeration or feedwater preheating. Alternative HRSG configurations include firetube boilers and natural forced circulation or once-through watertube boilers. Unfired heat recovery boilers and conventional fired boilers are similar in many respects. The principal difference is that exhaust gases are generally of much lower temperature, as compared with the combustion gases in a fired boiler, resulting in a far lower rate of transfer by radiant heat.

Representative examples of heat recovery boilers applied to reciprocating engine and gas turbine systems are provided in Figures 8-11 through 8-14. Figures 8-11 and 8-12 include heat recovery boilers applied to reciprocating engines. Figure 8-13 shows a two-pressure heat recovery boiler applied in a 45 MW combined-cycle cogeneration plant featuring a gas turbine and steam turbine. Figure 8-14 shows a multiple-pressure heat recovery boiler featured in a 150 MW plant.

Firetube Boilers

Exhaust gas firetube boilers route the exhaust gas through a bank of tubes as water is boiled in the surrounding chamber. Design and operation is much the same as with fuel-fired firetube boilers. Figure 8-15 illustrates the exhaust flow path through a single-pass firetube heat recovery boiler. Figure 8-16 illustrates a similar unit featuring two passes.

Due to their relatively large size, weight, and other design limitations, firetube boilers are generally limited to pressures below 250 psig (18.3 bar) and outputs of under 50,000 lbm (23,000 kg) per hour and do not offer superheat capabilities. They can be equipped with economizers to improve efficiency. They are generally less expensive and require less headroom than watertube boilers of comparable capacity, especially in smaller capacities.

Firetube boilers produce an additional benefit of sound attenuation, which results from the high-pressure drop through the fire tubes. For this reason, they are sometimes referred to as heat recovery silencers. This provides an economic benefit in reciprocating engine applications, in that muffler requirements for sound attenuation are reduced or, in some cases, eliminated.

Firetube hot water boilers are often included in small packaged reciprocating engine-driven systems. Hot water or steam units are sometimes, but not usually, used for small gas turbine applications. In addition to the disadvantage of size and weight, firetube boilers tend to produce high turbine

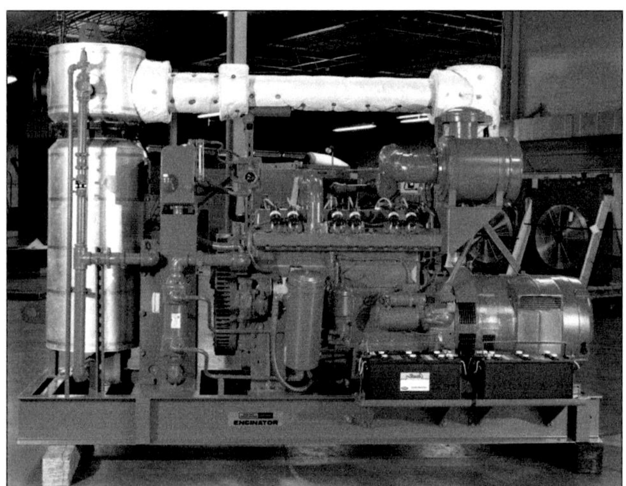

Fig. 8-11 Heat Recovery Boiler Applied to Reciprocating Engine. Source: Vaporphase by Engineering Controls, Inc.

Fig. 8-12 Jacket Water and Exhaust Heat Recovery Boiler Applied to a 150 kW Reciprocating Engine-Generator Set. Source: Waukesha Engine Division

Fig. 8-13 HRSG featured in 45 MW Combined-Cycle Plant. Source: Deltac

exhaust back-pressure and usually require an induced draft fan. Figure 8-17 shows a vertical packaged HRSG designed to produce low-pressure steam from reciprocating engine coolant system and exhaust heat. The flow paths of exhaust and jacket water are labeled.

Figure 8-18 shows three packaged jacket water and exhaust heat recovery silencers designed to generate 15 psig (2 bar) steam from 450 kW ebullient-cooled, gas-fired, reciprocating engine-generator sets. At full load, each gas engine has the following recovery:

Jacket Water			
(ebullient-cooled)	–	1,614,000 Btu/h	(1,703 MJ/h)
Exhaust	–	859,000 Btu/h(906 MJ/h)	
Total	–	2,473,000 Btu/h	(2,609 MJ/h)

As indicated in Figure 8-19, firetube units may also be equipped with supplementary firing. The separated chambers allow for firing the forced draft burner, without

Fig. 8-14 Multiple-Pressure HRSG Featured in 150 MW System. Source: Nooter/Eriksen

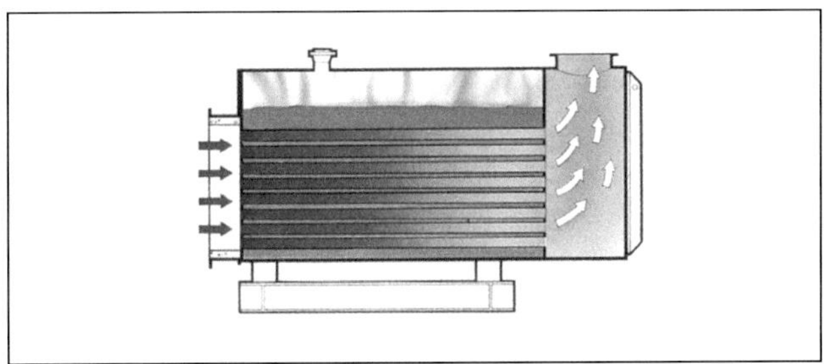

Fig. 8-15 Single-Pass Firetube Heat Recovery Boiler. Source: Superior Boiler Works

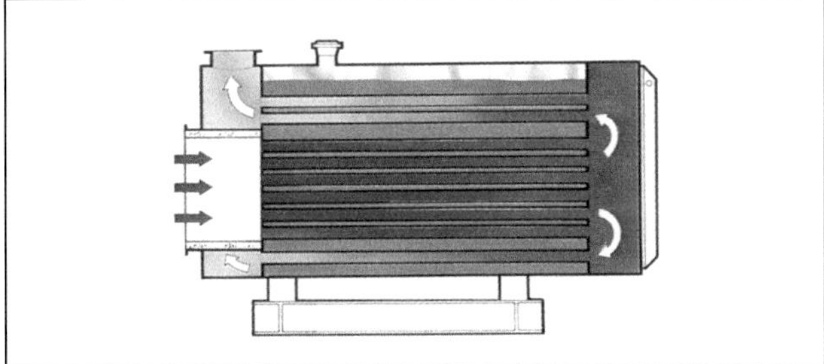

Fig. 8-16 Two-Pass Firetube Heat Recovery Boiler. Source: Superior Boiler Works

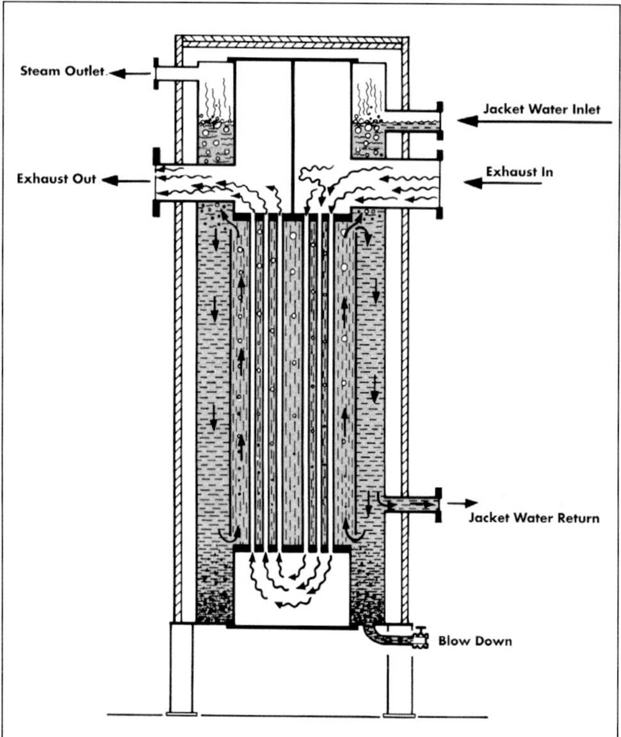

Fig. 8-17 Packaged Heat Recovery Low-Pressure Steam Generator.
Source: Vaporphase by Engineering Controls, Inc.

Fig. 8-18 Packaged Exhaust and Coolant System Recovery Silencers, Producing 15-psig Steam from Ebullient-Cooled Reciprocating Engines.
Source: Vaporphase by Engineering Controls. Inc.

Fig. 8-19 Dual-Chamber Firetube Heat Recovery Boiler Featuring Supplementary Firing Capability.
Source: Superior Boiler Works

Fig. 8-20 Supplementary Gas-Fired Firetube-Type HRSG Producing 140-psig Steam with Heat from Reciprocating Engine.
Source: Vaporphase by Engineering Controls, Inc.

running the heat recovery induced draft fan, or for the simultaneous firing of exhaust gas and the supplementary forced draft burner.

Figure 8-20 shows a reciprocating engine cogeneration system featuring two gas-fired heat recovery silencers (firetube HRSGs). Each HRSG can produce up to 10,000 lbm (4,500 kg) per hour of steam at 140 psig (10.7 bar), using heat recovered from the each of the two 1,928 kW dual-fuel reciprocating engine-generator sets.

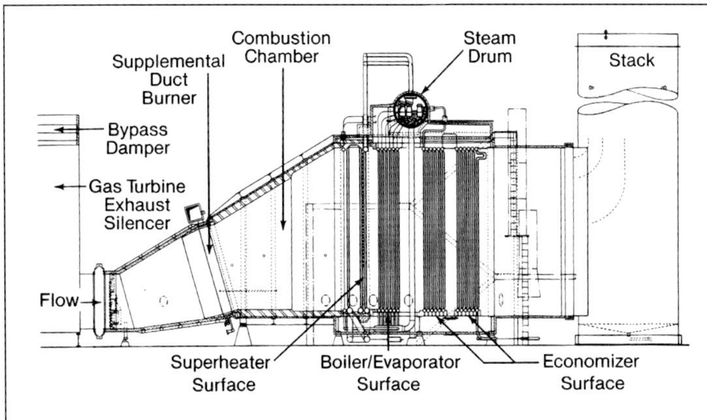

Fig. 8-21 Sectional View of Large HRSG Arrangement. Source: Babcock & Wilcox

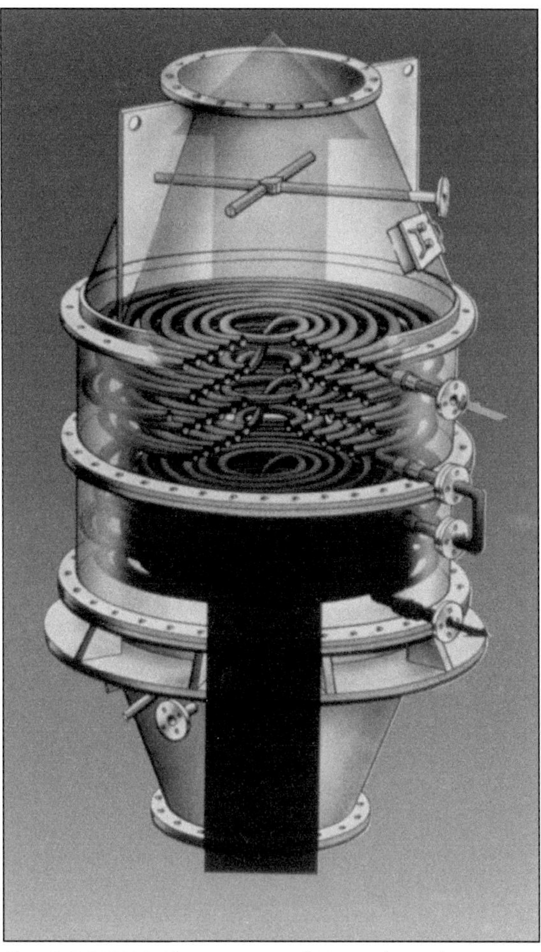

Fig. 8-24 HRSG Flow Diagram of Monotube Coil Design. Source: Clayton Industries

Fig. 8-22 Large Modular Natural Circulation HRSG. Source: Babcock & Wilcox

Fig. 8-23 Fin Tubes Used in Watertube Boiler. Source: Nooter/Eriksen

WATERTUBE BOILERS

As discussed in Chapter 7, watertube boilers are primarily employed in very large applications. Figure 8-21 shows a sectional view of a large HRSG arrangement, and Figure 8-22 shows a large modular HRSG featuring natural circulation design. Fins (Figure 8-23) are often used to increase heat transfer rates and reduce the total amount of surface area required.

Figure 8-24 is an HRSG flow diagram showing counter-flow through smooth monotube coil design. The coil tube is wound in a spiral pattern arranged to control velocities of the boiler gases. Figure 8-25 shows a five-section unit featuring this design, with a side inlet, top outlet configuration.

As in conventional-fired boilers, HRSGs can include economizers downstream of the boiler section to preheat feedwater to near saturation temperature. Superheaters may also be used, performing much the same function in conventional-fired or unfired heat recovery boilers.

Figure 8-26 shows four HRSGs applied with twin-pack gas turbine systems. The twin-pack systems each feature two FT8 aeroderivative gas turbines, with a combined power output of about 50,000 kW. The HRSGs are highly effective triple-pressure units that supply high-pressure steam at 1,005

Fig. 8-25 Five Section HRSG Featuring Monotube Coil Design. Source: Clayton Industries

Fig. 8-26 Four Triple-Pressure HRSG Applied with Twin-Pack Gas Turbines. Source: Nooter/Eriksen

psig/805°F (70 bar/429°C), intermediate-pressure steam at 190 psig/550°F (14 bar/288°C), and low-pressure steam at 24 psig/266°F (2.7 bar/130°C). Figure 8-27 shows a labeled cutaway illustration of one of these units. Features of this unit include carbon monoxide (CO) and selective catalytic reduction (SCR) catalysts, integral deaerator, induced draft fan, and stack damper.

The combined exhaust of the two gas turbines manifolded into a single HRSG can generate from 100,000 to 190,000 lbm (45,000 to 86,000 kg) per hour of steam, depending on ambient conditions. Supplementary firing can increase steam production by as much as 50%. Figure 8-28 shows expected steam generation performance curves for the twin-pack system based on a simple single-pressure level HRSG.

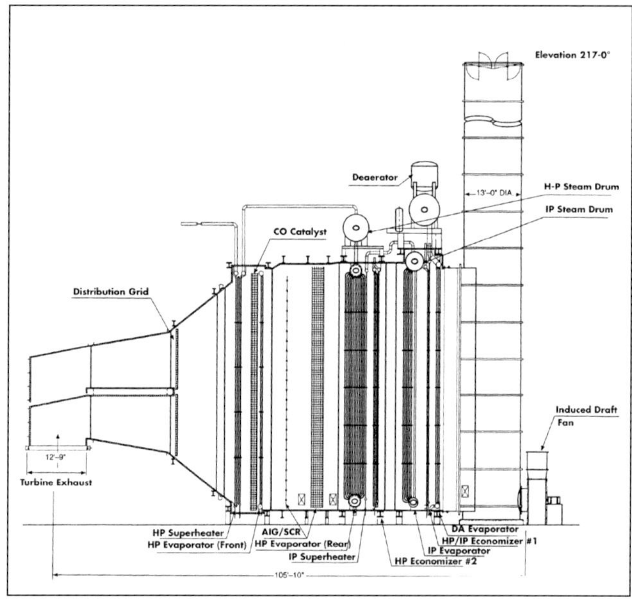

Fig. 8-27 Labeled Cutaway Illustration of Triple-Pressure HRSG with Integral Deaerator and CO and SCR Catalysts. Source: Nooter/Eriksen

Once-Through HRSG

The once-through HRSG is an emerging technology that consists of one or more serpentine circuits encompassing the economizer, boiler, and superheater sections. This eliminates the need for steam drums, level controls, blowdown, bypass controls, recirculation systems, and active water treatment. Also, the HRSG can be operated dry. These HRSGs can be constructed of a corrosion resistant material that permits lower stack temperatures.

Dry operation of the heat recovery boiler at turbine exhaust temperature is necessary for systems without a gas bypass to permit gas turbine operation whenever the boiler is inoperative. Some designs use carbon steel tubes and are rated at up to 6,000 hours dry operation at 900°F (482°C). Dry operation is possible for longer periods at reduced temperature.

Figure 8-29 is a flow diagram of a once-through steam generator showing the temperatures of the exhaust gas and the water/steam as they progress through the unit. Water and steam flow through continuous serpentine tube circuits that are countercurrent to the hot gas flow.

Figure 8-30 depicts a representative boiler performance profile for a large gas turbine application featuring steam injection. The plot shows gas and waterside temperatures, relative heat duties, and selective catalytic reduction (SCR) operating windows. Boiler duty sections are shown with

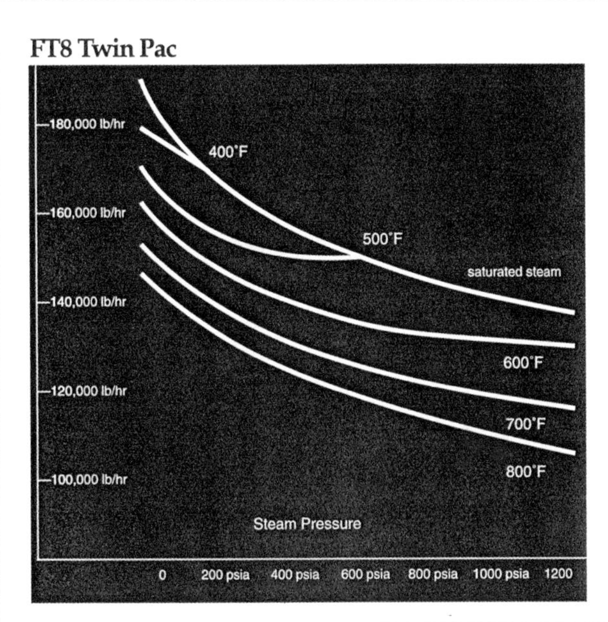

Fig. 8-28 Expected Steam Generation Performance Curves for FT 8 Twin Pack System with Single-Pressure HRSG.
Source: United Technologies Turbo Power Division

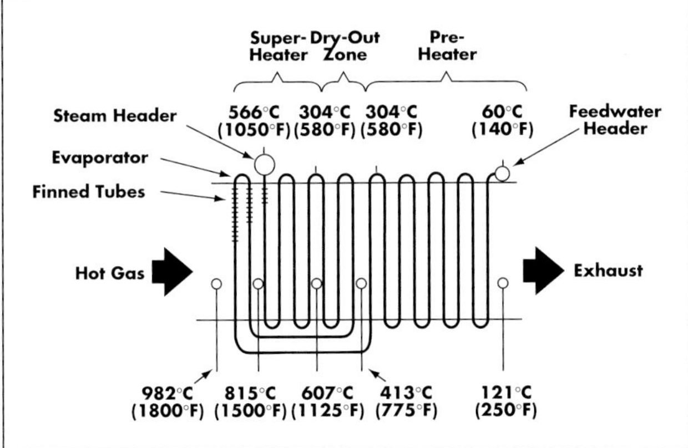

Fig. 8-29 Once-Through Steam Generator Flow Diagram.
Source: Solar Turbines

areas in proportion to their heat duty.

The primary disadvantage of once-through HRSG's is that they are not cost competitive on small scale cogeneration applications generating less than 40,000 lbm/h of steam and the units require high quality boiler feed water that usually require a demineralizer or reverse osmosis system to treat make-up water.

Supplementary-Fired Systems

Supplementary firing is the use of a burner in the sides of the duct upstream of the HRSG to raise the temperature of the entering gas stream. This is most commonly applied in

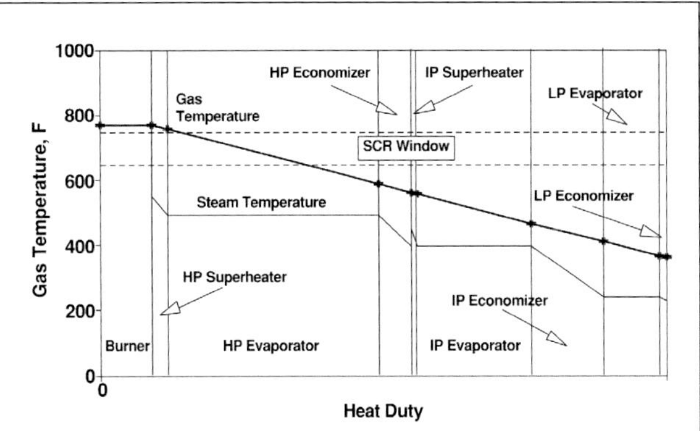

Fig. 8-30 HRSG Temperature Profile for Gas Turbine STIG-Cycle Application. Source: Cogen Designs, Inc.

gas turbine applications where the oxygen-rich (15 to 18%) exhaust can provide efficient combustion. Supplementary firing can also be applied with certain reciprocating engines, though on a more limited basis due to a lower oxygen content in the exhaust. The typical exhaust temperature from a turbine is in the range of 875 to 1,000°F (470 to 540°C). Duct firing can increase the temperature significantly, allowing increased steam production and higher temperatures and pressures. Figure 8-31 shows a 225 million Btu (237,375 MJ) per hour in-line natural gas duct burner. This unit has been applied for supplementary firing of exhaust from a 40 MW gas turbine into an HRSG.

Figure 8-32 shows the placement of the duct burner between the gas turbine and the HRSG. Efficiency is increased with supplemental firing because almost every Btu (kJ) of burner fuel is converted to useful thermal energy. This is because the mass flow and final temperature of the exhaust remain almost constant during supplementary firing. The increased temperature difference across the HRSG results in more heat recovered per lbm (kg) of exhaust gas.

With duct firing, the exhaust gas entering the HRSG can be elevated to as high as 1,800°F (980°C), given the typical operating limits of the ducting materials. Increased gas stream temperatures ranging up to nearly 3,000°F (1,650°C) can be achieved by adding a radiant heat section to the system. This section transfers radiant energy from the burner flame to water contained in a membrane on the outer wall of the gas duct. Elevated steam temperatures increase the efficiency of the steam cycle. However, steam temperatures exceeding 1,000 or 1,100°F (540 to 590°C) may require an upgrade in material quality, which will increase the cost of the equipment considerably.

Fig 8-33 shows a large horizontal-gas-flow HRSG with supplementary firing. Units with horizontal gas flow use

Fig. 8-31 225 MMBtu/h In-Line Duct Burner Applied for Supplementary Firing of Gas Turbine Exhaust. Source: Coen Company

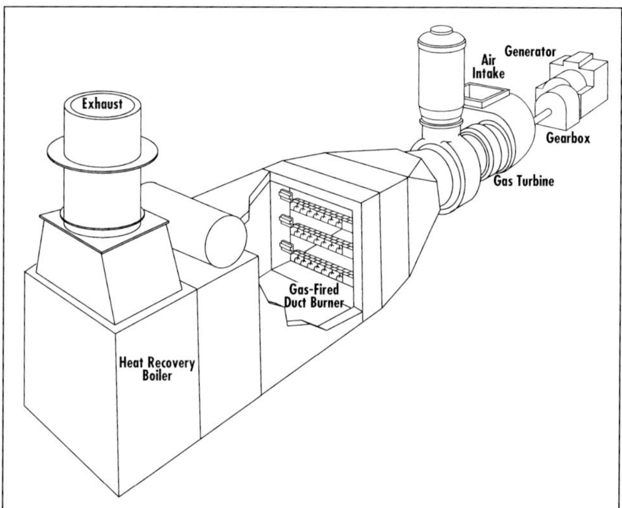

Fig. 8-32 Illustration Showing Placement of Duct Burner Between Gas Turbine and HRSG. Source: Coen Company

vertical tubes connected to headers at the top and bottom, with natural circulation.

Assuming a high oxygen content, in excess of 15%, a portion of the exhaust gas can alternatively be used for combustion air in a conventional-fired boiler. Exhaust gas temperatures can thereby be raised to about 3,000°F (1,650°C), greatly increasing steam production and allowing higher temperature and pressure. As the rate of supplementary firing increases above the level of available oxygen for combustion, the overall process becomes less thermally efficient because fresh combustion air must be introduced.

Figure 8-34 shows a large capacity boiler designed for supplementary firing in conjunction with a gas-turbine combined cycle. This unit is designed and sized similar to units using outside air through forced-draft fans. The stack temperature can be dropped economically to within 100°F (56°C) of the incoming feedwater temperature. Since the boiler is sized for a flue gas weight based on fresh-air firing, a portion of the gas turbine exhaust is bypassed to a separate steam-generating bank.

Figure 8-35 provides an example of the impact of supplementary firing and operation at reduced gas turbine load on HRSG temperature profiles. Boiler duty sections are shown, with areas in proportion to their heat duty. Also shown are the operating temperatures for SCR systems. In this HRSG design, an SCR system would be placed within the LP Evaporator section of the boiler and would be operable over the full range of load conditions.

Figures 8-36 and 8-37 provide comparative illustrations of two multiple pressure HRSGs. One is unfired, the other is supplementary-fired. Temperature distribution profiles are shown for both the gas and steam/water. Notice the elevated gas temperature resulting from supplementary firing.

Figure 8-38 shows one of two supplementary gas-fired triple-pressure HRSGs applied to a gas turbine in a 100 MW plant at an industrial facility. Figure 8-39 shows a labeled cutaway schematic of this unit. The HRSG produces high-pressure steam at 1,290 psig/925°F (90 bar/496°C), intermediate-pressure steam at 290 psig/430°F (21bar/221°C), and low-pressure saturated steam at 10 psig (1.7 bar).

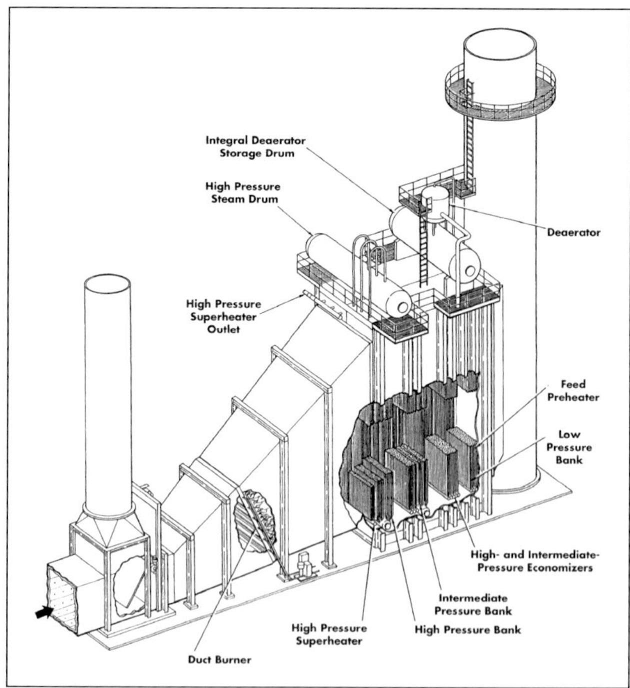

Fig. 8-33 Horizontal-Gas-Flow HRSG with Supplementary Duct Firing. Source: Combustion Engineering, Inc., Reprinted with permission from Combustion Fossil Power, 4th Ed., 1991

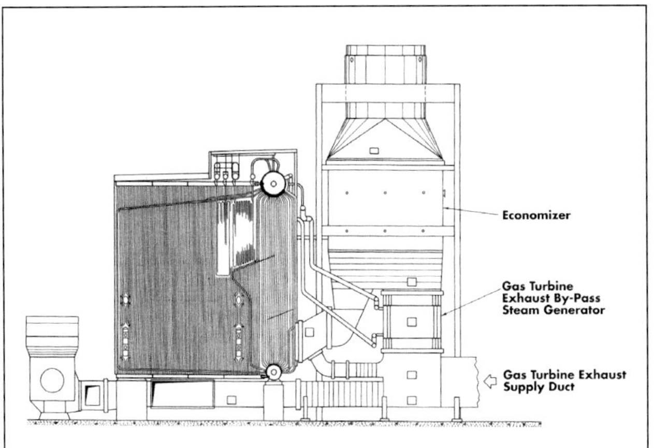

Fig. 8-34 Boiler Designed for Supplementary Firing in Conjunction with Gas-Turbine Combined Cycle. Source: Combustion Engineering, Inc., Reprinted with permission from Combustion Fossil Power, 4th Ed., 1991

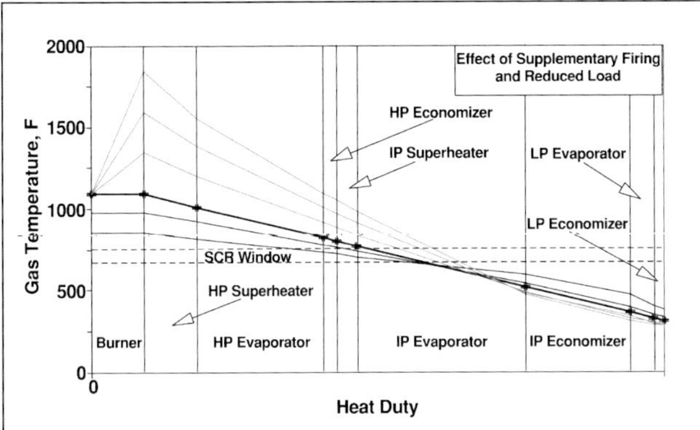

Fig. 8-35 Temperature Impact of Supplementary Firing Reduced Gas Turbine Load on HRSG. Source: Cogen Designs, Inc.

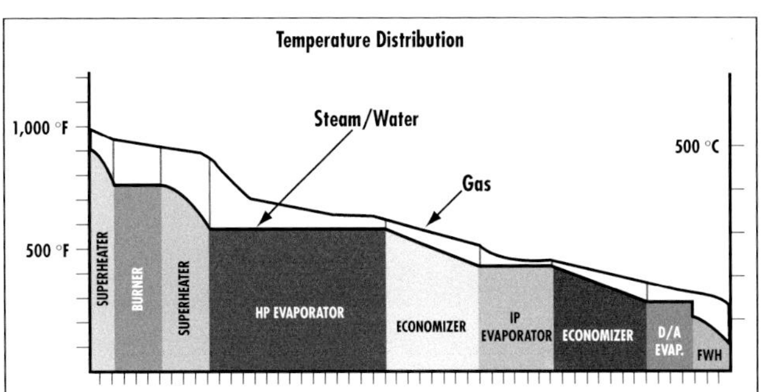

Fig. 8-36 Sectional Schematic of Unfired HRSG with Temperature Distribution Profile. Source: Nooter/Eriksen

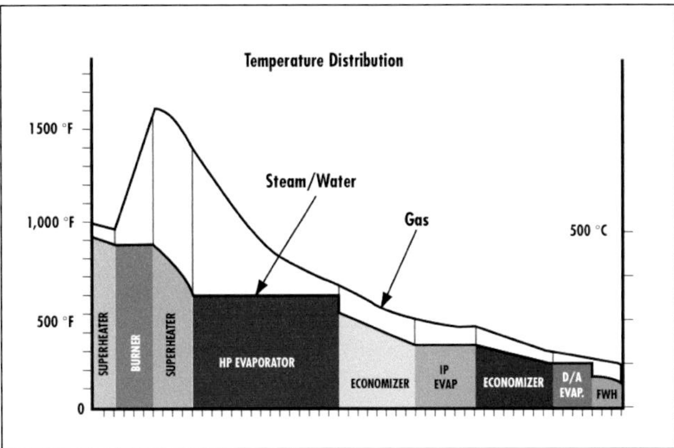

Fig. 8-37 Sectional Schematic of Supplementary-Fired HRSG with Temperature Distribution Profile. Source: Nooter/Eriksen

Factors Affecting Heat Recovery Efficiency

In all closed-loop heat exchangers, the heating gas or liquid is maintained throughout the process at a temperature greater than that of the heated fluid. This difference is often referred to as the **approach temperature**. While it is possible to bring the temperatures of the two flows down very close to equilibrium, it is not usually practical to do so. The principle of diminishing returns dictates that as the approach temperature falls, the added investment and space requirements for increased heat exchange surface area and effectiveness becomes cost-prohibitive. Further, because water evaporates at constant temperature while the exhaust gas stream continues to decrease in temperature while it cools, a temperature difference between the two streams is unavoidable.

Pinch point refers to the difference between the gas temperature leaving the evaporator and the saturation

Fig. 8-38 Supplementary Gas-Fired Triple-Pressure HRSG Applied to Gas Turbine in 100 MW Plant at Industrial Facility. Source: Nooter/Eriksen

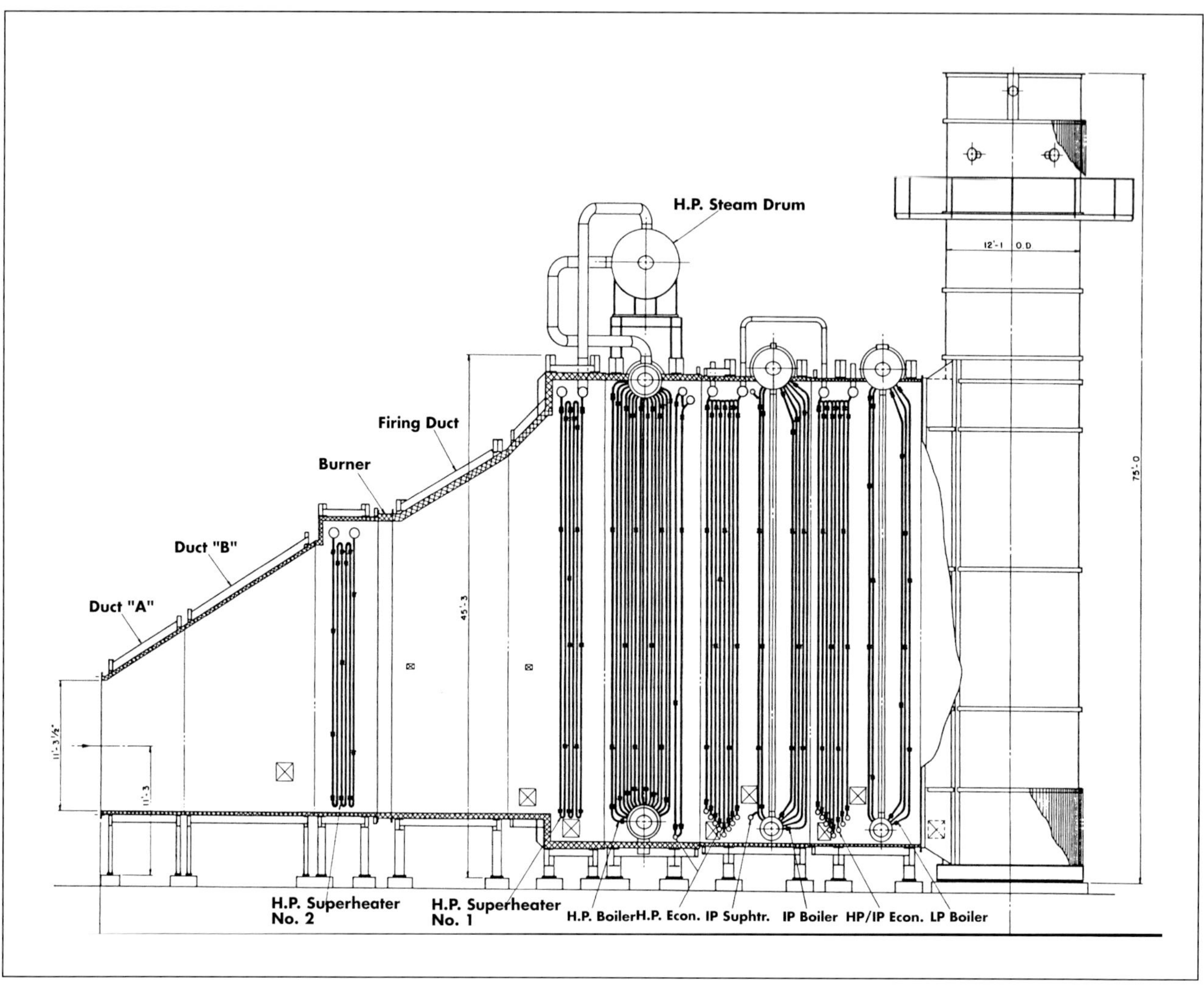

Fig. 8-39 Labeled Cutaway View of Supplementary-Fired HRSG Applied to Gas Turbine Exhaust.
Source: Nooter/Eriksen

temperature of the water in boiler applications. Small pinch point and superheater approach temperatures result in larger heat transfer surfaces and higher capital cost. Typical pinch point differentials range from 15 to 60°F (8 to 33°C). Generally, an HRSG with a pinch point in the range of 15 to 25°F (8 to 14°C) will have about 50% more surface in the evaporating section than a unit with a pinch point in the range of 40 to 50°F (22 to 28°C). Superheater approach temperature differentials range from 35 to 60°F (19 to 33°C). The economizer approach temperature is typically set to avoid economizer steaming at the design point. Typical economizer approach temperature differentials range from 10 to 30°F (6 to 17°C).

Tables 8-1 and 8-2 show the effect of pinch point and pressure on HRSG performance. Both are based on performance for a gas turbine-generator set that produces 3,621 kW, consuming 47.50 million Btu (50,100 MJ) per hour of fuel and discharging 141,734 lbm (64,300 kg) per hour exhaust gas at 948°F (510°C). Table 8-1 shows the effect of pinch point on HRSG performance for a system requiring 250 psig (18.3 bar) saturated steam. Notice that as pinch point decreases, steam production and system thermal efficiency increase. Table 8-2 shows the effect of pressure on HRSG performance for a given pinch point of 40°F (22°C). Notice that as required steam pressure increases, recoverable thermal energy, steam production rate, and thermal efficiency all decrease, while final exhaust stack temperature rises.

Fouling factors must be considered in all heat exchanger selections. Fouling reduces the efficiency of heat transfer because it reduces the overall heat transfer coefficient. Usual causes include: external tube deposits, such as oil

Table 8-1 Effect of Pinch Point on HRSG Performance

Pinch Point (°F)	Heat Transferred (MMBtu/h)	Steam Produced (lbm/h)	Thermal Efficiency (%)	Stack Temperature (°F)
40	21.58	21,117	71.4	338
30	22.00	21,538	72.3	327
20	22.44	21,958	73.3	314

Source: Solar Turbines

Table 8-2 Effect of Pressure on HRSG Performance

Pressure (psig)	Heat Transferred (MMBtu/h)	Steam Produced (lbm/h)	Thermal Efficiency (%)	Stack Temperature (°F)
100	22.60	22,383	73.6	310
250	21.58	21,117	71.4	338
500	20.52	20,018	69.2	368
750	19.82	19,431	67.7	388

Source: Solar Turbines

films or solids from a liquid or gaseous stream; internal tube-surface scaling due to precipitation of solid compounds from solution; or corrosion of surfaces, external or internal. To minimize fouling, natural gas exhaust is preferred. Using gas rather than oil will allow greater heat transfer efficiency and smaller heat exchange surface requirements.

Another critical parameter, in heat recovery system design is the exit temperature of the exhaust gas as it enters the stack and the potential for corrosion of the heat exchange surfaces. The limiting factor is the temperature at which, combustion product acids and water condense onto metal surfaces, causing corrosion.

The **acid dewpoint**, the temperature at which acids in the exhaust gas will begin to condense, is a function of the amount of sulfur in the fuel and, subsequently, in the exhaust gas. Exhaust streams from natural gas firing have a lower dewpoint than exhaust from fuels containing sulfur, allowing improved heat recovery efficiency.

In an HRSG, corrosion is a major concern in the economizer section, where exhaust gases are at their coolest prior to exiting the stack. Attention must be paid to both the cooling gas and the feedwater temperatures. Because the temperature of the tube metal tends to be near that of the feedwater at the inlet, it is advisable to maintain feedwater temperatures above 212°F (100°C) to avoid dewpoint precipitation. Deaerators operating at pressures above 5 psig (1.4 bar), which corresponds to a saturation temperature of 228°F (109°C), often provide for this.

The potential for **water dewpoint** corrosion increases when water or steam is injected into a gas turbine for emissions control or, in the case of steam, for increased power output and thermal fuel efficiency. Injection results in a higher dewpoint due to the moisture content of the exhaust stream.

The size of the heat recovery unit also depends on the pressure drop on both the gas and liquid side. The pressure drop of an exhaust gas system includes the exhaust gas ducting, heat recovery unit, auxiliary silencers, and outlet duct to the stack. System designs typically allocate as high of a pressure drop as possible at the heat recovery unit in order to minimize size and cost. Typical HRSG back-pressures are 10 to 15 in. wg (1.9 to 2.8 cm Hg).

Excess pressure drop can impose a high back-pressure on prime mover systems, resulting in loss of power and efficiency, as well as overheating. In relation to gas turbines, back-pressure reduces the pressure ratio across the turbine section and, therefore, the power output of the turbine. The turbine will lose about 0.25% of power per inch of water back pressure.

The pressure and temperature of the steam being generated are limiting factors in exhaust gas temperature and in resulting heat recovery efficiency. Figure 8-40 is an example of the impact of final stack temperature on overall system efficiency.

Performance Correction Factors

Figure 8-41 shows the steam production capability for the exhaust of a 17,000 hp (12,500 kW) single-shaft gas turbine at full load under ISO conditions with both an unfired and fired HRSG. These curves cover a wide range of steam pressures and temperatures typical of many industrial applications. The full load gas turbine heat rate in this example is about 9,000 Btu/hp-h (7,000 kJ/kWh) and the exhaust flow rate is about 385,000 lbm (175,000 kg) per hour at a temperature of about 1,017°F (547°C). The HRSG pinch point temperature difference is 27°F (15°C) and both inlet and exhaust losses are assumed to be 3 in. wg (0.6 cm Hg). With supplementary firing, the after-burning firing temperature is assumed to be about 1,470°F (800°C).

Correction factors must be applied to rated exhaust flow and temperature to accurately establish steam production capability under actual site conditions over the full range of operating loads. Manufacturers provide correction factors for elevation, compressor inlet temperature, inlet and exhaust pressure losses, water or steam injection, and part-

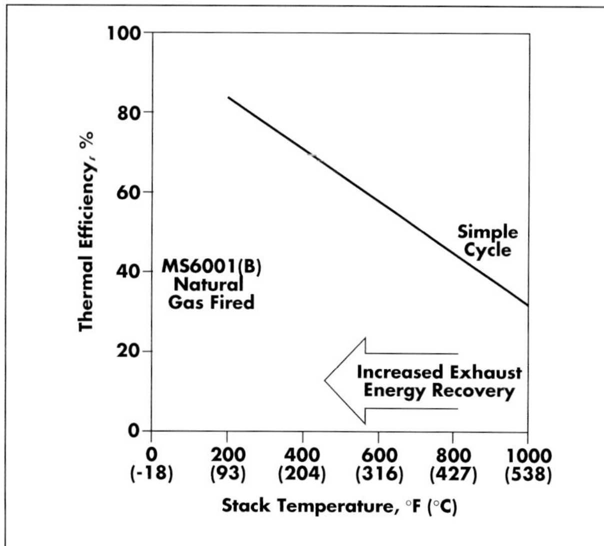

Fig. 8-40 Gas Turbine System Overall Thermal Efficiency vs. Stack Temperature. Source: General Electric Company

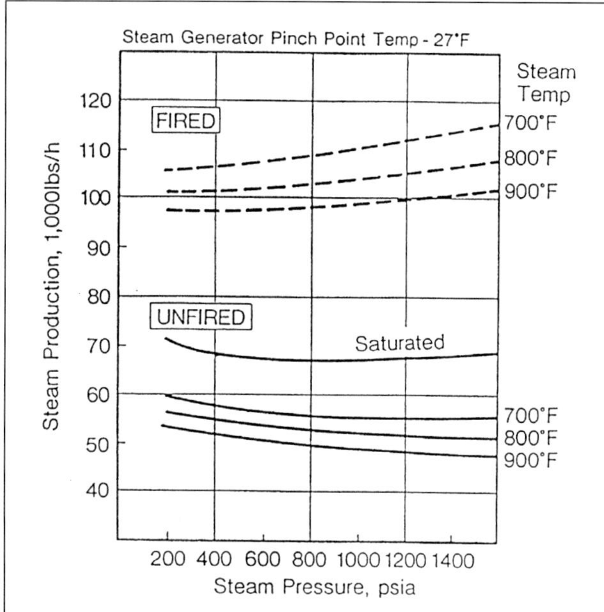

Fig. 8-41 Full-Load Steam Production Capability for Exhaust of 17,000 hp Single-Shaft Gas Turbine at ISO Conditions. Source: United States Turbine Corp./Mitsubishi Heavy Industries, Ltd.

load operation.

The following example lists correction factors for the gas turbine unit described above (Note that while there is a degree in similarity between gas turbines, correction factors will vary from one model to another.):

1. If site elevation was 1,400 ft (430 m) above sea level, the correction factor for exhaust flow is about 0.95. Thus, the exhaust flow would be adjusted by multiplying the exhaust flow of 385,000 lbm (175,000 kg) per hour at ISO conditions by 0.95, which would yield a corrected exhaust flow rate of about 366,000 lbm (166,000 kg) per hour. In this example, exhaust temperature is not affected by elevation.
2. If actual site ambient temperature was 80°F (27°C), the correction factor for exhaust flow would be about 0.95 and the correction factor for exhaust temperature would be about 1.02. Thus, the rated temperature of 1,017°F (547°C) would be corrected to 1,037°F (558°C) and the exhaust mass flow would be corrected to 347,700 lbm (157,715 kg) per hour.
3. The temperature correction factor for inlet pressure drop is to add 6.7°F (3.7°C) for every 10 in. (25 cm) of water excess loss in inlet above the base condition of 3 in. (8 cm) of water. At 2 in. (5 cm) of excess loss, the temperature correction would be 1.34°F (0.74°C), yielding a corrected exhaust temperature of about 1,038.3°F (559.1°C). Correspondingly, the correction factor for exhaust mass flow at 10 in. (25 cm) of excess loss in inlet is about 0.975, which would yield a corrected exhaust mass flow of 339,008 lbm (153,773 kg) per hour.
4. The temperature correction factor for exhaust pressure drop is to add 1.04°F (0.58°C) for every inch of water excess loss in exhaust, above the base condition of 3 in. (7.6 cm) of water. At 8 in. (20 cm) of excess loss, the temperature correction would be 5.36°F (2.98°C), yielding a corrected exhaust temperature of about 1,044.2°F (562.3°C). No exhaust flow correction is required.
5. If steam injection was applied at a rate of 1 lbm per lbm of fuel, the exhaust mass flow correction factor would be about 1.02, yielding a corrected exhaust mass flow of 345,788 lbm (156,848 kg) per hour. The temperature correction factor would be about 1.01, yielding a corrected exhaust temperature of about 1,054.6°F (568.1°C).
6. Finally, exhaust mass flow and temperature must be corrected for part-load operation at each load point over the operating regime. Exhaust mass flow tends to remain constant under part-load operation of single-shaft gas turbines and decrease with multi-shaft turbines. When inlet guide vanes (IGVs) are used with single-shaft turbines, exhaust mass flow is reduced as load is decreased over the upper part of the operating range. Flow tends to remain constant below 60 to 80% of full load, depending on ambient temperature. In this example, IGVs are used to maintain a constant exhaust temperature down to about 60% of full load. At that operating point, the correction factor for exhaust mass flow is about 0.79. At

90% of full load, for example, the correction factor for exhaust mass flow is about 0.95.

Summary equations that would be used to apply the correction factors discussed above are as follows:

$$\text{Exhaust mass flow} = EMF_{ISO} \times EF_1 \times EF_2 \times EF_3 \times EF_5 \times EF_6 \quad (8\text{-}7)$$

$$\text{Exhaust temperature} = (ET_{ISO} \times ET_2 \times ET_5 \times ET_6) + ET_3 + ET_4 \quad (8\text{-}8)$$

Where subscripts:

ISO = At ISO rated conditions
1 = Correction factor for elevation
2 = Correction factor for compressor inlet temperature
3 = Correction factor for excess inlet losses
4 = Correction factor for excess exhaust losses
5 = Correction factor for steam or water injection
6 = Correction factor for part load operation

Heat Recovery at Multiple Steam Pressures

Heat recovery with single-pressure HRSGs becomes limited at higher pressures due to higher saturation temperatures. In some cases, multiple-pressure HRSGs may be used to serve process loads or gas turbine steam injection requirements, improving overall efficiency. One to four separate pressure sections may be used, each with superheater, boiler, and economizer sections. Multiple-pressure operation is also effective for increasing combined-cycle system efficiency where admission steam turbines include high-pressure and low-pressure inlets.

Figure 8-42 shows a large-capacity, two-pressure HRSG applied in a 220 MW gas-turbine combined-cycle cogeneration plant. High-pressure steam is generated at 1,500 psig/1,000°F (104.5 bar/538°C) and low-pressure steam is generated at 223 psig (16.4 bar). Extraction steam is exported to a nearby fertilizer plant and supplementary firing is used to increase steam turbine generating capacity when gas turbine output falls off during high ambient temperature conditions.

Figure 8-43 shows a triple-pressure HRSG applied in a 41 MW gas turbine cogeneration plant that serves a university hospital campus. High-pressure extraction steam is used for campus heat distribution and air conditioning cold generation. Intermediate-pressure steam, generated at 85 psig (6.9 bar), is used for heating and cooling the inlet air to the gas turbine, and low-pressure steam is generated at 25 psig (2.7 bar) for selected heating loads. Supplementary firing allows steam output to be varied as required.

Fig. 8-42 Large Capacity Two-Pressure HRSG Applied in 220 MW Combined-Cycle Cogeneration Plant. Source: Deltak

Figure 8-44 shows the temperature profile of an unfired HRSG featuring three operating pressure levels. Illustrated is the distribution of heat exchanger sections and the associated temperature differences between exhaust gas and water and steam temperatures. This shows the manner in which heat-absorbing sections operating at certain temperature are located in the gas stream to minimize the amount of heat transfer surface required. There are ten discrete heat exchange sections distributed based on the gas temperature available and the fluid temperature requirement. Note that the high-pressure economizer is divided into three separate sections to provide appropriate temperature zones for the intermediate-pressure superheater, evaporator, and economizer and that the intermediate pressure-generating bank and economizer sections are intermeshed with sections of the high-pressure economizer to optimize performance.

Condensing Heat Exchangers

Most fuels contain small amounts of moisture that is vaporized during the combustion process. Furthermore, most fossil fuels contain hydrogen, which forms additional water vapor when the hydrogen in the fuel combines with the oxygen in the combustion. For example, when natural gas is burned, approximately 10% of the gas energy content is lost because of the water vapor in the flue gas. A condensing economizer allows the recovery of this "lost" heat by cooling the flue gas to the dew point; the typical dew point for natural gas applications is 135 °F (57 °F). When this water vapor is condensed, the latent heat is recovered. This latent heat can be recovered and used to heat low temperature sinks such as incoming combustion air, make up water,

Fig. 8-43 Triple-Pressure HRSG Applied in 41 MW Gas Turbine Cogeneration Plant Serving a University/Hospital Campus. Source: Deltak

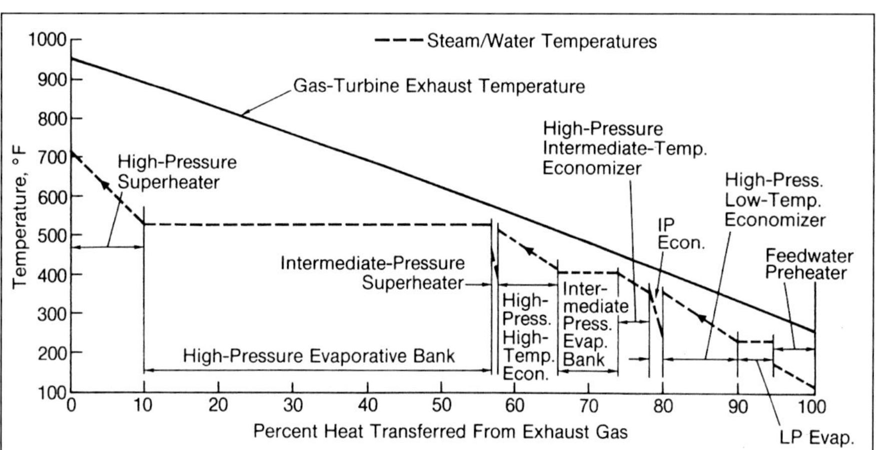

Fig. 8-44 Temperature Profile of Triple-Pressure Unfired HRSG. Source: Combustion Engineering, Inc. Reprinted with Permission from Combustion Fossil Power, 4th Ed., 1991.

or domestic water. The most important consideration for selecting a condensing economizer is to ensure that a proper heat sink is available.

There are several different types of condensing economizers currently available, including direct contact, indirect contact, open and closed loop; the selection depends on the particular application. While the use of a condensing economizer can yield significant energy savings, the following factors and trade-offs must be considered.

- Since cooling the flue gas to the dew point forms acids, the materials of construction must be carefully selected to avoid corrosion. While this is easily accounted for in the construction of new economizers, care must be exercised on retrofits of existing systems.
- Condensing economizers are not recommended for fuels with sulfur or chlorine because of the high probability of sulfuric and hydrochloric acid formation.
- Cooler flue gas may have the unintended effect of reducing the draft of the exhaust stack or impacting other downstream systems, depending on the design of the system (i.e. natural draft or balanced draft).

Section III

Prime Mover Technologies

Chapter Nine
Reciprocating Engines

Reciprocating engines are the dominant type of internal combustion engines (engines in which the fuel is burned within the working cylinder) and are the most commonly used prime movers. Reciprocating engines serve innumerable vehicular and equipment applications, as well as stationary applications that can range in capacity from a few hp (or kW) to about 90,000 hp (67,000 kW).

Heavy-duty reciprocating engines are extremely efficient and reliable power producers with the ability to generate usable thermal energy. Reciprocating engines generally provide the highest simple-cycle full-load and part-load thermal efficiency among comparably sized prime movers.

Reciprocating Engine Types

Spark- and Self-ignited Reciprocating Engines

There are two general types of reciprocating internal combustion engines:

- **Spark-ignited** engines operate on the **Otto cycle** and use gaseous or readily vaporized liquid fuels such as natural gas, gasoline, propane and various biomass-type and manufactured gases.
- **Self-ignited**, or **Diesel**, engines use liquid fuels and achieve ignition through the heat of compression. Diesel engines operate on the full range of liquid petroleum fuels, both distillate and residual. Some Diesel engines are modified to operate as dual-fuel engines using a mixture of liquid and gaseous fuels that are ignited by a compression-ignited pilot oil charge.

Two- and Four-Stroke Cycle Reciprocating Engines

The majority of reciprocating engines operate on what is known as the **four-stroke cycle.** In this cycle, power is generated through a series of four combustion process stages: air intake, compression, power, and exhaust. Two revolutions of the crankshaft occur in each four-stroke cycle (one power stroke every two crankshaft revolutions).

Figure 9-1 illustrates the piston location within the cylinder at the completion of each of the four combustion process strokes. From right to left, the four-stroke process operates as follows:

1. **The intake stroke** starts with the piston at top dead center (TDC), which is the upper most part of its stroke, and ends with the piston at bottom dead center (BDC), which is the end of the downward stroke. As the piston moves downward, it creates a partial vacuum in the cylinder. The intake valve is open and air, or a mixture of air and vaporized fuel, is drawn into or injected into the cylinder from the intake manifold past the open intake valve.
2. **The compression stroke** starts with the piston at BDC. Both the intake and exhaust valves are closed and remain so as the piston moves upward. The air (compression-ignition Diesel cycle) or air-fuel mixture (spark-ignition Otto cycle) is confined within the cylinder and is compressed to a small fraction of its initial volume. Toward the end of the compression stroke, combustion is initiated and the cylinder pressure begins to rise more rapidly.
3. **The power (or expansion) stroke** starts with the piston at or near TDC again and ends at BDC. The high-temperature, high-pressure gases resulting from fuel combustion push the piston down and force the crank to rotate. As the piston approaches BDC, the exhaust valve opens to initiate the exhaust process and drop the cylinder pressure to close to the exhaust pressure level.
4. **The exhaust stroke** starts with the piston at BDC. The exhaust valve remains open and the remaining burned gases exit the cylinder. The process continues first because the cylinder pressure may be significantly higher than the exhaust pressure, and then because the upward motion of the piston sweeps the now fully expanded gases through the exhaust valve and into the exhaust manifold. As the piston again reaches TDC, the exhaust valve closes and the cycle repeats.

In a **two-stroke cycle** engine, there are no separate intake and exhaust process strokes. Intake and exhaust begin near BDC during the last part of the power stroke and end shortly after the beginning of the compression stroke. Most modern designs feature some type of external scavenge blower. In simpler types, ports in the cylinder liner, opened and closed by the piston movement, control the exhaust and inlet flows. The two-stroke cycle requires

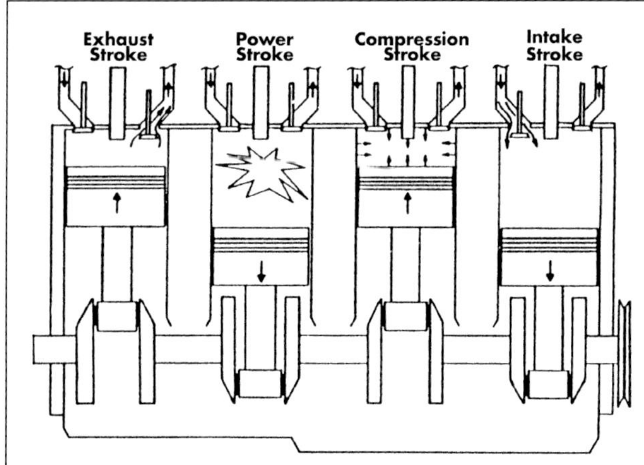

Fig. 9-1 Illustration of Four-Stroke-Cycle Reciprocating Engine Combustion Process Cycle.

only one revolution of the crankshaft, yielding one power stroke per revolution. The basic two-stroke process operates as follows:

1. **The compression stroke** starts by closing the inlet and exhaust ports and then compresses the cylinder contents. As the piston approaches TDC, combustion is initiated and pressure increases.
2. **The power (or expansion) stroke** is similar to that in the four-stroke cycle until the piston approaches BDC. At that point, the exhaust ports open and the intake ports are uncovered.

Engine Aspiration

The intake manifold distributes the air (self-ignited Diesel cycle) or the air-fuel mixture (spark-ignited Otto cycle) to the various cylinders of a multi-cylinder reciprocating engine. There are two types of engine aspiration:

- **Natural aspiration** draws combustion air into the piston cylinder at atmospheric pressure. Gas/fuel need only be supplied at low pressure (less than 1 psig).
- In **charged aspiration**, an air blower or mechanical compressor is used to pressurize the air or air-fuel mixture before it is inducted into the cylinder. When driven by an engine auxiliary output shaft or a separate driver, the device is known as a **supercharger**. When driven by an exhaust-powered turbine, the device is known as a **turbocharger**. In some cases, an engine may feature both a supercharger and a turbocharger. Because compressing the air or air-fuel mixture increases its temperature, an aftercooler (or intercooler) is used to cool the heated charge to further increase its density.

Charging increases the power output from a given cylinder size by increasing the amount of oxygen available for combustion. Thus, for a given engine size, weight, displacement, and piston speed, significantly more power is produced when supercharging or turbocharging is used. This increase in power density greatly reduces engine capital cost per hp (kW). It also, generally, increases engine volumetric efficiency.

OPERATING CYCLES

The working fluid (air or air-fuel mixture) in an actual reciprocating engine does not go through a complete cycle in the engine, meaning it is not returned to its original condition. Instead, the engine actually operates on the so-called open cycle. To analyze the engine, however, it is convenient to devise a closed cycle that approximates the open cycle. This approximation is known as the air-standard cycle.

In practice, engines operating on Otto and Diesel cycles do not duplicate the cycle or approach the theoretical efficiencies of these ideal cycles. In fact, in many modern engine designs, the characteristics of Otto and Diesel engines have become more similar to each other. It is still, however, useful to consider the air-standard cycles as a starting point for understanding the basic principals of operation. Following are descriptions of the Otto and Diesel cycles.

OTTO CYCLE

Figure 9-2 provides representations of the air-standard, or ideal, Otto cycle in both pressure/volume (P-V) and temperature/entropy (T-s) ordinates. The ideal Otto cycle consists of an isentropic compression (A-B), followed by constant-volume combustion (B-C), an isentropic expansion from which work is extracted (C-D), and, finally, a reversible constant-volume rejection of heat (D-A).

Specific heat (c) is the energy, as heat, transferred during a process per unit mass flow ($\dot{m}$) of fluid (working fuel),

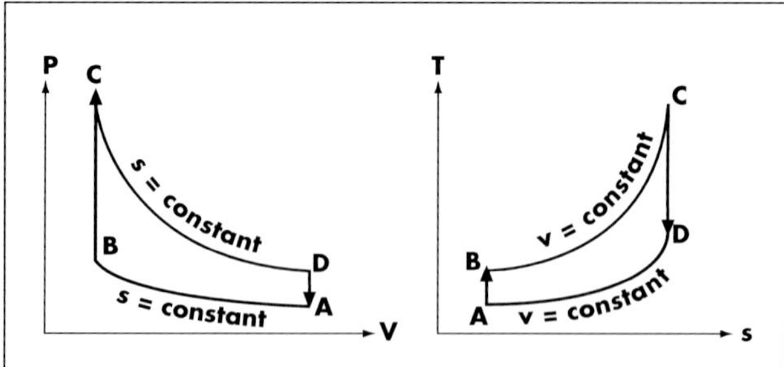

Fig. 9-2 The Air-Standard Otto Cycle.

divided by the corresponding change of temperature (T) of the fluid (working fuel) in the process (subscripts v and p denote specific heat at constant volume or constant pressure, respectively). In the ideal cycle, the heat added (Q_{in}) along the path B to C is thus:

$$Q_{in} = \dot{m}c_v\,(T_C - T_B) \qquad (9\text{-}1)$$

The heat rejected along the path D to A is:

$$Q_{out} = \dot{m}c_v\,(T_D - T_A) \qquad (9\text{-}2)$$

Thus, the net work (W_{net}) of the cycle is:

$$W_{net} = Q_{in} - Q_{out} = \dot{m}c_v\,[(T_C - T_B) - (T_D - T_A)] \qquad (9\text{-}3)$$

and the thermal efficiency (η_{th}) of the cycle is:

$$\eta_{th} = \frac{Q_{in} - Q_{out}}{Q_{in}} = 1 - \frac{Q_{out}}{Q_{in}} = 1 - \frac{\dot{m}c_v\,(T_D - T_A)}{\dot{m}c_v\,(T_C - T_B)} = 1 - \frac{T_A}{T_B}\,\frac{(T_D/T_A - 1)}{(T_C/T_B - 1)} \qquad (9\text{-}4)$$

Note that this is a mathematical description of the ideal constant-volume air cycle in which the properties of air are assumed to be consistent throughout the cycle. The Otto cycle, however, calls for the compression of an air-fuel mixture. Ideal thermal efficiency can therefore be expressed as a function of the compression ratio.

Compression ratio (r_c) expresses the relationship of the maximum volume of an engine cylinder (V_T) (with the piston at the bottom of its stroke) to the minimum, or clearance, volume (V_C), of the cylinder (with the piston at the top of its stroke):

$$r_c = \frac{\textit{maximum cylinder volume}}{\textit{minimum cylinder volume}} = \frac{V_d + V_c}{V_c} \qquad (9\text{-}5)$$

In the ideal Otto cycle,

$$r_c = V_A = V_D \qquad (9\text{-}6)$$

and thermal efficiency can be expressed as:

$$\eta_{th} = 1 - (r_c)^{1-k} = 1 - \frac{1}{r_c^{\,k-1}} \qquad (9\text{-}7)$$

where k is the ratio of specific heat at constant volume and at constant pressure. Assuming k to be constant, ideal cycle thermal efficiency for the Otto cycle is solely dependent on the compression ratio. For example, with k held constant at a value of 1.4, a compression ratio of 11:1 would yield the following ideal thermal efficiency:

$$\eta_{th} = 1 - \left(\frac{1}{r_c}\right)^{k-1} = 1 - \left(\frac{1}{11.0}\right)^{0.4} = 0.617 \; (i.e., 61.7\%)$$

Diesel Cycle

Figure 9-3 represents the air-standard Diesel cycle (also known as the compression-ignition cycle) in both P-V and T-s ordinates. The ideal Diesel cycle consists of an isentropic compression of the air after it has been inducted into the cylinder (A-B), followed by the injection of fuel and combustion at constant pressure (B-C), then an isentropic expansion from which work is extracted (C-D), and a reversible constant-volume rejection of heat (D-A).

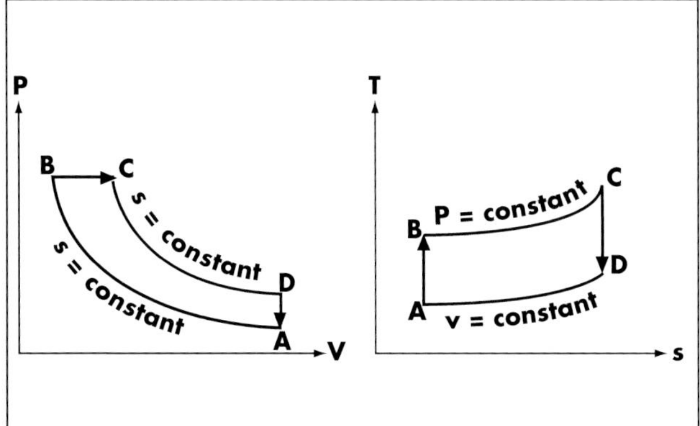

Fig. 9-3 The Air-Standard Diesel Cycle.

In the ideal Diesel cycle, the heat added along the constant-pressure path from B to C is:

$$Q_{in} = \dot{m}c_P\,(T_C - T_B) \qquad (9\text{-}8)$$

The heat rejected along the constant-volume path from D to A is:

$$Q_{out} = \dot{m}c_v\,(T_D - T_A) \qquad (9\text{-}9)$$

Thus, the net work of the cycle is:

$$W_{net} = Q_{in} - Q_{out} = \dot{m}[c_p(T_C - T_B) - c_v(T_D - T_A)] \qquad (9\text{-}10)$$

and the thermal efficiency of the cycle is:

$$\eta_{th} = \frac{Q_{in} - Q_{out}}{Q_{in}} = 1 - \frac{Q_{out}}{Q_{in}} = 1 - \frac{\dot{m}c_v\,(T_D - T_A)}{\dot{m}c_p\,(T_C - T_B)} = 1 - \frac{T_A}{kT_B}\,\frac{(T_D/T_A - 1)}{(T_C/T_B - 1)} \tag{9-11}$$

Whereas in the ideal Otto Cycle thermal efficiency is a function of compression ratio, the same is not true for the Diesel cycle. In the Otto cycle, the isentropic compression and expansion ratios are equal. As shown in Figure 9-3, in the Diesel cycle, the isentropic compression ratio is greater than the isentropic expansion ratio. The compression and expansion ratios are expressed as:

$$\textit{compression ratio } r_c = \frac{V_A}{V_B} \tag{9-12}$$

and

$$\textit{expansion ratio } r_e = \frac{V_D}{V_C} \tag{9-13}$$

Note that while physically, the expansion ratio is essentially the same as the compression ratio, the focus here is on the adiabatic portion of the expansion ratio. The ratio of compression to expansion is known as the cut-off ratio ($r_{c.o.}$) and is expressed as:

$$r_{c.o.} = \frac{T_C}{T_B} = \frac{V_C}{V_B} = \frac{r_c}{r_e} \tag{9-14}$$

If these relationships are used in Equation 9-11, the equation for thermal efficiency becomes:

$$\eta_{th} = 1 - \frac{1}{k}\left[\frac{(r_c/r_e)^k - 1}{(r_c/r_e - 1)(r_c^{\,k-1})}\right] = 1 - \left(\frac{1}{r_c}\right)^{k-1}\left[\frac{(r_{c.o.})^k - 1}{k\,(r_{c.o.} - 1)}\right] \tag{9-15}$$

Thus, in the ideal Diesel cycle, efficiency is a function of the compression ratio and the expansion ratio. For example, with k held constant at a value of 1.4, a compression ratio of 16:1, with a cut-off ratio of 2:1, would yield the following ideal thermal efficiency:

$$\eta_{th} = 1 - \left(\frac{1}{r_c}\right)^{k-1}\left[\frac{(r_{c.o.})^k - 1}{k\,(r_{c.o.} - 1)}\right] = 1 - \left(\frac{1}{16}\right)^{0.4}\left[\frac{(2^{1.4} - 1)}{1.4}\right]$$
$$= 0.614 \text{ (i.e., 61.4\%)}$$

Actual Cycle Performance

There are dramatic differences between ideal cycle efficiencies and those achieved in practice. In addition to the fact that engines do not actually operate on these ideal cycles, there are several causes of irreversibility that occur in practical applications. Following are some of the reasons for the differences in ideal and actual cycle performance:

1. The specific heat content of the actual gases increases with a rise in temperature.
2. Dissociation of combustion products occurs.
3. Combustion may not be complete due to incomplete air-fuel mixing and/or lack of sufficient oxygen.
4. Combustion does not occur instantaneously, so constant volume combustion is an ideal representation of the actual combustion event. In the actual combustion process, combustion commences prior to TDC and continues after TDC. Because combustion continues after TDC when the cylinder volume is much greater than the clearance volume, actual peak-pressure value is lower than ideal, resulting in less expansion. After peak pressure, expansion stroke pressure is greater than ideal because less work has been extracted from the cylinder gases than in the ideal cycle.
5. During the inlet and exhaust strokes, there is a pressure drop through the valves and a certain amount of work is required to charge the cylinder with air and exhaust the combustion products.
6. Actual compression ratios are less than the nominal value because of late intake valve or port closing.
7. Exhaust blowdown losses occur in actual cycles because the exhaust valve is opened well before BDC to reduce pressure during the first part of the exhaust stroke in four-stroke-cycle engines and to allow time for scavenging in two-stroke-cycle engines. The gas pressure at the end of the power (expansion) stroke is, therefore, reduced, resulting in a decrease in expansion stroke work transfer.
8. There is considerable heat transfer between the burned gases in the cylinder and the cylinder walls. This heat transfer causes the gas pressure in the actual cycle to be less than ideal as volume increases.
9. Cylinder gas leakage occurs. As cylinder pressure increases, gas flows into numerous crevices, such as the region between the piston rings. Some of the gas flows into the crankcase and is lost from the cycle. Some of the gas flow returns, but has been cooled

by heat transfer in the crevices. Both effects reduce cylinder pressure and, therefore, expansion work relative to ideal-cycle values.

10. Engine aspiration effectiveness is less than ideal because the cylinders are not completely filled with fresh air with each intake stroke and exhaust gases are not completely removed with each exhaust stroke. The faster the engine runs, the less time there is to fill the cylinder on each intake stroke and, therefore, the engine has less ability to rid exhaust gases and take in fresh air.
11. There is irreversibility associated with all real processes (e.g., pressure and temperature changes).

In practice, thermal fuel efficiency of an engine is the degree to which the engine is successful in converting the energy of the fuel into usable mechanical energy. It is that part of the heat energy in the cylinder that forces the pistons to move, resulting in crankshaft rotation.

Compression ratio is still a leading indicator of engine thermal efficiency as well as power density. Thermal efficiency improves with higher compression ratios because the combustion gases are able to expand further in the expansion or power stroke, allowing more heat energy to be transformed into mechanical energy. Despite all of the practical limitations, reciprocating engines generally offer the highest simple-cycle thermal fuel efficiency of all prime movers of comparable capacity. Thermal fuel efficiencies typically range from 28 to 48% (LHV basis), with both large capacity Diesel engines and natural gas engines (larger than 4 MW) at the highest end of the efficiency range.

Comparison of Otto- and Diesel-Cycle Engines

Both spark-ignited, Otto-cycle engines and self-ignited, Diesel-cycle engines are available as two-stroke and four-stroke designs and with natural or charged aspiration. In the Otto-type engine combustion process cycle, the engine compresses an air-fuel mixture in a cylinder. The mixture is generally ignited in the cylinder by a spark at or near the completion of the compression stroke. In the Diesel cycle, air only (not an air-fuel mixture) is compressed in the compression stroke. The temperature of the air within the cylinder rises to the auto-ignition temperature of the Diesel fuel prior to the end of the compression stroke. Ignition occurs when the fuel is injected, at high-pressure, into the cylinder starting at or near TDC and continues during most of the combustion (power) stroke.

From the ideal cycle P-V diagrams shown above, it can be seen that the isentropic compression and expansion ratios are equal in the Otto cycle, but compression ratio is greater than the expansion ratio in the Diesel cycle. From the T-s diagrams, it can also be seen that the ideal Otto cycle has higher efficiency and higher work area than the Diesel cycle. Less heat is rejected with the same amount of heat input and the same compression ratio. For the same amount of heat rejected, then, the Otto cycle would produce more work. In the example calculations provided above, the ideal thermal efficiency for an Otto cycle with a compression ratio of 11:1 was essentially the same as that of a Diesel cycle with a compression ratio of 16:1.

Thus, in theory, the Otto cycle is more efficient than the Diesel cycle for engines operating with the same compression ratio. However, in practice, Diesel engines achieve significantly greater thermal efficiencies than Otto engines due to the ability to operate with higher compression ratios and at higher peak pressures.

Traditional high efficiency low-speed Otto-cycle engines have achieved a maximum thermal fuel efficiency of about 41% on a LHV basis (this corresponds to about 37% on a HHV basis with natural gas). A Diesel-cycle engine of similar capacity and operating speed may achieve a thermal fuel efficiency of about 46% on a LHV basis (or 44% on a HHV basis using Diesel oil).

Many manufacturers are now offering natural gas fueled stationary Otto-cycle engines with thermal fuel efficiencies above 43% on a LHV basis. These higher efficiencies are achieved by improved programmable logic control (PLC) based combustion controls for lean burn firing, utilizing the Miller-cycle variation of the Otto-cycle and in one case adding a second stage of turbo charging. Turbo chargers have become a standard component on almost all of these newer units.

The newer higher efficient models with lean burn firing utilize PLC control for precisely controlling the ignition timing and fuel flow. PLC controlled fuel valves have been added upstream of the mechanically operated cylinder valves. These units may also have a pre-chamber where pilot fuel is pre-ignited by the spark plug at each cylinder. Typically, the fuel flow to the pre-chamber is also precisely metered by both a PLC controlled electronic control valve and a mechanically operated valve. This pre-ignition process is effective in lowering the overall cylinder chamber temperature and the resultant NO_X emission rates. Many manufacturer's are now guaranteeing NO_X emission rates down to 0.6 gr/bhp and lower for these new lean burn high efficient natural gas fueled models.

Recently, several engine manufacturers have reported thermal fuel efficiencies above 46% on a LHV basis. These large ultra-high efficient engines are offered in both medium speed (720-rpm) and high speed (1,500-rpm) depending on the manufacturer. Several of these ultra-high efficient models will require emissions control technology to lower NO_X emission rates for application in the United States.

With an Otto engine, a combustible air-fuel mixture is compressed and heated in the engine's cylinders. Fuel is mixed with the air prior to the compression stroke of the cycle. Ignition occurs due to an externally timed electrical spark. Since the fuel-air mixture is compressed, it is necessary to use a volatile gaseous or readily vaporized fuel that can be distributed uniformly into the incoming air at relatively low pressure. This fuel-air mixture can prematurely ignite during the compression stroke if the cylinder pressure and/or temperature become too high. Compression ratios and mean effective pressures of Otto-cycle engines are, therefore, limited by two types of detonation considerations: surface ignition and combustion knock.

The problems of surface ignition and knock generally do not exist in the Diesel engine, because air only is compressed during the compression stroke. Since there is no fuel present during the compression stroke, much higher compression ratios can effectively be achieved. Whereas the maximum compression ratios of Otto engines are limited to about 12.5:1 or 13:1, compression ratios for Diesel engines may exceed 20:1.

Another difference is that Otto-cycle engines are governed by throttling the charge while Diesel engine are governed by varying the amount of fuel injected into the cylinders. In Otto-cycle engines, a throttle is required for part-load operation. Throttling increases pumping work under partial loads and, therefore, decreases efficiency. Diesel engines do not require a throttle and, therefore, achieve better part-load efficiencies than do Otto-cycle engines.

The typical thermal fuel efficiency advantage of Diesel engines versus traditional spark-ignited Otto-cycle engines will range from 10 to 20%. However this gap in efficiencies closed significantly from 2000-2010. Otto-cycle engines will typically have 20 to 25% lower power output and correspondingly greater heat rejection than the same size Diesel engine at the same operating conditions.

Despite the simple-cycle thermal fuel efficiency disadvantage, Otto-cycle engines offer several advantages with respect to Diesel engines. The higher compression ratio and combustion pressure in Diesel engines produce increased engine stress, necessitating a heavier and sturdier design. The abuses of oil contamination and soot deposits from combustion also increase maintenance requirements and costs.

Of great significance in the current energy market is the need to control environmentally harmful air emissions. Otto-cycle engines operating on fuels such as natural gas produce lower emission rates in many regulated pollutants than do Diesel-cycle engines operating on liquid fuels. This environmental efficiency advantage often translates into capital and operating cost advantages and often allows natural gas-fired Otto-cycle engines to be more easily permitted for stationary applications.

In applications that employ heat recovery (cogeneration cycles) in facilities with sufficiently large heat requirements, the higher rate of heat rejection of the Otto-cycle engine compensates to some extent for the lower simple-cycle thermal fuel efficiency compared with Diesel-cycle engines. Since more heat is rejected from the simple-cycle, more heat is available to be recovered. Also, due to the composition of the exhaust gases, heat recovery systems can often extract more heat from Otto-cycle engine exhaust by cooling the exhaust gases to a lower final exit temperature than can be practically achieved with liquid fuel-fired Diesel engine exhaust.

A critical parameter in exhaust gas heat recovery system design is the exit temperature of the exhaust gas as it enters the stack. The limiting factor is the temperature at which the heat exchanger surfaces reach a point where combustion product acids and water precipitate onto the metal surfaces, causing corrosion. The acid dew point is a function of the amount of sulfur in the fuel and, subsequently, in the exhaust gas. Exhaust streams from natural gas firing, therefore, may be cooled lower than exhaust from sulfur-bearing liquid fuels.

The combined impact of higher heat rejection rate and the lower heat recovery system exit temperature can result in a significant heat recovery efficiency advantage for gas-fired Otto engines versus oil-fired Diesel engines. Therefore, with cogeneration-type applications, overall thermal efficiency may be fairly similar, though the constituent components of power output and recovered heat will be different in each case.

Dual-Fuel Engines

Dual-fuel engines are Diesel-type engines that are capable of operating on natural gas and other gaseous fuels, as well as on liquid fuels. The approach can provide fuel source and price flexibility, as well as some of the advantages of both alternatives.

All current designs feature combustion initiated by

self-ignited pilot oil, with remaining combustion energy provided by natural gas. Thermal fuel efficiencies well in excess of 40% (LHV basis) have been achieved while operating predominantly on natural gas, with NO_X emission levels of 1 gram/bhp-h (1.3 gram/kWh) or less.

There are two general, currently available dual-fuel engine design types. One type features direct-injection of highly compressed gas into the cylinder for Diesel cycle-type combustion. This type allows for up to 95% gas firing at full load, with 5% pilot oil. The second type locates the gas valve in the intake manifold with the air-gas mixture being compressed and then ignited by the compression-ignited pilot oil. This type allows for up to 99% gas firing at full load with 1% pilot oil.

The ability to vary the amount of gas used allows for the use of maximum gas based on fuel pricing and/or air emissions control strategies. The ability to purchase natural gas on an interruptible (non-firm) basis, based on the availability of an alternative fuel, generally results in a much lower gas price. Operation with maximum use of natural gas may also allow for flexibility in meeting air emissions requirements. For example, some NO_X emissions control regulations require lower NO_X emissions levels in the summer months when ground-level ozone problems are more severe. The ability to operate almost exclusively on natural gas during the summer months can enable the engine to meet strict permitting requirements, while still maintaining the ability to operate on liquid fuel during other months of the year.

Engine Components and Operation

The four-stroke combustion process cycle produces power in the following manner. The piston compresses air (compression-ignited Diesel-cycle engines) or an air-fuel mixture (spark-ignited Otto-cycle engine), fuel is burned within the cylinder, expanding combustion gases exert force on the piston and this force is transferred, through a connecting rod, to produce rotation in a crankshaft. Following is a discussion of how reciprocating engines operate, with a focus on several basic engine components and design features.

Engine Frame

The engine frame structure includes all fixed parts that hold the engine together. Its function is to support and align the moving parts, while resisting the forces imposed by the engine's operation. It also supports auxiliaries and provides jackets and passages for cooling water, a sump for lubricating oil, and a protective enclosure for all of these parts.

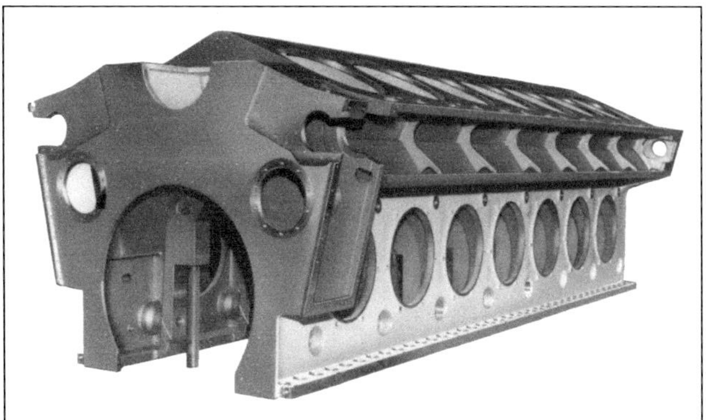

Fig. 9-4 Center-Frame for Large Capacity 14-Cylinder, V-Type Engine. Source: Fairbanks Morse Engine Div.

For stationary engines, which rest on a substantial foundation, two-piece construction is common. The lower section, or bedplate, forms a base, supports the main bearings, encloses the lower part of the crankcase, and forms a sump for lubricating oil. The upper section, or center-frame, includes the upper part of the crankcase and the cylinder block in which the cylinders are supported. Automotive type engines have a one-piece cylinder block and are typically constructed of cast iron. Figure 9-4 shows the center-frame of a large capacity 14-cylinder, V-type, four-stroke-cycle engine.

Figures 9-5 and 9-6 show the frame box and bedplate for a large capacity 6-cylinder, in-line two-stroke-cycle

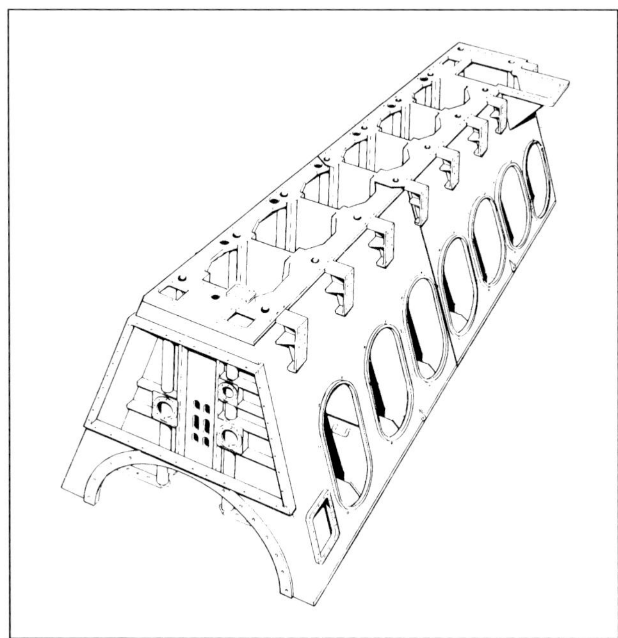

Fig. 9-5 Frame Box for Large Capacity Two-Stroke-Cycle, In-Line Diesel Engine. Source: MAN B&W

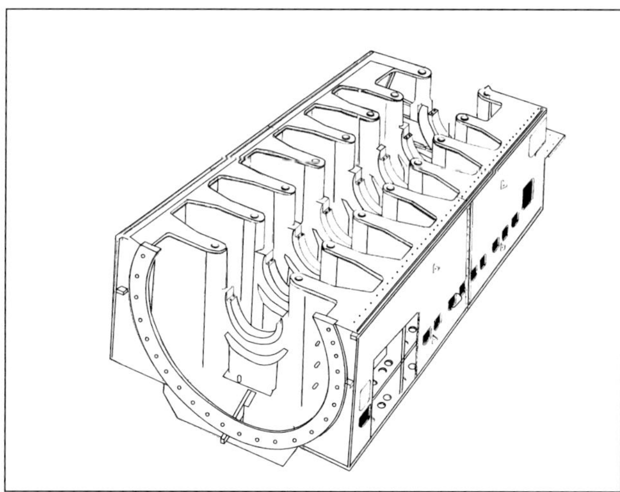

Fig. 9-6 Bedplate for Large Capacity Two-Stroke-Cycle, In-Line, Diesel Engine. Source: MAN B&W

Diesel engine. The frame box is of welded construction featuring a hinged door for access to crankcase components. The bedplate is built-up of longitudinal side girders and welded cross girders with cast steel bearing supports. Stay bolts connect the bedplate, the frame box, and the cylinder frame to form a rigid unit.

Cylinders and Pistons

Cylinders are chambers located inside the engine in which air or an air-fuel mixture is compressed, the fuel is ignited, and the power is produced. Figure 9-7 shows a single engine cylinder liner and water jacket. Liners are referred to as wet or dry, depending on whether the sleeve is in direct contact with the cooling water. The cylinder liners are inserted in the large circular holes in the cylinder block.

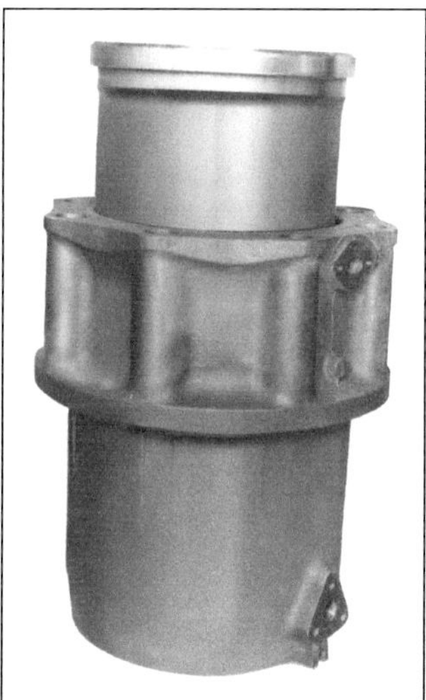

Fig. 9-7 Engine Cylinder Liner and Water Jacket.
Source: Fairbanks Morse Engine Div.

The cylinder head (or heads) forms the top or lid to seal the cylinders. Cylinder heads are typically made of cast iron or aluminum and must be strong and rigid to distribute the gas forces acting on the head as uniformly as possible through the engine block. The cylinder head contains the spark plug or fuel injector and, in overhead valve engines, parts of the valve mechanisms.

Figure 9-8 is a cross-section illustration of a cylinder head for a four-stroke-cycle Diesel engine. The cylinder head is designed to withstand operation with combustion pressures of up to 2,610 psi (180 bar). The fuel injector is flanked by the valve assemblies on either side. As shown in Figure 9-9, studs at the top of the frame are used to fasten the cylinder head to the frame. The cylinder head closes the top end of the cylinder so as to make a confined space in which to compress the air or air-fuel mixture and to confine the gases while they are burning and expanding.

Multi-cylinder engines can include in-line, V, flat, and radial cylinder arrangements. In-line arrangements are the simplest. If the engine has more than eight cylinders, it becomes difficult to make a sufficiently rigid frame and crankshaft with an in-line arrangement. Also, the engine becomes quite long and takes up considerable space. The V-type arrangement, with two connecting rods attached to each crankpin, reduces length and makes the frame and crankshaft stiffer. Typically, the angle between the two banks of cylinders in a V-type arrangement is between 40 and 75 degrees.

Variations on the V-type arrangement are the flat arrangement, in which the banks are on a 180 degree angle, and the radial arrangement. An additional long-standing design is the opposed piston type, which features two pistons facing each other in the same cylinder and two crankshafts. A given engine model will feature numerous

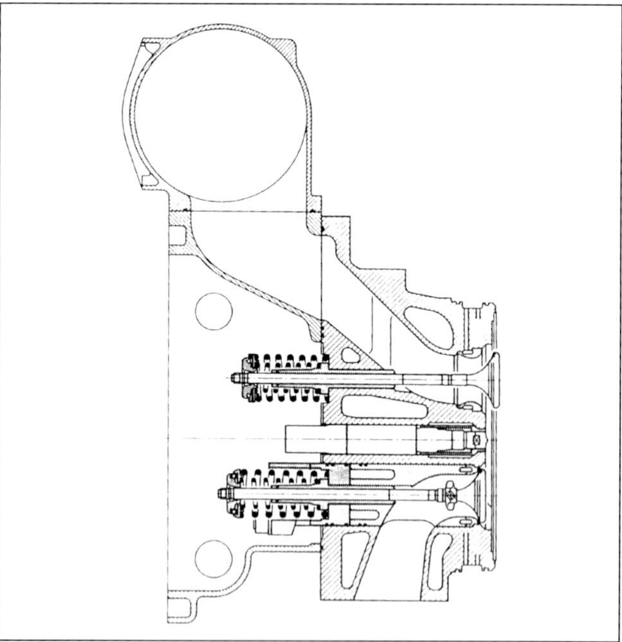

Fig. 9-8 Cross-Sectional Illustration of Four-Stroke-Cycle Diesel Engine Cylinder Head. Source: MAN B&W

sub-models that are differentiated by the number of cylinders, i.e., I6, I8, V8, V12, V16, etc.

The piston slides up and down within the cylinder and serves to seal the cylinder, compress the air charge or air-fuel mixture charge, resist the pressure of the gases while they are burning and expanding, and transmit the combustion-generated gas pressure to the crank pin via the connecting rod. Pistons may be made of cast iron, aluminum, steel, or a combination. Pistons are cooled by circulating lubricating oil (or, in some cases, cooling water) through the cavities or spaces in the piston — the method varying with different designs. Oil spray is also a common cooling method.

The piston is fitted with rings, which ride in grooves cut in the piston head to seal against gas leakage and control oil flow. The compression rings make the piston and the cylinder walls airtight by sealing the space between the piston and the liner. Oil rings, which are located below the compression rings, prevent surplus oil from being carried up into the combustion chamber where it would burn incompletely and form carbon. They are designed to scrape off, on the down-stroke, most of the lubricating oil splashed into the cylinder and return it to the crankcase and ride over the remaining oil film on the way up. The crankcase must be ventilated to remove gases that blow by the piston rings to prevent pressure build-up.

Fig. 9-9 Studs at Top of Frame Used to Fasten Cylinder Head to Frame of Very Large Capacity V-Type Engine.
Source: Fairbanks Morse Engine Div.

Figure 9-10 shows a piston made of high-tensile steel, with piston pin and rings, and Figure 9-11 illustrates a piston designed for a four-stroke-cycle Diesel engine. This composite piston consists of a forged steel crown, designed to withstand typical firing pressures of about

Fig. 9-10 Piston with Piston Pin and Rings. Source: Wartsila Diesel

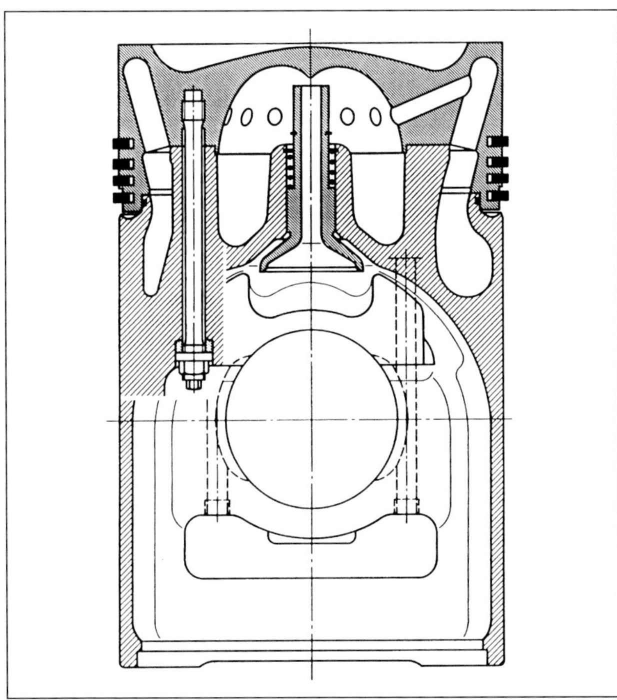

Fig. 9-11 Diesel Engine Piston Illustration. Source: MAN B&W

2,600 psi (180 bar), and a nodular cast piston skirt. This piston design features three compression rings and one oil scraper ring.

The piston pin, or wrist pin, is the link between the connecting rod and the piston. The skirt is the portion of the piston that extends below the piston pin and serves as a guide for the piston and connecting rod. A sometimes-used alternative design to a piston pin is a circular ball joint that allows the piston to rotate.

Cylinder displacement is the product of stroke (piston travel from BDC to TDC) and the cross-sectional area of the cylinder. The diameter of an engine cylinder is called the bore. Total engine displacement is equal to the displacement of one cylinder times the number of cylinders in the engine. It is calculated as follows:

$$(\pi r^2) \times (stroke) \times (number\ of\ cylinders) \quad (9\text{-}16)$$

Therefore, a 16-cylinder engine with a 13.5 in. bore and a 16.5 in. stroke would have a displacement of:

$$(\pi 6.75^2) \times (16.5) \times (16) = 37{,}789\ in^3 \text{ or } 21.9 \text{ cf}$$

In SI terms, the bore of this 16-cylinder engine would be 343 mm and the stroke would be about 419 mm. The displacement would be:

$$(\pi 171.5^2) \times (419.5) \times (16) = 619{,}458\ cm^3 \text{ or } 619 \text{ liters}$$

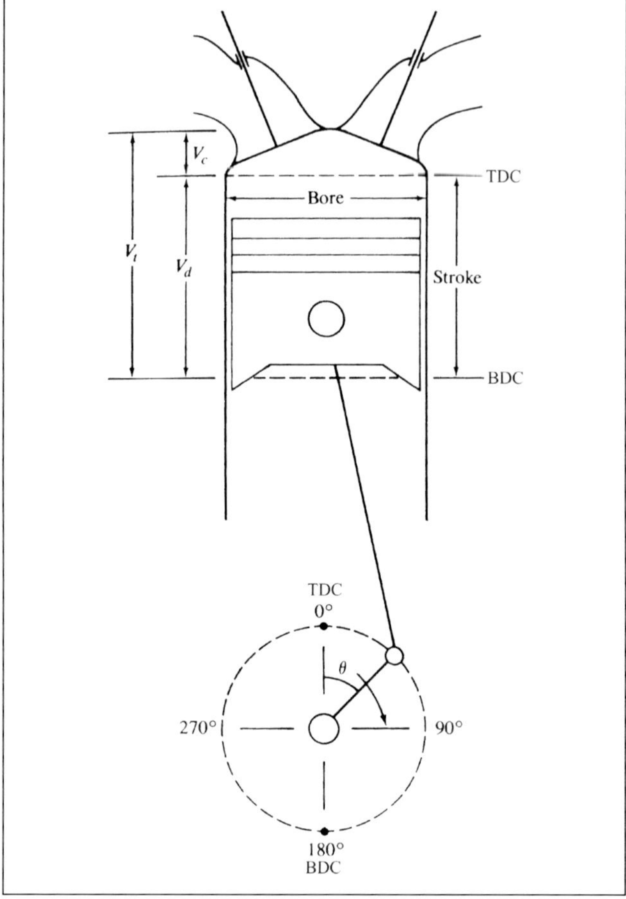

Fig. 9-12 Illustration Showing Engine Cylinder Geometry.

Figure 9-12 illustrates the measurements of bore and stroke, as well as engine displacement and crank angle. The crank angle is commonly used to refer to the crank and piston position with respect to TDC and BDC. As shown, at TDC, the crank angle is 0 degrees and at BDC, it is 180 degrees. Engine events are commonly described with respect to crank angle. The timing of fuel injection, for example, is commonly expressed as occurring at a certain number of degrees of crank angle before or after TDC.

Figures 9-13 and 9-14 are cross-sectional illustrations of large capacity, four-stroke-cycle Diesel engines. Both use the same cylinder cover, piston, cylinder liner, and connecting rod, and produce a cylinder output of about 1,300 hp (975 kW). Figure 9-13 features a V-type frame design and Figure 9-14 features an in-line frame design. In each engine, the cylinder bore is 18.9 in. (480 mm) and the piston stroke is 23.6 in. (600 mm). Mean piston speed varies from 32.8 to 29.5 ft/s (10.0 to 9.0 m/s), depending on engine rotational speed.

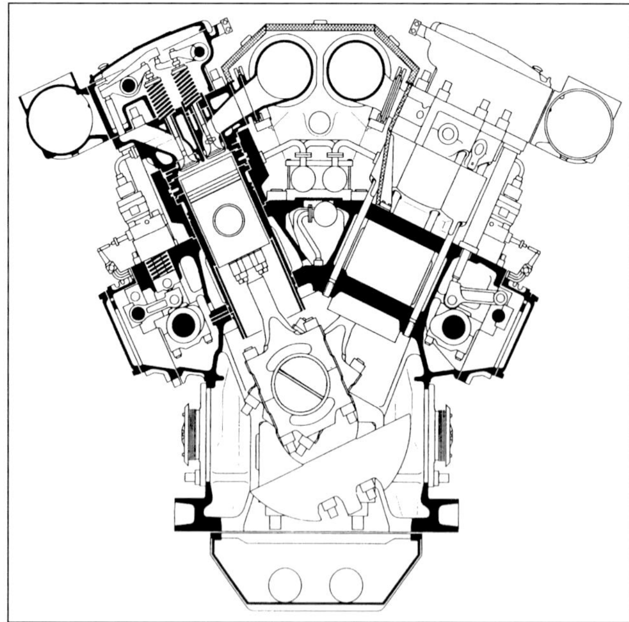
Fig. 9-13 Cross-Sectional Illustration of Four-Stroke-Cycle, V-type Engine. Source: MAN B&W

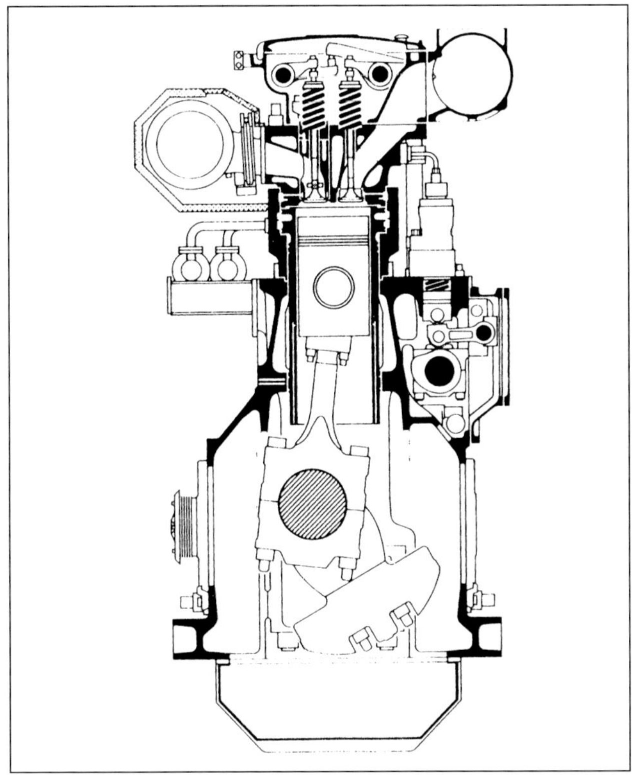

Fig. 9-14 Cross-Sectional Illustration of Four-Stroke-Cycle In-Line Engine. Source: MAN B&W

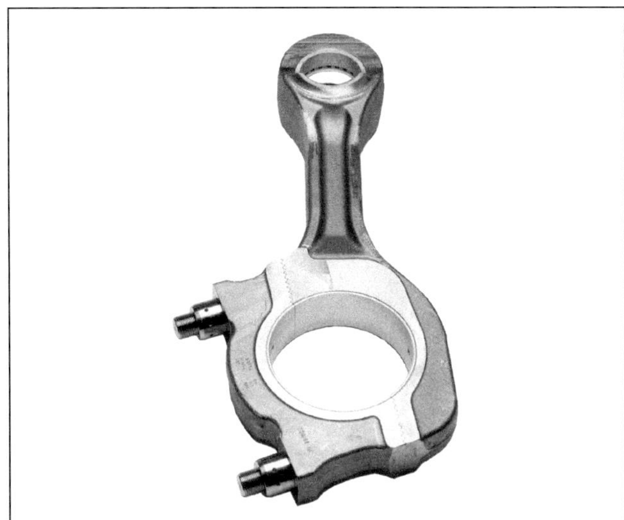

Fig. 9-15 Connecting Rod. Source: Wartsila Diesel

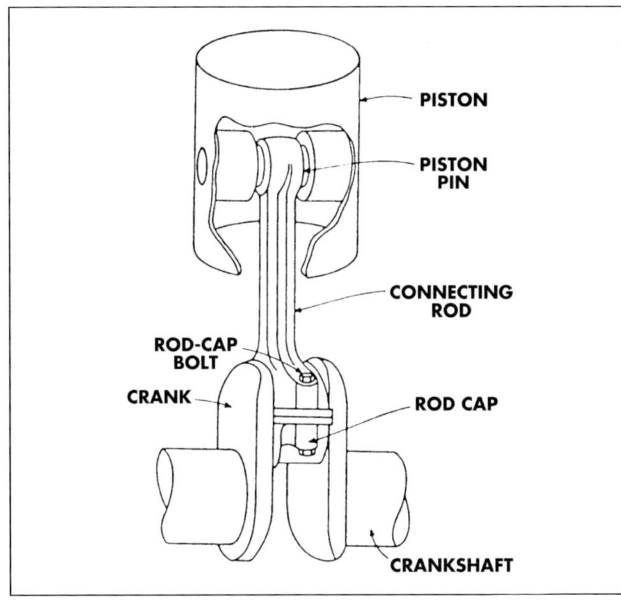

Fig. 9-16 Piston and Connecting Rod Assembly Attached to the Crankpin on a Crankshaft. Source: Waukesha Engine Div.

Converting Piston Reciprocating Motion to Shaft Circular Motion

To produce usable engine power output, the force of the reciprocating motion of the piston must be converted to the force of rotational motion of a shaft. A connecting rod (Figure 9-15) is a bar, or strut, with a bearing at each end. Its purpose is to transmit force in either direction between the piston and the crank on the crankshaft of an engine.

The reciprocating motion of the piston and its connecting rod is thus converted into rotating motion of the crankshaft, which is used to drive the load (i.e., generator, compressor, pump, etc.), as well as various engine components. Figure 9-16 illustrates a piston and connecting rod assembly attached to the crankpin on a crankshaft. The crankshaft, which is usually a steel forging, is made of a series of cranks. In the case of in-line engines, there is one crank for each cylinder. With V-type engines, each crank will serve a pair of cylinders.

The crankshaft is made up of a series of bearing surfaces called journals. Figures 9-17 and 9-18 show crankshafts of four-stroke-cycle engines. The crankshaft is housed within the crankcase and is supported in the cylinder block by means of bearings at each of the main bearing journals. Attached to the crankshaft in most engines is a heavy wheel or disc known as a flywheel. The flywheel helps to ensure that the crankshaft turns smoothly by evening out the power pulses from each cylinder.

Work and Power Expressions

There are several ways to refer to the work done by an engine. Engine power, or the rate at which work is done, is expressed as hp or kW and thus is a function of mean effective pressure (mep), the displacement of the engine, and the speed of the engine. The following are several

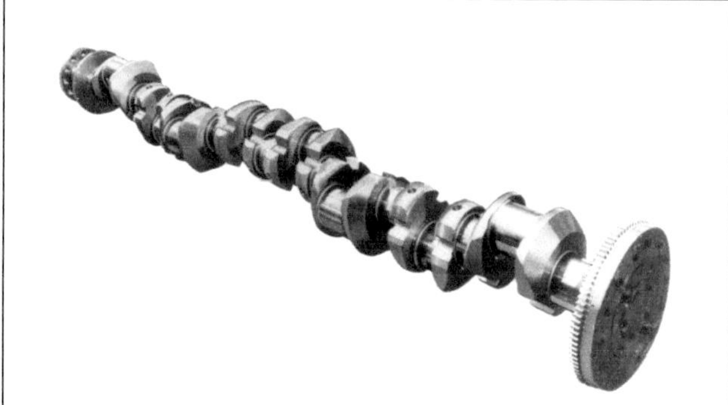

Fig. 9-17 Crankshaft of Four-Stroke-Cycle Engine.
Source: Fairbanks Morse Engine Div.

Fig. 9-18 Crankshaft for Very Large Capacity Reciprocating Engine.
Source: Fairbanks Morse Engine Div.

concepts and definitions common to discussions of reciprocating engine power and work. They build the basic concepts presented in Chapter 2.

Engine power (P) is measured by the product of a force (F) and the rate at which it moves, or the distance through which that force travels per unit of time. It can be expressed as:

$$P = (F)(2\pi r)(N) \qquad (9\text{-}17)$$

Where:
r = Effective length of a brake lever
N = Crankshaft rotational speed

In English system units, 1 hp is defined as the power needed to raise 550 lbm through a height of 1 ft in 1 second (550 ft-lbf/sec). This is equivalent to 33,000 ft-lbf/min or 745.701 (746) watts. Thus, when power is expressed in hp, force in lbf, length in ft, and crankshaft rotational speed in rpm, Equation 9-17 becomes:

$$hp = \frac{(F)(2\pi r)(N)}{33{,}000} = \frac{(F)(r)(N)}{5{,}252} \qquad (9\text{-}18)$$

In SI units, when power is expressed in kW, force in Newton (N), length in meters (m), and crankshaft rotational speed in rpm, while noting that 1 kW is equivalent to 60,000 N-m/min, Equation 9-17 becomes:

$$kW = \frac{(F)(2\pi r)(N)}{60{,}000} = \frac{(F)(r)(N)}{9{,}549} \qquad (9\text{-}19)$$

Actual net power available at the engine's crankshaft is called the power output or brake power and is commonly expressed in the English system as **brake hp** (bhp). This term is derived from the fact that the power output of an engine can be measured by absorbing the power with a brake.

Torque (T) is a measure of the force of rotation. It consists of the product of the force applied to a lever and the perpendicular distance from the line of action of the force to the axis of rotation. It is most commonly expressed in lbf-ft or N-m. For a given torque, there are numerous combinations of amount of force and length of a lever arm. The lever arm is the distance between the centers of the crankpin and the main journal and is, therefore, a radius (r). Torque can thus be expressed as F x r. If, for example, in English units, the force on the crankpin is 500 lbf and the lever arm (length of crank) is 2 ft, then the torque rotating the crankshaft is 1,000 ft-lbf.

While power is the rate at which an engine does work, torque is the capacity of an engine to do work. Torque does not vary in proportion to speed of the engine as does brake power, but depends primarily on volumetric efficiency and friction losses. Substituting torque for the product of force and radius (Fr) in the equation for power yields the following relationship between power and torque.

$$P = (T)(2\pi)(N) \qquad (9\text{-}20)$$

Mean effective pressure (mep) is the work produced by a cycle divided by the volume swept out by the piston in the working stroke and is expressed as:

$$mep = \frac{W}{V_{swept}} \qquad (9\text{-}21)$$

where W represents the cyclic work and V_{swept} is the swept volume ($V_A - V_B$), with the distinction that the two-stroke cycle requires one revolution and the four-stroke cycle requires two revolutions of the crankshaft.

Mep is the constant averaged pressure on the piston over the length of the piston stroke. It reduces the varying pressure on the piston to a single averaged value. Mep is useful in evaluating the ability to produce power. Since the product of piston travel and piston area is piston volume (displacement), the relative power-producing capability of two pistons can be established by comparing the product of mep times displacement in each case. The force on the piston is equal to the pressure times the area (A). Since mep is equivalent to a constant force on the piston over the length of the piston stroke, the work done per power stroke is F times L, where L is the length of the stroke. Thus, for a given displacement, mep is useful in evaluating an engine's ability to produce power due to the following relationship:

$$\textit{Engine power} = \textit{displacement} \times \textit{speed} \times \textit{mep} \qquad (9\text{-}22)$$

In accordance with Equations 9-20 and 9-22, if the displacement of the engine is doubled, torque also doubles.

Mep cannot be measured directly, but it can be calculated from the power equation if the power output, displacement (LA) and speed are known. The relationship of mep to power is shown by rearranging Equation 9-22 as follows:

$$mep = (P)(LA)(n) \qquad (9\text{-}23)$$

In English units, when mep is in psi, power is in hp, LA in cf, and n is in power strokes per minute (in a four-stroke-cycle engine, n = N/2; in a two-cycle engine, n = N, where N is the crankshaft rotational speed, in rpm), Equation 9-23 becomes:

$$mep = \frac{(P)(LA)(n)}{33{,}000} \qquad (9\text{-}24)$$

Note that when LA is expressed in in^3, 33,000 becomes 396,000.

In SI units, when mep is in kPa, power is in kW, LA is in liters (l), and n is in power strokes per minute, Equation 9-23 becomes:

$$mep = \frac{(P)(LA)(n)}{60{,}000} \qquad (9\text{-}25)$$

Given the relationship between power and torque, the relationship of mep and engine torque can be expressed, in English units, as:

$$mep = \frac{(LA)(T)}{150.8} \qquad (9\text{-}26)$$

where mep is in psi, torque in lbf-ft, and LA in in^3 for a four-stroke cycle. For a two-stroke cycle, 150.8 would be replaced with 75.4.

In SI units, the relationship of mep and engine torque can be expressed as:

$$mep = \frac{(LA)(T)}{12.56} \qquad (9\text{-}27)$$

where mep is in kPa, torque in N-m, and LA in l for a four-stroke cycle. For a two-stroke cycle, 12.56 would be replaced with 6.28.

Bmep is a variable independent of the capacity of the engine. As shown by Equation 9-24, for a given set of engine operating conditions, the torque that is developed is proportional to the brake mep. Typically, torque and bmep curves peak at about half that of brake power. Since brake power is proportional to the product of torque and speed, and torque is controlled by the capacity of the engine, higher brake power arises from higher speed. However, with increased speed, friction losses increase at a faster rate than does power, thereby resulting in a relative decrease in mechanical efficiency. Thus, while power increases with increased speed, it does so at a decreasing rate.

Typically, reciprocating engine bmep values will range from 100 to 250 psi (6.9 to 17.2 bar), depending on engine type, design, aspiration type, and operating condition. Generally, bmep will be greater with supercharged/turbocharged engines than with naturally aspirated engines. Bmep values are highest at the speed where maximum torque is achieved and somewhat lower at maximum rated power.

The following example illustrates the measurement of actual brake power and torque for a four-stroke-cycle engine with the following characteristics:

bmep = 200 psi
LA = 5,000 in^3
N = 900 rpm

$$\textit{Engine brake power} = \frac{(200 \times 5{,}000 \times 900)}{(396{,}000 \times 2)} = 1{,}136\ bhp$$

$$\textit{Engine torque} = \frac{(5{,}252 \times 1{,}136)}{900} = 6{,}629\ lbf\text{-}ft$$

Engine Aspiration Systems (Intake and Exhaust)

The engine air intake system supplies clean air for combustion. The exhaust system forces exhaust gases remaining from the previous power stroke from the combustion chamber.

Valve assemblies open and close the intake and exhaust ports that connect the intake and exhaust manifolds to the combustion chamber. Valves, which are usually made from forged alloy steel, are subjected to the direct pressure and temperature of the combustion occurring within the cylinders. The valve stem moves in a valve guide, which can be an integral part of the cylinder head or may be a separate unit pressed into the head.

In a four-stroke-cycle engine, about two-thirds of the way through the power (expansion) stroke, the exhaust valve starts to open. The exhaust gas flows through the valve into the exhaust port and manifold until the cylinder pressure and the exhaust pressure reach equilibrium. This piston then displaces, or sweeps, the gases from the cylinder into the manifold during the exhaust stroke. The intake valve opens just before TDC, while the exhaust valve remains open until, or just after, TDC.

Figure 9-19 illustrates the admission process for an air-fuel mixture in a spark-ignited engine, detailing the location of the spark plug between the intake and exhaust valves. In a Diesel engine, a fuel pump-nozzle assembly is used to atomize and inject the fuel into the combustion chamber after air compression is complete.

The camshaft and drive assembly operate the opening and closing of valves at a controlled rate of speed, as well as at a precise time in relation to piston position. Camshafts, which are usually made of cast iron or forged steel, perform valve actuation with one cam for each valve on the camshaft. Valves are closed by springs and opened by cam lobes. When the camshaft is in the engine frame and the valves are overhead, a push rod is used to transmit motion of the cam and lifter to a rocker arm on the cylinder head that opens the valves. In the case of overhead camshafts, the cams may operate directly on the followers or may first operate on a rocker lever. Figures 9-20 through 9-22 illustrate camshaft and valve opening mechanisms.

Valve timing refers to the adjustment of valves to open and close at the proper time for smooth and efficient engine operation. The valve actuating mechanisms are adjusted so that the valves open and close a designated number of degrees before and after the piston has reached TDC or BDC. If timing is inaccurate in a given cylinder, it will adversely effect power production and extraction.

Timing gears located at one end of the engine drive the camshaft and other components using power supplied by the crankshaft. In smaller applications, a chain- or cog belt-driven camshaft may be used. Figure 9-23 illustrates use of a chain drive applied to a large long-stroke engine to allow the camshaft to be located high up on the engine. Such engines are provided with a hydraulic chain tightener, which automatically maintains the tension of the chain throughout its service life.

During the intake and exhaust strokes of a four-stroke-cycle engine, the engine essentially acts as an air or fuel-air pump. In two-stroke-cycle engines, which include only the compression and power strokes, a separate pump or com-

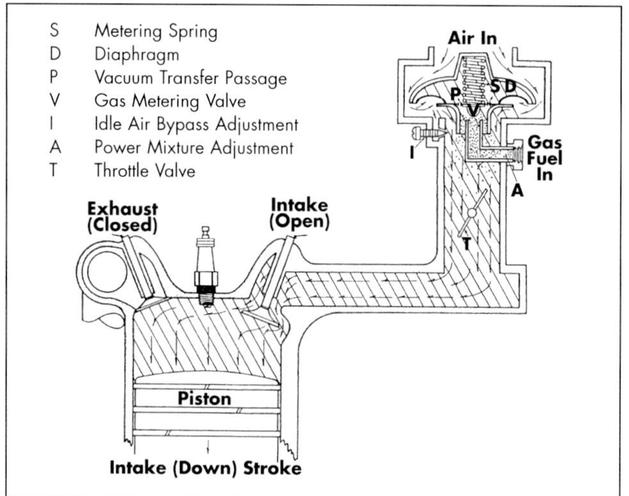

Fig. 9-19 Illustration of Air-Fuel Mixture Admission in Four-Stroke-Cycle Spark-Ignited Gas Engine. Source: Waukesha Engine Div.

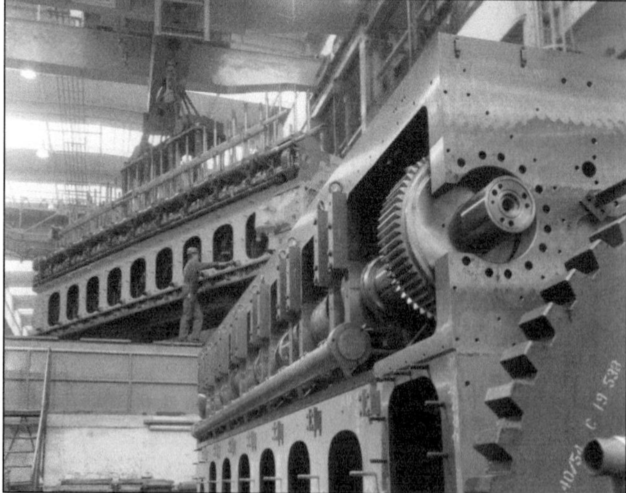

Fig. 9-20 Gear-Driven Camshafts on 17,000 hp (12,700 kW) In-Line Four-Stroke-Cycle Engine. Source: MAN B&W

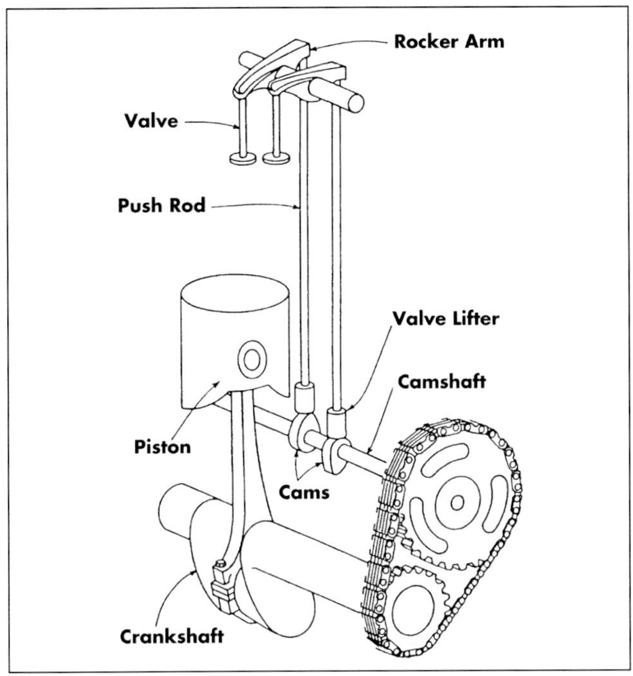

Fig. 9-21 Essential Moving Parts of Valve-Operating Mechanism for One Cylinder. Source: Waukesha Engine Div.

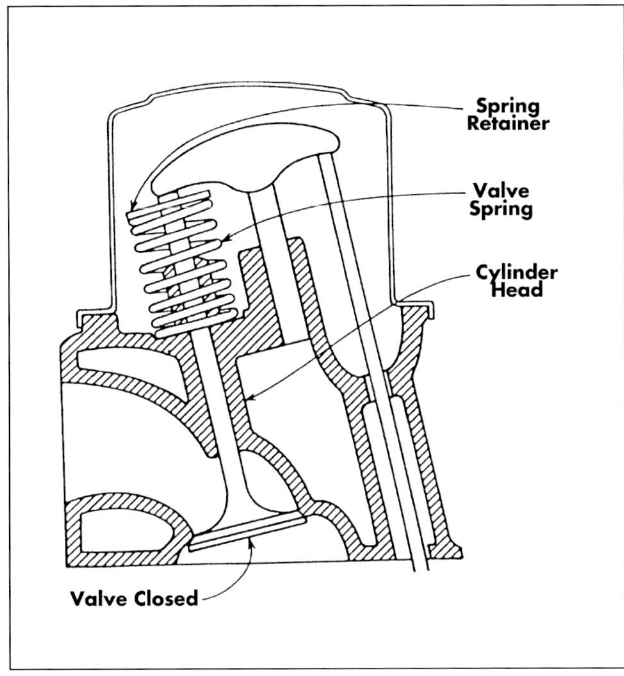

Fig. 9-22 Illustration of Valve and Valve Spring. Source: Waukesha Engine Div.

pressor may be used.

The operation of filling and clearing the cylinder in two-stroke engines is called scavenging. A series of ports or openings is arranged around the cylinder in such a position that the ports are open when the piston is at the bottom of its stroke. Scavenging arrangements are classified as cross-, loop-, and uniflow-scavenging, depending on the location and orientation of the scavenging ports. Cross- and loop-scavenging systems use exhaust and inlet ports in the cylinder walls, uncovered by the piston as it approaches. Uniflow systems may use inlet ports with exhaust valves in the cylinder head or inlet and exhaust ports with opposed pistons. Figure 9-24 illustrates a hydraulically operated exhaust valve assembly for a two-stroke-cycle engine.

With a typical uniflow design, exhaust valves open when the piston is more than half way down and a blowdown (or free exhaust) process begins. As the piston continues toward BDC, the scavenging ports open. Exhaust flow continues toward the exhaust valves, which now have a large open area. When the cylinder pressure falls below the inlet pressure, air enters the cylinder and the scavenging process begins. Flow continues while

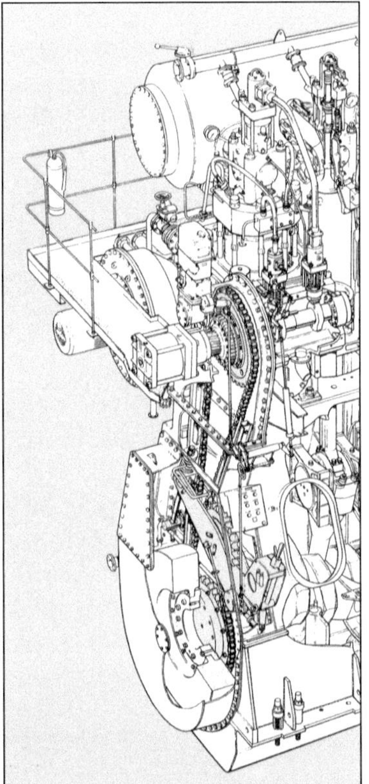

Fig. 9-23 Chain-Driven Camshaft on Large Long-Stroke Engine. Source: MAN B&W

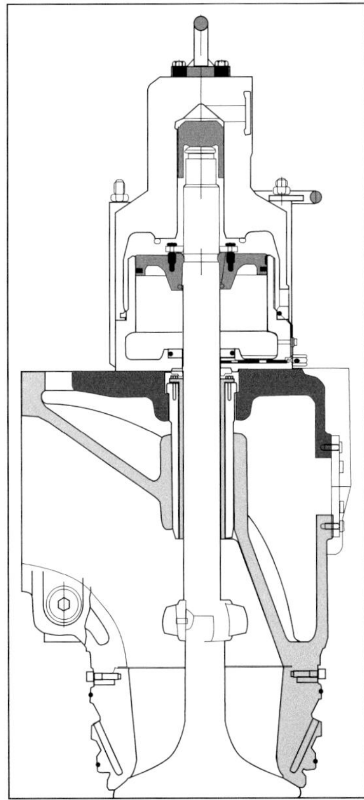

Fig. 9-24 Exhaust Valve Assembly for Two-Stroke-Cycle Engine. Source: MAN B&W

inlet pressure exceeds cylinder pressure. As cylinder pressure exceeds exhaust pressure, the fresh charge displaces the burned gases.

Volumetric efficiency is a term used to express the effectiveness of an engine intake system's induction process. Volumetric efficiency of a four-cycle, naturally aspirated engine is the ratio of the actual volume of air (stated in terms of standard temperature and pressure) taken into the engine cylinder during the intake stroke to the piston displacement. It essentially measures the weight of air actually in the cylinder compared to the weight of an equivalent volume of free air.

Engine power capacity is directly related to volumetric efficiency. As the weight of air in the cylinder increases, more fuel can be burned, producing more power. For two-stroke-cycle engines, the term **scavenge efficiency** is used to describe how thoroughly the burned gases are removed and the cylinder is filled with fresh air. The term volumetric efficiency as defined above is not applied to supercharged or turbocharged engines.

Regardless of the engine type, thermal fuel efficiency is impacted by how thoroughly the products of combustion are swept out from the cylinder and a fresh air charge is admitted. Lower-speed engines generally can achieve greater fuel efficiency because there is more time in each cycle for these events to occur. Engine designs that optimize mass flow through the engine by minimizing restrictions on both the air intake and exhaust sides are also able to achieve greater fuel efficiency.

SUPERCHARGERS AND TURBOCHARGERS

The higher the air pressure (or charge density), the greater the amount of air and oxygen that can be provided to the cylinder and the greater the amount of fuel that is combusted. When the air-charging device is driven mechanically by an accessory shaft from the engine, it is called a supercharger. When the device is driven by the exhaust gases, it is called a turbocharger.

Turbocharging and supercharging have become extremely common in today's high-performance engines. Capacity is commonly increased by 35 to 100% without increasing engine size, resulting in increased power density (increased power ratio per unit of space). However, the increase in power density does produce additional stress on engines, causing greater wear and shorter engine life.

Turbocharging/supercharging is an important part of the movement toward leaner air-fuel mixtures, designed to achieve low NO_X emissions. As engines decrease the total fuel in the air-fuel mixture, a lower-flame temperature is achieved, which reduces NO_X emission levels. It also reduces power output. Increasing the air-charge density compensates for this, allowing engines to maintain power density while operating under very lean combustion conditions.

Common positive displacement charging devices used as superchargers include sliding-vane or rotary compressors and roots blowers. Centrifugal compressors are continuous flow devices, which are well suited for the high-speed operation achieved with an exhaust-driven radial- or axial-flow turbine. Figures 9-25 and 9-26 illustrate two turbocharger turbine designs. Figure 9-27 is a charge-air schematic for a four-stroke-cycle natural gas-fired, spark-ignited engine.

The energy to drive the turbocharger comes from

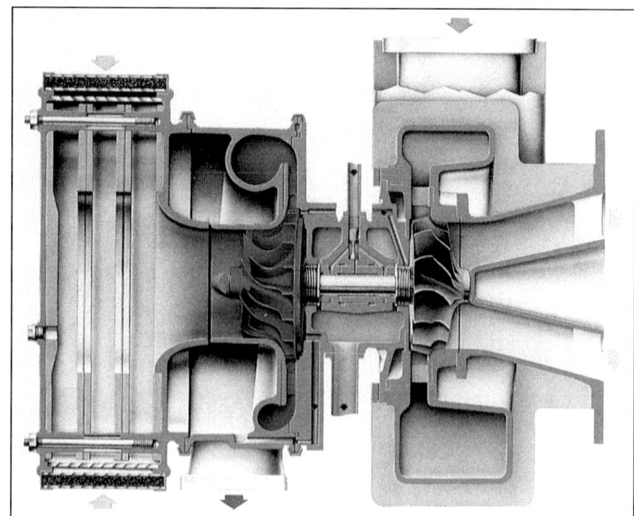

Fig. 9-25 Radial Turbocharger Design. Source: MAN B&W

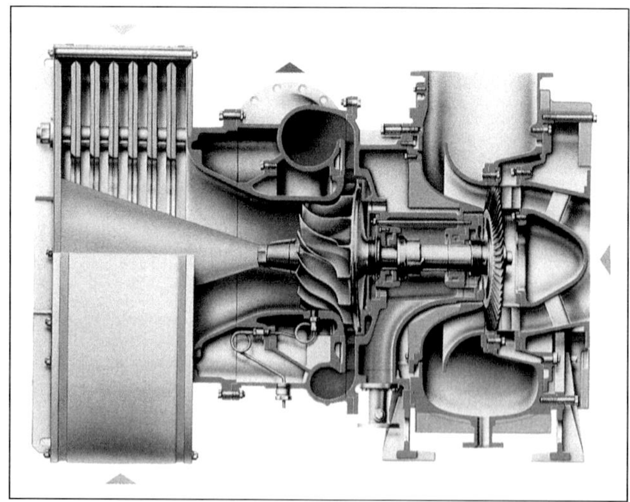

Fig. 9-26 Axial Turbocharger Design. Source: MAN B&W

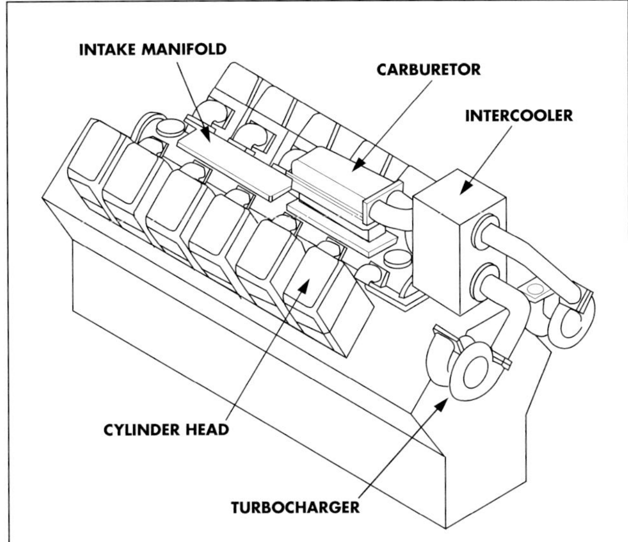

Fig. 9-27 Charge-Air Schematic for Four-Stroke-Cycle Spark-Ignited Engine. Source: Waukesha Engine Div.

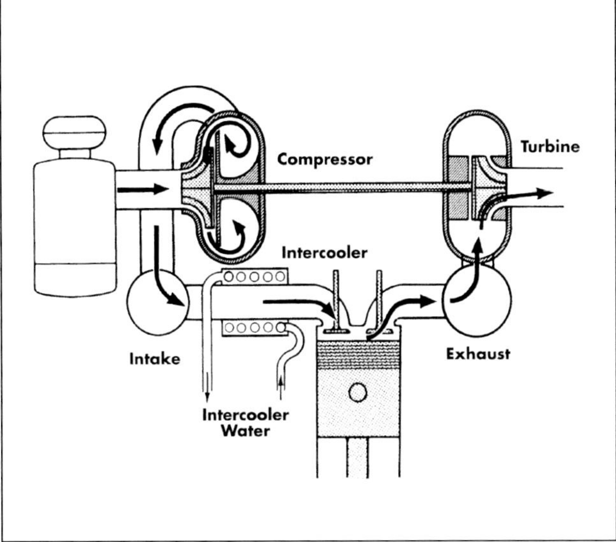

Fig. 9-28 Schematic Illustration of Turbocharger with Intercooler, on Four-Stroke-Cycle Engine. Source: Waukesha Engine Division

the blow-down energy in the engine exhaust. The potential exhaust gas energy available to a turbocharger turbine placed in the exhaust stream is called the blow-down energy because it represents the combustion products being blown-down from cylinder pressure at the point where the exhaust valve opens to atmospheric pressure.

Charge cooling with a heat exchanger, commonly referred to as an intercooler or aftercooler, prior to entry to the cylinder, can be used to further increase the air or air-fuel mixture density. By reducing charge pressure and temperature in spark-ignited engines, higher charge pressure and/or compression ratios can be used.

Figure 9-28 is a schematic drawing of a turbocharger on a four-stroke-cycle engine. The flow path of charge air can be seen from the compressor section through the intercooler and intake manifold, through the intake valve to the cylinder. In the exhaust path, the combustion gases exit the cylinder through the exhaust valve, pass through the exhaust manifold, then to the turbine section before being exhausted to atmosphere. Also shown is the intercooler water circuit.

There are numerous other designs and configurations used for charging air. Some designs feature two blowers in series or parallel. Figure 9-29 shows four configurations used for increasing charge density and optimizing volumetric efficiency with turbochargers and superchargers on a two-stroke-cycle opposed piston engine, which uses two pistons in each cylinder.

Figure 9-29(a) features a mechanical blower or supercharger only. Figure 9-29(d) features a turbocharger only. Figure 9-29(b) features both a turbocharger and supercharger in parallel and Figure 9-29(c) features a turbocharger and supercharger in series. In the series configuration, air is drawn into the turbocharger, where it is compressed and discharged through a cooler to the engine-driven blower. This second-stage blower, operating at a low pressure ratio, discharges the air directly into the engine intake manifold and then to the individual cylinder ports. The inlet air both scavenges the cylinder and supplies a sufficient air charge for proper combustion. The blower provides particularly good response during engine starting and to sudden load changes.

Fuel Delivery and Combustion Systems

Specific fuel consumption (sfc) and thermal fuel efficiency are based on the power delivered by an engine and the fuel energy consumed. Sfc is the ratio of the mass flow of fuel ($\dot{m}_f$), typically expressed as an amount of fuel, in number of lbm, gallons or cf used by the engine, per hour, to the power (P) produced or delivered by the engine (typically expressed as hp or kW). It is expressed as follows:

$$sfc = \frac{\dot{m}_f}{P} \qquad (9\text{-}28)$$

When it is based on the brake power delivered, it is termed brake sfc (bsfc).

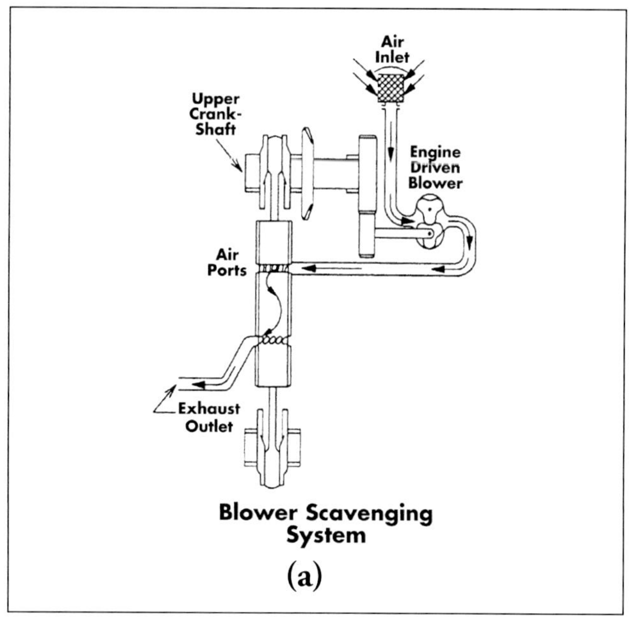

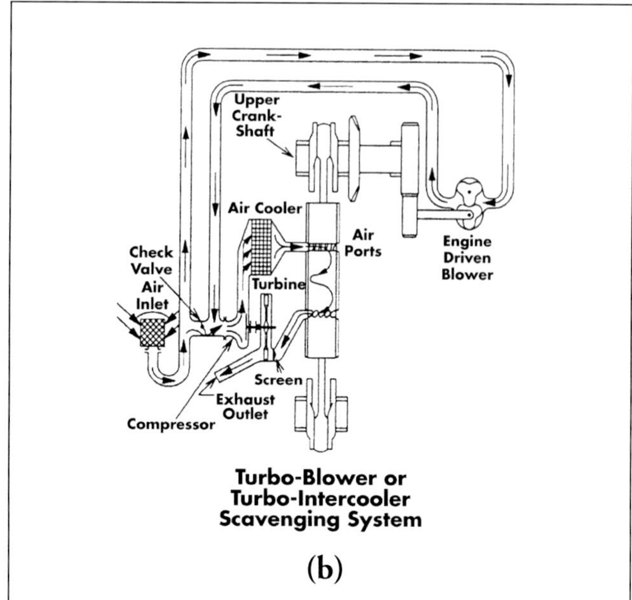

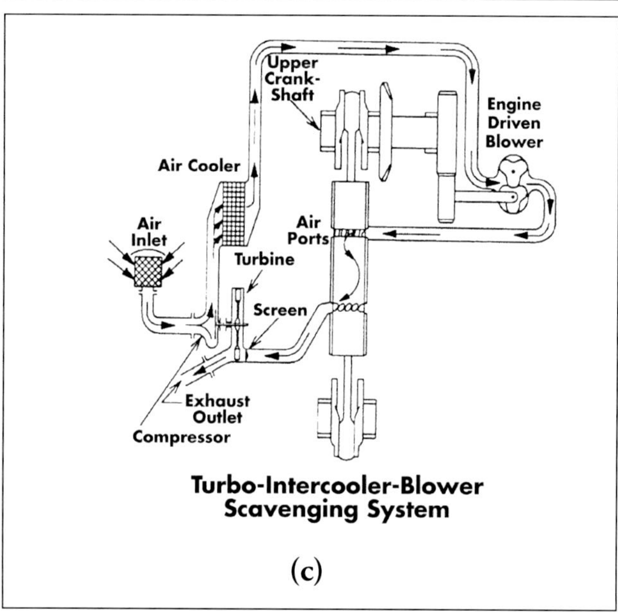

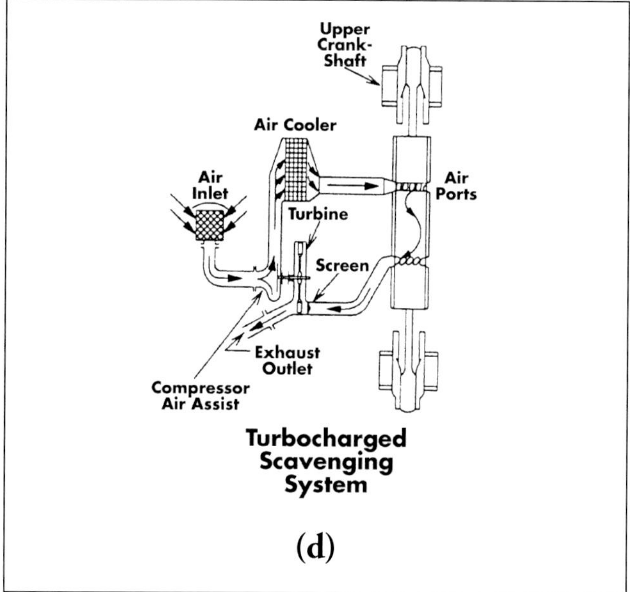

Fig. 9-29a-d Various Configurations for Increasing Charge Density. Source: Fairbanks Morse Engine Div.

Engine thermal fuel efficiency is the ratio of the work produced per cycle to the amount of fuel energy supplied per cycle that can be released by combustion. The fuel energy supplied that can be released by combustion is the product of the mass flow of fuel supplied to the engine and the heating value or energy intensity of the fuel. This may be expressed on a higher heating value (HHV) or lower heating value (LHV) basis.

Air-fuel ratios (and exhaust gas recirculation where applicable) are designed to achieve a specified power and torque output at a given speed and to optimize reliability, efficiency, and emissions control. Microprocessor control is now commonly used to carefully regulate air-fuel ratios. Some engine control systems feature air-fuel ratio control for each cylinder. Electronic control of ignition timing and fuel injection rates is also used.

Primary air emissions of concern in reciprocating internal combustion engines include NO_X, carbon monoxide (CO), hydrocarbons (HC)/organic compounds, and particulate matter (PM). CO emissions, which depend primarily on the air-fuel ratio, are minimal in both lean-burn spark-ignited engines and Diesel engines. CO emissions are relatively high in rich-burn spark-ignited engines. To minimize NO_X emissions, combustion system designs promote thorough mixing of air and fuel, lower peak combustion temperatures, reduced residence time at combustion temperature, and, in the case of dual-fuel engines, reduced pilot fuel use.

Depending on the air-fuel ratio used, engines may be classified as lean-burn, stoichiometric, or rich-burn. The ratio of oxygen (or air) to fuel that produces perfect combustion is referred to as the **stoichiometric air-fuel ratio.** If more oxygen (excess air) is used in the combustion process cycle, the engine is referred to as a **lean-burn** engine. If fuel intake is equivalent to or greater than the stoichiometric quantity, the engine is referred to as a **rich-burn** engine. Diesel engines always operate with lean air-fuel ratios and under part-load, operate with extremely lean air-fuel ratios. Spark-ignited engines may be designed to operate over a wide range of air-fuel ratios, from rich to very lean.

Combustion chambers include open chamber and pre-chamber designs. In open chamber designs, all of the charge is contained in a single space. In pre-chamber designs, combustion is initiated in a pre-combustion chamber or ignition cell that typically comprises from 1 to 6% of clearance volume (combustion chamber volume when the piston is at TDC). Figure 9-30 illustrates open chamber and pre-chamber configurations.

Pre-chamber combustion allows for an increase in the air-fuel ratio to levels exceeding that achievable with open chamber lean-burn engines. At the higher end of lean combustion in conventional spark-ignited engines, misfire may occur. To ensure proper combustion, a small volume of fuel-rich mixture (below the stoichiometric level) is burned in the pre-chamber. This fuel mixture may be spark- or compression-ignited.

Figure 9-31 illustrates a cylinder head igniter and shows the admission valve assembly, spark plug holder, ignition sleeve, and pre-chamber. Figure 9-32 details the pre-chamber assembly. Figure 9-33 is a cutaway illustration of a four-stroke-cycle spark-ignited, twin air-valve, carbureted gas engine. It details cylinder head igniter, admission valve, pre-chamber, and cylindrical combustion bowl on the piston head.

Very lean combustion in spark-ignited design can result in slow burning combustion and a reduction in

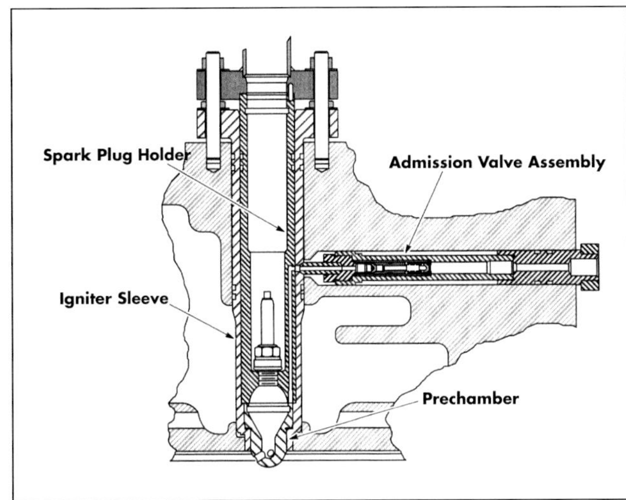

Fig. 9-31 Cylinder Head Igniter. Source: Waukesha Engine Div.

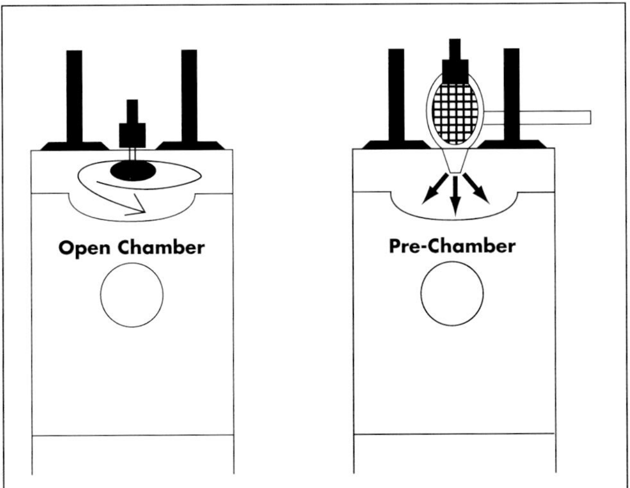

Fig. 9-30 Comparative Illustration of Open and Pre-Chamber Configurations. Source: Waukesha Engine Div.

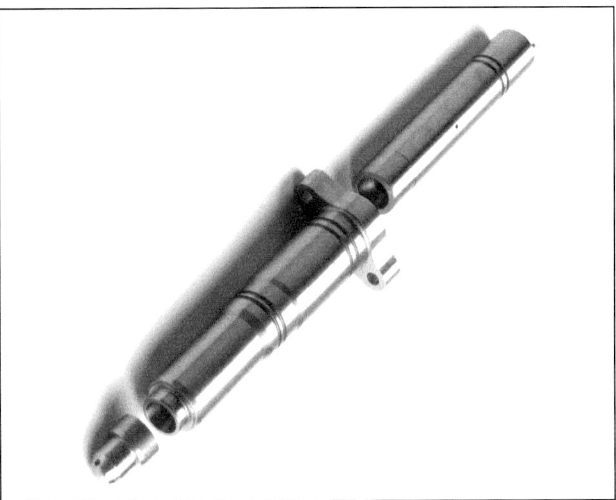

Fig. 9-32 Pre-Chamber Assembly. Source: Waukesha Engine Div.

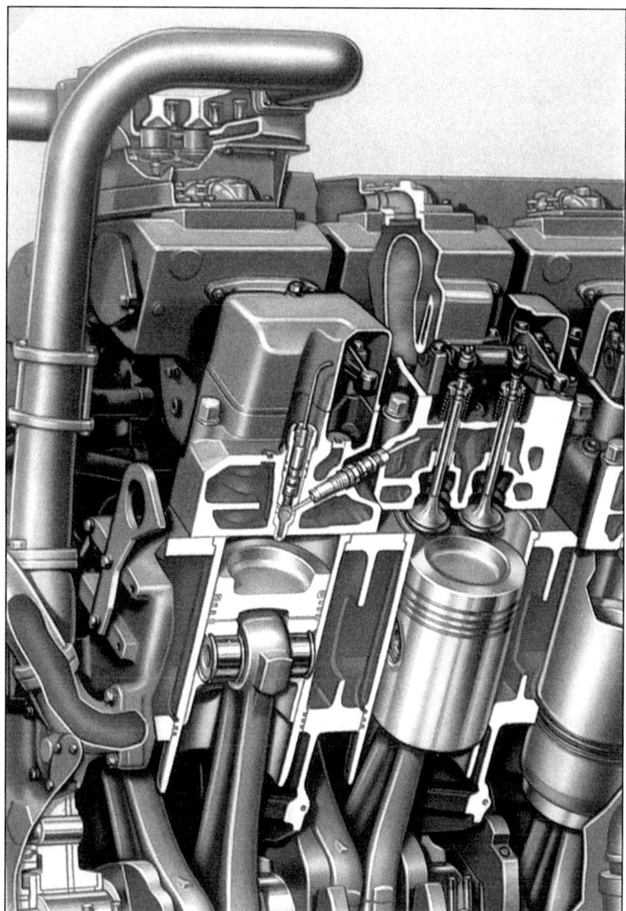

Fig. 9-33 Cutaway Illustration of Pre-Chamber, Spark-Ignited Engine, Detailing Combustion System Components.
Source: Waukesha Engine Div.

Fig. 9-34 Piston Head with Specially Shaped Combustion Bowl to Increase Turbulence. Source: Waukesha Engine Div.

engine capacity and fuel efficiency. To optimize performance, it is necessary to achieve thorough mixing of fuel and air in order to limit peak flame temperatures while achieving the highest possible flame front velocity.

To enhance mixing and burn rate, combustion chambers are designed for a high degree of turbulence. Figure 9-34 shows a piston head with a uniquely shaped combustion bowl, also designed to promote greater turbulence. Swirling flow about the cylinder axis is used to promote more rapid mixing and combustion in both compression ignition engines and spark-ignited engines, and to improve scavenging in two-stroke engines. Swirl may be generated by discharging the flow tangentially toward the cylinder wall, where it is deflected sideways and downward in a swirling motion. Another way is by forcing the flow to rotate about the valve axis before it enters the cylinder. Figure 9-35 shows a swirl flow cylinder head design for a spark-ignited engine.

Fig. 9-35 Swirl Flow Cylinder Head Design.
Source: Warstila Diesel

Combustion Control in Spark-Ignited Engines

Combustion occurs in a spark-ignited Otto-cycle engine when the ignition system provides a spark at a specific time (close to the end of the compression stroke) to ignite the compressed air-fuel mixture. The spark is generated by a spark plug located in the cylinder's combustion chamber. The high-temperature plasma kernel created by the spark develops into a self-sustaining and propagating flame front. This creates intense heat and pressure within the cylinder to force the piston downward in the power stroke.

A key variable in emissions control is air-fuel ratio. A long-term trend in stationary industrial spark-ignition engines has been toward lean-burn designs to reduce CO and HC emissions. Excess air provides for more complete combustion, thereby minimizing CO emissions. Excess air also allows unburned hydrocarbons, from cylinder crevices and cylinder walls, to be oxidized by the oxygen-rich gases present during the expansion and exhaust strokes. With older generation lean-burn engines operating at a lambda (λ) of about 1.1 or greater, CO and HC emissions were dramatically reduced. However, NO_X emissions were not effectively controlled. Therefore, the trend has been toward engines designed to operate with extremely lean air-fuel ratios, precisely controlled to minimize NO_X as well as CO and HC emissions while maintaining engine power density and high thermal fuel efficiency and while avoiding misfire. If the mixture is too lean, combustion quality becomes poor, HC emissions rise sharply, and engine operation becomes erratic.

One method of NO_X emissions control is the mixing of a fraction of exhaust gas with the air-fuel mixture. This process, referred to as exhaust gas recirculation (EGR), is effective under partial load conditions, though the amount of recycled gas that can be used is limited.

The most effective method of air emission control in engines operating near the stoichiometric air-fuel ratio is the use of three-way catalysts. These systems can be effectively applied to control NO_X as well as CO and HC emissions in both stationary and vehicular applications. To be effective, three-way catalyst systems require the engine air-fuel ratio to be maintained close to stoichiometric through the use of an oxygen sensor in the exhaust. While newly developed lean-burn engines can often be permitted without the need for exhaust emission treatment systems, rich-burn engines commonly require them. When lean-burn engines do require exhaust emission treatment, more expensive selective catalyst reduction (SCR) systems are required for the oxygen-rich exhaust.

Gas-fired spark-ignited engine carburetors provide gas pressure regulation and air-fuel mixing. Using an arrangement of a diaphragm and springs, the pressure regulator controls the pressure of the gas entering the mixer, where air is added to provide the desired air-fuel ratio. The air-fuel ratio of a spark-ignited engine should vary as a function of the load under normal steady-running conditions. The conventional carburetor uses the rate of airflow to the engine as the major variable to control air-fuel ratio. In many modern system designs, additional controls sensitive to variables such as torque output, pressure in inlet manifold, air density, temperature, or detonation are used as additional controlling elements.

The upper right hand corner of Figure 9-19 illustrates a carburetor for a spark-ignited gas engine and details the air and gas entry locations. Figure 9-36 shows a carburetor, revealing the air-gas metering valves in the upper air inlet housing. Airflow is controlled by a throttle downstream of the venturi. The gas inlet valve is shown on the side.

Although a carburetor is usually the simplest and least expensive device for fuel metering in a spark-ignited engine, increasingly, the trend has been toward fuel injection systems. The control precision achieved with fuel injection is a key element in today's high-performance, low-emission engines. Fuel is typically injected before or after the air charger into the intake manifold or intake valve ports. There are several different types of injection systems, including mechanical and electronic designs that may include individual cylinder injectors or a single fuel injector.

Electronic control systems commonly use injectors at each inlet port. Monitoring of engine speed, crank position, spark timing, air-fuel ratio, jacket temperature, air-inlet temperature, airflow, and throttle position is used to provide the optimal combination of fuel rate and emissions charac-

Fig. 9-36 Carburetor with Air-Fuel Metering Valves Shown in Upper Air Inlet Housing.
Source: Waukesha Engine Div.

teristics for each operating condition. Figure 9-37 is a block diagram of an integrated system for speed, air-fuel, and ignition timing control for spark-ignited gas-fueled engines. The exhaust sensor is usually an oxygen sensor, though a temperature sensor or NO_X sensor may also be applied.

Applied ignition system designs feature distributor ignition, magneto ignition, and electronic (solid state) ignition. Distributor ignition systems use a battery and a single ignition coil (transformer) to provide spark current. Most of the larger current gas engines are offered with either a magneto or electronic ignition system, both of which use one coil for every cylinder. The magneto generates its own power, while electronic systems require a battery. Figure 9-38 shows electronic ignition components.

The ignition system must provide sufficient voltage across the spark plug electrodes to ignite the air-fuel mixture under the full range of operating conditions. As the air-fuel mixture becomes increasingly lean, higher energy is required for effective flame propagation.

The ignition system also has to be timed so as to ensure that the high-tension spark occurs at the proper instant within each cylinder. Since combustion does not take place instantaneously, the spark should occur before the end of the compression stroke. This is known as spark advance, meaning the spark occurs before the piston is at TDC. Generally, the spark setting is dependent on the combustion time, which varies with cylinder design and operating conditions. Some engine cylinder designs feature use of two strategically placed spark plugs.

A turbulent flame develops from the spark discharge, propagates across the air-fuel mixture, and extinguishes at the combustion chamber wall. Typically, the duration of the burning process is about 40 to 60 crank angle degrees. The rate at which the air-fuel mixture burns increases from a low value, immediately following the spark discharge, to a maximum value, about halfway through the burning process, and then to close to zero as the burning process concludes.

For each engine design under given conditions, there is particular spark timing, which gives maximum engine torque. If the start of the combustion process is progressively advanced before TDC, the compression stroke work (from the piston to the cylinder gases) transfer increases. If the end of the combustion process is progressively delayed (or retarded), the peak cylinder pressure occurs later in the expansion stroke and is reduced in magnitude, resulting in reduced expansion stroke work.

A retarded spark **(timing retard)** is used as a means of controlling detonation and NO_X and hydrocarbon emission formation. Retarded timing results in an increase in exhaust temperature and a decrease in the optimal power and thermal fuel efficiency. Timing retard must, therefore, be precisely controlled to avoid poor combustion and the potential for misfiring. As speed increases, or as load and intake manifold pressure are decreased, the spark must be advanced to maintain optimum timing.

Surface ignition is the ignition of the air-fuel charge by overheated valves or spark plugs or by any other hot spot in the combustion chamber. When surface igni-

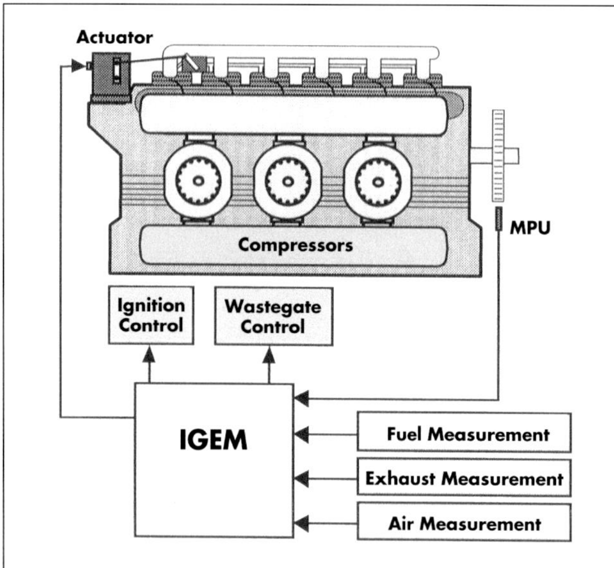

Fig. 9-37 Block Diagram for Integrated Gas Engine Manager System. Source: Woodward Governor

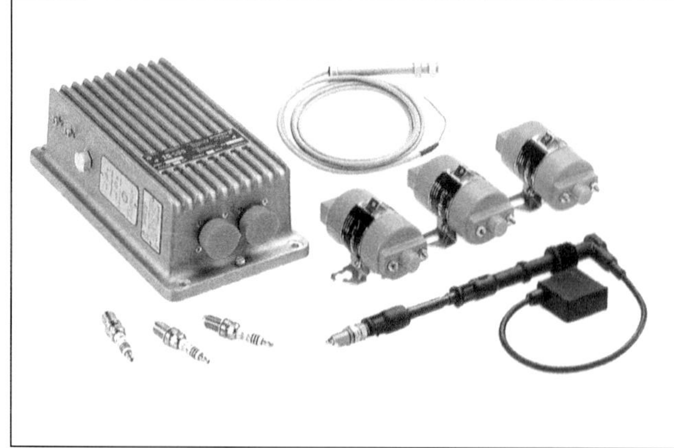

Fig. 9-38 Electronic Ignition Components for Spark-Ignited Engine. Source: Wartsila Diesel

tion occurs before the spark plug ignites the charge, it is referred to as pre-ignition. When it occurs after normal ignition, it is referred to as post-ignition. Pre-ignition, which is more common, can be very destructive to an engine, causing overheating and eventual failure of components.

The components that can cause pre-ignition are those that are least well-cooled and where deposits build up, such as spark plugs and exhaust valves. Surface ignition can be avoided through proper engine design and appropriate attention to fuel and lubricating oil quality.

A critical concern with spark-ignited engines is the avoidance of **knock**. As the flame propagates across the combustion chamber, the unburned mixture ahead of the flame, called the end gas, is compressed, causing its pressure, temperature, and density to increase. Some of the end-gas air-fuel mixture may undergo chemical reactions prior to normal combustion, the products of which may auto-ignite. This spontaneous ignition causes a rapid release of energy, which, in turn, causes high-frequency pressure oscillations inside the cylinder that produce a sharp metallic noise, or knock. Knock is avoided when the flame front consumes the end gas before these reactions have time to cause the air-fuel mixture to auto-ignite.

Knock primarily occurs under wide-open throttle operation. The tendency to knock depends on engine design and operating variables that influence end-gas temperature, pressure, and the time spent at high values of each. End-gas temperature, and therefore the tendency to knock, can be reduced by decreasing the inlet air temperature and retarding the spark timing. However, with any given engine design, the tendency to knock is directly dependent on the anti-knock quality of the fuel.

There is a wide variation in the ability to resist knock among hydrocarbon fuels, depending on molecular size and structure. The compression ratio at which, under specified operating conditions, a specific fuel will exhibit knock is known as the critical compression ratio. The fuel **octane number**, which will vary slightly depending on the rating method used, defines its ability to resist knock. The critical compression ratio of a fuel increases with increased octane number.

Methane, the primary constituent of natural gas, has a very high octane number of 120. Most pipeline natural gas will have an octane number of slightly less than this due to the other lower-octane constituents. Commercial gasoline will vary in octane number depending on its formulation and the types of chemical additives used to boost its octane number. Oxygenates such as methanol, ethanol, and methyl tertiary butyl ether (MBTE) offer good anti-knock blending characteristics. Because the octane number of gasoline is considerably lower than that of natural gas, engines operating on gasoline are limited to a lower compression ratio than those operating on natural gas.

The octane number requirement of an engine is defined as the minimum fuel octane number that will resist knock throughout its speed and load range. In addition to fuel composition, this is influenced by an assortment of engine design and operating variables. For example, octane number requirements tend to decrease when ignition timing is retarded, inlet air density and temperature are decreased, humidity is increased, and when engine load is reduced. Engine octane number requirement is also highest with operation slightly rich of the stoichiometric ratio. Movements toward leaner or richer mixtures tend to reduce the octane number requirement and, therefore, increase the allowable compression ratio.

A similar index used for measuring knock resistance characteristics of gaseous fuels is the **methane number.** This index establishes pure methane, which is highly knock resistant, with a methane number of 100. Hydrogen, which is very prone to knocking, represents the zero point on the methane index. Natural gases usually contain not only methane, but higher valence hydrocarbons as well, such as butane or propane with methane numbers of about 10 and 35, respectively. Other inert components, such as N_2 and CO_2, raise the antiknock rating, sometimes resulting in methane numbers higher than 100.

Combustion Control in Diesel Engines

Combustion occurs in a compression-ignition Diesel-cycle engine when fuel is injected into the hot compressed air charge in the cylinder. Diesel engine fuel injection systems are required to atomize and distribute fuel in the combustion chamber, as well as precisely control the rate of injection based on operating parameters.

During the compression stroke, air is compressed to about 600 psi (41.4 bar) at a temperature of about 1,000°F (538°C). Fuel-injection timing controls the crank angle at which combustion starts. If the fuel is injected or ignited too early, compression will not be at the maximum and ignition will be delayed. If fuel is injected too late, the piston will be past TDC and power output will be reduced.

Since injection commences just before combustion starts, there is no knock limit and higher compression ratios can be used. Also, since engine torque is controlled by varying the amount of fuel injected per cycle, with air-flow essentially unchanged, the engine can be operated unthrottled. Thus, pumping energy is low and part-load performance is superior to spark-ignited engines.

In a pump-injection system, the fuel pump forces the fuel under high pressure through the fuel line and nozzle. Each cylinder has its own fuel pump and each pump is operated by a separate cam on the camshaft. Fuel-injection nozzles may be hydraulically operated by the pressure provided by the fuel pump.

Combustion in compression-ignition engines occurs in stages. Liquid fuel must first be brought to vapor form before it will start to burn. Compression of the air charge in the cylinder provides heat for vaporizing the atomized liquid fuel. Following the initiation of fuel injection, there is a delay during which some oxidation takes place, but no appreciable pressure rise occurs. Next, rapid fuel burning begins, followed by a period of constant or controlled pressure rise. After injection is completed, after-burning of the heavier fuel molecules occurs while the piston is rapidly moving down. All burning must be completed prior to the exhaust valve opening to limit smoke in the exhaust.

Ignition delay is the time (or crank angle) interval between the start of injection and the start of combustion. Minimizing ignition delay through component design is important for optimizing thermal fuel efficiency and smoothness of operation, and for minimizing misfire and smoke emissions. The ignition characteristics of fuel also affect ignition delay and, therefore, engine performance. The ignition quality of a fuel is defined by its **cetane number**, which is determined by comparing the ignition delay of the fuel with that of primary reference fuel mixtures in standardized tests. Fuel with a higher cetane number produces a shorter ignition delay.

Air emissions concerns with Diesel engines focus on NO_X, HC, and particulate matter. Despite a higher cost, Diesel fuel and other distillate oils are generally used instead of residual oils due to the superior air emissions characteristics. The two major causes of HC emissions in Diesel engines are lean combustion and under-mixing of fuel, which leaves the fuel injector nozzle at low velocity late in the combustion process. HC emissions can also result from fuel droplets trapped in crevices that are not burned and from premature cooling of the fuel charge before it is completely burned.

Both particulate matter and NO_X emissions are, in large part, due to fuel composition, resulting from the carbon and nitrogen content, respectively. Diesel engine particulates consist primarily of combustion-generated soot on which some organic compounds have been absorbed. Due to the small particle size involved, some type of filter is the most effective trapping measure. NO_X emissions result from both the fuel nitrogen content (fuel-NO_X) and the combustion process (thermal NO_X). The Diesel cycle combustion process is inherently more disposed to thermal NO_X production than the Otto Cycle, in which thorough mixing of air and fuel initiates in the compression stroke. Retarded injection timing is used to reduce NO_X emissions (with a modest reduction in efficiency), though an increase in particulate emissions can occur.

While there are many methods used to partially limit NO_X emissions in Diesel engines, levels of control required to meet more stringent emissions standards can generally only be achieved with the use of costly SCR systems. Another alternative to particulate matter and NO_X control is dual-fuel operation, with natural gas used as the primary fuel.

Combustion Control in Dual-Fuel Engines

In dual-fuel engines, combustion is initiated with compression ignition of injected pilot oil. Significantly more energy is released from compression-ignited pilot oil than a spark, allowing for faster and more complete combustion of gaseous fuel. This allows for operation at higher compression ratios and mep than would be practical with a standard Otto-cycle type gas engine. Current dual-fuel combustion system designs achieve nearly the same cylinder power output and thermal fuel efficiency operating on natural gas as is achievable with standard Diesel engine operation on liquid fuel, while adding the benefits of strict air emissions control and fuel purchase flexibility. Applied designs have achieved thermal fuel efficiencies in excess of 40% (LHV basis) with operation on natural gas with NO_X emissions rates of 1 gram/bhp-h (1.3 gram/kWh) or less.

The methane number of the gaseous fuel used is an important consideration with dual-fuel engines. Engine ratings may be based on a methane number of 100 and require modification if fuel characteristics are significantly different.

Design options include compression of air only in the cylinder with injection of separately compressed gas or compression of a gas-fuel mixture. Figure 9-39 provides a cross-section view of a dual-fuel combustion system featuring injection of compressed natural gas. Highlighted are the fuel injection valve, gas fuel actuator, and control oil pump. This gas-Diesel type design uses a combined oil/gas injection valve and functions very much like a tra-

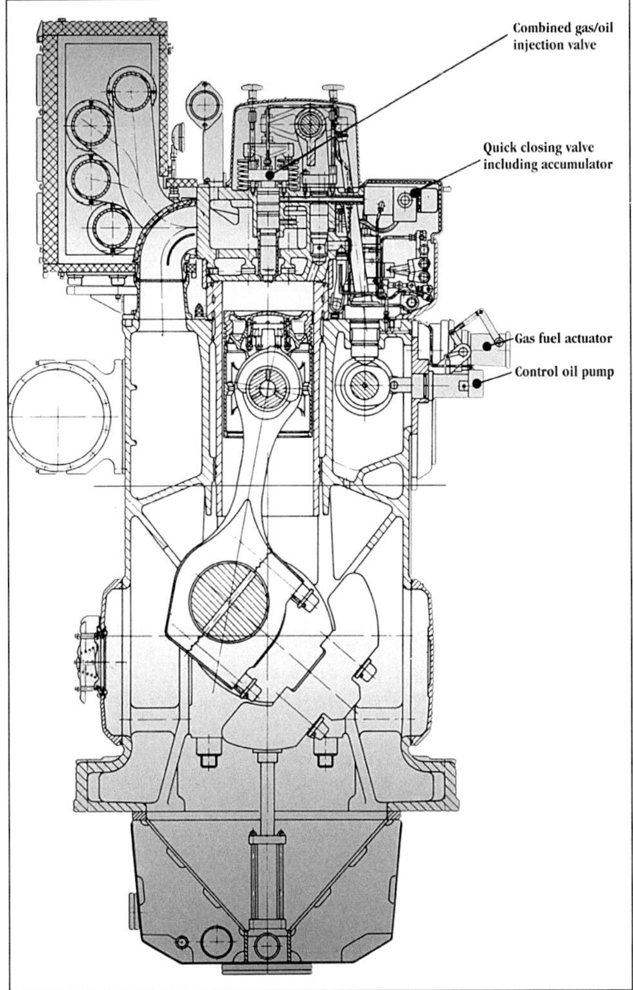

Fig. 9-39 Cross Sectional View of Dual-Fuel Engine Showing Combined Oil/Gas Injection Valve. Source: Warstila Diesel

ditional Diesel-cycle engine. Under start-up and low-load operation, the system operates exclusively on liquid fuel. As the load reaches about 35% of full load, gas-injection may be activated and the engine can operate with 5% pilot oil, with gas providing the balance of the fuel requirement.

Figure 9-40 illustrates a similar dual-fuel Diesel engine fuel system design that features high-pressure gas injection. Shown on the right is the high-pressure pump, which incorporates one element for injection of pilot fuel and a second element in the control oil system for opening the gas fuel spindle. Required gas pressure to enter the pressurized combustion chamber is about 3,600 psi (250 bar). This necessitates the use of an external gas compressor module for independent gas compression.

As opposed to the injection of high-pressure gas after air compression is complete, an alternative dual-fuel engine design features location of the gas valve in the intake manifold. The gas valves are hydraulically actuated and gas volume is controlled by varying the valve opening duration. Varying air-fuel ratios are achieved by bypassing air from the turbocharger compressor to the suction side. This arrangement promotes thorough mixing at the gas valve and achieves homogeneity of the mixture for very lean combustion and minimal NO_X production.

Figure 9-41 is a cross-sectional view of a cylinder, detailing the dual-fuel elements. Shown are the header locations and gas valve. Figure 9-42 is a diagram of the cylinder head, detailing the pre-combustion chamber with

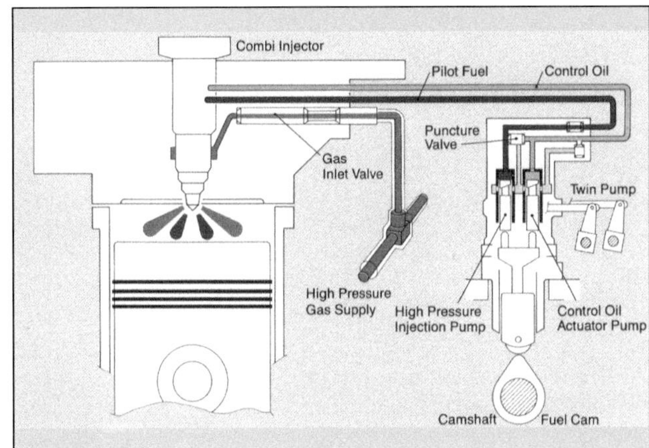

Fig. 9-40 Dual-fuel Diesel Engine Fuel-System Illustration, Showing High-Pressure Gas Injection. Source: MAN B&W

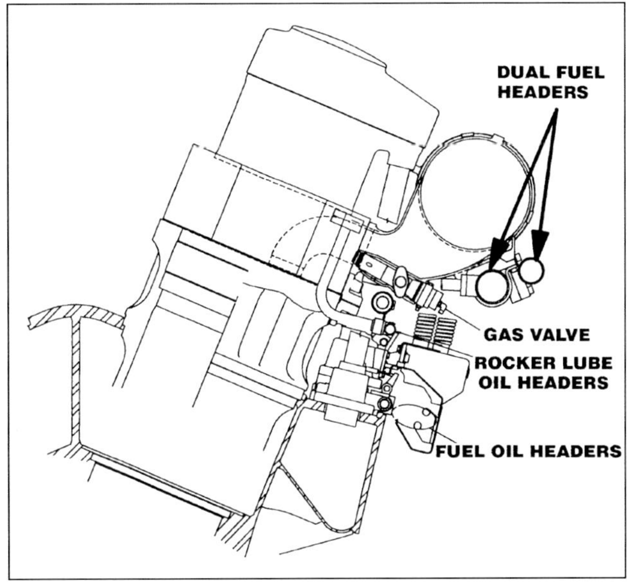

Fig. 9-41 Cross-Sectional View of Cylinder Detailing Dual-Fuel Elements. Source: Fairbanks Morse Engine Div.

pilot nozzle and the Diesel nozzle. This pre-ignition chamber design requires only 1% pilot oil, allowing for operation with as much as 99% gas.

Figure 9-43 illustrates the indicated pressure characteristic with respect to crank angle of combustion with full-load operation on 99% natural gas with pre-chamber injection of 1% pilot oil. The first pressure peak is associated with pre-chamber combustion at the end of the compression stroke.

Current industry research and development efforts are focused on development of direct-injected gas technology, which will allow for Diesel-cycle engines to operate without use of pilot oil, while matching the power and thermal efficiency of equivalent sized liquid-fuel engines. Due to high ignition temperature requirements, the direct-injected gas engine will require some type of ignition assist at typical Diesel engine compression ratios. An alternative to pilot Diesel injection may be a glow plug ignition assist system, which provides a sufficiently hot surface for ignition.

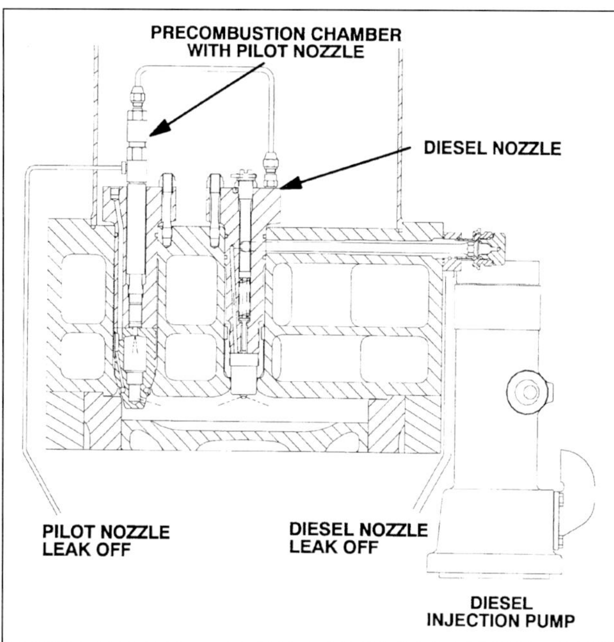

Fig. 9-42 Cylinder Head Diagram Detailing Pre-Combustion Chamber with Pilot Nozzle. Source: Fairbanks Morse Engine Div.

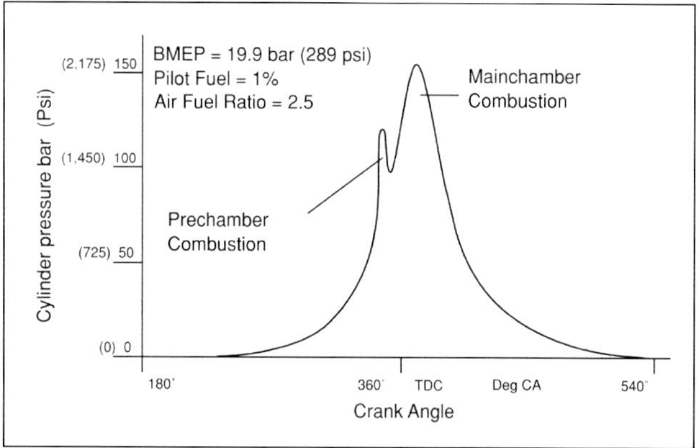

Fig. 9-43 Indicated Pressure Characteristics of Dual-Fuel Operation with 1% Pilot Oil. Source: MAN B&W

Engine Coolant Systems and Heat Recovery

A reciprocating engine requires a coolant system to extract heat from the engine and reject the heat to an external device such as a radiator, cooling tower, or heat recovery system. Peak gas temperatures in internal combustion reciprocating engine cylinders are on the order of several thousand degrees F. Maximum metal temperatures for the inside of the combustion chamber space are limited to much lower values and the gas-side surface of the cylinder wall must be kept below about 350°F (177°C) to prevent deterioration of lubricating oil film.

Coolant systems include cooling circuits for engine jacket water, lubricating oil, and, where applicable, charge air. Figure 9-44 shows the jacket and auxiliary water connections for a four-stroke-cycle spark-ignited engine. This system features a gear-driven jacket water pump, which is mounted on the front of the engine directly above the lubricating oil pump. Both the crankcase and cylinder heads contain passageways, which comprise the jacket water system. Cooling water exits the heat transfer device and enters the jacket water pump, from which it is directed

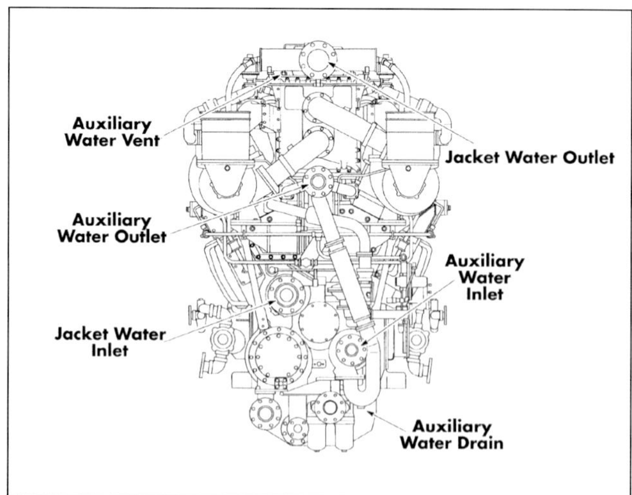

Fig. 9-44 Jacket and Auxiliary Water Connections for a Spark-Ignited Engine. Source: Waukesha Engine Div.

to the header of the water supply manifold. It circulates around the cylinder liners and then passes up into the cylinder head, through water jumpers, and flows around the valve seats.

Figure 9-45 illustrates a cylinder liner/water guide assembly. Exiting the cylinder head, the cooling water flows into the exhaust manifold above each cylinder bank. It then passes to a collection manifold that routes it to the water outlet header and back to the heat transfer device.

Much of the heat of friction due to metal surfaces in relative motion, with a lubricant in between, passes to the lubricating oil system, which serves the additional function of component cooling. The oil pump draws the oil from the sump and delivers it through the filter and strainers to the internal oil distribution system. Various distribution lines and spray nozzles are used to lubricate and provide cooling to components throughout the engine. Figure 9-46 shows a typical shell-and-tube oil cooler for an industrial-grade engine.

The heat balance of a reciprocating engine includes useful work, exhaust heat, heat to lubricating oil and cooling jackets, heat from charge air, and radiation losses. Nearly half (jacket water plus exhaust) of a moderately efficient, small capacity reciprocating engine's energy input can be recovered as useful thermal energy. Engines with higher thermal efficiencies have lower mass flow and heat rejection and, therefore, produce less recoverable heat. Values will vary widely depending on specific engine designs and operating conditions, so examples should not be considered as typical or representative.

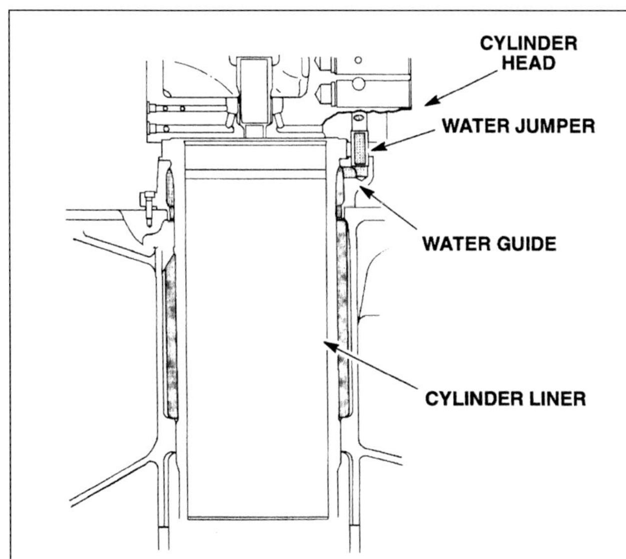

Fig. 9-45 Illustration of Cylinder Liner/Water Guide Assembly. Source: Waukesha Engines Div.

Recoverable heat outputs can be classified as either high temperature or low temperature:

- Part of the high-temperature heat contained in engine exhaust can be recovered for process heating or to generate high-pressure steam by means of an exhaust gas boiler. Small systems may combine exhaust gas and coolant system heat recovery to produce low-grade heat, or omit exhaust gas heat recovery altogether.
- Low-temperature heat, recovered from water jacket, lubricating oil, valve cage, and charge-air cooling systems, can be used for hot water or air heating or to generate low-pressure steam. Nearly all of the cooling system heat can be recovered. Heat is usually recovered by forced circulation systems, which produce hot water at temperatures up to 250°F (121°C). Low-pressure steam of about 15 psig (about 2 bar) can be raised by either flashing high-temperature hot water from forced circulation systems or through ebullient cooling.

Note that published heat balance data, found in manufacturers' technical data books, are usually calculated at conventional jacket water outlet temperatures. These values must be adjusted if the jacket water outlet temperature is elevated above the rated temperature. For example, with increased jacket water outlet temperatures, the heat rejection to the jacket water will decrease and the heat rejection to lube oil, radiation, and exhaust will increase. Heat to radiation increases due to the increased skin temperature of the engine. Correction factors for heat balance are provided by the manufacturer.

A water-to-water heat exchanger may be used to isolate the jacket water coolant circuit and recover engine heat for use in process thermal requirements. This provides protection for the engine in the event of a loss of coolant in the proccss loop and also provides a means for matching the

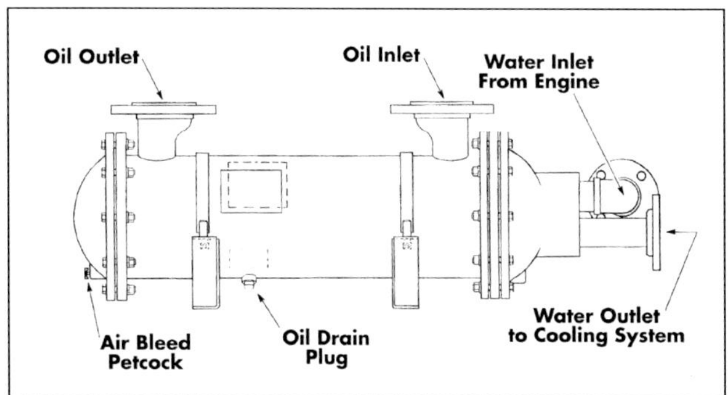

Fig. 9-46 Representative Oil Cooler for Stationary Industrial-Grade Engine. Source: Waukesha Engine Div.

process temperatures and flow rates of the jacket water system. An additional barrier should also be used when heating potable water. A supplemental or waste heat rejection transfer device may be included in the process loop (i.e., remote fan-cooled radiator or cooling tower) to allow for continued engine operation when there is little or no process heat requirement. The auxiliary water circuit for lube oil and intercooler may require a separate source of cooling water since operating temperatures are below the minimum required jacket water temperature.

The amount of heat that can be effectively recovered from the exhaust gas is limited by efficiency losses due to fouling of heat exchange materials. As noted previously, natural gas generally permits more effective heat recovery. Use of low-ash or ashless lubricating oil in Diesel and gas engines significantly reduces the fouling of exhaust heat exchangers. Refer to Chapter 8 for detail on heat recovery.

Instrumentation And Control Systems

The basic control functions and devices used for controlling prime movers are discussed in Chapter 13. While control systems are generally becoming increasingly complex, each engine system will feature, at a minimum, a basic instrument panel and control system. Typical instrument panels will provide gauges such as the following for quick inspection of operating conditions:

- Intake manifold pressure and temperature gauge
- Lube oil pressure and temperature gauge
- Exhaust temperature pyrometer
- Jacket water temperature gauge
- Digital tachometer

Electronics has brought an added dimension to reciprocating engine control, allowing for complete control of all mechanical and auxiliary systems. In many cases, engines can operate unattended, with control and monitoring from remote locations. Capabilities exist for extensive information retrieval, handling, and analysis.

Microprocessor-based programmable logic controllers are used to manage engine load, air-fuel ratios, charge-air density, fuel injection, and ignition timing. The precise regulation that is achievable allows for operation with extremely lean air-fuel mixtures to achieve air emissions control without misfiring and with a minimal compromise of power producing capacity and thermal fuel efficiency over a full range of operating loads. It also allows stoichiometric or rich-burn engines to achieve the precise air-fuel ratios required for effective three-way catalyst system operation.

Cylinder operation may be regulated either globally or individually. Figure 9-47 illustrates the operation of a closed-loop, power cylinder balancing system for large-bore, spark-ignited engines featuring electronic fuel injection. With this design, a combustion analysis system and auto-balancing controller receive pressure information from in-cylinder sensors. The system then determines whether to increase or decrease the gas flow to each cylinder using electronic fuel injection control to make the adjustments automatically.

In heat recovery applications, important heat recovery parameters, such as exhaust mass flow and temperature or charge-cooler temperatures, may be integrated into the engine control system. In dual-fuel engines, electronic control systems allow for uninterrupted, automatic changeover from liquid fuel to gaseous fuel operation.

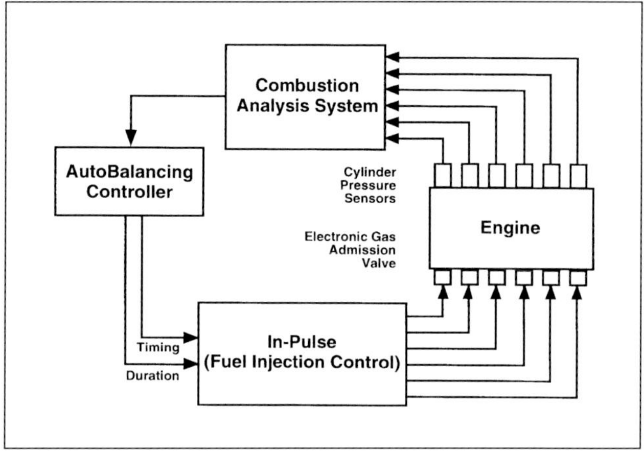

Fig. 9-47 Closed-Loop, Power Cylinder Balancing System, with Electronic Fuel Injection. Source: Woodward Governor

Physical Size, Sound, and Vibration

Weight and physical dimensions are often important considerations, particularly in retrofit applications. Reciprocating engines are relatively large and heavy compared with other prime movers. While dimensions vary considerably, Figure 9-48 provides representative examples of orders of magnitude for typical bare engines.

Reciprocating engines, like other technologies relying on combustion, generate a full range of sound from low to high frequency. Low frequencies are usually attributed to the muffled combustion within the cylinders and exhaust pulsations. High frequencies are often associated turbochargers. Ambient sound levels associated with a reciprocating engine are high when the engine is not located within a sound attenuated enclosure. Such acoustic enclosures are typically designed to reduce the magnitude of sound waves down to approximately 60 to 90 dba at 3 ft (1 m) on smaller engines. Sound problems can be

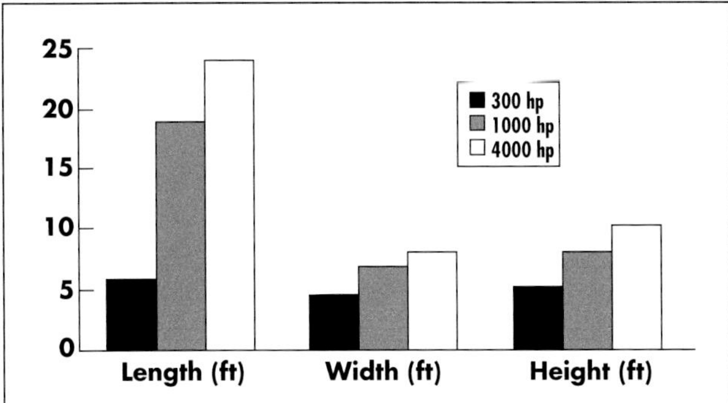

Fig. 9-48 Representative Dimensions of Engines by Capacity.

resolved on larger engines when the surrounding building is designed to acoustically absorb or negate these sound waves. Exhaust sound attenuation is commonly achieved by a quality muffler. Moderate sound attenuation can be achieved with vibration isolators, baffles, and use of absorption materials. However, when a significant level of sound attenuation is required, single- or double-walled enclosures may be required.

Reciprocating engines require an extensive design for vibration isolation. When a reciprocating engine is combined with driven equipment, a torsional analysis must be performed to ensure that the characteristics of the combined system do not exceed the limits in the equipment's operating range. Generally, when one or more harmonic damper devices are used, systems will be well below acceptable levels required for coupling to electric generators, compressors, pumps, and other end-use equipment.

Reciprocating Engine Selection

Reciprocating engine selection for any given application includes consideration of various trade-offs between quality, simple-cycle thermal fuel efficiency (at partial and full loads), heat recovery efficiency, capacity, maintainability, availability, durability, and first cost. The ultimate goal is to produce the best life-cycle cost that meets the requirements of the application. To the extent possible, factors such as the value of reliability should be quantified as a cost factor. If this is not possible, then such factors must be somehow used to qualify the results of the life-cycle analysis. For example, comparison of gas-fired spark-ignited Otto engines with liquid-fuel-fired compression-ignited Diesel engines often matches the benefits of the simple-cycle thermal fuel efficiency and power density advantage of the Diesel engine against the environmental and heat recovery advantages of the spark ignited engine.

Engine quality is independent of the operating characteristics of the engine, and must be judged relative to the specific engine type and cost. If a customer pays a premium for a product, as in the case of a low-speed engine, yet does not receive the benefits associated with such a purchase (i.e., high efficiency and long component life), the quality of the engine is low. If a customer pays a relatively low price for a high-speed engine that lasts for many years in standby duty and starts reliably every time, the quality of the engine is high.

Comparisons of different capacity engines will usually show that there is an increase in maximum bmep and a decrease in brake-specific fuel consumption with increased size. There is also a decrease in mean piston speed at maximum power as engine size increases. For engines of comparable capacity operating under the same conditions, lower engine speed and higher bmep for a given compression ratio usually produce higher thermal fuel efficiency. Lower rpm, lower piston speed, and lower bmep usually result in increased engine component life and reduced maintenance costs.

Supercharging or turbocharging increases bmep (and engine stress) and generally results in slightly higher thermal fuel efficiency. The higher power density associated with charged engines usually makes their first cost more attractive. Intercooling/aftercooling further increases inlet air density and, in spark-ignited engines, allows for greater air density within the knock avoidance region.

A major factor in the engine selection process is air emissions control. In addition to comparing the published air emissions rates of the various regulated pollutants, one should consider the methods used to achieve these emissions rates and the impact of these control methods on other performance characteristics. For example, a spark-ignited engine that achieves NO_X and hydrocarbon emissions control through a superior combustion chamber design and air-fuel ratio control system would be more favorable than one that achieves such control with retarded timing, since the latter method could adversely affect capacity, efficiency, and long-term maintenance cost.

Higher speed engines (above 1,200 rpm) offer weight, size, and first-cost advantages over lower speed engines. They do, however, require more maintenance and have a shorter service life. High-speed engines are often used for stand-by operation or less frequent duty applications where first cost is more critical than efficiency and maintenance cost. Applications that call for workhorse engines and a high level of reliability will use lower rpm engines.

Both piston speed and engine rotating speed should be considered in engine selection. Piston speed is commonly expressed in ft/s (or m/s), while rotational speed refers to crankshaft speed and is commonly expressed in rpm. While rotating speed is most commonly referred to when classifying engines, it does not take into consideration the size and piston stroke of an engine. A long-stroke engine with a low rotating speed may have a high piston speed. On the other hand, a high-speed, short-stroke engine may have a relatively low piston speed.

Application size, of course, greatly affects engine selection. Smaller systems of up to several hundred hp (200 kW) are usually pre-packaged and require minimal site engineering. They are designed around higher rpm engines, typically 1,200 to 3,600 rpm. Constant-speed base-load operations, such as electric generation, use engines rated at 1,800 rpm or less, while variable speed mechanical drive applications can run up to 3,600 rpm and higher. Mid-sized applications, from 300 to 1,500 hp (225 to 1,100 kW), usually require more site engineering, but are still fairly standardized. Engine speed ranges from 780 to 1,800 rpm, with 1,200 to 1,800 rpm engines more often used for intermittent duty. Large applications, ranging from 1,000 to 10,000 hp (750 to 7,500 kW), require substantial site engineering and generally use lower rpm engines ranging from 1,200 rpm down to below 300 rpm. Engines with capacities in the tens of thousands of hp (or kW) may operate at speeds of 100 rpm or even lower.

Increasingly, however, the gap between the performance, durability, and reliability of the low-speed and medium-speed engines is closing. In many applications, it is becoming increasingly difficult to justify the higher capital cost associated with the slower speed machines on the basis of economics, especially when these economics are based on simple payback. Design improvements over the past decade have improved performance, reliability, and maintainability of higher speed industrial and automotive derivative engines substantially.

Figure 9-49 shows three performance curves for a typical spark-ignition engine, highlighting the impact of speed on capacity and heat rate for a given engine model. Notice that as engine speed is reduced, so is the heat rate and maximum achievable capacity. For example, at 1,500 rpm, the maximum achievable generator output is 400 kW. Operation at 1,200 rpm produces a maximum generator output of 300 kW and does so with a significantly lower heat rate (or brake specific fuel consumption) than would be achieved at a higher operating speed.

The quantity and quality of the recoverable heat will also be a function of the engine speed and the percent of full load capacity at which it is operating. Generally, increasing load increases exhaust mass flow and temperature. Increasing speed also tends to raise gas temperature.

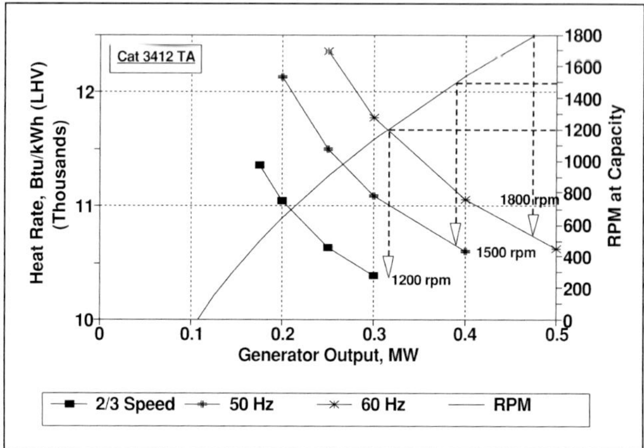

Fig. 9-49 Spark Engine Performance. Source: Cogen Designs, Inc.

To accurately determine heat recovery potential over a range of operation, one must utilize curves such as those shown in Figures 9-50 and 9-51. Figure 9-50 shows exhaust temperature and flow vs. generator output at three different engine speeds. Figure 9-51 shows jacket water heat rejection, in MBtu/min, as a function of speed and capacity.

Engine Rating Standards and Performance Adjustment Factors

An accurate assessment of performance requires consideration of the standards and variables used in manufacturers' published ratings. Figure 9-52 is an engine rating and fuel consumption performance curve for an 8-cylinder Waukesha engine. The engine is a turbocharged, intercooled four-stroke-cycle lean-burn unit, featuring pre-chamber combustion design and four valves per cylinder. Compression ratio is 9:1 and displacement is 8,699 in^3 (142.5 liter). The performance curves are based on continuous duty rating, which is defined as the highest load and speed that can be applied 24 hours per day, 365 days per year, except for normal maintenance. It is also based on 10% overload rating, which is allowed 2 hours per 24 hours.

The figure graphs bsfc, or fuel consumption per bhp-h output, versus load for three operating speeds: 800, 900, and 1,000 rpm, for operation with a 32:1 air-fuel ratio. Notice the decrease in capacity and increase in thermal fuel efficiency as speed is reduced: capacity is reduced from 2,090 bhp (1,558 kW) at 1,000 rpm to 1,670 bhp (1,245

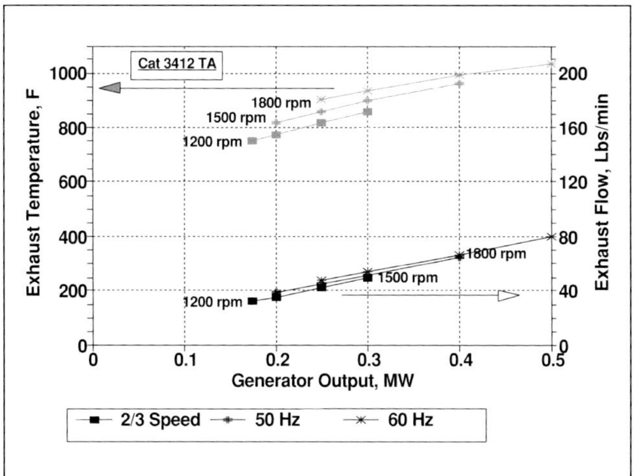

Fig. 9-50 Exhaust Flow and Condition as a Function of Speed and Capacity. Source: Cogen Designs, Inc.

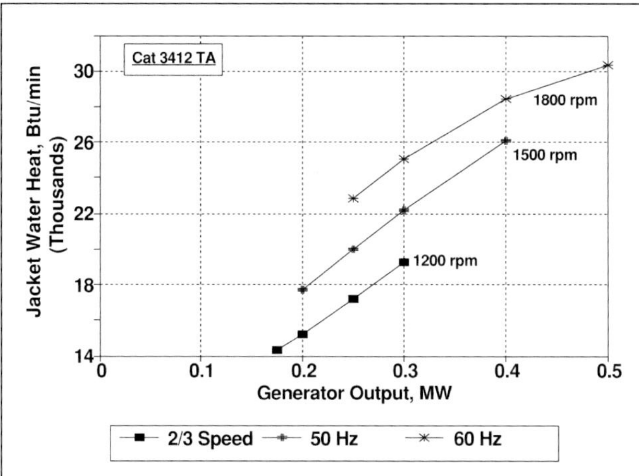

Fig. 9-51 Jacket Water Duty as a Function of Speed and Capacity. Source: Cogen Designs, Inc.

kW) at 800 rpm, while the full load heat rate is reduced from 6,556 Btu/bhp-h (9,851 kJ/kWh) at 1,000rpm to 6,369 Btu/bhp-h (8,793 kJ/kWh) at 800 rpm.

In the United States, **standard conditions** referenced for engines are 14.696 psia (101.325 kPa) and 60°F (16°C). Fuel flow, in standard cubic feet (scf), is referenced to a gas at standard conditions. New ISO standard conditions applicable to the example in Figure 9-52 are: 29.54 in. Hg (100 kPa) barometric pressure, 77°F (25°C) ambient and induction air temperature, and 30% relative humidity (1 kPa/0.3 in. Hg. water vapor pressure). Fuel is specified as dry natural gas with 900 Btu/ft^3 (33.5 J/cm^3) on an LHV basis and 118 octane rating. The performance curves are based on operation at 180°F (82°C) jacket water outlet with 130°F (54°C) intercooler water at a bmep of 190 psi

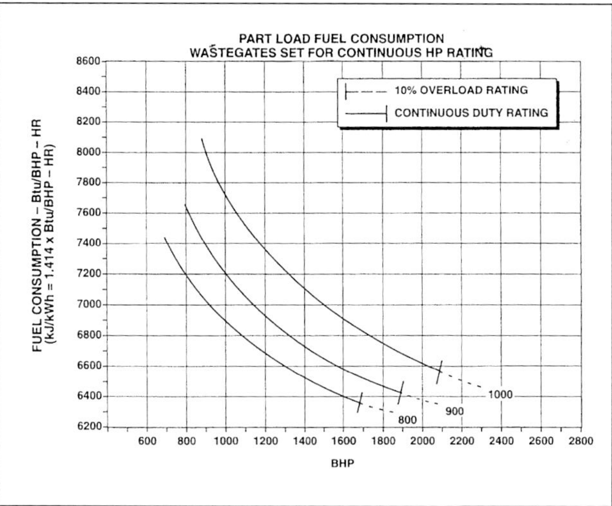

Fig. 9-52 Performance Curve for Spark-Ignited Gas Engine. Source: Waukesha Engine Div.

(13.1 Bar).

Based on the standard conditions used in establishing performance ratings, adjustment factors may be applied to calibrate the ratings for operation at conditions other than those specified. In the equipment specification package, heat balance data is provided for a range of characteristic factors, including:

- Speed (rpm)
- Power (bhp or kW)
- Bsfc (Btu/bhp-hr or kJ/kWh)
- Fuel consumption (Btu/hr or kW)
- Heat to jacket water (Btu/hr or kW)
- Heat to lube oil (Btu/hr or kW)
- Heat to intercooler (Btu/hr or kW)
- Heat to radiation (Btu/hr or kW)
- Total energy in exhaust (Btu/hr or kW)
- Exhaust temperature after turbine (+/–50°F or 30°C)
- Induction airflow (scfm or nm^3/h)
- Exhaust gas flow (lbm/hr or kg/h)

Following are examples of the impact of changes in the heat balance values resulting from changes in one variable:

- At 1,000 rpm, 180°F (82°C) jacket water temperature, and 130°F (54°C) intercooler water temperature, the continuous power is 2,090 bhp (1,558 kW) at a bmep of 190 psi (13.1 bar). By decreasing the intercooler water temperature to 90°F (32°C), the continuous power rating increases to 2,200 bhp (1,640 kW) at a bmep of 200 psi (13.8 bar). Bsfc increases only slightly.

- With the intercooler water maintained at 90°F (32°C) and changing jacket water temperature to 250°F (121°C), capacity at 1,000 rpm remains constant. Heat rejection to jacket water is reduced 18%, heat rejection to lube oil increases 29%, and heat loss to radiation increases 55%.
- At 209 bmep (14.4 bar), corresponding to the 10% overload rating, capacity increases to 2,299 bhp (1,714 kW), while bsfc decreases by about 1.5%.

Table 9-1 lists adjustment factors for selected Waukesha engines based on altitude and temperature. In this case, all natural gas engine ratings are based on a fuel of 900 Btu/ft^3 (35.3 MJ/m^3) SLHV, 119 octane (per ASTM D-2700 test method). All Diesel engine ratings are based on a #2-D fuel of 18,400 Btu/lbm (42.8 kJ/g) LHV. Ratings are based on ISO 3046/1-1986 with mechanical efficiency of 90% and Tcra (clause 10.1) as specified limited to +/–10°F (5°C). Ratings are valid for SAE J1349, BS 5514, DIN 6271, and AP 17B-11C standard atmospheric conditions.

The table lists adjustment values in two categories: Intermittent/Standby and ISO Standard/Prime Power. The **Intermittent Service** rating is defined as the highest load and speed that can be applied in variable speed mechanical system applications only. Operation at this rating is limited to a maximum of 3,500 hours per year. The **Standby Service** rating applies to those systems used as a secondary source of electric power. This rating is the output the engine will produce continuously (no overload), 24 hours per day for the duration of the prime power source outage. The **ISO Standard Power/Continuous Power** rating is defined as the highest load and speed, which can be on a continuous basis, year-round. It is permissible to operate the engine at up to 10% overload, or maximum load indicated by the intermittent rating, whichever is lower for two hours in every 24 hour period.

Tables 9-2 and 9-3 show the manufacturer's heat balance tables for a Fairbanks Morse Diesel engine designed for dual-fuel and Diesel operating mode, respectively. The heat balances are based on maintaining the jacket water out temperature at 175°F (+/–5 degrees), the lubricating oil inlet temperature at 135°F, and the intake airflow is 90°F.

Performance curves are particularly useful for predicting performance for operation under variable load and speed. The bsfc of a reciprocating engine operating at constant speed increases with decreasing load. If power requirements vary, variable speed operation can be a big advantage. Thermal efficiency levels will remain close to full-load ratings in the middle and upper ranges of a load curve. Efficiency drops off at the lower end of the load curve, but not nearly as quickly as with constant speed operation.

When a gear box is used, efficiency is reduced by 2% to 5%. In engine generator applications, efficiency is also reduced by an additional 1.5% to 4% as a result of generator losses. Usually, when output is specified in terms of the driven load, i.e., generator output in kW, generator and gearing losses are included in the performance data.

Another factor to consider when evaluating system performance is the auxiliary component requirement, or parasitic loads. For example, high-compression engines must have fuel delivered at a pressure higher than the combustion chamber pressure at the time of injection. For example, given a delivered gas pressure of 50 psig (4.5 bar) (at the gas meter) for a 7,000 hp (5,200 kW) engine requiring a delivery pressure of 1,600 psig (111 bar), the parasitic load for gas compression is greater than 4% of the engine output.

ENGINE MAINTENANCE, RELIABILITY AND LIFE

While first-cost and thermal efficiency are often the driving force behind investment decisions, engine life, reliability, and maintenance requirements are critical factors that greatly affect life-cycle economics.

Although there are exceptions to every rule, lower rotational speed, mep, and compression ratios for similar type engines generally result in longer engine life and a lower maintenance requirement. Higher speed engines accumulate more operating strokes, which means faster wear. Higher bmep and compression ratio mean higher firing pressures and temperatures, also promoting wear. Output and efficiency enhancement features, such as turbocharging or supercharging, tend to produce greater stresses and result in greater wear and shorter engine life.

The benefits of lower first cost and somewhat greater thermal efficiency of oil-fired engines come at the expense of higher maintenance requirements. This is due to deposit build-ups, oil contamination, clogging of injectors and nozzles, and higher fouling factors in exhaust and heat recovery components.

The type of operating duty is also a critical factor. In particular, frequent starting promotes for engine wear and engines running close to maximum power rating are subject to added wear. Engines used for standby or emergency use will experience a greater degree of wear, per hour of run time, than engines used for regular duty because they operate at a higher power rating, are subject to more starts and stops, and typically operate at higher speeds.

Table 9-1 Manufacturer's Performance Adjustment Factors

Selected Type and Model	Altitude and Temperature Adjustment	Intermittant Standby	ISO Standby Prime Power
Turbocharged & Intercooled	Deduct 2.0% for each 1,000′ above:	1,500′	3,000′
All VHP GSI and DSI,	Deduct 1.0% for each 10°F above:	100°F	100°F
All VSG	Deduct 2.0% for each 1,000′ above:	1,500′	1,500′
All VHP GL	Deduct 1.0% for each 10°F above:	85°F	100°F
Turbocharged Only	Deduct 3.0% for each 1,000′ above:	1,500′	3,000′
All VHP Diesel	Deduct 1.0% for each 10°F above:	85°F	110°F
Naturally Aspirated			
VHP Natural Gas and Diesel	Deduct 3.0% for each 1,000′ above:	500′	1,500′
All VSG; All VGF Nat. Gas	Deduct 1.0% for each 10°F above:	85°F	100°F

* Included in the specification is a footnote that these altitude and temperature adjustments are meant to be a guide only and cannot be applied without limit.

Table 9-2 Manufacturer's Heat Balance Table for Dual-Fuel Engine in Dual-Fuel Mode (Values in Btu/bhp-h)

% of Full Load	50%	75%	100%	110%
Thermal Input	6,980	6,250	5,900	5,930
Power	2,545	2,545	2,545	2,545
Jacket Water	875	800	740	720
Lube Oil	330	290	275	275
Air Cooler	205	380	500	550
Exhaust	2,840	2,085	1,730	1,720
Radiation	185	150	130	120

Table 9-3 Manufacturer's Heat Balance Table for Dual-Fuel Engine in Diesel Mode (Values in Btu/bhp-h)

% of Full Load	50%	75%	100%	110%
Thermal Input	6,370	6,190	6,260	6,280
Power	2,545	2,545	2,545	2,545
Jacket Water	880	830	820	820
Lube Oil	330	310	310	310
Air Cooler	240	415	575	610
Exhaust	2,220	1,950	1,880	1,870
Radiation	175	140	130	125

Derating an engine below its prime power rating produces less wear, greater reliability, and longer engine life. Because energy savings are based on output, derating reduces short-term savings, although the strategy can improve life-cycle economics.

A thorough routine or preventative maintenance program should always be required. Routine maintenance includes filters, oil, oil sampling, water treatment, water-side and combustion-side deposit buildup inspections, plugs, and wires or injectors. Simple actions, such as diligent replacement of filters, can have a tremendous affect on extending long-term maintenance intervals and engine life.

Recent technology with automotive derivative engines has extended routine maintenance intervals to about 2,000 operating hours. This is due to the use of automatic valve lash adjustment (hydraulic lifters), automatic oil level control with large supplemental, circulating oil sumps, and other components such as long-life spark plugs.

Long-term (overhaul) maintenance includes items such as valves and piston rings and partial rebuilding. The long-term maintenance intervals range widely between 7,500 and 50,000 hours. Complete rebuilding, which includes component replacement such as heads, pistons, liners, and bearings, may be required after two long-term maintenance intervals (15,000 to 100,000 hours). The high-end numbers only apply to very large, low-rpm engines. For medium-speed industrial engines, the time between minor overhauls ranges from 15,000 to 20,000 hours and the time between major overhauls typically ranges from 30,000 to 40,000 hours.

It may be more economical to replace some low-cost, high-production automotive derivative engines rather than rebuild them. This disposable interchangeable engine strategy offers good life-cycle economics by limiting first cost. It can also reduce down-time versus rebuilding because systematic engine change-out can be done relatively quickly.

Life-cycle maintenance costs, TBO (time between overhaul), and engine life are determined by the trade-offs between the three critical factors: engine characteristics, operating duty, and routine maintenance procedures. Complete life-cycle operations and maintenance (O&M) costs will vary widely, ranging from $0.004 to $0.020/hp-h ($0.003 to $0.015/kWh). The low-end range of life-cycle O&M costs is for the very large, low-rpm engines. O&M costs for most mid-sized, medium-speed engines are $0.010/hp-h ($0.0075/kWh) or less. A difference of $0.01/hp-h or more over the life of the engine can have a profound effect on long-term project economic performance.

Complete service contracts for all maintenance, including engine overhaul and/or replacement, are readily available from project developers, manufacturers, or local contractors. Service contracts can be designed for complete cradle-to-grave service or designed to cover specific functions. They can be based on a set annual fee, tied to output, or set on an as-needed time and materials basis. In-house or non-contracted O&M costs are typically lower on the front end and higher in the later years and can often produce better life-cycle economics.

The importance of good maintenance cannot be overstated. While significant, the cost of maintenance is relatively small when weighed against overall cost savings potential. The key is to ensure that the engine remains reliable and productivity and savings potential is protected.

Fig. 9-53 Waukesha APG-2000 Engine-Generator Source: Waukesha Engine Div.

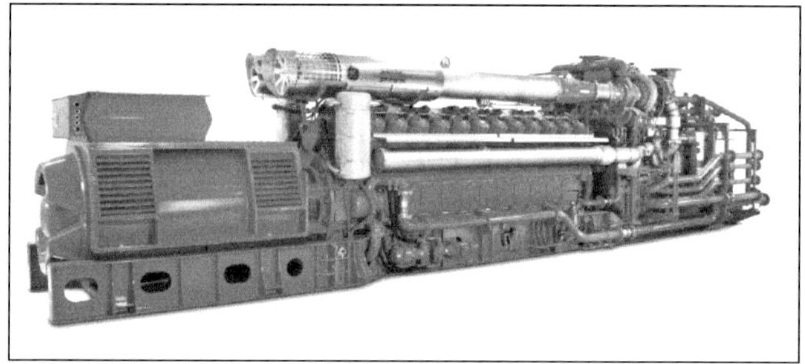

Fig. 9-54 GE/Jenbacher J624 Engine-Generator: Source: GE/Jenbacher.

Currently Available Engine Designs

Figures 9-55 through 9-59 are a series of cutaway illustrations of reciprocating engines, designed to show the interrelationships between the various main engine components. Improvements in reciprocating engine technology increased the electrical generating efficiencies for stationary natural gas fueled Otto-cycle units significantly from 1995 to 2010. Most manufactures now offer several models with thermal fuel efficiencies in excess of 40% on a LHV basis, and several manufactures off units with electrical generating efficiencies in excess of 46%.

Figure 9-53 is a Waukesha APG2000 unit rated for 1,860 kW at 4,800 volts, which achieves a 42% thermal fuel efficiency. This unit operates at 1,200-rpm and can limit NO_X emissions to 0.6 gr/bhp.

Figure 9-54 is a GE/Jenbacher J624 unit that utilizes a two-stage turbo-charger with an intercooler in between to achieve 46.5% electrical generating efficiency. This 4,400 kW unit operates at 1,500-rpm for 50-Hz installations and uses a gear box for 60-Hz applications.

Figure 9-55 is a cross-sectional illustration of a Waukesha over-head valve, four-stroke-cycle spark-ignited engine. The piston is at TDC, revealing the clearance volume for the combustion chamber. The spark plug location is shown just above the combustion chamber and just below the valve assembly. Within the crankcase, the crankpin and crankshaft are shown at the bottom of the connecting rod. The intake and exhaust manifolds are shown to the upper left. The relationship of the camshaft to the crankshaft and to the valve assembly via the push rod is also shown.

Figure 9-56 shows a Caterpillar G3500 spark-ignited gas engine. The right and left hand views are cut away at different depths to reveal different aspects of the various components. On the top right-hand side of the cylinder assembly is the ignition transformer, residing just below the value cover. The small pocket seen through the cutaway piston on the right, just below the spark plug, is the combustion chamber. Lubricating oil spray and pathways are shown around the connecting rod. Housed to the far right of the connecting rod is the control module, which features integral detonation-sensitive timing and ignition diagnostics. The left-hand side view illustrates the

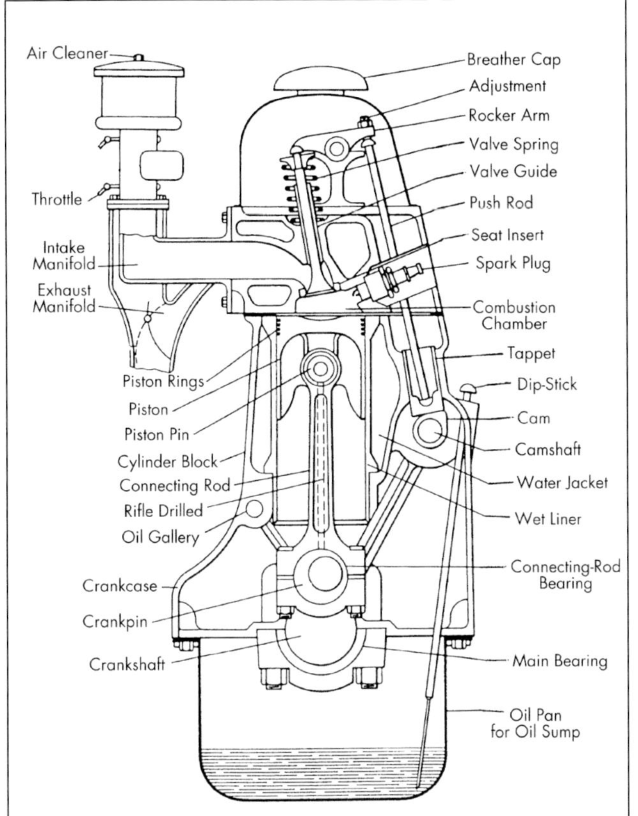

Fig. 9-55 Cross-Sectional Illustration of Over-Head Valve, Four-Stroke-Cycle, Spark-Ignited Engine. Source: Waukesha Engine Div.

Fig. 9-56 Cutaway View of Caterpillar Spark-Ignited Gas Engine. Source: Caterpillar Engine Div.

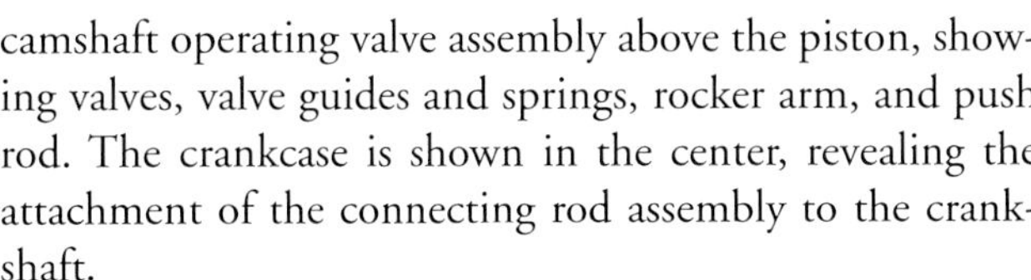

camshaft operating valve assembly above the piston, showing valves, valve guides and springs, rocker arm, and push rod. The crankcase is shown in the center, revealing the attachment of the connecting rod assembly to the crankshaft.

Figure 9-57 shows cutaway illustrations of a Waukesha in-line, six-cylinder, spark-ignited engine showing cylinder head and liner, piston, valve, connecting rod, crankshaft, crankcase, and camshaft. The engine has a compression ratio of 11:1 and a displacement of 1,096 in^3 (18 liter), with a bore and stroke of 5.98 in. (152 mm) and 6.50 in. (165 mm), respectively. At 1,800 rpm, this engine has a continuous duty rating of 375 hp (280 kW).

Figure 9-58 shows a cutaway view of a basic four-cycle Diesel engine that is also applied for dual-fuel operation. Figure 9-59 is a cross-sectional view of a four-cycle Diesel engine frame. This heavy-duty engine design features a one-piece frame. Thick partitions connect to the central deck and carry the main bearings and crank. Long studs thread into the central deck and extend up through the top of the cylinder head. This Diesel engine model is used for both straight Diesel fuel operation and for dual-fuel operation.

Figures 9-60 through 9-66 illustrate several natural gas spark-ignited engines and Figures 9-67 through 9-69 illustrate dual-fuel engines. These are essentially Diesel engines modified to operate on both liquid and gaseous fuel. Various design features are highlighted to provide an overview of currently available technologies.

Figure 9-60 shows a Waukesha 16-cylinder, spark-ignited, four-stroke-cycle gas engine. This engine features two turbochargers with two-pass, air-to-water intercooling and is designed for very lean combustion. The engine features a compression ratio of 11:1 and a total displacement of 2,924 in^3 (48 liter), with a bore and stroke of 5.98 in. (152 mm) and 6.50 in. (165 mm), respectively. Its continuous capacity ratings range from 710 bhp (530 kW) at 1,200 rpm to 1,065 bhp (795 kW) with intercooler inlet water temperature of 130°F (54°C). At full rated design capacity, the heat rate is about 7,100 Btu/bhp-h (10,039 kJ/kWh) on a LHV basis, which equates to a thermal fuel efficiency of about 36%. When operat-

Fig. 9-57 Cutaway Illustrations of Waukesha In-Line, Six-Cylinder, Spark-Ignited Engine. Source: Waukesha Engine Div.

DRIVE END

Fairbanks Morse
Engine Division
Colt Industries

1. Crankshaft
2. Counterweights
3. Connecting rod
4. Pistons
5. Cylinder liners
6. Hydraulic nut assembly
7. Bearing saddle stud
8. Main bearing saddle
9. Tie bolt
10. Cylinder block
11. Crankcase inspection cover
12. Lube oil header
13. Camshaft
14. Cooling water inlet cover
15. Cylinder water jacket
16. Fuel injection pumps
17. Tappet housing
18. Fuel oil pipe housing
19. Fuel oil leakage detectors
20. Vibration damper housing
21. Barring device
22. Air start distributor
23. Fuel headers
24. Air intake manifolds
25. Valve cover
26. Valve housing
27. Push rods
28. Fuel injection pump control rod
29. Cylinder head
30. Rocker arms
31. Air start header
32. Exhaust valves
33. Safety valve
34. Intake valves
35. Fuel injection nozzle
36. Air start check valve
37. Exhaust manifold
38. Cooling water outlet header
39. Intercoolers
40. Turbochargers
41. Water pump
42. Lube oil outlet
43. Lube oil pan
44. Thrust bearing housing
45. Camshaft thrust bearing cover
46. Idler gear
47. Pump drive gear (lube & fuel)
48. Lube pump mounting
49. Fuel oil pump
50. Camshaft flexible drive gear
51. Pilot air pipes

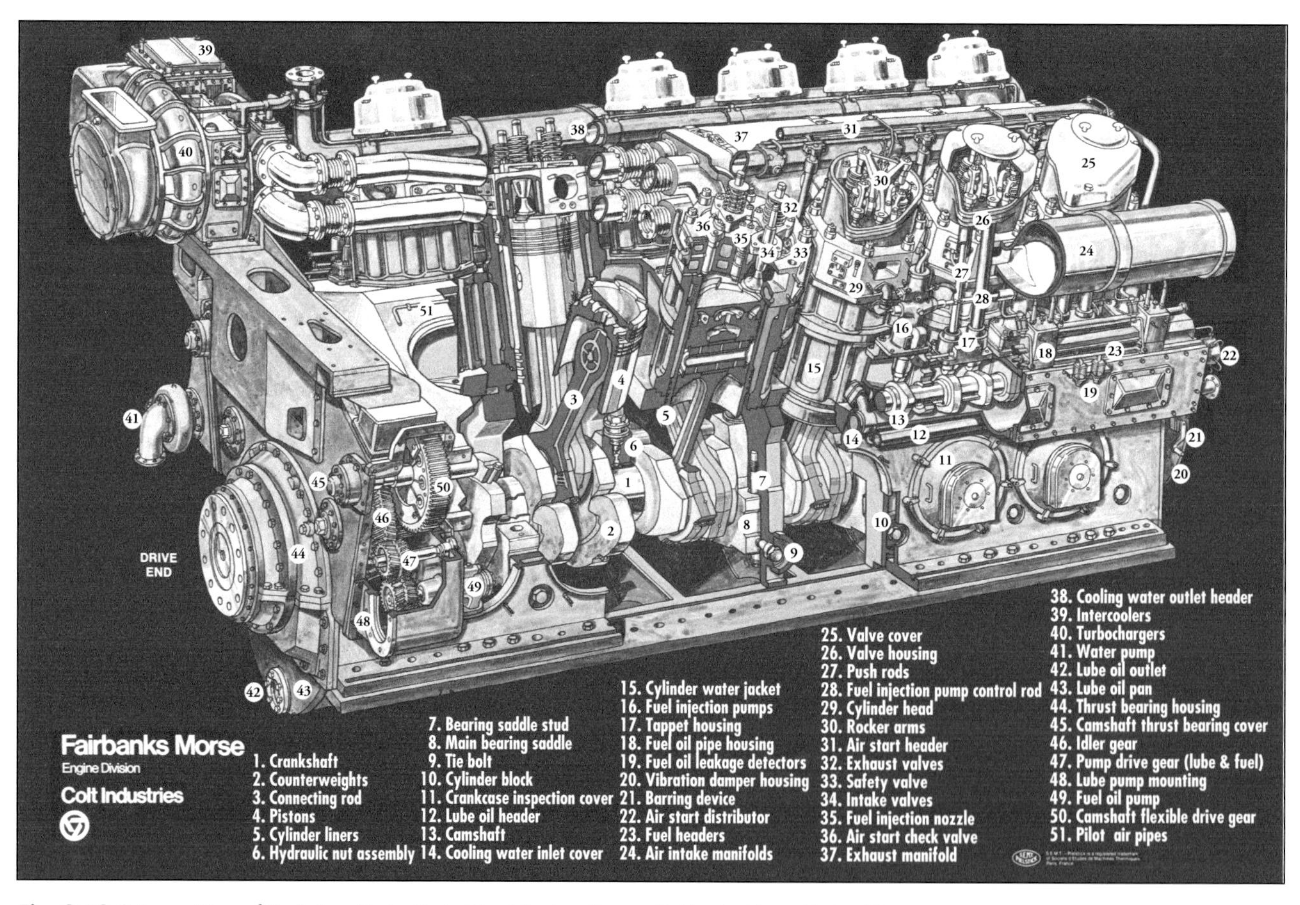

Fig. 9-58 Cutaway View of Basic Four-Stroke-Cycle Diesel Engine. Source: Fairbanks Morse Engine Div.

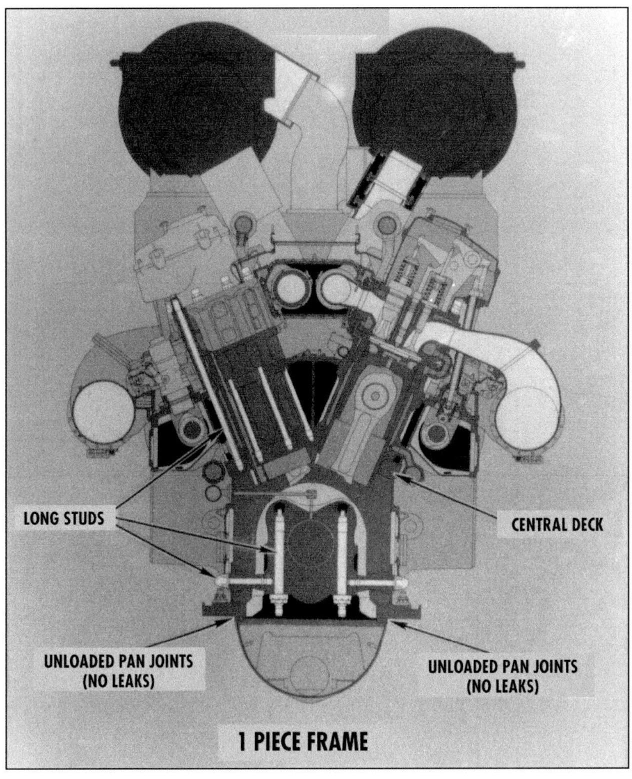

Fig. 9-59 Cross-Sectional View of Four-Stroke-Cycle Diesel Engine Frame. Source: Fairbanks Morse Engine Div.

ing at 75% of full load, at a constant speed of 1,800 rpm, the heat rate increases to about 7,450 Btu/bhp-h (10,534 kJ/kWh). When speed is decreased to 1,400 rpm, while maintaining rated power output, the heat rate is reduced to about 6,900 Btu/bhp-h (9,756 kJ/kWh) on a LHV basis.

Figure 9-61 shows a Waukesha 12-cylinder, spark-ignited, turbocharged, intercooled, four-stroke-cycle gas engine. This engine, designed for constant speed operation, features a compression ratio of 9:1 and a total displacement of 10,784 in^3 (177 liter), with a bore and stroke of 9.84 in. (250 mm) and 11.81 in. (3,000 mm), respectively. The lean burn, pre-chamber design is operated with an air-fuel ratio of 28:1 (λ = 1.74) and achieves typical NO_X and CO exhaust emissions of about 1.5 gram/bhp-h (2.0 gram/kWh) and 2.25 gram/bhp-h (3.0 gram/kWh), respectively. When set for operation at 750 rpm, the engine has a continuous power output capacity of 2,042 bhp (1,523 kW) and produces a torque of 14,299 lbf-ft (19,386 N-m). At full load, the engine operates with a bmep of 200 psi (13.8 bar) and has a heat rate of 7,000 Btu/bhp-h (9,905 kJ/kWh) on a LHV basis, for a thermal fuel efficiency of 36.4%. The exhaust mass flow at this rating is 20,440 lbm/h (9,271 kg/h) at a temperature of 749°F (399°C). When set for operation at 1,000 rpm, the full-load continuous power output rating increases to 2,723 bhp (2,030 kW). At this rating, the heat rate increases to 7,105 Btu/bhp-h (10,054 kJ/kWh), with corresponding increases in both exhaust mass flow and its temperature.

Fig. 9-60 Waukesha 16-Cylinder, Spark-Ignited, Four-Stroke-Cycle Gas Engine. Source: Waukesha Engine Div.

Fig. 9-61 Waukesha 12-Cylinder, Spark-Ignited, Turbocharged, Intercooled, Four-Stroke-Cycle Gas Engine. Source: Waukesha Engine Div.

Figure 9-62 shows a Fairbanks Morse eight-cylinder, two-stroke-cycle, spark-ignited, opposed-piston engine, which is described above, applied in an engine-generator set. The unit produces 1,710 kW of electric power output at a design rated speed of 900 rpm. The engine itself is 16 ft (4.9 m) long by 8 ft (2.4 m) wide and high. The engine-generator set is 23 ft (7 m) long and weighs 70,000 lbm (31,750 kg). This unit features a pre-combustion chamber with an independent injection of gas to the ignition cell. Electronic controls vary the timing and air-fuel ratio to achieve reliable ignition and optimal performance across the entire operating range. This unit operates at a heat rate of about 6,600 Btu/bhp-h (9,340 kJ/kWh) on a LHV basis, with a NO_X emissions rate of about 1 gram/bhp-h (1.3 gram/kWh).

Fig. 9-63 Wartsila 16-Cylinder, Turbocharged, Spark-Ignited Gas Engine-Generator Set. Source: Wartsila Diesel

Figure 9-63 shows a Wartsila 16-cylinder, turbocharged, spark-ignited, gas engine-generator set. The engine features a pre-combustion chamber and operates with a 2.3 excess air ratio at a low temperature to minimize NO_X formation. At 900 rpm (60 Hz) operation, the generator set output is 2,530 kW. The engine has a compression ratio of 11:1 and operates at a bmep of 218 psi (15 bar). The engine design features a 9.8 in. (250 mm) bore with an 11.8 in. (300 mm) stroke and a mean piston speed of 29.5 to 32.8 ft/s (9 to 10 m/s). The engine-generator set dimensions are about 29 ft (8.8 m) long by 7 ft (2.1 m) wide by 11 ft (3.4 m) high. The engine weight is about 22 tons (20,000 kg) and the entire engine generator-set weight is 43 tons (39,000 kg). The heat rate at full load is 8,137 Btu/kWh (8,584 kJ/kWh) on a LHV basis, which corresponds to a simple-cycle thermal fuel efficiency of 41.9% (LHV basis) and 37.8% (HHV basis), excluding minor deductions for auxiliaries and parasitics.

Figure 9-64 is a large-capacity, low-speed, Cooper-Bessemer 16-cylinder engine with turbocharging and pre-chamber combustion. The engine has a displacement of 66,419 in^3 (1,088 liter), with a bore and stroke of 15.5 in. (394 mm) and 22 in. (559 mm), respectively. The engine has a length of 29.5 ft (9 m) and a height of 14 ft (4.3 m). The floor-line width is about 6.5 ft (2 m) and the overhead width, including turbochargers, is about 15 ft (4.6 m). When operating at 400 rpm, engine power output capacity is about 6,800 bhp (5,070 kW) at a relatively low bmep of 207 psi (14.3 bar). This engine achieves a heat rate of about 6,200 Btu/bhp-h (8,740 kJ/kWh), which corresponds to a simple-cycle thermal fuel efficiency of about 41% (HHV basis) and 37% (LHV basis).

Fig. 9-62 Fairbanks Morse, Eight-Cylinder, Two-Stroke-Cycle Spark-Ignited Opposed Piston Engine. Source: Fairbanks Morse Engine Div.

Fig. 9-64 Low-speed, Cooper-Bessemer 16-Cylinder, Turbocharged Engine with Pre-Chamber Combustion.
Source: Cooper Energy Services, Cooper Cameron Corp.

Figure 9-65 is a Hercules, small-capacity, industrial spark-ignition engine. This 4-cylinder, overhead valve unit features a compression ratio of 7.5:1 and, with minor modification, is capable of operation on natural gas, gasoline, and certain other liquefied petroleum fuel. Engine displacement is 163 in^3 (2.7 liter) with a bore and stroke of 4.0 (101 mm) and 3.25 in. (83 mm), respectively. The engine features an overhead valve design with gear-driven camshaft, distributor, and governor. When operating on natural gas, a maximum continuous-duty capacity of about 44 hp (32 kW) is produced at 2,800 rpm with a heat rate of about 8,800 Btu/bhp-h (12,450 kJ/kWh) on a LHV basis. At 1,800 rpm, continuous-duty power is about 31 hp (23 kW) and torque output is about 84 lbf-ft (114 N-m). At this rating, the heat rate is about 7,865 Btu/bhp-h (11,130 kJ/kWh) on a LHV basis, corresponding to a simple-cycle thermal fuel efficiency of 32.4% (LHV basis) and 29.1% (HHV basis). Engine dimensions are 30 in. (76 cm) long by 24 in. (61 cm) high by 21 in. (53 cm) wide. Basic engine weight is 420 lbm (191 kg). This model has been commonly applied to small mobile and stationary vapor compression systems. Larger, similar models have been commonly applied to electric generators, industrial air compressors, and on-highway vehicles.

Fig. 9-65 Hercules, Small-Capacity, Industrial Spark-Ignition Engine. Source: Hercules Engine Company

Figure 9-66 shows two Caterpillar spark-ignited, four-stroke-cycle, gas engines applied in tandem. The combination is used to produce 1,600 kW electric output from a common generator. Each of these 12-cylinder, turbocharged engines has a full-load capacity of about 1,100 bhp (820 kW) with a compression ratio of 11:1. Total piston displacement for each engine is 4,210 in^3 (67.4 liter), with a bore and stroke of 6.7 in. (170 mm) and 7.5 in. (190 mm), respectively. These engines are designed for operation at 1,200 rpm and achieve a heat rate of 7,000 Btu/bhp-h (9,900 kJ/kWh), for a simple-cycle thermal fuel efficiency of 36.4% (LHV basis) and 32.7% (HHV basis). The exhaust gas flow rate for each engine is 5,647cfm (158m^3/min) at a temperature of 770°F (410°C).

Figure 9-67 is a Wartsila 18-cylinder engine generator set. This 6,425 kW unit includes a dual-fuel, compression-ignition Diesel engine with a high-pressure gas injection system. Engine displacement is 27,531 in^3 (451 liter) with a bore and stroke of 12.6 in. (320 mm) and 13.8 in. (350 mm), respectively. The compression ratio of this engine is 13:1 and the bmep rating ranges from 348 psi (24 bar) to 316 psi (21.8 bar). At 720 rpm, the engine develops 8,928 bhp (6,660 kW), with a mean piston speed of about 28 ft/s (8.6 m/s). In normal operation, pilot oil, corresponding to 5% of the design fuel oil consumption rating, is injected prior to the main injection. Start-up and low-load operation are always on Diesel oil. When operating in maximum gas mode, the main injection is pure gas at a pressure of about 3,600 psi (250 bar). The compressed

Fig. 9-66 Two Caterpillar Spark-Ignited, Turbocharged, Four-Stroke-Cycle Gas Engines Applied in Tandem. Source: Caterpillar Engine Div.

gaseous fuel is transported from a gas compressor, which is normally located outside the engine room, to the injection valve on the engine through a double-walled pipe system. Both Diesel oil and gaseous fuel are injected through a combined injection valve. The heat rate for this unit at full load operation with 95% gas admission is 7,469 Btu/kWh (7,880 kJ/kWh). This consists of 379 Btu/kWh (400 kJ/kWh) oil input plus 7,090 Btu/kWh (7,480 kJ/kWh) gas input. The combined simple-cycle thermal fuel efficiency is about 45.7% on a LHV basis or about 41.2% on a HHV basis. These performance figures do not include deducts for engine-driven pumps or gas compression (roughly 5% of shaft power output), which will result in lower net efficiency.

Figure 9-68 is a Fairbanks Morse 6-cylinder, two-stroke-cycle, compression-ignited, dual-fuel opposed piston engine. This engine-generator set produces 1,500 kW at a design rated speed of 900 rpm. Dual-fuel operation is accomplished using pilot fuel oil as the ignition source. The engine is fitted with duplex fuel injection pumps. The pilot fuel pump delivers a constant amount of pilot fuel during each stroke. When up to speed and with sufficient fuel oil and lubricating oil operating pressures, the engine may be switched over to gas operation. The amount of gas admitted to the cylinders is determined by the action of a governor-controlled throttle valve. Ignition timing is maintained by the pilot fuel system. The standard dual-fuel model achieves a heat rate of about 6,100 Btu/bhp-h (8,630 kJ/kWh) on a LHV basis, with a NO_X emissions rate of about 4 gram/bhp-h (5.4 gram/kWh). The low-emissions model achieves a NO_X emissions rate of about 1 gram/bhp-h (1.3 gram/kWh), but does so at a slightly higher heat rate of about 6,250 Btu/bhp-h (8,810 kJ/kWh).

Figure 9-69 is a Fairbanks Morse Colt-Pielstick 18-cylinder, turbocharged, dual-fuel, compression-ignition four-stroke-cycle engine. When applied to power generation, this dual-turbocharged engine produces 7,823 kW of generator output at a rotational speed of 514 rpm. The engine is 32 ft (9.8 m) long and 12 ft (3.7 meters) wide. The entire generator-set package is 43 ft (13.1m) long and weighs 99 tons (89,000 kg). This engine features the gas valve located in the intake manifold and admits gas at the beginning of the intake stroke. As opposed to high-pressure in-cylinder injection, this engine requires a far more modest inlet gas pressure of 60 psi (4.2 bar). The engine control system varies the air-fuel ratios and achieves very thorough mixing, allowing for combustion of a highly lean mixture. The engine achieves a full-load heat rate of 5,900 Btu/bhp-h (8,349 kJ/kWh) and a NO_X emissions rate of 1 gram/bhp-h (1.3 gram/kWh) or less when operating on 99% gas in dual-fuel mode. The simple-cycle thermal fuel efficiency in dual-fuel mode is about 43.1% (LHV basis) and 38.9% (HHV basis). In Diesel operation, the full-load heat rate is 6,260 Btu/bhp-h (8,858 kJ/kWh) with NO_X emissions of 9 gram/bhp-h (12 gram/kWh). Both dual-fuel and Diesel mode operation ratings are achieved at a bmep of 282 psi (19.6 bar). The exhaust gas stack temperature is 750°F (399°C) in dual-fuel mode and 720°F (382°C) in Diesel mode.

Note that while the high-pressure fuel injection type engine shown in Figure 9-67 shows a superior heat rate, when adjustments are made for the higher auxiliary power requirements, its performance is similar to that of the dual-fuel engines that admit lower pressure gas into the intake manifold, such as the unit shown in Figure 9-69.

Fig. 9-67 Wartsila 18-Cylinder, Duel-fuel, Diesel Engine, Featuring High-Pressure Gas Injection. Source: Wartsila Diesel

Fig. 9-68 Fairbanks Morse Six-Cylinder, Two-Stroke-Cycle, Compression-Ignited, Dual-Fuel Opposed Piston Engine. Source: Fairbanks Morse Engine Div.

Fig. 9-69 Fairbanks Morse Colt-Pielstick, 18-Cylinder Turbocharged, Dual-Fuel, Compression-ignition, Four-Stroke-Cycle Engine. Source: Fairbanks Morse Engine Div.

MILLER CYCLE

The Miller-cycle is an adaptation of a four stroke Otto-cycle that has gained popularity and is in early commercialization for stationary natural gas-fired engines and more recently on a wider scale for the automobile industry. In the Miller-cycle, the intake valve remains open as the compression stroke begins. In effect, the compression stroke is two discrete cycles: the initial portion when the intake valve is open and final portion when the intake valve is closed. As the piston initially moves upwards in what is traditionally the compression stroke, the charge is partially expelled back out the still-open intake valve. By leaving the intake valve open and allowing a portion of the air out during the compression stroke the compression ratio in the cylinder is reduced as well as the resultant shaft power requirements.

Charge air is partially lost during the initial stroke, which would normally reduce engine power. A supercharger with an intercooler is used to boost power output and lower charge air temperature. The air is cooled by the intercooler between the compression processes. The lower intake charge temperature, combined with the lower compression of the intake stroke, yields a lower final charge temperature than would be obtained by simply increasing the compression of the piston. This lower charge air temperature provides several benefits:

- Ignition timing can be advanced beyond what is normally allowed before the onset of detonation, thus increasing thermal efficiency
- An overall increase in the compression ratio (supercharger + compression stroke) is achieved thus increasing both power output and thermal efficiency
- NO_X emissions are reduced

SPLIT CYCLE ENGINES

Split-cycle engines separate the four strokes of intake, compression, power, and exhaust into two separate but paired cylinders. The first cylinder is used for intake and compression. The compressed air is then transferred through a crossover passage from the compression cylinder into the second cylinder, where combustion and exhaust occur. A split-cycle engine is really an air compressor on one side with a combustion chamber on the other.

This type of cycle has been around for a long time. Previous Split-cycle engines appeared as early as 1914. Many different split-cycle configurations have since been developed; however, none has matched the efficiency or performance of conventional engines, due to two major problems — poor breathing (volumetric efficiency) and low thermal efficiency.

Recent R&D developments are underway to try to solve these problems. An engine, shown in Figure 9-70, known as the Scuderi Engine, is currently in prototype testing. A single cylinder prototype of this engine is currently undergoing bench testing to create engine performance maps. Production versions are still at least several years away.

The Scuderi Engine design addresses the breathing problem by reducing the clearance between the piston and the cylinder head to less than 1 mm. With outwardly opening valves that enable the piston to move very close to the cylinder head without the interference of the valves, almost 100 percent of the compressed air can be transferred from the compression cylinder into the crossover passage, eliminating the breathing problems associated with previous split-cycle engines.

To address the efficiency problem, the Scuderi engine fires after top dead center (ATDC). Although considered bad practice in conventional engine design, firing ATDC in a split-cycle arrangement eliminates losses from recompressing the gas. Firing ATDC is possible because of high pressure air in the transfer passage and high turbulence in the power cylinder. The result is very rapid atomization of the air/fuel mixture, creating a fast flame speed or combustion rate. Computer simulations predict efficiency gains of 5-10% over conventional engine for the naturally aspired engine and 10-20% improvements over conventional engines for the turbocharged version. In addition, computer simulations by an independent research organization predicted that NO_X emissions would be reduced by 80% compared to a conventional engine.

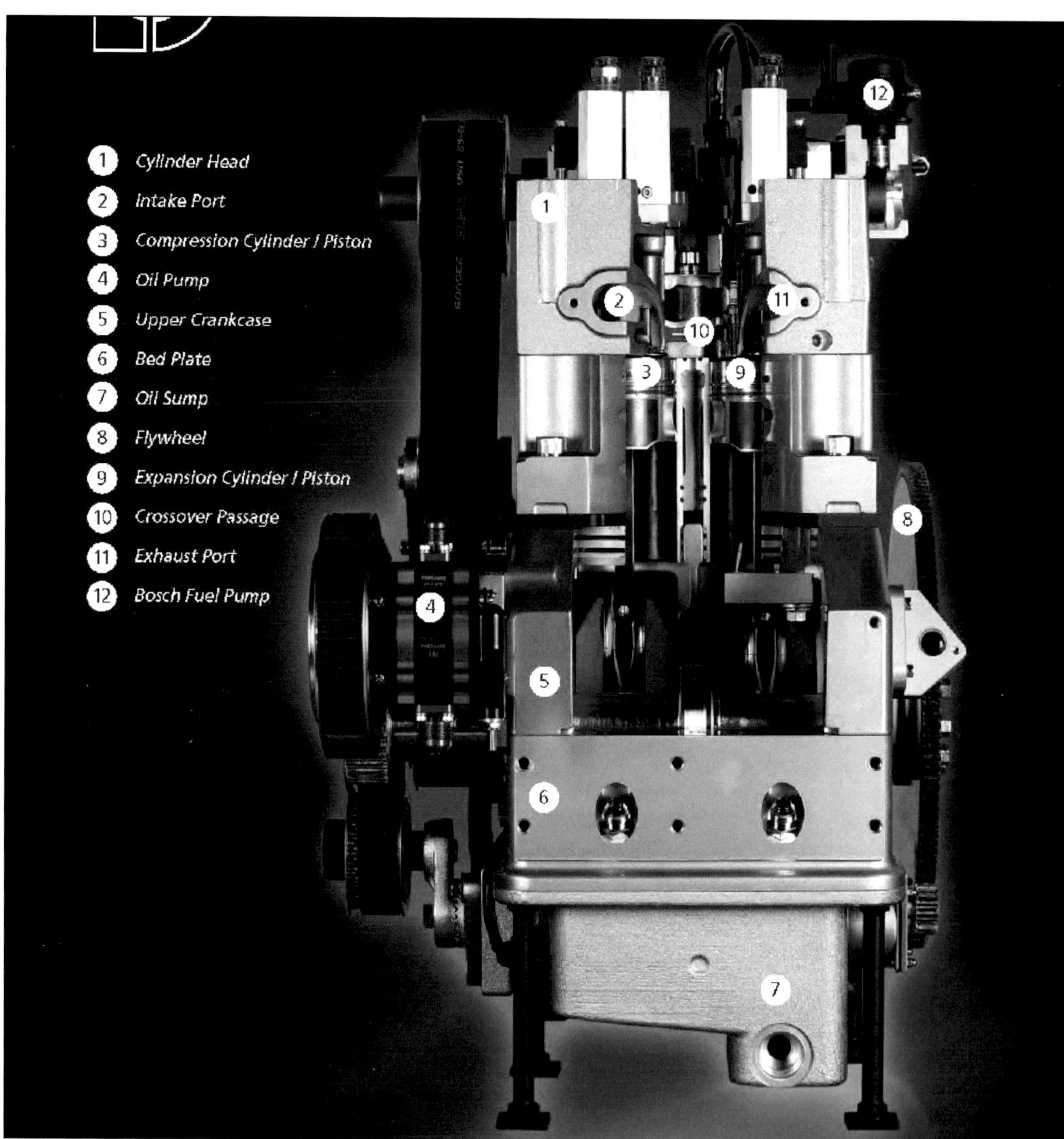

Fig. 9-70 Scuderi Engine Schematic.
Source: The Scuderi Group

Chapter Ten
Combustion Gas Turbines

Combustion gas turbines feature compact, lightweight designs and offer simplicity, versatility, high power-density, and excellent reliability. They are known for smooth, clean-burning, reliable operation, as well as easy installation, minimal vibration, long life, and low maintenance. Gas turbine models typically range in capacity from a few hundred hp (or kW) to hundreds of thousands of hp (or kW), though recently, small-capacity units of under 50 hp (37 kW) have been commercially introduced.

Most gas turbines in operation for industrial applications use natural gas as the primary fuel. Combustion gas turbines can however, subject to manufacturer approval, use a wide variety of fuels, including liquefied petroleum gas (LPG), alcohol, kerosene (jet fuel), propane, coal-derived gas, naphtha, distillate oils, some crude oils, and some residual oils. Manufacturers offer specifications to define minimum fuel quality requirements for their machines.

Combustion Gas Turbine Types

The workings of a gas turbine are similar to that of a reciprocating engine in that both feature induction, compression, combustion, expansion, and exhaust. However, in a gas turbine, combustion occurs continuously as a fluid dynamic process at a relatively low constant pressure, while in a reciprocating engine, combustion occurs as a batch process at higher, fluctuating pressures.

Open and Closed Cycles

Combustion gas turbines operate on the **Brayton,** or Joule, cycle. Gas-turbine systems feature open or closed cycles. **Open-cycle** combustion gas turbines are configured to continuously induce and compress atmospheric air, mix and combust fuel, extract mechanical energy from the products of combustion to drive the compressor and payload, and exhaust the spent products of combustion to atmosphere. The open Brayton cycle is more common than the closed Brayton cycle in gas-turbine mechanical drive, electric power generation, and cogeneration systems.

Closed-cycle combustion gas turbines continuously recycle their working fluids. Energy requirements for closed cycles are derived from near-atmospheric combustion of fuel at nearly theoretical air-fuel ratios. Energy released is transferred to the working fluid by a heat exchanger and products of combustion are discharged to atmosphere.

Advantages of the closed cycle include:

- Clean, consistent working fluid
- Ability to operate on a wide range of fuels
- Ability to control fluid density and pressure that offers a method of output control
- Marginally reduced altitude impact on system capacity
- Wider range of efficient operating capacity
- Decreased requirements for sound attenuation equipment in the fluid path
- Choice of working fluid for optimum physical and chemical properties

Disadvantages of the closed cycle include:

- Cost and size of combustion equipment and heat exchangers
- Cost of ducting and seals to contain the working fluid
- Employment of potentially toxic material as working fluids
- Cooling is required before fluid re-enters the compressor
- Temperature limitations imposed by heat exchangers
- Absence of significant commercial experience in North America

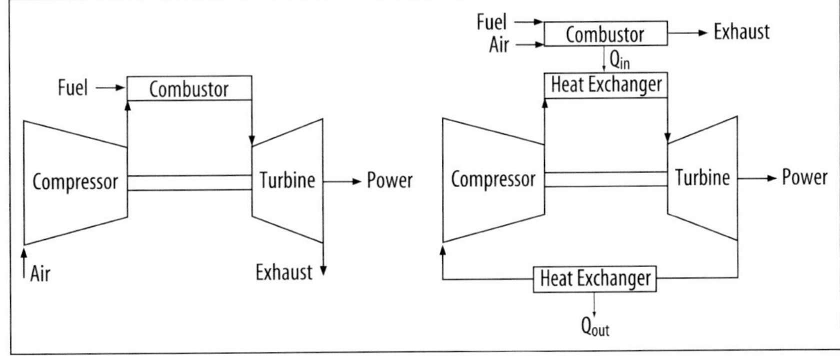

Fig. 10-1 Open and Closed Brayton Cycle.

Given the limited application of closed-cycle gas turbine systems, this chapter focuses on open-cycle systems.

Aeroderivative and Industrial Types

Aeroderivative and industrial classifications are commonly used to differentiate between available gas turbine model types, and there is a long standing debate between manufacturers as to which design is superior for stationary applications. In fact, both types share technologies for cycles, compressor and turbine aerodynamic designs, combustion and emission control systems, and much of the best available metallurgy. For any given stationary application, available models must be evaluated individually to determine comparative economic performance.

Defining characteristics of **aeroderivative gas turbines** include:

- Adaptation from aircraft gas turbines to industrial applications
- Compact dimensions and light weight, permitting factory packaging for minimum field installation time and cost
- Use of light, thin, super alloy materials for outer turbine and compressor cases
- Low thermal capacity, allowing them to endure rapid load changes and numerous start-up/shut-down events
- Relatively high simple-cycle thermal fuel efficiency
- High inlet gas fuel pressure requirement, requiring greater fuel compressor power
- Compact and lightweight combustor design, offering limited design flexibility for alternative fuels and emissions control modifications
- Low moment of inertia (flywheel effect), which allows for quick acceleration, with low starting torque and power requirements; as a result, faster control system response to changing load is required
- The application of ball and roller shaft bearings for radial and thrust loads and the use of more costly high-temperature synthetic lubricating oils
- More frequent maintenance and major overhauls (in some cases, off-site prime mover repair can limit outage duration)
- Pressure ratios are usually above 15:1.

Defining characteristics of **industrial gas turbines** include:

- Use of thick, low-alloy materials for outer turbine and compressor cases (hence, the alternative designation of heavy frame turbines)
- Restricted rate of load increase in order to prevent excessive thermal stress and resulting impact on maintenance and repair requirement
- High moment of inertia, which requires longer acceleration time and high starting torque and power requirements, but allows for stable operation with more time for the control system to make corrections under changing load
- For large-capacity units, components shipped to site by rail for field assembly and erection (factory packaging is generally limited to small-capacity units)
- Combustor design not constrained by weight or space limitations, offering more flexibility for designs aimed at alternative fuels and new emissions control technologies
- Long-standing experience using natural gas
- Use of long-life tilting pad journal shaft bearings and low-cost mineral lubricating oils
- Relatively infrequent maintenance and major overhauls
- Need for in-situ maintenance and major repair of large-capacity units, which reduces unit availability (for some smaller units, maintenance can be done by engine exchange)
- Existing market domination in unit capacities above 67,000 hp (50,000 kW); industrial designs are also more prevalent under 27,000 hp (20,000 kW)
- Pressure ratios are usually below 20:1, though large-capacity units may have pressure ratios as high as 30:1.

The General Electric (GE) LMS100 represents a hybrid design utilizing technologies from both GE's industrial (Frame) and aeroderivative gas turbine units. This new 134,000 hp (100,000 kW) unit achieves a 46% simple cycle efficiency (LHV) using intercooling between compression stages and a 42:1 cycle pressure ratio.

Operating Cycle

While actual operation differs from the ideal cycle, and actual performance is dramatically lower than the ideal cycle indicates, it is useful to review the basic principles and performance equations associated with the ideal

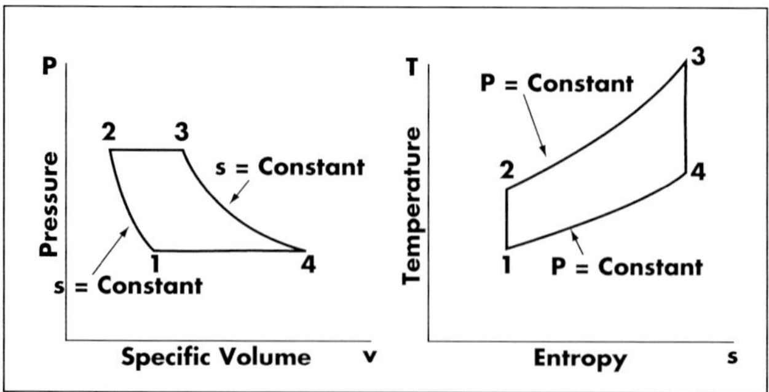

Fig. 10-2 The Air Standard Brayton Cycle.

cycle. The Air Standard (or ideal) Brayton cycle is shown in Figure 10-2.

In the ideal cycle, gas is isentropically compressed (Path 1-2) and heat is added at constant pressure (Path 2-3). The gas then undergoes an isentropic expansion to its initial pressure (Path 3-4) and heat is rejected at constant pressure (Path 4-1).

In a gas turbine, the lower pressure at State 1 represents atmospheric pressure and the upper pressure at State 2 represents the pressure after compression of the air by the compressor. The addition of heat between States 2 and 3 occurs in the combustor. Work is then derived from the expansion of the hot combustion gases from States 3 and 4.

In the ideal cycle, the heat input (Q_{in}) in the constant-pressure process (Path 2-3) per unit mass is the product of the mass flow rate ($\dot{m}$), the specific heat (c_p), and the corresponding change of temperature (T) of the fluid (working fuel) in the process. It is expressed as:

$$Q_{in} = \dot{m}c_p\,(T_3 - T_2) \qquad (10\text{-}1)$$

For the constant-pressure heat rejection (Q_{out}) part of the cycle (Path 4-1), the heat output per unit mass is:

$$Q_{out} = \dot{m}c_p\,(T_4 - T_1) \qquad (10\text{-}2)$$

The net work produced (W) is:

$$W = Q_{in} - Q_{out} = \dot{m}[c_p(T_3 - T_2) - c_p(T_4 - T_1)] \qquad (10\text{-}3)$$

Assuming that the specific heats are constant, and neglecting chemical changes that occur during combustion, the ideal thermal efficiency ($\eta_{Brayton}$) can be expressed as:

$$\eta_{Brayton} = \frac{(Q_{in} - Q_{out})}{Q_{in}} = \frac{c_p\,(T_3 - T_2) - c_p\,(T_4 - T_1)}{c_p\,(T_3 - T_2)} = 1 - \frac{(T_4 - T_1)}{(T_3 - T_2)} \qquad (10\text{-}4)$$

As can be seen from the P-V diagram, $p_2 = p_3$ and $p_1 = p_4$ and the pressure ratio (r_p) can be expressed as:

$$r_p = \frac{p_3}{p_4} = \frac{p_2}{p_1} \qquad (10\text{-}5)$$

In the ideal cycle, which has no internal losses, the thermal efficiency can also be shown to depend solely on the pressure ratio. Thus, cycle efficiency can also be expressed as:

$$\eta_{Brayton} = 1 - \left(\frac{p_1}{p_2}\right)^{(k-1)/k} = 1 - \left(\frac{1}{r_p}\right)^{(k-1)/k} \qquad (10\text{-}6)$$

where k is the ratio of specific heats (c_p/c_v) at constant pressure and constant volume.

Expressed as a function of volume (V) and compression ratio (r_c), the thermal efficiency equation is:

$$\eta_{Brayton} = 1 - \left(\frac{v_1}{v_2}\right)^{1-k} = 1 - r_c^{(1-k)} \qquad (10\text{-}7)$$

These equations show that the ideal efficiency of the Brayton cycle, which is a constant-pressure combustion cycle with complete expansion, is identical to the ideal efficiency of the Otto cycle, which is a constant-volume combustion cycle with incomplete expansion. Thus, as with the Otto cycle, the theoretical efficiency depends only on the compression ratio or pressure ratio.

In practice, the operation of the gas-turbine engine differs from the ideal Brayton cycle because of irreversibilities in the compressor and turbine, such as friction in the bearings, and pressure drop in the flow passages and combustion chamber. A distinguishing feature of the Brayton cycle is the large amount of compressor work (also called back work) compared to turbine work. The compressor might require 40 to 80% of the output of the turbine. This is considerably higher than the compression work required by reciprocating engines, which are more efficient compressors and expanders. This high-compressor energy requirement is particularly important when actual gas turbine cycle efficiency is considered, because overall efficiency drops rapidly with a decrease in the efficiencies of the compressor and turbine.

The effect of these inefficiencies on actual gas turbine cycle performance can be seen in Figure 10-3. This T-s diagram shows the effect of the irreversible aspects of the real

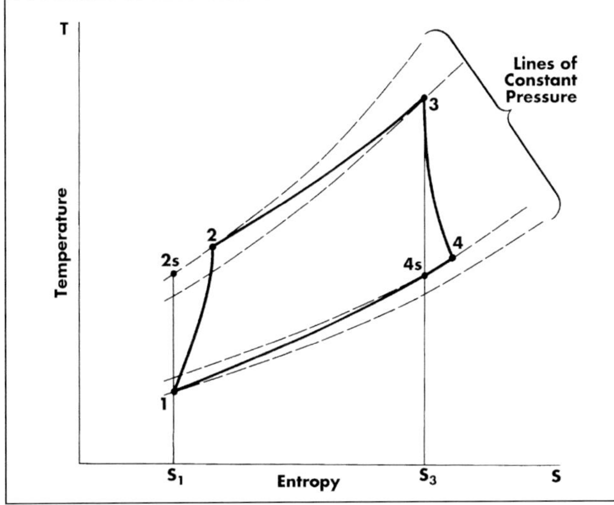

Fig. 10-3 T-s Diagram of Actual Gas Turbine System. Source: Babcock and Wilcox

gas turbine cycle. An isentropic compression would attain the Point 2s, whereas the real compressor attains the pressure P_2 with an entropy corresponding to Point 2. The turbine expansion lines on the diagram for P_2 and P_3 illustrate the effect of pressure losses in the combustor and connecting piping. The deviation of the process between Points 4 and 1 from a constant-pressure process illustrates the effect of compressor inlet and turbine exhaust pressure losses on cycle efficiency.

Accounting for practical application inefficiencies, an equation for actual efficiency becomes:

$$\eta_{actual} = \frac{\eta_t W_t - \frac{W_c}{\eta_c}}{Q_{in}} = \frac{W_{ta} - W_{ca}}{Q_{in}} \tag{10-8}$$

where subscripts c and t indicate ideal compressor and turbine, respectively, and ca and ta indicate actual compressor and turbine work, respectively.

The actual work requirement of the compressor as a function of temperature (T) or enthalpy (h) change can be expressed as:

$$W_{ca} = \dot{m}c_p(T_{2'} - T_1) = \dot{m}(h_{2'} - h_1) \tag{10-9}$$

where $T_{2'}$ and $h_{2'}$ represent the actual temperature and enthalpy, respectively. Actual compressor efficiency, in relation to the isentropic process, can be expressed as:

$$\eta_{ca} = \frac{h_2 - h_1}{h_{2'} - h_1} \tag{10-10}$$

Correspondingly, actual turbine work can be expressed as:

$$W_{ta} = \dot{m}c_p(T_3 - T_{4'}) = \dot{m}(h_3 - h_{4'}) \tag{10-11}$$

Turbine efficiency, in relation to the isentropic process, can be expressed as:

$$\eta_{ta} = \frac{h_3 - h_{4'}}{h_3 - h_4} \tag{10-12}$$

The actual enthalpy exiting the combustor section is equal to the entry enthalpy (from compressor discharge), plus the heat added due to combustion (Q). Thus,

$$\dot{m}c_p T_3 = \dot{m}_a c_p T_{2'} + Q \tag{10-13}$$

Since the mass of fuel is negligible compared with the mass of air, $\dot{m}$ is approximately equal to $\dot{m}_a$. Therefore, the heat added (Q) can be represented as:

$$Q \cong (h_3 - h_{2'})\dot{m}_a \tag{10-14}$$

Thus, accounting for combustor efficiency (η_{comb}), the actual efficiency of a gas turbine can then be expressed as:

$$\eta_{th} = \frac{(W_{ta} - W_{ca})}{\dot{m}_f(LHV)\eta_{comb}} = \frac{[(h_2 - h_1) - (h_3 - h_4)]}{(h_3 - h_2)} \tag{10-15}$$

Gas Turbine Components and Operation

The basic elements of a combustion gas turbine are the **compressor, combustor,** and **turbine**. Compressor and turbine sections are composed of one or more rows of bladed wheels called stages. The function of each stage is to increase the pressure (compressor) or decrease the pressure (turbine) of the working fluid. Rotating stages are often separated by stationary blades. Depending on the specific design, blades are referred to as buckets, vanes, or nozzles. Blades and buckets refer to the rotating parts, while vanes and nozzles refer to the stationary parts. The connection of the blade to the wheel is called a root.

In the most basic open cycle, the compressor rotates, drawing in ambient air, compressing it, and steadily forcing it into the combustor. The temperature of the air rises due to the compression. Fuel is burned with the compressed air in the combustor. As the fuel burns, it causes the gas temperature to rise quickly in an essentially constant-pressure process. The high-energy combustion gases expand in the turbine section and the temperature and pressure of the expanding gases drop as their energy is converted into mechanical work. This conversion involves two steps. In the nozzles section, the hot gases expand and a portion of the thermal energy is converted into kinetic energy. A portion of this kinetic energy is then transferred to the rotating buckets and converted to work.

Single-shaft gas turbine units feature direct mechanical coupling between power turbine stages and compressor drive turbine stages so that all rotating elements operate at the same speed. Single-shaft models have been applied to power generation and other suitable constant speed applications where rotating mass inertia is important for operating stability. Most large industrial turbines are of single-shaft design, as illustrated in Figure 10-4.

Multi-shaft designs (or **free turbines**, as they are sometimes called) feature two or more shafts that operate at independent speeds. In the two-shaft design shown in

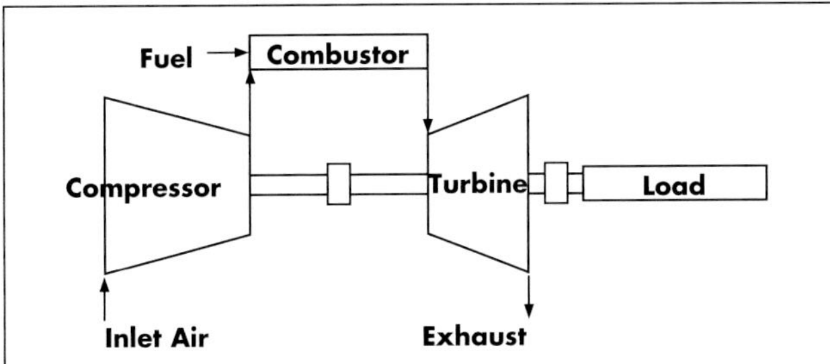

Fig. 10-4 Simple-Cycle, Single-Shaft Gas Turbine.

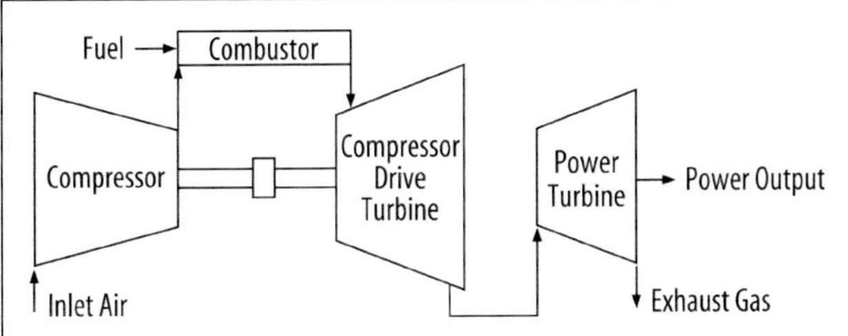

Fig. 10-5 Two-Shaft Gas Turbine with Free Power Turbine Stage.

Figure 10-5, one of the turbines is dedicated to driving the compressor, while the other drives the load. In a three-shaft design, there may be two compressor-driven turbines for high- and low-pressure compressors. Multiple shaft turbines offer a greater potential compressor surge margin and require less energy for starting. They also allow the compressor to operate at its most efficient speed while the power turbine speed varies with the driven load, eliminating the need for gearboxes and associated losses. As a result, multi-shaft turbines offer lower part-load heat rates (higher efficiency) than single-shaft units. Multi-shaft turbines are preferred for mechanical drive applications because of low starting torque and increased operational flexibility. Most aeroderivative type turbines are of the multi-shaft design.

The major components of commercially available gas turbines are discussed below.

COMPRESSOR

In the compressor section, air is compressed by several orders of magnitude in preparation for the combustion process. Since the compressor uses 65 to 75% of total power output, this component has significant impact on overall cycle efficiency. Sophisticated aerodynamic design is required to minimize the work required to compress the air.

Small-capacity units may include **centrifugal compressors**, while units of 4,000 hp (3,000 kW) or larger usually include **axial compressors**. Axial-flow compressors include multiple sets of airfoil rotor blades and stationary blade passages. Figure 10-6 shows an axial compressor rotor for a twelve-stage compressor. The rotor is an assembly of disks held together by a single bolt between the forward and aft stub shafts. The rotor assembly is supported at either end by tilting pad journal bearings. Figure 10-7 shows a radial (centrifugal) compressor design. Air enters the centrifugal compressor at the center of a rotating bladed impeller. Rotation imparts a centrifugal force on the air, driving it radially outward at very high speed into a stationary diffuser. Figures 10-8 through 10-11 show compressor designs for several different gas turbines.

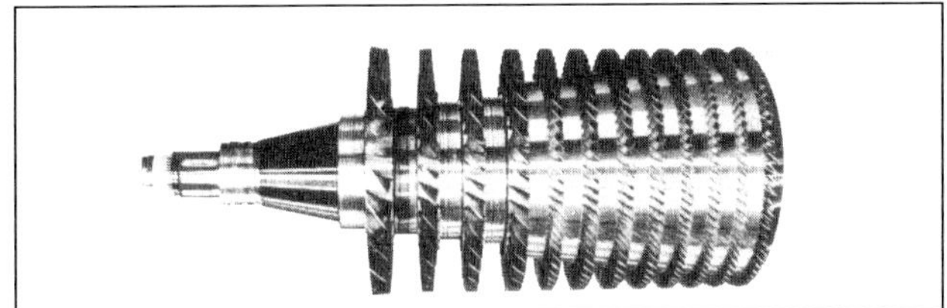

Fig. 10-6 View of a 12-Stage Axial Compressor Rotor.
Source: Solar Turbines

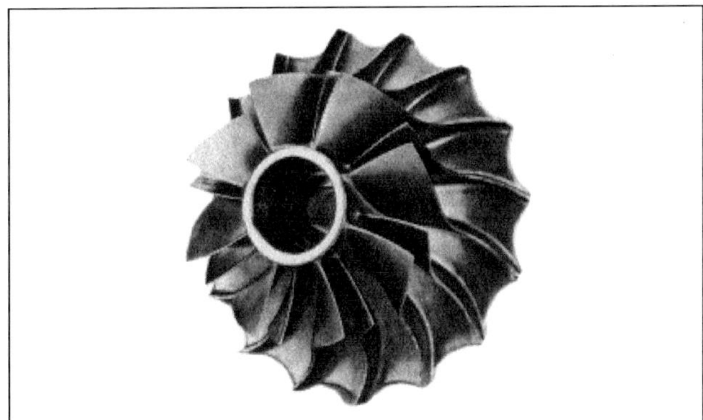

Fig. 10-7 Radial Compressor Design.
Source: Solar Turbines

Fig. 10-8 View of Gas Turbine with Compressor Casing Half Removed.
Source: Solar Turbines

Compressor designs may also include variable inlet guide vanes (IGVs) and multi-stage compression with intercooling. IGVs can be modulated to reduce mass flow under part loads, allowing a higher exhaust temperature to be maintained. The vanes are also used to restrict airflow for quick acceleration on starting. Figure 10-12 shows the air inlet casing and IGVs for an industrial gas turbine with a capacity of about 12,000 hp (9,000 kW).

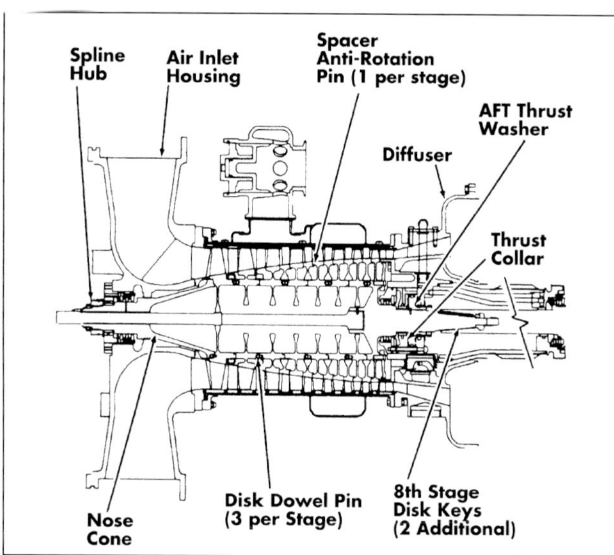

Fig. 10-9 Illustration of Compressor Drive Assembly. Source: Solar Turbines

Fig. 10-10 Rotor with 15 Stages of Compressor Blading in Large-Capacity Turbine. Source: Siemens Power Corp.

Fig. 10-11 Close-Up View of Compressor Blade Design. Source: ABB

Fig. 10-12 Industrial Gas Turbine Inlet Casing with Variable IGVs. Source: MAN GHH

Figure 10-13 illustrates a compressor with 10 axial stages followed by a centrifugal stage. The rotor is supported by two tilting-pad bearings, and a thrust bearing and counter thrust bearing, located in the front-end bearing housing. An epicyclic gear in front of the compressor transmits the necessary starting power to the turbine and the power required for driving ancillaries, such as lube oil pumps. An intercooler (heat exchanger) is used to decrease the work of compression by cooling air in the middle of its compression cycle. Intercooling is discussed later in this chapter.

Combustor

Compressed air and fuel are delivered continuously to the combustor, where a portion of the air is mixed with fuel in the combustion primary zone. Remaining compressed air is mixed with the products of combustion to reduce exiting fluid temperature to acceptable levels.

Several alternative configurations are used in applied combustor designs. **Can** configurations are designed with one fuel nozzle in each burner can. **Annular** configurations

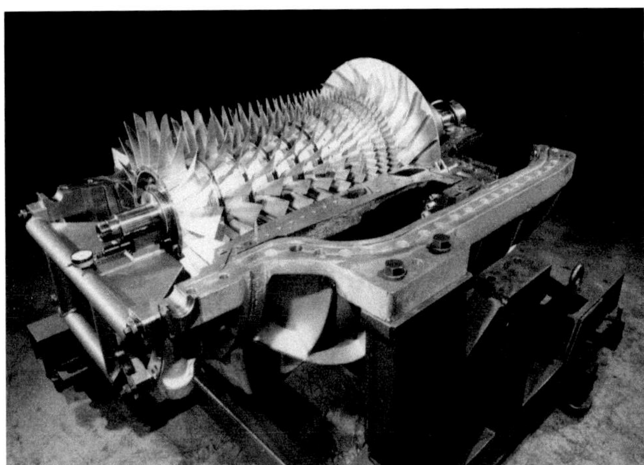

Fig. 10-13 Air Compressor with Ten Axial Stages Followed by One Centrifugal Stage. Source: MAN GHH

include one large combustor with multiple burners. While can combustors can be removed and replaced with limited disassembly, annular combustors require the power turbine and gas producer modules to be removed for maintenance. However, annular combustors have been the design choice among recent generations of high-performance gas turbines due to more efficient space utilization and superior performance and durability at high firing temperatures.

Another combustor design features one or more external cylindrical or silo combustors. The external design places the combustor at right angles to the normal flow of gases from the combustor to the turbine section. Because the combustor geometry is not constrained by other parts of the system, it can be adjusted to accommodate a wide range of fuel types and emission control strategies. Figure 10-14 is a labeled illustration of a single-silo, dry low-NO_X combustor for a 50 MW turbine.

Over the past decade, a great focus of combustor design efforts has been on development and improvement of dry low-NO_X combustion technology as a means of complying with strict air emission regulations while maintaining fuel efficiency and limiting the production of other harmful air pollutants, such as carbon monoxide. NO_X production is mostly a function of fuel nitrogen content (fuel NO_X), and combustion temperature, residence time, and pressure level (thermal NO_X). To limit thermal NO_X formation, combustor and burner designs have focused on minimizing peak flame temperature and residence time.

Peak flame temperature reduction has long been accomplished through wet injection with water or steam, which remains an important technology. However, due to cost and efficiency concerns, the design trend has been toward dry low-NO_X combustion technology. Minimum peak-temperature residence time is limited by the need for complete combustion to reduce emission of carbon monoxide and unburned organic materials. Thus, the design trend has been toward premixing of fuel and air through premix burners (as opposed to diffusion in which mixing and combustion take place simultaneously). Design limitations exist with premixing, however, due to flame instability with very lean fuel-air mixtures and under low part-load operation.

Figure 10-15 shows a dry low-NO_X combustor featuring an all-welded steel design and 18 burners of vortex break down type (or double cone). The nozzles are

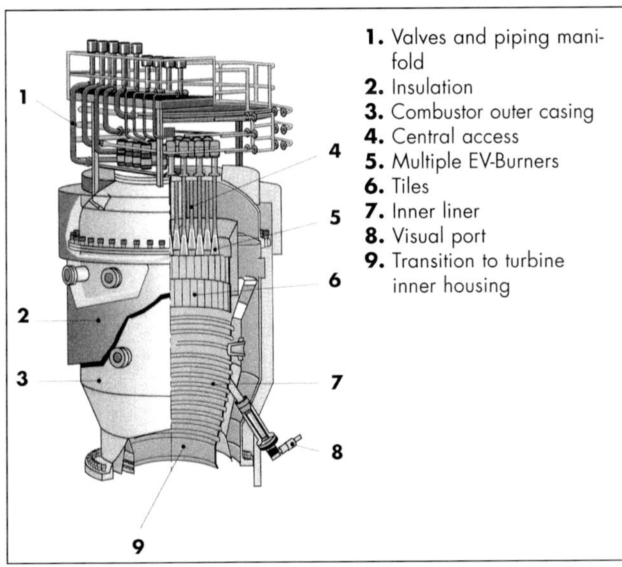

Fig. 10-14 Design Detail for Dry Low-NO_X Silo Combustor. Source: ABB

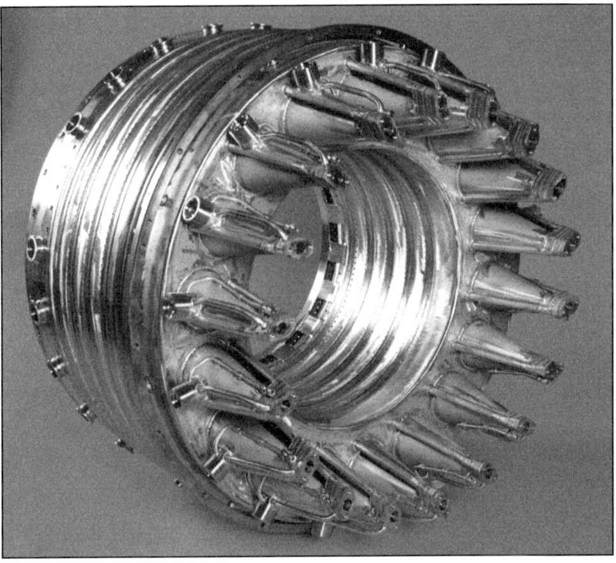

Fig. 10-15 Dry Low-NO_X Combustor. Source: ABB STAL

dual-fuel type for gas and liquid. In the rear end, openings for dilution bypass air are seen. Figure 10-16 illustrates a dual-fuel, double-cone burner featuring vortex breakdown design. This unit is reportedly capable of achieving NO_X emission levels of under 10 parts per million (ppm) or 20 mg/m^3. [Note that all emissions values given in this chapter are based on dry exhaust gas with 15% O_2 by volume.]

Figure 10-17 shows a gas turbine combustion liner and cap. The design provides for six fuel nozzles, which are mounted directly on the combustion end cover. The dry low-NO_X combustor design features a combination of lean diffusion flames, staged combustion, and lean premixed combustion. The combustion ignition system uses spark plugs, cross-fire tubes, and flame detectors. Ignition in one chamber produces a pressure rise, which forces hot gases through the cross-fire tubes, propagating ignition to all other chambers within one second.

Figure 10-18 shows assembly of a combustion chamber module used in the aeroderivative FT8 gas turbine. The streamline combustion chamber consists of nine separate chambers on the periphery and is designed for operation on liquid and/or gaseous fuels at any time. Figure 10-19 shows the dry low NO_X annular combustion liner for a small-capacity gas turbine. The sheet metal liner maintains low metal temperatures through film cooling, using a small fraction of the compressor discharge air.

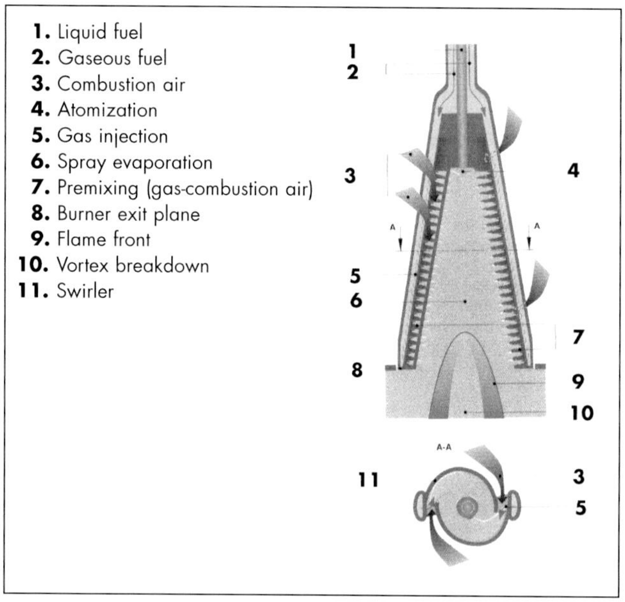

1. Liquid fuel
2. Gaseous fuel
3. Combustion air
4. Atomization
5. Gas injection
6. Spray evaporation
7. Premixing (gas-combustion air)
8. Burner exit plane
9. Flame front
10. Vortex breakdown
11. Swirler

Fig. 10-16 Dual-Fuel, Double-Cone Burner Illustration. Source: ABB

Fig. 10-18 Assembly of Combustion Chamber for Aeroderivative Gas Turbine. Source: United Technologies Turbo Power Division and MAN GHH

Fig. 10-17 Gas Turbine Combustion Liner and Cap. Source: General Electric Company

Fig. 10-19 Annular Combustion Liner for Small-Capacity Gas Turbine. Source: Solar Turbines

Figure 10-20 shows a Siemens hybrid burner for natural gas and fuel oil premix operation for a large-capacity industrial gas turbine. Two swirlers are used for the airflow, with the outer diagonal swirler supplying about 90% of the air. For gas operation, the unit has three separate nozzle systems: the diffusion burner, the premix burner, and the pilot burner nozzles. The diffusion burner is used under low-load conditions to maintain flame stability. The premix burner operates over a range from full load down to about 45% of full load with the assistance of a stabilizing pilot flame. Liquid fuel for premix operation enters through a ring header and nozzles located immediately downstream of the diagonal swirlers. The fuel is atomized into small droplets that evaporate due to the high air temperature.

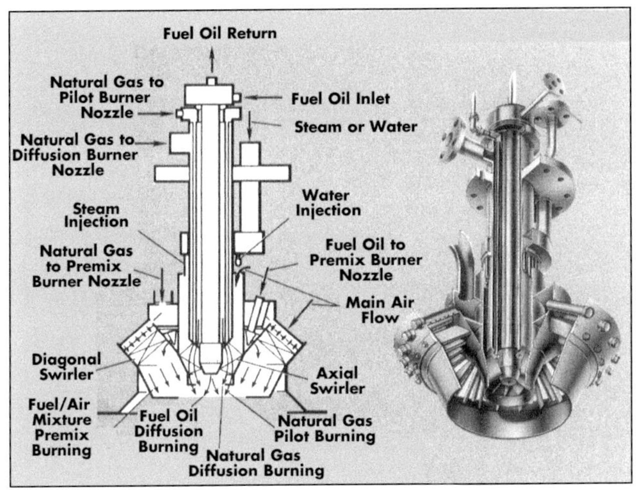

Fig. 10-20 Hybrid Burner for Natural Gas and Fuel Oil Premix Operation in Large-capacity Gas Turbine. Source: Siemens Power Corp.

Figure 10-21 shows the burner ring combustor for the system described above and its location in the gas turbine. The combustor consists of three casing sections forming an inner cone and two outer half shells. The shell is protected from exposure to high temperature with internal heat pads, which are cooled with compressor discharge air. As shown, the combustor is located between the compressor section on the left and the power turbine section on the right.

This same hybrid burner technology is available in a much smaller capacity 12,000 hp (9,000 kW) industrial turbine. The hybrid burner and large-volume flame tube with ceramic lining is used with two relatively small silo-type external combustion chambers in a V arrangement. Manufacturers' tests indicate that NO_X emissions levels of 9 ppm (18 mg/m^3) and CO emissions levels of under 4 ppm (8 mg/m^3) are achieved with premix opera-

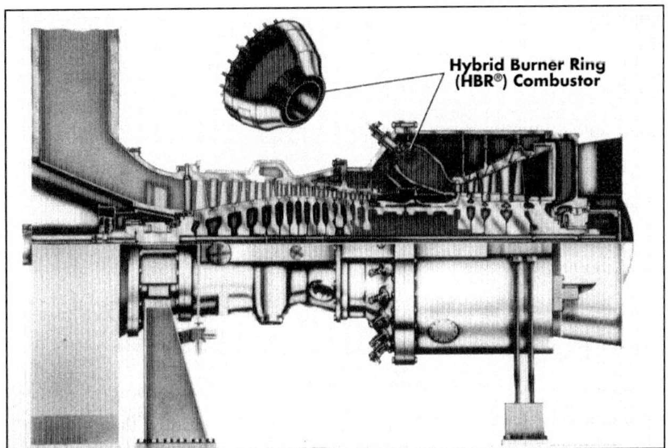

Fig. 10-21 Hybrid Burner Ring Combustor and its Location in the Gas Turbine. Source: Siemens Power Corp.

tion over most of the operating load range. NO_X emissions levels are slightly higher at full load, and both NO_X and CO emission levels increase dramatically at about 30% of full load, as the burner operates in diffusion mode, with maximum dilution air.

Figure 10-22 illustrates the combustor design applied to a sequential combustion process for a large-capacity industrial gas turbine. In the sequential, or reheat type, combustion process, there are two annular combustion chambers. Fuel is injected at the same time into both chambers. A high amount of excess air is premixed with fuel prior to combustion. Compressed air is fed into the double-cone burner on the right to create a homogeneous, lean fuel/air mixture. This burner is referred to as an EV burner. The mixture spirals through the burner and forms a vortex. The vortex flow breaks down as it exits the EV burner into the EV combustion chamber, shown on the right, forming a recirculation zone. The mixture is then ignited and a single low-temperature halo flame is produced.

The recirculation zone allows the flame to stabilize within the chamber, avoiding contact with and degradation of material surfaces. Hot gas exits the first chamber,

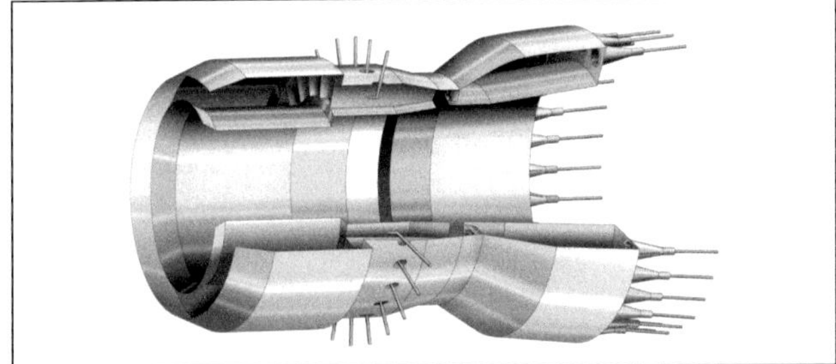

Fig. 10-22 Sequential Combustion System Featuring Two Annular Combustion Chambers. Source: ABB

moving through the first turbine stage before entering the second chamber, which is called the SEV combustor, shown on the left. Carrier air is injected at the fuel lance to lower the temperature of the hot gas moving through the burner so that spontaneous combustion is delayed. Therefore, ignition takes place as soon as the fuel reaches its self-ignition temperature. This occurs after it has moved beyond components into free space. Subsequently, the hot gas continues its steady flow through the low-pressure turbine.

Given the direct relationship between NO_X production and temperature, the low flame temperature achieved with thorough pre-mixing and the high amount of excess air used by the sequential combustion system results in a low level of NO_X. Additionally, in the second combustion chamber, where about 40% of the fuel is burned, the incoming hot gas has considerably lower oxygen content than normal air. This results in less oxygen being available for NO_X formation. NO_X emissions levels are reported to be below 15 ppm (30 mg/m^3) and single digit ppm values are anticipated.

Turbine

In the turbine section, work is extracted from the high-pressure, high-temperature working fluid as it expands to atmospheric pressure. Thus, hot gas energy is converted into mechanical energy to drive both the air compressor and the load. Figure 10-23 shows a gas duct, which transfers the hot gases from the flame tubes to the inlet of the high-pressure turbine.

Fig. 10-23 Single-Piece Super-Alloy Gas Duct Used to Transfer Hot Gas to High-Pressure Turbine. Source: MAN GHH

Applied turbine section designs include radial inflow and reaction configurations. Some small-capacity units may feature the radial inflow design, but most units use multi-stage reaction turbines. In multi-shaft (free turbine) configurations, turbines dedicated to driving compressors are referred to as compressor drive turbines or gas producer turbines. Turbines dedicated to driving external loads such as generators, pumps, or process compressors are referred to as free turbines or power turbines.

Figure 10-24 shows a power turbine module. The gas generator can be seen in the background. Figure 10-25 shows a high-pressure turbine, featuring two axial-flow stages, that is used to drive the compressor of an industrial gas turbine. Figure 10-26 illustrates the turbine section design for a large-capacity, heavy-duty industrial turbine. Figure 10-27 shows the turbine section of a heavy-duty gas turbine. Figure 10-28 details the vanes and blades of a high-pressure turbine. Figure 10-29 shows a turbine rotor with four-stage blading for a large-capacity gas turbine. The turbine blading has been designed to minimize any turbulent boundary zone and to provide a low entry velocity at the leading edge pressure side of the blades with continuous increase of the flow acceleration.

Cooling Systems

High-performance gas turbine models feature turbine inlet gas operating temperatures that exceed allowable operating metal temperatures. Hence, carefully controlled cooling is essential for the gas turbine's hot section. Temperature limits are maintained by cooling air, diverted from the compressor and passed through critical parts to carry excess heat into the turbine exhaust stream. The use of air for cooling, however, may involve a turbine efficiency penalty.

Figure 10-30 shows the hot section of an industrial gas turbine. Generally, at maximum cooling flow rates, basic convective cooling can limit turbine rotor inlet gas temperature to a maximum of about 2,050°F (1,121°C). To counterbalance the high heat transfer at the leading edge, more advanced cooling techniques, such as film cooling for

Fig. 10-24 Power Turbine Module in Foreground; Gas Generator in Background. Source: ABB STAL

Fig. 10-27 Turbine Section on Large Heavy-Duty Gas Turbine. Source: General Electric Company

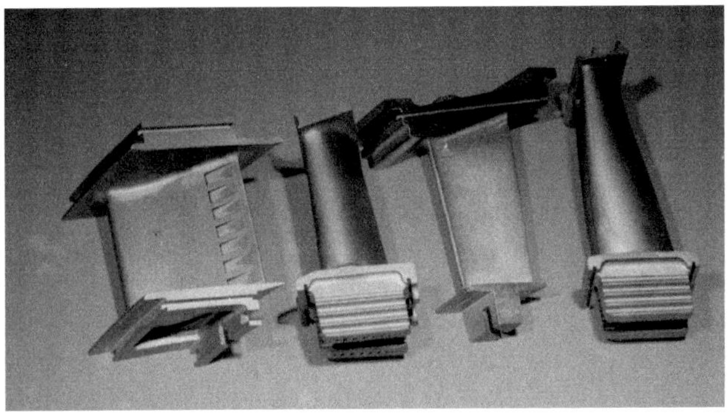

Fig. 10-28 Vanes and Blades of High-Pressure Compressor-Drive Turbine. Source: MAN GHH

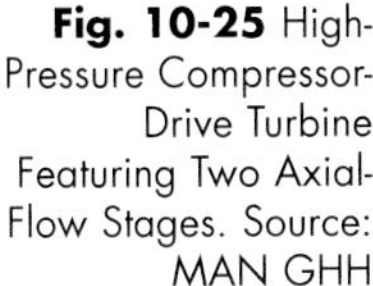

Fig. 10-25 High-Pressure Compressor-Drive Turbine Featuring Two Axial-Flow Stages. Source: MAN GHH

Fig. 10-29 Turbine Rotor with Four-Stage Blading for Large-Capacity Gas Turbine. Source: Siemens Power Corporation

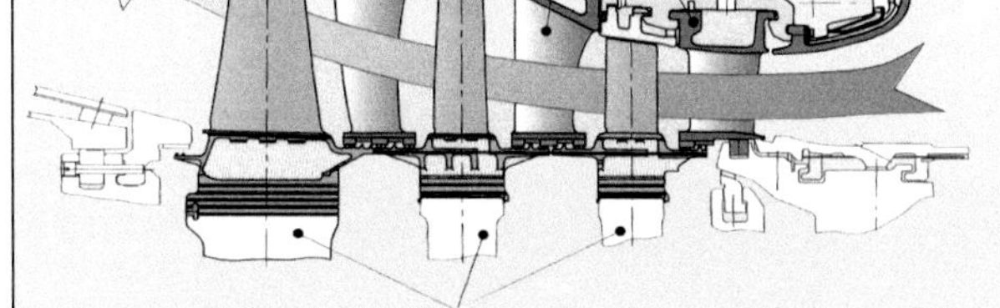

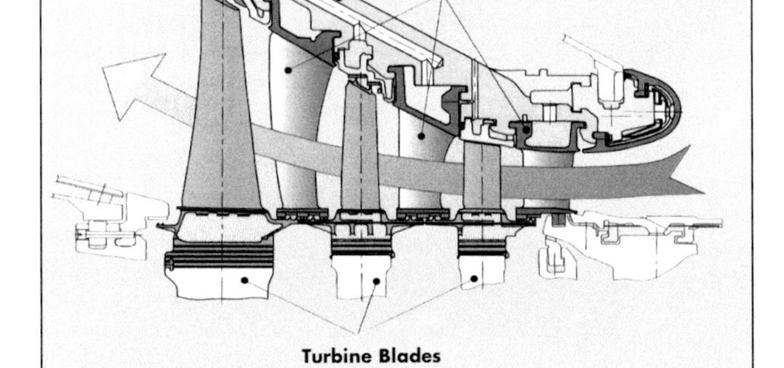

Fig. 10-26 Turbine Section Design for Large-Capacity Heavy-Duty Turbine. Source: ABB

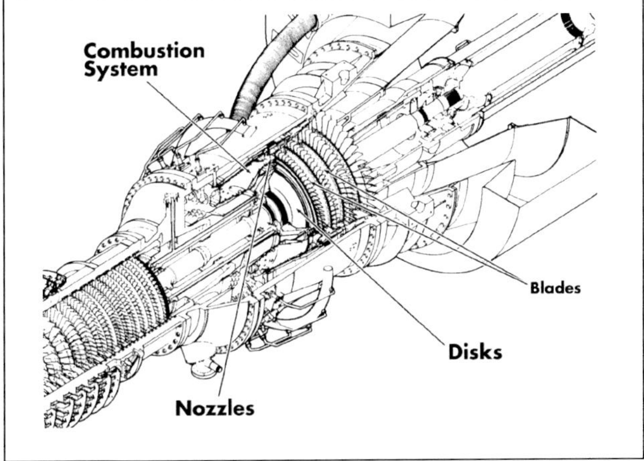

Fig. 10-30 Industrial Gas Turbine Hot Section. Source: Solar Turbines

the entire airfoil, have been applied. Internal airfoil cooling is accomplished by introducing cooling air at the root, or tip, of the air foil. It is then discharged at the opposite end of the blade through the trailing edge or ejecting holes. Currently available cooling systems have brought the high-end limit for turbine inlet gas temperature to about 2,400°F (1,316°C).

Figure 10-31 illustrates the cooling process and shows the flow of cooling air for vane and blade rows of a four-stage power turbine section of a large-capacity, heavy-duty industrial gas turbine. With the exception of the last stage rotating blades, all turbine stationary and rotating blades are air-cooled. Cooling air is provided at different pressure and temperature levels from compressor extraction to optimize cooling effect and thermal performance.

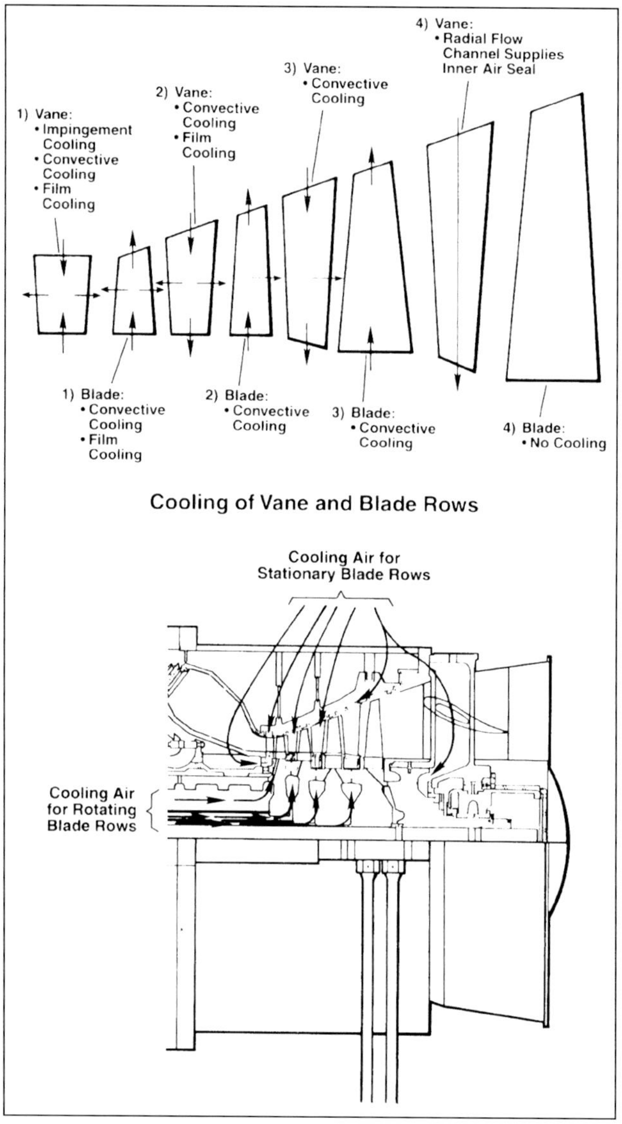

Figure 10-32 shows a typical turbine blade cooling passage for a small-capacity industrial gas turbine. Figure 10-33 is a gas turbine bucket design showing air-cooled passages. In this turbine, the first- and second-stage buckets, as well as all three nozzles, are air-cooled. The first-stage bucket is convectively cooled by means of an aircraft-derived serpentine arrangement. Cooling air exits through axial airways located on the bucket's trailing edge and sidewalls for film cooling.

Figure 10-34 illustrates airfoil-cooling techniques using combinations of impingement, convection, and film cooling. Figure 10-35 illustrates a typical nozzle vane pitch cross-section with lines of constant temperature superimposed upon it. Design considerations focus not only on maximum temperature, but also on thermal gradients in the material, since the thermal stress associated with these gradients causes fatigue damage.

Speed Matching Gears

Applied designs include a variety of gear configurations. The two most common configurations are **parallel-offset** and **epicyclic** designs. Parallel-offset gears require larger skids for equipment mounting, but otherwise provide reliable,

Fig. 10-31 Cooling Process for Four-Stage Power Turbine Section of Large-Capacity Gas Turbine. Source: Siemens Power Corp.

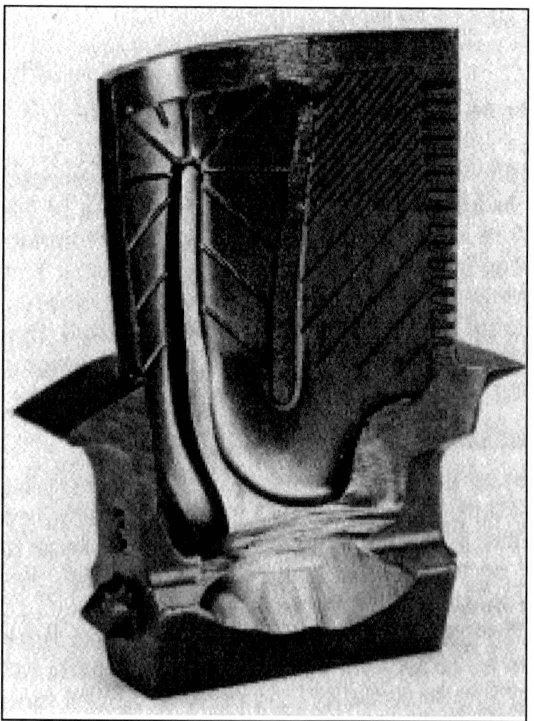

Fig. 10-32 Typical Turbine Blade Cooling Passage for Small-Capacity Turbine. Source: Solar Turbines

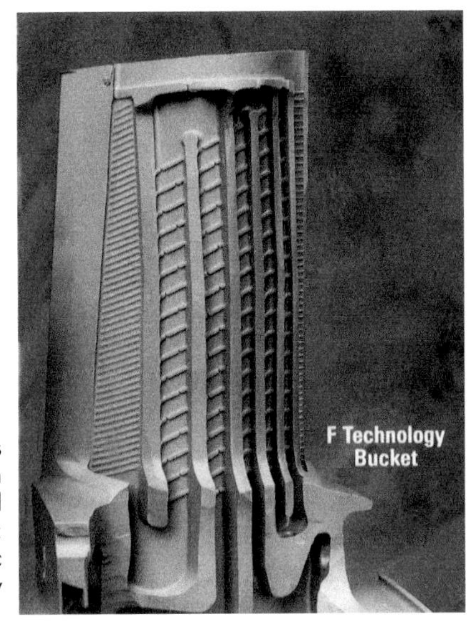

Fig. 10-33 Gas Turbine Bucket Design Showing Air-Cooled Passages. Source: General Electric Company

economic service and easy maintenance. Epicyclic gears permit gas turbine shafting to operate on a common centerline with its driven load and have lower gear mechanical losses and less oil flow requirement than a parallel shaft design. Epicyclic gears are often recommended for larger ratios where double reduction parallel offset gears would otherwise be required. Larger gas turbine models, which rotate at low enough speed, are designed for direct shaft coupling to 50 and 60 Hz power generators to eliminate losses associated with speed changing gears.

Figure 10-36 shows an epicyclic gear unit. The unit is designed for continuous-duty operation at output speeds of 1,800 rpm for 60 Hz service and 1,500 rpm for 50 Hz service. This gearbox unit has built-in accessory pads that drive the starter, lube pump, and liquid fuel pump (when required). Figure 10-37 shows a reduction drive assembly.

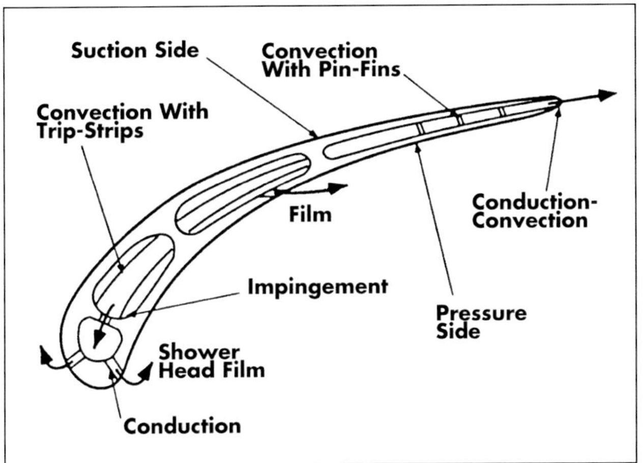

Fig. 10-34 Air Foil Cooling Techniques Using Combinations of Impingement, Convection and Film Cooling. Source: Solar Turbines

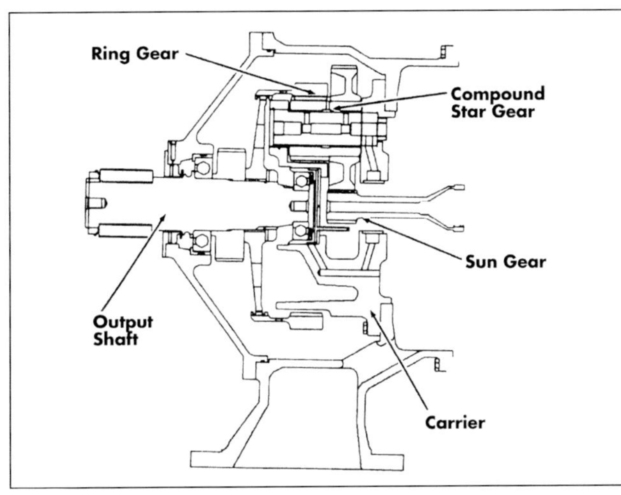

Fig. 10-36 Epicyclic Gear Unit. Source: Solar Turbines

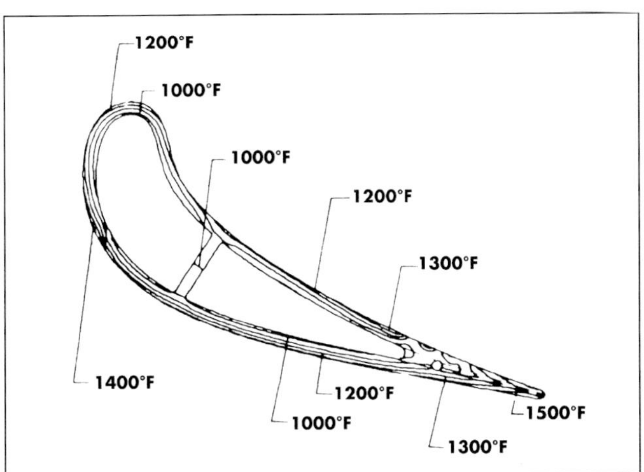

Fig. 10-35 Illustration of Typical Nozzle Vane Showing Isotherms. Source: General Electric Company

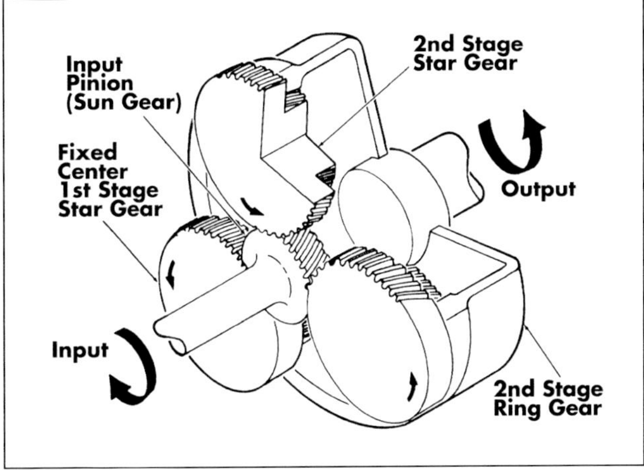

Fig. 10-37 Reduction Drive Assembly. Source: Solar Turbines

CONTROL SYSTEMS

Modern combustion gas turbines are equipped with programmable logic controllers (PLCs) using proprietary software. The system provides required control sequences and monitors the functions required to safely start, load, and operate the unit. The control system warns of out-of-specification conditions and shuts down the unit when more serious out-of-specification conditions are detected. Gas turbines generally require more complex control systems than do other prime movers, particularly during start-up, when a rapid sequence of events occurs. Normally, packaged combustion gas turbine control systems have sufficient capacity to monitor and control the load coupled to turbine output shafting.

Figure 10-38 shows a typical gas turbine control panel. Figure 10-39 shows a layout illustration of a control system for a gas turbine applied for electric power generation on a university campus. The control system features a 32-bit processor, with completely automated start-up and extensive monitoring, logging, and diagnostic capability. The system incorporates triple-monitoring redundant CPUs to provide two-out-of-three voting at every level of the trip string to avoid nuisance trips. Figure 10-40 shows a typical graphics display of the host computer, which provides plant operators with a full spectrum of real-time turbine information.

Fig. 10-38 Typical Gas Turbine Control Panel. Source: MAN GHH

Fuel Supply Pressure

Gas turbine fuel pressure requirements with natural gas can usually be estimated by multiplying the applicable compression ratio by site barometric pressure in psia and adding 100 psi (6.9 bar) to allow for fuel governing system pressure drop. A range of 150 to 750 psig (10.3 to 51.7 bar) has been established for existing commercial models. Some newly introduced, very small capacity turbines can operate at pressures as low as 5 psig (0.35 bar).

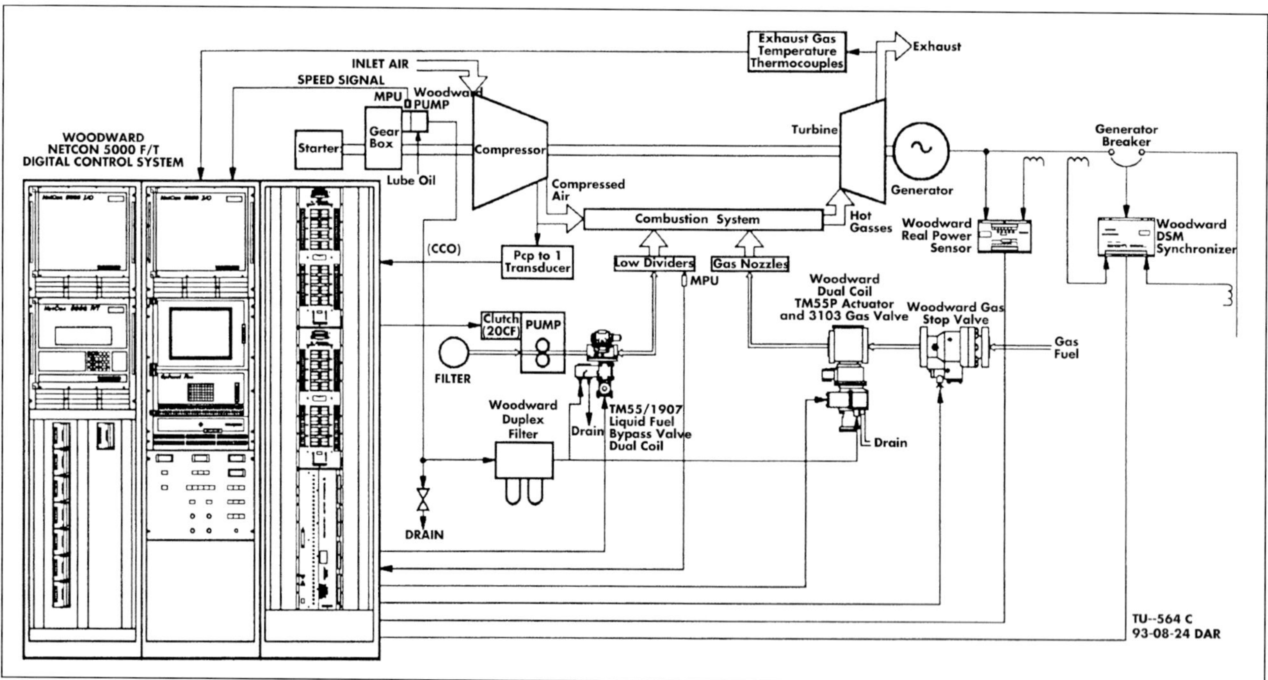

Fig. 10-39 Layout Illustration of a Control System for a Gas Turbine Applied for Electric Power Generation. Source: Woodward Governor

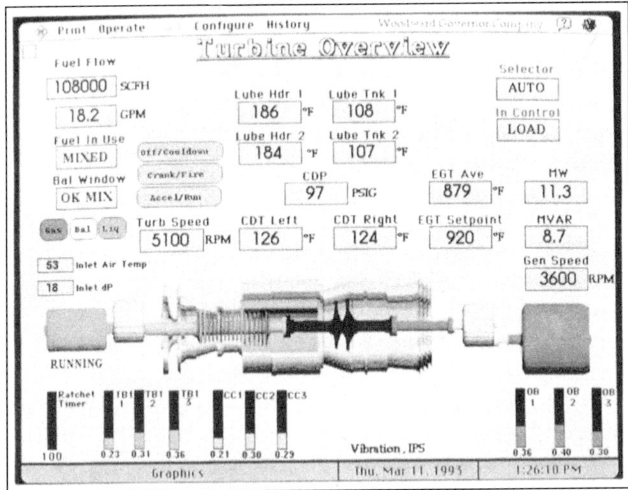

Fig. 10-40 Typical Graphics Display of Real-Time Turbine Information on Host Computer. Source: Woodward Governor

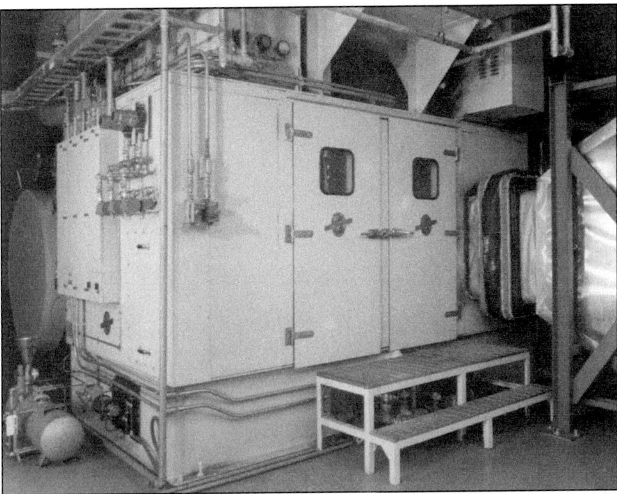

Fig. 10-42 Packaged Enclosure for 5,000 hp Turbine Applied for Mechanical Drive Service. Source: Stewart and Stevenson

Minimum fuel pressure available at-site is a critical consideration factor for gas turbine applications. Fuel compression is expensive in terms of first cost, maintenance cost, and parasitic power consumption, so a less thermally efficient turbine of lower compression ratio may be more attractive when the economics of fuel compression are factored in.

Enclosures & Sound Attenuation

Gas turbine machinery generates sound at the compressor air inlet, outer turbine cases, turbine exhaust, and ancillary equipment, including cooling fans, external pumps, power generators and power transformers. Sound attenuation is available at several levels of control from packagers of gas turbine systems and is usually offered in terms of a guaranteed sound power or sound pressure level at a specified distance from the turbine enclosure. Generally, sound levels of 82 to 85 dba at 3 ft (1 m) distance can be assured with normally available silencers and packaged sound attenuation. Figure 10-41 shows a packaged enclosure for a 40 MW LM6000 gas turbine-generator set. Figure 10-42 shows an indoor packaged acoustical enclosure for a gas turbine with a capacity of about 5,000 hp (3,700 kW) applied for mechanical drive service on a university campus.

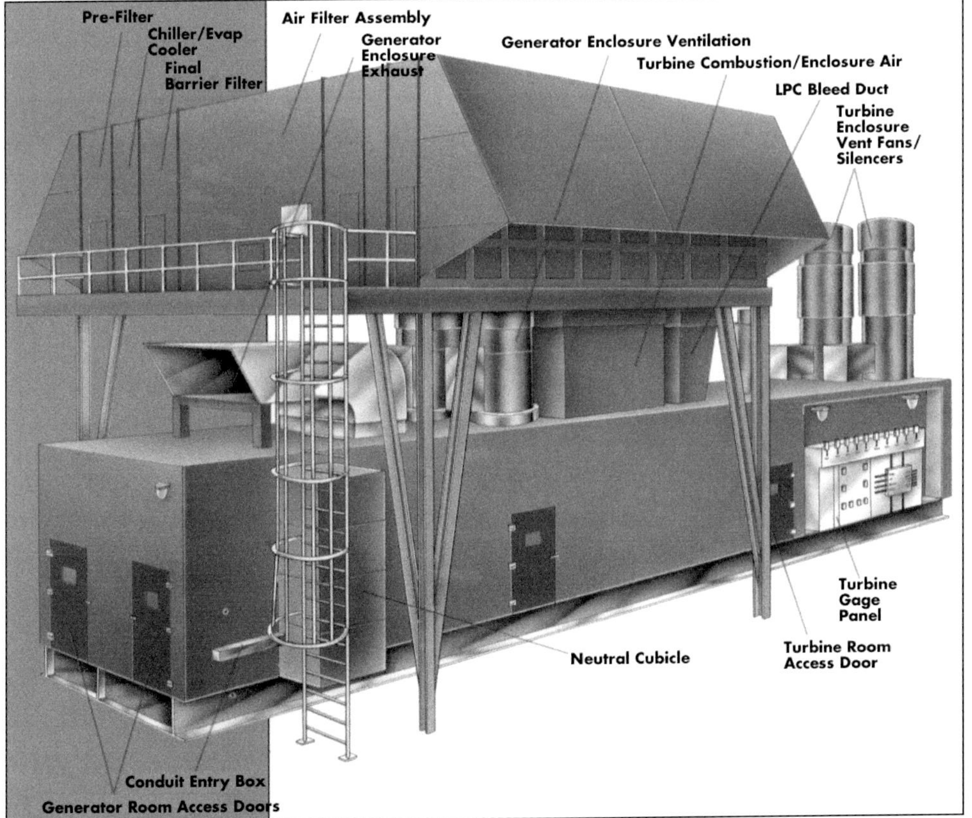

Fig. 10-41 Illustration of Packaged Enclosure for 40 MW Gas Turbine-Generator Set. Source: General Electric Co.

Gas Turbine Performance

Gas turbine thermal efficiencies (shaft power) range from 15% to more than 45% (LHV). A simplified heat balance diagram, showing average annual performance for a 4.7 MW gas turbine cogeneration system, is shown in Figure 10-43.

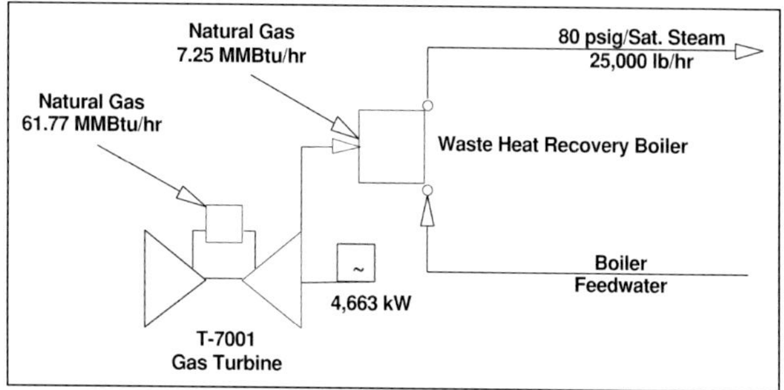

Fig. 10-43 Simplified Heat Balance Diagram Showing Average Annual Performance for 4.7 MW Gas Turbine Cogeneration System. Source: Cogen Designs, Inc.

As demonstrated above in the discussion of the Brayton cycle, in practice, gas turbine thermal efficiency is a function of the pressure ratio, turbine inlet temperature, and component efficiency. Pressure ratio is, in part, a function of the number of stages of compression and available compressor drive power. Larger turbines can be more thermally efficient because they can accommodate more stages. The higher thermal efficiency of large turbines derives from the lower surface-to-volume ratio (resulting in lower friction losses) and from the reduced losses at blade tips and roots.

Gas turbine simple-cycle thermal (electric generation) efficiency and total efficiency (including exhaust heat recovery down to 300°F/149°C) is shown as a function of capacity in Figure 10-44. Depicted are the power generation thermal fuel efficiency and the thermal energy producing efficiency of more than 70 gas turbine cogeneration systems. Figure 10-45 depicts gas turbine heat rate as a function of capacity (in kW) for industrial, aeroderivative, and steam injection (STIG) cycle models.

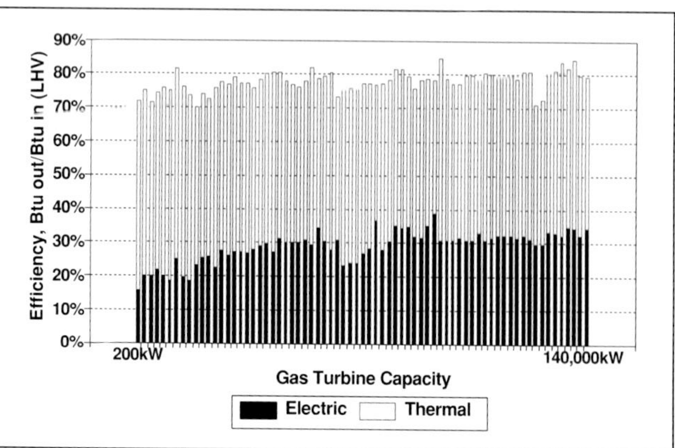

Fig. 10-44 Simple-Cycle and Total Efficiency as a Function of Gas Turbine Cogeneration System Capacity. Source: Cogen Designs, Inc.

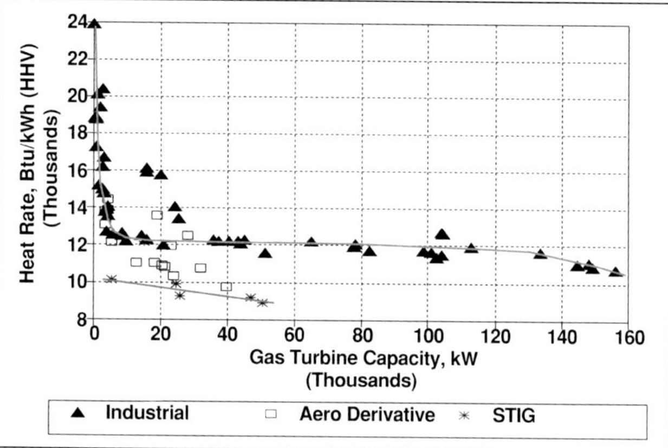

Fig 10-45 Heat Rate vs. Capacity for Industrial, Aeroderivative and Steam Injection Models. Source: Cogen Designs, Inc.

Figure 10-46 shows a plot of output and thermal efficiency for different gas turbine firing temperatures and various pressure ratios. The higher the output per unit mass of airflow, the smaller the gas turbine required for the same

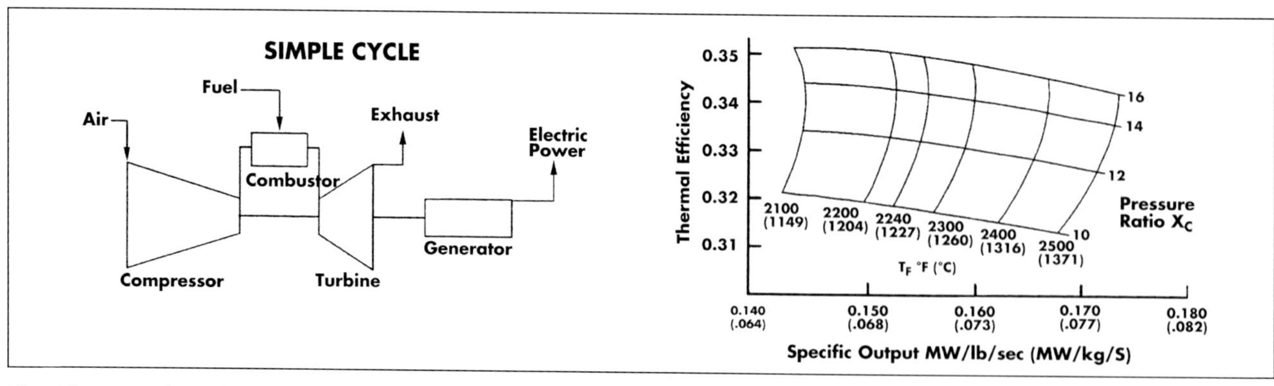

Fig. 10-46 Simple Cycle Gas Turbine Thermodynamics. Source: General Electric Company

output power. Note the importance of cycle-pressure ratio on output and thermal efficiency, and that at each firing temperature, maximum efficiency occurs at a pressure ratio other than that of maximum output. Also note that the pressure ratio resulting in maximum output and maximum efficiency changes with firing temperature, and the higher the pressure ratio, the greater the beneficial impact of increased firing temperature.

Gas turbine thermal efficiency is reduced under part-load operation. Figure 10-47 provides a representative sample of fuel rate versus load for applied turbine designs featuring single-shaft, two-shaft, and single-shaft with IGVs. The IGVs restrict mass flow under part-load operation, thereby improving thermal efficiency relative to the standard single-shaft design. Notice that two-shaft design offers the highest part-load thermal efficiency, increasingly so as loads are reduced.

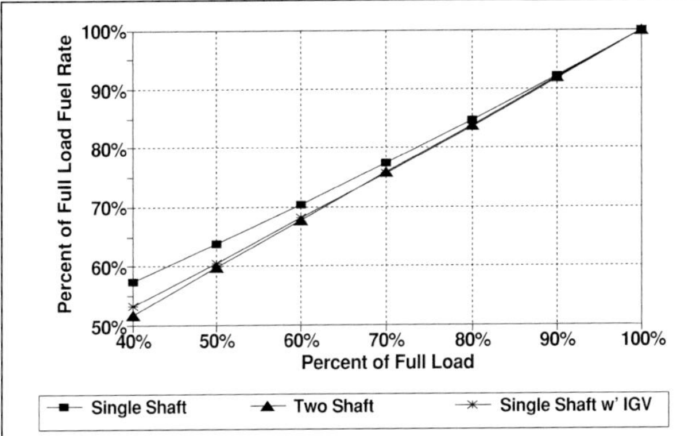

Fig. 10-47 Gas Turbine Fuel Rate vs. Load.
Source: Cogen Designs, Inc.

While in the ideal Brayton cycle, efficiency improves with increased pressure ratio, in actual practice, the effect of increased pressure ratios is linked to turbine inlet temperatures. For each inlet temperature, there is a certain pressure ratio that results in maximum thermal efficiency. Beyond this point, efficiency begins to decrease with increasing pressure ratios. Maximum pressure ratios and inlet temperatures are limited by current technological constraints of material temperature tolerance and cooling system effectiveness.

Turbine efficiency is also a function of turbine airflow design and turbine inlet temperature. Most of the efficiency losses in the turbine section are caused by flow conditions in the blade-to-blade channel of the rotating blades. A critical efficiency issue is the need to balance the impact on efficiency of higher-pressure ratios and firing temperatures with the need for increased cooling airflow.

As firing temperatures increase, some or all of the stationary nozzles and rotating blades must be cooled. As pressure ratios increase, the temperature of the cooling air typically bled from the compressor also increases. For modern industrial turbines, cooling airflow is about 12 to 15% of the total compressor flow. Since allowable component metal temperatures are limited to a maximum of about 1,750°F (954°C), increased cycle efficiency requires some form of turbine part cooling. Compressor air, diverted for cooling turbine parts, is expensive in terms of the energy required to deliver it. The impact of cooling is doubly felt because this air makes no contribution to the combustion or dilution processes in the cycle.

Impact of Ambient Conditions

Fuel pressure available and site ambient temperature and ambient pressure conditions are critical factors in gas turbine performance, more so than with internal combustion reciprocating engines or steam turbines. Since the turbine is a fixed displacement machine, air mass flow varies directly as a function of air density. Factors that can affect overall performance are altitude, ambient temperature and humidity, inlet duct loss, outlet duct loss, and heat recovery device-related exhaust gas back-pressure. Gas turbine power and thermal energy output are adjusted for air density variations from rating conditions and for inlet and exhaust losses.

Ambient air density decreases as altitude increases. This, in turn, decreases mass flow, fuel flow, and capacity of the gas turbine by roughly 3.5% per 1,000 ft (305 m) above sea level. Figure 10-48 is a representative altitude power correction factor curve for a gas turbine in HP/HP versus

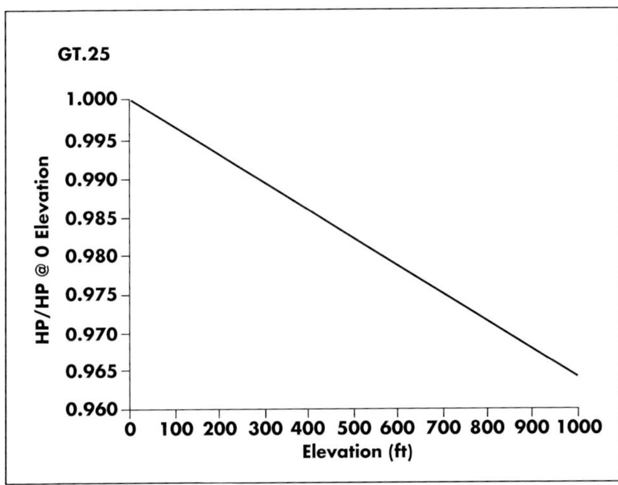

Fig. 10-48 Representative Altitude Power Correction Factor Curve.
Source: Solar Turbines

elevation. At 0 elevation, the HP/HP of 1.0 indicates full capacity. Consistent with the rule of thumb, the power correction of 0.965 HP/HP (or kW/kW) at 1,000 ft (305 m) elevation, indicates a capacity reduction of 3.5%. Altitude impact on exhaust gas temperature is not significant, because turbine inlet temperature is usually the governing factor for full-load operation and is usually held constant. Heat rate variations with altitude are not usually significant, because power and fuel flow variations are roughly proportional as compressor inlet air density varies.

The effect of increased ambient temperature can be dramatic, due in part to decreased air density and the impact of higher fluid temperatures. The compressor absorbs more power when inlet air is hot, leaving a smaller percentage of total power available for the output shaft. A general rule of thumb is that power output will be reduced by about 0.5% per degree F (0.9% per degree C) of increase in ambient temperature. Figure 10-49 is a representative ambient temperature power correction factor curve for a gas turbine in HP/HP at 59°F (15°C) versus ambient temperature. A power correction factor of 1.00 indicates the design nominal power rating based on an ambient temperature of 59°F (15°C). Consistent with the rule of thumb, a 40°F (22°C) increase in ambient temperature to 99°F (37°C) shows a power correction factor of 0.80, indicating a 20% reduction in power output.

Heat rate is also affected slightly as a result of increasing compressor work with increasing inlet air temperature. These levels will vary depending on turbine controls. Figure 10-50 shows relative efficiency as a function of ambient temperature for a gas turbine with a design-rated efficiency of 34.2% (LHV). Notice the modest increase in efficiency as ambient temperature falls and the larger decrease in efficiency as ambient temperature rises above the design point.

Exhaust losses include the pressure drop through the exhaust stack, silencers, and any heat recovery equipment that creates a back-pressure on the turbine. Back-pressure reduces the pressure ratio across the turbine wheels and, therefore, the power output of the turbine. The turbine will lose approximately 0.25% of power per in. H_2O (0.187cm Hg) back-pressure. Heat rate will also increase by about the same percentage.

Inlet depression, or loss, is the pressure drop resulting from restrictions that occur as the outside air passes through the inlet plenum, filter, and silencers. Inlet depression is more problematic than back-pressure because the turbine loses power, both because the inlet air density is reduced and because relative back-pressure is increased. Power loss is approximately 0.5% per in. H_2O (0.187 cm Hg) inlet depression and heat rate will increase by about 0.17% per in. H_2O (0.187 cm Hg) inlet depression. Turbine engine suppliers provide inlet and exhaust correction factors applicable to each design. Figure 10-51 shows representative power correction factor curves for exhaust and inlet loss in HP/HP versus pressure loss.

Inlet depression and exhaust gas back-pressure are under the designer's control, and restrictions can be minimized in some cases. Inlet air-cooling can also compensate for higher ambient temperatures.

Reducing the compressor inlet air temperature can be accomplished using an evaporative cooler or chilled water coil downstream of inlet filters. Care must be taken to prevent condensation or carryover of water to avoid compressor fouling. Evaporative cooling allows inlet air temperature to

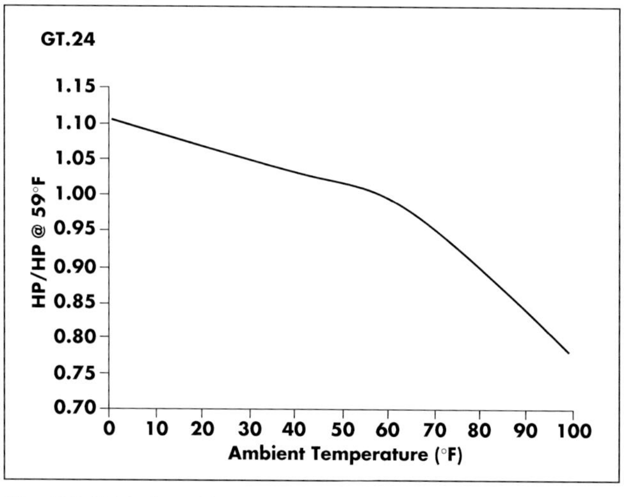

Fig. 10-49 Ambient Temperature Power Correction Factor Curve. Source: Solar Turbines

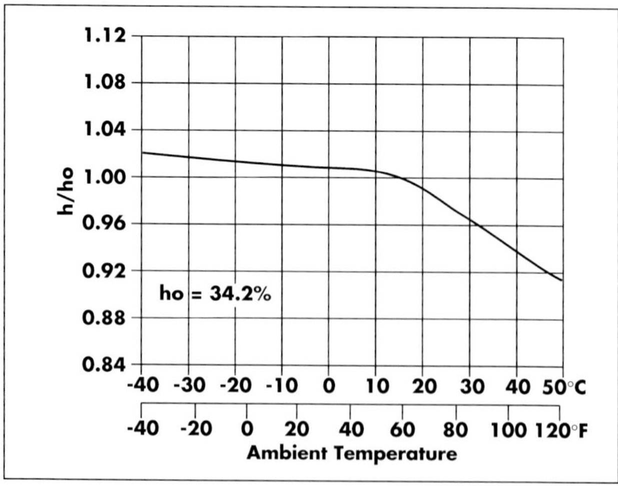

Fig. 10-50 Relative Efficiency as a Function of Ambient Temperature for Representative Gas Turbine. Source: ABB Stal

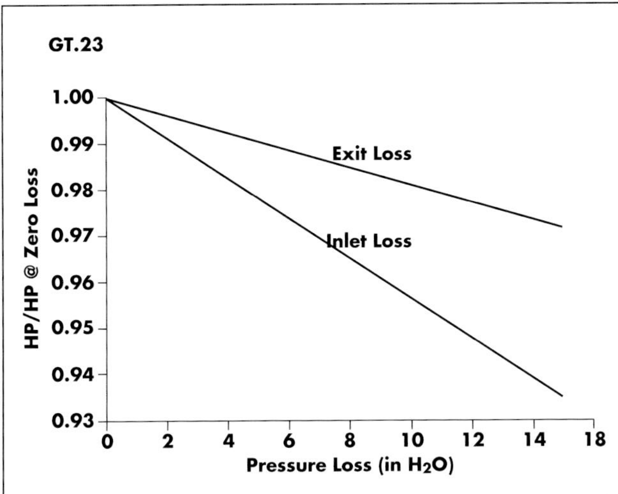

Fig. 10-51 Representative Inlet and Exit Pressure Loss Power Correction Factor Curves. Source: Solar Turbines

approach the ambient wet-bulb temperature, which is significantly below ambient dry-bulb during periods of moderate or low humidity. Thus, the biggest gains from evaporative cooling are achieved in hot, dry climates. Chillers are not limited by ambient wet-bulb temperature and allow a greater inlet air temperature reduction. The effectiveness of a chiller versus an evaporative cooler must be weighed against the capital and operating cost premium of the chiller system. Often, an appropriate option is to use recovered heat to generate chilled water for inlet cooling purposes. In general, if recovered heat is used for inlet cooling, nearly half the cooling capacity is recovered through increased thermal output.

Recently developed water spray injection technology is another means of compensating for capacity and thermal efficiency loss due to elevated ambient temperature. Spray inlet cooling increases mass flow as spray injection nozzles direct water droplets to the compressor inlets, cooling the air during the compression process. While this enhances capacity (and exhaust energy flow) under all conditions, it is particularly effective under very high ambient temperatures, producing a capacity increase of as much as 30% versus the normal derated performance at 90°F (32°C), while also modestly improving heat rate.

Humidity in the inlet air also increases the heat rate and decreases efficiency. Generally, turbine heat rate increases by about 1.0% for every weight percent of moisture in the inlet air. [Psychrometric tables or charts can be used to determine weight percent.]

In addition to these factors, power take-offs, such as gas compressors, oil pumps, hydraulic systems, and fuel pumps, reduce available power output. Devices may utilize gas turbine shaft power or be electrically driven.

As a means of comparison, a set of standard conditions has been established by the ISO. These standards, which have been adopted by most gas turbine manufacturers, are:

Ambient temperature	=	59°F (15°C)
Altitude	=	Sea level – 0 ft. (0 m)
Ambient pressure	=	29.92 in. Hg (101.3 kPa)
Relative humidity	=	60%
Inlet and exhaust losses	=	0 in. H_2O (0 cm Hg)

Based on these ISO standard conditions, correction factors are established based on physical mass flow properties to determine performance for conditions other than ISO, such as those shown in Table 10-1. Based on the specific conditions for a given application, these correction factors can be used to determine power output and heat rate. This will be apparent from manufacturer's data.

The example below demonstrates the method of calculating gas turbine performance at site conditions based on manufacturers' ISO-based ratings and correction factors for conditions other than standard.

Table 10-1

Gas turbine ISO power	=	10,000 hp (7,456 kW)
Gas turbine ISO heat rate	=	7,770 Btu/hp-h (10,992 kJ/kWh) LHV Basis
Altitude (above sea level)	=	500 ft (152 m)
Ambient temperature	=	69°F (21°C)
Inlet depression	=	4 in. H_2O (0.747 cm Hg)
Exhaust back-pressure	=	2 in. H_2O (0.374 cm Hg)
Relative humidity	=	60%

Using Figure 10-52, the correction factors are:

Condition	Power	Heat Rate
Altitude	0.983	N/A
Ambient temperature	0.956	1.015
Inlet depression	0.984	1.007
Exhaust back-pressure	0.997	1.003

The corrected power output and heat rate at the site conditions, respectively, would be:

Power = 10,000 x 0.983 x 0.956 x 0.984 x 0.997 = 9,219 hp (6,873 kW)

Heat rate = 7,770 x 1.015 x 1.007 x 1.003 =7,966 Btu/hp-h (11,269 kJ/kWh)

The curves in Figure 10-52 show the nominal rated performance for a continuous-duty, ISO-matched, gas-

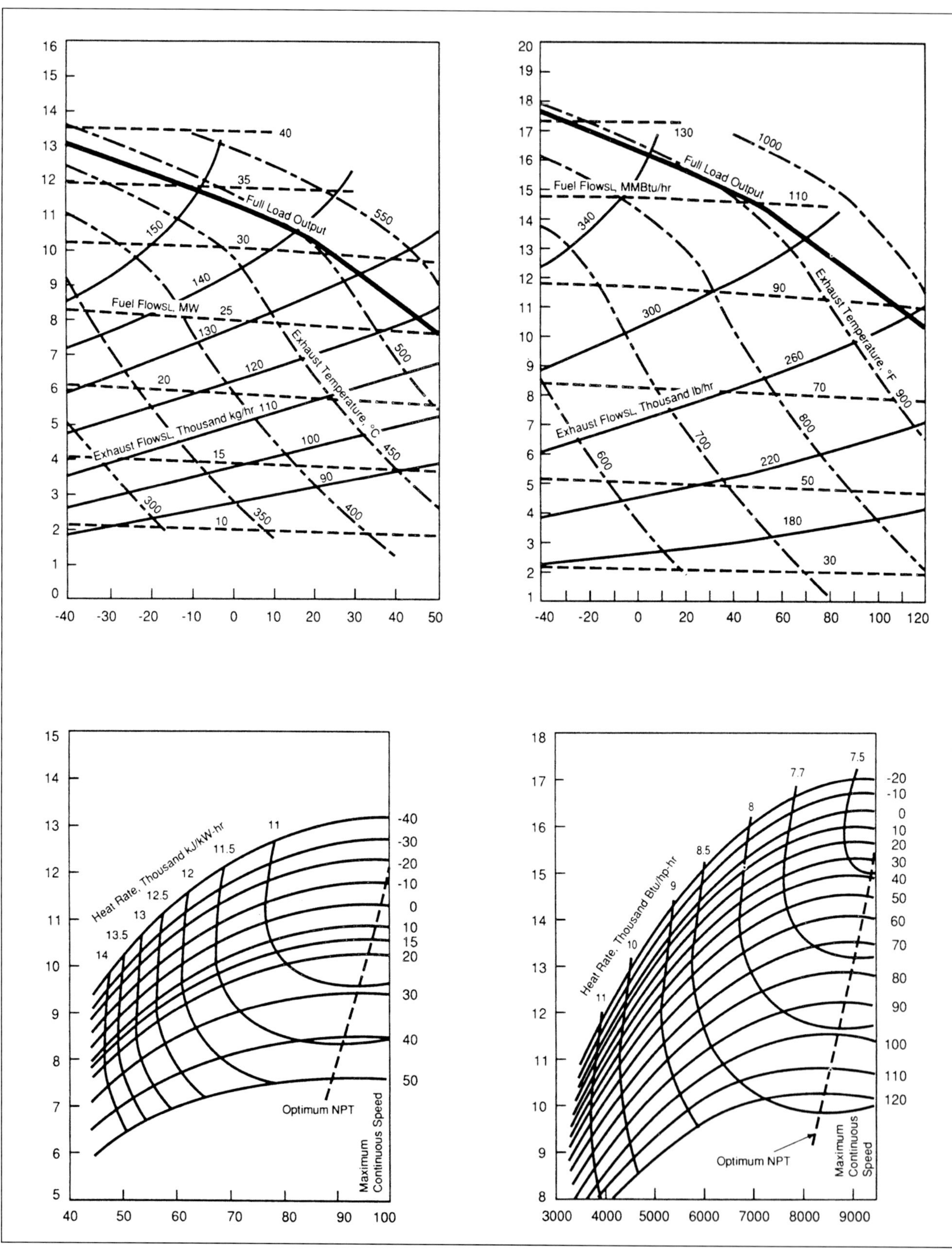

Fig. 10-52 Nominal Rated Performance for Continuous Duty, ISO Matched Gas Fueled, Gas Turbine. Source: Solar Turbines

fueled Mars T-14000 gas turbine mechanical drive or compressor set in English and SI system values. The top two sets of curves relate output power, fuel flow, exhaust temperature, and exhaust flow to inlet air temperature at optimum power turbine speed. The bottom two sets of curves relate output power and heat rate to inlet air temperature at various power turbine speeds. This gas turbine, shown in Figure 10-53, features a 15-stage axial compressor that achieves a compression ratio of 16:1 at a design rated speed of 10,780 rpm. The gas producer turbine operates at the same design speed and features two stages. The power turbine also has two stages and operates at a maximum continuous speed of 9,500 rpm.

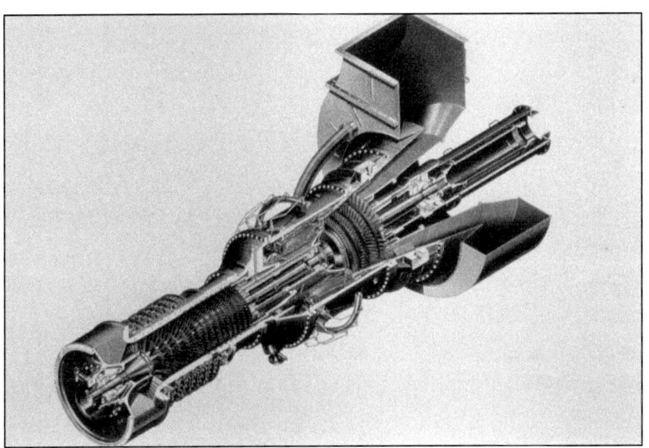

Fig. 10-53 Mars T-14,000 Gas turbine. Source: Solar Turbines

Combustion Gas Turbine Exhaust and Heat Recovery

Gas turbine advantages over reciprocating engines include the quality and capacity of recoverable rejected heat. While generally offering somewhat lower simple-cycle thermal fuel efficiency than reciprocating engines, particularly under partial loads, gas turbines produce more recoverable heat. Moreover, while industrial reciprocating engines deliver a portion of their recoverable heat in the form of low-grade heat rejection (from jacket water, lubricating oil, and intercooler heat rejection), all of the recoverable heat produced by a gas turbine is in the form of high-temperature exhaust. Thus, while only the exhaust gas from a reciprocating engine can be used to generate high-pressure steam, all of the recoverable heat from a gas turbine can be used for this purpose.

Gas turbines also generally produce exhaust gases with somewhat higher oxygen content than reciprocating engines, though some lean-burn reciprocating engines also produce high-oxygen content exhaust. Gas turbines draw in three to four times the air required for combustion, with some of the air used for turbine cooling. Cooling air mixed with combustion gas exits as a clean, dry, oxygen-rich stream at temperatures of approximating 1,000°F (490°C). Supplementary firing with duct-burners can increase heat recovery capacity and efficiency by increasing the temperature of the exhaust.

Gas Turbine Exhaust Characteristics Under Part-load Operation

Reducing fuel input to the gas turbine under part-load operation causes a reduction in exhaust gas temperature, particularly with single-shaft turbines, which maintain constant exhaust flow. Some applied turbine designs feature IGVs, which reduce mass flow under part loads, thereby maintaining relatively higher exhaust temperatures.

Figures 10-54, 10-55, and 10-56 provide representative samples of gas turbine exhaust performance characteristics as a function of percent of full-load operation for applied turbine designs featuring single-shaft, two-shaft, and single-shaft with IGV. Figure 10-54 depicts recoverable exhaust energy versus load. Figures 10-55 and 10-56 depict exhaust flow versus load and exhaust temperature versus load, respectively.

Figure 10-54 indicates a fairly linear relationship for all three applied designs, with the percent of full-load recoverable exhaust energy being slightly greater than the corresponding percent of full-load. Notice in Figure 10-55 that exhaust flow remains constant with the single-shaft design. The two-shaft design shows a fairly linear relationship, though exhaust flow is disproportionately high throughout the entire range. The single-shaft design featuring IGVs shows a fairly linear relationship down to 80% of full-load operation and remains constant at lower loads.

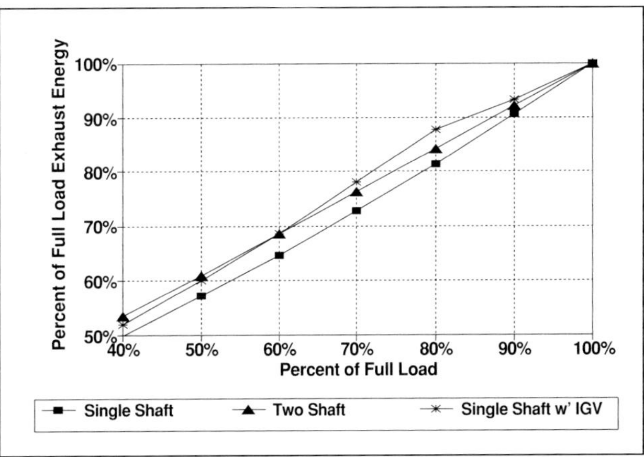

Fig. 10-54 Recoverable Gas Turbine Exhaust Energy vs. Load. Source: Cogen Designs, Inc.

Notice the relationship between the exhaust temperature in Figure 10-55 and the exhaust flow in Figure 10-56.

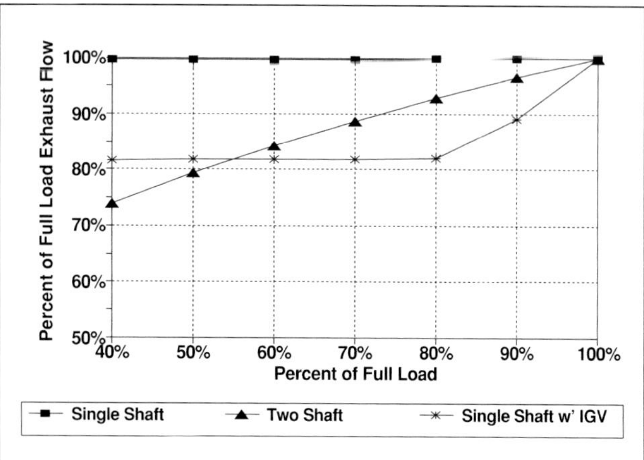

Fig. 10-55 Gas Turbine Exhaust Flow vs. Load. Source: Cogen Designs, Inc.

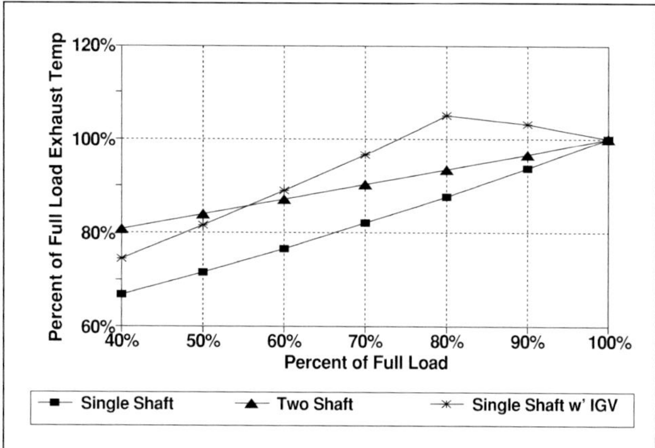

Fig. 10-56 Gas Turbine Exhaust Temperature vs. Load. Source: Cogen Designs, Inc.

Cycle Modifications & Enhancements

In addition to improved high-temperature materials, there are several variations to the gas turbine cycle that can be used to increase efficiency and unit capacity:

Gas turbine combined cycle: In this cycle, superheated steam generated from turbine exhaust in a heat recovery steam generator (HRSG) is supplied to a steam turbine to produce additional power. Typically, power output and overall thermal efficiency can be increased by up to 50%. Combined-cycles are discussed in detail in Chapter 12.

Steam (or water) injected gas turbine cycle: As discussed, the largest share of energy collected by the turbine section is used by the primary air compressor, leaving a relatively small fraction of the total available for export work. Because of this, relatively small changes in the ratio of compressor work to export work result in significant changes in the useful turbine output. One way to change this balance is to make the mass flow passing through the turbine greater than that passing through the compressor. For example, if the compressor takes two-thirds of the work done by the turbine and an increase in turbine mass flow increases overall power production by 10%, the effective increase in export work is 30%. This effect can be seen when steam is injected into the turbine combustor for emissions control. The small increase in exhaust mass flow increases power output of the turbine, usually by 3 to 10%.

Some gas turbine designs lend themselves to modifications that allow injection of significantly greater quantities of steam (or water) specifically for power augmentation. The result is that the gas turbine functions in a somewhat similar fashion to a combined cycle, which uses both gas and steam turbines to produce power. This process is frequently referred to by trade names as the Cheng™ or STIG™ cycle. Examples of these turbines and their relative power increases with steam injection include the Kawasaki GPCC15 (1.3-2.3 MW), Allison 501 KH (3.6-5.4 MW), the Mitsubishi MF-11AB (6-11 MW), and the General Electric LM2500 (22-28 MW). Depending on the turbine design, power augmentation can increase capacity by up to 50% while also increasing thermal efficiency. Steam-injection cycles are discussed in detail in Chapter 12.

Air bottoming cycle: In this cycle, hot gas turbine exhaust (or other process exhaust) is passed through an air-to-air heat exchanger. The heated air is then used to drive an air power turbine. Combined, the power output and efficiency of this type of combined cycle is about 30% greater than the Brayton cycle. While efficiency and total power output are less than with a conventional combined cycle, the use of a bottoming-cycle air turbine is simpler and less costly.

Regenerative cycle: A regenerator or recuperator is a heat exchanger that transfers heat from turbine exhaust to compressor discharge air prior to combustion. Recovered heat displaces a portion of fuel that would otherwise be required. Regeneration usually decreases unit capacity (as a result of heat exchanger pressure drops), but enhances cycle efficiency.

Regeneration causes the exhaust gases ultimately leaving the turbine to be much cooler than they would otherwise be. The reduced energy and temperature of the exhaust stream results in decreased heat recovery potential. Whereas the typical gas turbine exhaust temperature is about 1,000°F (538°C), the ultimate exhaust from a recuperative turbine is about 600°F (316°C). Both the quantity of recoverable heat and the maximum achievable temperature and pressure of heat recovery generated steam are greatly reduced. The simple open-cycle gas turbine cycle with regenerator is shown schematically in Figure 10-57 and in T-s diagram in Figure 10-58.

For the ideal cycle with regeneration, thermal efficiency depends not only on pressure ratio, but on the ratio of the minimum to maximum temperature as well. Figure 10-59 illustrates the variation of efficiency with cycle pressure ratio for two values of turbine inlet temperature (T_{T4}), and several values of recuperator effectiveness $\mathcal{E}_r$. The curves labeled $\mathcal{E}_r$ correspond to the overall efficiency of a simple-cycle gas turbine. As pressure ratios are increased for a given turbine inlet temperature, the compressor outlet temperature is increased. When it is equal to the turbine outlet temperature, there is no heat to return (recuperate) back to the air. This is where the crossover point occurs. Operating conditions to the left of this point (i.e., moderate pressure ratios) allow for efficiency enhancement to be achieved with the use of a recuperator.

The decision to use a regenerator largely depends on whether or not heat recovery is being used and, if so, on the relative values of increased power output versus decreased thermal energy output. Figure 10-60 is a 4.6-MW Solar

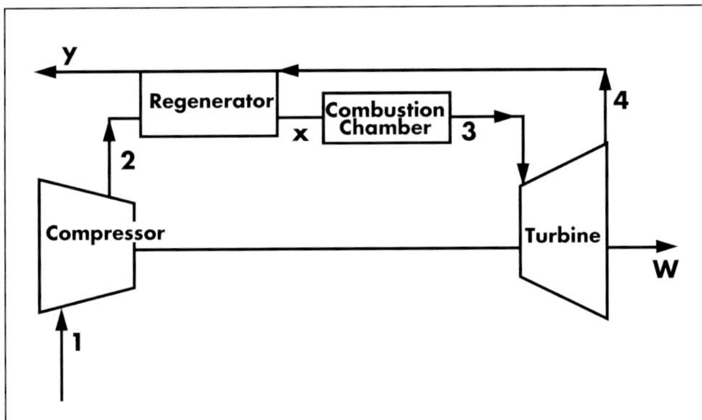

Fig. 10-57 Simple Gas Turbine Cycle with Regenerator.

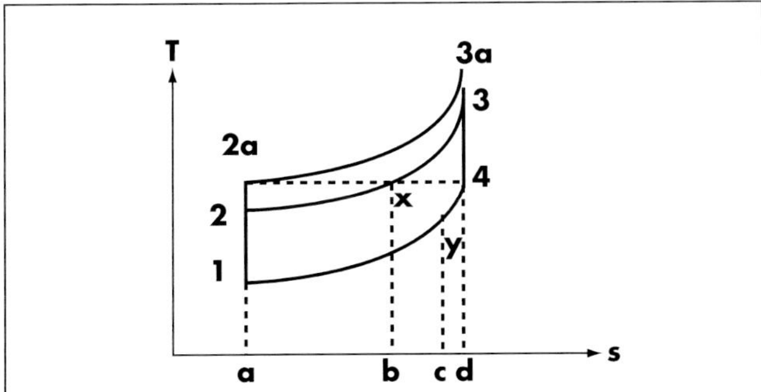

Fig. 10-58 Ideal Regenerator Cycle.

Mercury 50 unit that has a recuperator as part of the standard package that reduces the exhaust temperature to 710 F (377 C). This recuperator decreases the heat rate to 8,863 BTU/kWh (9,351 KJ/kWh) by pre-heating the combustion air after the compression process but before the combustor.

Reheat Cycle: The reheat cycle uses an additional combustor or reheat element in which additional fuel is combusted using the oxygen present in the exhaust gas. The reheat cycle increases the efficiency of the turbine by increasing the average temperature of the gases doing

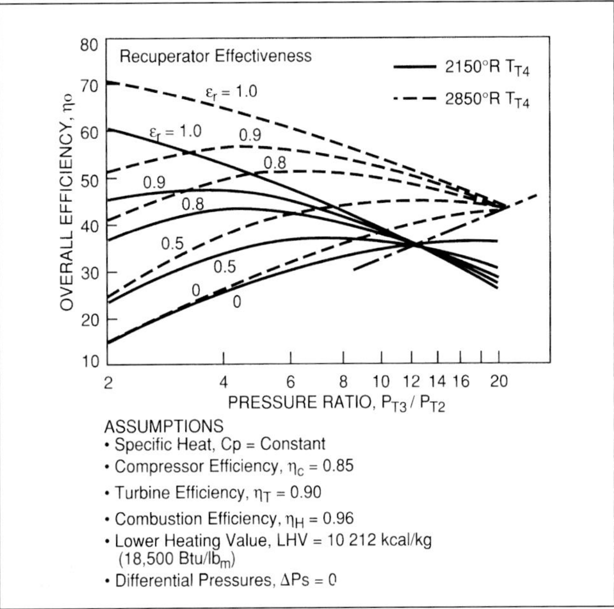

Fig. 10-59 Recuperated Gas Turbine Efficiency.
Source: Solar Turbines

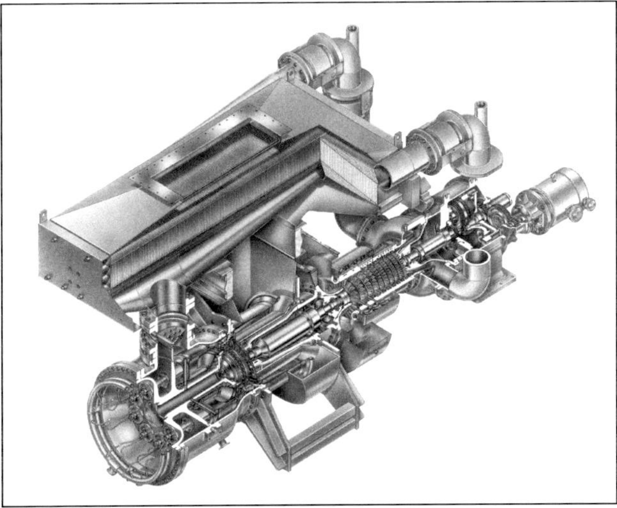

Fig. 10-60 Solar Mercury 50 Gas Turbine with standard recuperator.
Source: Solar Turbines.

expansion work in the turbine section.

In the reheat cycle, combustion gases are only partially expanded in the power turbine, or gas generator. At some intermediate pressure, these gases are exhausted from the turbine and diverted to the reheater, where additional fuel is burned to increase the inlet temperature to its original temperature. The elevated temperature gas, which is at a pressure between atmospheric and the compressor exit pressure, is then passed on to the second section of the expansion turbine, where it produces additional work.

Intercooling cycle: As shown in Figure 10-61, with intercooling, the compressor is split into two or more sections to decrease the average temperature of the air in the compressor. An intercooler is simply a heat exchanger through which air exiting the low-pressure compressor passes before entering the high-pressure compressor. Typically, compressor discharge temperatures are about 900 to 1,000°F (482 to 538°C). By cooling the air at an intermediate point in the compression cycle, the outlet temperature is reduced by a few hundred degrees, air density is increased, and compression power is reduced.

Intercooling is most effective for high-pressure ratio compressors, because compressor input power depends on pressure ratio. Since intercooling also lowers compressor discharge temperature, the potential capacity of a gas turbine with intercooling is greater than the same unit without intercooling. Potential unit capacity with intercooling is higher because compressor work is reduced and the lower compressor discharge temperature allows more energy to be introduced in the combustor for a given maximum turbine inlet temperature. A small additional benefit may be derived because cooling air for turbine blades and vanes is cooler from the intercooled compressor and part cooling is potentially more effective.

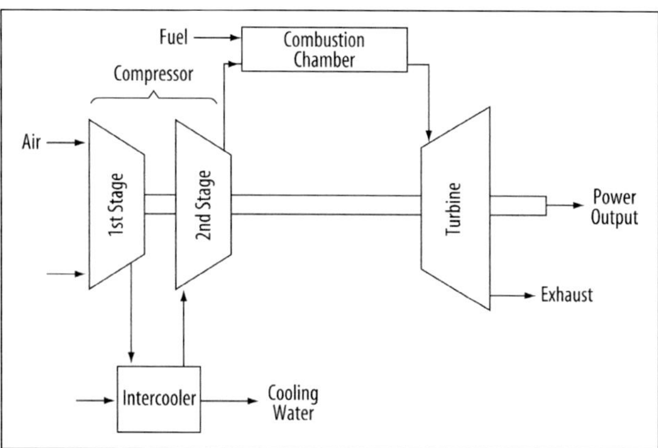

Fig. 10-61 Simplified Diagram of Gas Turbine Compressor Intercooling Cycle.

The General Electric (GE) LMS100 represents a hybrid design utilizing technologies from both GE's industrial (Frame) and aeroderivative gas turbine units. This new 134,000 hp (100,000 kW) unit achieves a 46% simple cycle efficiency (LHV) using intercooling between compression stages and a 42:1 cycle pressure ratio.

Maintenance

When applied, maintained, and properly operated, gas turbine technology offers high reliability and low life-cycle maintenance costs. Successful operation requires hands-on understanding of model characteristics, monitoring systems, important operating limits and ranges, safety hazards, tolerance for starting transients, lubrication requirements, fuel and air quality requirements, capacity, cooling requirements, and control interactions. The following points are significant:

- Rapid thermal transients associated with start-up and rapid load increases and decreases are life-reducing events for gas turbines and should be avoided to the extent possible. Good operations and planning can minimize the need for thermal transients associated with unnecessary starts and stops.
- Fuel quality must be maintained to assure maximum life and performance from any gas turbine. Equipment damage and degraded performance may result from a number of causes, including poor quality fuel.
- Performance indicators that require shutdown include off-line water washing, fire detection, fuel leaks, lubricating oil leaks, control system malfunction, low operating fluid levels, and driven load hazards.
- Monitoring the quality of lubricating oil and management of injection water quality are needed, along with appropriate corrective action when problems arise.
- Provisioning and inventory maintenance of spare parts needed to support the operation are a significant part of good operations management.

Turbine suppliers offer training and instruction manuals with correct operating procedures. Third-party firms also offer services that include system operation, usually for a fee, a share of the savings they generate, or both.

Viable model rating and service life depend heavily on site conditions, including compliance with air and fuel quality specifications. Fuel- and air-born alkali metals in trace quantities with small amounts of fuel-born sulfur are particularly destructive to the high-temperature alloy materials used for combustion and hot turbine parts. This so-called **hot corrosion** can render a 30,000-hour life turbine unserviceable in less than 2,000 operating hours. Selection of pipeline quality, natural gas fuel limits the potential for fuel-bound sulfur and reduces or eliminates the impact of alkali metals.

Air-born sodium (in common salt) and other alkali metals can be controlled by application of high-performance primary air filters with moisture barriers as required. Further protection is offered by properly scheduled off-line water washing of the compressor section to periodically remove contamination not prevented by the primary air filters.

Axial compressors tend to separate contaminants from primary air and store previously dissolved solids as a coating in the compressor air path. This separation results when evaporation liberates dissolved solids from water droplets in the primary inlet air as temperature rises in the compression process. Evaporated water vapor continues with primary air, while solids settle and coat compressor surfaces. As the process continues, the coating thickness increases until it can no longer adhere to air path surfaces and chunks are reintrained with primary airflow. Concentrations of sodium and calcium are introduced with primary air into the combustor, slugging the hot gas path with vaporous alkali sulfates. Vaporous contaminants condense in the turbine as work extraction lowers gas temperatures, and the resulting globs impact and stick on stator and blade surfaces, starting the irreversible hot corrosion process.

Almost all distillate fuels contain trace quantities of sulfur. Further, improper handling after refining can contaminate distillates with alkali metals as well. Barge transport in salt water and fuel contact with transport tankers, storage tanks, and piping with residual or crude oil residues are

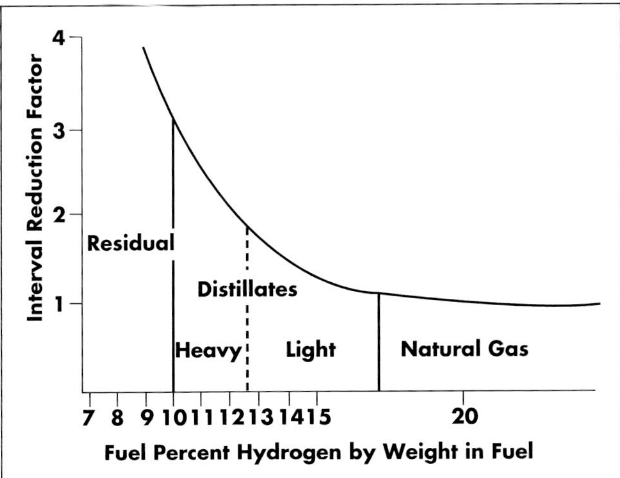

Fig. 10-62 Estimated Effect of Fuel Type on Gas Turbine Maintenance. Source: General Electric Company

examples of fuel quality contamination sources. Appropriate remediation steps can eliminate these sources of contamination. When heavy fuel oil is used, Vanadium is also a concern and usually requires treatment with additives.

Figure 10-62 is an estimate of the effect of fuel type on gas turbine maintenance. Natural gas is considered the optimum fuel and establishes a baseline maintenance factor of 1. Heavier hydrocarbon fuels show a maintenance factor ranging from 3 to 4, due largely to a frequently contained corrosive element and a higher amount of radiant thermal energy, which results in a subsequent reduction in combustion hardware life. Distillate fuels are shown to have maintenance factors ranging from the baseline up to as high as 3, depending largely on contaminants picked up during transportation.

Three basic steps can prevent hot corrosion:

1. Select a turbine model with the best available alloys and coating, as demonstrated by successful operation in field applications at the capacity and efficiency required to meet economic performance expectations.

2. Analyze the installation site to determine which hazards exist.

3. Select the design features required to produce economical operation at site conditions by eliminating or reducing hazards to acceptable limits.

Specific practices for preventing hot corrosion and preserving performance include:

- Careful study of model capacity ratings. Some manufacturers have recognized the benefits of natural gas firing by increasing their model capacity ratings by several percent for gas fuel operation. These higher ratings offer more capacity, higher thermal efficiency, and lower first cost per hp (kW) without increasing the cost per hp-h (kWh) for maintenance, increasing maintenance outages, or decreasing expected turbine life.
- For operation on distillate fuels, application of filter systems featuring coalescing stages followed by 1 to 3 micron particulate control stages for the removal of water-born alkali metals.
- Compressor water washing, which has been demonstrated in field experience to be useful in recovery of performance lost to compressor fouling. On-line water washing benefits applications where economics demand maximum on-line service time and maximum life-cycle efficiency. Alternatively, off-line water washing, which removes alkali coating and carries it out of the machine in solution through cleaning solution drains, offers superior performance recovery.

Gas turbines can operate for years without a single interruption of service. They are, however, typically shutdown every three to six months for a few hours for detailed inspection and routine maintenance. This includes inspection of fuel nozzles, thermocouples, and safety and control calibrations, as well as checking of the plenum drain valve and cleaning air/oil separator.

Monthly inspections that do not require shutdown are also required. These inspections include checking fittings and oil levels and inspecting for leaks, checking, cleaning, or replacing filters, and checking all instrumentation. Daily inspections (quick visual inspection and reading of instruments) and periodic vibration testing should also be done.

Time between overhaul (TBO) is related to the quality of the preventative maintenance program, the type and quality of fuel used, the number of starts, the percent of full power rating at which the engine is operated and the quality of the turbine components, design, and construction. The TBO may range from 15,000 hours to more than 100,000 hours.

Routine maintenance costs are very low. Repair and overhaul, however, can be costly. Typically, the life-cycle maintenance cost will range from $0.002 to $0.005 per hp-h ($0.0007 to $0.013 per kWh). Complete service contracts, including turbine engine replacement or overhaul insurance, can limit a facility's responsibility to visual inspection and recording of control and performance data. Remote monitoring can further reduce in-house responsibilities.

Currently Available Combustion Gas Turbine Designs

Following are descriptions of several gas turbines covering a wide range of capacities and designs and highlighting some basic design and performance characteristics. Power output capacity and thermal fuel efficiency curves and figures are presented for several models.

Figure 10-63 is a disassembled view, and 10-64 is an assembled cutaway view of a GE LM2500 aeroderivative gas turbine with a design capacity of about 32,000 hp (23,860 kW). This unit features a 16-stage axial-design compressor driven by a two-stage, high-pressure turbine, and produces a pressure ratio of 18:1. Power is extracted through a radial drive shaft at the forward end of the compressor. Output power is transmitted to the load by means of a coupling shaft on the aft end of the power turbine rotor shaft. The bare gas turbine is 14.1 ft (4.3 m) long and 5.0 ft (1.5 m) high and weighs only 10,300 lbm (4,672 kg). The low rotor inertia, inherent in this aeroderivative design, results in low-starting torque and power requirements, which reduces the size of the associated starting system (e.g., electric motor, pneumatic media storage, or engine-driven hydraulic system). The starting torque for this unit is less than 750 ft-lbm (1,017 N-m). The unit's power turbine operates at either 3,000 or 3,600 rpm, allowing for direct, gearless coupling to 50 Hz and 60 Hz generators.

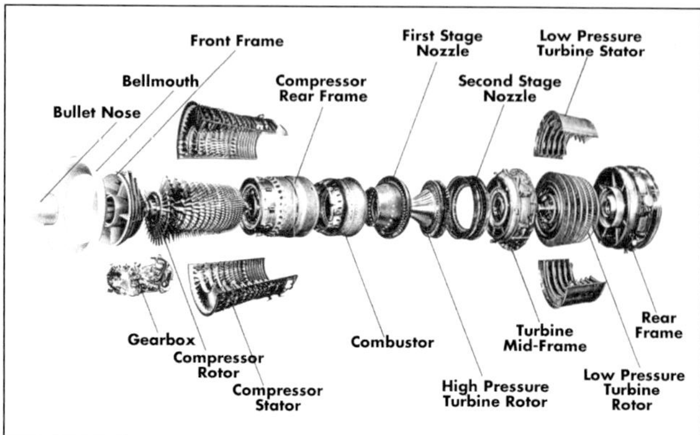

Fig. 10-63 Disassembled View of LM2500 Aeroderivative Gas Turbine. Source: General Electric Company

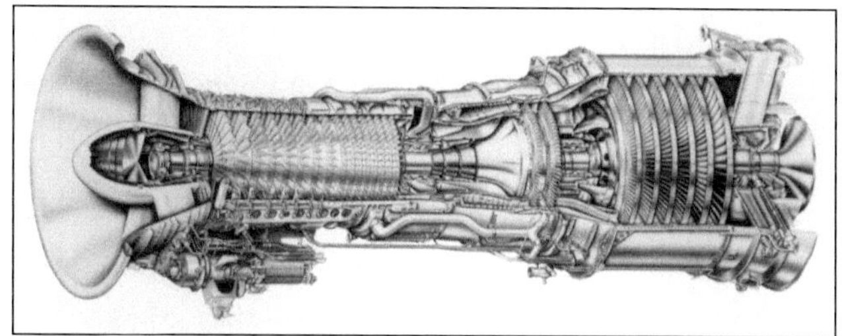

Fig. 10-64 Cutaway Illustration of LM2500 Aeroderivative Gas Turbine. Source: General Electric Company

Figure 10-65 shows a rotor of a large-capacity industrial turbine undergoing dynamic full-speed balancing and overspeed testing. The blading of the 16-stage compressor is shown on the right and the blading for the four-stage turbine is shown on the left. Figure 10-66 is an open-case model showing location of the rotor, which is perpendicular to the silo combustor.

Figure 10-67 is an X-ray picture of an ABB GT10 gas turbine with a design-rated capacity of about 35,000 hp (26 MW). For a heavy-duty industrial turbine, this unit offers a relatively lightweight and compact design. The unit features a 10-stage compressor (on the right), a dry low-NO_X combustor, and the two-stage compressor turbine. The two-stage power turbine (on the left) drives the load.

Figure 10-68 is a labeled cutaway illustration of a

Fig. 10-65 Rotor Undergoing Balancing and Testing with Compressor Blading on the Right and Turbine Blading on the Left. Source: ABB

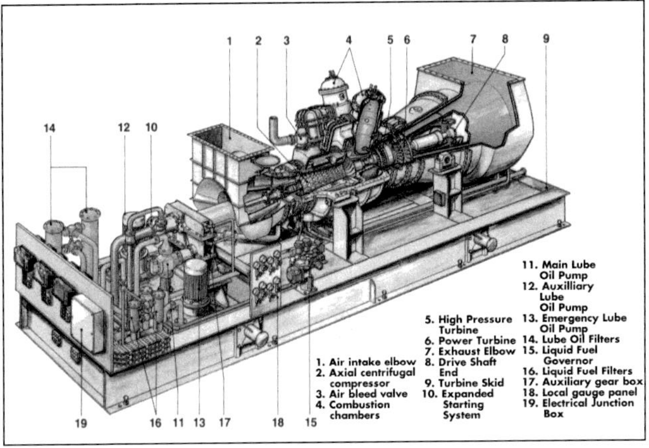

Fig. 10-68 Cutaway Illustration of MAN GHH Industrial Gas Turbine. Source: MAN GHH

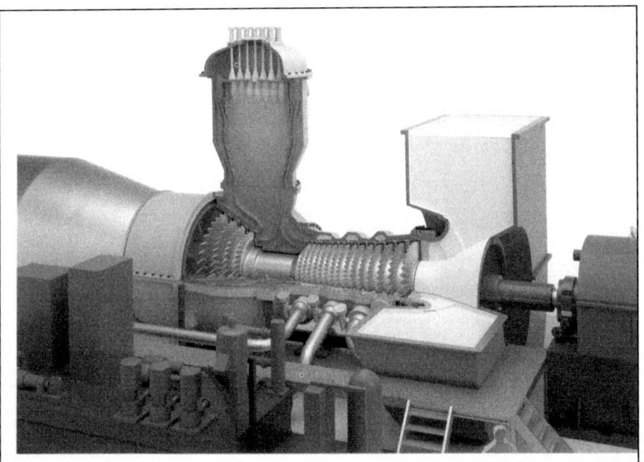

Fig. 10-66 Open-Case Model Showing Location of Rotor and Silo Combustor on Heavy-Duty Industrial Gas Turbine. Source: ABB

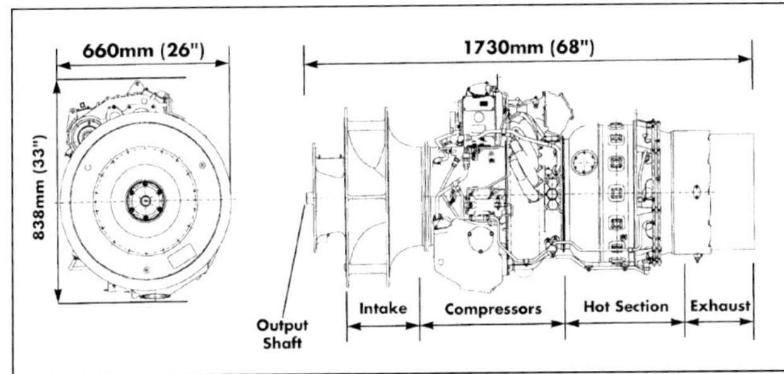

Fig. 10-69 Cutaway Views of ST18 2,660 hp (1,984 kW) Aeroderivative Gas Turbine. Source: United Technologies Pratt and Whitney Canada

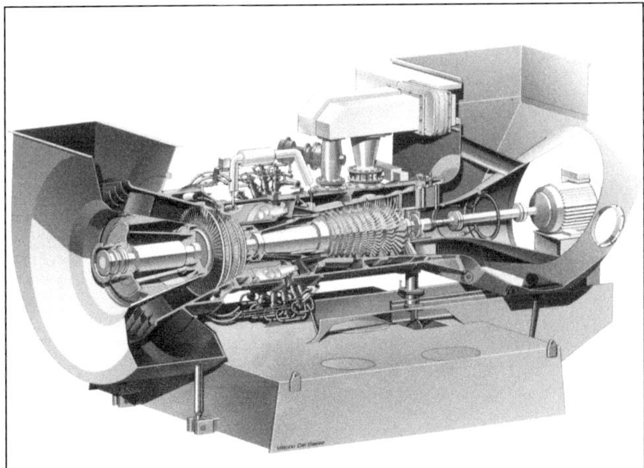

Fig. 10-67 X-ray Picture of 35,000 hp (26 MW) Gas Turbine. Source: ABB Stal

skid-mounted, packaged MAN GHH industrial gas turbine, which has a power capacity of about 12,400 hp (9,250 kW). The standardized, skid-mounted design of ancillary equipment is well-suited for mechanical drive service application in remote locations with hostile environments.

Figure 10-69 shows cutaway views of the ST18 aeroderivative gas turbine. At a dry weight of 772 lbm (350 kg), this unit achieves a baseload capacity rating of about 2,660 hp (1,984 kW) at 19,000 rpm, with a thermal fuel efficiency of about 30% (LHV basis) under uninstalled, ISO sea-level conditions. The turbomachinery comprises three independent rotating assemblies mounted on concentric shafts. The compressor is made up of two centrifugal impellers, each driven by a single-stage turbine. Power is extracted by a two-stage turbine and delivered to the inlet end of the engine, as shown. The combustion system comprises a reverse-flow annular combustion chamber, 14 fuel nozzles, and two spark igniters.

Figure 10-70 shows a two-shaft aeroderivative Allison

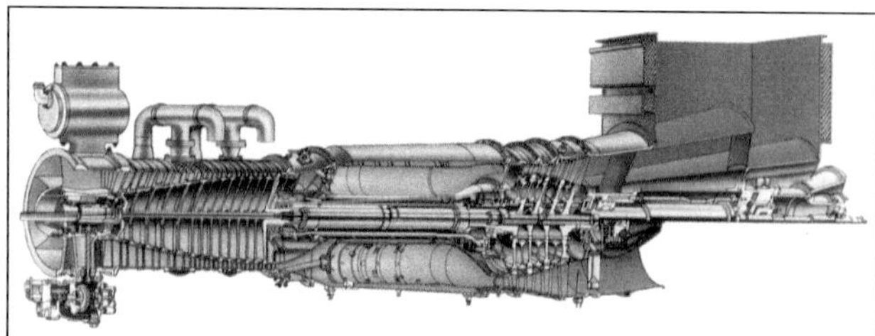

Fig. 10-70 Two-Shaft Aeroderivative Allison Gas Turbine. Source: Stewart and Stevenson

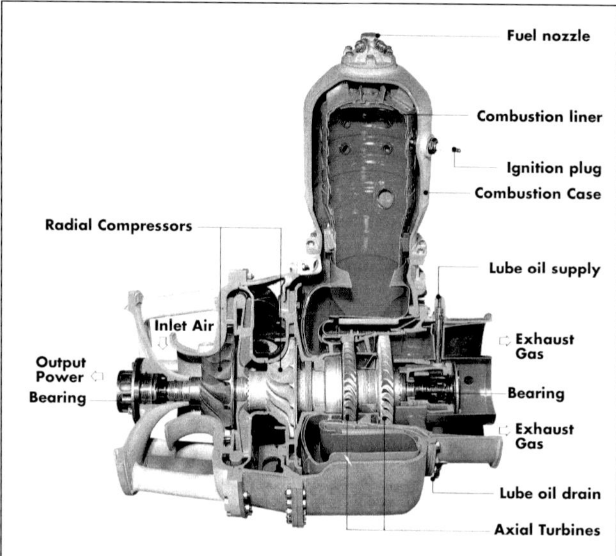

Fig. 10-71 Small-Capacity Kawasaki Turbine Showing Radial Compressors and Axial Turbines. Source: United States Turbine Corp.

gas turbine with a capacity of about 5,000 hp (3,700 kW). This unit features a modular design in which major subassemblies are interchangeable. The mechanical dimensions for this gas turbine are 15 ft (4.6 m) long by 8 ft (2.4 m) wide by 9 ft (2.7 m) high. Figure 10-71 shows a small-capacity Kawasaki aeroderivative industrial gas turbine. Radial compressors are shown on the right and axial turbines are shown on the left. The combustion case is in the center with the fuel nozzle shown at the very top.

Figure 10-72 is a labeled cross-sectional illustration of a single-shaft Solar Saturn gas turbine featuring cold-end drive; it is suitable for generator sets operating at constant speed. On the left of the annual combustor is the eight-stage axial compressor. To the right of the annular combustor section is the three-stage axial turbine. When operated at design-rated speed of 22,300 rpm, the pressure ratio is 6.5:1 at a mass flow rate of 13.5 lbm/sec (6.1 kg/sec) and the unit output is about 1,600 hp (1,200 kW). This unit features an epicyclic reduction drive (shown on the left),

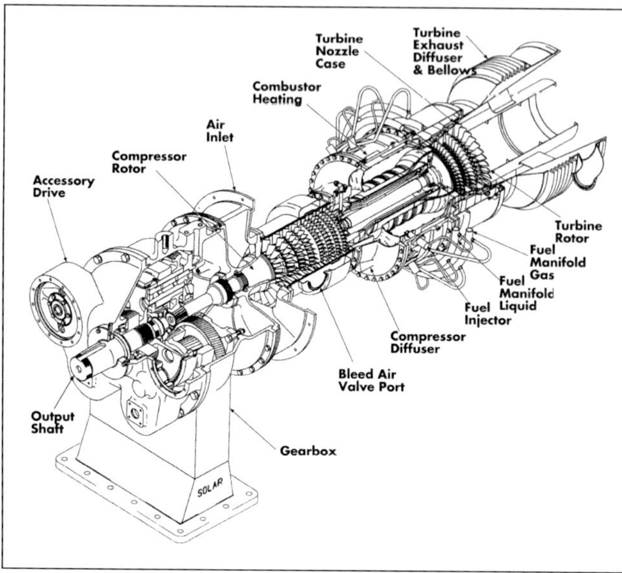

Fig. 10-72 Cross-Sectional View of Small-Capacity Single-Shaft Gas Turbine. Source: Solar Turbines

which provides an overall reduction of about 15:1 at 1,500 rpm and about 12.5:1 at 1,800 rpm output. Performance curves for this unit when applied to electric power generation are shown in English and SI units in Figures 10-73a and b, respectively.

Figure 10-74 is a labeled cross-sectional illustration of a two-shaft Solar Centaur gas turbine with a design full-load capacity of about 4,500 hp (3,350 kW). The turbine section, which features three axial stages, is shown on the right side. In this configuration, the first two turbine stages drive the compressor only. Gases leaving the gas producer turbine flow through the power turbine, forming a fluid coupling. Energy is absorbed by the power turbine and transferred to the output shaft. Turbine cooling is accomplished with internally bled compressor air. A speed governor on the shaft regulates the speed of the gas producer section to control the power level. The power turbine runs at a speed dependent only upon the load, making it suitable for driving equipment such as centrifugal compressors and pumps.

On the left of Figure 10-74 is the 11-stage compressor which operates at a speed of about 15,000 rpm. The compressor section features variable guide vane assemblies, compressor case, stator assemblies, and rotor assembly. The inlet guide vanes are modulated depending on the speed of the gas producer section. For starting, the vanes are in the closed position, restricting the flow of incoming air (bypassing a portion of the airflow to the exhaust collector), allowing for quick acceleration with a relatively low turbine rotor inlet temperature, while preventing the

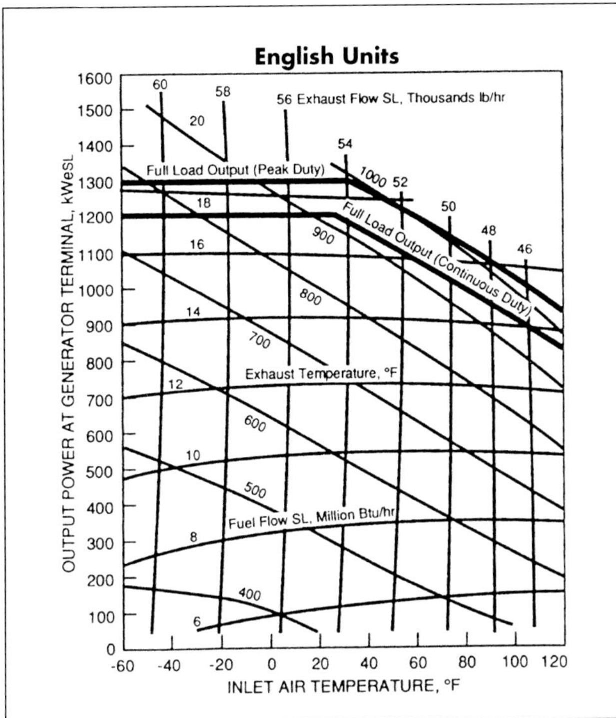

Fig. 10-73a Performance Curves for Single-Shaft Solar Saturn Gas Turbine Applied to Electric Power Generation (English Units). Source: Solar Turbines

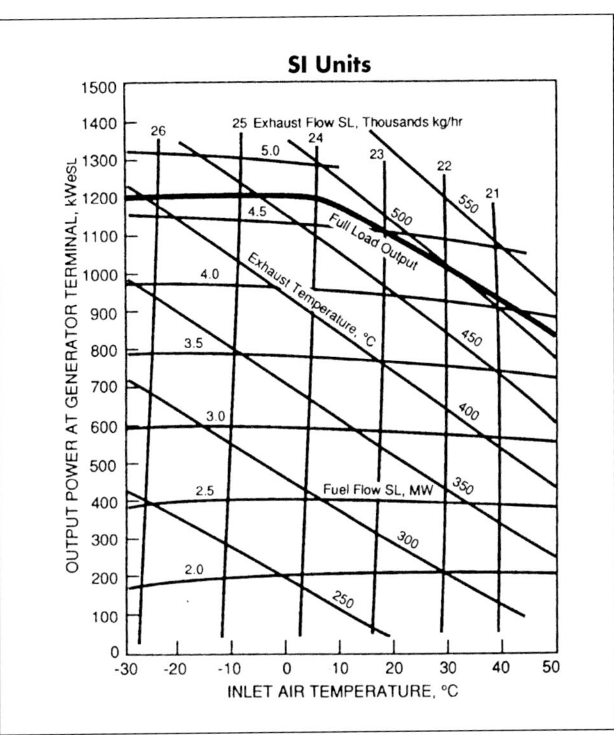

Fig. 10-73b Performance Curves for Single-Shaft Solar Saturn Gas Turbine Applied to Electric Power Generation (SI Units). Source: Solar Turbines

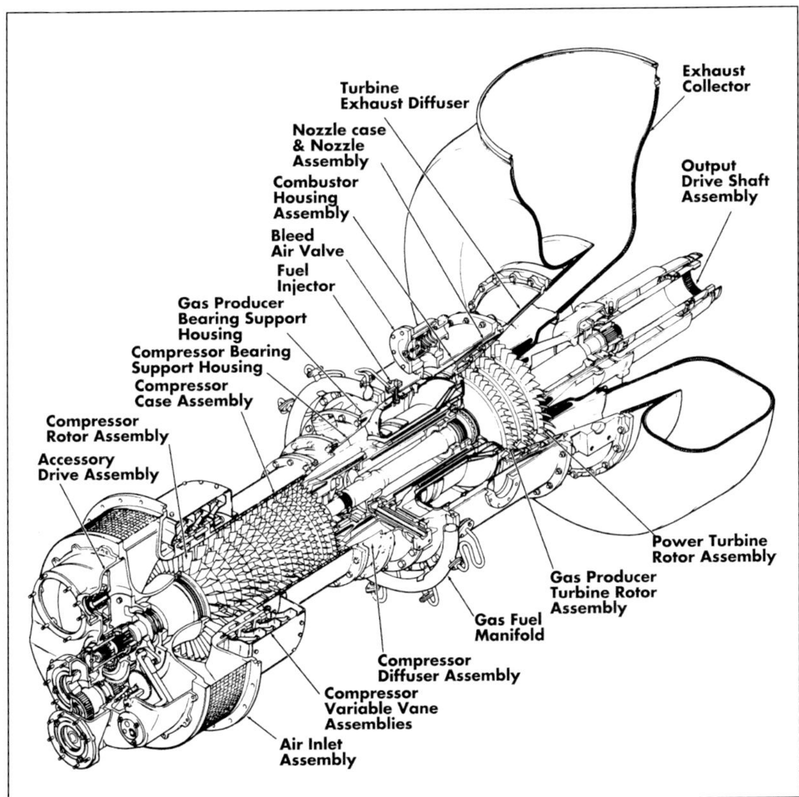

Fig. 10-74 Cross-Sectional View of Two-Shaft Gas Turbine. Source: Solar Turbines

compressor from reaching the surge limit. The annular-type combustor is shown in the center. The combustor section includes the combustor case, combustion liner, and fuel injectors. This unit achieves a heat rate of about 9,100 Btu/hp-h (12,877 kJ/kWh) on an LHV basis, for a simple-cycle thermal fuel efficiency of about 28% (LHV).

Figure 10-75 shows a Solar Titan 250 dual-shaft gas turbine designed for power generation, with an ISO baseload capacity rating of about 29,160 hp (21,745 kW). The heat rate with natural gas operation under ISO conditions with no inlet and or outlet losses and power output measured at the generator terminals is 8,775 Btu/kWh (9,260 kJ/kWh), which corresponds to a thermal fuel efficiency of 38.9% (LHV basis). The unit features a 16-stage axial compressor, which achieves a pressure ratio of 24:1. The unit has an IGV and 5 VGV's for surge control at part speed and mass flow control at rated speed. The combustor chamber is an annular type with torch igniters. The turbine section features a

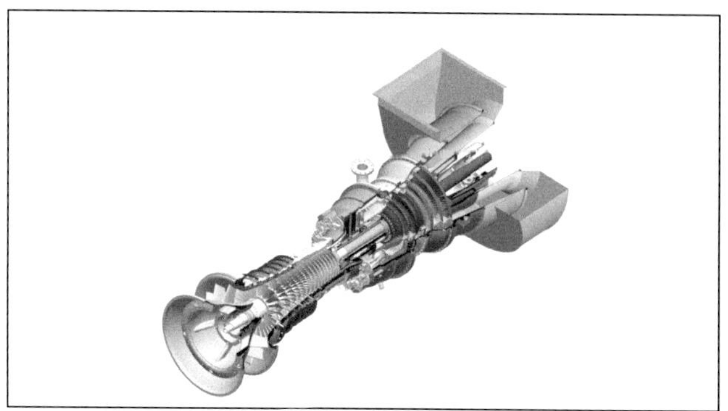

Fig. 10-75 Cross-Sectional View of Two-Shaft Gas Turbine (2). Source: Solar Turbines.

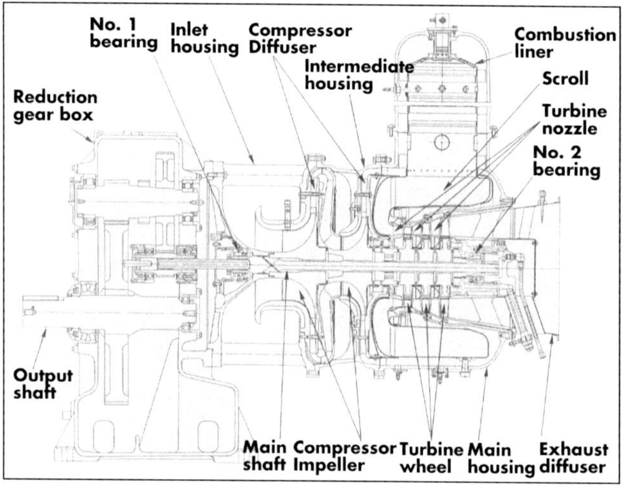

Fig. 10-76 Cutaway Illustration of 940 hp (700 kW) Gas Turbine. Source: Kawasaki Heavy Industries, LTD.

two-stage gas turbine (compressor power) and three-stage power turbine. The turbine exhaust temperature is 865°F (465°C). The dry weight of the gas turbine and generator package is 285,800 lbm (129,635 kg), with a length of 59.7 ft (18.1 m) and a width of 11.1 ft (3.4 m).

Figure 10-76 shows a cutaway illustration of A Kawasaki S2A-01 gas turbine. At ISO conditions, with no inlet or outlet losses, the peak capacity is about 940 hp (700 kW). The heat rate under these conditions, with natural gas operation, is about 11,590 Btu/hp-h (16,400 kJ/kWh) for a simple-cycle efficiency of about 21.9% (LHV basis). This unit features a two-stage centrifugal compressor, which produces a pressure ratio of 9.0:1 at a design speed of 31,500 rpm. The mass flow is 10.4 lbm (4.7 kg) per second with a turbine inlet temperature of 1,706°F (930°C) and an exhaust gas temperature of 923°F (495°C). The combustor is a single-can type with one fuel nozzle and spark plug ignition. The uncooled three-stage axial turbine also has a design speed of 31,500 rpm. The dry

Fig. 10-77 Single-Shaft 9,800 hp (7,300 kW) Gas Turbine Designed for Electric Power Generation. Source: Kawasaki Heavy Industries, LTD.

weight of the unit is 2,866 lbm (1,300 kg), with a length of 6.6 ft (2.0 m), a width of 3.9 ft (1.2 m), and a height of 4.3 ft (1.3 m).

Figure 10-77 shows a M7A-03D Kawasaki single-shaft gas turbine designed for power generation, with an ISO baseload capacity rating of about 9,700 hp (7,250 kW). The heat rate with natural gas operation under ISO conditions with 3.9 inch WG (.98 kPa) inlet and 13.8 inch WG (3.43kPa) outlet losses and power output measured at the generator terminals is 10,435 Btu/kWh (11,010 kJ/kWh), which corresponds to a thermal fuel efficiency of 32.7% (LHV basis). The unit features a 11-stage axial compressor, which achieves a pressure ratio of 13:1 at a design speed of 13,790 rpm. The first three stators and the IGV are variable for surge control at part speed and mass flow control at rated speed. The combustor features six cans with six fuel nozzles and two igniters. The turbine section features a four-stage axial turbine, with air-cooled blades and vanes used for the first and second stages. The turbine exhaust temperatures is 970°F (521°C). The dry weight of the gas turbine and generator package is 25,353 lbm (11,500 kg), with a length of 19.7 ft (6 m) and a width of 6 ft (1.85 m).

When applied to electric power cogeneration, with operation on natural gas and steam injection used for NO_X control at a rate of 150% of fuel flow, the ISO rated baseload capacity of the system is about 5,500 kW. At that rating, the simple-cycle electric generation efficiency, inclusive of generator and inlet and outlet losses, is about 28%. The high exhaust temperature allows for a high rate of heat recovery, resulting in high net power generating efficiency.

Figure 10-78 is a cutaway illustration of a GE LM1600 gas turbine. This turbine consists of a dual rotor gas generator and a power turbine. The gas generator, which operates at a compression ratio of 22:1, consists of a three-stage low-pressure compressor and a seven-stage variable-geometry high-pressure compressor. The annular venturi swirler

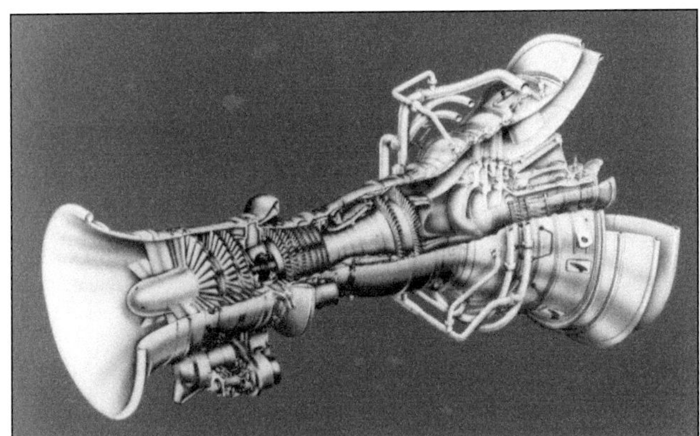

Fig. 10-78 Cutaway Illustration of GE LM1600 Aeroderivative Gas Turbine. Source: General Electric Company

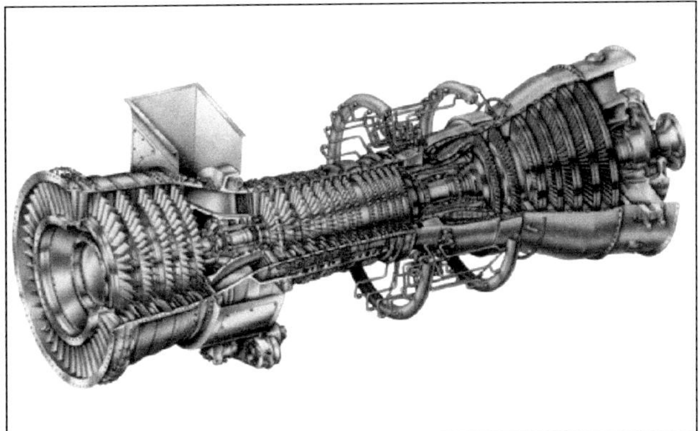

Fig. 10-79 Cutaway Illustration of GE LM6000. Source: General Electric Company

combustor is fitted with 18 fuel nozzles. There is a single-stage high-pressure turbine and a single-stage low-pressure turbine. The low-pressure compressor is driven by the low-pressure turbine using a concentric drive shaft through the high-pressure rotor. The high-pressure rotor is formed by the high-pressure compressor driven by the high-pressure turbine. Both the high-pressure and low-pressure turbine nozzles and blades are air-cooled and are coated to improve resistance to oxidation, erosion, and corrosion. The two-stage power turbine is attached to the gas generator by a transition duct that also serves to direct the exhaust gases from the gas generator into the first stage of the nozzle of the power turbine. An engine-mounted accessory drive gearbox is provided for starting the unit and driving accessories critical to operation.

Output power is transmitted to the load by a coupling adapter on the aft end of the power turbine rotor shaft. In mechanical drive service, the LM1600 has a design output of about 19,100 hp (14.3 MW), with a heat rate of 7,020 Btu/hp-h (9,932 kJ/kWh) on an LHV basis and a simple-cycle thermal fuel efficiency of 36.3% (LHV basis). Exhaust flow is 104 lbm/s (47.3 kg/s) at a temperature of 915°F (490°C). For generator drive applications, the power turbine is designed to operate at constant speed of 7,900 rpm, over the engine operating range, and provides an electrical output of 14,250 kW at a heat rate of 9,414 Btu/kWh (10,548 kJ/kWh) on an LHV basis. The turbine dimensions are 26 ft (7.9 m) long by 12 ft (3.7 m) wide by 13 ft (4.0 m) high, with a baseplate foundation load weight of 130,000 lbm (58,967 kg).

Figure 10-79 is a cutaway illustration of the GE LM6000. This unit, which is based on the GE CF6-80C2 aircraft engine, has the ability to drive the load from either the cold or hot end of the machine and does not require a separate aerodynamically coupled power turbine. It offers a capacity in excess of 56,000 hp (42 MW) with a thermal efficiency of greater than 40%. More than 500 of these units have been manufactured to date.

Figure 10-80 is a cutaway illustration of the aeroderivative FT8 gas turbine. This unit is designed with various power turbine optimization speeds, depending on the application. Power turbines are optimized for 3,000 rpm (50 Hz), and 3,600 rpm (60 Hz) for electric power generation applications. For mechanical service drive applications, a higher-speed power turbine is optimized to run at 5,000 to 5,500 rpm, with the ability to vary speed from 2,500 to 5,750 rpm. Both the generator set and mechanical drive configurations have an option for clockwise or counter-clockwise rotation of the output shaft to match the driven load.

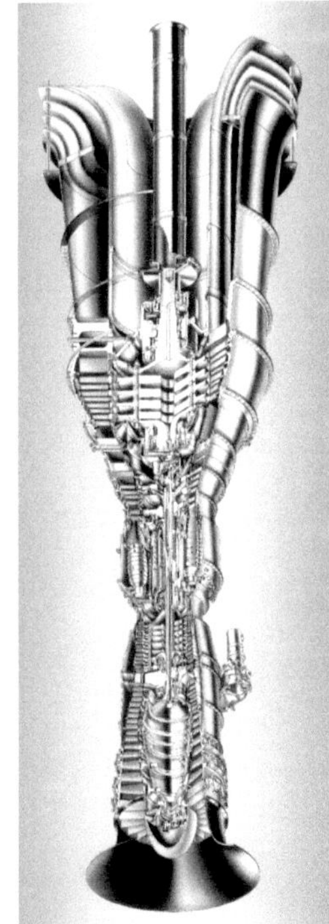

Fig. 10-80 Cutaway Illustration of FT8 Aeroderivative Gas Turbine. Source: United Technologies Turbo Power Div.

Figures 10-81 through 10-83 show performance characteristics of the FT8 when applied to electric power generation. Figure 10-81 shows power output versus ambient temperature. Included with the FT Power PAC is rating data for the twin-pack. The twin-pack configuration has two gas turbines that drive opposite ends of a single generator, for a combined electric power output of 51,100 kW. Power output is flat-rated at temperatures below 50°F (10°C), with a corresponding improvement in heat rate as shown in Figure

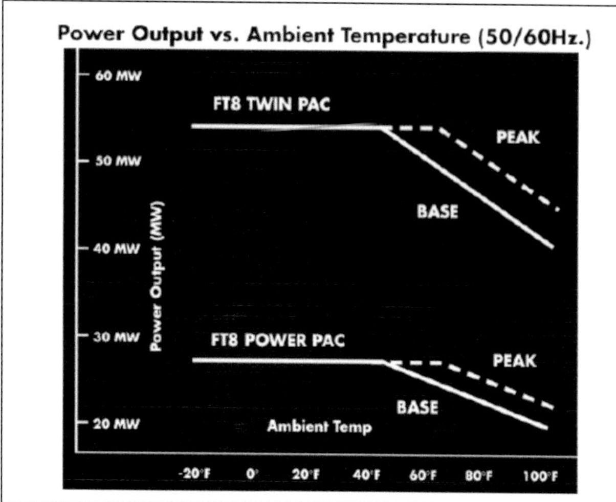

Fig. 10-81 Power Output vs. Ambient Temperature for FT8 Power PAC and Twin Pack Generator-Sets. Source: United Technologies Turbo Power Div.

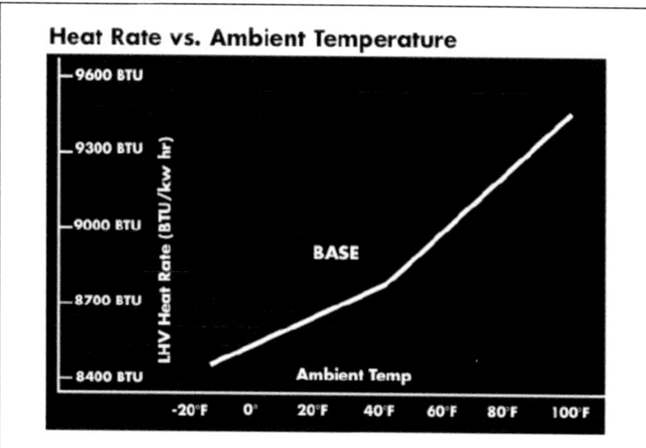

Fig. 10-82 Heat Rate (LHV Basis) vs. Ambient Temperature for FT8 Generator-Set. Source: United Technologies Turbo Power Div.

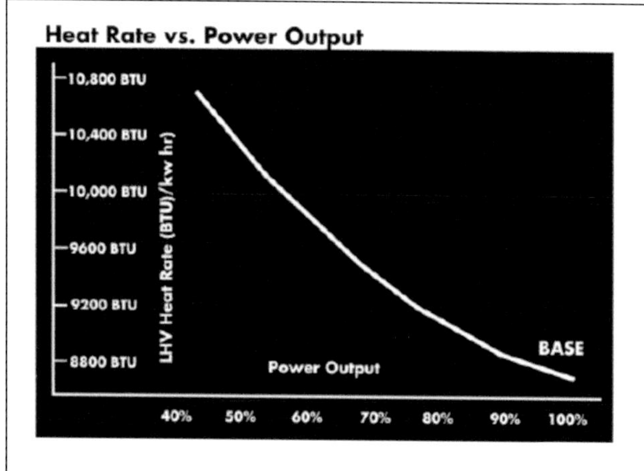

Fig. 10-83 Heat Rate (LHV Basis) vs. Power Output for FT8 Generator-Set. Source: United Technologies Turbo Power Div.

10-82. At sea level, without inlet or outlet losses, and without water or steam for NO_X, the single unit is rated at 25,420 kW ISO base and 8,950 Btu/kWh heat rate (LHV) or 38.1% simple-cycle efficiency. This includes an electric generator efficiency of 97.3%. Figure 10-83 shows heat rate versus power output. The two-shaft design allows for relatively good part-load thermal fuel efficiency. At 60% of full load, for example, there is an increase in heat rate of less than 15%.

Figure 10-84 shows expected mechanical drive performance in shaft output and heat rate versus power turbine speed. Capacity and heat rate, which are represented in kW, can be compared to the performance figures shown for the generator-set, with the primary differences being increased power turbine speed at full load and absence of generator losses. Figure 10-85 shows expected mechanical drive service heat rate versus ambient temperature, at ISO conditions, with no inlet or exhaust losses, at a power turbine speed of 5,000 rpm.

Figure 10-86 is an open-case model of an ABB GT24 industrial gas turbine showing location of the two annular chambers used in the low-NO_X sequential combustion system described above. Both combustion chambers operate with extremely lean fuel mixtures, allowing the unit to achieve a NO_X emissions rate conservatively estimated at

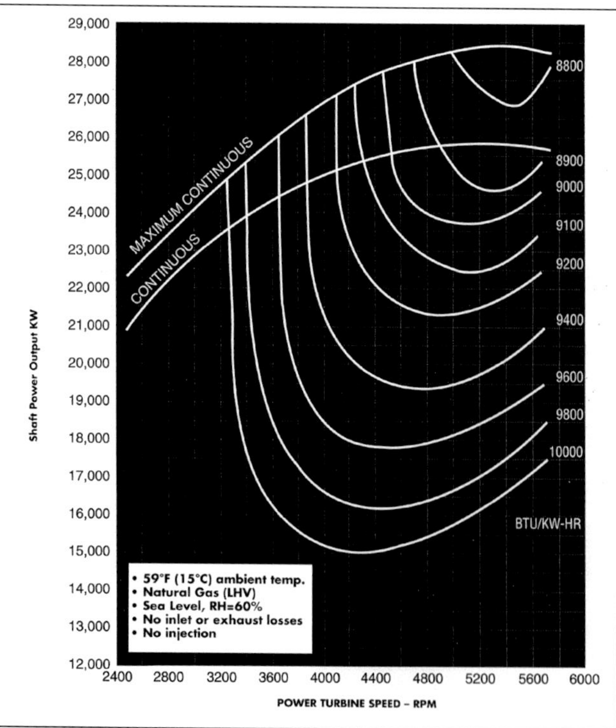

Fig. 10-84 Shaft Output and Heat Rate (LHV basis) vs. Power Turbine Speed for FT8 Applied for Mechanical Drive Service. Source: United Technologies Turbo Power Div.

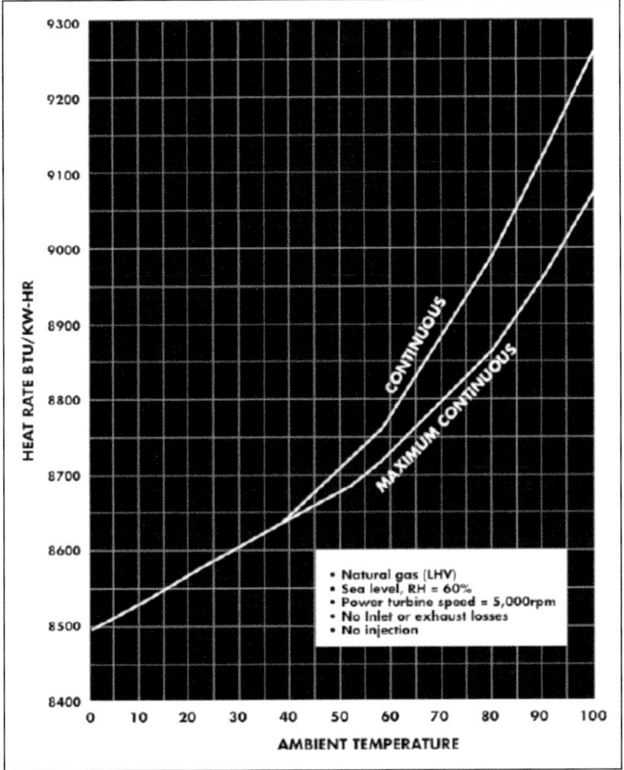

Fig. 10-85 Heat Rate (LHV Basis) vs. Ambient Temp. for FT8 Applied for Mechanical Drive Service. Source: United Technologies Turbo Power Div.

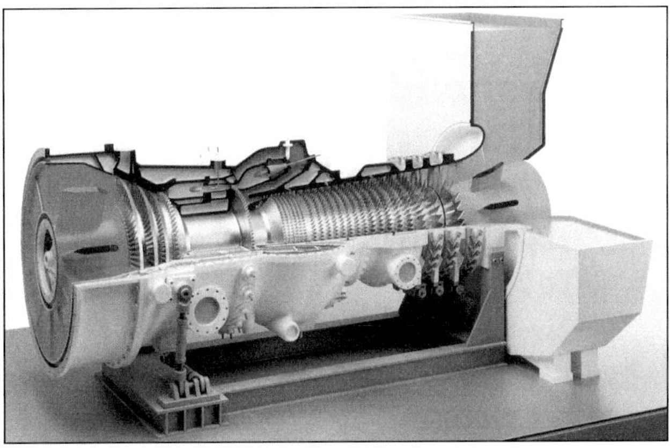

Fig. 10-86 Open-Case Model of ABB GT24 Showing Sequential Combustion System. Source: ABB

Fig. 10-87 Assembly of Very Large Capacity, Heavy-Duty Industrial Gas Turbine. Source: Siemens Power Corp.

well under 25 ppm when operating on natural gas and 42 ppm when operating on No. 2 oil (@ 15% O_2). The turbine inlet temperature, based on ISO conditions, is 2,291°F (1,255°C), enabling the electric generation set to achieve a simple-cycle fuel efficiency of 37.5% on an LHV basis, with a heat rate of 9,003 Btu/kWh (9,497 kJ/kWh). Mean temperature of blade materials, a key indicator for reliability and lifetime, as well as any of the hot-gas sections, are maintained at a maximum of 1,652°F (900°C).

At 60 Hz, design generator output is 166 MW. The unit operates at a shaft speed of only 3,600 rpm and achieves a compressor pressure ratio of 30:1. The full-load exhaust mass flow is 833 lbm/s (378 kg/s) at a temperature of 1,130°F (610°C), which is well-suited for combined-cycle operation. Off-design operation is controlled by adjustment of the first three variable stator rows. The exhaust temperature of the first combustor remains unchanged down to about 25% of full load. In this operating mode, the second turbine stage experiences only small thermal transients when the gas turbine is started. At the moment of ignition, it has already been preheated to approximately 70% of the full-load inlet temperature. The dimensions of the gas turbine are 34 ft (10.4 m) long by 13 ft (4.0 m) wide by 15 ft (4.6 m) high and the dry weight is about 197 tons (178.7 metric tons).

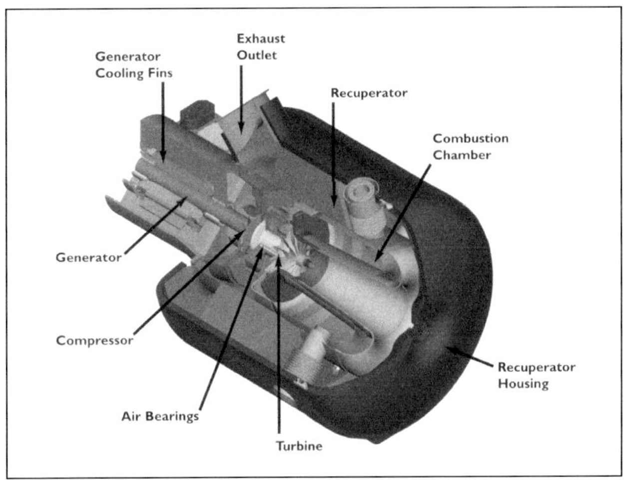

Fig. 10-88 Cutaway Illustration of 28 kW, Packaged "Micro" Gas Turbine-Generator Set. Source: Capstone Turbine Corporation

The contrast of Figures 10-87 and 10-88 provide insight into the wide range of types and capacities of gas turbines commercially available for field applications. Figure 10-87 shows the assembly of a very large capacity, heavy-duty industrial gas turbine. Figure 10-88 provides a cutaway illustration of a very small-capacity unit, termed a micro turbine, which has been recently introduced to the market. This packaged, recuperative turbine-generator unit, which produces an electrical output of just under 30 kW, has a weight of only 1,082 lbm (490 kg) and a length of about 6 ft (1.8 m). Manufacturer's specifications indicate that it can operate on a variety of fuels, with operating pressures ranging as low as 5 psig (0.35 bar), NO_X emissions of 9 ppmv, and sound emissions of 65 dBA at 33 ft (10 m). All rotating components are mounted on a single shaft that rotates at up to 96,000 rpm at full load. The air-cooled generator is designed for operation at 50 or 60 Hz. When operating at a gas pressure of 55 psig (3.9 bar), under ISO conditions, stated thermal efficiency is 26% (LHV) with a total exhaust energy of 277,000 Btu/h (295,000 kJ/h) at a temperature of 520°F (271°C). The objective of market introduction of micro-type turbines such as this one is to compete with small-capacity packaged reciprocating engines. Hence, there is a large focus on compactness of size, ease of installation, and reliability. Such units are also targeted for use in vehicular applications.

Gas Turbine Research & Development

Gas turbine research and development (R&D) efforts are currently focused on the continued development of alternative cycles, including regeneration, intercooling, power augmentation with steam injection, and NO_X controls without injection of steam or water. Component development technology has centered on ceramics, advanced alloys for high temperature, coatings, efficiency enhancements in compressor and turbine designs, and control technology.

Because specific capacity, pressure ratios, and cycle efficiency are closely tied to increased firing temperatures, materials technology has become a limiting factor. Additional aggravating factors include combustion hot spots, fuel and air purity, and the mechanical strength of high-temperature materials to withstand rotational and other mechanical stresses concurrently with high temperatures at the target operation temperature. Primary focus for new technology development is combustor and hot gas path parts associated with the turbine.

At turbine inlet temperatures of about 2,800°F (1,538°C) and a pressure ratio of about 50:1, potential simple-cycle efficiency could approach 50%. However, material and cooling technology is currently limited to about 2,300°F (1,260°C), with component cooling to maintain metal temperatures well below this. Without component cooling, conventional materials cannot withstand long-term operating temperatures exceeding 1,800 or 1,900°F (982 or 1,038°C). With state-of-the-art cooling techniques and other material treatment, achievable inlet temperatures are about 2,300°F (1,260°C). Current R&D is aimed at up to 2,600°F (1,427°C). The relationship between inlet temperature and pressure ratio to efficiency is shown in Figure 10-89.

Increasing material strength and operating temperature tolerance of turbine components can improve cycle efficiency in two ways. First, firing temperatures can be increased. Second, cooling air requirements can be reduced, decreasing parasitic loads and increasing net capacity.

Currently available advanced cooling systems have brought the high-end limit for turbine inlet temperature to about 2,350°F (1,288°C). New developments in turbine design are targeting parts in the path of the hot gas with blade cooling and improved air delivery systems and airfoil internal passage design.

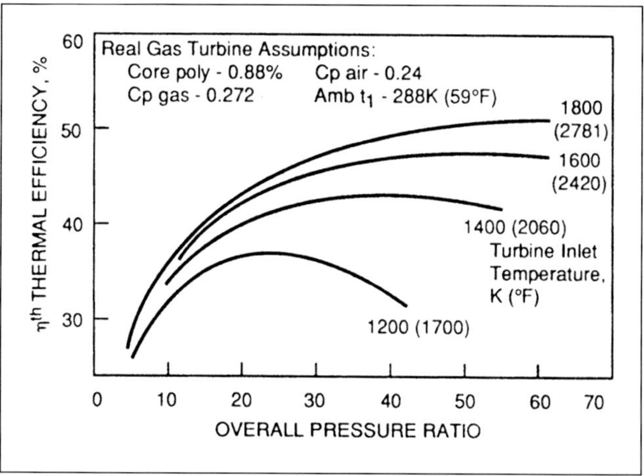

Fig. 10-89 Effect of Turbine Inlet Temperature and Pressure Ratio on Thermal Efficiency. Source: Solar Turbines

Materials Development

A comparison of conventional and future advanced materials for gas turbine design and the potential advantages of advanced materials are shown in Tables 10-2 and 10-3.

Nickel-based and cobalt-based superalloys have allowed for greatly increased temperatures and produced high resistance to oxidation and corrosion. Superalloys show great strength and high resistance to oxidation at elevated temperatures as a result of the interaction

Table 10-2 Conventional and Future Advanced Materials for Gas Turbine Design

Material	1995 Use Temp. °C (°F)	2015 Use Temp. °C (°F)	Density, g/cm3
Aluminum*	204 (400)		2.8
Organic composites	427 (800)	427 (800)	1.8
Aluminum metal matrix composites		538 (1000)	2.8
Titanium*	454 (850)	538 (1000)	4.5
Titanium aluminide		760 (1400)	4.1
Titanium and titanium aluminide metal matrix composites		870 (1600)	3.9
Nickel alloys*	982 (1800)		8.0
Single-crystal nickel*	1120 (2050)		9.0
Nickel aluminide intermetallic composites		1320 (2400)	5.2
Silicon carbide, silicon nitride ceramics*	1400 (2550)		3.2
Ceramic matrix composites*		1540 (2800)	3.0

* Current properties

Source: Solar Turbines

between the alloy's composition and structure. Various metallurgical and ceramic coatings are also used to provide a surface that is highly resistant to degradation by corrosion and erosion resulting from poor air or fuel quality.

Significant R&D has also been focused on high-temperature ceramics as structural materials for hot section components such as rotor blades, vanes, and combustor liners. Ceramics have the potential to advance gas turbine inlet temperatures beyond metal alloys. Various types of composites are being developed for different components. These include composites with metallic, glassy, and organic materials. Ceramic coatings are being developed to serve as thermal barriers for metallic hot section components.

Advanced ceramics have fairly simple chemistries. Generally, they consist of elements such as silicon, carbon, aluminum, oxygen, and nitrogen, in network arrangements held together by strong chemical bonds. Other elements used in certain compositions are titanium, barium, lead, and zirconium. Advanced ceramics may be classified as either monolithic ceramics or ceramic matrix composites.

Metal matrix composites have properties imparted to the metal by the non-metallic reinforcements used. The cross-linking of dissimilar materials creates new materials with unique properties that are both very strong and very light.

Metal matrix composites and intermetallic composites are not expected to have the resistance to extremely high temperatures of ceramics. However, they are potentially useful materials for intermediate temperature applications in compressor blades and downstream turbine components such as disks, blades, casings and shafts. Titanium-based metal matrix composites and reinforced titanium aluminides are seen as excellent potential structural materials.

Engineered plastics also have potential uses as materials for lower-temperature component applications such as compressor air-inlet ducting and filter housings where corrosion resistance is of primary importance. The strong damping characteristics of these materials may also allow them to be used as materials for vibration control, and their low density and durability make them potential candidates to replace metallic components for various types of seals and bushings. While material costs are expected

Table 10-3 Potential Advantages of Advanced Materials

Material	To Replace	Component	Advantage
Organic composites	Aluminum, stainless steels	Compressor blades, housings	Weight reduction, increased damping, corrosion resistance
Titanium aluminide intermetallics	Steels, nickel alloys	Compressor cases and blades, LP turbine blades	Lightweight, high strength
Nickel aluminides	Nickel superalloys	Turbine disks, blades	Lightweight, oxidation resistance
Ceramics	Nickel and cobalt-based	Blades, vanes, combustor liners, transition pieces, fuel injector tips, tubular materials	Durability, oxidation resistance, high-temperature strength, lightweight, reduced cost
Ceramic matrix	Nickel-based superralloys	Combustor liners, ceramic recuperator headers	Oxidation resistance, high-temperature strength, lightweight

Source: Solar Turbines

to remain high, the potential to manufacture composites into complex shapes could result in lower production costs compared with machined metallic assemblies.

Humid Air Turbine Cycle

The Humid Air Turbine (HAT) cycle is currently under development and is expected to be in production shortly. In the HAT cycle, exhaust heat from the gas turbine is used to heat and humidify the combustion air. The HAT cycle will operate with intercooling and high-pressure ratios and is expected to offer a thermal efficiency of about 47% (LHV). The HAT cycle is being designed to operate with natural gas or fuel from a coal gasifier.

Ericsson Cycle

One cycle that may, in the future, offer the potential of efficiency approaching 50% is the Ericsson cycle. This cycle combines the features of regeneration, reheating, and intercooling along with steam/fuel reformation. In the ideal Ericsson cycle, both isothermal compression and expansion are combined with regeneration. Cycle efficiency is improved versus the Brayton cycle because isothermal compression requires only about 75% of the energy (per unit of air) required by adiabatic compression. Isothermal expansion is also more efficient than adiabatic expansion.

Figure 10-90 shows a typical arrangement of this potential future gas turbine system. In the ideal cycle, it is assumed that intercooling is used between every blade row of the compressor and that reheat is used between each turbine section. In a simplified version of the cycle, multi-staged compression with intercooling is used to cool the air prior to entering the regenerator. The energy in the steam produced from the cooling water in the intercooler (heat exchanger) is then used in reforming the fuel. This form of heat recovery increases thermal efficiency by returning the sensible heat of the air to the cycle as part of the fuel energy.

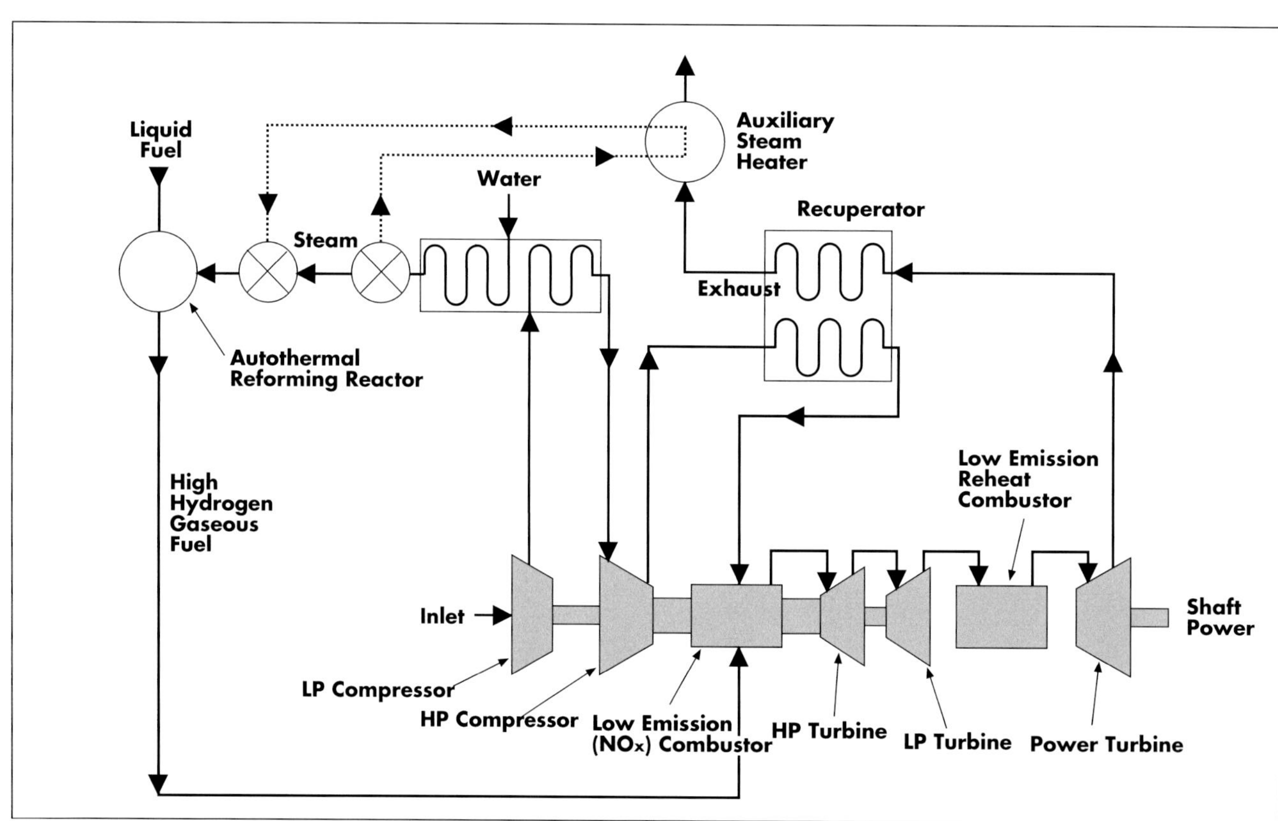

Fig. 10-90 Arrangement of Potential Future High-Efficiency Gas Turbine System. Source: Solar Turbines

Chapter Eleven

Steam Turbines

Steam turbines extract heat from steam and transform it into mechanical work by expanding the steam from high to low pressure. These machines have been a mainstay in industry throughout the past century and are available in virtually any capacity, from a few hp (kW) to several hundred thousand hp (kW).

Small- and intermediate-sized steam turbines are used for a wide range of applications, including generators, compressors, and pumps. When coupled with gears, they are used to drive fans, reciprocating compressors, and other classes of low-speed machinery. The largest turbine applications are generator drives in utility and other central power stations.

Since steam can be generated with any type of fuel and, in some cases, with recovered heat, steam energy can sometimes be produced at a very low cost. Industrial steam turbines can potentially be applied with any high-pressure steam system, and low-pressure turbines are available for special applications. The result is low-to-moderate cost power generation with high reliability, low maintenance (when added to existing steam plants), and extremely long life.

Steam Turbine Power Cycles

Steam turbines operate on the **Rankine Cycle**. The simple ideal Rankine cycle, which is described in detail later in the chapter, can be reduced to four processes:

1. Boiler feedwater (condensate) is pressurized and injected into a boiler.

2. Water is heated and evaporated in the boiler. The resulting steam may be superheated to increase its enthalpy and reduce moisture.

3. Steam is expanded in the turbine to a lower pressure. A small portion of steam thermal energy is converted to kinetic energy that is used to drive a rotating load.

4. Steam is condensed by a cooling medium in the condenser. In a back-pressure turbine, exhaust steam is delivered to a remote heating load, where condensation occurs.

A steam turbine is considered part of a cogeneration system when an application involves the sequential use of a single source of energy for both the generation of power and useful thermal energy output. As with other prime mover systems, these applications are broadly classified as either topping or bottoming cycles.

A **topping cycle** uses a back-pressure or extraction turbine as a pressure-reducing valve. As high-pressure steam is expanded to a lower pressure, shaft power is generated by the turbine at a minimal cost, as there is only a slight drop in the enthalpy content of the steam (usually less than 10%). This application is ideal for facilities that require both high- and low-pressure steam simultaneously.

A **bottoming cycle** uses excess steam, discharged from a high-pressure process, to generate shaft power. Bottoming cycles are also used for applications which discharge high-temperature exhaust gas, convert it to steam in a heat recovery steam generator (HRSG), and pass it through a steam turbine.

Traditional **combined cycles,** which integrate a topping cycle and a bottoming cycle into one sequential process, use recovered heat (generally from a combustion engine's exhaust) to produce additional shaft power in a steam turbine. When a condensing turbine is used, it functions as a bottoming cycle. In a cogeneration type combined cycle, a back-pressure or extraction turbine may be used with low-pressure steam passed on to a thermal load. In this application, the steam turbine may be classified as a second topping cycle or an intermediate cycle.

Shaft power can be extracted from virtually any pressure drop. Some multi-stage turbines are designed to operate with great efficiency at very low steam pressures. In general, however, condensing turbines are most cost-effective at steam pressures above 100 psig (7.9 bar) and back-pressure turbines should operate with at least a 4:1 pressure drop ratio (ratio of absolute inlet to exhaust pressure). When rejected steam or heat recovery-generated steam is available at lower pressures, bottoming cycle steam turbine applications may still be used. However, for most industrial size turbines, the size and cost of the turbine outweigh the value of available power output in low-pressure applications.

Equipment Classifications

Steam turbines are classified according to their fundamental operating principles, some of which are:

- Number of stages: single- or multi-stage
- Number of valves: single- or multi-valve
- Turbine stage design class: impulse or reaction

- Steam supply: saturated or superheated; single or multiple pressure
- Steam exhaust conditions: condensing, non-condensing, automatic extraction, mixed pressure, regenerative extraction, or reheat
- Casing or shaft arrangement: single casing, tandem compound, or cross compound
- Number of parallel exhaust stages: double flow or triple flow
- Direction of turbine steam flow: axial flow, radial flow, or tangential flow
- Type of driven apparatus: mechanical drive or generator drive

In a **single-stage turbine**, steam is accelerated through a nozzle or cascade of stationary nozzles and guided into the rotating buckets on the turbine wheel to produce power. A single pressure drop occurs between the inlet to the nozzle and the exit for the last row of blades. Single-stage turbines are usually limited to sizes of a few thousand hp (kW) or less, although larger capacities are available in special designs. Mechanical efficiency will generally range from 30% to 60%. In single-stage design, the emphasis is on simplicity, dependability, and low first cost.

Figure 11-1 is a cutaway illustration of a horizontal single-stage steam turbine. Notice the single disc wheel keyed to the shaft. Figure 11-2 shows the rotor assembly for this turbine. This assembly consists of a contoured, single-disc, two-row wheel. This is commonly referred to as a Curtis stage design.

A **multi-stage turbine** includes two or more stages in

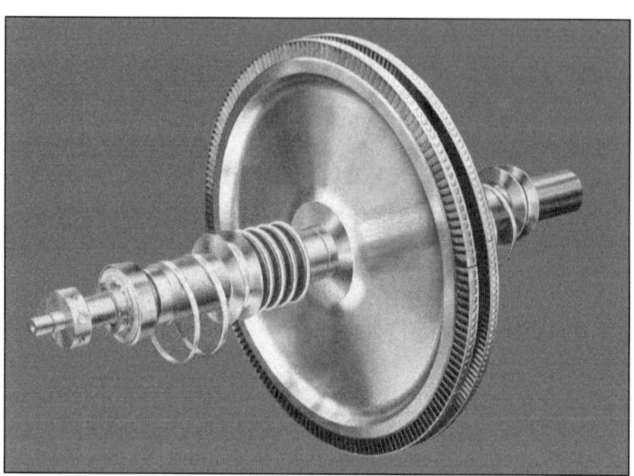

Fig. 11-2 Single-Stage Steam Turbine Rotor Assembly. Source: Tuthill Corp. Coppus Turbine Div.

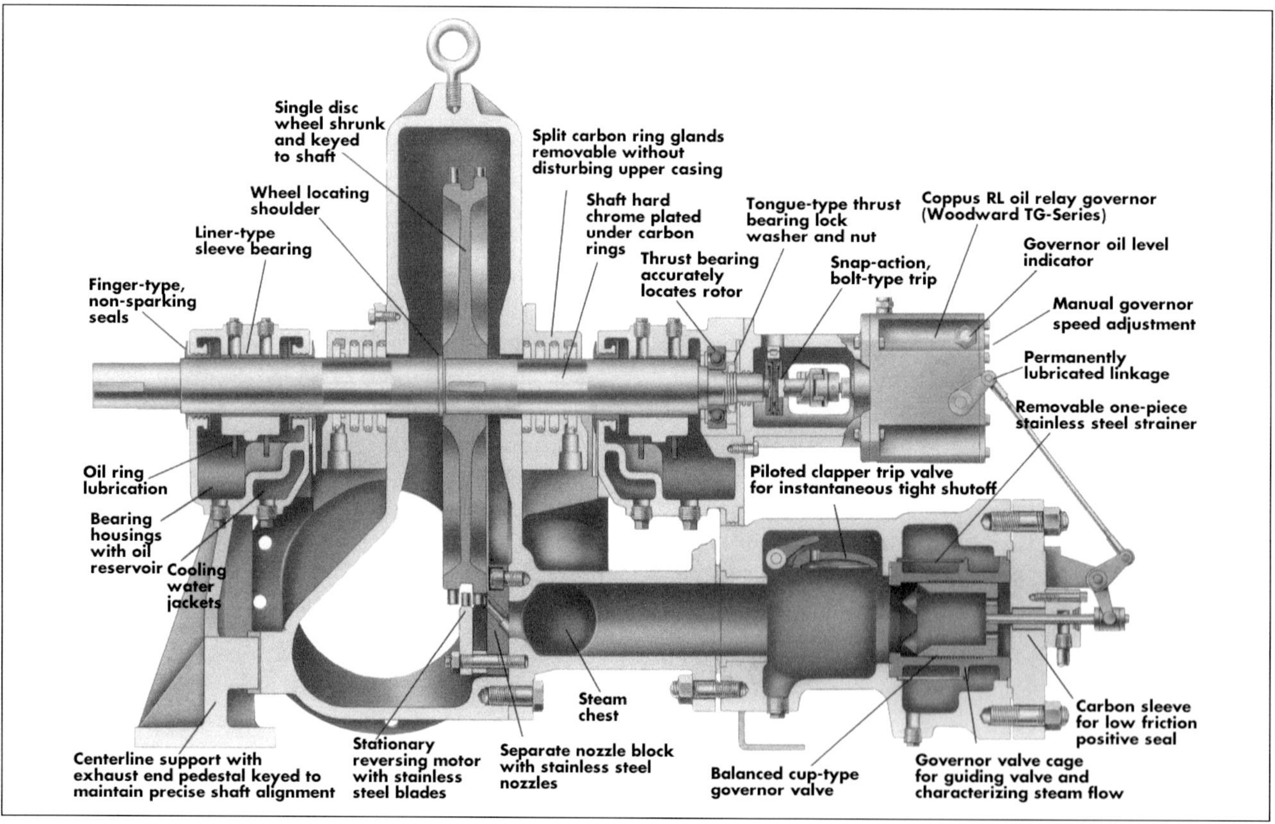

Fig. 11-1 Cut-Away Illustration of Single-Stage Steam Turbine. Source: Tuthill Corp. Coppus Turbine Div.

a single case, each providing a portion of the total pressure drop. Multi-stage turbines utilize either a Curtis or Rateau first stage, followed by one or more Rateau stages (the steam path design is discussed below). A diaphragm is used to divide stages and carries the stationary nozzle elements that are used to accelerate the steam into the next stage.

Multi-stage turbines are more efficient than single-stage turbines and also more complex and costly. First cost can be several times that of a single-stage unit, depending on the number of stages and other parameters. Multi-stage turbines will generally have mechanical efficiencies between 50 and 80%. Factors that differentiate performance between single- and multi-stage turbines are discussed later in this chapter.

Figure 11-3 is a cutaway illustration of a multi-stage steam turbine with the casing half removed. This unit can achieve capacities of up to 5,500 hp (4,100 kW), with inlet steam pressures ranging up to 400 psi (28 bar). Figure 11-4 is an open-case view of a 15-stage steam turbine with a capacity range of up to 15,000 hp (11 MW).

Fig. 11-4 Open-Case View of 15-Stage Steam Turbine with Capacity Range of up to 15,000 hp (11 MW). Source: Tuthill Corp., Murray Turbomachinery Div.

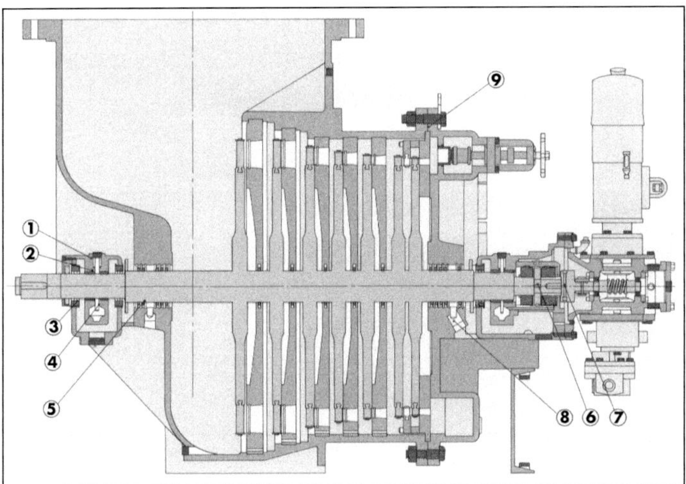

Fig. 11-3 Cutaway Illustration of Multi-Stage Steam Turbine. Source: Tuthill Corp., Murray Turbomachinery Div.

1. Exhaust and steam end bearing covers
2. Steel-backed journal bearings
3. Bronze labyrinth type oil seals
4. Pressure lubrication
5. Segmented carbon ring gland and interstage seals
6. Ball thrust bearings
7. Pin-type overspeed trip
8. Gland leak-offs
9. Casing splits

When the turbine speed cannot be matched directly to equipment speed in a given application (or if operation at that speed would be inefficient), gearing is required. When the optimum speed for the turbine is greater than that for of the driven equipment, the first cost and the efficiency loss of a reduction gear is matched against the efficiency gain of optimizing the speed of both the turbine and the driven equipment.

Applied Technology Types

Industrial steam turbine applications are generally classified as either condensing or non-condensing (back-pressure). A third classification is an extraction turbine, which can include elements of both.

Condensing turbines operate with an exhaust pressure less than atmospheric (this is a vacuum pressure). Because of the very low exhaust pressure, the pressure drop through the turbine is greater and more energy is extracted from each lbm (kg) of steam. Standard condensing turbines may be of straight flow (Figure 11-5a) or dual-flow opposed exhaust design (Figure 11-5b); the later being used to minimize stresses and optimize efficiency in larger units. The con-

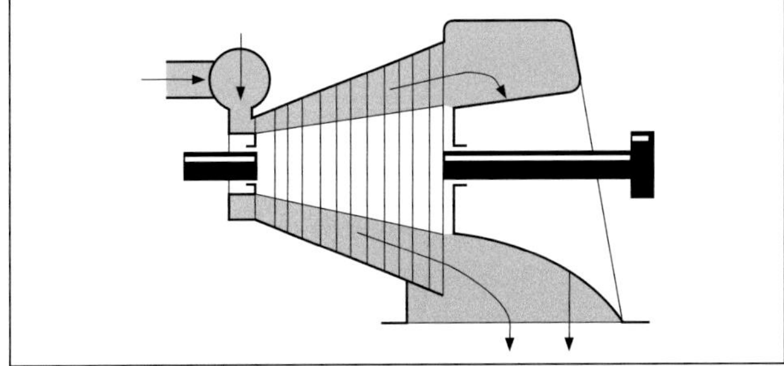

Figure 11-5a Straight Flow Condensing Turbine.

Figures 11-5a — 11-5i; adapted from graphics provided by Tuthill Corp., Murray Turbomachinery Div.

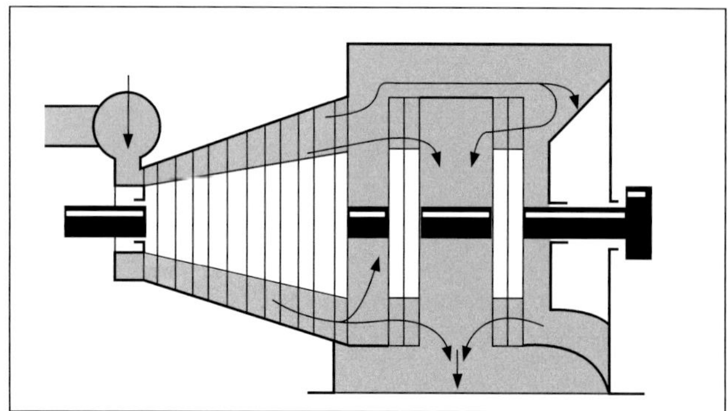

Figure 11-5b Dual-Flow Opposed Exhaust Condensing Turbine.

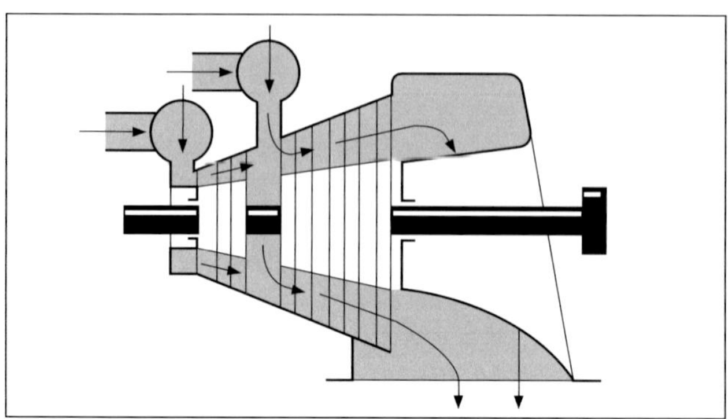

Figure 11-5e Controlled Induction Condensing Turbine.

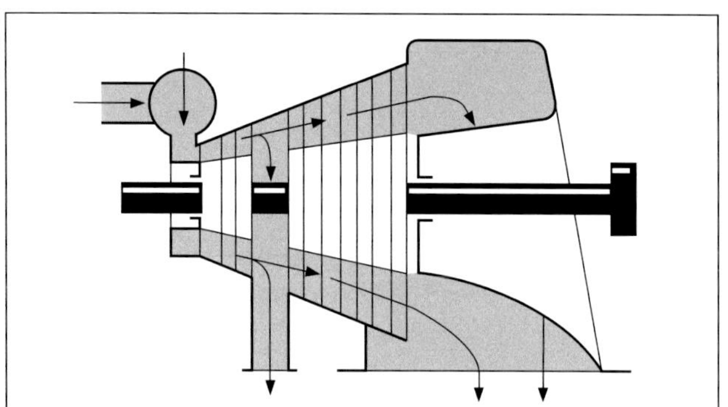

Figure 11-5c Non-Automatic Extraction Condensing Turbine.

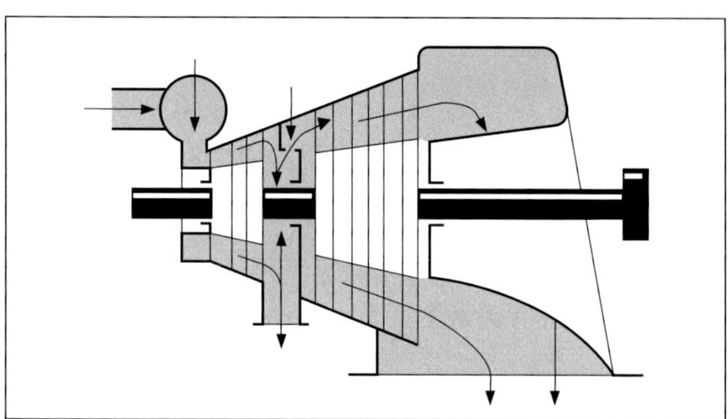

Figure 11-5f Controlled Extraction/Admission Condensing Turbine.

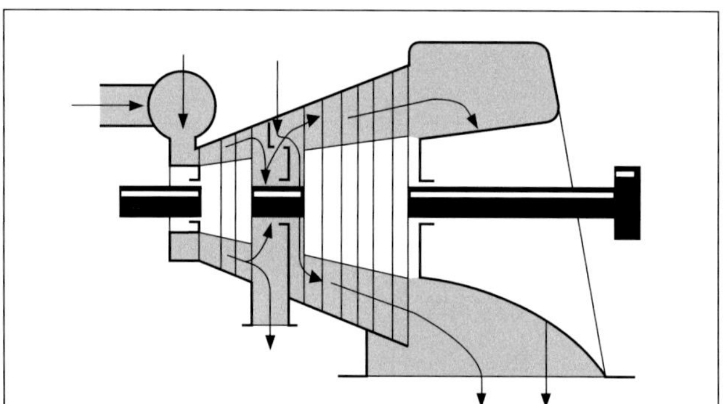

Figure 11-5d Controlled Extraction Condensing Turbine.

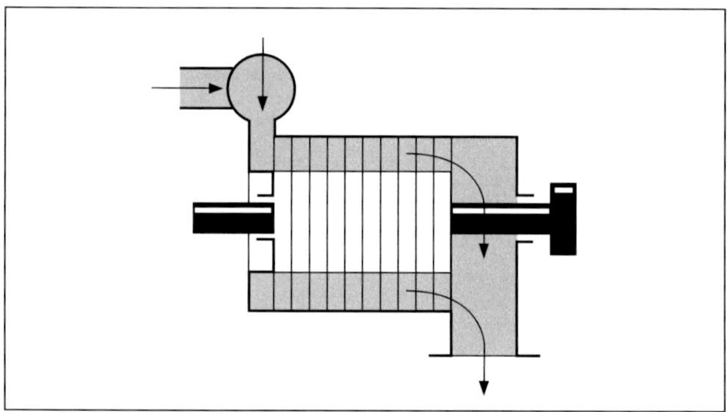

Figure 11-5g Straight Flow Back-Pressure Turbine.

denser can be either air- or water-cooled. In water-cooled systems, the cooling water exiting the condenser can sometimes be used for process or space heating hot water loads.

Because the unused steam energy is rejected to atmosphere and is, therefore, wasted, condensing turbines are usually designed with several stages to maximize efficiency. This design does, however, add to the cost. The auxiliary equipment required for handling and condensing the turbine exhaust also drives up capital and operating cost.

Non-condensing (back-pressure) turbines (Figure 11-5g) operate with an exhaust pressure equal to or in excess of atmospheric pressure. Exhaust steam is used for heating, process hot water, or other purposes. Back-pressure turbines may be of single- or multi-stage design. Because all of the unused steam in the power generation process is passed on to process application and, therefore, not wasted, mechanical efficiency is not a major concern. In some cases, where low-pressure steam applications are limited, multi-stage turbines are used to maximize power output.

An **extraction turbine** (Figures 11-5c, d and h) is a multi-stage machine with the added design feature of having one or more outlets to allow intermediate pressure steam (intermediate between inlet pressure and exhaust pressure) to be withdrawn. Automatic-extraction turbines are generally designed in three configurations: single, double, and triple, depending on the number of pressures at which steam is extracted. Adding automatic stages adds significantly to cost and complexity. As a result, the single-automatic extraction design is the most common.

Extraction turbines are generally designed for applications in which there is a need for discharged steam at different pressures or when there are varying low-pressure steam process requirements. While back-pressure turbines are somewhat inflexible, since shaft power and process steam requirements must be closely matched, extraction turbines can cope with changes in these variables and satisfy requirements over a broad range. If the mass flow requirement of process steam is reduced, an extraction turbine can pass additional steam through the low-pressure stage to maintain shaft power output.

Admission (or induction) turbines (Figure 11-5e) include steam input at two or more pressures at two or more locations. This configuration can allow byproducts from in-house processes to be used to increase turbine power output. Extraction and admission features can also be combined into a single unit knows as an extraction/induction turbine (Figures 11-5f and i).

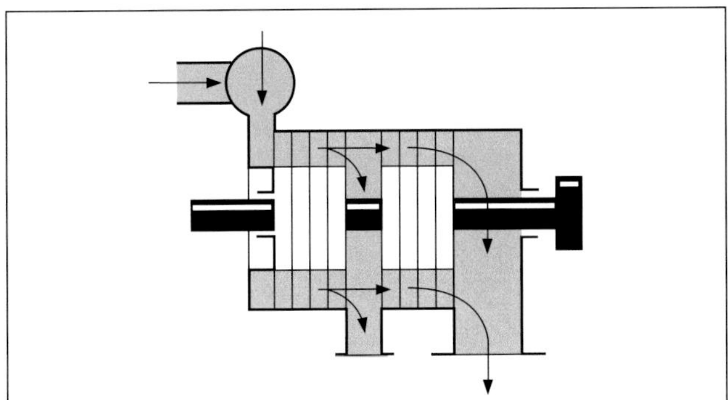

Figure 11-5h Non-Automatic Extraction Back-Pressure Turbine.

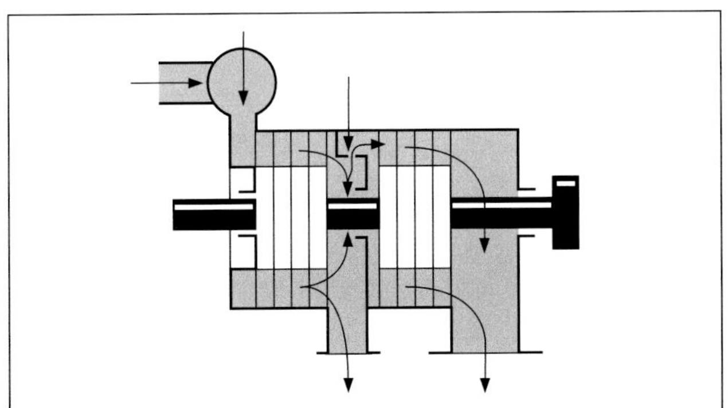

Figure 11-5i Controlled Extraction/Admission Back-Pressure Turbine.

Extraction and admission turbines may be either controlled (Figures 11-5d, e, f, and i) or uncontrolled (Figures 11-5c and h). A non-controlled extraction or admission steam turbine has one or more openings in the turbine casing for extraction or admission of steam, but does not have means for controlling the pressure of the induced or extracted steam.

The pressure at the bleed opening of a **non-automatic extraction (or bleeder) turbine** (Figure 11-5c) is proportional to the reduction in flow through the stages following the bleed. Under low-load conditions, the pressure in each stage will decrease. Therefore, the bleed opening must be located in a stage where the pressure at minimum turbine load will be adequate for the desired extraction load. At full power, the pressure in the bleed stage may exceed the desired pressure and a pressure-regulating valve must be used. In order to prevent steam from entering the turbine through the bleed opening, a non-return valve is also required.

An **automatic-extraction turbine** (Figure 11-5d) is designed to hold pressure in the extraction line constant, regardless of load, by regulating the flow of the steam to the turbine stages following the extraction or admission opening(s). Automatic extraction turbines do not experience the power loss associated with non-automatic units that require a pressure-reducing valve. Automatic extraction turbines can be designed to accommodate extraction flows from zero to about 9% of inlet flow. A small portion must always be maintained to flow through the low-pressure stages of the turbine to carry away the heat generated by windage. Generally, automatic units, which are far more costly than non-automatic units, are used if the extraction or admission steam flow is greater than 15 or 20% of the total steam flow.

Often, two or more types of turbines are applied to a specific application. For example, condensing and back-pressure turbines can be used in combination, rather than depending on a single extraction turbine. Extraction-admission turbines can also be used to balance out variable low-pressure steam conditions through the ability to either admit or extract steam, depending on operating conditions.

Industrial steam turbines ordinarily include a single casing with all rotating blades attached to one shaft and steam flow in one direction. Very large units often employ a **tandem-compound** design. With this design, steam enters the first section, or high-pressure end, and expands to an intermediate pressure. The steam is then transferred through a crossover pipe to the low-pressure end. Both the high- and low-pressure ends are on the same shaft. Typically, the low-pressure end of the turbine is a single-casing section with double flow, although, in some cases, units may have two low-pressure casings that produce triple flow.

Very large tandem-compound units have a high-pressure element, an intermediate (or reheat) element, and a triple-flow, low-pressure element. **Cross-compound** units are similar to tandem-compound units, except that the high- and low-pressure ends are not on the same shaft. When used as turbine-generators, each end is usually connected to a separate generator. When used as a mechanical drive, two ends may be connected to the same shaft by reduction gears.

Cycle Enhancements

The **reheat cycle** is a modification of the simple Rankine cycle, which was developed to take advantage of the increased efficiency available with higher pressures while avoiding excessive moisture in the low-pressure stages of the turbine. The steam, after partial expansion, is withdrawn, superheated at constant pressure in the boiler, and returned to the turbine. This approach is used to improve performance in utility and other large plants with multiple staged turbines.

Reheating also improves the quality (i.e., reduces the moisture content) of exhaust steam. As discussed below, if steam quality falls below about 90% for an extended period of time, blade erosion and reduced efficiency result. By incorporating a reheat cycle, optimal boiler and condenser pressures can be obtained without concerns about low-quality turbine exhaust. Figure 11-6 is a schematic representation of a reheat cycle and Figure 11-7 shows the cycle in a temperature-entropy (T-s) diagram.

Another method used to increase the thermal efficiency of a steam power plant is a **regeneration cycle.**

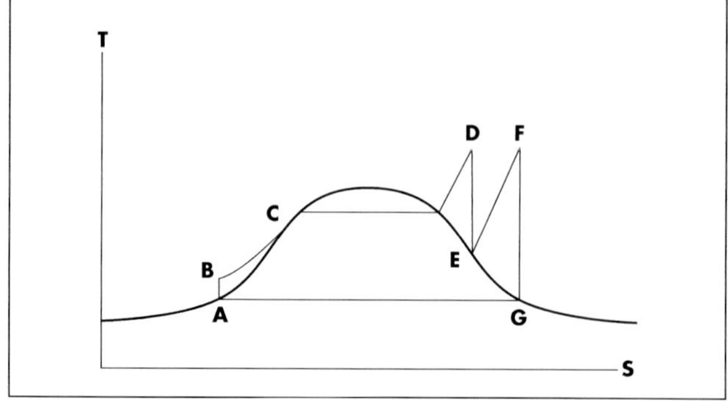

Fig. 11-7 Temperature-Entropy (T-s) Diagram of Steam Turbine Reheat Cycle.

With regeneration, steam is extracted from the turbine after it has partly expanded and used in a boiler feedwater heater. This produces a higher average temperature of heat addition (i.e., higher average boiler temperature). Since ideal efficiency is proportional to $1 - T_{out}/T_{in}$, the increase in average temperature of heat addition produces an overall increase in power plant efficiency.

Another way to look at this is, since the fluid temperature entering the boiler is higher, less energy is required to vaporize and superheat the steam and, therefore, less fuel is needed. However, the turbine also develops less work since some of the steam is bled from the second stage. Operating conditions, therefore, must be chosen so that the decrease in net work is more than compensated for by boiler fuel savings. A single regenerative heater can improve overall cycle efficiency by 2% or more. Figure 11-8 shows the flow loop of a regenerative cycle and Figure 11-9 shows the cycle in a T-s diagram.

In the figures, steam enters the turbine at state E. After expansion, to an intermediate state F, some of the steam is extracted and enters the feedwater heater. The steam that is not extracted is expanded to state G and condensed in the condenser. This condensate is pumped into the feed-

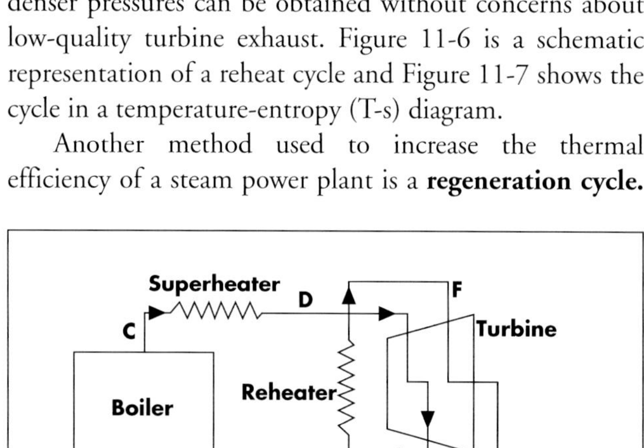

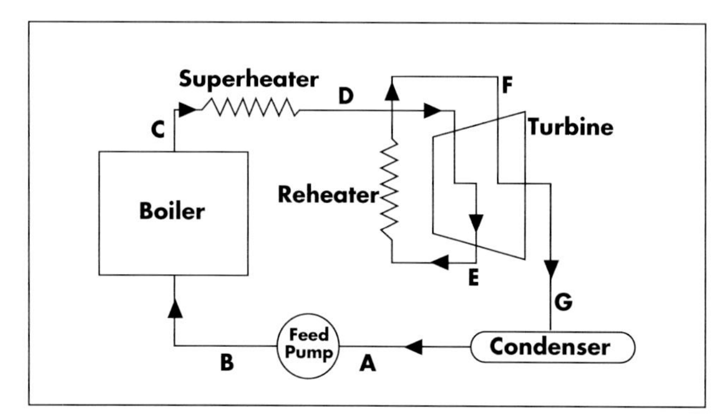

Fig. 11-6 Schematic Representation of Steam Turbine Reheat Cycle.

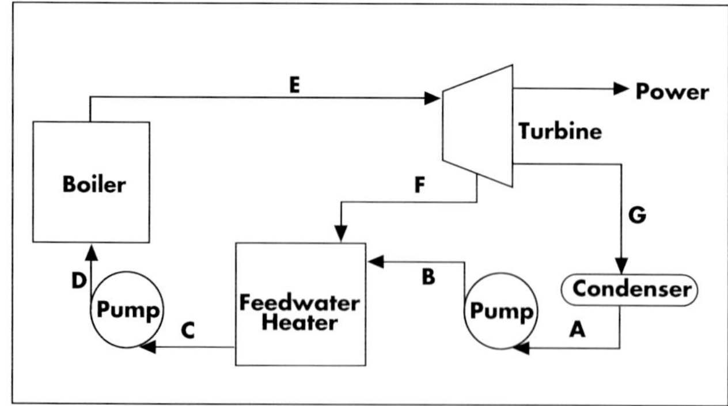

Fig. 11-8 Steam Turbine Regenerative Cycle Flow Loop.

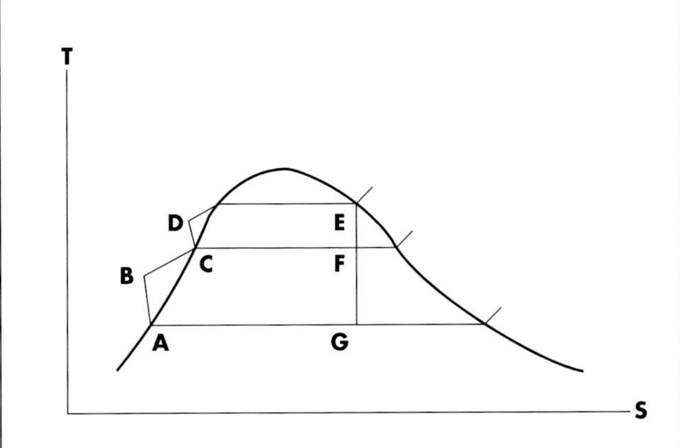

Fig. 11-9 Regenerative Cycle T-s Diagram (Non-Superheat, Non-Reheat).

water heater, where it mixes with the steam extracted from the turbine. The proportion of steam extracted is just sufficient to cause the liquid leaving the feedwater heater to be saturated at state C. Since condensate is pumped to an intermediate pressure corresponding to state F, a second-stage feedwater pump is required to achieve boiler pressure.

Types of Service Applications

Industrial steam turbines may be applied in **generator** or **mechanical drive** applications. Generator drives can be direct-drive applications, but more often the turbine drives the generator through a speed reduction gear, which allows the turbine to be designed for optimum speed and efficiency. Figures 11-10 and 11-11 show steam turbine-generator set applications.

Mechanical drives can be applied in numerous applications, including compressors, fans, blowers, pumps, mills, crushers, cutters, and line shafts. Depending on requirements, mechanical drive applications can operate at rotational speeds that are considerably higher than would be found with generators. Figure 11-12 shows a steam turbine mechanical drive application.

Steam turbines can be connected to drive two pieces of equipment, such as an air compressor and a generator. The turbine drives one or both machines, depending on low-pressure steam load. Figure 11-13 shows a dual-drive steam turbine featuring a double-ended shaft.

Fig. 11-10 Steam Turbine-Generator Set. Source: Dresser Rand

Fig. 11-11 Three, Multi-Valve, Condensing Steam Turbine-Generator Sets. Source: Dresser Rand

Fig. 11-12 Single-Stage Back-Pressure Steam Turbine Applied to Mechanical Drive Service. Source: Tuthill Corp., Coppus Turbine Div.

Fig. 11-13 Dual-Drive Steam Turbine Featuring Double-Ended Shaft. Source: Tuthill Corp., Coppus Turbine Div.

STEAM TURBINE COMPONENTS AND OPERATION

Steam Turbine Blades and Flow Path Design

The shape and arrangement of nozzles (stationary blades) and moving blades is perhaps the most important element of steam turbine design. It is the shape of the blades that determines the form of the flow passages and the energy transferred from the fluid (steam) to the rotor. A sufficient number of blades must be provided to ensure well-defined flow passages; yet too many blades may result in high resistance to flow. Since the arrangement of the flow passages affects manufacturing costs and maintenance requirements, a balance must be struck between operating efficiency and cost. Following is a discussion of the two basic design classes: impulse and reaction. The designs are compared in Figures 11-14 and 11-15.

In the **impulse turbine,** the pressure drop for the entire stage takes place across the stationary nozzle. Impulse nozzles orient the steam so it flows in well-formed high-speed jets. The pressure drop accelerates the steam to high velocity, with velocity related to the square root of the enthalpy drop. Steam energy is transferred to the rotor entirely by the steam jets striking the moving blades. Impulse turbines, which are somewhat similar to waterwheels, are far more rugged and durable than reaction turbines, making them preferable for most applications. The principle of the impulse turbine is shown graphically in Figure 11-16. The velocity ratio (v_r), which is the ratio of blade velocity (v_b) to the steam jet velocity (v_j), is the key consideration in optimizing impulse turbine performance.

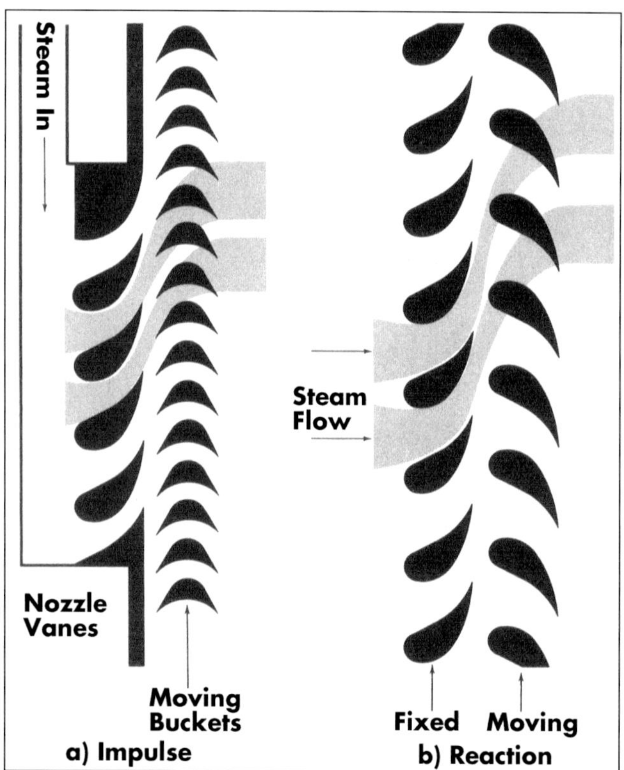

Fig. 11-14 Simple Impulse and Reaction Stage Diagrams.
Source: Babcock & Wilcox

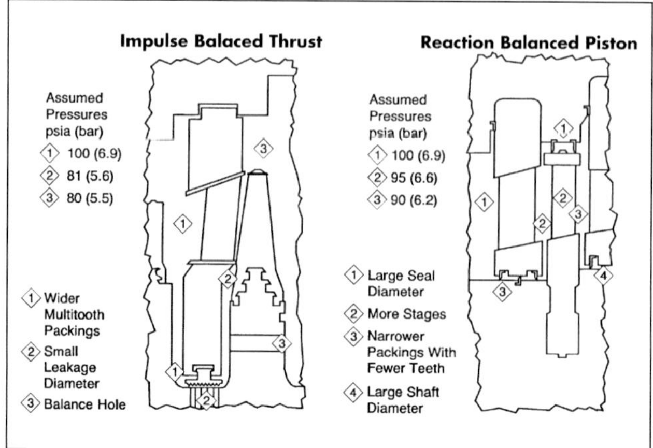

Fig. 11-15 Comparison of Impulse and Reaction Turbines.
Source: General Electric Company

$$v_r = \frac{v_b}{v_j} \tag{11-1}$$

Impulse turbines have two types of staging: pressure-compounded (Rateau) staging and velocity-compounded (Curtis) staging. A **Rateau stage** uses only a single row of moving blades (one row of buckets per stage). It is used in multi-stage turbines where a series of stages are included together in one casing. Optimum performance is achieved when the redirecting action of the blade causes the steam to come to a complete stop, with all of the kinetic energy of the steam jet transferred to the blade.

A **Curtis stage** consists of two rows of moving blades. Stationary nozzles direct the steam against the first row, and then reversing blades redirect it to the second row. The stationary blade absorbs no energy and redirects the jet 180 degrees so it can enter the second row of moving blades at twice the speed of the moving blade. This second moving blade absorbs the remaining kinetic energy in the steam jet, resulting in a steam velocity close to zero at the blade exit.

The design allows small wheel diameters and tip speeds, fewer stages, and a shorter, more rugged turbine. Curtis stages are only used in single-stage turbines and in the first stage of multi-stage turbines. Due to packaging considerations, relatively large angles must be used in the

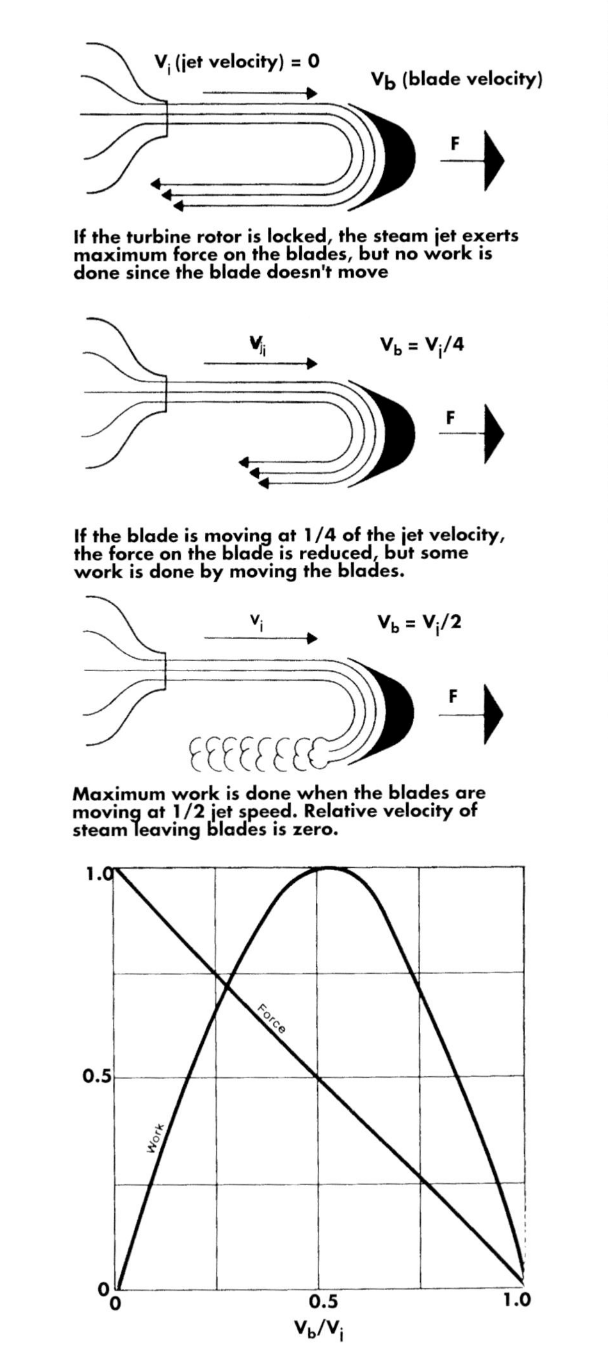

Fig. 11-16 Graphical Representation of Impulse Principle.
Source: Elliott Company

reversing blades and the second row of moving blades in a Curtis stage. Curtis stages are, therefore, generally less efficient than Rateau stages.

Diagram efficiency (η_D) is the portion of the steam jet's kinetic energy that is captured by the blade. Since kinetic energy is proportional to velocity squared, diagram efficiency can be expressed as:

$$\eta_D = \frac{v_1^2 - v_2^2}{v_1^2} \qquad (11\text{-}2)$$

Where:
v_1 = velocity of steam entering blade
v_2 = velocity of steam leaving blade

Ideal diagram efficiency will be achieved when v_b/v_j = 0.25 for a Curtis stage and when v_b/v_j = 0.5 for a Rateau stage. In practice, however, all of the kinetic energy in the steam jets cannot be captured. This occurs for several reasons. First, there are friction losses. Second, steam cannot come to a complete stop after leaving the blades, but must keep flowing in order to exit the turbine. Third, to keep the nozzles from being hit by the blades, they must direct steam at an angle. The greater the angle, the lower the portion of steam jet kinetic energy that can be transferred to the blades.

In many multi-stage turbine applications, turbine speed (rpm) is fixed by the speed of the driven equipment. The main design variable is the speed of the steam jets leaving the nozzles in each stage. Breaking up the total enthalpy drop optimizes velocity ratios across each turbine stage to produce an optimum steam jet velocity. The optimal number of stages is determined by the enthalpy drop that produces the optimal velocity ratio per stage. A higher steam jet velocity will produce a lower velocity ratio and therefore lower efficiency. However, since this results in a larger enthalpy drop across each stage, it takes fewer stages to achieve the full enthalpy drop across the turbine, thereby reducing production costs.

Due to first stage pressure limitations in multi-stage turbines, it is sometimes necessary to take a large enthalpy drop that results in the steam jet velocity being greater than four times blade velocity. The resulting velocity ratio of less than 0.25 makes the Curtis stage potentially more efficient than the Rateau stage. Also, the larger enthalpy drop required to produce sufficient velocity for the Curtis stage reduces the number of stages required. If the velocity ratio of the first (Curtis) stage is set at about half that used in the Rateau stages, the Curtis stage will consume as much of total enthalpy as four Rateau stages, due to the squared relationship of energy and velocity. This can reduce turbine construction costs in comparison to a design using only Rateau stages.

Reaction turbines make use of the reaction force produced as steam accelerates through the nozzles created by the blades themselves. In the reaction turbine, stages are composed of a stationary and a rotating row of blades, with the steam expanding in both the rows. The moving blades are designed to use the steam jet energy of the stationary blades and to act as nozzles themselves.

In a pure reaction stage, the stationary blade passage serves to increase the velocity of the steam and direct it at the moving blades. Reaction turbine blades are shaped so that the area between two adjacent blades of the same row will form a converging nozzle. This produces a pressure drop and, therefore, an increase in the relative velocity across each row. The pressure drop across these moving nozzles produces a reaction force that supplements the steam jet force across the stationary blades.

The diagram efficiency curve for reaction blading is flat at the peak, which allows optimal efficiency to be maintained over a wide range of velocity ratios. Ideal diagram efficiency is achieved with a velocity ratio that is twice that of Rateau blading and four times that of Curtis blading. For a 50% reaction design, optimum efficiency occurs at an ideal velocity ratio of 0.707. Therefore, for stages operating at the same diameter and stage pressure drop, the peak efficiency for a reaction stage would occur at a blade velocity, or wheel speed, considerably higher than for an impulse stage.

Enthalpy differentials for each stage of a reaction turbine are usually lower than those for impulse stages. Thus, reaction turbines require more stages. Reaction stages are generally more efficient than impulse stages, with reduced friction losses due to lower flow velocities. However, reaction turbines are more complicated and fragile due to tighter clearances. They are used in constant speed applications when efficiency is of critical importance. The greater number of stages allows for less space per stage and generally less rugged construction.

In practice, stages may be constructed with varying degrees of reaction and impulse blading and turbines may include various combinations of reaction, Curtis and Rateau stages. Turbine blades may be classified according to their degree of reaction or the ratio of the enthalpy drop in the moving blades to the enthalpy drop in the entire stage. Zero reaction would be the case with a pure impulse stage. When there is an equal enthalpy drop in the stationary and moving rows, the blades are 50% reaction.

One design used to balance capital cost and efficiency in larger turbines combines the impulse and reaction principles by employing a Curtis or Rateau stage as the first stage and using reaction stages for the remainder of the expansion. The Curtis stages reduce the pressure of the steam to a moderate level and produce a high proportion of output work, while the more efficient Rateau or reaction stages absorb the balance of the energy available.

Governors

Turbine shaft speed depends on the driven load and the amount of steam flowing into the steam chest. If the load on the shaft increases, at constant steam flow, the turbine speed decreases. Alternatively, if a constant load is maintained as steam flow increases, shaft speed will also increase. Mechanical energy output from a steam turbine is controlled by a **governor,** which operates a control valve to admit the correct amount of steam.

The mass flow rate of steam through a turbine is determined by the product of the absolute pressure of the steam entering the first stage nozzles and the throat area of the first stage nozzles. Figure 11-17 shows the effect of throttling on steam turbine operation in enthalpy vs. entropy (h-s). As throttling is increased, the steam energy available to do work in the turbine is reduced.

Steam turbines can operate at speeds ranging from 1,800 rpm to more than 14,000 rpm. Optimal speed for most industrial turbines is commonly in the range of 3,600 to 5,600 rpm. Typically, single-stage turbines will operate down to 50% of rated speed. Multi-staged turbines are often limited to 70% or above due to the length and weight of the shaft.

As turbines move away from design speed, they lose efficiency. Figure 11-18 shows steam flow versus hp for the

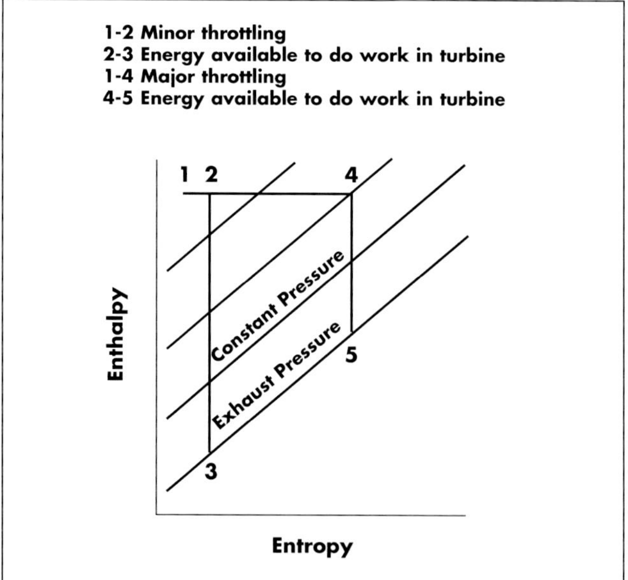

Fig. 11-17 Effect of Throttling on Steam Turbine Operation. Source: Tuthill Corp., Murray Turbomachinery Div.

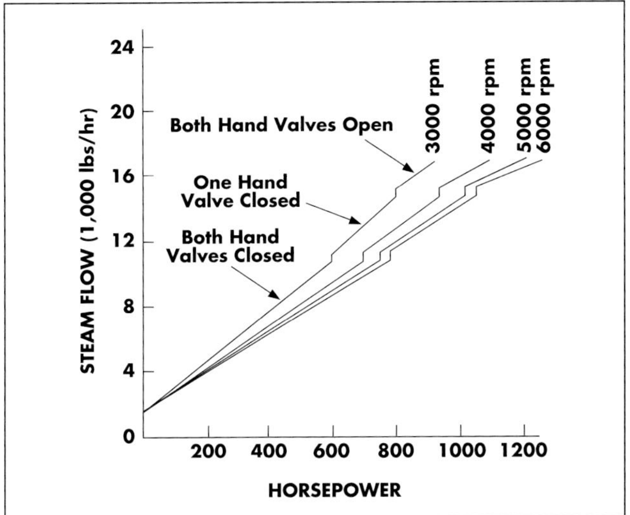

Fig. 11-18 Steam Flow vs. Power for Same Turbine Operating at Various Speeds. Source: Tuthill Corp., Murray Turbomachinery Div.

same representative steam turbine at various turbine speeds. Indications are provided as to when hand valves should be opened and closed to provide maximum efficiency.

Mechanical governors are used in the most basic control systems. In mechanical governors, shaft speed is sensed by a fly-ball mechanism and hydraulic relays provide input to the control valves. The speeder spring exerts a force on the thrust bearing, while the flyweights exert an opposing force proportional to the rotational speed of the turbine. Depending on the speeder spring tension and the speed of the turbine, an equilibrium point is established. Figure 11-19 is a cutaway illustration detailing a steam turbine governor valve.

Electronic governor systems are somewhat more sophisticated than mechanical governors and use speed pick-ups and electronic circuit boards instead of fly-balls and hydraulic relays. In the control process, sensors measure operating parameters of the turbine. A governor compares signals from the sensors with selected set points and produces a signal to the valve actuators that position the valves directly or indirectly.

The most advanced control systems use microprocessor-based digital logic and can provide an abundance of diagnostic information to the operator. Microprocessor-based systems also allow for the use of two governors, providing 100% emergency backup.

Toe piece
Snap ring
Stainless steel washer
Carbon ring
Stainless steel valve stem
Carbon sleeve for low friction positive seal
Stainless steel seats
End cap with low point drain
Trunnion block
Compression spring
Stainless steel seal block
Intermediate leakoff
Valve bonnet
Low-force, characterized V-ported governor valve
Valve body
Governor Valve

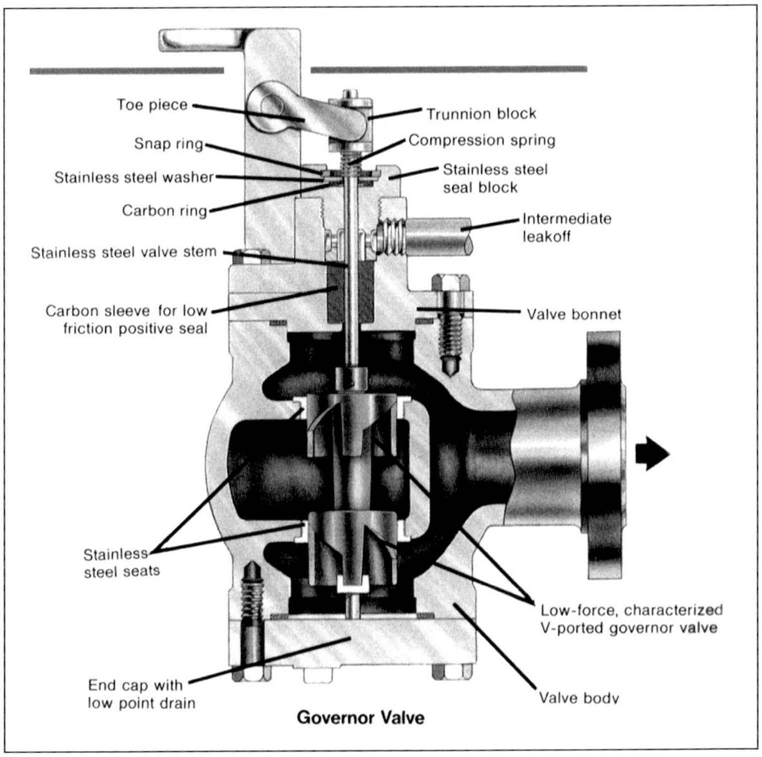

Fig. 11-19 Cutaway Illustration of Steam Turbine Governor Valve. Source: Tuthill Corp., Coppus Turbine Div.

Control Valves

Inlet control valves regulate steam flow and pressure to provide the appropriate power and speed. Throttling, which occurs across the control valve(s), reduces the thermal performance of the turbine. The efficiency loss will be a function of the control valve design and overall turbine pressure ratio. For this reason, multiple valves allow for more efficient operation under partial loads than a single valve. Turbines with larger pressure ratios usually experience smaller efficiency losses for a given amount of throttling than turbines with smaller pressure ratios.

The term hand valves is commonly used to refer to valves for individual nozzle sets. Hand values are often used to reduce throttling losses under partial loads and can improve efficiency in systems that include a single modulating control valve. As the load on the turbine is reduced, one or more hand valves can be closed to reduce throttling loss by closing off nozzle compartments in the first stage of the turbine. With reduced nozzle area, the governor must increase the ring pressure to allow the turbine to carry the load. The increased ring pressure results in more energy per lbm (or kg) of steam, thereby reducing the steam flow required to do the required amount of work. In order to bring the turbine back up to full load, the hand valve must be reopened.

Valves may be operated manually or actuated by remote or automatic control. Figure 11-20 shows a single-stage steam turbine model featuring multiple-hand valves. These nozzle hand valves provide control for part load, overload, and minimum inlet/maximum back-pressure operation. Figure 11-21 shows a representative steam turbine performance curve in steam flow versus ring pressure. Performance is indicated for turbine operation with zero, one, and two hand valves closed.

Multi-valve turbines are an alternative to hand valves for improving efficiency at reduced load. Several valves exist, each of which controls steam flow to an individual nozzle compartment. The governor linkage is configured to sequentially open and close these valves to minimize throttling loss. At any time, only one valve is throttling and the other values are either fully open or closed.

Figure 11-22 compares part-load performance of a single-valve turbine to that of a multi-valve turbine in steam flow versus power output. Notice the incremental impact of closing one and two hand valves as compared to the impact of sequential opening and closing of valves in the multi-valve configuration.

Fig. 11-20 Multiple Hand Valves on Single-Stage Steam Turbine. Source: Tuthill Corp., Coppus Turbine Div.

Figure 11-23 shows a multi-valve, controlled extraction turbine with the casing cover removed to reveal the rotor. This turbine was built to drive a generator and is rated at 5,000 kW at 9,483/1,800 rpm, with an inlet steam condition of 600 psig/750°F (42.4 bar/399°C). The turbine has an inlet steam flow of 158,000 lbm/h (71,668 kg/h) and a controlled extraction opening at 200 psig (14.8 bar), with an extraction steam flow of 125,000 lbm/h (56,669 kg/h).

Another type of valve is the **trip (or stop) valve.** Trip valves are used to shut off the supply of steam to the turbine in response to a shutdown (or trip) signal, indicating that an out-of-specification condition has developed. Typically, a spring-loaded trip valve is bolted to the inlet flange of the turbine and latched in the open position when the turbine is to run. A solenoid trip can also be used to trip the turbine in response to an electrical signal. As an option, a separate combination trip-and-throttle valve may be used to close off the steam flow in response to a shutdown (or trip) signal. The trip-and-throttle valve can be used to modulate steam flow during start-up as the turbine is being brought up to speed. It may be manually or hydraulically positioned from zero lift to 100% lift.

Figure 11-24 illustrates a governor assembly and overspeed trip and valve. On the left, the governor assembly is

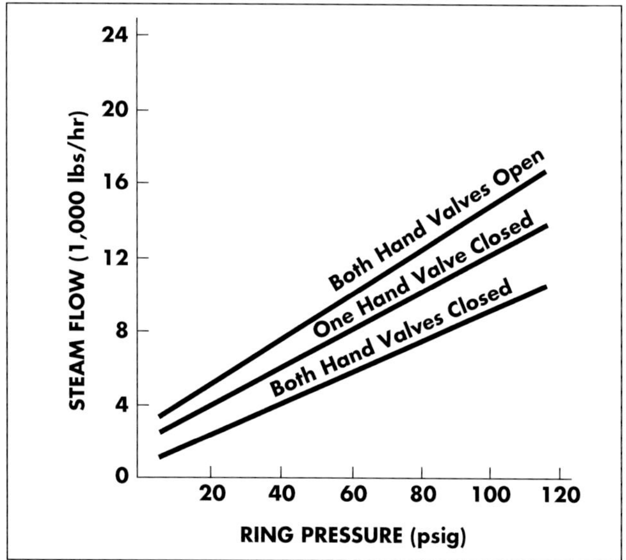

Fig. 11-21 Representative Steam Turbine Performance Curve Showing Impact of Closing Hand Valves. Source: Tuthill Corp., Murray Turbomachinery Div.

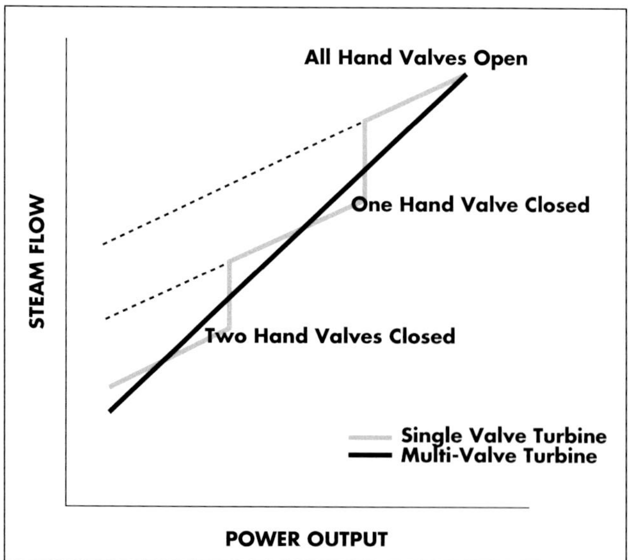

Fig. 11-22 Comparison of Part-Load Performance of Single-Valve and Multi-Valve Turbines. Source: Tuthill Corp., Murray Turbomachinery Div.

Fig. 11-23 Multi-Valve, Controlled Extraction Steam Turbine Designed to Drive Electric Generator. Source: Tuthill Corp., Murray Turbomachinery Div.

shown with steam chest detail shown below. The overspeed trip and valve is detailed on the right. The trip valve can be reset against full line pressure by using an auxiliary resetting lever. The trip setting is adjustable without lifting the bearing case cap or disturbing the linkage

Control Variables

Speed control is the primary control loop for all steam turbines, since without it, the turbine cannot be started or operated safely. While one steam control device can control only one parameter, such as speed, application flexibility is increased when the controlled parameter's set point is influenced as a function of a secondary parameter, such as exhaust pressure. For example, exhaust pressure can be used to bias the speed set point in order to maintain a desired exhaust set point.

In electric generation applications, it is often desirable to control the turbine for load rather than speed. For condensing turbines and admission turbines, load control is also usually the primary parameter. In the case of back-pressure and extraction turbines, load control may be secondary.

Typical driven equipment and/or process control parameters include pressure, flow, temperature, and speed. Often, it may be desirable to reduce turbine speed at low load to improve the efficiency of the driven equipment (a fan or centrifugal compressor, for example), even if the turbine itself would operate more efficiently at a higher speed. In such cases, the reduced power requirement of the driven equipment more than offsets the negative impact of reduced turbine efficiency.

In addition to speed and load, other important control parameters sometimes include the following:

- Inlet pressure
- Extraction pressure
- Induction pressure
- Exhaust pressure

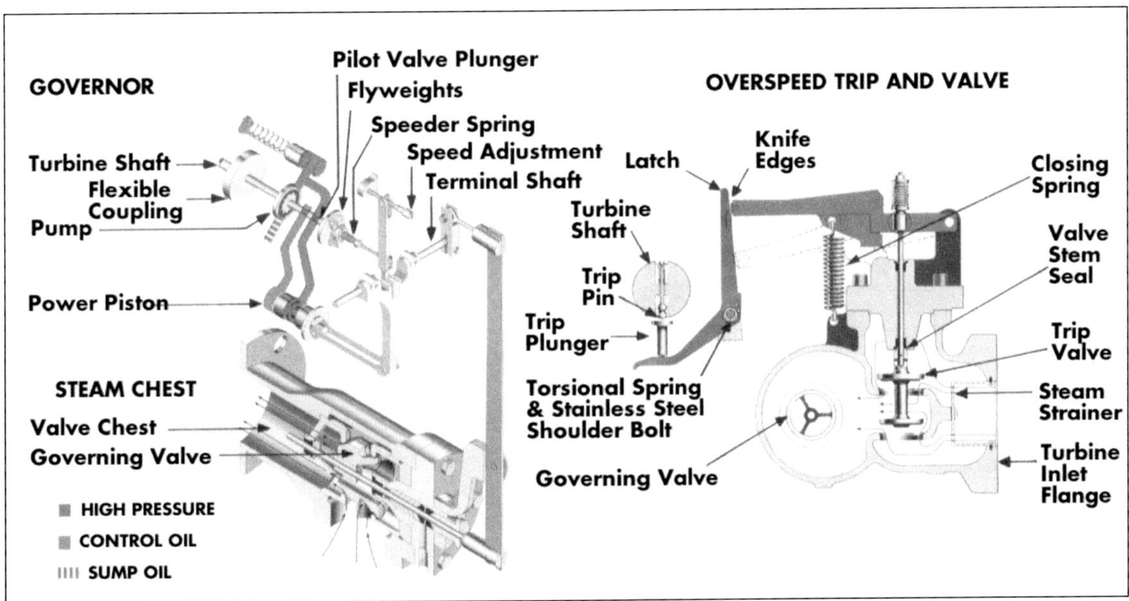

Fig. 11-24 Illustration of Governor Assembly and Overspeed Trip and Valve. Source: Elliott Company

Inlet pressure is controlled by manipulating the inlet governor valve(s). To control inlet pressure, a bias must be applied to the speed control loop that relates actual inlet pressure to set point values. That is, after comparing a pressure signal to a reference signal, the required speed reference is generated. The corrective signal repositions the actuator controlling the inlet steam valve. To promote stability and keep the turbine from reacting to minute changes, a dead band is established in the control system to provide a window within which no corrective action is taken.

Exhaust pressure control is used with back-pressure turbines when the exhaust steam is being used in a process that requires steam pressure control. Exhaust pressure is controlled by manipulating the inlet governor valve(s) in a manner analogous to that described above.

Extraction valves regulate flow to downstream stages of the turbine in order to maintain constant back-pressure at the exhaust. With extraction added to speed/load as a control parameter, the governor must control (ratio) the inlet valves and the extraction valves in such a manner that both speed/load control and extraction pressure are held at desired levels.

Induction (admission) control valves operate on a similar principle. Induction valves control flow into the downstream stages of the turbine, while maintaining constant pressure at the induction opening. As with extraction turbines, the control system must ratio the positioning of the inlet valve and the induction valve to maintain stable set point control.

For every application, a steam map, or envelope, must be established that provides a description of the operating range of the turbine and, therefore, establishes boundaries. Boundary lines include a maximum throttle (inlet) flow, maximum power, maximum exhaust flow, zero extraction flow, minimum power, and minimum exhaust. Within these boundaries are the combination of throttle and extraction flow that result in various power output levels with safe and stable operating conditions.

The minimum exhaust flow is usually of critical importance in order to avoid overheating the exhaust section. In addition, admission turbines must assure a minimum throttle flow so that enough steam flows through the inlet section to cool the forward section of the turbine rotor.

Figure 11-25 shows a schematic diagram of a digital control system for an automatic extraction steam turbine. The system features a 32-bit microprocessor-based digital control. Typical control functions include turbine speed control, load control/limiting, inlet and exhaust pressure

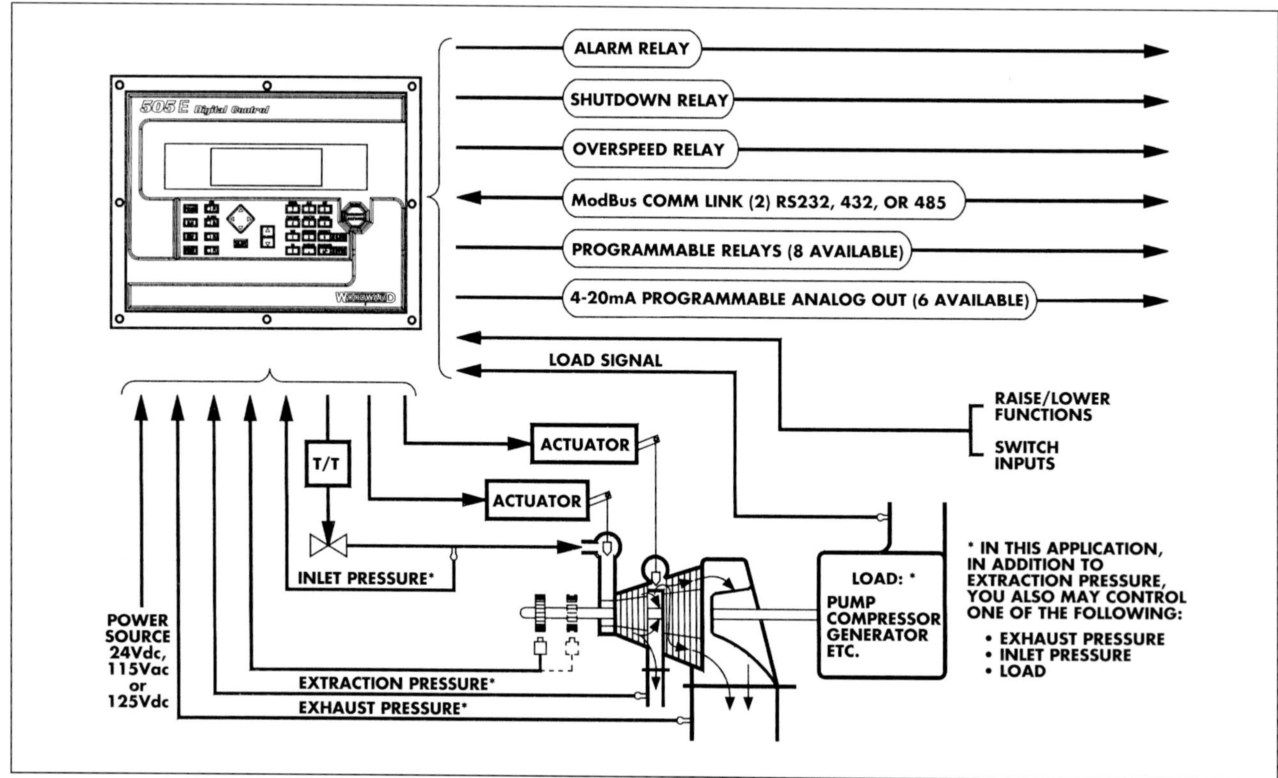

Fig. 11-25 Schematic Diagram of Digital Control System Configuration for Automatic Extraction Turbine. Source: Woodward Governor Co.

control/limiting, import/export control/limiting, and isochronous load sharing. As shown, in this application, two parameters are controlled. In addition to extraction pressure, load, exhaust pressure, or inlet pressure may be controlled.

Instrumentation

There are many types of instrumentation that can be used depending on the type of turbine application. Typical instruments used include:

- Steam pressure gauges for supply exhaust and nozzle ring or first stage pressure
- Steam temperature gauges for turbines operating with superheated steam
- Oil pressure gauges for lube and control oil
- Oil cooler temperature gauges
- Vibration detectors
- Bearing temperature gauges

Condensing Process

In condensing turbines, steam expands to a deep vacuum with the specific volume of the steam at exhaust equal to about 1,000 times its inlet volume. Steam is then condensed or changed to liquid form in the following stage, where thermal energy is extracted. Condensing turbines are exhausted to a vessel, or condenser, in which thermal energy is rapidly and continuously removed through direct or indirect contact with cooling water.

Heat rejection systems are generally classified as once-through (open) or closed systems. In once-through systems, water is withdrawn from a body of water, such as a river or ocean, and pumped through the condenser where its temperature increases by about 15 to 20°F (8.3 to 11.1°C). The warmer water is then discharged back to the source.

Due to environmental regulations and inaccessibility to acceptable heat sinks, closed-loop cooling towers are usually used. In wet cooling tower systems, water that has been circulated through the condenser is then circulated through the tower where the heat is rejected to atmosphere, mainly by evaporation. Refer to chapter 36 for a detailed discussion on cooling towers and related heat rejection systems.

After condensing, liquid water occupies less than 0.1% of the volume of an equal weight of saturated steam at atmospheric pressure. This change in volume creates a partial vacuum at the turbine exhaust.

To maintain this vacuum, the products of the condensing process must be continuously removed from the condenser. These products include the water resulting from the condensation of the steam, as well as other gases carried into the condenser with the exhaust steam. These products result from leakage in the exhaust line or from the presence of gases in the feedwater originally supplied to the boiler. If circulating coolant water is introduced directly into the exhaust steam, the water and the gases, which it contains, must also be removed from the condenser.

If the condenser were free of air, the pressure created would be the saturation pressure equivalent to the steam temperature. If, for example, the steam temperature were 80°F (26.7°C), the saturation pressure would be 0.5069 psia (3.495 kPa). This corresponds to 1.04 in. (2.64 cm) HgA, or 28.88 in. (73.35 cm) Hg vacuum. Final steam temperature is a function of the cooling water supply temperature and the amount of cooling water supplied per unit mass — lbm or kg — of steam condensed.

Condensing pressure for a surface condenser cooled with water from a cooling tower is estimated by computing the sum of four temperatures:

A. Ambient (outdoor) wet-bulb temperature

B. Cooling tower approach temperature (the temperature difference between cooling tower water and outdoor wet-bulb)

C. Temperature rise in the surface condenser

D. Condenser approach temperature (the temperature difference between saturated steam temperature and cooling water)

Consider the following example using typical values for a conventional cooling tower surface condenser application:

A. Design wet-bulb temperature = 78°F (26°C)

B. Tower approach to design wet-bulb = 7°F (4°C)

C. Temperature rise in surface condenser = 20°F (11°C)

D. Condenser approach temperature = 9°F (5°C)

The saturation temperature of the turbine exhaust steam is:

$$A + B + C + D = 78 + 7 + 20 + 9 = 114°F\ (46°C)$$

From steam tables, a 114°F (46°C) saturation temperature occurs at 2.91 in. (7.39 cm) HgA, or 27.01 in. (68.61 cm) Hg vacuum. If the design wet-bulb temperature were higher, either the condenser pressure would have to be increased or one or both approach temperatures decreased. Decreasing the approach temperatures requires a corresponding increase in condenser size and cost.

Figure 11-26 shows typical steam turbine condensing pressure versus saturation temperature and required design ambient wet-bulb temperatures. For a given

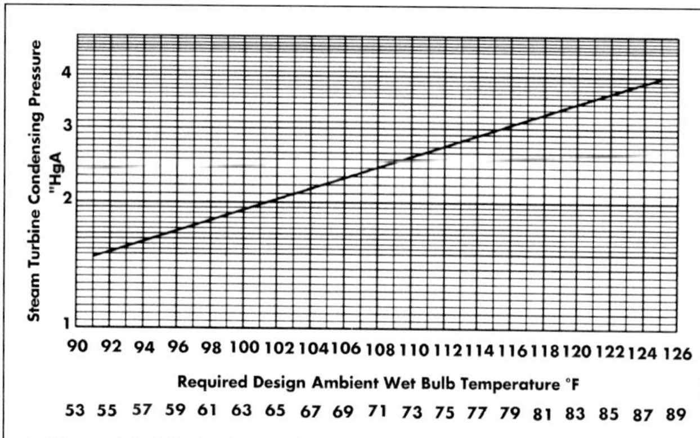

Fig. 11-26 Condensing Pressure vs. Saturation and Required Design Wet-Bulb Temperatures. Source: Tuthill Corp., Murray Turbomachinery Division

application, both the selected cooling tower and surface condenser supplier must confirm these approximations.

Water-cooled **surface condensers** are the most common types of technology applied to steam turbine condensing systems. Alternatives include direct contact condensers, which spray cooling water directly into the steam flow, and air-cooled condensers, which may be applied when cooling water is not readily available or cooling tower operation is not acceptable. The surface condenser serves three important functions: 1) providing a low pressure at the turbine exhaust to maximize plant system efficiency; 2) isolating the high purity water in the closed system to minimize water treatment costs; and 3) minimizing condensate flow to reduce corrosion potential to system components. The condenser also serves as a collection point for all condensate drains.

Surface condensers are usually shell-and-tube heat exchangers with exhaust steam condensing on the shell side and cooling water flowing in one, two, or four passes inside the shell. Most small and intermediate capacity surface condensers are cylindrical, while very large capacity units are rectangular. Figure 11-27 shows a single shell, single pass steam surface condenser for very large capacity applications. As shown, condensate is collected in the bottom of the condenser in the hot well for return to feedwater heaters. Figure 11-28 is an intermediate-capacity shell-and-tube surface condenser that serves a 4,500 kW steam turbine generator set. This two-pass tube-side, one-pass shell-side heat exchanger is designed to condense 49,000 lbm/h (22,226 kg/h) at 2.5 in. HgA (8.5 kPa) when provided with 5,200 gpm (328 L/s) of cooling water.

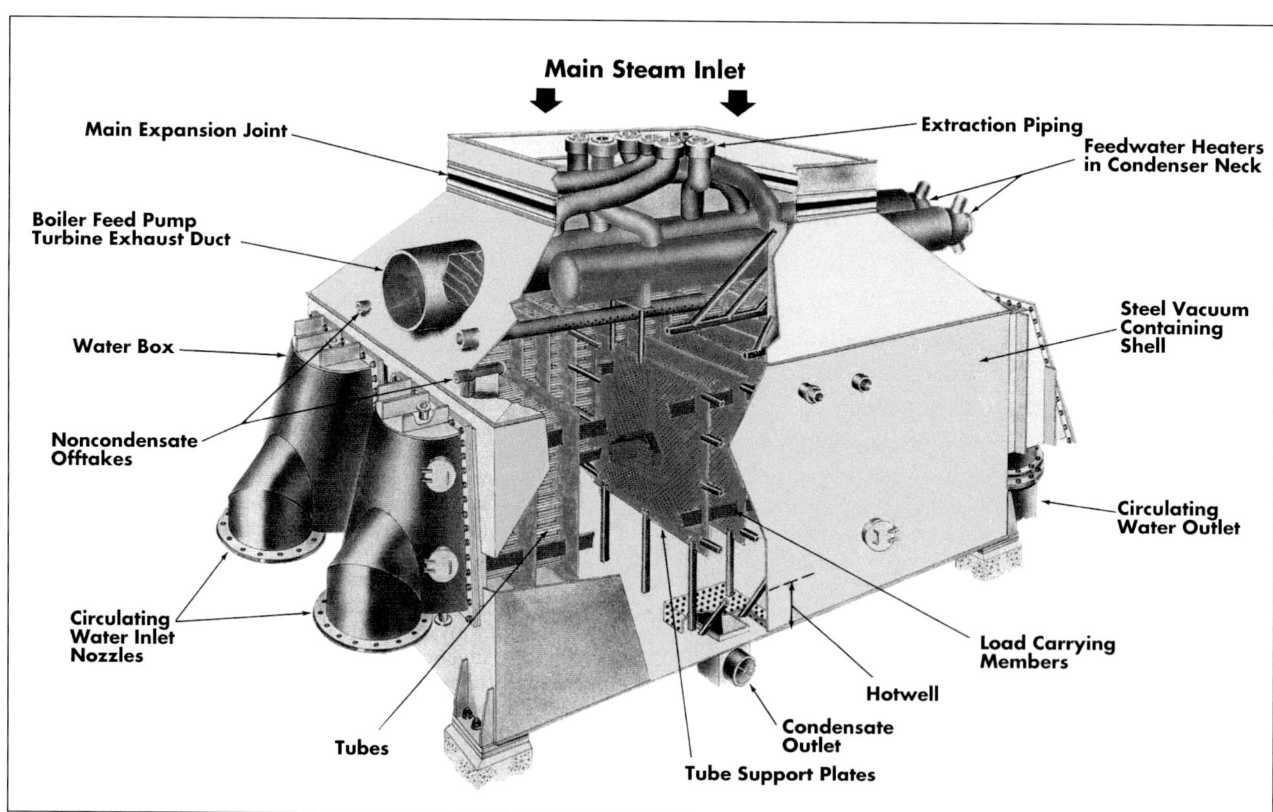

Fig. 11-27 Large Capacity, Central Power-Plant Type Surface Condenser. Source: Babcock & Wilcox

Fig. 11-28 Shell-and-Tube Surface Condenser Serving 4,500 kW Condensing Unit.
Source: Tuthill Corp., Murray Turbomachinery Div.

STEAM TURBINE POWER CYCLE OPERATION AND PERFORMANCE

Ideal Rankine Cycle

The energy available per unit mass (e.g., lbm or kg) of steam flowing through a steam turbine is a function of the turbine pressure ratio (ratio of absolute inlet pressure to absolute exhaust pressure) and the inlet temperature. The maximum ideal thermal efficiency achievable with a steam turbine cycle is, therefore, the difference between the inlet and exhaust energies (or enthalpies). This efficiency is based on the cyclic process being reversible and adiabatic (isentropic).

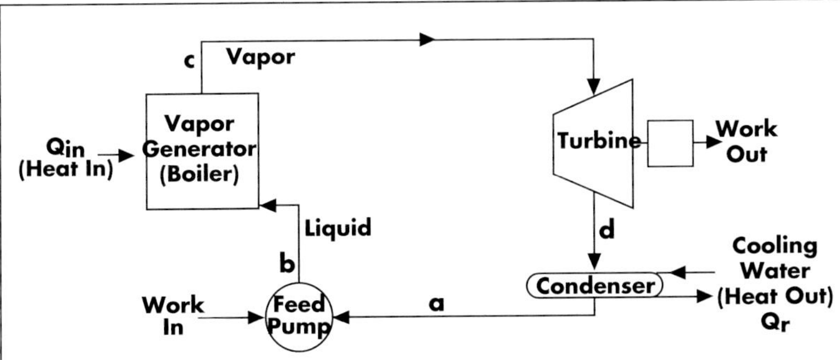

Fig. 11-29 Elements of the Simple Rankine Cycle.

As shown in Figure 11-29, the simple ideal Rankine cycle can be reduced to four processes:

1. Process A-B: Reversible adiabatic pumping of return condensate into the pressurized boiler. For this process:

$$\textit{Pump work} = (h_b - h_a) \qquad (11\text{-}3)$$

2. Process B-C: Constant-pressure heat transfer, resulting in steam generation from return condensate. For this process:

$$\textit{Heat added in boiler} = (h_c - h_b) \qquad (11\text{-}4)$$

3. Process C-D: Reversible adiabatic expansion of the steam in the turbine from a saturated vapor state to the condenser pressure. For this process:

$$\textit{Turbine work} = (h_c - h_d) \qquad (11\text{-}5)$$

4. Process D-A: Constant-pressure heat transfer resulting in condensation of steam in the condenser. For this process:

$$\textit{Heat rejected in condenser} = (h_d - h_a) \qquad (11\text{-}6)$$

Figure 11-30 shows schematic representa-tions of pressure-volume (p-v), temperature-entropy (T-s), and a **Mollier diagram** for the ideal Rankine cycle.

In the pressure-volume (p-v) diagram, the the two constant-pressure phases of admission (4-1) and exhaust (2-3) are connected by an isentropic-expansion phase (1-2). The shaded area 4-1-2-3 represents the work of the cycle.

The temperature-entropy (T-s) diagram shows the properties for liquid, wet vapor, and superheat, as taken from the steam tables. The Rankine cycle can be superimposed, as shown. Underneath the curve lies the vapor dome in which a liquid vapor mixture exists. The upper of the parallel lines (c-d) represents the path of heat addition at constant temperature as high-pressure liquid is vaporized to steam. The line (d-1) leading from the vapor dome to point 1 represents superheating of steam. The isentropic expansion in the steam turbine is the vertical line 1-2. The bottom line (2-a) represents the condensing path in which heat is rejected as low-pressure steam is condensed to a liquid. The isentropic compression in the feed pump is the vertical line a-b. Compression phase is usually negligible

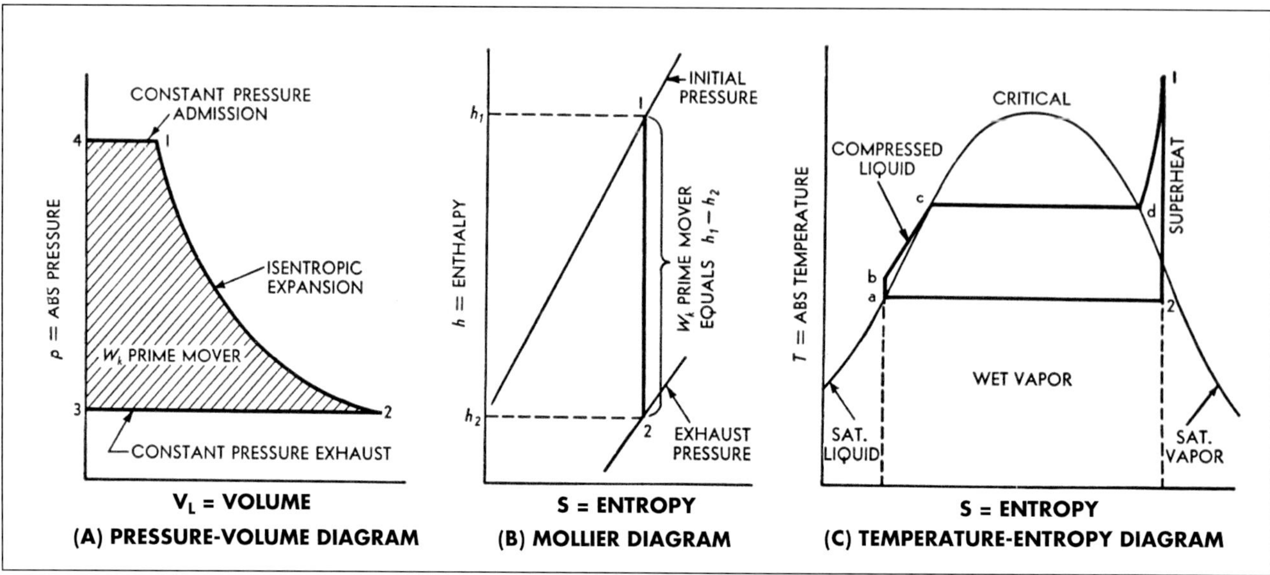

Fig. 11-30 Pressure-Volume, Mollier and Temperature-Entropy Diagrams for Ideal Rankine Cycle. Source: Babcock & Wilcox

and points a and b can, therefore, be considered as a single point, saturated liquid at the exhaust pressure. All of the necessary values for calculating power output and efficiency can be taken from standard steam tables.

Another way to identify the values required for calculating power output and efficiency from a given set of conditions is to use a Mollier (or h-s) diagram to represent the thermodynamic properties of steam. The ordinate (h) is enthalpy and the abscissa (s) is entropy. The available energy that can be converted into work is shown on the Mollier diagram as a vertical line $h_c - h_d$.

Ideal simple Rankine cycle thermal efficiency is represented by:

$$\eta_{th} = \frac{\textit{Turbine work out} - \textit{Pump work}}{\textit{Heat in}} = \frac{(h_c - h_d) - (h_b - h_a)}{(h_c - h_b)} \quad (11\text{-}7)$$

Pump work is often left out of this equation*, and thus:

$$\eta_{th} = \frac{\textit{Turbine work out}}{\textit{Heat in}} = \frac{(h_c - h_d)}{(h_c - h_b)} \quad (11\text{-}8)$$

* At lower pressures, the pump work is negligible, but at higher pressures, it may play a somewhat larger role. In a 100 MW regeneration cycle, for example, the total pump work per unit of mass passing through the first stage turbine is 3.73 Btu/lbm (8.68 kJ/kg) compared to turbine work output of 422.5 Btu/lbm (982.7 kJ/kg) and heat input of 1134.4 Btu/lbm (2,638.6 kJ/kg) into the steam generator. Therefore, the pump work is 0.3% of the heat input and 0.9% of the turbine work.

The **theoretical water rate (TWR)**, or **steam rate (TSR)**, is the amount of steam required by the turbine, at ideal expansion, to produce a given unit of work output. It is expressed, in English system units (refer to conversion tables for SI unit equivalents), as:

$$\textit{TWR (or TSR)} = \frac{2{,}545}{(h_c - h_d)} \quad (11\text{-}9)$$

Where TWR is in lbm/hp-h and $(h_c - h_d)$ is in Btu/lbm. This equation, expressed in kWh output, is:

$$TWR = \frac{3{,}413}{(h_c - h_d)} \quad (11\text{-}9a)$$

Where, TWR (or TSR) is in lbm/kWh and $(h_c - h_d)$ is in Btu/lbm.

Ideal heat rate for a condensing turbine is the product of TWR and boiler heat input. For a back-pressure turbine, where the steam evaporation process is debited to process, heat rate is simply the TWR times the heat drop. Thus:

$$\textit{Ideal heat rate (condensing turbine)} = TWR(h_c - h_b) \quad (11\text{-}10)$$

$$\textit{Ideal heat rate (back-pressure turbine)} = TWR(h_c - h_d) \quad (11\text{-}11)$$

Ideal cycle power output is determined by the steam flow rate times the heat drop. It is expressed in terms of hp as:

$$W = \frac{\dot{m}\,(h_c - h_d)}{2{,}545} \qquad (11\text{-}12)$$

Where:

$\dot{m}$ = steam flow rate, in lbm/h
$h_c - h_d$ = heat drop, in Btu/lbm
W = power output, in hp

Similarly, in terms of kW, it is expressed as:

$$W = \frac{\dot{m}\,(h_c - h_d)}{3{,}413} \qquad (11\text{-}12a)$$

Figure 11-31 provides a series of diagrams showing the effect of changes in steam temperature, steam pressure, and exhaust pressure on ideal Rankine Cycle performance.

Actual Rankine Cycle

Ideal simple Rankine cycle efficiency assumes isentropic expansion through the turbine or expansion that is carried out reversibly and adiabatically. This means that all enthalpy between the turbine inlet steam h_c and exhaust steam h_d is converted to work output at maximum possible thermal efficiency.

As with any power cycle, actual cycle efficiency is less than ideal. Under actual conditions, inefficiencies occur and the expansion process is not carried out at constant entropy. The factors affecting the cycle efficiency include throttling losses, friction between the steam and the turbine, bearing losses, and steam leakage. In addition to staging, there is generally a strong correlation between capacity and turbine mechanical efficiency. Figure 11-32 shows the relationship between design capacity and turbine mechanical efficiency.

In evaluating the relative performance of a steam turbine, the critical measure is the **actual Rankine efficiency.** This is the ratio of shaft work delivered by the turbine to the shaft work that would be developed by an ideal turbine expanding the steam from the same throttle conditions to the same exhaust pressure.

Thus, the actual turbine efficiency, η_T, is

$$\eta_T = \frac{(h_c - h_d)_{actual}}{(h_c - h_d)_{ideal}} \qquad (11\text{-}13)$$

The steam tables or Mollier chart can be used to determine the difference in throttle and exhaust enthalpy or the heat per lbm (or kg) of steam actually converted into shaft work. Actual expansion is not represented by a vertical (constant entropy) line on a Mollier diagram, but a line sloping downwards to the right.

Actual water rate is calculated as follows:

$$AWR = \frac{TWR}{\eta_T} = \frac{2{,}545}{(h_c - h_d)_{actual}} \qquad (11\text{-}14)$$

Actual water rate is affected by several factors, including:

A. Base efficiency — determined by inlet steam condition and output power.
B. Speed correction factor — determined by the speed and the basic pitch diameter of the turbine wheel; this factor is also used to plot efficiency under variable speed operation.
C. Superheat correction factor — determined by subtracting the saturation temperature of the steam from the actual inlet temperature.
D. Back-pressure correction factor — determined by the exhaust pressure.
E. Efficiency loss correction factor — determined by losses that result from reduction or step-up gears and generator operation, where applicable.

Thus:

$$AWR = \frac{TWR}{A \times B \times C \times D \times E} \qquad (11\text{-}15)$$

In calculating heat rate, the amount and temperature of condensate returned to the boiler, the make-up water, and the boiler efficiency must be considered. The actual heat rate is expressed as:

$$\textit{Heat rate (condensing turbine)} = AWR\,(h_c - h_b)_{actual} \qquad (11\text{-}16)$$

$$\textit{Heat rate (back-pressure turbine)} = AWR\,(h_c - h_d)_{actual} \qquad (11\text{-}17)$$

Actual power output in terms of hp is expressed as:

$$W_a = \frac{\dot{m}\eta_T\,(h_c - h_d)_{ideal}}{2{,}545} = W_a = \frac{\dot{m}\,(h_c - h_d)_{actual}}{2{,}545} \qquad (11\text{-}18)$$

Where:

W_a = power output, in hp
$(h_c - h_d)_{actual}$ = actual heat drop, in Btu/lbm.

Similarly, actual power output (W_a) in terms of kW is expressed as:

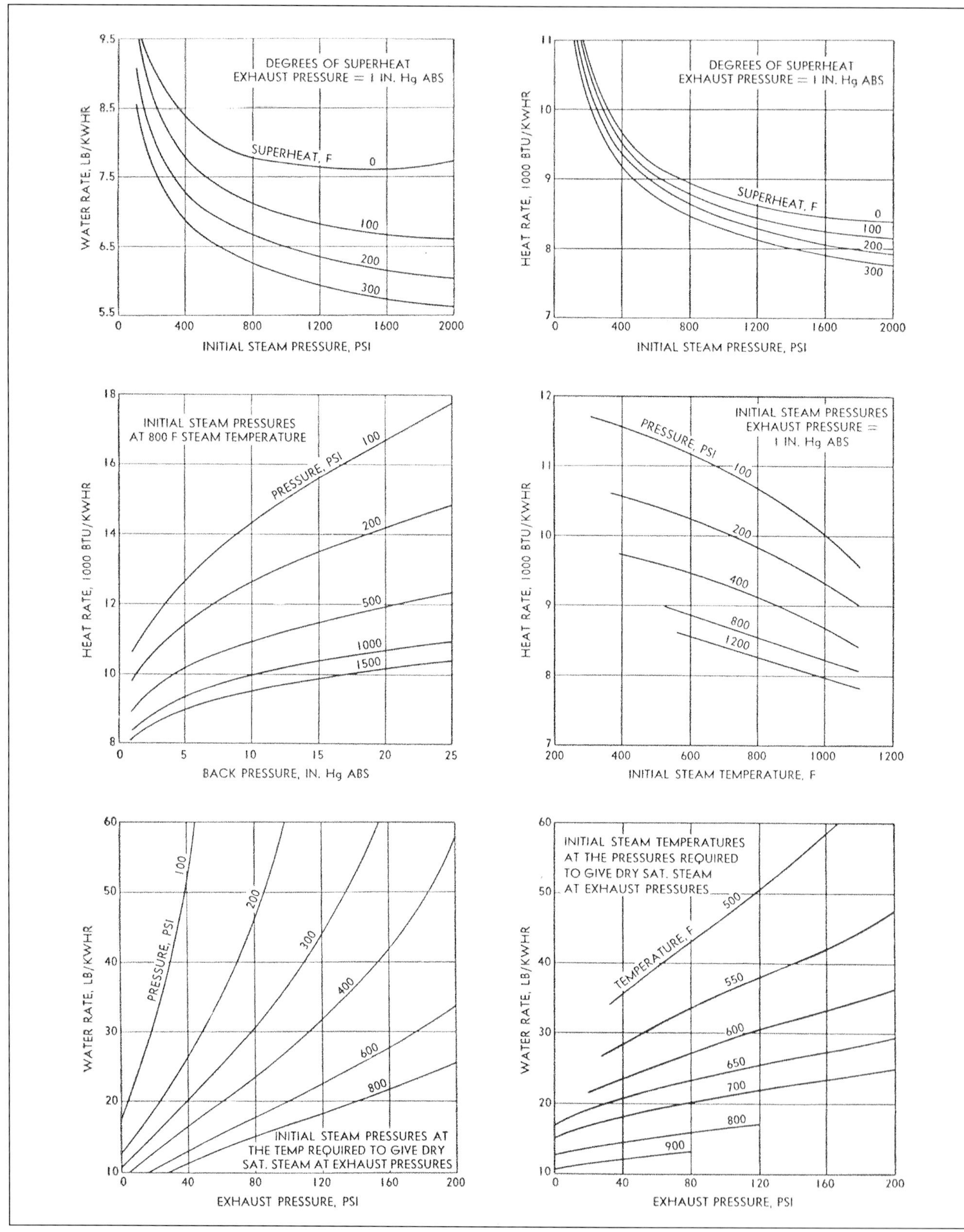

Fig. 11-31 Effect of Changes in Steam Temperature, Pressure, and Exhaust Pressure on Ideal Rankine Cycle. Source: Babcock & Wilcox

$$W_a = \frac{\dot{m}\eta_T (h_c - h_d)\ ideal}{3{,}413} = W_a = \frac{\dot{m}\,(h_c - h_d)\ actual}{3{,}413} \quad (11\text{-}18a)$$

Figures 11-33a and b are useful nomegraphs for determining approximate turbine efficiency when hp (kW), speed, and steam conditions are known. In Figure 11-33a, which has English unit values, a 25,000 hp, 5,000 rpm turbine using steam at 600 psig/750°F exhausting to 4 in. HgA has an approximate efficiency of 77%. Similarly, in Figure 11-33b, which has SI unit values, an 18,500 kW, 5,000 rpm turbine using steam at 40 bar/400°C exhausting to 150 millibar also has an approximate efficiency of 77%.

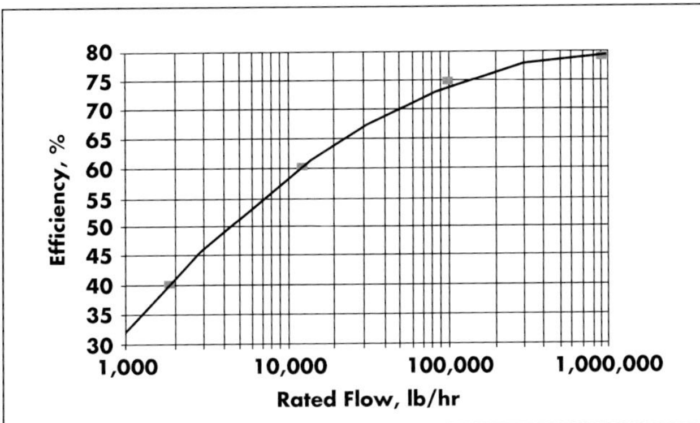

Fig. 11-32 Effect of Design Capacity on Turbine Mechanical Efficiency. Source: Cogen Designs, Inc.

Condensing vs Back-pressure Turbine Performance

Figure 11-34 is a Mollier chart with cycle efficiencies of 40, 50, and 60% shown. As illustrated, lower efficiency results in reduced heat drop, less work done, and higher final enthalpy. The term "heat absorbed" refers to the energy in the steam that is actually converted to work. In a condensing turbine, the heat that is not absorbed goes on to be condensed and represents wasted energy. In a non-condensing, (i.e., back-pressure) turbine, the heat that is not absorbed is used effectively in a downstream process.

The condensing cycle is a closed Rankine cycle in that the working fluid repeatedly executes the cycle processes. Non-condensing cycles may be viewed as open cycles in that the fluid exits the cycle and may or may not be returned from process loads.

Figure 11-35 illustrates the difference between a condensing and non-condensing Rankine cycle. Both cycles are shown with non-ideal expansion processes, each with

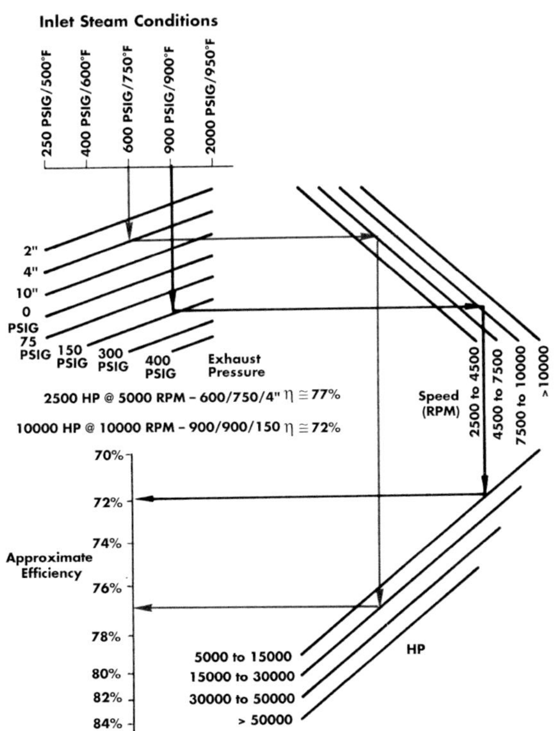

Fig. 11-33a Approximate Steam Turbine Efficiency Nomograph (English Units). Source: Elliott Company

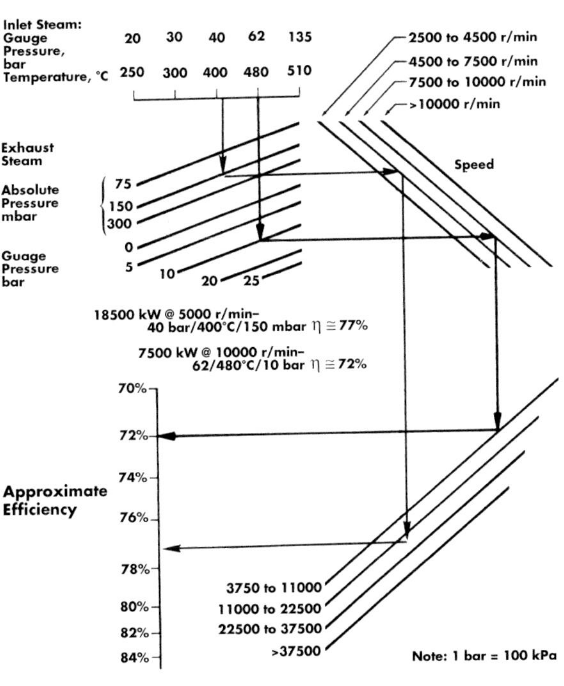

Fig. 11-33b Approximate Steam Turbine Efficiency Nomograph (SI Units). Source: Elliott Company

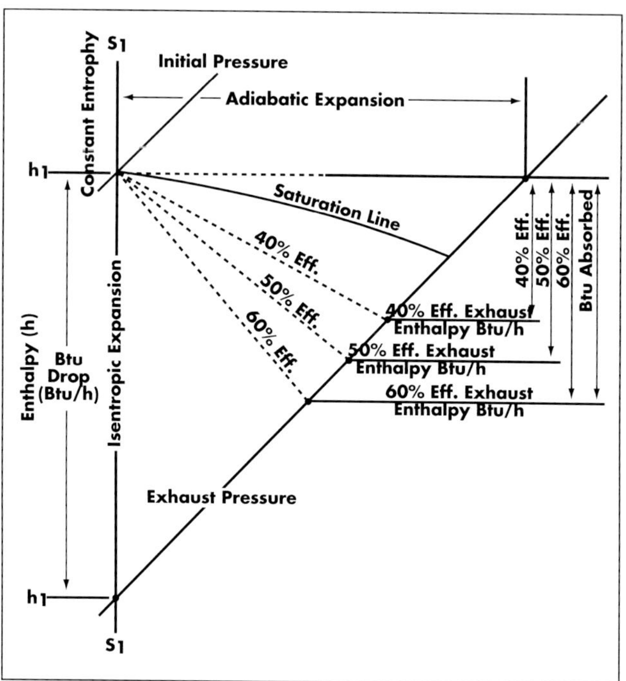

Fig. 11-34 Simplified Mollier Chart Indicating Representative Cycle Efficiencies.

cycle is indicated by the larger shaded region, indicating greater heat rejection.

Total heat rates for condensing steam turbines, including boiler efficiency losses, typically range from 7,500 to 15,000 Btu/hp-h (10,612 to 21,222 kJ/kWh). This correlates to thermal efficiencies ranging from 34 to 17% (HHV). Other than the 15 to 20% boiler efficiency losses and various minor losses in the turbine, the rest of the energy — the latent heat removed from the steam in the condenser — is lost to atmosphere.

When the condenser is replaced by a process thermal load, as in a topping cycle, the net heat rate required to produce shaft power is on the order of 2,800 Btu/hp-h (3,962 kJ/kWh), or greater than 90% thermal efficiency (HHV). Excluding boiler losses, the net thermal (Rankine) efficiency will always approach 100% because all of the heat that is not converted into useful work is simply passed on to meet process heating loads.

Figure 11-36 shows turbine efficiency for a condensing induction/extraction turbine with a variable medium-pressure extraction. Steam can either be extracted or induced through the extraction port. The percent generating effi-

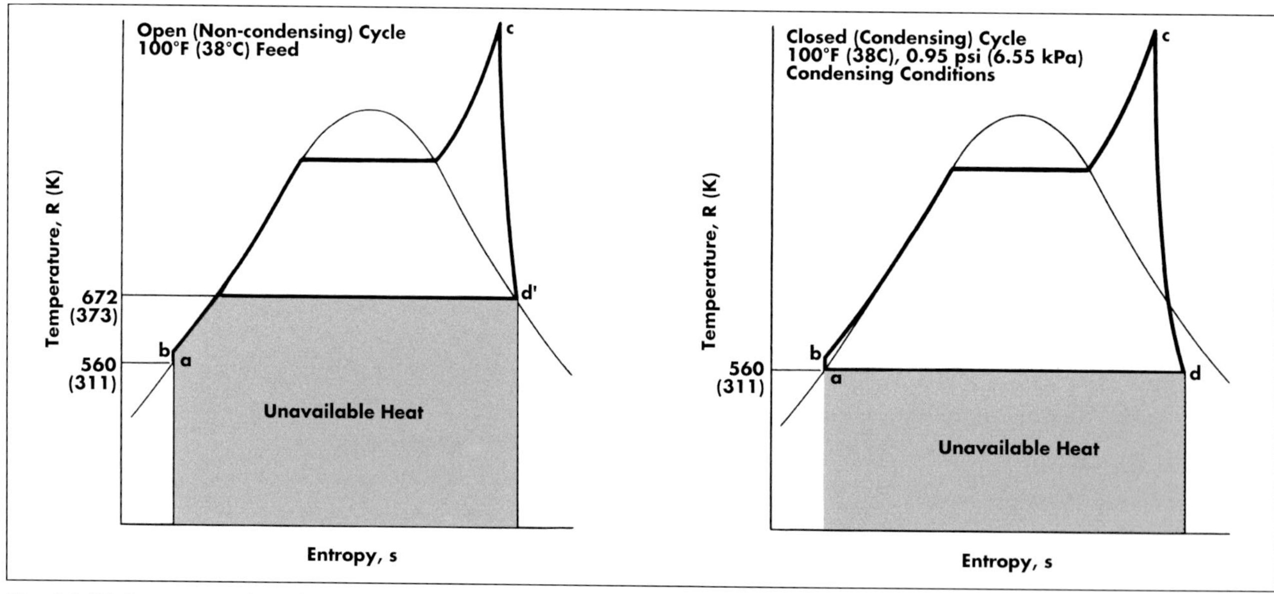

Fig. 11-35 Comparison of Condensing and Non-Condensing Rankine Cycles with T-s Diagrams. Source: Babcock & Wilcox

the same work and heat quantities involved. Liquid compression occurs between points a and b and heat is added between points b and c. Irreversible expansion takes place from c to d' in the non-condensing cycle and from c to d in the condensing cycle. Heat is rejected from points d' to a and d to a; the shaded areas are proportional to the amount of rejected heat. The lower efficiency of the non-condensing

ciency represents the ratio of actual efficiency to theoretical efficiency at each throttle flow rate. The curves indicate how efficiencies vary depending on the total steam entering the turbine (throttle steam) and induction or extraction. A dashed line shows the maximum extraction limit. Figure 11-37, which is similar to Figure 11-36, shows turbine efficiency for a back-pressure turbine with extraction.

Figures 11-38 and 11-39 are actual performance curves based on manufacturer's data. Notice the difference in power production for condensing and back-pressure units for a given throttle flow and extraction flow.

PART LOAD PERFORMANCE

Steam turbines maintain relatively good performance during off-design or part-load operation. Figures 11-40 and 11-41 provide representative curves showing off-design performance. As indicated, steam flow is roughly proportional to power output at most loads and efficiency is largely unchanged. Note that the minimum steam flow required to keep the machine operating is about 10%.

Figure 11-42 is a representative curve to approximate steam rates for turbines operating at part-load and -speed. Consider, for example, a turbine rated for full-load opera-

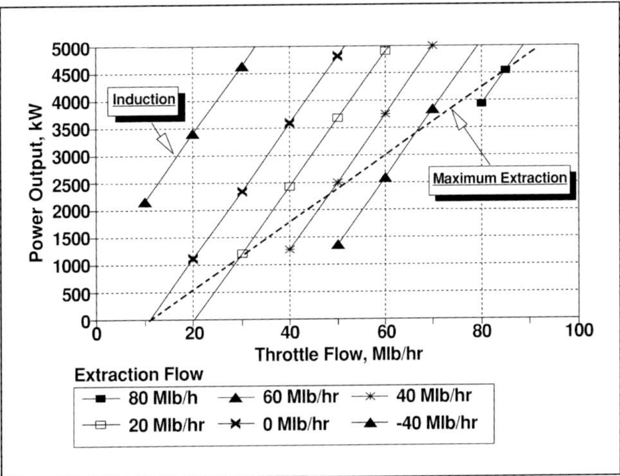

Fig. 11-38 Condensing Induction/Extraction Turbine, Power Output vs. Throttle-Flow for Various Extraction (and Induction) Flows. Source: Cogen Designs, Inc.

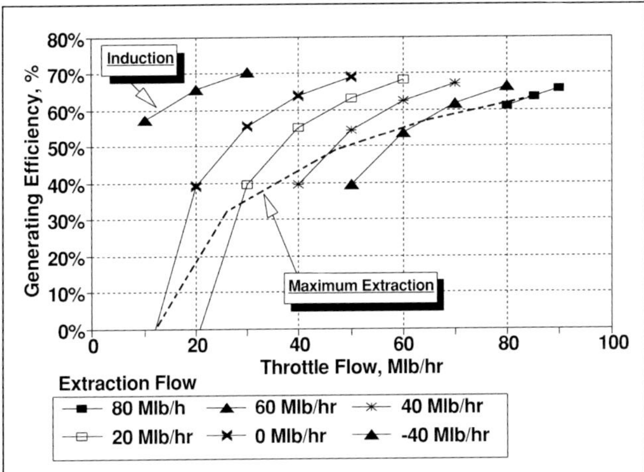

Fig. 11-36 Percent of Theoretical Efficiency vs. Throttle Flow for Condensing Induction/Extraction Turbine. Source: Cogen Designs, Inc.

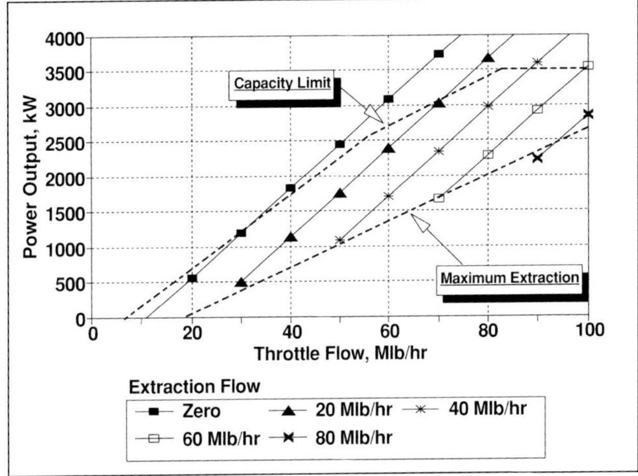

Fig. 11-39 Back-Pressure Turbine with Extraction, Power Output vs. Throttle-Flow for Various Extraction Flows. Source: Cogen Designs, Inc.

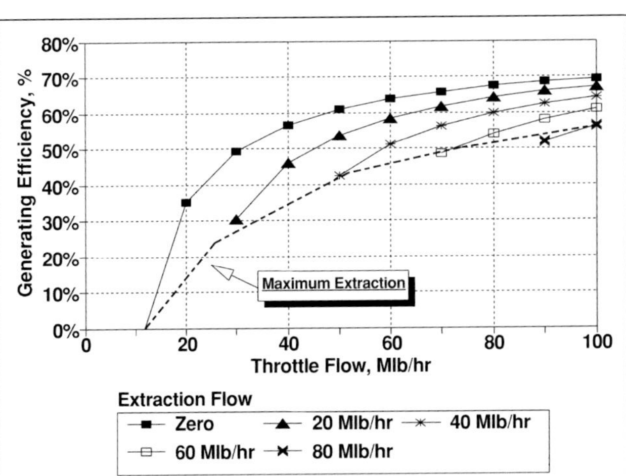

Fig. 11-37 Percent of Theoretical Efficiency vs. Throttle Flow for Back-Pressure Turbine with Extraction. Source: Cogen Designs, Inc.

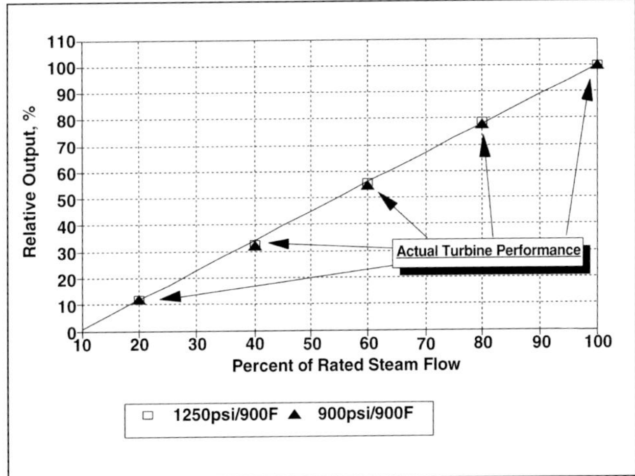

Fig. 11-40 Steam Turbine Off-Design Performance, Capacity vs. Steam Flow. Source: Cogen Designs, Inc.

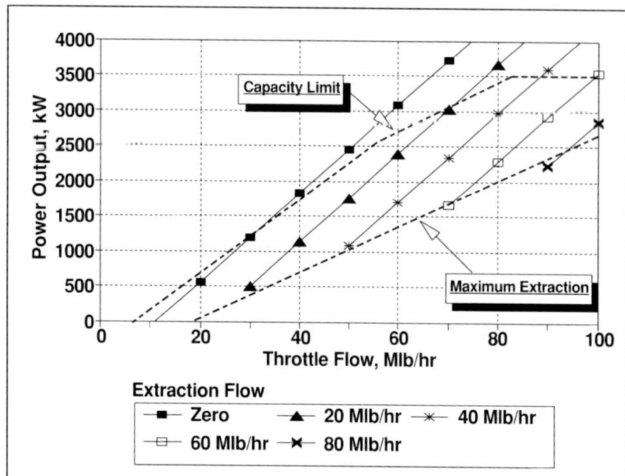

Fig. 11-41 Steam Turbine Off-Design Performance, Efficiency vs. Steam Flow. Source: Cogen Designs, Inc.

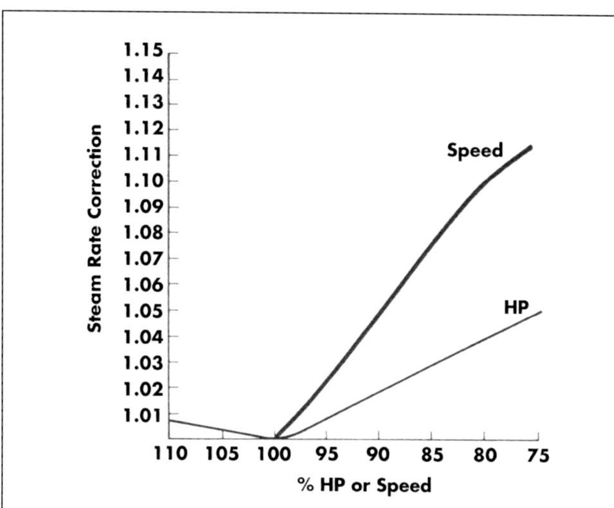

Fig. 11-42 Representative Part-Load/Speed Correction Curves. Source: Elliott Company

tion at 25,000 hp (18,640 kW) at 5,000 rpm. Off-design operation is 20,000 hp (14,912 kW), which is at 80% of full load, and 4,500 rpm (90%). From the curve, the power correction is 1.04 and the rpm correction is 1.05. The total correction is the product of the two correction factors (1.04 x 1.05), which is 1.09. The part-load steam rate is the product of the full-load steam rate and the correction factor. If the full-load steam rate is 7.40 lbm/hp-h (4.50 kg/kWh), the part-load steam rate would be 7.40 lbm/hp-h x 1.09 = 8.06 lbm/hp-h (4.50 kg/kWh x 1.09 = 4.91 kg/kWh).

While the steam turbine itself operates slightly less efficiently at reduced speed, the driven equipment may operate at improved efficiency at reduced speed under part load. Centrifugal compressors, pumps, and fans show this characteristic, as illustrated in Figure 11-43.

Figure 11-44 shows steam flow versus load for various steam turbines. Notice that while multi-stage turbines are more efficient at full load than single-stage units, they lose efficiency more rapidly as load moves away from the design point. This occurs because the nozzle area of a single-stage turbine can be adjusted to change flow throughout the turbine without affecting the velocity ratio. This cannot be accomplished in a multi-stage turbine.

In multi-stage turbines, hand valves or sequentially opening valves can only adjust the nozzle area in the first turbine stage. When some first stage nozzles are closed under partial pressure, pressure in all stages except the last one goes down, causing the steam jet velocity to be higher in the first stage and lower in the last stage. This reduces first stage

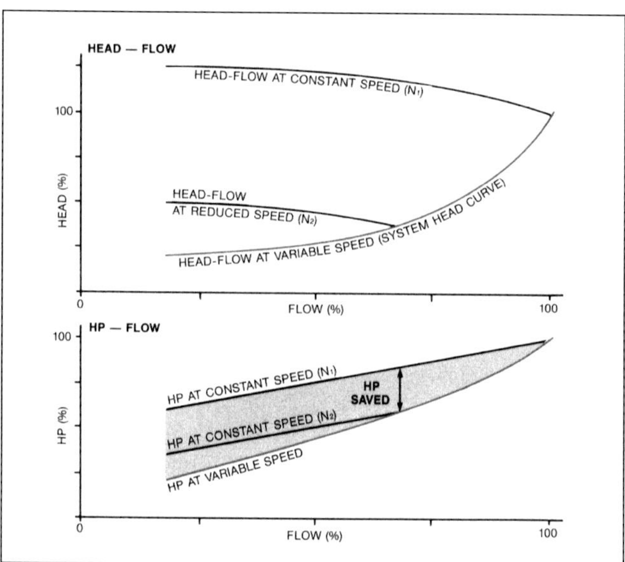

Fig. 11-43 Variable Speed Operation at Part-Load Conditions. Source: Tuthill Corp., Murray Turbomachinery Division

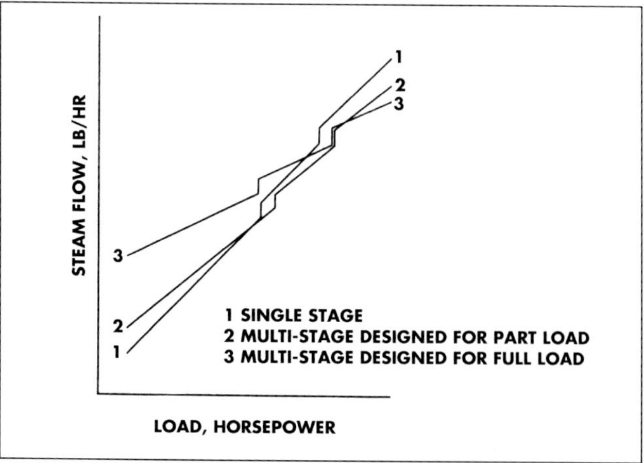

Fig. 11-44 Steam Flow vs. Load for Various Steam Turbines. Source: Tuthill Corp., Murray Turbomachinery Division

velocity ratio and increases last stage velocity ratio.

However, as shown in Figure 11-44, multi-stage turbines can be designed to improve efficiency at part load at the expense of full-load efficiency. Generally, the lowest operating costs will be achieved if the turbine is designed to optimize steam rate at the normal or most common load experienced, with allowances made for operation at other load points.

Factors Affecting Single- and Multi-Stage Turbine Performance

Factors that differentiate performance between single- and multi-stage turbines include the effects of reheating between stages, steam quality, and nozzle changes at partial load or overload.

A **reheating effect** occurs between stages in a multi-stage turbine. With each stage, the steam energy in the enthalpy drop that is not converted to power remains in the steam so it can be used by subsequent stages. The portion of the available energy remaining in the fluid is termed the reheat. The reheat in a given stage is available to do work in the succeeding stage, except for the last stage. Therefore, more power is produced because the multi-stage turbine makes more efficient use of the available steam energy.

There is also a reheating effect on the exhaust steam of single-stage turbines. This is due to **windage**, or resistance losses resulting from the blades and discs passing through exhaust steam and the kinetic energy remaining in the steam leaving the blades.

The steam **quality** or moisture level is an important consideration for any turbine, affecting both performance and maintenance requirements. Steam quality is a measure of the amount of vapor in the two-phase liquid-vapor mixture and represented as a percentage ranging from 0 to 100%:

$$Quality = x = \frac{mass_{vapor}}{(mass_{liquid} + mass_{vapor})} \qquad (11\text{-}19)$$

In all condensing and most non-condensing turbine applications, as the steam expands through the turbine, steam properties cross the saturation line. Seen on a T-s diagram, the process moves into the "vapor dome," the area under the saturation curve where liquid is present and, therefore, the steam has a lower enthalpy per lbm (hg) than the dry saturated steam. Figure 11-45 shows T-s diagrams using wet, dry saturated, and superheated steam.

If steam quality falls below 90 or 95% in the last low-pressure stage(s) of a steam turbine, liquid droplets can erode the turbine blades. Efficiency is also degraded by moisture because the presence of water drops increases friction losses in the steam itself and, since water drops tend to move more slowly than vapor, they strike the rotor blades at unfavorable velocities and exert a braking effect. Integral moisture separators and other design features can reduce the problem and, in multi-stage turbines, the reheating effect between stages can reduce moisture content. By superheating steam at the boiler, the quality of expanded steam in later turbine stages is improved.

Steam Turbine Performance Examples

Two sets of examples are provided to demonstrate the principles of steam turbine operation. The examples demonstrate actual and ideal cycle efficiencies for both condensing and back-pressure turbines. They also include a comparison of the thermodynamics associated with using a back-pressure turbine in place of a pressure reducing valve (PRV).

In all cases, available enthalpy drop and the turbine mechanical efficiency determine the amount of power that can be produced. It is clear that the higher pressure and temperature steam condition in Examples 4 to 6 produce greater simple-cycle efficiency and significantly more power output for a given steam flow. In the case of the back-pressure turbine, the ideal and actual net thermal efficiency in both examples are shown to be approximately equal to the boiler thermal fuel efficiency of 85%, confirming the high efficiency achieved with a topping cycle.

Values are given in English system units. Refer to Conversion Tables in Chapter 6 for SI unit equivalents.

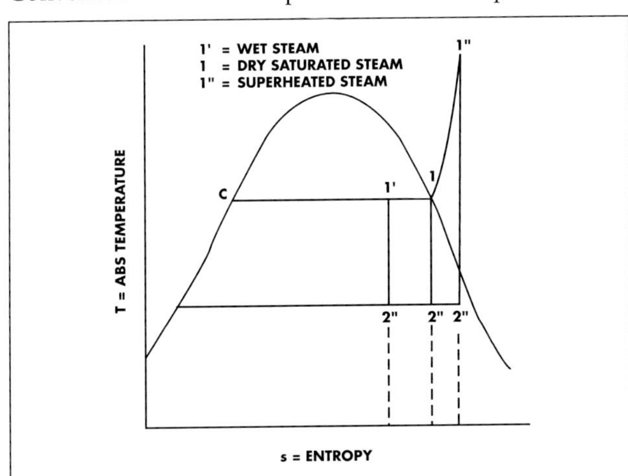

Fig. 11-45 Temperature-Entropy Diagram of Rankine Cycle for Wet, Dry Saturated and Superheated Steam. Source: Babcock & Wilcox

Examples Based on 250 Psig Saturated Steam

Examples 1 through 3, which are illustrated in Figures 11-46 through 11-49, are based on the following conditions:

Inlet condition (h_c)	=	1,201.7 Btu/lbm (250 psig – saturated steam)
Exhaust condition (h_d) (condensing example)	=	867.7 Btu/lbm – ideal (3 in. HgA)*
Exhaust condition (h_d) (back-pressure examples)	=	1,056.3 Btu/lbm – ideal (25 psig)
Turbine efficiency (η_T)	=	70% (condensing example)
Turbine efficiency (η_T)	=	50% (back-pressure examples)
Boiler efficiency (η_B)	=	85%
Steam (mass) flow ($\dot{m}$)	=	20,000 lbm/h
Condensate (h_a)	=	83 Btu/lbm (condensing example)
Condensate (h_a)	=	180 Btu/lbm (back-pressure examples)
Makeup water	=	6% (18 Btu/lbm)
Boiler feedwater (h_b)	=	79.1 Btu/lbm (condensing example)
	=	170.3 Btu/lbm (back-pressure examples)

*3 in. HgA refers to the distance a given amount of pressure can force a mercury column to rise. "A" refers to absolute pressure taken relative to 0 pressure. 1 in. HgA is approximately equal to 0.49 psia (or 3.387 kPa). When absolute pressure is lower than atmospheric pressure (29.92 in. Hg), it is referred to as vacuum pressure. 3 in. HgA is approximately equal to 27 in. Hg vacuum (actually 26.92 in.) relative to atmospheric pressure.

Notes:

1. h_b-h_a represents the difference in enthalpy between condensate and boiler feed temperature resulting from makeup water; it is not pump work (W).
2. Inlet conditions are based on saturated steam at 250 psig (406°F).
3. In the condensing turbine example, the critical determinants of thermal efficiency and output are the heat drop and turbine efficiency. Additional examples are provided to demonstrate the effects of increasing or decreasing these factors.
4. In the back-pressure turbine examples, turbine efficiency does not affect thermal efficiency because all heat that is not converted to useful work is passed onto process. However, turbine efficiency is still a critical factor because it will determine how much of the steam energy is converted to work output and how much is passed to process. Additional examples are included to demonstrate the effects of increasing turbine efficiency and heat drop.
5. In the back-pressure turbine examples, when steam inlet condition (potential heat drop) is increased (higher pressures or superheating temperatures), the value of replacing pressure reducing valves with back-pressure steam turbines is increased.
6. In both the ideal and actual cycle examples, the values of 79.1 Btu/lbm and 170.3 Btu/lbm are used for h_b in the condensing and back-pressure examples, respectively. The ideal cycle refers to the turbine, not the entire system. Thus, the actual need for make-up water is considered in the ideal cycle examples.
7. Note that the symbol h_d' is often used to indicate actual turbine exhaust condition as opposed to ideal.
8. Note that boiler efficiency losses are considered in both the ideal and actual Rankine cycle examples.

Example 1 – Condensing Steam Turbine

A. Ideal Rankine Cycle

$$\eta_{th} = \frac{(h_c - h_d)}{(h_c - h_b)} = \frac{(1{,}201.7 - 867.7)}{(1{,}201.7 - 79.1)} = \frac{334.0}{1122.6} \times 100\% = 29.75\%$$

Ideal heat drop $= h_c - h_d = 1{,}201.7 - 867.7 = 334$ *Btu/lb*

$$TWR = \frac{2{,}545 \text{ Btu/hp-h}}{(h_c - h_d)} = \frac{2{,}545 \text{ Btu/hp-h}}{334 \text{ Btu/lbm}} = 7.62 \text{lbm/hp-h}$$

$$\text{Heat rate} = TWR \frac{(h_c - h_b)}{\eta_B} = 7.62 \text{ lbm/h} \frac{(1{,}201.7 - 79.1) \text{ Btu/lbm}}{0.85} = 10{,}064 \text{ Btu/hp-h}$$

$$\text{Power Output} = \dot{m} \frac{(h_c - h_d)}{2{,}545 \text{ Btu/hp-h}} = 20{,}000 \text{ lbm/h} \frac{334 \text{ Btu/lbm}}{2{,}545 \text{ Btu/hp-h}} = 2{,}625 \text{ hp}$$

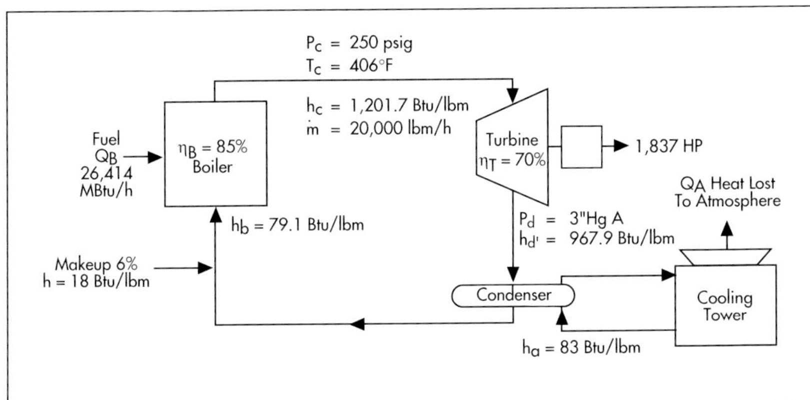

Fig. 11-46 Condensing Turbine Operation.

B. Actual Conditions

Actual heat drop $= (h_c - h_d)_{actual} = \eta_T (h_c - h_d)_{ideal} = 0.7(1{,}201.7 - 867.7) = 233.8$ *Btu/lbm*

$$\eta_R = \frac{(h_c - h_d)_{actual}}{(h_c - h_b)} = \frac{(233.8)}{(1{,}201.7 - 79.1)} \times 100\% = 20.38\%$$

$$\eta_{System} = \frac{(h_c - h_d)_{actual}}{(h_c - h_b)/\eta_B} = \frac{(233.8)}{(1{,}122.6/0.85)} \times 100\% = 17.70\%$$

Actual $h_d = h_c - \eta_T (h_c - h_d)_{ideal} = 1.201.7 - 233.8 = 967.9$ *Btu/lbm*

$$AWR = \frac{2{,}545 \text{ Btu/hp-h}}{(h_c - h_d)_{actual}} = \frac{2{,}545 \text{ Btu/hp-h}}{233.8 \text{ Btu/lbm}} = 10.89 \text{ lbm/hp-h}$$

$$\text{Heat rate} = AWR \frac{(h_c - h_b)}{\eta_B} = 10.89 \text{ lbm/h} \frac{1{,}122.6 \text{ Btu/lbm}}{0.85} = 14{,}382 \text{ Btu/hp-h}$$

$$\text{Power output} = \dot{m}\eta_T \frac{(h_c - h_d)_{ideal}}{2{,}545 \text{ Btu/hp-h}} = (20{,}000 \text{ lbm/h} \times 0.70) \frac{334 \text{ Btu/lbm}}{2{,}545 \text{ Btu/hp-h}} = 1{,}837 \text{ hp}$$

The condenser duty (P) $= \dot{m} (h_d - h_a)_{actual} = 20{,}000$ *lbm/h* $(967.9 - 83.0)$ *Btu/lbm* $= 17{,}698$ *MBtu/h*

Example 2 – Back-Pressure Steam Turbine

A. Ideal Rankine Cycle

$$\eta_{th} = \frac{(h_c - h_d)}{(h_c - h_b)} = \frac{(1{,}201.7 - 1{,}056.3)}{(1{,}201.7 - 170.3)} = \frac{145.4}{1031.4} \times 100\% = 14.09\%$$

Ideal heat drop $= h_c - h_d = 1{,}201.7 - 1{,}056.3 = 145.4$ Btu/lbm

$$TWR = \frac{2{,}545 \text{ Btu/hp-h}}{(h_c - h_d)} = \frac{2{,}545 \text{ Btu/hp-h}}{145.4 \text{ Btu/lbm}} = 17.50 \text{ lbm/hp-h}$$

$$\text{Heat rate} = TWR\,\frac{(h_c - h_d)}{\eta_B} = 17.50 \text{ lbm/hp-h}\,\frac{(1{,}201.7 - 170.3) \text{ Btu/lbm}}{0.85} = 21{,}235 \text{ Btu/hp-h}$$

$$\text{Net heat rate} = TWR\,\frac{(h_c - h_b)}{\eta_B} = 17.50 \text{ lbm/hp-h}\,\frac{145.4 \text{ Btu/lbm}}{0.85} = 2{,}994 \text{ Btu/hp-h}$$

$$\eta_{net} = \frac{\text{Work output}}{\text{Net heat rate}} = \frac{2{,}545}{2{,}994} \times 100\% = 85\%$$

$$\text{Power output} = \dot{m}\,\frac{(h_c - h_d)}{2{,}545 \text{ Btu/hp-h}} = 20{,}000 \text{ lbm/h}\,\frac{145.4 \text{ Btu/lbm}}{2{,}545 \text{ Btu/hp-h}} = 1{,}143 \text{ hp}$$

Net thermal output to process $= \dot{m}\,(h_d - h_a)_{ideal} = 20{,}000$ lbm/h $(1{,}056.3 - 180.0)$ Btu/lbm $= 17{,}526$ MBtu/h

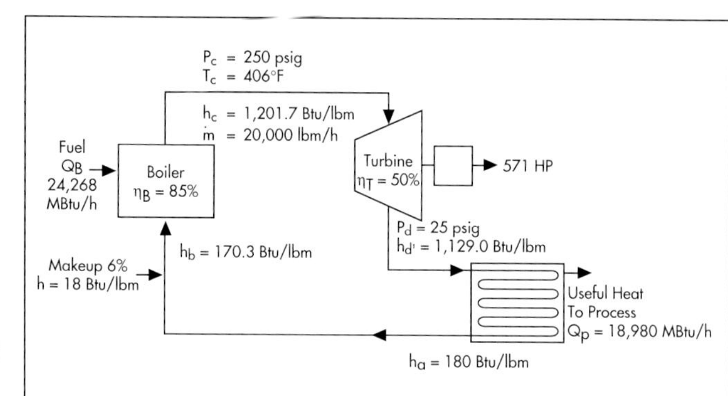

Fig. 11-47 Back-Pressure Turbine Operation.

B. Actual Conditions

Actual heat drop $= (h_c - h_d)_{actual} = \eta_T (h_c - h_d)_{ideal} =$
$0.5(1{,}201.7 - 1{,}056.3) = 72.7$ Btu/lbm

$$\eta_R = \frac{(h_c - h_d)_{actual}}{(h_c - h_b)} = \frac{(72.7)}{(1{,}201.7 - 170.3)} \times 100\% = 7.04\%$$

Actual $h_d = h_c - \eta_T (h_c - h_d)_{ideal} = 1.201.7 - 72.7$
$= 1{,}129$ Btu/lbm

$$AWR = \frac{2{,}545 \text{ Btu/hp-h}}{(h_c - h_d)_{actual}} = \frac{2{,}545 \text{ Btu/hp-h}}{72.7 \text{ Btu/lbm}} = 35.0 \text{ lbm/hp-h}$$

$$\text{Net heat rate} = AWR\,\frac{(h_c - h_d)_{actual}}{\eta_B} = 35.0 \text{ lbm/hp-h}\,\frac{72.7 \text{ Btu/lbm}}{0.85} = 2{,}994 \text{ Btu/hp-h}$$

$$\eta_{net} = \frac{\text{Work output}}{\text{Net heat rate}} = \frac{2{,}545}{2{,}994} \times 100\% = 85\%$$

$$\text{Power output} = \dot{m}\eta_T\,\frac{(h_c - h_d)_{ideal}}{2{,}545 \text{ Btu/hp-h}} = (20{,}000 \text{ lbm/h} \times 0.50)\,\frac{145.4 \text{ Btu/lbm}}{2{,}545 \text{ Btu/hp-h}} = 571 \text{ hp}$$

Net thermal output to process $= \dot{m}\,(h_d - h_a)_{actual} = 20{,}000$ lbm/h $(1{,}129.0 - 180.0)$ Btu/lbm $= 18{,}980$ MBtu/h

Example 3 – Pressure Reducing Valve vs. Back-Pressure Turbine (Figure 11-49)

A. Boiler and PRV Operation

Thermal output to process $= \dot{m}(h_c - h_a) = 20{,}000\ lbm/h\ (1{,}201.7 - 180.0)\ Btu/lbm = 20{,}434\ MBtu/h$

$$Input\ to\ boiler = \dot{m}\,\frac{(h_c - h_b)}{\eta_B} = 20{,}000\ lbm/h\ \frac{(1{,}201.7 - 170.3)\ Btu/lbm}{0.85} = 24{,}268\ MBtu/h$$

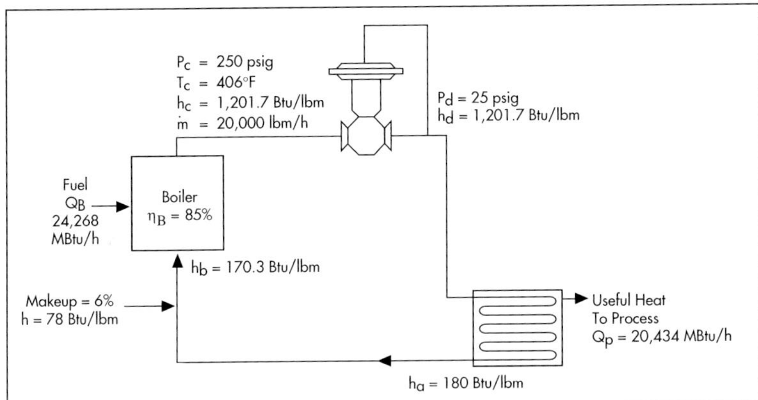

Fig. 11-48 PRV Operation.

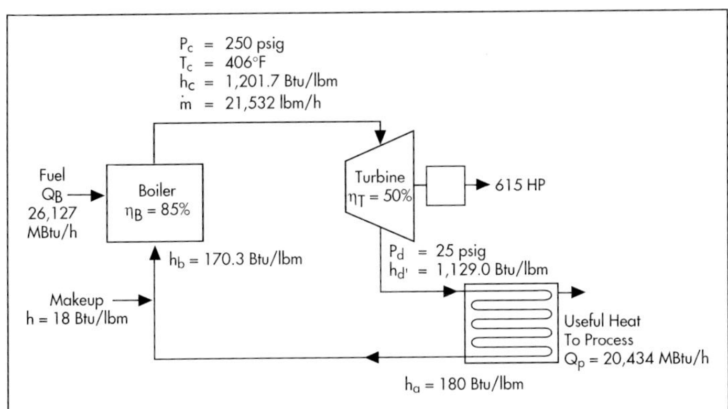

Fig. 11-49 Back-Pressure Turbine Operation vs. PRV.

B. Back-Pressure Steam Turbine Operation

$$Net\ steam\ input\ required = \frac{Net\ input\ to\ process\ with\ PRV}{(h_d - h_a)_{actual}} = \frac{20{,}434\ MBtu/h}{(1{,}129.0 - 180.0)\ Btu/lbm} = 21{,}532\ lbm/h$$

$$Power\ output = \dot{m}\,\frac{(h_c - h_d)_{actual}}{2{,}545\ Btu/hp\text{-}h} = 21{,}532\ lbm/h \times \frac{72.7\ Btu/h}{2{,}545\ Btu/hp\text{-}h} = 615\ hp$$

$$Input\ to\ boiler = \dot{m}\,\frac{(h_c - h_b)}{h_B} = 21{,}532\ lbm/h\ \frac{(1{,}201.7 - 170.3)\ Btu/lbm}{0.85} = 26{,}127\ MBtu/h$$

Conclusions (Examples 1-3)

Example 1 demonstrates how irreversibilites within the turbine can greatly reduce net power output and efficiency. In this case, ideal efficiency of 29.75% is reduced to 20.83% under actual conditions and net power output is reduced from 2,625 to 1,837 hp.

If the 70% efficient turbine in this example were replaced with a 50% efficient turbine, actual power output and overall system efficiency (including boiler losses) drop to 1,312 hp and 12.64%, respectively. On the other hand, if the inlet condition is increased to a higher pressure and temperature, or if the 250-psig steam is superheated, shaft power output and overall thermal efficiency would increase. Superheating the steam increases the Rankine cycle efficiency because the average temperature at which heat is transferred to the steam is increased. Note also that when steam is superheated, the quality of the steam leaving the turbine improves.

In Example 2, ideal back-pressure turbine efficiency of 14.09% is reduced to 7.04% under actual conditions and net power output is reduced from 1,143 to 571 hp. However, because the back-pressure turbine exhausts all unused energy to process, the net efficiency remains at almost 100% (neglecting boiler losses and small parasitic losses), regardless of the actual cycle efficiency. If the back-pressure steam load were large in relation to required power, a lower efficiency turbine would likely be most cost-effective. Conversely, if the back-pressure steam load is small, it may be more important to more fully utilize the power producing potential of each lbm (kg) of steam with a high-efficiency turbine.

As shown in Example 3, use of a back-pressure turbine in place of a PRV requires an added fuel energy input of 1,859 MBtu/h to produce 615 hp of power and deliver the equivalent amount of useful heat to process. This tradeoff would usually provide significant net energy cost savings at conventional fuel and electricity prices.

While the values will differ, similar observations and conclusions can be drawn from Examples 4 through 6 which follow.

Examples Based on 600 Psig, 750°F Superheated Steam

Examples 4 through 6, which are illustrated in Figures 11-50 through 11-53, are based on the following conditions:

Inlet condition (h_c)	=	600 psig, 750°F — (1,380 Btu/lbm)
Exhaust condition (h_d) (condensing example)	=	3 in. HgA (918 Btu/lbm — ideal)
Exhaust condition (h_d) (back-pressure examples)	=	80 psig (1,187 Btu/lbm — ideal)
Turbine efficiency (η_T)	=	70% (condensing example)
Turbine efficiency (η_T)	=	50% (back-pressure examples)
Boiler efficiency (η_B)	=	85%
Steam (mass) flow ($\dot{m}$)	=	40,000 lbm/h
Condensate (h_a)	=	83 Btu/lbm (condensing example)
Condensate (h_a)	=	180 Btu/lbm (back-pressure examples)
Makeup water	=	6% (18 Btu/lbm)
Boiler feedwater (h_b)	=	79 Btu/lbm (condensing example)
	=	170 Btu/lbm (back-pressure examples)

Note:
In these examples, enthalpy per lbm of steam is rounded to the nearest Btu and cycle efficiencies include consideration of makeup water requirements. Footnotes listed under previous examples (1 through 3) based on operation with 250-psig steam apply to these examples (4 through 6) as well.

Example 4: Condensing Steam Turbine

A. Ideal Rankine Cycle

$$\eta_{th} = \frac{(h_c - h_d)}{(h_c - h_b)} = \frac{(1{,}380 - 918)}{(1{,}380 - 79)} = \frac{462}{1301} \times 100\% = 35.5\%$$

Ideal heat drop = hc – hd = 1,380 – 918 = 462 Btu/lbm

$$TWR = \frac{2{,}545 \text{ Btu/hp-h}}{(h_c - h_d)} = \frac{2{,}545 \text{ Btu/hp-h}}{462 \text{ Btu/lbm}} = 5.51 \text{ lbm/hp-h}$$

$$\text{Heat rate} = TWR \frac{(h_c - h_b)}{\eta_B} = (5.51 \text{ lbm/hp-h}) \frac{(1{,}380 - 79) \text{ Btu/lbm}}{0.85} = 8{,}434 \text{ Btu/hp-h}$$

$$\text{Power output} = \dot{m} \frac{(h_c - h_d)}{2{,}545 \text{ Btu/hp-h}} = (40{,}000 \text{ lbm/h}) \frac{462 \text{ Btu/lbm}}{2{,}545 \text{ Btu/hp-h}} = 7{,}260 \text{ hp}$$

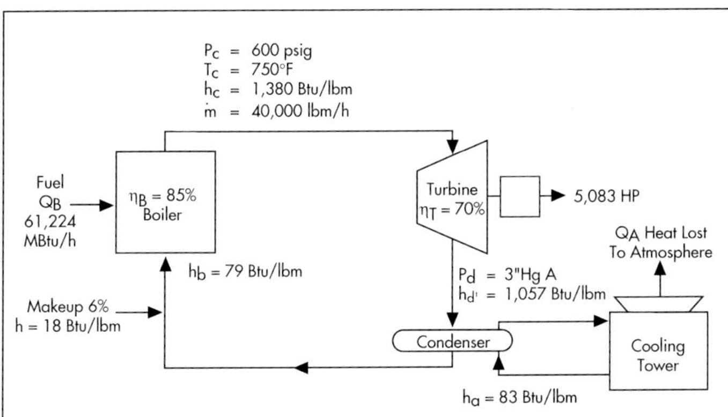

Fig. 11-50 Condensing Turbine Operation.

B. Actual Conditions

$$\text{Actual heat drop} = (h_c - h_d)_{actual} = \eta_T (h_c - h_d)_{ideal} = 0.70(1{,}380 - 918) = 323 \text{ Btu/lbm}$$

$$\eta_{th} = \frac{(h_c - h_d)_{actual}}{(h_c - h_b)} = \frac{(323)}{(1{,}380 - 79)} \times 100\% = 24.8\%$$

$$\eta_{System} = \frac{(h_c - h_d)_{actual}}{(h_c - h_b)/\eta_B} = \frac{(323)}{(1{,}301 - 0.85)} \times 100\% = 21.1\%$$

$$\text{Actual } h_d = h_c - \eta_T (h_c - h_d)_{ideal} = 1{,}380 - 323 = 1{,}057 \text{ Btu/lbm}$$

$$AWR = \frac{2{,}545 \text{ Btu/hp-h}}{(h_c - h_d)_{actual}} = \frac{2{,}545 \text{ Btu/hp-h}}{323 \text{ Btu/lbm}} = 7.88 \text{ lbm/hp-h}$$

$$\text{Heat rate} = AWR \frac{(h_c - h_b)}{\eta_B} = 7.88 \text{ lbm/hp-h} \frac{1{,}301 \text{ Btu/lbm}}{0.85} = 12{,}061 \text{ Btu/hp-h}$$

$$\text{Power output} = \dot{m}\eta_T \frac{(h_c - h_d)_{ideal}}{2{,}545 \text{ Btu/hp-h}} = (40{,}000 \text{ lbm/h} \times 0.70) \frac{462 \text{ Btu/lbm}}{2{,}545 \text{ Btu/hp-h}} = 5{,}083 \text{ hp}$$

$$\text{Condenser Duty } (P) = \dot{m}(h_d - h_a)_{actual} = 40{,}000 \text{ lbm/h } (1{,}057 - 83) \text{ Btu/lbm} = 38{,}960 \text{ MBtu/h}$$

Example 5 – Back-Pressure Steam Turbine

A. Ideal Rankine Cycle

$$\eta_{th} = \frac{(h_c - h_d)}{(h_c - h_b)} = \frac{(1{,}380 - 1{,}187)}{(1{,}380 - 170)} = \frac{193}{1{,}201} \ x\ 100\% = 16.1\%$$

$$\textit{Heat drop} = h_c - h_d = 1{,}380 - 1{,}187 = 193\ Btu/lbm$$

$$TWR = \frac{2{,}545\ Btu/hp\text{-}h}{(h_c - h_d)} = \frac{2{,}545\ Btu/hp\text{-}h}{193\ Btu/lbm} = 13.2\ lbm/hp\text{-}h$$

$$\textit{Heat rate} = TWR\ \frac{(h_c - h_d)}{\eta_B} = 13.2\ lbm/hp\text{-}h\ \frac{(1{,}380 - 170)\ Btu/lbm}{0.85} = 18{,}791\ Btu/hp\text{-}h$$

$$\textit{Net heat rate} = TWR\ \frac{(h_c - h_d)}{\eta_B} = 13.2\ lbm/hp\text{-}h\ \frac{193\ Btu/lbm}{0.85} = 2{,}997\ Btu/hp\text{-}h$$

$$\eta_{net} = \frac{\textit{Work output}}{\textit{Net heat rate}} = \frac{2{,}545}{2{,}994}\ x\ 100\% = 85\%$$

$$\textit{Power output} = \dot{m}\ \frac{(h_c - h_d)}{2{,}545\ Btu/hp\text{-}h} = 40{,}000\ lbm/h\ \frac{193\ Btu/lbm}{2{,}545\ Btu/hp\text{-}h} = 3{,}032\ hp$$

Net thermal output to process
$= \dot{m}\,(h_d - h_a)_{ideal} = 40{,}000\ lbm/h\ (1{,}187 - 180)\ Btu/lbm$
$= 40{,}280\ MBtu/h$

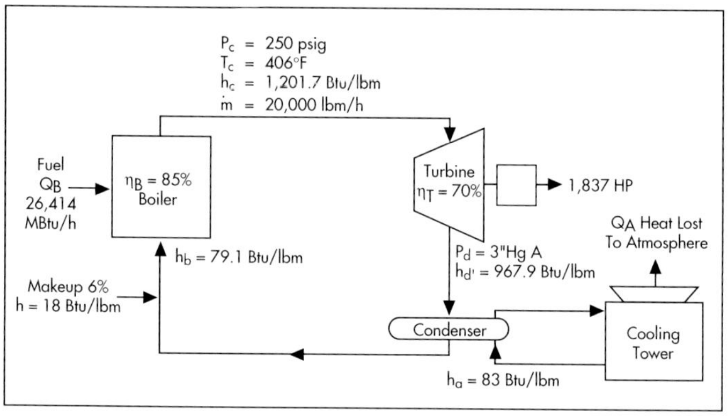

Fig. 11-51 Back-Pressure Turbine Operation.

B. Actual Conditions

Actual heat drop $= (h_c - h_d)_{actual} = \eta_T (h_c - h_d)_{ideal}$
$= 0.70(1{,}380 - 1{,}187) = 96\ Btu/lbm$

$$\eta_{th} = \frac{(h_c - h_d)_{actual}}{(h_c - h_b)} = \frac{96}{(1{,}380 - 79)}\ x\ 100\% = 7.4\%$$

$$\textit{Actual } h_d = h_c - \eta_T (h_c - h_d)_{ideal} = 1{,}380 - 96 = 1{,}284\ Btu/lbm$$

$$AWR = \frac{2{,}545\ Btu/hp\text{-}h}{(h_c - h_d)_{actual}} = \frac{2{,}545\ Btu/hp\text{-}h}{96\ Btu/lbm} = 26.5\ lbm/hp\text{-}h$$

$$\textit{Net heat rate} = AWR\ \frac{(h_c - h_d)_{actual}}{\eta_B} = 26.5\ lbm/hp\text{-}h\ \frac{96\ Btu/lbm}{0.85} = 2{,}993\ Btu/hp\text{-}h$$

$$\eta_{net} = \frac{\textit{Work output}}{\textit{Net heat rate}} = \frac{2{,}545}{2{,}993}\ x\ 100\% = 85\%$$

$$\textit{Power output} = \dot{m}\eta_T\ \frac{(h_c - h_d)_{ideal}}{2{,}545\ Btu/hp\text{-}h} = (40{,}000\ lbm/h\ x\ 0.50)\ \frac{193\ Btu/lbm}{2{,}545\ Btu/hp\text{-}h} = 1{,}517\ hp$$

Net thermal output to process $= \dot{m}\,(h_d - h_a)_{actual} = 40{,}000\ lbm/h\ (1{,}284 - 180)\ Btu/lbm = 44{,}160\ MBtu/h$

Example 6 – Pressure Reducing Valve vs. Back-Pressure Turbine (Figure 11-53)

A. Boiler and PRV Operation

Net input to process $= \dot{m}\,(h_c - h_a) = 40{,}000\ lbm/h\ (1{,}380 - 180)\ Btu/lbm\ \ 48{,}000\ MBtu/h$

$$\text{Input to boiler} = \dot{m}\,\frac{(h_c - h_b)}{\eta_B} = 40{,}000\ lbm/h\ \frac{(1{,}380 - 170)\ Btu/lbm}{0.85} = 56{,}941\ MBtu/h$$

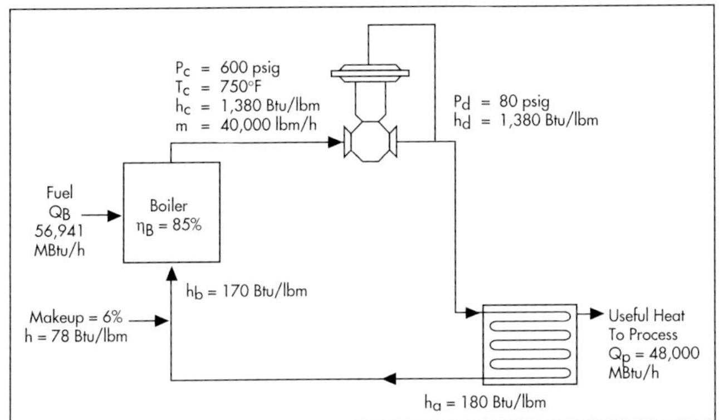

Fig. 11-52 PRV Operation.

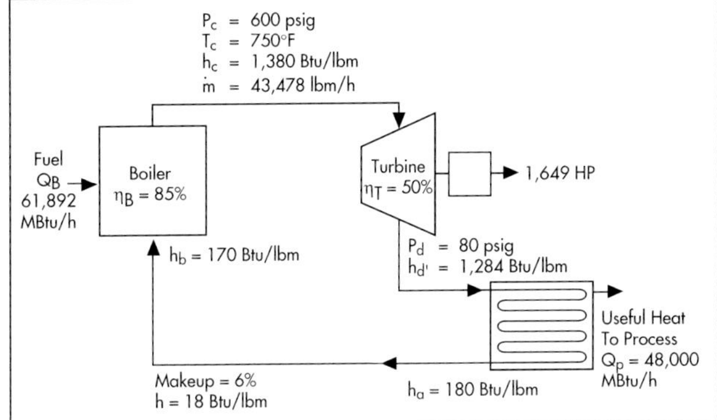

Fig. 11-53 Back-Pressure Turbine Operation vs. PRV.

B. Back-Pressure Steam Turbine Operation

Net steam input required

$$= \frac{\text{Net input to process with PRV}}{(h_d - h_a)_{actual}} = \frac{48{,}000\ MBtu/h}{(1{,}284 - 180)\ Btu/lbm} = 43{,}478\ lbm/h$$

$$\text{Power output} = \dot{m}\,\frac{(h_c - h_d)_{actual}}{2{,}545\ Btu/hp\text{-}h} = 43{,}478\ lbm/h\ \frac{96.5\ Btu/lbm}{2{,}545\ Btu/hp\text{-}h} = 1{,}649\ hp$$

$$\text{Input to boiler} = \dot{m}\,\frac{(h_c - h_b)}{h_B} = 43{,}478\ lbm/h\ \frac{(1{,}380 - 170)\ Btu/lbm}{0.85} = 61{,}892\ MBtu/h$$

Added fuel energy input $= 61{,}892\ MBtu/h - 56{,}941\ MBtu/h = 4{,}951\ MBtu/h$

Steam Turbine Maintenance, Reliability, and Service Life

Steam turbines offer great reliability, ease of maintenance, and long service life. Excluding consideration of the steam generation and distribution systems, steam turbines usually have relatively few maintenance requirements. This is due to simplicity of design, minimal moving parts, construction with durable materials, and operation at fairly low speeds. Turbines can operate for decades without requiring major overhaul and can last for 50 years with proper maintenance. When a high-pressure steam system is required for other purposes, steam turbines can be effectively applied with a relatively low incremental maintenance cost.

Causes of steam turbine performance deterioration include leakage, control damage, solid particle erosion, moisture erosion, steam path deposits, and foreign object damage. The quality of the turbine components and construction and the type of design are all important, particularly with respect to the ability of the blades to withstand the abuses of wet steam. More complex multi-staged turbines require greater care and attention than rugged single-stage turbines. Turbines designed with small clearances, as is common with reaction turbines, also require greater maintenance. Simple design produces greater reliability, less maintenance, and longer service life.

Routine procedures include monthly visual inspections that consist primarily of inspecting for oil or steam leaks and reviewing monitoring data. Any problem will show up in pressure and temperature readings or in measurement of vibrations.

Annual inspections and maintenance procedures include the following:

- Pull the casing to check the turbine wheel for blade wear
- Check nozzle alignment to ensure that steam is not cutting the turbine blades; re-balance and remove scale as needed
- Open packing box to check condition of the valve packing and spindle
- Inspect shaft seals for wear
- Check thrust bearing end play
- Inspect bearing clearance and end play
- Inspect gear tooth wear pattern, if applicable
- Check foundation
- Check and re-calibrate gauges
- Remove and clean steam strainer
- Drain water and clean foreign material from oil reservoir
- Drain oil from governor and flush clean, if applicable
- Drain small quantity of oil from system and conduct an oil analysis; change oil as needed and, if applicable, change filter element at same time
- Inspect grease in ball or roller bearing housings and renew if necessary

Premium turbines, with oil lubricated bearings, will typically run 3 to 5 years without requiring anything but the most minor adjustments: oil and some repacking. Provided there are no easily detected problems, such as blade wear, vibration, steam leaks, or contaminated oil, the turbine should continue to operate for years with no need for repair or lengthy interruption of operation.

As noted, steam quality is an important consideration for any turbine system. Moisture can erode the turbine blades, increasing maintenance requirements and decreasing efficiency.

Steam supply to the turbine always has some impurities in the form of various organic and inorganic compounds. As steam expansion proceeds and pressure and temperature decrease through the various stages, these impurities are likely to deposit on the turbine blades. This causes a reduction in flow area and a roughing of the passage walls, reducing capacity and efficiency.

Special design features are available to minimize erosion. Integral moisture separators or stainless steel moisture shields can be used. Steam purity indicators and modern day water purifiers can provide adequate means of control.

Currently Available Steam Turbine Designs

Following are illustrations and descriptions of several steam turbines and components, including a range of capacities and designs.

Figure 11-54 is an open-case view of a single-stage Elliot turbine available in capacities up to about 3,500 hp (2,600 kW). These units range in size up to about 5 ft (1.5 m) long and 3.5 ft (1 m) wide and high, with a weight of about 2,500 lbm (1,100 kg). The turbine has an operating speed range of between 1,000 and 7,000 rpm. To the far right are lubricated steel governor linkage pins connecting to a Woodward governor. To the left of the governor assembly is the rotor locating bearing, which was shown above. To the right of the turbine wheel are the carbon sealing rings. The three segments of each ring are held in place by springs, which prevent the seals from rotating. These units are commonly used to drive pumps, blowers, fans, compressors, and electric generators.

Fig. 11-54 Open-Case View of Single-Stage Steam Turbine. Source: Elliott Company

Figure 11-55 is a closed-case view of a horizontally split, single-stage steam turbine. The inlet and exhaust connections are shown in the lower half casing. The inlet connection on the left is differentiated from the exhaust connection by its smaller diameter. The upper half casing can be removed to permit easy access to internal components for inspection and routine maintenance.

Figure 11-56 is labeled a cutaway illustration of a vertical single-stage steam turbine. Vertical designs allow for easy mounting on vertical drive equipment. The compact configuration and small footprint is well suited for limited space and in-line installations. A vertical single-stage turbine is shown in Figure 11-57.

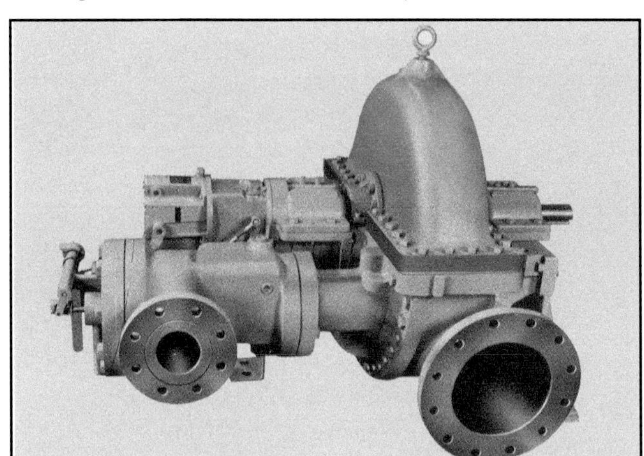

Fig. 11-55 Closed-Case View of Horizontally Split Single-Stage Steam Turbine. Source: Tuthill Corp., Coppus Turbine Div.

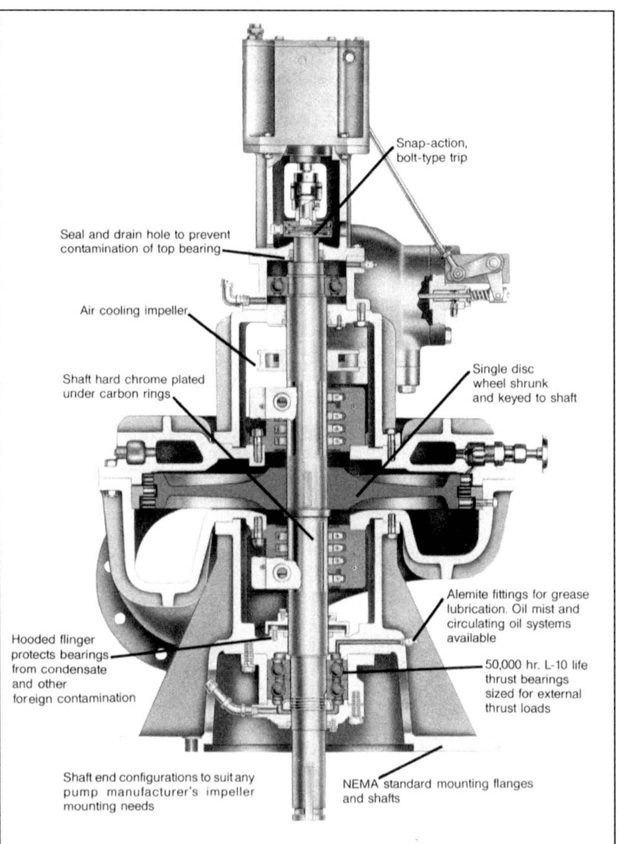

Fig. 11-56 Cutaway Illustration of Vertical Single-Stage Steam Turbine. Source: Tuthill Corp., Coppus Turbine Div.

Fig. 11-57 Vertical Single-Stage Steam Turbine. Source: Tuthill Corp., Coppus Turbine Div.

Figure 11-58 shows a standard multi-stage turbine with a capacity range of up to 7,000 hp (5,200 kW) at a speed of 8,000 rpm, with steam conditions of 700 psi (48 bar) and 825°F (441°C). Figure 11-59 is an open-

Fig. 11-58 Multi-Stage Steam Turbine with Capacity Range of up to 7,000 hp (5,200 kW). Source: Dresser-Rand

low-pressure rotor featuring extremely long exhaust blades.

Figure 11-62 shows a detailed cross-sectional view of three steam turbine designs: straight non-condensing, single automatic extraction/admission condensing, and double automatic extraction/admission condensing. The two units shown on the bottom are both double automatic extraction units.

Figure 11-63 shows a low-pressure condensing steam turbine driving a compressor. The turbine is rated at 997 hp (743 kW) at 3,980 rpm with steam conditions of 2 psig (115 kPa), dry and saturated, to 4 in. HgA (14 kPa). This turbine was designed to operate efficiently under such low steam inlet pressure and achieves a full-load steam rate of 25.8 lbm/hp-h (8.7 kg/kW).

Fig. 11-59 Open-Case View Revealing Rotor of Extraction/Induction Steam Turbine. Source: Tuthill Corp., Murray Turbomachinery Div.

case view revealing the rotor of an intermediate capacity multi-stage extraction/induction steam turbine. This turbine frame can accommodate up to 15 stages with a maximum power range of up to 15,000 hp (11 MW) and a maximum speed range of up to 15,000 rpm. Maximum inlet steam conditions range up to 900 psi (62 bar) and 900°F (482°C). Maximum non-automatic (bleed) extraction exhaust pressures range up to 400 psig (28 bar).

Figure 11-60 is an overhead open-case view of a 42,000 hp (31 MW) high-pressure extraction steam turbine. Figure 11-61 shows a large capacity steam turbine

Fig. 11-60 Overhead Open-Case View of 42,000 hp (31 MW) High-Pressure Extraction Turbine. Source: Dresser-Rand

Because turbines are custom built, efficiency and operating characteristics can be optimized for each application. Once a unit is designed, it may be pre-packaged to reduce erection time at the site. Figure 11-64 shows a high-pressure steam turbine unit in the 70,000 hp (52 MW) capacity range being prepackaged on its base frame.

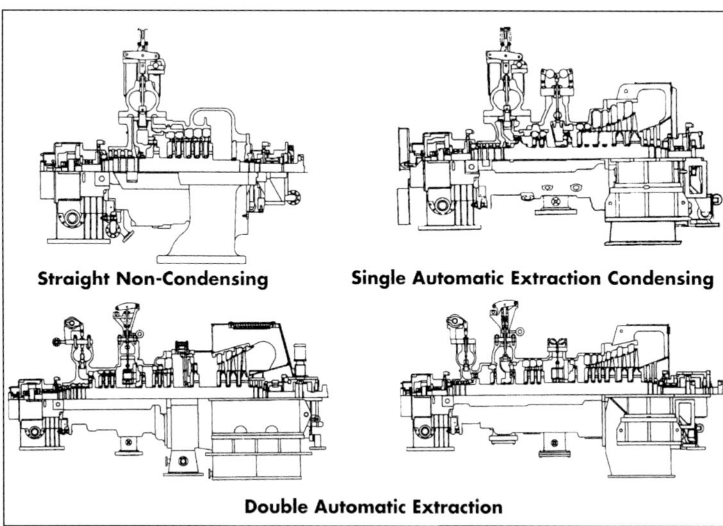

Fig. 11-62 Steam Turbine Design Types. Source: General Electric Company

Fig. 11-61 Large Capacity Low-Pressure Steam Turbine Rotor. Source: ABB Stal

Fig. 11-64 Packaged High-Pressure Steam Turbine in the 70,000 hp (52 MW) Capacity Range. Source: ABB Stal

Fig. 11-63 997 hp (743 kW) Condensing Steam Turbine Operating with 2 psig (115 kPa) Inlet Steam Condition. Source: Tuthill Corp., Murray Turbomachinery Div.

CHAPTER TWELVE

Combined and Steam Injection Cycles

Combined cycles and **steam injection cycles** are important enhancements to conventional prime mover simple cycles. They can conserve energy by converting rejected heat into additional power production and they reduce air emissions per unit of power generated. Whereas cogeneration cycles involve the use of recovered heat to serve thermal processes, combined and steam injection cycles use recovered heat in the form of high-pressure steam to produce additional power. As such, the techniques are important alternatives in applications where process uses for cogenerated thermal energy are either not available or somewhat limited.

A combined cycle is the sequential linking of any topping and bottoming cycle, or two simple cycles. Typically, a gas turbine or reciprocating engine is used to generate shaft power at the top of the cycle, with steam generated from turbine or engine exhaust heat. The steam is then passed through a steam turbine to generate additional power at the bottom of the cycle.

With gas-turbine systems, an alternative to adding a bottoming cycle is to use the steam injection cycle (also known by trade names as STIG™ or Cheng™ cycles), in which recovered heat is used to generate additional power by injecting steam directly into the gas turbine. It is similar in concept to a combined cycle in that increased mass flow is passed through a turbine to produce more power. The difference is that the increased mass flow is injected into the same (gas) turbine (operating on an open Brayton cycle), as opposed to a different (steam) turbine (operating on a closed Rankine cycle).

COMBINED-CYCLE SYSTEMS

Figures 12-1a and b show a simple combined-cycle system featuring a gas turbine-generator with a heat recovery steam generator (HRSG) and a steam turbine-generator, condenser, and auxiliary system. In this case, the HRSG is the heat exchanger that links the two cycles together by transferring the exhaust energy from the gas-turbine topping cycle to the steam-turbine bottoming cycle.

The benefit of combining these cycles is that combustion engine cycles are well suited for high-temperature operation, and the rejected heat of those cycles, still at a relatively elevated temperature, is well suited for the steam-turbine cycle. A fundamental limitation on thermal efficiency of combustion engine cycles is the high exit temperature, typically in the range of 900 to 1,200°F (482 to 649°C). The exhaust stream has a high energy level that is wasted if heat recovery applications are not available.

Generally, the basic gas turbine design (i.e., no after-coolers or recuperators) provides a good exhaust temperature for steam turbine power generating efficiency. Increases in exhaust gas temperatures of a few hundred degrees, via supplementary firing, may further optimize HRSG and steam turbine performance. These factors allow the use of basic gas turbines in high thermal-efficiency combined-cycle plants.

Thermal efficiency of a combined-cycle plant, assuming no firing in the HRSG, is the ratio of total power output (combustion engine and steam turbine) to the combustion engine energy input. This can be expressed as:

$$\eta_{th} = \frac{(W_{CE} + W_{ST})}{Q_{CE}} \qquad (12\text{-}1)$$

Where:

W_{CE} = Work produced by combustion engine (e.g., gas turbine or reciprocating engine)
W_{ST} = Work produced by steam turbine
Q_{CE} = Heat energy input to combustion engine

If additional heat is added via after-firing or supplemental firing in the HRSG, thermal efficiency can be expressed as:

$$\eta_{th} = \frac{(W_{CE} + W_{ST})}{(Q_{CE} + Q_{HRSG})} \qquad (12\text{-}2)$$

In a cogeneration application in which a portion of the recovered heat is passed on to process (Q_P), either directly from the HRSG or via extraction from the steam turbine, efficiency can be expressed as:

$$\eta_{th} = \frac{(W_{CE} + W_{ST} + Q_P)}{Q_{Total\ input}} \qquad (12\text{-}3)$$

The net fuel rate, or FCP equation, (developed in Chapter 2) applied to a combined-cycle cogeneration type system can be expressed as:

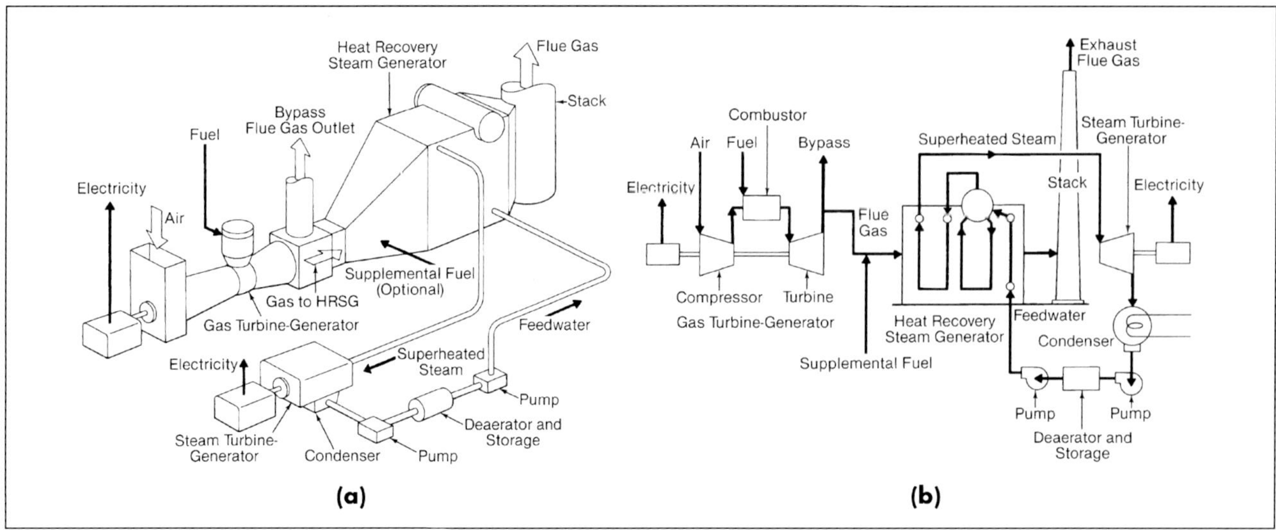

Figs. 12-1a and b Simplified Combined-Cycle Schematics. Source: Babcock & Wilcox

$$FCP = \frac{(Total\ fuel - Fuel\ credit)}{(W_{CE} + W_{ST} - W_{AUX})} \qquad (12\text{-}4)$$

Where:

Fuel credit = The avoided fuel use associated with the recovery of energy from the steam turbine

W_{Aux} = Auxiliary power requirements

Cost chargeable-to-power (CCP) in combined-cycle cogeneration applications can be expressed as:

$$CCP = \frac{Total\ energy\ cost - Energy\ cost\ credit + O\ \&\ M\ cost}{TC\ power\ output + BC\ power\ output - Aux.\ power\ input} \qquad (12\text{-}5)$$

Where: TC is topping cycle and BC is bottoming cycle.

Figure 12-2 shows thermal efficiency and specific output of gas-turbine combined cycles at different firing temperatures and various pressure ratios. In contrast to simple-cycle efficiency, where high pressure ratios are desirable, this figure indicates that more moderate pressure ratios may produce higher combined-cycle efficiency. Gas turbines featuring higher pressures can still produce very high efficiencies in combined-cycle operation, depending on the design and system configuration.

A significant difference between simple-cycle gas-turbine operation and combined-cycle operation is the effect of increased ambient air temperature. As shown in Chapter 10, increased ambient air temperature derates gas turbine output and thermal efficiency due to reduced density and, therefore, reduced mass flow and pressure ratio. However, in the combined cycle, the increase in ambient air temperature raises the temperature of the exhaust gas stream and, therefore, increases the thermal efficiency of the steam process. This lessens the relative

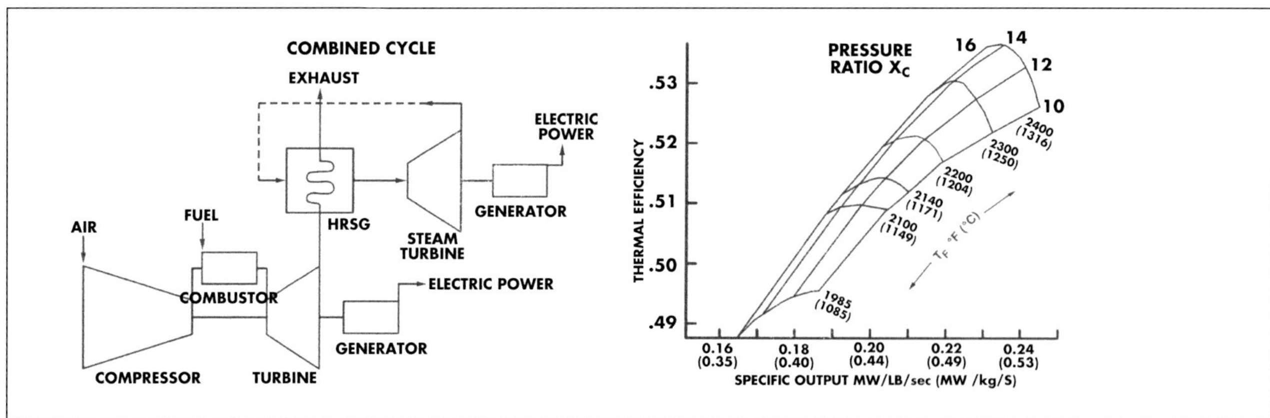

Fig. 12-2 Combined Cycle Gas Turbine Thermodynamics. Source: General Electric Company

importance of ambient air temperature and somewhat reduces the need for inlet air-cooling.

The most basic configuration for a combined-cycle plant is a single-pressure system. This typically consists of a gas turbine (or reciprocating engine), a single-pressure HRSG, and a condensing turbine. Multiple gas turbines and, in some cases, multiple steam turbines may be used.

A typical 100 MW (100,000 kW) combined-cycle system might consist of 68 MW output from the gas turbine, 33 MW output from the steam turbine, and 1.5 MW auxiliary input requirement. If the heat rate of the gas turbine is about 11,010 Btu/kWh (11,611 kJ/kWh), which corresponds to a thermal fuel efficiency of 31% (LHV basis), then the entire process has a heat rate of 7,518 Btu/kWh (7,929 kJ/kWh). The thermal efficiency of this combined-cycle plant, in which work rate is expressed in kW and energy use in Btu, is:

In the most basic design, extraction steam from the steam turbine is used to provide feedwater preheating. Overall thermal efficiency can be improved by a few percent by using a preheating loop with the lowest temperature HRSG exhaust gas. This improves heat recovery effectiveness and slightly raises power output by increasing steam generation capacity.

With a reciprocating engine-based combined-cycle system, similar overall thermal fuel efficiencies can be achieved as with gas turbine systems, though the reciprocating engine contributes a higher percentage of overall power output. Natural gas spark-ignited, Otto-cycle engines are particularly well suited for combined-cycle operation due to a high exhaust temperature and quantity relative to Diesel-cycle engines, though Diesel-cycle and dual-fuel engines are also effectively applied in combined-cycle applications.

Figure 12-3 is a line diagram of a combined-cycle system featuring a gas-fired reciprocating engine and steam turbine. All exhaust gas heat can be used for power generation at the turbine. Overall system thermal efficiency is further optimized when engine coolant heat is used for pre-heating, deaerating, etc., in the steam cycle or for space heating, absorption cooling, or process use in a cogeneration-type application.

With a spark-ignited gas engine, featuring a simple-cycle efficiency of 39% (LHV basis) for example, exhaust temperatures in excess of 750°F (399°C) can be produced. At an exhaust temperature of 788°F (420°C), for example, superheated steam at about 145 psig (11 bar) and 715°F (380°C) can be generated, allowing for a thermally efficient steam cycle. Engine coolant heat recovery can be used for feedwater heating to further enhance overall system performance.

In a gas turbine-based combined-cycle system (without supplementary firing), the ratio of power output of the gas turbine to the steam turbine may be 2:1. Alternatively, with a reciprocating engine-based system, the ratio of power output of the reciprocating engine to the steam turbine may be 4:1 or 5:1. Overall, however, similar combined-cycle thermal fuel efficiencies can be achieved due to increased heat recovery steam generation from gas turbine exhaust.

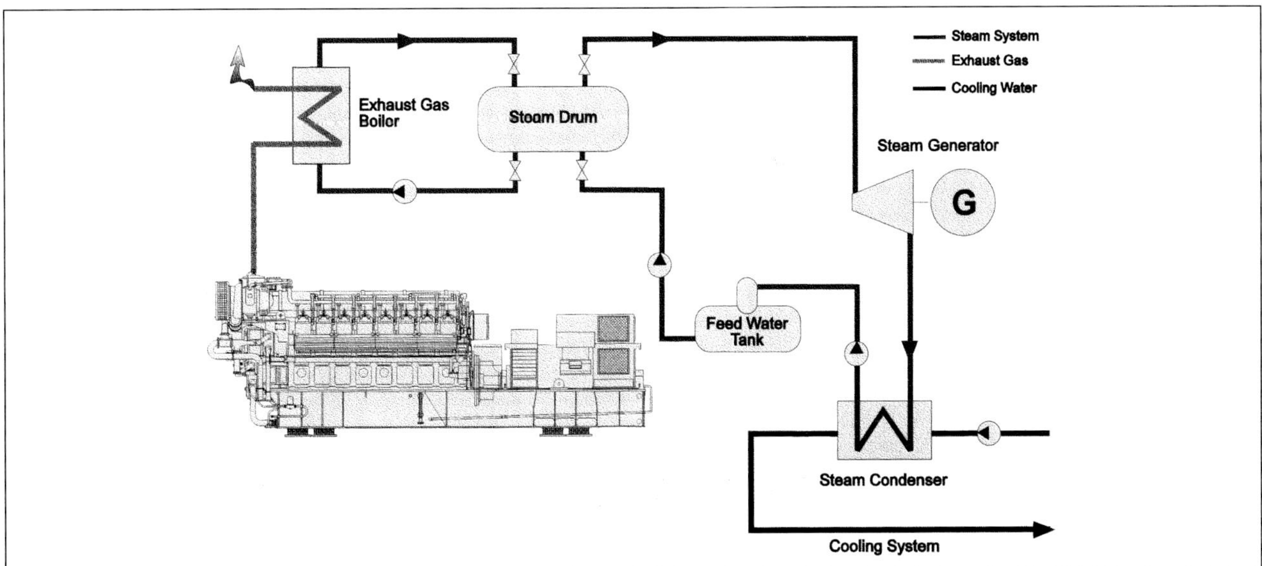

Fig. 12-3 Line Diagram of Combined-Cycle System Featuring Spark-Ignited Reciprocating Engine. Source: Wartsila Diesel

Multiple-Pressure Systems

Steam turbine Rankine efficiency is improved when the heat drop across the turbine is increased. Therefore, since the exit condition is set by ambient temperature (and cost considerations related to the condenser, final pressure, etc.), efficiency can be improved by operating the steam generator at higher temperature and pressure. With combined-cycle operation, a balance must be struck between gas turbine (or reciprocating engine), HRSG, and steam turbine efficiency.

As HRSG temperature and pressure are increased, the rate of exhaust gas heat energy utilization is decreased. Conversely, lower steam pressure and lower exhaust gas exit temperature increase the overall utilization of the exhaust gas heat. An effective way to incorporate higher steam pressures and temperatures into combined-cycle systems is to use a **multi-pressure HRSG**. This requires a steam turbine with two or more steam admissions: one at high pressure and one or more at lower pressures. These configurations allow for greater thermodynamic efficiency of the steam turbine because higher-pressure steam can be used in the first stage without compromising HRSG efficiency.

For example, with a multiple-pressure system that produces 39 MW from the steam turbine, overall net thermal efficiency would be increased from 45.4% single-pressure efficiency to 48.1% (LHV) and heat rate would be reduced to 7,096 Btu/kWh (7,483 kJ/kWh). The thermal efficiency of this combined-cycle plant is:

$$FCP = \frac{(68 MW_{GT} \times 11.3 MMBtu/MWh) - (360 MMBtu/h \div 0.83\eta_B)}{(68 MW_{GT} + 12 MW_{ST} - 0.75 MW_{AUX})}$$

$$= 4{,}223\ MBtu/MW$$

Today's larger capacity combined-cycles featuring high-efficiency gas turbines and triple-pressure steam cycles can achieve total thermal fuel efficiencies in excess of 55% (LHV basis). Figures 12-4 and 12-5 show combined-cycle power generation systems with capacities of about 250 MW. Both feature gas turbines, triple-pressure steam cycles, and single-shaft power blocks. Both systems guarantee NO_X emissions of below 25 ppm when operating on natural gas and are believed capable of achieving single-digit ppm NO_X emissions levels.

Figure 12-4 shows a labeled layout of a 251 MW system featuring a 160 MW ABB sequential combustion gas turbine, with dual combustion chambers. The gas turbine features a 30:1 pressure ratio, which is unusually high for such an efficient combined-cycle system. The steam turbine is part of the single power train, with the common generator located in the middle. The three variable stators at the compressor inlet allow the gas turbine to be operated with a relatively flat efficiency curve in the part-load range in combined-cycle operation. By adjusting the stators, the mass flow is reduced linearly to 60% of the full-load figure, allowing the turbine exhaust to be maintained at almost its design point of about 1,130°F (610°C).

Figure 12-5 shows a basic heat balance for a 254 MW system featuring a Siemens 170 MW gas turbine. The common hydrogen-cooled generator is solidly coupled to the gas turbine, with a synchronous clutch used for the steam turbine connection. The two-casing steam turbine features a high-pressure turbine and a combined intermediate- and low-pressure turbine with an axial exhaust to the condenser. The gas turbine features a pressure ratio of 16.6:1 and an exhaust flow of 3,600,000 lbm/h (454 kg/sec) at a temperature of 1,004°F (562°C).

Combined-Cycle Cogeneration Systems

If a given facility has a large low-pressure steam load (and some high-pressure steam load), it may make sense to employ a combined cycle that provides usable thermal energy to process. In this case, high-pressure steam is made in the HRSG, and a back-pressure or extraction steam turbine functions as a pressure-reducing valve (PRV), in addition to producing shaft power. Even if steam of pressures up to a few hundred psi (20 bar) are required for process, the gas turbine exhaust (often with supplementary firing) can normally make sufficiently high pressure steam that an intermediate steam turbine stage can be used. Figure 12-6 is a schematic representation of a combined-cycle cogeneration system featuring a gas turbine, HRSG, and steam turbine.

There are many variations on the cogeneration design with combined-cycle plants. Some divert a portion of the thermal energy directly from the HRSG over a wide range of pressures. Systems may feature gas turbines or reciprocating engines with a back-pressure or extraction steam turbine serving process loads. A combination of condensing and back-pressure turbines can also be used. Multistage extraction turbines are often used, because they can exhaust steam at varying pressures and quantities, while condensing the rest of the steam in the power generation process. These types of configurations are discussed in greater detail in the chapters in Section VI covering electric power cogeneration applications.

In addition to electric power generation, the steam turbine or combustion engine can be used for mechanical

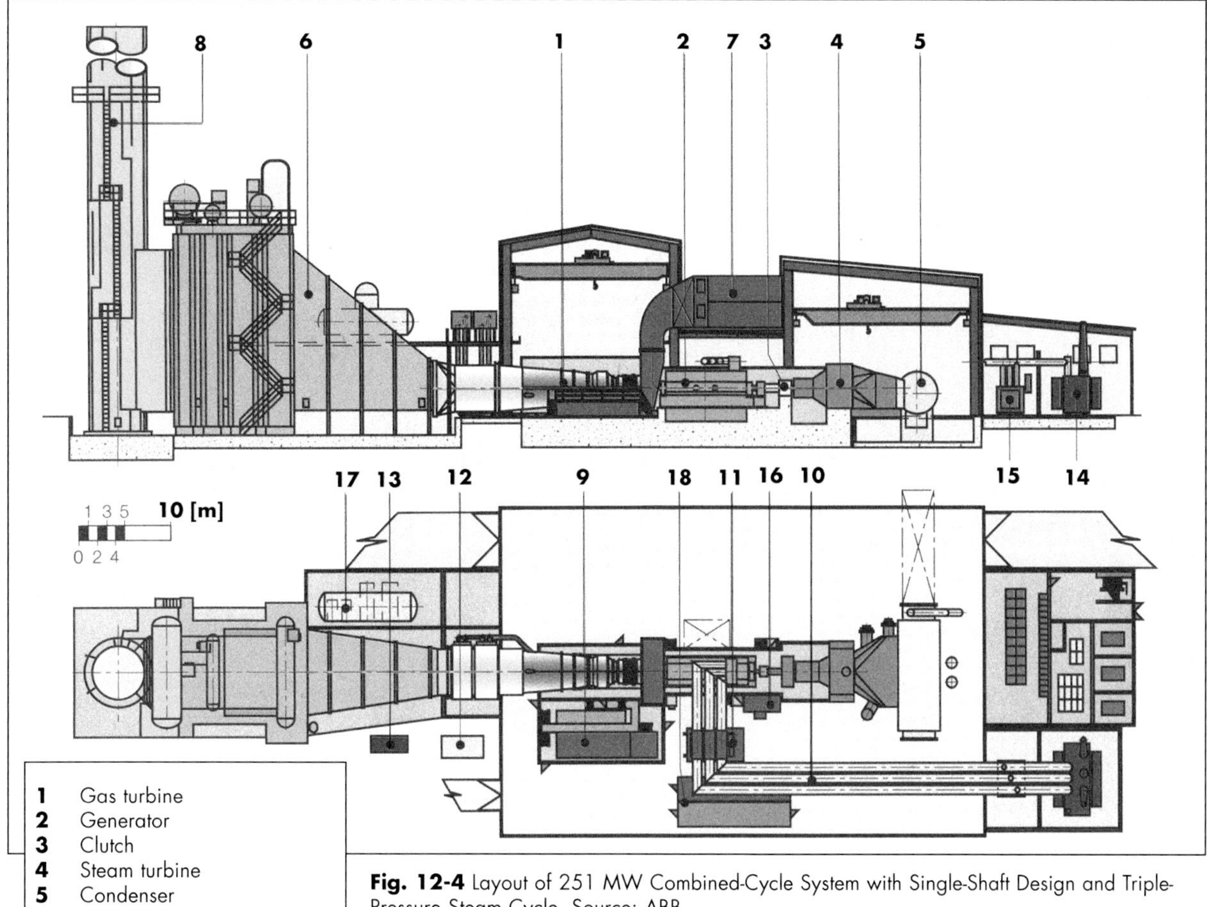

1 Gas turbine
2 Generator
3 Clutch
4 Steam turbine
5 Condenser
6 HRSG
7 Air intake
8 Stack
9 Auxiliary block, GT
10 Generator bus duct
11 Generator breaker
12 Fuel oil block
13 Fuel gas block
14 Main transformer
15 Auxiliary transformer
16 Lube oil system
17 Feedwater tank
18 Electric and Control modules

Fig. 12-4 Layout of 251 MW Combined-Cycle System with Single-Shaft Design and Triple-Pressure Steam Cycle. Source: ABB

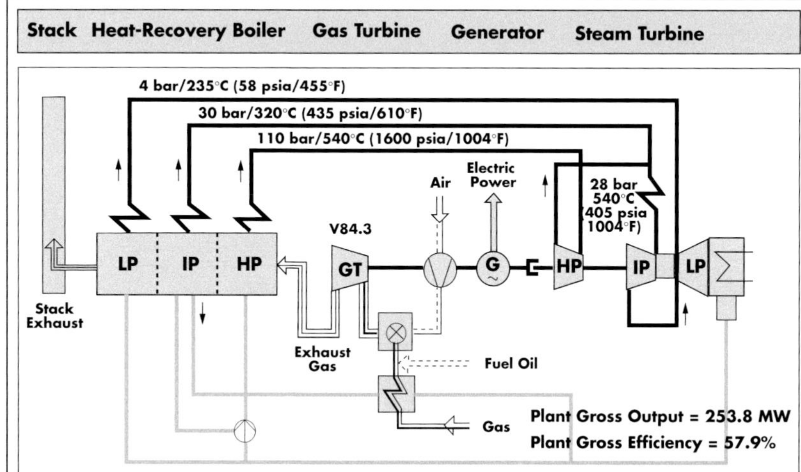

Fig. 12-5 Basic Heat Balance for 254 MW Combined-Cycle System with Single-Shaft Design and Triple-Pressure Steam Cycle. Source: Siemens Power Corp.

drive service. This is more prevalent in the smaller capacity applications. Many facility plants or district heating/cooling plants that have been built around the use of steam turbines to drive chillers, pumps, etc., have been retrofitted with gas turbines upstream of the existing plant and have become combined-cycle plants. Building summer load with condensing steam turbine-driven chillers, or a combination of back-pressure driven chillers and absorption chillers, is a way to level loads and create a more optimal combined-cycle load profile.

Because a condensing turbine produces more power than a non-condensing turbine, a combined-cycle cogeneration configuration using a non-condensing turbine has a lower power generating thermal efficiency than a condensing application. However, the total system efficiency and fuel chargeable-to-power (FCP) will be higher when steam energy that is passed on to process is credited to the energy input in the calculations.

While an extremely efficient combined-cycle plant can reach thermal efficiencies in excess of 55% (LHV), combined-cycle cogeneration plants can achieve net thermal fuel efficiencies higher than 80%. Consider the system described in the previous examples, except that 12 MW is produced from the steam turbine (with a reduced auxiliary requirement) and 360,000 MBtu/h (379,764 MJ/h) is passed on to process. In this case, the net heat rate, or FCP, is calculated as follows:

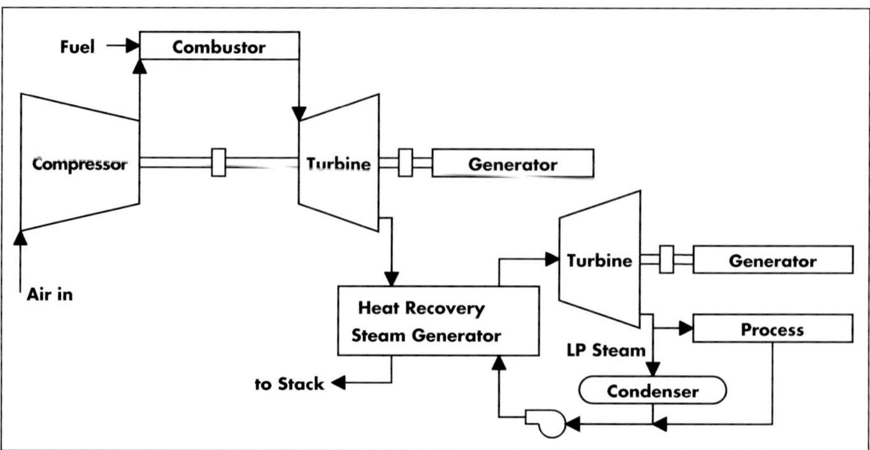

Fig. 12-6 Combined-Cycle Cogeneration System.

$$\eta_{th} = \frac{(68MW_{GT} + 39MW_{ST} - 1.5MW_{AUX}) \times 3{,}413 MBtu/MWh}{11{,}010\ MBtu/MWh_{GT} \times 68MW_{GT}} \times 100\% = 48.1\%$$

In this example the 0.83 value represents a boiler efficiency of 83 percent for the plant's fuel-fired boilers.

The relative values of achieving more power output or more thermal output will be a function of real market costs for power and fuel, availability of condenser cooling water as well as the particular plant configuration. While the net thermal fuel efficiency in the example above is 81% and FCP is 4,223 Btu/kWh (4,455 kJ/kWh), the simple power generating efficiency is only 35% and the heat rate is 9,696 Btu/kWh (10,228 kJ/kWh).

Figure 12-7 shows, schematically, a large 199 MW combined-cycle cogeneration system using a three-pressure waste heat boiler (WHB) at pressures of 1400, 450, and 5 psig (97.6, 32.0, and 1.4 bar), with gas-turbine steam injection. In this case, the steam turbine comprises 20% of the total system electrical generating capacity.

COMBINED-CYCLE APPLICATIONS

The primary use of conventional combined-cycle generation in the United States today is in medium- and large-scale power plants ranging from 100 MW to greater than 1,000 MW. In most discussions of future utility, non-utility, or independent power producer power plants, the gas-powered combined-cycle plant is considered state-of-the-art technology. This is due to the high efficiency of combined-cycle plants versus conventional power plants and the reduced societal cost resulting from pollution and other damaging externalities. In particular, low emission characteristics are typically associated with the types of fuels used in combined cycle plants (i.e., natural gas and lighter oils).

Prior to the 1990's, combustion turbines were used by electric utilities, mostly as peaking plants. Applications were designed for a limited number of run hours due to the higher cost of gas and oil fuels compared with coal, and to the relative inefficiency of simple-cycle combustion turbines. Simple-cycle combustion gas turbines were not operationally

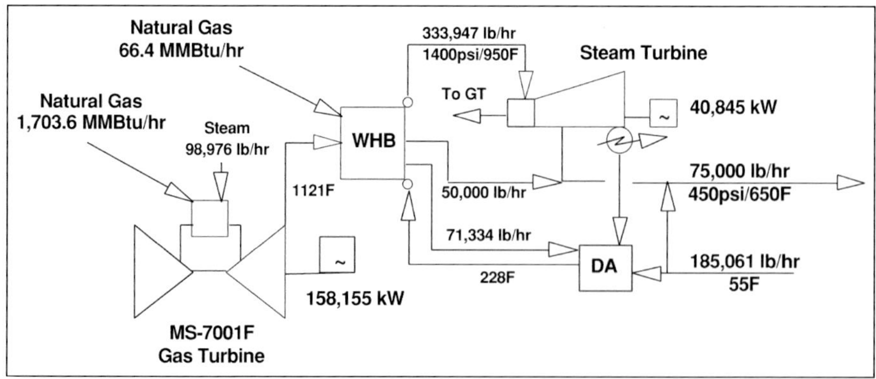

Fig. 12-7 199 MW Combined-Cycle Cogeneration Plant. Source: Cogen Designs, Inc.

cost-competitive with large coal plants for serving baseloads.

However, simple-cycle gas turbines continue to serve as effective peaking plants because, with limited hours of use, fuel cost does not significantly impact life-cycle costs. The low installed cost, easy dispatchability, and short construction lead time were the dominant factors leading to their selection for peaking plant applications.

Over the past couple of decades, power plant construction by **independent power producers (IPPs)** in the United States has been dominated by the following configurations:

1) Cogeneration plants in which recovered heat served some type of industrial steam host or a district heating/cooling system
2) Various waste-to-energy type plants
3) Combined-cycle power plants

With these technologies, IPPs have consistently demonstrated superior cost-efficiency in power production compared with utility generation plants.

Currently, the climate in the power generation industry is one of free-market competition. This climate is partially the result of advances in smaller-scale power production technology that reversed the economy-of-scale tradition that led utilities to build larger and larger plants. The low capital cost and high thermal fuel efficiency of combined-cycle plants, plus changes in federal legislation and the overall energy market, have all contributed to the current conditions.

As discussed in Section V, the Energy Policy Act of 1992 (EPAct 92) infuses competition into the electric generation industry by unbundling the electric generation function from the transmission and distribution functions. Furthermore, EPAct 92 imposes regulatory pressure for utility least-cost planning, including consideration of environmental and other external factors in the utility resource evaluation process. As a result, utilities are moving swiftly in the use of combined-cycle power plants to serve their current and future capacity needs in a cost-competitive and socially beneficial manner. The trend toward combined-cycle plants has been further fueled by the planning concept of building smaller modular plants of 50 to 300 MW, rather than traditional, much larger steam plants.

The primary factors contributing to combined-cycle market penetration are:

- Low air emissions
- High thermal efficiency
- Dispatchability
- Reliability
- Low construction cost
- Lower risk
- Short construction lead time

In addition to central power plant applications, combined-cycle power plants can be effective in capacities well below 100 MW. In fact, depending on improvements in small steam turbine performance and installed costs, variations on the combined-cycle theme may have implications for systems ranging down to less than 10 MW. Reciprocating engine-based combined-cycle systems and combined-cycle cogeneration systems also merit consideration for industrial plant operations.

Steam Injection Cycles

Steam injection-cycle systems augment gas turbine power output by increasing mass flow. As discussed in Chapter 10, the gas turbine compressor section typically requires about two-thirds of turbine power output, and simply increasing the size of the system increases power output only one-third as fast as mass flow. However, increasing mass flow downstream of the compressor by direct injection of high- or intermediate-pressure steam produces a direct increase in power output.

Figure 12-8 shows steam injection locations for the GE LM2500 STIG™-cycle gas turbine. Figure 12-9 is a steam balance and system layout diagram of a system featuring the LM5000, showing HRSG functionality and steam injection locations.

Steam injection-cycle systems follow operating strategies that are similar to combined-cycle systems. Recovered heat-generated steam is superheated and injected through the

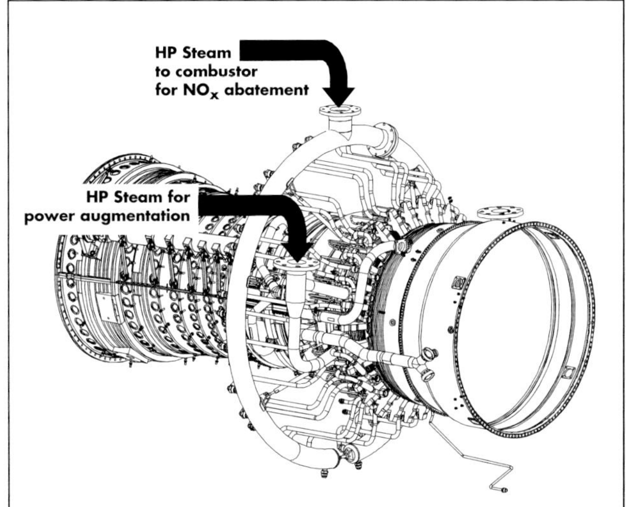

Fig. 12-8 Steam Injection Ports for LM2500 Gas Turbine. Source: General Electric Company

turbine, where the added mass flow can increase power output by as much as 50%. As with combined-cycle operation, steam injection dramatically lowers the overall power cycle heat rate. Table 12-1 shows the performance enhancement achieved with two GE STIG™ models. Table 12-2 shows the steam injection capability for various GE models, showing the steam flows and enhanced capacities and thermal efficiencies.

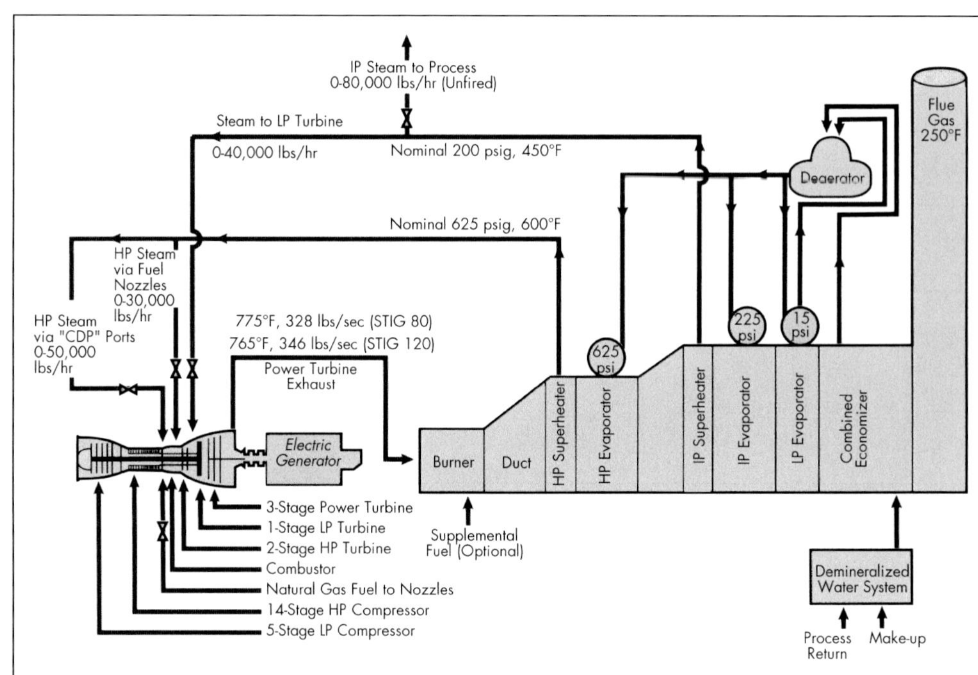

Fig. 12-9 Heat Balance and System Layout Diagram of STIG™- Cycle System Featuring LM5000 Gas Turbine. Source: Stewart and Stevenson

As shown in Figure 12-7, steam injection can also be used in gas turbines serving combined-cycle systems. Usually, the resulting increase in the gas turbine output is greater than the corresponding decrease in steam turbine output. However, there will also usually be a slight decrease in overall thermal efficiency. When limited steam injection is provided to reduce NO_X emissions, the changes in performance are fairly small. Chapter 17 provides a discussion of steam injection for NO_X control.

Standard Base Load, Sea Level, 60% RH, – Natural Gas – 60 Hertz – 4 in. (102mm) Inlet/10 in. (254mm) Exhaust Loss – Average Engine at the Generator Terminals*

Model	Dry Rating (MWe)	% Thermal Efficiency (LHV)	STIG Rating (MWe)	% Thermal Efficiency (LHV)
LM1600	13.3	35	16	37
LM2500	22.2	35	27.4	39

*3% margin on Eff. Included

Table 12-1 Performance Enhancement Achieved with Two GE Steam Injection Models. Source: General Electric Company

Standard Base Load, Sea Level, 60% RH, – Natural Gas – 60 Hertz – 4 in. (102mm) Inlet/10 in. (254mm) Exhaust Loss – 25 PPM NO_X

Model	Rating (MWe)*	% Thermal Efficiency (LHV)*	Steam Flows – lb/hr (kg/hr) Fuel Nozzle	Compressor Discharge
LM1600	16	37	11540 (5235)	9840 (4463)
LM2500	27.4	39	18300 (8301)	31700 (14379)
LM2500+	32.5	40	23700 (10750)	
LM6000	42.3	41.1	28720 (13027)	

*Average Engine at Generator Terminals
(2.5% on LM1600 Gen, 2.0% on all others Gen, 1.5% GB included)

Table 12-2 Capacity, Performance and Steam Flows of Various GE Steam Injection Cycle Models: Source: General Electric Company.

Steam injection can provide the flexibility to operate effectively under an extremely wide range of varying load and energy cost conditions. Steam injection in combined-cycle systems can reduce condenser load, allowing steam turbine and auxiliary equipment capacities to be reduced. Major disadvantages are that more water is required because steam is used in an open cycle, extensive boiler water treatment is required to avoid fouling of the gas turbine, and there is a tendency for increased production of CO emissions.

Some turbines can accept both high- and low-pressure steam, while others can accept only high-pressure steam. Steam can be injected at multiple pressures and temperatures, at multiple points in the cycle, and at varying rates, depending on power and thermal load requirements. Since steam from HRSGs can produce pressures well above those needed for injection, steam turbine extraction steam or multiple pressure HRSGs can be used.

For many stationary gas turbines, operation with steam injection rates that are equal to more than a few percent of the total air mass flow requires modifications, particularly in the compressor. Aeroderivative gas turbines are generally more compatible with large-scale steam injection.

Three common types of steam injection are NO_X control steam, compressor discharge port (CDP) steam, and low-pressure (LP) steam. NO_X control steam may be injected into the fuel, combustion air, or directly into the combustion chamber. High-pressure CDP steam is injected at a pressure above the compressor discharge pressure, depending on the pressure ratio. In practice, high-pressure (CDP) steam produces roughly three to four times more power augmentation than does low-pressure steam, since it passes through both the compressor drive turbine and the power turbine (see Fig. 10-5 in Chapter 10). Low-pressure steam is injected into the power turbine only and may be introduced at different points in the turbine section.

Figure 12-10 shows capacity versus steam injection for NO_X, CDP, and LP steam for a steam injection gas turbine. In this graphic, NO_X control steam is the first to be introduced. It is added up to about 30,000 lbm/h (13,600 kg/h). CDP steam is then added up to about 32,000 lbm/h (14,500 kg/h). At that point, capacity has increased to about 45 MW and LP steam can be added, along with some additional CDP steam. This brings the maximum capacity of the turbine to about 50.5 MW at a total injection rate of 120 Mlbm/h (54,400 kg/h).

The top line in the figure (filled squares) is the sum of the NO_X (32 Mlbm/h), CDP (45 Mlbm/h), and LP (43 Mlbm/h) steam. The incremental power production is fairly linear with steam injection, but the slope of the curve depends on the type of steam introduced. Notice NO_X and CDP steam show far greater capacity enhancement and efficiency than LP steam. In this case, the LP steam is introduced part way down the power turbine. Since it is not condensed, it is used inefficiently, relative to the performance achievable in a condensing steam turbine.

Figure 12-11 shows the capacity of this same turbine as a function of steam injection rate and ambient temperature. More steam always produces more power, but the increase is greatest around 59°F (15°C). The curves labeled 42 ppm and 25 ppm NO_X represent the steam injection rates that achieve these levels of NO_X in the exhaust gases. The other curves represent the steam injection rates in Mlbm/h, of high-pressure (625 psig) and lower-pressure (225 psig) steam.

The performance of an Allison 501 KH Cheng cycle system is illustrated in Figure 12-12. In this example, the gas turbine produces 3,587 kW at sea level and 59°F (15°C), with a fuel input of 50.7 MMBtu/h (53,478 MJ/h) on an HHV basis. Steam production, at a condition of 450

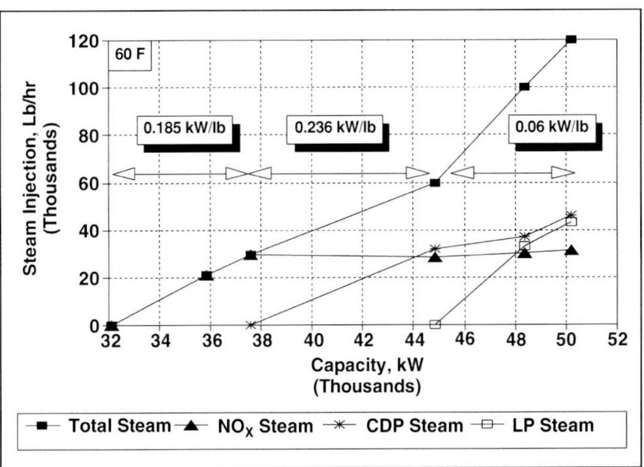

Fig. 12-10 Representative Steam Injection Performance Curves. Source: Cogen Designs, Inc.

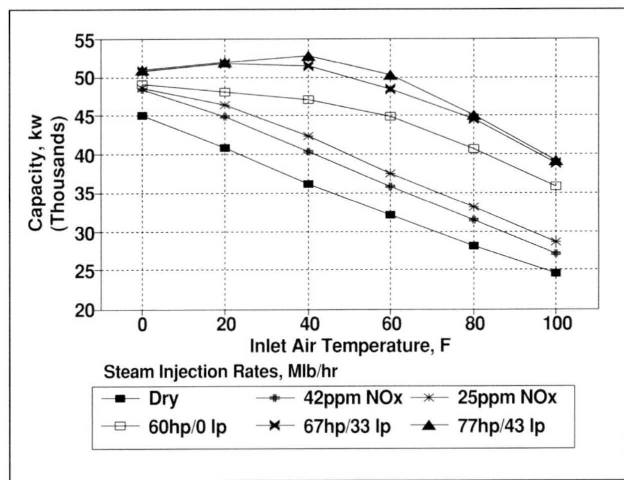

Fig. 12-11 Capacity as a Function of Steam Injection Rate and Ambient Temperature. Source: Cogen Designs, Inc.

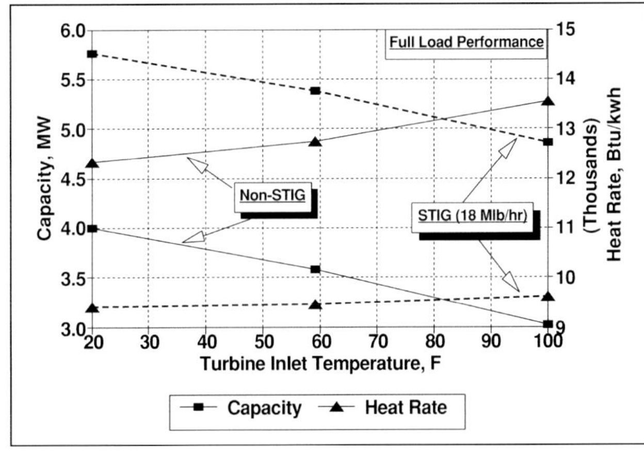

Fig. 12-12 Capacity and Heat Rate of Allison 501 KH vs. Ambient Temperature at Various Steam Injection Rates. Source: Cogen Designs, Inc.

psig/550°F (32 bar/288°C), is 21,340 lbm/h (9,679 kg/h) and the simple-cycle heat rate is 14,142 Btu/kWh (14,917 kJ/kWh). The FCP is 7,736 Btu/kWh (8,160 kJ/kWh).

As shown, injecting 18,000 lbm/h (8,165 kg/h) of steam increases power output to 5,374 kW and the fuel input rate to 56.5 MMBtu/h (59,600 MJ/h), for a heat rate of 10,513 Btu/kWh (11,089 kJ/kWh). If the maximum desired power output was constant at 3,587 kW, any excess steam could be injected back into the turbine, resulting in reduced turbine fuel requirements.

Alternatively, 21,340 lbm/h (9,680 kg/h) steam production and 3,587 kW power output can also be accomplished with 36.8 MMBtu/h (38,817 MJ/h) gas turbine fuel input, 10.6 MMBtu/h (11,181 MJ/h) HRSG supplemental firing, and 18,000 lbm/h (8,165 kg/h) steam injection. The ability to produce the same power output at two different firing rates provides great operational flexibility, but makes steam injection cycle optimization analytically challenging.

Figure 12-13 shows a typical arrangement of a packaged steam injection-cycle system. Shown are locations of the gas turbine generator set enclosure and the HRSG, including the location of the natural gas duct burner. Notice the narrow footprint as compared with the length and height of the entire system.

Figure 12-14 shows a steam injection-cycle package applied in a university medical center. The system features two Allison 501-KH Cheng-cycle steam-injected gas turbines, Coen duct burners, Abco high-pressure HRSGs, and Ideal 4,160 volt electric generators with Westech parallel shaft gearboxes. The system generates high-pressure steam for both steam injection and process use, and provides 80% of the electrical and 100% of the steam requirements of the medical center.

Supplementary-Fired Systems

As discussed in Chapter 8, supplementary firing of gas turbine or reciprocating engine exhaust can be used to efficiently increase thermal output. Because a relatively small portion of the oxygen in an open gas cycle is used for combustion, the remainder can be used for supplementary firing in the exhaust duct or the HRSG.

Increasing exhaust gas temperatures can improve steam turbine cycle efficiencies, depending on engine and

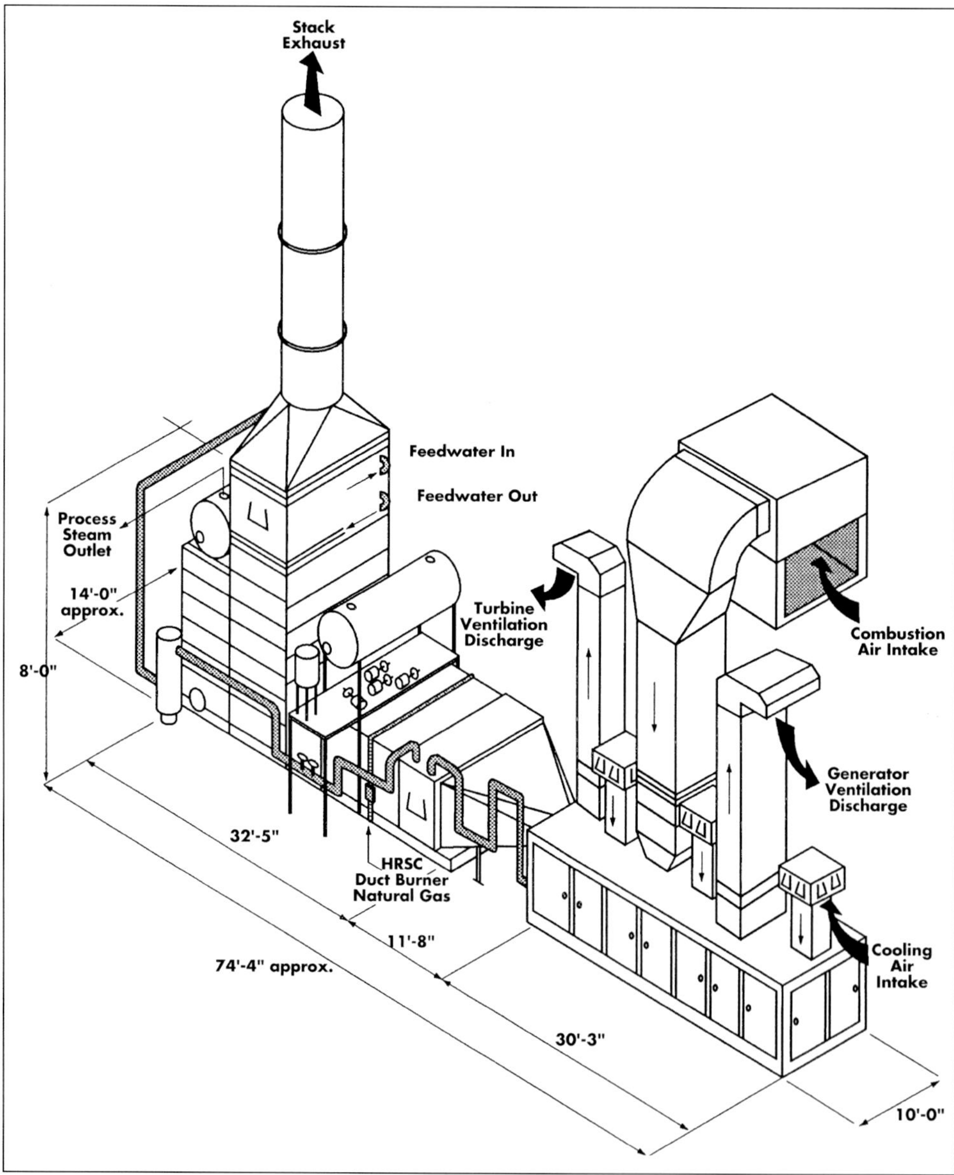

Fig. 12-13 Typical Arrangement of Cheng Cycle Package. Source: United States Turbine Corp.

Fig. 12-14 Cheng Cycle Package Applied in University Medical Center. Source: United States Turbine Corp.

system type. Also, since the capacity of the gas turbine or reciprocating engine can be reduced relative to total plant output, overall project costs may be reduced.

Another important potential benefit of supplementary firing is that it can allow a facility to be more flexible in responding to varying load conditions. In plants that either buy or sell power under differentiated electric rate periods, it may be advantageous to increase power output with supplementary firing in the higher cost rate periods. In cogeneration-type plants that have varying thermal loads, the gas turbine (or reciprocating engine) exhaust can be baseloaded and supplementary firing can be used to meet thermal load peaks. The supplementary firing HRSG boiler efficiency is approximately 89% on a HHV basis, which offers a substantial improvement over a typical auxiliary fired boiler efficiency of 85%.

Supplementary-fired units can be classified into two broad categories: those with limited supplementary firing and those with maximum supplementary firing. Limited supplementary firing is typically used to heat the exhaust gases to 1,200 to 1,600°F (649 to 927°C). At temperatures above 1,600°F (760°C), special HRSG designs are required using water wall construction and some type of combustion chamber cooling may also be required. The efficiency of steam production increases until fresh air must be added to increase oxygen for efficient combustion. Natural gas is particularly attractive for limited supplementary firing because it can be burned easily without combustion chamber cooling.

There are several conditions in which maximum-fired systems may be cost-effective. The installation of a gas turbine in a conventional utility or other large plant might be compatible with this practice. If costs for supplementary-fired fuels are lower than gas turbine fuels, as would be the case with coal, it could be economical to use a relatively small gas turbine and introduce the gas turbine exhaust directly into the coal-fired boiler (hot windbox firing) instead of a conventional HRSG. Industrial plants that seek to match power output to internal power requirements might also use maximum supplementaryfiring, passing a large portion of the thermal output to process. However, thermal efficiency is reduced due to the additional combustion air requirement and the plant will consume more resources, including peripheral resources such as water. Pollution is increased due to the use of more polluting fuel and because of the higher heat rates.

Figure 12-15 shows a gas turbine applied in a combined-cycle cogeneration system featuring a fired boiler. The gas turbine produces only 12% of the total system output in the form of power output and functions as an add-on to the boiler-fired steam-turbine cogeneration system. In this application, heat recovered from the gas turbine exhaust is used for boiler water preheating.

Spray Injection Technology

Whereas historically water injection was used only for NO_X emissions control, water spray injection is now used for capacity and performance enhancement. This is accomplished by injecting a mist of atomized water into the high-pressure (and in some cases also the low-pressure) compressor inlets. Spray intercooling increases mass flow by cooling the air during the compression process.

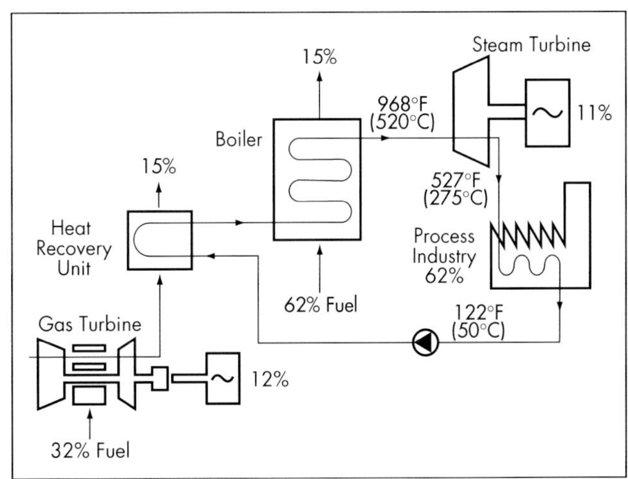

Fig. 12-15 Gas Turbine Applied to Combined-Cycle System with Fired Boiler.

The LM6000 gas turbine, shown in Chapter 10, has been effectively configured for spray injection. This LM6000 Spray Inter-cooled Turbine (Sprint™) uses bleed air from the 8th stage of the high-pressure compressor to atomize water into tiny droplets, from injection through nozzles, as a mist. Manufacturer's data indicates an increase of more than 8% in power output at 59°F (15°C) with virtually no change in heat rate and a 30% increase at 90°F (32°C) ambient condition, with a heat rate improvement of 2%. Exhaust mass flows are increased slightly and temperature is elevated, allowing for increased heat recovery and, therefore, greater total thermal efficiency.

Comparison of Combined-Cycle and Steam Injection Systems

Conventional combined cycles and steam injection systems offer similar performance and operating flexibility in responding to power and thermal load variations. Supplementary firing can be used with either type of system. Critical differences are:

- Combined-cycle systems generally have a higher capital cost than steam injection-cycle systems due to the addition of the steam turbine, condenser, and cooling tower. They also usually offer superior heat rates due to the low exhaust pressure achieved in a condensing steam turbine (a steam injection-cycle gas turbine exhausts at atmospheric pressure). Combined-cycle systems also do not require abundant use of highly treated water because steam is condensed in a closed cycle. However, moderate amounts of low quality water must be available for the cooling tower make-up water. Generally, combined cycles are suited for high-load factor applications where water and fuel costs are high. An additional consideration is that the steam turbine can run independent of the prime mover, possibly reducing back-up requirements for facilities that have boilers for steam generation.
- Typically, steam injection-cycle systems produce greater capacity for a given turbine model without supplementary firing and does so at a lower capital cost. The steam injection-cycle system is also a less-complex system with lower space requirements. They do, however, require far more water and an extensive water treatment program due to steam loss in turbine exhaust. Steam injection provides some NO_X emissions control, while the gas turbine in the combined-cycle system may require additional investment for NO_X control. Steam injection- cycle systems are better suited for low load factor applications, particularly where peak electric rates are high.

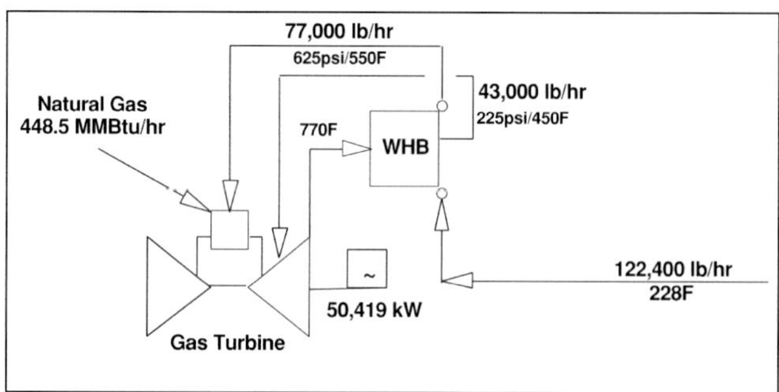

Fig. 12-16 Heat and Material Balance for 50.4 MW Steam Injection System. Source: Cogen Designs, Inc.

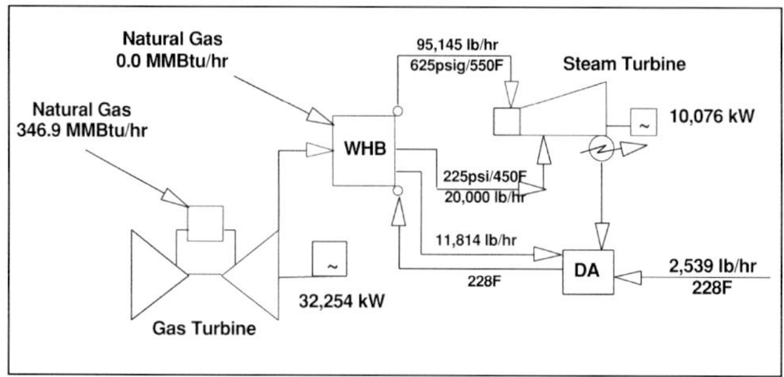

Fig. 12-17 Heat and Material Balance for 42.3 MW Combined-Cycle System. Source: Cogen Designs, Inc.

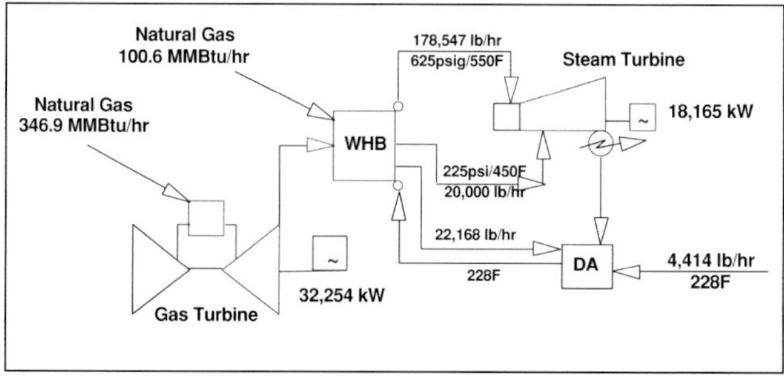

Fig. 12-18 Heat and Material Balance for 50.4 MW Combined-Cycle System with Supplementary Firing. Source: Cogen Designs, Inc.

Figures 12-16, 12-17, and 12-18 provide comparisons of similar steam injection and combined-cycle systems. Figure 12-16 is a heat and material balance schematic of a steam injection-cycle system. Fuel input is 448.5 MMBtu/h (473,078 MJ/h), producing 50.4 MW of power at a heat rate of 8,895 Btu/kWh (9,382 kJ/kWh) on an HHV basis.

Figure 12-17 is a combined-cycle system featuring the same gas turbine in which both high- and low-pressure steam is sent to a steam turbine. Without steam injection, the turbine has a capacity of about 32.3 MW and a fuel input requirement of 346.9 MMBtu/h (365,910 MJ/h) on an HHV basis. The steam turbine adds an additional 10 MW of capacity, reducing the overall heat rate to 8,196 Btu/kWh (8,645 kJ/kWh). Comparison of these two designs shows greater capacity with the steam injection cycle, but a superior heat rate with the combined cycle.

Figure 12-18 shows a combined cycle system configuration featuring the same gas turbine, but using supplementary firing to match the 50.4 MW capacity of the steam injection system. In this case, the supplementary firing results in a heat rate of 8,876 Btu/kWh (9,362 kJ/kWh), which is higher than the unfired combined-cycle, but still marginally superior to the steam injection-cycle system. However, if a larger, more costly gas turbine is used, the total capacity of the steam injection-cycle system could be matched without supplementary firing, at a superior heat rate of under 8,200 Btu/kWh (8,650 kJ/kWh).

Both combined-cycle and steam injection-cycle systems can be viewed as adding peak shaving capability into an otherwise baseloaded electric cogeneration system. Varying electric and thermal loads, time-of-use and real-time pricing, and seasonal fuel cost fluctuations all figure heavily into cogeneration project economics. Both systems provide flexibility to operate effectively under a wide range of load and energy cost conditions.

An ideal operating strategy uses either a combined-cycle or steam injection system with supplemental firing or a fuel-fired boiler. Gas turbine capacity is selected for baseload electric and steam loads. Supplementary firing is used as steam load increases, and combined-cycle or steam injection- cycle systems are brought on-line as electric load increases. By varying the levels of secondary combustion and secondary electric generation, a wide range of loads can be met. Microprocessor control is of great value in these types of applications.

This simplified comparison shows the typical trade-off of lower capital cost for the steam injection system versus lower operating cost for the combined-cycle system. Conditions of high peak electric rates, inexpensive water, and lower load factor (low utilization of required capacity) would favor the steam injection cycle. Conditions of higher load factor and water costs and, to some extent, higher fuel costs would favor the combined-cycle.

Prime mover control systems regulate input of energy to the prime mover and the conversion of this energy into power in a safe and efficient manner. Regardless of whether the prime mover drives a generator to produce electric power or drives a mechanical load, such as a compressor or pump, energy flows into the prime mover and the prime mover converts it into **developed power** (P_D) and **developed torque** (T_D). The driven load exerts a **load power** (P_L) and **load torque** (T_L) in the opposite direction of the developed power. Torque is a measure of force of rotation and power, which is the rate of doing work, is equal to torque times the speed of rotation. Therefore, if speed is held constant, torque and power are proportional.

There are two general modes of prime mover operation:

- **Steady state** refers to the condition in which P_D and T_D match P_L and T_L. This could be a no load, full load, or an intermediate-load condition. Under steady-state conditions, no acceleration or change in speed occurs.
- **Transient** refers to the condition in which P_D and T_D do not match P_L and T_L. When, for example, the load (T_L) is suddenly decreased, T_D will momentarily remain unchanged. The excess T_D immediately produces an acceleration of the prime mover and the speed increases.

The essential role of prime mover control is to continue to operate the prime mover safely in a steady-state mode, while responding to the transient situation as fast as possible with a minimum of instability. An **isochronous** control system holds rotational speed constant under varying load conditions. When a control system allows a change in operating speed that is inversely proportional to a given change in load, it is said to be a **droop** system.

CONTROL GOVERNORS

A **governor** is a device that controls the speed, or some other parameter, of a prime mover. All prime movers are rotating machines that derive their power from the flow of some energy input, such as fuel or steam. Most commonly, prime mover speed is controlled by varying the flow of the energy input.

In all governing systems, at least two components are required: a speed sensing element and a device to operate the energy input valve. In the simplest governors, these may be one and the same. However, where considerable force is required, an additional device, i.e., a servo motor, is required to change the position of the **energy-input valve**. Control valve characteristics and rangeability may vary, for example, from 100 to 20% (5:1 turn down ratio), while the pressure drop in the system rises from 5 to 80%.

The rangeability can be expressed in the form of an equation as:

$$R = (Q_1/Q_2)(\Delta P_2/\Delta P_1)^{0.5} \tag{13-1}$$

Where:
R = Rangeability
Q_1 = Valve initial flow, percentage of total flow
Q_2 = Valve final flow, percentage of total flow
P_1 = Initial pressure drop across valve, percentage of total pressure drop
P_2 = Percentage of final pressure drop across valve
In this case, the rangeability would be calculated as:

$$R = (100/20)(80/5)^{0.5} = 20$$

Three characteristics in valve design and manufacturing are:

1. Linear Design: Typical flow rangeability is between 12-1, equal stem movement for equal flow change.
2. Equal-Percentage: Typical flow rangeability of 30-1 to 50-1, equal stem movement for equal percentage flow.
3. On-Off: Linear for first 25% of travel and on/off after that point, same as linear up to on/off range.

Governors sense actual speed, compare it to a reference setting, and generate an error or difference signal. Its output adjusts the prime mover energy input until the actual speed returns to the required speed, at which P_D and T_D are again equal to P_L and T_L. The basic process involved with speed control with a transient in a closed- loop system can be summarized as follows:

1. Change in T_L (transient), causing

2. Change in speed, which is sensed by the governor, causing
3. Change in governor output, causing
4. Change in energy input valve position, causing
5. Change in T_D, causing
6. Return to steady state at governed speed

This is a closed control loop, which is central to all prime mover control systems. Figure 13-1 illustrates closed-loop control.

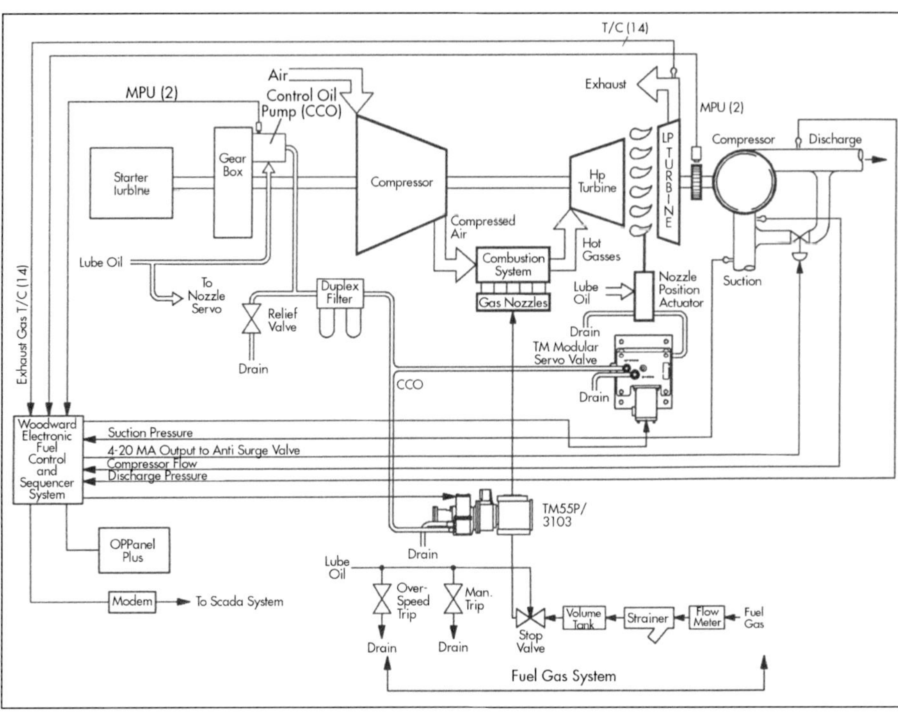

Fig. 13-1 Closed-Loop Control. Source: Woodward Governor

Governors may feature mechanical or electronic designs. In mechanical governors, shaft speed is sensed by a fly-ball mechanism that exerts a force proportional to the rotational speed of the prime mover. Hydraulic relays provide input to the control valves.

Figure 13-2 shows a governor used for Diesel or spark-ignition engines and steam turbines in mechanical drive service. The unit features an internal oil pump, relief-valve, and accumulator system for controlling governor operating pressure. A centrifugal fly-weight-head and pilot-valve assembly controls oil flow to and from the governor assembly. A power cylinder (servo motor) positions the fuel rack, fuel valve, or steam valve of the engine or turbine. A pneumatically-operated bellows mechanism sets governor speed, with a knob for manual speed adjustment. An adjustable needle valve and spring-loaded buffer compensation system provide governor stability.

In an electronic governor, sensors called magnetic pickups measure operating parameters of the prime mover. The governor compares signals from the sensors with selected set points and provides a signal to the valve **actuators.** An actuator converts the output signal from the electronic control to a mechanical movement that positions the throttle valve, steam valve, or fuel rack of the prime mover. Actuators may receive the power to move their output shafts in the form of hydraulic pressure or mechanical drive from the prime mover, or from an independent electrical or hydraulic source.

Figure 13-3 shows a proportional hydraulic actuator used for positioning steam and fuel-control valves requiring high-force linear input. Figure 13-4 is a cutaway illustration showing the actuator components. A torque-motor servo valve is energized by the electronic control to generate a pressure differential applied to the ends of a second-stage spool valve. Supply pressure is regulated by the spool valve to move a double-acting servo piston and provide one linear inch of output shaft travel. Internal mechanical feedback varies the location of the actuator shaft in direct proportion to the input-current signal.

Fig. 13-2 Governor Used for Controlling Speed of Prime Movers Driving Pumps and Compressors. Source: Woodward Governor

Figure 13-5 shows a gas valve actuator combination designed for use with electronic controls for industrial gas turbines. Its open-loop positioning characteristic

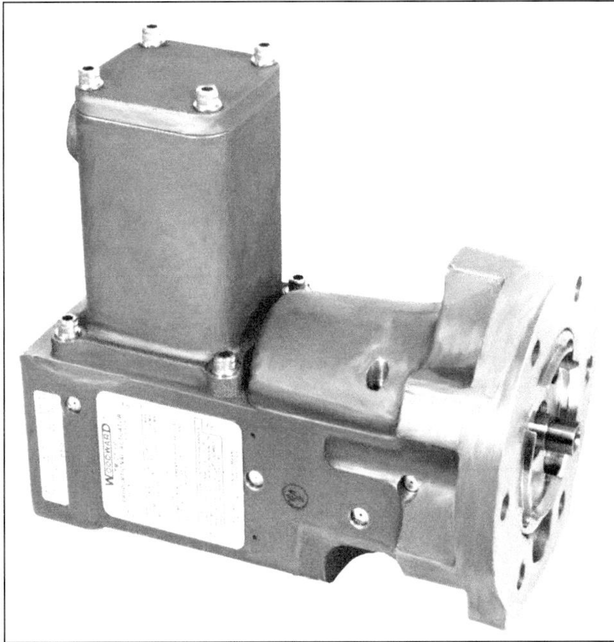

Fig. 13-3 Proportional Electro-Hydraulic Actuator Used for Steam and Fuel Control Valves. Source: Woodward Governor

Fig. 13-5 Gas Valve Actuator Combination Designed to Control Gas Turbines Used for Industrial Applications. Source: Woodward Governor

provides consistent valve positions from idle to full load.

The output signal from an electric controller usually includes proportional, integrating, and derivative response components. Proportional response is an output signal that changes in proportion to the error signal, or difference between speed set point and measured speed. Proportional control always results in some offset from a given set point. Integrating response eliminates the offset by considering the length of time that a given error has existed. Derivative response uses the rate of change of the error to produce a dampening signal that minimizes overshoot and oscillation around the set point value.

Mechanical governors use proportional and integrating response in varying amounts to produce the net output signal. Electronic controls include all three responses in varying combinations and are often referred to as **PID type control**.

Deadband is the amount of change a control system will allow in a controlled parameter before it responds with a correction. The smaller the deadband, the more accurately a parameter is controlled. **Limiting** a parameter allows it to vary within a specified range. A parameter may be limited in its maximum value, minimum value, or both.

Figure 13-6 shows a digital

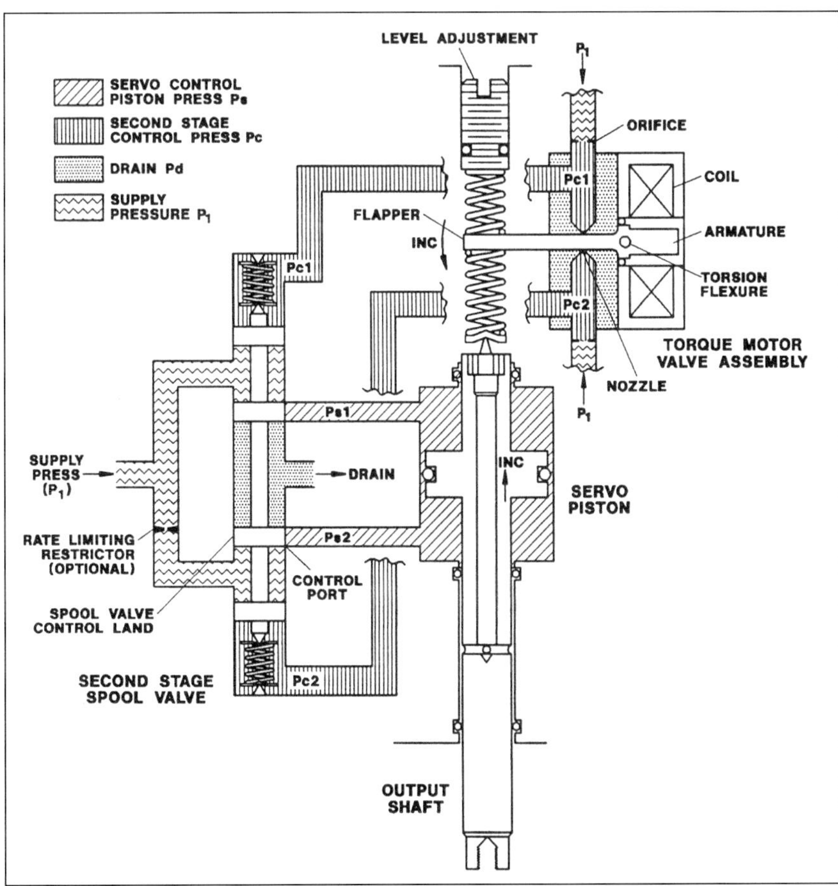

Fig. 13-4 Cutaway Illustration of Actuator Components. Source: Woodward Governor

electric-powered governor system designed for reciprocating engines. This system includes a 16-bit microprocessor control and a limited-angle rotational actuator. Two complete control programs allow for operation on two different fuels, or for alternate parallel or isolated generating service. Adjustment is done through a handheld programmer or a host computer. This unit provides 2.0 lbm-ft (2.7 N-m) of torque in steady state and 5.24 ft-lbm (0.725 kg-m) of work over 75 degrees of rotation.

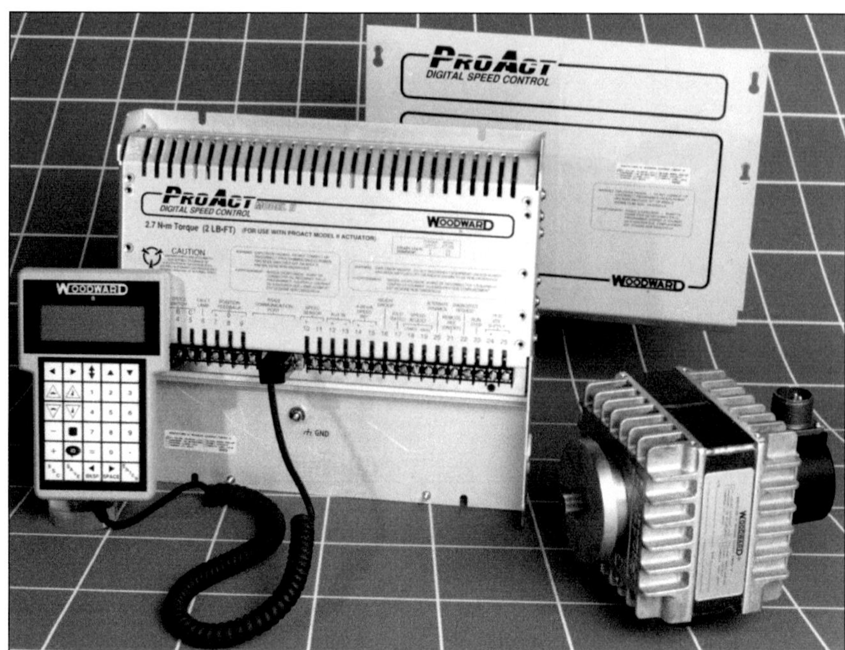

Fig. 13-6 Digital Electric Powered Governor System. Source: Woodward Governor

Sequencers or sequencer/monitors are programmable, microprocessor-based units that monitor several parameters of a prime mover or load and perform procedures such as starting, stopping, loading, or unloading. When a sequencer is teamed with a separate electronic speed control, complete automatic control may be achieved.

Older electronic control systems were the analog type, using varying voltages or currents proportional to the value of some parameter. The most advanced control systems use microprocessor-based digital logic. Digital logic allows the control system to perform PID type control, as well as limiting, sequencing, and monitoring. Performance calculations allow dynamic control strategies to be used. Remote system monitoring and control is possible via telecommunications networks.

The ability to control regulated air emissions from prime movers and related equipment is an important consideration in the selection and operation of the control system. Control systems must respond to feedback on emissions production parameters and adjust operations accordingly. Air-fuel ratios, for example, must be continuously adjusted on engines, based on such feedback, in order to stay within acceptable emissions limits. Many larger capacity systems require continuous emissions monitoring systems (CEMs).

Control Parameters

Prime movers can be controlled in a variety of ways, depending on the requirements of the specific application. Control options can be categorized into one of the following general areas:

- Speed control
- Load control
- Prime mover parameters
- Drive equipment parameters
- Process parameters

Regardless of the application, speed and/or load control are required. Speed control is the primary control loop for most prime movers. Speed control is used, for example, in mechanical drive applications where output power depends directly on running speed.

In most electric generation applications, however, it is desirable to control the prime mover for load rather than speed. When the generator is feeding electricity into a grid system, the prime mover rotational speed is locked to the frequency of the grid once its generator has been synchronized to the grid and the breaker is closed. However, the load (kW) can be controlled by raising or lowering the speed reference to vary the energy input to the prime mover. With speed held constant, a change in energy input will produce a corresponding change in output power.

In generator applications, the governor system may include an automatic synchronizer to adjust prime mover speed to match bus frequency and adjust generator phase to match bus phase and provide voltage matching. Import/export controls allow the governor to regulate the flow of power between the utility and the connected system and to limit output so that it does not exceed the generator rating. This would be complimented by the use of Load Shed system to reduce load when the utility source is not present. The Load Control section of Chapter 27 provides further discussion on the control of generators and use of load shedding systems.

Driven equipment and process control parameters can include pressure, flow, temperature, level, or speed. Gas turbines frequently are controlled on the basis of exhaust temperature. A second example is the control of a reciprocating engine- or turbine-driven compressor pumping gas through a pipeline, where pipeline pressure may be the controlling parameter.

When based on driven equipment or process parameters, prime mover control may be accomplished in one of two ways:

1. Adjusting the actuator output signal can control parameters that are directly related to prime mover rotational speed. Typically, this would be limited to process parameters such as pressure or flow. Figure 13-7 illustrates this process for a steam turbine. As shown, the process control and the speed control functions both reside within the prime mover control. Each receives an input signal, compares it with the reference set point, and generates a corrective signal. The two controllers then compete for control of the actuator through a low-signal selector (LSS). The lowest corrective signal is selected and used to position the energy-input valve.
2. A remote speed set point can be used to control the prime mover. As shown in Figure 13-8, a process controller (or some remote control system) performs all reference calculations and sends a corrective signal through the remote speed set point to adjust the speed reference.

Control algorithms are based on performance characteristics and limitations of the prime mover and the driven equipment. With steam turbines, for example, it is sometimes beneficial to maintain full-load design speed and limit output by throttling the driven load through its own control mechanisms. When driving equipment such as pumps, fans, and positive displacement type compressors, speed control may be effective, where the equipment performance gain will more than offset performance loss at the steam turbine. With a reciprocating engine, performance of both the engine and the driven equipment is enhanced through speed control, resulting in excellent part-load performance over a wide operating range.

With centrifugal compressors, the advantage of variable speed operation is not as significant. Although they can operate at variable speed, the turn-down range is limited because a minimum mass flow is required to prevent surge or back-flow in the compressor. In such cases, it may be advantageous to maintain full-load design speed or to use speed control over a more limited operating range.

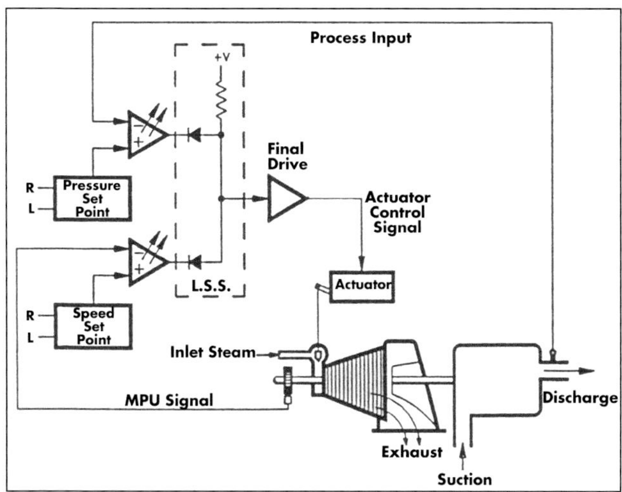

Fig. 13-7 Low-Signal Selector.

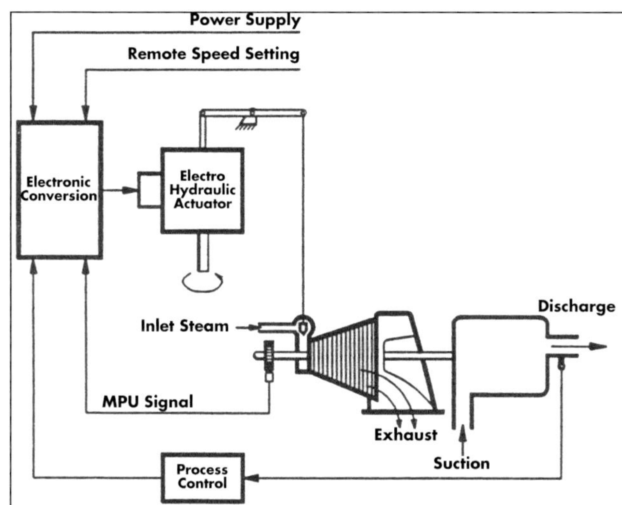

Fig. 13-8 Remote Process Control.

For large steam turbine systems, where most or all of the steam generation is dedicated to the turbine, both turbine and boiler must be controlled by the same system. Since the boiler cannot respond rapidly, the steam turbine valves modulate for initial load response at constant throttle pressure. The control system then acts to restore the proper rate of fuel flow to the boiler.

Control Lags and Governor Designs

All control systems include time lags. In particular, a time lag inherent to prime movers results from the relationship between speed and acceleration. A torque imbalance is proportional to acceleration, not speed, although speed is the desired control parameter. Since acceleration is the first derivative of speed (rate of change of speed), acceleration at any time is the angle of the tangent of the speed curve at that point. When speed is at a minimum, acceleration is zero. When acceleration is maximum, speed reaches its mid level, and when speed is at its maximum, acceleration returns to zero. Therefore, there is a 90 degree phase difference in time between acceleration reaching maximum and speed reaching maximum.

For similar reasons, when a servo is added to a governor system to position the energy input valve, there can be a 90 degree lag between governor speed and energy input valve position. Added to the phase shift described above, this results in a total phase shift for the prime mover system of 180 degrees. This means that the energy-input valve is closing to its minimum position while prime mover speed is increasing to its maximum. As soon as a change in load occurs, resulting in a transient condition, control becomes unstable, oscillating between maximum and minimum.

In practice, because other factors introduce additional phase shifts within the prime mover, the governor design must achieve an actual phase lag well below 90 degrees to ensure stable operation.

Figure 13-9 illustrates the workings of a **droop governor.** The unit has a feedback mechanism that reduces the tension in the speed reference spring (and decreases the speed setting) when the servo moves in the increase direction. The reverse occurs when the servo moves to the decrease direction. The feedback lever reduces the speed reference before the deviation reaches its maximum, reducing the phase shift in the governor as required. This results in speed eventually returning to steady state.

After a transient caused by an increase in load, the power piston has to take a position that increases the energy input. The feedback lever will take a new position where the speed reference is lower. Therefore, the actual speed where the prime mover settles for steady state after the transient will be lower than before the load increase. This characteristic is illustrated by the droop curve shown in Figure 13-10. As shown, the governor achieves stable operation during and after a transient by adding a lead factor in the form of a change in speed setting. This is known as permanent droop.

The ability to return to original speed after a change in load and maintain constant steady state speed regardless of load is called isochronous speed control. Figure 13-11 illustrates the workings of an isochronous compensated governor. This unit features a hydraulic feedback between the servo and the speed setting. When a load increase or decrease occurs, a temporary change in speed setting is produced, similar to the basic droop system. However, this temporary droop leaks away across a needle valve, which

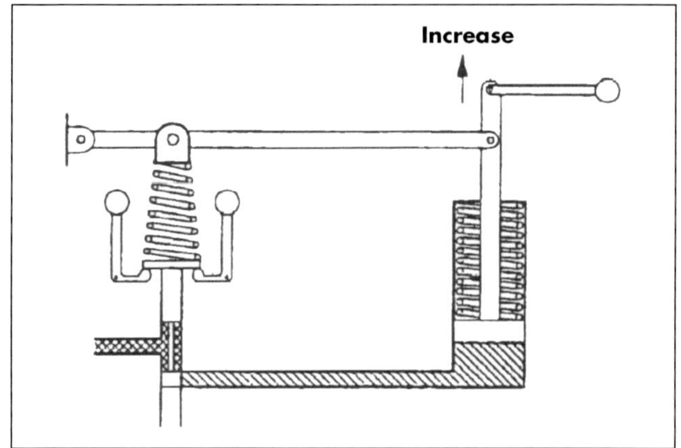

Fig. 13-9 Illustration Detailing Workings of Droop Governor.

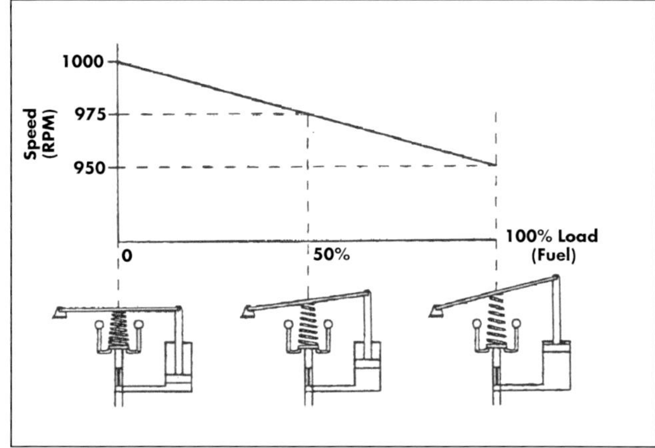

Fig. 13-10 Illustration of Droop Curve and Corresponding Speed Settings.

equalizes pressure between the two compensation pistons, and the governor returns to its original speed setting. This results in isochronous operation at constant steady speed.

Droop is a straight-line function, defined as the decrease in speed for a given increase in load:

$$\%\ Droop = \frac{Speed\ at\ no\ load - Speed\ at\ full\ load}{Speed\ at\ full\ load} \times 100 \qquad (13\text{-}2)$$

A typical droop governor lowers the speed reference from 3 to 5% of the reference speed over the full range of engine output. Thus, a 3% droop governor with a reference speed of 1,854 rpm at no fuel would have a reference speed of 1,800 rpm at maximum fuel. All electronic controls have circuits that provide a form of temporary droop by adjusting the change in actuator position according to speed. More complex hydraulic governors have adjustable droop. Figures 13-12 and 13-13 illustrate various types of

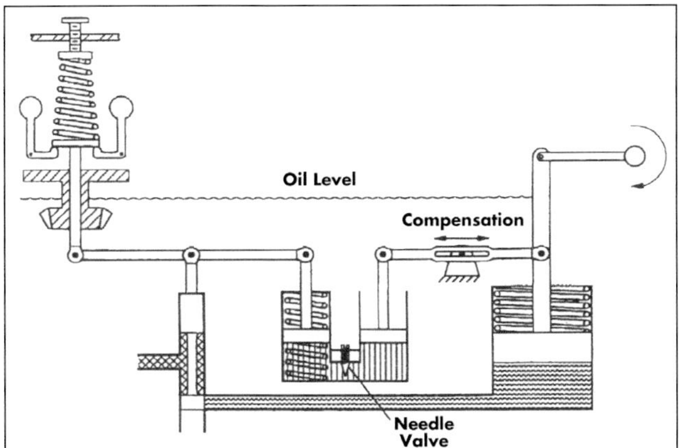

Fig. 13-11 Illustration Detailing Workings of Isochronous Compensated Governor.

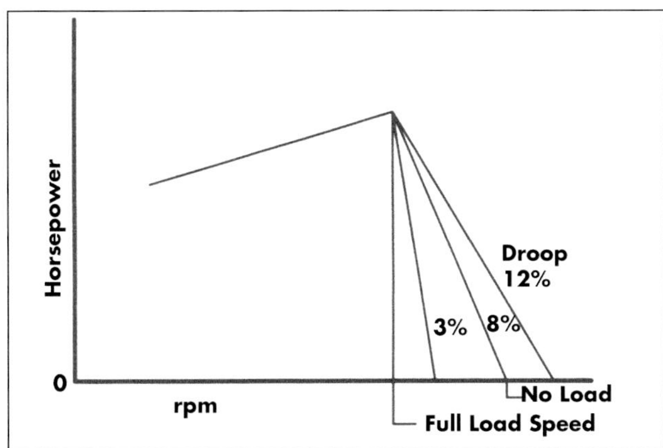

Fig. 13-13 Speed Droop Illustrated for Mechanical Drive Applications.

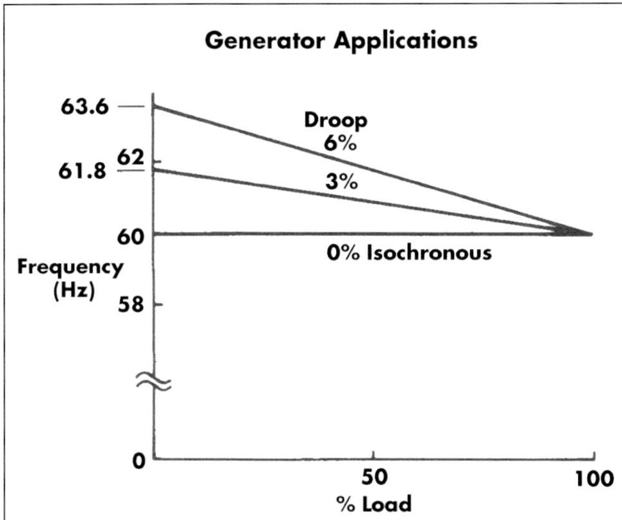

Fig. 13-12 Speed Droop Illustrated for Generator Applications.

droop for generator and mechanical drive applications.

Figure 13-14 is a governor used for numerous types of small-capacity prime movers that can operate isochronously or with droop. This type of governor may be applied to control engines or turbines driving generators, pumps, or compressors. Stability for isochronous control is achieved through pressure compensation. The temperature-compensated speeder spring minimizes speed drift caused by temperature change. Droop may be adjusted from 0 to 7% with a permanent movement inside the cover or with externally adjusted droop. This governor provides hydraulic-powered travel in the increase-fuel/steam direction, with a return spring.

An isolated prime mover generator can operate in isochronous mode, changing speeds only temporarily in response to changes in load. The system can also operate in droop if a lower speed is permissible under loaded conditions. If, however, the generator is connected to an electric grid, the grid will determine the frequency. Should the governor speed reference be less than the utility frequency, power in the utility bus will flow to the generator and motor the unit. If the

Fig. 13-14 Speed Droop Governor for Controlling Speed of Small Prime Movers. Source: Woodward Governor

governor speed is even fractionally higher than the frequency of a utility bus, load will go to maximum in a futile attempt to increase the bus speed.

Isochronous governor control is thus impractical when paralleling a utility. In such cases, droop is effective. When operating in parallel with a grid, the load on the prime mover is determined by the reference speed setting on the droop governor. Increasing the speed setting cannot cause a change in speed, which is locked in, but will cause a change in the amount of load the prime mover is carrying.

Droop may also be used to parallel multiple prime movers on an isolated bus. If all prime movers are operating in droop and variations in speed are acceptable, the voltage of the bus will vary with a change in load. Multiple prime movers can also be paralleled on an isolated bus with one unit in isochronous operation and remaining units operating with droop. These systems will be able to maintain a constant speed as long as the isochronous prime mover is capable of accommodating any load changes.

In addition to the in-phase lags described above, there are numerous additional lags in response. These can vary widely, depending on prime mover type and design. Lags due to scan intervals and other factors are even present in digital controls.

Three types of delays are associated with gas and steam turbines. Manifold lag is the time elapsed between energy inlet valve actuation and appearance of the new energy level at the turbine. For a gas turbine, the manifold lag depends on piping between the fuel valve and the combustors. Similarly, for a steam turbine, the manifold lag depends on piping between the steam valve and the first turbine stage. Turbine lag includes the time delay before the prime mover operates on a new power level. For a steam turbine, this lag is small because steam moves rapidly across the turbine stages. Gas turbines also experience combustor lag, which is the time it takes to establish a new flame pattern.

With reciprocating engines, time lags include the dead time between the charging of the cylinder with fuel and its conversion to torque and the time required for all of the cylinders to be firing at the new level. The dead time varies greatly and is least in Diesel-cycle engines and greatest in four-stroke Otto-cycle engines. Dead time decreases as engine speed increases. Natural gas and gasoline Otto-cycle engines also experience manifold lag due to the dead time associated with charging the manifold. Under large load changes, turbochargers also contribute to time lag.

In all prime movers, transient behavior is affected by the polar moment of inertia, or flywheel effect. For a given torque difference, acceleration is inversely proportional to the system's moment of inertia. With twice the moment of inertia, the system will accelerate only half as fast. With more inertia, the governor has more time to make corrections. However, under large load swings, it will also take more time for corrections to take effect.

Examples of Prime Mover Control Systems

Figure 13-15 is a block diagram of a Woodward Governor comprehensive multi-engine control system, a family of modules that include speed control, load control, engine control, monitoring, and sequencing. This system may be used with gas, Diesel, and dual-fuel engines in mechanical drive and power generation applications. The digital control application program is created in

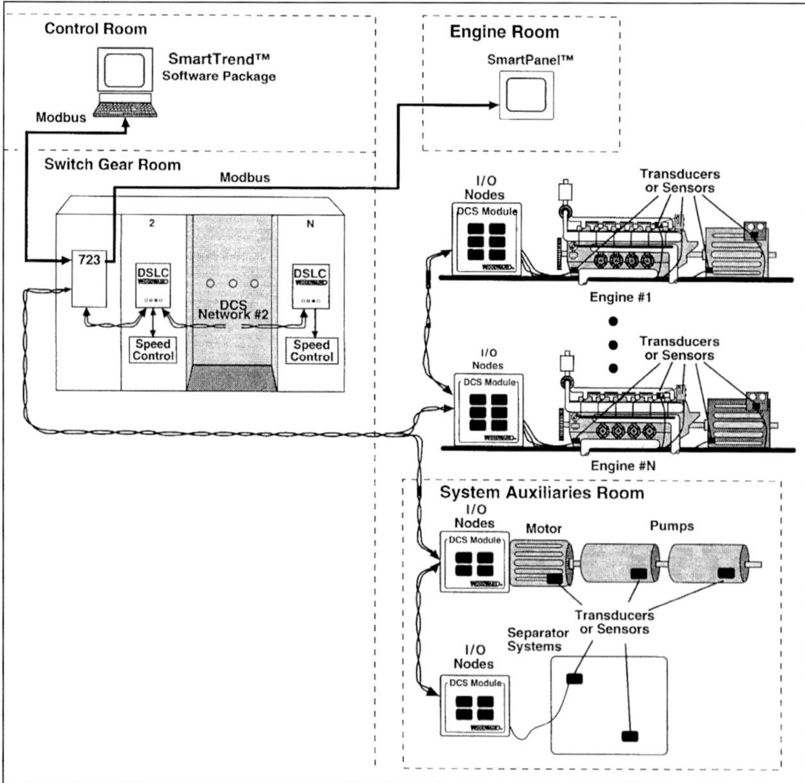

Fig. 13-15 Block Diagram of Comprehensive Multi-Engine System Digital Control System. Source: Woodward Governor

application block language software (ABLS), designed to be programmed by non-software engineers. The graphical application programmer (GAP) allows the ABLS to be programmed graphically, and it automatically creates the documentation for the software.

Figure 13-16 shows a representative block diagram illustrating the control systems applied to a power generation system featuring an extraction steam turbine. The control system has a full range of turbine-control functions, including conditional sequencing based on the information it receives through its discrete and analog functions. The system receives redundant input signals from all three CPUs and, based on a comparison of data and a two-out-of-three voting system, generates outputs to the active coil of a dual-coil actuator for turbine control.

Figure 13-17 shows a representative diagram of a control system for a gas turbine applied to mechanical drive service in a gas compression station. This system provides fuel control, automatic start/stop sequencing, temperature sensing and limiting, speed control, nozzle control, alarming functions, and compressor surge control. All of the control information is available through modem for remote control of operations and troubleshooting.

Figure 13-18 shows a comprehensive integrated boiler/turbine-generator control system. This system would be applied in large capacity plants, such as central utility plants, to coordinate the boiler and turbine-generator for fast and efficient response to automatic load dispatch controls. In its basic form, the system consists of ratio controls that monitor pairs of controlled inputs, such as boiler energy input to generator energy output and fuel flow to feedwater flow.

Vendors of gas turbine and reciprocating engine generator sets typically offer complete controls systems with off-the-shelve or proprietary programmable logic controllers. Figure 13-19 shows a typical schematic of a gas turbine generator control schematic offered by the prime mover vendor.

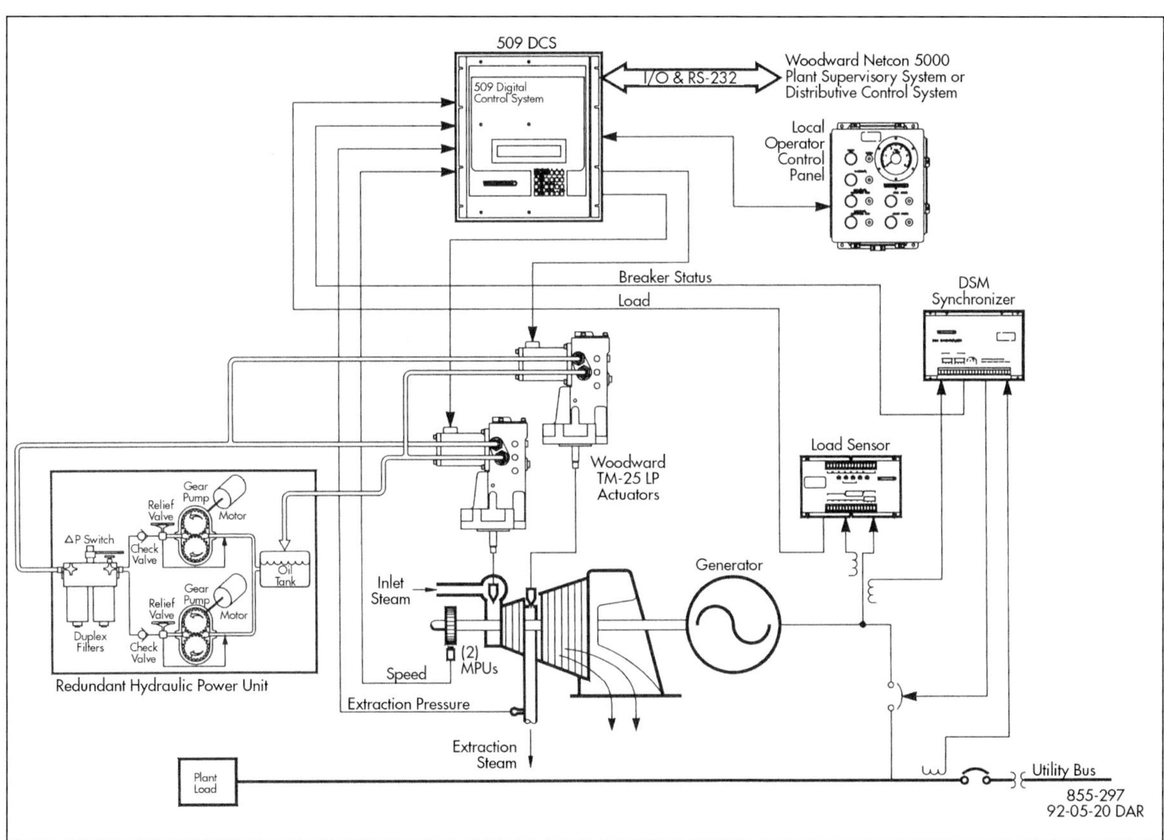

Fig. 13-16 Representative Block Diagram of a Steam Turbine-Generator Set Control System. Source: Woodward Governor

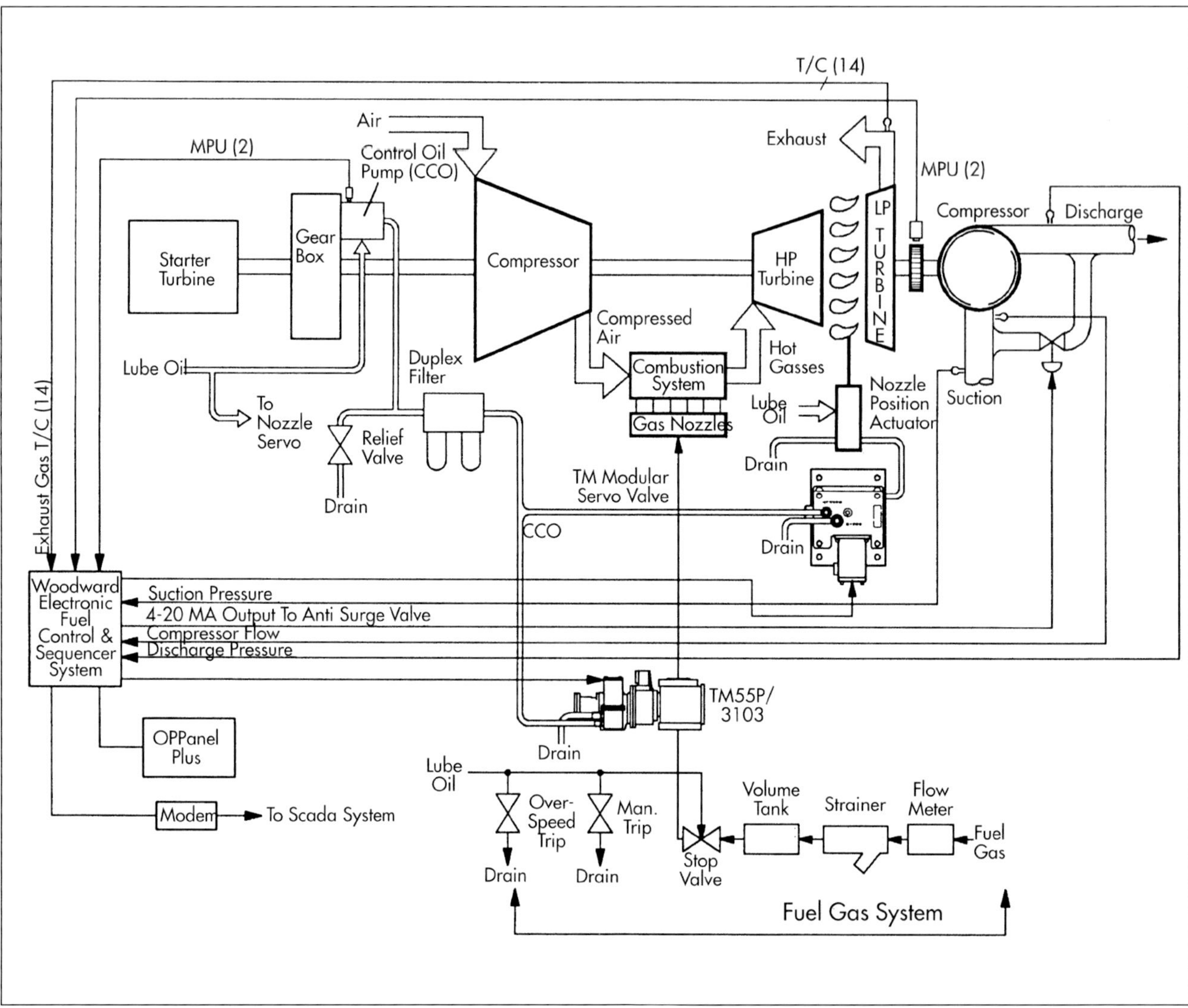

Fig. 13-17 Diagram of Control System for a Gas Turbine Applied to Mechanical Drive Service. Source: Woodward Governor

Maximum and Minimum Load Limits
Maximum Rate of Change Limit
Unit Frequency Error
Unit Load Demand Runbacks
Load Dispatch System
A
T
Unit Load Master
Megawatt Load Control
Unit Frequency Correction
Operational Limitations
MW Load Demand
MW Load Demand
Throttle Pressure Deviation
Boiler Transient Correction
MW Error
Boiler Demand Calibration
A T Boiler Master
Boiler Demand
Boiler Demand
Anticipating Signals
Firing Rate Transient Correction
Superheated Steam Temperature Error
Firing Rate Demand Calibration
A T Firing Rate Master
Feedwater Flow Error
Firing Rate Demand Limiting Control
Firing Rate Demand
Air Flow Demand
Reheat Temperature Error
Gas Flow Distribution Demand Calibration
Minimum Flow
Gas Tempering Program
Gas Recirculation Program
Minimum Flow
Throttle Pressure Program Development
MW Error
Boiler Turbine Coordination Control
Throttle Pressure Set Point
Oxygen Error
Air Flow Demand Calibration
Air Flow Error
Minimum Air Flow
Fuel Flow Error
Air Flow Demand Limiting Control
Firing Rate Demand Limiting Control
Total Gas Recirculation Flow (Fan Amperes)
Throttle Pressure
Throttle Pressure Control
Feedwater Flow
Pressure Compensated Drum Level
Feedwater Flow Control
Steam Flow
Fuel Flow Demand
Fuel Flow
Fuel Flow Control
Temperature Compensated Air Flow
Air Flow Demand
Air Flow Control
Gas Recirculation Flow Control
Maximum Gas Fan Amperes
Throttle Pressure Error
Feedwater Flow Error
Fuel Flow Error
Air Flow Error
Gas Fan Ampere Error
A T Turbine Master
Boiler Feed Pump Flow Control
Fuel Flow Control
Forced Draft Fans Flow Control
Gas Tempering Damper Positioning Control
Gas Recirculating Fans
Gas Recirculation Damper Positioning Control
Turbine Control Valves
Feedwater Flow (Drum Level) Control
Firing Rate Control
Gas Flow Distribution Control
Turbine-Generator Control

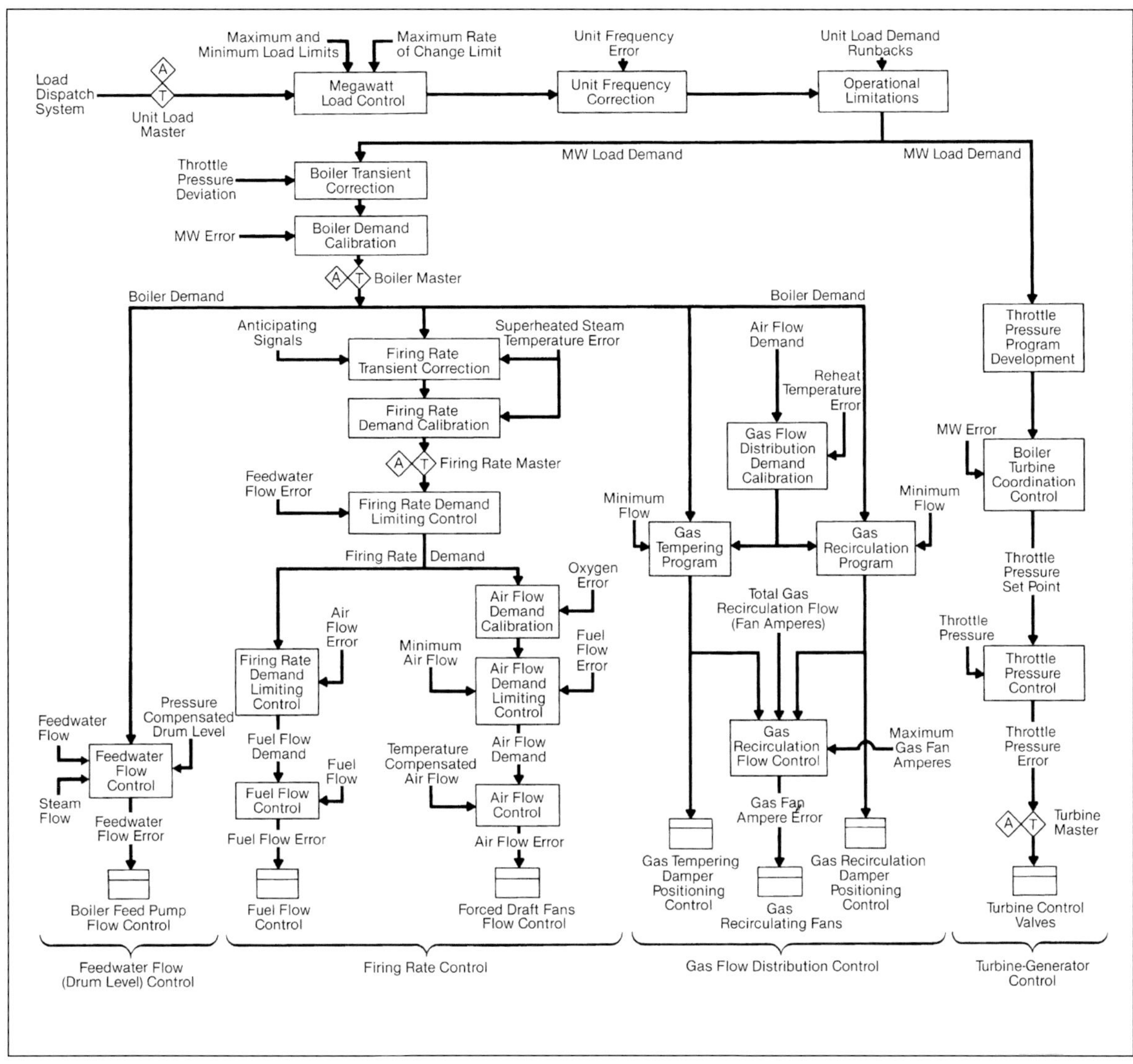

Fig. 13-18 Integrated Boiler Turbine-Generator Control System. Source: Babcock & Wilcox

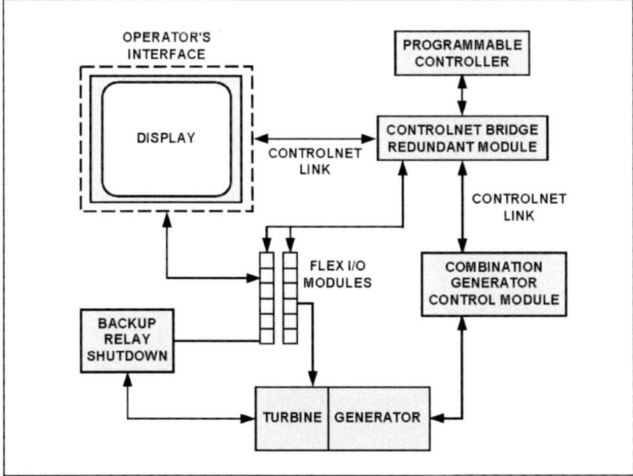

Fig. 13-19 Turbine Control System. Source: Solar Turbines

CHAPTER FOURTEEN

Renewable & Alternative Power Technologies

In addition to traditional forms of power generation that rely on finite reserves of fossil fuels, there are three basic categories of renewable resources that can be used to generate power. Renewable resources are those that are replenished within the ecosystem (wind, water) or whose source is so plentiful that it can be considered an infinite supply.

One category uses renewables as a prime mover combustion fuel source. Examples include landfill-derived methane and wood gasifiers. Another category uses renewable energy sources to raise sufficient thermal energy to drive conventional prime movers. This includes the use of renewable sources such as biomass and geothermal energy to produce steam for power production from a conventional or modified steam turbine. It also includes the use of developing solar thermal technologies in the same manner to raise high-pressure steam or in some cases, such as the prototype parabolic dish technology, to generate thermal energy in the form of a heated transfer fluid to drive a reciprocating engine or gas turbine, operating on a Stirling or Brayton cycle. These renewable sources and technologies for thermal generation that can be used for power generation are discussed in detail in Chapters 5 and 7. The third category uses renewable energy more directly to produce power. This involves the use of prime movers such as water and wind turbines. It also includes the use of photovoltaic technology, which converts solar radiation directly into electricity.

This chapter presents information on three renewable technology applications: hydro, wind, and photovoltaic. A fourth technology in this chapter is the fuel cell. While they involve the use of a renewable energy source (hydrogen), fuel cells currently must be powered by conventional fossil fuel sources such as natural gas to generate the hydrogen. While this does not satisfy the definition of a renewable technology, it is often categorized with renewable sources of clean or "green" power, since it produces such low levels of pollution relative to other fossil fuel powered technologies.

HYDROPOWER

Hydropower is an extension of the sun-driven hydrologic cycle, which gives the earth its renewable water supply. Atmospheric water reaches the earth's surface as precipitation. Some of this water evaporates, but much of it either percolates in the soil or becomes surface runoff. Water from rain and melting snow eventually reaches ponds, lakes, reservoirs, or oceans where evaporation is constantly occurring.

The use of hydropower for productive mechanical drive service applications dates back more than 2,000 years to Egypt, Rome, and Greece. There are numerous documented water wheel applications for grinding wheat and corn, with the technology changing and improving over time. These initial devices tapped the energy of running or falling water by means of a set of paddles mounted around a wheel. The force of the moving water was exerted against the paddles, and the consequent rotation of the wheel was transmitted to machinery via a shaft, either directly or through a crude set of gears for mechanical advantage.

All watermills used the same principle — running water was used to turn a wheel, which through a system of cogs, gears, shafts, and pulleys, could be used to power a range of equipment. Waterwheels were classified according to the point at which the water entered them. The earliest designs were undershot wheels, which had paddles around their circumference. The water passes beneath the wheel, so its speed is dictated by the speed of the stream. Using this principle, tide mills operated on the rising tide, allowing water to flow and be impounded into a millpond through a lock gate. As the tide ebbed, there was sufficient head to turn the heavy waterwheel. This relatively inefficient undershot design was largely replaced in the Middle Ages by the designs in which the water was introduced through a chute either halfway up (breastshot) or all the way up (overshot). Buckets built into the wheel structure filled with water, so it was turned not only by the speed of the water but also by its weight, thereby improving the efficiency. These were especially beneficial in hilly terrain where a large head (elevation) was available. Storage ponds were often built to guarantee a more reliable and steady source of water supply. By the mid 10th century, more than 5,000 waterwheels are reported to have been in operation in England alone.

The combination of waterwheel and transmission linkage, often including gearing, was developed and applied in the Middle Ages. Hydropower continued as a mainstay technology for providing mechanical power and later played an important role in colonial United States and in the industrial revolution for milling, textile manufacture, paper mills, irrigation, etc. In northern climates, where waterwheels were subject to freezing, they were enclosed in sheds with stoves used to provide heat. The early waterwheels were built from wood. These were subject to rapid wear and needed relatively frequent overhauls and rebuilds. By the mid-18th century, cast iron gradually began to take over from wood. In the 1840s, for example, new fabrication techniques, including the use of strong and durable wrought iron, enabled the replacement of several breast wheels, that had been in operation since 1801 at the famous Harpers Ferry Armory, with large 15 ft (4.6 m) diameter iron overshot wheels.

In the 19th century, a series of advancements produced modern, powerful hydraulic (water) turbines, many designs of which endure today. As with steam turbines, there are 2 basic types of modern water turbines: reaction and impulse. Each type has several sub-types featuring design variations within the same basic concepts.

In 1827, Benoit Fourneyron introduced a reaction turbine that channeled water through an enclosed chamber fitted with an inner ring of fixed guide blades. These guide blades deflected the water outward against the moving vanes of a runner (or waterwheel). The vanes of this outer runner were curved in the opposite direction from the fixed inner guide blades, reversing the direction of water flow within the device and creating a reactive force. In 1844, Uriah Boyden patented an improved outward flow turbine featuring a conical approach passage, giving the incoming water a gradually increasing velocity and a spiral motion that corresponded to the direction of the motion of the wheel. It also featured improved guide vanes that directed the flow of water through the wheel passages more efficiently. In 1845, James Francis conducted tests that showed 88% of the energy available in the falling water was converted into power. This was a significant improvement to the 60 to 75% efficiency achieved by the previous breast-type water wheels, an impressive achievement even by today's standards. In 1849, Francis further perfected the reaction turbine with a design that endures today. In fact, the terms "reaction" and "Francis" turbine are now used interchangeably.

In 1862, James Leffel patented the Double Turbine Waterwheel. This departed significantly from previous designs by combining two wheels in a single case. The upper wheel was comprised of inward-flow buckets, while the lower wheel had axial-flow buckets that curved inward and downward. This mixed flow design resulted in a longer, narrower, and faster turbine that operated very efficiently under a variety of water conditions.

In 1890, Lester Pelton designed an impulse turbine, also called the Pelton wheel. A high-pressure jet of water directed against buckets, on the rim of the wheel, turns this turbine. A few decades later, Forest Nagler developed a type of reaction turbine called the propeller turbine. It has fewer blades and larger spaces between blades than the other turbine designs. This design helps reduce the chance that the turbine will be damaged by debris in the water that passes through it.

The evolution of hydropower applications includes the taming of Niagara Falls with the first of the major American hydroelectric sites for power generation. These first plants were direct current stations built to power arc and incandescent lighting just before the start of the 20th century. With the advent of the electric motor, the market transformed from more localized mechanical drive plants to centralized hydroelectric stations, and the use of hydropower increased significantly. Backed by enabling legislation to build plants on federal land and the desire to control water resources and use them for irrigation, several massive depression-era dams and hydropower plants were built. By the 1940s, hydropower provided about one-third of the electric energy in the United States.

Since this zenith, its market share has dwindled significantly. However, hydropower still provides for approximately 10% of the nation's electricity production and remains a very valuable and cost-effective renewable resource. While not without environmental issues and negative impacts, hydropower stands alone as the most widely applied renewable technology. Currently, hydro-power provides the vast majority of the electricity produced from renewable technologies in the United States and in the world.

Historically, hydroelectric generation for power sales and transmission over interconnected power grid was often a by-product of water development associated with large dam projects. Internal mechanical equipment drive and pumping applications for irrigation were the targeted uses of hydropower at many of the dams. Power sales were viewed as an added return on investment, one that could reduce the cost recovery burden on irrigation water users alone. Over time the availability of low-cost power in the regions served by hydropower plants supported large-scale growth of industry and associated expansion of populations, notably in the western United States. In response, many plants were up-rated with the installation of additional electric generation capacity to serve these expanded loads.

Technology and Design

Hydropower plants can be located on rivers, streams, and canals. For reliable water supply, dams are commonly required. Dams store water for later release for irrigation, domestic and commercial/industrial use, and power generation. The reservoir acts much like a battery, storing water to be released as needed to generate power. The dam creates a head or height from which water flows. As illustrated in Figure 14-1, water flows from behind the dam, down the feeder canal from the intake, to the forebay, which is a tank that holds water between the feeder and the penstock. It must be deep enough to ensure that the penstock inlet is completely submerged so air is excluded from the power equipment. The penstock is a pipe connecting the forebay to the powerhouse. It pressurizes the water and is usually made of steel or high-density plastic capable of withstanding high pressures. The turbine-generator set(s) and associated power producing and control devices are located in the powerhouse, which stores and protects the equipment. The rushing water from the penstock simply pushes the turbine blades, imparting a force on the blades that turns the rotor. The rotor can drive a variety of mechanical devices; in the vast majority of cases, these are electric generators. The tailwater is the natural stream immediately downstream of the dam and the afterbay or tailrace is the flow of water out of the powerhouse back into the stream.

The waterwheel or hydraulic turbine is at the heart of the power plant. The most efficient of the traditional waterwheels is the overshot wheel. The water is supplied to the top of the wheel, where it is deposited into buckets, which then fall due to the weight of the water, thus turning the wheel. The overshot wheel uses the weight of the water for a full 180 degrees, giving it potential efficiency of as high as 65%. The wheels have rotational speeds ranging from 6 to 20 rpm, depending on their size. The main disadvantage of the waterwheel for most modern applications is its slow speed. Since electric generators and other modern machinery require significantly higher rotational speeds, and given the high torque of the wheel, expensive gearing must be added for such applications. Another disadvantage is the manufacturing cost for these relatively massive wheels, though maintenance requirements are generally fairly low.

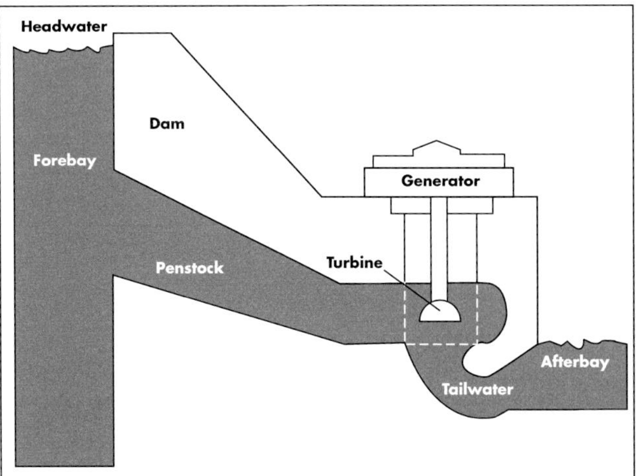

Fig. 14-1 Labeled Illustration of Dam and Hydropower Plant. Source: Bureau of Reclamation

The two major classes of modern hydraulic turbines are the impulse and reaction designs. Hydraulic conditions at dams, waterfalls, and river runs determine in large part the type of turbine that will be most effective (refer to chapter 11 for additional detail on reaction and impulse design principles).

Impulse Turbines

Impulse turbines tend to be employed for water resources with high heads and low flow rates. They extract energy from water by first converting the head into kinetic energy. This is accomplished by passing the water through a nozzle that discharges a jet into the air. This jet is directed onto buckets that are fixed on the rim of the runner (rotor) and formed to remove the maximum velocity from the water. The most widely used kind of impulse turbine is the **Pelton turbine** (or wheel), in which each bucket is divided in the center by a double-curved wall so that the jet of water is split upon hitting the bucket and diverted to either side, transmitting a maximum amount of energy to the turbine. Figure 14-2 is a cutaway illustration of an impulse turbine. Figure 14-3 shows a very large capacity Pelton waterwheel undergoing a service inspection at the Hoover Dam.

These turbines will typically exhibit mechanical efficiencies ranging from 80 to 90%, increasing as the size of the wheel increases. For maximum efficiency, the runner tip speed should equal about one-half the striking jet velocity. High velocities are achieved when the ratio of the diameter of the wheel and the diameter to the center of the nozzles is relatively small. Efficiencies in excess of 90% have been reported for operation at between 60 and 80% of full load. The power of a given wheel can be increased by using more than one jet. Two-jet arrangements are

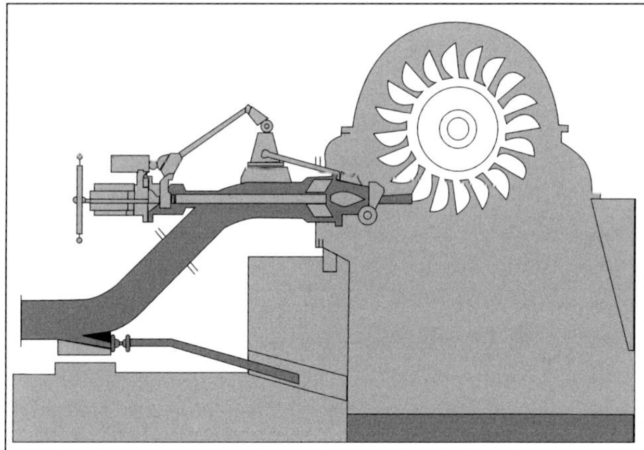

Fig. 14-2 Cutaway Illustration of Impulse Turbines.
Source: Bureau of Reclamation

common for horizontal shafts. Sometimes two separate runners are mounted on one shaft driving a single electric generator. Vertical-shaft units may have four or more separate jets.

Another type of impulse turbine is the **Turgo**, invented by Eric Crewdson in 1920. The jet impinges at an oblique angle on the runner from one side and continues in a single path, discharging at the other side of the runner. This type of turbine has been used in medium-sized units for moderately high head applications. Its design allows it to be relatively small, reducing capital cost, and operate at high speeds, reducing gearing cost requirements. Another type of impulse turbine design is the **cross-flow**, which is one of the simplest turbines. It is a radial impulse turbine that uses a rectangular water inlet. It has a drum-shaped runner consisting of parallel discs connected together near their rims by a series of curved blades. The water flow strikes the turbine blades as it enters and exits the turbine. The cross-flow design can be effectively used for very low to medium head and flow applications and can achieve an efficiency of 80% for small capacity units and as high as 88% for larger units.

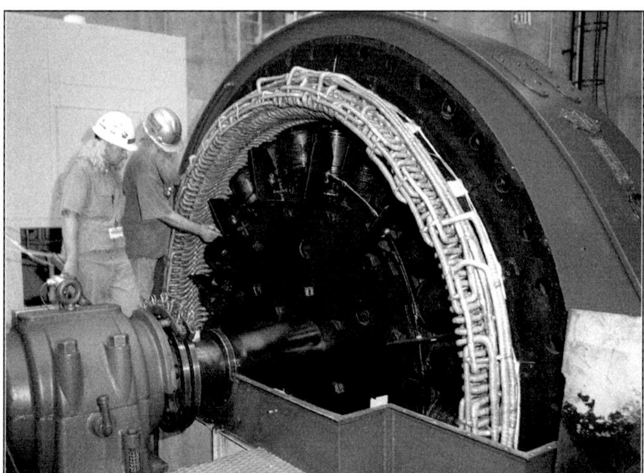

Fig. 14-3 Large Pelton Waterwheel Undergoing Service Inspection.
Source: Bureau of Reclamation

One of the simplest and most widely utilized of the impulse turbines is the **Banki Turbine**. Originally thought to be designed by A.G.M. Michell in England, it was introduced to America by a technical document written by Donat Banki. The turbine is manufactured widely throughout the world, with thousands of small generating units in operation utilizing heads of 15 to 100 ft (5 to 30 m). The turbine consists of a nozzle and turbine runner made up of two discs connected by a series of curved blades. The blades are placed within the outer edge of the discs and generally are one-fifth as long as the full diameter of the discs. The rectangular nozzle directs the jet of water at about 16 degrees across the width and at the outer edge of the curved blades in the turbine. The water flows across the blades to their inner edge and then flies across the central core of the turbine, where the water then hits the underside of the blades on the opposite side. In commercially manufactured models, the system allows up to 63% of the kinetic energy of the falling water to be absorbed in the first pass and close to 21% on the second pass, for a total conversion efficiency of about 84%.

Reaction Turbines

Reaction turbines are widely applied over a large range of heads of water with moderate or high flow rates. They achieve rotation primarily through the reactive force produced by the acceleration of water in the runner, rather than in the supply nozzles, as in the case of impulse turbines. While the exact manner in which this acceleration occurs varies depending on the specific design, in all designs, a fraction of the hydraulic pressure is first converted into velocity in the passage of the water through the inlet structure, which consists of a spiral casing and a gate device that leads to the runner. The energy from the water is transformed into mechanical energy in a single-stage runner (i.e., only one wheel of buckets or blades) that absorbs the full energy of the water. These may be horizontal or vertical and operate with the turbine wheel completely submerged in order to reduce turbulence. They may also be designed with adjustable blades that allow the blade angle to be changed to increase the turbine's efficiency. Figure 14-4 is a cutaway illustration of a reaction turbine.

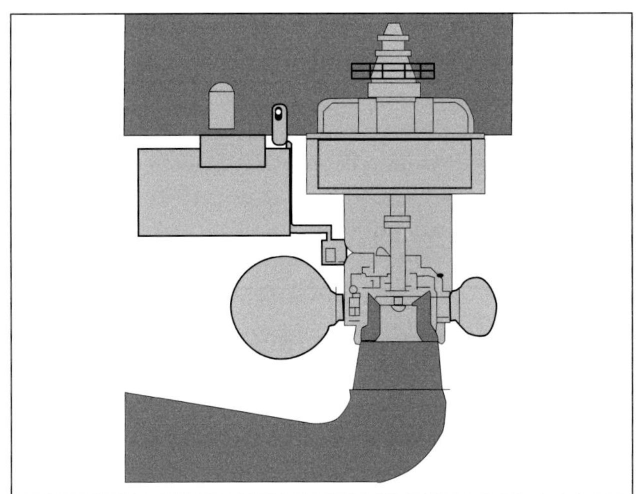

Fig. 14-4 Cutaway Illustration of Reaction Turbine. Source: Bureau of Reclamation

Due to the wide range of design options available, reaction turbines can be used over a larger range of heads and flow rates than impulse turbines. Reaction turbines typically have a spiral inlet casing that includes control gates to regulate water flow. In the inlet, a fraction of the potential energy of the water may be converted to kinetic energy as the flow accelerates. The water energy is subsequently extracted in the rotor.

There are four major kinds of reaction turbines in wide use: the **Francis**, **Deriaz**, **Kaplan**, and **propeller type**. In fixed blade propeller and adjustable blade Kaplan turbines, there is essentially an axial flow through the machine. The Francis and Deriaz type turbines use a mixed flow, where the water enters radially inward and discharges axially. Runner blades on Francis and propeller turbines consist of fixed blading, while in Kaplan and Deriaz turbines, the blades can be rotated about their axis, which is at a right angle to the main shaft.

Francis turbines are perhaps used most extensively because of their suitability with the widest range of heads, typically from 10 to 2,000 ft (3 to 600 m). They are particularly prevalent in medium-head applications ranging from 400 to 1,000 ft (120 to 300 m). They can have either horizontal or vertical shafts. Vertical shafts are generally used for machines with diameters of about 7 ft (2 m) or more. Vertical shaft machines are usually smaller than horizontal units for a given capacity, permit greater submergence of the runner with a minimum of deep excavation, and allow good maintenance accessibility for the tip-mounted generator. Horizontal shaft units are more compact for smaller sizes and allow easier access to the turbine, although generator access becomes more difficult as size increases. The most common form of Francis turbine has a welded, or cast-steel, spiral casing. The casing distributes water evenly to all of the inlet gates. The multiple gates operate from fully closed to wide open, depending on the power output desired. Most are driven by a common regulating speed ring that is rotated by one or two oil-pressure servomotors controlled by the speed governor. The number of blades can vary from 7 to 19. On installations that have low head and large flow, the turbine can be mounted in an open camber where water is directed onto the runners by adjustable guide vanes. By placing the turbine higher than the tailwater, a suction head can be created. Figure 14-5 shows numerous very large capacity Francis turbines in service at the Hoover Dam hydropower plant.

Fig. 14-5 Large Francis Turbines in Service at Hoover Dam Hydropower Plant. Source: Bureau of Reclamation

Propeller turbines are used extensively in North America, where low heads and large flow rates are common. The resulting design generally features large diameters and slow rotational speeds. The energy conversion efficiency of a conventional propeller turbine decreases rapidly once the turbine load drops below 75% of full load. Losses can be reduced by varying the inlet blade angle of the runner to match the runner inlet conditions more accurately with the water velocity for a given flow. In a Kaplan turbine-generator set, a servomotor adjusts angles and inlet flows to match electric load while keeping the main shaft with its direct-coupled generator rotating at constant speed. Runners with 4 to 6 blades are common, though more blades may be used for high heads. Figure 14-6 shows a sectional view of a Kaplan turbine-generator set with labeled components.

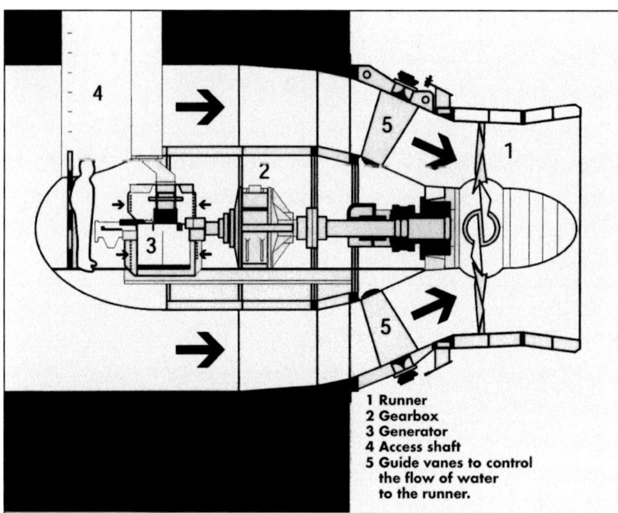

Fig. 14-6 Sectional View of Kaplan Turbine Generator with Labeled Components. Source: CADDET Renewable IEA/OECD

System Performance

The output of a hydropower plant will be a function of the volume of water released or discharged and the vertical distance the water falls (the head). The turbine design requirements will largely be a function of the head, discharge rate, and required rotational speed of the driven device.

Turbine power (P_T) requirements are determined as:

$$P_T = Q \times H \times \eta_T \times c \qquad (14\text{-}1)$$

Where:

Q = Flow rate

H = Head

η_T = Mechanical efficiency of the turbine (and generator if applicable)

c = Constant (the product of the density of water and acceleration due to gravity)

The mechanical efficiency of the turbine itself will vary. Older large capacity designs still in operation may have mechanical efficiencies below 50%. Very small capacity newer units may also have mechanical efficiencies well below 50%. On the other hand, some large capacity modern turbine designs have demonstrated mechanical efficiencies in excess of 90%. When assessing mechanical efficiency, however, one must also consider a number of other friction losses in the system, as well as mechanical losses associated with the driven device (e.g., electric generator efficiency losses). Ideal power can be converted into actual power by including all friction losses or efficiency factors in Equation 14-1.

The power produced by water depends upon the water's weight and head (height of fall). Assuming c equals 1 (at zero altitude and standard atmospheric conditions), each ft^3 (0.03 m^3) of water has a mass of 62.4 lbm (28.3 kg). For example, a column of water with a 1 ft^2 (0.3 m^2) cross-sectional area and 10 ft (3 m) height would contain 10 ft^3 (0.28 m^3) of water. It would impart a force of 624 (10 x 62.4) lbf (283 kg) on each ft^2 (or m^2) of turbine blade.

In English system units, the power potential of a water stream, in hp, is found by dividing the result of Equation 14-1 (with flow measured in ft^3/sec, head measured in ft, and standard density of water at 62.4 lbm/ft^3) by 550 ft-lbm/sec. For example, with a head of 100 ft (30.5 m) and a flow rate of 1,000 ft^3/sec (28.3 m^3/sec), an 85% mechanically efficient unit would develop (100 x 1,000 x 0.85 x 62.4 ÷ 550) 9,650 hp. To convert to kW output, hp can be multiplied by 0.746 kW/hp. This would yield 7,195 kW developed. Alternatively, when expressed in SI units, where head is rated in m and flow rate in m^3/sec, Equation 14-1 can be multiplied by 9.81 to yield kW developed. Again, this would yield (30.5 x 28.3 x 0.85 x 9.81) 7,195 kW.

An important factor in initial turbine selection is a ratio of design variables, termed the power specific speed (N). In U.S. design practice, it is expressed as:

$$N = \frac{nP^{1/2}}{H^{5/4}} \qquad (14\text{-}2)$$

Where:

n = Rotational speed in revolutions per minute

P = Power output in hp

H = The head of water in ft

Turbine types can be classified by their specific speed, which applies at the point of maximum efficiency. If N ranges from 1 to 20, corresponding to high heads and low rotational speeds, impulse turbines are commonly selected. For N between 10 and 90, Francis type turbines are commonly selected, with slow-running, near-radial units selected for the lower N values and more rapidly rotating mixed-flow runners for higher N values. Deriaz turbines may be selected for N of up to 110, while for N values ranging from 70 to a maximum of 260, propeller or Kaplan turbines may be selected.

Based on Equation 14-2, a turbine designed to deliver 100,000 hp (74.6 MW) with a head of 40 ft (12.2 m) operating at 72 rpm would have a specific speed of

226. This, for example, might suggest selection of a propeller or Kaplan type turbine. Based on Equation 14-1, at a turbine mechanical efficiency of 90%, the flow rate would be about 24,500 ft^3/sec (694 m^3/sec) and the runner diameter would likely exceed 30 ft (10 m). This typifies the very large sizes required for high-power, low-head installations and the low rotational speed at which these turbines must operate to stay within an appropriate specific speed range.

Applications

Two of the largest and most famous hydropower plants in the world are found at the Hoover and Grand Coulee Dams. These marvels of engineering, construction, and sheer human effort, which were built more than 60 years ago, are testaments to viability of hydropower as a cost-effective, reliable, renewable source of electric energy production, with extremely long service life.

The largest hydropower plant in the United States (though about half the capacity of the largest in the world) is the enormous Grand Coulee Dam in Washington Sate. Figure 14-7 provides an aerial view of the dam and power plant. It has an installed capacity of about 6,800 MW, with 27 Francis turbine-driven units ranging in capacity from 10 to 800 MW. The plant also includes an additional 300 MW of pump-generator capacity. With nearly 12 million cubic yards (9.2 million m^3) of concrete, the dam is said to be the largest concrete structure ever built. With a rated head of 330 ft (100 m), this barricade, which raises the water 350 ft (107 m) above the old riverbed, is 5,233 ft (1,595 m) long and 550 ft (168 m) high. The average water release from the dam is 110,000 ft^3/sec (3,115 m^3/sec). Figure 14-8 provides a cross-sectional view of the dam and third powerhouse, drawn to scale. Notice the massive penstock required to provide the necessary water flow to the turbine-generators.

Fig. 14-7 Aerial View of Grand Coulee Dam and Power Plant. Source: Bureau of Reclamation

After operating for some 60 years, with several expansions and renovations along the way, this hydroelectric facility remains highly cost-effective. Annual operations and maintenance (O&M) costs, which represent the majority of the operating costs, range between 2 and 3 cents per kWh produced and, of course, there is no real fuel-cost component. The Grand Coulee Dam project was undertaken as part of the Bureau of Reclamation's Columbia Basin Project, which, in addition to providing hydropower production, was intended to provide flood control, irrigation, recreation, stream flows, and fish and wildlife benefits. FDR Lake, behind the dam, is 151 miles (252 km) long, with over 5 million acre-ft (6.2 billion m^3) of active storage. Additionally, 550,000 acres (2,225 km^2) of irrigation is provided to the Columbia Basin.

The average annual power generation is about 21 billion kWh. This corresponds to a load factor of about 35%. Since Grand Coulee has such a large generation capacity rela-

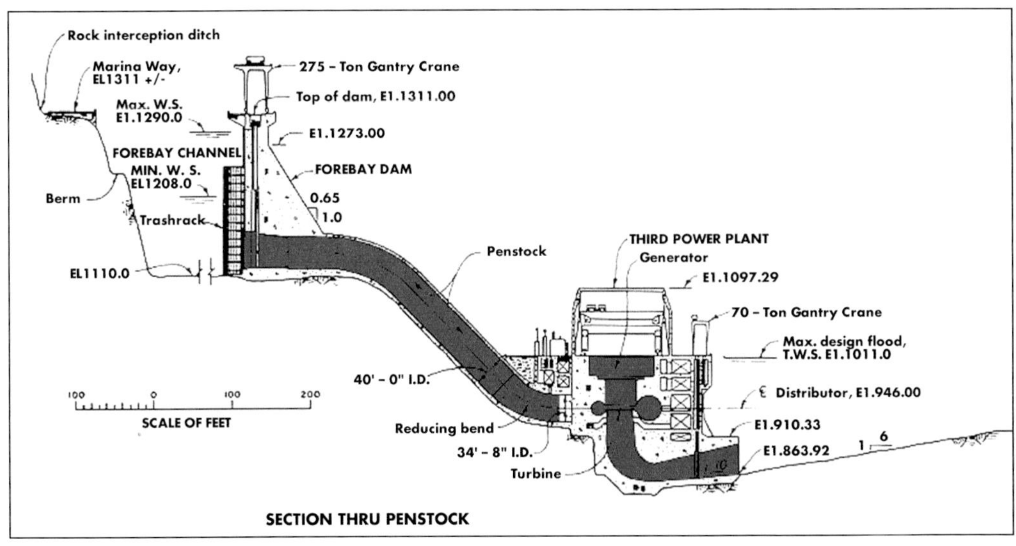

Fig. 14-8 Cross-Sectional View of Grand Coulee Dam and Third Power House Drawn to Scale. Source: Bureau of Reclamation

tive to baseloads connected to the distribution system, Bonneville Power Authority (BPA) varies the output considerably by time-of-use, creating a peaking load profile. In addition to daily output variation, monthly variation is also considerable, ranging from a typical year output of 1.5 million MWh in October to just under 2.5 million MWh in June. The actual availability of the plant has varied from about 80 to 90% in each of the last 10 years.

While less than one-third the capacity of the Grand Coulee, The Hoover Dam, pictured in Figure 14-9, is itself a massive concrete thick-arch structure that is 726 ft (221 m) high and 1,244 ft (379 m) long at the crest, with a power generation capa-city of more than 2,000 MW with 19 generating units. The Hoover power plant, located in the Black Canyon of the Colorado River, was developed with similar water management objectives as the Grand Coulee, using electricity sales as a means of making the project self-supporting and financially solvent. Floodwaters of the Colorado River are impounded by the Hoover Dam and released in response to downstream water orders. The quantity of water available for release through the power plant is, in part, based upon the water orders. Water for generation is conveyed through four penstocks from four intake structures immediately upstream and contiguous to the dam. Spillway structures use 16 ft (5 m) by 100 ft (30 m) drum gates, which provide for an additional 16 vertical ft (5 m) of storage capacity in Lake Mead, the reservoir impounded upstream of the dam. Lake Mead is the largest reservoir in the United States with a capacity of 28.5 million acre-feet (35.2 billion m^3), a length of 110 miles (177 km), a shoreline of 550 miles (885 km), a maximum depth of 500 ft (152 m), and a surface area of 157,000 acres (635 km^2). At 576 ft (176 m), the rated head of this plant is fairly high, almost double that of the Grand Coulee.

Fig. 14-9 Aerial View of Hoover Dam and Power Plant.
Source: Bureau of Reclamation

The ten-year average plant output is greater than 4.5 million MWh annually and has reached as high as 5.8 million MWh. O&M cost is fairly high for a large hydropower plant, ranging from 4 to 6 cents per kWh. While some plants are even higher, many medium- and large-capacity plants have demonstrated significantly lower long-term average O&M costs. Availability of the Hoover plant is very high, at an average of nearly 90% from year to year. Overall load factor on all installed capacity is fairly low at about 30%, with significant seasonal variation. Monthly electricity output generally varies from 300,000 MWh in winter to 500,000 MWh in spring.

While the famous massive hydropower plants draw the most attention, the majority of hydropower plants are much smaller. Many are similar in concept and design, but at a scale of one tenth or one hundredth the size. Projects of much smaller capacity and with rated heads of about 65 ft (20 m) or less are termed **low-head plants.** These can still be sizable systems of 20 MW or more, though many are under 1 MW of capacity. Low-head dams may often be located closer to where the real electric loads are, reducing the power lost in transmission. They may also be designed as **run-of-the river plants,** which use power in the river water as it passes through the plant without causing appreciable change in the river flow. These systems generally impound very little water and, in some cases, do not require a dam or reservoir. This reduces the likelihood of water quality changes, such as higher temperature, lower oxygen, increased phosphorus and nitrogen, and increased siltation.

Figure 14-10 shows the Boise River Diversion Dam on the Boise River, about 7 miles southeast of Boise, Idaho. It is a 68 ft (21 m) high rubble-concrete, weir type structure with a hydraulic height of 39 ft (12 m). It was originally built to provide irrigation and to supply power for the construction of the nearby Arrowrock Dam. The power plant, shown in Figure 14-11, consists of three vertical 500 kW electric generation units. The plant began operation in 1912 and retains many unique engineering features of the era, including double Francis turbines, wooden turbine bearings, and belt-driven auxiliary systems. When operating, the plant supplies power to irrigation loads and delivers any surplus power to the BPA for marketing and distribution to regional industries and municipalities. The power plant has been placed in standby mode, but is presently targeted for rebuilding. It will be up-rated to 2.1 MW (710 kW each unit) and special care will be taken to restore its historical significance. As part of the rebuilding effort, the plant will be automated and controlled as part of the southern Idaho automation system, which will remotely control several southern Idaho power plants.

Fig. 14-10 Boise River Diversion Low-Head Dam. Source: Bureau of Reclamation

On the other end of the spectrum from the Grand Coulee are projects such as the Lewiston Dam and power plant (Figure 14-12), a 40 year old, 350 kW plant on the Trinity River in California. Like the Grand Coulee and Hoover Dams, it was designed with similar multiple purposes. However, in stark contrast, this dam stands only 91 ft (28 m) tall, with a rated head of about 60 ft (18 m), and has a water release discharge rate of 350 ft^3/sec (10 m^3/sec). Given its age and small capacity, annual O&M costs, which have ranged widely from as low as 3 cents to as high as 8 cents per kWh over the past decade, are considerably higher on average than what is now generally achieved in larger, more modern plants. An impressive statistic of this plant, however, is that both annual load factor and availability factor have been very high, exceeding 90% and even approaching 100% in many years.

Load factor variation in hydropower plants that have high availability factors may result from seasonal variations in available water flow, permit limitations, and obligated downstream releases. In addition to the need to regulate water flow in the river, storage is intentionally used for enhanced reliability and for shaping of electricity output to match system load requirements. In some cases, hydropower plants are purposely operated as peaking plants and this may be enhanced by pump storage systems. Still, some plants, including many small-capacity, low-head plants, show annual load factors in excess of 90%.

In assessing a potential hydropower application, it is necessary to evaluate the source to determine how much water can be delivered to the turbine and how high the water source is above the turbine. The length of the penstock or

Fig. 14-11 Three Vertical 500 kW Turbine Generator Units Installed in 1912. Source: Bureau of Reclamation

Fig. 14-12 Lewiston Dam and Small Capacity Low-Head (350 kW) Power Plant. Source: Bureau of Reclamation

feed-in pipe must be considered to calculate friction losses. It is also necessary to distinguish between gross and net head. Gross head is the vertical height from the tailwater level to the intake level when the turbine is stopped. Net head is the head available to drive the turbine when it is running at normal full load, with all losses, such as pipe friction and rise in tail water level, taken into account. In the case of an impulse turbine, the height of the jet center above the tailwater level must also be deducted. With a long pipeline on a high head scheme, or with a restricted tailrace on a low head scheme, the difference between gross and net heads may be as much as 20%. Initial gross head and flow assessment can be done through direct measurement or from previously established sources and maps. For proper equipment selection and system design, it is important to know not only average stream flow, but also minimums and maximums to be expected. Normalized hourly or daily flow data is necessary to make an accurate assessment of expected annual power production and to project economic performance.

Beyond the initial assessment of approximate system capacity and power production, is the extensive environmental assessment. This aspect of project development can take several years and will greatly influence not only if a project can be implemented, but the design constraints under which the project must be developed.

Pumped Storage

Pumped storage is an important part of many hydropower plants and can also be used as a storage vehicle in conventional plants built in suitable locations. Its serves as a means of keeping water in reserve for peak period power demands. With pumped storage, water is pumped to a storage pool above the power plant (often to an upper reservoir) at a time when demand for, or price of, energy is relatively low. The reservoir acts much like a battery, storing energy in the form of elevated water. The water is then allowed to flow back through the turbine at times when demand, or price, is high. Such arrangements make good use of the hydropower plant's ability to start up quickly and make rapid adjustments in output, helping to balance supply (available water power) and demand (net system output). Even though these systems do not generate additional electricity (there actually is a reduction in total energy output due to losses in pump and turbine operation, as well as in the electric motor and generator), pumped hydro-storage often becomes economical when compared with the cost of constructing additional facilities for meeting peak power demands.

Modern pumped storage units in the United States normally use reversible-pump turbines that can be run in one direction as pumps or in the other direction as turbines. These are coupled to reversible electric motor/generators. The motor drives the pump during the storage portion of the cycle, while the generator produces electricity during discharge from the upper reservoir. Most reversible-pump turbines are of the Francis type. These are more costly than standard units due to the need for added hydraulic and electrical control equipment, but are less costly than completely separate pump-motor and turbine-generator assemblies with dual water passages. Reversible single-stage pumps have proven economical in applications with heads exceeding 1,000 ft (300 m) and are capable of operating at even higher heads. For medium heads, Deriaz turbines have been successfully applied. They allow ready adjustment of the runner-blade angles to match the opposite requirements of pumping and power generation and can vary the pumping load. They also allow for easy pump start-up with minimum load while the unit is submerged in water.

Pumped storage has become widespread in the United States and abroad. The largest plant is located in Bath County, Virginia, where six pump-turbines operate with enough storage to generate at a capacity of 2,100 MW for an 11 hour period. If a suitable reservoir area can be identified or constructed, excess power from a non-hydropower plant (e.g., a conventional fossil fuel-fired plant, a wind turbine plant, etc.) can be used to pump water into storage for later release through the turbine.

One driver behind the application of pumped storage with non-hydropower plants is to limit the construction cost for the installation of larger conventional generation facilities to meet peak load demand. Another is to maximize run time on lower cost baseload plants. Pumped storage may allow such plants to operate at full capacity during off-peak demand periods. These same concepts may be applied to other renewable electric generation technology applications, such as wind and photovoltaic systems.

Tidal Power

In addition to traditional hydropower plants located in rivers and streams, it is also possible to use tidal power for hydroelectric generation. In areas where the normal tide runs high, water can be allowed to flow into a dam-controlled basin during high tide and discharged during low tide to produce intermittent power. One such plant is located in France on the estuary of the Rance River in Brittany. There, a reservoir has been created by a barrage 2.4 miles (4 km) inland from the river mouth to make use of tides ranging from about 11 to 44 ft (3.4 to 13.4 m). The power station is equipped with 24 reversible bulb- type propeller turbines coupled to reversible motor/generators, each having a capacity of 10 MW. Pumped storage is used if the tidal outflow through the plant falls below peak power demands. The use of reversible turbines (a series of fixed and rotating blades) permits the tidal flow to work in both directions, from the sea to the tidal basin on the flood and on the ebb from the basin to the sea.

Although large amounts of power are available from the tides in favorable locations, this power is intermittent and varies with the seasons. However, while these are difficult, expensive applications and viable sites are limited, the potential does exist for application in locations where tides are extremely high. One site of consideration for future application is the Bay of Fundy, where the tidal range reaches more than (49 ft) 15 m.

Environmental Issues and Future Markets

While most of the major sites are already exploited, there remains significant potential for growth of hydropower. Many existing sites are still capable of being expanded. In its report "Assessment of Waterpower Potential and Development Needs" the Electric Power Research Institute (EPRI) estimated that power generation capacity could be increased by 23,000 MW by 2025 by increasing conventional hydropower (10,000 MW), new hydrokinetic sources (5,000 MW), and from ocean wave energy (10,000 MW). This assessment is consistent with DoE's Hydropower Program identification of nearly 6,000 sites that could produce as much as 30,000 MW of additional hydropower capacity. Moreover, hydropower is only generated at a very small percentage of the nation's dams. Of a reported 75,000 to 80,000 dams in the United States, only 3% are used for hydroelectric generation. The National Hydropower Association reports that in the 1980s, 91% of the 765 federally licensed hydro projects did not require a new dam and that more than 20,000 MW of new capacity could be installed in the United States without the construction of a new dam.

Given the climate of rising energy prices and air emission regulations, there is a strong market growth potential for hydropower in large, medium, and even very small applications. The National Hydropower Association estimates that in 2004, the 268 million MWH of hydropower generated in the United States helped avoid more than 160 million tons of carbon emissions. Despite this benefit, there remains long-standing environmental and land-use issues associated with hydropower that must continuously be resolved.

The majority of concerns regarding hydropower focus on the environment. While there are many advantages of hydropower dams, such as provision for irrigation, recreation, and flood control, when wild rivers are dammed up, ecosystems both up and down the river change. Downstream, water flow rates decline and water levels change, affecting habitats for many animals. Every case is different, but with such a drastic change in the environment, it is likely that some species or environmental element will be impacted. The most obvious impact of hydropower dams is the upstream flooding of vast land areas, much of which may have been previously forested or used for agriculture. This can also eliminate land from native peoples and destroy rare ecosystems. Further, as the biomass in this land is submerged, it decomposes, adding additional carbon emissions to the environment. An extremely large project could flood thousands, and even tens of thousands, of square miles of land. Even if the opposition can be overcome or an equitable settlement reached, there often remains significant cost associated with relocation and compensation to affected individuals.

Fish that populate these areas often get stuck in the penstock, which leads them to almost certain death in the turbines. Screens, as shown in Figure 14-13, can greatly limit this and modern turbines can be designed to allow

Fig. 14-13 Large Fish Screen at Hydropower Plant.
Source: Bureau of Reclamation

fish passage to further reduce death rate. Damming a river can alter the amount and quality of water in the river downstream of the dam. It can also prevent fish from migrating upstream to spawn. Setting minimum downstream flow requirements and employing fish ladders can mitigate these impacts. However, screens and fish ladders add significantly to capital costs, negatively impacting project cost-effectiveness.

Silt can slowly fill up a reservoir, decreasing the amount of water that can be stored. This also can deprive the river downstream of silt, which fertilizes the river's flood plain during high water periods. Bacteria present in decaying vegetation can also react with mercury that may be present in rocks underlying a reservoir, into a form that is water-soluble. This mercury can accumulate in the bodies of fish and pose a health hazard to those who depend on these fish for food. Dams are also susceptible to growth of other forms of hazardous bacteria. Continual efforts are being made to reduce these potential negative impacts, with a strong focus on reducing fish mortality rates and maintaining downstream dissolved oxygen at proper levels.

Still, in certain limited cases, facilities located on or near rivers can effectively apply hydropower systems for mechanical drive or electricity production services. Today, standardized micro-turbine systems are available for applications ranging from a few MW down to a few kW. These can be applied in run-of-the-river type systems that use the power in river water as it passes through the plant without causing significant change in river flow. Hence, given unique access to a flowing river, hydraulic turbines merit consideration as a prime mover system of choice for a commercial, industrial, or institutional facility. In consideration of potentially negative environmental impact, new hydropower facilities can be designed with the power plants placed underground, and selective withdrawal systems can be used to control the water temperature released from the dam.

Larger projects will continue to find many challenges in overcoming environmental and land-use issues. However, they also will continue to enjoy the benefit of economies of scale and federal agency support. Small projects will be aided by the availability of standardized designs for micro-turbine applications and the ability to work through various siting issues more quickly than with larger projects. The development and permitting process can be extremely laborious and costly and, therefore, risky. However, there have been so many successful ventures of all capacity ranges that market confidence is solid.

Wind Energy Conversion Systems

Similar in many ways to hydropower, wind is one of the oldest sources of power production, having been used for more than 1,000 years throughout the world. Early windmills were designed with a vertical shaft with paddle-like sails radiating outward and were located in a building with diametrically opposed openings for the inlet and outlet of the wind. Most of these wind devices were patterned after early waterwheels. Over the next several centuries, numerous advancements were made for grain milling as wood-framed sails were located above the millstone (instead of a waterwheel located below) to drive the grindstone through a set of gears. This allowed for the widespread development of milling in areas that did not have suitable streams for waterpower.

The first type of windmill to be widely adopted in Europe was the post-mill, in which the entire body of the mill pivoted on a post and could be turned to face the sails into the wind. During the 14th century, the tower-mill type of construction was adopted, in which the body of the mill remains stationary, with only the cap moving to turn the sails into the wind. With this design, a horizontal shaft protrudes from the cap, or upper portion, of the mill building. Generally, most windmills featured 4 to 8 wind sails, each about 10 to 30 ft (3 to 9 m) in length, radiating from the shaft. The wood frames of the sails were covered with canvas or fitted with wood shutters. The power of the turning shaft was transmitted through a system of gears and shafts down to the mill machinery at the base of the building.

With the coming of the industrial revolution, windmills became increasingly large and metal replaced wood components. Cast-iron drives were introduced in the mid-18th century. Fantail rudders were added to steer the wheels into the wind to optimize efficiency. In the 19th century, there were more than 9,000 windmills in operation in Holland alone. In addition to milling grain, windmills served a variety of uses, including other types of grinding, pressing, and sawing processes, as well as pumping water to drain lowland areas.

As airplane technology evolved, research into airfoil design was readily adapted into propeller blade design for wind turbines. By the early part of the 20th century, thousands of windmills were scattered across the United States, serving farms and other rural areas for pumping and other mechanical applications, and later for powering small electric generators. Figures 14-14 and 14-15 show traditional type (Aermotor) multi-bladed windmills in a farm setting. Generally used for water pump-

Fig. 14-14 Traditional Type Multi-Bladed Aermotor Windmill in Farm Setting. Source: U.S. Department of Agriculture

Fig. 14-15 Traditional Type Multi-Bladed Aermotor Windmill on Wood Frame. Source: Jim Green, U.S. DoE/NREL

ing, the blades or sails are mounted at an oblique angle on the horizontal shaft. The fantail rudder steers the bladed wheel into the wind. This same basic design has endured for more than a century and is still applied today. In the 1920s and 1930s, there were some 300 reported companies producing small wind turbines throughout the world, with more than 100 of them located in North America. Two of the largest producers were Jacobs Manufacturing and the Windcharger Company, which reported to have produced up to 1,000 units per day. Many of the turbines sold during this period were 200 W units that powered radios.

Concurrently, by the turn of the century, wind turbines were being applied in Denmark for electric generation. In the United States, however, with the rural electrification movement that brought grid-connected power to many of these areas, windmill applications declined significantly. While there were experimental large grid-connected wind turbine electric generators installed (e.g., 1.25 MW Smith Putnum unit at Grandpa's Knob, Vermont), the vast majority of manufacturers in the United States ceased production by the 1950s and the technology laid largely dormant for several decades thereafter.

During the 1980s, wind-energy conversion systems (WECS), as they became known, experienced a resurgence due to public policy encouragement through tax incentives and other subsidies, environmental regulations that limited the use of fossil fuel, and the expected long-term trend of rising energy prices. In the United States, most notably California, there was a tremendous surge of investment in wind turbine applications during that period. A decade later, this was followed by market stagnation. The emerging utility industry deregulation left great uncertainty in the market, and projections of stable or even falling electric rates furthered the stagnation.

In the late 1980s and early 1990s, total new WECS capacity installations ranged from between 250 and 500 MW per year. Since then, the United States market has experienced a surge in wind project development, with the first decade of the 21st century experiencing exponential growth. More than 10,000 MW of wind power was installed in 2009 alone, far exceeding any prior year. This growth was fueled by rising energy prices, continuing difficulties in permitting new fossil fuel plants, and increased public support for "green power". This sharp rise is consistent with worldwide increases in wind power project development. As shown in Figure 14-16, the cumulative United States wind power capacity is now over 35,000 MW, and electricity generated from wind, is estimated at 48 million MWh, just over 1% of the nation's electricity supply.

In the United States, Texas and California continue to be the leaders in implementing WECS technology.

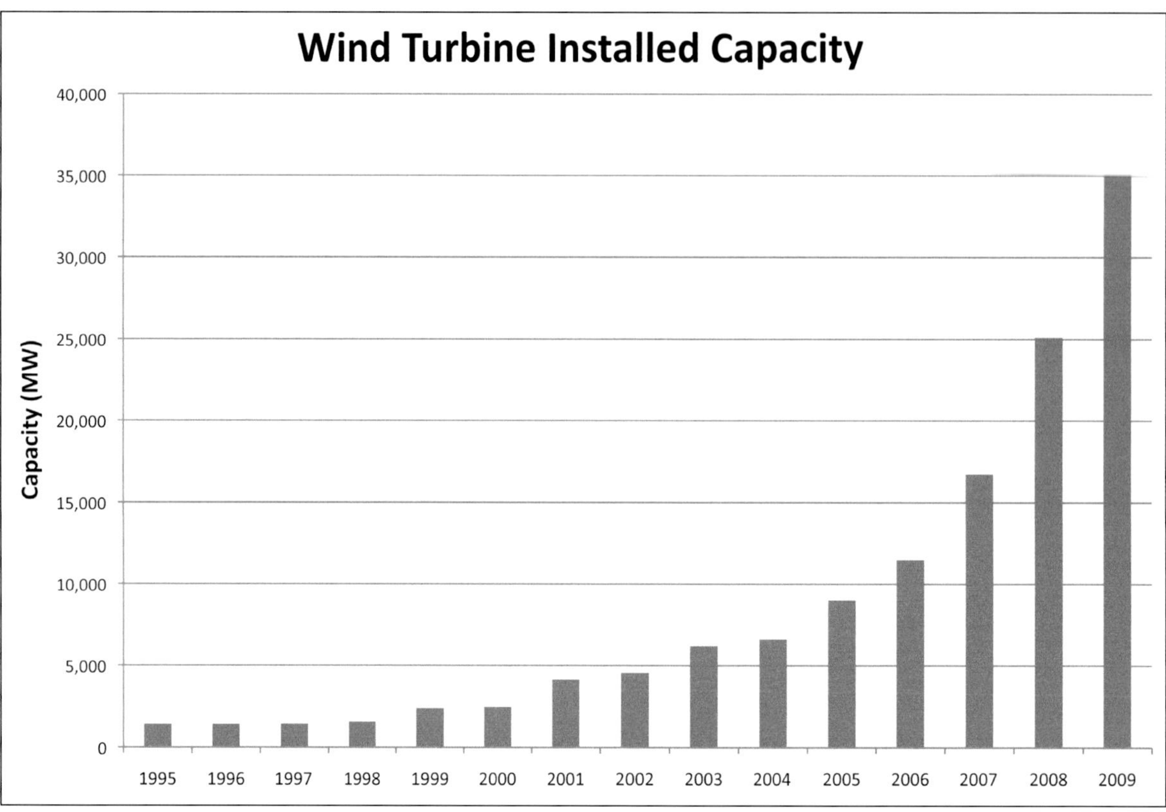

Figure 14-16 Wind Turbine Cumulative Capacity by Year. Source: American Wind Energy Association

This may be more due to environmental issues, high energy rates, and state support for wind systems than to the state's wind energy resources. In fact, in a 1991 study conducted by Pacific Northwest National Laboratory, California ranked only 17th among states in wind energy potential and available windy land area. States such as North Dakota, Kansas, South Dakota, Montana, Nebraska, Wyoming, Oklahoma, and Minnesota are all considered to have 10 to 20 times the wind energy potential of Texas and California, and the recent surge in wind turbine projects extended into many of these other states with high wind-power potential. While land cost is relatively low in these states, market growth limitations include lower energy costs, less stringent environmental regulations, and line loss inefficiencies associated with serving distant loads, given the lack of population density. Concepts such as using wind power to produce hydrogen (carrier energy) supplies from water, via electrolysis, which can then be exported to regions with high energy costs and stringent environmental regulations, continue to merit consideration. However, while logistically feasible, project economics are not currently sufficiently compelling to drive market development.

The U.S. Department of Energy has set lofty goals for the contribution that wind power can potentially make to United States electric generation. Their optimistic outlook is supported by improvements in technology, increased manufacturer market entry, and favorable public policy initiatives. Excluding land from wind energy development for environmental and land-use considerations, the amount of windy land suitable for wind power generation using today's available technology is about 6% of the total land area in the contiguous United States. The associated electric power generation potential, inclusive of practical system efficiencies, is 500 GW, which is many times the country's electric capacity requirement. This could be even higher with future generation technology that can more effectively exploit wind potential at lower wind speeds.

Although the nation's wind potential is very large, only part of it can be exploited economically. Market growth will likely depend on a combination of factors, including continued wind technology improvements and production cost reductions, market energy rates, environmental constraints on competing technologies, and aggressive public policy support. In Denmark, through a national emphasis on the proliferation of wind power, the country currently produces more than 18% of its electricity requirements from wind power.

Technology

Turbine performance continues to improve as turbine size, tower height and reliability have steadily increased. These factors, coupled with an increase in funding for research and development and component testing, have helped to enhance turbine capacity factor, defined as the ratio of the actual energy produced in a given period, to the hypothetical maximum possible (i.e., running full time at rated capacity power output).

Over the past decade, the average turbine capacity has continued to rise, with multi-megawatt machines becoming the norm, and a number of 3 MW turbines being installed on 328 foot (100 meter) towers.

The average capacity of wind turbines installed is currently approximately 1.7 MW and with a variety of models in the 2 to 3 MW range entering the market, average turbine capacity is likely to continue to increase.

Hub heights of installed turbines currenlty range from 148 to 344 feet (45 to 105 meters), rotor diameters from 187 to 325 feet (57 to 99 meters). Turbine performance has improved consistently with capacity factors of less than 30% in late 1990s increasing to an average of 35% for new projects, with some projects achieving more than 45%. As technical performance has continued to improve, annual operation, maintenance and repair (OM&R) requirements have also decreased, adding further to improved project economic performance.

Policies

A range of policies are under consideration in the United States that would further encourage investments in wind energy sources. These include:

- A national renewable electricity standard (RES) with a target of generating at least 25% of the nation's electricity from renewables by 2025, and a near-term target of 10% by 2012
- A high-voltage interstate transmission "superhighway" to tap the nation's vast renewable energy sources
- National climate change legislation that recognizes the value of wind power in reducing greenhouse gas emissions

WECS Technology

The prime mover in a modern WECS is the wind turbine. Wind turbine operation may be viewed simply as opposite that of a propeller type fan. Instead of using power to increase air velocity (produce wind) with a fan, wind turbines use wind to produce power. The wind imparts a force that turns the rotor blades, which spins a shaft that drives a generator or other mechanical device. Wind turbines thus change the horizontal movement of the wind into a rotational force turning a shaft.

There are two general classifications of wind turbines: **horizontal-axis wind turbines (HAWTs)** and **vertical-axis wind turbines (VAWTs)**. Figure 14-17 provides a side-by-side comparison of these two basic designs. As shown in Figure 14-18, the design of HAWTs is similar to historic

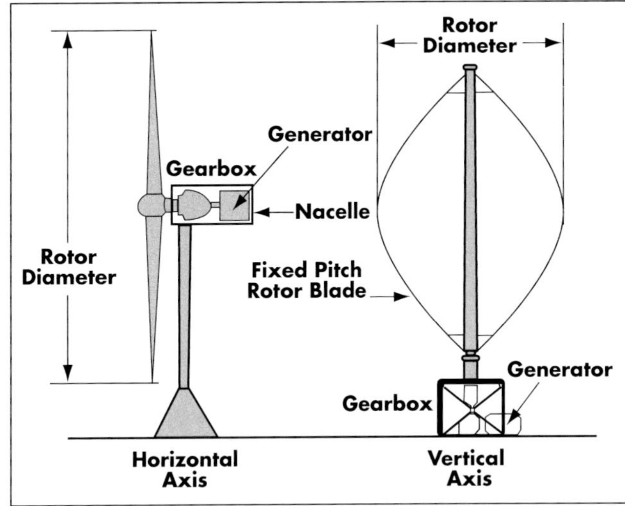

Fig. 14-17 Side-by-Side Comparison of Horizontal and Vertical Axis Turbines.

Fig. 14-18 Three-Blade HAWT. DoE/NREL. Source: Tom Hall, DoE/NREL

windmills, with a propeller type rotor on the axis. The axis of rotation is horizontal and roughly parallel to the wind stream. The blades on the horizontal drive shaft may be facing into the wind (an upwind turbine) or the wind may hit the supporting tower first (a downwind turbine).

The general momentum theory behind the operation of a conventional HAWT involves the extraction of energy from the air passing through it by reducing its pressure. This is much the same in principle as a steam or gas turbine. In this case, the wind is obstructed by the turbine, causing the air stream to slow and the upstream air pressure to increase. On the downstream side, pressure is lower, and atmospheric pressure is recovered by further slowing of the stream.

The traditional farm windmill features a large number of blades and typically operates at a relatively slow speed. It offers good starting torque, but suffers from lower efficiency and limited energy conversion potential. More modern two- and three-blade HAWTs run at higher tip speed ratios and provide superior power conversion ratios. These turbines generally operate with a constant blade pitch (or orientation) angle, but are often equipped with pitch-change mechanisms used to feather the blades in extreme wind conditions. In some cases, the blade pitch is continuously controlled to maintain constant speed and target power output. HAWTs now predominate in the market, largely because of their higher efficiency, which has been continually advanced due to extensive improvements in rotor design.

Fig. 14-19 Darrieus Rotor Type VAWT. DoE/NREL. Source: Scandia National Laborarories, U.S. DoE/NREL

As shown in Figure 14-19, VAWTs have an axis of rotation that is vertical with respect to the ground and roughly perpendicular to the wind stream. VAWTs are somewhat like classical waterwheels in that the fluid arrives at a right angle to the rotation axis of the wheel. The blades are typically long, curved, and attached to the tower at the top and bottom. While they are generally less efficient than HAWTs, they offer strength and simplicity of design and the benefit of having all of the drive train and generator parts at ground level.

The most prominent VAWTs, named after their designers, are the Savonius, used primarily for pumping, and the Darrieus, a higher-speed machine that can be used for grid-connected electric generation. The Darrieus rotor design (shown in Figure 14-19) typically features two or three C-shaped rotor blades that make it resemble an eggbeater. The blades are high-performance airfoils, formed into a modest curve to minimize the bending stresses in the blades and allow efficient operation at high speeds. This is a lift-based design in which each blade experiences maximum lift (torque) only twice per revolution, making for a high torque (and power) sinusoidal output similar to the cranking on a bicycle.

Another lift-type VAWT is called the Cycloturbine. It features straight vertical axis blades called Giromills and a wind vane to mechanically orient a blade pitch change mechanism to maximize performance. The Cycloturbine design is similar to the Darrieus rotor, except that its airfoils are straight and the blade pitch is continually changed during rotation to maximize wind force.

Lift, Drag, and Stall

There are two basic aerodynamic design concepts applied to power WECS: lift and drag. Some VAWTs use drag-based designs. Drag, which is essentially air resistance, will normally increase if the area facing the direction of motion increases. Drag-based designs work like a paddle used to propel a canoe through the water — the wind simply pushes the blades. Lift-based designs, which predominate, function much the same as airplane wings.

The reason an airplane can fly is because air flowing across the upper surface of the wing moves faster than on the lower surface. The pressure is, therefore, lower on the upper surface, creating an upward lift force that is perpendicular to the direction of the airflow. Lift-based wind turbines are configured to use this lift force as their driving energy. Since the blades are affixed to the rotor, which is secured to a stationary tower, the lift force on the blade turns the rotor producing rotational force on the shaft to drive mechanical devices.

The amount of lift force achieved for a given wind speed and rotor blade size and type will largely be a function of the angle of the blade in relation to the general direction of the airflow (also known as the angle of attack). Degradation of lift force will occur if the blade is not completely even and smooth. Dents and even dead insect build-up will reduce lift and cause the air to whirl around in an irregular vortex, a condition knows as turbulence. This also occurs if the angle of attack is too great. If the turbulence becomes high enough, the low pressure on the upper surface and the resulting lift are not sufficient to turn the rotor. This condition is known as stall.

The phenomenon of lift and stall are key control elements for wind turbines. Turbines may be designed for pitch control, stall control, or a combination of both. In the case of pitch-controlled turbines, when power output exceeds a preset limit, the power controller signals the blade pitch mechanism to pitch the rotor blades a number of degrees out of the wind to reduce lift and thus the energy conversion rate. When power output falls below the limit, the blades are pitched fully into the wind. Passive stall-controlled wind turbines have a fixed blade angle. The rotor blade profile is designed so that when the wind speed is too high, it creates turbulence on the side of the rotor blade that is not facing the wind. This causes a gradual stall effect that prevents the lifting force of the blades from acting on the rotor. Such systems are essentially self-correcting, allowing for simple, low-cost control, and avoiding the need for moving parts on the rotor. An active stall control is similar to a pitched control system, except that when the high load limit is approached, it pitches the blades to make the angle of attack of the rotor blades steeper, which causes the blades to go into a deeper stall.

WECS Components

Figure 14-20 provides an exploded cutaway illustration of a pitch-controlled upwind HAWT. While there are many different designs and configurations, this figure illustrates most of the basic components found in a modern wind turbine applied to electric generation. Following is summary detail on each of the main components.

Nacelle: The nacelle sits atop the tower and includes the gearbox, low- and high-speed shafts, generator, controller, and brake. A cover protects the components inside the nacelle and the rotor attached to it. It may also include an air- or water-cooling unit to cool the electric generator and gearbox oil. Some nacelles are large enough for a technician to stand inside while working. Figure 14-21 is a cutaway illustration of the nacelle of a 1.3 MW Nordex HAWT and Figure 14-22 is an open-case field view of a 1.5 MW Micon unit under construction.

Rotor/Blades: Most HAWTs have either two or three blades, though some designs feature as few as 1 or as many as 5 or more blades. Wind blowing over the blades

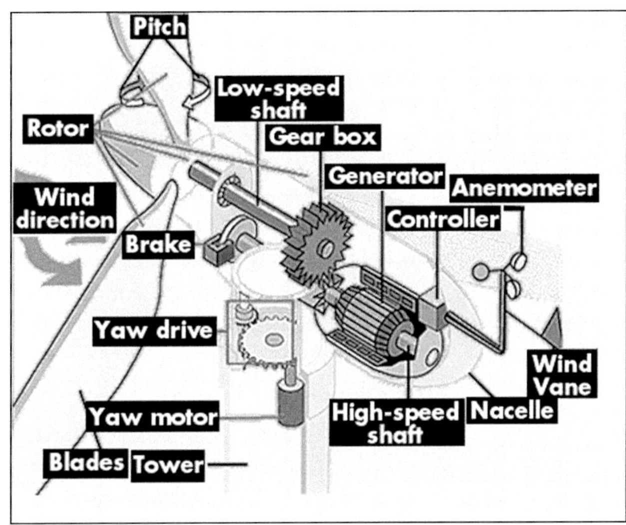

Fig. 14-20 Labeled Illustration of Modern Pitch-Controlled, Three-Blade Upwind Turbine Applied to Electric Generation. Source: U.S. DoE

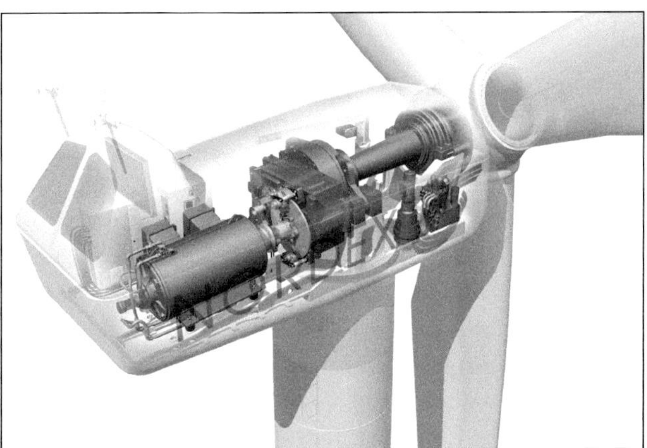

Fig. 14-21 Cutaway Illustration of the Nacelle of a 1.3 MW HAWT. Source: Nordex

Fig. 14-22 Open-Case Field View of the Nacelle of 1.5 MW HAWT Under Construction. Source: NEG Micon

Fig. 14-23 Large-Capacity Turbine Blade Fabrication. Source: NEG Micon

causes them to lift and rotate. Since the speed at which the blades move at the tip will be greatest and decrease to zero at the hub, blades on larger turbines are twisted so as to achieve an optimal angle of attack all along the length of the blade. Blades are usually made of fiberglass, polyester, or epoxy resins. Some have wood cores. In pitch-controlled turbines, blades are turned, or pitched, out of the wind to keep the rotor from turning in winds that are too high for safe operation or too low to produce electricity. The blades and the hub together are called the rotor. The hub of the rotor is attached to the low-speed shaft. The hub may be attached to the drive shaft or a gear. Upwind machines have their rotor in front of the tower (wind hits the rotor before the tower) and downwind machines have the reverse arrangement. Figure 14-23 shows the final stages of turbine blade fabrication for a large Micon unit and Figure 14-24 shows a blade undergoing a modal test. The blade is instrumented with accelerometers, which feed the frequency of oscillation into a computer upon being struck with a calibrated instrumental hammer. Figure 14-25 shows an innovative five-bladed rotor on an 80 kW turbine.

Drive Train: The rotor typically turns the low-speed shaft at a range anywhere from 15 to 60 rpm. A high-speed shaft drives the connected mechanical device (e.g., generator). A disc brake, which can be applied mechanically, electrically, or hydraulically, stops the rotor in emergencies and during equipment servicing. Gears connect the low-speed, high-torque shaft to the high-speed, low-torque shaft and increase the rotational speeds. For electric generators, very high gear ratios are required (e.g., from 20:1 to 80:1 to drive a 6-pole generator, or even greater for a higher speed generator). For certain mechanical drive

Fig. 14-24 Turbine Blade Undergoing a Modal Test at the NWTC. Source: NEG Micon

Fig. 14-25 Innovative Five-Blade 80 kW HAWT. Source: Paul Migliore, DoE/NREL

applications, lower gear ratios can be used, or in some cases, gearing can be eliminated. The gearbox is a costly (and heavy) part of the wind turbine. Currently, R&D efforts are focused on the use of direct-drive generators that operate at lower rotational speeds and do not need gearboxes. However, the tradeoff for reduced or eliminated gearbox cost and size is increased generator cost and size. Figure 14-26 shows the main (low-speed) shaft of a large-capacity HAWT.

Generator: This is usually a standard induction generator that produces 50 or 60 Hz ac electricity. Direct-drive and/or variable speed generators are in prototype and can be used to eliminate gearing and optimize turbine energy capture. There are also units that utilize two generators for different wind speed levels or one generator with bi-level output capability.

Controller: The controller receives a signal from the anemometer, at which point is starts the machine at wind speeds sufficient to drive the load and shuts off the machine at wind speeds that can damage the system. With a wind turbine, wind speed and force cannot be regulated like fuel or steam input to an engine. Hence, the turbine must be adjusted in response not only to load, but to wind speed. Modern wind turbine control systems may monitor in excess of 100 parameters, including rotor and driven device speed, air speed, direction and temperature, rotor blade pitch angle, yaw angle, vibrations, hydraulic pressure, gearbox bearing, and generator winding temperature.

Yaw Drive: Upwind turbines face into the wind. The yaw drive is used to keep the rotor facing into the wind as the wind direction changes. Downwind turbines do not require a yaw drive, since the wind blows the rotor downwind. A wind turbine is said to have a yaw error if the rotor is not perpendicular to the wind, meaning a reduced amount of the available wind energy is running through the rotor area. In addition, turbines that run with a yaw error are more subject to fatigue loads. Hence, most HAWTs use yaw motors to keep the turbine blades directly into the wind. The yaw mechanism is typically activated by an electronic controller that checks the position of the wind vane, which measures wind direction, several times per second. A yaw brake is commonly used whenever the yaw mechanism is unused.

Fig. 14-26 Main (Low-Speed) Shaft of a Large-Capacity HAWT. Source: NEG Micon

Tower: The tower carries the nacelle and the rotor. Towers may be made from tubular steel, steel lattice, or concrete. Towers usually are fabricated in sections, with flanges at either end, and bolted together on the site. They are conical, with their diameter increasing toward the base to increase their strength, while minimizing overall material requirement. For smaller systems, guyed singular poles or even a tripod arrangement may be used. Because wind speed increases with height, taller towers enable turbines to capture more energy and generate more power. Obviously the tower must at least be tall enough for blade clearance above the ground. Beyond that, a cost-effectiveness analysis must be performed to determine optimal tower height based on the gains in average wind speed achieved with additional height. In areas that have high roughness factors of the terrain (high roughness will more greatly reduce wind speed closer to the ground) or other obstructions, the tendency will be to increase tower height. It may also be beneficial to have sufficient height to level out the wind speed experienced over the full length of the blade. If there is too much variation, it produces greater bending of the blade when its position is perpendicular to the ground. This will increase fatigue loads on the turbine, which will negatively impact long-term maintenance and repair costs, as well as effective service life.

Anemometer: The anemometer measures the wind speed and transmits wind speed data to the controller. Cup-type anemometers are the most common. They have a vertical axis and three cups, which capture the wind. The wind speed can be calculated as the number of rpm is registered electronically. Other, less common types of anemometers include those that use propellers instead of cups and ultrasonic or laser anemometers, which detect the phase shifting of sound or coherent light reflected from the air molecules. New filtering techniques have been developed to enhance data gathered by the more inexpensive cup anemometers, making them almost as useful as the more sophisticated but costly types.

PERFORMANCE

The power available from wind, which is air in motion, can be expressed as:

$$P = 1/2\rho v^3 A \quad (14\text{-}3)$$

Where:

P = Wind power

ρ = Air density

v = Wind speed

A = Area (rotor swept area exposed to the wind), or $\pi/4D^2$, where D is the rotor diameter

In English system units, when density is expressed in lbm/ft^3, wind speed in ft/s and area in ft^2, wind power will be expressed in ft-lbf/sec (when divided by the gravitational constant of 32.17 ft/sec^2). This product can be converted to hp by dividing by 550 ft-lbf/sec. In SI units, with density expressed in kg/m^3, wind speed in m/s, and area in m^2, power will be Watts.

By rearranging Equation 14-3 for power per unit rotor area, under standard conditions, a wind speed of 40 ft/s (12 m/s), or about 27 mph, will have the potential power of about 0.135 hp/ft^2 (1.06 kW/m^2). Due to the cubic relationship between wind velocity and power, a 50% reduction in wind speed to 20 ft/s (6 m/sec) would reduce the potential wind power to 0.0164 hp/ft^2 (0.132 kW/m^2), an 8-fold reduction. In addition to the cubic relationship with wind velocity, the theoretical power obtainable varies as the square of the rotor diameter. For example, a WECS with a rotor diameter of 100 ft (30 m) will have a wind swept area of about 7,850 ft^2 (700 m^2). Since the rotor area increases with the square of the diameter, increasing the diameter to 200 ft (60 m) will increase the rotor area to about 31,500 ft^2 (2,800 m^2), for a four-fold increase in available power.

The power output from a WECS will be a function of the wind power available, the power coefficient, the efficiencies of the mechanical interface, and, in the case of electric generation, the generator efficiency and the efficiency of any power-conditioning equipment. Building on Equation 14-3, the electric power output from a WECS can be expressed as:

$$P_e = 1/2\rho v^3 A \eta_g \eta_m \eta_p C_p \quad (14\text{-}4)$$

Where:

P_e = Turbine electric power output

η_g = Generator efficiency

η_m = Mechanical interface efficiency

η_p = Power-conditioning equipment efficiency

C_p = Power coefficient

In order for the rotor to capture the wind's kinetic energy and convert it to rotational energy, the rotor must slow the wind. In contrast to the available power given by Equation 14-3, actual power is a function of the difference in wind velocity upstream and downstream of the rotor. For all of the potential wind power to be extracted, the exit speed would have to be zero. This is obviously impossible, since it would mean that the air could not leave the turbine and no additional air could enter the rotor. In accordance with Betz's Law, the power coefficient reflects the fact that not all of the wind's kinetic energy can be extracted by a WECS, because a wind turbine will deflect the wind, even before the wind reaches the rotor plane, and because there must be a finite velocity of air leaving the blades.

It can be shown from fluid dynamics that the maximum actual power that can be extracted from the wind by the rotor is equal to the mass flow rate times the drop in the wind speed squared. The ratio between this and the power of undisturbed wind shown in Equation 14-3 establishes the power coefficient. When plotted for the full range of potential ratios between leaving and entering wind velocities (v_2/v_1), the theoretical maximum value for the power coefficient, C_p, is 16/27, or 0.59. This maximum occurs at a v_2/v_1 ratio of 1:3. This is an optimum design parameter, since the maximum power coefficient will be lower for all other velocity ratios, either higher or lower. Accordingly, this sets the upper limit of WECS conversion efficiency, with no other system losses. In reality, the maximum efficiency obtainable with a highly efficient turbine operating under optimal conditions (which occurs when the propeller tip speed is between five and six times the wind velocity) is less than 50% and will generally be substantially lower.

When accounting for the power coefficient and all other normal system losses, the overall average conversion efficiency will typically range from 20 to 35%. The efficiency for a given unit will vary over its operating regime, depending largely on wind speed. At certain points in the operating regime, the unit may reach or exceed an overall efficiency of 45%. Conversely, under non-optimum conditions, this same unit may experience an efficiency of less than 20%. Generally, for a given rotor speed, the power coefficient, and overall system efficiency, drops as wind velocity decreases.

Since the majority of power produced over the course of the year will be at wind speeds above the annual mean (due to the cubed relationship of speed and power potential), systems may be designed to optimize efficiency under high wind speed conditions, at the expense of sacrificing efficiency under lower wind speeds. Alternatively, efficiency may be sacrificed at steady-state design load to minimize wear and lower long-term maintenance and repair requirements and perhaps to reduce sound emissions.

Usually, large HAWTs are designed to start running at wind speeds of 7 to 11 mph or 10 to 16 ft/sec (3 to 5 m/s). This is called the "cut-in" wind speed. The wind turbine will be programmed to stop at high wind speeds, typically above 56 mph or 82 ft/sec (25 m/s), to avoid damaging the turbine or its surroundings. The stop wind speed is called the "cut-out" wind speed.

Performance Example

Consider the manufacturer's performance data for a 2 MW turbine. The three 262 ft (80 m) diameter blades produce a swept area of 54,110 ft^2 (5,027 m^2) and operate at a design rotational speed of 16.7 rpm. This relatively low rotational speed at the nominal wind speed requirement for full-load operation enables the system to minimize sound emissions as well as long-term stress and wear. It is pitch-controlled and its air brake features 3 separate pitch cylinders. The cut-in and cut-out speeds are 13 ft/s (4 m/s) and 82 ft/s (25 m/s), respectively, with a nominal speed rating of 49 ft/s (15 m/s) for design operation at 2 MW output. Its generator operates at 690 volts and maintains its nominal 2 MW output over a rotational speed range of 1,090 to 2,300 rpm for 60 Hz operation and 905 to 1,915 rpm for 50 Hz operation. The nacelle weighs about 61 tons (56 tonnes) and the rotor weighs about 34 tons (31 tonnes). Total weight varies, depending on tower height, from about 225 to 350 tons (205 to 318 tonnes) at heights ranging from 200 to 330 ft (60 to 100 m).

The unit is designed to maintain a constant 2 MW output over a wide range of operating conditions. Its microprocessor-based control system allows the rotational speed of both the rotor and the generator to vary by about 60%. This minimizes both unwanted fluctuations in the power supplied to the grid and in material stresses on vital parts of the construction. Its control system also allows for rotational speed to be regulated by sound emission limits. For example, at a distance of 1,110 ft (340 m), its sound emissions can be reduced from 44.9 to 40.4 dba by reducing its blade tip speed.

Its power curve reveals a typical sharp increase in power output as wind speed increases, from under 200 kW at 16 ft/s (5 m/s) to about 1,200 kW at 33 ft/s (10 m/s), and then locking in at 2 MW at 49 ft/s (15 m/s) for all wind speeds up to the cut-out speed. Using Equation 14-3, it can be shown that the wind power potential for this unit (based on standard atmospheric conditions at sea level) is about 3 MW at a wind speed of 33 ft/s (10 m/s) and 10 MW at 49 ft/s (15 m/s). This indicates that overall efficiency, based on Equation 14-4, ranges from 40% to about 20% at its design nominal wind speed and power output rating. Actual field measurements reveal the system's ability to maintain its constant 2 MW output, with wind speeds varying between 49 and 82 ft/s (15 and 25 m/s) by varying the pitch-angle between 16 and 24 degrees and the generator speed between 1,400 and 2,000 rpm.

APPLICATION ASSESSMENT

A preliminary indication of an area's wind energy potential can be based on its rated wind power class. The Wind Energy Resource Atlas of the United States can be used to identify the areas potentially suitable for wind energy applications. More specific data can also be obtained for localized areas. The wind resource maps indicate the resource potential in terms of wind power classes in the United States (as shown in Table 14-1), ranging from Class 1 (the lowest) to Class 7 (the highest in the US). Each class represents a range of mean wind power density (in units of W/m^2) or equivalent mean wind speed at the specified height(s) above ground. Wind speed can go up to Class 10 in some areas of the world, such as off the coast of Tierra Del Fuego. Areas designated Classes 4 or greater are considered suitable for application with today's wind turbine technology. Power Class 3 areas may be suitable for some applications and, as technology continues to advance, may see more widespread applications. Class 2 areas are marginal and Class 1 areas unsuitable for wind energy development. The vast majority of the wind potential in Class 4 or greater is found in the Central and Midwestern states, though most states have some areas with such potential. In addition to wind resource, the economic viability of wind power will vary from region to region, depending on numerous factors, including production/demand match (seasonal and daily), market electricity costs, transmission and access constraints, public acceptance, and public policy support.

A wind energy database containing detailed wind statistics for 975 stations in the United States produced specifically for use in wind energy applications can be obtained from the National Climatic Data Center. More accurate prediction for site-specific applications must be based on a wind power distribution analysis. Somewhat similar to bin temperature analysis used for predicting heat or cooling loads, this shows the amount of time the wind blows at various wind speeds in an average year.

Average wind speed data, however, can produce misleading information, given the cubed relationship of wind speed to power. During periods when the wind is below or slightly above the cut-in speed, minimal power will be produced. During periods when the wind speed is very high, power production will be extremely high, and during periods when the wind speed is above the cut-out speed, no power will be produced. Hence, two sites with the same average wind speed can produce very different actual capacity factors. Absent precise hourly wind-speed data for a given site, there are two common wind distributions used: the Weibull distribution and the Rayleigh distribution (which is thought to be more accurate at sites with high average wind speeds and low levels of potential turbulence caused by buildings and rough ground). These are statistical tools that produce probability density distributions from average wind speed data.

The wind power density figures developed for the wind resource maps incorporate in a single number the combined effect of the frequency distribution of wind speeds and the dependence of the wind power on air density and on the cube of the wind speed. Table 14-1 assumes a Rayleigh distribution of wind speeds and air density at sea level. The decrease in air density with altitude means that a higher average wind speed is required to achieve a given wind power density. To obtain the same wind power density, the wind speed must be about 1% higher than shown in the table for every 1,000 ft (304 m) of elevation above sea level.

Even with these tools, the best method of determining annual average wind power density is site-specific measurements conducted at the various heights at which the wind turbine may be located. Table 14-2 provides data from three locations that indicate identical mean

Annual Avg. Site	Annual Avg. Wind Speed m/s (mph)	Wind Power Density W/m^2 (hp/ft^2)	Power Class at 33 ft (10 m)
Culebra, Puerto Rico	6.3 (14)	220 (0.027)	4
Tiana Beach, New York	6.3 (14)	285 (0.036)	5
San Gorgonio, California	6.3 (14)	365 (0.045)	6

Table 14-2 Comparison of Annual Average Wind Power at Three Sites with Identical Wind Speeds. Source: DoE, Pacific Northwest National Laboratory (Battelle) Classes of Wind Energy for the World Resource Maps

Wind Power Class	10 m (33 ft)		50 m (164 ft)[(a)]	
	Wind Power Density W/m^2 (hp/ft^2)	Speed[(b)] m/s (mph)	Wind Power Density W/m^2 (hp/ft^2)	Speed[(b)] m/s (mph)
	0	0	0	0
1	100 (0.013)	4.4 (9.8)	200 (0.025)	5.6 (12.5)
2	150 (0.019)	5. (11.5)	300 (0.037)	6.4 (14.3)
3	200 (0.025)	5.6 (12.5)	400 (0.050)	7.0 (15.7)
4	250 (0.031)	6.0 (13.4)	500 (0.062)	7.5 (16.8)
5	300 (0.037)	6.4 (14.3)	600 (0.075)	8.0 (17.9)
6	400 (0.050)	7.0 (15.7)	800 (0.10)	8.8 (19.7)
7	1,000 (0.13)	9.4 (21.1)	2,000 (0.25)	11.9 (26.6)

(a) Vertical extrapolation of wind speed based on the 1/7 power law.

(b) Mean wind speed is based on Rayleigh speed distribution of equivalent mean wind power density. Wind speed is for standard sea-level conditions. To maintain the same power density, speed increases 3% per 1,000 m (5% per 5,000 ft) elevation.

NOTE: Each wind power class should span two power densities. For example, Wind Power Class 3 represents the Wind Power Density range between 150 W/m^2 and 200 W/m^2. The offset cells in the first column attempt to illustrate this concept.

Table 14-1 Classes of Wind Power Density at 33 ft (10 m) and 164 ft (50 m).
Source: DoE, Pacific Northwest National Laboratory (Battelle) Classes of Wind Energy for the World Resource Maps

wind speeds at a height of 33 ft (10 m). However, the actual wind power density, based on the frequency distribution of the wind speeds, is substantially different for the three locations, such that each location has a different wind power class. What can be seen in these extreme cases is that the use of only the mean wind speed and the Rayleigh distribution provides a much lower estimate than the actual power density. Another example is a site near Ellensburg, Washington, that has a mean annual wind speed of 17 ft/s (5.2 m/s), which is Class 3 wind power (0.02 hp/ft^2 or 160 W/m^2) if the Rayleigh distribution is applied. However, because the wind at this site is steady and does not fluctuate as predicted by the Raleigh distribution, the actual wind power is Class 6 (0.04 hp/ft^2, or 320 W/m^2), or twice that estimated by the Rayleigh distribution.

Moreover, local terrain features can cause the mean wind energy to vary considerably over short distances, especially in coastal, hilly, and mountainous areas. These variations cannot be ascertained from the wind resource maps. The wind resource atlas indicates areas where high wind resource is possible, but does not provide detail on variability on a local scale.

The intent of WECS feasibility analysis is to be able to assign hours per year to discrete wind speeds experienced at discrete tower heights in discrete locations to model the annual energy output. Given the cubed relationship of wind speed to power output, this is very important, especially in locations where wind speed varies considerably. Consider the example of one hour of operation at a wind speed of 40 ft/s (12 m/s) and one hour with no wind versus two hours of operation at a steady wind speed of 20 ft/s (6 m/sec). While the average wind speed will be the same for both scenarios, the first case would have a theoretical wind power potential for the two hours of about 0.135 hp-h/ft^2 (1.06 kWh/m^2), while the second scenario would have a two hour wind power potential of only 0.0328 hp-h/ft^2 (0.264 kWh/m^2), a four-fold difference. The prevailing general rule is that the average of the cubes is significantly greater than the cube of the average. Even more discrete analysis would be required to match hourly wind speeds to specific times of use to properly evaluate electric outputs differently based on varying time-sensitive electricity market values.

If a turbine has a high cut-in speed, it will only produce its rated power at very high wind speeds, and its rated power does not indicate how much power it will produce over the entire year's operating regime. The expected power output over the year will be based on the turbine's capacity (or load) factor. As with any other prime mover, this is the actual annual power output versus the potential output if the machine operated at its rated power output for the entire year. While units may run 80% or even more hours in the year, they generally do not operate at full load for that many hours. Typical capacity factors for successful projects range from 20 to 40%, though in some applications it can be considerably higher. This critical factor is largely a function of wind availability and, as a general rule, higher availability translates into better project economic performance.

However, a higher capacity factor does not always mean a more economical application. For example, for a given rotor size, increasing the capacity of an electric generator will lower the annual capacity factor for a modest cost premium. Despite a reduced capacity factor, the system may be able to better exploit high wind conditions and produce more power over the year, for a better return on investment despite the lower capacity factor. This is, however, more of a design issue and should not undermine the general notion that higher capacity factor generally does correlate to superior project return on investment.

The key element to a successful WECS application is location. The location will determine wind availability in terms of wind speed and hours of duration at each wind speed. It will also have a great influence on the market value of the WECS electricity produced. In addition to the many grid-connected wind farms that are selected based on location suitability, there are many successful WECS on islands or other remote sites, such as pumping applications at many farms, where access to the interconnected grid is unavailable or uneconomical. In these locations, simple-cycle Diesel generators, with relatively high fuel delivery costs, typically establish the life-cycle cost baseline. Other attractive locations are those in the proximity of high regional electric rates and/or difficult air permitting restrictions. In such locations, even relatively modest wind speeds may produce an economical project.

Still, perhaps the single most critical factor remains wind resource availability. The best sites are generally those with few obstructions in all directions, notably:

- Flat expanses, including long valleys extending down from mountain ranges and high elevation plains and plateaus.
- Areas near bodies of water, including coastal sites and large lakes.
- Mountainous areas, including exposed ridges and

summits, as well as gaps, passes, and gorges in areas of frequent strong pressure gradients.

Flat expanses are commonly the windiest, since there is less frictional drag to slow wind speed than over rugged terrain. A study conducted by Pacific Northwest National Laboratory estimated that the wind energy potential of 12 states in the midsection of the country alone, inclusive of practical WECS efficiencies, is sufficient to produce nearly four times the amount of electricity consumed in the United States. This is far more than any other area of the country. In fact, North Dakota alone is estimated to have enough potential energy from windy areas of Class 4 and higher to supply one third of the electricity consumption of the 48 contiguous states. The regions covered by this 12-state area exhibit vast areas with topography of flat plains to rolling hills and uplands, resulting from glaciation, and with flat areas that are the beds of ancient lakes. Consequently, a large portion of the land area is well exposed to the wind and wind resources of Class 3 to 6 predominate in this region. The study indicated a wide variation in wind class between 33 and 164 ft (10 and 50 m), so careful tower height selection is particularly important in these regions.

Figure 14-27, shows a few of the WECS installed at a 25 MW wind farm located on a flat expanse in Southwestern Minnesota. Here, farmers generate additional income by leasing their land for the wind farm and the developers exploit the favorable Midwestern winds to generate revenue under a 25-year contract that pays an estimated $0.05/kWh generated.

Sites near bodies of water also generally experience high wind velocity caused by the temperature difference between the air over the water and the air over the land. Generally, since land is warmer than water during the day and cooler at night, the air moves from the cooler water toward the warmer land by day and in the opposite direction by night. Exposed coastal areas in the Northeast, from Maine to New Jersey, have good wind potential. The exposed Atlantic coastal and offshore islands of the Northeast are primarily Class 4, 5, and 6. Class 4 is found immediately along the coast, while Class 6 exists along the outer capes and islands, such as Cape Cod and Nantucket Island.

Fig. 14-27 A Few of the Zond Wind Turbines Installed at a 25 MW Wind Farm Located on a Flat Expanse in Southwestern Minnesota. Source: Warren, Gretz, DoE/NREL

In the Northwest, exposed coastal areas of Oregon and Washington southward to northern California have Class 4 or higher wind resources. Class 4 or higher wind resource may also be found over much of the Great Lakes. Class 3 wind resource can be found along exposed coastal areas, from Delaware to North Carolina and much of the California and the Texas coastal areas. High wind resource (up to Class 7) occurs over the Aleutian Islands, much of the coastal areas of northern and western Alaska, offshore islands in the Bering Sea, and the Gulf of Alaska. In Hawaii, interactions between prevailing trade winds and island topography produce areas with wind power as high as Class 6 in numerous locations. In Puerto Rico and some other Caribbean Islands, trade winds produce Classes 3 and 4 along certain coastlines. Figure 14-28 shows a wind farm located by the water and Figure 14-29 shows one turbine of a wind farm located off shore the coast of Sweden.

Mountainous areas often have wind patterns similar to those near a body of water. So-called valley wind is produced, as solar impact on mountain slopes is warmer than surrounding valley air, which forces the wind up the mountain. At night, the reverse occurs as cooler mountain air sinks along the slope, producing a mountain breeze. These conditions tend to produce localized higher winds. Still, careful site selection is important, since hills and rugged terrain can cause wind obstruction in one location, while nearby produce a venturi effect with converging winds, creating very favorable siting locations. Turbulence caused by obstructions not only reduces available wind speeds, but also produces stress on the WECS, increasing maintenance requirements and shortening service life. Siting, of course, can be particularly difficult in mountain terrain, either eliminating project options or driving up project construction and long-term maintenance costs.

In the Pacific Northwest and down through northern California, there are numerous mountain range breaches that provide a low-elevation connection between continental air masses east of the range and the maritime air of the Pacific coast. For example, particularly strong pressure gradients develop along the Cascades and force the air to flow rapidly eastward in the summer and westward in the winter. Wind resources of Classes 3 to 6 have been measured at numerous sites throughout the range. East of the Rocky Mountains, the convergence of the plains and the mountains causes gradients that produce very high seasonal winds, with resources reaching Class 6 in winter. For example, areas with up to Class 6 wind resource are found in several valley wind corridors in southwestern Montana. However, lower winds of Class 3 are commonly experienced through other parts of the year, limiting potential annual availability factors. As with the Cascades, gaps in the Rocky Mountains provide a channeling effect that allows prevailing westerly and southwesterly winds to blow with little resistance across relatively high plains and uplands of southern Wyoming. This produces perhaps the largest region of non-mountainous terrain in the west with a high wind energy resource, with areas ranging from Class 4 to 6 at heights to 164 ft (50 m). Eastern mountain ranges also have good wind potential, with the highest ratings in the Appalachian Mountain summits from Pennsylvania to Maine.

Figure 14-30 shows most of the 550 kW Zond wind turbines of the 6 MW Green Mountain power plant in Vermont. The turbines feature full-span, variable-pitch blades and were designed to ensure reliable cold-weather operation. This includes:

- specially coated black blades, which absorb solar energy to reduce icing;
- heaters for the control systems;
- synthetic gearbox oil and hydraulic fluids;
- a large nacelle to protect maintenance workers from inclement weather; and
- a 131 ft (40 m) freestanding tubular tower that provides access to the turbine top through an internal ladder.

Fig. 14-28 A Wind Farm Located by Body of Water. Source: NEG Micon

Fig. 14-29 One Turbine of an Offshore Wind Farm. Source: NEG Micon

Fig. 14-30 6 MW Wind Farm Located in Mountainous Terrain in Vermont. Source: Green Mountain Power Corporation and DoE/NREL

Numerous studies were first conducted to ensure that there would be no harmful environmental impact. Construction crews cut as few trees as possible when building the wind farm and have since allowed the plants and brush growth to return as close as possible to their natural state. The turbines in this challenging location were installed for a reported price of about \$1,800/kW of capacity and are maintained under long-term contract at a reported rate of just over \$0.01/kWh produced. The plant has operated with a very high availability factor and achieves an annual load factor of about 27%, producing about 14,000 MWh per year.

With location being perhaps the preeminent factor, large capacity wind farms are more commonly located in areas where wind conditions are most favorable, as opposed to where the loads are. There are, however, limits to this strategy, since the power generated must be either used on site, distributed locally, or transmitted to locations with sufficient load requirements and/or power market value. Transmission system constraints and line losses then become limiting factors. Hence, there are numerous other factors beyond wind resource that drive a successful WECS project. Regions with higher cost electricity, stringent air permitting regulations, and public policy support experience further enhanced project economics. Huge wind farms have been developed in such areas, making use of not only the favorable wind resource, but of regional economic, environmental, and political conditions, as well as the economies of scale that can be achieved in equipment acquisition, installation, and operations and maintenance with so many units in great proximity. Figure 14-31 shows a wind farm at Altamont Pass in California where clusters of turbines are scattered to the horizon.

Fig. 14-31 Clusters of Wind Farms at Altamont Pass in California. Source: Warren Gretz, DoE/NREL

Fig. 14-32 Massive Wind Farm in Palm Springs at Sunset. Source: Warren Gretz, DoE/NREL

Fig. 14-33 One of the Many Wind Farms Located in the Tehachapi, California, Area. Source: Warren Gretz, DoE/NREL

Locations with favorable circumstances, from the northern Great Plains down to Texas, and certainly in California, have become sites for the implementation of numerous massive wind farms, such as the one in Palm Springs shown in Figure 14-32. Figure 14-33 shows one of the many wind farms located in the Tehachapi, California, area. The Tehachapi Pass, near Mojave, is a wind corridor where winds are funneled from the San Joaquin Valley into the Mojave Desert. Areas of Class 6 annual average wind resource are indicated by new site data in the Tehachapi Pass vicinity. This area has become one of the largest and most diverse wind farm installations in the world.

During the 1980s, some 6,000 turbines were installed, featuring nearly every first-generation WECS on the market at that time. Since then, numerous additions have been made, with an emphasis on installing newly developed larger capacity units. One new project, Oak Creek Phase II, added 23.1 MW of capacity from 33 Micon turbines. These units have a hub height of 180 ft (55 m) and a blade length of 77 ft (23.5 m) and enjoy a mean wind speed of 20.9 mph (9.5 m/s). In designing such large-scale projects, attention and performance consideration must be given to the shadowing effect of one turbine on the next one downwind.

In addition to the vast wind farms, smaller projects designed to meet facility electricity needs may also be cost-effective in such locations, especially when the avoided retail cost of electricity is substantially higher than the cost of selling generated electricity on the wholesale market. Much like other on-site power generation alternatives, system capacity is based on integrated analysis of facility loads and wind availability factors to select the ideal configuration.

Numerous small islands have enjoyed good success in blending WECS with conventional Diesel engine generation plants in achieving superior life-cycle energy cost profiles. Figure 14-34 shows the wind turbines and access road at a naval facility on San Clemente Island off of the coast of California. The WECS consists of three Micon turbines, each with a rotor area of about 7,535 ft^2 (700 m^2) and a rated output of 225 kW. The wind turbines start producing power at approximately 9 mph (4 m/s) and continue producing power up to 56 mph (25 m/s). The WECS were installed to provide electrical energy and emission savings by reducing the use of Diesel fuel to the maximum extent possible.

Fig. 14-34 Wind Turbines Used along with Diesel Engine-Generators at a Naval Facility on San Clemente Island.
Source: Warren Gretz, DoE/NREL

While Diesel engine-generators are mostly commonly used in large mixed (hybrid) systems, either to share load with the WECS or as back-up, other technologies such as battery storage may be used. Batteries serve to stabilize power fluctuations from the turbine and store excess electricity produced for later use. Battery banks most frequently use lead-acid batteries and range in capacity from 1 to 3 days' supply. Small capacity WECS will often rely solely on batteries for storage/back-up, while larger systems more often rely on Diesel engine back-up, or a combination of engines and batteries.

Fig. 14-35 Small 10 kW WECS Intended for Automatic Operation in Adverse Weather Conditions.
Source: Bergey Windpower Co., Inc., DoE/NREL

Figure 14-35 shows a very small 10 kW Bergey Windpower WECS. It is a three-blade, direct-drive, upwind unit with passive blade-pitch control. To achieve over-speed protection, the 23 ft (7 m) diameter rotor yaws out of the wind. Its simple rugged design is intended for high reliability, low maintenance, and automatic operation in adverse weather conditions. In remote locations, it can charge batteries for standalone applications.

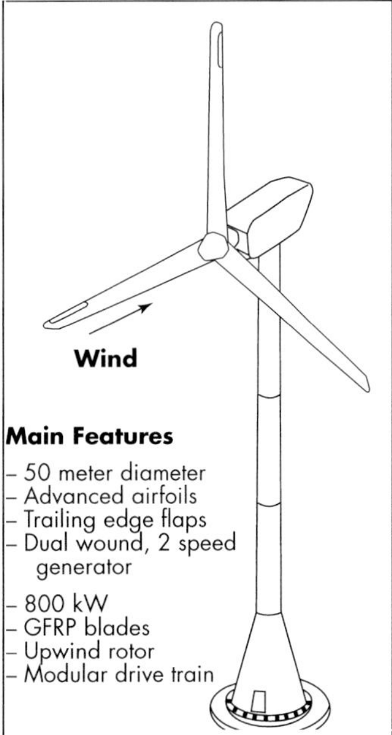

Fig. 14-36 Advanced Three-Blade HAWT Design Concept.
Source: DoE/NREL

Research, Development, and Market Penetration Potential

The objective of WECS research and development efforts is to reduce production cost, increase overall efficiency, and improve reliability and service life. There has been great advancement over the past two decades in the sophistication of analytical methods used as the basis of design. This has resulted in improved engineering methods for modeling of turbulent inflow, structural dynamic response, and power train and control systems. Complex computer programs allow designers to try out advanced components in different configurations prior to fabrication and testing. Development advancements continue on a wide range of design types, including both HAWTs and VAWTs, and in capacities ranging from a few kW to several MW.

Continuing efforts are being made to design blades to capture more energy from a given wind speed. These use lightweight, low-stiffness composite materials with more aerodynamic designs to improve energy capture, reduce production costs, and enable them to survive gusty winds. Rotor hubs are being made with increased flexibility, allowing increased rotor efficiency, while minimizing damaging drive train and structural loads. Tower designs are focused on increasing flexibility to reduce weight and increase allowable heights to place the turbine higher, where the wind is stronger and more energy is available.

Direct-drive, low-speed generator and power train designs are intended to allow the turbine to produce power at variable rotor speeds, increasing efficiency and ease of control, while reducing or eliminating gearing requirements. Several prototype units under development feature a variable speed, direct drive generator that spins at the same rate as the turbine rotor, completely eliminating transmission and gearing requirements. The variable speed feature allows the turbine rotor to increase or decrease speed in response to changes in wind velocity, increasing energy capture and reducing wear. A power electronics unit is being introduced to convert the variable frequency, ac produced by the generator into constant frequency, 50 or 60 cycle ac suitable for grid connection.

Following are illustrations of some advanced turbine concepts as envisioned by NREL. While these are not yet designed, researchers expect final configurations to be similar to those shown. Also shown are several advanced wind turbines currently in field operation.

Figure 14-36 is a HAWT conceptual design featuring three 164 ft (50 m) diameter blades with advanced airfoils and trailing edge flaps for over-speed control. It has a dual-wound, two-speed 800 kW generator with an upwind rotor and modular drive train. Figure 14-37 is a recently introduced Atlantic Orient advanced three-blade, 50 kW HAWT. Its rotor is 49 ft (15 m) in diameter. This downwind, stall-regulated turbine features passive yaw control, wood epoxy composite blades, incorporating NREL-designed airfoils, aerodynamic tip brakes, and an electrodynamic brake. This turbine was designed to be well-suited for remote standalone type applications. The integrated drive train eliminates many critical bolted joints found in conventional turbine designs. It also weighs less than conventional drive trains

Fig. 14-37 AOC Advanced 50 kW Three-Blade HAWT. Source: Warren Gretz, DoE/NREL

and eliminates maintenance-prone couplings between the gearbox and generator. It also features an optional yaw damper, which is a passive hydraulic system that limits yaw rates (and gyroscopic loads) and is specified for turbulent wind sites.

Figure 14-38 is an advanced VAWT conceptual design with three composite pultruded, 148 ft (45 m) diameter blades with advanced air foils and boundary layer control. It has a 1 MW generator with variable speed linear drive and no gearbox. This design is intended to build on the

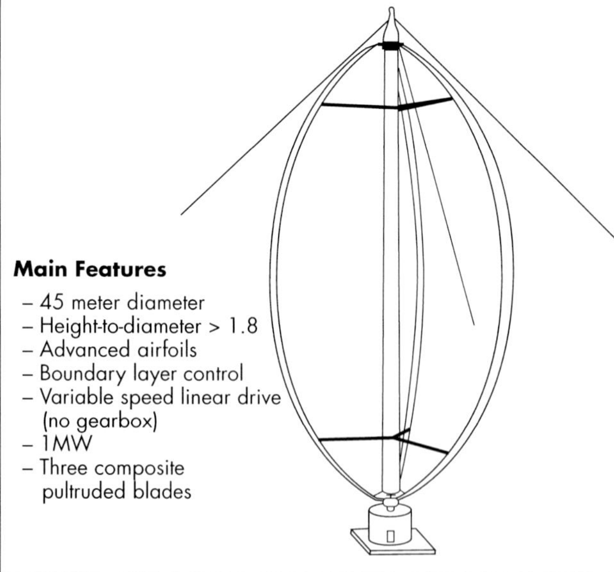

Fig. 14-38 Advanced VAWT Design Concept. Source: DoE/NREL

inherent advantages of previous Darrieus type turbines, which include simplicity of design and ground-mounted drive train and generator. This promotes easy access and eliminates weight-bearing requirements. One Darrieus type VAWT currently under commercial development is a 300 kW FloWind unit featuring three blades. The 56 to 69 ft (17 to 21 m) rotors will incorporate advanced airfoils and have a height-to-diameter ratio of 2.78, producing a narrow profile. This will enable lower production cost and the option to retrofit existing units to increase their power rating. The blades incorporate natural laminar flow airfoils developed at Sandia National Laboratories to keep air flowing smoothly over the blades and increase energy capture. They are being manufactured using an automated pultrusion technique in which fiber-resin blades are pulled through a die.

Figure 14-39 is a two-blade, lightweight teetered turbine conceptual design. The hybrid composite blades are 164 ft (50 m) in diameter with advanced airfoils. This stall-controlled downwind rotor model is intended for variable speed direct drive and has actively controlled ailerons for power clipping in response to gusts. Figure 14-40 is a leading edge, two-bladed, 250 kW WECS prototype that is built upon a 100 kW model developed in the 1980s. This upwind turbine uses aircraft-style airleron controls that deploy in high winds to smooth out energy spikes. Its design allows the rotor to teeter about the hub to minimize stresses through the machine, enabling a lighter-weight design and reduced wear. Another leading edge, two-bladed prototype is the 275 kW AWT-26. Figure 14-41 provides a closed-case illustration of this unit. This is a downwind, stall-regulated, free-yaw machine designed with a two-bladed teetered rotor. Its 85 ft (26 m) rotor incorporates aerodynamically efficient wood-composite

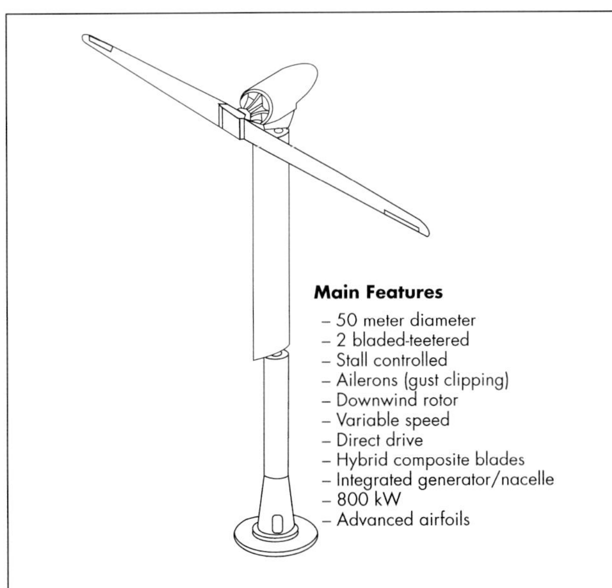

Fig. 14-39 Advanced Light Weight Teetered Two-Blade HAWT Design Concept. Source: DoE/NREL

Fig. 14-40 Advanced North Wind 250 kW Two-Blade Teetered HAWT in Prototype Field-Testing.
Source: New World Power Technology Co., DoE/NREL

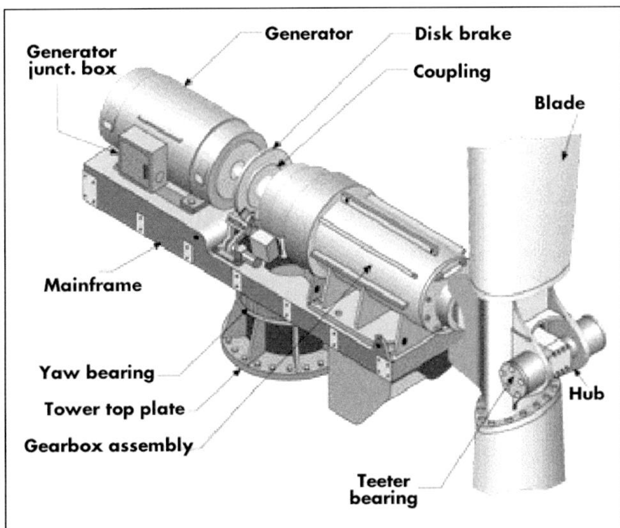

Fig. 14-41 Closed-Case View of 275 kW AWT-26 Advanced Two-Blade HAWT. Source: DoE/NREL-NWTC

blades using NREL-designed airfoils to promote extremely high wind capture. It also features an electromagnet-controlled high-speed shaft brake and aerodynamic top vanes, which serve as a fail-safe (emergency brake) and as an active brake for normal shutdown operation.

Market Penetration

Despite the vast wind resource potential in the United States and throughout the world, wind turbines have had limited market penetration due to relatively high capital costs and the need to often site them in remote locations in order to find access to adequate wind resource for a sufficient number of operating hours. Other factors have been uncertainty about long-term maintainability, with constant risk of damage due to severe weather conditions, and the long-term impact of high vibration levels and alternating stresses. Still the wind power market continues to grow. In fact DOE estimated that in 2007, wind accounted for 35 percent of all new generating capacity in the United States. This suggests that the combination of newer technology, rising energy rates and a demand for emission reductions have helped wind energy to overcome some of the barriers to market penetration that have previously stunted development. As the market continues to mature, more and more solid fleet performance data will be available on the current generation of advanced turbines. With this would also come an even greater market presence of a wind turbine service industry.

Continual advancements in design and manufacturing, as well as increased production quantities, have led to reductions in capital cost relative to other prime mover alternatives. Design advances have also resulted in improved power conversion rates, allowing for application in lower speed wind conditions and increasing the return on investment through increased output.

Given the economies of scale achieved with a very large wind farm, combined with industry advancements in cost-effective production and system reliability, it is believed that wind turbine capacity can be installed at below $800/kW and maintained at $0.01/kWh over the life of the equipment. While a conventional fossil fuel-based power plant might have a similar or even higher capital cost, it will also generally have a far higher load factor, as it is not dependent on wind variability. From this perspective, the relative capital cost per kWh for WECS will remain high. Counterbalancing this, of course, in the life cycle cost equation is the absence of a fuel cost component and associated air pollution control costs.

Photovoltaic Systems

The energy of the sunlight falling on the earth in just one day is the equivalent of all of the energy used worldwide in roughly a quarter of a century. In addition to thermal energy generation, electrical energy can be produced directly from this solar radiation by photovoltaic (aka PV or solar) cells. Unlike batteries or fuel cells, PV cells do not rely on chemical reactions to produce electricity, but simply convert sunlight into electrical energy.

PVs are made from semiconductor materials, most commonly silicon, which is the second most abundant element in the earth's crust and the same material used for computer chips. When combined with one or more other materials, it exhibits unique electric properties when exposed to sunlight. Electrons are excited by the light and move through the semiconductor. This is known as the photovoltaic effect and results in the production of dc electricity.

With the advent of the space program, PV cells made from semiconductor-grade silicon became the favored power source due to their ability to operate without a stored fuel source, relatively small size and weight, simplicity, and reliability of operation. The cost to manufacture PV systems was very high, but this was of minor concern for such operations given the advantages. As a consequence, most space satellites are solar cell-powered. As with so many other technologies, collateral benefits result from adapting this technology to terrestrial applications. In the 1970s, with the disruption of oil supplies to the industrialized world, along with growing environmental concerns about pollution caused by combustion of fossil fuels, interest in commercialization of PVs began. Since that time, researchers have made great strides in reducing production costs while increasing efficiency and reliability. Today, while production costs are still too high to achieve wide-scale market penetration, PVs are more widely used than ever before and their cost-effectiveness is continually increasing. PVs have no fuel requirement and operation is environmentally benign. The only negative environmental impact associated with PVs is the energy consumed and potentially toxic chemicals used during their manufacture. With no moving parts, they offer high reliability with long service life. They are modular and portable, can be installed quickly with respect to a conventional power plant, and can be easily expanded.

PV Effect

The PV effect is a process in which two dissimilar materials in close contact act as an electric cell when struck by light or other radiant energy. An electromotive force is generated by absorption of radiation, meaning a current will flow across the junction of these two dissimilar materials when light falls upon it. Sunlight photons are absorbed into an electron within the semiconductor material forming the cell. A positive-negative (p-n) junction within the cell ensures that these light-excited electrons flow unidirectionally to the cell terminals, and thus through the electrical load that connects them.

The PV effect was first recorded by Edmund Becquerel in 1839, when he noted the appearance of a voltage when illuminating two identical electrodes in a weak conducting solution. In the 1880s, Selenium PV cells were built that converted light in the visible spectrum into electricity with demonstrated conversion efficiencies of 1 to 2%. The first practical PV cells were made out of crystalline silicon in 1954 by Bell Laboratories. These initially demonstrated conversion efficiencies of about 4%, but with subsequent advancements achieved 11% efficiency.

Underlying this effect is photoionization or the production of an equal number of positive and negative charges. One of both charges can then migrate to a region in which charge separation can occur. This charge separation normally happens at a potential barrier between two layers of a solid material.

In crystals such as silicon or germanium, electrons are normally not free to move from atom to atom within the crystal. Light striking these crystals provides the energy needed to free some electrons from their bound condition. Free electrons cross the junction between two dissimilar crystals more easily in one direction than in the other, giving one side of the junction a negative charge and, hence, a negative voltage with respect to the other side.

PV System Characteristics

Materials and Cell Design

The basic PV building block is the PV cell. It is referred to as a cell because it produces dc electricity like a battery. A single PV cell is a thin semiconductor wafer, most commonly made of highly purified silicon. Groups of cells are joined together to form a PV module (or prefabricated panel) and modules may be connected into an array. The modules and arrays can provide electricity in virtually any quantity, starting at a few milliwatts to power a calculator and ranging up to power plant proportion.

The wafer is doped on one side with atoms that produce a surplus of electrons and on the other side with atoms that produce a deficit of electrons. This establishes a voltage difference between the two sides of the wafer. Metallic contacts are made to both sides of the wafer. When the wafer is bombarded by photons from solar radiation, electrons are knocked off the silicon atoms and drawn to one side of the wafer by the voltage difference and can flow through an external circuit attached to the metal contacts on each side of the wafer.

A PV wafer is typically made of silicon (which is in column IV of the periodic table with an atomic number of 14) and has a diamond crystal structure. It has 14 electrons arranged in 3 different shells. Its 10 core electrons are tightly bonded to the nucleus of the atom. The 4 valence electrons in the outer shell are covalently bonded to 4 neighboring atoms, forming its crystalline structure. At normal atmospheric temperatures, it is a very modest conductor. Its crystalline structure does not permit electrons to move about as freely as the electrons in a good conductor, such as copper.

By substituting, or doping, the host material with other atoms from columns III and V of the periodic table, its conductivity can be altered significantly. Doping with column V impurities (e.g., phosphorous, arsenic, or antimony) completes the covalent bond and leaves an additional loosely bound electron forming what is known as an n-type semiconductor. Since column V atoms have 5 electrons in their outer shell instead of 4, when these atoms bond with the silicon atoms, they have one extra electron that is more loosely held in place (only by a proton in the nucleus). This makes the material a far better conductor than pure silicon. When energy is applied to the altered substance, electrons are far more easily knocked loose from their orbit. As each electron is knocked loose, a hole is left behind where an electron could bond. The free electrons are called free carriers and can carry electrical current. The designation as an n-type semiconductor is for negative, due to the prevalence of free electrons.

Doping with column III impurities (e.g., boron, aluminum, gallium, or indium) leaves the covalent bond deficient of one electron or with a hole, forming what is known as a p-type semiconductor. Hence, in the n-type semiconductor, there are electrons looking for holes, and in the p-type, there are holes looking for electrons.

When these two types of conductors are joined, a p-n junction is formed, which is like a large diode area. Whenever different materials are placed in contact, an electric field exists at the interface. This field will exert a force on the electrons and cause current to flow whenever there are free electrons present. The concentration of electrons is greater on the n side than on the p side and the concentration of holes is greater on the p side than on the n side. The formation of the junction instantaneously allows the positive and negative electric charges to redistribute, establishing a built-in internal electric field. Equilibrium is established as an internal electric field builds up to the direction that opposes further flow of electrons from the n region and holes from the p region. If contacts are made with the two ends of the p-n junction and a voltage is applied (i.e., light impinging on the junction device), the state of equilibrium is disturbed and current will again begin to flow and a voltage will be established at the terminals, producing usable power. Figure 14-42 illustrates the electron distribution in the two materials and at the p-n junction.

Figure 14-43 illustrates the layers that form the basic construction of a typical crystalline silicon PV cell. As sunlight impinges on the top surface of the PC cell, some of the light is reflected off of the cell's grid structure and some is reflected by the surface of the cell. Antireflection coatings and texturing of the silicon surface can help to minimize surface reflection losses and promote the transmission of this light into the energy conversion layers below. Two additional electrical contact layers are also necessary. The electrical contact layer on the face of the cell where light enters is made of metallic material typically organized in a grid pattern. Since it is generally not transparent, it must have sufficient spacing so as not to cover the entire face of the cell and block light. The back electric contact layer is made of metal and functions as an electrical contact and covers the entire back surface of the cell.

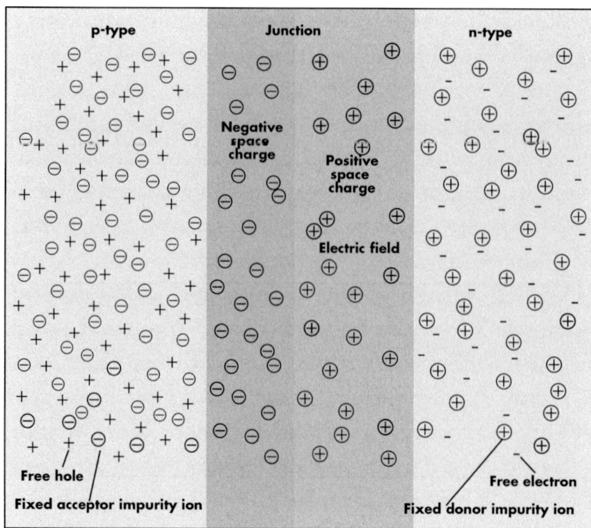

Fig. 14-42 Electron Distribution in N-Type and P-Type Materials and at the P-N Junction. Source: Jack Stone, NREL

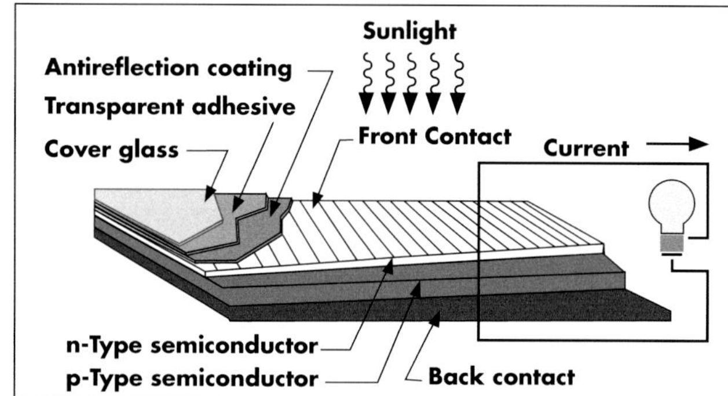

Fig. 14-43 Labeled Illustration of PV Cell Design. Source: NREL

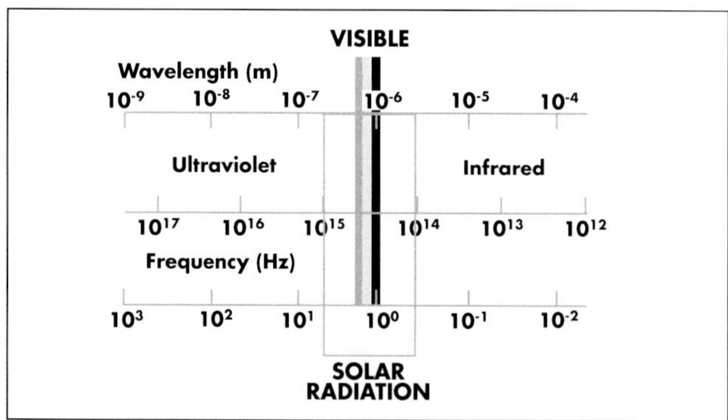

Fig. 14-44 Solar Wavelength, Frequency, and Photon Energy. Source: NREL

Absorption Material Selection

The sun emits virtually all of its radiation energy in a spectrum of wavelengths that range from about 7 x 10^{-7} to 13 x 10^{-6} ft (2 x 10^{-7} to 4 x 10^{-6} m). The majority of this energy is in the visible region. As shown in Figure 14-44, each wavelength corresponds to a frequency and an energy — the shorter the wavelength, the higher the frequency and the greater the energy, expressed in electron volts (eV). Only a certain amount of energy is required to knock an electron loose for a given material. This is called the band gap energy.

The photons in some of the light (long wavelength infrared) that fall upon the cell do not have the threshold energy needed to free electrons from the silicon atoms and pass through the cell without interacting. Concurrently, the photons in some of the light (short wavelength ultraviolet) have more than enough energy to create the electron hole pairs. The excess energy transferred to the charge carriers is dissipated as heat. About 40% of the incident light energy is absorbed and effectively used in freeing electrons from silicon atoms so that they can wander in the crystal lattice. The energy of the electrons increases from the ground state energy to an excited energy state.

In this exited state, electrons (or photons) are no longer associated with specific atoms in the absorber material, but are free to move. These energetic free electrons are then forced in the direction of the built-in electric field and collected by the electrical contact layers for use in an external circuit where they perform useful work.

Semiconductors are selected as absorber materials since they are a strong absorber of electromagnetic radiation in the visible range of wavelengths. Semiconductors in thicknesses of 0.004 in. (0.01 cm) or less can absorb all incident visible light. Band gap energy level is an important characteristic in material selection. The optimal balance between current and voltage is achieved at a band gap energy of about 1.4 eV. While the band gap of silicon, at 1.12 eV (which can effectively use wavelengths in the range of about 0.4 to 1.1 microns), is not ideal from this perspective, its natural abundance continues to make it the dominant material used.

In addition to silicon, other semiconductor materials that may be employed as absorbers include gallium arsenide, indium phosphide, copper indium diselendide, cadmium telluride, and titanium dioxide. The materials used for the junction-forming layers need only be dissimilar, and, to carry the electric current, they must be conductors. These may be the same materials used to produce diodes and transistors of solid-state electronics and microelectronics, which share the same basic technology as PV cells.

One way to increase overall efficiency is to create multiple-junction cells. The material with the highest band gap energy can be used on the surface, absorbing high-energy photons, while the lower-energy photons are absorbed by the lower band gap materials below.

System Types

In addition to the various layers, PV cell construction includes a transparent cover plate to protect the cell from the outside environment. PV modules, or panels, are then made by connecting numerous cells in series and in parallel to achieve useful levels of current and voltage. They are then mounted in a frame. These panels may stand alone, or can be electrically integrated into an array or subarray. They are built up and wired such that the panels can be disconnected without disabling or radically altering the array's current output. Figure 14-45 illustrates a PV solar array with an exploded view of the inner workings of a PV cell.

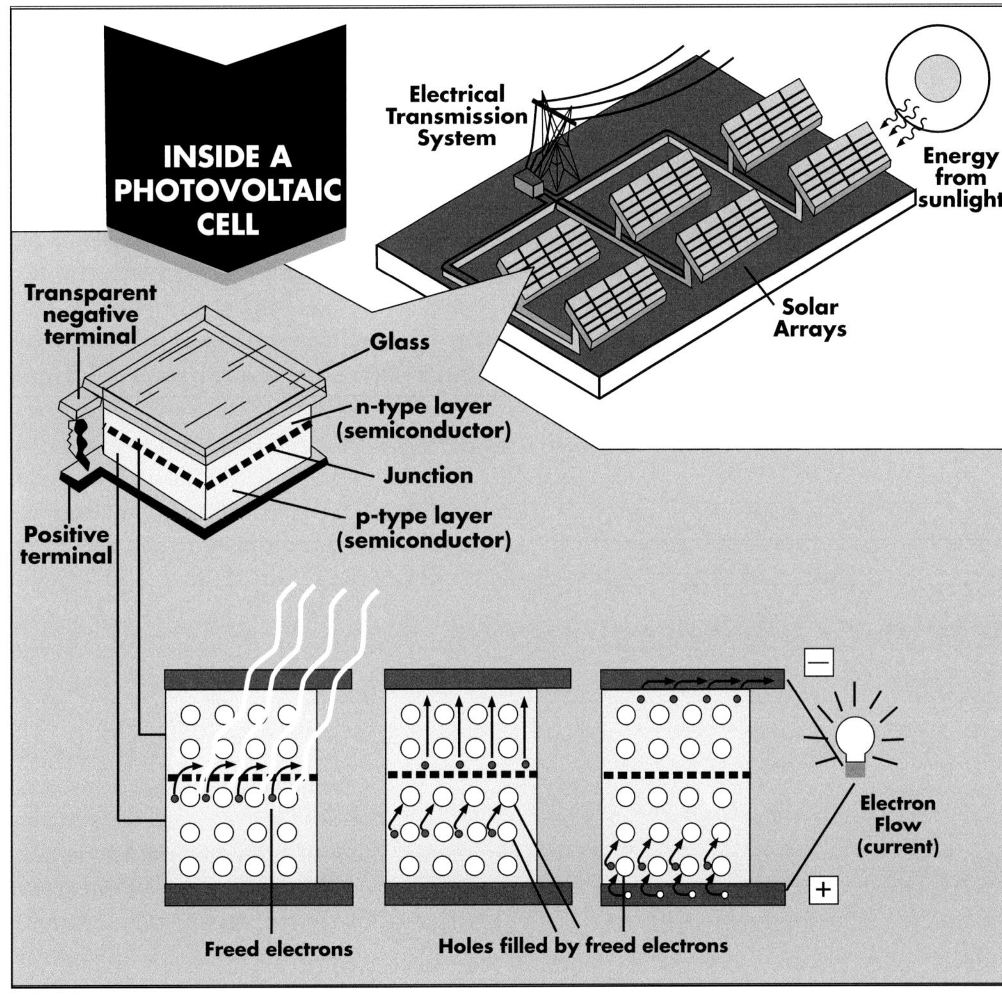

Fig. 14-45 PV Solar Array Illustration with Exploded View of the Inner Workings of a PV Cell. Source: DoE/NREL

For ac applications, power is transmitted through inverters and, in many cases, step-up transformers to increase the voltage output to a useable level. In addition to converting dc to ac power, the inverter limits current and voltage to maximize power output and, in grid-connected applications, matches the converted ac electricity to the utility's network, while providing necessary safeguards. Single or multiple inverters may be applied, usually requiring a compromise between cost, efficiency, and reliability. Figure 14-46 shows the electronic circuiting of a 300 kW inverter for a PV system.

There are three basic categories of PV systems: **flat plate collectors**, **thin film systems**, and **concentrator systems**. Each category has several types or subcategories.

Crystalline silicon flat plate collectors are the most developed and prevalent types applied today. These include single-crystal silicon and polycrystalline silicon, which is either grown or cast from molten silicon and later sliced into its cell size. They are then assembled on a flat surface into modules that can be fixed in place at an appropriate angle toward the sun. Alternatively, they can be mounted upon moveable devices that track the sun as it moves through the sky and position the modules so that they are always at an optimal position to maximize exposure to direct sunlight. The original and still most common semi-conducting material used for PV cells is single crystal silicon. This is one of the most efficient materials, converting up to 23% of incoming solar energy into electricity. It is also quite durable and has a long, proven service life in space applications. However, production costs for growing large crystals into thin, 0.004 to 0.012 in. (0.1 to 0.3 mm) wafers is high. This fact has driven researchers to explore a variety of alternative materials in an effort to find the optimal combination of cost, durability, and efficiency.

Fig. 14-46 Electronic Circuiting of a 300 kW Inverter for a PV System. Source: Jim Yost, DoE/NREL

Semiconductor materials vary in their ability to efficiently absorb sunlight. Materials such as gallium arsenide and amorphous silicon are extremely efficient absorbers. Hence, thinner flexible layers of such materials can produce amounts of power similar to a thicker rigid layer of material, such as crystalline silicon. Thin film systems are inherently less costly to produce than crystalline silicon, but still do not produce the same overall efficiency. In the thin film approach, a thin layer of the photovoltaically active material is deposited onto a supporting substrate, such as glass or metal. This greatly reduces the semiconductor material content of the finished product (using up to several hundred times less material). Commercial production of thin film materials can also be done at high throughput, since the module instead of the individual cell becomes the standard unit of production (a unit some 100 times larger). Since the thickness of the semiconductor material required may only be of the order of 1 micron, almost any semiconductor is inexpensive enough to be a candidate for use in the cell (silicon is one of the few that is inexpensive enough to be used as a self-supporting wafer based cell).

Many semiconductors have been investigated with several thin-film technologies that are now the focus of commercial development. Amorphous silicon, like the kind found in calculators, has been on the market for a decade. While production cost is relatively low, so are its efficiency and expected service life. Figure 14-47 shows rigid and flexible PV modules. The flexible module on the right is an amorphous silicon alloy thin-film system that has reportedly demonstrated an efficiency of 10.2%. Another type is thin films of polycrystalline silicon, which is very similar to the material that is already dominating the commercial market.

Silicon PV cell performance continues to improve due to both enhanced methods of manufacturing and a wider spectrum of material selection. An improvement in the performance of a monocrystalline silicon cell by the University of New South Wales (UNSW) to 24.4% conversion efficiency was measured at Sandia National Laboratories. Sandia also measured the performance of

Fig. 14-47 Comparison of Flexible and Rigid PV Modules. Source: United Solar Systems Corp., DoE/NREL

multicrystalline (large-grained polycrystalline) silicon cells at 19.8%.

Numerous materials other than silicon are viable for PV systems, and some have more optimal band gap energy and, therefore, higher theoretical energy conversion efficiency than silicon. Thin-film systems under development with materials such as copper indium diselenide and cadmium telluride have demonstrated very high efficiency. Despite potential difficulties in manufacturing, the compound semiconductor, copper indiumdiselenide, has produced the highest laboratory performance (reportedly 17% efficiency) for thin-film cells. Cadmium telluride is favorable from a manufacturing perspective, but its use of a heavy metal (cadmium) is viewed as a potential environmental hazard.

Studies have indicated that as material thickness is decreased, so is the absorbed usable photon fraction. Hence, while much less expensive, thin-film efficiencies will be lower. It is thus particularly important to enhance the optical path length with appropriate surface coatings or texturing to cause multiple passes of the light in the thin structure.

Concentrator systems employ a lens or reflector to concentrate sunlight on the PV cell, similar to the approach used with advanced solar thermal systems (see Chapter 7). Magnification ratios can range from 10 to 1,000 times. They can be produced less expensively than either of the other system types due to the reduced amount of PV material. However, they can only use direct sunlight (not daylight), so they must track the sun precisely and do not work when it is cloudy. The objective of these systems is to offset the additional cost for lenses, reflectors, and sun-following trackers with less PV material and higher efficiencies. Special silicon cells capable of withstanding increased light levels have been developed in an attempt to reduce overall costs in collector systems. Efficiencies in excess of 30% have been reached with the use of solar concentrators.

While concentrator systems offer perhaps the highest efficiencies, they are more limited in solar-spectrum utilization. Since they use only the direct-normal component of the sunlight, which is smaller than the total amount available to flat plates, they are best-suited for areas such as the southwestern deserts, where skies are clear and there is limited atmospheric interference.

As shown in Figures 14-48a, b, and c, the PV array may be either fixed or sun tracking, with one or two axes of rotation. Depending on the location, PV systems with two-axis tracking may receive 25 to 40% more global solar radiation annually than fixed-tilt systems. Using computer simulation, Fig. 14-49 shows how the output of two PV systems could be added to a utility's generation mix to help meet peak electric demand in summer. The fixed-tilt array faces south and is tilted from horizontal at an angle equal to the site's latitude. The more complex, expensive, and maintenance-intensive tracking array uses motors and gear drives to point the array at the sun throughout the day and to make seasonal adjustments as the solar altitude angles vary.

Performance

The efficiency of PV solar cells is a dimensionless value that indicates its effectiveness at converting solar energy into electricity. It measures energy output as a percentage of the solar radiation striking the surface of the cells. Rated efficiency is measured in a laboratory at 77°F (25°C) at an incidence angle of zero (beam radiation normal to the surface of the PV module) and with a radiation source that produces an energy and wavelength pattern that closely matches sunlight. The irradiance produced during testing is generally 316 Btu/ft^2 (1,000 W/m^2). Under these conditions, typical solar cell efficiencies range from 10 to 18%, though newer models can reach efficiencies of 24% or higher. These tests are completed in a controlled environment and should only be used for

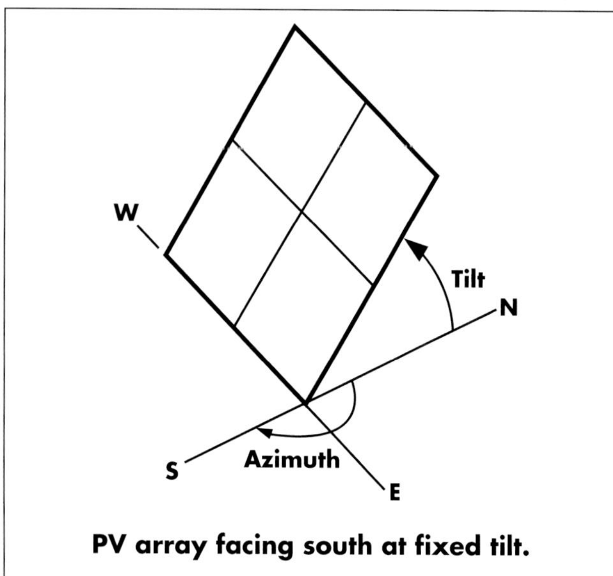

Fig. 14-48a Fixed Tilt Array. Source: PV Watts, DoE/NREL

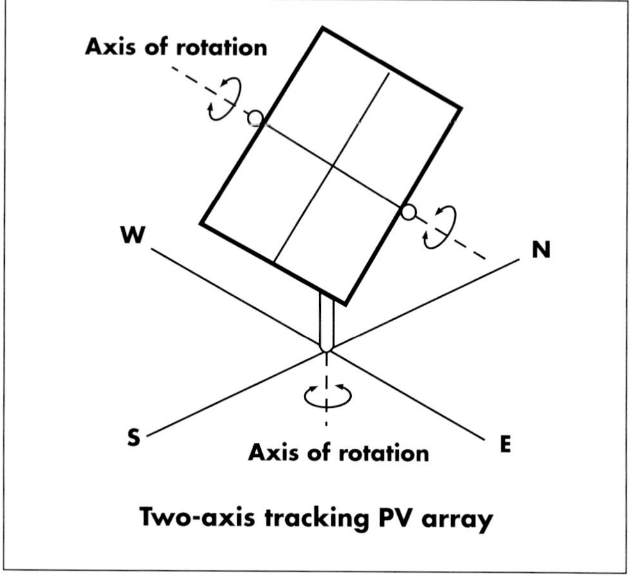

Fig. 14-48c Dual-Axis Tracking. Source: PV Watts, DoE/NREL

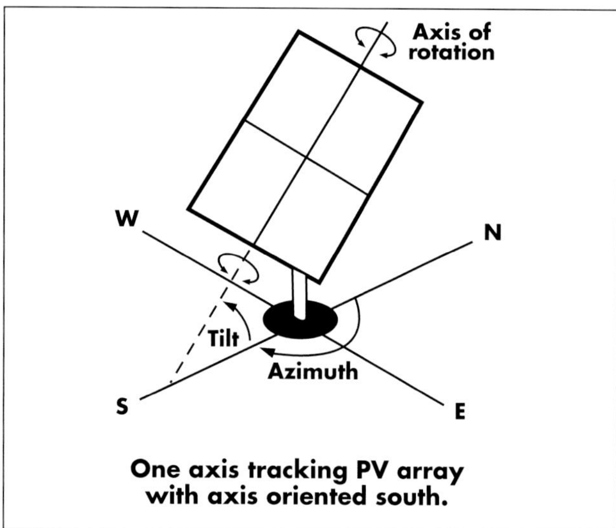

Fig. 14-48b Single-Axis Tracking. Source: PV Watts, DoE/NREL

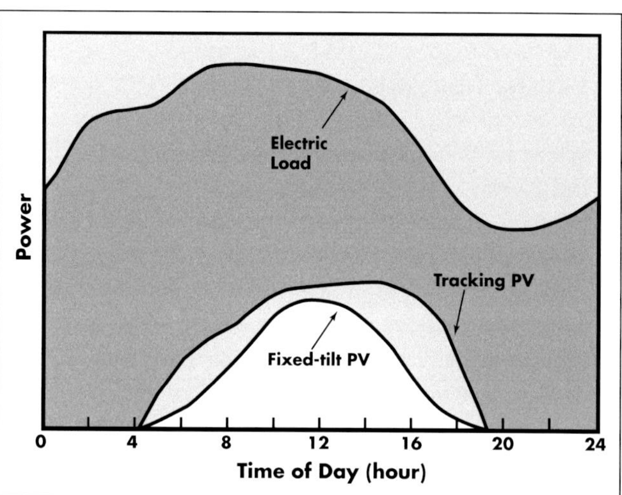

Fig. 14-49 Output Load Shape of Fixed-Tilt and Tracking PV Systems. Source: DoE/NREL

comparison between solar cell models. In the actual site application, unlike laboratory conditions, the temperature, the angle of incidence between the sun's radiation and the surface of the solar cell, and the amount of atmospheric interference of one type or another changes throughout the day. To get an estimate of the power output of the solar cell under actual site conditions, the following calculation is used to account for actual cell efficiency, available radiation, and PV configuration:

$$P = \eta_{PV} \, x \, S \, x \cos A \qquad (14\text{-}5)$$

Where:

P = PV power output (in Watts per ft^2 or m^2)

η_{PV} = Solar cell efficiency

S = Solar radiation available on a horizontal surface (in Watt per ft^2 or m^2)

A = Angle of incidence, between the sun's radiation and the normal to the PV surface

Efficiency

Published solar cell efficiencies depend on the materials and manufacturing processes used. There is a correlation between conversion efficiency and the operating temperature of a solar cell. In general, efficiency decreases by about 0.5% for every 1.8°F (1°C) rise. For example, a solar cell with a 15% conversion efficiency at 77°F (25°C) degrades to 10% efficiency when the operating temperature is 95°F (35°C). This is why solar cells typically perform better in cooler climates (assuming similar levels of available solar radiation) than in very hot ones. This de-rating factor should be based on the actual cell temperature, which may be slightly higher than ambient temperature. In addition, the efficiency can degrade over time. For example, the guaranteed efficiency of one solar cell manufacturer drops from 90% of rated efficiency to 80% of rated efficiency after 12 years. The conversion efficiency may also be lower under conditions of relatively low irradiance.

The published efficiency may show both dc and ac output. For ac output, inverter losses must be included in the efficiency term or added to the performance equation as an additional efficiency term as shown in Equation 14-6.

Available Solar Radiation

Quantification of available solar radiation begins with the extraterrestrial radiation from the sun (outside the atmosphere). This quantity, known as the Solar Constant (Gsc), has been established empirically to be 428 Btuh/ft^2 (1,353 W/m^2) per hour. This value varies somewhat with sunspot activity and earth-sun distance, but the variation is negligible. Before the solar radiation can be utilized by a PV cell, however, it is degraded by atmospheric diffusion and cloud cover, which are much more significant and must be accounted for. Values for actual horizontal radiation can be obtained for many locations in various formats, e.g., hourly values, daily means, monthly averages.

Databases containing solar irradiation and insolation values are compiled and published by the National Oceanic and Atmospheric Association (NOAA), the National Renewable Energy Laboratory (NREL), the World Meteorological Organization, and others. Irradiation is the instantaneous quantity of solar radiation received on a surface. This is expressed in Watt per unit area (e.g., W per ft^2 or m^2). Insolation refers to the average amount of solar radiation received on a surface during the entire day. This is expressed in W-h or kWh per unit area (e.g., kWh per ft^2 or m^2). It is essentially the summation of the irradiation on an object throughout a day. The overall worldwide daily average on an optimally sited collector is 0.465 kWh/ft^2 (5.01 kWh/m^2). The databases can be accessed directly from these organizations, as well as from their Web sites. Because weather patterns vary, database values will be better indicators of long-term performance than for a particular day or month or year. Monthly variation in solar radiation actually received may vary by 30%, while yearly variation may be 10%.

Incidence Angle

The angle of incidence varies with season and time of day. Maximum solar radiation is received when the sun is directly overhead, so that the incidence angle is closest to 0 degrees and "cos A" in Equation 14-5 is unity. At a given time of day, this happens when solar altitude angles are highest, which means that radiation is highest in the summer and lowest in the winter. Available radiation also peaks at noon and decreases at earlier or later times as the sun altitude angle decreases. The computation for angle incidence involves the use of a complex geometric equation that takes into account the sun's declination angle (for seasonal variation), the solar hour angle (for time of day), the latitude of the site where the solar cell is located, and the tilt and azimuth angles of the collector, which can also vary if a tracking device is used. Various nomographs and tables are available in solar literature to aid the calculation of this parameter. Some sources combine both horizontal radiation and incidence angle calculations into one look-up table for radiation available for common collector configurations.

As an example, at 10:00 am on June 21, at 40 degrees N latitude, there is 80 W/ft^2 (861 W/m^2) of direct solar irradiation available. The solar cell proposed for use is a high-density semi-crystal type with a published conversion efficiency of 15%. The solar panel is tilted at 45 degrees and the temperature is 77°F (25°C). Using Equation 14-5, the estimated output is:

$$15\% \times 9\text{-}9\ W/ft^2 \times \cos(45°) = 8.5\ W/ft^2\ (91.5\ W/m^2)$$

Thus, the output of a 2 ft (0.6 m) wide by 4 ft (1.3 m) long (8 ft^2 or 0.74 m^2) PV module under these conditions would be approximately 68 Watt. A summation for each hour is required to calculate the total daily electric output of the solar cell.

As an additional step in calculating the useful output, other system losses may need to be applied. If an inverter is used to convert dc power to ac, or batteries are used to store solar energy for re-release during nighttime or cloudy periods, then the cell output must be multiplied by these conversion efficiencies to get net useful output:

$$\text{Net Useful Output } (W/ft^2 \text{ or } W/m^2) = P \times \eta_{inverter} \times \eta_{storage} \quad (14\text{-}6)$$

Where:
P = PV output calculated from Eqn. 14-5
$\eta_{inverter}$ = Inverter efficiency (energy output/energy input)
$\eta_{storage}$ = Battery efficiency (energy output/energy input)

Other factors that will degrade system performance and, therefore, merit consideration are shading from obstacles (and even one PV module on another) and electrical losses due to wiring, module mismatches, and inverter performance below rated levels. Latitude-specific sun charts can be plotted to predict how buildings, trees, hills, and other obstacles will affect the amount of sunlight falling upon the PV system at a particular location at discrete time intervals throughout the year. These data can be entered into computer simulations programs to better refine insolation prediction.

Consider the example of a carefully monitored 60 kW grid-connected PCV system at an office building in southern Sweden situated at 56.6°N and 14.1°E. The system consists of two parts: the larger is a roof installation with four rows of multi-crystalline silicon modules and the smaller is an installation with facade integrated amorphous silicon thin-film modules. There are 58 SMA inverters, each with a rated output of 0.85 kW at 125-250 V_{in} dc, that connect the system to the grid. Table 14-3 provides a comparison of the accumulated energy from the 4,069 ft^2 (378 m^2), 49.5 kW rated roof plant between October 1997 and December 2000 with the measured annual solar insolation.

	Annual Insolation on roof (40° tilt) kWh/ft² (kWh/m²)	**Annual Energy Produced (kWh)**
1998	86.1 (927)	34,515
1999	94.9 (1,021)	38,592
2000	93.7 (1,009)	38,953

Table 14-3 Test Results for Available Radiation vs. Actual Production. Source: CADDET Renewable IEA/OECD Technical Brochure No. 142

From this and the other monitoring data taken, the overall performance ratio (PR) was determined to be 76% for the roof-mounted system. This PR represents the amount of actual power output achieved as a percentage of the maximum possible power output if the solar cells produced their rated power for the given amount of measured annual solar insolation. A PR of 100% means that the system would operate the entire period at standard test conditions and with an inverter efficiency of 100%, so that the solar cell conversion efficiency is the only factor affecting output. Note that in some cases, the inverter efficiency may already be included in the manufacturer's performance test data and would not be factored into the PR computation. In this case, dividing the annual kWh production by the product of the annual insolation on the roof per unit area and the area of the modules yields an average field-application efficiency of about 10%. Dividing this by a PR of 76% would indicate that the design rated performance for the modules (before consideration of inverter and other application-related losses) is about 13%.

The 24% overall system losses were due to inverter losses and local conditions. The principal element was inverter losses of 14%. An additional 3% loss was attributed to shadowing from adjacent rows of modules and from trees and buildings. The remaining 7% was attributed to accumulation of dirt and snow, light reflection from module surface at small incident solar angles, derating of conversion efficiencies at actual solar-cell temperatures in excess of 77°F (25°C), lower conversion efficiency at lower solar irradiance levels, mismatch in output between modules in subsystems, and Ohmic losses in cables.

As opposed to solar thermal technology applications, only direct sunlight (or daylight), not diffuse or reflected light, will cause the PV effect. Hence, ideal locations are those with little cloud cover. PVs can operate in daylight without sunshine, though at a reduced rate.

Applications

There are a wide variety of PV system types and structural applications. In most cases, the basic requirements for successful PV systems are that they be durable, waterproof, fire-restrictive, and aesthetically acceptable. The most common system applications involve the use of flat plate arrays either as stand-alone entities or as integrated building components. These may be fixed arrays or use single- or dual-axis tracking.

An ever-increasing number of structural applications are being installed, including many that are both aesthetically and functionally integrated into building structures. For example, a newly developed PV technology allows application of PV cells as roofing material. Roof tiles interconnect and can be installed over an existing roof or as a new roof. These may be designed as exchangeable PV shingles, which are flat panels that can be easily laid on and taken off a roof, or pre-fabricated roof panels that are roof-integrated modules composed of a panel that takes the place of roof sheathing and solar roof tiles. Thin-film technologies offer great application flexibility and can serve as windows, atriums, and awnings.

Three basic PV system application options are:

- **Stand-alone** systems that are not connected to the grid. These may use a battery for storage and may power dc devices directly or ac devices through an inverter.
- **Grid-connected** systems that provide ac power and are tied directly to the utility grid, or tied into the grid indirectly through the facility's internal electrical distribution system.
- **Hybrid** systems that may be tied into a dc or ac circuit along with other power producing devices, such as Diesel engine or wind power systems.

With both stand-alone and grid-connected systems, PVs are sometimes integrated with uninterruptible power supply (UPS) systems. These systems feature a PV module connected to a battery and, for some systems, to the electric grid. The battery stores power for use when the sun is not shining or during a utility outage. A charge controller regulates the flow of current to and from the battery subsystem to protect the battery from overcharge and over discharge. While this is similar in some respects to standard UPS and/or Diesel back-up systems, the PV has the advantage of being able to run as long as the sun shines.

In states with utility regulation that allows or requires net metering, a grid-connected PV system serves as an unobtrusive power plant that can provide power both to the facility and into the grid. Power fed into the grid allows the customer to either receive payment from the utility at the wholesale rate, or simply reduces consumption, lowering the bill at the retail rate. The latter situation typically applies as long as the customer is still a net consumer, not a producer of electricity for a given billing period.

Options for applying PV systems range from residential systems and small systems that power discrete equipment in remote locations, to larger utility power plant applications of several MW and larger. A 6.5 MW PV array has recently been retired after 20 years of operation at Carissa Plains, California. By PV system standards, this was quite large, but very small relative to the capacity of most conventional power plants. However, the potential for larger and larger central PV power plants certainly exists.

Relatively large-scale utility PV power plants consisting of many arrays installed together can prove useful to utilities. They can be built much faster than conventional power plants, which can be important under conditions of capacity shortages. In California, for example, the ability to add capacity quickly, with the added benefit of avoiding permitting and emissions trading to install new capacity, is of great importance and can make the more costly PV system a good tactical choice. For example, a 750 kW PV system currently being installed at a Naval facility in San Diego was developed and designed in just a few months and is expected to be constructed and operational in less than six months.

Figure 14-50 shows a traditional PV system array. This is a 500 kW system that has been in operation since 1993. Note the build-up of the array, with 3 rows of 12 PV cells (totaling 36) in each panel, which is connected to a series of panels, making up modules that are lined up in rows of 8. Figure 14-51 shows a grid-connected 300 kW facility located near Austin, Texas. Notice the conventional power plant in the rear. Figure 14-52 shows a 2

Fig. 14-50 Traditional 500 kW PV System Array. Source: Terry O'Rourke, DoE/NREL

Fig. 14-51 Grid-Connected 300 kW Facility Sited Near Conventional Power Plant. Source: 3 M Corporation, DoE/NREL

MW PV power plant operated by Sacramento Municipal Utility District. The 1,600 modules are spread across an 8,094 m^2 field in this very sunny region. Behind the PV system is a nuclear power plant. The utility has opted for nuclear plant decomissioning and has invested in expanded decentralized generation and/or in using renewable resources. This PV system has been expanded over time, with new arrays being added periodically.

In addition to the many standalone power station type applications, many PV systems are installed directly on a facility roof or integrated into the building construction. Figure 14-53 shows a roof-mounted PV system at a furniture factory in Massachusetts. Figure 14-54 shows a 372 panel system that is integrated into the roof of a downtown office building in Boston, Massachusetts. Figure 14-55 shows the installation of PV roof shingles. These newly developed triple-junction amorphous silicon shingles mount directly on the roof structure and take the place of asphalt shingles. Each shingle produces 17 Watt under full design conditions.

Figure 14-56 shows the final construction of a PV

Fig. 14-52 2 MW PV Power Plant Operated by Sacramento Municipal Utility District, Sited Near Decommissioned Nuclear Power Plant. Source: DoE/NREL

Fig. 14-53 Roof-Mounted PV System at a Furniture Factory in Massachusetts. Source: Bill Eager, DoE/NREL

Fig. 14-54 372 Panel PV System Integrated into Roof in a Downtown Boston Office Building. Source: Roman Piaskoski, DoE/NREL

Fig. 14-55 Installation of Newly Developed Triple-Junction Amorphous Silicon PV Roof Shingles. Source: Warren Gretz, Doe/NREL

system that was installed at the Olympic swimming complex in Atlanta, Georgia. A 340 kW rooftop PV array was installed on the main structure and a custom arched glass PV canopy was designed for the entrance to the complex. The rooftop PV system employs 2,832 Solarex 120-W PV modules mounted above the steel roof deck to allow for the free flow of cooling air below the array. A central Kenetech 300 kW dc-to-ac inverter feeds three-phase power into the campus utility grid. The total array has an area of about 27,900 ft^2 (3,000 m^2). The entry canopy features special large-area 250 kW Solarex PV modules that have a clear back skin to allow light transmission between the individual crystalline cells. Each PV module has its own integrated dc-to-ac micro inverter, developed by Advanced Energy Systems to deliver 60 Hz ac power directly to the building complex. Figure 14-57 shows an artist's rendition of a 48-story skyscraper that uses thin-film PV panels to replace traditional glass cladding material. The PV curtain wall extends from the 35th to the 48th floor on the south and east walls of the building.

While currently not cost-effective in most conventional markets, PVs have for some time proven useful and cost-effective in certain niche markets and for some specific applications. PVs have been effectively applied in numerous remote applications where utility interconnection was not feasible or too costly. In some cases, these applications were utility-sponsored where facilities in their service territory had isolated, low-energy, or low-revenue loads, or where line expansion was not feasible.

PVs have also been successfully applied for applications such as communications, warning signals, sectionalizing switches, cathodic protection, lighting, monitoring, and battery charging:

Fig. 14-56 Final Construction of PV System at the Olympic Swimming Complex in Atlanta, Georgia.
Source: Solar Design Associates, Inc., DoE/NREL

Fig. 14-57 Artist's rendition of Skyscraper Using Thin-Film PV Panels in Place of Traditional Glass Cladding Material.
Source: Kiss + Cathcart – Architects, DoE/NREL

- PV system battery charging has been applied at an electric vehicle recharging station in Southern Florida. The system is grid-connected and can sell power when vehicle-charging load is below system output.
- PV systems have been applied for certain sectionaliz-

ing switches on the Public Service Company of New Mexico's transmission line. Even though conventional power was available, the cost of transformers, surge arrestors, switches, and rectifiers to make the dc required for switch operation was greater than the cost of the installed PV system.

- Cathodic protection of pipelines is an economical and effective application of PVs to solve corrosion problems. The solution is an electric current from a PV source to counteract the natural corrosive currents generated around buried metallic devices.

PV systems have commonly been combined with other technologies at isolated sites in establishing a village energy concept by which multiple systems are combined to produce the optimal blend of life-cycle energy costs and reliability. This includes use with conventional Diesel engine power systems or, in some cases, with other renewable systems, such as WECS to eliminate the need for fuel or electricity acquisition and delivery. In one interesting application at a Tokyo Japan, water re-use promotion center, a desalination system uses both PVs and solar thermal applications for the membrane distillation process. The system produces fresh water by passing sea water through a membrane module. Heat to warm the seawater is provided by a solar collector and the pumps are powered by PV panels so that the system can be installed at sites without any other electricity source or fuel supply.

PVs may soon be used to generate hydrogen through the electrolysis of water. This hydrogen can be stored and ultimately used for normal combustion (as a cleaner source than fossil fuels) or to generate power in a fuel cell.

Costs and Market Penetration Issues

Similar to wind power resources, the availability of solar energy is orders of magnitude greater than the country's current electricity requirements. In fact, it is estimated that PV modules covering just a few tenths of a percent of the land in the United States could supply for all of the nation's electricity requirements. As conversion efficiencies improve, this number will continue to get smaller. Perhaps, as capacity costs continue to come down while conventional energy costs rise, such concepts will be viewed as more realistic. However, large-scale market penetration has been difficult to date.

Some historic problems with PVs were lack of standardization, lack of third-party certification, and poor warranty coverage. Many of these hurdles are now eroding. PV systems now meet most Underwriters Laboratories (UL), National Electric Code (NEC), and Institute of Electrical and Electronic Engineers (IEEE) applicable standards. Manufacturers and vendors are now also commonly offering long-term warrantees and guaranteed service contracts.

Through combined efforts of industry and the DoE, PV system costs have been reduced by 300% since 1982, but still remain far too high to be market competitive on a wide scale. An installation cost study was performed by the Solar Electric Power Association on systems ranging in capacity from 70 to 200 kW. For installation occurring between 1998 and 1999, they reported average total installation costs of under $8.50/W, with the module and inverters representing 56% and 9% of the total system cost, respectively. The study also documented a reduction in installed cost of some 14% over a two year period alone. Capital costs for larger systems may be somewhat lower, given greater economies of scale. In many locations, various state, agency, or utility grants are in place offering as much as $4.50/W as an incentive to promote this technology. Even with such lucrative support, capital cost per unit of capacity output remains the most significant obstacle to market growth. Constant improvements in production cost and system efficiency are driving capacity costs down each year, with the trend expected to continue.

The elegance and simplicity of PV system operation is quite compelling to both energy engineers and environmentalists. This derives from the fact that the system has no major moving integral parts, it requires only modest life-cycle operations, maintenance, and repair costs, and it experiences only modest performance degradation over its service life. This, combined with no fuel requirements and no environmentally harmful emissions, makes PVs a potentially favored and desired technology.

Fuel Cells

The fuel cell is a developing technology that offers the potential for high electric generation efficiency and very low environmentally harmful emissions. Unlike conventional power generation systems that rely on combustion or other sources of heat energy, a fuel cell converts chemical energy to dc electricity directly, via electrochemical reactions. The dc electricity can then be converted to ac electricity using solid state inverters.

Similar to a solar cell, the fuel cell may be thought of as a battery that is continually fed with fresh chemical energy and does not run down or require recharging. The process converts a fuel such as natural gas to hydrogen-rich gas through catalytic reaction with steam. The hydrogen-rich gas is reacted, or oxidized, with oxygen from the air to produce electricity and thermal energy. The process is the reverse of hydrolysis of water, in which a direct current is passed through water, decomposing the water to hydrogen and oxygen.

Currently, fuel cells are usually not commercially cost-competitive, absent of support in the form of federal or state grants, and there is still only a limited track record of proven reliability. There are, however, some near-term niche markets. One is for applications requiring high-quality power, or power conditioning for computers and other sensitive electronic equipment. Currently, this market is served by UPS systems. In these applications, the value of useful thermal energy provided by the fuel cell and the avoidance of required investment in and operations and maintenance of standby generators and UPS systems can offset some of the life-cycle cost premium.

To date, there has been widespread federal government support for R&D and market entry due to the upside potential for high-efficiency and environmentally benign operation. As the technology continues to develop, production and O&M costs will likely continue to decline.

Fuel Cell System Design

Operating Principle

Fuel cell systems have three main sections:

1. The **fuel processing section**, is where a steam reformer and a shift converter or reactor, are used to convert fuel into the hydrogen-rich gas required to sustain an electrochemical reaction in the power section. The steam reformer converts the fuel gas (e.g., methane) to hydrogen (H_2) and carbon monoxide (CO) over a nickel catalyst by the reaction:

$$CH_4 + H_2O \rightarrow CO + 3H_2 \tag{14-7}$$

In some fuel cell systems, in the reactor, the CO is reacted with steam over an iron-chrome catalyst at a lower temperature to produce more hydrogen and reduce the CO content to less than 1 ppm:

$$CO + H_2O \rightarrow CO_2 + H_2 \tag{14-8}$$

A variety of gaseous and liquid fuels, including gas from coal, can be used depending on the design of the fuel processing section.

2. The **power section** (or **stack**), is where H_2 and O_2 are combined to generate dc power, thermal energy, and steam. Stacks typically contain several hundred identical electrochemical cells, each consisting of a porous anode (a negative electrode), a porous cathode (a positive electrode), and an electrolyte. H_2-rich gas (e.g., reformed natural gas) is passed over the surface of the anode while air flows over the cathode surface. In some designs, cooling coils are dispersed throughout the stack to remove excess heat.

The H_2-rich fuel is oxidized at the anode to form hydrogen ions and release electrons. The hydrogen ions move from the anode through the electrolyte solution to the cathode. The electrons flow through an external electrical circuit, generating dc output from the power section. Oxygen is consumed in a reduction reaction at the cathode of each cell, reacting with hydrogen to form water.

Separator plates between cells separate the fuel gas and air from adjacent cells and provide channels for the flow of fuel and oxygen and an electrical conduction path from cell to cell. Each ft^2 (m^2) of electrochemical cell typically generates 100 (1,077) Watt or more of dc power at an electrical potential of less than 1 volt. Stacks are constructed with a specific number of cells and cell area in order to obtain the desired voltage and current flow.

3. The **power conditioning section**, is where solid-state inverters convert the dc power output from the power section to ac power output at a given voltage and frequency.

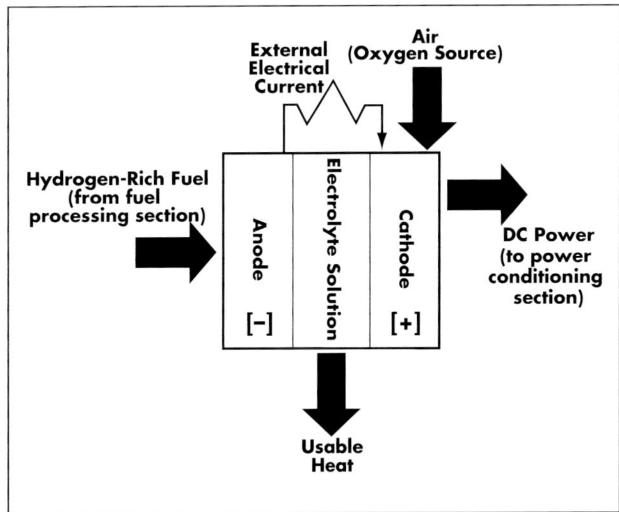

Fig. 14-58 Schematic Diagram of Fuel Cell.

Figure 14-58 is a schematic diagram and Figure 14-59 is a functional diagram of a fuel cell.

Phosphoric Acid Fuel Cell (PAFC)

R&D on the use of phosphoric acid as an electrolyte for fuel cells began in the United States in the early 1960s. A field test program of 40 kW units was completed in 1986 and the system was commercialized in 1992 at a 200 kW capacity. The 200 kW unit weighs about 40,000 lbm (18,000 kg) and is 18 ft (5.5 m) long by 10 ft (3.0 m) high by 10 ft (3.0 m) wide. Performance of these units shows reasonable reliability, long life, and good efficiency. In addition, an 11 MW unit has been demonstrated for utility applications.

The PAFC produces low-grade thermal energy in the form of hot water and has an electric production efficiency range of 36 to 40%. Relatively low operating temperatures of 300 to 410°F (149 to 210°C) have made development easier and maintenance lower, but also limit applications for using thermal energy output. When rejected heat can be effectively used, overall efficiencies of up to 80% are achievable. The 200 kW unit can supply more than 750,000 Btu/h (791,000 kJ/h) of hot water or low-pressure steam at temperatures of up to about 250°F (121°C).

Figure 14-60 illustrates a current cell design used in a PAFC. This design uses ribbed substrates as supports for the electrodes. These substrates are composed of porous graphic material that holds the excess acid inventory. The electrodes themselves are graphic structures with a Teflon-bonded platinum catalyst layer. The electrolyte is a porous, non-conductive matrix that acts as a sponge to contain the phosphoric acid.

Figure 14-61 shows a field installation of a 200 kW PAFC unit at a small commercial building. Figure 14-62 is a cutaway illustration of this packaged unit, revealing the position of the major components and the process flows through the system.

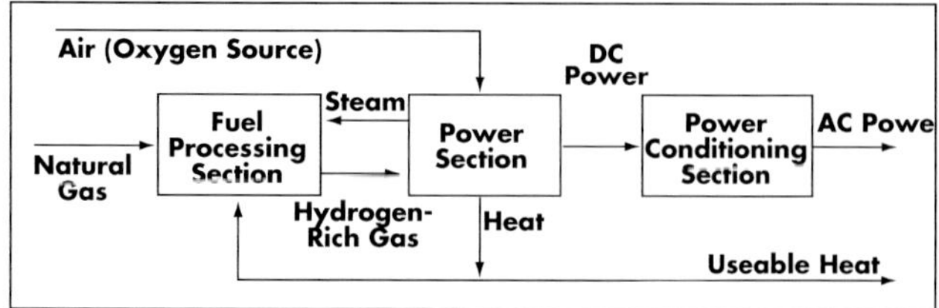

Fig. 14-59 Functional Diagram of Fuel Cell.

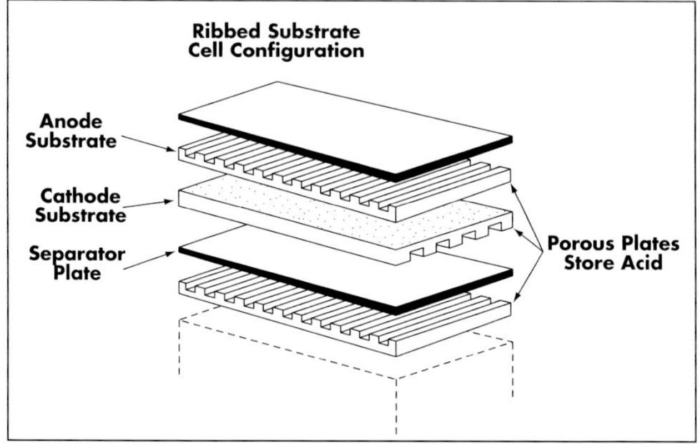

Fig. 14-60 Illustration of a Current Cell Design Employed with PAFC unit. Source: GRI

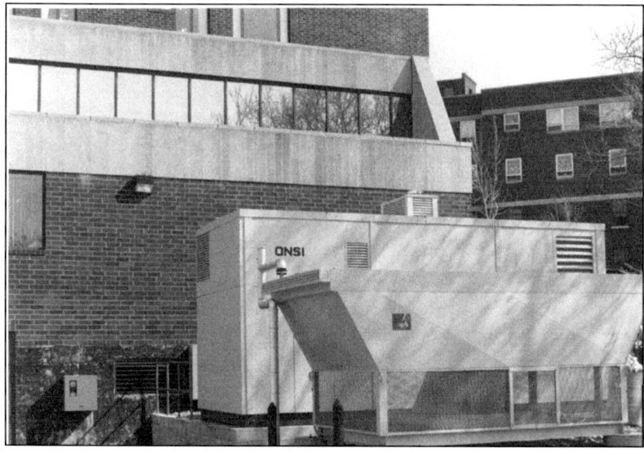

Fig. 14-61 200 kW ONSI PAFC Unit Installation. Source: Equitable Gas Co. and ONSI

Molten Carbonate Fuel Cell (MCFC)

Molten carbonate electrolyte development is proceeding to commercial system development with prototype capacities of 2 MW and smaller. Commercialization is currently under way.

MCFC design studies predict electric generation efficiencies ranging from 50 to 60% when operating on natural gas and just under 50% when operating on coal gas. Operating temperatures of approximately 1,200°F (650°C) allow for in-situ reforming of methane to hydrogen-rich gas directly without catalytic fuel processing. The higher operating temperature also provides high-grade thermal energy (steam) output.

Figure 14-63 is a schematic illustration of the chemical processes involved in electric production with an MCFC. Figure 14-64 shows the field installation of a 250 kW MCFC power plant used in a demonstration project. Figure 14-65 is a cutaway illustration of a conceptual design for a larger capacity natural gas-operated MCFC unit. The design goals for a 1,060 kW fuel cell power module are dimensions of 50 ft (15.2 m) long by 15 ft (4.6 m) wide by 14 ft (4.3 m) high. Target fuel efficiency is 54%, with NO_X emissions of under 1 ppmv (at 15% O_2) and sound emissions of 78 dba at a distance of 3 ft (0.9 m).

Solid Oxide Fuel Cell (SOFC)

SOFCs use ceramic materials such as zirconia as electrolytes. Cells are being developed in two geometric configurations: tubular and planar. The ceramic materials permit operation at approximately 1,800°F (980°C), at which temperature cells are capable of in-situ reforming of methane to hydrogen-rich gas directly without catalytic fuel processing. Long-term expectations of thermal efficiency range as high as 60%.

Commercialization of tubular SOFC generating equipment is currently under way. Planar SOFC, which is expected to provide higher power density, is considered to be a longer-term prospect. Figure 14-66 illustrates an SOFC fuel cell of tubular geometry. Figure 14-67 illustrates the operating principle of the SOFC fuel cell.

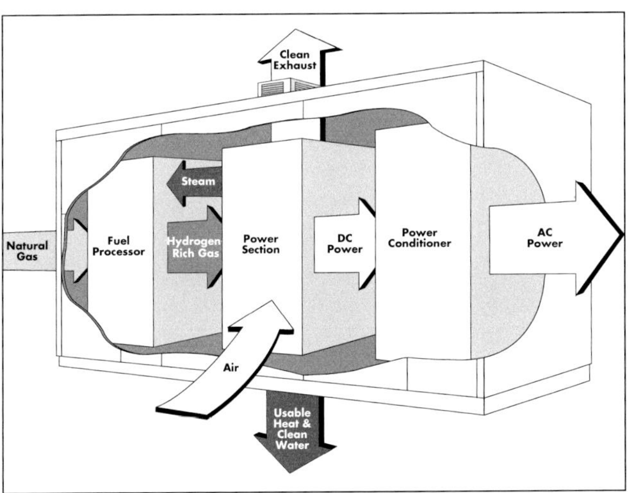

Fig. 14-62 Cutaway Illustration of PAFC Unit. Source: ONSI

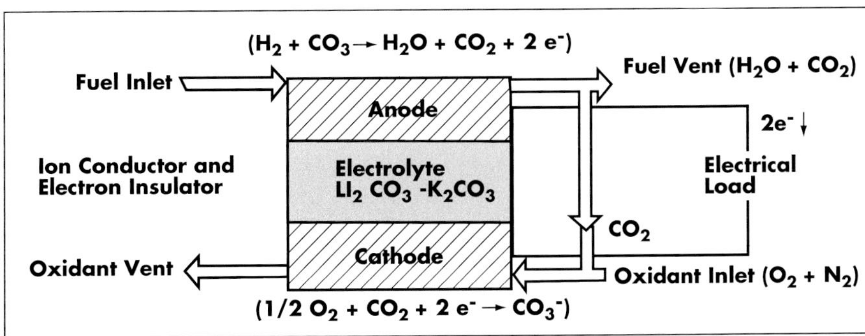

Fig. 14-63 Schematic of MCFC Chemical Processes. Source: GRI

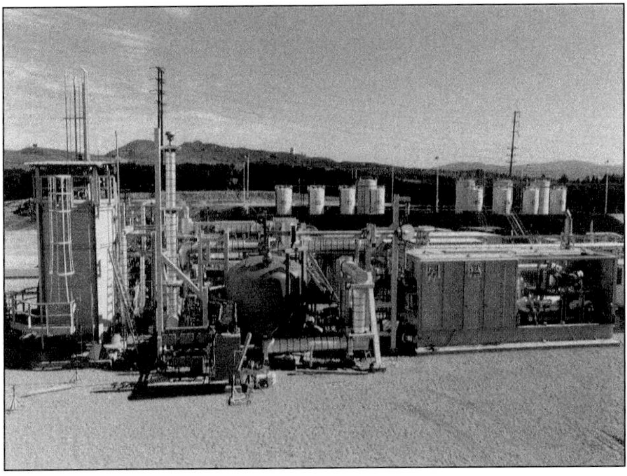

Fig. 14-64 Field Installation of 250 kW MCFC Demonstration Project. Source: Stewart and Stevenson

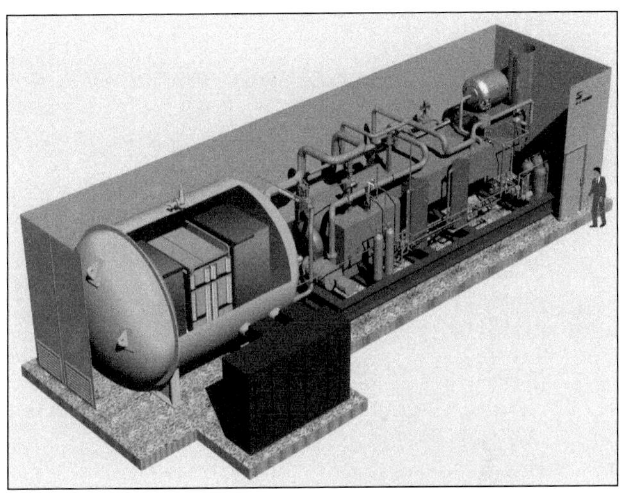

Fig. 14-65 Cutaway Illustration of Design Concept for Larger Capacity MCFC Module. Source: Stewart and Stevenson

Developing Fuel Cell Technologies

While the current generation of fuel cells struggles for market penetration, new technologies and advancements of current technology remain at the forefront of the industry. This includes the use of different electrolyte materials and new fuel cell designs.

Alkaline fuel cells have long been used for space applications. These cells are said to be capable of electric power generation efficiencies of up to 70%. They use alkaline potassium hydroxide as the electrolyte. Until recently they were consid-

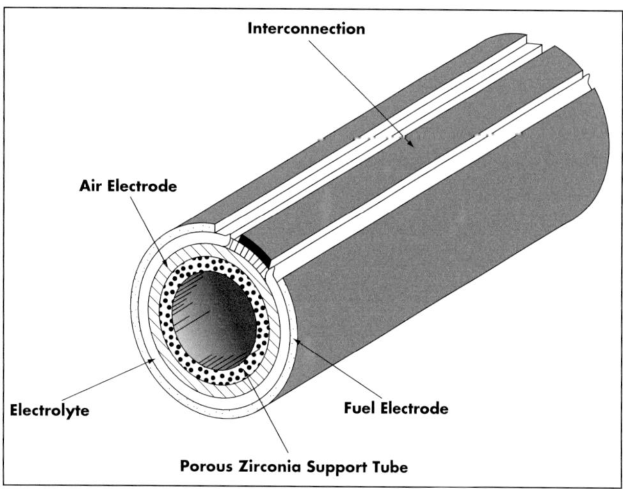

Fig. 14-66 Illustration of SOFC of Tubular Design. Source: Westinghouse

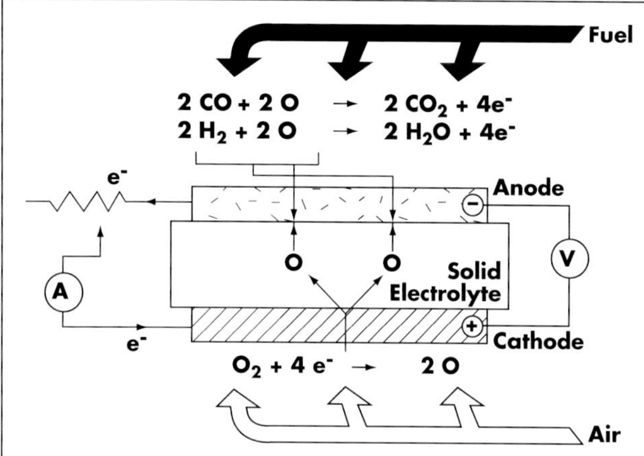

Fig. 14-67 Illustration of SOFC Operating Principal. Source: GRI

ered too costly for non-space applications. However, their proven application reliability and high thermal efficiency have drawn attention for development of stationary and vehicular land applications.

Proton Exchange Membrane (PEM) fuel cells operate at relatively low temperatures of about 200°F (93°C) and exhibit several times the power density of other fuel cell designs. They can vary their output quickly, making them suitable for vehicular and dedicated-load application. The proton exchange membrane is a plastic sheet that allows hydrogen ions to pass through it. The membrane is coated on both sides with highly dispersed metal alloy catalyst particles (mostly platinum). Hydrogen is fed to the anode side of the cell where the catalyst facilitates the release of electrons, as hydrogen atoms become ions (i.e., protons). The electrons in the form of electric current can be utilized before they return to the cathode side of the fuel cell where oxygen has been fed. Simultaneously, the protons diffuse through the membrane to the cathode side, where the hydrogen atom is recombined and reacted with oxygen to produce water, thus completing the process.

Direct Methanol fuel cells are similar to PEM cells in that they use a polymer membrane as the electrolyte. In these cells, the anode catalyst itself draws hydrogen from the liquid methanol, eliminating the need for a fuel reformer. Power generation efficiencies of up to 40% are targeted with operation at between 120 to 190°F (49 to 88°C). Higher conversion efficiencies can be achieved at elevated operating temperatures.

Regenerative fuel cells, a design currently under development, feature a closed-loop design, in which water is separated into hydrogen and oxygen by a solar-powered electrolyser. The hydrogen and oxygen are then fed into the fuel cell to generate the standard outputs of electricity, heat and water. The water is then recirculated back to the solar-powered electrolyser and the process is again initiated.

Direct Hydrogen Use

Several fuel cell field applications use hydrogen as their direct fuel source without an integral reformer to convert fossil fuel (e.g., natural gas) feedstock into hydrogen-rich gas. This process can be accomplished elsewhere, with hydrogen gas being delivered for direct use in the fuel cell. This can completely eliminate carbon dioxide emissions at the point of use, making it attractive for vehicular operation in urban areas. For example, a PEM fuel cell, powered directly by hydrogen stored on board as a compressed gas, has been successfully applied in a city transit bus in Vancouver. The very high power density makes it compact and light enough for practical vehicular use. Hydrogen's relatively low energy density by volume and the fact that it is lighter than air, somewhat complicates storage. An alternative under consideration is the use of hydrogen-rich fuel feedstock such as methanol with conversion occurring on board.

In addition to reforming, hydrogen can also be produced by electrolysis of water, with the electricity derived from conventional or other renewable sources. Experimental methods to produce hydrogen include photoelectrolysis, which uses sunlight to split the water molecules, and other reforming processes using various types of biomass as feeder fuels. Hydrogen produced as a by-product of industrial process, is an attractive source in that it requires no additional energy input and produces no additional carbon dioxide. In such applications, strict control of hydrogen purity is a critical element.

Costs and Market Penetration Issues

Similar to PVs, the largest barrier to wide-scale market penetration for fuel cells is high capital cost requirement. Unproven long-term reliability of most designs and relatively high life-cycle maintenance, repair and overhaul costs, notably the cost of stack replacement, are also significant obstacles. With power generation efficiencies that are currently comparable to conventional prime movers applied, either as simple or cogeneration cycles, and installed costs that can be double or even higher per unit of capacity, project economic performance will likely be inferior to conventional prime mover technology applications. Mitigating conditions include availability of financial incentives from utilities and federal and state agencies (notably the DoE) and unique application conditions under which fuel cells can provide significant collateral value to a facility.

Fuel cells provide the advantages of high-quality power output, quiet operation and very low environmentally harmful air emissions. Modular configurations and non-combustion/low emissions characteristics make siting and permitting much easier than fuel-burning plants.

Fuel cells also offer the potential for very high fuel efficiency in electric generation since power is generated directly and not subject to Carnot cycle limitations. High conversion efficiencies can also be obtained in smaller capacity modular units, where thermal efficiency is lower in smaller capacity conventional prime movers. While still not proven over time, fuel cells do have the potential for very high reliability, low routine maintenance, unattended operation and long periods between outages. Stack replacement periods (which are somewhat analogous to time between overhaul for conventional prime movers) continues to be extended with technology advances, but will likely remain a significant life-cycle cost.

While fuel cells already have a somewhat established niche market, wider spread market penetration is expected to increase over time, largely from decreased capital costs and increased electric conversion efficiencies. Capital cost reductions are expected to result from technology advancements and increased production volumes. Conversion efficiencies are expected to continue to increase with the potential to reach a level that is nearly double the simple cycle thermal efficiency of most conventional prime movers. Hence, despite limited current market penetration, the potential for electric generation efficiencies of as high as 70%, with relatively benign air emissions characteristics, may make fuel cells a compelling attractive near-term technology.

Continuing Fuel Cell Developments

Since the publication of the first edition of this book, fuel cells have continued to hold much promise and see much R&D attention. As such they remain a compelling attractive technology option. However, they still have not achieved significant market penetration. As the projected costs for high-volume manufacturing of automotive-type fuel cells has come down significantly, from $275/kW in 2002 to $61/kW in 2009, most of the emphasis has been on the transportation sector. This may also be an indication that fuel cells for power generation could one day become competitive with conventional sources.

Some of the progress has been a result of promotion by the U.S. Department of Energy. The Department's Efficiency and Renewable Energy division has been active in promoting research and development, through its Fuel Cell Technologies Program, supporting activities that have produced the following results:

- Doubled the durability of fuel cell systems in vehicles operating under real-world conditions
- Demonstrated fuel cell membrane electrode assembly durability of more than 7,300 hours in single-cell laboratory tests, exceeding DoE's 2015 target of 5,000 hours
- Reduced hydrogen production costs from both renewable resources and natural gas (hydrogen can now be produced by distributed reforming of natural gas at a projected high-volume cost of $3.00/gallon gasoline equivalent)
- Identified several new materials that show 50 percent improvement in on-board hydrogen storage capacity
- Established the International Partnership for the Hydrogen Economy (IPHE) among 16 countries and the European Commission to foster international cooperation on R&D, common codes and standards, and information sharing on infrastructure development through 30 collaborative projects

Commercially Availabale Fuel Cells

A decade ago, there were only a handful of fuel cells on the market. In 2010, the U.S. Fuel Cell Council lists 47 models as commercially available. However, most of these are designed for transportation, small portable power generation applications, or battery replacement. Of the 47 models, only 18, shown in Table 14-4 are configured for stationary power generation, either as emergency backup or as a primary power source. Moreover, many of these are small units, appropriate only as power sources for residential or small commercial applications. Still, the product list is expanding, and the future may see more penetration of this technology if the hydrogen infrastructure to support the market is developed. However, while this technology continues to see significant R&D attention and holds promise as a renewable energy source, wide spread use of fuel cells for utility-scale power production, or even large facility central power generation using stationary fuel cells is still not yet commercially viable, due to high first costs and ongoing OM&R costs.

Manufacturer	Product	Application	Type	Size
Ballard	Mark 1030	Cogeneration	PEM	1.3 kW
	Mark 1020 ACS	Back-up power / Light Mobility	PEM	300 - 5000 W
Fuel Cell Energy	DFC 300MA	Stationary	MCFC	300 kW
	DFC 1500MA	Stationary	MCFC	1200 kW
	DFC 3000	Stationary	MCFC	2400 kW
Hydrogenics	HyPM XR Power Modules	Stationary	PEM	4, 8, 12 kW
	HyPM XR (DC)	Stationary	PEM	4, 8, 16 kW
	FCXR Cabinet	Stationary	PEM	10, 20, 30 kW
	FCXR System	Stationary	PEM	150 kW
IdaTech	ElectraGen™ 3XTR	Backup	PEM - Liquid Fuel	3 kW
	ElectraGen™ 5XTR	Backup	PEM - Liquid Fuel	5 kW
	ElectraGen™ 3 XTi	Backup	PEM - Liquid Fuel	3 kW
	ElectraGen™ 5 XTi	Backup	PEM - Liquid Fuel	5 kW
Nuvera	Forza Industrial Power	Large Industrial / Stationary	PEM	250 kW
Relion	GenCore® 5T Series	Backup - Telecom	PEM	5 kW
	GenCore® 5U Series	Backup - Utilities	PEM	5 kW
UTC Power	PureCell® System Model 400	Stationary	PAFC	400 kW
	PureCell® System Model 5	Backup	PEM	5 kW

Table 14-4 Commercially Available Fuel Cells. Source: U.S. Fuel Cell Council

PART 2
Operating Environment

Section IV

Environmental Considerations

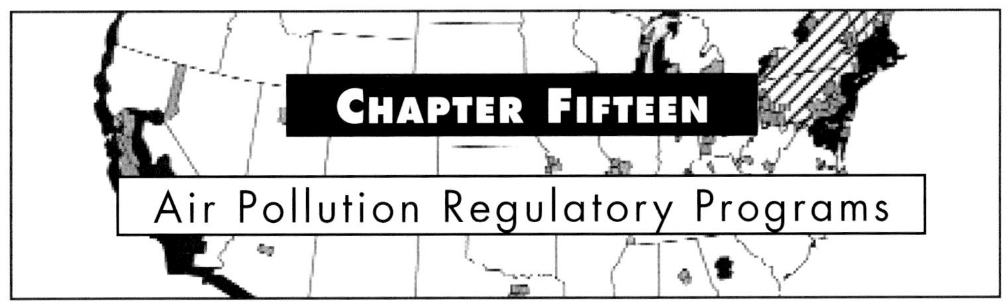

United States air pollution regulatory programs are designed to protect public health and prevent significant environmental damage. The **Clean Air Act (the Act)** first passed in 1970 and the **Clean Air Act Amendments of 1990 (CAAA)** are the main drivers behind most air pollution programs today. This chapter focuses on the Act, the CAAA, and the specific pollution control programs within them, which are of special importance to the technologies in this book. In addition, there is a discussion of emerging programs to address climate change at the end of this chapter.

CLEAN AIR ACT AND AMENDMENTS

The Act establishes the nation's air regulatory priorities and a basic regulatory framework. Although the Act has been amended several times over the years, the CAAA significantly expanded its scope to include control of acid rain, ozone-depleting chemicals (ODCs), and certain toxic air pollutants. Further, the CAAA established an operating permit program for large stationary sources. The CAAA also expanded requirements for the control of criteria pollutants, which are addressed through the National Ambient Air Quality Standards (NAAQS) including emissions standards for motor vehicles and combustion of transportation fuels.

The CAAA introduced the approach of allowing states and emissions sources to determine the best and most cost-effective emissions control options for their specific circumstances, including the use of clean fuels and alternative control measures such as emissions trading. This flexibility is intended to provide sources with opportunities to minimize compliance costs, and thereby encourage full and perhaps early compliance with environmental standards.

Finally, the CAAA lays out ambitious timelines for achieving these goals.

Since 1990 there have been minor changes to the ACT. Today the CAAA is organized into six titles as follows:

Title I—Pollution Prevention and Control
- Part A: Air quality and emission limitations
  - Criteria for establishing the national primary and secondary ambient air quality standards (NAAQS)
  - State Implementation Plan (SIP) requirements
  - Standards of performance for new stationary sources
  - Hazardous air pollutant standards
  - Federal enforcement
- Part B – Ozone Protection (this has been replaced by Title VI)
- Part C – Prevention of significant deterioration (PSD) of air quality
  - Clean air
  - Visibility protection
- Part D – Plan requirements for nonattainment areas
  - General provisions
  - Ozone nonattainment areas
  - Carbon monoxide nonattainment areas
  - Particulate matter nonattainment areas
  - Nonattainment areas for sulfur oxides, nitrogen dioxide, or lead
  - Savings provisions

Title II—Emission Standards for Moving Sources
- Part A – Motor vehicle emission and fuel standards
- Part B – Aircraft emission standards
- Part C – Clean fuel vehicles

Title III—General Provisions

Title IV—Noise Pollution

Title IV-A—Acid Deposition Control
- Mandates phased reductions of electric utility SO_2 and NO_X emissions nationwide
- Establishes a national emissions allowance trading program for SO_2

Title V—Permits
- Establishes a federal operating permit program to be implemented by states with federal oversight

Title VI—Stratospheric Ozone Protection
- Identifies stratospheric ozone depleting chemicals, such as CFCs and halons

- Establishes a schedule for phasing out the manufacture and import of these materials

The CAAA requires each state to develop a State Implementation Plan, or SIP, that lays out its plan for implementing the CAAA requirements. SIP development involves a lengthy planning process that results in a blueprint for reaching the air quality standards. It establishes milestones that can monitor a state's progress toward or compliance with the Act.

Specific Air Pollution Control Programs in the CAAA

Title I: Air Quality and Emissions Limits

Achieving the NAAQS across the nation is a major focus under Title I. The NAAQS set allowable limits for the presence of common pollutants in the ambient air. Currently, there are six identified pollutants, often called "criteria pollutants" because health-based criteria are used to set the standards. They are particle pollution (particulate matter including both PM-10 and PM 2.5), ground level ozone (including the precursors NO_X and VOC), NO_X, SO_2, CO and lead. The standards are developed in two phases: primary standards are set to protect human health and secondary standards are set to protect environmental health.

Under the Act, EPA must designate all areas that do not meet the NAAQS as nonattainment areas. States with nonattainment areas must develop State Implementation Plans (SIPs) that include the following:

- Certain mandatory limits affecting existing sources.
- Comprehensive inventories of actual emissions, and assessment of the impact of planned controls.
- Rules for permits for new and modified major stationary sources.
- Enforceable emissions limits and control measures (e.g., fees, marketable permits) and timetables for attainment of the NAAQS.
- Contingency measures if areas fail to make "reasonable further progress" (RFP) or attain the NAAQS on time.

If a state does not meet its attainment or progress requirements, the EPA can intercede and impose sanctions, such as reducing or eliminating federal highway funds or requiring even more stringent emissions requirements for new source construction or modification. Furthermore, if a state fails to develop a SIP that meets the requirements of the law, the EPA is required to intercede by developing its own federal implementation plan (FIP).

A FIP would include all the provisions that would be required in a SIP; however, the EPA would have the responsibility of developing the necessary regulations and enforcing the FIP in the state once it is final. Three air districts in California, for example, have had FIPs proposed to address the failure of the air districts to fulfill their obligations. The EPA maintains a current list of nonattainment areas for ozone, PM, CO, SO_2 and lead by county, for each state.

Ozone Nonattainment

Ground level ozone, otherwise known as smog, is considered to be the most widespread urban air pollution problem today. It is a highly reactive form of oxygen (O_3) that results from a complex set of photochemical reactions that occur when favorable conditions exist in the ambient air. These conditions are based on factors such as temperature and the presence of ozone precursors such as NO_X and VOC.

Ozone nonattainment areas are classified based on the severity of the problem. Each classification triggers various requirements that the area must comply with to have an EPA-approved SIP. The ozone standards were revised in 1997 and then EPA proposed further strengthening in 2008. The new 2008 standard met with controversy and as of January 2010, EPA had further extended the deadline for determining nonattainment with the 2008 standard. Because the level of stringency for the most severe areas is significantly greater than that for the less affected areas, the timelines for attainment are adjusted by attainment status level. As shown in Table 15-1, typically those areas with the most extensive requirements have longer to meet them, however, they must demonstrate each year that they are making reasonable further progress toward meeting the standards. Similar programs exist for areas that fail to meet the NAAQS for carbon monoxide and particulate matter.

As a state achieves the NAAQS standard, it submits a petition to EPA requesting redesignation.

Mandatory Measures for Ozone Nonattainment Areas

Emissions come from a variety of sources, but are generally categorized as mobile (from a vehicle), stationary (from a discrete non-moving source), or biogenic (naturally occurring emissions). Stationary sources include major, minor and non-point sources. **Major sources** are typically the focus of regulation because they are large,

Classification	Maximum Attainment Dates
Extreme	June 2024
Severe 17	June 2021
Severe 15	June 2019
Serious	June 2013
Moderate	June 2010
Marginal	June 2007
Subpart 1	June 2009
Subpart 1 Overwhelming Transport	June 2009

Table 15-1. Design Values and Attainment Dates for Ozone Nonattainment Area Classifications. Source: EPA

easily identifiable, and can be measured and monitored. Pollution control at these sources is often considered to be cost-effective in comparison to the cost of regulating a multitude of smaller, less identifiable sources that cannot be measured or monitored. Accordingly, under the Act, "major" sources of precursors trigger retrofit requirements and specific permitting rules. Mobile sources, however, represent an increasing portion of the inventory of emissions and are becoming a focus of renewed attention from regulators.

The Act defines major sources based on the **potential to emit (PTE)**. The PTE is considered to be the amount of pollution a source could generate if it were run all the time without any pollution controls. This term is described in more detail in the next chapter. The Act lowers the threshold for definition as a major source in the more severe nonattainment areas. For example, the threshold in Extreme nonattainment areas is the potential to emit 10 tons per year (tpy) of NO_X or VOC. In Marginal and Moderate nonattainment areas, the threshold remains the same at a PTE greater than 100 tpy of VOC and/or NO_X. The Act also defines modifications to a source as major if they result in a net increase in emissions that exceed specific thresholds for each nonattainment classification.

The Act requires that new major stationary sources and major modifications at existing major stationary sources of NO_X and VOC be offset by reductions either from existing sources or from offsets purchased from another source within the same nonattainment area. The offset ratio varies based on the area's nonattainment classification.

For example, a facility with a PTE of 100 tpy that is trying to locate or make a major modification in an area classified as Extreme must reduce or purchase 150 tpy of pollution offsets before it can locate the new facility or make the modifications.

VOC Controls

VOCs contain combinations of carbon and hydrogen and may also include traces of oxygen, sulfur, nitrogen, and halogens, such as fluorine and chlorine (methane, ethane, and few other compounds are excluded from the definition of VOCs). VOC emissions result from incomplete combustion and vaporization from substances such as paint, fuel, solvents and other coatings that use an organic carrier. Typically, major sources of VOCs include synthetic chemical manufacturers, coaters, and other industrial processes. Non-major stationary sources include gasoline service stations, degreasers, and consumer commercial solvents users.

The Act developed a concept known as **Reasonable Available Control Technology (RACT)** that called on all major sources and other sources subject to EPA **Control Technology Guidelines (CTGs)** to implement some reasonable level of emissions control. Since that time, the RACT standards have been ratcheted down to reflect changes in pollution control cost-effectiveness and revised health standards.

The RACT standards in the Act now include more than one dozen additional source categories to the list of 28 CTG source categories established prior to the CAAA. These guidelines define the RACT level that must be applied to existing facilities emitting certain threshold quantities of VOC emissions.

CTGs contain information on both the economic and technological feasibility of available control techniques. The CTGs describe the emissions source category and cover options for emissions reduction. In most cases, the CTG considers add-on control equipment, material substitution, process changes, and good housekeeping procedures. Most SIPs define RACT for CTG-affected facilities as the level of control specified in the CTG. All CTG-affected sources must comply with CTG regulation. In addition, major sources that are not covered by CTGs but are in a nonattainment area are required to implement a RACT standard that was negotiated by the state, with EPA oversight, through the SIP process.

The CAAA required that individual states amend their SIPs to require RACT for major VOC sources in Moderate nonattainment or higher designated areas or in the Ozone Transport Region. States are also required to go beyond RACT standards by requiring additional controls on VOC sources (and requiring controls on smaller sources) if necessary to meet attainment targets.

NO_X Controls

NO_X emissions, generally in the form NO_2, are generated as a byproduct of fossil fuel combustion. NO_X control is a key focus of the CAAA. Standards for new major sources and major modifications at existing sources are extremely stringent in all areas, including those designated as "in attainment." Major sources of NO_X emissions in ozone nonattainment areas with a Moderate or higher classification or in the Ozone Transport Region were required to achieve RACT by May 31, 1995.

Both atmospheric conditions and the ratio of VOC and NO_X concentrations affect ozone formation. In some areas, airshed modeling shows that NO_X reductions actually enhance the formation of ozone rather than diminish it.

In areas in which NO_X contributes to ozone formation, states must adopt NO_X control rules, similar to VOC rules, for major sources of NO_X. There are two types of rules: one set is aimed at new sources and the other is aimed at existing sources. New sources get controlled through New Source Review, while existing sources are controlled through RACT.

For most nonattainment areas, primarily Severe or Serious areas, NO_X reductions beyond RACT will be necessary to achieve the ozone standard throughout the region. Further, because ozone forms in the ambient air, it can travel via air currents for as much as 500 miles and/or three weeks after it is formed. As a result, areas such as Maine with very few sources of ozone precursors, are known to exceed the ozone standard with alarming frequency. To address this, the CAAA established the Northeast Ozone Transport Region (OTR), which extends from Northern Virginia to Maine along the eastern seaboard, as a means of developing a regional solution to the ozone problem. The CAAA allows the EPA to establish other OTRs if necessary.

To facilitate these reductions, EPA established the NO_X Budget Trading Program (NBP) in 1998. This program provides a cap and trade mechanism to reduce NO_X emissions from power plants and other large stationary sources throughout the ozone regional transport regions. This program is being used by all 20 eastern states impacted by ozone transport. Under terms of the agreement, the overall budget has been set as a percentage of the 1990 baseline emissions. Each state is allocating its budget as it sees fit. Typically, budget sources include utilities and large industrial facilities that generate at least 25 MW of power. All sources must hold enough allowances to cover their NO_X emissions from May through September.

In many regions, mobile sources are considered to be part of a facility's emissions inventory. While this adds to total emissions, it also provides another means of reductions that can be used as offsets for other facility emissions.

CO Nonattainment

Carbon monoxide (CO) is the product of incomplete combustion of carbon-based fuel, such as gasoline. CO impairs the oxygen-carrying capacity of blood. The NAAQS is designed to protect against any harmful effect of CO. Roughly three quarters of the nation's CO emissions come from mobile and small area sources such as home heating equipment and engines.

EPA established a one-hour standard of 35 ppm and an eight-hour standard of 9 ppm. As with ozone, the classification is used to set the attainment deadline and certain mandatory control measures. As of 2010, the only area designated as nonattainment for CO is Las Vegas, Nevada.

Control technologies for stationary sources include improved combustion control and capture systems that reuse CO-rich exhaust gases as fuel and afterburners or oxidation catalysts to destroy CO in exhaust streams. An effective means of reducing CO is switching to cleaner fuels, notably natural gas for stationary sources. Motor vehicle controls are similar to those for VOC as both are the product of incomplete combustion. Switching to alternative fuels and gasoline containing oxygen, as well as adopting more stringent automobile inspection and maintenance (I&M) programs, are key CO control measures.

Particulate Nonattainment

Particulate matter (PM) are small substances that are light enough to become airborne. They include smoke from burning coal, automobile exhaust, dust from industrial processes, and natural particles such as pollen. Particles are also formed in the air through reactions of NO_X, SO_X, VOC and ozone. It can contain metals, nitrates, sulfates, chlorides, fluorides, hydrocarbons, carbon silicates, and oxides. It can be corrosive, toxic to plants and animals, and harmful to humans. At high concentrations, PM can impair respiratory functions, reduce visibility, and discolor the air.

PM is categorized according to its size. The finer the particle, the more damaging it is to human health. Generally, EPA is concerned about those particles that are smaller than 10 micrometers in diameter. Particles are that are sized between 2.5 micrometers and 10 micrometers are known as "inhalable coarse particles" or PM-10 and can be commonly found near roadways and industrial operations. Particles that are 2.5 micrometers or smaller are known as "fine particles" or PM-2.5 and can be emitted directly from sources such emission stacks and automobiles, or they can be formed by reactions in the air.

The primary and secondary NAAQS for both PM-10 and PM-2.5 are the same. For PM-10, the standard is set at 150 micrograms per cubic meter over a 24-hour timeframe. The EPA established the PM-2.5 standards in 1997 and strengthened them in 2006, when a legal challenge ensued. For PM-2.5 there is an annual (arithmetic mean) standards set at 15 micrograms per cubic meter, and 24-hour standard set at 35 micrograms per cubic meter. These two approaches to the measurement yield different non-attainment results.

As of 2010, there were roughly 45 areas designated as nonattainment for PM-10. There are roughly 31 areas designated as nonattainment for the 2006 PM-2.5 standard and 39 areas designated nonattainment for the 1997 PM-2.5 standard.

Each SIP must require the implementation of control measures necessary to attain the PM-10 and PM-2.5 standards. These can include a combination of measures addressing stack emissions of PM-10, or fugitive dust. The Act also allows the EPA to address precursors, such as NO_X, VOC, and SO_2 in SIPs for nonattainment areas. Sources emitting precursors may also be subject to controls even though they are not directly subject to PM-10 or PM-2.5 regulation.

There are various control options used for limiting PM-10 and PM-2.5 from point sources, such as electrostatic precipitators for coal plants and switching to natural gas (which produces about one-tenth the PM emissions of fuel oil and less than one-hundredth of the emissions from uncontrolled burning of coal). As a result, add-on PM-10 controls are not used with gas firing.

Hazardous Air Pollutants

The Act established the National Emissions Standards for Hazardous Air Pollutants (NESHAPs) to regulate air pollutants designated by the EPA as hazardous. The EPA was directed to identify hazardous air pollutants (HAPs) that might have health effects, and then establish control standards for selected significant sources of these pollutants, restricting emissions to a level that produces no adverse health effects. This approach, however, was difficult to implement. Scientific controversy surrounding the identification of a "safe" exposure level made it very difficult for the EPA to write its regulations. Between 1970 and 1990, the EPA was only able to establish NESHAPs for certain sources of radon, beryllium, mercury, vinyl chloride, radionuclides, benzene, asbestos, and arsenic. Congress developed a new approach in the CAAA and shifted the regulatory emphasis from a risk-based approach to a technology-based approach for controlling air toxics.

Under the CAAA, technology-based control requirements are to be established to restrict the emissions of certain HAPs from certain categories of stationary sources. Only after these technical requirements are set will the EPA consider the significance of any remaining health risk.

The EPA is required to establish MACT standards for selected categories of industrial facilities that emit the listed HAP. The EPA identified the source categories since the Act passed. Table 15-2 lists some of the source categories and their subcategories that are relevant to fuel users and producers.

The MACT standards will be promulgated through successive regulations over ten years. The first standards were to have been promulgated for 40 source categories by November 1992, but some were delayed. The EPA did complete the Hazardous Organic NESHAP (HON), which covers many sources in the synthetic organic chemicals manufacturing industry, as well as a NESHAP for coke oven batteries. Standards have also been developed for HAP for benzene emissions. The development of these standards began before passage of the CAAA, but it has been drawn into the HAP process.

According to the EPA's final MACT promulgation schedule, which was issued on December 3, 1993, the next group of MACT standards to be set were to cover mostly dry cleaning operations. Additional groups of standards

Source Categories	MACT Schedule Date
Petroleum and Natural Gas Production and Refining:	
Oil and Natural Gas Production	11/15/1997
Petroleum Refineries	11/15/1997
Fuel Combustion:	
Engine Test Facilities	11/15/2000
Industrial Boilers	11/15/2000
Institutional/Commercial Boilers	11/15/2000
Process Heaters	11/15/2000
Stationary Internal Combustion Engines	11/15/2000

Table 15-2. Example MACT Categories Relating to Fuel Production & Use. Source: 57 Federal Register 44147, September 24, 1992

were to be adopted by a specified timetable, and all standards were to be promulgated by 2000. After each standard is promulgated, affected new sources must comply immediately and existing sources must achieve compliance within three years. In addition, MACT requirements can be triggered earlier by new major HAP sources under state/federal operating permitting programs. Lastly, if the EPA fails to meet the regulatory schedules, states must set MACT limits in the EPA's place. Should the states start developing MACT standards, they must work closely with the EPA to ensure that the state standard will not be invalidated by the EPA when it establishes its own MACT standards.

MACT Standards

The Act establishes criteria for setting technology-based MACT emissions limits. The MACT standards are based on the average emissions limitation achieved by the best performing 12% of existing sources in a source category or subcategory. If there are fewer than 30 sources in a category, the standard can be based on the average emissions limit achieved by the best performing five sources.

The Act also lists a number of control options that must be considered when establishing MACT, including:

- Reduction of emissions of pollutants through process changes, substitution of material, or other modifications
- Enclosure of systems or processes to eliminate emissions
- Collection, capture, or treatment of pollutants when released
- The design of equipment and operational standards

The Act provides incentives for sources to make early reductions in HAP emissions before the MACT standards are established. An existing source may obtain a six-year extension from compliance with MACT if it achieves a 90% reduction in emissions (95% if the emissions are particulates) prior to the proposal of the applicable MACT standards. Facilities will have to consider technical, regulatory, and economic factors before deciding to undertake the early reduction option.

Accidental Release Requirements

In addition to the MACT programs, the Act requires the EPA to develop regulations to address the potential accidental release of hazardous substances. The goal of these regulations is to focus on chemicals that may pose a significant hazard to the public should an accident occur, to prevent their accidental release, and to minimize the consequences of such releases. In October 1993, the EPA proposed a rule that would require the development and implementation of risk management plans (RMP) by sources that process or store more than a designated amount of a regulated substance. The RMP would include a program to prevent accidental releases and a response program to address accidental releases if they occur. The rule was finalized on August 19, 1996.

In January 1993, the EPA finalized the list of substances of concern and threshold quantities that will determine whether a facility must comply with the accidental release requirements. The EPA's final list includes 100 "acutely toxic" chemicals and 62 "flammable" gases and liquids. This rule addresses only the list of regulated substances and threshold quantities; the regulations for prevention and detection of accidental releases, including the finalization of requirements for the development of risk management plans, will be published at a later date.

State Air Toxics Programs

In the 1980s, when the EPA was making little progress in controlling air toxics, several states moved ahead in developing their own air toxics control programs. Some state programs are well established, while others are still under development or are very simple. Most state programs focus on new sources and address a list of chemicals similar to, but different from, the HAP list. Technology standards similar to MACT and/or ambient air impact limits are typically required. For ambient impact limits, the source must be controlled sufficiently so that modeled downwind air quality levels do not exceed the state's ambient limits for each toxic.

Title II: Mobile Source Regulations

Automotive emissions account for almost one half the emissions of VOCs and NO_X and up to 90% of the CO emissions in urban areas. Title II includes a number of significant amendments to the EPA's existing authority to regulate motor vehicle emissions and motor vehicle fuels, as well as several new programs designed to supplant the new nonattainment and air toxic programs:

- The EPA set more stringent tail pipe standards for hydrocarbons, CO, NO_X, and particulates to be phased in during the 1990s. The EPA is required to study whether even tighter Tier II standards are needed, technologically feasible, and economical. If the EPA determined by 1999 that lower standards are warranted, the standards will be cut in half beginning with 2004 model year vehicles.

- Stringent control of "evaporative emissions" were to be implemented by the mid-1990s. Devices that trap gasoline vapors from the engine and fuel system had to be improved. In addition, systems to capture gasoline vapors during refueling are required at service stations in Moderate or worse ozone nonattainment areas.
- The establishment of quality standards for fuel used for motor vehicle reformulated gasoline.
- The CAAA also calls for several programs to promote the introduction of clean fuels other than gasoline and diesel. Several of these programs may apply to businesses as opposed simply to car owners:
  - Fleets (ten or more vehicles in 22 urban areas in those Serious or worse areas for ozone) must purchase vehicles meeting tight emissions limits that may only be met by alternative fuels, although alternatives are not required. The phase-in began with 30% of the new fleet purchases of passenger automobiles and light vans by 1998 and increased to 50% in 1999 and 70% in 2000. For heavy-duty vehicles, the phase-in requirement is a constant 50% of new purchases starting in 1998. A fleet emission-trading program must also be put into effect within each nonattainment area.
  - A clean-fuel vehicle pilot program had to be established in California (the state with the greatest smog problem), requiring annual production and sale of 150,000 clean-fuel vehicles beginning in 1996 and increasing to 300,000 vehicles per year in 1999 and thereafter. This program differs from the overall program because credits are available to the manufacturers for producing more clean vehicles rather than to the fleet owners.

Title IV: Acid Rain Program

The **acid rain** program was developed to reduce acid deposition and its impact on public health and the environment. Acid rain occurs when SO_2 and NO_X are emitted into the atmosphere, combine with water molecules, and return to earth as acidic rain, fog, or snow. Acid rain is harmful to forests, lakes, and buildings. Electric generation plants have been targeted as the leading cause of acid rain.

At the time of the CAAA, approximately 20 million tpy of both SO_2 and NO_X were produced annually in this country, primarily from coal- and oil-fired electric generating plant. The CAAA established a two-phased goal that required a permanent 50% reduction (ten million tpy) in SO_2 and a two million tpy reduction of NO_X by year 2000. SO_2 emissions are regulated using a nationwide tradable "allowance" system. NO_X emissions are limited based on unit-specific achievable NO_X levels set by the EPA (this level sometimes is the same as the RACT level discussed under Title I). In 2009, the combined annual emissions were roughly 7.6 million tons, well below the established goals.

Specific exemptions from the acid deposition program requirements are provided for cogeneration facilities. If a cogeneration unit supplies less than one-third of its potential output, or less than 25 MW, to an electric utility, then it is exempt from the requirements of the program. A cogeneration unit is also exempt if there was a sales agreement between the unit and a utility in effect at the date of enactment of the CAAA, November 1990.

This aggressive program does not cover industrial facilities, but they may elect to opt-in to the SO_2 emissions market allowance under the Act. An industrial source can choose to have an emissions allowance allocation established based on its recent SO_2 emissions. If a source is then able to reduce its SO_2 emissions, it can sell its excess allowances to utilities. There is also a possibility that the EPA will, in the future, extend the program to industrial boilers that use a substantial portion of their total fuel use in order to produce electricity.

Control Levels and Schedules

SO_2 control is in two phases:

- Phase I of the acid rain program (from 1/1995-to-1/2000) requires those producing more than 100 MW (the highest pollution producing 110 power plants) to limit emissions to a level equivalent to 2.5 lbm of SO_2 per MMBtu of fuel input.
- Phase II (effective 1/2000) requires plants producing 25 MW or more to achieve an emissions level of 1.2 lbm SO_2 per MMBtu. Electric power plants must also install continuous emissions monitoring devices to verify compliance.

The law sets an indefinite long-term cap on SO_2 emissions at 8.9 million tpy (a 56% overall reduction) starting in 2000. There is, however, some flexibility. Facilities are allowed to choose the most practical and cost-effective option. They may install scrubbers, switch to low-sulfur fuels, shut down an older plant and replace it with a cleaner plant, or purchase credits for reductions achieved at other plants across the country.

SO_2 Allowance System

Rather than impose specific control requirements, the acid rain program establishes a phased-in "allowance system." SO_2 emissions from a source are limited to an allowance level. An allowance is defined as the authorization to emit one ton of SO_2. Generally, allowances will correspond to the two-phased emissions level requirements. The average fuel consumption rate for the years 1985 to 1987 was used to set the baseline.

Initially, each source is given a certain amount of pollution allowances associated with the baseline SO_2 product of its units. The source owners can buy and sell allowances, provided they have enough each year to cover these emissions. The system allows for temporary fluctuations in emissions levels during a given year as long as the total tons of SO_2 emitted during the year do not exceed the allowance level for the source. If the allowance is not met, the source is subject to a $2,000 per ton excess emissions fee (indexed over time) and is required to offset the excess emissions in the following year.

If a source cannot meet its allowance level, it must acquire allowances from other sources that are able to reduce emissions rates below their allowance level. The allowances are a free market commodity that is regulated by the EPA. They can be used, sold, bought, banked, or traded. The CAAA also provides that the EPA will auction off allowances to small generators that may be unable to obtain allowances on their own. The system allows the nation to reach its goals by providing incentives for over-controlling where possible, while allowing other sources to comply even if controlling emissions down to the required level is technically or economically unfeasible.

Control technologies for SO_2, such as flue gas desulfurization (scrubbers), can be very expensive, often as much as one-quarter of the cost of a new plant. An alternative, however, is the conversion to lower sulfur fuels, including lower sulfur coal, lower sulfur fuel oils, and natural gas. Conversion from coal to lower sulfur oil can reduce emissions by more than one half. Conversion to natural gas reduces SO_2 emissions to minuscule levels.

Since 1995, an additional 182 units opted into Phase I requirements as either substitution or compensating units, bringing the total participants to 445 units at 110 electric utility plants. The 445 units successfully met their compliance obligations for 1995. Emissions were reduced 39% below the allowable emissions level required by the CAAA. Nearly all Phase I units implemented the continuous emissions monitoring (CEM) and submitted CEM certification test results by the deadline.

NO_X Requirements

In addition to requiring reductions in SO_2 emissions, the acid deposition control program established a goal of reducing NO_X by two million tons from 1980 levels. The NO_X control program for acid deposition was implemented in two phases. Phase I applied to facilities with wall- or tangentially-fired boilers and required them to reduce NO_X emissions to 0.5 lbm/MMBtu or 0.45 lbm/MMBtu, respectively, by January 1, 1996. These requirements applied to approximately 170 units. Phase Two began in 2000; it added roughly 430 new Group I boilers and established Group 2 boilers including those using cell-burner technology, cyclone boilers, wet-bottom boilers, and other types of coal-fired boilers. By 2008, NO_X emissions were reduced by 5.1 million tons, which was more than two times the emission reduction goal.

Title VI: Stratospheric Ozone Protection

Title VI mandates a phase-out in the production of halogenated carbon compounds. These chemicals are used widely in fire protection, refrigeration, air conditioning, and industrial cleaning. The CAAA required the EPA to list all regulated substances and their ozone-depleting potential, atmospheric lifetimes, and global warming potential.

Pollutants and Schedules for Phase-out

In 1987, most industrialized countries agreed, under the **Montreal Protocol**, to revert to 1986 production and consumption levels of certain ozone-depleting substances. The Montreal Protocol was later amended to specify a phase-out schedule for the ozone-depleting chemicals that were subsequently accelerated.

The CAAA divided the ozone-depleting chemicals into two classes. Class I chemicals include specified **chlorofluorocarbons (CFCs)**, halons, 1,1,1 trichloroethane (or methyl chloroform), and carbon tetrachloride. Class II chemicals are mostly **hydrochlorofluorocarbons (HCFCs)**.

Class I Ozone-Depleting Substances

CFCs are the primary component of many common electric air conditioning refrigerants. These refrigerants are commonly referred to by their trade name and listed by trade number. CFC-based refrigerants, such as CFC-11 and CFC-12, were the leading refrigerants for most large systems. Because of their role in the destruction of stratospheric ozone, the EPA and the United Nations Environment Programme (UNEP) have restricted the use and production of CFCs. Venting of refrigerants to the environment became illegal in 1992. The final regulatory requirements limiting the production and consumption of ozone-depleting substances were issued by the EPA in December, 1993 (58 FR 65018 and 58 FR 69235). Consistent with the Montreal Protocol and other international treaties, the regulations required a complete phase-out of CFCs by 1996. Since that date, the only available CFCs are those that have been reclaimed from equipment that has either been retired or converted to a non-CFC refrigerant. Two major types of CFC replacements are HCFCs and hydro (HFCs).

HFCs differ from CFCs and HCFCs in that they contain no ozone depleting chlorine and have lower global warming potential (GWP) than CFCs. As a result, there are no current phase-out dates or production bans for HFCs, nor is one anticipated. The most prominent HFC, R-134a, was developed to replace CFC refrigerant R-12, which is used in equipment such as positive pressure centrifugal compressors, home refrigerators, and automobile air conditioners. HCFCs are addressed below in the discussion of Class II Ozone-Depleting Substances.

Halon, another Class I substance, is used in fire suppression and control equipment and was scheduled for phase-out by 1994. Halon production and importation in the United States ended on December 31 1993. In 1996, 1,1,1 trichloroethane or methyl chloroform, which was used extensively as a cleaning solvent, was phased out. Carbon tetrachloride, which is used primarily as a feedstock in the production of CFC-11 and CFC-12, was also phased-out in 1996.

Class II Ozone-Depleting Substances

HCFCs comprise most of the Class II substances and are used throughout the industry as substitutes for CFCs. Because HCFCs have hydrogen atoms included in their molecular structure, they have shorter atmospheric lifetimes and do less damage to the ozone than do CFCs. HCFC-22 is the most common refrigerant used in residential air conditioners. The CAAA freezes production and consumption of HCFC-22 by January 1, 2020. HCFC-123 has recently become a commonly used replacement for CFC-11. The law freezes productions of HCFCs by 2015 and eliminates their use entirely by 2030. HCFCs are referred to as bridge refrigerants because they will remain in use for several decades until current and newly developed alternatives take over the market.

Recycling and Labeling Requirements

The use of ozone-depleting substances will continue even though the manufacture of CFCs has stopped. Under the Act, the EPA must promulgate regulations that maximize the recapture and recycling of ozone-depleting substances. In 1992, the EPA issued final regulations on the servicing of motor vehicle air conditioners, which gave equipment standards for the mandatory recycling and/or reclamation of the ozone depleting substances.

The EPA was also required to promulgate rules to regulate the use and disposal of ozone-depleting substances used in appliances and industrial process refrigeration by 1992. The EPA was late in developing these rules, however, and finalized them in 1993. The regulations prohibit anyone who maintains, services, repairs, or disposes of appliances from venting or releasing substances such as refrigerants. The regulations also provide for a reduction in the use and emissions of CFCs and halons by maximizing the recycling of these substances.

The Act also establishes labeling requirements for containers of products containing CFCs. The labels must identify the particular substances and state that the substances harm public health and the environment by destroying stratospheric ozone.

Addressing Climate Change

Climate change has been at the center of divisive but prominent policy debate for more than two decades in the U.S. highlighted by the U.S. involvement in the UN's Framework Convention on Climate Change (UNFCCC) and the subsequent Kyoto Protocol. An important facet of this debate focuses on whether carbon dioxide can be considered a pollutant under the Act and hence should be – or can be – regulated by EPA.

EPA has taken several steps that begin to answer this question. In 2008 it released an Advanced Notice of Proposed Rulemaking (ANPR) regarding "Regulating Greenhouse Gas Emissions under the Clean Air Act." This document was developed as part of the response to a lawsuit brought by the state of Massachusetts in which the court found that the Act did authorize EPA to regulate tailpipe greenhouse gas emissions if the agency determines that they cause or contribute to air pollution that may endanger public health. In 2009, EPA signed two findings. First, that six greenhouse gases, including carbon dioxide, threaten the public health and welfare (an endangerment finding under Section 202(a) of the Act). Second, that the combined emissions from new motor vehicles and engines contributes to the greenhouse gas pollution that threatens the public health and welfare (a cause or contribute finding under section 202(a) of the Act). While these findings do not impose carbon dioxide emission reductions, they do support EPA's efforts to develop greenhouse gas emission standards for light-duty vehicles.

Further, in 2010 EPA promulgated the first part of a rule requiring all sources of greenhouse gas emissions that emit more than 25,000 tons per year to report their greenhouse gas emissions. Also in 2010, EPA issued the Final Tailoring Rule. This rule sets greenhouse gas emission thresholds that would trigger PSD or Title V Operating Permits for new and existing industrial sources. It is called a tailoring rule because it limits the program to certain stationary sources based on criteria that include the largest stationary sources in the country.

While EPA has been moving deliberately on climate change, many states have taken a lead role in developing policies and programs that will reduce emissions. More than half the states are involved in one of three major regional greenhouse gas cap and trade initiatives, the Western Climate Initiative, the Midwest Greenhouse Gas Reduction Accord, and the Regional Greenhouse Gas Initiative. In addition, Florida has adopted its own cap and trade initiative. Almost 30 states had adopted a renewable portfolio standard as of early 2010, setting requirements for the amount of power generated from renewable energy that must be delivered in the state. And more than 20 states had adopted statewide emission targets or goals by early 2010.

While there are no formal greenhouse gas reduction requirements facing large stationary sources today, the anticipation of such requirements in the future is making it imprudent to develop new facilities without including an assessment of potential future obligations. This kind of analysis is being increasingly done by potential sources of investment and the SEC is considering reporting rules to help investors consider the regulatory risk.

THE ROLE OF FEDERAL AND STATE/LOCAL GOVERNMENTS

Although the Act is a federal law, it is designed to be primarily implemented by state environmental regulators through SIPs, with the approval and oversight of the U.S. EPA. For many parts of the Act, particularly those involving NAAQS plans and permitting, the U.S. EPA provides regulatory and technical guidance to states as they adopt air pollution control rules. The EPA then approves and oversees the implementation of those rules by the state.

U.S. EPA Organization and Role

The U.S. EPA is divided into several groups, each with its own role in accomplishing the requirements of the Act. In general, EPA headquarters in Washington, DC, addresses broad policy development; it sets basic air pollution control goals and policy objectives. The staffs responsible for acid rain and ozone depletion are located in the Washington office.

The U.S. EPA has two other program offices, located in Ann Arbor, MI, and Durham, NC. The Durham office, often referred to as Research Triangle Park or RTP, issues specific guidance memos and regulations for air quality attainment and toxics programs, except those addressing mobile sources. RTP also creates and manages relevant air emissions control information, such as databases containing permitted information for facilities throughout the country. The U.S. EPA-Ann Arbor focuses on mobile source issues.

EPA Regional Offices

There are ten regional offices, each located within the geographical region that it manages (see Table 15-3). These offices are largely responsible for the day-to-day interaction with state and local agency officials. Each has the same basic responsibility to work with states in implementing the Act, but specific policies and level of involvement in a state's regulatory efforts vary from region to region. The EPA regional offices conduct detailed reviews of many state and local air regulatory decisions, such as permitting decisions for large sources, and they also assess the adequacy of rules adopted by the state to implement the Act.

State and Local Agency Roles

For the most part, states are required to meet or exceed federal environmental regulations. Each state has its own environmental agency that develops and implements pollution control requirements for that state through the development of a SIP.

SIPs contain all of the rules adopted by state and local agencies to implement the CAAA. To the extent that rules go beyond the requirements of the CAAA in terms of both

Regional Office and Location	States Within Region
EPA Region I Boston, MA	Connecticut, Maine, Massachusetts, New Hampshire, Rhode Island, Vermont
EPA Region II New York, NY	New Jersey, New York, Puerto Rico, Virgin Islands
EPA Region III Philadelphia, PA	Delaware, District of Columbia, Maryland, Pennsylvania, Virginia, West Virginia
EPA Region IV Atlanta, GA	Alabama, Florida, Georgia, Kentucky, Mississippi, North Carolina, South Carolina, Tennessee
EPA Region V Chicago, IL	Illinois, Indiana, Michigan, Minnesota, Ohio, Wisconsin
EPA Region VI Dallas, TX	Arkansas, Louisiana, New Mexico, Oklahoma, Texas
EPA Region VII Kansas City, KS	Iowa, Kansas, Missouri, Nebraska
EPA Region VIII Denver, CO	Colorado, Montana, North Dakota, South Dakota, Utah, Wyoming
EPA Region IX San Francisco, CA	Arizona, California, Hawaii, Nevada, Guam, American Samoa, Northern Mariana Islands, Micronesia, Marshall Islands, Palau
EPA Region X Seattle, WA	Alaska, Idaho, Oregon, Washington

Table 15-3 U.S. EPA Regional Offices

stringency and scope, these rules must also be included in the SIP. After state rules required by the Act are adopted, the state or local agency implements and enforces all rules, though the U.S. EPA approves and oversees the implementation of those rules. In most cases, the state staff is the primary contact for permit negotiations and all compliance issues.

In some states, there are also local agencies or regional offices that have varying degrees of authority in implementing air pollution rules. For example, California has divided its state into several districts, each with its own air quality management department that develops and enforces rules for its area. Some states have also established local agencies for large cities. The local agency, however, must still report to the state agency.

Multi-State Associations

Air pollution tends to be an airshed problem that knows no legal or geographic boundaries and, as such, is often a multi-state or regional problem. In order to address these concerns, multi-state or regional associations of state agencies have also been established to develop common and consistent strategies for air pollution control within a relevant geographic region. The Northeast Ozone Transport Region is an example of such an organization. Pollution problems affecting the Grand Canyon are handled by the Grand Canyon Western Visibility Transport Commission (GCWVTC).

In addition to the multi-state associations prescribed in the CAAA, other multi-state associations have formed. For example, the State and Territorial Air Pollution Program Administrators and the Association of Local Air Pollution Control Officials (STAPPA/ALAPCO) is the national association of state and local air quality control officials in the states and territories and over 165 metropolitan areas throughout the United States. It provides assistance and leadership for state efforts on certain regulatory requirements. Another example is the Northeast States for Coordinated Air Use Management (NESCAUM), a group composed of agency air officials from eight northeastern states in the OTR. These multi-state groups work to develop commonly needed technical information, such as airshed modeling or risk assessment, to develop regulatory approaches.

Sources For Additional Information

There are a number of sources that can provide additional current information on environmental legislation. These include:

- **Applicable State Regulations** — Copies of state regulations can be obtained by contacting the state environmental agency. Specific questions regarding the regulation's content can often be asked of state environmental staff.
- **Code of Federal Regulations (CFR)** — The CFR contains all federal regulations. Environmental regulations are contained in Volume 40 of the CFR, which is referred to as 40 CFR. Copies of CFRs can be ordered through various publication services. Additionally, local university libraries and government agencies may have copies on hand for use by the public.
- **EPA-Sponsored On-Line Information System** — EPA's Office of Air Quality Planning and Standards (OAQPS) has established a Technology Transfer Network (TTN), which is an on-line bulletin board containing information on air quality regulations and guidance.

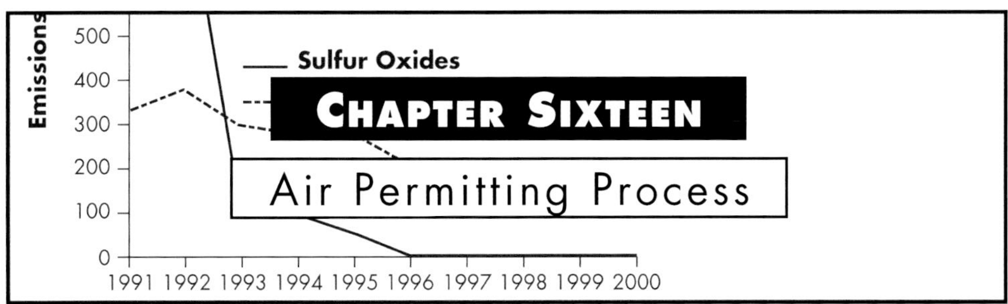

CHAPTER SIXTEEN

Air Permitting Process

Permitting is a fundamental requirement for most large stationary emissions sources. It is important to make sure that an emissions source receives the necessary permits for siting, construction/modification, and operation, because the penalties for failure to do so are severe. This chapter reviews the basic air-permitting requirements that apply to stationary sources and offers some permitting strategies. However, specific details in the permitting requirements vary by state and can change frequently. It is, therefore, necessary to check with the state environmental agency to find the most recent requirements that apply to sources in question.

There are two general types of permits under state and federal air regulations:

- **Permits to construct**, which allow the construction or modification of air pollution sources and are generally referred to as a new source or pre-construction permit.
- **Permits to operate**, which authorize the operation of air emissions sources.

Within each broad category, there are several types of reviews and approvals. A primary purpose of the pre-construction permitting process is to ensure that emissions control is consistent with the need to make steady progress toward achieving and maintaining federal and state air quality standards. Operating permits provide regulators with a vehicle for tracking compliance for the operation of existing equipment. Sources are categorized as either minor or major based, generally, on their potential to emit and face different permitting requirements accordingly.

Potential to Emit

Potential to emit (PTE) is the critical factor in determining permitting status as either a major or a minor source. PTE is the maximum emissions rate for a pollutant based on the maximum feasible operating rate of full capacity for 24 hours per day, 365 days per year. The basic formula for determining PTE is:

$$PTE = hr/day \times day/yr \times Firing\ rate_{Max} \times Emissions\ rate = Annual\ emissions \quad (16\text{-}1)$$

If, for example, the NO_X emissions rate is specified as 0.10 lbm NO_X/MMBtu and the firing rate is 60 MMBtu/h, the PTE NO_X, expressed in ton/yr (tpy), would be calculated, based on Equation 16-1, as:

$$PTE = 24\ hr/day \times 365\ day/yr \times 1\ ton/2{,}000\ lbm \times 60\ MMBtu/hr \times 0.10\ lbm\ NO_X/MMBtu = 28.9\ tons\ NO_X/yr$$

To express PTE in terms of daily output in lbm of NO_X, Equation 16-1 is modified to:

$$PTE = 24\ hr \times lbm\ NO_X/hr = lbm\ NO_X/day \quad (16\text{-}2)$$

A source wishing to limit its PTE may only do so by implementing emissions control technologies and/or adopting operational limits under a federally enforceable permit (e.g., a permit recognized by EPA as federally enforceable, and incorporated into an approved State Implementation Plan). The term "federally enforceable" means that the federal government has the ability to step in and enforce the permit or program. Some state permit programs either do not conform to federal programs or specifically do not recognize EPA's right to enforce the program and are therefore considered not to be federally enforceable.

PTE is almost always greater than the actual annual emissions rate, because a facility rarely operates at a 100% load factor on its full fuel-burning equipment capacity. A facility with fuel-burning equipment that operates at less than 100% load factor may reduce their recognized PTE by taking a **federally enforceable restriction (FER)**. By reducing their PTE below the major source threshold (see Chapter 15), the facility may avoid some of the permitting requirements. The FER typically requires a facility to install accurate fuel metering and maintain fuel use records.

PRE-CONSTRUCTION PERMITTING OVERVIEW

Pre-construction permitting is generally intended to ensure that local air quality goals are maintained and that any new or modified source has an acceptable level of emissions control built into it. The pollutants considered and technical criteria for permit approval vary with the size and type of source and the location's ambient air quality. In most instances, it is illegal to initiate construction of an air emissions source without obtaining the applicable pre-construction permit(s).

There are four types of pre-construction permitting review that a source might incur and all or any of them may apply to each source:

1. State-level minor source review
2. Federal Attainment Area Prevention of Significant Deterioration (PSD) review
3. Federal Nonattainment Area New Source Review (NSR)
4. The relatively new federal review for Hazardous Air Pollutant (HAP) requirements to determine if specific limits on HAPs must be established

All four reviews are normally carried out by the local or state agency, even those based on federal law. However, the U.S. EPA may be directly involved if a state has not been officially delegated the authority to implement the Clean Air Act (the Act) permitting provisions.

- **State/minor:** Most states have state-level permit programs, analogous to the federal NSR program, to address the construction and installation of minor sources and modifications that would not otherwise trigger federal NSR requirements. More than 90% of all pre-construction permits in the United States only go through state agency review with no direct involvement of the Act requirements.
- **Prevention of Significant Deterioration:** Areas that are in attainment with the NAAQS are required to conduct PSD reviews to assure that emissions from new or modified major stationary sources will not cause nonattainment. The definition of "major" varies with the pollutant, but is generally 100 or 250 tpy, depending upon the source type. PSD regulations limit air quality degradation below certain prescribed amounts, but do allow some margin for economic growth. Most PSD permits are issued by states. However, there are still some states that have not met the EPA's permit issuing criteria. In these states, the EPA plays varying roles in permit review, depending on the degree to which it has delegated permitting authority.
- **Nonattainment New Source Review:** Areas that are in nonattainment with the NAAQS are required to conduct NSR reviews to limit emissions growth at new major sources. Typically, the threshold level for defining a major source or a major modification are lower in nonattainment areas and range from 10 to 100 tpy. Permits for new major sources in nonattainment areas are issued by respective state environmental protection regulators.
- **HAPs Review, MACT:** Under the Act, EPA has proposed maximum available control technology (MACT) standards for many of the 189 identified HAPs [see Chapter 15]. As proposed, the Act also sets HAPs requirements related to the construction of new sources (both major and non-major) as well as existing sources. Under this provision, the definition of "major" is 10 tpy of any one HAP or 25 tpy of all HAPs now regulated. Existing sources and new non-major sources (referred to as "Area" sources by EPA) are expected to have less stringent emission limitations than major sources. Construction or modification of a major or area source of HAPs is prohibited prior to a state determination that the MACT standard for that source will be met. As currently proposed, existing sources will have several years to demonstrate compliance with the MACT standards.

Operating Permit Overview

Historically, operating permits issued by states were used to track sources, collect fees or confirm compliance requirements for a newly constructed source. They did not impose new requirements and were generally not federally enforceable. The Clean Air Act Amendments of 1990 (CAAA) created a new federally enforceable operating permit program that brings all control requirements under one federally enforceable permit. The operating permit is designed to be the key enforcement tool to be used by EPA in overseeing implementation of the Act.

The new federally enforceable operating permit program covers both new and existing major sources. Following EPA guidelines, all states and local permitting agencies that were required to develop operating permit regulations have submitted them to EPA. To date, 50 state programs and all 60 local programs have been approved. Under the program, annual fees of at least $25/ton of regulated pollutant must be imposed to cover the state's implementation costs, unless a state is able to show that a smaller fee is adequate to fund this program.

All air pollution control requirements applicable to an industrial source will be incorporated into the federal operating permit. The permit provisions will impose detailed requirements, including emissions limitations, schedules of compliance and monitoring requirements. Increased emphasis will be placed on the self-reporting of excesses or violations by a source owner.

Pre-construction Permitting: State Permit Programs

Every state has its own pre-construction permitting program that often goes beyond the federal program in both breadth and scope. They cover major and minor sources and often other pollutants in addition to the six criteria pollutants covered in Title 1 of the CAAA.

Examples of "minor" sources covered by state programs include: small commercial or industrial boilers, engines or heaters; small coating operations; new emissions control equipment; and minor changes to an already permitted major sources. Relatively minor modifications to a permitted source may require the source owner to file for a new minor source permit or permit amendment. Examples of new construction and modifications that may trigger minor source review include:

- Building any new facility with emissions sources
- Addition of new emitting equipment (boiler, coating operation, process unit) at an existing facility
- Installation of a pollution-control device
- Modification of a source in any way that increases the source's potential to emit regulated emissions.
- Change is a unit's fuel firing capability (from oil to oil and gas; from coal to gas, etc.)
- Modifications, repair or reconstruction of the source to replace components such that the new components exceed 50 percent of the fixed capital cost that would be required to construct an entirely new comparable facility.
- Significant change in raw materials (chemicals, solvents, inks) used in existing equipment

The criteria determining whether a new construction or modification needs to go through a minor source permit review vary by state. Some states have detailed listings of exemptions, while others decide exemptions on a case-by-case basis. Typical exemptions that are used in a number of states include:

- Routine maintenance and repair
- Welding operations
- Small gas-fired heaters
- Water cooling towers
- Blast cleaning devices
- Domestic heating and cooling
- Small engines
- Laboratory equipment

Permitting Process

Although the permitting process varies from state to state, typical steps include:

1. **Permit to Construct Application** — submitted by the applicant prior to installation, construction, or operation of new equipment.
2. **Review** — conducted by the state regulatory agency to check for completeness.
3. **Processing** — done by the state regulatory agency to determine that the proposed equipment installation will be built and operated in accordance with all necessary regulations. Permit conditions are set and public review may occur. If there is no compliance problem, the agency issues a permit to construct.
4. **Permit to Construct** — authorizes the applicant to install the equipment. It also can serve as a temporary permit to operate until a permit to operate is issued.
5. **Inspection** — is conducted after the equipment has been installed and before operation. The purpose of the inspection is to determine if the equipment was built and installed as indicated in the application and permit to construct. In addition, the inspection confirms that the equipment operates in compliance with state and federal regulations. This may involve emissions testing and other types of actual measurements.

Typical Pre-construction Permitting Requirements

Most states have standardized permit applications that are used by the applicant to convey the needed information to the state. Required information includes a detailed description of the new source or modification, as well as additional information such as emissions estimates and proposed control strategies.

Typically, states require minor source permit applicants to estimate the emissions resulting from the new construction or modification. States often require that the applicant detail the expected physical and chemical characteristics of the air contaminant stream, both before and after the installation of any emissions control equipment. This information may include emissions rate, concentration, exhaust gas volume and temperature, and exhaust stack locations and characteristics. Maximum annual and hourly emissions representations are required, which are subsequently used to set permit limits.

The applicant is required to select and propose emissions control equipment for all new or modified sources. Details of the equipment, including description, design parameters, and expected performance, are normally required. If wastes will be generated from using the pollution control equipment, the applicant may need to describe the intended disposal method to the state. Data on performance of alternative controls may also be required. The applicant may also need to provide cost data for the controls to show the state that more stringent controls are not cost-effective for the proposed application.

Some states require that an applicant perform air quality impact modeling to demonstrate the effect of the source's air emissions on the local air quality. States use this modeling to verify that the potential emissions are below prescribed air quality standards.

Representative Information Requirements

Representative information requirements for a state pre-construction permit for fuel-burning equipment are indicated in Figure 16-1. While the precise requirements will vary from state to state, the exhibit shows eight major information categories.

A project overview narrative is required to summarize the scope of the project as it relates to emissions generation, with background information on the location, the need for the project, and the pollution control strategies to be employed. Generally, the submission will be based primarily on a completed application form, which is obtained from the state environmental protection office. Supporting data for numerical entries in this form should be provided wherever necessary.

As Figure 16-1 indicates, site, engineering, and process flow drawings are typically required with a pre-construction permit application, as is engineering specification for fuel-burning equipment. Certified vendor pollution generation performance data also may be required. While available from qualified vendors, meaningful certifications may be difficult to obtain. An alternative approach is to require performance data, and also state performance requirements in the product specification. Penalty clauses for not meeting the specified requirements can serve two important purposes. They offer assurance to the state reviewer and the owner that the performance levels will be met, and place performance responsibility on the vendor/manufacturer.

Examples of supporting data formats to consider are presented in Figures 16-2 through 16-5. Figure 16-2 provides supporting data for projected fuel use estimates. Note that the data addresses several future years and indicates changes in fuel source use. Figure 16-3 indicates typical emissions factors that are used to calculate total emissions generation rates. Standard reference sources or boiler test data are typically used. If available, actual test data is preferred. Figure 16-4 indicates future trends for pollutant generation from the facility. Note that in this case, despite a proposed increase in steam generation capacity and projected increasing annual loads, pollution generation rates are projected to decrease significantly. This decrease is shown graphically in Figure 16-5 for SO_X and NO_X pollutants.

Frequently, air dispersion modeling is required in conjunction with permit applications. Air dispersion models are designed to simulate the dynamic dispersion of the pollutants into the atmosphere for varying climactic conditions in the proximity of the site being evaluated. Of particular importance are indicators of the pollution

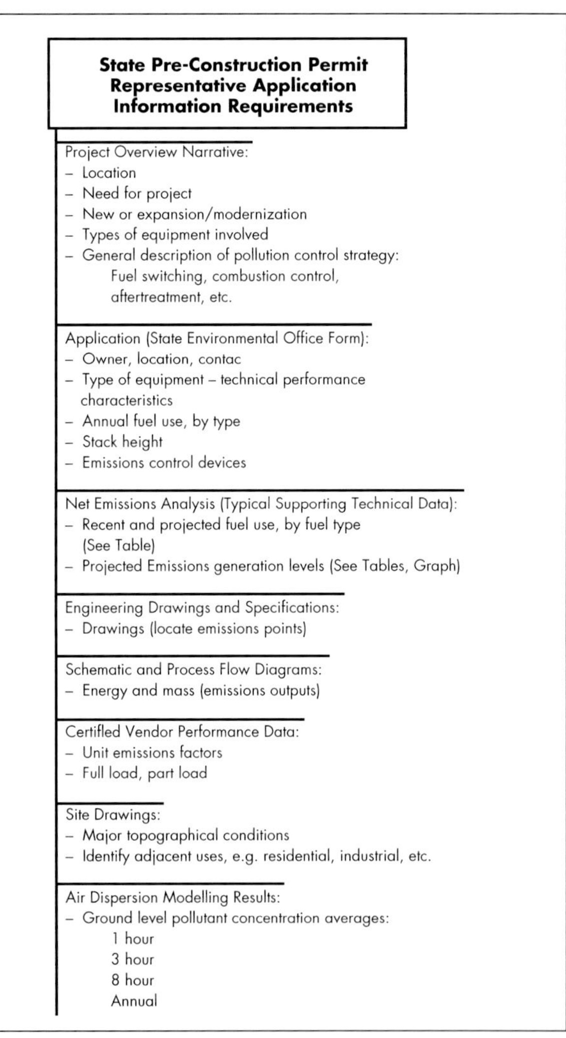

Fig. 16-1 Representative Pre-construction Permit Filing Requirements Summary.

YEAR	BOILER PLANT ACTIVITY	STEAM GENERATION PEAK (KPPH)	STEAM GENERATION ANNUAL (10^6 PPY)	EXISTING BURNERS NAT. GAS (10^9 BPY)	EXISTING BURNERS RES. OIL (10^9 BPY)	NEW BURNERS DIST. OIL (10-^9 BPY)	NEW BURNERS NAT. GAS (10^9 BPY)	TOTAL FUEL USAGE (10^9 BPY)
				ANNUAL FUEL USAGE				
1991	NONE (EXISTING SYSTEM)	415	1,182	0	1,478	0	NA	1,478
1992	NONE (EXISTING SYSTEM)	465	1,324	0	1,665	0	NA	1,665
1993	BOILER C & D: GAS SERVICE	510	1,452	907	0	908	NA	1,815
1994	BOILER B: FGR & NEW BURNER	540	1,538	769	0	769	385	1,923
1995	NEW BOILER E	560	1,595	798	0	398	798	1,994
1996	BOILER A&C: FGR	570	1,623	406	0	0	1,662	2,068
1997	NO CHANGE	590	1,680	420	0	0	1,680	2,100
1998	BOILER D: FGR & NEW BURNER	625	1,780	0	0	0	2,225	2,225
1999	NEW BOILER F	625	1,780	0	0	0	2,225	2,225
2000	NO CHANGE	680	1,936	0	0	0	2,420	2,420

Fig. 16-2 Example Supporting Data for Projected Fuel Use Estimates. Source: TA Engineering.

CRITERIA POLLUTANT	RESIDUAL OIL EMISSIONS FACTOR (LB/1000 Gal)	RESIDUAL OIL EMISSIONS FACTOR (LB/MMBtu)	DISTILLATE OIL EMISSIONS FACTOR (LB/lOOO Gal)	DISTILLATE OIL EMISSIONS FACTOR (LB/MMBtu)	NATURAL GAS EMISSIONS FACTOR (LB/1000 Gal)	NATURAL GAS EMISSIONS FACTOR (LB/MMBtu)
CARBON MONOXIDE	5.0	0.033	5.0	0.036	40.0	0.039
SULFUR OXIDES	157.0	1.026	35.5	0.254	0.6	0.001
PARTICULATE MATTER	13.0	0.085	2.0	0.014	3.0	0.003
VOCs	1.0	0.006	0.3	0.002	1.7	0.002
NITROGEN OXIDES	67.0	0.438	20.0	0.142	550.0	0.534
NITROGEN OXIDES (With FGR)	38.3	0.25	20.0	0.142	103.0	0.1

Fig. 16-3 Typical Emissions Factors Used to Calculate Emissions Rates. Source: TA Engineering.

CRITERIA POLLUTANT	EMISSION FACTOR (LB/MMBtu)	8,760 ANNUAL OPERATING HOURS (100% of YEAR)	7,008 ANNUAL OPERATING HOURS (80% of YEAR)	5,256 ANNUAL OPERATING HOURS (60% of YEAR)	3,504 ANNUAL OPERATING HOURS (40% of YEAR)	1,752 ANNUAL OPERATING HOURS (20% of YEAR)
		ANNUAL EMISSIONS (TONS PER YEAR AT FULL LOAD)				
CARBON MONOXIDE	0.039	41.9	33.5	25.1	16.7	8.4
SULFUR OXIDES	0.001	1.1	0.9	0.7	0.4	0.2
PARTICULATE MATTER	0.003	3.2	2.6	1.9	1.3	0.6
VOCs	0.002	2.1	1.7	1.3	0.9	0.4
NITROGEN OXIDES:						
STANDARD BURNER	0.3	321.9	257.6	193.1	128.8	64.4
With BURNER RECIRC.	0.2	214.6	171.6	128.8	85.8	42.9
With FGR	0.1	107.3	85.8	64.4	42.9	21.5

Fig. 16-4 Example Estimates for Future Facility Emissions Generation. Source: TA Engineering.

concentration at user-selected ground points. In addition to overall peak pollution levels, average concentrations during periods of one hour, three hours, eight hours, twenty-four hours, and annually are determined. Figure 16-6 provides an example of a summary of air dispersion modeling results.

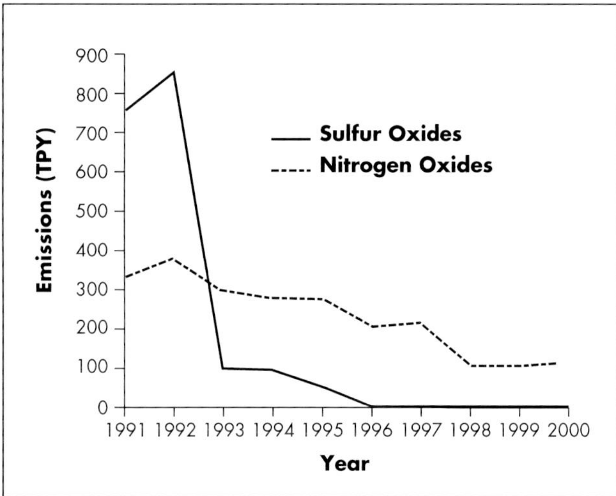

Fig. 16-5 Estimated (Hypothetical) Annual Boiler Plant Emissions. Source: TA Engineering

PERMIT CONDITIONS

Once a state and applicant have reached agreement on permit conditions, the state will issue a permit and the applicant is required to conform to all of the permit's provisions. Conditions are placed in permits in order to limit, or require some type of action by the source facility. These limits or actions may include:

- Compliance with specific regulations
- Keeping emissions at prescribed levels
- Limiting processing rates, throughput, and hours to match the representations in the permit application
- Limiting toxic emissions
- Requiring proper operation of control devices
- Establishment of monitoring record keeping and reporting mechanisms

Permit modifications are required any time a change is made to operations such that the permit conditions or other information listed on the permit are no longer accurate. This would include such changes as increasing operations beyond the condition limits or a change in ownership.

	EMISSION LEVELS (MICROGRAMS/CUBIC METER)									
	DISTILLATE FUEL					NATURAL GAS				
MODEL INPUT	SO_2	NO_X	Particulates (PM-10)	VOC	CO (Note 1)	SO_2	NO_X	Particulates (PM-10)	VOC	CO (Note 1)
EXISTING STACK HEIGHT										
1 -HR AVG.	NA	NA	NA	NA	NA	NA	NA	NA	NA	NA
3-HR AVG.	2,832	NA	NA	22	NA	7.7	NA	NA	0.06	NA
8-HR AVG.	NA	NA	NA	NA	NA	NA	NA	NA	NA	NA
24-HR AVG.	986	NA	54	NA	NA	3.5	NA	0.19	NA	NA
ANNUAL AVG.	106	112	5.8	NA	NA	0.31	142	0.02	NA	NA
NEW GEP STACK HEIGHT										
1-HR AVG.	NA	NA	NA	NA	NA	NA	NA	NA	NA	NA
3-HR AVG.	41	NA	NA	0.3	NA	0.14	NA	NA	0	NA
8-HR AVG.	NA	NA	NA	NA	NA	NA	NA	NA	NA	NA
24-HR AVG.	76	NA	0.44	NA	NA	0.03	NA	0	NA	NA
ANNUAL AVG.	0.32	0.34	0.02	NA	NA	0	0.48	0	NA	NA
NAAQS TARGET LEVELS										
1-HRAVG.	NA	NA	NA	NA	NA	NA	NA	NA	NA	NA
3-HR AVG.	1,300	NA	NA	160	NA	1,300	NA	NA	160	NA
8-HR AVG.	NA	NA	NA	NA	NA	NA	NA	NA	NA	NA
24-HR AVG.	365	NA	150	NA	NA	365	NA	150	NA	NA
ANNUAL AVG.	80	100	50	NA	NA	80	100	50	NA	NA

Notes:
1. CO Levels were determined elsewhere to be below NAAQS limits, so were not modeled.
2. A value of 0.00 indicates a concentration of less than 0.01 ug/m^3.
3. GEP = General Engineering Practice.

Fig. 16-6 Example of a Summary of Air Dispersion Modeling Results. Source: TA Engineering

PRE-CONSTRUCTION PERMITTING: PSD PERMITS

The PSD review process only applies to a new facility that has sufficient emissions to be considered "major" for a pollutant in an attainment area, or a significant modification of an existing major source in such an area. The steps in a PSD review are similar to, but more elaborate than, those for state minor source review and are standardized because they originate in the Act.

A source in an attainment area is considered major if it falls within one of the 28 categories listed in the Act and has the PTE of 100 tpy of a regulated pollutant, or, if unlisted, has the PTE of 250 tpy of any regulated pollutant. If the source meets the major source emissions threshold for any one regulated pollutant, it is considered a major source for that pollutant and PSD review is triggered. In addition, PSD review is triggered by any change to any existing major source that constitutes a "major modification" because it has an emissions increase of any regulated pollutant greater than those shown in Table 16-1.

Table 16-1 PSD Significant Increase

Pollutant	Emissions Rates (tpy)
Sulfur Dioxide (SO_2)	40
Nitrogen oxides (NO_X)	40
Volatile Organic Compounds (VOC)	40
Carbon Monoxide (CO)	100
Particulate matter (PM-10)	15

The assessment of the net change may also account for emissions decreases. For replacement of equipment, the net emissions change is estimated as the difference between the PTE for the proposed new unit and the historical actual emissions from the unit being shut down. In order to take credit for past decreases, all contemporaneous increases and decreases must be aggregated. A netting analysis will extend to a historical period of five years.

The following procedures are used to determine the net emissions change at a source:

1. Determine emissions increases resulting only from the project in question. If they do not trigger a major source or major modification threshold, the project generally can be permitted without PSD review. If they are in excess of the major modification threshold, further analysis is needed.
2. Determine the relevant time period.
3. Determine which equipment has or will experience a creditable increase or decrease in emissions during the relevant period.
4. Determine, on a pollutant-by-pollutant basis, the amount of each emissions increase and decrease.
5. Aggregate all contemporaneous and creditable increases and decreases to determine if a significant net emissions increase will occur.

After a regulatory analysis has been performed to determine applicability, the PSD permit application process generally consists of the following:

- Baseline air quality monitoring for the site.
- A case-by-case Best Available Control Technology (BACT) analysis, which considers all technically feasible air emissions control options with respect to economic, energy, and environmental impact.
- An air quality impact analysis through modeling to determine if the emissions from the proposed source would cause or contribute to exceeding an applicable PSD air quality increment.
- An assessment of the direct and indirect effects on the environment.

BACT Requirements

A BACT analysis is used to determine the control strategy required for each regulated pollutant emitted in significant amounts from a major new source or major modification. The CAAA defines BACT as "an emissions limit based on the maximum degree of emissions reductions for each pollutant, which the permitting authority determines on a case-by-case basis, taking into account economic, energy, and environmental impacts, is achievable for such facility through the application of production process and available methods, systems, and techniques."

BACT is not a control technology, but an emissions limit that is based on available integral and/or add-on control technology. It is not a static standard. Each new control technology introduced for a similar source and demonstrating a new lower level of emissions must be considered in subsequent BACT analyses.

For a technology to be considered as BACT, it must be technically feasible. This can mean that it has been used in other similar applications or is proven on a pilot scale. BACT compliance may include the use of alternate equipment, so that if a different type of equipment can do the job, it can also be used. Alternative fuel source usage may also be considered as part of a BACT determination or may even be BACT.

A key factor in the BACT determination is that it considers economic factors. This allows individual states some flexibility in permitting. BACT in one state may be considered too economically harmful in another. EPA oversight does provide for some degree of consistency, however.

The current application of BACT analyses is based on what is termed a "top-down" approach. This starts with the most stringent technology that has been applied elsewhere to the same or similar emissions source category and provides a basis for either accepting it as BACT or rejecting it in favor of another (equal or less stringent) technology. This top-down approach, coupled with cost-effectiveness evaluation of each technology, helps to determine which technology must be installed.

The basic steps in a top-down BACT analysis are:

1. Identify all potential control technologies applicable to each pollutant subject to PSD. This includes any technique that has been applied to a similar source. Also, consult the local air agency and obtain any BACT determinations for the type of equipment.
2. Evaluate the technical feasibility of the identified alternatives and reject those that are demonstrably unfeasible based on a documented engineering evaluation or on chemical or physical principles.
3. Assess and document the emissions limit achievable with each technically feasible alternative, based on the operating conditions of the unit under review and rank the alternatives.
4. Evaluate the top-ranked (most stringent) technically feasible control option to determine its economic, energy, and environmental impact. If it is proposed, the analysis is virtually complete. If it is rejected, the second-ranked control option is evaluated and so on until an acceptable option is proposed.
5. Evaluate any collateral air toxic effects of the proposed control option. If the air toxic impacts are sufficient to reject the option, Step 4 is repeated.

These steps must be completed as part of the permit application. The following is a list of the typical technologies that must be considered in BACT review for non-utility equipment:

Controls to be Considered for Boilers

- Particulates — Baghouses and/or electrostatic precipitators or the use of natural gas, which has inherently low particulates
- CO and VOC — Good combustion control
- SO_X — Lower-sulfur fuel or add-on scrubbers (dry or wet with various chemistries)
- NO_X — Combustion control (low excess air, overfire air, staged combustion), flue gas recirculation, selective catalytic reduction (SCR), selective non-catalytic reduction (SNCR)

Controls to be Considered for Gas Turbines

- CO, VOC, Particulates — Good combustion control
- SO_X — Low-sulfur fuel is normally used
- NO_X — Low-NO_X combustor design, water or steam injection, SCR

Controls to be Considered For Reciprocating Engines

- VOC, CO — Air/fuel adjustment, timing, add-on oxidation catalyst
- Particulates — Fuel quality, combustion control, traps
- SO_X — Low-sulfur fuel is normally used
- NO_X — Air/fuel adjustment, ignition retard, lean-burn combustion, SCR, SNCR, or non-selective catalytic reduction (NSCR)

BACT Cost-Effectiveness Test

Cost-effectiveness is a critical factor in the selection process. A typical BACT cost-effectiveness analysis should include the following steps:

1. Calculate the annual emissions reductions that will occur as a result of installing the control technology under evaluation. The annual emissions reductions may be calculated in several ways. It may be based on the daily emissions reduction achieved by the BACT method times 365 days per year. If the permit applicant is willing to accept a permit that has conditions to limit operation to fewer hours than this, adjustments are made to the calculation. For new equipment, the amount of pollutant removed will be the difference between maximum uncontrolled and maximum controlled emissions.
2. Calculate an equivalent annual cost from the capital cost. This may be done with the following equation:

$$A = P[i(1+i)^n] / [(1+i)^n - 1] \tag{16-3}$$

Where:

A = Equivalent annual capital cost of the control equipment
P = The control equipment installed cost

i = Interest rate (set at the prevailing applicable rate)
n = Equipment life (this may be fixed or set on a case-by-case basis)

Assuming an installed cost of $300,000, an interest rate of 10%, and an equipment life of 10 years, the equation would simplify to:

$$A = \$300{,}000 \times 0.1627 = \$48{,}810$$

3. Add to this figure the annual operating cost related to the control equipment. This may include labor, fuel, maintenance, other utilities or resources, and ongoing capital cost requirements.
4. Divide this annualized cost by the annual emissions reduction achieved and compare that value to the state's threshold value of annualized cost-effectiveness per ton of each regulated pollutant removed. For example:
In this case, if the annual operating costs were $20,000 and 10 tons were removed, the value would be:

$$\$68{,}810/10 = \$6{,}810/\text{ton removed}$$

If this value is lower than the value established by the control agency for annual cost-effectiveness per ton, the measure is considered acceptable as BACT under consideration of economic factors. If it is not, the same calculations are performed on the next (less costly) available control measure on the top-down list.

Air Quality Modeling and Air Quality Increments

An air quality analysis is a critical element of the PSD review process. In order to obtain a permit, an applicant must demonstrate that emissions from the proposed new or modified source will not violate the NAAQS or allowable PSD increments. The increments for each regulated pollutant are shown in Table 16-2. These are short-term increments that are not to be exceeded more than once per year. Increments vary with local classification. Class I covers certain pristine areas. All other areas are Class II. Class III treatment is an option to states that is not in use.

To conduct the air quality analysis, the permitting authority can require that the permit applicant monitor meteorological conditions and air quality at the selected site. Since this can take up to one year, this should be the first topic addressed when a PSD triggering project is under consideration.

Table 16-2 PSD Increments (in g/m3)

Sulfur dioxide	Class I Areas	Class II Areas	Class III Areas
Annual	2	20	40
24-hour	5	91	182
3-hour	25	512	700
PM-10			
Annual	4	17	34
24-hour	8	30	60
Nitrogen dioxide			
Annual	2.5	25	50

An air quality analysis is required for each pollutant subject to PSD review. The screening level analysis uses dispersion modeling techniques to predict ground level concentrations of each pollutant from the proposed new or modified source. The analysis compares the predicted concentrations to the significant impact levels (NAAQS and PSD increments). The complexity of the screening varies depending on the number of pollutants that must be analyzed and the proximity of the source to other major sources.

If the initial dispersion modeling shows that the impact from the proposed source for any regulated pollutant is above the NAAQS and PSD increments, more detailed and representative analysis may be required. This involves more elaborate modeling and may include collecting site specific meteorological data.

The **NAAQS analysis** consists of the following steps:

1. Define the significant impact area of the proposed new or modified major source for each pollutant subject to review.
2. Construct an inventory of all existing and proposed sources of emissions in the total screening area.
3. Determine the ambient air concentrations of each pollutant from monitoring data, excluding emissions from the proposed source.
4. Model the proposed source in conjunction with every identified source in the emissions inventory.
5. Determine the predicted total air quality impact by adding the total modeled concentrations to the existing monitored air quality data.
6. Compare the predicted total air quality impact to the NAAQS for each pollutant to determine compliance.

The **PSD increment analysis** consists of the following steps:

1. Define the significant impact area of the proposed

new or modified major source for each pollutant subject to PSD review. This is the same process as used in the NAAQS analysis.

2. Construct an inventory of all sites established after the baseline date (the data when the first PSD source was established in the area).
3. Determine the total predicted increment concentration from the proposed source or modification inclusive of other increment-consuming sources.
4. Compare the total increment concentration to the allowable PSD increment to determine compliance with the increments.

To obtain a permit, the predicted impact must be below the NAAQS and the allowable PSD increments. In addition, all pollutants subject to PSD review require additional impact analyses. These include:

- A growth analysis
- A soils and vegetation analysis
- A visibility impairment analysis

The incremental degradation that is allowed in an attainment area varies with the classification of attainment area. The incremental degradation allowed in Class I areas, which are normally national parks or national monuments, is smaller when compared to the remaining attainment areas.

In 2010, the EPA announced new one-hour average NAAQS for nitrogen dioxide (NO_2) and for sulfur dioxide (SO_2). The new NAAQS became effective for NO_2 in April 2010 and for SO_2 in June 2010.

Table 16-3 Major Source and Modification Thresholds for Nonattainment Areas

Non-Attainment Classification	POTENTIAL EMISSIONS (tpy) Major Source	Major Modification
Carbon Monoxide:		
Moderate	100	100
Serious	50	50
PM-10:		
Moderate	100	15
Serious	70	15
Ozone (VOC/NO_X):		
Marginal/Moderate	100	40
Ozone Transport Region	50 (100 for NO_X)	40
Serious	50	25
Severe	25	25
Extreme	10	Any increase

Pre-construction Permitting — Federal New Source Review Permits

Federal nonattainment NSR, like PSD, is normally implemented under EPA-approved state regulations with EPA oversight. Nonattainment NSR applies to new major sources of the nonattainment pollutants or pollutant precursors in the nonattainment area and major modifications (with significant net increases) at major sources of those pollutants. The analytical methods used to make these determinations are similar to those under PSD. Major source thresholds, however, are lower and vary with the severity of the nonattainment classification. Those thresholds are shown in Table 16-3.

The permitting process is similar to the process for PSD in several ways. The permit applicant may not, however, have to do extensive air quality modeling and the issue of baseline ambient monitoring does not arise. The control technology determination is somewhat different, as described below, but the applicant is still responsible for doing the necessary investigation and analysis of control alternatives. Lastly, as with PSD, a federal nonattainment NSR permit must go through a formal public review process.

The three key requirements associated with federal nonattainment NSR are:

1. The new facility or equipment use air emissions controls that meet the definition of the Lowest Achievable Emissions Rate (LAER).
2. The increase in air emissions is offset by reductions in actual emissions from other sources in the immediate area of air quality impact.
3. The source owner certifies that all other major sources in the state that are under the owner's control are in compliance with all requirements of the Act.

Lowest Achievable Emissions Rate (LAER)

EPA regulations define LAER as, "...that rate of emissions which reflects:

A. The most stringent emissions limitation which is contained in the implementation plan of any State for such class or category of source, unless the owner or operator of the proposed source demonstrates that such limitations are not achievable, or
B. The most stringent emissions limitation which is achieved in practice by such class or category of source, whichever is more stringent."

Further:

"In no event shall the application of this term permit a proposed new or modified source to emit any pollutant in excess of the amount allowable under applicable new source standards of performance."

The key distinction between BACT and LAER is that BACT takes into consideration cost and other environmental impacts while LAER does not. Therefore, LAER is nearly always more stringent or as stringent as BACT. As with BACT, technical feasibility and reliability are considered in a LAER determination.

Offsets

Offset requirements are often impediments to a nonattainment area project. Offsetting emissions must be based on actual emissions. They must go beyond control requirements already in the state rules and must be greater than 1 to 1. As shown in Table 16-4, the offset ratio varies with the local air quality classification and can involve another source owner. The source of the offsets will be given an enforceable permit to ensure that the offsetting reductions continue.

The last requirement for this permit is the compliance certification. A permit for a new source will not be issued if there is a compliance problem with an existing one. If a compliance problem exists, a consent order, variance, and/or fine would need to be settled before the new project permit could be issued.

Table 16-4 Offset Ratio for Major Sources or Modifications in Ozone Nonattainment Areas

Nonattainment Classification	Offset Ratio
Marginal	1.10 to 1
Moderate	1.15 to 1
Ozone Transport Region	1.15 to 1
Serious	1.20 to 1
Severe	1.30 to 1
Extreme	1.50 to 1

New Source Review Reform Activities

The federal and state processes for reviewing and approving equipment and process changes affecting a facility's air emissions have proven to be environmentally beneficial, but very complex and time consuming. As a result, EPA and several states, in response to industry concerns, are trying to improve the process, making it easier and quicker to meet permitting requirements while maintaining the key benefits of the pre-construction review process.

In 1994, the EPA issued guidance to state agencies indicating that certain types of pollution control projects could be excluded from federal major source NSR requirements. This guidance extends a similar policy established in 1992 for electric utilities to all of industry. The policy has no effect on state permitting requirements that are not derived from the CAAA. The exclusion may be used regardless of category. Low-NO_X burners, for example, might qualify, but the permitting agency must approve the exclusion.

Under the EPA's new guidance, both add-on emissions control projects and fuel switches to less polluting fuels can be excluded from NSR as "pollution control projects." The eligible projects include:

- The installation of conventional and advanced flue gas desulfurization and sorbent injection for SO_2.
- Electrostatic precipitators, baghouses, high-efficiency multi-clones, and scrubbers for particulate or other pollutants.
- FGR, low-NO_X burners, SNCR, and SCR for NO_X.
- Regenerative thermal oxidizers (RTO), catalytic oxidizers, condensers, thermal incinerators, flares, and carbon absorbers for VOC and toxic air pollutants.

Projects undertaken to accommodate switching to an inherently less polluting fuel such as natural gas also qualify for exclusion. In some instances, where the emissions unit's capability would otherwise be impaired as a result of the fuel switch, this may involve certain necessary changes to the pollution-generating equipment to maintain the normal operating capability of the unit at the time of the project. Those would also be excluded.

Permitting authorities may also apply the new EPA exclusion to switches to inherently less polluting raw materials and processes and certain other types of "pollution prevention" projects. For instance, many coating users may be making switches to water-based or powder paint application systems as a strategy for meeting RACT or switching to a non-toxic VOC to comply with MACT requirements.

Permitting authorities would be allowed to consider excluding raw material substitutions, process changes, and other pollution prevention strategies where the pollution control aspects of the project are clearly evident and the project will result in substantial emissions reductions per unit of output for one or more pollutants. In judging whether a pollution prevention project can be considered for exclusions as a pollution control project, permitting authorities would also consider as a relevant factor whether a project is being undertaken to bring a source into compliance on Act requirement. For more information on the flexibility afforded to pollution prevention projects, review the documentation on the EPA's website.

Federal NSR Reform Effort

In 1996, the EPA issued its NSR rule-making package. Its key features are summarized below:

- New criteria for evaluating the effect of operational changes on actual emissions. The rule allows states to determine whether or not NSR applies to modifications at existing facilities based on more realistic projections of actual emissions changes.
- Greater flexibility to facilities that are already well-controlled. For example, simplified and less restrictive applicability tests are allowed for modifications at facilities that have already installed state-of-the-art control technology.
- Pollution prevention exclusion. Projects that reduce or prevent pollution can be excluded from NSR, significantly reducing the administrative burden on facilities.
- Greater flexibility to set an emissions limit or cap, known as plant-wide applicability limits. Any changes or modifications can be made without triggering NSR as long as the emissions cap is not exceeded. This provision affords participating facilities much greater flexibility in meeting their changing operating and production demands.
- Exemption for certain pollution control projects that substitute less environmentally harmful compounds for stratospheric ozone-depleting substances.
- Reduction in administrative burden for affected facilities in determining the required control technology, BACT or LAER, under NSR.
- Greater flexibility to states in determining control technology requirements, while ensuring that the most effective control technologies are considered.
- Improvement in the process, consistent with the Act, for environmental protection of important natural areas, such as national parks and certain wilderness areas (the so-called "Class I" areas).
- Allowance of a broader range of preliminary construction activities at existing facilities to proceed before the issuance of a NSR permit.
- Lifting of certain restrictions on the use of emissions reductions that result from shutdown facilities as NSR offsets in nonattainment areas that lack EPA-approved attainment demonstrations.

In particular, industry is provided opportunities to exclude from NSR projects that are very low in emissions or are intended to cut emissions. New flexible plant-wide limits are allowed and special consideration is given when an undemonstrated technology is being required.

While the EPA has embarked on a formal process to reform federal major source NSR, several states are making efforts to simplify their pre-construction review processes. Specific reforms being pursued in several states include:

- Providing training and technical assistance, particularly to small businesses.
- Extending the number and types of exclusions from review.
- Creating standardized permits for small and common types of sources that can be issued without extensive permit applications and reviews.
- Trying to integrate pre-construction and operating permit reviews into a single process.

Operating Permits: State Programs

Prior to the CAAA, there was no federally enforceable operating permit program. If a source was required to obtain an operating permit, it was pursuant to a state program. Some states require facilities to obtain a state operating permit, separate from any pre-construction permit. State operating permit programs differ from state to state, although in general the requirements of the program are limited. Often, no additional operating requirements, other than those contained in the pre-construction permit, are set.

State operating permits have not tended to be a critical element of the permitting process and were established in the past as a means to accomplish the following:

- Verification of pre-construction limits, used as a confirmation that the conditions agreed upon in the construction permit have been met.
- Generation of revenue, used by some states that require annual fees to be submitted at the time of an annual state operating permit renewal.
- Inventory of sources, enables states to perform an annual inventory of the type of sources and equipment in operation.
- Annual emissions report, used as a means to review

the annual emissions of a source.

While states may use the information gathered by their operating permit programs for other uses, the scope and purpose of the state operating permit program are limited when compared to the federally enforceable operating permit program required by the Act.

Operating Permits: The Federal Program

The CAAA requires each major source of air pollutants to obtain a federal operating permit. The renewable permit incorporates all applicable conditions under one federally enforceable permit. Since the federal operating permit program does not create new emissions limits or requirements, it is considered a mechanism by which the EPA can enforce the requirements of the Act and provide the facility management and the public with a single document covering all air regulations and conditions affecting that facility.

The federal operating permit program applies, regardless of whether an area is in attainment or not. It applies to all sources that are major for any criteria pollutant or HAP and sources covered by federally enforceable NSPS or State RACT limitations.

States are responsible for adopting rules to implement this program, EPA approves those rules, and then the state implements the program. Applications for permits must be submitted to the state within one year of EPA approval of the state's program. In addition to EPA and state review, the Act also provides for the opportunity of public participation in the permitting process. All applications and supporting documents must be made available to the public for review. The permitting agency must also provide an opportunity for public comment and a formal hearing to address the application. Once the EPA, the applicable states, and the public have all reviewed the application and the permitting agency has addressed any comments, a final permit may be issued.

States are also required to collect an annual fee of at least $30/ton (1994 dollars) of regulated pollutants, unless a state is able to show that a smaller fee would be adequate for the state to fund its permitting effort. The Act requires that the fees be used to cover the cost of developing and implementing the operating permit program.

Monitoring Requirements

Under the Act and in concert with the operating permit program, the EPA has the authority to require expensive enhanced monitoring for determining compliance with the permit conditions. Enhanced monitoring may include continuous emissions monitoring systems (CEMS) or continuous emissions rate monitoring systems (CERMS) instead of periodic stack tests or simple emissions calculations, which has often been sufficient for compliance monitoring to date. CEM systems use techniques such as infrared or chemiluminescent analysis to measure emissions of NO_X, SO_2, CO, CO_2, particulates, and other regulated air pollutants.

Figure 16-7 is a representative example of the basic CEM process using dry extraction. As outlined in the exhibit, a sample probe is used to draw out stack gases on a continuous basis (except for short purge cycles) and filter out particulate material. The sample is pumped from the probe to the rest of the CEM via the sample transport line. The sample gas is hot and wet, and thus contains a certain amount of water vapor.

So that condensation does not occur in the sample transport line, the line is heated along its whole length so that the sample gas may be delivered unchanged to sample conditioning. In sample conditioning, the sample is chilled, condensing out the moisture. The now dry gas sample is filtered again and passed through a manifold to the appropriate device for sample analysis.

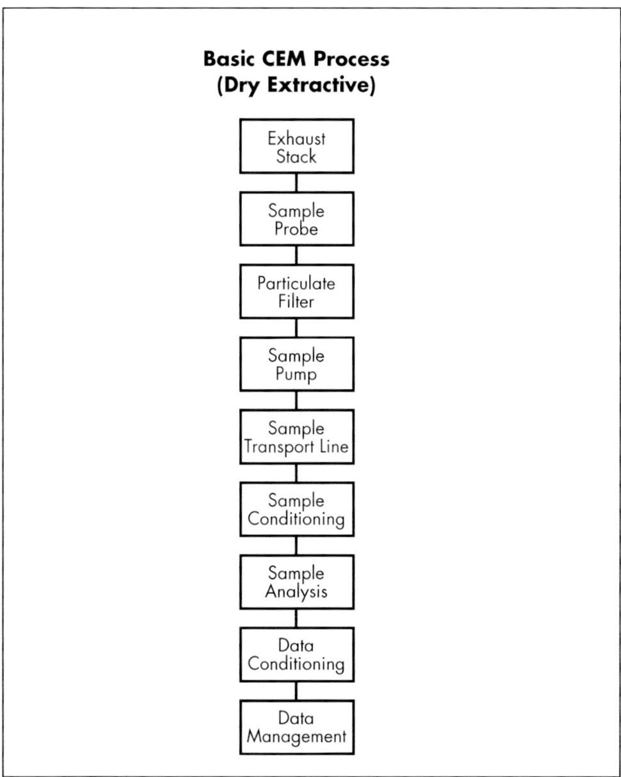

Fig. 16-7 Basic CEM Process (Dry Extractive).

The concentration of the substance of interest (e.g., NO_X, CO, etc.) in the gas sample is measured and sent in the form of an analog electric signal for data conditioning. The analog signals from the analyzers are translated into a form acceptable for computer communication (e.g., the RS-232 serial communications protocol) and sent to the central computer for data management. The information received from the data conditioner is converted to the appropriate units, validated, and reported in the required format.

CEMS may not be required if alternative methods are available that provide sufficiently reliable and timely information for determining compliance. This has been an area of debate in evolving regulations. The high cost of CEMS can have significant detrimental effects on project economics for small and intermediate sized systems. A lower cost alternative to CEMS is the use of predictive (or parametric) emissions monitoring (PEM) systems. These use predictive rather than actual-measured emissions monitoring based on combustion, scientific computer modeling, and confirmation by actual tests.

Compliance Schedule

As part of the permit application, applicants need to include a schedule of compliance. As part of that schedule, the facility will need to report, to the permitting agency every 6 months, its progress in meeting the conditions contained in the permit.

Under the CAAA, although this part of the operating program has created some controversy, an approved operating permit program must provide for certain changes within a permitted facility without requiring a permit revision. The changes may not be modifications under the ozone nonattainment requirements of the Act and they may not exceed the total emissions or emissions rates allowable under the permit. The facility must provide EPA and the permitting agency with written notification at least 7 days before the change, or a shorter time for emergencies.

HAP Permitting

The EPA has promulgated or proposed MACT standards for certain new and existing source categories that emit the listed 189 HAPs. The Act requires new and modified sources of HAPs to comply with MACT standards on a case-by-case basis and includes provisions to address pre-construction permitting of such facilities.

In the case of construction or reconstruction of a major source of HAPs, case-by-case MACT is immediately triggered. In the case of a modification of a major source of HAPs, case-by-case MACT standards are only triggered if the modification will increase emissions over a *de minimis* threshold level. Additional analysis on the part of a facility will be necessary to determine whether the emissions from the modification triggers case-by-case MACT. The facility also has the opportunity to "net" out of MACT requirements if it is able to offset its modification's emissions elsewhere in the facility, and thereby not exceed the *de minimis* threshold. If the facility has determined that the new construction or modification will trigger case-by-case MACT, however, it will need to negotiate a standard with the permitting agency.

In negotiating a MACT standard, the owner or operator of the facility will need to present what it considers should be MACT for its facility to the state permitting agency. Determining this technology-based standard will require analysis of the emissions sources, uncontrolled emissions, controlled emissions, and any applicable control technologies.

The state may request EPA oversight during the negotiation of a case-by-case MACT standard. Once a state permitting agency is prepared to propose a MACT standard for a new source or modification, the state must provide an opportunity for a public hearing and comment. After addressing any comments, the state agency may then send the proposed MACT standard to the EPA for review. If the EPA finds that the standard does not meet the requirements of the Act, it will disapprove the application and the state will need to amend the proposed standard. If EPA approves the MACT standard, the state may then promulgate it.

This permit negotiation, review, and approval process is required for each case-by-case MACT standard and is separate from other CAAA permit application and review requirements. Facilities may, however, have the opportunity to "merge" these permits in order to streamline the permitting process.

Merging Permits

Since the provisions of a federal operating permit program require every major source of air pollution to obtain an operating permit that incorporates all applicable requirements under the Act, it creates the potential for overlap between all three permit programs. Currently, the permit processes for federal operating permit revisions, case-by-case MACT, and NSR are separate and require the facility to meet unique notification and public comment obligations. The timing of the respective permit processes may not be compatible, resulting in lengthy delays in permit issuance and perhaps requiring the facility to undergo several public, state, and EPA review and comment periods for the same permit action.

The provisions for the federal operating permit program provide an option for states to merge the respective NSR and/or case-by-case pre-construction permit processes with the federal operating permit revision requirements and streamline the three separate processes. This option would allow states to "enhance" the NSR or case-by-case MACT processes to include the requirements of the federal operating permit so that facilities would be able to satisfy all permit revision obligations in one comprehensive process. Currently, most states do not have "merged" programs in place and it is not clear how many will pursue this option.

Permitting Strategies

Securing air permits can be a costly and time-consuming process. It is important to assess which permit requirements apply, opportunities to legally avoid them, how to go through the permit review quickly, and ideal permit conditions. The permit process is typically the time during which investment decisions for controls and operating conditions are made. Since one objective should be minimizing the need to revise a permit, these are important decisions.

Establishing a "Minor" Source Designation

The EPA has adopted a conservative approach for determining if a source is major. For example, a small backup generator or boiler may never emit more than 19 tpy of NO_X, but has a PTE of more than 100 tpy. Figure 16-8 provides a representative example of the impact operating hours on annual NO_X emissions. Assuming a 25 tpy target threshold, the parallel lines indicate allowable annual operating hours versus gas- and oil-fired boiler capacity with various levels of NO_X control. This graphic highlights the relevance of differentiating between the PTE and actual emissions.

The EPA will lower its estimate of a source's PTE if the source is constrained by a federally enforceable limit on its operation. The equipment owner, for example, could request a permit limit on hours of operation or the amount of fuel used and keep the allowable emissions below the major threshold and avoid many complex requirements. Because of the importance of this legitimate route to limiting a source owner's obligations, the

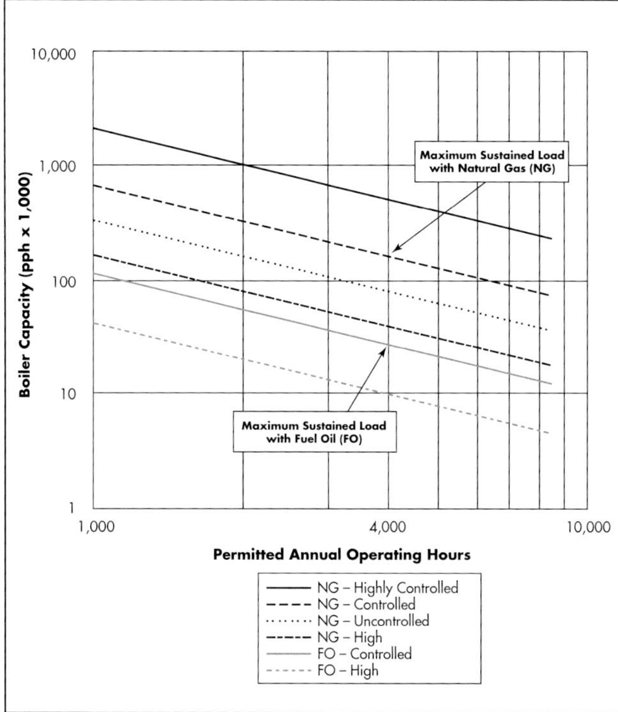

Fig. 16-8 Impact of Hours of Operation vs. Capacity on Actual Annual Emissions.

EPA and states are now developing mechanisms to establish what the EPA calls "synthetic minor" enforceable limits.

Netting Out of Federal NSR or MACT

Even if one has a major source and is modifying or expanding it, it is possible to "net out" of the key federal requirements. This is done by matching any increase in emissions with a creditable decrease within the same facility. Since the federal requirements are time consuming and expensive, at the outset of evaluating a project at a facility that is already major, it is important to explore what can be done to net out. Options include:

- Taking enforceable limits on the new or modified unit to limit its potential increase.
- Quantifying and receiving full credit for reductions due to equipment being removed or used less after the change. This may also require that the state sets enforceable limits on old units.
- Identifying voluntary controls for the new units or old units at the same site that can be made part of the overall project so that the project has no significant emissions increase.

As an example, if a large industrial facility, already

major for NO_X, were to seek a permit for a new boiler with potential emissions of 60 tpy in a Moderate ozone nonattainment area, LAER and offsets would likely be required. However, to ensure that no significant net increase would exist to trigger LAER and offsets, the permit applicant could:

- Switch an existing unit to a low-NO_X fuel.
- Retrofit added NO_X controls to existing units.
- Retire an existing NO_X source at the same site.
- Limit the use of the new unit or an existing unit (hours, fuel use, emissions).
- Voluntarily install advanced controls on the new unit

Obtaining Offsets

If a facility goes through Federal major source nonattainment NSR, the permit applicant must provide offsetting emissions reductions. Finding and acquiring offsets can be very difficult. Assuming it is not possible to reduce emissions at sources under the same ownership, there are generally three options:

1. Obtain the rights to "banked" reductions. Some states maintain emissions reduction banks where source owners who control or shutdown sources are allowed to preserve emissions reduction credits for future use. If the state has a bank, credits may exist in the bank. The owner has the right to set the price for those credits.
2. Conduct a private transaction. In areas without banks or where banked credits cannot be purchased economically, one must track down local source owners who may be willing to reduce emissions for a fee. The state must be involved in deciding how those reductions will be enforced and determine the amount of offsets that can be generated.
3. Obtain credits from the government. In some areas, to encourage new industry, states create a growth increment of allowable emissions in the state attainment plan. That increment is then available for the state to use to cover offset requirements for attractive new businesses or to maintain existing businesses.

Obtaining Appropriate Permit Conditions

A key part of any permitting process is the set of conditions in the permit designed to ensure that the source emissions are consistent with current rules, with the representations in the permit application, and with the control technology determination made as a part of the permit review. Those permit conditions are critical to the source in several ways. The most obvious is that they are legally enforceable and must be met regularly. Also, those limits will influence how often the permit must be amended when subsequent process changes occur. Practical and flexible permit conditions that meet the government's need for accountability are considered to be integral to a successful permitting effort. A few key considerations when negotiating permit conditions are:

- Avoid overly detailed limits, particularly when they have no environmental consequence. For example, a permit specifying a certain raw material may preclude the use of another raw material at a later time.
- Avoid redundant limits. States are inclined to set emissions limits (short- and long-term), as well as limits on process throughput and hours of operation and raw material type.
- Aggregate limits provide more flexibility than separate unit-by-unit limits. For similar operations, it may be easier to meet a single overall limit on a consistent basis than a limit on each unit. This provides flexibility to operate one unit more when other units are out of service or constrained.
- Clarify the basis for demonstrating compliance with a limit. Often, permits set limits with no description of how to show compliance. To the extent there is any ambiguity, it is best to spell out the compliance basis in the permit.

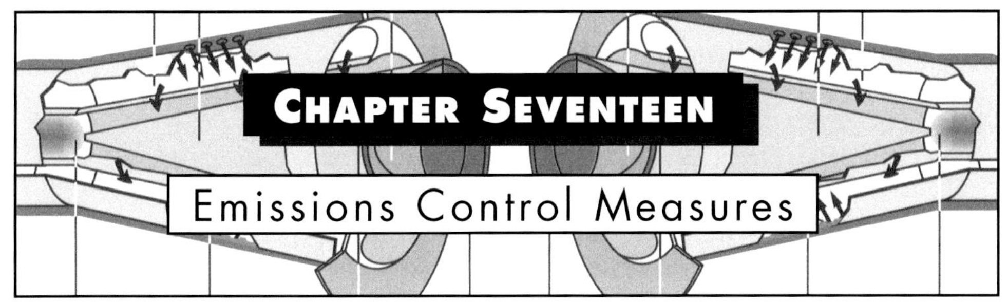

Chapter Seventeen

Emissions Control Measures

Federal and state regulations require emissions source owners to minimize or reduce air emissions under a number of regulatory circumstances. Four basic types of regulatory structures govern emissions control efforts. As shown in Table 17-1, a source may be affected by more than one of these. Selection of a technology and its cost-effectiveness are largely functions of the following factors:

- Type of fuel burning process. Major categories include boilers, combustion gas turbines, internal combustion reciprocating engines, reheat furnaces, glass melting furnaces, kilns, and other industrial processes. Specific control options are applicable to each.
- Capacity. In each category of fuel-burning equipment, capacity sub-categories are generally considered. Typically, a capacity-related distinction is made between industrial (and commercial) and utility equipment when emissions rules are established. Retrofits are typically more cost-effective in larger capacities.
- Load factor. Actual pollution output depends on emissions rates under various load conditions. The cost-effectiveness of emissions controls improves when run time and load factor are high.
- Type of fuel. A major factor in emissions levels for NO_X, SO_2, and PM, as well as other pollutants, is the chemical composition of the fuel. Critical factors in NO_X and SO_2 emissions are the levels of fuel-bound nitrogen and sulfur. Nitrogen- and sulfur-intensive fuels, such as coal and residual oils, need to achieve greater reductions in order to burn as cleanly as light oils and natural gas.

Table 17-1 Types of Control Requirements Leading to Use of Air Emissions Controls

Type of Control Requirement	Specific Examples
Technology Standards — A source is required to use a prescribed control technology, an equivalent control technology, or a more effective control technology.	• Required use of low-NO_X burners on some NO_X sources in specific ozone nonattainment areas • BACT is predetermined as an emissions level achieved by a specific technology for some new source permit decisions • MACT standards
Emissions Rate Limits — A source is required to meet an emissions limit, which can be an exhaust concentration or an emissions rate limit, e.g., parts per million (ppm), grams per hp–h (g/hp–h), or lbm per million Btu (lbm/MMBtu).	• LAER determinations and most BACT decisions under prevention of significant deterioration (PSD) requirements
Performance Standards – Sources or a group of sources, are given a performance target, typically in the form of emission budgets expressed in a mass quantity of emissions (e.g. tons). Sources are allowed to make reductions onsite or to exchange a portion of the budget (e.g. allowances) with another eligible source. Overall, a net emission budget or cap is met even though the distribution of reductions may be determined by the market. Increasingly this is being used in electric generation markets to facilitate the development of renewable energy. Companies must meet a performance standard established as a percentage of their output being supplied by renewable energy.	• SO_2 allowance program for utilities • South Coast Air Quality Management District's "Reclaim" program for NO_X/SO_X • Federal restrictions on ozone depleting chemicals (e.g., CFCs) • The OTC NO_X Budget Program • Renewable portfolio standards
Air Quality Impact Restrictions — Emissions must be controlled to a level that shows compliance with a downwind air quality concentration goal. Source location and stack height can influence compliance.	• State air toxics requirements • PSD increments protection

Concentrations								Emission – Factors				
Units	ppm (wet)	ppm (dry)	ppm (15% O_2)	ppm (13% O_2)	ppm (5% O_2)	ppm (0% O_2)	g/Nm3	g/BHPh	g/kWh	g/MJ input	g/kg fuel	kg/h
NO_X	1570	1660	1362	1820	3652	4767	3.41	13.70	18.63	2.49	99.9	352.1
CO	57	60	49	66	132	172	0.08	0.30	0.41	0.05	2.2	7.7
HC (as CH_4)	284	300	246	329	660	861	0.22	0.86	1.17	0.16	6.3	22.2
SO_X	516	545	447	598	1199	1565	1.56	6.26	8.52	1.14	45.7	161.0
O_2	13.0	13.7	15.0	13.0	5.0	0	196	790	1070	142.9	5740	20225
CO_2	5.2	5.6	4.6	6.1	12.3	16	108	440	590	78.9	3170	11188
H_2O	5.4	0	0	0	0	0	0	180	250	33.6	1350	4747
Particles	–	–	–	–	–	–	0.12	0.48	0.66	0.09	3.52	12.1
Fuel oil sulphur: 2.25%												

Table 17-2 Different Ways of Stating the Same Emissions from a Low-Speed Diesel Engine. Source: MAN B&W

- Equipment-specific factors. These include factors such as age, design, and logistics.
- Target level of emissions. Required percent reductions or allowable levels for equipment vary depending on regulatory distinctions, as discussed in the previous two chapters. Required control technology will depend on the project and attainment area in which it is located.
- Existing control measures. Many of the control methods described in this chapter can be used in conjunction with others already in place. It should be noted, however, that the effect of controls is not necessarily additive.

Pollutant Units Of Measurement

Emissions measurements and standards may be expressed in a variety of ways. They are generally listed as dry values with a correction factor for oxygen content because added air dilutes the emissions without actually reducing the amount emitted. The concentration of emissions in stack gases are usually corrected to a standard amount of dilution, expressed as excess oxygen content on a dry basis, as if the moisture in the flue gas were removed. Boilers typically use a correction factor to 3% O_2. Combustion gas turbines and reciprocating engines, which have significantly higher dilution levels than boilers, typically use a correction factor to 15% O_2, though sometimes other correction factors, such as 5% O_2, are used.

Table 17-2 shows different ways of stating emissions for an uncontrolled low-speed Diesel engine. The data represents the calculated exhaust gas composition at an ambient temperature of 77°F (25°C) and relative humidity of 50%. Unit values are broken into two categories — concentrations and emissions factors. Common methods of calculating emissions rates are indicated in Table 17-3.

The advantage of relating pollutants to output (i.e., shaft power) rather than input is that output takes into account efficiency. If, for example, two units consume the same amount of fuel, but one produces more useful output, the more efficient unit would be allowed a higher level of pollutant per energy unit consumed relative to the less efficient unit. The same applies to methods used to express process emissions factors. Nevertheless, this is not a common practice for boilers or process heaters.

Parts per million (ppm) is a measure of concentration, usually by volume. To express emissions in units of output and time, ppm must be converted to mass (weight). The conversion depends on units of measurement, molecular weight of the pollutant, variation of flow with time, and the basis on which the numbers are expressed (dry or wet, corrected for oxygen or uncorrected). The configuration of pollutant monitoring systems will dictate the conversions required to generate the corrected values from as measured raw data.

The formula used to correct to a fixed reference oxygen content when expressing NO_X concentrations is:

$$NO_X\,(ppm\ at\ 15\%\ O_2) = NO_X\,(ppm\ at\ X\%\ O_2)\ x\ \left(\frac{20.9-15}{20.9-X}\right) \tag{17-1}$$

Table 17-3 Common Methods and Units of Representation for Emissions Rates

Description	Example
Pollutant per energy unit consumed by heat value or fuel unit	lbm/MMBtu, kg/MJ, lbm/gal oil, or g/kg fuel
Pollutant per energy unit generated (typically used for combustion engines)	lbm/bhp-h, g/hp-h, or g/kWh
Pollutant per unit of product, also known as process emissions factor	lbm/ton of cement or kg/metric ton of glass pulled
Pollutant per unit volume of exhaust	ppm or g/Nm3
Pollutant mass emissions	lbm/h or kg/h

If, for example, the measured NO_X level reading is 20 ppm at 17% O_2, the NO_X level corrected to 15% O_2 would be:

$$20\ ppm\ NO_X\ (at\ 17\%\ O_2)\ x \left(\frac{20.9-15}{20.9-17}\right) = 30\ ppm\ NO_X\ (at\ 15\%\ O_2)$$

The generic conversion of ppm to lbm/MMBtu for NO_X is:

$$NO_X\ (lbm/MMBtu) = NO_X\ (ppm\ at\ X\%\ O_2)\ x \left(\frac{20.9-15}{20.9-X}\right) (F_D\ x\ 1.17\ x\ 10^{-7}) \qquad (17\text{-}2)$$

Where:

F_D for natural gas = 8,740 dscf/MMBtu at standard conditions

F_D for oil = 9,220 dscf/MMBtu at standard conditions

F_D for wood = 9,280 dscf/MMBtu at standard conditions

F_D for coal = 9,820 to 10,140 dscf/MMBtu at standard conditions (depending on coal type)

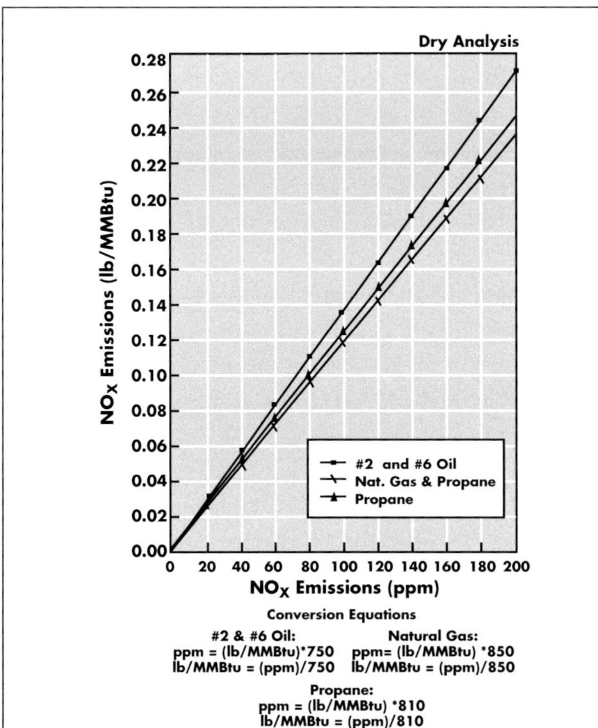

Fig. 17-1 NO_X Emissions Conversion Curves (15% excess air). Source: Cleaver Brooks

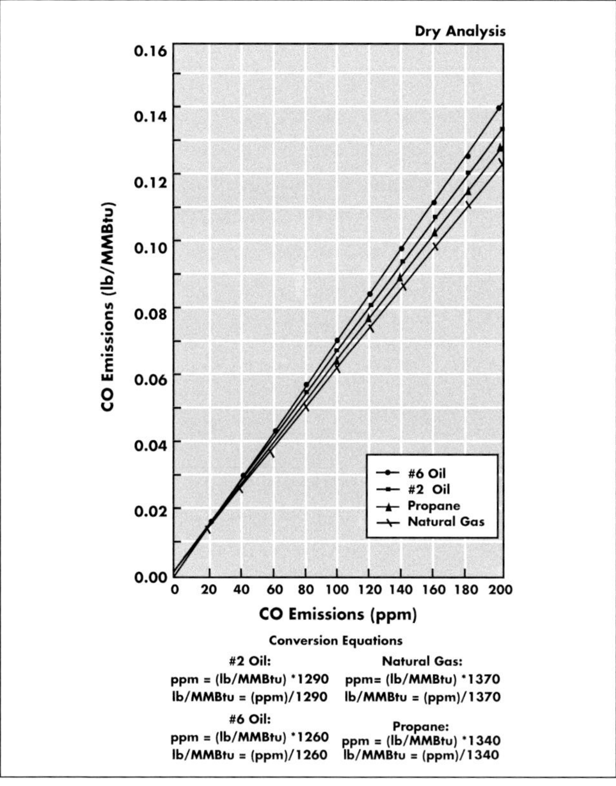

Fig. 17-2 CO Emissions Conversion Curves (15% excess air). Source: Cleaver Brooks

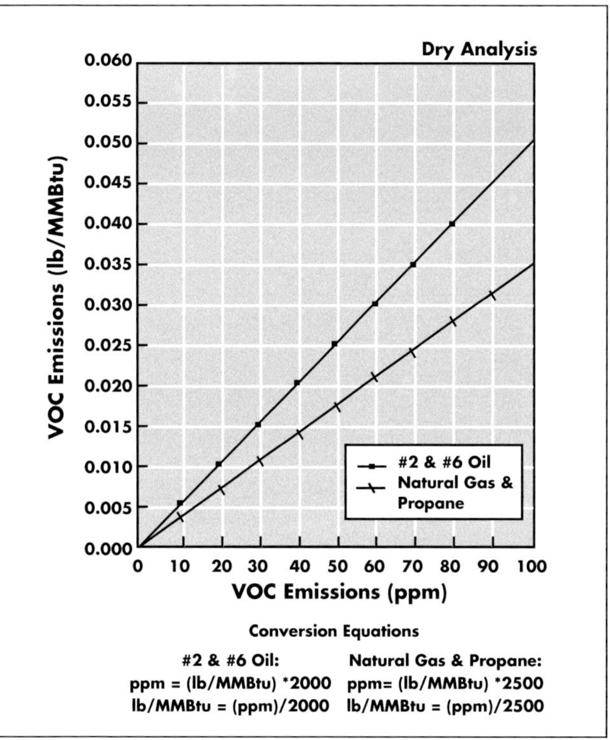

Fig. 17-3 VOC Emissions Conversion Curves (15% excess air). Source: Cleaver Brooks

Figures 17-1 through 17-4 are representative emissions conversion curves for NO_X, CO, VOC, and SO_X. Conversion data is provided for ppm and lbm/MMBtu. Values in these curves are based on dry analysis of boiler operation 15% excess air. Conversion curves and equations are provided for natural gas, propane, and fuel oil.

NO_X FORMATION AND CONTROL TECHNOLOGIES

NO_X Formation

During combustion, NO_X is formed by three fundamentally different mechanisms: thermal NO_X, fuel NO_X, and prompt NO_X formation.

Thermal NO_X results from the thermal fixation of molecular nitrogen and oxygen in the combustion air. Its rate of formation is extremely sensitive to local flame temperature and, to a lesser extent, to local oxygen concentrations. Virtually all thermal NO_X is formed in the region of the flame where the temperature is highest. Maximum thermal NO_X production occurs at a slightly lean fuel-to-air ratio due to the excess availability of oxygen for reaction within the hot flame zone. Control of local flame fuel-to-air ratio is critical in achieving reductions in thermal NO_X.

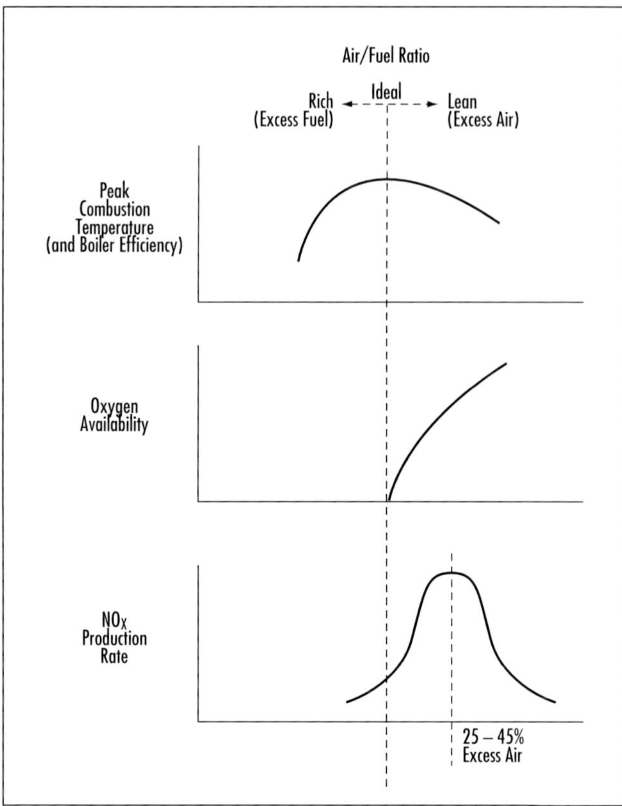

Fig. 17-5 NO_X Production in Boilers as Influenced by Excess Air Availability. Source: TA Engineering

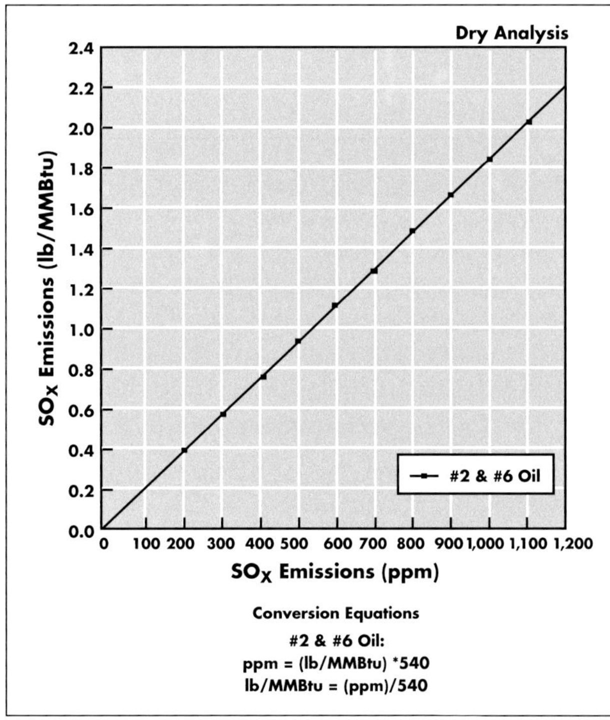

Fig. 17-4 SO_X Emissions Conversion Curves (15% excess air). Source: Cleaver Brooks

Figure 17-5 illustrates NO_X production in boilers as influenced by excess air availability. The three graphs show peak combustion temperature, boiler efficiency, oxygen availability, and NO_X production as a function of air-to-fuel ratio.

Feasible control mechanisms available for reducing the formation of thermal NO_X are:

1. Reduction of local oxygen concentrations at peak temperature
2. Reduction of the residence time at peak temperature
3. Reduction of peak temperature

Fuel NO_X results from the oxidation of organically bound nitrogen in fuels such as coal and heavy oil. Fuel NO_X formation, like thermal NO_X formation, is dominated by local combustion conditions. Its formation rate is strongly affected by the rate of mixing of the fuel and air in general and by the local oxygen concentration in particular. Typically, the exhaust gas NO_X concentration resulting from the oxidation of fuel nitrogen is a fraction of the level that would result from complete oxidation of all nitrogen in the fuel. Figure 17-6 shows the effects of fuel-bound nitrogen on NO_X emissions for fuel oils.

Although fuel-bound nitrogen occurs in coal and petroleum fuels, the nitrogen-containing compounds in petroleum tend to concentrate in the heavy residual and asphalt fractions upon distillation. Therefore, fuel NO_X formation is of importance primarily in residual oil and coal firing. Little or no fuel NO_X formation is observed when burning natural gas and distillate oil.

As with thermal NO_X, controlling excess oxygen is an important part of controlling fuel NO_X formation. A common control strategy involves introducing the fuel with a sub-stoichiometric amount of air (i.e., a "rich" fuel-to-air ratio). In this situation, fuel-bound nitrogen is released in a reducing atmosphere as molecular nitrogen (N_2) rather than being oxidized to NO_X. The balance of the combustion air enters above or around the rich flame in order to complete combustion.

Prompt NO_X is produced by the formation of intermediate hydrogen cyanide (HCN) via the reaction of nitrogen radicals and hydrocarbons in the fuel, and the oxidation of the HCN to NO. The formation of prompt NO_X has a weak temperature dependence and a short lifetime of several microseconds. It is only significant in very fuel-rich flames, which are inherently low-NO_X emitters.

Table 17-4 lists typical uncontrolled NO_X emissions from watertube boilers operating on natural gas, coal, and various grades of oil. Table 17-5 lists emissions from combustion gas turbine operation on natural gas and distillate fuel oil.

Table 17-4 Typical Uncontrolled NO_X Emissions from Watertube Boilers

Fuel	Boiler Type and Capacity	NO_X Emissions (lbm/MMBtu)
Natural Gas:	Packaged (50 MMBtu/hr)	0.14
	Field-Erected (150 MMBtu/hr)	0.23
Distillate Oil:	Packaged (50 MMBtu/hr)	0.13
	Field-Erected (150 MMBtu/hr)	0.21
Residual Oil:	Packaged (50 MMBtu/hr)	0.36
	Field-Erected (150 MMBtu/hr)	0.38
Crude Oil:	Enhanced Oil Recovery Boiler	0.46
Coal:	Wall-Fired	0.69
	Tangentially Fired	0.61
	Spreader Stoker	0.53
	Overfeed Stoker	0.29

Source: EPA

Table 17-5 Uncontrolled NOx Emissions from Combustion Turbines

Fuel Type	lbm/MMBtu
Natural Gas	0.10 - 0.32
Distillate Oil	0.24 - 0.88

Source: EPA

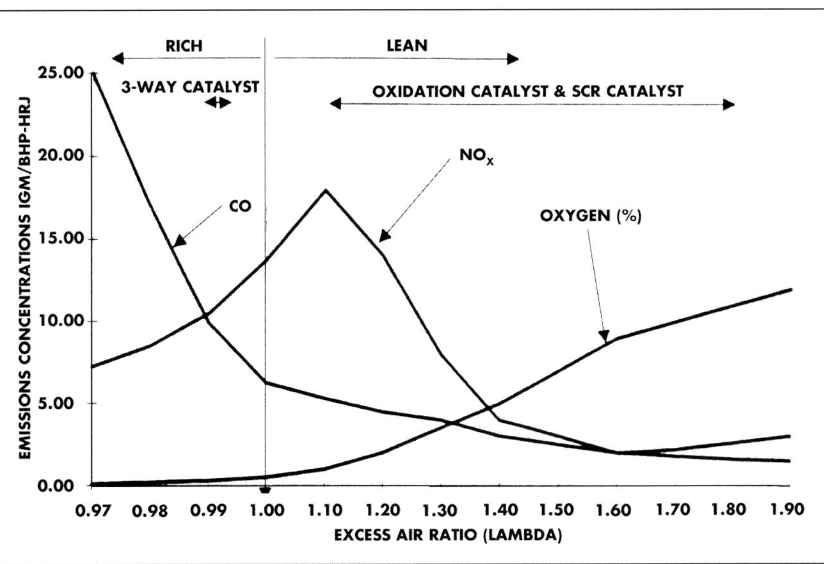

Fig. 17-6 Effects of Fuel-Bound Nitrogen on NO_X Emissions for Fuel Oils. Source: Cleaver Brooks

NO_X Control By Fuel Modification

Fuel switching can result in a reduction in NO_X when the alternate fuel has a lower nitrogen content, burns at a lower temperature, or is easier to control. Natural gas or distillate oils tend to result in lower NO_X emissions than coal or heavy fuel oils. Dual-fuel capability can allow natural gas use in selected periods. For example, some emissions control regulations require lower NO_X emissions levels in summer months when ground level ozone is most problematic. Dual-fuel boilers and gas turbines can operate exclusively on natural gas and dual-fuel Diesel engines, which feature injection of natural gas, can be operated at the highest gas injection levels (about 95 to 99% of the total fuel input), during such periods.

The use of fuel additives has been considered for reducing NO_X formation during fuel burning. An investigation of fuel additives used in a high-pressure gas turbine annular combustor indicated that transition metals added to Jet A fuel as organometallic compounds could reduce NO_X emissions by as much as 30%, with manganese, iron, cobalt, and copper the most effective. However, the study concluded that the resulting pollutants and operational problems would probably not warrant its use.

Fuel denitrification of coal or heavy oils could, in principle, be used to control fuel NO_X formation. The most likely application would be to supplement combustion modifications implemented for thermal NO_X control. Currently, denitrification is used to remove other pollutants, such as in oil desulfurization and chemical cleaning or solvent refining of coal for ash and sulfur removal. The low denitrification efficiency and high cost of these processes currently render them unattractive as emissions control technologies.

NO_X Control By Modification of Combustion Operating Conditions

For all conventional combustion processes, some excess air is required to ensure that all fuel molecules are oxidized. In **low excess air (LEA) firing**, less oxygen is supplied to the combustor than normal, reducing fuel nitrogen conversion to NO_X. LEA may be used with all fossil fuels as a primary NO_X control method or in combination with other NO_X controls.

Increased emissions of CO and smoke and a reduction in flame stability and fuel efficiency occur when excess air is reduced below the required minimum threshold. Adjustment of air registers, fuel injector positions, and over-fire air dampers can reduce required excess air while maintaining adequate air-to-fuel distribution. However, LEA controls require closer operator attention to ensure safe operation. Continuous LEA operations require the use of continuous oxygen (and preferably CO) monitoring and accurate and sensitive air and fuel flow controls.

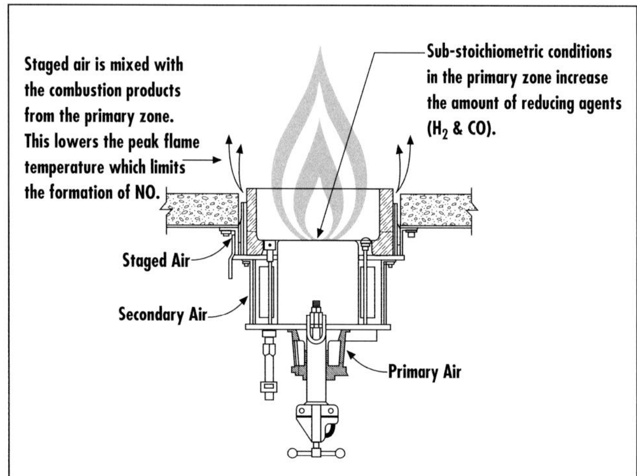

Fig. 17-7 Illustrative Example of Staged Combustion Process.

Alternatively, **lean combustion** (providing large quantities of excess air) can significantly reduce combustion temperature, resulting in much lower NO_X formation, as well as a reduction in CO emissions. With off-stoichiometric or **staged combustion methods**, initial combustion occurs in a primary, fuel-rich combustion zone and is completed in a second fuel-lean zone. Sub-stoichiometric oxygen introduced as primary combustion air into the high temperature, fuel-rich zone reduces fuel NO_X and thermal NO_X formation. Combustion in the secondary zone occurs at lower temperature, reducing thermal NO_X formation. This approach can be used for combustion of all fossil fuels. Figure 17-7 provides a simplified illustration of the staged combustion process.

The staged combustion concept is used in low-NO_X burners, as well as the following operational modifications:

- In **biased burner firing (BBF)**, the low rows of burners are fired more fuel-rich than the upper rows of burners. This may be accomplished by maintaining normal air distribution to the burners while adjusting fuel flow.
- **Burners out of service (BOOS)** combustion uses individual burners or rows of burners to admit air only with active (fuel-admitting) burners firing more fuel-rich than normal.
- In **overfire air (OFA)** combustion, the burners are fired more fuel-rich than normal, while the remaining combustion air is admitted through overfire air ports or an idle top row of burners.

Flue Gas Recirculation (FGR), or **Exhaust Gas Recirculation (EGR)**, is based on recycling a portion of flue gas back to the primary combustion zone. This system reduces NO_X formation by two mechanisms. First, heating in the primary combustion zone of the inert combustion products contained in the recycled fuel gas lowers the peak flame temperature. Second, FGR reduces thermal NO_X formation to some extent by lowering the oxygen concentration in the primary flame zone.

Recycled exhaust gas may be pre-mixed with the combustion air or injected directly into the flame zone. Direct injection allows for more precise control of the amount and location of FGR. In general, increasing the rate of recirculation results in a corresponding reduction in NO_X formation. Practical limits are dictated by equipment size, flame stability, and minimum oxygen requirements.

The use of FGR has several limitations. First, the

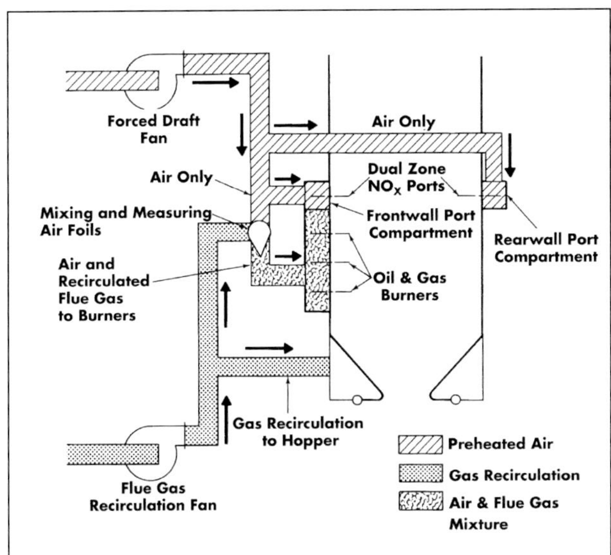

Fig. 17-8 Flue Gas Recirculation Low-NO_X System for Oil and Gas Firing. Source: Babcock and Wilcox

decrease in flame temperature alters the distribution of heat and lowers fuel efficiency. Second, because FGR reduces only thermal NO_X, the technique is applied primarily to natural gas or distillate oil. Third, FGR can be costly to retrofit because of required duct work, furnace penetrations, added fan capacity, and possible physical obstructions.

Figure 17-8 is a diagram of a large gas or oil boiler featuring FGR. Flue gas is introduced in the sides of the secondary air metering foils and exits through slots downstream of the air measurement traps. This method provides thorough mixing of flue gas and combustion air ahead of the burners and does not affect operation of the air metering foils. The required gas recirculation fan must be capable of overcoming losses through the flues, ducts, mixing devices, and the burners themselves. Figure 17-9 shows two profile views of an FGR system, integrated in an industrial watertube boiler. Figure 17-10 shows two

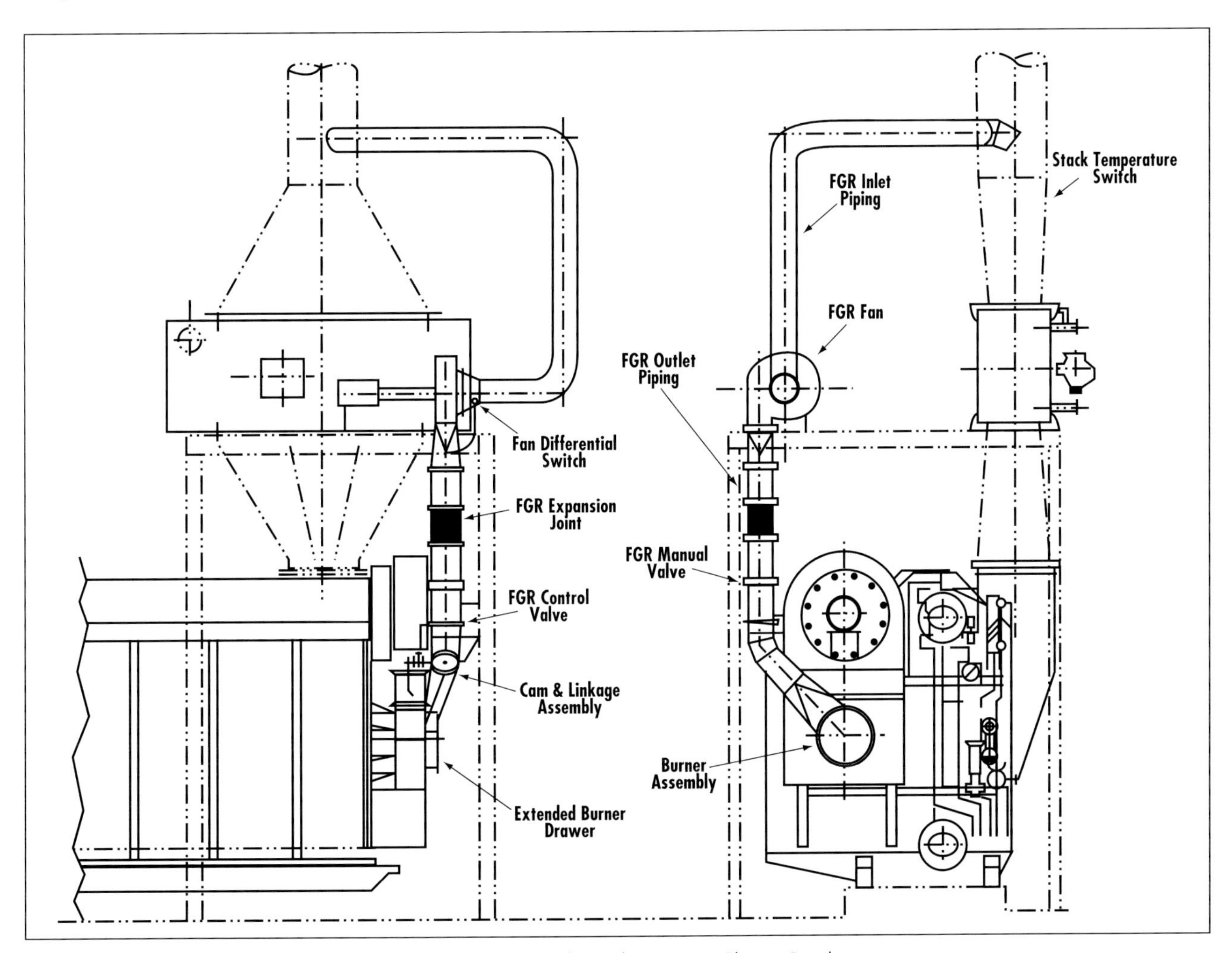

Fig. 17-9 Two Profile Views of FGR System in Industrial, Watertube Boiler. Source: Cleaver Brooks

500 boiler horsepower (BHP) industrial firetube boilers featuring FGR systems. Notice the ducted flue gas from the exhaust stack to the controlled burner. Each of these boilers produce 17,250 lbm/h (7,825 kg/h) of saturated steam at 120 psi (8.3 bar).

Table 17-6 shows NO_X and CO emissions for industrial, gas-fired firetube boilers. Values are included for uncontrolled emissions and for representative boiler models featuring FGR. Values are listed in both ppm on a dry basis (ppmd), corrected to 3% O_2 (@ 3% O_2), and in lbm/MMBtu of natural gas input.

Fig. 17-10 Two 500 BHP Industrial Firetube Boilers Featuring FGR Systems. Source: Superior Boiler Works, Inc.

Tables 17-7 and 17-8 show NO_X and CO for similar boilers operated on No. 2 oil and No. 6 oil, respectively. A wide range of emissions levels are typical of oil-fired boilers, depending on fuel nitrogen content, equipment type, and operating factors.

Reduced Air Preheat (RAP) is implemented by bypassing all or a fraction of the exhaust gas around an existing combustion air preheater. Reducing combustion air preheat lowers the primary combustion zone peak temperature and reduces thermal NO_X formation. Although NO_X emissions decrease significantly with reduced combustion air temperature, a significant loss in thermal efficiency

Table 17-6 Emissions Characteristics for Uncontrolled and FGR-Equipped Natural Gas-Fired Firetube Boilers

		Uncontrolled		System 1		System 2	
Pollutant	**Units**	**Typical**	**Range**	**Typical**	**Range**	**Typical**	**Range**
NO_X	ppmd @3% O_2	85	50-110	25	<30	15	<20
	lbm/MMBtu	0.1	0.06-0.13	0.03	<0.036	0.018	<0.024
CO	ppmd @3% O_2	200	30-275	100	10-130	100	25-200
	lbm/MMBtu	0.15	0.022-0.33	0.075	0.007-0.095	0.075	0.018-0.15

Source: Cleaver Brooks

Table 17-7 Emissions Characteristics for Uncontrolled and FGR-Equipped No. 2 Oil-Fired Firetube Boilers

		Uncontrolled		Equipped with FGR	
Pollutant	**Units**	**Typical**	**Range**	**Typical**	**Range**
NO_X	ppmd @3% O_2	150	115-175	115	90-140
	lbm/MMBtu	0.2	0.15-0.23	0.15	0.12-0.19
CO	ppmd @3% O_2	40	20-90	40	20-90
	lbm/MMBtu	0.031	0.016-0.07	0.031	0.016-0.07

Source: Cleaver Brooks

Table 17-8 Emissions Characteristics for Uncontrolled and FGR-Equipped No. 6 Oil-Fired Firetube Boilers

		Uncontrolled		Equipped with FGR	
Pollutant	**Units**	**Typical**	**Range**	**Typical**	**Range**
NO_X	ppmd @3% O_2	350	*	300	*
	lbm/MMBtu	0.47	*	0.4	*
CO	ppmd @3% O_2	40	20-90	40	20-90
	lbm/MMBtu	0.031	0.016-0.07	0.31	0.016-0.07

Source: Cleaver Brooks
*Range cannot be accurately predicted.

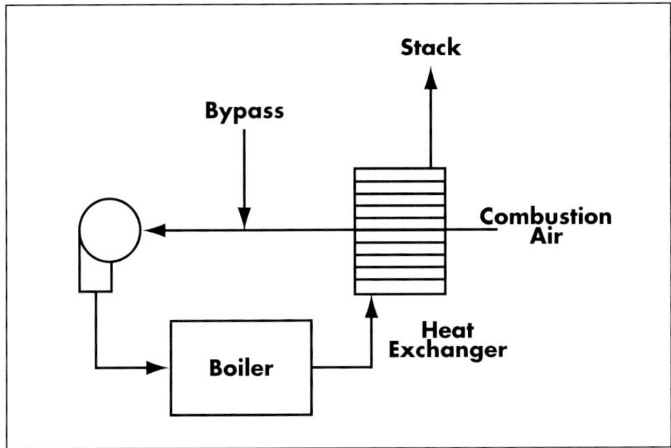

Fig. 17-11 Line Diagram of Reduced Air Preheat (RAP) Process.

will occur as a consequence of bypassing the air preheaters. Installing an economizer to preheat feedwater or condensate return, or enlarging the surface area of existing economizers, can partially recover the efficiency loss. RAP is illustrated in Figure 17-11.

Reburn, also referred to as in-furnace NO_X reduction or staged fuel injection, is the only NO_X control approach implemented in the furnace zone (i.e., the post-combustion, pre-convection section). In reburn, the products of combustion pass through a secondary flame, or fuel-rich combustion zone, downstream of the burner (primary combustion zone). Sufficient air is then supplied further downstream to complete the oxidation process.

Although coal or oil can be used for reburning, natural gas is best suited for this purpose because it contains no fuel-bound nitrogen, ash, sulfur, or other substances that cause pollution or create corrosive compounds in the boiler. In addition, natural gas mixes readily and reacts quickly with the flue gases and does not require pulverizers, preheating, or other forms of preparation. To date, gas reburn has shown no significant adverse effects on boiler thermal performance. As indicated in Figure 17-12, the process includes the following:

- In the primary combustion zone, original burners fired by coal, oil, or gas are turned down by 10 to 20%. The burners or cyclones may be operated at the lowest excess air consistent with normal commercial operation to minimize NO_X formation and to provide appropriate conditions for reburning.
- In the gas reburning zone, natural gas (between 10 and 20% of boiler heat input) is injected above the primary combustion zone. This creates a fuel-rich region where hydrocarbon radicals react with NO_X to form molecular nitrogen. The natural gas may be mixed with recirculated flue gases prior to injection to promote better mixing within the boiler. Gas reburning injectors require new boiler-wall penetrations on most units.
- A separate overfire air system redirects air from the primary combustion zone to a burnout zone downstream of the gas reburning reaction zone to ensure complete combustion of unreacted fuel and combustible gases. This separate overfire air system requires boiler penetrations and ducting.

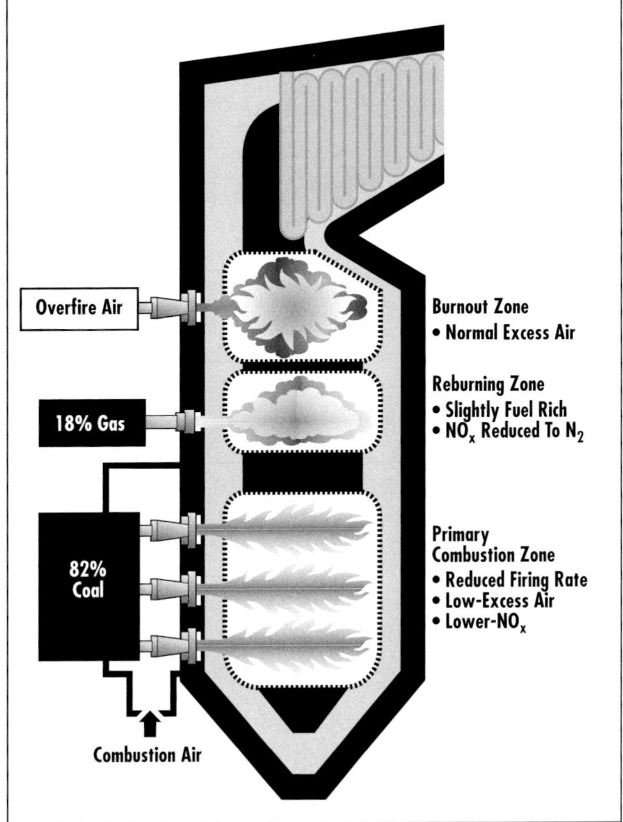

Fig. 17-12 Functional Diagram of Natural Gas Reburning Process. Source: GRI

Gas reburn can be applied to all types of utility boilers fired by coal (including wet-bottom and cyclone units), oil, or natural gas. The key requirement is adequate height above the main firing zone for reburning and burnout residence times. Boiler control system upgrades are usually included so that new parameters can be programmed into the boiler system for safe start-up, shut-down, and trip conditions. These control changes have been reviewed and approved by major boiler insurers.

Steam or water injection into the combustion zone can decrease flame temperature, thereby reducing the formation of thermal NO_X. Because steam and water act as a thermal ballast, it is important that the ballast reach the primary flame zone. To accomplish this, the ballast may be injected into the fuel, combustion air, or directly into the combustion chamber.

Water injection is sometimes preferred over steam due to its availability, lower cost, and higher heat absorbing capacity. However, water injection may exhibit some undesirable effects, including decreased thermal efficiency, increased maintenance costs, and reduced service life in gas turbines. This technology exhibits high operating costs chargeable to emissions control, with a fuel and efficiency penalty typically about 10% for utility boilers and about 1% for gas turbines. Water injection has not, therefore, gained much acceptance as a NO_X reduction technique for stationary combustion equipment, except for gas turbine applications where on-site steam is not available. Fuel-water emulsion (FWE) is a process that involves adding water to the liquid fuel used in stationary and marine Diesel engines. FWE reduces the peak temperature in the vicinity of the fuel droplets, and thus limits the formation of NO_X.

Gas- or coal-fired boilers that are equipped for standby oil firing with steam atomization already have a simple means for steam injection. Other installations may require a developmental program to determine the required degree of atomization and mixing with the flame, the optimum point of injection, and the quantities of water or steam necessary to achieve the desired effect.

When steam is injected into a gas turbine's combustor for emissions control, the small increase in exhaust mass flow increases the power output, usually by 3 to 10%. Some gas turbine designs lend themselves to modifications that allow injection of significantly greater quantities of steam specifically for power augmentation, with capacity increases up to 50%. Refer back to Figure 12-12, which shows gas turbine capacity, in MW, as a function of inlet air temperature and steam injection rates, in Mlbm/h.

Water- or steam-to-fuel ratio is a key variable affecting NO_X control in gas turbines. Typical steam-injection ratios are between 0.5 and 2.0 lbm steam per lbm fuel; water-injection ratios are generally below 1.0 lbm water per lbm fuel. Caution must be used with wet injection ratios, since higher ratios can increase CO and VOC emissions and reduce fuel efficiency. The quality of the water or steam is also an important consideration, since impurities may damage the combustion section and the turbine blades. The type of water treatment required will vary depending on the turbine design and the quality of the raw water, but can become a significant cost and reliability factor. Water treatment systems can include coagulation, filtration, absorption, ion exchange, reverse osmosis, or demineralization.

Burner Modifications for NO_X Control

Burner design influences the amount of NO_X formed during the combustion process and **low-NO_X burners** have been used or tested in a variety of boiler and process heating applications. Figure 17-13 is a high-capacity low-NO_X burner for gas and oil firing. The fuel elements are housed in a central flame stabilizer to improve flame stability and turndown while separating the fuel elements from the combustion air. Combustion air is regulated by dual air zones with multistage adjustable swirl vanes. Figure 17-14 illustrates a dual-fuel, staged low-NO_X burner designed to achieve low-NO_X operation without the use of flue gas recirculation. The burner, which may be applied in single or multiple units, features a rugged design with no moving parts and wide stability limits.

Figure 17-15 shows typical NO_X emissions rates for natural gas-fired industrial boilers in lbm/MMBtu. Emissions rates are shown for uncontrolled operation with low-NO_X burners and with low-NO_X burners plus 10% and 20% FGR.

Staged air burners are two-stage combustion burners that are fired fuel-rich in the primary zone (first stage). They increase flame length, delay completion of combustion, and limit peak flame temperature. The reduced temperature in the primary combustion zone inhibits NO_X formation. Staged air burners generally lengthen the flame and they are thus limited to installations large enough to avoid impingement. The installation of replacement burners may require substantial changes in burner hardware, including air registers, air baffles and vanes, fuel injectors,

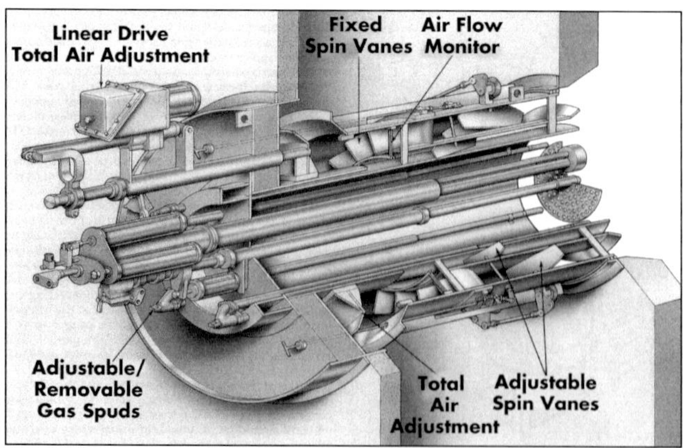

Fig. 17-13 Low-NO_X Burner for Gas and Oil Firing in Very Large Capacity Boiler. Source: Babcock and Wilcox

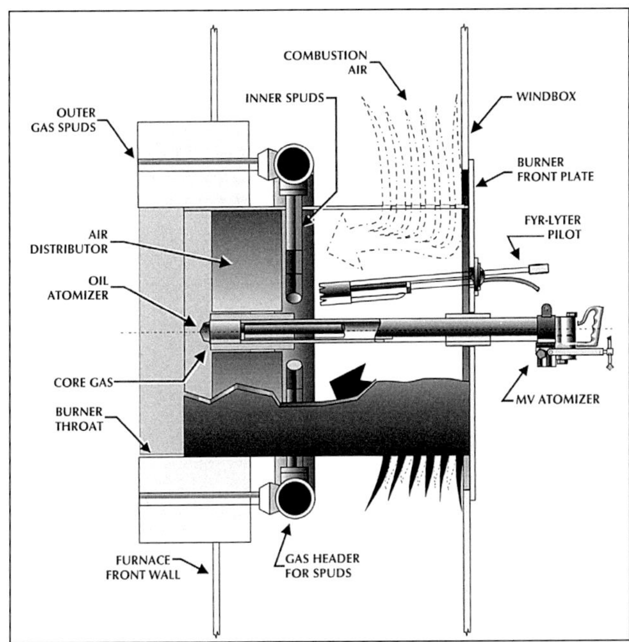

Fig. 17-14 Dual-Fuel, Staged Low-NO_X Burner.
Source: Coen Company

and throat design. Existing burners can incorporate staged air burner features by modifying fuel injection patterns, installing airflow baffles, or re-shaping the burner throat. Staged air burners can be used for all fuel types.

Figure 17-16 shows a packaged industrial watertube boiler featuring a cyclonic burner and two-staged combustion. In the first stage, a fuel-rich mixture is injected tangentially at high velocity into the cylindrical combustion chamber. The swirling flow pattern adds a significant convective component to the radiant heat transfer from the flame and limits flame temperature. In the second stage, additional air is introduced to complete combustion. A steam-injection system further reduces NO_X without significantly affecting the unit's fuel efficiency. NO_X emissions levels of 30 ppm and CO emissions levels of 50 ppm have been reported while maintaining a fuel efficiency of 84% or greater.

Staged fuel burners also use two-stage combustion, but mix a portion of the fuel and all of the air in the primary combustion zone. The high level of excess air greatly lowers the peak flame temperature achieved in the primary combustion zone. The secondary fuel is injected through nozzles, which are positioned around the perimeter of the burner. Because of its high velocity, the fuel gas entrains furnace gases and mixes them with first-stage combustion products, simulating flue gas recirculation and cooling the flame. The staged fuel burner can be operated with lower excess air than the staged air burner due to the mixing caused by high-pressure second-stage fuel injection. An additional advantage of the staged fuel burner is its compact flame. Unlike staged air burners, staged fuel burners are only designed for gas firing.

Burner spacing is a significant design element in NO_X control. The interaction between closely spaced burners, especially in the center of multiple-burner installations, increases flame temperature at these locations. Therefore, in most new utility boiler designs, vertical and horizontal burner spacing has been widened to provide more cooling

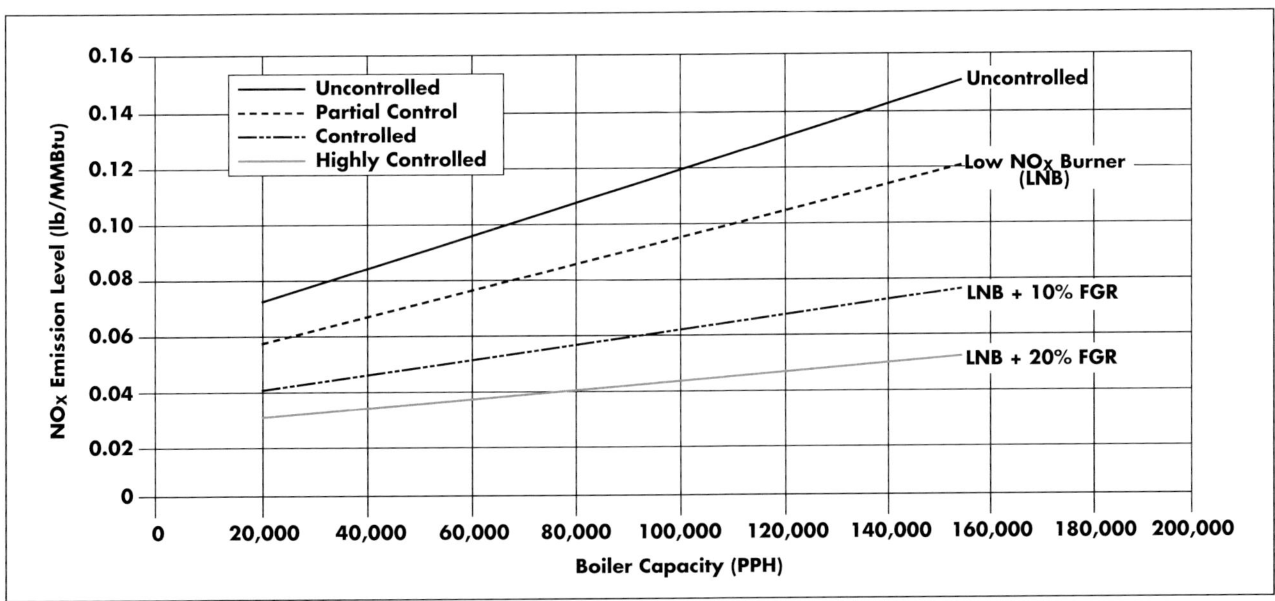

Fig. 17-15 Typical NO_X Emissions Rates for Gas-Fired Industrial Boilers with Various Control Technologies.

of the burner zone area.

Thermal NO_X formation generally increases as the heat release rate or combustion intensity increases. Load reduction, or derating, as a NO_X control option is applicable to all fuel types in existing units. In new installations, furnace enclosures are built to allow sufficient time for complete combustion with slower and more complete heat release rates.

Fig. 17-16 Packaged, Industrial Watertube Boiler Featuring Cyclonic Burner and Two-Staged Combustion. Source: Donlee Technologies

Reduced firing rates can lead to several operational problems. The reduced mass flow can cause improper fuel-air mixing during combustion, creating CO and soot emissions and reducing fuel efficiency. Increased excess air levels may improve mixing, but can further reduce efficiency and increase fuel NO_X generation. When the combustion unit is designed for a reduced heat release rate, the problems associated with derating are largely avoided.

Catalytic combustion is combustion occurring close to a solid surface that has a special catalyst coating. The catalyst accelerates the rate of a chemical reaction and can itself be recovered unchanged at the end of the reaction. As a result, reactions occur faster and with lower energy requirement. Catalytic combustion can be effective in reducing NO_X emissions, as well as emissions of CO and unburned hydrocarbons, by allowing combustion to occur at lower temperatures. However, at present, this control option has limited applicability due to catalyst degradation at temperatures above 1,830°F (1,000°C).

Reciprocating Engine Modifications

The differences between Diesel-cycle and Otto-cycle engines (described in Chapter 9) affect the production and control of NO_X. While the Diesel cycle is characterized by compression of air only with heat addition at constant pressure, the Otto cycle includes compression of an air/fuel mixture with heat addition at constant volume. Compression ignited (Diesel-cycle) engines produce high localized flame temperatures and, therefore, higher thermal NO_X emissions than lean-burn spark-ignited (Otto-cycle) engines. However, dual-fuel compression-ignited engines using a mixture of liquid and gaseous fuels can achieve significantly lower NO_X emissions than liquid fuel engines. Rich-burn (below the stoichiometric level) spark-ignited engines generally produce moderate levels of NO_X emissions, as well as high levels of CO and HC emissions.

Very lean mixtures generally cause slow burning combustion and can reduce bmep and, therefore, thermal fuel efficiency and capacity. To minimize thermal fuel efficiency losses, it is necessary to achieve thorough fuel-air mixing to limit peak flame temperatures and increase flame speed. Combustion chambers are designed for a high degree of turbulence. Swirl is used to promote more rapid mixing between the inducted air charge and injected fuel in compression ignition engines and is also used to speed up the combustion process in spark-ignited engines.

Figure 17-17 compares exhaust NO_X output to air-to-fuel ratio for a natural gas-fired spark-ignition engine. The graph shows NO_X emissions in ppmd (@ 15% O_2) versus air-to-fuel ratio and Lambda (λ). With rich combustion (left side of graph), NO_X decreases due to lower combustion temperatures and lack of oxygen. With lean combustion (right side of graph), NO_X initially reaches a peak because combustion temperature remains high and there is an abundance of oxygen. In establishing optimal air-to-fuel ratios, a balance must be achieved between numerous variables, including NO_X,

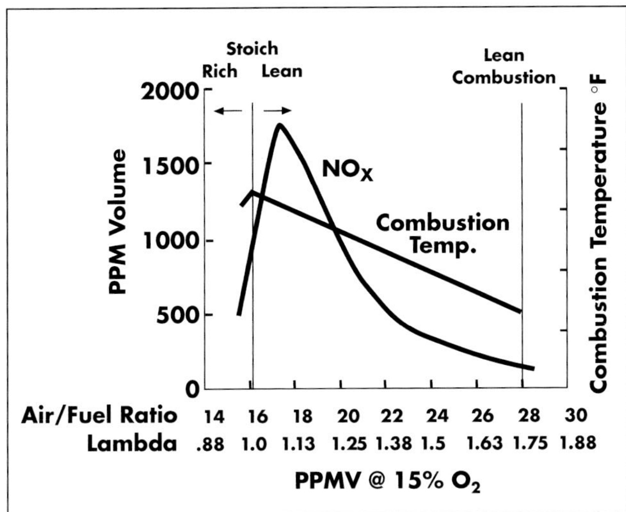

Fig. 17-17 NO_X Emissions of Natural Gas-Fired Spark-Ignition Engine as a Function of Air-to-Fuel Ratio.
Source: Waukesha Engine Div.

CO, and HC emissions, as well as bmep and thermal fuel efficiency. In spark-ignition engines, care must be taken to avoid both knocking limits and incomplete combustion. Figure 17-18 illustrates the relationship of these variables. As can be seen, there is a narrow band between knocking limits and incomplete combustion, where NO_X formation is lowest, fuel efficiency and specific load are greatest, and CO and HC emissions are relatively low.

Several methods are used for precise air-to-fuel adjustment. In one cogeneration system, the microprocessor controller considers engine output, air-to-fuel mixture pressure (boost pressure), and mixture temperature after

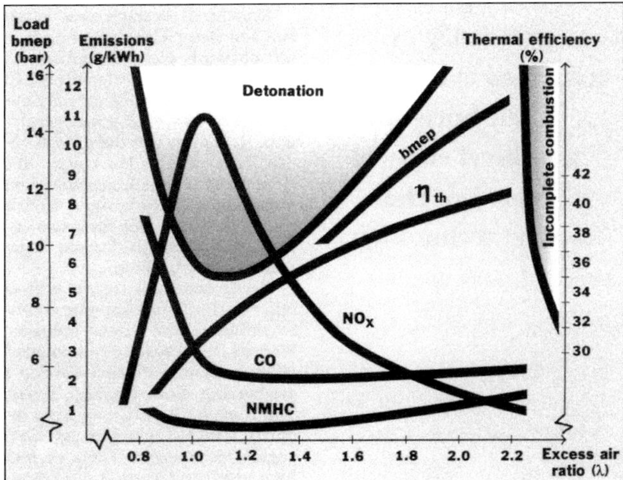

Fig. 17-18 Relationship Curves for Various Key Factors in Optimizing Air-to-Fuel Ratios for Spark-Ignited Engines.
Source: Wartsila Diesel

the intercooler. The signal output by the controller moves the adjusting cone of the air-to-gas mixer into the desired position to set the required ratio. This system is reportedly capable of achieving NO_X emissions levels of 45 ppmd (@ 15% O_2), or about 0.5 g/bhp-h (0.7 g/kWh). In injection-type reciprocating engines, including all Diesel and many dual-fuel and natural gas engines, the air-to-fuel ratio can be adjusted for each cylinder. A precise, homogenous combustion mixture is necessary in all cylinders in order to minimize NO_X emissions while avoiding misfire and detonation operating regions.

Ignition in a normally adjusted reciprocating engine is set to occur shortly before the piston reaches its uppermost position (top dead center, or TDC). At TDC, the air or air-to-fuel mixture is at maximum compression and power output and fuel consumption are optimum. **Ignition timing retard** is a NO_X control technique that causes more of the combustion to occur during the expansion stroke, thus lowering peak temperature, pressure, and residence time. Typical retard values range from 2 to 6 degrees, depending on the engine. Some increase in CO emissions and a reduction in fuel efficiency generally does occur.

Pre-stratified charge (PSC) combustion is a retrofit system that has been applied to small and mid-sized 4-stroke-cycle, carbureted, rich-burn reciprocating engines. Controlled amounts of air are introduced into the intake manifold before the fresh air-to-fuel mixture, causing stratification that decreases the combustion temperature. The sequence must be precisely controlled to prevent any mixing of the dilution layer and the fresh mix. If the two layers have time to mix, misfires can occur. Engines require an increase in intake air capacity via charging or derating of maximum power output when this system is used.

Pre-combustion chamber systems, which are described in detail in Chapter 9, are commonly used with newer lean-burn spark-ignition and certain dual-fuel compression-ignition designs. They allow the engine's main combustion chamber to operate at higher air-to-fuel ratios than open chamber lean-burn engines. To ensure proper combustion, a small volume of fuel-rich mixture (below the stoichiometric level) is burned in a pre-combustion chamber or ignition cell, which comprises about 1 to 5% of the clearance volume located in the cylinder head. This fuel mixture may be spark or compression ignited. The flame from the cell reaches into the main combustion chamber and ignites the remaining lean mixture. This technology, which is also sometimes referred to as torch ignition or stratified charge, is more expensive than open chamber design, but produces far lower NO_X levels than traditional lean-burn technology.

Figure 17-19 illustrates a pre-chamber design for a spark-ignited engine and Figure 17-20 illustrates the injection of pilot fuel into the pre-chamber of a dual fuel compression-ignition engine. Reportedly, NO_X emissions levels of under 0.5 g/hp-h (0.75 g/kWh) can be achieved along with CO and non-methane HC (NMHC) emissions of under 1.5 g/hp-h and 0.75 g/hp-h (2.0 g/kWh and 1.0 g/kWh), respectively, while maintaining high simple-cycle thermal fuel efficiency.

A key factor in dual-fuel engine designs is the ability to minimize the amount of pilot oil used. Figure 17-21 shows the influence of the pilot fuel quantity on pollutant emissions and thermal fuel efficiency, with direct pilot fuel injection. While thermal fuel efficiency and CO and HC emissions remain fairly constant from 1 to 3.5% pilot fuel proportion, NO_X emissions vary greatly. In the figure, the optimal pilot fuel proportion is indicated at about 1%.

Gas Turbine Dry Combustion Technology

Dry low-NO_X (DLN) combustion technology utilizing lean pre-mixed combustion continues to be the primary trend in development efforts for combustion gas turbines. Premixing marks a change from traditional diffusion flames, in which mixing and combustion take place simultaneously with high temperature peaks and short residence time. NO_X emissions levels to less than 5 ppmd (@ 15% O_2) in several systems have been reported, though 9 to 20 ppmd is typically the guaranteed level when combusting natural gas. There are numerous dry combustion designs. Three challenges associated with very lean combustors are:

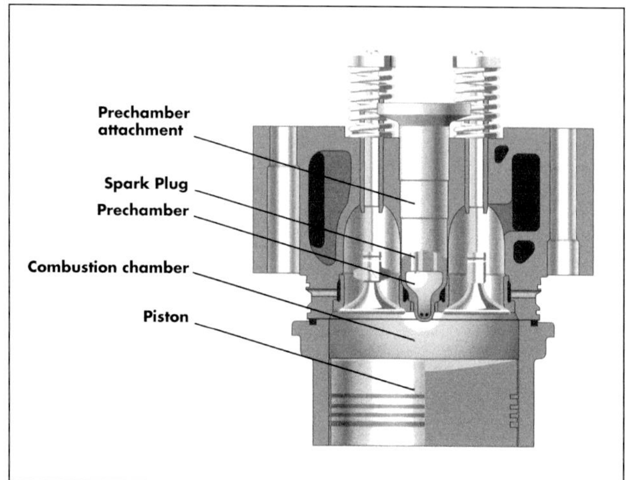

Fig. 17-19 Illustration of Pre-Chamber Design for Spark-Ignited Reciprocating Engine. Source: MAN B&W

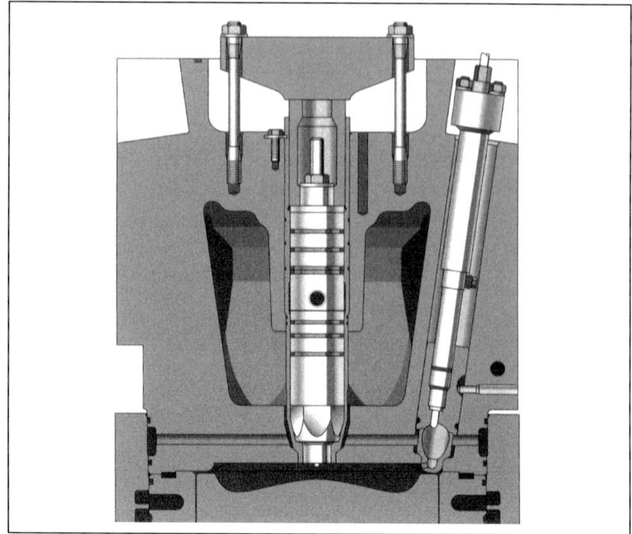

Fig. 17-20 Illustration of Pre-Chamber with Respect to the Main Combustion Chamber on Dual-Fuel Compression-Ignition Reciprocating Engine. Source: MAN B&W

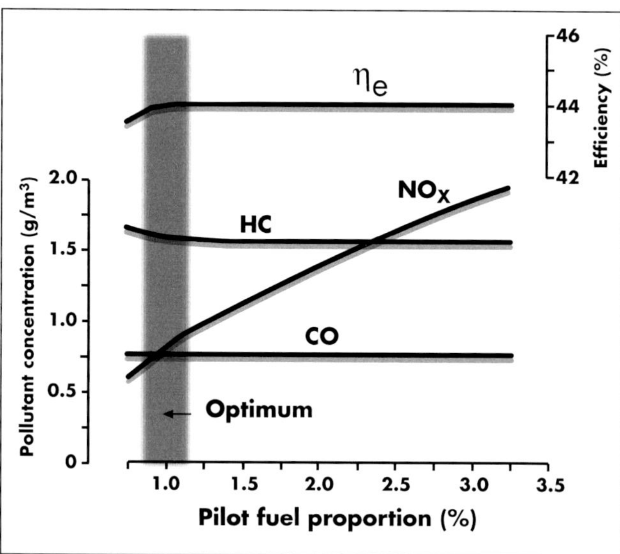

Fig. 17-21 Influence of the Pilot Fuel Quantity on Pollutant Emissions and Thermal Fuel Efficiency with Direct Pilot Fuel Injection. Source: MAN B&W

- The need to maintain flame stability at the design operating point.
- Precise turndown capability over the required load range.
- Balancing NO_X reductions with CO (and VOC) increases that occur with low flame temperature and short residence time.

Figure 17-22 illustrates a basic pre-mixing design. This is accomplished by operating the combustor primary zone at increased airflow and lower average temperature. Dilution zone airflow is reduced to keep total combus-

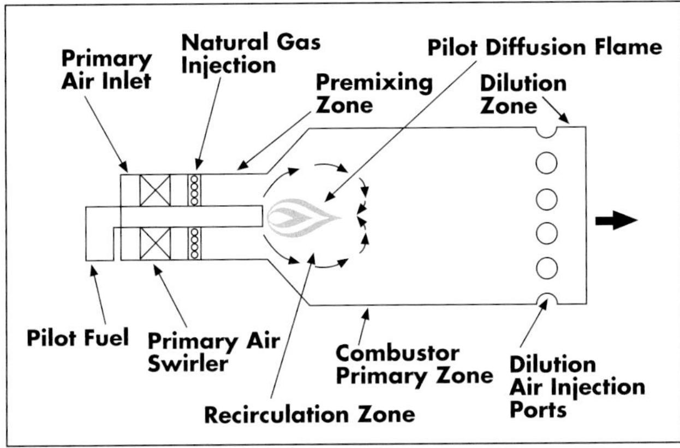

Fig. 17-22 Lean Pre-Mixed Combustor Concept. Source: Solar Turbines

Fig. 17-23 Lean Pre-Mixed Gas Turbine Low-NO_X Combustor Liner. Source: Solar Turbines

Fig. 17-24 12,400 hp (9,250 kW) Heavy-Duty Industrial Gas Turbine with Dual Silo-Type DLN Combustors. Source: MAN GHH

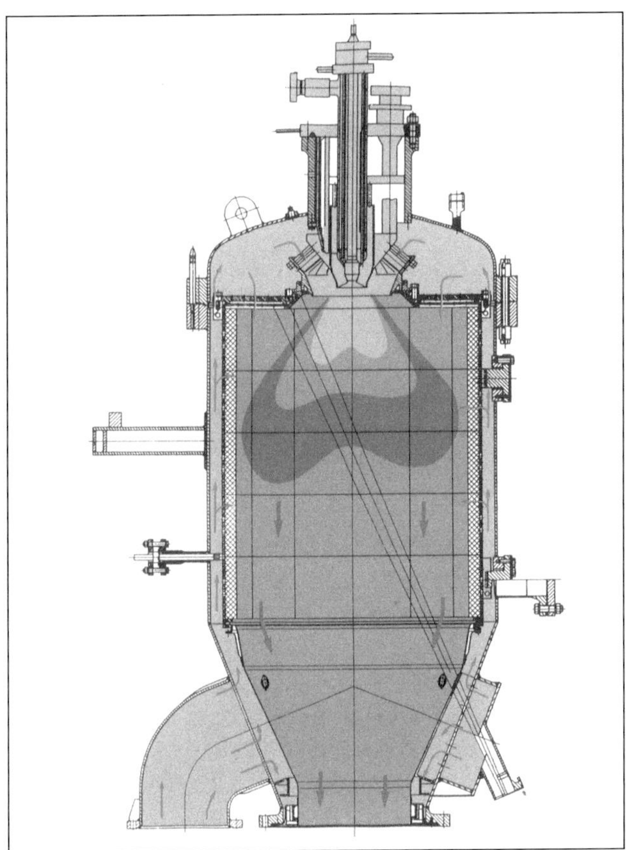

Fig. 17-25 Cutaway Illustration of Silo Combustion Chamber with Hybrid Burner. Source: MAN GHH

tor airflow and exit temperature unchanged. Gas turbine output and heat rate are therefore not affected. To provide a uniform flame temperature and avoid hot spots, the fuel and primary zone air are pre-mixed upstream. Figure 17-23 shows a lean pre-mixed gas turbine combustor liner for a 5,000 hp (3,700 kW) capacity unit. The sheet metal liner is similar to conventional combustor liners in geometry and construction. Low metal temperatures are maintained through film-cooling, using a small fraction of the compressor discharge air.

Figure 17-24 shows a 12,400 hp (9,250 kW) industrial gas turbine with dual silo-type DLN combustors. Figure 17-25 is a cutaway illustration of this silo-type combustor. Air at about 660°F (350°C) enters the lower combustion chamber area, feeding the burner and cooling the mixing tube, flame tube shell, ceramic holder, and flame tube end. Pre-mix burners are subject to limited operating ranges, since pre-mix flames are extinguished under conditions with high air-to-fuel ratios. In these particular units, a stabilizing pilot flame allows pre-mix operation over a range of 40 to 100% of full load with compressor inlet guide vanes. The burner operates in diffusion mode on start-up and under very low load conditions. In the diffusion mode, mixing and combustion take place simultaneously, producing high temperature peaks and correspondingly high levels of NO_X and CO.

Figure 17-26 shows NO_X and CO emissions data for operation on natural gas, in ppmd (@ 15% O_2), versus generator output. Curves are provided for varying levels of pilot fuel use. By minimizing pilot fuel, units are reportedly able to achieve NO_X emissions levels of 10 ppmvd or less through much of the operating regime. Figure 17-27 shows similar emissions data for operation on fuel oil.

Figure 17-28 illustrates a dual-fuel, double-cone DLN burner that has been applied in several combustor designs, achieving near single-digit ppm levels in NO_X. Compressed air and fuel are separately injected into the double-cone burner, where a homogenous lean fuel-to-air mixture is formed and spiraled to form a vortex.

One recent application of this burner is in a sequential combustion system design for a large capacity industrial turbine, illustrated in Figure 17-29. This design, which is discussed in Chapter 10, uses two separate annular combustors in what may be viewed as a type of reheat process. After adding about 60% of rated fuel input in the combustion chamber on the right, the gas expands through a high-pressure turbine stage where its pressure is reduced from about 450 psig (30 bar) to 225 psig (15 bar). The gas then enters into the second (SEV) combustion chamber at about 1,830°F (1,000°C) and self-ignites before expanding through the low-pressure turbine section. In a conventional combustor, self-ignition is avoided because it can result in combustor damage and high emissions. In this process, however, sufficient mixing is achieved, aided by the vortex flow from the burners, to eliminate flame temperature peaks. The process is aided by the lower O_2 content in the second combustion chamber, which inhibits NO_X formation. Typically, at start-up and idle, fuel is injected only into the primary zone.

As shown in Figure 17-30, there are two basic staged combustor designs: fuel-staged and air-staged. These designs may also be combined. In the simplest fuel-staged design, fuel flow is divided between two flame zones so that the amount of fuel fed to a stage is matched to the amount of available air at each operating condition. In an

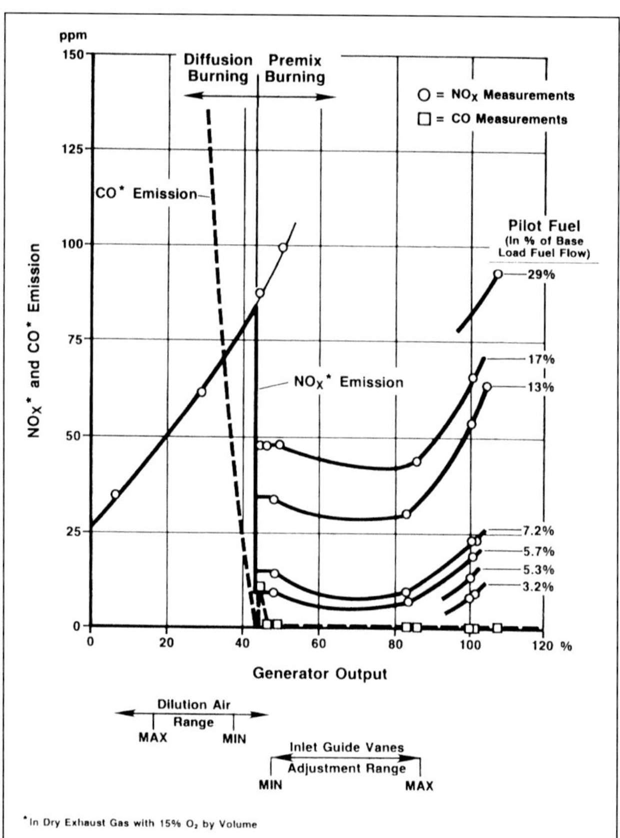

Fig. 17-26 Emissions Data for DLN Combustor Operating on Natural Gas. Source: Siemens Power Corp.

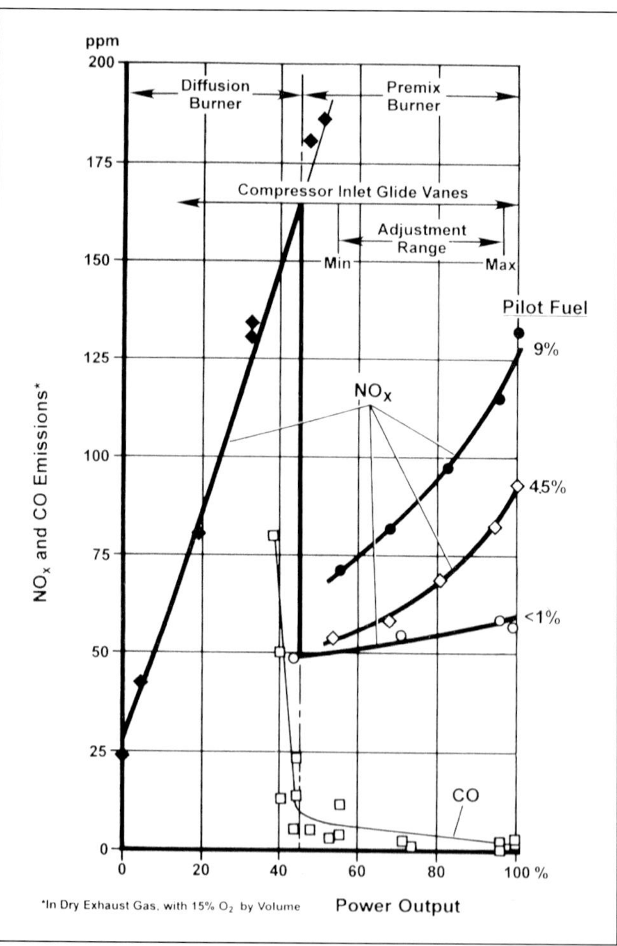

Fig. 17-27 Emissions Data for DLN Combustor Operating on Fuel Oil. Source: Siemens Power Corp.

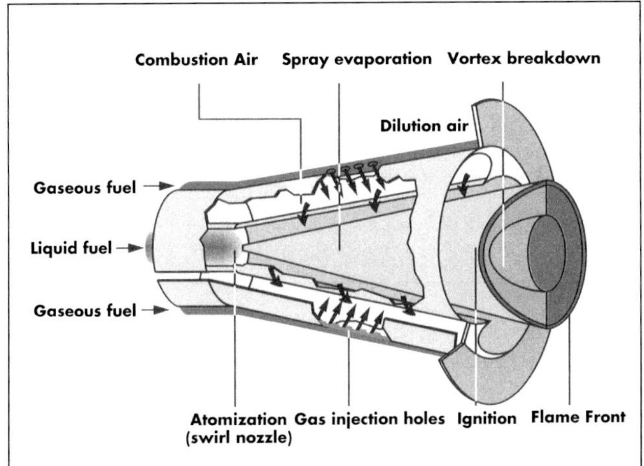

Fig. 17-28 Dual-Fuel, Double-Cone Burner Illustration. Source: ABB

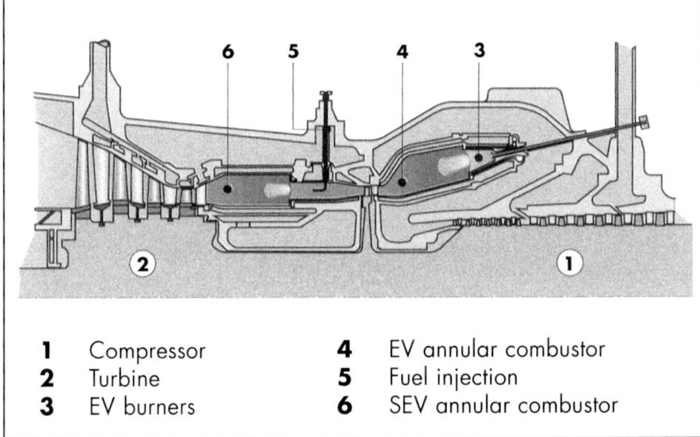

1	Compressor	**4**	EV annular combustor
2	Turbine	**5**	Fuel injection
3	EV burners	**6**	SEV annular combustor

Fig. 17-29 Sequential Combustion System Design. Source: ABB

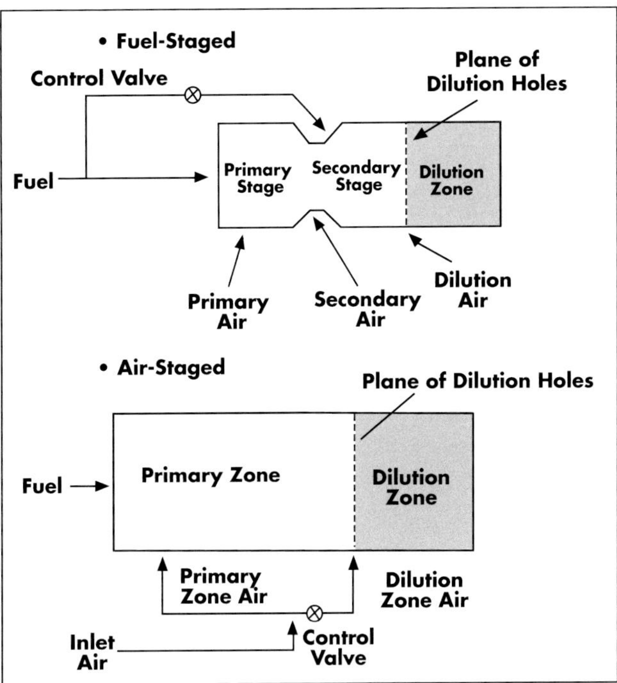

Fig. 17-30 Staged Combustors. Source: General Electric Company

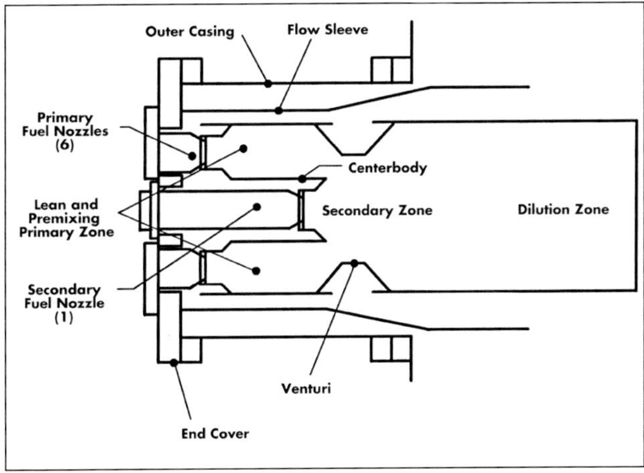

Fig. 17-31 DLN Combustor. Source: General Electric Company

air-staged design, a fraction of the airflow from the flame zone is diverted to the dilution zone under low loads. A currently used two-stage pre-mixed DLN combustor is shown in Figure 17-31. As shown in Figure 17-32, this combustor has four distinct fuel-staged operating modes.

Exhaust Gas Treatment — NO_X Controls

Exhaust gas treatment technologies can be used as the sole basis of control, or with reductions achieved upstream by combustion operation or equipment modifications. Exhaust gas treatment systems are classified as selective or non-selective, depending on whether they selectively reduce NO_X or simultaneously reduce NO_X, unburned hydrocarbons, and CO.

Selective catalytic reduction (SCR) is used for NO_X reduction in exhaust gas from systems operating with a λ of greater than 1.0, meaning to the lean side of stoichiometric combustion. SCR systems typically use ammonia or urea as a reducing agent/chemical to selectively reduce NO_X with NH_3 to N_2 plus H_2O vapor.

Typical reactions with ammonia are:

$$4NO + 4NH_3 + O_2 \rightarrow 4N_2 + 6H_2O$$
$$NO + NO_2 + 2NH_3 \rightarrow 2N_2 + 3H_2O$$

Typical reactions with urea are:

$$2NO + (NH_2)_2CO + 1/2O_2 \rightarrow 2N_2 + 2H_2O + CO_2$$
$$6NO_2 + 4[(NH_2)_2CO] + 4H_2O \rightarrow 7N_2 + 12H_2O + 4CO_2$$

The main variable affecting NO_X reduction is temperature and, for base metal catalysts, optimum performance is exhibited within a fairly narrow temperature range. Below this range, catalyst activity is greatly reduced and unreacted ammonia slips through. Above this range, ammonia begins to oxidize to form additional NO_X.

The catalyst grid can be of base metals or zeolite, as described below:

- Titanium/Tungsten/Vanadium (Ti/W/V) catalysts can operate within a temperature range of about 580 to 790°F (304 to 421°C). Ti/W/V is a more active catalyst than zeolite and, therefore, requires less volume per unit reduction in NO_X.
- Titanium/Vanadium/Platinum (Ti/V/Pt) catalysts can operate within a temperature range of about 340 to 380°F (171 to 193°C). While Ti/V/Pt has a limited operating range, it can be applied in the cold section of heat recovery units, allowing for easier installation. Additionally, the platinum provides some CO and hydrocarbon reduction.
- Zeolites, which are naturally occurring or synthetically manufactured alumino-silicate crystals, are referred to as molecular sieves. The micropore structure of the zeolites is about 3-10 angstroms in size and one gram of zeolite catalyst has a micro-pore surface reaction area of over 3,000 sq.ft. (278 sq.m.). Zeolites can operate within a temperature range of about 550 to 970°F (288 to 521°C), allowing flexibility in locating the catalyst (e.g., upstream of an HRSG). They have a relatively high tolerance to poisons. Zeolite has an additional advantage in that the spent catalyst grid may not be a hazardous waste. Figure 17-33 shows a zeolite/ceramic molecular sieve catalyst.

Poisoning elements, such as chloric compounds and heavy metals, introduced by lube oils and dirty fuels tend to react with metal catalyst surfaces, causing them to become non-reactive. Metals such as vanadium are susceptible to masking, or coating, of the catalyst due to reactions associated with fuel-bound sulfur. Metal catalysts also tend to burn, or cinder, at temperatures in excess of 800 or 850°F (427 or 454°C), causing damage to surface or substrate.

The design of each SCR system is somewhat unit-specific, depending on space constraints, location of equipment, catalyst operating temperature requirements, fuel, and cost. Figure 17-34 shows three potential options for how the SCR system can fit into the boiler system. System (a) is an example of a preferred system for new boilers or

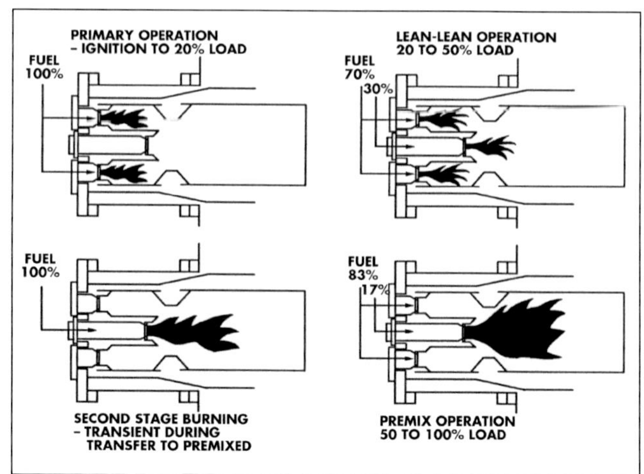

Fig. 17-32 Fuel-Staged DLN Combustor Operating Modes. Source: General Electric Company

Fig. 17-33 Zeolite/Ceramic Molecular Sieve Catalyst. Source: Environmental Emissions Systems, Inc./Steuler

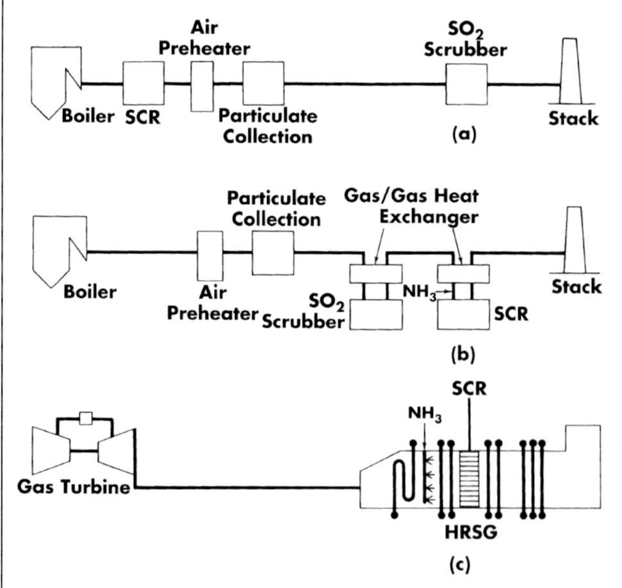

Fig. 17-34 Examples of Application of SCR Technology Placement Configurations. Source: Babcock &Wilcox

when space is available in retrofit applications. System (b) is more costly, requiring additional heat exchangers and an external heat source, and would be used on a retrofit application with limited space. System (c) is an example of SCR technology in a combined-cycle system with placement in the HRSG.

Figure 17-35 is an SCR system process flow diagram for a gas turbine equipped with HRSG. Figure 17-36 is a functional diagram of a cogeneration plant with two prime movers (gas turbines or reciprocating engines) featuring an SCR system. Shown are muffler/sound attenuation units, ammonia injection systems (controlled by residual NO_X measurement), the reactors with SCR (which may also include oxidation catalysts downstream), the central CEM and operating control system, and the ammonia storage tank. This system utilizes zeolite composite extruded honeycomb modules.

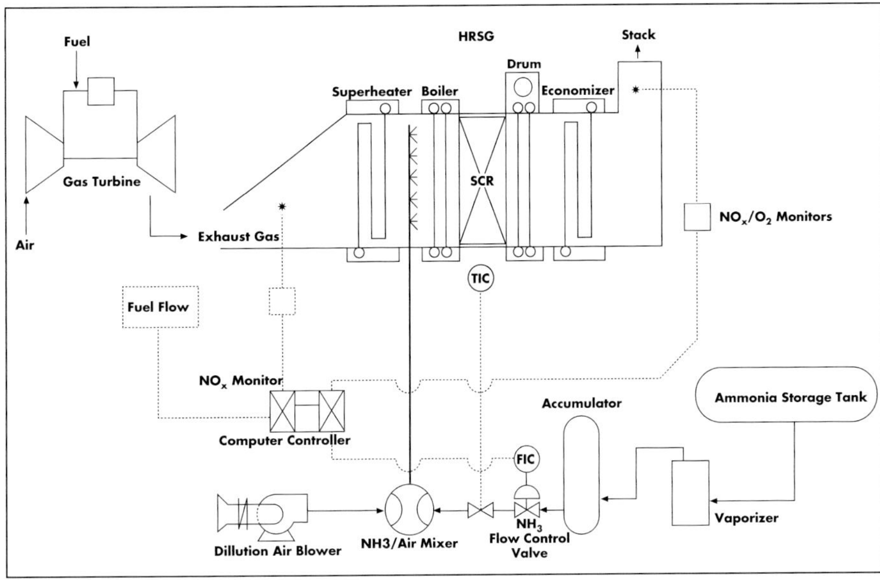

Fig. 17-35 SCR System Process Flow Diagram for Gas Turbine/HRSG.

Figure 17-37 is a cutaway diagram of the reactor/converter. The system operates as follows:

- NH_3 is injected into the waste gas stream in front of the reactor when the exhaust gases are in the proper temperature range. Below 572°F (300°C), NH_3 may form ammonia-disulfate with sulfur bearing fuel. Above 950°F (510°C), NH_3 thermally oxidizes excessively, yielding more NO_X. At 970°F (521°C), about 15% excess NH_3 is consumed.
- The rate of NH_3 injection is based on feedback signals from continuous NO_X measurements (CEM system). The system draws off a small quantity of waste gas downstream of the reactor. The sample is filtered and dried and sent to NO_X, CO, and O_2 analyzers. These data are relayed to the process controller. The controller activates a motor-driven valve on the NH_3 injection panel.
- Aqueous ammonia, 25.0 to 29.4% ammonia in water, or 33 to 40% urea, which is the reducing agent, is supplied by a metering pump through a feeder line from a non-pressurized stainless steel tank. A small sulfuric

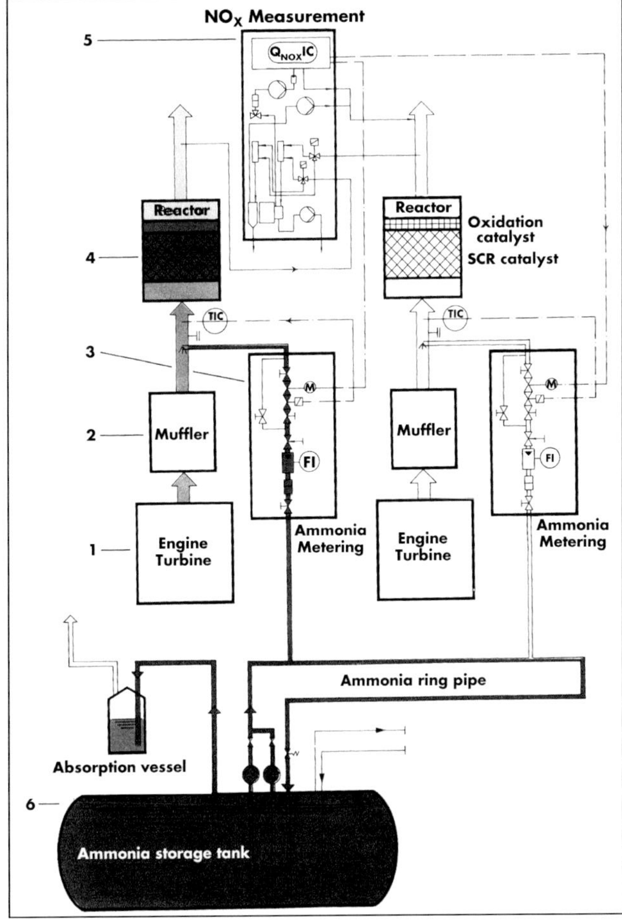

Fig. 17-36 Functionality Diagram of Cogeneration Plant with Two Prime Movers Featuring an SCR System.
Source: Environmental Emissions Systems, Inc./Steuler

1 Engine/turbine, single or multiple units
2 Muffler/sound attenuation unit
3 Ammonia injection systems
4 CER-NO_X reactor
5 Central, CEM system
6 Ammonia storage tank

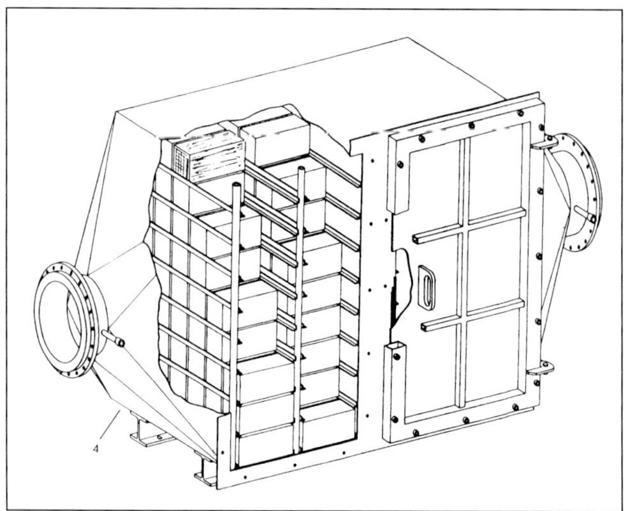

Fig. 17-37 Cutaway View of Reactor/Converter Featuring Zeolite/ Ceramic SCR. Source: Environmental Emissions Systems, Inc./Steuler

acid scrubber is used to eliminate ammonia vapors created during thermal venting of the tank.

- Both NO_X and NH_3 are absorbed into the micropores of the catalyst. An exothermic reaction in the zeolite micropore structure increases pressure from which the reaction products (N_2 and H_20 vapor) are forcefully expelled from the micropores, causing a self-cleaning action of the catalyst surface.

Figure 17-38 is a diagram of a ceramic molecular sieve catalyst (SCR) NO_X abatement system installed with a reciprocating engine cogeneration system in an industrial facility. When operating in dual-fuel mode (94% natural gas and 6% Diesel fuel), the exhaust gas flow is 71,910 lbm/h (32,575 kg/h) at a temperature of about 875°F (468°C). A minimum of 90% NO_X reduction has been achieved, with actual reductions from 46.0 to 3.2 lbm/h (20.9 to 1.5 kg/h). When operating in liquid fuel mode, the exhaust gas flow is 80,320 lbm/h (36,384 kg/h) at a temperature of 815°F (430°C). A minimum of 90% NO_X reduction has also been achieved in this mode, with actual reductions from 77.0 to 5.5 lbm/h (34.9 to 3.2 kg/h). Operation is guaranteed by the manufacturer with less than 10 ppm ammonia slip, and actual slip estimated at about 3 ppm. Since installation in 1988, the NO_X reduction capability of the system was increased to about 95% by increasing the amount of catalyst.

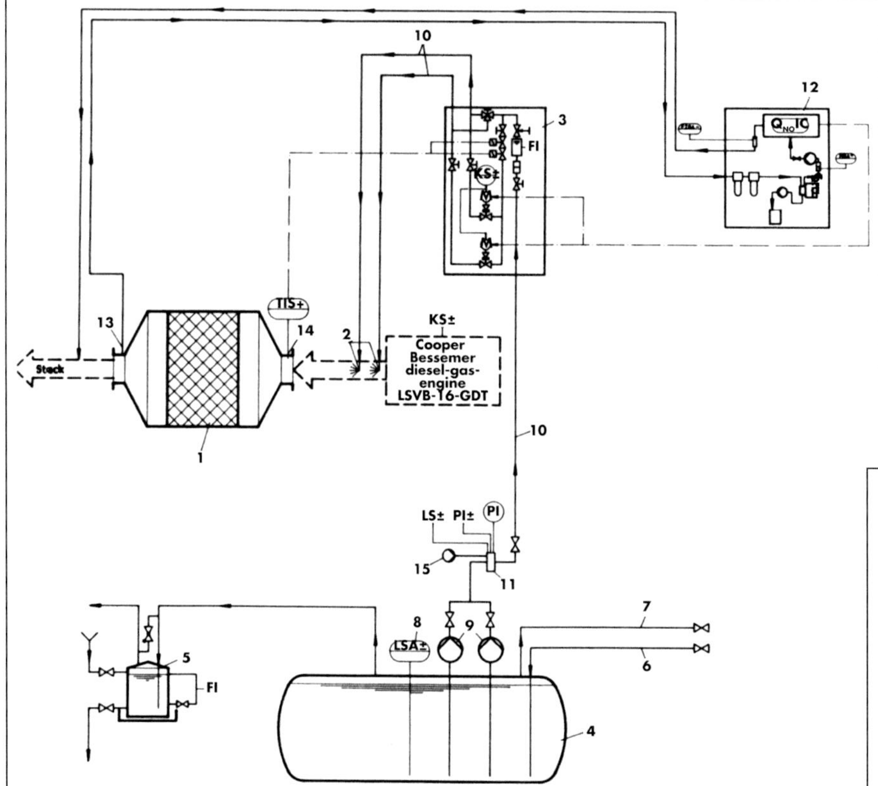

Figure 17-39 illustrates an exhaust gas treatment system for a Jenbacher lean-burn reciprocating engine featuring urea ($(NH_2)_2CO$) as the SCR reducing agent. The system includes a catalytic converter, a spraying device for urea, a urea tank, and controls. The converter itself includes individual honeycomb elements, with a final row that can function as an oxidation catalytic converter. An exactly dosed amount of urea (40% aque-

1 Reactor housing with SCR catalyst
2 NH_3 injection nozzles
3 NH_3 control panel
4 NH_3 storage tank (15,000 U.S. gallons)
5 Absorption vessel/vapor scrubber
6 NH_3 filling piping
7 Vent piping/vapor recycling
8 Level indicator
9 NH_3 supply pump
10 NH_3 supply piping
11 Air Chamber
12 NO measuring and control
13 Gas sampling probe
14 Temperature control
15 Pressure control

Fig. 17-38 Functionality Diagram of SCR NO_X Abatement System Applied to Dual-Fuel Reciprocating Engine Cogeneration System. Source: Environmental Emissions Systems, Inc./Steuler

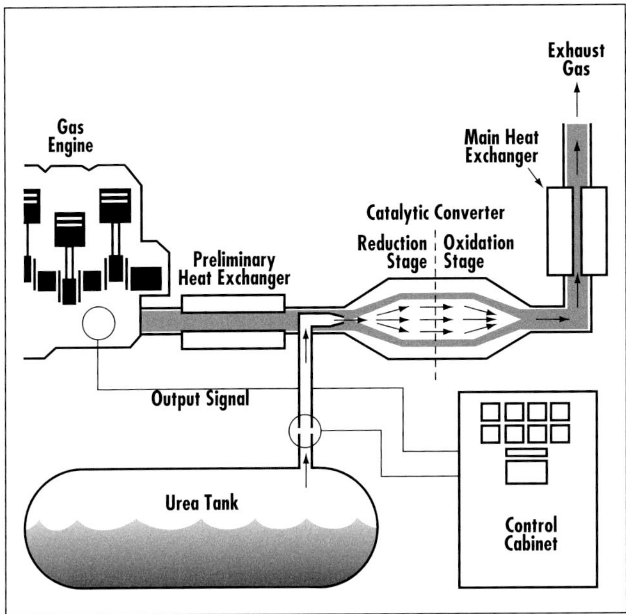

Fig. 17-39 Reciprocating Engine Exhaust Gas System. Source: Jenbacher Energies

ous solution) is spray-injected into the exhaust gas. Figure 17-40 illustrates the basic chemical reactions associated with this reciprocating engine exhaust gas treatment system.

Non-Selective Catalytic Reduction (NSCR) is used for NO_X reduction in exhaust gas from systems operating with a λ of equal to, or less than, 1.0, meaning to the rich side of stoichiometric combustion. In NSCR systems, NO_X is reduced in the presence of a catalyst by CO and HC (or VOC) in the exhaust gas, forming N_2, H_2O, and CO_2. Currently, some engine manufacturers report NO_X values as low as 0.5 g/hp-h (0.7 g/kWh). The remaining CO and HC in the exhaust gas can be oxidized to CO_2 and H_2O with oxidation catalysts. Suitable catalytic materials include platinum, palladium, and rhodium. These metals have relatively low

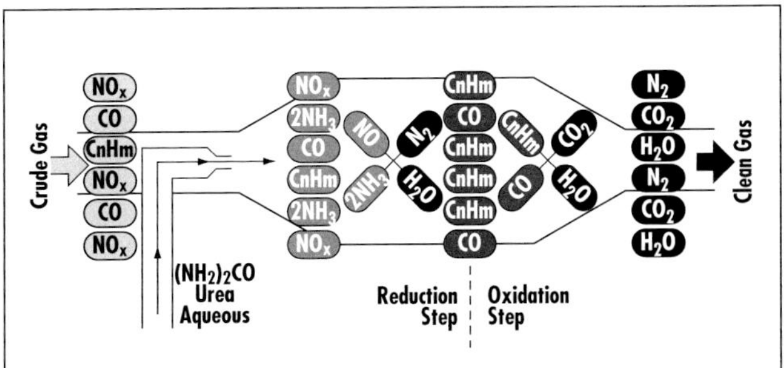

Fig. 17-40 Chemical Reactions of Reciprocating Engine Exhaust Gas Treatment System. Source: Jenbacher Energies

inlet temperature requirements and provide some poison resistance. As previously noted, use of certain oil additives (e.g., phosphorous or zinc) may result in catalyst poisoning or masking, although, in many cases, these deactivating agents may be chemically washed from the catalyst to restore activity.

Figure 17-41 shows an application of a NO_X, CO, and VOC/NMHC abatement system at a natural gas compressor station. The compressor station features four 8-cylinder 2-stroke-cycle Jenbacher natural gas-fired spark-ignition engines with a capacity of 670 hp (500 kW) each. The exhaust gas flow rate is 4,120 scfm (7,000 Nm^3/h) with a temperature range of 518 to 734°F (270 to 390°C) and an oxygen concentration level of 16% by volume. Emissions control is achieved with a (SCR) NO_X and (NSCR) CO/NMHC abatement system, with the NSCR oxidation catalyst downstream of the all-zeolite (SCR) catalyst in the same reactor housing. The total pressure drop across the catalyst bed is 4.7 in. water column (1.2 kPa). An 88% NO_X reduction is accomplished with reductions of 310, 373, and 1,000 ppmvd at O_2 levels of 16, 15, and 5%, respectively. A 68% CO reduction is accomplished with reductions of 154, 186, and 500 ppmvd at O_2 levels of 16, 15, and 5%, respectively. An estimated 10% NMHC reduction is accomplished with reductions of 15, 18, and

Fig. 17-41 Application of NO_X, CO, and VOC/NMHC Abatement System for Gas-Fired, Spark-Ignited Engines at Gas Compressor Station. Source: Environmental Emissions Systems, Inc./Steuler

48 ppmvd at O_2 levels of 16, 15, and 5%, respectively.

Since with NSCR systems, the engine air-to-fuel ratio will be at or close to stoichiometric, both NO_X reduction and CO and HC oxidation can be done in a single catalyst bed. In a **three-way catalyst system**, CO, HC, and hydrogen are oxygen acceptors. NO_X and molecular oxygen are oxygen donors. When donors and acceptors are balanced, the catalyst simultaneously reduces, NO_X, CO, and VOC. These catalysts have been applied to natural gas-fired reciprocating engines since 1962 and, since 1981, most passenger cars sold in the U.S. have employed three-way catalyst systems. These catalysts require close control of engine operation at very close to stoichiometric conditions through the use of feedback carburetors or fuel injection systems that are controlled by a reference signal from an exhaust gas oxygen sensor.

The chemical reactions that occur simultaneously across an NSCR three-way catalyst include, among others, the following:

$$CO + 1/2\ O_2 \rightarrow CO_2$$

$$H_2 + 1/2\ O_2 \rightarrow H_2O$$

$$HC + O_2 \rightarrow CO_2 + H_2O$$

$$NO_X + CO \rightarrow CO_2 + N_2$$

$$NO_X + H_2 \rightarrow H_2O + N_2$$

$$NO_X + HC \rightarrow N_2 + H_2O + CO_2$$

Figure 17-42 shows a catalytic converter unit used for stationary reciprocating engine applications. The steel catalyst housing frame can be used with either three-way or oxidation catalyst systems. The catalyst element shown on top of the housing can be quickly removed and replaced. Figure 17-43 is a schematic diagram indicating the converter and sensor location in the exhaust gas flue.

Figure 17-44 shows representative uncontrolled emissions for a standard 4-stroke-cycle rich-burn engine. At the crossover air-to-fuel ratio point, the emissions are around 10 g/hp-h (13 g/kWh) of both NO_X and CO. Figure 16-45 is a general performance diagram for a three-way catalyst system. Such a system will typically reduce NO_X and CO emissions by 85 to 95% based on exhaust temperature, air-to-fuel ratio set point, and catalyst volume.

There are several commercially available **selective non-catalytic reduction (SNCR)** (also called thermal NO_X reduction) systems that selectively reduce NO_X without catalysts.

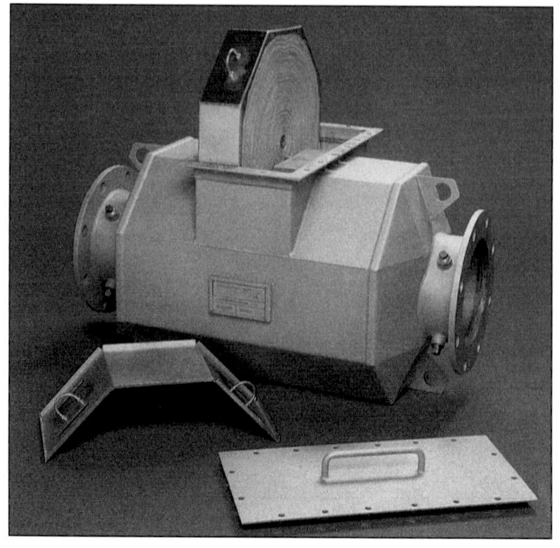

Fig. 17-42 Catalytic Converter Unit Used for Three-Way or Oxidation Catalyst Systems. Source: Miratech Corp.

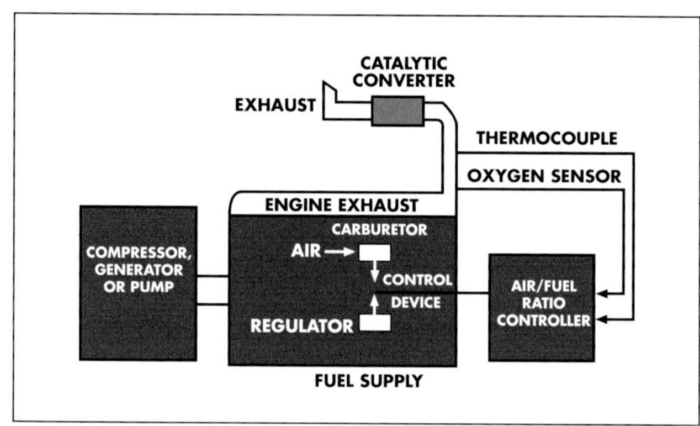

Fig. 17-43 Diagram Indicating Converter and Sensor Location in the Exhaust Gas Flue. Source: Miratech Corp.

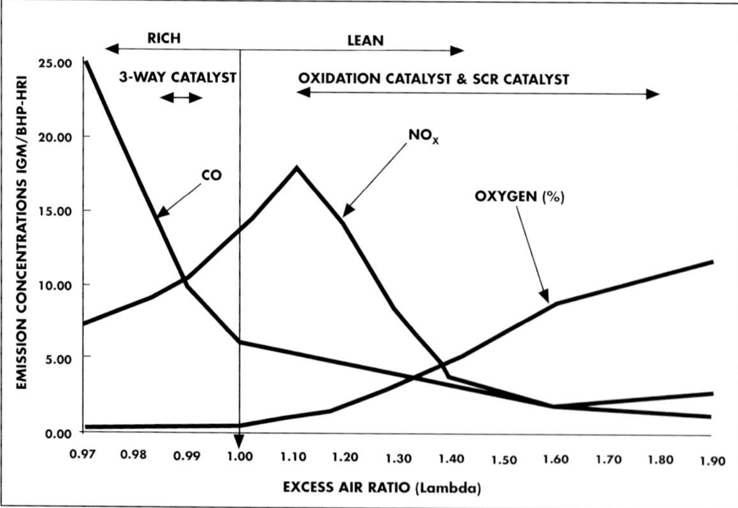

Fig. 17-44 Representative Uncontrolled Emissions for Four Stroke-Cycle Rich-Burn Engine. Source: Miratech Corp.

In some systems, gaseous ammonia (NH_3) is injected into flue gas to reduce NO_X to N_2. The chemical reaction mechanism is similar to that of SCR NO_X abatement systems using ammonia or urea. However, since no catalyst is present, reactions occur at a much higher temperature of about 1,300 to 1,400°F (700 to 760°C) and consumption of ammonia/urea is significantly higher than with SCR systems. The narrow temperature window for controlling reactions presents difficulties, particularly when loads vary. Additives have been developed to expand this temperature window.

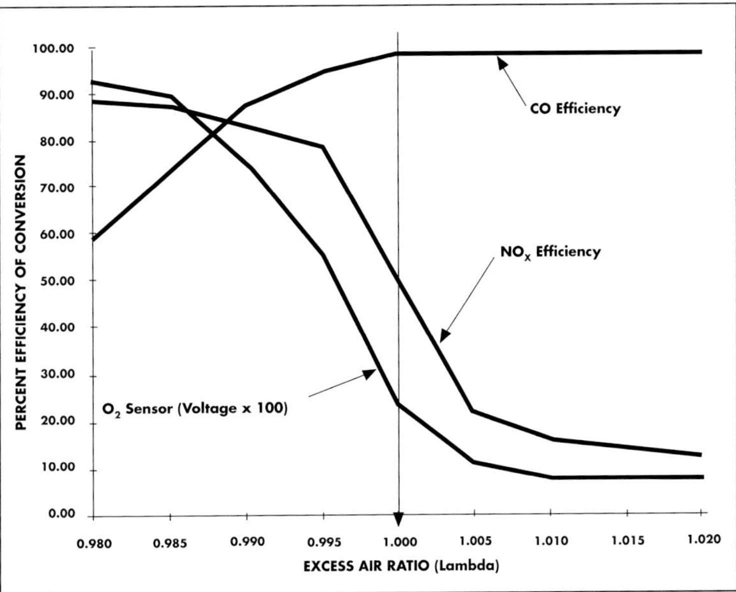

Fig. 17-45 Performance Curves for Three-Way Catalyst System. Source: Miratech Corp.

Summary of NO_X Emissions Control Technologies

Tables 17-9, 17-10, and 17-11, based on EPA data, indicate the applicability and effectiveness of various options for controlling NO_X emissions from commercial and industrial (C&I) boilers, reciprocating engines, and gas turbines, respectively.

CO and HC/VOC Controls

CO is normally created when carbon-containing fuels are burned in insufficient air or where the normal combustion process is halted before completion. CO production mechanisms include the presence of cold surfaces in the flame volume where the flame is quenched below the kinetic reaction temperature, direct leakage and quenching of partially reacted gases, and poor fuel-to-air mixing and subsequent quenching. CO emissions also depend on fuel type and can vary substantially, even in otherwise properly operating systems. In boilers, for example, observed CO emissions levels can be 10 ppm or below, or can exceed 300 ppm, and are often associated with severe NO_X control.

Control technologies for stationary sources include: improved combustion control, capture systems that reuse CO-rich exhaust gases as fuel, and afterburners or oxidation catalysts to destroy CO in exhaust streams. Fuel switching to cleaner fuels, notably natural gas, is also a commonly

Table 17-9 Summary of Potential NO_X Emissions Reductions Possible with Various Control Options for C&I Boilers

	Potential NO_X Reduction (%)							
	Coal-Fired Boilers		Oil-Fired Boilers			Gas-Fired Boilers		
	Watertube		Watertube		Firetube	Watertube		Firetube
	PC	Stoker	Field-Erected	Packaged		Field-Erected	Packaged	
Low Excess Air	5-30	5-30	5-25	5-25	5-25	5-35	5-35	5-35
Burners Out of Service	10-30	N/A	10-30	N/A	N/A	10-30	N/A	N/A
Overfire Air	15-30	0-30	25	20-40	N/A	35	20-40	N/A
Low-NO_X Burners	50	N/A	45	45	45	55	50	50
Radiant Burners	N/A	N/A	N/A	N/A	N/A	90	80	70-80
Flue Gas Recirculation	N/A	20-45	15-30	15-30	15-30	50-65	50-65	50-65
Water Injection	N/A	N/A	15-30	15-35	15-35	25-50	25-50	25-50
Natural Gas Reburn	60	N/A	80	N/A	N/A	N/A	N/A	N/A
Selective Catalytic Reduction	80-90	80-90	80-90	80-90	N/A	80-90	80-90	N/A
Selective Noncatalytic Reduction	30-70	30-70	30-70	30-70	N/A	30-60	30-60	N/A
Fuel Switching	65	N/A	40	60	40-65	N/A	N/A	N/A

Source: EPA/State and Territorial Air Pollution Program Administrators and the Association of Local Air Pollution Control Officials

Table 17-10 Potential Emissions Reductions from Reciprocating Internal Combustion Engines

	Potential NO_X Reduction (%)			
	Rich-Burn, Gas SI	**Lean-Burn, Gas SI**	**Lean-Burn, Diesel**	**Lean-Burn, Dual Fuel**
Air-to-Fuel Adjustment	10-40	5-30	N/A	N/A
Low Emissions Combustion	70-90	80-93	N/A	60-80
Ignition Timing Retard	0-40	0-20	20-30	20-30
Prestratified Charge	80-90	N/A	N/A	N/A
NSCR	90-98	N/A	N/A	N/A
SCR	N/A	90	80-90	80-90

Source: EPA/State and Territorial Air Pollution Program Administrators and the Association of Local Air Pollution Control Officials

Table 17-11 Potential Emissions Reductions from Gas Turbines

Control Technology	% Emissions Reduction
Water/Steam Injection	70-90
DLN Combustors	60-90
SCR	90

Source: EPA/State and Territorial Air Pollution Program Administrators and the Association of Local Air Pollution Control Officials

used option for stationary sources.

While CO can be reduced with control modifications, such modifications are often at odds with those required for NO_X control. Methods that treat exhaust gases directly (e.g., SCR and SNCR) do not exhibit this effect. In areas where CO is strictly controlled, NO_X reduction methods using combustion modification techniques may be limited by CO control considerations or require oxidation catalysts to reduce VOC emissions.

VOC emissions result from incomplete combustion and vaporization. There is considerable overlap between emissions classified as HC or VOC, since VOC are compounds containing combinations of carbon and hydrogen. However, they may also contain oxygen, sulfur, nitrogen, and halogens like fluorine and chlorine. Methane, ethane, and a few other compounds are excluded from definitions of VOC, giving rise to the term non-methane HC (NMHC).

Industrial VOC emissions are dominated by three types of sources: 1) incomplete combustion of fossil fuels; 2) solvent emissions resulting from coating and printing; and 3) organic emissions resulting from the handling and manufacture of petroleum products, chemicals, and chemically derived products. While all combustion sources may emit some VOC emissions as a result of incomplete combustion, internal combustion engines typically have the highest VOC levels and are, therefore, a primary focus of VOC control efforts. For the purpose of the discussion on emissions resulting from fossil fuel emissions, VOCs are addressed as HCs. To some extent, NO_X control methods can have deleterious side effects on HC emissions levels, and thus require oxidation catalysts.

As noted previously, oxidation catalysts are typically made of platinum (Pt), palladium (Pd), rhodium (Rho), or Pt/Pd/Rho catalyst formulae. CO emissions are effectively controlled from 450 to 1,250°F (230 to 670°C). HC control is dependent on the HC species. Typically, exhaust gas temperatures from 750 to 1,250°F (400 to 670°C) are sufficient for effective control.

Reciprocating Engine Emissions Controls

A number of emissions reduction technologies can be applied to new or existing engine designs, including lean-burn operation and add-on catalyst, with and without supplemental air. Figure 17-46 shows NO_X, CO, and NMHC (VOC) production for natural gas-fired reciprocating engines as a function of air-to-fuel ratio. Engines used for small utility applications are calibrated at very rich air-to-fuel ratios (typically ~12:1). Shifting to a leaner calibration can produce very large reductions in engine-out HC and CO emissions. Leaner operation, however, also tends to increase engine-out NO_X emissions so that the net change in combined HC and NO_X emissions may not be significant.

Exhaust treatment catalysts have been used on passenger cars since the mid-1970s to reduce emissions. They can apply in many instances to stationary engines as well. Oxidation catalysts are used to reduce HC and CO. This requires a net oxidizing exhaust environment, which is achieved through the use of lean engine calibrations or supplemental air injected into the exhaust manifolds.

The air-to-fuel ratio set point for operating in the catalyst window for post-combustion controls in stationary stoichiometric engines is very narrow. Air-to-fuel ratio, which depends on intake air density, temperature,

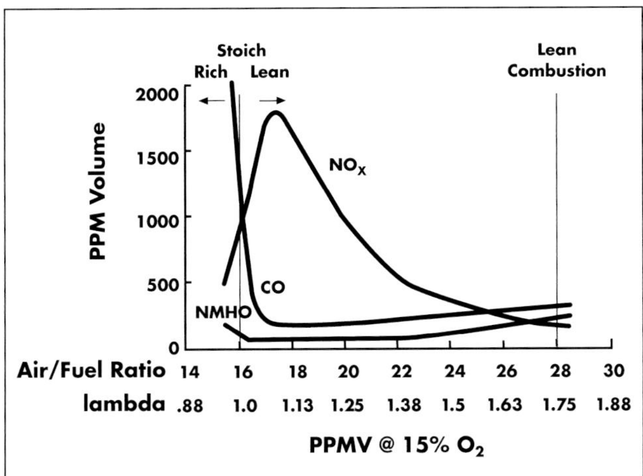

Fig. 17-46 Summary Illustration of NO_X, CO, and NMHC Emissions for Natural Gas-Fired Reciprocating Engines.
Source: Waukesha Engine Div.

fuel pressure, fuel flow, fuel heat value, and carburetor adjustment, must be precisely controlled. Figure 17-47 illustrates a simplified control system.

There are several ways to achieve the desired air-to-fuel set point for maximum catalyst performance. Typically, an oxygen sensor is placed in the exhaust stream. The sensor compares ambient oxygen on one side of the sensor with the exhaust gas sample on the other side. Some controllers also place a thermocouple next to the oxygen sensor for control limiting and temperature compensation due to oxygen. Through micro-circuit or microprocessor action, an output signal is generated to interface with a number of end devices. On charged reciprocating engines, it is common to control the waste gas, air bypass loop, or fuel regulator. With naturally aspirated engines, biasing the fuel regulator or some form of flow restriction is used. When an air-to-fuel ratio controller and catalyst are applied to a stoichiometric engine, standard emissions reduction efficiencies of 90% NO_X, 85% CO, and 50% HC are typically achieved.

Particulate Matter (PM) Control Technologies

As with other pollutants, fuel choice is an important determinant of particulate (PM) emissions. While uncontrolled particulate emissions for natural gas are negligible, particulates must be controlled for most other fuels, especially heavy oil and coal. Combustion control can improve particulate levels by reducing the level of unburned carbon. However, after-treatment, add-on controls are common for coal, oil, and many industrial material handling operations. The post-combustion control of particulate emissions from combustion sources and control of particulates from point sources in general can be accomplished by using one or more of the following devices:

- Electrostatic precipitators (ESP), which have no adverse effect on combustion system performance,

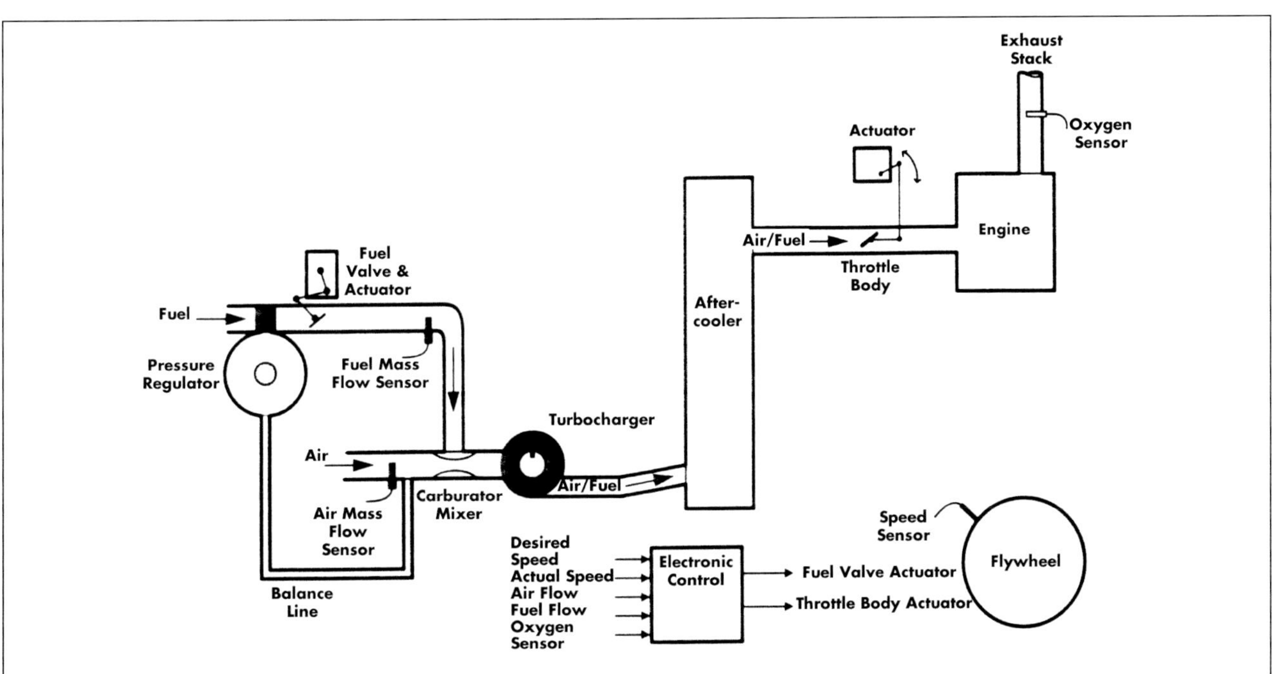

Fig. 17-47 Simplified Example of Air-to-Fuel Ratio Control. Source: Caterpillar Engine Div.

remove particles by charging them so that they are attracted to a collection surface.

- Fabric filters (or baghouses), which consist of a number of filtering elements (bags) along with a bag cleaning system contained in a main shell structure incorporating dust hoppers.
- Wet scrubbers, which include venture and flooded disc scrubbers, tray or tower units, turbulent contact absorbers, and high-pressure spray impingement scrubbers, are applicable for PM control, as well as for SO_2 control on coal-fired plants. A disadvantage is the stringent disposal requirements for the resulting wet sludge.
- Cyclone or multi-clone collectors are referred to as mechanical collectors because they do not rely on electrical, liquid, or barrier principles for removal of PM from a gas stream. The collection efficiency depends strongly on the effective aerodynamic particle diameter. These are often used as precollectors upstream of other controls because they are relatively ineffective for collection of the finer PM-10 and are not effective over wide load ranges.
- Side-stream separators combine a multi-cyclone and a small pulse-jet baghouse to more efficiently collect small diameter particles that are difficult to capture by a mechanical collector alone.
- In internal combustion engine applications using liquid fuel, zeolite SCR NO_X abatement catalysts have achieved up to 40% soot/particulate reduction. As the soot, most of which is carbon, passes through the catalyst, it is converted to CO_2.

Filterable particulate emissions can be controlled to various levels by all of these devices. Cyclones, ESPs, and fabric filters have little effect on measured condensable particulate matter (CPM) because they are generally operated at temperatures above the point at which most CPM remains vaporized and could pass through the control device. Wet scrubbers, however, reduce the gas stream temperature so they can remove some of the CPM. Side-stream separators are typically used on small stoker boilers as a lower cost (less effective) control method than ESP or baghouse.

SO_2 Control Technologies

Under atmospheric conditions, SO_2 is a reactive acrid gas, which can be rapidly assimilated back to the environment. However, the combustion of fossil fuels, notably coal and oil containing sulfur, in which large quantities of SO_2 are emitted at point source locations, results in environmental damage. In addition to power plant operation, certain industrial operations (refineries and smelters) can also be significant sources of SO_2 emissions.

During combustion, almost all of the sulfur in the fuel will oxidize to gaseous sulfur oxides (SO_X). The principal oxidizing reaction leads to the formation of sulfur dioxide (SO_2). Additionally, lesser quantities of sulfur trioxide (SO_3) are formed during combustion. While most emissions control standards target SO_2, some reduction in SO_3 occurs as well.

As with particulates, fuel choice is a major determinant of SO_2 emissions. Natural gas has negligible sulfur content, while coal and residual oil have relatively high sulfur levels. Figure 17-48 shows the effects of fuel sulfur content on SO_X emissions for fuel oils. Sulfur content of coals also vary widely. While somewhat more costly, the use of low-sulfur coal is one control option.

Fuel desulfurization is particularly cost-effective for oil-based fuels (distillates and residual fuel oil). The primary mechanism for desulfurization at petroleum refineries is hydrotreating in which hydrogen reacts with fuel-bound sulfur to produce H_2S and a lower sulfur oil. Coal can also be cleaned (desulfurized) in a mechanical process in which sulfur and ash components are separated from a ground coal based on differences in particle density.

Post-combustion flue gas desulfurization (FGD) uses an alkaline reagent to absorb SO_2 in the flue gas and produce sodium or calcium sulfate or sulfite compounds. These solid compounds are then removed in downstream particulate control devices. FGD technologies are categorized as wet, semi-dry, or dry, depending on the state of the reagent as it

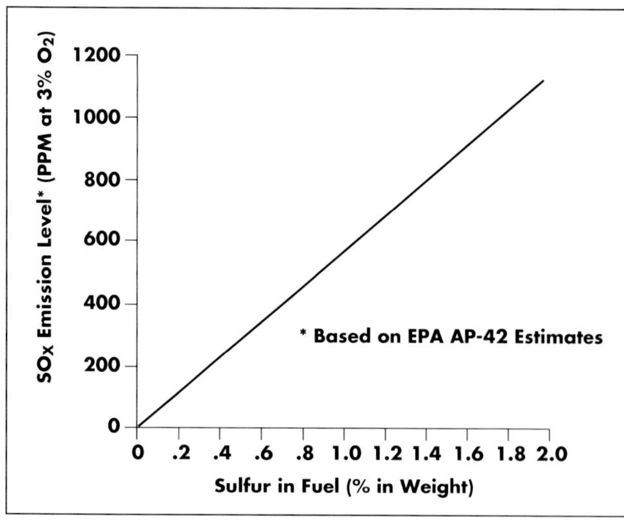

Fig. 17-48 Effects of Fuel Sulfur Content on SO_X Emissions for Fuel Oils. Source: Cleaver Brooks

leaves the absorber vessel. These processes are either regenerable (i.e., the reagent material can be treated and reused) or are non-regenerable when waste streams are de-watered and discarded.

Currently, wet systems are the most common. Wet systems generally use alkali slurries as the SO_2 absorbent medium and can be designed to remove more than 90% of incoming SO_2. Lime/limestone scrubbers, sodium scrubbers, and dual alkali scrubbing are among the commercially proven wet FGD systems. The effectiveness of these devices depends on system design and operating conditions.

The lime and limestone scrubbing process uses a slurry of calcium oxide or limestone to absorb SO_2 in a wet scrubber. Control efficiencies in excess of 91% for lime and 94% for limestone over extended periods have been demonstrated. The process produces a calcium sulfite and calcium sulfate mixture, which is precipitated in a holding tank. The slurry is recycled to the scrubber to absorb additional SO_2, while a slip stream from the holding tank is sent to a solid-liquid separator to remove precipitated solids. The waste solids, typically 35 to 70 weight-percent solids, are generally disposed of by ponding or landfill.

Sodium scrubbing processes generally employ a wet scrubbing solution of sodium hydroxide (NaOH) or sodium carbonate (Na_2CO_3) to absorb SO_2 from the flue gas. Sodium scrubbers are generally limited to smaller sources because of high reagent costs; however, these systems have been installed on industrial boilers of up to 430 MMBtu/h (453,400 MJ/h) input. SO_2 removal efficiencies of up to 96% have been demonstrated.

The dual alkali system uses a clear sodium alkali solution for SO_2 removal, followed by a regeneration step using lime or limestone to recover the sodium alkali and produce calcium sulfite and sulfate sludge. Most of the effluent from the sodium scrubber is recycled back to the scrubber, but a slipstream is withdrawn and reacts with lime or limestone in a regeneration reactor and thickener where the solids are concentrated. The overflow is sent back to the system, while the remainder is further concentrated in a vacuum filter or other device to about 50% solids content. The solids are washed to recover soluble sodium compounds which are returned to the scrubber. Performance data indicate average SO_2 removal efficiencies of 90 to 96%. However, operating histories have indicated that average system reliability is only 90%.

Spray drying is used in a dry scrubbing approach to FGD. The technology is best suited for low- to medium-sulfur coals with sulfur contents of up to 3%, but may be applied to higher sulfur-content coals. A solution or slurry of alkaline material is sprayed into a reaction vessel as a fine mist and contacted with the flue gas for a relatively long period (5 to 10 seconds). The slurry is dried by the hot flue gas to about 1% free moisture and reacts with SO_2 in the flue gas to form sulfite and sulfate salts. The solids entrained in the flue gas are carried out of the dryer to a particulate control device, such as an ESP or baghouse. Vendors have offered commercial guarantees of up to 90% capture on low-sulfur (less than 2%) coal.

A number of dry and wet sorbent injection technologies are under development to capture SO_2 in the furnace, the boiler sections, or ductwork downstream of the boiler. These technologies are generally designed for retrofit applications and are suited for coal combustion sources requiring SO_2 removal efficiencies of between 25 to 50%. There are commercial applications of furnace sorbent injection in Europe.

Wet regenerable FGD processes are attractive because they can have the potential for better than 90% sulfur removal efficiency, have minimal waste-water discharges, and produce salable sulfur product. Some of the current non-regenerable calcium based processes can, however, be operated to produce a salable gypsum end-product.

CO_2 Capture and Storage Technologies

Efforts are increasing in both the United States and around the world to implement programs to address climate change. Many scientists argue that significant reductions in CO_2 (order of magnitude of 50% below current trajectories) are necessary by 2050 and that total emissions globally will need to be further reduced by the end of the century. In order to achieve this scale of reductions from large stationary sources, there has been considerable research and development of a promising suite of technologies for carbon dioxide capture and storage (CCS).

CCS is a three-step process that includes the capture of CO_2 from power plants or industrial sources, transport of the captured CO_2 (usually in pipelines), and storage of that CO_2 in suitable geologic reservoirs. Technologies exist for all three components of CCS, but they have not yet been deployed at the scale necessary to help achieve GHG reduction targets. Cost estimates of current technology for CCS in power production range between \$60 and \$114 per metric ton of CO_2 avoided depending on the power plant type. Approximately 70–90 percent of that cost is associated with capture and compression.

There are three emerging technologies for capturing CO_2 from large stationary sources such as coal-fired boilers: post-combustion capture, pre-combustion capture, and oxy-fuel combustion.

In post-combustion capture, CO_2 is separated from the flue gas after combustion has taken place. Several methods are being tested for capture including use of an amine solution to facilitate chemical absorption, use of solvents or sorbents to stimulate physical adsorption, use of membranes, and cryogenic separation. Currently, the use of amine solutions is most common, but the technique is not applied widely. In addition to requiring the addition of a capture system, most post-combustion capture processes require the flue gases to be heavily scrubbed prior to treatment for separation. This imposes a space penalty and increases the cost.

Pre-combustion capture is typically associated with Integrated Gasification with Combined Cycle (IGCC) combustion in which coal is gasified into a synthetic gas or hydrogen. During gasification, CO_2 is removed using a solvent. The remaining synthetic gas or hydrogen is combusted. IGCC units are routinely used in the petrochemical industry; they are currently being developed for large scale, base-load power generation applications.

In oxy-fuel processes, combustion takes place in an oxygen-rich environment. This increases combustion efficiency and reduces the formation of NO_X, leaving a gas stream that is primarily CO_2 and a small amount of other constituents that can be removed.

Once CO_2 is captured, it is dehydrated and compressed for transport through a pipeline to a suitable storage location. These locations are deep geologic formations that include a reservoir of high porosity and high permeability rock such as sandstone, and a cap of low permeability, low-porosity rock such as shale. A deep injection well is drilled through the cap into the reservoir and the CO_2 is injected into the rock. This process is much like the reverse of oil production and draws on this industry for technology. The U.S. Department of Energy National Carbon Sequestration Atlas has developed a preliminary estimate of total pore volume in the United States that suggests there is enough geologic capacity to store several hundred years of emissions from stationary sources. Ultimately, very detailed site characterization is required to prove that a location is suitable for long-term storage.

The components of CO_2 capture and storage are being deployed at commercial scale and in several places there are efforts underway to integrate these components. Today, more than 35 million tons of carbon dioxide per year are transported to oil fields throughout the United States and injected into mature oil fields to enhance the recovery of oil that could otherwise not be produced.

Hazardous Air Pollutant (HAP) Control Technologies

There are currently 189 chemicals or groups of chemicals being regulated by the EPA as HAPs, many of which are also being regulated at the state level as air toxics. Most HAPs are either gaseous organics or particulates and, as such, the VOC and particulate controls described in the previous sections apply. Some HAPs are inorganic acids and bases that are most easily controlled with a wet scrubber.

Process and raw material choice can influence HAP emissions. For example, an organic HAP used as a solvent might be replaced by a physically similar but unlisted organic. EPA is just beginning to set HAP requirements, so expected levels of control and control cost are difficult to predict. Table 17-12 provides a general overview of HAP control options.

Economics of Emissions Control Technologies

Cost-effectiveness evaluation is used to determine the best way to meet emissions control requirements and is also often integral to the regulatory permitting process as a means of determining which control technologies will be required. An economic test, or screening process, is used to determine if an emissions control technology is beyond the scope of a particular project. Cost-effectiveness is measured in terms of controls cost (dollars) per air emissions reduced (tons). The general equation used is:

$$\textit{Cost effectiveness} = \frac{\textit{Cost of emissions controls}}{\textit{Tons of emissions reduced}} \qquad (17\text{-}3)$$

Cost-effectiveness values are important for two reasons. First, they allow the evaluator to compare alternatives on an equivalent basis. Second, they allow air agencies to establish cost-effectiveness thresholds. If the value of a technology to be implemented exceeds the cost-effectiveness limits established by the local air agency, the facility petitioning

Table 17-12 HAP Control Options

Pollutant Type	Control Methods
Acids and bases	Wet scrubber or dry scrubber with neutralization
Particulates (metals)	ESPs, bagfilters, scrubbers
Hydrocarbons	Containment, alternative materials and processes, incineration, condensation, adsorption
Mercury	Injection of Powdered Activated Carbon with Baghouse

for a permit can avoid that technology and employ the next most stringent. As discussed in the previous chapter, cost-effectiveness tests are used to determine if a particular emissions control technology should be required to comply with RACT or BACT regulations.

In conducting a cost-effectiveness evaluation, the costs to be considered include both capital and operating costs. Capital cost, usually stated per unit of capacity (e.g., \$/hp or \$/MMBtu input), include not only the equipment cost, including taxes, shipping, and auxiliary components, but the direct and indirect cost for installation as well. Direct installation costs include items such as foundations and supports, rigging, electrical and mechanical trade work, insulation, and painting. Indirect costs include engineering, performance testing, commissioning, and contingencies. Annual operating costs include direct costs for utilities (electricity, fuel, steam, water, and compressed air), waste treatment and disposal, maintenance and repair materials, and operating, supervisory, and maintenance labor, as well as indirect costs for overhead, property taxes, insurance, and administrative charges.

Example

Consider the case of a facility seeking to meet NO_X RACT requirements and evaluating the potential for installation of an SCR system on its boiler. Currently, the boiler is an uncontrolled source with a heat input rating of 350 MMBtu/h (369,000 MJ/h) operating continuously throughout the year. The boiler has never been source-tested and review of AP-42 factors indicates that the uncontrolled emissions rate is 0.130 lbm of NO_X/MMBtu (0.056 g/MJ) natural gas consumed. [Note that AP-42 factors often establish worst case uncontrolled emissions rates.]

The budget estimate for turn-key installation of the SCR system is \$533,000, with vendor warrantee of an emissions rate reduction down to 0.0195 lbm NO_X/MMBtu (0.0083 g/MJ) consumed of natural gas. The expected effective life of the SCR is 10 years and, assuming proper maintenance, the catalyst panels will require replacement every four years at a cost of \$75,000 per changeout. Based on vendor data, the estimated annual operating cost for the SCR will be \$33,000 in constant dollars. The time-valued cost of money is set at 8% in the facility's capital budgeting analysis.

All cost factors are expressed on an annual basis using the following formula:

$$\text{Total annualized costs} = \text{Capital costs} + \text{Replacement costs} + \text{O \& M costs} \qquad (17\text{-}4)$$

$$\text{Catalyst replacement present value} = \$75{,}000 \times 0.7350 + \$75{,}000 \times 0.5403 = \$95{,}648$$

$$\text{Catalyst replacement annualized cost} = \$95{,}648 \times 0.1490 = \$14{,}252$$

$$\text{Installed SCR annualized cost} = \$533{,}000 \times 0.1490 = \$79{,}417$$

$$\text{Total annualized costs} = \$79{,}417 + \$14{,}252 + \$33{,}000 = \$126{,}669$$

Where: the discount factors (at 8%) for years 4 and 8, are 0.7350 and 0.5403, respectively, and the 10-year annualized cost factor is 0.1490.

Annual NO_X emissions reductions can then be determined using the formula:

$$\text{Tons of } NO_X \text{ reduced/yr} = \left(\frac{\text{lbm } NO_{X\,uncontrolled}}{\text{MMBtu}} - \frac{\text{lbm } NO_{X\,controlled}}{\text{MMBtu}}\right) \frac{\text{hrs}}{\text{day}} \times \frac{\text{days}}{\text{yr}} \times \frac{1 \text{ ton}}{2{,}000 \text{ lbm}} \times \frac{\text{MMBtu}}{\text{hr}} \qquad (17\text{-}5)$$

Thus,

$$\frac{(0.130 - 0.0195)\text{lbm } NO_X}{\text{MMBtu}} \times \frac{8{,}760 \text{ hrs}}{\text{yr}} \times \frac{1 \text{ ton}}{2{,}000 \text{ lbm}} \times \frac{350 \text{ MMBtu}}{\text{hr}}$$

$$= 169.3 \text{ tons}_{NO_X} (153{,}700 \text{ kg}_{NO_X}) \text{ reduced/yr}$$

Cost-effectiveness is determined by dividing the annualized cost of control by the tons per year (tpy), or ton-years (ton-y), of NO_X reduced. Based on Equation 17-3, this would be:

$$\text{Cost effectiveness} = \frac{\text{Annualized costs}}{\text{Annual tons reduced}} = \frac{\$126{,}669}{169.3} = \$750/\text{ton/yr } (\$0.83/\text{kg/yr})$$

Variations in Cost-Effectiveness Methodologies

There is considerable variation from region to region in both the methods used to determine cost-effectiveness and in the cost per ton or kg of a particular emissions that is considered to be cost-effective. While cost-effectiveness values are continually subject to change, levels in excess of $15,000/ton ($16.5/kg), adjusted to current dollars, of NO_X, VOC, and SO_X have been applied for BACT in the South Coast Air Quality Management District (SCAQMD), which has the most stringent regulations in the United States.

There is also ongoing debate as to the most appropriate methods to be used for calculating cost-effectiveness. The method used above is a levelized cash flow (LCF) method, which determines the average annual cost by multiplying the control equipment capital cost by a capital recovery factor, and adding it to all other direct and indirect costs. An alternative method is the discounted cash flow (DCF) method. This calculates the present value of the control costs over the life of the equipment by adding the capital cost to the present value of all annual costs over the life of the equipment and applying a uniform interest rate that is independent of the inflation rate. Equipment life is typically assumed to be 10 years, unless a shorter period can be justified. One advantage of the DCF method is that it can more easily take into account annual operating and maintenance costs that are not constant, emissions reductions that vary with time, and capital costs that may occur after the first year.

Another area of debate for RACT compliance is whether or not cost-effectiveness evaluation should be based on standards for uncontrolled emissions or on actual test data for the specific system in question. If, in the previous example, other control measures had already been implemented, then test data would be used instead of specified uncontrolled emissions rate standards such as AP-42, depending on prevailing regulations. If, in the previous example, the facility had already installed control technology which achieved a NO_X emissions rate of 0.060 lbm/MMBtu (0.029 g/MJ), the annual NO_X emissions reduction resulting from additional control would be 62.1 tons (56,336 kg) and the cost-effectiveness would be $2,040/ton-y ($2.25/kg-y).

Similarly, there is also ongoing debate regarding the appropriateness of using marginal or average cost-effectiveness in a top-down BACT approach. In a top-down average cost approach, the control method with the highest emissions reduction is evaluated for cost-effectiveness ($/ton reduced) based on an uncontrolled baseline condition. If the method is judged to be cost-effective, then it is selected. In a marginal, or incremental, approach, the difference in cost and emissions reduction between one alternative and the next best alternative are compared. This yields an evaluation of the incremental benefit of employing successively more-effective control methods. In many cases, control methods that would pass based on the average cost approach would fail based on the marginal approach.

Consider, for example, a scenario in which uncontrolled emissions for a new fuel burning system would be 100 tpy (90,700 kg-y). If the highest emissions reduction method could reduce annual emissions to 10 tons (9,070 kg) at a cost of $900,000, the average cost-effectiveness would be $10,000/ton ($11.0/kg). If the next highest emissions reduction method could reduce annual emissions to 30 tons (27,200 kg) at a cost of $350,000, the average cost-effectiveness would be $5,000/ton ($5.50/kg). When compared with the next best method, the cost-effectiveness of the highest emissions reduction method using the marginal approach would be $27,500/ton ($30.3/kg), based on achieving an incremental reduction of 20 tons with an incremental investment of $550,000. If the cost-effectiveness threshold was $20,000/ton ($22.0/kg), the highest reduction method would be required under the average approach, but would not be required under the marginal approach.

Representative Costs for Emissions Controls

Capital costs for control measures vary considerably with system capacity, with per unit costs generally highest for the smallest units. Figure 17-49 provides an example of the relationship of capacity to capital cost. First cost of low-NO_X burners, with and without FGR, are shown as a function of boiler capacity in lbm/h (PPH).

Tables 17-13 through 17-16 provide representative control cost data for boilers, combustion gas turbines, and reciprocating engines. Capital cost, annual cost, and cost-effectiveness values are provided for different capacities, fuel types, and operating duties. Table 17-13 provides representative emissions control costs for oil- and gas-fired industrial and commercial boilers. Table 17-14 provides representative emissions control costs for tech-

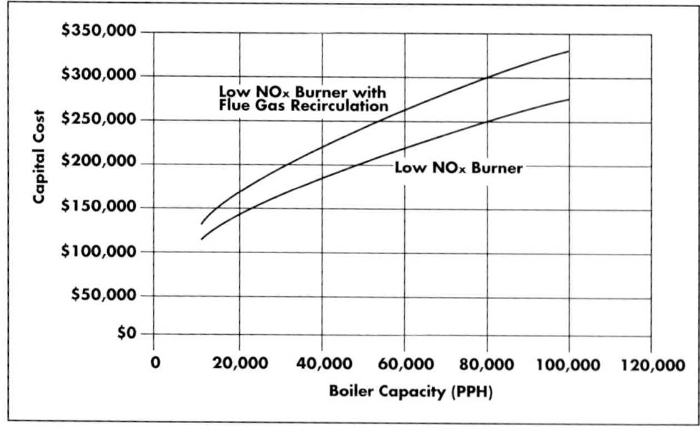

Fig. 17-49 First-Cost of Low-NO_X Burners with and without FGR as a Function of Boiler Capacity.

Control	Residual Oil Capital Cost ($/MMBtu/hr)	Residual Oil Annual Cost ($/MMBtu/hr)	Residual Oil Cost Effectiveness ($/ton)	Distillate Oil Capital Cost ($/MMBtu/hr)	Distillate Oil Annual Cost ($/MMBtu/hr)	Distillate Oil Cost Effectiveness ($/ton)	Natural Gas Capital Cost ($/MMBtu/hr)	Natural Gas Annual Cost ($/MMBtu/hr)	Natural Gas Cost Effectiveness ($/ton)
Firetube (10 MMBtu/hr)									
LEA[2]	2500	372	2280-4570	2500	270	3020-6040	2500	387	7360-14700
LNB[3]	5850	1190	2910-3640	5850	1190	5310-6640	5850	1190	9030-11300
RB[4]		See note 5 below			See note 5 below		3600	1060	5020-5730
FGR[3]	6110	1480	6040-12100	6110	1480	11000-22000	6110	1480	9360-11200
WI[2]		See note 2 below		2500	744	5550-8330	2500	627	5960-7950
Packaged Watertube (50 MMBtu/hr)									
LEA[2,6]	500	-33	<0	500	-136	<0	500	-18	<0
LNB[3]	2320	470	990-1240	2320	470	2750-3440	2320	470	2560-3200
RB[4]		See note 5 below			See note 5 below		6730	1960	6670-7630
FGR[3]	4160	1000	3530-7060	4160	1000	9780-19600	4460	100	4540-5450
WI[2]	500[2]			500	338	3900-4950	500	221	1500-2000
SCR[2]	6420	1560	2070-2360	6420	1510	5200-5890	6420	1510	4830-5480
SNCR[2]	3300	1040	2190-2740	3300	662	5040-6310	3300	869	4720-5910
Field-Erected Watertube (150 MMBtu/hr)									
LEA[2,6]	167	-101	<0	167	-203	<0	167	-86	<0
BOOS[2]	167	101	400-680	167	152	750-1250	167	94	620-1030
LNB[3]	1200	243	600-750	1200	243	600-750	1200	243	800-1010
RB[4]		See note 2 below			See note 2 below		6520	1900	3500-3940
FGR[3]		505	1690-3370	2070	505	2060-4130	2070	505	1390-1670
WI[2]		See note 2 below		167	271	1110-1660	167	154	640-850
SCR[2]	3770	1030	1290-1480	3770	1020	1560-1780	3770	996	2060-2350
SNCR[2]	3300	1050	2100-2630	3300	997	2450-3060	3300	937	3100-3880

1 Annual cost and cost effectiveness for 0.6 capacity factor; costs in 1993 dollars.
2 Source: EPA, March 1994.
3 Source: CARB, April 29, 1987.
4 Source: Santa Barbara County, December 1991
5 Not applicable, or not available.
6 Improved energy efficiency achieved using low excess air results in lower fuel costs and thus a net return on control strategy.

Table 17-13 Representative Emissions Control Costs for Oil- and Gas-Fired Industrial and Commercial Boilers.
Source: State and Territorial Air Pollution Program Administrators and the Association of Local Air Pollution Control Officials

Technology	Engine Size (hp)	RICH-BURN Total Cap. Cost ($)	RICH-BURN Annual Cost ($)	RICH-BURN Cost Effectiveness ($/ton)	LEAN-BURN Total Cap. Cost ($)	LEAN-BURN Ann. Cost ($)	LEAN-BURN Cost Effectiveness ($/ton)
Air/Fuel Adjustment[2]	250	11,000	6,000	580-870	74,000	26,000	3510-4680
	1000	16,000	15,000	350-520	78,000	31,000	1060-1420
	4000	25,000	45,000	270-400	94,000	53,000	450-600
Low Emission Combustion[2]	250	400,000	130,000	4500-5010	400,000	130,000	3970-4460
	1000	670,000	220,000	1850-2090	670,000	220,000	1610-1820
	4000	1,720,000	560,000	1190-1340	1,720,000	550,000	1030-1150
Ignition Timing Retard[2]	250	12,000	6,000	680-1130	12,000	5,000	980-4930
	1000	16,000	13,000	370-610	16,000	11,000	490-1470
	4000	25,000	38,000	270-450	25,000	30,000	340-1020
Prestratified Charge[2]	250	62,000	84,000	2670-3000		See note 3 below	
	1000	130,000	110,000	880-990		See note 3 below	
	4000	170,000	130,000	260-300		See note 3 below	
Non-Selective Catalytic Reduction[2]	250	20,000	10,000	290-310		See note 3 below	
	1000	42,000	27,000	200-220		See note 3 below	
	4000	130,000	96,000	180-190		See note 3 below	
Selective Catalytic Reduction[2]	250		See note 2 below		310,000	140,000	4280-4810
	1000		See note 2 below		340,000	180,000	1320-1490
	4000		See note 2 below		470,000	310,000	580-660

Engine Type[1]

1 Costs are estimated for engines running 8000 hours per year and are in 1993 dollars.
2 Source: EPA, July 1993.
3 Not Applicable.
4 Source: A.D. Little, September 1992.

Table 17-14 Representative Emissions Control Costs for Rich-Burn and Lean-Burn, Spark-Ignited Reciprocating Engines
Source: State and Territorial Air Pollution Program Administrators and the Association of Local Air Pollution Control Officials.

nologies applicable to both rich-burn and lean-burn, spark-ignited reciprocating engines. Table 17-15 provides similar data for Diesel- and dual-fuel-fired compression ignition reciprocating engines. Table 17-16 provides representative emissions control cost data for gas-fired and oil-fired combustion gas turbines.

The data contained in these tables should be viewed cautiously, since actual control costs vary widely depending on many application-specific factors. Moreover, due to the recently intensified focus on emissions control, control costs are subject to rapid change as new technologies are developed and existing ones refined. These factors, as well as economies of scale achieved in both manufacturing and field application of control technologies, have generally resulted in decreasing control costs, particularly for smaller capacity systems. Prime examples are SCR and DLN combustion systems for gas turbines. With DLN combustion systems, production costs are relatively low, but there is still a significant capital recovery cost associated with R&D development costs. Additionally, DLN combustors were first only available for very large capacity units. They are now becoming increasingly available for smaller capacity units, with production costs continually decreasing. While SCR systems were first applied in the United States on a very limited basis and at a very high cost, technology advancements and application experience have lowered costs.

Representative cost-effectiveness data should only be used as a general guide and not substituted for technology- and application-specific evaluation. This is particularly important in the field of emissions control, due not only to the rapid changes in the technology, but to the lack of standardization of performance measurement and reporting practices and the variability in field testing conditions. Manufacturer and vendor performance warrantees, with documented standards and testing tolerances, should be considered essential in the purchase and installation of control technologies.

		Fuel					
		DIESEL[1]			**DUAL FUEL[1]**		
Technology	**Engine Size (hp)**	**Total Capital Cost ($)**	**Annual Cost ($)**	**Cost Effectiveness ($/ton)**	**Total Capital Cost ($)**	**Annual Cost ($)**	**Cost Effectiveness ($/ton)**
Low Emission Combustion[2]	250	See note 3 below			520,000	170,000	11,370-12,990
	1000	See note 3 below			860,000	280,000	4650-5310
	4000	See note 3 below			2,210,000	710,000	2960-3390
Ignition Timing Retard[2]	250	12,000	6,000	760-1140	12,000	5,000	950-1420
	1000	16,000	13,000	420-630	16,000	11,000	470-700
	4000	25,000	40,000	310-470	25,000	29,000	320-480
Selective Catalytic Reduction[4]	250	190,000	99,000	4170-4690	190,000	98,000	5800-6530
	1000	250,000	140,000	1460-1640	250,000	130,000	1970-2210
	4000	510,000	300,000	780-880	510,000	270,000	1010-1140

1 Costs are estimated for engines running 8000 hours per year and are in 1993 dollars.
2 Source: EPA, July 1993.
3 Not Applicable.
4 Source: EPA, July 1993. Capital Costs corrected to remove double-counting of direct installation costs in EPA ACT.

Table 17-15 Representative Emissions Control Costs for Diesel- and Dual-Fuel-Fired, Compression-Ignition Reciprocating Engines.
Source: State and Territorial Air Pollution Program Administrators and the Association of Local Air Pollution Control Officials.

			Gas-Fired[1]			**Oil-Fired[1]**	
Technology	**Unit Size (MW) and Operation**	**Total Cap. Cost ($)**	**Annual Cost ($/year)**	**Cost Effectiveness ($/ton)**	**Total Cap. Cost ($)**	**Annual Cost ($/year)**	**Cost Effectiveness ($/ton)**
Water Injection[2]	5, continuous[3]	544,000	165,000	1390-1780	570,000	195,000	1000-1300
	25, continuous	1,140,000	408,000	690-880	1,210,000	547,000	560-710
	100, continuous	2,560,000	1,180,000	500-640	2,800,000	1,720,000	440-560
	25, peaking[4]	1,140,000	248,000	1670-2150	1,210,000	292,000	1190-1520
	100, peaking	2,560,000	624,000	1050-1350	2,800,000	786,000	800-1020
Steam Injection[2]	5, continuous	710,000	185,000	1560-2000	745,000	200,000	1010-1300
	25, continuous	161,000	448,000	760-970	1,730,000	514,000	520-670
	100, continuous	3,900,000	1,250,000	520-670	4,230,000	1,490,000	380-480
	25, peaking	1,610,000	319,000	2150-2760	1,730,000	350,000	1520-1820
	100, peaking	3,900,000	813,000	1370-1760	4,321,000	917,000	930-1190
Low NO_X Combustor[2]	5, continuous	482,000	63,400	530-800		See note 5 below	
	25, continuous	1,100,000	145,000	240-370		See note 5 below	
	100, continuous	2,400,000	316,000	130-200		See note 5 below	
	25, peaking	1,100,000	258,000	980-1470		See note 5 below	
	100, peaking	2,400,000	316,000	530-800		See note 5 below	
SCR[6]	5, continuous	572,000	258,000	2180-2450	572,000	274,000	1390-1560
	25, continuous	1,540,000	732,000	1230-1390	1,544,000	812,000	820-920
	100, continuous	3,300,000	2,190,000	920-1030	3,302,000	2,500,000	630-710
	25, peaking	1,540,000	517,000	3480-3920	1,540,000	537,000	2170-2440
	100, peaking	3,300,000	1,430,000	2400-2700	3,300,000	1,510,000	1530-1720

1 Costs in 1993 dollars.
2 Source: EPA, January 1993.
3 Continuous turbines operate 8000 hours per year.
4 Peaking turbines operate 2000 hours per year.
5 Not applicable.
6 Costs derived from Environex, 1991 and EPA, January 1993.

Table 17-16 Representative Emissions Control Costs for Oil- and Gas-Fired Combustion Gas Turbines.
Source: State and Territorial Air Pollution Program Administrators and the Association of Local Air Pollution Control Officials.

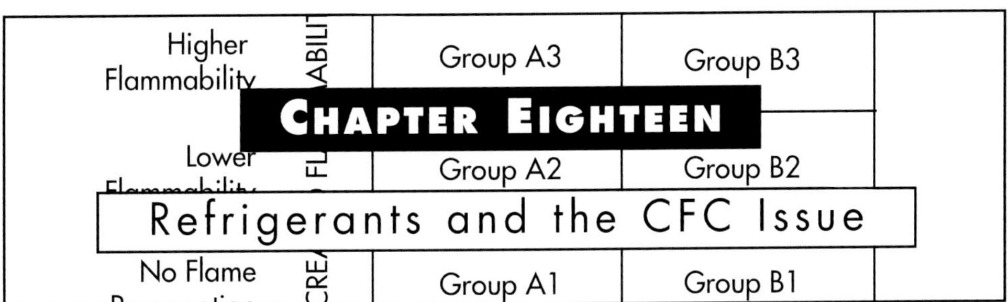

Chapter Eighteen

Refrigerants and the CFC Issue

The refrigeration industry is currently faced with two critical issues: the continuing rise in cooling related electric costs and the impact of environmental regulations on refrigerant production and usage. Three primary concerns driving the environmental issues are ozone depletion, global warming, and safety. The major regulatory drivers impacting the industry are:

- Reducing refrigerant emissions to the lowest allowable levels
- Increasing energy efficiency to the highest possible levels

This chapter provides an overview of commonly used refrigerants and current regulations related to safety and CFC phaseout. Possible retrofit alternatives are also discussed.

Refrigerants

Refrigerants are used as heat exchange media in refrigerant cycles to absorb and reject heat. They usually accomplish this as a result of phase changes, which occur through evaporation and condensation processes. To be useful, a refrigerant must satisfy numerous requirements. Chemical stability, specific heat, and latent heat of vaporization are important, as well as cost, availability, flammability, toxicity, and other factors. In a given application, the refrigerant used must be compatible with equipment, materials and operating characteristics. Finally, a refrigerant must be environmentally benign. Amongst the dozens of commercially available refrigerants, the need to satisfy a wide range of cost, safety, and thermodynamic requirements generally results in compromise.

The most commonly used refrigerants are ammonia, water, and a group of compounds containing fluorocarbons. They are commonly referred to by commercial trade names and their refrigerant, or R, numbers:

- **Ammonia (NH_3)**, or R-717, has been in use the longest of the widely used refrigerants. Ammonia has a boiling point of about -28°F (-33°C), a freezing point of about -108°F (-78°C), and a liquid-specific gravity of 0.684 at standard atmospheric pressure. It is highly toxic in high concentrations and flammable and several building code restrictions apply. However, ammonia remains a valuable refrigerant when handled properly due to its low cost, high efficiency, low volumetric displacement and low weight of liquid circulated per ton or kW_r of refrigeration. In addition to its excellent heat transfer qualities, ammonia has the ability to operate over a wide range of temperatures — often to -40°F (-40°C) or lower. An ammonia-water solution is the refrigerant used in certain absorption cycle refrigeration machines.
- **Distilled water (H_2O)**, or R-718, is the most commonly used refrigerant in large absorption chillers. It has a very large latent heat of vaporization — 1,070 Btu/lbm at 40°F (2,488 kJ/kg at 4°C). It is also stable, non-toxic, readily available, and relatively inexpensive. Its principle limitations are that it is only capable of producing cooling at temperatures above 40°F (4°C) and must operate under a vacuum. In addition to the refrigerant, absorption refrigeration requires a second fluid, an absorbent. Most absorption equipment currently on the market in larger capacities utilize lithium bromide (LiBr) in solution with water. LiBr is a non-toxic salt that dissolves in the water vapor it absorbs.
- **Chlorofluorocarbons (CFCs)** are a series of extremely stable fluorinated hydrocarbon compounds using ethane and methane as bases. The commonly used chlorofluorocarbons are trichloromonofluoromethane (CFC-11 or R-11), CCl_3F, which has a boiling point of about 75°F (24°C) at standard atmospheric pressure, and dichlorodifluoromethane (CFC-12 or R-12), CCl_2F_2, which has a boiling point of about -22°F (-30°C) at standard atmospheric pressure.
- **Hydrochlorofluorocarbons (HCFCs)** are similar but less stable than CFCs due to the presence of at least one hydrogen atom. The commonly used HCFCs are chlorodifluoromethane (HCFC-22 or R-22), $CHClF_2$, which has a boiling point of about -41°F (-41°C) at standard atmospheric pressure and diclorotrifluoroethane (HCFC-123 or R-123), $CHCl_2CF_3$, which has a boiling point of about 81°F (27°C) at standard atmospheric pressure.

- **Hydrofluorocarbons (HFCs)** differ from CFCs and HCFC due to the absence of any chlorine atoms. The most common HFC is tetrafluoroethane (HFC-134a or R-134a), CF_3CH_2F, which has a boiling point of about -15°F (-26°C) at standard atmospheric pressure.
- **Azeotropes** are mixtures of halocarbon compounds. A common azeotrope is R-500, which is a mixture of CFC-12 and HFC-152a. It has a boiling point of -27°F (-33°C) at standard atmospheric pressure. Considered a near-azeotrope, R-410a is a 50/50 mixture of HFC-32 and HFC-125 that is widely used.

The CFC Phase-Out

Current Regulations

CFCs have been used throughout this century for many applications, including refrigerants, solvents, and propellants. They are non-toxic, long-lived, and non-flammable. While the chemical composition varies with each CFC, they are all made from a combination of only three atoms: chlorine, fluorine, and carbon. CFC-12, for example, is made from two chlorine atoms, two fluorine atoms, and one carbon atom.

Once released to atmosphere, they persist long enough to slowly diffuse to the upper atmosphere (or stratosphere), where they are decomposed by ultraviolet radiation in gas-phase reactions. Released chlorine atoms then catalyze reactions which result in the decomposition of the ozone (O_3) present at this level. The concern over CFCs is that the massive quantities of CFCs being released to the stratosphere each year and the longevity of the compounds, has resulted in an ozone destruction rate that is far greater than the rate at which ozone is naturally created. While there has been continued scientific debate over the last two decades as to the seriousness of the situation, there is a widespread belief that ozone is being destroyed at a potentially dangerous rate.

Under federal mandate by the CAAA and the 1992 revisions to the Montreal Protocol, firm phaseout schedules are in place for production of chlorofluorocarbon (CFC) and hydrochlorofluorocarbons (HCFC) refrigerants. The final regulatory requirements limiting the production and consumption of ozone-depleting substances were issued by EPA on December 10, 1993, (58 FR 65018) and December 30, 1993, (58 FR 69235). Consistent with the Montreal Protocol and other international treaties, the regulations required a complete phase-out of CFC production by 1996. This phaseout included CFC-11 (R-11) and CFC-12 (R-12), which were the primary refrigerants used in centrifugal chillers and many supermarket refrigerators.

Prior to the Montreal Protocol, annual work production of CFCs was about 2.2 billion lbm (1 billion kg) per year, with about 40% of that produced in the United States. At that point, the cost of CFCs was about $1.00/lbm ($2.20/kg). This plan has driven up the price of available CFC-based refrigerants dramatically and is expected to continue the trend as supplies diminish. In accordance with the Energy Policy Act of 1992 (EPAct 92), the excise tax on CFC-based refrigerants increased from $3.35/lbm ($7.39/kg) as of 1993 to $7.15/lbm ($15.77/kg) 1999. It is likely that CFC-11, CFC-12, and CFC-500 chillers will begin to be left without replacement and service refrigerants, though stockpiles of recycled CFC-based refrigerant remain available to some extent. Already, costs for CFC-based refrigerants are well over $10/lbm ($22/kg) and are expected to continue to escalate at dramatic rates every year.

The Copenhagen Amendments also devised a phase-out schedule for HCFCs (Hydrochlorofluorocarbons). The molecular composition of HCFCs is similar to CFCs, with the distinguishing characteristic of being far less stable. Therefore, the impact on ozone depletion and global warming occurs at a slower rate than with CFCs. The schedule required consumption to level off in 1996 to a cap level based on 1989 consumption figures. At that point, HCFC use was limited to the sum of 3.1% of the 1989 CFC use, plus 100% of HCFC use in 1989, and is weighted by ozone depletion potentials. Future HCFC production limits, as a percent of the cap, are listed by year in Table 18-1.

Title VI of the CAAA also contains provisions for elimination of CFCs. It identifies stratospheric ozone depleting chemicals, such as CFCs, and establishes a schedule for the phase-out of the manufacture and import of these materials.

Table 18-1 Montreal Protocol/Copenhagen Agreement HCFC Production Caps

Date	Production Level
2004	65%
2010	35%
2015	10%
2020	0.5%
2030	0.0%

While operating on a separate parallel track, the CAAA establishes procedures and phase-out schedules that are roughly consistent with the international movement started under the Montreal Protocol. Under Title VI, ozone-depleting chemicals are divided into two classes. Class I chemicals include specified CFCs, halons, 1,1,1 trichloroethane (or methyl chloroform), and carbon tetrachloride. Class II chemicals are mostly HCFCs.

Class I chemicals are restricted in both production and use. Venting of refrigerants to the environment became illegal in 1992. Production of Class II substances is to be frozen by 2015 and their use eliminated by 2030. The CAAA also establishes a recycling program for CFCs used in air conditioning and refrigeration equipment. EPA has since promulgated regulations that maximize the recapture and recycling of ozone depleting substances. The regulations prohibit anyone who maintains, services, repairs, or disposes of appliances from venting or releasing substances such as refrigerants.

Table 18-2 shows the ozone depletion potential (ODP), global warming potential (GWP), and atmospheric lifetime (ATL) of various CFC, HCFC, and HFC refrigerants.

Alternative Refrigerants

Both low-ozone depleting and non-ozone depleting refrigerants are used as alternatives to CFCs. Low-ozone depleting refrigerants include HCFC refrigerants such as HCFC-22 and HCFC-123, which are referred to as **bridge refrigerants** because of the longer timetable governing their own phase-out. Other alternatives for vapor compression-based refrigeration systems include HFCs, notably HFC-134a, which has no chlorine atoms and no ozone depleting characteristics, and ammonia. Other alternatives include use of the two absorption-cycle refrigerants: water and ammonia-water solution.

Table 18-2 Ozone Depletion Potential, Global Warming Potential, and Atmospheric Lifetime of Various CFC, HCFC and HFC Refrigerants

Refrigerant	ODP	GWP	ATL
CFC-11	1.00	1.00	60
CFC-12	0.93	3.20	120
CFC-113	0.83	1.45	100
CFC-114	0.71	4.95	240
CFC-115	0.38	10.60	540
CFC/HCFC-500 (blend)	0.70	2.26	
HCFC-22	0.05	0.40	18
HCFC-123	0.02	0.02	2
HCFC-124	0.02	0.11	8
HCFC-141b	0.02	0.10	8
HCFC-142b	0.05	0.40	22
HFC-125	0.00	0.63	30
HFC-134a	0.00	0.31	18
HFC-143a	0.00	0.80	47
HFC-32/HFC-125 (blend)	0.00	1890	
Ammonia-717	0.00	0.00	(2 wks)

Source: UNEP/WMO

(Note that values for ODP, GWP, and ATL vary among published sources.)

Refrigerants such as HCFC-22, HCFC-123, and HFC-134a and other newly developed refrigerant blends can replace CFCs in original or slightly modified equipment. HCFC-22 is a refrigerant commonly used in systems with reciprocating and screw compressors. Production of large capacity systems that use HCFC-22 has also increased rapidly over the past several years. Some centrifugal compressor-based systems are designed to use HCFC-22, HCFC-123, or HFC-134a. Lower pressure centrifugal compressors are now replacing R-11 (CFC-11) with HCFC-123. Mid-sized systems previously operating on R-12 (CFC-12) now generally use HFC-134a.

There are no perfect drop-in replacements refrigerants and retrofit applications are generally more problematic than new equipment applications. Many systems, particularly hermetic designs, will be costly to retrofit, requiring new gaskets, seals, and possibly motors. Others simply cannot be retrofitted. However, with newly developed optimizing equipment and some hardware modifications, many systems can be retrofitted with a relatively straightforward, but lengthy (2 to 4 week), procedure. This is often done during scheduled overhauls.

The result of using replacement refrigerants is generally a loss of efficiency and cooling capacity output. This may be somewhat counteracted by efficiency and capacity increases achieved during traditional overhaul procedures. To compensate for losses, new equipment utilizing improved heat transfer surfaces and more efficient compressors have been developed. Computer models of engineering conversion techniques are also used to design optimal conversion systems which minimize losses.

When HFC-134a is used as a substitute for CFC-12, significant capacity reductions may result unless a gear set (or, in some cases, a motor change) is included as part of the retrofit. When replacing CFC-500, no hardware changes are necessary, but capacity can be reduced by 8 to 10%. HCFC-123 can be used to replace CFC-11, but may experience efficiency losses of up to 5% and capacity losses of up to 15%. Losses can be minimized by varying the impeller size or by over-sizing the compressor.

Retrofitting can be quite expensive and must be weighed against the cost of equipment replacement. In a chiller retrofit, typical conversion costs range from 30 to 80% of the cost of replacement. Replacement is more likely to be selected when existing equipment is fairly old, has already experienced significant efficiency losses, or where additional capacity would be needed to offset expected losses from refrigerant conversion.

The current trend is toward centrifugal compressors using R-22, R-123, or R-134a and replacing CFC refrigerants such as R-11, R-12, R-113, R-114, and R-500. Screw and reciprocating compressors are moving toward (or staying with) R-22, R-134a, and ammonia. One of the results of the new regulatory environment has been increased investment in absorption and desiccant technology development. This should lead to improved performance and lower capital cost for these systems. Tremendous investments have been made to optimize performance of traditional CFC-based systems during the past several decades and additional improvements are expected as a result of research and development efforts on alternative refrigerants. Other potential refrigerants are also currently under development and testing.

Carbon Dioxide (CO_2) is a component of our atmosphere that has no ozone depletion potential and insignificant global warming potential, so CO_2 has no regulatory liability, as do HFCs. There is no need to account for the amount used, and it does not need to be reclaimed. Other principal benefits of CO_2 are that it is a natural substance, it is relatively inexpensive (slightly over $1/lb), readily available, not poisonous in any common concentration, and nonflammable. The grade used must be dry, but it can presently be obtained in 99.9-percent purity from companies that supply welding gases, with a 20-pound tank selling for about $21. Characteristics of CO_2, compared with those of common alternatives, R-134a and R-404A, are shown in Table 18-3.

CO_2 is not new to refrigeration. Its use began in the mid-nineteenth century and steadily increased, reaching a peak in the 1920s. Its use declined with the introduction of CFCs that operated at much lower pressures. Use of CO_2 continued, but chiefly in cascade systems for industrial and process applications. Recently, strong interest has been shown in CO_2 as a refrigerant by vending machine and light-commercial refrigeration equipment manufacturers. There are also possibilities for other applications, such as residential air conditioning. The major challenges in CO_2 refrigeration involve the relatively high working pressures, since the supercritical portion of the transcritical cycle takes place above 1,067 psia (73.6 Bar).

Currently, CO_2 is becoming a popular choice for car air conditioning systems and supermarket refrigeration systems in Europe. However, engineering challenges remain. These includedesign improvements to increase cost-effectiveness, efficiency, and reliability of system that must accommodate the unique characteristics of CO_2, most significantly, five times the typical system operating pressure and a low critical temperature that requires cooling a supercritical fluid rather than condensing a two-phase mixture. As these challenges continue to be overcome, this refrigerant may see even more widespread commercialization.

Refrigerant Handling and Safety

ASHRAE Standard 34-1992, "Number Designation and Safety Classification of Refrigerants," and Standard 15-1994, "Safety Code for Mechanical Refrigeration," have been recently revised to address the classification and safety requirements of alternative refrigerants. With proper installation and the incorporation of a leak detection system with proper ventilation, all alternatives can be used safely.

Standard 34-1992 explains the format for refrigerant numbering and defines Safety Group Classifications for refrigerants according to their toxicity and flammability. Toxicity classifications are based on the Threshold Limit Value-Time Weighted Average (TLV-TWA). This is defined in Standard 34-1992 as "... the time-weighted average concentrations for a normal 8-hour workday and a 40-hour work week, to which nearly all workers may be repeatedly exposed, day after day without adverse effect."

Properties	Refrigerant		
	R-134a	R-404A	CO_2
Natural Substance	No	No	Yes
Ozone Depletion Potential	0	0	0
Global Warming Potential	1300	3260	1
Critical Point	590 psia; 214°F	541 psia 161°F	1,067 psia 88°F
Triple Point	0.058 psia -153.4°F	0.406 psia -148°F	75.1 psia -69.9°F
Flammable or explosive	No	No	No
Toxic	No	No	No

Table 18-3 Properties of CO_2 (R-744) compared with those of R-134a and R-404A.

As shown in Figure 18-1, refrigerants are assigned to one of two classes (A or B) based on the following criteria:

- "**Class A** signifies refrigerants for which toxicity has not been identified at concentrations less than or equal to 400 ppm, based on data used to determine TLV-TWA or consistent indices."
- "**Class B** signifies refrigerants for which there is evidence of toxicity at concentrations below 400 ppm, based on data used to determine TLV-TWA or consistent indices."

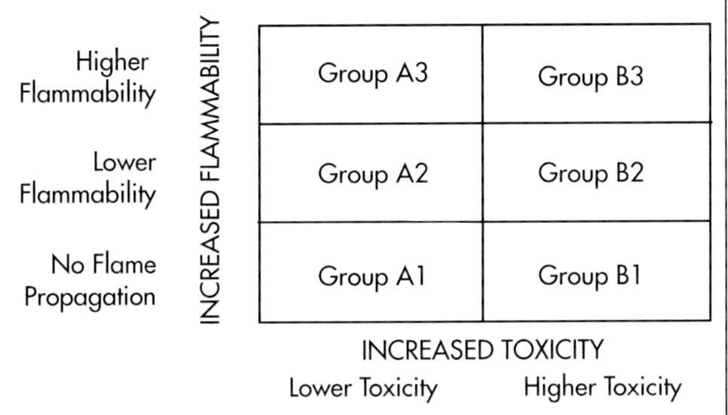

Figure 18-1 ASHRAE Standard 34-1992 Refrigerant Safety Classifications.

Additionally, refrigerants in each class are assigned a number used to designate flammability. Flame propagation and lower flammability limits (LFL) for each category are also defined in Standard 34-1992.

In low-pressure systems, HCFC-123 is the common replacement for CFC-11 used in centrifugal chillers and exhibits similar pressure-temperature characteristics. According to the standard, it is rated as having higher toxicity and is not flammable, placing it in a B1 refrigerant class.

In medium-pressure systems, HFC-134a is being used to replace CFC-12 in centrifugal, screw, and reciprocating compression-based systems. HFC-134a has a lower toxicity rating and is not flammable. Therefore, it is classified as an A1 refrigerant and is considered very safe.

In high-pressure systems, HCFC-22 is the primary refrigerant and is used for many applications in reciprocating, screw, and gear-driven centrifugal chillers. It is classified as an A1 refrigerant and is considered safe.

Ammonia, which is classified as a B2 refrigerant, is flammable, toxic in high concentrations, and corrosive and must be used carefully. Precautions include leak detection devices, even though its strong odor serves as a good warning sign. It can be smelled at levels of over 100 times less than when it is harmful. Most construction codes prohibit the use of ammonia in systems where ammonia may leak into a public occupancy area. Ammonia cannot be used with copper and zinc. Instead, more costly, lower heat transfer capability piping, such as aluminum or steel-welded schedule 40 or 80, must be used. However, because ammonia has better thermal characteristics than halocarbons, smaller pipe sizes are required.

ASHRAE Standard 15-1994 requires that "each machinery room shall contain a detector, located in an area where refrigerant from a leak will concentrate, which shall activate an alarm and mechanical ventilation ..." capable of removing the accumulation of refrigerant to the outdoors at the calculated airflow rate as listed in Standard 8.13.5. The detector shall be located in areas wherever refrigerant vapor from a leak will be concentrated to provide warning at a concentration not exceeding the refrigerant(s) TLV-TWA. For the machinery equipment rooms, a permanently mounted, continuously operating sensor shall be installed to provide long-term monitoring to assure occupant safety and refrigerant protection.

Section V

Utility Industry and Energy Rates

CHAPTER NINETEEN

Natural Gas Industry Overview

Natural gas is an abundant natural resource that can be produced and delivered at prices that are competitive with other available energy resources. Additionally, it is a comparatively clean-burning fuel with less environmental impact than many of the alternative fuels. Given the national focus on protecting the environment, reducing vulnerability to foreign oil markets, increasing use of domestic resources, and improving life-cycle energy-use efficiency, natural gas has become a critical resource in high demand.

Figure 19-1 shows the typical annual gas flow in the United States. As can be seen from this figure, approximately 22.9 trillion cubic feet (Tcf) of gas is withdrawn from gas wells in a typical year. Of this, 20.2 Tcf is consumed by end-users, including 5.0 Tcf by residential consumers, 2.91 Tcf by commercial consumers, 9.0 Tcf by industrial consumers, 0.6 Tcf by pipelines, and 2.7 Tcf by electric generation consumers.

Since the mid-1980s, the natural gas industry has busily experienced a continuing number of fundamental structural changes that have affected the business relationships among different industry entities. The evolving nature of the industry is depicted by Figure 19-2. As this exhibit shows, the natural gas industry includes the following basic components:

- **Producers:** Those who explore for new reserves and drill and develop gas wells.
- **Pipelines:** Those who are principally responsible for moving gas from producers to either local distribution companies (LDCs) or end-use consumers. Under current regulations, the LDCs, the ultimate end-users, or a gas marketer/broker can own the gas moved by the pipelines.
- **Gas marketers/brokers:** Those who purchase gas from producers or pipeline companies and sell that gas to LDCs or end-users.
- **LDCs:** Those companies that are primary distributors of gas, moving gas from pipelines to end-users. As is the case with the pipelines, the LDC, the end-users, or the gas marketer/broker can own the gas moved over the LDC's network.
- **Electric utility gas consumers:** One category of major end-users that use natural gas in electric generating facilities.
- **Residential, commercial, institutional, and industrial customers:** End-users that purchase gas from providers, LDCs, or gas marketers/brokers.

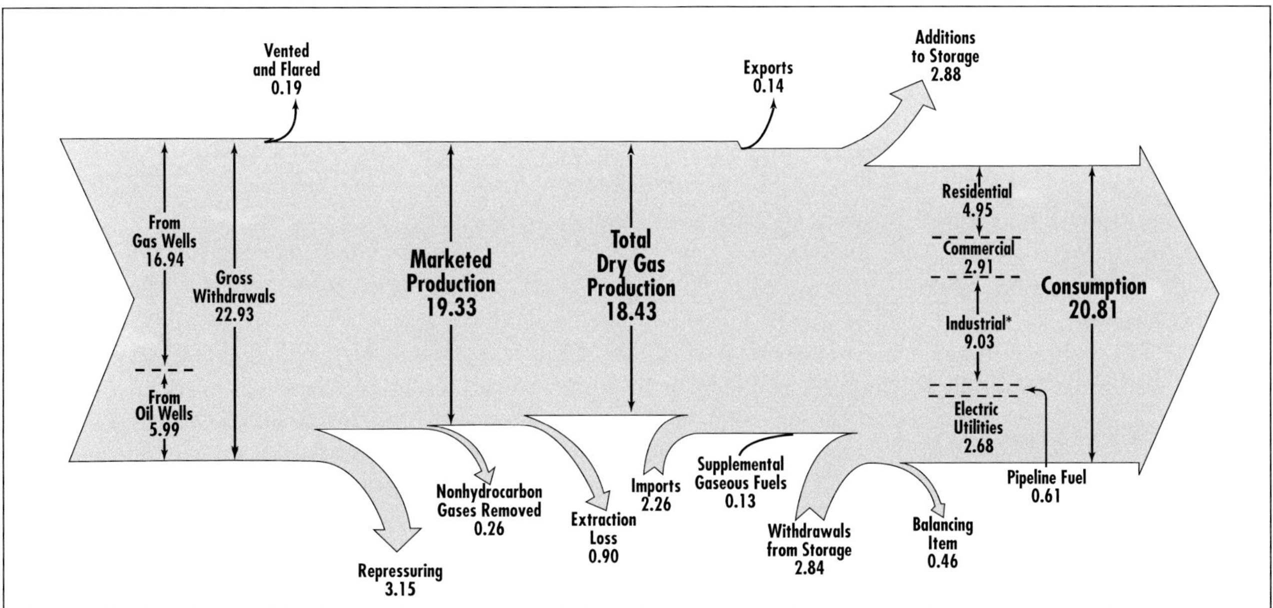

Fig. 19-1 Typical Annual Gas Flow (in Tcf) in the United States. Source: EIA/DoE

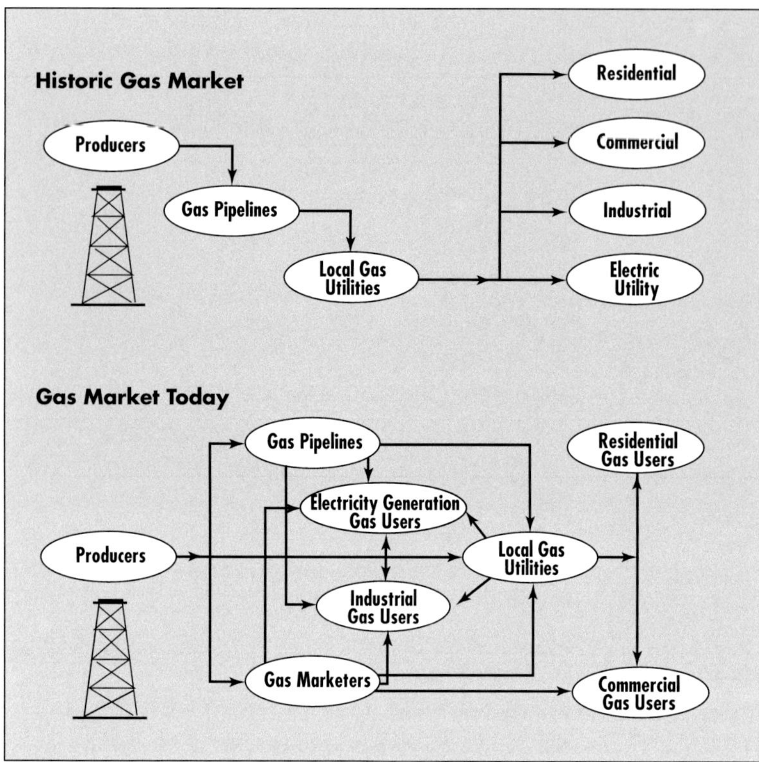

Fig. 19-2 Comparison of Historic and Current Gas Markets.

Historical Perspective

Over the past two decades, the natural gas industry has been transformed from a completely regulated industry to one in which market forces play a dominant role. Gas wellhead price deregulation, open access to pipeline transmission services, and comprehensive rate unbundling are three major shifts in federal policy designed to replace direct regulation with market forces.

In recent history, however, the gas industry was governed primarily by regulation. Passage of the Natural Gas Act of 1938 allowed the Federal Power Commission (FPC), now the **Federal Energy Regulatory Commission (FERC)**, to regulate the price interstate pipelines could charge distribution companies for gas. In 1954, FPC authority was expanded to include jurisdiction over the price producers could charge interstate pipelines. This ruling effectively brought all gas produced and sold in interstate commerce under federal regulation, allowing only intrastate gas to remain free of wellhead price control. As a result, growing gas demand produced higher intrastate gas prices because these gas prices were based on free-market conditions, not an artificially determined, regulated price level. For the interstate markets, it was truly a buyer's market during that time.

By the late 1960s, funding and development costs for gas had begun to exceed the artificially controlled price levels. As one would expect, the artificially low prices took a natural toll on industry investment and supply. For instance, in 1968, 19 Tcf of natural gas was consumed, but only 13 Tcf was found. By 1973, the gap between production and consumption was much wider: 22 Tcf was consumed while only 6 Tcf was found. Then, during the severe winter of 1976-1977, widespread shortages of natural gas developed in the Northeast and Midwest. While these shortages were most profoundly manifested in curtailments and interruptions, they did far greater long-term damage to the industry, as many would not trust future gas deliverability for years to come.

In 1978, Congress and President Jimmy Carter made deregulation of the natural gas industry a part of the National Energy Policy Act. They realized that pricing natural gas below its free market value distorted the market and prevented producers from obtaining an adequate return on their investment, thereby ultimately threatening consistent, reliable supply of natural gas to the nation. Thus, the 1978 Natural Gas Policy Act instigated the national-level decontrol of natural gas at the wellhead. Again, as expected, this decontrol has resulted in a dramatic increase in the responsiveness of gas exploration and production activity to standard market signals.

FERC, however, continues to regulate the interstate pipeline system. But, it has facilitated the unbundling of pipeline transportation and open access to pipeline transportation through FERC Orders Nos. 380, 436, 500, and 636. Along with wellhead deregulation, these orders have resulted in lower gas costs, as well as innovative and expanded gas services.

FERC Orders Nos. 380, 436, and 500 served to abolish minimum LDC commodity bill provisions in interstate pipeline sales tariffs and allowed LDCs and other pipeline customers to purchase gas from third parties. Order No. 436 also established the requirement for interstate pipelines to provide nondiscriminatory interruptible transportation and provided guidelines for partial automatic recovery of pipeline transition costs.

FERC Order No. 636 focuses on comprehensive

unbundling of merchant and transportation services. Traditionally, interstate gas pipelines have combined, or bundled, both the merchant and transportation functions and linked gas producers with downstream markets. Since November 1993, however, the situation has changed. Although pipelines, through their marketing affiliates, may sell unbundled gas just as any other marketer, the vast majority of gas moved on interstate pipelines today is owned and managed by others.

Today LDCs can competitively seek the lowest priced wellhead gas for both long-term (firm) and short-term (spot market) contracts, pay the pipelines only for transportation services, and then deliver gas to customers on either a firm or interruptible basis. Larger customers may also purchase their own gas, directly from the wellhead or through market brokers, and use both the pipelines and LDCs as vehicles for transporting the gas to their facility. Facilities with competitive energy alternatives and favorable utilization load profiles, resulting from extensive base or off-season usage, can benefit greatly from the new era of deregulated gas markets.

Natural Gas Industry Restructuring — Unbundling of 636

When FERC issued Order No. 636, known as the "Restructuring Rule," in April 1992, the Order was designed to improve the industry's ability to compete effectively for new markets by allowing more efficient use of the interstate natural gas transmission system. Its purpose was to increase competition among gas sellers and diminish the perceived monopolistic market power of pipeline companies.

This deregulation initiative shifted the gas purchasing responsibility from the pipeline to the LDC or end-user. These gas purchasing entities, which have traditionally relied on the pipelines for gas supply purchase logistics as part of the bundled service package, now must contract for these services separately or improve their own gas supply market intelligence so they can participate successfully in this new deregulated market.

A shift of regulatory responsibilities has also resulted from the Restructuring Rule. State regulators have taken on an increased oversight responsibility for gas supply related matters and have seen the need to work more closely with LDCs to evaluate the risks, pricing features, and other particulars of future natural gas supplies.

The restructuring under Order No. 636 has provided the natural gas industry with increased flexibility to respond to new marketing opportunities. Other details of the major provisions of Order No. 636 include:

- Pipeline companies that provided bundled city-gate firm sales service on May 18, 1992, now must offer a new "no-notice" firm transportation service (i.e., service that does not require advance notice by the shipper).
- Pipeline companies must provide open-access transportation that is equal in quality for all gas supplies, whether purchased from their pipeline company or not.
- Pipeline companies must provide customers with open access to storage.
- Tariff provisions cannot inhibit the development of market centers or production pooling areas. [Market centers allow diverse customer demands to combine as a single load. Pooling areas allow diverse supplies to combine as a single source.]
- Two new generic capacity-assignment mechanisms were established. One authorized and required pipeline companies to provide firm shippers on downstream pipelines with access to capacity on upstream pipelines held by the downstream pipeline. The other authorized a reallocation mechanism so that firm shippers can release unwanted capacity to those that want it. These reassignments must first be posted on an electronic bulletin board.
- All fixed costs associated with pipeline transportation service are now recovered through a capacity-reservation fee, i.e., a straight fixed-variable (SFV) rate design. Previously, the commonly used modified fixed-variable (MFV) rate design allocated certain fixed costs to the volumetric charge. As a result of this change, the cost of peaking service has been increased significantly. In addition to SFV, other rate designs may be adopted for very limited special circumstances primarily associated with small municipal customers.
- Pipeline companies were required to use various rate-making techniques to mitigate "significant" changes in revenue responsibility to any customer class. If any customer class experienced a change in revenue responsibility that exceeded 10% after mitigation, pipeline companies were required to phase in the increase over a four year period. This phase-in period was viewed as a temporary measure, serving to soften the sudden cost increases associated with the rate-design change.
- Pipeline companies were allowed to abandon sales and interruptible transportation service to any exist-

ing customer upon expiration or termination of the contract without seeking case-by-case approval from FERC. Service under firm transportation contracts for one year or less could also be abandoned. Under longer-term contracts, such service could be abandoned only if the existing customer failed to match the offer for the capacity made by another potential customer.

- Firm shippers must have flexibility in changing receipt and delivery points.

To the average customer, these changes within the natural gas industry may seem very technical. The impact on the industry, however, has been far reaching:

- Merger and acquisition activity among LDCs has increased as competition and the pressure from state regulators for lower gas costs increases.
- While FERC Order No. 436 caused a great deal of industrial end-use gas to move from LDC sales to LDC transportation service, Order No. 636 has led LDCs to offer services that entice industrial transportation customers back to sales services.
- Many new pipeline, pipeline interconnection, and storage projects have emerged.
- The availability and dependability of natural gas to all sectors of the consuming public has been enhanced.
- Natural gas has become even more competitive with alternative fuels in the industrial sector.

Natural Gas Loads

Since the nation requires more natural gas (and other fuels) in the winter, the laws of supply and demand usually drive up costs in the winter. In addition, the coldest months set the peak requirements for production and distribution facilities and, therefore, most of the fixed costs for the entire industry. Allocation of these fixed costs to natural gas purchased in this period also drives up the cost of winter gas.

Conversely, in the non-heating season months, when demand is lower, natural gas is less costly. In addition, because fixed costs are mostly allocated to winter gas supplies, non-winter gas gets somewhat of a free ride and is tied primarily to a variable cost already driven down by lower demand. To the extent that some fixed costs can be recovered through (or spread over) non-winter gas sales, LDCs and pipelines benefit by building their system load factor by increasing non-winter sales. LDCs in regions with moderate climates or where industrial process demand greatly exceeds heating demand will typically have higher load factors and experience a lower differential between winter and non-winter costs. The national effects of supply and demand as a whole, however, will still affect costs in these regions.

Integrated Resource Planning

Another aspect of natural gas industry regulation is the integrated resource planning (IRP) process. In the electric industry overview discussion in Chapter 20, the evolution of the IRP process is presented, as well as standard industry cost and benefit evaluation techniques. Following is a brief discussion of IRP in the gas industry.

Decades of regulated pricing at the wellhead caused market aberrations, which in turn led to a natural gas shortage in the mid-1970s. Natural gas utilities had to constrain their marketing efforts and began to sign desperate contracts with pipeline suppliers to ensure that they would be able to meet their obligation to serve. Rapidly rising natural gas prices finally led to federal price deregulation phased in through 1985.

The natural gas shortage and resulting escalating prices led to significant voluntary conservation efforts on the part of customers, and the beginning of government- and gas utility-sponsored conservation programs. Additionally, more stringent energy efficiency appliance and equipment standards have led to significant natural conservation as individuals and businesses replace old equipment.

Gas utilities also began to negotiate interruptible gas sales for large industrial customers as a way to help ensure that they could meet peak winter demands. While conservation was not yet being specifically targeted by utilities and state **public utility commissions (PUCs)**, it had become a supply planning consideration. The move toward interruptible contracts, however, was a significant step toward proactive supply planning through use of a **demand-side management (DSM)** tool. In other words, interruptible service has historically been, and continues to be, a DSM tool used by LDCs throughout the country.

Natural gas utility-sponsored DSM programs are fairly similar to those for electric utilities in terms of the basic load shaping strategies. These programs enable the gas utility to delay the purchase or building of low-load factor and, therefore, relatively high-cost peaking and storage supplies or facilities that are needed only for short duration. In addition, they serve to avoid gas purchases and costs associated with gathering, transmission, distribution, customer accounts, and general plant and other non-gas costs.

Load leveling programs, such as high-efficiency furnaces, enable the gas utility to get maximum use of its system delivery capabilities and contract reservation charges. Load building programs, such as those that encourage the use of gas-fired cooling technologies, are one way to achieve higher load factors with little or no incremental system cost. Load reduction or peak shaving programs generally involve interruption capability in exchange for reduced rates to customers that possess alternative fuel use capability. The goals for each of these load shape objectives are as follows:

- **Conservation:** The promotion of decreased gas use, primarily during peak periods, to reduce peak-day requirements and ensure adequate reserve margins. Examples include installation of building envelope insulation, high-efficiency burners, and heat recovery systems.
- **Load building:** The promotion of increasingly efficient use of existing capacity that can reduce rates by spreading fixed costs over greater sales units. Examples include fuel switching, natural gas vehicles, fuel cells, and application of prime mover technology for mechanical drive and electric generation.
- **Seasonal load reduction:** Changes in load shape that yield increased gas utility seasonal control over the level and/or timing of demand. Examples include high-efficiency furnaces and water heaters and gas heat pumps.
- **Valley filling:** The promotion of technologies that increase off-peak sales to improve utility load factor and generate additional revenue without increasing peak-day requirements. Examples include application of gas-fired cooling and dehumidification technologies and gas-fired electric peak shaving generators.
- **Peak clipping:** Options that reduce peak demand without adversely affecting off-peak sales. This involves storage deliverability service.
- **Peak load shifting:** The shifting of on-peak demand and energy use to off-peak periods, resulting in lower peak-day requirements and higher load factors. This involves various types of storage capacity.

On the supply side, gas utility IRP differs fundamentally from electric utility IRP. The native generation and off-system supply purchases of the electric utility are replaced with a variety of gas purchase, transportation, and storage contracts. Peaking facilities, such as liquefied natural gas (LNG) plants or propane-air injection facilities, may supplement these contracts. The chosen mix of these supply options is intended to meet the peak daily demand requirements of the system at lower cost.

Historically, LDCs evaluated opportunities to add load based on whether the added load would provide sufficient revenues over time to cover the cost to serve them and provide a fair rate of return. Hurdle rates, or rate-of-return tests, were established to determine if and how much a customer would have to contribute toward the cost of gas hook-up and associated construction required to provide gas service.

When considering the addition of a new resource that would allow it to serve new customer loads, the LDC would consider the change in average gas costs that would result to existing firm service rate payers. It would also consider the margins that would be earned on the new sales that would be used to reduce the revenue requirement of existing rate payers. These considerations assured that rate payers would not bear an increase in total revenue requirement as a result of the addition of new resources.

In natural gas IRP, this same logic is applied to consideration of all resource-related investments, be they supply-side options (e.g., contracting for new gas supplies, building new pipelines, or brokering capacity) or demand-side options (e.g., conservation or load building). Within an overall societal framework, gas utility IRP considerations go beyond revenue requirements. They seek to determine which combination of resource and marketing options provides the greatest benefits to customers, the gas utility, and society.

Gas utility IRP must still consider the commodity side for its native load. However, with the movement toward free market competition in gas commodity sales and the unbundling of services, the role of gas utility IRP will continue to shift away from the commodity side and focus more strongly on gas distribution services. With this shift, LDC-sponsored DSM programs have been de-emphasized. Since the LDC must now compete in a relatively free market for commodity sales, the obligation to include the commodity-side impact in IRP evaluation has also been reduced or eliminated. The result is that the cost-justification for administering DSM programs has been greatly diminished.

The transmission and distribution segments of both the gas and electric industries are similar in that transmission investments exhibit substantial economies of scale, and distribution in both industries has monopolistic features. However, the natural gas industry differs from the electric industry because gas can be more easily stored than electricity. With regard to IRP, the ability to easily store gas and to broker capacity has a significant impact on an LDC's marginal costs.

Reliability planning is also different for gas and electric utilities. LDCs plan for their own reliability, while electric utilities depend more on regional power pool planning since their transmission and distribution are interconnected with other electric utility systems. In addition, regulators have historically placed a higher priority on electric service reliability than on gas.

One important factor in the development of gas utility IRP has been the need to make quick decisions about new supply resources without the benefit of knowing when the next opportunity will occur. However, today's unbundled, competitive market for gas supplies, transportation, and capacity brokering has provided increased flexibility in terms of optimizing both short- and long-term resource portfolios.

As with the electric industry, the role of intensive IRP is seen as somewhat less critical given the replacement of regulatory control with competitive market forces. LDCs are obligated to make prudent resource acquisition decisions in order to remain competitive and financially viable. Still, the transmission and distribution of natural gas shall remain a heavily regulated area with limited access to market entry. Hence, IRP methods should still be required to overcome market imperfections brought about by regulation.

The Current Natural Gas Market

Natural gas is sold and/or distributed to consumers by the LDCs. LDC natural gas rates are determined through a cost-of-service regulatory process under the jurisdiction of a PUC. Increasingly, though, consumers may now also purchase gas through a variety of other means — usually a gas broker — and have it transported to the city gate of the LDC. The city gate is the delivery point at which the LDC takes control of gas delivered by pipelines and moves it through its distribution system to the customers in its franchise area. The price paid for the gas is market-based and the price paid for its transportation through the interstate pipeline system is regulated by FERC. Currently, the LDCs have exclusive franchise rights for sale of distribution services for delivering the gas from its city gate to the customer's gas service meter.

Gas flows from the wellheads scattered across the country to these gas meters through a massive distribution system that includes interstate and intrastate pipeline and local distribution networks. Most of these networks are woven with cross-connections forming a reliable patchwork of very large centralized trunk lines, with successively smaller branches.

Gas flows to and from massive hubs, or distribution centers, and is priced to consumers based on its cost at the hub plus a transportation differential from the hub to the city gate. When customers select a forward fixed price, the price is built up using the following three components:

1. The NYMEX natural gas futures contract price for the forward month(s), whose delivery location is at Henry Hub in Louisiana. [While there is an additional, smaller futures trading location at the Waha Hub in West Texas, the overwhelming reference point used to price gas from California to Maine is the NYMEX contract at Henry Hub.]
2. A basis differential from the Henry Hub to a local market hub.
3. The transportation differential from the market hub to the city gate.

Whether it is bought by the LDC for sale or by a broker arranging the sale for a customer, pricing is based on the delivery of a given volume at a given time to a given location with a given level of reliability.

Most consumers are not involved throughout the entire process. They most commonly purchase gas from the LDC, based on a price at their meter. Some, usually larger customers, may purchase gas that is delivered to the LDC's city gate by a broker. Either the broker negotiates its ultimate delivery to the facility meter with the LDC or the consumer may play a more direct role in this process. Still, only a few, usually the very largest, customers play an active role in all of the numerous transactions that occur along the way from the wellhead to the burner tip.

Mentioned above, gas, like many other commodities, is sold on the NYMEX futures exchange. It can be bought on a month-by-month basis, or in strips for a given volume at a specific monthly price for a year or several years, with specific volumes locked in for each month. A basis is added to the NYMEX price to establish a price at a given local market hub. Although the basis is traded in an over-the-counter market, the natural gas basis and NYMEX contract together are one of the most volatile and heavily traded products in U.S. energy markets.

In addition to these costs, various other transaction fees are incurred along the way as gas moves from system to system on its way to the city gate of the LDC.

For most process businesses, a known, or at least fairly regular, quantity of gas is required every day with known maintenance downtimes and the occasional forced shut down. Most customers, however, have heat sensitive loads that are driven by unpredictable weather. While these customers must have gas as it is needed, it makes it very difficult, and hence expensive, to bring the right amount of gas everyday to the city gate.

Even process oriented facilities experience some variability in their gas use. This may result from the variation in their production levels that depend on market conditions and/or a portion of their load being weather sensitive. The activity surrounding this unpredictability is referred to as "balancing" and usually involves local gas storage and propane-air or LNG injections. The cost of the facilities required in balancing this load is an additional charge that is passed on to the consumer.

Customers that have on-site backup fuels (e.g., fuel oil, propane, etc.) can save money by purchasing their gas and its transportation on an interruptible, rather than firm, basis. For many facilities in distant regions, such as the far northeast, firm delivery is not an option; there simply is not enough pipeline and storage capacity to service the winter requirements of everyone. For others, though, strategically paying for only seasonal off-peak transportation and gas can produce a substantial savings bucket to pay for alternate fuels and fuel-use equipment.

Natural gas is purchased on a firm or interruptible service basis or as a combination of these services. Firm sales are assured, i.e., the broker and/or LDC takes on the obligation to provide customers with as much gas as is required on a year-round basis.

While the commodity may be available on a firm basis, it is necessary to reserve room on the pipelines and local distribution system to ensure firm delivery. Interruptible delivery service, which requires little or no contribution to pipeline demand charges, involves the sale of gas, generally at a lower rate than firm service, when gas is available.

The fixed costs that must be recovered are based on the overhead of the companies involved and include the distribution system, storage facilities, and service and corporate facilities. The variable costs that must be recovered are those associated with purchasing natural gas supplies and transporting them through the vast national pipeline transmission system, and the annual storage activity by the LDC.

Rates are reflective of the degree to which fixed and variable costs are affected by the type of load served and service provided:

- **Firm sales and transportation services** are the most expensive, because to fulfill the obligation of firm service, the broker and/or LDC must enter into long-term guaranteed supply-purchase contracts. In many cases, they are also responsible for storage practices to keep extra gas available in the event of an unusual winter cold spell or some type of supply interruption.

  Because firm-service obligations set the level of peak-distribution capacity that must be built and maintained, they have the greatest effect on capital expenditure requirements. Firm service also may include uninterrupted transportation service for third-party purchased gas or backup for facilities using alternative energy sources.
- **Interruptible sales and transportation services** are less costly because the broker and/or LDC is unencumbered by the obligation of firm service, and they are required to sell or transport gas only when gas or capacity is available. Availability may result from excess gas from firm supply contracts or purchased gas from the spot market. Interruptible transportation service is provided whenever the pipeline and/or LDC have capacity available on their distribution system.

  Customers contracting for such services have the ability to withstand supply interruptions, typically through the use of alternative energy sources. There are also various levels of interruptible service. In many cases, agreements are made that call for gas service to be interrupted (either by the pipeline or LDC) for a given number of days in the year. This may range from a few weeks to several months, depending on the location of the facility and the price paid.

One of the important industry trends is the switch toward straight cost-of-service pricing for local distribution of transported gas. In some areas, distribution tariffs are negotiated based on competitive alternatives, much like interruptible gas service. This presents certain difficulties for gas marketers and end-users because the LDC ultimately controls the final price through its ability to vary the price of distribution between the LDC's city gate and the burner tip, regardless of the price paid for gas delivered to the city gate. For example, by charging a high price for distributing independently purchased gas from the city gate to the customer's facility, an LDC can make it more attractive for the consumer to purchase gas directly from the LDC. However, the trend continues toward local distribution tariffs that are cost-of-service based for both interruptible and firm distribution services.

VOLATILITY IN THE NATURAL GAS MARKET

One challenge in evaluating the economics of projects that utilize natural gas is choosing a representative gas rate. This has been especially difficult in the last decade since the natural gas market has been extremely volatile due to variations in supply and demand caused by weather, utility power plant dispatching decisions, fluctuations in storage levels, constraints in pipeline delivery capacity, and pipeline operational difficulties. In addition to these factors, the natural gas is a competitive fuel to other common fossil fuels, most notable heating oil, so the price of gas may tend to track oil prices, which can also be a volatile market.

As shown in Figure 19-3, the price of natural at both the wellhead and the city gate has varied by more than $7.00 per MCF, with prices doubling and even tripling within a period of a few years.

One way to address this volatility is to use long-term average utility rates for analyzing the cost of gas, along with a reasonable estimate of escalation for future price to account for inflation and future price impacts such as the consumer price index (CPI). Alternatively, forecast indices more tailored to energy pricing can be used, such as those provided for short term outlooks by the Energy Information Administration (EIA) or for longer term horizons by the National Institute of Standards and Technology (NIST). Securing multi-year supply contracts for all or a portion of the fuel requirement is one way of mitigating price volatility risk. Another option is to consider multiple energydual fuel capability where it is logistically feasible.

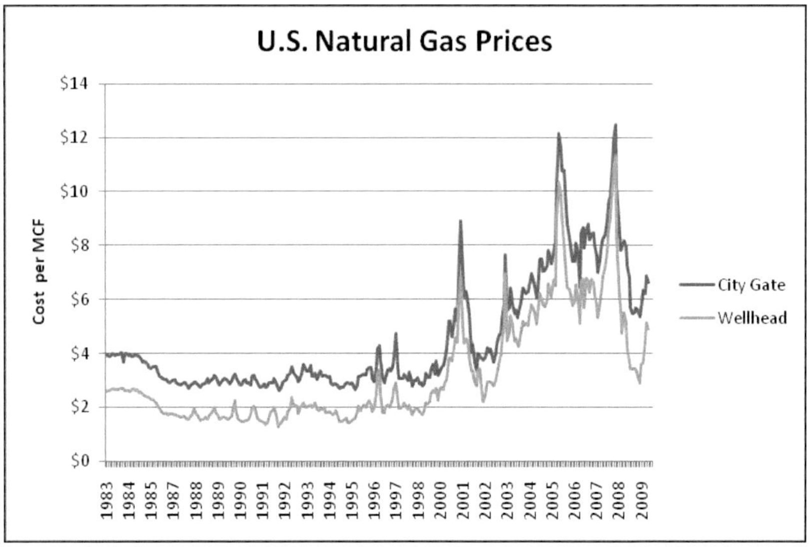

Although different energy sources tend to track each other in price somewhat, this still provides flexibility and cost advantage in energy source procurement. This might be in the form of a dual fuel (natural gas and oil) burner or a propane air system to allow interruption of natural gas supplies. In any event, it always makes sense to perform sensitivity analysis to determined the impact of utility rate changes. This is discussed further in Chapter 41.

SELECTING THE BEST RATE

For consumers, selecting the best natural gas rate entails matching available rate options with the requirements of their specific facility and the alternative energy services that are available. These factors may affect a facility's decision to locate in a given service territory and what type of goods and services it will provide. Once located, these factors will dynamically affect ongoing capital and operational decisions.

While more costly, standard firm rates may often be most attractive for many customers, because the gas volumes are guaranteed and the additional costs and complexity of maintaining an alternative energy source capability are eliminated. Limited hours of operation and the use of high-efficiency equipment and/or waste heat recovery all reduce the impact of higher-cost gas. Certain types of firm rates, such as demand/commodity or end-use specific, can be highly attractive when there is a strong match with the operating characteristics of the application.

Non-firm service, or interruptible, rates generally require additional capital costs associated with installing equipment that can use alternative energy sources. The gas cost differential, however, may be highly compelling.

Cost savings associated with non-firm service rates, with alternative fuels such as oil or propane, can greatly enhance dual-fuel, direct-fired equipment options. Interruptible rates with electricity as a backup can sometimes allow for lower-cost gas without the investment in dual-fuel capabil-

Figure 19-3 Natural Gas Price History Source: Energy Information Administration

ity. This applies to multiple-unit systems, cogeneration, and peak shaving applications. **Propane-air systems** are sometimes used to provide an alternative fuel source for gas-fired equipment during periods of interruption. These systems provide a propane-air mix (which is sometimes blended with natural gas) that is adjusted to match the heat content and distribution characteristics of natural gas. As such, it can be distributed through the gas piping at a facility and used as an energy source for gas-fired equipment, thereby eliminating the need to install dual-fuel operating capability on each gas-consuming device.

Risk Management

The most dramatic change in energy procurement management over the next few years may be the impact of price risk management. Where customers were previously passive recipients of the price risk management exercised by the LDC on their behalf — in effect, no risk management — nearly all customers will have the ability to manage their individual price risks in the future. While the utility essentially pays, and charges, whatever market prices are when the gas is bought, customers can easily lock in a fixed price for future periods by simply asking a broker for a quote. If agreed to, the broker hedges the transaction in its book of futures contracts and supply portfolio. If not agreed to, the customer can call the broker back for another quote should the futures prices fall, paying market prices until prices fall sufficiently or until the customer is fearful of forward prices rising higher.

Note that while risk management is typically thought of as only something that someone with very large energy positions might engage, this is no longer true. Some small heat-sensitive businesses and even residential customers already have this ability through brokers and this is expected to be a growing trend as the market continues to mature.

Further, while this may seem like a cumbersome addition to the manager's energy procurement effort, the risk management ability will provide invaluable efficiencies. Certain technology application projects may make sense only on the condition that the future gas or electric price holds; and the project economics fall apart with other future price constellations. With risk management, one can easily lock in the forward price and initiate the project without price risk. Alternatively, if the prices do not support sufficient economic performance for the project to be viable, one can safely put the project on hold until the forward prices support the project.

Summary

The industry trend in the sale of natural gas to end-users is one of increased competition and increased customer choice of services. Natural gas commodity and delivery services will be marketed as unbundled components or as rebundled components in an economically attractive and convenient package of services. The LDC will retain the primary responsibility for distribution of gas from the city gate to the customer's facility. This distribution will be available as a regulated unbundled service. LDCs may or may not continue to be the dominant seller of gas to its customers. Third-party brokering of gas is expected to continue its growth as a mainstream service and will likely penetrate even to the residential level. As a result of gas sales competition, commodity pricing is expected to be highly competitive, with modest profit margins achievable by the seller.

Commodity gas and delivery services will continue to be available on a firm and non-firm basis, or various combinations of both. Pipeline transportation rates and LDC distribution rates will remain regulated and will generally be based on the cost-of-service. Customers will be marketed to by pipelines, LDC, gas brokers, and other energy services companies, each striving to provide the best combination of price and customized services to meet specific requirements.

Chapter Twenty

Electric Industry Overview

Electricity is consumed by virtually every residential, commercial, industrial, and institutional facility in the United States. As illustrated in Figure 20-1, powered by coal, natural gas, petroleum, nuclear, hydro, and geo-thermal energy, some 30 quadrillion Btu (31.7 trillion MJ) are consumed in providing 3 billion MWh per year of electric power to the nation's electricity consumers.

The past half century witnessed a profound growth in electricity use. During the 1940s and 1950s, a unified electric grid reached the far corners of the nation, while electricity costs continually declined. But, during the mid-1970s, the high cost of building new large centralized power plants and the continuing growth of peak demand and oil prices produced substantial increases in the price of electricity.

Electricity costs today vary widely across the country, though some stabilization and cost efficiency has been achieved as a result of utility interactivity on a unified electric grid, price structures designed to direct the market away from peak usage and toward conservation, integrated resource planning (IRP) activities, and the infusion of increased competition in the electric generation market. Environmental considerations and growth in **non-utility generation (NUG)**, both fueled by federal legislation, have continued to have their impact in redefining the electric industry.

The trend toward deregulation in the electric industry has and will continue to have a profound impact on electricity consumers, as well as on electric utilities and various types of non-utility power producers. The NUG sector has evolved considerably over the past two decades as evolving legislation and regulation have infused competition into the electric generation portion of the industry. Whereas once electric utilities had nearly exclusive rights to generate and sell electricity as a sole-source provider, the long-term trend has been toward utilities only having such rights over the transmission and distribu-

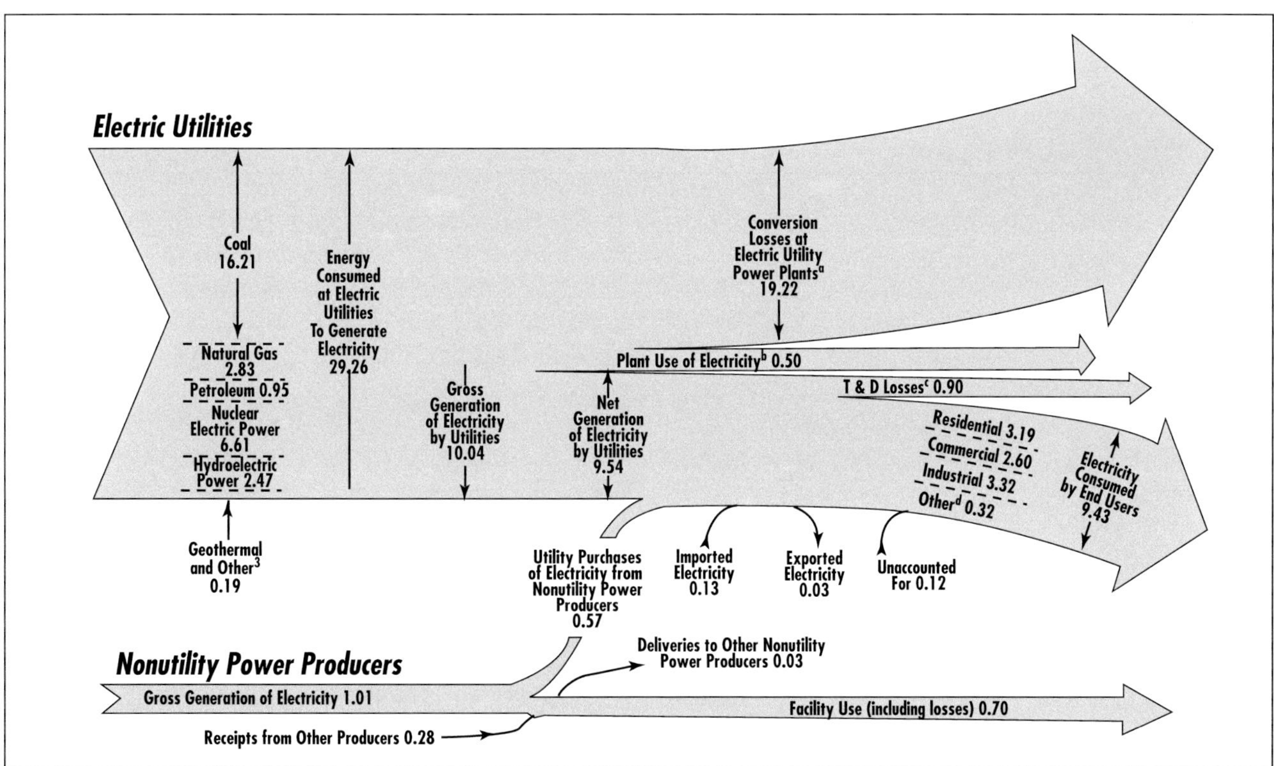

Fig. 20-1 Typical Annual Electricity Flow (in Quadrillion Btu) in the United States. Source: EIA/DoE

tion (T&D) of electricity — while becoming only one of numerous competitors in the area of power generation and sale. To some extent this trend has stalled in recent years and, in fact, several states have taken steps to put more regulations back in place to mitigate risk of outages or cost destabilization (such as was experienced in California and other states).

Despite the emergence of competition in the market, electricity prices have not plummeted to levels predicted by some, as stranded investment and recovery of other such costs have found their way into transmission and distribution pricing. Retail wheeling has not taken hold in the market, so it has not been a panacea either. The current view is that tremendous capital investment will be required to upgrade and modernize the distribution grid, which would also tend to hamper price reductions. Finally, higher fuel and other commodity costs and more stringent environmental regulations, notably for greenhouse gas (GHG) regulation are additional factors indicating the potential for significant increase in electricity prices.

Historical Perspective

The prolific growth of the electric utility industry during the middle portion of the last century was fueled by its ability to offset capital investments by gains in operational efficiency and market expansion. Early on, technology, market growth, and economies of scale combined to cause utility marginal costs (the cost of providing a new kWh) to be lower than average embedded utility costs (the average cost of providing a current kWh). By the 1960s, straight-slope growth projections had become standard in most supply planning models. In the 1960s, nuclear power gave rise to the dream of electricity "too cheap to meter." With these developments came a dramatic growth in electric demand and a pronounced market dominance of electric heat appliances and motor-driven technologies.

In the mid-1970s, when electric powered systems were at their zenith in market share, the electric industry was hit by the first of several price shocks. First, the energy crisis created by the Organization of Petroleum Exporting Countries (OPEC) oil embargoes during the late 1970s took a great toll on electricity generation costs. Because the overall electric production/distribution process is only about 30 or 35% efficient, skyrocketing fuel costs contributed to rising electricity costs.

Second was the exponential rise in the cost of building new power plants. Complications arose in many nuclear power plant projects and much publicity surrounded the accompanying construction delays and dramatic cost escalations. For many utilities, the high cost of including nuclear generation plants into the electric rate base had a dramatic effect on electricity rates. However, due to all the disallowances of cost recovery by the utility through rates, this impact has been somewhat dampened.

Conventional plant construction costs also increased significantly during this period. This was due primarily to siting difficulties, environmental protection requirements, inflation, and other market factors.

Despite national attention on conservation and alternative technologies, electric demand continued to increase. Electricity prices and load growth, however, stabilized, and in some cases declined, during the 1990s. On average, the annual rate of increase in electric demand has slowed from about 7% in the 1970s to about 2%, or slightly lower today, though some regions of the country are still experiencing sustained, and even dramatic growth.

More recently, electric utilities and the cost of their services to customers have been characterized by diversity. Rates have been somewhat stable, due in part to increased interactivity among utilities, improved long-term resource planning activities, and the infusion of competition into the generation market. However, where excess capacity and low fuel costs previously kept prices relatively low, these trends have reversed and are beginning to manifest themselves in rising rates. In addition to the significant increases in natural gas and oil, coal has been following these fuels and will continue to be impacted by the need to contend with GHG regulations.

The financial condition of electric utilities today and the cost of electricity for their consumers vary widely. Some utilities have too little capacity, while some have too much, though the condition of vast excess capacity is no longer prevalent. Hence, while there has been some convergence of price, consumer prices vary by hundreds of percent across the country.

Future Perspective and Smart Grid

It appears likely that the electric industry is moving toward a time when customers will pay for electricity and transmission on a real time basis. How soon that will happen and the benefits and costs that will arise from such a "smart grid" has been the topic of much discussion and has already attracted billions of dollars of investment. Simply put, a smart grid is one that allows two-way communications between customers and utilities. Like the Internet, the benefit is not the actual ability to communicate but rather the tasks that are capable of being accomplished due to myriad potential applications that arise. Cost savings and reliability improvement are two of the major benefits that are expected from the smart grid.

The electric utility system in the United States is very expensive, and yet operates at less than sixty percent capacity on average. In addition, it is estimated that between 2010 and 2030 over $1 trillion dollars will need to be invested in it to accommodate growth and replace aging equipment. Using two-way communications, utilities hope to encourage customers to use energy more efficiently and lower their loads during peak time. In addition, there is hope that the smart grid will allow customers to tap into small renewable and conventional generating systems located at customer homes and businesses, and will allow utilities to access the energy stored in the batteries of plug in hybrid vehicles and understand the power flows on their systems in much greater detail. Such abilities would allow utilities to stretch their existing systems and could save their customers tens of billions of dollars in new construction and operating costs. Furthermore, the ability to sample electricity thirty times per second, which is possible with current technology, at millions of data points throughout the system, would allow utilities to much more precisely understand and control power flows, thereby preventing outages or reducing the duration of an outage.

The smart grid is in its infancy. Many utilities across the country have been awarded several billion dollars of federal funding and are in the process of designing and implementing pilot programs to test technology, applications for the technology and customer reactions. When and what the smart grid ends up looking like in ten years is very difficult to predict. However, it seems inevitable that some of the smart grid features and benefits described above will be realized, the limits of which nobody really knows as technology continues its rapid advancement.

Public Utilities Regulatory Policy Act

The landmark **Public Utilities Regulatory Policy Act (PURPA) of 1978**, passed under the National Energy Act, was a response to market uncertainties that started a long, slow trend toward the break-up of the vertical integration of electric utilities by opening up generation to **qualifying facilities (QFs)**, which were **cogenerators** and other small **independent power producers (IPPs)**.

PURPA established a class of QF self-generators that were exempt from regulation under the Public Utilities Holding Company Act (PUHCA) and most sections of the Federal Power Act (FPA), Securities Exchange Commission (SEC), and state utility regulations. While QFs could be small power production facilities using renewable resources, they could also be cogeneration facilities using oil, gas, or coal and meeting the applicable operating and efficiency standards. However, the total ownership of the facility by one or more investor-owned electric utilities (IOUs) was limited to 50%.

One of the principal goals of PURPA was to promote energy independence following the oil embargo and energy crises of the 1970s. In this, the methods encouraged by PURPA included conservation of electric energy, efficient use of fuel and utility generation plants, and equitable pricing for non-utility generated electricity. PURPA led to a series of orders from FERC that required electric utilities to encourage — or at least not discourage — cogeneration and small power production.

Before PURPA, utilities could refuse to serve facilities that generated their own power. Instead of being able to size on-site generation systems for optimal energy efficiency and economic performance, a facility often had to provide for all of its power requirements all of the time. Further, redundant equipment had to be installed because the utility would not provide back-up power. These systems were referred to as total energy systems because they provided for all of the facility's energy requirements. Opportunities for self-generation, however, were extremely limited as the need for redundancy and complete power independence severely affected potential project economic performance.

Elements of PURPA

PURPA legislation encouraged on-site generation systems, principally by improving project economics. To some extent, it also encouraged electric utilities to improve their own system efficiency and use least-cost planning to remain competitive. PURPA required utilities to:

- Purchase electricity from on-site generating facilities at a just and reasonable price.
- Sell electricity to such producers at rates available to customers with similar load characteristics.
- Allow all QFs to interconnect.

FERC instituted a series of regulations to implement this legislation, which went into effect in 1981. In addition to mandating non-discriminatory practices, FERC regulations eliminated the cumbersome petition process for IPPs and required state regulatory authorities and utilities to establish reasonable interconnection standards based on system safety and reliability.

PURPA QF Criteria

The FERC rules implementing PURPA defined cogeneration as the combined production of electric power and useful thermal energy by sequential use of energy from one source of fuel. FERC prescribed three criteria that must be met by a qualifying cogenerator. The qualification test includes an ownership standard, an operating standard, and an efficiency standard. Under the ownership standard, electric utilities may participate in joint ventures, but may own no more than 50% equity in a QF. Refer to Chapter 21 for additional detail on FERC QF criteria for cogenerators.

Small power producers are defined as facilities with a capacity of 80 MW or less that use biomass, waste, or renewable resources to produce electricity. These facilities include solar, wind, geothermal, hydroelectric, biomass, and municipal solid waste systems. These systems must have a maximum fossil fuel input of 25% of the total input. All small power producer facilities of 30 MW or less are exempt from FPA, PUHCA, SEC, and state utility regulations. Only biomass and geothermal systems, however, can qualify at capacities above 30 MW. FERC rules, issued in 1981, permitted unlimited utility ownership of geothermal facilities.

Experience Under PURPA

The QF concept was developed from a 1970s emphasis on conservation, and, therefore, it does not necessarily recognize the benefits associated with peak-shaving, load shifting, and capital cost avoidance. Although it is now generally recognized that strategic peak shaving may, at times, be as beneficial to society as cogeneration, the statutory definitions and requirements for QF status remain in effect. Utilities are thus still permitted to discriminate against non-qualifying facilities, although, for the most part, regulatory authorities have strongly encouraged utilities to extend the same non-discriminatory practices.

FERC has given state utility regulatory bodies a great deal of leeway in promulgating regulations to support the PURPA precepts. Some state agencies have taken aggressive action to force utilities into decentralized power generation. Others have let the utilities set their own policies, provided those policies were not discriminatory. Some utilities have even led the way in making the process work.

Since PURPA, the NUG industry has consisted of QFs and IPPs. IPPs are non-utility, or non-utility-based, generators that do not receive PURPA protection because they do not cogenerate, do not use renewables, or are too large. However, these facilities can sell power on the wholesale or retail market, either within a franchise area or outside of it using wheeling.

While NUG generation increased in market share throughout the 1980s, the exportation of power from NUGs was largely dependent on the price paid for power by a utility, inclusive of transmission costs when applicable. The result was limited competition in the industry and continued utility investment in expensive power plants that, when added to the rate base, were driving up marginal costs.

Difficulties in purchasing power for partial-requirement on-site electric generators also limited the growth of on-site generation, thereby limiting competition in the generation market. This trend led regulators to set in place mechanisms that would encourage electric utilities to optimize resource selections through **supply-side management (SSM)**.

Concurrently, the trend of rising electric growth and marginal cost led regulators to set in place **demand-side management (DSM)**, i.e., mechanisms that would encourage consumers to implement conservation and other load management measures that reduce the need for additional electric utility generation and transmission capacity. When combined into one overall planning process, these two mechanisms came to be known as integrated resource planning (IRP).

IRP

IRP or least cost-planning, as it is sometimes referred to, is a process used by utilities to assess supply-side and demand-side resources based on consistent planning assumptions. The purpose of IRP was to overcome market imperfections in the highly regulated electric industry through development of a reliable resource mix that satisfied consumers' energy service needs and allowed for the utility to meet these needs at a reasonable total cost. IRP became a tool used in the regulation of public utilities to guide utility investment and operating decisions in a direction that is consistent with public policy objectives.

During the 1970s, the historical factors described above acted in many cases to shift utility marginal costs to levels above average costs. Thus, building new power plants would lead to rate increases for consumers. High marginal costs, in combination with the OPEC oil embargo, the Carter administration's conservation push, and the nuclear plant construction and budget overrun problems, created strong incentives to implement electric DSM. Cost-effective DSM programs could delay or obviate the need for new power plant construction. Utility-sponsored DSM programs are designed to stimulate consumer efforts to make capital expenditures that result in electricity being used more efficiently or not at all, or to level load by reducing peak demands and filling off-peak valleys in the system load profile. DSM activities are planned interventions by utilities to improve the efficiency with which their customers use energy.

Thus, the resource selection process began to evolve in the early 1980s with consideration of demand options in addition to supply alternatives. While conservation had little impact on supply planning techniques beyond lowering the straight-line slope of growth, the introduction of electric cogeneration that came next began to create changes in the supply market.

Cogenerated electricity was seen by regulators as another low-cost alternative to potentially costly new central power plant construction. Because of their resistance to purchase cogenerated electricity, electric utilities were eventually required by federal law to purchase cogenerated electricity from QFs at the utility's avoided cost. The process of determining avoided costs for cogeneration payments contributed greatly to the sophistication of the utility's SSM processes and gave regulators a new tool with which to measure utility planning efficiency.

Regulators targeted electric and gas utilities as key players to overcome the many market impediments to energy cost-optimization. Since electric and gas utilities are regulated monopolies, they were considered ideal vehicles for conveying incentives and disincentives to counter market force distortions. By spreading their investments over the broad base of utility customers, utilities could provide incentives to customers to install energy efficient equipment. The result was anticipated to be a net positive benefit for the utility, its customers, and society as a whole, if IRP was done properly.

Mandated conservation or DSM programs were the first step toward implementing utility IRP in most cases. DSM options, viewed as viable alternatives to supply, produce reductions in projected peak demand and annual loads that positively affect supply planning. IRP requires the determination and selection of the optimal mix between supply- and demand-side resources, based on criteria such as cost, risk, reliability, and environmental impact. The perspectives used in this analysis are usually associated with the utility, its ratepayers, DSM program participants, and society at large.

This least-cost planning process compares the utility's avoided generation and transmission cost with all available power purchasing options, as well as available DSM options. A basic tenant of IRP is that utilities should select the most economical resources available, subject to reliability considerations. So, for example, if a NUG can sell power to the utility at a price lower than the utility's avoided cost, that power should be purchased and the utility should forego the construction of its own facilities or generate a correspondingly lesser amount. If, on the other hand, a DSM option reduces supply requirements at a cost to the utility that is less than its avoided cost, the utility should implement that option.

DSM Options

Figure 20-2 shows representative load shape objectives of utilities that underlie utility-sponsored DSM

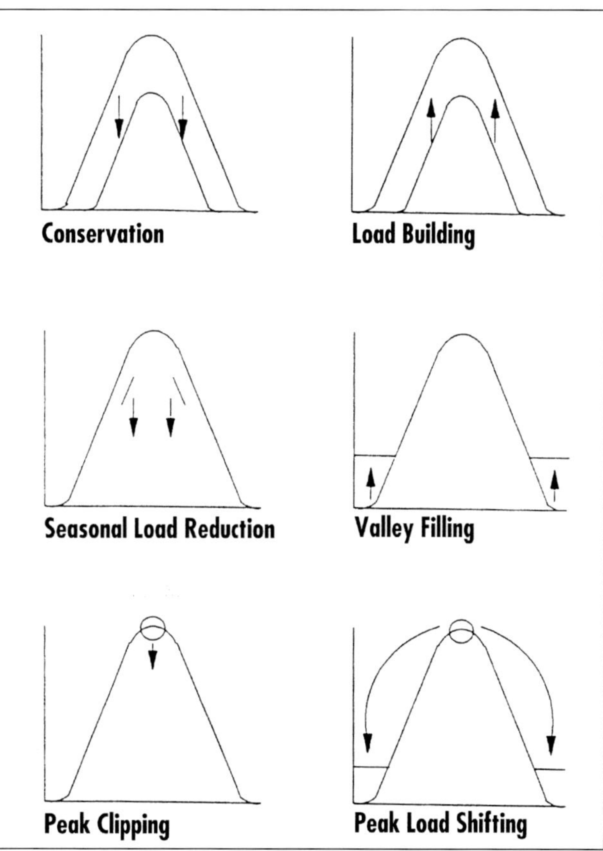

Fig. 20-2 Representative Load Shape Objectives from Utility IRP. Source: GRI

programs. These programs are designed to produce certain load shapes as a means of accomplishing their objectives. Following are descriptions of the main categories of electric utility DSM programs.

- **Conservation programs:** The promotion of decreased electricity usage throughout the year, usually with some commensurate reduction in peak period demand. Reducing the electricity consumed is accomplished by improving the efficiency of energy-using devices. Examples include the installation of high-efficiency lighting, motors, and electric chillers, as well as the installation of insulation to reduce heat gain and, therefore, cooling load.
- **Load leveling programs:** The adjustment of electric load according to the cost of serving it by time-of-use. Reducing peak seasonal or daily demands, for example, generally impacts directly on the need to build new transmission and distribution lines and peaking capacity (e.g., power plants) and, therefore, can result in significant savings. The five general types of load leveling programs are:
  - **Load building programs** that involve increasing electric loads. An example is the use of heat pumps to build winter electric load for summer peaking electric utilities, although this may also increase summer peak loads.
  - **Valley filling programs** that build load but target increases in load in daily or seasonal off-peak periods. Valley filling programs build load to improve system efficiency without the need for additional utility investment. An example is electric space heating equipment that does not increase peak loads for summer peaking utilities.
  - **Peak clipping/load shedding programs** that reduce peak seasonal, daily, or hourly demand in order to defer or avoid the need to add new generation, transmission, and/or distribution capacity. Examples are duty-cycle or load shedding energy management systems and interruptible electric rate programs.
  - **Peak shifting programs** that retain electric load, but move it from peak demand periods to lower demand or off-peak periods. Examples are programs that involve a change in production schedules or the use of thermal energy storage (TES) systems that allow for cooling systems to operate off-peak and store ice or chilled water for use during peak periods.
  - **Seasonal load reduction programs** that changes load shape, yielding increased electric utility seasonal control over the level and/or timing of demand. Examples include high-efficiency cooling equipment and fuel switching for seasonal end-uses such as cooling.
- **Fuel** (or **energy source) switching programs:** These involve encouraging customers to improve energy service efficiency by selecting an alternative fuel (or energy source). These fuel choice strategies are currently most commonly used by combined electric and gas utilities seeking to lower electric peak demand while filling a gas throughput valley. In some cases, fuel switching programs result through unilateral or joint electric and gas utility planning encouraged by state regulatory commissions. While listed separately, this category may be considered part of peak clipping, seasonal load reduction, or conservation.

The choice of load shaping strategy for a utility will depend upon its unique load characteristics. The utility's overall load factor and peaking times, as well as local transmission and distribution line loading will determine whether it will need new baseload power capacity or peaking capacity and where it may seek to do its load building or load reductions.

Effects of Electric Utility DSM Programs

A common effect of utility DSM program implementation is that customers' electric bills decrease, thereby also reducing the utility's revenue. This loss of revenues means lower contributions to fixed costs, or utility overhead. If this loss (and all other related utility DSM expenditures) is not fully compensated by the decrease in fixed cost requirements, rates must be increased to generate sufficient revenues to cover these costs.

Rates increase more from DSM than supply options whenever: 1) the marginal cost of supply is lower than average cost; or 2) the marginal cost of supply is greater than the average cost, but the cost of DSM is greater than the difference between the marginal cost of supply and the average cost.

From a macro standpoint, society still benefits from cost-effective DSM programs because the bill reductions are larger than the resulting rate increases; however, whether individual customers actually save money depends largely upon whether they participate in the DSM programs. So called "non-participants" will commonly see their bills rise because they are paying higher rates and receiving no reduction in the amount of electricity they use. As is explained in more detail below, the various states differ on the extent to which the "non-participant's" interest should be taken into account when evaluating whether DSM programs should be implemented.

Economic Evaluation of DSM Programs

Cost-benefit analyses of DSM programs are typically viewed from several perspectives. In order to evaluate the cost-effectiveness of DSM options as alternatives to supply options, an industry standard set of cost-benefit tests were developed. All DSM program costs and the associated energy, demand, and other benefits are summarized and compared with supply-based avoided costs through application of these tests.

Utilities, as part of their IRP processes, evaluate programs and individual measures on the basis of benefit-to-cost ratio (BCR) or NPV. The BCR is the present value of the program benefits derived over a defined period and divided by the present value of the program costs over the same time period. An NPV or net benefit (NB) analysis using present value benefits (i.e., discounted to reflect the time value of money) less discounted costs can show negative, zero, or positive net benefits. A ratio of benefits to costs that exceeds 1.0 indicates more benefits than costs, from the particular perspective being considered, and is used as a benchmark — or minimum threshold — in screening DSM programs.

BCR and NB are essentially accounting systems for benefits and costs. The values of each depend not only on all of the conditions involved in a particular action, but on the frame of reference, i.e., who or what bears the cost and reaps the benefit. Some actions might prove to have net benefits from all perspectives, some from just one. There are five standard industry benefit cost tests commonly used to measure the cost-effectiveness of DSM programs. These are based upon the California PUC's and Energy Commission's Standard Practice Manual. These five DSM tests are:

1. **Participant test.** This test measures DSM program cost-effectiveness from the perspective of those customers participating in the programs. The NB to participants equals the reduction in utility bills due to savings, plus the incentives or rebates from the utility, minus their cost to implement the measure.
2. **Ratepayers impact measure (RIM test).** This test measures what happens to customer bills or rates due to changes in utility revenues and operating costs caused by a DSM program. It quantifies the impact of implementing a DSM program on non-participants and includes the impact of lost marginal revenues resulting from the implementation of a DSM option. When the utility's rates and program costs and incentives paid to the participants are above its avoided costs, which is almost always the case, DSM programs will be deemed to have a negative net benefit and will not be implemented.
3. **Utility cost test.** The NB of a DSM program to the utility is based on the total costs incurred by the utility, including incentives, and excludes any net costs incurred by the participant. It is a measure of the total revenue requirements impact of a particular DSM program and equals avoided utility supply costs, minus utility DSM program administration costs, minus total costs to implement the measure.
4. **Total resource cost (TRC) test.** This test measures the NB of a DSM program from the combined perspectives of the utility, all utility customers, and all persons in a state or region (total resource). This test measures the total cost of a DSM program, including costs incurred by both the utility and the participants, compared with the avoided capacity- and energy-related costs of a traditional supply-side resource. This test does not include utility incentives or lost revenues, which are typically viewed as being transfer payments between the utility and participants or non-participants.
5. **Societal test.** This test is similar to the TRC test, except that it also includes externalities, environmental or otherwise, as a benefit. Externalities are benefits or costs resulting from the production, distribution, or consumption of goods that are not reflected in the price. These include environmental costs imposed on human health, quality of life, and the health of other species, and the ecosystem as a whole. Also included are environmental compliance costs borne by the utility or energy user. Other more far-reaching concepts include consideration of additional economic and social costs, such as impact on employment, risk of fuel interruption, and national security.

Table 20-1 provides a summary of the benefit and cost components that make up each of the five standard cost-effectiveness tests for load-reducing DSM programs. In the case of load-increasing programs, some benefits become costs (e.g., utility supply and capacity cost savings become incremental costs) and some costs become benefits (e.g., utility lost revenues become increased revenues). For each test perspective, a BCR can be computed

Table 20-1 Summary of Benefit/Cost Components In Five Standard Tests [1]

Benefit or Cost Component	Participant	Utility Cost	RIM	TRC	Societal
DSM measure cost (paid by participant)	Cost	—	—	Cost	Cost
Incentives from utility to customer	Benefit	Cost	Cost	—	—
Utility program administration cost	—	Cost	Cost	Cost	Cost
Utility lost revenue/ customer energy bill reduction	Benefit	—	Cost	—	—
Utility supply and capacity cost savings	—	Benefit	Benefit	Benefit	Benefit
Externality impact	—	—	—	—	Benefit/Cost

[1] The information above is for load-reducing DSM Programs

from the parameters described that allows for the comparison and ranking of DSM programs. The ratios are formed from the calculation of the individual perspective benefits and costs, and are represented as positive values between 0 and infinity. Theoretically, the higher the value, the better the option ranks against supply. An option fails the test if its ratio is less than 1.

In most cases, a proposed DSM program will pass one or more of the standard cost-effectiveness tests, but not all. Thus, an important issue is the relative weight given to the alternative test. Most utilities and PUCs have placed primary weight to either the TRC or RIM test. Of course, a proposed DSM program needs to pass the participant test, otherwise there will not be anyone participating in the program.

Supply-Side Management Options

SSM options for an electric utility include its own generating resources and the potential to purchase power from other utility networks or from IPPs and cogenerators. The price of ongoing power purchases from IPPs and cogenerators, as defined by PURPA, was originally based on the utility's avoided cost. Purchases from all other power producers, including non-qualifying facility cogenerators and electric wholesale generators (EWGs), are based on competitive bidding to meet additional utility-generating resource requirements. Many states now use competitive bidding for selection of all supply resources.

Competitive Bidding

Primarily because of substantial forecasting problems, alternatives to simple formula valuations for purchased power have been developed over the past decade. During this period, numerous state jurisdictions have found that rigid projections for the value of purchased power were wrong and, in many cases, too high. Moreover, many contracts established in earlier periods were later found to be over-priced as a result of incorrect assumptions on demand growth and fuel pricing. For these reasons, utility-forecasted avoided cost is often regarded somewhat critically by state regulators.

To avoid forecasting problems, competitive bidding was an attempt to establish a competitive "avoided-cost" level against which utilities could measure their own cost-to-build capacity. Cogeneration groups originally argued that such bidding was contrary to PURPA, because, in effect, what was paid was lower than actual utility constructed generation costs. However, competitive bidding became an integral part of utility IRP programs and was mandated by many utility commissions.

While the bidding process solved many of the forecasting problems, by itself, it ignored the significance of externalities. Therefore, another factor that was to be considered as part of the competitive bidding process in some states was externality costs. For example, in some states, less-polluting power projects and projects employing renewable energy sources were to be given priority treatment in the bidding queue, or have utility capacity requirements set aside or reserved for their use.

IRP SUMMARY

While the IRP process was having a relatively positive impact on the electric utility industry throughout the 1980s and early 1990s, its value diminished as energy costs fell and utilities grappled with excess generating capacity. More recently, rising oil and natural gas costs, tighter utility generation and transmission capacity, and concerns with climate change and energy security have revitalized the IRP process in general and DSM and renewable electric generation technologies in particular. In fact, states such as California are now using a legally mandated "loading order" that requires the utility to use all cost effective energy efficiency, demand management, and renewable energy options before it is permitted to purchase any conventionally generated electricity.

CURRENT STATUS OF ELECTRIC INDUSTRY REGULATION

While the competitive bidding process promoted competition for generation markets, it was still subject to the judgment of electric utilities as to how to evaluate the cost impact of purchased power versus utility-owned generation capacity. While subject to regulatory scrutiny, the evaluation process was still somewhat subjective, given all of the nuances associated with the evaluation process.

In many cases, there is still some disagreement about the degree to which capacity value should be included in avoided cost determination. Some utilities reason that so long as they do not need power in the present, the only value of purchased power in the short term is its commodity value. Others reason that NUG capacity is not dependable enough. Non-firm purchases are thus seen as having no capacity value, because the utility must still maintain its own capacity to ensure delivery of power to consumers.

Reliability remains a critical concern. In some cases, contracts are structured with performance incentives, penalties, or takeover provisions to promote delivery performance. While not always the case, capacity value should be treated equally for NUG purchases and DSM programs. Moreover, current data indicates that new NUG cogeneration and combined-cycle plants are every bit as reliable, if not more so, than electric utility-owned plants.

Since most of the newer contracts require NUG plants to be on-line to receive payment, utility rate payers often have better protection than they would with utility-owned plants. NUG plants must be reliable to ensure their own profitability. Also, because most NUG plants are far smaller than most utility-owned plants, overall system reliability is higher because of the diversity of multiple modular plants. Fixed price contracts also protect customers from construction cost overruns and the impact of fuel price escalation.

WHEELING AND OPEN ACCESS TO TRANSMISSION SERVICES

One of the main goals of the Energy Policy Act (EPAct) of 1992 was to increase competition in the wholesale electric generation market. Historically, electric utilities have been vertically integrated. PURPA began a long and continuing trend of breaking such vertical integration by opening up the generation function to NUGs and providing for non-discriminatory access to transmission systems, subject to rules currently being developed by FERC.

EPAct 92 redefines the domain of regulated utilities in the functions of transportation and distribution, while infusing more competition into the generation function and wholesale merchant function. It widely opens the generation function to NUGs and allows them greater access to transmission capacity.

The transmission function is still largely recognized as needing regulation. The image of multiple sets of competing power lines does not have the same appeal, from an economic or societal perspective, as multiple sets of decentralized power generators. In light of the growing decentralization of power generation, however, the future of the transmission function must be one of regulated open access.

Current conditions of excess capacity in certain areas, wide cost differences between utilities, and the growth of NUG capacity have made wheeling and transmission access focal points of the electric industry. Open access to transmission services refers to making a utility's transmission lines available at just and reasonable rates to other utilities and non-utilities, on a non-discriminatory basis, for buying and selling electricity. **Wheeling** is a process in which a utility transmits power for others when it is neither the generator nor purchaser of that power. In the process, the utility accepts a certain amount of power at one point and delivers a similar amount of power at another point, less system losses.

Wheeling may be conducted on a retail or wholesale basis. Wholesale wheeling refers to the transmission of power for wholesale sale in interstate commerce. Retail wheeling refers to transmission of power for the purpose of retail (end-use) sales. The various legislation and rule makings have firmly required open, non-prejudicial access to transmission services for the purpose of wholesale purchase and sales activities.

After decades, wheeling has remained a hotly contested issue. Due to the complexity of transmission systems and the need for reliability and safety, it is commonly agreed that the utility, or a designated **independent system operator (ISO)**, must control the actual operation of the transmission systems. If other utility or non-utility sources of supply produce lower-cost electricity, a utility's customers should have access to it as long as there is fair treatment for all rate payers.

ISO Functions

The purpose of an ISO is to manage and set the rules for accessing a regional transmission system. The ISO insures that a transmission system is reliable and eliminates any conflict arising from the fact that power pool members are owners of their own generation capacity. eliminate any conflict arising from the fact that power pool members are owners of their own generation capacity.

Historically, a principle benefit of competitive power-sourcing options with commensurate access to the regional transmission system that was customers could reduce their electric bills by contracting with non-utility power producers. However, it should be noted that such a system works especially well when marginal power costs are low. When marginal costs are high, due to high fuel costs and/or the high cost of new generation, the lowest rates will most likely be those provided by utilities because they are providing power from older, depreciated power plants. A customer that leaves the utility's system to purchase power on the outside market may find it difficult and expensive to return to utility purchased power rates if the market rises suddenly.

Wheeling promotes cost equalization by taking advantage of non-utility power plants and regional price discrepancies, thereby reducing average regional cost differences. It allows areas in need of additional capacity to use abundant capacity in other areas instead of building additional capacity. It also allows areas subject to high-cost power to gain access to supply from areas with lower-cost power.

Pricing

The Federal Power Act gives FERC jurisdiction over the rates charged for interstate wheeling. Rule making has required that such rates or charges be "just and reasonable." These rates must permit the wheeling utility to recover all of its costs, including variable costs and costs of facility additions required for the ordered access. The rates must pass on all transmission access costs to the requesting party and not to the utility's native load customers. State regulators will also play a role, because they can affect the ability of utilities to retain profits from such transactions.

Two critical factors that affect transmission service availability are the adequate supply of transmission facilities and transmission service pricing. The adequacy of transmission facilities greatly affects price. For a deliverability-constrained system, transmission can be quite costly due to the need for substantial capital investment. For a supply constrained system, transmission costs can be inexpensive relative to supply costs.

Setting pricing for wheeling services can be a complex task, particularly because costs will vary with each transaction. The cost to the utility for wheeling depends on several factors, including the system characteristics of the utility, the type of transmission services provided, and the relative locations of the buyer and the seller. However, average cost rate-making methodologies, based on previously published cost and capability data, may be sufficient for establishing transmission prices.

Capital cost requirements for providing wheeling also vary widely. In cases where most of the required transmission facilities are in place and there is adequate capacity available, costs would be based on average embedded cost rate-making. However, in some cases, incremental costs can be substantial. In order to carry increased loads on its transmission system and maintain reliability, a utility may have to improve substations, switchgear, and various other types of protection and monitoring devices. It may also have to add reactive power compensation equipment. The most costly situation is when new lines must be constructed or new circuits added to existing lines to increase transmission capacity. With increased transmission and an increased number of transactions, additional operations and maintenance costs are also incurred.

Identifying the specific operating costs associated with wheeling is also complicated. Electricity flows along the path of least resistance, not necessarily along the most direct path. Furthermore, wheeling can result in increased line losses. The relationship between current flow through a conductor and resistance is exponential. Thus, as current increases, losses increase dramatically. Therefore, in order to deliver a given amount of power to the purchaser, either the seller may have to provide more power than is being purchased, or the wheeling utility will have to generate additional power to compensate for the losses. Accurate identification of these incremental losses requires extensive study, which adds further expense to transmission costs.

In many cases, the wheeling utility will have to provide additional reactive power to support the transmission. In these cases, the wheeling utility incurs added costs for equipment, such as capacitors and, if generators must be run, some additional fuel and operating expenses.

Transmission service may be provided on a firm or non-firm basis. Generally, firm wheeling service means that the transmission capacity is available and guaranteed and that service may only be interrupted, in the same percentage amount as native load requirements are interrupted, for special circumstances, such as for an emergency, maintenance, or violation of the contract. Non-firm service generally provides no commitment for capacity. Service may, therefore, be interrupted for any number of reasons.

An additional cost associated with wheeling is the utility's opportunity cost. When committing its transmission lines to firm service wheeling, the utility may have to pass up opportunities to purchase low-cost power to displace its own higher cost power, to sell its own generated power, or contract for more lucrative wheeling arrangements in the future.

In equilibrium, it is likely that, as in the natural gas industry, all pricing will be done on a spot basis, with specialized contracts and hedging used to achieve maximum benefits. Buyers could be individual facilities or collective buying pools. The distribution of power would then be handled by ISOs responsible for operating power T&D pools.

Wholesale Wheeling

Wholesale wheeling covers supplier access to transmission. Supplier access allows for utility-to-utility transactions, as well as the purchase of power by a utility from a NUG not located in its service territory. Constraints to such access include the wheeling utility's system reliability, its own economic dispatch, and the needs of its core customers. Increased access to transmission for the purpose of wholesale wheeling will allow more utilities, and especially NUGs to participate in this already established practice.

One of the beneficiaries of increased wholesale wheeling may be very small electric distribution companies and municipal systems. Currently, they do not command as strong a market position in wheeling as the large utilities to purchase wheeled power from IPPs.

Ultimately, the impact of open access to transmission service for the purpose of wholesale wheeling on NUGs is that prices paid for power will not be controlled solely by the local utility. In cases in which the local utility has a surplus of capacity, the value of supply may be far greater to another utility. In addition, if the local utility wishes to purchase supply, it must now compete for it and may have to pay a higher price.

The rationale for open access is precisely the same efficiency argument used in leading the gas industry to its current structure. That argument is that there are always efficiency gains to be had, as long as the marginal costs are not equal to marginal benefits at every point in time for each location on the grid. Further, the only way to promote this equality is to eliminate actual barriers to this solution and allow traders to competitively determine the value of capacity and the value of wheeling throughout the system.

Still, there are many limits to wholesale wheeling. To make a project more profitable, the price that another utility offers must be greater than the cost that the local utility would pay, plus the cost of transmission service. If multiple utilities must wheel power to reach a given buyer, the combined wheeling charges can be quite substantial. As a result, increased wholesale wheeling will not result in complete supply cost equalization across the country. It is expected, however, that it will advance the trend toward such cost equalization. It will also focus the role of utilities on T&D, rather than on supply.

Retail Wheeling

Retail wheeling allows customers access to the grid to purchase power from suppliers other than the local utility. While this is fundamentally a bypass of the traditional utility's generation franchise monopoly, the utility retains its natural monopoly in T&D and distributes power on behalf of its shippers. Retail wheeling transactions essentially involve the purchase of unregulated power from any seller other than the customer's local distribution utility.

Retail wheeling can and has posed several problems. It adds a tremendous amount of complexity to the operation of utility distribution systems and may increase system operating costs. In addition to the issue of system reliability, the reliability of supply is also a concern. In the case of sales to a utility, the level of reliability of supply varies widely. While many NUGs are highly reliable, few provide the full range of service that a utility does, such as supplementary, backup, and maintenance services. Single system producers will often be unable to fully serve all of the requirements of full service customers. Utilities may also come to play a role as a supply partner or as a vendor of supplementary and backup service to customers purchasing primary power via wheeling.

Much, if not all, of the responsibility for managing these added complexities will fall upon a network of ISOs. Generally, reliability will be forced into the system and the ISO will manage the scheduling of power resources and loads subject to the power reliability constraints. While the addition of specialized ISO functions may eventually prove to be cost-efficient, the dramatically increased transmission activities and number of transactions is expected to add some to the cost of service.

In regions where retail power costs are high relative to the marginal cost of power, retail wheeling may have a more significant impact. The most likely candidates to purchase retail power are large commercial, industrial, and institutional customers with high load factors. In cases where a supplier is already wheeling power to a given utility, there would be no effect on the cost of supply or generation efficiency, but only a change in the wholesale revenues received by the affected utility.

With all of the legislative and regulatory pressure in favor of retail unbundling, however, it should be noted that it is not at all obvious that customers will be enticed to switch to a non-utility marketer. Even in cases where the prospective savings are the highest, there may be no significant savings over the near term as utility commissions allow IOUs to continue charging for the **"stranded costs"** of now non-competitive generation facilities. Where the plants may have been previously approved by the commission and are now non-competitive, most commissions are still allowing partial- or full-cost recovery for those plants as customers purchase elsewhere. As a result, the actual savings and incentive to switch are substantially reduced.

Despite all foreseeable obstacles and disadvantages, retail wheeling may further promote the role of the low-cost producer and seller in the electric market. It may maximize the development of cost-efficient projects and promote cost minimization throughout the industry as a result of widespread, vigorous competition. However, while many states have allowed retailed wheeling over the past decade, retail wheeling has not played a dominant role in the market to date as so many other factors have come to impact ultimate consumer price and decision making.

SUMMARY

An outgrowth of the self-generation and wheeling market is the breakdown of vertical integration in cost allocation. T&D charges are being detailed explicitly in many rate designs. This is part of the overall process to more clearly define cost causation and to apply these principles to cost recovery with greater precision.

The infusion of competition into the electric generation industry is also being fueled by the wholesale (or off-system) sale of power among utilities and the establishment of opportunities for non-regulated EWGs and independent power marketers. The overall goal of various legislation and rule making over the past two decades has been to allow enhanced competition.

While still not widespread, in the future, widespread retail wheeling may dramatically increase competition in the generation market by greatly expanding the opportunities for power sales. By increasing the number of potential buyers, generators can seek the best price for the power they provide, while buyers can use the competition for sales to their advantage by negotiating for lower rates. Over the past decade consumers ranging from residential to some large corporations have entered into agreements for buying and selling electricity with some benefit achieved. In some cases both consumer and the providers have experienced negative impacts.

It should also be noted that if the "smart grid" evolves as planned, it will allow much smaller renewable and other generation (often referred to as micro-generators) to be integrated into the utility system and to be called on during system peaks. In fact, businesses are already being rapidly developed that will install, maintain and even own these micro-generators located at customer facilities.

Overall, the trend in the electric industry toward deregulation and open access to transmission services should ultimately lead to a condition under which the sale and purchase of power is driven by free market competition. However, while deregulation and open access have progressed a fair distance, currently, they have not become a national phenomenon as predicted by many. In fact, in some states, the policies are being reversed because of failure of how deregulation was implemented.

Regardless of whether regulators allow retail wheeling in their states, the future of electricity prices appears to be one of stratification. The movement toward time-of-use (TOU) rates and other pricing mechanisms, such as real-time pricing (RTP), segments usage into discrete intervals and produces charges that more closely reflect- the utility's cost to serve. Peak (hour and season) costs may experience a far greater growth rate than the average cost of electricity. Likewise, off-peak (hour and season) costs may grow at rates lower than the average, or even decrease over time. Facilities with competitive alternatives and/or favorable load profiles should experience lower cost increases than the average, and facilities lacking alternatives and/or with unfavorable load profiles should experience greater increases.

Competition generally will have a positive effect on the market by forcing competing sources to be more cost-efficient. There remains, however, the potential for market distortions due to the hybrid nature of mixing competition into a regulated industry.

The operative terms for the future electric market are choice and diversity. Facilities that can dynamically control and vary their energy demands through system and source diversity, and/or self-generate electricity, will be more likely to be positioned to take advantage of inter-energy source pricing competition and to be most cost-competitive. Facilities with flexible load shapes and the ability to shift energy use among different time periods, and those that can actively switch between multiple energy sources, will have the opportunity to become active participants in energy-use decisions rather than simply passive energy consumers. Rates for large commercial, industrial, and institutional (CI&I) customers will likely increase less than those of residential and small CI&I customers due to concerns about competition.

Many facilities are increasing their competitive purchasing positions through careful planning of system utilization and the installation of multi-energy-source technologies. Prime mover-driven and heat-cycle equipment, such as absorption cooling, can be installed to allow a facility to operate its electrical demand at near 100 percent load factor (a flat purchased electric load profile), or with an inverted load curve in which peak demand is lower than off-peak demand. Facilities designed along these lines can purchase a varied mix of energy sources in a dynamic, least-cost usage strategy. This can be done through an emphasis on use of off-peak power, or through negotiated competitive energy rates based on the strength of alternative options. These options include both on-site technology applications and alternative retail purchasing options.

RTP type rates may become more common, allowing for purchase of power during specific days, hours, or minutes on a discrete, dynamic basis. These discrete, dynamic pricing mechanisms are anticipated to eventually drive the market as increased competition for retail markets takes hold. Though electricity may still be sold in bundled blocks, the underlying driver will be competition based on instantaneous availability of competitively priced power on the open market.

Finally, after decades of movement toward increased competition utility involvement in all phases of generation, transmission and distribution remain prevalent. This industry has not experienced the dramatic changes of other deregulated industries such as communication, but does continue to evolve. The infusion of current and anticipated environmental regulations, notably those related to climate change, remains uncertain but looms as a critical and potentially dominant factor. While still a minor portion of the overall electricity mix, renewable energy has continued to be infused on utility and NUG generation and retail consumer level as does the trend toward increased conservation and efficiency. As evidenced by public policy support such as the provision of grants and rebates or legislative and regulatory mandates, the trend toward increasing reliance upon DSM and renewables continues and is believed by many to be the road to the future.

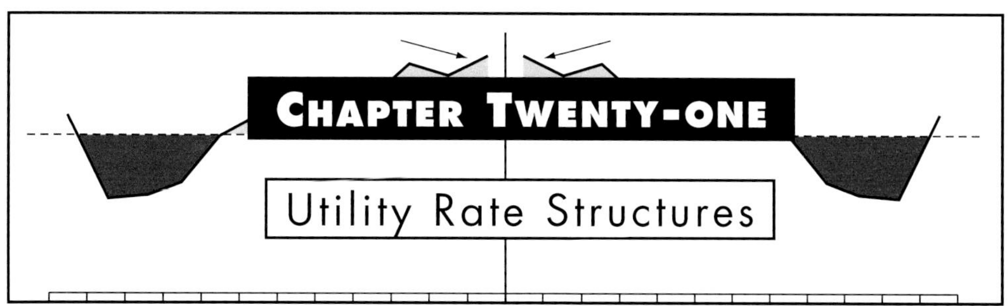

Chapter Twenty-one

Utility Rate Structures

The recent trend in the electric and gas industries has been one of deregulation, market competition, unbundling of services, and increased consumer choice in energy services. Historically, natural gas companies mostly sold and delivered purchased gas to customers at a price determined by a regulated rate structure that included a bundled host of services. Electric utilities sold electricity that they had generated or purchased under similar arrangements.

While far from fully evolved, the current trend is toward a condition under which the local utility's prime function is the transmission and/or distribution of natural gas or electricity, but not necessarily the sale of the energy commodity. The purchase and sale of natural gas and electricity is becoming more of an independent market function in which the utilities are but one broker among many. It is very conceivable that over the next decade, this trend will be all encompassing, inclusive of the residential market, with newly developed rate structures that are compatible with the evolving conditions surrounding consumer transactions.

Though the trend continues, utilities still provide both sales and distribution functions to most customers. Natural gas and electric utility rates are determined through a regulatory process under the jurisdiction of a **public utility commission (PUC)**. The utility periodically applies for a rate case hearing to request rate adjustments in response to changes in economic factors, such as supply, demand, distribution, and market forces that affect the utility's cost to serve.

Utility Rate Design Overview

Rates are designed so that the utility can recover sufficient funds to cover costs and, in the case of investor-owned utilities (IOUs), have an opportunity to earn a a reasonable return on investment for its stockholders. The PUC first determines the projected level of allowable investment by the utility in its facilities (referred to as the rate base) and its projected level of operating expenses. It then determines the rate of return the utility may earn on its investment and adds to it the operating expenses. This sets the total revenue requirement, which is equal to the total cost to serve. Rates are then designed to allocate revenue requirements among the various customer classes and subcategories within the classes.

Typically, customer classes include residential, commercial, industrial, institutional, and municipal. They may be subdivided into many rate classes or lumped together into broader rate classes. Commercial and industrial (C&I) as well as institutional (CI&I) customers, for example, are commonly lumped together.

General rate classes include categories such as residential, small CI&I, and large CI&I. Residential rate classes are typically limited to single-family dwellings and to multi-family dwellings metered separately from one another. Master-metered multi-family dwellings can either be treated as a separate rate class or as part of a commercial rate class.

Two common distinctions between customer classes are demand and the voltage at which service is taken, although distinctions are made between CI&I classes based on other criteria as well. Additional rate classes sometimes exist for cogenerators, non-utility generators (NUGs), and other end use- or equipment-specific categories. Another major distinction is between firm and non-firm service. Many utilities are also able to offer special contracts that are customer-specific.

In assigning rate classes, utilities attempt to determine commonalties among customers within a class. Contribution to a probability of a peak, for example, is an extremely important factor.

Types of Gas and Electric Utility Loads

Consumer electric and gas loads may be characterized as peak, off-peak, baseload, and seasonal. Each is rated differently by the utilities, produces different costs to the end-user, and has different implications for the system.

Peak and Off-Peak Loads

The broadest and most critical distinction is between peak and off-peak loads. Peak loads are those that occur during the time periods that the utility must supply the maximum amount of energy and, therefore, use the highest amount of system capacity. Increases in peak load are closely tied to the need for investment in facility expansion. For electric utilities, peak loads often necessitate the use of the less fuel-efficient peaking plants. During peak periods, the rate of transmission and distribution system line losses also increases, further adding to supply requirements. For these reasons, peak capacity is the most expensive to purchase and carries with it the burden of increased capital cost and decreased fuel and delivery system efficiency. For gas utilities, peak loads that must be served on a firm basis determines the capacity requirements of the local distribution system piping network and the amount of gas supply and transmission capacity that must be reserved on interstate pipelines and/or the amount of gas storage capacity required.

Off-peak loads are those that occur during the time of day, week, month, or year when the utility uses a relatively small amount of its total system capacity. Consequently, off-peak loads do not contribute to the need for facility expansion. For electric utilities, off-peak loads are usually served by the most efficient electric generation plants and distributed with relatively low line losses. For gas utilities, off-peak loads are generally served by a low-cost source of gas supply. For all of these reasons, off-peak capacity is the least expensive to purchase and carries with it the benefits of operating cost-efficiency and minimized capital cost impact.

In addition, many utilities further segment loads into intermediate categories, sometimes referred to as shoulder- or intermediate-peak periods. These loads and their effects lie between the peak and off-peak loads.

Baseloads

Baseloads, from an end-user perspective, are those that occur all of the time (i.e., a 100% load factor). The quantity that can be referred to as baseload is the baseload rate times every hour in a given period (i.e., day, month, year). Baseloads are served, in theory, by the portion of the utility's capacity that is used constantly. Baseloads are often considered optimal loads because the fixed costs, associated with capacity-related investments, are spread over the maximum potential units of sales, and, in the case of electric utilities, the life-cycle costs of highly efficient power plants are at the lowest possible rate.

In many rate structures, baseloads carry a higher weighted average cost than off-peak loads. This is because off-peak load is a component of baseload that is separated from the peak component. Capital investment-related costs are frequently stripped away and added to the peak constituent. Variable operating costs, which are typically lower than the average due to increased efficiency, are broken out and assigned incrementally to off-peak load. Baseload costs are thus a composite of a fixed load experienced continuously in both periods.

Different rate structures separate or stratify costs to varying degrees between peak and off-peak. Traditional rate structures tend to have a relatively low differentiation of costs in different rate periods. More progressive rate structures, on the other hand, have a wider range of prices. They attempt to assign costs to a greater number of finite blocks of usage based on their varying impact on capital and operating costs, with the result that the cost of a kilowatt-hour (kWh) of electricity or thousand cubic feet (Mcf) of gas at any given time may be far different than the average cost.

Seasonal Loads

Utility systems are generally either summer peakers or winter peakers, referring to the season during which their peak-capacity requirements are highest. Most of the nation's electric utilities are summer peakers, because summer cooling loads outweigh the winter-heat-load component. In most regions of the country, heating loads are mostly served by direct on-site fuel use, such as gas or oil.

Even though fuel costs (driven by supply and demand) are even greater in the winter, the capacity-related capital cost component and the inefficiency of electric generation peaking plants produce a greater effect on seasonal electric costs than do their fuel cost component. Summer peaking utilities, therefore, will often construct seasonally differentiated rates that are higher in the summer than in the winter.

Not all electric utility systems are summer peakers, however. Some are winter peakers, some are balanced, and some (due to a high growth rate) set a new peak every season. Winter peakers are found in certain northern regions where cooling loads are modest and/or electric heat is predominant. Winter peakers or balanced systems may also be found in moderate climatic regions that are more conducive to the use of electric heat. Balanced systems are more likely for utilities with a heavy industrial base in which temperature-related end uses are minor, compared with the baseload process end uses.

Almost all gas utilities are winter peakers, due to the preponderance of space heating loads. Northern climate local distribution companies (LDCs) tend to have the most dramatic winter peaking load profiles. LDCs that serve a significant industrial sector and/or provide large quantities of gas to electric generation plants tend to have more balanced loads, as do LDCs in warmer climates.

Load Factor and Usage Profile

The load factor is one of the most significant elements of a customer usage profile and of a utility's operating profile. For the utility system, it is the measurement of actual energy output over a period versus potential output over that period, based on the full capacity of the system. If, for example, an electric utility has 10,000 megawatt (MW) of total capacity and generates 120,000 MWh over a 24 hour period, its load factor is 50% [120,000 MWh/(10,000 MW x 24 hours)]. If this utility had a 100% load factor, it could generate the same 120,000 MWh over 24 hours with only 5,000 MW of capacity. Hence, an incremental load with a 100% load factor will produce a significantly lower revenue recovery requirement than one with a 50% load factor because the capital cost associated with the additional 5,000 MW of capacity is eliminated.

If, for example, a gas LDC has a maximum daily distribution capacity of 100,000 Mcf per day (Mcfd) and 100% annual load factor, it would distribute 36.5 million Mcf (100,000 Mcfd x 365 days) per year. If, in actuality, it distributed only 14.6 million Mcf per year, the annual distribution system load factor would be 40% (14.6 million Mcf/36.5 million Mcf).

A load factor of 100% is the theoretical ideal operating state for utilities. In this ideal state, the utility's investment in capacity is spread over the maximum amount of potential output, and the revenue requirement per unit (kWh or Mcf) sold is minimized. In reality, however, a 100% load factor is not an attainable goal. Some margin of excess supply, transmission, and distribution capacity is necessary for maintenance and emergency backup. It is also not possible to balance consumer loads perfectly.

There are, however, some conditions that produce utility load profiles approaching a 100% load factor. Electric utilities, for example, that operate as part of regional power pools and/or have sufficiently low operating costs to allow for export of all excess capacity can approach a 100% load factor. Utilities that are capacity constrained may also need to operate all of their plants at 100% load factor, and purchase power to meet the rest of their load requirements. Gas utilities that predominantly serve heavy industrial loads, for instance, may have a high load factor.

In order to improve system load factor, utilities often seek to market electricity or gas in low-load periods at a relatively low cost. They may also provide incentives, through various conservation and load management programs, to promote elimination or shifting of peak loads.

The utility considers load factor both on a discrete minute-by-minute basis and an hourly, daily, weekly, monthly, and yearly basis. Other common measurement periods are seasons, normal workdays, workweeks, and weekends. Sophisticated dispatch modeling is used to determine when to bring additional capacity (i.e., electric generation plants or gas storage) on- and off-line in response to load fluctuations.

Electric utility systems (or regional power pools) have a dispatch stack order. Generating stations are arranged in the order in which they will be brought on- and off-line in response to changing load requirements. Typically, plants in the stack are classified as baseload, intermediate, and peak-load plants. Baseload plants are generally the most cost-efficient to operate, or have the lowest operating cost, including fuel costs per unit output. Intermediate (or swing-load) plants run much of the time, often under varying loads. Peak-load or peaking plants are typically the least costly plants to construct, but also the least cost-efficient to operate. LDCs also have a type of dispatch order. Supply sources, including storage reserves and liquid natural gas facilities, are arranged in the order in which they will be used.

Load factor is one of the most important determinants of the cost to serve a given facility. Consider two electric utility customers that have the same monthly load of 72,000 kWh. One of the customers uses 100 kWh every hour of the month (720 hours). Under a typical demand/commodity type rate, this customer would have a peak demand of 100 kW and a monthly load factor of 100%. The other customer uses 75 kWh every hour of the month, except for one hour every day, when a certain process requires 600 kWh. This customer would have a peak demand of 600 kW and a monthly load factor of 17% [72,000 kWh/(600 kW x 720 hours)]. While, in this example, the monthly usage is exactly the same for both customers, the cost to serve the customer with the 17% load factor would likely be several times that of the customer with the 100% load factor. To serve the customer with the 17% load factor, the utility would have to reserve 600 kW in generation, transmission, and distribution capacity as opposed to 100 kW for the other customer. If both customers paid the same amount for electricity on a per kWh basis, much of the charges paid by the customer with the 100% load factor would be to support the investment required to serve the other customer.

Now consider a third customer that uses 72,000 kWh per month. This customer operates a night and weekend shift factory that uses 200 kWh every hour for half of the hours every month (360 hours). Therefore, this customer has a maximum demand requirement of 200 kW and a monthly load factor of 50% [72,000 kWh/(200 kW x 720 hours)]. In this case, since all of the load occurs in the utility's off-peak period when load requirements are low, the utility does not require any additional capacity to serve this customer, except for the local wires and transformers connected directly to the facility. Additionally, this customer's loads can be served by the utility's most efficient generation plant. When the power is required by the customer (i.e., the specific usage profile), in this example, it is even less costly to serve the off-peak customer with the 50% load factor than it is to serve the customer with the 100% load factor.

A parallel example can be drawn with three gas utility customers that all consume the same amount of gas on an annual basis. The first customer has a 17% load factor based on a winter heating load requirement. The second customer has a 100% load factor based on a continuous industrial process requirement. The third customer has a 50% load factor, based on a continuous non-winter month gas-fired cooling load requirement. Similar to the three electric utility customers, in this hypothetical example, the customer with the 17% load factor will likely be the most costly to serve and the customer with the 50% load factor will likely be the least costly to serve, with the 100% load factor customer falling somewhere in the middle.

The conclusion that can be drawn from these examples of customers with the same exact daily, monthly, or annual usage is that the load factor and, even more importantly, the specific load profile are key determinants of the actual cost of service. The main reason is the impact of the load profile on fixed cost requirements and how fixed costs are recovered with respect to overall usage requirements.

In addition to load profile, the magnitude of overall load is an important cost factor. There is an economy of scale in serving customers that requires large amounts of energy. Costs of running lines or pipes to a facility, installing service and meters, billing every month, and providing other goods and services are less when averaged over a large, rather than a small, volume.

Allocation of Costs

Utility allocation of fixed costs and the associated setting of rate levels is a process that is partly scientific and partly subjective. One major consideration in cost allocation and rate design is the desire for rate stability. All usage charges should contribute to fixed costs in some way, and demand charges should serve to mitigate against significant distortions in cost allocation resulting from varying load profiles.

If a utility placed all capital cost recovery in a single period of use, such as peak summer, the rate might be too unstable. The market might overreact with a rush toward solutions, such as peak shaving and, in short order, cost recovery would be insufficient. If all usage in that single usage period disappeared, the capital costs would not disappear. The utility system would still have to be financially supported. Of course, loads do not disappear all at once. Hence, as they change, the utility rate structures must change with them.

Fixed cost-related charges are neither arbitrary nor capricious, but neither are they a perfect scientific cost allocator. The underlying theory behind demand charges and other fixed cost rate components is that they better approximate real costs. If, for example, all customer electric bills under demand rates were broken down into incremental usage costs, the cost per kWh might range from $0.02 to $0.90, or even higher, depending on the severity of the impact of demand charges. This wide range of price differentiation is said to allow rates to more closely represent the utility's actual cost to serve. With the advent of advanced metering and the "smart grid", such determinations can be made dynamically, based on actual events, as opposed to predicted events.

Clearly, the movement in today's energy market is toward more precise cost allocation strategies. Factors that have contributed to this movement include:

- The need to protect utility revenues by having rates that are competitive with energy alternatives. More competitive, diversified rates have been constructed by electric and gas utilities to protect revenues and maximize sales during periods when electricity or gas is less costly.
- The desire of utilities to influence customer usage patterns in order to create higher system load factors. Price differentiation directs equipment investment choices and operating schedules toward customer load profiles that improve system load factor and lower the average cost to serve. If all similar energy units cost the same, load growth would tend to move away from a high load factor as the prevailing forces of weather and the normal work week direct loads to peak periods. Cost differentiation provides incentives to counter these forces through careful end use planning.

- The directive from regulators to redesign cost allocation processes and to avoid the need for additional capacity. Careful review of utility management decisions by regulators has directed utilities away from the business of building new plants much in advance of load growth.

In addition to least-cost planning, regulators have demanded more precise cost allocation in an effort to make rates fairer. Regulators have also provided utilities with incentives for activities other than construction. A preferred rate of return for investments in conservation and load growth reduction activity is an example. Many demand-side management (DSM) programs involve shifting usage from the utility peak period to an off-peak period. In order to make such programs effective, utilities must have rate structures that charge according to when energy is used instead of just how much energy is used.

The imperative of the regulated least-cost planning activities has increasingly been replaced by competitive market-based imperatives. As a result, utility rate design is driven more and more by competition and less by static regulatory planning.

On the other end of the spectrum from a single fixed-price usage charge is a completely market-driven discrete usage charge that changes from hour to hour, or even minute to minute. In this scenario, gas and electricity usage charges continually vary with the price being established on almost an instantaneous basis. This price must be market competitive and must reflect all embedded fixed and variable costs. This type of pricing would be applied to both the commodity side and to the transmission and distribution side, either separately or as a bundled price.

While the market is still not fully mature enough or unencumbered by imperfection to function with such pricing, the technology of metering and transmitting gas and electricity is approaching the level of sophistication necessary for such a market. Moreover, the forces of competition are moving the market in this direction.

However, today's market does not yet function in this manner. Instead, some utility rates are designed to approximate this type of instantaneous, or real-time pricing (RTP). To do so, rate designs use a mix of various rate components. They can be classified into several general categories on the basis of energy-use characteristics. Each category may constitute one rate or several categories may be aggregated to determine a rate.

Billing Factors

Utilities use a number of billing factors in their rate structures. Some of these factors are clearly stated in all utility bills, some are considered in every transaction but not necessarily identified separately in all bills. Others are considered and applied, depending on the particular characteristics of the utility or the customer class under a given tariff. The most common billing factors are:

- **Basic, or customer, service charge.** A fixed charge is assessed to each customer based on costs related to connection, metering, billing, service maintenance, etc. Typically, basic service charges are greater for rates designed to serve large users because of the greater cost associated with larger pipes, regulators, and metering equipment. Facilities with multiple services under the same rate may have summarized billing with only one basic service charge. Facilities with multiple services under different rates will often pay a basic service charge for service under each rate.
- **Minimum charge.** This is the lowest bill a customer is required to pay for service on a given rate schedule during each billing period, regardless of actual usage. In most cases, this is equal to the customer charge. However, for larger customers, it can include other charges, such as demand charges or charges established by individual contracts between the customer and the utility.
- **Commodity charge.** This is a charge based on energy usage or the number of energy units actually consumed by the facility during each billing period. Commodity charges generally include the utility's incremental operating costs, plus some contribution to fixed costs.
  - Common natural gas billing units are: cf (cubic feet), Ccf (hundred cubic feet), Mcf (thousand cubic feet) and therm (100,000 Btu or 105,480 kJ). While a therm is a specific quantity of energy, a cubic foot of gas has varying Btu or kJ levels. Typically, 1 cf of pipeline quality natural gas contains about 1,000 Btu, but this can vary by several percent, depending on the specific source of natural gas.
  - Common electricity billing units are: kilowatt-hours (kWh) and kilovolt-amperes (kVA).
- **Demand, or maximum level of service, charge.** This charge is based on a customer's peak rate of consumption and takes into account the util-

ity's required investment needed to serve that load. Measurement of demand, or rate-of-use, is somewhat like a car speedometer. However, a demand meter does not measure rate-of-use instantaneously, but averages it over a discrete utility-selected interval. This interval, or "demand window," for electric utilities is usually 15 or 30 minutes. The interval for gas utilities is usually the highest daily consumption or, in some cases, it may be measured as the peak usage for one month. Demand can be measured in multiple-use periods, and charges can be differentiated for each use period.

Demand charges are assessed in several different ways. They may vary by season, TOU, and level of use:

- **Seasonal variation** produces greater charges for peak demand during months in which the utility experiences its highest peak demand. With gas utilities, demand charges are often applied only in winter months.
- **TOU variation** can be applied in several different ways. In many cases, demand is only measured by electric utilities during a peak period. In other cases, demand is measured during several different periods, such as peak, shoulder, and off-peak. The charge per unit of demand varies with each period, with peak charges being the highest.
- **Level-of-use variation** produces different charges for increasing blocks, or steps, of demand. For example, the first 500 kW or 500 Ccf of demand will be billed at one rate, the next 1,000 kW or 1,000 Ccf of demand billed at another rate, and so on.
- **Minimum demand charges** are sometimes set at a given level for each rate tariff. In some cases, they may be based on a percentage of the customer's connected load or main transformer or gas meter size.

• **Ratchet adjustments.** A demand ratchet adjustment sets minimum monthly billable demand at a certain percentage (usually 60 to 100%) of the highest month's peak demand in a given preceding period (typically 11 or 12 months). When the utility has a severe summer or winter peaking system, the ratchet may be based on the highest month's peak demand only in the peaking season. The minimum monthly demand charge assures the utility of cost recovery for investment in peak capacity, even if a customer does not require that peak capacity in a given month. In some cases, this preceding period may be as long as several years or for the life of a contract. Ratchets serve as a mechanism to approximate the true cost of service and to distribute that cost impact on the customer over several months.

Consider the example of an electric utility with an 80% demand ratchet adjustment. If one month had a peak demand of 2,000 kW and the next 11 months had a peak demand of 1,000 kW, each of the proceeding 11 months' demand charges would be based on 80% of the 2,000 kW figure, or 1,600 kW. The result would be an additional annual billable 600 kW per month, totaling 6,600 kW additional billable kW over the next 11 months. Figure 21-1 provides an example of the impact of an 80% ratchet on an 11 month rolling basis.

The reasoning behind ratchet adjustments is that monthly demand charges alone do not sufficiently compensate the utility for the impact of a customer who sets a large peak once per year rather than every month. The utility must have sufficient capacity to meet the annual highest peak hour of demand at all times and must incur the cost of building or securing contracts to purchase the needed capacity. A customer that does not need that peak demand every month must still pay for a large portion of the demand in order to reserve it for the peak month.

• **Energy adjustment charge (EAC).** This mechanism is designed to pass increasing or decreasing fuel costs per billable unit directly to the customer. EACs may vary on a monthly, quarterly, or annual basis. It is normally based on a 12 month rolling average of fuel

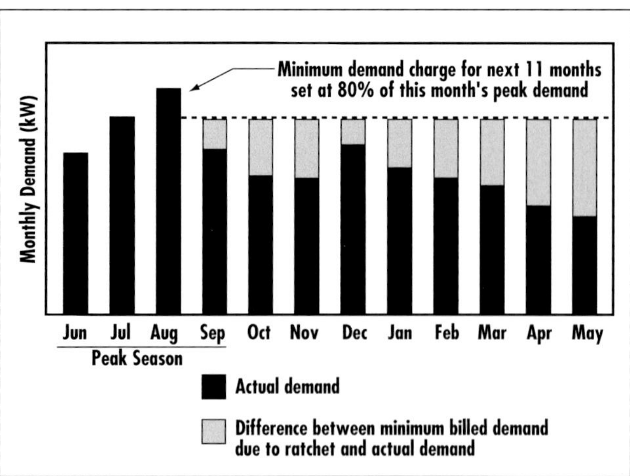

Fig. 21-1 Impact of 80% Demand Ratchet.

costs and is used to maintain rate stability between rate-case proceedings. Common terms used to express EAC are: fuel cost adjustment (FCA), fuel adjustment charge (FAC), purchased power adjustment (PPA), purchased gas adjustment (PGA), and levelized gas adjustment (LGA). Figure 21-2 is an illustration of a fuel adjustment charge to monthly electricity charges.

- **Other adjustment charges.** A utility bill may also include additional adjustment charges. Many electric companies have a nuclear capacity adjustment or nuclear plant decommissioning charge. Some companies also have conservation adjustment mechanisms or sales adjustments. Gas utilities have shrinkage or retainage charges. These are applied as a percentage increase to transported gas volumes at delivery points for transportation customers. This charge reflects lost or unaccounted for gas volumes that arise from the transportation of gas over the LDC's pipeline network.
- **Taxes and fees.** These are added charges that the utility bears as operating expenses and passes on to individual customers. These may include gross receipt taxes or sales taxes. They may also include surcharges for special regulatory commission approved programs that are added to the bill. Such charges may be shown on the bill as itemized charges, or they may be embedded in the derivation of other utility charges and are thus not easily spotted on the bill.

Rate Design Strategies

The purpose of rate design is to recover revenue requirements associated with the costs incurred by utilities to provide services to the customer classes, while also recognizing the different energy use profiles of customers between and within the customer classes. Rate design also sends price signals to consumers and can influence energy-use decision makers to favor more cost efficient energy use profiles. Price signals mean that a particular utility's rate design makes it clear when it is more or less costly to purchase a unit of energy.

If all customers had the same usage pattern, fixed and variable costs could be readily integrated into one simple cost recovery mechanism — a usage charge. To determine an average price per kWh or Mcf, the total revenue requirement would be divided by all of the kWh or Mcf to be sold. In reality, however, customers do not have the same usage patterns and, as demonstrated in the above load factor examples, distinctions must be made as to how varying customer load patterns affect system cost.

Block Rates

Block rates are rates in which the charges for a unit of service vary with consumption. The billing period's consumption levels within a rate are often broken down into blocks, or steps, with different charges for each block. There are several different common types of block rates:

- **Declining block rates.** Historically, the most common type of rates, declining block rates have lower usage charges as levels of consumption increase. The reasoning behind these rates is that increasing customer usage reduces the cost to serve the customer on a per-unit basis. These rates have been phased out in many utilities because they are believed to encourage greater consumption and discourage conservation. Figure 21-3 illustrates a natural gas declining block service rate.
- **Increasing, or inverted, block rates.** These rates charge more per unit as levels of consumption increase. They are relatively uncommon with natural gas service, but they do show up with electric

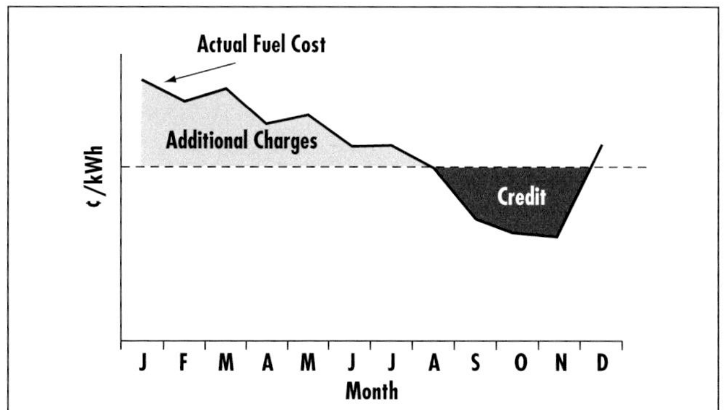

Fig. 21-2 Representative Illustration of FAC Charge to Monthly Electricity Charges.

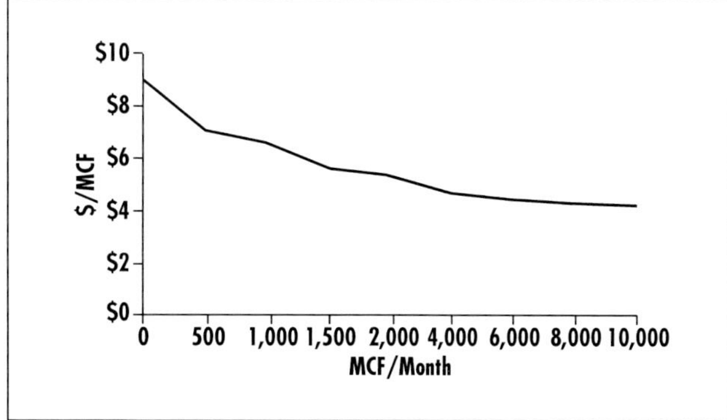

Fig. 21-3 Representative Natural Gas Declining Block Service Rate.

services. They are used by utilities that are supply-constrained, or used for conservation purposes to discourage increased consumption. Figure 21-4 illustrates a natural gas inverted block service rate.

- **Sliding block rates.** These rates use a peak demand value times a multiplier to determine the first-step size. They encourage greater load factors or, in other words, more constant levels of energy usage by the customer. This type of rate is common today in the electric industry for C&I customers. Figure 21-5 illustrates a natural gas sliding block service rate.

In some cases the utilities charge the same unit price regardless of the level of consumption, meaning there is only one block. This is somewhat common for residential customers.

Many utilities use a combination of block-pricing structures. They are often used to reflect a seasonal energy-cost differential. A utility may use an inverted block rate in the season of highest consumption and a flat or declining rate in the off-peak season.

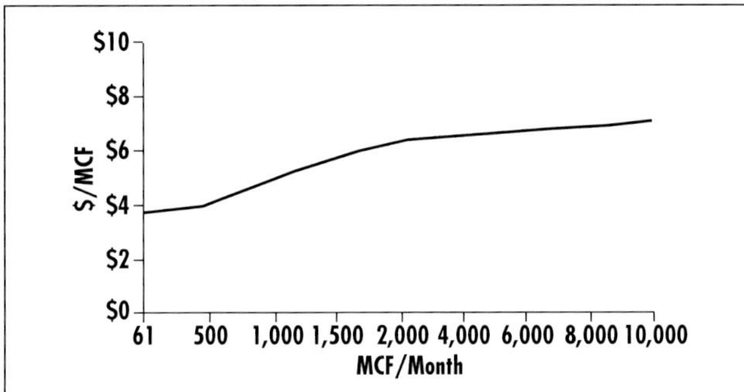

Fig. 21-4 Representative Natural Gas Inverted Block Service Rate.

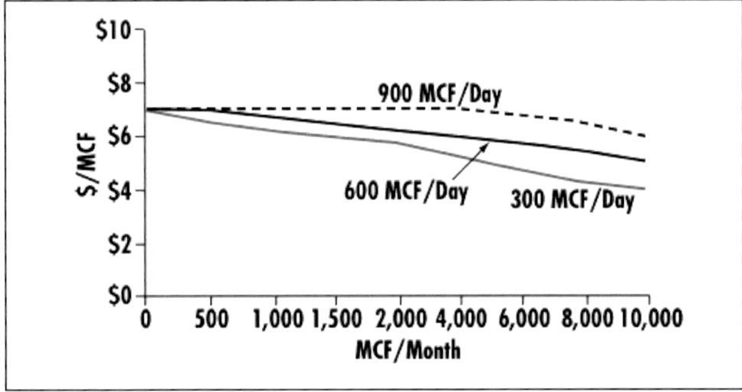

Fig. 21-5 Representative Natural Gas Sliding Block Service Rate.

Demand/Commodity Rates

These rates consist of two basic components: a demand charge and a commodity charge. The demand charge is typically based on the peak hourly or daily volume in the billing period. The commodity charge is based on the units of energy consumed. In many cases, a ratchet penalty is used. As previously discussed, the ratchet penalty is typically based on a certain percentage of the peak demand incurred within the previous year on an 11 or 12 month rolling basis. This type of rate somewhat more accurately reflects the cost to serve than flat commodity rates because it takes into account peak-day requirements. Such rates encourage customers to maintain a high load factor.

In many cases, a sliding block rate structure is used to blend demand and usage charges into stepped rates based on utilization (or load) factors. Usage is billed at varying rates based on hours of use of demand. For example, a 30 day billing cycle has 720 hours. If the peak demand on an electric utility bill is 1,000 kW, a 100% load factor would represent 72,000 kWh, or 720 hours of use of demand. A 25% load factor would represent 18,000 kWh, or 180 hours of use of demand. This type of rate schedule breaks the hours of the month into different blocks. With a declining block rate structure, in each subsequent block or hours of use of demand, the rate is lower. For example, the first 180 hours of use of equivalent full-load demand (18,000 kWh) would be billed at a certain rate per kWh. The next 180 hours of use of demand (from 18,000 kWh to 36,000 kWh) would be billed at a lower rate per kWh, and so on. This type of rate rewards high load factor customers. If the customer's peak demand increases, more kWh are billed at the higher rate and vice versa.

Time-of-Use Rates

TOU rate structures are often designed to differentiate among months of the year (seasonal rates), days of the week, or hours of the day. These rates assign greater costs to peak usage periods, to discourage consumption, and lower costs to off-peak periods. Commodity cost differentiation may be augmented by peak demand charge differentiation. These TOU rates are more reflective of the cost to serve than standard block rates and provide price signals that direct consumers toward off-season or off-time usage. The level of differentiation will often be greater for utilities with poor annual load factors, such as those with loads that are predominantly heating or cooling. Typically, higher summer rates will be used by electric utilities and higher winter rates will be used by gas utilities.

- **Seasonally Differentiated Rates.** These are a type of TOU rate designed to assign greater cost to consumption, on a per-unit basis, for corresponding rate blocks in peak usage months. Commodity cost differentiation may be augmented by peak season demand charges and ratchets. These rates provide price signals that influence the consumers toward off-season usage.

- **Off-season rates.** These services are provided to customers in specific non-peak months. They are designed for customers who do not require gas service in peak months, but do not have the alternative energy sources typically required for standard interruptible sales service. Similar to interruptible rates, which are discussed later, these rates provide lower cost gas, because they do not contribute to the high cost of maintaining peak facilities. These rates may also be attractive to customers that have alternative energy sources, but cannot use them in certain months because they must meet seasonal emissions standards.

End Use Rates

End use rates are designed to provide incentives to customers to install and operate specific equipment that uses the utility's energy. This usually includes space heating, water heating, and cooling equipment. These rates generally have some basic equipment requirements and often need separate metering. Specialized heating rates are generally used by electric utilities. These are sometimes combined into an all-electric rate that is offered to customers using electric heating, water heating, air conditioning, cooking, etc. Specialized cooling equipment rates are more commonly offered by gas utilities.

End uses selected for specific rates can often be placed in given rate periods or seasons. In some cases, end use rates are combined with TOU rates. The lowest rates typically are offered for end uses employed during the off-peak season or off-peak time of day or week. More moderate rates may be offered for baseload process end uses, such as water heating, cogeneration, or year-round industrial processes.

Many PUCs prefer that price signal tools be applied uniformly to usage characteristics rather than to specific end uses. However, some utilities feel that their typical rates already accomplish this and that further distinctions are needed to attract (or discourage) certain loads.

Non-Firm Service Rates

Non-firm rates include interruptible rates, standby rates, and load management rates. With multiple energy source options increasingly available, non-firm service is becoming increasingly popular:

- **Interruptible rates.** These rates, which are commonly offered to CI&I customers, are designed for customers whose entire load, or large blocks of load, can be dropped at virtually any time. The primary benefit to utilities is that it reduces the need to guarantee service during peak demand periods. The utility can pass on savings, which come predominantly from reduced fixed-capacity costs, to customers.

  Perhaps more important are the demand reduction incentives that electric utilities are currently offering customers who cut their demand during times of peak usage. These programs provide customers with cash payments based on the frequency and amount of the demand reduction. An indication of the growing importance of these programs is the recent formation of companies that will aggregate demand reductions for many customers and take care of the administrative duties necessary to secure payment. Interruptible service may involve commodity or distribution components.

  There are many conditions under which a facility can reduce its demand. Examples include:

  - Customers with dual-fuel burners capable of operation on natural gas and alternative fuels, such as propane or oil, or distribution systems supplied by propane-air mixtures can easily withstand interruption of natural gas service by switching over to their alternative burner fuel.
  - Customers with on-site electric power generation capacity can go off line and generate their own power to serve the entire facility or selected circuits within the facility.
  - Customers with equipment that can operate on either electricity or fuel or steam, can simply use the non-electrical equipment during periods of interruption. This would be the case with a dual-drive mechanical service device that had both an electric motor drive and a prime mover drive, or with mixed energy source (hybrid) multiple unit systems that feature both electrical and non-electrical powered units.

  Electric utility interruptible rates are often referred to as utility-controlled peak shaving. Customers are required to reduce their demand on the utility system completely, or to some predetermined level when necessary. Contracts may be designed with a set rate-break on demand or usage, a flat annual or monthly fee paid by the utility, or a specific payment rate for each period of interruption. Significant penalties for failure to interrupt are also common.

LDC interruptible rates are typically based on the customer's ability to use an alternative fuel such as oil or propane with the natural gas commodity charge set or negotiated based on the price of the alternative fuel. Natural gas distribution services (in cases where a customer purchases gas from a seller other than the LDC) may be purchased on a non-firm basis. Pricing by the LDC may be based on competitive alternatives similar to interruptible commodity pricing or may be offered at a firm cost-of-service based price.

- **Load control, or load management, rates.** These rates are a type of interruptible rate that gives the utility direct control over specific loads (sometimes via radio wave signals) during system peak periods. This strategy is most commonly used by electric utilities for residential water heating and air conditioning customers, but can be used for commercial, industrial, and agricultural customers. Load control is intended to minimize customer inconvenience by selecting loads that can most easily be eliminated or cycled.
- **Standby rates.** Utilities offer these rates to customers that have their own source of energy but require service from the utility on an intermittent basis. These rates are more common to electric utilities and may be categorized as: maintenance rates for power supplied by the utility when the customer prearranges downtime for generation equipment maintenance; supplementary rates used by self-generators that regularly require additional power from the utility; and backup rates used by self-generators in the event of an unexpected system outage.

Negotiated and Specialty Rates

Negotiated rates may be cost-based, designed to allow the utility to compete with alternatives, used to support economic development or business recovery, or implemented to permit unique arrangements, such as interruptible service.

Cost-based negotiated rates are designed for customers whose usage and characteristics vary considerably from the average of the rate class or have realistic competitive alternatives to the utility offering the rates. Most commonly, this is only done with large CI&I customers. In many cases, the utility has the ability to negotiate rates down to a level equal to or, more commonly, slightly above its short-term marginal cost. The regulatory justification is that the rest of the utility's customers will benefit from such contracts as long as the negotiated rate charged to the particular customer covers the incremental variable cost of service and provides some contribution to fixed costs. These rates provide the utility with a maximum degree of flexibility to market their product to customers with special needs or with competitive alternatives. Sometimes, particularly when the contract period is longer than five years, these rates may be designed to recover long-term marginal costs. With these rates, a larger minimum demand charge is required.

- **Economic development rates.** Utilities commonly offer these rates to provide economic incentives for businesses to locate or expand into their home service territory and/or into economically depressed areas. They are often based on a schedule in which rates are initially discounted and then phased into a standard rate over a period of several years.
- **Business retention rates.** These rates are designed to retain customers with either competitive options from other energy sources, self-generation capabilities, or an interest in moving to another state or service territory. Business recovery rates are designed to retain customers in financial difficulty.
- **Conservation and load management rates.** Many utilities offer these rates to customers that meet certain equipment or building envelope thermal efficiency standards and/or operating standards. These rates may be designed with a simple percentage rate break on usage based on achieving a certain level of conservation and efficiency. They may also involve a reduction in charges based on some type of load-control incentive, or they may be based on a combination of both. Rate design also may include incentive mechanisms for shifting load from peak to off-peak periods.
- **Special contract rates.** In cases where it is in the best interest of all parties (i.e., the utility, the customer, and the rest of the utility's customers) and where the unique conditions of the situation cannot be met under standard rules, utilities may develop special contracts with individual customers. Examples are a very large cogeneration application or a customer who makes year-round third-party gas purchases and is willing to make volumes available to the utility

for resale during peak periods. These contracts often require individual PUC approval, which can be a lengthy process.

Examples of other types of specialty rates are compressed natural gas rates for natural gas-fueled vehicles and rates that support the introduction of new technologies.

Competitive Energy Rates

Competitive energy rates give utilities the maximum flexibility to sell power or natural gas in competitive situations. For example, customers considering a gas or steam technology application, such as cogeneration, are often presented with some type of competitive energy rate by their electric utility in an effort to keep the full load on the utility system. Competitive pricing may be offered down to some small level above the incremental avoided cost.

This pricing structure is also used for peak usage by utilities with some excess system capacity. There is some concern that this practice skews market choices, keeps load at the expense of conservation opportunities, discounts opportunity costs, and, in the long run, leads to additional capacity needs at the expense of the rest of the rate payers.

Wholesale (Off-System) Sales

Another area for incremental-cost commodity sales is on the wholesale market. This market, while often less stable than on-system sales, can be very profitable for utilities if the plant supplying power or the reserved gas pipeline capacity is in the rate base. A utility with excess capacity in a given period can sell gas or electricity to other utilities or, in some cases, to customers outside of their service territory. The prevailing logic is that excess capacity should be marketed whenever possible as long as variable costs are recovered and some contribution, however small, is made toward fixed cost recovery. With the potential for capacity release, however, LDCs can instead sell excess capacity rights to others rather than use the capacity for the purpose of off-system sales.

Real Time Pricing

Real-time pricing (RTP) is an emerging utility rate strategy that goes a step further than demand charges and varying usage charges in allowing for differentiation of costs that better reflect the utility's actual incremental cost. RTP rates typically do not have a demand component. Instead, kWh, or Mcf, consumption is priced by the hour. For example, an hourly RTP structure may charge $0.90/kWh at noon on a Wednesday in August and only $0.02/kWh at midnight on a Sunday in March. Currently, RTP is being used by numerous electric utilities.

The theory behind RTP is that if customers are told in advance of the utility's anticipated system and price conditions, customer demand will respond most directly to price changes. That is, a decrease in consumption as the price rises and an increase in consumption as the price falls. Typically, customers are given a schedule of hourly prices one day in advance. In some cases, the cost per kWh is fixed for several categories (i.e., off-peak, utility-peak, or regional power pool-peak), but the hours during which they are applied are varied and communicated by the utility to customers on an hourly, daily, or weekly basis.

The procedures used to design rates that differentiate usage and demand charges by time and season of use come close to approximating what the utility determines is proper hourly cost allocation. RTP accomplishes this with greater certainty and fewer complications. Taking the example of a TOU rate with a peak demand period of the typical 9-to-5 workweek, a peak demand may be set at 9 a.m. by a facility. This peak may have no impact on the utility peak, but is charged as if it were set at the utility system peak hour. With RTP, if in fact this peak had no impact on the utility peak, it would be priced at a far lower level than a peak that did have an actual impact on the utility peak.

The concept behind RTP is that pricing reflects real, almost instantaneous, market conditions instead of predicted market conditions. While TOU rate blocks are bins which approximate what actual costs are in different periods, RTP more closely reflects the actual value of electricity (or gas) at any given point in time. Non-RTP rates are, therefore, based on probability of occurrence, rather than the occurrence itself.

Another type of pricing that more closely represents real events is ambient temperature based pricing. For example, when the outside temperature falls below 30 or 20°F (-1 or -7°C), natural gas pricing could automatically shift to a higher rate. Currently, there are many interruptible gas rates that base interruptions on temperature. A similar strategy could be employed for summer electricity pricing, based on rising temperature. While this is not an instantaneous pricing mechanism, it is one based on real events as opposed to predicted events. When the event occurs, the pricing schedule is in effect.

Electric utility dispatch modeling has become

an increasingly precise process. Utilities can identify where each incremental kWh comes from and its value. Large facilities can then perform the same modeling of in-house usage. Furthermore, the cost of telemetry is decreasing, while capabilities are increasing. As alternative electricity purchase options become available, RTP may become a mainstream sales pricing tool. Consumers may elect to purchase certain blocks from the utility, generate certain blocks on site, and purchase certain blocks from other sources through retail wheeling, all based on real-time price signals. With the advent of the smart grid discussed in the previous chapter, the potential for optimizing these approaches may be dramatically enhanced.

Rate Riders

In additional to various rate design options, rate riders are special charges or programs integrated into rate schedules that modify the structure based on specific customer qualifications. Riders are used to account for unique conditions or to give the utility added flexibility to apply rates without dozens of additional tariffs. Riders may include: negotiated competitive-energy riders, interruptible riders, standby riders, buy-back riders, conservation and other load-control riders, end use riders, and other types of discounts, such as an electric utility discount for customers receiving service at a voltage above the standard voltage.

Common Natural Gas Rate Schedules

Actual natural gas rate schedules include many of the same rate-design strategies previously mentioned. These rates are offered to customers who meet specific criteria. Commonly used natural gas rate schedules are:

- **Firm sales rates.** These are the highest priority of service offered by LDCs. Gas is made available throughout the year, on an uninterrupted basis, to serve customers' needs. Typically, there are several categories of firm sales rates, such as residential, small commercial, and large CI&I. Rate design may include block rates, seasonally differentiated rates, demand charges, etc.
- **Firm transportation rates.** These provide uninterrupted transportation service, through the LDC's distribution system, of natural gas purchased directly by the customer. The LDC takes on the obligation to deliver to the customers' facilities gas that has been delivered to the LDC's gate station by an interstate pipeline.
- **Dual-fuel firm rates.** These provide service to customers who have the option of using an alternative fuel source, but who, at any time, may request firm delivery of gas from the LDC. Since the LDC must stand ready to serve, and may have significant investment in, distribution facilities, supply contracts, or storage capacity, it may require purchase of some level of guaranteed volume or include a demand-charge component. Because this type of service has the potential to negatively affect the LDC's load factor if the customer only uses gas services during peak periods, the rate may be expensive.
- **Interruptible sales rates.** A utility can sell to its interruptible sales customers spot market gas or excess gas that it purchased as a reserve for its firm sales service customers. Selling gas on a commodity basis only allows gas to be priced competitively with oil, propane, and other energy sources. Therefore, customers benefit from lower costs.

  Prices are often negotiated competitively and indexed each month to the alternative fuel or energy source (e.g., No. 6 oil, No. 4 oil, No. 2 oil, propane, or electricity). A rate offering, commonly known as the standard offer, may be made to the entire group with the same energy-source alternative. Individual customers with better purchasing capabilities may negotiate price and be permitted to lock in a rate for a longer or shorter period of time. Prices may also vary with notice period. The shorter the notice period the lower the gas cost.

  The LDC may have the ability to negotiate downward to a certain floor. In many cases, the floor is set a few cents above the LDC's actual supply cost. This competitive approach is accepted by PUCs because it holds down overall rates by maintaining gas sales that would otherwise be lost to alternative fuels. Rate-case proceedings may include an agreement that a large portion of the marginal revenues from these sales flow back and reduce the rates of firm-rate customers, often via the PGA. In a sense, this flow-back of revenues provides compensation for the use of facilities (fixed costs) that is amortized through cost recovery via firm rate charges.
- **Interruptible transportation rates.** These services are similar to interruptible sales rates, except they relate solely to distribution rather than to both sales

and distribution. The LDC's ability and/or need to interrupt is tied to local distribution constraints, not to supply constraints.

- **Cogeneration, air conditioning, and other end use rates.** These services are offered for specific equipment applications and, because they have a predictable effect on the LDC's load factor and cost, they are typically grouped under individual rate schedules. These rates are designed to be attractive to customers because they are more closely tailored to actual energy use profiles of the applications, often improving LDC load factor.

Common Electric Rate Schedules

Electric rates use a mix of various rate components, and rate schedules include one or more of the rate design strategies discussed earlier. These components typically consist of an energy charge, a demand charge, a ratchet clause, a fuel adjustment charge, surcharges for factors such as conservation or nuclear plant decommissioning, power factor charges, and taxes. Often, rates are offered to customers who meet specific criteria, such as type of facility or type of equipment used. Some of the most common non-residential electric rate schedules include:

- **General service (GS) rates.** GS rates are typically used by most small commercial customers. Rate design may include block rates, seasonally differentiated rates, and demand charges. Generally, these rates place a greater emphasis on usage than on demand and are less differentiated than large customer power rates. In some cases, these rates are available without demand metering, using higher usage charges instead. Typically, availability of GS rates is limited to residential customers or customers whose demand does not exceed a particular specified level.
- **General service TOU (GST) rates.** GST rates are generally used by small C&I customers with multiple shift operations. They commonly consist of peak and off-peak periods and often register peak demand only during the peak period. They are not time-differentiated as much as are large customer power TOU rates, but they do allow customers to benefit from lower costs for extended use in off-peak periods. They also are often used by customers with electric heat or some type of thermal storage. Rate design may include block rates and seasonally differentiated rates. This rate is now commonly offered for residential customers as well.
- **General service heating (GSH) rates.** GSH rates are common end use-specific rates. Typically, they are general service rates that are available only to electric heating customers whose heating related usage makes up a certain minimum portion of total usage. They are often used by summer-peaking utilities to build winter load. They have a greater degree of seasonal differentiation than standard general service rates, with depressed winter rates compensating for extensive usage. They also may have a TOU component to allow for the use of domestic hot water or heating thermal storage.
- **Street lighting (SL) rates.** SL rates are end use-specific, typically offered to states, cities, or other municipalities, and sometimes to large campus-type facilities. The rate often includes an amortization of the first cost of the street light, ongoing maintenance on the lights, as well as the cost of the electricity used by the lights.
- **Large power (LP) rates.** LP rates are the traditional rates offered to larger CI&I customers and virtually always include monthly or fixed-contract demand charges and may use rate blocks and seasonal differentiation. Typically, the design is somewhat similar to general service rates, except that there is usually increased emphasis on demand charges.
- **Large power TOU (LPT) rates.** These rates are becoming more predominant for large CI&I customers. Typically, they consist of two, three, or four rate periods, such as peak, shoulder, or off-peak. They may register demand only during peak or all rate periods or have fixed-contract demand charges. Off-peak usage may be handled with varied charges or as peak usage with charges in the off-peak periods only for demand in excess of peak demand. Usage charges are varied by rate period. Rate design may include block rates and seasonally differentiated rates.

  These rates are more stratified than traditional large power rates. They are advantageous for facilities with high load factors and extended hours of operation, which can offset costly peak usage with inexpensive off-peak usage. They also are attractive for facilities that use thermal storage or some type of peak-shaving technology.
- **Real-time pricing (RTP) rates.** Typically, RTP rates do not have a demand component. Instead, kWh may be priced by the individual hour or, in some cases, charges may be fixed, but the hours in which

different charges are applied will vary. In either case, the utility communicates these varying costs or hours of application on an hourly, daily, or weekly basis to customers. Often, these rates are used by facilities with alternative energy sources in place or with the ability to shed loads on a regular basis. RTP rates may become increasingly common as electric rates become even more sensitive to market competition. As opposed to demand-based TOU rates, RTP rates are thought to more closely reflect the actual discrete price of power at a given hour or even minute.

- **Transmission (T) rates.** T rates are typically offered to large customers who take power at a high voltage level. T service has several advantages. First, electricity from a transmission line is usually highly reliable. Transmission lines are the backbone of the utility system and they receive high maintenance and restoration priority. Secondly, T rates are less expensive than other large customer rates because they do not include allocated costs associated with the utility distribution system. Finally, in those states that allow open access, being directly linked to a transmission line makes it easier to purchase power from sources other than the local utility. Over the long term, it is anticipated that the advent of retail wheeling and the further unbundling of electric rates will result in transmission/distribution rates being available to all customer classes.
- **Interruptible rates (IR).** These rates and rate riders are designed for customers that have blocks of load (or all of their load) that can be dropped at any (or almost any) time. Commonly, this includes the use of standby generation as a load-shedding technology. Interruptible rates may be designed with a set rate break on demand or usage, or may consist of a flat annual or monthly fee paid by the utility to the customer with a specific payment rate for each period of interruption. Rate design includes different steps, or levels, of availability, with 100% availability receiving the most beneficial treatment. Rates also vary with notice period. The shorter the notice needed for an interruption, the higher the rate discount.

  Further refinement of interruptible rates involves differentiation of services between non-firm electricity sales and non-firm transmission/distribution services. Facilities with on-site energy alternatives can benefit from the ability to withstand sales and transmission interruptions. Facilities with alternative electricity purchase options, via retail wheeling, can benefit from the ability to withstand sales interruptions, but may still require firm transmission/distribution services.
- **Qualifying Facility (QF) rates and rate riders.** Many utilities have special QF rates or rate riders for self-generators. In some cases, these are elective rates (or riders), while in other cases, they are required. These riders often include rate designs that emphasize high peak demand charges, such as TOU rates. These special QF rates also usually include mandatory provisions that require the self-generator to purchase a form of backup services to provide a payment stream to the host utility for providing a reliability service by virtue of the physical connection. The provision of backup service is generally desired by the customer to prevent interruptions. The charge for backup is supported by the argument that a rate recovery mechanism is needed to prevent self-generating facilities from taking advantage of utility capacity supported by other rate-paying classes, in the event of an outage of self-generation equipment, or during periods of planned system maintenance or for purchasing supplementary power.

  Rate restrictions are often put in place to limit rate options available to self-generators. A self-generator forced to be on a highly demand-sensitive rate with a ratchet penalty clause could end up with nearly a full year's worth of demand charges for a single outage. This is often considered excessive recovery. On the other hand, a self-generator allowed to be on a low-demand, highly usage-sensitive rate may pay a minimal amount, which is often considered insufficient cost recovery.
- **Standby rates.** These rates are offered to self-generators requiring power from the utility when their own or alternative energy supply is inadequate or unavailable. Standby rates are often riders which affect several rates offered by a utility. Many utilities have special rate recovery treatment for providing standby (QF backup) power to self-generators when their own or alternative supply is inadequate or unavailable. Standby charges are a type of demand (or insurance) charge paid to the utility to reserve replacement capacity and energy if a system failure or normal maintenance interval takes the on-site generator out of service. These standby rates may be offered as separate rates or rate riders.

There are three general types of standby rates offered by electric utilities: **backup rates, maintenance rates,** and **supplementary rates**. Backup and maintenance rates are offered to provide power when a self-generator's system is fully or partially out of service. Maintenance rates are offered for use during pre-arranged downtime and backup rates are offered for unanticipated downtime. Supplementary rates offer power for regular use and are offered to partial-requirements customers that may require purchased power in addition to their own self-generated power.

Standby rates for backup and maintenance are usually based on a monthly charge per kW of capacity reserved. There is also a commodity charge for actual energy usage during the down-time period. These standby demand charges are less costly per kW than the actual demand charges on a given full service rate, but are paid for on a take-or-pay basis, regardless of whether additional power is ever required.

Maintenance service rates provide convenience in that they allow self-generators to perform routine service and overhaul during peak demand setting periods. An alternative is to perform maintenance in off-peak periods. However, this is not always possible, particularly for lengthy overhauls. Backup service rates are somewhat like an insurance policy. By purchasing this capacity insurance for a given amount of kW on a monthly basis, self-generators avoid ratcheting and/or full demand charges that might otherwise result from outages.

Consider an example in which the full service rate demand charge is $18/kW/month and the standby charge is $8/kW. The facility pays this charge regardless of whether backup power is used. In this example, the annual fee of $96/kW would be a wise investment only if the facility experienced peak setting outages more than 5 months per year, since 5 months' demand charges would only cost $90/kW.

The cost of standby service varies widely. Some utilities require self-generators to purchase standby insurance. Other utilities offer it as an option, while some have no provision for standby power at all. The alternative is the use of standard rates. Key questions in cost allocation are: "What are true costs?", and "What is a fair and reasonable price for such capacity insurance?"

The logic behind this particular cost allocation is that the utility must stand ready to serve these loads when needed. Cost allocation is based on a determination of the impact on generation and distribution capacity requirements. The cost-of-service analyses take into account all self-generators and the real probabilities of peak demand impact resulting from random system outages. If, for example, each of 100 self-generators were to set a peak once a year at different times, what would the real impact be on the capacity requirements of the utility?

For cases in which standby service is not mandatory, but offered as an option, customers must make the determination whether to take this type of insurance or take their chances on standard rates. Customers may also elect to secure standby power for a portion, rather than all, of their self-generation load.

In cases where standby charges are mandatory and very high, the cost may be sufficient to make projects uneconomical. In some cases, a change to mandatory requirements, resulting from rate case proceedings years after a system has been installed, may provide sufficient incentive to abandon a project due to the evaporation of savings critical to successful economic operation.

One hypothetical example of such prohibitive effects is a system with three generation units with required standby charges for the full connected load at 66% of the standard demand charge. In this case, the cost of standby service is equivalent to the system operating on a standard rate and experiencing the highly unlikely occurrence of a peak-setting outage in every single month for two of the three units. Add to this the potential of being forced onto an uneconomical rate, and a self-generator could end up with no savings at all. While this example is extreme, it helps to explain why self-generation has been underdeveloped in certain utility service territories.

RTP rate structures may offer an effective means of allocating costs for standby power. RTP is an attempt to reflect short-term costs so that consumers may make short-term purchase decisions. These same varying short-term prices could be made available to QFs. For example, one rate structure on file with a state commission provides payment for a QF's energy sales at the corresponding marginal cost (i.e., system lambda, $/MWh) of power the host utility experiences. In California, a forecast of marginal costs are the primary input in determining an RTP pricing structure for as-available energy.

MEASURING ELECTRIC DEMAND

The measurement of demand is fundamental to most electric rate structures. It is a tool that allows a utility to differentiate capital cost requirements for serving customers of varying usage patterns. By measuring and billing for demand, the utility can assign costs more fairly to customers.

As opposed to gas utilities, the demand interval for electric utilities is extremely short, usually 15 or 30 minutes. It is not an instantaneous measurement, but an average of discrete measurements over time. If, for example, a facility experienced a rate-of-use pattern of 200 kW for 5 minutes, 300 kW for 5 minutes, and 700 kW for five minutes, a 15-minute demand interval meter would register an average demand of 400 kW. The demand recording meter would log 400 kW and reset only when a higher level of demand was reached. Sometimes, utilities set peak demand by averaging peaks of a few demand intervals over the billing period.

Many customers use demand monitoring and load-shedding techniques to minimize the impact of peak demand billing. These facilities often attempt to synchronize their operations with the utility demand interval and use intermittent load shedding to reduce their average rate-of-use during intervals when a large surge of power is required for a period less than the full demand interval.

From the utility perspective, this load shedding technique may partially defeat the purpose behind demand metering, which is to charge for peak capacity requirements. A sliding demand interval is sometimes used to more accurately measure the impact of peak demand. With a sliding demand interval, for example, a 15-minute demand interval may be broken into smaller intervals of 5 minutes. These smaller intervals are averaged and then added together, as in the previous example, to set the peak demand for the entire 15-minute interval.

In some cases, utilities simply use smaller demand intervals, starting as low as 5 minutes. More common, however, is the use of the typical 15 or 30 minute interval with a clause in the rate schedule that states that in cases of rapidly fluctuating loads or other special conditions in which the established demand measurement time interval does not equitably compensate the utility, demand may be based on the peak for a shorter period.

Traditional rates often use only one peak demand measurement for billing. Some rates call for demand measurement only in certain peak periods. The rationale is that individual facility peaks in utility off-peak periods have no real impact on capacity requirements. TOU rates, however, measure peak demand in several periods. Demand charges may be set at a different rate for each period. For example, peak demand might be billed at \$20/kW during peak periods and \$5/kW during off-peak periods. In some cases, off-peak period peak demand may only be billed for the portion that exceeds peak period peak demand. This is referred to as excess demand billing.

Integrating Power Factor into Demand Billing

Utility generation is measured in volt-amperes (VA), or apparent power, while most customer meters are measured in watts (W), or real power. In alternating current (ac) circuits, watts (power) are equal to volts (potential) times amps (current) only when the wave-forms of voltage and amperage are in phase. This is an ideal condition that does not exist in electric distribution systems. Many types of equipment, such as induction motors, require more apparent power than the amount of real power consumed because their inductive impedance causes current and voltage to be out of phase.

The difference between apparent power (VA) and real power (W) is called volt-amperes reactive (VAR). This is the component of VA that circulates back and forth between the utility and the equipment, but is not consumed by the load. It is, however, partially consumed by distribution losses.

Power factor (PF) is the ratio of W to VA. A facility with a PF of 0.75, requiring the same wattage as another facility with a PF of 1.0, for example, will be more costly to serve, because the utility will require one-third more system capacity (1.00 W/0.75 PF = 1.33 VA) to serve the facility with the PF of 0.75.

To more accurately allocate capacity costs through demand charges, some utilities measure and bill demand charges based on kVA rather than kW. This shifts the cost for maintaining non-productive capacity, or a PF of less than 1.0, to the customer and acts as an added incentive to improve the facility's PF.

Many utilities simply institute a penalty for a lagging PF. For every increment below a required minimum PF, a charge is levied against the facility. The minimum allowable PF is typically in the range of 0.80 to 0.90. Many utilities establish this penalty on the rate schedule, but often do not invoke it.

Another way utilities establish a PF penalty is to

specify a maximum free kVAR as a percentage of the maximum kW of demand. Utilities then bill for all metered kVAR above this level. There are several other ways to build PF into billing, such as increasing the peak demand measured by a certain percent for every percent the PF is below a specified level.

Many utilities currently do not penalize for lagging PF, and many that do only impose modest penalties. In those cases, from the customer's perspective, the advantages of a higher PF and the benefits of investment in capacitors and other higher PF equipment are savings from reduced internal line losses, down-sized equipment, and avoided billing penalties. To encourage such customer actions, utilities attempt to set PF penalties at levels that will have sufficient economic impact. Refer to Chapter 24 for a detailed technical description of PF.

Metering Point and Transformer Ownership

Another factor that is often an element in electric utility rate structures is metering point and transformer ownership. Utility distribution voltage is almost always greater than the voltage required inside a facility. The main transformer brings the voltage down to a suitable level for the service entrance at the facility. Typically, the utility owns the transformer and meters usage on the customer (low-voltage or secondary) side of the meter.

These factors will figure into rate design. If the customer owns the transformer, the utility saves on capital and maintenance costs and can pass those savings on to the customer. If power is metered on the high side of the transformer, the utility saves on transformer-related power losses and can pass those savings on to the customer to compensate for losses now occurring on the facility side of the meter.

Separation of Commodity and Distribution Function

As discussed in the two previous chapters, over time, the forces of deregulation are moving the jurisdiction of regulated utility rate structures toward the transmission and distribution functions and away from the commodity sale function, which is falling under the control of free market forces. However, in today's market, the rate structures presented above still widely apply as the utilities continue to sell a bundled commodity to their customers or sell unbundled services, notably transmission and distribution under regulated cost-of-service based pricing.

IPPs and cogenerators have long since made the market more competitive. This has helped or hurt utilities depending on whether they were sellers or buyers. With the advent of electricity wheeling and gas brokering, with open access to transmission and distribution service, opportunities for utilities to make off-system sales have become even more prevalent.

From the consumer perspective, the trend of competitive pricing can be very attractive. Currently, competitive pricing is most beneficial to large consumers, notably those with favorable load profiles or competitive options. Such customers often face a win-win situation in which they install operating cost-reducing alternatives or reap savings from a competitive energy rate break or, in some cases, both.

Historically, it was anticipated that an increasing number of consumers would be the beneficiaries of competitive pricing in those states that allow open access. There was also the concern that embedded high cost of utility capacity would drive up prices to customers stranded without strong competitive options. Ironically, the high cost embedded utility capacity costs of ten years ago are now much more competitive given the recent fossil fuel price spikes and volatility.

Summary

The concepts behind the design of natural gas and electric utility rate structures are extremely similar in that they are developed through the same regulatory process. Both have been greatly affected by the movement toward deregulation and are moving more toward market-based competitive rate structures. While the pace of change varies from state to state and utility to utility, the commodity component is being separated from the transmission and distribution component, necessitating significant changes in cost recovery and, therefore, rate design.

Finally, the trends toward some variation on the concept of RTP is apparent, as is the trend toward increased customer choice in the selection of utility rates and energy services in general. Still, the basic concepts of fixed and variable cost recovery shall continue to apply. An understanding of these concepts allows gas and electricity consumers to understand utility rates and select rates and services that most effectively meet their energy use needs.

Chapter Twenty-Two

Utility Bill Analysis

The ability to calculate utility bill impacts is fundamental to the evaluation of any energy system. An effective method for determining the energy operating cost impact of a proposed option is to compare the calculated annual utility bills with and without the option in place. The difference represents the energy operating cost impact associated with the option. To do so, one must carefully identify not only the aggregate change in fuel or electricity usage associated with the proposed option, but when the changes occur with respect to the various time periods and other billing factors integral to the rate schedule.

The average unit cost for utilities, defined as the total annual cost divided by the annual consumption, is a tempting simplification of utility rate structures. It allows easy calculation of energy cost impact from changes in energy usage patterns. In some cases, this simplistic approach yields accurate answers. More typically, however, application of average unit costs gives results that are inaccurate and can at times be very misleading.

For any proposed technology application that would change the number of fuel or electricity units consumed, the cost impact can rarely be accurately determined by merely multiplying the change in units by the average cost per unit for the total facility usage. This is because incremental costs are usually quite different from average costs. With most currently available utility rates, the addition or subtraction of usage during peak periods may have several times the cost impact as the addition or subtraction of the same amount of usage in off-peak periods. Therefore, one must consider the weighted average cost of the increase or decrease in consumption units associated with a proposed application.

Calculating Utility Bills

Utility bills can typically be broken down into the following basic components:

- A customer or minimum service charge
- An energy or commodity charge
- A demand or maximum level of service charge
- Power factor penalties
- Adjustments such as taxes levied by state, county, and city authorities
- Surcharges or credits associated with specific orders established by various regulatory authorities
- Fuel cost adjustments that reconcile the actual cost of fuel used or delivered by the utility, with the estimated cost used in the most recent rate proceeding to set the energy or commodity charge

These basic components are expressed and calculated in many ways using various units of measure. Combined, they comprise the total utility bill, with each contributing in different ways to the weighted average cost. To calculate a utility bill, one must carefully read the rate tariff inclusive of all rate riders and adjustment clauses. One must also know the current values for items that vary, such as fuel adjustment charges. Based on this information, one should be able to calculate the utility bill exactly. If such calculations do not equal the utility bill exactly, either a mistake has been made in the computation or a piece of information is missing.

Utility rate spreadsheets are useful in performing such computations. Once a spreadsheet is built, it can be used to calculate costs for any usage pattern under a given rate or set of rates. It can also be used to quickly calculate cost savings from energy efficiency improvements.

Typical Gas Bill Calculation

The following is a sample utility bill calculation for a given natural gas usage profile for natural gas service. This is a typical declining block rate structure where dif-

Monthly service charge: $80.00

Minimum monthly charge = The service charge

Commodity rate:

Block	**Ccf**	**Per Ccf**
First	10,000	$0.4374
Next	20,000	$0.4222
All Over	30,000	$0.4100

Purchased gas adjustment (PGA):	($0.0097)
Demand-side management (DSM) charge:	$0.0020

State tax: 4.3%
City tax: 1.5%

Fig. 22-1 Sample Natural Gas Rate Tariff.

ferent levels of usage are billed at different unit costs. In this case, the billing unit is 100 cubic feet (Ccf) of natural gas. Ccf is a commonly used volume of gas for billing purposes. Other commonly used billing units are 1,000 cf (Mcf), therm (100,000 Btu), and million Btu (MMBtu). The rate schedule is shown in Figure 22-1. Refer to Chapter 5 for detail on the energy or heat content of natural gas billing units. Assuming that the gas usage for the month was 47,500 Ccf, the bill would be calculated as follows:

The commodity rate is first adjusted to account for the purchased gas adjustment (PGA) and the DSM surcharge. To each rate block, \$0.0097 is subtracted to account for the PGA and \$0.0020 is added to the commodity rate to account for the DSM surcharge. The resulting net commodity rate is:

Block	Ccf	Per Ccf
First	10,000	\$0.4297
Next	20,000	\$0.4145
All Over	30,000	\$0.4023

Therefore,

For the first 10,000 Ccf, the charge is:	\$4,297.00
For the next 20,000 Ccf, the charge is:	\$8,290.00
For the final 17,500 Ccf, the charge is:	\$7,040.25
The total commodity charge is:	\$19,627.25
Add the service charge:	\$80.00
The total pre-tax utility bill is:	\$19,707.25
Add state and city taxes of 5.8%:	\$1,143.02
Total bill for the month is:	\$20,850.27

Note that if the customer used no gas during the billing month, the bill would have been only the service charge of \$80.00, plus the state and city taxes, for a total monthly bill of \$84.64.

	Summer Period*	Other Periods
Monthly service charge	\$71.29	\$71.29
Demand charge	\$10.02/kW	\$8.53/kW
Energy charge	\$0.05890/kW	\$0.05242/kWh

Minimum monthly charge: the customer charge plus \$4.57 per kW of the highest billing demand established during the 12 months ending with the current month.

Fuel adjustment charge (FAC):	(\$0.00123 per kWh)
DSM program surcharge:	\$0.00074 per kWh
State tax:	4.3%
City tax:	1.5%

*Summer Period is defined as the billing months of June, July, August, and September. All other billing months are defined as "Other Periods."

Fig. 22-2 Sample Electric Rate Tariff.

Typical Electric Bill Calculation

Figure 22-2 is a sample electric rate tariff for a seasonally differentiated electric rate. In this example, assume that the customer's electricity usage in August was 200,000 kWh and the peak demand was 1,655 kW. The energy rate is adjusted to account for the FAC and DSM surcharge. To the base rate of \$0.05890 per kWh, \$0.00123 is subtracted to account for the fuel adjustment and \$0.00074 per kWh is added to account for the surcharge. The resulting net energy charge is \$0.05841. Note that the FAC varies each month and can be either positive or negative.

The monthly charge for energy is therefore:

200,000 kWh x \$0.05841 per kWh =	\$11,682.00
The monthly charge for demand is:	
1,655 kW x \$10.02 per kW =	\$16,583.10
Add the service charge:	\$71.29
The resulting total pre-tax utility bill is:	\$28,336.39
Add state and city taxes of 5.8%:	\$1,643.51
The total bill for the month is:	\$29,979.90

If the customer had used no electricity during the billing month, and the highest demand in the previous 11 months is assumed to be 1,890 kW, the pre-tax bill would be only the service charge of \$71.29, plus a minimum (demand ratchet) charge of:

1,890 kW x \$4.57 per kW = \$8,637.30

for a total of \$8,708.59. Adding on the state and city tax of 5.8% results in a final bill of \$9,213.69 for the month. This extreme example illustrates the importance of accounting for all elements of the rate structure. Had the demand charge been ignored, the bill calculation would have been grossly underestimated.

Determining the Weighted Average Cost of Power

In the following pages, three electric rate examples are discussed. To keep the analysis manageable, certain billing factors, such as customer charges, taxes, and power factor, have been excluded. These examples, which represent typical industrial, institutional, and large commercial electric rate structures, clearly demonstrate the relationship between varying consumption load profiles and electricity costs. They are based on the following three rate structure types:

- Rate 1: A seasonal time-of-use (TOU) rate
- Rate 2: A conventional seasonal (CONV) rate
- Rate 3: A four-tier seasonal real-time pricing (RTP) rate

The three rates presented here are representative of current rates in many parts of the country (between $0.05 and $0.06 per kWh for baseloaded usage, inclusive of demand charges). However, they should not be used to evaluate specific technology applications. One must always use the rates charged by the local utility. It must be noted that electric rates vary dramatically across the country. In fact, neighboring utilities in the same state often have significant differences between the types of rates offered and rate levels. These differences will likely increase as the utility industry continues to undergo restructuring. While all three rates have fairly similar costs for baseloaded usage, they have very different structures.

Rate 2 is referred to as a conventional electric rate because it has historically been the most common type of rate. It is, however, being increasingly replaced by TOU-differentiated rates designed to send market price signals that shape consumer usage patterns and better reflect the cost to serve. Because the usage charge per kWh does not vary with time of use and because peak demand charges are more moderate, Rate 2 price signals do not strongly drive usage away from peak periods or attract usage in off-peak periods to the extent TOU rates do.

In the two standard (i.e., Rate 1 and 2) rates, a demand charge combines generation, transmission, and distribution system capacity charges, although each is charged separately in many rate structures. Many TOU rate structures will use varying demand charges in each rate period. In addition to peak demand charges, this TOU rate charges for excess demand in off-peak periods. The peak usage charges also include some allocation for capacity costs. But, in many rate structures, costs between peak and off-peak usage are not nearly so differentiated. In those rate structures, a larger portion of the various capacity costs are embedded in demand charges. Rates 1 and 2 have demand ratchets, which can only be set in summer months, as part of their seasonal differentiation. Many rate structures do not have ratchets and some have ratchets that can be set in any month. The RTP rate combines all capacity and commodity costs into usage charges, differentiated by four rate periods.

Weighted Average Cost for Representative Operating Load Profiles

The simple average price of electricity or gas for a given facility can be calculated by dividing the annual cost by the annual usage in billing units (e.g., kWh or Ccf). This yields an average cost per kWh or Ccf. However, average cost calculations provide limited and often misleading information about the actual incremental cost of a particular end-use or load profile. The weighted average cost for specific usage profiles may vary dramatically. In fact, it could be several times greater with one rate structure compared with another.

To demonstrate this important concept, a table reflecting the price of purchasing electricity under various usage profiles is presented for each of the three example rates. Each of these tables lists ten different power usage profiles that might be associated with usage of a certain device or, perhaps, the usage of an entire facility.

The individual profiles in Tables 22-1 through 22-3 show the annual usage, in kWh, for each profile, the weighted average incremental cost of a kWh, and the annual cost of consuming power under specific load profiles for a theoretical 1 kW device. Explanation of how the various profiles in Tables 22-1 through 22-3 are calculated and how they relate to various types of usage follows the three electric rate examples.

While different rate designs result in widely varied costs under different usage profiles, the weighted average cost for the baseloaded kW usually is fairly similar for a given utility's cost structure. The rate structures and costs used in these three rate examples could all realistically be offered by one utility. The baseloaded cost per kW of capacity requirement is based on continuous usage every hour of the year, with the total usage being 8,760 kWh per kW of demand. This type of usage is shown for each rate example as Profile 8. This particular profile is illustrated graphically in Figure 22-3. As shown, 1 full kWh is consumed in each of the 24 hours in each day in

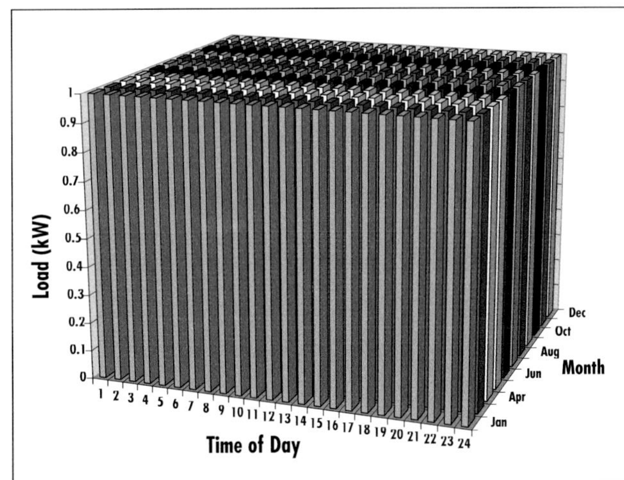

Fig. 22-3 Rate Structure Profile 8, the Baseloaded kW.

each of the 12 months of the year, producing a volume of 100% of usage for one kW of demand. Hence, the terms baseloaded kW and 100% load factor are applied. A decrease in usage volume per kW of demand corresponds to a decrease in load factor. The weighted average cost per kWh for the baseloaded kW is $0.0600 in the CONV rate, $0.056 in the TOU rate, and $0.056 in the RTP rate. Since the weighted average cost for baseloaded usage is close, comparison of these three rates clearly demonstrates the cost impact of rate design on various types of usage patterns.

Profile 4 in each of the rates is based on a total annual usage of only 1,400 kWh for the 1kW device. All of this usage is in the peak and shoulder rate periods during the four ratchet-setting summer months. As a result, in each of the three rates, the weighted average cost per kWh is significantly higher than the weighted average cost of the baseload usage associated with Profile 8, which also includes off-peak usage. The weighted average cost is so much higher because it is more expensive to provide power during peak periods than off-peak periods. This is reflected in the rate structures, though to varying degrees. The baseload usage profile of Profile 8 blends this high-cost peak usage with low-cost off-peak usage.

The annual usage for Profile 4 is illustrated graphically in Figure 22-4. Note that usage is only shown during the four summer months and during hours 6 through 21 of each day, which correspond to the peak and shoulder periods (6 a.m. through 9 p.m.) from Monday through Friday. Hence, this figure only represents the usage during the normal five-day workweek.

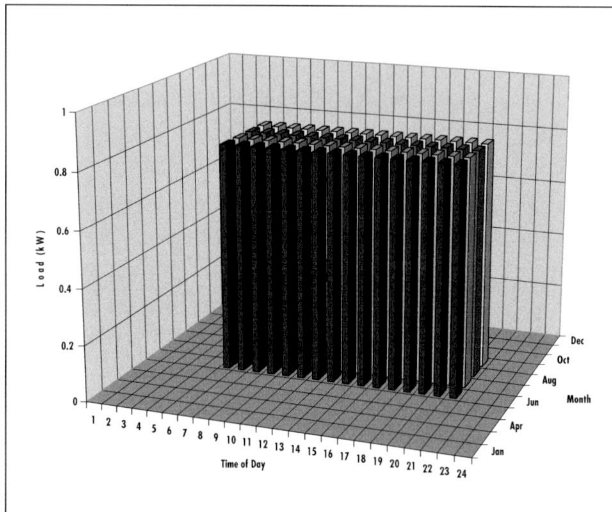

Fig. 22-4 Rate Structure Profile 4. Peak and Shoulder Summer Usage Monday through Friday.

In comparison to Profile 8, which shows an annual consumption volume of 8,760 kWh, Profile 4 only shows a volume of 1,400 kWh for the same 1 kW of peak demand. As will be shown below in the computations provided in the detailed discussion of each rate profile, the load factor for Profile 4 is only 16% since only 1,400 of a possible 8,760 kWh are consumed over the course of the year.

In contrast, Profile 10 is based on a total annual usage of 6,674 kWh, with all usage in the off-peak and shoulder rate periods. As a result, in each of the rates, the weighted average cost per kWh is even lower than the weighted average cost of the baseload usage profile (Profile 8). In this case, since it is far less costly to provide electricity in the off-peak period, the result is a very low weighted average cost. The utility now has this extra capacity available, during the peak periods, that it can sell at the much higher rate to balance the sale of this low-cost usage.

The ten profiles listed in each of Tables 22-1 through 22-3 were created to match the hours in each of the rate periods specific to the TOU rate structure. To allow for a reasonable basis of comparison, the same rate periods have been imposed on the rate structure used in the RTP rate. While the RTP rate has been calibrated for this purpose, it would also be appropriate to view these rate structures with respect to their own natural rate blocks.

For example, the RTP rate has a total annual base period, or lowest cost rate block of 3,020 hours per year, while in the TOU rate there are 4,588 annual hours in the off-peak rate block. To make the RTP rate compatible with this load profile, it was assumed that the 4,588 annual hours would be composed of 3,020 hours from the base period, with the remaining 1,568 hours assigned to the intermediate, or next lowest, cost rate period. The rest of the profiles for the RTP rate were calibrated to the TOU rate structure in a similar manner.

The table (i.e. Table 22-1, 22-2, and 22-3) accompanying each rate provides a calculation of the annual electric bill that would result from each of the ten usage profiles under each of the rate structures. Following the three rate descriptions are explanations for each of the profiles.

ELECTRIC RATE 1 (TOU)

Electric rate 1 is a seasonal TOU rate. The basic tariff is summarized in Figure 22-5. Under this rate, shoulder and off-peak demand charges are assessed only for demand levels that exceed that of the peak periods. For example, if summer peak usage was 1,000 kW and shoulder usage was 1,500 kW, the demand charge would be:

(1,000 kW x $12.00) + (500 kW x $6.00) = $15,000 per mo.

Electric Rate 1 (TOU) Seasonal, TOU Electric Rate	Summer (4 months)	Non-summer (8 months)
	Demand Charges	($/kW)
Peak	$12.00	$8.00
Shoulder Excess	$6.00	$4.00
Off-Peak Excess	$3.00	$3.00
	Energy Charges	($/kWh)
Peak	$0.068	$0.058
Shoulder	$0.058	$0.048
Off-peak	$0.032	$0.032

Figure 22-5 Electric Rate 1 Tariff Summary.

Ratchet Adjustment

Under this rate, if one month had a peak demand of 2,000 kW and the next 11 months had a peak demand of 1,000 kW, each of those next 11 months' demand charges would be based on 80% of the 2,000 kW figure, or 1,600 kW. The impact over the course of a year on the customer's bills would be an additional 600 kW in monthly billable kW charge over the actual demand. Over a period of one year, the customer would pay for 6,600 kW in additional demand charges. This is calculated as:

[(2,000 kW x 0.80) – 1,000 kW] x 11 months = 6,600 kW

Specific Hours of Operation

Peak:

10:00 a.m. — 6:00 p.m., Monday — Friday
4 summer months at 40 hr/wk (700 hr)
8 non-summer months at 40 hr/wk (1,386 hr)

Shoulder:

6:00 a.m. — 10:00 a.m. and 6:00 p.m. — 10:00 p.m., Monday — Friday
4 summer months at 40 hr/wk (700 hr)
8 non-summer months at 40 hr/wk (1,386 hr)

Off-peak:

10:00 p.m. — 6:00 a.m., Monday — Friday, all day Saturday and Sunday
4 summer months at 88 hr/wk (1,540 hr)
8 non-summer months at 88 hr/wk (3,048 hr)

ELECTRIC RATE 2 (CONV)

Electric rate 2 is a conventional seasonally differentiated CI&I rate. The basic tariff is summarized in Figure 22-6. Specific hours of operation are:

Summer—4 months (17.50 weeks totaling 2,940 hours)
Non-summer—8 months (34.64 weeks totaling 5,820 hours)

Electric Rate 2 (TOU) Seasonal, TOU Electric Rate	
Summer (4 months)	Non-summer (8 months)
Demand Charges	($/kW)
$10.00	$8.00
Energy Charges	($/kWh)
$0.051	$0.046

Figure 22-6 Electric Rate 2 Tariff Summary.

ELECTRIC RATE 3 (RTP)

Electric Rate 3 is a simplified real-time-pricing rate. There are several ways in which this developing rate is offered by electric utilities to customers. In one common approach, the pricing is established per rate block based on an analysis of the utility's costs associated with the dispatch of various generation stations in the stack. Each day, the utility informs the customer of which hours will be applied to each rate block.

In this case (Figure 22-7), four blocks have been assigned: base, intermediate, peak, and power pool peak. The power pool peak refers to costs incurred as a result of the utility requiring peak power from the pool. Since each hour of the year is assigned to a rate block based on actual (real time) dynamic conditions, there is no established schedule. A good approximation can be made based on experience, however. For the purpose of demonstrating the workings of this rate, hours have been assigned for winter (36.64 weeks) and summer (17.50 weeks) to each

Rate Block	Unit Cost ($/kWh)	Summer Hours	Non-summer Hours	Total Hours
1. Base	$0.025	700	2,320	3,020
2. Intermediate	$0.038	1,200	3,200	4,400
3. Peak	$0.170	1,000	300	1,300
4. Power pool peak	$0.740	40	0	40
Total		2,940	5,820	8,760

Fig. 22-7 RTP Rate Pricing Blocks.

of the four rate blocks. Figure 22-8 shows the maximum cost in each rate block for a baseloaded kW. Figure 22-9 shows the annual hours of operating and cost per kWh for various combinations of rate blocks.

Electric Rate Comparisons and Conclusions

The period during which power is used affects operating costs as much as, or more than, the amount of power used. This concept becomes critical in energy use planning as price differentiation by time of use increases.

While accountants may look at electric operating costs in terms of the average cost per kWh, energy use planners must look at the incremental costs of individual end uses and various consumption profiles to understand price impact. Energy planners audit facilities to develop incremental cost-usage profiles associated with individual equipment, systems, and activities. These audits are done in much the same manner as the ten profiles presented in the preceding pages.

Planners look at seasonal end uses, such as cooling, and understand that the relevant weighted average cost per kWh may be several times greater than the facility's

Rate Block	Summer	Non-Summer	Annual
1. Base	$17.50	$58.00	$75.50
2. Intermediate	$45.60	$121.60	$167.20
3. Peak	$170.00	$51.00	$221.00
4. Power pool peak	$29.60	$0.00	$29.60
Total	$262.70	$230.60	$493.30

Fig. 22-8 Maximum Cost per Rate Block.

Annual and Average Costs			
Rate Block	Hours	Ann. Cost ($)	Weighted Avg. Cost ($/kWh)
3+4 Summer	1,040	$199.60	$0.192
2+3+4 Summer	2,240	$245.20	$0.109
1+2+3+4 Summer	2,940	$262.70	$0.089
1+2 Annual	7,420	$242.70	$0.033
1+2+3 Annual	8,720	$463.70	$0.053
3+4 Annual	1,340	$250.60	$0.187
2+3+4 Annual	5,740	$417.80	$0.073
1+2+3+4 Annual	8,760	$493.30	$0.056

Fig. 22-9 Operating Hours and Cost per Rate Block Combinations.

overall average cost per kWh, especially if a ratchet adjustment is in effect. They look at baseloaded operations and consumption blocks and understand that the costs may be lower than the facility's average cost. They look at identical devices, in a multiple-unit system, that run the same amount of hours per year, and they understand that if they operate with different load profiles, their operating cost may be dramatically different.

Evolving RTP electric rate structures may extend this discrete differentiation to every hour of the year, or perhaps even every minute. An important benefit of RTP rates is that one peak hour of extraordinary usage might not have the dramatic cost impact that a rate with a high demand charge and ratchet adjustment would have. This type of rate flexibility is well suited for electricity purchase strategies that involve a mix of on-site power generation or purchase of non-utility-generated power along with the purchase of utility provided power. In the event of retail purchase of non-utility-generated electricity, traditional, TOU or RTP type rate structures may be applied to transmission and distribution services, while some type of RTP structure would be applied to the usage for the purpose of commodity transaction.

With this understanding, energy planners develop strategies to minimize operating costs and optimize productivity. Efficiency improvements and alternative energy source options are considered with respect to these incremental costs. Self-generation and electricity displacement strategies should be evaluated in the same manner. Electric cost savings opportunities should be considered on the basis of incremental cost per kWh, as well as the total usage. Elimination of 1 kW of low-load factor usage should produce larger cost savings per kWh than the elimination of 1 kW of high-load factor usage, even though it may not produce greater aggregate energy savings.

Following are explanations for each of the profiles and a discussion of the type of equipment usage or facility characteristics that would result in each profile. Also included are examples of how load factor, annual cost, and weighted average cost were calculated.

Determining the Weighted Average Cost for Various Load Profiles

For individual equipment or an entire facility operating with any one of the load profiles presented in Tables 22-1 through 22-3, the total annual usage and cost are based on the total input power (kW) of the equipment (or the connected load of the facility) times the usage and cost of 1 kW as presented in each table entry. In all cases, it is

Table 22-1 Billing Effect of 1 kW, With Usage Under Different Usage Profiles Operating on Electric Rate 1 (TOU)

Profile No.	Period of Use	Annual kWh (kWh)	Average Cost ($/kWh)	Annual Cost ($)	Annual LF (%)
1.	1 ratchet setting kWh per summer month	4	$24.870	$99	0.1
2.	BL summer peak (no ratchet set)	700	$ 0.137	$96	8.0
3.	50% LF summer peak (w/ratchet)	350	$ 0.351	$123	4.0
4.	BL summer peak and shoulder (w/ratchet)	1,400	$ 0.134	$187	16.0
5.	6 month cooling profile (w/ratchet)	1,870	$ 0.103	$193	21.3
6.	12 month cooling profile	3,379	$ 0.077	$260	38.6
7.	50% LF 12 months peak	1,043	$ 0.169	$176	11.9
8.	BL 12 months all rate periods	8,760	$ 0.056	$494	100.0
9.	MU 12 months all rate periods	4,755	$ 0.068	$321	54.3
10.	BL 12 months off-peak and shoulder (no peak demand)	6,674	$ 0.038	$254	76.2

Notes:
Load Factor (LF): Ratio of actual use vs. maximum potential use in all or certain rate periods.
Baseload (BL): 100% load factor, or the maximum use in rate period(s).
Mixed Use (MU): Usage based on 80% peak; 60% shoulder and 40% off-peak usage.
Cooling Profile: Demand based on 100% in 2 summer months, 85% in 2 summer months, and 60% in non-summer months. Summer usage based on 80% peak, 60% shoulder, and 30% off-peak. Non-summer usage based on 48% peak, 36% shoulder, and 18% off-peak.

Table 22-2 Billing Effect of 1 kW, With Usage Under Different Usage Profiles Operating on Electric Rate 2 (CONV)

Profile No.	Period of Use	Annual kWh (kWh)	Average Cost ($/kWh)	Annual Cost ($)	Annual LF (%)
1.	1 ratchet setting kWh per summer month	4	$22.850	$91	0.1
2.	BL summer peak (no ratchet set)	700	$ 0.108	$76	8.0
3.	50% LF summer peak (w/ratchet)	350	$ 0.312	$109	4.0
4.	BL summer peak and shoulder (w/ratchet)	1,400	$ 0.116	$163	16.0
5.	6 month cooling profile all rate periods (w/ratchet)	1,870	$ 0.097	$181	21.3
6.	12 month cooling profile all rate periods	3,379	$ 0.075	$252	38.6
7.	50% LF 12 months peak	1,043	$ 0.147	$154	11.9
8.	BL 12 months all rate periods	8,760	$ 0.060	$522	100.0
9.	MU 12 months all rate periods	4,755	$ 0.066	$315	54.3
10.	BL 12 months off-peak	6,674	$ 0.048	$318	76.2

assumed that the facility has only one billing meter that measures consumption and demand of all connected loads.

The annual and weighted average costs per kWh for the ten sample load profiles under each of the example utility rates are summarized in Figures 22-10 and 22-11. Note that because of the extremely low load factor for Profile 1, TOU and CONV average costs shown in Figure 22-11 are off the scale (>$20/kWh).

Explanation of Ten Sample Load Profiles

Profile 1 is based on a device rated at 1 kW, operated only one hour during each summer month (June, July, August, and September) when the entire facility is already operating at its highest peak demand level. This added load increases the peak demand level for the month by 1 kW. Therefore, there is a peak demand charge for that 1 kW in each of the 4 summer months. It also adds 1 kW to any applicable peak demand ratchet level. Under Rate 1,

Table 22-3 Billing Effect of 1 kW, With Usage Under Different Usage Profiles Operating on Electric Rate 3(RTP)

Profile No.	Period of Use	Annual kWh (kWh)	Average Cost ($/kWh)	Annual Cost ($)	Annual LF (%)
1.	1 kWh per summer month (peak)	4	$ 0.750	$ 3	0.1
2.	BL summer (peak)	700	$ 0.201	$ 141	8.0
3.	50% LF summer (peak)	350	$ 0.234	$ 82	4.0
4.	BL summer (peak)	1,400	$ 0.152	$ 213	16.0
5.	6 month cooling profile	1,870	$ 0.102	$ 190	21.3
6.	12 month cooling profile	3,379	$ 0.074	$ 249	38.6
7.	50% LF 12 months (peak)	1,043	$ 0.142	$ 148	11.9
8.	BL 12 months	8,760	$ 0.056	$ 493	100.0
9.	MU 12 months	4,755	$ 0.066	$ 316	54.3
10.	BL 12 months (base)	6,674	$ 0.039	$ 259	76.2

Notes:
Peak indicates that kWh are first charged to the highest rate block and then successively to lower rate blocks.
Base indicates that kWh are first charged to the lowest rate block and then successively to higher rate blocks.
Cooling profile kWh hours are allocated between rate blocks to correspond to the allocations used in the TOU rate examples.
Mixed-use profile kWh are allocated between rate blocks to correspond to the allocations used in the TOU rate example.

for instance, the effect of this 1 kW is a monthly demand charge based on 0.8 kW in each of those 8 non-summer months.

Thus, the 4 kWh produce a total annual billable demand charge based on 10.4 kW. In addition, there is a usage charge for the 4 kWh totaling $0.27. Based on Electric Rate 1, the annual cost for using the 4 kWh is:

$$(1\ kW \times \$12/kW \times 4) + (0.8\ kW \times \$8/kW \times 8) + (4\ kWh \times \$0.068/kWh) = \$99.47$$

The weighted average cost per kWh is:

$$\$99.47/4\ kWh = \$24.87/kWh$$

If that same 1 kW device is operated for 4 hours in the off-peak rate period, then no demand charges apply and the consumption charges are lower. The total annual cost of the 4 kWh is only $0.13, and the weighted average cost of 1 kWh is $0.032/kWh. While the first profile is an extreme example, it demonstrates how significant the impact of demand charges and ratchet adjustments can be in a given rate structure.

A large chiller used for space cooling, for example, may operate at peak capacity only a few hours during the entire year. In many cases, peak cooling demand coincides with the facility's maximum electric use period and establishes not only a demand peak for the month, but also an increased ratchet demand level for the year. If, at the hour of maximum monthly electric usage (the peak demand hour), the chiller consumes 1 extra kWh to satisfy load, that 1 kWh will increase the billing peak demand by 1 kW.

Taking the most extreme case of Profile 1 in Rate 1, the use of only 4 kWh at the peak demand period of the peak demand month would cost $99. This could actually be the case with a testing lab or a university that

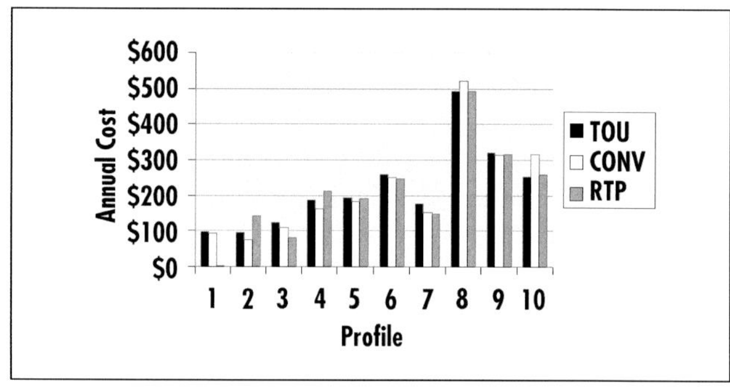

Figure 22-10 Annual Operating Cost for Each Load Profile.

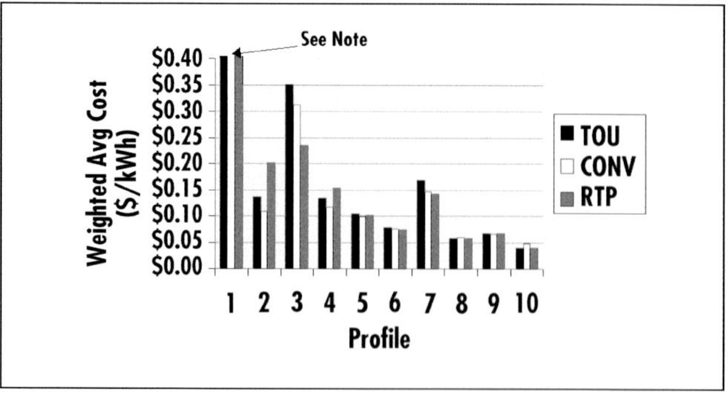

Figure 22-11 Weighted Average Cost ($/kWh) for Each Load Profile.

holds commencement in the summer and experiences an extraordinary peak load on only one day per year. If the facility is able to somehow shed these 4 kWh at the same time that the chiller or testing equipment consumes these ratchet-setting kWh, it would save \$99.

At first glance, the idea of 4 kWh costing \$99 might seem absurd. However, an understanding of how electric rate structures operate shows that the incremental cost per kWh can actually vary by several thousand percent. In fact, with a typical demand window of 15 minutes, this equipment need only set a peak for 15 minutes, consuming only 0.25 kWh to cause the facility to endure the full 10.4 kW of annual demand charge. Under the RTP rate, no such dramatic costs would be incurred. Since there is no demand charge impact, the costs only consist of the peak cost per kWh times the hours of use. In this example, the most expensive kWh of the year would cost \$0.74 and the 4 kWh would, therefore, cost \$2.96. However, it is important to note that in a real-time market, the cost for these 4 kWh could be quite a bit higher, conceivably as high as \$99, though perhaps not likely.

Profile 2 is based on a device rated at 1 kW and operated as a baseload in all 4 summer month peak hours. Under Rate 1 (the TOU example consisting of 700 hours per year based on a 40 hour per week rate period extending for 17.5 weeks), the annual cost for using those 700 kWh is:

$$(1\ kW \times \$12/kW \times 4) + (700\ kWh \times \$0.068/kWh) = \$96$$

The weighted average cost per kWh is:

$$\$96/700\ kWh = \$0.137/kWh$$

The annual load factor is:

$$\frac{700\ hours}{8760\ hours} = 8\%$$

In this scenario, demand charges total \$48, slightly more than half of the annual cost. It is assumed in this case that peak demand is sufficiently high in the winter months so that there is no ratchet in effect. Based on Rate 1, if, for example, an electric motor with an input power rating of 100 kW operates at full load in each of the 700 hours in this profile (represented in Table 22-1, Profile 2), the annual cost would be \$9,600. This can also be calculated by multiplying the input power (100 kW) by the total full-load hours of operation (700 hours) by the weighted average cost per kWh of \$0.137.

Under the RTP rate, assuming that the 700 hours corresponding to the TOU rate summer peak would be composed of the full 40 hours of the power pool peak block and 660 hours of the peak block, the total cost would be \$141 and the weighted average cost would be \$0.20/kWh. Based on this example, operating the same 100 kW electric motor during the same 700 hours would result in an annual cost of \$14,350.

In **Profile 3** the 1 kW device operates with a 50% load factor during the 40 hour per week peak rate period over the 4 summer months rather than at a 100% load factor. This results in a usage of 350 kWh over the 700 total hours in this rate period and a total annual load factor of 4%, rather than 8% with Profile 2. Usage over the 17.5 week summer period and annual load factor (LF), respectively, are calculated as follows:

$$(1\ kW \times 17.5\ wks) \times (40\ hrs/wk) \times (0.5\ LF) = 350\ kWh$$

$$\frac{700\ hrs \times 0.5\ LF}{8760\ hrs} = 4\%$$

Using the TOU rate, the total annual cost is reduced compared with Profile 2 due to reduced usage. Therefore, demand charges as a percent of the total cost increase. In Profile 3 (with ratchet adjustment), demand charges represent 84% of the total cost. The impact of reduced usage with constant demand charges is an increased weighted average cost per kWh. The weighted average cost per kWh for Profile 3 increases to \$0.351.

For this profile with only summer peak usage, the traditional non-time-differentiated rate is less costly than the TOU rate, and the RTP rate is the lowest, with a weighted average cost of \$0.234. The reason is that with such a low load factor load of 4%, the impact of demand charges, inclusive of ratchets, drives up costs dramatically for the other rate structures.

Profiles 2 and 3 are realistic examples of the cost of cooling equipment operation during peak periods. In many cases, these operating profiles are only a portion of a cooling unit's total energy usage. In other cases, they may represent the total operation. In facilities with multiple cooling units, one unit is often predominantly used as a peaking unit. Peak cooling loads often correspond to the TOU peak electric rate period (10:00 a.m. to 6:00 p.m., Monday through Friday) due to ambient temperature profiles, increased productivity, and increased internal gains from people and equipment. In single shift C&I operations, the peak period coincides with most of the operating hours of the facility. In those cases, profiles such as Profiles 2 and 3 may also be representative.

Electric usage with profiles of the type listed in Profiles 1 though 3 are often targeted for elimination or reduction by various load shedding or alternative energy source technologies. Peak-shaving generators and fuel- or steam-powered cooling are two commonly applied technologies for eliminating these blocks of electric usage.

Profile 4 is similar to Profiles 2 and 3, in that it reflects the higher cost of power resulting from seasonal differentiation and significant demand charges. With 100% usage in Summer Peak and Shoulder periods, this profile has greater usage, with a load factor of 16%, as compared with 4% in Profile 3. Under the TOU rate, this produces a weighted average cost per kWh of $0.134, as compared to $0.351 for the lower load factor usage of Profile 3.

In this profile, demand charges represent a lower percentage of the total cost than in Profile 3 because each unit of demand is spread over a greater usage base. However, demand charges still represent a significant portion of the total cost. Profile 4 also includes the effect of demand ratchets. Extending usage to the shoulder periods in this profile partially integrates lower-cost power into the profiles and results in increased load factor and decreased weighted average cost.

Under the RTP rate, assuming the 1,400 hours in Profile 4 would include 40 power pool peak hours, 1,000 peak hours, and the balance of 360 hours intermediate, the annual cost and weighted average cost, respectively, are:

$$(40\ kW \times \$0.74) + (1000\ kWh \times \$0.17) + (360\ kWh \times \$0.038) = \$213.28$$

$$\frac{\$213.28}{1{,}400\ kWh} = \$0.152/kWh$$

Profiles 5 and 6 refer to mixed use cooling season profiles. While Profiles 2 through 4 are all based on a 1 kW device running either all of the time or with a 50% load factor in a given rate period, Profiles 5 and 6 are based on a 1 kW cooling device running with a load that varies between each month and rate period. These profiles were designed to represent typical space cooling loads served by an electric vapor compression system.

During the 4 summer months, it is assumed that the equipment operates with a load factor of 80% peak, 60% shoulder, and 30% off-peak, based on the TOU rate structures. In July and August, it is assumed that the full 1 kW peak demand is set. In June and September, it is assumed that the peak demand impact of the 1 kW equipment is 0.85 kW. In the non-summer months, it is assumed that the equipment operates with a load factor of 48% peak, 36% shoulder, and 9% off-peak and the demand impact is 0.60 kW in each month. These profiles were then calibrated and applied to the CONV and RTP rate. The difference between the two profiles is that Profile 5 is based on 6 months of operation and profile 6 is based on year-round operation. Notice that profile 6 has a far lower weighted average cost per kWh than Profile 5 as more lower cost non-summer usage is blended in and there is no ratchet impact.

Profile 7 has peak usage every week of the year and demand charges in every month. This type of profile might be targeted for elimination with peak shaving power generation or replacement of baseloaded electric-driven equipment with fuel- or steam-driven equipment operated in the peak periods.

Profile 7 is representative of a load profile that might result from electric-driven equipment operated with a 50% load factor during the peak period only. Under the TOU rate, this corresponds to 20 hours of operation per week. There is a significant increase in weighted average cost for Profile 7 for all rate types. This is due to the fact that demand charges remain the same, but are spread over only half the usage as the load factor is only 11.9%. Under the RTP rate, the weighted average cost is relatively high because as load factor is decreased, a greater percentage of the usage is assumed to fall in the highest cost rate block.

Profile 8 (baseload 12 months, all rate periods) represents the baseloaded kW, or 1 kW baseloaded every hour of the year. The annual usage of 8,760 kWh has an annual load factor of 100%.

With this profile, the fixed investment in electric generation and distribution capacity is spread over the maximum possible annual usage. As shown in Profiles 1 through 7, different rate structures result in widely varied costs under different usage profiles. However, the weighted average costs for the baseloaded kW usually are fairly similar for different rates under a given utility's cost structure. The rate structures and costs used in the TOU, CONV and RTP rates could realistically be offered by one utility.

While with the profiles with lower load factor, peak

usage produced much higher costs for the RTP rate, compared with the CONV rate, the higher load factor off-peak usage produced much lower costs with the RTP rate. These opposing trends are roughly canceled out with continuous baseload usage, though traditional rate structures, such as Rate 2, commonly produce slightly higher baseloaded kWh costs. Thus, a three-shift facility with a very high load factor would choose the TOU or RTP rate structures.

The annually baseloaded kW is the load profile most often targeted for prime mover-driven power generation and mechanical service applications that employ heat recovery (cogeneration cycles). The weighted average cost per kWh of the baseloaded kW is often the benchmark for determining the feasibility of such applications. While the cost per kWh is lower than most of the other load profile entries, the total annual cost is the greatest.

Profile 9 represents a profile that might result from annual operation of individual electric motors or other process equipment. It might also result from the combined operation of multiple equipment in a facility. Facilities rarely have absolutely flat loads. Therefore, the baseload, or 100% load factor load profile, is not necessarily representative of the weighted average cost of power. Some type of mixed use profile, such as Profile 9, is generally more representative of the weighted average cost. This is an aggregate of varying individual components, such as lights and motors, each with a different load profile.

Sometimes, equipment does operate with a load factor of 100%, either in a specific rate period or in all rate periods. More often, equipment operates under varying load or under full load for intermittent periods. Profile 9 is an example of such operation. The load factor of this profile is about half that of Profile 8. The total annual cost is lower due to significantly lower use, but is greater than half the cost of Profile 8, because the weighted average cost per kWh is greater. This is due to the greater relative impact of demand charges (the dollar value of which remains the same as in Profile 8).

Profile 10 represents extensive off-peak non-demand setting usage. It has no demand charge at all. Notice that with no demand charges, the weighted average costs are lower for the TOU rate than the CONV rate. This is a result of the price signals of the TOU differentiated rates, which greatly emphasize usage in off-peak and shoulder periods — which the traditional rate does not do. Also note that only a lower cost excess demand charge can be assessed to these usage profiles in the TOU rate example, while a full demand peak can be set in the CONV rate. This would increase costs still further. Under the RTP rate, these average weighted costs are produced by blending the lowest cost rate block with the successively higher cost rate blocks.

As Profile 10 shows, TOU rates offer relatively low-cost power for about three quarters of the hours of the year. When an excess demand charge is assessed to shoulder usage, or if standard shoulder-period demand charges are in effect, the costs will be slightly greater, although still significantly lower than during the costly peak period.

The RTP rate offers relatively low-cost power for about 85% of the annual hours. The weighted average annual cost for the 3,020 hours of base block usage and 4,400 hours of intermediate block usage is only $0.033/kWh. This is balanced by a much higher unit cost in the peak and power pool peak blocks, which comprise about 15% of the total annual hours, but slightly more than half the total annual cost. The weighted average cost is only a usage cost, but it also reflects imbedded fixed costs, a large portion of which are included in the demand charges associated with the other two rates. Table 22-4 provides a comparison of Profiles 7, 8, and 10 for the CONV (2) and RTP (3) rates.

Rate No.	Profile No.	Usage (kWh)	Annual LF (%)	Avg. Cost per kWh	Total Cost
2	7	1,043	12	$0.147	$154
2	10	6,674	76	$0.048	$318
2	8	8,760	100	$0.060	$522
3	7	1,043	12	$0.142	$148
3	10	6,674	76	$0.039	$259
3	8	8,760	100	$0.056	$493

Table 22-4 Comparison of Profiles 7, 8, and 10 for CONV and RTP Rates.

Notice that the sum of total costs for Profiles 7 and 10 approximately equal Profile 8, the annually baseloaded kW. The total annual costs of Profiles 7 and 10 are somewhat close, compared with the stark contrast in usage. The weighted average cost per kWh for Profile 7 is about three times that of Profile 10.

Matching Energy Technology Alternatives with Electric Rates and Usage Profiles

As Tables 22-1 through 22-3 demonstrate, the weighted average cost of a kWh is highly variable. Different technology applications must be matched with different cost scenarios and electric load profiles, as the following discussion points out. The aggregate energy cost for any particular application is the real determinant of operating cost savings potential. While savings of $0.30 or $0.60 per kWh are attractive targets, savings must accrue over enough hours to provide sufficient payback on the investment in alternative energy equipment. Baseloaded cogeneration applications, for example, emphasize overall thermal fuel efficiency, effective heat recovery, and durability. These systems will save less per hour of operation than other technology applications in many cases, but will accrue savings over a greater number of operating hours. Conversely, peak shaving applications may run fewer hours, but target the most expensive electricity usage. In these applications, thermal fuel efficiency is generally less of a concern than low capital costs, because the key is elimination of low load factor high-cost power purchases.

Following is a brief discussion of the fuel- and steam-powered technology applications that should be considered for use in eliminating electricity purchases associated with the 10 representative load profiles presented in Tables 22-1 through 22-3.

Electric Peak Shaving Generation Applications

The primary focus here is on peak demand and usage charge savings. Thus, the emphasis is on eliminating costly peak demand charges resulting from poor load factor or, in the case of RTP rates, eliminating the usage in the highest cost rate blocks. As shown above in the RTP rate, the total annual cost is about the same for a load with a 15% load factor occurring in the highest cost rate blocks as a load with an 85% load factor occurring in the lowest cost rate blocks. Since peak shaving applications will have relatively low annual hours of operation, low capital cost is emphasized more heavily than optimum thermal efficiency and simple energy costs, and heat recovery is not commonly used. Representative load profiles that might be targeted for peak shaving generation include Profiles 1, 2, 3, and 7. Other potential technology applications include load shedding control, battery storage and thermal energy storage (TES).

Electric Cogeneration Applications

The primary focus of power generation applications that employ heat recovery is on overall system thermal fuel efficiency and durability, since equipment run times may range from several thousand hours to continuous operation. The object is to minimize the use of purchased electricity while simultaneously eliminating other internal fuel usage via heat recovery. The 8,760 hour load profile associated with Profile 8 is ideal for cogeneration in most cases. Other high load factor load profiles, such as Profile 9, may also be targeted for elimination with on-site application of cogeneration technologies. In some cases, notably with highly stratified rate structures, it may not be economical to generate power on site in periods when the lowest cost power is available from the utility. In such cases, load profiles with somewhat less than 100% load factor may be appropriately targeted for elimination. Other potential technology applications include combined cycles and steam injection cycles, along with a host of other standard baseload energy conservation measures.

Single-Unit, Year-Round Mechanical Drive Applications

The primary focus here is on satisfying end use requirements with a single unit that has the lowest life cycle cost. Single-unit systems may operate several thousand hours or more annually. In some cases, prime mover-driven systems using heat recovery may be less costly to operate than electric-driven systems in all use periods. In other cases, prime mover-driven systems may be more costly in some periods (i.e., off-peak), but less costly to operate on the average (mixed use) due to significant savings in peak and shoulder periods. System efficiency and durability are particularly emphasized for applications with significant run-time requirement. Heat recovery will be increasingly cost-effective with increased hours of operation. In single or two-shift operations, the unit may operate only in costly peak periods or in peak and shoulder rate periods. Representative load profiles that might be targeted for elimination with single-unit prime-mover mechanical drive systems include Profiles 8 and 9. Other potential technology applications include building and process automation systems and variable volume distribution systems (i.e., air, water, steam, etc.).

Multiple-Unit Mechanical Drive Mixed (Hybrid) System Applications

The primary focus here is overall system optimization with use of electric units in off-peak periods and non-electric units during peak and shoulder periods, or a variation with some baseloading of equipment. Equipment may operate anywhere from several hundred hours to several thousand hours annually. Thermal fuel efficiency (and heat recovery potential) of non-baseloaded individual units may be sacrificed for lower capital costs.

A cogeneration cycle unit may be baseloaded with

electric units used for the remaining off-peak load and non-electric units used for the remaining peak and shoulder loads. Alternatively, an electric unit may be baseloaded and non-electric units used for peaking duty only. In one- or two-shift, five-day operations, a single unit might experience a similar operating profile as would a peaking unit in a three-shift, seven-day operation. Profile 7 might be targeted for elimination with use of a prime mover-driven mechanical drive system. Other potential technology applications include building and process automation, peak shaving, load shedding control and variable volume distribution systems (i.e., air, water, steam, etc.).

Single-Unit Seasonal Cooling Applications

The primary focus here is satisfying cooling requirements with a single unit with the lowest life cycle cost. Equipment may operate anywhere from several hundred hours to a few thousand hours annually. Heat recovery is a viable option, but has less impact than in year-round, single-unit applications, because operating hours are typically lower. Heat recovery becomes more important with higher hours of operation and fuel costs. In single- or two-shift operations, the unit may operate predominantly in costly peak periods or peak and shoulder rate periods. Representative load profiles that might be targeted for elimination with use of a non-electric-driven cooling systems include Profiles 2, 3, 4, 5, and 6. Other potential technology applications include peak shaving, load shedding control, battery storage and TES.

Multiple-Unit Cooling (Hybrid) Applications

The primary focus here is to minimize system operating costs with use of electric units in off-peak periods and non-electric units in peak and shoulder periods, or a variation with some baseloading of equipment. Electric peak demand charges are a critical consideration. Equipment may operate anywhere from two hundred to several thousand hours annually. Thermal efficiency (and heat recovery potential) on non-baseloaded individual units may be sacrificed for lower capital costs. This is similar to strategies for year-round, multiple-unit mechanical drive systems. However, optimization strategies will differ somewhat due to the greater electric unit costs with seasonal pricing and ratchet potential. In one- or two-shift, five-day operations, a single unit might experience a similar operating profile as would a peaking unit in a three-shift, seven-day operation. Representative load profiles that might be targeted for elimination with use of non-electric-driven cooling equipment as part of a mixed system application include Profiles 2, 3, and 4. Other potential technology applications include peak shaving, load shedding control, battery storage and TES.

PART 3
Applications

Section VI

Localized Electric Generation

Chapter Twenty-three

Localized Electric Generation Applications Overview

Localized on-site electric power generation was once the dominant source of electricity in the United States and a driving force in the later stages of the industrial revolution. During much of the early and mid 20th century, there was a gradual long-term trend away from localized electric generation and toward the use of electricity generated at central utility stations and distributed over a nationally linked power grid. However, over the past two decades, non-utility generation (NUG) in the form of on-site electric cogeneration and independent power generation has re-emerged as a driving force in the evolution of electric industry deregulation.

NUG market share is now significant and is rapidly growing with the trend toward free market competition in electric generation. Regulated public electric utilities will likely continue to maintain somewhat exclusive rights and obligations for the transmission and distribution (T&D) of electricity. However, they will also likely become only one of numerous competitors in the area of power generation and sale.

The ability of independent power producers (IPP) and local facilities with on-site generators to sell electric power as the low-cost producer has been enhanced by federally mandated open access to transmission services for wholesale wheeling. This has occurred in stages. In the period following the Public Utilities Regulatory Policies Act of 1978 (PURPA), power sales could be made to the local electric utility at either its avoided cost or through competitive bidding. After the Energy Policy Act of 1992 (EPAct 92), open access was provided to transmission services for wholesale sales in interstate commerce. On the horizon and already in effect in certain states, retail wheeling allows NUGs open access for retail — or end use — sales. In certain areas of the country, Independent System Operators (ISO's), which manage the transmission of power in independent regions such as New England, actively solicit competitive power supply pricing from IPPs. The competitive power supply procurements are for both long-term and short term (day ahead) contracts for supply of power to a region.

Still, the long-term impact of industry deregulation on on-site generators is far from clear. On one hand, on-site generators will have increased flexibility in system design and operating strategies. The cost of standby, or backup, power is no longer a major economic obstacle to on-site generation projects as in the past. No longer bound only to one buyer — the local electric utility, the ability to sell excess power in a competitive market should also enhance project economics for many on-site generators.

On the other hand, competition has and should continue to reduce the cost of electricity in general. Large consumers, particularly those with high-load factor electricity requirements and those with on-site generation options, will continue to enjoy low-cost power. This means that while on-site generators will have free market opportunity to sell excess power, they will do so in a highly competitive environment.

Inside-the-fence electric generation (localized generation for on-site use) will still maintain certain advantages. One advantage is the avoidance of T&D costs for purchased electricity. Even with cost-of-service based rates, T&D services remain a significant cost factor — one that may vary widely depending on a facility's usage pattern, physical location on the distribution system, and the extent to which utility stranded investments in generation capacity are recovered through T&D tariffs. Another advantage is the potential operating cost savings that can be achieved through application of baseloaded cogeneration or strategic deployment of generation capacity during periods when prices are at their highest. Even if rarely used, on-site stand-by generation capacity can be used to support low-cost interruptible power purchases. Finally, on-site generation is a potentially profitable internal investment opportunity.

NUG Classifications

The two traditional classifications of NUGs have been Qualifying Facilities (QF), as defined by the Federal Energy Regulatory Commission (FERC), and IPPs. A third classification, established by EPAct 92, is Exempt Wholesale Generators (EWGs). Some additional classifications are partial requirement (inside-the-fence) generators and independent self-generators.

Qualifying Facilities, as established by PURPA and defined by the FERC, are small power producers or cogenerators. Small power producers are defined as facilities that use biomass, waste, or renewable resources to generate electricity. They are limited by size and ownership standards. Cogenerators are facilities that produce electric power and useful thermal energy by the sequential use of energy from one source of fuel. They, too, are limited by ownership, operating, and efficiency standards.

QFs may generate power, interconnect with the utility grid, and purchase power from the utility grid, all with non-discriminatory treatment. QFs can sell power to the local utility, to another utility, or, in some cases, to another facility on a retail basis. QFs have open access to utility transmission services for wholesale wheeling and, where allowed by individual states, for retail wheeling at non-discriminatory transmission rates.

Independent power producers (IPPs) may or may not be cogenerators. Their decision to maximize power production or to recover thermal energy for process type use is application specific. Like QFs, IPPs have certain ownership restrictions, but they are not given all of the benefits and protections provided to QFs, such as the right to purchase power from the local utility and the right to sell to the utility at the utility's avoided cost. Like QFs, IPPs can sell power to the local utility, to another utility via wholesale wheeling, or, in some cases, to another facility via retail wheeling, with the same right of open access to transmission services. An IPP may also use a portion of the generated power on site.

Exempt wholesale generators (EWGs) are similar to IPPs, except that they are exempt from certain ownership restrictions that apply to QFs and IPPs. Both QFs and IPPs are exempt from several Public Utilities Holding Company Act (PUHCA) and Federal Power Act (FPA) regulations if they meet ownership restrictions limiting public utility company participation. EPAct 92 amends PUHCA and FPA to allow public utilities to own or operate all or part of an EWG. Like QFs and IPPs, EWGs can sell power on a wholesale basis, but unlike QFs and IPPs, EWGs cannot participate in any form of retail wheeling.

Partial requirement (inside-the-fence) generators generate power for internal use, as well as purchase additional power from the utility grid. They do not, however, meet QF certification criteria. Often, these are peak-shaving systems, which eliminate the most costly blocks of purchased electricity. The utility is not obligated to interconnect or supply power to these facilities at non-discriminatory rates, but it may do so anyway as part of a negotiated agreement. If the self-generating facility has other competitive options, such as generating all of its own power or becoming a QF, and the utility values the facility's remaining load, it may be willing to sell power to the partial requirements customer. With the advent of retail wheeling, partial requirement customers may find access to less costly supplemental power and may, at times, export power to other retail consumers.

In addition, there are utility-sponsored peak-shaving programs by which the utility controls, via contract or incentive mechanism, the customer's use of on-site generation equipment. Utility incentives might include rebates, rate reductions, or a payment arrangement based on the customer's level of availability to interrupt service. The utility may also provide similar payments for the customer to export power to the grid upon request. Generally, actual hours of operation are limited and, in some cases, emergency generators are used. These programs allow the utility to reduce capacity requirements or to retain load by providing a customer with an economic alternative to expanded self-generation.

Independent self-generators operate total energy plants that are isolated from the utility grid. This category was more common prior to the 1978 PURPA legislation when utilities could deny service to self-generators. Today, total energy plants are most often used at facilities in remote locations or extremely large facilities operating large power plants. These facilities might also be cogenerators or other IPPs that export power. Total energy plants may also be economical for facilities with good heat/electric balance, or special power reliability requirements.

Cogeneration Systems

When on-site generation or mechanical drive systems provide recovered heat for thermal applications, they are referred to as cogeneration or combined heat and power (CHP) systems. Simple-cycle power production without the use of recovered heat — whether from a central electric-utility plant or from an on-site prime mover — is typically only 25 to 40% thermally efficient. The wasted fuel energy and unnecessary emission of combustion products and heat have a strong negative environmental and economic impact.

In contrast, electric cogeneration systems offer overall thermal efficiencies as high as 80 or even 90%. What turns the relatively inefficient simple-cycle power generation process into the highly efficient cogeneration process is the use of rejected heat in the form of steam, hot water, or heated air for thermal process applications.

Facilities that have year-round electric and thermal energy requirements are good candidates for application of electric cogeneration systems. Applications may range from a few kW packaged units that serve apartment buildings, health clubs, etc., to applications of more than 100 MW. Industrial and manufacturing facilities, refineries, hospitals, prisons, military facilities, hotels, restaurants, universities, district heating/cooling systems, and other commercial and institutional facilities can often capitalize on the high thermal efficiency of applied cogeneration cycles.

In most applications, facilities use electricity producing cogeneration systems to replace a portion of their electric load, while continuing to purchase electricity from the utility grid. Systems are designed for maximum cost-effectiveness by matching the outputs to the base-load requirements of the facility. Both fuel and electric utility systems remain in place to supplement and back up the cogeneration plant. Alternatively, in applications with large thermal loads, facilities may produce excess power for sale as described above.

Currently, on-site electric cogeneration is encouraged by some electric utilities because it helps to relieve electric capacity constraints, eliminates the need to build expensive new central power plants, and reduces utility emissions of regulated pollutants. Moreover, in today's deregulated market environment, many utilities are seeking joint ventures and other financial arrangements with cogenerating facilities.

The electric cogeneration market has become increasingly sophisticated over the past two decades. Experience in markets of all sizes has led to reduced capital costs and construction lead times and increased thermal efficiency and reliability. A focus on the stationary market by the reciprocating engine and gas turbine industry has led to the development of product lines with vastly improved environmental and other performance characteristics. Advances in reciprocating engine and gas turbine emissions control have been dramatic. The continued development of low-emission, high-efficiency dual-fuel engine technology has also contributed greatly to market development.

Packaged systems ranging in size from a few kW to several hundred kW are becoming financially attractive for certain applications. Packaged systems require far less site and system engineering, and smaller systems can often be installed in a few days. While economies of scale, especially capital and maintenance costs, work against the small end of the market, economies of mass production and installation standardization work for it.

Guaranteed service and insurance contracts by manufacturers, vendors, and energy services companies have taken much of the risk out of electric cogeneration investments for host facilities, allowing customers to reap economic benefits without adding responsibilities to in-house staff. While this increases operating costs, it opens the market to many facilities for which taking maintenance and overhaul responsibilities had been a prohibitive factor.

PURPA QF Criteria for Cogeneration

FERC rules implementing PURPA define cogeneration as the combined production of electric power and useful thermal energy by sequential use of energy from one source of fuel. A topping cycle first uses thermal energy to produce electricity and then uses the remaining energy for thermal process. In a bottoming cycle, the process is reversed.

FERC prescribed three criteria that must be met by a qualifying cogenerator. The qualification test includes an ownership standard, an operating standard, and an efficiency standard:

- **Ownership standard.** The owner of a qualifying cogeneration facility cannot be primarily engaged in the generation or sale of electric power, other than the electric power solely cogenerated. Electric utilities may participate in joint ventures, but may own no more than 50% equity in a QF.
- **Operating standard.** For topping-cycle cogeneration facilities, the useful thermal energy output of the facility during any calendar year can be no less than 5% of the total energy output, but for bottoming-cycle cogeneration facilities, no operating standard was prescribed.
- **Efficiency standard.** Efficiency standards for topping cycles are as follows:
  - If useful thermal energy is in the range of 5 to 15%, then the sum of the useful e°lectric output plus half of the useful thermal output must be greater than or equal to 42.5% of the total energy input; or
  - If useful thermal energy is greater than 15%, then the sum of the useful electric output plus half of the useful thermal output must be greater than or equal to 45% of the total energy input.

Although there is no minimum thermal output required for a bottoming cycle, the annual useful power output must be at least 45% of the energy input from the fuel used for supplementary firing to the thermal energy cycle before it enters the electricity generating cycle. Individual states are allowed to require greater thermal utilization levels. In many cases, higher total efficiency or waste heat utilization levels are required by states to encourage greater overall efficiency.

QF Certification

QF certification can be accomplished by either self-certification or FERC certification:

- To self-certify as a cogenerator, the owner or operator must provide FERC with an application that includes basic information describing the facility, the energy sources, the capacity, and the percentage of ownership. It must also include sufficient information to ensure that all applicable operating and efficiency standards are met.
- To attain FERC certification, the application must also include a detailed description of the system, the initial date of installation, and a notice for publication in the Federal Register.

Within 90 days of filing an application, FERC will issue an order permitting or denying the applications or setting the matter for hearing. If no order is issued within this period, certification is deemed to be granted. The advantage of self-certification is that it is a faster and easier process. The disadvantage is that the certification status may be less secure since FERC has not firmly endorsed it. As a result, lending institutions may be more reluctant to provide project financing and local utilities might consider challenging the validity of the certification.

Types of Electric Cogeneration Applications

Five very general categories of electric cogeneration applications are outlined below (and summarized in Table 23-1) in order of decreasing size.

1. **Cogeneration projects of several hundred MW at large facilities that sell either all or a large percentage of their electric output to an electric utility or other party**. These can be facility-owned plants in which the thermal loads support more generation than is required in-house, or third-party-owned systems that sell steam, electricity, and/or other energy products to the host facility. Third-party IPPs are often interested in finding a host for thermal output in order to meet the PURPA QF requirements or to improve project economic performance.
2. **Intermediate-to-large industrial, institutional, or commercial facilities in which all, or the majority, of the electric and thermal energy produced is consumed by the facility**. Often, the local host facility sells a portion of the electric output through the utility system or purchases a portion of their required power to balance electric and thermal loads. Host facilities may sell steam or chilled water to a neighboring facility and, where allowable, sell electricity to a neighboring facility or more distant facility through retail wheeling. Third-party ownership or some type of performance contracting is common in this category.

   Cogenerators in these first two categories often do not meet the FERC QF certification efficiency standards because their electric output exceeds that allowed by the available thermal load. Various types of prime movers are applied in these two categories. When high-pressure steam requirements are very high, gas turbines and steam turbines are commonly applied, either individually or in combined-cycle arrangements.
3. **Intermediate and large institutional and commercial facilities with lower thermal loads that purchase a significant part of their electricity from the utility.** Some of these facilities may also export a portion of their on-site generation output for sale at different times of the day. These facilities often select cogeneration systems with a high ratio of electric to thermal output, notably reciprocating engine-driven systems. To be economical, these cogeneration systems must be well matched to the facility's base thermal loads. In some cases, peak-shaving generators that do not use heat recovery are integrated with baseload cogeneration units.
4. **Small and intermediate facilities with cogeneration units that are small relative to their overall electric load.** In these facilities, systems are usually designed strictly around thermal load characteristics. In most cases, system capacity is limited to the baseload thermal energy requirement and electric generation capacity is much lower than the facility's requirements. These facilities import the bulk of their electricity from the utility grid.
5. **Small commercial facilities that produce more power than they need in some or all periods and export power under some type of standard offer or net billing arrangement.** In some states, regulations are in place that establish set prices for exported electricity from small units, usually 50 to 100 kW. Net billing arrangements are sometimes set up so that all power exported simply turns the meter backwards. These cogeneration systems, which are usually pack-

Table 23-1 Summary Table of Cogeneration Applications

Typical Characteristics	Type 1	Type 2	Type 3	Type 4	Type 5
Equipment size	>100MW	10–100MW	<20MW	<5MW	<100kW
Host use of electric output	Minor	Mixed	Major	Major	Major
Primary use of thermal output by host	Major/ Mixed	Major/ Mixed	Major	All	All
Equipment ownership	3rd Party	3rd Party/ Self	Self/ 3rd Party	Self/ 3rd Party	Self

aged units, are operated continuously to match the facility's thermal energy baseload, generally in the form of hot water. During night or weekend periods, the facility's electricity requirements may be extremely low. The ability to automatically export power may allow these units to operate economically during off-peak electric periods, even though the sales value of generated power is low.

Modular Cogeneration Systems

Modular, or packaged, cogeneration systems can be integrated into a central plant or serve individual thermal end uses at a facility. Reciprocating engines, micro gas turbines, and small back-pressure steam turbines are often installed to match a specific thermal load. Systems ranging from about 30 kW up to several hundred kW are commonly used to provide domestic hot water or boiler feedwater heating, for example. Figure 23-1 shows a factory-assembled gas-fired reciprocating engine cogeneration module. This module features an internal control and switchgear panel, a heat recovery hot water boiler, and an absorption chiller/heater.

There are several advantages to packaged systems. Capital costs are reduced and, in many cases, operation and maintenance are simplified. Standardized production and simplified installation requirements reduce initial costs. Modules can be inspected and tested in advance, and installation time is minimized. Electric interconnection is standardized and may be pre-approved by the electric utility and microprocessor controls can be integral to the system. With reciprocating engine-generator sets, the heat recovery and rejection system can be prepackaged as well.

A packaged 60 kW reciprocating engine-driven unit, for example, can produce about 400 MBtu/h (422 MJ/h) of hot water at up to 220°F (104°C). The unit would typically operate continuously, displacing a small portion of the facility's electric and thermal load. Typical installed costs range from $3,000 to $4,000 per kW, with smaller units at the higher end of the cost range. Installation cost and operations and maintenance costs per kW can be lower for larger systems.

Small packaged steam turbine generators find application at facilities with high-pressure steam distribution systems. For example, it may be cost-effective to install back-pressure or extraction steam turbines at the point of use of lower pressure steam. In such cases, the turbine produces useful power while functioning as a pressure reducing station. While this strategy is more commonly used for mechanical drive service applications, it can also be effective with electric generation applications.

District Heating and Cooling Systems

District heating and cooling (DHC) is the distribution of thermal energy from a central-source energy center to consumers in the surrounding area. DHC systems are used to serve large campus-type facilities, airports, and other multi-building facilities. They are also used to serve retail markets with sales to individual buildings

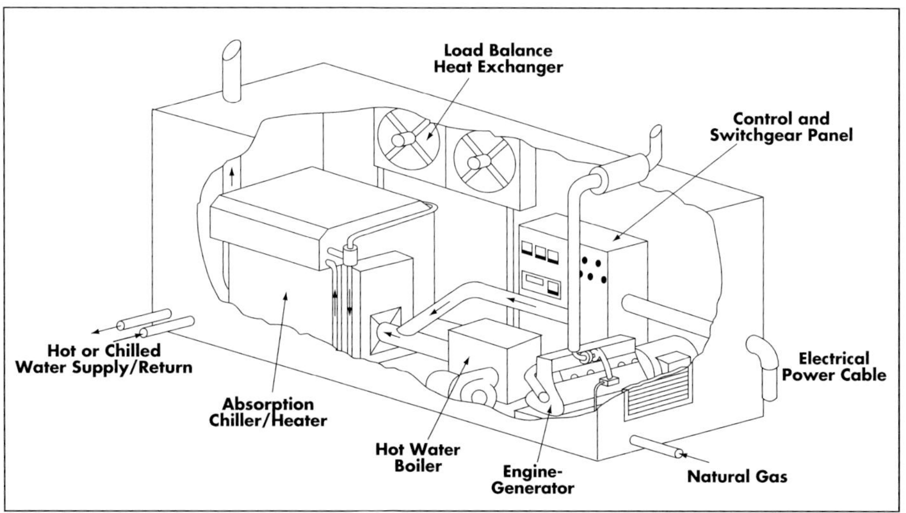

Fig. 23-1 Factory-Assembled Cogeneration Module with Integral Controls and Heat Recovery Powered Chiller/Heater.

or complexes in metropolitan areas and even suburban residential areas. In a district system, the energy center produces steam or hot water and chilled water, which is piped to buildings distributed around the site.

District heating has been used since the 1880s, but district cooling did not become popular until the 1940s and 50s. Prior to this, typical once-through cooling systems used well water pumped from large aquifers below the city to meet cooling needs. Since many district systems still only provide steam, facilities may use purchased steam to power turbine-driven or absorption chillers.

DHC systems have an efficiency advantage due to economies of scale. They also have a cost advantage in the ability to purchase low-cost fuel, and they offer high reliability, sometimes in excess of 99%. However, since DHC systems may replace small, uncontrolled sources of air pollution with one fully controlled source, they may accrue additional costs in meeting applicable environmental standards. A further disadvantage is the losses associated with the transmission of energy from the energy center to remote customers, particularly where older, direct burial systems are used. In some cases, makeup water (and water treatment) requirements and their costs can be substantial.

Electric cogeneration can be effectively applied in DHC systems as described above under Category 1, where the DHC has numerous host facilities for its thermal output. Generated power can be used for energy center mechanical drive services, such as chillers, pumps, and fans, and, in single-entity campuses, electricity can often be distributed without violating utility franchise rights or power sales contracts. Alternatively, DHC systems can potentially be a low-cost power producer and, therefore, sell power to the local utility or wheel it to another buyer in a competitive power sales market.

Figure 23-2 shows a cogeneration retrofit to an existing thermal production plant (heat pumps and gas- and oil-fired boilers) that serves a district system. Electricity generated in the plant is used for operating the heat pumps, while the surplus is exported to the external grid. The cogeneration system features an ABB STAL GT10 gas turbine equipped with a dry low-NO_X burner. Electrical power output is 21.9 MW with a simple-cycle electrical efficiency of 32% (LHV basis). Thermal output is about 129 MMBtuh (136,000 MJ/h), bringing the total system thermal fuel efficiency up to 86.8% (LHV basis). Achieved NO_X emissions levels are below 25 ppmv (corrected to 15% O_2, dry, volumetric basis).

To avoid risk, minimize capital investment, reduce maintenance, and save space, many small- and intermediate-sized facilities choose to connect with a district system. To avoid becoming a captive market to a non-regulated seller, they may retain their ability to cost-effectively install their own systems. In many cases, customers have existing equipment that can be recommissioned if needed. Otherwise, to avoid vulnerability to price increases, many buyers try to establish a long-term contract prior to committing to a district system.

Matching Capacity With Loads

Electric cogeneration is most frequently applied in baseload or nearly baseloaded operations. Key elements that make on-site electric cogeneration applications economical are heat recovery, high load factor, and high prevailing utility energy rates. The objective of baseloaded cogeneration is to eliminate as many purchased kWh as possible or sell as much power as possible, while maximizing the use of recovered heat.

Baseload cogeneration systems that operate continuously or for most of the year tend to require high thermal fuel efficiency, as well as equipment durability, operating and maintenance efficiency, and strict emissions control. Effective use of recovered heat is a critical element in achieving overall thermal fuel efficiency. In most northern climate areas, a facility's thermal requirements are usually substan-

Fig. 23-2 Cogeneration Plant Retrofit to Existing Thermal Production Plant Serving DHC System. Source: ABB STAL

tially greater in winter than in summer. Consequently, most systems are sized based on the summer load. In some cases, facilities apply steam turbine-driven chillers or steam absorption chillers to balance annual loads.

Annual load factor, thermal requirements, and power-to-heat ratio are among the most important factors affecting cogeneration potential. Load factor refers to the number of equivalent full-load hours per year that the equipment operates versus the total number of hours in a year and, therefore, the amount of cost savings that can be generated. The **power-to-heat ratio** is the energy used as power, in Btu or kJ (i.e., in kW times 3,413 Btu/kW), divided by the energy used as heat (in Btu or kJ).

Figure 23-3 presents a sample collection of facilities showing power-to-heat ratio versus process temperature requirement. A high power-to-heat ratio indicates suitability for a system with a high simple-cycle thermal efficiency. For example, the combination of low process temperature requirement of 200°F (93°C) and high power-to-heat ratio, 0.6 and 1.7 for the primary metals plant and chemical plant, respectively, are good matches for a reciprocating engine-driven system with high simple-cycle thermal efficiency.

On the other hand, a low power-to-heat ratio indicates suitability for a system with lower simple-cycle thermal efficiency or an application that features exportation of on-site generated power. For example, the two chemical plants with high process temperature requirements and power-to-heat ratios of about 0.1 are within the reasonable supplementary firing range of a gas turbine. The two chemical plants with power-to-heat ratios of about 0.05 are outside the normal capacity for a gas turbine. This type of screening would indicate that export of power might be appropriate for this application or that a back-pressure steam turbine, with relatively low simple-cycle thermal efficiency might be a good match.

A very high power-to-heat ratio, such as that shown for the chemical plant with a ratio of 3.5, would indicate limited cogeneration application potential. Combined-cycle systems, peak-shaving systems, or cogeneration systems that use only partial heat recovery would be appropriate considerations.

The following alternative sizing arrangements are often considered:

1. **Sizing based on minimum internal thermal and electric loads.** All offset electricity is valued at the facility's retail price, which is generally higher than the export sales prices, and the system is baseloaded for maximum run time at full load. As shown in Figure 23-4, all electricity requirements exceeding the baseload are met with purchased power. These plants achieve typical overall thermal efficiencies in the range of 70 to 85%. When displaced boiler losses are considered, the actual fuel-chargeable-to-power (FCP) will be even lower. Most utilities require that the facility purchase a minimum amount of electric power at all times if the facility does not have a contract to export power. This is to ensure that no power is exported if the facility's internal load drops suddenly and the on-site generator is slow to react and lower the level of generation. For example, California Utilities Rule 21, which governs the import and export of power for on-site generation, requires that the on-site generator be tripped if the imported amount of utility power drops below 5% of the generator capacity.
2. **Sizing based on thermal load and selling excess electrical output.** In these facilities, the thermal baseload can support thermally efficient generation of more

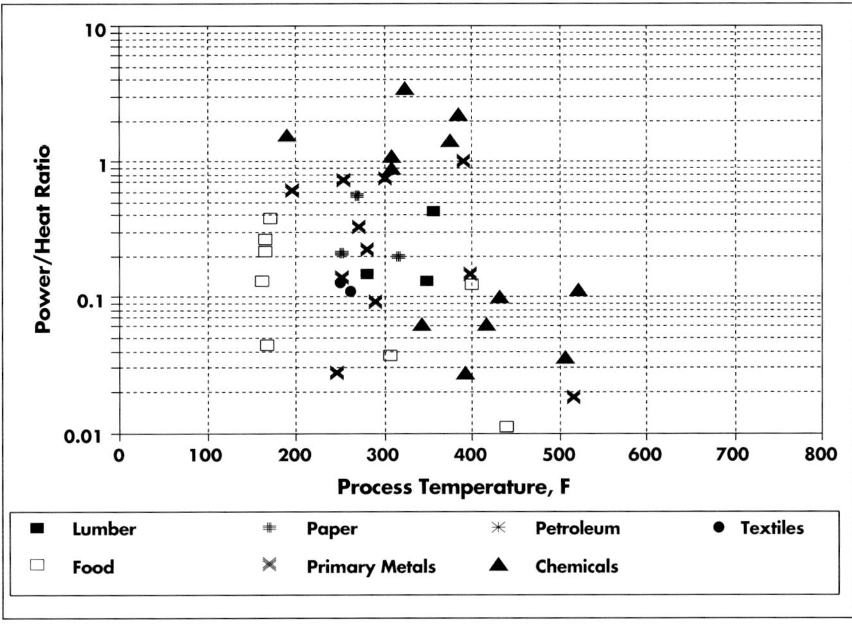

Fig. 23-3 Industrial Load Characteristics Showing Power-to-Heat Ratio vs. Process Temperature in °F. Source: Cogen Designs, Inc.

power than the facility requires. The larger capacity system generally results in lower capital and maintenance costs per installed kW and greater thermal fuel efficiency. This benefit is balanced by the lower value of electric output resulting from the portion that is exported. An example is shown in Figure 23-5. In this case, the cogeneration system is matched to the base-load thermal requirements. Excess power is produced and exported from midnight until 8 a.m., while purchased power is required during most other hours.

3. **Sizing to maximize electric production.** In some cases, these systems are designed to provide just enough thermal output to process applications to stay within the QF thermal use requirements. These are often combined-cycle plants with some type of fixed or variable steam extraction pressure and flow rate.

There are many tradeoffs for a developer choosing between becoming a QF cogenerator (and using or selling thermal energy) and constructing a non-cogeneration combined-cycle plant on an independent site. Where a thermal host is not needed, plants can be sited in better locations relative to land cost, permitting, and electric sales opportunities. Currently available modular combined-cycle power plants with capacities up to a few hundred MW can achieve thermal fuel efficiencies in excess of 55% (LHV basis), making them an attractive alternative to cogeneration. The cogeneration-cycle systems, however, still maintain a thermal fuel efficiency advantage, which improves resource use and the ability to control regulated environmental emissions.

Figure 23-6 depicts the operation of an electric cogeneration system with a full-load capacity that can meet 75% of the facility's peak electric demand. The system operates at full capacity for 10 hours per day on weekdays and tracks the facility's demand for 6 hours per day. The system is shut down overnight due to the combination of low electricity cost in the off-peak period and low thermal requirements. This strategy of dynamic operation depending upon hourly economic considerations is often referred to as economic dispatch.

Internal-Use Generation vs. Power Exportation

Internal-use electric generation is the simplest and most economical generation strategy. Systems designed for sell-back or wheeling involve additional engineering, economic, regulatory, and legal issues. Power exportation,

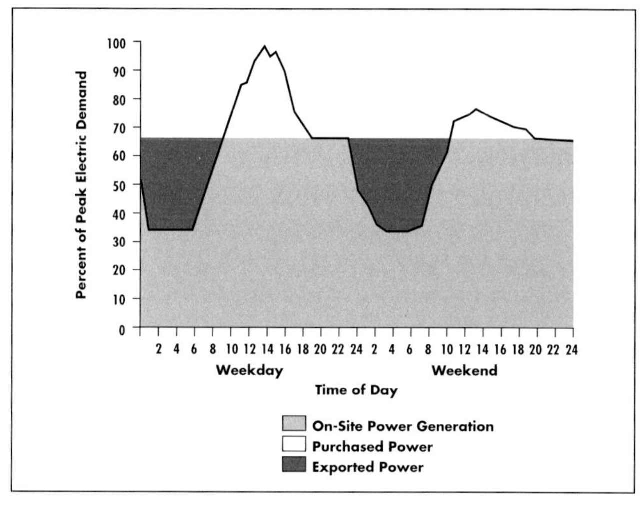

Fig. 23-5 Cogeneration System Matched to Thermal Baseload Requirement.

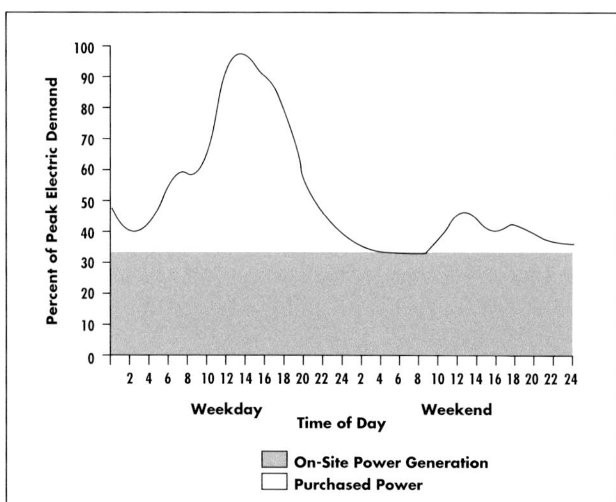

Fig. 23-4 On-Site Power Production vs. Purchased Power for Electrically Baseloaded Plant.

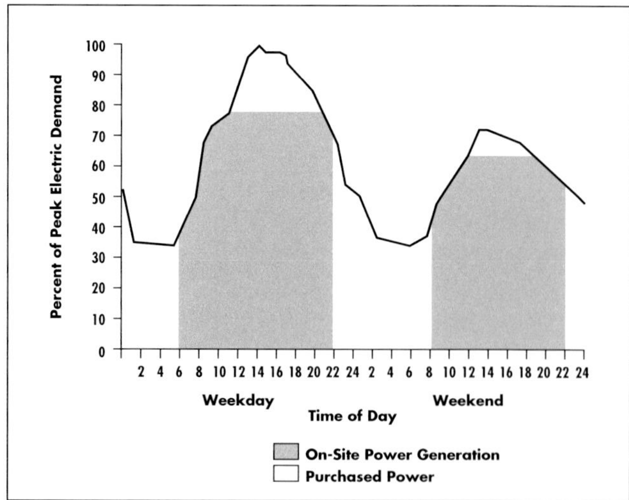

Fig. 23-6 Economically Dispatched Cogeneration System Operation.

however, is integral to many large generation projects. It is fundamental to IPPs that produce power solely for the grid. It is also an important option for cogenerators with sufficient thermal loads to support power generation in excess of their own requirements. The following items must be considered:

- Many utilities have set low values for buy-back power. Even with retail wheeling options, transmission and other costs reduce the value of exported power versus the retail value of internally used power. If power must be wheeled over several utility systems to reach a buyer, the combined wheeling charges could be cost-prohibitive.
- Engineering, operational, system protection, and monitoring requirements are greater when power is exported. Utilities often require expensive and time consuming impact studies to evaluate if the export of power will have a detrimental effect on the utility electrical distribution system. The installed cost per kW of a generation system designed for exporting power is generally greater than that of a system of similar capacity designed for in-house use only because of utility electrical interconnect requirements.
- Interconnection and power sales contract negotiations, legal matters, and other associated activities represent non-value-added costs that may delay projects and make smaller projects uneconomical.

Despite these obstacles, power exportation continues to grow, serving the interest of individual business, the public, and many utilities. Access to wheeling markets can allow self-generators to exact a higher price for exported power than they would if the local utility were the only potential buyer. The long-term logic for efficient, economic, and environmentally sound integrated resource planning is to match thermal loads with power production. This is not a hard and fast rule, however, since, in some cases, a larger facility without cogeneration can still achieve cost-effective performance levels.

As a simple guideline in the near term, smaller projects are most economical when sized for in-house usage only, unless encouraged by the state and local utility. Adding a few hundred kW to a system to export power may simply not be worth the cost and effort. Economies of scale for projects designed to export large quantities of power are usually needed to absorb the added complications and cost.

Facilities with base steam loads far in excess of their corresponding electric requirements may find power sales opportunities via wholesale or retail wheeling extremely beneficial, particularly if the power is wheeled, with modest transmission costs, to a nearby retail customer. Abundant power sales opportunities, however, will encourage competition, which will depress retail prices. Large cogenerators may have the opportunity to be low cost producers if they are fuel efficient and if they can minimize capital, maintenance, and environmental related costs.

Key elements to consider when assessing the likelihood of establishing a successful power sales agreement include:

- How are avoided costs established by the potential buyer, and what are they? Is the buyer capacity constrained, and is capacity value considered in contract costs? Is a competitive bidding procedure required for the generation project or by the potential buyer?
- Is retail wheeling an available option? Is there a need for power on the local grid or at some location that will be accessible? Is there available transmission service capacity and, if so, what are the rates for firm and non-firm service?
- What is the current criteria attainment status under the Clean Air Act Amendments and what impact will permitting and emissions control have on project costs? Are emissions reduction benefits reflected in the price paid for power by the purchasing utility?
- What are the short- and long-term market price projections for the region and local area? Is it possible to enter into a long-term firm sales contract? Is it possible to secure a long-term fuel purchase contract? Can fuel purchase escalations be indexed to power sales revenues?

There are three basic arrangements for power exportation:

1. **Sale of excess power only.** Local generators produce a fixed output and either purchase power or export power, depending on the level of varying facility loads. Non-firm sales generally pay much less than firm contracts, but utilities may still be required to purchase all power at avoided costs. In some cases, the ability to sell excess power can improve cogeneration system load factor and help project economics.
2. **Sale of a contracted amount of power.** The facility takes on an obligation to provide a specified amount of power to the grid. This is usually done at facilities with fairly constant loads that are significantly below generating capacity. The resulting firm sales contract

is more attractive than a non-firm contract. In some cases, purchased power may be used to meet demand peaks of limited duration.

3. **Sell-all/buy-all.** A facility enters into an agreement with the utility or a third party to export all of the power generated on-site and purchase all of the power required by the facility. This arrangement would occur when the value of exported power is greater than the price of purchased power, or when the host facility only has a marginal electricity requirement compared with the amount it can generate.

A facility that can create a favorable load factor and reduce the average price of purchased electricity to a low level may command a higher price on the export market by contracting for full output sale rather than varying excess output sale. Because avoided costs vary widely among utilities, it may be lucrative to buy from one utility and sell to another. The local utility can benefit from a partnership arrangement or from the combined revenues of selling power to the facility and of transmission charges for wheeling.

One additional consideration in exporting all of the power generated on-site is the loss of the power factor improvement that a synchronous generator can provide. In the sell-all/buy-all arrangement, the generator injects valuable kilovolt-amperes reactive (kVARs) into the utility system. This improves the utility network power factor, but is of no benefit to the host facility. A facility that is billed in kVA, or is subject to a power factor penalty, thus has added incentive to use power generated in-house. The trade-off of such benefits are items that may be negotiated as part of a power sales agreement.

Cogeneration System Case Study

Pictured in Figures 23-7 and 23-8 is the United Technologies Pratt and Whitney Plant in East Hartford, CT, that features an FT8 26 MW gas turbine cogeneration system operating on natural gas, with aviation kerosene as the backup fuel. At sea level, without inlet or

Fig. 23-7 Aerial View of 26 MW Gas Turbine Cogeneration Plant. Source: United Technologies Turbo Power Division

Fig. 23-8 View of Rooftop Cogeneration Plant Installation. Source: United Technologies Turbo Power Division

outlet losses, and without water or steam for NO_X emissions control, the single unit is rated at 25,420 kW (ISO base) and 8,950 Btu (9,442 kJ) per kWh heat rate (LHV basis) for a 38.1% simple-cycle efficiency. This includes an electric generator efficiency of 97.3%.

A 1,200 hp (900 kW) reciprocating compressor elevates 35 psi (2.4 bar) gas, supplied to the plant to the required 435 psi (30 bar) fuel gas pressure. This system features a 31 MVA, 13,800 volt, enclosed water-to-air-cooled electric generator and uses a compressed air starting system. The plant is equipped with a two-level emissions control system: water injection and selective catalytic reduction (SCR). The SCR module is located in the HRSG as required to match catalyst operating temperature requirements. Stack emissions are monitored by a continuous emissions monitoring (CEM) system and ammonia injection in the SCR is automatically controlled based on actual turbine exhaust. The total cogeneration plant occupies about 14,000 ft^2 (1,300 m^2). Figure 23-9 is a schematic drawing of the installed FT8 cogeneration system.

Figures 23-10 and 23-11 are curves of annual capacity demand versus total hours of occurrence for steam and electricity. Figure 23-12 shows the cogeneration plant fit with the annual steam duration curve. The plant requires 225 psig (16.5 bar) superheated steam at 460°F (238°C) to match the conditions in their main distribution header. Steam is generated in the HRSG at 250 psig (18.3 bar) at 510°F (266°C) and desuperheated to 465°F (241°C).

Notice that the steam capacity of 85,000 lbm/h (38,500 kg/h) almost never matches the facility's actual steam demand. Conventional boiler operation is required for more than half of annual hours and excess capacity is available for the remainder of the year. Still, this is a reasonable fit because it features relatively little excess steam capacity.

Figure 23-13 shows the cogeneration plant fit with the electrical load duration curve. The facility's electric demand varies from a peak of 45 MW on a typical winter weekday to a minimum of 17 MW on a typical spring

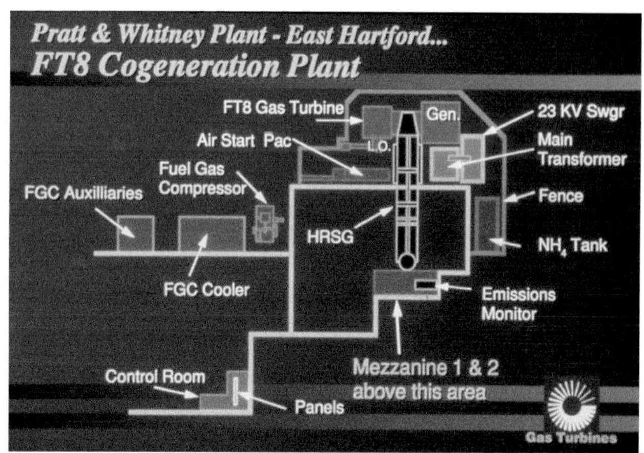

Fig. 23-9 Schematic Drawing of Installed Gas Turbine Cogeneration System. Source: United Technologies Turbo Power Division

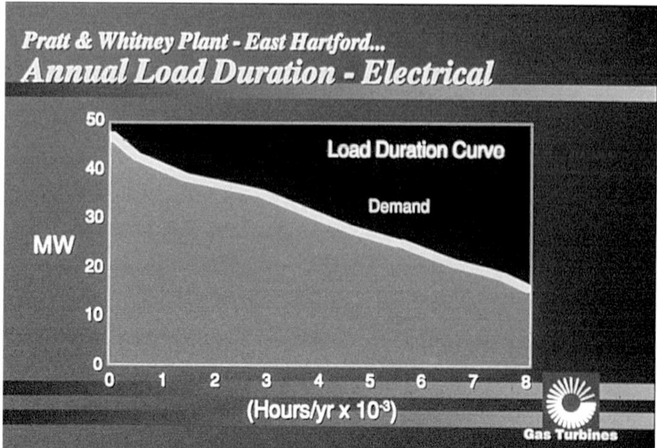

Fig. 23-11 Cogeneration System Annual Electric Duration Curve. Source: United Technologies Turbo Power Division

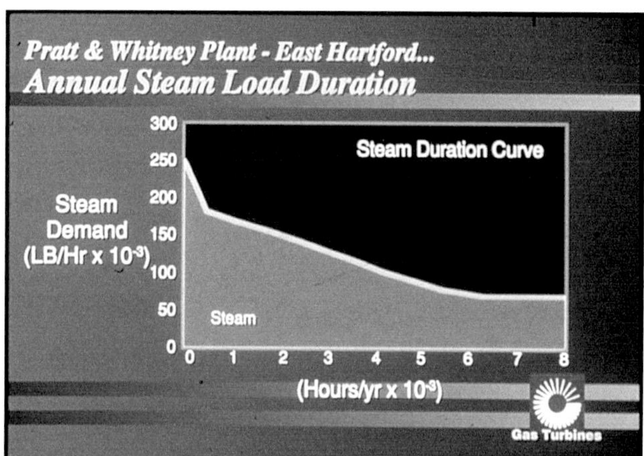

Fig. 23-10 Cogeneration System Annual Steam Duration Curve. Source: United Technologies Turbo Power Division

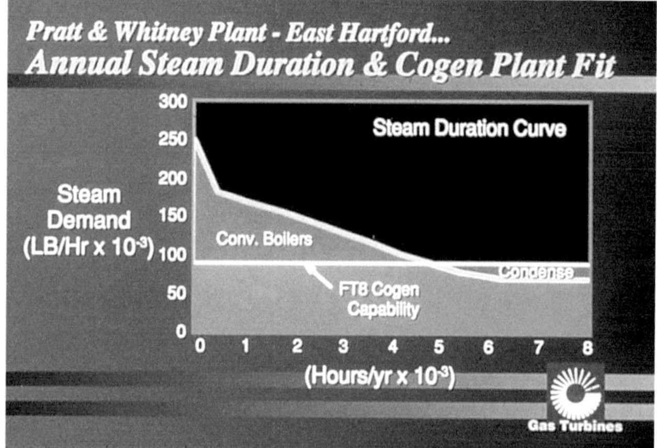

Fig. 23-12 Annual Steam Duration and Cogeneration System Fit. Source: United Technologies Turbo Power Division

weekend night. Notice again that the cogeneration system electrical capacity almost never matches the electric demand, with electricity purchases required for more than half of annual hours and excess capacity available for power sales during some of the year. There is relatively little excess capacity, which must be sold at a very low price, and most of the facility's purchased electricity requirements are eliminated. A larger unit would allow increased power sales, a configuration that might be desirable in the future, with access to retail wheeling markets.

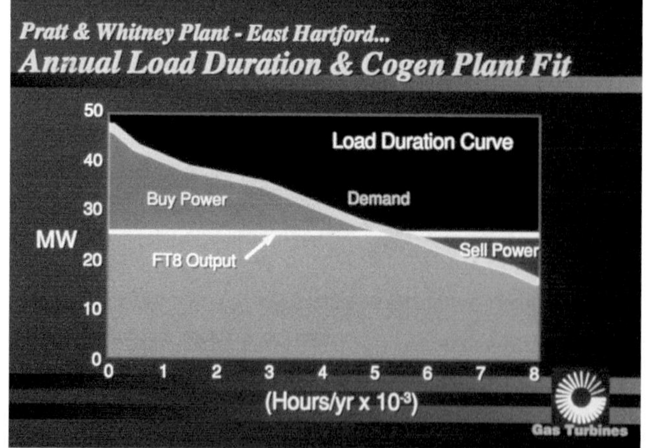

Fig. 23-13 Annual Electric Load Duration and Cogeneration System Fit. Source: United Technologies Turbo Power Division

Peak-Shaving/Load Shaping Generation

Most facilities experience spikes in their electricity demand because of occupancy, seasonal electric heating or cooling loads, or production requirements. Peak demand charges and time-of-use (TOU) rates can make the cost of relatively few peak period kWh extremely expensive.

Facilities often develop strategies designed to reduce demand peaks. These include shutting off equipment during peak-setting time, shifting entire operations to off-peak periods, the use of gas- or steam-driven cooling, thermal storage during off-peak periods, or battery electric storage.

On-site electric generation is another potentially cost-effective solution. This may be applied independently or in conjunction with other load shedding applications. Figure 23-14 shows an example of a peak-shaving generation load profile. In this case, electric power production represents a very small part of the facility's usage requirements, but about 40% of its billable peak demand.

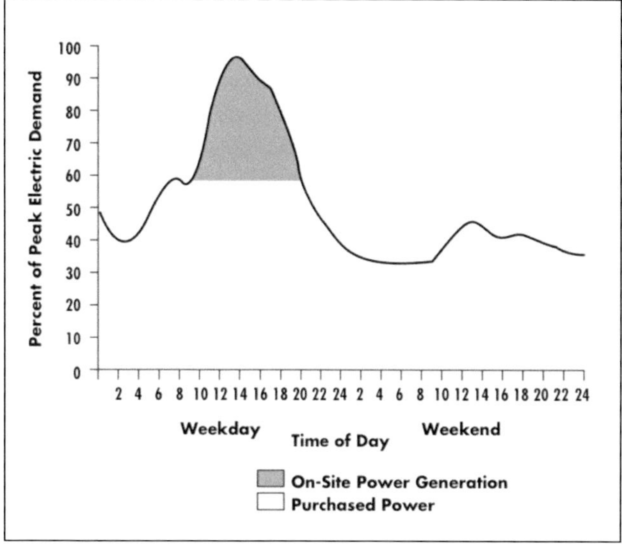

Fig. 23-14 Electric Peak-Shaving Generation Profile.

Typically, with peak-shaving generation systems, low first-cost is emphasized more than thermal fuel efficiency. In most cases, heat recovery is impractical due to insufficient thermal load or because of limited hours of operation. In some cases, however, a minimal amount of heat recovery is driven by the need to satisfy QF requirements, thereby ensuring that the local electric utility must continue to provide non-discriminatory services.

In some cases, facilities may also be able to use the peak-shaving generation equipment as a backup power source for critical electric equipment. Figure 23-15 shows two skid-mounted 12-cylinder dual-fuel reciprocating engine-generator sets used to peak shave and provide emergency stand-by power for a hospital.

Fig. 23-15 Two Skidded Dual-Fuel Reciprocating Engine Generator Sets used to Peak Shave and Provide Emergency Stand-by Power for a Hospital. Source: Fairbanks Morse Engine Division

Many on-site power generation systems include, in addition to baseloaded electric cogeneration systems, low capital cost electric generator sets that do not use heat recovery and are not designed for continuous operation. These units may be used as backup capacity to the primary electric cogeneration systems and may also, in some cases, be used for electric peak-shaving.

Standby Generator Applications

Many electric utilities and ISOs have peak-shaving, or interruptible service, programs in which facilities are encouraged to install standby generators. This is usually done with an installation incentive or a flat annual fee based on capacity. Customers may generate power during utility-designated periods, or use other methods to reduce their own power requirements. Many utility programs have different levels of payment or compensation based on availability.

Utility-controlled peak-shaving can allow utilities to improve load factor and reduce capacity requirements with minimal loss of sales volumes. They also sometimes serve as competitive alternatives to strategies designed by customers to reduce peak demand charges at their own facilities. Commonly, utility peak-shaving programs result in far fewer generator run-hours than in-house peak-shaving programs, meaning there is less loss of sales. In some cases, these utility controlled peak-shaving programs are packaged with other competitive energy rate offerings to provide sufficient incentive for a facility to forego an alternative, more extensive on-site self-generation option.

Where permitting regulations allow and expected run time is very low, emergency generators may be used in utility peak-shaving programs. Generator sets rated for emergency duty are less expensive than prime power units and they may either already exist on-site or be required anyway. However, in many states, on-site generation equipment may only be exempt from air emissions control regulations as long as the sole purpose, without exception, is emergency service.

An emerging application of standby generation systems is to support interruptible power service purchased on the retail wheeling market. A facility with dispatchable on-site generation capacity can lessen or eliminate the need to pay for firm purchased power capacity. This enables the facility to purchase low-cost commodity power, and, during periods when capacity is not available or when costs are high, the on-site system can be used. In the evolving deregulated power market, access to dispatchable electric generation capacity is becoming an increasingly valuable asset.

Interconnecting Peak-Shaving Generators

Often, when a facility decides to install standby generation capacity, it also considers expanding the project scope to include peak-shaving generation. Added capital cost may be incurred in selecting a more durable or fuel-efficient system when both interconnected operation and isolated standby duty are required.

Standby generating systems commonly operate at 480 volts and are coupled to critical loads via a transfer switch that locks out the utility feeder. This causes an interruption of power to the critical loads, since one power source must be dropped out before the other is picked up. Furthermore, this type of internal electrical distribution usually limits the ability of the emergency generator to peak shave load equivalent to its capacity since the connected loads on the emergency buss are typically much less than the generator output. In cases where interruption cannot be tolerated, facilities may have multiple feeders from the utility network. In these cases, where required interconnection is in place, the system can also be used for demand limiting purposes. To ensure required standby availability, the additional duty may warrant consideration of a higher quality, more durable engine-generator system.

If the generator cannot support the entire facility load, an operational strategy must be in place to allow the facility to restart or continue operation after the main breaker has opened. A typical sequence is to open the generator circuit breaker, shed non-essential loads, and then reconnect critical loads back to the generator. With the use of high-speed relays and a suitable plan, this process may be accomplished without interrupting generator output. Alternatively the generator circuit breaker can be tied directly into the feeder that supplies power for critical loads. The utility tie-breaker then will function as the feeder circuit breaker for non-critical loads.

Peak Shaving/Load Shaping Case Study

In 2000, a military facility in Georgia was on an electric rate with Georgia Power (GP), under which the electricity costs were calculated from demand and energy charges that are determined by a customer base load (CBL). The CBL is an hour-by-hour demand load profile at the facility's substation, metered at half-hour intervals. The facility's CBL was originally established by contract from the 1994 demand profile and adjusted once for a lighting efficiency upgrade project. The full value of the CBL is charged at a conventional TOU rate, regardless of actual usage. All energy consumed above the CBL is billed under an RTP rate. If demand falls below the CBL, the bill is credited at the RTP rate for that time interval. The RTP, a function of GP's generation and power purchase contracts, is calculated continually and announced an "hour ahead" of taking effect.

Since converting to the CBL-based rate in 1994, peak summer demand at the facility has declined from 27.5 to 24.9 MW, while the average peak winter (non-cooling) demand has remained relatively constant. Since, most of the time, RTP rates are less than half the TOU rate, there is substantial benefit from reducing the CBL and thereby shifting kWh from TOU to RTP.

The facility leases nine 1.5 MW packaged 700 rpm EMD Diesel engine generator sets built and previously used by the military during the 1960s. Three units are wired to one control trailer to form an independently operating 4.5 MW plant. The plant capacity is presently three of those sets, for a total of 13.5 MW. The facility pays a fixed annual lease cost plus a variable cost per engine operating hour that includes accrual for major overhaul at 16,000 hours. While these engine generator sets are thermally inefficient by today's standards and also have fairly high regulated air pollutant emissions levels, they are renowned for their reliability, a critical element for the facility.

The generators have historically run 200 to 300 hours per year, all on summer afternoons and evenings. Most of the revenue from this operation came from complying with interruptible service (IS) dispatch requirements from GP, while most of the operating hours occurred during peak shaving when RTP costs were high. The IS rate rider requires the facility to reduce demand to their IS threshold within 30 minutes of notice from GP. The levels and the potential for credit are agreed upon in advance under a 3-year contract with GP. The IS load is defined as the difference between the maximum CBL demand during the IS peak period and the kW amount of the contracted IS threshold. Up until 2001, the IS threshold was set at 12.5 MW with the IS peak established at 23.8 MW. Credit is reduced if the CBL load profile does not exceed the required 600 hours use of demand (or HUD, defined as the total monthly kWh divided by the peak demand) in the contracted CBL profile. When the HUD is below 600 hours, the incremental IS credit is reduced by a load factor. Additionally, excessive failure to respond to curtailments within the allotted dispatch time may forfeit the credit for the year. In addition to IS-related savings, the units have been operated whenever the RTP rose above $0.08/kWh. This is the approximate threshold at which the operating costs, inclusive of O&M accruals, are lower than the RTP market price for generating power.

Peak Shaving/Load Shaping Strategy

During 2000, the facility contracted with an energy services company (ESCo), to implement a peak shaving/load shaping program under a master Energy Savings Performance Contract (ESPC). An objective of this program was to ensure reliable long-term operation of these engine-generators as emergency backup units for the facility's critical loads. Hence, the key was to utilize savings generated by a more aggressive operating strategy to fund overhauls of the engines, switchgear, and other related auxiliary equipment and provide a more rigorous ongoing preventative maintenance program. An additional objective was to generate excess savings that could be used to leverage upgrade of other energy infrastructure equipment at the facility. Under the contract, the ESCo would implement and maintain all the measures at their own expense and be compensated by the facility out of a portion of the annual savings generated. In this manner, the facility could accomplish its mission objectives with no upfront capital investment.

The primary basis of savings for this strategy was to take advantage of the RTP component of the facility's utility bill. Under the applicable GP rate guideline, for a facility to have an RTP component, a CBL must be established. While the existing CBL provided some RTP operating benefits, a lower CBL would provide additional opportunities to generate further RTP savings. Hence, this load shaping program targeted an additional adjustment to the CBL. This required a new contract rate with GP and one year of operation to establish a new CBL. After implementation of the new rate, one year of proven operation of several load management strategies was required to reduce the peak electrical demand for calendar year 2001.

After detailed analysis of all electric rate options and load shape strategies, it was determined that it would be most advantageous to lower the CBL during the more expensive peak period, 2:00 p.m. to 7:00 p.m. summer weekdays. In addition to meeting the peak period requirements, operation would be required for all GP peak hours, 7:00 a.m. to 10:00 p.m. weekdays. Per GP's guidelines, this could be accomplished by going off of the RTP-type rate during 2001. This would remove the CBL and the facility would be charged only at the traditional TOU rate. After 2001, the facility could return to service under the RTP rate with a newly established CBL based on the 2001 actual load. If the actual load in 2001 could be driven down by on-site electric generation and other load management systems during peak periods, then the CBL could be set at these lower load levels for 2002 and beyond. This strategy would create significant recurring annual savings. It was also determined that additional savings could be achieved each year by setting more aggressive targets for the IS program and through more aggressive electric generation during RTP periods when the price was above the break-even threshold.

Three basic load management strategies were developed to achieve program objectives.

1. Run the generators and manage load aggressively during high priced summer peak hours in 2001 on the new rate tariff. This would shift kWh from the CBL price to the RTP during the critical afternoon hours in 2002 and beyond when the facility switches back to the original rate. The 2001 load, which was driven down by management of the generator plants and other load control systems (involving duty cycling of air conditioning systems and interruption of other non-critical systems) will become the new CBL beginning in 2002. Dramatic savings are achievable in that removing peak kWh from the CBL drops the average price from $0.25 to under $0.06 per kWh during selected periods.
2. Once the new CBL is set and the facility is back on the original rate tariff in 2002, the generators and chilled water storage will be tactically dispatched to

displace or shift load when the RTP is higher than the break-even threshold, which will vary depending on fuel cost. As the price of fuel drops, the generators can be run cost-effectively at lower RTP prices and vice versa.

3. Establish more aggressive IS targets. By lowering the target from 12.5 to 10.0 MW or even less, more IS capacity credits would be achieved, producing greater annual savings. Also, by reshaping the load profile, the HUD-related credit reduction would be greatly reduced or eliminated.

This strategy carried considerable risk in that once the new CBL was established, it would remain in place indefinitely. One significant outage of peak shaving capacity during a 30 minute period could result in setting a higher than planned CBL. This would result in a shortfall of the expected cost savings, which were to be used to pay for the system overhauls and other energy infrastructure improvements. Additionally, more risk was assumed by targeting the lower IS threshold since penalties for missing the target are more severe than the benefits of hitting it. To mitigate these risks, major system overhauls were designed to ensure generation system reliability. Detailed automated operating protocols were established for monitoring, control, and dispatch of the generation system and all other load shedding systems. In addition to the system automation, well-trained, highly focused system operators were required to ensure smooth, reliable operation and respond quickly to any system failure or emergency conditions.

Another challenge was to work within the facility's air permit limits, while substantially increasing generator operation during 2001. The facility is located in an EPA Attainment Area. It emits about 100 tons (91 tonnes) per year of a potential 417 tons (378 tonnes) per year of regulated pollutants. The Georgia State Air Quality Division allows for the operation of the Diesel engine-generators for 500 hours per year under the Title V Major Source Operating Permit. To ensure that the generators could run enough to meet load management goals, the facility filed for an air quality construction permit on the innovative basis of kWh generated (in lieu of the normal potential-to-emit basis). Permitted output was designed to avoid triggering Prevention of Significant Deterioration (PSD) status that would have delayed the project for months. While this provided more headroom for emissions in 2001, it would require careful management of the generator output so that the target load profile shape would be achieved without hitting the strictly enforced emissions cap allowed under the permit.

System Upgrades

Prior to commencing operation for testing and fine-tuning in May 2001 and active operation in June 2001, the ESCo established a system upgrade scope of work to improve reliability. The following improvements were made prior to the critical summer operating period:

- Overhauled and upgraded the 13.5 MW Diesel engine-generation plant.
- Installed 30,000 gallons (113,500 liters) of additional fuel storage and redundant fuel pumping capability.
- Installed backup transformers and a 1.8 MW standby generator at the facility Hospital and added a communication system to the central operating station (the Hospital on-site electric generator was also included in this load shedding strategy, so a temporary backup generator was installed to increase reliability).
- Expanded the load shedding control system to include 2.5 MW of routine load shedding circuits and several additional MW of potential load shed circuits to be used only in the event of short-term emergency outage of a significant portion of the engine-generator capacity.
- Upgraded the software, hardware, and communications of the enormous central energy management system (UMCS).
- Developed and implemented an automated operational protocol and manual backup protocol and conducted extensive training of all operators.

Results

On January 1, 2001, the facility changed to the TOU rate as a new two-meter load group consisting of the electric meter at the Hospital (located at and paid for by the facility) and the facility's main electric meter. This combined load group was then aggressively managed to shape a favorable purchased power load profile. On January 1, 2002, the facility will return to the RTP rate with the actual 2001 load profile as the new CBL.

Figure 23-16 documents the power usage for one day at the facility. It shows the hourly kW of actual demand, old and new CBL, and the on-site generation system output. Notice that the shape and area of the generation output closely matches the difference between the actual demand and the new CBL curves. Using this same basic daily load shape strategy of producing up to 8 MW on-site, 3 to 4 MW was shifted from TOU to RTP permanently. Since the old CBL was lower than the actual demand (this had been previously negotiated), new additional cost savings are only achieved based on the reduction from the old to the new CBL. Yet, the peak shaving profile, which is the effective combined impact of the on-site generation and any other load shedding, had to work against the actual higher demand level producing up to 8 MW in order to achieve the 3 to 4 MW reduction objective.

Through careful focus on the objective, the program exceeded the original target load management goal by 700 kW. This was accomplished by keeping the local electric generation and load shedding systems in operation or fully available every minute on weekdays from 7:00 a.m. to 10:00 p.m., June through September 2001. Through extensive monitoring and forward modeling of operations, the air permit limitations were managed to ensure that the engines could continue to operate as needed throughout the summer. Actual kWh generated this summer came within 2% of the permit limit.

The ESCo will now continue to operate and maintain the Diesel engine plant, a new chilled water storage system, load shedding control, and the UMCS system on a year-round basis for the duration of a 20 year contract. Long-term annual net savings from this measure will be in excess of $800,000, net of all OM&R costs. This will be sufficient to pay for all of the system upgrade work performed in 2001, as well as some additional energy infrastructure improvements desired by the facility. While there remains some ongoing risk and uncertainty about the exact long-term savings in each year due to

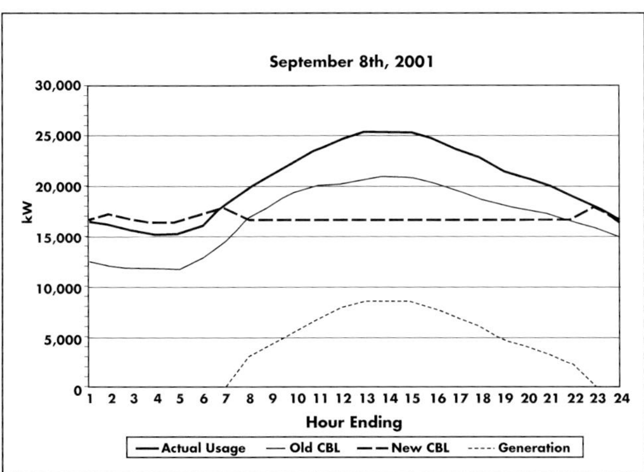

Fig. 23-16 Single-Day Power Usage Profile Showing Hourly kW of Actual Demand, Old and New CBL, and On-Site Generation System Output.

RTP and IS operation, the vast majority of savings have been assured with the achievement of the new CBL.

Electric Rate Selection Strategies

Partial requirement self-generators purchase electricity from the utility for supplemental and backup services. There are often several electric rate options available to these customers, and rate selection can have a significant impact on project economics.

The generation system may be a major factor in rate choice. A demand-based rate presents the potential for substantial savings, but also the risk of penalties for an outage at a critical time. A peak-shaving system will likely target such a rate, while a baseload cogeneration system might be better suited for a rate with a higher energy cost component. System reliability and redundancy will be important considerations.

If the local utility has reasonable QF rate options or demonstrates support of self-generators through cooperation or incentive programs, the choices of rates and system configuration are made easier. Some utilities offer customized rates equal to the project's cost of power to avoid bypass. If power is available from sellers other than the local utility via retail wheeling, prices for supplementary and standby power may be more attractive than in cases in which the local utility is the only retail seller. Even with retail wheeling options, utility rate options for T&D services remain a key factor.

The situation is more difficult if the utility offers either standard firm service rate options or a somewhat economically unattractive QF rate option. For instance, when supplementary or backup power is provided under typical demand/commodity rates, potential demand charges must be weighed against the cost of providing required reliability. This is especially critical if the facility has a ratcheted demand charge, since a failure that occurs even once during a peak demand period would negate any demand savings from the on-site generator for the remaining 11 months.

Three strategies are commonly used to mitigate the impact of supplemental electricity cost:

1. **Multiple generating units.** In a single-unit system, an outage could bring financial exposure to the full impact of demand charges and penalties. The probability of multiple units experiencing simultaneous outages is less likely, hence providing a strong measure of security. An effective strategy is to purchase utility standby rate insurance for the capacity of only one of the units, with the confidence that the prob-

abilities of more than one unit being down in a peak setting period is unlikely. Service and overhaul scheduling is also easier.

2. **Redundant generating capacity.** If, for example, an operating strategy is developed for a 6 MW load, this strategy might call for the installation of three 3-MW units. Standby insurance would likely not be required and project life would be extended because run-time is reduced on each of the units. Alternatively, redundancy can be provided by a lower cost backup unit. The prime units would then always be used except during an outage or maintenance interval when the standby generator is brought into service. In some cases, the standby units could also be operated for electric peak-shaving or export power duty.
3. **Electric load-shedding programs.** These may be automatically triggered in the event of a generator outage. Non-critical loads that can be shed to reduce the peak demand that would be incurred during a generator outage are identified. These loads can be tied into a single circuit under automatic control.

The various strategies, including purchase of standby insurance from the utility for a portion of the connected generator load (if available), take into account the probability of an outage and the potential financial exposure it represents. If demand charges are not high relative to energy (or commodity) charges, the impact of an outage will generally not be great. The project feasibility analysis should include a loss of savings resulting from the predicted number of outages.

With the availability of real-time pricing (RTP) or the ability to purchase NUG power, host facility on-site generation strategies are beginning to change. In some cases, RTP can have a detrimental effect on self-generation economics. RTP often lends itself to peak-shaving or more limited use of baseload cogeneration due to extremely low electricity costs in certain off-peak periods.

When operating on a demand sensitive rate, the facility would design the system to eliminate as many peak kW as possible. This includes the kW that may occur for only a few hours per month. With RTP, that incremental kW becomes less important; $0.50/kWh for a few kWh per month, for example, is small compared with $20/kW charged monthly for demand. Because the customer focus shifts away from elimination of peak demand charges and toward elimination of peak purchased kWh, RTP may result in the installation of a somewhat smaller generation system.

Renewable and Alternative Electric Generation Options

In addition to the three main types of prime movers previously discussed and the basic configurations (i.e., simple-cycle, cogeneration, STIG, and combined-cycle options), a range of other alternatives should be considered for electric power generation applications. Three categories of renewable and/or alternative energy options, each discussed below, are:

1. **Alternative energy sources** that can directly or indirectly provide driving energy to the traditional three prime movers (e.g., solid waste, biomass, and waste- and recovery-derived gases and geothermal energy).
2. **Alternative electric generator driver technologies** (e.g., wind and water turbines).
3. **Alternative (non-power cycle) electric power production technologies** (e.g., photovoltaics and fuel cells).

Several of these options are cost-competitive with conventional technology application options and certain others are being made cost-competitive with financial support from various federal and state agencies and local utilities. The renewable technology options, namely those that use geothermal, wind, water, and solar energy sources, have no fuel-cost component. Given the recent rise in world energy costs, following a long period of flat and even declining prices, life-cycle costs of renewables have improved relative to traditional fossil fuel-driven applications. Many states have recently required their utilities to meet renewable energy portfolio requirements and/or offer Renewable Energy Credits (RECs) to subsidize electrical power generated from renewable sources. In some states the RECs increase the value of power generated from a renewable source by as much as $0.04/kWh. Still, in some cases, this is not sufficient to overcome high initial capital and/or ongoing operations and maintenance cost premiums.

Technology applications involving solar, wind, and water resources do not produce air emissions and therefore enjoy the benefit of avoiding air permitting and emissions control expenses and, in some cases, of creating marketable air emissions credits. In more stringent air permitting locations, such as California, this can enhance the relative financial performance of these options versus fuel-burning options.

While each of these alternative technologies is limited for wide-scale application in CI&I facilities, each offers advantages, which, under the proper set of conditions, can make them life-cycle cost competitive. Moreover, increased fuel and environmental control costs for conventional fuel-burning technologies and continued advances in design for improved production effectiveness and conversion efficiency provide the expectation that these "green" technologies will continue to penetrate the energy market place at an increasing rate.

Alternative Energy Sources that Power Traditional Prime Movers

There is a wide range of low-cost or no-cost energy sources that can be used to produce high-pressure steam or synthetic (by-product) fuel gases. Refuse (mass-burn or refuse-derived), wood, and biomass, as well as coal, are fuels that can be burned to provide low-cost steam. Many of these energy sources can also be gasified to produce a wide variety of clean-burning, transportable fuel gases that can be burned in a boiler to produce steam or used directly in a combustion engine. These include coal and biomass-derived gases, plus a wide range of industrial process recovery type gases, as well as bio-gases produced from anaerobic digestion by methane-producing organisms such as sanitary landfill and sewage or sludge digester gases. This category also includes no-cost geothermal sources that can be used either directly or indirectly (through a flash evaporator) and concentrating thermal solar systems that can provide high-pressure steam to drive a steam turbine generator set. Chapter 5 provides detail on the energy value and composition of these sources and Chapter 7 provides detail on their use for steam generation or gasification.

Whereas geothermal and solar energy can be used for a variety of thermal applications, electric generation is generally limited to abundant high-temperature sources. Geothermal energy sufficient for electric generation application is limited to specific regions (e.g., the Western United States), where high-temperature sources can be economically accessed. Historically, the extraction and use of geothermal energy for electric generation systems has focused on sources that are dry or vapor-dominated and wet or liquid-dominated (partially flashed steam from hot water at or near saturation temperature). High-temperature geothermal sources are classified as being in excess of 302°F (150°C). Bottom-hole temperatures as high as 800°F (427°C) have been reported at a depth of 8,100 ft (2,469 m). Figure 23-17 is a process flow diagram of a geothermal power plant operating on a single-flash cycle. Hot water flows through a pipeline from the production well to a cyclone separator where brine and steam are separated. Brine is then sent to an injection well and returned to the geothermal reservoir. Steam moves through insulated lines to a scrubber, where condensate and impurities are removed. Steam then enters a condensing steam turbine, which operates as it would in a conventional electric generation plant application. Excess fluid is then returned to the geothermal reservoir via injection.

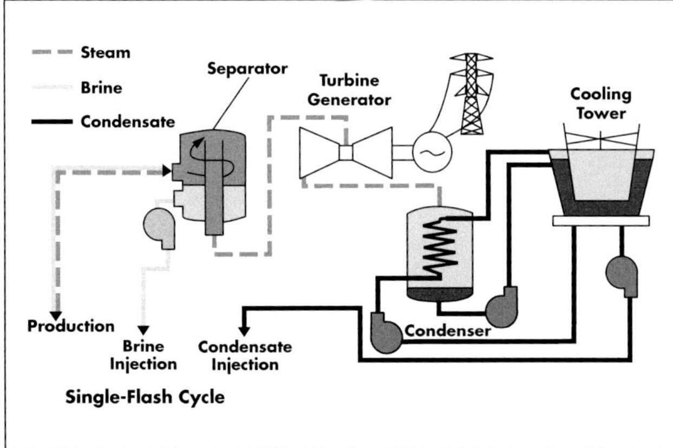

Fig. 23-17 Single-Flash Cycle Geothermal Power Plant Layout Illustration.

High-temperature solar thermal applications feature heliostat, parabolic dish, or trough type collectors that intensely concentrate solar energy, allowing for generation of extremely high fluid temperatures sufficient to produce high-pressure steam. Figure 23-18 is a process-flow diagram that is representative of most parabolic trough solar electric generation plants currently in operation. The collector field features several parallel rows of single-axis (east-to-west) tracking parabolic trough collectors aligned on a north-south horizontal axis. A heat transfer fluid (HTF) is heated as it circulates through the receiver and returns to a series of a heat exchangers in the power block where the fluid is used to generate high-pressure superheated steam. The steam is fed to a conventional reheat steam turbine-generator and then condensed and returned to the heat exchangers. Under design conditions, the plant can operate at full rated power output using solar energy alone. This can reach as high as 10 to 12 hours per day in the summer. The system is configured with an optional fuel-fired HTF heater situated in parallel with the solar field, or a fuel-fired steam boiler/reheater located in parallel with the solar heat exchangers. These can provide backup fossil fuel capability that can be used to supplement the solar output during periods of low solar radiation or provide full rated output dur-

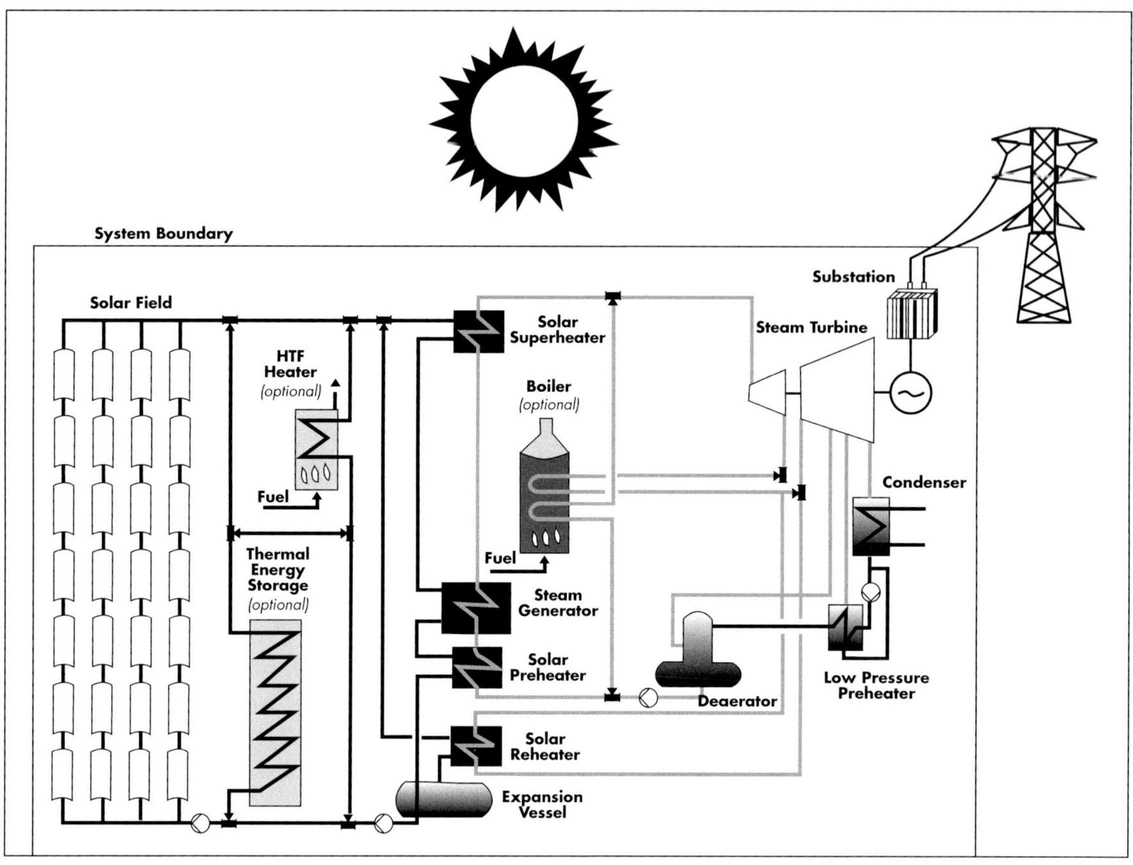

Figure 23-18 Process Flow Diagram for Parabolic Trough Solar Electric Generation Plant, with Thermal Storage and Fossil-Fuel Steam Generation Capability Options. Source: DoE/EREN

ing nighttime periods. Also shown is an optional thermal energy storage system for additional supplemental or backup energy capacity for generating high-pressure steam.

Alternative Electric Generator Driver Options

Today's wind energy conversion systems (WECS) and water (hydraulic) turbines are modern versions of traditional prime movers that date back more than a thousand years. They are also similar to today's conventional prime mover technologies (steam turbines and combustion engines) in that they generate shaft power that can drive a conventional electric generator. The main differences are that they are highly site-specific, limited to locations with sufficient wind and/or water characteristics, and they rely on cost- and pollution-free renewable resources.

Hydropower is by far the dominant renewable resource used for electricity generation, dwarfing all others by comparison. It provides nearly 20% of the world's electricity, 10% of the electricity in the United States, and the majority of the electricity in some countries, such as Canada or New Zealand. Today, most hydroelectric facilities are owned and operated by major utilities, regional power authorities, and IPPs that locate facilities at remote sites and feed electricity to the grid. Opportunities for new major project initiatives are somewhat limited by environmental and other political and economic obstacles. However, numerous opportunities remain in many of the tens of thousands of existing dams to install hydroelectric facilities, or to increase the current capacity of existing plants. These can be designed (at a significant cost premium) to minimize detrimental environmental impacts.

Electricity was once commonly produced at individual CI&I facilities using hydropower. While such applications are now far less common, there are many application opportunities for localized hydro-power plants that serve individual facilities, groups of facilities, or municipalities. These include less invasive low-head dam-type plants and run-of-the-river type systems that require little or no water impoundment and do not cause an appreciable change in river flow.

Like hydropower, WECS have moved from local mechanical service applications for pumping or grain production processes to utility or IPP electric power generation. After laying dormant for nearly one-half of the past century in the United States, WECS have experienced a series of resurgences over the past several decades and have in many respects taken hold as an accepted mainstream technology. Today, several thousand MW of capacity are being installed each year, largely in Europe and the United States, mostly in large wind farm applications located at windy, somewhat isolated, sites where their tall towers are unobtrusive and accessible to steady wind flow at sufficient speed and frequency to be cost-effective. However, they have also been effectively applied on a smaller scale at individual facility sites with exposure to favorable wind conditions.

While grid-connected systems typically require wind speeds of about 12 or 13 miles per hour (5.4 to 5.8 m/s) for operation, smaller non-grid connected electric generator applications can effectively operate at wind speeds as low as 7 or 8 miles per hour (3.1 to 3.6 m/s). Given their free energy source characteristic, WECS can be life cycle cost-competitive in today's market at locations with steady minimum wind speed. In areas where it is difficult and expensive to permit new combustion sources, WECS are further advantaged versus traditional fuel/steam-driven prime mover systems. They also can be particularly attractive in remote locations where grid connection is either costly or inaccessible and/or where fuel delivery is costly. Chapter 14 provides extensive detail on the applied wind and hydro technologies, including numerous examples of system applications. Chapter 26 provides additional detail and examples of wind and water electric generation applications.

Another somewhat unique renewable electric generator driver technology is the heat engine powered by solar thermal energy. The solar dish engine system (discussed in Chapter 7) uses an engine that converts heat to power. It functions in a manner similar to conventional engines, i.e., compressing a lower temperature fluid, heating the compressed working fluid, and then expanding it through a turbine or with a piston to produce shaft power, which can drive a traditional electric generator. The solar dish system uses a mirror array to reflect and concentrate incoming direct normal solar insolation to a receiver in order to achieve the high temperatures required to efficiently convert heat to work. This requires that the dish track the sun in two axes. The concentrated solar radiation is absorbed by the receiver and transferred to the heat engine. A number of thermodynamic cycles and working fluids can be used for such applications. To date, prototype development efforts have focused on the Stirling (reciprocating engine) and open Brayton (gas turbine) cycles. These heat engines can be supplemented with a conventional fuel burner to allow operation during cloudy weather and at night. Both engine types utilize recuperators to recover and reuse heat to improve thermal efficiency.

Alternative Electric Power Production Options

Unlike conventional electric power generation systems that operate on power cycles, photovoltaic (PV) and fuel cell systems rely on photoelectric and electrochemical reactions, respectively, to produce dc electricity. Their output can then be converted, through an inverter, to ac electricity. While still extremely capital-cost intensive, PVs have continued to decrease in production cost and increase in electric generation efficiency over the past two decades — a trend that is expected to continue. For maximum effectiveness and associated financial returns, they must be sited in locations with continuous or near-continuous daytime sunlight.

Whereas PV cells rely exclusively on renewable solar energy, which is cost and pollution free, fuel cells operate on hydrogen. While a potentially renewable resource, current technology limits hydrogen derivation to fuel-or electricity-consuming processes, such as reforming of fossil fuels or electrolysis. In certain cases, hydrogen can be recovered directly from industrial processes. As opposed to burning this hydrogen to produce steam for conventional steam turbine-driven electric generation at thermal efficiencies of under 30%, the hydrogen can be used directly in a fuel cell to produce electricity at higher generation efficiencies along with recovered heat. Regardless of the hydrogen source, fuel cells are considered less polluting than traditional fuel-driven power generation systems.

The cost of fuel cells has also continued to decrease due to increased production levels and technoogy advancements, while their thermal efficiency has increased. Because fuel cells are not power-cycle limited, it is believed that over time, the thermal efficiency of commercially available fuel cells will surpass that of traditional combustion and steam-driven prime movers. Still, both capital and long-term maintenance costs (largely the periodic cost of stack replacement) remain significant obstacles to wide-scale market penetration. To date, fuel cells have proven effective in "clean power" applications, a niche market in which the avoidance of power conditioners, uninterrupted power supply (UPS) systems, and standby generators combine to offset much of the first-cost premium.

Figure 23-19 is a process flow diagram of a 100 kW phosphoric acid fuel cell (PACF) installed at a 300-room hotel. The hotel uses the lower temperature heat to preheat domestic hot water and the higher temperature heat to power an absorption chiller-heater and for direct space and hot water heating. Since electric demand dips below the 100 kW output level at certain periods, the interconnected system is equipped with an isolated operation prevention system and is able to transmit the reverse current of surplus power back into the grid. The fuel cell unit is specified to produce an output of 100 kW of 3-phase, 210 v, 60 Hz electricity with a generation efficiency of just under 40% (LLV basis). Thermal output includes 164 MBtu/h (48 kW or 172 MJ/h) of high-temperature hot water at 185 to 194°F (85 to 90°C) and 259 MBtu/h (76 kW or 272 MJ/h) of low-temperature hot water at 104 to 122°F (40 to 50°C). Assuming all of this heat can be successfully recovered and used, the overall thermal efficiency of this system can reach 89%. Absent the low-temperature output (which is difficult to fully utilize), this figure drops to about 60%. The unit is 12.5 ft (3.8 m) long, 7.2 ft (2.2 m) wide, and 9.2 ft (2.8 m) high and weighs about 13 tons (12 tonnes). Its NO_X emissions concentration is rated at 5 ppm or less (at 0% O_2) and its sound emissions are rated at 65 dBa at 3 ft (1 m) distance.

Compared with hydropower, and even with wind power, these electric power generation technologies have experienced extremely small market penetration to date and, except in very specialized market niches, are generally considered not cost-competitive as power producing alternatives, due primarily to very high initial capital costs. Both, however, continue to hold great promise for the future.

Availability and Reliability

With many electric generation applications there is a strong requirement to reliably operate on specified load profiles. Project economic performance may be vulnerable to costly financial impact from outages (e.g., exposure to demand charge penalties, or loss of business production). Other facilities cannot tolerate outages due to the need to support various types of mission-critical and even life-sustaining operations. System availability and reliability are therefore important considerations.

Availability of electric generation systems is largely limited by equipment performance reliability and input energy source availability. While all equipment is subject to mechanical or electric system failure, fleet data will indicate that some systems and components have proven more reliable than others, making reliability planning a manageable task. Many of the newer alternative electric generation technologies have also proven reliable. However, since the design of many of these technologies continues to change and evolve rapidly, it is difficult to develop long-term, proven fleet experience in operations, maintenance, repair, and overhaul. This and the lack of

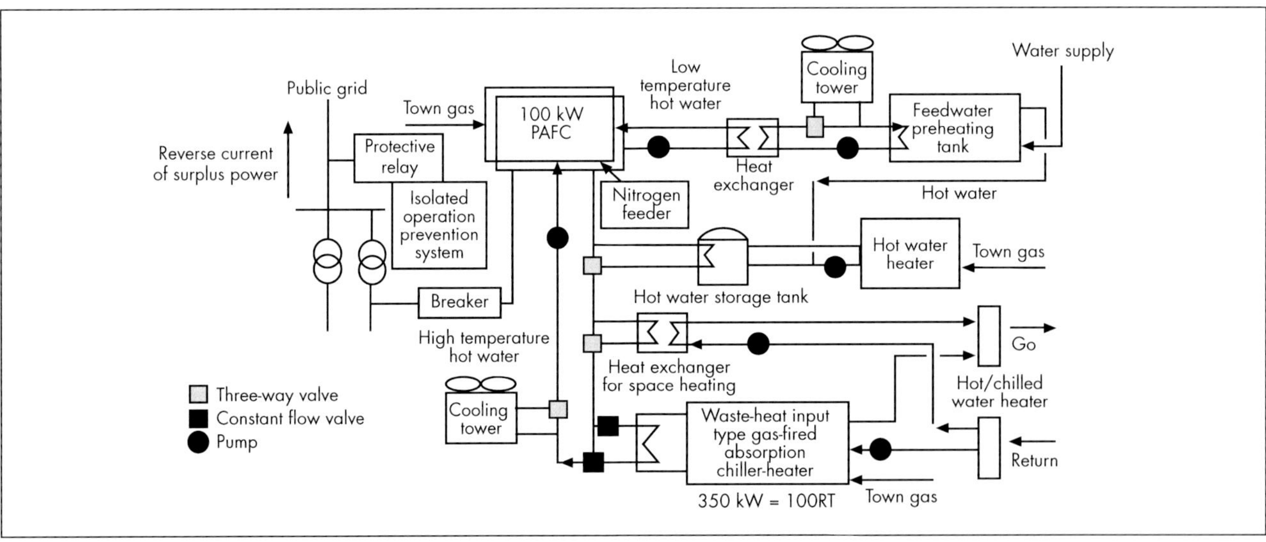

Fig. 23-19 Layout Illustration of 100 kW PACF Hotel Fuel Cell Application. Source: CADDET Energy Efficiency IEA/OECD

market penetration have limited component standardization. Also, a widespread service industry to support ongoing maintenance, repair, and replacement over a broad geographical area is also often lacking.

Energy source interruption can result from market curtailments (e.g., interruptible natural gas service contracts), delivery system physical failure or a variety of force majeure conditions. With certain renewable resources, availability is limited due to dependence on energy sources subject to variation (e.g., sun and wind) as opposed to easily stored and metered fossil fuels. For example, with PVs and solar thermal steam generator systems, availability is limited to about 25 or 35%, regardless of equipment reliability, even in the most optimal solar energy locations. With WECS, availability can be much higher, but can still be subject to wide variability and unpredictability of prevailing wind speeds. Hydropower plants may have very high availability, but, like wind, can vary depending on site-specific factors, such as seasonal or daily water level variations.

Selection of equipment with a proven record of high reliability and redundant capacity, along with judicious operations and maintenance practices are key elements to meeting availability and reliability challenges. Additional options include the application of various types of energy storage systems and selection of mixed technology (hybrid) systems.

Storage Systems

There are a variety of storage systems that can enhance the reliability, availability, and annual load factor, and balance the output characteristics of many electric generation systems. These include batteries, pumped water, flywheels, superconducting magnets, compressed air or other fluids to store electro/mechanical energy, and thermal storage systems to allow for steam production and subsequent electric generation. Since electric generation system output must be matched with facility load requirements or TOU related power sales market factors, storage may be used not only to mitigate availability risk but to allow for generation of electricity in one (off-peak) period for use or sale in another (peak) period.

While still quite expensive, **electric battery storage** is a simple technology application that can store power for dispatch when the power generation system cannot provide enough electric power to meet its end-use or power sales requirement. Similar to a fuel cell, a battery (or electric cell) utilizes the potential difference that exists between different elements. When two different elements are immersed in electrolyte, an electromotive force (emf) exists, tending to send current within the cell from the negative pole to the positive. The poles, or electrodes, of a battery form the junction with the external circuit. If the external circuit is closed, current flows from the battery at the positive electrode (or anode) and enters the battery at the negative electrode (or cathode).

In a primary battery, the chemically reacting parts require renewal. In a secondary (or storage) battery, the electrochemical processes are reversible to a high degree and the chemically reacting parts are restored, after partial or complete discharge, by reversing the direction of current through the battery. This highly reversible electrolytic action makes storage batteries effective for storing and discharging electrical energy in a highly controlled manner.

Limitations of battery use include their relatively low energy density, which dictates large size/weight requirements, their efficiency losses, and their high capital cost. However, advances in storage battery materials, driven largely by the perceived strong market potential in electric power applications (for peak shaving and backup power), continue to improve the economic performance of battery use for electric energy storage.

Flywheel technology involves the storing of mechanical energy as a wheel winds up through some system of gears and then delivering rotational energy until friction dissipates it. Flywheel energy storage is not a new technology. In fact, it is used in a diverse range of industrial applications, such as punch presses and high field pulsed magnets. Flywheels offer the benefit of high reliability, low maintenance, long service life, and virtually no environmental issues. Its stored energy density is, however, relatively low. In addition, its practical use has been limited to areas where the energy holding time is short because of the excessive rotational loss caused by bearings and windage.

Market penetration will depend on the ability to maximize storage, minimize friction losses, and establish effective system interface and control components. Significant R & D efforts are ongoing with this promising technology, with focus on increasing the energy density through work on high-strength, high-speed flywheels and expanding the holding time through development of a very low friction magnetic bearing. Most commercial flywheels are made of metal and rotate at relatively low speeds to maintain the tensile stresses in the flywheel within reasonable limits. Since the stored energy is proportional to the mass of the flywheel and to the square of the rotational speed, increases in rotational speed yield a large benefit in energy density. Advanced flywheels employ materials such as fiber-reinforced plastics that can be engineered to have very high strength in the radial direction, thus permitting higher operational speeds. The use of superconducting magnetic bearings is seen as a means to greatly reduce the long-term energy storage loss in the bearings, increasing the overall, or round-trip, efficiency of the storage process.

Another developing technology is **superconducting magnetic energy storage (SMES)**. In SMES, electric energy is stored by circulating current in a superconducting coil, or inductor. If the coil were wound using conventional wire, such as copper, the magnetic energy would be dissipated as heat due to the wire's resistance to the flow of current. However, if the wire is superconducting (no resistance), then energy can be stored in a persistent mode, almost indefinitely, until required. Because no conversion of energy to other forms is involved (e.g., mechanical or chemical), round-trip efficiency can be very high. SMES can respond very rapidly to dump or absorb power from the grid or local distribution network, limited only by the switching time of the solid-state components doing the dc/ac conversion and connecting the coil to the grid.

One favored superconductor for storage applications is a niobium-titanium alloy, which needs to be kept at liquid helium temperature in order to superconduct. Higher-temperature superconductors (i.e., those that operate at liquid nitrogen temperature or above) are not considered technically advanced enough to be considered for large-scale application. Market development continues based on the premise that once SMES has established a foothold in the utility market based on conventional superconductors, high-temperature superconductors can be introduced to reduce capital and operating costs once their physical characteristics have improved and the manufacturing processes are more mature.

Pumped hydro storage is a proven technology application for storing energy and increasing availability of electric generation systems, including both large-scale hydropower and other conventional central utility power plants. Two reservoirs at different elevations are required. When water is released from the upper reservoir, mechanical energy is produced by the down-flow through high-pressure shafts linked to turbines. The turbines drive generators to produce electricity. Water can be pumped back up to the upper reservoir by linking a pump shaft to the turbine shaft and using an electric motor to drive the pump. During off-peak periods when excess generation capacity is available, the surplus power is used to pump water into these specially designed reservoirs. Then, during periods of peak demand when more power is required by the system and/or during periods when availability is low or zero, the water is allowed to flow down again to generate electricity. Given a compatible location, pumped storage systems are relatively efficient and can be economical. While these are most commonly applied to hydropower or other major central generation plants, pumped storage systems can also be effectively used with other technology applications, such as a wind farm serving an isolated facility, assuming an appropriate reservoir site can be identified in the location of the project.

Modern pumped storage units in the United States normally use reversible pump-turbines that can be run in one direction as pumps and in the other direction as turbines. These are coupled to reversible electric motor/generators. The motor drives the pump during the storage portion of the cycle, while the generator produces electricity during discharge from the upper reservoir.

Somewhat analogous to pumped water storage, **compressed air storage** has been proven on a limited basis. Air is compressed at high pressure and stored in a large tank or underground cavity. While this can be effective, significant inefficiencies arise due to the energy required to cool the air as it is compressed and energy required (usually from fuel) to expand the cool air taken from storage as it enters the turbine. It is possible, however, to recycle the waste heat from the compression stage and use it to reheat the air during the expansion stage, thereby reducing overall system losses.

Solar thermal applications can use the above technologies to store their electric power output or can use a **thermal storage system** to store their thermal energy output for subsequent steam production and, in turn, electric power generation. Application for batteries or pumped storage require additional steam and electric generator capacity to be installed to produce excess power. Thermal storage systems only require excess solar thermal concentrator/receiver capacity.

In order to produce sufficiently high pressure/temperature steam for optimal steam turbine operation, a relatively high-temperature storage system is necessary. While high-temperature hot water can be effectively used, the space requirements are enormous. There are also a variety of phase-change materials, notably various salts, that are being used for high-temperature thermal storage. Figure 7-48 in Chapter 7 shows a high-temperature storage system using molten nitrate salt as the storage medium, which is part of the Solar Two Power Tower project. Liquid salt is pumped from a "cold" storage tank through the receiver, where it is heated and then pumped to a "hot" tank for storage. When steam is needed, hot salt is pumped to a steam generation system. From the steam generator, the salt is returned to the cold tank, where it is stored and eventually reheated in the receiver. While some types of phase-change salts are reasonably cost-effective for traditional thermal energy storage (TES) systems, these systems operate with higher hot and cold storage temperatures of 1,050°F (565°C) and 545°F (285°C), respectively. At a minimum, the salt must be maintained above its 464°F (240°C) melting phase point. Such systems can more than double the availability of a thermal solar system (from about 25% to more than 60% in some cases), but remain quite costly and still unproven for wide-scale market application.

Hybrid Systems

As either an alternative or compliment to storage, mixed (hybrid) systems can provide an extra measure of availability and reliability insurance for electric generation systems. This may involve the mixing of different types of prime movers (e.g., a gas turbine cogeneration system with a reciprocating engine back-up), additional equipment that can provide an alternative means to produce and deliver high-pressure steam, or the use of different types of equipment that rely on different energy sources.

Hybrid systems are fairly common among renewable technology applications, notably those providing on-site generation at their host facilities, as opposed to IPP and utility plants that simply feed the grid when electric power is available. Hybrid renewable and alternative energy systems may be considered in two categories. Systems that rely on solar thermal or geothermal energy may be configured with conventional fuel-fired systems to generate high-pressure steam. Other renewable systems will integrate conventional fuel-fired prime movers, the most common being the Diesel reciprocating engine-generator set. When gaseous or readily vaporized fuel is available, spark-ignited engines or gas turbines provide a similar function. While these units may be designed for simple-, cogeneration, or combined-cycle operation, their critical function is to supplement and balance the power output of these alternative systems or provide backup operation insurance.

In addition to these more conventional backups, it is not uncommon for alternative energy systems to be paired together, such as PV and wind systems, in arrangements to extend availability and reliability with two distinct renewable systems that have different operating regimes driven by availability of the energy source. Still greater availability will be achieved with the addition of more conventional backup/supplemental systems.

Some hybrid systems combine a number of electric production and storage components to meet the energy demand of a given facility or community. This is commonly referred to as the "village power" concept. PVs, engine-generators, wind turbines, small hydro plants, and any other source of electrical energy can be added as needed to meet energy demand and fit the local geographical and temporal characteristics, while limiting dependence on delivered outside conventional energy sources.

Localized On-Site Power Generation Plant Layout Examples

Figures 23-20 through 23-29 are examples of on-site generation plant layouts. Each of these plants require adequate space for the various equipment systems, as well as expected operations and service requirements. Space considerations include prime mover dimensions, arrangement of accessories, and passage for operation, maintenance, and crane work. The height of the machinery space depends on the headroom required for component removal, crane size, and ventilation systems and, where applicable, arrangement of the cooling water systems. Basements are beneficial for housing accessories and running cables and pipes, and can also be used to draw in ventilation air.

Figure 23-20 is a layout for a large-capacity single reciprocating engine electric generation plant. Air intake and exhaust flow paths are shown along with the location of the ventilation fan. Notice the multiple floor levels on both sides of the engine room and the exterior installation of the coolant water systems. Excluded from this plant layout are provisions for exhaust gas heat recovery and air emissions treatment. Figure 23-21 illustrates these provisions. Notice the basement location of the exhaust gas catalyst, silencer, and heat exchanger, along with coolant water heat exchanger.

Figure 23-22 shows the mechanical room of a 5,650 kW cogeneration plant featuring five dual-fuel reciprocating Diesel engines, each with an electrical output of 1,130 kW and a thermal output of 4.5 MMBtu/h (4,748 MJ/h). Notice the headroom and floor space clearances, which allow access for efficient operations, maintenance, and equipment removal. Figure 23-23 shows the mechanical

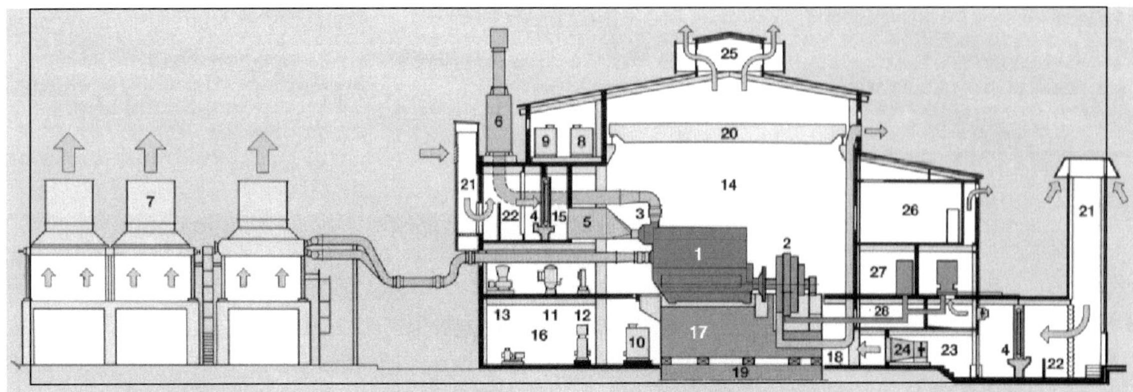

Fig. 23-20 Layout Diagram for Large-Capacity Single Reciprocating Engine Electric Generation Plant. Source: MAN B&W

1	Reciprocating engine	**16**	Basement
2	Electric generator	**17**	Engine gen-set block
3	Turbocharger	**18**	Spring isolators
4	Rotary air filter	**19**	Engine gen-set foundation
5	Air intake silencer	**20**	Crane
6	Exhaust silencer	**21**	Air intake tower
7	Radiator cooling plant	**22**	Baffle plate
8	Oil tank	**23**	Room for pressure fan
9	Oil tank	**24**	Ventilation pressure fan
10	Lube oil service tank	**25**	Extract ventilation
11	Lube oil automatic filter	**26**	Control room, office, etc.
12	Lube oil indicator filter	**27**	Switchgear, battery, transformer
13	Lube oil centrifuge	**28**	Cable floor
14	Engine room		
15	Equipment annex		

Fig. 23-22 View of Mechanical Room Housing Five 1,130 kW Dual-Fuel Reciprocating Engine Cogeneration Units. Source: MAN B&W

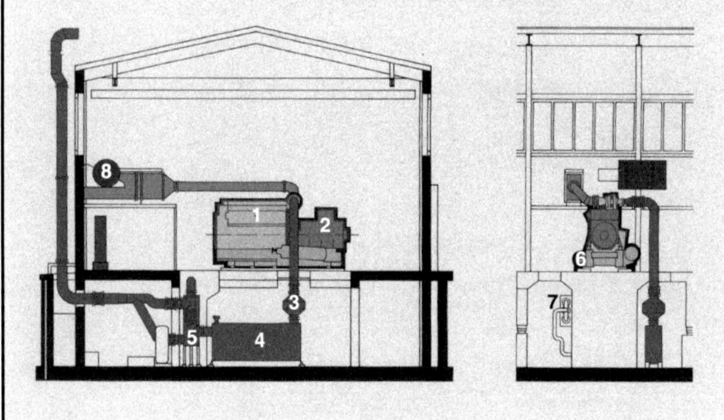

Fig. 23-21 Layout Diagram of Provisions for Reciprocating Engine Cogeneration Plant. Source: MAN B&W

1 Reciprocating engine
2 Electric generator
3 Exhaust gas catalyst system
4 Exhaust gas silencer
5 Exhaust gas heat exchanger
6 Lube oil heat exchanger
7 Cooling water heat exchanger
8 Fuel service tank

room for a 14 MW municipal power plant featuring two 18-cylinder dual-fuel reciprocating engine-generators.

Figure 23-24 shows the mechanical room for a gas-turbine cogeneration system with a capacity of about 9 MW. With headroom requirements established by HRSG, air intake, and exhaust duct configurations, floor space requirements are minimized by using multiple level structures to house auxiliary components. Use of sound-attenuating acoustical enclosures and intake and exhaust silencers minimize overall sound emission levels and allow lighter wall materials to be used for the mechanical room structure.

Figure 23-25 shows the layout of a condensing extraction steam turbine cogeneration system. Boiler-generated high-pressure steam is passed through the steam turbine-

Fig. 23-23 Mechanical Room for 14 MW Municipal Power Plant Featuring Two 18-Cylinder, Dual-Fuel Engine Generators. Source: Fairbanks Morse Engine Division

generator set and extracted at various points for preheater, deaerator, and facility heating services. The balance of the steam passing through the turbine is condensed and returned, along with condensate from the other thermal processes at the site.

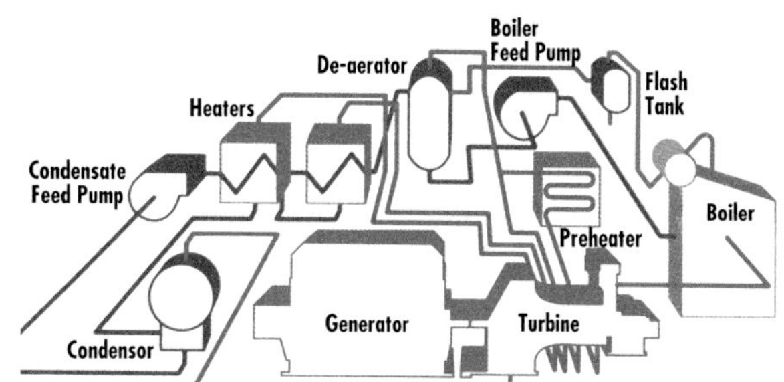

Fig. 23-25 Layout Illustration for Condensing Extraction Steam Turbine Cogeneration System. Source: Elliot Company

Figure 23-26 is a flow diagram for a condensing extraction steam turbine-generator set applied in a combined-cycle system. Heat recovery-generated steam serves both the steam turbine-generator and an absorption chiller. Uncontrolled extraction steam is used at the deaerator.

Figure 23-27 shows a system flow diagram (top) and a plant arrangement diagram (bottom) for a cogeneration

Fig. 23-24 View of Mechanical Room Housing 9 MW Gas-Turbine Cogeneration System. Source: Solar Turbines

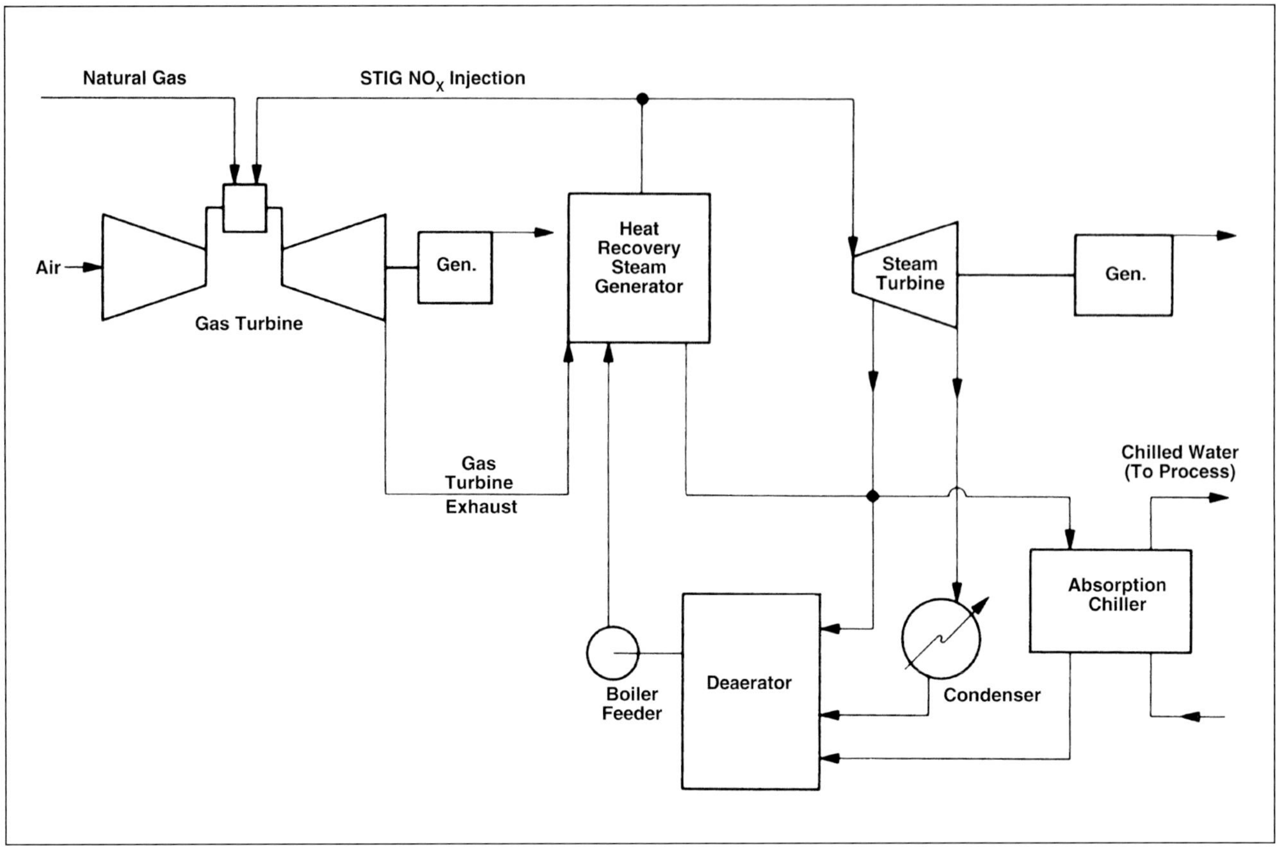

Fig. 23-26 Flow Diagram for Condensing Extraction Steam Turbine Generator Set Applied in Combined-Cycle System.
Source: Tuthill Corp. Murray Turbomachinery Division

plant featuring three small capacity gas turbines, two of which are designed for STIG-cycle operation. Two of the gas turbines are dual-fuel capable for natural gas and Diesel fuel. The other unit operates only on natural gas. The total steam production capacity of the three natural-circulation, watertube boilers is about 29,000 lbm/h (13,000 kg/h) at a pressure of 230 psig (16.9 bar). Depending on operating load requirements, all or part of the steam is distributed through the main header to a district heating plant and a newspaper plant. Electric output capacity ranges from 4.0 MW with no steam injection to 6.1 MW with full steam injection. With steam injection of about 5,000 lbm/h (2,300 kg/h), a NO_X emissions level of 26 ppmvd (corrected to 15% O_2) is achieved when operating on natural gas. With full steam injection, the cogeneration plant provides 7,850 lbm/h (3,560 kg/h) of steam to process.

Shown on the top of the flow diagram in Figure 23-27 are the fuel storage tank, gas compressor station, and cooling water and water treatment stations. Beneath the three turbine-generator and HRSG units, the electric circuitry is shown. Also shown is the air compressor station, required for pneumatic starting. The plant arrangement diagram shows the physical layout of the main components. Dimensions are 195 ft (59.4 m) by 53 ft (16.2 m). Notice that to reduce the length of the plant, one of the cogeneration units is set perpendicular to the other two.

Figure 23-28 shows the layout for a 100 MW combined-cycle plant featuring six dual-fuel reciprocating Diesel engines and one steam turbine. The plant is divided into two separate three-by-three engine rooms, with a middle section for common equipment. Each of the engine rooms has a natural gas compression system, as required for high-pressure Diesel-cycle injection of gas. The middle section includes components of the high- and low-voltage electrical systems, the steam turbine, and central cooling and water systems. This arrangement minimizes plant size and reduces piping and cabling distances. The reciprocating engines are enclosed in separate engine rooms to shut in the sound close to the sources and allow for lighter wall materials to be used in the rest of the plant. The plant is equipped with small overhead cranes in each engine room and at key places to meet

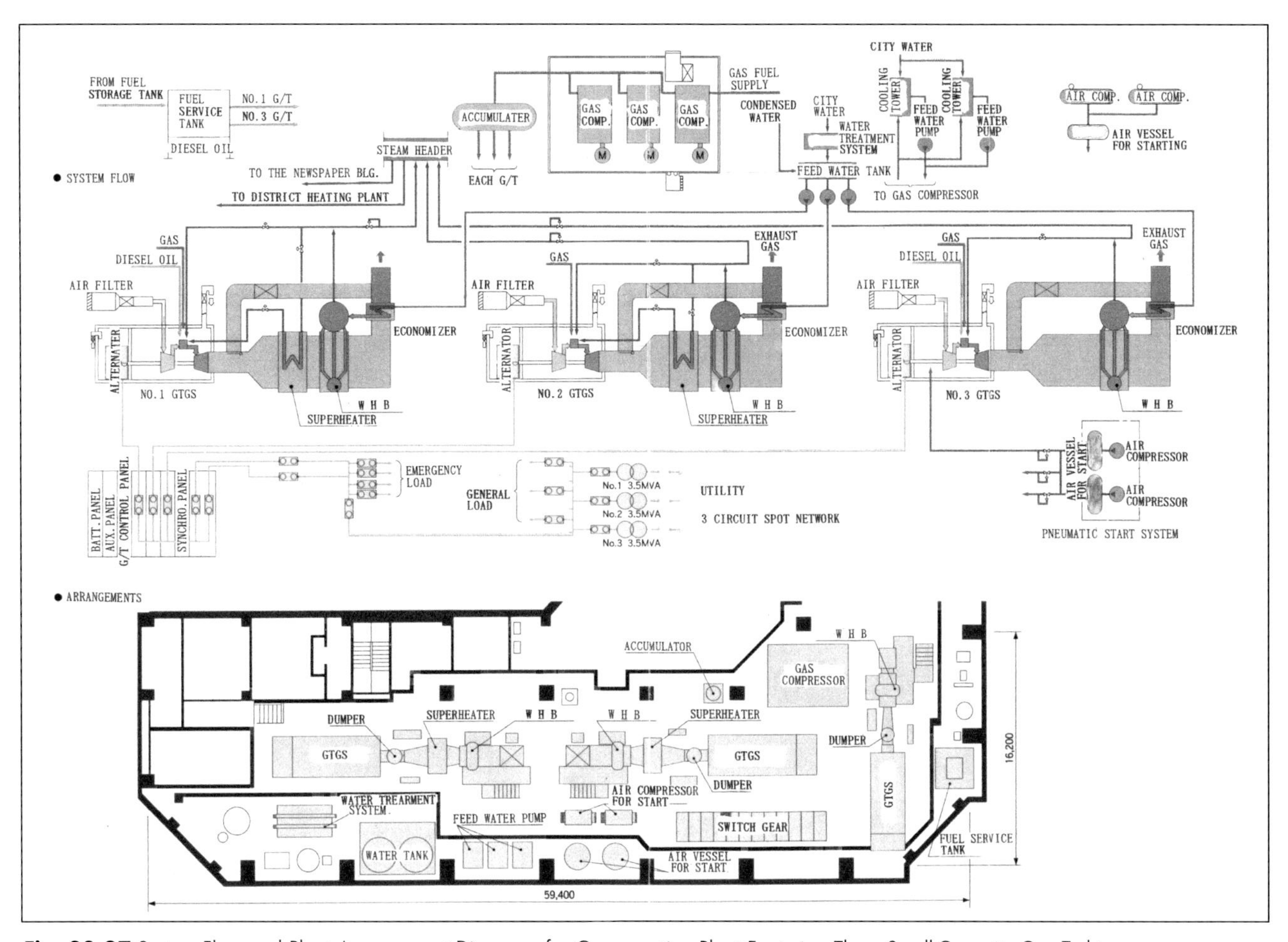

Fig. 23-27 System Flow and Plant Arrangement Diagrams for Cogeneration Plant Featuring Three Small-Capacity Gas Turbines. Source: Kawasaki Heavy Industries, LTD.

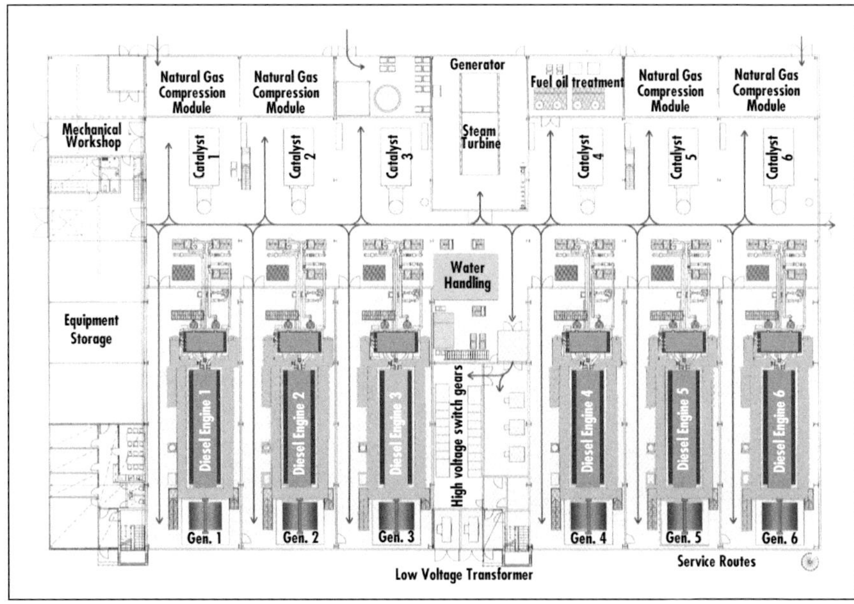

Fig. 23-28 Layout for 100 MW Combined-Cycle Plant Featuring Six Dual-Fuel Reciprocating Engines and One Steam Turbine. Source: Wartsila Diesel

lifting needs during service. Service routes, which have been designed to allow clearance for forklift trucks, are shown.

It should be noted that on-site electric cogeneration plants need not be considered as a dominant physical feature. They can be housed within a facility or out of view from the public, with sound and air emissions effectively controlled. Figure 23-29 shows a health sciences center facility that contains a 68 MW combined-cycle cogeneration system. The cogeneration plant features a 40 MW LM6000 gas turbine generator-set. Heat recovery-generated steam is used to power a 28 MW steam turbine that serves

Fig. 23-29 68 MW Combined-Cycle Cogeneration System Inconspicuously Housed in Health Sciences Center. Source: Stewart and Stevenson

the heating and process steam loads.

As an alternative to mechanical room installations, on-site generation systems may be installed outdoors with components protected by covers and various types of packaged housings. Figure 23-30 is a very compact gas turbine cogeneration plant installation, directly outside of an industrial facility featuring a 2,300 kW Kawasaki Cheng-cycle system.

Figure 23-31 shows another outdoor cogeneration plant. In the center is the packaged 20 MW LM2500. On the left, perpendicular to the generator set package, are the HRSG and auxiliaries. The electrical system components are shown on the right.

Figure 23-32 is a labeled diagram of a gas-turbine cogeneration system with HRSG. Figures 23-33 a and b compare layouts of a large-capacity simple-cycle gas turbine plant (a) with a combined-cycle system (b) featuring the same gas turbine.

Fig. 23-30 2,300 kW Kawasaki Cheng-Cycle Gas Turbine Installation at an Industrial Facility. Source: Kawasaki Heavy Industries, LTD.

Fig. 23-31 Outdoor Installation of 20 MW LM2500 Gas-Turbine Industrial Cogeneration Combined Cycle Plant. Source: General Electric Company

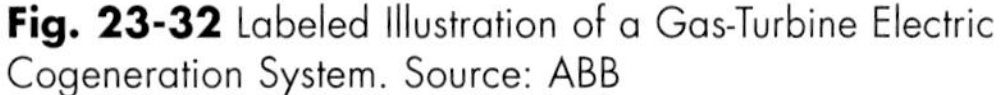

Fig. 23-32 Labeled Illustration of a Gas-Turbine Electric Cogeneration System. Source: ABB

1 HRSG
2 Bypass stack (optional)
3 Exhaust diffuser
4 Thermal block
5 Combustor block
6 Air intake system
7 Generator/Exciter block
8 Auxiliary block
9 Fuel pump block
10 Gas control block

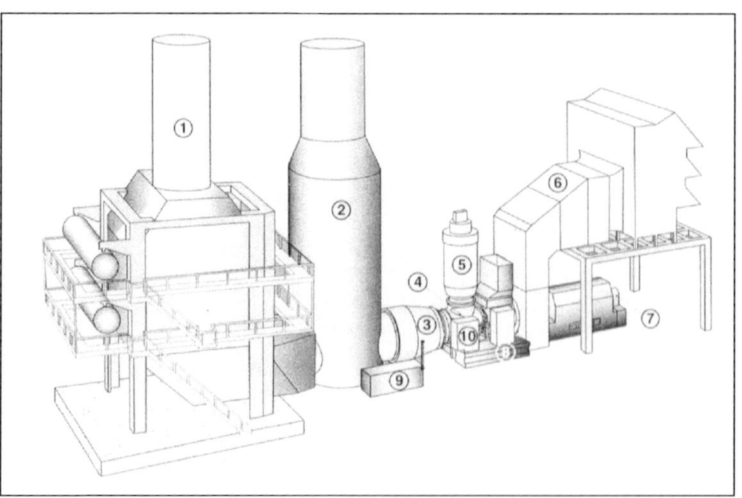

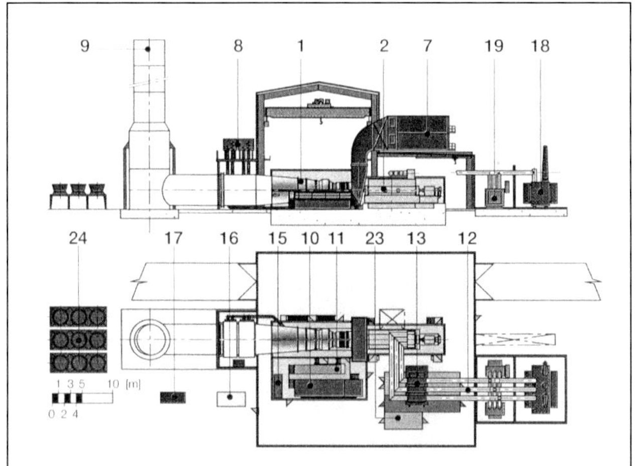

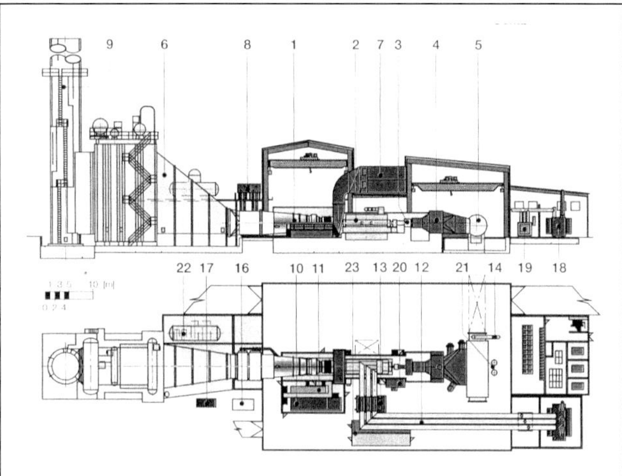

Fig. 23-33a and b Layout Comparisons of Simple-Cycle and Combined-Cycle Gas-Turbine Electric Generation Systems. Source: ABB

1	Gas turbine	**9**	Stack	**17**	Fuel gas block
2	Generator	**10**	Auxiliary block GT	**18**	Main transformer
3	Clutch	**11**	Control valve block	**19**	Auxiliary transformer
4	Steam turbine	**12**	Generator bus duct	**20**	Lube oil - ST
5	Condenser	**13**	Generator breaker	**21**	Condenser pump
6	HRSG	**14**	CW pipe	**22**	Feedwater tank
7	Air intake block	**15**	NO_X water injector	**23**	Control modules
8	Rotor air coolers	**16**	Fuel oil block	**24**	Coolers

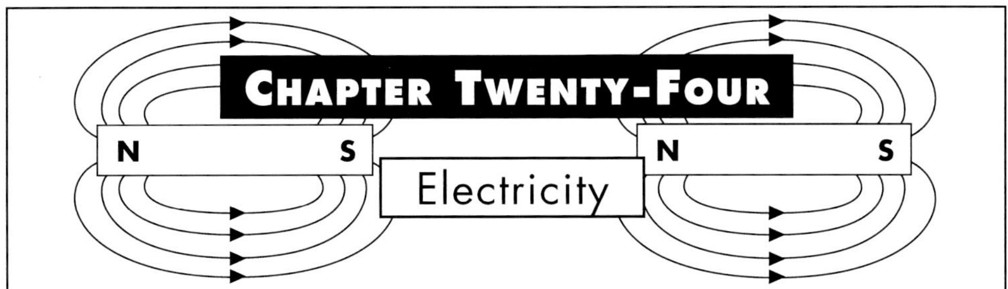

A significant portion of energy used on a daily basis is in the form of electricity, a secondary energy source. The bulk of electricity production occurs in large central power plants using fuel as the primary energy source. Switchgears and transformers control the delivery of the electricity from the central power plants to the consumers through a network of transmission lines, which interconnect nearly every utility service territory. While still a minority contributor, a growing percentage of the total electricity generated in this country comes from small and large independent power producers and cogenerators.

This chapter presents basic electricity definitions and theory that are necessary for understanding the various technologies and applications presented throughout this section.

DEFINITIONS

Electric current (or conventional current) is the movement of free electrons within an electrical conductor caused by a difference of electric potential (voltage) between the ends of the conductor. When charges flow, they can be positive, negative, or both. It is a convention to choose the direction of current to be in the direction of flow of positive charges. However, in a conductor, such as copper, the current is due to the motion of the negatively charged electrons. Therefore, with an ordinary conductor, such as copper wire, it is customary to refer to the direction of current as being opposite the flow of electrons. However, this convention is not always used. In some cases, current is considered to be in the direction of electron flow. The direction of the flow of electric current (motion of positive charges) is always from positive to negative, while the physical motion of electrons (negative charges) is from negative to positive.

Ampere (I), or **amp,** is the unit of measure of electric current. It is the quantity of 6.24×10^{18} electrons flowing past a point per second. Ampacity is the current-carrying capacity of conductors. The Coulomb is the quantity of electricity transported in one second by a current of one ampere. Sources such as the National Electrical Code provide tables indicating the maximum allowable current for any given size conductor and insulation system.

Electromotive force (EMF) is the force that causes current to flow within a conductor. This force, which can be considered electric pressure, is commonly referred to as electric potential.

Volt (V) is the unit of measure of EMF. Voltage is the potential between two charges, or two points in a circuit, or the electrical pressure in the electric system. The greater the voltage (pressure), the greater the flow.

Watt (W) is the common measurement of electrical power. One horsepower (hp) of mechanical work is equal to 745.7 W of electric power. Electric power, measured in watts, is the product of volts times amperes (rate of motion). For single-phase circuits, the relationship is expressed as:

$$P = EI \cos \theta \qquad (24\text{-}1)$$

Where:

P	=	Power in watts
E	=	EMF in volts
I	=	Current flow in amperes
$\cos \theta$	=	Power factor

Conductor is any material, such as silver, copper, or aluminum, that has a very low resistance to the flow of electric current. It can refer to a wire, coil, cable, bus bar, or any other object used to carry electric current.

Circuit is a conductor or a series of conductors through which an electric current flows. A circuit may have one or more electric components, known as circuit elements. Electric current will only flow in a closed or continuous circuit, on a continuous, non-interrupted path back to the source.

Cycle is the complete pattern of a single wave form of alternating current (see Figure 24-5).

Frequency is the number of cycles per second, specified in Hertz (Hz), where one Hz equals one cycle per second. The most typical frequency for alternating current in the United States is 60 Hz, although many other countries use 50 Hz.

Harmonics are wave forms (in both voltage and current) with frequencies that are multiples of the fundamental wave (60 Hz or 50 Hz). Harmonic distortion is a measure of distortion caused by devices such as inverters, rectifiers, transformers, and arc furnaces that distort the sinusoidal wave form, a wave in the form of a sine wave. A sine wave, which is shown in Figure 24-5, is a wave form that represents periodic oscillations.

Ohm (Ω) is the unit of measure of the resistance to flow of electric current in a circuit. One ohm is the value of resistance through which a potential difference of 1 volt will maintain a current flow of 1 ampere.

Ohm's law states that voltage in a circuit is equal to the current flow times the resistance in the circuit:

$$E = IR \qquad (24\text{-}2)$$

Where:

E = EMF in volts
I = Current in amperes
R = Resistance in ohms

Figure 24-1 illustrates relationships developed using Ohm's law. Included are relationships between power in watts, current flow in amps, electric potential in volts, and resistance in ohms.

Reactance is the measure (in ohms) of opposition to the flow of alternating current. Capacitive reactance is the opposition offered by capacitors and inductive reactance is the opposition offered by an inductive load.

Magnetism concerns magnetic fields and their effects on materials. A magnet establishes a field around itself that can be graphically represented by directed lines, i.e., magnetic lines of force (or flux). When an electric current passes through a conductor, a magnetic field surrounds the conductor. This is **electromagnetism**.

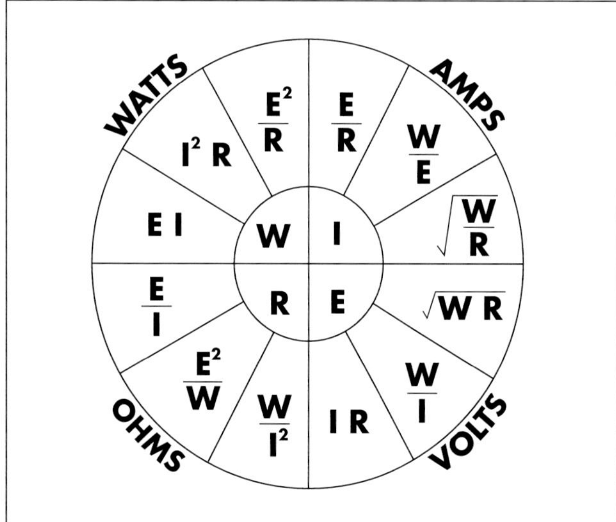

Fig. 24-1 Ohm's Law Relationships.

When a conductor is wound into a coil, the field around each turn in the coil interacts with the other turns. The net result is a large number of long parallel lines of flux running down the axis of the coil and creating a large field around the coil. Such a coil is known as an electromagnet. As shown graphically in Figure 24-2, lines of force, or magnetic flux, are considered to leave the magnet's north (N) pole, travel externally, and re-enter the magnet at its south (S) pole.

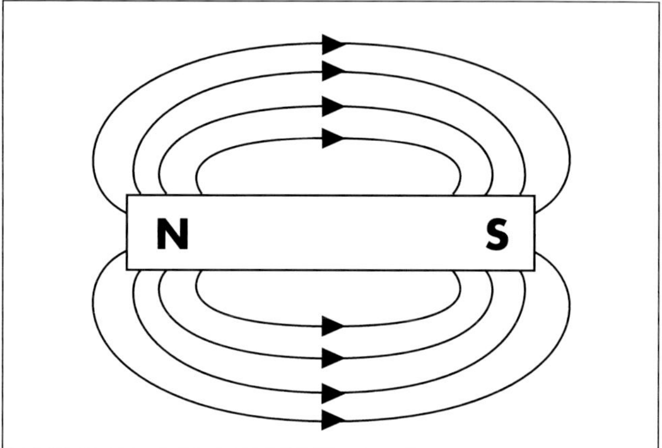

Fig. 24-2 Magnetic Flux.

According to Faraday's Laws, when a conductor (or coil) is moved through a magnetic field, a current is produced in the conductor. This current is said to be induced. Thus, the operative principle for electric motor, generator, and transformer action is called **electromagnetic induction**.

Direct and Alternating Current

There are two broad classifications of electric power: direct current and alternating current. The main difference is in the flow of current:

Direct current (dc) is an electric current that flows in one direction only. Flow can either be steady, as shown in Figure 24-3, or variable, as shown in Figure 24-4. Figure 24-3 shows a dc current that is very tightly controlled at a constant rate. Figure 24-4 shows a more typical dc current, one that exhibits some fluctuation.

Alternating current (ac) is an electric current that flows first in one direction for a period of time, and then in the reverse direction for a period of time. It is also constantly changing in magnitude. Ac current builds from zero to a maximum in the positive direction, reduces back to zero, then builds to a maximum in the negative direction and again reduces to zero.

Fig. 24-3 Pure dc.

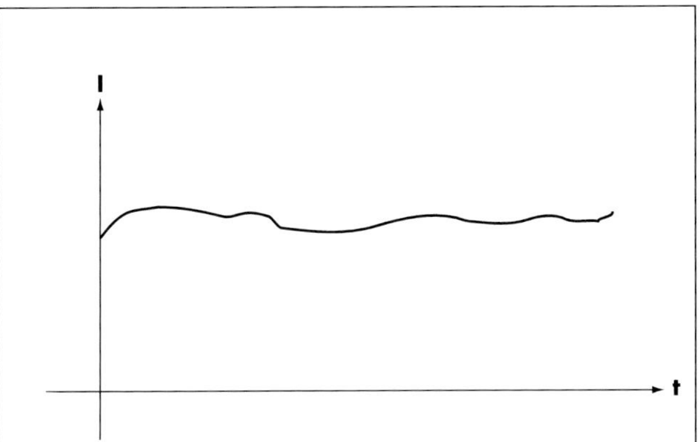

Fig. 24-4 General dc.

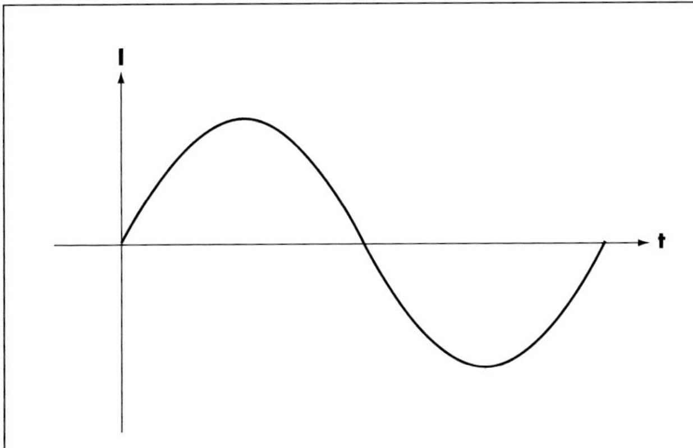

Fig. 24-5 Plot of an ac Current.

As shown graphically in Figure 24-5, plotted, the shape of ac flow is that of a wave. The common ac wave shape is the sine wave. The complete pattern of zero to maximum positive, to zero, down to maximum negative and back to zero is termed a cycle. The time needed to complete one cycle is termed the period of the wave.

The three basic circuit elements, or types of load, that compose an ac electric circuit are: resistance, inductance, and capacitance. Resistance consumes real power, while inductors and capacitors are reactive elements that only store and discharge energy.

Resistance (R) in a circuit is any element that consumes real power in the form of heat. R, measured in Ω, is the friction of the electrons flowing in a conductor. It is a physical property of the conductor wire used in a distribution system and results in a loss of power. Resistance causes a voltage drop in conductors and power-using devices. Power loss in a conductor from resistance is equal to the product of the current squared times the resistance of the conductor. The power loss, which is manifest as heat, is the limiting factor for allowable current.

The flow of electric current is similar to the flow of fluid in a pipe. Voltage is like the operating pressure, amperage the flow rate, and resistance the friction. The resistance to current flow is similar to the friction that impedes the flow of the fluid. Like friction, resistance generates heat when current flows through substances. In some cases, such as resistance heating, this phenomenon is advantageous as it provides a useful output.

Conductors are materials, such as copper, that have a very low value of resistance to current flow, thus generating a minimum of heat (power loss). Resistors are materials that conduct current, but have a significant value of resistance to current flow. They generate considerable quantities of heat energy.

Resistance is expressed as:

$$R = \frac{E}{I} = \left(\frac{P}{I}\right)\left(\frac{1}{I}\right) = \frac{P}{I^2} \qquad (24\text{-}3)$$

Where:

R = Resistance in ohms
E = Voltage in volts
P = Power in watts
I = Current in amperes

Power loss due to resistance is expressed as:

$$\textit{Power Loss (Watts)} = \textit{Current}^2 \times \textit{Resistance} = I^2R \qquad (24\text{-}4)$$

In accordance with Ohm's Law (I = E/R), the greater the resistance, the lower the current flow. For example, in a basic dc circuit with a 5 ohm resistor, the supply of 10 V will result in a current flow of 2 amps (2 = 10/5). If the voltage across the circuit is reduced to 0 (the circuit is open), then the current flow will be 0. If the voltage is doubled to 20, the current flow will be doubled to 4 amps.

In any given circuit, resistance is usually given in a measurement of ohms per foot (meter) for the material and size of the conductor. It is measured as the length of the conductor times the resistance per foot (meter). In many cases, the effects of resistance are relatively minor. However, at large sites with long-load leads, resistance can result in a significant reduction in voltage — or line voltage drop — that can seriously affect the performance of the electrical devices comprising the load.

Actual drop in voltage because of the resistance of the conductors is directly proportional to current flow (E = IR). Voltage drop will be maximum at full load and minimum at no load. In order to compensate for voltage drop, upstream voltage can be increased to a level that will result in the desired voltage at the load-drawing devices.

Inductance occurs when a circuit is closed, current starts to flow in a conductor coil, and lines of magnetic flux establish a magnetic field around the coil. Until the current flow has reached full load and established a complete field, the flux lines are cut by the turns in the conductor, inducing a current in the coil itself. When a circuit is opened, the opposite action occurs. For a brief period after the current flow ceases, the field is still in the process of collapsing. During this period, flux lines again cut the conductors and current is induced in the coil until the field is fully collapsed.

In a dc circuit, induction of current in the coil only occurs when power is being applied or removed. When current flow is constant, there is no increase or decrease in the magnetic field to cause induction of current in the coil.

In an ac circuit, inductance can be of great significance. The flow of current, in a wave form, is constantly changing direction and magnitude. This means that a magnetic field is constantly in the process of forming or collapsing. When a coil is part of an ac circuit, current is always being induced in the coil as long as the circuit is receiving power.

The two factors that affect induction of current in a coil are (a) the number of turns of conductor wire in the coil and (b) the number of lines of magnetic flux being cut per unit time. This rate of change of current flow is affected by the peak value of current in the coil and the frequency (number of complete electric rotations per second, generally 60 or 50 Hz) of the ac applied to the device.

In an ac circuit, as current rises in the cycle, the magnetic field grows stronger with more and more energy being stored in the form of magnetism. As current decreases in the cycle, the magnetic energy returns to the circuit and works to prevent the change in current that produced the magnetism.

An inductor in a circuit creates a hindrance to current flow because the induced current opposes the flow of current. This is called inductive reactance and forms a type of inertia that causes the magnetizing current to be out of phase with the driving voltage. Inductance causes the magnetic field current to lag behind the driving voltage and produces an effective loss of apparent power.

In an ac circuit with a purely inductive load imposed on it, voltage is at peak value when current is at zero, and voltage is at zero when current is at peak value. During the second and fourth quarters of the cycle, the voltage and current waves are of opposite sign (polarity). The zero point of current occurs 90 degrees after the zero point of voltage: the current wave lags the voltage wave by exactly 90 degrees. Inductive loads are, therefore, referred to as lagging loads.

Inductive reactance (X_L) is measured in ohms. While voltage across a resistance is equal to current times the resistance (E = IR), voltage across an inductor is equal to the current times the inductive reactance ($E = IX_L$). Inductive reactance, like resistance, tends to limit the magnitude of current flow in an ac circuit. For any given voltage across an inductance, the higher the X_L, the lower the current flow. Unlike resistance, however, X_L varies directly with frequency. If, for example, at constant voltage, frequency is reduced from 60 Hz to 50 Hz, the current flow will increase to 120% of the value at 60 Hz.

Capacitance is a quantity that defines the ability of a capacitor — or electron collector — to hold a charge. While inductance is related to current and magnetism, capacitance is related to voltage. Capacitors, sometimes referred to as condensers, are circuit elements that have the ability to store up an electric charge. Its capacity is a function of the areas of the conducting plates, the thickness of the insulating material, and the impressed voltage. With an inductance in a dc circuit, voltage is induced only when there is a change in current flow. With a capacitor, however, current will only flow when there is a change in voltage.

Inductance and capacitance both occur only when there is a change in rate of current flow and in voltage, respectively. Capacitance, thus, has the same significance as inductance in an ac circuit.

In an ac circuit, both capacitance and inductance

hinder current flow. However, while inductance hinders flow by opposing it, capacitance hinders flow by storing it. Capacitive reactance (X_C), measured in ohms, is the hindrance to current caused by the presence of capacitance in an ac circuit. The voltage across a capacitor in an ac circuit is equal to the current times the capacitive reactance ($E = IX_C$).

As voltage rises in the cycle, greater volumes of electrons are impressed and stored on the plates of a capacitor up to the peak voltage. As voltage drops (the second quarter of the first half of the sine wave), the electrons are returned to the circuit. As voltage changes polarity (second half of the sine wave), the electron charges on the plates change polarity and work to prevent a change in the voltage.

This forms a type of inertia, causing the voltage to be out of phase with, and lag behind, the current. It is the opposite effect of inductance and causes an equally effective loss of apparent power. A leading current is the current flowing in a circuit that is mostly capacitive. If a circuit contains only capacitance, the current leads the driving voltage by 90°. Figure 24-6 shows a plot of voltage versus current for both a pure capacitance circuit and a pure inductance circuit.

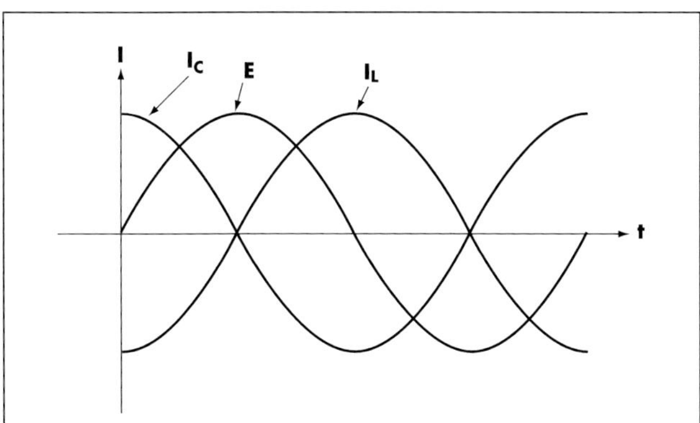

Fig. 24-6 Plot of Voltage vs. Current.

Where:

I_C = Current for purely capacitive circuit
I_L = Current for purely inductive circuit
E = Voltage

Capacitive reactance depends on rate of voltage change, in volts per second. Like inductive reactance, capacitive reactance is affected by frequency and tends to limit the magnitude of current flow. That is, the greater the capacitive reactance, the lower the magnitude of current flow. The critical difference is that while inductive reactance is directly proportional to frequency, capacitive reactance is inversely proportional to frequency. If frequency is increased, the value of capacitive reactance is reduced. For example, if the frequency in an ac circuit containing capacitive reactance is increased from 50 Hz to 60 Hz, the magnitude of current flow will increase by 120% of the 50 Hz value.

Impedance (Z) is the sum of total hindrance to current flow. The equation for impedance, measured in ohms, is:

$$Z = \frac{E}{I} \qquad (24\text{-}5)$$

Impedance is the total opposition or hindrance (resistance and reactance) of a circuit to the flow of alternating current at a given frequency. In alternating current, voltage drop in a conductor is equal to current times impedance. In direct current, since there is no effect of capacitance and inductance, voltage drop is equal to current times resistance.

There is always some amount of resistance in all circuits. In ac circuits, one or both of the other circuit elements — inductive and capacitive reactance — are often present. Impedance in an ac circuit, therefore, is a function of one or more of the following equations, depending on the circuit elements involved:

Resistive elements: $E = IR$ (24-6)
Inductive elements: $E = IX_L$ (24-7)
Capacitive elements: $E = IX_C$ (24-8)

To calculate the total hindrance to current flow in a circuit, the opposite (offsetting) effects of X_L and X_C must be considered. The relevant factor is the net effect of these two phenomena. To obtain the net total of these two reactances, one is subtracted from the other. With resistance, inductance, and capacitance in series, the equation for impedance is:

$$Z = [R^2 + (X_C - X_L)^2]^{1/2} \qquad (24\text{-}9)$$

In ac circuits, where resistance is only one of three possible forms of hindrance to current flow, Z must be substituted for R to validate Ohm's Law. The formula of E = IR, therefore, becomes E = IZ.

Electrical Power

As discussed in Chapter 2, power (P) is the rate of doing work, expressed as the product of force and rate of displacement. Power can be expressed in various interchangeable units, such as hp and W. Neglecting efficiency losses, an electric motor requires approximately 746 W or 0.746 kilowatt (kW) of electric power to accomplish 1 hp of mechanical work.

An alternative equation can be developed from the basic power equation P = EI to express power consumed in resistance. If, for example, a resistance load had 10 V across it and a 2 amp current flowing through it, the power consumed would be 20 W (20 = 10 x 2). Since voltage equals current flow times resistance (E = IR), IR can be substituted for voltage, yielding the following equation:

$$P = (IR)I = I^2R \qquad (24\text{-}10)$$

If the resistance in the above example was 5 ohms, the power consumed would also be 20 W ($20 = 2^2 \times 5$).

The relationship of electric power input to actual power output is primarily a function of electric motor efficiency. Motor efficiencies typically range from 75 to 98%, depending on factors such as size and material quality. This corresponds to actual power requirements ranging from 761 to 1,066 W per hp (1.02 to 1.33 W per W).

ac Electrical Power

In the plot of a dc circuit or a perfectly in-phase ac circuit, the relationship of voltage times current flow always produces a positive value or zero. In the dc circuit, this is because voltage and current flow always have a positive value. In a perfectly in-phase ac circuit, the values of voltage and current are always in the same direction at the same instant, as indicated by the corresponding sine waves on the power curve. The periods of positive voltage and positive current and the periods of negative voltage and negative current correspond perfectly and pass through zero at exactly the same instant. When multiplied, they always produce a positive value or zero, because a negative times a negative equals a positive.

In a purely inductive ac circuit in which the wave forms are 90 degrees out of phase, the values of voltage and current are sometimes of the same sign and sometimes of opposite signs. They are never zero at the same time. Multiplying voltage times current at any instant produces values that can be positive, negative, or zero.

During periods of positive power (voltage and current are either both positive or both negative), power is stored in the inductor. This power is returned to the source during periods of negative power (when the signs of voltage and current are opposite). Because the periods of positive and negative power are identical, the sum total of positive and negative power is zero. Therefore, even though there is current and voltage in the circuit, no power is being consumed.

The product of voltage at an alternator's terminals times the current flowing at the terminals will have a given value in a purely inductive circuit of volt-ampere (VA). This only indicates, however, that the alternator is delivering apparent power to a load, while the load is consuming no real power.

Exactly the same phenomenon occurs in a purely capacitive ac circuit. Here, too, the wave forms of voltage and current are 90 degrees out of phase — in this case, current leading voltage. The 90 degree separation of voltage and current causes electric power to be delivered to the capacitor and then returned to the source from the capacitor without the process consuming real power.

Analysis of either purely inductive or capacitive ac circuits reveals that where current is displaced from voltage by 90 degrees (either lagging or leading by one-quarter cycle), no real power is consumed by the load. However, while no power is being consumed, the voltage and current do represent a load on the alternator which can be measured in VA. This apparent power load on the alternator would be referred to as reactive power (or watt-less power).

In reality, all of the conductors in the coil and the circuit have resistance. Circuits are never purely inductive or capacitive. Given resistance plus inductive and/or capacitive reactance in a circuit, an angle of displacement (either lagging or leading) between the voltage and current waves results in greater than zero and less than 90 degrees. The predominant reactance determines whether the angular displacement will be either leading or lagging. This angle of phase displacement is expressed as θ.

Inductive and capacitive reactance in unequal amounts will result in a certain amount of reactive power, expressed as volt-amperes reactive (VAR), which will exceed the amount of apparent power — VA. The inductive reactance of a motor's stator windings creates an angle of separation between the current and voltage sine waves. Every ac induction motor draws more apparent power than it consumes real power, unless the amount of inductive reactance is countered by an equal amount of capacitive reactance.

Apparent power, which is the product of voltage and current in an ac circuit without the angle of phase separation, is simply expressed as:

$$VA = EI \qquad (24\text{-}11)$$

Where:
VA = Apparent power
E = Voltage in volts
I = Current flow in amperes

Real Power in ac-Resistive Circuits

In dc circuits, for any fixed load, the values of current and voltage are constant. The product of the two at any instant will be the same as at any other instant. In ac circuits, however, the product of these two values is constantly changing, as shown in Figure 24-7. Power calculations over the entire cycle are the total of all the instantaneous products of voltage and current. Thus, the power formula is:

$$P = E\ (effective) \times I\ (effective) \qquad (24\text{-}12)$$

In a purely resistive ac circuit, the sine waves of voltage and current are perfectly in phase; that is, the angle of phase displacement or separation is 0 degrees. Under full load in a purely resistive circuit, therefore, real power and apparent power are equal. Thus:

$$P = W = VA = EI \qquad (24\text{-}12a)$$

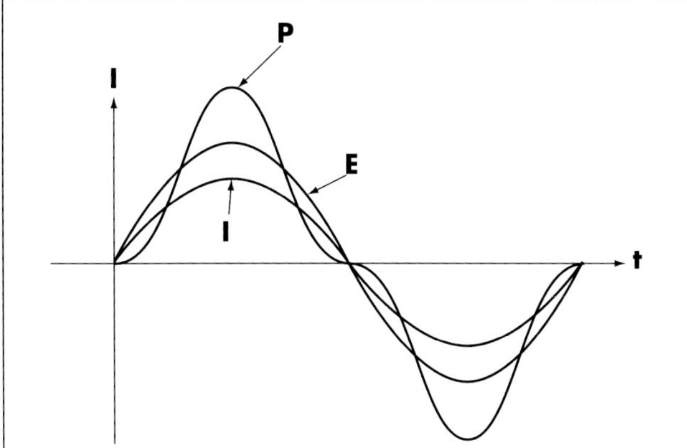

Fig. 24-7 Plot of Relationship of Voltage, Current, and Power in a Resistive ac Circuit.

Where:
E = Voltage
I = Current
P = Power in watts

The Power Triangle

The relationship of real power, apparent power, and reactive power can be understood through the trigonometric relationships of a right, or power, triangle. The power triangle consists of the following three components:

1. **VA:** Volt-amperes — or **apparent power** — expresses power as the simple product of voltage, in volts, and current flow, in amperes, without regard to the phase relationship of the current and the voltage. A kilovolt-ampere (kVA) equals 1,000 VA.
2. **W:** Watt — or **true power** — measures power corrected for the degree to which voltage and current are out of phase. It is the product of current times the voltage across the resistance. If voltage and current are in phase, W equals VA. If they are at all out of phase, W is less than VA. A kW equals 1,000 W.
3. **VAR:** Volt-amperes reactive — imaginary or **reactive power** — measures the product of the voltage across and the current flowing through a reactance (inductive, capacitive, or both). It is the component of total current that is 90 degrees out of phase with voltage. Reactive power must flow from a generator, but none of this reactive power is actually expended. A kilovar (kVAR) equals 1,000 VAR.

The power triangle, shown graphically in Figure 24-8, consists of these three values arranged in a right triangle. VA is the hypotenuse. The angle between the hypotenuse and the W vector is the displacement between line voltage and current. From the power triangle, it is seen that W is, being the adjacent side, the cosine of the angle. VAR, being the side opposite, is the sine of the angle. The angle indicates how far current is lagging or leading voltage. In the case of inductance, VAR expresses the current lagging behind the driving voltage. In the case of capacitance, it represents the current leading ahead of driving voltage. The equation for real power is:

$$VA \cos \theta = W \qquad (24\text{-}13)$$

The cosine of 0 degrees is 1 and the cosine of 90 degrees is 0. At 90 degrees of angular-phase displacement (a purely reactive circuit), real power is 0. At 0 degrees of

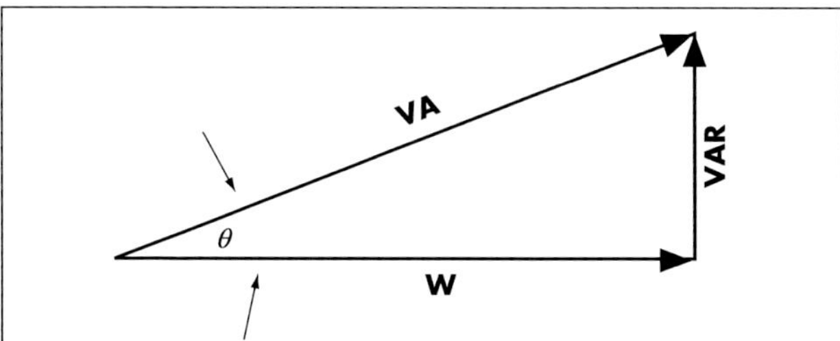

Fig. 24-8 Power Triangle.

angular-phase displacement (purely resistive), real power equals apparent power and reactive power is 0. Thus:

$$\theta = 90° \rightarrow W = VA \cos 90° = VA\ (0) = 0$$
$$\theta = 0° \rightarrow W = VA \cos 0° = VA\ (1) = VA$$
$$\theta = 30° \rightarrow W = VA \cos 30° = VA\ (0.866) = 0.866VA$$

The equation for VAR is:

$$VA \sin \theta = VAR \qquad (24\text{-}14)$$

The sine of 90 degrees is 1 and the sine of 0 degrees is 0. Thus, at 90 degrees of angular phase displacement (no real power), apparent power (VA) equals reactive power (VAR). Thus:

$$\theta = 90° \rightarrow VAR = VA \sin 90° = VA\ (1) = VA$$
$$\theta = 0° \rightarrow VAR = VA \sin 0° = VA\ (0) = 0$$
$$\theta = 30° \rightarrow VAR = VA \sin 60° = VA\ (0.5) = 0.5VA$$

These relationships can also be expressed as a function of the Pythagorean theorem: $A^2 + B^2 = C^2$. VA is C (the hypotenuse) and A and B (the legs of the triangle) are W and VAR. Thus:

$$VA = (W^2 + VAR^2)^{1/2} \qquad (24\text{-}15)$$
$$W = (VA^2 - VAR^2)^{1/2} \qquad (24\text{-}16)$$
$$VAR = (VA^2 - W^2)^{1/2} \qquad (24\text{-}17)$$

Power Factor

Power factor (PF) is a common measure of this angle of displacement of current from voltage. The cosine of the angle of current lag behind or lead ahead of voltage is the power factor of the circuit. This is quantified as the ratio of kW to kVA. The greater the phase displacement, the lower the power factor. Power factor is expressed as:

$$PF = \frac{Real\ power}{Apparent\ power} = \frac{W}{VA} \qquad (24\text{-}18)$$

Because power factor is a numerical expression of the angle-of-phase relationship, it can also be expressed as:

$$PF = \cos\theta = \frac{W}{VA} \qquad (24\text{-}19)$$

The leading contributor to low power factor in most facilities is the standard ac induction motor. Full-load power factor ratings typically range from 0.75 for small motors to 0.90 for large motors. Any other equipment with magnetic coils and fields, such as induction furnaces, also have power factors of less than unity, or 1.

Power factor varies with the load at any point in time. Most facilities have power factors ranging from 75 to 95%. When there is more motor power on line and less lighting, for example, power factor will be lower. Power factor will sometimes decrease with the load on a given motor. In general, for open drip-proof motors, power factor decreases as the percent full-load on the motor decreases. Because VAR from inductance and capacitance are in opposite directions, capacitors can be added to a line to cancel inductive VAR and improve power factor.

Purchased power from most utilities is measured in kW. A lower power factor penalizes the utility because it requires more current-carrying capacity in the generator for a given real-power demand, indicated by kW billing, and for resistance losses in the conductors on their system. This is why utilities often require a minimum power factor (ranging from 80 to 90%) from customers. Some utilities include power factor penalties as part of demand charges. Some utilities simply bill in kVA. Both strategies shift the burden of power factor correction onto the customer. Utilities that do not account for power factor simply compensate through the capacity and energy components of the rate structure. When electricity is generated on-site, poor power factor reduces the effective output of the generation equipment by reducing the generator efficiency.

Another advantage of improving power factor is that reduction of reactive current may allow for added load (growth) on the electric system without the need for upsizing transformers, main feeders, and bus ducts. Operating-cost savings potential also lies in the reduction of resistive power loss due to current flow in the conductors. All of these cases may provide sufficient impetus to cost-effectively maximize power factor.

A **capacitor** compensates for the inductive magnetizing current locally by producing capacitance kVARs that cancel out the inductance kVARs. Capacitors typically come in various voltages in discrete increments of 30 kVAR. Number and total capacity is determined by economic analysis. They can be installed together at any point past the utility metering device, in groups connected at the electrical center of power feeders or on individual motor terminals. Ideally, they should be as close to the load as economically possible.

Another way to improve power factor is with the use of synchronous motors. These motors can be designed with unity or leading power factor. When operating at a leading power factor, these motors are referred to as synchronous condensers. The leading power factor can be adjusted to balance the effect of the lagging power factor of induction motors. This practice, however, is quite costly relative to installing capacitors.

The benefits of improved power factor can be summarized as follows:

- Reduction in size requirements and capital cost for transformers, main feeders, and bus ducts.
- Reduced power losses.
- Reduced loads on transformers and motors, resulting in longer life and lower failure rates.
- Reduced rates where electricity is billed in kVA and elimination of power factor penalties where they exist.
- Reduction in voltage drop at the facility, which tends to improve the overall voltage profile.

Reactive Power and On-Site Generation

Power factor becomes more complicated with the interconnection of on-site generators. The power output of a generator is controlled by varying the torque applied by the prime mover to its shaft. The kVAR output is controlled by varying the field of excitation. If an on-site generator is overexcited, it produces kVAR as well as kW. This kVAR flows into the facility's load to provide excitation current, reducing the amount of kVAR the loads draw from the utility system. This serves the same function as installing capacitors to correct facility power factor.

Interconnected on-site generators that operate in parallel with the utility can use voltage regulators as power factor correction devices to constantly adjust their excitation according to the power factor. The excitation level is trimmed to control the flow of kVARs from the generator so that the generator's power factor rating is not exceeded.

Single- and Three-Phase ac Power

Single-phase ac consists of either a single voltage or two voltages in series, with exactly the same phase relationship. Three-phase power, which is the most common form of power used by industrial facilities, consists of three separate voltages spaced 120 electrical degrees apart using three lines. This is shown graphically in Figure 24-9. No two phases are ever at zero voltage at the same time, and no two phases are ever at peak voltage at the same time. Also, when any one phase is at zero voltage, the other two phases are of opposite sign (polarity) and are at 86.6% of peak voltage.

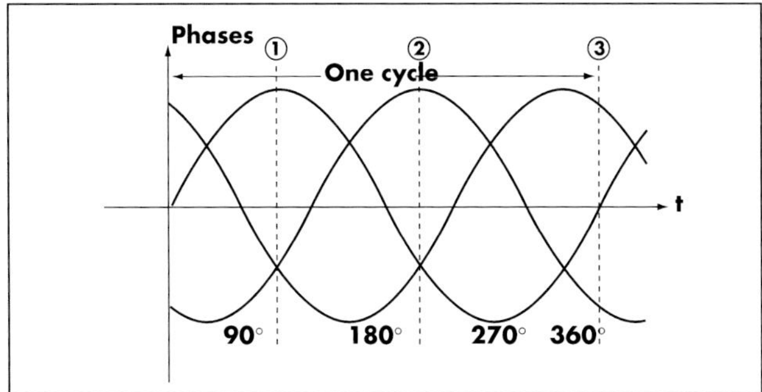

Fig. 24-9 Plot of Three-Phase Power.

The advantages of three-phase versus single-phase power are:

- It requires less cable to supply power at a given voltage.
- It produces a magnetic field of constant density that rotates at the line frequency, thus reducing the complexity of motor construction.
- It provides constant instantaneous output of the alternator if each of the phases has an identical load.

When calculating the relationship between three-phase power, voltage, and amperage, a phase factor of 1.732 (square root of 3) is used.

Basic ac power formulas are:

Term	Single-Phase Formula	Three-Phase Formula	Equation
I	$\frac{746\ hp}{(E)(Eff)(PF)}$	$\frac{746\ hp}{(1.732)(E)(Eff)(PF)}$	(24-20)
I	$\frac{1{,}000\ kW}{(E)(PF)}$	$\frac{1{,}000\ kW}{(1.732)(E)(PF)}$	(24-21)
I	$\frac{1{,}000\ kVA}{E}$	$\frac{1{,}000\ kVA}{(1.732)(E)}$	(24-22)
kW	$\frac{(I)(E)(PF)}{1{,}000}$	$\frac{(I)(E)(PF)(1.732)}{(1.000)}$	(24-23)
kVA	$\frac{(I)(E)}{1{,}000}$	$\frac{(1.732)(I)(E)}{1.000}$	(24-24)
hp	$\frac{(I)(E)(Eff)(PF)}{746}$	$\frac{(1.732)(I)(E)(Z)(Eff)(PF)}{746}$	(24-25)

Where:

I = Amperes per phase
E = Volts per phase
Z = Impedance per phase
Eff = Efficiency of hp output generating device
PF = Power factor

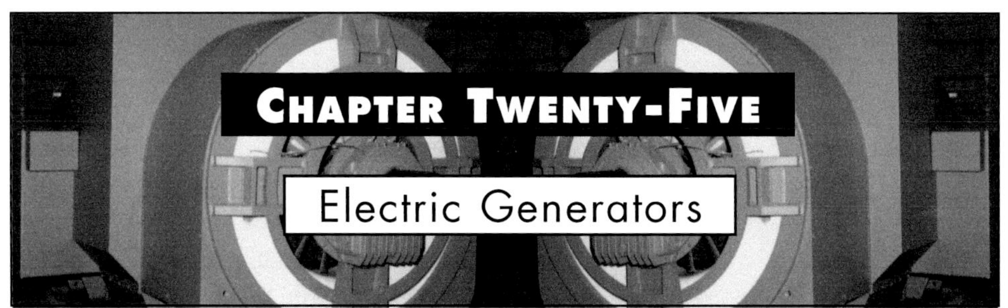

Chapter Twenty-Five

Electric Generators

In its simplest form, an electric generator consists of a loop of wire rotated in a magnetic field by an external force. As the loop rotates, an **electromotive force (emf)** is induced and, if an external circuit exists, a current is produced.

Most electric generation systems produce three-phase power by driving an alternating current (ac) generator with a prime mover. An **ac generator**, or **alternator**, as it is commonly referred to, is an electric machine that converts rotating mechanical energy into ac power.

There are two types of generators typically used for ac power production. **Induction generators** are used for parallel operation with an existing ac power source. Induction generators typically range in output from a few kW up to about 1,000 kW, though much larger capacity units do exist. The interconnection of the induction generator with the utility-derived power system is typically quite simple.

Synchronous generators are capable of operating independently. From the largest central utility generating station to the smallest emergency or standby power system, this is the most common type of generator in service today. Interconnection with another ac power source requires careful planning and design. Protection at the interconnection point consists of at least a synchronizing device to assure that the two sources are synchronized before they are connected together. After connection, controls and protective devices are required to assure continued safe operation for the equipment and the power system.

Following are descriptions of both types of generator, as well as descriptions of single-and three-phase power generation, generator selection, and control.

Induction Generators

An induction generator is a comparatively simple and low-cost machine. The stator can be single- or three-phase and the rotor can be of squirrel-cage or wound design. Figures 25-1 and 25-2 are induction machines that may be used as a motor or generator.

Synchronous speed is a function of frequency and the number of poles that constitute the magnetic field. It is determined as follows:

$$\text{Synchronous speed (RPM)} = \frac{120f}{P} \qquad (25\text{-}1)$$

Where:
f = Frequency in Hertz
P = Number of poles

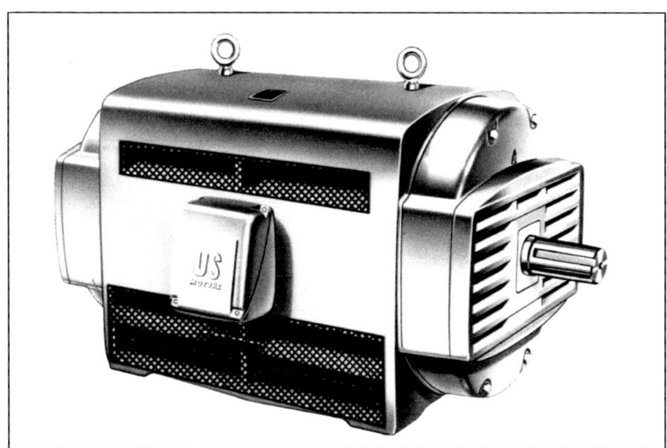

Fig. 25-1 Induction Machine. Source: United States Motors

An induction machine serves as a generator when it is driven above its synchronous speed. When it is operating below its synchronous speed, it serves as a motor. When operating as a motor (below synchronous speed), the induction machine absorbs current and produces torque (at a positive value). When operating as a generator, the machine does the opposite, absorbing torque and producing current.

Fig. 25-2 Induction Machine.
Source: Tuthill Corp. Murray Turbomachinery Division

The speed of an induction motor decreases below its synchronous speed as the load increases. The difference between synchronous speed and the motor's actual operating speed is referred to as **slip**, and is calculated as follows:

$$Slip = \frac{(Synchronous\ speed - Actual\ speed)}{Synchronous\ speed} \quad (25\text{-}2)$$

To generate power, an induction machine is driven by a prime mover above its synchronous speed. Circulating currents are induced in the rotor bars by the reactive current of the grid or other power source. The magnetic field interactions between the rotor and stator are converted into electric power that flows into the grid. The faster the generator is driven, the more power generated.

Induction generators do not have exciters or voltage regulators. They require an external source of reactive power for excitation, typically provided by the electric grid. Excitation can also be provided from other external power sources, such as capacitors or local synchronous generators connected to the system bus.

Induction generators are much more simple to operate and control than synchronous units. Protective devices required for a typical induction generator installation include:

- Over/under voltage
- Over/under frequency
- Overspeed
- Overload
- Reverse power

Synchronous Generators

A synchronous generator (or alternator) is the machine used to generate most of the ac power used today. A three-phase generator consists of a stator with a three-phase winding and a salient-pole rotor, which carries a dc field. The rotor (rotating winding or element) is driven by a prime mover at synchronous speed. As the magnetic flux of the dc-field winding crosses the stator windings, a three-phase voltage and current is induced.

Most synchronous generators are **revolving field generators** in which an armature, or assembly of coils placed in slots of a laminated steel core, is held stationary in a rotating field. Ac power is induced (generated) in the stationary winding of the machine.

In a less common arrangement, the **revolving armature generator**, the field is held stationary and an ac voltage is induced in the revolving armature winding. Ac power is supplied to an external load through slip rings and brushes that maintain a sliding electrical contact between rotating and stationary components.

Figure 25-3 is a 400 kW synchronous ac generator with permanent magnet generator excitation and 12-lead reconnectable design. Figure 25-4 is a cutaway view of a 35 kW synchronous generator suitable for 50/60 Hz operation at any voltage through 480 V.

Fig. 25-3 400 kW Synchronous ac Generator.
Source: Marathon Electric

An **exciter** supplies dc current to the field windings of the alternator, as illustrated in Figure 25-5. **Static exciters** are external solid state devices with no rotating parts. When the main alternator is used for exciter power, a battery or other external dc power source is usually required for initial field excitation and voltage buildup.

Fig. 25-4 Cutaway View of 35 kW Synchronous ac Generator.
Source: Marathon Electric

Rotating exciters are internal devices that include a stationary field and a small three-phase rotating armature on the generator shaft. The output of this armature is rectified and fed to the rotating main field of the generator.

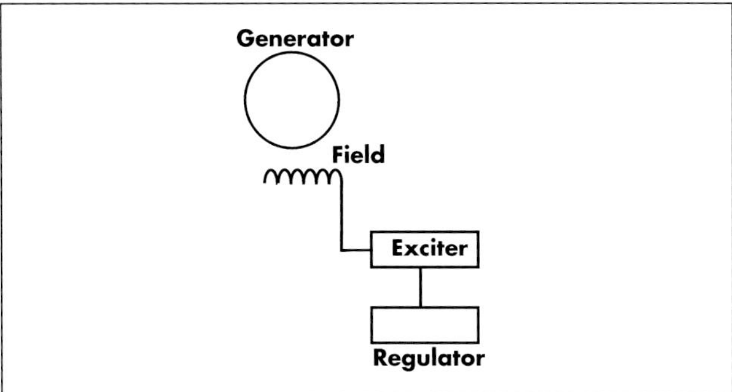

Fig. 25-5 Illustration of Exciter Operation.

Modern generators use solid-state rectifiers and no longer require slip rings and brushes for external excitation. There are two basic types of brushless generators. In the recent past, most generators of the brushless design depended upon residual magnetism in the main field to produce a small voltage upon initial starting. Today, many generators are **permanent magnet generators (PMG)** containing a separate alternator for the voltage regulator power supply. In this design, the voltage regulator is unaffected under heavy load and during transient load changes. PMGs typically provide better motor starting and short circuit performance.

Precise speed control is required to match the frequency of the utility grid or to maintain the required frequency in isolated operation. The most common frequency used in North America is 60 Hertz (Hz). In Europe, 50 Hz is used. These frequencies have been selected as norms to allow economical mass production of equipment. There are, however, a number of applications for which equipment standards are based on other frequencies.

The frequency produced by any ac generator is a function of two factors:

1. **The number of field pole pairs.** The number of pairs of poles that construct the field determines the number of electrical cycles produced by each revolution. A two-pole field produces one complete cycle per revolution, a four-pole field produces two cycles per revolution, a six-pole field produces three cycles, etc.
2. **Rotational speed (RPM).** For a given number of field poles, the frequency produced will be directly proportional to rotational speed.

Frequency can be calculated as follows:

$$Frequency = \frac{(Speed \times Number\ of\ poles)}{120} \qquad (25\text{-}3)$$

Where:
Frequency is measured in Hz (cycles per second)
Speed is measured in revolutions per minute (RPM)

For example, to produce a frequency of 60 Hz, a two-pole alternator would be driven at 3,600 RPM, a four-pole alternator at 1,800 RPM, a six-pole alternator at 1,200 RPM, etc. To produce 50 Hz, a two-pole alternator would be driven at 3,000 RPM.

Single- and Three-Phase Generators

Single-phase ac power consists of either a single voltage or two voltages in series with exactly the same phase relationship. NEMA lists standard single-phase voltages for 60 Hz systems as 120 volts, 240 volts, or a combination of 120 and 240 volts. There are three basic types of single-phase generators:

- **Single-voltage, two-lead generators.** These usually have a single circuit armature with two load leads and are designed to produce a single-output voltage.
- **Three-load lead, dual-voltage generators.** These are similar to the two-load lead design, except that a center-tapped lead is brought out to allow a dual single-phase voltage output.
- **Four-load lead reconnectable generators.** These are the most versatile and commonly used single-phase generators. They have a two-circuit armature winding, with each coil group designed for 120 volt output. Low voltage is produced by connecting the coils in parallel, and high voltage is produced by connecting them in series. The series configuration allows 240 volt, two-wire service. A third output lead can be used for three-wire, 120/240 volt service.

Three-phase generator armatures are constructed so there are three separate independent phase coil groups spaced 120 degrees apart. Each phase may be composed of one or two major coil groups (single or two-circuit design). There are two basic methods of connection for three-phase armatures: wye- (or star-) connected and delta-connected.

In the **wye** method of armature connection, as illustrated in Figure 25-6, the finish leads of each of the three-phase coil groups are tied together to form a neutral point. This is the wye, or star, point. The start leads then become the three output load leads. Sometimes, a fourth, or neutral, lead is affixed to the neutral point and brought out for external connection.

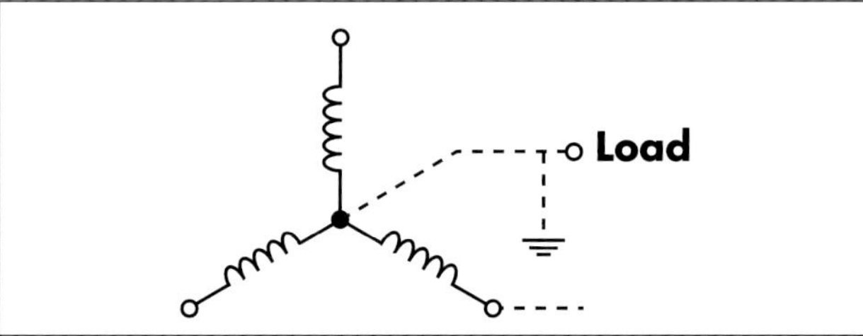

Fig. 25-6 Wye (or Star) Connection.

The voltage generated in each phase of a wye-connected generator is termed line-to-neutral, or phase voltage. The voltage measured between any two output load leads is termed line-to-line, or phase-to-phase, and is equal phase voltage times the square root of three (1.732). A wye-connected 480 volt generator (480 volts phase-to-phase) can also provide 277 volts (phase-to-neutral). Most generators are wound with wye connections to minimize voltage across winding insulation.

Alternatively, on **delta**-connected generators (Figure 25-7), the phase coil groups are configured in a triangle. The delta connection is made by joining the start lead of one phase coil group to the finish lead of the phase coil group to its left, and so on. An output lead is affixed to each corner of the triangle. On a three-lead delta-connected unit, any single point can be grounded.

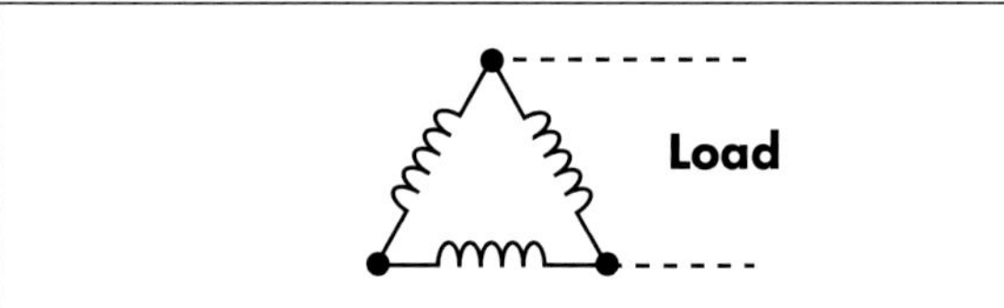

Fig. 25-7 Delta Connection.

Wye and delta configurations are used to provide a range of output voltage combinations. Typical generator terminal voltages for 60 Hz service are shown in Table 25-1.

Low Voltage	Medium Voltage
208Y/120	2,400 Δ
240 Δ	4,160Y/2,400
440 Δ	12,470Y/7,200
480Y/277	13,200Y/7,600
600 Δ	13,800Y/8,000
600Y/347	

Table 25-1 Typical Generator Voltages for 60 Hz Service.

Low voltage is found in smaller generators and is common for customer delivery. Medium voltage would usually be found in utility distribution systems and in larger generators. High voltage is usually reserved only for major transmission systems and can range anywhere from 30,000 to 1 million volts.

Generator Controls

Whenever a prime mover and generator are operated in parallel on the utility-derived power system, it is necessary to provide controls to assure that equipment is operated within its ratings. The role of the governor is to maintain the frequency within some specified range in the steady state, as load on the prime mover varies. Modern isochronous governors operate prime movers in response to load demand with no droop in frequency. Prime mover governors and controls are described in Chapter 13.

Generators are designed to produce a specific voltage under fixed conditions of field excitation, speed, and load. Any change in the connected load without a change in the alternator's excitation will cause the generator's voltage to change.

Voltage regulation is defined as the difference between the steady-state voltage under no load and the steady-state voltage at full-load output, expressed as a percentage of the full-load voltage. Most alternators control field strength to regulate voltage and have the ability to compensate for factors that affect voltage. Figure 25-8 shows a typical digital voltage regulator.

There are two types of voltage regulated generators:

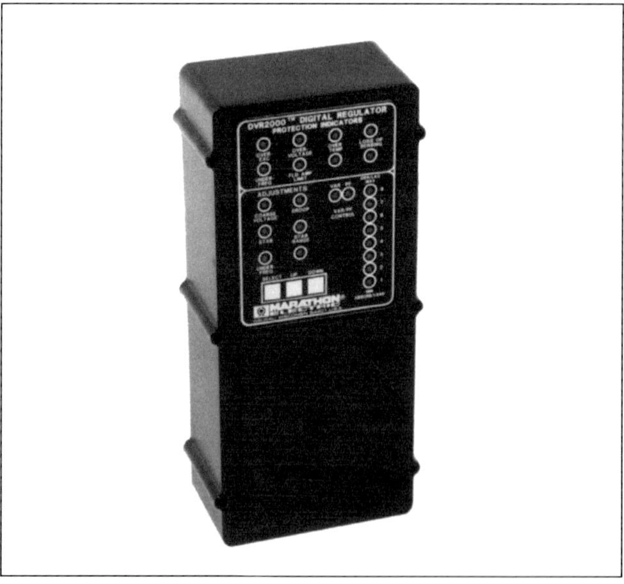

Fig. 25-8 Typical Digital Voltage Regulator.
Source: Marathon Electric

externally regulated and self- (or inherently) regulated. **Externally regulated** generators include devices that sense the output voltage and adjust the exciter output to maintain output voltage at the required set point. **Self-voltage-regulated generators** are simpler devices that use current transformers to compensate excitation based on load current.

In addition to load, the major factors affecting voltage regulation are speed and winding temperature. Rotational speed depends on the prime mover and its speed regulating system. Generators are inclined to vary voltage directly with the speed at which they are rotated and this can be problematic with self-voltage-regulated alternators. Generators equipped with external voltage regulators will provide better control due to their accuracy in compensating for speed variations.

Winding temperature depends on load current. The increased resistance from no load (with cold windings) to full load causes the output voltage to drop gradually as the windings warm up. This drop in voltage is referred to as temperature drift. Externally regulated units can compensate for this drift.

Load and power factor control are required for parallel operation with the utility grid. Load controllers output a proportional signal to the prime mover governor based on generator power output. A VAR/PF (or power factor) controller provides input to the voltage regulator to meet load requirements.

Generator Selection

Listed below are some of the main considerations involved in generator selection.

- **Synchronous versus induction.** Synchronous generators are slightly more efficient than induction units and may be used for any capacity application and any type of load. They can operate independently of the grid to provide power during standby or isolated operation. Induction generators are used for smaller applications (usually under 1,000 kW), where isolated operation is not required. They are less expensive than synchronous generators in smaller capacity ranges, have a more simple construction, and are easier to install and connect to the grid because they do not require synchronizing controls. Induction generators tend to place a high parasitic kVAR load on the utility-derived system and, in most cases, will require PF correction.
- **Voltage selection.** Generator voltage is determined by the voltage at the point of interconnection with the system. If the alternator operates in an isolated mode only (no parallel operation with the utility), then the economical choice would be to generate at the load utilization voltage. If parallel operation with the utility-derived system is contemplated, it may be desirable to operate at the delivery voltage to avoid the equipment cost and energy loss of voltage transformation. Other considerations that could require a voltage transformer between the generator and the utility interconnection are the need to isolate the generator from utility voltage fluctuations and higher frequency harmonics. Other considerations could include cable and switchgear costs or line losses.
- **Power ratings.** Prime mover/generator systems are rated in terms of standby power and prime power (or continuous duty). Standby ratings are designated for a brief and infrequent operating regime and the standby power rating will usually be greater than the prime power rating.
- **Insulating material temperature rating.** This will largely depend on the operating duty. Cost is considered with respect to the type of duty and the expected operating life and reliability of the unit.
- **Power factor (PF).** Generators are rated by their total current-carrying capacity in kVA and most three-phase units are rated at a minimum allowable PF of 0.8. With increased load PF, more output power will be delivered per kVA. Required kVA for a given load can be calculated as follows:

$$kVA = \frac{kW}{PF} \tag{25-4}$$

It can be seen that when voltage is held constant as PF decreases, more current is required to produce the same value of kW. As current flow through the winding increases, voltage drop increases and more power is dissipated in the form of heat. Operation of the generator outside its PF rating can cause overheating of its rotor or stator. This is typically not a problem for isolated operation, but merits special attention during parallel operation with the utility. In this case, a PF controller is required to monitor the flow of reactive power out of the generator and to make the necessary corrections to keep the generator within its power factor rating.

- **Transient loads.** If generators operate isolated from the electric grid, and motors constitute a large percentage of the connected load, the effect of motor starting will be a large drop in voltage, which may cause light flickering and instability on equipment with limited voltage variation tolerance. For induction motors, starting current is usually about six times that of running current (for high-efficiency motors, this can be 8 to 10 times).

- **Generator efficiency.** The power input requirements from the prime mover will be a function of the kW output requirement of the load and the alternator efficiency. Prime-mover power output (alternator input) requirements, expressed in hp, are calculated as follows:

$$hp = \frac{kW}{Generator\ efficiency \times 0.746} \qquad (25\text{-}5)$$

Prime movers are sometimes rated in kW or metric hp. The SI unit of kW refers to mechanical rotating power. This is different than electric power output from the alternator. A distinction is made between kW mechanical (kW_m) and kW electrical (kW_e). The relationship is calculated as follows:

$$kW_e = kW_m \times Generator\ efficiency \qquad (25\text{-}6)$$

Typically, generators have efficiencies ranging from 92 to 97%.

Temperature Effects, Insulation, and Cooling

Heat is the primary source of wear and eventual failure of rotating electric equipment. NEMA *Standard MG-1* defines ambient temperature as the temperature of the cooling air as it enters the ventilating openings of the unit. Alternators are rated for a maximum temperature rise above the NEMA ambient temperature standard of 104°F (40°C). Since operating temperature is the sum of ambient temperature and temperature rise, generators must be derated if ambient temperature exceeds the standard.

A further distinction is made by the type of operating duty. NEMA duty-cycle ratings are based on two classifications: continuous duty and standby duty.

- **Continuous duty.** This type of duty assumes continuous full-load operation. The allowable rated temperature rise ranges from about 140 to 300°F (60 to 149°C), depending on the class of insulation. Units that periodically run on overload (above rated temperature limits) will experience greater wear and be subject to premature failure.
- **Standby duty.** Allowable rated temperature rise for standby alternators may exceed continuous ratings by about 75 to 95°F (24 to 35°C), depending on the insulation class. Operation at elevated standby temperatures causes the insulation to wear at a rate four to eight times (per NEMA, MG-1, part 22.85) that of continuous-duty ratings. The increased wear is less a concern for standby units due to very low hours of operation.

Along with insulation type, generator cooling has a great impact on generator life. Generally, air cooling is used for small- and medium-capacity generators. Hydrogen and liquid cooling systems are used for medium- and large-capacity generators.

Figures 25-9 and 25-10 illustrate two types of air-cooled designs. Figure 25-9 is an open-ventilated cooling design schematic. Air is drawn through the generator by fans on the rotor and exhausted back to the surround-

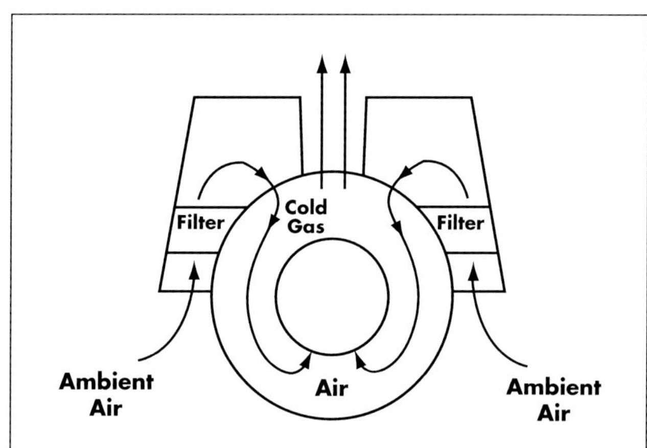

Fig. 25-9 Open-Ventilated Generator Schematic. Source: General Electric Company

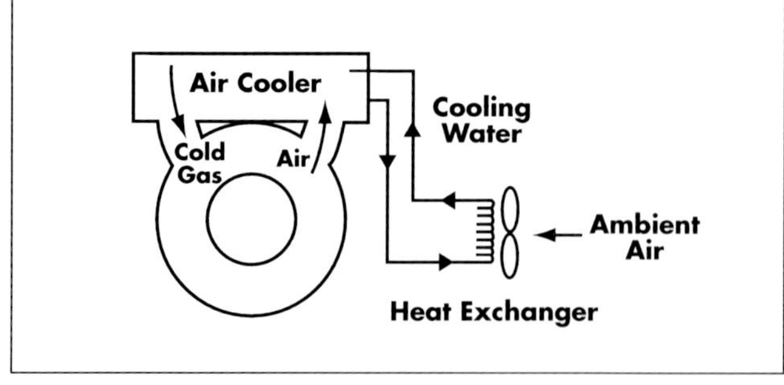

Fig. 25-10 TEWAC Generator Schematic. Source: General Electric Company

Fig. 25-11 Open-Case View Showing Design of Air-Cooled Small-Capacity Generator. Source: Kato Engineering/Reliance Electric

ing space outside the frame. Figure 25-10 is a totally enclosed water-to-air cooling (TEWAC) schematic. With this design, the generator is enclosed to prevent dirt and moisture from entering and water-to-air heat exchangers are used.

Hydrogen cooling has been applied to generators in a wide range of capacities, ranging from below 20 MVA to above 1,000 MVA, though its use is more common in capacities above 100 MVA. Hydrogen is an excellent cooling medium for generators due to its low density and high thermal conductivity relative to air. Historically, hydrogen cooling has resulted in higher generator efficiency and lower size than air cooling. These advantages are achieved, however, at a higher capital cost and with more complex installation, operation, and maintenance requirements. Water cooling is only used on very large capacity utility-grade generators. While providing the advantage of a vastly superior heat removal capability, the capital cost and operating complexity requirements make water cooling prohibitive in smaller capacity units.

Figure 25-11 provides an open-case view of a small-capacity air-cooled generator, revealing the air vents. Figure 25-12 shows a 3,500 kW, 4,160 Volt (1,800 rpm) generator featuring formed coils with mica insulation and epoxy insulation of all windings. The machine has a dual-end vent design to maximize air cooling effectiveness.

Figure 25-13 provides a cutaway view of a large-capacity TEWAC generator. The closed air circuit inside the generator performs all of the necessary cooling for the stator and rotor. Figure 25-14 shows a 300 MVA air-cooled generator designed for application with a 360 MW gas turbine generator set. The demonstrated mechanical efficiency of greater than 98% has previously only been achieved by hydrogen-cooled machines.

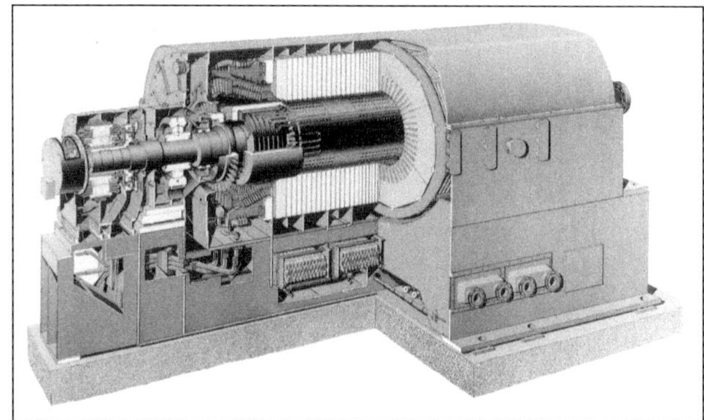

Fig. 25-13 Cutaway View of Large-Capacity TEWAC Generator. Source: ABB

Fig. 25-12 3,500 kW Air-Cooled Generator with Dual Vent Design. Source: Kato Engineering/Reliance Electric

Fig. 25-14 300 MVA Air-Cooled Generator. Source: ABB

HARMONICS

Harmonics are the multiples of the fundamental waveform frequency being produced by the generator and the load. Normally, since generators are magnetically symmetrical, only odd harmonics are significant. For example, a 60 Hz generated waveform will contain the 60 Hz fundamental, and may contain a 180 Hz 3rd harmonic, a 300 Hz 5th harmonic, a 420 Hz 7th harmonic, etc. Generally, the magnitude of the harmonic decreases with increasing harmonic order. Harmonics of lower order (i.e. the 3rd, 5th, and 7th, are the most significant).

As with motors, harmonics produce undesirable effects in the generator. Harmonic currents cause heat to be generated in the winding, core, and rotor and essentially act as derating factors since generator ratings are limited by allowable temperature rise. Harmonics also cause sine wave distortion. The distortion increases with increased harmonic content in the generated wave and, when severe, can cause voltage regulator sensing difficulties and inaccurate instrument readings.

The choice of coil pitch significantly impacts harmonic generation. As noted previously, in a generator, the magnetic field induces voltage in coils placed in slots in the armature. Maximum voltage is produced when the span of each of these coils is exactly equal to the span of the north and south field poles. This positioning is referred to as full pitch. Most generators are fractional pitch windings. For example, a 2/3 pitch eliminates most 3rd harmonics, while a 4/5 pitch eliminates most 5th harmonics. An example of a design that minimizes harmonics of lower order is a 5/6 pitch and a three-wire Delta connection. With a 5/6 pitch, 5th and 7th harmonics are low and, with the three-wire Delta connection, 3rd harmonic currents have no path.

Chapter Twenty-Six

Generator Driver Applications and Selection

Driver selection is one of the key elements in the development of an electric generation system. Once a general capacity range is identified, electric and thermal load characteristics are matched against available driver options and comparisons are made between general prime mover types (e.g., reciprocating engine, gas turbine, steam turbine) and sub-types within each category. Consideration must also be given to the energy sources available for each technology option, since thermal efficiency must be translated into operating costs. This includes traditional fossil fuels as well as a wide range of alternative energy sources such as recovered heat, biomass refuse, synthetic gases and solar. It also includes consideration of alternative generator drivers such as wind and water turbines.

One of the more commonly performed comparative analyses is between reciprocating engines and gas turbines. During preliminary driver selection for a cogeneration application, a reciprocating engine will be credited with providing greater power output per unit of available thermal load and higher thermal efficiency. While not always the case, reciprocating engine-generator sets are also usually assumed to incur a lower capital cost and offer better part-load thermal efficiency than gas turbines of similar capacity. Thus, for screening of applications that are not continuous and that require extensive part-load operation, reciprocating engine-generator sets are often first considered.

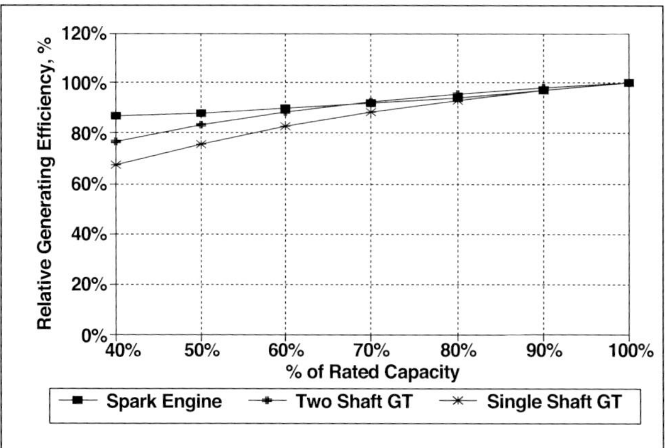

Fig. 26-1 Relative Effect of Part-Load Operation on Generating Efficiency. Source: Cogen Designs, Inc.

Figure 26-1 shows the relative effect of part-load operation on generating efficiency for a spark-ignition reciprocating engine, a two-shaft and a single-shaft combustion gas turbine. These curves reveal the benefits achieved by maintaining prime mover systems at full load. Notice that below 80% of rated capacity, the single-shaft turbine's relative generating efficiency starts to fall significantly, as compared with the reciprocating engine and two-shaft turbine. At 60% of rated capacity, the performance of the two-shaft turbine falls off relative to the reciprocating engine. A Diesel or dual-fuel engine will show even better part-load performance than the spark-ignited gas engine depicted in this graphic.

Gas turbines will more likely show superior economic performance during preliminary screening when thermal load requirements are high relative to electric loads and when higher temperature thermal output is required. When used with heat recovery-powered absorption or steam turbine-driven chillers for cooling applications, recovered heat from a combustion gas turbine will be more effective than heat recovered from a reciprocating engine system. This is the result of both the higher quantity of recoverable heat and the ability to power double-effect absorption or steam turbine-driven vapor compression systems with all of its thermal output.

For example, a moderately thermally efficient 2,000 kW reciprocating engine-driven electric cogeneration unit directly connected to a customized heat recovery absorption chiller/heater can provide about 8.5 MMBtu/h (8,959 MJ/h) of heating, or 650 tons (2,300 kW_r) of cooling capacity, or a combination of these heating and cooling outputs. A typical 2,000 kW gas turbine, however, can produce about 13 MMBtu/h (13,700 MJ/h) of heating, or 1,300 tons (4,600 kW_r) of cooling output.

For electric generation applications with more limited thermal loads or with widely varying thermal loads, both combined-cycle and steam injection-cycle systems will commonly be considered. These options provide operating flexibility in that heat recovery steam can serve either power generation or thermal loads, depending on demand. Supplementary firing can be used with either type of system. The steam injection cycle typically produces greater capacity for a given turbine model without supplementary firing and does so at a lower capital cost and with lower space requirements.

On the other hand, the combined-cycle system often achieves a superior heat rate and does not require abundant water use because its steam is condensed in a closed steam-cycle system. Thus, a simplified comparison shows the typical trade-off of lower capital cost for the steam injection-cycle system versus lower operating cost for the combined-cycle system. Generally, conditions of high peak electric rates, inexpensive water, and low load factor (i.e., full capacity is not required all of the time) tend to favor the steam injection cycle, while conditions of high load factor and high water and fuel costs tend to favor the combined-cycle system.

Once preliminary comparisons are made to match available alternatives to a given application, a more detailed analysis is required to determine if the project is economically feasible and to limit the driver options to a small set of candidates. At that point, a very detailed technical and financial analysis will be required before a prudent final decision can be made.

This chapter summarizes factors relating to the application of the main prime mover types in electric power generation systems. A number of detailed examples and illustrations are provided to show the effect of driver and heat recovery options in simple-, cogeneration-, combined-, and STIG-cycle applications.

Reciprocating Engine-Driven Electric Generation Systems

Internal combustion reciprocating engines are commonly used to provide standby (or emergency) power, as well as primary power. They offer many attractive features, including leadership in small unit, simple-cycle thermal fuel efficiency.

There are many types of reciprocating engine-driven electric generation systems with capacities ranging from 1 to 70,000 kW. Defining engine characteristics can include:

- Otto, Diesel or other cycles.
- Two- or four-stroke cycle.
- Atmospheric pressures or supercharged/turbocharged elevated pressures.
- Fuels used: natural gas, liquid natural gas (LNG) or landfill gas, light oil or heavy oil, or combinations of these fuels.
- Operation on variable stoichiometric mixtures (rich or lean burn) and injection methods for air emissions control.

Other distinctions include compression ratios, mean effective pressures (meps), emissions controls, coolant systems, combustion sequences, and a full array of component types.

Reciprocating Engine Characteristics

In commercial stationary power applications, reciprocating engines are most commonly found in the 50 to 10,000 kW capacity range. Two of the most prominent characteristics of reciprocating engines are their high simple-cycle thermal fuel efficiency and their large size and weight. Reciprocating engines are capable of generating electricity at thermal efficiencies of 25 to 45% (LHV basis) over a wide range of engine capacities. They are especially efficient relative to other prime movers in smaller capacities.

Reciprocating engine-generator sets are also relatively large and heavy compared with gas turbines and steam turbines. In small- and medium-capacity ranges, reciprocating engines are commonly available as factory-packaged power generation systems with air emissions and sound attenuation options, guaranteed performance, and factory testing. The packaged system offers low installation costs.

Reciprocating engines have their most successful applications where electric load is large in proportion to thermal load and where simple-cycle thermal fuel efficiency is particularly important to economic performance. They are attractive for cogeneration applications that use lower- temperature recoverable heat. Due to their high part-load efficiency relative to other prime movers, reciprocating engines are well-suited for applications with extensive part-load operation.

Engine-generator systems that do not use recoverable heat are well-suited for peak-shaving generation applications. In some instances, the high simple-cycle thermal fuel efficiency and reliability of large-capacity, low-speed reciprocating engines allows for successful application as prime power units, even without the benefit of heat recovery. Figure 26-2 shows a 179 MW power station featuring 15 nine-cylinder, in-line, 400 rpm Diesel engine-generators.

Figure 26-3 is an energy flow diagram for a small-capacity, spark-ignited, gas-fired reciprocating engine electric cogeneration system and Figure 26-4 is an energy flow diagram for a small-capacity, heavy oil-fired Diesel

Fig. 26-2 179 MW Power Station Featuring 15 Low-Speed Diesel Engines. Source: MAN B&W

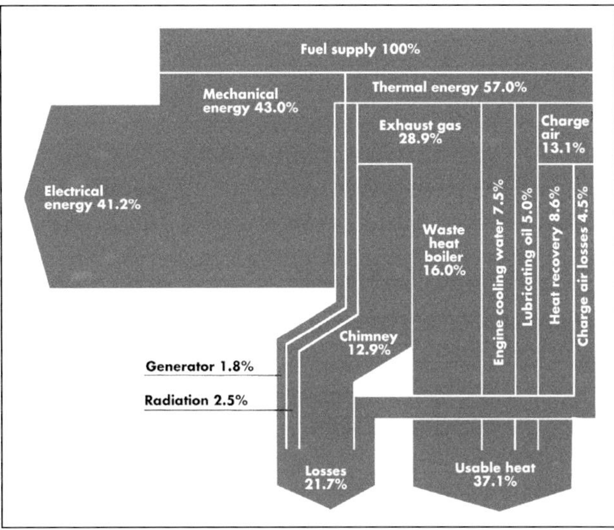

Fig. 26-4 Energy Flow Diagram for a Small-Capacity, Heavy Oil-Fired Diesel Reciprocating Engine Cogeneration System.
Source: MAN B&W

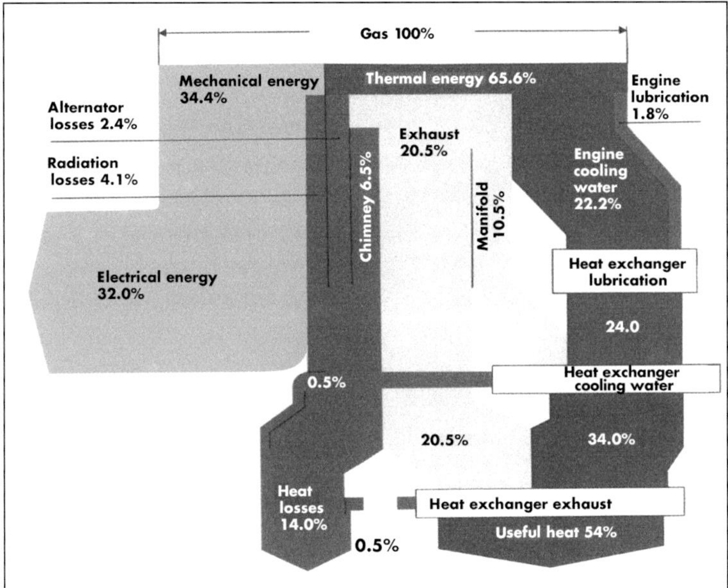

Fig. 26-3 Energy Flow Diagram for a Small-Capacity, Spark-Ignited Reciprocating Engine Cogeneration System. Source: MAN B&W

engine. Notice in these examples that the Diesel engine system produces a greater simple-cycle thermal fuel efficiency and the spark-ignited engine produces a greater total thermal fuel efficiency.

Sound and Vibration

Reciprocating engines, like other technologies relying on combustion, generate a full range of sound emissions from low to high frequency. The low frequencies are usually attributed to unbalanced forces (i.e., structure-borne noise), combustion exhaust, and other mechanical sources, such as valve trains, pumps, fans, and other ancillary equipment. High frequencies are often associated with turbochargers. Most smaller systems are available with acoustic enclosures that are typically designed to attenuate sound emissions to approximately 60 to 90 dba at a distance of 3 ft (1 m). In some applications, sound attenuation may not be required. Many reciprocating engine-generator sets achieve under 100 dba at a distance of 6 or 7 ft (2 m) without enclosure. With larger systems, the surrounding building features can be designed for sound attenuation.

Another common concern is the level of vibration associated with reciprocating engines. Torsional and lateral vibration may be eliminated or reduced to safe levels by appropriate shaft coupling applications and rotor balancing combined with appropriate provisions to isolate vibration from surrounding structures.

Operations, Maintenance and Reliability

One of the primary drawbacks of most reciprocating engines is their relatively costly, and frequent requirement for, maintenance. The abuses of friction and internal heat take a significant toll, because reciprocating engines have so many moving parts.

As detailed in Chapter 9, maintenance requirements of a reciprocating engine are fairly predictable. The maintenance intervals will be a function of the engine design, load characteristics, fuel quality, air quality, and type of duty cycle. Generally, lower rotational speed, piston speed, mep, and compression ratios will all result in longer relative engine life and a lower maintenance requirement. Likewise, the number of starts and stops the engine experiences will impact maintenance cost and frequency.

Routine maintenance includes filters, oil, oil sampling, water treatment, water-side and combustion-side deposit buildup inspections, plugs, and wires or injectors. Long-term (or overhaul) maintenance intervals for items such as valves, piston rings, and partial rebuilding range from 7,500 and 50,000 hours. Complete rebuilding intervals, including the replacement of components such as heads, pistons, liners, and bearings, typically range from 15,000 to 100,000 hours. Very large capacity, low-speed engines are at the high-end of the ranges provided for maintenance and rebuilding intervals and medium-speed industrial engines are in the middle of these ranges. Smaller capacity automotive derivative engines, which are at the low end of these ranges, may be designed for replacement rather than rebuilding.

Typically, single-engine electric generation applications achieve availabilities of 85 to 95%. [Availability level refers to the ratio of the amount of time a system is able to operate and perform its duty to the amount of time it is, or may be, called upon to perform.] Multiple engine systems that include some level of redundancy can achieve availability levels as high as 99%. Extremely large, very low speed reciprocating engines can also sometimes demonstrate availability levels as high as 99% and exhibit very low maintenance requirements. Since reliability is strongly linked to maintenance practices, well-implemented preventative maintenance programs generally improve availability and avoid exposure to downtime during inopportune periods.

Use of Rejected Heat

Economic returns from cogeneration projects are closely tied to baseload operation and effective use of available thermal output. In most cogeneration applications, systems run all or most of the time and it is important that the heat recovery equipment and operating strategy are carefully chosen to match system capacity with the facility thermal and electric load characteristics.

Figure 26-5 shows the various heat recovery circuits for a reciprocating engine cogeneration system. Many reciprocating engines are used to provide hot water via a series of heat exchangers that recover engine jacket water, oil cooler, turbo intercooler, and exhaust gas heat. However, there are possible alternative heat recovery systems that can be used to match thermal outputs to facility thermal characteristics.

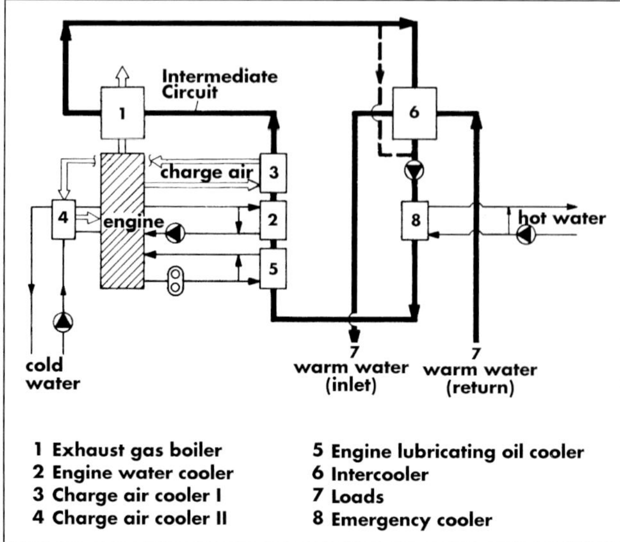

Fig. 26-5 Heat Recovery Circuits for Reciprocating Engine Cogeneration System. Source: MAN B&W

The reciprocating engine coolant system can alternatively produce hot water, low-pressure steam (via ebullient or high-temperature forced circulation cooling), or heated air (using a radiator). The exhaust gases can produce high- or low-pressure steam.

Heat recovery-generated low-pressure steam can be extremely useful. Many industrial facilities have low-pressure steam loads, such as water and solution heating, deaeration, space heating, absorption cooling, and desiccant drying. Figure 26-6 shows an ebullient-cooled, 800 kW

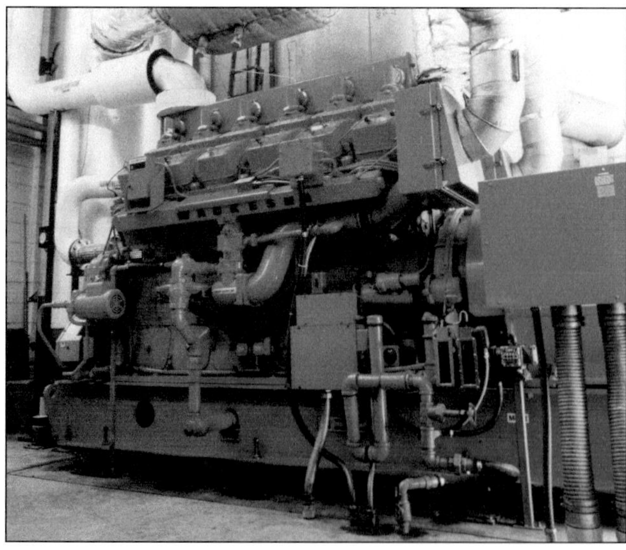

Fig. 26-6 800 kW Ebullient-Cooled Gas-Fired Reciprocating Engine Cogeneration System. Source: Waukesha Engine Division

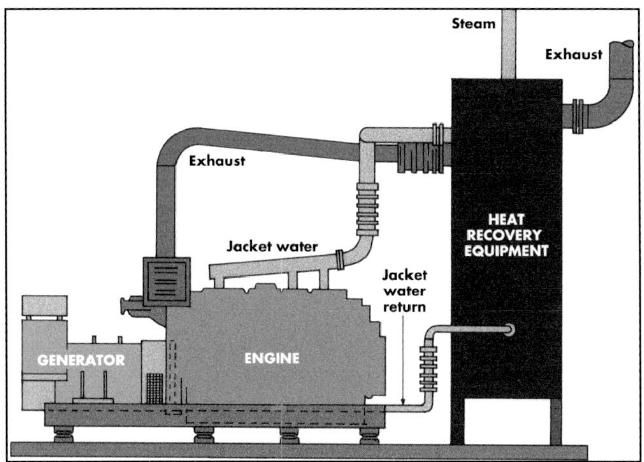

Fig. 26-7 Schematic Diagram of 800 kW Ebullient-Cooled Reciprocating Engine Cogeneration System. Source: Waukesha Engine Division

gas-fired reciprocating engine cogeneration system applied in an industrial facility. The heat recovery system produces 5,000 lbm/h (2,300 kg/h) of low-pressure steam for process solution tank heating, space heating, and for driving a 200 ton (700 kW_r) single-stage absorption chiller. Figure 26-7 is a schematic diagram of this system.

Figure 26-8 shows a flow diagram for a 1,558 kW reciprocating engine-driven cogeneration system with a fully packaged exhaust gas heat recovery silencer. Figure 26-9 shows a heat balance diagram of a 600 kW reciprocating engine system applied to cogeneration service. Notice that the engine jacket is combined with the exhaust heat in the waste heat boiler (WHP) to generate 10 psig (0.7 bar) steam and the oil and aftercooler heat is used separately for a lower temperature application.

Engine coolant and exhaust gas can also be coupled with a heat recovery absorption chiller/heater. Absorption chillers are especially useful in cogeneration applications where a year-round thermal load is required. Figure 26-10

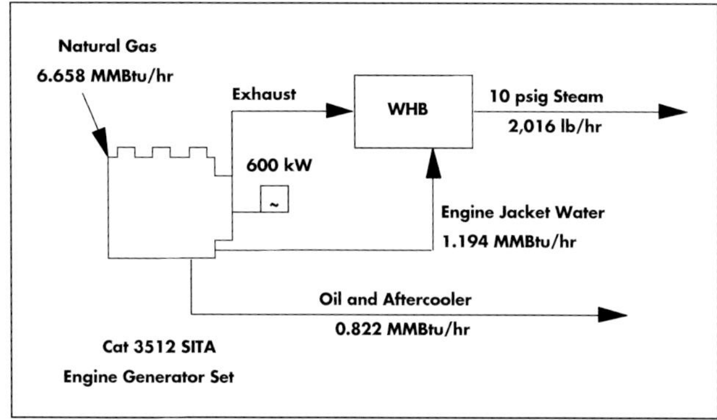

Fig. 26-9 Heat Balance Diagram of 600 kW Reciprocating Engine Cogeneration System. Source: Cogen Designs, Inc.

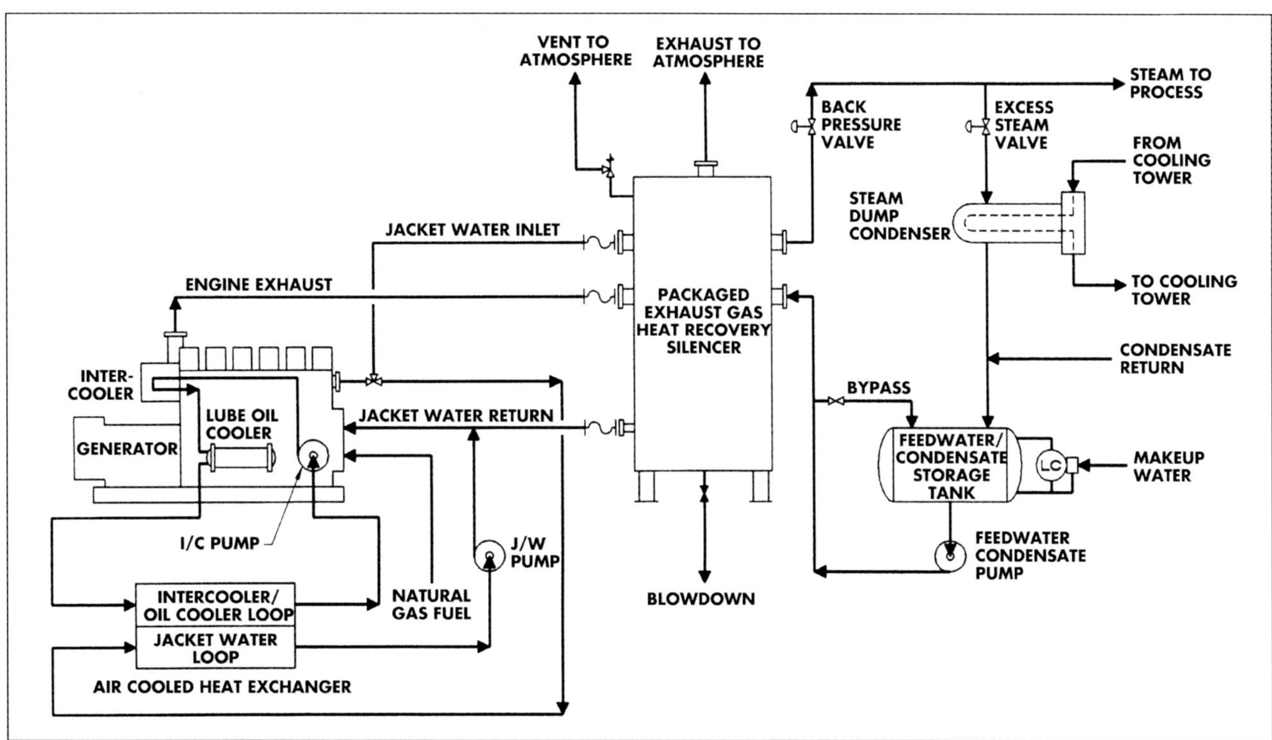

Fig. 26-8 Flow Diagram for 1,558 kW Reciprocating Engine-Driven Cogeneration System with Packaged Exhaust Gas Heat Recovery Silencer. Source: United States Turbine Corp.

illustrates the integration of a hot water-driven absorption chiller with a natural gas-fired reciprocating engine electric cogeneration system.

RECIPROCATING ENGINE APPLICATION EXAMPLES

Figures 26-11 through 26-18 provide examples of reciprocating engine electric power generation applications.

- Figure 26-11 shows three 2,500 kW, gas-fired, spark-ignition reciprocating engine-generator sets. These units operate at 900 rpm at 60 Hz (1,000 rpm at 50 Hz) at an mep of about 218 psi (15 bar), with an ISO rated heat rate of 8,742 Btu/kWh (9,223 kJ/kWh), for a simple-cycle thermal fuel efficiency of 39% (LHV basis). These generator sets weigh about 43 tons (39,000 kg), with dimensions of 29 ft (8.8 m) long, 6.7 ft (2.0 m) wide, and 10.5 ft (3.2 m) high.
- Figure 26-12 shows a dual-fuel compression ignition reciprocating engine-generator set applied in a cogeneration application. The engine and generator are mounted on a common modular frame, along with the heat exchangers for cooling water, lube oil, and charge air. The exhaust gas heat exchanger is installed separately. This seven-cylinder in-line engine produces 630 kW at an operating speed of 1,000 rpm.
- Figures 26-13 and 26-14 show skid-mounted reciprocating engine-generator sets. The skid-mounted package eliminates the need to build a heavy concrete structure to support the engine and generator. The skid also provides support for the essential electrical, mechanical, and piping systems. The complete unit includes factory engineering, fuel, lubrication, and cooling systems, engine control panel, generator, and exciter. These units may be equipped with a radiator or heat exchanger mounted on the skid. The skid may also be built with a complete housing.
- Figure 26-15 illustrates a 30 kW factory-assembled, gas-fired, spark-ignited automotive-derivative reciprocating engine cogeneration module. This module features an induction generator with integral switchgear for parallel operation and microprocessor controls. Module dimensions are about 7 ft (2.1 m) long by 2.5 ft (0.8 m) wide by 4.5 ft (1.4 m) high, and the weight is 2,600 lbm (1,180 kg). Acoustic levels are rated at 70 dba at a distance of 20 ft (6 m). The generator set achieves a simple-cycle thermal fuel

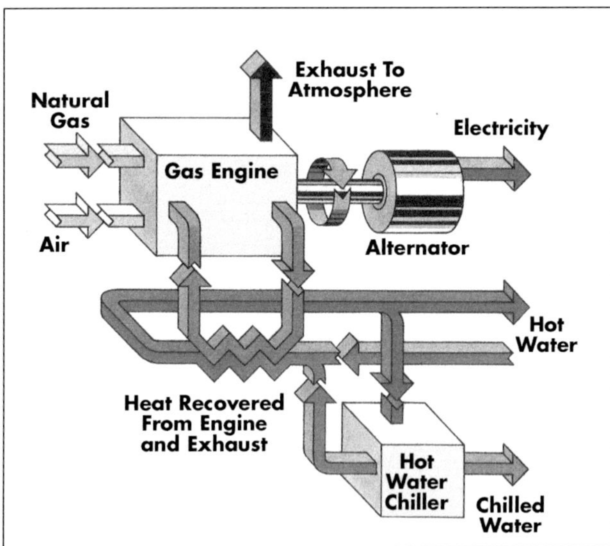

Fig. 26-10 Integration of Hot Water-Driven Absorption Chiller with Reciprocating Engine Cogeneration System. Source: British Gas

Fig. 26-11 Three 2,500 kW, Gas-fired, Spark-Ignited Reciprocating Engine-Generator Sets. Source: Wartsila Diesel

Fig. 26-12 Dual-Fuel Compression-Ignition Reciprocating Engine in Cogeneration Application. Source: MAN B&W

Fig. 26-13 Skid-Mounted, Spark-Ignited Reciprocating Engine-Generator Set. Source: Caterpillar Engine Division

efficiency of about 25% (HHV basis) and produces about 200,000 Btu/h (211,000 kJ/h) of hot water at temperatures of up to 230°F (110°C).

- Figure 26-16 shows a 649 kW Jenbacher natural gas-fired reciprocating engine-generator set applied to cogeneration service. This 16-cylinder engine features a compression ratio of 12.5:1, a bmep of 203 psi (14 bar) at the continuous power rating, and a nominal speed of 1,500 rpm. The engine achieves a thermal fuel efficiency of 37.4% (LHV basis). Exhaust gas flow rate is 8,584 lbm/h (3,894 kg/h) at a temperature of 959°F (515°C). Nearly 50% of the input fuel energy is recoverable from intercooler, oil cooler, and jacket water heat exchangers, with a maximum jacket water temperature of 194°F (90°C) and exhaust gas, at 248°F (120°C). Uncontrolled NO_X and CO emissions are reported to be 46 ppm (0.54 g/bhp-h) and 90 ppm (0.64 g/bhp-h), respectively (@ 15% O_2). The engine weight (dry) is about 9,000 lbm (4,000 kg) and the total system weighs about 24,000 lbm (11,000 kg) with dimensions of about 21 ft (6.4 m) long by 7 ft (2.1 m) wide and high. The 4,000 lbm (1,800 kg) generator operates at 1,800 rpm with an efficiency of 96.6% at a power factor (PF) of 1.0 and 95.8% at a PF of 0.8. The gear box efficiency is 98.7% with a reduction gear ratio of 1:1.2.
- Figure 26-17 shows one of two Waukesha 1,100 kW turbo-charged, intercooled, dual-fuel reciprocating engine cogeneration units applied at a sanitation station. Operating at 1,200 rpm synchronous speed, the twelve-cylinder engines run on natural gas or digester gas to provide about 2,100 kW of 4,160 volt, 60 Hz,

Fig. 26-14 Large-Capacity Packaged Reciprocating Engine-Generator Set with all Components Mounted on a Common Skid. Source: Fairbanks Morse Engine Division

Fig. 26-15 30 kW Factory-Assembled Automotive Derivative Reciprocating Engine Cogeneration Module. Source: Tecogen

Fig. 26-16 649 kW Natural Gas-Fired Reciprocating Engine-Generator Set. Source: Jenbacher Energies

Fig. 26-17 One of Two Dual-Fuel (Natural Gas and Digestor Gas) Reciprocating Engine Cogeneration Units at a Sanitation Station. Source: Waukesha Engine Division

three-phase power. The jacket water, at a temperature of 210°F (99°C), and the exhaust gas, at a mass flow rate of about 10,000 lbm/h (4,500 kg/h) and a temperature of 1,100°F (600°C), pass through separate heat exchangers that heat water in a separate closed loop to 180°F (82°C). This hot water is used to help the anaerobic digester process produce methane gas, which is then used to fuel the engines. The digester gas flows from a large storage sphere through water traps and filters before being delivered to the engines at a pressure of about 25 psi (1.7 bar). Automatic controls are used to shut off the digester gas supply and open the natural gas supply when storage sphere pressure falls below 35 psi (2.4 bar) and to switch back to digester gas once a pressure of 70 psi (4.8 bar) is achieved.

Fig. 26-18 Mechanical Room View of Four Reciprocating Engine-Generators Applied in 5,500 kW Cogeneration Application. Source: Jenbacher Energies

The engine-generators use two 24 volt dc electric starting motors and are controlled by an electronic governor through a mounted electric actuator and magnetic pickup. The engines feature air-to-water intercooler and dry type, wastegate-controlled turbochargers. These V-type engines each have a piston displacement of 7,040 in^3 (115 liters) and a weight of about 36,000 lbm (16,300 kg), with dimensions of about 18 ft (5.5 m) long by 7 ft (2.1 m) wide and 9 ft (2.7 m) high.

- Figure 26-18 shows a 5,550 kW cogeneration system at a New England college, featuring four gas-fired, turbocharged, aftercooled, 1,500 rpm reciprocating engines. Each unit drives a synchronous generator through a gear at 1,800 rpm, producing 1,388 kW of electric output at 4,160 volts. Each unit can produce 2,000 lbm/h (907 kg/h) of steam at 125 psig (8.6 bar) in a vertical firetube heat recovery boiler/silencer.

Engine exhaust gas is cooled from 860 to 500°F (460 to 260°C) in the heat recovery steam generator (HRSG) and then passes through an economizer where it is cooled to a final discharge temperature of 350°F (177°C). The economizer boosts the engine cooling water loop to 214°F (101°C) for use in a single-stage absorption chiller. The chiller is derated to operate with a hot water supply of 214°F (101°C). Prior to entering the economizer, the cooling water passes through three heat exchangers on each engine skid to provide engine cooling.

Upon leaving the economizer, the cooling water provides thermal energy to the absorption chiller, an oil storage tank heater, a domestic water heater, and a boiler feedwater heater. Any excess heat can be rejected via a bank of air-cooled radiators on the roof of the plant. The engines use a low-NO_X combustion system that produces NO_X emissions of 0.015 ounces (0.43 grams) per kWh. CO and NMHCs are controlled by a two-way catalytic converter.

Electrical output of the engines is controlled by an import/export controller that tracks the campus electric load. The engines are automatically started and stopped in response to changes in the total electric load. The engine cooling water loop is controlled by a central digital control system that also controls the entire chiller plant. Steam is produced by the cogeneration system on a baseload basis, with the main boilers tracking the campus steam load.

Normal operation features three of the engines on-line, with the fourth available for standby duty. In the event of a utility electrical problem, the intertie with the electric grid is opened and the entire campus is powered by the cogeneration plant. The plant can also export up to 1 MW of power to the utility when required. The thermodynamic heat balance for the three engines operating at full-load is shown in Table 26-1.

Table 26-1 Cogeneration System Heat Balance

	MMBtu/h	(MJ/h)	% Total Input
Fuel input	39.18	(41,335)	100
System usable outputs:			
Electric output (4,161 kW)	14.20	(14,981)	36.2
Steam output	6.00	(6,330)	15.3
Economizer	4.05	(4,273)	10.3
Jacket water	9.09	(9,590)	23.2
Total usable energy	33.34	(35,174)	85.1
System losses:			
Exhaust gas loss	3.24	(3,418)	8.2
Engine radiation	1.90	(2,005)	4.9
Generator losses	0.70	(739)	1.8
Total losses	5.84	(6,161)	14.9
Gross heat rate (no heat recovery)	9,416 Btu/kWh (9,934 kJ/kWh)		
Net heat rate (full heat recovery)	4,816 Btu/kWh (5,081 kJ/kWh)		
PURPA Efficiency	60.6%		

Combustion Gas Turbine-Driven Electric Generation Systems

Stationary combustion gas turbine engines have been prominent in large-scale power cogeneration over the past two decades. Recent advances in emissions control, production cost reductions, and ease of operation have allowed combustion gas turbines to achieve further market penetration in the medium- and large-scale commercial peaking and cogeneration industries, where they have become cost-competitive with conventional steam power plants. They have also had significant market penetration in capacities ranging down to 1 MW.

Commercially available gas turbine engine electric generation systems range from 30 kW to more than 100,000 kW, though capacities of below 400 kW are still uncommon. Designs and configurations include:

- Open or closed cycle.
- Aeroderivative or industrial design.
- Single-shaft or multi-shaft design; alternative compressor and combustor types.
- Fuels used: natural gas, LPG, alcohol, kerosene (jet fuel), coal-derived gas, refinery gas, landfill gas, naphtha, distillate oils, crude oils, and residual oils.
- Emissions control applications, i.e., wet injection, dry low-NO_X (DLN) combustors, and selective catalytic reduction (SCR) systems.
- Capacity enhancements based on injection of steam or water.

Other distinctions include number of stages, pressure ratios, operating pressures, emissions controls, coolant systems, combustion sequences, and materials used. Cycle enhancement techniques include intercooling, reheat, and regeneration (recuperation).

Combustion Gas Turbine Characteristics

Gas turbines may be characterized by the following qualities:

- High power density resulting from high rotating speeds, materials derived from flight technology, and absence of reciprocating parts.
- High mass flow and exhaust temperatures that allow for generation of large quantities of steam or process heat.
- High level of availability and low frequency and cost of maintenance. While this is not always true, it is commonly found to be the case in most capacity ranges.

Gas turbine electric generation shows simple-, combined- and cogeneration-cycle efficiencies that are equal to or higher than conventional coal power plants. Large- and medium-capacity units in heat recovery applications exceed the thermal efficiency of the best available utility power generation cycle. Smaller gas turbine systems offer lower simple-cycle thermal efficiencies than reciprocating engines of similar capacity, but offer a greater proportion of recoverable heat.

Combustion gas turbines in a wide range of capacities are available as factory-packaged power generation systems with emissions and sound attenuation options, guaranteed performance, and factory testing. The packaged system offers relatively low installation costs and

Fig. 26-19 Assembly of Small-Capacity Packaged Gas Turbine Generator Set. Source: Kawasaki Heavy Industries, LTD

can minimize space requirements. Figure 26-19 shows the assembly of a small-capacity packaged Kawasaki gas turbine generator set. Larger cogeneration systems with HRSGs can be delivered to the site as separate standardized assembled components.

Commonly, gas turbine electric generation systems are applied in facilities that feature large baseload, high- or intermediate-pressure steam loads, where thermal loads are relatively large in proportion to electric loads. In these applications, somewhat lower simple-cycle thermal fuel efficiency is offset by increased heat recovery. However, their high power density, high level of reliability, and the potential to recover high-temperature heat also make gas turbines attractive for a wide range of other electric generation applications, including stand-by power service, small capacity cogeneration, and stand-alone prime power service in remote locations.

Operation, Maintenance, and Reliability

When applied properly, maintained well, supplied with clean fuel and air, and operated and monitored properly, gas turbine electric generation systems offer higher reliability and significantly lower life-cycle maintenance costs than most available alternatives.

Preventative maintenance procedures typically include a shutdown every 3 to 6 months for detailed inspection and routine maintenance. Monthly inspections that do not require shutdowns are also required to check fittings, filters, oil levels, and instrumentation and to perform cleaning. Control of fuel quality and control of airborne particles with high-performance primary air filters are critical routine procedures. Time between overhaul (TBO) may range from 15,000 to 100,000 running hours, depending on the quality of applied equipment, type of duty cycle, load characteristics, and the quality of the preventative maintenance program.

Gas turbine generation systems typically offer high on-line availability. Availability factors of 95% and higher are common. In most electric generation applications, this is critical to the realization of savings and/or fulfillment of power sale obligations. Key elements of high availability include routine maintenance, provisioning and inventory maintenance of spare parts, and careful planning for major overhauls.

Use of Rejected Heat

Gas turbine systems provide most or all of their recoverable waste heat in exhaust gas at between 850 and 1,100°F (450 and 600°C). This valuable, high-temperature heat can be used in a wide range of applications, including direct contact process heating and indirect heating/steam production. Watertube boilers, including once-through designs, are used almost exclusively. Depending on the turbine's simple-cycle thermal efficiency and the temperature and pressure of the steam being generated, a heat-recovered unfired gas turbine system will typically produce 4 to 7 lbm (1.8 to 3.2 kg) of steam per kWh generated.

The thermal output of gas turbine systems can be easily augmented through the use of supplemental firing. This refers to the use of a duct burner or a register burner in the hot gas duct upstream (i.e., ahead) of the HRSG to raise the temperature of the exhaust gas. This process increases steam mass flow (the amount of steam generated) and may improve steam conditions (temperature and pressure) as well. With duct firing, the exhaust temperature can be elevated to as high as 1,800°F (982°C) and temperatures exceeding 2,200°F (1,478°C) may be accomplished by register burners in boilers equipped with radiant heat transfer. Figure 26-20 shows the typical heat recovery performance curves for the FT8 gas turbine with a single-pressure level HRSG.

A **regenerator**, or **recuperator**, is a heat exchanger that transfers heat from turbine exhaust gas to compressor discharge air prior to fuel combustion, displacing a portion of fuel heat input that would otherwise be required. Unit capacity (as a result of heat exchanger pressure drops) is usually decreased somewhat, but cycle thermal fuel efficiency is increased. Reduced exhaust temperatures limit the quality and quantity potential for exhaust gas heat recovery.

Recovered heat can be effectively used to drive either single- or double-stage absorption chillers. Usually, this is accomplished with heat recovery-generated steam, though turbine exhaust can also be used directly with a custom-designed heat recovery absorption chiller/heater to produce chilled and hot water. Figure 26-21 illustrates the integration of a steam-driven absorption chiller with a gas turbine cogeneration system.

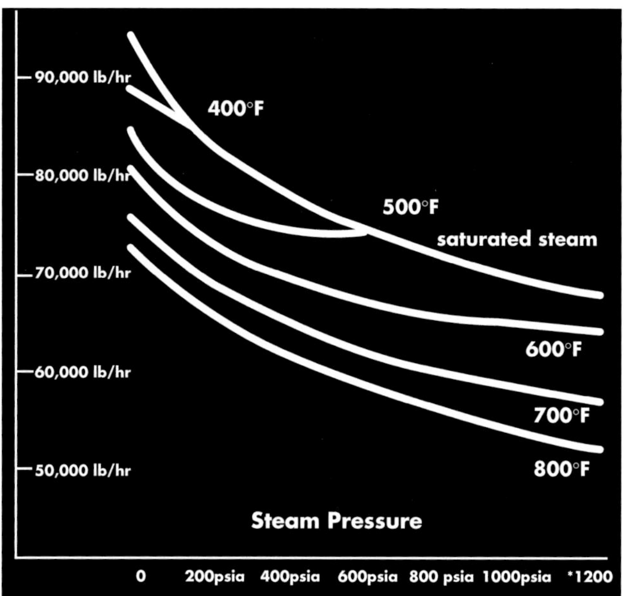

Fig. 26-20 Typical Heat Recovery Performance Curves for FT8 with Single-Pressure HRSG. Source: United Technologies Turbo Power Div.

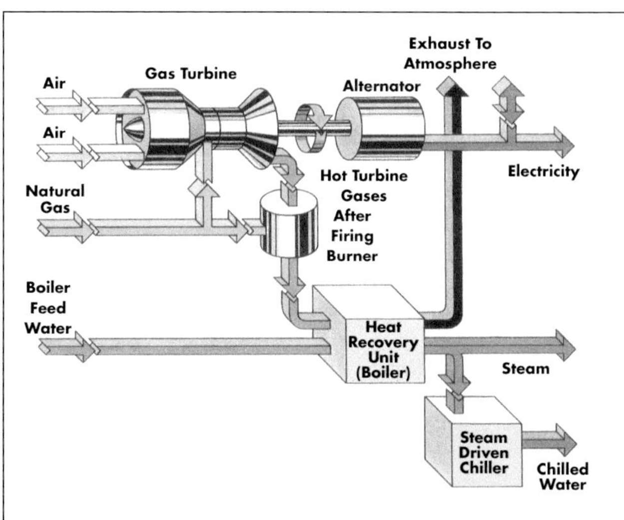

Fig. 26-21 Integration of Steam Driven Absorption Chiller with a Gas Turbine Cogeneration System. Source: British Gas

Gas Turbine Application Examples

Figures 26-22 to 26-27 provide examples of gas turbine electric power generation applications.

- Figure 26-22 shows a 1,050 kW packaged gas turbine generator set. This unit features a Kawasaki gas turbine and a Kato generator, with an off-skid-mounted electrohydraulic starting system. Exhaust heat is used to produce steam to serve thermal load requirements at a university campus.
- Figure 26-23 shows a skid-mounted Kawasaki M7A-01 gas turbine generator set designed for cogeneration-cycle operation. With operation on natural gas and steam injection used for NO_X control, at a rate of 150% of fuel flow, the rated baseload capacity of this generator set under ISO conditions is 5,480 kW. The twelve-stage axial compressor, which achieves a pressure ratio of 12.7:1, and the four-stage axial turbine both operate at 14,000 rpm. Inclusive of

Fig. 26-22 1,050 kW Packaged Gas Turbine-Driven Cogeneration System Applied at a University. Source: United States Turbine Corp.

Fig. 26-23 Skid-Mounted Gas Turbine Generator Set Designed for Cogeneration Duty. Source: Kawasaki Heavy Industries, LTD.

assumed inlet and outlet losses and a generator efficiency of 96%, the rated simple-cycle electric generation efficiency is 28.2% (LHV basis). The module is configured with a 250 kW ac motor with torque converter for starting and a single-stage parallel shaft gear unit with a gearbox shaft output speed of 1,800 rpm for 60 Hz and 1,500 rpm for 50 Hz electric service. The brushless synchronous generator has a maximum capacity of 7,059 kVA with a PF of 0.85. A 160 hp (120 kW) single-stage screw compressor is used to deliver natural gas to the combustor at a pressure of 256 psig (18.7 bar).

This unit can be packaged for cogeneration applications in a prefabricated, self-standing, steel, sound-attenuating enclosure that achieves a sound emission level of 85 dba at 3.3 ft (1 m). While the gas turbine itself only weighs about 10,000 lbm (4,500 kg), the entire generator set package weighs about 160,000 lbm (73,000 kg), with dimensions of 36 ft (11 m) long by 12.5 ft (3.8 m) wide by 24 ft (7.4 m) high, including the intake air filter.

- Figure 26-24 is a system flow chart of a cogeneration application featuring a dual-fuel-fired M7A-01 gas turbine. The relatively high exhaust temperature of about 1,000°F (540°C) allows for a high rate of heat recovery (about 49% of the thermal fuel energy input) from the 2 million ft³/h (60,000 m³/h) exhaust gas flow. About 30,000 lbm/h (13,000 kg/h) of saturated steam can be generated at a pressure of 240 psig (17.6 bar), with a feedwater (economizer inlet) temperature of 140°F (60°C). Steam flow paths are shown for steam injection and process use. About 14% of the steam output is used for NO_X emission abatement, with the rest available for process use. The system flow chart shows both natural gas and oil fuel delivery systems. The cooling circuit, served by a cooling tower, and the feedwater circuit are also shown.
- Figure 26-25 is a model of an ABB 83 MW dual-fuel gas turbine electric-generation system module. This heavy-duty industrial generator set features a silo-type dry low-NO_X combustor which, when equipped with

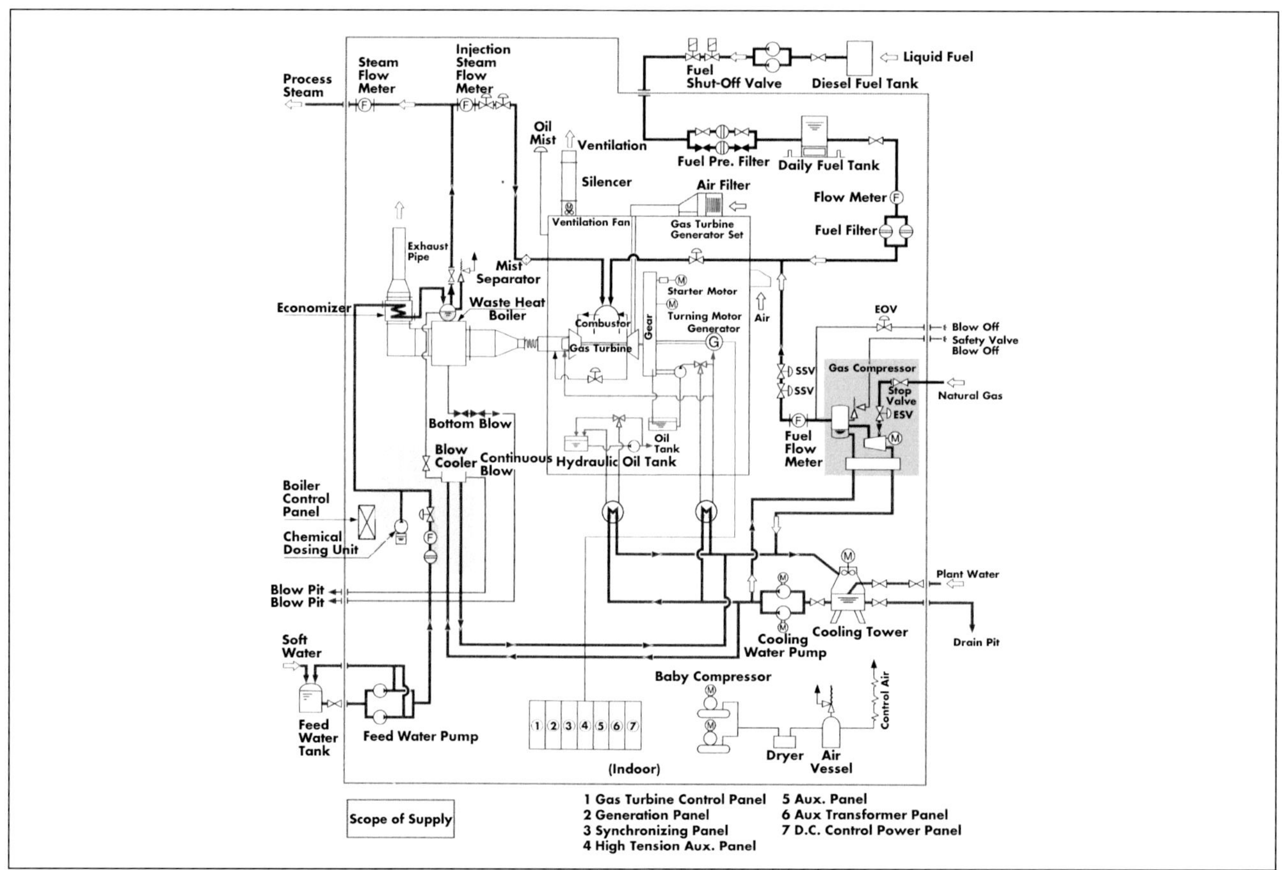

Fig. 26-24 System Flow Chart of a Cogeneration Application Featuring a Dual-Fuel-Fired Gas Turbine. Source: Kawasaki Heavy Industries, LTD

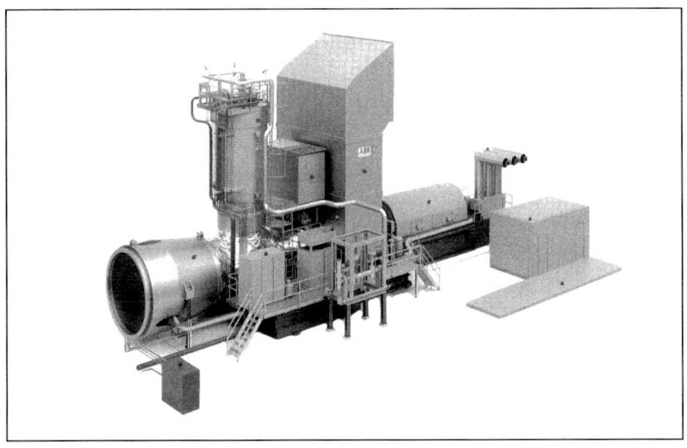

Fig. 26-25 Model of an 83 MW Heavy-Duty Industrial Gas Turbine Electric Generation System Module. Source: ABB

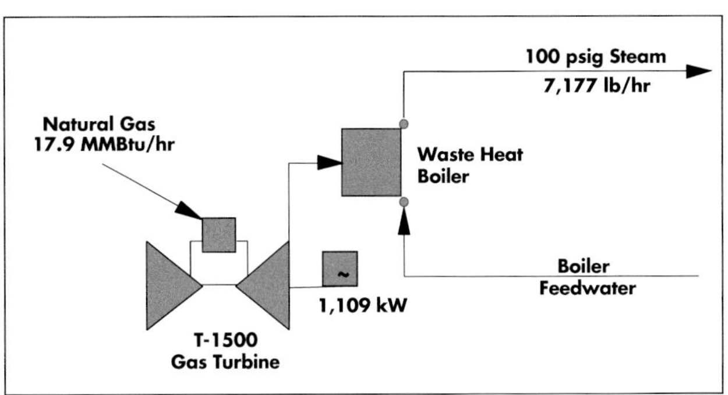

Fig. 26-26 Heat and Material Balance for 1,100 kW Gas Turbine-Driven Electric Cogeneration System. Source: Cogen Designs, Inc.

36 EV burners, can reportedly achieve NO_X emissions levels of 9 ppmvd (corrected to 15% O_2) with operation on natural gas. The full load simple-cycle heat rate for this unit is 10,400 Btu/kWh (10,972 kJ/kWh) on an LHV basis. The unit features a five-stage turbine, as well as an eighteen-stage compressor that achieves a pressure ratio of 13.3:1. The exhaust mass flow is 699 lbm/s (317 kg/s) at a temperature of 941°F (505°C). The generator is driven at a shaft speed of 3,600 rpm and produces a nominal output of 137 MVA at 60 Hz with a PF of 0.8, a rated voltage of 13.8 kV, and a rated current of 5,732 amps. The gas turbine module weighs 242,550 lbm (110,000 kg) with dimensions of 25 ft (7.5 m) long by 12 ft (3.5 m) wide by 13 ft (4 m) high. The generator exciter module weighs 277,830 lbm (126,000 kg) with dimensions of 26 ft (8 m) long by 11 ft (3 m) wide by 12 ft (3.5 m) high. All auxiliaries, control system, and electrical equipment are consolidated into one package. The combined weight of the packaged generator set module is 604,000 lbm (274,000 kg).

- Figure 26-26 shows a heat and material balance for 1,100 kW gas turbine-driven cogeneration system, on an HHV basis, featuring a single-pressure HRSG applied in a hospital. In this system, all recovered heat is delivered to the hospital's steam loop in the form of 100 psig (7.9 bar) steam. Deaerator steam would be taken from the 100 psig (7.9 bar) line through a PRV.
- Figure 26-27 shows a preliminary heat balance diagram for a 3,967 kW gas turbine cogeneration system applied at an industrial site. On an LHV basis, 46.56 MMBtu/h (49,121 MJ/h) is supplied to the turbine's combustor at a pressure of 225 psig (16.5 bar) via the fuel gas compressor.

An additional 48 MMBtu/h (50,640 MJ/h) can be introduced through the duct burner for supplementary firing. This raises the exhaust gas temperature

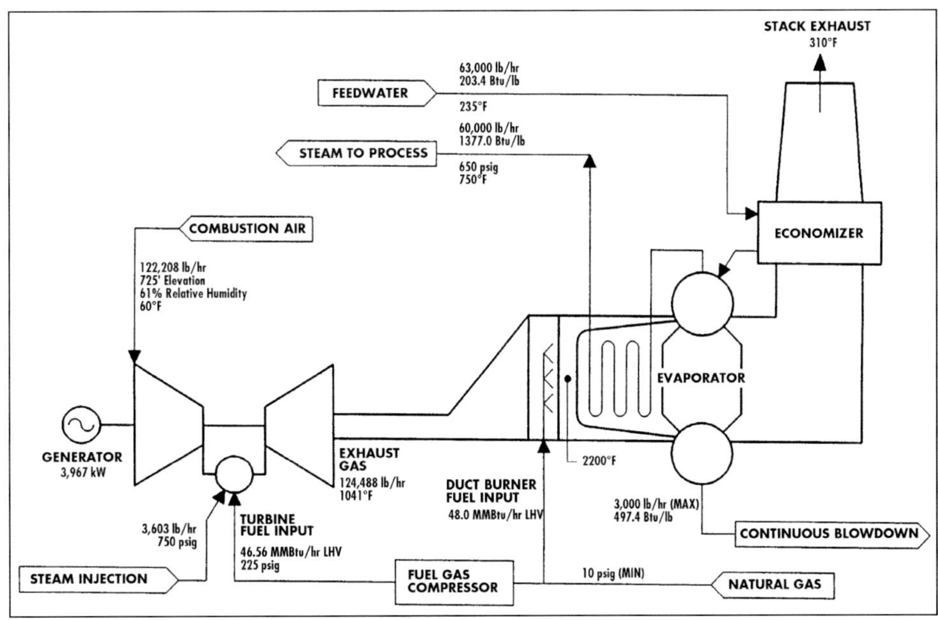

Fig. 26-27 Preliminary Heat Balance Diagram for Gas Turbine Cogeneration System Applied at Industrial Site. Source: United States Turbine Corp.

from 1,041 to 2,200°F (561 to 1,204°C). With a final stack exhaust temperature of 310°F (154°C), the total steam production comes up to 60,000 lbm/h (27,200 kg/h) at a condition of 650 psig/750°F (45.8 bar/ 399°C). Com-bustion air requirements and actual site conditions that will impact gas turbine performance are also shown.

Steam Injection-Cycle Systems

Steam and water injection (wet injection) are used to reduce NO_X emissions in gas turbines. With the use of steam injection in the simple-cycle gas turbine configuration, power output can be increased by as much as 50% and thermal efficiencies of 45% or more (LHV basis) can be achieved. The trade-off is the decrease in the amount of steam available for thermal processes.

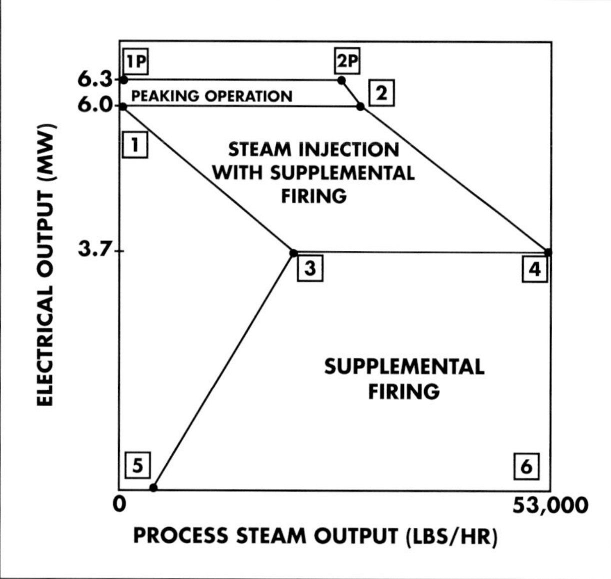

Fig. 26-28 Cheng- (STIG-) Cycle Operating Range Example Based on Allison Gas Turbine. Source: United States Turbine Corp.

For many stationary turbines, some modifications are required for operation with steam injection rates that are equal to more than a few percent of the total air mass flow.

Steam Injection-Cycle Application Examples

Figures 26-28 through 26-30 provide examples of steam injection-cycle applications.

- Figure 26-28 shows the operating range in electric output versus process steam output for a Cheng™- (or STIG™-) cycle system featuring an Allison gas turbine. The system can operate in steam- and electric-following load, steam-following mode, or maximum power output mode. Operating points are summarized in Table 26-2.

  Steam- and electric-following load matches facility steam and electric load requirements simultaneously at any point within its operating regime. This mode is particularly effective when power sales opportunities are limited, or when the plant is operating in isolation from the utility grid. The steam-following mode consists of operating along Line 1-3-4. Facility steam demand determines the operating point, and electric generation is allowed to follow accordingly. Steam injection takes place only as necessary to efficiently meet low points in steam demand. When there is minimal steam demand, the unit operates at Point 1 and produces about 6,000 kW and 3,497 lbm/h (1,586 kg/h) of steam. Power output is reduced to about 3,600 kW at Point 3 as about 24,000 lbm/h (10,900 kg/h)

Table 26-2 Nominal Operating Performance Specifications for Cheng-Cycle Operation on Gas Fuel

Operating points	1	2	3	4	5	6	1P	2P
Power output – (kW from generator)	5,997	5,997	3,601	3,601	0	0	6,256	6,256
Process steam output – (lbm/h)	3,497	40,285	23,947	44,938	6,458	52,739	6,288	40,203
Turbine fuel – (MMBtu/h – LHV)	55	55	47	15	15	15	58	58
Duct burner fuel – (MMBtu/h – LHV)	0	40	0	24	0	53	0	38
Injection steam – (lbm/h)	21,600	21,600	0	0	0	0	21,600	21,600
Temperature – (°F)	997	997	1,040	1,040	510	510	1,040	1,040
Mass flow rate – (lbm/sec)	41	41	35	35	35	35	41	41

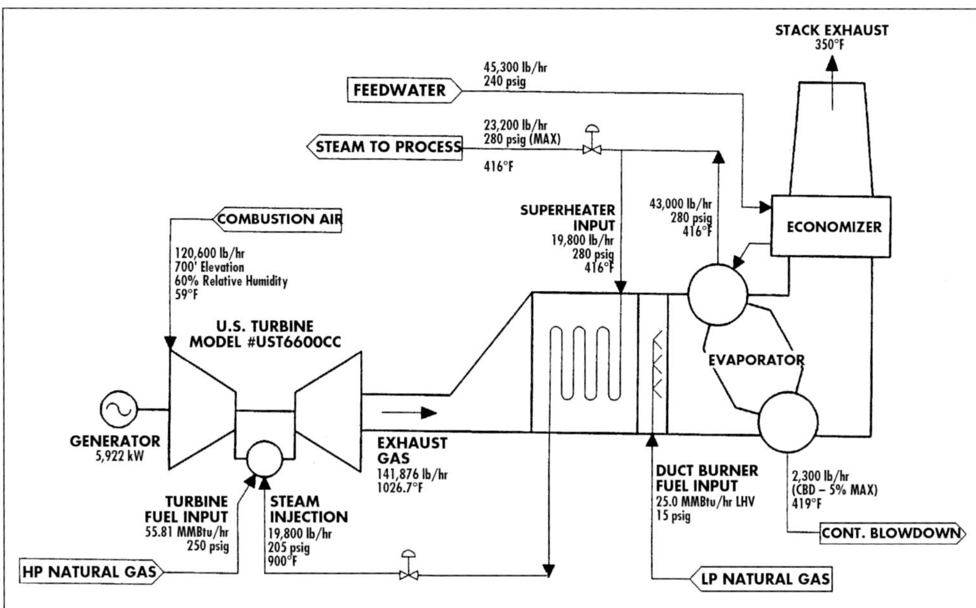

Fig. 26-29 Heat Balance Diagram for 5,922 kW Gas Turbine STIG-Cycle Cogeneration System. Source: United States Turbine Corp.

of steam is used to meet facility requirements. When steam demand exceeds this amount, power output remains constant and supplementary firing is used to provide a maximum output of about 45,000 lbm/h (20,400 kg/h). Maximum power output occurs along Line 1P-2P or in the areas immediately below it. This peaking mode is recommended for a maximum of about 1,250 hours per year and is typically used for peak shaving during the highest electric rate periods.

- Figure 26-29 shows a heat balance diagram for a gas turbine steam injection-cycle plant with a maximum electrical output of 5,922 kW. This is achieved with a maximum energy input to the gas turbine combustor of 55.81 MMBtu/h (58,880 MJ/h) of natural gas at a pressure of 250 psig (18.3 bar) and 19,800 lbm/h (8,981 kg/h) of steam. 43,000 lbm/h (19,500 kg/h) of steam at a condition of 280 psig/416°F (20.3 bar/213°C) is produced from a combination of 141,876 lbm/h (64,354 kg/h) of turbine exhaust, at a temperature of 1,027°F (553°C), and 25 MMBtu/h (26,375 MJ/h) of supplementary-fired natural gas. Under maximum steam injection conditions, 19,800 lbm/h (8,980 kg/h) of steam is diverted from the main steam header to the superheater section of the HRSG and then passed to the turbine combustor at a condition of 205 psig/900°F (15.2 bar/482°C).
- Figure 26-30 shows a STIG™-cycle system fully packaged for an outdoor installation, with the electric transformers and power distribution grid shown in the foreground. To the right of the package are the air filters. The middle section contains the gas turbine with the lube oil coolers and ventilation fans resting above it outside of the enclosure. The generator is housed on the left, with the exhaust ventilation duct above it. Excluding the HRSG, the dimensions for this 405,000 lbm (184,000 kg) packaged system, inclusive of air filter, are 75 ft (22.9 m) long by 21 ft (6.4 m) wide by 30 ft (9.1 m) high.

This system features a 22 MW GE LM2500 gas turbine with a simple-cycle ISO design condition heat rate of 9,280 Btu/kWh (9,790 kJ/kWh) on an LHV basis. The exhaust mass flow of the STIG™ unit is 167 lbm/s (76 kg/s) at a temperature of 926°F (497°C), producing an HRSG output of up to 75,000 lbm/h

Fig. 26-30 28 MW Packaged STIG-Cycle System. Source: Stewart and Stevenson

(34,000 kg/h) of steam at a pressure of 550 psi (37.9 bar). A minimum of 18,000 lbm/h (8,200 kg/h) of steam at 525 psi (36.2 bar) is injected for NO_X emissions suppression. Up to 50,000 lbm/h (23,000 kg/h) of high-pressure steam can be injected via compressor discharge ports and fuel nozzles to augment power output to a maximum of 28 MW. The remaining steam is used for process applications. With full steam injection, the STIG-cycle ISO rated heat rate is 8,325 Btu/kWh (8,783 kJ/kWh). As noted in chapter 12, the Brush electric generator is direct-driven by the hot end of this two-shaft machine at 3,600 rpm and produces 60 Hz ac power at 13,800 volts.

Steam Turbine-Driven Electric Generation Systems

Steam turbine systems are available in virtually any capacity, ranging from only a few kW to several hundred thousand kW. Steam turbines are the dominant technology for very large scale centralized electric generation in large utility plants. Steam turbine electric generation is potentially viable at any facility that has a high-pressure steam system. Low-pressure steam turbines are also available for special applications. Since steam can be generated with virtually any type of fuel, in some cases, where recovered heat, biomass, refuse, coal, geothermal or even solar energy can be used, steam energy can potentially be produced at a very low cost. The result can be low-cost power generation.

There is a variety of steam turbine types and configurations including:

- Single-stage or multi-stage designs.
- Steam exhaust conditions, including condensing, non-condensing, automatic extraction, mixed pressure, and others.
- Mechanical, electronic, or digital governors.
- Single- or multi-valve control.

Characteristics of Steam Turbine Systems

Steam turbines are characterized by customized design and selection, simplicity, reliability, and relatively low maintenance requirements. Steam turbines are well-suited for constant speed, baseload electric generation applications. Because turbines are custom-built, efficiency and operating characteristics can be optimized for each application.

At a given steam flow and heat drop, turbine mechanical efficiency will be selected based on the incremental capital cost for the more mechanically efficient turbine and the incremental value of increased power output. For example, the capital cost of a 60% efficient multi-stage turbine may be several times that of a 30% efficient single-stage unit. If the value of power is very great, investment in a more efficient turbine may be cost-justified.

Topping-cycle applications using single-stage back-pressure turbines are far less costly than condensing turbines (which are usually multi-stage) and do not require the space and cost associated with condensing equipment. Existing pressure reducing valves (PRVs) are common targets for installation of topping turbines in retrofit applications.

The high net thermal efficiency of topping-cycle applications makes them highly competitive with any form of electric generation. The key factor is the availability of a continuous low-pressure heat sink. A 1,000 kW topping turbine would typically serve a low-pressure steam load of 30,000 to 40,000 lbm/h (14,000 to 18,000 kg/h). As the heat sink becomes smaller, so does the return on investment.

An **extraction turbine** is a multi-stage unit having one or more outlets to allow intermediate-pressure steam to be withdrawn. Extraction turbines are generally designed for applications in which there is a need for steam at different pressures or where there are varying low-pressure process steam requirements. Commonly, though not always, extraction turbines are condensing turbines. Fixed extraction turbines may sometimes be used as an alternative for two smaller turbines (one back-pressure and one condensing). Varying extraction turbines can function as either condensing or non-condensing units in different periods or as both simultaneously, depending on load.

A **bottoming cycle** can be used for industry applications that discharge high-pressure steam or waste heat. Admission turbines operating on a bottoming cycle can utilize steam as available at multiple pressures.

While the steam generation source is commonly an independent boiler system, the turbine generator sets are often delivered to the site in compact skid-mounted packages, factory-designed and -assembled for the application. Larger systems may be assembled on site, though various components may be prepackaged.

Steam Turbine System Performance

The thermal efficiency of a steam turbine system varies far more than with reciprocating and gas turbine engines and is mostly a function of turbine design and steam supply and exhaust conditions. Turbine mechanical efficiency typically ranges from 30 to 60% for single-stage turbines and 50 to 85% for multi-stage turbines. System simple-cycle electric generation (thermal) efficiency typically ranges from 20 to 30% for condensing systems and 5 to 15% for non-condensing systems.

Tables 26-3 and 26-4 show representative steam rates for condensing steam turbines, in lbm/kWh (kg/kWh), under various inlet conditions, each at an exhaust condition of 3 in. HgA (10.2 kPa), in English and SI units, respectively. Included are theoretical steam rates and actual steam rates (excluding generator losses) with various steam turbine mechanical efficiencies. Tables 25-5 and 25-6 show representative heat rates (HHV basis) in English and SI units, respectively. Heat rates are calculated under the assumption of 83 Btu/lbm (193 kJ/kg) hotwell enthalpy and 83% boiler efficiency.

The distinguishing performance feature of non-condensing (topping-cycle cogeneration) applications is that while simple-cycle thermal efficiency is low, net thermal efficiency approaches 100% (excluding boiler

Table 26-3 Representative Condensing Turbine Steam Rates Under Various Inlet Conditions and Steam Turbine Mechanical Efficiencies (English Units)

Inlet Condition	Exhaust Condition	Theoretical Steam Rate (lbm/kWh)	Steam Rate $\eta_{Turbine} = 75\%$ (lbm/kWh)	Steam Rate $\eta_{Turbine} = 65\%$ (lbm/kWh)	Steam Rate $\eta_{Turbine} = 55\%$ (lbm/kWh)
150 psig/D&S	3″ HgA	11.2	14.9	17.2	20.4
250 psig/450°F	3″ HgA	9.8	13.0	15.0	17.8
400 psig/550°F	3″ HgA	8.8	11.7	13.5	16.0
600 psig/750°F	3″ HgA	7.4	9.9	11.4	13.5
900 psig/900°F	3″ HgA	6.5	8.7	10.0	11.8
1,500 psig/900°F	3″ HgA	6.3	8.4	9.7	11.5

Table 26-4 Representative Condensing Turbine Steam Rates Under Various Inlet Conditions and Steam Turbine Mechanical Efficiencies (SI Units)

Inlet Condition	Exhaust Condition	Theoretical Steam Rate (kg/kWh)	Steam Rate $\eta_{Turbine} = 75\%$ (kg/kWh)	Steam Rate $\eta_{Turbine} = 65\%$ (kg/kWh)	Steam Rate $\eta_{Turbine} = 55\%$ (kg/kWh)
11.4 bar/D&S	10.2 kPa	5.1	6.8	7.8	9.3
18.3 bar/232°C	10.2 kPa	4.4	5.9	6.8	8.1
28.6 bar/288°C	10.2 kPa	4.0	5.3	6.2	7.3
42.4 bar/399°C	10.2 kPa	3.3	4.4	5.2	6.1
63.0 bar/755°C	10.2 kPa	3.0	3.9	4.5	5.4
104.5 bar/755°C	10.2 kPa	2.9	3.8	4.4	5.3

Table 26-5 Representative Condensing Turbine Heat Rates Under Various Steam Inlet Conditions and Turbine Mechanical Efficiencies (English units)

Inlet Condition	Steam Enthalpy (Btu/lbm)	Theoretical Heat Rate (Btu/kWh)	Heat Rate $\eta_{Turbine} = 75\%$ (Btu/kWh)	Heat Rate $\eta_{Turbine} = 65\%$ (Btu/kWh)	Heat Rate $\eta_{Turbine} = 55\%$ (Btu/kWh)
150 psig/D&S	1,196	15,019	19,980	23,065	27,355
250 psig/450°F	1,231	13,536	18,046	20,823	24,609
400 psig/550°F	1,276	12,649	16,817	19,404	22,998
600 psig/750°F	1,379	11,556	15,469	17,800	21,079
900 psig/900°F	1,459	10,776	14,423	16,678	19,562
1,500 psig/900°F	1,429	10,217	13,623	15,719	18,576

losses) because virtually all unused heat energy (in the form of lower pressure steam) is passed directly to process. The topping-cycle turbine is, therefore, an extremely efficient power generator. While the energy input to the topping-cycle steam turbine, per unit of power output, is far greater than with a reciprocating or combustion turbine engine, net thermal efficiency, assuming effective use of all low-pressure output steam, is generally equal to or greater than either of the engine types.

Table 26-7 shows representative steam rates, in lbm/kWh, for non-condensing turbine operation under various inlet and exhaust conditions. Included are theoretical steam rates and steam rates for turbines with mechanical efficiencies of 35%, 50%, and 60%. Table 26-8 shows the steam rates from Table 26-7 in SI units.

Figure 26-31 represents performance of a large industrial condensing turbine with a variable medium-pressure extraction. This turbine's power output (in thousand kW, or MW) varies depending on the total steam entering the turbine (throttle steam) and the amount of steam that leaves via the extraction port. The straight parallel lines show how power output varies with these two steam flow rates. Note that the turbine has several physical limits: the maximum throttle (or inlet) flow, the maximum exhaust flow, and the minimum exhaust flow.

Maximum throttle flow of about 200,000 lbm/h (90,000 kg/h) occurs between 8.3 and 12.0 MW. The maximum exhaust flow, which is the ability of the condenser to accept steam and maintain constant condenser pressure, is about 90,000 lbm/h (41,000 kg/h). During conditions when no extraction steam is being provided, power production is reduced to 9.5 MW. To produce 11 MW, 40,000 lbm/h (18,000 kg/h) must be extracted, meaning the throttle flow increases to 130,000 lbm/h

Table 26-6 Representative Condensing Turbine Heat Rates Under Various Steam Inlet Conditions and Turbine Mechanical Efficiencies (SI units)

Inlet Condition	Steam Enthalpy (kJ/kg)	Theoretical Heat Rate (kJ/kWh)	Heat Rate $\eta_{Turbine}$ = 75% (kJ/kWh)	Heat Rate $\eta_{Turbine}$ = 65% (kJ/kWh)	Heat Rate $\eta_{Turbine}$ = 55% (kJ/kWh)
11.4 bar/D&S	2,781	15,935	21,247	24,516	28,973
18.3 bar/232°C	2,863	14,284	19,046	21,975	25,971
28.6 bar/288°C	2,967	13,421	17,895	20,648	24,402
42.4 bar/399°C	3,207	12,149	16,199	18,691	22,090
63.0 bar/755°C	3,393	11,492	15,323	17,680	20,894
104.5 bar/755°C	3,323	10,896	14,473	16,662	19,839

Table 26-7 Representative Steam Rates for Non-Condensing Steam Turbines Under Various Inlet and Exhaust Conditions and Turbine Mechanical Efficiencies (English Units)

Inlet Condition	Exhaust Condition	Theoretical Steam Rate (lbm/kWh)	Steam Rate $\eta_{Turbine}$ = 60% (lbm/kWh)	Steam Rate $\eta_{Turbine}$ = 50% (lbm/kWh)	Steam Rate $\eta_{Turbine}$ = 35% (lbm/kWh)
250 psig/D&S	15 psig	23.4	39.1	46.9	67.0
250 psig/550°F	15 psig	18.2	30.3	36.4	52.0
	50 psig	26.8	44.7	53.6	76.6
400 psig/750°F	50 psig	17.6	29.3	35.2	60.3
	100 psig	23.9	39.8	47.8	68.3
600 psig/750°F	50 psig	15.4	25.7	20.8	44.0
	100 psig	19.4	32.3	38.8	55.4
900 psig/900°F	50 psig	12.1	20.2	24.2	34.6
	100 psig	14.5	24.2	29.0	41.4
	200 psig	19.5	32.5	39.0	55.7
1,500 psig/900°F	50 psig	10.9	18.2	21.8	31.1
	200 psig	15.8	26.3	31.6	45.1
	400 psig	22.3	37.2	44.6	63.7

Table 26-8 Representative Steam Rates for Non-Condensing Steam Turbines Under Various Inlet and Exhaust Conditions and Turbine Mechanical Efficiencies (SI units)

Inlet Condition	Exhaust Condition	Theoretical Steam Rate (kg/kWh)	Steam Rate $\eta_{Turbine}$ = 60% (kg/kWh)	Steam Rate $\eta_{Turbine}$ = 50% (kg/kWh)	Steam Rate $\eta_{Turbine}$ = 35% (kg/kWh)
18.3 bar/D&S	2.0 bar	10.6	17.8	21.3	30.4
	4.5 bar	13.8	23.0	27.6	39.5
18.3 bar/288°C	2.0 bar	8.3	13.8	16.5	23.7
	4.5 bar	12.2	20.2	24.3	34.7
28.6 bar/399°C	4.5 bar	8.0	13.3	15.9	22.7
	7.9 bar	8.8	18.1	21.6	30.9
42.4 bar/399°C	4.5 bar	7.0	11.7	14.0	20.0
	7.9 bar	8.8	14.7	17.6	25.2
63.1 bar/755°C	4.5 bar	5.5	9.1	10.9	15.6
	7.9 bar	6.6	10.9	13.1	18.8
	14.8 bar	8.8	14.7	17.6	25.2
104.5 bar/755°C	4.5 bar	4.9	8.3	9.9	14.1
	14.8 bar	7.2	11.9	14.3	20.5
	28.6 bar	10.1	16.9	20.2	28.9

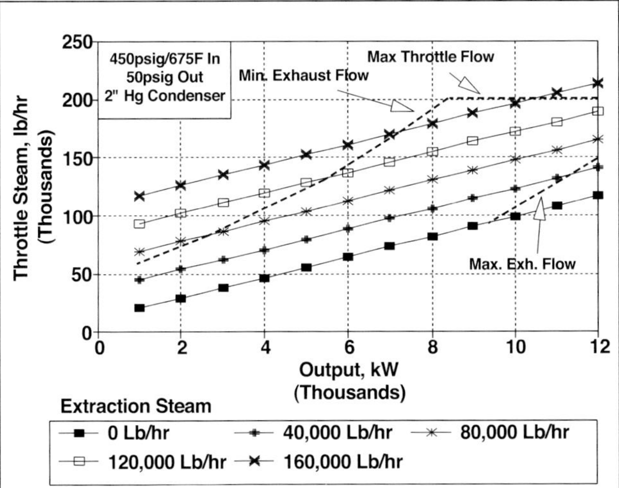

Fig. 26-31 Condensing Extraction Steam Turbine Performance in Output vs. Throttle Flow for Various Extraction Steam Rates. Source: Cogen Designs, Inc.

(59,000 kg/h). Also, a minimum steam flow of about 10,000 lbm/h (4,500 kg/h) to the condensing section is required for component cooling. If 80,000 lbm/h (36,000 kg/h) of extraction steam were required, a minimum of 3 MW of power output would have to be produced and the throttle flow would have to be at least 90,000 lbm/h (41,000 kg/h).

Figure 26-32 represents performance of an industrial back-pressure turbine. The percent generating efficiency represents the ratio of actual efficiency to theoretical efficiency at each throttle flow rate. The curves indicate how efficiencies and electric power output vary depending on the throttle steam flow. Starting at the throttle flow requirement of 10,000 lbm/h (4,500 kg/h) to produce 200 kW, the steam flow curve indicates a linear relationship between throttle flow and power output. Thus, steam flow requirements can be estimated using a minimum throttle flow plus a linear function.

Operations, Maintenance and Reliability

Steam turbines have low maintenance requirements and can sometimes operate for decades without major

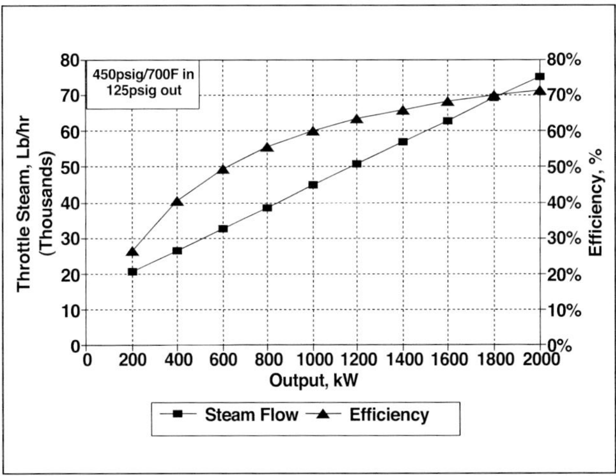

Fig. 26-32 Back-Pressure Steam Turbine Performance in Output vs. Throttle Flow for Various Extraction Steam Rates. Source: Cogen Designs, Inc.

overhaul because of simple design, few moving parts, construction with durable materials, and operation at relatively low speeds. A turbine generator set can be mounted and shipped on a common baseplate.

Maintenance requirements will vary depending on the quality of the turbine components and construction and type of design. Multi-stage turbines, particularly those designed with small clearances, will require more maintenance. Condensing turbine systems have auxiliary components, including surface condensers and cooling towers, that require added maintenance.

Routine procedures include monthly visual inspections — primarily looking for oil or steam leaks and reviewing monitoring device data. Any problem will show up in pressure and temperature readings or in the measurement of vibrations.

A key element of successful long-term steam turbine operation is the ability to control steam purity and the ability of the blades to withstand the abuses of wet steam and deposits. To minimize erosion, integral moisture separators and stainless steel moisture shields can be used.

Steam turbines are highly reliable and commonly offer availability factors ranging from 95 to 99%. Their high level of reliability makes them well-suited for critical applications and as back-up power generation for critical systems.

Steam Turbine Application Examples

Figures 26-33 through 26-38 are examples of steam turbine-driven electric generation system applications.

- Figures 26-33 and 26-34 show skid-mounted packaged steam turbine generator sets. Figure 26-33 shows a single-stage steam turbine generator package for applications of up to 2,000 kW. Figure 26-34 shows a skid-mounted multi-stage turbine generator set with the full range of accessories, including integral speed reduction gear.
- Figure 26-35 shows two 12,500 kW multi-valve extraction non-condensing steam turbine generator sets in a university application. These cogeneration units operate between 400 psig/750°F (28.6 bar/399°C) and 15 psig (2 bar), with extraction at 60 psig (5.2 bar). High turbine mechanical efficiency and a relatively large pressure/temperature drop allow these steam turbine generator sets to achieve a high power-to-steam flow ratio over the full range of extraction steam conditions. In choosing the multi-valve extraction steam turbine, the focus in this application is on optimizing power output through conservative exploitation of the low-pressure steam load or heat sink.
- Figure 26-36 shows a multi-valve condensing steam turbine generator set installed in a Midwest energy recovery plant. The turbine generator rating is

Fig. 26-33 Skid-Mounted, Packaged Single-Stage Steam Turbine Generator Set. Source: The Elliot Company

Fig. 26-34 Skid-Mounted, Packaged Multi-Stage Steam Turbine Generator Set with Speed Reduction Gear. Source: Tuthill Corp. Murray Turbomachinery Div.

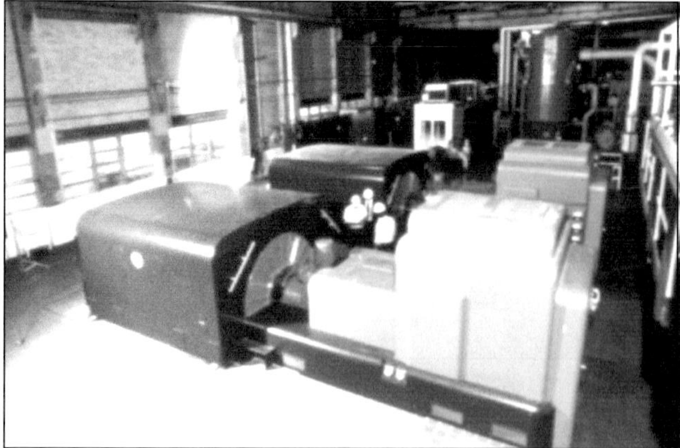

Fig. 26-35 Two 12,500 kW Multi-Valve, Extraction, Non-Condensing Steam Turbine Generator Sets in University Application. Source: Dresser-Rand

6,327 kW at 4,670 rpm with steam conditions of 500 psig/650°F (35.5 bar/343°C) to 3 in. HgA (10.2 kPa). The generator is driven at a speed of 1,800 rpm through a reduction gear. This bottoming-cycle application makes use of heat recovery-generated steam from a relatively high-temperature source. Figure 26-37 is a performance curve for this axial flow impulse type steam turbine.

- Figure 26-38 is an example of an application featuring two very different steam turbines. The system on the right features a multi-valve back-pressure unit rated at 1,962 kW at 7,500/1,800 rpm with steam conditions of 600 psig/650°F (42.4 bar/343°C) to 60 psig (5.2 bar). The system on the left features a condensing steam turbine rated at 2,217 kW at 5,300/1,800 rpm with steam conditions of 60 psig/323°F (5.2 bar/162°C) to 4 in. HgA (13.5 kPa).

Fig. 26-36 Multi-Valve 6,327 kW Condensing Steam Turbine Generator Set. Source: Tuthill Corp. Murray Turbomachinery Div.

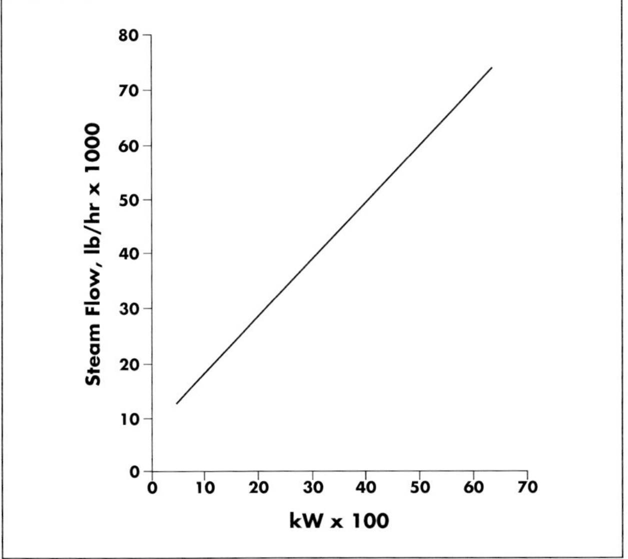

Fig. 26-37 Performance Curve for 6,327 kW Condensing Steam Turbine Generator Set. Source: Tuthill Corp. Murray Turbomachinery Div.

Drivers for Combined-Cycle Electric Generation Systems

Combined-cycle electric generation systems typically consist of one or more combustion engines (most commonly gas turbines), an HRSG, and a steam turbine. Combined-cycle systems may be designed as cogeneration systems or as power generation systems without additional heat recovery. In combined-cycle cogeneration, a portion of the heat recovered from the combustion-engine topping cycle is used to meet thermal process loads. For a cogeneration system to be classified as a Qualified Facility (QF), a minimum portion — 5 or 15% — of the thermal output must be passed on to a thermal process. Combined-cycle cogeneration applications feature back-pressure or extraction condensing turbines that discharge steam at pressures above atmospheric to process.

The primary application for conventional combined-cycle generation today is in medium- and larger-scale power generation projects. However, combined-cycle power plants can also be effective in cogeneration applications as low as 5 or 10 MW. Variations on the combined-cycle theme are used in which the steam turbine (and, in some cases, the combustion engine as well) is used for mechanical drive service instead of electric generation. While gas turbines are most common in combined-cycle applications in the United States, reciprocating engines are also sometimes applied.

Fig. 26-38 Steam Turbine System Application Featuring One Back-Pressure and One Condensing Turbine Generator Set. Source: Tuthill Corp. Murray Turbomachinery Div.

If a facility has an extensive low-pressure steam load (and some high-pressure steam load), it often makes sense to install not only a gas turbine, but also a steam turbine as an intermediate-stage power producer. High-pressure steam is made in the HRSG, and the back-pressure or extraction steam turbine functions as a PRV, while producing shaft power for electric generation (or mechanical drive) service.

Steam pressure and quality requirements are one potential difference between conventional cogeneration and combined-cycle operation. The need for high-pressure superheated steam to drive the steam turbine can result in a two- or even three-fold increase in the cost of the heat recovery steam generator. This may be cost-justified when the value of the power produced substantially exceeds the value of the potential thermal energy output or when there is insufficient thermal load to support a simple-cycle turbine-based cogeneration plant.

Combined-Cycle Cogeneration Plant Configurations

The typical combined-cycle cogeneration plant uses back-pressure or extraction turbine steam output to meet thermal load requirements. In some cases, all of the exhaust gas-generated steam is passed on to thermal processes and in other cases just a portion, depending on the load characteristics and on the relative values of producing added power or thermal energy. With a back-pressure turbine, power generation is fixed by steam flow.

Similar to the turbine arrangement shown in Figure 26-38, some combined-cycle configurations include one condensing turbine and one back-pressure turbine sized to match the process load. For facilities that experience varying thermal load requirements, the use of a single multi-stage extraction/condensing turbine adds a large degree of operating flexibility. Compared with the use of back-pressure steam turbines, however, condensing and extraction turbines add significantly to overall system capital cost.

Many existing facilities that are built around steam turbine operation can be re-configured to operate as combined-cycle plants. By strategically building steam turbine loads, combined-cycle operation can be made increasingly compatible and financially attractive, even on a relatively small scale (below 10 MW). For facilities already configured with topping-, condensing-, or bottoming-cycle steam turbine operation, the addition of a gas turbine can be viewed as an alternative upstream heat source that also produces power.

Supplementary firing of combustion engine exhaust can be used to efficiently increase thermal output and control steam condition. Because a relatively small portion of the oxygen in an open gas cycle is used for combustion, the remainder, which provides oxygen-rich exhaust, can be used for supplementary firing in the exhaust duct or the HRSG. In some cases, reciprocating engine exhaust gases can also be supplementary-fired.

An important potential benefit of supplementary firing is that it can allow a facility to be more flexible in responding to varying load conditions. For combined-cycle plants that either buy or sell power with differentiated cost rate periods, it may be advantageous to increase steam turbine power output with supplementary firing in the higher cost rate periods. In cogeneration type plants that have varying thermal loads, the combustion engine exhaust can be baseloaded and supplementary firing can be used to meet all additional thermal loads. In many cases, dynamic operating strategies are developed to use supplementary firing to meet either peak electric or thermal loads. Similar supplementary firing strategies may also be employed with STIG-cycle systems.

Usually, the combustion engine, HRSG, and steam turbine are housed in a centralized plant, though sometimes the steam turbine is housed separately and used downstream of the steam distribution system as a pressure reducing station. Combined-cycle systems are also available as prepackaged modules, with some configurations featuring a single common shaft.

Combined-Cycle System Performance

While a large capacity combined-cycle plant can exceed thermal fuel efficiencies of 50% (LHV), cogeneration combined-cycle plants can reach overall thermal efficiencies of 80 or 90%, with net thermal fuel efficiencies in excess of 65%. While the heat rate of a very efficient conventional combined-cycle plant might be as low as 7,000 Btu/kWh (7,385 kJ/kWh), the net heat rate of the cogeneration combined-cycle plant may be 5,000 Btu/kWh (5,275 kJ/kWh).

Multiple-pressure HRSGs can improve overall combined-cycle system thermal fuel efficiency when used in conjunction with steam turbines that include two or more admissions. In cogeneration applications, the lower- pressure steam can sometimes be used directly to serve thermal processes.

Overall combined-cycle thermal fuel efficiencies may be unchanged when reciprocating engines are used as the driver for the topping cycle component.

However, the reciprocating engine will provide a greater portion of the total system power output. It is also necessary to find a productive process use for the lower-temperature outputs from the reciprocating engine's coolant system.

Combined-Cycle Application Examples

Figures 26-39 and 26-40 provide examples of combined-cycle applications. Figure 26-39 shows a relatively small system with a gas turbine driver and Figure 26-40 shows an application using reciprocating engine drivers (refer to Chapter 12 for additional examples of combined-cycle systems).

- Figure 26-39 shows a heat and material balance for a 9 MW combined-cycle cogeneration system applied in a paper mill, featuring a two-pressure boiler (450/5 psig) with a 150 psig steam turbine discharge pressure and water injection for NO_X abatement. Notice that the non-condensing steam turbine produces only about 14% of the system's total capacity.
- Figure 26-40 is a heat balance diagram for a 45 MW combined-cycle system featuring four 6.42 MW dual-fuel-fired reciprocating engines and one 20.7 MW steam turbine. In this application, about 46% of the total energy input goes directly to the fuel-fired HRSG. More than half of the engine coolant system heat is used for preheating condensate through a heat exchanger located upstream of the deaerator. The remaining engine heat is dissipated in a jacket cooler.

Comparison of Cogeneration and Combined-Cycle Alternatives

Figures 26-41 through 26-45 illustrate five different cogeneration and combined-cycle arrangements that could be considered alternatives for a given application. Figure 26-46 is an energy map showing thermodynamic performance for each of the arrangements and Table 26-9 compares their performance. Three performance factors are compared in the table:

1. Overall thermal fuel efficiency based on the sum of the electrical output and the thermal output divided by the fuel input.
2. Net electric generation thermal efficiency based on a fuel credit for steam with an assumed displaced fired boiler fuel efficiency of 90%.
3. Electrical-to-thermal ratio (ETR) based on the heat value of electrical production versus steam production.

The five alternative configurations are described as follows:

1. Figure 26-41 illustrates the baseline arrangement featuring a 10 MW gas turbine with HRSG and water injection for NO_X emissions control in a simple cogeneration cycle. The baseline arrangement produces about 48,000 lbm/h (22,000 kg/h) of saturated steam at 150 psig (11.4 bar). The range of electric and steam output is shown in the lower left corner of the energy map in Figure 26-46. As Table 26-9 indicates, the overall thermal fuel efficiency of this system is 76%. The net electric generation efficiency is 60%,

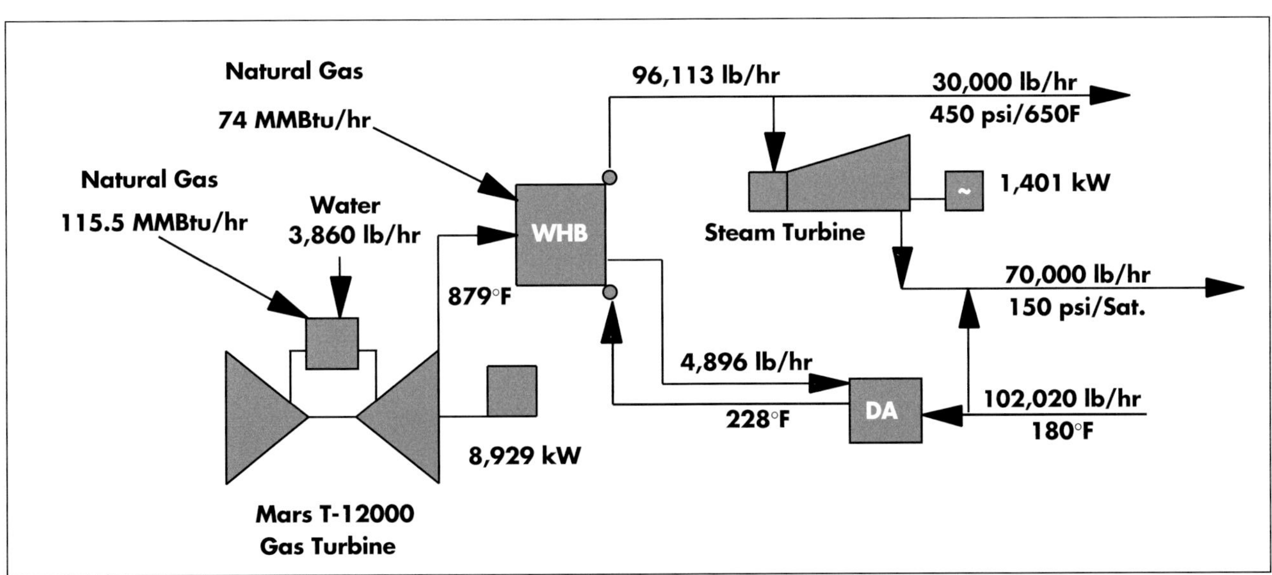

Fig. 26-39 9 MW Combined-Cycle Cogeneration System. Source: Cogen Designs, Inc.

Net Power Output:	45.0 MW
Gross Heat Rate (LHV):	8150 Btui/kWh
Net Heat Rate (LHV):	8400 Btu/kWh
Ambient Conditions:	90°F / 60% R.H.

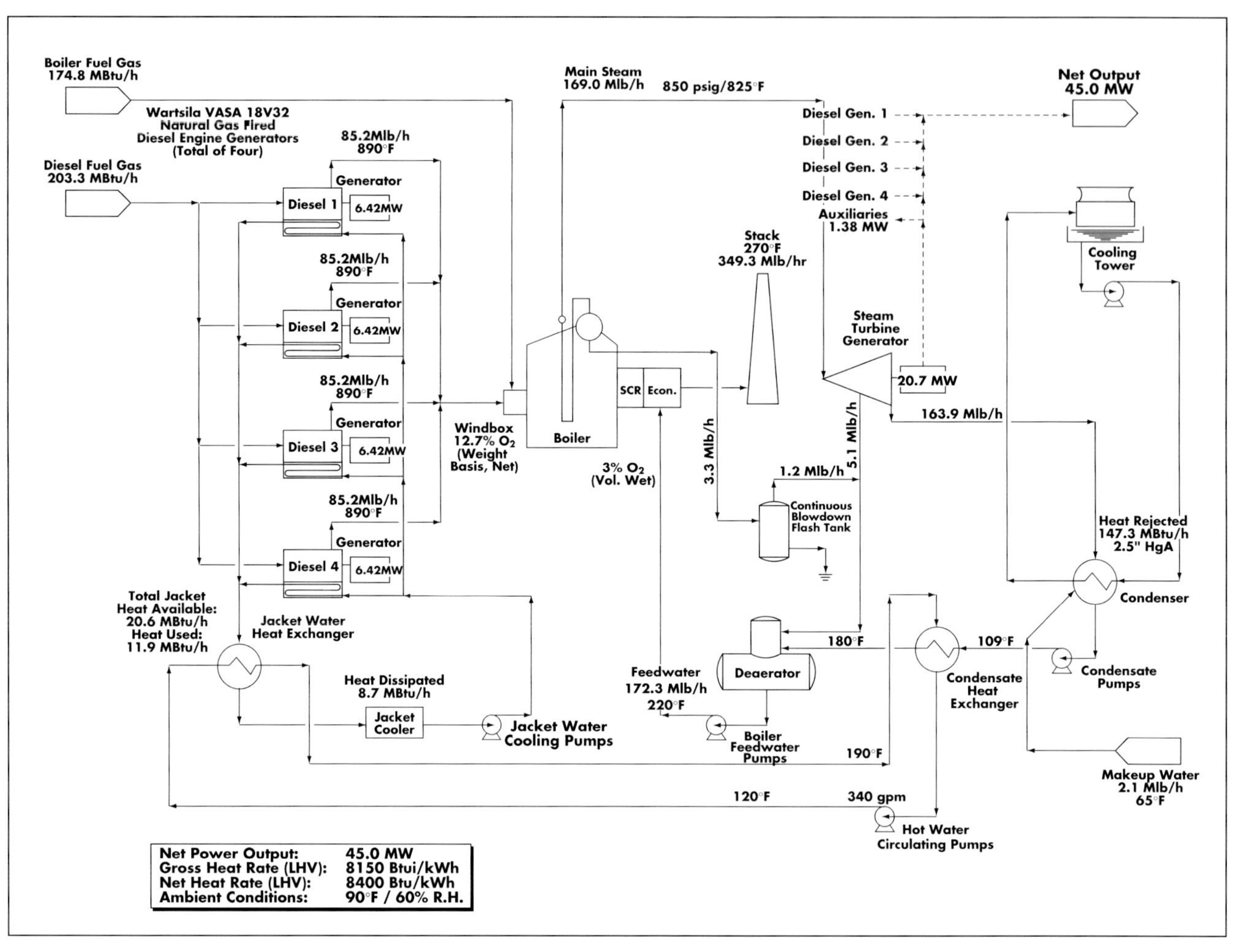

Fig. 26-40 Heat Balance Diagram of 45 MW Combined-Cycle System Featuring Four Dual-Fuel-Fired Reciprocating Engines and One Steam Turbine. Source: Wartsila Diesel

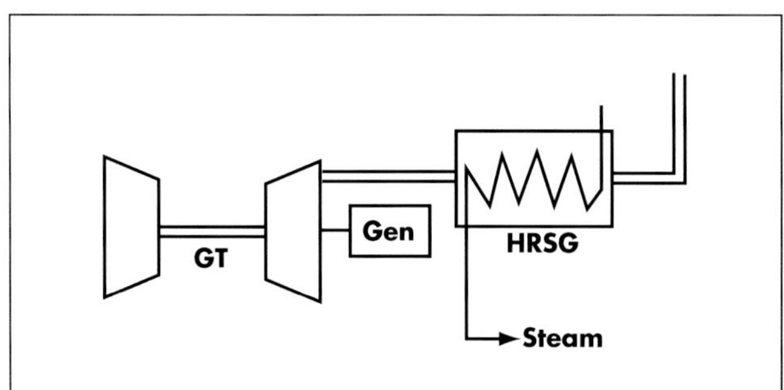

Fig. 26-41 Conventional Simple-Cycle Cogeneration System. Source: Solar Turbines

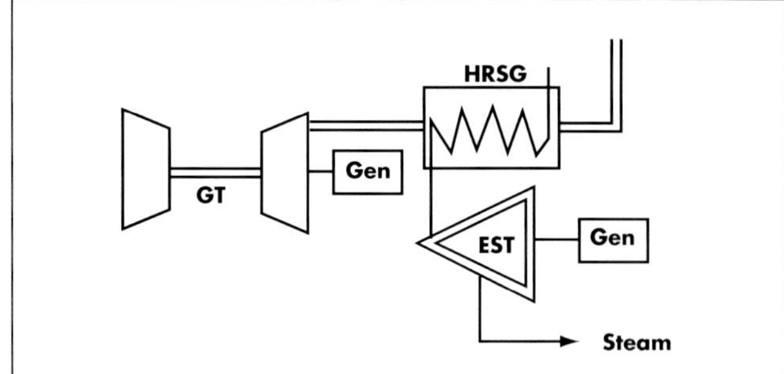

Fig. 26-42 Conventional Combined-Cycle Cogeneration System. Source: Solar Turbines

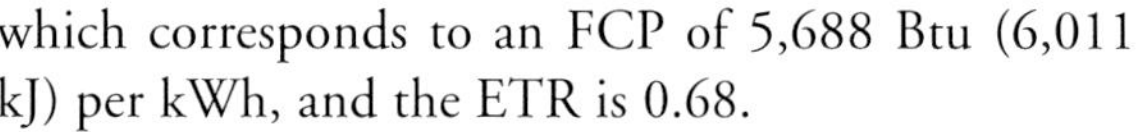

which corresponds to an FCP of 5,688 Btu (6,011 kJ) per kWh, and the ETR is 0.68.

2. Figure 26-42 illustrates a conventional combined-cycle cogeneration system featuring the addition of an extraction steam turbine. With this arrangement, the HRSG is designed to produce superheated steam at a condition of 600 psig/800°F (42.4 bar/427°C). The operating range for this arrangement, with varying amounts of extraction steam being passed on to process, is shown near the bottom center of the ener-

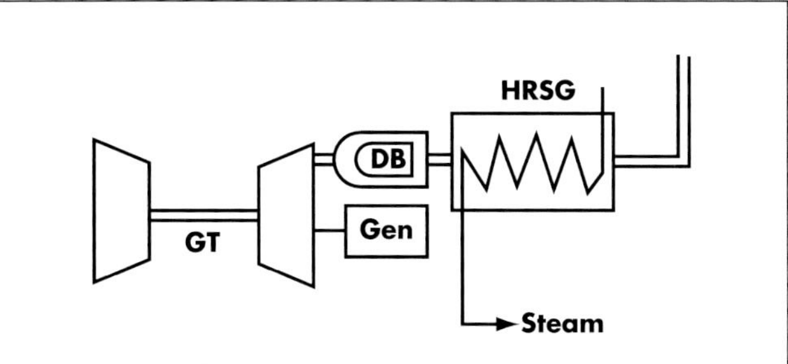

Fig. 26-43 Simple-Cycle Cogeneration System with Supplementary Firing. Source: Solar Turbines

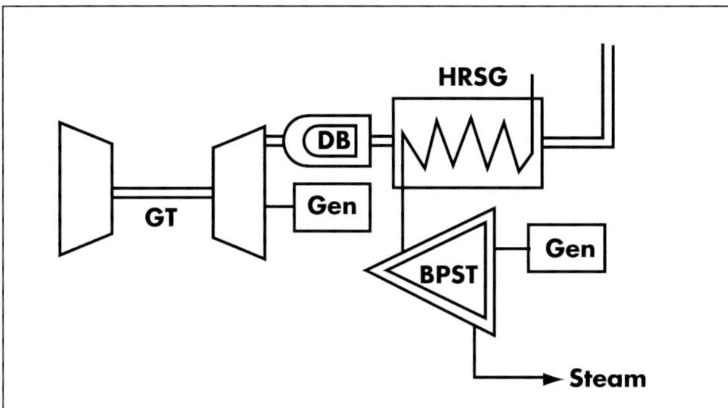

Fig. 26-44 Advanced Combined-Cycle Cogeneration System with Back-Pressure Steam Turbine and Supplementary Firing. Source: Solar Turbines

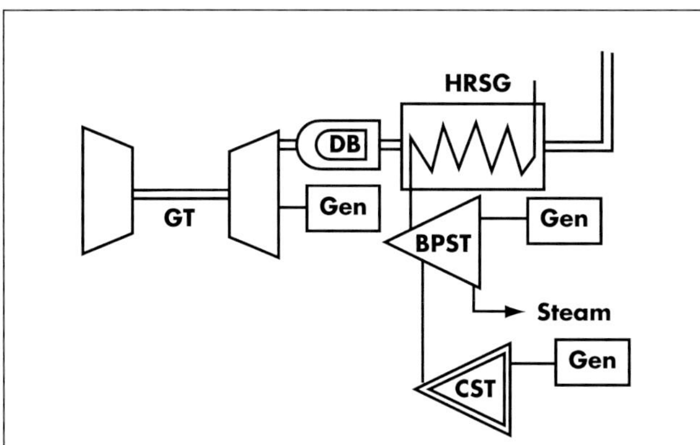

Fig. 26-45 Advanced Combined-Cycle Cogeneration System with Back-Pressure and Condensing Steam Turbines and Supplementary Firing. Source: Solar Turbines

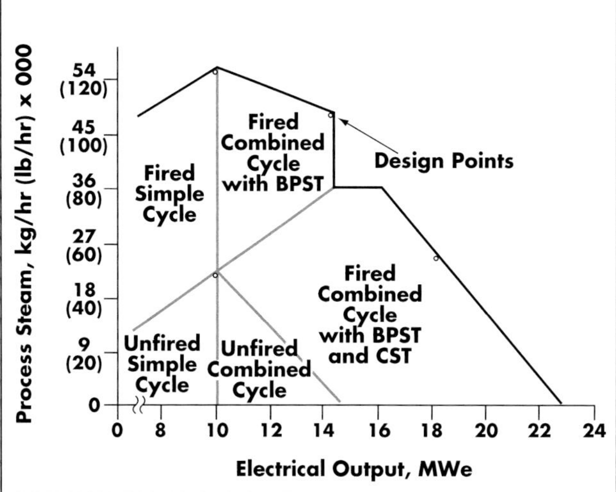

Fig. 26-46 Energy Performance Map for 10 MW Gas Turbine with Various System Arrangements. Source: Solar Turbines

System Arrangement	Fuel Efficiency %	Net Electrical Efficiency %	Electrical-to-Thermal Ratio
Simple Cycle	76	60	0.68
Combined Cycle with Extraction Steam Turbine	58	38	1.88
Supplementary-Fired Simple Cycle with Duct Burner	85	79	0.26
Supplementary-Fired Combined Cycle with duct Burner and BPST	84	79	0.43
Supplementary-Fired Combined Cycle with Duct Burner, BPST, and Condensing Steam Turbine	62	49	1.09

Table 26-9 Performance Comparison for Five System Arrangements. Source: Solar Turbines

gy map in the region labeled unfired combined-cycle. When operating in full condensing mode, the combined-cycle system can produce more than 14 MW of power. At the selected comparative design point for Table 26-9, 50% of the steam flow, or 21,000 lbm/h (9,500 kg/h), is extracted and the total system electric output is 12.6 MW. At this design point, both the overall thermal fuel efficiency and the net electrical efficiency are the lowest of all the arrangements considered, while at 1.88, the ETR is the highest.

3. Figure 26-43 illustrates the addition of a duct burner to the simple-cycle cogeneration system for the purpose of increasing steam production. With an upgraded HRSG, the steam production can be increased to 126,000 lbm/h (57,000 kg/h), with supplementary firing increasing the exhaust temperature from 920 to 1,800°F (493 to 982°C). Compared with the baseline system, this arrangement provides higher overall thermal fuel efficiency (85%) and net electrical efficiency (79%), with an FCP of 4,321 Btu (4,558 kJ) per kWh. This is because all fuel energy added to the duct burner produces useful steam, with the stack temperature from the HRSG remaining con-

stant. However, at 0.26, the ETR is the lowest of all arrangements considered, due to the increase in steam production with no corresponding increase in electric output. Similar to the use of only a BPST electric generation system, despite the low ETR, if there is sufficient process steam requirement, this operating arrangement is highly economical as indicated by the high net electrical efficiency. However, capital costs are increased due to the addition of the duct burner and the need for an upgraded HRSG.

4. Figure 26-44 illustrates a combined-cycle system arrangement featuring supplementary firing and the use of a BPST. Under the steam conditions used in the conventional combined-cycle arrangement example, the steam turbine adds about 3 MW of power output to the system. However, in this example, steam conditions are increased significantly, compared with current practice for smaller systems, to 1,250 psig/1,050°F (87.2 bar/566°C), and a maximum system electrical output of 14.5 MW is achieved by this advanced cycle. This is shown on the upper-middle portion of the energy map. As shown in Table 26-9, due to the elevated steam conditions and the inherent high net electrical efficiency of back-pressure steam turbine (BPST) operation, this arrangement produces similar overall thermal fuel efficiency and net electrical efficiency to the conventional cogeneration system with supplementary firing. However, the ETR is increased due to the additional power provided by the BPST. While operation is very economical, capital costs are increased further due to the addition of the steam turbine and additional upgrades required for operation at elevated steam temperature and pressure conditions.

5. Figure 26-45 illustrates the addition of a low-pressure condensing turbine to the arrangement shown in Figure 26-44. With the combination of back-pressure and condensing turbines, the steam flow to process can be modulated between 0 and 110,000 lbm/h (50,000 kg/h). With no extraction steam, the electrical output of the full condensing advanced combined-cycle, at the elevated pressure and temperature conditions shown in Figure 26-45, is 22.7 MW. At the selected comparative design point, 50% of the steam is extracted for process, with the remainder passing through the condensing turbine, for a combined electrical output of 18.4 MW. The operating range of this arrangement is shown on the bottom right of the energy map.

   At the selected design point, overall thermal fuel efficiency (62%) and net electrical efficiency (49%) are sacrificed for an increase in ETR (1.09). Compared with the conventional combined-cycle system, this arrangement produces a significant increase in net electrical efficiency due to supplementary firing and higher steam turbine inlet pressure and temperature. While the ETR is lower, overall electric output is greater and operating flexibility is greatly enhanced.

Renewable Generator Driver Options

Beyond the conventional fuel- and steam-powered forms of localized electricity generation, there are several types of renewable resource and alternative technologies that can be practically applied to the same ends. These are summarized in Chapter 23. In addition to the three conventional electric generator driver technologies addressed previously in this chapter, two other prime movers of long standing, water (hydraulic) and wind turbines, merit consideration as generator driver application options, where respective site-specific conditions of prevalent water and wind energy resources exist. Hydroelectric power has long been one of the leading electricity producers in the world, while wind energy conversion systems (WECS) are a far smaller, but growing, contributor to the world electricity production energy mix.

Hydraulic Turbines

The two major classes of modern hydraulic turbines are the impulse and reaction types. Hydraulic conditions at dams, waterfalls, and river runs determine, in large part, the class and type of turbine that will be most effective for a given application.

- **Impulse turbines**, which achieve rotation through the acceleration of water through supply nozzles, tend to be employed for water resources with high heads and low flow rates. The most widely used type of impulse turbine is the Pelton turbine (or wheel), in which each bucket is divided in the center by a double-curved wall so that the jet of water is split upon hitting the bucket and diverted to either side, transmitting a maximum amount of energy to the turbine. These turbines will typically exhibit mechanical efficiencies ranging from 80 to 90%, increasing as the size of the wheel increases. Other types of impulse designs include the Turgo and Banki turbines.
- **Reaction turbines** are widely applied over a large range of heads of water with moderate or high flow

rates. Rotation is achieved primarily through the reactive force produced by the acceleration of water in the runner. Reaction turbines typically have a spiral inlet casing that includes control gates to regulate water flow. These may be horizontal or vertical types and operate with the turbine wheel completely submerged in order to reduce turbulence. They may also be designed with adjustable blades that allow the blade angle to be changed to increase the turbine's efficiency. The four major types of reaction turbines in wide use are the Francis, Deriaz, Kaplan, and propeller type. Fixed blade propeller and adjustable blade Kaplan turbines feature an axial flow through the machine, while the Francis and Deriaz type turbines use a mixed flow, where the water enters radially inward and discharges axially. Francis turbines are most commonly used, due to their suitability with the widest range of heads, ranging from 10 to 2,000 ft (3 to 600 m) and are most commonly applied under medium-head applications ranging from 400 to 1,000 ft (120 to 300 m).

The output and design specifications of a hydraulic turbine applied to electric power generation will be a function of the volume of water discharged, the vertical distance the water falls (the head), and the required rotational speed of the generator. The electric conversion efficiency of different turbine types varies considerably. Older large-capacity designs still in operation may have mechanical efficiencies below 50%, as do certain newer very small capacity units. Conversely, several large capacity modern turbine designs have demonstrated mechanical efficiencies in excess of 90%.

Chapter 14 presents details on several hydroelectric plants ranging from the massive 6,800 MW Grand Coulee Dam to the 350 kW Lewiston Dam project. Individual hydraulic turbines are available in capacities ranging from a few kW to more than 800 MW. Applications may be classified as high head or low head, with low head referring to smaller capacity plants with rated heads of about 65 ft (20 m) or less. Very low head systems may also feature run-of-the- river designs, which use power in the river water as it passes through the plant without requiring water impoundment and without causing appreciable change in the river flow.

In assessing a potential hydropower application, it is necessary to evaluate the source to determine how much water can be delivered to the turbine and how high the water source is above the turbine. However, beyond the initial assessment of approximate system capacity and power production is the extensive and often lengthy environmental assessment and permitting process. While very large new projects are becoming increasingly uncommon due to the fact that many ideal sites have already been developed or cannot be developed due to environmental or other land use obstacles, numerous opportunities for smaller capacity applications remain in many regions of the country and the world. Very small projects are further aided by the availability of standardized designs for microturbine applications and the ability to work through various siting issues more quickly than with larger projects.

Wind Turbines

The prime mover in a modern WECS is the wind turbine. Electricity is produced as the wind imparts a force that turns the rotor blades, which spins a shaft that drives an electric generator. Turbines are classified as either horizontal-axis wind turbines (HAWTs) or vertical-axis wind turbines (VAWTs).

- The design of HAWTs, which dominate the WECS electric generation market, is similar to historic windmills, with a propeller type rotor on the axis. The axis of rotation is horizontal and roughly parallel to the wind stream. The blades on the horizontal drive shaft may be facing into the wind (an upwind turbine) or the wind may hit the supporting tower first (a downwind turbine). A yaw drive is used with upwind turbines to keep the rotor facing into the wind as the wind direction changes. Downwind turbines do not require a yaw drive, since the wind blows the rotor downwind. More modern two- and three-blade HAWTs run at high tip speed ratios that allow them to achieve very high electric power conversion ratios. These turbines generally operate with a constant blade pitch (or orientation) angle, but can be designed so that the blade pitch is continuously controlled to maintain constant speed and target power output.
- VAWTs, which are far less common for modern electric generation applications, feature an axis of rotation that is vertical to the ground and roughly perpendicular to the wind stream. They are similar to traditional waterwheels, with the fluid arriving at a right angle to the rotation axis. While less efficient than HAWTs, they feature a simple, durable design and have the advantage of having all of the drive train and generator components at ground level.

Wind turbines usually use a standard induction generator that produces 50 or 60 Hz ac electricity. Direct-drive and/or variable speed generators are in prototype and can be used to eliminate gearing and optimize turbine energy capture. There are also units that utilize two generators for different wind speed levels or one two-speed generator with bi-level output capability. Turbines may be designed for pitch control, stall control, or a combination of both.

As detailed in Chapter 14, the electric power output from a WECS will be a function of the wind power available, the power coefficient, the efficiencies of the mechanical interface, the generator efficiency and the efficiency of any power-conditioning equipment. When accounting for the power coefficient and all other normal system losses, the overall average conversion efficiency will typically range from 20 to 35%. Electric conversion efficiency will vary depending on turbine type and design specifications. The efficiency of a specific unit will also vary widely depending largely on wind speed at any given time. At certain points in the operating regime, a given turbine may achieve an electric conversion efficiency of 45% or greater, while under non-optimum conditions, efficiency of the same machine may be reduced to well under 20%.

Chapter 14 presents detail on several WECS with individual turbine-generator capacities ranging from 10 kW to 2 MW and applications featuring individual turbines at specific sites, as well as large wind farms with more than 100 integrated wind turbines.

A preliminary indication of an area's potential for a successful WECS project can be based on its rated wind power class. The *Wind Energy Resource Atlas of the United States* can be used to identify the areas potentially suitable for wind energy applications. More specific data can also be obtained for localized weather stations and/or long-term monitoring at specific sites. Usually, wind turbine-driven electric generators are designed to start running at cut-in wind speeds of 7 to 8 mph (3.1 to 3.6 m/s) for non-grid connected operation and 12 to 13 mph (4.3 to 5.8 m/s) for most grid-connected applications. To protect the wind turbine from damage, the cut-out speed will typically be set at about 56 mph (25 m/s). Available wind at these required speeds usually limits capacity (or annual load) factors to 20 to 40% at most good locations, though in some applications it can be considerably higher.

Renewable Electric Generation Application Examples

Figure 26-47 shows the main components of the nacelle of a 600 kW up-wind, stall-regulated HAWT that is one of 40 WECS installed at a 24 MW grid-connected inland wind farm in Denmark in 1995. The turbines have a 3-blade rotor that is 141 ft (43 m) in diameter with a swept area of 15,630 ft^2 (1,452 m^2). The units are mounted on tubular towers at a hub height of 150 ft (46 m). Each turbine drives a two-speed (1,000/1,500-rpm), 3-phase asynchronous induction generator. The cut-in speed for electricity production is 7.8 mph (3.5 m/s). At

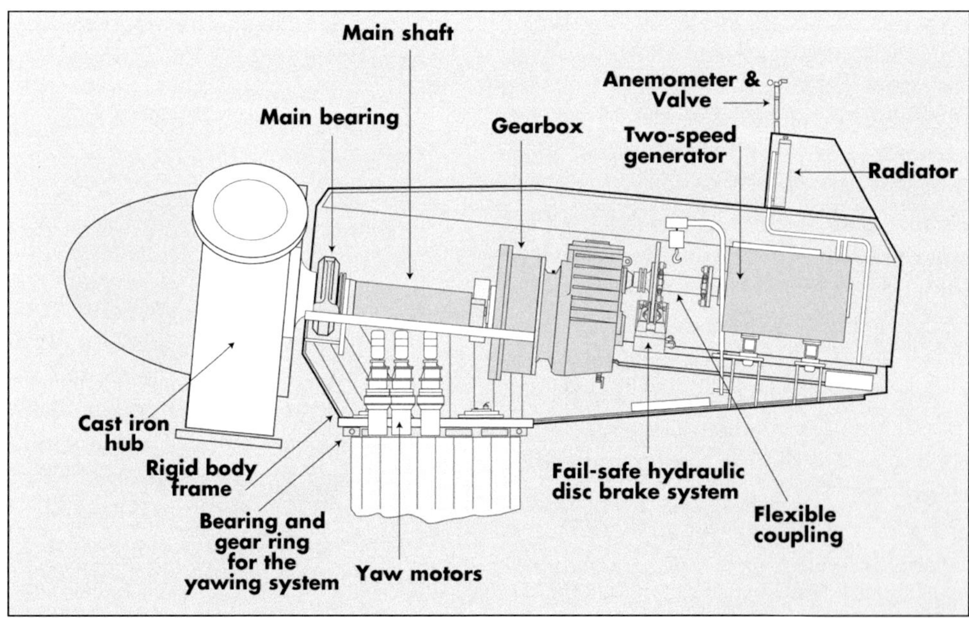

Fig. 26-47 Main Components of the Nacelle of a 600 kW Up-Wind, Stall-Regulated HAWT. Source: CADDET Renewable IEA/OECD and Micon A/S

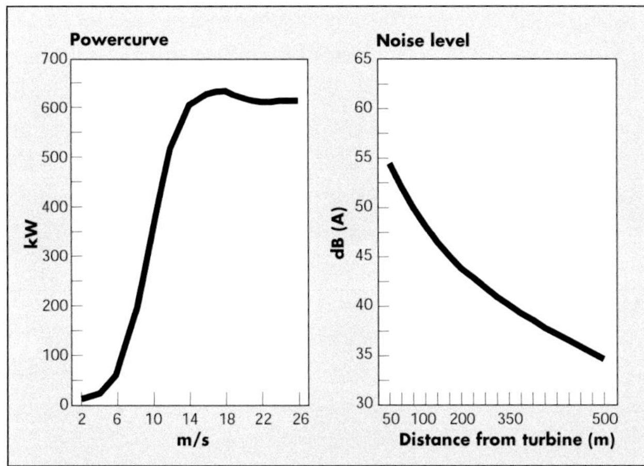

Fig. 26-48 Power and Sound Emission Curves for Wind Turbine Shown in Figure 26-47. Source: CADDET Renewable IEA/OECD and Micon A/S

a wind speed of about 16 mph (7.1 m/s), the generator output reaches 150 kW. Above that speed, the generator automatically switches to high-speed operation, reaching full capacity of 600 kW at a wind speed of about 31 mph (14 m/s). Output remains roughly constant up to the cut-out speed of 56 mph (25 m/s). Figure 26-48 provides power and sound emission curves for this turbine model.

After an initial period of gaining commercial operating experience, the 24 MW wind farm is reported to produce about 60,000 MWh annually, which correlates to an annual load factor of 28.5%. On a basis of annual electricity production, this would be equivalent to the output of a baseloaded 7.6 MW conventional fossil fuel- or steam-powered prime mover-driven system, operating with a load factor of 90%, or a three-unit system of 10.5 MW, with one redundant generator set, and an annual availability for two units of 98%. The total capital cost (subject to variation based on exchange rates) was reported to be about $18 million, for a unit cost of $750/kW. A comparison with conventional utility electric generation systems in the region shows a reported avoidance of 165 tons (150 tonnes) of SO_2, 49,600 tons (45,000 tonnes) of CO_2, 143 tons (130 tonnes) of NO_X, 6.6 tons (6 tonnes) of fine dust, and 3,090 tons (2,800 tonnes) of flyash and cinders.

Table 26-10 provides a summary of a hydroelectric technology application that is comparable in terms of total annual electricity production to the above wind farm. The project includes five small hydroelectric units installed by the city of Boulder, Colorado, within its municipal water supply system, with a total rated capacity of 4.1 MW. Two larger units, that will add another 7 MW to their total capacity, are currently being installed. These plants use the excess pressure in the city's water supply, which flows from the North Boulder Creek

Name of hydro station	Rated capacity (kW)	Type of turbine	Type of generator	Water source	In-service date	Annual generation (kWh)	Annual revenue (US $)	Construction cost (US $)
Maxwell	70	Reaction	Induction	Treated	Mar. 1985	513,382	20,573	110,000 1
Kohler	136	Reaction	Induction	Treated	Nov. 1985	598,889	32,903	280,000 2
Orodell	180	Reaction	Induction	Treated	Sep. 1987	823,022	19,507	540,000
Sunshine	800	Reaction	Induction	Treated	Sep. 1987	4,151,523	177,833	1,100,000
Betasso	2,900	Impulse	Synchronous	Raw	Dec. 1987	9,056,116	539,419	3,200,000
Silver Lake	3,500	Impulse	Synchronous	Raw	May 2000	17,000,000 *	500,000 *	4,426,503 *
Lakewood	3,500	Impulse	Synchronous	Raw	Late 2001	19,621,000 *	800,000 *	2,174,000 *
Total	11,086					51,763,932	2,090,235	11,731,503

* Estimated

1 Hydroelectric portion only; the total cost of the Maxwell Pump Generation station was $300,000

2 Hydroelectric portion only; the total cost of the Kohler Pump Generation station was $526,000.

Table 26-10 Technology, Production, and Capital Cost Summary for Municipal Hydroelectric Facilities.
Source: CADDET Renewable IEA/OECD

watershed at an altitude that is 4,699 ft (1,400 m) above the city. In the four smaller sites, which range from 70 to 800 kW, reaction Francis turbine-driven induction generators were used. These were selected for applications in which pressure-relief valves release some of the pressure in the pipelines, such as along treated-water lines for delivery to the customer. The reaction turbines help maintain constant pressure in the line at the outlet. These induction generators can also act as pumps for use in emergencies. Elsewhere in the system, such as at the water treatment plant where pipelines discharge into reservoirs at atmospheric pressure, impulse-type Pelton turbines are being used with synchronous generators for supplemental power.

This project did take several years to receive final approvals, but since that time has produced solid operating results. The total reported installation cost of the five completed and operating plants was about $1,280/kW, with simple paybacks estimated between 10 and 15 years. The cost for connection to the electricity grid was relatively low in these cases, since the plants were all located close to existing electricity substations. The electricity produced is sold to the local electric utility at an average rate of $0.05/kWh. At $942/kW, the larger units have a lower projected capital cost than the smaller capacity units and a far shorter simple payback. Combined, the capital cost for the entire project is expected to total $11.7 million, or $1,058/kW, with an estimated combined simple payback of 5.6 years. Once the additional two plants are commissioned, expected electricity production is 51,800 MWh/yr. The environmental impact is reported to be quite low, since most of the water supply facilities were already in place. Boulder also maintains a minimum flow of water in Boulder Creek set by the state wildlife board to protect local flora and fauna.

In comparison with the 24 MW wind farm, this 11.1 MW project will produce about 86% as much electricity annually. This is based on having a considerably higher annual load factor of 53%. Whereas the first 5 installed units operated with an annual load factor of 42% and showed a capital cost of about $0.35 per annual kWh output (or $0.0175/kWh when amortized over 20 years on a non-discounted basis), the overall project shows a capital cost of about $0.25 per annual kWh output. This is comparatively similar to the wind farm, which shows a capital cost of about $0.30 per annual kWh output. These figures are considerably higher than can be achieved with conventional, higher load factor, fuel-fired prime mover systems. A disadvantage resulting from relatively low load factor is the requirement for far more capacity to be installed than with conventional fuel-fired plants. Also, in many cases, availability cannot be guaranteed given the reliance on potentially intermittent conditions of available water and wind. Additionally, in order to provide a specified quantity of electricity at a specified time, to eliminate demand charges, support facility demand requirements, or fulfill a firm sales agreement, additional investment in storage capacity or a conventional backup system (e.g., Diesel engine generator set) is required. The compensation for these limitations, though, is the life cycle cost advantage of fuel- and emission-free electricity generation.

Generator Driver Selection Process

Given the wide array of driver and system configuration options, it is necessary to carefully evaluate and define the baseline conditions at the facility and perform a detailed electric generation feasibility analysis that compares all potentially viable system application options. A pre-screening process should be used to limit the number of options to be studied in detail. A rigorous technical and economic performance analysis will then be required to make a reliable final financial investment decision.

An integrated approach that considers all other potential conservation and load impact applications interactively is strongly recommended. To the extent that other load and cost reduction opportunities show superior economic performance to that of the electric generation project options, they should generally be considered first. As such, the generation system options would then be considered based on the new operating load and cost profiles that would result from implementation of the other measures.

Additionally, as with most system application analyses, the analysis results will be largely dependent on site-specific conditions. When comparative economic performance results are close for two or more options, it is clear that modest changes in site-conditions, operating loads, or other factors like forward energy price projections, could alter system selection decisions. For this reason, once a base integrated analysis is completed, sensitivity analyses should be performed to evaluate the degree to which such modest changes would impact expected economic performance.

Chapter 41 provides a detailed description of a technical study approach that may be applied to evaluation of cogeneration system feasibility. Following is an example of a detailed electric cogeneration system feasibility analysis, which focuses largely on driver selection.

Electric Cogeneration Feasibility and Driver Selection Example

The following example describes a study that was conducted to determine the potential for on-site cogeneration at a Massachusetts manufacturing facility. This study investigates the feasibility of installing and operating two different types of electric cogeneration driver technologies at the facility: a reciprocating engine-based system and a gas turbine-based system.

The study includes a discussion of current energy consumption, rate structures, and costs, as well as the existing facilities devoted to meeting electrical and steam loads within the complex. Following the discussion of existing conditions are descriptions of the cogeneration systems considered for this facility and the operations and economic simulation used to estimate annual operating costs.

Utility Service Analysis

Natural Gas Service

Facility engineering staff has developed plans to install a high-pressure gas line that will connect the boiler plant with a nearby high-pressure natural gas line as a separate project. This connection will allow the facility to receive gas at approximately 200 psig (14.8 bar) and will eliminate or significantly reduce the need to boost the gas pressure for use in a gas turbine cogeneration system, if applicable.

The projected average natural gas cost was determined to be $3.25 per MMBtu. This study uses this flat rate for calculations of all fuel costs, including cogeneration system and boiler fuel.

Electric Utility Interconnection

Utility electrical service originates on the western side of the complex, entering the site through two 23 kV, 2,000 amp feeders. These lines continue through individual 1,200 amp breakers to two main substation transformers. These units each have a maximum rating of 30 MVa (with cooling fans on high speed), reducing the 23 kV to 13.8 kV. Each main transformer feeds nine 1,750 kVa substations, reducing the service from 13.8 kV to 480 Vac. As a backup precaution, there is a 2,000 amp tie-breaker that will allow each of these secondary substations to be serviced by either of the main transformers in the event of a main transformer failure. There is at least one completely empty cubicle that was designated for connection to the proposed cogeneration system. Utility meters are installed on the 23 kV side of the main transformers.

Electric Utility Rate Structure

Electricity is purchased from the local utility under a firm industrial time-of-use (TOU) rate. The rate has several adjustments, including a fuel adjustment and a credit for high-voltage delivery. There are two defined TOU rate periods. The peak period is defined as 8:00 a.m. to 9:00 p.m., Monday through Friday, excluding holidays, and the off-peak period is defined as all other hours, including all weekend days and the nine observed holidays. The monthly demand charge is based on either the highest 15 minute kW demand occurring during the peak period, or 90% of the highest 15 minute kVa demand occurring during the peak period, whichever is higher.

Effective rates and adjustments are shown in Table 26-11. The demand charge shown has been reduced by $0.45 per kW, because all meters are installed on the 23 kV side of the transformers. All charges have been reduced by 1% to account for the high-voltage metering adjustment. Energy charges include a fuel adjustment charge (fac) of $0.02501 per kWh.

It is worth noting that current utility rates are considerably higher than those used in this example. Both gas and electricity costs have gone up since this example was developed. Gas prices are almost three times higher (over $9.00 per MMBtu national average versus $3.25 per MMBtu used in this example). Electricity costs have doubled (over $0.10/kWh national average versus a blended rate of approximately $0.053/kWh in the example).

Since natural gas, the input fuel for the cogeneration system in the above example, has increased more than electricity, the net effect would be a lower "spark spread" and poorer economic performance for this sample project. However, using national average pricing would not be appropriate for an economic analysis of an actual project – local utility tariffs should be used for a true picture. Moreover, a sensitivity analysis (similar to that discussed in the section on System Simulation And Economic Analysis later in this Chapter) would be useful in testing the impact of variations in both gas and electricity pricing to see what combinations of the two yield acceptable project paybacks.

This latter point applies to many of the examples of project evaluations that appear throughout this book, not just in this Chapter. It important to gain an understanding of the sensitivities to analysis assumptions such as utility rates that can change over time, and the relationships between those rates and other economic variables. With this approach, the project evaluation will be flexible enough to accommodate these variations, and will not seek "canned" solutions that may prove wrong under changing conditions.

RATE I	PEAK	OFF-PEAK
Customer Charge	$59.61/mo	–
Energy Charge (includes all adjustments)	$0.06832/kWh	$0.04258/kWh
Demand Charge	$11.68/kW	–

Table 26-11 Electric Rate Detail.

Electrical Loads

The facility provided the study team with historical electrical demand data covering an entire calendar year. These data included the total electricity consumed during each 4 hour period of the year. Figure 26-49 shows how the maximum, minimum, and average electrical consumption within each month varies over the course of the year. Note the significant drop during the month of July, resulting from an annual two-week plant shutdown.

The study team used these values to develop cogeneration plant configurations that would satisfy a substantial quantity of the facility's electrical requirements while maintaining high utilization factors for the selected equipment.

Steam Service

Boiler Plant

The four existing Cleaver-Brooks model DLD-76 watertube boilers each have a design rating of 60,000 lbm/h (27,216 kg/h) at 150 psig (11.4 bar). The pressure vessel design rating is 260 psig (18.9 bar). Each boiler includes a dual-fuel, steam-atomized burner (natural gas/No. 6 oil) manufactured by The Engineer Company. Partial flue gas recirculation is utilized to reduce air emissions levels using a 30 hp (22 kW) blower on each boiler to return gas from the breeching to the burner.

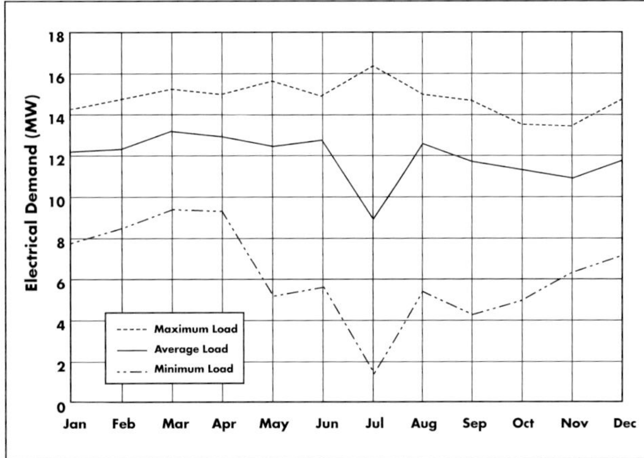

Fig. 26-49 Electric Load Variation Curves.

Forced-draft fans for Boilers 1, 2, and 4 are driven by electric motors with 60 hp (45 kW) ratings, while the fan on Boiler 3 is driven by a 53 hp (40 kW) Carling backpressure steam turbine exhausting at 5 psig (1.4 bar) into the deaerating feed-water heater. Exhaust gas economizers provide heat to a closed feedwater heater downstream of the deaerating unit.

A total of four feedwater pumps serve the boilers. Each pump can serve any boiler and is rated to deliver 215 gpm (814 lpm) at 550 ft (164 m) of head. Makeup water is drawn from the adjacent pond, with water treatment provided by a bank of zeolite softeners.

Other auxiliary equipment includes soot blowers for use while burning fuel oil and fuel oil tank heaters used to maintain oil storage temperature within the range required for transport. All boilers and much of the auxiliary equipment is controlled by a computerized Bailey Infi 90 system.

Steam Distribution System

Steam from each boiler steam drum discharges into a common 18 in. (46 cm) header inside the boiler plant. Export steam to production buildings is carried by an 18 in. (46 cm) main that leaves the boiler plant at roughly 135 psig (10.3 bar) and splits to serve the various buildings.

Pressure reducing stations are located around the production building. There is an existing flange and 16 in. (41 cm) isolation valve designated for interconnection with the proposed cogeneration facility. The proposed cogeneration configuration and pricing assume a single interconnection between the cogeneration heat recovery boilers and the 18 in. (43 cm) header through this valve.

Historical Steam Loads

The existing permit from the State Department of Air Quality (DAQ) allows only three boilers to operate at any given time, providing a total steam capacity of 180,000 lbm/h (81,647 kg/h). Of the total steam produced, a portion is required for loads within the boiler plant. The boiler plant consumes between 1,000 and 15,000 lbm/h (454 and 6,804 kg/h).

Under current load conditions, a single boiler easily meets all export steam requirements. This typically results in either Boiler 1 or 2 operating, because there is insufficient load on the deaerator to utilize all of the exhaust steam from the turbine on the forced draft fan for Boiler 3. When loads are sufficient, the use of the steam turbine drive produces significant operating cost savings versus the electric motor drives.

The facility provided steam consumption data for an entire year, including measurements of total steam exported during each 4 hour period. Figure 26-50 shows the maximum, average, and minimum steam exported for each month. Again, the two-week shutdown during July is evident.

It can also be seen that there was a strong seasonal variation in steam consumption. Facility staff expect this seasonal pattern to continue in the future. The analysis of cogeneration operations will consider the cyclic nature of the steam loads by simulating the operation of the cogeneration plant for each 4 hour interval.

Boiler Plant Efficiency

To determine the net fuel efficiency of the overall plant, the study team analyzed boiler plant records of fuel consumption and exported steam. The ratio of steam production to total boiler fuel consumption (lbm/gal) is commonly used to calculate efficiency, but the measurement of export steam in this case included fixed steam loads and losses (e.g., boiler blowdown and soot blower steam).

The study team performed a linear regression analysis, breaking the data into three linear sections. This process allows for a better approximation of efficiency over the wide fluctuation in export steam load than would result from a single linear function. Figure 26-51 shows the results of this analysis.

The efficiency versus load curve shows thousand lbm (Mlbm) of steam per gallon of fuel over the normal range of daily steam loads. The cogeneration simulation uses the relationship, indicated by the solid line, to calculate the total fuel consumed within the boiler plant to meet any steam load not met by the cogeneration system.

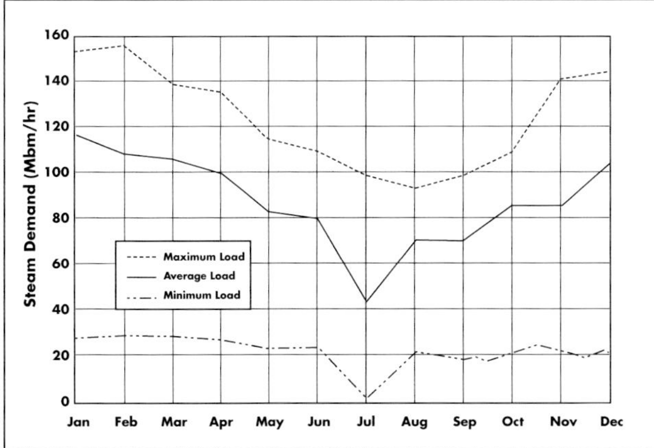

Fig. 26-50 Steam Exportation Curves.

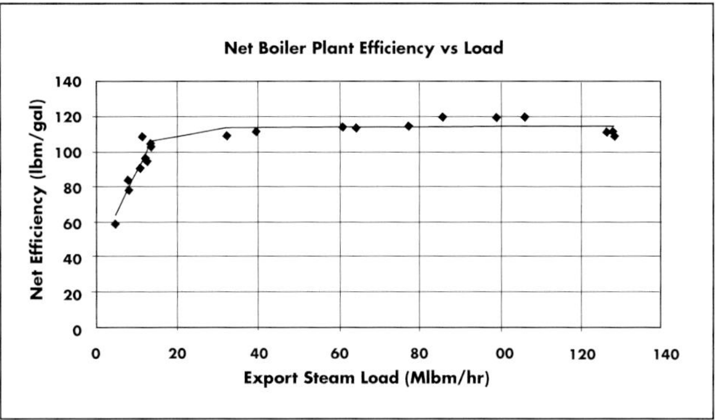

Fig. 26-51 Fuel-Efficiency vs. Steam Load Curves.

COGENERATION SYSTEM OPERATIONS ANALYSIS

The study team performed a simulation of utility costs under three different plant configurations to determine which had the lowest overall operating cost.

CASE	COGENERATION CONFIGURATION	ELECTRICITY SOURCE	STEAM SOURCE
Base	—	1) Electric Utility	1) Boiler Plant
GT-1	Three 4.8 MW Gas Turbines and 24,700 lbm/h HRSGs	1) Cogeneration 2) Electric Utility	1) Cogeneration 2) Boiler Plant
RE-1	Two 6.6 MW Reciprocating Engines and 8,500 lbm/h HRSGs	1) Cogeneration 2) Electric Utility	1) Cogeneration 2) Boiler Plant

Table 26-12 Three Plant Options.

The base case represents the operation of the existing facilities, while the other two cases represent the potential cogeneration configurations described below.

As the first step in configuring cogeneration plants for this site, the study team performed a simplified screening analysis of numerous potential systems and configuration options. A total of six potential system options passed the initial screening analysis. The study team then spoke with six manufacturers that produced these systems. In order to obtain the most appropriate cogeneration plant configurations for this project, they developed an outline of plant requirements and constraints that was sent to each of the manufacturers, as follows:

1. Electrical:
   - Total production of 10 to 15 MW using 2 to 4 prime movers (i.e., generator sets).
   - Tie into existing switchgear facility at 13.8 kV using existing empty cubicle.

2. Thermal:
   - Generate steam at 135 psig (10.3 bar) and discharge into header in boiler plant.
   - Historical steam loads average roughly 90,000 lbm/h (40,800 kg/h), with minimum values of 20,000 to 30,000 lbm/h (9,000 to 13,600 kg/h).
3. Fuel:
   - Natural gas available at 200 psig (14.8 bar) for 365 days per year; the facility has requested that the system not require a booster compressor.
4. Controls:
   - System should track electrical load (prevent export to utility grid). Existing steam and electrical loads track each other well.
5. Physical:
   - Existing cogeneration space includes two adjacent areas: a 60 ft (18 m) by 50 ft (15 m) area with no height limit and a 60 ft (18 m) by 80 ft (24 m) area with an 18 ft (5 m) height limit.

As a result of these specifications and discussions with suppliers, this list was narrowed to three prime mover configurations that would meet the above specification, including one reciprocating engine system and two gas turbine-based systems. This list was ultimately reduced through a second screening analysis to one for each prime mover technology by selecting the systems offering superior performance and lower capital and maintenance cost. Following are detailed descriptions of the two final alternatives.

COGENERATION SYSTEM PLANT CONFIGURATION: CASE GT-1

Equipment Configuration

This cogeneration plant consists of three gas turbine generator sets, associated heat recovery boilers, electrical switchgear, and auxiliary equipment from Solar Turbines. Total plant capacity includes 14.3 MW of electricity at 60°F (15.6°C) outside temperature and 74,100 lbm/h (33,611 kg/h) of 150 psig (11.4 bar) steam. This configuration does not include supplemental firing of HRSGs.

The plant will be controlled to track the facility's electrical demand and produce as much of that demand as the plant can provide. The simulation will start one gas turbine as soon as the electrical load exceeds the minimum operating level for one turbine (50% of capacity) and will start additional turbines as the load exceeds the capacity of the operating turbines. The simulation assumes that all operating engines share the total load equally.

Cogenerated steam production roughly tracks electrical production, resulting in the need to have standby steam-generating capacity. For this reason, the simulation assumes that one of the existing Cleaver-Brooks boilers will operate to maintain temperature for all hours of the year.

Maintenance Requirements

Solar Turbines reached agreement with the facility to offer a five year extended service program for the plant that will cover all scheduled and emergency maintenance and repairs. The cost for the program is $3.50 per MWh generated, with a guarantee of 95% availability for each of the turbines. This program includes eight preventive maintenance visits per turbine per year and unlimited trouble calls. Scheduled maintenance includes a minor service every six month period requiring two to three days of outage per turbine and an annual major overhaul requiring three to five days of downtime per turbine.

The operations simulation has accounted for these scheduled service intervals by making one turbine unavailable for these periods. In order to minimize the impact of this work on the operating cost savings, it was assumed that this work will be scheduled during the July plant shutdown and during minimum load periods in January.

Based on the above unit cost for the maintenance program and the projected total power generation of 98,616 MWh, the annual maintenance cost to the facility for this plant was estimated as:

98,616 MWh x $3.50 per MWh = $345,156/year

In addition to this figure, the following two maintenance costs were applied to each year of GT-1 system operation:

1. Maintenance of boilers, steam system, and related equipment totals $50,000 per year.
2. Four operators for 24 hour per day coverage, at $70,000 each, totals $280,000 per year.

Note that a standalone plant would require more than four operators to provide 24 hour operating coverage. In this case, this study considers the net staff requirement increase over existing boiler plant staff. This assumes that there will be overlap in coverage and responsibilities between boiler plant operations and cogeneration plant operations, and is consistent with the staffing plan prepared by the facility.

Plant Performance

Solar Turbines provided plant operations and performance data. The power output from a gas turbine is highly dependent upon the temperature of the combustion air. For this reason, the performance of the plant has been modeled using two independent variables: outdoor air temperature and load level. For a given outdoor air temperature, the plant will have a capacity three times that of a single gas turbine unit.

Steam generation capacity is a function of exhaust flow, exhaust temperature, and steam pressure. Because the plant controller will operate the gas turbine to generate as much power as the facility needs, there may be periods when the steam capacity of the heat recovery boilers exceeds the total boiler load. Under these conditions, exhaust gas would be diverted around the heat recovery boiler, thereby eliminating the need to vent steam to atmosphere. The simulation is capable of detecting and allowing for this condition.

Table 26-13 shows a subset of the performance data for this cogeneration plant configuration, including electrical and steam production, as well as fuel consumption values for the plant as a function of ambient temperature. These values are used by the operations simulation discussed below.

Project Cost

The study team developed a project cost estimate to implement this cogeneration configuration, including the cogeneration equipment described above, extension of the existing boiler plant building structure over the cogeneration system, construction, and system startup. The total cost to complete this project was estimated at $13,950,000, as summarized in Table 26-14.

COGENERATION SYSTEM PLANT CONFIGURATION: CASE RE-1

Equipment Configuration

This system is based on the installation of two gas-fired reciprocating engine-generators, heat recovery boilers, hot water heat exchangers, electrical switchgear, and auxiliary equipment from Coltec Industries/Fairbanks Morse Engine Division. Total plant capacity includes 13.06 MW of electricity, 13,800 lbm/h (6,260 kg/h) of 135 psig (10.3 bar) steam, and 21.8 MMBtu/h (22,988 MJ/h) of hot water at 180°F (82°C).

Switchgear will be installed to control the engine-generators, with capacity control to limit electrical production to the facility's electrical demand. The generators produce power at 13.8 kV, so that this can be tied directly into the existing facility switchgear without the need for a transformer.

Thermal energy from the engines will be used to offset steam heat in two areas. First, a new heat exchanger

Ambient Temp	Electric Capacity			Fuel Rate (MMBtu/MWh)						Steam Capacity (Mlb/MWh)					
	Gross	Loss	Net	Load Level						Load Level					
°F	MW	MW	MW	100%	90%	80%	70%	60%	50%	100%	90%	80%	70%	60%	50%
50	4.94	0.02	4.92	12.72	13.55	14.38	15.21	16.04	16.87	4.23	4.73	5.23	5.73	6.23	6.73
51	4.92	0.02	4.90	12.73	13.56	14.39	15.23	16.06	16.89	4.23	4.73	5.23	5.73	6.23	6.73
52	4.90	0.02	4.88	12.74	13.58	14.41	15.25	16.08	16.91	4.23	4.73	5.23	5.73	6.23	6.73
53	4.89	0.02	4.87	12.76	13.59	14.43	15.26	16.10	16.94	4.23	4.73	5.23	5.72	6.22	6.72
54	4.87	0.02	4.85	12.77	13.61	14.45	15.28	16.12	16.96	4.22	4.72	5.22	5.72	6.22	6.72
55	4.85	0.02	4.83	12.79	13.62	14.46	15.30	16.14	16.98	4.22	4.72	5.22	5.72	6.22	6.72
56	4.83	0.02	4.81	12.80	13.64	14.48	15.32	16.16	17.00	4.22	4.72	5.22	5.72	6.22	6.72
57	4.81	0.02	4.79	12.81	13.66	14.50	15.34	16.18	17.03	4.21	4.71	5.22	5.72	6.22	6.72
58	4.79	0.02	4.77	12.83	13.67	14.52	15.36	16.21	17.05	4.21	4.71	5.21	5.71	6.22	6.72
59	4.77	0.02	4.75	12.85	13.69	14.53	15.38	16.23	17.07	4.21	4.71	5.21	5.71	6.21	6.71
60	4.75	0.02	4.73	12.85	13.70	14.55	15.40	16.25	17.10	4.21	4.71	5.21	5.71	6.21	6.71

Table 26-13 Subset of Performance Data for Gas Turbine Cogeneration System.

Item	Cost
Engine Generators	$9,978,000
Electrical Switchgear	$390,000
Building	$701,000
Design/Construction Services	$270,000
Overhead and Profit	$2,214,000
Contingency	$398,000
Total	$13,950,000

Table 26-14 Capital Cost Summary for Gas Turbine Cogeneration System.

will be installed to use the hot water to preheat boiler makeup water from approximately 70°F (21°C) up to 170°F (77°C). For this analysis, the magnitude of this load was estimated by assuming this temperature change on a flow equivalent to 25% of the steam flow (corresponding to a makeup rate of 25%). Any additional heating requirements will be met by a combination of the existing deaerating feedwater heater and the economizer feedwater heater.

The other thermal load for hot water from the engines is process hot water load within the manufacturing plant. This analysis was based on a load of 4 MMBtu/h (4,200 MJ/h) for heating reject process water from an existing holding tank from an initial temperature of 120°F (49°C) up to approximately 170°F (77°C). Currently, this water is heated to 170°F (77°C) by steam heat exchangers. Included in the cost estimate is a hot water piping system consisting of 1,500 ft (457 m) of insulated 6 in. (15 cm) piping, with all associated heat exchangers, pumps, controls, and miscellaneous equipment. During any period when the process load is insufficient to provide proper heat rejection for the hot water loop, two waste heat radiators for each engine will maintain return water temperature within acceptable limits.

Maintenance Requirements

Coltec Industries estimated the total scheduled maintenance cost of this plant at $2.90 per MWh. This price was developed assuming the plant operates 8,000 hours per year at 100% load, and includes a breakdown of 30% parts and 70% labor. Additional operating costs include lube oil consumption, estimated at $0.32 per MWh. Based on the simulation projections of 99,312 MWh per year, annual maintenance cost was estimated as follows:

99,312 MWh x ($2.90 + $0.32) per MWh = $319,785/yr

Based on the facility's plan to hire a staff of four individuals to operate and maintain the cogeneration facility, most of the labor component of the above estimate can be re-assigned from the maintenance budget to the staffing budget. In addition to this figure, the same two maintenance costs used for the GT-1 system were applied to each year of RE-1 system operation:

1. Maintenance of boilers, steam system, and related equipment totals $50,000 per year.
2. Four operators for 24 hour per day coverage, at $70,000 each, totals $280,000 per year.

Plant Performance

Plant performance data was provided by Coltec Industries for the above system. Unlike the combustion turbines, reciprocating engine performance is largely independent of ambient air temperature — combustion air temperature is regulated by a heating system to within a narrow range. For this reason, the table of plant operations included for this system shows overall plant performance as a function of load only. Engine loading can vary within the range of 50 to 100%. When the load exceeds the capacity of one engine-generator, a second engine-generator is started and the two share the load equally.

As discussed above, hot water produced by the cogeneration system will be utilized by both the boiler plant and the manufacturing process. For clarity, the simulation considers the thermal output of the engines in terms of equivalent steam generation at 1,000 Btu/lbm (2,325 kJ/kg). Table 26-15 shows a representative portion of the performance data for this cogeneration plant configuration, including electrical and equivalent steam production, as well as fuel consumption values for the plant. These values are used by the operations simulation.

Project Cost

The study team developed a project cost estimate to implement this cogeneration configuration, including the cogeneration equipment described above, extension of the existing boiler plant building structure over the cogeneration system, construction, and system startup. The cost estimate also includes the hot water distribution and makeup water heating systems necessary to utilize available recovered heat in place of steam. The total cost to complete this project was estimated at $15,320,000, as summarized in Table 26-16.

System Simulation and Economic Analysis

The study team developed a simulation of the facility's loads, boiler plant, electric utility service, and cogeneration systems to compare the annual operating performance and costs of each case. This simulation begins with the historical electrical and steam loads and determines which equipment is available to meet those loads. In the base case, all steam loads are met by the existing boiler plant and all electrical loads are met by the electric utility. In the cogeneration cases, the cogeneration equipment will carry all loads up to operating capacity, with the excess met by the boiler plant and electric utility.

Ambient Temp	Electric Capacity			Fuel Rate (MMBtu/MWh)						Steam Capacity (Mlb/MWh)					
	Gross	Loss	Net	Load Level						Load Level					
°F	MW	MW	MW	100%	90%	80%	70%	60%	50%	100%	90%	80%	70%	60%	50%
50	6.53	0.13	6.40	9.28	9.42	9.57	9.84	10.23	10.63	2.73	2.76	2.79	2.82	2.85	2.88
60	6.53	0.13	6.40	9.28	9.42	9.57	9.84	10.23	10.63	2.73	2.76	2.79	2.82	2.85	2.88

Table 26-15 Selected Performance Data for Reciprocating Engine Cogeneration Plant Configuration.

Item	Cost
Engine Generators	$10,820,000
Electrical Switchgear	$320,000
Building	$701,000
Hot Water System	$288,000
Design/Construction Services	$330,000
Overhead and Profit	$2,426,000
Contingency	$437,000
Total	$15,320,000

Table 26-16 Capital Cost Summary for Reciprocating Engine Cogeneration System.

The simulation is structured to sequentially step through each 4 hour interval for a calendar year. For example, Table 26-17 is an excerpt from the simulation showing the time, temperature, and facility loads for January 1 and 2. The column headings show how each of these sources will operate to satisfy facility loads.

The simulation calculates total monthly consumption for each source of energy and the cost to provide this service is calculated based on boiler efficiency, projected fuel costs, and the electric utility rate structure. The end result of this simulation for each case is an estimate of the total cost to operate the available equipment to meet facility loads. By comparing the total operating cost for each cogeneration case with the base case, the analysis determines the net projected savings.

The study team assumed that the facility will opt to have one of the existing boilers kept warm for standby purposes at all times when operating cogeneration systems. For any period where the cogeneration system is providing 100% of the export steam demand, the simulation imposes a minimum steam load on the existing boiler plant of 2,000 lbm/h (907 kg/h).

Simulation Results

Tables 26-18 through 26-20 show the annual cost data for each area of plant operations modeled by the simulation. There is one table for the base case and one for each of the cogeneration plant configurations.

Table 26-21 summarizes the project capital cost and annual operating costs and savings for each of the two cogeneration configurations, as well as the base case representing existing conditions. These values show the gas turbine system to offer slightly superior potential economic performance based on simple payback. This is due to a combination of reasons, including:

- Smaller unit size of gas turbines and their resulting ability to operate at lower loads that will leave the larger reciprocating engines idle.
- Higher utilization of recovered thermal energy from the gas turbine cogeneration plant. The reciprocating engine system is limited by the lack of low-temperature thermal loads, resulting in the need to dump heat using a radiator under certain conditions.

The gas turbine cogeneration system evaluated does provide a good match of electrical and thermal capacities for the projected facility loads. In the absence of a more sizable low-temperature load for the reciprocating system, the

		FACILITY LOADS		COGEN PLANT			ELECTRIC UTILITY			BOILER PLANT	
Date/Time	Temp °F	Elec MW	Steam Mlb/hr	Elec MW	Steam Mlb/hr	Gas MMBtu	Demand MW	Peak MWh	Off-peak MWh	Steam Mlb/hr	Gas MMBtu
1/1 at 02:00	19	7.3	107.7	6.4	13.3	218	0.9		3.5	94.4	795
1/1 at 06:00	24	7.3	107.0	6.4	13.3	218	0.9		3.5	93.7	791
1/1 at 10:00	37	7.3	114.5	6.4	13.5	218	0.9	3.5		101.0	529
1/1 at 14:00	42	7.5	118.4	6.4	13.6	218	1.1	4.2		104.8	549
1/1 at 18:00	37	8.6	116.3	6.4	13.5	218	2.2	8.6		102.7	538
1/1 at 22:00	27	8.7	110.9	6.4	13.4	218	2.3	2.3	7.0	97.5	511
1/2 at 02:00	22	8.7	110.6	6.4	13.4	218	2.3		9.3	97.2	509

Table 26-17 Excerpt from 4-Hour Simulation Model for Base Case.

		ELECTRIC UTILITY				COGENERATION SYSTEM							BOILER PLANT				
		Demand	Peak	Off-peak	Billing	Electrical		Steam		Maint	Natural Gas		Steam		NOx	Natural Gas	
Month	num	MW	MWh	MWh	Cost	MW1	MWh	Mlb/hr1	Mlb	$	MMBtu	Cost	Mlb/hr1	Mlb	lb	MMBtu	Cost
Jan	1	13.4	3,600	4,900	$ 454,812	0.0	-	0.0	-	$ -	-	$ -	176.1	99,986	128	130,417	$ 423,855
Feb	2	13.9	3,171	4,660	$ 415,274	0.0	-	0.0	-	$ -	-	$ -	179.4	84,053	-	109,730	$ 356,622
Mar	3	14.3	3,752	5,486	$ 490,117	0.0	-	0.0	-	$ -	-	$ -	159.5	90,914	-	118,739	$ 385,901
Apr	4	14.1	3,648	5,088	$ 466,091	0.0	-	0.0	-	$ -	-	$ -	156.2	83,023	-	108,510	$ 352,657
May	5	14.7	3,542	5,166	$ 462,197	0.0	-	0.0	-	$ -	-	$ -	131.8	71,132	-	93,168	$ 302,798
Jun	6	14.0	3,751	4,867	$ 463,733	0.0	-	0.0	-	$ -	-	$ -	126.2	66,225	-	86,786	$ 282,053
Jul	7	15.5	2,804	3,414	$ 337,128	0.0	-	0.0	-	$ -	-	$ -	114.1	37,498	-	50,104	$ 162,838
Aug	8	14.1	3,579	5,249	$ 468,255	0.0	-	0.0	-	$ -	-	$ -	107.4	61,022	-	80,071	$ 260,232
Sep	9	13.8	3,454	4,484	$ 427,106	0.0	-	0.0	-	$ -	-	$ -	114.1	58,426	-	76,671	$ 249,182
Oct	10	12.7	3,333	4,607	$ 424,122	0.0	-	0.0	-	$ -	-	$ -	126.2	73,508	-	96,255	$ 312,830
Nov	11	12.6	2,945	4,442	$ 390,542	0.0	-	0.0	-	$ -	-	$ -	162.8	71,205	-	93,192	$ 302,873
Dec	12	13.9	3,469	4,785	$ 440,999	0.0	-	0.0	-	$ -	-	$ -	166.1	89,635	-	117,083	$ 380,520
TOTAL			41,048	57,147	$5,240,376		-		-	$ -	-	$ -		886,628	128	1,160,726	$3,772,360

Table 26-18 Annual Cost Data for Base Case.

		ELECTRIC UTILITY				COGENERATION SYSTEM							BOILER PLANT				
		Demand	Peak	Off-peak	Billing	Electrical		Steam		Maint	Natural Gas		Steam		NOx	Natural Gas	
Month	num	MW	MWh	MWh	Cost	MW1	MWh	Mlb/hr1	Mlb	$	MMBtu	Cost	Mlb/hr1	Mlb	lb	MMBtu	Cost
Jan	1	2.7	484	282	$ 45,149	13.3	7,734	70.0	33,876	$ 54,570	97,116	$ 315,626	130.2	66,165	97	86,784	$ 282,046
Feb	2	2.6	253	368	$ 33,075	13.9	7,209	71.9	32,005	$ 52,732	91,850	$ 298,514	127.3	52,107	-	68,457	$ 222,486
Mar	3	2.7	130	578	$ 33,603	14.3	8,529	71.4	38,211	$ 57,352	110,975	$ 360,668	114.1	52,783	-	69,408	$ 225,576
Apr	4	2.6	147	509	$ 31,850	14.1	8,079	69.9	36,206	$ 55,777	106,534	$ 346,237	106.2	46,881	-	61,745	$ 200,670
May	5	2.4	79	219	$ 14,824	14.0	8,410	66.9	37,156	$ 56,935	112,837	$ 366,719	91.8	34,048	-	45,109	$ 146,604
Jun	6	2.2	57	67	$ 6,801	13.6	8,495	64.7	37,297	$ 57,231	115,388	$ 375,009	71.7	28,992	-	38,537	$ 125,246
Jul	7	2.7	255	359	$ 32,756	14.3	5,604	60.4	22,136	$ 47,114	76,537	$ 248,745	61.5	15,806	-	21,913	$ 71,216
Aug	8	2.4	73	143	$ 11,159	14.0	8,612	63.1	36,882	$ 57,643	116,680	$ 379,208	60.7	24,219	-	32,423	$ 105,374
Sep	9	2.5	90	339	$ 20,670	13.7	7,509	65.3	33,170	$ 53,780	101,294	$ 329,207	65.8	25,318	-	33,887	$ 110,134
Oct	10	2.5	305	432	$ 39,273	12.7	7,204	62.3	31,461	$ 52,716	94,356	$ 306,656	86.1	42,080	-	55,633	$ 180,807
Nov	11	2.5	282	280	$ 31,269	12.6	6,825	65.6	30,128	$ 51,387	88,927	$ 289,012	120.7	41,165	-	54,375	$ 176,719
Dec	12	2.7	140	497	$ 30,862	13.9	7,617	69.8	34,648	$ 54,159	98,681	$ 320,713	120.9	55,035	-	72,363	$ 235,178
TOTAL			2,295	4,073	$ 331,291		91,827		403,176	$ 651,395	1,211,174	$ 3,936,314		484,599	97	640,633	$ 2,082,057

Table 26-19 Annual Cost Data for Case GT-1.

		ELECTRIC UTILITY				COGENERATION SYSTEM							BOILER PLANT				
		Demand	Peak	Off-peak	Billing	Electrical		Steam		Maint	Natural Gas		Steam		NOx	Natural Gas	
MONTH	num	MW	MWh	MWh	Cost	MW1	MWh	Mlb/hr1	Mlb	$	MMBtu	Cost	Mlb/hr1	Mlb	lb	MMBtu	Cost
Jan	1	3.2	37	248	$ 13,202	12.8	8,215	21.5	14,049	$ 53,951	78,672	$ 255,683	154.9	85,936	110	112,324	$ 365,053
Feb	2	3.1	87	118	$ 11,030	12.8	7,626	21.8	12,752	$ 52,057	73,065	$ 237,462	157.7	71,301	-	93,281	$ 303,162
Mar	3	3.2	138	113	$ 14,330	12.8	8,987	21.3	14,601	$ 56,437	85,637	$ 278,319	138.2	76,313	-	99,908	$ 324,702
Apr	4	3.2	76	80	$ 8,721	12.8	8,580	21.1	13,896	$ 55,126	81,947	$ 266,328	135.5	69,127	-	90,559	$ 294,318
May	5	3.2	186	188	$ 20,827	12.8	8,334	20.4	13,419	$ 54,336	79,294	$ 257,706	112.1	57,713	-	75,804	$ 246,362
Jun	6	2.9	137	135	$ 15,158	12.8	8,347	20.4	13,205	$ 54,377	79,294	$ 257,704	105.8	53,021	-	69,688	$ 226,487
Jul	7	3.1	137	200	$ 17,941	12.8	5,881	20.1	9,340	$ 46,435	57,188	$ 185,860	94.0	28,456	-	38,535	$ 125,238
Aug	8	2.9	113	188	$ 15,837	12.8	8,527	19.9	13,365	$ 54,956	81,158	$ 263,764	87.7	47,657	-	62,784	$ 204,049
Sep	9	3.2	51	84	$ 7,167	12.8	7,802	20.0	12,485	$ 52,624	74,753	$ 242,946	94.6	45,941	-	60,535	$ 196,737
Oct	10	3.2	-	100	$ 4,347	12.7	7,841	20.0	13,035	$ 52,747	75,676	$ 245,947	107.1	60,473	-	79,380	$ 257,986
Nov	11	3.1	95	247	$ 17,129	12.6	7,044	20.9	12,038	$ 50,182	67,832	$ 220,454	142.4	59,168	-	77,646	$ 252,349
Dec	12	3.2	110	279	$ 19,495	12.8	7,865	21.4	13,417	$ 52,827	75,171	$ 244,304	144.7	76,218	-	99,769	$ 324,248
TOTAL			1,167	1,979	$ 165,184		95,048		155,602	$ 636,055	909,686	$ 2,956,479		731,324	110	960,213	$ 3,120,691

Table 26-20 Annual Cost Data for Case RE-1.

Case	Project Cost	Annual Operating Cost	Net Savings	Simple Payback
Base	–	$9,012,736	–	–
GT-1	$13,950,000	$7,001,057	$2,011,679	6.9 yr
RE-1	$15,320,000	$6,878,409	$2,134,327	7.2 yr

Table 26-21 Economic Performance Summary.

gas turbine-based system will continue to offer better load matching and project economics for the facility.

It is interesting to note that in this case, two general rules of thumb are broken. While gas turbine systems are generally assumed to have a higher capital cost and lower maintenance cost than reciprocating engines, the reverse was true in this case. The gas turbine system proved to be less costly on a capital cost basis and have a slightly higher annual maintenance cost. This underscores the important point that rules of thumb cannot be relied upon when making significant investment decisions.

Moreover, as with most system application analyses, the results are largely dependent on site-specific conditions. When comparative economic performance results are close for two or more options, it is clear that modest changes in site conditions, operating loads, or other factors could alter system selection decisions. For this reason, sensitivity analyses are often performed to evaluate the degree to which such modest changes would impact expected economic performance. Since energy conservation or load reduction can also affect central plant requirements, integrated studies that interactively consider all opportunities at a facility are a preferred analytical approach.

Following are two sensitivity analyses that demonstrate different results. The first shows how results would be different for a facility with somewhat varying conditions. The second shows how the expected implementation of interactive measures at this facility would alter the study results.

Sensitivity Analysis 1

The major factor impeding the economic performance of the reciprocating engine configuration in this potential application is the cost and inefficiency of utilizing the low-grade portion of its thermal energy. In this particular facility, the lack of a sufficient nearby hot water load increased the capital budget for this case by requiring the installation of a heat recovery hot water distribution system.

Alternately, in this case, the high-pressure natural gas line available to the project provides substantial benefit to the gas turbine configuration. In another location, where such facilities were not available, the gas turbine cogeneration plant option would be burdened with the capital and operating cost of a gas compressor to raise the pressure to the 200 psig required by the DLN combustor of the gas turbine.

As a test of sensitivity, we will remove the hot water distribution system from the RE-1 plant configuration and require the addition of a booster gas compressor for each of the gas turbines included in the GT-1 plant configuration. This sensitivity assumes that gas is available to the facility at approximately 75 psig, the minimum gas pressure required by the reciprocating engines in the RE-1 configuration. For the GT-1 case, the simulation assigned an additional parasitic load of 75 kW for primary gas compression on each operating turbine and a total installed cost of $675,000 for the three compressors. This results in an increase in both capital cost and annual operating cost for the gas turbine system. Table 26-22 summarizes the economic performance for each option that results from running the system simulation for each of these revised cases. Notice that in this scenario, the reciprocating engine-based system now has a shorter simple payback than the gas turbine-based system.

	Project Cost	Annual Operating Cost	Net Savings	Simple Payback
Base	–	$9,012,736	–	–
GT-1	$14,625,000	$7,071,535	$1,941,327	7.5 yr
RE-1	$14,960,000	$6,878,409	$2,134,327	7.0 yr

Table 26-22 Economic Performance Summary Under Sensitivity Analysis 1.

Sensitivity Analysis 2

To the extent that other load and cost reduction opportunities show superior economic performance to that of the electric cogeneration project options, they should generally be considered first. This concept is commonly referred to as an integrated approach to energy use optimization. As such, the cogeneration system options would then be considered based on the new operating load and cost profiles that would result from implementation of the other measures.

This sensitivity analysis considers the interactive impact of one such measure. At the facility, there is a highly cost-effective option that would improve the energy and water use efficiency of a dying process through application of thermal and water recovery technology. The simple payback for this technology application is 2.0 years, making it extremely attractive to the facility. Assuming that this measure would be installed, the potential to use the recovered low-temperature heat from the reciprocating engine in the form of hot water would be greatly reduced, thereby reducing the economic performance of that system option. The gas turbine system would be affected, but to a much smaller degree, since during most operating hours the total steam load remains in excess of the gas turbine plant production.

Starting with the first sensitivity analysis that shows superior economic performance for the reciprocating engine system option, the simulation was run again, this time based on the new load profile that would result from the implementation of the thermal and water recovery measure. Changes modeled by the simulation include a reduction of 15,000 lbm/h of hot water heating steam load during most hours and the elimination of the heat recovery hot water distribution system from Case RE-1. The resultant cogeneration system economic performance analysis is summarized in Table 26-23.

These results show that, as expected, the reduction in process hot water load will have a greater impact on the reciprocating engine configuration than on the gas turbine configuration. As such, under these site-specific conditions, the gas turbine system now shows superior economic performance.

Conclusions

The key conclusions to be drawn from the previous analytical examples are that:

1. A pre-screening process should be used to limit the number of options to be studied in detail.
2. A rigorous technical and economic performance analysis will be required to make a reliable financial investment decision.
3. Rules of thumb should not bias the selection process.
4. Site-specific conditions must be carefully identified and applied to system performance evaluation.
5. Technology applications should be viewed within an integrated approach that considers the interactive effects of other viable facility improvement options.

Case	Project Cost	Annual Operating Cost	Net Savings	Simple Payback
Base	-	$8,469,718	-	-
GT-1	$13,950,000	$6,506,568	$1,963,150	7.1 yr
RE-1	$14,960,000	$6,570,592	$1,899,126	7.9 yr

Table 26-23 Economic Performance Summary Under Sensitivity Analysis 2.

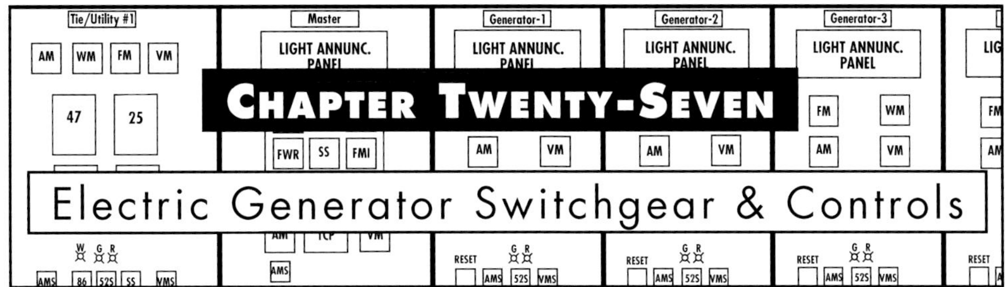

Chapter Twenty-Seven

Electric Generator Switchgear & Controls

Switchgear is an assembly of switching, interrupting, regulating and protective devices, control logic circuitry, instrumentation, and metering devices required for the safe distribution of electric power. With localized on-site electric generation applications, generator switchgear manages one or more generators in a prearranged control strategy to safely meet the needs of the host facility and, where applicable, the requirements of the utility grid. Figure 27-1 shows switchgear for an on-site generator application.

Switchgear applications may be classified as **automatic transfer systems** (for isolated operation), **paralleling systems** (for isolated operation with utility backup), and **utility paralleling systems.** Regardless of voltage class or system complexity, switchgear performs three basic functions: protection, control, and monitoring of electric generation and distribution systems.

Generator systems are generally classified by voltage. For low-voltage systems (600 volts or less), typical voltages are 480Y/277, 208Y/120, or 208, 240, or 480 volt, three-wire. In smaller facilities of this type, the generators are likely to be low-voltage machines. Medium-voltage systems are typically 5 kV (2,400 or 4,160 volts) or 15 kV (13,800 volts). In large facilities or single generator units with capacities of several MW, a medium-voltage design would typically be used. In equivalent multi-generator systems, either low- or medium-voltage switchgear could be used, depending on facility requirements. These values are for 60 Hz; there are corresponding voltages for 50 Hz. Higher voltages are generally restricted to major transmission systems.

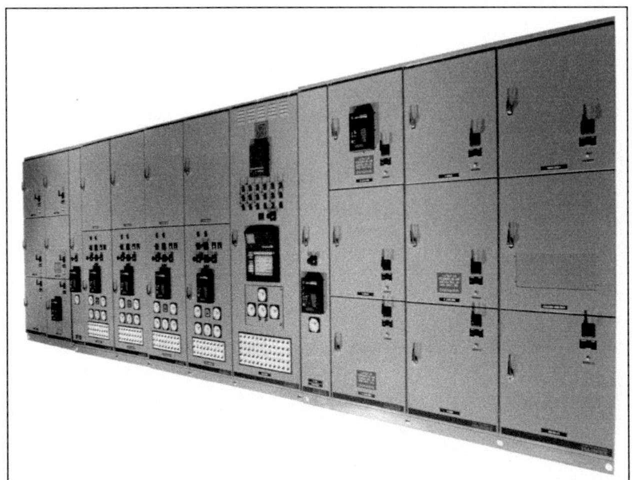

Fig. 27-1 Switchgear for an On-Site Electric Generation Application. Source: Zenith Controls

Switchgear characteristics depend on service voltage, continuous current rating, fault-duty ratings, and installation environment. The following are common types of switchgear:

- **Low-voltage metal-enclosed switchgear** (ANSI/IEEE C37.20.1). A single enclosure contains power switching or interrupting devices with bus bars, control, and other auxiliary devices. Bus bars are either copper or aluminum. Low-voltage power circuit breakers are contained in individual compartments and can be controlled by integral or remote devices.
- **Metal-clad switchgear** (ANSI/IEEE C37.20.2). This type of switchgear differs from the low-voltage type in that the main switching and interrupting device must be removable. The major components are completely enclosed by grounded metal barriers and are isolated from all instruments, relays, control devices, etc. Bus bars for medium voltage are completely insulated. Medium-voltage circuit breakers (either 5 kV or 15 kV class) may be air or vacuum type.
- **Metal-enclosed interrupter switchgear** (ANSI/IEEE C37.20.3). Used above 1,000 volts, circuit breaker, power fuses, bare bus, instrument transformers, and control wiring are completely enclosed with sheet metal. The breakers and power fuses may be stationary or drawout type.

Generator controls and switchgear are generally integrated into one system, or "line-up," which is usually compartmentalized by function. Figure 27-2 illustrates switchgear for a local on-site electric generation system featuring four generator sets.

A typical paralleling system consists of the following sections:

- **Generator control sections.** A separate cubicle is usually provided to control each generator set and synchronize the unit to a common bus. Typically, the cubicle will include metering devices and door-mounted switches, along with a status and alarm panel. Mounted inside

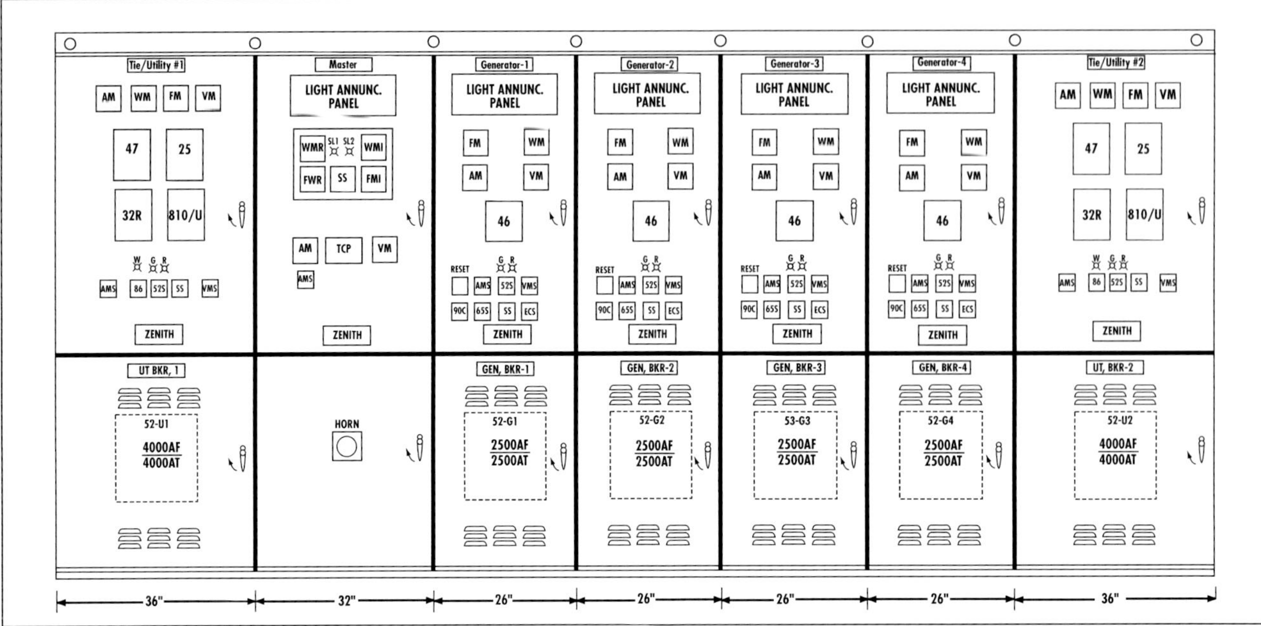

Fig. 27-2 Switchgear for Localized Electric Generation System Featuring Four Generator Sets. Source: Zenith Controls

the enclosure are the automatic starting module, engine governor, automatic synchronizer, protective relays, and breakers. The rear of the enclosure includes the system bus, generator potential, and current transformers. Main generator cables may enter the top or bottom of the cubicle. For medium-voltage switchgear, the breaker and bus section is usually separate from the control section.

- **Master control section.** The master cubicle houses controllers for engine generator set starting, loading, running, and load shedding. Both automatic and manual controls are provided. A synchronizing panel is usually provided to assist an operator in the manual mode.
- **Utility control section.** This section may contain the utility breaker, as well as automatic synchronizer, import/export controller (if used), and necessary control circuitry. Utility metering, switches, and protective relays are mounted on the door.
- **Distribution.** Distribution equipment includes circuit breakers, fuses, automatic transfer switches, and other devices that electrically link the power source to building loads.

Protection

Discriminative protection is required in all power systems to quickly isolate faulty circuits, while maintaining continuity of service to unfaulted circuits. The protective relaying and control system must be designed to function automatically in a safe, logical sequence to protect generators, the utility system, and host facility equipment. If the fault remains connected, the whole system may be in jeopardy. Protective systems can be complex, especially if the generators operate in parallel with the utility.

The switchgear contains circuits that sense operating conditions and act to prevent damage to the prime mover. For example, in the case of a reciprocating engine, the circuits would maintain surveillance on such parameters as oil pressure, water temperature, oil level, water level, exhaust temperature, engine speed, and direction of power flow. When any of these parameters goes outside a preset limit, these circuits would initiate immediate disconnect and shutdown of the prime mover. Some of these circuits have two-stage limits. The first stage would cause a warning to be sent to the operator, allowing for sufficient time to intervene before the second stage is reached so that unnecessary service interruptions can be avoided.

Circuits are included for protection of on-site loads and conductors as well. All electric power systems include overcurrent protection of a power feeder at the point where that feeder receives its power. The trip element continuously measures the amount of current flowing in the conductor. When that current exceeds a preset value for a specific time, the circuit breaker opens to disconnect the feeder from its source.

When a generator operates in parallel with a utility-derived power source, the switchgear must incorporate protective circuits that separate the two power sources upon occurrence of an abnormal operating condition. Because the power bus is common to the generator and the utility system, when these systems are operating in parallel, voltage and frequency protective circuits are ineffective. Typically, current and power flow directions and magnitudes are the best means of detecting faults and initiating protective actions.

Most utility system faults are cleared with automatic breaker operation, and the facility system must be protected against the effects of automatic closing. When a very large facility generator remains connected at the time the utility breaker is attempting to reclose, the generator can be connected out of synchronization, with damaging overcurrents. If there is more than one utility feeder, the controls must ensure that utility tie breakers do not disconnect the generator and reconnect it in an out-of-phase condition. Bus-tie circuit breakers are often used when there is more than one utility feeder.

For synchronous generators, the prime mover governor is controlled to maintain power flow from the generator into the power system and the voltage regulator is controlled to maintain proper VAR sharing. Protective relaying detects changes in current or power magnitude and direction, voltage, frequency, and system impedance. In most cases, when these changes exceed predetermined limits, the relays send a signal to trip the main breaker and disconnect the generating system from the utility system.

Somewhat less protective relaying is required for induction generators. In smaller systems, undervoltage and overvoltage protection may be adequate, though additional relays, such as under- and over-frequency, are sometimes required.

Figure 27-3 is a simplified one-line diagram of a system having the on-site generators capable of paralleling with utility. The diagram shows the utility circuit breaker 52-U, the generator circuit breakers, and the building distribution system. For the generator to parallel with the utility system, relaying must be used to trip the utility breaker under loss of power or fault conditions. Figure 27-4 is a one-line diagram illustrating commonly used protective relays for protection and control of the utility breaker. Figure 27-5 is a one-line diagram illustrating typical protective relays used on a medium voltage generator.

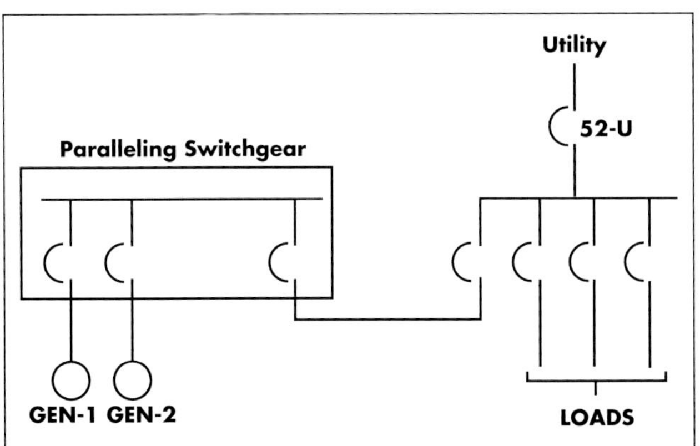

Fig. 27-3 Line diagram Showing Circuit Breakers on a System with Localized Generators Capable of Paralleling with the Utility Grid. Source: Zenith Controls

Circuit Breakers

A circuit breaker is defined by NEMA Standards as a "device designed to open and close a circuit by non-automatic means and to open the circuit automatically on a predetermined overcurrent when properly applied within its

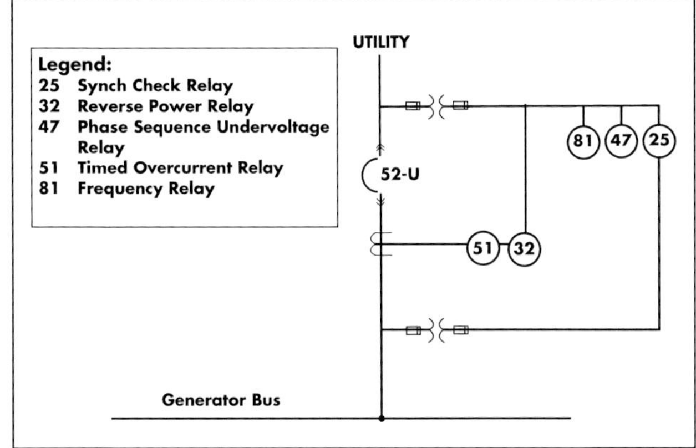

Fig. 27-4 Line Diagram Illustrating Commonly Used Protective Relays for Protection and Control of Utility Breaker. Source: Zenith Controls

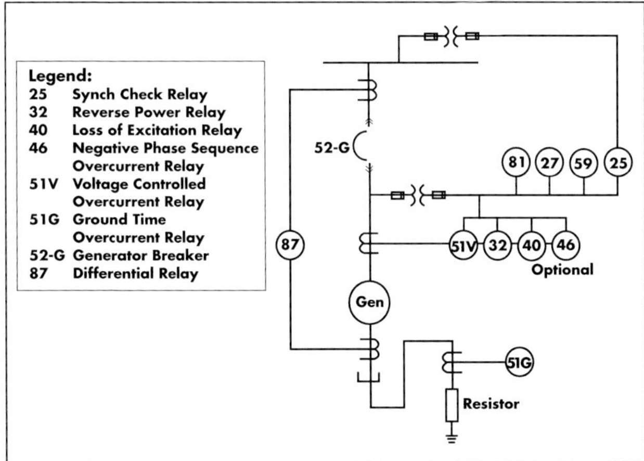

Fig. 27-5 Line Diagram Illustrating Typical Protective Relays Used on a Medium-Voltage Generator. Source: Zenith Controls

rating". Circuit breakers are switches that open in response to a prolonged overload or short-circuit current. The contacts are mechanically closed against the action of a heavy spring and held closed by a latch. Circuit breakers can be manually or automatically opened (tripped) by external signals, typically from some type of protective relay.

Molded case circuit breakers are designed to provide circuit protection from overcurrent for low-voltage distribution systems. High-voltage circuit breakers may be: airblast type, in which the arc is extinguished by a strong blast of air through an orifice across the arc; oil-type, which uses oil to extinguish the arc; or vacuum-type, in which contacts are separated in a vacuum.

Trip elements are either thermomagnetic or solid-state. Thermomagnetic (or electro-mechanical) trips, which are the industry standard in molded case breakers, use two components: bimetals and electromagnets. The bimetals are heated by the current flowing through the breaker. When overloading the breaker, the bimetal will bend and push the trip bar, causing the breaker to trip. Solid-state trips use current transformers and solid-state circuitry for reliability, accuracy, and repeatability. Once the tripping point is set, it remains constant at the set point.

An arc extinguisher confines, divides, and extinguishes the arc drawn between the breaker contacts each time the breaker interrupts current. A terminal connector joins a circuit breaker to a desired power source and load. Connection methods include bus bars, panel board straps, plug-in adapters, and reconnected studs.

Breaker selection is based on several site-specific parameters, including the following:

- **Interrupting capacity** is the maximum fault current the breaker can interrupt without damaging itself. The interrupting capacity of a breaker must be equal to or greater than the amount of fault current that can be delivered at the point in the system where the breaker is applied. This value can be in terms of symmetrical or asymmetrical, momentary or clearing current, as specified.
- **Continuous current rating** is the current the breaker will carry in the ambient temperature for which it is calibrated.
- **Voltage rating** is the maximum voltage that can be applied across its terminals.
- **Frequency rating** is the maximum frequency that can be applied without derating. High-frequency applications often require that the breaker be specially calibrated or derated.
- **External conditions** are the various (unusual) external conditions that the breaker may be subject to (temperature, altitude, moisture, etc.).

When a circuit breaker opens (while current is flowing), an arc is drawn out between the contacts (for a few cycles) until the breaker action extinguishes the arc. Trip elements must have several protective modes, typically classified as follows:

- Long time functions allow overload currents of a low value to flow for several tens of seconds. This allows motors to be started without causing feeder disconnect.
- Short time functions allow higher currents to flow for up to 0.5 seconds. This allows transformers to be magnetized, for example.
- Instantaneous functions will cause the feeder to be disconnected immediately, isolating a short circuit in a feeder circuit before it takes the source down. Ground circuit protection will cause the feeder to be disconnected when excessive current flows from the feeder into grounded equipment.

In low-voltage circuit breakers, the trip elements are integrally mounted. In medium-voltage breakers, trip functions are provided by separately mounted protective relays. In addition to overcurrent, protective relays are available to protect against: unacceptable conditions in current flow direction, balance, and sequence; power flow for direction and magnitude; voltage and frequency for over- and under-conditions; voltage for balance and phase rotation; and other parameters.

SYNCHRONIZATION

Synchronization is a primary control function for operation of localized on-site generators in parallel with one another or with a utility-derived service. This is accomplished by matching the output voltage wave form of one ac electrical generator with the voltage wave form of another ac system. Paralleling ac generators can be likened to a car entering a highway which needs to synchronize its speed and position with the other cars.

There are two basic processes used for paralleling generator sets to a bus:

- **Sequential paralleling.** The generator sets are connected to the bus in a predetermined order. The lead generator set is connected first to the bus. When the second generator set is required and ready to be connected, it is synchronized to the previous generator and placed on the bus. This system is not the preferred approach because it delays power to critical loads. It is, however, an economical system because only one synchronizing device (automatic synchronizer) is used for all generators on the bus.

- **Random paralleling.** The generator sets are all started at the same time and the first generator to reach nominal voltage and frequency will be closed to the dead bus (in emergency systems). Dead bus logic is usually employed to prevent the possibility of connecting more than one generator to the dead bus at the same time. As the remaining generators approach proper operating parameters, their automatic synchronizers bring them into synchronism with the now live bus and cause their generator circuit breakers to close. This method is the most commonly used; it allows the generators to connect to the bus in the fastest time and lessens the possibility of failure of all generators to connect to the bus if the synchronizer fails, as may be the case in sequential paralleling.

For two systems to be synchronized, the following variables must be matched:

- The number of phases in each system
- The direction of rotation of these phases
- The voltage magnitude of the two systems
- The frequencies of the two systems
- The phase angle of the voltage waves of the two systems

The first two conditions are determined when the equipment is specified and installed. As illustrated in Figure 27-6, paralleling two ac generators requires that voltage, frequency, and phase angle be matched within close tolerances.

With synchronous generators, voltage, frequency, and phase must be matched each time before the breakers are closed. The voltage regulator controls generator output voltage by changing its excitation voltage. If two synchronous generators of unequal voltage are paralleled, the difference in voltage results in reactive currents and lowered system efficiency. When a generator is paralleled to a utility bus of unequal voltage, the generator will operate at varying power factor.

The frequency of a paralleling generator must be nearly the same as the utility system, usually within 0.2%. With a synchronous generator, the frequency match is normally accomplished by controlling prime mover speed.

The phase relationship between the voltages usually must be within 10 degrees. With synchronous generators, phase matching — like frequency matching — is accomplished by controlling prime mover speed. Synchroscopes and synchronizing lamps are used to measure the phase angle between two sources.

Paralleling is more easily accomplished with induction generators. No voltage regulator is needed because output voltage and phase angle will automatically match the system supplying its field voltage. Frequency is determined automatically by the field voltage of the utility system, but not until the tie-breaker is closed. Thus, the generator must be kept close to synchronous speed prior to breaker closure.

Synchronizing can be done manually or automatically. For the manual process, indicating lights, a synchroscope, a synch-check relay, or a paralleling phase switch may be used. Increasingly, however, manual systems are giving way to automatic synchronizing systems. Automatic synchronizers monitor phase voltage of an off-line generator and the voltage of the same phases of the active bus. The synchronizer compares the frequency and phase of these two voltages and sends a correction signal to the governor controlling the prime mover of the oncoming generator. When the outputs of the two systems are matched in frequency and phase, the synchronizer issues a breaker-closing signal to the tie-breaker, thereby paralleling the two systems. Figure 27-7 is a functional block diagram of automatic synchronizer operation.

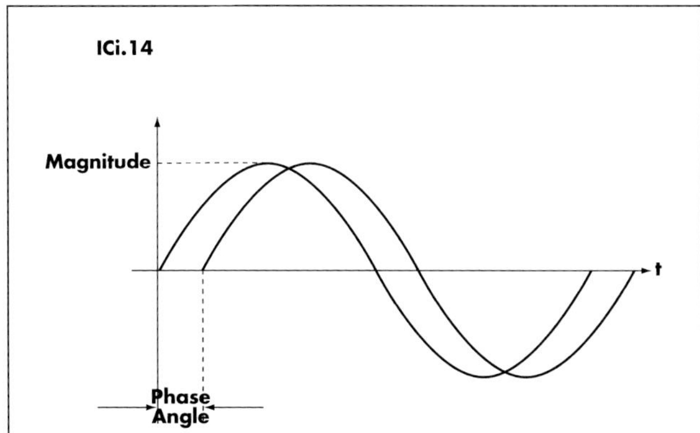

Fig. 27-6 Synchronizing Two ac Generators.

Load Control

The most simple, low-cost system for load transfer between generating systems is an **open-transition system**. With open-transition transfer, the load is disconnected completely from one power source before being connected to the other. This causes a brief loss of power during transfer, with resulting surge when the load is reconnected to a power source. This disturbance is referred to as a **bump**.

A **closed-transition system** allows load transfer between generators and the utility system without inducing a surge. Closed-transition transfer thus has the advantage

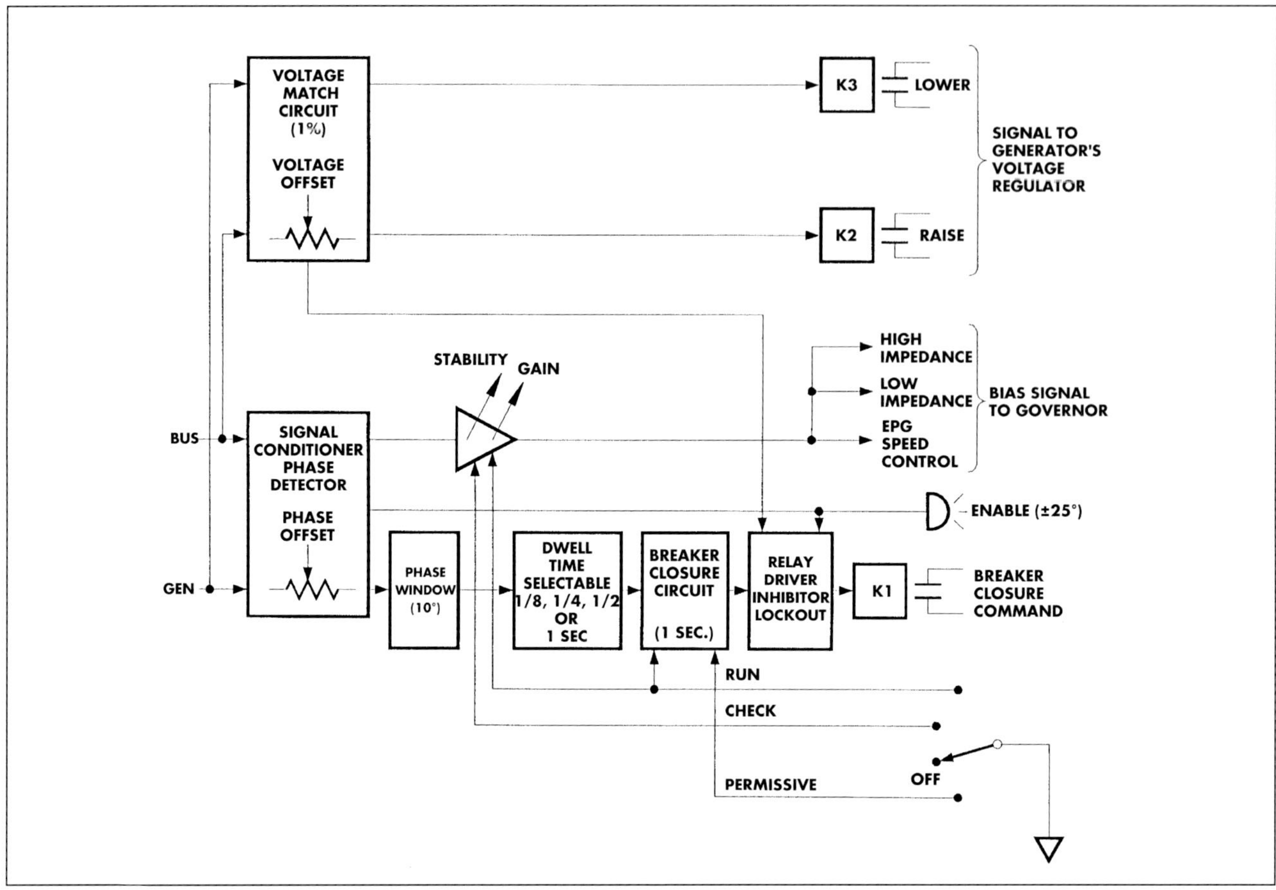

Fig. 27-7 Schematic Representations of Paralleling Switchgear System. Source: Woodward Governor

of a **bumpless** transfer between sources. Closed-transition transfer operates as follows:

- To initiate on-site generation, the generators start and parallel with the utility each through its own synchronizer. As the generators connect to the bus, the load will be transferred from the utility-derived source to the generators. When the transfer is complete, the utility breaker is opened.
- To return to utility derived power, a plant synchronizer adjusts the frequency and phase angle of running generators to match the utility system. When the generator bus and utility are in synch, the utility breaker will be closed and the load will be transferred back to utility. When the transfer is complete, the generators will be tripped off-line.

Unlike the load transfer systems described above, import and export control systems are designed to allow for sharing of loads between the on-site generation system and the utility system. An **import control system** will limit the amount of utility-derived power by adjusting generator output. In a peak demand limiting application, the generators are started and paralleled with the utility when demand reaches a predetermined value. As the load fluctuates, utility power is limited to the demand set point by increasing or decreasing generator loading.

A minimum import level of utility derived power is frequently established by the utility to prevent exporting of power from the on-site generation system. Sudden load decreases of a site electrical load can cause a temporary export of excess power to the utility. Export of power to the utility may not be allowed per the interconnection agreement and may trip a reverse power relay, if one is present. The minimum import level is also important to maintain synchronization of the utility to the plant, enabling a quick response in an event where the local generation source suddenly goes offline.

Consider a scenario in which the total facility load fluctuates between 1,000 and 3,000 kW, there are two on-site generators, each with a maximum capacity of 1,000 kW, and the import level is set at 500 kW. Table 27-1 shows the system response to different load conditions:

Table 27-1 Examples of System Response to Different Load Conditions

Load Condition	Load Value (kW)	Generator Load (kW)	Utility Load (kW) (Import = 500 kW)
1	600	100	500
2	1,000	500	500
3	2,500	2,000	500
4	400	0	400
5	3,000	2,000	1,000

In Load Conditions 1, 2, and 3, generator output fluctuates with the load to maintain the import set point. Under Load Condition 4, the entire load is served by utility-derived power. Under Load Condition 5, the local on-site generators reach their full capacity and excess load must be shed at the facility to return to within the import set point.

Import/export (I/E) systems are the most flexible systems for local on-site cogeneration and peak shaving applications. Since facility loads, the cost of imported power, and the value of exported power all vary, I/E control logic can be used to optimize system economic performance.

Process I/E control systems can be used to control import or export power, or to control thermal process parameters such as exhaust pressure or temperature. Figure 27-8 is a system diagram of exhaust pressure/import control for a non-condensing steam turbine topping-cycle application. The turbine generator will be controlled at the lowest power level required to maintain exhaust pressure or import power set point. In Figure 27-9, the turbine generator will be controlled at a power level based on inlet pressure and power export set point.

Figure 27-10 is a block diagram for a control system that can be used with single or multiple generators sets. The system, which includes electrically powered governors, operates in zero power transfer mode for bumpless utility transfer. In addition to the actuator and speed control, this system requires generator load sensors, automatic generator loading controls (AGLCs), synchronizers (SPM-A unit and master), an automatic power transfer load control (APTL), and a watt transducer.

In the figure, the AGLC is designed to be used with load-sharing and speed control to automatically control the loading and unloading of the generator set. Bumpless transfer is accomplished when paralleling the generator set to a load-sharing system. The AGLC activates when the generator-breaker auxiliary contacts close. It compares the existing potential on the load-sharing lines of the oncoming generator with the potential on the system load-sharing lines. Starting at the existing load level, the generator load is ramped at a preset rate until the voltage on the unit load-sharing lines matches the voltage on the system load-sharing lines. When the voltages match, a relay energizes to connect the unit load-sharing lines to the system load-sharing lines and to isolate the load ramp signal. The unload sequence is initiated by momentarily opening the unload contacts. The AGLC de-energizes the load-sharing relay to separate the generator load-sharing lines from the system load-sharing lines.

The APTL control is used to eliminate power bumps or surges at the moment of connecting or separating the

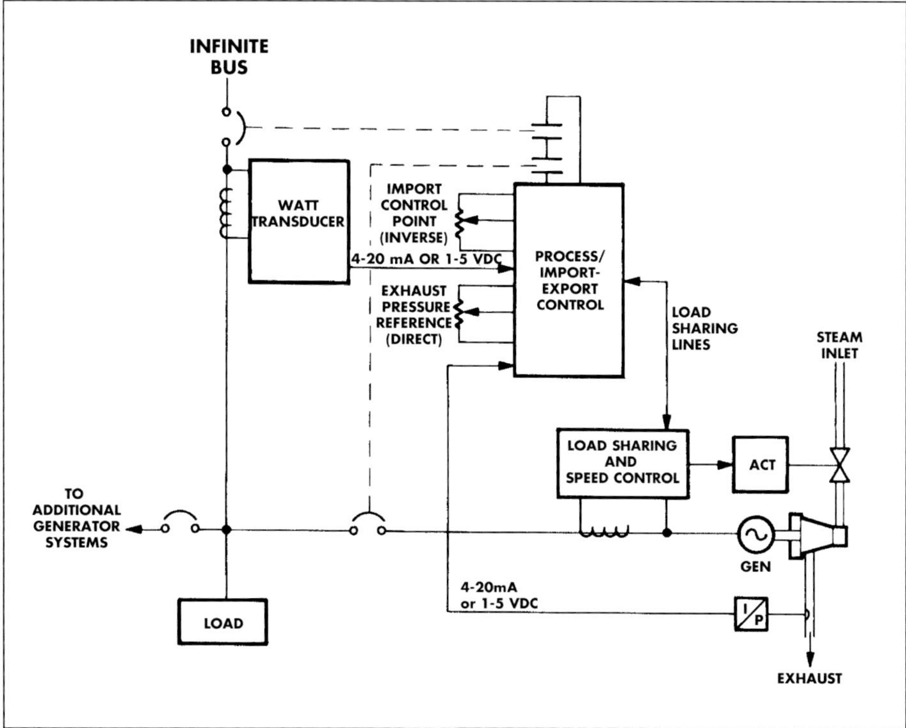

Fig. 27-8 System Diagram of Exhaust Pressure/Import Control. Source: Woodward Governor

on-site generator from the utility system. The APTL tracks voltage on the load-sharing lines of the utility system and, when the on-site system is paralleled with the utility system, biases the lines with a voltage that holds the on-site system to the same power level as that being generated at the moment of paralleling.

The SPM-A is a phase-locked-loop synchronizer that biases the speed of an off-line generator set so that the frequency and phase match those of the bus. It then automatically issues a contact closure signal to close the circuit breaker when phase and frequency are matched within the limits for a specified minimum match-up time. The synchronizer is pow-

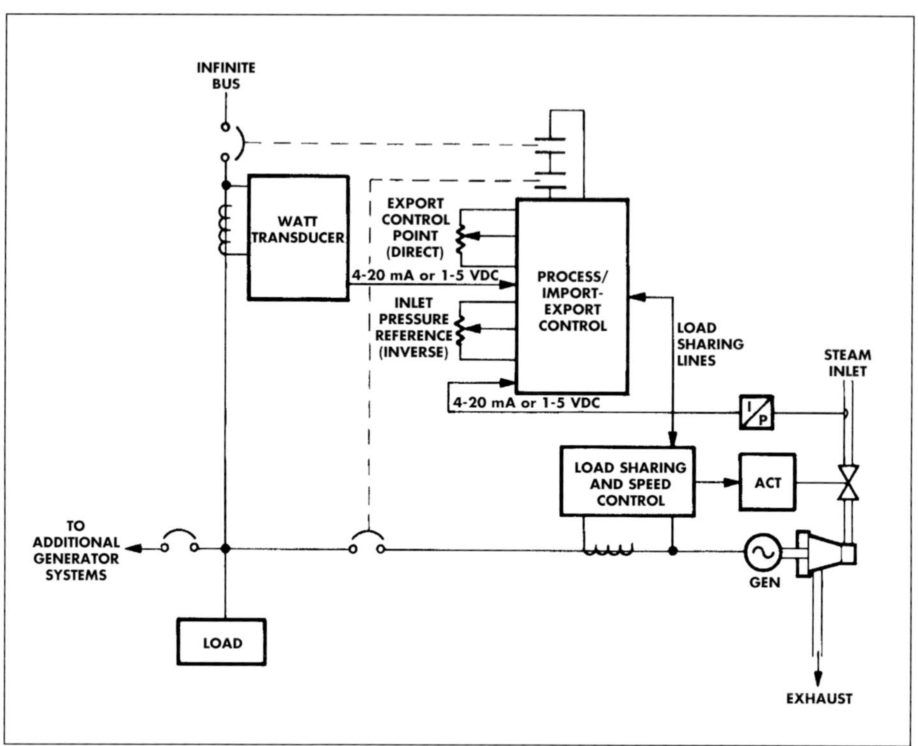

Fig. 27-9 System Diagram of Inlet Pressure/Export Control. Source: Woodward Governor

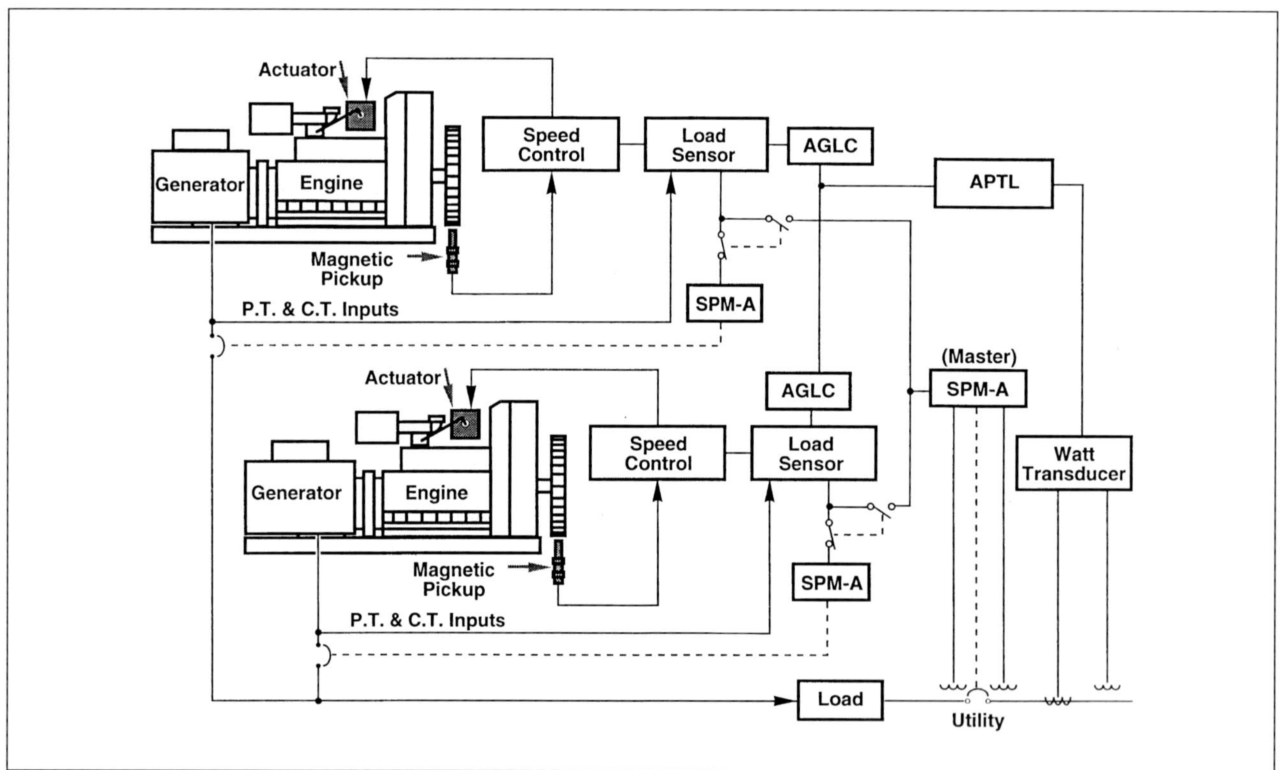

Fig. 27-10 Arrangement Diagram for Single or Multiple Units Operating in Zero Power Transfer Mode for Bumpless Utility Bus Transfer. Source: Woodward Governor

ered by voltage supply connections to the generator potential transformers (PTs).

The speed control system uses a speed-sensing magnetic pick-up to sense prime mover speed and convert it to an ac signal or proportionate frequency. A frequency-to-voltage converter receives the ac signal and changes it to a proportional dc voltage. A speed-reference circuit generates a dc reference voltage with which the speed-signal voltage is compared. The actuator responds to the signal from the control amplifier by repositioning the fuel or steam rack until the speed-signal voltage and the reference voltage are equal.

Figure 27-11 is a block diagram for a control system featuring electro-hydraulic governors that can be used for single or multiple generator sets operating in I/E mode with the utility bus. The controls are fairly similar to those shown in Figure 27-10, except that this system uses a common speed and load-sharing control and the APTL is replaced with a process I/E control.

Figure 27-12 is a diagram of a typical combination generator control module which can be used for load-sharing across multiple units. Units like these are typically employed in PLC based control systems to provide generator load control directly interfaced to electrical PT's and CT's for gas turbine control.

Figure 27-13 illustrates an instrument and controls architecture for a typical gas turbine cogeneration facility. Depending on the value of the electricity and the amount of thermal loads, the gas turbine may be controlled to maximize electrical output with the HRSG bypass damper and duct-burner controlled independently to maintain steam header pressure. Long-term capture and study of the data generated on transient response and thermal efficiency can suggest adjustments in overall plant performance, which will in turn lead to enhanced return on investment.

Monitoring and Metering Devices

Monitoring requirements for electric generators include acquisition and recording of a variety of data, including fuel input and power output, as well as the functioning of all control and protection devices. Monitoring systems can also include processing of data to provide reports on instantaneous and historical system operation.

Monitoring systems may range from basic metering to elaborate computer control systems. In the most basic case, meters are required to indicate individual and collective voltage, current, frequency, kW, kVAR, perhaps some key temperatures and flows, and operating hours. Strip-chart recorders can be used, but are rarely seen in modern plant

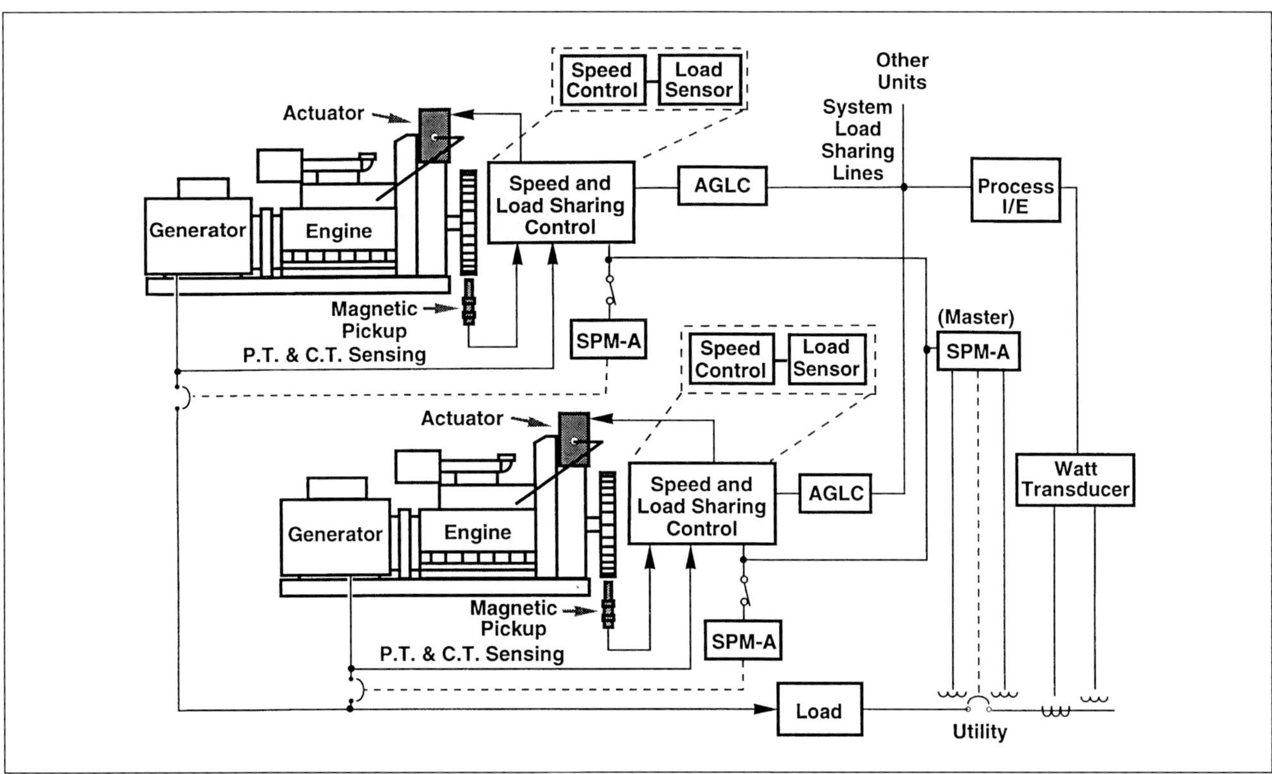

Fig. 27-11 Arrangement Diagram for Single or Multiple Units Importing and/or Exporting to the Utility Bus.
Source: Woodward Governor

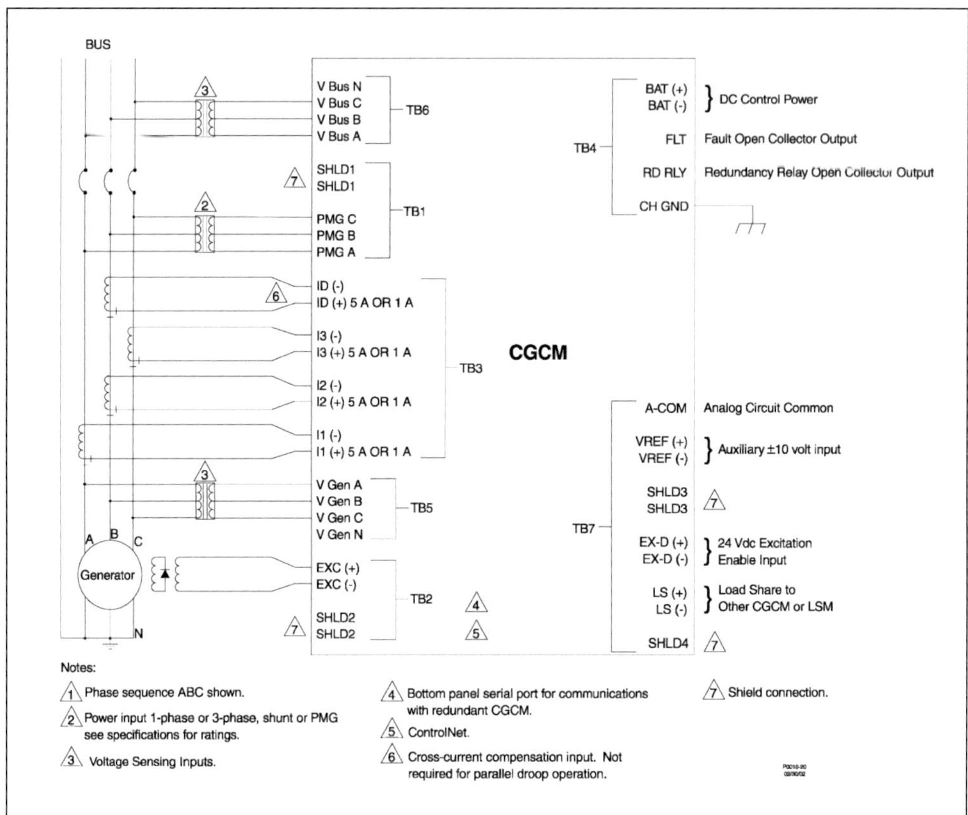

Figure 27-12 Arrangement Diagram for Single or Multiple Units Importing and/or Exporting to the Utility Bus. Source: Waldron Engineering & Construction

design. Data loggers and Graphical (Data Acquisition & Recorders (DAQs) are increasingly more common. When the plant is interconnected with the utility grid, the level of required sophistication is dramatically increased and a more elaborate system will be required. Computer-based systems for monitoring, control, and protection can be housed on-site or connected from a remote location.

Each type of prime mover, generator, and heat recovery system will require its own set of monitoring, control, and protection devices. These devices may be isolated in stand-alone consoles or integrated into the prime mover system, the switchgear system, or an overall computer control network. In small plants, the isolated skid approach is used, and the only network of communications between skids is via human operator, whose attention is usually requested by visual/audio signaling from these isolated systems. Typically an "islands of automation" concept, where each skid is allowed it own control system around a central system, is employed in plants of moderate to large size. A limited set of virtual telemetry and/or direct wired signals is handled from each skid to the center control system or Balance of Plant (BOP). Skid–to-Skid "bridges" (i.e. direct communications not thru the BOP) is limited to direct wired signals required for closed loop control. All other telemetry is moved to and from the central process controller. Smaller auxiliaries (Cooling tower, Deaerator, Condensate tank) will be implemented as direct-wired controls to the BOP.

Monitoring provides essential feedback information for servicing and calibrating the various controls of the switchgear system. Ammeters, voltmeters, and VAR meters are used in the initial calibration of the voltage regulation and VAR/PF controller. Frequency and kW meters are used in the initial calibration of the governor and loading controls for proper load sharing. The readings from these meters are commonly logged and analyzed for trends. These trends indicate the need for maintenance as it arises.

Support for control systems is an important consideration. Systems must be periodically reviewed, adjusted, updated, calibrated, etc. This support may come from inhouse expertise or through a vendor contract. In either case, it must be accounted for as part of the long-term investment decision. Sophisticated control systems are only useful when they are properly designed and maintained and when the operators are properly trained to use them.

The local utility or independent system operator (ISO) of the utility system selects metering arrange-

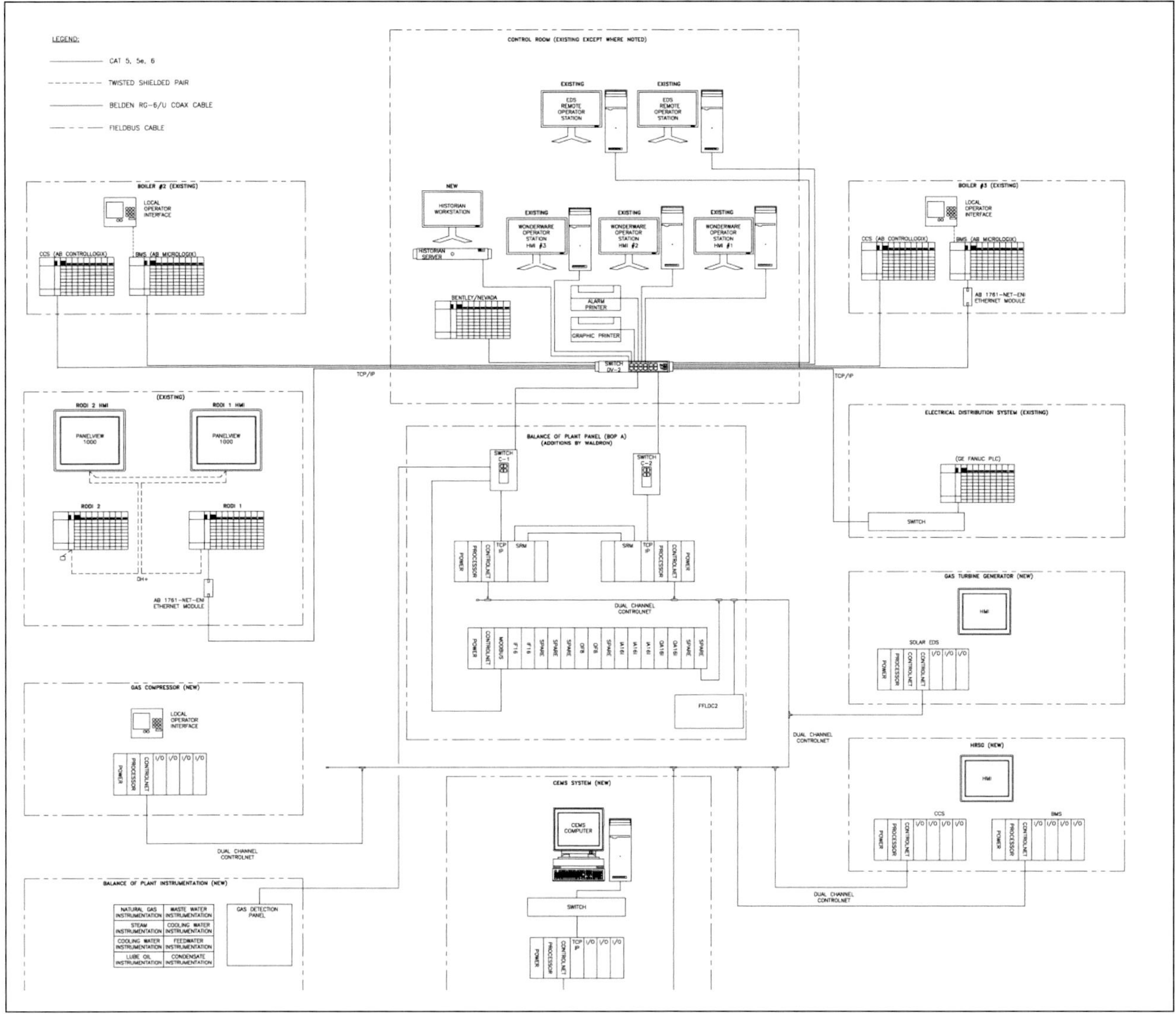

Figure 27-13 Typical Gas Turbine Cogeneration Facility Control System. Source: Waldron Engineering & Construction

ments and types of metering devices, typically reviewing the following:

- **kW and kVA demand.** This allows the utility or ISO to determine if primary (high-voltage) or secondary (low-voltage) metering is required.
- **Rate or load shedding considerations.** Time-of-use or seasonal rates, for example, require meters capable of recording use and demand in various rate periods. Special programs may require remote control of generators or load-shedding devices and recording of such events.
- **Export capability and backup requirement.** Remote monitoring of power "in flow" and "out flow" may be required. The level of back-up required for the host facility and the requirements of station service power by the generation facility may affect metering requirements.

Facilities with generators that do not export power may often use standard meters with demand registers. Some utilities will have special additional metering requirements, such as detents, which are devices that prevent the meter from rotating backwards.

If the system is intended to export power, an "export" kWh meter with detent and demand register will be required to record the amount of energy delivered to the utility system and peak period output, if the host facility has a power sales or wheeling agreement. A meter on output of the generating plant may also be required for the purposes of assessing utility stand-by charges.

Electric Devices and Functions for Switchgear Apparatus

The American National Standards Institute (ANSI) designates a number to each device function used in electrical drawings. Generator diagrams use switchgear device function numbers to indicate the required device. Some of the more common devices are listed below. For each number, there are several subcategories marked by a letter which more specifically defines the function. Multi-functional electronic relay systems are commonly used today and perform many of the relay and device functions noted below with a single digital system. Many utilities require a redundant multi-functional relay system to enhance the reliability of the protective system.

21 Distance relay functions when the circuit admittance, impedance, or reactance increases or decreases beyond predetermined limits.

25 Synchronizing relay operates when two ac circuits are within the desired limits of frequency, phase angle, or voltage to permit or cause the paralleling of these two circuits.

27 Undervoltage relay functions on a given value of undervoltage.

32 Directional power relay functions on a desired value of power flow in a given direction, or upon reverse power.

40 Loss of excitation (field) relay functions on a given or abnormally low value or failure of machine field current, or on an excessive value of the reactive component of armature current in an ac machine indicating abnormally low field excitation.

43 Manual transfer or **selector device** transfers control of switching equipment or other devices.

46 Negative (current) phase-sequence or **phase-balance relay** functions when the polyphase currents are of reverse-phase sequence, or when the polyphase currents are unbalanced or contain excessive negative phase-sequenced components.

47 Phase-sequence voltage relay functions on a predetermined value of polyphase voltage in the desired phase sequence.

49 Machine thermal relay monitors generator stator windings temperature and initiates shutdown when a predetermined limit is reached.

50 Instantaneous overcurrent relay functions on excessive current, or on an excessive rate of current rise, thus indicating a fault in the apparatus of the circuit being protected.

51 Time overcurrent relay has either a definite or inverse time characteristic to avoid trips due to instantaneous current spikes.

52 Circuit breaker is used to close and interrupt an ac power circuit under normal operating conditions or to interrupt the circuit under fault or emergency conditions.

59 Overvoltage relay functions on a given value of overvoltage.

65 Governor regulates fuel or steam to the prime mover to maintain and return to a steady-state operating status.

65C Governor control biases governor operation to match desired system performance, usually power output.

67 Directional overcurrent relay functions on a desired value of ac overcurrent flowing in a predetermined direction.

81 Frequency relay (U — under; O — over) controls the automatic closing and reclosing of a dc circuit interrupter, generally in response to load circuit conditions.

86 Locking-out relay shuts down and holds a piece of equipment out of service on the occurrence of abnormal conditions.

87 Differential current relay provides primary protection of the generator from internal faults. It can also be used to protect a transformer or to detect bus faults.

90 Voltage regulator operates to regulate the voltage output of a generator to match a desired strategy.

90C kVAR/power factor controller regulates the reactive current output flows to match a desired strategy.

Switchgear and Control System Application Examples

The following one-line diagrams are samples of basic control and protection requirements for several on-site electric generation applications. No one methodology can be taken as standard. The preferred methodology in a given application depends on site-specific factors, as well as preferences of the utility or ISO and the design engineer.

Example 1

Figure 27-14 is a single-line diagram of a relatively simple switchgear and control system serving an induction generator. Upon detection of a fault, circuit breaker trip elements will disconnect and isolate faulted conductors and equipment. Two types of protection are shown:

1. The directional over-power relay will initiate disconnect and shut down the prime mover generator if power flows from the bus to the generator, indicating prime mover or fuel system malfunction.
2. The definite time over-current relay would initiate disconnect of the prime mover generator set from the bus if the current into or out of the generator exceeded some multiple of generator rating, 2 to 3 times for 0.1 seconds or longer. This would indicate a bus disturbance of an unknown nature to the systems, but for which power source segregation is an appropriate first step in system protection.

A typical operating strategy for this system would begin with the decision to generate power by either a controller or an operator. The following sequence would occur:

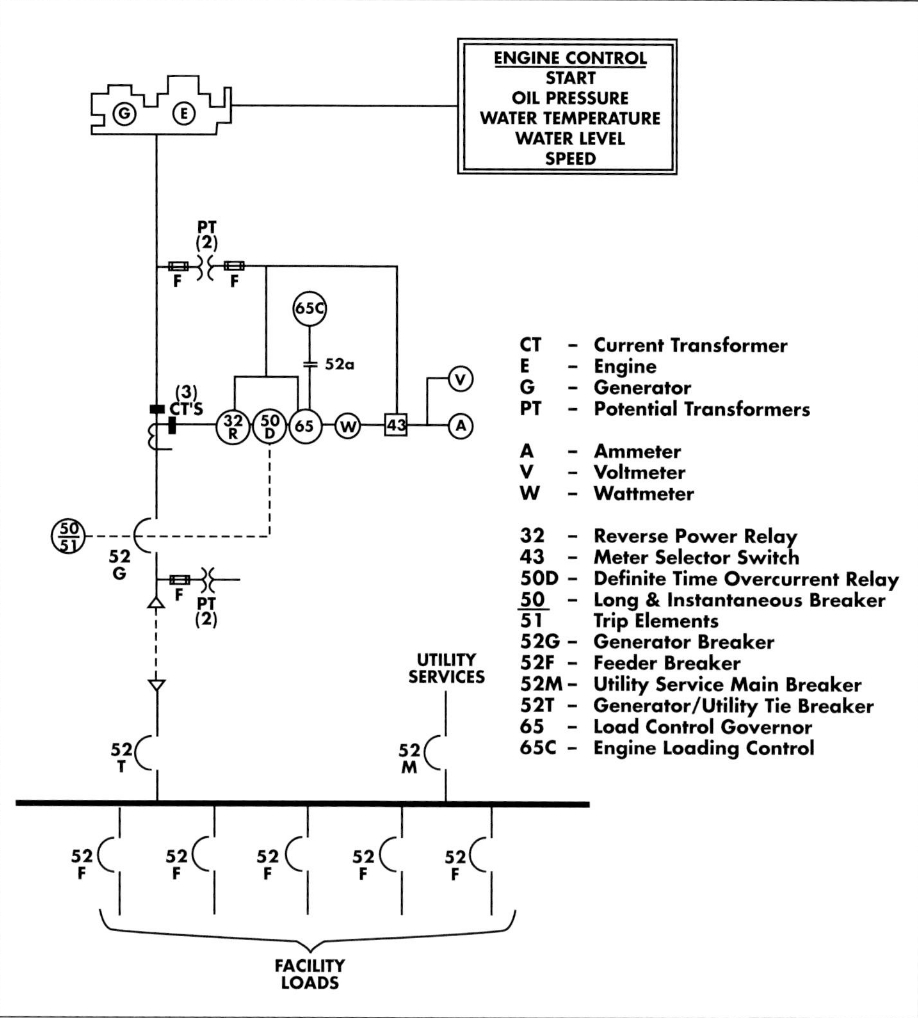

Fig. 27-14 Line Diagram of Simple Switchgear and Controls for Induction Generator. Source: Automatic Switch Company

1. The prime mover controls initiate: startup, fuel (or steam) turn-on, and fuel (or steam) setting through the governor.
2. The governor goes to full-on until the prime mover speed approaches the steady-state speed. When this speed is achieved, the governor meters fuel (or steam) to keep the speed constant.
3. At the appropriate time, the generator breaker is closed and the generator becomes excited from the line.
4. Simultaneous with breaker closure, the loading control causes the governor to increase fuel (or steam) to the prime mover. Since the generator is now paralleled with the bus, the prime mover does not increase speed, but produces shaft power, which the generator converts to kW.
5. The loading control measures the generator output through the CTs and PTs. Load control is maintained as long as the generator is paralleled with the utility-derived service.
6. When it is no longer desired to operate the plant, the generator breaker is opened and the shutdown sequence is initiated.

Example 2

Figure 27-15 is a simple line diagram of a typical central plant switchgear configuration for parallel operation of on-site generation with a utility-derived service. This would be found in large cogeneration applications serving industrial or institutional sites.

In this example, there are two utility service connections and an on-site generator system composed of four prime mover generator sets. The on-site power system can provide either peaking power, or operate as a baseloaded cogenerator. In either case, the generators are intended to operate in parallel with the utility system for extended periods. These generators are synchronous machines.

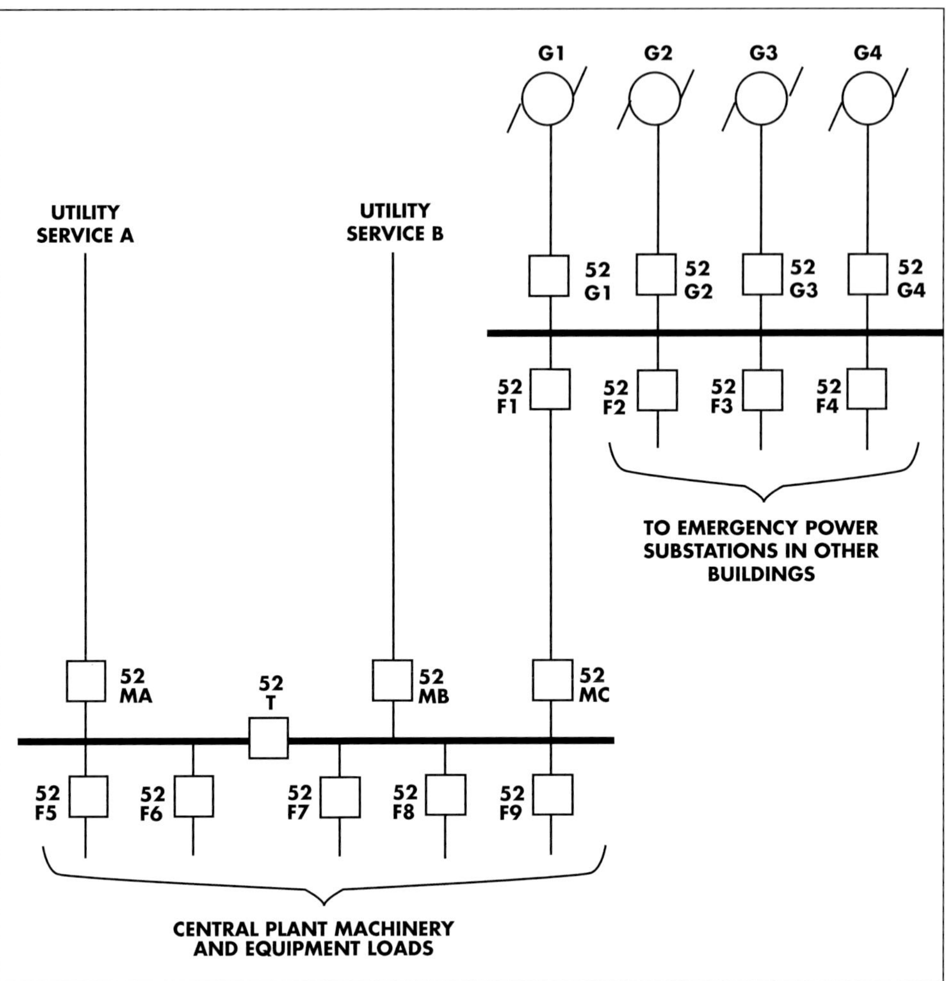

Fig. 27-15 Line Diagram of Integrated Generator Switchgear in Peaking or Baseload Application with Provisions for Emergency Power Service.
Source: Automatic Switch Company

The two utility services feed a switchgear bus through two main breakers. This switchgear bus is split in the middle to provides for isolation of the two utility services. Standard operation for this type of switchgear, commonly referred to as a doubled-ended switchgear, is for only two of the main devices to be closed at any given time. Either both mains or one main and the tie can be closed. [Closure of all three would expose the switchgear bus to too much fault current from the utility service and interfere with the coordination of the utility protective relaying system.] The generators are paralleled on a switchgear bus that is tied to the main switchgear though feed breakers 52F1 and 52MC.

Systems installed for baseload cogeneration can also serve as emergency power systems. Referring to Figure 27-15, note that there are additional feeder breakers on the generator switchgear bus. These feeders could provide emergency power to other buildings for emergency purposes. When the utility service is disrupted, the generators would start automatically and feed power to these emergency loads. Emergency power can also be fed to the switchgear bus of the central plant by opening both utility mains and closing the generator main, 52MC.

Example 3

Figure 27-16 illustrates suitable protective relay schemes for the utility and generator mains. In comparison with the previous example, there are some additional devices included in the generator main circuit. Devices 50/51, 32R, 65, and 65C and meters W, V, and A serve the same functions as in the previous example. Because this is a synchronous generator, it generates voltage on its own and does not require an energized bus. Device 59G is a ground overvoltage relay. Its purpose is to sense when there is a ground on one of the phases and to enunciate it. The resistor in parallel with the 59G limits ground current to or from the generator when it is in parallel with the utility system.

Devices 27 and 81 check the generator voltage and

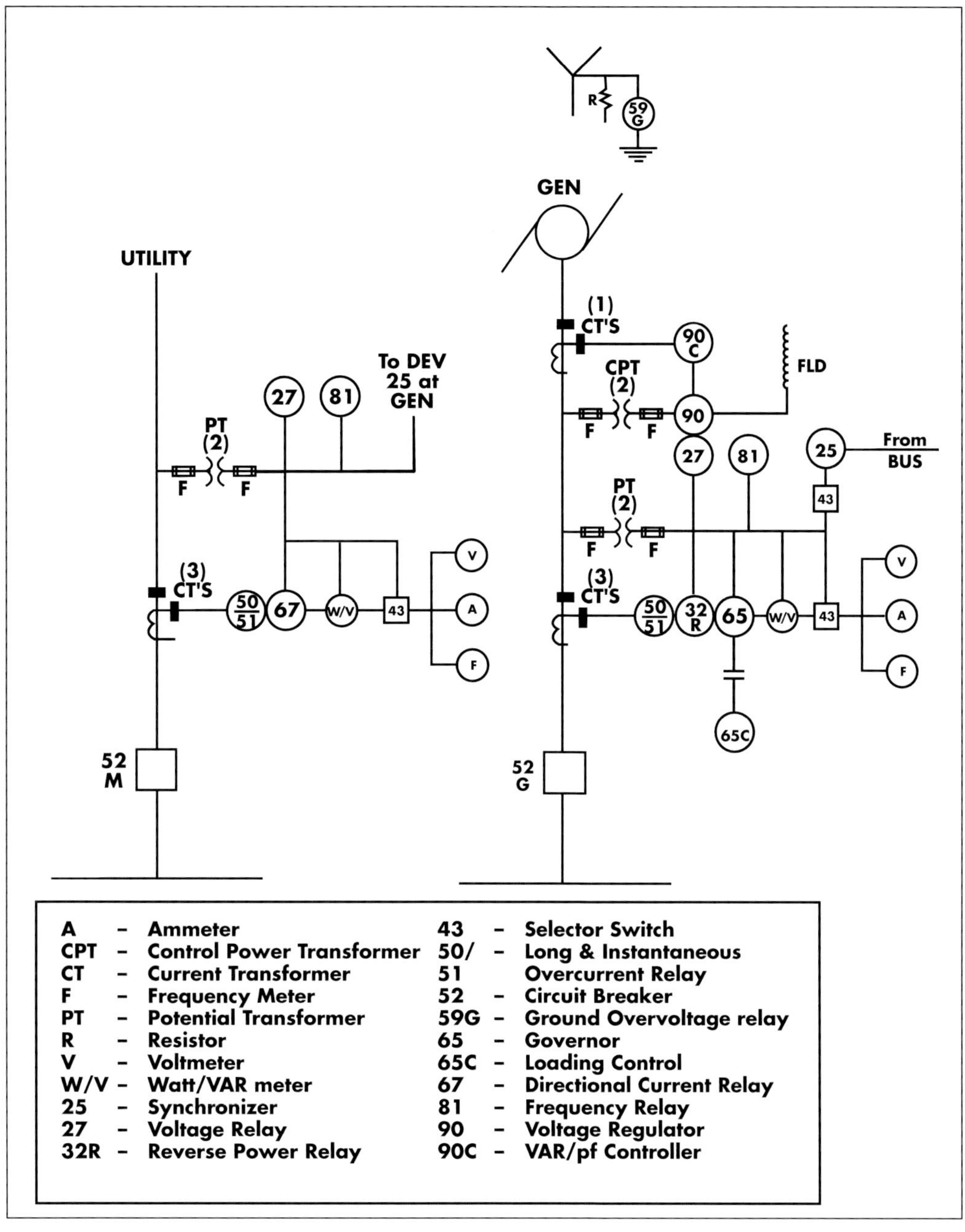

Fig. 27-16 Protective Relay Scheme for Parallel Operation of Generator and Utility-Derived Power Sources. Source: Automatic Switch Company

frequency before initiating synchronizing operations. The synchronizer, Device 25, compares the voltage and frequency of the generator with that of the bus. This synchronizer takes control of the voltage regulator and governor to cause the generator output to match the voltage and frequency of the bus. The synchronizer then matches the phase angle of the generator to the bus and initiates closing of the general breaker.

As is the case with the induction generator (Example 1), the loading control causes the governor to increase fuel (or steam) to the prime mover so that electrical energy is delivered to the bus. Similarly, the VAR/PF controller causes that voltage regulator to increase excitation to the generator field windings so that the generator produces the proper VARS to run at the desired power factor.

Protective relaying at the utility main provides voltage

and frequency functions. These assure that the utility services are at nominal values prior to an attempt to parallel it with the generator bus. This would be required, for example, if the facility were running on the generators alone following a utility outage.

The control strategy for this type of system can operate any number of engine generators. In the cogeneration mode, sets are added as the demand for thermal energy increases. In the peaking mode, sets are added to the bus when demand increases above its set point. The principle difference between the two strategies is that in the case of the baseloaded cogenerator, electric power can be fed back to the utility service. In the case of the peaking operation, no power is fed back to the utility.

For the cogeneration case, Device 67 would be set for a known value of current flowing back into the utility service. When current flow into the utility from the generators exceeds that value, the 52MC breaker is tripped. In the case of the peaking operation, Device 67 is set to trip on any current flow back into the utility service.

Example 4

An emergency generator control switchgear system is designed to automatically supply and control emergency power for distribution to building loads in the event of a normal power failure. Users of emergency standby power systems include hospitals, data centers, and other facilities with critical loads. Standby power systems can also be used for electric peak-shaving applications.

When a facility uses on-site generation exclusively as a backup for prime power provided by the utility grid, a relatively simple power transfer system is required. A group of loads, typically critical or life safety, are segregated and serviced by one or more automatic transfer switches. During normal operation, the local utility grid provides power. Upon a failure of utility power, backup generators are automatically started and connected to the loads. Figure 27-17 is a schematic representation of this system.

Per NFPA 110, system design must ensure that power is being supplied at rated voltage and frequency to emergency loads within 10 seconds of a loss of normal power. The generator sets are started by their solid-state cranking control and then accelerated to rated speed. The circuit breaker of the first engine generator to achieve 90% of nominal voltage and frequency then closes to the dead bus and loads are transferred in order of priority. Dead bus closing circuits prevent simultaneous closing of two or more generator breakers.

As the remaining generators approach proper operating parameters, their automatic synchronizers bring them in line with the generator bus and cause their circuit breakers to close. Generators that do not reach synchronism within a predetermined time will cause a "Fail to Synchronize" alarm. When the generator breaker closes, the generator synchronizer will relinquish control of the generator speed to the electronic load-sharing governors.

The operating sequences of automatic transfer switches and distribution equipment are controlled by the load sequence controller (LSC) located in the master control cubicle. The LSC monitors generator circuit breaker position to sense the number of generators on the bus. When the first generator closes to the bus, the LSC adds the first priority load by operating transfer switches that have been programmed as Priority 1. The controller will continue to monitor emergency bus capacity and add load in the preprogrammed sequence until all distribution equipment has transferred to the emergency bus. The LSC logic is designed to ensure that load is not added to the emergency bus until sufficient generator capacity is available.

Generator optimizing logic allows the operator to maintain equal running hours among all generators. The optimizing logic will remove generators from the bus as the load decreases. Upon return of normal power, the transfer switches will transfer back to the normal power source and remove the generator start signals, causing the generators to trip off line and shut down after a cool-down period.

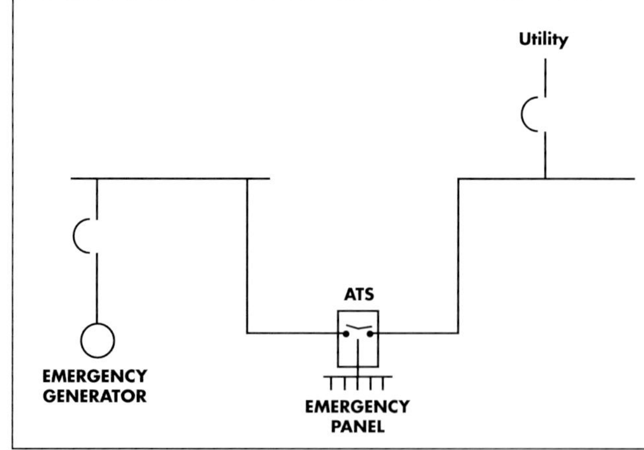

Fig. 27-17 Line Diagram of Automatic Transfer System for Standby Generation. Source: Zenith Controls

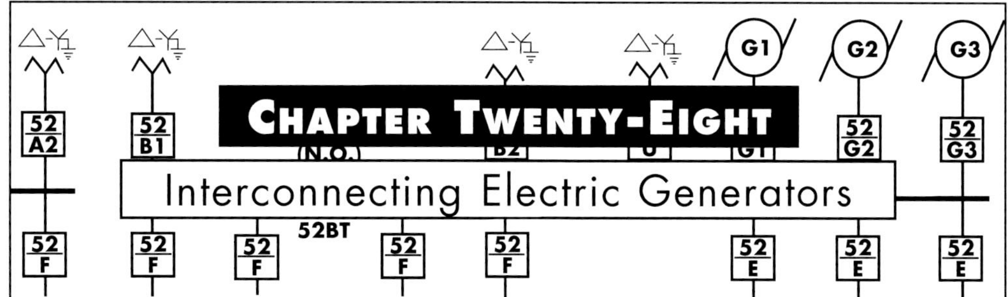

Chapter Twenty-Eight

Interconnecting Electric Generators

There are three general ways localized on-site generation can be applied in relation to the utility grid:

- **Isolated operation.** As illustrated in Figure 28-1, a stand-alone system can serve designated facility loads independent of the utility. These systems typically require off-line backup generators, as well as on-line reserve capacity, to provide sufficient power during load swings and scheduled or unscheduled outages. High installation and operating costs result.
- **Isolated operation with utility system backup.** This arrangement, illustrated in Figure 28-2, includes two variations. With open transition transfer, the load is disconnected completely from one power source before being connected to the other. This causes a brief power loss and resultant power surge, or bump. With closed-transition transfer, the transfer switches may overlap or parallel between the two sources, requiring actual synchronization of the two power sources during a period of 100 milliseconds to 60 seconds, depending on the utility and their allowable configurations. While more costly than open-transition transfer systems, closed transition has the advantage of a bumpless transfer and it is not as expensive as a fully parallel system.
- **Parallel operation with the utility system.** As illustrated in Figures 28-3 and 28-4, paralleling controls are configured in one of two ways. They may track the facility's electric load and adjust the output of the generators accordingly, with excess required power being furnished from the utility. Or, they may operate the generators at full load and sell excess power to the utility or another buyer. Parallel operation, which is the primary focus of this chapter, provides the greatest degree of flexibility and generally the best opportunity to optimize economic performance.

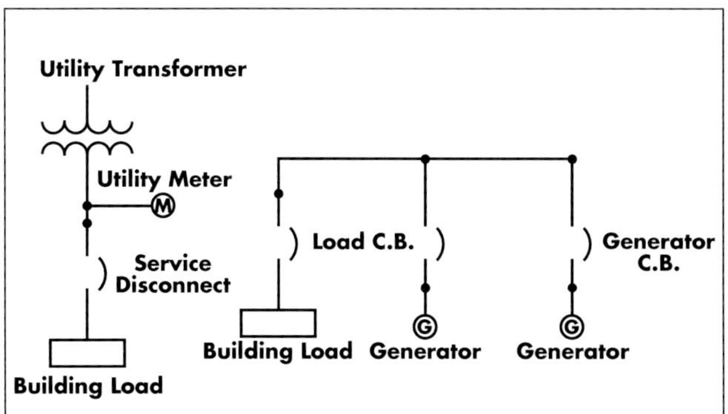

Fig. 28-1 Example of Connection for Isolated Operation (No Interconnection with Utility System).

Many localized on-site generation plants are designed to operate in parallel with the utility system. This approach

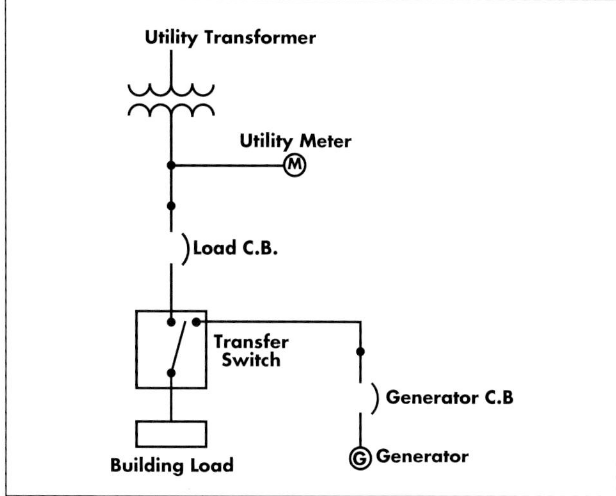

Fig. 28-2 Example of Connection for Isolated Operation with Utility System Backup (or Standby Generator Operation).

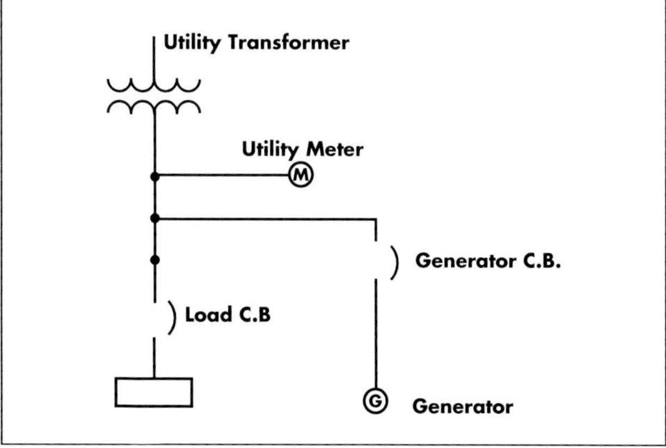

Fig. 28-3 Example of Connection for Parallel Operation with Utility System Not Capable of Standby Operation.

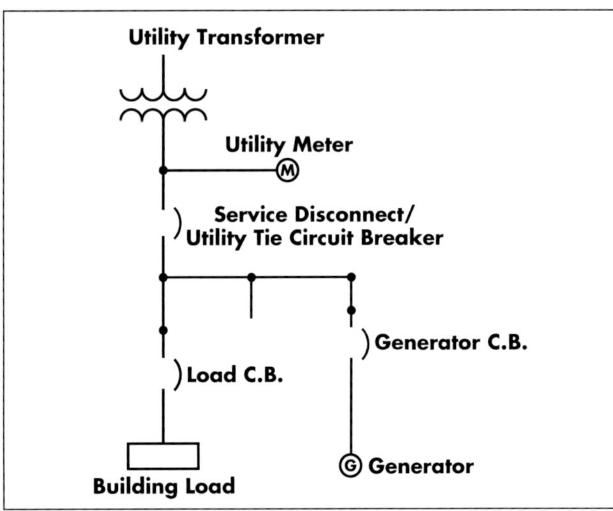

Fig. 28-4 Example of Connection for Parallel Operation with Utility System Capable of Standby Operation.

provides instantaneous backup for the host facility and allows optimum sizing of the on-site generation system in relation to facility thermal loads or peak-shaving requirements. The flexibility to purchase supplemental power and to sell excess power often provides a significant thermodynamic and economic advantage to on-site generators.

With the advantages of parallel operation, however, come added system complexity and cost. Interconnection requires that the host facility take necessary precautions to protect its own generation system, the utility system, and other facilities on the system that are served by utility-derived power. Protective relaying, as required in guidelines issued by the local utility, must be incorporated so as to ensure that immediate separation of the two electric power sources can be accomplished under possible fault conditions.

Interconnection is often a difficult issue, because there is a need to balance the host facility's interconnection cost and the utility's need to provide reliable power to its customers while maintaining its own safety requirements. Detailed discussion and negotiation is often required between the local generator and the utility to reach agreement on the components and operation of an interconnected system.

The standard approach to operational difficulty by the utility or an independent system operator (ISO) is simply to drop a generator off line as soon as any problem is detected. Except during peak periods, this is easy for the utility system, due to massive redundancy. In contrast, the on-site generator's goal is to maintain uninterrupted operation and, in some cases, to fulfill power exporting obligations or to avoid backup charges. For this reason, it is often difficult to reach agreement on the required relays and where relay trip limits should be set. A lengthy negotiating process must sometimes be endured to satisfy the needs of both parties.

On-site generators that are large relative to the capacity of the utility feeder present more complex coordination issues and likely will require a more extensive protection scheme. The construction of the relays is also often a contested area. Some utilities require electromechanical rather than more modern (and less maintenance intensive) solid-state type. The cost of these more sophisticated relay schemes usually must be absorbed by the local generating facility and can be prohibitive, particularly for small systems of a few MW or less.

Due to all of the protective devices required by utilities for interconnected systems, it is increasingly common for main facility breakers to open for no apparent reason. For example, voltage surges on the utility line are often sufficient to trip a voltage relay and disconnect a facility. From the utility or ISO perspective, this is considered normal operation. From the facility perspective, it is a problematic interruption that is often considered overprotection (and in some cases, regarded as harassment).

By far the greatest concern of all utilities is human safety, particularly for line workers who are exposed to possible electrocution if a power line is not locked out and dead. Strict operating procedures are in place to prevent such an occurrence, and interconnected generators must be included as part of the overall safety procedure to ensure that a facility does not accidentally re-energize a line while work is being done on it. Utilities often insist on the right to inspect an on-site generation facility to ensure that proper procedures are followed. The partnership between the utility or ISO and host facility is never more important than in cases involving line worker protection and other catastrophic potential.

The knowledge and experience of the utility or ISO line engineer is a good resource and should be considered as such by the local on-site generator. Legitimate concerns of the utility or ISO must also be respected and addressed with great care and responsibility. Conversely, the need to eliminate unnecessary power interruptions and facility investment in superfluous protection equipment must also be respected. For these reasons, circuit breaker systems must be carefully designed, extensively tested upon initial installation, and then periodically tested to ensure continued proper functioning.

Connecting with the Utility System Grid

For a self-generator to connect to an electric system served by the utility or ISO, several key factors, such as point of connection, grounding, protective relaying, and system isolation, must be addressed. Essential concerns include:

- **Safety.** Protection of customers, line workers, and the general public is paramount.
- **Quality of power.** The ability of the utility system is designed to assure that a certain level or range of power quality (i.e., proper voltage, frequency, etc.) to customers is not compromised.
- **Equipment protection.** The formidable fault currents available from electric utility systems always pose a risk to connected equipment and on-site generators, like other major equipment at the host facility requiring protection.
- **Unencumbered system control by the utility.** The utility must be able to control and respond to all events (i.e., faults, etc.) that interrupt operation or present risks, including network subsection isolation and dynamic repair functions.

Interconnection requirements for parallel operation can vary widely, depending on a number of factors. The capacity of the generating plant (kVA or kW) compared with the host facility's distribution system and the utility substation feeder will be major determinants of system requirements and cost. Most utilities place a maximum limit on the rating of a generator that can be connected to their distribution systems.

Power exportation adds complexity and cost, particularly if the utility's local distribution system is inadequate to handle additional exported power. While the utility will generally make the needed modifications, the cost usually must be absorbed by the self-generating facility. Inter- connection for small induction generators is usually simpler than for larger synchronous units.

Generator voltage is also an important factor in terms of in-house distribution and power exportation. Voltage selection for on-site generation at existing facilities depends on the existing distribution system and any reasonable retrofits that can be made. In some cases, existing transformers may require replacement or major modification, or the distribution system may require modification. At higher voltages, current is lower and generator internal losses are less. At low voltage, a higher ampacity interconnect is required and the system may not be able to export power without significant voltage variances. For export, generator voltage must be raised to account for transformer impedance and voltage drop, and the resulting overvoltage may have an undesirable effect on plant equipment. For an on-site generating plant of 2,000 kVA (or slightly larger), 480 volts may be well matched. Above that, consideration should be given to using higher voltages.

Voltage regulation is the primary means for utility network feeder system control. In contrast, current and power factor (PF) depend on customer loads and frequency is set by central utility prime movers. Utility-supplied voltage is rarely constant and typically varies by 5%, depending on the size of the load and the distance to the nearest substation. When an on-site generator is large with respect to the capacity of the utility intertie, it may be preferable to let the on-site generator set the facility voltage rather than the network.

Voltage is adjusted by the utility using load tap changers or induction voltage regulators. A **load tap changer** is a motor-driven switching device that can adjust transformer ratio in response to voltage variation. A **voltage regulator** is an automatic transformer connected in series with the main transformer. Capacitors that can be switched in and out of the circuit are used to control voltage changes resulting from PF variations. Figure 28-5 illustrates a primary distribution feeder with step voltage regulator and voltage profile.

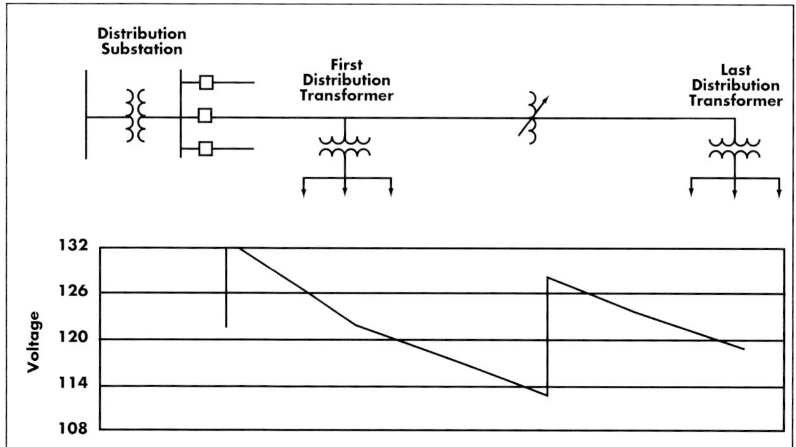

Fig. 28-5 Primary Distribution Feeder with Step Voltage and Voltage Profile.

By controlling on-site generator excitation with a PF or kVAR controller, the utility network PF can be improved and better regulated. Over-excitement produces excess kVARs, thereby reducing the kVARs drawn from the utility. Because PF is improved, the utility has an incentive to cooperate with the on-site generator. A communication link can be installed to vary excitation with respect to utility network load changes. Figures 28-6 and 28-7 are block diagrams of a self-excited and a separately excited synchronous generator, respectively.

Protective devices — typically combinations of **relays** and **circuit breakers** — guard the power system against damage. Relays sense electrical parameters of voltage, current, power (real, apparent, and reactive), frequency, and phase angle. The relays respond to over, under, and differential limits, direction of flow, sequence of phases and currents, etc., with or without time functions to permit coordination with other devices.

The basic principle of device coordination is to identify and isolate only the faulted circuit, leaving the rest of the distribution system operating. When a disturbance occurs in the system, the cause is typically immediately indeterminate and the optimum approach is to isolate the power sources and let their respective relaying schemes segregate the affected circuits.

Given the need to customize each protection system, the selection and application of protective devices to power systems is often referred to as the "art of protective relaying." The protection scheme used at a given installation will depend on the physical installation, available fault currents, distribution voltage, and skill of the designer.

A protective relaying and control system is required for any type of utility tie. The system must function automatically in a safe, logical sequence. Certain manual controls will also be required for system testing, calibration, and maintenance procedures.

A primary safety requirement for interconnection is safe access to the circuits for maintenance personnel.

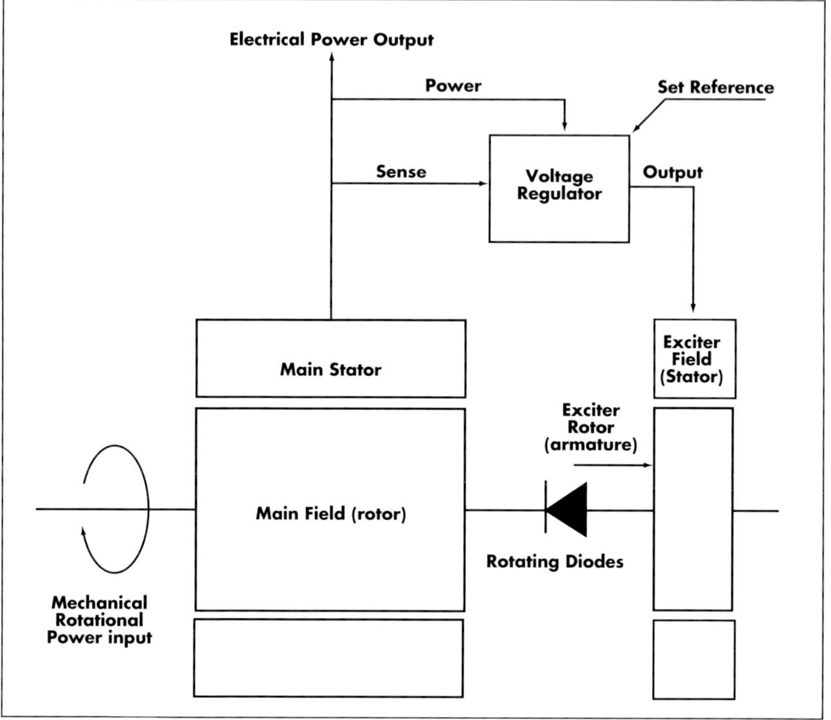

Fig. 28-6 Block Diagram of Self-Excited Synchronous Generator.

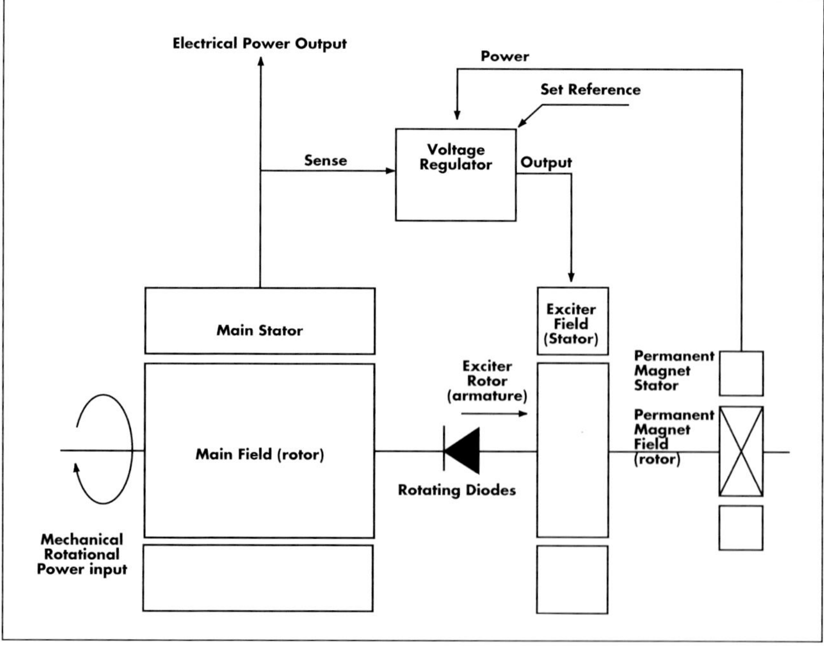

Fig. 28-7 Block Diagram of Separately Excited Synchronous Generator.

There must always be a locking provision on the tie-point disconnect that is accessible to utility or ISO company service personnel. The local on-site generator must be able to be separated from the utility grid immediately upon an unsafe power system disturbance.

When on-site generators are connected to the utility grid, they must be prevented from reclosing to a dead bus. This prevents generators from backfeeding the utility system when it is down. Additionally, when utility service is restored, it may not be synchronized with the on-site system. Therefore, it must be ensured that the two systems remain isolated. When the utility system re-establishes stable conditions, the generators can then be synchronized and paralleled with the grid.

TRANSFORMERS

Host facilities are connected to the utility-derived power supply via transformers, which reduce utility distribution or transmission voltage to a usable level and provide isolation and phase matching. When facilities have on-site generation systems, the transformer matches the generator voltage with the utility voltage while limiting the effects of transients on the generator.

For retail power purchase arrangements, decisions must often be made regarding transformer ownership and metering point. Often, the utility owns and maintains the main transformer and the metering point for billing is on the secondary (or low-voltage) side of the transformer. In this arrangement, the utility absorbs capital and maintenance costs as well as the power losses through the transformer. Facilities are often afforded the opportunity to own and maintain the transformer and take power metered at primary voltage. There are various methods by which the utility provides incentives or compensation to such facilities. This usually involves some type of rate discount calculated as a percentage reduction of use or cost.

There are several different transformer types, designated according to the connections on the primary and secondary side. The critical difference between wye- and delta-connected windings is that wye is generally grounded and delta is not:

- **Wye/delta-connected transformers** (with the ungrounded delta on the facility side), as shown in Figure 28-8, are the easiest to use for interconnected on-site generating facilities. These systems do not typically have the neutral resistance grounded on the utility side. Primary ground faults cleared on the utility side require immediate generator shutdown to prevent generation into open utility lines.
- **Delta/wye-connected transformers,** as shown in Figure 28-9, are the most common service transformer connections. Many generator neutrals are solidly grounded, especially in low voltage (<600 V) systems. Island operation often makes this the preferred configuration.
- **Wye/wye-connected transformers,** as shown in Figure 28-10, can be problematic. In addition to coupling harmonics (particularly third harmonics), this transformer connection complicates system protection. However, this configuration is becoming increasingly popular as most utilities require the generators to be a source of ground current for a ground on the utility side. Generator pitch can be specified to reduce harmonic concerns.
- **Delta/delta-connected transformers** with no grounding, as shown in Figure 28-11, is relatively

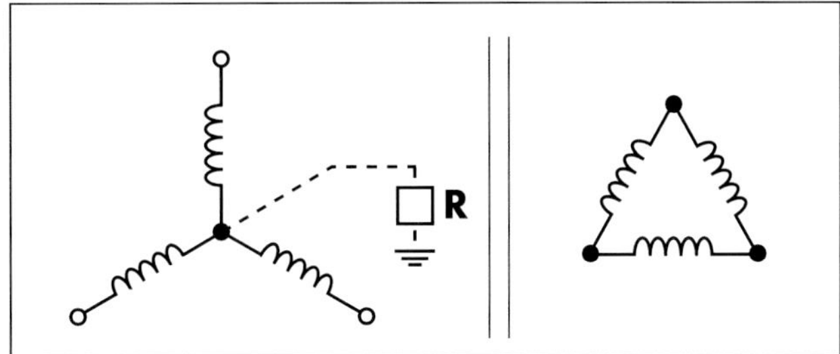

Fig. 28-8 Wye/Delta Connection.

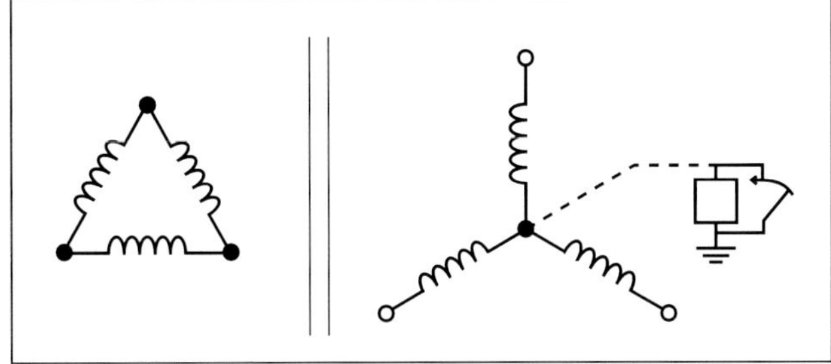

Fig. 28-9 Delta/Wye Connection.

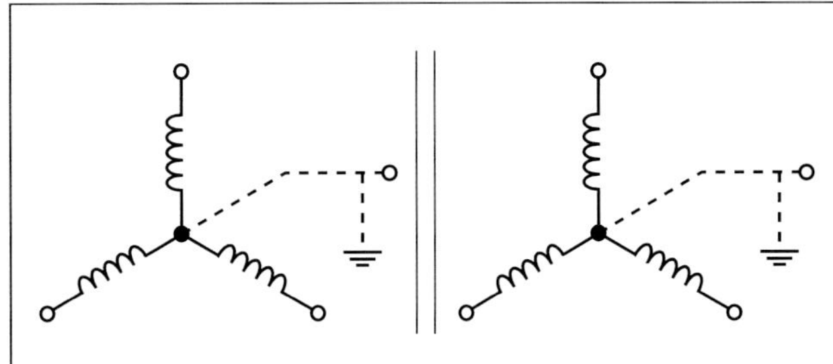

Fig. 28-10 Wye/Wye Connection.

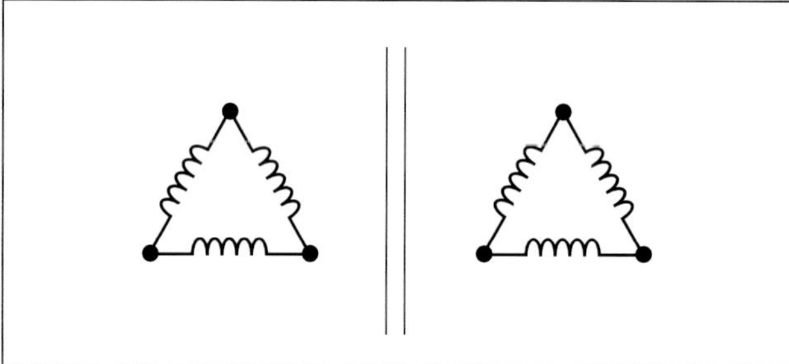

Fig. 28-11 Delta/Delta Connection.

uncommon but detection of ground faults can be accomplished with the addition of ground over voltage relaying using potential transformers connected line to ground primary and open corner delta secondary, referred to as a 3VO scheme.

The delta-grounded system eliminates harmonic problems resulting from on-site generation. Ac generators produce slightly distorted waveforms, and the harmonic currents in a grounded system may cause repeated ground-fault interruptions resulting from incorrect detection by overcurrent relays. However, because circulating currents cause internal heating, ac generators are usually supplied with wye-connected windings. The grounded systems must either be replaced or modified to avoid problems.

FAULT CONTROL

A **fault** is any disruption in the normal operation of an electric power system. A **system,** or **primary, fault** is an abnormality that involves, or is the result of, failure by primary equipment, including generators, transformers, busbars, overhead lines, and cables and all other items of the plant that operate at power system voltage. A system fault requires the disconnection of the affected equipment from the system by the tripping of associated circuit breakers.

System faults are caused mainly by pollution resulting from deposited soot or cement dust in industrial areas and salt deposited by wind-borne sea spray in coastal areas. Other causes of system faults on overhead lines include: birds, aircraft, lightning, fog, ice and snow loading, broken insulators, abnormal loading in machines, cables and transformers, failure of solid insulation due to moisture; mechanical damage; accidental contact with earth or earthed screens; and flashover in air caused by overvoltage or abnormal loading.

A **non-system fault** is any incorrect circuit breaker operation from a cause other than electrical equipment malfunction. Non-system faults may result from defects in protection (incorrect setting or faulty or incorrect connection) or from human error in testing or maintenance work.

Transients are temporary disruptions from normal steady-state conditions. On utility network feeders, transients can result from sudden load changes, starting of very large motors, circuit breakers interrupting large currents, or disturbances from lightning strikes. On the utility side, transients are subdued by station arresters, lightning arresters, or surge arresters, plus surge capacitors. On the facility side, transients are controlled through the use of lower voltage motor starting and sequential procedures for adding loads. Because paralleling an on-site generator out of phase establishes a large transient, synchronization and generator-loading procedures are critical. Utilities often request that generators ramp export power on and off to minimize transients.

There are many types of faults, as summarized below and illustrated in Figures 28-12 through 28-14:

Short-circuited phase faults (Figure 28-12)

- Three-phase fault clear of ground (earth)
- Three-phase-to-ground (earth) fault
- Phase-to-phase fault
- Single-phase-to-ground (earth) fault
- Two-phase-to-ground (earth) fault
- Phase-to-phase plus single-phase-to-ground (earth) fault

Open-circuited phase faults (Figure 28-13)

- Single-phase open-circuit
- Two-phase open-circuit
- Three-phase open-circuit

Winding faults (Figure 28-14)

- Winding-to-ground (earth) short-circuit
- Winding-to-winding short-circuit
- Short-circuited turns
- Open-circuited winding

Fault detection and interruption is one of the primary functions of utility distribution system operations. Utility protection systems are usually designed around a single power source (the utility substation) using a complex system that includes overcurrent, undervoltage, and impedance monitoring relays. When an on-site generator

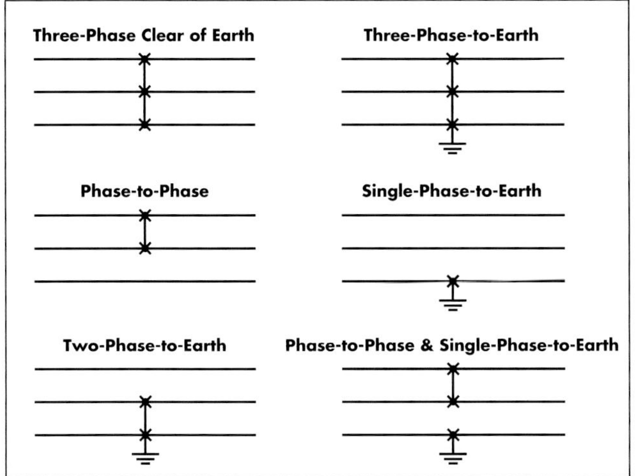

Fig. 28-12 Short-Circuited Phase Faults.

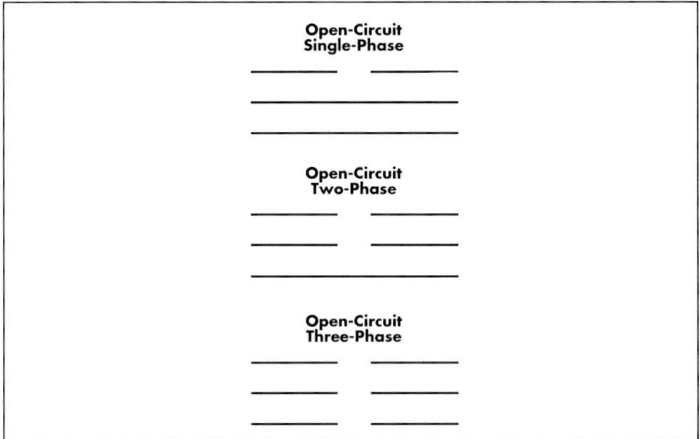

Fig. 28-13 Open-Circuited Phase Faults.

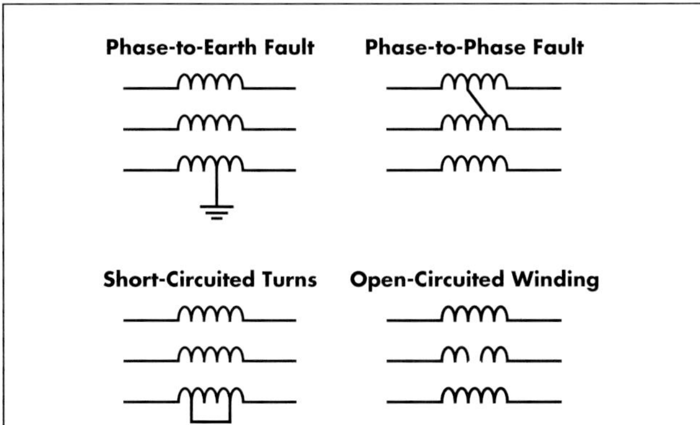

Fig. 28-14 Winding Faults.

is connected with the network, a complete review, resetting, and possible enhancement of the protection system is required.

The type of devices required to protect against feeding a fault depends on the location and type of fault and the kind of generator involved. Some types of faults are easily detected and some are not. For instance, a line-to-line short-circuit fault results in a dramatic increase in current, making detection easy.

A line-to-ground fault, which is common, is not always easy to detect. When the wire connects to ground, the line impedance drops, but not necessarily to zero. The change in current and voltage levels may sometimes be too modest to detect without grounded systems. This type of fault detection capability can be established through the connection of the distribution transformer grounded wye/delta, with measurement of the neutral current (Figure 28-15), or the connection of three potential transformers, connected wye/open delta (Figure 28-16), to the transformer primary (measuring voltage across the open delta).

Generally, when there is a fault or the line is de-energized, the voltage drop is quite significant. Therefore, voltage relays without great sensitivity are sufficient to provide adequate protection. Upstream faults (faults that occur between the facility and the network substation) can be sensed by undervoltage relays. For instance, if the utility trips off the line, the undervoltage relay at the facility will respond.

Downstream faults, which are not always easily detected, often result in both the facility and the utility feeding the fault. In some cases, a synchronous generator may carry the load and feed a dead line. Current, voltage, and frequency relays are used to sense this condition,

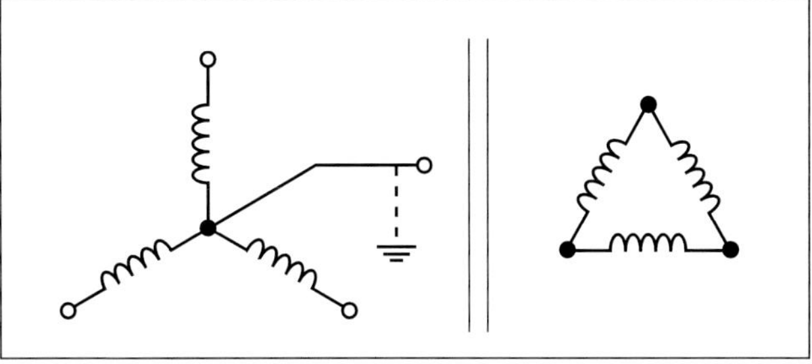

Fig. 28-15 Grounded Wye/Delta.

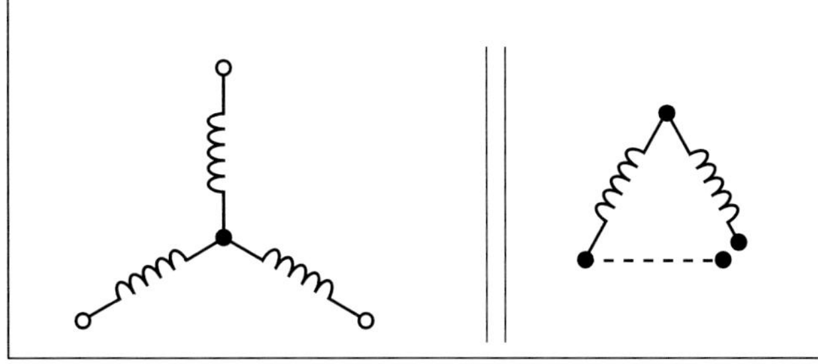

Fig. 28-16 Grounded Wye/Open Delta.

depending on possible scenarios.

Reclosing breakers are automatic devices much like circuit breakers that are used to time and interrupt faulted circuits and to re-energize the line by reclosing to restore service. Because most faults are temporary (lasting only a few cycles or seconds), reclosers momentarily trip open and closed, after which most faults will be cleared. Reclosers are set for one or several reclosing operations. Typically, with multiple reclosing settings, the initial response is instantaneous, lasting a couple of cycles. If the fault is not cleared, subsequent responses will be time delayed (for lower levels of current) to allow for a slightly longer interruption in an effort to clear the fault. Finally, if the fault is not cleared, the recloser will lock open. Figure 28-17 illustrates a typical recloser operating sequence.

Sectionalizers are used in conjunction with a circuit breaker or recloser to isolate more permanent faults. After a circuit breaker or recloser opens, sectionalizers open, isolating the fault and allowing the breaker or recloser to close and restore normal load current to downstream circuits. If a recloser is used and set for pre-set multiple operations, the sectionalizer will count the operations and lock open after the second-to-last operation to allow the recloser to close.

Several problems may occur when a utility system breaker opens and the local on-site generator is isolated from the network. If the generator is undersized relative to the connected load, it will be overloaded and begin to drop voltage and frequency. If the generator is oversized relative to the remaining load (meaning that it had been exporting power), it may be subject to overspeeding. In either case, the customer's main breaker must trip whenever utility power is disconnected. Under-frequency or over-frequency relays (set at 1 Hz +/-), undervoltage and overvoltage relays, and directional current relays can be used.

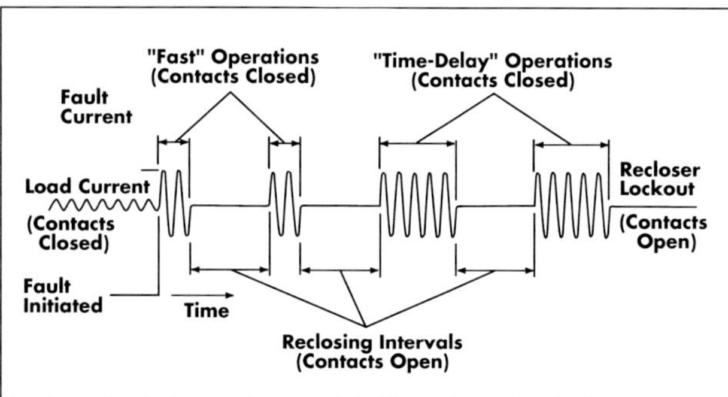

Fig. 28-17 Typical Recloser Operating Sequence.

When generators are being used for isolated operation, a synchronizer is used at the main breaker to allow reconnection without stopping. The facility breaker cannot be reclosed onto a dead or live unsynchronized utility line. Reconnection of the generator should only occur when trip conditions have been absent for a period of time greater than the longest reclose time for the main breaker at the feeding substation. A synchronizing breaker should only close when the frequency difference across the breaker is within the generator tolerance level. If an automatic synchronizer is not used, a synchronism check relay is required to prevent manual reclosing out of synchronism. A synchronizer is used at the generator breakers to allow the generators to be brought off and on without interruption of utility service.

Design of a facility's distribution net- work includes analysis of the available short-circuit current at various points in the network. Switchgear must be specified with sufficient short-circuit capacity. While an on-site generator may decrease the actual load drawn from the utility, it will usually increase the available short-circuit current at the facility, which may necessitate the use of heavier switchgear.

When an on-site generator and the utility network feed the same bus, the short-circuit current available from each must be added together to determine the capacity requirements of the circuit breakers. When several buses are used for in-house distribution, short-circuit current considerations limit the ability to transfer loads from one source to another in the event of a failure. Reactors are sometimes used to tie facility buses together to allow transfer of power from one bus to another without increasing short-circuit current in the event of a failure.

Representative Protection Configurations for Parallel Operation

The previous discussion has presented some of the relay functions. The following discussion will center on a series of examples. These examples assume that on-site generation is to be operated in parallel with the utility-derived service. They represent a medium- and a low-voltage system.

With the medium-voltage system, it is assumed that the facility has two utility-derived feeds and a number of on-site generators. The generators serve as emergency power for the facility as well. Whether these generators are run occasionally as peak-shaving units or continuously as baseload units, the relay scheme would remain the same. The fact that parallel operation is to last for more than 1 minute dictates the need for a coordinated protective scheme. If the parallel operation was solely for the purpose of closed transition transfer of the load from one source to the other, a very simple scheme could be used because the operation would typically last less than 30 seconds.

Medium Voltage Examples

Figure 28-18 represents a facility with two service feeders from the utility and a number of on-site generators. This example is representative of a campus with several buildings. The switchgear would be medium-voltage, 5 or 15 kV, and is divided into two line-ups: utility and generator. The utility switchgear is a typical double-ended design with a tie. It can be operated with any two of the main tie breakers closed at any time. All three breakers are never closed simultaneously to prevent feed-through of fault current from one utility feeder to the other.

In this example, 52A and 52T are closed and 52B is open. When the generators are to operate in parallel with the utility, 52T1 and 52T2 are closed and the selected generators are then started. When synchronized with the utility service, their respective breakers are closed. When operation is completed, the generator breakers are opened and 52T1 is opened. The feeders from the utility switchgear provide normal power to the campus and the generator switchgear provides emergency power to these buildings.

Figure 28-19 represents a minimum protective relay scheme for the utility main breakers shown in Figure 28-18. The transformers are connected and, to limit ground fault current, the neutral of the secondary winding is low-resistance grounded. A current relay is connected to a current transformer in series with the grounding resistor and has the designation of 50G. Upon establishing a ground fault, the 50G trips the associated main. This relay will typically have a time delay of 12 to 18 cycles to permit a downstream device to detect and respond to the fault by opening the breaker closest to the fault. This coordination avoids shutdown of the entire distribution system for faults down stream.

Each main breaker would have the same protective devices. Device 27/59 allows closure of the main when the utility voltage is within acceptable limits. Most utility companies require a frequency relay at the connection point, as well, to ensure that the on-site generation will not change the frequency of the utility grid. However,

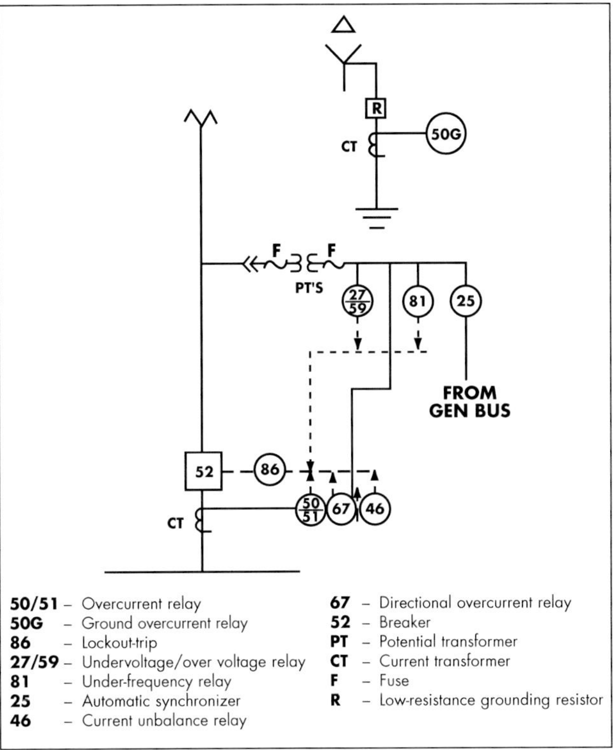

Fig. 28-19 Minimum Protective Relay Scheme for Utility System Main Breakers for Figure 28-18 Circuit.
Source: Automatic Switch Company

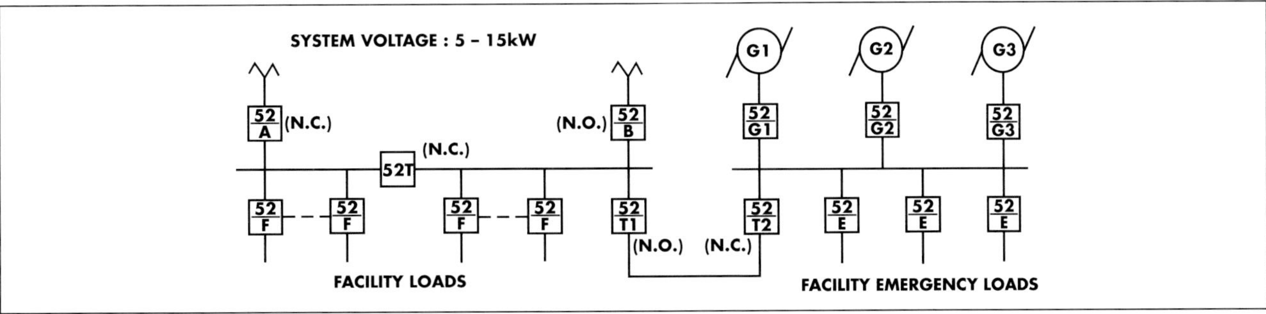

Fig. 28-18 Line Diagram for Parallel Operation of Local On-Site Generation with Utility-Derived Service at Medium Voltage.
Source: Automatic Switch Company

this relay serves as backup protection for isolating the generators from the bus to prevent overload when the utility system has an upstream failure.

Device 50/51 is an overcurrent relay that opens the main for various overcurrents. The 51 includes a delay operation relay that allows brief in-rush currents for motor starting, etc., while the 50 is an instantaneous relay that acts at much higher current levels to respond to fault currents. Device 46 acts as a current balance and phase sequence relay. Device 67 is a direction overcurrent relay and would typically be set at the full-load current rating of the generators where normal operation includes export power. If planned operation does not include export, the overcurrent relay setting would be low to operate immediately upon detection of flow into the utility system. Unlike the other relays, Device 67 would trip 52T1 to isolate the generators from the utility.

Where the on-site generation capacity is small (less than 50%) compared with the utility grid on the load side of the nearest utility sectionalizer or recloser, the relaying illustrated here could be sufficient. Where the local on-site generation is equal to or larger than that section of the grid, further study would be needed to determine the extent of required protective relaying.

Figure 28-20 represents a protective scheme for the generators in this power system. These generators will typically be six lead machines connected in wye. Where the generator rating is small compared with the utility service at the point of interface, it is common practice to high-resistance ground the center point of the wye. This prevents the disconnect and shutdown of the generator on a perceived fault when a ground fault occurs. For large generation systems, the typical configuration is unit transformer isolation of the generator at the point of connection to the utility-derived service.

Devices 27 and 81, voltage and frequency relays, monitor the operating generator as a precondition to initiating synchronizing of the generator to the bus. Because the generator is a limited source, comparatively soft bus, the trip settings of the relays will use somewhat longer time curves than would be used on a stiffer utility-derived service. The automatic synchronizer, Device 25, will match the generator voltage and frequency and phase angle to the bus. When synchronism is achieved, the synchronizer initiates the closure of the paralleling breaker, in this instance, the generator breaker.

Device 46 is a current balance and sequence relay that protects against unbalanced loading of the generator. Device 32 is a directional power relay that will disconnect the generator upon detection of power flow into the generator, indicative of a loss of power by the prime mover. Device 65 is the load-sensing portion of the prime mover governor used to control power output. Device 40 is a loss of excitation relay. Devices 90 and 90C are the generator voltage regulator and excitation control, respectively. The excitation control biases excitation to maintain loading within the generator VA rating. Device 86 is an electrically tripped lockout that requires manual reset. It would be tripped by the protective devices in the relay scheme.

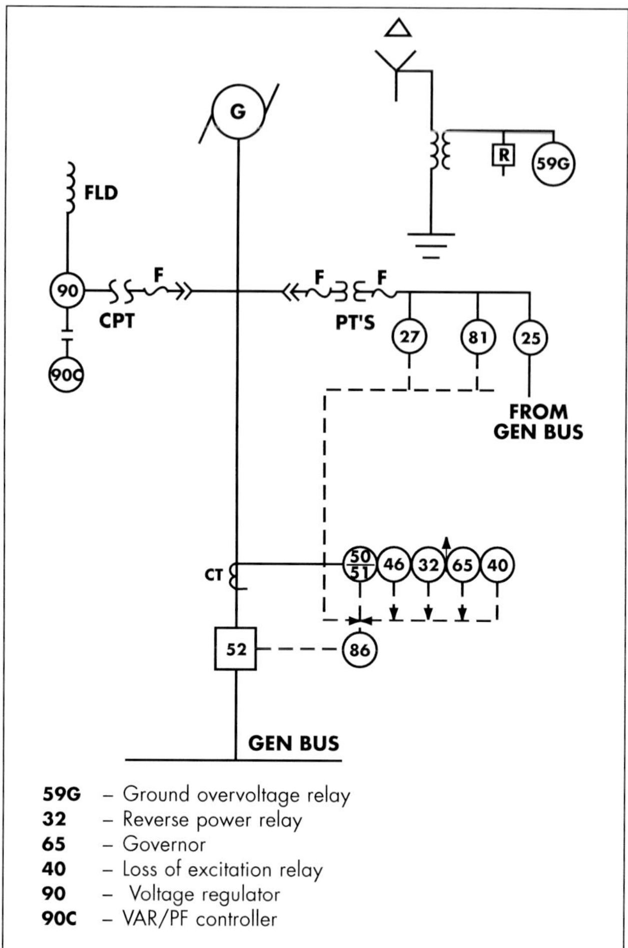

Fig. 28-20 Minimum Protective Relay Scheme for Generator Main Breakers for Figure 28-18 Circuit.
Source: Automatic Switch Company

Low Voltage Examples

Figures 28-21 and 28-22 show two basic manners in which a low-voltage generation system may be intertied with a utility-derived service. In Figure 28-21, the in-house generation is tied directly to the service switchgear with the generator wye point solidly connected to the ser-

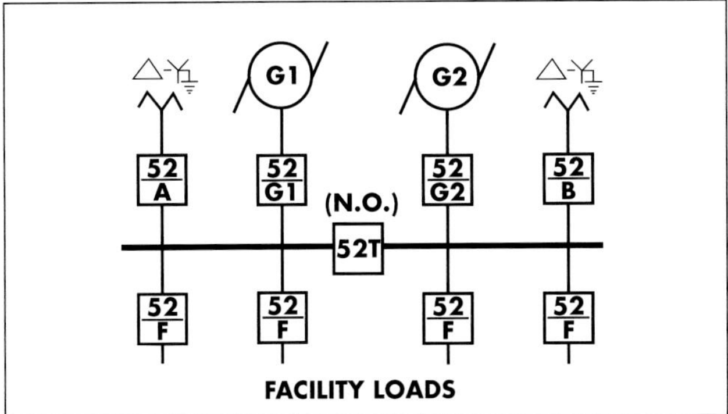

Fig. 28-21 Line Diagram for Parallel Operation of Local On-Site Generators with Utility Service at Low Voltage for Medium-Sized Facility. Source: Automatic Switch Company

vice neutral. The generators would have the same protective relay scheme as discussed for the medium- voltage case.

In Figure 28-22, generation is connected to the utility grid as a unit generator. This scheme is used where energy is sold to or wheeled by the serving utility. The low-voltage winding would be delta-connected and the medium- voltage winding would be wye-connected. Protective relaying for both the utility tie and generator circuits would be similar to those illustrated in Figures 28-19 and 28-20. The treatment of the unit transformer of Figure 28-22 would be the reverse of Figure 12-19.

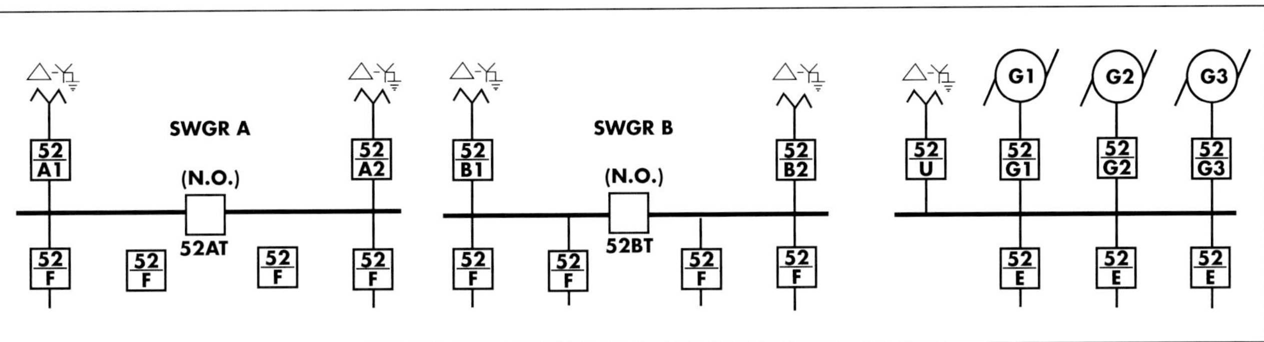

Fig. 28-22 Line Diagram for Parallel Operation of Local On-Site Generators with Utility Service at Low Voltage for Large-Sized Facility. Source: Automatic Switch Company

Section VII

Mechanical Drive Services

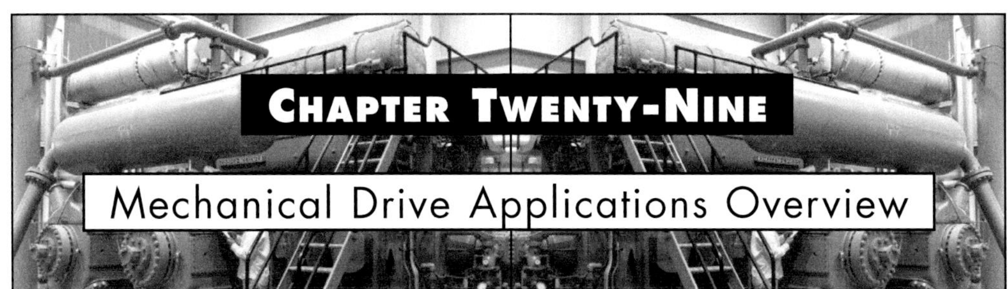

Chapter Twenty-Nine

Mechanical Drive Applications Overview

Technologies that provide mechanical drive service include a broad range of options for application, under a wide range of loads at constant or variable speed. Largely due to their low first-cost, relative ease of installation, simplicity of operation, and low maintenance requirement, electric motors continue to serve most mechanical drive applications. However, the thermal efficiency of reciprocating engines and gas turbines and the availability of packaged systems, combined with higher electric costs, can make prime mover technologies very competitive in some cases. In particular, cogeneration-cycles can be cost-effective where on-site thermal loads can be effectively served by recovered heat. Mechanical equipment may also be pneumatically or hydraulically driven or, in certain circumstances, be shaft-driven by alternative prime movers such as wind energy systems and water turbines.

This chapter includes a brief discussion of torque and power relationships among various load types, followed by a description of electric motor technology and comparison with the main categories of applied prime movers. A number of application considerations are then described.

Theoretical Torque and Power Formulae

Following are basic formulae used in estimating power and torque requirements of rotating equipment. Actual requirements are always greater due to friction, windage, or other machine factors.

Power (P) is equal to the net force (F) produced times the distance through which that force travels per unit of time:

$$P = (F)(2\pi r)(N) \qquad (29\text{-}1)$$

Where:

r = Radius

N = Rotational speed

In English system units, when power is expressed in hp, force in lbm-force (lbf), length in feet (ft), and rotational speed in rpm, Equation 29-1 becomes:

$$hp = \frac{(F)(2\pi r)(N)}{33{,}000} = \frac{(F)(r)(N)}{5{,}252} \qquad (29\text{-}2)$$

In SI units, when power is expressed in kW, force in Newton (N), length in meters (m), and rotational speed in rpm, while noting that one kW is equivalent to 60,000 N-m/min, Equation 29-1 becomes:

$$kW = \frac{(F)(2\pi r)(N)}{60{,}000} = \frac{(F)(r)(N)}{9{,}549} \qquad (29\text{-}3)$$

Torque is equal to the product of the force applied to a lever and the perpendicular distance from the line of action of the force to the axis of rotation:

$$T=(F)(r) \qquad (29\text{-}4)$$

In English system units, when torque is expressed in lbm-ft, power in hp, and rotational speed in rpm, torque can be expressed as:

$$T = \frac{(P)(5{,}252)}{N} \qquad (29\text{-}5)$$

In SI units, when torque is expressed in N-m, power in kW, and rotational speed in rpm, Equation 29-3 becomes:

$$T = \frac{(P)(9{,}549)}{N} \qquad (29\text{-}6)$$

Types of Mechanical Service Loads

Torque and power requirements are critical concerns in selecting a driver for a given mechanical load. Loads can be classified as follows:

- **Constant torque loads** include positive displacement pumps and compressors, conveyors, and friction loads such as extruders. The torque demanded by the load is constant throughout the speed range. Power requirements can be variable or constant, though usually they will be variable. Figure 29-1 shows torque and power versus speed curves for a constant torque load. In this case, power, which is directly proportional to speed, varies with the load requirement.
- **Variable torque/variable power loads** are typical of centrifugal fans, pumps, and compressors. Figure 29-2 shows the idealized relationship between torque and power versus speed. As the curves indicate, torque

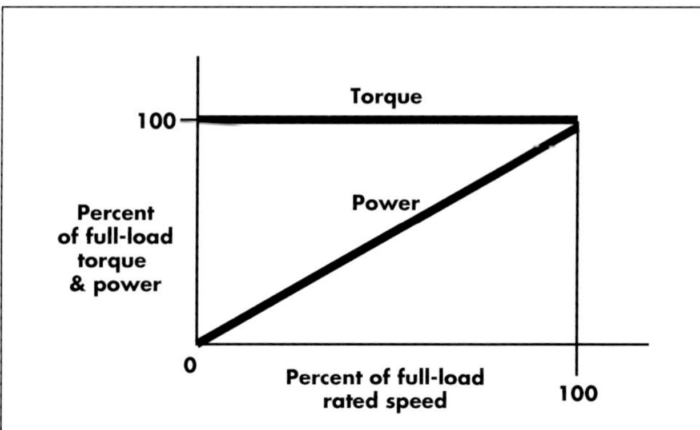

Fig. 29-1 Torque and Power vs. Speed Curves for Constant-Torque Load.

is directly proportional to a mathematical power of speed, usually speed squared (speed^2), and power is typically proportional to speed cubed (speed^3). Note that these are somewhat ideal relationships. Actual variations in power and torque will occur according to the characteristics of the system and the control scheme used. Usually, high starting torque is required for quick acceleration.

- **Variable torque/constant power loads** are typical of center-driven winders and machine tool spindles.
- **Higher inertial loads** are usually associated with machines, such as punch presses and centrifuges, using flywheels that supply most of the operating energy.
- **Shock loads** are experienced by a system subject to intermittent high peaks. This type of load is present in conveyors, crushers, winches, cranes, and vehicular systems. Drives applied to these types of loads must be able to effectively manage loads ranging from a small fraction to several hundred percent of nominal rated load.

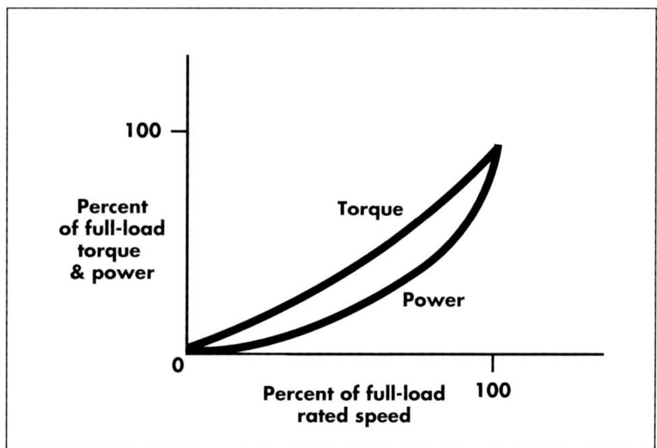

Fig. 29-2 Torque and Power vs. Speed Curves for Variable-Torque Load.

Electric Motor Drive Technology

Electric motors are efficient, relatively low-cost drivers for virtually any drive application. First cost, matched capacity, convenience, and simplicity of installation and operation make electric motors the most common choice for driving equipment, particularly in smaller capacities.

Types of Motors

Alternating current (ac) synchronous motors transform ac electric power into mechanical torque using field magnets excited with direct current (dc). They are generally used for large applications, or where maintaining precise speed and control is essential. Synchronous motors are seldom selected for small applications or for operation above 1,800 rpm due to the high cost of rotor construction. As application speed requirements decrease and load factor increases, synchronous motors become more economical in smaller capacities. For very large applications of a few thousand hp (or kW) at 3,000 to 3,600 rpm, synchronous motors may be used with step-up gears. Figure 29-3 shows a synchronous motor-driven process compressor.

Synchronous motors offer controllable power factor (PF) and very high operating efficiencies. Those sized above a few thousand hp (or kW) can reach efficiencies in excess of 97%. Because they can operate with a leading PF, synchronous motors can provide the added benefit of PF improvement. The ability to control the resistance of the rotor can allow for easier starting and reduced system voltage drop. They are not well suited for starting high inertial loads due to their tendency to overheat, and can also present operating difficulties by pulling out of synchronism during voltage dips.

For smaller applications, **ac induction motors** are almost always selected. When an induction motor (Figure

Fig. 29-3 Synchronous Motor-Driven Process Compressor. Source: Compressed Air and Gas Institute

Fig. 29-4 Induction Motor. Source: United States Motors

29-4) is driven by an electric current at a speed less than the synchronous speed, current flow in the stator windings induces current flow in conductors contained in the rotor. The magnetic attraction between the rotor and stator produces torque. The difference between normal operating speed and synchronous speed that induces current to flow in the rotor is called slip.

When an induction machine is driven by a prime mover above its synchronous speed, circulating currents are induced in the rotor bars by the reactive current of the power source and the magnetic field interactions between the rotor and stator are converted into electric power. Thus, induction machines can function as either motors or generators.

Induction motors are most commonly applied in capacities ranging from fractional hp (or kW) to several hundred hp (or kW), and range in efficiency from less than 70% for small motors to more than 96% for large units. One disadvantage of induction motors is that they take lagging current, producing PFs from 0.75 to 0.90 at full load, and substantially lower under part load. Induction motors also draw high starting instantaneous current, typically producing a system voltage drop. Once started, induction motors are very stable and able to withstand voltage drops of up to 25%.

Prior to the advent of solid state variable frequency drives (VFDs), **direct current motors** were used for variable speed applications. DC motors provide ease of control and comparatively high operating efficiency. Basic types of dc motors include shunt-wound, series-wound, and compound-wound, with each selected based on speed and torque characteristic of the driven load.

Shunt (or parallel) motors are best applied to controlled speed operations, since speed varies only slightly with load. In a series motor, the armature and field are in series. Speed is almost inversely proportional to the current and torque varies as the square of the current. Therefore, an increase in current produces a larger proportionate increase in torque, making series design well adapted for applications that require large starting torque such as cranes, hoists, and traction work. If the load is removed from a series motor, it can accelerate to a destructive speed. It is, therefore, common for series motors to have additional parallel windings that serve to limit the maximum speed.

The mixing of parallel and series windings results in a compound motor design. Compound motors feature higher starting torque than shunt motors, but poorer speed regulation. They are well suited for applications in which large and intermittent increases in torque occur as with punches and rolling mills.

In shunt-wound motors, speed is controlled by adjusting the armature voltage from zero to rated nameplate voltage (while applying full rated motor field voltage). The motors can provide a constant torque capability through nearly the entire controllable speed range, with power increasing from 0 to 100%. With some dc shunt motor designs, by weakening the motor field, speed can be further increased up to four or five times base speed. Over this speed range, the dc motor can demonstrate a constant power characteristic where torque decreases and is inversely proportional to speed. In a series-wound motor, speed can be controlled for any load by varying the voltage. Figure 29-5 shows a dc motor-driven process compressor.

Fig. 29-5 DC Motor-Driven Process Compressor. Source: Compressed Air and Gas Institute

Motor Ratings and Efficiency

The most common design for motors of up to a few hundred hp is 460 volt, 3 phase Delta. For larger motors, 2.3 kV or 4.16 kV are more common. For very large motors, 7,500 hp (5,600 kW) or more, 13.2 kV is typical.

The National Electrical Manufacturers Association

(NEMA) provides both manufacturers and motor users with guidelines for the standardization of motor dimensions and performance. Internationally, the commonly used rating system for motors is provided by the International Electro-technical Commission (IEC), a European standards organization. The IEC introduced a new standard in 2008 relating to energy efficient motors, standard IEC 60034-30. The standard defines new efficiency classes for motors and harmonizes the currently different requirements for induction motor efficiency levels around the world. The standard defines three International Efficiency (IE) classes for single speed, three phase, cage induction motors: premium efficiency (IE3); high efficiency (IE2), and standard efficiency (IE1). The standard also envisions a class for super premium efficiency (IE4) motors. IE4 motors are not yet commercially available. There are numerous differences in the design philosophies between the European and United States rating standards and care must be taken in application of IEC-rated devices to NEMA rated systems and vice versa.

The Consortium for Energy Efficiency (CEE) and the National Electrical Manufacturers Association have developed a joint specification defining a "premium" efficiency motor. Motors meeting this minimum specification are eligible to carry the NEMA Premium®, designation and exceed the EPAct minimum efficiencies in each covered category. The NEMA Premium motors specification covers single-speed, polyphase, 1-500 hp, and 2-, 4- or 6-pole squirrel cage induction, NEMA design A or B, continuous rated motors. The standard contains a table specifying nominal and minimum efficiency standards for motors. Additionally, there is a standard NEMA table for motor manufacturers that appears in the National Electric Code (NEC), providing the minimum and maximum kVA/hp for a motor on starting. Motor starting torque may be less than 100% or greater than 400% of the full-load torque, depending on the motor design. When voltage is applied to a motor at rest, the motor appears to be in the stalled rotor condition and will typically draw from four to six times its full load running current with a PF in the range of 0.15-0.20 and starting kW of 0.9-1.2 times the full load.

When operating on a utility derived power system, motor starting is typically not problematic except when many motors are started simultaneously. However, when operated on an on-site generator derived system, the starting inrush could stall the generator set and will at least tend to cause voltage and frequency transients. Overcurrent protective devices must be selected with the ability to avoid unnecessary tripping during normal starting inrush while providing adequate protection.

Where applicable, staggered starting or reduced voltage starting are commonly employed to minimize the effects of inrush currents. Three basic types of reduced voltage starting are: Star-Delta starting, auto-transformer starting, and soft starting. Star-Delta starting arranges the motor stator windings in a Star (Wye) configuration upon initial application of terminal voltage and then, at a given motor speed, reconnects the stator in the Delta configuration. Auto-transformer starting is similar, except that voltage is controlled by the auto-transformer. Soft starting uses gated SCR switching to control starting inrush, and is a feature of ac adjustable speed drives.

Full-load motor rating is based on maximum winding temperature, which may be determined from NEMA *Standard MG-1,* based on rated ambient temperature and the insulation rating (i.e., NEMA designation A, B, F, and H). Service factor is the maximum overload that can be applied without exceeding the temperature limitation of the winding insulation.

Electric service quality is always an important consideration for motor operation. NEMA specifies that motors must provide satisfactory performance when the motor voltage supply is within 10% of the motor nameplate rating. At the end of this limit, motors will generally experience lower efficiency and potential failure. As voltage decreases, motor current must increase, resulting in increased heating of the motor winding. Motors are highly susceptible to damage from voltage imbalances. A voltage imbalance of 3%, for example, may cause a rise in motor winding temperatures by as much as 25%. Motor protection includes phase current-sensing thermal devices that heat at the same rate as the motor and can take the motor off-line before unstable operating conditions can cause damage.

Input power requirement of a motor is determined by dividing the power output by motor efficiency. When power output is expressed in hp, conversion to kW is achieved by multiplying the hp by 0.746 kW/hp as follows:

$$kW = \frac{hp \times 0.746\ kW/hp}{\eta_{motor}} \tag{29-7}$$

Where η_{motor} is expressed as a decimal.

The relation is expressed in SI units as follows:

$$kW_e = \frac{kW_m}{\eta_{motor}} \tag{29-8}$$

Where subscripts e and m refer to electric power input and mechanical power output, respectively.

Premium-efficiency motors typically cost 20 to 30% more than standard-efficiency motors, depending on capacity and speed. The cost premium is largely the result of the use of more and better materials. Lamination material is a higher grade, higher cost steel. Typical standard-efficiency motors use low-carbon steel laminations, while premium-efficiency motors typically use high-grade silicon steel laminations that are thinner but have lower electrical losses. There are also more laminations and the rotor and stator core are lengthened. In addition, the laminations slots are larger in premium-efficiency motors, so more larger- diameter copper can be used in the windings. Friction and windage losses are reduced in premium-efficiency designs. Premium-efficiency motors tend to run at lower operating temperatures, resulting in longer life for lubricants, bearings, and motor insulation. They also generally operate at higher PFs.

Figure 29-6 provides a representative efficiency comparison for standard- and premium-efficiency motors operating at or near full load. Two caveats apply to the application of premium-efficiency motors, particularly in retrofit applications. First, performance of all motors is significantly reduced at low load and full load comparisons may not apply. Second, premium-efficiency motors have reduced slip (e.g., operate at higher rpm than standard-efficiency units.) In a retrofit, a slight increase in speed of the driven equipment will result in a large increase in motor output power requirement due to the cubed relationship of speed and load for centrifugal equipment. In some cases, increased motor rpm may be compensated by using variable-pitch sheaves, VFDs, or impeller trimming to maintain optimum mechanical efficiency.

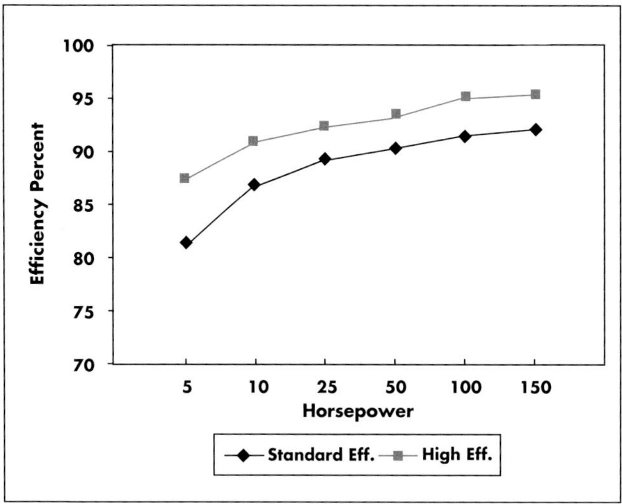

Fig. 29-6 Efficiency Comparison by Full-Load Capacity for Standard- and Premium-Efficiency Motors.

Variable Speed Control

Variable speed control can be accomplished using dc motors or multiple-speed ac motors or by equipping ac motors with variable frequency drives (VFDs). DC motor drives provide excellent speed control, but are costly and becoming increasingly less common. Induction motors are inherently constant-speed machines and speed can be changed only by changing the frequency, number of pole pairs, or slip. Multiple-speed ac motors are a low cost option for achieving some degree of speed control by varying the number of pole pairs used. Commonly, solid state devices are used to vary input power frequency to standard ac induction motors. These drives are referred to as variable speed drives (VSDs), adjustable speed drives (ASDs), or VFDs.

Variable speed ac drive technology has evolved considerably over the past decade and has been applied in a wide range of mechanical drive service applications. Unlike prime mover drive installations, electric VFDs are easily retrofitted to existing motor-driven equipment. VFDs, although available above 1,000 hp (750 kW), are particularly applicable to smaller loads for which the cost and complexity of prime mover drives are prohibitive. Motors driven by VFDs can slow to about 40% of full speed before motor cooling becomes a problem, well below the 50 to 60% lower limit for most prime movers.

Given the efficiency benefit of variable speed operation for many mechanical drive applications, VFDs can produce dramatic reductions in electricity use. Often, however, operating cost savings are limited because peak electrical demand is not substantially reduced, or may be slightly increased. Motors driven by VFDs can use slightly more power than constant-speed motors when operating at full speed due to the efficiency of the VFD itself, which is about 96%. Economic performance of VFD applications is, therefore, enhanced by electric rate structures that feature higher energy charges and more moderate demand charges. It is also enhanced when applied to loads that vary significantly, with substantial operating hours at low load levels. The cost of VFDs has declined to the point that, with moderate electric commodity (usage) cost structures, the efficiency benefits potentially achievable with variable speed operation will commonly produce good economic returns on the relatively modest cost premium for VFDs.

Prime Mover Drive Technologies

Prime movers have long been successfully applied to a wide range of mechanical drive services. Steam engines and steam and water turbines, for example, have been used since the industrial revolution and reciprocating engines have been applied for almost a century. Combustion gas turbines, introduced in the 1940s, have seen extensive development for use in mechanical drive service applications.

Common to both reciprocating engine and gas turbine drives is the ability to recover rejected heat, improving overall economic performance by allowing displacement of additional thermal energy purchases. The following sample calculation compares energy input or heat rate for simple-cycle power generation to the net energy input, or fuel-chargeable-to-power (FCP), for cogenerated power, when 45% of the original energy input can be recovered and effectively used:

Simple-cycle operation

Prime mover total fuel input	=	10 MMBtu/h (10,540 MJ/h)
Simple-cycle power production	=	1,300 hp (970 kW)
Simple-cycle heat rate	=	7,690 Btu/hp-h (10,880 kJ/kWh)

Fuel credit due to heat recovery

Heat recovered from prime mover	=	4.5 MMBtu/h (4,743 MJ/h)
Efficiency of displaced heating system	=	84%
Displaced purchased energy	=	$\frac{4.5 \text{ MMBtu/h}}{0.84}$ = 5.36 MMBtu/h (5,649 MJ/h)

Cogeneration-cycle operation

Net prime mover fuel requirement	=	10 – 5.36 = 4.64 MMBtu/h (4,891 MJ/h)
FCP (or net heat rate)	=	3,569 Btu/hp-h (5,042 kJ/kW)

What can be seen from this example is that the incremental fuel requirement for cogenerated mechanical power production, referred to as FCP or net heat rate, may be half that of the simple-cycle. The calculation illustrates the potential thermodynamic advantage of on-site prime movers over purchased (simple-cycle) electricity. The major prime mover types used in on-site mechanical drive applications are discussed below.

Reciprocating Engines

Reciprocating engines have been applied to numerous types of mechanical drive applications, including irrigation water pumping, mobile air compressors, refrigeration compressors, and gas compressors for the oil and gas industry. One hurdle that historically inhibited wide use of reciprocating engine drives has been the need for customized designs, particularly the need for torsional-lateral critical analysis to assure the absence of destructive modes of vibration at or near operating speeds. However, pre-engineered, packaged gas engine-driven equipment is available from several manufacturers. Standardization and market entry of leading manufacturers have produced lower costs and more widespread market recognition.

Engine types and operation are discussed in Chapter 9. A principle characteristic of reciprocating engines is their relatively high simple-cycle thermal fuel efficiency, particularly in smaller capacities. Another key characteristic is good part-load performance, achieved over much of the operating range with variable speed operation. Disadvantages include relatively large size and weight and relatively high maintenance requirements.

Typical life-cycle operations and maintenance (O&M) costs range from \$0.005 to \$0.015/hp-h (0.007 to \$0.020/kWh). Heat recovery options include generation of hot water or low-pressure steam from engine coolant systems. Higher-pressure steam can be produced from engine exhaust. Coolant system and exhaust heat recovery are commonly combined to produce one output. In applications that require limited peak load operation, a reciprocating engine-driven unit can sometimes be sized for 90% of prime capacity rating, reducing first cost. If a purchaser specifies Diesel Engine Manufacturers Association (DEMA) ratings, equipment will provide 110% of rated load for two hours per day without impact on engine life or maintenance cost.

Typical simple-cycle thermal efficiency for reciprocating engine equipment drive systems ranges from 25 to 45% (LHV basis), with small capacity, higher speed, spark-ignited engines at the lower end and large capacity, slow speed, Diesel engines at the higher end of the range. This thermal fuel efficiency range equates to simple-cycle heat rates of 5,650 to 10,200 Btu/hp-h (8,000 to 14,400 kJ/kWh) on a LHV basis. Typical recovered heat potential ranges from 30 to 45% of the total energy input, depending on simple-cycle efficiency and the temperature requirement of recovered heat. When fuel use displaced by recovered heat is considered, net heat rates, or FCPs, range from 4,500 to 3,500 Btu/hp-h (6,400 to 4,900 kJ/kWh).

Sound emissions can be attenuated to approximately 60 to 90 dba at a 3-ft (0.9-m) distance with acoustic enclosures. On some units, sound emissions are below 100 dba at a 7-ft (2.1-m) distance without enclosures. NO_X emissions can be controlled down to well below 1 gram/hp-h (1.3 grams/kWh) on many new gas-fired, lean-burn technology models and even lower with the use of exhaust gas treatment.

Figures 29-7 through 29-12 shows several different types of reciprocating engine mechanical drive applications. Figures 29-9, -10, and -11 are different types of gas and air compressor drive applications. Figure 29-10 is a somewhat unique technology application, in which the compressor is integral to the engine, with both compressor and engine on the same crankshaft. Figures 29-7 and 29-12 show pump and blower applications, respectively, and Figure 29-8 is a scrap metal shredder drive application.

Fig. 29-7 Two Air-Cooled Reciprocating Engines Driving Centrifugal Pumps at 900 rpm. Source: Waukesha Engine Division

Fig. 29-8 Reciprocating Engine Driving a Car Shredder at a Scrap Metal Plant. Source: Waukesha Engine Division

Fig. 29-9 3,300 hp (2,460 kW) Gas-Fired Reciprocating Engine Gas Compressor Drive. Source: Fairbanks Morse Engine Division

Gas Turbines

Combustion gas turbine capacities generally start at about 500 hp (375 kW) and range up to several hundred thousand hp (or kW). "Micro" turbine technology is commercially available; applications of 30 hp (22.5 kW) or even lower may be economically viable. Although their use has historically been confined to large facilities with significant steam loads or to applications in remote locations, recent technology improvements, reduced manufacturing costs, and the ability to control air emissions to extremely low levels have made gas turbines increasingly attractive for other mechanical drive applications.

Gas turbine types and operation are discussed in Chapter 10. Distinguishing characteristics include good reliability and high power density (low specific weight per unit of output capacity). Because of high exhaust mass flow rate at temperatures of about 900 to 1000°F (480 to 540°C), high temperature heat recovery is possible, allowing, for example, recovery of half or more of the total energy input in the form of high-pressure steam. Maintenance frequency is relatively low and typical life-cycle O&M costs range from \$0.004 to \$0.012/hp-h (\$0.005 to \$0.016/kWh).

Fig. 29-10 Integral Gas-Engine Compressor, with Engine and Compressor on Same Crankshaft.
Source: Cooper Energy Services, Cooper Cameron Corp.

Fig. 29-11 Reciprocating Engine Driving a Manufacturing Plant Air Compressor. Source: Waukesha Engine Division

Fig. 29-12 Reciprocating Engine Driving a Blower for Aeration in Waste-Water Treatment. Source: Waukesha Engine Division

Typically, single-shaft turbines may be effectively applied to full-capacity baseloaded applications. However, they do not operate efficiently at variable speeds and are inefficient at part loads. With variable speed operating capability, multi-shaft turbines are more frequently applied to mechanical drive services with significant part-load operational requirements. Multi-shaft units are also advantageous because they can develop the high torque required to quickly accelerate compressors and other driven equipment to full operating speeds.

Augmentation by supplementary firing in the oxygen-rich exhaust can significantly enhance the quantity and temperature of the exhaust gas stream. Steam injection can also be used for power augmentation. Because heat recovery-generated steam can be split between process and power augmentation service, steam injection cycles accommodate wide variations in host demands for thermal and mechanical energy. When thermal loads are small relative to mechanical load, combined cycle, steam injection (STIG) cycle, or recuperation can be used to increase power generation efficiency. Capacity enhancements to gas turbines are discussed in Chapters 10 and 12.

Generally, sound emissions can be attenuated to 80 to 90 dba at a 3-ft (0.9-m) distance with normally available silencers and packaged sound attenuation. NO_X emissions can be controlled to a range of 30 to 9 ppmv on models with low-NO_X combustors, and even lower with the use of exhaust gas treatment.

Typical simple-cycle thermal fuel efficiency for combustion gas turbine mechanical drive systems range from 17 to 35 (LHV basis), resulting in simple-cycle heat rates of 15,000 to 7,000 Btu/hp-h (20,000 to 10,000 kJ/kWh). With use of steam injection or recuperation, thermal fuel efficiencies of greater than 40% can be achieved.

Typical recovered heat potential ranges from 45 to 60% of the total energy input, depending on simple-cycle efficiency and temperature of the recovered heat. When fuel use displaced by recovered heat is considered, net heat rates, or FCPs, range from 4,800 to 3,300 Btu/hp-h (6,800 to 4,700 kJ/kWh).

Figures 29-13 through 29-16 show different types of gas turbine engine mechanical drive applications. Figure 29-13 is a cross-sectional illustration of a two-shaft unit and Figure 29-14 shows a two-shaft turbine applied to compressor drive service. Figure 29-15 shows a gas turbine applied to mechanical drive service in a water injection plant and Figure 29-16 shows a recuperator applied to a gas turbine mechanical drive for power augmentation.

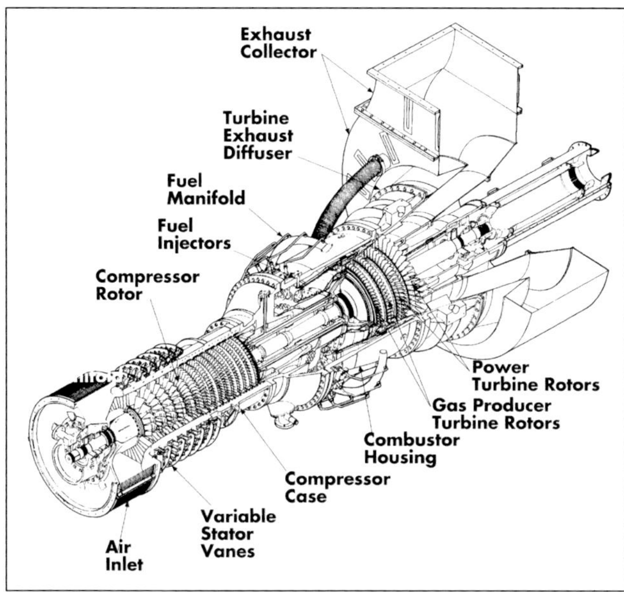

Fig. 29-13 Cross-Sectional Illustration of Two-Shaft Gas Turbine. Source: Solar Turbines

Fig. 29-14 Two-Shaft Gas Turbine Driving Propane Compressor. Source: Solar Turbines

Fig. 29-15 Heavy-Duty Industrial Gas Turbine Applied in Water Injection Plant. Source: MAN GHH

Fig. 29-16 Recuperated Gas Turbine Compressor Set. Source: Solar Turbines

STEAM TURBINES

Steam turbines are available in virtually any capacity and are very well suited for most mechanical drive applications. They offer good reliability, low maintenance requirements, and long service life, in addition to providing precise control with variable speed capability and excellent overload capability. While operating speeds in excess of 10,000 rpm are common, optimal operating speeds for many turbine models range from 3,600 to 5,500 rpm, which is compatible with the speed ranges of many types of equipment, such as centrifugal air and refrigeration compressors. Steam turbines generally require less floor space and foundation than reciprocating engines or gas turbines (with heat recovery), and can therefore be more easily fitted in many applications.

Steam turbine characteristics and operation are discussed in Chapter 11. The distinguishing features of condensing turbines are reasonably good thermal efficiency and a need for both a high-pressure steam system and a condensing system. They usually feature a high mechanical efficiency, multi-stage design. The distinguishing feature of non-condensing (topping cycle) back-pressure turbines is that while simple-cycle thermal efficiency is very low, net thermal efficiency approaches 100% (excluding boiler losses) because almost all unused heat energy (in the form of lower-pressure steam) is passed directly to process. They usually feature a low or moderate mechanical efficiency, single-stage design, but are none-the-less very efficient shaft power producers due to their high net thermal efficiency. Extraction steam turbines are multistage machines that combine these elements, with reasonably good thermal efficiency for the portion of the steam that is condensed and high net thermal efficiency for the portion of the steam that is extracted as low-pressure steam for thermal applications.

In facilities with high-pressure steam systems and abundant low-pressure steam loads, topping-cycle applications using back-pressure steam turbines are often highly cost-effective. When low-pressure steam loads are limited, high-efficiency multi-staged full condensing or extraction turbines can be used. Multi-stage back-pressure turbines can also be used under conditions of moderate low-pressure steam loads. Back-pressure and extraction turbines can also be designed to operate under varying intake and discharge pressures.

While shaft power can be extracted from virtually any pressure drop, condensing turbines generally become cost-effective when boiler-generated steam pressures are at least 100 psig (7.9 bar). Steam turbines are sometimes used as part of a mixed system hybrid configuration in which the turbine-driven equipment is run during peak periods and electric driven-units run during off-peak periods. An electric generator can be attached to the turbine shaft along with the driven equipment, allowing two modes of operation.

Multi-stage turbines require more maintenance than turbines with single-stage design. Condensing turbine systems have auxiliary components such as surface condensers and cooling towers that require maintenance. In all cases, key considerations include control over steam purity and protection from erosion due to moisture and impurity deposits. Typical life-cycle O&M costs range from $0.003 to $0.007/hp-h ($0.004 to $0.010/kWh).

With steam turbines, system thermal efficiency varies far more than with reciprocating and gas turbine engines and is mostly a function of turbine design and steam supply and exhaust conditions (that is, the enthalpy, or heat, drop through the turbine). Turbine mechanical efficiency typically ranges from 30 to 60% for single-stage turbines and from 50 to 80% for multi-stage turbines. System simple-cycle thermal efficiency typically ranges from 20 to 30% for condensing systems and from 5 to 15% for non-condensing systems. Assuming total steam flow and heat drop are fixed by given conditions, turbine mechanical efficiency selection should be based on the incremental capital cost for the more efficient turbine and the incremental value of increased power output.

A multi-stage condensing turbine can generate shaft power at a typical steam rate of 7 to 13 lbm/hp-h (4 to 8 kg/kWh). Accounting for boiler efficiency losses, this equates to heat rates ranging from 12,000 to 15,000 Btu/hp-h (15,000 to 21,000 kJ/kWh) on an HHV basis.

Table 29-1 shows representative steam rates, in lbm/hp-h, under various inlet conditions, each at an exhaust condition of 3 in. HgA (10.2 kPa). Included are theoretical steam rates and actual steam rates with various steam turbine mechanical efficiencies. Note that D&S refers to dry, saturated steam. Table 29-2 shows the same steam rates from Table 29-1 in SI units. Table 29-3 shows representative heat rates, in Btu/hp-h (HHV basis), under the same set of conditions shown in Table 29-1. Table 29-4 shows the same steam rates from Table 29-3 in SI units. Heat rates are calculated under the assumption of 83 Btu/lbm (193 kJ/kg) hotwell enthalpy and 83% boiler efficiency in each case.

In topping-cycle applications, steam rates typically range from 20 to 55 lbm/hp-h (12 to 34 kg/kWh). Since they use only a small portion of the energy flowing through the turbine, topping-cycle applications depend on a large downstream thermal load. Since net efficiency for a topping turbine will approach 100%, total system efficiency will be a function of boiler steam generation and net heat rate or energy chargeable-to-power (ECP) will range from 3,000 to 3,500 Btu/hp-h (4,200 to 5,000 kJ/kWh).

Table 29-5 shows representative steam rates, in lbm/hp-h, for non-condensing turbine operation under various inlet and exhaust conditions. Included are theoretical steam rates and steam rates for various turbine mechanical efficiencies. Table 29-6 shows the same steam rates from Table 29-5 in SI units.

Figures 29-17 through 29-22 show several types of steam turbine mechanical drive applications. Figure 29-17 shows a sugar mill in which multi-stage steam turbines are used to drive two cane knife and six mill drives

Table 29-1 Condensing Turbine Steam Rates Under Various Inlet Conditions and Turbine Mechanical Efficiencies (English Units)

Inlet Condition	Exhaust Condition	Theoretical Steam Rate (lbm/hp-h)	Steam Rate $\eta_{Turbine}$ = 75% (lbm/hp-h)	Steam Rate $\eta_{Turbine}$ = 65% (lbm/hp-h)	Steam Rate $\eta_{Turbine}$ = 55% (lbm/hp-h)
150 psig/D&S	3" HgA	9.4	11.2	12.9	15.3
250 psig/450°F	3" HgA	7.3	9.7	11.2	13.3
400 psig/550°F	3" HgA	6.6	8.8	10.2	12.0
600 psig/750°F	3" HgA	5.5	7.3	8.5	10.0
900 psig/900°F	3" HgA	4.9	6.5	7.5	8.9

Table 29-2 Condensing Turbine Steam Rates Under Various Inlet Conditions and Turbine Mechanical Efficiencies (SI Units)

Inlet Condition	Exhaust Condition	Theoretical Steam Rate (kg/kWh_m)	Steam Rate $\eta_{Turbine}$ = 75% (kg/kWh_m)	Steam Rate $\eta_{Turbine}$ = 65% (kg/kWh_m)	Steam Rate $\eta_{Turbine}$ = 55% (kg/kWh_m)
11.4 bar/D&S	10.2 kPa	5.1	6.8	7.8	9.3
18.3 bar/232°C	10.2 kPa	4.4	5.9	6.8	8.1
28.6 bar/288°C	10.2 kPa	4.0	5.3	6.2	7.3
42.4 bar/399°C	10.2 kPa	3.3	4.4	5.2	6.1
63.0 bar/755°C	10.2 kPa	3.0	3.9	4.5	5.4

Table 29-3 Representative Condensing Turbine Heat Rates Under Various Steam Inlet Conditions and Turbine Mechanical Efficiencies (English Units)

Inlet Condition	Steam Enthalpy (Btu/lbm)	Theoretical Heat Rate (Btu/hp-h)	Heat Rate $\eta_{Turbine}$ = 75% (Btu/hp-h)	Heat Rate $\eta_{Turbine}$ = 65% (Btu/hp-h)	Heat Rate $\eta_{Turbine}$ = 55% (Btu/hp-h)
150 psig/D&S	1,196	11,264	15,018	17,298	20,517
250 psig/450°F	1,231	10,097	13,462	15,534	18,358
400 psig/550°F	1,276	9,487	12,649	14,661	17,248
600 psig/750°F	1,379	8,588	11,399	13,272	15,614
900 psig/900°F	1,459	8,123	10,775	12,433	14,754

Fig. 29-17 Multi-Stage Steam Turbines Applied at Sugar Mill for Cane Knife and Mill Drives. Source: Tuthill Corp. Murray Turbomachinery Div.

through reduction gears. These units, which have been in operation for 25 years, feature oil relay governing systems and produce 200% torque at stall.

Figures 29-18 through 29-20 show various steam turbine process compressor drives. Note the large 45,000 hp (33,500 kW) capacity of the system in Figure 29-20.

Figure 29-21 is a steam turbine-driven centrifugal steam compressor applied in a mechanical vapor recompression evaporation process in a brewery. An illustrative line-drawing of system operation is shown in Figure 29-22. This single-stage compressor produces 75,000 ft^3/m (2,100 m^3/m) of compressed steam. In contrast to the mechanical drive system shown in Figure 29-18, in which the gear box is shown in the drive train, this unit is driven directly by the steam turbine without an integral gear box. The steam turbine is directly connected to the impeller of the compressor and compressor speed is matched to that of the turbine.

Table 29-4 Representative Condensing Turbine Heat Rates Under Various Steam Inlet Conditions and Turbine Mechanical Efficiencies (SI Units)

Inlet Condition	Steam Enthalpy (kJ/kg)	Theoretical Heat Rate (kJ/kWh_m)	Heat Rate $\eta_{Turbine}$ = 75% (kJ/kWh_m)	Heat Rate $\eta_{Turbine}$ = 65% (kJ/kWh_m)	Heat Rate $\eta_{Turbine}$ = 55% (kJ/kWh_m)
11.4 bar/D&S	2,781	15,935	21,247	24,516	28,973
18.3 bar/232°C	2,863	14,284	19,046	21,975	25,971
28.6 bar/288°C	2,967	13,421	17,895	20,648	24,402
42.4 bar/399°C	3,207	12,149	16,199	18,691	22,090
63.0 bar/755°C	3,393	11,492	15,323	17,680	20,894

Table 29-5 Representative Steam Rates Under Various Inlet and Exhaust Conditions and Turbine Mechanical Efficiencies (English Units)

Inlet Condition	Exhaust Condition	Theoretical Steam Rate (lbm/hp-h)	Steam Rate $\eta_{Turbine}$ = 60% (lbm/hp-h)	Steam Rate $\eta_{Turbine}$ = 50% (lbm/hp-h)	Steam Rate $\eta_{Turbine}$ = 35% (lbm/hp-h)
250 psig/D&S	15 psig	17.5	29.2	35.0	50.0
	50 psig	22.7	37.8	45.4	64.9
250 psig/550°F	15 psig	13.6	22.7	27.2	39.0
	50 psig	20.0	33.3	40.0	57.1
400 psig/750°F	50 psig	13.1	21.8	26.2	37.4
	100 psig	17.8	29.7	35.6	50.9
600 psig/750°F	50 psig	11.5	19.2	23.0	32.9
	100 psig	14.5	24.2	29.0	41.4
900 psig/900°F	50 psig	9.0	15.0	18.0	25.7
	100 psig	10.8	18.0	21.6	30.9
	200 psig	14.5	24.2	29.0	41.4

Table 29-6 Representative Steam Rates Under Various Inlet and Exhaust Conditions and Turbine Mechanical Efficiencies (SI Units)

Inlet Condition	Exhaust Condition	Theoretical Steam Rate (kg/kWh_m)	Steam Rate $\eta_{Turbine}$ = 60% (kg/kWh_m)	Steam Rate $\eta_{Turbine}$ = 50% (kg/kWh_m)	Steam Rate $\eta_{Turbine}$ = 35% (kg/kWh_m)
18.3 bar/D&S	2.0 bar	10.6	17.8	21.3	30.4
	4.5 bar	13.8	23.0	27.6	39.5
18.3 bar/288°C	2.0 bar	8.3	13.8	16.5	23.7
	4.5 bar	12.2	20.2	24.3	34.7
28.6 bar/399°C	4.5 bar	8.0	13.3	15.9	22.7
	7.9 bar	10.8	18.1	21.6	30.9
42.4 bar/399°C	4.5 bar	7.0	11.7	14.0	20.0
	7.9 bar	8.8	14.7	17.6	25.2
63.0 bar/755°C	4.5 bar	5.5	9.1	10.9	15.6
	7.9 bar	6.6	10.9	13.1	18.8
	14.8 bar	8.8	14.7	17.6	25.2

Mechanical Driver Part-Load Performance

Many mechanical system drivers operate under varying load conditions. For a given process, power requirement generally decreases as output decreases. Usually, this relationship is not linear; as the load changes, the efficiency of the driven equipment changes. The motor or prime mover also has a power requirement that varies according to load, and efficiency tradeoffs may be made in system selection. Efficiency at full load, for example,

Fig. 29-18 Steam Turbine-Driven Process Compressor, with Gear Box in Drive Train. Source: Compressed Air and Gas Institute

Fig. 29-19 Multi-Stage Steam Turbine Process Compressor Drive. Source: Dresser-Rand

Fig. 29-20 45,000 hp (33,500 kW_m) Steam Turbine Process Gas Compressor Drive. Source: Dresser-Rand

Fig. 29-21 75,000 CFM Steam Turbine Driven Centrifugal Steam Compressor to be Applied in Mechanical Vapor Recompression Process. Source: Centrifugal Compressor Division, Ingersoll-Rand Company

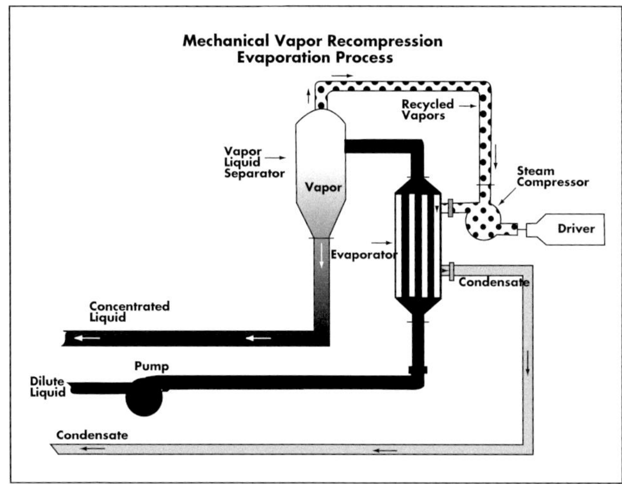

Fig. 29-22 Line Drawing Illustrating Operation of Steam Turbine Driven Mechanical Vapor Recompression Process. Source: Centrifugal Compressor Division, Ingersoll-Rand Company

may be compromised to allow optimal efficiency to be achieved at a commonly experienced part-load condition. Optimum application efficiency is achieved with the combination of driver, driven machine, and auxiliary components that produce the best weighted average performance over the full annual operating regime.

Figures 29-23 and 29-24 illustrate the performance of two systems operating at varying load. In each case, the driven machine's performance is represented in hp-h per unit output and the driver performance in energy units per hp-h. The combined performance of driver and driven machine is expressed as the **system performance ratio (SPR)**, which is the energy input required per unit of

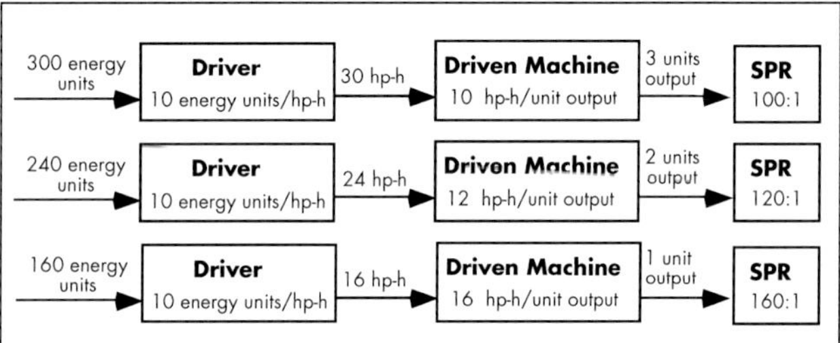

Fig. 29-23 System 1 SPR.

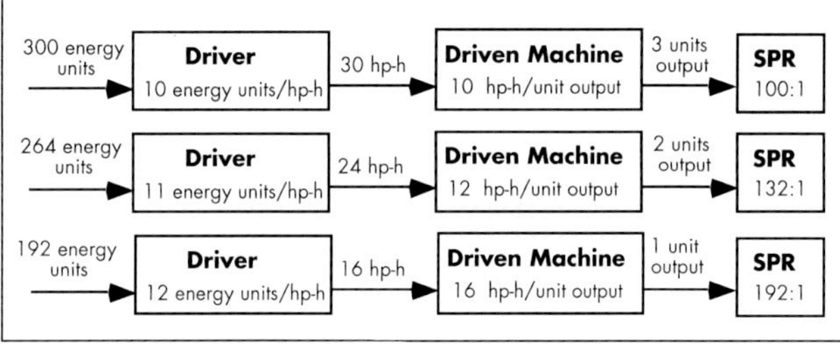

Fig. 29-24 System 2 SPR.

product output. The SPR serves as a benchmark by which full- and part-load performance can be compared for various system options.

For both System 1 and System 2, the output at full load is 3 units of product. At this level, the driven machines require 10 hp-h per unit of output. The driver requires 10 units of energy per hp-h. Given the full-load energy input requirement of 300 energy units and the full-load system output of 3 units of product, the SPR is 100:1.

In System 1, the driver performance remains constant while the performance of the driven machine degrades significantly under part-load operation. At two-thirds of full load the driven machine requires 12 hp-h per unit of output. This necessitates an input of 240 energy units to produce two units of product, bringing the SPR to 120:1. At one-third output, performance of the driven machine degrades further, resulting in an SPR of 160:1.

System 2 experiences the same performance degradation for the driven machine. However, in System 2, driver performance also degrades at part-load. To produce two units of output, the driver operates at 80% of full load and requires 11 energy units input per hp-h output. This increases the SPR to 132:1, significantly higher than System 1. At one-third output, SPR is 192:1.

Systems 1 and 2 are both representative examples of systems designed for constant speed operation. It is assumed that the driven machine uses some type of throttling to reduce system output and does so with degraded performance. Traditionally, throttling has been used to control flow with devices such as control valves on pumps, inlet and outlet dampers on fans, and air inlet throttling or cylinder unloading on positive displacement compressors.

In System 1, driver performance remains constant at all three operating load levels, similar to the performance of an electric motor or a steam turbine. In System 2, the driver performance degrades with reduced load, similar to a reciprocating engine or gas turbine operating at constant speed.

Many types of driven equipment show improved part-load efficiency when speed is reduced. Electric motor VSDs and a number of prime mover designs can also maintain efficient operation over a wide range of capacities by operating at variable speed. The performance of both the driven equipment and the driver may vary as speed is reduced. In some cases, the effect will be complementary, with the performance of both driver and driven equipment improving. In other cases, one or both components may operate at constant or reduced efficiency as speed is reduced.

For example, in the case of a reciprocating engine driving a pump, as the load decreases to about 60% of full load, the efficiency of both the engine and the pump will continue to improve. Below that point, engine speed can no longer be effectively reduced, although pumping efficiency would continue to improve at lower speeds.

Figures 29-25 and 29-26 are representative examples of reciprocating engine performance under variable speed operation. Figure 29-25 compares simple-cycle heat rates, in Btu/bhp (HHV basis), over operating loads ranging down to 25% of full load for both constant (C) speed and optimal (O) variable speed operation. Figure 29-26 shows recoverable heat available for the same engine when operating at variable speed.

In steam turbine-driven pump applications, for example, the efficiency of the pump increases as speed is reduced, but the efficiency of the steam turbine is reduced. While the steam turbine provides excellent control over a very wide speed range, the benefits of reduced speed operation of driver and driven machine are not complimentary as with reciprocating engines.

A typical control sequence for prime movers applied to mechanical drive services is to first reduce speed, then use standard part-load control mechanisms to modulate the driven machine. Minimum speed will be set either by limitations of the prime mover or the driven equipment.

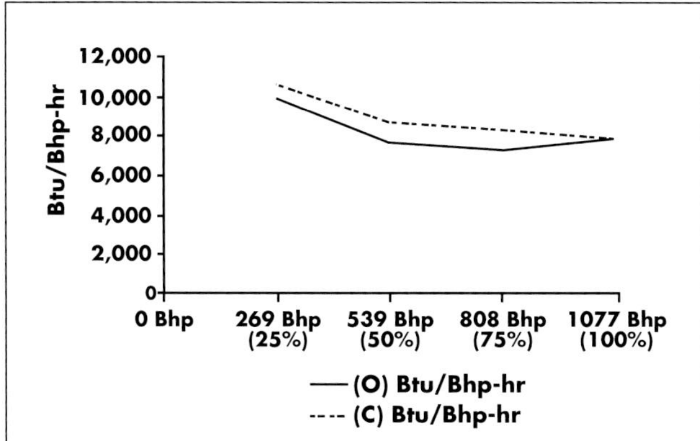

Fig. 29-25 Reciprocating Engine Performance at Constant vs. Variable Speed Operation (on HHV Basis).

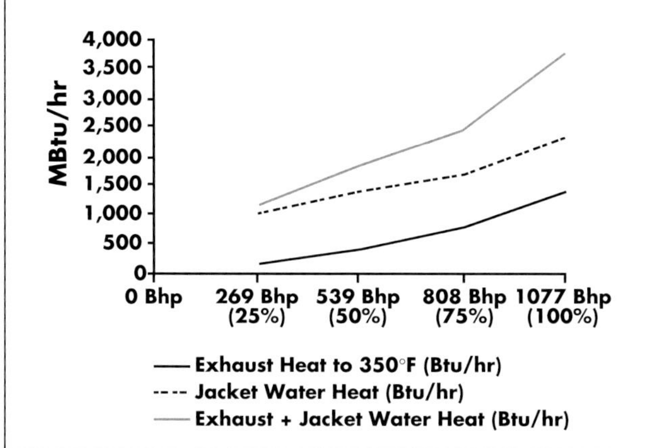

Fig. 29-26 Recoverable Heat from Reciprocating Engine Operating under Variable Speed.

Figures 29-27 and 29-28 show SPRs for two mechanical drive applications using speed control for part-load operation. In Systems 3 and 4, speed control allows the driven machine to maintain constant performance (10 hp-h per unit of product) at two-thirds of full load. At one-third of full load, it is assumed that driver speed is at minimum and throttling increases the power requirement to 12 hp-h per unit of product output.

In System 3, driver thermal efficiency improves at two-thirds load to 9 energy units input per hp-h, producing an SPR of 90:1. Driver efficiency degrades to 11 energy units input per hp-h at one-third of full load, producing an SPR of 132:1. This type of performance could be expected with a reciprocating engine operating under variable speed. Compared with System 2, which is assumed to have a similar driver and driven machine, the SPRs for part-load operation using speed control are superior by a significant margin.

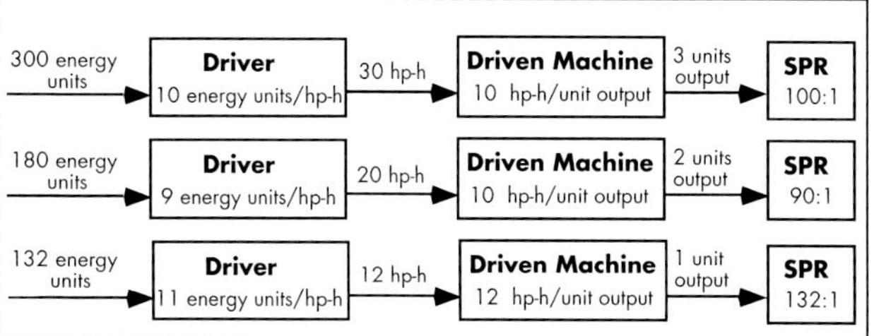

Fig. 29-27 System 3 SPR.

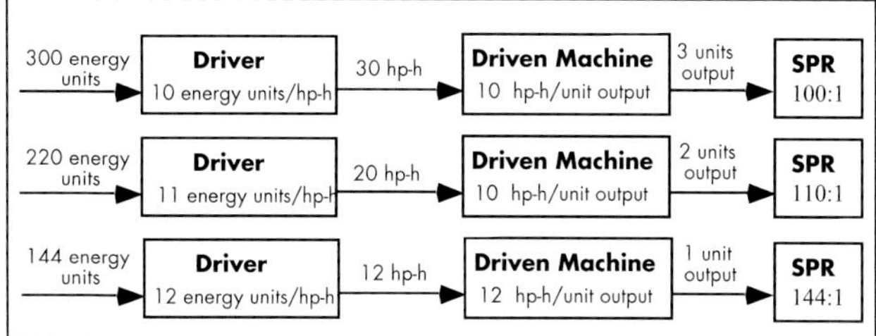

Fig. 29-28 System 4 SPR.

In System 4, driver thermal efficiency gradually decreases as speed is reduced, typical for a steam turbine. However, compared with constant speed operation in System 1, this performance loss is more than offset by the performance improvement in the driven machine. Hence, the resultant SPRs for both part-load conditions are superior to that achieved at constant speed. Operation at two-thirds and one-third of full load produces SPRs of 110:1 and 144:1, respectively, compared with 120:1 and 160:1 in System 1

Figure 29-29 illustrates the affinity laws for centrifugal loads, showing percent of full-load power when speed is reduced proportionately with flow. Input power

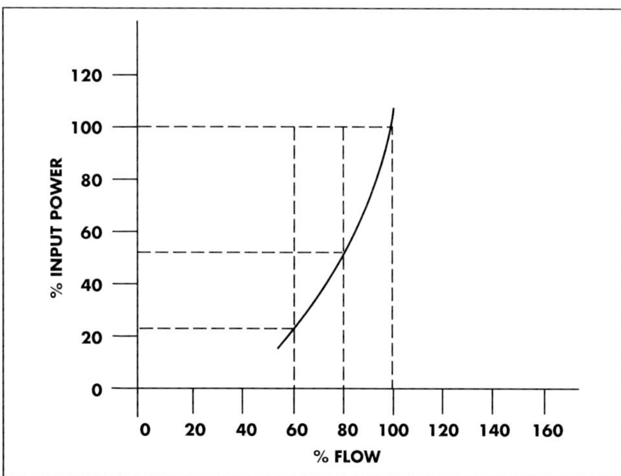

Fig. 29-29 Illustration of Affinity Laws for Centrifugal Loads, Showing Input Power Requirement vs. Flow Rate, Assuming Flow and Speed are Directly Related.

requirements have a cubed relationship with flow rate, while flow rate varies directly with rotational speed. Application of variable speed control can offer dramatic energy savings in centrifugal fan, pump, or compressor applications where the above relationships apply. For example, at 65% speed, a centrifugal pump may require as little as 27% of full-load input power.

Figures 29-30 and 29-31 show SPRs for two different systems in which the driven machines follow the affinity laws somewhat. As with Systems 3 and 4, it is assumed that at one-third of full-load system output, minimum driver speed limitations necessitate the use of throttling in addition to speed control. In System 5 and 6, speed control results in improved performance of the driven machine under part-load operation. At two-thirds of full-load output, there is a complimentary relationship between driver and driven machine performance. At one-third of full-load output, the trend reverses for the driver. Still, system SPR improves as a result of operation at reduced speed. While System 6 includes a driver performance penalty for operation at reduced speed, it is more than offset by the dramatic improvement in the performance of the driven machine.

Application Considerations and Capital Cost

In comparative analyses of driver options, the lowest first cost option serves as a base case to which alternatives can be compared on a life-cycle economic basis. For most

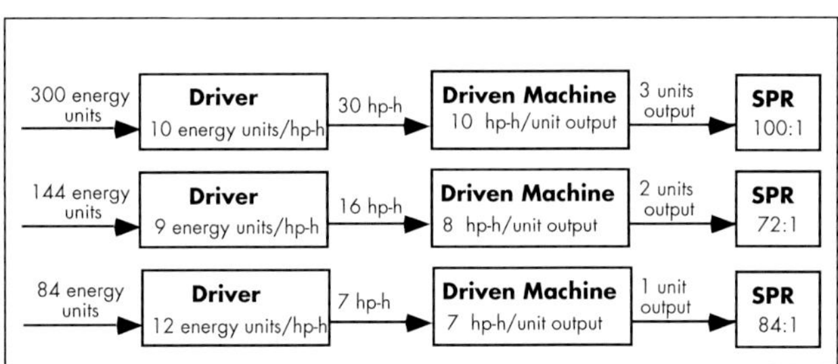

Fig. 29-30 System 5 SPR.

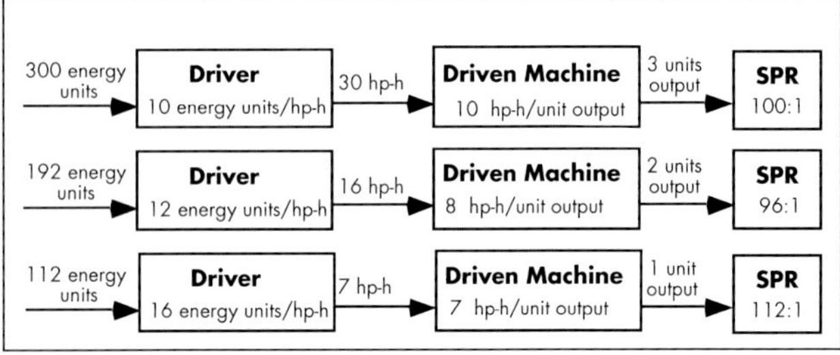

Fig. 29-31 System 6 SPR.

conventional applications, particularly those of small or moderate capacity, the base case option will be a conventional electric induction motor.

During preliminary screening, a few key factors, such as torque and power requirements, installation logistics, and energy prices, should be considered. These will often rule out several options. Logistical obstacles, such as clearance space, weight limitations, installation requirements for long exhaust gas venting or piping runs, or the need to modify existing systems to operate on recovered thermal energy, will often indicate poor economic potential before a more detailed analysis is undertaken.

The capital and operating costs of the remaining options are evaluated in the context of a rigorous life-cycle analysis to determine the system that will produce the best economic performance. A good representation of the average loading, average performance under that loading, and actual hours of operation should be used in the evaluation process, as well as consideration of peak loads.

In retrofit applications, load requirements can sometimes be estimated by metering the existing mechanical driver and the output of the driven equipment during operation. In some cases, metering data indicates that a lower capacity drive may actually be required. Metering should include measurement of all key operational parameters. With an electric motor, for example, this would include the motor's rpm, voltage, amperage, power factor, and total power usage.

In general, when energy costs are high, operating efficiency may be a primary consideration. When electricity costs are high and fuel costs are low, prime mover applications become increasingly more attractive. When considering potential application of cogeneration-cycle technologies, concurrence of thermal loads with system mechanical output requirements becomes a critical consideration. The temperature and form of the recovered heat must be compatible with on-site requirements and it must also be economically delivered to the point of use. If large pipe runs or other significant modifications or equipment additions are necessary, for example, increased capital costs may render the option uneconomical.

Another factor to consider is the load characteristic. For constant loads, the full-load operating characteristic of the driver is a principle concern. For varying loads, both the full-load and part-load characteristics of the driver should be considered. The impact of variable speed operation is thus an important factor.

VFDs enable electric induction motors to achieve

relatively constant efficiency from full load all the way down to their minimal operating speed at about 40% of full load. In the case of a reciprocating engine, it has been shown that there may be a complimentary effect between driver and driven machine performance under certain part-load conditions. On the other hand, steam turbine performance will suffer during operation at reduced speed. In the case of a gas turbine, the economies of scale in cost and thermal efficiency that are achieved with larger capacity systems are an important consideration. While variable speed operation with a multi-shaft turbine will produce better performance than constant speed operation, thermal efficiency will still suffer under part-load operation.

Such deficiencies may be addressed by using a fixed speed prime mover to drive an electric generator and using the generated electricity to power electric motor-driven machines at variable speed using VFDs. This approach can be particularly effective when there are multiple mechanical drive service applications under consideration. Economies of scale in thermal efficiency and superior overall load factor may be more readily achievable with one large-capacity on-site electric generation system providing power to numerous motor drives, as opposed to numerous prime mover mechanical drive systems. In such cases, the comparative advantages of VSD technology can be effectively exploited in multiple decentralized applications, while a facility can concurrently reap the benefits of on-site power generation.

Mixed (Hybrid) Mechanical Drive Systems

Mixed or hybrid systems that integrate prime mover-driven equipment with electric motor-driven equipment are common in modern industrial operations. They provide standby for critical loads during an electric power outage and allow for operating costs to be optimized. The use of a dual-shaft drive allows either a prime mover or an electric motor to drive the same piece of equipment. The equipment can be sized for the power output of both the motor and the prime mover combined, or for either one separately. Some typical configurations are:

- Prime mover-driven units used during expensive peak electric rate periods and electric-driven units used during inexpensive off-peak electric periods. The prime mover-driven units may or may not be equipped with heat recovery.
- Prime-mover driven units sized for and operated to match thermal loads served through heat recovery. Units can be baseloaded or designed for thermal load following. In some cases, the prime-mover units will be operated in peak electric rate periods, even if potentially recoverable rejected heat exceeds the thermal load requirement and must be vented.
- Prime mover-driven units used for peak-shaving duty only. Electric units may be baseloaded and the prime mover-driven units used to meet additional peak load.

For example, a plant with a critical 750 hp (560 kW) air compressor load may install as many as four units of 250 hp (186 kW) each, including an extra unit for backup capacity. With an all-electric system, the facility may also install standby electric generation capacity to support one or more of the units in the event of an electrical outage. By installing one or more 250 hp (186 kW) prime mover-driven units, the required standby generation capacity can be eliminated, possibly offsetting some or all of the incremental capital cost of the prime mover system. Under each operating condition and rate period, the most economical combination of units can be operated. Use of the electric units would be maximized under lower cost off-peak electric rates and minimized during more costly peak rates.

Capital Cost Considerations

In relatively straightforward retrofit applications (e.g., pumps and fans), the cost of replacing a conventional induction motor will typically range from $60 to $100 per installed hp of motor capacity ($80 to $135 per kW). Power factor correction devices, starters, and other components may add as much as $102/hp ($15/kW) to the retrofit installation. In new installations, the cost may be double or triple that, depending on the cost of electrical and controls installation. Premium-efficiency induction motors will add from 20 to 30% to the installed cost and synchronous motors will add 70 to 100%. VFDs will add $100 to $300 per hp ($134 to $400 per kW), depending on electrical and controls requirements, with larger capacity motors being on the lower end of the cost range. The relatively low capital cost as compared with prime mover drives, along with installation logistics and maintenance cost advantages make VFDs the logical base case for many VSD applications, against which prime mover applications may be compared. For more complex applications (e.g., compressors), retrofit costs can be considerably higher and in some cases, will not be feasible, necessitating the purchase of an entirely new system. Additionally, for such applications, more complex controls and programming will commonly be required, adding additional costs to the overall project.

The capital cost premium for prime mover-driven mechanical systems versus the conventional electric motor base case typically ranges from $300 to $1,000 per hp ($400 to $1,300 per kW) of installed capacity. Back-pressure steam turbines are the least expensive prime mover alternatives, followed by moderate-speed reciprocating engines and condensing steam turbines (assuming a high-pressure steam system is already in place). At the high end of the range are gas turbines and low-speed reciprocating engines. For all systems, the cost premium per unit of capacity is less as system capacity is increased. The costs associated with gas, exhaust, and rejected heat piping for reciprocating engine- and gas turbine-driven units, or steam boiler plants and piping for steam turbine-driven units, may add to the premium. Except for very large steam turbine plants, it is generally considered cost-prohibitive to build or greatly modify a steam generation plant for the sole purpose of using steam turbines.

With cogeneration-cycle applications, heat recovery can add $60 to $250/hp ($80 to $335/kW) to the cost premium for a combustion engine drive. On the low end of this cost range are simple engine coolant heat recovery systems. On the high end are more complex systems, such as multiple-pressure HRSGs used with gas turbine drives. Significant additional costs may be incurred to interface heat recovery systems with existing thermal loads. This may involve long-pipe runs, the use of storage systems, or replacement of heat exchangers or downstream HVAC coils. It may even involve the conversion of an entire process to accommodate a lower temperature energy source.

When an existing facility does not have sufficient electric service capacity available, the cost of providing additional electric service to the facility can vary widely. Where service expansion is required, the required costs can often offset a portion of the capital cost differential between a prime mover-driven unit and an electric unit.

Another factor that sometimes comes into play is the need for standby electric generation capacity and uninterrupted power systems (UPS) to support critical equipment loads. In these cases, the combined cost of the electric service, standby generator, and UPS to support electric motor drives may be equal to, or even greater than, the cost of a prime mover-driven unit. In some cases, an electric generator may be mounted in line with mechanical equipment (e.g., air compressor or pump) on a single prime mover-driven system. This allows the prime mover to serve a dual function of providing emergency power to the facility for other loads in the event of a utility outage.

Beyond the drive technology application itself, many systems will require significant modification to enable beneficial variable speed/variable flow operation. For example, a variable speed/variable flow pumping application may require change-out of 3-way control valves to 2-way valves throughout the distribution system. In the case of boiler or chiller system, the application may require installation of additional pumps and piping to create a primary/secondary system that enables constant volume circulation through the boiler or chiller, with variable volume circulation throughout the distribution system. Hence, while the cost-premium for a VFD may be quite modest, the overall capital cost of the project can be substantial. Still, such applications may prove to be financially attractive.

After gathering the appropriate data and limiting the options and potential operating strategies down to a few, through the screening process, a detailed study may be performed. It is important to conduct such studies within the context of overall facility system optimization planning and, therefore, interactively consider other potentially cost-effective improvement options, as well as facility long-term goals and objectives. Refer to the various chapters in Section IX for detail discussions on performing integrated technical and financial project performance analyses and on project implementation.

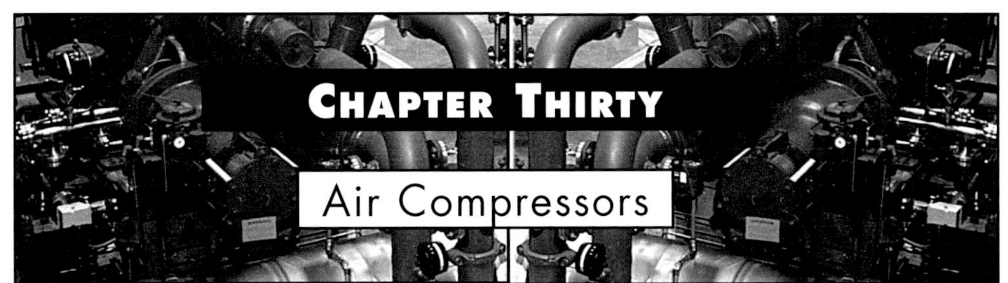

Chapter Thirty

Air Compressors

Pneumatic tools and controls served by a central air station have many advantages in industrial operations. They feature high power density, reliability, and precise control under partial loads and may be used steadily for long periods without overheating. Air motors are widely applied in operations that involve flammable or explosive liquids, vapors, or dust. They may be operated in corrosive, hot, or wet environments, frequently stopped and started, and operated at varying speeds without damage.

Pneumatically actuated equipment such as clamps, presses, feeders, and conveyors are commonly applied in production lines, as are pneumatic conveyers. Other common pneumatic applications include drilling, hammering, blast cleaning, spraying, automatic packaging, hoisting, cutting, stapling, sanding, grinding, plastic molding, liquid agitation, and fuel atomization.

Due to the wide range of applications for compressed air, air compressors typically account for a large part of an industrial facility's energy operating budget. The proper application of compressor technology is extremely important, since annual compressor operating costs are often greater than their total capital investment cost.

Compressor Types

Air compressors can produce pressures ranging from slightly above atmospheric to more than 60,000 psi (4,000 bar), although most industrial applications use pressures of around 100 psig (7.9 bar). There are two general methods used to compress gaseous matter:

1. Positive displacement compressors compress air (or other gases) by admitting successive volumes of air into a closed space and then decreasing the volume. Reciprocating and rotary screw compressors operate on this principle.
2. Dynamic compressors are machines in which air or gas is compressed by the mechanical action of rotating vanes or impellers imparting velocity and pressure to the air or gas. Dynamic compressors include axial and centrifugal types.

The basic concept behind the dynamic or centrifugal compressor principle can be easily explained with an analogy of a person whirling a ball attached to the end of a string. The ball pulls outward on the string due to the action of centrifugal force, and the pull is increased by a heavier ball, a longer string, or a faster rotation. As shown in Figure 30-1, the effect is the same if the ball is replaced with a molecule of gas, the string is replaced with an impeller, and the person is replaced with a mechanical driver. The centrifugal force imparted to the gas molecule will fling it outward, compressing it into the narrower impeller passageway. The larger the diameter of the impeller, the heavier the molecular weight of the gas, or the greater the speed rotation, the greater the pressure produced.

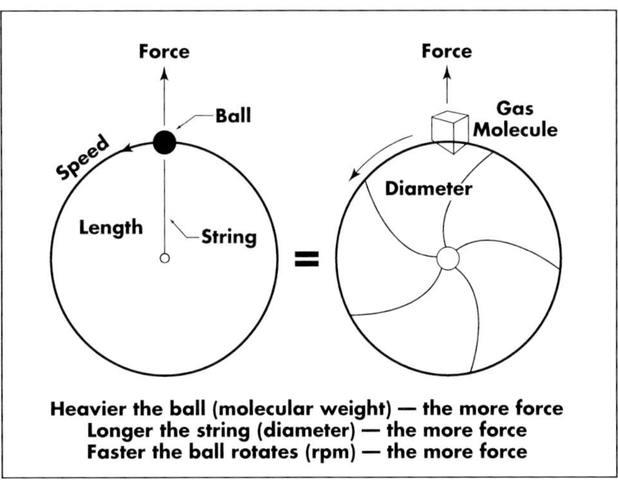

Fig. 30-1 Centrifugal Principle Illustration. Source: Carrier Corporation

In contrast, in positive displacement machines, a finite amount of the working fluid (e.g., air) is positively moved through the machine. Compression occurs as the machine encloses a finite volume of gas and reduces the internal volume of the compression chamber. The batch process characteristic allows positive displacement compressors to achieve higher pressures and respond to load change more effectively than centrifugal compressors. In particular, positive displacement compressors are more compatible with speed control as a means of controlling capacity under part-load conditions.

Rotary screw and reciprocating (piston) compressors are both positive displacement machines, which trap a finite volume of gas, compress it, and transports it to discharge. Dynamic machines, such as axial and centrifugal compressors, use rotary action to exert a torque on the gas to transport it and change its kinetic energy, without positive displacement.

Whereas positive displacement compressors are essentially constant volume, variable pressure machines, centrifugal compressors are essentially variable flow, constant pressure machines. Figure 30-2 provides a comparison of performance characteristics of centrifugal versus reciprocating compressors. Line JK represents the constant flow, variable pressure performance of the reciprocating compressor. Due to the decrease in volumetric efficiency at increasing pressures, the compressor will actually have a sloping characteristic, as shown by Line JL. Line FM represents the variable flow, constant pressure performance of the centrifugal compressor. Due to internal losses, the compressor characteristic is not a straight line, but is similar to Line FG.

Generally, positive displacement units are selected for smaller volumes of gas and higher pressure ratios. Dynamic machines are selected for higher volumes of gas and smaller pressure ratios. As such, centrifugal compressors are most common in large capacities, except when high pressure ratios are required.

Figure 30-3 shows the basic types of compressors discussed in this chapter.

Reciprocating Compressors

Reciprocating compressors compress gas with piston

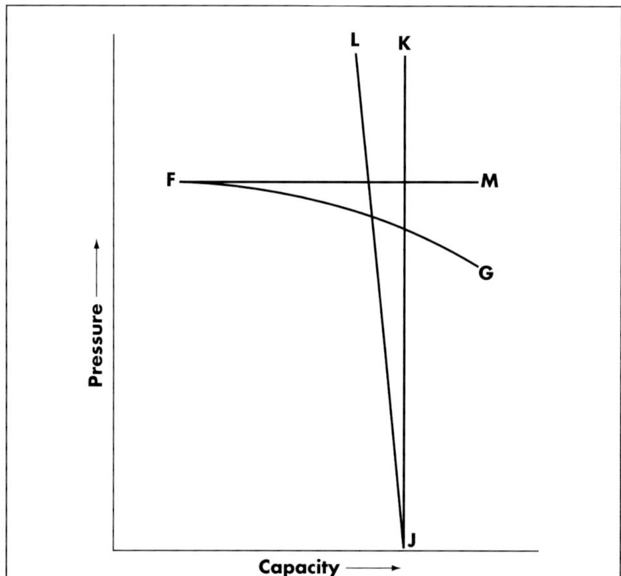

Fig. 30-2 Performance Characteristics of Centrifugal vs. Reciprocating Compressors. Source: Compressed Air and Gas Institute

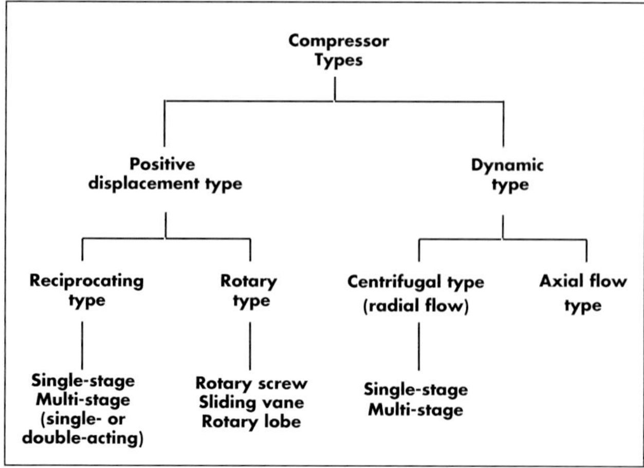

Fig. 30-3 Common Compressor Types.

action. They offer multi-stage and high-pressure capability and can range in capacity from fractional hp (or kW) to several thousand hp (or kW), although most applications are below 250 hp (185 kW). Pressures range from low vacuum at suction to more than 10,000 psi (700 bar). Reciprocating compressors are categorized as either single-acting or double-acting.

In **single-acting compressors**, air is compressed only on the upstroke of the piston. They may feature single or multiple cylinders, with single-stage or multi-stage compression. Single-stage units are typically rated at discharge pressures of 25 to 100 psig (2.7 to 7.9 bar). Two-stage units are typically rated at discharge pressures of 100 to 250 psig (7.9 to 18.3 bar), though designs for higher pressure are not uncommon.

Air-cooled units reject the heat of compression from cylinders, heads, and intercoolers to cooling air driven from the compressor fan. Liquid-cooled compressors have jacketed heads, cylinders, and intercoolers through which the heat of compression is rejected to the circulating coolant.

Figure 30-4 is an illustration of a single-acting reciprocating compressor, which reveals the characteristic automotive-type pistons, driven through connecting rods from the crankshaft, with compression taking place on the top of the pistons on each revolution of the crankshaft. Pistons typically use heat-resistant, nonmetallic guides and piston rings. Figure 30-5 is an illustration of a two-stage compressor with a liquid-cooled intercooler.

In **double-acting compressors**, air is compressed on both the upstroke and downstroke of the piston. The double-acting piston is driven by a piston rod extending through a packing gland to a crosshead, which is driv-

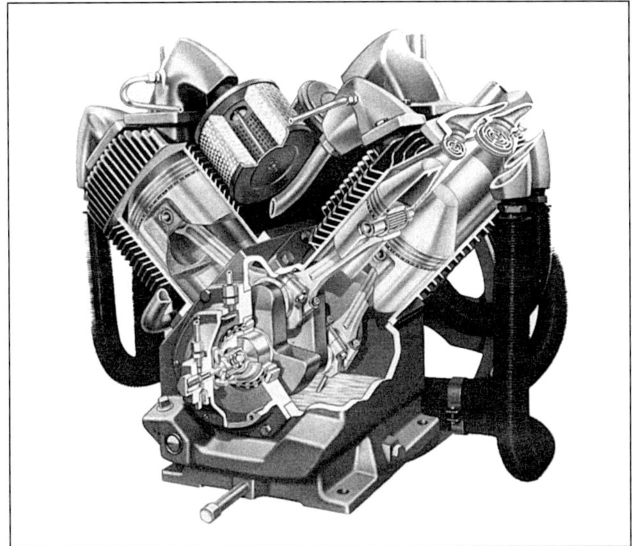

Fig. 30-4 Single-Acting, Reciprocating Compressor with Automotive-Type Skirted Pistons. Source: Compressed Air and Gas Institute

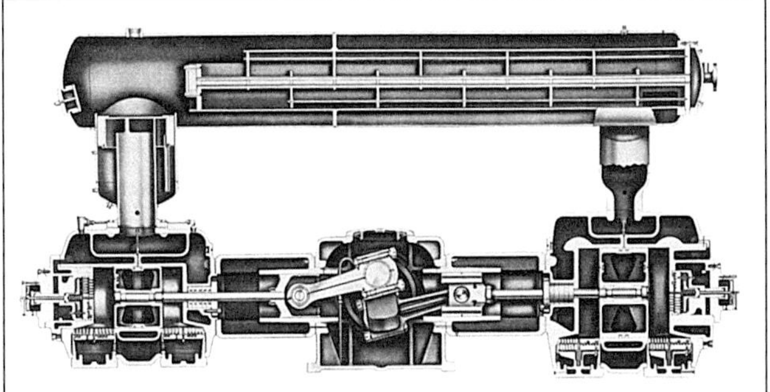

Fig. 30-5 Two-Stage Reciprocating Compressor with Liquid Cooled Intercooler. Source: Compressed Air and Gas Institute

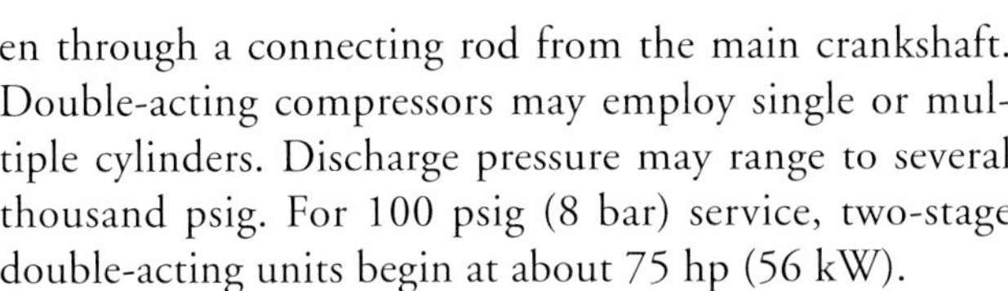

en through a connecting rod from the main crankshaft. Double-acting compressors may employ single or multiple cylinders. Discharge pressure may range to several thousand psig. For 100 psig (8 bar) service, two-stage double-acting units begin at about 75 hp (56 kW).

Typically, double-acting compressors are used for heavy-duty continuous service and employ cooling water-jacketed cylinders and heads. Figure 30-6 illustrates a double-acting compressor cylinder separated from the compressor frame by a structural member called a distance piece. Figure 30-7 shows a 125 hp (93 kW), two-stage double-acting reciprocating compressor installed in a paper mill.

Reciprocating compressors are essentially constant capacity, variable pressure machines. Capacity regulation can be achieved with one or more of the following methods:

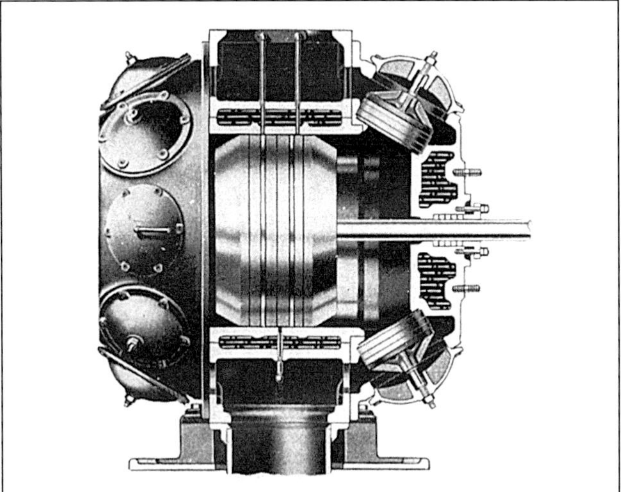

Fig. 30-6 Double-Acting Reciprocating Compressor Cylinder Separated from the Frame by a Distance Piece. Source: Compressed Air and Gas Institute

Fig. 30-7 125 hp (93 kW) Two-Stage Double-Acting Reciprocating Air Compressor. Source: Compressed Air and Gas Institute

- Automatic stop-start control by means of a pressure-actuated switch. This method is typically used when compressed air demand is light and intermittent.
- With constant-speed control, unloading can be accomplished in several different ways. Two common methods are inlet valve unloaders and clearance unloaders. Inlet valve unloaders mechanically hold the cylinder inlet valve open, thereby preventing compression. Clearance unloaders open pockets or small reservoirs, which increases the clearance volume of the cylinder, thereby reducing the volume

of air being compressed. For higher capacity motor-driven reciprocating compressors, this process can be accomplished in discrete steps varying from full load to no load. This process is referred to as step control.

- With variable-speed control, efficient unloading can be accomplished through much of the operating regime. By varying speed in response to changes in air demand, the compressor will operate at near 100% of full-load volumetric efficiency. The ability to reduce speed will be limited by the driver's capability. Once the lower speed limit of the driver is reached, one of the previously mentioned compressor control methods will be required to further reduce output.

All of these control systems utilize an air discharge pressure set point that actuates a pressure-sensor or pilot. A falling pressure indicates that air is being used faster than it is being compressed and that more air is required. A rising pressure indicates that more air is being compressed than is being used and that less air is required.

With prime mover drives, speed control is a common method of varying reciprocating compressor capacity. In these cases, the regulator actuates the fuel- or steam-admission governor valve on the driver to control the speed. Electric motor-driven compressors generally operate at constant speed, although it is possible to apply variable frequency drives (VFDs) to achieve variable speed operation. On reciprocating compressors of small and intermediate capacities, both constant-speed control and automatic start-stop control are typically used.

Reciprocating compressors typically require constant oil feed to the cylinders. Because this oil can contaminate the air stream, heavy-duty filtration is required to keep the oil out of the compressed air system. Oil-free reciprocating units, which are more costly, are used for applications requiring high-purity, non-oil contaminated compressed air.

Advantages of reciprocating compressors include their high efficiency and excellent part-load performance, as well as the large range of available sizes and pressures. Disadvantages include relatively high maintenance requirements and downtime because of their many moving parts and the vibration and stress caused by their reciprocating action. Reciprocating units are also physically larger and more expensive than alternative compressor types in larger capacities. A substantial foundation is required for their support and to dampen the dynamic loading. Pulsation is inherent in reciprocating compressors because suction and discharge valves are open during only part of the stroke.

Rotary Screw Compressors

In a rotary screw compressor, air enters the compressor and is trapped between mating male and female rotors and compressed to the required discharge pressure. In the basic single-stage design, the compressor consists of a pair of rotors meshing in a one-piece, dual-bore cylinder. Oil lubricates, seals, and cools the compressor. Oil-flooded rotary screw compressors are typically available in capacities ranging from 25 to 3,000 cubic feet per minute (cfm) or 0.7 to 85 cubic meters per minute (m^3/m) at pressures up to 600 psig (42 bar). Most units are ported for a pressure of 100 psig (7.9 bar), but are suitable for operation between 50 and 200 psig (4.5 to 14.8 bar) without large losses of efficiency. Single-, two-, and three-stage designs are available.

Figure 30-8 is an illustration of a rotary helical-screw compressor. In a single-stage unit, the air inlet is usually located at the top of the cylinder. The cylinder provides air inlet passages, oil injection points, compression zone, and discharge port. The male rotor typically has four helical lobes that are spaced 90 degrees apart. The female rotor has corresponding helical grooves, usually six, spaced 60 degrees apart. Typically, the male rotor is driven directly or indirectly by an electric motor or prime mover and the female rotor is driven by the male rotor. The thrust bearings (at the discharge end) take the rotor axial thrust and carry radial loads. The floating bearings on the opposite end allow for unequal thermal expansion of the rotor and cylinder.

Figure 30-9 illustrates the compression cycle for a single-stage helical screw-type compressor. With oil injected rotary screw compressors, air is drawn into the cavity between the main rotor lobes and secondary rotor grooves. As they continue to rotate, the rotor lobes pass

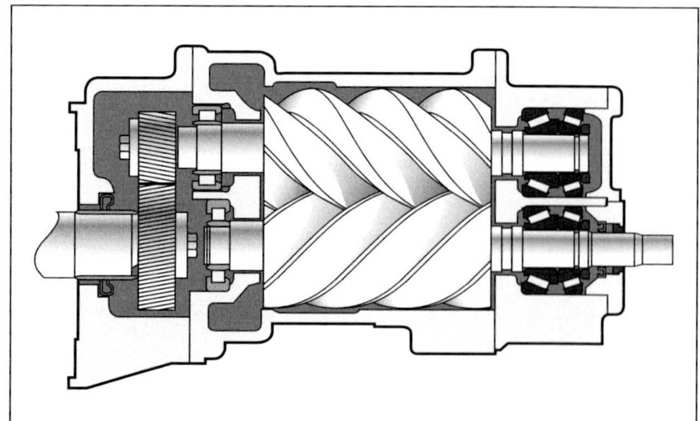

Fig. 30-8 Rotary Helical Screw Compressor Showing Thrust-Carrying Roller Bearings at One End and Floating Bearings at the Other End. Source: Compressed Air and Gas Institute

the edges of the inlet ports, trapping the air in a cell that forms between the rotor cavities and the cylinder wall. Continued rotation causes the main rotor lobe to roll into the secondary rotor groove, reducing volume and thereby raising pressure. After the cell is closed to the inlet, oil is injected to seal the clearances and remove heat. Compression ceases when the rotor lobes pass the edge of the discharge port and release the compressed air/oil mixture.

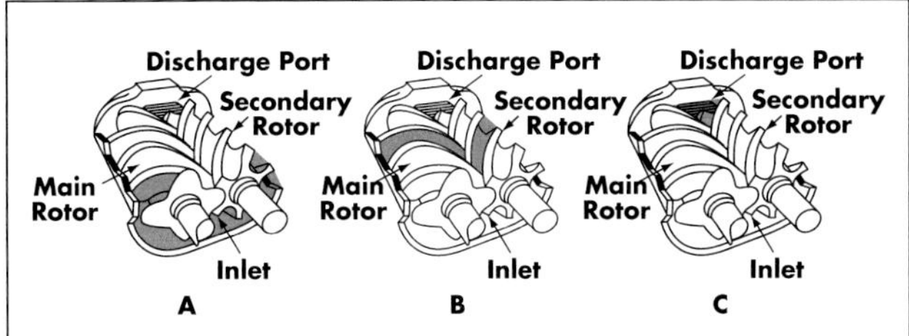

Fig. 30-9 Compression Cycle Illustration for a Single-State Helical Screw Compressor. Source: Compressed Air and Gas Institute

As the mixture passes to the oil reservoir, velocity change and impingement cause much of the oil to fall from the air. The air then passes through a separation device which removes most of the rest of the oil. Oil carryover to discharged air typically ranges from 0.002 to 0.005 ounces/cf (2.09 to 5.22 ml/m^3). An effective oil filtering system is also required to protect bearings and rotating elements.

Oil-free or dry screw compressors are also available, but are much more expensive and less efficient. In an oil-free screw compressor, the rotors have to be geared so that they do not touch. There is also no oil to seal the rotor tips or absorb the heat of compression. Oil-free screw compressors are typically two-stage units.

Capacity modulation can be affected in several ways:

- Air inlet throttling utilizes an inlet air valve that modulates in response to pressure sensing controls. A typical inlet throttled screw compressor operating at 70% capacity will require nearly 90% full-load power (22% efficiency loss).
- On-line/off-line control cycles the inlet air valve between the fully-open and fully-closed position. This method is more efficient than air inlet throttling, but results in pressure swings and usually requires a discharge air receiver. It can also produce increased axial stress on bearings.
- Geometry control uses a turn- or slide-valve to change the geometry within the compression chamber, changing the effective length of the rotors.
- Speed control can be used to efficiently reduce airflow capacity. Like all positive displacement machines, screw compressors continue to operate near full-load volumetric efficiency when speed is reduced. Input power requirement is reduced roughly in proportion to airflow. Since the ability to reduce speed is limited by the driver's capability, it is usually necessary to rely on an additional control method for operation under low-load conditions.

Rotary screw compressors now dominate the mid-sized compressed air market. They are characterized by low vibration and simple foundation requirements, broad pressure and capacity ranges, low maintenance, and long service life. A critical advantage of the rotary screw compressor over the reciprocating compressor is reduced maintenance requirements. While reciprocating units will typically require minor overhaul (e.g., valves) every 8,000 hours, screw compressor intervals may be 20,000 to 40,000 hours. Screw compressors are also smaller, quieter, and less expensive than reciprocating units in most mid-range and larger capacities and use less oil.

Screw compressors are usually less efficient than reciprocating units of comparable capacity, particularly under part-load conditions. In many applications, compressors operate at less than full load all or most of the time. As noted above, operating cost can exceed capital investment cost by several times during the life of the compressor, and the inability of constant-speed screw compressors to unload as efficiently as reciprocating units can thus be a significant disadvantage in many applications.

Centrifugal Compressors

The compression element of a centrifugal compressor is the impeller or wheel. As the impeller rotates, the air caught between the blades is forced to move outward at increasing speed, away from the eye of the impeller. As the air exits the impeller, it enters the diffuser and scroll volute, where the velocity of the air (kinetic energy) is converted to pressure (static energy). Smooth, continuous flow is established as the motion of the air away from the eye causes a low-pressure area that draws more air into the impeller. Figures 30-10a and 30-10b show the impeller, diffuser, and scroll volute of a centrifugal compressor.

Centrifugal compressors range in capacity from about 125 hp (93 kW) to more than 10,000 hp (7,500 kW). Figure 30-11 shows a three-stage centrifugal compressor rated at 7,250 cfm (205 m^3/m). This unit operates at a

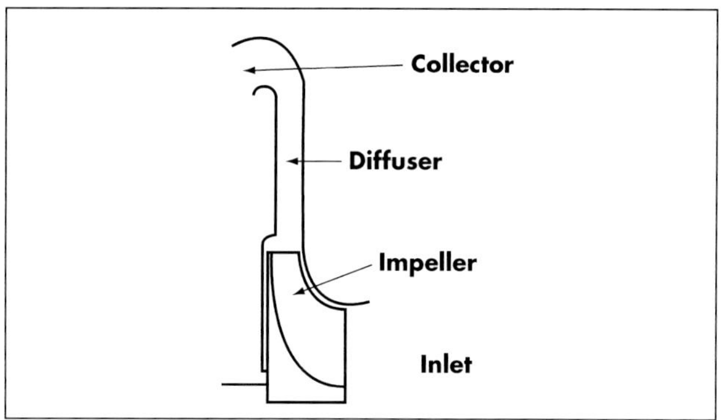

Fig. 30-10a Impeller, Diffuser, and Scroll Volute of Centrifugal Compressor. Source: The Elliot Company

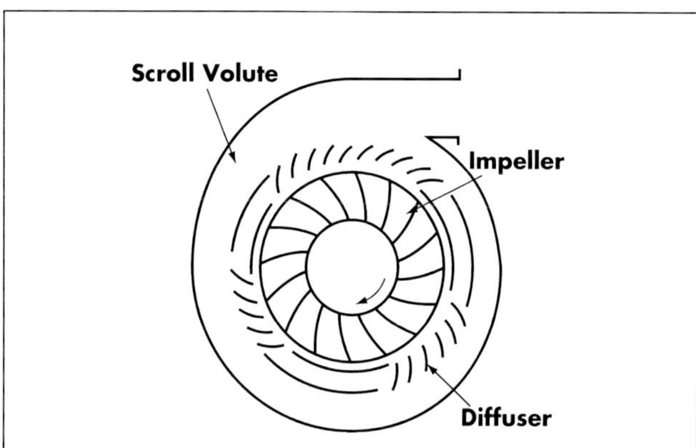

Fig. 30-10b Impeller, Diffuser, and Scroll Volute of Centrifugal Compressor. Source: The Elliot Company

discharge pressure of 124 psia (8.6 bar) and requires a 1,700 hp (1,270 kW) driver. In a multi-stage compressor, as shown in Figures 30-12 and 30-13, air is ducted from the scroll volute through interstage piping to the first intercooler, then to the second stage impeller, again through a diffuser and scroll volute to the second intercooler. Air from the second intercooler moves through a third impeller, diffuser, and volute where it reaches the final discharge pressure.

Centrifugal compressor capacity, or mass flow rate, is a function of volume flow and air density. Because air density is inversely proportional to absolute temperature, mass flow is reduced at higher temperatures. Impellers must be selected to deliver the required flow at the highest anticipated operating temperature. Altitude and the humidity of inlet air are also important variables. As discussed later in the chapter, moisture is condensed out of compressed air in intercoolers and aftercoolers. When inlet air is very humid, more moisture is condensed and mass flow rate is reduced.

Figure 30-14 is a representative centrifugal compres-

Fig. 30-11 Centrifugal Compressor. Source: Compressed Air and Gas Institute

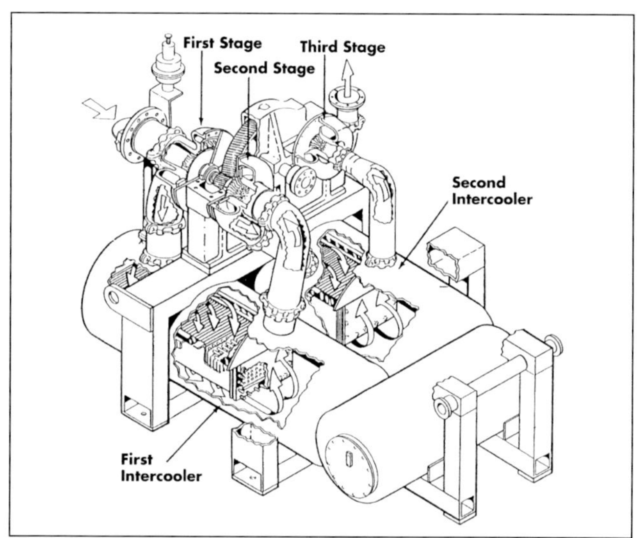

Fig. 30-12 Cutaway Drawing of Multi-stage Centrifugal Compressor. Source: The Elliot Company

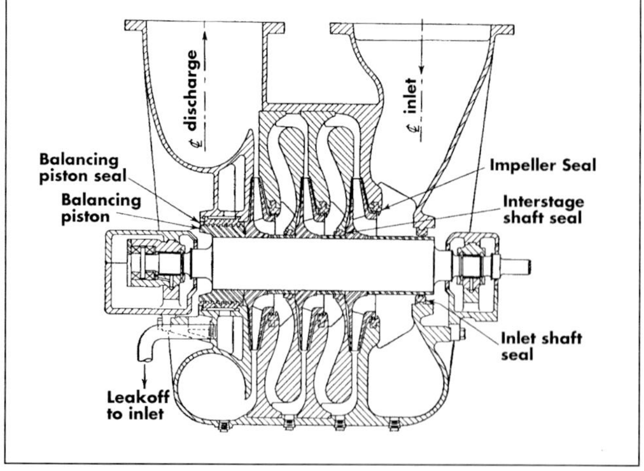

Fig. 30-13 Vertical Section Drawing Showing Typical Multi-stage Centrifugal Compressor. Source: Compressed Air and Gas Institute

sor performance curve in discharge pressure versus flow. Air delivery rises with decreasing pressure until the air velocity in the compressor reaches the speed of sound. At this point, flow is said to be choked, or stonewalled, because any further reduction in system pressure will not result in increased air delivery. This is shown on the lower right region of the curve. Figure 30-15 shows the effect of inlet air temperature on centrifugal compressor performance. Notice that mass flow and power requirement rise with decreasing temperature.

All dynamic compressors have a surge limit or minimum flow point below which the performance of the compressor is unstable. Operation below this point results in pulsations in pressure flow, which could become severe enough to cause damage. The intersection of the surge line and the sloping performance curve represents the maximum discharge pressure of the compressor. As system pressure increases, the compressor delivers less air until the system resistance is matched. This continues until the compressor is unable to maintain a steady flow of air and backflow, or surge occurs from the system through the compressor. The surge limit in a given application is a function of the compressor type, pressure ratio, inlet temperature, gas properties, blade angle, and operating speed.

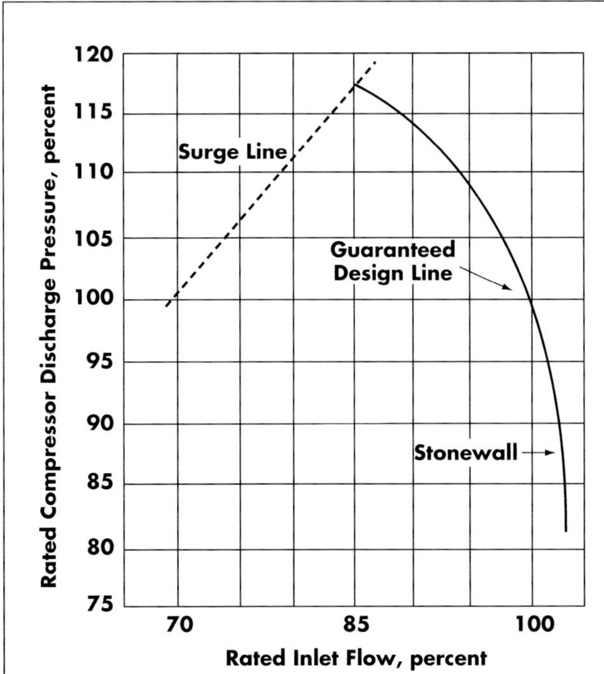

Fig. 30-14 Representative Centrifugal Compressor Performance Curve. Source: Compressed Air and Gas Institute

Since centrifugal compressors are dynamic machines, flow is much more sensitive than with positive displacement machines. Therefore, more sophisticated control logic is required. Capacity modulation is achieved via the following:

- Inlet throttling is often used, particularly in smaller capacities.
- Adjustable guide vanes allow for precise control of air inlet flow and, therefore, reduce the risk of surge. Adjustable vanes also reduce power requirement under partial loads.
- Speed control, achieved with a variable speed prime mover or electric driver, also provides some degree of output capacity control. Unlike the positive displacement compressors, however, centrifugal compressors achieve a more limited benefit through variable speed operation due to the minimum flow rate that is required to prevent surge. Figure 30-16 shows typical performance under variable speed control.

A major advantage of centrifugal compressors is that they provide oil-free air and use oil only in the gearbox. They offer simplicity of design, few moving parts, large clearances, and minimal vibration. This results in high reliability and low maintenance requirements. Their efficiency increases with size and surpasses even reciprocating compressors at full-load operation in the larger capacities. However, like screw compressors, centrifugal compressors exhibit poor part-load performance and are best suited for baseload duty. In smaller capacities, they are less efficient and more costly than screw compressors and offer lower pressure rise per stage.

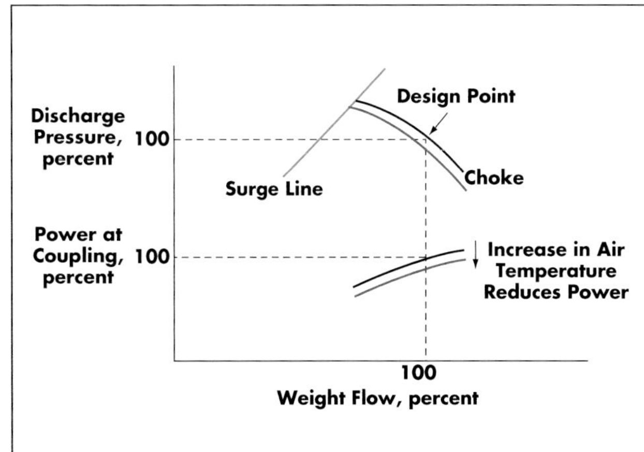

Fig. 30-15 Effect of Inlet Air Temperature on Centrifugal Compressor Performance. Source: Compressed Air and Gas Institute

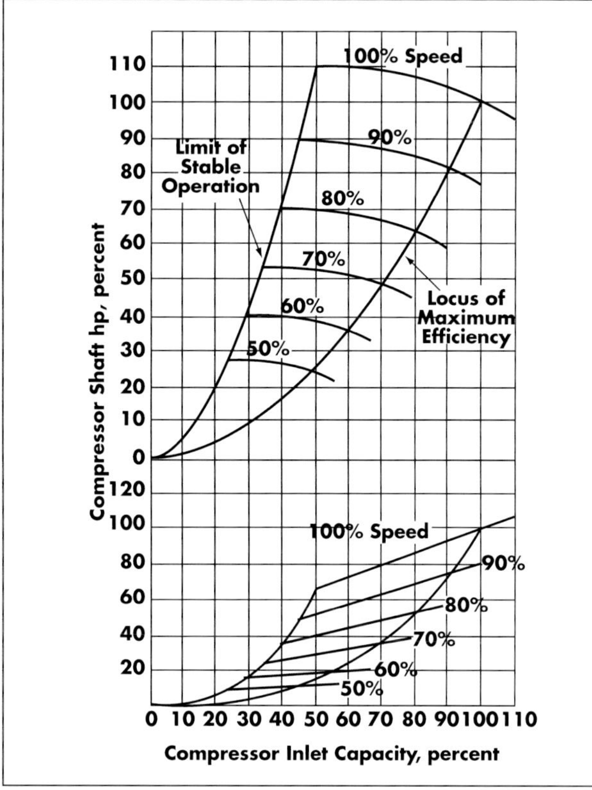

Fig. 30-16 Typical Performance Curves for Centrifugal Compressors. Source: Compressed Air and Gas Institute

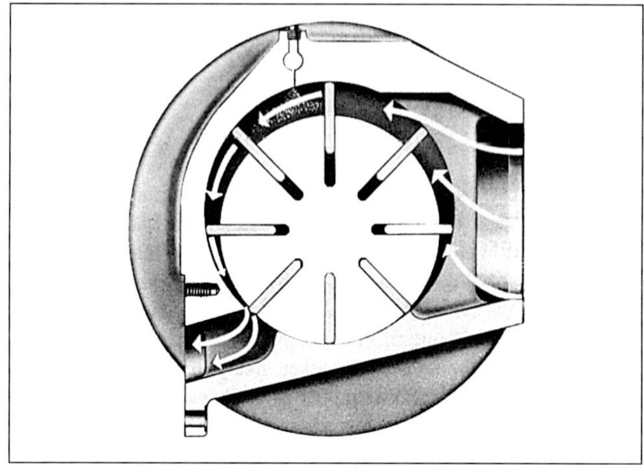

Fig. 30-17 Illustration of Oil-Flooded Sliding Vane Rotary Compressor. Source: Compressed Air and Gas Institute

Fig. 30-18 Axial Flow Compressor. Source: Compressed Air and Gas Institute

OTHER COMPRESSOR TYPES

Sliding vane rotary compressors include a radially slotted rotor that is offset in its housing. Sliding vanes held in the slots trap air as the rotor turns and provides compression as the volume between the vanes changes due to the rotor offset in the housing. Figure 30-17 is an illustration of an oil-flooded sliding vane rotary compressor that demonstrates the compression cycle. Vane compressors offer moderate pressure capability. Over the past few decades, the use of sliding vane compressors has dropped substantially, to the point where they are now rather uncommon.

Oil-free rotary-lobe compressors feature rotating lobes that have an intermeshing profile. Each stage consists of two rotors held within a figure eight-shaped cylinder. As the two rotors intermesh, compression takes place around the perimeter of the rotor as opposed to along the axis.

In **axial compressors**, a series of axial blades draws in, compresses, and discharges air, continuing along the axis of the compressor. Stationary axial compressors are the type used in very large combustion gas turbines. Axial compressors typically start in capacity at several thousand hp (or kW) and are usually larger than 10,000 hp (7,500 kW). Figure 30-18 shows an axial flow compressor used for blast furnace duty in a steel mill. Figure 30-19 shows typical performance curves for axial compressors, including variable speed control.

COMPRESSOR AUXILIARIES

Important compressor plant auxiliaries include air receivers, inlet air filters, intercoolers, aftercoolers, and air dryers designed to remove unwanted contaminant's and moisture from the compressed air stream.

An **air receiver**, typically used to reduce the impact of pulsations from reciprocating compressors, can also be used to provide storage. Storage capacity serves many functions. For example, it can limit distribution pressure swings and reduce problems associated with frequent compressor cycling or capacity modulation. Adequate storage allows compressor plants to respond to high demand occurring over very short periods. Storage can

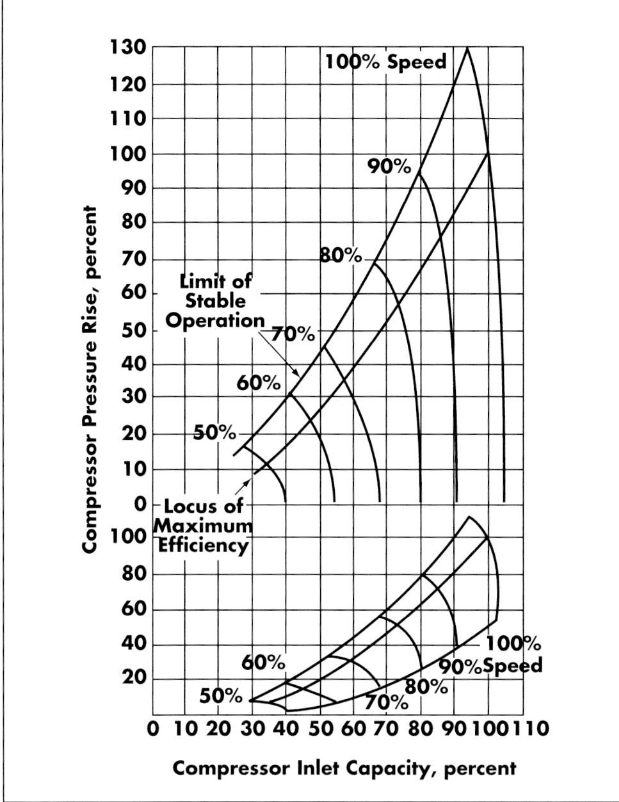

Fig. 30-19 Typical Performance Curves for Axial Compressors. Source: Compressed Air and Gas Institute

allow on-line compressors to run closer to full load, improving energy efficiency. This is particularly beneficial with constant speed-operated screw and centrifugal compressors.

The time interval during which a receiver can supply air without excessive drop in pressure is:

$$t = \frac{V(p_1 - p_2)}{qp_0} \tag{30-1}$$

Where:

t = Time in minutes
p_1 = Initial receiver pressure
p_2 = Final receiver pressure
p_0 = Atmospheric pressure
q = Air requirement (flow rate of free air per minute)
V = Receiver capacity

In most compressed air applications, the presence of oil or particles of grit or scale cannot be tolerated. **Inlet air filters** help prevent dirt or foreign matter from entering the compressor and possibly causing damage. Three basic types of filters are used:

- Viscous-impingement or oil-wetted filters include layers of wire mesh, screen, or fibrous filter pads contained in a canister or panel. Debris in the air drawn through the filter medium is retained on the oil-wetted surface. These are relatively low-cost, inefficient filters that are not recommended for non-lubricated compressors or for duty where heavy atmospheric contamination exists.
- Oil bath filters first draw the unfiltered air through an oil reservoir; this is followed by a screen mesh or other medium that scrubs out the oil and dirt particles on the oiled element surfaces. These are more efficient than viscous-impingement types, but also have a higher pressure drop. Due to the oil entrainment in the filtered air, they are also not suitable for non-lubricated compressors.
- Dry-type filters, consisting of densely spaced material that block particle penetration, are usually the most efficient filter option and the best option for non-lubricated compressors. The finer the filter element, the greater the efficiency and the higher the pressure drop. Routine inspection is required to ensure that the filter does not become clogged, possibly allowing a rupture that releases retained debris into the air system. Some elements may be cleaned by blowing them out or scrubbing them. Other less expensive paper-treated elements are disposable types.

The oil bath and dry type are available as filter silencers that also provide sound suppressing characteristics. With viscous-impingement filters, an additional silencing device, such as a pulsation damper, is usually used.

Air always contains some moisture in a vapor state. The maximum amount of moisture that a given volume of air can hold is dependent upon temperature and pressure. As temperature increases, the air is able to hold more moisture, and as pressure increases, the air is able to hold less moisture. For example, in English units, air at 70°F, 50% RH, holds about 4 grains/cf of moisture (1 lbm of water is equal to 7,000 grains). At 100% RH, the air holds about 8 grains/cf. At that point, the air is said to be saturated, i.e., it cannot hold any more moisture. If the absolute pressure is doubled (at constant temperature), the volume is reduced and the air will retain 4 grains/cf in the form of vapor and drop out 4 grains/cf in the form of liquid.

The dewpoint is the temperature at which condensate will begin to form if the air is cooled at constant pressure. Pressurized warm air typically leaves the compressor under saturated conditions. As the air cools in the distribution system, water will begin to condense. Condensed water vapor can have a number of deleterious effects in compressed air systems: causing corrosion, promoting wear, washing out lubricants from pneumatic devices, causing control instruments to malfunction, and freezing in exposed lines during cold weather.

Aftercoolers are air- or water-cooled tube bundles designed to cool the compressed air exiting the compressor. Separators then remove the condensed moisture from the air through mechanical or cyclonic action. The droplets are discharged from the system by a **moisture trap**. There are many types of traps available, such as float, inverted bucket, and solenoid valve with timer. Figure 30-20 is a representative illustration of a compressor arrangement showing the piping to a receiver with an aftercooler.

Air dryers are used to further reduce moisture. Representative flow diagrams for the following types of dryer are shown in Figure 30-21.

- Refrigerated air dryers chill the air to remove moisture by use of refrigerant-to-air heat exchangers. The air is then often reheated to near room temperature. These are relatively low cost, low maintenance units that provide constant dewpoint. The limitation is that very low dewpoints (below 32°F, 0°C) are not achievable, to avoid restriction in the dryer due to ice buildup.
- Deliquescent air dryers use a hygroscopic desiccant material having a high affinity for water. Systems consist of a large vessel filled with a desiccant such as lithium chloride or calcium chloride. The desiccant removes and dissolves water vapor from the compressed air. A primary advantage of these systems is their low first cost. Disadvantages are limited suppression of dewpoint, the need to replace the desiccant a few times per year, the need for a downstream filtering system, and the tendency to cake and channel, thus reducing the effectiveness of the units.
- Regenerative desiccant air dryers consist of two or more towers. Dry desiccant absorbs moisture from the compressed air flowing through one of the towers, while the other tower is being regenerated (dried) using heat or purge air. After the desiccant in the operating tower becomes saturated, the compressed air is diverted to the second tower and the first one is regenerated. Advantages of regenerative desiccant systems are the ability to achieve extremely low dewpoints, moderate operating costs, and no requirement to drain water. Heat-driven systems may be direct-fired with natural gas, or use steam, hot air, or

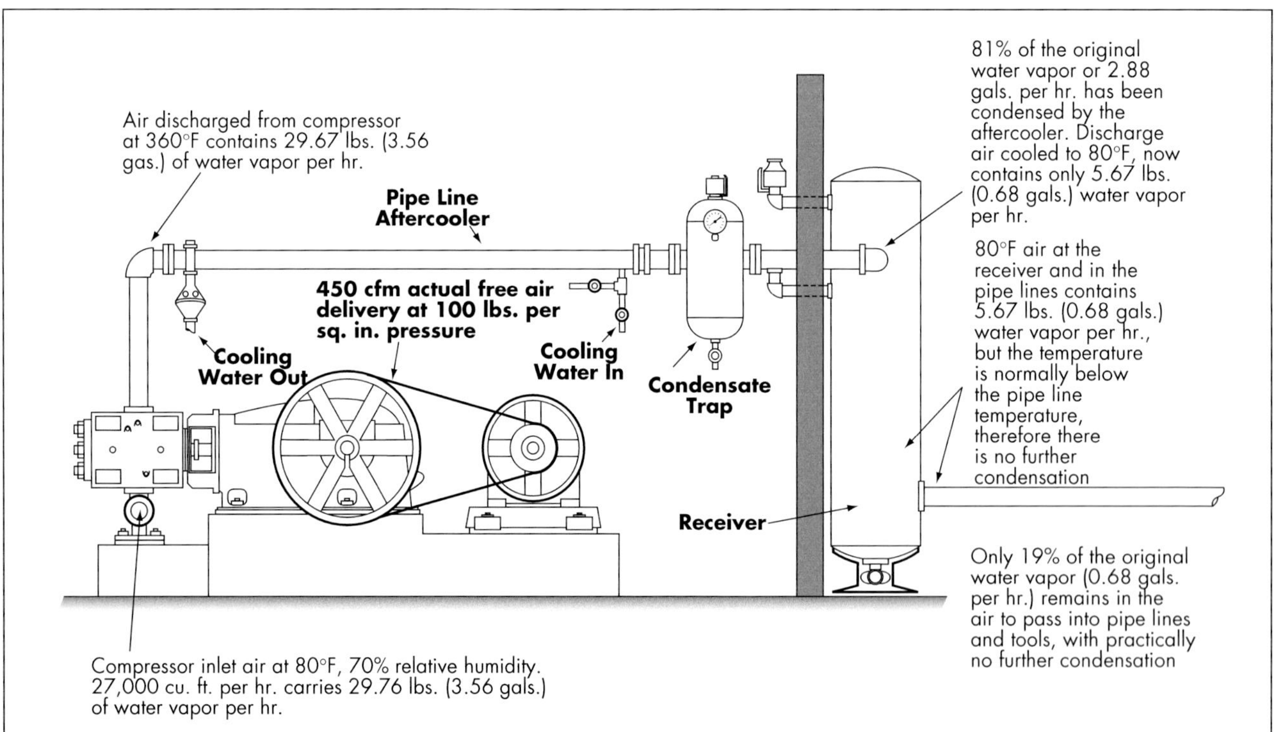

Fig. 30-20 Representative Compressor Arrangement Showing Piping to Receiver with Aftercooler. Source: Compressed Air and Gas Institute

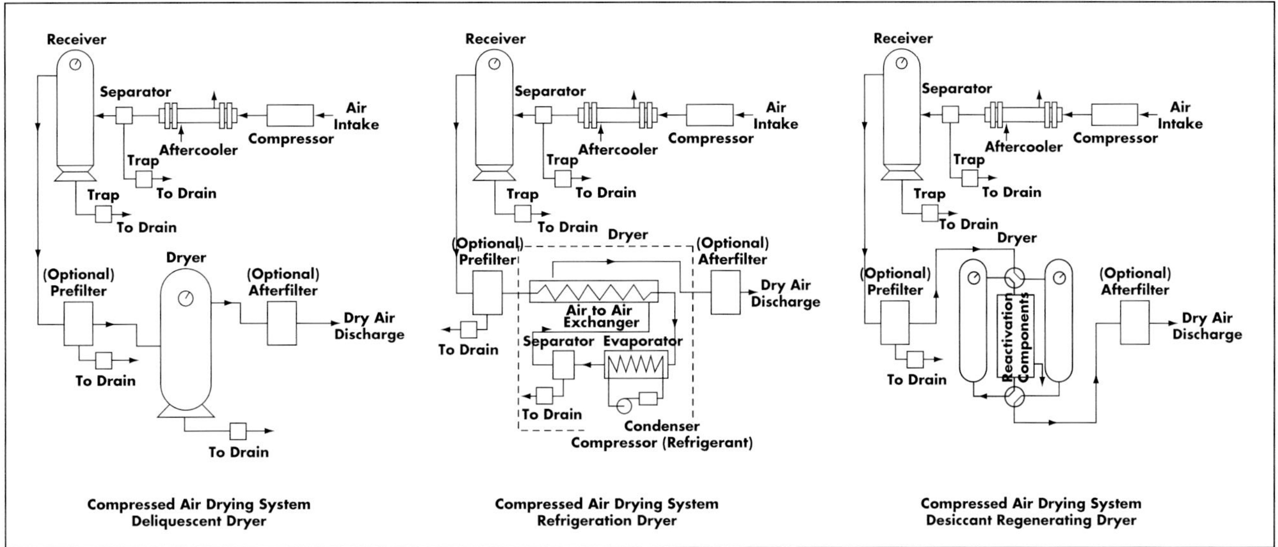

Fig. 30-21 Typical Flow Diagrams for the Three Basic Types of Air Drying Systems. Source: Compressed Air and Gas Institute

hot water. Recovered heat from prime mover cycles or other applications can be effectively utilized in these systems. Disadvantages include high first cost and the potential for oil aerosols to coat the desiccant materials, rendering them ineffective. Desiccant drying technology is discussed in detail in Chapter 39.

Multi-staging, or the connection in series of two or more identical compressors, is often used to limit the pressure rise per stage based on the limitations of the machine type, i.e., axial thrust load in centrifugal compressors, piston rod stress in reciprocating compressors, and rotor deflection and thrust in rotary compressors. **Intercooling** is the removal of heat from the air between stages to maintain safe discharge temperatures and improve overall compression efficiency. Intercoolers are typically water-cooled tube bundles with an external means of heat rejection (e.g., cooling tower or once-through city water cooling). Ideal intercooling exists when the temperature of the air leaving the intercooler equals the temperature of the air at the intake of the first stage. In practice, the benefits of intercooling are also partially offset by the pressure drop across the cooler and by the mechanical losses consumed in driving the stages.

Air Compressor Operation

There are several commonly used standards in system engineering to define air. One definition for a standard cubic foot (scf) of air is the quantity of dry air needed to fill a volume of 1 cf (0.03 m^3) at 14.7 psia (101 kPa) pressure and 60°F (15.55°C) temperature. Another definition is based on air at 36% relative humidity (RH) filling the same volume at the same pressure, but at 68°F (20°C). In both cases, the air density has the value of 0.075 lbm/cf (1.2 kg/m^3).

Air services with atmospheric inlet should be specified for 100% RH. The water content must be added to the net dry air requirement of the process. Note that saturated air at 90°F (32.2°C) contains about 3% water vapor by weight.

In the United States, the Compressed Air and Gas Institute (CAGI) has selected as standard conditions 1 bar (14.5 psia or 100 kPa), 20°C (68°F), and 0% RH. The most commonly used units of flow are cf per minute (cfm or ft^3/m) or per hour (cfh or ft^3/h) and m^3 per minute (m^3/m) or per hour (m^3/h). Power is normally expressed as hp or kW. Work of compression may be expressed positively as work output or negatively as work input.

Inlet pressure is generally defined as the absolute total pressure existing at the intake flange of a compressor. Inlet temperature is the initial temperature at the intake flange. Discharge pressure is generally defined as the absolute total pressure at the compressor's discharge flange and is commonly referenced in psig. Discharge temperature is the total temperature at the discharge flange of the compressor.

Typically, compressors are analyzed using ideal gas law with an assumed constant specific heat. A compressibility factor (Z) is used for real gas deviations.

The isentropic work of compression for a real gas is expressed positively as:

$$W = p_1 V_1 \frac{k}{k-1} \left[r^{(k-1)/k} - 1 \right] \left(\frac{Z_1 + Z_2}{2Z_1} \right) \qquad (30\text{-}2)$$

Where:
W = Work done
p = Pressure at inlet
V = Volume at inlet
k = Ratio of specific heats
r = Compression ratio (p_2/p_1)
Z = Compressibility factor = $\left(\frac{PV}{RT}\right)$

and subscripts:

1 = Inlet conditions
2 = Discharge conditions

Compressibility factors become increasingly more critical for analysis of hydrocarbon gases and refrigerants (which generally deviate significantly from the ideal-gas laws) and for non-hydrocarbon gas at very high pressures (i.e., greater than 1,000 psig or 70 bar).

Power (P) requirement is a function of the work done and the adiabatic efficiency of the compressor (η_c). It can, therefore, be expressed as:

$$P = \frac{W}{\eta_c} \qquad (30\text{-}3)$$

or as:

$$P = \frac{p_1 V_1 k}{(k-1)} \left[r^{(k-1)/k} - 1 \right] \frac{1}{\eta_c} \left(\frac{Z_1 + Z_2}{2Z_1} \right) \qquad (30\text{-}4)$$

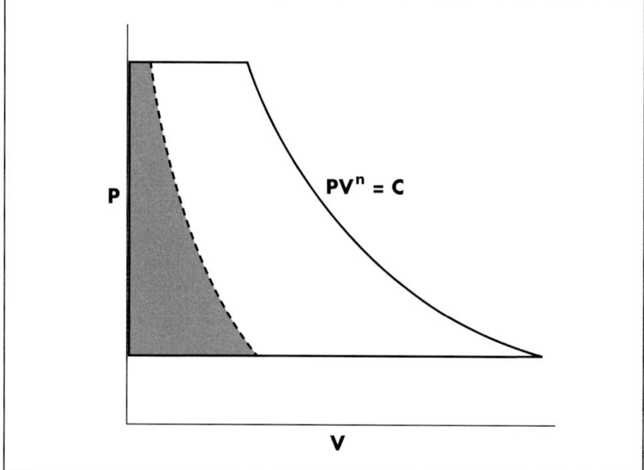

Fig. 30-22 Theoretical Pressure-Volume Diagram of Helical Screw-Type Compressor. Source: Compressed Air and Gas Institute

When power is expressed in hp, pressure in psia, and capacity in cfm, power can be expressed, based on Equation 30-4, as:

$$hp = \frac{144\, p_1 V_1 k}{33{,}000\,(k-1)} \left[r^{(k-1)/k} - 1 \right] \frac{1}{\eta_c} \left(\frac{Z_1 + Z_2}{2Z_1} \right)$$

With the volume capacity (V) expressed in m^3/s and the pressure (p) in Pascal, Equation 30-4 yields actual power in watts.

For a positive displacement compressor, at any given rate of flow, the ideal power is affected by the inlet pressure (P_1), the ratio of specific heats (k), and the compression ratio (r). The actual power requirement is increased as a result of losses through the intake and discharge valves (or ports). Additional losses result from turbulence in and leakage from the compression chamber and preheating of the inlet gas.

Figure 30-22 is a theoretical pressure-volume diagram of the helical screw-type compressor. The compression process approaches isothermal compression because of the cooling effect of the oil. The power and capacity loss can be represented by the shaded area, although leaking also occurs along the compression line.

Isentropic equations are often used so that isentropic work can be determined from thermodynamic reference charts or tables. However, isentropic, or reversible adiabatic, compression is ideal compression in which there are no internal energy losses due to friction, windage, or throttling and no heat transfer takes place. When internal compressor losses due to friction, windage, and throttling are considered, the process is known as irreversible adiabatic. This represents the actual work required to compress the gas. When thermal losses through the compressor casing are accounted for, the process is known as polytropic. Polytropic compression, therefore, represents the actual compression process.

For ideal gases with constant specific heats, the relationship of isentropic efficiency to polytropic efficiency can be expressed as:

$$\eta_c = \frac{(r^{(k-1)/k} - 1)}{(r^{(k-1)/(k\eta_p)} - 1)} \qquad (30\text{-}5)$$

Where:
η_c = Adiabatic efficiency
η_p = Polytropic efficiency

This relationship applies with reasonable accuracy

to real gases that do not deviate greatly from ideal gases because the real-gas errors cancel.

Adiabatic equations can be applied to a polytropic process by multiplying $\frac{k-1}{k}$ by $\frac{1}{\eta_p}$

This may be expressed as $\frac{(n-1)}{n}$, where n represents the polytropic coefficient, which can be determined if inlet and discharge temperature and pressures are known.

For positive displacement compressors in which velocities, turbulence, and slip are relatively low, power requirements can be reasonably approximated on an isentropic basis. However, dynamic compressors are more typically evaluated on a polytropic basis. Power, often referred to as gas power, on a volumetric flow basis is expressed as:

$$P = Q_1 p_1 \left[r^{(n-1)/n} - 1 \right] \left(\frac{n}{\eta_p (n-1)} \right) \left(\frac{Z_1 + Z_2}{2Z_1} \right) \quad (30\text{-}6)$$

Where:

Q_1 = Volumetric flow at inlet conditions

When power (P) is expressed in hp, volumetric flow (Q_1) in cfm, and inlet pressure (p) in psia, Equation 30-6 becomes:

$$hp = Q_1 p_1 \left[r^{(n-1)/n} - 1 \right] \left(\frac{n}{229\, \eta_p (n-1)} \right) \left(\frac{Z_1 + Z_2}{2Z_1} \right)$$

With the volumetric flow (Q_1) is expressed in m^3/s and pressure (p) in Pascal, Equation 30-6 yields gas power in watts.

As noted above, the calculated power is not the true power input to the compressor. A series of additional losses, such as bearing and seal losses, must be added into the gas power to determine the shaft power requirement. Driver efficiency losses, as discussed below, must also be considered in determining the total system energy input requirement.

Compressor Driver Options

Drivers produce the shaft rotational power required to operate the air compressor. Following is a discussion of the commonly used compressor drivers.

Electric induction motors are by far the most common drivers selected for stationary air compressor applications. Motors are typically the lowest first-cost driver option and offer simplicity of operation and minimal maintenance requirements. Motor-driven air compressors are abundantly available in all required capacity ranges, often as part of pre-engineered packaged systems. Recently, variable frequency drives (VFDs) have begun to be applied to some larger systems to improve part-load efficiency. Synchronous motors are also occasionally used in larger applications.

Reciprocating engines have a long history of driving air compressors. Screw compressors are somewhat commonly used with reciprocating engine drives due to their simplicity and durability under torsional stress. The typical control sequence for a reciprocating engine-driven rotary screw compressor is first to reduce speed, and then to use standard compressor part-load control mechanisms. This is beneficial because both the engine and the compressor will operate more efficiently under most partial- load conditions.

Pre-engineered, packaged, reciprocating engine-driven modules are now available from several leading manufacturers as a response to the increasingly high operating costs of electric units. The new modules use high-quality, heavy-duty industrial engines, which are compact and offer good predictability of operation. They are designed using components from standard industrial product lines to simplify maintenance requirements and limit down-time.

Off-the-shelf engine-driven screw compressors range in capacity between 50 hp (37 kW) and 500 hp (375 kW). Several manufacturers offer to package their units with reciprocating engines in capacities up to 1,000 hp (750 kW) or above. A typical packaged system will operate between 1,800 and 1,200 rpm in capacities above 200 hp (150 kW) and between 2,400 and 1,400 rpm in capacities below 200 hp (150 kW). Minimal constant efficiency idle speed will typically be reached at the 50 to 65% capacity range.

Reciprocating engines are sometimes packaged with reciprocating compressors, depending on required outlet pressures. This includes the under 50 psig (4.5 bar) market, as well as the over 200 psig (14.8 bar) market. Packaging of compressors with piston engines requires delicate torsional balancing, making conversions of existing reciprocating compressors more difficult. Because centrifugal compressors operate at very high speed and have sleeve bearings, they are more sensitive to torsional vibrations than screw compressors. For this reason and due to the need for gearing, reciprocating engines are not as commonly used with centrifugal compressors.

Useful system design options are available with current packaged units include:

- Heat recovery potential that may exceed 65% (LHV) by combining compressor heat rejection with engine heat rejection.
- An integral regenerative desiccant air dryer can be powered by engine exhaust.
- Compressors using half of the cylinders for power and half for air compression are available for applications ranging from about 20 to 375 hp (15 to 280 kW). The fuel consumption rate of a relatively small 130 cfm (3.7 m³/m) air compressor (e.g., 1,800 rpm with four power cylinders and a 36.75 hp rating) is about 9,500 Btu/hp-h (7,500 kJ/kWh) on an HHV basis.

Figures 30-23 and 30-24 show different packaged air compressors with reciprocating engine drivers. Figure 30-23 shows a 5,500 cfm (156 m³/m) Ingersol-Rand centrifugal air compressor designed for 100 psig (7.9 bar) service. The unit is driven by a 1,500 hp (1,120 kW) Caterpillar gas-fired reciprocating engine. Figure 30-24 shows a 585 cfm (17 m³/m) packaged air-cooled screw compressor system. This unit requires 133 hp (99 kW) for full-load operation at 100 psig (7.9 bar), which is provided by a Hercules engine at a fuel consumption rate of 8,200 Btu/hp-h (6,450 kJ/kWh) on an HHV basis.

Steam turbine drives are used for large screw and centrifugal compressors because of their mechanical compatibility, reliability, and precise control. They are more cost-effective in larger capacities. Some manufacturers offer packaged systems, although steam turbines are sometimes retrofitted to replace existing electric motor drives. A hybrid application can include a dual clutch on the same shaft as an electric motor. The benefits of turbine variable speed control are not as pronounced as with reciprocating engines, because steam turbine mechanical efficiency is reduced at low speed. Microprocessor controls can be used to optimize performance.

Back-pressure turbine-driven compressors have the potential to achieve extremely low operating costs — often as low as $0.010/hp-h ($0.014/kWh). This is dependent, of course, on the availability of a concurrent low-pressure steam requirement, or heat sink. Condensing turbines can be cost-effective under certain conditions, including

Fig. 30-23 Reciprocating Engine-Driven Centrifugal Air Compressor. Source: Ingersol Rand and Caterpillar Engine Div.

Fig. 30-24 585 cfm Packaged, Air-Cooled Reciprocating Engine-Driven Screw Compressor. Source: Dearing Compressor and Pump Co.

availability of low-cost fuel or heat-recovery generated steam, high cost electric rates, or avoidance of costly electric service upgrades.

Figure 30-25 shows a mixed (hybrid) system featuring two steam turbine-driven centrifugal air compressors and one electric motor-driven unit. Each unit is equipped with external coolers shown in the foreground. Each air compressor system requires 800 hp (596 kW) to produce 3,500 cfm (99 m^3/m) at 125 psig (9.6 bar). The steam turbine drivers are single-stage back-pressure machines that provide low-pressure steam to serve other processes at the facility.

Gas turbine compressor drives are generally only used for large industrial operations and remote field applications of at least 5,000 cfm (142 m^3/m), and more commonly at least 20,000 cfm (566 m^3/m). These applications involve centrifugal and axial air compressors and have driver requirements that start at several thousand hp (or kW). They may be specified where there is a large, constant high-pressure steam requirement, where electric capacity is limited or very costly, or when independent operating reliability is essential. They are very common for pipeline gas compressors, since they have no cooling or water requirements.

Matching Supply to Demand

Basic compressor selection factors normally include capacity, operating pressure, first cost, energy efficiency and cost, air purity requirements, installation location,

Fig. 30-25 Mixed (Hybrid) Compressed Air System Featuring Steam Turbine- and Electric Motor-Driven Centrifugal Compressors. Source: Centrifugal Compressor Division, Ingersoll-Rand Company

water availability and cost, noise constraints, and maintenance. Figure 30-26 shows the overlapping ranges of pressure and capacity typical of different types of compressors. Table 30-1 lists typical air compressor types and designs used in various capacity ranges for 100 psig (7.9 bar) air service. Under higher or lower operating pressures, selection criteria for each capacity range may differ.

Ideally, the supply side or compressed air capacity should be selected and configured after the demand side is controlled and optimized. The primary components of compressed air demand are real production requirements, artificial demand created by supply pressure that is greater than that required at the point of use, and air leaks. The lowest operating cost occurs when demand is minimized and precisely matched to supply. The demand side should be carefully audited, corrected, and optimized by minimizing leaks, providing manual controls, reducing distribution pressure, and making piping modifications to reduce pressure drop.

A primary reason for excessive operating costs is that air compressors are typically sized to meet peak-load requirements. A unit or system that meets the peak-load demand with optimal efficiency does not necessarily efficiently meet the more common part-load demand. It is often convenient to break down the load into two components: **baseload** is the portion of the load that is constant and **trim-load** is the additional varying component. In multiple-compressor systems, baseload may be served by units best suited for continuous operation at or near full load. Cycling trim-load compressors and storage can be used to efficiently meet temporary demands. Trim-load units should, therefore, offer superior part-load performance, precise supply air modulation, and rapid response.

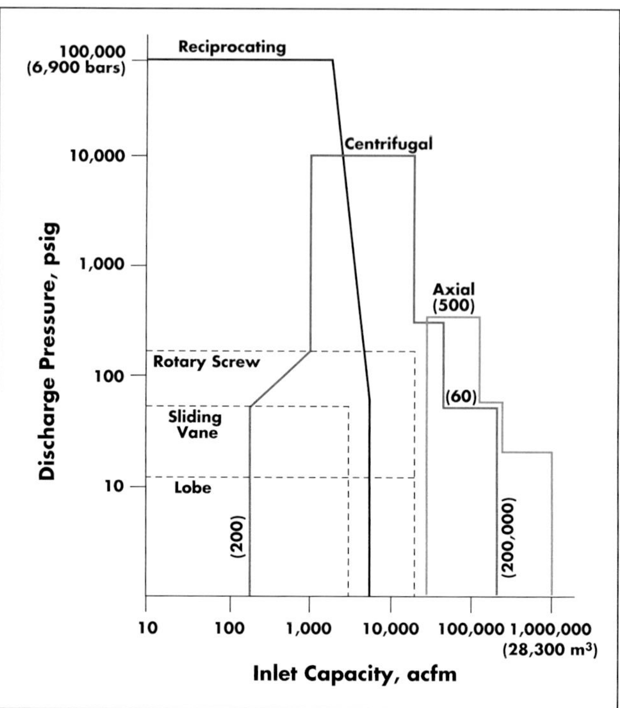

Fig. 30-26 Pressure-Capacity Chart Showing the Effective Ranges of Most Compressors. Source: Compressed Air and Gas Institute

In larger capacities, centrifugal compressors are often the baseload technology of choice. They offer good full-load efficiency in larger capacities and their reliability and low maintenance requirements are well suited for the long run-hours required of a baseload unit. In somewhat smaller capacity applications, screw compressors are used in baseload service. Prime mover-driven systems are potentially competitive alternatives to electric motor-driven systems when there is also a concurrent baseload demand for ther-

Table 30-1 Typical Air Compressors Used in Various Capacity Ranges [Based on 100 psig (7.9 bar) Operation]

Capacity range hp/cfm (kW/m³/m)	----------Reciprocating--------- Single acting	Double acting	Rotary positive displacement	Dynamic
1-30/3-120 (0.75-22/0.1-3)	One & Two stage	One stage	Rotary screw*	N/A
30-150/100-750 (22-112/3-21)	One & Two stage	Two stage	Rotary screw	N/A
150-300/600-1,600 (112-224/17-45)	One & Two stage	Two stage	Rotary screw	Centrifugal*
300-750/1,300-4,000 (224-559/37-113)	N/A	Two stage*	Rotary Screw	Centrifugal
>750/>3,400 (>559/>96)	N/A	Two stage*	Rotary Screw*	Centrifugal

*Not very common in capacity range due to relatively high capital cost and/or other logistical considerations.

mal outputs in the form of recovered heat.

The classic trim-load unit is a reciprocating compressor. Whether operated at fixed or variable speed, reciprocating compressors offer rapid response, superior part-load efficiency, and precise modulation. Rotary screw compressors operated with variable speed drivers may be a cost-effective alternative in some cases. The electric motor-driven reciprocating compressor has one distinct advantage over prime mover-driven trim units — reliable, fast response from a dead start. While some prime mover-driven units can provide quick start, this is not desirable on a regular basis. In addition, for reciprocating engine-driven systems, the combination of the rapid decline in engine performance and reduced compressor volumetric efficiency result in inefficient operation below 40% of full load.

In many smaller facilities, particularly those with single-shift operations, only one air compressor is used. Since most facilities face varying loads, a unit with good trim characteristics is often the most attractive. Constant speed electric motor-driven compressors are most common for such applications, although variable speed electric units and packaged reciprocating engine units can also be used to enhance part-load efficiency. In some cases, packaged reciprocating engine-driven units up to 400 hp (300 kW) can be cost-effective even without heat recovery because of high-cost peak electric rates during single-shift weekday operation.

In multiple compressor systems comprising similar units, it is common for both machines to share load equally, as demand requires each successive unit to be brought on line. Efficiency is often improved by using dissimilar units that feature required trim or baseload characteristics. Energy efficiency is optimized when all but one compressor is operating at full load and a relatively small trim unit is cycled or modulated as required.

For example, given a baseload of 3,400 cfm (96 m^3/m) and a peak load of 4,000 cfm (113 m^3/m), the baseload units can be sized at 3,000 cfm (85 m^3/m). A trim-load unit rated at 1,000 cfm (28 m^3/m) would always be loaded to 400 cfm, or 40%, maintaining reasonable performance under the base condition. As the typical load on the trim unit is between 400 and 1,000 cfm (11 and 28 m^3/m), the unit will operate at close to full-load efficiency all or most of the time, if operated at variable speed.

Another alternative is to split the trim-load service into two components: a bottom end and a top end. The bottom end consists of demand that is variable, but almost always present and can be met by a screw or reciprocating compressor driven by a reciprocating engine. The top end, or intermittent demand, is met by a cycling electric-driven reciprocating unit. This strategy provides for one quick-start unit, optimizes energy operating costs, and adds redundant capacity for increased reliability.

Most large facilities with multiple-shift operations are on some type of time-of-use (TOU) differentiated electric rate. Most are fairly stratified with peak period (5 day, daytime) electricity being relatively costly and off-peak period electricity being relatively inexpensive. A disadvantage of electric motor-driven trim compressors is that electric demand charges will apply to the peak load met by the unit, whenever this coincides with the peak demand setting period of the facility. Energy charges incurred during more typical part-load operation may be only a fraction of trim compressor operating cost during the utility peak periods.

TOU or the increasingly common real-time-pricing (RTP) electric rates may present opportunities for mixed, or hybrid, system configurations, featuring both electric and non-electric drivers. During peak periods, prime mover drivers may prove economical as baseloaded units, even without using recovered heat. One potentially cost-efficient strategy is to use a prime mover-driven screw or reciprocating compressor during peak and shoulder periods and an electric motor-driven unit during off-peak periods. Unless there is a concurrent load for the thermal output, a reciprocating engine-driven unit will generally not be as economical to operate as a baseloaded electric motor-driven unit during off-peak periods. Back-pressure turbine-driven units will almost always be economical (peak or off-peak), as long as there is full use of back-pressure steam.

The operating cost advantage of a hybrid system is often significant enough to offset the incremental cost of the prime mover-driven unit. Consider the following simplified example of matching system configuration to load:

Compressed air requirement during peak electric rate period:

- Baseload demand of 5,000 cfm (142 m^3/m)
- Peak demand of 6,000 cfm (170 m^3/m)

Compressed air requirement during off-peak electric rate period:

- Baseload demand of 4,000 cfm (113 m^3/m)
- Peak demand of 5,000 cfm (142 m^3/m)

Assume also a back-pressure steam turbine-driven centrifugal compressor can be designed to provide 4,000 cfm (113 m³/m) with a low-pressure steam output to match the facility's minimum requirement. One useful configuration would consist of a 4,000 cfm (113 m³/m) back-pressure steam-driven centrifugal compressors and a 2,000 cfm (57 m³/m) electric motor-driven reciprocating compressor. The steam turbine-driven unit would be baseloaded, with the electric motor-driven unit serving remaining peak baseloads and meeting peak demands.

A 2,000 cfm (57 m³/m) packaged reciprocating engine-driven screw compressor could be added to further reduce operating costs and provide some redundant capacity. It would replace the electric unit during peak periods and could be expected to operate efficiently at variable speed between 50 and 100% of full load. The electric unit would continue to operate and trim load off-peak. In the event that the largest unit, the steam turbine-driven machine, was out of service, the maximum capacity shortfall would be only 2,000 cfm (57 m³/m).

If expected low-pressure steam load is reduced during the off-peak period, perhaps by half, alternative configurations would be considered. One option would be to install a 2,000 cfm (57m³/m) steam turbine-driven unit and a 2,000 cfm (57 m³/m) baseload electric motor-driven unit. With the addition of two trim units (one engine-driven and one electric motor-driven), the total system capacity would be 8,000 cfm (226 m³/m), providing 100% redundancy when any one unit is out of service.

If in this example there was no steam demand, the following configuration might be selected:

- Baseload duty by a 4,000 cfm (113 m³/m) electric motor-driven centrifugal compressor;
- Peak trim by a 2,000 cfm (57 m³/m) reciprocating engine-driven screw compressor; and
- Off-peak trim by a 2,000 cfm (57 m³/m) electric motor-driven reciprocating compressor.

Alternative configurations would be considered based on maintenance and reliability concerns, required redundancy, energy prices, and available thermal loads. Because the lifecycle operating costs associated with compressed air plants are generally so much higher than initial capital costs, hybrid configurations will often show a relatively short payback when compared with least capital-cost alternatives.

Simplified Economic Analysis of Air Compressor Options

The following examples include simplified economic analyses of alternative air compressor systems that could be applied to an assumed load. Each example compares the application of a packaged variable-speed reciprocating engine-driven screw air compressor with electric motor-driven units that use inlet air throttling modulation or cycling control. The compressor systems under consideration have a design rating of 1,537 cfm (44.6 m³/m) at 125 psi (9.9 bar) service. Electricity and gas use under varying load are based on manufacturer's data. The examples consider:

- Simple payback on the incremental investment in a cycling control electric unit versus the lower cost modulation control electric unit.
- Simple payback on the incremental investment in the reciprocating engine unit, with variable speed and low-end modulation control.

The examples include simplified screening tools that can be applied to economic analysis of air compressor system options. They show how part-load performance can be used as the basis of the analysis and demonstrate the potential benefits achieved with variable speed control. The examples also demonstrate how economic performance is highly dependent on energy cost. Section IX provides a discussion on performing more detailed technical and economic project feasibility analyses.

Example 1

This example is based on a relatively high-cost electric rate that is summarized in Figure 30-27.

The first-cost premium for the electric unit with cycling control versus the base case electric unit with modulation control is $30,000. The first-cost premium for the gas-fired reciprocating engine unit is $130,000 versus the electric unit with modulation control and $160,000 versus the electric unit with cycling control.

	Summer 4 Months	Winter 8 Months	Annual Average
Demand, $/kW/Month	$ 18.00	$ 14.00	$ 15.33
On-peak, $/kWh	$ 0.0900	$0.0750	$ 0.0800
Shoulder, $/kWh	$ 0.0750	$0.0650	$ 0.0683
Off-peak, $/kWh	$ 0.0500	$0.0500	$ 0.0500

Fig. 30-27 Electric Rate Summary for Example 1.

Figure 30-28 shows the assumed load during annual hours in each of the electric rate periods. Figure 30-29 shows the percent of full-load power required for operation of each of the electric units under varying load. Input power and annual hours of operation are used to determine annual energy use and demand, with costs calculated at the assumed billing rate.

Figure 30-30 shows energy use and operating cost for the electric unit under modulation control. Energy use and operating cost for the electric unit under cycling control is shown in Figure 30-31. With an annual operating cost savings of $20,726, the incremental investment of $30,000 for cycling control is recovered in 1.45 years.

Figure 30-32 shows a similar analysis for the gas-fired reciprocating engine-driven system with the compressor operating in speed control through much of the operating range and modulation control under low load. The fuel required to drive the system is listed in the table and is supplied at a cost of $4.50/Mcf, independent of time of use.

The total cubic feet (CF) of gas requirement is tabulated for each annual block of operating hours. Annual operating costs are determined using a gas cost equivalent to $4.50/Mcf. The annual O&M cost, based on $3.80 per equivalent full-load hour (EFLH), is calculated as $21,584. Figure 30-33 shows the simple payback on the incremental investment in the reciprocating engine system versus the two electric compressor system options. A comparison of the engine option with the cycling control option shows that the additional incremental investment in the engine-driven system of $130,000 versus the cycling control system is returned in 2.24 years.

Load		Hours			
CFM	% Capacity	On-peak	Shoulder	Off-peak	Total
1,537	100%	125	75	200	400
1,383	90%	300	250	400	950
1,230	80%	450	400	700	1,550
1,076	70%	426	425	750	1,601
922	60%	400	410	800	1,610
769	50%	235	306	800	1,341
615	40%	100	100	480	680
461	30%	50	75	280	405
307	20%	-	45	110	155
154	10%	-	-	68	68
		2,086	2,086	4,588	8,760

Fig. 30-28 Loads by Hours and Electric Rate Periods.

Load	Electric Drive	
% Capacity	Modulation % Power consumed	Cycling % Power Consumed
100%	100%	100%
90%	97%	93%
80%	92%	85%
70%	88%	78%
60%	84%	70%
50%	80%	60%
40%	76%	52%
30%	72%	44%
20%	68%	36%
10%	64%	28%

Fig. 30-29 Percent Power vs. Percent Load for Electric Drive Options.

			Consumption kWh/Year				Cost $/Year			
% Capacity	Modulation % Power consumed	Power kW	On-peak	Shoulder	Off-peak	Total	On-peak	Shoulder	Off-peak	Total
100%	100%	300	37,500	22,500	60,000	120,000	$ 3,000	$ 1,537	$ 3,000	$ 7,537
90%	97%	291	87,300	72,750	116,400	276,450	$ 6,984	$ 4,969	$ 5,820	$ 17,773
80%	92%	276	124,200	110,400	193,200	427,800	$ 9,936	$ 7,540	$ 9,660	$ 27,136
70%	88%	264	112,464	112,200	198,000	422,664	$ 8,997	$ 7,663	$ 9,900	$ 26,560
60%	84%	252	100,800	103,320	201,600	405,720	$ 8,064	$ 7,057	$10,080	$ 25,201
50%	80%	240	56,400	73,440	192,000	321,840	$ 4,512	$ 5,016	$ 9,600	$ 19,128
40%	76%	228	22,800	22,800	109,440	155,040	$ 1,824	$ 1,557	$ 5,472	$ 8,853
30%	72%	216	10,800	16,200	60,480	87,480	$ 864	$ 1,106	$ 3,024	$ 4,994
20%	68%	204	–	9,180	22,440	31,620	$ –	$ 627	$ 1,122	$ 1,749
10%	64%	192	–	–	13,056	13,056	$ –	$ –	$ 653	$ 653
		Totals	**552,264**	**542,790**	**1,166,616**	**2,261,670**	**$44,181**	**$37,073**	**$58,331**	**$139,584**
				Demand	**300 kW x $15.33/kW/Month x 12 Month/Year**					**$ 55,188**
								Total Cost		**$194,772**

Fig. 30-30 Energy Usage and Cost for Modulation Control Unit.

% Capacity	Cycling % Power Consumed	Power kW	Consumption kWh/Year On-peak	Shoulder	Off-peak	Total	Cost $/Year On-peak	Shoulder	Off-peak	Total
100%	100%	300	37,500	22,500	60,000	120,000	$ 3,000	$ 1,537	$ 3,000	$ 7,537
90%	93%	279	83,700	69,750	111,600	265,050	$ 6,696	$ 4,764	$ 5,580	$ 17,040
80%	85%	255	114,750	102,000	178,500	395,250	$ 9,180	$ 6,967	$ 8,925	$ 25,072
70%	78%	234	99,684	99,450	175,500	374,634	$ 7,975	$ 6,792	$ 8,775	$ 23,542
60%	70%	210	84,000	86,100	168,000	338,100	$ 6,720	$ 5,881	$ 8,400	$ 21,001
50%	60%	180	42,300	55,080	144,000	241,380	$ 3,384	$ 3,762	$ 7,200	$ 14,346
40%	52%	156	15,600	15,600	74,880	106,080	$ 1,248	$ 1,065	$ 3,744	$ 6,057
30%	44%	132	6,600	9,900	36,960	53,460	$ 528	$ 676	$ 1,848	$ 3,052
20%	36%	108	–	4,860	11,880	16,740	$ –	$ 332	$ 594	$ 926
10%	28%	84	–	–	5,712	5,712	$ –	$ –	$ 286	$ 286
		Totals	**484,134**	**465,240**	**967,032**	**1,916,406**	**$38,731**	**$31,776**	**$48,352**	**$118,858**
			Demand	**300 kW x $15.33/kW/Month x 12 Month/Year**						**$ 55,188**
								Total Cost		**$174,046**
			Reduction in Cost as Compared to Modulation Control, $/Year							**$ 20,726**
			Incremental Capital Cost Over Modulation Control							**$ 30,000**
			Simple Payback Over Modulation Control, Years							**1.45**

Fig. 30-31 Energy Usage and Cost for Cycling Control and Payback vs. Modulation Unit.

	Fuel Req'd CFH	Operation Hours/Year	Fuel Use, CF/Year	Cost $/Year
100%	3,294	400	1,317,428	$ 5,928
90%	2,928	950	2,781,239	$ 12,516
80%	2,643	1,550	4,097,115	$ 18,437
70%	2,398	1,601	3,839,678	$ 17,279
60%	2,228	1,610	3,587,708	$ 16,145
50%	2,058	1,341	2,760,422	$ 12,422
40%	2,031	680	1,381,107	$ 6,215
30%	2,004	405	811,454	$ 3,652
20%	1,976	155	306,302	$ 1,378
10%	1,949	68	132,512	$ 596
		8,760	**21,014,964**	**$ 94,567**
Equivalent Full Load Hours/Year				**5,680**
O & M Cost @ $3.80/EFLH, /Year				**$ 21,584**

Fig. 30-32 Operating Cost for Reciprocating Engine Unit.

Example 2

This example is based on a lower-cost electric rate, summarized in Figure 30-34.

A first pass screening calculation revealed that under the lower electric rate, the reciprocating engine-driven air compressor could not be economically applied without benefit of heat recovery. Comparing Figure 30-35 (electric-driven unit with modulation) and Figure 30-36 (electric-driven unit with cycling control), the electric-driven unit featuring cycling control would be a likely choice.

Figure 30-37 considers the reciprocating engine sys-

	Summer 4 Months	Winter 8 Months	Annual Average
Demand, $/kW/Month	$ 9.00	$ 7.00	$ 7.67
On-peak, $/kWh	$ 0.0490	$ 0.0440	$ 0.0457
Shoulder, $/kWh	$ 0.0440	$ 0.0410	$ 0.0420
Off-peak, $/kWh	$ 0.0330	$ 0.0330	$ 0.0330

Fig. 30-34 Electric Rate Summary for Example 2.

Reduction in Cost as Compared with Electric Modulation Control, Per Year	**$ 78,621**
Incremental Capital Cost Over Electric Modulation Control	**$ 160,000**
Simple Payback Over Electric Modulation Control, Years	**2.03**
Reduction in Cost as Compared with Electric Cycling Control, Per Year	**$ 57,895**
Incremental Capital Cost over Electric Cycling Control	**$ 130,000**
Simple Payback Over Electric Cycling Control, Years,	**2.24**

Fig. 30-33 Payback for Reciprocating Engine Unit vs. Electric Unit Options.

tem with heat recovery. Listed natural gas consumption is based on the fuel-chargeable-to-power (FCP) as developed in Chapter 2, assuming heat recovery equivalent to 40% of engine fuel energy input. Displaced boiler efficiency is assumed to be 83% and gas cost is $4.50/Mcf. The O&M rate is increased to $4.20 per EFLH to reflect additional costs associated with the heat recovery system.

The first-cost premium for the reciprocating engine cogeneration system is $155,000 as compared with cycling control unit and $185,000 as compared with the modulation control unit. Results shown in Figure 30-38 indicate that with heat recovery, the engine option would produce a simple payback of 4.48 years versus electric unit with modulation, and a payback of 5.49 years versus the electric unit with cycling control.

			Consumption kWh/Year				Cost $/Year			
% Capacity	Modulation % Power Consumed	Power kW	On-peak	Shoulder	Off-peak	Total	On-peak	Shoulder	Off-peak	Total
100%	100%	300	37,500	22,500	60,000	120,000	$ 1,714	$ 945	$ 1,980	$ 4,639
90%	97%	291	87,300	72,750	116,400	276,450	$ 3,990	$ 3,056	$ 3,841	$ 10,886
80%	92%	276	124,200	110,400	193,200	427,800	$ 5,676	$ 4,637	$ 6,376	$ 16,688
70%	88%	264	112,464	112,200	198,000	422,664	$ 5,140	$ 4,712	$ 6,534	$ 16,386
60%	84%	252	100,800	103,320	201,600	405,720	$ 4,607	$ 4,339	$ 6,653	$ 15,599
50%	80%	240	56,400	73,440	192,000	321,840	$ 2,577	$ 3,084	$ 6,336	$ 11,998
40%	76%	228	22,800	22,800	109,440	155,040	$ 1,042	$ 958	$ 3,612	$ 5,611
30%	72%	216	10,800	16,200	60,480	87,480	$ 494	$ 680	$ 1,996	$ 3,170
20%	68%	204	–	9,180	22,440	31,620	$ –	$ 386	$ 741	$ 1,126
10%	64%	192	–	–	13,056	13,056	$ –	$ -	$ 431	$ 431
		Totals	**552,264**	**542,790**	**1,166,616**	**2,261,670**	**$25,238**	**$22,797**	**$38,498**	**$ 86,534**
			Demand	**300 kW x $7.67/kW/Month x 12 Month/Year**						**$ 27,612**
								Total Cost		**$114,146**

Fig. 30-35 Energy Usage and Cost for Modulation Control Unit.

			Consumption kWh/Year				Cost $/Year			
% Capacity	Cycling % Power Consumed	Power kW	On-peak	Shoulder	Off-peak	Total	On-peak	Shoulder	Off-peak	Total
100%	100%	300	37,500	22,500	60,000	120,000	$ 1,714	$ 945	$ 1,980	$ 4,639
90%	93%	279	83,700	69,750	111,600	265,050	$ 3,825	$ 2,930	$ 3,683	$ 10,437
80%	85%	255	114,750	102,000	178,500	395,250	$ 5,244	$ 4,284	$ 5,891	$ 15,419
70%	78%	234	99,684	99,450	175,500	374,634	$ 4,556	$ 4,177	$ 5,792	$ 14,524
60%	70%	210	84,000	86,100	168,000	338,100	$ 3,839	$ 3,616	$ 5,544	$ 12,999
50%	60%	180	42,300	55,080	144,000	241,380	$ 1,933	$ 2,313	$ 4,752	$ 8,998
40%	52%	156	15,600	15,600	74,880	106,080	$ 713	$ 655	$ 2,471	$ 3,839
30%	44%	132	6,600	9,900	36,960	53,460	$ 302	$ 416	$ 1,220	$ 1,937
20%	36%	108	–	4,860	11,880	16,740	$ –	$ 204	$ 392	$ 596
10%	28%	84	–	–	5,712	5,712	$ –	–	$ 188	$ 188
		TOTALS	**484,134**	**465,240**	**967,032**	**1,916,406**	**$ 22,125**	**$ 19,540**	**$31,912**	**$ 73,577**
				Demand 300 kW x $7.67/kW/Month x 12 Month/Year						**$ 27,612**
								Total Cost		**$101,189**
				Reduction in Cost as Compared with Modulating Control, $/year						**$ 12,957**
				Incremental Capital Cost Over Modulation Control						**$ 30,000**
				Simple Payback Over Modulation Control, Years						**2.31**

Fig. 30-36 Energy Usage and Cost for Cycling Control and Payback vs. Modulation Unit.

% Capacity	Fuel Req'd CFH	Fuel Equivalent Recovered as Heat, CFH	Net Fuel Charged-to-Power (FCP), CFH	Operation Hours/Year	Fuel Use, CF/Year	Cost $/Year
100%	3,294	1,587	1,706	400	682,523	$ 3,071
90%	2,928	1,411	1,517	950	1,440,883	$ 6,484
80%	2,643	1,274	1,369	1,550	2,122,602	$ 9,552
70%	2,398	1,156	1,242	1,601	1,989,231	$ 8,952
60%	2,228	1,074	1,154	1,610	1,858,692	$ 8,364
50%	2,058	992	1,066	1,341	1,430,098	$ 6,435
40%	2,031	979	1,052	680	715,513	$ 3,220
30%	2,004	966	1,038	405	420,392	$ 1,892
20%	1,976	952	1,024	155	158,686	$ 714
10%	1,949	939	1,010	68	68,651	$ 309
				8,760	**10,887,271**	**$ 48,993**
Equivalent Full Load Hours/Year				**5,680**		
O & M Cost @ $4.20/EFLH, /Year				**$ 23,856**		

Fig. 30-37 Energy Usage and Cost for Reciprocating Engine Cogeneration Option.

Reduction in Cost as Compared with Electric Modulating Control, $/year	**$ 41,298**
Incremental Capital Cost Over Electric Modulation Control	**$ 185,000**
Simple Payback Over Electric Modulation Control, Years	**4.48**
Reduction in Cost as Compared with Electric Cycling Control, $/year	**$ 28,341**
Incremental Capital Cost Over Electric Cycling Control	**$ 155,000**
Simple Payback Over Electric Cycling Control, Years	**5.49**

Fig. 30-38 Payback for Cogeneration Unit vs. Electric Unit Options.

Pumps are volumetric machines that cause a fluid, such as water, brine, oil, etc., to flow across an external resistance. Pumping energy depends on the volume of fluid moved, the resistance against which the pump works, and the machine efficiency. Pumps are used in almost every type of facility and are usually integral to most industrial process applications.

PUMP TYPES

Figure 31-1 provides a comprehensive listing by category and subcategories of types of pumps. Pumps are broadly classified as **positive-displacement** types (which include reciprocating and rotary pumps) and kinetic or **centrifugal** types.

- Pumps
  - Positive displacement
    - Reciprocating pumps
      - Steam — Horizontal / Vertical — Double-acting — Piston / Plunger — Simplex / Duplex
      - Power — Horizontal / Vertical — Single-acting / Double-acting — Piston / Plunger — Simplex / Duplex / Multiplex
      - Controlled volume — Horizontal / Vertical — Packed plunger / Packed piston / Diaphragm (Mechanically coupled / Hydraulically coupled) — Simplex / Duplex / Multiplex — Manual control / Automatic control
    - Blow case
    - Rotary pumps
      - Vane — Blade, bucket roller or slipper
      - Piston — Axial / Radial
      - Flexible member — Flexible tube / Flexible vane / Flexible liner
      - Lobe — Single / Multiple
      - Gear — External / Internal
      - Circumferential piston — Single / Multiple
      - Screw — Single / Multiple
  - Kinetic
    - Centrifugal 1)
      - Overhung impeller
        - Close-coupled Single & two stage — End suction (including submersibles) / In-line
        - Separately coupled Single & two stage — In-line / Frame mounted / Centering support AP1610 / Frame mounted ANSI B73.1 / Wet pit volute / Axial flow impeller (propeller) volute type (horiz. or vert.)
      - Impeller between bearings
        - Separately coupled Single stage — Axial (horiz.) split case / Radial (vert.) split case
        - Separately coupled Multistage — Axial (horiz.) split case / Radial (vert.) split case
      - Turbine type
        - Vertical type Single & Multistage — Deep well turbine (including submersibles) / Barrel or can pump / Short setting or close-coupled
        - Axial flow impeller (propeller) or mixed flow type (horizontal or vertical)
    - Regenerative turbine — Overhung impeller / Impeller between bearings — Single stage / Two stage
    - Special effect — Reversible centrifugal / Rotating casing (pitot)

1) Includes radial mixed flow and axial flow designs

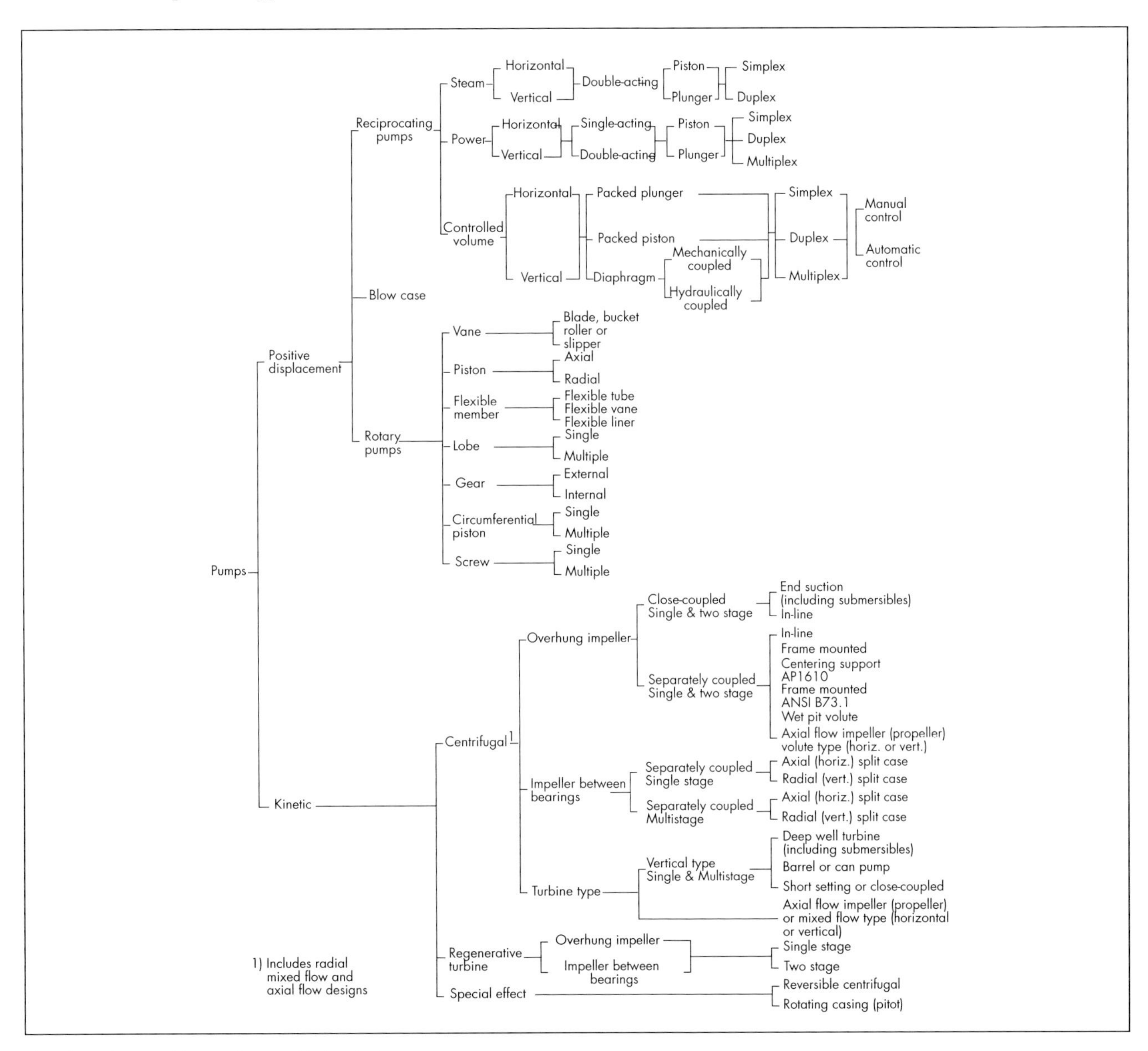

Fig. 31-1 Types of Pump. Source: Hydraulic Institute

Reciprocating piston pumps, or **power pumps**, are characterized by high capital cost and the ability to operate under extremely high pressure. At constant speed, a power pump delivers essentially the same flow at any pressure within the capability of the driver. Power pumps offer high full-load efficiencies and can provide constant delivery at varying pressures. Where applicable, flow rate can be varied by varying driver speed, bypassing the outlet of the pump back to the suction inlet, or varying the stroke length of the pump.

Power pumps include horizontal and vertical types, piston and plunger types, and single- or double-acting operation. Figure 31-2 illustrates a horizontal single-acting plunger type, Figure 31-3 illustrates a vertical single-acting plunger type, and Figure 31-4 illustrates a horizontal double-acting piston type.

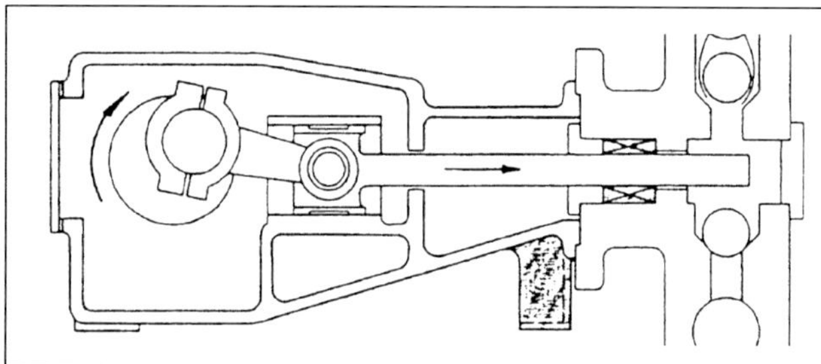

Fig. 31-2 Horizontal Single-Acting Plunger Power Pump. Source: Hydraulic Institute

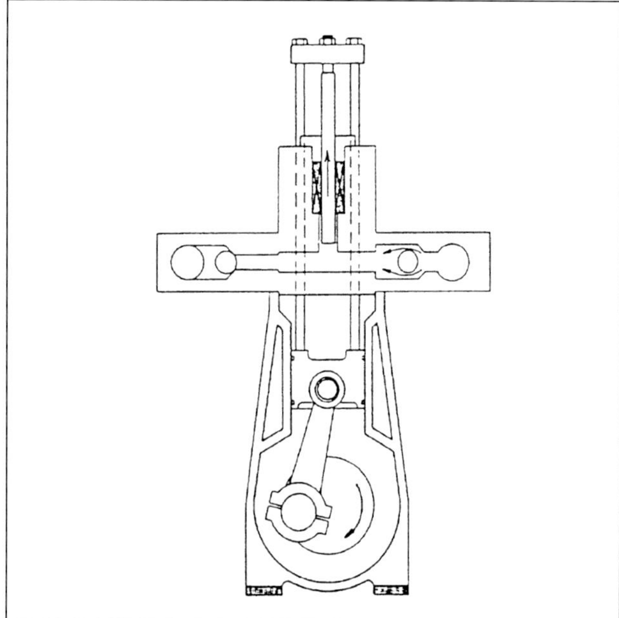

Fig. 31-3 Vertical Single-Acting Plunger Power Pump. Source: Hydraulic Institute

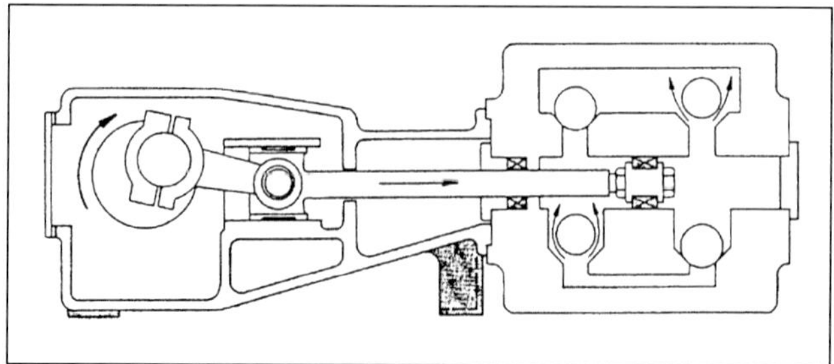

Fig. 31-4 Horizontal Double-Acting Piston Power Pump. Source: Hydraulic Institute

In a direct-acting steam pump, a reciprocating pump and a steam engine are built together as a unit. The steam piston is connected to the pump piston and provides power to drive the pump. Steam pumps, while quite inefficient, are known for reliability and long life. They are becoming increasingly less common, but can still be found in operation. Similar units may also be driven by compressed gases, such as air or natural gas. Figure 31-5 is a direct-acting, horizontal, double-acting, piston steam pump.

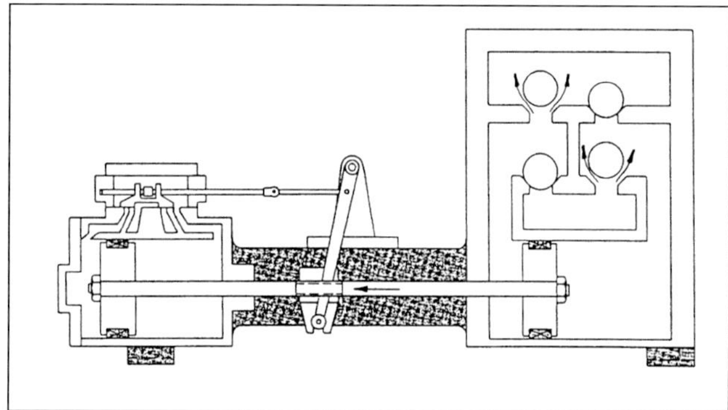

Fig. 31-5 Direct-Acting, Horizontal, Double-Acting, Piston Steam Pump. Source: Hydraulic Institute

Figures 31-6 through 31-8 show a variety of vintage pumps. Figure 31-6 shows a simplex valve pot piston pump. This 157 hp (117 kW) pump has a flow range between 0.5 and 450 gpm (1.9 and 1,705 L/m), with a maximum discharge pressure of 600 psi (42 bar). Figure 31-7 shows a forged steel simplex hydraulic plunger pump, built in 1938. This 185 hp (138 kW) pump can produce a maximum flow of 63 gpm (240 L/m) with a discharge pressure of up to 10,000 psi (690 bar). Figure 31-8 shows a 250 hp (186 kW) duplex valve pot piston pump. It has a maximum flow rate of 711 gpm (2,261 L/m) at a maximum discharge pressure of 600 psi (42 bar).

Rotary pumps are usually compact, light, simple, and low cost. While they can operate at pressures of up to 5,000 psig (346 bar), typical field applications range from

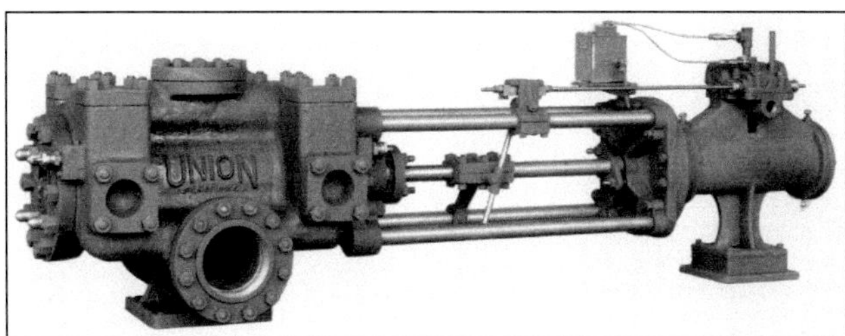

Fig. 31-6 Simplex Valve Pot Piston Pump.
Source: Union Pump Co.

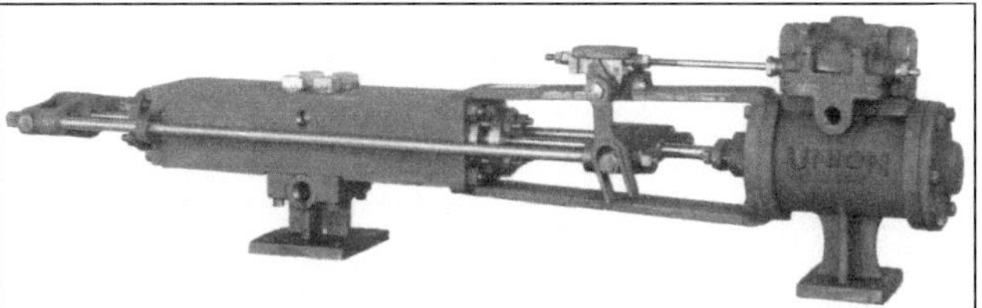

Fig. 31-7 Simplex Hydraulic Plunger Pump.
Source: Union Pump Co.

Fig. 31-8 Duplex Valve Pot Piston Pump.
Source: Union Pump Co.

25 to 500 psig (2.7 to 35.5 bar). Efficiencies are generally lower than those of power pumps, ranging from 60 to 85%. Rotary pumps are most commonly used for pumping oils or other liquids with high viscosity. There are several designs as listed below, all constant-capacity machines at a given speed.

- In screw type rotary pumps, fluid is carried in spaces between screw threads and is displaced axially as they mesh. Single screw pumps (Figure 31-9), which are commonly called **progressing cavity pumps**, have a rotor with external threads, which are eccentric to the axis of rotation. Screw and wheel pumps (Figure 31-10) use a plate wheel to seal the cavity. Multiple screw pumps (Figure 31-11) have multiple external screw threads and may be timed or untimed.
- Vane type pumps (Figures 31-12 and 31-13) include buckets, blades, rollers, or slippers that operate with a cam to draw fluid across the pump chamber. These pumps may be made with vanes in either the rotor or stator and with radial hydraulic forces on the rotor balanced or unbalanced. In the axial piston pump

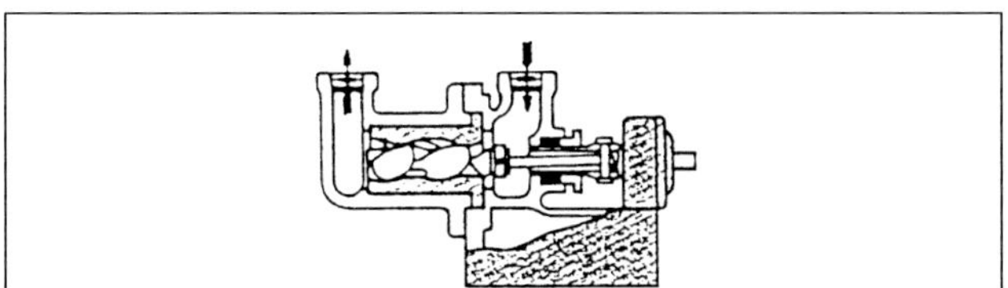

Fig. 31-9 Single Screw Pump (progressing cavity).
Source: Hydraulic Institute

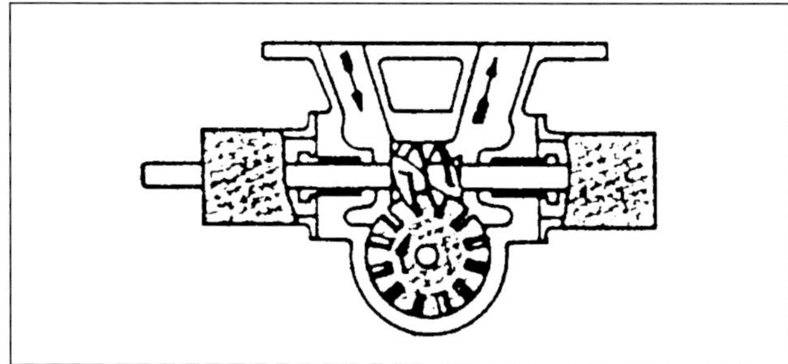

Fig. 31-10 Screw and Wheel Pump.
Source: Hydraulic Institute

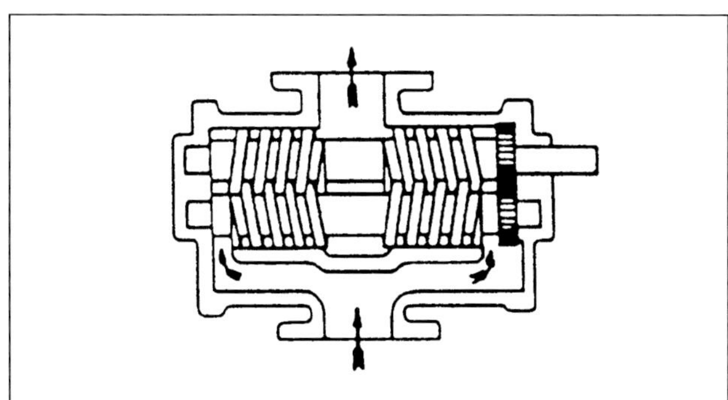

Fig. 31-11 Multiple (Two) Screw Pump.
Source: Hydraulic Institute

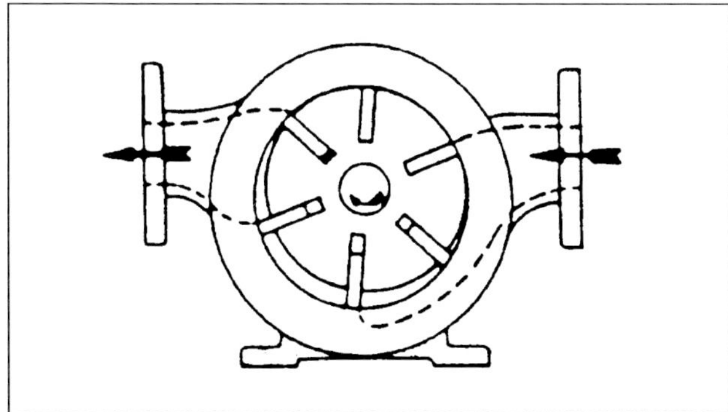

Fig. 31-12 Sliding Vane Pump.
Source: Hydraulic Institute

shown in Figure 31-14, valving is accomplished by rotation of the pistons and cylinders relative to the ports.

- In flexible member type pumps (Figure 31-15), the sealing action depends on the elasticity of a flexible tube, vane, or liner.
- In lobe type pumps (Figures 31-16 and 31-17), fluid is carried between rotor lobe surfaces and the pumping chamber, with the rotor surfaces providing continuos sealing.

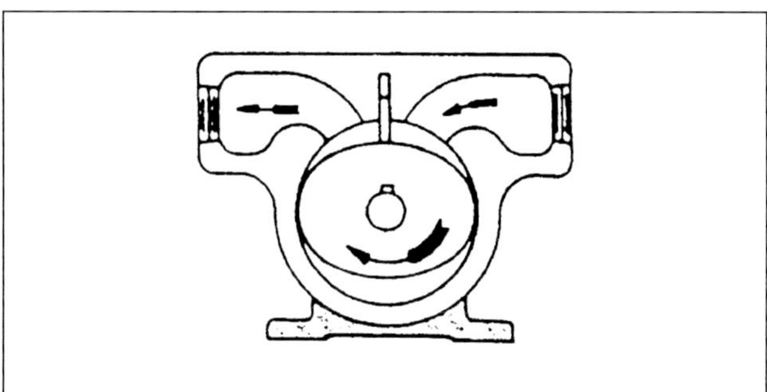

Fig. 31-13 External Vane Pump.
Source: Hydraulic Institute

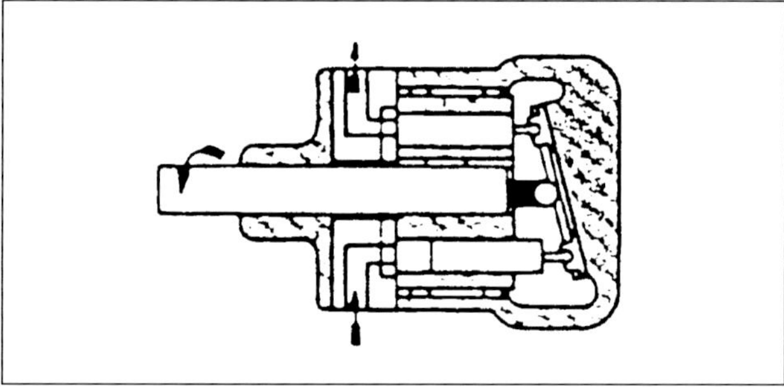

Fig. 31-14 Axial Piston Pump.
Source: Hydraulic Institute

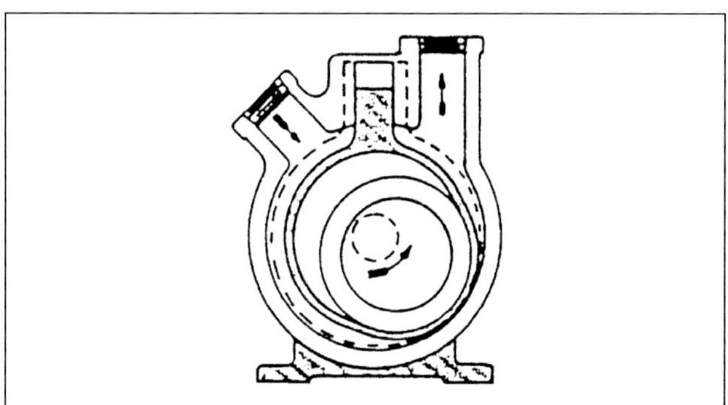

Fig. 31-15 Flexible Liner Pump.
Source: Hydraulic Institute

- In circumferential piston type pumps (Figure 31-18), fluid is carried in spaces between piston surfaces. There are no sealing contacts between rotor surfaces.
- In gear type pumps (Figures 31-19 and 31-20), fluid is carried between gear teeth and displaced when they mesh, with the surfaces of the rotors providing continuos sealing. Each rotor is capable of driving the other. Internal gear pumps have one rotor with internally cut gear teeth meshing with an externally cut gear.

Centrifugal pumps produce a pressure increase within a rapidly spinning vaned impeller by centrifugal action. Fluid enters the center of the impeller, called the eye, flows radially outward, and is discharged around the entire circumference into a casing. They typically operate at higher volume and lower head than positive displacement pumps, although higher head can be obtained by using multiple stages in series on a single shaft.

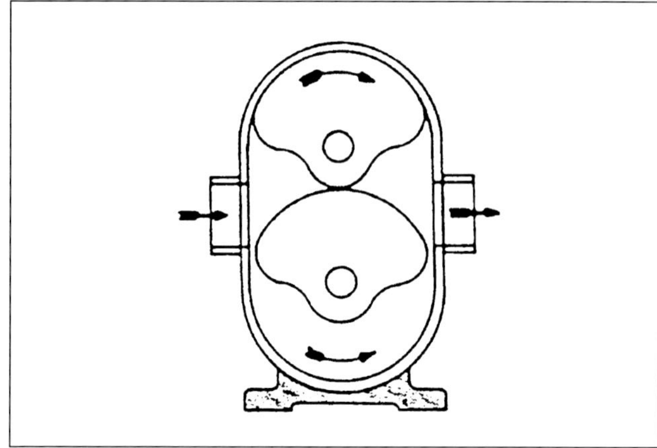

Fig. 31-16 Single Lobe Pump.
Source: Hydraulic Institute

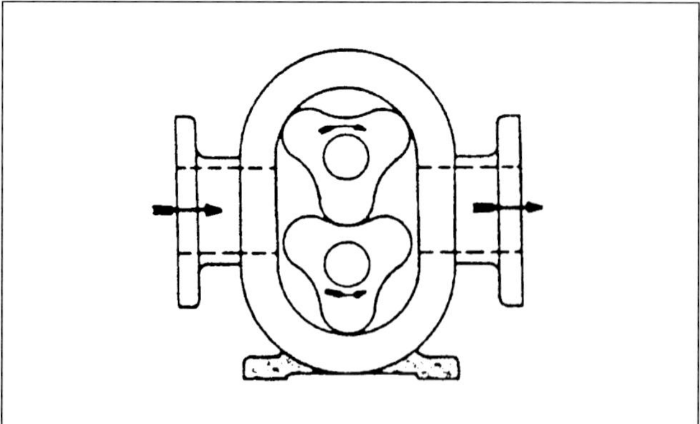

Fig. 31-17 Three-Lobe Pump.
Source: Hydraulic Institute

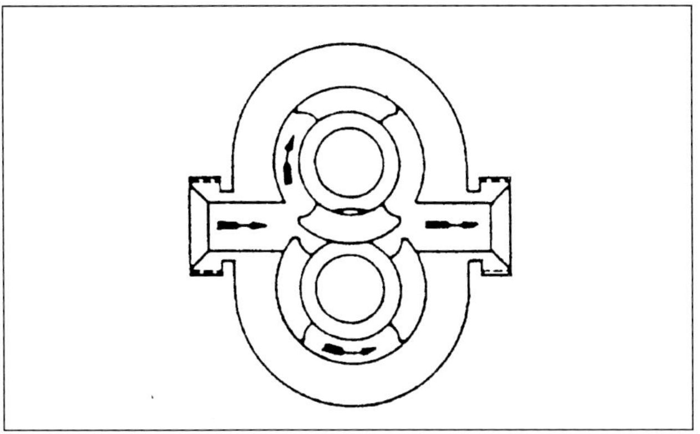

Fig. 31-18 Circumferential Piston Pump.
Source: Hydraulic Institute

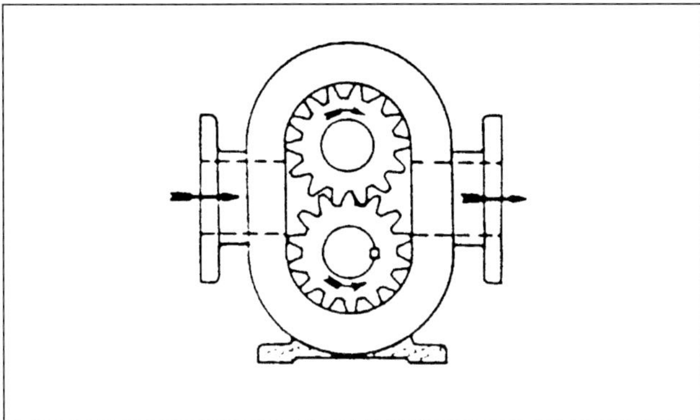

Fig. 31-19 External Gear Pump.
Source: Hydraulic Institute

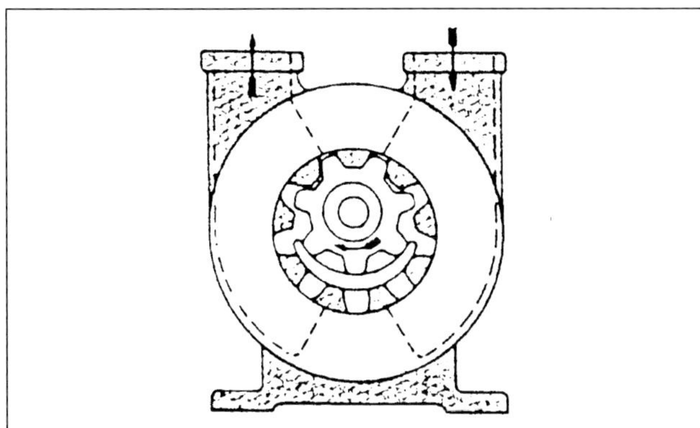

Fig. 31-20 Internal Gear Pump (with Crescent).
Source: Hydraulic Institute

Centrifugal pumps are classified according to impeller type and direction of flow as radial (centrifugal) flow, mixed flow, and axial flow. Radial flow pumps, as shown in Figure 31-21, usually have the lowest specific speed range of the three classes of centrifugal pumps. Figure 31-22 illustrates a mixed flow pump having a single inlet impeller with the flow entering axially and discharging in an axial and radial direction. Pumps of this type usually have an intermediate speed range. Figure 31-23 illustrates an axial flow pump, sometimes referred to as a propeller pump, having a single inlet impeller with the flow entering axially and discharging nearly axially. Axial flow pumps usually have a relatively high specific speed range.

Fig. 31-21 Radial Flow Pump.
Source: Hydraulic Institute

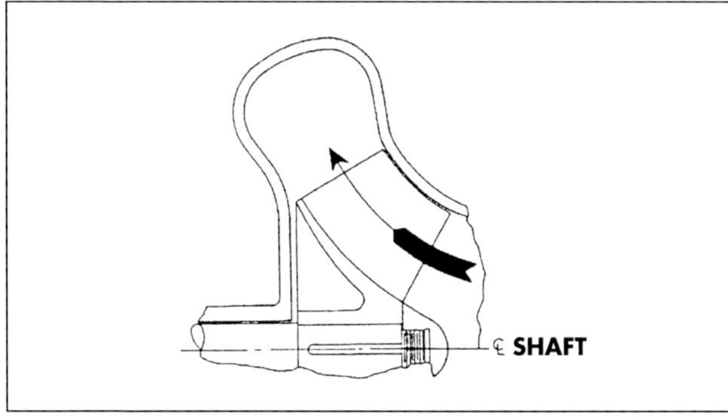

Fig. 31-22 Mixed Flow Pump.
Source: Hydraulic Institute

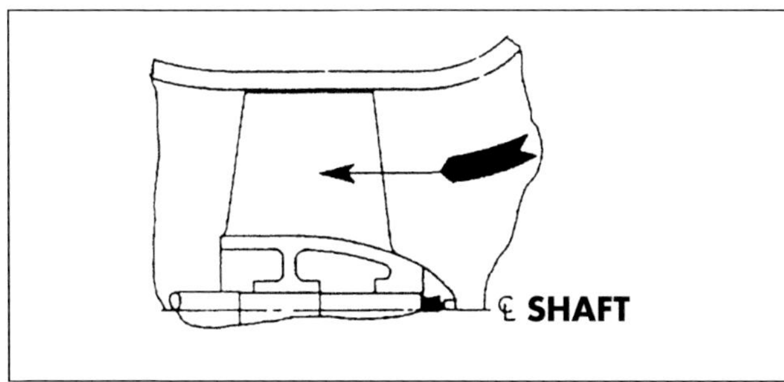

Fig. 31-23 Axial Flow Pump.
Source: Hydraulic Institute

Centrifugal pumps may also be classified as either volute pumps or diffuser pumps. In the volute pump, the impeller is surrounded by a spiral case, with the curved outer boundary called a volute. The absolute velocity of the fluid leaving the impeller is reduced in the volute casing, with a resulting increase in pressure. In the diffuser pump, the impeller is surrounded by diffuser vanes that provide gradually increasing passages to produce a gradual reduction in velocity.

Figures 31-24 and 31-25 are cutaway illustrations of centrifugal pumps. Figure 31-24 is a double-suction type single-stage axial (horizontal) split case. Figure 31-25 is a multistage radial (vertical) split case.

Figure 31-26 shows a radially split single-stage double-suction centrifugal pump with mechanical seals and bearing end caps not yet installed. On the right is the thrust end of the pump and on the left is the radial, or coupling, end showing the shaft extension. This pump is designed for a maximum flow rate of 1,850 gpm (7,002 L/m) against a head of 550 ft (16.4 bar), with a brake power requirement of 360 hp (268 kW). Figure 31-27 is a horizontally split heavy-duty multistage opposed impeller centrifugal pump driven by a 200 hp (149 kW) non-condensing steam turbine. The pump operates at 3,560 rpm and delivers 280 gpm (1,060 L/m) of 250°F (121°C) condensate in a high-pressure coolant application. Similar to a centrifugal pump is a turbine pump, which uses a regenerative effect to increase pressure at the periphery of a closed impeller inside a casing with swirl inducing chambers. Turbine pumps have been successfully employed in applications requiring high head pressures from a pump with small dimensions.

Pump Formulas and Affinity Laws

Energy must be converted to work if a liquid is to be moved and this is the function of a pump. The work may be lifting the liquid against gravity from one elevation to another or forcing it into a pressurized vessel. In addition to these fixed or static resistances to flow, there is also a dynamic component that varies with velocity of flow, i.e., frictional losses to be overcome in the pipe, fittings, heat exchangers, spray heads, and other equipment and materials. To make calculations easier, all of these quantities are reduced to a coherent format with the same dimensions, know as head.

Head (h) can be expressed in several ways. For example, the specific energy (energy per unit mass) imparted on the medium by the pump is a function of the pressure difference between the outlet and the inlet of the pump and the specific volume of the fluid pumped. Neglecting the velocity and elevation differences between the inlet and outlet of the pump, the head is expressed as the

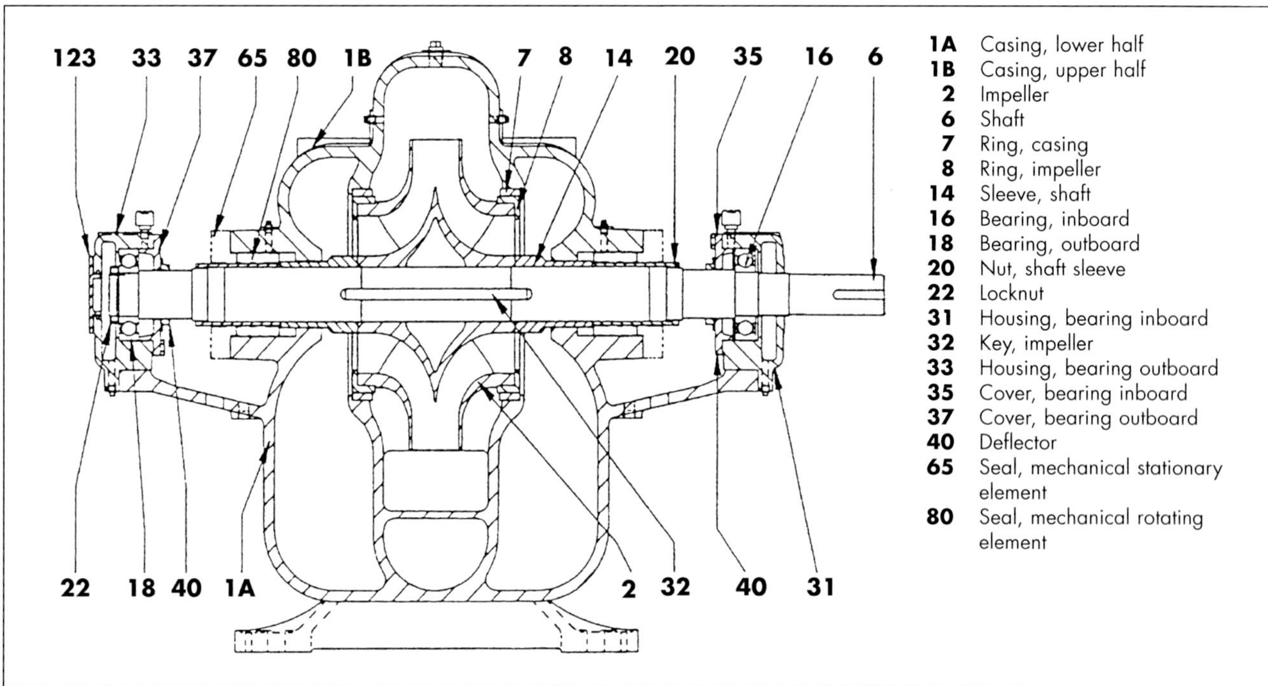

Fig. 31-24 Cutaway Illustration of Double-Suction Axial Centrifugal Pump. Source: Hydraulic Institute

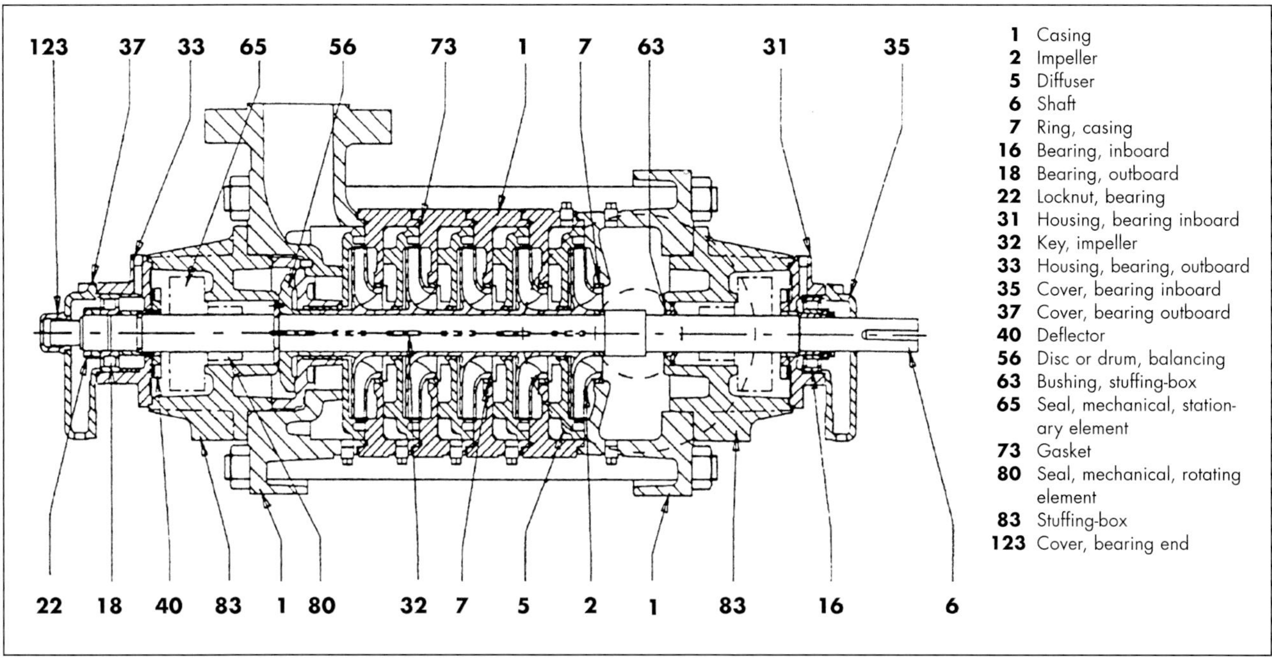

Fig. 31-25 Multistage (Radial) Centrifugal Pump. Source: Hydraulic Institute

Fig. 31-26 Radially Split Single-Stage Double-Suction Centrifugal Pump Showing Thrust and Coupling Ends. Source: Goulds Pumps Inc.

Fig. 31-27 Steam Turbine-Driven Horizontally Split Opposed Impeller Centrifugal Pump. Source: Goulds Pumps Inc.

length of a vertical column of the pumped fluid and calculated as:

$$h = \frac{(p_1 - p_2)}{\rho g} \qquad (31\text{-}1)$$

Where:

p_1 = Inlet pressure
p_2 = Outlet pressure
ρ = Fluid density
g = Acceleration due to gravity

The fluid flow, both liquid and gas, through pipe is impeded by frictional resistance that causes a pressure or head loss, which can be evaluated by application of pipe friction equations, which can be found in piping design handbooks (refer to Fanning's or the Darcy Weisbach equation and to Hazen and Williams formula). Consideration must also be given to the roughness of the pipe surface, as friction will vary widely depending on the type of material used. The head losses associated with flow through fittings can also be significant and will vary depending on the size and construction of the fitting. For example, a circular bend with corrugated inner radius will have a friction loss from 1.3 to 1.6 times greater than that of an equivalent smooth elbow or bend. As velocities increase, so do the dynamic losses and the power consumption. Roughness factors and friction losses in standard fittings are also listed in piping design handbooks.

Once the total head is determined, input power may be calculated for a known discharge flow rate as follows:

$$P = \gamma q h_T = G h_T \quad (31\text{-}2)$$

Where:
P = Power
q = Discharge rate
γ = Specific weight of fluid
h_T = Total head
G = γq = Mass rate of flow

In English system units, if q is in cf/s, γ is in lbm/cf, h_T is in ft of head, and G is in lbm/s, then the theoretical hp, also known as the hydraulic hp requirement, can be expressed as:

$$hyd\ hp = \frac{q\gamma h_T}{550} = \frac{G h_T}{550} \quad (31\text{-}3)$$

In SI units, when h_T is in kPa and G is in kg/s, the hydraulic power equation becomes:

$$hyd\ kW = \frac{G h_T}{1{,}000} \quad (31\text{-}3a)$$

When q is expressed in gpm, the hydraulic power equation becomes:

$$hyd\ hp = \frac{q h_T SG}{3{,}960} \quad (31\text{-}4)$$

Where:
SG = Fluid specific gravity

In SI units, if q is in L/m and h_T is in kPa, the hydraulic power equation becomes:

$$hyd\ kW = \frac{q h_T SG}{60{,}000} \quad (31\text{-}4a)$$

Considering total pump efficiency (η_p), the required power input in bhp that must be imparted to the shaft at a given flow (in gpm), at a specific gravity of the liquid being pumped and total head is:

$$bhp = \frac{q h_T SG}{3{,}960\,\eta_p} \quad (31\text{-}5)$$

In SI units, where flow is in L/m, the power input, expressed in bkW, becomes:

$$bkW = \frac{q h_T SG}{60{,}000\,\eta_p} \quad (31\text{-}5a)$$

To determine the capacity or energy input rating of the driver required to impart to the shaft of the pump the required power, the driver's efficiency (η_D) or ability to convert energy into shaft brake power, must also be considered.

Affinity Laws

Capacity (q) or rate of flow varies directly with the speed of the pump. Thus, the effect of a change of speed from N_1 to N_2 on pump capacity can be expressed as:

$$\frac{q_1}{q_2} = \frac{N_1}{N_2} \quad (31\text{-}6)$$

Total head (h_T) varies in proportion to the square of the flow rate and of the speed. Thus:

$$\frac{h_{T1}}{h_{T2}} = \left(\frac{q_1}{q_2}\right)^2 = \left(\frac{N_1}{N_2}\right)^2 \quad (31\text{-}7)$$

Hydraulic power input varies in proportion to the cube of the flow rate and of the speed. Thus:

$$\frac{hyd\,p_1}{hyd\,p_2} = \left(\frac{q_1}{q_2}\right)^3 = \left(\frac{N_1}{N_2}\right)^3 \quad (31\text{-}8)$$

and:

$$\frac{q_1 h_{T1}}{q_2 h_{T2}} = \left(\frac{q_1}{q_2}\right)^3 = \left(\frac{N_1}{N_2}\right)^3 \quad (31\text{-}9)$$

Pump Selection

Pump selection is based on required fluid flow rate and total head against which the pump must operate, including friction losses and net vertical lift. A selection is made from among alternatives to optimize hydraulic efficiency and meet other operating requirements. Figures 31-28 through 31-30 show three sets of performance curves for a selected pump, operating at speeds of 3,500, 1,700, and 1,170 rpm, respectively.

Table 31-1 provides data from a system curve showing the relation between head loss (HL) and fluid flow rate for a given piping system. As an example, for a required flow rate of 2,350 gpm (8,895 L/m) at 365 ft (111 m) head loss, an impeller size of 10.5 in. (26.7 cm) has been selected. On the 3,500 rpm pump curve shown in Figure 31-25, the brake power is calculated, based on Equation 31-5, as follows:

$$bhp = \frac{HL \, x \, gpm}{3{,}960 \, x \, \eta_p} = \frac{365 \, x \, 2{,}350}{3{,}960 \, x \, 0.83} = 261.0 \; bhp \; (194.6 \; kW)$$

The affinity laws can also be used to calculate performance of the selected pump at reduced speed. For example, at 1,750 rpm, flow rate would be calculated using Equation 31-6 as follows:

$$\frac{N_1}{N_2} = \frac{q_1}{q_2} = \frac{3500}{1{,}770} = \frac{2{,}350}{N_2}$$

From Equation 31-8, power requirement at 1,750 rpm would be:

$$\left(\frac{N_1}{N_2}\right)^3 = \frac{bhp_1}{bhp_2} = \left(\frac{3{,}500}{1770}\right)^3 = \frac{261}{bhp_2};$$

Thus, $bhp_2 = 33.7 \; bhp \; (25.1 \; kW)$

Table 31-1

HL (ft)	Flow (gpm)	HL (ft)	Flow (gpm)	HL (ft)	Flow (gpm)
447	2,600	180	1,650	37	750
413	2,500	159	1,550	28	650
381	2,400	139	1,450	20	550
350	2,300	120	1,350	13	450
306	2,150	103	1,250	8	350
278	2,050	87	1,150	4	250
251	1,950	73	1,050	1	150
226	1,850	60	950	0	50
202	1,750	48	850	0	0

Figure 31-31 is a generic representation of a speed versus power curve that indicates the potential benefits of variable speed operation. Table 31-2 provides a numerical summary of the relationship. As shown, a 20% reduction in speed (and flow) will result in a 50% reduction in required power. It is important to note, however, that the indicated power reduction may not be fully achievable in practice due to pressure losses in piping runs, throttling at individual loads, or hydraulic lift requirements.

Pump Driver Options

As with most mechanical drive applications, ac electric induction motors predominate, particularly in smaller

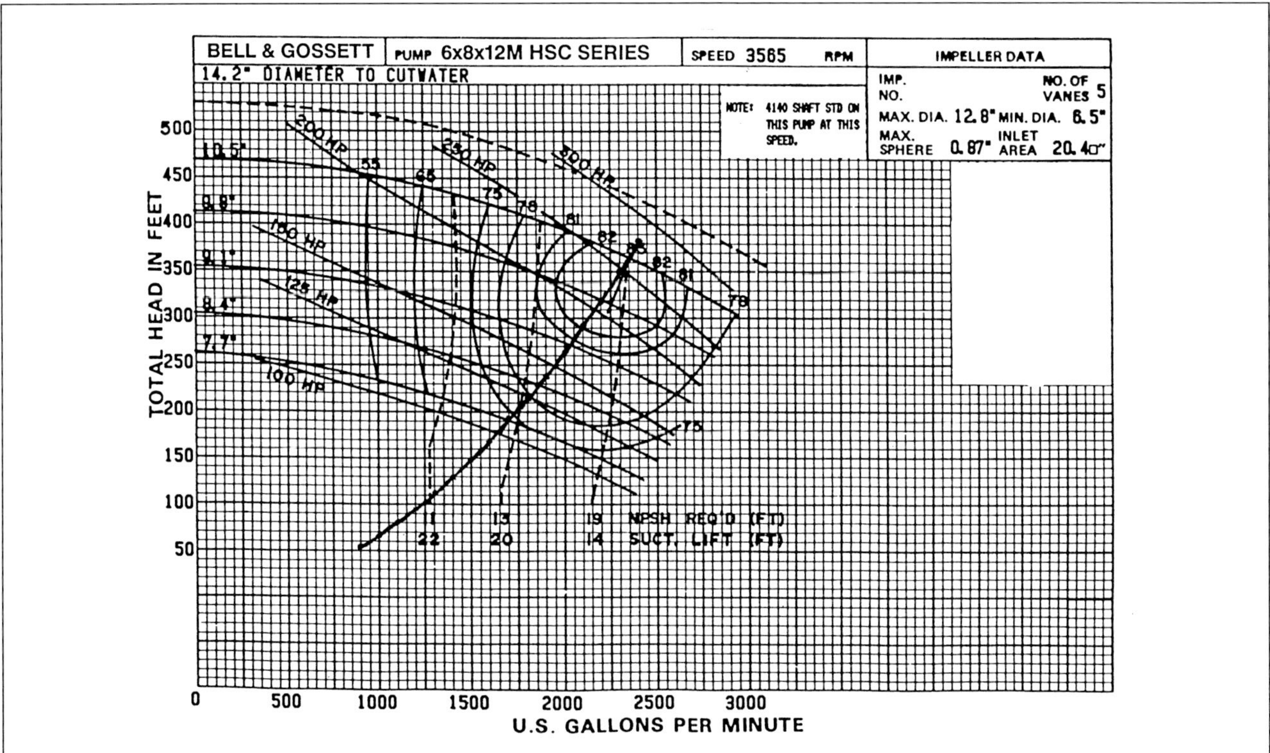

Fig. 31-28 3,500 rpm Pump Performance Curves. Source: Bell & Gossett

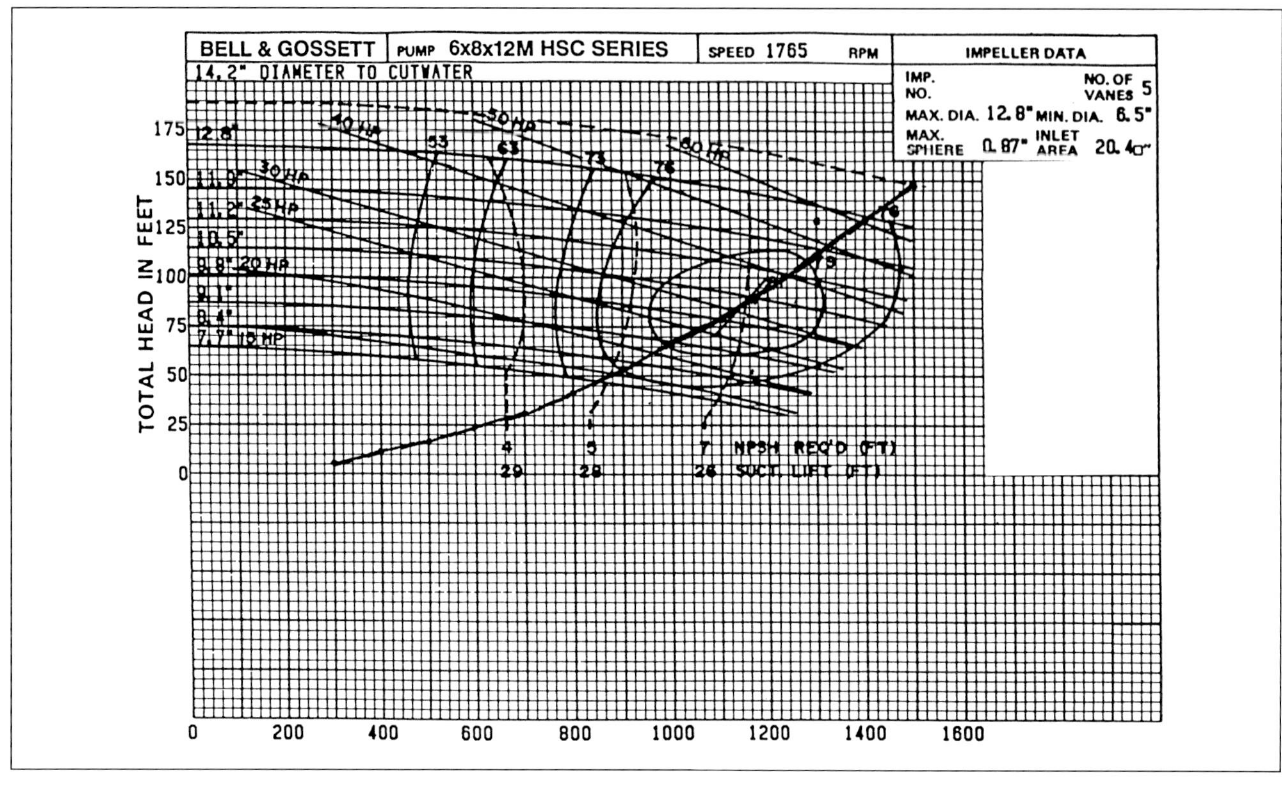

Fig. 31-29 1,700 rpm Pump Performance Curves. Source: Bell & Gossett

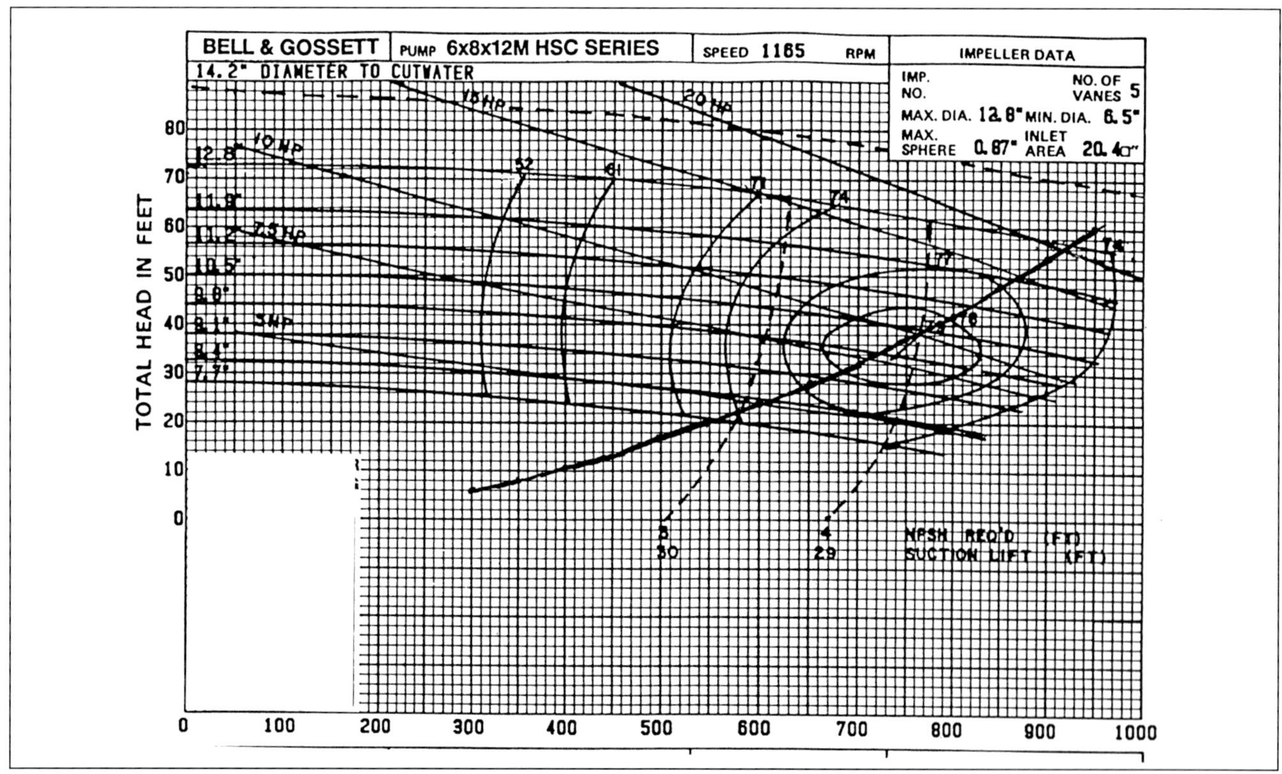

Fig. 31-30 1,170 rpm Pump Performance Curves. Source: Bell & Gossett

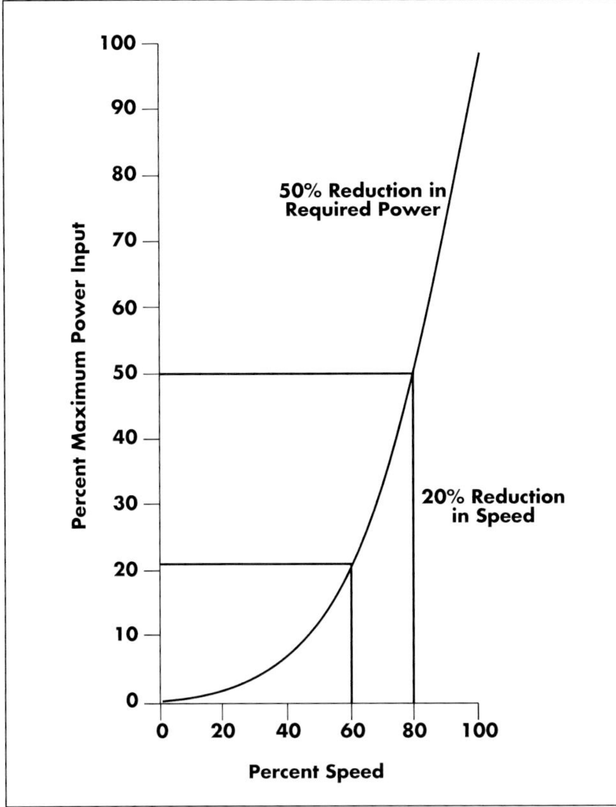

Fig. 31-31 Representative Speed vs. Power Curve.

Table 31-2
Numerical Description of Affinity Laws

Speed (%)	Flow (%)	Required Power (%)
100	100	100
90	90	73
80	80	50
70	70	34
60	60	22
50	50	13
40	40	6
30	30	3

capacities. Speed control can be accomplished with a dc motor, a multiple-speed ac motor, or by equipping an ac motor with a variable frequency drive (VFD). VFDs can control speed precisely over a wide range of operation, typically from 40 to 100% of full load. While far less costly and easier to install than prime mover drives, VFDs do not provide electrical demand savings during operation at full speed.

Under certain circumstances, investment in prime mover drives is justified on a life-cycle economic basis or by the need for backup in the event of an electric service outage. Along with the ability to provide speed control, prime mover drives provide electrical demand savings and the potential for heat recovery. Pumps may also be pneumatically or hydraulically driven. This can be effective, for example, when the pump location cannot support the weight or size of the driver. For remote applications, as still found in several rural farms, windmills are sometimes used to drive pumps. In other areas, water turbines are used. While wind and water turbine applications are not very common, they can be practical and cost-effective, especially when electric service is not readily available or cost-prohibitive.

Steam turbines have been commonly used for more than a century for driving pumps in various industrial process and central plant applications. Figure 31-32 shows a multistage steam turbine driving a decoking pump. Figure 31-33 shows a compact packaged steam turbine process pump drive rated at 1,750 hp (1,300 kW). The design includes a forced-feed lubrication system, a separate oil pump, twin coolers and filters, rotor vibration

Fig. 31-32 Multistage Steam Turbine in Decoking Pump Drive Application. Source: Dresser-Rand

Fig. 31-33 Packaged 1,750 hp Steam Turbine-Driven Process Pump Drive. Source: Tuthill Corp. Coppus Turbine Division

Fig. 31-34 Reciprocating Engine Pump Drives Applied in Flood Control Station. Source: Fairbanks Morse Engine Div.

and axial position monitoring system, and radial bearing temperature monitoring system.

Combustion engines are far less common for stationary pump applications, though they are used for remote field applications. Reciprocating engine-driven pumps are commonly applied in irrigation and district or pipeline pumping applications. In such cases, reciprocating engines offer relatively high simple-cycle thermal fuel efficiency and the ability to operate efficiently under variable load conditions. Figure 31-34 shows ten reciprocating engines used to drive large propeller pumps in a flood control station.

Fig. 31-35 Two-Shaft Gas Turbine Featured in Large Pumping Application. Source: Solar Turbines

Gas turbine drives are generally used only for very large capacity remote pumping applications. Figure 31-35 shows a two-shaft design applied in a pumping application requiring 10,000 hp (7,500 kW).

Pumps are generally not pre-packaged with prime movers because applications and equipment types are too diverse. There are three basic drive configurations: a single prime mover drive, a hybrid setup using one pump and two drivers on a shaft, and a hybrid setup using two driver/pump sets.

It may also be less costly to purchase a hybrid unit rather than two entirely separate pump/driver sets. Dual-shaft, or common single-shaft, pumps, which allow operation from two different types of drivers, are often used in prime mover drive applications. An electric motor can be included along with a prime mover drive, such as a back-pressure steam turbine. In the case of the failure of one driver, the other driver can easily take over. The electric drive could also be used when back-pressure steam loads are insufficient, or during inexpensive off-peak rate periods. When two pumps are desired to provide increased reliability, a hybrid system may be appropriate.

Boiler feedwater pumps are a particularly attractive back-pressure turbine application. As feedwater flow is reduced, turbine steam flow and available back-pressure steam are also reduced. Turbine discharge is used for feedwater heating, which varies in direct proportion to feedwater flow. Because feedwater pumps are critical for plant operation, multiple pumps or drivers are usually installed. Often, hybrid systems are used, with one or more electric motors and prime movers.

Figure 31-36 shows a boiler feed pump driven by a steam turbine. The 26 hp (19 kW) turbine serves a 40,000 lbm/h (18,000 kg/h) steam boiler and operates at 3,553 rpm, with an inlet steam condition of 195 psig (14.5 bar) dry saturated steam. Turbine exhaust steam at 5 psig (1.4 bar) feeds a deaerator tank.

Figure 31-37 shows a 375 hp (280 kW) vertical turbine pump driven by a gas-fired reciprocating engine. This system is part of four pumping stations that move water through 14 miles (22.5 kilometers) of pipeline from a desert to a New Mexico municipality. The pump specification required variable speed operation to meet the following head-capacity requirements:

- 1,600 gpm at 380 ft of head (6,056 L/m at 11.3 bar)

Fig. 31-36 26 hp Steam Turbine-Driven Boiler Feed Pump. Source: UTC Sikorsky Aircraft

- 2,900 gpm at 410 ft of head (10,977 L/m at 12.3 bar)
- 2,550 gpm at 490 ft of head (9,652 L/m at 14.6 bar)
- 2,070 gpm at 552 ft of head (7,835 L/m at 16.5 bar)

The system design called for maximum efficiency to be obtained when pumping 2,900 gpm (10,977 L/m) at 410 ft of head (12.3 bar) operating under a maximum ambient temperature condition of 100°F (38°C) at an elevation of about 6,000 ft (1,829 m). An exhaust silencer was also required, designed to maintain a 60 dba sound emission level at a distance of 25 ft (7.6 m). Figure 31-38 shows the performance curves for the featured engine over the range of operating speeds.

Fig. 31-37 375 hp Vertical Turbine Pump Driven by Reciprocating Engine in Pumping Station Application. Source: Waukesha Engine Div.

The driver is a turbocharged and intercooled, lean combustion, eight cylinder, four cycle, natural gas-fired engine with an 11:1 compression ratio. Operation is between 1,400 and 1,750 rpm to deliver between 205 and 375 hp (153 to 280 kW). Dimensions are 7.9 ft (2.4 m) long by 4.5 ft (1.4 m) wide by 5.6 ft (1.7 m) high, with a dry weight of 7,200 lbm (3,300 kg) and a piston displacement of 1,462 cubic inches (24 L). The engine is equipped with a 24 volt dc starting and charging system and necessary controls for automatic operation.

Simplified Economic Analysis of Pump Driver Options

Following are simplified comparative analyses for mechanical service application of three alternative pump drivers:

- Back-pressure steam turbine
- Natural gas spark-ignition industrial-grade reciprocating engine
- Variable frequency drive (VFD) electric motor

All three options are compared to a base case consisting of a constant speed electric motor. A natural gas cost of $4.00/Mcf is assumed for each of the fuel-driven options. Electricity costs are based on a time-of-use (TOU) rate with the following components:

Demand, $/kW/Month	$14.67
Peak, $/kWh	$0.0506
Shoulder, $/kWh	$0.0403
Off-peak, $/kWh	$0.0279

It is assumed that the pump would operate continuously throughout the year under varying flow requirements. Maximum flow condition is 2,230 gpm (8,441 L/M) against a total dynamic head (TDH) of 140 ft (43 m), and minimum required head is 30 ft (9 m). Pumping energy requirement, pump efficiency, and driver efficiency all vary with load and speed. Required flow rate during the various electric billing periods is shown in Figure 31-39.

Base Case Constant Speed Motor Option

The constant speed motor drives the pump at 1,750 rpm and flow control at the load is achieved by by-passing a varying fraction of the discharge flow. Input power is assumed constant at 74.3 kW, as indicated in Figure 31-40. Figure 31-41 shows the calculation of annual operating cost.

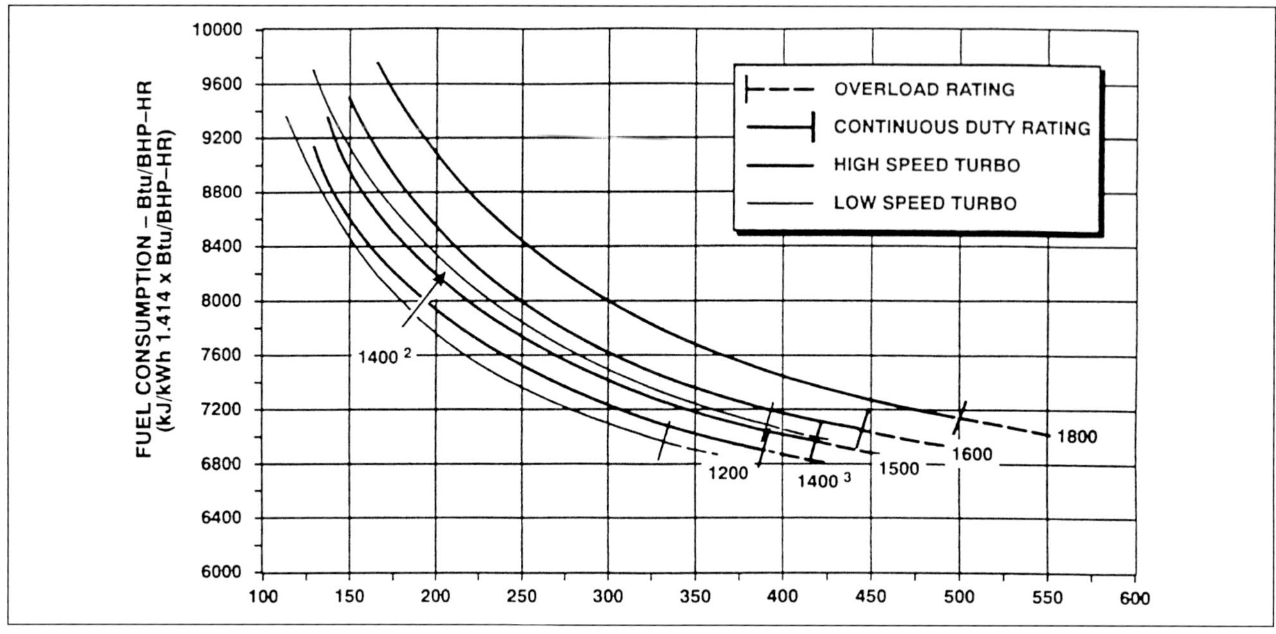

Fig. 31-38 Full- and Part-Load Performance Curves for Engine Pump Driver. Source: Waukesha Engine Division

Load		Hours			
Flow rate	% Capacity	On-Peak	Shoulder	Off-Peak	Total
2,230	100%	200	80	120	400
2,007	90%	350	300	300	950
1,784	80%	500	450	600	1,550
1,561	70%	500	480	620	1,600
1,338	60%	310	400	890	1,600
1,115	50%	190	250	910	1,350
892	40%	40	130	1,140	1,310
		2,090	**2,090**	**4,580**	**8,760**

Fig. 31-39 Flow Rate During Electric Billing Periods.

VFD Option

The first-cost premium for the VFD driver versus the constant speed motor was assumed to be $15,000, with no incremental O&M cost. Input power is indicated in Figure 31-42, based on representative pump curves and the pump affinity laws. An additional 5% peak power requirement was added to account for the power draw of the VFD itself. Operating costs are calculated in Figure 31-43. Notice that while total kWh usage is about 45% that of the constant speed electric motor, demand charges actually increase due to the power draw of the VFD itself. Simple payback versus the base case is 1.6 years, as shown in Figure 31-44.

Back-Pressure Steam Turbine Option

The first-cost premium for the back-pressure steam turbine driver versus the constant-speed motor was assumed to be $60,000, with an incremental O&M cost of $2,500. Steam boiler efficiency is 83% and it is assumed that all back-pressure steam is passed on to process. Energy requirement and operating cost are calculated in Figure 31-45. Notice the net steam energy requirement remains constant at 2,546 Btu/bhp, even though the turbine mechanical efficiency would decrease significantly under lower load at lower speed. This is due to the fact that all unused steam energy from the topping cycle is passed on to process. The simple payback for the steam turbine driver versus the constant-speed motor base case is 2.0 years, as shown in Figure 31-46.

% Capacity	Load (gpm)	Bypass (gpm)	TDH, (ft)	Pumpage hp	Pump Eff.	Pump hp	Motor Eff.	Input Power (kW)
100%	2,230	–	140	78.5	83%	94.6	95%	74.3
90%	2,007	223	140	78.5	83%	94.6	95%	74.3
80%	1,784	446	140	78.5	83%	94.6	95%	74.3
70%	1,561	669	140	78.5	83%	94.6	95%	74.3
60%	1,338	892	140	78.5	83%	94.6	95%	74.3
50%	1,115	1,115	140	78.5	83%	94.6	95%	74.3
40%	892	1,338	140	78.5	83%	94.6	95%	74.3

Fig. 31-40 Load and Power Requirement for Constant Speed Motor.

Reciprocating Engine Option

The first-cost premium of the

% Capacity	Power (kW)	Consumption kWh/Year On-Peak	Shoulder	Off-Peak	Total	Cost $/Year On-Peak	Shoulder	Off-Peak	Total
100%	74.3	14,858	5,943	8,915	29,716	$ 753	$ 240	$ 250	$ 1,242
90%	74.3	26,002	22,287	22,287	70,576	$ 1,317	$ 899	$ 624	$ 2,840
80%	74.3	37,145	33,431	44,574	115,151	$ 1,882	$ 1,348	$ 1,248	$ 4,478
70%	74.3	37,145	35,660	46,060	118,865	$ 1,882	$ 1,438	$ 1,290	$ 4,610
60%	74.3	23,030	29,716	66,119	118,865	$ 1,167	$ 1,199	$ 1,851	$ 4,217
50%	74.3	14,115	18,573	67,604	100,292	$ 715	$ 749	$ 1,893	$ 3,357
40%	74.3	2,972	9,658	84,691	97,321	$ 151	$ 390	$ 2,371	$ 2,911
	TOTALS	**155,267**	**155,267**	**340,251**	**650,786**	**$7,867**	**$6,262**	**$9,527**	**$23,656**
		Demand	**Max kW x $14.67/kW/Month x 12 Month/Year**						**$13,024**
							Total Cost		**$36,680**

Fig. 31-41 Operating Cost for Constant Speed Motor.

% Capacity	Load (gpm)	TDH (ft)	Pump Speed (rpm)	Pumpage hp	Pump Eff.	Pump bhp	Drive/ Motor Eff.	Input Power (kW)
100%	2,230	140	1,750	78.5	83%	94.6	90%	78.4
90%	2,007	119	1,610	60.1	83%	72.4	90%	60.0
80%	1,784	100	1,470	45.0	83%	54.3	90%	45.0
70%	1,561	84	1,330	32.9	83%	39.7	90%	32.9
60%	1,338	70	1,190	23.4	77%	30.4	90%	25.2
50%	1,115	58	1,050	16.1	70%	23.0	85%	20.2
40%	892	48	910	10.7	65%	16.4	80%	15.3

Fig. 31-42 Load and Power Requirement for VFD Motor Option.

% Capacity	Power (kW)	Consumption kWh/Year On-Peak	Shoulder	Off-Peak	Total	Cost $/Year On-Peak	Shoulder	Off-Peak	Total
100%	78.4	15,684	6,273	9,410	31,367	$ 795	$ 253	$ 263	$ 1,311
90%	60.0	21,014	18,012	18,012	57,038	$ 1,065	$ 726	$ 504	$ 2,296
80%	45.0	22,495	20,245	26,994	69,734	$ 1,140	$ 817	$ 756	$ 2,712
70%	32.9	16,448	15,790	20,396	52,634	$ 833	$ 637	$ 571	$ 2,041
60%	25.2	7,816	10,085	22,440	40,342	$ 396	$ 407	$ 628	$ 1,431
50%	20.2	3,841	5,054	18,398	27,294	$ 195	$ 204	$ 515	$ 914
40%	15.3	613	1,992	17,465	20,070	$ 31	$ 80	$ 489	$ 600
	Totals	**87,911**	**77,453**	**133,115**	**298,479**	**$4,454**	**$3,124**	**$3,727**	**$11,305**
			Demand	**Max kW x $14.67/kW/Month x 12 Month/Year**					**$13,802**
							Total Cost		**$25,107**

Fig. 31-43 Electric Usage and Cost Breakdown for VFD Motor Option.

Reduction in Cost: VFD Compared with Bypass Control, Per Year	$ 11,573
Incremental Capital Cost of VFD Over Bypass Control	$ 15,000
Simple Payback of VFD Over Bypass Control, Years	**1.6**

Fig. 31-44 Simple Payback on Investment in VFD Motor vs. Base Case.

reciprocating engine versus the constant-speed motor is assumed to be $70,000, with an annual incremental O&M cost of $0.01/bhp-h. It is assumed that no heat is recovered from the engine. Figure 31-47 calculates reciprocating engine energy requirement and operating cost.

In this case, the minimum flow and TDH is limited by the reciprocating engine minimum operating speed, which is assumed to be 50% of the design rating of 1,800 rpm. The simple payback versus the base case is shown in Figure 31-48 to be 3.6 years.

Driver Option Comparison

All three driver options show excellent economic performance in these simplified analyses. The VFD option produces a very short simple payback peri-

Capacity	Load (gpm)	TDH (ft)	Pump Speed (rpm)	Pump bhp	Turbine, Btu/bhp-h	Fuel Rate (Mcf/h)	Hours per Year	Fuel Use (Mcf/Yr)	Fuel Cost ($/Yr)
100%	2,230	140	1,750	94.6	2,546	0.282	400	113	$ 451
90%	2,007	119	1,610	72.4	2,546	0.216	950	205	$ 820
80%	1,784	100	1,470	54.3	2,546	0.162	1,550	251	$ 1,002
70%	1,561	84	1,330	39.7	2,546	0.118	1,600	189	$ 756
60%	1,338	70	1,190	30.4	2,546	0.091	1,600	145	$ 580
50%	1,115	58	1,050	23.0	2,546	0.069	1,350	93	$ 370
40%	892	48	910	16.4	2,546	0.049	1,310	64	$ 256
						Totals	8,760	1,059	$ 4,236
							Annual O&M Cost		**$ 2,500**
							Total Cost		**$ 6,736**

Fig. 31-45 Load and Operating Cost Breakdown for Steam Turbine Drive.

Reduction in Cost: Turbine Compared with Bypass Control, Per Year	$ 29,945
Incremental Capital Cost of Turbine Over Bypass Control	$ 60,000
Simple Payback of Turbine Over Bypass Control, Years	**2.0**
Reduction in Cost: Turbine Compared with VFD, Per Year	$ 18,371
Incremental Capital Cost of Turbine Over VFD	$ 45,000
Simple Payback of Turbine Over VFD, Years	**2.4**

Fig. 31-46 Paybacks for Steam Turbine Drive vs. Base Case and VFD Option.

od of 1.6 years, making it a highly attractive option. The back-pressure steam turbine, produces a 2.0 year payback versus the base case and a 2.4 year incremental payback versus the VFD option, while the reciprocating engine produces a 3.6 year payback versus the base case and a 7.1 year incremental payback versus the VFD option. These results indicate that the additional investment in the more costly prime mover drive options would merit serious consideration.

It is important to note that in these cases, project economics are particularly attractive because the pumping loads vary considerably and the demand component of the electric rate is relatively high. With an alternative electric rate in which the demand component were lower but the commodity cost (per kWh) were higher, the electric VFD motor would show even better economic performance. Under such a rate structure, the two prime mover options would show similar payback periods versus the base case option, but would show higher incremental payback periods versus the VFD option.

It is also important to note that the capital cost requirement for the two prime mover options are based on somewhat optimal installation conditions. Pricing is based on the assumption of a logistically straight-forward installation. Any additional system modifications

Capacity	Load (gpm)	TDH (ft)	Pump Speed (rpm)	Pump bhp	Engine, Btu/bhp-h	Fuel Rate (Mcf/h)	Hours per Year	Fuel Use (Mcf/Yr)	Fuel Cost ($/Yr)
100%	2,230	140	1,750	94.6	10,099	0.928	400	371	$ 1,484
90%	2,007	119	1,610	72.4	8,848	0.622	950	591	$ 2,365
80%	1,784	100	1,470	54.3	8,464	0.446	1,550	691	$ 2,765
70%	1,561	84	1,330	39.7	8,948	0.345	1,600	552	$ 2,207
60%	1,338	70	1,190	30.4	10,299	0.304	1,600	487	$ 1.947
50%	1,115	58	1,050	23.0	12,516	0.280	2,660	745	$ 2,978
						Totals	**8,760**	**3,436**	**$ 13,746**
							Annual O&M Cost		**$ 3,642**
							Total Cost		**$17,388**

Fig. 31-47 Load and Operating Cost Breakdown for Reciprocating Engine Drive.

Reduction in Cost: Engine Compared with Bypass Control, Per Year	$ 19,292
Incremental Capital Cost of Engine Over Bypass Control	$ 70,000
Simple Payback of Engine Over Bypass Control, Years	3.6
Reduction in Cost: Engine Compared with VFD, Per Year	$ 7,719
Incremental Capital Cost of Engine Over VFD	$ 55,000
Simple Payback of Engine Over VFD, Years	7.1

Fig. 31-48 Paybacks for Reciprocating Engine Drive vs. Base Case and VFD Option.

required for the installation would add to the capital cost and, therefore, increase the payback period.

In the case of the steam turbine driver, it is assumed that all of the rejected heat, in the form of low-pressure steam, can be effectively used without any significant system modifications. If all of the heat cannot be used during each period of operation, or if system modifications were required to allow for the use of this heat, such as a requirement for additional piping, economic performance would be decreased.

In the case of the reciprocating engine driver, no consideration has been given to heat recovery. If heat recovery was included in the analysis, the capital cost would be increased to account for the heat recovery system and the annual O&M cost would also be increased. However, the annual fuel consumption would be decreased by as much as half. The net impact would be improved economic performance, demonstrated by a shorter payback period on the incremental investment.

Still, while these are just simplified examples, based on only one set of assumptions, the general conclusion holds true that alternative driver options that feature variable speed control can produce attractive economics for pumping applications.

Consideration of Renewable Pump Driver Options

While rare in application for commercial, industrial and institutional facilities, both water and wind power systems merit consideration as pump drivers under certain circumstances. Wheras a century ago these were mainstream pump drivers, water and wind turbines are now predominantly used for electric generation only.

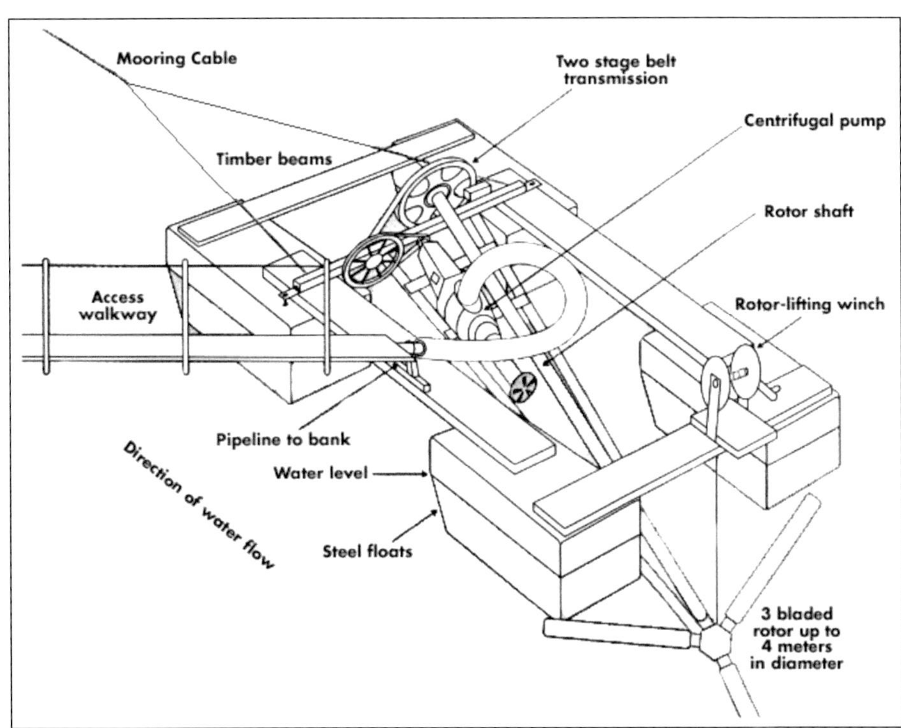

Figure 31-49. Water-Turbine Driven Pump Used to Provide Potable Water in Remote Location. Source: CADDET Renewable IEA/OECD

Water turbines are certainly effective pumps. Their mechanical efficiency in generating power from rushing water can be equally applied in reverse when used as pumps. In most cases, however, water turbine pumping applications are integrated into hydroelectric plants. Dedicated pump generators-sets are most commonly used, in which turbine-generated electricity is then used to power motor-driven pump sets. In pump storage applications, pumping is commonly accomplished by simply running the water turbine-generator set in reverse. In such cases, the generator operates as a motor and the water turbine as the pump.

Still there are some instances where hydropower plants operate exclusively to drive pumps.

Figure 31-50. Low-Wind Water-Pumping Windmill. Source: Peter Michaelis, DoE/National Renewable Energy Laboratory

These are usually in remote locations that do not require much power other than for pumping and have little or no access to electric transmission lines. Hence the electric generator and motor can be eliminated, reducing capital cost and additional mechanical efficiency losses associated with the generator. Figure 31-49 is a layout illustration of a water-turbine driven pumping system, used to provide potable water at a remote location in Sudan. This is a water current turbine that floats on a river with the rotor completely submerged. It is moored in a free stream to a post on one bank to minimize obstruction to river traffic. This run-of-the-river type system operates with no hydraulic head and, therefore, requires no dam or reservoir. The system uses an 11-ft (3.5-m) diameter, 3-blade Garman water current turbine with an inclined-axis rotor design. A two-stage belt transmission links the rotor to a centrifugal pump. The usable power is proportional to the rotor area and the cube of the water speed. The maximum pumping head for a single turbine set such as this one is about 80 ft (25 m). However, by installing turbines side-by-side with their pumps in series, higher heads can be generated.

Along with mill grinding, water pumping was historically among the most common applications for windmills. While in great decline following the national electrification movement of the early- and mid-20th Century, windmill water pumping applications were previously found scattered across thousands of farms throughout the United States. Today, most wind power applications are for electric generation, though numerous farm pumping applications can still be found. For grid-connected applications, minimum wind speeds of about 12 miles/hr (5.4 m/s) are required. Non-grid connected electric generation and other mechanical drive applications, however, can be effectively accomplished at lower wind speeds. Figure 31-50 shows a low-wind water-pumping windmill that is still frequently used for pumping water for livestock, irrigation, village water supplies, pond aeration and remote homes and farms where ordinary windmills cannot function because of low available wind speeds. This Oasis windmill pumps water in winds below 5 miles/hr (2.2 m/s), pulls it from very deep wells, and economically pumps large volumes of water.

As can be seen from these two examples, the use of wind or water power directly for pumping has become more of the exception to the rule as electric power generation has become the predominate application for these technologies. With available electricity from such applications, standard motor-driven pumps are used. However, given the appropriate set of circumstances, such technology applications are viable and can be practically applied.

Chapter Thirty-Two

Fans

Fans are volumetric machines that move quantities of air (or other gases), overcoming resistance to flow by supplying the energy necessary for continued motion. Fans are similar in many respects to pumps and compressors. All three are turbomachines that transfer energy to a flowing fluid. While fans are easily distinguished from pumps (which handle liquids), distinctions between fans and compressors are not so clearly defined. Historically, distinctions have been made on the basis of compression ratio or density change. For most purposes, low-pressure-rise machines may be classified as fans and high-pressure-rise machines may be classified as compressors. A broad distinction is that the function of a compressor is generally to increase pressure, whereas fans are generally used to propel the air or gas.

Fan Components

Fans typically include the following components:

- The **impeller** is the rotating element that transfers energy to the fluid. Impellers may be referred to as wheels, rotors, or, in certain designs, squirrel cages or propellers.
- The **blades**, or **vanes**, are the principal working surfaces of the impeller.
- The **housing**, also referred to as the casing or stator, is the stationary element that guides the air or gas across the impeller.
- The **inlet**, which may be referred to as the eye or suction, is the opening through which air enters the fan.
- The **outlet**, or **discharge**, is the opening through which air leaves the fan.
- **Stationary vanes** may be used to guide the flow. Vanes before the impeller are referred to as inlet-guide vanes. Vanes after the impeller are referred to as discharge-guide, or straightening, vanes.

Fan Types

Figure 32-1 illustrates the aerodynamic classification of fans. If fans are classified according to the direction of the flow through the impeller, there are four distinctive types: axial-flow and radial-flow fans (the types normally encountered in general service), mixed-flow fans, and cross-flow fans. Axial-flow fans produce flow in a direction parallel to the axis of the rotation, while radial-flow (or centrifugal) fans produce a flow that is parallel to the radius of rotation.

The **axial-flow fan** most commonly used is the disk or propeller fan, which consists of a propeller or a disk wheel placed within a ring casing or plate. It is the simplest type of fan and is best adapted to applications with high flow against low frictional resistance (through-wall building exhaust fans are an example). Figure 32-2 is a cutaway view illustration of an axial-flow fan.

When an axial-flow fan is placed in a cylindrical drum-type housing equipped with stationary directional vanes, it is called a **vane-axial-flow fan**. Vanes straighten discharge airflow that would otherwise follow a spiral path. If vanes are not part of the assembly, the fan

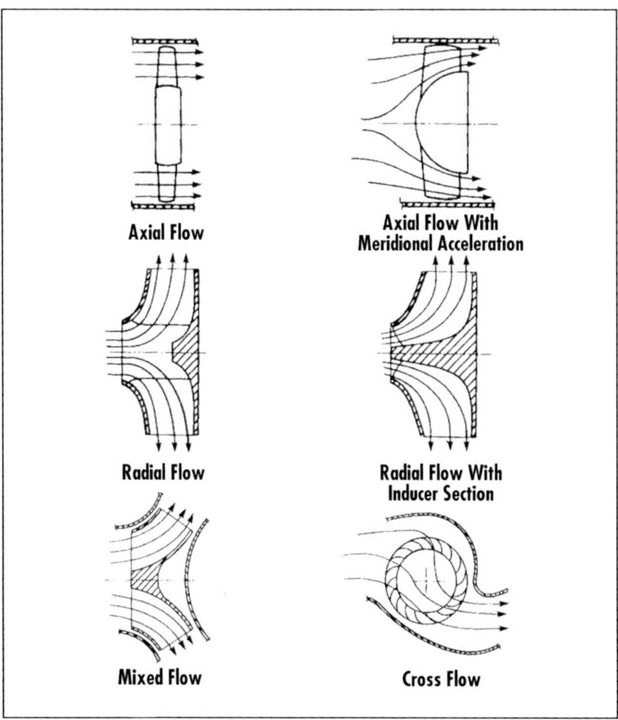

Fig. 32-1 Aerodynamic Classification of Fans. Source: Buffalo Forge, The Howden Fan Company

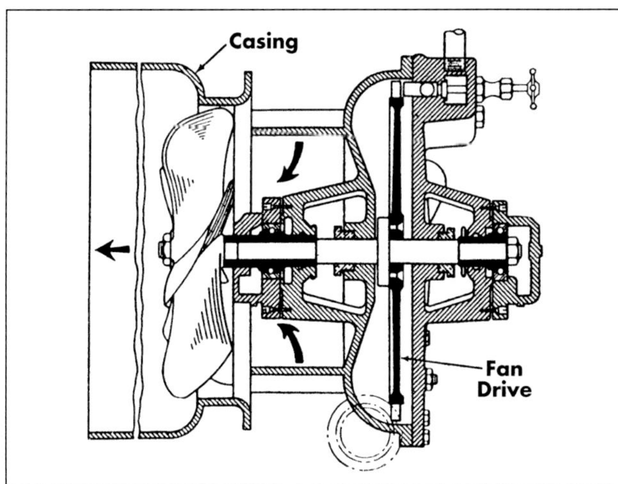

Fig. 32-2 Cutaway View Illustration of Axial-Flow Fan.
Source: Babcock and Wilcox

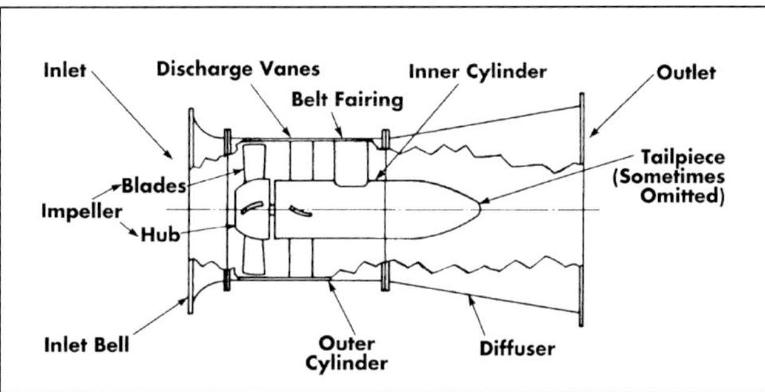

Fig. 32-3 Cutaway View of Vane-Axial Fan.
Source: Buffalo Forge, The Howden Fan Company

is called tube-axial. Axial-flow fans operate at good efficiencies and satisfactory sound emission levels if properly selected for air delivery and system resistance. Figure 32-3 shows a cutaway view of a vane-axial fan.

Radial or **centrifugal fans** are typically selected for airflows having higher frictional resistance. The impeller is mounted in a scroll-type housing, with the

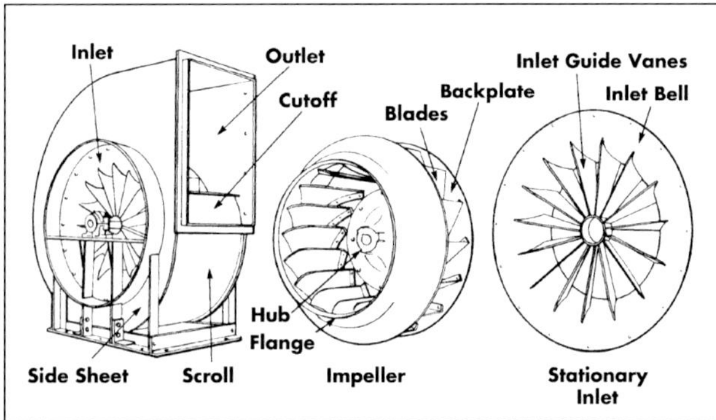

Fig. 32-4 Exploded View of Centrifugal Fan.
Source: Buffalo Forge, The Howden Fan Company

air entering the impeller parallel to the axis of rotation and leaving radially. The scroll formation of the fan housing converts the kinetic energy of the discharge air into potential energy in the form of static pressure. Figure 32-4 is an exploded-view of a centrifugal fan.

Figure 32-5 illustrates the various blade designs for centrifugal fans. Fans are classified according to blade configuration as follows:

- **Radial-blade** types can achieve high static pressure at high rotational speed, but have relatively low airflow capacity. These are rugged units that offer moderate efficiency.
- **Forward-curved blade** types develop high airflows at lower rotational speeds and are smaller and quieter. Input power requirement and delivered static pressure rise rapidly (allowing motor overloading) as airflow is increased under free-delivery conditions.
- **Backward-curved** or **backward-inclined blade** types typically develop lower airflows with higher delivered static pressure. These types have a non-overloading power characteristic, with input power and static pressure falling as airflow is increased under free-delivery conditions. Backward-curved fans operate at high efficiency, although with high sound emission levels. When the backward-curved blades are manufactured with an airfoil cross-section, there are measurable gains in stability, performance, and efficiency. Airfoil centrifugal fans are also relatively quiet.

Whether a forward- or backward-curved blade is used depends on sound emission limitations, efficiency, and desired performance characteristics. The maximum mechanical efficiencies obtainable for the listed fan types are not widely different, although there are marked differences in the way power and static pressure vary with changes in airflow. Centrifugal fans usually have a minimum sound emission level near the point of maximum efficiency. If a fan is noisy, it is often running at high speed to meet an excessive static pressure requirement or is undersized for the application.

Fan Formulas and Affinity Laws

In fan engineering, standard air is considered to be air with a density of 0.075 lbm/ft^3 when English units are used and 1.2 kg/m^3 when SI units are used. Commonly applied industry values for standard air are shown in Table 32-1. For moist air, it is common to

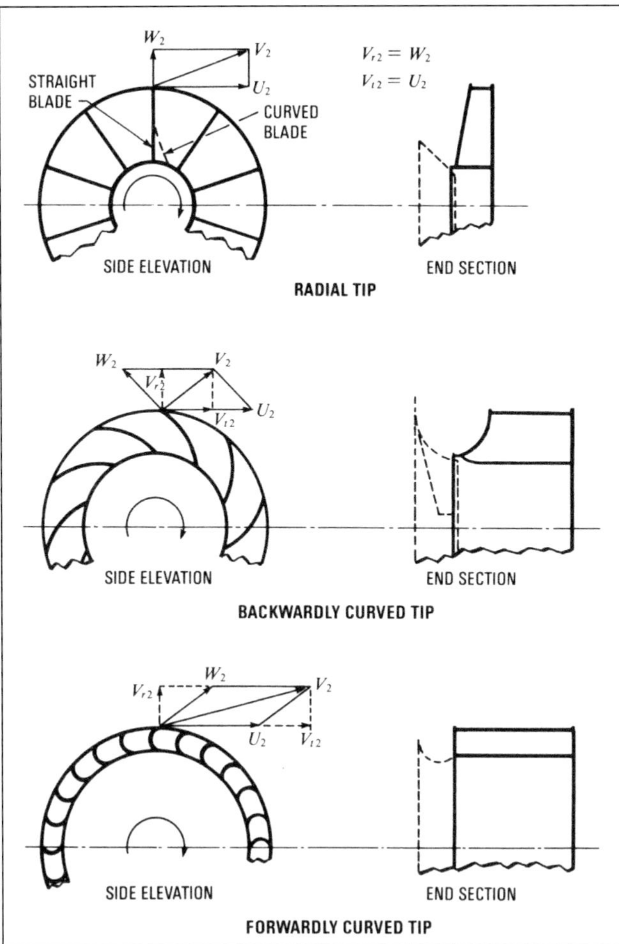

Fig. 32-5 Blade Design for Backward, Radial, and Forward Tips. Source: Buffalo Forge, The Howden Fan Company

Table 32-1 Commonly Applied Industry Values for Standard Air Condition

Property	English Units	SI Units
Dry Air Pressure	29.921 in. Hg	101.325 kPa
Temperature	70°F (529.7°R)	21°C (294.2°K)
Humidity	0 %	0 %
Density	0.075 lbm/ft^3	1.2 kg/m^3

assume a humidity of 50% and a temperature of 68°F (527.7°R) in English units and 20°C (293.2°K) in SI units.

Flow rates may be considered on a mass or volumetric basis. On a mass basis, the flow rate is the mass of the fluid passing through the fan per unit of time. Since air or gas is compressible, the volumetric flow rate will vary depending on the location at which it is measured. The volumetric flow rate at the fan inlet is equal to the mass flow rate divided by the gas density.

The same static and dynamic head or pressure components that exist with pumps are present in compressible flow applications. Pressure gauges are calibrated in inches (in.) of water (the pressure exerted by a column of water 1 in. high) or in. of mercury (Hg) for higher ranges at 60°F (15.55°C). One in. of water equals 0.036 psi and 1 in. of mercury equals 13.6 in. of water. In SI units, 1 millimeter (mm) of Hg equals 13.6 mm of water and 0.133 Pa.

Head (h) is the ratio of pressure (p) and fluid density (ρ), expressed as follows:

$$h = \frac{p}{\rho} \tag{32-1}$$

If fluids other than Hg or water are used for measuring pressure, the conversion is based on the relationship:

$$h_1 p_1 = h_2 p_2 \tag{32-2}$$

where the subscripts 1 and 2 denote the fluids under consideration.

Fan total pressure (P_t) is the difference between the total pressure at the fan outlet and the total pressure at the fan inlet. Total pressure is expressed as:

$$P_t = P_{t2} - P_{t1} \tag{32-3}$$

If the fan draws directly from atmosphere:

$$P_{t1} = 0$$

If the fan discharges directly to atmosphere:

$$P_{t2} = P_{v2}$$

Fan velocity pressure (P_v) is the pressure that corresponds to the average velocity at the fan outlet. This is expressed as:

$$P_v = \left(\frac{q_{v2}/A_2}{1{,}097}\right)^2 \rho_a \tag{32-4}$$

or, as:

$$P_v = \left(\frac{v}{1{,}097}\right)^2 \rho_a \tag{32-5}$$

Where:

P_v = Fan velocity pressure (in wg)
q_v = Volumetric flow rate at fan outlet (in cfm)
A = Flow area at fan outlet (in ft²)
ρ_a = Fan air density at fan outlet (in lbm/ft³)
v = Velocity (in ft/min)

In SI units, when P_v is in kPa, ρ_a is kg/m³ and v is in m/s, Equation 32-5 becomes:

$$P_v = \frac{\rho_a v^2}{2} \tag{32-5a}$$

For air at standard density, fan velocity pressure can be expressed as:

$$p_v = \left(\frac{v}{K}\right)^2 \tag{32-6}$$

where K is a dimensionless compressibility factor defined as the ratio of fan total pressure that would be developed with an incompressible fluid.

Fan static pressure (P_s) is the difference between fan total pressure and fan velocity pressure and, therefore, the difference between the static pressure at the fan outlet and the total pressure at the fan inlet. Thus:

$$P_s = P_t - P_v = P_{s2} - P_{t1} \tag{32-7}$$

Fan power output (H_O) is expressed as:

$$H_O = \frac{P_t \dot{q} K}{C_q} \tag{32-8}$$

Where:

q = Volumetric flow rate
p_t = Total fan pressure
K = Compressibility factor
C_q = Dimensional constant

When H_O is expressed in hp, P_t in in. wg, and q in cfm, C_q is 6,354. When H_O is expressed in kW, P_t in kPa, and q in m³/s, C_q is 1.0.

Fan power input (H_I) is a function of the fan power output and the fan total efficiency, η_t. This is expressed as:

$$H_I = \frac{P_t \dot{q} K}{\eta_t C_q} = \frac{H_O}{\eta_t} \tag{32-9}$$

Total input power to a fan will also depend on the driver's efficiency (η_D).

Fan total efficiency (η_t) is the ratio of fan power output to fan power input, considering losses due to skin friction, turbulence, leakage, and mechanical friction. From Equation 32-9, it is expressed as:

$$\eta_t = \frac{H_O}{H_I} \tag{32-10}$$

Fan static efficiency (η_s) is the product of total fan efficiency and the ratio of fan static pressure to fan total pressure. It is expressed as:

$$\eta_s = \frac{\eta_t P_s}{P_t} \tag{32-11}$$

The actual power transmitted to the fluid and the actual head developed will both differ from the ideal due to various losses that affect head, power, or flow rate. **Hydraulic efficiency** is the ratio of the actual head to the ideal head. Head losses depend on the design of flow passages and result from skin friction and change of direction or velocity in the machine. Friction losses generally vary as the square of the velocity and, therefore, of flow rate. **Volumetric efficiency** is the ratio of the net volume flow rate handled by the machine to the volume flow rate handed by the impeller. Leakage flow passes through the clearance spaces between the rotating and stationary parts to recirculate through the impeller. **Mechanical efficiency** is the ratio of the power transmitted to the fluid to the power that is applied to the shaft. Mechanical losses include power loss due to disk friction and other losses in bearings, seals, etc.

The term **slip** refers to the fact that the impeller does not develop the full ideal head, nor transmit the full ideal power because the fluid leaves the impeller at a small angle to the tangential direction. Slip differs from other losses in that it will occur even with an ideal fluid.

Figure 32-6 shows calculated centrifugal fan performance characteristics for radial, forward curved, and backward-inclined blades. Head (H) and power (P) are indicated versus flow rate ($\dot{Q}$). Ideal head and power are indicated with subscript E. For radial-tip fans, the theoretical power is directly proportional to the flow rate. However, for forwardly curved tips, the theoretical power rises much more rapidly and, for backwardly curved tips, much less rapidly, even to the point where it may fall off with increasing flow rate.

The fan laws summarized in Table 32-2 correspond to pump laws discussed in the previous chapter. Following are a series of examples designed to show the relationship of fan law variables.

1. Effect of Speed Variation

For a given fan size, duct system configuration, and air density:

- Airflow capacity varies directly as the speed ratio
- Pressure varies as the square of the speed ratio
- Input power varies as the cube of the speed ratio

For example, given a fan that delivers 25,000 cfm (708 m^3/m) at a static pressure of 2 in. H_2O (51 mm H_2O), speed of 900 rpm and draw of 15 bhp (11.2 kW), the capacity, static pressure, and power at 1,200 rpm are:

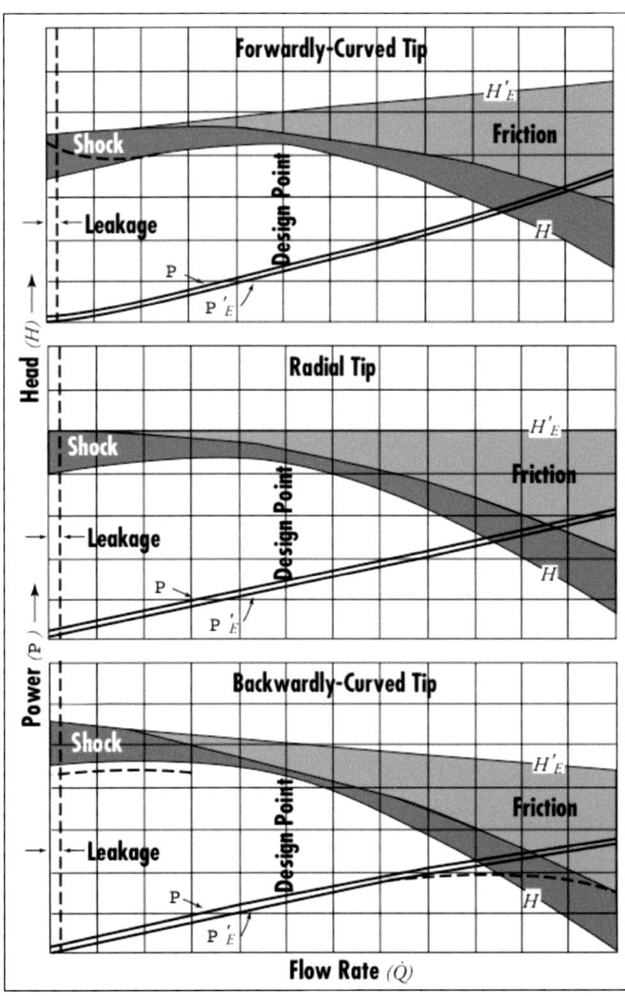

Fig. 32-6 Calculated Performance Characteristics for Three Centrifugal Fan Blade Designs. Source: Buffalo Forge, The Howden Fan Company

Capacity: $(25{,}000)(1{,}200/900)$ = 33,333 cfm (944 m^3/m)
Static pressure: $(2.0)\ (1{,}200/900)^2$ = 3.6 in. H_2O (91 mm H_2O)
Power: $(15)\ (1{,}200/900)^3$ = 35.6 bhp (26.5 kW)

2. Effect of Pressure Variation

For a given fan size, duct system configuration, and air density:

- Airflow capacity and speed vary as the square root of the pressure.
- Input power varies as the pressure ratio (3/2).

For example, given the fan in the example above, the capacity, speed, and power if the static pressure is increased to 3 in. H_2O (76 mm H_2O) are:

Capacity: $(25{,}000)(3/2)^{0.5}$ = 30,619 cfm (867 m^3/m)
Speed: $(900)\ (3/2)^{0.5}$ = 1,102 rpm
Power: $(15)\ (3/2)^{3/2}$ = 27.6 bhp (20.6 kW)

3. Effect of Fan Size Variation

For a constant pressure, density, point of rating, and fan type:

- Airflow capacity and input power vary as the square of the wheel diameter
- Speed varies inversely as the wheel diameter

For example, if the fan in the first example has a wheel diameter of 48 in. (122 cm), the capacity, pressure, and power for a similar fan with a wheel diameter of 60 in. (152 cm) running at the same rpm is:

Capacity: $(25{,}000)\ (1/1)\ (60/48)^3$ = 48,828 cfm (1,383 m^3/m)
Static Pressure: $(2.0)\ (1/1)^2\ (60/48)^2$ = 3.13 in. H_2O (79.5 mm H_2O)
Power: $(15)(48{,}828/25{,}000)(3.13/2.0)$ = $(15)\ (60/48)^5$ = 45.8 bhp (34.1 kW)

4. Effect of Simultaneous Fan Size and Speed Variation

For a constant pressure, density, point of rating, and fan type:

- Airflow capacity varies as the product of speed and the cube of the wheel diameter.
- Pressure varies as the product of the square of the speed and the square of the wheel diameter.
- Input power varies as the product of capacity and pressure or the cube of the speed and the fifth power of the wheel diameter.

5. Effect of Varying Inlet Air Density

For constant pressure, the speed, capacity, and input power vary inversely as the square root of the density,

Table 32-2 Summary of Fan Laws

Variable	When Speed Changes	When Density Changes	
Volume	Varies directly with speed ratio	Does not change	
	$q_2 = q_1 \left(\frac{N_2}{N_1}\right)$		(32-12)
Pressure	Varies with square of speed ratio	Varies directly with density ratio	
	$P_2 = P_1 \left(\frac{N_2}{N_1}\right)^2$	$P_2 = P_1 \left(\frac{\rho_2}{\rho_1}\right)$	(32-13)
Power of speed ratio	Varies with cube density ratio	Varies directly with	
	$H_2 = H_1 \left(\frac{N_2}{N_1}\right)^3$	$H_2 = H_1 \left(\frac{\rho_2}{\rho_1}\right)$	(32-14)

Where:
N = Speed of rotation
H = Fan power input
r = Fan air density
q = Fan capacity
P = Fan pressure

i.e., inversely as the square root of the barometric pressure and directly as the square root of the absolute temperature.

For constant capacity and speed, the input power and pressure vary directly with the air density, i.e., directly as the barometric pressure and inversely as the absolute temperature.

Consider, for example, a fan selected to handle 25,000 cfm (708 m^3/m) at 68°F (20°C) air at 2 in. (51 mm) static pressure, and requiring 15 bhp (11.2 kW). If the speed remains the same but the air temperature drops to 32°F (0°C), the static pressure and hp are:

Static Pressure: (2.0) (459.67 + 68) / (459.67 + 32)
= 2.15 in. H_2O (55 mm H_2O)

Power: (15)(459.67 + 68) / (459.67 + 32)
= 16.1 bhp (12.0 kW)

Fan Selection and Control

Many fan designs are designated based on their typical field of application. Examples include:

- **Ventilating fans**, are typically designed for clean-air service at normal temperatures. Heavy-duty ventilating fan designs may be selected for application under severe conditions.
- **Industrial exhausters** are typically designed with a focus on simplicity, durability, and the ability to withstand corrosive environments.
- **Pressure blowers** are designed to withstand the high tip speeds required to develop high pressures.
- **Mechanical draft fans** are designed for the capacities, temperatures, and environmental conditions encountered in a variety of combustion systems.

The fan specification should provide the supplier with necessary information regarding performance, service, arrangements, duct layout, size of the connecting ductwork, and any unique site-specific conditions. The fan output must be sufficient to overcome the losses caused by the various system elements. In an open system, differences in kinetic energy between inlet and outlet must also be considered. The effects of leakage, heat transfer, pressurization, and other fans operating in the system must be taken into account. Factors that affect air or gas density, such as barometric pressure, temperature, and relative humidity, must be considered, as well as any entrained material such as dust.

Fans can operate at any flow rate, from zero to maximum (free air delivery). Performance of a fan in a given application depends on the system characteristics and the fan pressure characteristics. Fan performance rating is a procedure, based on the fans laws, that permits the determination of variables such as flow rate, specific output, gas density, fan size, speed, input power, sound power level, and efficiency. Figure 32-7 shows a typical base performance test curve that relates several performance characteristics to airflow rate. For a given fan, performance values can be changed by varying the operating speed (rpm) to yield a family of curves. Figure 32-8 shows a family of fan performance curves over an operating range of 860 to 1,160 rpm. The curves show static pressure and shaft hp versus flow output.

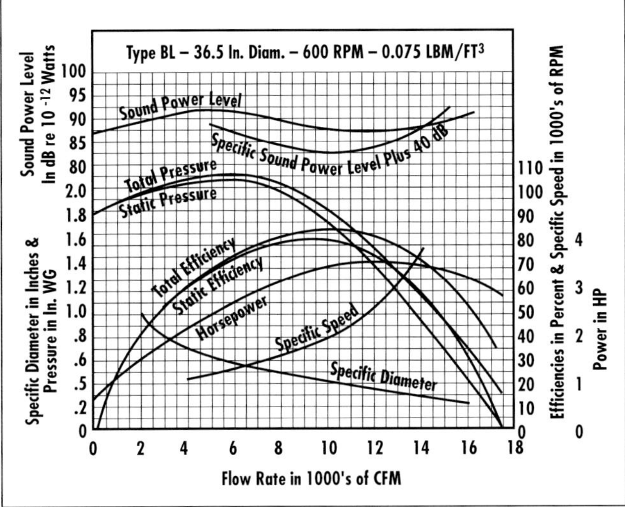

Fig. 32-7 Typical Fan Test Curves.
Source: Buffalo Forge, The Howden Fan Company

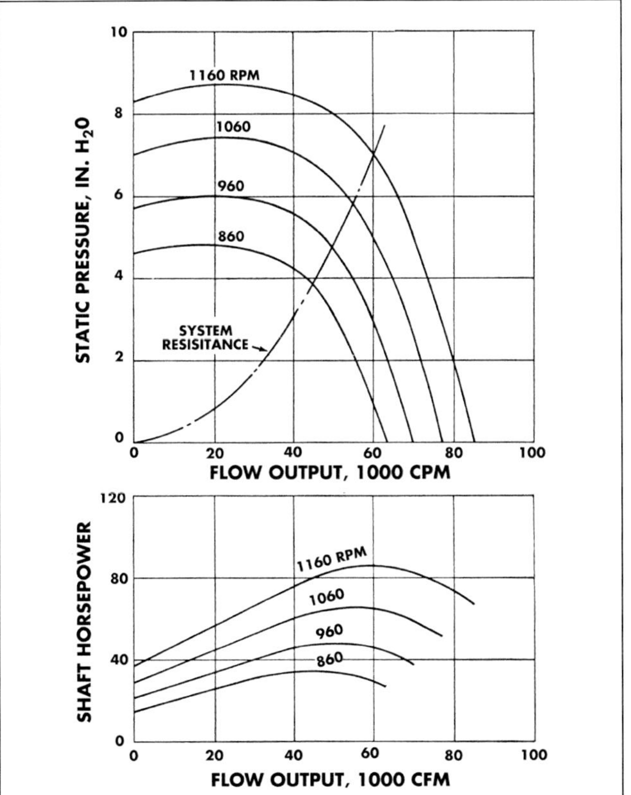

Fig. 32-8 Family of Fan Performance Curves. Source: Babcock and Wilcox

Fan capacity must often be controlled to meet varying system airflow requirements. The impact on performance of alternative control methods can be determined using the fan laws. Following is a discussion of various methods of control, along with an illustration of power requirement versus flow under varying load conditions.

- **Outlet damper control** is a throttling method that decreases air volume at constant speed by adding some resistance to the system. The maximum turndown rate is determined by the leakage rate at the fully closed position, although, in some cases, fan operation may be unstable under very low load. As shown in Figure 32-9, some power reduction accompanies the capacity reduction as the damper position is changed from wide-open to 3/4 to 1/2 to 1/4 open. Note that this is only true if the basic power characteristic of the fan is of the type shown which has a positive performance slope. In some cases, particularly with axial-flow and propeller fans, the fan's power characteristic will have a negative slope over part of the range and there will be a power increase rather than a power reduction if the fan is throttled with an outlet damper. Fan performance at 100% capacity is typically reduced slightly due to pressure losses across the dampers in the wide-open position.
- **Inlet box damper (IBD) control** decreases air volume at constant speed with a resistance effect, as

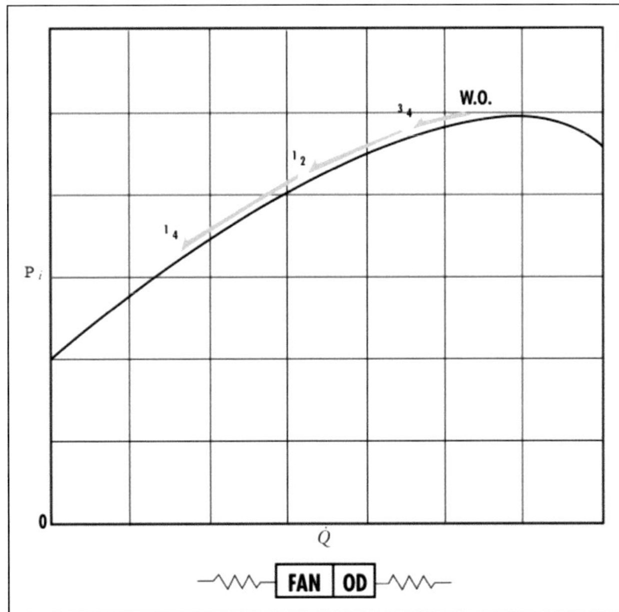

Fig. 32-9 Power Savings Potential with Outlet Damper Control.
Source: Buffalo Forge, The Howden Fan Company

well as a flow-modifying effect. Like outlet dampers, the maximum turndown rate is limited by leakage through the damper at the fully closed position. As shown in Figure 32-10, some power reduction can be achieved when inlet box dampers are applied to a fan with a positively sloping power characteristic. The spin produced by the IBD also provides a greater reduction in input power than is achieved using outlet box dampers.

- **Variable inlet vanes (VIV)** are designed to spin the air in the direction of fan rotation, providing better fan efficiency at lower flow rates than simple damper control. Like inlet box dampers, VIVs can be considered to have a resistance effect as well as a flow-modifying effect, although VIVs are more effective as they are usually placed close to the impeller. As shown in Figure 32-11, VIVs usually save more power than do IBDs.
- **Variable blade pitch control** is used for axial-flow fans. Flow is controlled at constant speed by varying the blade angle of the fan to create the optimal aerodynamic configuration at each point of operation. As the blade angle is adjusted from minimum to maximum, the change in flow is nearly linear. Variation in pitch can be achieved in various ways, including manually and automatically using pneumatic or hydraulic systems. For a properly designed fan with variable-pitch capability, the power savings can be significant, as indicated by Figure 32-12.
- **Variable speed operation** is usually the most efficient means of capacity control for centrifugal fans. As indicated by Figure 32-13, power savings are significant and predictable according to the fan laws, and stable operation occurs at all speed settings for the given system. Lower sound levels are also achieved under partial load. The turndown ratio will

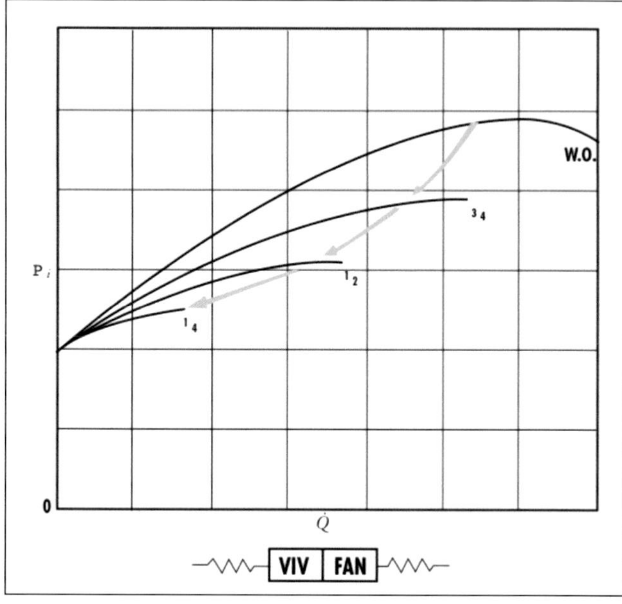

Fig. 32-11 Power Savings Potential with Variable Inlet Vane Control. Source: Buffalo Forge, The Howden Fan Company

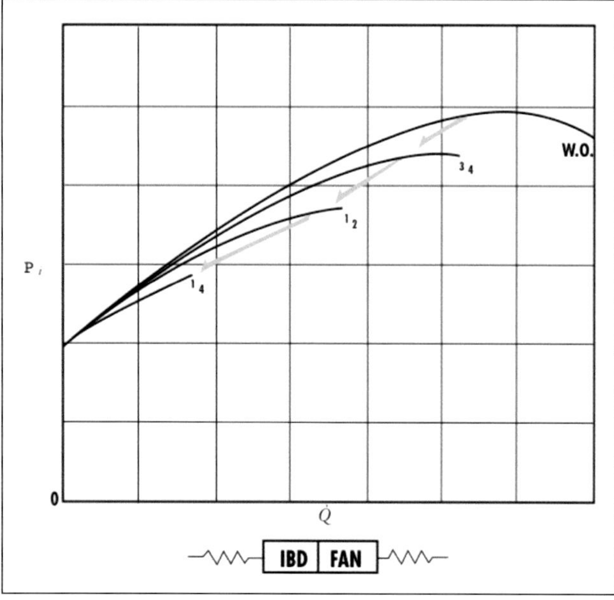

Fig. 32-10 Power Savings Potential with Inlet Box Damper Control. Source: Buffalo Forge, The Howden Fan Company

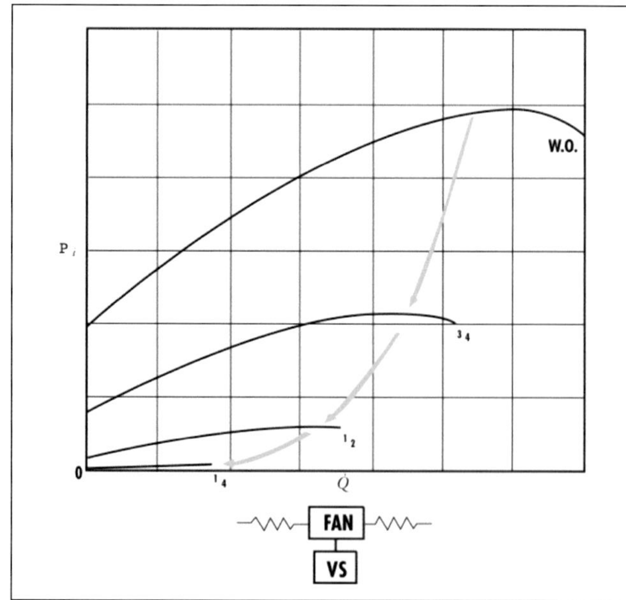

Fig. 32-12 Power Savings Potential with Variable Blade Pitch Control. Source: Buffalo Forge, The Howden Fan Company

be limited by the minimum speed that the variable-speed driver can achieve. The full-speed curve will be identical to the fan curve without the variable speed device if no slip occurs, although a small power loss typically occurs in the variable speed device.

Figure 32-14 compares various methods of capacity control. The potential impact of slip losses on power savings is indicated by the difference between the variable speed curves with subscripts i and o. Figure 32-15 illustrates the power savings that can be achieved using multiple fans to vary airflow capacity. Only limited control is possible when two fans are used; however, dampers, vanes, pitch, or speed control can also be used for one or both fans.

Fan Driver Options

As with pumps and most other mechanical drive equipment, ac electric induction motors predominate as fan drivers, and several effective control methods can be used with standard constant-speed electric motors. Speed control can also be achieved using variable-pitch belt drives, hydrokinetic and hydroviscous fluid drives, multiple-speed ac motors, dc drives, ac variable frequency drives (VFDs), and the three main classes of prime movers — reciprocating engines, gas turbines, and steam turbines.

Fans are generally not pre-packaged with prime movers because applications and equipment types are too diverse. However, once the requirements of the application are identified, packaging a fan with a prime mover is relatively simple and inexpensive. Similar to pumps, there are three common configurations: a single prime mover drive, a mixed system (or hybrid) set up using one fan

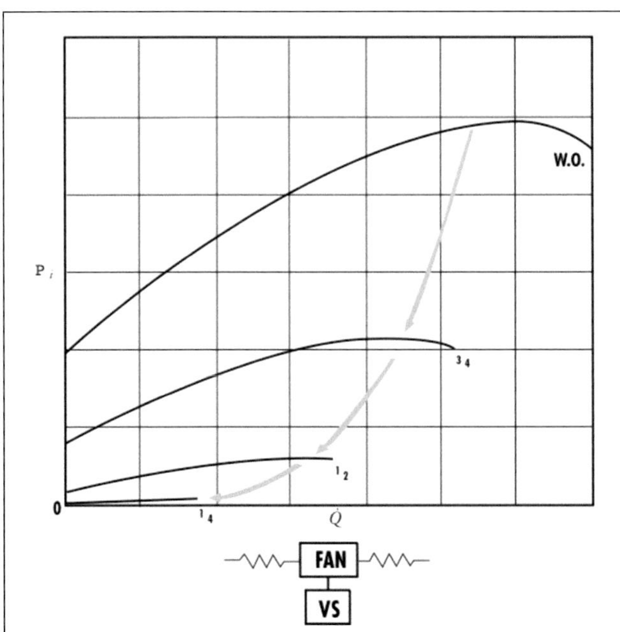

Fig. 32-13 Power Savings Potential with Variable Speed Control. Source: Buffalo Forge, The Howden Fan Company

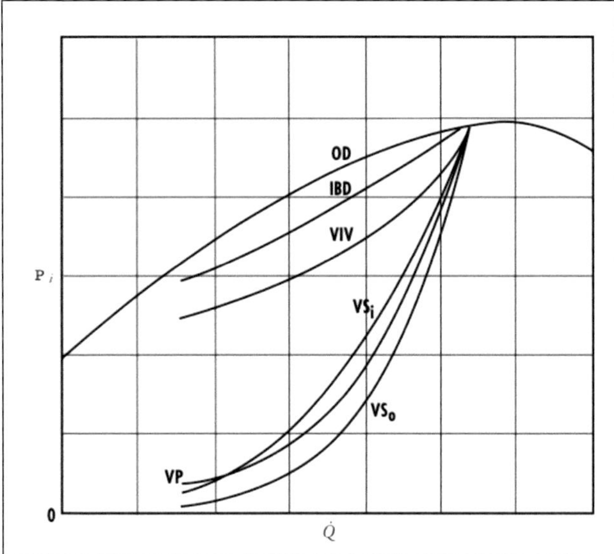

Fig. 32-14 Comparison of Power Savings Potential with Various Methods of Control. Source: Buffalo Forge, The Howden Fan Company

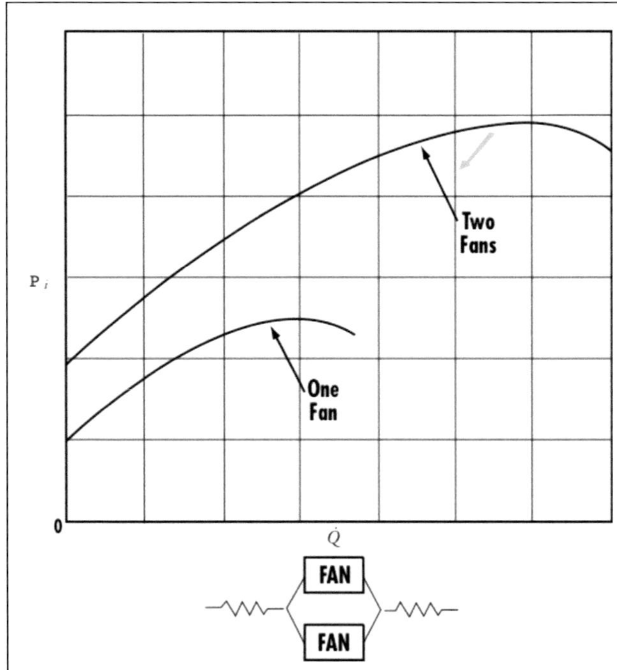

Fig. 32-15 Comparison of Power Savings Potential with Use of Multiple Parallel Fans as a Means of Control.
Source: Buffalo Forge, The Howden Fan Company

and two drivers on a shaft, and a hybrid setup using two driver/fan sets. Following are a few examples of common applications for prime mover-driven fans.

Back-pressure steam turbines are economical for a variety of industrial applications and are often used as drivers for large capacity boiler fans. They are commonly applied in mixed system configurations, along with an electric motor drive.

Boilers are controlled by varying combustion air and fuel flow. The lower the airflow, the lower the firing rate or the greater the boiler turndown. Fan variable inlet vanes tend to flutter at low flow rates, and can thus limit boiler turndown in constant-speed applications. The combination of variable speed and inlet vanes may allow for a lower minimum firing rate. The use of back-pressure steam turbine-driven fans is somewhat common for large boiler operations, with the back-pressure steam being used for feedwater heating. As airflow requirements are reduced at lower firing rates, deaerator and feedwater heating requirements are also reduced, often producing a system that is practically self-governing.

Because boiler fans are critical for plant operation and because an independent power source may be needed for plant start-up, hybrid systems are often used. As with pumps, a convenient and cost-effective arrangement is to have a fan with a double ended shaft — a steam turbine on one side and an electric motor on the other.

Figures 32-16 and 32-17 show dual-drive boiler fans featuring steam turbine and electric motor drives. Both fans serve a 40,000 lbm/h (18,000 kg/h) steam boiler. Figure 32-16 shows a dual-drive, forced draft boiler fan driven at 1,750 rpm by either a 17 hp (13 kW) back-pressure steam turbine or a 20 hp (15 kW) electric motor. Figure 32-17 shows an induced-draft boiler fan driven by a 26 hp (19 kW) steam turbine. The turbine operates at 4,000 rpm with an inlet steam pressure of 195 psig (14.5 bar). The turbine exhausts steam at 5 psig (1.4 bar), which feeds a deaerator tank. The alternative drive for this fan is a 30 hp (22 kW) electric motor.

Though sometimes used for industrial process applications, combustion engines are mostly used only in remote field applications or for reliability as a non-electric mechanical energy source for critical applications. Figure 32-18 shows a reciprocating engine driving a blower to aerate digester gas basins at a pollution control plant.

Prime movers are often used for emergency exhaust fans and other critical fan applications. Figure 32-19 shows a natural gas-fired reciprocating engine-driven cooling fan for a fiberglass furnace used in making building insulation. The engine is rated at 197 hp (147 kW) at 1,600 rpm. When operating at 1,600 rpm, the fan produces 6,650 cfm (188 m^3/m) of 100°F (38°C) air with 75 ft (23 m) H_2O static pressure and a specific density of 0.692 at an elevation of 700 ft (213 m).

Fig. 32-16 Forced Draft Fan Featuring Steam Turbine and Electric Motor Drive. Source: UTC Sikorsky Aircraft

Fig. 32-17 Induced Draft Fan Featuring Steam Turbine and Electric Motor Drive. Source: UTC Sikorsky Aircraft

Consider a mixed system fan drive in a waste incinerator application. The system includes an induced draft fan driven by both an electric motor and a non-condensing steam turbine. The 12-blade fan has a rotor diameter of 46.5 in. (118.1 cm) and a working blade tip width at the outside of 7.75 in. (19.7 cm). At a condition of standard air density and 70°F (21°C), the fan will produce 21,782 cfm (617 m^3/m) against a static pressure of 34 in. (864 mm) H_2O operating at 1,700 rpm and will require 159 brake hp (119 kW). At the normal operating temperature of 350°F (177°C), the air density is

Fig. 32-18 Reciprocating Engine Driving Blower at 1,000 RPM to Aerate Basins at Pollution Control Plant. Source: Waukesha Engine Division

Fig. 32-19 Natural Gas-Fired Reciprocating Engine-Driven Cooling Fan. Source: Courtesy of Robinson Industries

0.0477 lbm/ft^3 (0.76 kg/m^3) and the fan will produce 22,000 cfm (623 m^3/m) against a static pressure of 22 in. (559 mm) H_2O operating at 1,700 rpm and will require 103 brake hp (77 kW).

Starting is accomplished with the damper closed by the 125 hp (93 kW) motor which operates at 1,800 rpm. As the air temperature increases during startup, the density and static pressure are reduced to the point that the fan can achieve full-load output with the 125 hp (93 kW) motor driver. When the operating temperature reaches 350°F (177°C), the fan can develop 22,000 cfm (623 m^3/m) when driven by the lower hp steam turbine. The turbine operates with a normal inlet steam pressure of 100 psig (7.9 bar) and an exhaust pressure of 5 psig (1.4 bar).

Figure 32-20 shows the assembly of a large-capacity forced draft fan designed to be driven by either an electric motor or a steam turbine. The fan is designed to produce 62,188 cfm (1,761 m^3/m) against a static pressure of 37 in. (940 mm) H_2O when operating at 1,480 rpm and will require 443 hp (336 kW). The electric motor drive is already mounted on the right side of the skid. The steam turbine drive will be mounted on the left side.

Fig. 32-20 62,188 CFM Mixed System Forced Draft Fan. Source: Buffalo Forge, The Howden Fan Company

Section VIII

Refrigeration and Air Conditioning

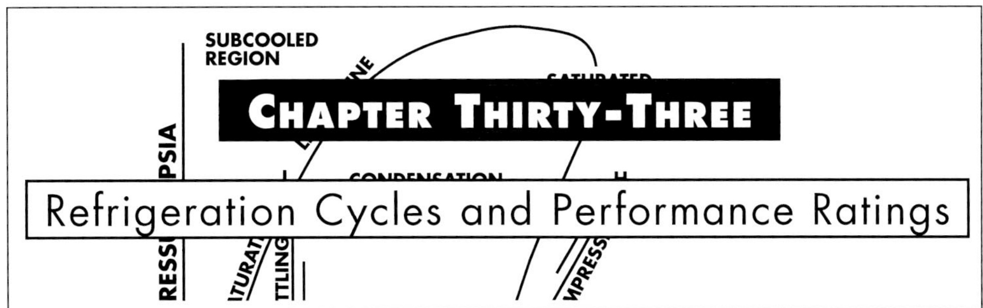

REFRIGERATION CYCLES

The basic refrigeration cycle includes a circulating refrigerant that undergoes phase changes in which it absorbs or rejects heat. Thermal energy passes from a heat source into low-pressure liquid refrigerant, causing it to evaporate. Heat is then rejected from the refrigerant vapor as it condenses at high pressure and temperature. In vapor compression cycles, mechanical energy is input to provide the compression needed at the refrigerant condenser and a throttling device maintains low-pressure conditions at the evaporator. Absorption cycles are similar to vapor compression cycles systems, except instead of compression, they rely on application of heat and the strong chemical affinity of an absorbent for the refrigerant to drive the cycle.

In practical application, heat is transferred from the air-conditioned space or process heat source directly to the refrigerant or indirectly through a secondary cooling fluid such as brine or chilled water. The refrigerant gives up its heat directly to atmosphere or to a cooling medium such as condenser water, which then transfers the heat to a heat rejection devise such as an outdoor cooling tower.

In the ideal heat engine cycle described in Chapter 2, work is produced as heat passes from a higher to a lower temperature. In a refrigeration cycle, work is input to transfer heat from a lower to a higher temperature. The ideal refrigeration cycle can thus be represented by a reverse of the Carnot cycle used to model heat engine performance. Figure 33-1 is a temperature vs. entropy (Ts) diagram of the reversed Carnot cycle. In the reverse cycle, work is added to the system and heat is removed from the lower temperature region (T_L). The energy rejected to the higher temperature region (T_H) equals the work performed by the compressor plus the thermal energy removed from the heat source.

In actual operation of mechanical compression systems, the heat rejected from the cycle is equal to the refrigeration effect plus the driving energy to the compressor. In actual absorption cycles, the heat rejected from the cycle is equal to the refrigeration effect plus the generator input.

In Figure 33-1:

Q_{out} = Heat rejected from the cycle = area 3-4-1-a-b-2-3

Q_{in} = Heat added to the cycle (heat removed) = area 1-2-b-a-1

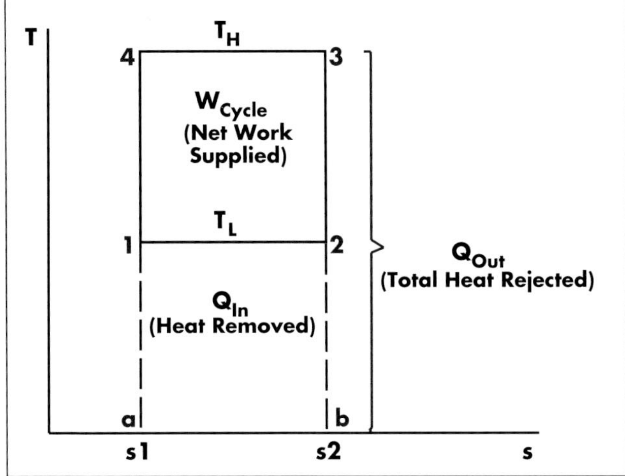

Fig. 33-1 Ts Diagram of Reversed Carnot Cycle.

W_{cycle} = Net work supplied to the cycle = $Q_{out} - Q_{in}$ = area 1-2-3-4-1

As can be seen from the Ts diagram, in the reversed cycle, area 1-2-3-4 measures the net work supplied to the cycle and area 1-2-b-a represents the heat removed from the colder region. Note that points a and b are at absolute zero. The cycle processes described above are represented diagrammatically as a vapor refrigeration cycle in Figure 33-2. It operates in the following manner:

- Refrigerant enters the evaporator at State 1 as a two-phase liquid-vapor mixture. In the evaporator, heat from the conditioned space or process load is transferred to the refrigerant.
- Evaporation of refrigerant (States 1 to 2) occurs as heat transfer from the conditioned space or process load causes the refrigerant to undergo a phase change from liquid to vapor. This process occurs at a constant temperature T_L and constant pressure.
- The refrigerant is then compressed isentropically between States 2 and 3 and changes from a low-pressure vapor to a high-pressure vapor at State 3. The compression process causes the pressure to increase along with a corresponding temperature increase from T_L to T_H.
- The refrigerant then passes through the condenser/subcooler (States 3 to 4) where it changes from a superheated vapor to a subcooled liquid as heat is

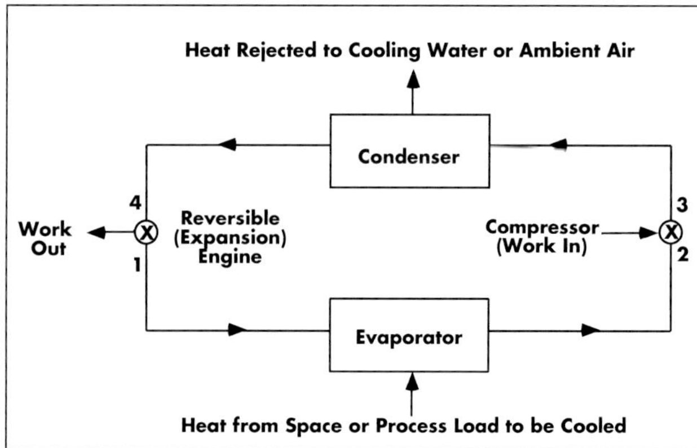

Fig. 33-2 Diagram of Basic Refrigeration Cycle Processes.

rejected to the warm air region. This process occurs at a constant high temperature and pressure.

- Between States 4 and 1, the refrigerant undergoes an ideal expansion process in which the pressure is reduced and the refrigerant temperature returns to T_L. An ideal or reversible engine is used for the expansion process.

Measuring Refrigeration Cycle Performance

Refrigeration Units of Measure

Quantitative measures of heat energy are typically the British Thermal Unit (Btu), Calorie, or kilowatt-hour (kWh). As defined, the removal of one Btu of heat from one lbm of water at 61°F will lower the temperature of the water by 1°F. The heat flow rate, or amount of heat that flows from one substance to another within a given time period, is commonly measured in Btu/h or kW.

In U.S. customary refrigeration engineering units, the standard ton of refrigeration (sometimes referred to as the standard refrigerant ton or RT) derives from the fact that the latent heat of fusion of ice is approximately 144 Btu/lbm (93 W-h/kg), which is 288,000 Btu (84.4 kWh) per ton. A refrigeration machine operating at one ton of capacity is absorbing energy at a rate equal to that which would exist if one ton of ice were melting in the refrigerated space every 24 hours. Thus:

1 ton refrigeration = (2,000 lbm ice/day) (144 Btu/lbm) /(24 hr/day) = 12,000 Btu/h

Note that in SI units, commonly used in refrigeration system engineering, the equivalent for one standard ton (12,000 Btu/h) is expressed as 12,658 kJ/h or 3.516 kW_r.

The kW shall be used as the SI measurement unit throughout this Section, with the following subscripts: m for mechanical power input, e for electric power input, h for heat or fuel energy input, and r for refrigeration output.

Coefficient of Performance (COP)

Heat engine thermal efficiency considers the amount of work that is produced from a given heat supply. With the refrigeration cycle, performance measurement — termed coefficient of performance (COP) — considers the amount of work required to remove a given amount of heat from the low-temperature source. Performance is optimized when input work is minimized at a given load.

Based on Figure 33-1, refrigeration cycle COP can be described as the ratio of the refrigeration effect (the amount of heat the system removes from the conditioned space or process load, Q_{in}) to work input (the net amount of energy transferred into the system to accomplish the refrigeration effect, W_{cycle}). Thus, the COP is expressed as:

$$COP = \frac{Refrigeration\ effect}{Work\ input} = \frac{Q_{in}}{W_{cycle}} \quad (33\text{-}1)$$

Where:
W_{cycle} = Net work required to drive refrigerator ($Q_{out} - Q_{in}$).

If we let Q_{in} = 1 ton of refrigeration, then equation (33-1) leads to the following convenient relationships:

$$hp/ton = \frac{(12{,}000)}{(COP)(2{,}545)} = \frac{4.72}{COP} \quad (33\text{-}2)$$

and

$$kW/ton = \frac{(12{,}000)}{(COP)(3{,}413)} = \frac{3.51}{COP} \quad (33\text{-}3)$$

In SI units, these expressions are more easily simplified when both energy input and refrigeration effect are expressed in kW. Thus:

$$kW_{input}/kW_r = \frac{1}{COP} \quad (33\text{-}4)$$

and

$$COP = kW_r/kW_{input} \quad (33\text{-}5)$$

For heat-driven refrigeration cycles, or, more generally, for Carnot refrigeration cycles, COP can be expressed as:

$$COP = \frac{Refrigeration\ effect}{Energy\ input} = \frac{Q_{in}}{Q_{out} - Q_{in}} \quad (33\text{-}6)$$

Where:

Q_{in} = Heat input to the cycle (this is the heat removed from the conditioned space or process load, i.e., refrigeration effect)

Q_{out} = Heat rejected by the cycle to the surrounding atmosphere

Referring back to Figure 33-1,

$$COP = \frac{area\ (3\text{-}b\text{-}a\text{-}4\text{-}3)}{area\ (1\text{-}2\text{-}3\text{-}4\text{-}1)} = \frac{Refrigeration\ effect}{Work\ added} \quad (33\text{-}7)$$

and

$$COP_{Carnot} = \frac{T_2(s_2 - s_1)}{(T_3 - T_2)(s_2 - s_1)} = \frac{T_2}{(T_3 - T_2)} \quad (33\text{-}8)$$

Where:

T_1 = T_2 = Low temperature (T_L)

T_4 = T_3 = High temperature (T_H)

s_1 = s_4 = Entropy during expansion process

s_2 = s_3 = Entropy during compression process

Consider, for example, a Carnot refrigeration cycle that is used to maintain a space temperature of 40°F and rejects heat at a temperature of 90°F. To calculate the Carnot COP, the temperatures would first be converted to degrees Rankine by adding 460° to each. COP would then be calculated as follows:

$$COP_{Carnot} = \frac{500}{(550 - 500)} = 10$$

Since one ton of refrigeration is equal to 12,000 Btu/h and one hp is equal to 2,545 Btu/h, the energy required to produce one ton of refrigeration, expressed in hp/ton, as shown in Equation 33-2, is 4.72/COP. In SI units, this would be expressed as:

$$\frac{1}{COP} = \frac{kW_m}{kW_r} \quad (33\text{-}9)$$

If, in the previous example where the COP was found to be 10, the heat removal rate was 12,000 Btu/h, the work required, based on Equation 33-2, would be calculated as:

$$\frac{4.72\ hp/ton}{10} = 0.472\ hp/ton$$

Alternatively, the work rate required per ton, in Btu/h, could be calculated as:

$$\frac{12{,}000\ Btu/h}{10} = 1{,}200\ Btu/h$$

The total heat rejected from the refrigeration cycle (Q_{out}) equals the rate of heat removal (or load) plus the work input (vapor compression cycle) or high-temperature heat input (absorption cycle) computed as follows:

$$Q_{out} = Q_{refrig} + (W_{in}\ or\ Q_{in}) \quad (33\text{-}10)$$

In the above example, the total heat rejected from the cycle would be 12,000 Btu/h (3.515 kW_r), which was the heat removal rate, plus 1,200 Btu/h (0.352 kW_r), which was the work energy required, for a total rate of 13,200 Btu/h (3.867 kW_r).

Energy Efficiency Ratio

Energy Efficiency Ratio (EER) is a term used in the United States to quantify the efficiency of electric-driven refrigeration systems. EER is defined as:

$$EER = \frac{Refrigerant\ effect\ (in\ Btu/h)}{Work\ input\ (in\ watts)} \quad (33\text{-}11)$$

$$= \frac{Q_{in}}{W_{cycle}}\ (Btu/watt\text{-}h)$$

This leads to the following convenient relationships:

$$COP = \frac{EER}{3.413} \quad (33\text{-}12)$$

and

$$kW/ton = \frac{(12{,}000)}{(EER)(1{,}000)} = \frac{12}{EER} \quad (33\text{-}13)$$

Pressure-Enthalpy Chart

Figure 33-3 illustrates a mechanically driven refrigeration cycle on a pressure versus enthalpy (p-h) chart. The p-h chart is divided into three general areas by the saturated liquid line and the saturated vapor line. The area to the left of the saturated liquid line is called the subcooled region, the area to the right of the saturated vapor line is called the superheated region, and the area between the saturated liquid and saturated vapor lines is called the liquid vapor mixture, or wet, region. Due to the shape of this region, it is also sometimes called the vapor dome. If, for example, refrigerant at point A on the saturated liquid line absorbs heat with no change in pressure, evaporation will take place and its enthalpy will increase. Evaporation would be complete at point B on the saturated vapor line. Any additional heat absorbed at constant pressure would move the refrigerant into the superheat region, as shown by point C.

In the figure, arrows indicate the four parts of the refrigeration cycle described below:

- Compression occurs from F to H. Notice that the heat content of the vapor increases and that the mechanical work expended on the refrigerant vapor to increase its pressure also superheats it. In the ideal cycle, compression occurs at constant entropy.
- Heat rejection occurs in the condenser from H to J. First, superheat is removed and then condensing occurs. In practice, there will be some pressure drop between H and J and the heat rejection process extends further to the left than point J into the subcooled region as some subcooling occurs in the condenser.
- Expansion, or throttling, occurs from J to D as refrigerant pressure is reduced from condenser pressure to evaporator pressure. The vertical line indicates that this occurs at constant enthalpy.
- In the evaporation process, the refrigerant is shown to enter the evaporator as a liquid-vapor mixture (inside the vapor dome). As the refrigerant flows through the evaporator, it absorbs heat and is completely vaporized at point E. In practice, some superheating takes place in the evaporator and some as the refrigerant flows to the compressor. This superheating is indicated from E to F.

System Energy Input

In the ideal case of isentropic compression, it is assumed that there is no heat transfer between the vapor being compressed and the surroundings, and that the compression process is frictionless. The actual power that must be applied to the shaft of a mechanical compression system (the brake power) is a function of the compressor isentropic efficiency, which is expressed as:

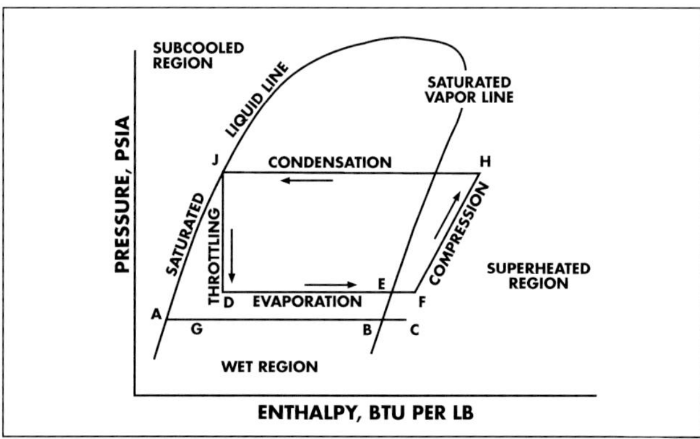

Fig. 33-3 Plot of Refrigeration Cycle Phases on p-h Chart. Source: The Trane Company

$$\eta_{compressor} = \frac{Ideal\ power}{Brake\ power} \quad (33\text{-}14)$$

For electric motor-driven systems, input power depends on motor efficiency as well as refrigeration and compressor efficiency. In English units, input power would be expressed as:

$$kW = \frac{(hp_{compressor} \times 0.746\ kW/hp)}{(\eta_{motor}/100)} \quad (33\text{-}15)$$

In SI units, this would be expressed as:

$$kW_e = \frac{kW_m}{(\eta_{motor}/100)} \quad (33\text{-}16)$$

For a combustion engine-driven refrigeration system, specific fuel consumption (SFC) of the prime mover would be expressed in English units as:

$$SFC\ (Btu/h) = \frac{(hp_{compressor} \times 2{,}545\ Btu/hp\text{-}h)}{(\eta_{prime\ mover}/100)} \quad (33\text{-}17)$$

In SI units, this would be expressed as:

$$SFC\ (kW_h) = \frac{(kW_m)}{(\eta_{prime\ mover}/100)} \quad (33\text{-}18)$$

It is important to identify the assumptions used in calculating COP and fuel usage. With direct fuel-fired equipment, COP and energy usage may be based on either lower heating value (LHV) or higher heating value (HHV). [Refer to Chapter 5 for detail on heating value of fuel.]

For prime mover-driven refrigeration systems operating on cogeneration cycles, recovered energy must be considered. When heat recovery from a combustion engine is used in a refrigeration system, Equation 33-1 becomes:

$$COP_{net} = \frac{Refrigeration\ effect}{Energy\ input - Energy\ displaced} = \frac{Refrigeration\ effect}{Net\ energy\ input} \quad (33\text{-}19)$$

In English units, the net energy input can be expressed as:

$$Btu_{net}/ton\text{-}h = Btu_{input}/ton\text{-}h - \frac{Btu_{recovered}/ton\text{-}h}{\eta_{boiler}} \quad (33\text{-}20)$$

and in SI units as:

$$kWh_{net}/kWh_r = kWh_{h\ input}/kWh_r - \frac{kWh_{h\ recovered}/kWh_r}{\eta_{boiler}} \quad (33\text{-}21)$$

In cogeneration systems using a back-pressure or extraction steam turbine, the energy input is the fuel input to the boiler, and steam at the turbine outlet provides recovered heat. A method of expressing net energy use for a steam turbine is to subtract the total enthalpy (expressed in Btu or kWh_r) at the turbine outlet from the total enthalpy at the turbine inlet, and divide by the boiler efficiency. If recovered heat is used to generate additional cooling via an absorption chiller or a bottoming cycle steam turbine, the additional cooling output potential is calculated as:

$$Additional\ capacity = \frac{Heat\ recovered}{Energy\ input\ requirement} \quad (33\text{-}22)$$

Combining the output of both systems, the energy input rate can be expressed as:

$$\begin{matrix}Total\ energy\\ input\ rate\end{matrix} = \frac{Energy\ input}{(Primary\ capacity + Additional\ capacity)} \quad (33\text{-}23)$$

COP can then be expressed as:

$$COP = \frac{Combined\ refrigerant\ effect}{Energy\ input} \quad (33\text{-}24)$$

Refer to Chapter 2 for a detailed discussion on cogeneration cycle thermodynamic performance expressions. This includes calculations applicable to quantifying the impact of recovered heat on net and overall system performance.

Equipment Rating Standards

Standard Conditions

Refrigeration and air conditioning (AC) equipment manufacturers provide energy usage data, based on equipment tests under specified conditions. This data typically includes efficiency ratings, fuel consumption, and heat rate. Performance is typically expressed in terms such as COP, EER, Btu/ton-h, kW/ton, hp/ton, kW_m/kW_r, or kJ/kWh_r. Also included are water pressure drops, fouling factors, and chilled and condenser water temperatures at which the performance data was measured. To effectively use this data to evaluate the anticipated performance of equipment for a given project, the conditions under which they were calculated must be specified. Often, they must be adjusted to reflect performance under anticipated operating conditions.

Examples of questions that should be considered when evaluating equipment ratings include:

- What ambient conditions and supply temperatures are the ratings based on?
- Do capacity and performance values include required auxiliary equipment?
- What are the allowable tolerances in the conditions used for performance rating evaluation and certification?
- Are the values used for thermal fuel efficiency and heat rate given as HHV or LHV?
- In the case of a steam- or hot water-driven system, has boiler efficiency been considered?

To allow comparison of competing equipment designs for a given application, manufacturers have adopted sets of standard conditions established by organizations such as the Air Conditioning and Refrigeration Institute (ARI) and the American Society of Heating, Refrigeration, and Air Conditioning Engineers (ASHRAE). These published standards are the basis for a certification program in which ratings are verified by testing. Ratings include data such as capacity in tons (or kW_r), energy input, and performance under part load.

In 1998, ARI updated its standards for water chilling packages. The list of standard rating conditions, as published in *Standard for Water Chilling Packages Using the Vapor Compression Cycle — ARI Standard 550/590*, is available through the ARI website and summarized in Table 33-1.

In addition, recommendations are provided for ranges to be used for applications at other than standard rating conditions. For example, performance curves can be provided for equipment operation at not only 85.0°F (29.44°C) entering condenser water and 44°F (6.7°C) leaving chilled water, but under a wide range of condenser and chilled water temperatures. This allows for discrete operating cost analyses to be performed using modeling of expected system performance under the varying expected load and ambient condition that will be experienced in actual operation.

Given that cooling and refrigeration equipment more often than not operates at off-design conditions, specifying engineers require data on performance over a wide range of conditions to properly evaluate the expected

	Water-Cooled	Evaporatively-Cooled	Air-Cooled
Condenser Water			
Entering Flow Rate	85°F [29.4°C] 3.0 gpm/ton [0.054 l/s per kW]		
Condenser Fouling Factor Allowance			
Water-Side	0.00025 h • ft² • °F/Btu [0.000044 m² • °C/W]		
Air-Side		0.000 h • ft² • °F/Btu [0.000 m² • °C/W]	0.000 h • ft² • °F/Btu [0.000 m² • °C/W]
Entering Air			
Dry Bulb			95°F [35.0°C]
Wet Bulb		75°F [23.9°C]	
Evaporator Water			
Leaving Flow Rate		44°F [6.7°C] 2.4 gpm/ton [0.043 L/s per kW]	
Evaporator Fouling Factor Allowance			
Water-Side		0.0001 h • ft² °F/Btu [0.000018m² • °C/W]	
		Condenserless	
		Water or Evaporatively Cooled	Air-Cooled
Saturated Discharge		105°F [40.6°C]	125°F [51.7°C]
Liquid Refrigerant		98°F [36.6°C]	105°F [40.6°C]
Barometric Pressure – 29.92 in. of Hg [101 kPa]			

Table 33-1 Standard Rating Conditions from ARI Standard 550/590-1988. Source: ARI

performance and energy usage under the intended application. Since fouling will inevitably occur, it is important to evaluate performance based on expected fouling factors. Manufacturers can test new machines with assumed fouling factors by adjusting a parameter, such as leaving fluid temperature, to a level that corresponds with the specified fouling factor.

IPLV AND NPLV

The concepts of integrated part-load value (IPLV) and non-standard part-load value (NPLV) were developed to standardize estimates of annualized average performance. These criteria define a load level, a percentage of time at each load level, and the available condenser water (or air) temperature at each loading to provide a single annualized performance figure. IPLV references ARI Standard Rating Conditions and NPLV references ARI Selected Application Rating Conditions. The time weighting values used for IPLV and NPLV are:

$$IPLV \text{ or } NPLV\ COP = 0.01A + 0.42B + 0.45C + 0.12D \tag{33-25}$$

Where:
A = COP at 100%
B = COP at 75%
C = COP at 50%
D = COP at 25%

When performance is expressed in kW/ton or hp/ton, the IPLV or NPLV can be expressed as:

$$kW/ton \text{ or } hp/ton = \frac{1}{\frac{0.01}{A} + \frac{0.42}{B} + \frac{0.45}{C} + \frac{0.12}{D}} \tag{33-26}$$

Where:
A = kW or hp at 100%
B = kW or hp at 75%
C = kW or hp at 50%
D = kW or hp at 25%

In computing COP or thermodynamic performance at the four specified load levels, the standard conditions remain constant, with the exception of condensing temperatures. Fluid or air temperatures entering the condenser are varied linearly between designated values at 0 and 100% load. ARI 550/590-1998 standards for rating performance at varying condensing temperatures associated with part-load conditions are summarized in Table 33-2.

An accurate evaluation of system performance under varying load conditions serves two purposes: sizing

% LOAD	1998 STANDARD WC °F (°C) ECWT	AC °F (°C) EDB	EC °F (°C) EWB
100%	85 (29.4)	95 (35.0)	75 (23.9)
75%	75 (23.9)	80 (26.7)	68.7 (20.4)
50%	65 (18.3)	65 (18.3)	62.5 (17.0)
25%	65 (18.3)	55 (12.8)	56.2 (13.5)

WC = water-cooled ECWT = entering condenser water temperature
AC = air-cooled EDB = entering air dry bulb temperature
EC = evaporative cooled EWB = entering air wet bulb temperature

Table 33-2 Condensing Temperatures used for Part Load Ratings from ARI Standard 550/590-1988. Source: ARI

equipment and performing economic analyses of operating costs. The IPLV standard is useful because it provides a uniform set of conditions for system comparisons. Still, it is necessary to appreciate its limitations and understand how to predict actual application performance more precisely. The ARI standards are based on a weighted average of building types and average weather data from 29 cities across the United States using the ASHRAE temperature bin method. Operating hours are based on the weighted average of various operations with and without the use of an economizer.

It is important to note that the latest ARI standard reflects significant changes from its previous 1992 standard. Perhaps the most significant change is the adjustment of the weighting of part-load points used in calculation of IPLV and NPLV. For example, whereas the previous standard used 17% weighting for full-load operation, the current standard uses only 1% and a greater distribution at the part-load levels. As another example, whereas the previous standard was based on the profile of a single building located in Atlanta, Georgia, the current standard is based on weighted national averages for load and weather conditions. While the new standard may be more representative of conditions nationally, it is still based on the assumption of single-chiller operation and assumed load and weather conditions.

To generate more application-specific representative data, the time-weighing values of the NPLV can be adjusted by the design engineer to reflect the impact of local weather conditions, multiple chiller installations, or internal cooling loads. The NPLV rating method also allows the design engineer to adjust water temperature, flow rates, cooling tower performance, and fouling factors for a particular application.

In a multiple-chiller system, for example, it is possible for one chiller to be fully loaded and one only partially loaded with a common condenser water supply. The assumption that the part-load chiller is operating under a lower condenser temperature than the fully loaded unit would be inaccurate in this case. Analysis should also include energy usage requirements for auxiliary equipment.

% Load	ECWT (°F)	Tons	kW/ton	COP	Weighting	Weighted Values
100	85.00	1,000	0.60	5.86	0.01	0.017
75	78.75	750	0.55	6.39	0.42	0.764
50	72.50	500	0.57	6.17	0.45	0.789
25	66.25	250	0.73	4.82	0.12	0.164
Total Weighted Value (Ton/kW)						1.734
IPLV kW/ton = 1/1.722						0.58
IPLV COP						6.10

Table 33-3 IPLV Rating Example for Electric-Driven Centrifugal Chiller.

% Load	ECWT (°F)	Tons	hp	Btu/hp-h (fuel)	Btu/ton-h (fuel)	COP	Weighting
100	85.00	1,000	770	7,871	6,061	1.98	0.01
75	78.75	750	548	7,495	5,476	2.19	0.42
50	72.50	500	375	8,040	6,030	1.99	0.45
25	66.25	250	330	10,024	9,986	1.20	0.12
Btu/ton-h APLV fuel input							6,272
IPLV COP							1.98

Table 33-4 IPLV Rating Example for Reciprocating Engine-Driven Centrifugal Chiller.

% Load	ECWT (°F)	Tons	Btu/ton-h (steam)	Btu/ton-h (fuel)	COP steam	COP overall	Weighting
100	85.00	1,000	10,084	12,121	1.19	0.99	0.01
75	78.75	750	9,302	11,215	1.29	1.07	0.42
50	72.50	500	8,696	10,435	1.38	1.15	0.45
25	66.25	250	10,084	12,121	1.19	0.99	0.12
IPLV COP steam							1.32
IPLV COP overall (@ 83% Boiler efficiency)							1.10

Table 33-5 IPLV Rating Example for Double-Effect Steam Absorption Chiller.

% Load	ECWT (°F)	Tons	Btu/ ton-h	Btu/ton-h credit	Btu/ton-h (net)	COP (net)	Weighting
100	85.00	1,000	6,061	2,921	3,140	3.82	0.01
75	78.75	750	4,476	2,639	2,837	4.23	0.42
50	72.50	500	6,030	2,906	3,124	3.84	0.45
25	66.25	250	9,986	4,813	5,173	2.32	0.12
Btu/ton-h APLV fuel input							3,250
IPLV COP net							3.82

Table 33-6 IPLV Rating Example for Reciprocating Engine-Driven Centrifugal Chiller With Heat Recovery.

Examples

IPLV calculations are presented below for three different types of 1,000 ton (3,515 kW_r) chillers. Tables 33-3 through 33-5 are based on an electric motor-driven centrifugal chiller, a spark-ignited reciprocating engine-driven centrifugal chiller, and a double-effect, steam-powered absorption chiller, respectively.

IPLV and NPLV can be used to express performance for cogeneration-cycle systems as well. Table 33-6 shows the IPLV ratings for the engine-driven system referenced in Table 33-4 with the addition of heat recovery. In this case, it is assumed that 40% of the energy input (HHV basis) is recoverable and that the displaced energy conversion process efficiency is 83%. Notice that the net fuel energy input is almost cut in half and the net COP is almost doubled when heat recovery is included in this example.

CERTIFICATION OF PERFORMANCE

To comply with ARI certification requirements, published ratings for all cooling and refrigeration systems shall include the standard rating. For water chilling packages, this corresponds to the applicable standard rating condition shown in Table 33-1. Part-load rating is intended to permit the development of part-load performance data over a range of operating conditions that reflect "real world" operation, since cooling equipment rarely operates at design, or rated, conditions. Part-load rating points shall be presented as IPLV, NPLV, or separate part-load data points suitable for calculating IPLV or NPLV. Typically, this involves calculation of performance at 100%, 75%, 50%, and 25% of full load. If, due to capacity control limitations, the unit cannot be operated at any of these points, then the unit can be operated at other load points and performance can be determined by plotting the performance versus the percent load using straight line segments to connect the actual performance points.

As with any testing and certification program, a key element in interpreting published data is allowable tolerances. For a specifying engineer concerned with equipment sizing and/or prediction of actual energy usage, the allowable tolerances will indicate the range of potential actual capacities and performance that can be expected from a given model, expressed as percentage deviation from the published performance data. ARI's allowable test tolerance, in percent, is calculated in U.S. standard units (using °F) as:

$$\%Tolerance = 10.5 - (0.07 \times \%FL) + \left(\frac{1{,}500}{DT_{FL} \times \%FL}\right) \tag{33-27}$$

In SI units (using °C), it is calculated as:

$$\%Tolerance = 10.5 - (0.07 \times \%FL) + \left(\frac{833.3}{DT_{FL} \times \%FL}\right) \tag{33-28}$$

Where:

FL = Full load

DT_{FL} = Difference between entering and leaving fluid (i.e., chilled water) temperature at full load, in °F (°C)

Consider, for example, a chilled water unit specified to operate with a typical 10°F differential between entering and leaving chilled water temperature at full load. The full load and 25% of full load allowable tolerances would be calculated, respectively, as:

$$10.5 - (0.07 \times 100\%) + \left(\frac{1{,}500}{10 \times 100\%}\right) = 5\%$$

and

$$10.5 - (0.07 \times 25\%) + \left(\frac{1{,}500}{10 \times 25\%}\right) = 14.8\%$$

For the 1,000 ton (3,516 kW_r) electric-driven chiller represented in Table 33-3, rated at 0.60 kW/ton (0.171 kW_e/kW_r) at full load, the test tolerance of 5% would allow the chiller performance to be as high as 0.63 kW/ton (0.179 kW_e/kW_r) and the chiller capacity to be as low as 950 tons (3,340 kW_r). At 25% of full load, where the chiller is rated at 0.73 kW/ton (0.208 kW_e/kW_r), the test tolerance of 14.8% would allow the chiller performance to be as high as 0.84 kW/ton (0.239 kW_e/kW_r).

The allowable tolerance on IPLV and NPLV are to be determined in standard U.S. and SI units, respectively, as follows:

$$\%Tolerance = 6.5 + \frac{35}{DT_{FL}} \quad for\ DT_{FL}\ in\ °F \tag{33-29}$$

and

$$\%Tolerance = 6.5 + \frac{19.4}{DT_{FL}} \quad for\ DT_{FL}\ in\ °C \tag{33-30}$$

For the above example of a 10°F (5.5°C) temperature differential, this yields a tolerance of 10%. The test tolerance would allow for the chiller rated at an IPLV of 0.580 kW/ton (0.170 kW_e/kW_r) to operate at 0.638 kW/ton (0.187 kW_e/kW_r). Correspondingly, recalculating the IPLV by substituting the maximum tolerance performance rating for each of the four load levels individually in Table 33-3 also yields an IPLV performance value of 0.638 kW/ton (0.187 kW_e/kW_r).

In addition to these allowable performance result tolerances, one must also consider the impact of tolerances on other test variables. For example, chilled and condenser water flow rates and temperatures may deviate by up to 5% and 0.5°F (0.3°C), respectively. Correspondingly, for air-cooled condensers, the average entering air dry-bulb temperature to the condenser may vary by up to 1°F (0.6°C). As discussed below, varying evaporator and condenser temperature will impact performance. If, for example, the leaving chilled water temperature is measured at an increased level of 0.5°F (0.3°C) and the entering condenser water temperature at the same level of decrease, the measured performance of the unit will be improved by 1% or more.

Allowances such as these can be added to the allowable performance tolerances listed above to describe the actual full deviation that could be experienced in the worst case scenario and still meet certification requirements. This means that performance at the target temperatures could actually deviate by more than the allowable performance tolerance in the test result. The actual total IPLV deviation then might be 11 or 12%, as opposed to the 10% tolerance. In this example, the actual normalized IPLV performance might be 0.650 kW/ton (0.190 kW_e/kW_r), as opposed to the published rate of 0.58 kW/ton (0.170 kW_e/kW_r).

It is also important for the specifying engineer to consider the impact of these tolerances on auxiliary equipment for sizing and energy usage analysis. As another example, if the test flow rate is increased by the allowable 5%, the capacity of the unit will appear greater by this amount. However, in order to achieve that capacity, the water pressure drop and expected auxiliary pumping requirements will be elevated above the level that would be calculated based on the published performance rates. Other subtle differences may also be found. For example, if the performance rating does increase, so does the heat of compression and, in turn, the heat rejection duty imposed on the condenser. Hence, slightly more fan energy or cooling tower water usage requirement can be expected. While these all may be modest changes, if all of the deviations occur in the same direction, the additive impact can be significant.

All of these factors should be carefully considered when specifying equipment and performing life-cycle cost analyses. It may be prudent to confer with the manufacturer and request zero tolerance data on performance, indicating the worst case scenario that would still meet certification standards. It should be understood that the manufacturers will have their own actual fleet manufacturing tolerances, which may differ from the certification tolerances. Therefore, they may seek to ensure that the equipment they produce does not fall below the tolerances and will establish their published ratings accordingly. In summary, while certification tolerances may be fairly significant, the certification process allows for uniformity in the industry. The tolerance specifications also allow the specifying engineer to understand the deviations that may result and plan accordingly.

Effect of Varying Evaporator and Condensing Temperature

While air conditioning systems operate with a refrigerant evaporating temperature of about 35 to 45°F (1 to 7°C) to produce 44°F (7°C) chilled water or 55°F (13°C) supply air, process refrigeration systems often operate at much lower temperatures. Industrial fluid chillers can supply glycol mixtures, brines, or other specialty process fluids as low as -60°F (-51°C), with refrigerant evaporating temperatures 5 to 10°F (3 to 6°C) lower than leaving fluid temperatures. In most cases, the power requirements per ton (or kW_r) are higher for process refrigeration applications, sometimes dramatically. For example, under design conditions for a conventional refrigeration system, assuming 105°F (41°C) condensing temperature, performance figures at corresponding evaporating temperatures may be similar to the values shown in Table 33-7.

To achieve a large differential between evaporating and condensing temperature, process refrigeration equipment and refrigerant choices frequently differ from

Evaporating Temperature	Representative Power Requirement
40°F (4°C)	0.80 hp/ton (0.17 kW_m/kW_r)
0°F (-18°C)	1.66 hp/ton (0.35 kW_m/kW_r)
-40°F (-40°C)	3.59 hp/ton (0.76 kW_m/kW_r)

Table 33-7 Relationship of Evaporating Temperature to Power Requirement.

those selected for air conditioning duty. Commonly used process systems include screw compressors and two-stage compound reciprocating compressors. Commonly used refrigerants include ammonia and HCFC-22. While uncommon for air conditioning applications, the thermodynamic characteristics of ammonia (NH_3), or R-717, make it attractive for lower temperature refrigeration applications. It has a boiling point of about -28°F (-33°C) and a freezing point of about -108°F (-78°C).

Figures 33-4 and 33-5 show the effect of suction temperature and condensing temperature on capacity and performance for a vapor compression system featuring a screw compressor. As suction (evaporating) temperature is increased, the mass flow through the compressor increases due to a decrease in refrigerant-specific volume. Concurrently, the compressor's volumetric efficiency increases due to the decrease in compression ratio at constant condensing temperature.

In Figure 33-4, for example, at a suction temperature of 40°F (5°C), system capacity is about 600 tons (2,100 kW_r), while at a suction temperature of 5°F (-15°C), the capacity is only 280 tons (985 kW_r). Notice also that the power requirement at 40°F (5°C) is about 560 bhp (418 kW_m), while at 5°F (-15°C) it is 460 bhp (343 kW_m). Both capacity and COP increase at higher suction temperatures. Also notice that while total power requirement decreases with decreasing suction temperature, power requirement per unit output increases dramatically from 0.93 bhp/ton (0.20 kW_m/kW_r) at 40°F (5°C) to 1.64 bhp/ton (0.35 kW_m/kW_r) at 5°F (-15°C).

Figure 33-5 illustrates the effect of condensing temperature. For example, at a condensing temperature of 77°F (25°C), the capacity is 470 tons (1,653 kW_r), while at 131°F (55°C), the capacity is 363 tons (1,276 kW_r). At 77°F (25°C), the full-load power requirement is 503 bhp (375 kW_m), while at 131°F (55°C), the full-load power requirement is 667 bhp (497 kW_m). Both capacity and COP decrease at higher condensing temperatures.

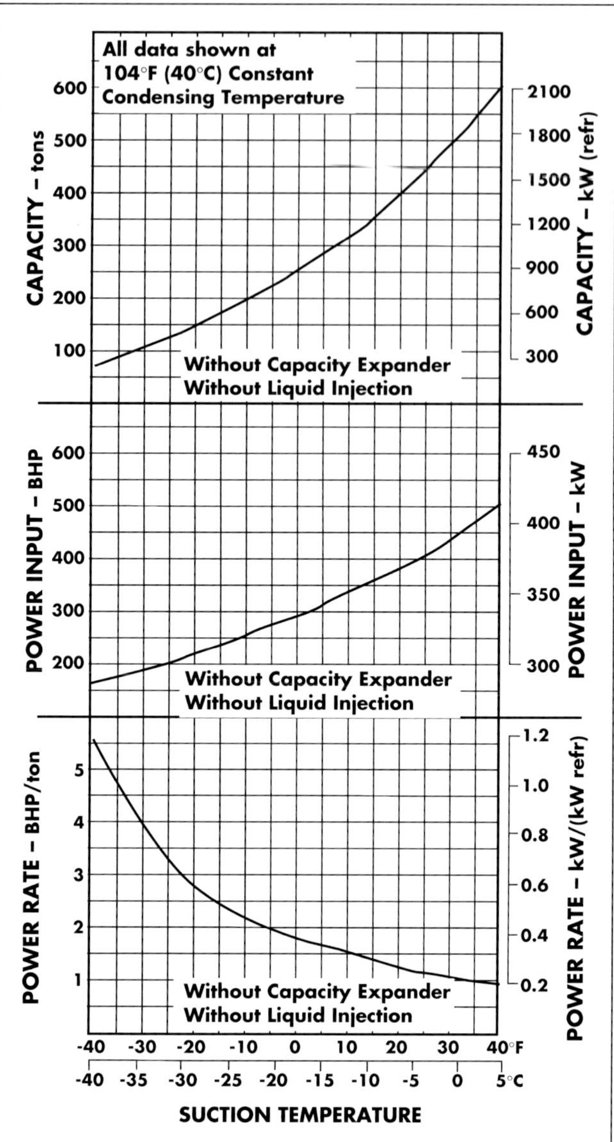

Fig. 33-4 Suction Temperature Effect on Performance. Source: Carrier Corp.

Source Versus End-Use Efficiency

Equipment performance rating standards are based on end-use efficiency. To establish a basis of comparison for overall thermodynamic performance (and environmental impact), it is necessary to understand the relationship between end-use and source thermodynamic efficiency as it applies to performance rating of cooling and refrigeration systems.

Electric-driven equipment typically exhibits COPs that are 2 to 4 times greater than comparable fuel- or heat-driven equipment. However, to calculate the actual fuel-related, or source energy COP, of an electric-driven unit, the efficiency of the electric generation and transmission and distribution (T&D) processes must be considered. As a starting point, the relationship between end-use COP and EER (Btu/watt-h) was previously derived and shown in Equation 33-12 as:

In considering source energy utilization, the relation can be expressed as:

$$COP = \frac{EER}{3{,}413}$$

$$COP = \frac{EER}{\textit{Power plant heat rate in Btu/watt-h}} \quad (33\text{-}31)$$

In SI units, EER would be replaced with watt-h of refrigerant output divided by watt-h of electricity input, and power plant heat rate would be expressed in watt-h heat input divided by watt-h electric output.

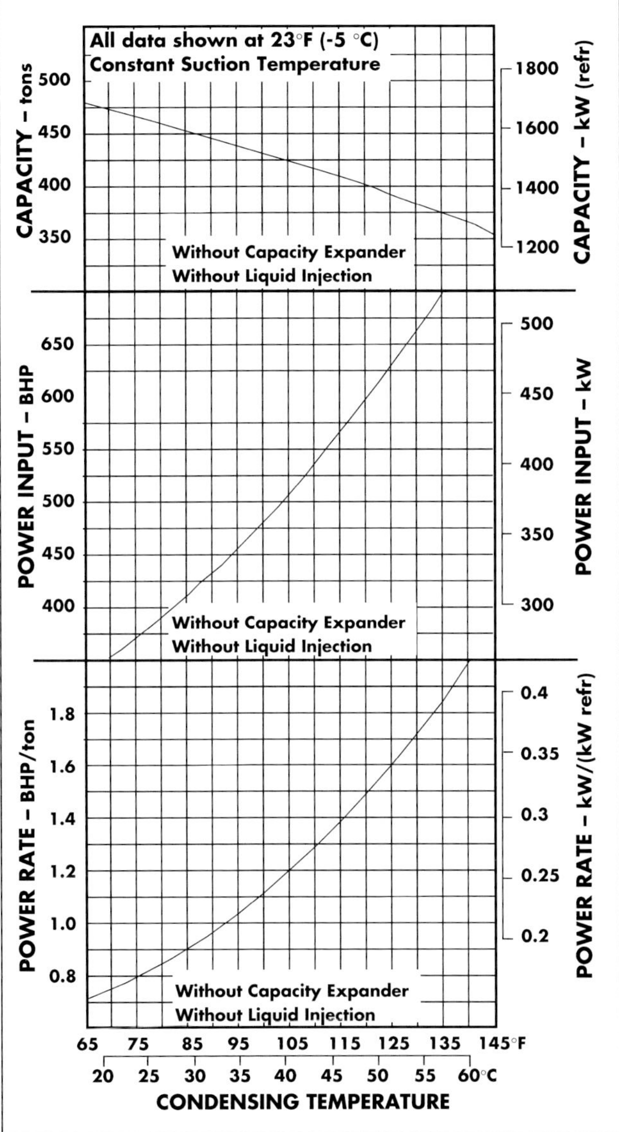

Fig. 33-5 Condensing Temperature Effect on Performance. Source: Carrier Corp.

Examples

Consider an electric-driven centrifugal chiller rated at 20 EER. The end-use COP would be calculated as:

$$\textit{End-use COP} = \frac{EER}{3.413} = \frac{20}{3.413} = 5.86$$

This same chiller unit, when driven by a prime mover, operating on a simple-cycle, with a thermodynamic efficiency of 32% (HHV basis), might have an end-use COP rating of 2.0. In this case, the end-use rating is the same as the source COP rating, since the fuel energy is being consumed on site.

As a basis of comparison, the source COP of the electric-driven unit is computed for three alternative central electric generation systems that could realistically be serving power to this unit.

Case A. An intermediate-load power (i.e., oil-fired steam) plant may have a cooling season heat rate of 10,000 Btu/kWh (2.93 kWh_h/kWh_e) and experience T&D losses of 7.5%, resulting in a total heat rate of just under 10,750 Btu/kWh (3.15 kWh_h/kWh_e). The source COP for the 20 EER unit described above would be calculated as:

$$\textit{Source COP} = \frac{\textit{20 Btu/watt-h}}{\textit{10.750 Btu/watt-h}} = 1.86$$

Case B. A peaking plant (i.e., simple-cycle gas turbine) may have a heat rate in excess of 14,000 Btu/kWh (4.10 kWh_h/kWh_e) and experience T&D losses of 12% (line losses are generally at their highest in peak conditions), resulting in a total heat rate for a delivered kW of 15,680 Btu/kWh (4.59 kWh_h/kWh_e). The source COP for the same 20 EER unit would be calculated as:

$$\textit{Source COP} = \frac{\textit{20 Btu/watt-h}}{\textit{15.680 Btu/watt-h}} = 1.28$$

Case C. An efficient baseload (i.e., coal) plant serving an off-peak cooling load may have a heat rate of 8,500 Btu/kWh (2.49 kWh_h/kWh_e) and experience T&D losses of 3.5% (line losses are generally at their lowest in off-peak conditions), resulting in a total heat rate of 8,798 Btu/kWh (2.58 kWh_h/kWh_e). The source COP for the same 20 EER unit would be calculated as:

$$\textit{Source COP} = \frac{\textit{20 Btu/watt-h}}{\textit{8.798 Btu/watt-h}} = 2.27$$

In these three examples, the source energy usage was shown to range from 2.5 to 4.5 times that of the end-use energy usage. On the basis of an end-use COP computation, the prime mover-driven unit appears to have grossly inferior thermodynamic performance when compared with the electric-driven unit. However, on the basis of a source COP computation, the two drive options are roughly comparable, varying by degree as opposed to by orders of magnitude. What can be concluded from this analysis of source versus end-use COP for electrically driven equipment is that the source results, or true thermodynamic performance, are quite comparable with on-site fuel- or steam-powered systems. The ensuing chapters in this section apply the standards and concepts presented in this chapter to a variety of cooling, air conditioning, and refrigeration technologies and applications.

Chapter Thirty-Four

Psychrometrics

Refrigeration is the process of cooling below ambient environmental temperature. Air conditioning (AC) is the process of treating air to simultaneously control its temperature, humidity, cleanliness, and distribution. Psychrometrics is the science of air/water vapor mixtures. It is used to study air at its various stages in the AC process and determine how the air moves from one state to another. The AC process changes the psychrometric condition of air and may involve cooling, heating, humidification, or dehumidification. This chapter provides a basic overview of psychrometrics as the underlying thermodynamic process involved in the application of certain technologies addressed in ensuing chapters in this Section. Examples include condensing methods, cooling towers, and desiccant dehumidification.

Figure 34-1 is a psychrometric representation of the basic AC processes. The arrows point to the direction on the chart that is followed to track the changing condition of the air as each of the processes occur. Directions A, C, E, and G are the four basic elemental AC processes: humidifying, heating, dehumidifying, and cooling, respectively. These four lines form four quadrants much like the four directions on a compass. Any point lying within these quadrants represents a combination of processes. For example, just as a point lying between north and east is in a northeast quadrant, any point lying between A and C represents heating and humidifying combined.

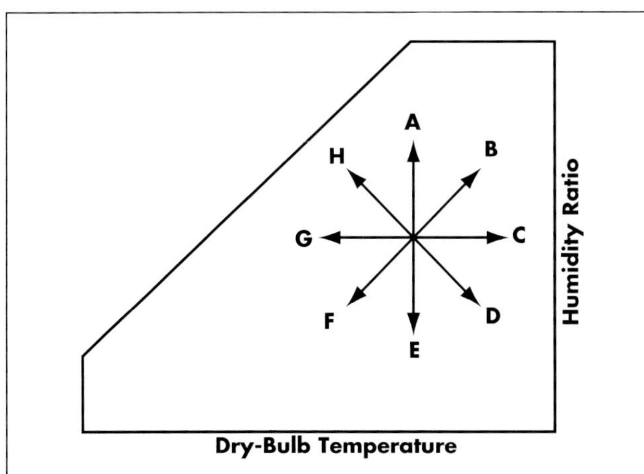

Fig. 34-1 Psychrometric Representation of Basic AC Processes.

Where:
A = Humidifying only
B = Heating and humidifying
C = Sensible heating only
D = Heating and dehumidifying (or chemical dehumidifying)
E = Dehumidifying only
F = Cooling and dehumidifying
G = Sensible cooling only
H = Evaporative cooling only

For the purposes of HVAC engineering, all aspects of these processes must be considered. Of primary interest for the ensuing chapters in this section are the cooling and dehumidification processes.

Air-Water Vapor Mixtures

Air is a mixture of many gases, predominantly nitrogen and oxygen. Atmospheric air also contains water vapor. The **dry-bulb (db) temperature** of an air stream is the temperature of the mixture of air and water vapor at rest. Moist air is said to be saturated when it can coexist in equilibrium with an associated condensed moisture phase.

Wet-bulb (wb) temperature is the temperature that will be reached when air and water are mixed with no transfer of heat from the outside. By evaporating, the water cools itself and the surrounding air until saturation is reached and the water can no longer provide additional cooling. While the db temperature indicates the actual temperature of the air, the wb temperature indicates the temperature that can be reached by evaporating enough water into the air to make it fully saturated.

The relationship between db and wb temperature is useful because if both are known, all other properties of the air-water mixture can be read directly from a **psychrometric chart**. The concept can be understood through a comparison with db temperature determination. Consider the case of two thermometers placed in a moving stream of gas (i.e., air), with the bulb of one covered with a linen, wet with distilled water. The covered thermometer functions as a wb thermometer. Water will evaporate from the wick due to heat transfer to the air at the db temperature and due to radiant heat transfer from the surroundings. The water vapor is diffused from the wick to the surrounding air-water vapor mixture until the evaporation rate reaches equilibrium with the drying capacity of air. At this point, the temperature reached by the wick-thermometer is called the wb temperature.

The wb thermometer will register a lower temperature than the db thermometer as long as evaporation continues. When the wb temperature is less than the db, the air is only partially saturated. The difference between db and wb temperature for a particular state is known as the wb depression. The greater the RH, the smaller the differential between db and wb temperatures, i.e., the wb depression. The maximum depression for a given db temperature will occur when the gas is dry. Zero depression will be observed at saturation, i.e., at 100% RH. Wb and db temperature will be equal because the gas is completely saturated.

When the cooling of an air stream proceeds at constant total pressure, the partial pressures of the constituents remain constant until the saturation state of the vapor is reached and condensation of the vapor occurs. **Dewpoint**, or condensation temperature, is the temperature at which condensation begins when the moist air mixture is cooled at constant pressure. The dewpoint is equal to the saturation temperature. When air is saturated, it has reached a condition of 100% RH.

Relative humidity (RH or **ϕ)** is the ratio of the mole fraction of water vapor in the mixture to the mole fraction of water vapor in saturated air at the same db temperature and barometric pressure. It is also the ratio of the pressure of the vapor in the air to the vapor pressure at saturation corresponding to the temperature. In simpler terms, RH expresses the moisture content of air as a percentage of what it can hold when air is fully saturated. This maximum value increases as the temperature increases, reaching its maximum at 100% RH (saturation).

The humidity ratio (ω), or specific humidity, of an air-water mixture is the ratio of the mass of the water vapor to the mass of dry air in the mixture. It is essentially a comparison of the weight of all of the water molecules to the weight of all of the air molecules in a lbm (or kg) of a mixture. Thus, the humidity ratio may be expressed as:

$$\omega = \frac{M_v}{M_a} \qquad (34\text{-}1)$$

Where:
M_v = Mass of water vapor
M_a = Mass of dry air

A typical humidity ratio is 0.00785 lbm (kg) of water per lbm (kg) of dry air. That is the amount of water in air at 70°F (21°C) and 50% RH at standard atmospheric pressure. In order to work in whole numbers, AC engineers commonly use the English system of measurement, which defines a lbm as consisting of 7,000 grains. So a humidity ratio of 0.00785 lbm of water/lbm of dry air is converted to whole numbers by multiplying by 7,000 grains/lbm of water. Thus,

$$0.00785\ lbm_{water}/lbm_{air} \times 7{,}000\ grains/lbm_{water}$$
$$= 55\ grains/lbm_{air}$$

For conversions to SI units, there are 15,432 grains/kg, so a humidity ratio of 0.00785 would be equivalent to 121 grains/kg_{air}. For example, assume that air at 90°F (32°C) and 70% RH is cooled and dehumidified so that the final state is 80°F (27°C) and 40% RH. The amount of water per lbm (kg) of dry air (ω) is reduced from 0.0214 to 0.0087. The 0.0127 lbm (kg) of water removed per lbm (kg) of dry air corresponds to (7,000 x 0.0127) 88.9 grains/lbm (196 grains/kg) of dry air.

The air cooling process consists of removing sensible and latent heat, so that the enthalpy, or the total energy, in the air is reduced. When air is hot, its enthalpy is high. When air is moist, its enthalpy is also high because additional heat was required to evaporate moisture into the air. The total enthalpy (h_m) of a mixture is equal to the enthalpy of the dry (h_a) air plus the enthalpy of the water vapor (h_v) times the humidity ratio (ω). Thus,

$$h_m = h_a + h_v\omega \qquad (34\text{-}2)$$

The amount of heat that must be removed to make this change, or enthalpy reduction, may be expressed as:

$$h_m = (M_{air} \times SH \times \Delta T) + (M_{H_2O} \times SH \times \Delta T)$$
$$+ (M_{condensate} \times heat\ of\ vaporization) \qquad (34\text{-}3)$$

Where:
M = Mass in lbm (kg)
SH = Specific heat in Btu/lbm • °F (kJ/kg • °C)

The enthalpy of the air is expressed as the number of Btu/lbm (kJ/kg or kW/kg) of dry air. Typical values range between 0 Btu/lbm (0 kJ/kg) at 0°F (-18°C) if the air is perfectly dry and 63 Btu/lbm (147 kJ/kg) if air is saturated at 95°F (35°C).

The heating and cooling requirements of AC loads are characterized by their sensible and latent components. The ratio of the sensible load component to the total heat load is sometimes referred to as the sensible heat fraction (SHF) or sensible heat ratio (SHR). The ratio of the latent component to the total load is sometimes referred to as the latent heat fraction (LHF) or latent heat ratio (LHR).

Generally, the more water vapor that must be removed from the air stream to be conditioned, the more work a mechanical refrigeration compressor (or absorption chiller) must do to achieve a low enough cooling coil temperature to condense the water vapor. The cooling coil must be much colder simply to lower the (sensible) temperature of the air to the desired level. Thus, just as with other lower temperature refrigeration applications, the compressor must operate at a higher compression ratio, requiring more shaft power per ton (kW_r or kJ/h), if there is a high LHR or if a low RH is desired.

The Psychrometric Chart

The psychrometric chart is a plot of the properties of atmospheric air. It graphically illustrates the relationship between db temperature, wb temperature, RH, humidity ratio, and enthalpy. It is essentially a graphic representation of the condition of air (air-water vapor mixture) at each point in the AC process. It relates db temperature to absolute moisture content of the air and includes all of the possible combinations of temperature, moisture content, density, and heat content properties that can occur in air. The chart can be used to make calculations to determine the sensible and latent loads associated with HVAC equipment processes. Figure 34-2 is a psychrometric chart in English units. The numbered lines or scales highlight the functionality of the chart. Figure 34-3 is a psychrometric chart in SI units.

Figure 34-4 is a basic sketch of a psychrometric chart; the scales and lines of the chart are highlighted as they relate to one specific set of air conditions. Included are db and wb temperatures, RH, specific humidity, vapor pressure, dewpoint temperature, and enthalpy. The darkened circle pinpoints the location of the specific condition and the intersecting horizontal and vertical lines show the scales from which the condition can be identified. If any two properties of an air mixture are known, the chart allows a quick determination of all of its other properties.

Following are a series of skeleton psychrometric charts. Each highlights particular scales and lines associated with each one of the seven conditions shown jointly in Figure 34-4.

Figure 34-5 shows db temperature lines, i.e., the temperature of air measured on a standard thermometer. These are shown as straight vertical lines. The scale called db temperature is laid out horizontally at the bottom of

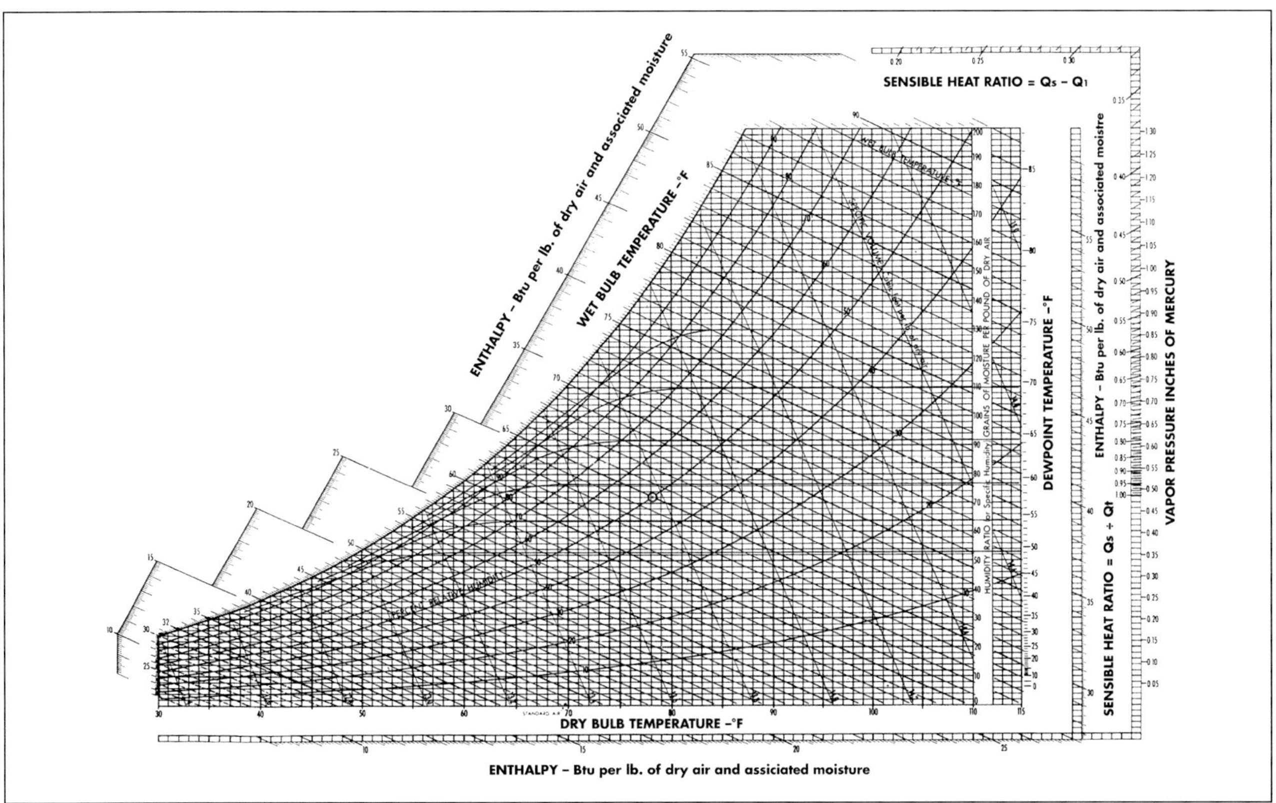

Fig. 34-2 Psychrometric Chart (English units). Source: The Trane Company

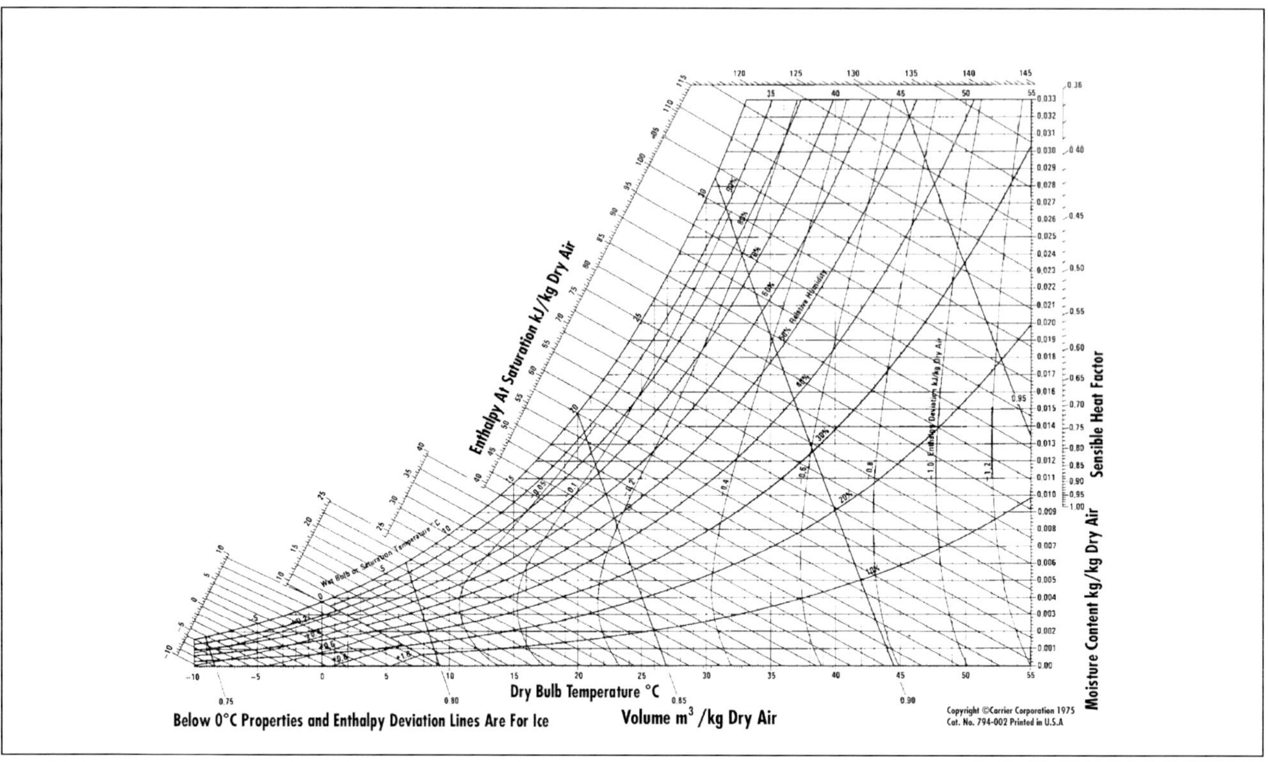

Fig. 34-3 Psychrometric Chart (SI units). Source: Carrier Corp.

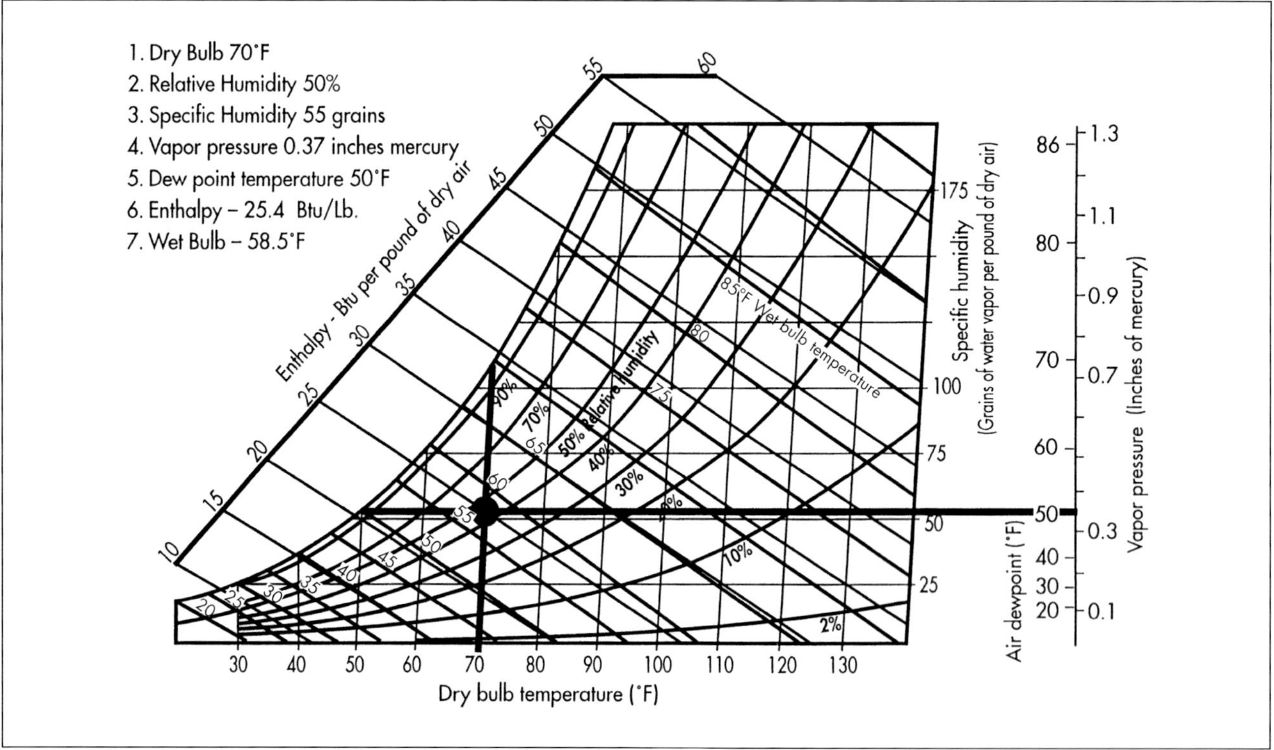

Fig. 34-4 Sketch of Psychrometric Chart Highlighting One Specific Set of Conditions.

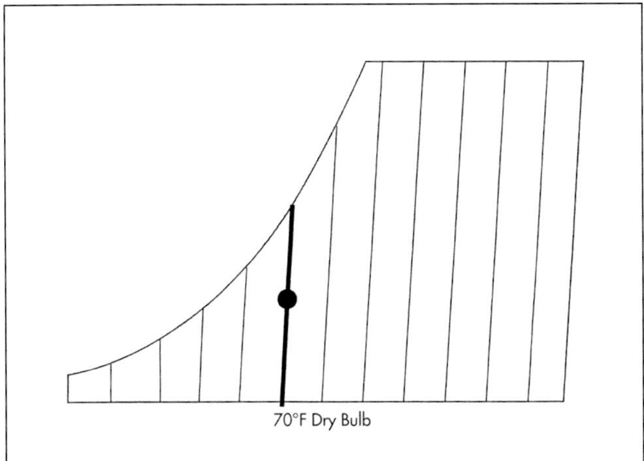

Fig. 34-5 Dry-Bulb Temperature Lines.

the chart in °F, with the incremental lines extending vertically. In SI unit charts, the scale is in °C.

Figure 34-6 shows RH (ϕ) lines. The RH lines are curved and the values appear in increments of 10%, representing the degree of saturation, or the ratio of the pressure of the vapor (p_v) to the pressure of the vapor at saturation (p_{vs}). At 100% RH, db temperature is equal to wb temperature. As RH decreases, the wb temperature becomes lower than the equivalent db temperature.

Figure 34-7 shows specific humidity, or humidity ratio (ω or HR), lines. These are shown as straight horizontal lines that are perpendicular to db lines. The scale is in grains of moisture per lbm of dry air (grains/lbm) and typically ranges from 0 to about 200. In SI unit charts, the corresponding scale is kg/kg of dry air and typically ranges from 0 to 0.033. For example, to determine the grains/lbm of dry air removed in conditioning air from 90°F (32°C) and 70% RH to 80°F (27°C) and 40% RH, proceed vertically from 90°F (32°C) db to the 70% RH curve. The intersection is at 150 grains (0.0214 kg/kg), while 80°F (27°C) and 40% RH intersect at 61 grains (0.0087 kg/kg). The difference is 90 grains (0.0128 kg/kg).

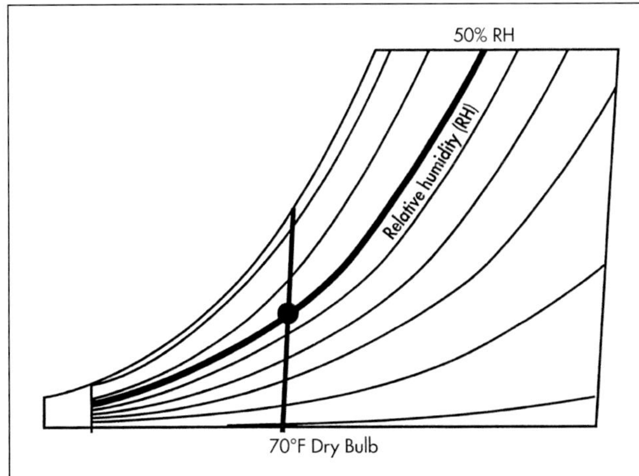

Fig. 34-6 Relative Humidity Lines.

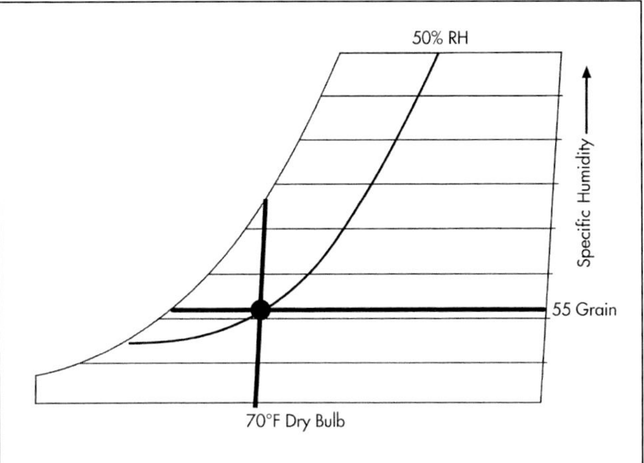

Fig. 34-7 Specific Humidity (Humidity Ratio) Lines.

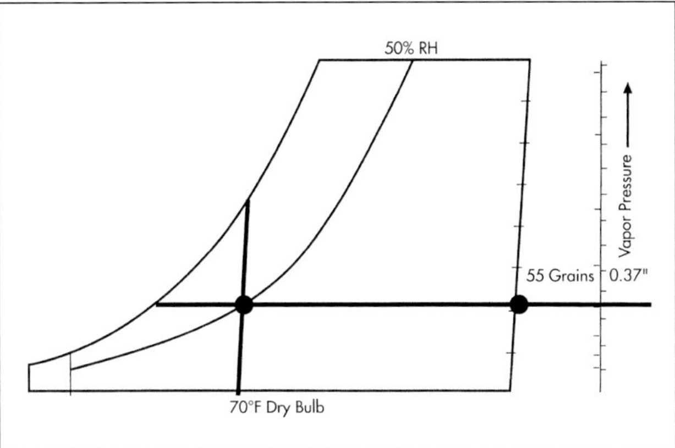

Fig. 34-8 Vapor Pressure Scale.

Figure 34-8 shows the vapor pressure (P_v or P_w) scale. This measures the pressure exerted by water vapor in the air. The scale, when included, is typically on the far right of the chart and the unit of measurement is in-Hg abs (cm-Hg abs or Pascal).

Figure 34-9 shows the dewpoint temperature scale. Dewpoint temperature lines run horizontally, like the grains of moisture lines, with a scale that typically ranges from 20 to 90°F (-7 to 32°C). To determine the dewpoint temperature of air at 80°F (27°C) and 50% RH, for example, start at the db temperature of 80°F (27°C) and proceed vertically to the 50% RH curve. From this intersection, proceed horizontally to the saturation curve (100% RH). This yields a dewpoint temperature of 60°F (16°C).

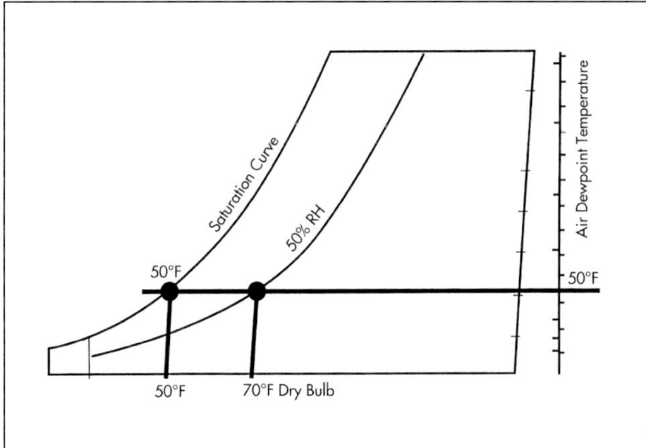

Fig. 34-9 Dewpoint Temperature Scale.

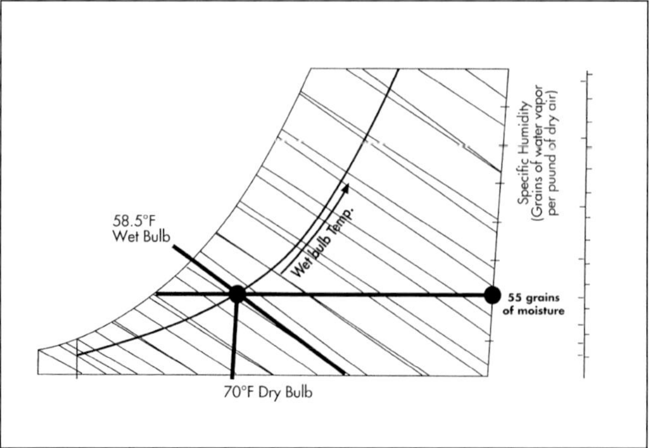

Fig. 34-11 Wet-Bulb Temperature Lines.

Figure 34-10 shows the enthalpy (h) scale. It is typically on the left, but sometimes on both sides, of the chart. Typically, the scale ranges from about 5 to 60 Btu/lbm (12 to 140 kJ/kg). Commonly, the wb lines are used to represent enthalpy lines. While these are not exactly the same, they are generally considered to be close enough for practical purposes.

Figure 34-11 shows wb, or saturation, temperature lines. These indicate the temperature of air above 32°F (0°C). Below 32°F (0°C), temperatures are measured on a wb thermometer on which the water in the wick has frozen to ice. Note that the slope of the wb lines change below 32°F (0°C). The scale, in °F, is the curved line (with a slope of about 30 degrees) at the left edge of the chart.

Additional Features of Psychrometric Charts

Following are some additional lines and scales commonly found on psychrometric charts:

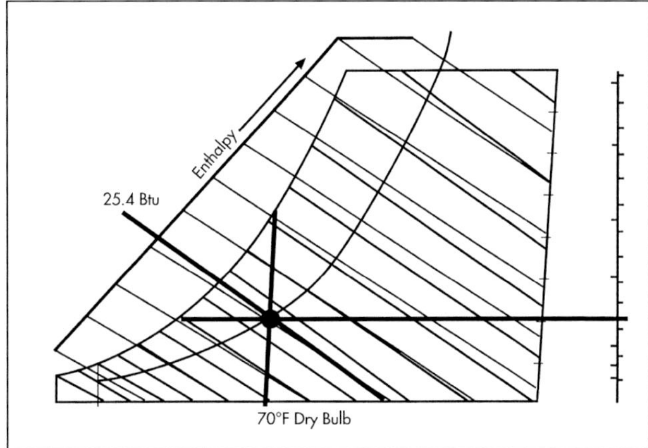

Fig. 34-10 Enthalpy Scale.

- Specific volume (v) lines. Specific volume is the volume that a unit mass of air occupies. It is the reciprocal of the density of air. Specific volume lines run almost diagonally from the upper left to the lower right of the chart, with values listed on each line. Values falling between each line can be found through interpolation by placing a straight edge over the point of intersection and paralleling the specific volume lines. The unit of measurement is ft^3 (m^3) of the air mixture per lbm (kg) of dry air, with typical values ranging from 12.5 to 14.5 (0.75 to 0.95) in standard charts.
- Sensible heat factor (SHF), or ratio, scale. This is useful when plotting some AC processes, such as cooling and dehumidification, and helps to determine the required supply air conditions. The scale typically ranges from 0.35 to 1.00, or 35 to 100%. This percentage represents the sensible work. The remaining percentage would be latent heat.
- Enthalpy deviation lines. These are used to correct enthalpy readings when extreme accuracy is required. Enthalpy is the total heat in the air at 100% saturation. If the air is not completely saturated, slight error is present in the enthalpy reading. The scale typically ranges from –0.02 to –0.30 Btu (–0.02 to –0.32 kJ). This is the amount of heat that would be subtracted from the enthalpy reading.

Cooling and Dehumidifying Air

The process of sensible cooling is represented on the psychrometric chart by a horizontal line extended to the saturation line. If air is cooled sensibly, it changes in its db and wb temperature, RH, and total heat, but does not change in its moisture content, dewpoint temperature, or vapor pressure. For example, to determine how much heat must be removed to sensibly cool air having a db temperature of 80°F (27°C) and 50% RH to 50°F (10°C), identify the humidity ratio and enthalpy at 80°F (27°C) and 50% RH. This is 26 grains and 23.4 Btu (0.0037 kg/kg and 54.4 kJ), respectively, per lbm (kg) of dry air. Proceed horizontally at a constant humidity ratio to 50°F (27°C), at which point enthalpy is 16.1 Btu/lbm (37.4 kJ/kg) of dry air. The difference, or sensible heat removal requirement, is 7.3 Btu/lbm (17.0 kJ/kg).

Until the air temperature reaches its dewpoint, all of the cooling is sensible. For example, an air stream at 70°F (21°C) and 50% RH can be sensibly cooled to 51°F (11°C) before any moisture is removed. At 51°F (11°C), it is saturated (100% RH). If it is cooled further, its moisture will begin to condense out of the air.

The db temperature and absolute humidity of the air stream and the cooling coil surface temperature determine sensible and latent cooling. If the cooling surface temperature is below the initial dewpoint temperature, this process can be portrayed on the chart as a straight line extending from the initial condition to the surface temperature on the saturation curve. During this process, the db and wb temperature, moisture content, vapor pressure of the moisture, and total heat all decrease. The amount of moisture removed depends on how cold the air can be chilled. The lower the temperature, the drier the air.

The cooling/dehumidification process is illustrated diagramatically and the process air path is drawn on a psychrometric chart in Figure 34-12. In this example, air is cooled from 70 to 45°F (21 to 7°C) and moisture level is reduced from 56 to 44 grains/lbm (0.008 to 0.006 kg/kg).

To meet the specified temperature and humidity set points in a given space, the sensible and latent removal capacity of the refrigeration system must be equal to the corresponding fractions of the cooling load. In situations in which the LHF is much greater than the SHF, or where a low dewpoint temperature is desired, excess sensible cooling capacity must be designed into the system.

In order to satisfy the latent cooling requirement, the air stream must be cooled below the dewpoint temperature, or excess air must be introduced and cooled. When a system is designed to produce low humidity levels, the air stream must be reheated before being discharged to the space.

The required coil temperature depends on the humidity level desired. If latent load is high, design options to consider are:

- Use of a deeper coil.
- Use of lower temperature to dry part of the air and then mix.
- Use of some other special design.

An inherent inefficiency in conventional AC systems is the need to overcool in order to achieve a low dewpoint. For each degree of dewpoint temperature that the air stream must be cooled beyond the point it satisfies the sensible cooling requirement, the system must be overdesigned, driving up system cost and driving down system performance. This means that there is more total cooling capacity requirement than there is total cooling load. This additional energy use is further increased by the energy required for reheating. For each required Btu (kJ or kWh) of sensible overcooling, there is a corresponding Btu (kJ or kWh) required for reheating.

At lower temperatures, moisture removal by cooling

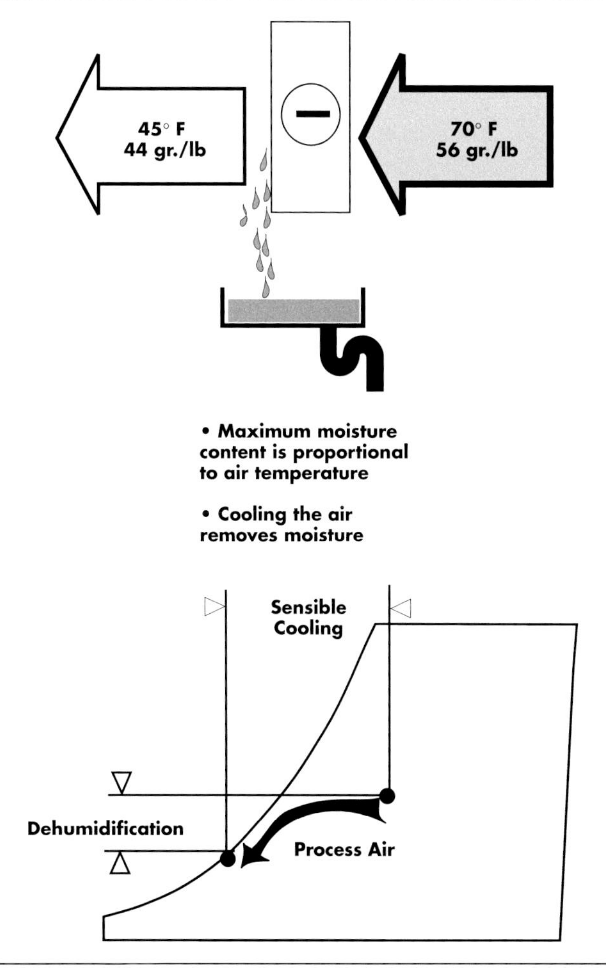

Fig. 34-12 Cooling/Dehumidification Process.
Source: Munters Cargocaire and Mason Grant Company

is less efficient. Also, the rate of sensible overcooling continually increases with each declining degree of dewpoint temperature. The corresponding rate of reheat requirement increases proportionally with the rate of sensible overcooling.

When cooling coil surfaces are held below the freezing point, frost will develop on the coil. This places a limit on how low conventional AC systems can reduce dewpoint temperature. The frost insulates the coil and it becomes physically clogged, reducing airflow and heat transfer efficiency. Defrost equipment must be designed into the system in order to eliminate ice build-up, further adding to the energy inefficiency. Thus, designing a system to achieve low humidity levels results in increased capital and operating costs for excess cooling capacity, reheat, and defrosting equipment.

While beyond the scope of this discussion, it is important to note that for refrigeration freezing processes, the psychrometric process extends to the product freezing range. Product temperature is brought down to the freezing range (sensible heat removal above freezing) and then water in the product changes to ice while the temperature remains constant (latent heat removal). Product temperature is then lowered below the freezing point to the ultimate storage temperature (sensible heat removal below freezing).

STANDARD AIR CONSTANTS

Standard air is defined as dry air at 70°F (21°C) and 29.92 in. Hg column (101.325 kPa). The air constants below apply specifically to standard air and can be used in most AC calculations. If very precise data are required, the constants below can be adjusted to correspond to actual dry air conditions. Note that in English system units, the weight of air may be expressed in lbm or lb, which both imply lb-mass (lb is more commonly used in HVAC engineering than lbm). Note also that ft^3 is also commonly expressed as cf and cf per minute is commonly expressed as cfm.

Specific volume of standard air = 13.33 ft^3/lbm (0.833 m^3/kg)

Density of standard air = 0.075 lbm/ft^3 (1.2 kg/m^3)

Specific heat of standard air = 0.24 Btu/lbm • °F (1.00kJ/kg • °C or 0.278 Watt-h/kg • °C)

Average latent heat of water vapor = 1,054 Btu/lbm (2,451 kJ/kg or 681 W-h/kg)

Use of these standard air constants is often sufficient in performing HVAC engineering calculations. Such calculations are, however, limited to the accuracy of the assumption of standard air conditions. To refine the accuracy of such calculations, one must adjust factors such as air density (and specific volume), specific heat, and latent heat. These can be found in psychrometric charts designed for non-standard conditions, including charts for high and low temperatures and barometric pressures.

Heat Extraction-Evaporators, Chilled Water, Economizers and Thermal Storage

The object of a refrigeration cycle is to produce a cooling effect, to maintain environmental or process conditions at desired levels. In the cycle, heat is extracted from the space or medium being cooled to the refrigerant in the evaporator. This is sometimes referred to as the cold-, low- (pressure) or suction-side of the cycle. The evaporator may provide direct or indirect heat extraction from a space or process. For large facilities and large processes, brine or chilled water usually serve as the indirect cooling medium.

This chapter focuses on the evaporators and chilled water systems, with emphasis on chilled water distribution systems and design strategies. Also included is a discussion of other types of cooling optimization strategies including water-side economizer cycles, to obtain "free" cooling effect when ambient conditions permit, and thermal energy storage to reduce peak electric demand and/or balance system operation.

LIQUID COOLERS

In the refrigeration process, heat is transferred from the medium being cooled to the refrigerant as it changes phase in the **evaporator**. A refrigerant-to-air, or **direct expansion (DX)**, evaporator is typically used with smaller vapor compression systems with a higher-pressure refrigerant, such as HCFC-22, CFC-12, and HFC-134a. The DX evaporator shown in Figure 35-1 is a finned-tube coil over which air is passed and thereby cooled.

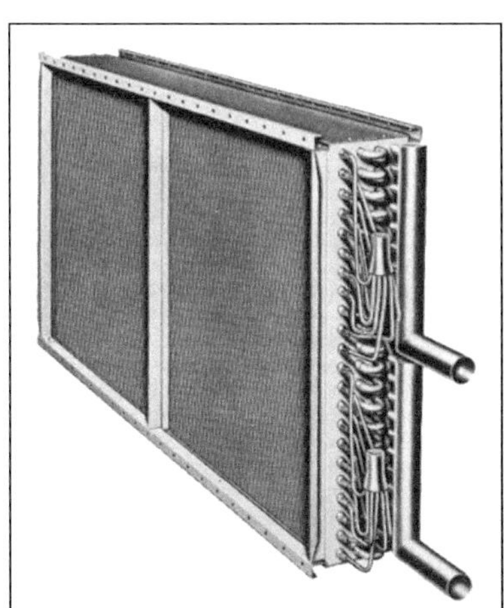

Fig. 35-1 Illustration of Fin-Tube Evaporator. Source: The Trane Company

Large central air conditioning or process cooling systems with multiple terminal units use a **brine** or **chilled water** cooling system because it is impractical to circulate large volumes of refrigerant throughout a facility. In this case, the chilled water serves as an indirect or secondary cooling medium that circulates between the loads and a central refrigerant-to-liquid evaporator, or **cooler**. Upon exiting the cooler, the chilled water is distributed to coils housed in air handling units (AHUs) or to other air conditioning or process heat exchangers.

The liquid cooler is a shell-and-tube heat exchanger of either the flooded or dry type. In a flooded cooler, the refrigerant is vaporized on the outside of bare or augmented surface tubes that are submerged in evaporating liquid refrigerant within a closed shell. Refrigerant may be metered by a float valve or orifice with flooded systems. Evaporators are said to be dry when a portion of the evaporator area is used for superheating the refrigerant. Refrigerant flow to the evaporator is controlled by a

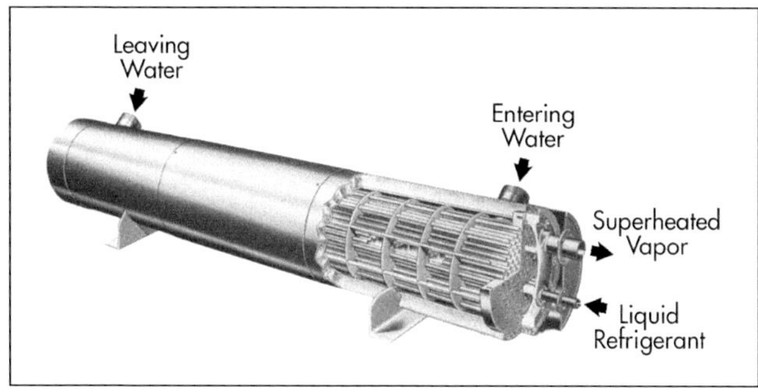

Fig. 35-2 Direct-Expansion Shell-and-Tube. Source: The Trane Company

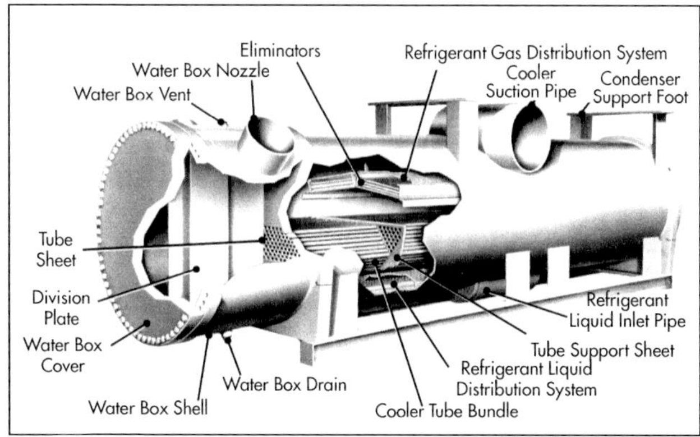

Fig. 35-3 Cooler (Evaporator) for Centrifugal Chiller. Source: Carrier Corp.

thermal expansion valve in response to superheat in the return line from the evaporator coil. Typically, this is set at about 10°F (6°C) of superheat to assure gas inlet to the compressor to avoid compressor damage due to liquid carryover. Figure 35-2 shows a direct-expansion shell-and-tube evaporator.

Figure 35-3 is a cutaway schematic of an evaporator (or cooler) used with a centrifugal chiller. Notice the eliminator which, in this case, is a series of parallel plates used to trap liquid droplets ahead of the compressor. Also notice the marine water box, which is sometimes used to allow for easier inspection and cleaning of heat exchanger tubes.

Chilled Water Temperature and Flow

Chilled water supply and return temperature are a function of the cooling load and flow rate, commonly measured in gallons per minute (gpm) or liters per minute (lpm). The temperature differential (ΔT) between supply and return temperature can be calculated in English units as:

$$\Delta T(F^\circ) = \frac{(Btu/h)_{cooling}}{8.33\ lbm/gal \times 60\ min/h \times gpm \times SG \times C_p} \qquad (35\text{-}1)$$

and in SI units as:

$$\Delta T(C^\circ) = \frac{kW_r}{1\ kg/liter \times 60\ min/h \times lpm \times SG \times C_p} \qquad (35\text{-}2)$$

Where:

SG = Specific gravity of fluid, which is assumed to be 1

C_p = Specific heat of fluid in Btu/lbm • °F, kJ/kg • °C or kWh/kg • °C [it is assumed to be 1 Btu/lbm • °F (4.187 kJ/kg • °C or 1.16 Watt-h/kg • °C)]

Based on a ΔT of 10°F, required flow rate is determined to be 2.4 gpm/ton (2.6 lpm/kW_r) as follows:

$$\frac{(12{,}000\ Btu/h)_{cooling/ton}}{8.33\ lbm/gal \times 60\ min/h \times 10^\circ F \times 1 \times 1} = 2.4\ gpm/ton\ (2.6\ lpm/kW_r)$$

Increasing the ΔT to 15°F (8.3°C), for example, would result in a reduced flow rate of 1.6 gpm/ton (1.7 lpm/kW_r).

If the flow rate and supply temperature are held constant in an operating system, the return temperature will fall as facility load is reduced. Under part-load conditions, there is an opportunity to conserve energy by allowing the supply water temperature to increase. This reduces the energy input requirement for a given load level by minimizing the pressure differential between the evaporator and the condenser. This is called **chilled water reset.**

In an air conditioning application, the ability to reset chilled water supply temperature is limited by the AHU's capacity to control humidity at the higher water temperature. In a large, diverse facility, available reset will be limited by the AHU or zone with the most severe cooling or dehumidification requirement.

Chilled water supply temperatures can be adjusted manually or automatically. Commonly, higher supply temperatures are used in spring and fall months, when both the temperature and the humidity of outside air are generally reduced. Automatic controls can provide a varying reset based on specified conditions, such as AHU coil valve position, outside air temperature and humidity, supply air, or indoor conditions.

Chilled Water Piping and Distribution Systems

The most common multiple chiller systems involve parallel piping configurations designed for constant flow. Chilled water flows through all chillers whenever the system is operating and an identical ΔT is maintained across each machine. Typically, three-way control valves are used at terminal units to maintain constant system flow by bypassing coils during periods of part-load. There are several limitations with this configuration:

- Pump energy is wasted due to system operation at constant full flow.
- Cooler tubes are subject to unnecessary wear and fouling.
- Supply temperature control is limited as supply water from operating and off-line chillers is mixed.

Consider the example of two identical chillers designed to provide supply water at 44°F (6.7°C) with a 10°F (5.5°C) temperature rise. Under typical operating conditions, at 40% of design load, one chiller is operating at 80% capacity with 44°F (6.7°C) supply temperature and an 8°F (4.4°C) temperature rise. Due to mixing, the minimum supply temperature leaving the two-chiller plant is 48°F (8.9°C), which may not provide adequate dehumidification in all areas of the facility. Energy consumption is also penalized, because chilled water reset is not possible for the operating chiller. Reducing supply temperature, by operating both chillers at 40% load, could further penalize energy consumption due to the need for added condenser water pumping and, possibly, hot gas bypass to allow chiller operation at low-load.

Inactive chillers can be isolated when individual chilled water pumps are provided for each unit. A common parallel chiller configuration includes the use of a bypass line, with differential pressure control designed to maintain constant flow to operating chillers under varying load conditions. Two-way control valves can be used with this approach.

When two chillers are connected in series, as shown in Figure 35-4, a constant supply temperature can be maintained, even when only one chiller is operating. Overall refrigeration energy efficiency is improved, because the first chiller is able to operate at a higher leaving water temperature.

Consider operation of a mixed (or hybrid) system that includes an absorption chiller and a vapor compression chiller. Because the COP of the absorption chiller is improved more than that of the vapor compression chiller at high evaporating temperature, the absorption unit is placed first in sequence. Also, since absorption chillers are limited to a minimum supply temperature of about 40°F (4.4°C), lower supply temperatures can be obtained by locating the vapor compression chiller downstream of the absorption unit.

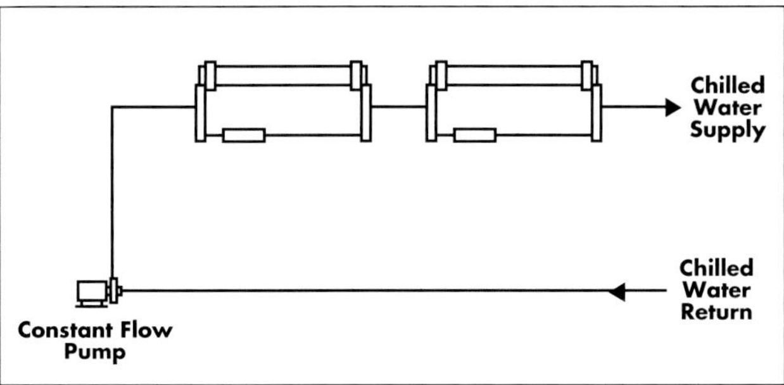

Fig. 35-4 Series Flow Configuration.

In addition to thermodynamic performance considerations, relative chiller loading may be varied based on other factors, such as prevailing time-of-use energy rates. For example, during peak electric rate periods, the temperature differential across the absorption chiller could be increased to reduce load on the electric-driven chiller. Absorption chiller load could then be decreased during off-peak periods when electricity prices are low.

A disadvantage of the series arrangement is that the system experiences head loss (and tube wear) through off-line chillers under part-load conditions. As with the parallel arrangement, series systems can be adapted for variable flow with a bypass that maintains constant flow through the chillers.

PRIMARY/SECONDARY DISTRIBUTION SYSTEMS

In a typical **primary/secondary system**, there are two separate pumping loops: a primary loop for the chillers and a secondary loop for distribution to various facility loads. The chillers, operating on the primary system, can be cycled, staged, and loaded in the most efficient configuration. The primary circulator pumps are matched to the individual flow and head of each chiller, and when a chiller in a multiple unit configuration is inactive, its pump is off. The secondary distribution system operates at variable flow based on load requirements. Typically, multiple variable speed pumps are used under differential pressure control.

Primary/secondary piping arrangements increase capital costs, but can often produce significant pumping energy savings, particularly in systems that experience a wide range of load variation. A number of variations can be used for multiple chiller systems.

- In the most basic arrangement, shown in Figure 35-5, a bypass line decouples the chiller (primary) flow from the distribution system (secondary) flow. The bypass line is located so that all chillers operate at the same discharge and return temperature. When the secondary system pump is at part-load, flow is reduced and the bypass line allows a portion of the primary flow to bypass the secondary system. Chillers share load equally and constant supply temperature can be maintained.
- A modification of this basic arrangement, shown in Figure 35-6, locates the bypass line so as to create a lead chiller, which can be fully loaded before the second chiller is brought on-line. This arrangement may be preferable when there is an operational advantage to using a designated chiller to meet base loads.
- A more sophisticated arrangement, shown in Figure 35-7, locates the bypass between two chillers. A valve is installed between each chiller and the bypass, allowing either chiller to operate as lead unit. This arrangement can be particularly effective with mixed chiller systems. Consider, for example, a hybrid system consisting of an absorption chiller and an electric chiller, with the absorption unit providing baseload cooling during peak electrical rate periods and the electric chiller providing baseload cooling during off-peak periods.

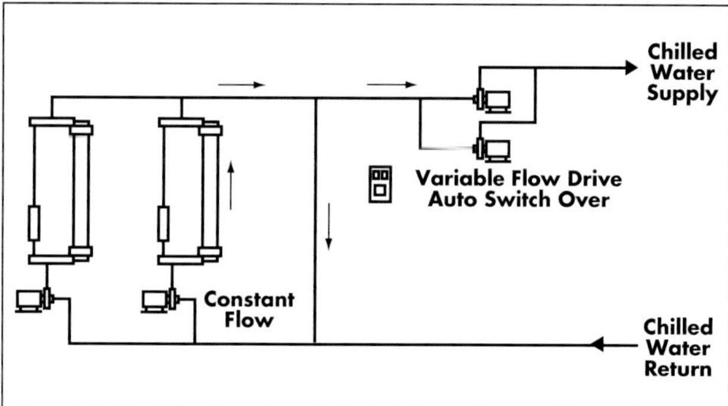

Fig. 35-5 Basic Primary/Secondary Configuration.

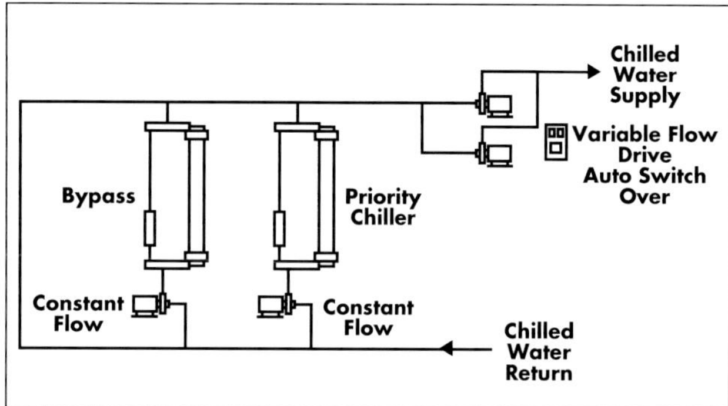

Fig. 35-6 Primary/Secondary Priority Loading Configuration.

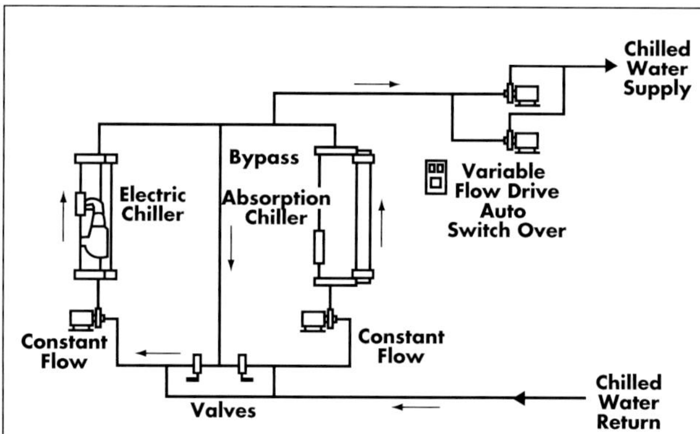

Fig. 35-7 Primary/Secondary Variable Priority Loading Configuration.

Historically, flow variation through chillers was severely limited by design parameters. Hence, primary/secondary systems were the only reasonable alternative available for capturing pump energy savings through variable system flow. However, current designs do allow for a relatively large variation in flow through chillers. This allows a single (primary) pump to operate with a variable flow rate, enabling the capture of significant pump energy savings. While this configuration does not offer the full range of flow variation achievable with a primary/secondary system, it does offer the potential for energy savings to be achieved at a far lower capital cost.

Water-Side Economizer Cycle (Free Cooling) Applications

Free cooling, or **economizer cycle**, applications take advantage of favorable outdoor temperatures to meet air conditioning or process cooling load requirements. **Water-side free cooling** involves the use of heat rejection systems to produce chilled water without operation of refrigeration systems.

In many localities, depending on the particular cooling process, there are periods during the course of the year when ambient wet-bulb temperature will be sufficiently depressed to permit the cooling tower (or other heat rejection system) to produce a low enough water temperature to satisfy the facility cooling requirement. Supply water temperature can also usually be raised during free cooling operation in air conditioning applications, because overall load, and particularly dehumidification load, is reduced as a result of cool and/or dry ambient air conditions.

Direct water-side free cooling, illustrated in Figure 35-8, bypasses the chiller completely and allows condenser water to flow directly to the load. Typically, the chilled water pump is bypassed and only the condenser pump is used. It is critical to ensure that the raw condenser water from the tower or body of water does not contaminate the relatively clean chilled water circuit. This system is often referred to as a strainer cycle, indicating the need for a strainer or filter. Cooling tower freeze protection is also a concern, and variable-speed fans as well as basin

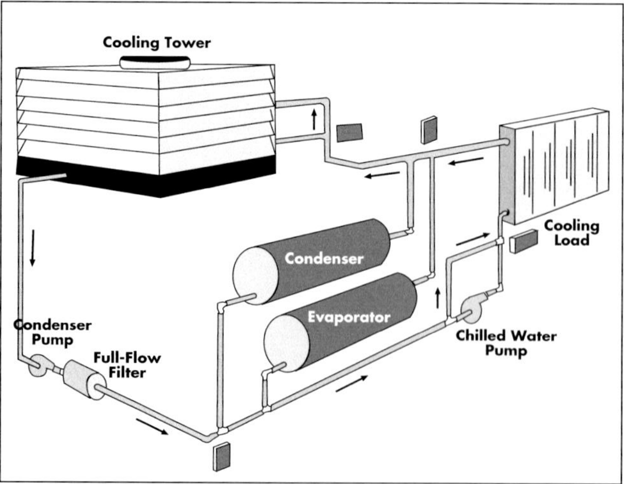

Fig. 35-8 Direct Water-Side Free Cooling Configuration. Source: The Trane Company

heating are often required. While it has proven effective in a number of field applications, the stringent operating requirements and risks associated with this approach have limited its use.

Indirect water-side free cooling, illustrated in Figure 35-9, is a common approach that is accomplished by the inclusion of a heat exchanger to separate the heat rejection and chilled water circuits. Available free cooling is somewhat reduced due to the temperature differential across the heat exchanger, although efficient plate-and-frame units are generally used. The pumping energy requirement is also greater with indirect free cooling, since both condenser water and chilled water pumps are required and the heat exchanger presents additional head to be overcome by the pumps.

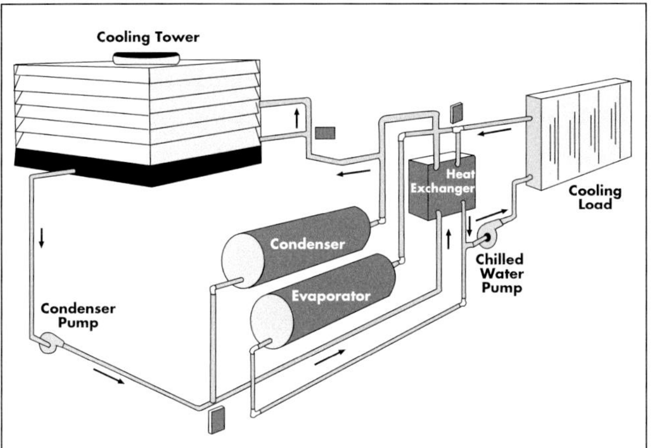

Fig. 35-9 Indirect Free Cooling Configuration, with Heat Exchanger. Source: The Trane Company

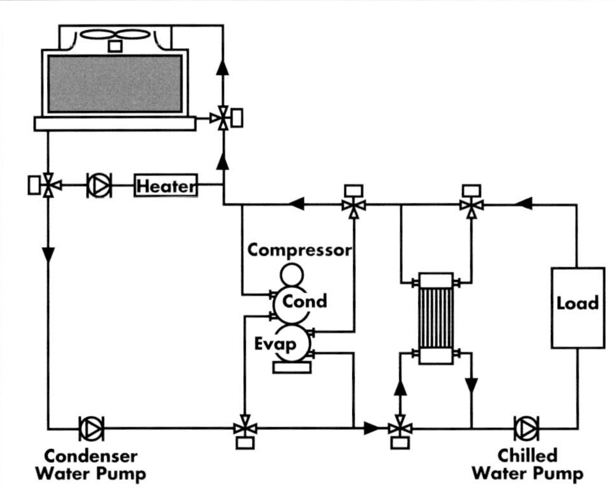

Fig. 35-10 Schematic Representation of Indirect Water-Side Free-Cooling System.

The transition between chiller and free-cooling operation deserves attention in any application. In temperate climate regions, cooling tower water temperatures in the range of 40 to 45°F (4 to 7°C) are required for free cooling, while the minimum acceptable chiller condenser water temperature is often above 55°F (13°C). When chillers are started, controls are required to maintain condensing temperature above this lower limit as the temperature of the cooling tower circuit rises. A multiple cooling tower array can allow cells that operate on free cooling to be isolated from cells that provide heat rejection in standard operation. Figure 35-10 is a schematic representation of a vapor compression system designed to operate as an isolated indirect free cooling loop.

An additional water-side free cooling method, known as thermal-siphon or refrigerant migration, is accomplished by allowing the chiller itself to function as a heat exchanger. When the condenser water is at a lower temperature than the chilled water, it is sometimes possible for the refrigerant to reject heat absorbed from the chilled water to the condenser water without mechanical compression. A small refrigerant pump may be used. Care must be taken when switching to normal compressor operation, since condenser water must be brought up to safe operating temperatures.

THERMAL ENERGY STORAGE

Thermal energy storage (TES) involves the capture of thermal energy generated during one time period for later use during another. TES has been used for decades in one form or another for space and process heating and cooling applications. There are many types of TES applications. Examples include storage of solar heat for night heating, storage of heat recovery-generated hot water for use as needed for fluctuating loads, storage of ice in winter for use in summer, and storage of a cooling medium generated by an electric chiller during off-peak periods for use during peak periods.

TES may involve sensible or latent storage. Sensible storage is accomplished by raising or lowering the temperature of storage media, such as water or solids, without producing a phase change. Latent storage is accomplished by a phase change of the storage media, typically from liquid to solid, such as water/ice or salt hydrates.

For heating applications, water, oils, phase change media (such as eutectic salts), and solids (such as rock or brick) are most commonly used. For cooling applications, water and phase change media, including both ice and eutectic salts, are typically used. Other latent-heat types of cool storage media include carbon dioxide (CO_2), which is used for applications such as low-temperature food freezing, and liquid nitrogen, which is used for cryogenic operation conditions.

Refrigeration-Cycle TES Technology Overview

For refrigeration and air conditioning applications, all TES systems operate on the same fundamental concept: the cooling equipment produces a refrigeration effect (direct or indirect), either to meet the load or to be added to storage. The storage system either accepts excess cooling capacity or supplies it to meet the load. The load may be served directly by the cooling system or the storage system. Figure 35-11 is a simplified schematic representation of a basic TES system. Descriptions of the three most common cool storage systems follow, and a fourth is briefly mentioned.

Liquid Systems

Liquid (water) storage is usually the simplest for heating and cooling applications. In cooling applications, liquid storage may be applied with a wider range of refrigeration systems than can ice systems, including LiBr-type absorption chillers. The high specific heat of water, which is approximately 1 Btu/lbm • °F (1.16 Wh/kg • °K), makes it well suited for TES. Above ground tanks are commonly made of steel and buried tanks are commonly made of concrete.

The principle advantages of chilled water storage systems are simplicity and the ability to operate chillers at or close to normal supply (suction) temperatures. The

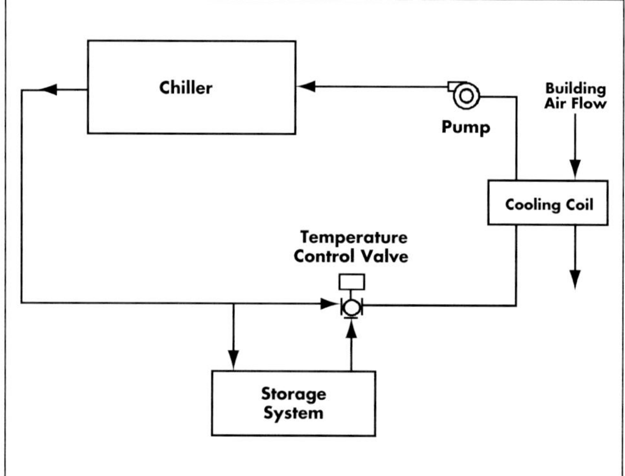

Fig. 35-11 Schematic Representation of Basic TES System.

ability to operate at relatively high supply temperatures results in greater chiller performance.

A principle disadvantage of water storage is the immense storage equipment size and space requirement. Given the temperature differential range between stored and supplied chilled water, which is limited to 10 or 20°F (6 or 11°C), chilled water storage requires a very large amount of storage volume. With hot water storage systems, however, a far greater temperature differential can be used, resulting in a lesser volume requirement for equivalent amounts of stored energy.

Another disadvantage of water storage systems is the tendency for mixing or temperature blending of the water returning to storage and the colder stored water. There are several techniques to limit mixing, although they add to system cost. Water storage systems are also subject to problems associated with lower than desired return water temperatures and the potential for leaks in large concrete installations. They also require chemical treatment and filtering systems.

Ice Systems

The high heat of fusion of ice — 144 Btu/lbm (0.093 kWh/kg) — makes it an excellent storage medium. Three common types of ice storage systems are ice builder systems, ice harvesting systems, and ice slurry systems.

- **Ice builder systems** may operate with DX systems or brine (typically 25% ethylene glycol/75% water) chiller systems. The refrigerant or brine coils, which are submerged in a water storage tank, freeze the water in their immediate vicinity. The water then melts when warm return water is introduced to the tank. Primary concerns with any type of refrigerant coil system are 1) minimizing bridging of ice between coils to promote water circulation and increase exposed ice surface area and 2) creation of the minimal necessary ice thickness to reduce compressor loading during the build cycle. Typical storage systems use plastic containers filled with deionized water and an ice nucleating agent placed in a steel, concrete, fiberglass, or polyurethane tank. The inventory of available ice can be determined by measuring the water volume and adjusting for the expansion to ice (about 9%).
- **Ice harvesting systems** consist of an ice producing section and an ice/water storage section. Ice is accumulated on the outside of the evaporator in the ice producing section and then drops to storage as hot gas is passed through the evaporator to break the bond

between the ice and the evaporator wall. Typical cycle intervals are about 30 minutes. Since the ice floats on water and does not completely displace its own volume, inventory of available ice must be determined by water conductivity or heat balance methods.

- **Ice slurry systems** use a brine solution, which is cooled to its freezing point where ice crystals form in its fluid film. The slurry portion is pumped to storage, where it forms a floating porous ice pack, and the remaining concentrated solution is re-cooled. The process is continuous, with the brine solution continually circulating through the evaporator until the desired stored capacity is reached as indicated by temperature measurement of the solution in storage at equilibrium conditions. An advantage of this system is that no defrost cycle is required.

A primary advantage of ice versus water storage is that size requirements are much lower. Storage space requirements for ice systems are generally less than one fifth those of chilled water systems. Ice systems are also less subject to problems associated with lower than desired return water temperatures.

A major disadvantage of ice systems is decreased cooling system performance due to the need to produce the lower refrigerant (suction) temperatures required for ice making. As discussed in Chapter 37, the performance of a vapor compression system depends on the saturated suction and discharge pressures of the compressor. The suction temperature is a function of the inlet and outlet supply temperature and the discharge is a function of ambient conditions. As the entering water temperature drops, the capacity of the system drops and the temperature differential increases. This results in a requirement for more energy input per ton-h (or kWh_r) output. Another disadvantage is that ice storage tends to present more complications than water storage with respect to generation, storage, discharge, and control.

Eutectic Salts

Eutectic salts change phase at various temperatures. The most commonly used eutectic salt consists primarily of sodium sulfate, which has a heat of fusion of 41 Btu/lbm (0.026 kWh/kg) and freezes and melts at about 47°F (8°C). Other types of eutectic salts may have a higher heat of fusion. Systems typically consist of containers stacked in a tank with small clearances for chilled water to circulate. A principal advantage is that cool storage can be generated with typical chilled water temperatures in excess of 40°F (4°C). Their advantage versus ice is that they can operate more efficiently because they do not have to produce such low temperatures. Their advantage over water is that they require half of the space (significantly more space than ice). Their primary disadvantage is significantly higher capital cost than conventional water or ice storage systems.

Carbon Dioxide (CO_2)

Although not common, CO_2 systems are sometimes useful for very cold applications, such as commercial food freezing. CO_2 has a triple thermodynamic equilibrium point of –70°F (–57°C) at about 60 psig (5.2 bar), which makes it a somewhat unique storage medium because it is used as a solid, liquid, or gas. It can be used as a vapor compression refrigerant, a liquid refrigerant, and a solid storage medium. Though not as dense a storage medium as ice, CO_2 typically requires less than one-third the storage space of chilled water. The advantage of a CO_2 system is its compatibility with very low temperature refrigeration applications. The disadvantages are complexity of operation and significantly higher capital cost than conventional water or ice storage systems.

TES Applications

TES applications can be classified into four basic categories: load shifting, downsizing equipment, intermittent use to improve system performance, and excess capacity storage.

Load Shifting

Load shifting involves generating thermal energy during periods when energy costs are lower and storing it for use when energy costs are higher. This type of application is used almost exclusively with electric motor-driven vapor compression systems. The basic concept is to generate thermal energy in the form of ice, chilled water, or other media during off-peak periods when electricity costs are low, and use it during peak periods when electricity costs are high. This can be particularly effective in eliminating peak electric demand charges.

Over the past two decades, these applications were a major focus of electric utility demand-side management (DSM) and marketing programs. The impetus for the customer is operating cost savings resulting from replacing the use of high-cost peak cooling season electric usage with low-cost off-peak usage. The impetus for the electric utility has been to limit utility system peak demand and/or to retain loads that might otherwise be lost to non-electric technologies. Given the reduced emphasis on

DSM, due to the evolving process of deregulation and/or the condition of excess generation capacity in many regions, utility support for this technology application has recently declined. However, with the trend toward increasingly stratified electricity rates, load shifting remains an effective technology application for many facilities, even absent additional utility incentives.

Systems may be designed for full storage — where all of the peak cooling requirements are served by storage; or partial storage — where only a portion of the peak cooling load is met by storage, with the rest served directly by the refrigeration-cycle machine. Frequently, partial storage applications are more cost-effective because of their lower capital cost. With full storage, more refrigeration machine and/or storage capacity is required.

In a typical partial storage system for a host facility with a one-shift operation, the refrigeration machine operates at part load during peak periods and at full load during off-peak periods to generate storage. In most cases, it is necessary for the host facility to be on a time-of-use (TOU) electric rate for electric-driven thermal storage to be cost-effective. However, for the remainder of the electric loads, more traditional, less time-differentiated rates are preferable for one-shift operations, since they cannot fully benefit from extensive low-cost off-peak usage. If a rate change is necessary, the facility must consider any negative economic impact on the remaining electricity purchases that result.

Facilities with three-shift operations, such as hospitals and industrial plants, can effectively use storage if their off-peak cooling requirement is significantly lower than their peak requirement. In cases where such facilities have high load factors and require similar or only slightly lower levels of cooling during off-peak rate periods, excess refrigeration machine capacity may be required. This strategy may prove financially acceptable if redundant capacity is desired and considered an added benefit.

The large reliance on the specific characteristics of the electric rate structure can be an application-limiting factor. In cases where the rate structure features either off-peak demand billing or where a demand charge is assessed to any portion of off-peak demand that exceeds the peak demand, the economic benefits of the load shift may be somewhat limited. Moreover, should the rate structure change, which is always a possibility given the dynamics of today's rapidly changing electricity market environment, long-term returns on the investment could be diminished since system selection and operations are so closely dependent on the specific (predictable) rate structure. It may therefore be beneficial to consider a long-term agreement on the rate structure with the utility. Given these risks, load-shifting TES systems still offer a good measure of operating flexibility, allowing facilities to dispatch storage and limit the use of electricity during the most costly periods of the day. With the advent of real-time-pricing (RTP) type rates, the ability to reduce electricity consumption during periods of very high market pricing provides a strong measure of financial security, though electricity market volatility does make system selection and operational strategy a difficult challenge.

Figure 35-12 shows a 3.3 million gallon stratified water thermal storage system tank used at a Veteran's Medical Center to balance energy demand during peak load periods. With a 12°F (5.6°C) temperature rise, a system of this capacity can store about 27,500 ton-h (97,000 kWh_r) of cooling capacity. This could eliminate the operation of a 5,500 ton (19,300 kW_r) chiller over a 5 hour daily peak period or the same capacity chiller operating over an 8 hour peak period with a 62.5% load factor.

Downsizing Equipment

Given the short duration of design cooling load conditions, the limited use of storage can typically reduce refrigeration machine capacity requirement by providing cooling during those limited peak hours. Thus, the refrigeration machine and system auxiliaries, such as the cooling tower and condenser pump, can be sized at a capacity lower than that of a facility's full peak requirement. The stored cold thermal energy, generated during off-peak load periods, can then be used during peak load conditions to make up the capacity deficit. With the trend toward market-based RTP electric rates, the availability of a limited amount of storage capacity may prove economical by providing the ability to eliminate the few most costly kWh purchased on a given day. As with other

Fig. 35-12 3.3 Million Gallon Stratified Thermal Energy Storage Water Tank. Source: Carter & Burgess, DoE/NREL

load shifting strategies, it is, however, difficult to predict the optimal amount of storage capacity.

This strategy can be employed with virtually any type of refrigeration system. It can also be used when load growth has exceeded the current chiller or boiler capacity, or both. Storage can be employed permanently, or as a stop-gap measure until load growth is significant enough to warrant investment in added capacity.

Intermittent Use to Improve System Performance

Storage of thermal energy for intermittent use can improve performance of systems that would otherwise cycle frequently or operate often under very low loads. In both heating and cooling system applications, equipment cycling and/or operation under very low loads often results in decreased overall system performance.

In cases where equipment does cycle on and off due to low load conditions, cycling frequency can be reduced as run time is extended to charge the storage system. In other cases, limited cycling can be induced in conjunction with storage as an alternative to constant low load operation. The system generates cold (or hot) water and then cycles off for a lengthy period, which can improve efficiency and limit auxiliary equipment run time. Equipment with significant auxiliary usage, such as single-stage absorption systems, can be run at full load for shorter intervals rather than at part load for longer intervals. Further, as peak demand setting periods approach, the unit can be cycled off to limit peak demand.

Limited storage can also be used to balance a closed-loop water-source heat pump system when heating and cooling loads are not exactly coincidental. A relatively small tank can provide significant benefits. Another application is for cooling units (or boilers) that must be operated to serve only a small load for a given period of the day on a consistent basis. Under this circumstance, it may be effective to charge storage during normal (higher load) operation and shut off the cooling unit (or boiler), allowing storage to serve the small (off-peak) load.

Excess Capacity Storage

Storage of available thermal energy in excess of the amount that can be used during a given time period for use during periods when there is sufficient demand may allow this energy to be used rather than wasted. One of the most common uses of TES is to capture available thermal energy during periods when supply is greater than demand, which often is the case in heat recovery applications. The stored thermal energy is then metered in accordance with demand and, therefore, not wasted. This strategy is commonly used with solar heating systems and combustion engine applications that recover heat in the form of hot water. An example is a cogeneration-cycle system that operates to meet an electric or mechanical load during a low thermal load requirement period. While it may not be economical to operate the system and reject all of the heat, it may be economical to operate the system and store recovered heat.

While storage is typically in the form of hot water for these types of applications, there are cases in which it makes sense to use cool storage produced by heat recovery-driven cooling equipment. When heat recovery is in the form of steam, it may be effective to use the steam in an absorption chiller or steam turbine-driven chiller to generate cool storage, which can then be used to serve cooling loads as they occur. While typically a greater amount of energy can be stored in the form of hot water than chilled water, generating cool storage may also permit equipment downsizing, as described above.

Evaluation of TES Application Opportunities

The economic performance analysis of potential TES applications should compare the cost of the storage equipment, minus any avoided capital costs, and the cost of generating or capturing the stored thermal energy with the alternative purchase and/or generation of that energy during the appropriate use period. Following are several factors that should be included in the evaluation of TES application opportunities.

Operating at lower suction temperatures to produce colder temperatures as required with many TES systems reduces cooling unit performance and capacity and, therefore, increases energy usage. This is most significant with ice storage systems. Conversely, in some cases, during the off-peak periods in which the system operates to charge the storage tanks, ambient conditions may be cooler and or less humid. This allows for a lower condensing temperature and, therefore, an efficiency gain.

Operation with reduced chilled water distribution temperatures can provide capital and operating cost benefits. In new construction, opportunities should be evaluated for designing a lower temperature chilled water distribution system. Capital cost savings may be achieved with smaller capacity pipes, ducts, pumps, and fans. Operating cost savings may be achieved through reduced pump and fan energy requirements. Caution should be taken in retrofit applications to account for all redesign costs, including, where applicable, fan, pump, pipe, and duct modifications and the installation of pipe and duct insulation. Partial storage systems often allow for the purchase of less refrigeration machine (i.e., chiller) capacity and associated heat rejection equipment. If, for example, the peak load is 6,000 tons (21,000 kW_r) and this can be satisfied with a 3,000 ton (10,500 kW_r) chiller plus storage, the savings on the additional 3,000 tons of chiller capacity and associated condenser pump, fan, and cooling tower capacity can offset much, if not all, of the thermal storage system capital cost.

Counterbalancing the prospective benefits of potentially reduced distribution system capital and operating costs are capital costs associated with added storage system pumping, water treatment and filtering costs, and maintenance requirements. The opportunity cost of allocating space for storage and the cost of any additional landscaping or structural modifications should also be considered. Ice storage systems are fairly large; cold water systems are extremely large. In the case of large concrete water storage systems, consideration should be given to the potential for leaks. The use of at least two partitions should be considered to allow for simultaneous operation and repair.

Sizing of storage capacity and predicting how much to store each day is difficult as a result of continually changing load requirements. If storage is undersized, the refrigeration machine may have to operate under peak conditions. If an electric-driven refrigeration machine is brought on line to serve the unintended need for peak operation, an electric demand may be set (under a demand sensitive electric rate), thus reducing the savings associated with load shifting. If a system is oversized, the refrigeration machine may operate inefficiently under very low loads.

Factors such as plant diversity and tenant use can further complicate prediction of required storage capacity. As with other peak-shaving technologies, it should also be assumed that, periodically, the system will be out of service during peak demand periods. In the case of electric-driven systems operating under a load-shifting strategy, this may also necessitate peak refrigeration machine usage and could result in the incidence of peak demand charges.

As noted above, another risk associated with electric load shifting applications is that the system design and selection of the optimal amount of storage is so highly dependent on the specific electric rate structure. A system design that produces optimal economic performance under one rate structure may produce inferior economic performance under another. The dramatic changes occurring in the electricity industry, therefore, present uncertainty risk to load-shifting applications. Still, the application may remain very useful if electric rates experience further stratification between peak and off-peak periods, or move toward RTP. While perhaps more limited storage capacity will be required and the optimal amount may be difficult to determine, the ability to eliminate the most costly electricity purchases may become even more valuable over the next decade.

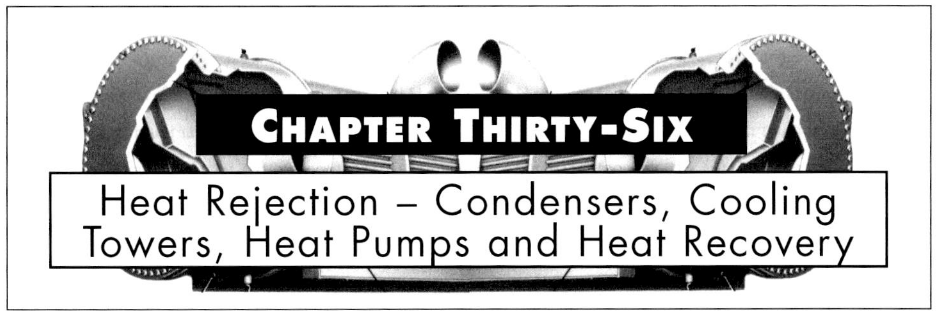

Heat Rejection – Condensers, Cooling Towers, Heat Pumps and Heat Recovery

Heat absorbed in the refrigeration cycle, including both the heat extracted in the evaporator and the energy used to drive the cycle (i.e., work of compression), must be rejected to the outside environment. The heat exchanger that rejects heat from condensing refrigerant to an external medium is known as the condenser. This is sometimes referred to as the hot-, high- (pressure) or discharge-side of the cycle. The heat rejection may occur directly from the condenser to ambient air, as is the case with an air-cooled condenser, or indirectly through a cooling tower or dry cooler.

For large facilities, cooling towers are often used to reject heat absorbed in the condenser to the ambient environment. They use the evaporative cooling process discussed in Chapter 34 and condenser water as the heat rejection medium.

The condenser is a heat source, and this heat must be rejected in order to complete the refrigeration cycle. When there is no practical use for the condenser heat, either because of condition (i.e., temperature or quantity) or the time when it is available, it must be rejected to the ambient environment. When this condenser heat is productively used, the refrigeration system may be considered a heat pump.

Refrigeration-cycle systems can be, in fact, very efficient heaters, sometimes generating four to five times the energy in the form of heat as they require as input. Some refrigeration-cycle systems reverse their cycles to deliver either heating or cooling, while others are used only for heating. Some can do both simultaneously, by directing the rejected heat to a productive use while in the cooling mode. These systems may be considered a type of heat recovery heat pump. They recover refrigeration cycle-generated heat, normally rejected to the outside environment, for productive use, such as hot water for reheat or domestic use.

This chapter provides a detailed discussion of refrigeration cycle heat rejection processes. This includes a presentation of key system components, namely condensers and cooling towers, with details on both energy and water use associated with their operation. This is followed by a discussion of heat pumps and heat recovery systems that allow heat rejected from refrigeration cycles to be used productively.

CONDENSERS

EQUIPMENT CLASSIFICATIONS

Air-cooled condensers, comprising a finned-tube coil located in a stream of ambient air, are generally the least expensive type of condenser. When ambient dry-bulb temperatures are sufficiently low, air-cooled condensers can allow the cooling system to operate efficiently. However, when ambient temperature is high, the resulting high condensing temperature can significantly reduce refrigeration system performance. This is particularly problematic for facilities, operating under demand-sensitive electric rates, that have electric-driven units serving air conditioning applications. This is because peak electric load usually occurs under design conditions at high outdoor ambient temperatures.

Applications for air-cooled condensers are also limited by size. In mechanical compression systems, maximum capacity is generally about 400 tons (1,400 kW_r). When used for absorption chillers, they are typically limited to very small applications due to the high rate of heat rejection of absorption cycles.

Air-cooled condensers require less routine maintenance than alternative types. However, significant reductions in performance often occur over time due to fouling of coils exposed to unfiltered outdoor air, and service life is typically shorter than with water-cooled systems due to coil degradation. Air-cooled systems have no water treatment requirements and low incidence of freeze-ups during cold weather operation temperatures.

Figure 36-1 illustrates an air-cooled condenser featuring dual fans. Capacity is controlled by modulating the airflow rate across the coil based on condensing pressure. One fan is cycled while airflow across the other is controlled by damper modulation, according to the following control sequence: at 170 psi (11.7 bar) condensing pressure, the damper blades begin to open; at 225 psi (15.5 bar), the controlled fan is started; and at 255 psi (17.6 bar), the damper blades are fully open. The process

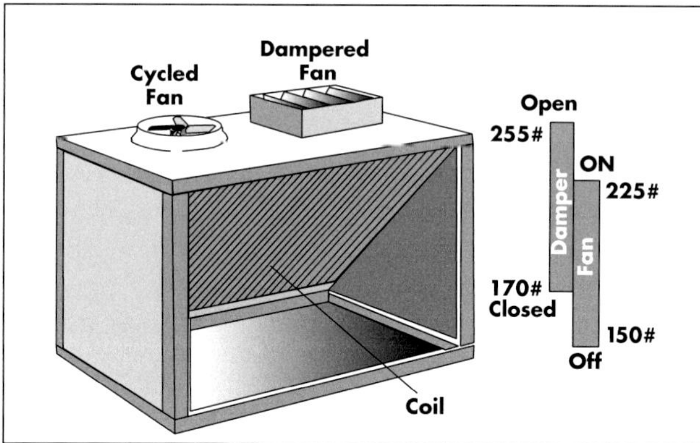

Fig. 36-1 Illustration of Dual-Fan, Air-Cooled Condenser.
Source: The Trane Company

is reversed as condensing temperature falls, with the lead fan stopped at 150 psi (10.3 bar).

Water-cooled condensers are classified as either shell-and-tube or tube-in-tube. In both, warm refrigerant vapor enters the top of the condenser and flows over the outside of the water tubes. As heat is transferred to the water, the refrigerant condenses and falls to the bottom of the condenser. A water regulating valve sometimes controls water flow in response to refrigerant pressure. A cooling tower is the most commonly used device to cool the leaving condenser water in recirculating systems. Once-through cooling can be provided by city water or water pumped from another external source.

Figure 36-2 is a cutaway drawing of a water-cooled condenser used with a centrifugal chiller system. The refrigerant is forced to travel a serpentine course on the shell side counter to the flow of the water in the tubes. Notice the marine water box and the thermal economizer, or subcooler, a circuit in which some of the tubes in the first water pass are below refrigerant liquid level. In vapor compression systems, subcooling increases cycle efficiency by reducing total flow of refrigerant through the system. In low-temperature applications, subcoolers are generally not used, due to the possibility of water freezing in the subcooler tubes on failure of the condenser water pump.

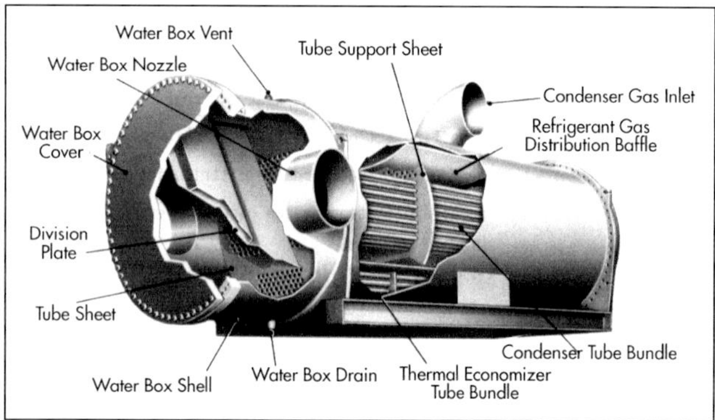

Fig. 36-2 Condenser for Centrifugal Chiller.
Source: Carrier Corp.

Evaporative condensers (Figure 36-3) are similar to air-cooled units, but include a water spray to cool the external surfaces of condenser coils and reduce condensing temperature. Ambient air passing over the coils evaporates a small amount of water, absorbing heat from the air stream and refrigerant coils. Condensing refrigerant drains into a liquid receiver at the bottom of the condenser. Non-evaporated water drains into a sump at the bottom of the unit and is then pumped along with required makeup water to spray nozzles above the coils.

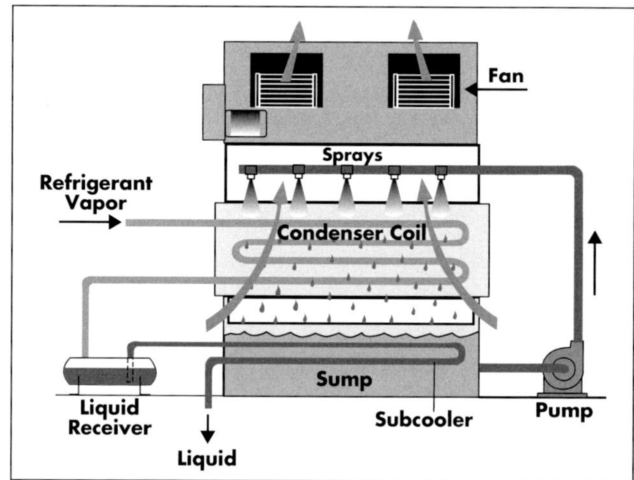

Fig. 36-3 Illustration of Evaporative Condenser.
Source: The Trane Company

Condensing Method Selection

The choice of condensing methods is influenced by several factors, including water quality, water acquisition and treatment cost, energy cost, system capacity and load factor, sound emissions and space considerations, and ambient air quality. Table 36-1 shows typical condensing temperatures achievable under design conditions with alternative condensing methods, in °F (°C).

Condenser Load

Refrigeration system heat rejection, typically expressed in Btu/ton-h (kWh_h/kWh_r), depends on the amount of heat absorbed (or the refrigeration effect) and the thermal or mechanical energy input to the refrigeration cycle. For vapor compression systems, the driving

**Table 36-1 Condensing Temperature, °F (°C) vs. Condensing Media
Source: Carrier Corp.**

Condensing Media	Inlet Temperature °F (°C)	Temperature Rise °F (°C)	Outlet Temperature °F (°C)	Leaving Differential °F (°C)	Condensing Temperature °F (°C)
City	75 (24)	20 (11)	95 (35)	10 (6)	105 (41)
Water	80 (27)	20 (11)	100 (38)	10 (6)	110 (44)
Cooling Tower					
75-78°F (24-25.5°C) WB	85 (29)	10 (6)	95 (35)	10 (6)	105 (41)
Evaporative Condenser					
75-78°F (24-25.5°C) WB	- -	- -	105 (41)		
78-80°F (25.5-27°C) WB	- -	- -	110 (44)		
Air-Cooled	95 (36)	13 (7)	108 (43)	12 (7)	120 (50)
	110 (44)	13 (7)	123 (51)	12 (7)	135 (58)

energy to the compressor can be supplied by an electric motor or a prime mover. In a hermetically sealed electric-drive unit (motor contained in compressor housing), the heat of compression as well as motor heat losses are included in the condenser load. If, for example, a hermetic unit requires a driving energy input of 0.63 kW/ton (0.18 kW_e/kW_r), the total heat rejection in English units would be:

0.63 kW/ton x 3,413 Btu/kWh + 12,000 Btu/ton-h = 14,151 Btu/ton-h

In SI units, the heat rejection would be:

$$0.18\ kWh_e/kWh_r + 1\ kWh_h/kWh_e = 1.18\ kWh_h/kWh_r$$

where the full electric energy input, expressed in kWh_e, is converted to rejected heat, expressed as kWh_h.

In an open-drive unit, the heat associated with motor efficiency losses is dissipated to ambient air and is not rejected directly to the condenser. For example, if the motor efficiency is 93% (0.93), the total heat rejection in English units would be:

0.63 kW/ton x 0.93 x 3,413 Btu/kWh + 12,000 Btu/ton-h = 14,000 Btu/ton-h

In SI units, the heat rejection would be:

$$0.18\ kWh_e/kWh_r \times 0.93 + 1\ kWh_h/kWh_e = 1.17\ kWh_h/kWh_r$$

In the case of a reciprocating engine-driven compressor, an additional heat rejection load may be present. When the heat that must be rejected from the engine cooling system cannot be used productively, it may be rejected by integrating it into the cooling tower loop or dissipating it separately through a dump radiator.

In the case of a condensing steam turbine, the cooling load required for steam condensation may also be passed to the refrigeration condenser water loop. Water-cooled surface condensers are the most commonly applied technology. Alternatives include direct contact condensers that spray cooling water directly into the steam flow, and air-cooled condensers, which may be applied when cooling water is not readily available or cooling tower operation is not acceptable.

In absorption chillers, the heat energy is input to the generator cycle in the form of direct-fired fuel, steam, hot water, or high-temperature exhaust gas. The full energy input minus the energy of exhausted combustion gas or condensate return must be removed from the tube bundle to the cooling tower or other heat sink. Total absorption heat rejection will vary widely, depending on the performance of the cycle. Typically, input energy ranges from 9,000 to 18,000 Btu/ton-h (0.75 to 1.50 kWh_h/kWh_r), bringing total heat rejection (condenser load) to between 21,000 and 30,000 Btu/ton-h (1.75 to 2.50 kWh_h/kWh_r).

If, for example, the total generator energy input to a direct-fired, double-effect absorption chiller is 12,000 Btu/ton-h (1.0 kWh_h/kWh_r) and the fuel combustion efficiency is 82%, total system heat rejection would be:

12,000 Btu/ton-h x 0.82 + 12,000 Btu/ton-h = 21,840 Btu/ton-h

In SI units, the heat rejection would be:

$$1.0\ kWh_h/kWh_r \times 0.82 + 1\ kWh_h/kWh_r = 1.82\ kWh_h/kWh_r$$

Cooling Towers

Equipment Classifications

Evaporative cooling towers are the most commonly applied heat rejection technology for large refrigeration-cycle systems. In a typical cooling tower, a pump delivers water to the top of the tower where it is sprayed, or cascaded, over a series of baffles and agitated by contact with moving air. As a small portion of the water evaporates, it cools the air and the remaining water. The water then collects in a sump at the bottom of the tower and returns to the refrigeration or process heat exchanger for reuse. Often, fan controls regulate airflow to maintain leaving water at the required temperature set point under varying load and ambient conditions. A controlled source of makeup water is also required to offset the evaporation and other losses. Figure 36-4 illustrates basic cooling tower operation.

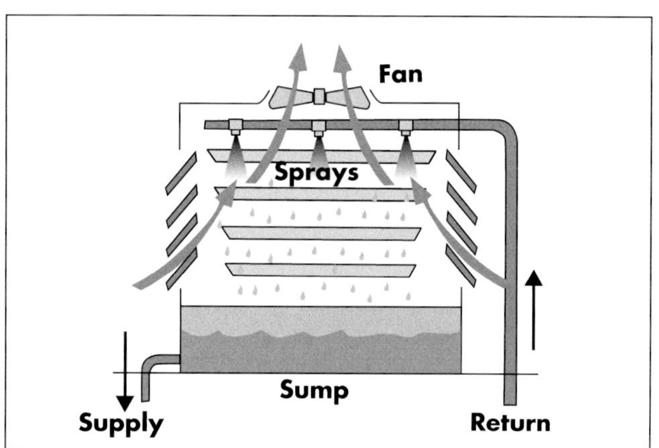

Fig. 36-4 Basic Cooling Tower Operation Illustration. Source: The Trane Company

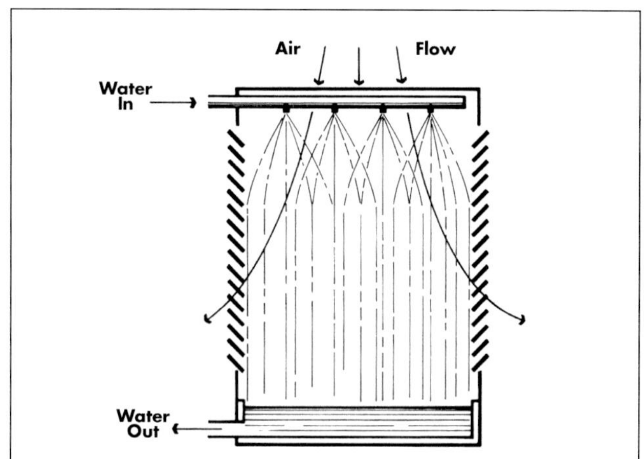

Fig. 36-5 Atmospheric Cooling Tower. Source: Marley Tower Company

Cooling towers may be designed to use **natural draft** or **mechanical draft**. Small, inexpensive natural draft towers (Figure 36-5) induce airflow using pressurized water spray. Larger units, such as the hyperbolic natural draft towers (Figures 36-6 and 36-7), sometimes used for large electric generation facilities, produce airflow as a result of the density differential between the heated air inside the stack and the cooler air outside the tower.

Mechanical draft towers use one or more fans to produce the required airflow. Mechanical draft units, the most common systems used in air conditioning and process refrigeration applications, allow stability in changing conditions and provide a means of regulating airflow to respond to varying load conditions. Two mechanical draft configurations exist:

1. In **forced draft towers** (Figure 36-8), the fan is

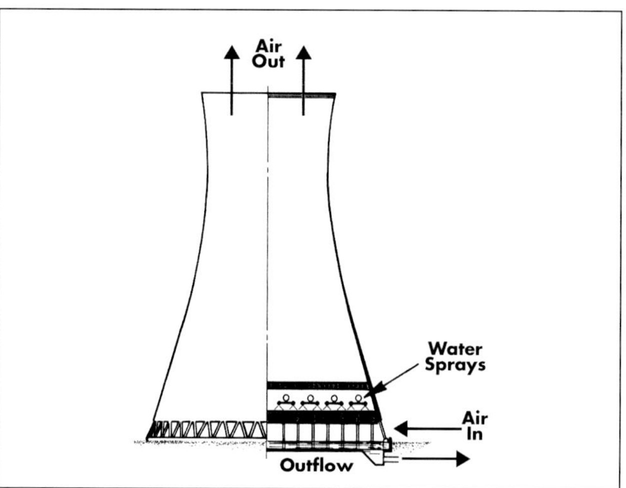

Fig. 36-6 Hyperbolic Counterflow Natural-Draft Type Tower. Source: Marley Tower Company

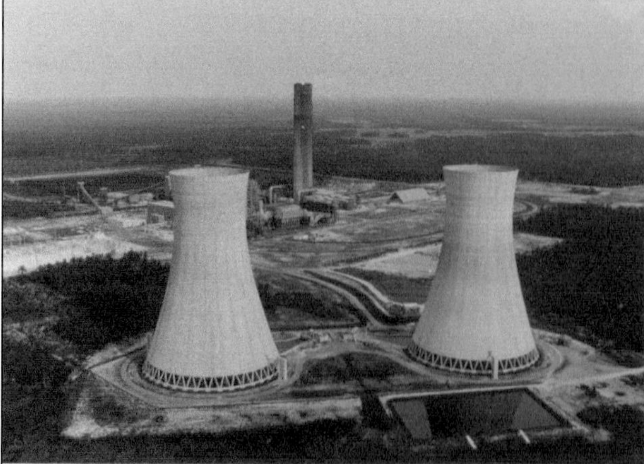

Fig. 36-7 Hyperbolic Natural Draft Cooling Tower Serving Large Power Plant. Source: Marley Tower Company

located in the ambient air stream entering the tower. Forced draft towers are characterized by high entrance velocities and low exit air velocities. They are susceptible to recirculation, in which a portion of the tower's moist discharge air re-enters the tower, elevating the average entering wet-bulb temperature. They are also susceptible to icing when moving air is laden with either natural or recirculated moisture. Typically, centrifugal fans are used. While these usually require more power than propeller fans, they are able to operate against the high static pressure associated with ductwork and can, therefore, be installed indoors or within enclosures that can provide sufficient separation between intake and discharge locations to minimize recirculation. They also offer low sound emission levels.

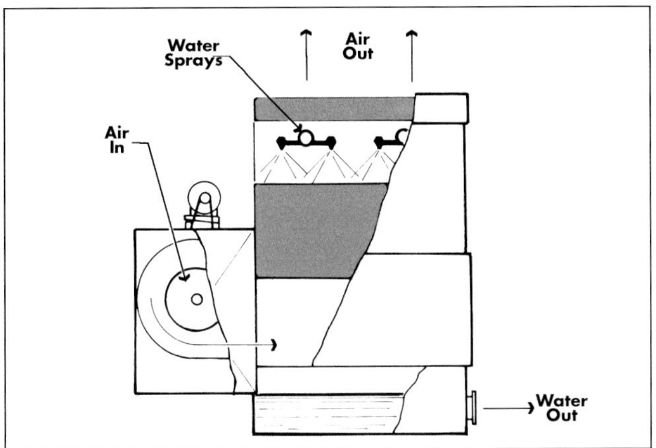

Fig. 36-8 Forced Draft Cooling Tower.
Source: Marley Tower Company

2. In **induced draft towers** (Figure 36-9), the fan is located in the exiting air stream. The air discharge velocity is several times greater than the air entrance velocity, limiting the potential for recirculation and icing. Axial fans are typically used and power requirements are less than with forced draft towers. The units exhibit higher sound emission levels than forced draft towers in most applications and sound emission reduction is difficult due to the fan location.

Hybrid draft, or fan assisted, towers use a combination of natural and mechanical draft. They are designed to minimize the power required for air movement while limiting the overall size and cost of the tower. Often, the fan is only required during peak periods.

Cooling towers may also be classified as **crossflow** (Figures 36-10 and 36-11) or **counterflow** (Figure 36-12). In crossflow towers, air flows horizontally across

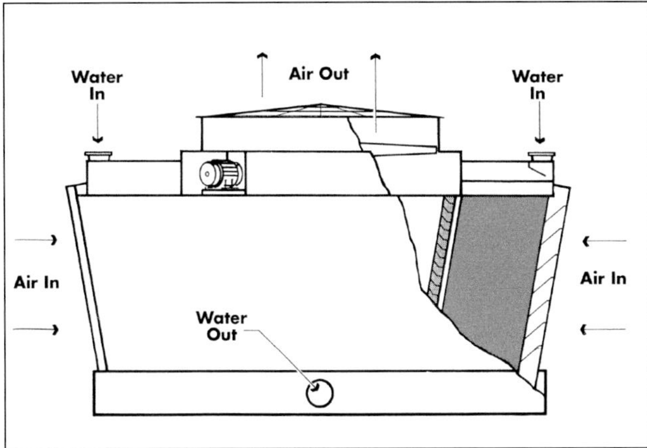

Fig. 36-9 Induced Draft Cooling Tower.
Source: Marley Tower Company

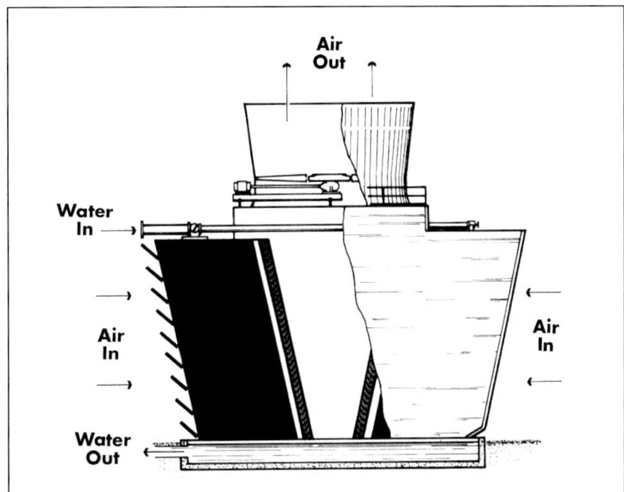

Fig. 36-10 Crossflow Cooling Tower.
Source: Marley Tower Company

Fig. 36-11 Crossflow Cooling Tower Featuring PVC Film Fill.
Source: Marley Tower Company

the downward fall of water. In a typical crossflow unit, water is delivered to elevated basins and distributed to heat transfer surfaces (or fill) through metering orifices in the floor of the basins. In counterflow towers, water is sprayed downward over the fill, while airflow is upward, with a forced draft or induced draft fan. In smaller units, the counterflow design requires more pumping head and fan power, as well as greater tower height, than the crossflow design.

Towers may be field-erected or factory-assembled.

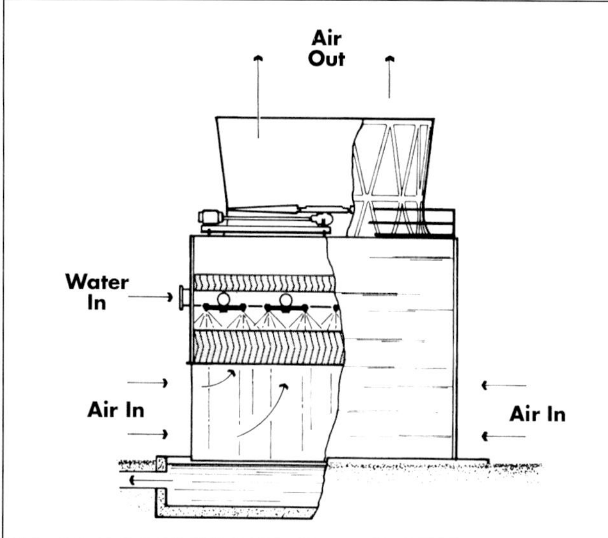

Fig. 36-12 Counterflow Cooling Tower.
Source: Marley Tower Company

Larger towers are commonly shipped with prefabricated components for field erection. Factory-assembled towers are shipped either intact in smaller sizes or in as few sections as possible, depending on the mode of transportation. Figure 36-13 shows three packaged cooling towers featuring a low-profile design.

For a given capacity requirement, tower selection is based on several factors, including:

- Logistical factors, such as available space, structural support enclosure requirements, and aesthetics.
- Climate and annual operating duty, including freeze protection if needed.
- Permitting limitations, including sound emissions levels and proximity to other equipment or structures.
- Economic performance analyses, focusing on the balance between capital cost and operating cost among available options.

Fig. 36-13 Three Packaged Cooling Towers.
Source: Marley Cooling Tower Company

Cooling Tower Sizing and Performance

The cooling effect of evaporation is about 1,000 Btu/lbm (0.65 kWh_h/kg) of water evaporated. Cooling towers optimize evaporation by providing the maximum transient water surface and airflow rate. Figure 36-14 shows a typical cooling tower performance curve, graphing leaving (cold) water temperature versus wet-bulb temperature. Figures 36-15 and 36-16 illustrate the process. In this example, the wet-bulb temperature remains unchanged across the cooling tower at 65°F (18°C), while the dry-bulb temperature drops from 100 to 70°F (38 to 21°C). Water temperature falls from 95 to 85°F (35 to 29°C) across the tower.

The design conditions used in specifying a cooling tower include water flow rate, hot (entering) water temperature, cold (leaving) water temperature, and ambient wet-bulb temperature. The load served at a given flow

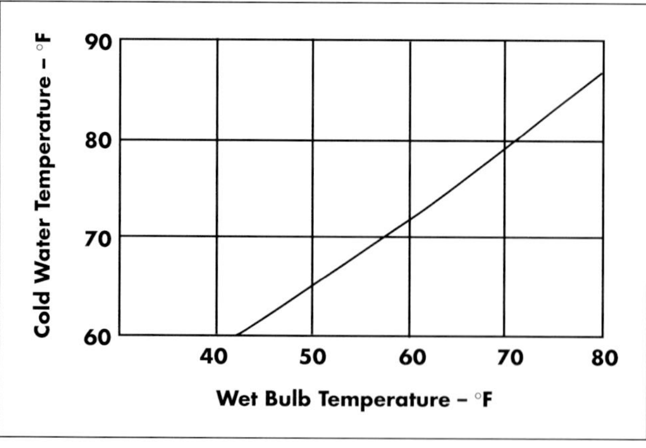

Fig. 36-14 Typical Cooling Tower Performance Curve — Cold Water Temperature vs. Wet-Bulb Temperature.

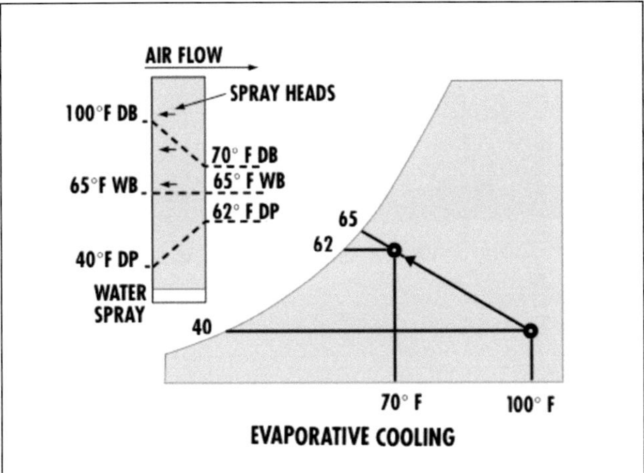

Fig. 36-15 Evaporative Cooling Process. Source: Carrier Corp.

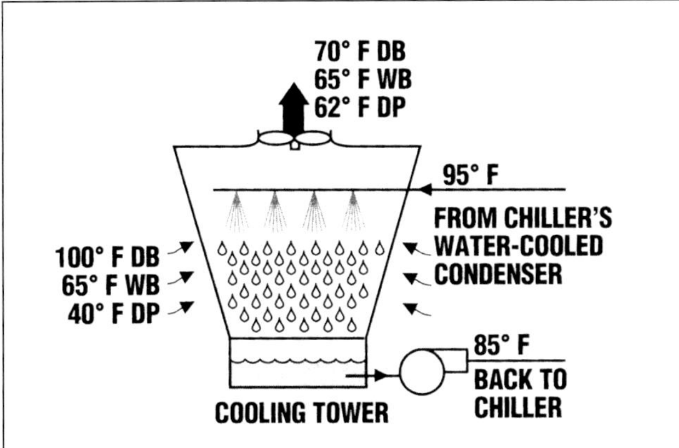

Fig. 36-16 Cooling Tower Process. Source: Carrier Corp.

rate and temperature differential is determined using the following equation in English units:

$$gpm = \frac{Btu/h}{8.33\ lbm/gal\ x\ 60\ min/h\ x\ \Delta T\ x\ SG\ x\ C_p} \quad (36\text{-}1)$$

and in SI units:

$$lpm = \frac{kW_r}{1\ kg/liter\ x\ 60\ min/h\ x\ \Delta T\ x\ SG\ x\ C_p} \quad (36\text{-}2)$$

Where:

SG = Specific gravity of fluid

C_p = Specific heat of fluid, in Btu/lbm • °F, kJ/kg • °C, or kWh_h/kg • °C

Possible temperature and flow rate combinations are dictated by the load and by hydraulic and temperature requirements of the process being served. As the temperature difference is increased and flow rate reduced, tower size and cost are generally reduced. Trade-offs between cooling tower and refrigeration (or process) system performance are commonly considered.

A common design for conventional electric-driven refrigeration machines is based on a 10°F (6°C) differential, with 85°F (29°C) entering and 95°F (35°C) leaving condenser water. Required flow rate is 3 gpm/ton (3.2 lpm/kW_r) to reject about 15,000 Btuh/ton (1.25 kW_h/kW_r). A greater water temperature differential is sometimes used to reduce condenser water pumping and tower capacity requirements, although refrigeration system performance is reduced somewhat as a result.

A 15°F (9°C) differential, from 85 to 100°F (29 to 38°C), is often used with absorption chiller systems, due to their very high heat rejection rate. For example, when a double-effect absorption chiller is designed for a 10°F differential, required condenser water flow rate is typically 4.5 gpm/ton (4.8 lpm/kW_r) to reject about 22,000 Btuh/ton (1.83 kW_h/kW_r). Using a 15°F (9°C) differential allows the absorption unit to operate at 3 gpm/ton (3.2 lpm/kW_r). This can be particularly valuable in retrofit applications where vapor compression systems are being replaced with absorption systems, because it may eliminate the need to replace the existing condenser water pumps.

As shown in Figure 36-17, the **approach** is the difference between the cold water temperature and the ambient wet-bulb temperature. The approach is set by the size and effectiveness of the tower relative to the total heat load, with larger, more-efficient towers able to operate with a smaller approach. Figure 36-18 shows the effect of the chosen approach on tower size at fixed head, load, water flow rate, and wet-bulb temperature. The relationship of required cooling tower fan power to ambient wet-bulb temperature is shown in Figure 36-19. Figure 36-20 illustrates the effect of fan operation on tower performance for a three-cell tower. As ambient wet-bulb temperature increases, successive fans must be brought on-line.

There are several ways to improve cooling tower performance. One method is to increase the height of the tower, thereby increasing the contact time between air and water. Contact time can also be increased by using a fill that impedes the progress of the falling water, although increased fan power would typically result.

Water Treatment

Cooling towers are very efficient air washers — the air exiting the tower is much cleaner than the air entering the tower. Air washing, plus the presence of dissolved solids in entering makeup water, continuously contaminates

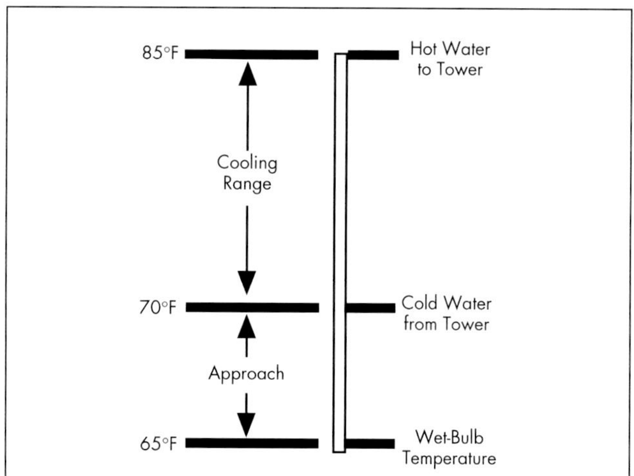

Fig. 36-17 Illustration of Approach Temperature.

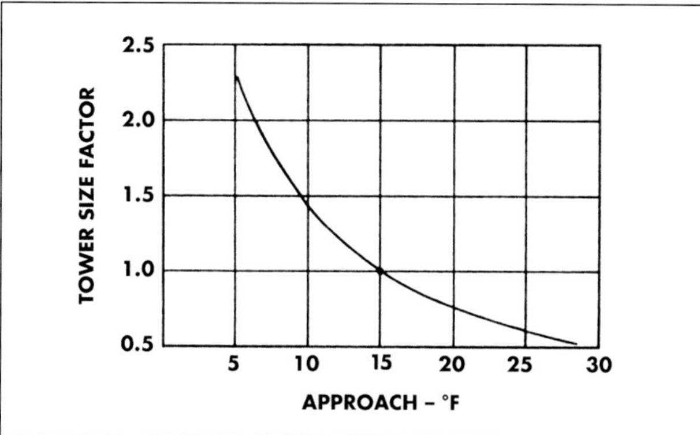

Fig. 36-18 Tower Size Factor vs. Approach. Source: Marley Tower Company

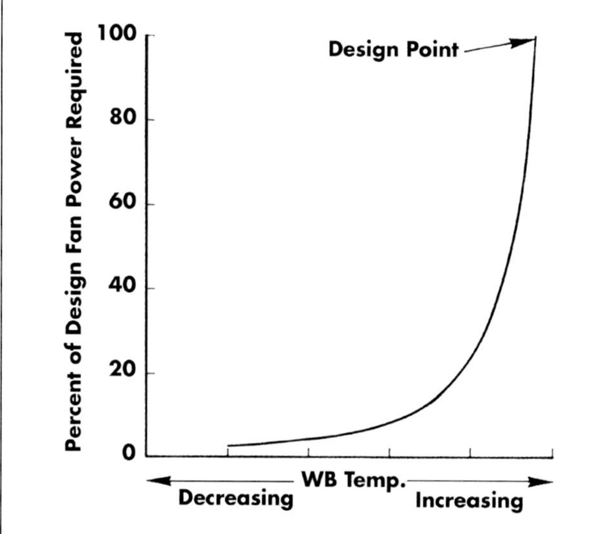

Fig. 36-19 Relationship of Fan Power Requirement to Wet-Bulb Temperature. Source: Marley Tower Company

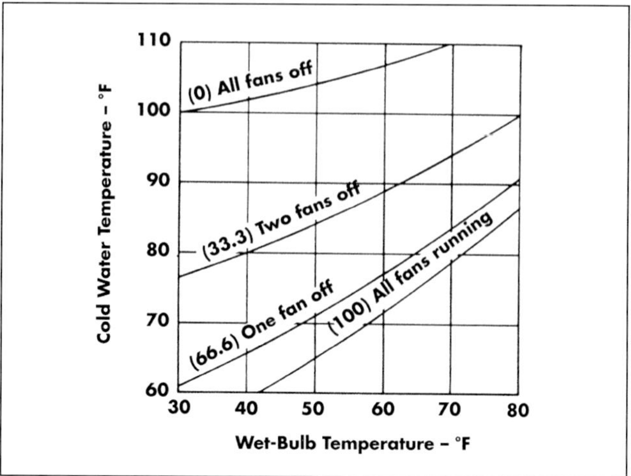

Fig. 36-20 Tower Performance as Function of Fan Power (with Single-Speed Motors). Source: Marley Tower Company

recirculated cooling tower water. Contaminant concentration levels are referred to as **total dissolved solids** (TDS).

Blowdown is the controlled discharge of a limited amount of tower water to regulate the concentration of impurities and is a necessary part of most cooling tower systems. In addition, chemical, electrostatic, or electronic water treatment is often required due to a number of potentially damaging conditions:

- Scale deposits occur due to the presence of calcium and other ions in cooling water. The maximum amount of calcium carbonate that can be held in solution depends on the water temperature and the amount of free carbon dioxide. An increase in water temperature or a decrease in free carbon dioxide beyond the point of equilibrium will result in scale deposition. Various chemical compounds are used to keep scale-forming solids in solution.
- Biological growth, such as slime and algae, may develop in towers and impede performance. Chlorine compounds are most commonly used as algaecides and slimicides. Because excess chlorine can damage wood and other organic tower materials, it is often used only intermittently as shock treatment.
- Corrosion can be a problem even with highly resistant metals. Corrosion may result from low PH, high oxygen content, carbon dioxide, or contact between dissimilar metals. An increase in the dissolved solids may promote corrosion, since electrolytic action increases with increased conductivity. Plastic components and an array of sophisticated metals are commonly used to avoid corrosion. Once present, various treatment compounds are used as inhibitors.

- Foaming can be caused by certain combinations of dissolved solids or other forms of contamination. If blowdown is insufficient for control, foam depressant chemicals must be added to the system.
- Suspended solids can best be removed by continuous filtration. Oils and fats, which will reduce tower performance and can even be a fire hazard, can be eliminated with the use of a skimmer. Many non-dissolved solids tend to settle out in low-velocity areas of the system, such as the tower basin. These areas can become a breeding ground for bacteria and can degrade tower performance and increase chemical treatment costs. A system designed to filter the cooling basin water at a rate of about 5% of the total circulating water can effectively control particulate levels.

Ozone systems are being increasingly used to reduce blowdown and chemical treatment requirements. Ozone is triatomic oxygen (O_3), which has very high oxidizing power. It is a colorless gas produced from oxygen in air by electric discharge. An ozone generator produces ozone, which an in-line mixer injects directly into the tower water system. Ozone is an effective method of controlling all microbiological organisms (including Legionella) and biofouling. By destroying slime-forming bacteria, calcium carbonate deposits are freed from heat exchange surfaces, allowing crystals to be deposited in the quiescent areas of the cooling tower's basin whenever solubility in water is exceeded. Ozone discharge itself does not produce any harmful by-products. However, due to capital cost considerations and some remaining uncertainty about effectiveness, ozone systems are still not yet widely applied.

Water Usage

The total water makeup rate (WMR) requirement for a cooling tower is the sum of losses from evaporation (E), blowdown (B), and drift (D), a term that refers to liquid droplets entrained in the exhaust air stream. All of the lost and removed water must be replenished with fresh makeup water.

Using 1,000 Btu/lbm (0.646 kWh_h/kg) as the amount of heat absorbed in the evaporation of water, evaporation rate, in gpm, can be calculated as:

$$E = \frac{(Tons \times Btu\ heat\ rejection\ per\ ton)}{(1{,}000\ Btu/lbm \times 8.34\ lbm/gal \times 60\ min/hr)} \quad (36\text{-}3)$$

In SI units, the evaporation rate, in lpm, would be:

$$E = \frac{(kW_r \times kWh_h/kWh_r)}{(0.646\ kWh_h/kg \times 1\ kg/l \times 60\ min/hr)}$$

If, for example, the heat rejection is 15,000 Btu/ton-h (1.25 kWh_h/kWh_r), the evaporation rate for a 4,000 ton (14,064 kW_r) system is:

$$E = \frac{(4{,}000\ tons \times 15{,}000\ Btu/ton)}{(1{,}000\ Btu/lbm \times 8.34\ lbm/gal \times 60\ min/hr)} = 120\ gpm\ (454\ lpm)$$

If the actual load is not known, the evaporation rate can be approximated based on the condenser water flow rate and temperature differential. A commonly used formula for estimating the evaporation rate, in English units, is:

$$E = flow\ rate \times \Delta T \times 0.0008 \quad (36\text{-}4)$$

Dissolved and suspended solids entering the system are not removed in the process of evaporation; therefore, the concentration of these impurities can rapidly increase in operating cooling towers. Blowdown requirements for a given system are based on maintaining the concentration of contaminants below maximum acceptable levels. The ratio of allowable concentration of impurities to the concentration existing in entering makeup water is referred to as cycles of concentration (C). Excluding consideration of drift, required blowdown rates can be determined by the following formula:

$$B = \frac{E}{C - 1} \quad (36\text{-}5)$$

Drift reduces the blowdown requirement because it continually takes with it the elevated levels of total dissolved solids (TDS). If not known, drift may be approximated, in English units, as:

$$D = flow\ rate \times 0.0002 \quad (36\text{-}6)$$

Including drift, the blowdown rate can be expressed as:

$$B = \frac{E - [(C\text{-}1) \times D]}{(C\text{-}1)} \quad (36\text{-}7)$$

If, for example, the flow rate is 10,000 gpm, the water temperature differential is 15°F, and the desired concentration ratio is 5, the required blowdown rate would be:

$$B = \frac{(10{,}000 \times 15 \times 0.0008) - [(5-1) \times (10{,}000 \times 0.0002)]}{(5\text{-}1)}$$

$$= 28 \text{ gpm } (106 \text{ lpm})$$

Blowdown requirements resulting from air washing are generally not calculated, but are determined empirically based on field monitoring results.

Annual water use can be estimated based on the WMR in gpm (lpm) and an estimate of equivalent full-load hours (EFLH). This is calculated as:

$$WMR \times 60 \text{ min/h} \times EFLH = \text{gallons/yr (liters/yr)} \quad (36\text{-}8)$$

In the example considered above, WMR is estimated at 152 gpm (575 lpm), including 120 gpm evaporation, 28 gpm blowdown, and 4 gpm drift. Assuming the system runs 1,500 EFLH per year, total water usage based on Equation 36-8, is:

$$\frac{152 \text{ gpm} \times 60 \text{ min/h} \times 1{,}500 \text{ EFLH}}{1{,}000 \text{ gal/kgal}}$$

$$= 13{,}680 \text{ kgal/yr } (52{,}460 \text{ kl/yr})$$

If the cost of water is $5.00 per kgal, the annual water cost is:

$$13{,}680 \text{ kgal} \times \$5.00/\text{kgal} = \$68{,}400$$

Water usage is often measured in hundred cubic feet (Ccf) or cubic meters (m^3) for water utility billing. Since there are approximately 748 gallons per Ccf, kgal can be converted to Ccf by dividing by 0.748. Thus:

$$\frac{kgal}{0.748} = Ccf \quad (36\text{-}9)$$

Commonly, water utility billing consists of a water charge and a sewer charge, with charges for both based on metered city water flow. A facility can sometimes petition for a sewer charge abatement when metering is installed for cooling tower makeup water, since evaporation does not add to city sewer load. Blowdown may or may not be metered in such cases.

Cooling Tower Freeze Protection

Freeze protection is essential for towers that operate in cold weather, particularly during free-cooling operation or low-load periods. Thin ice that forms on the louvers or intake structure presents no problems and can actually be beneficial by retarding airflow. However, significant ice formation on the fill or in tower support regions may cause serious structural damage. The potential for ice formation varies directly with the airflow rate through the tower and inversely with the heat load and with the amount of water flowing over the fill. Freeze protection can be achieved using air-side or water-side control, with heating elements and bypass systems used to protect towers during shut-down periods.

Mechanical draft towers can sometimes utilize air-side control, depending on the number of fans and the ability to vary fan speed. For example, multiple- or variable-speed fans can be automatically modulated to reduce the amount of cold air coming into contact with the circulating water. In a multi-cell cooling tower, freezing potential is reduced by running all fans at low speed rather than shutting off one or more cells. Natural draft towers do not have the potential for air-side control.

Large towers employ water-side control by increasing the concentration of flowing water nearest the tower air intakes. This places the greatest heat load closest to the coldest air. The resulting rapid increase in intake air temperature precludes freezing within the fill, and the flow of warm water protects the most vulnerable areas.

Heating systems and bypass circulation can offer protection when towers are shut down. Electric heaters, direct steam injection, or closed-loop steam or hot water systems are used with heating elements immersed in the cold water basin. Heating is not required when all tower water drains by gravity into an indoor sump. During startup, automatic controls allow water to be bypassed directly into the cold water basin until the required temperature is reached — usually about 80°F (30°C).

Heat Pumps and Heat Recovery Systems

All systems operating on refrigeration cycles are essentially pumping heat — they extract heat from one medium (a source) and transfer it, along with the driving energy input to the system, to another medium (a sink) at a higher temperature. In most cases, the primary purpose of such systems is to cool a space or process medium by removing heat. The heat is, thus, a waste product that must be rejected.

In cases where systems are specifically designed to absorb heat and transfer it to serve a heat requirement, they are commonly referred to as heat pumps. Heat pumps can provide heating only, simultaneous heating and cooling, or alternate heating and cooling.

Refrigeration cycle heat recovery systems recover the heat that would otherwise be rejected through the condenser system or heat sink, and transfer it to serve a heat load. These systems operate only when there is a simultaneous cooling load to serve.

Heat Pumps

There are numerous types of heat pump system designs and various heat source and heat sink media used. Some systems reverse their cycles to deliver both heating and cooling, while others are used only for heating. Common to all is the same basic operating principle — extract heat from one medium and deliver to another. As a means of understanding the basic operation of a heat pump cycle, consider the operation of a simple air-to-air reverse flow direct expansion (DX) heat pump.

As shown in Figure 36-21, with a basic DX type heat pump operating in the cooling mode, heat from the inside conditioned space is absorbed by the vaporizing refrigerant in the indoor coil. The heat-laden refrigerant is pumped outdoors by the compressor, where the heat is rejected by the outdoor coil. The refrigerant is then condensed and pumped back to the indoor coil, where the cycle is repeated.

As shown in Figure 36-22, when the unit is operating in heating mode, a reversing valve changes position to reverse the refrigerant flow. The heat is absorbed into the system by the outdoor coil as the low-pressure liquid refrigerant becomes a low-pressure vapor. It then flows to the compressor and is discharged to the indoor coil as a high-temperature vapor. The heat pick-up from the outdoor coil plus the heat of compression is released from

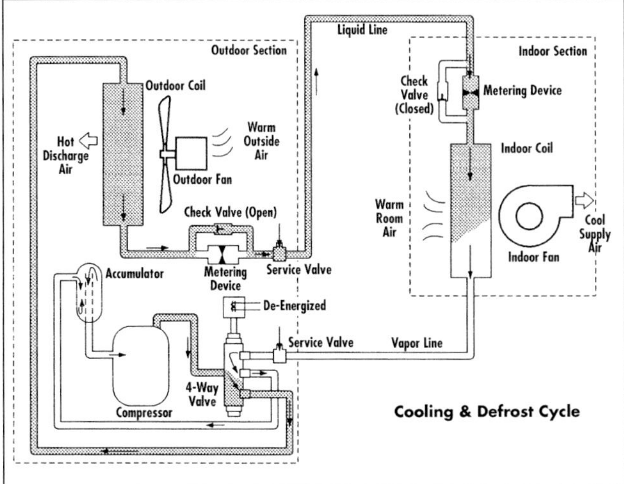

Fig. 36-21 Heat Pump Operating in Cooling Mode.
Source: Carrier Corp.

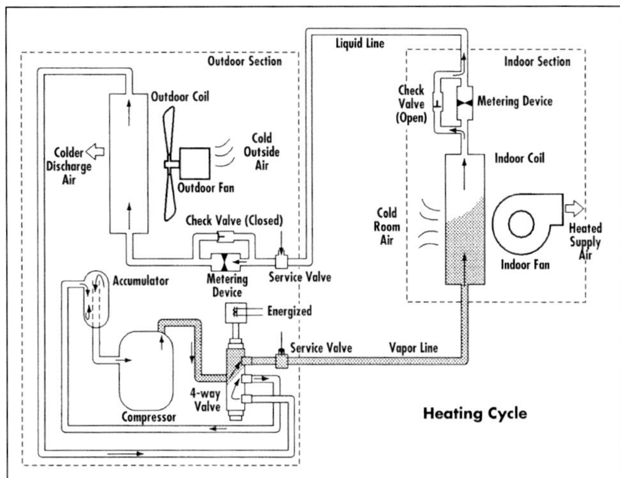

Fig. 36-22 Heat Pump Operating in Heating Mode.
Source: Carrier Corp.

the refrigerant in the indoor coil, directly or indirectly, to the conditioned space.

Types of Heat Source and Heat Sink Media

Numerous heat sources and distribution fluids can be used with heat pumps and there are many variations on the basic system design. Following are brief descriptions of commonly used heat source and heat sink media.

Air

Outdoor air represents a virtually unlimited heat source and heat sink medium for heat pumps and is widely used in small-capacity systems. Other sources of air for heat pumps include air from internal spaces and air from process or HVAC exhaust systems. Since air is a relatively inefficient heat transfer medium, a large surface area is required for heat exchange coils.

Since outside air temperature fluctuates so widely, set point design is a critical task. As the outdoor temperature decreases, heating capacity and efficiency of an air source heat pump decreases rapidly. Supplemental heating sources are generally required in colder climate regions. As outdoor air temperature increases, capacity and efficiency in cooling mode decreases. Air source heat pumps also require defrosting capability and are subject to degraded performance due to frost build-up. Exhaust air is usually a better heat source than outdoor air, but is usually an inferior heat sink due to relatively higher temperatures. It also has a low variation in temperature, which is advantageous. Disadvantages are limited load capability, compared with outside air, and, in cases where the exhaust stream contains dirt, grease, or other impurities, the potential for coil degradation.

Water

There are many types of water sources that can serve as heat source and heat sink for heat pumps, including various types of ground (well) water and surface water from lakes, ponds, oceans, and rivers. Other potentially usable sources include municipal water supply, cooling tower water, and various types of waste water or fluid, such as sanitary sewage or discarded process water.

Temperature will be a function of climate and source depth. Ground water is especially effective because of its relatively high and constant temperature. Heat exchangers can also be submerged in surface water sources, though the temperature drop across the evaporator in winter may need to be limited to prevent freeze-up. Water quality should be analyzed for the potential of scale formation and corrosion. In some cases, it may be necessary to separate the water fluid from the equipment with an additional heat exchanger. Water-to-refrigerant heat exchangers are usually DX or flooded-water cooler types.

Ground (Earth)

The ground is commonly used as a heat source and heat sink medium, with heat transfer occurring through buried coils. Heat can be transferred directly with the refrigerant in the buried coil or indirectly through a secondary loop in which heat is transferred between a circulating brine and the ground. While suitability as a heat source and heat sink varies depending on soil composition, the ground exhibits relatively low temperature variation.

Solar Energy

Solar energy can be an effective heat pump heat source. When available, it provides heat at a higher temperature than other conventional sources, which produces a greater heating COP. The advantage of using solar as a heat pump heat source instead of directly for heating is that collector efficiency and capacity is increased due to a lower temperature requirement.

Types of Heat Pumps

The air-to-air system with a refrigerant changeover system described above is one of several types of commercially available heat pumps. Following are descriptions of various heat pump system designs.

Air-to-Air

During heating, this type of system operates by removing heat from the outside air (heat source) by evaporating refrigerant in the outdoor coil. Refrigerant temperature in the coil is below the temperature of the outside air, which causes heat to flow from the air to the refrigerant. This heat is then transferred to the indoor space.

As an alternative to the refrigerant changeover design discussed and shown graphically above, the thermal cycle change can also be done with an air changeover. In this type of design, one heat exchanger coil is always the evaporator and the other is always the condenser. The positioning of the dampers causes the change from cooling to heating. Thus, instead of reversing the refrigerant flow, the indoor and outdoor airflows are redirected to accomplish the same ends. This is usually the least costly type of heat pump. As with any air-cooled system, heat transfer is not as efficient as with water systems. At low ambient air temperature conditions, less heat is available and/or energy must be supplied to defrost the coil.

While an air-to-air unit is in heating mode, the outdoor air passes over the outdoor coil. A typical temperature reduction of about 10°F (5.6°C) occurs as the outdoor air gives up heat to the refrigerant. Extended-surface forced-convection heat transfer coils are most commonly used to transfer heat between the air and the refrigerant. Typically, the surface area of the outdoor coil will be 50 to 100% larger than the indoor coil.

When the air temperature is reduced below the atmospheric freezing temperature of water, moisture contained in the air will freeze and form frost on the surface of the outdoor coil. This frost will continue to form, thereby increasingly degrading heat transfer performance. When the coil is sufficiently iced over, it will render the system completely ineffective unless some means of removing the ice is provided. There are numerous methods used to defrost the unit. One method is the use of a resistance heater on the outdoor coil to melt the ice. Another method is to reverse the refrigerant flow, thereby putting the unit back in the cooling mode. The outdoor coil again becomes the condenser and hot refrigerant gas melts the frost buildup. To maintain indoor and conditioned space temperature during this period, when the unit is temporarily operating in cooling mode, a supplemental heat source may be required.

Water-to-Water

This type of system uses water as the heat source and heat sink for both heating and cooling. Water, waste water, or process effluent streams serve to treat the outdoor coil and also to treat water at the indoor coil.

Heating-to-cooling changeover can be done in the refrigerant circuit or by switching in the water circuit. The water source may be directly admitted to the evaporator (similar to direct free cooling systems) or passed through a heat exchanger to avoid the potential for contamination (similar to indirect free cooling systems). Another alternative is to use a closed-circuit condenser water system. Water-to-water heat pumps are very efficient compared to other types, but have high capital cost and require a substantial amount of maintenance.

Water-to-Air

This type of system uses water in the outdoor coil as the heat source and air to treat the indoor coil to transmit heat to or from the conditioned space. The outdoor coil can be served by the same types of water sources as the water-to-water heat pump. Air passes through a DX indoor coil, where it is cooled and dehumidified. This type of unit has a lower capital cost than a water-to-water heat pump and the indoor coil requires less maintenance.

Air-to-Water

This type of system is the reverse of the water-to-air heat pump. Air is used to treat the outdoor coil, as in the air-to-air system, and water is used in the indoor coil.

Ground (Earth)-Coupled

These types of systems can use direct expansion of the refrigerant in a buried coil of five or more feet underground. More commonly, they are designed to operate indirectly through a secondary loop in which heat is transferred between a circulating brine and the ground. In either case, heat is extracted from the ground, which remains at a fairly constant temperature all year. These units are usually not highly efficient and can be costly to repair if leaks or other problems occur in the buried coil. However, significant maintenance, repair, and replacement savings can be achieved in cases where coil degradation associated with corrosive air or water conditions is avoided.

Solar-Assisted

These types of systems use relatively low temperature solar heat as the heat source. They may be configured as water-to-air or other types depending on the type of solar collector and HVAC distribution system used. Systems can be designed with the evaporator coils directly in the solar collector, or with a secondary medium, such as air or water, bringing the heat from the collector to the coil.

Water-Loop (Load Transfer) Systems

These types of systems are designed to transfer internal load from one zone to another, with multiple water-to-air heat pump units connected hydraulically with a common two-pipe system. Each unit uses conventional refrigeration cycle cooling, supplying air to the individual zone and rejecting the heat removed to the two-pipe system. When heat is required in certain zones, the individual units switch into heating mode, usually by means of a refrigerant-reversing valve.

Figure 36-23 shows a water source two-pipe heat pump loop system consisting of a series of independently controlled units served by a central boiler and cooling tower. The heat that is rejected to the water loop from spaces requiring cooling is transferred to spaces or zones requiring immediate or after-hours heating. Under peak cooling conditions, excess heat that is rejected to the loop can be used to serve additional loads, such as domestic water preheating, or can be stored for later use. Under peak heating conditions, the central water heater adds heat to the system whenever the water loop temperature falls below an established set point. Since the heat of compression contributes heat to the system, the boiler can be sized smaller than a conventional system.

Figure 36-24 shows a geothermal heat pump loop system application. Instead of the cooling tower, the earth serves as the heat sink for cooling, and instead of the boiler, the earth also serves as the heat source for heating.

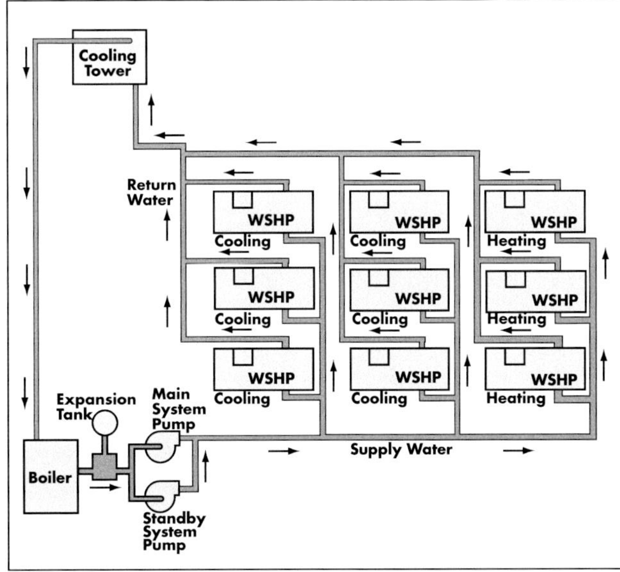

Fig. 36-23 Water Source Two-Pipe Heat Pump Loop System with Cooling Tower and Boiler. Source: The Trane Company

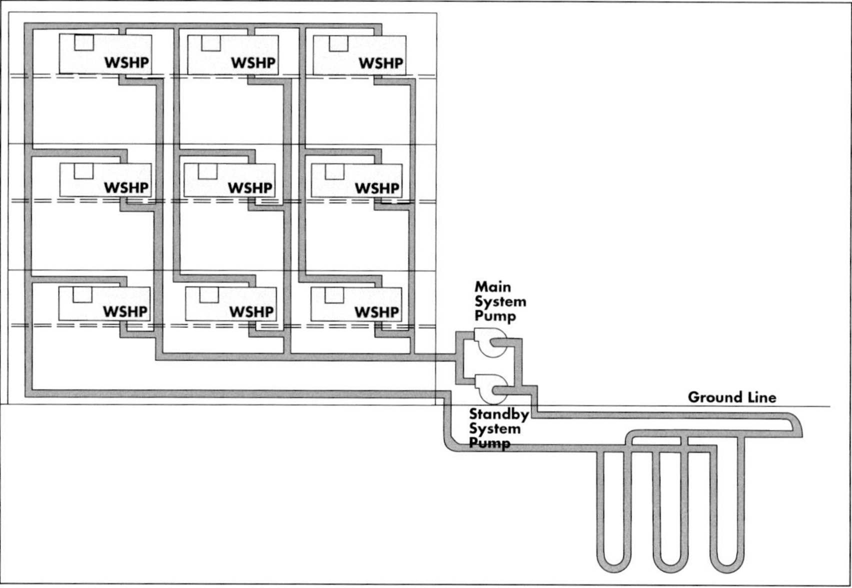

Fig. 36-24 Geothermal Heat Pump Loop System. Source: The Trane Company

The absorption cycle may also be used for heat pump applications. Currently, there are numerous types of absorption cycle heat pump systems under prototype development. Many of these are based on variations of the conventional LiBr-based absorption cycle or the NH_3-H_2O based GAX cycle, which are discussed in Chapter 38. While the basic GAX cycle is an excellent heat producer, it is not as thermally efficient in the cooling mode as other cooling system alternatives. The advanced GAX-cycle machines, however, are expected to improve the cooling-side efficiency. Under standard ARI conditions, this unit is expected to achieve a cooling COP of 0.95 and a heating COP of 1.55. These COP figures are particularly impressive because these are air source heat pumps operating through a much wider temperature range than can be achieved with conventional absorption cycle machines.

Absorption Cycle and Engine-Driven Heat Pumps

While almost all of the heat pumps currently in commercial operation use electric motor-driven vapor compression systems, the compressor itself is indifferent to its driver and may be driven with any type of prime mover. Figure 36-25 shows a series of small-capacity gas-fired reciprocating engine-driven heat pumps applied in a manufacturing plant. In contrast to traditional electric-driven heat pump systems, these units can use heat recovered from the engine to meet peak heating requirements instead of relying upon inefficient electric resistance heat. Figure 36-26 is a schematic diagram of an engine-driven heat pump. Included is a heat exchanger to provide for domestic hot water heating and a gas-fired auxiliary heater.

Fig. 36-25 Series of Small-Capacity Gas Engine-Driven Traithlon Heat Pumps. Source: York International and The American Gas Cooling Center

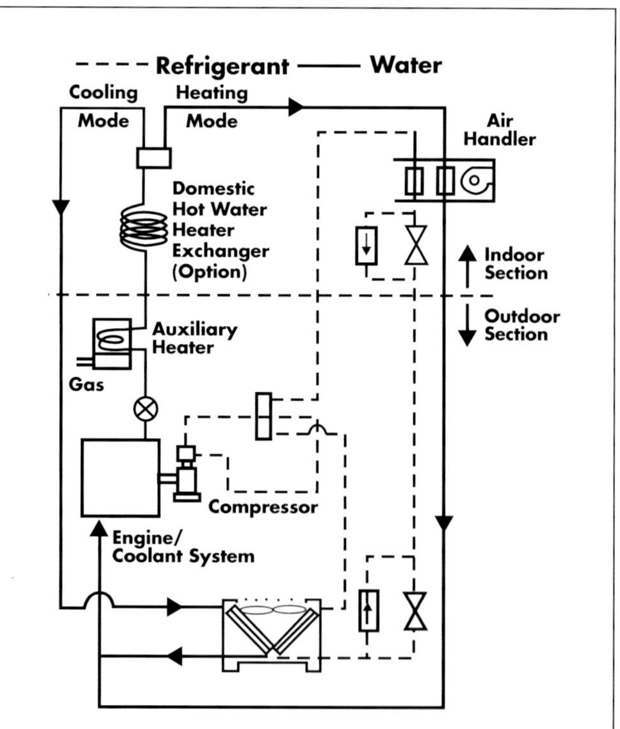

Fig. 36-26 Schematic Diagram of an Engine-Driven Heat Pump. Source: GRI

Heat Pump Performance

In contrast to the ideal COP for a conventional refrigeration cycle, which is expressed as:

$$COP_c = \frac{T_L}{(T_H - T_L)} \quad (36\text{-}10)$$

the ideal COP of a heat pump is expressed as:

$$COP_h = \frac{T_H}{(T_H - T_L)} \quad (36\text{-}11)$$

where all temperatures are in absolute units and subscripts h and c refer to heating and cooling COP, respectively.

Whereas, with the conventional refrigeration cycle, COP_c represents the ratio of cooling effect to energy input, with a heat pump cycle, COP_h represents the ratio of heat output to energy input. What can be seen from the ideal COP expression is that the heat pump can be a very efficient heater. The energy available for heating duty includes the heat normally rejected in conventional refrigeration cycles — the heat associated with the refrigeration effect, i.e., 12,000 Btu/ton-h (1 kWh_h/kWh_r) — plus the driving energy input that would normally be rejected through the condenser to the outside environment. The heat available (H_A) can thus be expressed as:

$$H_A = \textit{Refrigeration effect} + \textit{Driving energy input} \quad (36\text{-}12)$$

The driving energy input refers to the portion of the heat imparted on the system that must also be rejected to the condenser. In the case of vapor compression systems, it is the heat of compression. In absorption systems, it is the heat duty of the generator. This is roughly equivalent to the heat input that is not otherwise rejected (i.e., in the form of steam condensate, return water, or exhaust gas). The practical equation for heating COP is the H_A divided by the driving heat input (H_I).

Consider, for example, an electric-driven air-cooled vapor compression system with a 100,000 Btu/h (29.3 kW) heating capacity and an electricity input of 10 kW. In heating mode, the COP would be:

$$COP_h = \frac{100{,}000\ Btu/h}{10\ kW \times 3{,}413\ Btu/kWh} = 2.93$$

In SI units, this would be:

$$COP_h = \frac{29.3\ kW}{10\ kW} = 2.93$$

In practice, most heat pumps operate as efficient heaters in moderate temperatures, i.e., above 40°F (4.4°C). However, as the ambient temperature approaches freezing, capacity and efficiency fall off dramatically, and eventually the system must use an additional heat source to maintain heating capacity. To determine the seasonal heating COP, performance would have to be calculated over the full operating regime and then averaged. Generally, in smaller conventional systems, the supplemental heat source is electric resistance heating, which is considered highly inefficient. In electric resistance heating mode, the heating COP would approach 1.0. In some cases, conventional fuel-fired furnaces are used in a hybrid system configuration. These systems are both more thermally efficient in use of source energy and more costly to install than electric resistance supplemental systems.

When comparing performance for systems that use electricity as opposed to fuel or systems that use a combination of electricity and fuel, it is understood that a direct comparison of COPs can provide misleading results. More valid comparisons would be based on source energy usage (considering the source fuel efficiency of electricity production) or, more directly, on economic performance.

Another concern with air-to-air heat pumps is frosting. As discussed above, when ambient temperatures reach 40°F (4°C) or below, the temperature of the air passing over the coil is reduced below freezing, causing frost buildup on the surface of the outdoor coil. This necessitates the use of some type of defrosting method. Energy used for defrosting should also be considered in the determination of the seasonal heating COP.

Refrigeration Cycle Heat Recovery

Any type of refrigeration cycle system that transfers heat for beneficial purposes may be considered a heat pump. Refrigeration cycle systems that provide cooling but also recover heat rejected from the refrigerant to provide useful heating may be considered a type of heat recovery heat pump. These are differentiated from reverse cycle systems in that when they operate, they always provide cooling with conventional cycle operation. An additional feature is that they recover refrigeration cycle-generated heat, normally rejected to the outside environment, for productive use.

With refrigeration cycle heat recovery systems, there is a net gain in thermodynamic efficiency when the relatively low temperature — typically, but not always, lower than 130°F (54°C) — recovered heat can be used. Since there is 12,000 Btu/ton-h (1 kWh_h/kWh_r) of heat rejection, plus a portion of the cycle driving energy input to the system that also must be rejected, heat recovery can result in extremely efficient refrigeration cycle system energy utilization.

Heat recovery can be applied to almost any type of refrigeration cycle system, limited chiefly by the economic viability of using a relatively low-temperature heat source. The underlying theory is that the energy rejected from refrigeration cycles contains available thermal energy that can be recovered for productive use. Applications that heat or preheat relatively cold water (as in domestic water heating), for example, are more likely to be cost-effective. Principle obstacles are capital and operating costs associated with recovery of any relatively low-temperature energy stream, potentially negative impacts on cooling system performance resulting from installation and operation of heat recovery components, design complications associated with non-concurrence of cooling and heating loads, and potentially unstable chiller operation under low-load conditions.

Refrigeration cycle heat recovery systems typically use desuperheaters and condensers or auxiliary condensers. These are discussed below.

Desuperheaters, which are commonly used in smaller applications featuring air-cooled reciprocating and scroll compressors, function somewhat as auxiliary condensers. As shown in Figure 36-27, the desuperheater is usually connected to the system between the compressor discharge and the condenser. Heat is transferred from the hot refrigerant to the water being heated. The desuperheated refrigerant then flows through the condenser. The temperature of the water leaving the desuperheater is dependent on the cooling (refrigeration) loads. Typically, water temperature gain ranges from 5 to 40°F (3 to 22°C).

Two important application considerations are sizing and avoiding contamination of the heating loop by the refrigerant charge. If the desuperheater is undersized, it restricts the refrigerant charge line, causing the compressor to operate inefficiently against excessive refrigerant head pressure. An oversized unit may result in insufficient discharge refrigerant gas velocities, which in turn can result in oil trapping and insufficient oil to the compressor.

Condenser heat recovery systems are typically used for larger systems featuring centrifugal or screw compressors. These systems may feature either a single condenser or a dual-condenser configuration with a dedicated heat recovery condenser. In applications in which the heating load exceeds the cooling load at all times, the chiller essentially functions as a heat pump with all of the rejected heat from the condenser serving the heating load. In applications in which the cooling load exceeds the heat recovery requirement, typical heat rejection apparatus (i.e., a cooling tower or other heat sink) is required.

In the simplest arrangement, the condenser is sized for a relatively large peak temperature differential of 30°F (17°C) or more. The hot water exiting the condenser is passed through a filter to the heat recovery (heat exchanger) unit and then on to the cooling tower (heat sink). To limit heat exchanger fouling and the risk of contamination by cooling tower water (or other source, such as river water), a second separate condenser is commonly used, which establishes a separate heat recovery circuit. The separate condenser is sized in accordance with the portion of the available heat energy targeted for recovery.

Dual-condenser systems, shown in Figure 36-28, operate on the principle that refrigerant will migrate to the coldest point in the system. Raising the temperature

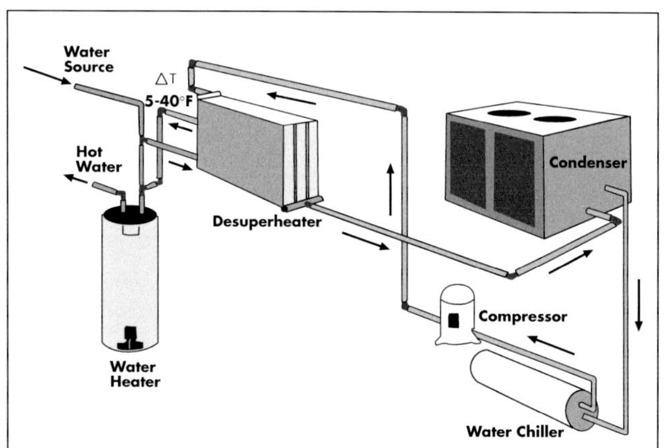

Fig. 36-27 Small-Capacity Heat Recovery System Featuring Desuperheater. Source: The Trane Company

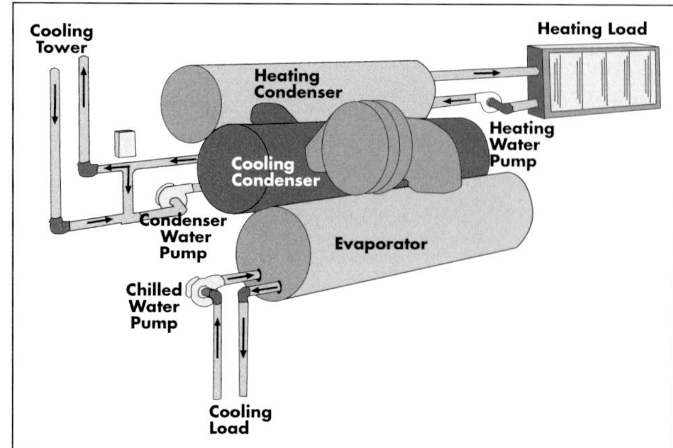

Fig. 36-28 Dual-Condenser Heat Recovery System. Source: The Trane Company

of one condenser thus forces heat rejection to the other. By modulating the flow to the heat recovery condenser or the temperature of the tower water loop, the temperature of the standard condenser and the heat rejection to the heat recovery condenser can be controlled. With centrifugal chillers, care must be taken to avoid unstable operation under low load conditions, given the high head pressure imposed on the system.

Heat Recovery System Performance

As with heat pumps, the energy available to a heat recovery system consists of the refrigeration (or cooling) effect, plus the driving energy input. In English units, when driver energy for a vapor compression system is measured in hp, the heat available from a system can be expressed as:

$$H_A = [12{,}000\ Btu/ton\text{-}h_{cooling\ effect} + hp/ton \times 2{,}545\ Btu/hp\text{-}h_{compression}] \times tons\ capacity \quad (36\text{-}13)$$

When driver energy is measured in kW, the equation becomes:

$$H_A = [12{,}000\ Btu/ton\text{-}h_{cooling\ effect} + (kW/ton \times 3{,}413\ Btu/kWh)_{compression}] \times tons\ capacity \quad (36\text{-}14)$$

In SI units, this can be expressed as:

$$H_A = [1\ kWh_h/kWh_{r\ cooling\ effect} + (kWh_m/kWh_r)_{compression}] \times kW_r\ capacity \quad (36\text{-}15)$$

In this equation, subscript h refers to the heat energy rejected and subscript m refers to mechanical energy input. It is understood that the mechanical energy input is converted to heat as a result of the compression process.

Total system performance includes the cooling benefit of the refrigeration effect and the heating benefit of the refrigeration effect plus the driving energy input. Combined, the total useful energy output usually reflects a very efficient use of the energy input. However, a principle disadvantage is the chiller capacity and efficiency penalty associated with the higher pressure differential against which the compressor must work. This particularly affects summer energy consumption, because the compressor speed is optimized for the high-head duty through selection of gear ratio and/or impeller duty. Annual performance is increasingly degraded by increasing elevated leaving condenser hot water temperature. This penalty increases as the hot water temperatures produced increase and may exceed 25% of the total chiller energy input. As a result, the energy saved by utilizing heat recovery is partially offset by the additional energy required per unit of cooling output.

Table 36-2 provides a representative comparison of two heat recovery system design set points. In winter, leaving condenser temperature is 100°F (38°C) in one unit and 105°F (41°C) in the other. In both cases, through high gear ratio, leaving condenser water temperature is 95°F (35°C) in summer.

This example is based on a constant leaving chilled water temperature of 44°F (7°C). In addition to minimizing leaving condenser water temperature, chilled water reset can be used to minimize performance degradation. This is accomplished by allowing the leaving chilled water temperature to be raised to the maximum temperature that will still satisfy the cooling load. This reduces peak compressor pressure requirements during winter operation (or cooler ambient temperature periods) when lower humidity levels and cooling capacity requirements allow for elevated (or reset) chilled water temperatures.

The economic performance of heat recovery systems is a function of the achievable net energy cost savings and the cost of the heat recovery equipment and any design modifications necessary for the distribution and use of relatively low-temperature hot water. As with any low-temperature heat recovery system, capital costs are negatively impacted by the need to increase distribution system size. If, for example, the hot water is used for space heating, the capital cost penalty includes larger size pipes, pumps, and coils. Pump energy use is also increased. Selection of distribution temperature must balance these factors with the desire to minimize leaving condenser water temperatures

Operating Mode	Leaving CWT (°F)	HP per Ton	Performance Penalty
Base	95	0.82	Base
100 °F Temp.			
Summer	95	0.86	4%
Winter	100	0.89	9%
105 °F Temp.			
Summer	95	0.88	7%
Winter	105	0.98	20%

Table 36-2 Comparison of Heat Recovery Design Set Points.

to minimize the system performance penalty.

A key operating factor to system economic performance is the concurrence of heating and cooling loads. When there is insufficient cooling load to support the heating requirement, a false cooling load must be applied to the chiller in order to generate the required heat rejection, rendering the system highly inefficient. An alternative to this operating mode is the provision of heat from a supplemental source. When operating in cooling only mode, with a lower pressure differential, the chillers can achieve full capacity and improved efficiency. However, efficiency will always remain lower than a comparable chiller system designed for a lower lift requirement.

A variation of this applied chiller heat-recovery technology involves the use of two interrelated systems. One system functions at typical temperatures and efficiencies with, for example, a condenser water rise of 85 to 95°F (29 to 35°C). In such a system, the 95°F (35°C) exiting condenser water bypasses the cooling tower and enters the second system where it is cooled by refrigerant and returned at 85°F (29°C). To accomplish this, the compressor work is relatively low. The refrigerant in the second system is then condensed, giving off its heat to the heating loop. Thus, 12,000 Btu/ton-h (1 kWh_h/kWh_r) plus the heat of compression from both systems' compressors are passed to the heating loop. When heating requirements are lower than the heat energy made available by the cooling process, it is bypassed to the tower.

Absorption Cycle Heat Recovery

Though not commonly used in heat recovery applications, absorption cycle systems provide abundant quantities of low-temperature heat. While a single-stage LiBr absorption chiller with a COP of 0.66 is considered a very inefficient refrigeration machine, it is an extremely efficient heat-producing machine. While 1 Btu (or kWh) of heat input produces only 0.66 Btu (or kWh) of refrigeration effect, it produces 1.66 Btu (or kWh) of rejected heat (this should not be confused with the absorption heating cycle). While more efficient in providing cooling than a single-effect absorption chiller, a double-effect LiBr absorption machine with a COP of 1.0 is also a good heat producer, as it rejects about 2.0 Btu (or kWh) of heat per Btu (or kWh) energy input. Hence, by combining the beneficial use of the cooling effect and the recovered heat, the absorption machines can operate at an extremely high performance rate.

However, as with vapor compressions systems, the heat product is at a relatively low temperature, which requires large pipe and pump sizes, and there is also an efficiency penalty for elevating the temperature of the recovered heat. Generally, the limit on using full heat rejection of absorption chilling machines is about 100°F (38°C). This is usually the maximum temperature of the combined condenser and absorber cooling water circuits. By targeting only the absorber cooling water circuit, heat can be recovered at substantially higher temperatures in LiBr machines. The amount of recoverable heat will be proportional to the driving energy input to the generator. With ammonia-based machines, heat recovery potential is greatly increased due to increased absorber operating temperatures. Due to the large temperature glide, absorber temperatures can exceed 180°F (82°C), allowing for relatively high temperature heat recovery. This opens up a wider range of useful applications and reduces pump and pipe size requirements.

Chapter Thirty-Seven

Vapor Compression-Cycle Systems

The basic vapor compression refrigeration cycle consists of the four processes discussed in Chapter 33 — expansion, evaporation, compression, and condensing. The ideal vapor compression refrigeration cycle is the reverse Rankine cycle (see Chapter 11). Whereas, in the Rankine cycle, vapor expansion produces work output, in the reverse Rankine cycle, vapor compression work produces refrigeration effect. This vapor compression work is done by refrigerant compressors that are driven by electric motors and fuel- or steam-powered prime movers.

This chapter describes the vapor compression cycle and types of equipment commonly used for its application, as well as applicable compressor types, refrigerants, system performance, and capacity control. It also includes a detailed discussion on vapor compressor drivers.

Vapor Compression-Cycle Technologies

There are four basic components used in the vapor compression cycle: a compressor, a condenser, an expansion valve (or throttling device), and an evaporator, each corresponding to one of the four cycle processes. The relationship of these components in the vapor compression cycle is shown in Figure 37-1.

In all practical applications, the actual vapor compression cycle performance is always lower than the ideal cycle. This is the result of several factors, including friction losses, heat exchanges between parts of the system, and pressure drops in suction and discharge lines. Additionally, the refrigerant vapor charge enters the compressor with a small amount of superheat (to avoid potential damage to the compressor through the entrance of slugs of liquid refrigerant that are not completely vaporized during compression), and the liquid refrigerant entering the expansion valve is usually subcooled a few degrees. Also, since compressor suction and discharge are actuated by pressure difference, the process requires the actual suction pressure inside the compressor to be slightly below that of the evaporator and the discharge pressure to be above that of the condenser.

Vapor compression cycle systems can be categorized in several ways, including by compressor type, condenser type, and evaporator type. Vapor compression systems consist of a compressor, driver, liquid cooler (evaporator), condenser(s), refrigerant control device, and a controller. They often include a receiver(s), oil separator, intercooler and/or subcooler accumulator, and, in some cases, dual liquid pumps. They may use air-cooled, water-cooled, or evaporative condensers and refrigerant-to-air or refrigerant-to-liquid evaporators.

As detailed in Chapter 35, since it is generally impractical to circulate large volumes of refrigerant throughout a facility, chillers that use refrigerant-to-liquid evaporators are almost always used for large central air conditioning systems with multiple terminal units and for small to large process loads. They are also sometimes used in smaller units. Vapor compression systems using refrigerant-to-air type evaporators are called **direct expansion (DX)** units. These systems provide direct cooling as the refrigerant absorbs heat directly from the medium (i.e., air) being cooled. DX systems typically range in capacity from fractional tonnage up to a few hundred tons (1,000 kW_r) and use high-pressure refrigerants such as HCFC-22, R-410A, a mixture of difluoromethane (CH_2F_2, called R-32), or pentafluoroethane (CHF_2CF_3, called R-125).

DX units have two basic configurations: packaged and condensing systems. **Packaged,** or **rooftop, units (RTUs)** are self-contained units with all four vapor compression components in one housing. They are typically roof-mounted and directly connected to ductwork or

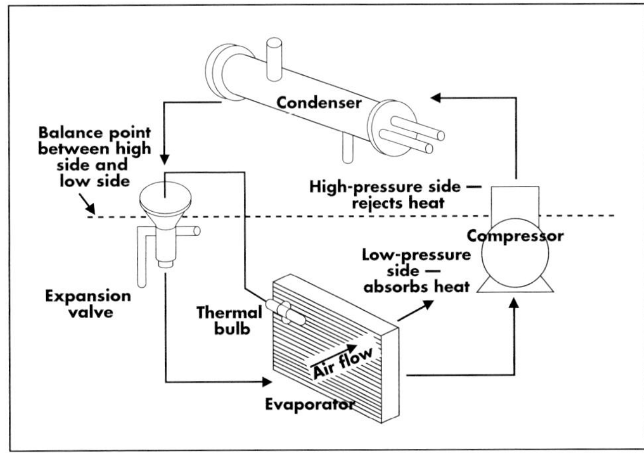

Fig. 37-1 Relationship of Vapor Compression Cycle Components.

to a diffuser and distribute chilled air directly to a conditioned space. These are air-cooled units that range in capacity from fractional tonnage window units to several hundred tons, with extensive ducting systems and controls. **Condensing,** or **split, systems** are units that consist of a compressor and driver with an interconnected air-cooled condenser mounted in a self-contained housing. Instead of direct ducting, a liquid and suction (vapor) line set sends and receives refrigerant from these components to an air handling unit (AHU) or, in some cases, multiple AHUs.

REFRIGERATION COMPRESSORS

In a refrigeration-cycle system, the compressor raises the pressure of the refrigerant gas from evaporator pressure to condenser pressure. The compressor delivers the refrigerant to the condenser at a pressure and temperature at which the condensing process can be readily accomplished. Since thermal losses through the compressor casing are usually small and cannot be measured accurately, they are usually neglected. Hence, both polytropic (input) work and irreversible adiabatic (actual gas) work may be considered the same.

There are several types of compressors and each type can be of hermetic or open-drive design:

Hermetic (or **closed-drive) compressors** (Figure 37-2) consist of an electric motor and compressor built into an integral housing using a common shaft and bearings. Since the motor is built into the housing, the windings run in a lower temperature environment that is cooled by the refrigerant. The motor, therefore, can be smaller than in other configurations. Typically, the motor is cooled with suction gas as it passes through the windings. Motors may also be cooled by liquid refrigerant and economizer gas. A penalty in thermal efficiency occurs as refrigeration capacity is used to absorb the motor heat. This does, however, eliminate the need to remove the motor heat from the mechanical room.

An advantage of hermetic compressors is that they eliminate the need for mounting the motor and aligning the couplings, and minimize other site work during installation. They also eliminate the potential for refrigerant leaks at the shaft seal.

Open-drive compressors (Figure 37-3) operate with an external or exposed electric motor or prime mover that drives the compressor. Large motors are common and driver heat gain is discharged into the surrounding atmosphere. In some cases, mechanical energy must be expended (i.e., a fan) to rid the mechanical room of excessive heat.

An advantage of the open-drive arrangement is that it offers slightly greater thermal efficiency than hermetic compressors because driver cooling with suction gas or water is eliminated, allowing more refrigerating effect to be applied to the actual load. Refrigeration changeover is also less cumbersome because the driver is open and accessible. Disadvantages include size and the potential leak path occurring at the shaft seal.

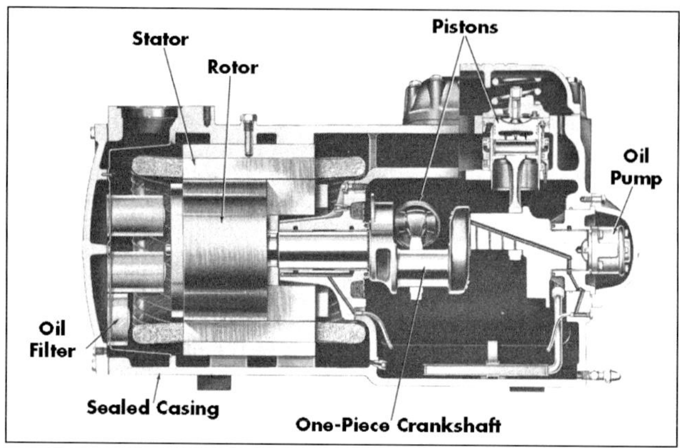

Fig. 37-2 Cutaway View of Hermetic Compressor.
Source: The Trane Company

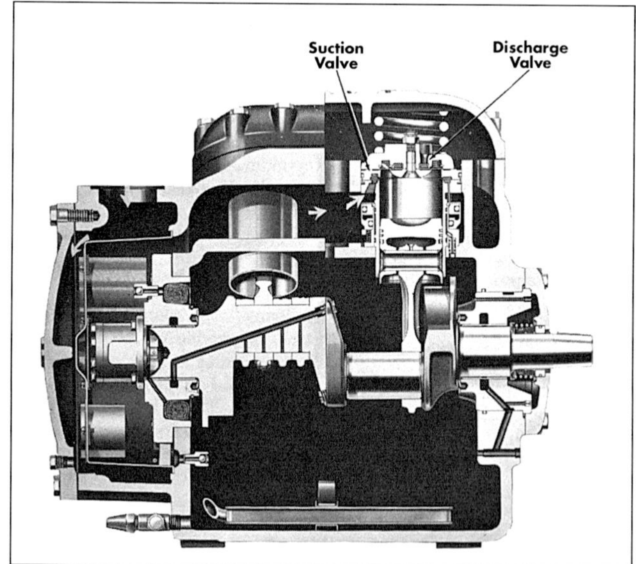

Fig. 37-3 Cutaway View of Open-Type Compressor.
Source: The Trane Company

COMPRESSOR DESIGN

The five basic refrigeration compressor designs are described below and compared in Table 37-1:

- **Rotary compressors** have an out-of-center rotor that rotates within a cylinder. Applied designs include use of a rolling piston or sliding vanes.

Characteristic	Rotary	Scroll	Reciprocating	Screw	Centrifugal
Typical	< 5 tons	5–30 tons	< 1–150 tons	100–750 tons	100–10,000 tons
Capacity Range	(< 18 kWr)	(18–105 kWr)	(< 1-525 kWr)	(350–2,600 kWr)	(350–35,000 kWr)
Maximum Capacity Range	5 tons (18 kWr)	10 tons (35 kWr)	400 tons (1,400 kWr)	2,000 tons (7,000 kWr)	20,000 tons (70,000 kWr)
Types	Sliding vane, rolling piston types, welded shell only	Compliant and non-compliant, welded shell only	Open, serviceable, and welded shell hermetics	Rolling rotor open and hermetic	Single and multi-stage open and hermetic
Displacement	Positive	Positive	Positive	Positive	Not positive
Capacity Control	Variable speed and on-off	Variable speed and on-off	Variable speed, on-off, and	Variable speed and intake slide valve cylinder unload	Variable speed and guide vanes

Table 37-1 Comparison of Currently Applied Compressor Technologies.

- **Scroll compressors** have two spiral-shaped parts; one remains fixed while the other orbits against it. Suction gas is brought in from the outer portion and compressed gas exits at the center of the scrolls.
- **Reciprocating compressors** have a piston(s) that travels back and forth in a cylinder(s).
- **Screw compressors** have two helical rotors, one male and the other female. As they turn, the rotors mesh.
- **Centrifugal compressors** have a high-speed impeller with numerous blades that rotate.

Rotary compressors are commonly used in appliances such as refrigerators or water coolers and for air conditioning and heat pump units of less than 5 tons (18 kW_r). The scroll compressor is typically used in applications of about 5 to 30 tons (18 to 105 kW_r). The three major compressor types, however, are the reciprocating, screw, and centrifugal, which are used for almost all larger-capacity applications. These three types, which are discussed in detail below, may be designed as single- or multiple-stage units, depending on the application requirements.

Reciprocating Compressors

Reciprocating compressors are used for applications with capacity requirements of up to 400 tons (1,400 kW_r), based on ARI standard conditions. They are generally used for smaller-capacity applications ranging from about 5 to 150 tons (18 to 525 kW_r), most commonly in air-cooled systems. The major components of a reciprocating compressor are the body, cylinder head, valve plates, and pistons driven by a connecting rod from the crankshaft. The design has many similarities to a reciprocating engine.

In the basic design, much like a reciprocating engine, the pistons provide compression by reducing the volume of refrigerant vapor in a cylinder. The cylinder head serves as a pressure plate to support and hold the valves and valve plate in position. It also provides the vapor passage into and out of the compressor. As the piston moves downward, it draws refrigerant vapor from the evaporator into the cylinder through an intake valve. As the piston moves upward, it compresses the vapor. The piston displacement is the maximum theoretical volume of the compressor. The swept volume is the actual volume of refrigerant vapor entering the cylinder and is usually less than the piston displacement due to the re-expansion of some refrigerant gases that remain in the cylinder (in the dead volume) after discharge. The piston is designed to come as close as possible to the cylinder head, without touching it, in order to press as much of the vapor into the high-pressure side as possible.

Figures 37-2 and 37-3 are cutaway illustrations of reciprocating compressors. The intake and compression strokes of a reciprocating compressor are illustrated in Figures 37-4 and 37-5, respectively. In multiple-cylinder units, the cylinders may be arranged in pairs with one cylinder head covering two cylinders. One or more openings are provided for access to the crankcase for assembly and overhaul.

In larger reciprocating compressors, oil and refrigerant mix continuously. In most compressors, the lubrication system is a positive displacement type pump that forces the oil to the components. Refrigeration oils are soluble in liquid refrigerant and generally mix completely at typical ambient temperatures. The term miscibility is used to describe a refrigerant's ability to mix with oil. Because oil is passed through the compressor cylinders to provide lubrication, a small amount of oil (about 1%) is constantly circulating with the refrigerant. In order to return oil and

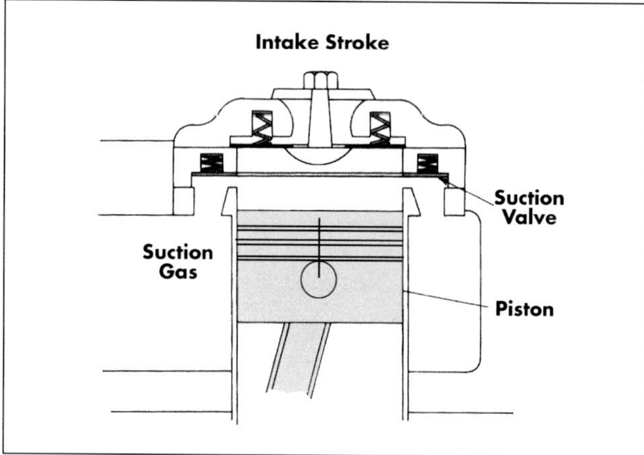

Fig. 37-4 Illustration of Intake Stroke of Reciprocating Compressor. Source: The Trane Company

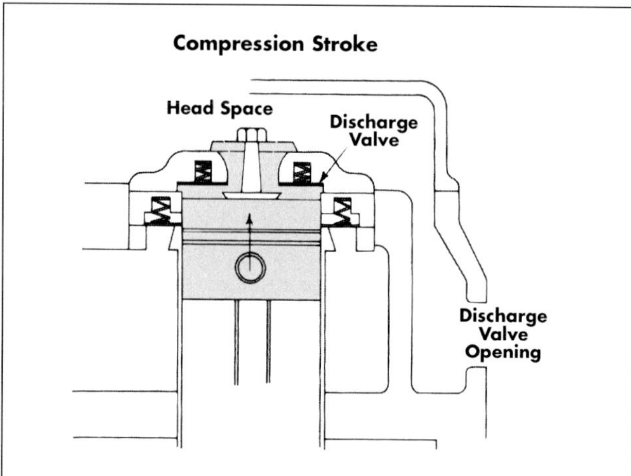

Fig. 37-5 Illustration of Compression Stroke of Reciprocating Compressor. Source: The Trane Company

refrigerant gas, velocities must be high enough to sweep the oil along. If gas velocities are insufficient, oil will tend to lie on the bottom of refrigerant tubing. This potential becomes more pronounced under lower evaporating temperatures, since the viscosity of oil increases as temperature decreases.

In hermetic designs, the motor is directly connected to the compressor. Both are sealed inside a housing to lubricate the compressor. To lubricate a small hermetic compressor, the return suction gas is fed into a hollow disk mounted on the motor compressor shaft. Centrifugal force throws the oil and liquid refrigerant to the outer rim of the disk, after which it flows over the motor windings. Only the vapor refrigerant remains at the center and is drawn into the cylinders of the compressor. In larger hermetic compressors, the lubricating oil pump is driven off of the crankshaft. The oil flows through the bearing and the unloaders, and, simultaneously, through the crankshaft to the rod bearing, up the connecting rod, out of the piston, onto the cylinder walls, then back to the oil pump.

In external- (or open-) drive designs, the compressor is bolted together with the crankshaft extending through the crankcase. The crankshaft can be driven by a flywheel (pulley) and belt or can be driven directly by a coupling.

Small compressors usually have fins cast with the cylinders to provide better air cooling. Larger compressors may have water jackets surrounding the cylinders for cooling. Some compressors have cylinder liners (or sleeves) that can be replaced when worn.

Compressor pressure may range as high as 600 psi (41 bar), depending on the type of refrigerant used. Typically, currently applied reciprocating compressors operate on refrigerants such as HCFC-22 (R-22), HFC 134a (R-134a), R-410a, and ammonia, which have high vapor densities and condensing pressures. Single-stage units are commonly applied to chilled water and other air conditioning applications. Two-stage compressors are used for low-temperature applications. Two stages of compression are achieved with booster and high-stage units, which can be two individual compressors or integral machines. Intercooling is generally used between stages.

Figure 37-6 illustrates a vapor compression cycle system with a reciprocating compressor. The two gauges on either side of the compressor are used to measure the suction, or low-side, pressure before the refrigerant reaches the compressor, and head, or high-side, pressure after the refrigerant is discharged from the compressor as a high-pressure vapor. An additional component, the receiver is added to the system in order to store liquid refrigerant and balance the system. Upon leaving the receiver, the liquid refrigerant passes through the expansion valve on the way to the evaporator. Whereas the compressor maintains a difference in pressure between the evaporator and the condenser, this difference can only be maintained with the use of the expansion valve, which separates the high-pressure part of the system from the low-pressure part. The valve is adjusted so that only just as much liquid can pass through it as can be vaporized in the evaporator, or cooling coil.

Screw Compressors

Screw (rotary positive displacement) compressors range in capacity from under 20 tons (70 kW_r) to about 1,800 tons (6,300 kW_r) under ARI standard conditions. Their rotation motion allows for smooth, quiet, almost vibration-free operation and low maintenance require-

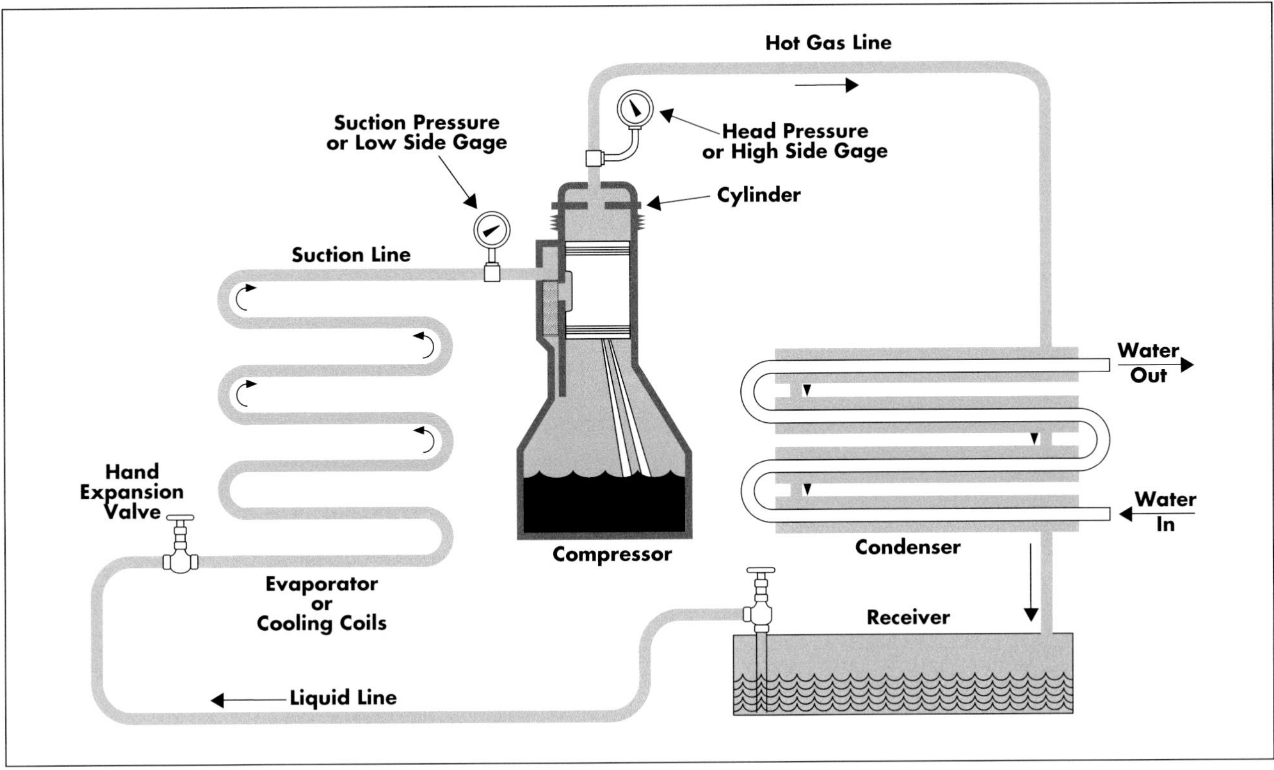

Fig. 37-6 Illustration of Vapor Compression Cycle System with a Reciprocating Compressor. Source: The Trane Company

ments, as compared with reciprocating compressors. The two main types of screw compressors used today are single screws and twin screws. Similar to reciprocating compressors, currently applied screw compressors typically operate on refrigerants such as HFC 134a and ammonia.

Screw compressors have meshing male and female helical rotors that compress refrigerant vapor. As the rotors turn and mesh, they trap the vapor that has been introduced by suction in the interlobe space. Typically, the male rotor is driven by the compressor driver and, in turn, drives the female rotor. The vapor is further compressed by continued rotation of the meshing rotors. The compression space becomes progressively smaller until the compressed refrigerant vapor is discharged toward the condenser as the interlobe space becomes exposed to an outlet port. The screw threads form the boundary separating several compression chambers that move down through the compressor at the same time. Twin screw compressors consist of two helically grooved rotors of the same shape, mounted in a housing with inlet and outlet ports. One or both rotors may be driven, depending on the specific design.

Figure 37-7 is a cutaway illustration of a screw compressor. Figure 37-8 illustrates the compression cycle of a typical screw compressor. On top, the lobes and grooves

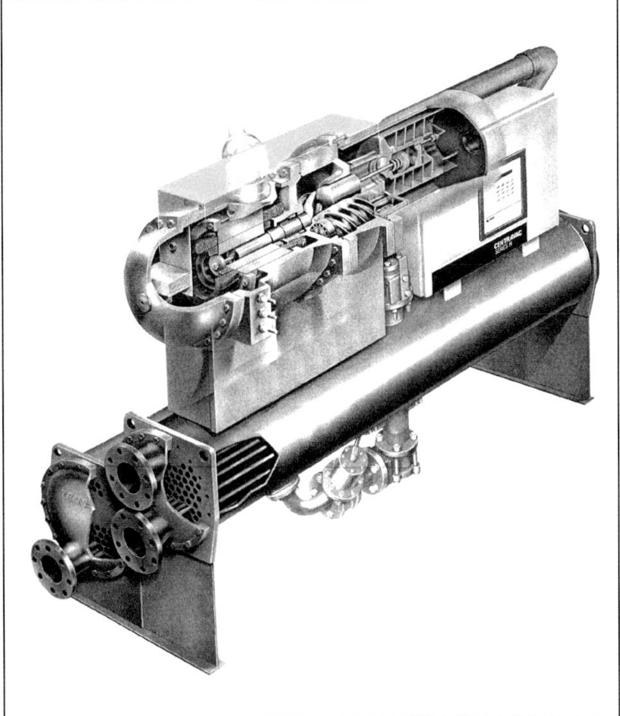

Fig. 37-7 Cutaway Illustration of a Chiller Featuring a Helical-Rotary Compressor with a Hermetic Motor.
Source: The Trane Company

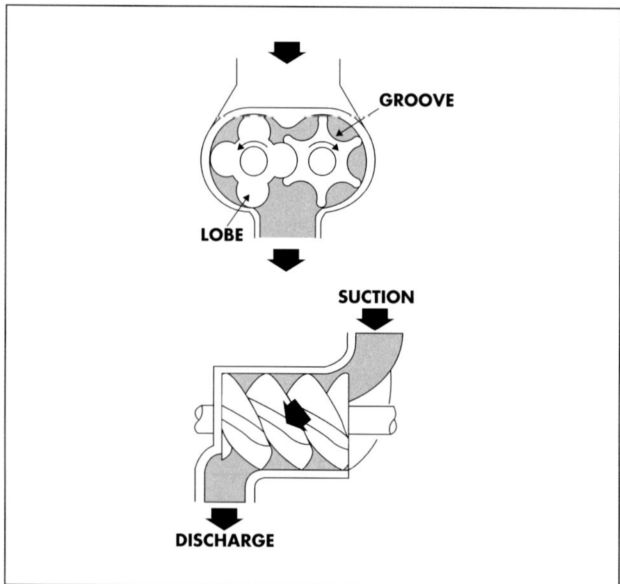

Fig. 37-8 Screw Compressor Compression Cycle Illustration.

are shown in the compressor housing. On the bottom, the refrigerant flow path is shown from suction to discharge. Notice the profile design is asymmetrical to reduce refrigerant leakage. This is a 4-lobe/6-flute configuration. For higher capacity, lower-pressure applications, a 3-lobe/4-flute configuration may be used. For lower capacity and higher pressure, a 6-lobe/8-flute configuration may be applied.

Figure 37-9 illustrates the refrigeration cycle for a chiller system featuring a screw compressor. Notice the change in refrigerant phase and pressure as it moves through the cycle, with high-pressure refrigerant exiting the screw compressor on the upper left before being condensed in the condenser on the lower right.

Many designs feature an oil injection system to seal the clearance between the rotors and the cylinder. An oil separator is used to remove oil from the refrigerant discharge. The oil absorbs heat, which limits refrigerant discharge temperature. The oil is then cooled through methods such as external heat exchange or injection of liquid refrigerant. Alternatively, oil injection-free screw compressors are also commonly used. Instead of injecting oil into the compression chamber, liquid refrigerant can be injected. This allows for the elimination of discharge oil separators and external coolers associated with oil injection operation.

Screw compressors are typically more expensive than reciprocating compressors in smaller capacities, but offer smoother, often less noisy operation and lower maintenance requirements. As capacity increases, capital cost decreases relative to reciprocating compressors. When comparing similar (load) capacity equipment, screw compressors are relatively smaller than their counterpart centrifugal or reciprocating compressors. Screw compressors are typically less thermally efficient than comparable capacity centrifugal compressors at full load, but, due to their positive displacement characteristics, offer equal or superior performance under part-load conditions.

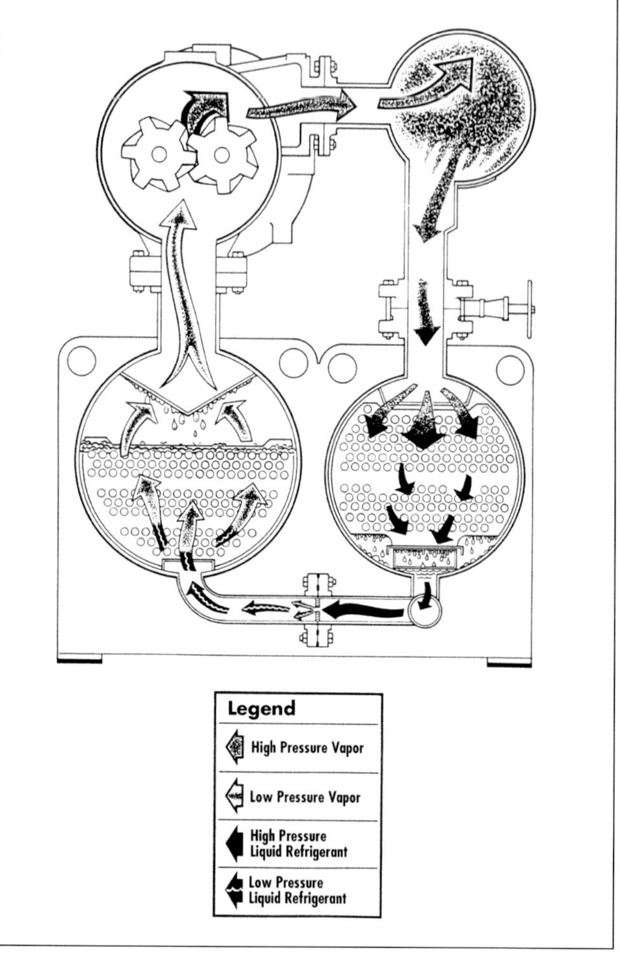

Fig. 37-9 Illustration of Refrigeration Cycle for a Screw Compressor-Based Chiller System. Source: York International

The typical operating speed for a screw compressor is about 3,600 rpm. They are, however, well suited for operation over a wide range of speeds. To avoid gearing requirements, they can sometimes be matched with the ideal operating speed of the driver. Screw compressor output capacity is reduced linearly with speed reduction. For example, operation at 1,800 rpm, which would be well suited for a reciprocating engine driver, would require a doubling of compressor size. Even though the added cost of the larger compressor is roughly equivalent to, or slightly greater than, the cost of the speed increaser, lower speed operation extends compressor life and reduces noise levels.

Centrifugal Compressors

Centrifugal compressors range in capacity from 50 to 10,000 tons (175 to 35,000 kW_r) under ARI standard conditions. Economic factors relating to manufacturing limitations and efficiency losses generally dictate that most applications start above 200 tons (700 kW_r). Custom-built units may exceed 20,000 tons (70,000 kW_r). They offer smooth, almost vibration-free operation, and, because they have so few moving parts, offer long service life with limited maintenance requirements.

Whereas reciprocating and screw compressors are positive displacement machines that physically squeeze gas with pistons or rotors to compress it, centrifugal compressors are variable displacement machines that impart centrifugal force to the gas to produce velocity energy that is converted to pressure. Centrifugal compressors operate at speeds of 3,000 to 20,000 rpm and produce compression with a high-speed impeller(s). Vapor is fed through the suction line into a housing near the center of the compressor. A disk with radial blades (impellers) spins rapidly in the housing.

Whereas the piston and the helical rotors are at the heart of the reciprocating and screw compressors, respectively, the impeller is at the heart of the centrifugal compressor. Figure 37-10 shows an impeller. The center, or eye, of the impeller is fitted with vanes that draw gas into radial passages, which are internal to the impeller body. The rotation of the impeller increases centrifugal force and moves the vapor to the outside of the impeller, where it slows down and expands into a diffuser tube, causing pressure to increase. Figure 37-11 shows an impeller, volute, and diffuser passage.

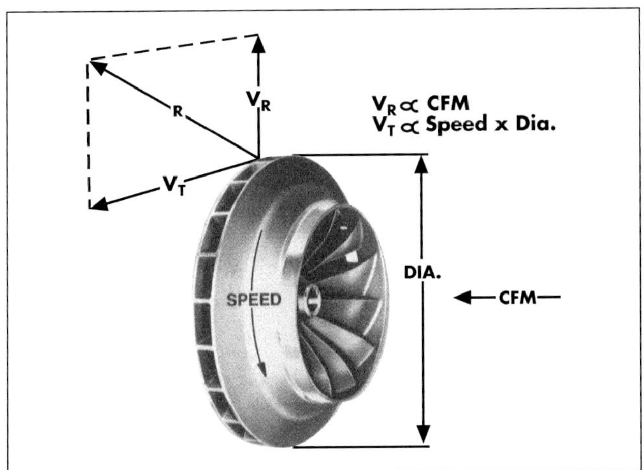

Fig. 37-10 Centrifugal Refrigeration Compressor Impeller. Source: The Trane Company

Figure 37-12 illustrates the refrigeration cycle for a chiller system featuring a centrifugal compressor. Notice the suction flow of low-pressure vapor into the eye of the impeller and the subsequent flow of high-pressure vapor outside of the impeller in the volute casing and down through the discharge baffle into the condenser. In this system, a subcooler is shown directly downstream of the primary condenser, and an oil cooler is shown directly upstream of the cooler.

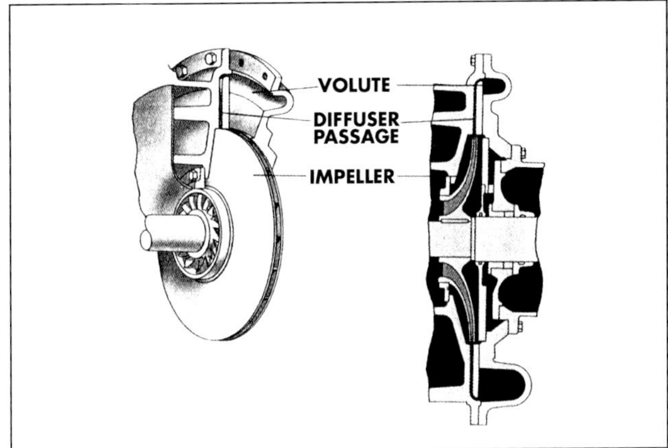

Fig. 37-11 Impeller, Volute, and Diffuser Passage in Centrifugal Compressor. Source: The Trane Company

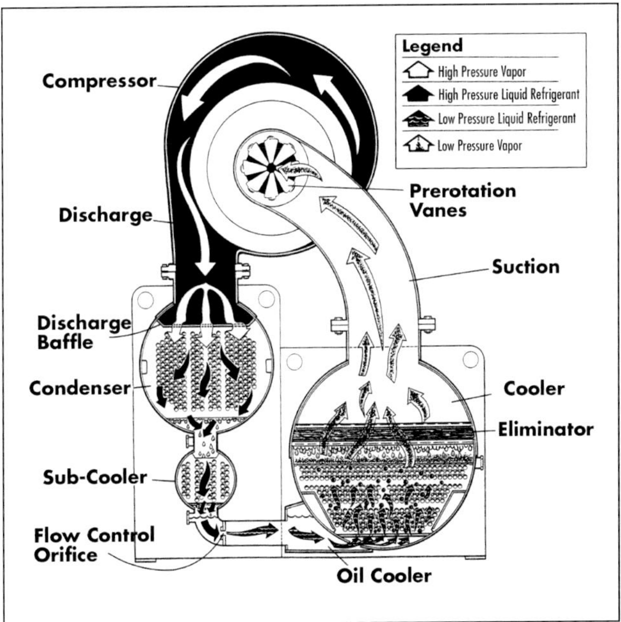

Fig. 37-12 Illustration of Refrigerant Flow in a Centrifugal Compressor-Based Chiller. Source: York International

The forces that act on the impeller can be broken down into two components: A radial velocity component, v_r, acts to move the gas away from the impeller in a radial direction. A tangential velocity component, v_t, acts to move the gas in the direction of impeller rotation. Both of these forces act to generate the resultant velocity vector, R, the length of which is proportional to the kinetic energy available for conversion to static pressure. For a given compressor, v_r is directly proportional to the mass flow of refrigerant handled, and v_t is proportional to the impeller diameter times the rotational speed. Thus, changes in speed or impeller diameter for a given mass flow rate will result in increased or decreased static pressure producing capability.

The process described above is called a stage of compression. A centrifugal compressor using one impeller is called a single-stage machine. Multi-stage units use a series of impellers and provide higher compression ratios because they increase vapor pressure with each stage. The limiting factors of compression are tip speed and the number of stages. Tip speed is a function of the impeller's size and rpm. Single-stage machines may be limited by their surge line and may have to run under hot gas bypass at loads less than 15 or 25% of full load. Most centrifugal compressors have internal oil pumps that are driven by the compressor shaft or by an internal motor.

Figure 37-13 is an open-case view of an open-drive, horizontally split casing centrifugal compressor revealing the impellers in this integral multi-stage compressor. On the far right is the mechanical contact seal for drive-line shaft connections. Figure 37-14 provides a cutaway view of this type of open-drive centrifugal compressor. Notice that the compressor shaft extends through the casing to facilitate connection to a number of different types of drives.

Figure 37-15 is a cutaway illustration of a three-stage centrifugal compressor of hermetic design. Figure 37-16 shows a large-capacity, multi-stage, open-drive centrifugal compressor. This unit has a cast-iron casing that is horizontally split for accessibility to internal components. This compressor can be designed with two to eight stages, depending on the intended application.

Many centrifugal compressors use lower-pressure refrigerants such as HCFC-123 (R-123), and the evaporator operates at below-atmospheric pressure. Although HCFC-123 is still in use, it is being phased out, and while

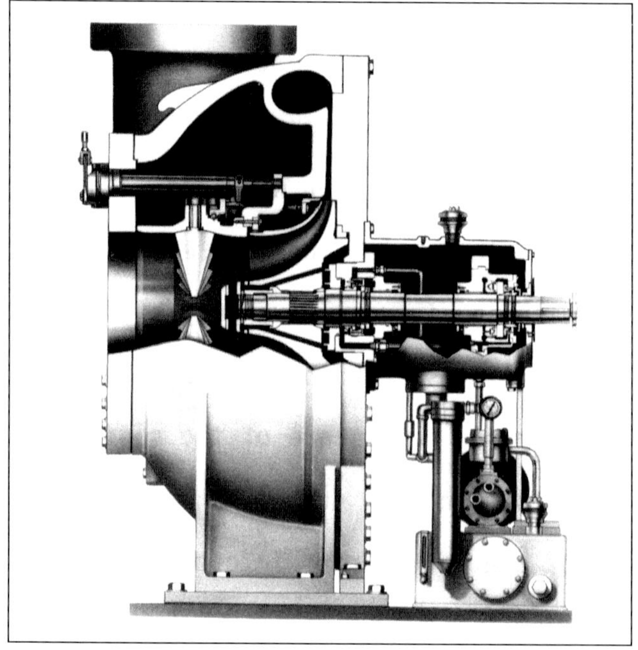

Fig. 37-14 Cutaway View of Open-Drive Centrifugal Compressor. Source: Carrier Corp.

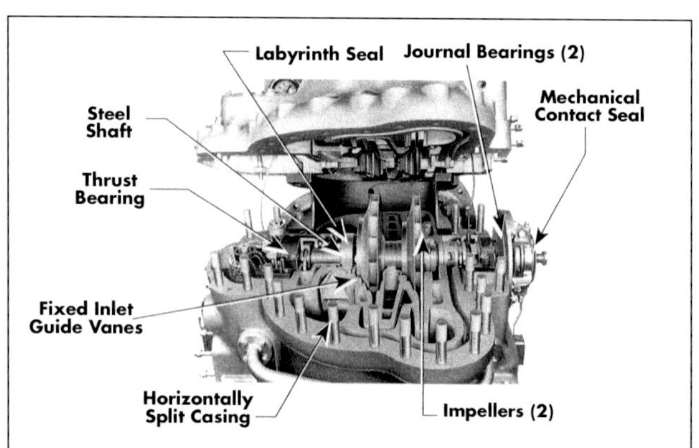

Fig. 37-13 Open-Case View of Open-Drive, Horizontally Split Casing Centrifugal Compressor. Source: Carrier Corp.

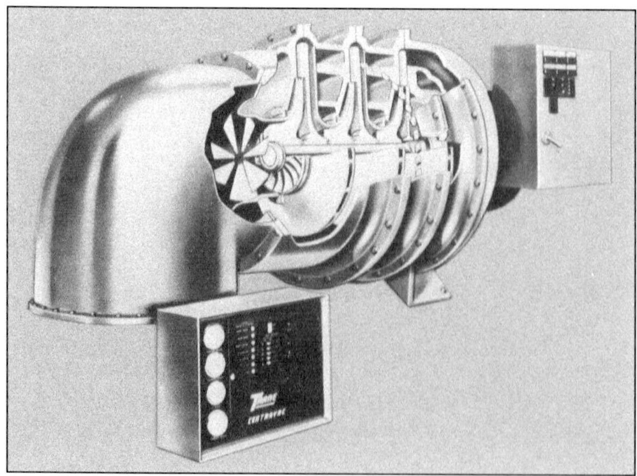

Fig. 37-15 Cutaway View of Three-Stage Centrifugal Compressor of Hermetic Design. Source: The Trane Company

Fig. 37-16 Large-Capacity, Open-Drive Multi-stage Centrifugal Compressor with Horizontally Split Cast-Iron Casing. Source: York International

there is currently no commercially available replacement, research continues to find alternatives. Larger units of several thousand tons and some of the newer smaller units are designed to operate with higher pressure refrigerants, such as HFC-134a and in some cases, HCFC-22 (R-22).

Negative- (low-) pressure systems operate under a vacuum, which draws leaks into the chiller during operation. Noncondensables, such as air and water, must be eliminated from the system by a purge unit. These noncondensables can cause problems, compromising chiller performance and causing unplanned shutdowns, if left untreated. Purge units regularly vent refrigerant to the atmosphere along with the noncondensables. These purge systems must be maintained, and current environmental regulations prohibit the direct venting of refrigerants. Also, negative-pressure chillers may require use of a vacuum prevention system that heats the refrigerant to increase the pressure and prevent air from entering the unit. This is intended to decrease the amount of purging required as a result of leaks into the system during shutdowns.

Compressor Capacity and Compression Ratio

The pressure change accomplished by the compressor, or compression ratio, of a compressor is a function of discharge and suction pressure. It may be expressed as a ratio of the absolute discharge pressure to the absolute suction pressure:

$$\text{Compression ratio} = \frac{\text{Discharge pressure}}{\text{Suction pressure}} \tag{37-1}$$

If, for example, the compressor discharge and suction pressures are 280 and 85 psia, respectively, the compression ratio would be:

$$\text{Compression ratio} = \frac{280\ psia}{85\ psia}$$

As suction pressure increases, compressor capacity increases. As discharge pressure increases, compressor capacity decreases somewhat. Power requirement increases with increasing compression ratio and with increasing discharge pressure for a given suction pressure. In a reciprocating compressor, for example, gas flows from the suction line into the space created by the downward stroke of the piston and is forced into the discharge line by the piston on the upward stroke. As the pressure in the suction line increases, the movement of the gas increases. Because of the higher pressure, the gas is more dense, and thus a greater mass of refrigerant can be compressed per unit of time. As the suction pressure decreases, the gas becomes less dense.

Excessively high compression ratios cause loss of energy and excessive superheating of discharge gas, which can result in compressor damage. For applications requiring high compression ratios, compression is, therefore, accomplished in stages. The discharge of the first compression stage becomes the suction of the next stage.

Suction and Condensing Temperature Effects

As discussed in Chapter 33, suction and condensing temperatures are critical factors that affect compressor capacity and system performance. As suction temperature is increased, the mass flow through the compressor increases because the specific volume decreases and, at constant condensing pressure, the compressor's volumetric efficiency increases. Thus, at constant condensing temperature, as the evaporator temperature increases, both compressor capacity and power requirements increase. However, the power requirement increases at a lower rate due to the lower pressure ratio that the compressor must develop. Figure 37-17 shows the variation of capacity and power requirements as a function of suction temperature for a reciprocating compressor. Notice that the refrigerating capacity increases rapidly as the evaporator temperature increases. In this example, the brake power of the compressor reaches a maximum when the evaporator temperature is 36°F (2°C) and decreases if the evaporator temperature is either higher or lower.

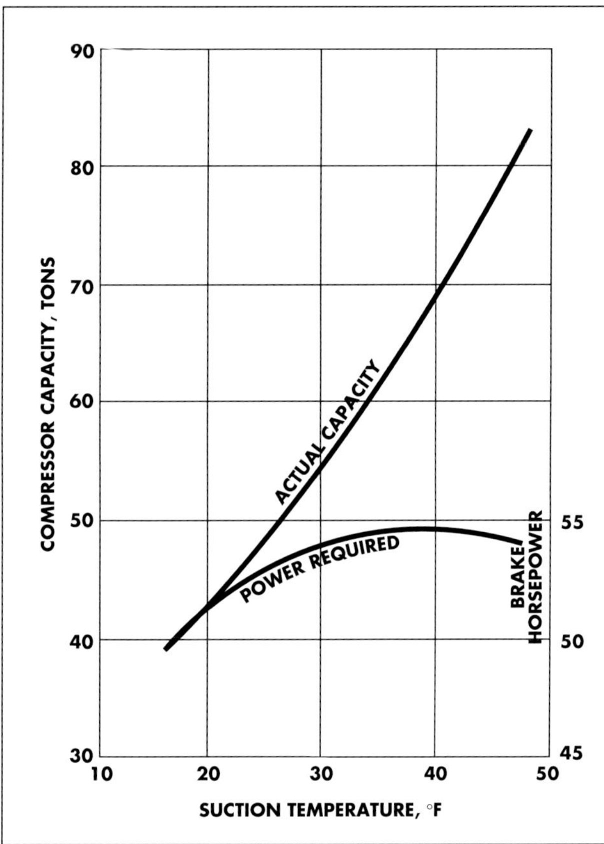

Fig. 37-17 Variation of Capacity and Brake Power with Suction Temperature. Source: The Trane Company

For any given evaporator temperature, the power required for a compressor will vary considerably with head pressure. A rise in condenser pressure produces a corresponding rise in condenser temperature, which then produces a corresponding rise in compressor power requirement.

REFRIGERANTS

A refrigerant is any fluid that, usually with a phase change, picks up heat at a low temperature and pressure and gives up heat at a higher temperature and pressure. There are numerous types of refrigerants used in vapor compression machines, each with a different set of characteristics. The type of refrigerant used will affect compressor capacity. Displacement of refrigerant in constant-speed, positive displacement compressors is constant. The capacity will vary, however, as a function of refrigerant density. Increased density at constant volume means greater mass flow and, therefore, greater capacity. It also means that more brake power is required to turn the compressor.

Refrigerant types, characteristics, and environmental factors are presented in Chapter 18. Factors to be considered in refrigerant selection for vapor compression systems include:

- Temperature-pressure effects (i.e., evaporating, condensing, and discharge temperature-pressure)
- Latent heat of vaporization
- Specific volume in vapor state
- Specific heat in liquid state
- Heat and power requirements
- Stability and life
- Safety ratings (i.e., toxicity, fire, and explosion tendency)
- Compatibility (i.e., non-corrosive, non-solvent)
- Leakage tendency and ease of detection
- Effect on climate change, or global warming potential
- Future availability and cost (e.g., phase-out of CFCs)

A critical vapor compressor system design factor is the refrigerant's temperature-pressure relationships. This dictates the minimal design working pressure for the system components. The difference between the intended condensing and suction temperatures affects the compressor size and drive power and should be considered in refrigerant selection.

There is a definite temperature-pressure relationship for all substances. Figure 37-18 is a simplified Mollier type diagram. Within the saturation region, there is an exclusive pressure value corresponding to each temperature. Outside this region, the pressure will vary depending on the quantity of superheat or subcooling involved. Figure 37-19 compares the temperature-pressure relationship at saturation for several refrigerants. Note that the apparent curvature changes result from the different scales used for pressure above and below standard atmospheric pressure of 29.92 in. mercury (Hg) column (101.325 kPa).

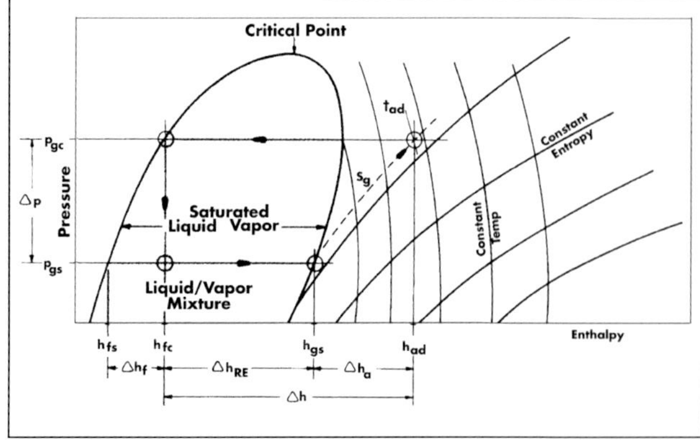

Fig. 37-18 Simplified Mollier Type Diagram. Source: Carrier Corp.

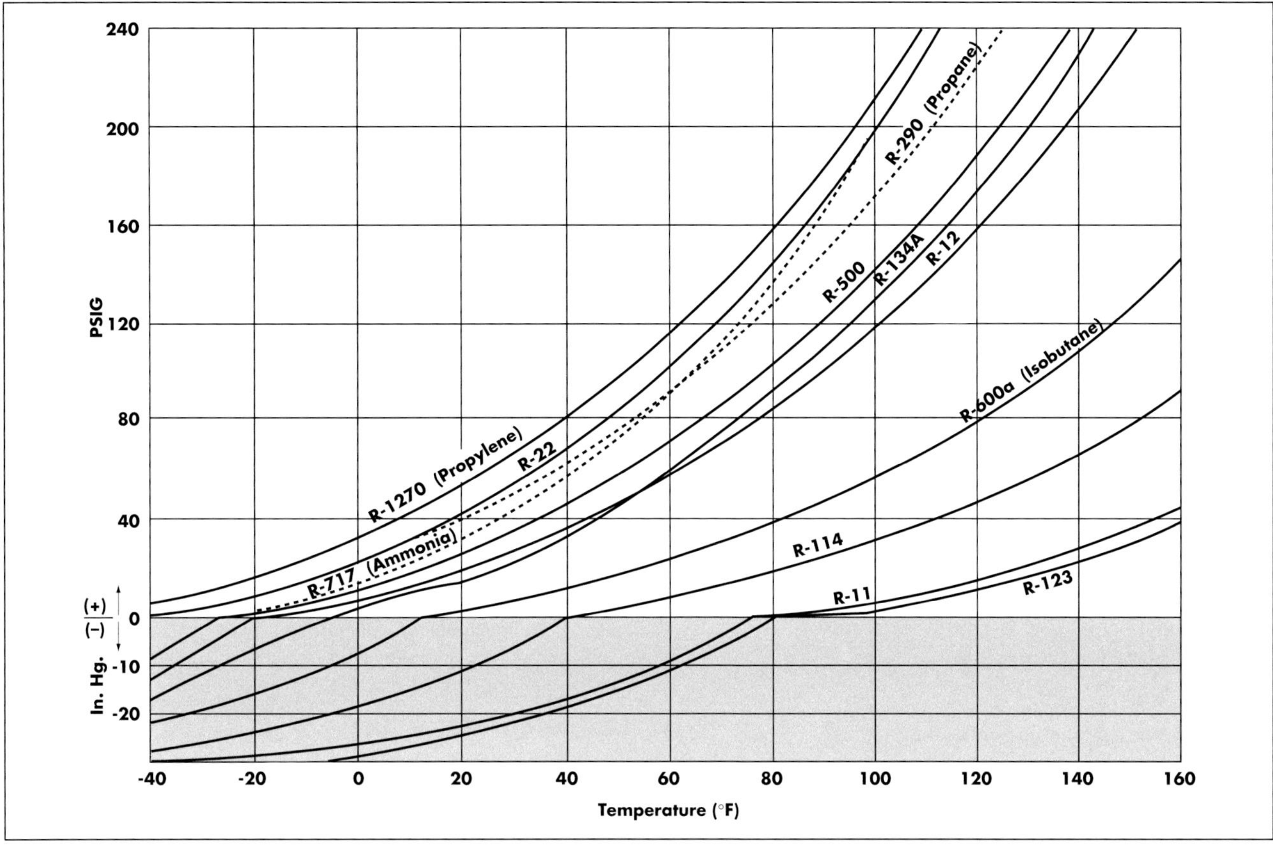

Fig. 37-19 Pressure-Temperature Relationships of Refrigerants. Source: Carrier Corp.

The amount of refrigeration effect produced by a refrigerant is directly related to its latent heat of vaporization (h_{fg}), which is the difference between the enthalpies of the saturated vapor and liquid conditions of a refrigerant at any given temperature. In theory, the larger the latent heat of vaporization, the greater the cooling effect. Table 37-2 provides a comparison of the latent heat of vaporization of representative refrigerants at various saturation temperatures.

Refrigerant Number	-20°F (-29°C)	0°F (-18°C)	20°F (-7°C)	40°F (4°C)
R-11	85.5 (55.2)	83.9 (54.2)	82.2 (53.1)	80.5 (52.1)
R-12	70.9 (45.8)	68.8 (44.4)	66.5 (43.0)	64.2 (41.5)
R-22	97.3 (62.9)	94.1 (60.8)	90.6 (58.5)	86.7 (56.0)
R-123	81.8 (52.8)	79.4 (51.3)	78.1 (50.4)	76.7 (49.5)
R-134a	93.6 (60.5)	90.7 (58.6)	87.5 (58.6)	84.0 (54.3)
R-717	583.6 (377.0)	586.9 (379.1)	553.1 (357.2)	536.2(346.4)

Table 37-2 Latent Heat of Vaporization at Various Saturation Temperatures in Btu/lbm (W-h/kg).

The specific volume of the refrigerant in its vapor state (V_g) will be a determinant of the volume of refrigerant that must be compressed. Therefore, lower specific volume is beneficial. Table 37-3 lists the specific volumes of saturated vapor for representative refrigerants at various temperatures.

Refrigerant Number	-20°F (-29°C)	0°F (-18°C)	20°F (-7°C)	40°F (4°C)
R-11	23.9 (1.492)	13.9 (0.878)	8.5 (0.531)	5.4 (0.337)
R-12	2.4 (0.150)	1.6 (0.100)	1.1 (0.069)	0.8 (0.050)
R-22	2.1 (0.131)	1.4 (0.087)	0.9 (0.056)	0.7 (0.044)
R-123	29.3 (18.9)	16.2 (10.5)	9.6 (6.2)	5.9 (3.8)
R-134a	3.5 (0.219)	2.2 (0.137)	1.4 (0.087)	1.0 (0.062)
R-717	14.7 (0.918)	9.1 (0.568)	5.9 (0.368)	4.0 (0.250)

Table 37-3 Specific Volume of Saturated Vapor at Various Temperatures in ft^3/lbm (m^3/kg).

Refrigerants R-11 and R-12, are being phased out, but are included in Tables 37-2 and 37-3 since there is still equipment in service that uses these refrigerants. R-123 (the "bridge" refrigerant, designed as a temporary replacement for R-11) is being phased out, but is still in production, so it is included as well. Refrigerant R-22 is also being phased out, in favor of refrigerants with lower climate change potential. One such refrigerant is Carbon Dioxide (CO_2). As noted in Chapter 18, this refrigerant has very low global warming potential, but also presents some challenges due to its operating conditions, compared with conventional refrigerants. It has five times the typical system operating pressure and a low critical temperature that requires cooling a supercritical fluid rather than condensing a two-phase mixture.

In practice, as the liquid refrigerant passing through the throttling device is cooled adiabatically to evaporator temperature, a portion is evaporated, or converted into "flash gas." This produces a reduction in the change of fluid enthalpy (Δh_f) in the latent heat of evaporation of the refrigerant. This can be calculated as:

$$\Delta h_f = (C_{pf})\,(T_c - T_s) \qquad (37\text{-}2)$$

Where:
C_{pf} = Specific heat of refrigerant in its liquid state
T_c = Condensing temperature
T_s = Evaporator temperature

Thus, the actual latent heat of evaporation available for cooling duty is lower than the ideal heat of vaporization. Therefore, it is beneficial for a refrigerant to have a low specific heat in its liquid state to minimize flash gas and, thus, the flow rate of refrigerant required for cooling duty.

The net refrigeration effect (Δh_{RE}), in Btu/lbm (kWh/kg), can be calculated as:

$$\Delta h_{RE} = \Delta h_{fgs} - \Delta h_f \qquad (37\text{-}3)$$

Where:
Δh_{fgs} = Δh_{fg} at the evaporator temperature (T_s)

Economizer Cycles

The primary method of improving vapor compression system cycle performance is increasing the refrigeration effect per unit mass of refrigerant by cooling the temperature of the liquid refrigerant prior to expansion to the primary evaporator. The colder the liquid refrigerant's temperature prior to flashing, the lower the mass flow and power required. Three methods for accomplishing this are: subcooling with a thermal economizer, use of an open economizer, or use of a closed economizer.

A thermal economizer, or sub-cooler, is a circuit in which some of the tubes in the first water pass of the condenser are below the refrigerant liquid level in the condenser. The refrigerant liquid level is maintained to cover the subcooler tubes in order to increase the cooling of liquid refrigerant. By reducing the required total mass flow of refrigerant through the system, brake power is reduced and cycle efficiency increased. In low-temperature applications, subcoolers are not used due to the risk of freezing water in the subcooler tubes in the event of a pump failure.

Open and closed economizers are commonly used for improving cycle efficiency in process refrigeration systems. An open, or flash, economizer is a tank that is vented to the inlet of one of the compression interstages. Liquid refrigerant is cooled prior to entering the cooler as some refrigerant is allowed to flash, cooling the remaining liquid to the saturation temperature corresponding to the inlet pressure of the compressor interstage. The closed economizer takes the liquid from the condenser and splits the flow into two streams. Most of the flow is directed through the tubes of a shell-and-tube heat exchanger, while the remainder goes to the shell side through a control value and is flashed to cool the refrigerant in the tubes. The gas generated is vented to the inlet of one of the compressor stages.

Compressor Part-Load Operation

Refrigeration systems must be sized for peak design conditions, even if these conditions occur only several hours in a year. Most of the time, conditions are in the middle regions of the cooling load curve. There are several factors that affect performance under varying loads, including driver efficiency, compressor volumetric efficiency, ambient temperature, the relationship between heat exchange surface area and load, and the characteristics of the refrigerant.

To a certain extent, compressors automatically reduce capacity as load decreases. This is commonly referred to as "riding with the load." While a compressor operating at constant speed without constricting refrigeration flow maintains constant volume flow, mass flow rate varies with operating conditions. It is the mass flow rate that determines the capacity to transfer heat. Mass flow, which is a function of volumetric flow and specific heat, can be expressed as:

$$\text{Mass flow rate} = \frac{\text{Volumetric flow rate}}{\text{Specific volume}} \qquad (37\text{-}4)$$

As loads decrease, less heat is absorbed into the evaporator, so less refrigeration is vaporized. The suction pressure drops, the saturation temperature of the refrigeration is reduced, and the specific volume increases. Since the volumetric flow is constant, the mass flow decreases, meaning less refrigeration mass is being pumped per unit of time. As a result, capacity decreases. As load is increased, the process is reversed and mass flow and, therefore, capacity, increases.

As suction pressure and temperature decrease, partial freezing of the moisture that collects on a refrigerant-to-air evaporator coil may result. The frost decreases the amount of air that passes through the coil, which lowers the suction pressure and temperature even further. This blocks airflow and can eventually damage the compressor.

In order to control operation at lower loads, compressors are designed to "unload" under partial-load conditions. Each compressor design has a different technique for unloading or reducing power draw as loads decrease. Being positive displacement machines, as cooling load requirements decrease, reciprocating and screw compressors reduce displacement. Centrifugal compressors, which are dynamic machines, reduce output by reducing the effect of centrifugal force.

As load decreases, heat dissipation requirements fall and suction pressure and temperature are usually lower than they are under fully loaded conditions. However, because the heat exchange surface area is constant, the heat exchanger is effectively oversized and, therefore, more efficient under partial load. If load falls as a result of decreased outside air temperature, condenser water temperature also falls. Hence, refrigeration cycle efficiency is improved. Counteracting these improvements is usually a reduction in compressor isentropic efficiency.

Typically, in the upper third of the load curve at constant speed, power requirement decreases and cycle efficiency increases as loads decrease. In this load range, the negative impact of compressor volumetric efficiency loss is usually more than compensated for by other efficiency gains. As loads fall further, system performance begins to decrease as the effect of degrading compressor volumetric efficiency begins to outweigh other system efficiency enhancement factors. In the bottom third of the load curve, system performance begins to fall off sharply. While decreasing ambient temperatures (assuming a relationship between load and ambient temperature exists in the particular application) continue to partially off-set compressor volumetric efficiency degradation under low loads, most systems have limitations as to the minimum acceptable condenser temperature.

Typically, as loads fall below 15% of full load, the power requirement rate may be as much as double the full-load design rating, even at reduced ambient temperature conditions. In this operating range, driver thermal efficiency degradation also becomes a contributing factor. For this reason, systems designed for frequent operation in lower load ranges may be designed with multiple compressors to reduce system performance degradation under low-load operation.

Reciprocating Compressor Capacity Control

Capacity is controlled in reciprocating compressors through cycling, cylinder unloading, speed control, or a combination of these techniques. Small reciprocating compressors often simply cycle on and off to meet varying load requirements — an inefficient process. Larger compressors, starting at 10 or 20 tons (35 to 70 kW_r), typically use cylinder unloading devices to reduce total displacement or compression action. Cylinder unloading helps to control the compressor pumping capacity as evaporator loads decrease by deactivating cylinders until only one, two, or three cylinders remain operative.

Cylinder unloading methods incorporate unloader assemblies, which are controlled directly by electronic controllers or by hydraulic devices operated by pressure or by electrically actuated valves attached to the cylinders. Figure 37-20 illustrates a pressure-actuated unloader. A bellows assembly senses suction pressure from within the compressor crankcase and operates a valving mechanism. As suction pressure in the crankcase

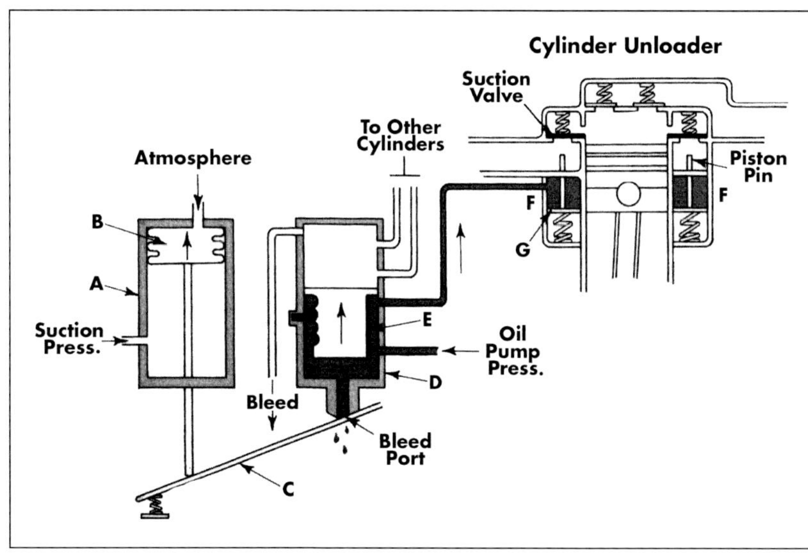

Fig. 37-20 Pressure-Actuated Unloader for Reciprocating Compressor.
Source: The Trane Company

rises, thus indicating an increasing load, pressurized oil is fed into the valving mechanism, causing the cylinder to become loaded. If a decreasing suction pressure is sensed, oil is bled from the unloader through the bleed port and back into the crankcase. When the oil pressure in the cylinder unloader is relieved, springs cause the piston to move upward and raise the suction valve off of its seat, thus preventing any gas from being pumped through the cylinder.

Hermetic motors use a valve to close the suction passage of the cylinders that are to be unloaded. Typically, there are one-to-four stepped unloading stages ranging down to 25% of full capacity. If a compressor is required to run at a system load less than the capacity that would be produced at the minimum stage of compressor unloading, the entire compressor must be cycled on and off. For increased load control, as well as system reliability, multiple smaller units are often packaged together for parallel operation. Compressor displacement may also be changed by adjusting its rotation speed. Reducing speed reduces displacement and, therefore, capacity and power requirement. Speed control is discussed below.

Screw Compressor Capacity Control

Capacity is controlled in screw compressors through variable compressor displacement modulation. This is accomplished by varying the speed, by using a moveable slide valve mounted near the suction area of the compressor casing, or both. With slide valve control, at part-load, the slide valve produces a slot that delays the point at which the suction entraps the vapor and compression begins. This causes a reduction in swept volume, and thus, compressor throughput. Units are also available with stepped unloading.

Typically, a capacity controller will monito chilled water temperature. Under decreasing load (i.e., increased return water temperature), a valve, actuated by a hydraulic or gas piston and cylinder assembly located in the compressor or by a positioning motor, will reduce the working length of the rotors and provide stepless capacity modulation down to about 10% of full load. While somewhat less efficient at full load than comparable capacity centrifugal compressors, their positive displacement characteristics allow screw compressors to unload more efficiently than centrifugal compressors, and, therefore, operate with relatively superior part-load performance. As with reciprocating compressors, reducing speed reduces displacement and, therefore, capacity and power requirement.

Centrifugal Compressor Capacity Control

Capacity can be controlled in centrifugal compressors by varying speed, bypassing hot gas, use of variable inlet guide vanes, or a combination of all of these techniques. The primary means of control for part-load operation of centrifugal compressors is variable inlet guide vane operation. Inlet guide vanes are mounted on the compressor casing just upstream of the impeller inlet and may be electrically, pneumatically, or hydraulically operated. Referring to the diagram of the impeller and volute casing in Figure 37-21, inlet vanes are located in front of the impeller at the suction vapor inlet.

The inlet, or pre-rotation, vanes direct the refrigerant vapor into the impeller. During start-up, the vanes are closed to reduce starting load. At the full-load condition, the vanes are wide open and parallel to the refrigerant flow. As loads decrease, a controller, which typically monitors leaving chilled water temperature, begins to close the vanes. This allows the vanes to adjust both the refrigerant gas quantity and the angle at which it enters the impeller. By adjusting the angle, the refrigerant fluid is spun in the same direction as the impeller, and the relative velocity differential of the fluid as it enters the eye of the impeller is reduced. With each new vane position, a new compressor performance characteristic is created to help stabilize the system under varying loads. Figure 37-22 shows pre-rotation vanes that are mounted on the suction end of a large-capacity, multi-stage centrifugal compressor. Pneumatic controls are used to automatically position the vanes to maintain the desired chilled water temperature.

As with other types of compressors, partial load operation of centrifugal vapor compressors is limited by the surge (or pulsation) point that occurs at reduced volume flow rates. **Surging** is a condition in which the machine rapidly alternates between shut-off and a near

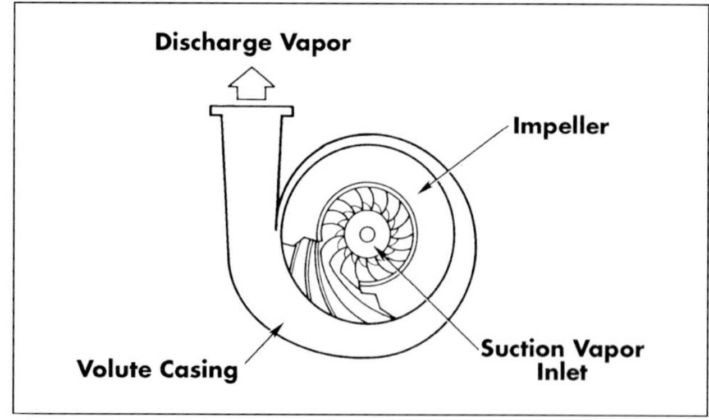

Fig. 37-21 Diagram of Impeller and Volute Casing Showing Suction Vapor Inlet. Source: Carrier Corp.

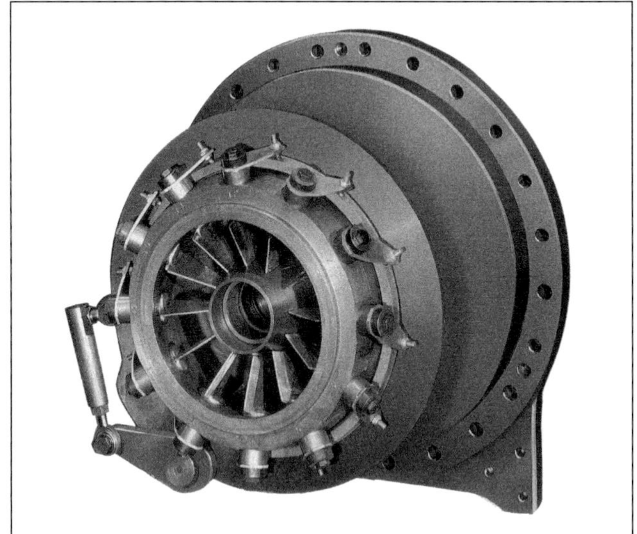

Fig. 37-22 Pre-rotation Vanes Used with Large-Capacity, Multi-stage Centrifugal Compressor. Source: York International

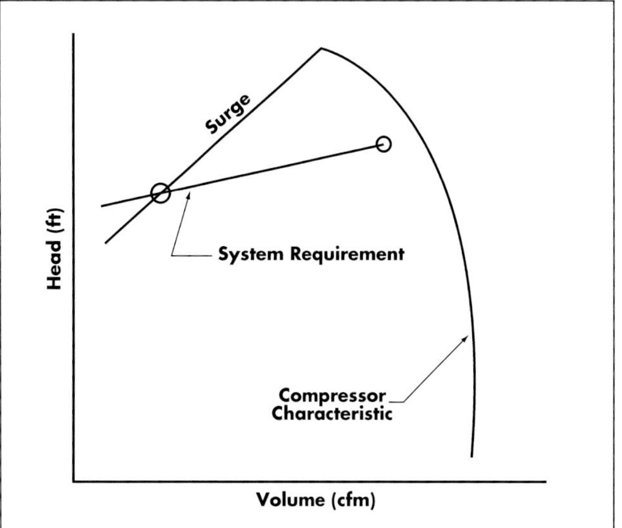

Fig. 37-23 Example Surge Line for Centrifugal Compressor. Source: Carrier Corp.

maximum flow rate. Capacities below the surge point are said to be in the unstable range.

Surges occur as load decreases beyond a given point, at which the pressure in the system is higher than at the compressor outlet. This causes flow to stop momentarily. Refrigerant flow continues to leave the casing of the compressor and system pressure decreases to a point at which the compressor again begins to deliver refrigerant vapor. However, the discharge pressure does not build up fast enough to deliver the near full-load volume of refrigerant. The surge is thus characterized by a rapid decrease in flow rate and is marked by a change in noise level of the compressor. This cycle repeats until the load increases sufficiently for the compressor to exit the surge region or until the head decreases.

Figure 37-23 shows the intersection of the surge and system requirement line for a given compressor characteristic, expressed in feet of head versus volume. To avoid surging under a given set of operating conditions, a certain minimum volume flow through the impeller must be maintained. If there is no partial load control, the surge point is typically about one-half to two-thirds of full-load capacity. In controlling centrifugal compressors under part load, the focus is on widening the stable operating range of the impeller as efficiently as possible.

Figure 37-24 provides representative performance characteristic curves for a single-stage centrifugal compressor operating at constant speed over a range of inlet vane positions. The pressure differential between the inlet (evaporator) and outlet (condenser) of the compressor are

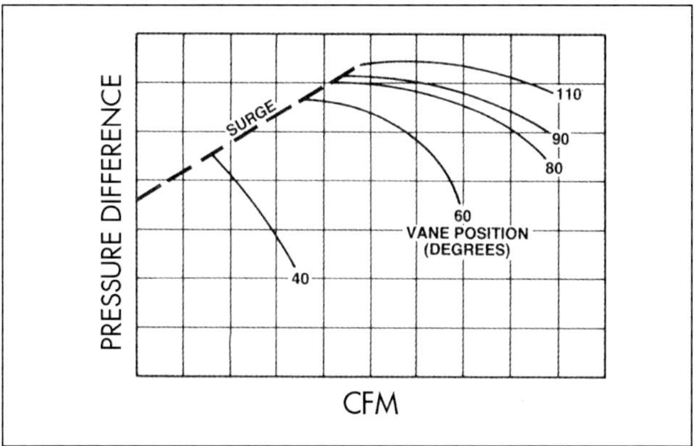

Fig. 37-24 Representative Single-Stage Centrifugal Compressor Performance Characteristics Curves Over a Range of Inlet Vane Positions. Source: The Trane Company

on the vertical axis and the compressor refrigerant flow rate capability at various pressure differences and vane positions are on the horizontal axis. The surge line represents the flow ratc and pressure difference conditions that produce surge. Any compressor operating point that falls to the right of this line is satisfactory from a stability standpoint.

The two principal concerns for operation at low load are compressor stability and volumetric efficiency. As the refrigerant flow is reduced in response to falling load, the inlet vanes are modulated. This stabilizes the evaporator pressure and temperature by balancing the compressor pumping capacity with the new gas flow rate. As the system load decreases, the reduction in refrigerant flow reduces the amount of heat transferred to the condenser.

This increases the heat transfer capability of the condenser relative to the load, meaning it is capable of condensing the refrigerant at a lower temperature and pressure. This reduces the difference between the stabilized evaporator pressure and the new condensing pressure, moving the operating point. With effective modulation, compressors can typically operate down to a minimum of 10 or 20% of design load, depending on design and operating factors. Below this level, a low-temperature control will stop the compressor and then restart it when a rise in temperature indicates the need for additional cooling. This process is highly undesirable, particularly for large machines. Typically, performance starts to fall off rapidly below 30 to 40% of full load. To extend the stable operating range, a false load could be put on the system through hot gas bypass. Alternatively, a design featuring multiple impellers or speed control could also be used to extend the operating range and improve compressor volumetric efficiency.

Hot Gas Bypass

Hot gas bypass is a technique to eliminate problems with systems that must be operated at evaporator loads below the minimum stage of compressor unloading. Hot gas bypass adds heat load to the system to stabilize suction pressure. It is used to eliminate surging and frost build-up on evaporator coils. Under very low loads, the relatively small mass flow of refrigerant, produced by the less than minimum acceptable load, can cause a gas-cooled motor to heat up and the suction and discharge line gas velocities to fall to a point at which they cannot move the lubricating oil to and from the compressor at a uniform rate.

With hot gas bypass, some of the hot refrigerant gas is diverted from the discharge line to the low-pressure side of the system. Instead of going to the condenser to release heat, the hot refrigerant gas returns uncooled to the system, thus adding heat load and producing a stable suction pressure and temperature when the actual system load is less than the minimum stage of compressor unloading. The passage of hot gas is regulated by a pressure-actuated or electronic modulating valve, which operates when the suction pressure falls to the valve setting. These valves are typically sized to bypass a heat load that is roughly equivalent to the minimum capacity of the compressor and can, therefore, maintain compressor stability down to actual cooling loads of almost zero.

The obvious disadvantage of hot gas bypass is that it is inherently inefficient, or even counter-efficient. Because the hot gas puts a false load on the machine, the compressor must do added work, but no real refrigeration benefits are provided. Hot gas bypass is, however, often an important feature because it eliminates cycling, which results in efficiency losses and accelerated system wear, and protects the compressors from damage under low loads.

Speed Control

All methods that reduce refrigerant flow reduce the amount of compressor work done under low loads. However, compressor volumetric efficiency will decrease as output goes down. Reducing the speed (rpm) of the driver and compressor allows the compressor to operate at part load more efficiently with a lower degree of reliance on compressor unloading techniques. Most applications that use speed control have prime movers as drivers, although variable frequency drive (VFD) electric motors are also used and are becoming increasingly popular, even on smaller capacity compressor motors.

Without using other compressor unloading techniques, capacity control can be achieved by reducing speed. Once the driver reaches the minimum (idle) speed level, for stable operation of the driver and/or compressor, compressor unloading is required to further reduce capacity. Once unloading is required, compressor isentropic efficiency begins to decrease. However, this occurs at a lower load level than would occur under constant speed operation. Therefore, overall part-load efficiency is greater with speed control. Under very low loads (bottom 20% of the load curve), all compressors (electric- or prime mover-driven) experience a significant reduction in volumetric efficiency. Reducing speed in this range can reduce compressor volumetric efficiency degradation to some extent. However, under these low load conditions, the driver itself will begin to experience a fall off in thermal efficiency as well. This is often far more pronounced with a prime mover than an electric motor.

The amount that speed can be reduced depends on both the characteristics of the compressor and the driver. Generally, the minimum speed is 50 to 70% of the full load design speed with centrifugal compressors. The minimum speed is 40 or 50% of the full load design speed for screw compressors and somewhat lower with reciprocating compressors.

Speed control with positive displacement compressors is a relatively simple process. As speed is reduced, compressor volumetric efficiency remains fairly constant and the refrigeration effect improves as a result of improved condenser operation and other related factors.

Speed control is more limited with centrifugal machines than with positive displacement machines, but it is still effective. The inherent characteristics of the compressor are that head drops with the square of the speed. Speed must, therefore, be carefully controlled with centrifugal compressors to avoid surging. In some cases, as load decreases, speed must actually be increased to avoid surging. Typically, speed control is used down to about one-half or two-thirds capacity. In order to optimize overall system efficiency, a complex control sequence is often required.

Figure 37-25 shows a centrifugal compressor's response to load change by using speed control. Consider this example in which the vertical line on the right represents 100% capacity, and the compressor is selected to operate at 120% of full load design speed. At Point A, the head output matches the head required. At Point B, the compressor requires 120% of the nominal power. As the load drops off to the vertical line on the left, the compressor speed is reduced. In this case, reducing speed will allow the compressor to match the reduced head and flow requirements of the heat exchangers down to about 40% of full load. Below 40%, the head required by the heat exchanger performance is represented by Point D, while the compressor is able to only produce head equivalent to Point C. This disparity will force the compressor into surge. To prevent surge, the compressor must operate at a load corresponding to Point E. This may be accomplished by using hot gas bypass and increasing the speed to 80% of full load design speed. Under this condition, the cooler handles the load corresponding to the vertical line at the left, while hot gas is passed from the condenser to the bottom of the cooler to make up the difference in flow from Points D and E.

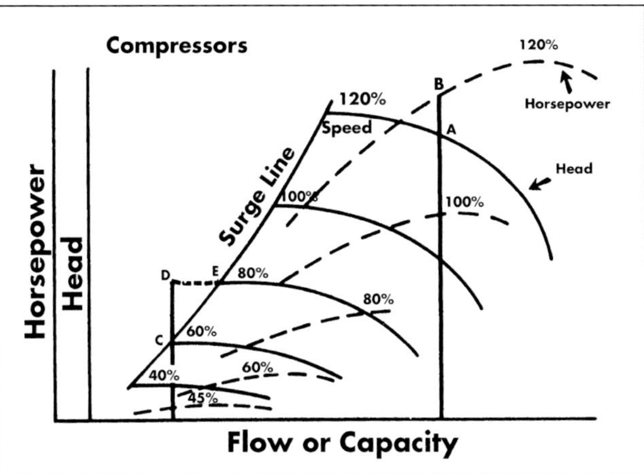

Fig. 37-25 Speed Control Effect on Centrifugal Compressor's Response to Load Change. Source: Carrier Corp.

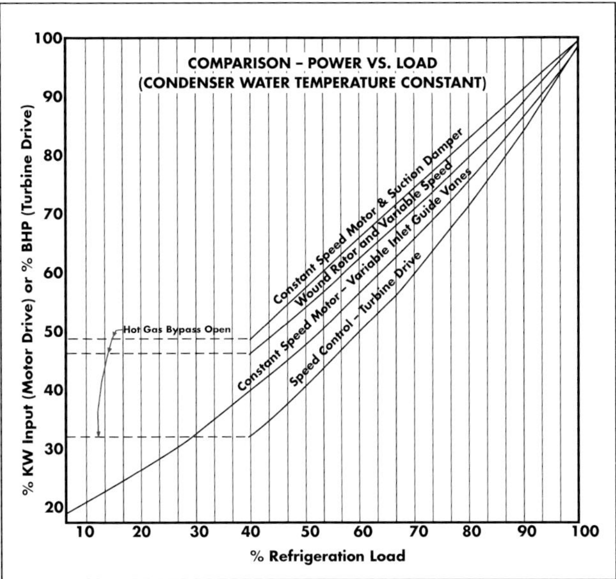

Fig. 37-26 Relative Performance of Various Centrifugal Compressor Control Methods. Source: Carrier Corp.

Speed control is a valuable efficiency tool, whether done simply or with complex integration of control functions. A computer module can be used to collect system operating conditions and process the information to optimize control of the driver speed and the compressor unloading. For each system, a control sequence must be developed to optimize performance under the entire range of operation. In either case, available microprocessor control technology allows for optimization of any type of system using speed control. Refer to Chapter 29 for additional detail on mechanical equipment system drivers and speed control.

Figure 37-26 shows the relative performance of various part-load control methods for a centrifugal compressor system. Performance is expressed as a percentage of full-load power versus a percentage of full-load refrigeration load. Power requirements are shown in kW_e for several electric motor-driven compressor options and in bhp for a steam turbine-driven compressor option. Among the options shown, speed control using a steam turbine shows the best performance down to about 40% of full load. Next to that is constant speed operation with inlet guide vanes. This methods offers control down to very light loads without requiring hot gas bypass for stable operation.

Metering (Throttling) Devices

Another critical component of a vapor compression system is the throttling (or expansion or metering) device. Just as the condenser and evaporator are counterparts, the throttling device is the counterpart of the compressor. The compressor and throttling device prepare the refrigerant for the phase change by manipulating the pressure and flow rate, while the evaporator and condenser actually cause the refrigerant to change phase through heat transfer processes. The throttling device is located at the inlet of the evaporator. Its purpose is to reduce the pressure of the refrigerant from condenser exit pressure to the evaporator inlet pressure, while regulating the mass flow of refrigerant.

The types of metering devices most commonly used are automatic expansion valves, thermostatic expansion valves, electronic expansion valves, capillary tubes, float valves, and fixed orifices. These are discussed below.

An **automatic expansion valve** is a pressure-actuated diaphragm needle valve used to maintain a constant pressure in the evaporator of a DX system. Typically, it is only used in small constant-load applications. Evaporator pressure acts on the lower side of the diaphragm and atmospheric and spring pressure act on the upper side. As evaporator pressure drops, indicating more cooling is required, the combined pressure of the atmosphere and the spring push the diaphragm down, opening the needle valve wider to allow more refrigerant flow to the evaporator from the liquid line. The result is increased evaporator pressure, which then pushes the diaphragm up to restrict refrigerant flow.

Thermostatic and electronic expansion valves are most commonly used in DX systems in capacities of up to 400 tons (1,400 kW_r). A **thermostatic expansion valve** (TXV), which is the most commonly used, regulates the amount of refrigerant entering the evaporator and maintains the refrigerant vapor leaving the evaporator at fairly constant temperature. Its design is similar to an automatic expansion valve, except that a temperature-sensing bulb responding to the load on the evaporator replaces the pressure of the spring and atmosphere.

The TXV meters refrigerant into the evaporator by measuring the condition of the vapor at the leaving side of the evaporator. Figure 37-27 illustrates expansion valve operation, based on measurement of the superheat of the refrigerant vapor in the evaporator suction line. Superheat is determined by comparing the suction line refrigerant pressure and temperature. A rise in evaporator temperature expands the fluid from the bulb through a capillary tube. This forces the diaphragm down. In the example shown, liquid enters the expansion valve at a condensing pressure of 196 psig (14.5 bar) and, with 5°F (3°C) subcooling, a temperature of 95°F (35°C). Liquid pressure drops to 69 psig (5.8 bar) immediately upon passage through the valve. The pressure reduction causes the liquid to boil, or flash. This heat absorbing process sensibly cools the liquid to the saturation temperature of 40°F (4°C), which corresponds to 69 psig (5.8 bar). The greatest limitation of a TXV is that its control can be unstable at low loads or under conditions of low-pressure differential across the valve.

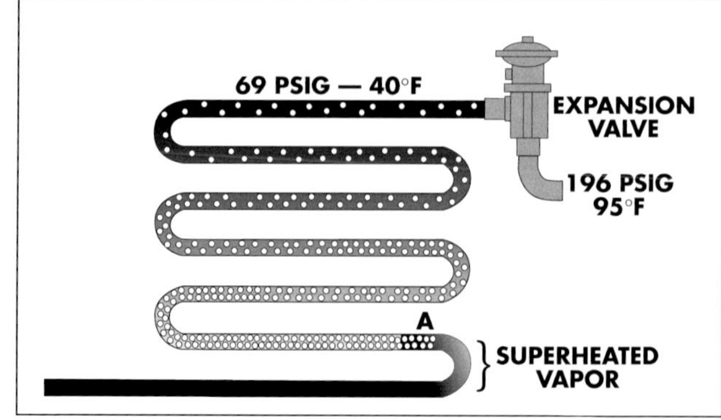

Fig. 37-27 Illustration of Expansion Valve Operation. Source: The Trane Company

An **electronic expansion valve** (EXV) performs the same function as a TXV, except it is operated with a microprocessor-based controller. The controller can be dedicated to the EXV or integrated into the microprocessor controlling the entire chiller operation. When integrated, EXV operation can improve low load and overall efficiency, based on the operating conditions of all chiller components.

A **capillary tube**, which is typically used in household refrigerators and small window air conditioning units, is simply tubing whose length and bore are precisely designed to control refrigerant flow to the evaporator. Pressure reduction between the condenser and the evaporator results from the friction loss, or pressure drop, in the long, small-diameter passage. One or more tubes are required, depending on the evaporator design. The use of capillary tubes is largely limited to pre-assembled units because the bore diameter and length are critical to its efficiency.

A **float valve** is used with flooded evaporators. Its assembly is a variable orifice device located in a chamber connected to the evaporator. It is actuated by a float immersed in a liquid container. A low-side float valve is operated by low-side level. It opens at a low level and closes as the liquid refrigerant, under pressure, forces the float to rise. A high-side float valve is operated by high-side level. The valve opens when the liquid level increases to the liquid level in the float chamber and admits liquid to the low side. Float valves maintain the evaporator refrigerant over a wide range of operating conditions and are commonly used in ammonia systems in industrial refrigeration applications and with large flooded evaporators. Typically, these are applied in chilled water systems with capacities in excess of 150 tons (500 kW_r).

Fixed orifice type metering devices are typically fixed hole plates, or a hole in a series of plates, in the liquid refrigerant line entering the evaporator. The amount of liquid refrigerant passing through the device is based on the level of refrigerant in the condenser line and the pressure differential between the evaporator and the condenser. Fixed orifices have no moving parts and are, therefore, less subject to wear than float valves. However, part-load performance is not as good. Similar to float valves, fixed orifices are typically applied in chilled water systems with capacities in excess of 150 tons (500 kW_r).

Vapor Compression System Drivers

Vapor compression system drivers, which produce shaft rotational power, are essentially the same as those used for air compressors and other mechanical drive services applications. As with air compressors, the vapor compressor is indifferent to the type of driver, except with respect to the operating speed. Its brake power requirement is matched by the torque and rotational speed at the output coupling of the drive shaft. The energy requirement of the driver will be a function of the compressor's brake power requirement and the performance of the driver in providing that power to the compressor at the particular operating speed, plus any gearing efficiency losses.

While there is some component variation, vapor compression systems are quite similar, regardless of the driver. This is illustrated by Figure 37-28, which shows two factory-assembled large-capacity compressor/driver packages. On the bottom is an electric motor-driven unit and on the top is a steam turbine-driven unit. Between the electric motor and the compressor is a speed increaser gear. With this particular unit, no gearing is required for the steam turbine driver.

Following is a comparative discussion of commonly used drivers for vapor compression systems — electric motors, reciprocating engines, gas turbines, and steam turbines. Given the commonality with other mechanical drive applications, the focus here is on driver characteristics as they are specifically applied to vapor compression systems. A review of Chapter 29, which provides greater detail on mechanical drive service for each driver type, is, therefore, recommended prior to proceeding with the following section. For additional detail on driver performance expressions, refer to Chapters 2 and 33. For additional detail on each of the prime mover types, refer to the applicable chapters in Section III.

Fig. 37-28 Comparison of Factory-Assembled Centrifugal Compressor Packages — Steam Turbine Drive on Top and Electric Motor Drive on Bottom. Source: York International

Electric Motor-Driven Systems

Electric motors are by far the most common drivers selected for vapor compression systems. They typically are the lowest first-cost driver option. They are also usually the smallest and easiest to install and offer simplicity of operation and minimal maintenance requirements.

Electric motors are abundantly available in capacities ranging from fractional hp (kW) to very large capacities, either as part of pre-engineered, pre-packaged vapor compression systems, or built-up, application-specific packaged systems. As with most mechanical drive service applications, ac induction motors predominate, particularly in smaller capacities. Synchronous motors, which offer the collateral benefit of power factor correction, are also sometimes used in larger applications. Both hermetic and open-drive motors are commonly applied, though hermetic units are more common, particularly in smaller capacities.

Most motors applied to vapor compression system drive service are operated at constant speed. However, variable frequency drives (VFDs) are applied to some systems (typically larger capacity systems, although they are increasingly being applied to smaller capacity equipment) to operate under variable speed capacity control with improved part load performance.

Performance

Typically, the full load power requirements for an electric motor-driven chiller system operating under standard ARI conditions range from 1.1 kW/ton (0.31 kW_e/kW_r) for small, air-cooled reciprocating compressor-based chiller systems to about 0.50 kW/ton (0.15 kW_e/kW_r) for large high-efficiency water-cooled centrifugal compressor-based systems. Small air-cooled DX units typically range from 1.6 to 1.0 kW/ton (0.46 to 0.28 kW_e/kW_r). Caution should be used when interpreting chiller performance ratings from manufacturers at ARI standard conditions. The rating system allows a considerable testing tolerance on chilled water temperature and condenser water temperature, and these variables can significantly affect the rating.

Full-load and part-load efficiencies have gradually improved throughout all system types and capacity ranges over the past two decades, resulting from modest improvements in motor efficiency, compressor efficiency, and controls and the aggressive use of increased heat exchanger surface areas and improved materials.

Application of premium-efficiency motors has become increasing common for vapor compression drive service. At a driver cost-premium of 20 to 30%, they offer typical efficiency advantages of 3 to 5% versus standard-efficiency models. Typical induction motor efficiencies range from about 70% for the smallest capacities to 96% for the largest capacities, with full-load power factors ranging from 70 to 90%, also increasing with respect to capacity. Synchronous motors can achieve efficiencies in excess of 97% in large capacities and can operate over a wide range of power factors, including leading power factors.

While adding to capital costs, speed control through application of VFDs provides significant efficiency benefits over much of the part-load operating range, as indicated by the previous discussion of speed control. Combined with lower condenser water temperatures, power requirements can reach as low as 0.35 kW/ton (0.10 kW_e/kW_r) under certain part-load conditions. VFDs do add a slight efficiency penalty at full-load operation as compared with standard constant speed motors, due to the power requirement of the VFD itself.

Application Examples

Figure 37-29 shows a chiller package with a hermetic motor-driven multi-stage centrifugal compressor. Figure 37-30 shows a large capacity packaged chiller with an open motor-driven centrifugal compressor. The motor drive is open drip-proof, squirrel cage induction type. The 60 Hz version operates at 3,570 rpm and the 50 Hz version at 2,975 rpm. This unit has a full load ARI performance rating of about 0.58 kW/ton (0.165 kW_e/kW_r), with an IPLV rating of about 0.56 kW/ton (0.159 kW_e/kW_r).

Figure 37-7 (shown earlier) is a cutaway illustration of a chiller featuring direct-drive of a rotary screw compressor with a hermetic electric motor. This unit has a full-load ARI performance rating of about 0.78 kW/ton (0.22 kW_e/kW_r), with an IPLV rating of about 0.70 kW/ton (0.20 kW_e/kW_r). Figure 37-31 is a chiller package featuring an open induction type motor driving a rotary screw compressor.

Figure 37-32 is a packaged chiller with a reciprocating compressor driven by a 1,750 rpm semi-hermetic electric motor. Under ARI design conditions, with a water-cooled condenser, this unit has a full load performance rating of about 0.84 kW/ton (0.24 kW_e/kW_r). With an air-cooled condenser, the full load rating, under ARI design conditions, increases to about 1.05 kW/ton (0.30 kW_e/kW_r).

RECIPROCATING ENGINE DRIVES

Vapor compression chiller systems driven by reciprocating engines generally range in capacity from 20 tons (70 kW_r) to several thousand tons (more than 10,000 kW_r). Basic engine type distinctions include: cycle (i.e., Otto or Diesel), aspiration (i.e., atmospheric or supercharged/turbocharged), fuel type, air-fuel ratio (i.e., lean or rich), operating pressures, coolant systems, and methods used to control emissions.

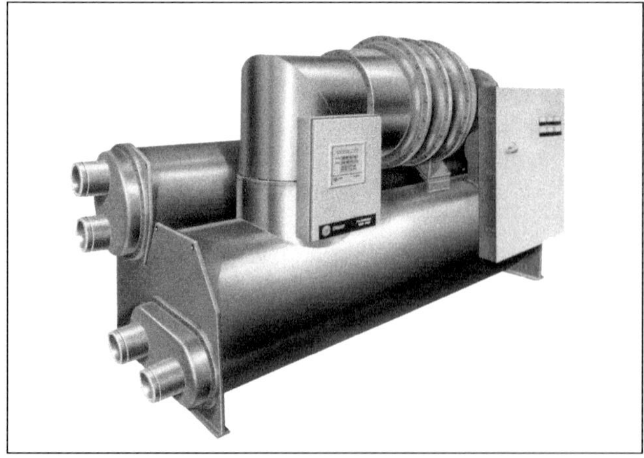

Fig. 37-29 Chiller Package with a Hermetic Motor-Driven Multi-stage Centrifugal Compressor. Source: The Trane Company

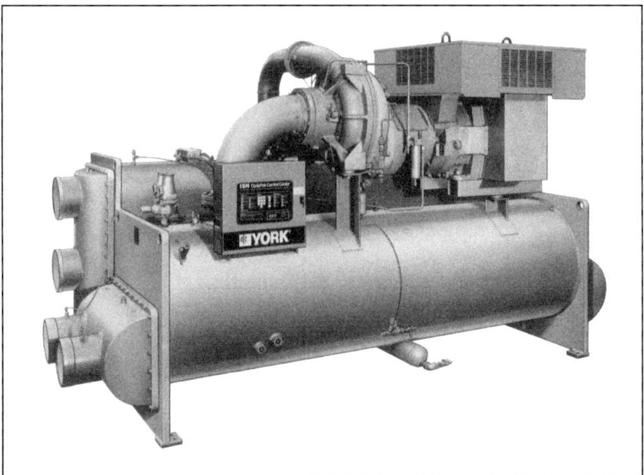

Fig. 37-30 Packaged Chiller with an Open Motor-Driven Centrifugal Compressor. Source: York International

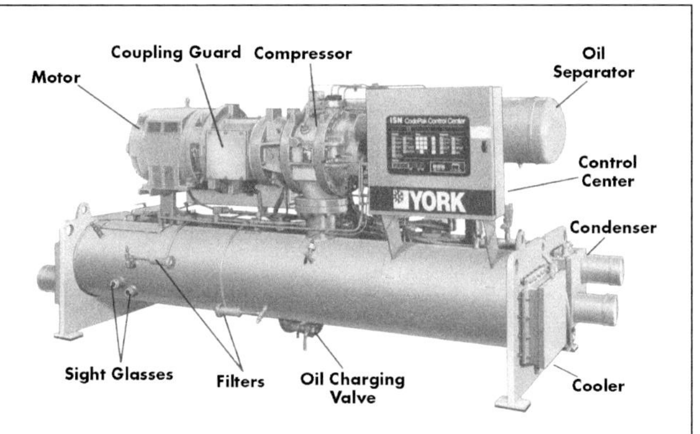

Fig. 37-31 Chiller Package Featuring an Open Induction Type Motor Driving a Rotary Screw Compressor. Source: York International

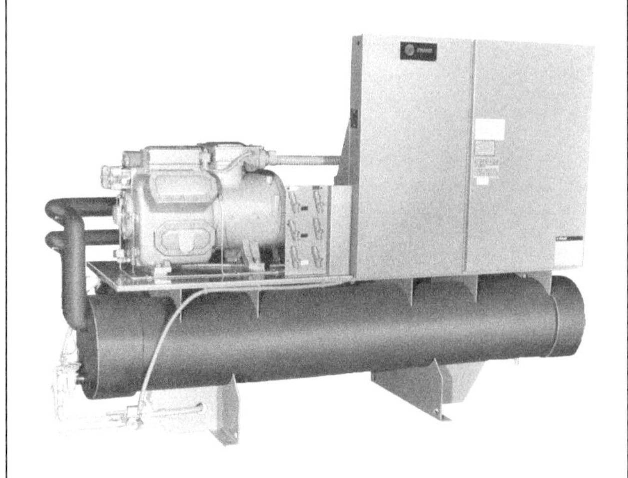

Fig. 37-32 Packaged Chiller with an Electric Motor-Driven Reciprocating Compressor. Source: The Trane Company

Reciprocating engine-driven chillers are much less common than electric-drive applications, but may be a viable cooling technology in cogeneration applications where engine heat recovery is beneficial. Figure 37-33 shows a 3,000 ton (10,500 kW_r) skid-mounted, packaged reciprocating engine-driven chiller. Figure 37-34 is a labeled illustration of a large-capacity packaged chiller featuring a gas-fired reciprocating engine-driven centrifugal compressor.

The most common reciprocating engines applied to larger vapor compressor drive service are those with maximum speeds ranging from 1,200 to 1,800 rpm. This speed is usually matched with compressors designed for operation at higher rpm, driving the compressor through a gear box. Higher speed automotive derivative engines can drive some higher speed compressors directly. Since reciprocating compressors operate at shaft speeds either the same or slightly lower than reciprocating engine speed, the gear box acts as an engine speed decreasing mechanism. With screw and centrifugal compressors, gearing is usually used to increase speed.

Fig. 37-33 3,000 Ton Skid-Mounted Packaged Reciprocating Engine-Driven Chiller with Centrifugal Compressor. Source: Caterpillar Engine Division and York International

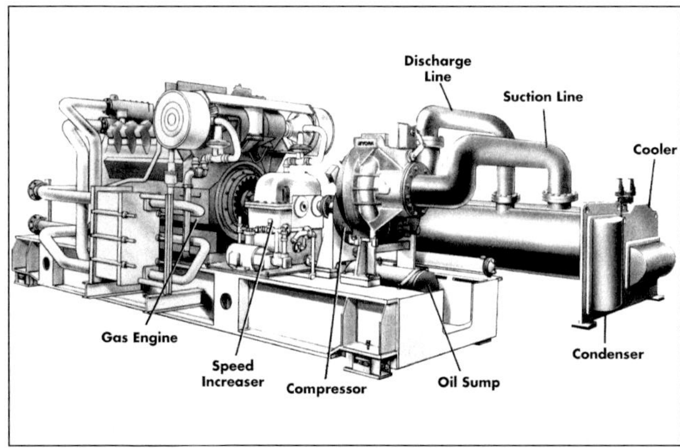

Fig. 37-34 Labeled Illustration of Large-Capacity Packaged Reciprocating Engine-Driven Chiller. Source: York International

Moderate- to high-speed engine selections are usually made for seasonal space conditioning applications, while low-speed engines are more commonly applied to base-loaded applications.

Since space conditioning loads will commonly hit an absolute peak for relatively few hours over the course of a cooling season, variable speed capability usually provides operational benefits. Since variable speed-operated engines will operate at less than their rated speed, wear is reduced and thermal efficiency is increased.

Additionally, such load profiles make it possible to sometimes select a system with a prime capacity rating of about 90% of the peak load requirement. The system can then operate in over-speed mode to serve the 100% load requirement, as required.

Base-loaded, year-round refrigeration operations and larger space conditioning applications with substantial annual operating hours often use lower speed, industrial grade engines. Increased thermal efficiency, reliability, and engine life and lower maintenance requirement become far more significant factors as hours of operation increase. Therefore, the life-cycle economics may favor the more expensive lower-speed industrial grade engines and other components for such applications.

Heat Recovery Options

Commonly, engine heat recovery is used for hot water applications, particularly in smaller capacity applications. However, high- or low-pressure steam can be generated using exhaust gas heat recovery systems and low-pressure steam can be generated using ebullient engine cooling or systems that flash high-temperature circulated cooling system water. For facilities that have thermal loads that can be served by recovered heat, these systems can operate on standard cogeneration cycles.

For facilities that do not have alternative uses for recoverable heat, chilling can improve overall system thermodynamic and economic performance. Either hot water or low-pressure steam can be used for add-on, single-stage absorption chilling. For example, for a 1,000 ton (3,500 kW_r) cooling application, an 850 ton (3,000 kW_r) reciprocating engine-driven chiller could be paired with a 150 ton (500 kW_r) absorption chiller that is powered by the combined engine and exhaust recovered heat. One thousand tons (3,500 kW_r) of cooling are provided with the fuel input required by an 850 ton (3,000 kW_r) engine-driven chiller. This configuration offers a similar capital cost to the 1,000 ton (3,500 kW_r) engine-driven chiller system. Alternatively, exhaust gas-generated high-pressure steam can be used to power a more efficient two-stage absorption chiller, and lower temperature recovered engine heat can be used for other purposes.

Figures 37-35 through 37-37 are schematic diagrams of heat recovery systems commonly applied to reciprocating engine-driven vapor compression systems. Figure 37-35 is a closed-loop cooling system, Figure 37-36 is an ebullient cooling system, and Figure 37-37 is a forced circulation/steam cooling system.

Performance

Typically, the full-load HHV energy input requirements for simple-cycle reciprocating engine-driven chillers, under standard ARI conditions, range from 10,000 Btuh/ton (0.83 kW_h/kW_r) or 1.2 COP for a small reciprocating compressor-based unit to 5,500 Btuh/ton (0.46 kW_h/kW_r) or 2.2 COP for a large-capacity, high-efficiency centrifugal compressor-based unit. When full heat recovery is employed, the net energy input requirements, inclusive of displaced boiler efficiency, range from about 5,000 Btuh/ton (0.41 kW_h/kW_r), or 2.4 COP_{net}, to 2,900Btuh/ton (0.24 kW_h/kW_r), or 4.1 COP_{net}.

Figure 37-38 provides performance detail on a 125 ton (440 kW_r) automotive derivative reciprocating

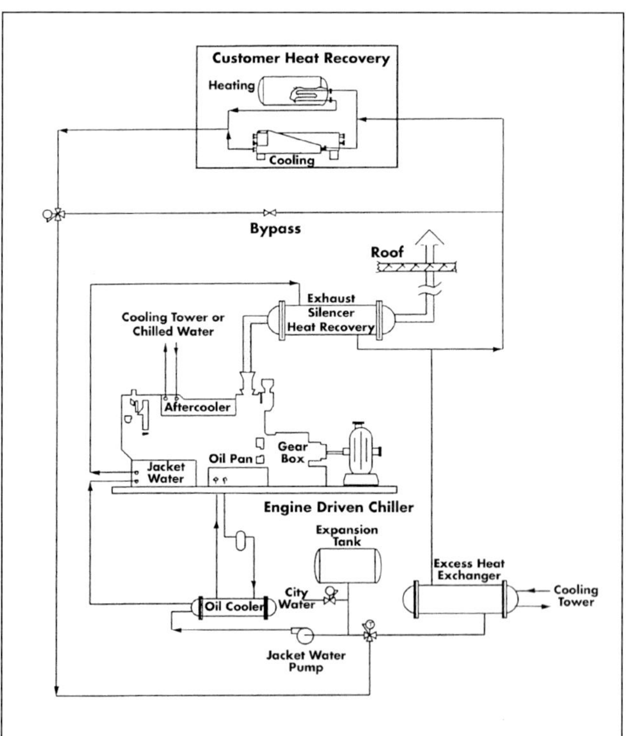

Fig. 37-35 Schematic Diagram of Closed-Loop Cooling Heat Recovery System for Reciprocating Engine-Driven Chiller. Source: Waukesha Engine Division and The American Gas Cooling Center

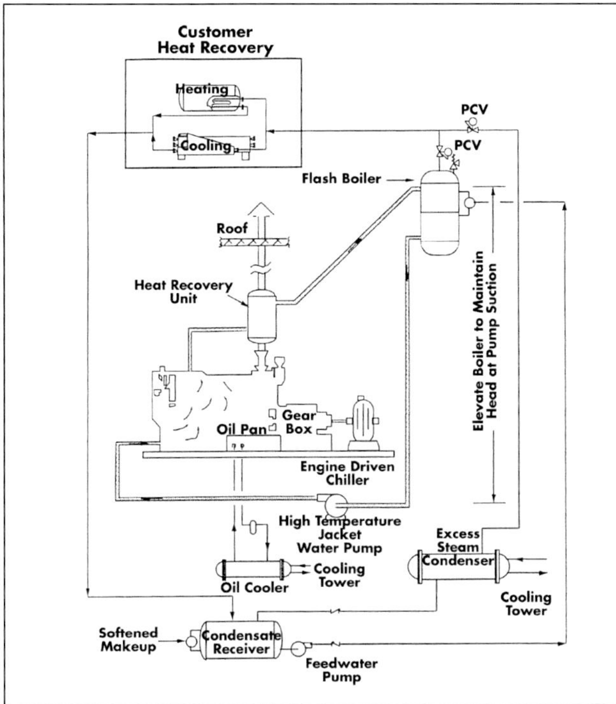

Fig. 37-36 Schematic Diagram of Ebullient Cooling Heat Recovery System for Reciprocating Engine-Driven Chiller. Source: Waukesha Engine Division and The American Gas Cooling Center

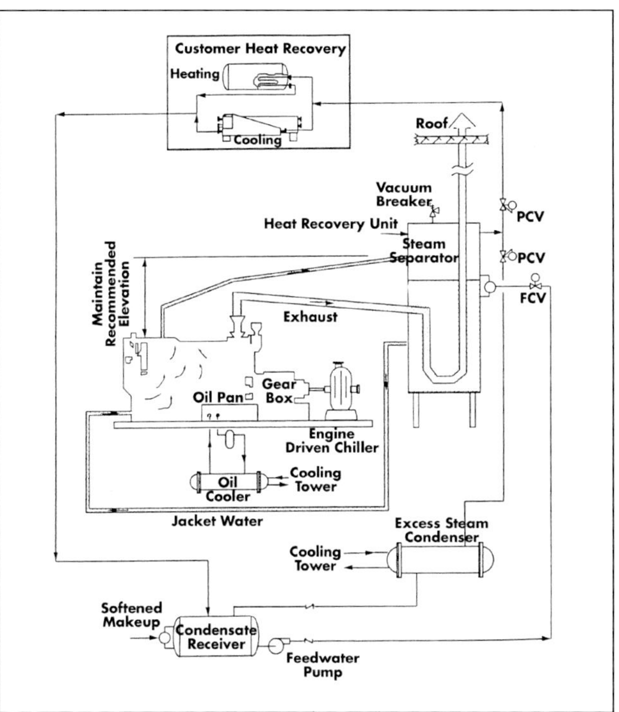

Fig. 37-37 Schematic Diagram of High-Temperature Forced Circulation/Steam Cooling Heat Recovery System for Reciprocating Engine-Driven Chiller. Source: Waukesha Engine Division and The American Gas Cooling Center

General

Evap LWT	Entering Condenser Water Temperature (°F)											
	75			80			85			90		
(°F)	Tons	Gas	HR	Tons	Gas	HR	Tons	Gas	HR	Tons	Gas	HR
40	123	1,005	294	120	1,058	310	117	1,112	325	113	113	339
42	128	1,005	294	124	1,056	309	121	1,107	324	117	117	339
44	133	1,001	293	129	1,049	307	125	1,120	328	122	122	338
45	136	1,003	294	132	1,051	307	128	1,100	322	124	124	338
48	143	997	291	139	1,046	306	135	1,096	320	131	131	337
50	149	993	291	144	1,043	305	140	1,093	320	136	136	335

ARI-550

Evap LWT	Percent of Full Load Output											
	25%			50%			75%			100%		
(°F)	Tons	Gas	HR	Tons	Gas	HR	Tons	Gas	HR	Tons	Gas	HR
44	31	350	102	63	360	105	94	671	196	125	1,120	328

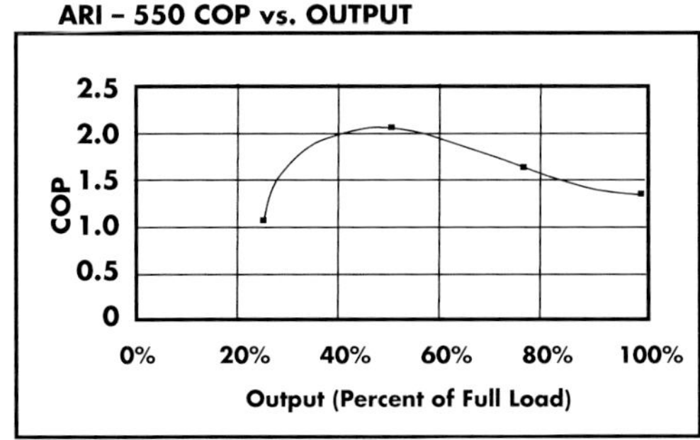

Full Load COP = 1.3

IPLV COP = 1.7

Fig. 37-38 Performance Data for 125 Ton Packaged, Gas-Fired, Automotive Derivative Engine-Driven Chiller with Screw Compressor. Source: Tecogen

engine-driven chiller featuring a screw compressor. The general data table shows the impact of varying evaporator and condenser water temperatures on system performance. At 44°F (7°C) leaving chilled water temperature and 85°F (29°C) condenser water temperature, the design full-load capacity rating is 125 tons (440 kW_r). The fuel rate is 1,120 scf of natural gas at an assumed HHV of 1,020 Btu/scf. Under the highlighted conditions, this indicates a full load fuel rate of 9,142 Btu/ton-h (0.762 kWh_h/kWh_r). Recoverable heat (HR) from the engine cooling system is listed in MBtu/h. In this case, the recoverable heat of 328 MBtu/h (96 kW_h) at a hot water temperature of 201°F (94°C) corresponds to 2,624 Btu/ton-h (0.219 kWh_h/kWh_r).

When combined with heat recovered from engine exhaust, the total system can produce about 4,500 Btu/ton-h (0.38 kWh_h/kWh_r) at a hot water temperature of 217°F (103°C). The ARI-550 table and COP versus output curve shows part-load performance with varying condenser water temperature. The ARI-based IPLV COP rating of 1.7, which is greatly enhanced by variable-speed operation and integral microprocessor controls, is significantly greater than the full load rating of 1.3. Acoustic levels are 96 dba at 3 ft (1 m), which can be reduced to 89 dba with an acoustical enclosure. This system also requires an electric input of 3 kW for parasitics.

Figure 37-39 through 37-41 are three types of performance curves for a 950 ton (3,340 kW_r) gas-fired reciprocating engine-driven chiller operating under standard ARI conditions. Figure 37-39 shows fuel input in Btu/ton-h (HHV basis) versus cooling load. Notice that with this particular configuration, performance is optimized at 75% of full load and begins to fall off rapidly when operating below 50% of full load. Figure 37-40 plots COP for this system. Also included is a curve indicating total COP for a piggyback unit featuring a recovered heat-powered absorption chiller. In this case, COP is calculated by comparing the fuel input to the combined output of the vapor compression chiller, plus the absorption chiller which operates on heat recovered from the engine.

Figure 37-41 plots the recoverable heat, in MBtu/h, versus cooling load for this system. Assuming a displaced boiler efficiency of 82%, at full load, the recovered heat would produce a fuel credit of about 3,100 Btu/ton-h (0.26 kWh_h/kWh_r). Subtracting that from the full load fuel input of about 6,200 Btu/ton-h (0.52 kWh_h/kWh_r)

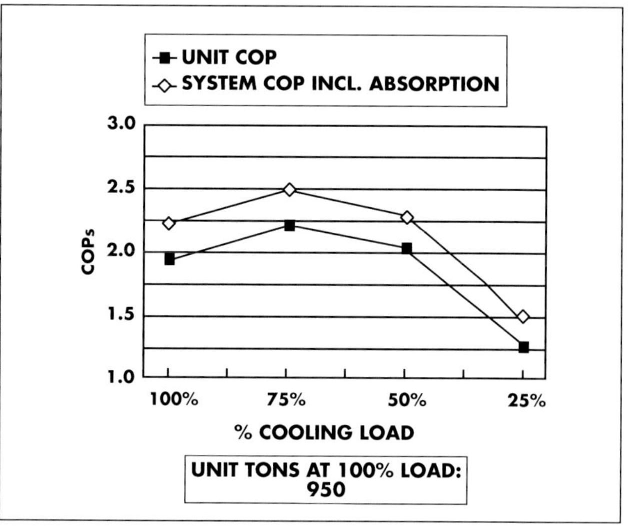

Fig. 37-40 COP vs. Cooling Load for 950 Ton Reciprocating Engine-Driven Chiller With and Without Recovered Heat Powered Absorption Chiller. Source: York International

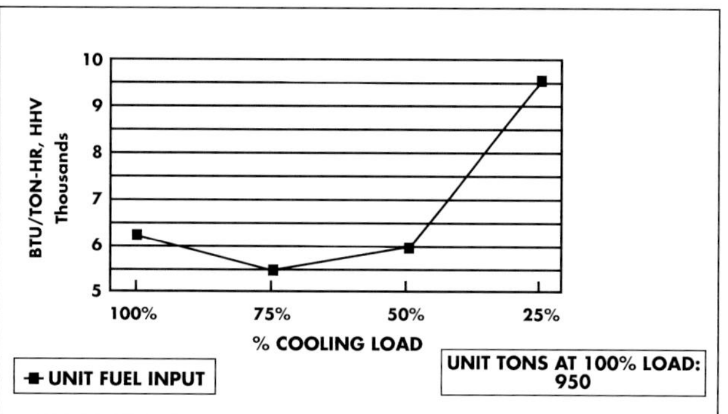

Fig. 37-39 Fuel Input vs. Cooling Load for 950 Ton Reciprocating Engine-Driven Chiller. Source: York International

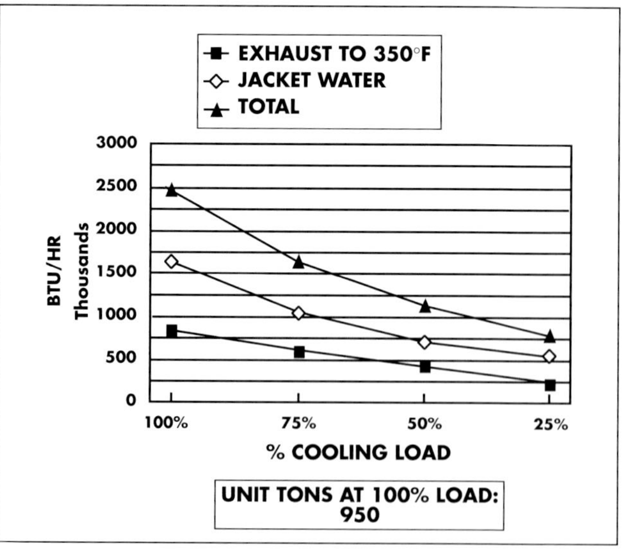

Fig. 37-41 Recoverable Heat vs. Cooling Load for 950 Ton Reciprocating Engine-Driven Chiller. Source: York International

yields a net fuel rate of about 3,100 Btu/ton-h (0.26 kWh_h/kWh_r), for a COP_{net} of 3.9. While this net COP is significantly higher than the total COP achieved with the add-on recovered heat-powered absorption chiller, the performance enhancement achieved with the absorption chiller feature would be an attractive option if there were limited opportunities to displace other facility fuel use with recovered heat.

Maintenance

Reciprocating engine-driven vapor compression systems have relatively high maintenance frequency requirements. In seasonal space conditioning applications, however, which typically operate fewer hours than other applications, maintenance frequency is somewhat less of a concern. As such, the lower maintenance and longer life benefits of the lower-speed, heavy-duty industrial-grade engines may not affect the economics sufficiently to offset higher first cost. Additionally, seasonal applications allow for off-season maintenance programs to be completed without the pressure to quickly bring a unit back on line.

Complete life-cycle OM&R costs for the engine drive will vary widely, ranging from \$0.005 to \$0.02 per ton-h (\$0.002 to \$0.006 per kWh_r). OM&R costs for most mid-sized, medium-speed engines will be typically less than \$0.010/ton-h (\$0.003/kWh_r). Complete service contracts are commonly procured for reciprocating engine-driven vapor compression systems to allow a facility to capture the benefits of the system without absorbing additional maintenance and repair responsibility and risk.

Application Examples

Reciprocating engines have been widely applied for driving vapor compression systems. Applications are fairly common with reciprocating, screw, and centrifugal compressors and range from small DX units to extremely large water-cooled systems. Figures 37-42 through 37-47 provide several application examples.

Figure 37-42 shows a 98 ton (345 kW_r), roof-mounted, air-cooled chiller system. This system features two packaged 49 ton (172 kW_r) Carrier reciprocating compressors, each driven by a Waukesha engine. The units, which are housed in weather-proof, sound-attenuating enclosures, have been applied for office building cooling in an inner city environment.

Figure 37-43 shows one of three 4,150 ton (14,500 kW_r) reciprocating engine-driven chillers that provides 40°F (4°C) water for air conditioning at a 6.5 million ft^2 (600,000 m^2) airport. Each chiller consists of a 2,585 hp (1,925 kW_m), turbocharged, lean-burn gas-fired Waukesha engine driving a York centrifugal compressor, operating on HCFC-22 refrigerant. Figure 37-44 shows the engine performance curves for this unit. Data is provided in Btu/bhp-h (LHV basis) for this engine-driven system, which operates at varying speed depending on the cooling load. At an extremely low power requirement of less than 0.65 hp/ton (0.14 kW_m/kW_r), this system can achieve a COP of 2.5 under optimal operating conditions.

Figure 37-45 shows a process cooling application at a winery featuring three gas-fired reciprocating engine-driven screw compressors with ammonia refrigerant. These units are connected to a flooded ammonia/glycol chiller that circulates glycol/water mixture through jacketed wine tanks for fermentation and cold stabiliza-

Fig. 37-42 Two 49 Ton Gas-Fired Reciprocating Engine-Driven, Air-Cooled Chillers with Reciprocating Compressors in Sound Attenuating Enclosures. Source: Alturdyne Energy Systems and The American Gas Cooling Center

Fig. 37-43 4,150 Ton Gas-Fired Reciprocating Engine-Driven Chiller with Centrifugal Compressor. Source: Waukesha Engine Division

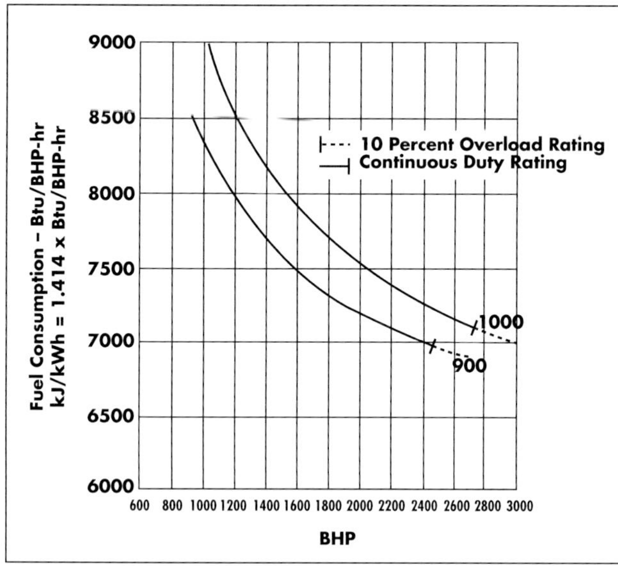

Fig. 37-44 Performance Curves for 2,585 hp Gas-Fired Reciprocating Engine Used to Drive a 4,000 Ton Chiller with a Centrifugal Compressor. Source: Waukesha Engine Division

Fig. 37-46 Open Enclosure View of Packaged 15 Ton, Gas-Fired Reciprocating Engine-Driven Condensing Unit. Source: Trico Energy Systems

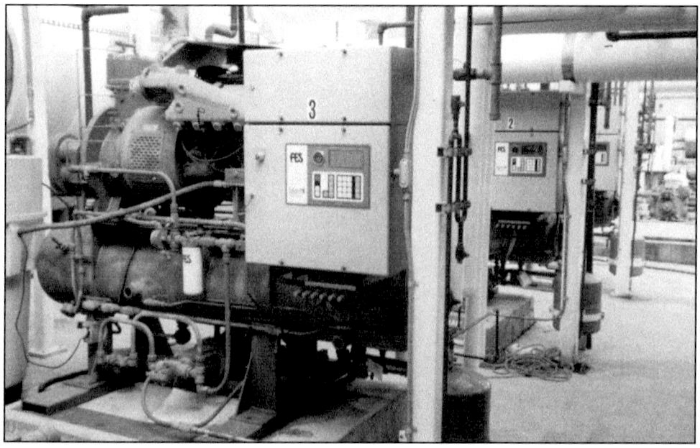

Fig. 37-45 Three Reciprocating Engine-Driven Ammonia Screw Compressors Applied for Process Cooling. Source: FES and The American Gas Cooling Center

Fig. 37-47 Reciprocating Engine Matched with 1,500 kW Electric Generator and 1,000 Ton Air-Cooled Chiller. Source: Carrier Corp. and Caterpillar Engine Division

tion and through air handlers to cool the barrel storage room. Recovered heat is used to replenish a 10,000 gallon (38,000 liter) water tank, which supplies hot water for cleaning and sanitation duty.

Figure 37-46 shows an open-enclosure view of a packaged roof-mounted 15 ton (52 kW_r) gas-engine condensing unit. This unit features a Thermo King air-cooled DX unit driven by a four-cylinder variable-speed Hercules industrial-grade reciprocating engine.

Figure 37-47 is a creative application featuring a single reciprocating engine matched with a 1,500 kW_e electric generator and a 1,000 ton (3,500 kW_r) air-cooled chiller. The engine is used to drive either load. While under standard ARI conditions, these components would be mismatched, with an air-cooled condenser operating under 115°F (46°C), desert conditions, both the centrifugal compressor and electric generator power requirements are equal.

Gas Turbines

Gas turbine-driven vapor compression systems are rare and are typically applied only in large capacities. While available in capacities of 20 tons (70 kW_r) or even less, they are uncommon below the 1,200 ton (4,200 kW_r) capacity range. Gas turbines are theoretically attractive vapor compression drivers because they provide precise control, large quantities of high-quality recoverable exhaust energy, high reliability, and the capability for effective air emissions control.

Gas turbines are usually matched with centrifugal compressors, but, in some cases, lower capacity units are matched with custom-designed screw compressors. While centrifugal compressors typically operate at high operational speeds, gas turbine speeds are usually even higher, necessitating a speed reducer (gear) to match driver and compressor shaft speeds. With some vary large capacity systems, centrifugal compressor speed may actually be matched for direct drive by the gas turbine without use of gearing.

Single-shaft gas turbines may be effectively applied to full-capacity, base-loaded applications where their poor part-load efficiency and inability to effect speed control are not detriments. For applications where part-load operating time is significant, multi-shaft turbines are well suited. As with reciprocating engines, their variable speed operating capability provides for complimentary part-load performance of both the turbine and the compressor. Multi-shaft units can also quickly develop high starting torque to accelerate compressors to design operating speeds.

Key characteristics that support, but are not necessarily essential for, economical application of gas turbine drivers are extensive annual cooling requirement with concurrent baseload thermal energy requirements and a focus on reliability and electricity independence.

A potential application drawback is the concurrence of high inlet temperature conditions with peak cooling capacity requirements. Due, in part, to decreased air density and the impact of higher fluid temperatures on adiabatic head, the turbine's compressor absorbs more power when inlet air is hotter, thereby reducing total power available to the shaft. This can result in the system output being at its lowest when load requirements are at their highest. Inlet cooling, which can be provided for by available chilled water, can cost-effectively cool the turbine air inlet to improve performance. While this can be included in the application design, there will be a capital and operating cost penalty.

Heat Recovery

High mass flow and exhaust temperature characteristics provide large quantities of potentially recoverable heat, which can produce strong economic performance when matched with a sufficient existing concurrent thermal load requirement. Augmentation by supplementary firing in the oxygen-rich exhaust can significantly increase exhaust gas temperature and, therefore, steam (or hot water) production.

Alternatively, recuperation can be used to transfer heat from turbine exhaust to turbine compressor discharge air prior to combustion, thereby displacing a portion of the fuel requirements. This results in a thermal efficiency increase and corresponding decrease in recoverable energy due to the reduction in exhaust gas temperature — typically to about 600°F (315°C), and also a slight decrease in capacity.

In addition to using recovered heat for non-cooling process applications, recovered heat can be used to provide additional cooling output in a piggyback arrangement. Options include:

- A double-effect steam absorption chiller (or customized directly coupled heat recovery powered unit).
- A hot water recovery boiler to power a single-effect absorption chiller.
- A steam turbine-driven chiller (either condensing or non-condensing turbines may be used to increase overall system cooling output).

Steam injection may also be used to enhance capacity and cycle efficiency. Capacity enhancement with steam injection- (STIG-) cycle operation will be equal to or greater than the combined-cycle configuration, while overall thermal efficiency will be slightly lower.

Performance

Typical simple-cycle HHV energy input requirements for gas turbine vapor compression drive systems, under ARI standard conditions, range from 11,500 Btuh/ton (0.96 kW_h/kW_r), or 1.04 COP for smaller capacity systems, to 6,300 Btuh/ton (0.53 kW_h/kW_r), or 1.9 COP for very large systems. Taking into account recoverable heat and displaced boiler efficiency losses, net energy input requirements range from 4,000 Btuh/ton (0.33 kW_h/kW_r), or 3.0 COP_{net}, to 2,600 Btuh/ton (0.22 kW_h/kW_r), or 4.6 COP_{net}.

When steam injection or recuperation is used, energy input requirements may reach slightly under 5,000 Btuh/ton (0.42 kW_h/kW_r), or 2.4 COP. When matched with steam turbine-driven systems or absorption chillers powered by recovered heat, combined COPs in excess of 2.0 are achievable.

Maintenance

As with other types of combustion gas turbine applications, gas turbine-driven vapor compression systems, when applied, maintained, and operated properly, offer high reliability and relatively low life-cycle maintenance costs. However, a high-quality control and monitoring system and a hands-on understanding of the specific system in operation are important elements in successful operation.

Typically, the life-cycle OM&R cost on the turbine drive will range from \$0.004 to \$0.012 per ton-h (\$0.001 to \$0.03 per kWh_r). Complete service contracts, including turbine engine replacement or overhaul insurance, are common. Third-party operation is somewhat common for larger systems.

Application Examples

Figure 37-48 shows a 6,100 ton (21,500 kW_r) gas turbine-driven chiller system that provides chilled water and steam for a campus-wide distribution system at a large university. The system features a 5,400 hp (4,000 kW_m) Allison gas turbine driving a York multi-stage, high-pressure centrifugal compressor operating on HCFC-22 refrigerant.

The turbine operates at a full load rated speed of 14,770 rpm to drive the centrifugal compressor at 13,600 rpm through a 1.086:1 gear reducer. Variable speed operation allows the system to operate efficiently under part-load conditions and minimizes the use of hot gas bypass under very low loads. The turbine exhaust is passed to an HRSG that generates 100,000 lbm/h (45,000 kg/h) of steam at a pressure of 150 psig (11.4 bar). The turbine exhaust heat is capable of generating about 25,000 lbm/h (11,250 kg/h) of steam, with the remainder of the heat to the HRSG provided by fired duct burners.

Figure 37-49 illustrates two gas turbine drive configurations. Both application options are used to achieve additional capacity through the use of recovered heat. The configuration on the top matches a gas turbine-driven centrifugal compressor with an absorption chiller. The configuration on the bottom is a traditional combined-cycle system matching a gas turbine-driven chiller with a steam turbine-driven chiller.

Figure 37-50 illustrates a system in which the gas turbine driver is paired in line with a smaller electric motor/generator. With this configuration, chiller capacity is matched with the combined shaft power output of both the turbine and the motor. This application configuration is used to extend operating hours and enhance system economics.

Under design cooling conditions, full turbine and motor power are used to drive the chiller. Under partial load, the turbine at full-load power is used alone to drive the chiller. As loads continue to fall, the turbine remains at full load, with part of the power output used to drive the generator. In this arrangement, the turbine remains at full load under most or all conditions and can operate for a

Fig. 37-48 6,100 Ton Gas Turbine-Driven Chiller System Applied at a University Campus. Source: Stewart and Stevenson

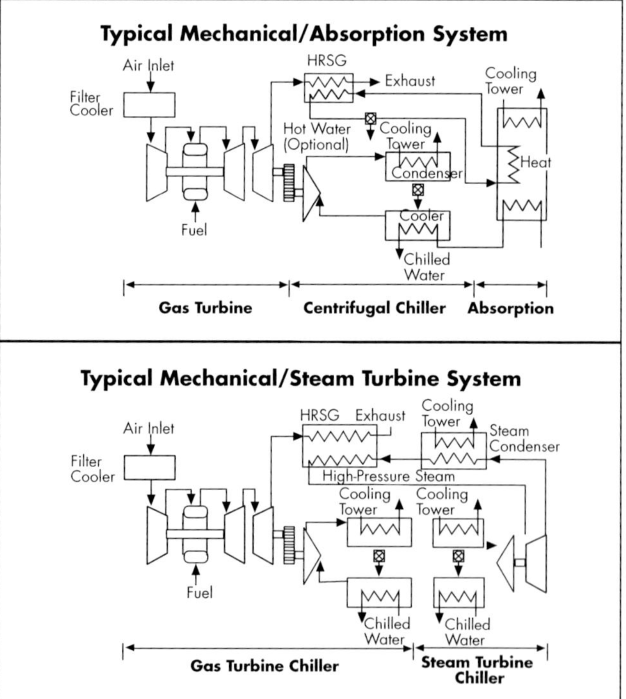

Fig. 37-49 Two Configurations Used with Gas-Turbine Systems for Achieving Additional Capacity Through Use of Recovered Heat. Source: Solar Turbines and The American Gas Cooling Center

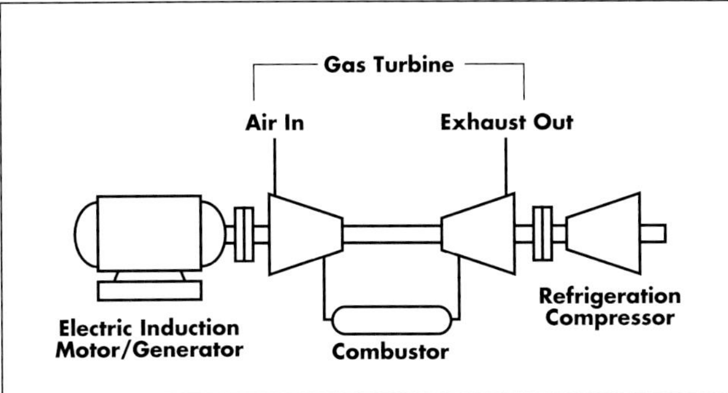

Fig. 37-50 Gas Turbine-Driven Refrigeration Compressor with Induction Motor/Generator.

Fig. 37-51 Large-Capacity Packaged Steam Turbine-Driven Centrifugal Compressor. Source: Carrier Corp.

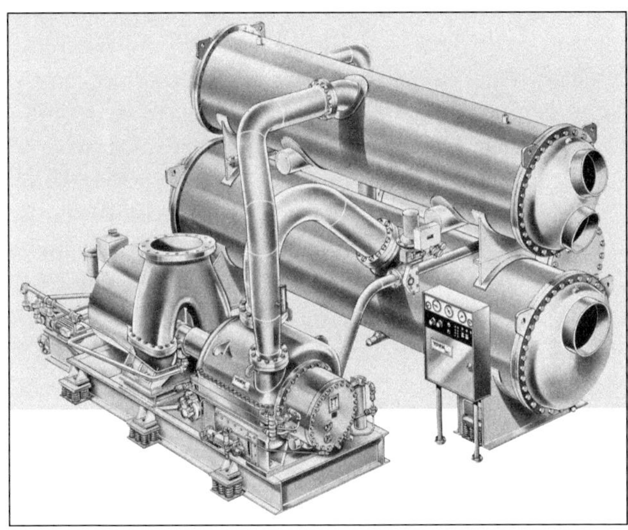

Fig. 37-52 Illustration of Large-Capacity Chiller with Factory-Assembled Steam Turbine-Driven Centrifugal Compressor Package. Source: York International

majority of the hours in the year under peak efficiency. The value of shaft power and heat recovery over 4,500 to 7,000 hours makes this type of chiller/cogeneration application very effective and economical. Factoring in reliability, low maintenance, and long service life makes the gas turbine-driven chiller a highly attractive alternative.

With this configuration component capacity (i.e., turbine and motor) is a function of the operating and load characteristics of the facility. In facilities with limited or no winter season cooling, the motor generator, for example, could be sized to match full turbine output capacity. It would serve as a complete electric cogeneration system in the winter and as a complete chilling cogeneration system in the summer. In swing seasons, it would serve as both. Direct inlet cooling from the chiller is also used to further increase total power output.

Steam Turbine Drives

Known for their reliability, efficiency, and long service life, steam turbine-driven chillers have been successfully applied for most of this century. Large-capacity steam turbine-driven chillers featuring centrifugal compressors, such as the ones shown in Figures 37-51 and 37-52, remain a commonly used technology for large industrial facilities, universities, and hospitals with high-pressure steam systems, as well as for district type heating and cooling plants.

Steam turbines have good speed compatibility with centrifugal and screw compressors. Typically, steam turbines operate best in the 3,000 to 6,000 rpm range, which encompasses the ideal speed ranges for many of these compressor designs. Though not as common, steam turbines have also been successfully applied to smaller capacity systems featuring reciprocating compressors.

Back-pressure, condensing, and extraction/admission turbines may all be effectively applied to vapor compression drive service. Central power house plants often use a combination of condensing and back-pressure turbines for district type chilled water and steam distribution systems. These plants maximize boiler system utilization by balancing winter heating loads with summer cooling loads. They also commonly use steam turbine-driven chilled and condenser water pumps, as well as boiler fans and feed water pumps.

When sufficient low-pressure steam loads can be served, back-pressure turbine-driven units are the lowest cost, highest net thermal efficiency option. When existing low-pressure steam loads are insufficient, non-condensing turbine-driven chillers can be used in conjunction with single-stage absorption chillers in a piggyback arrangement. By condensing steam in the absorption chiller, the need for a surface condenser is eliminated. The large discharge steam piping and valves required for condensing turbines are also eliminated. An example of such an application is where the turbine-driven chiller is load-matched with a year-round process cooling application. Discharged back-pressure steam could be used for heating in the winter and space cooling via the absorption chiller in the summer.

When load swings are more severe and/or less predictable, a multi-stage extraction/condensing turbine can be used to either provide increased cooling output or maintain cooling output with less steam input. Extraction turbines can discharge steam at varying pressures and can be set to discharge steam at different quantities and pressures simultaneously. While they are more expensive than standard multi-stage condensing turbines and far more expensive than simple single-stage back-pressure units, they provide optimal operating flexibility and economy.

Dual-shaft operations are sometimes used with a steam turbine and an electric motor, each driving the same chiller to allow for mixed system (hybrid) operation, standby security, and elimination of a need for a second chiller. An electric generator can be installed on the other end of the shaft connecting the steam turbine to the chiller. In these applications, the turbine can be used to generate electricity in non-cooling periods or during power outages.

Maintenance

As with other types of mechanical equipment drive applications, steam turbine-driven vapor compression systems require relatively low maintenance and, with proper preventative maintenance, will have a low incidence of forced downtime. Given proper control of steam purity and quality and reduced blade exposure to moisture, they can operate for decades without major overhaul.

Life-cycle OM&R costs for the turbine drive typically will range from \$0.002 to \$0.008 per ton-h (\$0.0006 to \$0.0023 per kWh_r). Routine maintenance is frequently performed in-house. Steam turbine overhaul and repair is commonly contracted out as needed. In seasonal air conditioning applications, turbine components can be sent out for repair and overhauls can be conducted in the winter months.

Performance

As with any steam turbine application, performance will largely be a function of the entering and exiting steam conditions and the mechanical efficiency of the turbine. For example, a highly efficient multi-stage turbine operating from 250 psig (18.3 bar) saturated steam into a deep vacuum can perform at a rate of 8.5 to 10 lbm/ton-h (1.1 to 1.3 kg/kWh_r). With superheated steam at 600 psig and 750°F (42.4 bar and 400°C), steam rates may range from 6 to 7 lbm/ton-h (0.8 to 0.9 kg/kWh_r).

As with any prime mover mechanical drive application, the back-pressure steam turbine is generally the most efficient and lowest operating cost technology that can be employed when sufficient low-pressure steam loads exist. The turbine acts as a steam pressure-reducing valve (PRV), providing cooling output with an HHV net fuel consumption rate as low as 2,250 Btu/ton-h (0.19 kWh_h/kWh_r) input to the boiler or a 5.3 COP_{net}. Moreover, there are no additional costs associated with steam condensing because all unused steam is passed into the process. This results in reduced operating costs associated with water usage, water treatment, and additional auxiliaries for steam condensing.

Turbine part-load efficiencies will be fairly similar for single-stage and multi-stage back-pressure or condensing units. Single-stage units can often operate at lower speeds than multi-stage. Due to the weight of the shaft and other factors, multi-stage units are more limited in speed ranges.

Whereas reciprocating engines gain in thermal efficiency at reduced speeds relative to constant speed operation at a given partial load, steam turbines tend to lose efficiency as speed is reduced. Still, modest reductions in speed can be used to improve part-load operation versus constant speed operation due to relatively greater improvements achieved in compressor volumetric efficiency.

Under part-load operation, the compressor will often be unloaded first, then speed will be reduced in the lower range of the load curve. Unloading the compressor in order to maintain higher turbine speed, or reducing speed to optimize compressor performance, are efficiency trade-offs. Modern microprocessor-based steam turbine modulators have been developed to further optimize variable speed performance.

Application Examples

Figure 37-53 shows a central plant application featuring two 5,000 ton (17,600 kW_r) steam turbine-driven chillers. In such large capacities, the capital cost premium for a steam turbine (versus an electric motor) is reduced on a relative basis due to economies of scale.

Figure 37-54 shows a steam turbine-driven chiller applied in a commercial building for space conditioning service. This system features a 2,500 hp (1,860 kW_m)

Fig. 37-53 Central Plant Application Featuring Two 5,000 Ton Steam Turbine-Driven Chillers. Source: Carrier Corp.

Fig. 37-54 Steam Turbine-Driven Chiller Applied in Commercial Building for Space Conditioning. Source: Tuthill Corp. Murray Turbomachinery Division.

multi-stage Murray steam turbine driving a centrifugal compressor. Under full-load conditions, the steam turbine operates at 5,000 rpm with a steam rate of 12.4 lbm/hp-h (7.5 kg/kWh$_m$). Steam enters the turbine at a condition of 125 psig (9.6 bar) dry and saturated and exhausts at a pressure of 3 in. HgA (10 kPa). Figure 37-55 is a performance curve for this steam turbine. Notice that performance remains fairly constant over most of the operating range.

Comparison of Driver Options

The economic performance of any applied driver-technology involves a number of factors. It is a function of the specific annual cooling load profile of the facility, the application-specific thermodynamic performance of each system under actual load conditions, long-term electricity and fuel cost rates as they apply to the specific load profile and time of use, installed costs for the complete system, and long-term OM&R costs. Careful consideration must also be given to the host facility's requirements for redundancy and reliability. First cost alone thus provides insufficient indication of economic performance. Instead, life cycle cost analysis with focus on total cost of service is required.

First cost, size, familiarity, simplicity of operation, and minimal service and maintenance requirements typically make electric motors the most common choice for vapor compression system applications. The capital cost of prime mover-driven vapor compression systems is usually significantly greater than that of conventional electric-driven systems, with the premium ranging anywhere

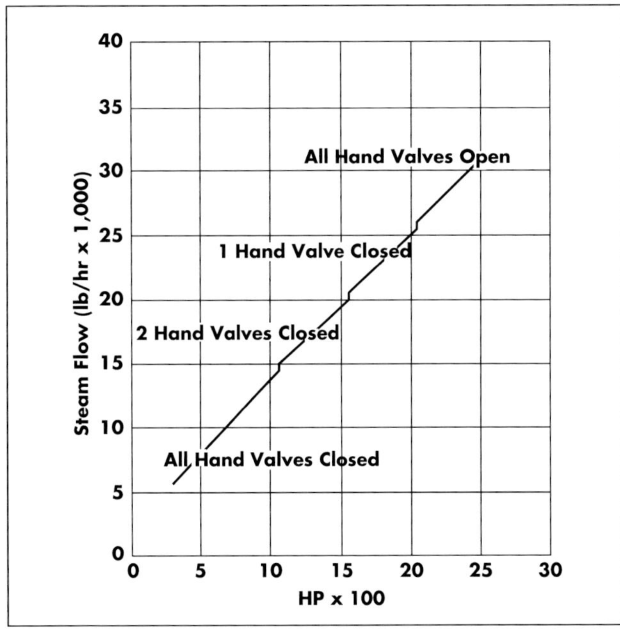

Fig. 37-55 Performance Curve for 2,500 hp Steam Turbine Applied with Centrifugal Compressor. Source: Tuthill Corp. Murray Turbomachinery Division.

from \$250 to \$1,000 per ton (\$65 to \$290 per kW$_r$) of installed cooling capacity. On the low end of this range are back-pressure steam turbines, in the middle are moderate-speed reciprocating engines and condensing steam turbines, and on the high end are gas turbines and low-speed reciprocating engines.

The capital cost differential would move higher or lower for each technology, depending on capacity and site-specific conditions. Generally, as chiller capacity is increased, the cost per ton (or kW_r) differential between prime mover-driven chillers and electric-driven chillers drops, since economics of scale are reached with the prime mover itself and the coupling of the driver to the compressor. However, this reduction in cost differential only occurs at very large capacities.

In new installations, the costs associated with gas or oil, exhaust, and waste heat piping for reciprocating engine- and gas turbine-driven units, or steam piping for steam turbine-driven units, will add to capital costs as will costs for air permitting and emissions control. Electric service (i.e., wiring, transformers, and switchgear), a large unit's motor starter, and the cost of extensive conduit runs will elevate the cost of an electric-motor drive unit. In cases where a facility would need to install additional electric service capacity to support the installation of a new electric system, the cost can be substantial. Depending on the logistics of the facility and the location and available utility substation capacity, these costs can often offset a portion or, in an extreme case, all of the installed capital cost premium for a prime mover drive.

Another potential cost factor is the need for standby electric generation capacity to support critical cooling loads. In these cases, the combined cost of the electric cooling equipment and the standby prime-mover generator set may be equal to or even greater than the cost of the prime mover-driven vapor compression system. In some cases, an electric generator may be mounted in line with the compressor on a prime mover-driven system. This allows the prime mover to serve the dual function of providing emergency power to the facility for other (non-cooling) loads in the event of a utility outage.

Operating cost generalizations are difficult to make given the wide range of fuel and electricity prices and differing utility rate structures found across the country. Fuel prices are an important factor with simple-cycle prime mover-driven systems. It is less critical for systems operating on cogeneration cycles, which will generally produce net energy operating costs of $0.013 to $0.04/ton-h ($0.004 to $0.011/$kW_r$). Even so, when electricity prices are extremely low, it is difficult to generate sufficient operating cost savings to overcome the capital cost differential on a life-cycle basis. The increased OM&R cost requirements also reduce savings potential.

Generally, the relationship between the facility load requirement curve and the electric rate structure are the key influences on operating costs. The trend toward higher peak electricity pricing, either on a commodity basis (such as the case with real-time pricing) or on the basis of high demand charges, places an operating cost premium on electric motor-driven units operating during peak periods. Given the common coincidental relationship between peak facility cooling requirements and high peak electricity pricing, peak or near-peak capacity operation for relatively few hours can dramatically drive up annual operating costs. This is, however, counterbalanced when systems also operate extensively in lower cost off-peak or intermediate rate periods.

Stratified electric rate structures also provide an opportunity to achieve operating cost optimization through use of mixed (hybrid) systems, featuring both electric motor- and prime mover-driven systems. For larger facilities, particularly those that require some measure of system redundancy, mixed systems allow for a variety of operating strategies driven by the dynamic changes in facility load and electricity pricing.

Chapter Thirty-Eight
Absorption Cooling Systems

The use of absorption cycle technology has varied since its introduction in the late 19th century, depending on the relative cost of fuel and electricity and on improvements to mechanical compression and absorption technology. In recent years, seasonal pricing of electricity and fuel, improvements in absorption technology, heat recovery applications, utility promotion, and environmental concerns over halocarbon-based refrigerants have driven a resurgence of absorption cooling applications.

In response to market demand, U.S. manufactures have all introduced expanded lines of direct-fired and steam-powered absorption chillers and chiller/heaters ranging from 3 tons (11 kW_r) to more than 1,500 tons (5,000 kW_r). This chapter presents detail on a wide range of absorption technologies and applications.

Absorption Refrigeration Cycle

Absorption cycles are similar to vapor compression cycle systems, but rely on the chemical affinity of an absorbent for the refrigerant to produce the cooling effect. The cycle is best understood by examining the functions of four major components: absorber, generator, condenser, and evaporator. The system is illustrated in Figure 38-1 and described as follows:

1. In the **absorber**, low-pressure refrigerant vapor is condensed and absorbed by the warm, concentrated absorbent solution. Absorption (mixing) is made possible because of the affinity between absorbent and refrigerant molecules. The heat of condensation and heat of absorption are removed from the absorber tube bundle by cooling water to an external heat rejection system. After the refrigerant vapor and the strong solution combine, the solution has a lower concentration and is referred to as a weak solution. Its pressure is then elevated as it is pumped into the generator.
2. In the **generator**, heat is introduced and the refrigerant and absorbent are separated (refrigerant is regenerated) by a boiling process. Hot concentrated solution is cooled by a heat exchanger using incoming weak solution as it leaves the generator and is throttled back to the absorber.
3. High-temperature/high-pressure refrigerant vapor flows to the **condenser** where it condenses to liquid as heat is rejected. Except in certain small capacity systems, water-cooled condensers are used.
4. Warm liquid refrigerant is then expanded through an orifice and enters the evaporator where it vapor-

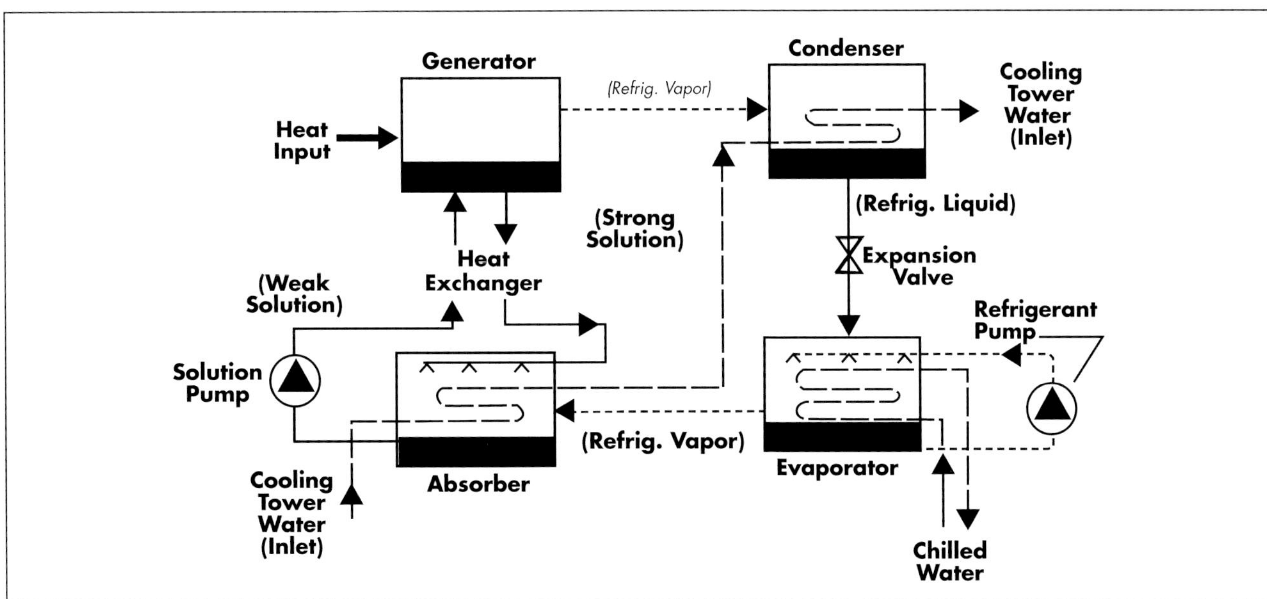

Fig. 38-1 Schematic Diagram Showing Basic Absorption Cycle Components. Source: The American Gas Cooling Center

izes at low temperature and pressure, drawing heat from an external, low-temperature source. In many absorption chillers, the refrigerant is pure water with evaporation occurring under a vacuum produced by the absorption process.

In comparison to the vapor compression cycle, three of the four main elements — condenser, evaporator and expansion valve — are essentially the same. The absorber and generator combine to serve a similar role as the compressor.

Absorption Chiller Technologies

Equipment Classifications

Currently, the two absorption cycles in widespread use are the lithium bromide (LiBr) cycle and ammonia-water (NH_3H_2O) cycle. In the LiBr cycle, water acts as the refrigerant and LiBr is the absorbent. In the ammonia-water cycle, an ammonia-water solution acts as the refrigerant with water as the absorbent. Most of the large capacity units on the market today use the LiBr cycle. The ammonia-water cycle is most commonly used in small-capacity direct fuel-fired single-effect units or larger capacity custom-designed systems for low-temperature industrial process applications.

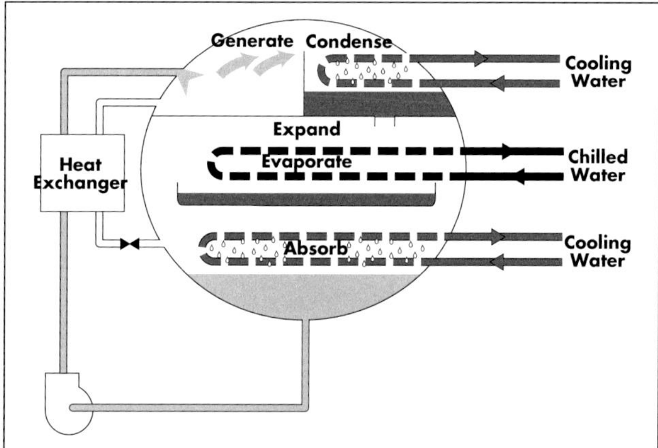

Fig. 38-2 Illustration of Single-Effect Absorption Cooling Cycle. Source: The American Gas Cooling Center

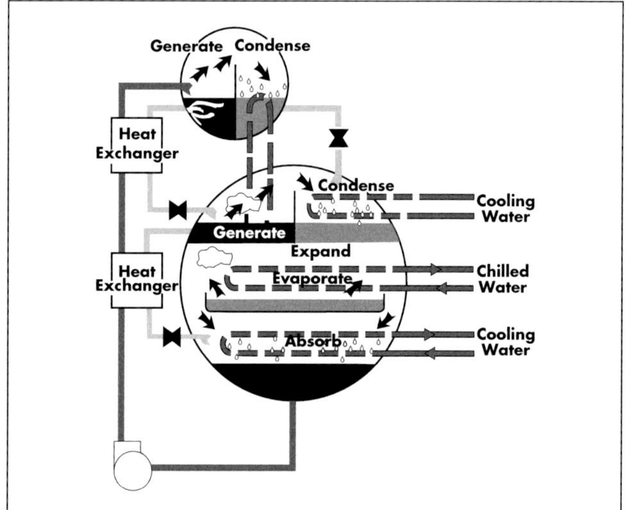

Fig. 38-3 Illustration of Double-Effect Absorption Cooling Cycle. Source: The American Gas Cooling Center

Two variations are used with LiBr cycle absorption chillers: single-effect or single-stage cycles (Figure 38-2) and double-effect or two-stage cycles (Figure 38-3). In two-stage designs, which are more thermally efficient, heat recovered from the first stage condensing process is used to boil additional refrigerant at a lower temperature in the second stage.

Absorption machines can also be classified by energy source:

- Direct-fired units, which have an integral burner assembly to combust fuels such as natural gas, propane, and light distillate oils.
- Indirect-fired units, which use steam or thermal fluid from an outside source to power the cycle.
- Exhaust gas or heat recovery units, which are custom-designed machines that can be directly connected to a heat source such as an engine.

Common characteristics of all currently available absorption machines are:

- Simplicity of design with few moving parts and operation at relatively low temperatures and pressures.
- Low electrical requirement.
- High rate of heat rejection, typically from about 21,000 Btu/ton-h (1.75 kWh_h/kWh_r) to 30,000 Btu/ton-h (2.50 kWh_h/kWh_r). This necessitates more cooling tower capacity with higher pump and fan energy use, relative to conventional electric-driven vapor compression systems.
- Large physical size and weight. The operating weight of a 100 ton (352 kW_r) unit, for example, is typically about 10,000 to 12,000 lbm (4,500 to 5,400 kg) with dimensions of 13 ft (4.0 m) long by 4.5 ft (1.4 m) wide by 7.5 ft (2.3 m) high. The operating weight of a 1,000 ton unit is typically about 65,000 to 75,000 lbm (29,000 to 34,000 kg), with dimensions of 25 ft (7.6 m) long by 8 ft (2.4 m) wide by 12 ft (3.7 m) high.
- Use of low-global warming and ozone-safe refrigerants.

Single-Effect LiBr-H_2O Cycle Absorption Systems

Low-pressure steam or hot water can be used as an energy source in single-effect systems. Typical temperature requirements range from 200 to 270°F (93 to 132°C). Steam-powered systems are generally designed for operation at pressures between 9 and 15 psig (1.6 to 2.0 bar). When hot water or steam temperatures are below design specifications, chiller capacity is reduced.

Although single-effect absorption technology is relatively thermally inefficient by today's standards, it is useful when steam costs are low or when recovered heat can be used. Under ARI conditions, full-load steam rates typically range from 18.3 to 19.0 lbm/ton-h (2.4 to 2.5 kg/kWh_r), corresponding to COPs between 0.65 and 0.69. At 80 to 87% boiler efficiency, gross heat input ranges from 20,000 to 23,000 Btu/ton-h (1.67 to 1.92 kWh_h/kWh_r) on an HHV basis, reducing COP to between 0.60 and 0.52.

Figure 38-4 is a cutaway illustration of a single-effect absorption chiller. Figure 38-5 shows a single-shell design and Figure 38-6 shows a two-shell design in which the upper shell houses the generator and condenser and the lower shell houses the absorber and evaporator. Representative part-load performance data is provided in Figure 38-7.

Power requirements for internal auxiliary components such as solution, refrigerant, and purge pumps and controls typically range from 0.01 to 0.04 kW/ton (0.003 to 0.01 kW_e/kW_r), with a minimum of 0.004 kW/ton (0.001 kW_e/kW_r) for some smaller machines. Heat rejection, which is close to 30,000 Btu/ton-h (2.5 kWh_h/kWh_r), requires cooling water flow rates of between 3.6 and 6.4 gpm/ton (3.9 to 6.9 lpm/kW_r).

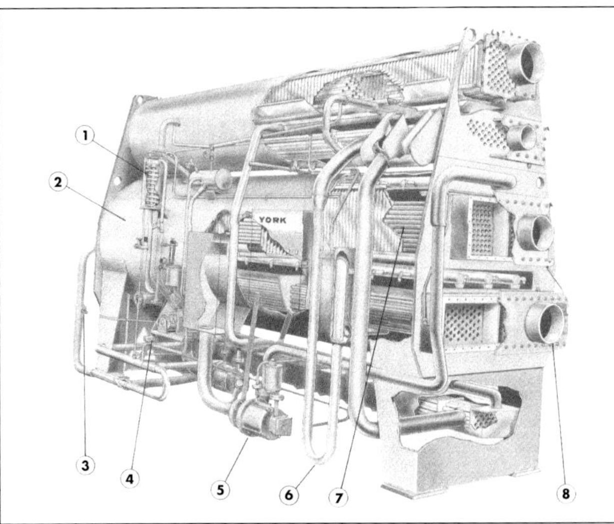

1. Purge system to expel non-condensable gases from units external purge chamber
2. Service access to controls and service points
3. Stablizer control to limit excessive cycling due to rapid reduction in load or condenser water temperature
4. Unloader control to maintain refrigerant level for proper pump operation during low-capacity and condenser-water conditions
5. Hermetic pumps
6. Decrystallization system to automatically correct minor crystallization
7. Double-walled evaporator
8. Water and generator pass heads

Fig. 38-4 Cutaway Illustration of Single-Effect LiBr Absorption Chiller. Source: York International

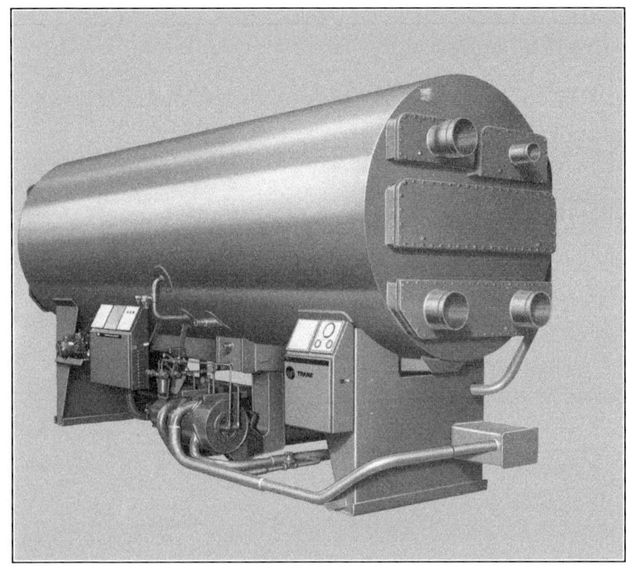

Fig. 38-5 Single-Effect LiBr Absorption Chiller with Single-Shell Design. Source: The Trane Company

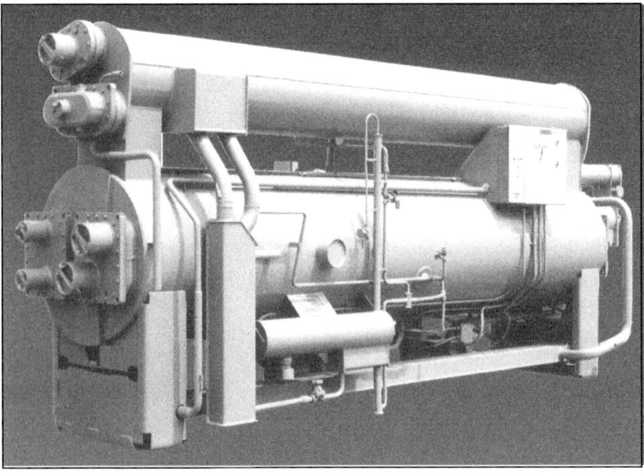

Fig. 38-6 Single-Effect Absorption Chiller with Two-Shell Design. Source: Carrier Corp.

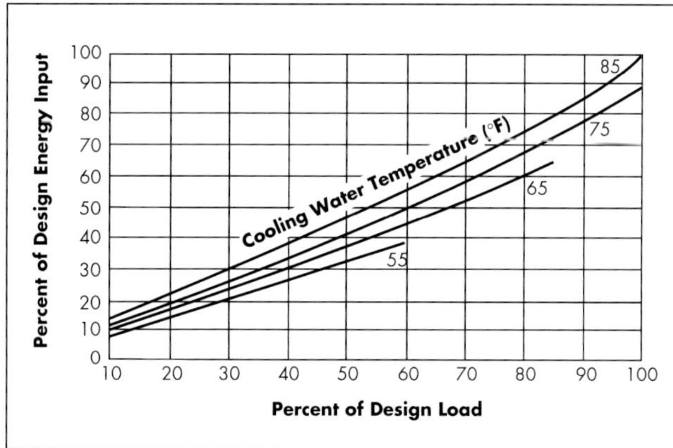

Fig. 38-7 Energy Input vs. Load at Various Cooling Water Temperatures for Single-Effect Absorption Chiller.
Source: The Trane Company

is used to boil additional water from the low-temperature, second-stage generator. In addition, a recuperative heat exchanger is used to recover heat from solution leaving the low-temperature generator. As a result of this double-effect, a thermal efficiency improvement of about 70% is achieved versus the single-stage cycle.

Operating temperature for double-effect units is about 370°F (188°C), corresponding to a saturated steam pressure of about 115 psig (8.9 bar). Typical full-load steam rates, under ARI standard conditions, range from 9.7 to 10 lbm/ton-h (1.25 to 1.29 kg/kWh_r), with corresponding COPs of 1.22 to 1.19. At 80 to 87% boiler efficiency, gross heat input ranges from 11,300 to 12,600 Btu/ton-h (0.94 to 1.05 kWh_h/kWh_r), reducing the range of COPs to 1.08 to 0.96. Direct-fired units typically consume between 11,800 and 13,400 Btu/ton-h (0.99 to 1.11 kWh_h/kWh_r), on an HHV basis, with COPs of 1.02 to 0.90. Figure 38-9 shows a double-effect absorption machine.

Double-effect design options include series flow, parallel flow, and reverse flow machines. With the series flow design, shown in Figure 38-10, a weak solution is pumped through low-temperature and high-temperature heat exchangers in series before entering the primary

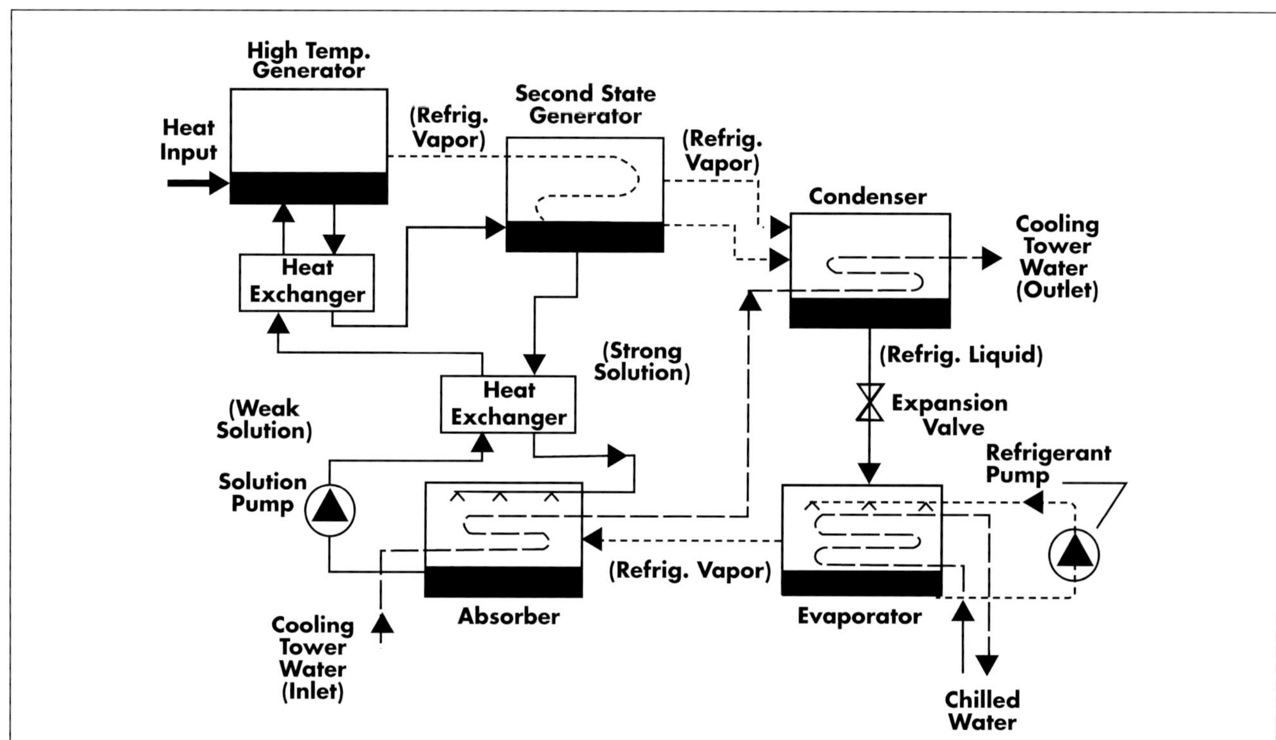

Fig. 38-8 Schematic Diagram Showing Components of an LiBr Double-Effect Absorption Machine.

Double Effect LiBr-H_2O Cycle Absorption Systems

Double-effect, or two-stage, absorption systems are shown schematically in Figure 38-8. The double-effect cycle uses a second generator, condenser, and heat exchanger that operate at higher temperature. Refrigerant vapor is recovered from the first-stage generator in the high-temperature condenser. The heat from this condensing process

Fig. 38-9 Large-Capacity, Double-Effect, Steam-Powered LiBr Absorption Chiller. Source: Carrier Corp.

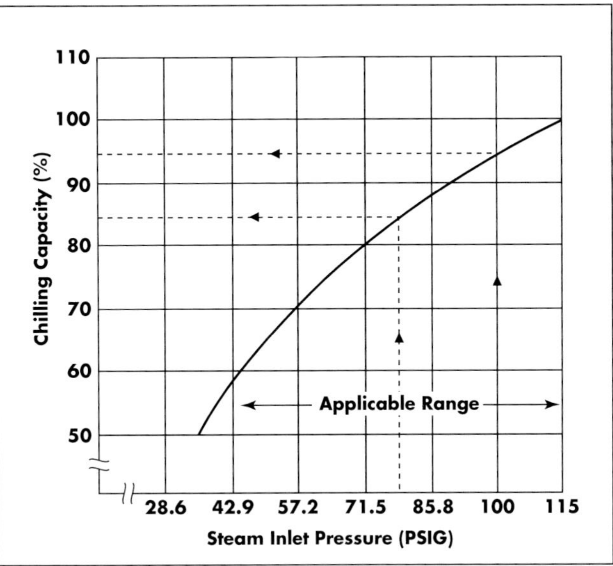

Fig. 38-11 Capacity Output vs. Steam Pressure for Double-Effect LiBr Absorption Machine. Source: York International

generator. In the parallel flow design, the solution stream leaving the low-temperature heat exchanger is split between the high-temperature heat exchanger and a secondary low-temperature generator. In the reverse flow design, weak solution is preheated by hot vapor from the first-stage generator before passing into the second-stage generator.

As with single-effect systems, capacity of double-effect units is reduced when operated at lower than design temperature. Figure 38-11 demonstrates typical capacity output versus steam pressure. Figure 38-12 shows

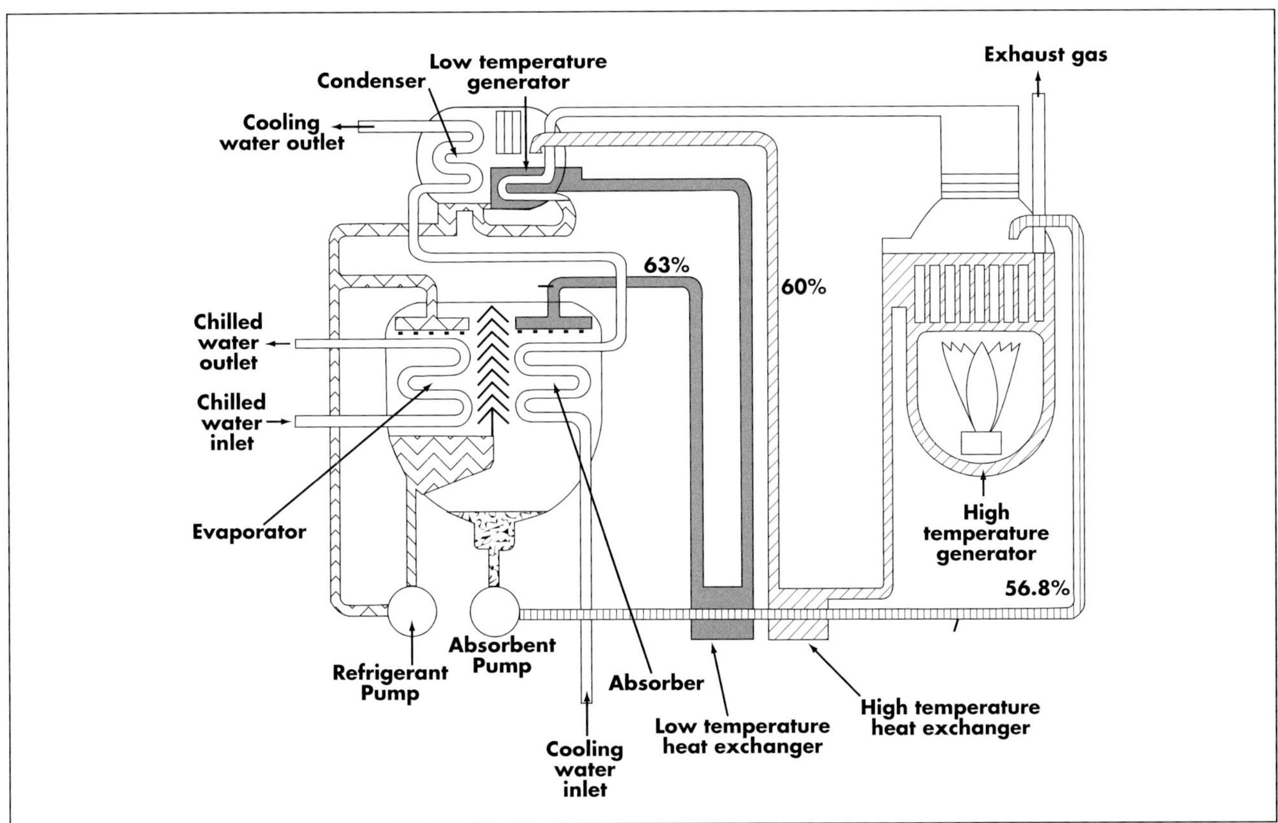

Fig. 38-10 Schematic Diagram of Series Flow Design. Source: Adapted from The American Gas Cooling Center

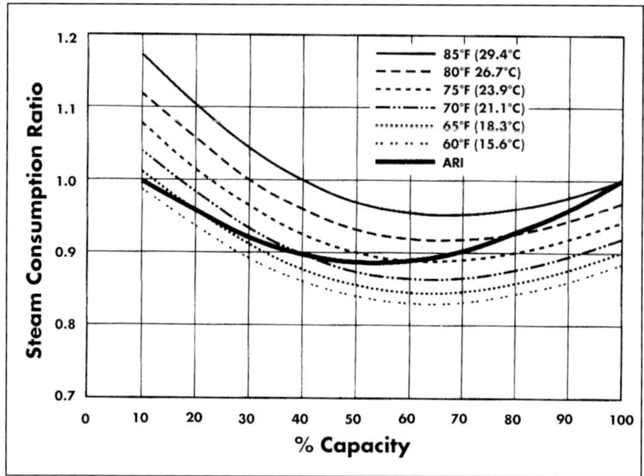

Fig. 38-12 Part-Load Performance Curves for a Double-Effect Steam LiBr Absorption Chiller. Source: Carrier Corp.

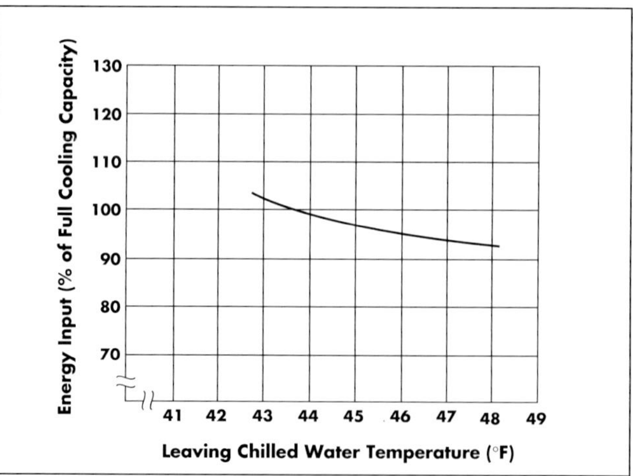

Fig. 38-13 Energy Input as a Function of Leaving Chilled Water Temperature for Double-Effect LiBr Absorption Chiller. Source: York International

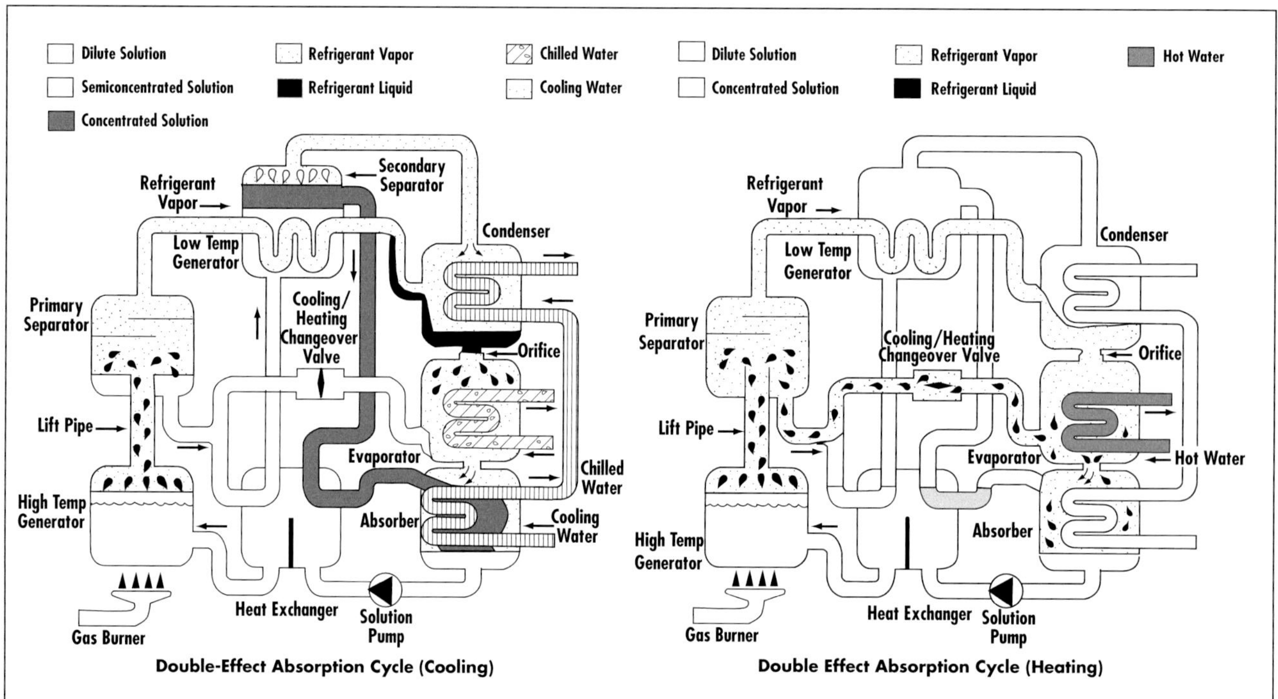

Fig. 38-14 Comparison of Absorption Cooling and Heating Cycles. Source: Adapted from American Yazaki Corp. and The American Gas Cooling Center

part-load performance curves for a double-effect unit. The curves depict steam consumption ratio (SCR) versus percent capacity at several condenser water temperatures. Figure 38-13 shows energy input, in percent of full-load cooling capacity, as a function of leaving chilled water temperature.

Internal electrical requirements range from 0.01 to 0.04 kW/ton (0.002 to 0.008 kW_e/kW_r), with a minimum of 0.004 kW/ton for some smaller machines. Heat rejection at typical rates of about 22,000 Btu/ton-h (1.83 kWh_h/kWh_r) may be achieved, with condenser water flow rates ranging from 3.0 to 4.5 gpm/ton (3.2 to 4.8 lpm/kW_r).

Absorption Chiller/Heaters

Absorption machines designed to provide both cooling and heating service are referred to as chiller/heaters. An advantage of the chiller/heater design is that it can either eliminate or reduce the requirement for a separate boiler, thus reducing capital cost and space requirements. Figure 38-14 provides an illustrative comparison of absorption cooling and heating cycles. The absorption heating cycle is illustrated in Figure 38-15 and described as follows:

1. Heating is enabled by opening a valve that allows hot refrigerant vapor to bypass the condenser and flow directly to the evaporator (or alternative auxiliary heat exchanger).
2. Condensing refrigerant heats the system water flowing through the evaporator tube bundle and overflows into the absorber where it dilutes the absorbent.
3. The weak absorbent is pumped from the absorber to the first- and second-stage generators. Strong absorbent is recovered in the direct-fired first-stage generator as combustion boils the entering dilute solution.
4. The strong absorbent liquid flows to the absorber and refrigerant vapor returns to the evaporator.

This sequence describes the operation of a double-effect LiBr absorption machine designed to produce 140°F (60°C) hot water. Available design alternatives are capable of producing higher temperatures of about 180°F (82°C).

Figure 38-16 is a schematic illustration of a direct-fired double-effect chiller/heater. Figures 38-17 and 38-18 show direct-fired double-effect chiller/heaters. Figure 38-17 shows a large-capacity system and Figure 38-18 shows a small-capacity system housed inside a cabinet.

Operation of double-effect direct-fired LiBr absorption machines designed for simultaneous heating and cooling basically follows the normal cooling cycle, with the exception that some of the refrigerant vapor produced in the first-stage generator is bled off into a hot water heat exchanger to meet heating load requirements. Because the vapor bypasses the second-stage generator, less cooling

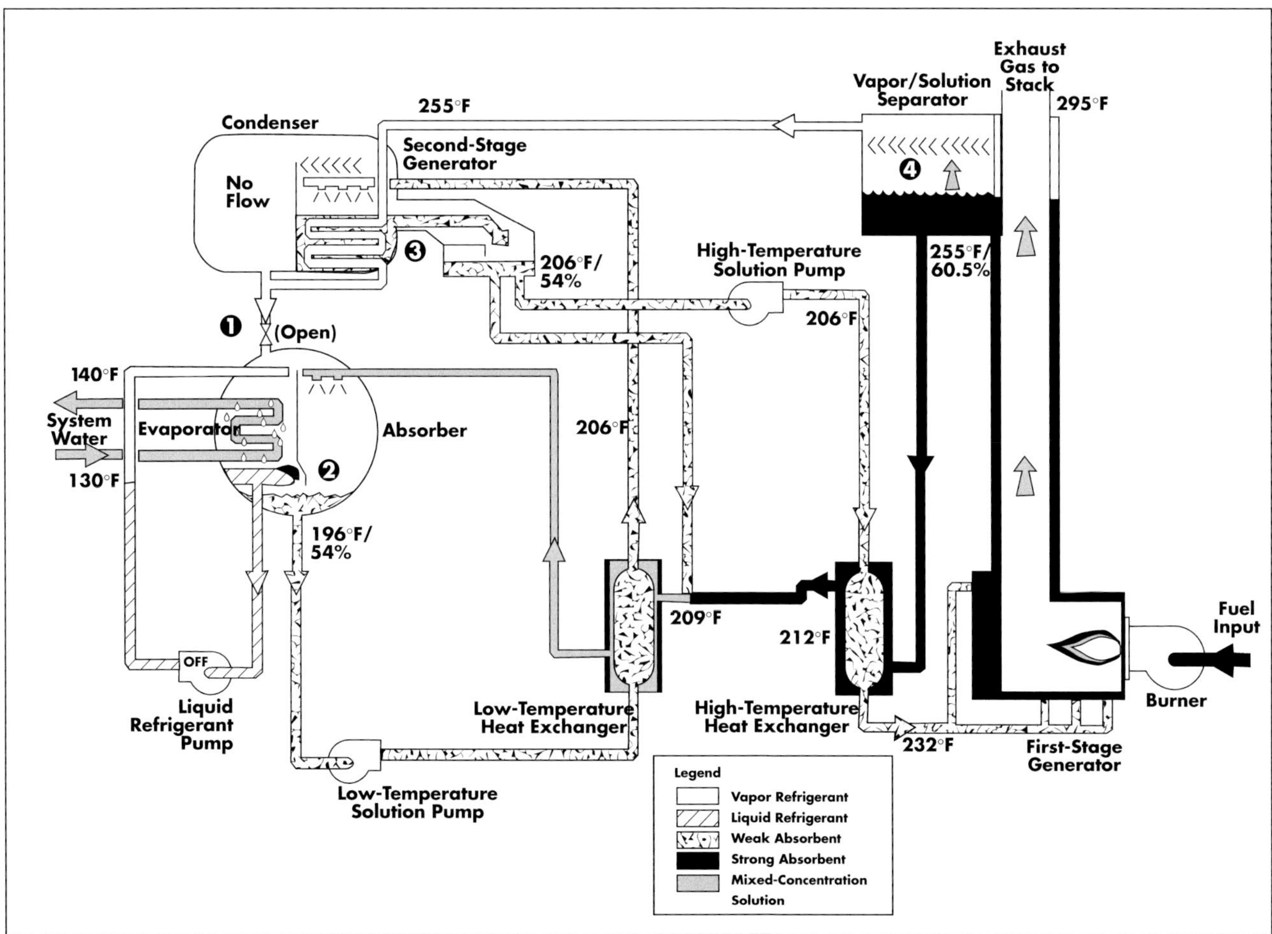

Fig. 38-15 Double-Effect Direct-Fired Absorption Heating-Only Cycle. Source: Adapted from The American Gas Cooling Center

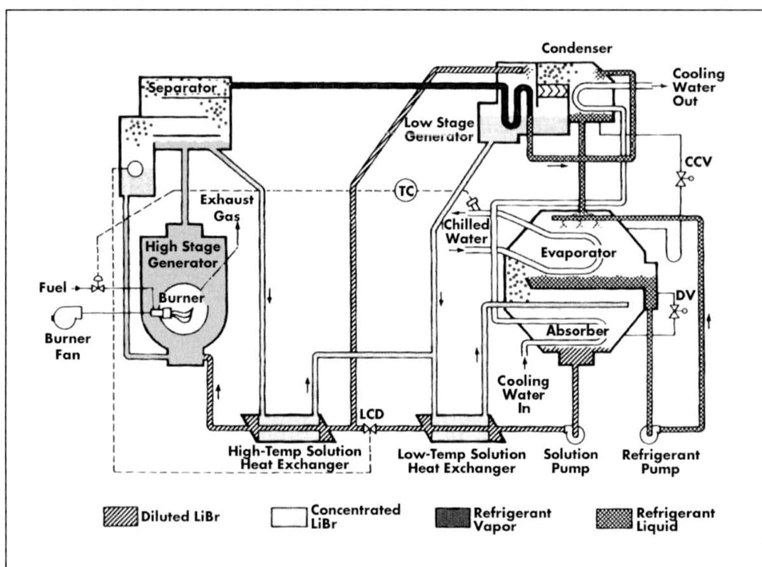

Fig. 38-16 Schematic Illustration of Direct-Fired Double-Effect LiBr Absorption Chiller/ Heater. Source: Carrier Corp. and The American Gas Cooling Center

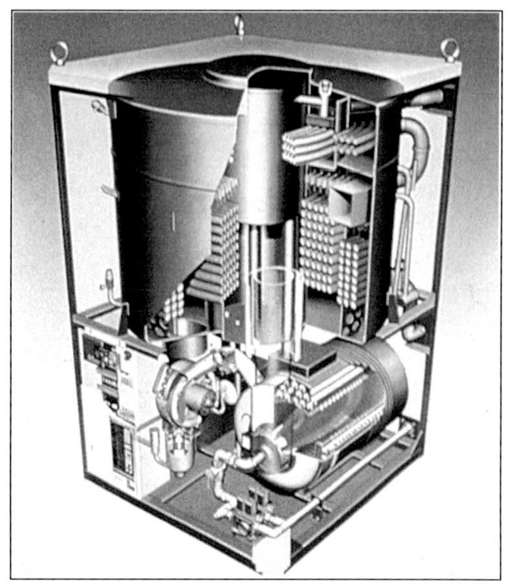

Fig. 38-18 Small-Capacity, Direct-Fired LiBr Absorption Chiller/Heater Inside Cabinet. Source: American Yazaki Corp.

capacity is available during simultaneous heating and cooling. Temperature controllers modulate the flow of refrigerant to maintain hot water and chilled water temperature set points. Either cooling or heating can be given priority, in case of unbalanced loads. Figure 38-19 shows how cooling capacity varies during simultaneous heating/cooling operation.

Heat Recovery Systems

Heat recovery or exhaust gas-fired chillers and chiller/heaters are double-effect machines that can be powered from a source providing clean, hot exhaust gases, such as a gas turbine or reciprocating engine. Direct coupling to a combustion engine can reduce installation cost and space requirements because a separate heat recovery boiler is not needed. They will, however, carry the premium cost associated with custom design as this is not a standardized item.

A 2,000 kW_e reciprocating engine-generator set can produce between 500 and 700 tons (1,760 to 2,460 kW_r) of absorption cooling under standard ARI conditions in a typical application, while a gas turbine of the same capacity can produce between 1,000 and 1,500 tons (3,500 to 5,200 kW_r) of cooling. Figures 38-20 and 38-21 show the available cooling and heating capacity, respectively, based on engine exhaust flow and temperature.

Fig. 38-17 Large-Capacity, Direct-Fired LiBr Absorption Chiller/ Heater. Source: Carrier Corp.

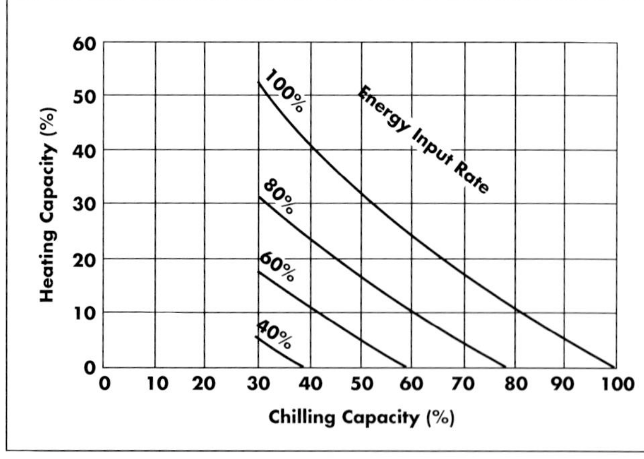

Fig. 38-19 Simultaneous Heating Capacity vs. Cooling Capacity. Source: York International

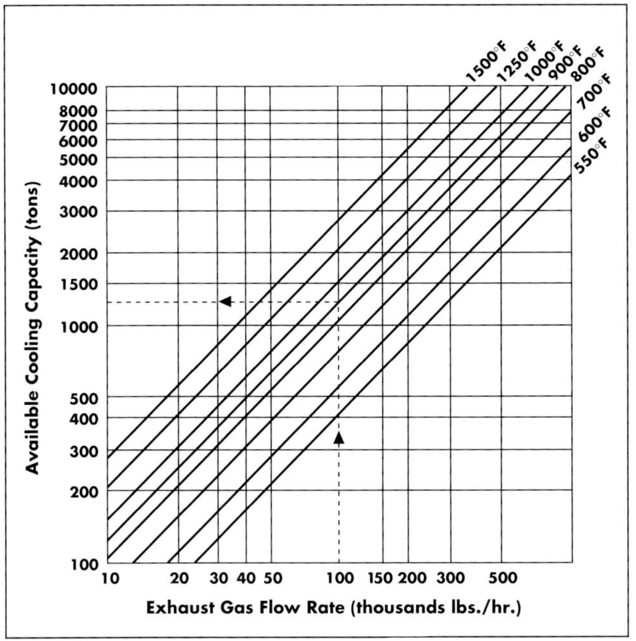

Fig. 38-20 Graph for Estimating Cooling Capacity Available vs. Engine Exhaust Rate and Temperature. Source: York International

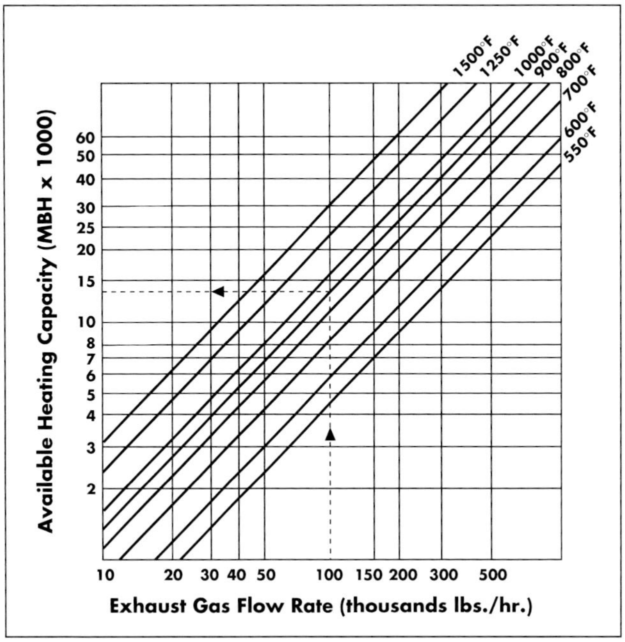

Fig. 38-21 Graph for Estimating Heating Capacity Available vs. Engine Exhaust Rate and Temperature. Source: York International

Triple-Effect Absorption Cycles

LiBr absorption machines using what is termed a triple-effect absorption cycle have been in prototype development for many years. These triple-effect absorption chillers would be direct-fired designs and provide an expected 50% thermal efficiency improvement over existing double-effect units. Units currently under development are expected to have full-load COPs in excess of 1.4 with IPLV COPs better than 1.6, under standard ARI conditions. Cooling tower and makeup water requirements will also be reduced due to the lower heat rejection associated with the lower cycle energy input requirement.

While these systems do not feature a distinct third stage, they increase the use of internally recovered heat to achieve high thermal efficiencies. One design under development consists of two cascaded single-effect chillers, one operating at conventional temperatures and the other at higher temperatures. The smaller high-stage is energized by direct fuel combustion, producing a generator temperature of 450°F (222°C). High-stage heat rejection at about 200°F (93°C) serves as the energy source for a conventional LiBr low-stage cycle. Figure 38-22 is a schematic diagram of one triple-effect cycle currently under development. At this time it is unclear when or if this technology will reach commercialization.

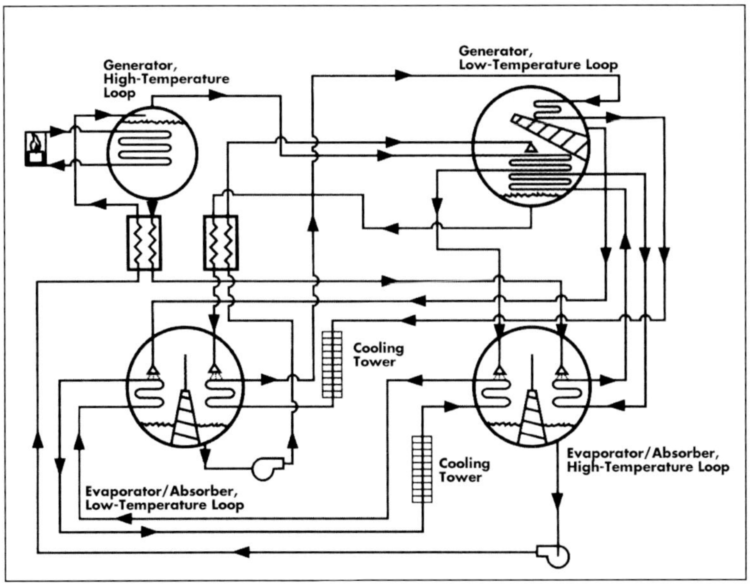

Fig. 38-22 Schematic Diagram of Triple-Effect Absorption Cycle System. Source: GRI

Ammonia-Water Absorption Cycles

Ammonia-water (NH_3-H_2O) based absorption machines have been used for most of this century, notably in residential refrigerators and air conditioners. Presently, the technology is used in custom-designed systems for low-temperature industrial applications and in small-capacity, direct-fired air-cooled units. Typical units are 3, 4, or 5 tons (10, 14, or 17 kW_r) capacity, with up to five units assembled on a single skid. Figure 38-23 shows a 50 ton (166 kW_r) system featuring two 25 ton (88 kW_r) factory-assembled modules. Larger capacity, higher efficiency models are becoming more common and are the subject of extensive R&D efforts.

In the ammonia-water cycle, water acts as the absorbent and the ammonia-water solution acts as the refrigerant. Both the refrigerant and the absorbent boil in the generator, and the regeneration of strong absorbent is a fractional distillation process. Figure 38-24 shows a schematic diagram of an ammonia-water absorption chiller. In this diagram, heat energy is designated Q, with the arrows showing the direction of flow. A refrigerant heat exchanger (RHX), also called a precooler, is commonly placed upstream of the evaporator to transfer heat to the absorber.

Currently, double-effect LiBr absorption machines achieve higher cooling cycle COPs than do ammonia-water units. There are, however, certain advantages offered by ammonia-water cycle systems. A limitation of LiBr machines is their inability to achieve evaporator temperatures below about 40°F (4.4°C). In contrast, the ammonia-water cycle allows extremely low evaporator temperatures to be achieved, and for very low applications, is competitive with vapor compression technology on a cost and efficiency basis. Such systems have been effectively applied for various process refrigeration applications, including ice storage.

The ammonia-based refrigerant also allows the cycle to be operated at condenser pressures of up to 300 psia (20.1 bar) and at evaporator pressures of about 70 psia (4.8 bar), allowing vessels to be under 6 in. (15 cm) diameter. The solution pumps in small-capacity units must be positive-displacement types. Also, due to the higher condenser pressures, ammonia-water systems can be more effectively air-cooled than LiBr-based systems, avoiding the capital cost and complexity associated with evaporative cooling.

GAX Cycles

Another promising area of absorption technology development involves the generator-absorber heat exchange (GAX) cycle. While this cycle has gone undeveloped since its invention at the beginning of this century, its simplicity and potential for adaptation to more efficient variations has led to recent intensive development activity. In the GAX cycle, generation (desorption) and absorption each occur over a large temperature glide (a wide concentration range), allowing for a temperature overlap between the hot end of the absorber and the cold end of the generator. In this overlap zone, heat can be transferred from the absorber to the generator, directly or through a circulating heat transfer fluid. As this temperature overlap is increased,

Fig. 38-23 Two 25-Ton Factory-Assembled, Direct-Fired, Air-cooled Ammonia Absorption Chiller Systems on Single Skids.
Source: Robur Corporation and The American Gas Cooling Center

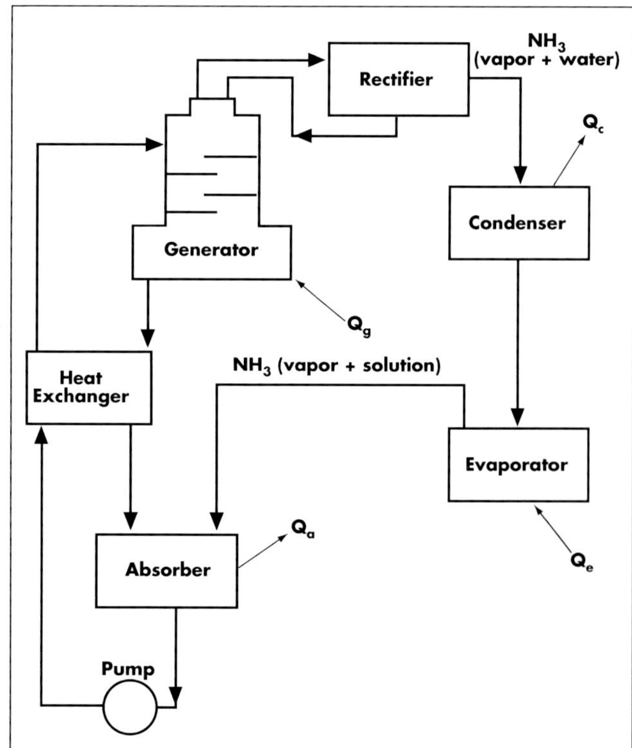

Fig. 38-24 Schematic Diagram of Ammonia-Water Absorption System.

the potential cycle COP is increased. In addition, heat can be recovered from the absorber section at temperatures of up to 180°F (82°C), making the cycle well-suited for heat pump application. This is one of the primary focuses of this developing technology.

The basic GAX-cycle system, shown in Figure 38-25, has two pressure levels and a single sorbent circulation path. There is no temperature overlap until the generator temperature increases to a point at which the temperature difference between the generator and the condenser is about 2.6 times the lift (temperature difference between the condenser and the evaporator). At full load, high COP can only be achieved when the input energy is significantly hotter than required for a conventional single-effect absorption machine. Under part load, at reduced lift, temperature overlap increases, improving COP.

In one design variation, the branched GAX cycle, solution flow through the temperature overlap portion of the absorber is increased in order to improve heat release. The increased sorbent flow is pumped to an appropriate midpoint in the generator, reducing flow through the cold end of the generator with a corresponding decrease in input heat requirement. In another variation, the vapor exchange (VX) GAX-cycle system, a third (intermediate) pressure level is introduced along with an auxiliary generator and absorber. External cooling is applied to the auxiliary absorber and a second GAX exchanger transfers otherwise wasted heat to the auxiliary generator. Figure 38-26 is a schematic diagram of a VX GAX-cycle system.

GAX-cycle systems are currently available for residential and light commercial applications with a COP of approximately 0.60. Target COPs for these air-cooled units range from 0.70 to 0.95 under ARI standard conditions.

Operation and Maintenance

Absorption chillers are characterized by their simple design with few moving parts and relatively low operating temperatures and pressures. Following is a description of some of the basic recommended O&M practices.

Regular O&M of the purge unit at manufacturer recommended intervals is required to maintain design pressure within the system. The purge pump oil level must be checked and changed as necessary, the pump motor bearings must be oiled, and the belt tension must be adjusted as needed. A monthly inspection of the non-condensable accumulation rate should also be performed.

Manufacturer's specifications should be followed for inspection of refrigerant and solution pump-motor assemblies. Typically, bearings and shaft seals should be inspected every three years and hermetic pumps every five years.

Absorption machines use corrosion inhibitors to protect internal components from corrosion and to limit the formation of non-condensable gas. They also use a heat-transfer additive to improve refrigeration capacity, particularly during initial start-up. The solution must be analyzed each cooling season, and more frequently when chillers operate year-round, to ensure that both inhibitor and additive levels are within acceptable limits and to determine when replenishment is required. Periodic analysis of refrigerant water is also recommended and manufacturer's procedures should be followed for treatment of evaporator and condenser water to ensure coverage by equipment warrantees.

Evaporator, condenser, and absorber tube surfaces require ongoing maintenance. The primary fouling deposits are scaling and sedimentation (or sludge). Scale

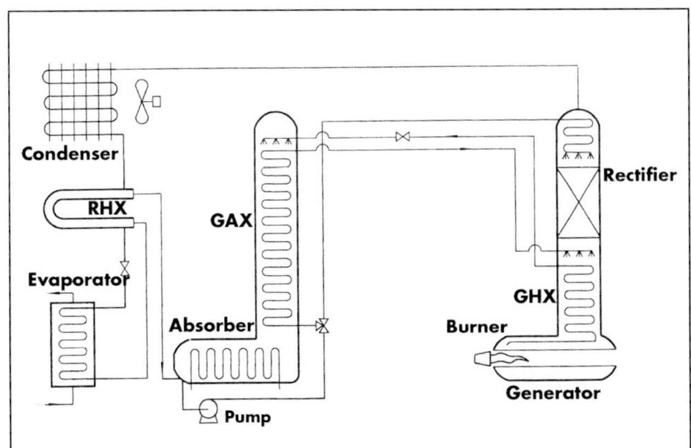

Fig. 38-25 Basic GAX-Cycle Schematic Diagram.
Source: Energy Concepts Co.

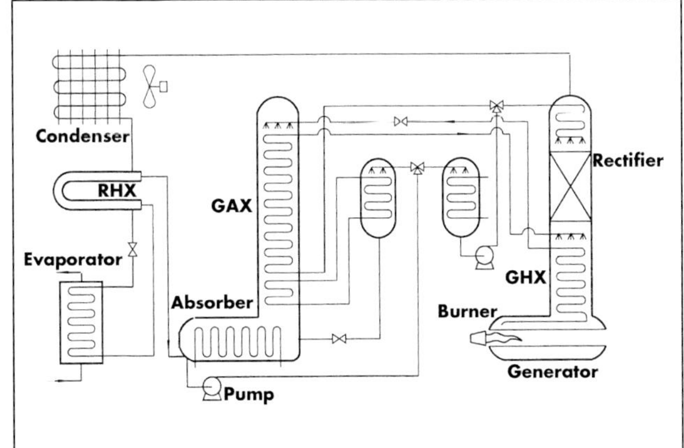

Fig. 38-26 VX GAX-Cycle Schematic Diagram.
Source: Energy Concepts Co.

deposits can be removed by chemical cleaning. Sludge deposits must be removed mechanically by removing the headers, loosening the deposits with a brush, and flushing the loosened materials from the tubes with water.

In order to minimize long-term maintenance costs, ready access should be provided to the tubes for cleaning and inspection. Marine water boxes are useful in this regard. Whereas standard water boxes require disassembly of the water piping and removal of the water box for tube access, marine boxes provide side connections for water piping and a removable cover on the end of the box for full access to the tubes.

For direct-fired units, the combustion air fan and motor must be inspected at least semiannually and lubricated annually. Exhaust flue, flue damper, and exhaust devices should be inspected semiannually or in accordance with the manufacturer's specifications. Exhaust flue gases should be analyzed for emissions and for combustion performance evaluation.

Combustion controls (i.e., gas/oil pressure and airflow switches, solenoid valves and proof-of-flame controls) should be checked regularly, typically every one to three months. The gas train valves and switches and, where applicable, oil pump and oil supply system should be inspected and tested regularly for leaks. If the equipment area is subject to adverse conditions, more frequent inspections are recommended.

Crystallization

Crystallization, or solidification, of the LiBr solution has been an important historical operating concern of absorption chiller operators. As the temperature of the solution decreases, the maximum possible concentration also decreases and it can begin to solidify and block piping in the machine. It is then necessary to heat the piping where the blockage has occurred in order to melt the crystals. Crystallization can result from condensing water temperatures becoming too low, from operation under very low load, or as a result of other occurrences, such as power failure or air leakage into the system.

Typically, condenser water is limited to a minimum temperature of about 55 to 60°F (13 to 16°C), with a bypass valve used to maintain this temperature limit. Most designs also include an overflow pipe, which allows heated solution to pass from the generator to the absorber section and raises the temperature via a heat exchanger to liquefy any crystallization. This configuration allows for automatic decrystallization whenever necessary.

An important improvement in absorption equipment is the development of various flow schemes that allow the equipment to operate at lower solution concentrations and temperatures, reducing the potential for crystallization. Modern microprocessor based controls have virtually eliminated crystallization nuisances, in part through the development of internal responses to rapidly changing external conditions.

Performance Comparison

Figure 38-27 compares conventional absorption machines including single-effect (1E) NH_3 and LiBr and double-effect (2E) LiBr machines, as well as developing GAX machines, including the basic GAX-cycle system and the VX and Branched (B) GAX-cycle systems. The assumed condenser temperature is 91°F (33°C) and lifts range from 126 to 18°F (70 to 10°C), corresponding to evaporator temperatures of between -40 and 41°F (-40 to 5°C). The following characteristics are evident in the figure:

- The COP of the conventional single-effect NH_3 machines remains fairly constant over the full range, indicating superior performance under high-lift conditions and relatively poor performance under low-lift conditions.
- NH_3 cycles can operate at very low evaporator temperatures, while the LiBr machines are limited to the higher temperature range.
- Relatively high prospective COPs are indicated for advanced GAX cycles, particularly as cycle lift is decreased. The VX GAX-cycle systems show the best COP under high-lift conditions, indicating attractiveness for industrial refrigeration applications.

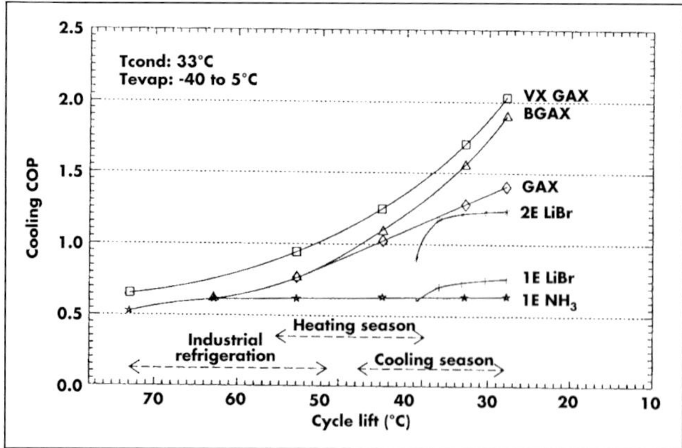

Fig. 38-27 Comparison of Absorption Cycle COPs.
Source: D.C. Erickson and M.V. Rane, Energy Concepts Co.

Absorption Technology Applications

Selection Criteria

Application feasibility of absorption technology depends largely on the relative cost of fuel and electricity. Where applicable, the combination of low summer fuel costs and high electricity costs may result in strong economic performance for absorption cycle systems versus conventional electric motor-driven refrigeration.

Availability of recovered heat can also favor application of absorption technology. In a cogeneration application, absorption refrigeration can use heat recovered from a high-temperature application, such as a prime mover applied to electric generation or mechanical drive refrigeration service.

Similar to prime mover vapor compression system driver options, which are discussed in the previous chapter, another factor that may support superior economic performance for a given absorption technology application is the condition of limited electric service capacity. This may be due to on-site logistics or unavailability of utility substation capacity.

Capital costs for single-effect absorption machines are comparable to electric motor-driven vapor compression systems of the same capacity and about half the cost of double-effect machines. While, historically, single-effect steam units were the only absorption machines available for central cooling plant applications, double-effect LiBr machines are now more commonly applied for applications that do not use recovered heat. The superior COP and reduced heat rejection of double-effect units often outweigh the capital cost advantage of the low-pressure single-stage steam units. Double-effect systems also require smaller pumps and cooling towers which reduces the capital cost impact.

Still, low-pressure steam (or hot water) units remain a viable technology. They can be cost-effectively applied when fuel (and make-up water) costs are low. More commonly, however, single-effect absorption chillers are used with cogeneration cycle applications and other applications using recovered heat.

Heat Recovery Applications

Often, a limiting factor to the economic performance of cogeneration-cycle technology applications is the availability of an adequately sized baseload requirement for the heat output. Often, the capacity of a cogeneration system will be limited by low cooling season thermal loads as compared with the heating season. Absorption cooling technology is thus commonly specified as a means to balance annual thermal loads, thereby improving cogeneration application project economics.

Double-effect LiBr absorption units requiring high-temperature heat can be effectively applied using exhaust from combustion engines or other processes or when steam turbine outlet pressures are sufficiently high (usually in excess of 85 psig/6.9 bar). However, low-temperature single-effect systems are more economical for many heat recovery applications. In some cases, the higher capital cost of high-temperature heat recovery equipment partially offsets the operating cost advantage of double-effect units.

In reciprocating engine applications, for example, only a portion of available waste heat is of high enough temperature for effective operation of a double-effect absorption machine. A single-effect system can be powered either by low-temperature waste heat from the engine's coolant system, by heat recovery from high-temperature exhaust, or a combination of both.

A non-condensing steam turbine-driven vapor compression system with a "piggybacked" single-stage absorption machine can sometimes be a cost-effective combination. In this arrangement, discharge steam from the turbine is condensed in the absorption chiller, avoiding capital costs of a condensing turbine system. Because most non-condensing steam turbine applications require minimum pressure ratios of around 4 or 5 to 1, most applications are not compatible with high-pressure absorption chillers, but are better matched with single-effect units using 15 psig (2 bar) steam.

Low-pressure steam absorption machines can also allow an intermediate, or second topping cycle to be used. For example, a gas turbine cogeneration unit producing high-pressure steam can discharge to a back-pressure steam turbine generating additional power and discharging low-pressure steam for absorption cooling in summer and space heating in winter.

Figure 38-28 shows four large-capacity absorption chillers powered by steam generated using recovered process heat. Figure 38-29 shows a 200 ton (700 kW_r) single-effect absorption chiller powered by low-pressure steam generated by recovered heat from a reciprocating engine.

Another effective use of recovered heat-powered absorption cooling technology is for gas turbine inlet cooling. As described in Chapter 10, the thermal performance of gas turbine systems is improved when low

Fig. 38-28 Absorption Chillers Powered by Heat Recovery-Generated Steam in Industrial Process Application. Source: The Trane Company

Fig. 38-29 Single-Effect Absorption Chiller Powered by Heat Recovered from a Reciprocating Engine. Source: Waukesha Engine Division

inlet air temperatures exist. As a rule of thumb, if recovered heat is used to power a source of inlet air cooling, nearly half of the cooling capacity is recovered through increased thermal output. Figure 38-30 illustrates this arrangement. Figure 38-31 shows a 1,400 ton (4,900 kW_r) single-stage absorption chiller that provides inlet air cooling for a 48 MW gas turbine-driven electric generation plant. Steam, at a pressure of about 15 psig (2 bar), is provided to the chiller from the low-pressure section of the heat recovery steam generator. With this arrangement, under peak ambient temperature conditions, power output is increased by about 10% versus what could be achieved with simple evaporative inlet cooling.

LOW-TEMPERATURE ABSORPTION APPLICATIONS

For low-temperature applications, ammonia-water absorption technology has long been an effective option. Some of the newly developed advanced cycles are already being applied. Recently, an absorption ice maker, based on a modified VX GAX-cycle system, was successfully demonstrated in a process refrigeration application. This system, shown in Figure 38-32, is powered by 160°F (71°C) hot water recovered from a reciprocating engine coolant system and produces 5°F (-15°C) refrigeration.

While excluded from consideration for very low-temperature systems, LiBr-based systems can be used in series with vapor compression chillers to produce chilled water somewhat below the unit's 40°F (4.4°C) lower limit. Chilled water at 38 or 39°F (3.3 to 3.9°C) may be required in some air conditioning applications, for example, with strict humidity requirements. Another strategy is the use of absorption chillers in conjunction with desiccant dehumidification. In this case, the desiccant system meets the latent (dehumidification) load, while the chilled water from the absorption unit meets the sensible load. Desiccant systems are discussed in Chapter 39.

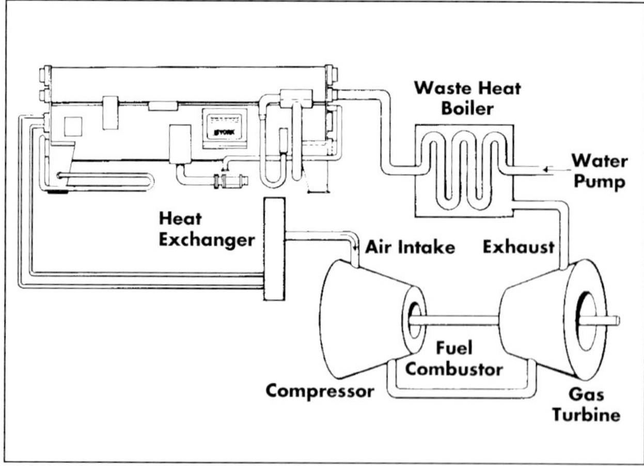

Fig. 38-30 Illustration of Absorption Chiller Powered by Heat Recovery Steam Applied to Gas Turbine Inlet Air Cooling. Source: York International

Fig. 38-31 1,400 Ton Single-Effect Absorption Chiller Operating on Heat Recovery-Generated Steam, Providing Inlet Air Cooling for 48 MW Gas Turbine-Generator Set. Source: The Trane Company and The American Gas Cooling Center

Fig. 38-32 12 Ton/Day Recovered Heat-Powered Absorption Ice Maker Operating on a Modified VX GAX Cycle.
Source: Energy Concepts Inc.

Chapter Thirty-Nine

Desiccant Dehumidification Technologies

Cooling loads have both sensible and latent components. Sensible cooling is the removal of (dry) heat. Latent cooling is a dehumidification process for moisture (wet heat) removal. Desiccant technologies can often be effectively applied to provide latent cooling, or dehumidification, due to the ability to alternately attract and reject moisture. In desiccant dehumidification applications, moisture in an airstream is removed by passing the air over a substance that absorbs moisture.

Desiccants are various solid or liquid substances that, due to their very strong affinity for moisture, attract water vapor directly from an airstream. Once the desiccant material has become saturated with moisture, the moisture can be driven off by applying heat to the desiccant. Heat can be applied by direct fuel-firing, steam, hot water, or waste heat. This process, which continually renews the desiccant's ability to collect moisture, is referred to as **regeneration**.

Desiccant dehumidification is a well-established technology for applications that call for low humidity levels. It provides for the ability to precisely control humidity levels in achieving very low dewpoint conditions. Mechanical (vapor compression) systems are challenged to economically achieve dewpoints below 40°F (4°C) — coil frosting and the need for large coil surface areas are key limitations. Also, they must often overcool an airstream to achieve the desired humidity level and then reheat the air to meet acceptable space temperature conditions. In addition to potential thermal performance benefits, desiccant air drying can also remove harmful airborne pollutants and bacteria, reduce equipment corrosion due to moisture and impurities, and protect product integrity.

Conventional and Desiccant Cooling

Figure 39-1 compares conventional and desiccant cooling processes using the psychrometric chart. As shown on the chart, Process 1-2 in conventional vapor compression cycle systems consists exclusively of sensible cooling. Dehumidification (latent cooling) begins at Point 2 after the airstream has been cooled to saturation. Additional sensible cooling also occurs in Process 2-3. Finally, once the desired dehumidification has been accomplished, reheating is often required (Process 3-4) to bring the discharge temperature up to the required level. In the desiccant system process, air entering the system is dehumidified and sensibly heated in Process 1-2. The air is then sensibly cooled, often with conventional vapor compression or absorption systems cooling, to the desired temperature indicated by Point 4.

In the traditional air conditioning process, dehumidification is accomplished by passing the airstream to be conditioned over a cooling coil surface that is cold enough to cause condensation. If the cooling coil did not have to dehumidify as well as sensibly cool the airstream, the refrigeration system could operate at higher (suction) temperatures and would, therefore, be more thermally efficient. In addition, properly dehumidified air may be too cold to maintain the desired temperature in the conditioned space. Expenditure of energy to reheat the airstream may be needed to maintain comfortable conditions.

Desiccant air drying is often a more effective way to reduce latent load than mechanical compression systems. Generally, the greater the requirement for dehumidification, the more thermally efficient desiccant systems are relative to conventional air conditioning (condensation) processes.

In general, desiccants tend to be more effective than mechanical compression systems when:

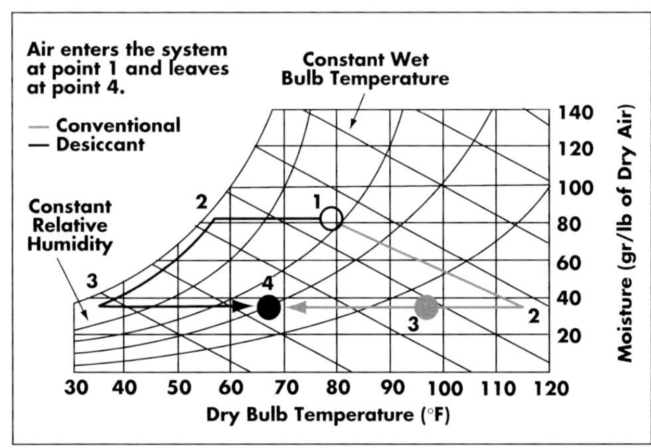

Fig. 39-1 Simplified Psychrometric Chart Comparison.

- Ratio of latent to sensible load is high.
- Low humidity levels are required (or desired).
- Electric rates are high relative to fuel or recovered heat is available.
- Mold or mildew problems exist in the ductwork or building.

Desiccant Air Drying Process

Desiccant dehumidifiers do not cool air to remove moisture. They attract moisture from air by creating an area of low vapor pressure at the surface of the desiccant. The pressure of the water in the air is higher, so the water vapor molecules move from the air to the desiccant and the air is dehumidified. As a wet desiccant is heated, its vapor pressure becomes high and it will give off moisture to the surrounding air. Water vapor moves back and forth from the air to the desiccant, depending on the vapor pressure difference. Figures 39-2a through 39-2c describe the process diagramatically.

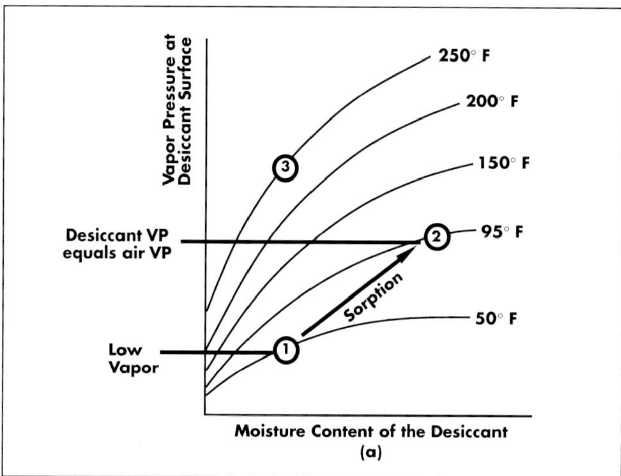

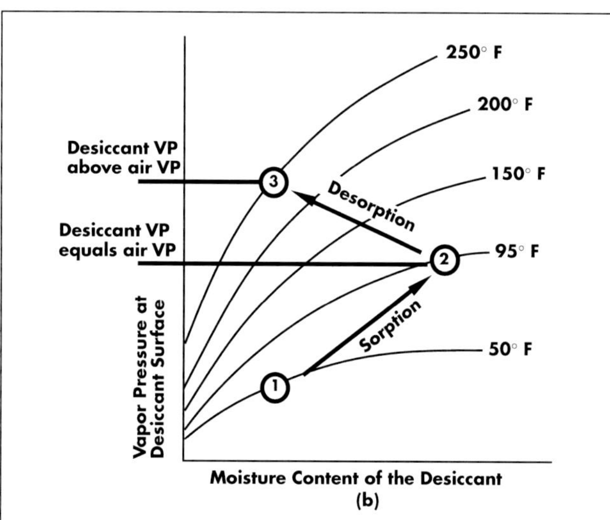

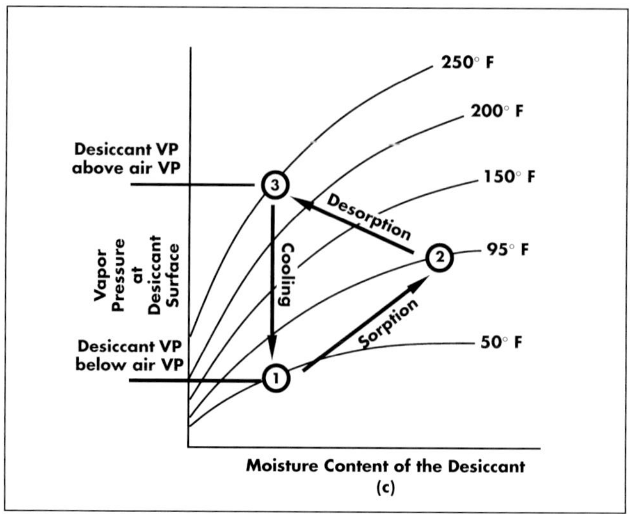

Figs. 39-2a, b, and c Diagrammatic Representations of Desiccant Process. Source: Munters Cargocaire and Mason Grant Company

The process of removing moisture from air with a desiccant is termed **sorption**. When water vapor is sorbed by a desiccant (sorbent), the water changes phase from vapor to sorbed liquid. This phase change releases energy, primarily the latent heat of condensation, plus a small additional amount (approximately 10 to 20%) due to the attraction between the sorbed water and the desiccant. The total heat released is referred to as the **heat of sorption**. The release of latent energy results in an increase in the temperature of the desiccant and the surrounding airstream.

The thermodynamics of the sorption process are similar to the reverse of what happens in an evaporative cooler. As moisture is removed from the airstream, its sensible temperature rises. The rise in air temperature is directly proportional to the amount of moisture removed. The dryer the air, the warmer it will be.

The critical element of the sorption process is reversibility. Desiccants are regenerated by being heated to temperatures above that in the sorption process, then placed in a different airstream. The high vapor pressure desiccant surface gives off, or **desorbs**, moisture to the air to equalize the vapor pressure differential. The dry, hot regenerated desiccant is then cooled to restore its low vapor pressure and is returned to the moist airstream. Typical regeneration temperatures range from 130 to 250°F (54 to 121°C). Thermal energy can be applied with direct firing of fuel or by steam, hot water, or waste heat. Figure 39-3 is a block diagram of the desiccant air conditioning process.

The sorption cycle has three parts:

1. Exposing the desiccant to the airstream to be conditioned when it is cool and dry (low vapor pressure) and can attract moisture.
2. Heating the desiccant to further increase its vapor pressure to a point higher than that of the discharge airstream so that moisture will move off of the surface to the air to equalize the pressure differential.
3. Cooling the desiccant to restore its low vapor pressure so that it can begin attracting moisture once again.

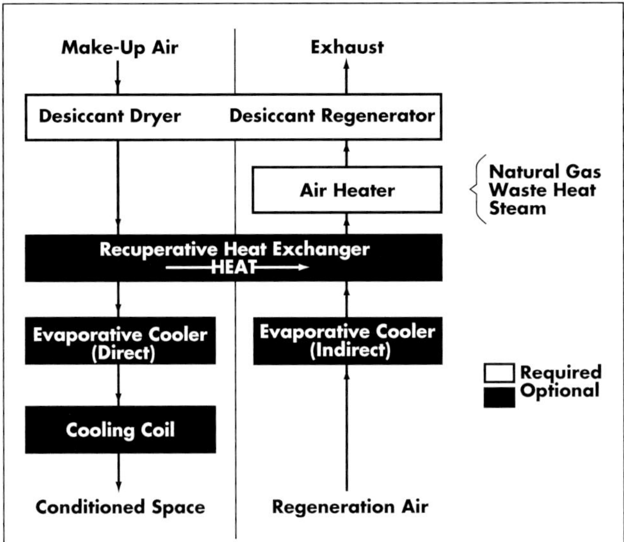

Fig. 39-3 Block Diagram of Desiccant Air Conditioning Process.

In the sorption process, the removal of moisture from the air results in a release of heat (heat of sorption) that raises the air temperature as the process converts latent cooling load to sensible cooling load. Figure 39-4 shows three sensible cooling options that can be incorporated in desiccant systems to provide sensible cooling. The recuperative heat exchanger transfers heat to the air used to regenerate the desiccant. This can be supplemented by or replaced with a conventional chiller. The direct evaporative cooler evaporates water into the warm, dry air to cool it. Usually, this process cannot provide the full measure of required sensible cooling. The indirect evaporative cooler evaporatively cools a separate airstream and uses it to cool the dry air through an air-to-air heat exchanger.

Desiccant Materials

There are many different types of solid and liquid desiccant materials. The common element is their capacity to collect moisture. Desiccants that collect moisture

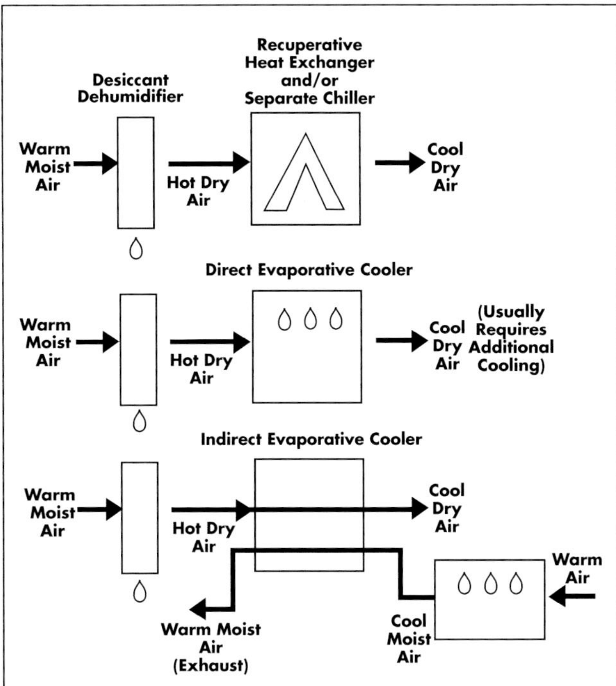

Fig. 39-4 Three Sensible Cooling Options for Use with Desiccant Systems.

through strong molecular attraction (much like a sponge) and hold it on the surface of the material are called **adsorbents**. These are mostly solid materials such as silica gel. Other desiccants, called **absorbents**, undergo a physical change as they collect moisture. These are usually liquids, or solids that readily become liquids as they absorb moisture. Following is detail on various types of solid adsorbents and liquid absorbents.

Solid Adsorbents

Solid desiccants are materials with a tremendous internal surface area per unit mass, such as silica gel, molecular sieves, activated alumina, or hygroscopic salts. The capacity of solid desiccants per unit mass, however, is usually less than that of liquids.

- **Silica gels** are formed by condensing soluble silicates from water or solvent solutions. They are relatively inexpensive and easy to customize. They are available in beads as large as 0.2 in (0.5 cm) in diameter or as fine powders.
- **Zeolites** are aluminosilicate minerals that occur in nature. They have an open crystalline lattice that attracts and traps molecules, such as water vapor, like objects in a cage. Particular atoms of the material determine the size of the openings or gaps in the cage-like structure, which in turn governs the size of

the molecules that can be adsorbed.

- **Synthetic zeolites (molecular sieves)** are manufactured crystalline aluminosilicates. By controlling the temperature of the thermal manufacturing process and the materials used, the zeolite's structure and surface characteristics can be closely controlled. While more costly than naturally occurring zeolites, this process can produce a more uniform product.
- **Activated alumina** are manufactured oxides and hydrides of aluminum. Their structures can be controlled by the gases, temperatures, and duration of the thermal manufacturing process.
- **Carbons** are usually used to adsorb gases other than water vapor, because they have an affinity for nonpolar molecules, such as organic solvents. Carbons have a large internal surface and very large capillaries.
- **Solid polymers** are long molecules twisted together like strands of string. The sodium ions in these molecules can each bind several water molecules, and the spaces between the strands can also contain condensed water. This polymer's capacity exceeds that of many solid adsorbents, particularly at high relative humidities.

Liquid Absorbents

Liquid desiccants have a vapor pressure lower than water at the same temperature. The air passing over the solution approaches this reduced vapor pressure and is dehumidified. If the vapor pressure of the solution is greater than that of the water, the air is humidified. At a given concentration and temperature, liquid desiccant solutions are in equilibrium with air at a fixed humidity. The vapor pressure of liquid absorption solution is directly proportional to its temperature and inversely proportional to its concentration. If concentration is reduced or the temperature is increased, the vapor pressure increases. If concentration is increased or temperature reduced, the vapor pressure decreases.

- **Hygroscopic salt solutions** are water solutions with salts such as lithium chloride (LiCl) or calcium chloride (CaCl). While more costly than CaCl, LiCl is more commonly used because it is more effective and less corrosive. At 40% concentration and 70°F (21°C), LiCl is in equilibrium with air at 19.3 grains/lbm (0.0027 kg/kg). The concentration is determined by the specific gravity. Concentration limits determine the maximum and minimum allowable solutions. If cooled sufficiently, LiCl will form ice at concentrations below 33%. At concentrations above 33% it will become supersaturated and form solid salt. Compared with its use as a solid, liquid LiCl has a far greater ability to hold water molecules. In a solid state, each LiCl molecule can hold two molecules of water. As a solution in equilibrium with air at a 90% RH, each molecule can hold 26 molecules of water.
- **Glycols** exhibit similar characteristics to hygroscopic salt solutions, except that they require much higher equilibrium concentrations and tend to evaporate. Typical equilibrium concentrations are greater than 90%, which is more than double that of hygroscopic salts. Systems, therefore, require far more solution given that there is only a small percent of water to work with. Also due to the tendency to evaporate, requiring frequent solution replacement, glycols are more commonly used for lower temperature applications when evaporation rates are reduced. Commonly used glycols are triethylene glycol and propylene glycol. Propylene glycol is commonly used in food processing and frost-free applications due to its low toxicity. In other types of applications, triethylene glycol is used due to its lower evaporation rate.

Types of Desiccant Dehumidification Systems

While there are several types and variations of desiccant air drying systems, five basic types are: liquid spray tower, solid tower, multiple vertical bed, rotating bed, and rotating wheel. Each has unique characteristics in terms of efficiency, complexity, cost, and dewpoint range. Selection will depend on the specific capacity and conditions of the application.

Liquid Spray Tower

Liquid desiccant systems operate on the principle of chemical absorption of water vapor from air. A representative liquid system featuring LiCl solution is shown schematically in Figure 39-5. In this system, air to be conditioned is cooled and dehumidified by contacting the desiccant (absorbent) solution in the conditioner. Heat extracted from the air by the desiccant solution is transferred to the coolant by continuously recirculating the solution through a heat exchanger. Coolant flow through the heat exchanger is modulated to control the amount of heat and moisture extracted from the air.

Figure 39-6 shows a small packaged LiCl liquid system with a conditioner airflow capability of up to 3,000 cfm (85 m^3/m). To remove the water extracted from the

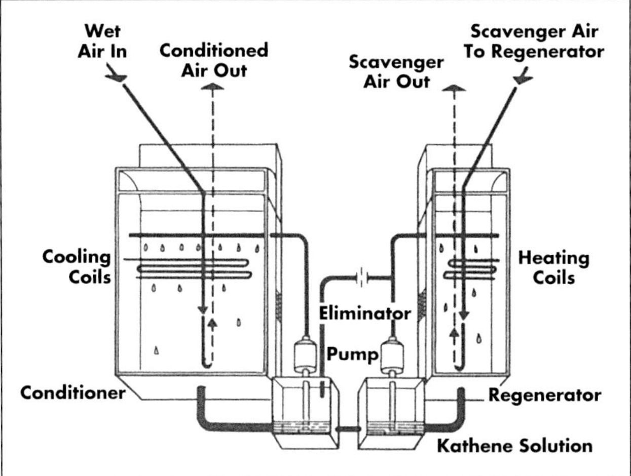

Fig. 39-5 Schematic Diagram of Liquid System. Source: Kathabar Systems

air and keep the desiccant solution at a fixed concentration, a small amount of solution is transferred from the conditioner to the regenerator. Heated desiccant solution contacts a scavenger airstream and is concentrated by evaporation of water from the solution to the air. Heat flow to the heat exchanger is modulated to match the moisture load in the conditioner. The amount of moisture

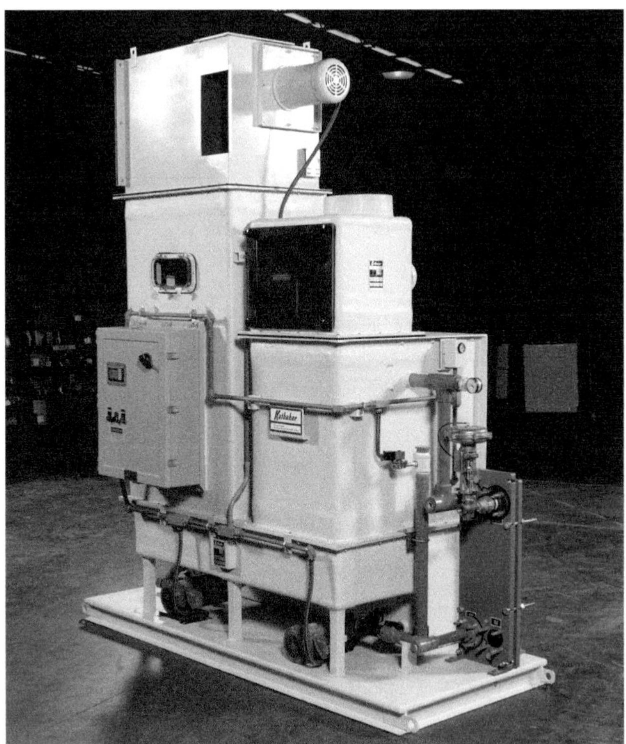

Fig. 39-6 Small Packaged LiCl Liquid System. Source: Kathabar Systems

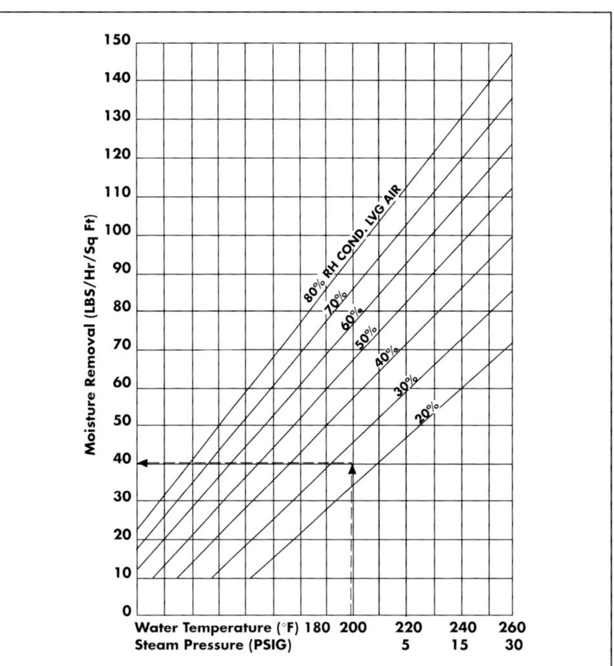

Fig. 39-7 Regenerator Capacity in Moisture Removed vs. Temperature/Pressure. Source: Kathabar Systems

absorbed in the conditioner is released in the regenerator to maintain solution concentration. A small amount of the concentrated solution is then transferred back to the condenser.

Liquid systems typically range in capacity from 1,000 to 85,000 cfm (28 to 2,400 m^3/m). RH levels ranging from 90 to 18% can be achieved by adjusting the solution concentration. Liquid systems are generally more expensive than solid systems, especially in smaller capacities. However, operating cost efficiency is generally equal to or better than any other desiccant system in this range. They are very controllable under part loads. They also tend to be more complex and require more maintenance.

Commercial coolants, well, river, and pond water, cooling tower water, and direct expansion systems may all be used for cooling. Heat can be supplied to the regenerator in many forms, including direct fuel-firing, steam, hot water, hot air, or exhaust gases. Regenerators can operate effectively with heat source temperatures of as low as 100°F (38°C), providing opportunities for effective use of recovered low-grade heat.

Figure 39-7 shows the regenerator capacity for a liquid system in moisture removal versus water temperature or steam pressure. As indicated on the performance curves, with 200°F (93°C) hot water and 25% RH air, regenerator capacity is 40 $lbm/h/ft^2$ (195 $kg/h/m^2$). Figure 39-8 is a nomograph used to calculate regenerator

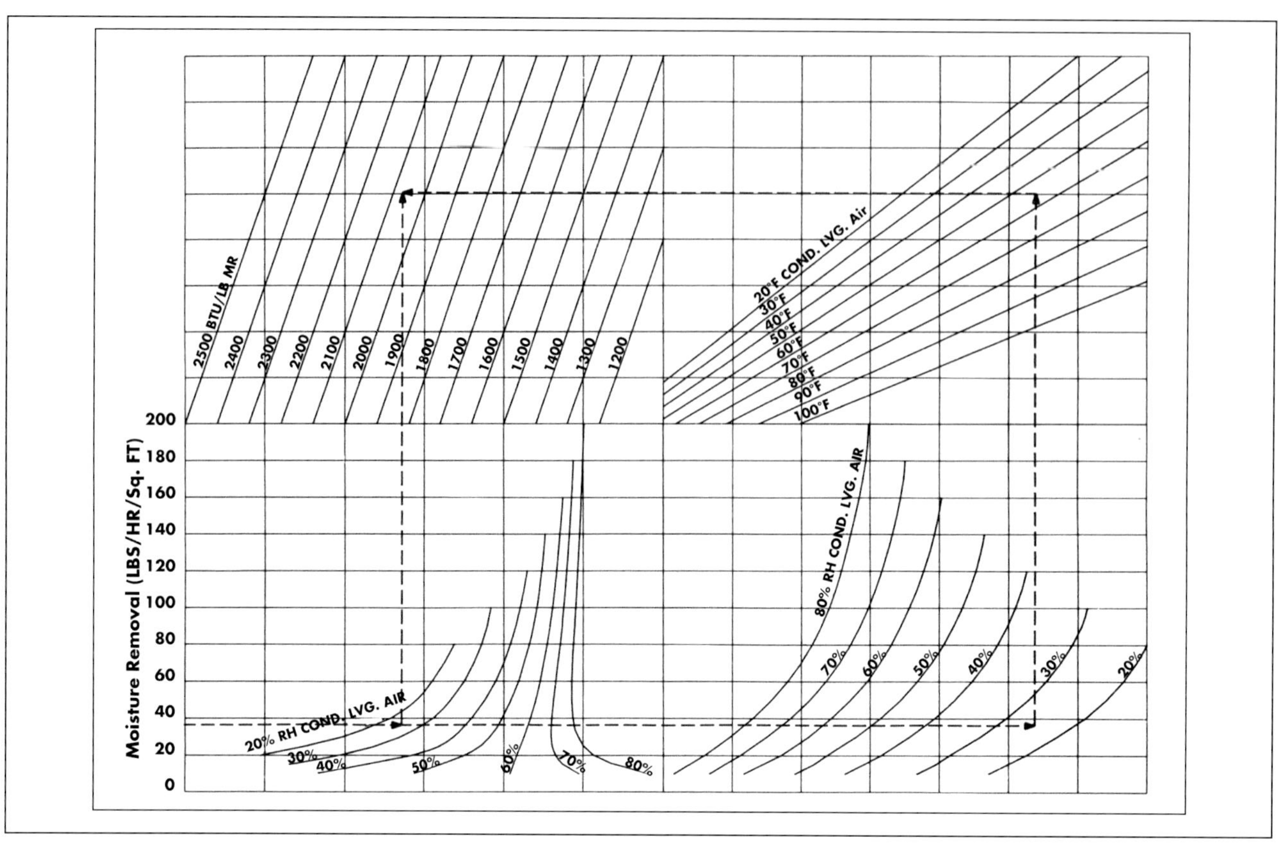

Fig. 39-8 Regenerator Heat Requirement Nomograph. Source: Kathabar Systems

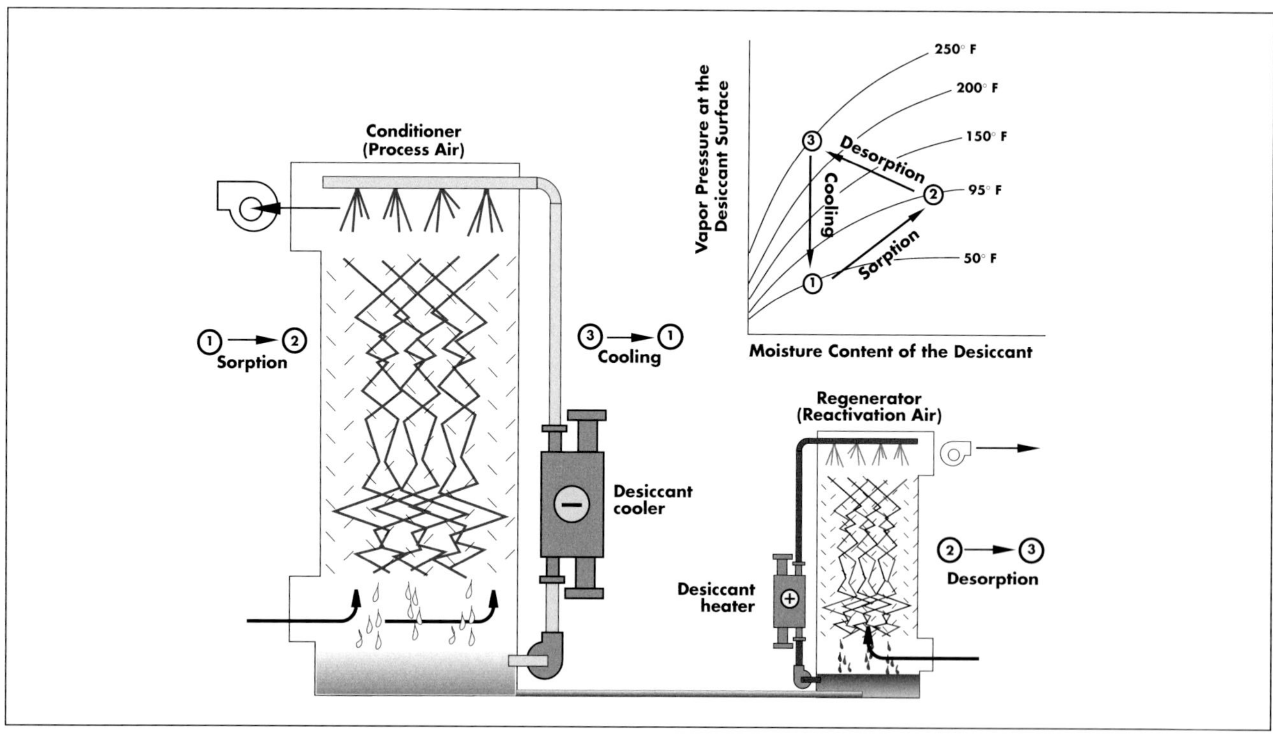

Fig. 39-9 Illustration and Equilibrium Diagram of Liquid Spray Tower Process. Source: Munters Cargocaire and Mason Grant Company

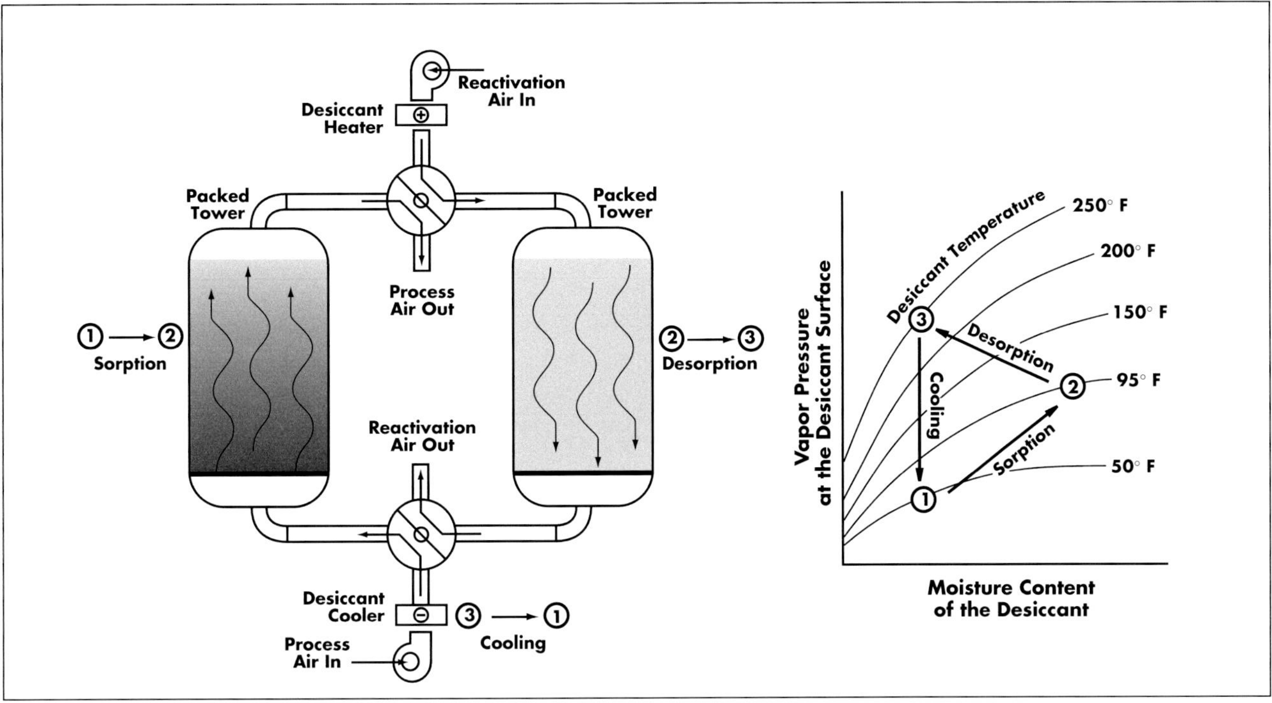

Fig. 39-10 Solid Packed Tower System Illustration with Equilibrium Diagram. Source: Munters Cargocaire and Mason Grant Company

heat requirements for a LiCl liquid system. In this example, the regenerator load of 35.8 lbm/h/ft^2 (175 kg/h/m^2) is located on the left-hand scale. A line is drawn through the two RH curves at the bottom of the chart at 25% RH. From the RH curves on the right, a line is extended upward from 25% RH to the 55°F (13°C) conditioner leaving temperature. Lines are extended from both the RH curves at the bottom left and the conditioner leaving temperatures. These lines intersect at 2,075 Btu/lbm (1.34 kWh/kg). Multiplying this by the total regenerator load in lbm/h (or kg/h) yields the total regenerator heat input requirement for a given application.

Figure 39-9 illustrates a representative liquid spray tower and shows the process graphically on an equilibrium chart. In the sorption process (1-2), the desiccant absorbs water in the conditioner, becoming warmer and rising in vapor pressure. At Point 2, the desiccant is in the sump. It has absorbed a relatively large amount of water and its surface vapor pressure is too high to attract more moisture. In the desorption process (2-3), the dilute desiccant passes through the heater and its vapor pressure rises. When it is sprayed into the reactivation air, the high pressure forces the water out of the desiccant and into the air. The desiccant returns from the regenerator to the sump dry and concentrated, though still warm and at a high vapor pressure. In the cooling process (3-1), part of the liquid is pulled out of the sump and circulated through a heat exchanger connected to a chiller or tower. It exits the process both dry and cool and is then circulated back through the conditioner to start the process again.

Solid Desiccant Tower (Mixed Beds)

Solid towers use desiccants, such as silica gel or molecular sieve, loaded into a vertical tower. Figure 39-10 illustrates a representative solid packed tower system and shows an equilibrium diagram. In the sorption process (1-2), the dry desiccant picks up moisture from the process air that flows through the tower. After the desiccant has become saturated with moisture, the process air is diverted to a second drying tower for the desorption process (2-3). The first tower is then heated and purged of its moisture with a small reactivation airstream. Hot reactivation air is used to heat the desiccant to raise its surface vapor pressure. Process air is used to remove the heat from the bed. When the hot desiccant must be cooled to lower its vapor pressure (3-1), the cool process air removes the heat from the bed.

Solid towers are not commonly used for ambient-pressure applications. Since the drying and reactivation occur in separate sealed compartments, they are especially effective for dehumidifying compressed air and pressurized process gases. They can be designed to achieve dew-

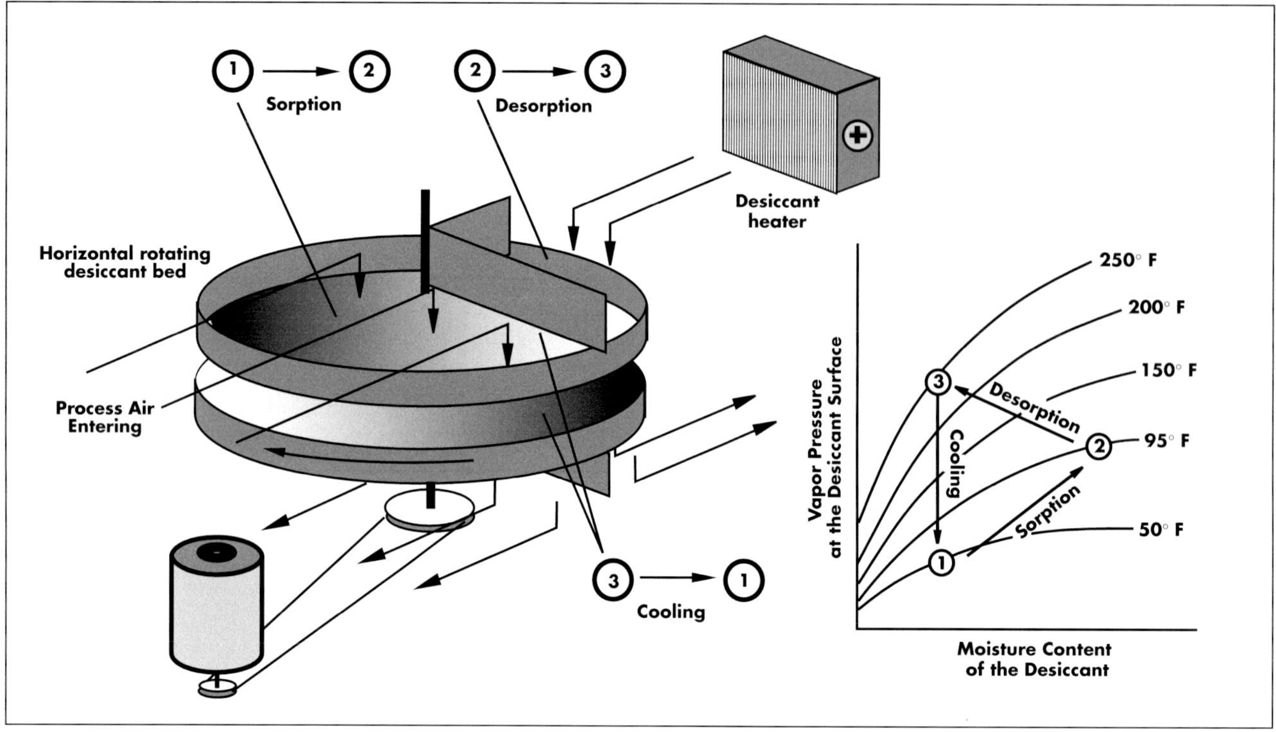

Fig. 39-11 Illustration and Equilibrium Diagram of Rotating Horizontal Bed System. Source: Munters Cargocaire and Mason Grant Company

points as low as -60°F (-51°C). These systems may range in capacity from as low as a few cfm to 5,000 cfm (1 to 140 m³/m).

The sorption process (1-2) occurs in one tower, while the desorption process (2-3) occurs in the other. The thermal energy that drives the desiccant cycle is added to the process by heating and cooling the reactivation and process airstreams.

The tower design can result in a changing outlet condition. As the desiccants begin to approach saturation, they do not dry the air as well as when they are fully reactivated. If this is not acceptable, the airflow must be changed to a fresh tower before the first tower becomes saturated. Tower systems tend to be large because air velocities are kept very low to avoid lifting and damaging the desiccant.

Rotating Beds

Rotary dehumidifiers have one or more beds with solid granular desiccants. The beds handle air to be dried and regeneration air simultaneously. The two airstreams are separated by seals to prevent mixing, and the beds are physically rotated within the casing to expose one portion of the desiccant to the process stream and one portion to the regeneration stream.

As the desiccant leaves the reactivation side, it is still warm. It is cooled by the process air during the first few degrees of rotation through the process side. The desiccant then dries the rest of the process air. As the desiccant is rotated into the hot reactivation air, it is heated, thus releasing moisture.

Typically, rotating bed systems range in capacity from 60 to 20,000 cfm (2 to 570 m³/m) and offer dewpoint temperatures ranging from -5 to 60°F (-21 to 16°C). They can range as low as -60°F (-51°C) using zeolites. These systems have the lowest capital costs of all desiccant systems. They have a simple design that is expandable and that can be easily repaired. In smaller systems, they are generally more cost-effective than liquid systems, since the low first cost tends to outweigh operating cost. However, the process and reactivation airflows are parallel rather than counterflow. This is done to minimize pressure differences and, therefore, leakage within the bed. This arrangement reduces the heating efficiency in reactivation, which raises operating costs above other types of desiccant systems.

Figure 39-11 illustrates a rotating horizontal bed system with an equilibrium chart. In the sorption process (1-2), the desiccant dries out the process air and picks up moisture. As the trays rotate into the hot reactivation air

(2-3), the desiccant is heated and releases moisture in the reactivation/desorption process. The desiccant exits dry, but still warm from the reactivation process. In the cooling process (3-1), the desiccant is cooled by the process air during the first few degrees of rotation through the process side. The dry cool desiccant is then ready to dry the rest of the process air.

Multiple Vertical Bed

Multiple vertical bed systems are somewhat of a hybrid between solid tower and rotating bed systems. They are well-suited for ambient pressure conditions, yet can achieve low dewpoints because there is minimal leakage. They use a circular carousel with many granular desiccant vertical beds that rotate between the process and reactivation airstreams. Typical capacity ranges for multiple vertical bed systems are from 500 cfm to 25,000 cfm (14 to 700 m^3/m). These systems are more complex and costly than rotating beds and require more maintenance. Figure 39-12 illustrates a multiple vertical bed desiccant system with an equilibrium diagram.

Rotating Wheel

The desiccant wheel functions much like the rotating bed. The difference is that the rotating bed type uses a wheel of granular desiccant in a packed bed and the wheel type uses a wheel of corrugated heat exchanger type surface. This results in different heat and mass transfer and pressure drop characteristics. The rotating desiccant wheel structure is lightweight and porous and can use several kinds of solid desiccants. The corrugated design combines high surface area with low total mass and presents minimal air pressure resistance, resulting in good efficiency. The surface can also be loaded with more than one type of desiccant in order to match very specific air treatment requirements.

Figure 39-13 provides an illustration and equilibrium diagram of a rotating desiccant wheel. In the sorption process (1-2), the desiccant picks up moisture as the process air flows through the flutes formed by the corrugations, becomes saturated, and its surface vapor pressure is elevated. As the wheel rotates into the reactivation airstream (2-3), the hot reactivation air causes its surface vapor pressure to rise farther, causing the desiccant to release its moisture in the desorption process. As the hot desiccant rotates into the process air (3-1), a small portion of the process air cools the desiccant, allowing it to collect more moisture from the rest of the process airstream.

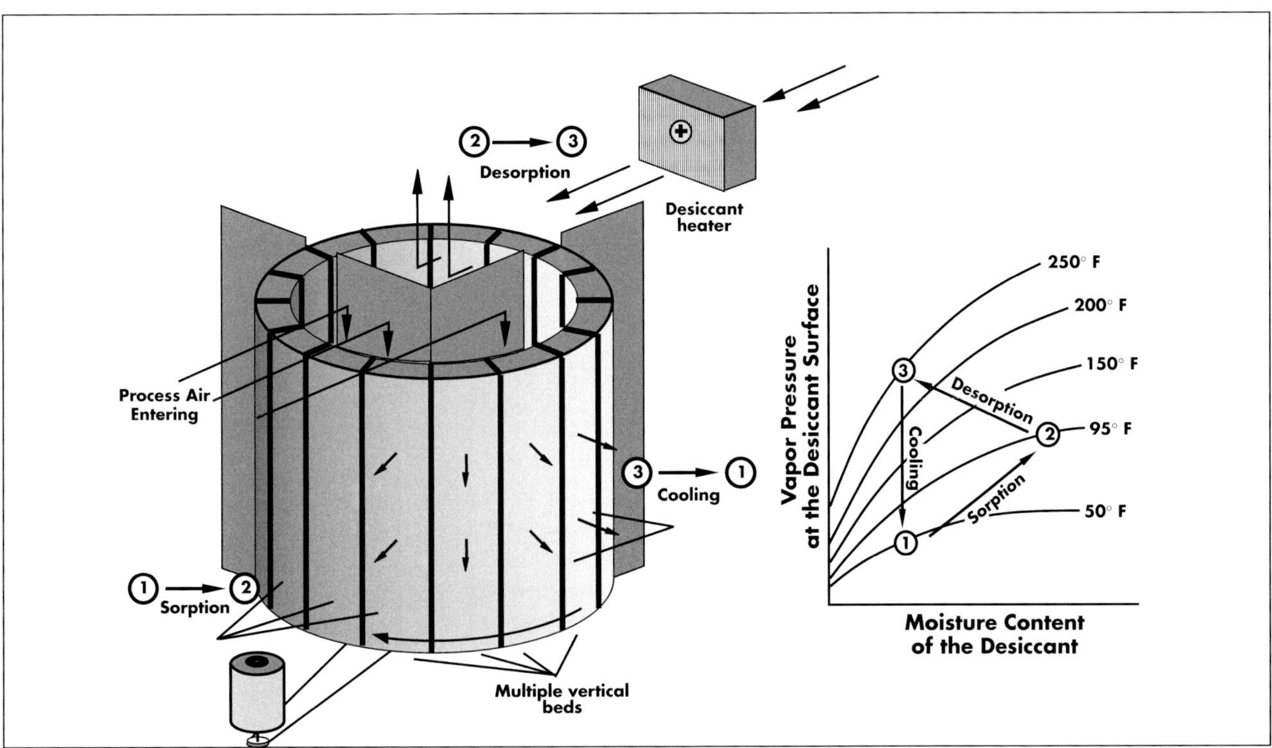

Fig. 39-12 Illustration of Multiple Vertical Bed System. Source: Munters Cargocaire and Mason Grant Company

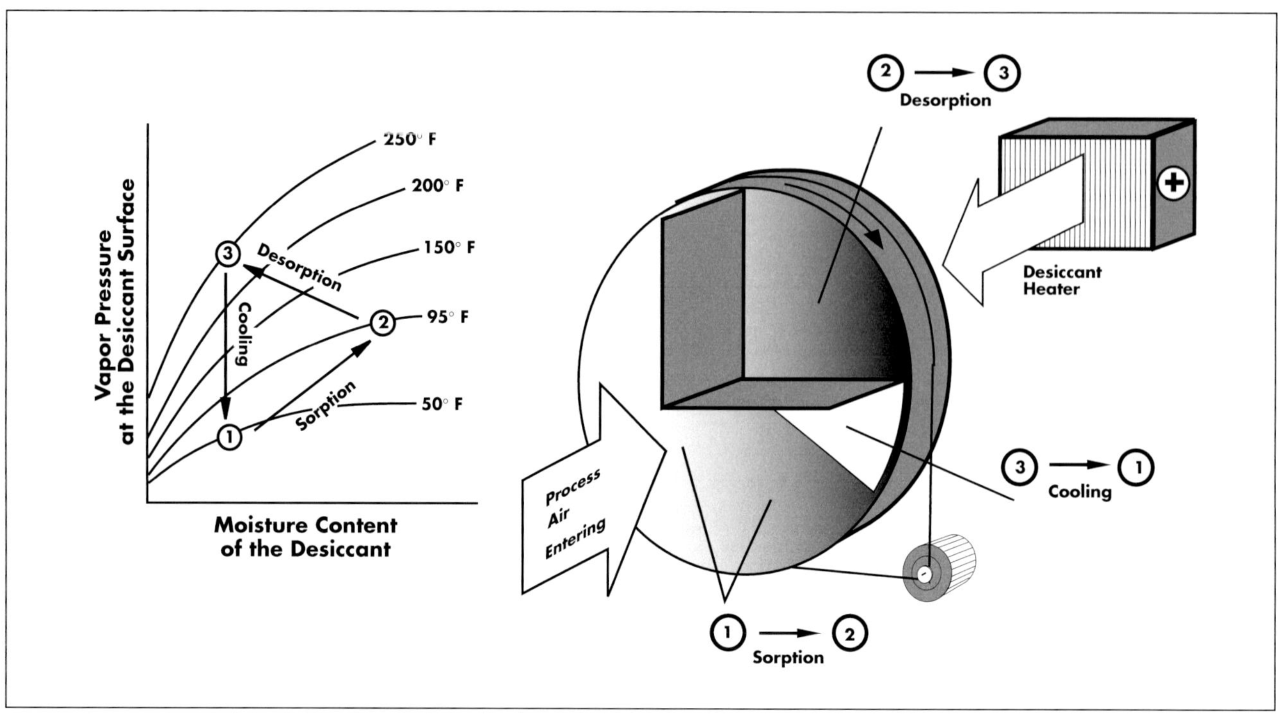

Fig. 39-13 Illustration of Rotating Wheel with Equilibrium Diagram. Source: Munters Cargocaire and Mason Grant Company

Rotating wheel systems are becoming increasingly more common as they serve all conditions down to -60°F (-51°C) dewpoint. They are similar in cost to multiple vertical beds and offer similar or superior efficiencies under most loads. They are generally less costly than liquid systems and dry tower systems down to -30°F (-34°C) dewpoint. When solid adsorbent, rather than absorbent, salts are used in the wheel, maintenance is potentially the lowest of all types of desiccant system. This technology is also becoming popular for use in conjunction with commercial space cooling systems.

Figure 39-14 shows a plan view and Figure 39-15 is

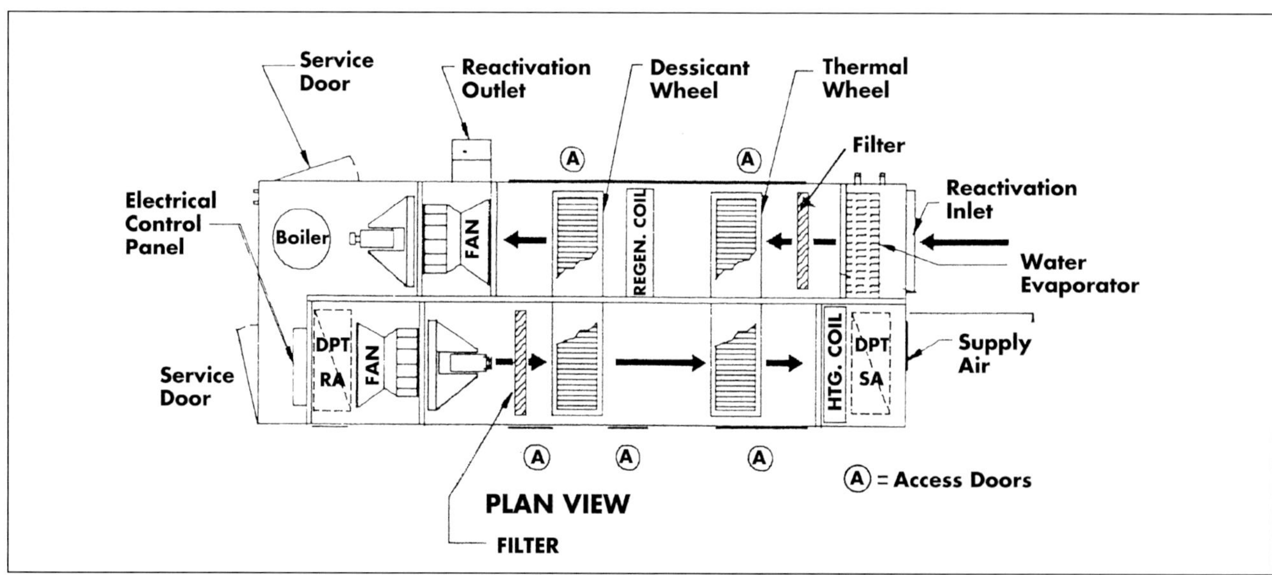

Fig. 39-14 Plan View of Double Rotating Wheel System. Source: ICC Technologies

Natural Gas Boiler or Steam Converter

Exhaust 100°F D.B.

Backward Curved Air Foil Fans

180°F D.B.

Regeneration Hot Water Coil 137°F D.B.

68°F D.B.

Evaporative Cooler

Outdoor or Return Air 95°F D.B./75°F W.B. 98 gr/lb

Desiccant Wheel 150°F D.B. 52 gr/lb

Outdoor or Building Exhaust Air 80°F D.B. 67°F W.B.

Thermal Wheel 81°F D.B. 52 gr/lb

Heating Hot Water Coil (for supplemental heat in winter)

Conditioned Supply Air 81°F D.B./62°F W.B. 52 gr/lb

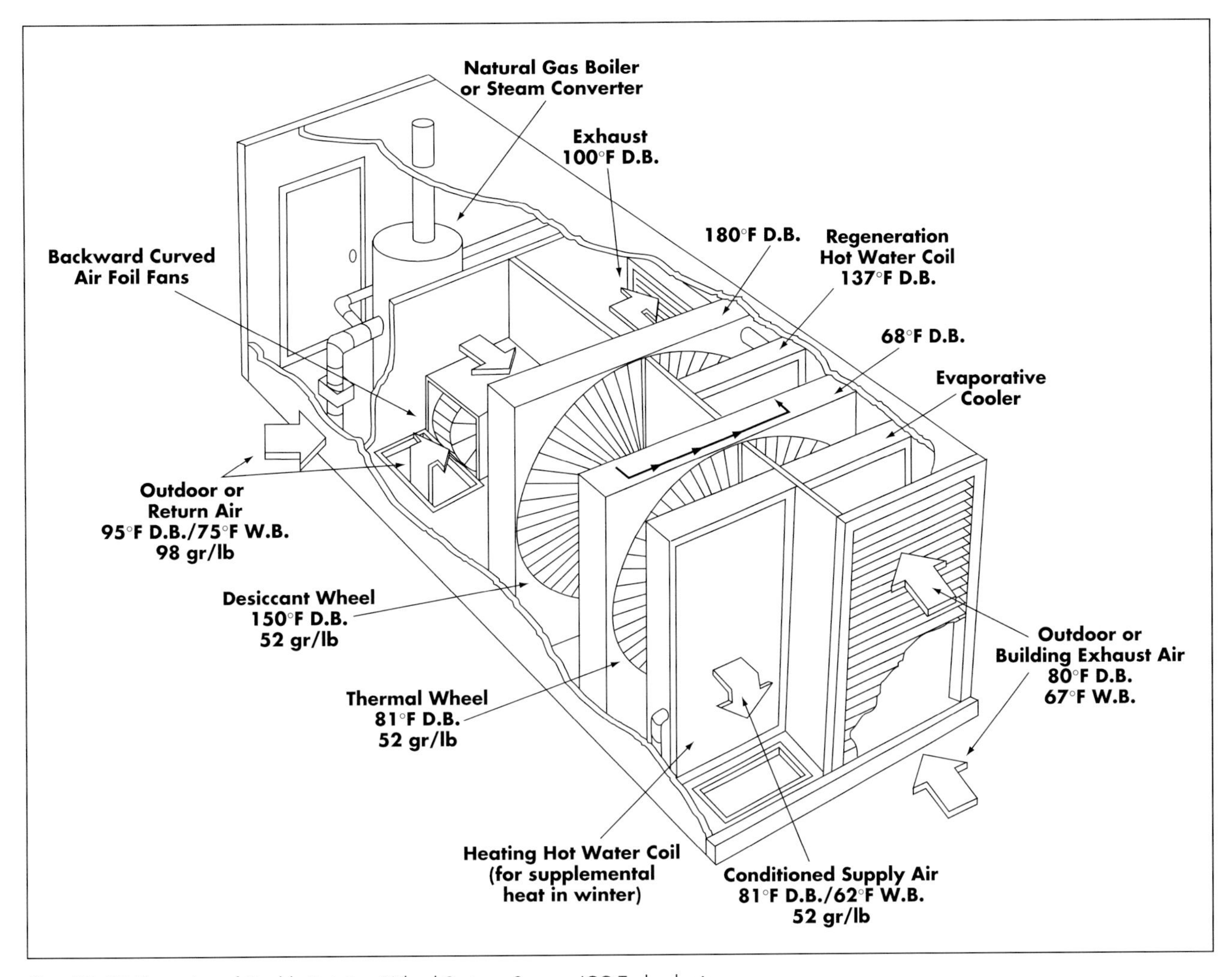

Fig. 39-15 Illustration of Double Rotating Wheel System. Source: ICC Technologies

a cutaway illustration of a double rotating wheel system, with representative air conditions shown. This system features two wheels. First, the air is dried in the desiccant wheel, then the air temperature is reduced in the second (thermal wheel) before it is delivered to the conditioned space. The second wheel, which is metal, serves as a heat exchanger. Most of the heat from the air leaving the desiccant wheel is removed to preheat the regeneration airstream, thereby reducing the external heat source requirement. As the wheel gives off heat to the incoming air on the regeneration side, it is cooled and rotates back to the conditioning side to extract heat from the hot dry airstream. As shown, outdoor or building exhaust air entering the regeneration side of the system first passes through an evaporative cooler, which cools the incoming air and raises its RH. The cooler incoming air serves to increase the efficiency of the thermal wheel. Figure 39-16

Fig. 39-16 Installation of Packaged Rooftop Unit Featuring Double Rotating Wheel System. Source: ICC Technologies

Fig. 39-17 Cut-Away View of Packaged Double Rotating Wheel System Showing Rotating Wheel Technology. Source: ICC Technologies

shows an installation of this technology in a packaged rooftop unit. Figure 39-17 is an open-case view, showing the rotating wheel.

Desiccant System Performance

Desiccant system performance depends on several factors. These include:

- The sorption/desorption characteristics of the particular desiccant used.
- The amount of desiccant exposed to the reactivation and process airstreams by the system.
- The process and reactivation air velocity through the desiccant.
- The process air moisture and temperature levels.
- The reactivation air moisture and temperature levels.

Each type of system will offer varied performance depending largely on the operating conditions. Some systems are more effective at higher dewpoint temperatures, while others perform better at very low dewpoint temperatures. The best desiccant for a particular application will depend on the range of water vapor pressures that will occur in the air, the temperature level of the available regeneration heat source, and the moisture sorption and desorption characteristics of the desiccant operating within those parameters.

The range of vapor pressures that may be encountered varies widely. For example, consider the range of vapor pressures for a constant dry-bulb (db) temperature of 70°F (21°C). At 100% RH, the dewpoint temperature is 70°F (21°C) and the corresponding vapor pressure is 0.74 in. Hg (2.5 kPa). At 10% RH, the dewpoint temperature is 12°F (-11°C) and the corresponding vapor pressure is 0.07 in. Hg (0.24 kPa). Each desiccant has a different sorption characteristic that affects system performance. With liquid absorption solutions, vapor pressure is directly proportional to its temperature and inversely proportional to its concentration.

A graphical representation of the relationship of moisture capacity to RH when desiccant and air are at the same temperature can be shown as a capacity isotherm. Figure 39-18 shows representative capacity isotherms of four desiccants at 77°F (25°C). Capacity is expressed as the weight of water retained as a percentage of the dry weight of the desiccant. As shown, at 20% RH, molecular sieve holds about 20%, while LiCl holds 35%. Note that large variations from these isotherms occur because manufacturers use different optimization methods for a given desiccant, depending on the particular application. Also note that these capacity figures assume that the des-

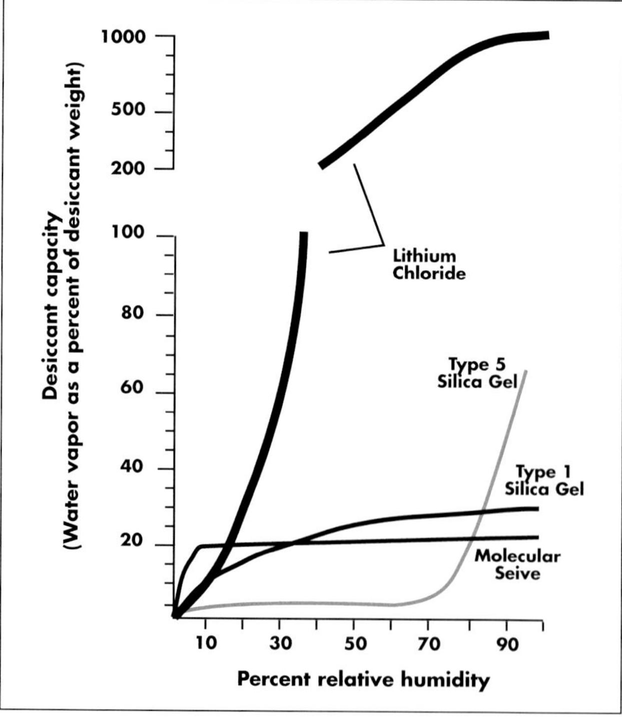

Fig. 39-18 Representative Capacities of Various Desiccants. Source: Munters Cargocaire and Mason Grant Company

iccant has enough time to absorb the moisture. In practice, capacities are lower due to the need to remove water from a fast-moving airstream.

Regeneration energy requirement consists of the sum of the heat required to raise the desiccant's vapor pressure higher than that of the surrounding air, plus the heat required to vaporize its moisture content, plus the heat from desorption of the water from the desiccant. For a given system, desiccant performance will be affected by the reactivation temperature level. Liquid desiccants, for example, can achieve a high level of performance at low reactivation temperatures, while the performance of solid desiccants improves considerably with increasing temperatures.

Consider the performance of the rotating wheel system. Figure 39-19 illustrates the changes in moisture and db temperature throughout the desiccant system on both the process air and reactivation air sides and shows the corresponding performance curves. On the process side, air enters at conditions of 70°F (21°C) and 56 gr/lbm (0.0080 kg/kg) at about 50% RH. The process air exiting the dehumidifier is both warmer and dryer at 109°F (43°C) and 13 gr/lbm (0.0018 kg/kg). On the reactivation side, outside air passes through a heater and proceeds to heat the desiccant in the wheel. As the desiccant gives up its moisture, the air becomes moist and cool. System performance is plotted graphically in Figure 39-20.

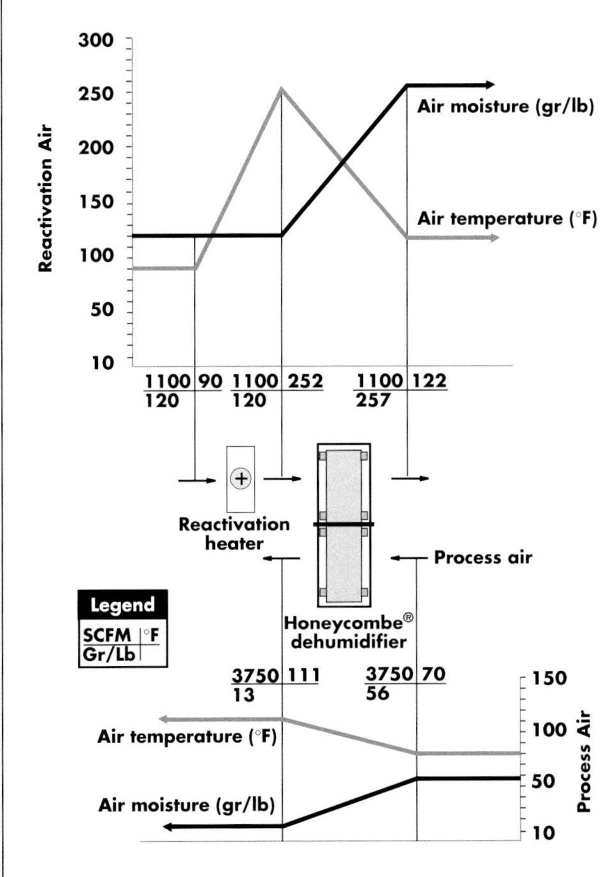

Fig. 39-19 Airflow Temperature and Humidity Change Diagram of Rotating Wheel System With Performance Curves. Source: Munters Cargocaire and Mason Grant Company

With this system, performance is optimized with high initial moisture in the process air, high reactivation air temperature, and low process air velocity:

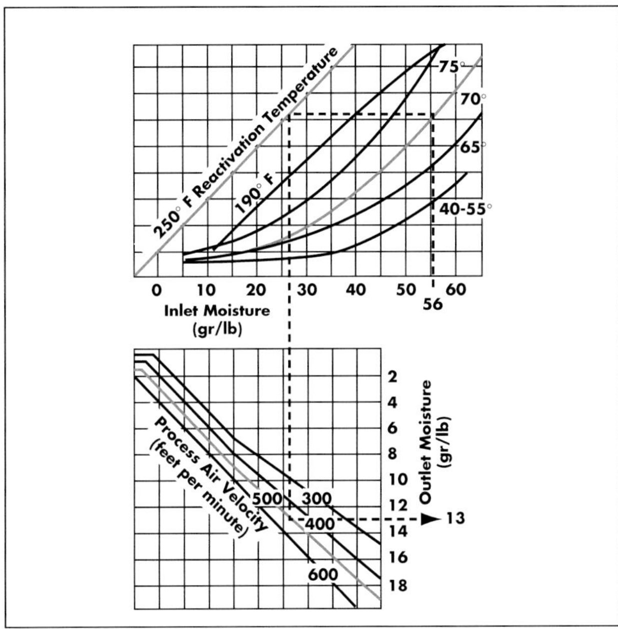

Fig. 39-20 Performance Plots for System Shown in Fig. 39-19. Source: Munters Cargocaire and Mason Grant Company

- As the moisture content of the entering process air is decreased, the final moisture level will also decrease, but so will the total amount of moisture removed.
- As the process inlet air temperature is reduced, the moisture leaving the process will also be reduced. This is due to the fact that performance is improved because the desiccant is cooler and, therefore, has a lower surface vapor pressure which allows it to attract more moisture.
- As the air velocity through the system decreases, so does the outlet moisture level. However, lower air velocity requires a larger, more costly system. In most cases, the trade-off is not economical, except when very low moisture levels are required.
- Generally, the hotter the desiccant, the more easily

it gives up moisture. The dryer the desiccant, the more moisture it can absorb when it rotates into the process airstream. Thus, performance increases with increased reactivation temperature.

Desiccant Applications

Generally, the potential benefits achieved with desiccant systems include energy savings, increased productivity, equipment and product protection, and improved environmental conditions. In industry, desiccant technology has been widely used for several decades for applications that require very low specific humidity — in the range of 45 gr/lbm (0.0064 kg/kg) and less — such as the case with clean rooms and other specialty operations. Given recent improvements in desiccant technologies, desiccant systems are being used more often and for more types of applications every year, inclusive of commercial space conditioning. Following are some common types of desiccant applications.

Condensation and Corrosion Prevention

Desiccant systems are used to protect heavy machinery from rusting and light equipment, such as computers and telecommunication gear, from microscopic level corrosion. Ferrous metals, such as iron and steel, can be protected by desiccants from moisture related corrosion. They are also used to eliminate sweating due to water vapor condensation on cold surfaces.

- Air drying is used to blanket the mold surface in plastic injection molding to protect the equipment and the product. It also allows for lower coolant temperatures, which reduces cycle time, thereby increasing productivity.
- Ice rink surface temperatures are colder than the dewpoint of the conditioned air, causing condensation that softens and distorts the ice surface. Desiccant systems can reduce operating costs and refrigeration capacity requirements, improve the quality of the ice and surrounding conditioned air, and eliminate corrosion of structural members.
- Hospital operating rooms are maintained at low temperatures to ensure surgeon comfort and productivity in spite of the heavy gowning necessary to avoid transmission of infectious diseases. Desiccant systems allow compliance with humidity specification required by licensing authorities at such low temperatures.
- Refrigerated warehouse loading docks can be hazardous for workers who load and unload trucks at high speed across floors that are slippery with condensation. Desiccant systems remove condensed moisture, providing safety while removing the latent load from the warehouse refrigeration system.
- Breweries and many other food processing facilities use coolant and process temperatures at or below freezing. Desiccant systems are used to prevent condensation, frost, and ice build-up and to prevent product degradation due to condensation or frost.
- In supermarkets (which are highly energy intensive facilities), desiccants eliminate sweating from display cases and products, reduce reheat and defrost requirements, and improve the efficiency of the mechanical compression refrigeration systems. Typically, desiccants are used to lower the specific humidity of the facility below the dewpoint temperature of the medium-temperature cases. This also serves to reduce the defrost cycle time on the low-temperature cases. Sophisticated systems can be built around the rotary wheel dehumidifier to take advantage of hot gas for winter heat and use a desuperheating coil to preheat regeneration air. Capital costs of supermarket desiccant systems are now comparable to conventional electric-driven vapor compression systems in new construction where the benefits of reduced ductwork, fan capacity, and refrigerated case requirements can be attained. Fan usage can also be reduced because the airflow requirements are reduced to handle only the sensible load.

Moisture Regain Prevention

Materials that are highly hygroscopic will absorb moisture under high RH conditions. This often results in damaged product and/or handling equipment. In these applications, desiccant systems may provide savings in productivity and product protection.

In controlled temperature and RH environments (such as process clean rooms), desiccant dehumidification is very commonly applied. The lower the RH requirement, the less efficient mechanical cooling is because supply air must be made extremely cold to purge the moisture and then often reheated.

- Pharmaceutical powders will easily absorb moisture, ruining the product, shortening shelf life, or making processing difficult. Clean rooms with desiccant dehumidifiers are used to prevent moisture absorption. RH down to 10% at room temperature is often required for granule/powder handling, proper compounding, warehousing, and transportation.

- Sugars, salts, and all types of candies are highly hygroscopic. Processing becomes extremely difficult when they absorb moisture and become sticky. Desiccant systems can improve processing efficiency and protect the products during storage. All types of powder, flour, vegetables, coffee, gum, and dozens of other products can be protected to improve processing time and/or storage.
- Desiccant systems are used in clean rooms for production of microcircuits. They prohibit moisture absorption by the polymers used to mask the circuit, which would interfere with the etching process that forms the circuit lines. Desiccants also speed up the curing of laminated circuit boards and protect the thin transparent films used for many applications from absorbing moisture and bubbling.
- Other industries that use desiccants to prevent moisture regain include chemical preparation, rubber and plastic fabrication, metals treatment, photographic production, wood processing, cosmetic manufacturing, printing, painting, and glass lamination.

Figure 39-21 shows a packaged 2,250 cfm (64 m^3/m) gas-fired desiccant dehumidification system that is commonly applied in moisture regain prevention applications. This unit can be installed directly in the product room and used to circulate air without special ductwork.

Product Drying

Desiccant systems can be effectively applied for product drying, regardless of the temperature, whenever there is a need for a low dewpoint. Additionally, they are effective for speeding product drying under conditions where temperature must be limited to 120°F (49°C) or less to avoid product damage.

Enzymes and other food preserving ingredients, such as yeast, are protected from damage by drying at low temperatures and low dewpoints. Dozens of food and drug products ranging from seeds and cereal to food powders are produced with higher quality when dried at low temperature and low humidity levels. Desiccants allow the drying process to be accelerated by establishing a deep wet-bulb (wb) depression. This refers to the difference between the db and wb temperatures of the air surrounding the product and the vapor pressure at the surface of the material being dried and the vapor pressure in the air.

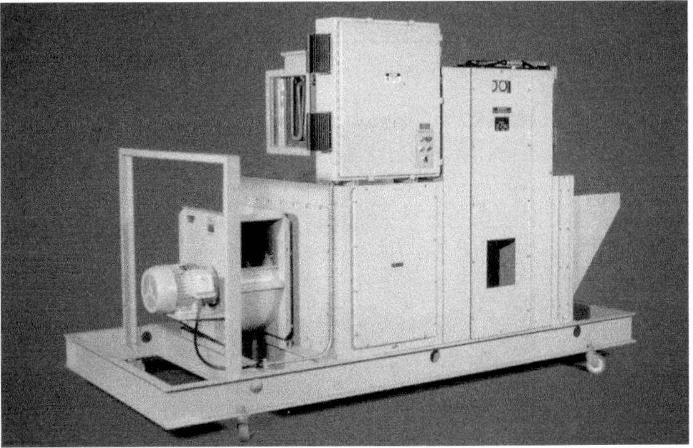

Fig. 39-21 2,250 cfm Packaged Gas-Fired Desiccant Dehumidification Unit Commonly Applied for Moisture Regain Prevention. Source: Munters Cargocaire

Compressed Air/Gas Drying

Desiccants are commonly used for drying compressed air. Moisture in a compressed air system can cause serious corrosion and wear of many types of equipment and can cause instruments and controls to malfunction. Drying of instrument air to a dewpoint of -40°F (-40°C) prevents condensation or freeze-up in the instrument control lines. Dry air prevents rusting of the air lines, which produces abrasive impurities and causes excessive wear on tools. Desiccant drying effectively solves these problems, while also removing dirt and oil.

Since the drying and reactivation occurs in separate sealed compartments, solid tower systems are especially effective for dehumidifying compressed air and pressurized process gases. They can be designed to achieve dewpoints as low as -60°F (-51°C) over a wide range of capacities.

Drying and cleaning of industrial gases or fuels, such as natural gas, is done before storage underground to ensure that valves and transmission lines do not freeze from condensed moisture during very cold weather. Other gases, such as liquid oxygen, hydrogen, and nitrogen, must also have a high degree of dryness.

Mixed HVAC System Applications

In certain cases, life-cycle costs for an HVAC system can be optimized with a combination of conventional refrigeration cycle-based cooling and desiccant dehumidification. In many space conditioning applications, desiccant systems can be integrated into the overall facility HVAC system, providing energy cost savings in addition to environmental health and building protection benefits. This allows for the use of lower capacity refrigeration equipment that can be sized to overcome only the facility's sensible loads.

A typical commercial application design consists of a desiccant system sized to deep-dry minimum outdoor ventilation air requirements. The dry air is ducted through the building and is dried to a point that satisfies the full latent load. This can be integrated into a down-sized (smaller capacity) chiller or DX system with fan coil air handling units (AHUs). This arrangement allows the refrigeration system to operate at the increased thermal efficiency allowed by higher suction temperatures. When airflow can be reduced, duct and fan capacities can be reduced, lowering associated capital and operating costs.

Determination of the economic viability of incorporating a desiccant unit into a mixed system will depend on many factors. Mixed system strategies are more likely to be cost-effective when large quantities of humid outside air are required due to high exhaust rates, or for facilities with high internal latent gains from crowds of people. As general rules of thumb, a mixed system featuring a desiccant unit may prove economical under any of the following circumstances:

- Conventional cooling alone cannot meet year-round dehumidification requirements.
- Electric power costs are high and/or thermal energy costs are low.
- Latent heat loads are high relative to sensible heat loads.
- Latent and sensible heat loads peak at different times.
- The system is required to operate dry with very low RH or no condensation on the cooling coils.

Given one or more of these circumstances, several mixed system configurations may be possible. Figure 39-22 illustrates four different configurations that can be used in mixed systems:

- In System 1, the desiccant is used only to dry precooled makeup air.
- In System 2, the desiccant is used to dry a blend of precooled makeup air and return air.
- In System 3, the desiccant is used to dry a blend of makeup air and return air which is precooled.

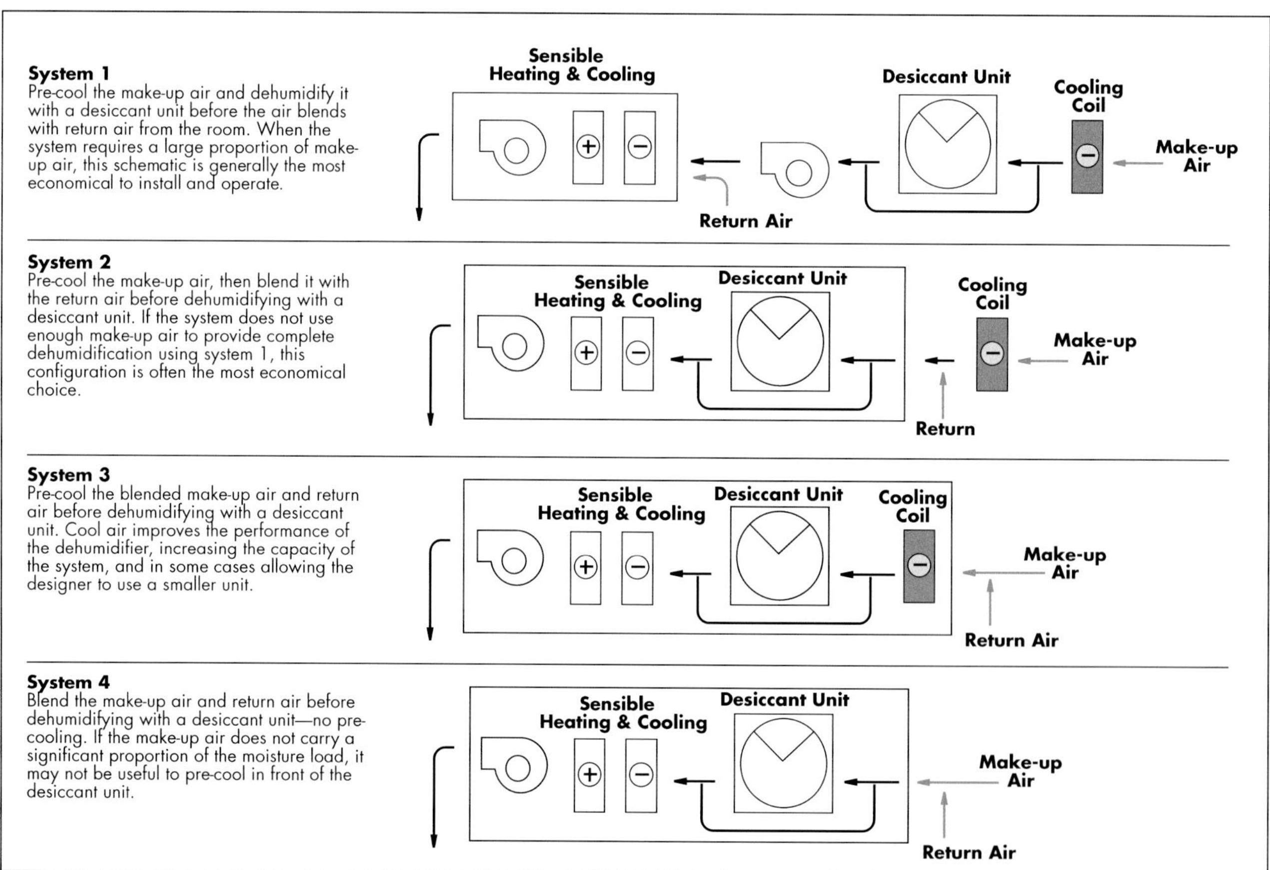

Figure 39-22 Representative Mixed System Configurations. Source: Munters Cargocaire and Mason Grant Company

- In System 4, the desiccant is used to dry a blend of makeup air and return air that is not precooled.

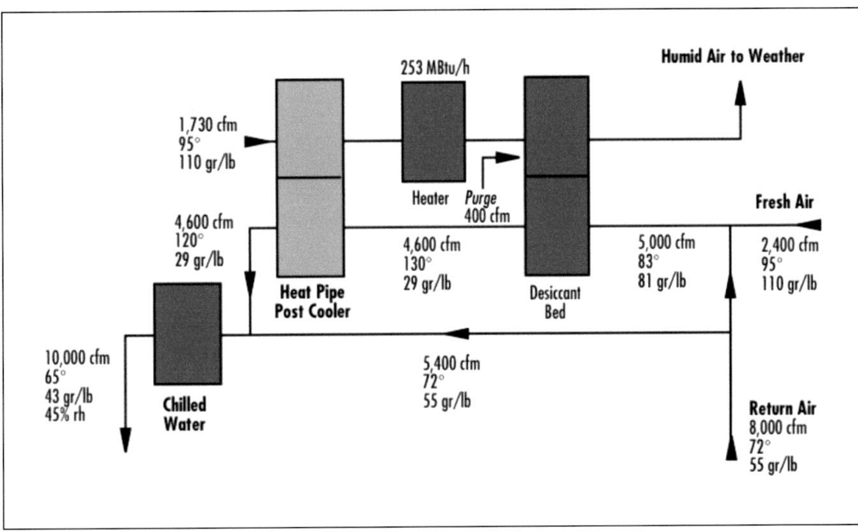

Fig. 39-24 Energy Balance Schematic of Desiccant Bed System for Operating Rooms. Source: Munters DryCool

Environmental Health and Building Protection

In addition to energy cost saving HVAC applications, desiccants can be used to reduce many biological hazards. Hospitals and health-care facilities are especially sensitive to such hazards. Commercial buildings, hotels, and other public facilities are also vulnerable. Buildings can have problems with microbiological contamination (i.e., mold, mildew, fungus, bacteria) in stagnant condensate pans and the insulation in ducts downstream of cooling coils.

In HVAC applications, desiccants keep evaporator coils dry to minimize the opportunity for microbiological growth on HVAC coils, drain pans, and duct work. Drying with desiccants up-stream of cooling coils can result in improved indoor air quality, reduced solvent emissions into the air, and protection from mold and mildew growing on building structures and furnishings.

Desiccant systems are sometimes used in hospital operating rooms to minimize moisture. Figure 39-23 shows a desiccant bed system used for hospital operating rooms. Figure 39-24 shows an energy balance schematic of this type of system. Liquid desiccant systems are sometimes used to remove bacteria, while providing dehumidification. This is accomplished by passing air directly through the system.

Fig. 39-23 Desiccant Bed System Applied in Hospital Operating Room. Source: Munters DryCool

When combined with the benefits of air quality and building protection, desiccant systems may be effectively applied at hotels and motels. Typically, room thermostats responding to the db temperature turn on the packaged terminal air conditioner only when the sensible temperature rises above the setpoint. Since the room is only occupied for a few hours during the day and the unit is sized for peak demand, the sensible load is quickly satisfied and the unit turns off, circumventing dehumidification and keeping the moisture load high. Moisture can continue to build up, creating mold growth and odor problems.

Desiccant Dehumidification Examples

Consider an industrial application in which pharmaceutical mixing, tableting, and packaging operations are carried out and require a moisture level of 10% RH and a temperature of about 70°F (21°C) in a 9,600 ft^2 (892 m^2) internally located room. To maintain these conditions, the absolute humidity is specified at a 13°F (-11°C) dewpoint, which corresponds to 11 gr/lbm (0.0015 kg/kg). The internal sensible heat gain is 41,000 Btu/h (12 kW) and the moisture load from people and door openings totals 105,000 gr/h (6.8 kg/h).

Due to exhaust requirements, 800 cfm (23 m^3/m) of makeup air is required at a design condition of 91°F (33°C), with a moisture level of 147 gr/lbm (0.0210 kg/

kg). The incoming air is precooled to 50°F (10°C) and 54 gr/lbm (0.0077 kg/kg), adding 154,800 gr/h (10.0 kg/h) to the moisture load, which must be removed by the dehumidifiers.

The airflow required to overcome the sensible heat load can be calculated as:

$$(cfm_s) = \frac{Q_s}{1.08(T_1 - T_2)} \qquad (39\text{-}1)$$

Where:
1.08 = Btu/h per cf of air per min. per °F

Thus,

$$\frac{41{,}000\ Btu/h}{1.08(70°F - 60°F)} = 3{,}796\ cfm \quad (107\ m^3/min)$$

Given an estimated return airflow rate of 3,000 cfm (85 m³/m) at 70°F (21°C), the blended supply air temperature is:

$$\frac{(3{,}000\ cfm \times 70°F) + (800\ cfm \times 50°F)}{3{,}800\ cfm} = 66°F\ (19°C)$$

and the blended moisture condition is:

$$\frac{(3{,}000\ cfm \times 11\ gr/lbm) + (800\ cfm \times 54\ gr/lbm)}{3{,}800\ cfm}$$
$$= 20\ gr/lbm\ (0.0028\ kg/kg)$$

The condition of the supply air required to remove the room moisture load of 105,000 gr/h (6.80 kg/h) with 3,800 cfm (108 m³/m) can be determined based on the following equation:

$$Capacity = Airflow \times Air\ Density \times Moisture\ Difference \qquad (39\text{-}2)$$

Given a return air moisture level of 11 gr/lbm (0.0015 kg/kg), the required moisture level of the supply air can be calculated from the following series of equations:

$$105{,}000\ gr/h = 3{,}800\ cfm \times 60\ min/h \times 0.075\ lbm/cf/min \times (11\ gr/lbm - x\ gr/lbm);$$

$$11\ gr/lbm - \frac{105{,}000\ gr/h}{17{,}100\ lbm/h} = 4.9\ gr/lbm\ (0.0007\ kg/kg)$$

Thus, the 3,800 cfm (108 m³/m) of supply air must be at a moisture level of 4.9 gr/lbm (0.0007 kg/kg). To allow for about 15% excess capacity, a desiccant unit with a capacity of about 120,000 gr/h (7.77 kg/h) is desired. From the performance curves for a rotary LiCl dehumidifier, it is shown that air entering the unit at 72°F (22°C) and 20 gr/lbm (0.0028 kg/kg) can exit between 1.5 and 5 gr/lbm (0.0002 and 0.0007 kg/kg), depending on the velocity of the air through the desiccant bed. A unit with 7.5 ft² (0.7 m²) of free face area will process 3,800 cfm (108 m³/m) with a velocity of 506 ft/min (154 m/min) so that air will leave the unit at a moisture level of 4 gr/lbm (0.0005 kg/kg). The total capacity of this unit would be 119,700 gr/h (7.76 kg/h).

This example is shown in Figure 39-25 and the performance is plotted in Figure 39-26. As shown, the precooling of the 800 cfm (23 m³/m) of outside air reduces the moisture level from 147 to 54 gr/lbm (0.0210 to 0.0077 kg/kg) and the temperature from 91 to 50°F (33 to 10°C). The outside air then blends with the return airflow to produce 3,800 cfm (108 m³/m) of supply air at a moisture level of 20 gr/lbm (0.0029 kg/kg) and a temperature of 66°F (19°C). This airstream exits the desiccant unit at 4 gr/lbm (0.0006 kg/kg) and 86°F (30°C).

It is important to note that had the outside air not been precooled, the desiccant unit specified would be about twice the capacity and the application would be less economical. Since conventional refrigeration cycle systems perform more effectively at high temperature and high moisture contents and desiccants are more effective at lower temperatures and drier conditions, combining the two technologies to dehumidify fresh air can be less costly

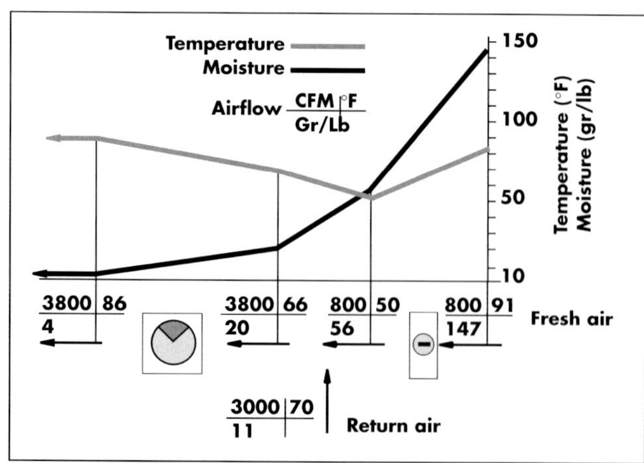

Fig. 39-25 Changes of Temperature and Moisture of Air Moving Through System. Source: Munters Cargocaire and Mason Grant Company

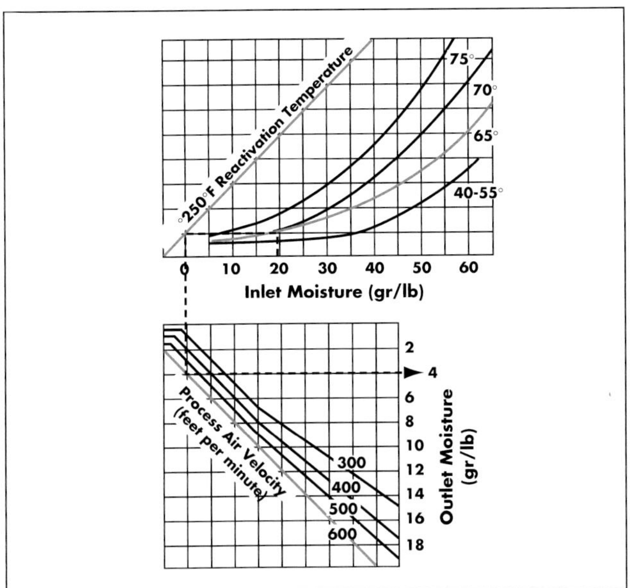

Fig. 39-26 Performance Plots for System Example. Source: Munters Cargocaire and Mason Grant Company

than using either technology exclusively.

Given the condition of the air exiting the desiccant unit of 4 gr/lbm (0.0006 kg/kg) and 86°F (30°C), the post-cooling requirement is calculated as:

$$\frac{3{,}800\ cfm \times 1.08 \times (86°F - 60°F)}{12{,}000\ Btu/ton} = 8.9\ tons\ (31\ kW_r)$$

Figure 39-27 is a schematic diagram of the entire HVAC system serving the conditioned space. The summary table lists values for airflow, temperature, and moisture at various locations. Note that the 800 cfm (22.6 m^3/m) of incoming outside air (Location A) balances 500 cfm of exhaust (14.2 m^3/m) at Location G and 300 cfm (8.5 m^3/m) of air leakage.

This system's controls are as follows:

- Room air pressure is controlled by a differential pressure controller, which automatically adjusts the fresh air intake damper. As air pressure exceeds the set differential, the damper begins to close to reduce incoming air.
- Dehumidifier capacity is controlled by a bypass damper that passes air around the system as humidity levels are less than design conditions. As a dewpoint controller signals a humidity change, the bypass damper is modulated.
- Desiccant system energy use is controlled by a temperature controller in the reactivation airstream. This minimizes energy use by reducing the energy of the reactivation heaters when the temperature rises above the set point of 120°F (49°C), which indicates that all of the moisture absorbed on the process air side has been desorbed in reactivation.
- Chilled water flow through the outside air precooling coil is controlled by a three-way valve, which responds to a controller located downstream of the coil that is set at 50°F (10°C). The post-cooling coil, which sets the final delivered air temperature, is controlled in the same manner, with the temperature sensor located in the conditioned space to maintain constant temperature. In a variable flow system, two-way valves could be used to accomplish this control function more economically.
- A supply air heating coil is used to provide heat whenever room temperature drops below 68°F (20°C). This is controlled by the same sensor that controls the post-cooling coil.

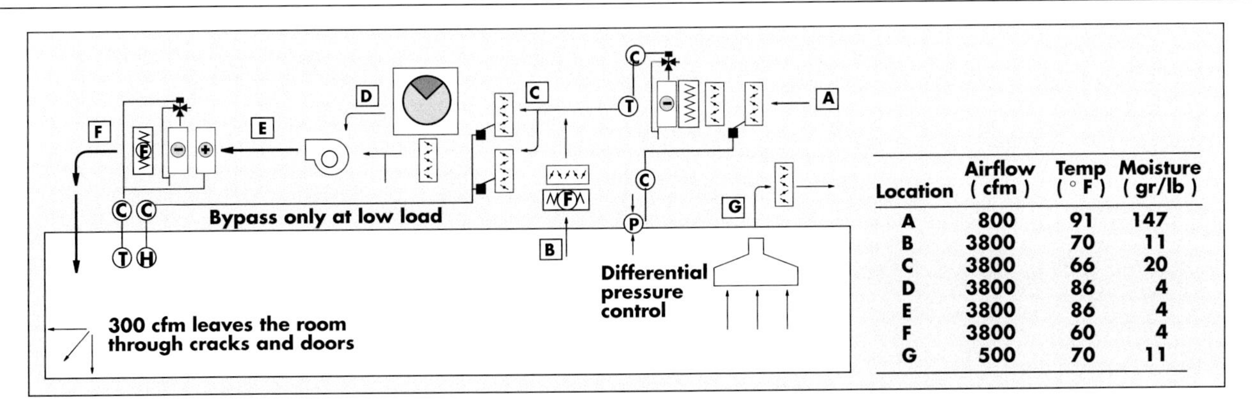

Location	Airflow (cfm)	Temp (°F)	Moisture (gr/lb)
A	800	91	147
B	3800	70	11
C	3800	66	20
D	3800	86	4
E	3800	86	4
F	3800	60	4
G	500	70	11

Fig. 39-27 Schematic Diagram of HVAC System. Source: Munters Cargocaire and Mason Grant Company

PART 4
Analysis and Implementation

Section IX

Integrated Approach to Facility Resource Optimization

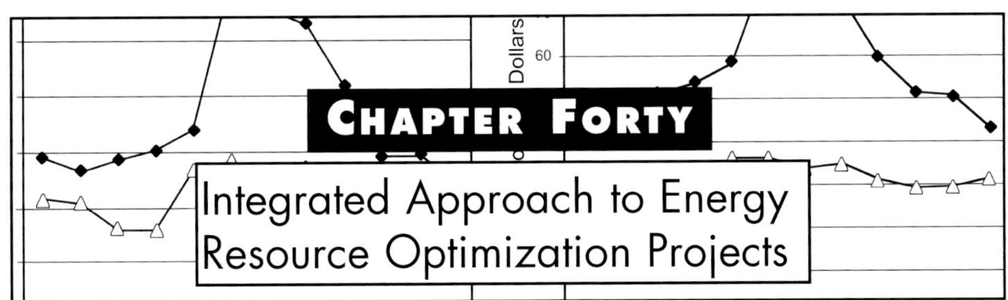

CHAPTER FORTY

Integrated Approach to Energy Resource Optimization Projects

An integrated approach to developing and implementing an energy resource optimization project involves evaluation of the entire facility rather than one specific application. The process is based on an interactive analysis of end-use, distribution system, central plant, and resource supply-side opportunities to find an optimal solution that takes advantage of the synergies between them. It includes, at a minimum, evaluation of electrical and thermal energy, water, solid waste and wastewater, and operations, maintenance, and repair (OM&R) activity. All results must also be tested against the facility's environmental, operational, strategic, and financial objectives.

With the integrated approach, a facility's resource use is viewed as both dynamic and interactive. Therefore, when considering energy supply, central plant, and distribution system improvement options, one must simultaneously consider the loads being served, such as comfort conditioning or process end-uses. The analysis must also consider upgrades based on the loads that would be prudently anticipated once cost-effective end-use improvements are implemented. Supply-side options, such as energy source switching, energy supplier switching, utility negotiations, load management, or on-site power generation, must also be considered interactively with respect to the anticipated central plant, distribution system, and end-use modifications. Finally, careful consideration must be given to the facility's overall thermal energy (heat) balance so that potential changes in one area are considered along with their impact on other areas.

For example, when lighting systems, envelope components, air handling systems, or process applications are upgraded, or compressed air use is optimized, then steam, hot water, chilled water, and compressed air requirements can be significantly altered. If peak loads for any of these end-uses change, then central plant systems of different capacities may be required and the economic performance of upgrading those central systems may be changed. Even if the peak loads remain the same, the load profile (i.e., the timing and duration of the end-use load events) may alter central plant performance. Conversely, when the central plant systems are upgraded, the incremental cost of steam, hot water, chilled water, and compressed air may be reduced, thereby reducing the savings potential of end-use improvements and sometimes eliminating certain measures that would have been cost-effective with the old, less efficient central plant systems. Supply-side improvements have a similar effect. If fuel or electricity can be purchased at a lower price, or if heat and power can be provided less expensively through an on-site cogeneration system, the economic performance of central plant and end-use upgrade programs will often be significantly altered.

The technical engineering analysis evaluates what will happen when systems are installed and/or operated in a different manner that will change resource use. The analysis is a creative, logical process involving accurate, careful characterization of actual operating thermal and electrical loads and other resource uses, both on a facility-wide and system-specific basis. In addition to implementation cost and savings evaluation, all other factors that will affect the installation and operation of various systems are considered. Systems and components are selected that exhibit the flexibility, efficiency, durability, and performance to meet all loads and satisfy the facility's mission requirements and goals. This selection is often an iterative approach as alternatives are tested against various (and often competing) constraints until an optimal solution is reached. Typically, the process is repeated at increasing levels of detail as more data is gathered and the accuracy of the cost and savings estimates increases. This avoids wasted time doing detailed evaluation of opportunities that do not meet the facility's technical or financial criteria.

While analytical in nature, a successful evaluation study involves a hands-on approach with extensive fieldwork, comprehensive direct system metering, and rigorous technical and financial analysis. While computer modeling is commonly used, its function is to compliment and support, not to replace, the primary hands-on fieldwork and direct metering activities. Qualifying the viability of technology applications begins with careful physical inspection of facilities, equipment, and systems. This involves not only observation and metering of centralized equipment, such as boilers, chillers, and air compressors, but also confirming field conditions in distribution systems and spaces. For example, the site survey often includes crawling through tunnels to inspect distribution piping size and condition and examining ceiling space to inspect air-side distribution systems, confirming as-built drawings, and determining if there is adequate space and access for modifications.

The analysis of potential energy and resource use optimization projects determines the cost to provide the same end-use products and services with and without a series of system improvements. This is based on an analytical framework that constructs a baseline against which the new systems can be compared. The baseline can be either current or recent average utility billing history, or it can be an adjusted baseline to account for changes that would have occurred in the absence of the energy cost reduction program. For example, a facility expansion, change in usage, or change in operation may be planned, which may justify an adjusted baseline to give a realistic prediction of true avoided energy and operating costs. The analysis is then re-run with appropriate changes to the model that affect all of the recommend changes. The difference between baseline and this predicted usage is the incremental effect resulting from the new system(s). Given the interactive effects between systems, all potentially beneficial project application options must be identified during the study process. Options are then mixed and matched and considered interactively until the optimal portfolio of measures is selected.

The purpose of an integrated incremental analysis is, therefore, to determine the incremental change in usage and cost of all affected resources with the implementation of a system improvement. In addition to the direct energy usage of a given system, resulting changes in usage and cost of associated energy systems and non-energy resources must be considered. This change in facility operating cost must then be weighed against the capital investment requirement in a time-valued life-cycle analysis that extends over the full term of the project life.

The financial analysis evaluates operating economics against current systems and/or all possible competitive alternatives aimed at the selection of the optimum program portfolio. This is performed within the context of all available project funding options and the host facility's investment criteria. It also includes sensitivity analyses whereby the robustness of the economic performance is tested against likely variation in system variables, such as energy cost and operation schedules.

Project investment decisions are based primarily on whether the expected cost savings, reduced OM&R, or increased economic performance resulting from the project meet or exceed an acceptable threshold level or produce sufficient cash-flows to justify the financial commitment. The financial analysis translates the technical analysis into time-sensitive dollar values. The capital and operating costs of existing systems are compared to those of the new systems. Should a project be required due to ongoing or planned facility or process expansion or imminent equipment failure, the lowest cost option which will satisfy the project requirement becomes the base case upon which each improvement option is tested for incremental cost and savings. The threshold level and evaluation methodology employed will vary among investors. However, all financial analyses should include consideration of tax treatment, financing, ongoing OM&R costs, and other time-valued costs in a life cycle economic format.

In addition to viewing all resources used by a facility, an integrated approach requires a big-picture view of the physical environment surrounding the facility and the legal and economic environment impacting the facility. One must therefore look beyond the confines of the facility, at neighboring resources, such as bodies of water that could provide cooling, and at neighboring facilities that could provide a source for either selling or purchasing chilled water, steam, electricity, or by-products. Additionally, groups of facilities that could be aggregated to purchase energy resources more cost-effectively should be considered.

Finally, consideration is given to prudently anticipated changes in legal, regulatory, and economic factors. These may include changes in environmental regulations and changes being brought about by the wide-scale restructuring of the utility industry. Restructuring is changing not only the cost of buying and selling energy resources, but also the load shapes and rate structures under which they are purchased and sold. This impacts the economic performance of various technology applications as well as the types of technologies and applications that should be considered. While it is not possible to accurately predict all future events and market changes, these factors should be carefully considered in the sensitivity analysis. Moreover, given the risk of change and market volatility, as well as the fast pace of technological developments, programs should be developed that have the flexibility to be adapted to a wide range of potential future conditions.

Once the facility has decided to proceed with an energy resource optimization project, the next step is project implementation. This begins with the selection and execution of implementation contract(s) and financing arrangements. Work may be executed under one master contract that includes all phases of implementation and may include the evaluation and study phase as well. A turnkey contract such as this is typically executed with an energy services company (ESCo) that is responsible for all aspects of project coordination and management. Alternatively, the project can be done through a series of contracts, each developed and signed prior to subsequent phases of implementation. Facilities may finance project implementation internally through capital budgeting or through the use of external funding, or a combination of both. Alternatively, projects may be implemented and then owned and operated by a third party.

The major implementation phases are design and construction. The design and construction approach will vary, based on the contractual approach. However, the end result should be the same regardless of the approach events — a functional project that meets the intent of the development documents. The design process and resulting documents will vary depending on whether work is to be performed under a traditional plan and specification format or a more flexible design-build type approach. Milestones for design review (e.g., 30% complete, 50% complete, etc.) should be mutually agreed upon in advance. The overall construction phase may be further divided into three main elements: procurement, installation, and commissioning. Following commissioning, when all systems have been tested for performance and conformance with the design intent, the project is closed out. Punch lists are completed and as-built documents and OM&R manuals are delivered, training of operations and maintenance staff is conducted, and the project is accepted by the facility. At this point, the operational phase begins.

During the operational phase, all systems must be effectively operated and maintained in order to ensure that the facility remains fully and safely operational, and that project economic performance is realized throughout the life of the installed systems. Ongoing observation, data acquisition, and analysis should serve as a basis for preventative or predictive maintenance programming, continual system optimization efforts, and identifying future cost reduction project opportunities. Post-installation activities may also include either one-time or ongoing savings metering and verification (M&V) processes.

There are numerous phases involved in the full process of identifying, developing, implementing, operating, and maintaining an integrated energy resource optimization project. This chapter provides an overview of the process, while subsequent chapters within this section describe these elements in detail.

Elements of an Integrated Program

Figure 40-1 is a process flow chart of the major steps involved in developing and implementing an integrated energy resource optimization project. This flow chart is based on using a performance-based design-build approach where the contractor develops a program, proposes it to a client facility, and then enters into a contract to execute the entire scope of work. This is one of several methods of development and implementation of such a program. In other cases, the process could be broken into several parts, where the client facility contracts for an independent engineering study, then for engineering design and implementation, each separately. Regardless of the approach, the process will include the same essential elements. Figure 40-2 provides a representative timeline of several central elements of the entire process in the form of a Gantt chart. The operational phase is not shown, since this is an ongoing effort for the duration of the project life. Expanded versions of this chart, showing critical path tasks, are provided below along the overview discussions of the development and implementation processes.

Project Development Phases

All integrated energy resource optimization projects must begin with some type of study that identifies the opportunities and quantifies them in such a manner that a scope of work can be developed and financial decisions made. The multi-phased integrated approach involves a multidisciplinary team effort designed to process a large quantity and variety of information. The team should consist of technical experts from analysis, design, construction, OM&R, and M&V disciplines. These technical disciplines are supplemented by financial, contractual, energy source procurement, and risk management expertise. The process requires this cross-functional team because the nature of the projects involves all of these disciplines and requires their specialization and the volume of information requires a team effort to process. Moreover, risk management dictates extensive cross-checking and internal quality control.

Energy efficiency opportunities may be categorized by their location in the energy conversion flow, e.g., end-use measures, distribution system measures, central plant measures, and utility supply-side options. Specialists are

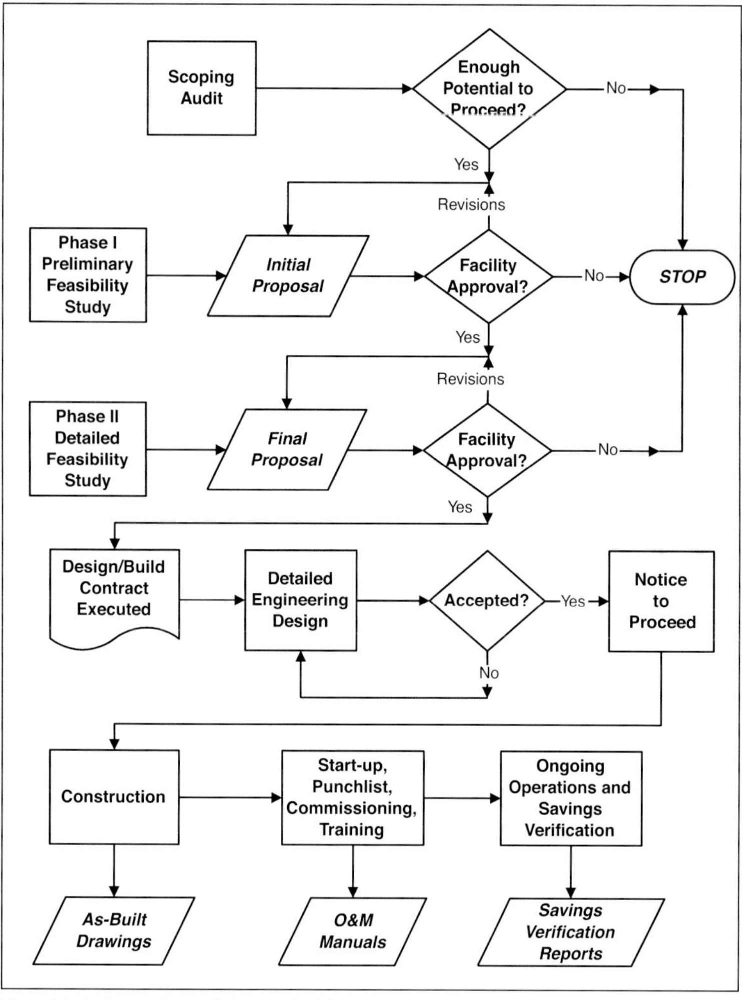

Fig. 40-1 Flow Chart of Design-Build Process.

assigned by technical discipline, such as lighting, HVAC, pumping systems, and central chillers/boilers, though in many cases, an individual may be assigned multiple tasks, depending on the skill sets possessed.

All utilities are evaluated, as well as energy-related operating costs, such as water usage, waste water, emissions, and solid waste. The utility supply scenario is reviewed in the context of the facility master plan and future opportunities that are prudently anticipated to arise from deregulation. Alternative supply and purchasing options are explored along with supply risk management and load management techniques that will enhance these options.

The development approach should intentionally be split into multiple phases to allow early high-level screening of opportunities to streamline subsequent efforts. During the early phases, the project scope and potential measures are reviewed by representatives of each discipline and/or experienced senior managers with cross-disciplinary expertise. This ensures practical and efficient review of all critical success factors early in the process.

In most cases, a formal two-phased study approach should be undertaken. Prior to this, an additional stage, or Scoping Audit, should be added to the process to initially verify if there is viable project potential. It can also serve to determine an overall plan of action. With the Scoping Audit (for early screening and master plan development) and the phased feasibility approach, the development team should make effective use of its professional resources and not waste the client facility management's time. As the process proceeds, results are carried forward, additional information is gathered, and the team consults regularly (internally and with facility representatives at key milestones) to continuously screen measures in increasing detail and refine them to match site conditions and facility needs. Interaction between measures (both physical effects and analytical impact) is

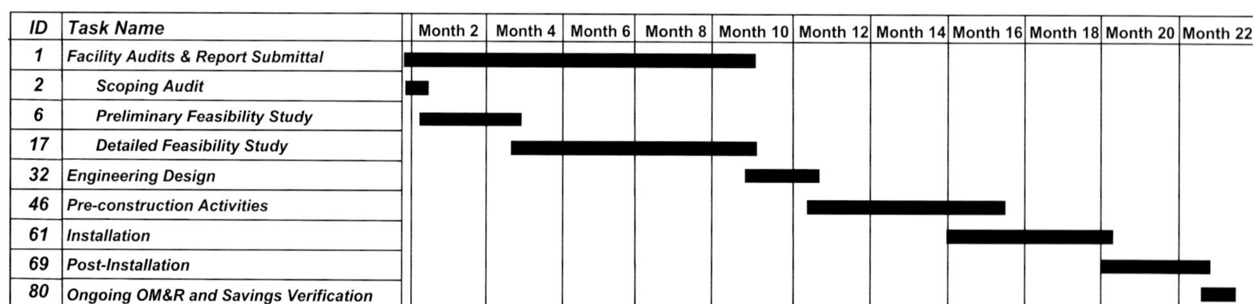

Fig. 40-2 Summary of Key Program Elements Shown in Gantt Chart Format.

given increasing attention so that an optimal package of measures is developed. The phased approach also affords the opportunity for multiple quality control review milestones. Internal review and sign-off are integrated with the project development work plan.

Figure 40-3 provides a representative timeline in the form of a Gantt chart, showing critical path tasks of the development process. The development tasks are organized under three major categories — Scoping Audit, Preliminary Feasibility Study, and Detailed Feasibility Study. The timeline shows about 9 months for the development process (ID 1-28) and an additional month (ID 29-31) for technical and financial review, and contract and finance negotiation and execution. This timeline allows for an extensive process, suitable for a large, complex project. This can be condensed considerably for smaller projects or where all parties agree to fast-track the process. Chapter 41 provides a description of the Scoping Audit and Project Development Plan, followed by a detailed step-by-step description of a two-phased preliminary and detailed feasibility study approach. These components are summarized below.

Scoping Audit

To identify the work required and properly allocate resources to the initiative, a project scope must be developed. Therefore, a multi-phased approach often initiates the process with a facility Scoping Audit, which is a low-cost, brief review to define the energy and resource savings potential and system infrastructure upgrade requirements.

The Scoping Audit includes a broad review of the major energy and resource consuming systems in the facility, without a lot of detail. It is developed based on a preliminary review of resource usage and cost records, a walk-through of the buildings, and review of the mechanical, electrical, and controls systems. The walk-through

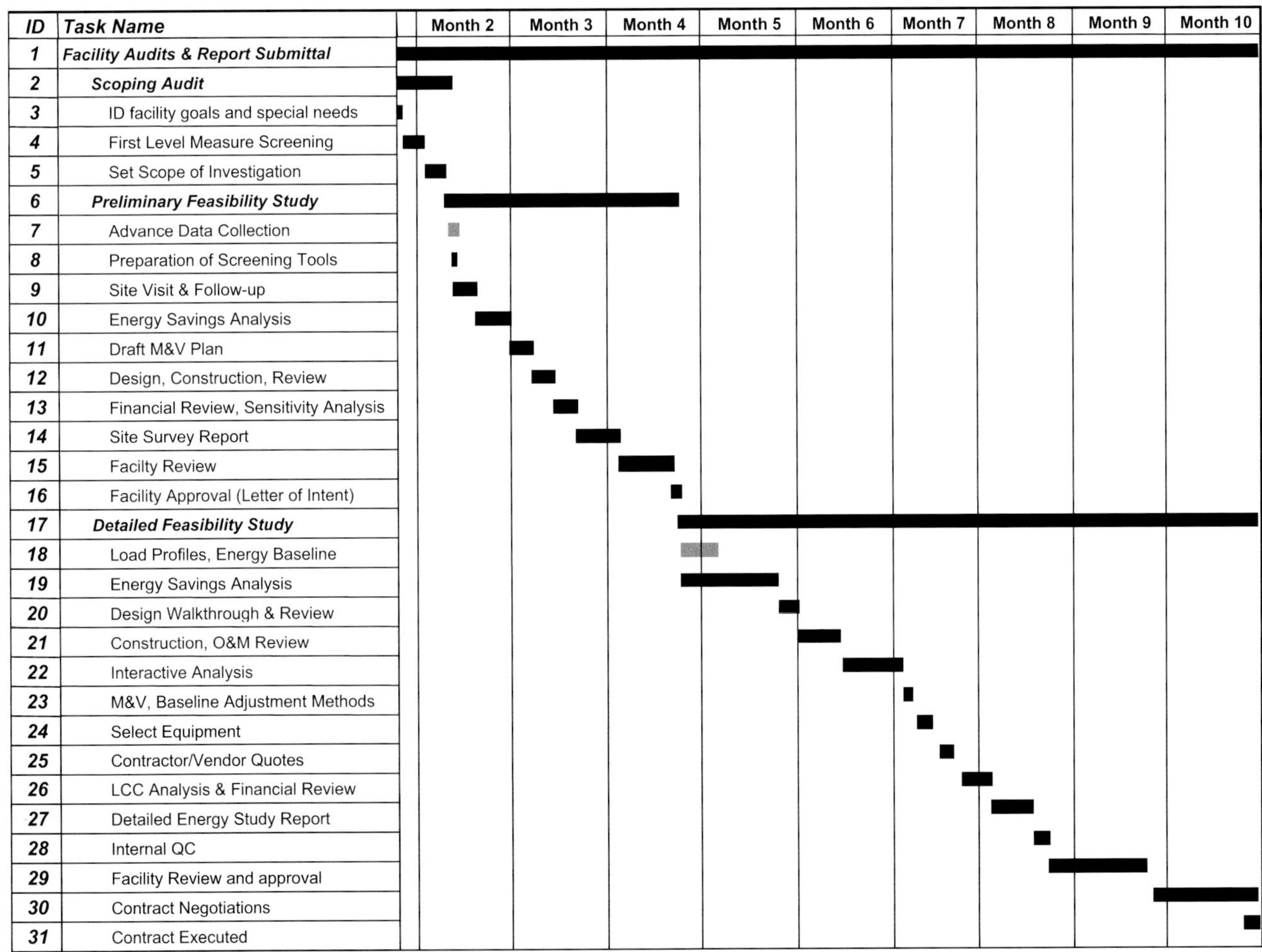

Fig. 40-3 Representative Critical Path Tasks of the Development Process in Gantt Chart Format.

may take one or two days and will include interviews with key operating personnel, general observation of site conditions and major systems, and identification of known problems for which solutions should be sought. Potential project measures are identified and rough estimates of installed costs and savings for major technology categories are prepared as an early indication of project potential.

Another key element of the Scoping Audit is the identification of facility goals and objectives. These are critical factors that must be identified early in the process. They become important screening criteria for review throughout the project development process.

Based on the results of the Scoping Audit, a brief report is developed to indicate the likely areas of focus, establish an overall order of magnitude of the project opportunities, and identify particular areas of expertise and time/effort required. At this point, an assigned management team reviews the results and determines whether to recommend that the facility proceed with a formal study. Feedback from the management team will be synthesized in making the final determination if there is sufficient benefit to be achieved through further pursuit of the project.

Project Development Plan

Following a positive recommendation to proceed, all project planning information, beginning with the Scoping Audit results, should be documented in a Project Development Plan. The plan is initiated during the Scoping Audit and is continually updated throughout the study phase to incorporate new information as it is compiled and as the scope of work and measure concepts evolve accordingly. The plan will identify the areas of the facility to be investigated, project objectives, goals, deliverables, and key global assumptions and constraints. It also establishes the project team, including representation from all necessary disciplines.

Additionally, the plan will identify the scope, objectives, and rationale behind the plan. In large facilities, for example, it is common to target only certain locations or types of systems initially as part of a multiple-project, long-term master plan. This allows for a timely, focused effort and a systematic approach to developing subsequent projects. It is still necessary to scope the entire facility to develop the master plan and make logical selections for each project effort. In smaller facilities, the entire facility is usually addressed from the onset.

The project plan will also identify the schedule and requirements of the host facility. Subject to feedback from facility management and subsequent modifications to the plan and its final acceptance, the study team will proceed with the next phase of the process, which is the Preliminary Feasibility (Phase I) Study. This same iterative process of recommendation, feedback, modification, proposal, and acceptance should be followed for this study phase and each subsequent phase to follow.

Preliminary Feasibility Study

When the Scoping Audit is complete, a more clearly defined group of measures has been established, and the commitment to proceed has been made, the study team conducts an initial site survey to refine the scope of the project and enhance the Project Development Plan, and then focus on the development of the initial integrated project. The study team will develop a full understanding of the existing conditions and begin to refine measures for inclusion in the project.

During the Preliminary Feasibility Study, the team continues to compile detailed information on the energy and other resource usage of equipment targeted for potential improvements. The team should construct a simple model of the facility's primary energy, distribution, and end-use systems, as well as the loads being served. They will gather data for a preliminary analysis of measure potential, locate metering and monitoring points, and identify areas of focus for subsequent phases. Long-term facility master planning objectives are incorporated, including utility supply optimization, capital equipment upgrade, and efficiency improvements. A list of potential measures is identified and screened, yielding a refined list of technically and economically feasible project opportunities. While this study phase represents only the preliminary phase of the process, quality work performed during this phase enables reliable preliminary decision making and will improve the accuracy and efficiency of the subsequent analytical and design phases.

The deliverable for this phase is a report that will include an economic summary of measures and alternatives recommended for further consideration and the key assumptions that must be confirmed. The report will be presented to facility management for their review to serve as a basis for discussion of further steps toward implementation. Key success factors will be identified during the site visits to eliminate wasted effort in subsequent phases and to expedite the entire process. This effort is far more robust than the Scoping Audit, but less than the detailed study to come. It may require several team members to spend a couple of weeks on site in total, and may take one to two months to complete.

Detailed Feasibility Study

At this point, the potentially viable measures have been selected during the Preliminary Feasibility Study and facility management has agreed to pursue those measures that withstand rigorous feasibility and cost-effectiveness screening. Upon receiving authorization to proceed with the target list of measures, the team can commence a comprehensive Detailed Feasibility Study. Much of the required data will have been gathered and analyzed using preliminary screening tools. During the study, which involves thorough site investigation, all remaining data will be identified and obtained. For each energy- and resource-using subsystem, the team will review all relevant facility documentation and establish all operating conditions.

This phase involves accurate, careful characterization of actual thermal and electrical loads. Qualifying the viability of a potential measure starts with a more thorough physical inspection of facilities, equipment, and systems. This involves more than observing equipment and collecting nameplate data. The team should inspect distribution systems to determine their capacity and condition, confirm as-built drawings, and determine if there is space and access for installing the new equipment. The team will select systems and components with the flexibility, efficiency, durability, and performance to meet individual or concurrent loads, and evaluate operating economics against current systems and/or possible competitive alternatives. Technology application concepts are also refined with sufficient design detail for investment grade construction cost estimates. The final result is a package of measures with reliable savings and accurate firm cost estimates incorporated into a life-cycle financial model and ready for implementation. Deliverables include firm energy (and other resource) and operational savings, an optional savings verification plan, conceptual design of each measure, final construction costs, and results from a financial model that shows the rate of return or cash flows for the life of the project. The Detailed Feasibility Study is sometimes referred to as an investment-grade audit, as it serves as the basis for overall project investment decisions.

Financial Analysis

As a final step in the development phase, rigorous financial analysis must be performed to evaluate potential capital commitments. There are a variety of capital budget analysis and presentation tools that may be employed. These include techniques such as:

- Simple payback
- Simple rate of return
- Discounted payback period
- Net present value (NPV)
- Internal rate of return (IRR)
- Savings investment ratio (SIR)

Prospective investors often expend great effort to properly evaluate capital investments. Often times, they will use a mixture of simple analytics, such as payback or substitute comparisons, and more complex tools, such as IRR and other cash analysis tools as well as price earnings (PE) and other net income-oriented ratio analysis tools. Each method has its strengths and weaknesses. A good investor will use all of these analytical tools, but will rely most heavily on previous investment experience in making a decision.

Life-cycle analyses evaluate the sum total of project incremental costs and benefits over the life of the project. In each year, different situations will occur, such as a significant expenditure for overhaul in a certain year. Also, energy and other resource costs are projected to change over time. Given a stream of expected annual cash flows, present value (PV) analysis is a means of equating an amount received or paid in the future in today's dollar. Virtually all sophisticated economic analyses use the basic concepts of PV to account for the time value of money over the life of a project. The value of a dollar saved in one year must be differentiated from a dollar saved in another based on the time value of money. Thus, all cash flows in every year of a project, whether costs, savings, or net cash flows, must be related to each other in a way that accounts for when they occur.

In performing time-valued project economic analyses, numerous factors must be considered on an annual basis. These factors include capital and interest, energy and other resource operating cost savings, OM&R costs, inflation, salvage value, replacement costs, disposal costs, property tax, insurance, deprecation, and other tax deductions. These factors, along with detail on the various financial analysis techniques, are presented in Chapter 42.

Contracting for Project Implementation

Once a determination to proceed with a project is made, the work must be contracted for. Under a design-build arrangement, such as that detailed in Figure 40-1, the contract would be executed after the detailed study phase and prior to the design stage. With other approaches, there may be a series of contracts issued for various stages of the work, such as with an engineering firm to perform the initial study, an A&E firm to perform the design engineering, and then a construction firm who is the winning bidder on the scope of work that was put out to bid. In all cases, the purpose of a contract is to detail the responsibilities of both the owner and contractor, including their rights and liabilities, throughout the execution of a project. A properly drawn contract, in which the scope of work and expectations of the owner and contractor are clearly defined, will minimize the risk to all parties of trouble during project execution. The same is true between a contractor and its subcontractors.

There are many contracting methods available to an owner, including cost plus, lump sum, unit price, turnkey, and performance-based, among others. Each method has its own benefits and liabilities that must be considered prior to selecting one for a project. Once selected, it is common to utilize prepared standard contracts for professional services that have been developed by the American Institute of Architects (AIA), which has been time-tested throughout the architecture, engineering, and construction industries. With certain performance-based contracting methods or energy services agreements involving third-party ownership, non-standard customized contracts may be used. While careful contract execution greatly reduces risks, it is still necessary to anticipate and deal with contract issues that may arise during the execution period. Chapter 43 provides detail on contract method options and key issues for consideration.

Project Funding

Facilities may choose to finance project implementation internally, through a capital budgeting process, through the use of external funding, or a combination of both. The options for funding capital projects range from unsecured balance sheet financing to complex collateralized structured project financing. Key factors that will influence the choice of options are the credit strength of the user/host and the distribution of risk between the owner and the source of capital funds. The stronger the credit profile of the user/host, the greater the financing options. The objective is to capture the cash flow generated by a project in a secure, segregated financial structure that is bankruptcy-proof and directs cash flow to the exclusive use of the projects.

Chapter 43 provides detail on the various financing options that relate to the economics of the assets being acquired. It includes a review of the determinants of available financing options and the advantages and disadvantages of each. It also includes a step-by-step approach that leads to the selection of a financing format and compares and contrasts sample financial offerings.

Implementation Phases

Once a determination has been made to proceed with a project and contracting formats have been selected and capital funding has been approved, the focus turns from development to implementation. Implementation begins with the engineering design phase and proceeds with construction. Once the project is commissioned and accepted by the host facility, it enters into the operations or service phase in which long-term OM&R activities and, in some cases, savings verification activities are carried out.

Figure 40-4 provides a representative timeline in the form of a Gantt chart showing critical path tasks of the implementation process. The tasks are organized under four major categories — Engineering Design (ID 32-45), Pre-Construction (ID 46-60), Installation (ID 61-68), and Post-Installation (ID 69-79). Facility review and contract negotiation and execution (shown as ID 29-31 in Figure 40-3) may be considered as either development or implementation tasks, or as individual tasks falling between the two processes. Ongoing savings verification (ID 80, shown in the project summary timeline in Figure 40-2) is an optional step that may be considered part of the on-going service phase activities. This sample implementation timeline shows about a year for the process. This would allow for the implementation of a sizable, multi-measure, integrated project of up to about $10 million. For considerably larger projects, a longer timeline of up to two years may be expected. For smaller projects, a condensed timeline of as few as six months may be practical.

Chapter 44 provides a description of the various implementation stages. It begins with a detailed description of the engineering design phase, including the preparation of design documents (drawings, specifications, and submittals) and development of an implementation plan for construction. It then proceeds with a discussion of the various construction phase elements. This includes detail on material procurement, the installation, startup, and commissioning components, and the site safety and quality assurance plans developed and implemented during the construction phase. This is followed by descriptions of the development and execution of training, long-term OM&R, and savings verification programs.

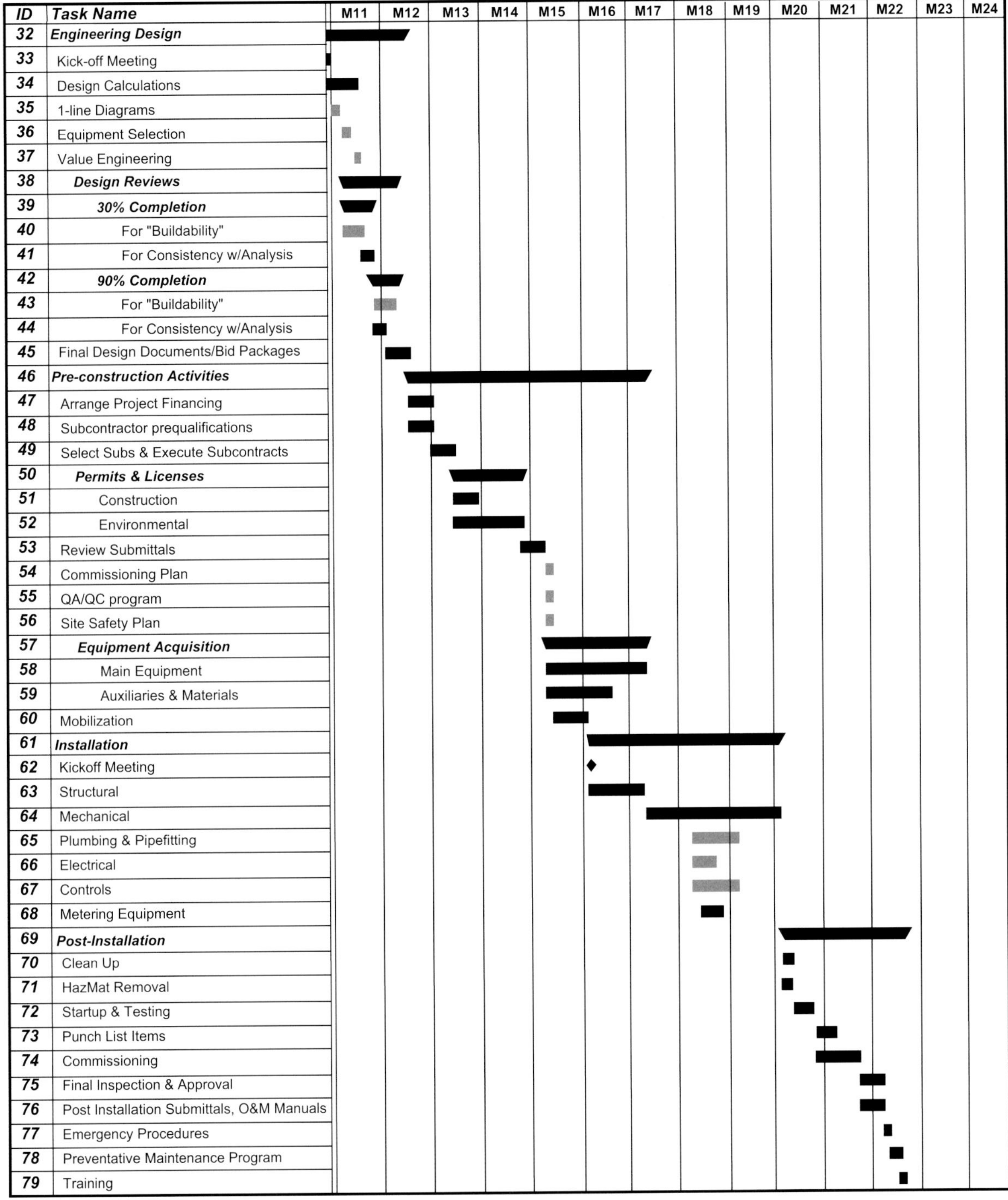

Fig. 40-4 Representative Critical Path Tasks of the Implementation Process in Gantt Chart Format.

Design Phase

Once a project has been approved, it is necessary to develop a comprehensive design that will serve as the basis of project implementation. This is so, whether contracted for separately, as in the case of a standard plan and specification arrangement, or as part of a turn-key design-build initiative. During this phase, the preliminary designs that were used as a basis of project cost estimating and other feasibility analyses are brought to completion over a series of phases. Final engineering analysis, such as zoning and sizing calculations, are performed and equipment selections are finalized. A design document package is produced that contains, at a minimum, all procedures, drawings, specifications, design analyses, cost estimates, and other documentation required to fully describe the proposed work to the facility for its review and ultimate sanctioning of project implementation. During this phase, a design-builder will also finalize the implementation plan for construction.

While this phase is led by engineering design, it is still a multidisciplinary effort to ensure smooth transfer of information. Assumptions and concepts are carried forward from the analysis phases to the design team, and later to the construction team. A sound design team will include individuals familiar with the original feasibility study to ensure that the design is consistent with all energy saving assumptions. Construction experts will also participate in design development, monitoring the design for value engineering opportunities and looking for more cost-effective equipment selections and layouts. In the case of a design-build arrangement, the construction group may also pre-order long lead-time items as needed to meet project timelines. Operations experts will lend their perspective on the OM&R aspects of the proposed measures during this phase.

Construction Phase

Common to all types of implementation and contractual approaches is the need for a construction methodology that incorporates standard industry practices for assuring proper coordination, communications, materials quality, construction methods, budget controls, code compliance, and safety practices. Appropriate measures must be taken to ensure that the measures are installed on time and within budget and are consistent with all previous analysis and design concepts. These planning, oversight, tracking, and control measures are the responsibility of the construction management team. They include:

- Pre-construction activities, such as planning, constructability review, and value engineering
- Cost estimating and document control at various stages of development
- Final equipment selection and material acquisition
- Scheduling and coordination of construction trades and activities, materials delivery, etc.
- Start-up and commissioning planning and oversight
- Site safety program development and management
- Quality assurance program development and management

The construction management team uses various control systems and tracking tools and checks to ensure that all subcontracted work is performed in a timely manner and meets high standards of quality. All construction team members' work must be closely coordinated. Projects should be delivered on time, within budget, and with a minimal amount of scope and cost changes. With a standard plan and specification approach, risks and responsibilities are more spread out between the facility owner, the design firm, and the construction firm. In the case of performance-based design-build type arrangements, many of the risks of project performance are shifted away from the owner and onto the contractor. With performance-based contracts, the achievement of actual verified savings becomes the engine that drives this construction process. The risk of a flawed design or substandard construction no longer lies primarily with the owner, but with the performance contractor.

Training and OM&R

Planning for short- and long-term OM&R and replacement activities should commence concurrent with the construction phase. For these tasks, an OM&R team can be formulated either by the host facility or the contractor who is to be responsible for providing such services. The primary responsibilities of the team leader are: development and execution of the routine operating and maintenance systems; definition and management of the repair and emergency response resources; sustaining a trained and supported OM&R site staff; collecting and analyzing performance data; and responding to client concerns.

The OM&R supervisor should develop manuals that cover the total operation of each measure being installed. The manuals will contain step-by-step methods and illustrations for operating systems and their components. They will detail the location, function, characteristics, and component arrangements and relationships. Decisions should be made as to the intended long-term OM&R approach (i.e., as needed, preventative, or predictive) and who will be responsible (contractor or host facility) for each component. Ideally, an optimally planned program should be developed for each installed measure that will clearly prescribe the recommended maintenance schedule and expected frequency. Emergency maintenance procedures and parts inventory requirements should also be included.

The OM&R supervisor will provide, or arrange for, comprehensive training on the operation, troubleshooting, maintenance, and repair of equipment and systems modified or installed under each measure. Instruction may include a classroom phase, as well as practical hands-on field training. For each measure, the OM&R manuals and supplemental materials should be included in the training program.

Baseline and Performance Measurement

An optional additional step in the implementation process is the development of a savings verification program. This can serve to document the success of the project over time in generating projected savings or meeting anticipated operational goals. In various types of performance-based contracts, savings verification programs establish the basis of payments made by the host to the project financier or prime contractor. Even when not a contract requirement, savings verification programs provide much meaningful data to the analyst and operators, and ultimately to the host/owner. They also provide feedback on the accuracy of the study originally used to justify the investment. Moreover, the system performance data gathered can also provide collateral benefit as part of preventative or predictive maintenance programs.

Generally, verification plans set up resource use evaluation procedures, typically using direct metering of energy and other resource input to the affected process or building area. Sometimes, the metering and data collection system is linked to the overall energy management control system and provides operating data that is used to optimize system performance and indicate when service is required.

The type of verification process used will vary widely. Savings verification programs are designed with specific M&V protocols. Protocol selection will depend on several factors, including project complexity, metering complexity, magnitude of costs and savings, the level of interactivity with other measures, and contractual allocation of risk associated with performance factors not controlled by the contractor. Also considered is the collateral value of the M&V installation with respect to other uses for the data and metering systems.

Case Study

Following is a case study of an integrated energy and resource usage and cost reduction project. It was developed and implemented by an energy services company specializing in performance contracting and energy infrastructure services. The host facility, a university medical center, had a critical need for major energy infrastructure improvements, including expansion of its emergency electric generation and chilled water production capacity. Hence, while the energy usage and cost reductions in and of themselves were important objectives, the cost savings produced a revenue stream sufficient to pay for all of the needed capital improvements.

The solution was found by leveraging a series of cost-effective energy efficiency measures and by optimizing the thermal balance of the central steam system.

In the first case, the university medical center desired expanded emergency electric generation and chilled water production capacity. The solution was found by leveraging a series of cost-effective energy efficiency measures and by optimizing the thermal balance of the central steam system. In the second case, the industrial-type laboratory facility desired a complete renovation of virtually all of its major energy systems and the ability to meet its air conditioning capacity requirements. The solution was found through energy source switching and a complete redesign of its air handling system, including the effective use of new control technology and energy recovery systems.

University of Massachusetts Medical Center Project

The University of Massachusetts (UMass) Medical Center, Hospital, and Research Laboratory recently implemented a $30 million integrated energy and resource efficiency and cost reduction program at its 2 million ft^2 (186,000 m^2) facility in Worcester, Massachusetts. The program was implemented under a 10-year energy services agreement that served to reduce energy usage and cost, while modernizing the facility's energy infrastructure and providing reliable standby electric service.

The completed integrated project included major upgrades to the facility's combined heat, power, and cooling plant. It also provided for new lighting and electric motors throughout the facility, upgrades to air handling equipment, and expansion and modernization of the facility's energy management control system (EMCS).

Facility Goals and Objectives

The facility's objectives were to reduce operating costs and generate annual savings sufficient to leverage numerous important energy infrastructure upgrades. These included the following:

- Upgrade and expand the 25-year old electric cogeneration system

- Secure the ability to reliably operate the facility under the emergency power system
- Expand chilled water system capacity
- Install new low air emission central boilers
- Modernize and expand the EMCS
- Optimize energy cost savings
- Implement the project without interrupting facility operations

Baseline Conditions

The existing central plant housed three steam watertube boilers that provided steam for heat, power, and cooling. Their original full capacity rating was 115,000 lbm/h (52,000 kg/h), but they were operated at up to 90,000 lbm/h (41,000 kg/h) at 250 psig/500°F (18 bar/260°C) and had limited environmental controls. The plant also included a cogeneration system consisting of two extraction/condensing turbine generators. The 250 psig (18 bar) header also supplied three 2,500 ton (8,790 kW_r) steam turbine-driven centrifugal chillers. Steam turbines were also used to drive boiler feed pumps and draft fans.

There was insufficient on-site electric generation capacity to support the entire facility in the event of a utility outage. There was also a deficiency in chilled water capacity, given the load growth experienced over the past decade, including the addition of a new research facility.

There was an old central pneumatic and partially upgraded DDC EMCS that provided some measure of control over the HVAC system. However, it required both upgrade and expansion. The central plant was largely dependent on manual or equipment-specific automated controls.

There were about 5,000 hp (3,728 kW_m) of standard-efficiency motors, none of which operated under variable speed. The lighting system was old and inefficient, with over-lighting in some locations and inadequate lighting in others.

Project Measures

To improve thermal efficiency, the steam system was upgraded to operate at a much higher initial pressure and temperature. Two of the three original boilers were targeted for decommissioning and two new dual-fuel 115,000 lbm/h (52,000 kg/h) Babcock and Wilcox watertube boilers with low-NO_X burners and flue gas recirculation were installed. The new units generate steam at 1,100 psig/850°F (77 bar/454°C) that is used to power a new Dresser-Rand 5 MW back-pressure steam turbine generator. The new turbine generator produces electricity at a voltage of 13.8 kV to match the plant's existing normal and emergency electrical distribution systems, thereby avoiding the additional expense of a new step-down transformer. The plant has black-start capability and can operate isolated from the grid if necessary. As an additional redundancy measure, a new Caterpillar 1,250 kW reciprocating Diesel engine generator was installed.

Also installed was a boiler plant master control (BPMC) system to oversee operation of the new boilers, turbine generator set, chiller, water treatment, condensate, etc. Header pressure controllers and other "off skid" items are controlled directly through the BPMC, as well as the supervisory and data sharing with the other controllers. The system serves to reduce operating and repair costs related to pressure problems and the amount of employee time required to monitor the systems. The BPMC system works in conjunction with the individual control system on the existing boilers, turbines, and chillers.

A newly installed medium-pressure steam header, which cross-connects to the existing header, delivers 250 psig/500°F (18 bar/260°C) steam, exhausted from the new turbine, to the two existing extraction/condensing turbine generators, three steam turbine-driven chillers, and turbine-driven boiler feed pumps and draft fans. Extraction steam at 50 psig (4.4 bar) is then used to power a new 5,000 ton (17,580 kW_r) turbine-driven York centrifugal chiller designed for reasonably efficient operation at this relatively low operating pressure. Instead of replacing the entire system with a larger turbine generator operating with one larger pressure drop, the retrofit design optimized capital costs by keeping the existing steam turbine generators in operation. The new unit captures the enthalpy drop (extra ability to do work) at the now higher work availability level resulting from the generation of higher temperature/pressure steam by the new boilers.

In addition to the major combined heat, power, and cooling project, the other six implemented measures provided a combination of infrastructure improvements and operating cost savings. Five of the six measures, described as follows, offered simple paybacks of less than five years.

- Lighting System Upgrade: The lighting system upgrade involved the installation of about 27,000 new fixtures. Significant savings were achieved as a result of the new fixtures, as well as the reduction in the total number of lamps and ballasts needed to provide proper light levels. The reduced electric demand and cooling requirement and the added heating requirement resulting from this measure were interactively considered in the savings analysis, heat balance, and capacity selection of the boilers, turbine-driven generator, and chillers.

- EMCS Upgrade: This measure involved the replacement of the old imprecise controls with state-of-the-art technology. This was achieved by overlaying the existing pneumatic system with direct digital controls (DDC) and expanding the number of control points at the facility by 65%. The reduced electric, heating, and cooling loads and altered load profiles were considered interactively with the other measures.
- Electric Motor Upgrade: This upgrade included the replacement of 134 motors with high-efficiency models. Due to the relatively large average capacity of the motors that were replaced, the total of which was 5,000 hp (3,728 kW_m), the simple payback for this measure was extremely low at 1.6 years.
- Small Chilled Water System Upgrade: This measure produced a good payback while providing for some air handling system improvements.
- Domestic and Process Water Improvements: This measure provided for new bathroom fixtures, a new reverse osmosis demineralized water system, and the redesign of cooling water flow for the refrigeration system.

The sixth measure involved the conversion to a variable air volume (VAV) system with the use of VFD technology for variable speed HVAC fan operation. The measure improved the indoor air quality and comfort conditions at the facility, but provided a modest economic return with a simple payback of 11 years.

Implementation of the overall project required careful planning to minimize and, in many cases, eliminate interruptions to facility operations. Boilers, turbines, and chillers were brought into the plant, staged, and installed while the existing plant was fully operational. Equipment cut-overs were carefully scheduled during off-peak periods to minimize the impact on the facility's critical care and emergency operations.

Results

Table 40-1 summarizes the capital costs, annual savings, and simple paybacks for the portfolio of measures installed as part of the program. It reflects the ability to implement a major facility upgrade by leveraging longer payback measures with shorter ones in developing an overall project with acceptable financial returns.

The new combined heat, power, and cooling plant has enabled UMass Medical Center to achieve its energy operating cost and infrastructure improvement objectives. Because of the critical nature of services provided to the campus, the primary and backup systems have been designed to provide uncompromised power reliability. The electric cogeneration plant now produces about 75% of the facility's annual electricity. The upgrade also improved system reliability and now enables the facility to provide emergency power sufficient to fully support the Hospital and Medical Center. The new chiller, along with the HVAC upgrades, improved the capacity, efficiency, and reliability of the central chilled water system. Overall, the increased capacity and improved system thermal efficiency of the cogeneration system produced energy

Measure Implemented	**Capital Cost ($)**	**Annual Savings ($)**	**Payback (yr)**
Chilled Water System Improvements	271,000	55,000	4.93
Energy Management Control System Upgrade	750,000	160,000	4.90
Air Handling System Upgrade/VAV Conversion	1,800,000	164,000	10.98
Domestic and Process Water Improvements	500,000	156,000	3.21
Premium-Efficiency Motor Upgrade	240,000	150,000	1.60
Lighting System Upgrades	3,000,000	650,000	4.62
Combined Heat, Power, and Cooling Cogeneration Plant	23,000,000	2,293,000	10.03
Total	**29,561,000**	**3,628,000**	**8.15**

Table 40-1 Cost, Savings, and Payback Summary of Integrated Program.
Source: Energy User News

savings of \$2.3 million per year.

The five short payback measures reduced energy consumption and provided several operational benefits. Combined, they generated \$1.17 million in annual savings at a cost of \$4.76 million, yielding a simple payback of 4.1 years. The robust economic performance of these measures was leveraged to balance the longer paybacks of two major infrastructure upgrades, i.e., the combined heat, power, and cooling measure, and an air handling system measure, both of which had simple paybacks in excess of 10 years. Like the cogeneration measure, the air handling measure provided important operational benefits. Where it might not have been judged economical and funded on its own, the leveraging effect of bundling a portfolio of longer and shorter payback measures into one integrated project allowed the facility to achieve their objective of providing the maximum energy savings and infrastructure benefits within a financeable framework.

❑ Electronic Ballasts
❑ Specular and Aluminum Reflectors
❑ T8 Lamps
❑ Street Light Replacements
❑ Task Lighting
❑ Occupancy Sensors
❑ Daylighting Controls
❑ LAN Systems/Network Interfacing
❑ Control Programming/Systems Control
❑ Enthalpy Control/Economizers
❑ Preventative/Predictive Maintenance

CHAPTER FORTY-ONE

Technical Analysis

All integrated energy and resource cost reduction programs must start with a study that identifies the opportunities and quantifies them in such a way that a scope of work can be developed and financial decisions can be made. Commonly, a two-phased study approach is used for this purpose. Prior to this, an additional third stage, or Scoping Audit, should be added to the process to initially verify if there is viable project potential. It serves to provide early screening of potential measures and leads to the determination of an overall plan of action.

Upon completion of the Scoping Audit, an initial group of measures to be proposed to the host facility as the core of the project has been established. This is reviewed by the facility for comment and acceptance. Upon acceptance, the facility commits to proceed with the next phase of evaluation, which is the Preliminary Feasibility Study. This study refines the proposed group of measures and critically tests their feasibility and economics. It includes a fairly comprehensive review of the entire facility. The study team develops a full understanding of existing conditions and formulates energy resource cost reduction and facility upgrade measures that will be proposed to the facility as the basis of the eventual implementation contract. This Preliminary Feasibility Study goes beyond the standard measure categories and includes any appropriate site-specific measures in the evaluation. Once these measures are modified as needed and accepted by the facility, the study team will complete the evaluation of cost and savings through a final "investment grade" Detailed Feasibility Study.

The technical study, or measure evaluation, process from conception to contract can be described as having three phases.

1. **Scoping Audit**: Identify a general approach to the facility and a core group of measures on which a resource efficiency and cost reduction program can be based so that intent can be established.
2. **Preliminary Feasibility Study**: Evaluate and refine the core group and any other measures to generate a reliable, quantified proposal so that a commitment can be established to proceed with a final investment-grade study.
3. **Detailed Feasibility Study**: Finalize all evaluation with detailed analysis and measurement to generate an investment-grade proposal so that capital funding commitments can be made and a contract can be executed to proceed with design or turnkey implementation.

This process is flexible and can be adjusted to accommodate special conditions. For example, results of previous energy audits may be used to streamline or even eliminate the Scoping Audit and Preliminary Energy Study phases.

These two analytical phases — the Preliminary Feasibility Study and the Detailed Feasibility Study — are to be executed sequentially. They should be based on the groundwork laid in the Scoping Audit phase and be in accordance with a Project Development Plan. This chapter presents step-by-step detail for executing a two-phased technical analysis for the purpose of developing an integrated energy and resource efficiency and cost reduction program. This is preceded by brief descriptions of the Scoping Audit and Project Development Plan processes.

The technical analysis study format is presented from the perspective of an independent analytical team performing the work as would be the case with a consulting engineering firm. In practice, such a team can be fully or partially staffed by internal personnel. The team may also be part of an energy services company (ESCo) offering turnkey design-build or performance contracting services. Regardless of the source of the study team, the same basic approach would apply.

OVERVIEW OF TECHNICAL ANALYSIS STUDY PROCESSES

While the technical analysis study process may consist of one, two, or even three phases, quality studies will all generally include the same basic components. These are:

- Compiling all data collected during the preliminary study phase.
- Conducting in-depth interviews with the facility energy manager, engineers, and maintenance and operating personnel to assess the operating characteristics of existing energy systems and system improvement goals.
- Reviewing facility plans for other planned non-energy

related capital improvements and renovations.

- Reviewing long-term fuel and electricity purchasing options available to the facility (this includes all viable power purchase and sales and fuel purchase options).
- On-site evaluation of all mechanical, electrical, and controls systems associated with targeted application opportunities.
- On-site direct metering.
- Organizing and analyzing all data collected.
- Computer modeling as appropriate both for specific measures and for facility-wide measure interactivity.
- Utility bill reconciliation analysis.
- Developing detailed equipment and installation budget cost estimates for each potential application.
- Developing budget estimates for long-term annual operations, maintenance, repair, and replacement and comparing with existing baseline annual costs.
- Financial analysis, including structuring of terms for payment to ensure the facility receives the optimum benefit of any grants or utility incentive payments to reduce the project implementation cost.
- Developing cost-effective savings verification protocols for all proposed measures.

Informational Requirements

While different types of information and different levels of accuracy will be required during different phases of the study process, the same basic types of information will be required to complete most integrated studies. Following is a representative listing of some of the basic information that is typically required.

Incremental economic values must be determined for items such as:

- Electricity
- Fuel
- Steam or other heat sources at each thermal level (temperature) required
- Purchased chilled water
- Water use
- Operations and maintenance
- Personnel requirements, including special training
- Environmental permitting and emissions controls costs
- Cost of standby electricity (in electric generation feasibility studies)
- Cost of capital/debt
- Cost of insurance
- Cost/value of required or avoided floor space
- Downtime costs

Long-term life-cycle factors must also be integrated into the economic analysis for items such as:

- Cost/value of electricity (internal use and, where applicable, power sales)
- Fuel supply cost and contract security
- Escalation of energy, water, and other resources costs and operations, maintenance, and repair costs (and contracts)
- Replacement costs
- Salvage value
- Performance degradation

Incremental capital investment costs to be integrated include:

- Engineering and planning costs
- Acquisition costs of systems to be installed
- Cost for modification of existing equipment
- Demolition and material disposal costs
- Utility program incentives or penalties
- Startup and debugging costs and cost of production downtime
- Permitting, development, legal, and other consulting fees

The selection and utilization of tools and the assembly of information are combined in the processes of analysis described in detail below.

Initial Development Steps

Scoping Audit

A Scoping Audit is a low-cost, brief review of a facility intended to define the energy and resource savings potential and system infrastructure upgrade requirements. It consists of a preliminary review of resource usage and cost records and a walk-through of the facility. The walk-through may take one or two days, depending on facility size and system complexity. It should include interviews with key facility engineering and operating personnel, general observation of site conditions and major systems, and identification of known problems for which solutions should be sought. Utility rates and supply contracts, operating conditions (e.g., equipment schedules and controls), and major equipment nameplate data are compiled. Local, state, and regional factors are also considered, such as environmental regulations, weather, gas and electric supply, transmission, and distribution conditions.

Figure 41-1 is a checklist of numerous efficiency and cost reduction measures that may be initially considered during the Scoping Audit, with the objective being to identify those measures that merit more detailed consideration. This list is by no means exhaustive, but provides a good start. Manufacturing and industrial process efficiency opportunities are mentioned, but a comprehensive list is not provided, as the range of process equipment and systems that may be encountered is so far reaching. Often, in addition to the core study team, experts are brought in to focus on specific processes that are beyond the expertise of the team members, though generally at a later phase in the study process.

As the process advances, results are carried forward, with additional information gathered, and more rigorous analysis is performed during each development phase. Because the scoping phase sets the stage for the remainder of the project, it is wise to dedicate senior staff to conduct these audits. They have the experience to work closely with facility representatives and develop a practical plan for project development that matches the special needs of the facility.

A key element of the Scoping Audit is the identification of facility goals and objectives. These are critical factors that must be identified early in the process. They become important screening criteria for review throughout the project development process. In total, the Scoping Audit allows the team to record the following facility needs and objectives through discussions with facility management:

- Long-range master plans for renovation, expansion, or new construction
- Cost reduction goals, financing preferences, investment thresholds, and risk tolerance
- Equipment and energy source preferences
- Design standards to maintain consistency with existing equipment and spare parts inventories
- Current operational and/or maintenance problems
- Equipment replacement needs dictated by age, capacity, or refrigerant phase-out plans
- Price and risk management objectives for the procurement of energy supplies
- Constraints such as reliability, safety, sound, aesthetics, historical preservation, and environmental regulations
- Inadequate comfort control or indoor air quality

At the conclusion of the Scoping Audit, a brief report should then be developed to indicate likely areas of focus and establish an overall order of magnitude of the project opportunities. The report should also indicate particular areas of expertise required for the study. Based on Scoping Audit results and the desired level of study certainty, the time, effort, and analyst expertise requirements can be established. Once a final budget is established and authorization to proceed is provided by the host facility, the senior member of the study team must make a realistic assessment of what can be accomplished and how, and then formulate a Project Development Plan.

Project Development Plan

The technical analysis process begins with the establishment of a Project Development Plan. This plan results from the development of an understanding of the goals and objectives of the host facility's senior management. These criteria act as needed guideposts throughout the study. Objectives may be quantitative attributes, such as percent energy cost reduction figures, replacement of old equipment, and expansion of system capacity. They may also be qualitative attributes, such as preferred fuel and technologies (e.g., renewable resources) and the desire to upgrade or replace certain systems based on potential savings or the desire to improve indoor air quality, working conditions, or overall facility reliability.

Constraints are limitations to the acceptance of certain equipment and system characteristics. Constraints may include factors such as reliability requirements, desired degree of maintenance sophistication, adaptability to other facility systems, sound emission levels, or aesthetics, or may be externally imposed, such as environmental regulations or air emission credit limitations. Constraints of both qualitative and logistical nature can be useful for setting boundaries and eliminating certain measures from further consideration.

The technical analysis study team should then articulate and prioritize project objectives, framed within the host facility's goals and constraints, thus establishing the criteria that identifies the most valued system characteristics. If available, facility long-range planning documents should be reviewed in the process. Clarity in this early task of setting objectives will greatly facilitate the efficiency of screening, selection, and the eventual ranking of project opportunities. This process, well executed, will identify the measures that meet the project's objectives and best fit the facility goals.

The Project Development Plan should be based on establishment of a budget for the study process. The budget will largely determine the comprehensiveness of the study and allow for designation of staffing assignments. The budgeting process should consider the study objectives, size, and complexity of the facility, potential for savings, and documentation and reporting requirements. If significant investment decisions are to be made, the budget should be adequate to support a process that will produce a high level of certainty in the study results. Typically, a quality study will cost, at a minimum, a couple percent of the ultimate project implementation budget. Assuming a relationship between cost and quality, higher quality studies will mitigate project financial investment risk and also lead to reduced costs in the design phase of the project.

The project plan should also indicate the schedule and requirements of the host facility. This will ensure that data is compiled and made available at the appropriate time and that access to the facility is properly arranged. Subject to feedback from facility management and subsequent modifications to the plan and its final acceptance, the study team will proceed with the next phase of the process, which is the Preliminary Feasibility (Phase I) Study.

Preliminary Feasibility Study – Phase 1

The Preliminary Feasibility Study is a comprehensive survey of the entire facility to develop a full understanding of existing conditions and identify resource cost reduction and facility upgrade opportunities that merit a more detailed evaluation. The study team should construct a simple, yet accurate, model of the facility's primary energy systems, distribution systems, end-use systems, and the loads being served. It should gather sufficient data for a preliminary analysis of savings potential and cost, as well as locate potential metering points and areas of focus for potential detailed study.

Quality work performed during this phase enables credible preliminary decision making and will improve the accuracy and efficiency of subsequent phases. Key factors identified early eliminate wasted effort in subsequent phases and expedite the entire process, while minimizing the effort and, therefore, cost of this study phase.

Lighting	**Energy Management System Upgrades**
❑ Electronic Ballasts ❑ Specular and Aluminum Reflectors ❑ T8 Lamps ❑ Compact Fluorescent Lamps ❑ LED Exit Sign Replacements ❑ Street Light Replacements ❑ Direct and Indirect HID Lighting ❑ Task Lighting ❑ New Fixture Layout Design ❑ Occupancy Sensors ❑ Daylighting Controls ❑ Dimming Controls ❑ Exterior Lighting Photoelectric Controls ❑ Timed Lighting Controls	❑ LAN Systems/Network Interfacing ❑ Control Programming/Systems Control ❑ Process Control ❑ Equipment Sequencing or Seasonal Operation ❑ Temperature and Relative Humidity Control ❑ Chiller/Boiler Optimization Control ❑ Enthalpy Control/Economizers ❑ Duty Cycling ❑ Load Shedding ❑ Remote Communication Control and Monitoring ❑ Preventative/Predictive Maintenance ❑ Equipment Performance and Energy-Use Monitoring ❑ Water and Airflow Measurement ❑ Continuous or Predictive Emissions Monitoring
Motors/Drives	**Compressed Air (and Gas) Systems**
❑ Electric Variable Frequency Drives ❑ Premium-Efficiency Motor Retrofit ❑ Downsizing Oversized Motors ❑ Motor Idle Elimination Control ❑ Motor Power Factor Control ❑ Synchronous Motor Power Factor Correction ❑ Fuel/Steam Prime Mover Mechanical Drives	❑ Air Compressor Replacement ❑ Hybrid (Mixed) Air Compressor Systems ❑ Compressed Air Leak Elimination Programs ❑ Compressed Air Automation Systems ❑ Compressed Air End-Use Optimization ❑ Air Motors ❑ Compressed Air Storage Systems ❑ Compressed Air Drying Equipment Upgrades
Electric Service Systems	**Building Envelope**
❑ Energy Efficient Transformer Upgrade ❑ Primary and Secondary Selective Systems ❑ Power Factor Correction ❑ Centralized Demand Limiting Controls ❑ Power Quality, Protection, and Coordination ❑ UPS Systems ❑ Energy Monitoring and Communications Systems ❑ Multiple Electric Service Systems ❑ Solid State DC Converters ❑ Load Shedding Switchgear	❑ Wall, Ceiling, and Floor Insulation ❑ Reglazing and Window Replacement ❑ Storm Windows ❑ Window Film Treatments ❑ Air Pressure Balancing ❑ Convection Loss Reduction/Air Sealing ❑ Vestibules/Air Curtains ❑ Roof Covers/Reflective Roof Services ❑ Vapor Barriers ❑ Thermal Shutters ❑ Interior/Exterior Shading

Fig. 41-1 Checklist of Potential Efficiency and Cost Saving Measures.

Steam and Chilled Water Distribution Systems	Air-Side Distribution Systems
❑ Steam Distribution Pressure Controls ❑ Central Steam/Chilled Water Distribution Repairs ❑ Steam Trap Repair/Replacement ❑ Humidification System Upgrades ❑ Steam/Chilled Water Pipe Insulation Upgrade ❑ Piping System Balancing ❑ Variable Speed Pumping ❑ Primary/Secondary Loops ❑ Heat Exchanger Upgrades ❑ Radiator Upgrades ❑ Condensate System Upgrades ❑ Decentralization/Centralization Programs ❑ District Heating/Cooling Systems ❑ Seasonal Central Plant Shutdowns/Remote Units ❑ Isolate Off-Line Systems	❑ DDC System Upgrade ❑ Variable Speed Drives on Fan Motors ❑ HVAC Air Distribution Modifications ❑ Constant Volume to Variable Volume Conversions ❑ Rezoning, Zone Control, and Air Balancing ❑ Air-Side Economizer Free Cooling Applications ❑ Temperature Set Point Adjustments, Setbacks ❑ Exhaust Hood Controls and Modification ❑ Reheat Controls and Load Reduction ❑ Air-to-Air Heat Exchangers for Energy Recovery ❑ Destratification Fans ❑ Humidification/Dehumidification ❑ Damper Controls and Upgrades ❑ Clean Room Environment Control ❑ Indoor Air Quality Improvements
Central Plant Supply-Side Systems	**Energy Procurement Supply-Side Opportunities**
❑ Cogeneration-Cycle Applications ❑ Combined-Cycle or STIG-Cycle Applications ❑ Peak-Shaving Generation Applications ❑ Demand Limiting Applications ❑ Fuel Cell Applications ❑ Solar Energy Heating/Cooling Applications ❑ Photovoltaic Applications ❑ Wind Energy Conversion Systems Applications ❑ Geothermal Applications ❑ Municipal Waste and Biomass Applications ❑ Electric Battery Storage	❑ Overall Commodity Risk Management Programs ❑ Pipeline Investment and Construction ❑ Utility Bypass ❑ Submetering of Individual Buildings ❑ Natural Gas Transportation Options ❑ Interruptible Gas Purchasing Options ❑ Electric Wheeling Options ❑ Power Sales Opportunities ❑ Negotiated Competitive Electric Rate Options ❑ Interruptible Electric Service Options
Boiler Plant Systems	**Chilled Water and Refrigeration Systems**
❑ Boiler Optimization/Replacement ❑ Burner Optimization/Replacement ❑ Oxygen and Excess Air Trim Controls ❑ Boiler Fuel Switching/Dual-Fuel Capability ❑ Boiler Heat Recovery ❑ Radiant Heat Systems ❑ Fluidized Bed Systems ❑ Condensing Furnaces ❑ Boiler Auxiliary System Optimization ❑ Boiler Temperature Reset ❑ Domestic Water Heating System Optimization ❑ Emissions Control ❑ Off-Line Boiler Isolation ❑ Automated Continuous Blowdown Control ❑ Blowdown Heat Recovery ❑ Condensate System Upgrade and Optimization ❑ Feedwater Delivery Improvements ❑ Water Treatment Optimization ❑ Thermal (Hot) Energy Storage	❑ Chiller Optimization ❑ Chiller Replacement ❑ Chiller Energy Source Switching ❑ Topping or Bottoming Cycle Chiller Applications ❑ Heat Recovery-Driven Chiller Applications ❑ Cooling Tower Repair/Optimization/Replacement ❑ Cooling Tower Ozone Water Treatment Systems ❑ Cooling Tower Variable Speed Fan Application ❑ Water-Side Free Cooling Applications ❑ DX Unit Replacement ❑ Direct Evaporative Cooling ❑ Refrigeration System Optimization ❑ Refrigeration System Replacement ❑ Off-Line Chiller Isolation ❑ Chilled Water/Condenser Water Reset ❑ Liquid Refrigerant Pumping ❑ Thermal (Cold) Storage ❑ Air, Water, Solar, and Geothermal Heat Pumps ❑ Desiccant Dehumidification ❑ Chilled Water System Heat Recovery
Water Resources	**Miscellaneous Technologies**
❑ Lavatory Fixture Upgrades ❑ Low-Flow Sink Aerators and Shower Heads ❑ Sink and Shower Timer Controls ❑ Water Coolers ❑ Elimination of Once-Through Cooling Systems ❑ Waste Water Recovery Systems ❑ Laundry System Upgrades ❑ Cooling Tower Control ❑ Irrigation Optimization/Xeriscape ❑ Gray Water Systems ❑ Reverse Osmosis Demineralizing Systems ❑ Eliminate Once-Through Cooling Systems	❑ Process Equipment Automation and Optimization ❑ Process Equipment Replacement ❑ Process Energy Recovery Systems ❑ Reduce Operating Hours for All Energy Systems ❑ Dishwashing Equipment and Practices ❑ Fuel Recovery Systems ❑ Laundry Heat Reclaim and Energy Source Switching ❑ Elevator Controls and Upgrades ❑ Incinerator Upgrades and Heat Recovery ❑ Swimming Pool Covers and Humidity Control ❑ Vending Machine Energy Reduction ❑ Plug Load Efficiency and Management Programs ❑ Alternative Fuel Vehicles

Fig. 41-1 (cont.) Checklist of Potential Efficiency and Cost Saving Measures.

Figure 41-2 summarizes the recommended approach to processing information during the preliminary study phase of the development process. The table includes a description of the information to be gathered, the work to be performed to process and analyze this information, and the results and/or deliverables produced by that work.

Description of Preliminary Feasibility Study

There are 9 main steps that should be executed during the Preliminary Feasibility Study phase. The following discussion describes each of these steps, including the methods, work effort, and results produced during each step of the Preliminary Feasibility Study.

Step 1. Advance Data Collection

Before the study team goes on-site, information should be collected to familiarize the team with the facility and provide guidelines for identifying measure opportunities. Advance data collection can typically be defined by four types of data:

A. Previously Completed Studies

While earlier studies may have a different focus, they often provide a source of raw and qualified data. Previous studies can be of particular value by providing information about energy system issues of special importance to the facility and by offering alternative perspectives emerging from differing analytical approaches. However, caution should be taken not to formulate pre-conceived notions prior to the analysts own site investigation.

B. Utility Information

The study team should secure monthly bills, including electricity, fuel, waste disposal, water, and sewer, for the most recent two year period. In addition to offering an aggregate picture of use and seasonal variation, a review of the bills may uncover inconsistencies, such as multiple billings or seemingly inappropriate rates. Further examination and clarification may reveal important limitations, such as an inability to combine electrical services or delivery capacity constraints. Computerized records of daily, hourly, or 15 minute demand and consumption may also be available from the electric or gas utility; they provide a deeper picture of load shape and variation patterns. The utility information compiled during this step provides the basis for on-going load analysis and aggregation work required to fully develop an integrated approach. Where available, internal submeter readings will be collected. These can provide load and consumption information for sub-units, such as kitchens, boilers, chillers, cooling towers, or process equipment.

C. Utility Rates

The study team should obtain copies of the current tariff books from the utilities serving the facility. In addition to the specific rate schedule that the facility currently operates on, all potentially applicable rate schedules should be identified and entered into computer spreadsheet format to simulate how utility bills are actually calculated. This will often be necessary to determine the real impact of modifications under consideration. Certain changes could affect the overall cost of fuel or electricity for the entire facility. The study team should also conduct an evaluation of utility rebates that may be available. Advanced knowledge of the mechanics and magnitude of rebate and incentive program offerings may increase the accuracy of the screening process and alter the focus of the study.

Assessments should also be made of the marginal costs of electricity, fuel, and steam from local utilities. On-site cogeneration, peak shaving, load shifting, gas transportation, and electric wheeling options should all be considered in the determination of prudently anticipated electric and fuel cost reductions. Cost reduction potential will be a function of the ability to negotiate lower rates based on the competitive options available to the facility or the savings achieved through self-generation (e.g., cogeneration, peak shaving, etc.), electric and gas brokering, or forms of utility bypass. Establishment of current and anticipated energy rates and alternative rate structures will establish a cost range within which potential measures can be screened.

Given the potential volatility in energy costs (or utility rates), it may be worthwhile to include a rate sensitivity analysis in the Preliminary Feasibility Study. The payback of each efficiency and cost reduction measure would be calculated under a range of possible energy rates, to determine the thresholds at which it becomes no longer economically viable. With energy analysis models, either computer simulations or spreadsheet calculations, this is a relatively easy and worthwhile exercise. It will result in an unbiased or, at least, less biased, study and will reduce the risk that future changes in the utility market place will invalidate the results of the study. This is especially important for measures that depend on the difference between energy rates, such as cogeneration,

Category	Information and Data Gathered	Scope of Work Performed	Results and Deliverables
Utility Information	Two-year billing history (annual consumption and demand for electricity, fuels, purchased utilities, water); Copies of actual bills; Utility rates; Copies of the current tariff books	Become familiar with usage patterns; Review current rates and service options; Assess marginal, incremental costs; Consider load aggregation supply-side options, such as real-time pricing, interruptible service, etc.	Provides basis for ongoing load analysis; Graphical load profiles, energy use indices (e.g., $/kWh, $/Btu, Btu/sq.ft.); Model to simulate actual bill calculation and accurately predict utility cost savings
Equipment	Nameplate data, operating and/or computer trend logs and/or maintenance contracts; Known maintenance or operational problems; Pressure, flow, voltage requirements	Inspect equipment and gather basic model and operating data; Assess condition, reliability, ability to meet loads	Retrofit, replacement, or control options for efficiency improvements consistent with an overall master plan for the facility; Load reduction opportunities
Facility Documentation	Site plans, distribution system schematics, mechanical equipment schedules, control system schematics; Interviews with facility technical, operations, and maintenance personnel; Previously completed studies	Gain general understanding of layout, configuration, and loads served; Begin to confirm accuracy of record drawings; Review and determine status of any applicable measures	List of all applicable measures and first level screening for technical and financial feasibility
Facility Needs	Field work logistics; Activity schedules; Points of contact for particular resources	Develop list of buildings to be surveyed, work out access and communications	Minimal disruptions to facility operations; Efficient use of staff time
Constraints	Long-range facility planning documents; HazMat regulations; Areas of concern	Brief review for consistency with proposed measures; Identify and record potential problems	Preparation for Detailed Study by identifying logistical and environmental barriers
Measure Savings	List of targeted measures from Scoping Audit; All survey data; Occupancy and activity schedules	Prepare measure- and site-specific screening tools; Correlation plots of loads vs. temperatures to give aggregated load profiles and seasonal operating efficiency	Quickly screen cost-effective and practical measures with simple analytical techniques; Can consider more alternatives (e.g., equipment options, fuel types)
Measure Costs	OM&R records; Gross cost indicators (capacity, floor area, etc.)	Establish current OM&R costs; Rule of thumb installed cost estimates	Capital cost and avoided maintenance cost estimates for new equipment
Financial and Contractual	Preliminary cost and savings estimates; Predictions of future market values	Parametric runs of financial model for sensitivity analysis	Identify technical and financial assumptions that most affect financial goals
Design Issues	Photographs and/or sketches	Establish physical limitations of retrofit construction	Space limitations, site code requirements, site logistics
Construction Issues	Construction-related site difficulties; Vendor quotes for major equipment; Facility preferences or recommended contractors	Assess rigging, piping, structural modifications, demolition, etc.	Modifiers to "rule of thumb" cost estimates
Operational Issues	Operating conditions (controls, operating schedules, comfort, and ventilation standards)	Interviews with facilities staff (all shifts), inspections, review of equipment logs	Establish clear baseline and confirm with facility management

Fig. 41-2 Summary of Work Effort – Preliminary Feasibility Study.

power generation, or fuel switching, since small changes in one or both utility rates can have a large impact on the project economic performance. Still, price uncertainty should not be considered a reason not to invest in energy efficiency projects, since the decision not to invest must be based on the same analysis as the decision to invest. The status quo simply carries forward the risk of higher operating costs versus the returns on a capital investment in energy efficiency.

D. Site Plans and Major System Drawings

Available copies of site plans of the entire facility should be obtained and studied for the purpose of identifying the location of each area to be addressed in the study. Distribution system and central plant schematic diagrams and equipment schedules should also be obtained to provide a general understanding of those systems. Particularly valuable before embarking upon a site visit are equipment schedules for energy intensive equipment. Reduced scale floor plans are often a useful aid in preliminary study work (for example, to keep notes on locations of equipment or to assist in compiling lighting inventories). If these are available, they should be collected and distributed to the study team in advance.

Step 2. Preparation of Screening Tools

Screening tools are standardized methods of analyzing opportunities using a mixture of site-specific parameters (such as utility rates) and typical performance characteristics to narrow the field of practical measures in an efficient but reliable manner. Such tools include spreadsheet rate programs capable of quickly determining incremental costs for any change in energy or resource consumption, lighting simulations, cost databases, manual nomographs, and application-specific spreadsheet programs. The rate programs should also be capable of producing various graphs and indices of usage profiles and unit costs. Usage profile graphics are very useful in assessing characteristics such as load factor and daily, weekly, and monthly usage variation. They can also produce meaningful indices, such as cost and usage per square foot and incremental costs (e.g., \$/kWh, \$/Btu). This type of analysis lends itself to comparison with a database of similar type and size facilities. For large enough studies, the team may develop new tools to address particular requirements of the host or to handle multiple applications of a single technology. In any case, the study team must tailor its screening tools to the applicable conditions of the facility.

Another aspect of the preparation is a marginal utility cost analysis, which will indicate the local utility's ability to flex price in competitive negotiations over cogeneration projects, peak shaving projects, fuel switching projects, and gas and electric brokering options. This will provide a sensitivity benchmark for measuring potential project economic performance. While all cost factors may not be available, a reasonably accurate assessment of utility marginal cost can be made from a review of public records. Utility decisions as to negotiated rate offerings are not necessarily driven by marginal cost alone. Therefore, the experience of other similar facilities in the same utility service territory can also provide valuable information on the anticipated behavior of a utility during such negotiations.

The goal is to establish a range between current and possible future energy costs for the screening process. Current rate structure alternatives and anticipated changes in rate structures must be considered as well, since savings is commonly influenced by time-specific, incremental energy costs, not average costs. The current movement toward rate stratification and real-time pricing in electricity markets are examples of key elements that will impact the selection of technology applications.

Energy rates (as well as rates for other resources, such as water and waste disposal) — both cost and rate structure — should not be considered static elements. Therefore, sensitivity analyses based on extensive research of both short- and long-term market trends is a very important part of the study process, particularly when large long-term investment decisions are to be made based on the study results.

Step 3. First Site Visit

While it is possible for documentation and operator information to provide great detail regarding the operation and layout of a facility's energy systems, it is necessary for the study team to become familiar with the site. Scheduling of the initial site visit should be made after the evaluation of the preliminary data collection has been completed. The site-survey data collection process should be separated into two parts: data retrieval and data validation.

Upon developing a prioritized list of buildings and systems to be evaluated in the preliminary study and reviewing available utility bills, the study team should proceed with a physical site visit. At the onset, the study team should get an overview of the physical layout of the facility, work out access and communication issues, and establish a base of operations. Team members should be given assignments for buildings (or building zones), systems, and other investigation tasks and a plan should be established for daily review of findings, assembly of data, identification of problems and missing information, and realignment of tasks.

During the course of the visit, whatever preliminary data was not procured in preparation for the site visit should be obtained. Making summary data assumptions too rapidly may cause time delays at a later date. It is, therefore, important to develop an early understanding of control systems, operating conditions, and schedules and to establish rapport with key facility staff.

Potentially hazardous materials, such as asbestos, ammonia, halogenated hydrocarbons, PCB ballasts, and lead piping should be identified with the aid of installation documentation and facility personnel. Potential type, location, and amount of hazardous material should be estimated and recorded during the site visit for further professional investigation should the project proceed. The study team should identify areas where cost reduction may be realized by repairing, modifying, or replacing equipment or installing new equipment and facilities. The study team should identify measures that do and do not warrant further efforts and whether or not a Detailed Feasibility Study is warranted to determine technical and/or cost feasibility.

A. Facility Equipment

The study team should review the operation, age, and condition of equipment on-site and, if available, historical and maintenance and operational records. This begins the development of perspective for the number and types of measures that may be applicable. Replacement for conservation purposes is always more financially acceptable if the equipment is a known maintenance problem or is scheduled for replacement. Determination of whether new equipment is required for replacement purposes, expansion, or simply for operating cost savings will impact the financial analysis of the potential application.

The initial equipment inspection should include observation of the conditions and operations of major systems and their primary components. Examples are:

- Lighting systems: Lamp/ballast types, condition, efficiency, light levels, color rendition, etc.
- Cooling towers: Capacity, cold basin water conductivity, current load, projected load, condition of structure, fill, fan drive, water temperature (fan) controller, water chemistry controller, and hot basin water distribution.
- Distribution systems: Chilled water, condenser water, hot water and steam, electricity, compressed air and distribution temperatures, pressures, volumes and voltages, and control schemes for each major system.
- Pumps: Service, basic type, design point (head and flow), observed pressure drop, fixed speed or variable speed, method of flow control for part-load conditions.
- Electric motors: Efficiencies, loads, service.
- Boiler, chiller, and compressed air systems: equipment condition, capacity vs. maximum and minimum loads, standby capacity, leaks and losses, environmental concerns, end-use load reduction opportunities, configuration vs. optimal resource use, backup and interchangeability, energy recovery, and thermal storage.
- Process systems: Equipment condition, method of operation, capacity vs. current and anticipated loads, end-use reduction opportunities, and fuel substitution.
- Control and monitoring systems: Type of system, active control enhancements and potential for expansion to control and/or monitor new systems, compliance with current codes and standards, and existing energy management strategies.
- Air handling systems: Type, capacity, age, typical operating settings and positions, condition, configuration, zoning, and ability to meet loads and maintain required environmental conditions.
- Unitary HVAC equipment: Equipment condition, capacity vs. maximum and minimum loads, environmental concerns, load reduction opportunities, configuration vs. optimal resource use, energy recovery, and control system improvements.

The make, model, and serial number of all primary equipment under study should be obtained during the site visit. Often, the serial number is most important for larger equipment, since most manufacturers are able to supply the original design specifications of a particular unit. Copies of representative on-site logs and records of checking and testing by maintenance personnel should be obtained and reviewed to gain understanding of operations and condition.

In addition to the equipment review, the study team should collect and review information on existing operations and maintenance contracts for the equipment. Compliance with current or pending equipment and operational standards and codes should also be considered. This may include, for example, air or effluent emissions, safety, and minimum outside air requirements.

B. Facility Site Plans

To the extent that they are available, accurate, and up-to-date, facility drawings of the current site configuration will provide the study team with a preliminary understanding of the systems located within the facility and how they were designed to interact. These drawings also provide a basis for determining the physical limitations of retrofit construction from mechanical, electrical, civil, and architectural perspectives. System drawings and schedules that were identified but not located during site visit preparation should be located at this time. Copies of central plant, distribution, and end-use mechanical and electrical drawings and schedules should be procured for detail of the layout of the facility and how its utilities are delivered.

As part of the site visit, the study team should confirm the accuracy of the record drawings in terms of actual physical layout and system configurations. Deviations from the record drawings are common and must be identified to eliminate erroneous conclusions by the study team. Open areas shown on drawings are often used by equipment installed subsequent to plan development and therefore, may not be available for expansion purposes. At this point, deviations should be identified from empirical observations or non-extensive testing procedures. Any deviations that cannot be reconciled should be targeted for more in-depth testing during the detailed study.

During the initial site visit, special attention will be paid to those factors that will limit or eliminate the feasibility of certain application options. Space limitations, site code requirements, and site logistics are often critical factors in the determination of system viability. Early elimination of system alternatives allows for a more streamlined study focus. A fairly extensive data retrieval procedure should be employed to gauge feasibility for the maximum number of system options.

C. Interviews with Operating Personnel

Extensive interviews with knowledgeable management and front-line engineering, operating, and maintenance personnel are a critical aspect of the study process. System operators can provide valuable and timesaving detail on how and when equipment operates, the control parameters used, equipment capabilities, known problems that require solutions, and how and where operating records are kept. The personnel that have daily responsibility for the facility's systems should always be considered an invaluable source of information. Their talent and knowledge should be outwardly appreciated and respected by the analyst. It is, however, important to corroborate information provided through interviews with several persons and through review of records. For multi-shift operations, it is important to interview personnel from each shift. It should not be assumed, for example, that day shift personnel have a full understanding of night shift procedures. If the analyst wants to know about space conditions or whether banks of lights and other equipment are shut-off at night, and this data is not clearly indicated by energy management system records, the best way to find out is to meet with night shift personnel during their shift.

The analyst should also seek recommendations from the operators, as well as information as to what obstacles may be encountered when considering system modifications and changes in operating procedures. Project measures that have been tried and failed in the past should be discussed with an attempt to understand why they were attempted and why they failed.

These interviews should continue throughout the study process. The analyst should continually review observations with operating personnel and seek to achieve a more comprehensive understanding of how things are done at the facility and why.

Step 4. Data Retrieval

In addition to monthly and annual load data, load characterization requires evaluation of electric demand records. Many facilities keep a logbook on a regular schedule, which can supply this data. Review of these records with the operators and, if possible, a walk-through of data collection procedures will provide the study team with a firm understanding of the numbers and their reliability. This data is often the only available long-term data on the facility. The team should also investigate what type of record keeping is done by operations and management for energy and water use and cost. Frequently, numerous in-house meters and settings are recorded and logged, either by shift or hour. These may include softener water flow, boiler firing rate, status of boilers and chillers, steam flow, generator output, various supply and return temperatures, fuel use, tower make-up, weather, etc. Other data to be collected includes steam balance, water balance, air balance, and other such reports, which may have been generated on a one-time or periodic basis.

Other forms of historical data include recording charts (such as strip or circular charts), computer operated recording forms, or direct digital control (DDC) systems. Computer systems generally allow for storage of trend logs, which can be downloaded, to electronic media or to hard copy. If trend logs are not stored, it may be possible to develop and produce them for the period of study. Also, the utility company is a useful source of historical consumption data. Both electric and gas companies often maintain long-term hourly records of large customers; they use this data for billing and planning purposes and will generally make it available to the customer.

It is important to note that any data obtained from logs or computer outputs will only be as accurate as the equipment (sensors) providing the data. Operators will often not know the accuracy of such equipment, but will depend on its repeatability. However, it is best to check current readings with good portable calibrated gauges. These readings can serve as a reference point for interpreting or adjusting previously collected data.

In order to ensure accurate readings, it is essential that proper measurement protocols be followed (e.g., the distance of flow measurement sampling sensors from elbows). The best way to check temperatures and pressures, for example, is with one of each type of probe. In most cases, it is the difference between two readings that is of importance, not necessarily the actual value. Using one gauge eliminates errors resulting from two gauges being calibrated differently. If two gauges are used, the location of the gauges should be reversed after a reading to make sure that the readings do not change.

It is important to know the magnitude of all loads in discrete time periods so they can be related to occupancy schedules, utility rate periods, and weather patterns. It is also necessary to identify peak capacity requirements and coincidence of peak demand.

Operations, maintenance and repair (OM&R) records are also important data. When replacing equipment, it is necessary to determine the costs that will be avoided so that it may be compared to the costs required for maintaining the new equipment. All related operations costs should also be considered, including in-house and contracted personnel costs and material requirements, such as chemicals used for boiler or cooling tower treatment.

During the site visit, it is often helpful to take photographs of all major systems, representative samples of smaller equipment, and facility layout and structural features. To check that identified application opportunities will be logistically feasible, it is necessary to identify the basic logistical components involved with any major system installations, including significant structural obstacles.

In addition to collecting system data, preliminary building envelope data for walls, windows, doors, roofs, etc., should also be taken, supported by photographs and/or sketches to facilitate identification of opportunities to make improvements to the thermal properties of the building envelope. An inherent benefit of the site visit is that a record of the conditions of the building envelope will be produced and may be used by the facility as a blueprint for future facility upgrades not necessarily included in the original scope of work.

Step 5. Data Validation

All data collected during the site visit should then be reviewed to identify any missing or inconsistent information that will need to be collected and validated during later follow-up visits. The subsequent data validation process will determine if the collected data is consistent with the findings of the study team and if the various data components are consistent with each other. Primary data should be entered into electronic format to allow for efficient review and preliminary analysis. Graphical plots of data, such as load vs. outside air temperatures, should be checked for correlation. Thermal (heat) balance and other operating standards should be reviewed for consistency with the site visit observations and with energy and other resource usage records.

The facility record drawings and system schedules should also be carefully reviewed at this time. Any discrepancies that cannot be resolved should be marked for further scrutiny and reviewed with operating personnel at the facility. At this point, an internal report should be generated to summarize the status of the preliminary study and identify the tasks to be completed during follow-up site visits.

As part of the continuing load analysis and aggregation effort, the magnitude of all loads will be developed in discrete time periods so that they can be related to occupancy schedules, utility rate periods, and weather patterns. Peak load requirements and coincidence of peak demand will also be studied. All of the log and record data will be processed into load profiles in electronic format for further graphical and statistical analysis. These profiles make a valuable planning tool for facility managers and the study team should make the data files available upon request.

Additionally, decisions about energy sources must be tied to the understanding of the facility's overall heat balance. This is particularly important when considering the application of cogeneration cycles or other heat recovery technologies. A decision as to whether or not to expand or reduce low-pressure steam usage or a decision to convert an electric heating application to a thermal application (or vice versa) will greatly impact the heat sink available for effective use of recovered heat. Hence, topping cycles and other processes that use energy at its highest level of availability (lowest entropy) must be considered interactively with bottoming cycles and processes that use energy at lower levels of availability (higher entropy).

A different type of example is a situation in which one is considering some type of end-use heating application (e.g., process use, water heating, etc.). If this location is being served by a central steam distribution system, converting from a steam-technology device to a direct fuel or electricity-driven device would lower the demand on the central steam system. In the case where consideration is being given to decentralize and eliminate the entire system or portions of the system, this may be an excellent opportunity. Conversely, if the system is to remain in place and is already underutilized, this conversion may adversely impact overall system efficiency by, for example, further reducing loading on boilers that are already experiencing efficiency degradation due to low-load operation.

Step 6. Follow-up Site Work

During follow-up site visits, the study team should bring a list of questions derived from the data validation process and subsequent internal reports, and research answers on-site. Interviews with operating personnel should be conducted to finalize understanding of system operations and to resolve any discrepancies that have been identified. Measurement and testing procedures should be employed as deemed necessary by the study team so that effective application opportunity screening can be performed.

At the conclusion of the follow-up site visit, a master list of application opportunities should be compiled. Each potential opportunity should be accompanied by the collected data that will be used for screening analysis.

Step 7. Application Opportunity Screening Process

From the master list of identified potential application opportunities (which can be based on the checklist of potential measures presented in Figure 41-1), different types of screening methodologies can be employed to qualify measures as eligible for inclusion in the detailed study. As part of the site visits, project implementation costs should also be estimated by qualified members of the study team. Once preliminary costs are established, the screening tool models developed in the preparation phase will be used to evaluate the economic performance potential of each application opportunity. Levels of certainty will vary amongst technologies and will depend on the quality of the data available during the site visits.

Lighting system upgrade applications, for example, can be assessed with a fairly high degree of accuracy during the preliminary study phase. Based on record drawings or site visit take-offs, fixture counts can be established along with existing wattage. A reasonable assessment of hours of operation can be identified through interviews with operating personnel. Based on these data and the difference in wattage between existing fixtures and logical replacements, monthly usage and demand savings can be determined with reasonable accuracy. Standard cost estimates for equipment and installation can also be done in a fairly accurate, yet cost-effective manner.

In contrast to lighting systems, central chilled water, steam, or hot water system upgrade programs present a broad range of complexities that tend to obviate accurate prediction of economic performance. Quantification of cost and savings is complicated by the variations in cooling or heating loads over time and the changing performance of chillers and auxiliary equipment with load and ambient conditions. With cooling systems featuring electric chillers, the structure of seasonal, time-differentiated demand-oriented utility rate schedules adds to the complication.

The central plant screening process includes utility bill analysis, which will often reveal a cooling or heating related load profile, review of operating records, and cooling or cooling degree day analysis. Discussion with operating personnel may also be extremely helpful. Ultimately, a gross estimate of effective full-load hours (EFLH) of operation can be generated and applied to an estimated average efficiency or COP for the system. With cooling systems, it is common to use the integrated part-load value (IPLV) to establish commodity usage. For electric chillers, peak demand, inclusive of ratchet effects, will significantly impact operating costs. An assessment of the peak demand on a monthly basis is therefore also important in developing reasonable estimates.

Alternatively, a preliminary model of the chiller loads, systems, and equipment can be developed using the many competent building simulation software packages available. While much of the detail (e.g., metering data) required to produce highly accurate results may not yet be available, default values can be used to allow for development of reasonable estimates without a great deal of time expenditure. If simulation modeling is ultimately to be used during the Phase 2 detailed study, it may be advantageous to begin using it at this stage, thereby allowing for continual refinement as more validated data becomes available.

With chiller replacements, there are also numerous options that should be initially considered, including a variety of electric chillers, fuel-fired, steam- or hot water-powered single-effect or double-effect absorption chillers, and a variety of prime mover-driven chillers, with and without use of heat recovery. Options for use of thermal storage should also be considered.

Project costing estimates should be developed with a similar level of accuracy as the savings estimates. Generalized equipment costs can be estimated based on rules of thumb for a given equipment type and capacity (e.g., $/ton, $/hp, etc.). Installation costs are modified for identified site difficulties, such as rigging, piping, and structural modifications. Preliminary estimates should also be made for projected OM&R savings and additional OM&R costs.

Step 8. Summary of Findings

Upon completion of the preliminary screening process, a summary table should be developed for each application opportunity. Generally, the table will include estimated capital costs and rebates, which will yield net capital costs, and estimated costs and savings for energy, OM&R, and other resource, which will yield net savings. From the net capital costs and savings figures, simple paybacks can be computed for each project opportunity.

Given the fast changing energy market, it is often useful to construct two sets of preliminary economic performance values: the first based on current conditions and energy supply costs and the second on rudimentary consideration of measure interactivity and future energy supply costs. The two sets of values will establish a range from which each opportunity can be judged.

Based on the results of the economic performance screening process, measure recommendations can be categorized as:

- Likely — appears to be desirable from an economic and practical standpoint and has been generally well received
- Marginal — may be deficient in one of the criteria categories, but is strong in the others
- Rejected — has had initial interest and support, but will clearly not pass one or more of the critical tests for implementation

This summary will be the tool used in final discussions with the host facility prior to determining the proposed portfolio of measures. The facility may have overriding interest to accept or reject individual measures in the recommendations. At this point, it is important to eliminate projects that fail preliminary screening in order to reduce the cost and time associated with the detailed study phase of the process. Application opportunities that are ranked as marginal can be further scrutinized with an additional round of screening in an effort to re-categorize them as either likely or rejected.

Once estimated energy savings, system life, construction costs and utility rebates, and energy price sensitivities are established, simplified life-cycle analyses will be performed as the final economic screening in the preliminary study phase. These financial analysis techniques are discussed in Chapter 42.

Step 9. Preliminary Study Report

Upon completion of the screening process and organization of the summary of findings, a preliminary study report should be delivered to management detailing the study team's findings and project recommendations. The report can be used to document all significant aspects of the initial study. This may include, but not be limited to, problems under investigation, personnel contacted during the site visits, relevant information obtained from facility personnel, availability and quality of as-built drawings, maintenance records, measurements taken, equipment inspection details, determination of problem areas, and recommendations to resolve the problems. Finally, it should include a detailed breakout of qualified application opportunities, inclusive of preliminary cost and savings projections.

The report should include estimates of the potential energy (e.g., Btu, kWh, kW, etc.), water, and cost savings that can be expected by implementing the recommended measures and the project implementation costs. Type, location, and amount of potentially hazardous materials (HazMat) should be documented, along with a proposed method of labeling, handling, removal, and disposal of the HazMat and associated costs. The report will include a narrative describing the proposed applications in concept, accompanied by photographs and preliminary sketches for the qualified opportunities, descriptions of the equipment to be supplied, and details and scope of work for the subsequent detailed study.

The report should also include a plan and budget for proceeding with the detailed study phase. A timeline should be established for the detailed study and study team members and duties should be assigned. A list should be prepared of all remaining data to be collected. A preliminary determination of metering points, procedures, and schedules for on-site activities should also be made.

Detailed Feasibility Study — Phase 2

The feedback and direction provided by host facility management after review of the initial study work will help guide the study team in refining the overall project scope and limiting the focus to those measures that the facility wishes to include in the final investment-grade study or proposal. The Detailed Feasibility Study is initiated following acceptance of the Preliminary Feasibility Study and authorization to proceed with this final development phase effort.

In the case of a standard plan and specification type approach to project implementation, the Detailed Feasibility Study would typically be performed by an engineering firm under contract with the host facility. It would serve first as a basis of making internal funding commitments and then as a basis of commissioning a design-engineering firm (sometimes the same firm) to develop the design and construction bid documents. In the case of many types of energy services arrangements, the Detailed Feasibility Study results in a firm proposal from an ESCo to perform a turnkey design-build project. Hence, the study must be highly detailed to allow the host facility to make a major investment commitment and to ensure that the ESCo has mitigated its own risk and can stand behind its guarantees of price and savings. Either way, the Detailed Feasibility Study should produce an investment-grade final report.

An update to the Project Development Plan is made at this stage. The Plan should now reflect the final package of measures and the associated work effort required to prepare detailed analysis of costs and savings.

Having more clearly defined the overall project scope and limited the focus to opportunities that have a realistic chance of being implemented, the detailed study phase proceeds with more rigorous on-site investigation, inclusive of direct system metering, and more in-depth, precise analysis. This is followed by the development of a preliminary design, a detailed scope of work (SOW), and firm construction pricing, along with finalized energy baselines and, where applicable, final OM&R and M&V Plans.

Overview

Following is a representative listing of some of the basic technical and financial information that will be required. This is proceeded by the discussion of the main analytical components of the Detailed Feasibility Study.

Technical Information

Technical information will be gathered, reviewed, and confirmed through observation and inspection. Missing data will be identified. Data will be used to develop engineering assumptions for analysis of energy savings. Facility management and operations information will become constraints to ensure that technology application concepts are compatible with standard facility practices and that recommended measures enhance, rather than burden, facility operations. Technical information resources include the following:

- Facility documentation will be reviewed and its accuracy confirmed.
- Operating conditions will be assessed through observation and interviews with facility staff.
- Operations management structure and practice will be reviewed.

Financial Information

Detailed financial information will be needed as input to the financial models. They will provide more parameters to predict life cycle costs and cash flows than are needed for simple payback evaluations. Financial data review involves the following:

- Incremental economic values must be determined.
- Long-term life-cycle factors must also be integrated into the economic analysis.
- Incremental capital investment costs must be integrated.

Figure 41-3 illustrates the depth of information gathered and processed during a Detailed Feasibility Study. It can serve as a checklist of information items to be acquired and reviewed.

Main Analytical Components

The three main analytical components of a comprehensive Detailed Feasibility Study are acquisition and analysis of metered and field-collected data, computer modeling, and utility bill reconciliation.

Technical Information	
Facility Documentation	❑ Mechanical, electrical, and architectural drawings; Site plans and floor plans ❑ Billing histories for all metered usage (including purchased CHW, steam, and water) ❑ Utility rates (copies of the current tariff books) ❑ Records from facility-owned submeters ❑ Demand profiles, if available (24 hour demands for selected day types) ❑ Equipment inventory (e.g., from PM programs); Nameplate and manufacturer's data ❑ Automation system documentation (points lists, manuals, diagrams, sequences) ❑ Previous submittal packages and equipment operating manuals ❑ Reports from previous studies ❑ Operating logs, EMS computer trend logs, maintenance records, balancing reports
Operating Conditions	❑ Space inventory by function, with location and floor area ❑ Environmental control standards (temperature/humidity, ventilation, lighting) ❑ Operating schedules (daily, weekly, seasonal hours of operation) for each space ❑ Known maintenance or operational problems; Critical deferred maintenance items ❑ Shortfalls in equipment capacity or distribution system bottlenecks ❑ HVAC control strategies (system operating schedules, setback, and reset schedules) ❑ Central equipment operations (sequencing of equipment and auxiliaries, fuel sources)
Operations Management	❑ Maintenance practices (standard preventative or predictive maintenance intervals) ❑ Work order scheduling systems; Staffing levels ❑ Service contracts with outside firms ❑ Long-range facility planning documents, such as a Master Plan ❑ Design standards (vendors and materials, labeling/tagging, controls compatibility) ❑ Applicable regulations and codes ❑ Operating budgets (utilities and maintenance)
Financial Information	
Incremental Operating Costs	❑ Electricity, natural gas, purchased chilled water, other fuels ❑ Steam or other heat sources at each thermal level (temperature) required ❑ Water use ❑ Environmental permitting and emissions controls costs ❑ Cost of standby electricity (in electric generation feasibility studies) ❑ Cost of capital/debt and cost of insurance ❑ Cost/value of required or avoided floor space ❑ Reliability and redundancy requirements and associated downtime costs
Capital Investment Costs	❑ Technical project support (engineering, planning, commissioning, M&V costs) ❑ Costs of systems to be installed with quotes from vendors and subcontractors ❑ Energy delivery infrastructure and generation equipment ❑ Turnkey construction costs with construction management, demolition, disposal, etc. ❑ Utility program incentives or penalties ❑ Startup and debugging costs and cost of production downtime ❑ Permitting, development, legal, and other consulting fees
Life-Cycle Cost Factors	❑ Cost/value of electricity (internal use and, where applicable, power sales) ❑ Natural gas and fuel supply cost, contract commitment, and contract security ❑ Escalation of energy, water, and OM&R and replacement costs (and contracts) ❑ Replacement costs and salvage value ❑ Performance degradation

Fig. 41-3 Checklist of Review Items for Detailed Feasibility Study.

Extensive direct system metering and recording activities have a firm scientific basis, but also involve hands-on activities. It involves the practical determination of what to meter, where to meter it, how to meter it, and for how long. It involves activities such as selecting and installing probes and data acquisition systems. Examples include taping voltage and current, installing temperature probes into pipes, setting airflow measurement devices into ducts, over registers, and into exhaust hoods, and measuring power and energy, fuel flow, etc.

Critical to the design and application of an effective metering process is the understanding of how the acquired data will be used. Temperatures, flows, and power requirements of equipment must be analyzed in such a manner that reveals the actual operating characteristics and resource usage of existing equipment and allows for proper sizing of new equipment, development of optimal operating strategies, and determination of associated costs and achievable savings. This process can also be extended to the development of baselines and system performance prediction tools to use for savings verification programs.

The measurement process, be it baseline or post-implementation, continues for a period long enough to encompass the normal variation of the significant factors, or independent variables, which determine the loads and operation of each system. When system-specific metering techniques are used, the senior analyst will make a determination as to the duration of the metering activities. For equipment serving loads that operate consistently over time, short-term metering will be most appropriate. For systems whose loads fluctuate, such as those affected by weather, data should include the key variables that are believed to impact load variation. This data will be gathered through intermediate or even long-term metering, depending on variations, the magnitude of the project, and the required degree of accuracy. The resultant data can be statistically analyzed to determine the effect of those independent variables on resource consumption and demand, deriving their coefficients in a multiple linear regression.

Since specific metered data is gathered during a given range of operating conditions, this analytical process allows for the development of an operating performance prediction tool to be used to reflect what the usage and demand, and therefore operating cost, would be under any given set of conditions. Examples of independent variables include weather conditions, occupancy, and process production quantity. From this, the baseline performance and usage for each system can be developed, which at any time is the original measured consumption given a set of values for the independent variables. The same process can be followed for establishing post-implementation performance.

Computer building load, system operation, and economic analysis simulation modeling using software such as DOE-2, TRACE©, or HAP© is also a valuable tool in the detailed study process. It allows for interactive analysis of multiple measures and simulation of the impact of proposed and actual changes in facility-wide systems and operations over time. It also allows for rapid testing of numerous potential options for each application type under varying conditions. Simulation modeling automatically accounts for measure interactivity, time-of-use utility rates, independent variables (such as weather), and the effect of a wide range of potential system optimization strategies. Simulations of conditions that can critically affect building and equipment loads (e.g., solar, partial shading, variable schedule-dependent activities, building mass, multiple HVAC optimization strategies, etc.) are straightforward with such modeling and can be arduous and less accurate with other methods. When based on the results of field inspection and calibrated with actual metered data, modeling will allow for consistently reproducible results of the effects of long-term system changes at the facility.

Utility bill reconciliation refers to the matching of analytical results, such as those provided by simulation software, with actual historical records. When adjusted for any given historical year's operation, utility rates, and weather, the baseline model should be able to reproduce the actual costs shown in the historical records. If the predicted and historical results agree (i.e., ± 5 to 10%) for a range of facility activity, the baseline model is validated. The analyst should seek to reconcile all utilities on a monthly, annual, and peak demand basis to ensure model validity.

When these three main analytical components are used together, the metered data and utility bill data provide a factual basis by which to calibrate the model so that results are fully grounded in reality. When so validated, the computer simulation model can then provide a high-powered tool capable of evaluating measures interactively to produce optimal system configurations and can rigorously analyze savings projections.

Figure 41-4 provides a summary of the work required for executing the Detailed Feasibility Study. The table includes the work performed to process and analyze the technical and financial information listed in Figure 41-3, and the results and/or deliverables of that work.

Description of Detailed Feasibility Study

There are 14 main steps that should be executed during the Detailed Feasibility Study phase. The following discussion describes each of these steps, including the methods, work effort, and results produced during each step of the Detailed Feasibility Study.

Step 1. Selecting the Appropriate Analytical Tools

The first step in the Detailed Feasibility Study is to select the appropriate analytical tools. While these may have been partially determined during the preliminary study phase, they must be reassessed based on the results

Category	Scope of Work Performed	Results and Deliverables
Utility Information	Graphical analysis of profiles for load management and self-generation or cogeneration opportunities; Rate review, fuel source assessment, supply options; Reconcile models to highest resolution data available, at least to monthly usage and peak demand	Fuel switching opportunities, alternative suppliers, alternative procurement strategies; Models are accurate and reflect actual usage patterns and end-use consumption
Equipment	Determine metering points, methods, and durations; Establish data analysis plan in advance (e.g., regression); Include auxiliaries (e.g., pumps); Establish equipment performance curves	Verified baseline usage encompassing normal load variations; Higher certainty in savings estimates
Facility Documentation	Verify accuracy of documentation; Refine measure concepts to reflect additional information	Improved accuracy; Better screening for feasibility; Time savings during later phases
Facility Needs	Accommodate all planned changes into measure analysis; Review recommended measures against these requirements	Load shape impacts will be included; measures may be altered
Constraints	Evaluate compliance impact on measures; Review measures for compatibility with end-use requirements	May uncover opportunities for beneficial changes in operating strategies (e.g., pressure reductions, distribution system changes)
Measure Savings	Determine optimal level of analysis effort; Develop energy analysis models (building simulations and/or custom applications); Calibrate to billing history; Supplement with metering	Accurate analysis of complex measures using validated model; Based on accurate load profiles; Includes interactive effects; Complete cost savings calculations using full rate structures
Measure Costs	Review construction conditions with estimators and/or subcontractors	Preliminary estimates (± 20%) for screening and refinement
Financial and Contractual	Run financial models with all first costs and recurring costs	Cash flow projections, including finance charges, M&V costs, OM&R savings; Construction financing, draw-down schedule
Design Issues	Review measure concepts for equipment locations, utility service, and other design feasibility issues; Review impact and test economics of emissions monitoring and controls	Measures can be designed to meet standards; Conceptual designs and equipment selections
Construction Issues	Review measure impact; Prepare installation and commissioning budgets	Detailed cost estimate with line item breakdown, materials, and labor by trade, plus other costs; Installation schedule
Operational Issues	Estimate current and future OM&R costs; Assess personnel requirements, including training	OM&R savings incorporated into life-cycle cost analysis; Preliminary training plan; Analysis reflects actual operating strategies (e.g., sequencing of chiller pumps)

Fig. 41-4 Summary of Work Effort for Detailed Feasibility Study.

of the Preliminary Feasibility Study. Given the complete range of analytical tools available (from simple spreadsheets to customized engineering calculations to detailed hourly simulation programs), the experience of the study team will be drawn upon to perform feasibility studies that produce efficient effort with accurate results. The level of analytical complexity and, therefore, the cost of that effort must be justified by the level of certainty achievable and the potential for savings at the facility.

In some cases, the full range of activities described above is necessary, given the magnitude and complexity of the application. In other cases, such comprehensive analysis is not cost-justified and less rigorous methodologies are appropriate. If the study process requires a high cost relative to the potential savings, it may be judged inappropriate on economic grounds. Instead, use of engineering estimates, stipulated hours of operation, and many of the techniques described in the preliminary study approach are indicated. There are some standardized, menu-driven study models that may provide the most cost-effective analysis method for the feasibility study for lower cost, lower savings items. These models may have pre-set fields for entry of site-specific inputs that are supported by a series of default assumptions. Similar proprietary models, developed by analysts, are also sometimes used. The more effective versions provide numerous default values that can be overridden or modified based on the level of detail available to the study team.

However, in the case of applications involving significant capital costs and savings potential, along with numerous site-specific considerations, the rigorous approach described above is usually quite cost-effective, with the results being the selection of the optimal system applications and minimization of technical and financial risk to all parties. The level of confidence achieved will also allow a smooth transition to the design and construction processes, given the wealth and accuracy of the known information.

A more rigorous approach may also be beneficial to M&V efforts, since the detailed performance models may be used to extrapolate short-term metered data to longer-term savings results. When warranted, a model such as DOE-2.1 may be selected in favor of other methods, such as temperature bin calculations, for calculating building energy performance because:

- It is a dynamic model.
- It accounts for seasonal solar loads and the associated effects of building features and complex exterior shading.
- The contribution to cooling load from changes in relative humidity are simply accounted for.
- Varying hourly loads due to occupants, lighting, and equipment is done automatically.
- It accounts for equipment performance at varying load and outdoor weather conditions, accounting for key factors, such as part-load performance and condenser relief at outdoor air temperatures lower than design.
- It facilitates quick review of daily, monthly, and annual load profiles and trends for any modeled element.
- It facilitates evaluation of multiple, interactive measures.

A complete cost/benefit evaluation will be performed to select the best tool for each proposed measure. Regardless of the type of tool selected, the analysis methodology is based on accepted industry standard engineering algorithms. These are well documented and can be made available to facility representatives for review.

Step 2. Load Development

Mixing and matching components and operating conditions to yield the highest possible system efficiency while providing for the necessary capacities is a creative process. It must, however, be based on accurate load characterization. Quantification of cost and savings is complicated by the structure of demand-oriented and time-of-use utility rate schedules, peak utility loads attributable to combinations of peak building loads and discretionary operational modes, variations in end-use demands over time, and the changing performance of equipment with load and ambient conditions.

Quantification of concurrent extreme loads provides a basis for definition of maximum system demand and operation ranges and, ultimately, a basis for equipment selection and operating cost analysis. Electric load characterization typically involves evaluation of utility demand interval records (usually 15 or 30 minutes) to determine electric load peaks, averages and minimums, and total annual consumption in kWh (per time-of-use billing period). Thermal load characterization typically involves evaluation of hourly loads (in Btu/h, kJ/h, or kWh/h) to define thermal energy demand maximum, minimum, average, and total annual consumption (in Btu, kJ, or kWh). This is of particular importance in the analysis of thermal storage systems, either cool storage (e.g., ice or chilled water) for space conditioning or heat storage (e.g., hot water) for domestic, process, or space conditioning. It is equally critical for evaluation of cogeneration system application alternatives that depend heavily on timing, or coincidence of electric and thermal load profiles.

If equipment operates at a constant load with a constant usage, load development is fairly straightforward. The load needs to be measured and the specific hours of operation need to be identified. In the case of lighting systems, for example, load development will require accurate fixture counts, grouped by lamp-ballast combination type and by common hours of usage. When hours of operation are not known or must be verified, relatively simple long-term metering devices can often be used. An example is a light-sensitive photocell with built-in microprocessor, which is an inexpensive self-contained device that measures the run-time of a lighting fixture. Another example is an activity logger, which adds the feature of occupancy sensing, allowing for the measurement of run-time during periods of occupancy or non-occupancy.

If the equipment operates intermittently or under varying loads, load development requires a more complex approach. Both the load served and resource usage requirements must be determined based on some type of simulation of actual operations or direct measurement. Note, however, that modeling or metering is complicated by the fact that systems may serve a given load (e.g., chilled water loop) with multiple chillers, pumps, and towers, or a combination of central and perimeter space conditioning.

Consider the analysis of an individual measure, such as an upgrade to existing chilled water, hot water, or steam systems. The first step is to determine the end-use load (e.g., chilled water, steam, or hot water) with a high degree of certainty and the energy required to serve the load on a short-term (instantaneous) and annual basis. The first line of questioning in the development of usage patterns determines the nature of the load:

- Is it seasonal and based mainly on weather conditions (i.e., temperature, humidity, and solar loads), as would be the case with most comfort cooling or heating applications?
- Is it process and based mainly on production volumes?
- Is it heavily influenced by daily, weekly, or seasonal operating schedules?
- Is it process and based on internal heat gain or loss?
- Is it some mix of the above?

The answers to these questions will enable the development of reliable modeling that will be used in both the detailed study, as well as the savings verification process.

Many types of loads are influenced greatly by ambient outdoor weather conditions. This includes several factors in addition to outside air temperature such as wet-bulb temperature, wind speed and direction, insolation, and time of year. In addition, there are other factors that are independent of, or only partially related to, ambient conditions. These may include baseload internal gains, time-of-day, day-of-week, time-of-week, occupancy patterns, and process activities. Depending on the season, internal gains may comprise a significant part of the overall cooling load, particularly in facilities that do not use a large degree of outside air makeup.

In considering all of these factors, the goal is to develop a prediction methodology, frequently in the form of a regression equation, that can be used on a set of data that is well defined and readily available. This usually limits the analyst to variables such as time, wet-bulb and/or dry-bulb temperatures, or process production rate.

Once a regression analysis is performed, it is valuable to plot a graph of the residuals obtained. This residual plot shows how far each predicted value, based on the regression equation, falls from the actual data. If an error pattern does occur, it would be necessary to define the regression criteria differently or possibly look for a different shape curve for the regression. Ideally, as discussed above, modeling software should be used to develop a complete facility load profile, which will then be calibrated based on the metered-based results, as well as historical operating and utility bill records.

Step 3. System Performance and Energy Usage

If there is an existing system serving the load, the amount of energy (or water, lubricating oil, etc.) currently used by that equipment must be determined using empirical or metered data. In the case of essentially non-varying systems, instantaneous metering can often be used to accurately determine system performance.

For systems that operate with varying input and/or output, longer-term monitoring of current performance is the preferred method. Metering should include not only the major use of energy (or other resources), but also the various auxiliary loads necessary to support and distribute to end-use devices, such as air handlers or perimeter radiation. Extending metering procedures to all or most of the multiple-system components will enable the analyst to determine which units are operating at each load condition and any performance deviations between individual units. The resultant system performance prediction tool must encompass all the numerous variables that affect system energy (or water) usage, including weather conditions, load requirements, and current operational schedules.

Once identified and plotted, total system performance data should be further separated between main equipment and auxiliary equipment usage. This process is complicated by various operating strategies that may be employed with multiple-unit systems. Interviews with operating personnel are also helpful in identifying the general operating strategy employed under various load conditions. An example would be the sequencing of chillers in a multiple chiller central cooling plant. Various options for chiller selection (capacity, type, energy source, etc.) may be available at any given loading and ambient condition. Selection and starting of pumps and cooling tower cells/fans adds another dimension.

Furthermore, the input data required to determine a chiller energy performance curve includes water temperatures in and out of the chillers, flow rates through the chillers, and part-load chiller energy consumption. Some other important factors relating to actual performance include entering condenser water temperature, condenser water flow rate (assuming a water-cooled system), and chiller system operating settings. In the existing systems, some of these factors, such as condenser water flow and chilled water supply temperature, may be fixed or may be varied where load and operational parameters are based on actual conditions. As part of a proposed technology application, many of these factors may be changed due to the use of different equipment and/or implementation of new operational optimizing strategies.

With a performance curve developed that includes corrections for part-load operation, as well as off-design chilled water use and condenser water temperatures, a basic spreadsheet format can be used to distribute actual equipment energy usage requirements under varying loads for varying periods. By assigning hours of use at each loading level, total annual energy usage can be calculated. In order to determine operating costs under varying energy rate periods, the hours need to be broken out separately for each billing time-of-use period. Total consumption, therefore, reflects the amount of energy consumed over all of the hours the system operates at each specified percentage of total capacity output.

Energy using systems under consideration will have different performance characteristics under varying loads and corresponding ambient conditions. For existing equipment, this can often, but not always, be determined from metered data. For example, if the system operates under variable speeds, there may be several combinations of speed and equipment modulation that will produce the same system results. A map consisting of a series of integrated performance curves of the driver and the driven equipment can be developed to identify the optimum combination for each level of output requirement under variable conditions.

Time at Temperature and Simulation Models

If a bin method is used, with regression analysis based on outside air temperature, the best and most readily available energy analysis methodology is to use historical or standard weather data. It is important that average weather data is not used exclusively, since it would dramatically lessen the effects of extreme weather. For example, the temperature may frequently reach 100°F (38°C) in areas where the average temperature is 75°F (24°C). In the case of a chilled water system analysis, for example, use of average figures would lower the calculated peaks in both chilled water requirements and chiller power.

The local weather data station, NOAA, ASHRAE, publishers of modeling software, and various equipment manufacturers are all potential sources of weather data (Typical Meteorological Year, Test Reference Year). In addition, many utilities now closely monitor weather in their service territory, which can be made available to customers. Typically, it is possible to obtain, at a minimum, 5°F (about 3°C) bin weather data showing the number of hours in each month during which the temperature is in the 5°F bin range. In the best case, this is further split to 1°F (or 0.6°C) increments and also provides the time-of-day of the temperature occurrence. The best available weather data should be used to generate predictable cooling and heating loads and relationships of system performance with varying loads. In addition to establishing the number of hours at each load with weather data, hours at load will be further differentiated by utility rate period when necessary.

Difficulties can arise because this use of bin data results in a static representation. Analysis is complicated by the lag in a building's thermal response to changing temperature conditions. Facilities with substantial mass may experience a thermal time lag of several hours. A peak temperature at 2:00 p.m., for example, could result in a peak cooling requirement at 4:00 p.m. Conversely, as outside air temperatures are reduced, peak cooling requirements may persist for several hours due to stored internal heat gains. Facilities with high ventilation rates will be less subject to such impacts, but most facilities will experience some thermal time-constant influence. Care must be taken to evaluate the regression data to establish a level of certainty in the results. It is often necessary to make an assessment of what additional independent variables, such as facility operational or production schedules, will be used as multiple regression coefficients. Also, unless the data is already separated, it is difficult to assign hours to discrete energy rate periods. Still, this format is very useful for energy savings verification purposes.

To perform more dynamic analyses, hourly computer simulation models are used. The building construction and system performance characteristics are input to the simulation model, which applies these characteristics to determine load conditions over every hour of the year. Time constants, thermal lag factors, operation and production schedules, and other influential features are modeled in such programs and the interactive effects of numerous measures are simultaneously considered. When considering multiple replacement options or other system improvements (e.g., chilled water reset), modeling, while generally more time-consuming, is quite valuable.

Auxiliary Equipment and Resource Use

The same evaluation procedure should be followed for all auxiliary equipment, such as pumps and fans that are integral to the system's operation, as well as other resources, such as water, sewer, and chemicals. It is usually necessary to treat auxiliary usage separately, because usage will not necessarily have a linear relationship with the main equipment. For example, a pump serving a compressor's coolant system may have constant usage as the system's overall load varies. Similarly, while the energy usage of a chiller, boiler, compressor, reciprocating engine, or turbine can vary significantly with the load placed on it, the auxiliary load will often remain fairly constant and/or may be affected by other factors. In the case of a cooling water system, the cooling tower fan will operate more in hotter and more humid weather to maintain temperature at a given cooling load. However, the condenser water circulating pump loading remains constant in most cases.

In the case of a chiller system, if the chilled water pumps have variable speed capability and air handling units are equipped with two-way valves, pump energy usage will vary significantly with load. It is important to note that while auxiliary usage can be considered separately, there is often an interactive effect between auxiliary equipment operation and the performance of the main equipment. For example, increased cooling tower fan usage can produce lower return condenser water temperature, which in turn can improve chiller performance.

The system or main and auxiliary component energy rate vs. load curves are the data necessary to determine the load imposed on the existing system for each outside air temperature (or other correlated variables). They also allow for the determination of the energy used by the total system, existing equipment, and their auxiliaries combined, to meet that load.

Step 4. Baseline Determination

Having developed the load shape profiles for the facility and the equipment performance characteristics of the existing systems, annual energy usage and peak demand should then be determined. This can be done with a bin-method analysis, computer simulation analysis, or both. If computer simulation modeling is used, it is critical that the assumptions used in the computer model are validated. Otherwise, the most high-powered system will still yield results that are inaccurate and of little use. The field metering and regression analysis, therefore, remain essential to the process. These provide either the direct input to the model or the method by which model assumptions are calibrated.

To determine the impact on peak capacity requirements and costs, the full energy usage rate will be identified for each hour (or smaller time increment). System or building-level results must be overlaid on the facility's total profile and differentiated by specific energy rate periods, as dictated by the loads served by a given utility meter. From this, net changes in total facility demand and the associated incremental cost can be determined. Critical to this determination is the understanding that the incremental peak demand impact is a function of coincidental peak of all loads connected to a billing meter. If, for example, the peak capacity requirement of the system occurs at a period other than that of the facility's peak, the incremental impact will be lower than 100% of the system's peak. If the billable demand in winter months is greater than the actual demand level due to an electric rate ratchet adjustment based on summer peak demand, then an increase in coincidental peak will not necessarily increase the billable demand in the winter months.

For cogeneration system applications, the coincidental thermal and power requirements are critical aspects of the analysis. Recovered heat will have value only if it can be used when it is produced or stored for effective later use. Hour-by-hour load analysis combined with the system's performance under each load condition is, therefore, the most reliable performance prediction tool. Refer to the cogeneration system and driver selection study provided at the conclusion of Chapter 26 for an example of such a baseline analysis.

The annual usage and demand for the existing system will become the baseline against which savings can be projected. The usage and demand of the systems under consideration for replacement or upgrade, or the overall facility usage and demand under consideration for on-site electric generation, will be compared to the baseline to determine the energy and demand savings projections.

The baseline operating cost can be determined by entering the usage and demand by month and rate period into a utility or fuel cost rate program. Energy rates can either be programmed directly into simulation modeling, or model outputs imported to a spreadsheet rate calculation program. In either case, the annual baseline costs for the facility should be reconciled with the actual billing records. Either the billing records should be normalized to analysis conditions, or analysis outputs adjusted to the conditions experienced during the actual billing periods. In either case, if the analysis output cannot be reconciled with actual billing histories, the entire process must be reviewed, with certain assumptions recalibrated as appropriate. The goal is to develop an analytical process that can deliver reliable and reproducible results. While it may not be possible to identify all of the specific factors that resulted in a particular year's energy costs (i.e., occupancy patterns, production schedules, etc.), if the monthly (or shorter duration) bills for energy and other resource costs for a given year or series of years cannot be closely reconciled with the analysis, the analysis results will generally be considered unreliable.

Step 5. Savings Analysis

Once the baseline has been determined, the same basic methodology, facility model, system model, or equipment model can then be used to determine the projected energy (and other resource) usage, demand, and operating cost for the proposed application options developed in the preliminary feasibility phase. The difference is that the performance characteristics of the proposed system(s) are substituted for the existing equipment, and changes in efficiency and operating conditions are included as part of each measure (e.g., equipment changes, space condition temperature setbacks, changes in outside air volumes, equipment modulation controls, etc.). In order to substitute the proposed equipment for the existing equipment, performance characteristics of the proposed equipment must be identified for all of the potential conditions to be experienced in the proposed operating regime. Whereas existing equipment performance can be based on actual field measurement, proposed equipment performance can usually be obtained by matching the data provided by the equipment manufacturer to the site-specific conditions.

Manufacturer's data should be obtained for performance under the full range of anticipated operating conditions. These performance data plots are particularly important for equipment such as chillers and gas turbines that experience widely varying performance under different loads and ambient air conditions. With a new chiller, for example, performance data should be provided for chiller loading ranging from 10 to 100% in 10% increments with condenser water temperature held constant at temperature increments from 85°F to 60°F (29°C to 16°C) in 5 degree increments. Varying condenser water temperatures, or condenser water relief, can also be modeled with the proper manufacturer's performance curves. If chilled water supply temperature is to be varied, it will be necessary to obtain performance data for operation under each anticipated chilled water supply temperature. Once the data is obtained, it will be a relatively simple interpolation process to determine the projected efficiency for any combination of load and operating conditions. These performance characteristics should be warranted by the manufacturer as part of the purchase agreement and confirmed by well-defined standards-organization testing data. Careful evaluation of the certified testing data is recommended to determine testing tolerance or margin of error allowed. It is sometimes necessary to de-rate the performance data to compensate for allowable test tolerances to ensure more accurate, risk-mitigated results. Refer to the discussion on equipment rating standards in Chapter 33 for an example of expected performance variation.

Annual operating costs for the proposed changes are determined in the same manner as with the baseline. The proposed new operating costs may be calculated for the entire facility with the new system included or on an incremental basis for a specific measure application. Operation cost savings projections are then computed by subtracting the proposed cost from the baseline costs, again on either a facility-wide or measure-specific basis.

The true measure of achievable savings is the incremental change in usage and cost that results from the installation of a given measure. The value of heat recovery, for example, will be measured by the cost avoided in using recovered thermal energy (or heat) for a specific purpose, as opposed to using another source of energy for the same purpose. Most commonly, recovered heat replaces thermal energy output from some type of fuel burning equipment, usually a boiler or furnace. In these cases, the value of recovered thermal energy is equivalent to the cost of fuel energy that would have otherwise been consumed. The amount of energy displaced by recovered heat is a function of the efficiency of the displaced energy conversion equipment. Refer to Chapters 2 and 8 for detail on methods used for accounting for value of recovered heat.

To properly evaluate thermal load requirements and fuel displacement potential, the thermal requirements of each end use must be accurately quantified. For example, steam may be required at numerous pressures. These requirements must be matched to available levels (for example, those that exist with extraction or back-pressure turbines).

In summary, the baseline development effort provides a mechanism by which to determine the energy or resource consumption and costs with the existing systems, given a set of independent variable conditions. This is based on actual operating measurements and data, in conjunction with equipment performance models. The facility system or equipment model permits the projection of the energy or resource consumption and costs with the proposed systems. This is based on actual equipment performance data where available or on engineering calculation. By comparing the two, the net benefit of the project will be determined.

Accounting for Measure Interactivity

Inherent interactions between all facility systems impacted by the potential project should be identified and accounted for in the study. This is imperative if accurate savings are to be calculated. In and of themselves, measures may result in increases and/or decreases in resource use in interrelated systems. For example, decreasing lighting energy will commonly both decrease cooling load and increase heating load requirements. With an integrated multiple technology project, proposed measures will most likely impact several other proposed measures, making the analysis more complicated.

Consider the example of two potential project recommendations, one which improves chiller efficiency and the other which reduces cooling load. If savings for the chiller efficiency improvement are calculated at the existing cooling load, the resultant savings will be greater than if calculated at the reduced cooling load. Alternatively, if the load reduction measure is considered with the existing chiller efficiency, the savings will be greater than if the chiller efficiency improvements are considered. If both measures are ultimately recommended, then the interactive savings will be calculated based on the chiller operating with improved efficiency and serving reduced load. Ignoring this interaction will result in overestimation of savings and, in some cases, over-sizing of new equipment. The EPA laboratory case study at the conclusion of the previous chapter is an example of this. Both heating and cooling peak demand and consumption loads were dramatically reduced. If one were to have analyzed both the chiller/heater measure and the AHU redesign measure separately, overall savings clearly would have been overstated. Moreover, far more new chiller/heater capacity would have been installed (based on the baseline loads) than was needed for the new, more efficient load profile, driving up capital costs unnecessarily and likely producing an inefficient, oversized central plant.

Similarly, the heat and steam use from a boiler plant presents similar trade-offs. For example, low-pressure steam or hot water used for process or feedwater heating can be replaced with recovered waste heat, supporting cogeneration system application potential. In the cogeneration study example provided at the conclusion of Chapter 26, the reciprocating engine generator set option relied on serving such a load with heat recovered from its cooling system. However, in the second sensitivity analysis, consideration was given to a very economical alternative process that would greatly reduce this thermal load. The result of the interactive analysis was that the expected economic performance of the reciprocating engine was reduced, making the gas turbine option more favorable.

Conservation, alternative process, or heat recovery options for low-pressure steam loads can also all have a profound impact on steam turbine topping cycle applications. Such options will often prove economical in and of themselves. However, they will also lower the turbine throughput and power generation output, representing a trade-off that might prove uneconomical when considered interactively. To properly analyze such situations, an overall facility thermal (or heat) balance must be developed and used to derive the fuel/electricity time-dependent economic benefits to the whole, rather than parts of, the facility.

Consider the actual case of a Connecticut manufacturer that sought to identify the most cost-effective strategy for adding chilled water and compressed air capacity to meet expanded load and production process requirements. In this case, the baseline heat balance reflected a modest non-heating season imbalance under which steam from an existing back-pressure steam turbine-driven chiller had to be vented to allow full-load operation under peak load conditions.

On the basis of the interactive technical and financial performance analysis, a back-pressure steam turbine-driven air compressor and a single-stage absorption chiller were selected. The exhaust steam was tied into the low-pressure steam distribution header and used primarily for heating in the winter and to serve the new single-stage steam absorption chiller in non-winter months. Functioning as a low-pressure steam heat sink, the added absorption chiller capacity served to balance the facility's seasonal steam loads, eliminating the need to vent excess steam, thereby allowing the existing turbine-driven chiller and the new turbine-driven compressor to operate optimally.

A separate, non-interactive analysis of the single-stage steam absorption chiller (inclusive of auxiliary components) demonstrated both higher capital and operating costs than the electric chiller option, showing it to be uneconomical. If a sufficient low-pressure steam load requirement had already existed at the facility, the investment in the topping-cycle turbine-driven air compressor would have been extremely economical, offering a simple payback of about one year. The extremely low operating cost of the steam turbine-driven air compressor results from its low actual fuel-chargeable-to-power (FCP), as all rejected steam energy in the form of back-pressure steam is credited against the total steam energy input to the turbine. In this case, the facility did not have a suitable low-pressure steam baseload in place, necessitating the installation of the absorption chiller at a capital and operating cost premium. Hence, the dramatic operating cost savings with back-pressure turbine operation were only made achievable with the additional heat sink in the form of the new absorption chiller. The interactive annual operating cost savings, inclusive of the disadvantage with the absorption chiller, showed this to be by far the lowest operating cost option and the most cost-effective option (even with its higher capital cost) on a life-cycle basis.

This project exemplifies optimization of heat and power resource use at a facility and the need for interactive analysis. It highlights, however, only one of the numerous interactive impact considerations that go into the development of an integrated energy and cost reduction project. In a case such as this, the selection of the exact capacity for both the air compressor and chiller should be made only after determining which end-use improvements are to be made. These load reductions will also result in correspondingly lower operational savings potential for the new air compressor and chiller.

A broader integrated project analysis would include a series of measures designed to reduce chilled water and compressed air end-use requirements and their interactive impact on the central plant project. It would also include other end-use measures, such as a lighting system upgrade. This would lower electricity demand requirements and chilled water requirements (by reducing internal gain), while raising winter steam heating requirements. Steam distribution system improvements, such as a steam trap program, would lower steam system requirements, perhaps limiting the extent to which the heat sink would be large enough to support operation of all of the topping cycle turbines. Other distribution system optimization measures, such as variable flow chilled water pumping, would also need to be analyzed concurrently. Finally, expansion of the facility's automated energy management system would have to be interactively analyzed. This would also likely serve to reduce steam and chilled water loads, redefining the capacity requirements. These are but several of the many end-use and distribution system measures that would be considered interactively with central plant measures in a fully comprehensive integrated energy study.

Another consideration is the interactive impact of such a project on the long-term potential for on-site power generation. In this case, the increased steam baseload has made this an attractive site for application of a 10 MW gas turbine electric cogeneration system. Prior to the installation, a smaller capacity gas turbine or a reciprocating engine-based system would have been a better match with the facility load profile. The interactive analysis of electric cogeneration system economic performance potential includes consideration of opportunities to further modify the facility's heat balance through conversion of several additional electric processes to steam processes. Also considered would be the interactive effects of various potential utility incentive offerings and alternative electricity and gas purchasing and risk management options.

To accomplish such interactive analyses with computer simulation modeling, the annual usage and demand (actual and billed for existing and proposed systems) is summarized automatically. If a series of proposed measures has been modeled, total changes in the facility will be reported on an interactive basis. To isolate the predicted savings from a given measure, an order must be established so that the impact of successive measures can be separated and evaluated on their own merit, while still considering their interactive impact on other systems. An effective way to accomplish this is to model groups of highly interactive measures together and determine the aggregate savings result. Then, in order to interactively determine the contribution of any single or subset of measures, the model is run with the single measure or subset removed. The difference in savings is then the incremental contribution of this single measure or subset of measures.

Step 6. Refinements to Selected Project Measures

Commonly, more than one type of system or system configuration and control strategy are considered in an effort to develop the most effective option for a given application. Each option must be modeled and then compared with the baseline. Once these savings are compared with the budget cost estimates for each option, the most appropriate option will be selected and proposed. Finally, as discussed above, measure impact must be determined on a system-specific basis, as well as on an interactive facility-wide basis.

Capacity factor and reliability of actual system operation must also be considered. Capacity or load factor is a term used to express average load as a percentage of full load in a given period. For example, if a system operates at full load for 360 hours in a month that has a total of 720 hours, the capacity factor is 50%. The efficient utilization of many systems rests largely on the capacity factor of the plant. Generally, the more hours the unit is able to run, the better the project economics. This is significantly affected by competing utility rate structures. Alternatively, the profitability of a peak shaving application depends more on strategic operation over relatively few hours. In this case, reliability is a critical factor because the system must operate consistently in peak electric demand periods in order to produce expected savings.

When evaluating project economics, unplanned forced outages and other planned outages will be considered. When planning for and evaluating a project, one must therefore consider the expected capacity factor of the plant and the reliability of the proposed system. This includes projecting savings based on less than optimum performance, or incorporating the cost of mitigation measures such as standby capacity.

Another part of energy system analysis for the facility will be to identify the reliability and redundancy requirements. This pertains not only to central systems, but also to auxiliary components such as pumps and fans. Having an extra engine on standby but only one fuel pump for both engines is an example of a vulnerability to interruption that is sometimes overlooked. If the fuel pump fails, neither engine will function. Hence, well conceived, redundancy considerations should figure largely in the equipment selection process inclusive of system auxiliary components. Installation of equipment redundancy allows for more creative operating strategies. However, it also increases project capital cost, so the benefits must be identified and carefully considered. If the intent of redundancy is exclusively to protect savings, it may not be valued as highly as if it is to meet mission critical requirements as well.

An important aspect of measure refinement is the assurance that the system output is compatible with the load requirements. Heat recovery systems are particularly sensitive to this, as discussed in the Savings Analysis step. When considering use of recovered heat, it is important to identify the thermal levels (temperature) of energy that can be produced (e.g., hot water or steam) and the amount of heat available at the required temperature. It is also important to consider any modifications that can be made to existing systems to allow them to use recovered energy in the form that it is available. For example, if recovered heat is to be used to serve a process that currently uses steam, either the recovered heat must be available as steam, or the recovered heat can be made available as hot water and the process converted to operate with hot water. Processes can also be converted to operate at lower temperatures. By serving a lower temperature process, the efficiency of the heat recovery process can often be increased.

The potential for conversion of process requirements becomes an important part of potential measure refinement. As part of the feasibility analysis, various options should be considered for improving the balance between summer and winter thermal loads. Where applicable, it may be economical to increase thermal loads through added baseload applications. The application of steam turbine-driven or absorption chillers as discussed above are two of several options.

Step 7. Consideration of Environmental and Other Code Compliance Impact

Integral to the analysis of any fuel burning system is consideration of air emissions. Other environmental considerations include water usage and disposal, chemical usage and disposal, asbestos and PCB (or other HazMat) removal and disposal, fuel storage, indoor air quality, sound emissions, and refrigerant usage and disposal. The installation of a natural gas pipeline is also subject to consideration of environmental impact and associated costs. All other prevailing code compliance regulations should also be considered and project strategies should be formulated that include the limitations imposed by these regulations. Often, a more efficient or lower first cost option will be eliminated from consideration based on compliance regulations.

Installation of new equipment, such as a gas turbine cogeneration system, will result in costs associated with emissions control technology and, in some cases, emissions offsets, and it may affect other equipment at the facility. It can add costs by subjecting the entire facility to certain compliance regulations or can lower costs, as might be the case in waste heat recovery displaced boiler operation. The analysis should consider what would have been the present or prudently anticipated environmental costs with and without the installation of the system under consideration.

As part of the Detailed Feasibility Study, for any potential technology application that involves the installation of fuel burning equipment such as boilers, process furnaces, direct-fired absorption chillers, gas turbines, or reciprocating engines, the analyst should determine emissions control requirements and their associated costs. Emissions control economics will vary with the source type and the degree of control needed and must also relate to the current air-permit conditions at the facility. Cost-effectiveness is an economic test, or screening process, to determine if an emissions control technology is beyond the scope of a particular project.

To determine the cost effectiveness of a control strategy option, one must first establish uncontrolled emissions rates (i.e., base case) prior to measure implementation. With existing equipment, this can be done through equipment metering or through the use of accepted emissions standards data, such as EPA's AP-42 factors. For new equipment, the manufacturer can provide certified emissions rates. All cost factors can be expressed on an annual basis by considering the capital cost, the component replacement costs and time intervals, and the associated OM&R costs. The cost-effectiveness can then be determined by dividing the annualized cost of control by the tons (or kg) of emissions per year reduced. In a number of cases, measures can be developed which both improve operating efficiency and satisfy environmental mandates. An example is the switch from CFC-based refrigerants in older, less efficient chillers and systems to non-ozone depleting refrigerants in new, high-efficiency chillers with well-designed operating and distribution systems.

The development of annualized costs is often sufficient for survey purposes. However, it is more effective to identify the year in which each cost will actually be incurred. Additionally, if continuous emissions monitoring (CEM) is required, the capital and annual operating costs must also be considered in the feasibility analysis.

Determination of cost-effectiveness values is important for two reasons. First, it will allow the project team to compare various options on an equivalent basis. Second, it may provide economic relief from installing technology that is cost-prohibitive in a given situation. Air agencies may establish cost-effectiveness permitting thresholds. If the value of a control technology to be implemented exceeds the cost-effectiveness limits established by the local air agency, the facility petitioning for a permit may be able to avoid that technology and employ the next most stringent control technology.

Additionally, the cost of environmental compliance may also be considered as an adjustment to the baseline operating cost in certain situations. Consider, for example, a building that is out of compliance with indoor air quality codes due to inoperable outside air dampers. A measure such as a VAV system will truly improve system efficiency. However, in the process, outside air levels will be brought up to code. This may result in excess energy being used vs. the original condition where that outside air was not being introduced and conditioned. In an absolute sense, this measure may actually increase energy usage. However, it may be appropriate to adjust the baseline calculation to reflect the energy that would have been used if the dampers were operable and brought into code compliance under the assumption that the codes need to be met regardless of whether the efficiency measure is implemented or not.

Finally, it is often worth quantifying the impact the project will have on climate change. In fact, in some cases, it could be a major motivation for pursuing energy efficiency, alternative generation options, and energy conservation initiatives. For each type of energy savings generated, the CO_2 emissions factors in Chapter Five can be utilized to convert energy savings into carbon reductions.

For fossil fuel reductions, this is a straightforward calculation, using a single conversion factor per Btu of fuel saved. For electricity, the factor must first be selected based on the region or state where the project is located, to account for the fuel mix used in that state's power plants. For indirect energy sources, such as purchased steam or purchased chilled water, the source energy mix for the plant that produces that energy stream must first be determined. For example, a central steam plant may utilize a mix of gas and oil, or a central chilled water plant may utilize electricity and gas (through steam-fired absorption chillers). In these cases, a weighted average emission factor can be derived from the mix of primary fuels used to generate each unit of purchased energy (e.g., lbm of steam or ton-hour of chilled water).

The results of the above carbon calculations are typically expressed in metric tons of CO_2 eliminated from the atmosphere each year (assuming that energy savings results are calculated as annual figures). Since this unit of measure may not be familiar to those who will evaluate the results, it may be worth converting metric tons of CO_2 into carbon offsets that anyone can relate to. Two common metrics are equivalent cars removed from the road, and equivalent trees planted.

The factor for equivalent cars removed is based on the average American car that gets 20.4 miles per gallon (8.7 km per liter), driven for an average of 11,720 miles (18,862 km) each year. With these assumptions, and the carbon content of gasoline at 8,890 metric Tons CO_2 per gallon, the conversion factor is 5.23 metric tons CO_2 equivalent per vehicle per year. So by dividing CO_2 reduction results by this number, climate change benefits from a project can be presented in terms of the equivalent number of cars that would have to be removed from the road each year to achieve the same result.

The factor for trees planted recognizes the fact that growing forests store carbon. Through the process of photosynthesis, trees remove CO_2 from the atmosphere and store it as cellulose, lignin, and other compounds. The rate of accumulation is equal to growth minus removals (i.e., harvest for the production of paper and wood) minus decomposition. In most United States forests, growth exceeds removals and decomposition, so there has been an overall increase in the amount of carbon stored nationally. The U.S. EPA estimates the annual average rate of carbon accumulation is 4.69 metric tons of CO_2 per acre of pine or fir forests. So, by dividing CO_2 reduction results by this number, climate change benefits from a project can be expressed in terms of equivalent acres of trees that would have to be planted to achieve the same result.

Step 8. Evaluating OM&R Requirements

With the installation of any energy system, there will be a change in OM&R requirements at the facility. This change may be minor or major. If new motors, pumps, or lighting fixtures are installed, there will usually be a relatively minor, predictable impact. Generally, these produce a slight decrease in OM&R requirements when an older, more maintenance-intensive inventory of equipment is replaced with newer, less maintenance-intensive equipment. The group replacement nature of this type of retrofit allows for more controlled and predictable ongoing maintenance practices, such as group relamping. They also generally produce an initial OM&R savings as lamps and ballasts or motors are replaced midway through their replacement cycle with all new equipment. In the case of installing a prime mover for power generation or mechanical drive service, there will generally be a major change in OM&R requirements. Additional operators may be required and additional preventative maintenance and repair services will be required. Examples of additional OM&R cost factors are lubricating oil consumption, chemical treatment and effluent disposal, and periodic overhaul.

In any case, the analyst should include all OM&R factors in the analysis, including anticipated future overhaul or replacement costs. OM&R requirements will be based on consideration of the intended operating duty, including not only hours of operation, but load variation and frequency of systems starting and stopping. In developing the integrated life-cycle analysis, these anticipated costs should be inserted in the years they are expected to be incurred. Long-term complete service contracts (if applicable) are included in the calculations as such. In addition to regular system maintenance costs, there may be other operation costs to be included, such as additional personnel required to operate the system and costs associated with the hiring of plant operators.

The incremental cost analysis of OM&R provides a more realistic and precise view of the actual cost. For example, compare a prime mover-driven equipment application and an electric motor-driven alternative. While the prime mover will likely have a far greater OM&R cost than the electric motor, motor maintenance and overhaul are not free. Therefore, the incremental cost consists of the OM&R cost of the prime mover minus the OM&R cost of the electric motor.

When analyzing multiple alternatives, the OM&R cost of each should be inserted into the analysis. When evaluating one alternative with respect to another, or a base case, the incremental OM&R cost can be determined by subtracting one from the other. In developing projections for OM&R costs, vendor quotes are very helpful. Even if work is to be done in-house, vendor quotes for operations, routine maintenance, and cradle-to-grave service options provide meaningful insight to market costs and consistency between options. For each system under consideration, the equipment vendor should be asked to provide a proposal for various short- and long-term OM&R options.

Fuel and power management services and their associated costs require additional analysis to accurately evaluate the benefits of pipeline installation, transportation or wheeling initiatives, and end-use and central plant technology conversions to alternate fuels. These service requirements include forecasting plant usage and market conditions, negotiations and procurement arrangements, execution of curtailment strategies, managing daily volumes and storage requirements, regulatory monitoring and invoice reconciliation. Whether the facility or an outside firm provides these services, the operational costs of the various supply options must be included for analysis in their development.

Step 9. Evaluating Utility Supply Options

Given the fast developing deregulated energy industry environment, alternative sources or arrangements for purchasing fuel and power are integral to the evaluation of long-term cost reduction opportunities in that they can significantly lower operating costs. Therefore, as part of the study process, the project team should investigate various options available for providing fully reliable fuel and electric power services to the facility at the lowest possible price. This includes evaluation of natural gas delivery infrastructure to provide gas purchase or transportation purchase, as well as retail wheeling and utility bypass, if appropriate. It also includes negotiations with the local utility, as well as any other potential power sellers on the grid.

This process is integral to the evaluation of on-site electric power generation options. Accurate and objective evaluation of the economic performance of on-site generation system applications will be integral to any electric rate negotiations with the local utility or other power supplier or purchaser. Competitive energy rates will be offered almost exclusively on the basis of verifiable competitive options available to the facility. Complete development of on-site generation system project economics therefore will serve as a basis of electric rate negotiation process.

It is important to separate the supply functions of capacity and commodity in this process. The ability to withstand interruptions to gas or electricity supply will usually allow significant reductions in operating costs. With fuel supply, this is commonly accomplished through multiple fuel source operating capability. With power purchases, this is usually accomplished with the ability to shed loads, either through a demand controls strategy or on-site generation. On-site generation capacity can, therefore, allow for elimination of capacity costs associated with the purchase of power.

Interactive impacts of alternative fuel and power purchase options will also be considered. Just as a reduction in load brought about by another measure can lower the economic performance of a given measure under consideration, the reduction in fuel or power cost brought about by an alternative purchasing arrangement can lower the economic performance of a potential technology application. Therefore, savings analyses must be performed with a sensitivity toward the cost of energy with alternative fuel and power purchase options in place including long-term escalation (or de-escalation) rates.

Step 10. Developing an Interactive Summary

With the integrated approach, the facility will be viewed as a whole whose sum total of resource use is both dynamic and interactive. Therefore, when considering energy supply, central plant, and distribution system improvement options, the analyst must consider all loads being served, such as comfort conditioning or process end-uses. This front-end analysis will then be upgraded based on the loads that would be prudently anticipated once cost-effective end-use improvements are implemented. Supply-side options, such as energy source switching, energy supplier switching, utility, negotiations, or on-site power generation, should be considered interactively with respect to the anticipated central plant, distribution system, and end-use modifications.

As indicated in the above discussion of measure interactivity, the project team should explore, for the purpose of understanding, the effect of each potential change on the rest of the facility. For example, when lighting systems, envelopes, air handling systems, or process applications are upgraded, or compressed air use is optimized, then steam, hot water, chilled water, and compressed air requirements can be significantly altered. Different capacity central plant systems may be required as a result, and the economic performance of upgrading those central systems may be changed as a result of the load changes. Conversely, when the central plant systems are upgraded, the cost of steam, hot water, chilled water, and compressed air is reduced, thereby reducing the savings potential of end-use improvements — sometimes eliminating certain improvements that would have been cost-effective with the old, less economically efficient central plant systems.

Supply-side improvements have a similar effect. If fuel or electricity can be purchased at a lower price, or if heat and power can be provided less expensively through an on-site cogeneration system, the economic performance of central plant and end-use upgrade programs will often be significantly altered. Given this potential for interactive impacts, all potentially beneficial project application options should be identified during the study process, inclusive of their interactive impact. Options should then be mixed and matched and considered interactively until the optimal portfolio of measures is selected. Therefore, it is necessary to develop a summary format that identifies the impact of individual measures, as well as the impact of the complete portfolio. The following is an example of a study report format that uses a series of linked spreadsheets to build an interactive survey of a portfolio of measures, many of which have interactive savings impact.

The process starts by building a base case with the total energy/fuel consumption and cost and OM&R cost in the scope of the analysis. The spreadsheet design allows for the comparison of the total electric and fuel consumption and costs (plus OM&R and other resources costs) before and after each proposed measure using the energy calculation model employed by the project team. This integrated analytic framework allows calculation of cost reduction potential for each individual measure while accounting for interactive effects and eliminating the potential for double counting.

1. The first section of the spreadsheet includes the historical energy (and other resource) usage, repriced at the most current energy/fuel rates, plus OM&R costs. This provides a baseline of costs.
2. The second section is constructed to perform computations, in energy units, of changes being made by the proposed project or series of projects. This includes both increases and decreases in consumption of all resources. OM&R changes are computed on a cost basis.
3. The third section then applies the net decrease (or increase) in each source of energy and other resources used to the base case consumption and recalculates the cost of the facility total energy usage, and adds in net changes in OM&R costs.

This process will then be replicated for each additional measure or group of measures by building upon what has become a revised base case. These can then be sorted in the order of measures most likely to be implemented based on cost-effectiveness or preference stated by the host facility. The spreadsheets will be repeated for each building and summed to provide a facility-wide scenario that reflects all applicable interactions. By comparing the final case to the original base case, the net resource and cost savings of the program is determined. The cost related to the impact on OM&R will be included along with energy and other resource cost savings, as well as the impact of alternative fuel and power purchasing options.

Step 11. Equipment Selection

Having analyzed the savings potential for various measures, and having made preliminary equipment and component selections, the next step is to develop conceptual designs and select specific equipment. This allows for preparation of budget estimates for development, installation, and commissioning of each measure under consideration. From this, final measure cost-effectiveness can be evaluated.

All equipment selected and/or specified and installed should satisfy the necessary requirements of the governing agencies, such as local and state codes, testing and approving agencies (UL, AGA, NFPA, ANSI, ASHRAE, ASME, etc.), and construction agencies (SMACNA, ASTM, NEC, NEMA, etc.) and should represent the highest standards of the industry. Equipment and systems should be selected that both minimize OM&R requirements and can be simply and effectively serviced. Equipment is also selected on the basis of environmental emissions characteristics, with emphasis on minimizing emission of pollutants.

It is generally most cost-efficient to not favor particular manufacturers in the equipment selection process. However, customer preferences or specific request should also be considered. Equipment selection should be based on the engineering specifications required for the best performance and on economic analysis of all available models that can meet the specifications. This also encourages vendors to be creative in matching the most cost-effective equipment option to the specified performance, or suggesting alternate performance parameters that may improve measure economics. This process allows the purchaser to optimize price performance, since project capital cost projections are relatively accurate and minimal contingency budgeting is required.

Another important selection criterion is technological obsolescence protection. It is prudent to specify only manufacturers that have demonstrated a corporate commitment to backward compatibility, i.e., that new systems will be able to operate with existing field equipment. An example is the specification of a facility EMS or system-specific DDC controls. Building automation controls used should support open protocols such as Building Automation and Control Network (BACNet) because they provide the ability to view and operate with data of different systems from a central operator's workstation, allow for interaction of data from different systems, and offer the ability to select the best products from various vendors to meet the facility's needs over the long term.

In addition, specified manufacturers should have a demonstrated commitment to supplying equipment repair and replacement parts for a minimum of ten years from the time the particular devise is no longer manufactured. This will avoid obsolescence of an otherwise usable devise due to lack of a proprietary component or part.

Step 12. Budget Estimates of Incremental Capital Cost

In the preliminary study phase, cost estimates will have been considered in order to screen and select several options from all possible choices. All cost components will have been developed as rough estimates, but at this point, it will be necessary to develop a firm set of installed costs for each measure still under consideration. Since the validity of the study depends as much on accurate implementation cost estimates as it does on accurate savings estimates, the preparation of budgets in the detailed study phase should be performed with the same rigor as savings estimating.

Preliminary conceptual drawings should be made for the various system options. These should include footprint and height requirements and anticipated layouts for piping, conduit, exhaust flues, pumps, and other auxiliaries. Clearances for maintenance access (e.g., pulling tubes) or code requirements (e.g., around electrical equipment) should be illustrated. Access locations for bringing in the equipment or disposing of existing equipment should be identified. Structural requirements should also be identified to ensure that the weight of the new equipment could be tolerated. Utility service availability and capacity should also be noted and included on the drawings.

In addition to the drawings, a comprehensive list of all known and potential cost line items should be made for each measure. In addition to the cost of all major and auxiliary components, this includes environmental compliance related equipment and procedures inclusive of permitting and testing costs. Asbestos and PCB removal and disposal costs should also be included. Examples of other costs include insurance, the cost of lost floor space, land acquisition, and facility construction. Applicable taxes, prevailing labor wage rates, and bonding requirements should also be considered. A separate list of potential scope items which may or may not have savings associated with them, but support the system and overall infrastructure improvements, should be included as well.

Once the master list is completed, budget pricing can be established based on detailed internal pricing procedures or through solicitation of contractor and vendor bids. This may be done through a formal RFP process or on a somewhat less formal basis. Regardless, it should be based on a bid package with a detailed SOW, supporting design documents, contract requirements and a bid form identifying line item cost. It should be facilitated with contractor walk-throughs of the facility. Following the contractors' site visit is a question and answer forum, which provides all bidders with follow-up information. If required, SOW addenda can be issued at this time.

In addition to the lump sum pricing required by the RFP, each prospective contractor should also submit the following items:

- Break out costs for each trade and/or task as outlined in the bid form.
- A brief summary of the sequence of construction, accompanied by a timeline for the proposed SOW.
- Identification of all subcontractors and either subcontractor pricing or an approximation of the percentage of work to be completed by subcontractors.
- The contractor's responsibilities and processes for commissioning should be outlined.
- All documents, including a Contractor Qualification Statement, in the bid package should be completed and signed.

After receiving bids, a series of clarifications will generally be necessary to ensure that all price items are covered. With this approach, pricing for all major items can then be based on actual quotations. These quotations will include all necessary equipment and service specifications and can then be used to confirm initial assumptions and estimates. For example, equipment weight and dimension data will confirm adequate access, space, and structural support where the equipment is to be installed; actual performance data will confirm assumptions used in the energy savings analysis. Bids must be carefully considered based on the quality and detail of the response, as well as the reputation and financial condition of the contractor/vendor. In addition, other bid evaluation considerations include their proposed timeline, bid exceptions, and value engineering opportunities. If one were to get a very low bid from one subcontractor and then, in turn, submit this as part of their own guaranteed firm fixed price offering, they would be exposed to liability if that contractor was unable or ultimately unwilling to execute the work at the stated price. Hence, adding in some cost contingency or using another higher bid price from a proven quality subcontractor might be a more prudent approach.

An alternative, or complimentary, step is conducting an internal pricing process, based on conventional pricing programs or reference sources such as R.S. Means. While some firms have the internal estimating capabilities to do so, others may use this only as a check on the bid pricing approach. Also, use of database information from previous jobs serves as a good reference for price checking, with the understanding that conditions will, of course, differ for each application and job site.

After the incremental project cost is totaled, any utility rebate and other type of available financial incentive will be subtracted to determine the net investment cost. Rebate potential will often depend on the specific equipment chosen. For example, there may be a dozen options for chiller makes and models. Each will have different efficiency and, therefore, may qualify for a different rebate.

Concurrent with the development of budget estimates is the establishment of an installation schedule. This can be included in the form of a Gantt chart. While this is important with all contract vehicles, it is particularly important with performance-based contracts since savings will not begin to accrue until the measure is commissioned and because construction financing typically is arranged with finance charges accruing on the money borrowed and expended as construction progresses. The timeline will be based on a critical path method wherein each construction element is identified and carefully planned for completion on the latest possible finish date to delay finance charges while still maintaining the construction schedule.

Once the budget estimates are completed, the final savings analysis can then be run for each measure alternative still under consideration. Certain attractive system alternatives or system enhancements may fail financial screening, but offer other qualitative benefits. These benefits will be identified so the options can be considered as alternates by the host facility. In cases where replaced existing equipment can be sold, its salvage value can be subtracted from the full cost of the new system. In cases where the existing equipment is in working order, but approaching the end of its useful life, considering the full cost of the new equipment would obscure the fact that replacement would be required at some point in the near future for reliable operation and, therefore, not reflect the full economic benefit of replacement. The decision to install new equipment immediately, as opposed to waiting until the end of the existing equipment's service life, will take into consideration the time-valued incremental investment cost associated with early retirement of the existing system, which will be less than the full project cost.

Thorough development of design concepts and installed cost estimates during the detailed study phase will generally reduce overall design costs and minimize instances where proposed system applications are later rejected by the design team, requiring re-analysis of system alternatives. Where appropriate, design engineers should be called upon to review preliminary design concepts to confirm logistical feasibility and identify any omitted cost factors. An example is review by a structural engineer of the weight-bearing potential of a particular location selected for the installation of heavy equipment. This review may show the location to be impractical on the basis of inadequate structural support or may identify structural reinforcement requirements that will be included in the budget estimate.

While these activities add to study costs, they often prove cost-effective as they can reduce design costs and minimize the need for subsequent follow-up studies. In cases where projects are proposed with firm or not-to-exceed pricing following the study phase, the involvement of professional design engineering and construction estimating in the process is considered essential. A turnkey design-build contractor or ESCo has the advantage of having all or most of the required expertise in-house, avoiding the need to subcontract design and construction review during this critical phase of the project.

Step 13. Life-Cycle Financial Analysis

To evaluate the sum total of project incremental costs and benefits over the life of the project for each potential system option, a cash flow analysis should be performed for every year of each potential measure's life. Total project cash flow will be determined for each year of the project life as the sum of all measures combined. Any alternate measures, measure groups, or equipment or supply options should be analyzed individually. Integrated programs may allow for combining certain measures that in and of themselves do not necessarily satisfy the financial screening requirements. This may enable projects that are desired by the host facility, such as central plant upgrades, building envelope architectural and structural improvements, or alternative technology applications with marginal project economics to be fully funded as part of the overall integrated cost reduction program. The cash flow presentation will show total cost savings and total performance payments and will break each into its component parts. The financial analysis will be formatted to meet the host facility's specifications. Life-cycle costing assumptions, such as escalation and discount factors, will be clearly stated so they can be reviewed.

The type of financial analysis performed and the results will be largely influenced by the types of contract and funding vehicles being contemplated for project implementation. The following two chapters present detail on contract types, financial analysis techniques, and project financing options. Chapter 42 provides an introduction to basic financial analysis techniques that are widely used for project evaluation. Chapter 43 focuses on the various standardized and customized performance-based contract vehicles available for implementing an integrated energy optimization project and presents details on various conventional and non-conventional financing options that may be used to fund project implementation. These contract and financing options and analytical methods should be considered in advance by the study team so that financial results can be analyzed and reported in a format that is compatible with that of the host facility.

Step 14. Final Report and/or Contract Proposal

At the conclusion of the Detailed Feasibility Study process, a final report should be delivered to management describing the study team's findings and project recommendations. The report can be used to document all significant aspects of the study. It should contain the results of the work of the project team, state the approach taken in developing each section of the proposal, discuss all assumptions, and provide details for each proposed measure with the associated SOW. The technical proposal should contain a detailed savings analysis for each proposed measure, with full documentation, along with the proposed M&V approach where applicable. The technical sections should document the technical analysis approach, including pertinent drawings, sketches, photographs and equipment catalog cuts, to the extent needed to define the means and method of each measure. It should also address project implementation, providing sufficient detail to present the design concepts and the approach to construction, startup, OM&R, and training. The report should also include the financial analysis as discussed in the previous step so that firm commitment decisions can be made with the study serving as the basis.

In the case of a turnkey design-build, performance contracting-based, or third-party ownership and operation offer, in addition to the study report, the contractor should prepare and deliver a firm price offer to the host facility. The price proposal can then provide the facility with an evaluation of the financial aspects of proposed measures from both the team's and the host facility's perspective, proposing the implementation costs and service phase costs throughout the contract term. The financial sections should include agreed upon financial analysis techniques and, in the case of a performance-based contract, should include the amount that the contractor will guarantee in savings, discriminating between each category of savings, including energy and demand, water, and avoided OM&R and capital renewal savings. Additionally, this section should include a schedule detailing any incentives to be received over the life of the measure and the effect of such incentives on the paydown of the principal investment. This should include a life-cycle cost analysis and the guaranteed price schedules indicating total savings, required payment to the contractor, and net savings to the host facility on an annual basis for the duration of the contract term.

The proposal should also include the proposed project management team, further detail on the proposed implementation plan, and a Gantt chart indicating the project implementation schedule. Chapter 40 provides an example of project implementation Gantt Chart. Finally, it should clearly define the roles of the contractor and the host facility during the implementation and service phases of the project.

If changes or adjustments are requested by the host facility, the contractor will make appropriate revisions and then typically submit a best and final offer (BAFO). Questions on the proposal can be answered in writing, but it is recommended that an oral presentation be made to properly review the proposal. At this presentation, the contractor can cover the results of the Detailed Feasibility Study, answer any questions, and provide an overview of the remaining steps to implementation.

Chapter Forty-Two

Evaluating Project Financial Potential

In order to receive funding, be it equity, debt, or a combination of both, energy resource optimization projects must be evaluated with respect to other project opportunities and uses of available funding. Prospective investors go to great lengths to properly evaluate capital investments. Often times, they will use a mixture of simple analytical tools, such as payback or substitute comparisons, and more complex tools, such as internal rate of return and other cash analysis tools, as well as price earnings (PE) and other net income oriented ratio analysis tools. All have their place and all have their weaknesses. A good investor will use many or even all of these analytical tools, while relying heavily on previous experience in investing to make a decision.

It is important to remember that the results of any analysis are only as valid as the assumptions on which they are based. Sophisticated analysis performed with gross estimates of costs and savings, or of risk or escalation rates, will likely not yield useful results. In making capital budgeting presentations, it is strongly recommended to properly categorize the low- and high-risk projects and assign higher return requirements to the riskier commitments. To do otherwise might allow the analysis results to direct all funds to the high return, high risk ventures.

This chapter focuses on the analytical tools used to evaluate capital commitments. It begins with definitions of the most commonly used tools in formal capital budget analysis and then proceeds with a detailed discussion of the most commonly used financial analysis techniques.

Commonly Used Tools

The tools most commonly used in formal capital budget analysis and presentation include:

- **Simple payback** analysis, which determines how many years of earnings (or net benefits) it will take to payback the investment.
- **Simple rate of return (ROR)** analysis, which determines the percentage or rate at which an investment will return.
- **Discounted payback period** analysis, which considers the time in which the discounted value of net benefits equals the initial investment.
- **Net present value (NPV)** analysis, or sum of the discounted net cash flows (DCFs) over the life of the project, which measures the amount by which a project's stream of benefits exceed all costs, including the cost (or opportunity cost) of capital. Generally, the discount rate applied to the stream of benefits is equal to the cost (or opportunity cost) of capital.
- **Internal rate of return (IRR)** analysis, which determines the discount rate for which NPV is equal to 0.
- **Savings investment ratio (SIR)** analysis, which compares benefits and costs by dividing the present value of returns by the present value of the investment.

Each of these methodologies is presented with the assumption that an initial investment is made in Year 0. This is the full investment minus any offset costs. In retrofit applications, if the replaced equipment can be sold, it may have a salvage value. This value, minus any depreciation impact or other selling costs, can be subtracted from the initial investment to arrive at a net investment. Similarly, assets at the end of a contract term may have value and could possibly be sold. Costs that are avoided as a result of the investment should also be considered. For example, if equipment overhaul is required to allow a facility to continue to use existing equipment, that cost may be subtracted from the initial investment in the new equipment for analysis purposes. If the investment allows a facility to avoid installing new equipment, such as a new electric service substation to support growth requirements elsewhere in the facility, that too may be subtracted from the initial investment. Difficulties arise when replaced equipment remains in place and adds backup in the form of system redundancy. In some cases, this can be quantified based on the cost of an avoided outage, while in other cases it is relegated to qualitative benefit status.

One should conduct a similar analysis for ongoing operations, maintenance, and repair (OM&R) expenses. If costs are avoided as a result of the installation of new equipment, the investment analysis should consider only incremental operations costs. Generally, a savings of $1 per year of OM&R expenses equates to about $5 to $8 in capital savings. Thus, emphasis should be placed on operating cost savings.

Payback and simple ROR methods are generally reserved for preliminary screening analyses. They are too crude for proper evaluation of capital commitments because they do not adequately consider the time value of money. They are important, however, as practical tools used to limit a wide range of options and determine, on a preliminary basis, project potential.

Consider, for example, life-cycle feasibility analyses, as discussed in the previous chapter, and the level of detail required for more intensive capital budget analyses. The most efficient approach to project financial analysis may be to eliminate many or even most of the potential options through simple payback or ROR analyses, and then to apply the more rigorous IRR or NPV analysis to the remaining few options. It is important to remember that the more high-powered analyses may be equally as misleading as the simple analyses if the input assumptions are not reasonably accurate.

Net benefits resulting from an investment is usually measured by the annual positive net cash flow from a project which is generally considered to be after-tax distributable cash. While most taxes are not paid at a project level (but at a corporate level), taxes are a cost of entering into the project and must be considered. Generally, the project is considered a full taxpayer by most corporations and tax benefits are assumed to be used in the year they are made available. The most commonly used tools in formal capital budget cash analysis and presentation include IRR, NPV, and its derivative, SIR, analyses. IRR is also a good tool for evaluating opportunities of different risk and for lease vs. buy (or various combinations of debt and equity) analysis. Other net income-oriented analyses are also used by some organizations, mostly publicly traded companies.

Financial Analysis Techniques

Simple payback is a commonly employed method that calculates the time required to recover an original investment through the net savings realized or net income derived from the investment. Simple payback is calculated as follows:

$$\text{Simple payback} = \frac{\text{Project investment cost}}{\text{Annual net positive cash flow}} \qquad (42\text{-}1)$$

If the project investment cost is \$100,000 and the annual net positive cash flow is \$25,000, the simple payback, based on Equation 42-1, is:

$$\frac{\$100{,}000}{\$25{,}000} = 4 \text{ years}$$

The advantages of simple payback analysis are that it requires simple computation and is easily conceptualized as an economic screening indicator. The two critical disadvantages are that it does not consider the time value of money or the effect of cash flow occurring after the payback period is considered (i.e., where cash flow is heavily weighted toward the out years). Because project service life is not considered, there is no way to differentiate between two investments that have the same payback but different useful service lives.

Simple ROR calculates the percentage or rate at which an investment is going to return. This is calculated in percent as:

$$\text{Simple rate of return} = \frac{\text{Annual net positive cash flow}}{\text{Project investment cost}} \times 100\% \qquad (42\text{-}2)$$

The simple ROR is the reciprocal of payback (and as such offers the same advantages and disadvantages), expressed as a percentage. Using the previous example, the simple ROR, in percent, based on Equation 42-2, is:

$$\frac{\$25{,}000}{\$100{,}000} \times 100\% = 25\%$$

Life-cycle analyses evaluate the sum total of project incremental costs and benefits over the life of the project. To accomplish this, a cash flow analysis must be performed for every year of the project's life. In each year, different things will happen. For example, a significant expenditure for overhaul may be expected in certain years. Also, energy and other resource costs are projected to change over time. Therefore, each year will have a different net benefit and the net benefits (or net revenues) of each year must be related to each other.

Present value (PV) or **present worth** analysis is a means of equating an amount received or paid in the future in today's dollars. Virtually all sophisticated economic analyses use the basic concepts of PV to account for the time value of money over the life of a project. Energy saving projects produce cost savings over varying number of years. The value of a dollar saved in Year 5 must be differentiated from a dollar saved in Year 1 based on the time value of money. So must the investment made at the beginning of the project in Year 0. Thus, all cash flow in every year of a project, whether costs, savings, or net cash flow, must be related to each other in a way that accounts for when they occur. This can be accomplished through PV analysis.

In some cases, there is a uniform flow of savings in each year. In other cases, there are unequal amounts of savings produced in each year. Given some anticipated **future value (FV)** or **future worth** of savings over a number of years, PV calculations are used to determine the dollar amount today (PV) that is equivalent to some anticipated FV amount. The equivalence depends on the rate of interest that can be earned on investments during the time period under consideration, or the discount rate.

The PV of a future sum of money over a number of periods at a given interest rate per period is calculated as:

$$PV = \frac{FV}{(1 + i)^n} \qquad (42\text{-}3)$$

Where:
i = Threshold interest rate or discount rate per period
n = Number of periods over which the value is calculated

PV may be considered the determination of the sum of money that would have to be invested at a given interest rate per year to yield a specified amount at a specified future date. For example, if the annual interest rate were 10%, the PV of $1,000 to be received at the end of a period of 10 years would be calculated as:

$$PV = \frac{\$1{,}000}{(1 + 0.1)^{10}} = \$386$$

Thus, given an interest or discount rate of 10%, were one to consider $1,000 of savings in Year 10 as equivalent to $1,000 in the present, conclusions drawn from the analysis would be erroneous. Conversely, the FV of a present sum of money (PV) over n periods at an interest rate i per period would be calculated as:

$$FV = P(1 + i)^n \qquad (42\text{-}4)$$

FV may be considered the determination of what a sum of money in the present would be worth if invested at a given interest rate per year at a specified future date. For example, if the annual interest rate is 10%, the future value at the end of Year 10 of $1,000 invested today would be calculated as:

$$FV = \$1{,}000\,(1 + 0.1)^{10} = \$2{,}594$$

Discounting

The process of calculating PV is called discounting. To determine the PV of future sums, a stream of annual net cash flows must be adjusted (or discounted) by multiplying each year's total by a PV factor (PVF). PVF is expressed as:

$$PVF = \frac{1}{(1 + i)^n} \qquad (42\text{-}5)$$

Where:
i = Threshold interest or discount rate per period
n = Period in which the future amount is earned

For any given discount rate, the PVF can be identified from a standard compound interest table or calculated. For example, given a discount rate of 10%, annual PVFs or discount factors over a three year period would be calculated, based on Equation 42-5, as:

$$Year\ 1 = \frac{1}{(1 + 0.10)^1} = \frac{1}{(1.10)^1} = 0.9091$$

$$Year\ 2 = \frac{1}{(1 + 0.10)^2} = \frac{1}{(1.10)^2} = 0.8264$$

$$Year\ 3 = \frac{1}{(1 + 0.10)^3} = \frac{1}{(1.10)^3} = 0.7513$$

Using these annual PVFs and an annual end-of-the-year positive net cash flow of $100,000, the PV of each of the cash flows over the three year period would be calculated as:

Year	PVF	Net Cash Flow	PV of Net Cash Flow
1	0.9091	$100,000	$90,910
2	0.8264	$100,000	$82,640
3	0.7513	$100,000	$75,130
Total			**$248,680**

If, in Year 0, the project required an investment of $200,000, the simple payback would be 2 years. Yet, when the cash flows are time-valued, the investment would not be recovered until sometime in Year 3. Where as a non-discounted method such as simple payback or ROR would assume the cumulative net cash flow to be $300,000 at the end of Year 3, the discounted analysis reveals a smaller PV sum of $248,680.

As indicated by the continually diminishing PVF, when applied over a longer term, for example 10 or 20 years, this gap between discounted and non-discounted cumulative net cash flow widens, showing non-discounted analysis methods to be crude and potentially misleading. Extending this analysis out to 15 years, for example, shows the PV of the cumulative net cash flow to be about \$760,000, or just about half of the non-discounted amount of \$1,500,000. If the initial investment were \$800,000 instead of \$200,000, the simple payback would be 8 years. However, if the useful life of the project were only 15 years, the discounted analysis would show the net annual cash flow to never fully pay back the investment.

The critical value of PV-type analysis is that by the process of discounting, a series of cash flows can be adjusted to determine their actual value in the present. Thus, an investor can view two opportunities, each perhaps with a different life cycle and cash flow shape, and conduct a more objective analysis as to which one, on the basis of the assumptions used, is better.

If cash flows are a fixed uniformed series of equal amounts in each year, the PV is the product of the cumulative PVF for the series and the payment amount. PVFs can be found in standard tables or determined using computer spreadsheet functions. PVF for a series of years can be calculated as:

$$PVF(i,n)_{ser} = \frac{(1+i)^n - 1}{i(1+i)^n} \qquad (42\text{-}6)$$

Using the previous example, the 3-year cumulative PVF is calculated as:

$$PVF(i,\ n)_{ser} = \frac{(1+0.1)^3 - 1}{0.1(1+0.1)^3} = \frac{0.331}{0.1331} = 2.4868$$

and the PV is:

$$(\$100{,}000)PVF\ (i,n)_{ser} = (\$100{,}000)(2.4868) = \$248{,}680$$

Capital recovery factor (CRF) is the reciprocal of the PVF. It is used to calculate future equal payments required to repay a PV of money over a specified number of periods at a given interest rate. This is used to determine mortgage payment and can be found in a table of annual capital recovery factors or determined using computer spreadsheet functions. CRF can be expressed as:

$$CRF(i,n) = \frac{i(1+i)^n}{(1+i)^n - 1} = \frac{i}{1-(1+i)^{-n}} \qquad (42\text{-}7)$$

If the \$200,000 investment required to generate the \$100,000 of annual savings, shown above, were to be financed over three years at a 10% interest rate, the CRF would be:

$$CRF(i,n) = \frac{0.1331}{0.331} = 0.4021$$

and the annual payment would be:

$$(\$200{,}000)CRF\ (i,n) = (\$200{,}000)(0.4021) = \$80{,}402$$

Since debt is a factor in most investments, discounting and PV analysis is a critical element of the evaluation process. Financing options are addressed further in the next chapter.

Consideration of Inflation

In addition to the time value of money, another factor to consider with respect to the return on a multi-year investment is inflation. Inflation reflects the rise in the actual cost of commodities over time. Different economic goods inflate at different rates. In an energy savings project, fuel, electricity, and maintenance, for example, may all inflate at different rates. In a multi-year cash flow analysis, each cost or savings line item can be inflated at a different rate. In many cases, net annual savings are simply inflated at an average rate.

In order to perform a PV calculation for a uniform series of annual cash flows, inflation should be taken into account when calculating the benefits, costs, and the discount rate or cost of capital. One way to do this is to use an effective interest rate that reflects the combination of inflation and interest rate. The effective interest rate can be calculated as:

$$i_{eff} = \frac{1+i}{1+inf} - 1 = \frac{i - inf}{1+inf} \qquad (42\text{-}8)$$

Where:

i = Interest rate

inf = Inflation rate

Given the fact that interest rates, energy costs, and other project cost factors are subject to unpredictable volatility, the concept of accounting for the effects of inflation should be viewed cautiously. An alternative approach is the use of sensitivity analyses for major cost factors. Sensitivity analyses may be designed, for example, to evaluate the impact on the investment if power or fuel costs double, or drop in half. The investor can attempt to assess the risk of such an occurrence and adjust the investment hurdle rate accordingly.

Time Value of System Replacement Costs

In most retrofit applications, the new equipment being installed replaces existing equipment that is in working order. If the replaced equipment can be sold, it has salvage value, which is deducted from the initial investment. Similarly, if the new equipment will have a value after the term of the contract (or finance period), it also has salvage value. However, it is important to consider the consequences of maintaining that equipment in place and not making the investment in the new system. To do this in a life-cycle analysis, one must consider as the base case (or baseline) keeping the unit in place over its useful life. If the existing equipment requires an overall, the overhaul is considered the initial Year 0 investment in the base case.

In the life-cycle analysis, one must consider the year in which the existing equipment would have to be replaced. If, for example, the unit had a potential useful life of five years, the base case option would be to keep the unit in place and replace it in five years. The life-cycle analysis would compare the annual cash flows of the base case and the alternative case — installing the new equipment in Year 0. This life-cycle cost can then be compared with investing in the more efficient replacement equipment in Year 0 and operating that unit in each of the proceeding years. This type of analysis is only possible when one considers annual cash flows over the life cycle of an investment, put into financial perspective based on the time value of money.

Types of Time Value Financial Analysis Techniques

In performing time-valued project economic analyses, numerous factors must be considered on an annual basis. These factors include capital and interest expense, energy operating cost savings, OM&R costs, salvage value, replacement costs, disposal costs, property tax, insurance, deprecation, and other tax deductions. Following are descriptions of several types of time-valued financial analysis techniques.

Discounted cash flows (DCFs) are a series of annual net cash flows, including all of the incremental economic factors of a project, translated into PV dollars in each year. The PVs listed previously in the three year series represent the annual DCFs. When considering various project options, applied discount rates may be varied based on an estimate of risk. The period in which the cumulative DCF equals zero, or the time-valued positive cash flow equals the initial investment in Year 0, is referred to as the discounted payback period.

Uniform annualized costs for a project are the uniform periodic, or average annual, cost over the project life or analysis period. To determine uniform annualized costs, one must first calculate the PV of all cost factors (i.e., net OM&R costs and depreciation) and then apply a capital recovery factor to the investment (minus salvage value) to determine equal payments over the analysis period.

The **NPV** measures all of the economic consequences of the project by determining the net gain or loss from a project at a given discount rate. NPV is the sum of the DCFs over the life of the project, or the cumulative DCF. This measures the amount by which a project's benefits exceed all costs, including the cost (or opportunity cost) of capital.

If a project is fully financed, there is no discounted payback period because there is no initial cash outlay in Year 0. However, the NPV can always be calculated, because it simply looks at net cash flow in each year. The payments made for debt and interest are accounted for in each year they are made and calculated into the cash flow of each year. The totals (net cash flow) for each year are then discounted to establish their PV and summed to determine the NPV.

IRR measures all economic consequences of a project by determining the discount rate as applied to the project's cash flow, which produces an NPV equal to 0. IRR can be hand-calculated using trial and error discounting calculation or performed easily using a programmed computer function. Consider a project requiring a $200,000 investment that produces net annual cash flow over three years of $105,000, $108,000, and $110,000. Using the trial and error approach, IRR may be determined as shown in Figure 42-1:

The IRR can be found by interpolating the NPVs between 25% and 30%:

$$IRR = 25\% + \frac{9{,}440}{(9{,}440 + 5{,}312)}\ (30\% - 25\%) = 28.2\%$$

NPV and IRR are generally calculated using spreadsheets with net DCFs over the life of the project. When considering multiple options, spreadsheets are generated for each option based on the complete operation or on the incremental changes vs. a base case.

The development of such spreadsheets allows for a complete life-cycle analysis of the investment. They contain a translation of what is expected to occur in each year of the project into a PV dollar figure. If, for example,

Year	Net CF	PVF (@25%)	DCF (@25%)	Cum. DCF (@25%)	PVF (@30%)	DCF (@30%)	Cum. DCF (@30%)
0	($200,000)	1.0000	($200,000)	($200,000)	1.0000	($200,000)	($200,000)
1	$105,000	0.8000	$84,000	($116,000)	0.7692	$80,766	($119,234)
2	$108,000	0.6400	$69,120	($46,880)	0.5912	$63,850	($55,384)
3	$110,000	0.5120	$56,320	$9,440	0.4552	$50,072	($5,312)

Fig. 42-1 IRR Determination Using Trial and Error Approach.

over the course of a 30 year project life, system overhaul is required every 10 years, that cost is plugged into Years 10 and 20.

Recurring costs such as replacement costs, overhaul costs, and other periodic costs are often included on an annualized basis or on a per unit output basis (i.e., $/hp-h or $/kWh). If these costs are to be set aside for the life of the project, then this method allows for more streamlined life-cycle analyses in which annual costs can be uniformly inflated over the life of the project. Typically uniform inflation rates are established for each major cost factor, such as electricity, fuel, and maintenance.

A more basic method simply applies one inflation rate to first year net savings in order to generate a stream of annual cash flows. Consider, for example, a $1,000,000 project that produces net annual savings, inclusive of all operating costs, of $125,000. Fifteen year cash flows are shown in Figure 42-2 using a uniform 2% inflation rate.

The pre-tax IRR for this investment works out to 11%, which assumes that the $1 million investment is made at the end of Year 0 and each annual cash flow is received in one payment at the end of Years 1 through 15. At a cost of capital, or discount rate, of 10%, the NPV over the 15 year project life works out to a modest $53,707. If, instead of 2%, a 4% inflation rate is used, the IRR and NPV work out to 12.8% and $168,311, respectively. If this cash flow stream were considered at a discount rate of 12%, the NPV would be reduced to $43,212, showing the powerful impact of discount rate consideration.

Another use of IRR analysis is to evaluate buy vs. lease (or finance) options. In such cases, the buy scenario, as shown in Figure 42-2, includes negative cash flow in Year 0, followed by positive cash flows over each year, ending perhaps with a salvage value at the end. The lease scenario will show annual cash flows, net of the lease (or finance) payments. By comparing the DCF or IRR of each scenario, the higher annual cash flows of the buy option, excluding Year 0, may be accurately matched against the lower annual cash flows of the lease option, which does not require a capital expenditure.

Commonly, a project will be funded with a combination of equity and debt. If, in the above example, the investor funded the project through a combination of 40% equity and 60% debt, at a rate of 9%, the pre-tax IRR would be increased from 11% to 13.5% and the NPV increased to $84,470. While this approach does produce a higher return on equity, it also comes with an incremental increase in the risk profile for the equity holder, as equity will be subordinate to debt service.

This is a very simplified overview of a complex issue. In most cases, an investor that performs a buy vs. lease (or finance) analysis must consider multiple issues in making such a decision. These include balance sheet, credit effects, top line and bottom line growth aspects, and benefits relating to book income (as IRR and NPV are strictly cash analyses tools). Such considerations are often specific to individual companies and are not discussed herein.

Limitations of NPV and IRR

When considering an investment, a positive NPV indicates that the investment is expected to produce a PV of all of the benefits, which outweighs the PV of all of the costs, given the cost of capital or the hurdle rate. However, difficulties may arise when comparing several investment options, each with a positive NPV. Also, because access to capital always has limitations, a consumer must often choose between several sound investment opportunities. While NPV does consider all parts of the investment over its economic life, including the time value of money,

YEAR	0	1	2	3	4	5	10	15
Annual cash flow	$(1,000,000)	$125,000	$127,500	$130,050	$132,651	$135,304	$149,387	$164,935

Fig. 42-2 15 Year Cash Flow on $1 Million Investment with $125,000 First Year Net Annual Savings and 2% Inflation Rate.

it depends on selection of an appropriate discount rate (the impact of which can be significant, as shown in the above example), which involves subjective judgment.

When considering two options with the same useful project lives and the same initial investment requirement, the one with the greater NPV would be the logical choice, assuming equal risk. However, when considering two projects that require different investment levels and/or have different useful project lives, additional information, such as what will be done after the life of the shorter project, would be required to make the best selection.

Consider two project options in which Option 1 had an NPV that was twice that of Option 2, but also required an investment that was three times greater. If the IRR of Option 2 was considerably greater than that of Option 1, Option 2 might be chosen despite having a lower NPV. It would tie up less capital (or available debt) and provide greater return per dollar invested. The evaluation of these investment options might also include an analysis of what else could be done with available capital if the lower cost investment is chosen. In this case, the combined NPV and IRR of Option 2 and another investment opportunity would be compared to Option 1. In some cases, the threshold discount rate might be greater for the larger investment. Additionally, the larger capital requirement may preclude another project and the profits or savings lost from that opportunity must be considered in the threshold discount rate.

Another factor that must be applied in capital budgeting analysis is risk rate. Simply put, if two options show the same NPV or IRR, and one is considered to have more risk, the other would be the logical choice. Consider an economical performance evaluation of a cogeneration system option with risk rate impact factors mitigated through manufacturer's warrantees and guaranteed OM&R contracts. Compare this option to capital allocations to a new product line. When using the NPV method, investors usually add a premium to the discount rate to account for the risk, which results in lower discounted values. The new product option would have to display a significantly higher IRR to compensate for the greater risk of under-performance or failure.

SIR is another investment analysis tool used for evaluating energy operating cost savings projects. It is a variation of the NPV method, computed by dividing the PV of returns by the PV of the investment. Whereas NPV combines all benefits and costs, SIR compares benefits and costs. If the resulting ratio is greater than 1, it implies that returns exceed the investment. The greater the SIR, the more desirable the project. SIR is computed as:

$$SIR = \frac{PV\ of\ returns}{PV\ of\ investment} \tag{42-9}$$

Whereas IRR does not indicate the amount of profit to be earned from a project, and NPV does not indicate the amount of investment required for a project, SIR considers both. A greater SIR indicates a greater value of NPV for each dollar invested. Therefore, when several projects are being considered with either limited access to capital (and/or debt) or limited applications, the projects with the highest SIR would be the ones that will maximize the NPV for each dollar invested.

If the values of assumptions in the financial analysis can vary widely, then the results will have a large uncertainty. In this case, it may be worthwhile to perform sensitivity analysis to understand the range of results that an investor could expect. The analysis could go further and apply a probability of occurrence of each scenario resulting in an expected value (e.g. NPV or IRR) rather than a single point estimate. Another option is to use more sophisticate methods like Monte Carlo simulation.

Depreciation

Depreciation of assets is a principal tax consideration in most projects. Depreciation may be considered from different perspectives. The perspectives used in capital budgeting are physical depreciation and tax depreciation as it affects cash flows.

- **Physical depreciation** is a loss of ability to produce future cash flows because of obsolescence or deterioration resulting from wear. As a result of time and use, the value of the asset erodes. Installed energy systems have varying useful lives. While performance may be maintained, the value of the system is decreasing because its useful service life is being used up. At the end of its useful life, the system may have no real value or may have some salvage value.

- **Accounting depreciation** is an internal method of tracking the life of an investment and recovering an initial investment in an asset so that funds will be available to replace it. In an accounting sense, the initial investment in the asset may be considered as a prepaid expense and the accounting depreciation is a method of charging against it over time. It is meant to reflect, in an accounting and fiscal sense, the effect of physical depreciation of the asset.

- **Tax depreciation** is a method of claiming non-cash flow expenses over the useful life of an asset to allow for the recovery of the initial investment so that it can be replaced when it is retired. This allows a portion of the returned earnings to be set aside, free of tax, to serve as a replacement fund.

For tax depreciation, the value of the asset is considered to be used up over a specific time frame in specific, not necessarily uniform, increments. There are several acceptable methods of depreciating assets. These include a straight line method, in which the loss of value is constant and directly proportional to the age of the asset, and various accelerated depreciation methods which allow for greater depreciation during the early years of the asset's useful life. The IRS sets guidelines for the useful life of most equipment. It is usually advantageous to accelerate tax depreciation as much as possible due to the time value of money.

After an asset has been fully depreciated from a tax perspective, it is considered to have no value. Often, however, a system will remain in operation long after it has been fully depreciated, meaning the physical depreciation period extends beyond the tax depreciation period. Any continued earnings are fully taxable since the full value of the asset has already been recovered. Capital gains and losses occur when assets are sold for more or less than their cost. If equipment is sold after it has been fully depreciated, its salvage value is fully taxable. As with all other project factors, tax rates and depreciation schedules will vary widely among different facilities.

Determining After-Tax Earnings

While some simplified project analyses may deal with gross operating profit or pre-tax revenues, the ultimate value of an investment is based on its anticipated net, or after-tax, financial impact on a cash, and sometimes, book basis. The following definitions are useful when considering net financial impact:

- **Gross operating profit** is the difference between total revenue and the sum of all operating costs. This is referred to in accounting terms as **earnings before interest, taxes, and depreciation and amortization (EBITDA)**.
- **Earnings before interest and taxes (EBIT)** is EBITDA minus book depreciation and amortization on the asset.
- **Pre-tax book income** is EBIT minus any interest on any project debt. This is also called **earnings before taxes (EBT)**.
- **Taxable income** is EBITDA minus interest, as well as tax depreciation and amortization on the asset. Often, taxable income will be less than EBT because tax depreciation is commonly accelerated in the early years, as compared to straight-line book depreciation.
- **Taxes** are equal to the product of the applicable federal, state, and local income tax rates and EBT. Current taxes are equal to the product of the applicable federal, state, and local income tax rates and taxable income. Deferred taxes are equal to the difference between taxes and current taxes. Current taxes are owed in the year the income is recognized, while deferred taxes accrue as a liability on the project's balance sheet and are paid during years in which taxable depreciation is less than book depreciation.
- **After-tax cash flow** is EBITDA minus any interest and principal on any project debt, as well as current taxes.
- **After-tax earnings**, or **net income**, is EBT minus taxes.

Summary

The financial analysis tools presented in this chapter provide a broad, but simplistic, introduction to the types of analyses required to make informed investment decisions. Greater detail is available in a wide number of standard accounting and financial analysis texts. Updated information is essential, since tax laws are continually changing. Beyond this basic understanding, no major investment should be made without the review and approval of knowledgeable financial experts.

The financial analyst must be familiar with the specific financial situation of the host facility/investor and must understand all of the key financial implications of the project under consideration, inclusive of all obligations and risks. The type of contract to be used for the execution of the work is a key consideration factor, as is the financial condition of the contractors and any potential partners. Finally, it is important to note that most projects are partially, or even fully, funded with debt. Hence, the type of financing instrument contemplated for the project is of principal importance to the financial analyst. The following chapter presents information on the various types of contracts and financing arrangements most commonly used for the implementation of energy resource optimization projects.

0	1	2	3	4	5	10	15
	$166,666					,181	$219,912
	($146,714)	($146,714)	($146,714)	($146,714)	($146,714)	($146,714)	($146,714)
	$19,952	$43,237	$69,923	$100,076	$133,767	$357,806	$681,515

Chapter Forty-three

Project Contracting And Financing Options

A capital project that is shown to be technically feasible and financially attractive must be contracted for and funded before it can be built. The process of determining what contracting and financing vehicles are to be used is somewhat interactive in that one will influence the other, i.e., certain contracting methods are more or less compatible with certain financing methods and vice versa. Hence, an understanding and evaluation of the wide range of available contract and finance methods is necessary to select the optimal combination of both for a particular type of project.

Once the decision has been made to proceed with a project and funding is confirmed to be available, the next step is for the parties to agree on a contracting method, and then execute contracts necessary to implement the project. The major hard asset related functions of project implementation are design and construction. However, depending on the type of contracting method, there may be one or multiple contracts that are required to be in place prior to the commencement of construction proper. The range of contract options consists of several standard methods, including cost plus, unit price, lump sum, and various turn-key, design-build type arrangements. There are also numerous performance-based options, which can be tied to the above, as well as outsourced third-party ownership and operation arrangements. Generally, the more active hands-on involvement the customer wants, the more likely a cost plus or other open book type methodology will be used. Conversely, a performance-based or outsourcing method will generally shift all or most of the risk of cost control, schedule, and delivery of products and services to the contractor, typically an energy services company (ESCo).

Some facilities choose to fund projects internally through a capital budgeting process, either on an all-equity or a corporate (or institutional) debt equity mix. For larger projects, some form of external funding is typically required. Generally, such debt is part of a project-specific collateralized structured project financings. In the case of many of the performance-based or third-party ownership arrangements, it is the contractor or ESCo that will seek to secure project financing for itself as opposed to for the host facility.

The key factor that influences the choice of financing options is the stability of operating cash flow from the project. This stability is based on: a) the structure of the off-take (revenue) and the fuel and service (expense) streams; and b) the underlying credit strength and historical performance of the counter-parties in each of the contracts. Generally, the credit strength of the off-taker(s) (the user/host and any other parties purchasing off-take) is of paramount importance, as the project is usually implemented on a site- or host-specific basis. Structure aside, the stronger the credit profile of the user/host, the greater the financing options. Structure cannot be ignored, however, as it determines the allocation of risk between the parties entering into a contractual relationship. In its best form, project finance captures the cash flow generated by a project in a secure, segregated legal and financial structure that is creditworthy by its own merits, bankruptcy proof to its ultimate owner, and allocates risk and directs cash flow to the exclusive benefit of the parties associated with the project.

This chapter focuses on contracting methodologies and associated financing options. It also includes a review of the determinants of available financing options, the advantages and disadvantages of each, and the selection of a financing format. The rationale that leads to a selection of a financing format is presented in five steps, which are discussed in detail. Three sample financial offerings are then compared and contrasted.

Project Contracting

The purpose of a contract is to detail the responsibilities of both the owner and contractor, including their rights and liabilities, throughout the execution of a project. A properly drawn contract, in which the scope of work and expectations of the owner and contractor are clearly defined, will minimize the risk to all parties of problems during project execution. The same is true between a contractor and its subcontractors.

In the case of traditional plan and specification contracting, it is usually necessary for an owner to contract with an architect and/or engineer to prepare the contract documents that will be the basis for a project. The same may be true with a design-build contractor that is acting as a prime contractor and seeking those services from an outside firm. The contract documents typically consist of three main components: the completed drawings, the construction specifications, and any other requirements or special conditions that illustrate contractual obligations of both parties. Other items identified include communication protocols, dispute resolution and payment methods.

There are many contracting methods available to an owner/host facility. Each method has its own advantages and disadvantages, all of which should be carefully considered prior to choosing one for a project. With many contracting methods, it is common to utilize prepared standard contracts for professional services that have been developed by the American Institute of Architects (AIA). These standard contracts have been developed and tested throughout the architecture, engineering, and construction industries and are common in versions appropriate for contracting with consulting firms for architectural or engineering services and versions for contractors and subcontractors. The AIA contracts also have developed requisitions for payment forms, which can simplify an owner's review of contractor invoices. With certain performance-based and other types of energy services contracting methods, non-standard customized contracts are commonly used.

CONTRACTING METHODS

Cost Plus

This contracting method can take one of several forms. Typically, it is one in which the contractor agrees to perform the work for direct costs plus a negotiated percentage mark up. The costs are considered to include all materials, equipment, and labor (often not including general supervisory or overhead staff labor). This method can be modified by agreeing upon a fixed fee in lieu of a percentage mark up. A simple form of this contract is also known as a **time and materials (T&M)** contract, in which the contractor passes through the cost of the material (sometimes with mark up, depending on the agreement) and bills the owner for labor provided at its current or negotiated billing rates.

A cost plus contract is a good option in time-sensitive, emergency, or other fast-track situations, when there is not adequate time to prepare complete plans and specifications, especially in the case of very small projects. It may also be a good option when there are many unforeseen conditions, such as tunneling or foundation work. The contractor, in this case, is responsible for maintaining clear and accurate records of project expenditures and providing those records to the owner for payment.

A disadvantage is that this contracting method does not allow for knowledge of the final cost of the project until the project is complete. Not knowing the final cost presents budgeting difficulties and the open-ended risk can negatively impact raising of project funding. Scheduling and other difficulties may also arise due to the uncertainty of the project completion date. These risks can be somewhat mitigated if a cost plus fixed fee arrangement is considered, in which the contractor is obligated to complete the work for the stipulated fee regardless of the final cost of the job. This arrangement will also often motivate the contractor to complete the project as expeditiously as possible, presumably to move its staff on to other projects.

Lump Sum (or Firm Fixed Price)

This contracting method is one in which the contractor agrees to perform a particular scope of work (SOW) for an amount proposed to the owner, typically in the form of a proposal or bid response to a set of plans and specifications. This is advantageous to the owner, as the owner will know exactly how much the project will cost from the outset, excluding any scope additions the owner requests throughout the construction process.

This contracting method places much of the risk on the contractor, as compensation will be based on the amount bid to complete the project, even if the contractor encounters trouble or cost overruns during the construction process. For that reason, it is important that the owner provide completed plans and specifications to the bidding contractor to allow for the development of a comprehensive price estimate. Risk for cost overruns and other liabilities may still reside with the owner (or owner's contracted engineer) for any errors or omissions in the plan and specification package.

Unless all, or at least most, of the details of the project are known and specified, this contracting method may not be appropriate. Unknown details, that are the responsibility of the contractor, will force the contractor to inflate the bid price in order to protect against serious unrecoverable cost overruns. However, this process can work well for the experienced and organized contractor, who can benefit from effective project management by value-engineering cost savings, thereby increasing effective profit margin. This method is generally not recommended for emergency or other time-sensitive situations in which complete plans are typically not developed, or in cases in which there are many other circumstances out of the contractor's control, such as weather hazards. A hybrid version of this approach is to use a lump sum contract with exceptions made for items known to be difficult to estimate in advance (e.g., a road or river crossing).

Unit Price

This contracting method involves the payment on the basis of cost per unit of service. This method is preferable when a project requires the provision of large quantities of similar items, such as stone, concrete, or earth. It can also be used for projects requiring the same type of materials as well as labor throughout the job, such as the replacement of standard light fixtures throughout a facility. This allows a contract to be executed without requiring final drawings and SOW to be completed. It also allows flexibility in adding or subtracting SOW from the original plan, based on unit price adjustments. This may require establishment of some minimum fixed cost base, or adjustment factors that ensure that fixed costs can be recovered if the SOW is reduced. Conversely, a unit price reduction may be appropriate once a certain quantity is exceeded.

A disadvantage to this contracting method is the large quantity of bookkeeping required. It may also not be compatible with highly diversified work, where there are many types of activities, each with a moderate number of units. Also, as with the cost plus method, the owner will not know the project cost until the end of the construction period. Finally, construction limits should be carefully considered when using this approach. Items such as lighting fixture quantities can be clearly estimated and documented. However, complications arise with items such as cubic yards (or meters) of earth excavation for a trench. In such cases the contractor can lay the trench walls back or dig them tight. Both approaches may be acceptable to get the job done but there will be a large difference in the trench excavation quantity. To mitigate against excessive quantities and therefore costs, the contract can provide for a quantity limit for which the contractor will be paid, with quantities beyond that limit absorbed by the contractor. Provisions may also be added to promote value engineering with incentives for successfully accomplishing the work with quantities below an established floor.

Turn-key

Also known as **engineer, procure, construct (EPC)** or **design-build (D/B)**, this contracting method can be used with firms that specialize in providing all components of project engineering and construction. The basis of the contract can be similar to any of the other contracting methods, such as lump sum or cost plus, but in this case one firm will provide all engineering and construction work.

This method can help to expedite a project for several reasons. Because completed plans and specifications are not necessary to allow contractors to bid on the project, the schedule of the project can be compressed by that bidding time that is no longer required. It is also possible for preliminary construction work to begin or long lead-time materials ordered, while final design documents are being developed. There is also one line of responsibility for both design and construction, reducing or eliminating the ability for the party to assign blame to another entity if problems occur.

This approach places the design risk on the contractor (builder). Not only will the contractor offer a firm bid price on the basis of an incomplete design (which will only be completed later by the contractor), but if the design is flawed, the contractor will bear the responsibility for all corrections, presuming the contractor had legitimate access to known existing conditions.

Sometimes, the parties will agree to a **gross maximum price (GMP)** for the project. The actual final price is then determined through some open-book type accounting/reporting from the contractor, but cannot exceed the initial GMP (absent SOW change). This approach mitigates owner risk by establishing a price ceiling. The contractor may seek some financial premium for offering this price ceiling in advance and may also carry some contingency in the price estimate for unforeseen cost exposure. Sometimes, the contracts are structured to provide incentive for the contractor to value-engineer cost reductions by sharing a portion of the underage from the GMP once final prices are trued up. The GMP provision can be applied not only to turn-key contracts, but to several of the other contracting methods as long as sufficient detail exists as to the overall SOW of the project.

Since the work is based on a proposal with incomplete design and specification, the contract must carefully specify the final conditions and minimal levels of performance that are to be achieved by the contractor. Also, with this approach, the owner is placing a great amount of the construction inspection, normally reserved for the design engineer, in the hands of the turn-key contractor. To mitigate risk of substandard or incomplete work, the owner must still maintain (or hire a separate contractor to perform) a vigilant oversight role. Also, to reduce the potential for future claims, special attention must be given to the contractual methodology to add or subtract SOW.

Performance Based

This contracting method places much of the financial risk of a successful project on the contractor. In this case, the contractor typically finances the implementation of a project, or helps the owner procure financing, with little or no upfront cost to the owner. The contractor commonly retains ownership of (or at least some responsibility for) the equipment or systems for a specified period of time, and the cost of the project is repaid, over the contract term, usually by proof of operation, provision of a commodity, or generation of savings.

Sometimes the projects will be paid for following completion and verification of performance. More commonly, however, projects are financed so that the owner can acquire goods and services without the capital usually required, and instead budget the payment to the contractor as an operating cost rather than an upfront capital expenditure. Projects are considered to be self-funding when the savings achieved are sufficient to make the annual finance and long-term operations, maintenance, and repair (OM&R) payments required. If project performance is guaranteed by the contractor, resulting in variability of payments by the owner, the owner may be able to recognize the transaction as **off-balance sheet financing**.

This contracting method is most commonly undertaken with an ESCo that specializes in performance-based contracting and has both the technical and financial capability to provide sufficient guarantees. The method for verifying performance or savings must also be clearly specified and understood by all parties, as this is the largest risk for both parties. Third-party financial institutions that may be providing the project capital to the contractor will also require security of a clear method to verify payment requests.

Performance-based contractors will also commonly assume some or all of the OM&R of the installed equipment or systems during the contract term. This has several benefits, including a reduction of the owner's routine OM&R duties, as well as confirmation for the contractor that the equipment will remain in good working order during the contract term. This ensures that the contractor will continue to receive payment for the work until the project cost has been fully amortized and/or the contract term has expired.

Performance based contracts need not be used only for integrated energy projects and may be used in cases where there are no (or little) savings being generated to fund the project. In such cases the performance element is used as a criteria to measure and ensure that the desired outcome is achieved, such as processing a certain number of units of product in a specified period of time at a specified quality.

Third-Party Ownership and Operation Agreement

In these types of arrangements, a third-party ESCo (or contractor/investor) purchases a facility's existing energy system assets and/or finances, designs, builds, owns, operates, and maintains an energy plant serving one or more host facilities that use the energy supplied. The ESCo may produce electricity, steam, hot water, or chilled water and dispatch them to the host. In cases where electricity is the only output, the agreement is called Power Purchase Agreement (PPA). The host generally makes a long-term commitment to purchase these forms of energy at some agreed upon rate, which usually has both the fixed and variable components necessary for capital construction and operation. Project Financing is then dependant on several factors including a) the off-takers' credit risk and energy consumption profile, b) the price at which the outputs will be sold in the future, and c) fuel price risk.

The structure of the arrangement may be somewhat similar to a performance-based or shared savings agreement, except that the host facility has no form (or need) of ownership and no responsibility for short- or long-term OM&R and future plant upgrade. Savings are usually realized in the form of a reduced purchase cost of energy resources. Often, the rate is tied to energy costs associated with alternative options or the original energy purchasing arrangement at the host facility prior to signing an agreement. The ability to provide energy outputs at a lower cost presumes more efficient use of energy and management resources at the central plant. It may also include arrangements for lower cost alternative energy supplies by ESCo and demand-side energy system upgrade projects within the facility.

There are many types of third-party owned and operated applications. Historically, the most common has been electric cogeneration plants. In such cases, the system is usually tied to the load characteristics of one major host. Electricity is sold to the host facility and excess power produced is sold to the local utility or is wheeled to another buyer. Output products such as steam, hot water, chilled water, and compressed air may be sold to the host and, in some cases, to other nearby facilities or an existing district heating/cooling plant.

In many cases, third-party ownership and operation is tied solely to one host and involves operation of all or most energy systems. In some cases, the host retains ownership of some of the existing equipment, but relinquishes operational responsibility to the contractor. In other cases, the contractor may purchase or design, build, own, operate and maintain all of the on-site energy system assets from the host. In many cases, these arrangements are made in conjunction with a major central plant upgrade.

If the thermal loads at the host facility allow for efficient cogeneration of more power than can be used by the host, the contractor may sell the excess power. With the advent of open access to transmission services for the purpose of wheeling power, sales opportunities are available if the third-party operator can become a low-cost power producer. With the advent of retail wheeling, power may then be sold to nearby local facilities somewhat similarly to steam or chilled water. There is, however, increased competition in the power sales market from other independent producers. As such, the best market for sale of power is usually the host, which will displace power it is purchasing at retail rates. Alternatively, the ESCo may choose not to generate power on site, but to negotiate for lower cost electricity purchases from the utility or from another power provider, or to install standby generation capacity to support purchase of low-cost power on an interruptible service basis.

Facilities will have varying degrees of expertise in operation of energy plants. Third-party firms specializing in ownership and operation of energy plants, however, have proven expertise and bring the resources to effectively design, build, operate, and manage all aspects of plant operation. Because they participate in multiple projects, such firms enjoy economies of scale. The cost of bringing this expertise into a facility-owned plant might otherwise be considered prohibitive. These firms can also absorb certain risks as part of their portfolio that might be prohibitive for a single facility.

Currently, there is a growing trend toward outsourcing of facility energy system operation and sale of energy system assets. ESCos are now offering such services to a wide variety of facilities, both with and without on-site electric cogeneration potential. Potential services range from central plant operation alone, to operation of HVAC, process distribution system, and other end-use equipment, to providing specified climate conditions (e.g., 70°F, 50% RH, 30 ft-candles). As a result of the ongoing deregulation process that is underway in the natural gas and electric industries, simultaneous to the outsourcing trend is a trend toward aggregation of load by ESCos. The aggregation of multiple host sites allows for energy commodity acquisition with leverage of greater buying power. Finally, this allows the host facility to more fully concentrate it financial and management resources on its core business or mission.

ANTICIPATING AND DEALING WITH CONTRACT PROBLEMS

Even with a well conceived and properly executed contract, each party must be prepared to address problems that can and often will arise during the execution of a project to ensure that the work is completed expeditiously and in accordance with the contract documents. Issues particular to project completion, payment of funds, liability, and warranty are typically detailed in the general conditions section of the project specification. These conditions should clearly delineate responsibility of the owner, contractor, and any subcontractors.

There are several layers of responsibility in the typical project; that between owner and engineer, owner and contractor, and contractor and subcontractor. Interwoven throughout the contractor and subcontractor relationships is also generally input and inspection by the engineer of record. Because of these layers of responsibility, groups that may not be party to a dispute can be unfairly affected by those disputes.

Withholding of Funds

With most contracting methods, the owner holds the largest clout, that being the funds for payment. With varying degrees of success, withholding payment for incomplete work may be the owner's most effective means of getting a contractor's attention. However, the effectiveness of this can largely depend on the amount of money still outstanding to be paid to the contractor, assuming that progress payments have been made during the course of the project.

As a project nears completion, the amount of money outstanding can be so small that a contractor may be willing to forfeit this payment in lieu of performing the project close-out, or addressing problems in equipment or system operation. Because of this, it is important for the owner to specify a retainage of the total project cost, expressed as a percentage, in the contract documents. This percentage is withheld from each progress payment and only paid after final project close-out and completion of punch list items. Retainage generally falls in a range of 5 to 15% of a contract amount and is meant to equate (roughly) to the contractor's profit. Thus, the contractor would not be inclined to walk away in the above example, as the owner would use the retainage to hire another contractor to complete the work. The original contractor would forfeit all or at least a substantial portion of its entire gross margin, as well as the margin on what work it failed to complete.

Another limiting factor is the owner's risk exposure for incomplete work. This could be loss of production time or inability to maintain space conditions or support critical mission activities or the loss of expected energy cost savings from an efficiency project. In such cases, withholding money does not necessarily result in getting the project completed on schedule. An extreme result of lack of contractor performance can be termination of the contract. This can be done for a number of reasons, including lack of performance, negligence, or failure to adhere to project rules and regulations.

It is, however, important to note that the contractor that has performed the work as planned and has completed all punch list items rightly deserves prompt payment of all outstanding monies owed, including retainage. In order to protect the contractor from undue withholding of final payment, it is important for the contractor to ensure that the contract documents call for payment of outstanding invoices if the owner does not respond to a final written request for payment within a specified period of time. If necessary, this response should take the form of an updated punch list of problems or issues that the owner feels have not been addressed. If no such issues exist, this clause will provide the contractor some legal weight against the owner to demand due funds.

Subcontracting Relationships

It is important for the owner to know in advance if a contractor plans to utilize subcontractors to perform portions of the work. The general conditions of the specifications should state that the owner has the right to approve (without consent being unreasonably withheld) the use of any subcontractors on the project. The owner should also be aware that problems between the contractor and the subcontractor can adversely affect the completion of the project.

In most instances, the contractor will not compensate its subcontractor until it has first received a progress payment from the owner. Therefore, in the instance that the owner delays payment to the contractor, it is possible that the subcontractor will also see a delay in payment from the contractor, even if the dispute does not involve the subcontractor's work. This can lead to a suspension of the work by the subcontractor, and a further delay of the project. Ultimately, the contractor has responsibility to perform the work according to the schedule, and the line of authority places the weight of contract enforcement with the owner and contractor.

It is good practice for large projects to require the contractor to provide a payment bond. This will help to ensure that the contractor is compensating its subcontractors for work they have performed on the project. The bond will also help to protect the owner from recourse that a subcontractor may take against the owner to demand payment. It is typical to require a payment bond in the amount of at least 50% of the contract price.

Protection from Non-Performance

A **performance bond** is a prudent means of protecting the owner from non-performance of the contractor. Should the contractor default on execution of the work, the owner will have the ability to complete the project using another contractor and the owner will be protected from the risk of cost overruns due to the need to bring a new contractor onboard mid-stream. This is, of course, provided the performance bond has been purchased in an appropriate amount. It is typical to require a performance bond in the amount of 100% of the contract price.

The contractor may, in turn, also require performance bonds from its subcontractors. It may also wish to assign other risk items that it has obligated itself to onto its subcontractors, including **liquidated damages** for schedules and even equipment performance.

The owner/host facility can also protect itself from financial loss by requiring the contractor to pay liquidated damages in the event of failure to meet performance standards or a contracted schedule. In addition to mitigating risk of substandard performance or non-performance, it can protect against project completion delays. The prime contractor may, in turn, also use liquidated damages provisions to protect itself from similar issues with its subcontractors. In some cases, delays can have a significant economic impact on the host facility or the prime contractor. As such, passing some or all of this risk (at least for critical path items) onto contractors and subcontractors in the form of liquidated damages, may be a prudent and reasonable approach.

In contracts in which liquidated damages are used as a method to compensate for a lack of performance (either technically or regarding schedule), the liquidated damages can be backed either by a performance bond, corporate guarantee of the contractor or its parent company, or a letter of credit. Generally, a direct pay letter of credit is the most secure vehicle to use, as it allows the owner immediate access to funds held at a bank in the event the project does not perform and an independent engineer certifies as such. Generally, liquidated damages are in the range of 20 to 40% of a contract amount (though in certain situations it may be greater), depending on the type of technology used and the track-record of the contractor.

Project Financing

Five Evaluation Steps

There are five basic steps in selecting a financing format. The first three steps profile the requirement and the last two match the financing to the profile. The steps are:

1. Evaluating the credit worthiness of the user/host facility.
2. Determining the amount and volatility of cash flow generated by the project under various economic conditions.
3. Identifying the tax considerations, risk implications, and compatibility with financial goals and strategies.
4. Matching the available financing formats to the cash flow and its stability, risk tolerance, and strategic goals.
5. Identifying the financing sources for the preferred financing format and selecting a source.

1. Credit Worthiness

In all cases, underlying credit governs the cost of the financing and the available financing options. How the market evaluates the credit of the host/user is critical to the efficiency of the financing.

Credit evaluation considers the credit contribution of the particular project and, more important, the credit of the host/user. Usually, an energy savings project will not be pivotal to the credit of any going concern. If it is, however, that concern's financing options become restricted to structures that protect the creditor from the general risks of the business and rely on the economics of the project. The usual circumstance is that the host/user has an established credit history and current risk profile and that the project's impact on the credit will be inconsequential. However, in certain cases, it may be acceptable to put in place a project with a host that many consider non-creditworthy. The beauty of structured finance lies in its ability to structure around what would otherwise not be acceptable from a pure corporate credit sense. There are many examples in which a project has a better credit than the party it sells the majority of its off-take to.

Essential elements of credit analysis are measurements of the adequacy of capital and cash flow and collateral value. Capital and cash flow are measured by financial statement analysis. Collateral value is determined by the recovery value of the assets securing the financing, usually, but not limited to, the equipment of the project.

The evaluation of established credits can be aided by independent credit rating agencies. These agencies include Moody's Investor Services, Standard & Poor's, D&B, and other industry and trade specific reporting and credit rating services. It is important to note that these credit rating agencies are not infallible. They have made some significant errors in judgment on both sides of the norm. They are a good yardstick, but must not be used as the sole determinant. In addition, an investor's ability to structure a deal in the manner most desired may be inversely proportional to the credit of the host. There are many cases in which poor deals have been done with stellar-credit hosts. One reason might be that the host takes advantage of its position and commoditizes the energy provider, making the provider look like any other vendor. The art of the deal is to structure a transaction that is a win-win situation for all parties.

Lenders and investors will rely upon the ratings of these independent agencies to provide initial screening and supplement that data point with their own evaluation of the host/user's industry, its market trend, current circumstances specific to both the customer and its sector, and an analysis of recent financial statements. This analysis will confirm or question the standing ratings of the rating agencies. Credit ratings tend to trail the events within an enterprise, so an analysis of financial statements, interviews with management, and checks within the industry or trade updates the rating and highlights any deterioration or improvement in the situation. For credits that are not covered and reported by the agencies, a special private review may be requested by the prospective lender/investor or, if the lender/investor has sophisticated credit analysis resources, it will categorize the credit using the framework of the rating agencies.

Credit ratings and credit history are the first determinant of the financing options available. Higher grade credits are immediately aware of their broad range of choice. Lower grade credits must look to a financing structure that secures itself by the cash flow generated by the project, the available collateral, and warranties and guarantees of the contractor and third-party investor/owner.

The value of equipment as collateral ranges from good to worthless. If the equipment is chattel, can be easily removed, has an established secondary market, and a perfected lien can be established, the credit can be enhanced by using it as collateral. Standby generators are generally viewed as good collateral. Circumstances which make equipment less attractive as collateral are when the equipment is real property, has a high labor cost in its installation, is difficult to remove, or cannot be attached with a clear lien. The acid test of collateral value is the answer to the question "what happens in bankruptcy scenario?" This is a special case, where projects are concerned, in which the host or off-taker "goes away." The ability of a project to compete in the open market is of paramount importance in this scenario.

The ratings, updated evaluation, and the additional security of collateral are the basis for the pricing of the financing. The perceived ranking of the host/user's ability to pay on a commonly understood quality spectrum becomes the determinate and justification for the magnitude of the fees, interest rate, and costs imbedded in the financing.

Within the banking and institutional lending community, there are prevailing funding rates relating to the grades of credit risk. While transaction size, investment liquidity, market recognition, and other factors affect pricing, the ranking of a credit within the "investment grade" range of risk determines the "spread" which the financing must yield over a transaction of comparable maturity or term that is deemed riskless. Risks that are determined to be less than "investment grade" are categorized as "high yield" (or the colloquial term "junk") credits and carry substantially higher financing costs, which reflect not only the market's risk appraisal, but the cost of more complex financing structures as well. If the owners are government entities (e.g. Federal, State, County), lenders pay special attention to non-appropriation language which specifies that the owner is not obligated to appropriate funds to pay for the debt service for the project in its annual budget. The lender might accept this type of language if they feel comfortable with the credit profile of the owner, their capacity and their history of making payments.

2. Project Cash Flow

The previous chapter discussed project cash flow and the various methods by which cash flow is used to evaluate the project's financial impact. The financial analysis tools presented in that chapter, such as NPV and DCF, consider present value (PV) of future cash flows over the life of a project at a given discount or cost of capital rate. These are tools of capital budgeting analysis. The tools of credit analysis are more pragmatic: they measure the host's ability to pay for the project.

For project financing, cash flow is usually expressed in monthly amounts, which can be targeted to offset projected financing costs. For any given financing vehicle, the financing rate will generally be lower for more credit worthy facilities. Also, a longer financing term will produce a lower monthly financing payment, although the yield will likely increase since the uncertainty of payment in out years would increase.

To produce positive cash flow, it is therefore necessary to extend the finance term to the point at which the combination of monthly financing and operating costs still produce net monthly savings. However, the length of the financing term may be limited by several factors, most importantly the useful life of the equipment and/or the life of the contract; the latter is usually the limiting factor. Generally, however, if the host/user can demonstrate credit worthiness and the project shows enough positive cash flow after operating costs, project financing can usually be obtained. A simplified expression of monthly project cash flow, excluding all non-cash charges, is:

Cash flow = Monthly incremental project savings –
Monthly incremental operating & financing costs

In the previous chapter, methods of calculating fixed payment requirements for financing were shown using a capital recovery factor (CRF). CRF may be calculated using Equation 42-7, found in CRF look-up tables, or derived from computer spreadsheet functions. Consider the example of a facility seeking to borrow $500,000 to finance a project. At a 9% simple annual interest rate, the monthly repayment amount would be $22,842 for 24 months, $15,900 for 36 months, and $12,443 for 48 months. Presuming total monthly incremental savings is an operating cost savings of $12,000 in energy usage and $1,500 in OM&R during the first four years, simple monthly cash flow calculations are as shown in Table 43-1.

Term	Energy Savings	+	Net OM&R Savings	-	Financing Costs	=	Cash Flow
24 months (negative)	$12,000	+	$1,500	-	$22,842	=	($9,342)
36 months (negative)	$12,000	+	$1,500	-	$15,900	=	($2,400)
48 months (positive)	$12,000	+	$1,500	-	$12,443	=	$1,057

Table 43-1 Monthly Cash Flow Analysis for Various Financing Terms at 9% Interest Rate.

As shown in Table 43-1, with a financing term of 48 months or longer, the project would show monthly positive cash flow. In a sense, the cost of project financing is buried in the existing expense budget. If a project can produce positive cash flow, a third-party financing format allows a facility to use dollars, itemized as expense, to help implement capital projects that might otherwise not capture limited capital investment budget dollars.

If, in this example, the simple annual interest rate were increased to 15%, the project would show negative cash flow, or shortfall, in each of the three finance term options considered in Table 43-1. As shown in Table 43-2, a longer finance term of 60 months would be required if the project is to yield a positive cash flow in each monthly payment period.

Term	Energy Savings	+	Net OM&R Savings	-	Financing Costs	=	Cash Flow
48 months (negative)	$12,000	+	$1,500	-	$13,915	=	($415)
60 months (positive)	$12,000	+	$1,500	-	$11,895	=	$1,605

Table 43-2 Monthly Cash Flow Analysis for Various Financing Terms at 15% Interest Rate.

It is important to note that extending the financing term usually increases the interest rate applied to the financing. If, in this example, the simple annual interest rate was 13% for a 36 month term, 14% for a 48 month term, and 15% for a 60 month term, the net cash flows would be as shown in Table 43-3.

Even though the simple annual interest rate increases with increasing term, net cash flow increases in the positive direction more rapidly. Thus, the 60 month term, which has a simple annual interest rate of 15%, shows positive cash flow, while the 36 month term, which has a lower simple annual interest rate of 13%, shows negative cash flow.

Projects that do not produce positive cash flow during the financing term may still be desirable and financeable, depending on the circumstances:

- Projects that are required by a facility, such as the replacement of a retired system, may produce a level of cost savings that only partially offsets the financing costs. The new system alternative with the lowest negative cash flow will generally be selected. If the facility is credit worthy, these types of projects are financeable, even though they are not self-funding. While savings may not be seen in existing operating budgets, an objective view recognizes the required replacement cost as part of the financial baseline.
- Positive cash flow financing enables the facility to cover finance costs with project savings, but may result in excess interest costs. Depending on the financing strategy of the facility, projects that produce negative cash flow during the financing term, but positive cash flow after the term is complete, may be desirable. These projects may still be considered as self-funding, as long as they produce a positive NPV over the life-cycle of the investment. Here again, the credit strength of the host/user must support the financing.

If, in the example shown in Table 43-3, one were to extend the cash flow analysis for the scenario in which the project is financed at 13% over 36 months to 60 months, it would produce a cumulative positive cash flow of $203,508. This is more than double the $96,300 of cumulative cash flow produced with the 60 month financing term arrangement. While these differences might support the notion that the shorter financing term is superior, not considered is that fact that with the shorter term, the project must endure several years of negative cash flows.

The above examples are simplistic and are meant to show only short-term on balance sheet projects. Numerous other factors come into play when energy outsourcing contracts (i.e., projects that provide services to

Term	Energy Savings	+	Net OM&R Savings	-	Financing Costs	=	Cash Flow
36 months (negative)	$12,000	+	$1,500	-	$16,847 (@13%)	=	($3,347)
48 months (negative)	$12,000	+	$1,500	-	$13,663 (@14%)	=	($163)
60 months (positive)	$12,000	+	$1,500	-	$13,663 (@15%)	=	$1,605

Table 43-3 Monthly Cash Flow Analysis at Various Financing Terms and Interest Rates.

a host customer, but are owned by other parties and are not treated by the host as a long-term capital obligation) are used.

3. Tax Position, Risk Adversity, and Financial Goals and Strategies

Financing strategies have different tax and accounting consequences. Because balance sheet analysis is used when obtaining credit, the tax and accounting consequences of any financial obligation may become an important part of the choice of financing vehicles.

In formulating a financing strategy, a primary determination is usually whether the project cost is treated as a capital budget item or an expense budget item in internal accounting. If it is to be treated as a capital budget item, various types of loan and lease/purchase options are considered. If it is to be treated as an expense budget item, rental or true lease agreements or various types of energy services agreements are considered. Consider the following examples:

- A rental agreement implies no ownership on behalf of the renter. The renter wishes to use equipment for a period of time, after which the owner may rent the equipment to someone else. The payment of an equipment rental becomes an expense item that is deducted from a company's profits before taxable income is determined.
- As opposed to a rental agreement, a loan agreement used to finance a project implies ownership transferring to the user or borrower. The asset can be depreciated over its useful economic life for both bookkeeping purposes and tax purposes, and the interest incurred in the transaction can be treated as an expense item.

In the previous two examples, ownership is fairly straightforward. There are, however, many financing agreements in which ownership determination is not so straightforward and may be subject to interpretation. For example, an operating lease, which is a long-term (up to 80% of the useful life of the asset) rental agreement with a fair market value purchase option may, in certain cases, be treated as an operating expense item to the lessee for book purposes, but as ownership for tax purposes.

These considerations also become important to institutions and corporations, which for organizational reasons, charter restrictions, or other internal reasons cannot or choose not to include the project in the capital budget. For example, a school board may be able to execute a lease agreement, but not increase the commitment to physical plant without voter approval. In such cases, the lease option or some variation becomes the choice financing format. Should a company have loan covenants with preexisting lenders, leaving an asset on or off the balance sheet often becomes a threshold decision whether or not to undertake a project. In other cases, even if the treatment results in an off-balance sheet treatment, some rating companies might include the project in their debt analysis of the owner. In all cases, accountants or tax counsel should be consulted to determine appropriate tax and accounting treatment for financing vehicles.

Optimize Depreciation Benefits

To optimize effectiveness of financing, the tax position of the user/host and that of lender/investor must be considered. If a facility cannot effectively utilize depreciation to reduce its tax position, it will likely seek to assign ownership to the lender/investor or another party. By assigning ownership to the party that can best utilize the depreciation, total financing costs are minimized. The desire to keep a project off of the balance sheet may lead a facility to accept a higher financing rate.

Risk Versus Reward

A facility with a high level of risk adversity may lean toward financing vehicles that are more costly, but provide installation cost or savings guarantees. Many host facilities wish to minimize all project responsibilities. These facilities will trade the opportunity for greater profits for fewer responsibilities, lower risk, and avoidance of ownership.

On the other side of the spectrum are facilities that seek the lowest-cost financing vehicle and assume the full range of ownership and operational risks and responsibilities. These facilities usually have a high degree of technical and financial expertise.

Terms of the Financing

Another factor is the length of the financing term. Many facilities will seek the longest possible term in order to maximize annual positive cash flow. Others will seek a short financing term, sometimes even with a negative cash flow, in exchange for reducing overall interest costs.

4. Financing Formats

Numerous financial vehicles may be used to finance energy projects, ranging from conventional bank loans to energy savings performance contracts (ESPCs). Each will differ in the way the payment is calculated, in tax consequences and accounting treatment, and in the amount of risk assumed by each party. As a general rule, the greater the risk assumed by the investor/lender, the greater the cost of funds will be.

Traditional and Non-Traditional Financing Sources

Traditional lending sources consider traditional risks using traditional financial vehicles, such as leases, loans, and mortgages. Conventional financial instruments usually have a payment schedule based on a pre-agreed term and interest rate. Interest rates may be fixed or variable, with both usually tied to one or more indices of money rates, which are uniformly reported by financial press. Floating rates generally key off of the London Interbank Offer Rate (LIBOR) and fixed rates off of U.S. Treasury Instruments, each with a term equal to the average life of the loan being considered. Conventional financing sources are risk adverse, meaning greater risk will demand greater reward in the form of higher financing rates. The cost of debt will be largely dependent on the amount of cash flow coverage the project provides and the strength of the balance sheet.

Conventional lenders may or may not factor in, understand, or care about the relative impact of project performance risk. The universe of banks that will even undertake any form of asset-backed or project financing is small. Some ESCos or project teams, which consist of a combination of an ESCo and a conventional lender, fully understand both credit and project performance risk and factor both into the ultimate cost of financing.

Non-traditional sources accept higher risk and may use non-traditional financing vehicles, such as shared savings and other energy service agreements. Other types of non-traditional sources include state government or utility-backed financing instruments. Financing is made available from these sources for projects in order to help accomplish other objectives, such as business retention.

Conventional Financing Vehicles

The following is a list of general conventional financing vehicles, with brief descriptions of each:

1. A **commercial loan** is a formal agreement in which a lender provides a borrower with funds for a stated purpose. It is backed by the full faith and credit of the borrower. Loans may be secured by the particular asset being purchased, and/or with additional assets of the company, such as accounts receivable, property, or other tangible assets.
2. A **conditional sales lease** (sometimes called **capital lease)** agreement is an agreement for financing the purchase of an asset in which the borrower is treated as the owner of the asset for federal income tax purposes, but does not become the legal owner of the asset until the terms and conditions of the agreement have been satisfied. This entitles the borrower to the tax benefits of ownership, such as depreciation.
3. An **operating lease** (also called true lease) is a lease that has the financial reporting characteristics of a usage agreement and also meets certain criteria established by the Financial Accounting Standards Board (FASB) and or the IRS. Generally, an operating lease is not required to be shown on the balance sheet of the lessee. Operating leases also include conditions in which the lessor has taken a significant residual position in the lease pricing and, therefore, must salvage the equipment for a certain value at the end of the lease term in order to earn its rate of return.

   A **true lease** is a long-term rental agreement that contains fair market value purchase options, which may be exercised by the lessee at the end of the lease term. Since it is unknown if the equipment will ultimately be purchased, rental payments are usually expensed for tax purposes.

Public Sector Financing Vehicles

The following is a list of general financing vehicles available for the public sector, with brief descriptions of each:

1. A **federal lease** is a financing alternative for federal government entities. The federal government usually provides the terms and conditions of their financing transactions and customarily contains termination-for-convenience clauses, which provides the ability to cancel the lease.
2. A **General Obligation (GO) Bond** is the traditional municipal bond vehicle. These bonds are backed by the full faith and credit of the state or local government. Unlike revenue bonds, repayment is not tied to user fees or other income derived from the use of the financed facility.
3. A **Private Activity Bond** is a municipal bond vehicle that allows for private ownership. While such bonds are similar to GO Bonds in that they are backed by the full faith and credit of the state or local government, repayment is tied to user fees or other income

derived from the use of the financed facility. Also, the private investor must obtain enough volume cap to cover the total amount of the tax-exempt bonds required. Volume cap is available each year and is allocated by the municipality amongst its various constituent needs, such as schools, low-income housing, waste management, etc. Thus, the politics of obtaining volume cap are not insignificant.

4. A **municipal lease** is a conditional sales type contract packaged in the form of a lease, which is available only to municipalities, states, counties, and certain special authorities. Tax-exempt interest accrues to the lessor, resulting in financing rates that are lower than conventional commercial lease rates. Since the lease structure can direct operating funds into capital purchases, a municipal lease provides a cost-effective alternative to floating a bond issue. Unlike a conventional lease, which is non-cancelable for the lease term, a municipal lease is usually subject to an annual approval of a budget appropriation clause to provide for the continuance of lease payment. This appropriations clause typically allows the municipality to return the leased asset to the lessor without penalty prior to the end of the lease term should funding become unavailable. The essential nature of the equipment is the lessor's protection against this termination provision (e.g., chillers installed in a municipal facility).

Non-Conventional Utility- or State-Sponsored Financing Vehicles

Generally, these types of financing arrangements are designed to support various utility or state objectives. Often, these instruments are similar to conventional financing vehicles and may be provided at below prevailing market rates and be available for host facilities with higher risk credit profiles.

- **State-sponsored financing vehicles** are often provided as part of state economic development efforts, usually through an agency designated as the Economic Development Authority. Loans are provided in order to retain or expand businesses currently operating in the state or to attract new business to the state. Business retention financing may be provided to important community employers that are either facing bankruptcy or being enticed to move their operations to another state. The benefit to the state is not based on the profitability of the financing, but on the future tax revenues or the improved economic vitality of the state that results from the project.
- **Utility-sponsored financing vehicles** are one of many types of incentives offered for energy system installations as part of utility demand side management (DSM) or marketing programs. Low-cost or even no-cost financing may be provided to customers installing energy systems that either strategically reduce or expand loads, depending on the utility's objectives. Marketing programs are somewhat similar to state-sponsored economic development financing programs in that they are designed to retain or expand loads of current customers or to attract new customers into the service territory. In many cases, customers must provide guarantees of a certain amount of usage as part of the agreement. DSM-based financing programs are designed to encourage various types of energy efficiency projects which serve the utility's integrated resource planning objectives. With the advent of utility deregulation, many of these programs have been cut back or eliminated.

Project Financing Enhancements

Borrowers, contractors, and lenders should always be up to date on the latest available incentives that would improve the economics of a project. Some of these are direct payments (e.g. grants, tax credits, rebates) provided by government entities to promote energy efficiency projects. Others are dependant on the outputs of the project including production tax credits. In some cases, market conditions might help the project by providing higher purchase rates for its output due to state requirements for renewable portfolio standards. Some of these benefits might only be available to the owner of the project which puts additional emphasis on the optimum ownership structure of the project.

Non-Conventional, Performance-Based Financing Vehicles

Guaranteed or shared savings performance-based financing provides low risk and responsibility to the host facility and higher risk and responsibility to the contractor/investor. Consequently, they provide lower net financial benefits to the host facility and greater financial reward to the contractor/investor. Performance-based contracting is a specialty field, which generally consists of large ESCos that have the full range of technical and financial resources to analyze, develop, design, install, maintain, and manage large projects. Performance-based contracting is also done by smaller design-build engineering firms in joint venture with firms that have substantial financial resources.

The following is a list of non-conventional performance-based financing vehicles, with brief descriptions of each:

- **Guaranteed** or **insured savings agreements** are programs that utilize the savings of energy cost generated by the energy efficiency equipment to service a third-party financing of the equipment. The financing entity may be the installing company or an institutional investor. The energy savings are guaranteed by either the installing company and/or an insurance program underwritten by a top rated national company. Guaranteed savings agreements call for a pre-arranged, one-time, or ongoing savings verification program.
- A **shared savings agreement** is an arrangement in which the contractor/investor installs the equipment and receives payments from the end-user/host facility based on the energy savings produced by the equipment. Shared savings agreements are sometimes considered to be off-balance sheet financing, which often makes them attractive.

Two general types of agreements are fixed and true shared savings contracts.

- **Fixed shared savings contracts** define the savings based on a one-time savings analysis. The savings-based payment is not subject to change following acceptance by the customer. This may be based on an analysis prepared prior to construction, a verification analysis at the time of project completion, or one year after project completion. These are somewhat similar to many standard guaranteed savings agreements.
- **True shared savings contracts** require the savings to be verified periodically (e.g., monthly or yearly) and the savings-based payment is subject to change based on the period measurement of savings.

To proceed with a shared savings agreement, the third-party, contractor/investor, or financier is required to provide a detailed program for verification and guarantee of the savings revenue stream. Funding for such a guarantee can be provided in the form of a letter of credit or by posting a performance bond for an amount that would equal the customer's share of the savings over the term of the contract.

One of the keys to ensuring optimization of these types of energy cost savings programs is effective projection and validation of savings to ensure maximum financial returns. In addition, if recommended measures are not properly designed, then the use of poor quality components, technical obsolescence, early retirement of equipment, and excessive maintenance can erode the savings stream over the life of the project.

For Public Sector owners, current enabling legislation in numerous States allows them to contract with an ESCo and fund the project using several alternatives (e.g. tax-exempt lease, capital funds, bonds). One benefit of this is that the customer is not obligated to select the contract purely based on low cost but being able to take into account other factors like value of proposed solution, the ESCo's experience, and the ESCo's credit risk.

Savings Verification Programs

Guaranteed/shared savings agreements are predicated on some type of agreed upon savings verification process. The type of verification processes vary widely, but all of them include the development of a baseline and a methodology for determining savings. Methods used for savings verification programs are discussed in Chapters 41 and 44. One-time verification agreements result in lower overall project annual costs because they eliminate management fees and analytical procedures required with ongoing savings verification programs, as well as higher costs associated with the ongoing risk on the part of the contractor/investor.

In a true shared savings arrangement, the contractor/investor will generally take on maintenance, and in some cases operational responsibilities. Since the contractor/investor must produce the amount of savings guaranteed over the life of the contract, it is necessary to have a sound element of control over equipment performance. This increases to project costs. The percentage of savings provided to the host facility will also be lower as a result of savings metering and verification (M&V) procedures incurred by the contractor. The M&V procedures used will vary depending on the specific technology application and the level of certainty agreed upon by the contractor/financier and host/customer. M&V procedures based on periodic metering or inspection and standardized engineering calculations are relatively inexpensive, while procedures based on continuous monitoring and sophisticated regression analysis can be very costly.

Whether shown in the analysis or not, the host facility will often benefit from additional savings resulting from the elimination of previous responsibilities. The actual cost of OM&R may increase or decrease. When new equipment replaces old equipment, there is the presumption that OM&R costs will decrease. However, in many cases it increases. If an electric cogeneration system is added to a facility and the existing equipment is kept in place, or if a more OM&R intensive technology is installed, such as an engine-driven chiller, the cost will likely increase. Still, the most accurate approach is to identify the baseline OM&R costs prior to the project and subtract the post-implementation OM&R costs, regardless of who performs the work, to yield the net change.

5. Compare Available Financing Options and Select a Source

Public Sector Facility Options

In today's market, the two most common financing vehicles sought by public sector facilities are municipal lease/purchase options and performance-based (or shared savings) options. These are commonly solicited in the form of complete project financial offerings through public notice requests for proposals (RFPs). In some cases, the RFPs specify the exact SOW and the specific financing vehicle sought. In other cases, the RFPs are general and contract/investors are requested to make their own recommendations as to the physical scope and/or the financing vehicle.

Many RFPs lend themselves to integrated projects in which shorter payback measures, such as lighting retrofits and high-efficiency motor programs, are blended with longer payback central plant modifications that might not be self-funding. In some cases, the utilities will play a significant role in the process through special programs designed to assist public sector facilities seeking energy efficiency improvements.

Both municipal lease/purchase and performance-based arrangements may allow projects to be self-funding and, therefore, do not require capital appropriations. The attractiveness of a performance-based energy savings program is that the third-party financier assumes the financial risk of all construction costs and only receives payment based on the energy savings generated. The attractiveness of a municipal lease/purchase option is that, for qualifying borrowers, it probably provides the least-cost financing. In addition to the low finance rate, advanced funding for all construction and implementation costs can be arranged with the tax-exempt funding source.

As opposed to a performance-based agreement, a tax-exempt lease purchase agreement is generally seen as a somewhat greater financial risk because it must generate sufficient energy savings to fully amortize the lease cost over the term of the financing program. However, mitigating risk is the substantially lower cost of financing and avoidance of additional fees, which results in significantly increased debt coverage. This may allow a project to show positive cash flow over the shorter financing term, while the performance-based contract term must be extended far longer to produce positive cash flow.

Despite the lower rate of return to the host facility over the life of the project, performance-based arrangements are becoming a major source of investment in the public sector. From an individual facility's perspective, the amount of savings returned is not always the overriding concern. Often, facilities are in need of a complete overhaul or significant equipment upgrade and also wish to downsize the scope of future internal OM&R responsibilities. Having a third party handle all installation and operating aspects of the project takes a tremendous burden off of the facility's management.

Private Sector Options

Projects in the private sector are usually not developed through an RFP process, but through the normal course of business. Project developers may be brought in by physical plant managers to make proposals to their own management or may be brought in by utility marketers or DSM program administrators. In some cases, senior management will bring in a contractor/investor with whom they have had a longstanding relationship.

Financing rates for both conventional and performance-based arrangements will vary considerably. Facilities that have very solid credit can achieve rates only slightly higher than those available through municipal lease/purchase arrangements, while less creditworthy facilities may face rates that are more than double.

As opposed to many public sector facilities, private sector facilities will typically seek greater financial returns on a project. While central plant upgrades and outsourcing of OM&R responsibilities are compelling benefits, the bottom line increase of profits is a more clear focus. This will often lead to the selection of a less comprehensive performance-based contracting arrangement, rather than a true shared savings agreement (i.e., one with a one-time savings verification guarantee). The appeal of a true shared savings arrangement is often the off-balance sheet finance structure. Additionally, more comprehensive services offered by the third-party contractor/investor allows the facility to more clearly focus on their own core business.

SIMPLIFIED FINANCING EXAMPLES

Following are a series of simplified examples of project financing for a $1,000,000 project that produces a first year savings of $166,666 with a simple payback of 6 years. In these examples, first year savings are projected over a 15 year project life at a 2% escalation rate. Construction financing is required over the course of one year and the cost of construction financing is rolled into the total amount financed. Due to the timing of construction and invoicing, it is assumed that construction financing is based on financing the full project cost over half of one year. At an interest rate of 10%, the total amount financed would be $1,050,000. Assuming a financing term of 10 years, the 15 year project cash flow would be as shown below in Table 43-4.

The assumptions in these examples are for illustration of project financing calculations. Since many financing parameters can vary over a wide range, change frequently, and can have a significant impact on cash flow, actual project financing calculations should use real, documented values for market interest rates, tax rates, energy rates, and inflation when modeling project financing. It may also be worthwhile to run sensitivity analyses, varying these parameters individually and in combination to test for threshold values where financing and cash flow conclusions might shift.

Notice that the project shows negative cash flow in the first three years, though ultimately a cumulative 15 year positive cash flow of $1,173,395. Despite the significant cumulative positive cash, a project such as this should be viewed cautiously for several reasons. First, almost all of the positive cash is achieved in years 10 through 15, which, when considered in present value dollars, would only be a fraction of the total amount. Second, the 2% escalation rate is somewhat speculative, though it is not considered overly aggressive. However, one must also consider that, absent of performance guarantees, savings tend to erode due to the equipment performance degradation over time. If, for example, no annual energy cost escalation was applied, the project would show negative cash flow throughout the 10 year finance term. Given the marginal NPV of this project under the proposed financing arrangement and the risks involved, the project would have a fairly high likelihood of being rejected.

Alternatively, in order to produce positive cash flow in every year following project implementation, the finance term might be extended to 15 years. Assuming the interest rate increases to 11%, the 15 year project cash flow would be as shown in Table 43-5.

Notice that with a 15 year term, even with a higher interest rate, the project now produces positive cash flow in every year, though the cumulative cash flow over the project life is reduced by about one-half million dollars. Cumulative project cash flow is reduced due to both the higher interest rate and the longer finance term. Still, in many cases, the prospects for a self-funding project that generates positive cash flow starting in the first year may be more attractive than the prospects for greater total cumulative positive cash flow. If analyzed with no energy

YEAR	0	1	2	3	4	5	10	15
Annual Energy Cost Savings		$166,666	$169,999	$173,399	$176,867	$180,405	$199,181	$219,912
Annual Financing Cost		($170,883)	($170,883)	($170,883)	($170,883)	($170,883)	($170,883)	
Annual Cash Flow		($4,217)	($884)	$2,516	$5,984	$9,522	$28,298	$219,912
Cumulative Cash Flow		($4,217)	($5,101)	($2,584)	$3,400	$12,922	$116,116	$1,173,395

Table 43-4 15 Year Cash Flow Analysis for $1 Million Project Financed at 10% over 10 Year Term.

YEAR	0	1	2	3	4	5	10	15
Annual Energy Cost Savings		$166,666	$169,999	$173,399	$176,867	$180,405	$199,181	$219,912
Annual Financing Cost		($146,714)	($146,714)	($146,714)	($146,714)	($146,714)	($146,714)	($146,714)
Annual Cash Flow		$19,952	$23,285	$26,685	$30,153	$33,691	$52,467	$73,198
Cumulative Cash Flow		$19,952	$43,237	$69,923	$100,076	$133,767	$357,806	$681,515

Table 43-5 15 Year Cash Flow Analysis for $1 Million Project Financed at 11% over 15 Year Term.

escalation rate, this project would still produce positive cash flow in every year. However, if analyzed with a 2% negative escalation (or savings erosion) rate, the project would produce negative cash flow by year eight. Again, this sensitivity analysis and the long-financing term requirement would indicate a marginal project.

Consider instead a slightly different set of project economics in which the same capital investment produced $200,000 in first year savings, for a simple payback of 5 instead of 6 years, and project financing was available at an interest rate of 9%. The 10 year cash flow analysis for this project, based on a more conservative negative energy savings escalation rate of 1% per year, is shown in Table 43-6.

Compared with the first example, the impact of slightly increased annual savings (a reduction of simple payback from 6 to 5 years and a slightly decreased interest rate) from 10 to 9% result in a proposed project that is far more attractive, especially given the more conservative treatment of savings escalation. Still, for many corporations, returns on such a project may remain below their investment threshold. Additionally, the certainty of the annual energy cost savings projection would remain a critical risk factor, even with the somewhat more conservative treatment.

An alternative approach would be to undertake the project in a performance-based agreement with an ESCo providing either project financing or turn-key delivery of the entire project. The cash flow analysis for this project, based on an annual payment to the ESCo of 90% of the annual savings, is shown in Table 43-7.

Compared with the conventional financing approach shown in Table 43-6, annual (non-discounted) cash flow to the host is reduce by $17,168 in the first year and $92,801 over the 10 year term of the shared savings contract. The reduction in net cash flow to the host results from the fact that, in addition to project credit risk, the ESCo is also taking on project performance risk, as well as project management and savings M&V costs. From the perspective of the host, project risk is dramatically reduced and the investment can generally be removed from their balance sheet, freeing up investment capital.

It is assumed in this example that the ESCo is a vertically integrated firm for which margins are achieved on placement of design-build services, as well as on project financing. It is also assumed in this case that the ESCo has access to lower cost capital than does the host and, due to its expertise in project analysis, and perceives the energy savings performance risk to be lower than does the host/client. In cases where the ESCo provides only project financing and performance guarantees and design-build services are provided by another third party, it is likely that in order to achieve targeted returns on investment, annual payments to the ESCo would be somewhat greater than the representative example shown in Table 43-7.

Table 43-8 provides a somewhat more detailed analysis of an energy cost savings project financed under a shared savings type arrangement. In this example, the

YEAR	0	1	2	3	4	5	10
Annual Energy Cost Savings		$200,000	$198,000	$196,020	$194,060	$192,119	$182,703
Annual Financing Cost		($162,832)	($162,832)	($162,832)	($162,832)	($162,832)	($162,832)
Annual Cash Flow		$37,168	$35,168	$33,188	$31,228	$29,287	$19,871
Cumulative Cash Flow		$37,168	$72,336	$105,524	$136,752	$166,039	$284,038

Table 43-6 10 Year Cash Flow Analysis for $1 Million Project Financed at 9% over 10 Year Term.

YEAR	0	1	2	3	4	5	10
Annual Energy Cost Savings		$200,000	$198,000	$196,020	$194,060	$192,119	$182,703
90% Savings to ESCo		($180,000)	($178,200)	($176,418)	($174,654)	($172,907)	($164,432)
10% Savings to Host		$20,000	$19,800	$19,602	$19,406	$19,212	$18,271
Cumulative Cash Flow		$20,000	$39,000	$59,402	$78,808	$98,020	$191,237

Table 43-7 10 Year Cash Flow Analysis Where ESCo Provides 90% of Annual Savings to Host Facility.

simple payback on investment is about 4 years when considering only energy savings and 5 years when considering net savings after OM&R and metering costs. In this case, the ESCo guarantees a 20% share of the savings to the host facility and bears the full cost of OM&R and project M&V. In many cases, the host's share will be allocated as a percentage of the savings after these costs are accounted for. It is assumed that the ESCo takes a 20% equity position in this project and finances the rest at an interest rate of 10%.

If the project performs as anticipated, the host will achieve a before tax net annual savings of $56,420 and the ESCo will earn a return of 19.70% on the equity investment. If the project only produces 90% of the projected savings, the host would achieve a slightly lower before tax net annual savings of $50,778, but the ESCo's return on investment would shrink to only 5.06%. If, on the other hand, the project over-performed by 10%, the host would achieve a slightly higher before tax net annual savings of $62,062, while the ESCo would earn a return of 32.18% on the investment. This large swing in the potential returns to the ESCo underscores the relative risk and rewards of such ventures. The relatively modest swings in net annual savings to the host underscores the relative stability of such arrangements.

Comparison of Conventional and Performance-Based Financing Options

Considering the previous project financing examples, it is clear that the conventional host-owned debt-financing option is potentially most lucrative to the host if the project performs as anticipated. In addition to the higher

Project Cost		**Project Financial Data**	
Construction	$1,100,000	Percent of Savings to Owner	20.00%
Development	$124,000	Debt Interest	10.00%
Less Rebates	($100,000)	Debt Percent	80.00%
Net Project Cost	$1,124,000	Investor IRR	19.70%
		OM&R and M&V Escalation	2%
		Energy Escalation	0%

YEAR	**0**	**1**	**2**	**3**	**4**	**5**	**15**
DEBT		$899,200	$870,899	$839,767	$805,523	$767,854	$107,474
SAVINGS							
Utility Cost Savings		$341,700	$341,700	$341,700	$341,700	$341,700	$341,700
Increased Utility Purchases		$59,600	$59,600	$59,600	$59,600	$59,600	$59,600
NET UTILITY SAVINGS		$282,100	$282,100	$282,100	$282,100	$282,100	$282,100
PROJECT OPERATING COSTS							
Portion of Savings to Owner		$56,420	$56,420	$56,420	$56,420	$56,420	$56,420
Project OM&R Costs		$44,960	$45,859	$46,776	$47,712	$48,666	$59,324
Project M&V Costs		$11,240	$11,465	$11,694	$11,928	$12,167	$14,831
TOTAL YEARLY COSTS		$112,620	$113,744	$114,890	$116,060	$117,253	$130,575
FINANCIAL COSTS							
Depreciation		$74,933	$74,933	$74,933	$74,933	$74,933	$74,933
Interest		$89,920	$87,090	$83,977	$80,552	$76,785	$10,747
Taxes at 43%		$1,989	$2,723	$3,569	$4,538	$5,645	$28,313
TOTAL FINANCIAL COSTS		$166,843	$164,746	$162,479	$160,024	$157,364	$113,994
CASH FLOW TO INVESTOR							
Net Cash from Project		$77,571	$78,543	$79,664	$80,949	$82,417	$112,465
Principal Payment		$28,301	$31,131	$34,244	$37,669	$41,436	$107,474
Capital Outlay		$(224,800)					
NET CASH OUTLAY		($175,531)	$47,412	$45,420	$43,280	$40,981	$4,991
CASH FLOW TO OWNER							
Pre Tax		$56,420	$56,420	$56,420	$56,420	$56,420	$56,420
After Tax		$32,159	$32,159	$32,159	$32,159	$32,159	$32,159

Table 43-8 Analysis of Energy Cost Savings Project Financed Under a Shared Savings Arrangement.

return potential, the host facility can utilize depreciation benefits and maintain the flexibility to change operations without the need to negotiate with a third party. However, all performance risk falls on the host/owner. The host/owner must also maintain sole responsibility for OM&R and for equipment and savings performance. Finally, financing is on the owner's balance sheet and may limit access to additional capital.

With virtually any type of performance-based/shared savings arrangement, the host will commonly pay a premium for the performance guarantee and laying off other project risks and responsibilities in the form of reduced potential positive cash flow. The magnitude of the premium will be proportional to the comprehensiveness of the guarantees and obligations committed to by the contractor/third party. With the single-year savings calculation type of performance-based/shared savings agreement, the ESCo is assured payments over the full term if first year performance is as anticipated. Under the yearly savings calculation, the ESCo must continually prove performance throughout the term. This adds further to actual project operating costs and performance risk. Conse-quently, the percent of operating energy savings and, therefore, the incentive offered to the host in the form of net savings will likely be lower.

However, the ability to enter into a more secure project arrangement and off-load operational responsibilities may be viewed as a major benefit to the host facility. The avoidance or capital and/or debt requirements may also be an attractive feature, though depreciation benefits cannot be utilized by the host/owner. Another potential disadvantage of these arrangements is that as operational changes become necessary, the owner may have to negotiate either operational changes or contract termination with the third party. There may also be potential discrepancies over savings between the host and the ESCo/third party. These risks can be mitigated through well-conceived and carefully reviewed contracts.

Integrating Performance Contracting and Conventional Project Financing

Several steps can be taken to add performance-based contracting benefits to a municipal lease/purchase arrangement or a conventional owner debt financing arrangement. To ensure increased profitability with the tax-exempt lease or other conventional financing agreement, a rigorous savings verification procedure should be put in place, as well as an ongoing OM&R program. These, along with other services and guarantees, can be bundled into the overall project financing package.

A performance contractor/manager may be selected to manage the entire project for a given fee or for a percentage of the total construction cost. The contrac-tor/manager can bid out the project on behalf of the facility through a series of performance-based contracts, requiring some type of savings guarantee as well as ongoing OM&R agreements. Financing packages can also be bid. In addition, the contractor/manager may also provide an extensive savings verification program and/or bring in a third-party savings insurance policy.

By bundling these features with the relatively low-cost debt, the project package more closely resembles a performance-based/shared savings package, though less comprehensive and less costly. Savings guarantees, however, will only be as secure as the stated obligations and the capabilities and balance sheet of the party offering the guarantee. Moreover, third-party savings insurance policies should be viewed cautiously, as they may be more limited than the guarantees that an ESCo implementing the turnkey work would be willing to offer.

Outsourcing Energy Systems to Third-Party Owners/Operators

As noted in the discussion of contracting method options, another type of project contractual arrangement is a third-party ownership and operation agreement. With this option, a third-party ESCo (or contractor/investor) purchases a facility's existing energy system assets and/or finances, designs, builds, owns, operates, and maintains an energy plant serving one or more host facilities that use the energy supplied under a long-term contract.

In this case, the host facility need not secure project funding, but instead generally makes a long-term commitment to purchase the various forms of energy at some agreed upon rate. The structure of the arrangement may be somewhat similar to a true shared savings agreement, except that the host facility has no form of (or need for) ownership and no responsibility for short- or long-term OM&R and future plant upgrades. Savings are usually realized in the form of a reduced purchase cost of energy resources. This savings stream can then be compared to the net cash-flow generated through the various other types of financing approaches. The ESCo/investor will still be interested in the credit worthiness of the host facility in much the same way as the financier, since the monthly or annual payments made by the host facility for the purchased energy must ultimately compensate for the repayment of the capital funding of the project.

Chapter Forty-Four

Program Implementation and Operation

When a host facility has decided to proceed with an energy resource optimization project, selected a contracting method, executed contracts, and secured project funding, the next step is implementation. The various contract vehicles presented in the previous chapter will greatly influence the exact method used for implementation. However, in all cases, the major implementation phases are design and construction. The overall construction phase may be further divided into three main elements: procurement, installation, and commissioning. Once systems are installed and commissioned, the implementation phase extends to the establishment of long-term operations, maintenance, and repair (OM&R) procedures and training. Post-installation activities include the ongoing execution of OM&R procedures and may also include either one-time or ongoing savings verification processes. This chapter presents a detailed approach to the project implementation and operation. While the process will vary somewhat depending on the contract format, the basic tenants of a sound implementation process will remain essentially the same.

Implementation Tasks

The main tasks of design, procurement, installation, commissioning, and OM&R may each be performed by separate entities, or may be provided as bundled services with two or more tasks being provided by one entity. In some cases, all five tasks may be provided by one entity. This is typical of turnkey performance contracting and own and operate arrangements that have become prevalent in the energy services industry.

The level of detail required for each of these main tasks will vary widely depending on the type of project being implemented and the contract method utilized. Projects may range from simple installation of a single device or a single system to highly complex installations involving multiple interactive systems. Installations may be performed in a single building or in multiple buildings served by centralized systems in a campus setting or spread out over a geographical area.

For more straightforward, single-technology installations, the design tasks may be relatively simple, supported by little or no technical and financial analysis. This is sometimes the case when equipment is replaced by similar or even the same equipment due to age or disrepair. In such cases, OM&R requirements will also remain essentially the same for the new equipment as they were for the previous equipment. Even in such cases, it is recommended that a technical analysis be commissioned to evaluate the potential for improving operations through the use of alternative technologies and/or operating procedures. Still, cost can be contained by limiting or simplifying these activities for smaller, less complex projects.

For larger, more complex projects involving the integration of multiple technology applications, each of the five main tasks represent significant endeavors that should be preceded by rigorous technical and financial analysis and implemented by highly skilled industry professionals. In such cases, it is often beneficial to also include the sixth task of savings verification. While potentially useful with all contracting arrangements, this task is an essential element for projects being implemented on a performance basis.

The following discussion provides overviews of the management approaches that can be employed to accomplish the tasks described above. These approaches are presented within the context of implementing large-scale, integrated projects involving multiple technology applications.

Overview of Potential Management Approaches

Project implementation management approaches will vary depending on the contract vehicle used. As described in Chapter 43, approaches will include traditionally bid plan and specification types, including cost plus, unit price, and lump sum and various turnkey, design-build, and performance-type or out-sourced own and operate arrangements.

Under the plan and specification system, a design engineering firm is selected by the host facility/owner to engineer the project and prepare detailed contract drawings and specifications. These drawings and specifications are then competitively bid, by multiple construction firms and subsequently used by the selected firm as the documentation basis to construct the project. In some cases, procurement of major equipment may be handled directly by the host facility and only installation and commissioning are bid out. In the design-build approach, an individual firm bids project design and the three main construction tasks of procurement, construction, and commissioning as a combined service. In some cases, a design-build firm may initiate the project by performing the original feasibility study as a means of placing (or securing work for) their design-build services. Energy services companies (ESCos) may also offer study and design-build services complimented by project financing, performance warrantees, and ongoing OM&R services.

A key benefit of procuring bundled services, inclusive of design and construction and, in some cases, OM&R, is that it can promote efficiency throughout implementation and continuity of service into the operational phase of the project. The Construction, Commissioning and OM&R Teams, for example, can participate throughout the design phase, establishing a firm understanding of all finite design detail. Construction Team members participating in the design process will provide value engineering, constructability evaluation, and will begin to develop the construction schedule and sequencing. Commissioning members will provide a review of design and ensure that all elements required for planned operation are included in the design. OM&R team members can provide very valuable design review by ensuring that the design provides for effective and efficient operations and maintenance over the life of the equipment. They can also begin ordering long-lead time equipment before all final documents are completed, thereby expediting project completion. The need to build what they specify and design, and, in turn, maintain what they design and build, places both the responsibility and the incentive for effective economic implementation more squarely on the single firm.

Risk Management

There are numerous design, construction, and operational risks that must be mitigated to ensure successful project performance. The design is subject to the conceptual and technical risk. Construction risks include unforeseen problems leading to cost escalation, project delays, poor project management, low quality contractor work, statutory problems, and improper commissioning. Operational risks include improper installation practices, commissioning and lack of proper long-term OM&R.

It is not uncommon for a design-engineering firm to use off-the-shelf designs from previous projects that are modified just enough to fit the application. Alternatively, design costs can be driven up by over-inclusive designs that exceed the reasonable requirements for successful implementation. There is also always the risk of a low quality design. This often results from a profit motive-driven rushed design process that does not produce the optimal product. Since a design-build firm must build what they design, there is adequate incentive to produce a sound, workable design to the level required for successful implementation without going too far. With performance-based contract arrangements, there is adequate incentive to produce an economical design, yet one that will provide for optimal technical and financial project performance. Performance-based or own and operate type contracts evaluate the long-term OM&R costs when selecting equipment. Commonly, they will select high-quality equipment based on life-cycle analysis to provide longevity, as well as sustained performance over the term of the contract, while not unduly burdening either the first cost or the long-term OM&R costs.

Poor project management can lead to a substandard quality project, cost overruns, or both. Oversight errors during the analysis or design phase often result in construction cost overruns, lower than expected project performance, or both. Cost escalation may also result from any type of project delay, including those caused by contentious relationships between the owner, design firm, and construction firm. Statutory problems, notably environmental permitting and zoning, can lead to project delays. In cases where equipment modification is required, there is risk of redesign costs and additional procurement requirements.

There is also the risk that under-bidding to win the award can result in substandard construction, the requirement for additional expenditures during construction, or legal remediation. Performance-based bundled service approaches transfer much of the combined project risk to the contractor. As a single entity, the contractor has the incentive and ability to use internal control mechanisms to mitigate risk. Conversely, there is inherent risk in the removal of the checks and balances provided by use of independent design and construction firms.

During initial startup, once installation is completed, improper commissioning can result in damaged equipment and additional project delays. Once fully operational, there remains the long-term risks associated with improper OM&R. Improper operation can result in poor project economic performance and present various system reliability and health and safety risks. Improper long-term maintenance can result in degradated system performance, unnecessary system down-time, and excessive component or system replacement requirements.

The risks associated with design, construction, and operations must be carefully assessed with both the traditional plan and specification approach and with the various types of performance-based and/or bundled services approach. These risks must be mitigated through careful oversight by the host facility and through strong contractual agreements that include various types of performance guarantees.

The performance contracting approach has become increasingly common in the energy field. Full service ESCos take responsibility for the project at the onset and are constrained through their energy savings and other performance guarantee obligations. This approach represents a major shift from the conventional method of constructing a capital improvement project. Performance contracting realigns these relationships by transferring much of the risk of project performance from the host/owner and to the performance contractor. The performance contractor incorporates the roles of design professional, construction contractor, and owner into a single entity. Instead of drawings and specifications, the achievement of actual verified savings or specific performance criteria becomes the engine that drives the construction process. The risk of flawed design or substandard construction no longer lies primarily with the owner, but with the performance contractor.

In the case where the host facility is outsourcing ownership and/or operation of energy (and other resource) systems to a third-party ESCo, the obligations for effective design, construction, and OM&R fall even more squarely on the ESCo. In these cases, the ESCo purchases or leases the host facility's existing energy system assets and/or installs new systems. The ESCo then operates and maintains these systems for the duration of the contract. Contracts can be performance-based or based on the sale of energy output to the host at an agreed upon rate. The latter arrangement places even more of the design, construction, and OM&R risk on the ESCo than the typical performance contract. In order to maximize profit margins, the ESCo must minimize life-cycle (implementation and ongoing OM&R) costs without compromising operational reliability.

Regardless of the approach taken, design, construction, and ongoing OM&R must be planned and executed by competent personnel whose objective is to provide effective, professional services by meeting the established quality, cost, and time goals established in the contract. Moreover, regardless of the approach, the host facility/owner, whether through internal staff or through trusted consulting professionals, should have a thorough understanding of the process and carefully scrutinize each aspect of project implementation. Even if the contractor is to own and operate installed equipment, it must be understood that the host facility's mission requirements remain the paramount consideration throughout the process. To the extent that a sense of trust and partnering can be developed between owner/host facility and contractor, projects will likely be implemented more smoothly, more quickly, and more economically. This is more likely to occur with turnkey design builders and ESCos, since they have total involvement in the project from inception to completion.

Before selecting a preferred contracting method or design approach, the owner/host facility must realistically evaluate its own capabilities and expertise to manage the process. The more complex the design and installation the more likely issues will arise. By attempting to reduce first cost project estimates and assume more risk early in the process, owners may subject themselves to unanticipated project cost increases and delays later on, when it is more difficult to contain the impacts on the project.

Following are overview descriptions of the basic project implementation elements described above.

Design Management Approach

The first main element of project implementation is the design phase. At a minimum, design work includes the preparation and development of all procedures, drawings, specifications, design analyses, cost estimates, and other documentation required to fully and completely describe the proposed work for its review and acceptance. To effectively accomplish the objectives of meeting the contractually established quality, cost, and time goals, the Design Team must balance design sensitivity, functional objectivity, and fiscal prudence.

The initial step is to establish design criteria. This is the foundation of formats, conventions, and global assumptions upon which the rest of the design work is built. At the onset of the design phase, the Design Team should review the final feasibility study report and conduct a project programming analysis, which includes a cursory evaluation describing and evaluating the various primary design parameters. At a minimum, this should include applicable facility design guidelines, standard load calculations, design space and process conditions, current life-cycle parameters, cost assembly formats and access, fuel source options, emissions requirements, operational requirements and time constraints. Applicable facility design guidelines must include mission considerations and reliability, which will be key decision factors related to design for implementation and redundancy requirements. This project programming analysis should be maintained through design option elimination until data yields parameters limiting the number of design options to a single viable final system selection.

Following concurrence on design criteria, a present conditions report should be prepared containing a description of all major systems and sub-systems that are anticipated to be affected. Although much of this work will have been completed during the study phases (described in Chapter 41), this process allows the Design Team to confirm and/or add to the database such that the capacities and equipment will be quantified and basic control strategies established for all systems.

The Design Team must evaluate the viability of the feasibility study conclusions and solutions and value engineer for practicality and cost-effectiveness. Involvement of designers in the analysis process generally proves to be cost-effective as it can eliminate certain design assumption errors that could result in underestimation of project costs or selection of systems or components whose installation is impractical or would present significant logistical difficulty.

The Design Team objectives should be to provide state-of-the art designs with selection of systems that feature optimal cost-performance resulting from efficiency, maintainability, reliability, and ease of operation. The Design Team must also be sensitive to operational or mission-sensitive requirements of the host facility and seek to solve potential problems with innovative methods. Some of the implementation work can be disruptive to occupants and process operations, no matter how well it is managed. An important design objective, therefore, is to implement the changes in a timely manner and with a minimum of activity interruptions. This can include, for example, measures to isolate airborne contamination that will be induced by the construction work from adjacent operating areas.

Another critical aspect is the development of implementation procedures that will avoid interruption to mission-sensitive systems that cannot withstand any interruption during construction. An example is the design of bypass valves and piping that allow critical systems to continue to operate while the new systems are installed. Often, with proper design and planning, the new system can be brought on-line with virtually no interruption to current operation.

Deliverables

There are several sets of design deliverables. Depending on the project-specific strategy, this could include two, three, or four sets, denoted by the level of completion at the time of the deliverable. Design deliverables may include a 30% schematic design development package, a 60% review package, a 95% review package, and a final design package.

After internal quality control review by the Design Team, each set of deliverables should be reviewed and approved by the party commissioning the Design Team. In the traditional plan and specification process, this would be a representative from the host facility/owner. With performance contracting or other design-build approaches, this would be the contractor/investor. Ideally, the package should also be reviewed by both the Study Team and the Construction Team. In all cases, the host facility must remain involved in the process and give approval to proceed at the completion of major milestones.

At about 30% completion, a preliminary schematic design development package should be submitted for review. Those documents should include a specific construction scope of work (SOW), all major equipment requirements and installation location, updated air/water riser diagrams, new equipment schedules, control SOW and DDC or PLC points list (where applicable), operational process flow diagram and an outline specification. In addition, where appropriate, a schematic floor plan of prototype occupancies should be developed indicating the recommended number of control zones. All plans, diagrams, and schedules should be compatible with the latest CAD-type software package and delivered in electronic and reproducible media. In essence, the 30% submittal should be sufficient for confirmation of feasibility study assumptions based on complete definition of all new systems, existing systems, modifications, and control strategies.

With many types of performance-based turnkey type projects, the 30% schematic design package will have been developed during the proposal stage and serve as the basis for the proposal price offering. This is typically performed at the contractor's risk. Beyond that point, if their proposal is accepted, the parties may enter into a full contract for design and implementation or an initial contract for the completion of the design and final pricing. Sometimes these services will be provided at the contractor's cost, with contract provisions for an upset fee to be paid to the contractor for design and development work done, beyond the initial proposal, in the event the project is not implemented or the owner wishes to take possession of the design and either perform the work themselves or bid the work to other contractors.

With a traditional design approach, upon approval of the 30% design development set, with comments, work can commence on the 60% submission. At this point, it is recommended to commission a value engineering review of the package by construction and long-term OM&R management professionals. These reviews should examine construction and OM&R procedures and requirements, materials specifications, recyclability, environmental impact, and maintainability. Also during this phase, the measurement and verification (M&V) components may be incorporated into the finial design and pricing package. Beyond the 60% review, no significant changes are possible without significant rework of detailing and plans.

At about 95% of the completed design and documents, a full draft of the design documents should be submitted for final review and comment. The results of the value engineering review (delivered as marked up drawings or notes) should be incorporated into the 95% submission or appended as special conditions.

Once the comments are resolved and changes incorporated, the final design documents should be issued. The final design drawings should contain the necessary details for all phases of the project and should be sufficient to attract competitive bids and afford a clear understanding of the job throughout construction. Final design specifications should be prepared as required to detail the project and supplement the drawings for the project. Additional documents to be prepared and submitted with the specifications may include a list of host facility-furnished equipment, lists of items to be removed during construction and their ultimate destination (e.g., recycle, dispose, etc.), list of proprietary items, statement of work sequence, and statement of construction duration.

The final design analysis should be submitted with the final drawings and specifications. A final cost estimate based on the final drawings and specifications should also be submitted. This cost estimate will be a corrected, expanded version of the cost estimates previously submitted. Other cost estimates to be provided include cost estimates for development and execution of an OM&R program, inclusive of OM&R manual development and training requirements. Where applicable, cost estimates should also be provided for savings verification equipment and services.

All cost estimates should be submitted under separate cover with the design. The life-cycle cost analysis, which was provided during the feasibility stage, will then be updated with design-specific data as required. In some cases, project costs and savings could change significantly. This might occur if the Design Team rules out the equipment selected by the Study Team due to practical design considerations, such as size or weight limitations. Unless a suitable alternate system had already been evaluated, both the engineering and financial study must be re-run based on the characteristics of the alternate system or systems. Cost estimates should also clearly identify elements of the design or implementation that may require a project contingency. Typical contingency items in this type of design would be for elements of the work that may be difficult or impossible to quantify at the time of the estimate, such as environmental and regulatory concerns, unforeseen conditions and hazardous waste disposal.

The schedules and processes described above are representative of the more formal, traditional approach to design. When operating under alternative design-build approaches, review requirements by the host facility are minimized, as the responsibility for a properly constructed project ultimately lies with the design-builder. In many cases, design is performed as an in-house function of the design-builder or ESCo and the review process tends to be more iterative.

With design-build approaches, construction drawings and specifications are often provided on drawing sheets, which are understood to be a design-build set. These are detailed to the minimum extent necessary to provide for a complete workable installation. Details and construction techniques typically understood by, and familiar to, experienced contractors, subcontractors, and skilled construction personnel are not necessarily shown. However, plan and layout drawings still must be complete and fully dimensioned, showing all major elements of construction and clearly describing the design intent.

In cases, where the design and study are provided by the same firm, this process is done internally. In cases where design and analysis are provided by separate firms, the process tends to be less fluid if iterations must pass back and forth between the two firms until a new final design and analysis is completed. In some cases, the design firm may perform the re-analysis function itself with either an independent analysis or by modifying the Study Team's original analysis. Final construction cost estimates may be provided by the design firm's construction estimators or performed by independent construction estimators. In cases where the design and construction are provided by the same firm, this process is done internally.

Upon completion and approval, the final design and final feasibility study will then form the basis of the Implementation Plan.

Quality Management

An important aspect of the design process is quality assurance (QA) and quality control (QC). QA is typically more programmatic in nature and may provide for review of the design and to ensure the design is consistent with the intent of the project and will meet its objectives. QC focuses on specific project deliverables such as material delivered to the project or weld inspections. Quality management should be an ongoing activity at each project team level. Design review presentations ensure the critical input of cross disciplines and the most senior engineers. This process should include representation from the Design Team and from the host facility's management team.

Ideally, every product that is issued should undergo peer review for engineering and editorial quality. Engineering review includes a critical analysis of scope, approach, assumptions, and calculations by an engineering peer who has, to date, not contributed in any substantial manner to the project. The designer and reviewer confer and work to reach consensus on the reviewer's commentary. The reviewer signs off formally on the document control routing sheet, which becomes a part of the project record.

At a minimum, where applicable, all project and associated equipment and facility installation designs should comply with the recommendations and standards of the following:

American Association of State Highway and Transportation Officials (AASHTO)
American Concrete Institute (ACI)
American Institute of Steel Construction (AISC)
American National Standards Institute (ANSI)
American Society for Civil Engineers (ASCE)
American Society for Heating, Refrigeration, and Air Conditioning Engineers (ASHRAE)
American Society for Mechanical Engineers (ASME)
American Society for Testing and Materials (ASTM)
American Water Works Association (AWWA)
Architectural and Engineering Instructions (AEI)
Environmental Protection Agency (EPA)
Factory Mutual (FM)
Federal, state, and installation environmental codes and regulations
Institute of Electrical and Electronic Engineers (IEEE)
Local utility standards
Metal Building Manufacturers Association (MBMA)
National Electric Code (NEC)
National Fire Protection Association (NFPA)
National Standard Plumbing Code (NSPC)
Occupational Safety and Health Act (OSHA)
State, local, and installation safety documents applicable to the project
Uniform Building Code (UBC)
Uniform Electrical Code (UEC)
Uniform Federal Accessibility (UFA)
Uniform Fire Code (UFC)
Uniform Mechanical Code (UMC)
Uniform Plumbing Code (UPC)

Note that this is only a partial representative listing. There are numerous other standards that may apply, depending on the type of equipment, facility, and processes involved. Another governing factor will be the type of business or institutional mission involved (e.g., hospital, military base, etc.). Each will usually have several additional standards that must be complied with. These standards should be identified at the onset of the design process and agreement should be reached with the host facility as to which standards apply and, in the case of conflict, which standard supersedes the others.

Design Document Control

Since several different teams and companies may be involved in the process, effective document control is a critical area. Document control should begin at the study phase with careful planning of what documents, both written and electronic, will be required throughout the project. The flow of these documents and responsibility for their generation and handling should, therefore, be established at the start of the project. Consistency is maintained throughout, with documentation from the study phase flowing into the design phase and construction phase through to commissioning and the ongoing OM&R phase.

Given the wide range of personnel that will require access to the electronic documents, various levels of access should be established. Design documents can be controlled by title (or file name) and revision level. Letter or number suffixes can be added to permanent file names to identify revisions. The Design Team Leader should maintain control over revision levels for drawings and specifications. Read-only access should be allowed, with write access granted only by document control and accompanied by a new revision level suffix. Field changes can be accumulated in a revision format and incorporated into the newest revision only when approved by the appropriate Design Team members. Once construction is completed, final record drawings will be established reflecting construction modifications. It is also advisable to keep an electronic record of each formal submission, including the proposal period drawings or initial 30% submission and each subsequent submission up to the 100% construction drawings and finally the as-built drawings. The drawings used for final contractor pricing (the pricing set) are particularly important in the event any disputes arise as to what firm price offerings were based upon. Field changes are sometimes paper copies, formally transferred into the finals after construction.

CONSTRUCTION MANAGEMENT APPROACH

Depending on the contract vehicle used and the types and make-up of the contractors involved, there are many ways to approach the three main elements of construction — procurement, installation, and commissioning — of integrated energy resource optimization projects. Rarely can complete Construction and Design Teams be fielded with in-house personnel, so even with turnkey services being provided by a single contractor, various subcontractor arrangements will generally come into play.

Under the traditionally bid plan and specification system, the host facility relies on design professionals to prepare contract drawings and specifications that detail the means by which a project is to be built. During the construction phase, the design professional stands watch over the construction contractor to assure that the work conforms to the construction documents and to assist with the selection of the best alternative if and when changes become necessary. While usually effective, construction and administration (C&A) can be extremely rigorous in its attempt to minimize the costs to the owner. Its strict adherence to design specifications can also limit the ability to take advantage of value engineering opportunities, or limit costs imposed by unforeseen items uncovered during construction.

A key element to this relationship is the qualifications of the bidder and the quality of the bid. In many cases, the most qualified bidder is not selected. For example, many government agencies or institutional facilities require "low-bid" contract awards. The low-bid award process does not truly evaluate a contractor's ability to complete the project as well as perform the work as originally intended by the design professional. In this environment, it can be difficult for the design professional to control the quality and conformance of a project. A contentious process of ongoing "price add" (in the form of change orders) negotiations and even work stoppage is not uncommon, especially in the case where the winning bidder has under-priced the project to ensure contract award with the expectation that price adds will follow as a means to make target margins. While this will not necessarily occur, it is a risk that should be carefully considered when a low-bid award process is used. When contractors are selected based on a "best value" selection criteria, some of this risk is mitigated, but of course still not eliminated. In a performance-based scenario, the design-build firm or ESCo has a lot more at risk themselves, as a firm-fixed price offering, and in some cases guaranteed equipment performance, drive the process to completion with fewer incidents of price add negotiations. Still, no process is completely immune from these risks.

It is also not uncommon for the relationships between the design professional, contractor, and owner/host to become polarized, characterized by competition and contention rather than cooperation. When this occurs, the construction contractor may act to maximize the value of its contract and the design firm may move to reduce its liability exposure and increase its fees. This course of events could leave the host exposed to costly change orders and poor workmanship. Several remedies and solutions to such potential problems are provided in the discussion on contract vehicles in Chapter 43.

Open lines of communication and careful documentation of all relevant details are often the best vehicles for avoiding such problems. The more interaction between all parties involved, the greater the chance of establishing harmony and commonalty of goals. By anticipating future problems and making the extra effort for explicit documentation, conflicts can also be minimized. Another tool is to have extensive involvement of qualified host facility representatives throughout each phase of the project. While adding a cost factor to the project, these representatives can provide continuity from one phase to another and often seek to solve controversy before it evolves into a major problem.

A rigorous bid review process, conducted by a knowledgeable management representative, can be very effective. The unit price contract approach described in Chapter 43 is another risk mitigation alternative. With this arrangement, when the Design Team finishes a set of documents, it is given to a professional estimator/bidder that establishes unit quantities for each aspect of the project. Contractors then bid unit prices only against an already established quantity. This process unbundles bid components allowing for detailed analysis and direct comparisons of bids. It also locks in the rates of the construction contractors, assuming items that are not unit-priced (e.g., wire nuts, pipe hangers, etc.) are included elsewhere in the price. Its use, however, is generally limited in that it may not be well-suited for diverse and complex projects and does not promote value engineering. Also, there is a tendency for contractors to constantly look for (and, if possible, over-price) items that the initial bid-developing estimator undercounted or omitted.

There are also several alternatives to the traditional plan and specification method that can be used to mitigate risk and improve project continuity. These include pre-selection of construction contractors, or the use of construction management, design-build or performance contracting services.

Pre-selecting the construction contractor prior to design completion can be quite beneficial. This is sometimes possible when there is a long-standing relationship between the host/owner and a construction firm, or when the selection process is based on firm qualifications as opposed to bid and specification. Under these scenarios, the Construction Team can participate during the design phase, allowing for ongoing design review and value engineering input from the Construction Team. Final price negotiation can be a problem since the work has essentially been pre-awarded without being based on a final price.

A construction management (CM) firm would join the process either during or prior to the start of the design process. In some cases, the CM firm may be brought in very early and assist in the selection of the design firm. In other cases, they assist with the selection of the construction contractor and subcontractors. CM firms can provide value engineering cost analysis and construction feasibility analysis. During the construction phase, CM firms can provide on-site management, contract administration, bid reviews and analyses, pre-qualification of contractors and subcontractors, and risk management services.

In essence, the CM firm becomes the agent of the owner/host and can assume some degree of risk, depending on the preference of the owner/host and the financial structure of the CM firm. CM firms can be asked to fully estimate the project and set a not-to-exceed (or gross maximum price) budget. In such cases, they assume a greater portion of the risk. Alternatively, the CM firm may simply manage the construction process under one of several financial arrangements, including a fixed fee, costs plus, or a percentage of the total project cost. The potential benefits of using a CM firm lie in the professional experience they bring to the project, the increased continuity achieved between design and construction phases, and the fact that they serve as agent for the host/owner. For host/owners that do not have in-house expertise, this arrangement can be very helpful. However, most commonly, the CM will not take on much risk and will be paid either a fixed fee, for their hours, or a percentage of the construction cost.

With performance-based contracting, construction performance is largely controlled through energy savings and other performance guarantee obligations, as opposed to owner or design firm watchdog functions. With this approach, the risks of flawed design or construction lie primarily with the performance contractor. The sole-source responsibility provides for a high level of accountability to the host facility, as well as optimal continuity of service. Moreover, with this arrangement, the contractor is driven to maximize project economic performance. Still, the owner or agent of the host facility must remain involved and exert sufficient control to ensure that their objectives are met.

Construction Approach Methodology

Regardless of the overall implementation approach taken, the construction approach methodology for procurement, installation, and commissioning should incorporate standard industry practices for assuring proper coordination, communications, materials quality, installation methods, budget controls, code compliance, safety practices, etc. All of the procedures to be followed should be established and documented in a procedures manual, or project manual, made available in electronic and hard copy. The manual will list all key personnel, along with their relationship to the project, the objectives of the project, established milestones and critical dates. It should include a system of communication, including the initiation, routing, and recording of all correspondence. The proper processing of contracts, field orders, payment requests, change orders, shop drawings, cut-sheets, and samples should all be explained. The time and place for job and project meetings should be established and responsibility for the preparation of agenda and the recording of minutes determined. The procedures manual should remain a flexible document that will take form according to the needs of the project and the management group.

Pre-Construction

To initiate the pre-construction activities, the Construction Team and their subcontractors, along with the principal Design Team members, should hold a detailed planning session to discuss all aspects of the project, from the design phase through final completion. At this session, the project schedule will be reviewed and modified as necessary, based on the current status of the project and the full range of planned activities culminating in commissioning and testing of all project components. Input from all principal team members, as well as subcontractors, should be required so that agreement can be reached on schedules, processes, and key milestones for all remaining project activities.

In a performance contracting arrangement, or in cases where the construction contractor or CM firm is selected during the early stages of the design phase, the Construction Team can be involved in the review and comment role throughout the design process. This allows for familiarization with not only the design, but also the concepts that went into the design as they relate to the energy savings and construction methodologies. The experience of the Construction Team representatives can be very helpful to the designers, allowing for immediate feedback on the implications of a particular design and potential problems related to constructability, as opposed to addressing such issues only after the construction phase has begun. Construction input is particularly valuable when contemplating the interconnection of new and old systems to minimize interruption in normal operations and services. Change orders, schedule and cost overruns, and unnecessary rework can be minimized or eliminated. Simultaneously, the Construction Team can get a head start in preparation for the construction phase by knowing the direction the Design Team is taking and the projected scope of the construction activities. They can also begin to solicit qualifications and pre-qualify potential contractors that may be invited to bid on the project.

As noted previously in the design approach discussion, during the design phase, the Construction Team can be involved in carefully reviewing the equipment and system selections, the methods of construction, and the scheduling implications of material and equipment selections as related to project costs and lead times. The assistance provided to the Design Team by the Construction Team in locating cost-effective and readily available materials and equipment is invaluable and continues throughout the design phase. Not only are construction costs reduced, but engineering design time and costs are also reduced. This type of interactivity is yet another benefit of having combined, or at least cooperative, design-build functionality.

Scheduling

All Construction Team participants should use an agreed upon scheduling software package. This can include off-the-shelf packages, as well as self-developed databases and spreadsheets. The programs should be used to carefully schedule and track the performance of projects. Scheduling should be based on the critical path method (CPM) so that timely purchase and delivery of material and equipment is ensured and that adequate personnel and resources are available as needed. Scheduling and frequent auditing for compliance should be a major focus of the construction management staff. With such procedures in place, deviations from the schedule will be quickly detected and swift corrective action can be taken as necessary to restore the schedule. Careful attention to scheduling allows for anticipation of delays and minimization of their effects.

Schedules are a critical component of a successful project and should be developed, reviewed and approved by professionals with sufficient experience to produce a realistic baseline schedule. While some elements of the project are independent of each other most have some interdependency. Interdependencies between one work element and others in the schedule are called predecessors (a prior element affects the timing) and successors (an element affects the timing of a later element). The link that is established between interdependent elements has a significant impact on properly indentifying and maintaining the CPM schedule. Throughout the project it is important to maintain the schedule accuracy by updating actual vs. planned activities. If a project is not being constructed according to plan, then a proper and accurate schedule will be the source document for determining the root cause of performance failures.

Construction Document Control

Document control during the construction phase includes control of all drafted, written, or printed material relating to the project. This includes all contract documents, memoranda, transmittals, letters, orders, authorizations, records, reports, meeting minutes, cost estimates, schedule printouts, verbal conversation documentation, daily work reports, safety inspections and accident reports, quality control reports, assignments, regulations, sketches, brochures, shop drawings, samples, and manufacturers' data. Design drawings and specifications should continue to be controlled by the Design Team, with field changes accumulated in a revision format.

Upon initiation of the project construction phase, the Construction Team should develop any additional documentation required for construction and maintain a filing system and listing of all anticipated documents. This listing will be continually updated to reflect the current status. The listing will include contract drawings, design specifications, shop drawings, equipment specifications, permitting and inspections, OM&R manuals, pre-functional testing, commissioning reports, training manuals and handouts, and any other information necessary for the timely and successful implementation of construction. By keeping this updated listing available, all of the various project team members and management representatives can more easily follow the construction process. Further, upon completion of construction, the current body of records will remain available for the ongoing training and operations and maintenance phases.

All critical correspondence should be catalogued and tracked, with a report generated and distributed to all appropriate parties. Report status tracking will ensure timely response on all necessary issues. To the full extent possible, standard forms should be used for all correspondence.

To ensure proper control of documents and proper correspondence procedures, a flowchart should be developed defining the relationships between the parties involved. For each major form or record, a separate flowchart may be required, since they may become hard to follow when too many elements are included. Use of flowcharts should be restricted to job-related orders and documentation where distribution is critical. A general procedure is sufficient for the processing of all other correspondence. Increasingly, electronic communication is being used for these purposes, ranging from group list e-mail to confidential web sites containing all project data, with various levels of clearance access provided as appropriate. Numerous software packages are commercially available.

Procurement

Procurement involves the selection and acquisition of all equipment and materials required for proper project installation. It also extends to the selection and contracting with subcontractors that will provide the direct labor and/or materials associated with specific aspects of the overall project.

Material procurement is the final step of the process initiated in Step 11 (Equipment Selection) of the Detailed Feasibility Study described in Chapter 41 and confirmed during the design phase described above. As noted, all equipment selected and/or specified and installed should satisfy the necessary requirements of the governing agencies, such as local and state codes, testing and approving agencies (UL, AGA, NFPA, ANSI, ASHRAE, ASME, etc.), and construction agencies (SMACNA, BOCA, ASTM, NEC, NEMA, etc.) and should represent the highest standards of the industry. Equipment selection should be based on the engineering specifications required for the best performance and economic analysis of all available models that can meet the specifications. Consideration should also be given to technological obsolescence protection, with specifications only for manufacturers that have demonstrated a corporate commitment to backward compatibility, i.e., that new systems will be able to operate with existing field equipment. The same procedures are followed, with the exception that more explicit detail regarding design, schedules, and contractual obligations is included in the specification. The manufacturer's or vendor's obligations for delivery, installation, and commissioning are also usually specified at this time. Obligations for OM&R are also sometimes included along with extended warrantees.

In some cases, the exact specification and type of materials will have already been established. In other cases, a final determination will not be made until final pricing is received between competing equipment. In the case where two different manufacturers' equipment will satisfy the design specification, selection may be made at this point based on price or a revised life-cycle analysis and some design detail may have to be added to account for the specific characteristics of the chosen equipment (i.e., size, weight, electrical connections, etc.). Delivery time for major components is also a significant factor in the selection of equipment due to the critical path nature of these components.

With major components or systems for which it is intended that the manufacturer or vendor provide full or partial OM&R work, it is best to negotiate all of the terms of the contract at this point in time. For long-term service-type arrangements, renewable contracts with annual escalation rates should be included. In this manner, the actual life-cycle purchase cost, inclusive of long-term OM&R, can be evaluated, allowing for the best purchase decision to be made.

To properly secure accurate pricing, a clear and concise design must be prepared and a comprehensive bid package developed. It is important that each bidding contractor is clear as to what they are bidding on and what their responsibilities will be. If changes are required, all addenda must be clearly dated and properly annotated. A comprehensive bid package includes a detailed SOW, a formal request for proposal, a contractor qualifications questionnaire, prior safety record information, a bid form outlining specific price breakouts, and any other requirements that will be placed on the contractor that will add cost to the project. It is important to include specifics, such as if the project is tax exempt or if the contractor will be required to provide a payment, performance or bid bond or specific insurance limits and requirements. Since such items will impact the overall pricing, they should be clearly defined to the bidding contractors up front. It is best to provide a copy in draft form of the final contract that the selected contractor will be asked to sign so that they can note any exceptions they will require during the bidding stage and also price in items that will be required for full contract compliance. Project scheduling should be carefully considered at this point.

The construction plan should include a critical path timeline that reflects scheduling conditions. A project that will allow for the building to be shut down with the contractor in complete control will be significantly less costly and take a far shorter time to execute than a situation where all work must be done during off hours or with zero impact on building activities. Numerous variations of approaches and timelines may be developed for the contractor in order to identify the impact on price. An economic balance can then be struck by the owner in a separate analysis of cost impact on their business/mission vs. construction cost impact.

As in the initial budgeting phase, multiple quotations should be solicited. The vendors that provided the most attractive offerings during the initial budget estimating phase should be invited to submit bids. This may be complimented by solicitation of additional bids, depending on the level of satisfaction with the original offerings. At this point, negotiation of price, equipment features, and services offered is appropriate. In some cases, there may be a second round of bid solicitation, referred to as "best and final offer," following the negotiation process, though negotiations may continue up until contracts are signed.

In some cases, more typically with single technology applications, major equipment may be purchased by the host facility and then provided to the contractor hired to perform the installation. As noted earlier in this Chapter, if an owner has the internal capabilities to perform and manage this effort it may present an opportunity to reduce the overall project cost. The potential for savings should be carefully weighed against the increased involvement and risk to the owner. For more complex integrated projects, the equipment along with the other required materials is purchased directly by the construction contractor. Similarly, the prime contractor may procure equipment and labor separately to avoid additional price mark-ups, or may find it more efficient to allow the subcontractor to procure, handle, and take responsibility for the materials they will be installing.

The process for establishing the SOW and selecting contractors is described in Step 12 (Budget Estimates of Incremental Capital Cost) of the Detailed Feasibility Study presented in Chapter 41. As with material selection, this process now moves to final selection and firm contractual commitments. Whether performed by the host facility, construction contractor, design-build contractor, performance contractor, or CM firm, it is necessary to have the estimating skill and experience to fully evaluate the bids of the subcontractors to not only ensure the lowest reasonable project costs possible, but to ensure that all work has been included and, that ultimately, the project will be successfully installed.

The process of pre-qualification and selection of contractors that will be invited to bid on the work is often started during the development process as preliminary bids are used to develop the initial project cost estimates. Pre-qualification may be based on a formal process, often referred to as a request for qualifications (RFQ), or based on discussions with known contractors in the area or contractors known for their qualifications for the type of work contemplated. Efforts should be made to identify contractors that are familiar with the site and its mission and working conditions as well as those that have performed well in the past for the host facility/owner and/or prime contractor. If possible, this pre-qualification process should start early in the development process so as to avoid lengthy delays after plans and specifications have been developed. In the case of design-build and other turnkey work, the prime contractor will have already been selected and will also likely be ready with its list of pre-qualified subcontractors. In the traditional plan and specification process, this can add significant time to the implementation process.

In making final contractor selections, some key questions to be answered include:

- Does the contractor have proven experience in the technology applications to be implemented?
- What is the contractor's current workload?
- Does the contractor have the internal resources to properly staff the project?
- Who, if any, are the contractor's proposed subcontractors?
- What is the company's credit rating?
- Can the contractor obtain a performance bond, and, if so, at what bonding rate?
- Does the contractor have a demonstrated safety record and proper safety methods and procedures in place?
- Has this firm included all of the pricing elements for this project in their bid?
- Has this firm successfully completed prior projects without claims, liens or similar issues?

Successful contract (and subcontract) negotiation is not a matter of securing the lowest stated price or shortest stated schedule for completion. Instead, it requires the use of skill, knowledge, and understanding of the project to obtain the best combination of price, schedule, proficiency, construction quality, safety, completeness of bid, past performance history, and follow-up service by the subcontractor.

INSTALLATION

During the installation phase, the equipment and materials that have been procured are fully installed and prepared for project commissioning. For most large projects, actual installation is accomplished through the use of various subcontractors with specialties in areas such as mechanical, electrical, structural, rigging, demolition, waste disposal, and general contracting. With the equipment and materials already selected and the SOW and contractual arrangements defined for each subcontractor, successful project installation is largely dependent on effective scheduling and site management.

Site Management

During the installation phase of all major construction projects, daily site management should be provided and documented by a Site Superintendent who is the senior on-site construction person responsible for all day-to-day functions on the job site. Some of the major functions are scheduling, contingency planning, securing all required permits and necessary approvals from the host facility, and ensuring compliance by all on-site personnel with all prevailing standards, regulations, and safety procedures. The Site Superintendent's (or Construction Manager's) management activities should be supplemented by assistants, trade superintendents, and a site safety and quality control officer. The Site Superintendent should be empowered to make all necessary job site decisions that require immediate response and should have control over all on-site construction personnel, either directly or indirectly through trade superintendents or foremen.

The Site Superintendent must provide clear direction through unambiguous communication with each trade subcontractor and all other site personnel. This includes negotiating the sequence and timing of milestones with each subcontractor in order to manage the critical flow path of deliveries, installation work, and commissioning. The Site Superintendent also has the responsibility to maintain a daily log of personnel on site, changes or directions issued, construction activity completed, and any constraint, such as delivery delays, weather, or labor actions. In addition to enforcing compliance with content and timing of subcontract provisions, the Site Superintendent is ultimately responsible for enforcing compliance with the site safety plan and all other established laws, regulations, and governing work practices, as well as industry standards of quality such as workmanship, completeness, material storage, and waste disposal.

Other site management responsibilities include approving payment requisitions in accordance with contract payment terms, accepting substantial completion of subcontractor's work, and developing or approving punch list items and requirements for final completion. Along with the OM&R Supervisor, the Site Superintendent should also have responsibility for coordinating testing commissioning and acceptance of work from all subcontractors, inclusive of enforcing the specific provisions and intent of subcontracts for commissioning.

Job Meetings and Minutes

There are several classifications of meetings that should be convened on a regular basis during the installation phase in order to manage properly and keep all interested parties currently informed. Executive meetings should be held regularly (e.g., on a monthly basis) to bring together the top-level management of the host/owner and the Design and Construction Team firms.

Regular job meetings, attended by representatives of each prime contractor and the Project Manager for the host/owner, should be held on a weekly or bi-weekly basis. In cases where a CM firm is used, the CM firm may serve as the Project Manager. In the case of performance contracting, the Project Manager will be an employee of the ESCo. Design group representatives, subcontractors, suppliers, and material persons will attend when required by the Project Manager. The Project Manager will chair this meeting, take minutes, and publish a conference memorandum which will be distributed to the home office, the design group, the owner/host, all prime contractors, and all others in attendance.

Daily meetings should be held with on site personnel from the respective trades performing the work. The Site Superintendent should chair these meetings and distribute daily minutes, or summarize verbally and include in the daily report. Special meetings for clarification, design planning, scheduling, coordination, crises, etc., are called as directed by either the Site Superintendent or the Project Manager and may be initiated by any of the interested parties.

The superintendents of the prime contractor and the major subcontractors should meet briefly each day either before the start of the day's work or at the conclusion. The purpose of this informal gathering is for each party to state their planned activity for the coming work day, to warn of any unusual event, to discuss anticipated deliveries, and other such matters in order to discover and avoid any unforeseen sources of conflict. Potential areas of cooperation such as sharing the use and cost of heavy equipment may be found that will facilitate the work of two or more trades. The Site Superintendent should assure that these meetings are brief, but effective.

COMMISSIONING

At the completion of the installation phase, and sometimes concurrent with the installation phase depending on the occupancy of the area served, the new or modified systems must be commissioned for service. The construction process is not complete until all systems have been fully commissioned and accepted by the host facility/owner. In essence, this is the phase of construction in which all design assumptions and calculations are verified through practical application of the installed equipment. While the specific procedure varies depending on the type of equipment and applications involved, the basic elements of the commissioning process apply to almost all types of projects. These basic elements are startup, testing, and operational service. These elements are followed by training of operational personnel and the development of long-term OM&R procedures.

Start-up

In preparation for start-up, a review of the manufacturer's recommendations for start-up is completed along with a review of the design parameters of the equipment in the specific installation. This preparation for startup is sometimes referred to as pre-functional testing. This review is conducted for the purpose of determining any operational parameters that could affect performance and/or safety of operation. This may include items such as system cleaning, flushing, initial fluid level and lubrication requirements, valve positions, wiring connections, minimum or maximum flow rates or pressure differentials, cleanliness of filters, and rotational direction. A list of each of these items is developed along with the parametric measurement values required during operation. It must be understood that improper installation procedures or mistakes made once the equipment is started up can be very costly and delay eventual project completion significantly. Operation without lubrication or coolant or mismatched power connections can have devastating effects. Even something as seemingly insignificant as one day of poor water treatment can have severe consequences. Hence, the need for extensive preparation and check-out, along with proper safety procedures, cannot be overstated.

Once the system is initially started, a check of each of these items is made to determine if the unit is performing as proposed in the design. Assuming the unit is within design parameters, the commissioning goes on to the next phase. If the unit is not within design parameters, additional external parameters may need to be measured prior to contacting the designer, thereby giving the designer a full review of all external items that could be affecting the unit's performance. The designer must then advise on the best course to proceed. Once corrective action is taken, the process is repeated until it is determined that all design parameters have been met. At this point, system testing can begin.

Testing

During execution of testing procedures, all of the operational parameters, such as head loss/pressure drop, motor amperage, temperature rise or fall, rotational speed, fuel or power usage, and any other parameters related to the operation and/or performance of the installed systems, are measured. For a controls system, this would include a complete point-by-point checkout, graphical interface review, sequence of operation verifications, historical trend log review and equipment responses to changing parameters. Depending on the proposed operation and complexity of the system, this may involve performance testing at numerous loading points over a period of time sufficient to confirm conformance to specifications. If the system is designed for constant-duty service, a one-time measurement may be acceptable. The results of these tests are then reviewed with the Design Engineer or independent third party engineer and, where applicable, the Savings Analyst to confirm consistency with original design and analysis assumptions.

If the installed systems required source emissions review, the regulatory agency will require specific testing be completed at the end of construction to verify the emissions are in compliance with approved limits. In plants that serve more critical operations and require high reliability, the owner should also require testing of the various failure modes covered by the design. Each of these tests will have a predetermined test protocol and expected performance requirement. If any of the tests fail, the engineer and contractor will determine the cause of failure, design and affect repairs before re-testing. These processes will be repeated until all tests have been successfully passed and the systems accepted.

Operational Service

Upon completion of all testing, the units are put into operational duty on a daily or as-needed basis. Operation is still monitored for conformance to specifications. Maintenance of any filters or fluids that may need frequent cleaning and/or changing due to construction debris is also reviewed during the initial phases.

For large and/or complex systems, a manufacturer's representative may be involved in the commissioning process to lend their expertise and confirm warranty parameters. The role of the manufacturer or vendor in the commissioning process is generally specified during the procurement process. In some cases, this will include almost full responsibility for system commissioning. The commissioning processes must, of course, still be monitored and confirmed by the Construction Manager and Design Engineer. The process is complete when they, and ultimately the owner, formally accept the project as complete.

While project acceptance indicates the successful completion of the construction phase, it is often necessary to properly train operating personnel and establish a long-term OM&R program prior to acceptance. This ensures a successful transition from construction to the operational phase.

QUALITY CONTROL

Quality control is a mechanism by which the quality of work performed is measured and monitored and any deficiencies are corrected. All implementation work should be executed under the auspices of a formal quality control (or assurance) program. Some companies establish their own quality control programs based on ISO 9000. This is a group of standards and guidelines for quality management and quality assurance. The series evolved from the International Organization for Standardization founded in Geneva, Switzerland, in 1946. Whether based on this or internal company programs, quality assurance involves a series of planned or systematic inspections, analyses, and actions required to provide adequate confidence that a product or service will satisfy its intended purpose.

For every project, a Quality Control Plan must be developed that establishes guidelines, procedures, and formats for monitoring and ensuring quality at all stages of implementation, from pre-construction activities through completion and commissioning. At each stage, results are compared to the plan, allowing corrective action to be taken early on. A feedback loop is established under which corrections are then evaluated to ensure that the ultimate quality standard has been achieved. The Quality Control Plan must be customized to match the scope and type of construction work anticipated for the project. It should be flexible enough to be modified, should the SOW change and require different quality control procedures. All personnel and subcontractors involved in the project must be committed to the plan for it to be effective.

The process must start with adequate job knowledge. This includes knowing the goals, objectives, and deliverables of the project. A performance standard is established so that job tasks are completed efficiently, while following a system of checks and balances to ensure proper execution at each stage. Since most projects inherently require a multi-functional effort, teamwork and communication are essential elements. The role of every member of the Quality Control Team is recognized as an important contribution to the overall goal of a quality installation.

Proper documentation throughout the process is also critical to demonstrating and controlling quality.

Routine inspections are carried out by the Quality Control Officer (QCO). This may be a full-time position at the job site for very large projects, or may be a duty performed by the Site Superintendent or Project Manager. Inspection items are established in a checklist included as part of the Quality Control Plan. A daily production inspection report should be maintained and serve as a primary vehicle for the quality control documentation process. This report should list daily (as well as longer duration) project information, as well as any safety and quality concerns.

In addition to routine independent inspections, the QCO should accompany all local authorities and inspectors when they perform their inspections and should accompany the Design Engineer for review of all work performed to ensure conformance to the design intent and specifications. The timing and frequency of each of these procedures will be dictated by the work scope and construction schedule.

All discrepancies found during inspections must be documented and a corrections punch list developed. These punch list items are then forwarded to the respective party responsible (e.g., subcontractor). The responsible party will correct discrepant work and notify the QCO and/or Site Superintendent when corrected work is ready for re-inspection. Subcontractors are to provide copies of all inspection reports, which will be complied and reviewed, with all records kept by the QCO and/or Site Superintendent. This process and feedback loop continues throughout the entire implementation process and does not conclude until all work has been completed and the quality of that work has been confirmed and documented.

Site Safety, Health, and Loss Prevention Programs

Fundamental elements in the construction process include a unilateral commitment that all work to be performed will be planned to minimize the potential for personal injury, property damage, and loss of productive time and a system of prompt detection and correction of unsafe practices and conditions will be implemented.

As a means of maintaining a safe working environment at the construction site and adhering to all required safety codes, a complete site safety, health, and loss prevention program should be instituted. Such programs serve to exercise all available means and methods of controlling and/or eliminating hazards and risks associated with construction, thus minimizing personal injury, conserving property, achieving greater efficiency, and reducing direct and indirect costs. The program should conform to all federal, state, and local health and safety requirements, including Occupational Safety and Health Administration (OSHA) requirements. The program should address those elements that are specific to this site and have the potential for negative impact on the safety and health of workers and other personnel on site.

Following are some key elements that should be included in the site safety, health, and loss prevention program:

- A review of the project to be constructed, the construction environment, and processes to be utilized, culminating in the development of a formal hazard review analysis
- A safety indoctrination program for staff, supervisory, and trade personnel
- A fire protection and prevention plan, including hot work permits
- Building floor plans and established evacuation plans
- Lock out, tag out procedures
- Educational programs
- Safety meetings
- First aid program and facilities
- Identification and monitoring of all potential hazardous conditions or practices
- Establishment of procedures for use of all mechanized construction equipment
- Provisions for a prompt response to safety recommendations, violations (if any), and complaints
- Thorough and effective accident investigation, reporting, and corrective action procedures

For very large projects, these elements should be monitored through a safety, health, and loss prevention checklist program administered by a site safety, health, and loss prevention officer (SSO). This officer should be empowered to have access to all necessary information and make necessary corrections as appropriate up to and including the cessation of all work until safe work practices are restored. As such, the SSO must have the complete cooperation of the entire Construction Team, inclusive of senior management, supervisors and all employees. In cases where the project magnitude does not justify the expense of a full-time SSO, the Site Superintendent or a principal assistant may fulfill this critical function, but under the supervision of the company's SSO.

OM&R Management Approach

Once a project is fully installed and all equipment has been commissioned, all of the new or modified systems must be maintained. All energy systems will require some degree of OM&R. Certain measures that may be installed, such as pipe or envelope insulation, require no operation. Nonetheless, even these must be maintained. The OM&R plan for each system installed will have been considered in the Detailed Feasibility Study. During this phase, anticipated future incremental OM&R costs are weighed against previous incremental OM&R costs as part of the economic performance analysis. These factors are once again considered during the final equipment selection process, since reliability and maintainability are important selection criteria.

First cost and efficiency considerations are often the driving force behind investment decisions, but reliability, maintenance requirements, and equipment life are important balancing factors that greatly affect project life-cycle economics. These factors are directly related to the equipment and quality of maintenance procedures selected.

Once systems have been purchased and all final detail for installation and subsequent operation is complete, it is an appropriate time to develop a complete OM&R program. When in-house personnel are to operate and/or maintain equipment, training programs are usually required.

When projects are implemented under performance-based or own and operate types of contracts, clear delineation of responsibility must be made between the contractor and host facility. If a contractor is responsible for equipment performance, it is logical to maintain with that the responsibility for equipment OM&R, or at least M&R. In this manner, the contractor can ensure equipment performance as opposed to risking dispute over who failed to execute the responsibilities that resulted in equipment non-performance. For such purposes, it is advisable to draw up two SOWs, one for the contractor and one for the host facility, listing in detail who is responsible for each OM&R activity and who has the liability for each potential out-of-parameter condition. With complete new system installation, it is sometimes easy for the contractor to clearly assume all responsibility for the life of the contract and this can be budgeted for in the original agreement. This process becomes further complicated in cases where some new equipment is installed in conjunction with older existing systems (e.g., new controls being installed with old dampers and actuators remaining in place). In this case, issues can easily arise as to who is responsible for routine problems, such as a no heat or cooling complaint. One way to mitigate this risk is to allow the contractor to take responsibility for the entire system, including the existing components, and to evaluate their condition and budget for their long-term OM&R along with the new equipment being installed.

Establishing an OM&R Program

To put an OM&R program into place, a staff member of the host facility or contractor responsible for OM&R should serve as the Team Leader. For very large projects, there may be a need to have separate teams for operations and maintenance, with a team leader for each. The primary responsibilities of the OM&R Team Leader will be the development and implementation of an OM&R and training plan. This may include completion and coordination of several or all of the following:

- Development of OM&R procedure manuals
- Development and execution of initial and ongoing training sessions
- Development of OM&R responsibility assignments (based on cost-effectiveness and customer preference)
- Procurement and execution of vendor service contracts
- Development and execution of ongoing quality control and site safety procedures
- Development and execution of final acceptance, testing, and start-up procedures for each newly installed system
- Development and execution of ongoing procedures for gathering of long-term metering data for maintenance, quality control, and manufacturer warranty verification
- Development and coordination of emergency service and repair services procedures

OM&R Training Manuals

Implementation of individual OM&R and training programs is typically the responsibility of the on-site OM&R Supervisor, who will report to the OM&R Team Leader. The OM&R Supervisor will develop operating manuals that comprehensively cover the total operation of each new system brought on line. The manuals will contain step-by-step methods for operating systems and individual components, detailing the location of the items, their function, and characteristics, as well as component relationships.

Maintenance manuals provide necessary component detail and illustration outlining arrangements and locations. The manuals should clearly prescribe the manufacturer's recommended schedule for preventative maintenance, seasonal maintenance requirements, and expected frequencies of maintenance. Emergency repair procedures should also be included.

Both operating and maintenance manuals should include all necessary manufacturer's details, service manuals, and parts lists. They should also include installation data provided by the Construction Team. The manuals should be bound and clearly marked, tabbed, and indexed and also be available electronically. In addition, condensed operating and maintenance instructions and diagrams should be strategically deployed in environmentally protected casings.

Electronic OM&R manuals are becoming increasingly popular, offering more comprehensive information and faster access. Several manufacturers have developed detailed OM&R software programs for their products. These include equipment drawing and photo libraries and hypertext-type features to call up detail on any of the items discussed in the general procedures. Some programs include features under which every part can be bar-coded with database reference to a manufacturer's part number. The OM&R Supervisor can organize the available software products for each system into one library, combined with project- and facility-specific files in order to construct a comprehensive OM&R resource.

The OM&R Supervisor should provide or arrange for comprehensive instruction on the safe operation, troubleshooting, maintenance, and repair of all equipment and systems modified or installed. The day-to-day operating personnel must thoroughly understand how the equipment works, how it should be operated, what can go wrong, and how to respond to all abnormal conditions. Instruction should include both classroom and practical application phases. The OM&R plans and manuals for each system, along with supplemental materials, should be used and instruction on their use provided. These elements can be supplemented with training video tapes (or CDs or DVDs), workbooks, computerized instruction programs, and on-line help services. Equipment manufactures should be brought into the process to impart their particular expertise to facility operating personnel. Off-site factory training courses are also recommended for specialized systems.

The time and effort associated with developing superior OM&R procedure manuals and training programs will usually pay long-term dividends. This is manifested in the avoidance of lost productivity, injury to personnel, accidental release of toxic substances, and a host of other potential problems that could arise absent strict adherence to proper procedures. Long-term maintenance and repair costs can also be minimized through early detection of component wear or system malfunction.

Types of Maintenance Strategies

The importance of proper maintenance for all types of systems cannot be overstated. No matter what type of equipment is selected, inadequate maintenance can lead to unreliable operation/production that can, in turn, result in excessive downtime and lost savings and productivity that cannot be made up.

A comprehensive maintenance program employing the practice of repair before failure generally represents a form of inexpensive insurance. There are three basic approaches to ongoing maintenance:

- **Unscheduled maintenance,** which is performed only when an incident occurs.
- **Scheduled,** or **preventative, maintenance,** which is performed on a schedule of periodic tasks designed to preclude failure; costs are a fairly predictable variable that can be included in the investment analysis.
- **Predictive,** or **on-condition, maintenance**, which is based on the equipment's condition; measurement of machine conditions are trended over time, allowing for prediction and avoidance of impending failure.

Generally, design and cost of a long-term maintenance program represents a trade-off between more frequent, less costly routine procedures and more costly, unscheduled procedures and long-term OM&R costs. Unscheduled maintenance strategies are often less costly when equipment is new. Over the long term, however, unscheduled maintenance strategies can result in longer downtime, and if system failures do occur, they are more likely to be serious. Often, more rigorous routine preventative maintenance programs produce lower life-cycle maintenance costs.

Preventive maintenance with long-term schedules allows maintenance costs to be more predictable. Components are serviced or replaced before failure occurs or performance degrades in order to increase system availability and efficiency. The downside is that certain procedures may be initiated before they are actually required. Over the life of the system, this may mean that the unit is taken out of service more often and cost outlays are greater than they would be with a predictive maintenance strategy.

In establishing preventative maintenance schedules, site-specific conditions of fuel, air, water, and other environmental factors should be considered. If, for example, the fuel is dirtier than is assumed in manufacturer specifications, schedules for inspecting, cleaning, and replacement might be shortened. Operational factors, such as loading and start/stop frequencies, should also be considered. In addition, schedules should be modified over time based on empirical observation.

Predictive maintenance results in a more finely tuned approach. Components are serviced or replaced most efficiently with respect to their actual condition. If a component demonstrates a field life 5,000 hours beyond the expected life, replacement based on scheduled maintenance wastes 5,000 hours worth of component life. If the component fails 5,000 hours earlier than its expected life, an unscheduled, and potentially inopportune, outage may occur. Processes such as oil analysis can be used to predict actual component wear and, therefore, actual replacement need rather than timed replacement need. The downside of predictive maintenance is the additional cost of monitoring equipment and procedures. To be effective, more data must be gathered and analyzed on a continual basis.

The OM&R plan should include detailed descriptions of systems, the specific requirements for system testing, and a listing of measurements to be taken with methods and means indicated. The testing and measurement process should continue throughout the life of the system to ensure that quality maintenance is performed and that preventative maintenance is effective. This process should also provide necessary data for documentation that equipment performance is in accordance with manufacturers' warrantees.

The OM&R plan should be extended beyond major system components to include auxiliary components such as fans, pumps, and coolers, as well as peripheral elements such as facility water quality control. This includes potable water, wash, rinse, and other process water, chilled water, condenser water, and various equipment cooling circuit water. Cooling towers are commonly the source of system coolant water. Careful water quality control is necessary not only to maintain system performance and limit deposits, but also to ensure health and safety, which can be at risk if proper biological controls are not in place. Of course, minimizing health and safety risk should be considered as the paramount OM&R objective, followed then by objectives such as maintaining system reliability, performance, and service life.

Microprocessor technology allows for close monitoring of equipment performance via remote data link. This technology provides a way to anticipate impending problems and correct them before an untimely outage occurs, and it extends the interval between scheduled maintenance procedures. In cases where metering and monitoring is used for system optimization or for ongoing savings verification processes, the same systems can often be used for predictive maintenance purposes and for other alarm functions. The computerized monitoring system can be linked with the OM&R resource library to provide immediate access to information required to make necessary corrections.

Often, effective strategies combine all three approaches. Unscheduled maintenance can be used for inexpensive, easily replaceable, or non-critical components. Scheduled maintenance can be used for components that are more costly or critical, or for those whose end of service life is predictable. Finally, predictive maintenance can be reserved for the most costly or critical components.

The benefits of well-conceived long-term maintenance programs include more efficient use of labor, greater productivity, more efficient equipment operations and fewer unplanned interruptions. It also allows materials and services to be priced out, bid, and contracted for in advance, instead of under the pressure of an emergency situation. Finally, costs can be predictably accrued, resulting in more effective budgeting and accounting.

Regardless of the strategy, project savings should be substantial enough to cover all maintenance and overhaul costs. If not, the project should not have been judged economical during the life-cycle analysis performed as part of the Detailed Feasibility Study and should only have been implemented on the basis of other benefits.

In-House Versus Contracted Services

Maintenance can be handled in-house or through a fixed-price, complete-service contract that can include repair, overhaul, and/or equipment replacement insurance, or through as-needed service by a time-and-materials service contract. Complete service contracts are often structured on a flat rate per machine run-hour over the life of the equipment, with inflation adjustments included. These contracts can take much or all of the responsibility of maintenance and repair from a facility, adding a degree of stability and security to the investment, although usually at a cost premium.

Service contracts can cover specified functions or can be completely comprehensive. Complete service contracts inclusive of all maintenance as well as overhaul and/or replacement are readily available from project developers, manufacturers, or local contractors. Service contracts can be designed for complete cradle-to-grave service, ideally allowing a facility to simply generate profits or execute their mission without having to absorb any maintenance responsibilities. These are, however, generally only practical for larger types of equipment or abundant quantities of similar smaller equipment.

In-house maintenance programs can often be handled least expensively if the facility staff has the technical expertise. In-house or as-needed maintenance tends to cost less in the early years of a project, when only routine maintenance is required. After a system has been in operation for several years, maintenance costs are higher due to increased incidence of repair and eventual overhaul and/or replacement requirements.

In general, in-house or as-needed maintenance may cost less over the life of the project than service contracts. This, however, is a higher risk option that could result in higher life-cycle costs. Facilities that have qualified technicians and experience in project maintenance will often opt for complete responsibility, which typically produces higher profitability. In such cases, even greater emphasis should be placed on the development of OM&R manuals and training programs. Facilities seeking to make secure financial returns on the project without added responsibilities will more likely contract for complete service. This also involves a risk analysis. When considering, for example, catastrophic failure of a major system, it may be difficult for a small facility to absorb the potential cost impact, and the incremental cost of insurance for such an event may be relatively high. A large facility with numerous sites might be better suited to absorb this risk. A large manufacturer with a large portfolio to spread the risk over, however, may be even better suited to take on this risk for, of course, a cost premium. A common trade-off between these strategies is for a facility to perform daily operations and routine maintenance, and then contract out for long-term service and overhaul or replacement.

Emergency Response and Spare Parts

Emergency response procedures to equipment failure should be established for all critical systems. For systems under long-term preventative maintenance contracts, emergency service will often be a component of the contract. Provisions can be made for various levels of response time, depending on how critical system operation is to the facility. Provisions for temporary service should also be included, as appropriate.

As noted in the prior discussion of the OM&R manuals, for each equipment component related to an installed system, the OM&R Supervisor should develop and maintain a spare parts list/database with electronic inventory capability. The database should detail spares for normal operations and specify all relevant tooling. As noted, there are numerous software packages commercially available that can add speed and sophistication to this process. Local sources of parts should also be maintained, with identification of those capable of responding within 24 hours of a call for service. Where appropriate, an inventory of spare parts should be maintained on site.

Establish Savings Verification Program

A recommended additional step in the implementation process for integrated energy resource optimization projects is the development of a savings verification program. Such programs are developed as a means of documenting the success of the project over time in generating projected resource savings or meeting anticipated economic and operational goals. These programs are often an integral part of contracts between facilities and project contractors to provide guarantees to the facility that proposed savings will be realized. In various types of performance-based contracts, savings verification programs establish the basis of payments made by the host to the project financier or prime contractor. The obligation to either prove that the savings have been achieved or make financial restitution shifts project risk to the contractor, providing the incentive to meet or exceed project goals through effective design, construction, and, where applicable, long-term OM&R.

Even when not required as part of a contract, savings verification programs provide much meaningful data to the analyst and operators, and ultimately to the host/owner. They provide feedback on the accuracy of the study originally used to justify the investment and they provide feedback on the ultimate investment decision. Additionally, system performance data gathered in the process can also be used collaterally as part of preventative or predictive maintenance programs and to confirm the status of manufacturer's performance warrantees. It can also be used as the basis of subsequent project development studies.

The fundamental question to be answered by savings verification is, "What would the energy (or other resource) consumption and cost have been, under today's conditions, had the new measure or group of measures not been installed?" The answer to this question is the energy baseline or, more accurately, the adjusted energy baseline; where the adjustment refers to changes in operation of the facility (e.g., occupancy, space condition requirements, etc.) not caused by the installation of the measure or group of measures.

In general, verification plans set up resource use evaluation procedures, usually characterized by direct metering of energy and other resource input to the affected process or building area. Whenever possible, meters are installed early in design (and sometimes during development), so they enhance the accuracy of projected savings as well as establish the baseline. Additional metering and data collection is usually installed during construction.

Where relevant to savings or cost, concurrent indirectly measured data such as occupancy, schedule, production (e.g., lbm of product produced per shift), climate (e.g., dry- and wet-bulb temperature, percent cloud cover, precipitation, wind conditions), and OM&R expense (e.g., softener salt consumed per shift or per lbm of production) should be tracked. Where OM&R material and labor is expected to change as a result of a measure, work orders or other documents applicable to the affected systems may also be tracked. The installed metering and data collection system may be maintained in place over the life of the project or removed after the host facility is satisfied that savings have been and will continue to be achieved. Sometimes, the metering and data collection system is linked to the overall control system and provides operating data that is used to optimize system performance and indicate when service is required. An example of this is provided in the final case study presented in Chapter 40.

General Verification Process Approaches

The type of verification process used will vary widely. Savings verification programs are designed with specific metering and verification (M&V) protocols. Protocol selection will depend on several factors, including project complexity, metering complexity, magnitude of costs and savings, level of interactivity with other measures, contractual allocation of risk associated with performance factors not controlled by the contractor, and collateral value of the M&V installation with respect to other uses for the data and metering systems.

The M&V process may, therefore, be based on engineering estimates, adjusted utility cost analysis, facility-wide metering, and/or equipment specific pre- and post-construction metering. More extensive programs add significantly to project costs, but also provide more security in verifying system performance, contractual guarantees, and/or in allowing for ongoing system modifications based on monitored performance data. With the continued advancements in direct digital control (DDC) and communications technology, extensive long-term system metering has become less costly, rendering this comprehensive approach increasingly economical.

Three general M&V approaches are:

1) Projects for which the potential to perform and generate savings needs to be verified, but the actual performance (savings) can be stipulated:

   This approach involves procedures for verifying that baseline conditions have been properly defined; the equipment and/or systems that were contracted to be installed have been installed; the installed equipment/systems meet and continue to meet the specification of the contract in terms of quantity, quality, and rating and continue to operate and perform in accordance with the specification in the contract and meet and continue to meet all functional tests.

2) Projects for which the potential to perform and generate savings needs to be verified and for which actual performance during the term of the contract needs to be measured:

   This approach includes the features of Item 1 above, plus procedures for verifying actual achieved energy (or other resource) savings during the term of the contract based on engineering calculations, with metering and monitoring of one or more variables. This approach may also include the use of statistically valid samples.

3) For projects consistent with Item 2 above, except that verification techniques involve utility whole-facility meter analysis and/or computer simulation calibrated with utility billing data.

Regardless of the approach, the starting point is the establishment of the baseline. Without this, there is the risk that energy and other resource use will not be allocated to the appropriate consumers, or that quantification, and hence savings projections, will be in error. The methods of baseline measurement cover the range from the simple and direct to the complex and derived. With all methods, however, the baseline can be conceptually reduced to the sum product of the following three variables:

1. Load or output (in tons cooling, Btuh heating, lumens lighting, tons of cement produced, lbm of laundry washed, etc.)
2. Efficiency (in kW/ton, Btuh fuel/Btuh heating, watt/lumen, kJh/ton of cement, lbm softener salt/lbm wash, etc.)
3. Time

Under the simplest conditions, where all three of these variables are constant, the baseline consumption can be measured directly. At the same time, the constancy can be confirmed by sampling the resource use over a brief time interval and recording the hours of operation with an hour meter.

On the other extreme, where all three variables are changing, establishing baseline consumption is a much more intensive and expensive process. In this case, it is necessary to determine explicitly the functional relationships between these variables and the independent variables that effect the change. Consequently, the baseline methodology requires a detailed understanding of system performance.

Baselines used for savings verification can be developed in the same manner as those used in savings analyses. This process is described in Chapter 41. One important potential difference from the savings analysis process is that the original system usage need not always be determined. Baseline adjustment can be avoided as long as the post-retrofit load is measured directly and the baseline efficiencies are established. With this approach, the savings can be defined as the product of the measured actual load and the difference in efficiency between the original or base system and the new system. In contrast, if the post-retrofit consumption is monitored as opposed to the load, then it is necessary to determine what part of the consumption reflects changes from the new system and what part is due to changes in the baseline conditions. For this, it is necessary to devise models that calculate consumption as a function of all the independent variables, such as temperature, production, occupancy, etc., and then adjust the baseline for these parametric changes.

Selecting the Appropriate M&V Approach

As indicated in the discussion above, the choice of M&V baseline strategy depends, to a large extent, on the nature of the variation of the loads and efficiency, the allocation of contractual risk, and the magnitude and complexity of the project. Within these constraints, a detailed understanding of the system performance can lead to significantly reduced costs with minimal increase in risk. The above post-retrofit load measurement, as opposed to consumption measurement, is a good example. This choice allows the avoidance of extensive computer modeling and baseline adjustment. On large projects, such as a central boiler or chiller renovation or the installation of a cogeneration plant, the magnitude of project cost and savings often justifies extensive metering. Careful selection of monitored variables and savings calculation methodology can, nevertheless, significantly reduce costs.

Another important decision is whether to employ a one-time or ongoing process. The one-time approach allows for a reasonably good confirmation that savings have been achieved, while avoiding the recurring expense of annual M&V activities. With this, however, the contractor does not bear the risk for degradation of performance over time resulting from normal wear or ineffective OM&R work.

When contractor payments are based on the results of an ongoing savings verification process, the contractor would have to maintain an element of control over ongoing OM&R in order to protect their payment stream. Annual M&V approaches are, therefore, only recommended as part of performance-based contracts when the contractor is also providing ongoing OM&R for the systems being installed. They may also be cost-justified in cases where the collateral value of the information provides operational and additional savings benefits to the host facility. This would be the case when the acquired data and analysis process is dovetailed with ongoing maintenance and performance optimization programs. Observation and analysis of operational trends often allow for equipment to be operating more efficiently and also provides useful data for ongoing preventative and predictive maintenance activities.

Summary

The implementation processes described in this chapter cover several of the main elements associated with management of project implementation and operations. However, there are numerous other aspects of the process, each of which must be carefully managed to ensure project success. These include resource procurement and management of the relationships with the various utilities serving the facility, environmental regulation compliance and project permitting, and management of a host of financial, legal, and contractual matters. Many of these subjects are covered in previous chapters and should be reviewed within the context of project implementation.

Management responsibility should be assigned for each of the major functional areas associated with project implementation, including design, construction, OM&R, training, M&V, legal/contractual, financial, environmental, and resource procurement. With the traditionally bid plan and specification system, the team leaders for each of these functional areas may reside in several different companies. When projects are implemented by an ESCo, all or most of these team leaders may reside in one firm and will have reporting responsibility to one overall project executive. In either case, the host facility must have its own management team in place to ensure project success. Each of these functional areas must be understood by the host facility's management team and carefully reviewed as the process unfolds. Independent consultants may be engaged to supplement the knowledge base of the management team as appropriate. Still, the owner or a senior manager at the host facility must be dedicated to the process and should have a clear understanding of the facility's mission and overall objectives as they relate to the project at hand.

Appendix

Conversion Factors

Table 1.1	Length
Table 1.2	Length, Small Units
Table 1.3	Area
Table 1.4	Volume
Table 1.5	Mass
Table 1.6	Force
Table 1.7	Pressure or Force per Unit Area
Table 1.8	Energy, Work and Heat
Table 1.9	Power or Rate of Doing Work
Table 1.10	Density (Mass/Volume)
Table 1.11	Thermal and Heat Transfer Parameters
Table 1.12	Multiplying Prefixes
Table 1.13	Selected Additional Conversion Factors

Source: Babcock and Wilcox

Table 1.1
Length

Starting With → Multiply By To Obtain ↓	Inch	Foot	Yard	Statute Mile	Millimeter	Centimeter	Meter	Kilometer
Inch	1	12	36	63.36×10^{3}	3.937×10^{-2}	0.3937	39.37	3.937×10^{4}
Foot	0.08333	1	3	5,280	3.281×10^{-3}	3.281×10^{-2}	3.281	3,281
Yard	0.02778	0.3333	1	1,760	1.094×10^{-3}	1.094×10^{-2}	1.093	1,094
Statute Mile	1.578×10^{-5}	1.894×10^{-4}	5.682×10^{-4}	1	6.214×10^{-7}	6.214×10^{-6}	6.214×10^{-4}	0.6214
Millimeter	25.40	304.8	914.4	$1,609 \times 10^{6}$	1	10.00	1,000	1.000×10^{6}
Centimeter	2.540	30.48	91.44	1.609×10^{5}	0.1000	1	100	1.000×10^{5}
Meter	0.0254	0.3048	0.9144	1,609	1.000×10^{-3}	0.010	1	1,000
Kilometer	2.540×10^{-5}	3.048×10^{-4}	9.144×10^{-4}	1.609	1.000×10^{-6}	1×10^{-5}	0.0010	1

Table 1.2
Length, Small Units

Starting With → Multiply By To Obtain ↓	Mil	Inch	Angstrom	Nanometer	Micron	Millimeter	Centimeter
Mils	1	1,000	3.937×10^{-6}	3.937×10^{-5}	3.937×10^{-2}	39.37	393.7
Inch	0.0010	1	3.937×10^{-9}	3.937×10^{-8}	3.937×10^{-5}	3.937×10^{-2}	0.3937
Angstrom	2.540×10^{5}	2.540×10^{8}	1	10.00	1.000×10^{4}	1.000×10^{7}	1.000×10^{8}
Nanometer	2.540×10^{4}	2.540×10^{7}	0.100	1	1,000	1.000×10^{6}	1.000×10^{7}
Micron	25.40	2.540×10^{4}	1.000×10^{-4}	1.000×10^{-3}	1	1,000	1.000×10^{4}
Millimeter	2.540×10^{-2}	25.40	1.000×10^{-7}	1.000×10^{-6}	1.000×10^{-3}	1	10
Centimeter	2.540×10^{-3}	2.54	1.000×10^{-8}	1.000×10^{-7}	1.000×10^{-4}	0.100	1

Table 1.3
Area

Starting With → Multiply By To Obtain ↓	Square Inch	Square Foot	Square Yard	Acre	Square Centimeter	Square Meter	Square Kilometer
Square Inch	1	144.0	1,296	6.273×10^{6}	0.1550	1550	1.550×10^{9}
Square Foot	6.944×10^{-3}	1	9.000	4.356×10^{4}	1.076×10^{-3}	10.76	1.076×10^{7}
Square Yard	7.716×10^{-4}	0.1111	1	4840	1.196×10^{-4}	1.196	1.196×10^{6}
Acre	1.594×10^{-7}	2.296×10^{-5}	2.066×10^{-4}	1	2.471×10^{-8}	2.471×10^{-4}	247.1
Square Mile	2.491×10^{-10}	3.587×10^{-8}	3.228×10^{-7}	1.563×10^{-3}	3.861×10^{-11}	3.861×10^{-7}	0.3861
Square Centimeter	6.452	929.0	8,361	4.047×10^{7}	1	$1\,000 \times 10^{4}$	1.000×10^{10}
Square Meter	6.452×10^{-4}	0.09290	0.8361	4,047	1.000×10^{-4}	1	1.000×10^{6}
Square Kilometer	6.452×10^{-10}	9.290×10^{-8}	8.361×10^{-7}	4.047×10^{-3}	1.000×10^{-10}	1.000×10^{-6}	1

Table 1.4
Volume

Starting With → Multiply By To Obtain ↓	Fluid Ounce	Quart	Gallon	Barrel (Petroleum)	Cubic Foot	Cubic Yard	Cubic Centi-Meter (milliliter)	Liter	Cubic Meter
Fluid Ounce	1	32.00	128.0	5376	957.5	2.585×10^{4}	3.381×10^{-2}	33.82	3.382×10^{4}
Quart	0.03125	1	4.000	168.0	29.92	807.9	1.057×10^{-3}	1.057	1,057
Gallon	7.813×10^{-3}	0.2500	1	42.00	7.481	202.0	2.642×10^{-4}	0.2642	264.2
Barrel (Petroleum)	1.860×10^{-4}	5.952×10^{-3}	0.0238	1	0.1781	4.809	6.290×10^{-6}	6.290×10^{-3}	6.290
Cubic Foot	1.044×10^{-3}	3.342×10^{-2}	0.1336	5.615	1	27.00	3.532×10^{-5}	3.532×10^{-2}	35.32
Cubic Yard	3.868×10^{-5}	1.238×10^{-3}	4.951×10^{-3}	0.2079	3.704×10^{-2}	1	1.308×10^{-6}	1.308×10^{-3}	1.308
Cubic Centimeter (Milliliter)	29.57	946.4	3785	1.590×10^{5}	2.832×10^{4}	7.646×10^{5}	1	1,000	1.000×10^{6}
Liter	2.957×10^{-2}	0.9464	3.785	159.0	28.32	764.6	1.000×10^{-3}	1	1,000
Cubic Meter	2.957×10^{-5}	9.464×10^{-4}	3.785×10^{-3}	0.1590	2.832×10^{-2}	0.7646	1.000×10^{-6}	1.000×10^{-3}	1

Table 1.5
Mass

Starting With → Multiply By To Obtain ↓	Pound	Short Ton	Long Ton	Gram	Kilogram	Metric Ton
Pound	1	2,000	2240	2.205×10^{-3}	2.205	2,205
Short Ton	5.000×10^{-4}	1	1.12	1.102×10^{-6}	1.102×10^{-3}	1.102
Long Ton	4.464×10^{-4}	0.8929	1	9.842×10^{-7}	9.842×10^{-4}	0.9842
Gram	453.6	9.072×10^{5}	1.016×10^{6}	1	1,000	1.000×10^{6}
Kilogram	0.4536	907.2	1016	1.000×10^{-3}	1	1,000
Metric Ton	4.536×10^{-4}	0.9072	1.016	1.000×10^{-6}	0.0010	1

Table 1.6
Force

Starting With → Multiply By To Obtain ↓	Newton	Grams	Kilograms	Pounds
Newton	1	9.807×10^{-3}	9.807	4.448
Grams (f)	102.0	1	1000	453.6
Kilograms (f)	0.1020	0.001	1	0.4536
Pounds (f)	0.2248	2.205×10^{-3}	2.205	1

Table 1.7
Pressure or Force per Unit Area

Starting With → Multiply By To Obtain ↓	Atmospheres	Cm of mercury at 0C	Inch of mercury at 0C	Inch of water at 4C	Feet of water at 4C	Kilograms per square meter	Pounds per square inch (abs)	Pascal	Bar
Atmospheres	1	1.316×10^{-2}	3.342×10^{-2}	2.458×10^{-3}	2.950×10^{-2}	9.678×10^{-5}	6.804×10^{-2}	9.689×10^{-6}	0.9869
Cm of mercury at 0C	76.00	1	2.540	0.1868	2.242	7.356×10^{-3}	5.171	7.501×10^{-4}	75.01
Inch of mercury at OC	29.92	0.3937	1	7.355×10^{-2}	0.8826	2.896×10^{-3}	2.036	2.953×10^{-4}	29.53
Inch of water at 4C	406.8	5.354	13.6	1	12	3.937×10^{-2}	27.68	4.015×10^{-3}	401.9
Feet of water at 4C	33.90	0.4460	1.133	8.333×10^{-2}	1	3.281×10^{-3}	2.307	3.345×10^{-4}	33.49
Kilograms per square meter	1.033×10^{4}	136.0	345.3	25.40	304.8	1	703.1	1.0197×10^{-1}	1.0197×10^{4}
Pounds per square inch (abs)	14.70	0.1934	0.4912	3.613×10^{-2}	0.4335	1.422×10^{-3}	1	1.450×10^{-4}	14.50
Pascal	1.01325×10^{5}	1.333	3,387.4	249.0	2,989	9.80665	6,895	1	1.000×10^{5}
Bar	1.013	1.333×10^{-2}	3.387×10^{-2}	2.49×10^{-3}	2.989×10^{-2}	9.80665×10^{-5}	6.895×10^{-2}	1×10^{-5}	1

PRESSURE is the force per unit area exerted on or by a fluid.
Absolute pressure is the sum of atmospheric and gauge pressure.
Standard atmospheric pressure is 14.696 lbf/in or 29.92 mm Hg.

Table 1.8
Energy, Work and Heat

Starting With → Multiply By To Obtain ↓	British thermal units	Centi-meter-grams	Foot-pounds	Horse-power-hours	Joules watt-second	Kilo-watt-hours	Kilogram meters	Watt-hours
British thermal units	1	9.297×10^{-8}	1.285×10^{-3}	2545	9.480×10^{-4}	3,414	9.297×10^{-3}	3.414
Centimeter-grams	1.076×10^{7}	1	1.383×10^{4}	2.737×10^{10}	1.020×10^{4}	3.671×10^{10}	10^{5}	3.671×10^{7}
Foot-pounds	778	7.233×10^{-5}	1	1.98×10^{6}	0.7376	2.655×10^{6}	7.233	2655
Horsepower-hours	3.929×10^{-4}	3.654×10^{-11}	5.050×10^{-7}	1	3.725×10^{-7}	1.341	3.653×10^{-6}	1.341×10^{-3}
Joules watt-second	1054.8	9.807×10^{-5}	1.356	2.684×10^{6}	1	3.6×10^{6}	9.807	3600
Kilowatt-hours	2.93×10^{-4}	2.724×10^{-11}	3.766×10^{-7}	0.7457	2.778×10^{-7}	1	2.724×10^{-6}	0.001
Kilogram-meters	107.6	10^{-5}	0.1383	2.737×10^{5}	0.1020	3.671×10^{5}	1	367.1
Watt-hours	0.293	2.724×10^{-8}	3.766×10^{-4}	745.7	2.778×10^{-4}	1000	2.724×10^{-3}	1

Table 1.9
Power or Rate of Doing Work

Starting With → Multiply By To Obtain ↓	Btu per minute	Foot-pounds per minute	Foot-pounds per second	Horse-power	Kilo-watts	Watts
Btu per minute	1	1.285×10^{-3}	7.716×10^{-2}	42.44	56.91	5.691×10^{-2}
Foot-pounds per minute	778	1	60	3.3×10^{4}	4.426×10^{4}	44.26
Foot-pounds per second	12.97	1.667×10^{-2}	1	550	737.6	0.7376
Horsepower	2.357×10^{-2}	3.030×10^{-5}	1.818×10^{-3}	1	1.341	1.341×10^{-3}
Kilowatts	1.757×10^{-2}	2.260×10^{-5}	1.356×10^{-3}	0.7457	1	10^{-3}
Watts	17.5725	2.260×10^{-2}	1.356	745.7	1000	1

Table 1.10
Density (Mass/Volume)

Starting With → Multiply By To Obtain ↓	Pounds per Cubic Foot	Pounds per Gallon (U.S. Liquid)	Grams per Cubic Centimeter	Kilograms per Liter	Kilograms per Cubic Meter
Pounds per Cubic Foot	1	7.481	62.43	62.43	6.243×10^{-2}
Pounds per Gallon (U.S. Liquid)	0.1337	1	8.345	8.345	8.345×10^{-3}
Grams per Cubic Centimeter	1.602×10^{-2}	0.1198	1	1.000	1.000×10^{-3}
Kilograms per Liter	1.602×10^{-2}	0.1198	1.000	1	1.000×10^{-3}
Kilograms per Cubic Meter	16.02	119.8	1,000	1,000	1

Table 1.11
Thermal and Heat Transfer Parameters

Thermal Conductivity

1 Btu/h ft F = 1 Btu/h ft R
= 12 Btu in/h ft^2 F
= 1.7307 W/m K

Heat Content

1 Btu/lb = 2.326 kJ/kg
1 Btu/Standard cubic foot = 0.039338 MJ/Nm^3

Heat Flux, or Heat Transfer Rate

1 Btu/h ft^2 = 3.1546 W/m^2

Volumetric Heat Release Rate

1 Btu/h ft^3 = 10.3497 W/m^3

Heat Transfer Coefficient or Conductance

1 Btu/h ft^2 F = 5.6784 W/m^2K

Table 1.12
Multiplying Prefixes

Multiplication Factors	Prefix	SI Symbol
1 000 000 000 000 = 10^{12}	tera	T
1 000 000 000 = 10^{9}	giga	G
1 000 000 = 10^{6}	mega	M
1 000 = 10^{3}	kilo	k
100 = 10^{2}	hecto*	h
10 = 10^{1}	deka*	da
0.1 = 10^{-1}	deci*	d
0.01 = 10^{-2}	centi*	c
0.001 = 10^{-3}	milli	m
0.000 001 = 10^{-6}	micro	μ
0.000 000 001 = 10^{-9}	nano	n
0.000 000 000 001 = 10^{-12}	pico	p
0.000 000 000 000 001 = 10^{-15}	femto	f
0.000 000 000 000 000 001 = 10^{-18}	atto	a

* To be avoided where possible.

Table 1.13
Selected Additional Conversion Factors

Length and area	
100 feet (ft)/minute (min)	= 0.508 meter (m)/ second(s)
1 square mile (mi^2)	= 640 acres
	= 259 hectares (ha)
1 m/s	= 196.9 ft/min
1 square kilometer (km^2)	= 100 ha
	= 0.3861 mi^2
1 ha	= 10,000 m^2
	= 2.471 acres
1 nautical mile	= 6080 ft
	= 1.853 km
1 nautical mile/hour (h)	= 1 knot
Weight	
1 pound (lb)	= 16 ounces (oz)
	= 7000 grains (gr)
	= 0.454 kilogram (kg)
1 oz	= 0.0625 lb
	= 28.35 grams (g)
1 grain (gr)	= 64.8 milligrams (mg)
	= 0.0023 oz
1 lb/ft	= 1.488 kg/m
1 g	= 1000 mg
	= 0.03527 oz
	= 15.43 grains
1 kg/m	= 0.672 lb/ft
Volume	
1 cubic foot (ft^3)	= 1728 cubic inches ($in.^3$)
1 $in.^3$	= 16,390 cubic millimeters (mm^3)
1 imperial gallon	= 277.4 $in.^3$
	= 4.55 liters (l)
1 U.S. gallon (gal)	= 0.833 imperial gallon
	= 3.785 l
	= 231 $in.^3$
1 U.S. barrel (bbl) (petroleum)	= 42 U.S. gallons
	= 35 imperial gallons
1 liter (1)	= 10^6 mm^3
	= 0.220 imperial gallon
	= 0.2642 U.S. gallon
	= 61.0 $in.^3$
1 board ft	= 12 in. x 12 in. x 1 in. thick
	= 144 $in.^3$
1 ft^3/min	= 1.699 m^3/h
1 m^3/h	= 0.589 ft^3/min
Temperature, measured	
F	= (C x 9/5) + 32
	= R − 459.67
C	= (F − 32) x 5/9
	= K − 273.15
Density	
1 ft^3/lb	= 0.0624 m^3/kg
1 grain/ft^3	= 2.288 g/m^3
1 grain/U.S. gallon	= 17.11 g/m^3
	= 17.11 mg/l
1 m^3/kg	= 16.02 ft^3/lb
1 g/m^3	= 0.437 grain/ft^3
	= 0.0584 grain/ U.S. gallon
1 g/l	= 58.4 grain/U.S. gallon
Water at 62F (16.7C)	
1 ft^3	= 62.3 lb
1 lb	= 0.01604 ft^3
1 U.S. gallon	= 8.33 lb
Water at 39.2F (4C), maximum density	
1 ft^3	= 62.4 lb
1 m^3	= 1000 kg
1 lb	= 0.01602 ft^3
1 l	= 1.0 kg
1 kg/m^3	= 1 g/l
	= 1 part per thousand
1 g/m^3	= 1 mg/l
	= 1 part per million (ppm)
Pressure	
1 atmosphere (metric)	= 98,066.5 pascals (Pa)
	= 1 kg force (kgf)/square centimeter (cm^2)
	= 10 m head of water
	= 14.22 lb/$in.^2$
1 lb/ft^2	= 47.88 Pa
	= 0.1924 in. of water
	= 4.88 kg/m^2
1 ton (t)/$in.^2$	= 13,789 kilopascals (kPa)
	= 1.406 kg/mm^2
1 in. head of water	= 5.20 lb/ft^2
1 m head of water	= 9806 Pa
	= 0.1 kg/cm^2
1 m head of mercury (Hg)	= 133.3 kPa
	= 1.360 kg/cm^2
	= 1333 millibars
1 kg/m^2	= 9.806 Pa
	= 1 mm head of water
	= 0.2048 lb/ft^2
1 kg/cm^2	= 98.066 kPa
	= 735.5 mm Hg
	= 0.981 bar
	= 14.22 lb/$in.^2$
1 kg/mm^2	= 9.8066 megapascals (MPa)
	= 0.711 t/in^2

In these conversions, inches and feet of water are measured at 62F (16.7C), millimeters and meters of water at 39.2F (4C), and inches, millimeters, and meters of mercury at 32F (0C).

INDEX

I